P9-DVF-722

Genomics and Clinical Medicine

OXFORD MONOGRAPHS ON MEDICAL GENETICS

16. C.R. Scriver and B. Childs: *Garrod's inborn factors in disease*
18. M. Baraitser: *The genetics of neurological disorders*
21. D. Warburton, J. Byrne, and N. Canki: *Chromosome anomalies and prenatal development: an atlas*
22. J.J. Nora, K. Berg, and A.H. Nora: *Cardiovascullar diseases: genetics, epidemiology, and prevention*
24. A.E.H. Emery: *Duchenne muscular dystrophy, second edition*
25. E.G.D. Tuddenham and D.N. Cooper: *The molecular genetics of haemostasis and its inherited disorders*
26. A. Boué: *Fetal medicine*
30. A.S. Teebi and T.I. Farag: *Genetic disorders among Arab populations*
31. M.M. Cohen, Jr.: *The child with multiple birth defects*
32. W.W. Weber: *Pharmacogenetics*
33. V.P. Sybert: *Genetic skin disorders*
34. M. Baraitser: *The genetics of neurological disorders, third edition*
35. H. Ostrer: *Non-mendelian genetics in humans*
36. E. Traboulsi: *Genetic diseases of the eye*
37. G.L. Semenza: *Transcription factors and human disease*
38. L. Pinsky, R.P. Erickson, and R.N. Schimke: *Genetic disorders of human sexual development*
39. R.E. Stevenson, C. E. Schwartz, and R.J. Schroer: *X-linked mental retardation*
40. M.J. Khoury, W. Burke, and E. Thomson: *Genetics and public health in the 21st century*
41. J. Weil: *Psychosocial genetic counseling*
42. R.J. Gorlin, M.M. Cohen, Jr., and R.C.M. Hennekam: *Syndromes of the head and neck, fourth edition*
43. M.M. Cohen, Jr., G. Neri, and R. Weksberg: *Overgrowth syndromes*
44. R.A. King, J.I. Rotter, and A.G. Motulsky: *The genetic basis of common diseases, second edition*
45. G.P. Bates, P.S. Harper, and L. Jones: *Huntington's disease, third edition*
46. R.J.M. Gardner and G.R. Sutherland: *Chromosome abnormalities and genetic counseling, third edition*
47. I.J. Holt: *Genetics of mitochondrial diseases*
48. F. Flinter, E. Maher, and A. Saggar-Malik: *The genetics of renal disease*
49. C.J. Epstein, R.P. Erickson, and A. Wynshaw-Boris: *Inborn errors of development*
50. H.V. Toriello, W. Reardon, and R.J. Gorlin: *Hereditary hearing loss and its syndromes, second edition*
51. P.S. Harper: *Landmarks in medical genetics*
52. R.E. Stevenson and J.G. Hall: *Human malformations and related anomalies, second edition*
53. D. Kumar and S.D. Weatherall: *Genomics and clinical medicine*

Genomics and Clinical Medicine

EDITED BY

Dhavendra Kumar, MD, FRCI, FACMG

Institute of Medical Genetics
University Hospital of Wales
Cardiff University
Cardiff, Wales
United Kingdom

ADVISORY EDITOR

Professor Sir David Weatherall, DM, FRS

Weatherall Institute of Molecular Medicine
John Radcliffe Hospital
Oxford University
Oxford, England
United Kingdom

OXFORD
UNIVERSITY PRESS

2008

UNIVERSITY PRESS

Oxford University Press, Inc., publishes works that further
Oxford University's objective of excellence
in research, scholarship, and education.

Oxford New York
Auckland Cape Town Dar es Salaam Hong Kong Karachi
Kuala Lumpur Madrid Melbourne Mexico City Nairobi
New Delhi Shanghai Taipei Toronto

With offices in
Argentina Austria Brazil Chile Czech Republic France Greece
Guatemala Hungary Italy Japan Poland Portugal Singapore
South Korea Switzerland Thailand Turkey Ukraine Vietnam

Published by Oxford University Press, Inc.
198 Madison Avenue, New York, New York 10016
www.oup.com

Library of Congress Cataloging-in-Publication Data

Genomics and clinical medicine / [edited by]
Dhavendra Kumar and Sir David Weatherall.
p. ; cm.
Includes bibliographical references and index.
ISBN 978-0-19-518813-4
1. Medical genetics. 2. Genomics. I. Kumar, Dhavendra. II. Weatherall, D. J.
[DNLM: 1. Genomics. 2. Clinical Medicine. 3. Genetic Predisposition to Disease.
4. Pharmacogenetics. QU 58.5 G3352 2007]
RB155.G464 2007
616′.042–dc22 2006034212

9 8 7 6 5 4 3 2 1

Printed in China
on acid-free paper

To Anju, who encourages me to write, and Ashish and Nikita, who make me proud

Foreword

A scant 20 years have passed since the word "genomics" was coined by Victor McKusick, Frank Ruddle, and Tom Roderick to describe a new discipline. The suffix of the word derives from the Greek *ome* meaning all, and aptly conveyed an intention to transition the study of heredity from a focus on single genes (genetics) to the more global perspective of all of the hereditary material. A proliferation of other "omics" disciplines has subsequently erupted—including proteomics, metabolomics, transcriptomics, glycomics, microbiomics, and many more.

However, genomics remains the foundation of the rest, reflecting as it does a comprehensive analysis of the deoxyribonucleic acid (DNA) instruction book. The success of the Human Genome Project has now laid that instruction book wide open. As a result, the life sciences have been catapulted forward, and biology has now taken its rightful place alongside physics and chemistry as a truly digital and quantitative science.

It is the application of genomics to medicine that carries its greatest promise of benefit to humankind. Thus, the publication of this first textbook of "Genomics and Clinical Medicine" marks a milestone, a coming of age. Here, in the early years of the third millennium, we can see the emerging outlines of a new synthesis of the noble tradition of the healing arts with an increasingly precise way of understanding the causes of disease, based on an understanding of the human genome.

For some in the clinical medicine community, however, this textbook may come as a surprise. After all, there are still many practicing physicians who would say they see no evidence of genetics or genomics as part of their daily medical practice. Surely, however, that reveals a problem with the successful communication of rapid new developments in this field, not the facts of the matter. For in these 41chapters, a vast array of genomic implications for nearly every condition that affects humankind is laid out in an elegant and comprehensive fashion.

The pace of progress in genomics has been astounding. Over just the last 15 years, largely as a consequence of the tools made available through the Human Genome Project, genes have been identified for more than 2000 inherited conditions. With recent rapid advances in the understanding of human genetic variation, the specific hereditary contributions to common diseases like diabetes, heart disease, cancer, and mental illness are emerging at an unprecedented rate. The very real possibility of offering individuals, who are currently healthy, a personalized prediction of future risks of illness is no longer a distant dream. And given that, many of the common disorders for which predictions are becoming possible are associated with proven means of reducing risk through diet, exercise, lifestyle change, medical surveillance, or pharmacotherapy, the real likelihood of widespread individualized programs of preventive medicine grows by the day. Similarly, the ability to make predictions about the possibility of a beneficial or undesirable response to drug therapy, the field of pharmacogenomics, is advancing rapidly, and will soon require healthcare providers to determine the genotype before writing the prescription, at least for certain drugs. Many of us predict that the complete genome sequence of an individual will become part of that person's medical record within approximately 10 years, at a cost of $1000 or less. And the therapeutics that we use in the future will likely be heavily dependent on an understanding of the genomic basis of illness, leading to interventions that are both more accurately targeted to the underlying problem and less likely to cause side effects.

All of these advances should be welcomed by anyone interested in the alleviation of human suffering. Yet, a number of major ethical, legal, and social challenges lie along the path if this vision is going to be realized. In the United States, for example, we still lack effective federal legislation to prevent discriminatory uses of predictive genetic information. Major challenges also lie ahead with regard to ensuring equitable access to new genomic technologies, especially as our medical care system seems to undervalue opportunities for preventive medicine, focusing instead on treating disease once it has already appeared. However, perhaps the greatest barrier, and the one which this book admirably seeks to address, is an educational one. Most members of the public are interested in genomics, but relatively unsure of the details. Seeking advice, they generally turn to their healthcare providers, but many of those professionals are poorly prepared to become practitioners of this new art. After all, most physicians have had little or no training in genetics or genomics, and will be hard pressed to quickly acquire the scientific principles, the medical knowledge, and the psychosocial skills that will be necessary for the successful introduction of genomic medicine. Busy practitioners will desperately need an authoritative source of information that includes both principles and specific applications. The introduction of this textbook, with its distinguished and authoritative list of contributors, thus arrives in the nick of time. Welcome to the genome era.

Francis S. Collins, MD, PhD
National Human Genome Research Institute
National Institutes of Health
Bethesda, MD, USA

Preface

Although the science of genetics is only 150 years old, genetics has been in existence since ancient time. The evolution and natural selection theories put forward by Charles Darwin had clear overtones reflected in some of our present-day concepts of the genetic basis of biological life. Mendel's laws of inheritance, and successive discoveries in various aspects of genetics, laid the foundations of a number of disciplines covering different areas within the science of genetics. Human genetics was no exception. However, this remained heavily shrouded under the dark clouds of the so-called Eugenics movement of the early twentieth century, when history recorded one of the worst practical applications of modern science on fellow human beings under the pretext of scientific research.

It has taken almost 60 years to arrive at our present state in the science of genetics. The future now appears bright, opening up many opportunities on the horizon. Clinical genetics is now a recognized medical specialty among several disciplines comprising the current spectrum of modern medicine. The basis of clinical genetics is grounded in the sound knowledge and understanding of medical genetics, which emerged as a spin-off of "Human Genetics."

Fifty years after the discovery of the double-helix structure of the deoxyribonucleic acid (DNA) molecule (Watson and Crick, 1953), characterization of the complete sequence and organization of the human genome was successfully accomplished (Lander et al., 2001; Venter et al., 2001). This major scientific achievement laid the foundation of "*human genomics*"; that section of the biological sciences, which studies variations, mutations, and functions of genes and controlling regions, and their implications on human variation, health, and disease. This is strengthened by developments in the other areas of genomics relating to bacteria, vectors, parasites, animals, and plants.

The identification of all human genes and their regulatory regions provides the essential framework in our understanding of the molecular basis of disease. This advance has also provided a firm foundation for the future development of genomic technologies that can be applied to modern medical science. Rapid developments in global gene analysis, gene product analysis, medical bioinformatics, and targeted molecular genetic testing are destined to change the practice of modern medicine. However, many practicing clinicians perceive developments in genomics as primarily confined to the research arena with little clinical applicability. DNA/ribonucleic acid (RNA)-based methods of disease susceptibility screening, molecular-based disease diagnosis and prognosis, and genomics-based therapeutic choices and prediction of treatment outcome are some of the key areas that are likely to influence the practice of modern clinical medicine.

Undoubtedly, the science of genomics holds tremendous potential for improving human health. The World Health Organization (WHO, 2002) has recently made several recommendations on the scope and application of genomics on global health. It is acknowledged that the information generated by genomics will provide major benefits in the prevention, diagnosis, and management of communicable and genetic diseases as well as other common medical diseases, including cardiovascular diseases, cancer, diabetes, and mental illnesses (Cardon and Bell, 2001). Together, these constitute the major health burden, as reflected in chronic ill health and mortality. In addition, a number of infectious diseases are associated with genomic mutations manifesting in the form of increased susceptibility, clinical severity, favorable and unfavorable response to antimicrobial therapy, and in conferring protection. It is possible that the protective effect of a microbial vaccine might be influenced by genomic variation.

The sequence of the entire human genome is now complete, but this is not limited to one individual alone. Each person carries a distinct sequence. The variation among all humans is reflected in variation within the human genome. The genomic variation between individuals, together with environmental factors probably determines the disease susceptibility and is important in drug efficacy and side effects (Chakravati, 2000; Holden, 2000). The key to genomic variation lies in finding single nucleotide polymorphisms (SNPs) and its use in disease association studies (Stephens et al., 2001). The positional cloning (identifying the gene by location followed by functional analysis) of the disease susceptibility loci will depend upon the successful application of haplotype associations. In addition, these will be important in clinical studies to find individuals in whom a drug is likely to be efficacious. The use of SNPs in pharmacogenetics is currently restricted to studying genes for drug-metabolizing enzymes, such as P450s, and variations in genes that target drug receptors. The newly emerging dynamic field of pharmacogenomics is an exciting application of genomic variation in drug discovery and drug development.

The recent cloning of real disease susceptibility genes for multifactorial diseases is encouraging, for example, the identification of *NOD2* as a susceptibility gene for Crohn's disease (Hugot et al., 2001; Ogura et al., 2001). This is a major development in understanding the pathophysiology of inflammatory bowel disease. Similar studies are likely to unravel the genetic mechanisms in other complex medical diseases. A comprehensive SNP map will allow the cloning of other susceptibility alleles. However, this will depend upon population sample and size, the method employed, linkage disequilibrium, or association studies rather than the technology used (Cardon and Bell, 2001). Some of the best genetic studies of this kind include susceptibility to infectious disease, for example, an association between chemokine receptors (CCR5) and human immunodeficiency virus (HIV) susceptibility, and between the bacterial transporter protein Nramp and resistance to macrophage-infecting bacteria such as *Mycobacterium tuberculosis*. Similarly, various alleles at the G6PDH locus determine malaria susceptibility (Tishkoff et al., 2001).

These kinds of studies and clinical applications of the resulting outcomes are not without ethical concerns. Some of the questions and concerns are related to ownership of the genes and freedom to use collected DNA for such studies. These are complex and emotional issues, especially when dealing with populations who may have been exploited or perceived to have been exploited. These issues should always be dealt with carefully under the statutory requirements and rules.

There has been a tremendous surge in various subspecialties and technologies with names ending in *-omics*. We are rapidly moving into the "*omics*" era. In addition to genomics, several new specialist fields with an "omics" suffix have recently appeared, for example, pharmacogenomics, nutrigenomics, metabonomics, transcriptomics, proteomics, microbiomics, glycomics, toxicogenomics, and many more. Some of these areas are included in this book. Whatever the basis of distinction might be, the driver of all these terms is *GENOMICS*—the study of genomes in its entirety.

Genomics is not just about genome sequencing. Apart from full-length complementary DNAs (cDNAs) and their sequences, copies of messenger RNAs (mRNAs) that actually exist and code for different proteins are probably more important. The study of proteins thus derived, falls within the broad field of proteomics, a likely outcome of functional genomics and probably a true companion to genomics. It is likely that eventually *proteomics* will have more practical applications in clinical medicine. This is rapidly moving ahead with the completion of the HapMap Project (International HapMap Consortium, 2005) and the future "functional-variant database," a natural outcome of the HapMap Project (Gibbs, 2005).

It is vital that existing gaps in our knowledge about various "omics" disciplines are filled to ensure efficient use of the valuable information emerging from research. It is also important that the gap between "genetic professionals" and the "primary-care community," and as well as the "public health community," is narrowed (Khoury et al., 2003). Integration of this knowledge in the medical education curriculum and the continued professional education programs is urgently required to ensure applications of genomics in the provision of healthcare.

During the last two decades, the practice of medical genetics or clinical genetics, has found its niche within the broad horizon of clinical medicine. Genetic services now constitute a small, but albeit important, component of modern medical practice and public health. Currently, genetic services focus on providing information on chromosomal and single-gene diseases with limited contribution to multifactorial/polygenic diseases. How would this then be different from genomics? Already, there is tremendous enthusiasm for the recently introduced term of "genomic medicine." In a primer on genomic medicine, Guttmacher and Collins (2002) viewed "genetics as the study of single genes and their effects" and genomics as "the study not just of single genes, but of the functions and interactions of all the genes in the genome." In simple terms, there is a quantitative difference between the two fields—the study of multiple genes as opposed to that of one gene. Thus, genetics can be seen as part of genomics! However, there is a qualitative difference between genetics and genomics in medical and health applications ranging from the concept of disease in genetics to the concept of information in genomics (Khoury et al., 2003).

The practice of medical genetics has traditionally focused on those conditions that result from specific alterations or mutations in single genes (e.g., inborn errors of metabolism, Duchenne muscular dystrophy, and Huntington's disease), whole or part of chromosomes (e.g., trisomy 21 in Down syndrome), or associated with congenital malformations and developmental disabilities. The existing model of medical genetic services for these conditions includes laboratory diagnosis, genetic counseling, and management. This is supported by public health measures to ensure delivery of genetic services and genetic screening (e.g., newborn screening or screening the high-risk population). On the other hand, the practice of genomics in medicine and public health will focus on information resulting from variation at one or multiple loci and strong interactions with environmental factors, for example, diet, drugs, infectious agents, chemicals, physical agents, and behavioral factors (Khoury et al., 2003).

What medical and public health applications could one foresee following the completion of human genome sequence in 2003? How could these be applied and delivered to the 95% of human diseases that do not fall under the rubric of genetic disorders? These are some of the likely questions related to genomic medicine. Medical and public health professionals urgently need to make the changes necessary to accommodate rapid identification and characterization of the numerous genomic variants at multiple loci, which increase or decrease the risks for various diseases, singly or in combination with other genes, and with various chemical, physical, infectious, pharmacologic, and social factors (Khoury, 1999). This genetic and genomic information is crucial in assessing the disease susceptibility among healthy individuals, and in personalized primary and secondary prevention planning. Collins and McKusick (2001) stated that, "By the year 2010, it is expected that predictive genetic tests will be available for as many as a dozen common conditions, allowing individuals who wish to know this information to learn their risks for which interventions are or will be available. Such interventions could take the form of medical surveillance, lifestyle modifications, diet, or drug therapy. Identification of persons at highest risk for colon cancer, for example, could lead to targeted efforts to provide colonoscopic screening to those individuals, with likelihood of preventing many premature deaths."

Personalized medicine will encompass not only common medical diseases, but could include a wide range of preventable diseases (www.genovations.com). Genetic testing for future disease susceptibility using multiple genomic variants will be possible and affordable with the application of "high-throughput" microarrays-based genetic testing (Collins and Guttmacher, 2001).

A wealth of information on genomics is rapidly being acquired with the potential for major impact on human health. However, this data and information is scattered throughout several scientific journals, reviews, and state-sponsored reports and bulletins. A clinician or health professional often has difficulty in accessing and assimilating this information for application in the medical and public health practice. More importantly, an inability to assimilate and interpret leads to frustration and avoidance of potentially useful information.

In view of the above developments and the rapidly increasing gulf in the available literature resource, the need for a dedicated book on genomic medicine was appreciated. It was obviously an impossible task for a single author. Several leading experts in different fields of the genome science and technology offered to contribute. The views and opinions reflected in individual chapters are largely influenced by their own experience, perception, and interpretation of the available data and information.

This book provides a wide coverage of the subject from the historical progress to general aspects of genomics and the describing in some detail the medical and health applications. Generally, all chapters follow a similar format and are written by experts in their respective field of research and clinical expertise. Each chapter provides a detailed and comprehensive account of the subject. However, it is likely that some gap might exist due to inevitable time lag between the time of writing and appearing in the print. This is due to rapid developments in each field. However, all efforts are being made to provide the reader core information on the basic principles, scientific facts, current and likely future applications, useful relevant references, and information on Internet-based resources that should be helpful in exploring the subject further.

It is hoped that this book will facilitate in acquiring factual information on genomics, developing concepts on the genomic basis of human disease, and in providing a practical base to enable an interested clinician and health professional to develop an understanding of applications of genomics in clinical medicine and health. It is aimed at a wide range of scientists, clinicians, and health professionals who are engaged in research, teaching, and training in medical and health applications of the genome-based science and technology.

Finally, the practice of Medicine is an art based on sound scientific principles. It would be appropriate to quote Sir William Osler's remarks, "*If there were no individual variability, medicine would have been science not an art.*" Genomics in this context provides the basis of individual variability and the modern postgenomic clinician will need to ensure that this is applied as an art.

Dhavendra Kumar MD, FRCPI, FACMG
Institute of Medical Genetics
University Hospital of Wales
Cardiff University, Cardiff, Wales, UK

References

Cardon, LR and Bell, JI (2001). Association study designs for complex diseases. *Nat RevGenet,* 2:91–99.

Chakravati, A (2000). To a future of genetic medicine. *Nature,* 409:822–823.

Collins, FS and Guttmacher, AE (2001). Genetics moves into medical mainstream. *JAMA,* 286:2322–2324

Collins, FS and McKusick, VA (2001). Implications of the Human Genome Project for medical science. *JAMA,* 285:540–544.

Genovations—the advent of truly personalized healthcare. Available: http://www.genovations.com [July 12, 2007].

Gibbs, R (2005). Deeper into the genome. *Nature,* 437:1233–1234.

Guttmacher, AE and Collins, FS (2002). Genomic medicine: a primer. *N Engl J Med,* 347:1512–1520.

Holden, AL (2000). The SNP consortium: a case study in large pharmaceutical company research and development collaboration. *J Com Biotech,* 6:320–324.

Hugot, JP, et al. (2001). Association of NOD2 leucine-rich variants with susceptibility to Crohn's disease. *Nature,* 411:599–603.

International HapMap Consortium (2005). A haplotype map of the human genome. *Nature,* 437:1299–1320.

Khoury, MJ (1999). Human genome epidemiology: translating advances in human genetics into population-based data for medicine and public health. *Genet Med,* 1:71–73.

Khoury, MJ, McCabe LL, and McCabe, ER (2003). Population screening in the age of genomic medicine. *N Engl J Med,* 348:50–58.

Lander, ES, et al. (2001). Initial sequencing and analysis of the human genome. International Human Genome Sequencing Consortium. *Nature,* 409:860–921.

Ogura, Y et al. (2001). A frameshift in NOD2 associated with susceptibility to Crohn's disease. *Nature,* 411:603–606.

Stephens, C et al. (2001). Haplotype variation and linkage disequilibrium in 313 human genes. *Science,* 293: 489–493.

Tishkoff, SA et al. (2001). Haplotype diversity and linkage disequilibrium at the human G6PDH: recent origin of alleles that confer malarial resistance. *Science,* 293:455–461.

Venter, JC et al. (2001). The sequence of the human genome. *Science,* 291:1304–1351.

Watson, JD and Crick, FHC (1953). Molecular structure of nucleic acids. *Nature,* 171:737–738.

World Health Organization (2002). Genomics and world health—report from the advisory committee on health research. WHO, Geneva.

Acknowledgments

Editing a multiauthor book was a huge challenge. This was impossible without the support of several people particularly when a large team of expert contributors was involved. Although my idea of producing a book on genomic medicine looked attractive, it raised several questions. As for me, this was indeed daunting but convincing enough to approach several authors and advisors. The proposal for this book was positively considered by several leading publishers and as well as few key experts in the fast-emerging virgin territory of genomic medicine. I will always remain truly grateful to all expert reviewers, contributors, as well as strong support from the Oxford University Press in bringing my dream to fruition.

Several eminent persons offered wise counsel during the early stages of book planning. Notable names include Prof. Sir David Weatherall (Oxford), Prof. John Bell (Oxford), Prof. Michael Patton (London), Prof. Michael Parker (Oxford), and Prof. Stuart Tanner (Sheffield). Their invaluable advice, support, and encouraging and positive responses were a good kick-start.

I was privileged to receive support from Prof. Francis Collins (Director, NHGRI, NIH, USA) on this idea for a comprehensive book dealing with genomic medicine. He very kindly agreed to write the foreword for the book. I was fortunate to have a regular flow of suggestions and reflections from some leading experts including Prof. Colin Munro (Glasgow), Prof. Peter Harper (Cardiff), Dr. Eli Hatchwell (New York), Prof. David Cooper (Cardiff), Prof. Angus Clarke (Cardiff), and Prof. Julian Sampson (Cardiff). I was extremely fortunate to receive continuous flow of expert editorial guidance and supervision from Prof. Sir David Weatherall. Special thanks are due to Mr. Bill Lamsback, the Senior Executive Editor at the New York office of the Oxford University Press, who believed in the book project from start and supported it throughout to its final production.

I am indebted to Dr. Annie Procter and Prof. Julian Sampson, both at the Institute of Medical Genetics, Cardiff, for providing me generous time, space, and essential resources in the completion of this book. Several of my colleagues in Cardiff spent hours reviewing the manuscript. I am particularly grateful to Drs. Annie Procter, Daniela Piltz, Mark Tein, Alex Murray, Mark Davies, Andrea Edwards, and Prof. Angus Clarke for critically reviewing some chapters. Other staff members were always enthusiastic, positive, and supportive of this project.

My special thanks are due to Ms. Cetra Hastings, who worked tirelessly as a secretary and offered valuable and constructive editorial advice, even while working away as an English language expert in Norway (originally from Oxford).

The book is dedicated to all lead authors and their team of coauthors who worked extremely hard in writing excellent chapters. The quality of the material presented in almost all chapters is exemplary and of very high standards. I will always remain grateful to all those whose name I might have omitted, but without their support and encouragement this book could not have been conceived, written, and produced.

Finally, this book could not have been completed without the affection and support of my family who stood beside me like a buttress till the final draft of the manuscript was mailed to the Oxford University Press.

Contents

Contributors

Dan E. Arking, PhD
Institute of Genetic Medicine
Johns Hopkins University School of Medicine
Baltimore, MD

Wadie F. Bahou, MD
Department of Hematology and Medicine
Chair—Program in Genetics
State University of New York
Stony Brook, NY

Michael D. Bates, MD, PhD
Division of Gastroenterology, Hepatology and Nutrition
Division of Developmental Biology
Cincinnati Children's Hospital Medical Center
Cincinnati, OH

Graeme C.M. Black, DPhil, FRCOphth
Academic Units of Medical Genetics and Ophthalmology
University of Manchester
Manchester, UK

David A. Burden, MD, FRCP
Department of Dermatology
Western Infirmary
Glasgow, Scotland

Alan Burnett
Department of Haematology and Laboratory Medicine
University Hospital of Wales
Cardiff University
Cardiff, UK

Angus J. Clarke, DM, FRCPCH
Institute of Medical Genetics
University Hospital of Wales
Cardiff, UK

William O.C. Cookson, MD, DPhil
Respiratory Genetics Unit
National Heart and Lung Institute
Imperial College
London, UK

Francis S. Collins, MD, PhD
Director
National Human Genome Research Institute
National Institutes of Health
Bethesda, MD

James J. Cox, BSc, PhD
Department of Medical Genetics
Cambridge Institute for Medical Research
Addenbrooke's Hospital
Cambridge, UK

Mark Davies, MBBCh, MRCP
Institute of Medical Genetics
University Hospital of Wales
Cardiff, UK

Campus Drie Eiken
Antwerp, Belgium

Ian Dunham, MA, DPhil
Wellcome Trust Sanger Institute
Wellcome Trust Genome Campus
Hinxton, Cambridge, UK

Christopher Everett, MRCP, PhD
Royal London Hospital
London, UK

Richard Festenstein, FRCP, PhD
MRC Clinical Sciences Centre
Faculty of Medicine
Imperial College London
London, UK

Ian Frayling, PhD, MRCPath
Director, All-Wales Genetics Laboratory Service
Institute of Medical Genetics
University Hospital of Wales
Cardiff, UK

Mark Gardiner, MD, FRCPCH
Department of Paediatrics and Child Health
Royal Free and University College
Medical School
University College London
London, UK

Christopher S. Garrard, FRCA, DPhil
Intensive Care Unit
John Radcliffe Hospital
Oxford, UK

Philip Gaughwin, PhD
Stem Cell and Developmental Biology
Genome Institute of Singapore
Singapore

Pille Harrison
University of Oxford Institute of Musculoskaeletal Sciences
Botnar Research Centre,
Headington,
Oxford UK

Eugene Healy, PhD
Division of Infection, Inflammation and Repair
University of Southampton
Southampton General Hospital
Southampton, UK

Charles Hinds, FRCP, FRCA
Intensive Care Medicine
West Smithfield
London, UK

Sahoko Ichihara, MD, PhD
Department of Human Functional Genomics
Life Science Research Center
Mie University
Tsu, Mie, Japan

Alan Irvine, MD, FRCPI, MRCP
Department of Paediatric Dermatology
Our Lady's Hospital for Sick Children
Crumlin
Dublin, Ireland

Sarra E. Jamieson, PhD
Genetics and Infection Laboratory
Cambridge Institute for Medical Research
Addenbrooke's Hospital
Cambridge, UK

Derek Jewell, DPhil, FRCP
Department of Gastroenterology
University of Oxford
Gastroenterology Unit
Radcliffe Infirmary
Oxford, UK

Jane Kaye
The Ethox Centre
Department of Public Health and Primary Care
University of Oxford
Oxford, UK

Julian C. Knight, DPhil, MRCP
Wellcome Trust Centre for Human Genetics
Roosevelt Drive, Headington
Oxford, UK

Michael R. Konikoff, MD
Division of Gastroenterology, Hepatology and Nutrition
Cincinnati Children's Hospital Medical Center
Cincinnati, OH

Dhavendra Kumar, MD, FACMG
Institute of Medical Genetics
University Hospital for Wales
Cardiff, UK

John T. Lear, MD, FRCP
Central Manchester Dermatology Centre
Manchester Royal Infirmary
Manchester, UK

Bing Lim, PhD
Stem Cell and Developmental Biology
Genome Institute of Singapore
Singapore
AND
Harvard Institutes of Medicine
Harvard Medical School
Boston, MA

Forbes Manson D.C.
Academic Unit of Eye and Vision Science
Manchester Royal Eye Hospital
School of Medicine and
Centre for Molecular Medicine
Faculty of Medical and Health Sciences
University of Manchester
Manchester, UK

Peter A. Maxwell, FRCP
Regional Nephrology Unit
Belfast City Hospital
Queens University
Belfast, N. Ireland, UK

Mark I. McCarthy, MD, FMedSci
Diabetes Research Laboratories
Oxford Centre for Diabetes, Endocrinology and Metabolism (OCDEM)
University of Oxford
Churchill Hospital
Old Road, Headington
Oxford, UK

Duncan McHale, MBBS, PhD, MRCP
Senior Director, Molecular Profiling
Worldwide Development
Pfizer Ltd
Sandwich
Kent, UK

John H. McVey, PhD
Haemostasis and Thrombosis
MRC Clinical Sciences Centre
Imperial College London
Hammersmith Hospital Campus
London, UK

Kenneth I. Mills, PhD, FRCPath
Experimental Haematology
Department of Haematology
School of Medicine, Queen's University
Belfast, N. Ireland, UK

Nilesh Morar, MBChB, FCDerm
Wellcome Trust Centre for Human Genetics
University of Oxford
Oxford, UK
AND
Molecular Genetics Unit
National Heart and Lung Institute
Imperial College
London, UK

Huw R. Morris, PhD, FRCP
Academic Department of Neurology
School of Medicine, Cardiff University
Cardiff, UK

Colin S. Munro, MD, FRCP
Alan Lyell Centre for Dermatology
Southern General Hospital
Glasgow, UK

Albert Ong, DPhil, FRCP
Academic Unit of Nephrology
The Kidney Institute
Northern General Hospital
University of Sheffield
Sheffield, UK

Michael Parker, PhD
The Ethox Centre
Department of Public Health and Primary Care
University of Oxford
Oxford, UK

Saad Pathan, DPhil
Department of Gastroenterology
Wellcome Trust Centre for Human Genetics
University of Oxford
Oxford, UK

Christopher S. Peacock, PhD
The Pathogen Genome Unit
Wellcome Trust Sanger Institute
Wellcome Trust Genome Campus
Hinxton, Cambridge, UK

Michelle Penny, PhD
Director, Oncology Molecular Profiling
Worldwide Development
Pfizer Inc.
New London, CT

Anna Ponnampalam
Center for Women's Health Research,
Department of Obstetrics and Gynaecology
Monash University
Melbourne, Australia

Don Powell, PhD
Wellcome Trust Sanger Institute
Wellcome Trust Genome Campus
Hinxton, Cambridge, UK

Lucy F. Raymond, DPhil, FRCP
Department of Medical Genetics
Cambridge Institute for Medical Research
Addenbrooke's Hospital
Cambridge, UK

Andrew P. Read, PhD, FAMSci
Academic Department of Medical Genetics
University of Manchester
St.Mary's Hospital
Manchester, UK

Peter A.W. Rogers
Center for Women's Health Research
Department of Obstetrics and Gynaecology
Monash University
Melbourne, Australia

Julian Sampson, DM, FRCP
Institute of Medical Genetics
University Hospital of Wales
Cardiff, UK

Andrew Singleton, PhD
Molecular Genetics Unit
National Institute on Aging
National Institutes of Health
Bethesda MD

Patrick J. Stover, PhD
Cornell University
Division of Nutritional Sciences
Ithaca, NY

Patrick F. Sullivan, MD
Departments of Genetics, Psychiatry, and Epidemiology
The University of North Carolina at Chapel Hill
Chapel Hill, NC

Masaharu Takemura, PhD
Department of Human Functional Genomics
Life Science Research Center
Mie University
Tsu, Mie, Japan

Wai-Leong Tam
Stem Cell and Developmental Biology
Genome Institute of Singapore
Singapore

Edward G.D. Tuddenham, MD
Haemophilia Centre and Haemostasis Unit
Royal Free Hospital
London, UK

Santiago Uribe Lewis, MD, PhD
MRC Clinical Sciences Centre
Faculty of Medicine, Imperial College London
London, UK

Guy van Camp, PhD
Department of Medical Genetics
University of Antwerp
Campus Drie Eiken
Antwerp, Belgium

Tineke C.T.M. Van der Pouw Kraan PhD
Department of Molecular Cell Biology
and Immunology
VU Medical Center
Amsterdam, The Netherlands

Lut van Laer, PhD
Department of Medical Genetics
University of Antwerp

Colin D. Veal, PhD
Department of Genetics
University of Leicester
Leicester, UK

Cornelis L. Verweij, PhD
Molecular Biology and Biochemistry
VU Medical Center
Department of Molecular Cell Biology and Immunology
Amsterdam, The Netherlands

Sir David J. Weatherall, DM, FRS
Weatherall Institute of Molecular Medicine
University of Oxford
John Radcliffe Hospital
Oxford, UK

Andrew R. Webster
Institute of Ophthalmology
University College London
London, UK

Gareth C. Weston
Center for Women's Health Research
Department of Obstetrics and Gynaecology
Monash University
Melbourne, Australia

Paul Wordsworth, DPhil, FRCP
Nuffield Department of Orthopaedic Surgery
Nuffield Orthopaedic Centre
Windmill Road
Headington, Oxford, UK

Yoshiji Yamada, MD, PhD
Department of Human Functional Genomics
Life Science Research Center
Mie University
Tsu, Mie, Japan

Part I

General Genomics

1

From Genes to Genomes: A Historical Perspective

Ian Dunham and Don Powell

You may or may not have noticed, but during the past 10 or so years, you have just experienced the genomics revolution. A precise start point is difficult to define, but for argument's sake we can identify the first publication of the genome sequence of a free-living organism (*Haemophilus influenzae*) (Fleischmann et al., 1995) as the harbinger of the flood of genomic information that would appear over the next 10 years. Notably this flood included the first complete sequences of a human chromosome (Dunham et al., 1999), the completed sequence of the human genome (IHGSC, 2004), and draft sequences of the mouse (Waterston et al., 2002) and rat (Gibbs et al., 2004) genomes. After the wealth of whole-genome sequences have come the beginnings of genome re-sequencing as human genomes have been sampled to generate single nucleotide polymorphisms (SNPs) (Sachidanandam et al., 2001) and these SNPs have been used to build maps of the sequence content of different individual human genomes (Altshuler et al., 2005). All of this information is freely available through public domain Web servers such as Ensembl (Hubbard et al., 2005), so any researcher with access to the Internet can interrogate the human genome. It is now hard to imagine a time when genome sequence information was not central to human genetics. However, looking back to the time before genome sequencing was a reality there were genuine questions as to whether we could afford or indeed needed the human genome sequence (Lewin, 1986). Thus it is worth looking back to see how we got to the position we are now in and what were the motivating factors behind the drive toward genome sequencing (Table 1–1).

The Classical Period of Human Genetics

Genetic disorders have been documented in humans since early times. For instance, sculptures found at Ain Ghazal in Jordan show examples of birth defects, and there is evidence in the Talmud for an understanding of the genetic basis of hemophilia (Nielsen, 2003). However, the study of genetics in relation to human traits began in earnest with the rediscovery of Mendel's work. In 1902, Archibald Garrod identified that alkaptonuria appeared to be inherited as a Mendelian trait and that it, along with albinism and cystinuria, was very likely an "individuality of metabolism" (Garrod, 1902), that is, an inborn error of metabolism (Garrod, 1923). Against the background of key developments in the field of genetics from plants, *Drosophila*, bacteria and viruses, Pauling et al. (1949) established that there were significant differences in the biochemical properties of the hemoglobin in normal and sickle cells, and thus that sickle cell anemia was a molecular disease. This was confirmed when Vernon Ingram utilized the new protein sequencing approaches of Fred Sanger to identify a single amino acid difference between normal and sickle cell hemoglobin, establishing the molecular basis of a human genetic disease for the first time (Ingram, 1957). This was a clear demonstration that human disease could be caused by mutation in a single gene, the paradigm that was to dominate study of genetic disease for the next 40 years.

Around the same time, evidence was emerging that larger-scale changes in genetic material could also lead to disease. Lejeune et al. (1959) and Jacobs et al. (1959), working independently, presented the first evidence linking an abnormal chromosome with a human disease by demonstrating an extra chromosome 21 in patients with Down syndrome. The following year Nowell and Hungerford (1960) in Philadelphia observed that chromosome 22 was abnormally short in the malignant cells of chronic myelogenous leukemia patients, showing for the first time an association between a chromosomal rearrangement and human disease. Thus it was clear that chromosomal scale abnormality could also play a role in disease, an observation that was later to play a pivotal role in disease gene mapping.

Over the next two decades the combined efforts of clinicians and cytogeneticists documented many hundreds of diseases with potential genetic causes. In many cases these were relatively rare diseases segregating in families with clear modes of inheritance. In others there might be an association between a chromosomal rearrangement and a disease or a particular type of tumor. These observations were elegantly summarized by Victor McKusick in the compendium *Mendelian Inheritance in Man* (1998). Hence by the time of the recombinant DNA revolution the stage was set for application of the new molecular techniques to identify the genes involved in many of these diseases.

The Advent of Human Molecular Genetics

Arthur Sturtevant, working in T.H. Morgan's laboratory, established the basis for genetic linkage mapping in *Drosophila* in 1913 (Sturtevant, 1913). The technique used genetic crosses to follow segregation of a phenotype with respect to other phenotypes and

Table 1–1 Major Landmarks in the History of Genetics and Genomics

Year	Event	Clinical Loci (No.)[a]	Genetic Map (No.)[b]	Physical Map (No.)[c]	DNA Sequence (Mb)[d]	DNA Sequence (human, Mb)[e]	Longest Contiguous Sequence (kb)[f]	References
1902	Garrod shows alkaptoneuria is a Mendelian trait							Garrod, 1902
1944	Avery, MacLeod, and McCarthy show DNA can transform bacteria							Avery et al., 1944
1952	Hershey and Chase show DNA, not protein, is hereditary material							Hershey and Chase, 1952
1953	Structure of DNA elucidated							Watson and Crick, 1953
1956	Human diploid chromosome complement established as 46							Tjio and Levan, 1956
1959	Lejeune demonstrates trisomy 21 in Down syndrome							Lejeune et al., 1959
1961	Genetic code proposed							Crick et al., 1961
1970	First restriction enzyme isolated							Smith and Wilcox, 1970
1971	First DNA sequence—10 bp from a bacterial virus						0.01	Wu and Taylor, 1971
1973	DNA cloning developed; first physical map (of SV40 virus)						0.05	Cloning: Cohen et al., 1973 SV40: Khoury et al., 1973
1976	First genome sequence—of RNA virus MS2						5.38	Fiers et al., 1976
1977	First DNA genome sequence—ΦX174							Sanger, Coulson, et al., 1977
1977	Dideoxy sequencing developed							Sanger, Nicklen, et al., 1977
1981	Mitochondrial genome						16.6	Anderson et al., 1981
1982	First whole-genome shotgun sequence				0.6		48.5	Sanger et al., 1982
1984	Huntington disease (HD) gene localized; Epstein–Barr virus genome sequenced				3		172.3	HD: Gusella et al., 1983 EBV: Baer et al., 1984
1985	Three proposals to sequence the human genome; polymerase chain reaction (PCR) developed				5			Genome: Watson and Cook-Deegan, 1991 PCR: Saiki et al., 1985
1986					10			
1987	First commercial DNA sequencing machine		403 Genetic markers		15			Donis-Keller et al., 1987
1988					23			
1989	CFTR gene—implicated in cystic fibrosis—isolated				34			Riordan et al., 1989
1990	US Department of Energy and National Institutes of Health present plans for HGP to Congress				49		229.3	HGP: http://www.genome.gov/10001477 CMV: Chee et al., 1990
1991					71			
1992	Sanger Center founded				101		315.3	*Saccharomyces cerevisiae* chromosome 3: Oliver et al., 1992
1993					157			

(Continued)

Table 1–1 Major Landmarks in the History of Genetics and Genomics *(Continued)*

Year	Event	Clinical Loci (No.)[a]	Genetic Map (No.)[b]	Physical Map (No.)[c]	DNA Sequence (Mb)[d]	DNA Sequence (human, Mb)[e]	Longest Contiguous Sequence (kb)[f]	References
1994					217		562.6	*S. cerevisiae* chromosome 8: Johnston et al., 1994
1995	*Haemophilus influenzae*—first genome of a free-living organism; entire map of human genome			15,086 RH/STS; 1 per 199 kb	384		1830	Fleischmann et al., 1995 Hudson et al., 1995
1996	Yeast genome; HGP data to be deposited in public databases 24 h after production		2032 Genethon SSR/ CEPH		651	259	3573	Transcript map: Schuler et al., 1996 *Synechocystis* sp. genome: Kaneko et al., 1996
1997	*Escherichia coli* genome				1160	551	4639	*E. coli* sequence: Blattner et al., 1997
1998	*Caenorhabditis elegans* genome				2008	1084	8774	*C. elegans.* Chromosome 5: *Science* 282: 2011
1998	Refined gene-based map	4325	8031 SSR/ CEPH	30,181				Broman et al., 1998. Deloukas et al., 1998.
1999	Human chromosome 22	5216			3841	2668	23,051	Dunham et al., 1999.
2000	Human draft sequence; Drosophila sequence; microRNA in human genomes; Human chromosome 21	6293			11,101	6702	25,491	Human: Hattori et al., 2000 Drosophila: microRNA: Pasquinelli et al., 2000 Hattori, 2000
2001	Stanford RH MAP; Draft human sequence published	7557		36,678	15,849	7995		Stanford map; Olivier et al., (2001). Lander et al., 2000.
2002	deCODE map; mouse draft genome sequence	8118	5136		28,507	9261		deCODE map: Kong et al., 2002 Mouse genome: Waterston et al., 2002
2003		8784			36,553	10,307	87,410	Human Chromosome 14: Heilig et al., 2003
2004	Finished human genome analysis published; chicken genome published				44,575	10,989		Human: International Human Genome Sequencing Consortium, 2004 Chicken: Hillier et al., 2004
2005	Chimpanzee genome published; HapMap published			1 million SNPS	116,106	11,223		Chimpanzee: The Chimpanzee Sequencing and Analysis Consortium, 2005 HapMap: Altshuler et al., 2005

[a]Number of clinically relevant loci mapped (from OMIM gene map loci). OMIM counts from Internet Archive Wayback Machine (http://web.archive.org/collections/web.html search term http://www.ncbi.nlm.nih.gov/Omim/).

[b]Number of genetic markers mapped.

[c]Number of sequence-based markers mapped.

[d]DNA sequence (Mbp) for all organisms reported by Genbank (http://www.ncbi.nlm.nih.gov/Genbank/genbankstats.html). Includes draft and redundant data (sequences generated more than one time).

[e]DNA sequence (Mbp) for human genome reported by Genbank (ftp://ftp.ncbi.nih.gov/genbank/release.notes/) at last release of each year. Includes draft and redundant data (sequences generated more than one time).

[f]Longest contiguous sequence reported by the Institute for Genomic Research (http://www.tigr.org/tdb/contig_list.shtml) and original papers.

HGP Proposal: http://www.genome.gov/10001477.

allowed the distance between the genes involved to be estimated. Essentially, the more frequently two mutations segregated together the closer they were. The first human gene maps were built using a combination of cytogenetics and somatic cell and biochemical genetics. Linkage analysis could also be performed in families to position disease genes relative to biochemical markers. By the time of the first workshop on human gene mapping, held at Yale in 1973, 83 markers had been placed on the map, although some chromosomes had no markers and some of the assignments made at the workshop later proved incorrect.

The process of placing genes on maps of the human genome was made dramatically easier by the developments in recombinant DNA technology that occurred over the course of the 1970s. Key technologies among the many that appeared at this time included the application of restriction enzymes and plasmid and phage vectors for molecular cloning (Maniatis et al., 1982), the Southern blotting technique, which allowed a single DNA species to be identified within a complex DNA sample (Southern, 1975), and DNA sequencing (Maxam and Gilbert, 1977; Sanger, Nicklen, et al., 1977). The availability of cloned human DNA fragments allowed direct mapping of genes and other markers on to human metaphase chromosomes after radiolabeling by in situ hybridization. Thus the DNA fragments could be positioned by observation of the site of radioactive decay on Giemsa-stained chromosomes. The subsequent refinement of using fluorescently labeled probes (fluorescent in situ hybridization—FISH) greatly facilitated application of this approach. In an alternative approach to building maps, Botstein et al. (1980) realized that the natural variations in DNA sequence between individuals could be detected by application of restriction enzymes and Southern blotting (so-called restriction enzyme fragment length polymorphisms—RFLPs; Fig. 1–1) and that these RFLPs could serve as markers in genetic linkage mapping in the same way as biochemical markers or observable phenotypes. This allowed much more rapid mapping of genes and markers and became the basis for the first extensive maps of the genome (Donis-Keller et al., 1987).

The new tools for genetic linkage mapping in humans allowed a more comprehensive approach to locating the genes involved in human disease. Researchers could apply linkage mapping to the families in which the disease was segregating to associate the disease with the new genetic markers. In some cases this was spectacularly successful, such as the early mapping of the location of the Huntington's disease gene to chromosome 4, band p16.3 (Gusella et al., 1983). Similarly, the cystic fibrosis locus was mapped by linkage analysis to the long arm of chromosome 7, band q31 (Wainwright et al., 1985). In other cases, such as Duchenne muscular dystrophy, cytogenetic abnormalities in patients would also play a role in pinning down the location of the disease gene (Kunkel et al., 1985; Ray et al., 1985). The linkage mapping approach was taken up by many human geneticists, and many disease loci were placed on the map. However, the RFLP approach was laborious and could not provide the density of markers that was needed to refine the location of disease genes to a small enough region to make cloning the gene a simple task. New markers for mapping were required. Alec Jeffreys identified a new class of polymorphism, in which tandem arrays of short sequence varied in number between individuals [variable number of tandem repeats—VNTRs (Fig. 1–1)]. VNTRs could be typed in linkage analysis but still required Southern blotting. However, with the advent of the polymerase chain reaction (PCR) (Saiki et al., 1985) it became possible to use other especially variable sequences as markers. Particularly important were so-called microsatellites or simple sequence repeats consisting of small blocks of 2–10 bp repeated in tandem arrays within our genome. As with VNTRs, these microsatellite blocks can vary in number between individuals, possibly as a consequence of errors in DNA replication. Their advantage over VNTRs is that they are simpler to analyze by PCR and more widely distributed, and microsatellites provided the next stage of linkage mapping (Weissenbach et al., 1992).

Once the location of a disease gene was established with respect to available genetic markers, the task of identifying the gene was not a trivial one. The cystic fibrosis gene was finally identified in 1989 after a heroic effort involving sequential cloning of DNA along the genome in the region of interest by chromosome walking and jumping (Riordan et al., 1989; Rommens et al., 1989). Between 1986 and 1995, more than 50 human disease genes were discovered by pure positional cloning (Strachan and Read, 1996). However, it has been variously estimated that the search for the cystic fibrosis gene required 10 years and up to $50 million, and many researchers began to realize that future gene searches would be greatly assisted if the maps were already in existence and cloned DNA was available on request. This centralization of the mapping activity would considerably improve efficiency. In this respect it is noteworthy that despite the genetic location of the Huntington's disease gene having been identified in 1983, the gene was finally cloned 10 years later after the involvement of more than 100 scientists (Huntington's Disease Collaborative Research Group, 1993). It was these considerations and the coming of truly high-throughput technology that drove the development of the project to sequence the human genome.

The Human Genome Project

Calls for a project to sequence the human genome had begun in the mid-1980s (Watson and Cook-Deegan, 1991). Robert Sinsheimer had conceived of a genome project to raise funds at University of California–Santa Cruz (UCSC) in 1984. In 1985 the Nobel Laureate Renato Dulbecco gave a lecture proposing to sequence the genome to understand the genetic basis of cancer (Dulbecco, 1986). Charles DeLisi, also in 1985, championed the genome sequencing project as a "big science" project within the Department of Energy that would fit with the department's long-standing interest in the genetic lesions caused by radiation. By 1988 sufficient political and scientific momentum had been gathered for money to be made available from the US government, and the United States took the lead in proposing and developing funding to set up the systems that would be needed to underpin an assault on the human genome. In 1989 the National Center for Human Genome Research (now the National Human Genome Research Institute) was established at the National Institutes of Health (NIH) to coordinate a distributed effort. Goals (reduction in sequence costs, maps of DNA clones ready to sequence, proof of principle using model organisms with smaller genomes) were established that were hugely challenging (National Human Genome Research Institute, 1990). At the beginning a key objective was to establish the tools by which a later assault on the genome sequence could be mounted.

Improving Genome Maps

The genetic maps that had been developed for disease gene mapping provided a framework for study of our genome, but the process was very demanding and uneven, and at the beginning of the 1990s the markers were too sparse to access the underlying DNA with ease. There was a real need for methods to localize markers rapidly onto chromosomes or sequences. The map of the human genome had too

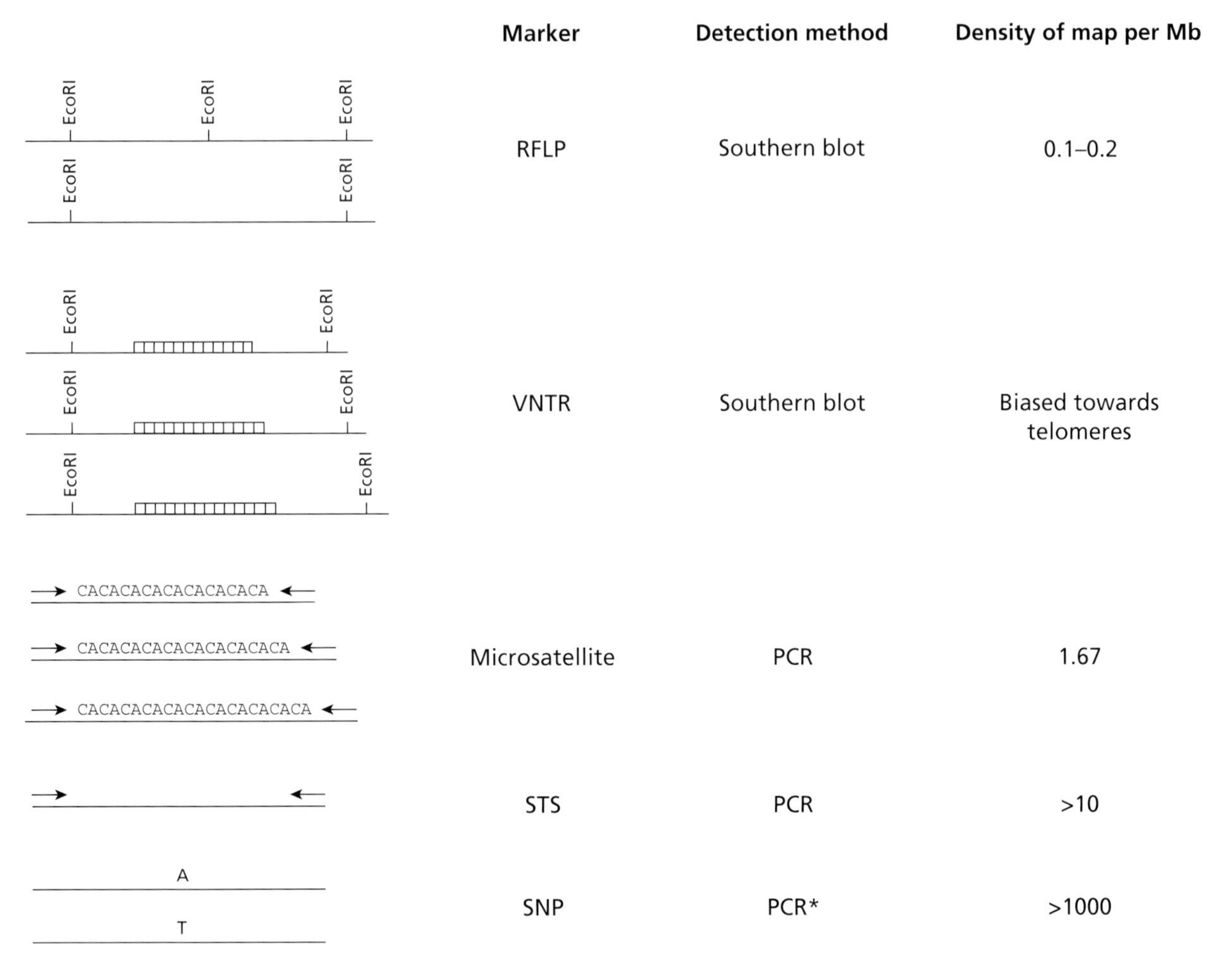

Figure 1–1 Types of marker used for genetic and physical maps of the human genome and their effective map densities. Horizontal lines indicate segments of DNA that are assayed with the particular marker class. Where more than one line is shown, the marker is polymorphic between genomes; that is, two or more alleles are shown. For restriction fragment length polymorphisms (RFLPs) the vertical ticks show the location of restriction enzyme sites (in this case for *Eco*RI) that may be lost or gained by sequence changes and thus can be recognized as polymorphic by digestion with the appropriate restriction enzyme. RFLPs usually have only two alleles. For variable numbers of tandem repeats (VNTRs) or minisatellites the open boxes indicate the occurrence of different numbers of repeated (tandem) copies of a core DNA sequence (16–72 bp) that is variable between alleles. Many alleles may be present in the population. For microsatellites, also known as simple sequence length polymorphisms, simple sequence repeats, or short tandem repeats, the variable short tandem repeat of 1–7 bp is indicated by the run of CA bases. Microsatellites usually have multiple alleles. The CA (or GT) dinucleotide repeat is the most common form in the human genome, although other tri- and tetra-nucleotide repeats are also suitable for use as genetic markers. For all polymerase chain reaction (PCR) amplifiable markers arrows indicate oligonucleotide PCR primers used to amplify the specific segment of interest. For single nucleotide polymorphisms (SNPs) the single nucleotide variable between alleles is indicated, in this case either A or T. Most SNPs are biallelic. SNPs are assayed by a number of approaches, most of which use PCR to some degree. For the map densities, the density of markers per Mb is given for a representative map as follows: RFLPs (Donis-Keller et al., 1987), microsatellites (Kong et al., 2002), sequence-tagged sites (Deloukas et al., 1998), SNPs (Altshuler et al., 2005).

few of the interesting places marked on it and too many empty regions. Several emerging molecular biological techniques provided the means to create higher-density maps Fig. 1–1). The development of microsatellite genetic markers that were assayable by PCR provided a much more efficient means to build dense genetic maps (Weissenbach et al., 1992). Other sequences could also be assigned to chromosomes to increase the density of markers available. Researchers around the world had isolated sections of complementary DNA (cDNA), either through cloning projects or in producing expressed sequence tags (ESTs), short regions of messenger RNA (mRNA) passed through high-throughput sequence analysis. Markers could be assigned locations either by association with known markers using radiation hybrid analysis (Schuler et al., 1996) or by hybridization of fluorescently labeled sequences to chromosomes (FISH). By 1998 more than 30,000 markers had been placed on the genome (Deloukas et al., 1998).

Despite the generation of high-resolution marker maps, a further step was needed before high-throughput sequencing of the genome would be possible. The DNA sequencing methodology required access to the genomic DNA in a suitable cloned form; that is, the genomic DNA between the markers would have to be cloned and organized in an additional map to make it accessible. The clone resources used by the HumaN Genome Project (HGP) have reflected the most efficient methods available at each stage. Early projects,

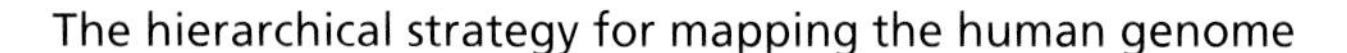

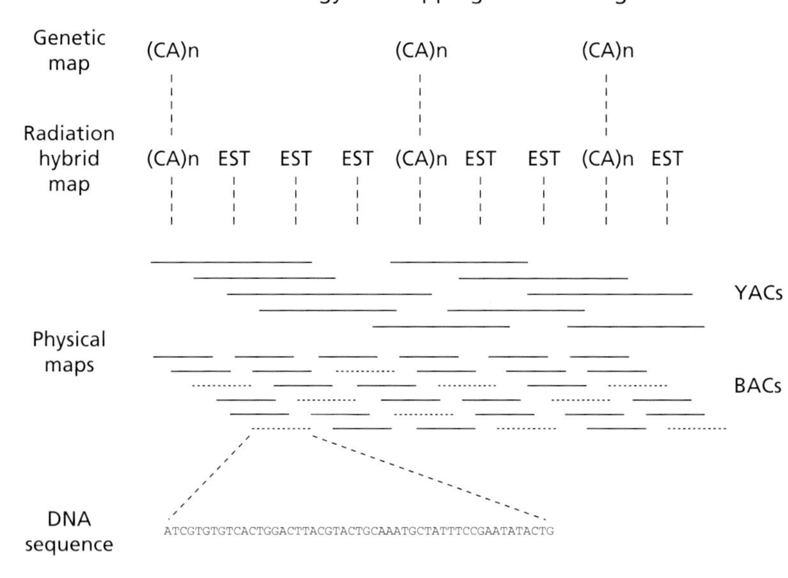

Figure 1–2 The hierarchical strategy for mapping the human genome. The human genome was mapped using a series of maps of increasing resolution to establish a framework organization of the genome (so-called top-down approach). Bacterial artificial chromosome (BAC) contigs assembled by bottom-up restriction enzyme fingerprinting methods could be assigned to landmarks in the framework to position them. As BAC contigs were built and extended, individual BAC clones sufficient to represent the underlying DNA were selected for sequencing (the tile path—dotted lines). EST, expressed sequence tag; YAC, yeast artificial chromosome.

such as the mapping of *Caenorhabditis elegans*, used libraries of cosmids, which hold approximately 40–50 kb. For the human genome, the small size of cosmids was too limiting. Early proposals sought to use yeast artificial chromosomes (YACs), which could hold almost 1 Mb of inserted DNA. In the event, YACs proved to be too unstable for many sequencing applications, rearranging or losing inserted DNA sequences, and the HGP settled on libraries of bacterial artificial chromosomes (BACs). BACs can accommodate 100–200 kb (typically approximately 140 kb) and are very stable. The intermediate goal of the HGP was to map BACs onto the human genome using the genetic and radiation hybrid maps as a scaffold to orient the map (Fig. 1–2).

The array of markers in these maps could be matched to each BAC by hybridization. Additional mapping was carried out by a restriction enzyme fingerprinting method. The BACs were digested with restriction enzymes and the fragments separated on gels, all using standardized techniques. Where BACs overlapped on the genome sequence, they would share fragments. To cover the human genome requires many thousands of clones, and the mapping was an enormous task. But automation and high-throughput labs accelerated to provide a set of BAC clones for the entire genome by 2000. The genome coverage was provided by 354,510 mapped BACs; from these a tiling path (Fig. 1–3) of 29,298 clones was eventually selected for sequencing (McPherson et al., 2001).

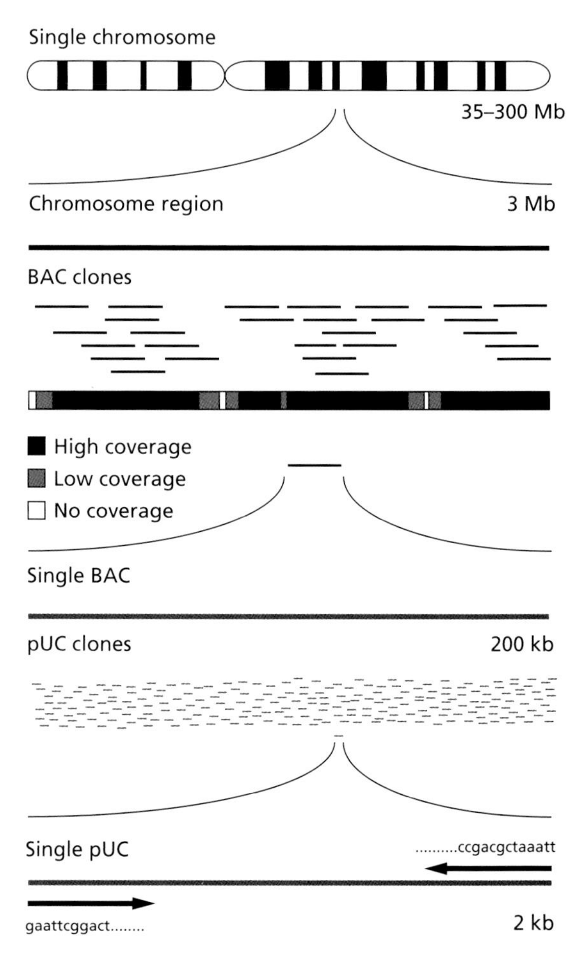

Figure 1–3 Bacterial artificial chromosome (BAC)–based DNA sequencing used by the International Human Genome Sequencing Consortium. See text for explanation.

The Development of Sequencing Technology

With the generation of maps and clones well in hand, the next step toward determining the genome sequence was the application of high-throughput sequencing methodologies. To understand the advances that were necessary to achieve this we need to examine the historical development of the sequencing method (Fig. 1–4). In the early 1970s, several groups approached the problem of how to determine the base sequence of DNA. Probably the most successful was that of Fred Sanger at the Laboratory of Molecular Biology, Cambridge, UK. Sanger had won a Nobel Prize in 1958 for developing methods to determine the sequence of the protein insulin. His group then devoted themselves to sequencing DNA: the Holy Grail of molecular biology. The first efforts were to determine the sequences of genomes that were tractable, such as viruses, which

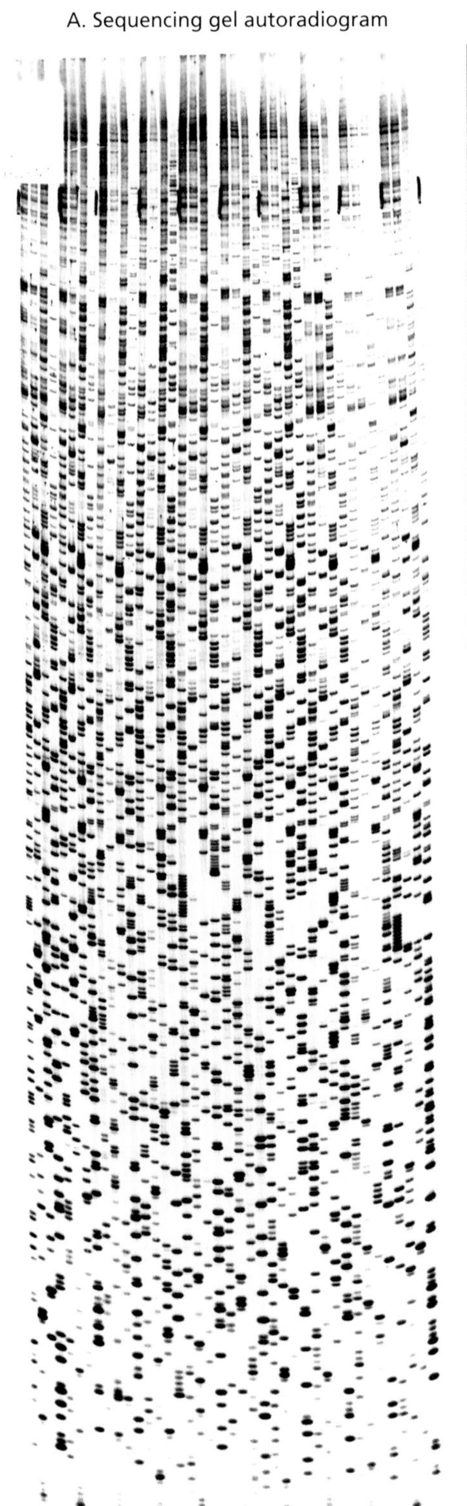

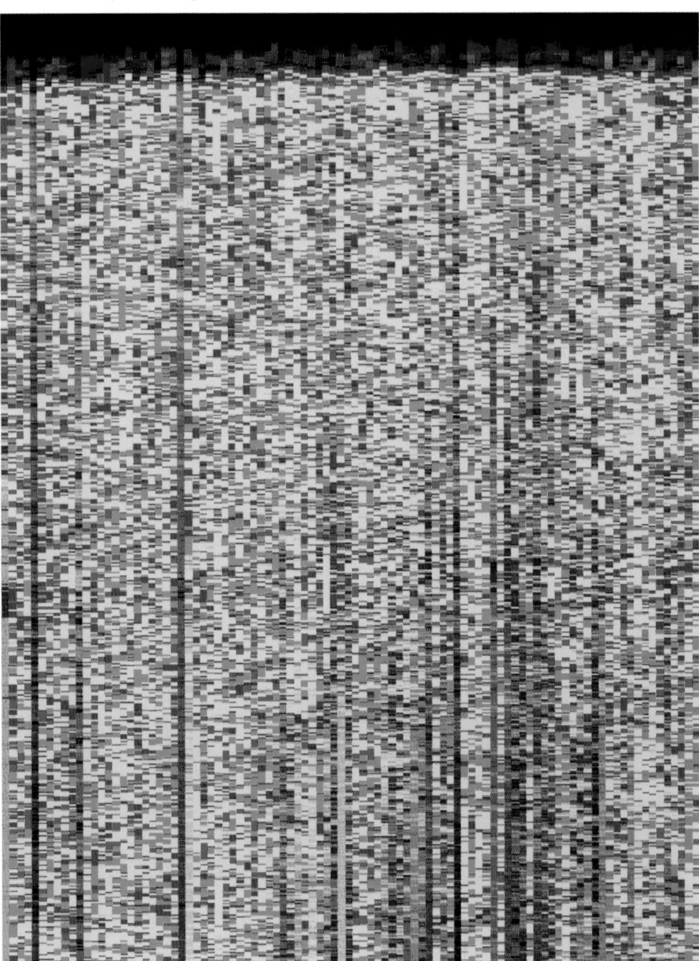

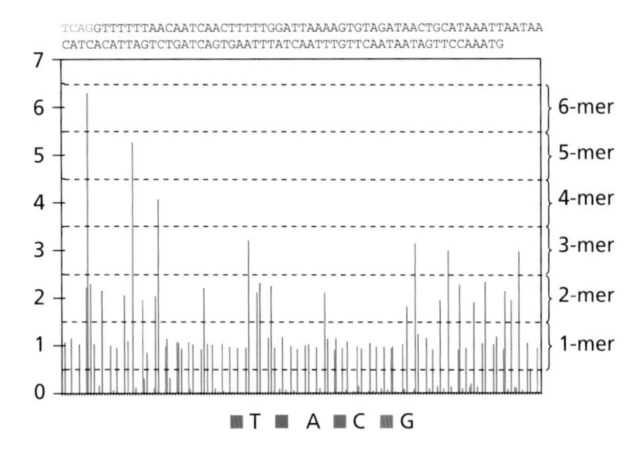

Figure 1–4 Evolution of DNA sequencing methodology. The figure illustrates the evolution of sequencing methodologies by the changing output from the DNA sequencing experiment. (*A*) Autoradiogram of a dideoxy sequencing gel. Four ^{35}S-labeled sequence reactions are prepared for each sample, one for each base. There are 10 reactions on this gel, each with approximately 250 bases, for a total of 2500 bases. Each gel and exposure takes approximately 3 days. (*B*) Output of an ABI3730 DNA sequencing machine. The 3730 analyzes 96 samples (each row in the figure is one sample) with the four dideoxynucleotides in the sequencing reaction labeled with a different fluorescent color. The display shown has approximately 500 bases in each row, for a total of 48,000 bases. Each run takes approximately 2 hours. (*C*) Output from a 454 Life Sciences sequencer. In this technology, chemiluminescence is measured as each base is synthesized on template strands in the presence of enzymes and luciferin. The sequencing is done in a massively parallel fashion and can generate 25 million bases in 4 hours (Margulies et al., 2005). (Credits: *A*, courtesy of Dr. Bart Barrell; *B*, Wellcome Trust Sanger Institute; *C*, Reprinted by permission from Macmillan Publishers Ltd.: Margulies et al. (2005), *Nature* 437: 376–380, copyright 2005.)

consisted of only a few thousand bases or base pairs. Sanger's early attempts to derive DNA sequence were founded on the principles used to define protein structure and RNA sequence. He had been hugely successful in the 1960s in sequencing RNA molecules. Indeed, the first genome sequenced was an RNA virus, MS2. His early methods used chromatography of partially degraded and synthesized DNA molecules. The crucial developments toward the final method of DNA sequencing were three. First, thin polyacrylamide gels were developed that gave good resolution of DNA fragments. Second, restriction enzymes allowed researchers to produce discretely sized fragments of larger DNA molecules. Third, and most important, was the production of dideoxynucleotides, which caused termination in a DNA synthesis reaction. Sanger, Nicklen, and Coulson (1977) published their dideoxy method in 1977. Almost 30 years later, essentially the same method provides the major output of sequencing centers, including the 60 million bases each day generated at the Wellcome Trust Sanger Institute.

To begin to attack genomes of greater complexity, improvements were needed in the methods of producing suitable samples, in the efficiency of the sequence reaction, in data input, and in software to analyze the data. For genomes larger than a few thousand base pairs, it remained very difficult to imagine how such long tracts of sequence could be broken up and managed. In 1981, Sanger and his colleagues produced the sequence of the human mitochondrial genome, some 16,000 bp (Anderson et al., 1981). In 1982, Sanger developed a procedure called the whole genome shotgun (WGS), in which an entire genome is broken into pieces, the sequence of each piece is acquired at random, and software (and the human brain and eyes) is used to assemble the overlapping sequence strings into one contiguous whole. Remarkably, Sanger's group sequenced the 48,000 bp of bacteriophage λ using this method (Sanger et al., 1982). Thus at the time that proposals to determine the 3 billion bp sequence of the human genome were being initiated, the longest DNA sequence determined was 172,281 bp for Epstein–Barr virus and the *total* length of sequences in GenBank was less than 50 million bp (Fig. 1–5). Each base cost approximately US$5–10 to produce, and a postdoctoral researcher might produce 2000 bases each day. The HGP was to be a leap in the dark.

The Move to Production DNA Sequencing

Even at the scale of the bacteriophage λ genome, sequencing was very much a cottage industry, in which four reactions (one for each base) incorporating radioactivity were carried out for each sample, and the reaction products were loaded by hand on to an acrylamide gel and visualized as exposed images on an autoradiogram of the gel after electrophoresis (Fig. 1–4). Each gel comprised perhaps 32 or 48 lanes (8–12 samples) and each run provided perhaps 200 bases, for a total of 1600–2400 bases. The whole process took several hours over a period of several days. For the genome project to succeed, the sequencing process needed to be industrialized. Two vital developments ensued. The first was to label the DNA not with radioisotope but with four differently colored fluorescent tags, one for each base (Smith et al., 1986). The four reactions could then be combined into one sample. Improvements in the chemistry of the fluorochromes meant the sensitivity grew rapidly. The second development was to separate the DNA sequence products not on a rectangular slab gel, but through a gel-containing capillary tube. Small-diameter tubes gave high resolution and prevented crossover from one sample to adjacent samples. The gel was a viscous polymer and, at the end of the run, the capillaries could be washed out and fresh polymer loaded. The output is collected as traces of fluorescence (Fig. 1–4).

By 1997, capillary machines that ran 96 samples at one time were available from several manufacturers. With an average of more than 400 bases per sample, they generated at least 38,400 bases per run. With eight runs per day, the potential output was more than 300,000 bases per machine per day. The advantages of capillary-based machines led to rapid replacement of gel-based protocols, especially in the larger sequencing centers. The impact of these advances on the pace of sequence collection is illustrated in Figure 1–5.

To move from genomes of 100,000 bp or so to the 3 billion bp of mammalian genomes was deemed to be too large a step in 1990. As mentioned above, the total length of sequences in GenBank at this time amounted to approximately 50 Mb from both genomic DNA and cDNA *of all species*. The HGP would move from bacterial and yeast genomes (less than 15 Mb) to the nematode *C. elegans* (approximately 100 million bp) to the human. The lessons in mapping, producing clone resources, assembling sequences, dealing with repetitive DNA sequences, and managing data and distributed results would be applied to the key target, the human genome sequence.

Generating High-Quality Sequences

The procedure for deriving sequences from BAC clones is illustrated in Figure 1–3. Each BAC is broken into fragments of 2000–4000 bp and cloned into a pUC vector. These pUC subclones are the resource for sequencing. Randomly selected clones are sequenced to generate predicted *sequence coverage* of 4 $\times$ —each base should on average be sequenced four times. This is the draft sequence of a genome. Typically it should produce more than 90% of the regions of the genome amenable to dideoxy sequencing and assembly to an accuracy of better than 99.9%.

The problem for large genomes—from bacterial to human—is that not all genomic regions are generated by this approach. Some regions are cloned poorly and underrepresented (clone gaps). Some regions sequence poorly, producing misreads, missing bases, or regions that will not find an overlap (sequence gaps). To produce a human genome sequence that gives clinical research an accurate and complete resource requires both high-quality maps and attention to the detail of the sequence in a process called sequence finishing. Finishing is the use of targeted biochemical reactions to close the gaps and to reduce the number of ambiguous bases. It is an intensive and demanding stage that requires skilled research teams, but it remains the only method to produce high-quality sequences. Clone gaps can be closed using alternative cloning methods or using PCR-based reactions to "bridge" the missing region. Sequence gaps can also be closed using PCR or, in some cases, adjusting the standardized high-throughput sequencing protocols.

The results of this process were first seen for a human chromosome with the publication in 1999 of the sequence of chromosome 22 (Dunham et al., 1999). Chromosome 22 in many ways typifies the problems of sequencing the human genome. Although it is relatively small, it contains large regions that are not amenable to dideoxy sequencing and assembly. These constitute at least 15 Mb and are largely repetitive sequences found on the short arm and around the tips of the chromosomes (telomeres) and the centromere (the region at the "waist" of the chromosome that plays a role in chromosome segregation during cell division). The region that was sequenced was approximately 33.4 Mb and contained 11 gaps that totaled 1 Mb. But the entire sequence of a chromosome gave us something much more—a first glimpse of the landscape of our genome, a view hidden from us before this. It told us the distribution of genes was uneven, that repetitive sequences were frequent but had characteristic patterns of distribution, and that large-scale duplications were much

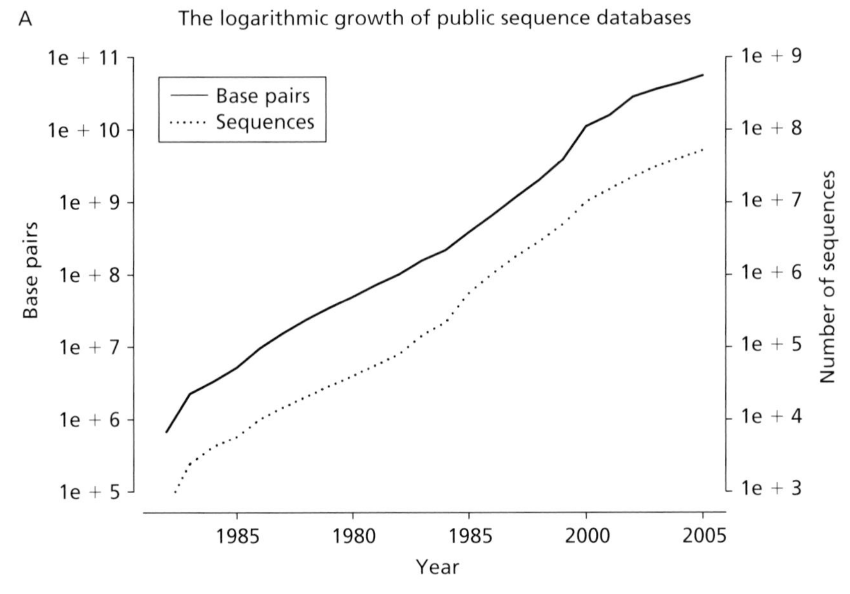

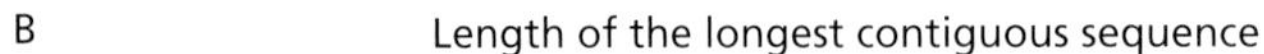

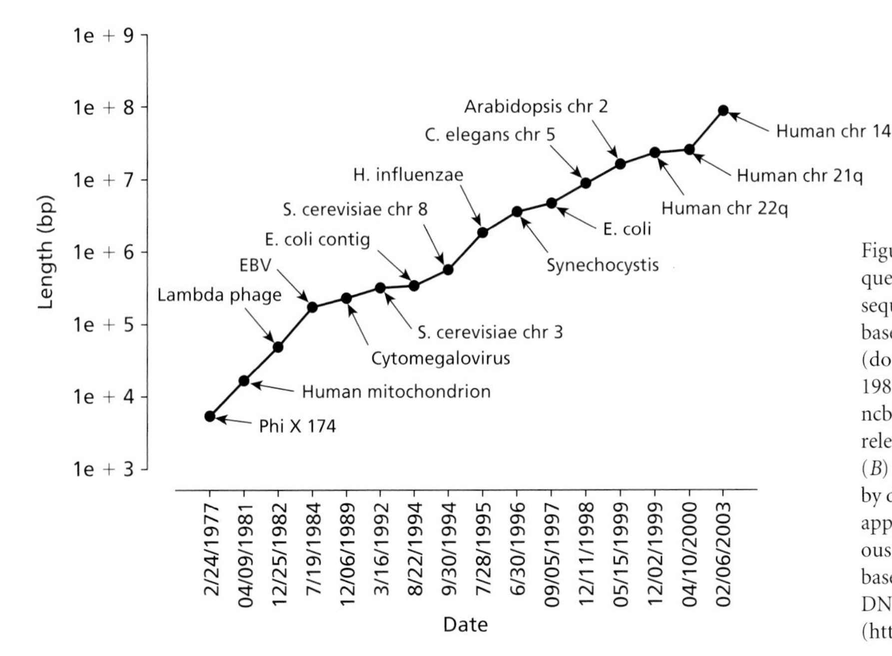

Figure 1–5 (*A*) The logarithmic growth of public sequence databases. The graph shows the growth in DNA sequences deposited in the public DNA sequence databases (GenBank/EMBL/DDBJ) by number of entries (dotted line) and total base pairs (solid line) between 1982 and the end of 2005. Data taken from http://www.ncbi.nlm.nih.gov/Genbank/genbankstats.html and the release notes for NCBI-GenBank Flat File Release 151.0. (*B*) The size of the largest contiguous finished sequence by date of sequence submission. The graph shows the approximately logarithmic increase in size of contiguous finished sequences available in the sequence databases. Data taken from "World's Longest Contiguous DNA Sequence" at the Institute of Genomic Research (http://www.tigr.org/tdb/contig_list.shtml).

more prevalent than previously thought. It was clear that much more remained to be discovered in the rest of the genome. The sequence of chromosome 22 led the way for new discoveries in genome biology.

The sequence of chromosome 21 followed in May 2000 (Hattori et al., 2000), and while the teams of the HGP sought to produce finished sequence of all chromosomes, an analysis of the draft of the human genome sequence was published in *Nature* in February 2001 (IHGSC, 2001). The draft human genome published in *Nature* in 2001 contained approximately 150,000 gaps. Although it was an intermediate resource and all the sequences and analyses had been freely available during production, the draft sequence paper is a remarkable testament to international cooperation, to the support of funding agencies, and to how limited our understanding of *genome* biology was before this.

The sequence of chromosome 22 had suggested that the complement of human genes might be less than half the widely touted value of 100,000: the draft sequence suggested a value nearer 30,000. The draft sequence also described the remarkable content of repetitive sequences in our genome and outlined the spread and, in many cases, "death" of movable DNA sequences. The overview paper and accompanying individual papers, which occupied an entire issue of *Nature*, told us how much we had learnt and how much remained for us to learn, to understand, and to use.

In 2004, an overview of the finished human genome sequence—consisting of 99% of euchromatic regions to an accuracy of better than 99.999%—was published (IHGSC, 2004). The finished sequence contained only 351 gaps, an improvement of 400-fold over the draft sequence. Annotation (see below) was markedly improved because the high-quality sequencing helped the teams to remove some ambiguities and to combine some gene predictions that were separated (by gaps) in the draft.

In 2006, the analysis papers for the last human chromosomes were published. As with the first human chromosome, the largest, chromosome 1, was sequenced by teams led by the Sanger Institute (Gregory et al., 2006). Chromosome 1 represents some 8% of the human genome, contains more than 3100 genes, and contains more than 350 mapped Mendelian disorders. With finished sequences for all human chromosomes to hand, the tasks are to translate those data and emerging knowledge into biological and biomedical advances. The first step in generating knowledge from genome data is to annotate the sequence. Many other drafts of large genomes (produced by WGS sequencing) have been produced and a smaller number, including the mouse and the zebrafish, will be finished to a standard similar to that of the human genome. These genome sequences play a vital role in annotation of the human genome (see below) as well as supporting biomedical research using these important model systems.

Making Information from Data—Databases and Annotation

Completion of the human genome sequence was a magnificent achievement by a dedicated group of genome researchers worldwide. The very act of sequencing the human genome also had the effect of moving attention from the study of single genes toward wider horizons. Genome sequence information combined with new technologies such as DNA microarrays suddenly allowed experiments to be performed with a genome-wide perspective. In addition, overshadowed in earlier times by the success with rare monogenic diseases, the complex multifactorial nature of common disease suddenly came back into focus. Almost as soon as the genome was sequenced it was clear that much more needed to be done.

Before these new horizons could be explored, it was also clear that the sequence needed to be interpreted and made accessible for researchers. DNA sequencing is not an end in itself. Biologists and medical researchers sequence DNA because it can provide baseline information about gene structure and a route to understanding biology and disease. For medical research, the interest lies largely in genes and the regions of the genome that control genetic activity. However, finding genes by sequence analysis alone is not a simple task (Hubbard et al., 2002). The signals that define the boundaries of genes are, for the most part, "consensus" sequences and are not founded in absolutely predictable base sequences that can be searched for by computer. Moreover, mature mRNA sequences in complex organisms are often derived from non-contiguous sequences in the genome. Hence, a protein of 600 amino acid residues, which could be encoded by an mRNA of 2000 nucleotides, might be specified by a gene that is 10-fold or more larger and split into 10–20 exons. Finding these snippets of a gene is tremendously demanding. Quality annotation requires more experiments and dedicated and trained researchers.

Three resources could be used to help annotation of the human genome (Fig. 1–6). First, the publicly available databases, such as GenBank, European Molecular Biology Laboratory (EMBL), and DNA Databank of Japan (DDBJ), contain many thousands of (mostly) single genes from organisms sequenced over the past 20 years. Second, several groups, including one funded by the HGP, had cloned and sequenced mRNAs in a high-throughput, random fashion to generate ESTs. As ESTs are derived from mRNA, they would

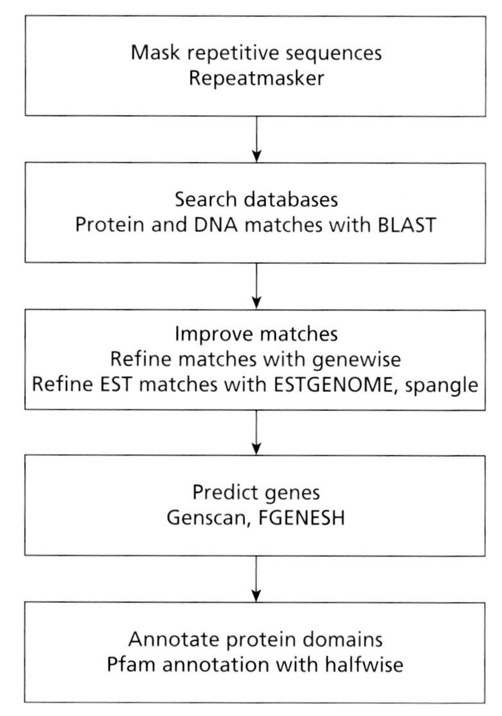

Figure 1–6 Automated gene prediction pipeline for Ensembl. An example workflow of annotation for genome sequences. Repetitive sequences are first removed from assembled DNA sequences. Gene prediction algorithms are not normally used as a first step for genomes where there are sufficient messenger RNA (mRNA) or expressed sequence tag (EST) data. However, they may be used as a final step to annotate regions of the genome where there is no mRNA or EST evidence: predicted exons are accepted only if they are supported by other experimental evidence, such as protein matches. For some genomes that lack significant amounts of mRNA or EST data, a larger proportion of the genes are built using gene predictors as a starting point. Although automated annotation provides rapid output, "curation" of those predictions still must be carried out. Accurate annotation, supported by complementary DNA evidence and inspection of each gene prediction, has been performed on the majority of human chromosomes and is contained in the Vertebrate Genome Annotation (VEGA) database: http://vega.sanger.ac.uk/index.html.

highlight expressed regions of the human genome. Third, whole genome sequences—especially the draft sequence of the mouse genome, produced in 2002 (Waterston et al., 2002)—could be compared directly with the human genome sequence to identify conserved regions including many genes. The comparison of the draft mouse genome sequence with the human led to the identification of 1200 genes in the human that had not been found by other methods.

The comparison of genome sequences (comparative genomics) can thus yield a rich harvest of understanding of our genome. Comparison of the human and chicken genomes [separated by 310 million years (My) of evolution] showed sequences were conserved within regions previously thought to be junk, suggesting they have an important role (Hillier et al., 2004). The chicken genome has also helped us identify genes that do not code for protein, but are transcribed into an "active" RNA. Such non-coding RNAs were first described in the nematode worm, *C. elegans* (Lee et al., 1993), but the number encoded in the genomes of complex organisms has been fully appreciated only in the past 5 years. These RNAs are now finding increasing importance in the study of cell biology and diseases such as cancer (He et al., 2005; Lu et al., 2005). In contrast, comparison of the chimpanzee and human genomes identifies regions that might be essential in the divergent paths of the two species during the past 6 My and that might throw light on interspecies differences such as human speech (Cheng et al., 2005; CSAC, 2005).

The logarithmic growth of genome sequence data (Fig. 1–5) and the increasing complexity of analysis have made data provision much more difficult. Although individual sequences are small files, the raw data are the images of traces of fluorescence from the sequencing machines and the interpretation of the sequence data through the annotation of gene structures, other features, and the overlaying of variations in sequence. All these produce enormous databases. For example, the raw sequence records (traces from sequencing machines) held at the Wellcome Trust Sanger Institute constitute one of the ten largest Unix databases in the world. Today, three main databases hold mirrored copies of all publicly available sequence data of large genomes: the US National Centre for Biotechnology Information (NCBI: http://www.ncbi.nlm.nih.gov/), the Ensembl server, co-directed by the Wellcome Trust Sanger Institute and the European Bioinformatics Institute (http://www.ensembl.org/index.html), and the DDBJ (http://www.ddbj.nig.ac.jp/). But the data are of use only if tools exist to analyze and search the sequences.

Genome "browsers," including Ensembl, the entry points to NCBI (such as Map Viewer), and the UCSC browser (http://genome.ucsc.edu/), provide such tools. Each can be customized to include as much or as little information in the display as required. These resources have grown swiftly in response to user demands and on the basis of a changing and improving set of genome sequences. For example, Ensembl grew from a few pages of genes predicted in 1998 from the unfinished human sequence (Fig. 1–7). By 1999, a scheme had been developed for automated annotation of the emerging human sequence and the database and browser were launched in January 2000 (Hubbard et al., 2002, 2005). By January 2006, the Ensembl site included 24 large genomes, annotation of genes, maps of putative regulatory regions and of variation, and links to OMIM (Online Mendelian Inheritance in Man: http://www.ncbi.nlm.nih.gov/omim/), as well as providing mechanisms for individual users to add their own data and make them available. In the week of January 15–22, 2006, the Ensembl site registered nearly 850,000 page impressions and downloads totaling 64 GB (Fig. 1–7). Thus the aim of sequencing the human genome was to provide the information to make biological and medical advances. This aim has now been realized, and the freely available tools and resources developed through the HGP are heralding completely new research avenues that were unimaginable a decade—and in some cases only 1 year—ago.

Conclusions

We have come a long way since the first suggestions of determining the human genome sequence. The first human genome sequence was, by the nature of its construction, a composite, representing at

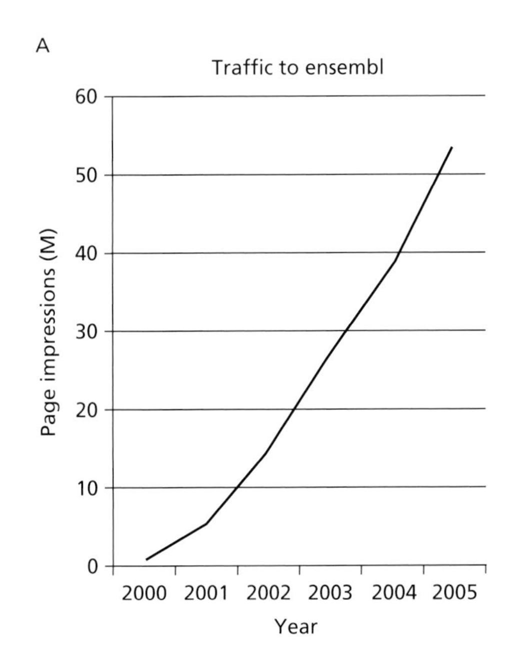

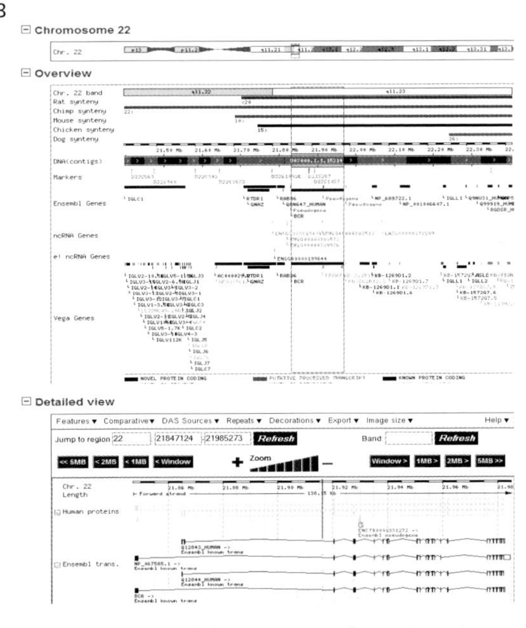

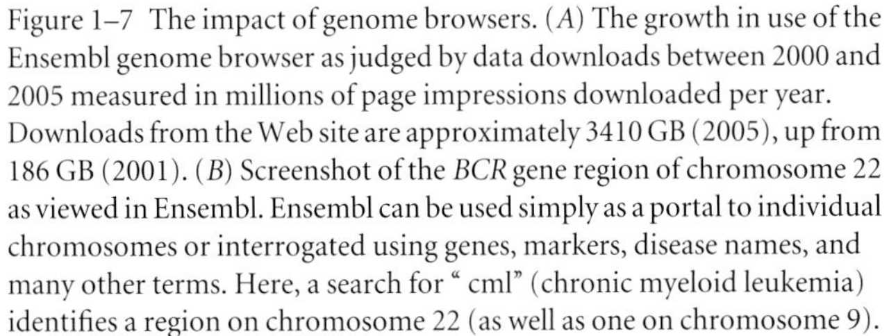

Figure 1–7 The impact of genome browsers. (*A*) The growth in use of the Ensembl genome browser as judged by data downloads between 2000 and 2005 measured in millions of page impressions downloaded per year. Downloads from the Web site are approximately 3410 GB (2005), up from 186 GB (2001). (*B*) Screenshot of the *BCR* gene region of chromosome 22 as viewed in Ensembl. Ensembl can be used simply as a portal to individual chromosomes or interrogated using genes, markers, disease names, and many other terms. Here, a search for " cml" (chronic myeloid leukemia) identifies a region on chromosome 22 (as well as one on chromosome 9). The display shows ("Chromosome 22," at top) an ideogram of chromosome 22, with the region of interest boxed. In the "Overview" (center panel), the relevant gene, *BCR*, is highlighted. The outline box indicates the view shown in the "Detailed view" (bottom panel), which gives detail of the *BCR* gene structure. An additional panel (indicted by the red line, not shown) gives the sequence of the *BCR* gene. The view can be customized to display only the features or evidence the researcher requires. In this example, synteny with other species, markers, and two categories of genes are displayed.

the same time the genomes of every human, but not of any one individual. Subsequent work has confirmed what was previously suspected, that each individual genome contains base differences at a rate of approximately 1 base in 1000 (Sachidanandam et al., 2001). However, these differences occur together in a limited number of combinations (or haplotypes), and the first maps of the genome incorporating the haplotypes have now been completed (Altshuler et al., 2005). The genome is more complex still in that both small (Dawson et al., 2001) and large insertions or deletions are common and an appreciable number of copy number changes are also present (Iafrate et al., 2004; Sebat et al., 2004). To fully explore the complexities of our genomes we need more genome sequences, and technology is now becoming available that might enable this to be a reality within just a few years (Margulies et al., 2005; Shendure et al., 2005).

But is there a need for more genome sequences? Human geneticists had originally envisaged the genome sequence as a tool with which to capture their favorite disease genes. This has come to pass, and many were cloned either as a by-product of the mapping and sequencing effort or subsequent to completion. Other rare disease genes remain to be mapped, and each case gives us exquisite insight into the function of single genes. But those studying complex disease where many genes may be involved as well as environmental components have found the path less easy. Realistically, only a handful of true "common disease genes" have been identified. In part, this may be due to the limited power of some of the early study designs, which is now being remedied with larger collections (http://www.wtccc.org.uk/; WT media release http://www.wellcome.ac.uk/doc_WTX026809.html), perhaps up to whole national or pan-national collections (e.g., UK Biobank: http://www.ukbiobank.ac.uk/). However, researchers now also argue that they need a deep understanding of the patterns of polymorphism in their cases and controls to unpick the complex genetic effects that may be present. Despite the advent of new high-throughput genotyping methods the only way to obtain this is ultimately to re-sequence the genome in each case or control.

The genome information has also changed our approach to studying biology. Comparative analysis (comparing different genome sequences) can now bring the power of evolution to bear on many aspects of the genome. New perspectives on scale and new technologies have opened up new avenues for exploration. For instance, we have now moved from the world where individual labs study a single gene or protein in great detail to one where single experiments can interrogate the whole genome. Similarly, that information can be made instantaneously available to the world research community over the Internet for further study. In the past, we thought of pathways of proteins interacting in a linear way in a defined process, a paradigm that originated in the study of metabolism. Now we can start to think of complex webs of proteins interacting with several or many partners within the cell (Vidal, 2001). Similarly, we can build up many layers of information from the expression level of an mRNA, through the location of its protein and its interaction partners, to the functional consequences of a knockout or knockdown. And again, by providing the information through databases that talk to each other, researchers are free to move from gene to gene. We have truly now gone from gene to genome.

Acknowledgments

The authors would like to acknowledge the efforts of scientists worldwide who have made the human genome sequence a reality. We would also like to thank Andrew Mungall for comments on the manuscript.

References

Altshuler, D, Brooks, LD, Chakravarti, A, Collins, FS, et al. (2005). A haplotype map of the human genome. *Nature*, 437(7063):1299–1320.

Anderson, S, Bankier, AT, Barrell, BG, de Bruijn, MH, et al. (1981). Sequence and organization of the human mitochondrial genome. *Nature*, 290 (5806):457–465.

Avery, T, MacLeod, CM, and McCarty, M (1944). Studies on the chemical nature of the substance inducing transformation of pneumococcal types: induction of transformation by a desoxyribonucleic acid fraction isolated from pneumococcus type III. *J Exp Med*, 79:137–158.

Baer, R, Bankier, AT, Biggin, MD, Deininger, PL, et al. (1984). DNA sequence and expression of the B95-8 Epstein–Barr virus genome. *Nature*, 310:207–211.

Blattner, FR, Plunkett, G III, Bloch, CA, Perna, NT, et al. (1997). The complete genome sequence of *Escherichia coli* K-12. *Science*, 277:1453–1474.

Botstein, D, White, RL, Skolnick, M, and Davis, RW (1980). Construction of a genetic linkage map in man using restriction fragment length polymorphisms. *Am J Hum Genet*, 32(3):314–331.

Broman, KW, Murray, JC, Sheffield, VC, White, RL, et al. (1998). Comprehensive human genetic maps: individual and sex-specific variation in recombination. *Am J Hum Genet*, 63:861–869.

Cheng, Z, Ventura, M, She, X, Khaitovich, P, et al. (2005). A genome-wide comparison of recent chimpanzee and human segmental duplications. *Nature*, 437(7055):88–93.

Chee, MS, Bankier, AT, Beck, S, Bohni, R, et al. (1990). Analysis of the protein-coding content of the sequence of human cytomegalovirus strain AD169. *Curr Top Microbiol Immunol*, 154:125–169.

Cloning: Cohen, SN, Chang, AC, Boyer, HW, Helling, RB. (1973). Construction of biologically functional bacterial plasmids in vitro. *Proc Natl Acad Sci USA*, 70:3240–3244.

Crick, FH, Barnett, L, Brenner, S, and Watts-Tobin, RJ (1961). General nature of the genetic code for proteins. *Nature*, 192:1227–1232.

CSAC (2005). Initial sequence of the chimpanzee genome and comparison with the human genome. *Nature*, 437(7055):69–87.

Dawson, E, Chen, Y, Hunt, S, Smink, LJ, et al. (2001). A SNP resource for human chromosome 22: extracting dense clusters of SNPs from the genomic sequence. *Genome Res*, 11(1):170–178.

Deloukas, P, Schuler, GD, Gyapay, G, Beasley, EM, et al. (1998). A physical map of 30,000 human genes. *Science*, 282(5389):744–746.

Donis-Keller, H, Green, P, Helms, C, Cartinhour, S, et al. (1987). A genetic linkage map of the human genome. *Cell*, 51(2):319–337.

Dulbecco, R (1986). A turning point in cancer research: sequencing the human genome. *Science*, 231(4742):1055–1056.

Dunham, I, Hunt, AR, Collins, JE, Bruskiewich, R, et al. (1999). The DNA sequence of human chromosome 22. *Nature*, 402(6761):489–495.

Fleischmann, RD, Adams, MD, White, O, Clayton, RA, et al. (1995). Whole-genome random sequencing and assembly of Haemophilus influenzae Rd. *Science*, 269(5223):496–512.

Fiers, WR, Contreras, R, Duerinck, F, Haegeman, G, et al. (1976). Complete nucleotide sequence of bacteriophage MS2 RNA: primary and secondary structure of the replicase gene. *Nature*, 260:500–507

Garrod, AE (1902). The incidence of alkaptonuria: a study in chemical individuality. *Lancet*, ii(3):1616–1620.

Garrod, AE (1923). *Inborn Errors of Metabolism*. London: Hodder & Stoughton.

Gibbs, RA, Weinstock, GM, Metzker, ML, Muzny, DM, et al. (2004). Genome sequence of the Brown Norway rat yields insights into mammalian evolution. *Nature*, 428(6982):493–521.

Gregory, S. G., Barlow, K. F., McLay, K. E., Kaul, R., et al. (2006). "The DNA sequence and biological annotation of human chromosome 1." Nature 441 (7091): 315–21.

Gusella, JF, Wexler, NS, Conneally, PM, Naylor, SL, et al. (1983). A polymorphic DNA marker genetically linked to Huntington's disease. *Nature*, 306 (5940):234–238.

Hattori, M, Fujiyama, A, Taylor, TD, Watanabe, H, et al. (2000). The DNA sequence of human chromosome 21. *Nature*, 405(6784):311–319.

He, L, Thomson, JM, Hemann, MT, Hernando-Monge, E, et al. (2005). A microRNA polycistron as a potential human oncogene. *Nature*, 435 (7043):828–833.

Heilig, R, Eckenberg, R, Petit, JL, Fonknechten, N, et al. (2003). The DNA sequence and analysis of human chromosome 14. *Nature*, 421:601–607.

Hershey, AD and Chase, M (1952). Independent functions of viral protein and nucleic acid in growth of bacteriophage. *J Gen Physiol*, 36:39–56.

Hillier, LW, Miller, W, Birney, E, Warren, W, et al. (2004). Sequence and comparative analysis of the chicken genome provide unique perspectives on vertebrate evolution. *Nature*, 432(7018):695–716.

Hubbard, T, Andrews, D, Caccamo, M, Cameron, G, et al. (2005). Ensembl 2005. *Nucleic Acids Res*, 33(Database issue):D447–D453.

Hubbard, T, Barker, D, Birney, E, Cameron, G, et al. (2002). The Ensembl genome database project. *Nucleic Acids Res*, 30(1):38–41.

Hudson, TJ, Stein, LD, Gerety, SS, Ma, J, et al. (1995). An STS-based map of the human genome. *Science*, 22:1945–1954.

Huntington's Disease Collaborative Research Group (1993). A novel gene containing a trinucleotide repeat that is expanded and unstable on Huntington's disease chromosomes. *Cell*, 72(6):971–983.

Iafrate, AJ, Feuk, L, Rivera, MN, Listewnik, ML, et al. (2004). Detection of large-scale variation in the human genome. *Nat Genet*, 36(9):949–951.

IHGSC (2001). Initial sequencing and analysis of the human genome. *Nature*, 409(6822):860–921.

IHGSC (2004). Finishing the euchromatic sequence of the human genome. *Nature*, 431(7011):931–945.

Ingram, VM (1957). Gene mutations in human haemoglobin: the chemical difference between normal and sickle cell haemoglobin. *Nature*, 180:326–328.

Jacobs, PA, Baikie, AG, Court Brown, WM and Strong, JA (1959). The somatic chromosomes in mongolism. *Lancet*, 1(7075):710.

Johnston, M, Andrews, S, Brinkman, R, Cooper, J, et al. (1994). Complete nucleotide sequence of *Saccharomyces cerevisiae* chromosome VIII. *Science*, 265:2077–2082.

Kaneko, T, Sato, S, Kotani, H, Tanaha, A, et al. (1996). Sequence analysis of the genome of the unicellular Cyanobacterium *Synechocystis* sp. strain PCC6803. II. Sequence determination of the entire genome and assignment of potential protein-coding regions. *DNA Res* 3:109–136.

Khoury, G, Martin, MA, Lee, TN, Danna, KJ, et al. (1973). A map of simian virus 40 transcription sites expressed in productively infected cells. *J Mol Biol*, 78:377–389.

Kong, A, Gudbjartsson, DF, Sainz, J, Jonsdottir, GM, et al. (2002). A high-resolution recombination map of the human genome. *Nat Genet*, 31 (3):241–247.

Kunkel, LM, Monaco, AP, Middlesworth, W, Ochs, HD, et al. (1985). Specific cloning of DNA fragments absent from the DNA of a male patient with an X chromosome deletion. *Proc Natl Acad Sci USA*, 82(14):4778–4782.

Lander, ES, Linton, LM, Birren, B, Nusbaum, C, et al. (2000). Initial sequencing and analysis of the human genome. *Nature*, 409:860–892.

Lee, RC, Feinbaum, RL, and Ambros, V. (1993). The *C. elegans* heterochronic gene lin-4 encodes small RNAs with antisense complementarity to lin-14. *Cell*, 75(5):843–854.

Lejeune, J, Turpin, R, and Gautier, M (1959). Mongolism; a chromosomal disease (trisomy). *Bull Acad Natl Med*, 143(11–12):256–265.

Lewin, R (1986). Proposal to sequence the human genome stirs debate. *Science*, 232(4758):1598–1600.

Lu, J, Getz, G, Miska, EA, Alvarez-Saavedra, E, et al. (2005). MicroRNA expression profiles classify human cancers. *Nature*, 435(7043):834–838.

Maniatis, T, Fritsch, EF, and Sambrook, J (1982). *Molecular Cloning: A Laboratory Manual*. New York: Cold Spring Harbor Laboratory Press.

Margulies, M, Egholm, M, Altman, WE, Attiya, S, et al. (2005). Genome sequencing in microfabricated high-density picolitre reactors. *Nature*, 437(7057):376–380.

Maxam, AM and Gilbert, W (1977). A new method for sequencing DNA. *Proc Natl Acad Sci USA*, 74(2):560–564.

McKusick, VA (1998). *Mendelian Inheritance in Man. A Catalog of Human Genes and Genetic Disorders*. Baltimore, MD: Johns Hopkins University Press.

McPherson, JD, Marra, M, Hillier, L, Waterston, RH, et al. (2001). A physical map of the human genome. *Nature*, 409(6822):934–941.

National Human Genome Research Institute (1990). Understanding our genetic inheritance. The United States Human Genome Project. The first five years: fiscal years 1991–1995. Available: http://www.genome.gov/10001477.

Nielsen, H (2003). Genetic disorders in history and prehistory. In DN Cooper (ed.), *Nature Encyclopedia of the Human Genome* (Vol. 2, pp. 815–818). London: Macmillan.

Nowell, PC and Hungerford, DA (1960). Chromosome studies on normal and leukemic human leukocytes. *J Natl Cancer Inst*, 25:85–109.

Olivier, M, Aggarwal, A, Allen, J, Almendras, AA, et al. (2001). A high-resolution radiation hybrid map of the human genome draft sequence. *Science*, 291:1298–1302.

Oliver, SG, van derAart, QJM, Agostoni-Carbone, ML, Aigle, M, et al. (1992). The complete DNA sequence of yeast chromosome III. *Nature*, 357:38–46.

Pasquinelli, AE, Reinhart, BJ, Slack, F, Martindale, MQ, et al. (2000). Conservation of the sequence and temporal expression of let-7 heterochronic regulatory RNA. *Nature*, 408:86–89.

Pauling, L, Itano, HA, Dinger, SJ and Wells, IC (1949). Sickle cell anemia, a molecular disease. *Science*, 110(2865):543–548.

Ray, PN, Belfall, B, Duff, C, Logan, C, et al. (1985). Cloning of the breakpoint of an X;21 translocation associated with Duchenne muscular dystrophy. *Nature*, 318(6047):672–675.

Riordan, JR, Rommens, JM, Kerem, B, Alon, N, et al. (1989). Identification of the cystic fibrosis gene: cloning and characterization of complementary DNA. *Science*, 245(4922):1066–1073.

Rommens, JM, Iannuzzi, MC, Kerem, B, Drumm, ML, et al. (1989). Identification of the cystic fibrosis gene: chromosome walking and jumping. *Science*, 245(4922):1059–1065.

Sachidanandam, R, Weissman, D, Schmidt, SC, Kakol, JM, et al. (2001). A map of human genome sequence variation containing 1.42 million single nucleotide polymorphisms. *Nature*, 409(6822):928–933.

Saiki, RK, Scharf, S, Faloona, F, Mullis, KB, et al. (1985). Enzymatic amplification of beta-globin genomic sequences and restriction site analysis for diagnosis of sickle cell anemia. *Science*, 230(4732):1350–1354.

Sanger, F, Coulson, AR, Friedmann, T, Air, GM, et al. (1977). The nucleotide sequence of bacteriophage phi-X174. *J Mol Biol*, 125:225–246.

Sanger, F, Coulson, AR, Hong, GF, Hill, DF, et al. (1982). Nucleotide sequence of bacteriophage lambda DNA. *J Mol Biol*, 162(4):729–773.

Sanger, F, Nicklen, S, and Coulson, AR (1977). DNA sequencing with chain-terminating inhibitors. *Proc Natl Acad Sci USA*, 74(12):5463–5467.

Schuler, GD, Boguski, MS, Stewart, EA, Stein, LD, et al. (1996). A gene map of the human genome. *Science*, 274(5287):540–546.

Sebat, J, Lakshmi, B, Troge, J, Alexander, J, et al. (2004). Large-scale copy number polymorphism in the human genome. *Science*, 305 (5683):525–528.

Shendure, J, Porreca, GJ, Reppas, NB, Lin, X, et al. (2005). Accurate multiplex polony sequencing of an evolved bacterial genome. *Science*, 309 (5741):1728–1732.

Smith, LM, Sanders, JZ, Kaiser, RJ, Hughes, P, et al. (1986). Fluorescence detection in automated DNA sequence analysis. *Nature*, 321 (6071):674–679.

Smith, HO and Wilcox, KW (1970). A restriction enzyme from *Haemophilus influenzae*: I. Purification and general properties. *J Mol Biol*, 51:379–391.

Southern, EM (1975). Detection of specific sequences among DNA fragments separated by gel electrophoresis. *J Mol Biol*, 98(3):503–517.

Strachan, T and Read, AP (1996). *Human Molecular Genetics*. New York: John Wiley.

Sturtevant, AH (1913). The linear arrangement of six sex-linked factors in Drosophila, as shown by their mode of association. *J Exp Zoo*, 14:43–59.

Tjio, JH and Levan, A (1956). The chromosome number of man. *Hereditas*, 42: l–6.

Vidal, M. (2001). A biological atlas of functional maps. *Cell*, 104(3):333–339.

Wainwright, BJ, Scambler, PJ, Schmidtke, J, Watson, EA, et al. (1985). Localization of cystic fibrosis locus to human chromosome 7cen-q22. *Nature*, 318 (6044):384–385.

Waterston, RH, Lindblad-Toh, K, Birney, E, Rogers, J, et al. (2002). Initial sequencing and comparative analysis of the mouse genome. *Nature*, 420 (6915):520–562.

Watson, JD and Cook-Deegan, RM (1991). Origins of the human genome project. *FASEB J*, 5(1):8–11.

Watson, JD and Crick, FHC (1953). A structure for deoxyribose nucleic acid. *Nature*, 171:737–738.

Weissenbach, J, Gyapay, G, Dib, C, Vignal, A, et al. (1992). A second-generation linkage map of the human genome. *Nature*, 359(6398):794–801.

Wu, R and Taylor, E (1971). Nucleotide sequence analysis of DNA. II. Complete nucleotide sequence of the cohesive ends of bacteriophage lambda DNA. *J Mol Biol*, 57:491–511.

2

The Human Genome: Structure and Organization

Andrew Read

Humans have two genomes, nuclear and mitochondrial. Normal cells contain two copies of the nuclear genome and a much larger but variable number of copies of the mitochondrial genome. The nuclear genome is approximately 2×10^5 times larger than the mitochondrial genome (3×10^9 vs. 16,659 bp), and contains more than 500 times the number of genes (approximately 24,000 vs. 37), including many required for mitochondrial functions. Understandably, the phrase "the human genome" normally refers to the nuclear genome. The structure and organization of the mitochondrial genome is described separately in the following section.

The Human Genome as Chromosomes

The nuclear genome is most readily seen when it is tightly packaged at the metaphase stage of mitotic cell division (Fig. 2–1). It is important to remember that this represents a highly abnormal state of the genome. At this stage in the life of the cell, the DNA has already been replicated, so that each chromosome consists of two identical sister chromatids, joined at the centromere. Thus, a mitotic cell contains four copies of the nuclear genome, and the tightly packaged DNA is largely inactive. In a more typical cell, each chromosome consists of a single highly extended chromatid, with regions of active and inactive DNA.

The 24 different chromosomes can be recognized by their size, the position of the centromere (Table 2–1), and the pattern of dark and light bands produced by laboratory manipulation. Most commonly, this manipulation consists of subjecting the chromosomes to a limited digestion with trypsin and then staining with Giemsa reagent. The bands so produced are called G-bands. A standard nomenclature of the G-bands is used to identify chromosomal locations (Fig. 2–2). G-banding reflects small differences in the GC content of the DNA at different chromosomal locations. The genome-wide GC content is 41%. Dark staining G-bands contain DNA with a slightly lower GC content (average 37%), whereas the light bands average 45% (International Human Gene Sequencing Consortium, 2001). These differences correlate with systematic variations in the distribution of genes and various classes of repeat element across the genome (see below). It is intriguing to speculate why such regularities across 1–5-Mb regions have persisted through evolution.

Chromosomes have two general functions. On a large scale, they are necessary to ensure accurate partitioning of the replicated DNA between daughter cells at mitosis and to allow the more complicated events of meiosis. For this purpose, the gene content and DNA sequence are irrelevant. All that matters is that each chromosome should be a stable DNA package with a single centromere and sufficient origins of replication. On the micro scale, chromosomes organize the DNA into domains that govern gene expression. This is a very important aspect of human genome organization, but as described in the following section, our understanding of this area is not yet very advanced.

Structure of Chromosomes

The stunning success of the Human Genome Project has had one small negative effect. By focusing attention on DNA sequences written as extended lines of A, G, C, and T characters, it encourages the unwary seeker to forget that our genome functions, not as extended naked DNA, but as highly folded coils and loops of chromatin, a DNA–protein complex. Watson and Crick showed that successive base pairs along the DNA double helix are 0.34 nm apart. This allows us to calculate that at metaphase, 1 metre (3×10^9 bp) of DNA is packed into 23 chromosomes with a total length of approximately 100 μm. During interphase, the DNA is more extended, but still highly organized. Proteins are the main packaging agents of chromatin, with some involvement of small RNA molecules. The structure and spatial organization of chromatin within the interphase nucleus are crucial determinants of the functioning of the genome. Current knowledge of these important matters is rudimentary, but we can recognize several levels of organization:

1. One hundred and sixty five base pairs of naked DNA wrap around an octamer of histone proteins (two molecules each of H2A, H2B, H3, and H4) to form a nucleosome. Successive nucleosomes are separated by 10–80 nucleotides (nt) of spacer DNA. Nucleosomes are relatively stable structures that nevertheless must permit polymerases and other progressive enzymes to move along a DNA strand. Adenosine triphosphate (ATP)-powered chromatin remodeling complexes of proteins assist in this process, and the DNA of active gene regulatory sequences is often relatively devoid of nucleosomes.
2. Adjacent nucleosomes are linked by H1 histones. They can associate tightly or loosely to form "open" or "closed" chromatin. This level of packing is determined by chemical modification of the histones and DNA. Human DNA is

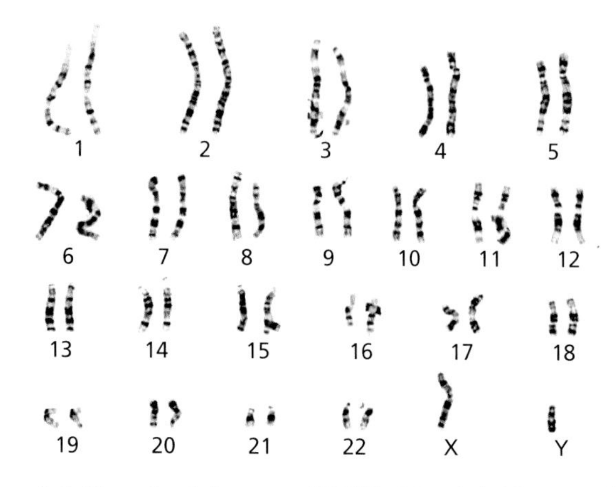

Figure 2–1 Normal male karyotype (46,XY), G-banded. (Courtesy of Dr. Lorraine Gaunt, Manchester, UK.)

modified by methylation of the 5-position of cytosine bases (see below). Histones are subject to a great variety of modifications, including methylation, acetylation, phosphorylation, and ubiquitination of specific residues. Taken together, these are thought to comprise a histone code that determines the activity of genes (Koch et al., 2007; reviewed by Felsenfeld and Groudine, 2003). Large tracts of closed, genetically inactive chromatin are called heterochromatin, whereas open, potentially active chromatin is euchromatin. The closed state can propagate along a fiber and may be controlled at the next level of organization.

3. The 30-nm basic chromatin fiber forms loops containing 20–100 kb of DNA, with the base of each loop anchored to a proteinaceous scaffold. Looping brings together DNA sequences that appear widely separated in the linear DNA sequence, allowing them and their bound proteins to interact.
4. In metaphase chromosomes, the loops and scaffold are compacted further to form the visible chromosome. This is an unusual and largely inert state of the genome, packaged up so that segregation of the two copies into the daughter cells can be carefully controlled. In the interphase nucleus, the extended loops occupy defined territories whose location (central vs. peripheral) and proximity to one another are believed to be important factors controlling gene expression, although there is currently little specific information on this.

Centromeres

A functioning chromosome must have one, and only one, centromere. This is the attachment site for the spindle fibers that pull chromatids apart during mitosis. Human centromeres are marked by long tracts of repetitive DNA. Tandem arrays of highly similar 171 bp α-satellite sequences are, in turn, assembled into higher-order structures (as not each repeat is perfect) from 200 to 9000 kb long. Sequencing long highly repetitive tracts of DNA is extraordinarily difficult, because it can be near-impossible to work out the correct assembly of the individual near-identical clones. Because of this difficulty, and because they are believed not to contain any genes, centromeres are not included in the current human genome sequence.

The DNA of centromeres is complexed to centromere protein A (CENPA), a variant histone H3 (Rudd and Willard, 2004) and adopts the compact structure of heterochromatin. The relation of the DNA sequence to centromere function is not simple. Some abnormal human chromosomes have functioning "neocentromeres" that lack the α-satellite arrays, whereas others have α-satellite arrays that do not function as centromeres. In budding yeast, the centromeres are only 125 bp long, yet they seem to function in the same way as human centromeres without needing the long tracts of repetitive DNA.

Table 2–1 Nomenclature of Chromosomes

Type of Chromosome	Centromere Position	Human Examples
Metacentric	Central	3, 16, 19, 20
Submetacentric	Internal but not central	1, 2, 4–12, 17, 18, X
Acrocentric	Close to, but not at, end	13, 14, 15, 21, 22, Y
Telocentric	At end	(No human examples)

A chromosome can be described as metacentric if its arms look roughly the same size under the microscope, even if they do not have precisely the same DNA content.

Telomeres

The ends of chromosomes require a special structure for two reasons:

- Because of the detailed enzymology of DNA replication, it is not possible to replicate the extreme 3′-end of a DNA strand. Each round of replication shortens a chromosome by 50–100 nt. Within the life of a multicellular organism, that would be tolerable, provided no vital gene is located near the end of any chromosome. However, over evolutionary time, it would be disastrous. Some special mechanism is needed to restore chromosome ends, at least once every generation of the whole organism.
- As part of their mechanism for repairing DNA damage, cells check for loose DNA ends and join together any that they find. Chromosome ends need protecting from this mechanism.

Telomeres of human chromosomes carry long arrays of tandemly repeated $(TTAGGG)_n$ sequence. These contract during successive rounds of somatic cell replication. Germ cells have a special RNA–protein complex, telomerase, that is able to restore telomeres to full length by nontemplated addition of the telomeric repeat (Hahn, 2005). Specific proteins bound to telomeres protect the DNA end, which is formed into a nonstandard looped structure that does not trigger the DNA damage response.

The Human Genome as DNA

The "finished" human genome sequence was published in 2004 (International Human Genome Sequencing Consortium, 2004). Table 2–2 shows the current best estimates of the size of each chromosome.

Uncertainties in the figures relate primarily to the centromeres and telomeres; there are also surprisingly large variations between individuals in the size of their chromosomes (see below). The genetic code uses 3 nt to encode each amino acid. Thus, we have enough DNA to encode 10^9 amino acids or several million proteins—vastly more than the actual number of genes. The gene counts in Table 2–2 are provisional, but no future adjustments can alter the fact that very little of our genome codes for protein—probably only 1.5%.

Why Do We Have So Much DNA?

The 98.5% of our DNA that is noncoding can be broadly described under three headings:

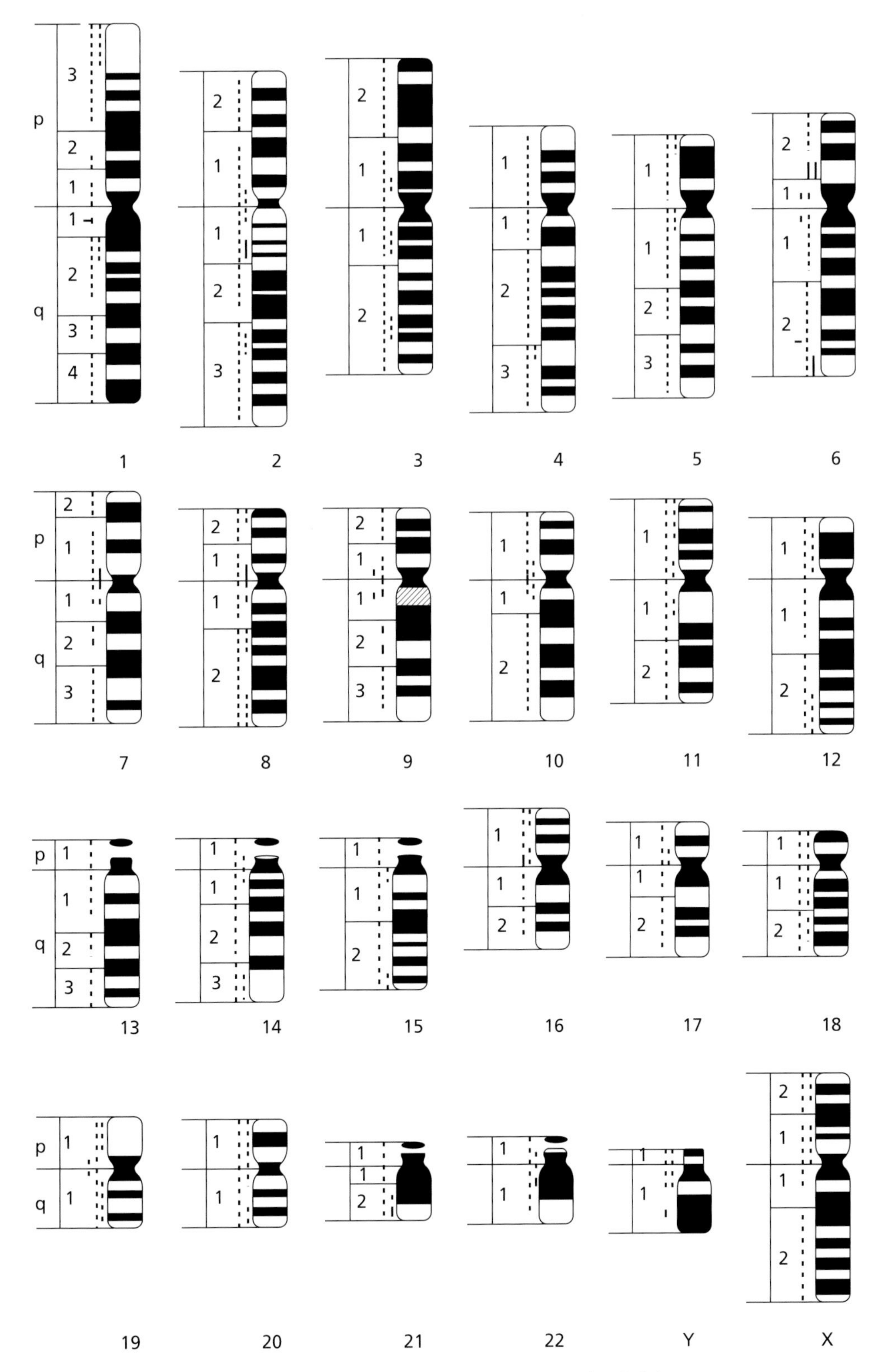

Figure 2–2 Standard cytogenetic nomenclature of human G-banded chromosomes.

- Intragenic noncoding DNA—as described in the following section, the intron–exon structure of genes means that most of the DNA within a gene is noncoding. In addition, if a gene is defined as a functional unit of DNA, it extends well beyond the coding sequence.
- Intergenic unique sequence DNA—genes are seldom densely packed along human chromosomes as they are in bacteria. Usually, there are quite large stretches of DNA between adjacent genes.
- Repetitive DNA—approximately 50% of the human sequence is present in more than one copy. Repetitive DNA is rare in

Table 2–2 DNA and Gene Content of Human Chromosomes

Chromosome	Length (bp)	Genes (Known + Novel)	Known Genes	Pseudogenes	Genes (Mb)
1	245,522,847	2281	1988	723	9.29
2	243,018,229	1482	1246	690	6.10
3	199,505,740	1168	1043	499	5.85
4	191,411,218	866	743	424	4.52
5	180,857,866	970	833	434	5.36
6	170,975,699	1152	1050	459	6.74
7	158,628,139	1116	916	454	7.04
8	146,274,826	794	692	364	5.43
9	138,429,268	919	778	309	6.64
10	135,413,628	862	730	317	6.37
11	134,452,384	1426	1264	359	10.61
12	132,449,811	1104	1009	396	8.34
13	114,142,980	399	318	213	3.50
14	106,368,585	733	646	260	6.89
15	100,338,915	766	589	308	7.63
16	88,827,254	957	839	216	10.77
17	78,774,742	1257	1104	305	15.96
18	76,117,153	322	267	168	4.23
19	63,811,651	1468	1337	164	23.01
20	62,435,964	631	592	180	10.11
21	46,944,323	271	243	87	5.77
22	49,554,710	552	471	129	11.14
X	154,824,264	931	766	380	6.01
Y	57,701,691	104	76	54	1.80
Total	3,076,781,887	22,531	19,540	7892	7.32

Source: Data from Ensembl Release 35, November 2, 2005.

coding sequences but ubiquitous in the human noncoding DNA, including in the introns of genes. It includes some functional sequences such as the ribosomal RNA genes, but the bulk of human repetitive DNA is apparently nonfunctional and much of it may be parasitic. Variation in the amount of repetitive DNA is the main reason why eukaryotic genomes vary so greatly in size despite varying relatively little in gene numbers (our genome is 200 times the size of the yeast genome but 1/200 of the size of the *Amoeba dubia* genome).

The Extended Gene

Human genes are much larger than their coding sequences. If we define a gene as a functional unit of DNA, extra elements include:

- 5′ and 3′ Untranslated regions—the AUG initiation codon of a messenger RNA (mRNA) is located some distance downstream of the 5′-end (the genome-wide average is 300 nt). The 5′ untranslated region binds the ribosomes, which slide along until they encounter an AUG codon embedded in a suitable context (a Kozak sequence, consensus GCCRCCAUGG, where R means a purine (A or G) and the initiator codon is underlined). At the 3′ end, the stop codon (UAA, UAG, or UGA) is usually several hundred bases or even kilobases upstream of the physical end of the mRNA. 3′ Untranslated sequences contain important elements that regulate the activity and turnover of mRNAs. Note that unlike introns and other regulatory elements (see below), the 5′ and 3′ untranslated regions form part of the mature mRNA.
- Introns—as with all higher organisms, most human genes consist of a number of exons separated by introns. Only the exons end up in the mature mRNA. Exons include both the coding sequence and the 5′ and 3′ untranslated regions. Introns are usually considerably larger than the exons they separate. Table 2–3 gives some statistics.
- Gene regulatory sequences—one reason for the increase in genome size as we move up the evolutionary scale is the greater complexity and sophistication of gene regulation in higher organisms. Table 2–4 lists some of the players in human gene regulation.

Most of these are quite poorly understood at present. The ENCODE Consortium of laboratories (ENCODE Project Consortium, 2007) is pursuing a large-scale effort to define the nature and action of these regulatory elements. Some genes that play critical roles in development, and that presumably need particularly extensive regulation, are located in "gene deserts" on chromosomes (defined as regions of 500 kb or greater containing no genes). This suggests that their regulation requires a large region free of competing or interfering transcription units and hints at a function for some of the large amount of intergenic DNA.

Table 2–3 Human Genome Statistics

Average exon number	9
Average exon size	145 bp[a]
Average intron size	3365 bp
Average size of transcription unit	27,000 bp
Average size of mRNA	2600 bp

Source: Data from International Human Genome Sequencing Consortium (2001).
[a]This is the average size of internal exons. The 3′ exon of a gene is often considerably larger than the internal exons. 3′ Untranslated regions average 770 bp. 5′ Untranslated regions are also relatively large, averaging 300 bp.

Functions of Noncoding DNA

In addition to the noncoding DNA of the extended gene as described above, we can identify several categories of functional noncoding sequence:

- Sequences required for chromosomal function—as described in the preceding section, chromosomes need centromeres and telomeres. Other sequences may be required as origins of replication, and for helping to regulate the behavior of chromosomes during mitosis and meiosis.
- Sequences encoding functional RNA molecules—in addition to protein-coding genes, some DNA sequences encode RNA molecules that play important roles in the cell. The short arms of the acrocentric chromosomes each carry approximately 50 copies of the genes encoding ribosomal RNA, in the form of tandemly repeated arrays. Four hundred and ninety seven genes encoding transfer RNAs (tRNAs) are scattered around the genome. Almost 100 genes encode small RNA molecules involved in splicing pre-mRNA. Many other species of small RNA are involved in chemically modifying ribosomal and tRNA and in regulating gene expression. As mentioned in the following section, the number of noncoding RNAs (ncRNAs) that can be identified experimentally is extremely large, but it is not clear how many of them are just nonfunctional by-products of poorly regulated transcription.
- Conserved noncoding sequences—in general, the 1.5% of our genome that codes for protein is well conserved between humans and related organisms, whereas the noncoding DNA is poorly conserved. However, 3%–8% (depending on the criteria used) of the noncoding DNA is strongly conserved. Bejerano et al. (2004) identified 481 segments, more than 200 bp long, that are 100% conserved between human, rat, and mouse. Most were also highly conserved in the chicken and puffer fish. They included 100% conserved sequences of 770 and 732 bp on the X chromosome in introns of the *POLA* gene, and a nearby 1046 bp region with only a single nucleotide change. Siepel et al. (2005) aligned human, mouse, rat, chicken, and puffer fish sequences to identify elements that were highly conserved, though not ultraconserved. These highly conserved elements comprise 0.14% of the human genome. Less than half these elements overlap exons of known protein-coding genes, and many are hundreds or even thousands of bases long. They are often close to genes, but not always, and when distant from genes they often lie within gene deserts. It is generally assumed that these sequences must have some important functions, most probably in gene regulation, for which the precise sequence is critical. However, currently, there is little insight into what function they might have. No known cellular process requires a sequence as long as a kilobase to be totally conserved. A radical alternative hypothesis is that these sequences happen to be subject to some unknown error-correcting process, such that though nonfunctional they are unable to mutate. A possible, though entirely hypothetical, mechanism might make long-lived transcripts and periodically thereafter compare the DNA sequence with the sequence of the transcripts, editing the genomic DNA as necessary to keep it identical to the transcripts.

Table 2–4 Regulatory Elements in the Human Genome

Element	Comments
Promoter	The 500 bp of DNA immediately upstream (5′) of a gene usually includes a number of different motifs that attract and bind the transcription factor proteins that recruit and assemble an RNA polymerase complex.
Enhancer	A sequence that increases transcription of the gene by binding proteins that help attach or activate the RNA polymerase. Enhancers may be upstream or downstream of the promoter, but are normally quite close, within a few hundred base pairs. Although present in every cell, they usually act only in specific tissues, presumably because the protein(s) they bind are present only in that tissue.
Silencer	Similar to an enhancer, but with a negative action.
Locus control region(LCR)	Sequences many kilobases upstream of a gene that are essential for expression of the gene. A single LCR may be needed for expression of a series of downstream genes (e.g., the α and β globin units). LCRs probably function by controlling the large-scale loop structure of the chromatin, through the proteins that bind to them.
Insulator (boundary element)	Sequences that limit the extent of influence of an LCR or other regulatory element. Insulators have to be located between the LCR and the DNA that they are protecting. Like LCRs, insulators probably affect the large-scale loop structure of the chromatin.

Repetitive DNA

At least half the human genome comprises sequences that are present in more than one copy. Repetitive DNA can be classified in many different ways; one basic distinction is between tandem and interspersed repeats. In tandem repeats, the same sequence occurs a number of times, one after another, at a particular chromosomal location. Interspersed repeats are present in a number of copies at different locations in the genome. The great majority of repetitive DNA is noncoding; exceptions to this include the tandemly repeated arrays of genes encoding ribosomal RNA, some small RNAs and olfactory receptor proteins, and a few interspersed or clustered multicopy genes such as the histone or ubiquitin gene families.

Tandem Repeats

The repeat unit in tandem repeats varies from a single nucleotide [poly(A) runs are particularly common] through short units such as the hexanucleotide telomeric repeat, to long units like the 171 bp α-satellite present at chromosomal centromeres. As mentioned in the following section, tandem repeat arrays are often polymorphic as regards the number of repeat units.

- Satellite DNA was so named because in early experiments using density gradient centrifugation of cellular DNA, it formed a satellite to the main peak of bulk DNA. There are several families of satellite DNA (α, β, satellite 1, etc.). All are mainly located at centromeres and heterochromatic regions of chromosomes, and comprise large arrays (up to several megabase pairs) of tandem repeats.
- Minisatellite DNA comprises tandem arrays, typically 1–20 kb long, of repeating units of a dozen or so nucleotides. Although distributed through the genome, they are particularly found near chromosome ends, proximal to the telomeric repeats. Minisatellites with the same repeat unit are often present at a number of different chromosomal locations—these are the basis of the original DNA fingerprinting technique of Jeffreys (1985).
- Microsatellites are short arrays, typically less than 100 bp, with repeat units mostly 1–5 nt. Approximately 3% of our genome consists of microsatellites, distributed randomly across all chromosomes.

Interspersed Repeats

Most interspersed repeats, comprising almost half the entire human genome, belong to four classes of semiautonomous elements that have, or once had, the ability to propagate themselves within a genome (International Human Genome Sequencing Consortium, 2001). These transposon-derived repeats can be seen as a sort of intracellular parasite. Whether they have any function useful to the host cell is much debated. The four classes are:

- Long interspersed nuclear elements (LINEs). Full-length LINEs are 6 kb long. There are several families, of which the L1 repeats are the commonest. They encode two proteins and have a promoter, enabling them to be transcribed. . One protein has endonuclease and reverse transcriptase activities and the other binds nucleic acids, but its exact function is unclear. The L1-encoded proteins preferentially associate with their encoding transcript to form a ribonucleoprotein particle, which is a proposed retrotransposition intermediate. They make DNA copies of the transcript and insert them at new locations in the genome (retrotransposition). Most L1 copies are truncated—probably only 80–100 copies are fully functional and the average LINE is only 900 bp long. The human genome contains an estimated 8,50,000 LINEs, comprising 21% of our total DNA.
- Short interspersed nuclear elements (SINEs). These are 300 bp long. They encode no proteins, but can be propagated by the LINE enzymes. Human SINEs can be grouped into the Alu, MIR, and Ther2 families. Approximately 1.5 million complete or partial copies are present, comprising 13% of our genome.
- Long terminal repeat retroposons are closely related to retroviruses, though lacking the envelope gene necessary for extracellular existence. Some 4,50,000 copies make up 8% of our genome. Again, many copies are defective.
- DNA transposons are virus-like entities that propagate by a cut-and-paste mechanism. At least seven major families are distinguishable in the human genome, totaling approximately 3,00,000 copies and 3% of the genome.

The distribution of LINEs and SINEs in the human genome is interesting. LINEs are four-fold, more common in the AT-rich DNA that forms the gene-poor dark G-bands on chromosomes, whereas SINEs show the opposite distribution. It is difficult to explain this without ascribing some beneficial function to the SINEs. Transposable elements are dangerous, disrupting the host sequence, and one would expect there to be selection against elements transposing into gene-rich regions.

A minor but interesting class of interspersed repeats are the pseudogenes. These are nonfunctional copies of active genes. Some pseudogenes contain the entire functional gene sequence, including introns, but carry mutations that disable them. Others (processed pseudogenes) lack introns and must have been produced by reverse transcription of an mRNA, probably using the LINE machinery. Processed pseudogenes necessarily lack a promoter, so cannot be expressed (unless by chance they land adjacent to an unrelated active promoter). Pseudogenes of both classes are frequently truncated. They provide a snapshot of the processes of gene duplication and divergence that underlie much of evolution. Pseudogenes also present a trap for attempts to compile catalogs of genes by analysis of genome sequences. As Table 2–2 shows, approximately a quarter of all gene-like sequences are pseudogenes, the majority of them processed.

Segmental Duplications

Approximately 5% of the human genome consists of long sequences, 1–400 kb in size that are present in two or more near-identical copies (Bailey et al., 2002). Such duplicons, with more than 90% sequence identity occur not only on every chromosome, mainly in the pericentromeric and subtelomeric regions, but also at other positions (Fig. 2–3). The copies may be repeated tandemly and spaced apart on the same chromosome or on different chromosomes. Mispaired recombination between syntenic duplicons that lie near together but not adjacent is a frequent cause of pathogenic chromosome abnormalities; common copy number polymorphisms (see below) are also associated with segmental duplications.

The Mitochondrial Genome

The mitochondrial genome is very different from the nuclear genome (Fig. 2–4; Table 2–5).

In many respects, it has more in common with bacterial genomes than the eukaryotic nuclear genome. This is consistent with the idea that mitochondria originated as endosymbiotic bacteria within some ancestral eukaryotic cell. If this theory is correct, then over the years the mitochondria have gradually transferred more and more of their functions to the nucleus. The great majority of mitochondrial proteins are now encoded by nuclear genes. Cells contain many mitochondria (typically 100–1000; maybe 1,00,000 in an oocyte) so that mitochondrial DNA (mtDNA) might be formally classified among the repetitive DNA in a cell. Although the mitochondrial genome is very small compared to its nuclear counterpart, because there are many copies mtDNA often makes up 1% or so of total cellular DNA.

As in bacteria, the mitochondrial genome is circular and closely packed with genes. There are no introns and little intergenic noncoding DNA. Some genes even overlap. In the nuclear genome, it is not uncommon for genes on opposite strands to overlap—Nusbaum et al. (2005) recorded 59 such pairs on chromosome 18—but in this case, genes on the same strand overlap, using the same template but read in different reading frames. Twenty-four of the thirty-seven genes specify functional RNAs (2 ribosomal RNAs and 22 tRNAs); the other 13 genes encode components of the electron transport pathway. A short segment of the genome is triple stranded.

This D-loop (displacement loop) is produced by replication forks overlapping as they travel in opposite directions round the circular DNA. The D-loop contains the only significant amount of noncoding DNA in the mitochondrial genome. Perhaps because of

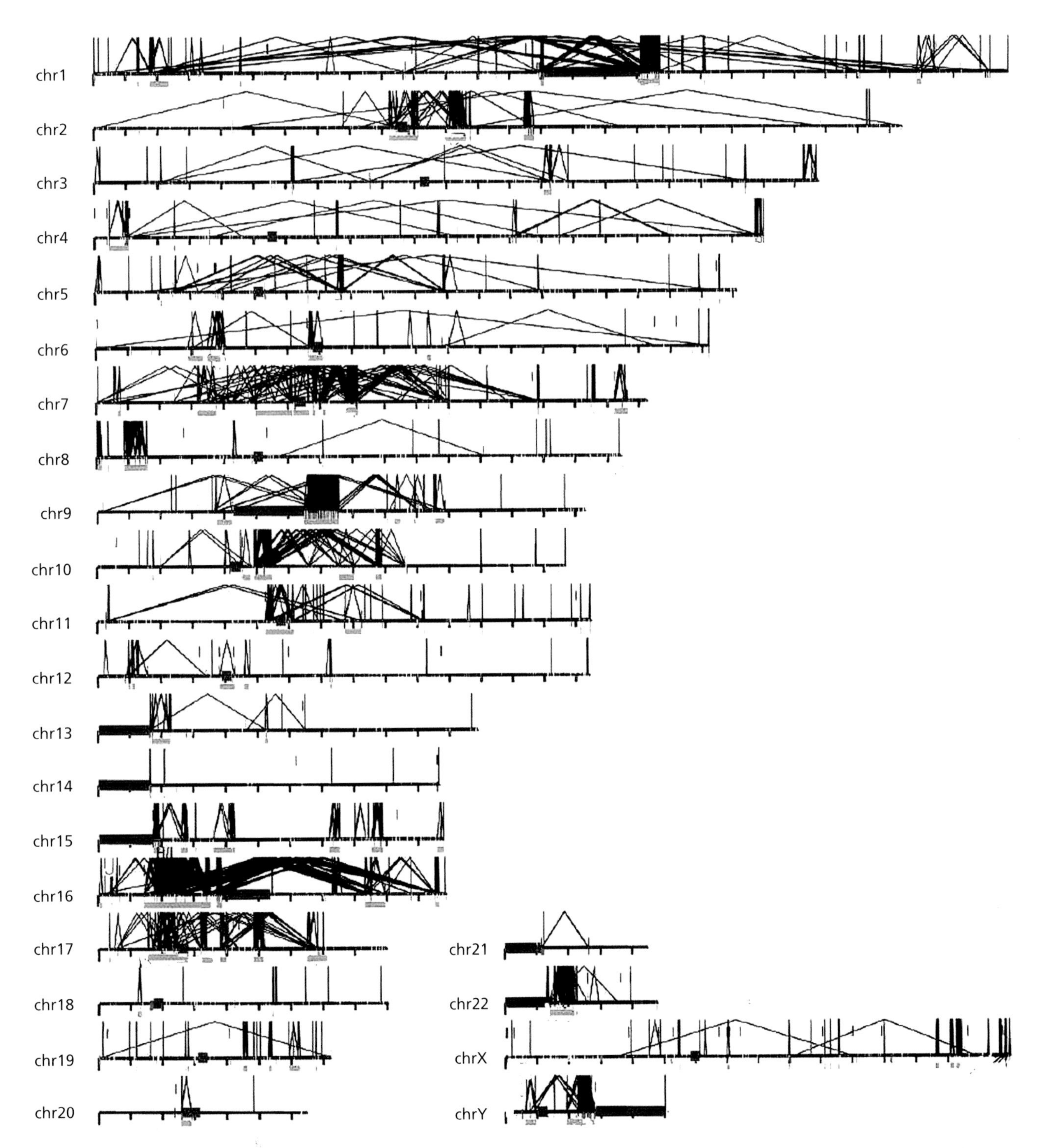

Figure 2–3 Segmental duplications in the human genome. Regions duplicated within a chromosome are connected by diagonal lines; other duplications are between different chromosomes. Bars below the line show regions of unique sequence that are flanked by duplicated segments. Such regions are hotspots for chromosomal rearrangements mediated by recombination between mispaired duplicons. All duplications shown are more than 10 kb long with more than 95% sequence identity. Thick bars on the chromosomes indicate areas of heterochromatin (mainly centromeres and acrocentric short arms) that have not been analyzed. [Adapted with permission from Bailey et al. (2002) *Science* 297:1005.]

this, it is the location of many of the DNA polymorphisms that are such useful tools for anthropologists researching the origins of human populations. Because there is no recombination among mtDNAs, complete haplotypes of polymorphisms are transmitted through the generations, modified only by recurrent mutation, making mtDNA a highly informative marker of ancestry, at least along the maternal line.

mtDNA replication and transcription use nuclear-encoded polymerases. Transcription proceeds in both directions round the circle. The initial products are two large multicistronic RNAs, which

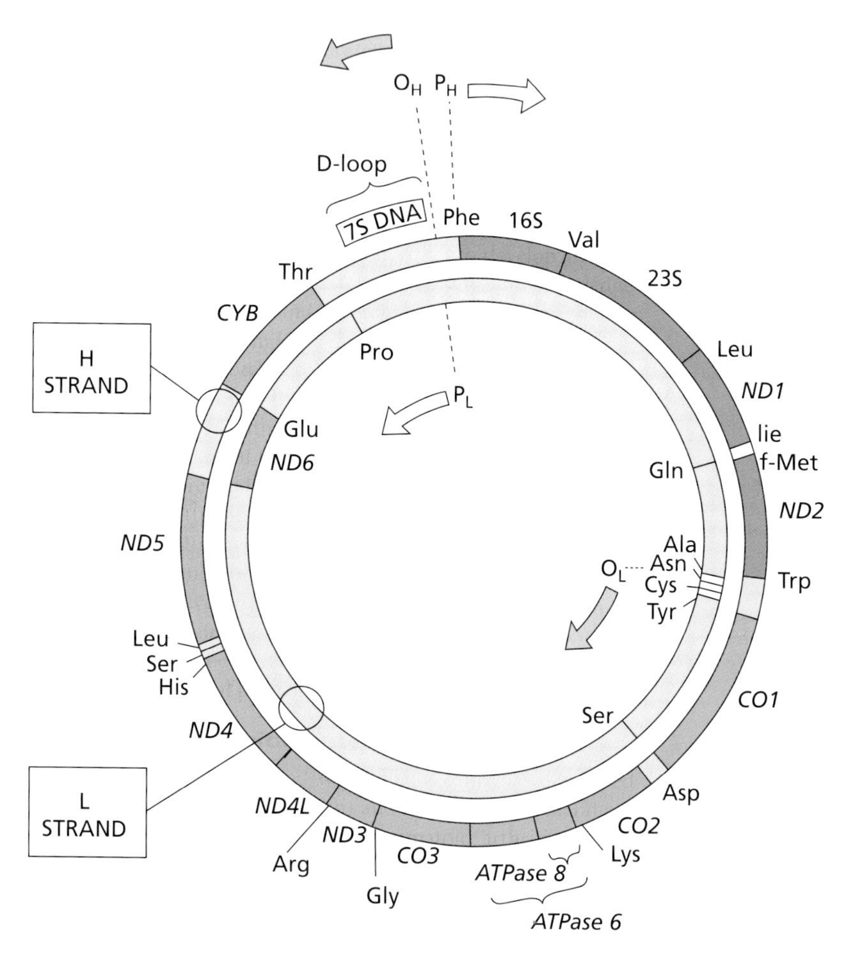

Figure 2–4 The mitochondrial genome. The heavy (H) and light (L) strands of the circular 16,659 bp double helix are shown. Protein-coding genes are shaded; transfer RNAs genes are shown as short lines with the name of the amino acid. There are no introns. O_R, O_L, and the heavy arrows indicate the origins and directions of replication of the two strands. P_R, P_L, and the light arrows show the promoters and the direction of transcription of the two multicistronic transcripts that are subsequently cleaved into individual mRNAs. [Adapted with permission from Fig. 9.1 of Strachan, T and Read, A (2004). *Human Molecular Genetics* (3rd edn), Garland.]

are subsequently cleaved to make the individual mRNAs. All the protein components of the translation machinery are nuclear encoded, but the tRNAs are exclusively mitochondrially encoded, and these use a coding scheme slightly different from the otherwise universal code. There are four stop codons, UAG, UAA, AGG, and AGA; UGA encodes tryptophan, and AUA specifies isoleucine, rather than arginine as normally. Presumably, with only 13 protein-coding genes, the mitochondrial system could tolerate mutations that modified the coding scheme in a way the main genome could not.

Table 2–5 Comparison of the Human Nuclear and Mitochondrial Genomes

	Nuclear Genome	Mitochondrial Genome
Size	3×10^9 bp	16,659 bp
Topology	23 linear molecules	1 circular molecule
No. of genes	Approximately 24,000	37
% coding sequence (incl. genes for functional RNAs)	Approximately 1.4%	93%
Average gene density	Approximately 1 per 125 kb (variable)	1 per 0.45 kb
Introns	Average 8 per gene (variable)	None
Repetitive DNA	Approximately 50%	None

Mutations in mtDNA are important causes of disease, and perhaps also of aging (Trifunovic et al., 2004). Phenotypes caused by variation in mtDNA are transmitted exclusively down the maternal line (matrilineal inheritance), but most genetic diseases where there is mitochondrial dysfunction are caused by mutations in nuclear-encoded genes, and so follow normal Mendelian patterns. As cells contain many copies of the mitochondrial genome, they can be heteroplasmic, containing a mix of different sequences. Unlike mosaicism for nuclear variants, heteroplasmy can be transmitted by a mother to her children.

Variation within the Human Genome

The human genome sequence as contained in the databases is a composite, made up of sequence derived from a number of individuals. Probably, nobody has "the human genome sequence" exactly as listed. Just as people are all different phenotypically, so everybody's genome sequence is unique. However, just as we are all recognizably humans, so sequence individuality comprises numerous small variations superimposed on a general standard pattern. It is convenient to divide the variants into large-scale and small-scale. These are described in more detail in Chapter 6; here we discuss their relevance to the structure of the human genome.

Large-Scale Variation

Until recently, it was generally supposed that most human genetic variation was on the 1–100 nt scale, and that the large-scale structure of the human genome was fixed. New techniques, such as comparative genomic hybridization (Speicher and Carter, 2005) that allow an individual's whole genome to be surveyed for large-scale copy number variants, have radically changed this view. It is now apparent that large-scale copy number polymorphisms are a major class of human variants.

Work is still at an early stage to define all the normal human large-scale variants. Sebat et al. (2004), Iafrate et al. (2004), and Sharp et al. (2005) identified many variants from surveys of limited numbers of individuals. Iafrate et al., for example, surveyed 55 individuals and found 255 genomic clones that were not present at the same dosage in all subjects. One hundred and two variants occurred in more than one individual and 24 were present in more than 10% of the samples. On an average, each person had 12.4 large-scale variants. Using a different technique that could identify rearrangements as small as 8 kb, Tuzun et al. (2005) identified 139 insertions, 102 deletions, and 56 inversion breakpoints in a comparison of just two individuals. A more recent study of 270 individuals by Redon et al. (2006) identified 1,447 "copy number variable regions," covering a total of 360 megabases (12% of the entire genome). Even this study underestimates the true extent of variation because the techniques used were insensitive to variants involving less than several kilobases of DNA. Databases of variants have been set up, for example, the Structural Variation Database (http://paralogy.gs.washington.edu/structuralvariation). Many rare variants are pathogenic, but until the range of normal variation is well documented, it is difficult to interpret laboratory findings on patients.

Small-scale Variation

There are three main types of small-scale variants:

- Variable numbers of tandem repeats (VNTRs, sometimes called simple sequence repeat polymorphisms, SSRPs);
- Insertion–deletion polymorphisms (indels);
- Single nucleotide polymorphisms (SNPs).

The main classes of tandem repeats (satellites, minisatellites, and microsatellites) have been described earlier. Inter-individual variation in the number of repeat units is common in all three classes, although by no means all tandem repeats are polymorphic. The variations in microsatellites arise by slippage when the DNA is replicated; with longer repeat units, variations are more likely to arise by recombination between misaligned repeats. Variations of both types have proved very useful as genetic markers. These can be used in genetic mapping, to track transmission of a particular chromosomal segment through a family, and in forensic work to determine whether two DNA samples come from the same person (e.g., a scene of crime sample and DNA of a suspect). Early work used minisatellite repeats that were characterized by Southern blotting. For both mapping and forensics, these have been superseded by microsatellites, characterized by polymerase chain reaction (PCR).

Microsatellites make up approximately 3% of the human genome (International Human Genome Sequencing Consortium, 2001). There is approximately one microsatellite every 2 kb. The most important, both in terms of their frequency and their use as genetic markers, are dinucleotide repeats; $(CA)_n$ and $(TA)_n$ are the most common. During the early phases of the Human Genome Project, much effort was devoted to defining and mapping over 10,000 microsatellites, predominantly $(CA)_n$ repeats, for use as markers. The less frequent tri- and tetra-nucleotide repeats are preferred, particularly for forensic work, because they are less prone to replication slippage during the PCR. Slippage makes most dinucleotide repeats appear as a little ladder of stutter bands after PCR amplification, which can be hard to read on a gel or sequencer trace.

Indels may be of any size. Large indels may be due to insertion or excision of a transposon. They are readily detected on a sequencer trace of a heterozygote: beyond the location of the variant, the trace will consist of two superimposed but shifted traces. Such variations are approximately 13 times less frequent than substitutions in the human noncoding DNA (and 25 times less frequent in coding sequences) (Ogurtsov et al., 2004).

Among SNPs, the subset that create or abolish a restriction enzyme recognition site were among the earliest DNA polymorphisms to be studied. For example, a GAATTC ↔ GATTTC SNP would abolish or create an *EcoRI* recognition site. Thus, when genomic DNA was digested with that enzyme, the pattern of fragments would differ according to the genotype at this SNP. The pattern could be studied by hybridizing a Southern blot to a probe containing or flanking the variable site, revealing a restriction fragment length polymorphism (RFLP). As genetic markers, RFLPs were superseded during the early 1990s by the more informative multiallele microsatellites. Recent developments have refocused attention on SNPs. Genetic association (as distinct from linkage) studies require a high density of markers that is achievable only with SNPs. The SNP Consortium has identified some 10 million SNPs and cataloged them in the public dbSNP database (www.ncbi.nlm.nih.gov/entrez/query.fcgi?db=snp.). Because they lend themselves to automated high-throughput genotyping, in a way that microsatellites do not, there is now a vast body of SNP genotype information, much of it public.

The Pattern of Variation across the Genome

The recent report of the HapMap Consortium (International HapMap Consortium, 2005) provides a picture in unprecedented detail of human small-scale genetic variation. Phase I of the HapMap project typed 1,007,329 SNPs in 269 DNA samples from people from four locations; Nigeria (Ibadan), USA (Utah), China (Beijing), and Japan (Tokyo). The aim was to characterize 1 SNP every 5 kb across the genome. Phase II of the project will extend this to a further 4.6 million SNPs. In addition, in Phase I, 10 regions, each 500 kb in length, were fully sequenced in 48 of the samples, and all SNPs identified in these regions were typed in all 269 samples. SNPs almost always have only two alleles (i.e., at a given location, there is a choice of two alternative nucleotides, but not three or four) and are best characterized by the minor allele frequency (MAF). Forty-six percent of the SNPs in the HapMap data had MAF less than 0.05, and 9% were seen in only a single individual. However, 90% of heterozygous sites in each individual were due to common SNPs (MAF > 0.05).

Thus, the pattern of variation is of a very low level of variation at the majority of nucleotide positions, but common variants at approximately 1 in every 300 nt. This pattern does not arise because those positions are intrinsically mutable; rather, it is a product of the evolutionary history of our species. When data on adjacent SNPs are combined, it becomes apparent that chromosomes are a mosaic of haplotype blocks (Fig. 2–5).

Blocks are statistical concepts, whose exact properties depend on the statistical criteria used, but they reflect a real and important structural feature of human genomes. Blocks average 5–15 kb in length (depending on the population and the statistical method used to define the blocks) and contain 20–70 SNPs on average. If a stretch of DNA contains 50 biallelic SNPs, there are 2^{50} possible haplotypes. However, in reality at any given location, there is a far more limited range of haplotypes. In the Phase I HapMap data, the average number of haplotypes per block ranged from 4.0 in the Chinese and Japanese sample to 5.6 in the Nigerians. In both cases, this is

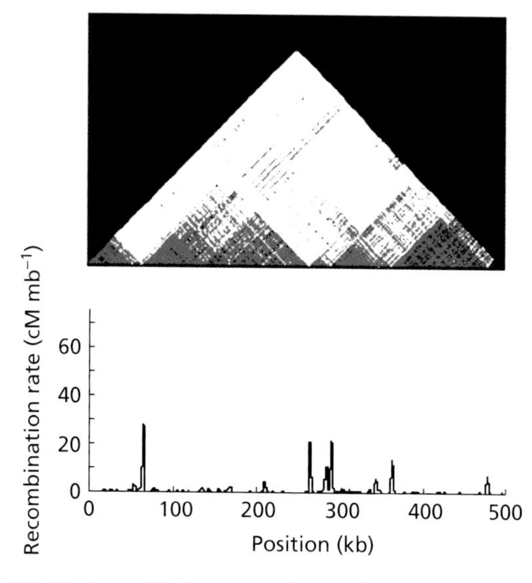

Figure 2–5 Haplotype block structure in a 500 kb region of chromosome 7q31. The triangular plot shows the pattern of linkage disequilibrium. The same map of single nucleotide polymorphisms (SNPs) is plotted along each diagonal axis. Shaded cells mark locations where the genotype at one SNP predicts the genotype at another SNP. The histogram at the bottom shows the distribution of recombination events across the region. It is evident that SNPs are correlated within a series of blocks, the boundaries of which tend to coincide with recombination hotspots. [HapMap project data on 90 white Caucasians from Utah, from International HapMap Consortium (2005) *Nature* 437:1307.]

vastly less than the number of possible haplotypes that could be generated by permutation of all the SNPs in a block.

This remarkable pattern can be explained if each block haplotype is derived from a single common ancestor. Thus, all humans inherit their DNA at any particular location from only a tiny handful of ancestors. More than 90% of all chromosomes are due to haplotypes with an MAF more than 0.05; the exceptions are likely the result of recent mutations that generated SNPs which have found their way into only a small number of descendant chromosomes. This does not mean that we are all descended from four or six cavemen. The four or six haplotypes in one block probably came from a different set of ancestors from the four or six haplotypes in the next block. The size of the blocks is a function of the number of generations between us and the common ancestor. With each round of sexual reproduction, recombination during meiosis fragments the chromosomes. African populations are older than others are, so there has been more fragmentation and haplotype blocks are smaller. The mtDNA and the Y chromosome (except for its tip) are not subject to recombination, and so each is a single large haplotype block.

Much of the research described in the rest of this book starts from this picture of our genomes as a mosaic of limited numbers of ancestral chromosomal segments. The common disease–common variant hypothesis assumes that most of the genetic determinants of susceptibility to common disease are ancient polymorphisms. At the relevant chromosomal location, susceptibility or resistance alleles should segregate with particular blocks. Thus, identifying them comes down to identifying blocks that are associated at the population level with susceptibility or resistance. It is not necessary to type every SNP in a block to distinguish the four or six alternatives at a particular location. Blocks can be defined by a small number of SNPs (tagging SNPs). The research protocol therefore comes down to typing individuals for sufficient tagging SNPs to define every block, and looking for associations. If blocks average 10 kb, and require three tagging SNPs each, this means typing for approximately 1 million SNPs in a case-control study large enough to detect whatever relative risk is ascribed to the factor (a relative risk of 1.1–2.0 would often be reasonable). Note that this protocol rests on the assumption that susceptibility factors are ancient common variants. It would fall down if most susceptibility were due to a heterogeneous set of recent mutations, as with most Mendelian diseases. A recent report from the Wellcome Trust Case Control Consortium (2007) has provided the first major systematic test of this strategy .

Annotating the Human Genome Sequence

The finished human genome sequence is only the raw material for the serious work of understanding what it does. A major goal is to have a complete catalog of genes. Genes can be identified in many ways apart from analysis of the DNA sequence. These various methods generate lists of genes that must ultimately all correlate with identified segments of the genome sequence.

- Pedigree analysis identifies genes as determinants of Mendelian characters (Huntington disease, blood group O, etc.). The corresponding DNA sequence can be identified only after the character has been mapped to a chromosomal location by linkage studies in large families, and then mutations have been demonstrated in a candidate gene at the appropriate location.
- Where biochemical investigation has provided some amino acid sequence of a protein, the corresponding gene sequence can be inferred using a table of the genetic code. The redundancy in the code (a given amino acid can usually be encoded by any of several codons) makes this process difficult—it is much more easily done in the other direction, starting from a DNA sequence.
- Sequencing of mRNA has identified most expressed sequences in the genome. For technical reasons, RNA is converted to a complementary DNA (cDNA) using a reverse transcriptase enzyme, before it is sequenced. Much of the data comprises short partial sequences, a few hundred nucleotides long (expressed sequence tags or ESTs), that are sequenced on an industrial scale. ESTs should be sufficiently unique to allow unambiguous alignment with genomic sequence, and may allow a more complete gene sequence to be deduced from overlapping ESTs. When cDNA and genomic sequences are aligned, the introns can be identified as stretches of genomic sequence that are missing from the cDNA.

Complementing these approaches, computer analysis of the genome sequence attempts to identify genes from first principles. Identifying short coding sequences scattered across long tracts of noncoding DNA is a very uncertain business, hence the provisional nature of gene counts. Very short exons, exons flanked by nonstandard splice sites and exons with unusual base compositions are particularly likely to be missed. Supporting evidence would include conservation of the sequence between humans and mice or other animals, particularly of the inferred amino acid sequence, and biochemical demonstration of the existence of a correctly spliced mRNA corresponding to a putative multi-exon gene.

Ab initio gene identification methods are particularly poor at identifying genes whose product is a functional RNA. This is

Table 2–6 References for Details on Annotation of Individual Human Chromosomes

Chromosome	Reference
1	Gregory et al. (2006)
2	Hillier et al. (2005)
3	Muzny et al. (2006)
4	Hillier et al. (2005)
5	Schmutz et al. (2004)
6	Mungall et al. (2003)
7	Hillier et al. (2003)
8	Nusbaum et al. (2006)
9	Humphray et al. (2004)
10	Deloukas et al. (2004)
11	Taylor et al. (2006)
12	Scherer et al. (2006)
13	Dunham et al. (2004)
14	Heilig et al. (2003)
15	Zody et al. (2006a)
16	Martin et al. (2004)
17	Zody et al. (2006b)
18	Nusbaum et al. (2005)
19	Grimwood et al. (2004)
20	Deloukas et al. (2001)
21	Hattori et al. (2000)
22	Dunham et al. (1999)
X	Ross et al. (2005)
Y	Skaletsky et al. (2003)

particularly frustrating given the recent explosion of interest in microRNAs (miRNAs) and the emerging consensus that RNA plays a much more important and varied part in the life of the cell than previously thought.

Detailed annotations of individual chromosomes have been published (Table 2–6); and the latest analyses for the entire genome are available through Ensembl (http://www.ensembl.org/) or UCSC (http://genome.ucsc.edu/) genome browsers.

The Functional Genome: The Transcriptome and Proteome

The transcriptome is the totality of transcribed sequences, and the proteome is the total complement of proteins. Both the human transcriptome and proteome are considerably larger than our total of 24,000 genes would suggest.

Transcription of a gene depends on assembling an initiation complex upstream of the gene. This includes the RNA polymerase, but also a whole suite of transcription factors and coactivators that provide the specificity and control of transcription. Sequences to be transcribed are identified in the DNA both by the large-scale structure (euchromatin vs. heterochromatin) and by specific small sequence motifs that bind transcription factor proteins. Whether such sequences are actively transcribed or not then depends on the availability of the necessary proteins, and the absence of inhibitory proteins, in the particular cell at that particular time. Transcripts often vary considerably in abundance between people, and much of this variation is heritable (Deutsch et al., 2005). Presumably, this variation explains much of human individuality.

The Puzzle of ncRNAs

Only approximately 1.5% of the human genome consists of protein-coding sequence. A further 27% is transcribed as introns or the 5′ and 3′ untranslated regions of genes. Thus, we might expect 28% of our genome to be transcribed, from one strand of the double helix. In fact, RNA molecules corresponding to a much larger fraction of the genome can be isolated experimentally (Huttenhofer et al., 2005). Some of these ncRNAs have been known for many years, notably the ribosomal RNAs and tRNAs. Other classes of functional small ncRNAs have proliferated over the past few years. These include the 60–300 nt small nucleolar RNAs and the 21–25 nt miRNAs and short interfering RNAs. Small RNAs are involved in the chemical modification of ribosomal and other RNAs, in splicing the introns out of gene transcripts and in regulation of gene expression. Some very large ncRNAs function in gene regulation as antisense RNAs. They are transcribed from the opposite DNA strand to a gene, and appear to compete with the gene promoter for transcription. Examples are the *SNRPN*, *AIR*, and *TSIX* transcripts. *XIST* is another large ncRNA that is part of the X-inactivation machinery.

In addition to these ncRNAs with known functions, a very large number of small and large ncRNAs can be isolated that have no known function. The number of such RNAs exceeds the number of mRNAs. Often transcripts corresponding to both strands of the DNA can be identified (ENCODE Project Consortium, 2007). It is a matter of considerable controversy whether these transcripts indicate a whole world of RNA function of which we are ignorant, or whether they show that transcription is a much more hit-and-miss affair than previously supposed.

CpG islands

A striking feature of human DNA is the scarcity of G nucleotides directly downstream (3′) of a C nucleotide. The overall GC content is 41%, so we might expect 4.2% (0.205 × 0.205) of all dinucleotides to be CG. The observed frequency is one-fifth of this. The explanation lies in DNA methylation. Most eukaryotes have DNA methylase enzymes that attach a methyl group to the 5-position of cytosines in CG (traditionally written CpG) sequences. MeCpG sequences recruit DNA binding proteins that, in turn, recruit histone deacetylases and lead to modification of the chromatin structure and silencing of the DNA. CpG methylation is a major mechanism of gene control in humans.

Methylated cytosines are mutational hotspots. All cytosines, methylated or not, have a tendency to deaminate spontaneously. Deamination of unmethylated cytosine produces uracil. Cells recognize this as an unnatural base in DNA and repair the damage by replacing uracils with cytosine. However, deamination of 5-methyl cytosine produces the natural DNA base thymine. Cells are therefore unable to recognize such events. Over evolutionary time, the majority of methylatable cytosines have mutated to thymine, hence the scarcity of CpG sequences in the human genome.

Although most of the human genome is depleted in CpG sequences, approximately 1% consists of "CpG islands" where the cytosines are usually unmethylated, and so have not been lost. CpG islands are associated with the 5′ ends of approximately 60% of human protein-coding genes, particularly the ubiquitously expressed "housekeeping" genes. Typically, they are a few hundred base pairs long, and lie immediately upstream of the gene, or overlap the first exon. Methylation of the island is associated with pathological gene silencing.

Alternative Splicing and the Proteome

Totaled over all cell types and all human development, humans have far more different proteins than different genes. Four main mechanisms account for this:

- The genes encoding immunoglobulins and T-cell receptors use special mechanisms of DNA splicing, recombination, and mutagenesis to produce a potentially infinite variety of antibodies and receptors (reviewed by Gellert, 2002).
- It is extremely common for a gene to have two or more alternative promoters. These may be differentially regulated, and produce transcripts with alternative first exons joined to common downstream exons. For example, the dystrophin gene has eight promoters, producing tissue-specific variants of the protein. The *CDKN2A* gene has two promoters; the 5′ exons produced from them are spliced to the downstream exons in different reading frames, so that the single gene encodes two proteins that have totally different amino acid sequences.
- A majority of all human genes produce more than one mature mRNA by alternative splicing of exons (Maniatis and Tasic, 2002). The ENCODE project (ENCODE Project Consortium, 2007) reported an average of 5.4 transcripts per gene. Common mechanisms involve differential incorporation or skipping of exons and use of alternative splice sites within an exon. Splice sites are not all the same. They are more or less active depending on the sequence context surrounding the invariant GT. AG splice signals, on the presence or absence nearby of enhancers or silencers of splicing (sequences that bind proteins, which help or hinder deployment of the spliceosome machine) and on the repertoire of the binding proteins (SR proteins) available at the time. Some alternative splicing is unquestionably functional, generating two or more functional proteins from a single gene in a controlled way. How much of the total is of this type, and how much reflects loose control of splicing, is not clear.
- After assembly on the ribosomes, polypeptide chains are often extensively modified to make the functional protein. Sometimes proteolysis produces several functional polypeptides (e.g., some peptide hormones). Many modifications are reversible—for example, reversible phosphorylation controls the activity of many signalling proteins—so that cells can contain multiple differentially active versions of the same protein.

Conclusion

The successful completion of the Human Genome Project ushered in a new era in human genetics. The finished sequence is complemented by a rapidly increasing number of genome sequences from other species. A major surprise has been the relatively low number of genes in our genome—not vastly more than that in the drosophila fruit fly (approximately 13,000) or the *Caenorhabditis elegans* worm (approximately 19,000). The greater complexity of mammals compared with these organisms has been accompanied, neither by a great increase in the number of genes nor by a significant increase in the number of different proteins produced per gene by alternative splicing, but by a vast increase in the amount of noncoding DNA with no known function. This hints at the existence of much more sophisticated systems for regulating gene expression, probably mediated by combinatorial binding of numerous proteins and small RNA molecules to some of the noncoding DNA, and maybe controlled by the local structure and organization of the chromatin.

A major interest is in how the genome varies between people. Techniques such as array comparative genomic hybridization are revealing a wholly unexpected degree of large-scale structural variation between normal individuals, as well as many pathogenic variants. The HapMap project is detailing a remarkable organization of chromosomes as mosaics of short ancestral conserved blocks, with all humans having only a tiny handful of ancestors for any particular block. Hopefully, cataloging these blocks will provide the key to identifying ancient common variants that predispose us to the many common diseases where there is evidence of genetic predisposition.

References

Bailey, JA, Gu, Z, Clark, RA, et al. (2002). Recent segmental duplications in the human genome. *Science*, 297:1003–1007.

Bejerano, G, Pheasant, M, Makunun, I, et al. (2004). Ultraconserved elements in the human genome. *Science*, 304:1321–1325.

Deloukas, P, Matthews, LH, Ashurst, J, et al. (2001). The DNA sequence and comparative analysis of human chromosome 20. *Nature*, 414:865–871.

Deloukas P, Earthrowl ME, Grafham DV et al. (2004). The DNA sequence and comparative analysis of human chromosome 10. *Nature*, 429:375–381.

Deutsch, S, Lyle, R, Dermitzakis, E, et al. (2005). Gene expression variation and expression quantitative trait mapping of human chromosome 21 genes. *Hum Mol Genet*, 14:3741–3749.

Dunham, I, Shimizu, N, Roe, BA, et al. (1999). The DNA sequence of human chromosome 22. *Nature*, 402:489–495.

Dunham A, Matthews LH, Burton J et al. (2004). The DNA sequence and analysis of human chromosome 13. *Nature*, 428:522–528.

ENCODE Project Consortium (2007). Identification and analysis of functional elements in 1% of the human genome by the ENCODE pilot project. *Nature*, 447:799–816.

Felsenfeld, G and Groudine, M (2003). Controlling the double helix. *Nature*, 421:448–453.

Gellert, M (2002). V(D)J recombination: RAG proteins, repair factors and regulation. *Annu Rev Biochem*, 71:101–132.

Gregory, SG, Barlow, KF, McLay, KE, et al. (2006). The DNA sequence and biological annotation of human chromosome 1. *Nature*, 441:315–320.

Grimwood, J, Gordon, LA, Olsen, A, et al. (2004). The DNA sequence and biology of human chromosome 19. *Nature*, 428:529–535.

Hahn, WC (2005). Telomere and telomerase dynamics in human cells. *Curr Mol Med*, 5:227–231.

Hattori, M, Fujiyama, A, Taylor, TD, et al. (2000). The DNA sequence of human chromosome 21. *Nature*, 405:311–319.

Heilig, R, Eckenberg, R, Petit, JL, et al. (2003). The DNA sequence and analysis of human chromosome 14. *Nature*, 421:601–607.

Hillier, LW, Fulton, RS, Fulton, LA, et al. (2003). The DNA sequence of human chromosome 7. *Nature*, 424:157–164.

Hillier, LW, Graves, TA, Fulton, RS, et al. (2005) Generation and annotation of the DNA sequences of human chromosomes 2 and 4. *Nature*, 434:724–731.

Humphray, SJ, Oliver, K, Hunt, AR, et al. (2004). DNA sequence and analysis of human chromosome 9. *Nature*, 429:369–374.

Huttenhofer, A, Schattner, P, and Polacek, N (2005). Non-coding RNAs: hope or hype? *Trends Genet*, 21:289–297.

Iafrate, AJ, Feuk, L, Rivera, MN, et al. (2004). Detection of large-scale variation in the human genome. *Nat Genet*, 36:949–951.

International HapMap Consortium (2005). A haplotype map of the human genome. *Nature*, 437:1299–1320.

International Human Genome Sequencing Consortium (2001). Initial sequencing and analysis of the human genome. *Nature*, 409:860–921.

International Human Genome Sequencing Consortium (2004). Finishing the euchromatic sequence of the human genome. *Nature,* 431:931–945.

Jeffreys, AJ, Wilson, V, and Thein, SL (1985). Individual-specific fingerprints of human DNA. *Nature,* 314:67–73.

Koch, CM, Andrews, RM, Flicek, P, et al. (2007). The landscape of histone modifications across 1% of the human genome in five cell lines. *Genome Res,* 17:691–707.

Maniatis, T and Tasic, B (2002) Alternative pre-mRNA splicing and proteome expansion in metazoans. *Nature,* 418:236–243.

Martin, J, Han, C, Gordon, LA, et al. (2004). The sequence and analysis of duplication-rich human chromosome 16. *Nature,* 432:988–994.

Mungall, AJ, Palmer, SA, Sims, SK, et al. (2003). The DNA sequence and analysis of human chromosome 6. *Nature,* 425:805–811.

Muzny, DM, Scherer, SE, Kaul, R, et al. (2006). The DNA sequence, annotation and analysis of human chromosome 3. *Nature,* 440:1194–1198.

Nusbaum, C, Zody, MC, Borowsky, ML, et al. (2005). DNA sequence and analysis of human chromosome 18. *Nature,* 437:551–555.

Nusbaum, C, Mikkelsen, TS, Zody, MC, et al. (2006). DNA sequence and analysis of human chromosome 8. *Nature,* 439:331–335.

Ogurtsov, AY, Sunyaev, S, and Kondrashov, AS (2004). Indel-based evolutionary distance and mouse–human divergence. *Genome Res,* 14:1610–1616.

Redon, R, Ishikawa, S, Fitch, KR, et al. (2006). Global variation in copy number in the human genome. *Nature,* 444:444–454.

Ross, MT, Grafham, DV, Coffey, AJ, et al. (2005). The DNA sequence of the human X chromosome. *Nature,* 434:325–337

Rudd, MK and Willard, HF (2004) Analysis of the centromeric regions of the human genome assembly. *Trends Genet,* 20:529–533.

Scherer, SE, Muzny, DM, Buhay, CJ, et al. (2006). The finished DNA sequence of human chromosome 12. *Nature,* 440:346–351.

Schmutz, J, Martin, T, Terry, A, et al. (2004). The DNA sequence and comparative analysis of human chromosome 5. *Nature,* 431:268–274.

Sebat, J, Lakshmi, B, Troge, J, et al. (2004). Large-scale copy number polymorphism in the human genome. *Science,* 305:525–528.

Sharp, AJ, Locke, DP, McGrath, SD, et al. (2005) Segmental duplications and copy-number variation in the human genome. *Am J Hum Genet,* 77:78–88.

Skaletsky, H, Kuroda-Kawaguchi, T, Minx, PJ, et al. (2003). The male-specific region of the human Y chromosome is a mosaic of discrete sequence classes. *Nature,* 423:825–837

Speicher, MR and Carter, NP (2005). The new cytogenetics: blurring the boundaries with molecular biology. *Nat Rev Genet,* 6:782–792.

Taylor, TD, Noguchi, H, Totoki, Y, et al. (2006). Human chromosome 11 DNA sequence and analysis including novel gene identification. *Nature,* 440:497–500.

Trifunovic, A, Wredenberg, A, Falkenberg, M, et al. (2004). Premature ageing in mice expressing defective mitochondrial DNA polymerase. *Nature,* 429:417–423.

Tuzun, E, Sharp, AJ, Bailey, JA, et al. (2005) Fine-scale structural variation of the human genome. *Nat Genet,* 37:727–732.

Wellcome Trust Case Control Consortium (2007). Genome-wide association study of 14,000 cases of seven common diseases and 3,000 controls. *Nature,* 447:661–683.

Zody, MC, Garber, M, Sharpe, T. et al. (2006a). Analysis of the DNA sequence and duplication history of human chromosome 15. *Nature,* 440:671–675.

Zody, MC, Garber, M, Adams, DJ, et al. (2006b). DNA sequence of human chromosome 17 and analysis of rearrangement in the human lineage. *Nature,* 440:1045–1049.

3

Human Functional Genomics and Proteomics

Yoshiji Yamada, Sahoko Ichihara, and Masaharu Takemura

Genomics offers a new opportunity to determine how diseases develop by allowing researchers to take advantage of experiments of nature and by providing them with a growing array of sophisticated tools with which to identify the molecular abnormalities that underlie disease processes (Guttmacher and Collins, 2002; Burke, 2003). As shown by the recent completion of the draft sequence (Lander et al., 2001; Venter et al., 2001), one characteristic of the human genome with both medical and societal relevance is that, on average, the DNA sequences of two unrelated individuals are more than 99.9% identical. However, this still means that the DNA sequences of two such individuals differ at more than 3 million base positions, given that the human genome consists of approximately 3 billion base pairs. This is commonly referred to as single nucleotide polymorphisms (SNPs). Intensive efforts are currently under way, in both the academic and commercial sectors, to catalog these differences and correlate them with phenotypic variations relevant to health and disease (Guttmacher and Collins, 2002).

Although most SNPs are silent, a small proportion of them affect the function or amount of the protein encoded by the gene in which they are located. Such variants can have physiological effects that are clinically important, such as giving rise to differences in the response to drugs or environmental factors or to differences in susceptibility or predisposition to disease (Burke, 2003). For example, many of the enzymes responsible for drug metabolism exist in variant forms, leading to differences among individuals in drug efficacy and in the risk of adverse effects (Evans and McLeod, 2003; Weinshilboum, 2003).

Recent advances in genomics (Guttmacher and Collins, 2003) have contributed to the development of gene expression profiling useful in assessing patient prognosis and guidance of therapy, for example, in breast cancer (van de Vijver et al., 2002). Other examples are genotyping either for stratification of patients according to risk for a disease such as myocardial infarction (Yamada et al., 2002), prediction of patient response to certain drugs such as antiepileptic agents (Siddiqui et al., 2003), developing new approaches to the design and implementation of new drug therapies, such as treatment with imatinib for hypereosinophilic syndrome (Cools et al., 2003), understanding of the role of specific genes in common conditions such as obesity (Branson et al., 2003; Farooqi et al., 2003), and identification of genes responsible for diseases such as progeria (Eriksson et al., 2003). Such advances also allowed the rapid identification of the pathogen responsible for severe acute respiratory syndrome, or SARS (Drosten et al., 2003; Ksiazek et al., 2003).

Studies of gene mutations have provided a new model of pathophysiology in which the molecular causes of diseases are illuminated by genetics. Evidence is now emerging of the complex gene–gene and gene–environment interactions that underlie many diseases, and the characterization of these interactions represents the next important step in genomic research (Burke, 2003). Unlike the gene mutations responsible for inherited diseases, which tend to be rare and result in clinically significant alteration of gene function, most common genetic variants cause relatively small changes in gene function. Efforts to understand the molecular mechanisms that underlie common complex diseases build on insights gained and strategies developed in the study of single-gene diseases. However, the scope of these efforts is immense and will require continuing development and improvement of molecular, biological, and informatics tools that will allow the simultaneous analysis of many genetic variants and environmental risk factors (Collins, 1999). In the future, genetic information will increasingly be used in population screening to determine individual susceptibility to common disorders, such as coronary heart disease, stroke, diabetes mellitus, allergy, and various cancers. Such screening will identify groups at risk and thereby facilitate the initiation either of primary prevention efforts, including changes in diet, exercise, and smoking habits, or of secondary prevention efforts such as early disease detection and pharmacological intervention (Khoury et al., 2003).

This chapter reviews recent advances in various basic and clinical aspects of human functional genomics and proteomics. The chapter also summarizes approaches to the genetic analysis of common complex diseases as well as the application of bioinformatics to systems biology.

Human Functional Genomics

Functional Genomics

Functional genomics is a systematic effort to understand the function of genes and gene products by high-throughput analysis of gene transcripts in a biological system (cell, tissue, or organism) using automated procedures that allow scale-up of experiments classically performed for single genes (Yaspo, 2001). Functional genomics can be conceptually divided into gene-driven and phenotype-driven approaches (see also Chapter 10). Gene-driven approaches rely on genomic information to identify, clone, and express genes as well as to characterize them at the molecular level. Phenotype-driven approaches rely on phenotypes either identified from random mutation screens or associated with naturally occurring gene variants, such as those responsible for mouse mutants or human diseases, to identify and clone the responsible genes without prior knowledge of the underlying molecular mechanisms (Yaspo, 2001). The tools of functional genomics have enabled systematic approaches to obtaining basic information for most genes in a genome, including when and where a gene is expressed, and what phenotype results if it is mutated as well as the localization of the gene product and the identity of other proteins with which it interacts (Steinmetz and Davis, 2004). Functional genomics aspires to answer such questions systematically for all genes in a genome, in contrast to conventional approaches that address one gene at a time.

DNA Microarray Analysis for Studies of Gene Expression

At the cellular level, any injurious stimulus triggers adaptive stress responses determined by quantitative and qualitative changes in signaling cascades that interact in complex, often redundant, manners (Chung et al., 2002). The nature of these complex interactions renders analysis at the level of the single gene (or gene product) insufficient to provide an adequate description of them. The integration of information from genome-wide differential expression profiling (at various time points and between phenotypes under different experimental conditions) with other types of genomic information will enable a better understanding of gene regulatory networks and their malfunction in disease (Podgoreanu and Schwinn, 2004). Microarray technologies have revolutionized the analysis of changes in gene expression associated with biological events (Cho et al., 2001) or with complex diseases (Levene et al., 2003) by allowing the simultaneous determination of the abundance of many thousands of gene transcripts in a single experiment.

There are two major technologies for DNA microarray analysis (Steinmetz and Davis, 2004). In the first type of DNA microarray platform, oligonucleotide probes are synthesized directly on the glass surface of the array (Fodor et al., 1991). Photolithography and photosensitive oligonucleotide synthesis chemistry are the most common processes used in the production of such microarrays and achieve a density of more than 1 million spots per square centimeter. In the second type of DNA microarray platform, represented mainly by complementary DNA (cDNA) microarrays, probe synthesis is separated from array manufacture (Schena et al., 1995; Ferguson et al., 1996). The probes are thus synthesized and then mechanically spotted or printed onto the glass or membrane surface in an array format, with a density of approximately 10,000 spots per square centimeter.

For microarray analysis, RNA is first extracted from a sample. Amplification of transcripts may be necessary if the RNA yield is low. Although linear amplification with acceptable fidelity is possible in general, this step has the potential to introduce artifactual distortion of the original gene expression pattern. The isolated RNA itself, cDNA copies, or amplified RNA is then labeled with a fluorophore or radioactive element before hybridization, either competitive or noncompetitive, with the microarray. Complementary sequences remain bound to the array as unbound sequences are washed off. Expressed genes are identified by the positions of the bound sequences on the array (Cook and Rosenzweig, 2002).

Gene expression profiling has been proposed as a general method for identifying candidate genes responsible for complex multifactorial diseases. Microarray technologies are also expected to identify potential new drug targets in various diseases (Podgoreanu and Schwinn, 2004). However, the transcriptome is not fully representative of the proteome, given that many transcripts are not actively targeted for translation (Abbott, 1999), as evidenced recently by the discovery of gene silencing by RNA interference (Lipardi et al., 2001). Furthermore, alternative RNA splicing as well as posttranslational modification of encoded proteins and their interactions with other proteins, all of which can affect the biological function of a gene, are not detected by DNA microarray analysis. Finally, integration and comparison of results obtained with microarrays remain problematic, especially with regard to data derived from different species or with different platforms. Despite these limitations, DNA microarray analysis is currently the most powerful tool available in functional genomics.

Clinical Application of DNA Microarray Analysis

The primary impact of DNA microarray analysis on medical practice is likely to result from the ability to obtain, from a single sample, multiparametric data sets on a far greater scale than previously possible. The simultaneous hybridization of analytes in a single sample to thousands of different specified targets on a microarray has become central to genomics research (Petricoin, Hackett, et al., 2002). Arrays of oligonucleotides or cDNAs have thus been used for expression profiling (Schena et al., 1995; DeRisi et al., 1996; Brown and Botstein, 1999; Golub et al., 1999; Alizadeh et al., 2000; Bittner et al., 2000; Hedenfalk et al., 2001; Pomeroy et al., 2002; Rosenwald et al., 2002; Shipp et al., 2002; van't Veer et al., 2002) or for genome-wide genotyping (Chee et al., 1996; Hacia et al., 1996; Mei et al., 2000; Cutler et al., 2001; Klein et al., 2005), and several potential clinical applications have begun to emerge. The resulting data can comprise thousands of individual measurements and provide an intricate and complex representation of the biological properties of a cell, tissue, or organ. Evaluation of quantitative changes in messenger RNA (mRNA) in clinical samples is likely to prove informative for:

- Assessment of predisposition to disease in subpopulations
- More precise diagnoses;
- Determination of disease severity and patient prognosis
- Selection of optimal therapies;
- Monitoring of a patient's response to treatment (especially in individuals with cancer);
- Identification of therapeutic target genes (Khan et al., 2001) (Fig. 3–1 and Box 3–1).

Box 3–1 Clinical Applications of DNA Microarrays.

- Simultaneous hybridization of analytes in a single sample to thousands of different specified targets on a DNA microarray
- Evaluation of quantitative changes in messenger RNA in clinical samples useful in assessing predisposition to disease
- Establishment of laboratory methods for high sensitivity and specificity
- Evaluation of disease severity and patient prognosis
- Selection of optimal therapies
- Identification of new therapeutic target genes
- Monitoring patient's response to treatment

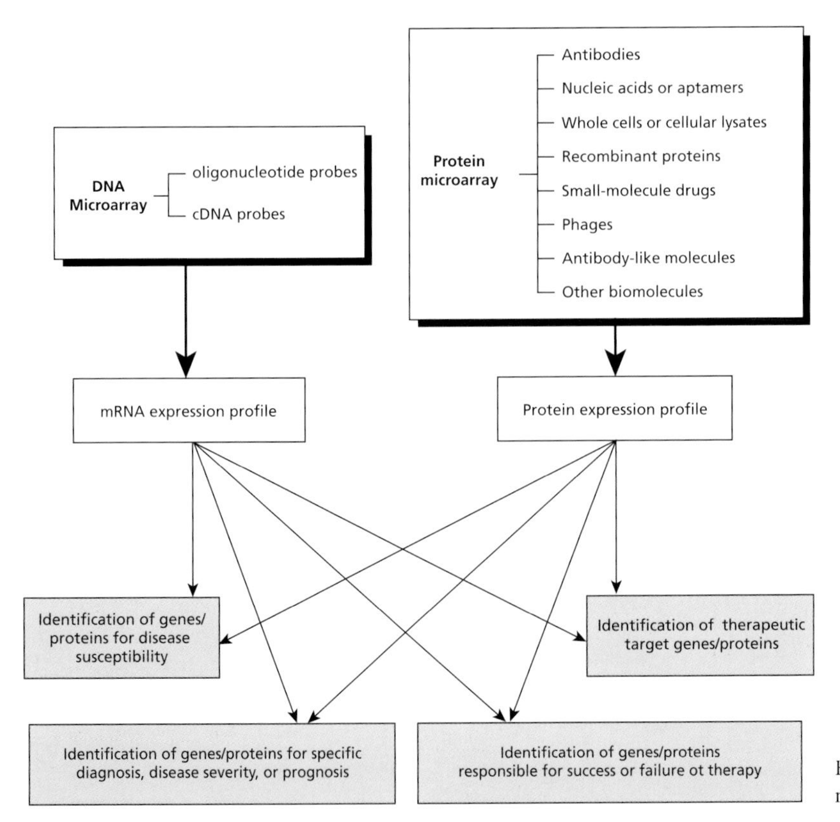

Figure 3–1 Clinical applications of DNA and protein microarrays.

Evidence from clinical studies increasingly suggests that transcript profiling is able to provide valuable insight into disease classification. Cancer has been a focus of initial transcript profiling studies of human disease because of its clinical importance, the availability of tissue samples, and the major role of genetic abnormalities in its pathogenesis (Cook and Rosenzweig, 2000). For example, gene expression profiling of histopathologically indistinguishable B cell lymphomas has revealed the existence of molecularly distinct disease subtypes that are associated with different prognoses (Alizadeh et al., 2000) and responses to treatment (Rosenwald et al., 2002). In a similar way, expression profiling has facilitated prognostic classification of renal cell carcinoma (Takahashi et al., 2001). Microarray analysis also identified previously unknown subtypes of human melanoma as well as a novel set of genes associated with malignant transformation (Bittner et al., 2000). Inhibition of the expression of one of these latter genes—that for the small GTPase RhoC—reduced the extent of metastasis in an animal model (Clark et al., 2000). Transcript profiling is thus able to provide not only useful clinical markers but also mechanistic insight pathogenesis and potential targets for intervention into disease.

DNA microarrays can be reliable tools, provided all nucleotide sequences that hybridize with high sensitivity and specificity are included to allow precise detection of intended targets. It is important to ensure that results are highly reproducible, and quality control and quality assurance systems are established (Petricoin, Hackett, et al., 2002). Determination of the appropriate level of analytical and biological validation needed for each medical application of DNA microarrays and the supporting computer-based bioinformatics systems (Eisen et al., 1998) raises new challenges for scientists, given that the benefits of genomics to clinical medicine and public health are potentially enormous.

Comparative Genomics for Studies of Human Disease Genes

Comparative genomics is an important component of functional genomics, because model organisms play a pivotal role in the functional characterization of genes and in the dissection of basic biochemical in vivo mechanisms (Yaspo, 2001). Almost 62% of known human disease proteins (mostly those for cancer or for neurological or metabolic conditions) and approximately 68% of cancer genes surveyed have counterparts in *Drosophila melanogaster* or *Caenorhabditis elegans* (Banfi et al., 1996; Rubin et al., 2000). The Sonic hedgehog signaling pathway, for example, is essential for embryonic patterning and neural development in species ranging from flies to vertebrates but is also involved in holoprosencephaly, a genetic defect that affects craniofacial development (Ming et al., 1998), as well as in several types of cancer (Dahmane et al., 1997). There is a core of approximately 1400 gene clusters with homologs present in yeast, worms, flies, and mammals (Lander et al., 2001). These core genes encode mostly proteins that participate in transcription, intermediary metabolism, cell cycle regulation, or protein transport or trafficking, indicating the essential nature of these cellular activities. Human diseases caused by defects in basic cellular processes are likely to involve enzymatic or morphogenetic pathways mediated by such core proteins. In contrast, cardiac, immunologic, and endocrine disease genes are underrepresented among genes conserved in invertebrates, reflecting the more specialized nature of these systems in vertebrates (Yaspo, 2001).

Determination of the function of a gene in an organism as a whole requires either the identification of individuals harboring mutations of the gene followed by analysis of the associated phenotype (gene-driven paradigm) or the identification of individuals with a mutant phenotype of interest followed by mapping of the mutated gene (phenotype-driven paradigm) (Yaspo, 2001). The systematic generation of mutant phenotypes in invertebrate and vertebrate models is an important aspect of this latter approach. It has been estimated that approximately one-third of all genes in the yeast, worm, fly, and mouse genomes are essential for viability (Miklos and Rubin, 1996). It is likely that lethal mutations localize to a substantial fraction of core-conserved genes. In drosophila, mutation of approximately 20% of genes generates an observable phenotype, whereas this figure is 50% in yeast (Miklos and Rubin, 1996). In mouse, gene targeting methods, despite their potential for revealing gene function in defined contexts and for rescuing lethal phenotypes, have hit at most a few thousand loci, representing only a small fraction of all genes. Systematic mutagenesis strategies and analysis by gene-driven or phenotype-driven approaches are thus required to create extensive collections of mutants in mammals. Phenotypes at the cellular or tissue level may also be examined and characterized by genomic and proteomic analyses (Yaspo, 2001).

Future Perspectives

DNA microarray analysis has made possible many new discoveries regarding the underlying mechanisms of human disease, most notably cancer. The ability of such analysis to examine gene expression at the whole-genome level will continue to be an important asset in research in both biology and medicine. Genome-scale microarrays will continue to be used for candidate gene selection generating interesting hypotheses, mechanistic investigations, and discovery of potential biomarkers. In the future, they will be used routinely in the selection, assessment, and quality control of drugs for pharmaceutical development as well as in disease diagnosis and for monitoring of desired and adverse outcomes of therapeutic interventions.

Human Functional Proteomics

Functional Proteomics

The Human Genome Project has provided the nucleotide sequence of the human genome as well as the localization of genes within the genome and the predicted amino acid sequences of encoded proteins (Lander et al., 2001; Venter et al., 2001). However, less than 5% of the human genome actually encodes proteins. Furthermore, determination of the nucleotide sequence of a genome reveals its potential for gene and protein expression but does not directly provide information on which proteins are expressed in the various types of cells in an organism or individual, at what level they are expressed, or whether they undergo posttranslational modification (Kramer and Cohen, 2004). In addition, there is a poor correlation between the abundance of an mRNA transcribed from genomic DNA and that of the protein translated from the mRNA, and the initial transcript can be spliced in various ways to yield different protein forms (Banks et al., 2000). The discipline of functional proteomics therefore aims to perform large-scale studies of gene expression at the protein level. Such studies are likely to lead to the discovery of new proteins, markers of disease, pathophysiological mechanisms, and targets for drug development (Pandey and Mann, 2000). Proteomics complements genome-based approaches by providing information on protein expression, turnover, localization, posttranslational modification, and interaction with other biological molecules, thus yielding new insights into complex cellular processes (Podgoreanu and Schwinn, 2004).

It is estimated that there are approximately six to seven times as many distinct proteins as there are genes in humans, in part owing to RNA splicing and exchange of various structural cassettes among genes during transcription. Proteins are not only more numerous than the genes that encode them, but they are also more structurally complex, comprising primary, secondary, tertiary, and quaternary structural elements. In addition, they collectively possess a great variety of biochemical activities that are highly dependent on structure. Proteins are also subject to a host of posttranslational modifications, including proteolytic cleavage, disulfide bond formation, phosphorylation, glycosylation, S-nitrosylation, fatty acylation, and oxidation (Loscalzo, 2003). Given that proteins are the functional output of the genome, proteomics provides information that is most immediately relevant to organismal and cell biology (Banks et al., 2000).

Strategies of Functional Proteomics

There are five basic elements to any proteomic analysis: sample acquisition, protein extraction, protein separation, protein sequence determination, and sequence comparison with reference databases for protein identification (Loscalzo, 2003). A variety of methods for proteomics research has recently been developed (de Hoog and Mann, 2004). The tools of proteomics include two-dimensional gel electrophoresis (Gygi et al., 2000); mass spectrometry (MS) (Aebersold and Mann, 2003); yeast two-hybrid analysis (Uetz et al., 2000); protein microarray analysis (Phizicky et al., 2003; Schweitzer et al., 2003; Zhu and Snyder, 2003; Ramachandran et al., 2004); structure-based methods such as x-ray crystallography, nuclear magnetic resonance spectroscopy, electron crystallography, and electron tomography (Sali et al., 2003); bioinformatics (Redfern et al., 2005); small-molecule screening (Kuruvilla et al., 2002; Winssinger et al., 2002); activity-based profiling (Adam et al., 2002; Speers and Cravatt, 2004); imaging (Huh et al., 2003; Phizicky et al., 2003); and reagents such as antibodies, cDNAs, sets of recombinant proteins, and small interfering RNA (siRNA) constructs. In the following sections, we describe two-dimensional gel electrophoresis, MS, and protein microarray analysis, all of which are central technologies of protein analysis and proteomics.

Two-dimensional Gel Electrophoresis

Two-dimensional gel electrophoresis is widely used to fractionate the numerous proteins present in a given cell type or tissue and to identify differentially expressed or modified proteins (Kramer and Cohen, 2004). In the standard format, two-dimensional gel electrophoresis allows the visualization of between 3000 and 10,000 proteins, depending on the method of spot detection (Issaq, 2001). This approach separates proteins on the basis of two distinct characteristics: size and charge. In the first dimension, proteins are separated from each other by isoelectric focusing according to a pH gradient formed in a cylindrical polyacrylamide or agarose gel. Each protein migrates in the applied electric field until it reaches the region of the gradient with a pH identical to its isoelectric point. The gel containing the separated proteins is then layered on top of a polyacrylamide slab gel in a buffer containing the detergent sodium dodecyl sulfate (SDS) for electrophoretic fractionation in the second dimension according to protein size. It is possible to analyze subpopulations of proteins, or to fine-tune the analysis, with the use of limited pH gradients (such as pH 7–9) or of second-dimension gels with different levels of cross-linking that allows resolution within a more limited size range. The protein spots present in the second-dimension gel after electrophoresis are commonly visualized by staining with the dye Coomassie blue or with silver. For higher-sensitivity detection, the gel can be stained with a fluorescent dye such as Sypro Ruby (Molecular Probes, Eugene, OR) or the proteins can be metabolically labeled with a radioactive isotope before sample preparation (Clarke et al., 2003). An important drawback of two-dimensional gel electrophoresis is that it does not provide a true representation of the proteome; it is biased against short-lived, low-abundance, hydrophobic, highly insoluble or basic, or very small or large proteins (Gorg et al., 2000). Recent improvements of the technique allow more reliable quantification of protein abundance with the use of an internal standard (difference gel electrophoresis) (Unlu et al., 1997) or better proteomic coverage with the use of narrow-range immobilized pH gradient strips after sample prefractionation (Gorg et al., 2002). An important resource for two-dimensional gel electrophoresis is the World-2DPAGE Index to 2D PAGE Databases and Services (http://au.expasy.org/ch2d/2d-index.html).

Mass Spectrometry

Two-dimensional gel electrophoresis coupled with MS is widely applied to identify the individual protein spots of interest on second-dimension gels (Kramer and Cohen, 2004). MS is a technique that measures the mass-to-charge ratio (m/z) of ions in the gas phase as well as the number of ions with each m/z value present in a sample. A mass spectrometer consists of two principal components, a system for sample introduction and ionization and the mass analyzer and detector. MS is an important analytic tool, because it provides both quantitative and qualitative information for an analyte of interest. The emergence of biological MS based on matrix-assisted laser desorption ionization (MALDI) (Karas and Hillenkamp, 1988) or electrospray ionization (ESI) (Fenn et al., 1989) has contributed substantially to advances in protein analysis and proteomics (Clarke et al., 2003). MALDI and ESI generate ions in the gas phase from large proteins without extensive fragmentation. MALDI produces ions by sublimation and ionization of proteins out of a dry crystalline matrix. Mass spectra generated with this type of ionization are relatively simple to interpret, because singly charged ions are the predominant product; thus, in most instances, each peak corresponds to an individual sample component. A variation of MALDI, in which the surface of the target is modified to mimic a chromatographic column, has allowed the laser desorption-based method to be used with more complex mixtures. With this technique, which is known as surface-enhanced laser desorption ionization (SELDI), it is possible to fractionate and characterize sample components directly at the point of application on the basis of their differential interaction with the surface (Merchant and Weinberger, 2000; Issaq et al., 2002). The ESI method of ionization produces ions from a solution, which makes it amenable to coupling with liquid-based separation technologies such as high-performance liquid chromatography or capillary electrophoresis. ESI mostly generates multiply charged ions, with the result that data handling is more complex than that for MALDI. However, the ability to couple the ionization source directly with a separation method allows the analysis of more complex mixtures by ESI-based with ESI MS (Kramer and Cohen, 2004).

The key characteristics of mass analyzers for proteomics are sensitivity, resolution, and accuracy (Aebersold and Mann, 2003). The major types of mass analyzer used for proteomics include quadrupole, ion trap, time-of-flight (TOF), and Fourier transform ion cyclotron (FT-MS) systems. Quadrupole and ion trap analyzers operate with the use of applied DC voltage to establish stable trajectories for ions to reach the detector. These analyzers are coupled with ESI for protein analysis, because ESI produces multiply charged ions; as z increases, the m/z ratio decreases and enters the useful range of the analyzers. A TOF analyzer is the simplest configuration; the time taken by an ion (in a vacuum) to reach the detector is proportional to the molecular weight of the protein fragment. TOF analyzers are more sensitive and accurate than are quadrupole or ion trap analyzers. An FT-MS instrument captures and analyzes ions in a high vacuum with a high-strength magnetic field. Quadrupole and TOF analyzers can be combined in various configurations to perform what is known as tandem MS (MS/MS) (Kramer and Cohen, 2004).

Protein Microarray Analysis

Protein microarrays provide information on the abundance or post-translational modification of proteins or on protein–protein interactions. Such information regarding the proteome of a given cell population can aid in the characterization of complex cellular circuit diagrams, revealing the networks of proteins whose expression or function is upregulated or down-regulated in association with health or disease. Protein microarrays are based on protein–protein, DNA–protein, RNA–protein, or protein–ligand binding (Espina et al., 2004). They can thus contain antibodies, aptamers (short protein-binding oligonueclotides of random sequence) or other nucleic acid molecules, whole cells or cell lysates, recombinant proteins, small-molecule drugs, phage-displayed proteins, or antibodies or antibody-like molecules (Figeys and Pinto, 2001; Sreekumar et al., 2001; Kuruvilla et al., 2002; Petricoin, Zoon, et al., 2002; Walter et al., 2002). Comparison of the protein profiles of samples from multiple patients or from the same patient at multiple time points during treatment is one important clinical application of protein microarray technology attempted to date (Espina et al., 2004) (Fig. 3–1).

Two general strategies for protein microarray analysis have been pursued. The first, abundance-based microarray analysis, seeks to measure the amounts of specific biomolecules with the use of analyte-specific reagents such as antibodies. The second, function-based microarray analysis, examines protein function in high-throughput screens by printing a collection of target proteins on the array surface and assessing either their interactions with other molecules or their biochemical activities (LaBaer and Ramachandran, 2005).

Abundance-based Protein Microarrays

There are two types of abundance-based microarrays, capture microarrays and reverse-phase protein blots (LaBaer and Ramachandran, 2005). Capture microarrays are generated by spotting specific capture molecules, including antibodies, aptamers, photoaptamers (aptamers that have the ability to form covalent bonds with their cognate proteins when they are electronically excited), or affibodies (in vitro selected binding protein), on the array surface in order to trap and assay their targets present within complex mixtures. These profiling arrays are directly analogous to DNA microarrays. Typically, the capture microarray is probed with a complex sample and the relative amounts of the targeted analytes are then determined by comparison with a reference sample (MacBeath, 2002; Zhu and Snyder, 2003). In contrast, reverse-phase protein blots are not true arrays in that they do not comprise an arrangement of known elements. Instead, these blots are produced by spotting the unknown experimental sample itself (or a series of experimental samples) on a membrane, which is then probed with analyte-specific reagents in a manner that is directly analogous to nucleic acid dot blots (Kononen et al., 1998; Paweletz et al., 2001). Both capture microarrays and reverse-phase protein blots rely largely on the availability of well-defined and highly analyte-specific reagents.

Function-based Protein Microarrays

Function-based microarrays are used either to examine the interactions of proteins with other proteins, nucleic acids, lipids, other biomolecules, and small molecules (MacBeath and Schreiber, 2000; Zhu et al., 2001; Ramachandran et al., 2004) or to evaluate enzyme activity or substrate specificity. Such microarrays are produced by printing the proteins of interest on the array with the use of methods designed to maintain their integrity and activity, allowing hundreds to thousands of target proteins to be screened simultaneously for function (Mitchell, 2002; Zhu and Snyder, 2003).

There are two basic types of function-based protein microarrays, referred to as protein-spotting microarrays and self-assembling microarrays (LaBaer and Ramachandran, 2005). For the preparation of protein-spotting microarrays, the various recombinant target proteins are expressed, purified, and then spotted onto the array surface. In general, proteins are affixed to the surface of such microarrays either by chemical linkage or with the use of peptide fusion tags (LaBaer and Ramachandran, 2005). In the chemical-linkage format, proteins are attached to a surface that displays functionalized groups, such as aldehydes, activated esters, or epoxy residues. Peptide fusion tags incorporated at the amino or carboxyl terminus of the recombinant proteins during their expression from modified DNA coding sequences mediate protein affixation through interaction with an affinity capture reagent on the microarray surface. Proteins are expressed and purified before affixation to the array or are expressed and captured simultaneously (Box 3–2).

Box 3–2 Applications of Protein Microarrays.

- Important tool for functional proteomics analysis, providing information that is not obtainable by DNA microarray analysis.
- Applicable to studies of protein–protein or protein–ligand interactions.
- Evaluation of kinase or other enzymatic activity.
- Detection of posttranslational modification of proteins.
- Early detection of disease on the basis of proteomic patterns.
- Personalized selection of therapeutic combinations.
- Real-time assessment of therapeutic efficacy and toxicity.

The difficulty in obtaining purified proteins has limited the number of studies performed with function-based protein microarrays. An alternative to protein-spotting microarrays avoids the need to purify proteins in advance and instead relies on the production of proteins on the microarray surface with a cell-free transcription and translation system (LaBaer and Ramachandran, 2005). In such self-assembling protein microarrays, also known as nucleic acid-programmable protein arrays (Ramachandran et al., 2004), full-length cDNA molecules, rather than purified proteins, are immobilized on the microarray surface and are expressed in situ with a mammalian cell-free expression system such as rabbit reticulocyte lysate. The cDNAs encode the proteins of interest with an attached fusion tag that is recognized by a capture molecule arrayed (together with the cDNA molecules) on the glass surface. The capture reaction then immobilizes the proteins on the surface in a microarrayed format.

Clinical Proteomics

Clinical proteomics is defined as the systematic, comprehensive, and large-scale identification of protein patterns associated with disease and the application of this knowledge to improve patient care and public health through better assessment of disease susceptibility, prevention of disease, selection of therapy on an individual basis, and monitoring of the response to treatment (Granger et al., 2004). The cause of most human diseases lies in the dysregulation of protein interactions. Proteins assemble into networks through a variety of protein–protein interactions and posttranslational modifications. In cells affected by disease, such protein networks are disrupted, deranged, or locally hyper- or hypo-active compared with those in healthy cells (Liotta et al., 2001). Pathogenic gene mutations as well as gene loss, duplication, or amplification result in the production of defective proteins or in protein excess or deficiency, and such changes are likely to lead to rewiring of the normal protein circuit diagram and an aberrant biological state. The goal of clinical proteomics is thus to characterize changes in protein networks and consequent alterations in information flow within cells, tissues, or organs of diseased individuals (Banks et al., 2000; Legrain et al., 2000; Liotta and Petricoin, 2000; Uetz et al., 2000; Blume-Jensen and Hunter, 2001; Ideker et al., 2001).

Clinical proteomics can be divided into three stages: (1) discovery and selection of protein patterns, (2) validation in large clinical data sets, and (3) translation into clinical practice. Advanced analytic technologies, bioinformatics, and carefully defined patient populations are important for the generation of high-quality clinical proteomic data sets. Whereas moderate numbers of well-characterized clinical samples are required for discovery of protein markers, large numbers of equally high-quality samples are needed to validate these markers and to develop them for routine clinical use. Although the most commonly collected clinical sample is blood, other body fluids such as urine and bronchoalveolar lavage fluid may provide additional important opportunities for clinical proteomics (Granger et al., 2004). Compared with nucleic acids, proteins show a much wider variation in physical properties and abundance. The concentrations of different proteins in plasma thus vary by up to 10 orders of magnitude, presenting additional technical challenges for the detection of minor components. The further development of clinical proteomics will thus require technology capable of measuring numerous candidate markers accurately in small volumes of many samples at reasonable cost (Granger et al., 2004).

Future Perspectives

The proteome is a dynamic collection of proteins that varies among individuals, among cell types, and among cells of the same type under different physiological or pathological conditions (Huber, 2003). Clinical proteomics is likely to directly change clinical practice by affecting critical elements of patient care and management, potentially allowing early detection of disease on the basis of proteomic patterns of body fluid samples, diagnosis based on proteomic signatures as a complement to histopathology, individualized selection of therapeutic combinations that best target the entire disease-specific protein network, real-time assessment of therapeutic efficacy and toxicity, and rational modulation of therapy, based on changes in the diseased protein network associated with drug resistance (Liotta et al., 2001). Functional proteomics will redefine biomedical research in the genome era.

Genetic Analysis of Complex Multifactorial Diseases

The Human Genome Project has made the sequence of all human genes available (Lander et al., 2001; Venter et al., 2001), and our knowledge of the variability of the genome has recently increased considerably. New technologies that allow high-throughput accurate genotyping at low cost as well as bioinformatics and statistical tools for dealing with the huge amount of data generated are also becoming available. Genetic analysis is a prerequisite for characterization of the etiology of complex multifactorial diseases, including coronary heart disease, stroke, and cancer, and such knowledge will likely lead to important advances in the diagnosis and treatment of such conditions. Comprehensive integration of genetic variability and accurate phenotype determination may be the only way to improve our understanding of the genetics of complex multifactorial diseases (Cambien, 2005).

Although monogenic disorders are uncommon, elucidation of the mechanisms by which single genes are able to cause disease has clarified our understanding of the molecular basis of human genetic diseases overall, including the more common disorders that are genetically complex. Such multifactorial, polygenic diseases characteristically arise from an interplay of many genetic variations that affect molecular and biochemical pathways and their interactions with environmental factors (Podgoreanu and Schwinn, 2004). According to the common-variant/common-disease hypothesis (Lander, 1996; Risch and Merikangas, 1996; Collins et al., 1997; Lohmueller et al., 2003), common functional allelic variations modulate individual susceptibility to common complex diseases as well as the manifestation and severity of such diseases and patient prognosis. Indeed, the major challenge in the genome era, to which much research effort is being devoted, is to connect the approximately 22,000 human genes to the genetic basis of complex polygenic diseases and to the integrated function of complex biological systems. The identification of SNPs and the construction of haplotype maps of the human genome will likely play essential roles in this effort (Podgoreanu and Schwinn, 2004).

Although most SNPs simply serve as genetic markers, a small proportion of functional SNPs will likely be found to have discernible phenotypes. Determination of the relations of SNPs to human phenotypes will accelerate the process of disease gene mapping and may ultimately allow the identification of patients likely to respond to specific therapies (Yoshioka and Lee, 2003). Gene polymorphisms that confer susceptibility to myocardial infarction have been recently identified. For example, through screening of 65,671 SNPs, Ozaki et al. (2002) found that polymorphisms in the lymphotoxin-a gene were associated with susceptibility to myocardial infarction in a large-scale case–control study (1133 subjects with myocardial infarction, 1878 controls). These researchers suggested that variants in the lymphotoxin-a gene are risk factors for myocardial infarction, implicating this gene in the pathogenesis of this condition. The same group also subsequently showed that an SNP in the galectin-2 gene is associated with the prevalence of myocardial infarction (Ozaki et al., 2004). Galectin-2 plays a role in the secretion of lymphotoxin-a, and the isolated SNP affects the transcriptional activity of the galectin-2 gene. Yamada et al. (2002) determined the genotypes for 112 polymorphisms of 71 candidate genes in 2819 subjects with myocardial infarction and 2242 controls. In this large-scale study, logistic regression analysis with adjustment for conventional risk factors revealed that the risk of myocardial infarction was associated with a 1019C→T (Pro319Ser) polymorphism of the connexin 37 gene in men and with a –668/4G→5G polymorphism in the plasminogen activator inhibitor type 1 gene and a –1171/5A→6A polymorphism in the stromelysin-1 gene in women.

The magnitude of disease risk conferred by each SNP is usually low, and thus disease depends on complex interactions of each SNP with other genes and with environmental factors (Komajda and Charron, 2001). In contrast to monogenic diseases, the development of common diseases will likely prove not to be predictable on the basis of an individual polymorphism alone. Predictions will require models that include the interplay of multiple polymorphisms, environmental factors, and gene–gene and gene–environment interactions (Yoshioka and Lee, 2003). The most effective strategies for disease prevention will likely result from a balanced integration of new genetic information into epidemiological studies (Willett, 2002).

Strategies for Identifying Disease Susceptibility Genes

There are two basic strategies for identifying genes that influence common diseases or other complex traits, the whole-genome approach and the candidate gene approach, both of which rely on linkage analysis and association studies (Fig. 3–2).

Linkage Analysis

Linkage analysis involves the proposition of a model to account for the pattern of inheritance of a phenotype observed in a pedigree. It determines whether the phenotypic locus is transmitted together with genetic markers of known chromosomal position. Linkage analysis is thus used to identify the chromosomal location of gene

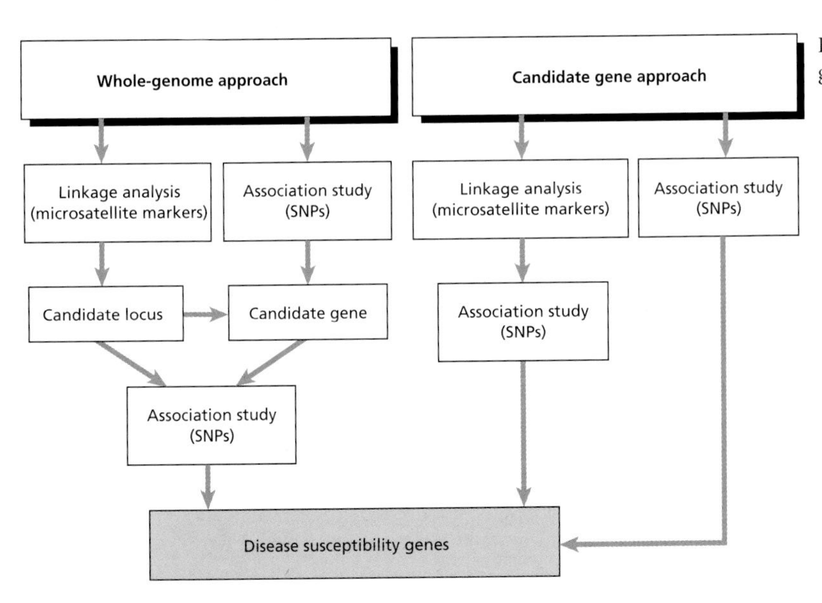

Figure 3–2 Strategies for identifying disease susceptibility genes.

variants related to a given disease. This approach has been applied successfully to map hundreds of genes for rare, monogenic disorders. However, the genetic contribution to complex disorders is likely to involve multiple susceptibility genes, each of only moderate importance at most, a situation that limits the application of linkage analysis (Podgoreanu and Schwinn, 2004). The nature of complex diseases precludes the use of extended families under the hypothesis that the same disease allele acts in most affected individuals throughout a pedigree. Rather, in such diseases, a multitude of genes with common alleles gives rise to an apparently chaotic pattern of heterogeneity within and between families. New linkage methods, such as affected sib-pair linkage analysis, are being developed and tested for complex disorders (Podgoreanu and Schwinn, 2004).

Association Studies

Association studies determine whether certain alleles occur at a frequency higher than that expected by chance in individuals with a particular phenotype. They examine the frequency of specific genetic variants in groups of unrelated individuals with disease and unaffected controls (Podgoreanu and Schwinn, 2004). Their higher statistical power (Risch and Merikangas, 1996) and the fact that they do not require family-based sample collection are the two main advantages of this approach over linkage analysis.

There are two types of association between gene polymorphisms and clinical phenotypes. The first and more persuasive type is based on a well-understood, genome-based pathophysiological mechanism that allows prediction of differences in clinical phenotype. The second type is identified without a clear pathophysiological mechanism and may require a more substantial data set. In both instances, the finding of an association needs to be replicated. Whether identified polymorphisms become reliable markers for disease or not will depend on several factors, most importantly on the extent to which they are able to predict clinical outcome (Petricoin, Hackett, et al., 2002).

One of the main weaknesses of association studies is that, unless the marker of interest is in linkage disequilibrium with the functional variant (or the marker allele is the actual functional variant), the analysis will not be able to distinguish them (Podgoreanu and Schwinn, 2004). Other known limitations of genetic association studies include potential spurious findings that result either from population admixture of different ethnic or genetic backgrounds in the case and control groups (Cardon and Palmer, 2003) or from multiple comparisons in the statistical analysis of association (Cardon and Bell, 2001). The most reliable method of identifying a true association between genetic polymorphisms and disease is replication of findings across different populations or related phenotypes (Tabor et al., 2002). Poor reproducibility in subsequent studies is one of the main criticisms of the association study (Hirschhorn et al., 2002).

Candidate Gene Approach

The candidate gene approach involves the direct examination of whether or not an individual gene or genes might contribute to the trait of interest. This strategy has been widely applied to analysis of the possible association between genetic variants and disease outcome, with genes selected on the basis of a priori hypothesis regarding their potential etiologic role (Tabor et al., 2002). It is characterized as a hypothesis-testing approach because of the biological observation supporting the proposed candidate gene (Podgoreanu and Schwinn, 2004). The candidate gene approach is not able, however, to identify disease-associated polymorphisms in unknown genes.

Whole-genome Approach

In the whole-genome screening approach, hundreds to hundreds of thousands of microsatellite DNA markers or SNPs uniformly distributed throughout the genome are used to identify genomic regions that harbor genes that influence the trait of interest with a detectable effect size. This is a hypothesis-generating approach, allowing the detection of previously unknown potential trait loci (Podgoreanu and Schwinn, 2004). For example, several whole-genome linkage analyses with families and sib pairs (Pajukanta et al., 2000; Francke et al., 2001; Broeckel et al., 2002; Harrap et al., 2002; Gretarsdottir et al., 2003; Hauser et al., 2004; Wang et al., 2004), and an association study with unrelated individuals (Ozaki et al., 2002) have been performed in search of genetic variations that contribute to cardiovascular diseases. A common variant of the complement factor H gene was also recently shown to be associated with age-related macular degeneration in a study based on the use of whole-genome SNP microarrays (Klein et al., 2005), which allowed the genotypes for 1,16,204 SNPs to be determined simultaneously.

The specific set of SNP alleles present on a single chromosome or region of a chromosome is referred to as a haplotype. Although, in theory, many haplotypes might exist for a given chromosomal region, recent studies suggest the existence of just several common haplotypes. The International HapMap Project (The International HapMap Consortium, 2003; http://www.hapmap.org/index.html/en) is an ongoing effort that aims to determine the common patterns of DNA sequence variations in the human genome by characterizing such variations, their frequencies, and correlations between them in DNA samples from populations with ancestry in regions of Africa, Asia, or Europe. The project will thus provide information that will allow the indirect association approach to be applied readily to any functional candidate gene in the genome, to any genomic region implicated by family-based linkage analysis, or, ultimately, to the entire genome for scans of disease risk factors. Establishment of a comprehensive haplotype map and identification of tag SNPs in each linkage disequilibrium block will result in a substantial reduction in the amount of genotyping required for such studies with little loss of information. The haplotype map will be an important tool with which to perform association studies, given that certain phenotypes may not be determined by the genotype of a single polymorphism but rather by a diplotype (a pair of haplotypes). In the near future, whole-genome association studies with sets of haplotype-tagging SNPs will constitute a powerful approach to identifying common genetic factors responsible for disease (Cambien, 2005) (Box 3–3).

Future Perspectives

Studies have indicated the existence of a substantial genetic component in complex multifactorial diseases. However, despite the identification of a variety of candidate genes related to these conditions, the replicability of such findings in most studies is poor, mainly because of the limited population size of the studies, the ethnic diversity of gene polymorphisms, complicating environmental factors and gene–environment interactions, and linkage disequilibrium with nearby genes. The development of personalized prevention of complex diseases will require the performance of large-scale whole-genome case–control association studies in various ethnic groups in order to identify the genes that confer susceptibility to such diseases. Large-scale population-based longitudinal studies will also be needed to confirm the relations between the genes and disease.

Box 3–3 Identification of Disease Susceptibility Genes.

- Linkage analysis involves the proposition of a model to account for the pattern of inheritance of a phenotype observed in a pedigree. It determines whether or not the phenotypic locus is transmitted together with genetic markers of known chromosomal position.
- Association studies determine whether a certain allele occurs or not at a frequency higher than that expected by chance in individuals with a particular phenotype.
- The candidate gene approach involves the direct examination of individual genes for a possible role in determination of the trait of interest.
- The whole-genome approach examines all genes systematically with panels of microsatellite markers or single nucleotide polymorphisms uniformly distributed throughout the genome.
- Both linkage and association approaches provide mathematical probability for existence of susceptibility genes or loci.
- Whole-genome association studies with sets of haplotype-tagging SNPs will constitute an important approach to identifying genetic factors for complex multifactorial diseases.

Bioinformatics and Systems Biology

Advances in molecular biology and related technologies have allowed the increasingly rapid sequencing of large portions of the genomes of several species, including the entire human genome. Popular sequence databases, such as GenBank (National Center for Biotechnology Information) and European Molecular Biology Laboratory (EMBL) have thus been growing at exponential rates. This deluge of data has necessitated the careful storage, organization, and indexing of sequence information. The application of information technology to biology has given rise to the field of bioinformatics, which encompasses the creation, storage, maintenance, and organization of databases of biological information, including nucleic acid and protein sequences, as well as the computational analysis of such information (http://biotech.icmb.utexas.edu/pages/bioinfo.html). Bioinformatics thus relies on techniques from applied mathematics, informatics, statistics, and computer science to solve biological problems. Major research efforts in this field are directed at sequence alignment, gene hunting, genome assembly, protein structure alignment, protein structure prediction, prediction of gene expression and protein–protein interactions, and the modeling of evolution (Fig. 3–3). The terms bioinformatics and computational biology are often used interchangeably, although the latter typically focuses on algorithm development and specific computational methods. A common aspect of bioinformatics and computational biology is the use of mathematical tools to extract useful information from noisy data produced by high-throughput biological techniques. Representative problems in computational biology include the assembly of high-quality DNA sequences generated by fragmentary shotgun DNA sequencing and the prediction of gene regulation from data obtained by DNA microarray analysis or MS (http://en.wikipedia.org/wiki/bioinformatics). Bioinformatics research tools and resources are available from the European Bioinformatics Institute (http://www.ebi.ac.uk/information) (Brooksbank et al., 2003; Brooksbank et al., 2005).

In parallel with such data-driven research, the new approach of model-driven research is being developed (Yao, 2002).

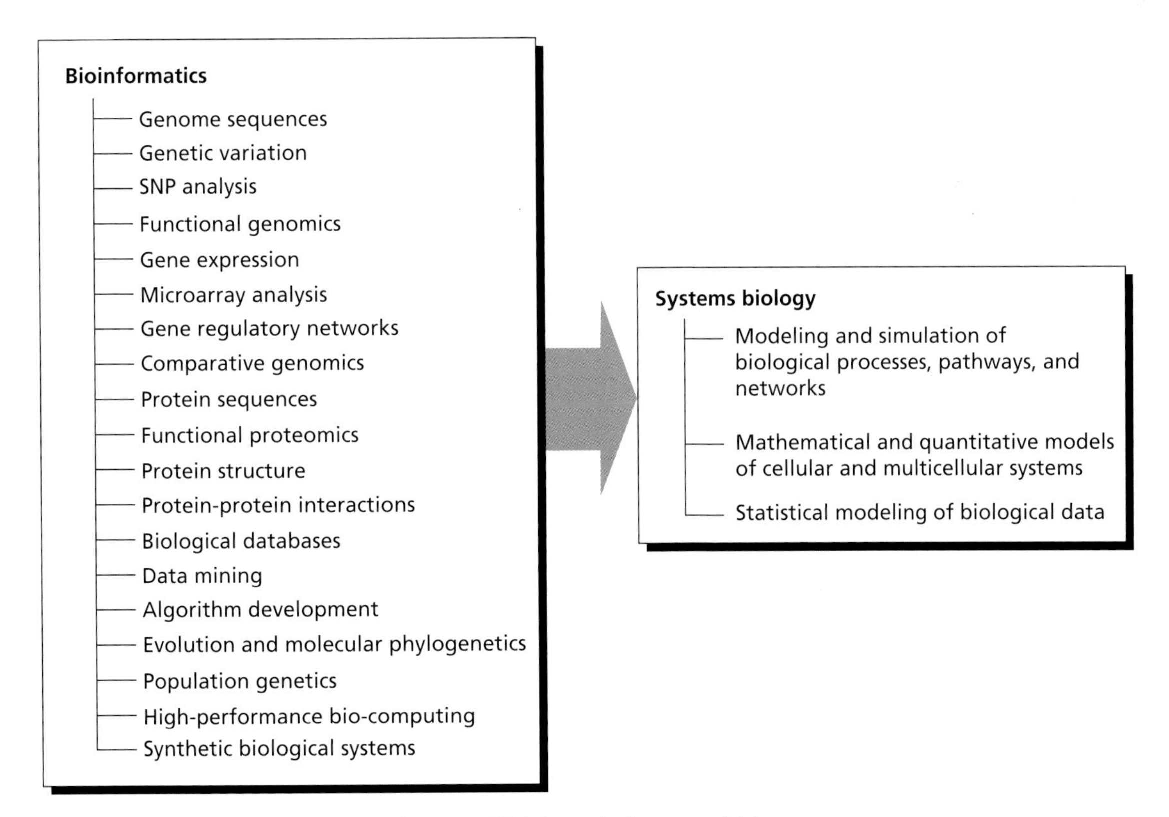

Figure 3–3 Bioinformatics for systems biology.

Model-driven research, also referred to as systems biology, sets up a biological model by combining knowledge of the system in question (e.g., a metabolic pathway, organelle, cell, physiological system, or organism) with related data and then simulates the behavior of the system in order to understand its underlying biological mechanisms (Fig. 3–3). Systems biology thus seeks to integrate data from high-throughput biological studies of the relations and interactions among various system components in order to develop an understandable model of the whole system. Beginning with knowledge of the complete set of genes and predicted proteins for an organism, systems biology relies on high-throughput techniques to measure changes in gene or protein expression in response to a given perturbation. A hypothesis is then proposed to explain the observed changes and such hypotheses form the basis for development of a mathematical model of the biological system. The model is subsequently used to predict how different changes will affect the behavior of the system (http://en.wikipedia.org/wiki/systems_biology). The ultimate goal is to develop a life simulator, which will require the step-by-step simulation of systems of increasing complexity, from the subcellular level to the cellular, tissue, organ, and organismal levels, as well as simulation of development, physiology, and pathology (Yao, 2002). The next section briefly describes three biological simulation systems of particular interest.

Simulation of Biological Systems

The company Entelos (http://www.entelos.com) has developed a virtual patient system (PhysioLab) and constructed virtual models of obesity, diabetes, asthma, and rheumatoid arthritis with this system. A top-down approach was adopted in the development of this system by first listing various factors that affect a particular disease and then breaking down these factors to various successive subsystems toward the metabolic pathway level. The system incorporates genetic, physiological, and lifestyle factors and permits researchers to investigate the underlying pathophysiology that causes seemingly similar patients to respond differently to the same drug therapy.

The Virtual Heart Project has been undertaken by several researchers (Noble, 2002). The virtual heart model contains more than 1 million cells or elements, each of which has complex internal biochemical reactions, and is governed by more than 30 million equations. The virtual heart was constructed initially at the cellular level but is currently incorporating the genetic level, for example, by reconstructing the effects of particular mutations in relevant genes such as those for sodium channel subunits. This model has immense potential for prediction of which drugs are likely to have serious cardiac side effects, a common cause of drug failure in late clinical trials and of withdrawal of drugs from the market. By changing a small number of ion channel-mediated currents to simulate the effects of a particular drug, it is possible to predict whether or not the drug will cause minor changes in the electrocardiogram that have been associated with sudden heart failure. This work forms part of the international collaboration of the Physiome Project (http://www.physiome.org).

The recent progress in large-scale genomics research has made it possible to measure gene expression at various developmental stages and under various mutational conditions on an unprecedented scale (Ko, 2001). Davidson et al. (2002) have described a genomic regulatory network for development. These researchers investigated a regulatory network that controls the specification of endoderm and mesoderm in the sea urchin embryo with the use of a large-scale perturbation analysis in combination with computational methodologies and incorporating genomic data, cis-regulatory analysis, and molecular embryology. With this approach, they were able to obtain various system-level insights into the developmental process (Box 3–4).

Box 3–4 Bioinformatics and Systems Biology.

- Bioinformatics encompasses the creation, storage, maintenance, and organization of databases of biological information, such as nucleic acid and protein sequences, as well as the computational analysis of such information.
- Systems biology is the study of living organisms in terms of their underlying network structure rather than simply with regard to their individual molecular components. A system can be anything from a gene regulatory network to a cell, a tissue, or an entire organism.
- Systems biology requires investigation of all interacting components simultaneously using high-throughput, quantitative technologies.
- Computational approaches are necessary to handle and interpret the volumes of data in understanding complex biological systems.
- Advanced bioinformatics is essential for the development of systems biology.

Clinical Application of Systems Biology

Systems biology, with its combination of computational, experimental, and observational approaches, is highly relevant to drug discovery and the optimization of medical treatment regimens for individual patients. Although the analysis of individual SNPs is expected to provide insight into individual genetic susceptibility to various pathological conditions, it may be impossible to identify such relations when complex interactions are involved (Kitano, 2002). Consider a hypothetical example in which an SNP of gene A is responsible for a certain disease. This relation may not be apparent, however, if circuits exist to compensate for the effect of the SNP. The SNP of gene A will thus be linked to disease susceptibility only if these compensatory circuits are disrupted for some reason. A more mechanistic, systems-based analysis will therefore be necessary to elucidate such complex situations involving multiple genes (Kitano, 2002). It has been predicted that computer simulation and analysis, together with bioinformatics approaches, will greatly increase the efficiency of drug discovery (Cascante, et al. 2002).

Future Perspectives

The rapid growth in the amount of biological information, including genome and protein sequences as well as data on the expression of genes, protein structure, and protein–protein interactions, will greatly increase our knowledge of biology and pathology. Construction of models to derive such knowledge from data for various integrated systems will be inevitable. Given that systems of living organisms possess features distinct from those of artificial systems in that adaptability, evolution, redundancy, robustness, and emergence combines to give rise to the former, the integration of life science and bioinformatics will be important to provide insight into complex living systems (Yao, 2002). The application of systems biology to model and simulate various systems and to visualize the results for a better understanding of life mechanisms will be an important aspect of functional genomics and proteomics. The application of systems biology to medical practice will also be important in the future. Drug discovery, the design of multiple drug therapies, and the elucidation of therapeutic gene circuits will be pursued with modern, complex, and precision engineering approaches through iterative cycles of hypothesis and simulation-driven research (Kitano, 2002).

Summary

The application of functional genomics and proteomics is expected to directly change clinical practice by affecting critical elements of patient care and management. Although the most immediate benefits from the simultaneous analysis of multiple genes or proteins in relation to disease are likely to be in diagnosis, classification, and prediction of treatment response and prognosis, identification of the specific molecular pathways responsible for a disease and of potential new therapeutic targets will lead to the rational design of new therapies. Comprehensive integration of genetic variability and accurate phenotypes will improve our understanding of the genetics of complex multifactorial diseases and facilitate the development of personalized medicine. Clinical genomics and proteomics as well as genetic analysis of complex diseases are thus likely to have important direct bedside applications.

References

Abbott, A (1999). A post-genomic challenge: learning to read patterns of protein synthesis. *Nature,* 402:715–720.

Adam, GC, Sorensen, EJ, and Cravatt, BF (2002). Chemical strategies for functional proteomics. *Mol Cell Proteomics,* 1:781–790.

Aebersold, R and Mann, M (2003). Mass spectrometry-based proteomics. *Nature,* 422:198–207.

Alizadeh, AA, Eisen, MB, Davis, RE, Ma, C, Lossos, IS, Rosenwald, A, et al. (2000). Distinct types of diffuse large B-cell lymphoma identified by gene expression profiling. *Nature,* 403:503–511.

Banfi, S, Borsani, G, Rossi, E, Bernard, L, Guffanti, A, Rubboli, F, et al. (1996). Identification and mapping of human cDNAs homologous to *Drosophila* mutant genes through EST database searching. *Nat Genet,* 13:167–174.

Banks, RE, Dunn, MJ, Hochstrasser, DF, Sanchez, J-C, Blackstock, W, Pappin, DJ, et al. (2000). Proteomics: new perspectives, new biomedical opportunities. *Lancet,* 356:1749–1756.

Bittner, M, Meltzer, P, Chen, Y, Jiang, Y, Seftor, E, Hendrix, M, et al. (2000). Molecular classification of cutaneous malignant melanoma by gene expression profiling. *Nature,* 406:536–540.

Blume-Jensen, P and Hunter, T (2001). Oncogenic kinase signalling. *Nature,* 411:355–365.

Branson, R, Potoczna, N, Kral, JG, Lentes, K-U, Hoehe, MR, Horber, FF (2003). Binge eating as a major phenotype of melanocortin 4 receptor gene mutations. *N Engl J Med,* 348:1096–1103.

Broeckel, U, Hengstenberg, C, Mayer, B, Holmer, S, Martin, LJ, Comuzzie, AG, et al. (2002). A comprehensive linkage analysis for myocardial infarction and its related risk factors. *Nat Genet,* 30:210–214.

Brooksbank, C, Cameron, G, and Thornton J (2005). The European Bioinformatics Institute's data resources: towards systems biology. *Nucleic Acids Res,* 33:D46–D53.

Brooksbank, C, Camon, E, Harris, MA, Magrane, M, Martin, MJ, Mulder, N, et al. (2003). The European Bioinformatics Institute's data resources. *Nucleic Acids Res,* 31:43–50.

Brown, PO and Botstein, D (1999). Exploring the new world of the genome with DNA microarrays. *Nat Genet,* 21:33–37.

Burke, W (2003). Genomics as a probe for disease biology. *N Engl J Med,* 349:969–974.

Cambien, F (2005). Genetic analysis and coronary heart disease. *Future Med,* 1:17–27.

Cardon, LR and Bell, JI (2001). Association study designs for complex diseases. *Nat Rev Genet,* 2:91–99.

Cardon, LR and Palmer, LJ (2003). Population stratification and spurious allelic association. *Lancet,* 361:598–604.

Cascante, M, Boros, LG, Comin-Anduix, B, de Atauri, P, Centelles, JJ, Lee, PW (2002). Metabolic control analysis in drug discovery and disease. *Nat Biotechnol,* 20:243–249.

Chee, M, Yang, R, Hubbell, E, Berno, A, Huang, XC, Stern, D, et al. (1996). Accessing genetic information with high-density DNA arrays. *Science,* 274:610–614.

Cho, RJ, Huang, M, Campbell, MJ, Dong, H, Steinmetz, L, Sapinoso, L, et al. (2001). Transcriptional regulation and function during the human cell cycle. *Nat Genet,* 27:48–54.

Chung, TP, Laramie, JM, Province, M, and Cobb, JP (2002). Functional genomics of critical illness and injury. *Crit Care Med,* 30:S51–S57.

Clark, EA, Golub, TR, Lander, ES, and Hynes, RO (2000). Genomic analysis of metastasis reveals an essential role for RhoC. *Nature,* 406:532–535.

Clarke, W, Zhang, Z, and Chan, DW (2003). The application of clinical proteomics to cancer and other diseases. *Clin Chem Lab Med,* 41:1562–1570.

Collins, FS (1999). Shattuck lecture—medical and societal consequences of the Human Genome Project. *N Engl J Med,* 341:28–37.

Collins, FS, Guyer, MS, and Charkravarti, A (1997). Variations on a theme: cataloging human DNA sequence variation. *Science,* 278:1580–1581.

Cook, SA and Rosenzweig, A (2002). DNA microarrays: implications for cardiovascular medicine. *Circ Res,* 91:559–561.

Cools, J, DeAngelo, DJ, Gotlib, J, Stover, EH, Legare, RD, Cortes, J, et al. (2003). A tyrosine kinase created by fusion of the PDGFRA and FIP1L1 genes as a therapeutic target of imatinib in idiopathic hypereosinophilic syndrome. *N Engl J Med,* 348:1201–1214.

Cutler, DJ, Zwick, ME, Carrasquillo, MM, Yohn, CT, Tobin, KP, Kashuk, C, et al. (2001). High-throughput variation detection and genotyping using microarrays. *Genome Res,* 11:1913–1925.

Dahmane, N, Lee, J, Robins, P, Heller, P, and Altaba, AR (1997). Activation of the transcription factor Gli1 and the sonic hedgehog signalling pathway in skin tumours. *Nature,* 389:876–881.

Davidson, EH, Rast, JP, Oliveri, P, Ransick, A, Calestani, C, Yuh, CH, et al. (2002). A genomic regulatory network for development. *Science,* 295:1669–1678.

de Hoog, CL and Mann, M (2004). Proteomics. *Annu Rev Genomics Hum Genet,* 5:267–293.

DeRisi, J, Penland, L, Brown, PO, Bittner, ML, Meltzer, PS, Ray, M, et al. (1996). Use of a cDNA microarray to analyse gene expression patterns in human cancer. *Nat Genet,* 14:457–460.

Drosten, C, Günther, S, Preiser, W, van der Werf, S, Brodt, HR, Becker S, et al. (2003). Identification of a novel coronavirus in patients with severe acute respiratory syndrome. *N Engl J Med,* 348:1967–1976.

Eisen, MB, Spellman, PT, Brown, PO, and Botstein, D (1998). Cluster analysis and display of genome-wide expression patterns. *Proc Natl Acad Sci USA,* 95:14863–14868.

Eriksson, M, Brown, WT, Gordon, LB, Glynn, MW, Singer, J, Scott, L, et al. (2003). Recurrent de novo point mutations in lamin A cause Hutchinson-Gilford progeria syndrome. *Nature,* 423:293–298.

Espina, V, Woodhouse, EC, Wulfkuhle, J, Asmussen, HD, Petricoin, EF III, and Liotta, LA (2004). Protein microarray detection strategies: focus on direct detection technologies. *J Immunol Methods,* 290:121–133.

Evans, WE and McLeod, HL (2003). Pharmacogenomics—drug disposition, drug targets, and side effects. *N Engl J Med,* 348:538–549.

Farooqi, IS, Keogh, JM, Yeo, GSH, Lank, EJ, Cheetham, E, and O'Rahilly, S (2003). Clinical spectrum of obesity and mutations in the melanocortin 4 receptor gene. *N Engl J Med,* 348:1085–1095.

Fenn, JB, Mann, M, Meng, CK, Wong, SF, and Whitehouse, CM (1989). Electrospray ionization for mass spectrometry of large biomolecules. *Science,* 246:64–71.

Ferguson, JA, Boles, TC, Adams, CP, and Walt, DR (1996). A fiber-optic DNA biosensor microarray for the analysis of gene expression. *Nat Biotechnol,* 14:1681–1684.

Figeys, D and Pinto, D (2001). Proteomics on a chip: promising developments. *Electrophoresis,* 22:208–216.

Fodor, SP, Read, JL, Pirrung, MC, Stryer, L, Lu, AT, and Solas, D (1991). Light-directed, spatially addressable parallel chemical synthesis. *Science,* 251:767–773.

Francke, S, Manraj, M, Lacquemant, C, Lecoeur, C, Leprêtre, F, Passa, P, et al. (2001). A genome-wide scan for coronary heart disease suggests in Indo-Mauritians a susceptibility locus on chromosome 16p13 and replicates linkage with the metabolic syndrome on 3q27. *Hum Mol Genet,* 10:2751–2765.

Golub, TR, Slonim, DK, Tamayo, P, Huard, C, Gaasenbeek, M, Mesirov, JP, et al. (1999). Molecular classification of cancer: class discovery and class prediction by gene expression monitoring. *Science,* 286:531–537.

Gorg, A, Boguth, G, Kopf, A, Reil, G, Parlar, H, and Weiss, W (2002). Sample prefractionation with Sephadex isoelectric focusing prior to narrow pH range two-dimensional gels. *Proteomics,* 2:1652–1657.

Gorg, A, Obermaier, C, Boguth, G, Harder, A, Scheibe, B, Wildgruber, R, et al. (2000). The current state of two-dimensional electrophoresis with immobilized pH gradients. *Electrophoresis,* 21:1037–1053.

Granger, CB, Van Eyk, JE, Mockrin, SC, and Anderson, NL (2004). National Heart, Lung, and Blood Institute Clinical Proteomics Working Group report. *Circulation,* 109:1697–1703.

Gretarsdottir, S, Thorleifsson, G, Reynisdottir, ST, Manolescu, A, Jonsdottir, S, Jonsdottir, T, et al. (2003). The gene encoding phosphodiesterase 4D confers risk of ischemic stroke. *Nat Genet,* 35:131–138.

Guttmacher, AE and Collins, FS (2002). Genomic medicine—a primer. *N Engl J Med,* 347:1512–1520.

Guttmacher, AE and Collins, FS (2003). Welcome to the genomic era. *N Engl J Med,* 349:996–998.

Gygi, SP, Corthals, GL, Zhang, Y, Rochon, Y, and Aebersold, R (2000). Evaluation of two-dimensional gel electrophoresis-based proteome analysis technology. *Proc Natl Acad Sci USA,* 97:9390–9395.

Hacia, JG, Brody, LC, Chee, MS, Fodor, SPA, and Collins, FS (1996). Detection of heterozygous mutations in *BRCA1* using high density oligonucleotide arrays and two-colour fluorescence analysis. *Nat Genet,* 14:444–447.

Harrap, SB, Zammit, KS, Wong, ZY, Williams, FM, Bahlo, M, Tonkin, AM, et al. (2002). Genome-wide linkage analysis of the acute coronary syndrome suggests a locus on chromosome 2. *Arterioscler Thromb Vasc Biol* 22:874–878.

Hauser, ER, Crossman, DC, Granger, CB, Haines, JL, Jones, CJ, Mooser, V, et al. (2004). A genomewide scan for early-onset coronary artery disease in 438 families: the GENECARD Study. *Am J Hum Genet,* 75:436–447.

Hedenfalk, I, Duggan, D, Chen, Y, Radmacher, M, Bittner, M, Simon, R, et al. (2001). Gene-expression profiles in hereditary breast cancer. *N Engl J Med,* 344:539–548.

Hirschhorn, JN, Lohmueller, K, Byrne, E, Hirschhorn, K (2002). A comprehensive review of genetic association studies. *Genet Med,* 4:45–61.

Huber, LA (2003). Is proteomics heading in the wrong direction? *Nat Rev Mol Cell Biol,* 4:74–80.

Huh, WK, Falvo, JV, Gerke, LC, Carroll, AS, Howson, RW, and Weissman, JS, et al. (2003). Global analysis of protein localization in budding yeast. *Nature,* 425:686–691.

Ideker, T, Thorsson, V, Ranish, JA, Christmas, R, Buhler, J, Eng, JK, et al. (2001). Integrated genomic and proteomic analyses of a systematically perturbed metabolic network. *Science,* 292:929–934.

Issaq, HJ (2001). The role of separation science in proteomics research. *Electrophoresis,* 22:3629–3638.

Issaq, HJ, Veenstra, TD, Conrads, TP, and Felschow, D (2002). The SELDI-TOF MS approach to proteomics: protein profiling and biomarker identification. *Biochem Biophys Res Commun,* 292:587–592.

Karas, M and Hillenkamp, F (1988). Laser desorption ionization of proteins with molecular masses exceeding 10,000 daltons. *Anal Chem,* 60:2299–2301.

Khan, J, Wei, JS, Ringner, M, Saal, LH, Ladanyi, M, Westermann, F, et al. (2001). Classification and diagnostic prediction of cancers using gene expression profiling and artificial neural networks. *Nat Med,* 7:673–679.

Khoury, MJ, McCabe, LL, and McCabe, ERB (2003). Population screening in the age of genomic medicine. *N Engl J Med,* 348:50–58.

Kitano, H (2002). Computational systems biology. *Nature,* 420:206–210.

Klein, RJ, Zeiss, C, Chew, EY, Tsai, JY, Sackler, RS, Haynes, C, et al. (2005). Complement factor H polymorphism in age-related macular degeneration. *Science,* 308:385–389.

Ko, MSH (2001). Embryogenomics: developmental biology meets genomics. *Trends Biotechnol,* 19:511–518.

Komajda, M and Charron, P. (2001). The heart of genomics. *Nat Med,* 7:287–288.

Kononen, J, Bubendorf, L, Kallioniemi, A, Barlund, M, Schraml, P, Leighton, S, et al. (1998). Tissue microarrays for high-throughput molecular profiling of tumor specimens. *Nat Med,* 4:844–847.

Kramer, R and Cohen, D (2004). Functional genomics to new drug targets. *Nat Rev Drug Discov,* 3:965–972.

Ksiazek, TG, Erdman, D, Goldsmith, CS, Zaki, SR, Peret, T, Emery, S, et al. (2003). A novel coronavirus associated with severe acute respiratory syndrome. *N Engl J Med,* 348:1953–1966.

Kuruvilla, FG, Shamji, AF, Sternson, SM, Hergenrother, PJ, and Schreiber, SL (2002). Dissecting glucose signalling with diversity-oriented synthesis and small-molecule microarrays. *Nature,* 416:653–657.

LaBaer, J and Ramachandran, N (2005). Protein microarrays as tools for functional proteomics. *Curr Opin Chem Biol,* 9:14–19.

Lander, ES (1996). The new genomics: global views of biology. *Science,* 274:536–539.

Lander, ES, Linton, LM, Birren, B, Nusbaum, C, Zody, MC, Baldwin, J, et al. (2001). Initial sequencing and analysis of the human genome. *Nature,* 409:860–921 (Erratum, *Nature,* 2001;411:720, 2001;412:565).

Legrain, P, Jestin, JL, and Schachter, V (2000). From the analysis of protein complexes to proteome-wide linkage maps. *Curr Opin Biotechnol,* 11:402–407.

Levene, AP, Morgan, GJ, and Davies, FE (2003). The use of genetic microarray analysis to classify and predict prognosis in haematological malignancies. *Clin Lab Haematol,* 25:209–220.

Liotta, LA, Kohn, EC, and Petricoin, EF (2001). Clinical proteomics: personalized molecular medicine. *JAMA,* 286:2211–2214.

Liotta, L and Petricoin, E (2000). Molecular profiling of human cancer. *Nat Rev Genet,* 1:48–56.

Lipardi, C, Wei, Q, and Paterson, BM (2001). RNAi as random degradative PCR: siRNA primers convert mRNA into dsRNAs that are degraded to generate new siRNAs. *Cell,* 107:297–307.

Lohmueller, KE, Pearce, CL, Pike, M, Lander, ES, and Hirschhorn, JN (2003). Meta-analysis of genetic association studies supports a contribution of common variants to susceptibility to common disease. *Nat Genet,* 33:177–182.

Loscalzo, J (2003). Proteomics in cardiovascular biology and medicine. *Circulation,* 108:380–383.

MacBeath, G (2002). Protein microarrays and proteomics. *Nat Genet,* 32:526–532.

MacBeath, G and Schreiber, S (2000). Printing proteins as microarrays for high-throughput function determination. *Science,* 289:1760–1763.

Mei, R, Galipeau, PC, Prass, C, Berno, A, Ghandour, G, Patil, N, et al. (2000). Genome-wide detection of allelic imbalance using human SNPs and high-density DNA arrays. *Genome Res,* 10:1126–1137.

Merchant, M and Weinberger, SR (2000). Recent advancements in surface-enhanced laser desorption/ionization-time of flight-mass spectrometry. *Electrophoresis,* 21:1164–1177.

Miklos, GL and Rubin, GM (1996). The role of the genome project in determining gene function: insights from model organisms. *Cell,* 86:521–529.

Ming, JE, Roessler, E, and Muenke, M (1998). Human developmental disorders and the Sonic hedgehog pathway. *Mol Med Today,* 4:343–349.

Mitchell, P (2002). A perspective on protein microarrays. *Nat Biotechnol,* 20:225–229.

Noble, D (2002). Modeling the heart—from genes to cells to the whole organ. *Science,* 295:1678–1682.

Ozaki, K, Inoue, K, Sato, H, Iida, A, Ohnishi, Y, Sekine, A, et al. (2004). Functional variation in *LGALS2* confers risk of myocardial infarction and regulates lymphotoxin-a secretion *in vitro. Nature,* 429:72–75.

Ozaki, K, Ohnishi, Y, Iida, A, Sekine, A, Yamada, R, Tsunoda, T, et al. (2002). Functional SNPs in the lymphotoxin-a gene that are associated with susceptibility to myocardial infarction. *Nat Genet,* 32:650–654.

Pajukanta, P, Cargill, M, Viitanen, L, Nuotio, I, Kareinen, A, Perola, M, et al. (2000). Two loci on chromosomes 2 and X for premature coronary heart disease identified in early- and late-settlement populations of Finland. *Am J Hum Genet,* 67:1481–1493.

Pandey, A and Mann, M (2000). Proteomics to study genes and genomes. *Nature,* 405:837–846.

Paweletz, CP, Charboneau, L, Bichsel, VE, Simone, NL, Chen, T, Gillespie, JW, et al. (2001). Reverse phase protein microarrays which capture disease progression show activation of pro-survival pathways at the cancer invasion front. *Oncogene,* 20:1981–1989.

Petricoin, EF, Hackett, JL, Lesko, LJ, Puri, RK, Gutman, SI, Chumakov, K, et al. (2002). Medical applications of microarray technologies: a regulatory science perspective. *Nat Genet,* 32:474–479.

Petricoin, EF, Zoon, KC, Kohn, EC, Barrett, JC, and Liotta, LA (2002). Clinical proteomics: translating benchside promise into bedside reality. *Nat Rev Drug Discov,* 1:683–695.

Phizicky, E, Bastiaens, PI, Zhu, H, Snyder, M, and Fields, S (2003). Protein analysis on a proteomic scale. *Nature,* 422:208–215.

Podgoreanu, MV and Schwinn, DA (2004). Genetics and the circulation. *Br J Anaesth,* 93:140–148.

Pomeroy, SL, Tamayo, P, Gaasenbeek, M, Sturla, LM, Angelo, M, McLaughlin, ME, et al. (2002). Prediction of central nervous system embryonal tumour outcome based on gene expression. *Nature,* 415:436–442.

Ramachandran, N, Hainsworth, E, Bhullar, B, Eisenstein, S, Rosen, B, Lau, AY, et al. (2004). Self-assembling protein microarrays. *Science,* 305:86–90.

Redfern, O, Grant, A, Maibaum, M, and Orengo, C (2005). Survey of current protein family databases and their application in comparative, structural and functional genomics. *J Chromatogr B Analyt Technol Biomed Life Sci,* 815:97–107.

Risch, N and Merikangas, K (1996). The future of genetic studies of complex human diseases. *Science,* 273:1516–1517.

Rosenwald, A, Wright, G, Chan, WC, Connors, JM, Campo, E, Fisher, RI, et al. (2002). The use of molecular profiling to predict survival after chemotherapy for diffuse large-B-cell lymphoma. *N Engl J Med,* 346:1937–1947.

Rubin, GH, Yandell, MD, Wortman, JR, Miklos, GLG, Nelson, CR, Hariharan, IK, et al. (2000). Comparative genomics of the eukaryotes. *Science,* 287:2204–2215.

Sali, A, Glaeser, R, Earnest, T, and Baumeister, W (2003). From words to literature in structural proteomics. *Nature,* 422:216–225.

Schena, M, Shalon, D, Davis, RW, and Brown, PO (1995). Quantitative monitoring of gene expression patterns with a complementary DNA microarray. *Science,* 270:467–470.

Schweitzer, B, Predki, P, and Snyder, M (2003). Microarrays to characterize protein interactions on a whole-proteome scale. *Proteomics,* 3:2190–2199.

Shipp, MA, Ross, KN, Tamayo, P, Weng, AP, Kutok, JL, Aguiar, RC, et al. (2002). Diffuse large B-cell lymphoma outcome prediction by gene-expression profiling and supervised machine learning. *Nat Med,* 8:68–74.

Siddiqui, A, Kerb, R, Weale, ME, Brinkmann, U, Smith, A, Goldstein, DB, et al. (2003). Association of multidrug resistance in epilepsy with a polymorphism in the drug-transporter gene *ABCB1*. *N Engl J Med,* 348:1442–1448.

Speers, AE and Cravatt, BF (2004). Chemical strategies for activity-based proteomics. *Chembiochem,* 5:41–47.

Sreekumar, A, Nyati, MK, Varambally, S, Barrette, TR, Ghosh, D, Lawrence, TS, et al. (2001). Profiling of cancer cells using protein microarrays: discovery of novel radiation-regulated proteins. *Cancer Res,* 61:7585–7593.

Steinmetz, LM and Davis, RW (2004). Maximizing the potential of functional genomics. *Nat Rev Genet,* 5:190–201.

Tabor, HK, Risch, NJ, and Myers, RM (2002). Opinion: Candidate-gene approaches for studying complex genetic traits: practical considerations. *Nat Rev Genet,* 3:391–397.

Takahashi, M, Rhodes, DR, Furge, KA, Kanayama, H, Kagawa, S, Haab, BB, et al. (2001). Gene expression profiling of clear cell renal cell carcinoma: gene identification and prognostic classification. *Proc Natl Acad Sci USA,* 98:9754–9759.

The International HapMap Consortium, (2003). The International HapMap Project. *Nature,* 426:789–796.

Uetz, P, Giot, L, Cagney, G, Mansfield, TA, Judson, RS, Knight, JR, et al. (2000). A comprehensive analysis of protein–protein interactions in *Saccharomyces cerevisiae*. *Nature,* 403:623–627.

Unlu, M, Morgan, ME, and Minden, JS (1997). Difference gel electrophoresis: a single gel method for detecting changes in protein extracts. *Electrophoresis,* 18:2071–2077.

van de Vijver, MJ, He, YD, van't Veer, LJ, Dai, H, Hart, AA, Voskuil, DW, et al. (2002). A gene-expression signature as a predictor of survival in breast cancer. *N Engl J Med,* 347:1999–2009.

van't Veer, LJ, Dai, H, van de Vijver, MJ, He, YD, Hart, AA, Mao, M, et al. (2002). Gene expression profiling predicts clinical outcome of breast cancer. *Nature,* 415:530–536.

Venter, JC, Adams, MD, Myers, EW, Li, PW, Mural, RJ, Sutton, GG, et al. (2001). The sequence of the human genome. *Science,* 291:1304–1351 (Erratum, *Science,* 2001;292:1838).

Walter, G, Bussow, K, Lueking, A, and Glokler, J (2002). High-throughput protein arrays: prospects for molecular diagnostics. *Trends Mol Med,* 8:250–253.

Wang, Q, Rao, S, Shen, GQ, Li, L, Moliterno, DJ, Newby, LK, et al. (2004). Premature myocardial infarction novel susceptibility locus on chromosome 1p34-36 identified by genome-wide linkage analysis. *Am J Hum Genet,* 74:262–271.

Weinshilboum, R (2003). Inheritance and drug response. *N Engl J Med,* 348:529–537.

Willett, WC (2002). Balancing life-style and genomics research for disease prevention. *Science,* 296:695–698.

Winssinger, N, Ficarro, S, Schultz, PG, and Harris, JL (2002). Profiling protein function with small molecule microarrays. *Proc Natl Acad Sci USA,* 99:11139–11144.

Yamada, Y, Izawa, H, Ichihara, S, Takatsu, F, Ishihara, H, Hirayama, H, et al. (2002). Prediction of the risk of myocardial infarction from polymorphisms in candidate genes. *N Engl J Med,* 347:1916–1923.

Yao, T (2002). Bioinformatics for the genomic sciences and towards systems biology. Japanese activities in the post-genome era. *Prog Biophys Mol Biol,* 80:23–42.

Yaspo, M-L (2001). Taking a functional genomics approach in molecular medicine. *Trends Mol Med,* 7:494–502.

Yoshioka, J and Lee, RT (2003). Cardiovascular genomics. *Cardiovasc Pathol,* 12:249–254.

Zhu, H, Bilgin, M, Bangham, R, Hall, D, Casamayor, A, Bertone, P, et al. (2001). Global analysis of protein activities using proteome chips. *Science,* 293:2101–2105.

Zhu, H and Snyder, M (2003). Protein chip technology. *Curr Opin Chem Biol,* 7:55–63.

4

Epigenomics and Human Disease

Santiago Uribe Lewis, Christopher Everett, and Richard Festenstein

The different cellular phenotypes that compose multicellular organisms are generated by the expression of housekeeping and cell-type-specific genes and by the repression of inappropriate ones. The pattern of gene expression that defines a cell type is termed the "epigenotype," which is established and maintained by "epigenetic" mechanisms that are able to govern gene expression regardless of the underlying genetic code (Holliday, 2005). Genomic imprinting, where genes are expressed from only one of the inherited parental alleles, represents a classical example of epigenetic gene regulation; memory of the expression state, presumably established during gametic meiosis, is thus transmitted to the zygote, maintained throughout embryonic and postembryonic development, and reestablished during gametogenesis of the newly formed organism in a sex-specific manner (Reik and Walter, 2001). It follows that epigenetic "plasticity" would enable pluripotent stem cells to give rise to a variety of epigenotypes. In contrast, lineage commitment of cells toward postmitotic differentiated epigenotypes requires a more stable epigenetic memory that confers their phenotypic identity.

Cells can acquire an epigenotype by modulating the availability of trans-acting factors to regulatory cis-acting genetic elements that specify gene activity or inactivity. Such availability can be controlled by the manner in which DNA is packaged as chromatin inside the nucleus. Silent genes may thus be packaged in "condensed" chromatin such as heterochromatin. In contrast, active genes may be packaged in "open" chromatin, termed euchromatin.

Chromatin is formed when approximately 147 bp of DNA associate with the histone octamer (two histone H3–H4 dimers plus two histone H2A–H2B dimers) to form the nucleosome (Kornberg and Lorch, 1999). Linker histones of the H1 class associate with the DNA between single nucleosomes, facilitating a higher level of organization, the so-called solenoid helical fibers (30-nm fibers) (Rattner and Hamkalo, 1979; Thoma et al., 1979). Chromatin architecture beyond the 30-nm fiber is less clear, but even higher-order folding (condensation) and unfolding (decondensation) are conceived as functionally relevant chromatin states.

Centromeric/pericentromeric heterochromatic regions are gathered together in the mammalian interphase nucleus as distinct domains known as centromeric clusters (Hsu et al., 1971). These clusters have characteristic biochemical features that help define heterochromatin: abundant repetitive DNA (satellite DNA) (Joseph et al., 1989), replication in mid to late S phase (Guenatri et al., 2004), histone hypoacetylation (Jeppesen et al., 1992), and specific methylation at lysine 9 of histone H3, the latter placed primarily by the histone methyltransferase Suv39h (Peters et al., 2001). These modifications, together with an RNA component (Maison et al., 2002) and possibly the RNA interference (RNAi) machinery (Fukagawa et al., 2004; Pal-Bhadra et al., 2004; Schramke et al., 2005), maintain the presence of heterochromatin protein 1 (HP1), a component of constitutive heterochromatin (Eissenberg and Elgin, 2000).

Previously regarded as a static, condensed structure, heterochromatin is now known to be a highly dynamic structure. Fluorescence recovery after photobleaching experiments using green fluorescent protein-tagged HP1 (HP1-GFP) revealed that HP1 is highly mobile at both heterochromatin and euchromatin (Cheutin et al., 2003; Festenstein et al., 2003; Schmiedeberg et al., 2004). The structural and functional integrity of heterochromatin is therefore maintained by dynamic processes. The dynamic exchange of HP1 indicates the presence of "windows of opportunity" for the binding of additional factors, and suggests that gene regulation in heterochromatin results from a dynamic competition between regulatory factors (Dillon and Festenstein, 2002).

Processes exist to allow changes in accessibility at both the chromatin fiber and the core DNA. Covalent modifications in DNA (cytosine methylation) and histones (lysine acetylation, methylation or ubiquitination, serine and threonine phosphorylation, and arginine methylation) modify the interaction between histones, DNA, and chromatin-binding factors (Fischle, Wang, and Allis, 2003; Peters and Schubeler, 2005), whereas nucleosome-remodeling factors modify core histone and DNA accessibility (Narlikar et al., 2002). It is very likely that these biochemical processes are coordinately regulated by an "epigenetic code" written as modifications in DNA and histones, and "read" by factors that specifically recognize single or combined modifications (Strahl and Allis, 2000; Turner, 2000; Narlikar et al., 2002). For example, histone acetylation, recognized by proteins with bromodomains (de la Cruz et al., 2005), as well as histone H3 lysine 4 methylation, recognized by WDR5-containing H3 lysine 4 methyltransferase complexes (Wysocka et al., 2005), are generally associated with "open" chromatin and gene expression (Jeppesen et al., 1992; Bernstein et al., 2002; Schubeler et al., 2004). On the other hand, histone H3 lysine 9 or 27 methylation, recognized by the chromodomains of HP1 and Polycomb proteins, respectively (Lachner et al., 2001; Fischle et al., 2003), and DNA methylation, recognized by methyl binding domain (MBD) proteins (Hermann et al., 2004), associate with "condensed" chromatin and gene repression (Bird and Wolffe, 1999; Lachner and Jenuwein, 2002; Dellino et al., 2004).

Methylated DNA, through MBDs, may target histone deacetylases (HDACs) (Nan et al., 1998; Ng et al., 1999) and H3 lysine 9 methyltransferases (e.g., Suv39h) to specific loci (Fuks, Hurd, Wolf, et al., 2003). In contrast, histone H3 lysine 9 methylation may bring about DNA methylation (by association with DNA de novo methyltransferases) (Fuks, Hurd, Deplus, et al., 2003) as well as HP1 (Aagaard et al., 1999; Lehnertz et al., 2003). Therefore, the DNA and histone modification machineries may crosstalk to generate condensed chromatin (Figs. 4–1A and B). DNA methylation and H3 lysine 9 methylation may, however, function independently (Lehnertz et al., 2003; Lewis et al., 2004) and thus confer different degrees of epigenetic plasticity. DNA methylation, thought to be more stable throughout meiosis and mitosis, may provide long-term (repressive) epigenetic memory, whereas histone modifications provide a more labile epigenetic mark.

The location where genes reside in the nucleus also provides a mechanism for epigenetic gene regulation. Genes may dynamically relocate to associate with nuclear structures rich in the transcription factors required for expression (Osborne et al., 2004), or be developmentally inactivated by their relocation to heterochromatic pericentromeric clusters (Brown et al., 1997). In the phenomenon of position effect variegation (PEV), genes are repressed in a proportion of the cells when abnormally juxtaposed to heterochromatin (Cattanach, 1974; Locke et al., 1988; Festenstein et al., 1996; Henikoff, 1996). The point that gene expression is affected by the placement of genes nearby heterochromatin has focused attention on the associations between genes and their native cis-acting regulatory elements. DNA sequences, such as locus control regions (Grosveld, 1999) or boundary elements/insulators (West and Fraser, 2005), are thought to regulate gene expression by establishing "permissive" chromatin domains (Festenstein and Kioussis, 2000), or by facilitating nuclear relocation and/or regulating interactions between gene promoters and enhancer or silencer elements (West and Fraser, 2005). Imprint control regions, typically found as DNA differentially methylated regions (DMRs) in imprinted gene clusters, are conceived as sites from which chromatin structures are propagated bidirectionally to control the expression of genes within imprinted domains (Soejima and Wagstaff, 2005).

Disease states with an epigenetic basis are classified into those where changes in chromatin structure at the deregulated gene(s) result from mutations in DNA sequences in the same chromosome, that is, in *cis*, and those where genetic mutations affect the genes that encode for factors that establish or maintain chromatin structures, that is, in *trans.* Table 4–1 illustrates the disease states with a confirmed or possible epigenetic basis, some of which we describe in greater detail in the text.

Genetic Mutations Affecting Epigenetic Regulation in Cis

Several human diseases are associated with the expansion of untranslated trinucleotide repeats, and their molecular pathogenesis may be mediated by the effects on chromatin packaging of nearby genes. These include Friedreich's ataxia (FRDA), myotonic dystrophy, and fragile X syndrome. Deletion of repetitive DNA may deregulate nearby genes via an epigenetic effect in facioscapulohumeral dystrophy. Nearby genes may also be deregulated by genetic mutations in imprint centers that control the expression of genes within imprinted gene domains. We have used the Beckwith–Wiedemann syndrome (BWS) to exemplify this mechanism (Table 4–1).

Friedreich's Ataxia

FRDA was described by Nicholaus Friedreich in 1863. It is the most common of the hereditary ataxias with a prevalence of 1 in 50,000. FRDA is an autosomal recessive disease with age of onset usually before the age of 25. Progressive ataxia, cardiomyopathy and associated diabetes are the core features (Harding, 1981). The most common genetic abnormality is a homozygous, expanded GAA trinucleotide repeat in the first intron of the *frataxin* gene located on chromosome 9q (Campuzano et al., 1996). Frataxin is a mitochondrial protein thought to be involved in iron metabolism [for a review, see Everett and Wood (2004), and references therein]. The correlation of the severity of certain clinical features and age of onset with the shorter of the two expanded repeats (Durr et al., 1996; Filla et al., 1996; Lamont et al., 1997; Monros et al., 1997; Montermini et al., 1997) may be explained by *frataxin* expression being inversely proportional to the length of the expanded GAA repeat, which is particularly true for smaller expansions (Campuzano et al., 1997). Therefore, residual expression from the allele with the shorter expansion may be important in modulating disease severity. Interestingly, atypical patients have GAA repeats of similar length to the patients with more classical features (Gellera et al., 1997; Geschwind et al., 1997; Montermini et al., 1997). Factors such as environment, modifier genes, and somatic mosaicism may play a role in such phenotypic variation.

Frataxin Gene Repression in FRDA

Abnormally expanded GAA repeats within the *frataxin* gene impair *frataxin* expression. Using a two-exon construct derived from the HIV *gp120* gene with 9-270 GAA repeats present in the only intron in transient transfection experiments, transcription and replication were found to be impaired in a GAA repeat length and orientation-dependent manner (Ohshima et al., 1998). In yeast, replication fork stalling occurs when the expanded GAA strand serves as the lagging strand template (Krasilnikova and Mirkin, 2004).

GAA repeats form abnormal secondary DNA structures, notably triplexes. Such structures may be responsible for impaired *frataxin* expression (Wells et al., 1988; Ohshima et al., 1996; Mariappan et al., 1999). GAA DNA triplexes formed by GAA are said to be "sticky" due to the propensity to self-associate, and "sticky" DNA may impair transcription (Sakamoto et al., 1999). GAA repeats caused pausing of RNA polymerase in vitro, due to the transient formation of an intramolecular DNA triplex initiated by RNA polymerase (Grabczyk and Usdin, 2000). In addition, "sticky" DNA impaired transcription in vitro by sequestering RNA polymerase to the "sticky" repeat (Sakamoto et al., 2001), which may also sequester transcription factors resulting in transcriptional repression. Interestingly, GAAGGA hexanucleotide repeats do not form triplexes, "sticky" DNA, or repress transcription (Ohshima et al., 1999; Sakamoto et al., 1999; Sakamoto et al., 2001). A pure GAA repeat seems necessary to cause disease as "imperfections" may interfere with the pathologic properties. Thus, inclusion of GGA imperfections in long GAA repeats reduced "sticky" DNA formation and relieved in vitro transcriptional repression (Sakamoto et al., 2001). It is likely that the abnormal DNA structure formed by GAA repeat expansions may serve to affect DNA packaging, and therefore chromatin configuration.

Recently, experiments in transgenic mice have shown that GAA repeat–induced gene silencing may be associated with further condensation of the chromatin packaging of DNA. A $(GAA)_{200}$ repeat

Table 4–1 Genetic mutations generating "epigenetic" disease in cis or trans

	Mutation	Disease	OMIM
In cis	Triplet repeat expansion	Friedreich ataxia	229300
		Myotonic dystrophy	160900
		Fragile X syndrome	158900
	Repeat contraction	Facioscapulohumeral dystrophy	309550
	Locus control region	β-Thalassemia	141900
	Imprint control region and/or chromatin boundary genetic or epigenetic mutation	Beckwith–Wiedemann syndrome	130650
	Imprint control region genetic or epigenetic mutation	Prader–Willi/Angelman syndrome	176270/105830
	Imprint control region genetic or epigenetic mutation	TNMD	601410
	Imprinted gene genetic and/or epigenetic mutation	AHO/PHP-Ia and PHP-Ib	103580 and 603233
In trans	DNMT3B, DNA methyltransferase	ICF syndrome	242860
	MECP2, methyl DNA binding protein	Rett syndrome	312750
	ATRX, chromatin remodeller	ATR-X syndrome	301040
	NSD1, histone methyl transferase	Sotos syndrome	117550
	RSK2, histone H3 (?) kinase	Coffin–Lowry syndrome	303600
	SMARCAL1, chromatin remodeller?	SIOD	242900
	CBP, CREB binding protein.	Rubinstein–Taybi syndrome	180849
	Hairless	Atrichia	209500
	Emerin	EDMD	310300
	Lamin B receptor	Pelger–Huet anomaly	169400

Abbreviations: OMIM, online Mendelian inheritance in man (www.ncbi.nlm.nih.gov/entrez/query.fcgi?db=OMIM); TNMD, transient neonatal diabetes mellitus; AHO/PHP-Ia/PHPIb, Albright hereditary osteodistrophy/pseudo-hypoparathyroidism type Ia/Ib; ICF, immunodeficiency, centromere instability, and facial anomalies; SIOD, Schimke immuno-osseous dysplasia; EDMD is X-linked Emery–Dreifuss muscular dystrophy (additional examples can be found in Hendrich and Bickmore, 2001; Bickmore and van der Maarel, 2003; Jiang, Bressler, et al., 2004; Robertson, 2005).

expansion was linked to a human CD2 (hCD2) reporter. The direct inhibitory transcriptional effects of GAA repeats on DNA transcription were excluded as the GAA repeat was linked to the 3′ untranscribed region of the *hCD2* transgene. In this transgenic mouse model, the *hCD2* reporter gene alone is sensitive to juxtaposition to constitutively tightly packaged DNA (heterochromatin), for example, centromeres, and results in variegated hCD2 expression on T cells, or PEV (Festenstein et al., 1996). In PEV, rather than gene expression being silenced in all cells, a proportion of cells become silenced with the remaining continuing to express. Linking a GAA repeat expansion to the *hCD2* transgene also resulted in PEV (Saveliev et al., 2003), but importantly this occurred even when the transgene was situated in regions of the chromosome that are usually loosely packaged in euchromatin, suggesting that the presence of GAA repeats induced chromatin condensation and heterochromatin formation. In T cells where *hCD2* was silenced, DNase I hypersensitive site analysis showed condensation of chromatin packaging at the promoter of the gene. This silencing was also modified by altering the dosage of HP1(an important component of heterochromatin (Saveliev et al., 2003). Similar condensation of chromatin packaging at the expanded GAA repeat, which may also affect chromatin packaging at nearby regulatory elements (e.g., the promoter), may occur in the mutant *frataxin* gene in FRDA (Figs. 4–1C and D). Interestingly, the phenotypic variation seen in FRDA could be due to variegation of *frataxin* gene expression. A longer repeat may induce "enhanced" variegation with few cells expressing frataxin and a more severe phenotype. Modifiers of epigenetic silencing such as HP1(may also play a role. Histone deacetylation is associated with condensed chromatin and gene silencing; and interestingly, upregulation of *frataxin* expression occurred with butyric acid (an HDAC inhibitor) treatment in cells lines containing constructs of the entire human FRDA functional sequence (Sarsero et al., 2003). Recently, a novel HDAC inhibitor has been shown to upregulate *frataxin* in lymphoid cells from patients with FRDA providing a potential therapy that is currently being investigated in mice (Festenstein, 2006; Herman et al., 2006).

In the GAA repeat *frataxin* knockin mouse, with 230 GAA repeats placed in the first intron of the *frataxin* gene, there was only a small reduction in *frataxin* expression in homozygous mice with no associated phenotype. Even when frataxin expression was reduced to 25%–30% of wild type in compound heterozygous mice carrying both a GAA repeat expansion and a non-functional *frataxin* allele, there was apparently no phenotype (Miranda et al., 2002). Frataxin levels of less than 25% are probably required for a phenotype to develop in mice. In FRDA lymphoblastoid cell lines, frataxin expression determined by immunoblotting ranged from 4% to 29% compared to normal (Campuzano et al., 1997). One explanation for the relatively small repression of *frataxin* observed in the GAA-knockin mice would be that a repeat expansion of 230 may not be sufficient to affect chromatin structures such that *frataxin* is not repressed enough to cause an observable phenotype.

Myotonic Dystrophy

Myotonic dystrophy is an autosomal dominant disease caused by an unstable CTG repeat expansion located in the 3′ untranslated region (UTR) of the *DMPK* gene on chromosome 19q (Brook et al., 1992; Fu et al., 1992; Mahadevan et al., 1992), and is the most common adult form of muscular dystrophy. It is a multisystem disorder typically causing myotonia, progressive muscular dystrophy, cataracts, cardiac conduction defects, and endocrine anomalies.

Although myotonic dystrophy is now thought to be largely due to abnormal nuclear CUG RNA processing and sequestration of nuclear proteins (Jiang, Mankodi, et al., 2004), the CTG expansion

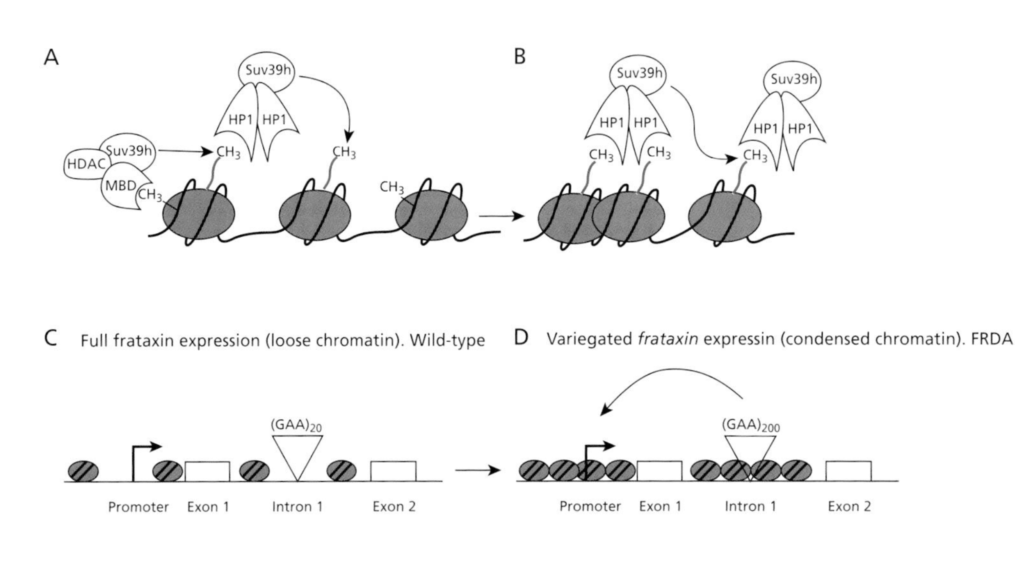

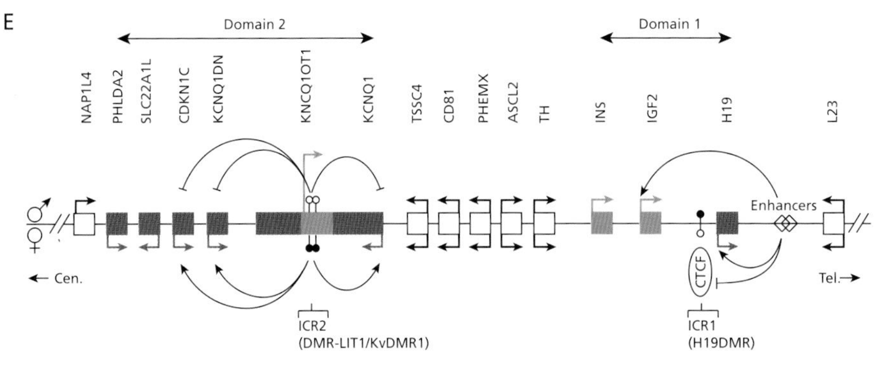

Figure 4–1 Chromatin and disease. (*A*) Schematic representation of chromatin condensation. DNA (thick black line) wraps around histone octamers (gray oval) to form nucleosomes. Methylated (CH_3) DNA is recognized by DNA MBD proteins that complex with histone deacetylases (HDAC) and histone H3 methyltransferases (e.g., Suv39h) to deacetylate histones, and specifically methylate lysine 9 of histone H3. A methylated lysine 9 of H3 is specifically recognized by heterochromatin protein 1 (HP1), which interacts with Suv39h and normally exists as a dimer. (*B*) HP1 dimers may link nucleosomes methylated at lysine 9 of H3 by Suv39h and facilitate chromatin compaction. (*C*) Potential mechanism underlying Friedreich's ataxia. In the presence of a normal GAA repeat in intron 1 of the *frataxin* gene, the region around the repeat and the nearby promoter may lie in loosely packaged chromatin. Therefore, the gene is actively expressed. (*D*) In the presence of an expanded GAA repeat within intron 1 of the *frataxin* gene, densely packaged heterochromatin may form. This may induce similar changes at the nearby promoter (arrow). Variegated expression of *frataxin* may then occur. The proportion of *frataxin*-positive cells may depend on GAA repeat length and the degree of heterochromatinization. In transgenic mice, an untranslated GAA repeat expansion linked to a *hCD2* reporter gene was associated with variegated expression and heterochromatin formation close to the repeat and at the more distant promoter (see text). (*E*) Schematic representation of the imprinted domains 1 and 2 in chromosome 11p15.5 associated with the Beckwith–Wiedemann syndrome. Domain 1 contains insulin (*INS*), insulin-like growth factor 2 (*IGF2*), and *H19*. A differentially methylated region (DMR) upstream of *H19* (H19 DMR) is the imprint control region 1 (ICR1) that, when bound by CTCF on the unmethylated maternal chromosome (open lollipop), acts as a chromatin boundary which prevents an *IGF2*-downstream enhancers (diamonds) interaction. Domain 2 contains the *KCNQ1OT1* gene expressed from the paternal allele, and the *PHLDA2*, *SLC22A1L*, *CDKN1C*, *KCNQ1DN*, and the *KCNQ1* genes expressed from the maternal allele. A DNA DMR at the 5′ end of *KCNQ1OT1* (DMR-LIT1 or KvDMR1; methylated on the maternal allele filled lollipop) is the imprint control region for domain 2 required for repression of genes in the paternal chromosome and expression of those in the maternal chromosome. Arrows indicate transcriptional orientation. Genes with two arrows represent non-imprinted genes (modified from Robertson, 2005).

also affects the chromatin packaging of DNA. For example, in fibroblasts from myotonic dystrophy patients, the presence of a CTG expansion in the *DMPK* gene is associated with the condensation of chromatin at the *six5* enhancer present in the 3′ region of the *DMPK* gene, rendering it inaccessible to transcription factors and causing downregulation of *six5* expression (Otten and Tapscott, 1995; Klesert et al., 1997). CTG repeats also efficiently recruit nucleosomes, the basic structural element of chromatin (Wang et al., 1994; Wang and Griffith, 1995). CTG expansions also behaved similarly in the hCD2 transgenic mouse model causing gene silencing and chromatin condensation (Saveliev et al., 2003). Some features of myotonic dystrophy (particularly cataract formation) may be secondary to *deregulation* of the *six5* gene lying nearby to the CTG repeat (Inukai et al., 2000; Klesert et al., 2000; Sarkar et al., 2000), and this deregulation may be secondary to CTG repeat--induced chromatin condensation. Other diseases where DNA repeats are found in UTRs, such as spinocerebellar ataxia type 10 (Matsuura et al., 2000) and myotonic dystrophy type 2 (Liquori et al., 2001), may share similar so-called epigenetic molecular pathogenic mechanisms.

Fragile X Syndrome

Fragile X syndrome is one of the most common forms of mental retardation affecting males. At least 95% of cases are caused by the expansion of an unstable CGG triplet repeat lying in the 5′ UTR of the *FMR1* gene on the X chromosome (Verkerk et al., 1991). Expansion of the CGG repeat to greater than 200 results in the syndrome. The *FMR1* gene product (FMRP) is involved in the ribosomal regulation of RNA translation, perhaps, via the RNAi pathway (Gatchel and Zoghbi, 2005). Epigenetic mechanisms, particularly DNA methylation, are thought to underlie *FMR1* silencing.

The CGG expansion causes CpG methylation within the tract and also of the nearby upstream promoter, associated with silencing of the gene (Pieretti et al., 1991; Sutcliffe et al., 1992; Hornstra et al., 1993). In support of the hypothesis that this DNA methylation causes the disease, there are cases within fragile X families of unaffected males with unmethylated expansions (Smeets et al., 1995; de Vries et al., 1996; Wang et al., 1996). In mice, disabling the *FMR1* gene causes "anxiety" and altered social behavior (Spencer et al., 2005). Further, treatment of fragile X cells with 5-azadeoxycytidine (5-azadC, an inhibitor of DNA methylation) resulted in transcriptional activation of *FMR1* (Coffee et al., 1999), and the responsiveness to 5-azadC was inversely proportional to CGG repeat length (Coffee et al., 2002). Treatment of the same lymphoblastoid fragile X cells with Trichostatin A (TSA, an HDAC inhibitor) also resulted in moderate transcriptional activation of *FMR1*, suggesting histones are deacetylated at the inactive promoter (Chiurazzi et al., 1999). Treatment of fragile X cells with HDAC inhibitors and 5-azadC synergistically activated transcription (Chiurazzi et al., 1999). Further, there was hypoacetylation of histone H3 and H4 at the 5′ region of *FMR1* in fragile X compared to normal cells (Coffee et al., 1999). Histone acetylation, particularly of histone H4, is also inversely proportional to CGG repeat length (Coffee et al., 2002). Demethylation at the *FMR1* promoter after 5-azadC treatment also resulted in histone hypoacetylation. After the withdrawal of 5-azadC treatment, methylation at the promoter resumed with associated hypoacetylation of histones H3 and H4, suggesting a direct link between the two processes (Coffee et al., 2002). Chromatin at the fragile X *FMR1* promoter was inaccessible to restriction enzyme digestion in disease when compared to normal controls, although histone deacetylation had no effect on this higher-order chromatin packaging (Coffee et al., 2002). Therefore, histone deacetylation at the promoter may be secondary to the putative primary silencing effect of CpG methylation. Furthermore, H3 lysine 9 methylation at this region is not reversed with 5-azadC, suggesting that histone and DNA methylation are not directly linked in silencing *FMR1* (Coffee et al., 2002).

Interestingly, in a cell line derived from an unaffected individual with 400 CGG repeats and normal *FMR1* messenger RNA (mRNA) levels, there was H3 lysine 4 methylation at the promoter and repeat, as well as histone H3 lysine 9 methylation and hypoacetylation in the promoter and exon 1 (Pietrobono et al., 2005). Thus, H3 lysine 9 methylation and hypoacetylation may occur in response to the presence of the CGG repeat but are insufficient to silence the gene, again suggesting that DNA methylation is an essential factor for gene silencing in this disease. Recently, further chromatin immunoprecipitation (ChIP) experiments at the *FMR1* promoter and within exons 1 and 16 have confirmed that basal histone acetylation and H3 lysine 4 methylation are much higher in wild-type controls, and that after 5-azadC treatment in fragile X cells, histone H3 acetylation, and H3 lysine 4 methylation levels rose and H3 lysine 9 methylation levels fell. In addition, it was suggested that H3 lysine 4 and DNA methylation were the main epigenetic marks in fragile X syndrome with histone acetylation being secondary, as the elevating histone acetylation levels without *FMR1* gene reactivation had no effect on H3 lysine 4 methylation (Tabolacci et al., 2005).

Interestingly, CGG expansions within the premutation range (60–200) are associated with a recently described disorder, fragile X–associated tremor ataxia syndrome, which is thought to be mediated by the overexpression of *FMR1* (Tassone, Hagerman, Chamberlain, et al., 2000; Tassone, Hagerman, Taylor, et al., 2000) and impaired FMRP protein translation (Primerano et al., 2002) with associated neuronal intranuclear inclusion formation and neurodegeneration (Hagerman et al., 2001; Greco et al., 2005).

Facioscapulohumeral Dystrophy

Facioscapulohumeral dystrophy (FSHD) is the third most common inherited muscular dystrophy, and is inherited in an autosomal dominant fashion. It is primarily a disease of skeletal muscle, but retinal telangiectasia, sensorineural deafness, cardiac arrhythmias, and mental retardation also occur (Tawil et al., 1998; Tawil, 2004). Its molecular genetic pathogenesis is also thought to be mediated by epigenetic mechanisms.

Contraction of a tandem array of 3.3 kb D4Z4 repeats lying in the subtelomeric region of chromosome 4q is associated with FSHD (Wijmenga et al., 1992; van Deutekom et al., 1993). Ninety five percent of FSHD cases are due to this deletion, and contraction to a threshold of less than 11 repeats results in disease (Wijmenga et al., 1992; Lunt, 1998). Generally, the shorter the remaining D4Z4 tandem array, the earlier the onset and more severe the disease (Tawil et al., 1996; Lunt, 1998). Haploinsufficiency of the entire chromosome 4q subtelomeric region, including the D4Z4 tandem repeats and nearby genes, does not cause FSHD, (Tupler et al., 1996) suggesting a gain of function underlying the pathogenesis of this disease. An open reading frame (*DUX4*) exists with D4Z4 (Gabriels et al., 1999), but it is not thought to be expressed (Yip and Picketts, 2003). It is hypothesized that D4Z4 repeats are heterochromatinized and that variable spreading of heterochromatin silences the nearby genes. Loss of the repeats may result in inappropriate derepression of these nearby genes (Fisher and Upadhyaya, 1997). Specific sites within D4Z4 repeats are hypomethylated in FSHD, suggesting that epigenetic mechanisms are important in its pathogenesis (van Overveld et al., 2003). However, ChIP analysis of the FSHD locus in human-rodent somatic cell hybrid cells has suggested that regions close to the D4Z4 repeats and of the two nearby genes (*FRG1* and *ANT1*) are not packaged in heterochromatin, having histones acetylated more in the pattern of euchromatin than heterochromatin (Jiang et al., 2003). *ANT1* itself is an attractive candidate as it has been associated with another myopathy (autosomal dominant chronic progressive external ophthalmoplegia) (Kaukonen et al., 2000), although the molecular pathogenesis of this disease is very different to FSHD, being characterized by large-scale mitochondrial DNA deletions. It is not clear how the overexpression of genes nearby to the D4Z4 tandem repeat causes FSHD. Interestingly, there are two allelic forms of the subtelomeric region of chromosome 4q (4qA and 4qB) due to a 10-kb polymorphic region distal to the D4Z4 repeats. FSHD only occurs with D4Z4 deletions in 4qA (Lemmers et al., 2002), although the mechanism for this is unknown.

Derepression of the *FRG2* gene may be fundamental to the molecular pathogenesis of FSHD. In muscle from FSHD patients the *FRG2* gene, which lies nearest to the D4Z4 repeats and is usually silenced, was overexpressed with levels inversely related to the number of repeats. *FRG2* was specifically overexpressed in FSHD and not in other hereditary myopathies, suggesting that this overexpression was not an effect of muscle dysfunction in general. This was a muscle-specific effect, as *FRG2* overexpression did not occur in lymphocytes from FSHD patients (Gabellini et al., 2002). However, in other experiments *FRG2* was not expressed in FSHD muscle biopsies, only becoming upregulated in differentiating cultured FSHD myoblasts (Rijkers et al., 2004). Interestingly, *FRG1* and *ANT1*, which also lie nearby to the D4Z4 region, and are usually expressed, may also become upregulated in FSHD (Gabellini et al., 2002) although this too has since been disputed (Jiang et al., 2003). A repressor multiprotein complex, D4Z4 recognition complex (DRC), was identified containing YY1 (a transcriptional repressor) that binds to an element in D4Z4. This complex proportionally repressed the transcription of a neomycin resistance gene in HeLa cells in transfection experiments, depending on the number of binding elements included in the construct. Reducing levels of the components of this complex in HeLa cells resulted in upregulation of *FRG2*. Thus, contraction of the D4Z4 tandem array in FSHD may result in less binding of the DRC, which may derepress nearby genes in cis, with their resultant over expression (Gabellini et al., 2002). Interestingly, *FRG2* has a homolog lying close to D4Z4 repeats in the subtelomeric region of chromosome 10q. *FRG2* expression in differentiating myoblasts originated from both *FRG2* homologs on chromosome 4 and 10. Thus, in addition to a putative change in chromatin status at 4qtel allowing upregulation of *FRG2* in cis, somatic pairing of subtelomeric D4Z4 repeats on chromosome 4 and 10 may allow for activation of the chromosome 10 *FRG2* homolog in trans, when chromosome 4 D4Z4 repeats are contracted (Rijkers et al., 2004). The finding that *FRG2* upregulation in FSHD may be partially due to overexpression from the gene on chromosome 10 may help to explain FSHD occurring in patients where an extensive chromosome 4 D4Z4 deletion also included *FRG2* (Lemmers et al., 2003). However, it remains unclear whether *FRG2* overexpression is of primary pathogenic importance in FSHD. In contrast, FRG1 overexpression in transgenic mice reproduces features of FSHD strongly implicating this overexpression of this gene in the disease pathogenesis (Gabellini et al., 2006).

Others have invoked a looping mechanism whereby D4Z4 repeats mediate long-range interactions with distant controlling elements (Jiang et al., 2003). Intriguingly, in transient transfection experiments in C2C12 myoblasts, D4Z4 repeats linked to a LacZ reporter had no effect on LacZ expression but caused defective differentiation of the myoblasts (Yip and Picketts, 2003). This trans effect may be due to competition for the DRC by the presence of additional D4Z4 repeats, resulting in *deregulation* of the endogenous FSHD region.

A further aspect of FSHD molecular pathogenesis may lie in aberrant interactions of the locus with the nuclear periphery, an established heterochromatic nuclear region. Indeed, the FSHD locus from both affected and unaffected individuals localized to the nuclear periphery (Masny et al., 2004). The interaction was not mediated by D4Z4 repeats, but by more proximal DNA sequences. D4Z4 contraction may result in an abnormal locus interaction with the nuclear envelope, possibly with resultant perturbed chromatin modification and abnormal interaction with transcription factors. This may have effects on other genes localized to this nuclear domain (Masny et al., 2004). This mechanism would allow for possible deregulatory effects on multiple putative genes located in cis or in trans to the mutant FSHD locus. Others have suggested that localization of the FSHD region to the nuclear rim may render it susceptible to telomere position effects (Tam et al., 2004)—common in yeast but rare in humans (Baur et al., 2001). There was no evidence for the dissociation of the mutant FSHD locus away from the nuclear periphery, which is a potential mechanism for aberrant upregulation of genes usually silenced by proximity to heterochromatin, in these experiments. It is possible that the DRC plays a role in localizing the FSHD locus to heterochromatin. Indeed, YY1 is known to interact with heterochromatin at γ satellite pericentromeric DNA (Shestakova et al., 2004).

The precise molecular pathogenic mechanisms that underlie FSHD remain to be elucidated. However, there is accumulating evidence that epigenetic mechanisms are fundamental to its pathogenesis.

Beckwith–Wiedemann Syndrome

The BWS is a clinically heterogeneous and multigenic disorder associated with genetic and epigenetic mutations that affect the expression of imprinted genes clustered at chromosome 11p15.5. The involvement of genomic imprinting in the phenotype was suggested by the preferential loss of maternal alleles in BWS-related tumors (Henry et al., 1991) and the maternal inheritance of the autosomal dominant forms of the condition (Moutou et al., 1992).

The present findings may include macrosomia (prenatal and/or postnatal overgrowth), hemihyperplasia, enlarged tongue (macroglossia), abdominal wall defects (exomphalos), embryonal tumors (most commonly Wilms' tumor), ear anomalies, visceromegaly, renal abnormalities, and neonatal hypoglycemia (Weksberg et al., 2003). Additional supportive findings may include polyhydramnios and prematurity, an enlarged placenta, cardiomegaly, hemangioma, cleft palate, advanced bone age and characteristic facies with midfacial hypoplasia and infraorbital creases.

The chromosome 11p15.5 cluster has two independently regulated domains of imprinted genes that are highly conserved between mouse and human. Domain 1 (Fig. 4–1E) contains the paternally expressed *IGF2* gene, and approximately 70 kb downstream is the maternally expressed *H19* gene. IGF2 is a fetal-specific growth factor, whereas *H19* encodes an apparently untranslated transcript. In mice, both *Igf2* and *H19* require a common set of downstream enhancers for expression (Leighton et al., 1995). Access to the enhancers by *Igf2* or *H19* is controlled by the imprint control region 1(ICR1; *H19*DMR) present in between the two genes (approximately 2 kb upstream of *H19*). In mice, ICR1 has been shown to act as a chromatin boundary element when bound by the zinc finger protein, CTCF (Bell and Felsenfeld, 2000). Binding of CTCF to the unmethylated DMR1 on the maternal allele facilitates *H19* promoter–enhancer interactions, but blocks *Igf2* promoter–enhancer interactions. CTCF cannot bind to methylated DNA. Therefore, the chromatin boundary is not functional in the paternally methylated allele, and *Igf2* is able to interact with the downstream enhancers (Lewis and Murrell, 2004). CTCF is also thought to protect the maternal allele from DNA methylation during embryogenesis (Fedoriw et al., 2004). Mutations that result in *IGF2* loss of imprinting (biallelic expression) such as microdeletions affecting CTCF-binding sites within the ICR1 (H19DMR) (Prawitt et al., 2005), which are thought to contribute but not to be sufficient to produce symptomatology, underscore the functional relevance of the ICR1 in the control of imprinted expression of domain 1 in humans.

Domain 2 or KIP2/LIT1 domain contains at least six imprinted genes (Fig. 4–1C). *KCNQ1OT1* (*LIT1*) is the only gene expressed exclusively from the paternal allele (Lee et al., 1999). Other imprinted genes in the region, *PHLDA2* (*IPL*), *SLC22A1L* (*IMPT1*), *CDKN1C* (*p57KIP2*), *KCNQ1DN*, and *KCNQ1* (*KvLQT1*), are expressed preferentially or exclusively from the maternal allele. A DMR at a CpG island 5′ of *KCNQ1OT1*, the DMR-*LIT1*, is the functional imprint control region for the KIP2/LIT1 domain that is methylated on the maternal allele. The mouse homolog of DMR-*LIT1* was shown to regulate the neighboring imprinted genes within the domain in cis

(Fitzpatrick et al., 2002). Deletion of mouse DMR–*LIT1* on the paternal chromosome resulted in biallelic expression of the genes normally silent on this allele. Similar results were obtained using human microcell hybrids; deletion of the paternal DMR-*LIT1* led to derepression of *CDKN1C*, *KCNQ1DN*, and *KCNQ1* from the paternal allele (Horike et al., 2000). The evidence therefore indicates that an unmethylated DMR-*LIT1* on the paternal allele acts in cis to silence maternal-specific genes on the paternal chromosome. However, in humans microdeletion of *LIT1*, including the DMR, showed no phenotype when inherited paternally but caused BWS and reduced *CDKN1C* expression when inherited maternally (Niemitz et al., 2004). This indicates that DMR-*LIT1* may be required for the activation of genes in the maternal allele. Detailed molecular analysis of modifications in DNA and histones at the DMR-*LIT1* (also called KvDMR1) in mouse (Lewis et al., 2004; Umlauf et al., 2004) and humans (Diaz-Meyer et al., 2003; Higashimoto et al., 2003; Murrell et al., 2004; Beatty et al., 2006) have now paved a way toward elucidating the mechanisms of KvDMR1 function.

Most cases of BWS are sporadic with no obvious genetic mutations, and are therefore thought to be epigenetic in nature. The most common epigenetic alteration associated with BWS (50% of cases) is the loss of methylation at KvDRM1 (Smilinich et al., 1999) associated with the loss of imprinting of *KCNQ1OT1* (Lee et al., 1999; Weksberg et al., 2001). Loss of imprinting of *IGF2*, without changes in DNA methylation at *H19*, is observed in 25%–50% of cases (Weksberg et al., 1993), and with concomitant *H19* hypermethylation in 2% of cases. Some cases where loss of imprinting for KvDMR1/*KCNQ1OT1* occurs also show loss of imprinting of the *IGF2* gene (Lee et al., 1999; Mitsuya et al., 1999), indicating a potential interaction between the two domains or perhaps a common trans-acting regulator.

Genetic alterations amid chromosome 11p15.5 in BWS have been characterized. These mutations may not always be mutually exclusive to the epigenetic alterations mentioned earlier. Sporadic paternal uniparental disomy and mutations in *CDKN1C* contribute to approximately 5%–20% of cases (Weksberg et al., 2003). Autosomal dominant mutations in *CDKN1C* are observed in 25% of cases (Weksberg et al., 2003). Sporadic chromosome 11p15 duplication, translocation, and inversion, which may also be recurrent in families, are observed in 1% of cases (Weksberg et al., 2003).

BWS also associates with mutations in genes outside chromosome 11p15.5, such as the *NSD1* gene on chromosome 5—encoding a histone methyltransferase [also mutated in Sotos syndrome, Online Mendelian Inheritance in Man (OMIM) 117550] (Baujat et al., 2004). Further, mouse Eed (a polycomb group protein) is essential for *Kcnq1* imprinting (Mager et al., 2003). These observations underscore the multigenic nature of BWS, and suggest that undetected mutations in factors potentially affecting chromatin structure in the 11p15.5 domains in trans may contribute to the epigenetic defects observed in sporadic cases of BWS.

Genetic Mutations Affecting Epigenetic Regulation in Trans

ICF Syndrome: The Importance of DNA Methylation

Immunodeficiency, Centromere Instability, and Facial Anomalies (ICF) syndrome is a rare autosomal recessive disorder in which patients display immunodeficiency, centromere instability (association, breakage, and stretching of the pericentromeric heterochromatin of chromosomes 1, 9, and 16), and facial anomalies. Mental retardation and developmental delay are also observed (Smeets et al., 1994). The disease maps to chromosome 20, and the gene encoding for the DNA de novo methyltransferase DNMT3B. Mutations in both the catalytic domain responsible for methyltransferase activity and the aminoterminal end of the protein, which is likely to be responsible for its targeting to pericentromeric sequences (Bachman et al., 2001; Chen et al., 2004), are present in ICF patients (Wijmenga et al., 2000; Shirohzu et al., 2002). Even though deletion of the Dnmt3b catalytic domain in mice results in perinatal lethality (Okano et al., 1999), missense mutations in this domain in ICF patients probably impair rather than completely abolish enzymatic activity (Wijmenga et al., 2000).

Reduced DNA methylation levels in ICF cells are not global but primarily observed in specific DNA sequences. As expected from the sites of chromosomal instability, there is hypomethylation of pericentromeric satellite 2 and 3 DNA in many cell types including lymphocytes (Jeanpierre et al., 1993). Undermethylation of sequences within the inactive X chromosome did not, however, affect the inactivation process, and sex-specific differences in disease severity are not observed. A whole genome scan to identify sequences that are consistently hypomethylated in lymphoblasts from ICF patients showed no global differences in methylation levels compared to controls, but detected undermethylation of two types of repeats, one of which was the D4Z4 repeats implicated in FSHD (Table 4–1 and see above) (Kondo et al., 2000).

Why should particular sequences be hypomethylated in ICF cells? One possibility is that these sequences are present at the chromosomal sites where the DNMT3B enzyme is normally localized. The subcellular localization of endogenous human DNMT3B has not yet been reported, but exogenously tagged murine Dnmt3b colocalizes with pericentric heterochromatin in some cell types (Bachman et al., 2001). The domain necessary for targeting DNMT3B to heterochromatin has not been determined, but it is likely to be at the aminoterminus where there are two domains commonly found in other chromatin-associated proteins. One of these is a PHD finger similar to that found in ATRX, a chromatin remodeling protein that is concentrated at sites of heterochromatin and repetitive DNA sequences (McDowell et al., 1999). There is also hypomethylation of specific repetitive DNA sequences in ATR-X patients (but hypermethylation of other sequences). However, there is no chromosome instability reported at the sites of hypomethylation in ATR-X cells, and little phenotypic overlap between ATR-X and ICF syndromes is observed (see ATR-X below). The other domain that may be involved in targeting DNMT3B is a PWWP (conserved Proline and Tryptophan) domain, which binds DNA (Qiu et al., 2002) and based on the similarity to tudor and chromodomains (Maurer-Stroh et al., 2003) may also recognize methylated proteins. A homozygous missense mutation in the PWWP domain of *DNMT3B* has been identified in ICF sibs resulting in a serine-to-proline change that may have a profound consequence on the mutant protein's structure (Shirohzu et al., 2002). In addition, the PWWP domain of murine Dnmt3b has been shown to bind DNA nonspecifically and to be required for targeting DNA methyltransferase activity to murine pericentric sequences (Chen et al., 2004).

In some ICF patients, no mutations in *DNMT3B* are detected. In these instances, it may be possible that catalytically inactive splice variants, overexpressed in disease, compete with endogenous DNMT3B for binding to pericentromeric heterochromatin sites. Interestingly, DNA hypomethylation of pericentromeric satellite regions occurs during human hepatocarcinogenesis associated with overexpression of the splice variant DNMT3B4 that lacks catalytic domains IX and X (Saito et al., 2002). An alternative possibility is that stimulatory factors normally required for DNMT3B activity are absent or depleted. In vitro experiments have shown that the catalytic activity of mouse Dnmt3b is stimulated by the presence of human DNMT3L (Suetake et al., 2004). It is, however, not known whether *DNMT3B* splice variants are overexpressed or DNMT3L absent or underexpressed in ICF cells.

Presumably, DNA hypomethylation in ICF syndrome leads to *deregulation* of the genes that perturb cranio-facial, cerebral, and immunological development. Recently, microarray analysis was used to identify genes with significantly altered mRNA levels in ICF lymphoblastoid cell lines as compared with controls (Ehrlich et al., 2001). Many of the genes identified have known roles in immune function that could account for the immunodeficiency consistently manifest in ICF syndrome. However, no alteration of DNA methylation was detected at the promoters of any of the deregulated genes tested (Ehrlich et al., 2001) consistent with the findings of the whole-genome scan (Kondo et al., 2000). Furthermore, none of these genes are located on chromosomes 1, 9, or 16. This raises the question of how hypomethylation of specific repetitive DNA sequences in ICF patients can lead to altered expression of genes located at distant genomic sites. One possibility is that the hypomethylation of satellite DNA alters their heterochromatin properties, and that it is the physical association of deregulated genes with these domains in the nucleus, which is aberrant in ICF cells (Bickmore and van der Maarel, 2003). Silenced genes in human B and T lymphocytes have been shown to colocalize with the domains of pericentromeric heterochromatin and to relocate away form these domains upon gene activation (Brown et al., 1999). It will be of interest to determine the subnuclear localization pattern of the deregulated genes in ICF cells compared to healthy controls.

Alternatively, the loss of DNA methylation at large arrays of satellite repeats may release or recruit protein complexes and affect the balance of regulatory complexes throughout the genome. Interestingly, ICF lymphoblastoid cell lines showed altered binding patterns of HP1 with the formation of large foci containing HP1 and components of promyelocytic (PML) nuclear bodies that colocalize with chromosome 1qh and 16qh DNA sequences (Luciani et al., 2005). The altered pattern was, however, only observed during the G_2 phase of the cell cycle and not in fibroblasts. These results show that binding of HP1 is not dependent on DNA methylation and also indicate that cell-type-specific defects in the timing of heterochromatin packaging may be a major determinant of chromosomal abnormalities and gene deregulation (Luciani et al., 2005).

Rett Syndrome: Failure to Read and/or Maintain Methylated DNA

The presence of DNA methylation can influence transcription (Kass et al., 1997). One mechanism through which this influence is achieved involves the recognition of methylated DNA by MBD proteins, which then recruit co-repressor complexes containing histone deacetylase and/or histone methyltransferase activity, as well as heterochromatic proteins, to repress transcription (Bird and Wolffe, 1999; Fujita et al., 2003; Fuks, Hurd, Wolf, et al., 2003). Once the methylated locus is recognized and repressed, MBDs may also serve to maintain the repressed state after cell division. MBDs such as MECP2 (methyl-CpG binding protein 2), MBD1, and MBD2 differentially recognize methylated DNA dependent on genomic context (Jorgensen et al., 2004; Klose et al., 2005) with distinct binding patterns in chromosomal DNA, and also differ in the identity of the co-repressor complex with which they interact (Nan et al., 1998; Ng et al., 2000). A methylated cytosine mark may thus be interpreted by distinct complexes with different downstream effects on gene regulation.

Classical Rett syndrome (RTT) is characterized by the progressive loss of intellectual functioning, fine and gross motor skills and communicative abilities, deceleration of head growth, and the development of stereotypic hand movements, occurring after a period of normal development from birth to 6 months of age. After initial regression, the condition stabilizes and patients usually survive into adulthood (Weaving et al., 2005). RTT results from sporadic mutations in *MECP2* that is encoded from Xq28 (Amir et al., 1999). MECP2 contains an aminoterminal MBD, a transcriptional repressor domain (TRD) that interacts with co-repressor complexes, and a carboxyterminal WW domain shown to interact with splicing factors (Kriaucionis and Bird, 2003; Buschdorf and Stratling, 2004). In addition, splice variants of MECP2A (*MeCP2β* in mice) and MECP2B (*MeCP2α* in mice) have been characterized, the latter shown to be more abundantly expressed in somatic tissues including brain (Kriaucionis and Bird, 2004; Mnatzakanian et al., 2004). Mutations in the *MECP2* gene are diverse. Missense mutations and those at the TRD correlate with milder phenotypes than nonsense/frameshift mutations or those at the MBD. Mutations in the carboxyterminus are also frequent. Comprehensive databases focusing on collecting *MECP2* mutations have been compiled (www.mecp2.org.uk and http://mecp2.chw.edu.au/).

RTT is almost exclusively a disease of females, because *MECP2* is X-linked and patients are heterozygous for the mutated allele usually inherited from the father. After random X chromosome inactivation, typically half of the cells express the wild-type allele and the other half a mutated *MECP2*. Cases of symptom-free female carriers of such *MECP2* mutations are very rare where skewing of X chromosome inactivation prevents expression of the mutant allele (Villard et al., 2000). Males carrying comparable *MECP2* mutations rarely survive beyond 2 years and have a different more severe phenotype than RTT (Villard et al., 2000; Ravn et al., 2003). However, rare cases of males with classical RTT indicate de novo mutations occurring early in development (Topcu et al., 2002). The majority of single nucleotide changes in *MECP2* are C to T transitions at CpG hotspots. Spontaneous deamination of 5-methylcytosine to thymine, responsible for approximately one-third of point mutations giving rise to human genetic disease (Cooper and Youssoufian, 1988), is therefore thought to give rise to the de novo mutations.

Phenotype–genotype correlation is conflictive in RTT. Aside from the absence of mutation in *MECP2* in 5%–10% of patients (Weaving et al., 2005), perhaps indicating failure to detect mutations, the diverse nature of the mutations (in the MBD, TRD or WW domain) may affect protein structure/function to different extents. Phenotypic variability may also be due to skewed X chromosome inactivation, genetic background, or unidentified environmental factors. In addition, *MECP2* mutations have been found in other disorders, including neonatal-onset encephalopathy, X-linked recessive mental retardation, autism, and some patients exhibiting the Angelman phenotype (Hammer et al., 2002).

Deletion of MeCP2 in mice generates features with remarkable similarity to those observed in RTT (Chen et al., 2001; Guy et al., 2001). Mice with the absence of MeCP2 have no apparent phenotype until approximately 6 weeks of age. There follows a period of rapid regression resulting in reduced spontaneous movement, clumsy gait, irregular breathing, hind limb clasping, and tremors. Rapid progression of symptoms leads to death at approximately 8 weeks of age, and detailed brain examination revealed reduced brain and neuron size (Chen et al., 2001; Guy et al., 2001). This indicates that, aside from neurodevelopmental defects, RTT is also a neurodegenerative disorder. Expression of *Mecp2* is ubiquitous in mice, rats, and humans although levels vary widely between tissues. In the brain, *Mecp2* is preferentially expressed in neurons but not in glia, and increased numbers of high-expressing neurons are observed postnatally (Caballero and Hendrich, 2005). Strikingly, conditional deletion of *Mecp2* in the brain showed the same phenotype as "whole animal" null mutants (Chen et al., 2001; Guy et al., 2001).

MeCP2 null mice showed no major differences in global transcription levels in a microarray analysis using cells from the cortex, forebrain, or hippocampus compared to wild type (Tudor et al., 2002), although some gene expression variation was observed in a small number of genes between the two genotypes (Tudor et al., 2002). It may be that transcription changes in the absence of MeCP2 occur in a small subset of cells undetectable by this microarray analysis approach, and that MeCP2 has a transcription-independent role in the brain, or that other MBDs compensate for the absence of MeCP2 (Tudor et al., 2002).

MeCP2/MECP2 target genes in mice/humans have started to be identified (Caballero and Hendrich, 2005). ChIP analysis identified MeCP2 binding within the mouse chromosome 6 imprinted domain where *Dlx5* and its non-imprinted neighbor *Dlx6* are present (Horike et al., 2005). *Dlx5* and *Dlx6*, whose proteins regulate neurotransmitter production (Stuhmer et al., 2002), were upregulated in MeCP2-deficient animals with accompanying changes in locus chromatin structure (Horike et al., 2005). Importantly, *DLX5* loss of imprinting was observed in three out of four RTT-derived lymphocyte cell lines (Horike et al., 2005). Other potential targets include the imprinted gene *UBE3A* (ubiquitin protein ligase E3A, OMIM 601623) (Samaco et al., 2005), the absence of which associates with the Angelman syndrome where *MECP2* mutations have been observed (Hammer et al., 2002), and brain-derived neurotrophic factor (*BDNF*) (Chen et al., 2003; Martinowich et al., 2003). Using a candidate gene approach, *Hairy2a* was identified as a clear target in *Xenopus laevis* (Stancheva et al., 2003).

ATR-X Syndrome: A Connection to Chromatin Remodeling

The α-thalassemia mental retardation syndrome/X-linked is another example of genetic mutation in a factor involved in chromatin organization affecting disease loci in trans. The *ATRX* gene at chromosome Xq13.3 encodes for an SNF2-like chromatin remodeling helicase. Functional domains of ATRX include a PHD zinc finger–like motif at its aminoterminus (homologous to the PHD motifs in DNMT3A and DNMT3B) and a helicase domain at its carboxyterminus. Mutations in ATRX are clustered in these two domains, and are thought to impair its nuclear localization, protein–protein interactions, or chromatin-remodeling functions. (Gibbons et al., 1997; McDowell et al., 1999; Cardoso et al., 2000; Tang et al., 2004).

Affected individuals have low levels of α-globin subunits that favor the formation of unstable β-globin tetramers which precipitate within erythrocytes causing varying degrees of haemolysis and splenomegaly. Affected males have relatively severe mental retardation together with facial and skeletal abnormalities, urogenital abnormalities, and microcephaly, whereas heterozygous females are usually asymptomatic (Hendrich and Bickmore, 2001; Ausio et al., 2003).

ATRX may be involved in gene activation, suggested by the reduced expression of the α-globin locus in ATR-X syndrome. This, however, does not explain the additional phenotypic traits observed, and presumably deregulation of many other loci gives rise to the complex phenotype. The observation that diverse DNA methylation defects (hypermethylation at DYZ2 Y-chromosome repeats and hypomethylation of ribosomal DNA) are present in disease (Gibbons et al., 2000) indicates that ATRX is able to regulate chromatin structure at several distinct loci. Direct (stimulatory) effects of ATRX upon loci are also indicated by its association with the transcriptional regulatory death-associated protein 6 (DAXX). The DAXX-ATRX containing complex, levels of which are reduced in ATRX patient cell lines, is able to remodel nucleosomes in vitro (Xue et al., 2003). Furthermore, ATRX associates with PML bodies, which are thought to function as regulatory (activator) factor reservoirs in the nucleus (Xue et al., 2003; Wang et al., 2004).

On the other hand, chromatin remodeling by ATRX may facilitate chromatin condensation and gene silencing. ATRX was found to associate with the histone methyltransferase Enhancer of Zeste Homolog 2 (EZH2) (Cardoso et al., 1998). EZH2 is part of the polycomb group repressor complexes (PRC) that methylate histone H3 lysine 27 (PRC2) and histone H1 lysine 26 (PRC3) (Kuzmichev et al., 2004). H3 lysine 27 methylation is a mark recognized by the chromodomain of the Polycomb protein (contained within the PRC1 complex) implicated in chromatin condensation and developmentally regulated gene silencing (Cao et al., 2002; Czermin et al., 2002; Kuzmichev et al., 2002). Methylation at H1 lysine 26 may also be involved in chromatin condensation, as HP1 has been shown to specifically interact with this mark (Daujat et al., 2005). These interactions reveal potential new pathways of gene regulation where the HP1 and the polycomb group silencing pathways may be synergistic. For example, a novel polycomb group complex, PRC4, was recently shown to arise and be upregulated upon oncogenic transformation (Kuzmichev et al., 2005). PRC4 has special activity toward histone H1b lysine 26, particular native complex subunit isoforms (Eed2) and the presence of HDACs. Excess of PRC4 might lead to increased lysine 26 methylation and subsequent recognition by HP1. Dependent on the genomic context and associated complexes, HP1 may either silence or derepress the affected loci (Hiragami and Festenstein, 2005). Furthermore, ATRX and HP1 have been shown to colocalize by immunofluorescence (McDowell et al., 1999) and to directly interact (Le Douarin et al., 1996; Kourmouli et al., 2005; Lechner et al., 2005). These interactions may generate regulatory complexes that directly affect chromatin condensation.

Moreover, mutations in ATRX are frequent at the PHD zinc finger motif, which is likely to be responsible for the targeting of ATRX to pericentromeric heterochromatin (McDowell et al., 1999). Inadequate targeting of catalytically active ATRX may therefore result in ectopic binding at loci deregulated in disease. Targets for ATRX clearly need to be determined. In addition, due to the predominant localization of ATRX to pericentromeric heterochromatin, ATRX may be indirectly regulating additional loci by modulating the nature of gene association with pericentromeric heterochromatin.

Conclusion

It is clear that previously mysterious aspects of gene regulation that can be grouped under the term "epigenetic" are finally yielding to molecular biology approaches and have revealed a new level of genome organization and regulation. It is hoped that the rapid increase in our understanding of the factors and processes involved in establishing and maintaining gene expression patterns will reveal potentially powerful new therapeutic avenues for an ever increasing number of human diseases.

References

Aagaard, L, Laible, G, Selenko, P, Schmid, M, Dorn, R, Schotta, G, et al. (1999). Functional mammalian homologs of the *Drosophila* PEV-modifier Su(var) 3-9 encode centromere-associated proteins which complex with the heterochromatin component M31. *EMBO J*, 18:1923–1938.

Amir, RE, Van den Veyver, IB, Wan, M, Tran, CQ, Francke, U, and Zoghbi, HY (1999). Rett syndrome is caused by mutations in X-linked MECP2, encoding methyl-CpG-binding protein 2. *Nat Genet*, 23:185–188.

Ausio, J, Levin, DB, De Amorim, GV, Bakker, S, and Macleod, PM (2003). Syndromes of disordered chromatin remodeling. *Clin Genet*, 64:83–95.

Bachman, KE, Rountree, MR, and Baylin, SB (2001). Dnmt3a and Dnmt3b are transcriptional repressors that exhibit unique localization properties to heterochromatin. *J Biol Chem*, 276:32282–32287.

Baujat, G, Rio, M, Rossignol, S, Sanlaville, D, Lyonnet, S, Le Merrer, M, et al. (2004). Paradoxical NSD1 mutations in Beckwith–Wiedemann syndrome and 11p15 anomalies in Sotos syndrome. *Am J Hum Genet*, 74:715–720.

Baur, JA, Ying, Z, Shay, JW, and Wright, WE (2001). Telomere position effect in human cells. *Science*, 292:2075–2077.

Beatty, L, Weksberg, R, and Sadowski, PD (2006). Detailed analysis of the methylation patterns of the KvDMR1 imprinting control region of human chromosome 11. *Genomics*, 87:46–56.

Bell, AC and Felsenfeld, G (2000). Methylation of a CTCF-dependent boundary controls imprinted expression of the Igf2 gene. *Nature*, 405:482–485.

Bernstein, BE, Humphrey, EL, Erlich, RL, Schneider, R, Bouman, P, Liu, JS, et al. (2002). Methylation of histone H3 Lys 4 in coding regions of active genes. *Proc Natl Acad Sci USA*, 99:8695–8700.

Bickmore, WA and van der Maarel, SM (2003). Perturbations of chromatin structure in human genetic disease: recent advances. *Hum Mol Genet*, 12 (Spec. No. 2) :R207–213.

Bird, AP and Wolffe, AP (1999). Methylation-induced repression—belts, braces, and chromatin. *Cell*, 99:451–454.

Brook, JD, McCurrach, ME, Harley, HG, Buckler, AJ, Church, D, Aburatani, H, et al. (1992). Molecular basis of myotonic dystrophy: expansion of a trinucleotide (CTG) repeat at the 3′ end of a transcript encoding a protein kinase family member. *Cell*, 69:385.

Brown, K, Guest, S, Smale, S, Hahm, K, Merkenschlager, M, and Fisher, A (1997). Association of transcriptionally silent genes with Ikaros complexes at centromeric heterochromatin. *Cell*, 91:845–854.

Brown, KE, Baxter, J, Graf, D, Merkenschlager, M, and Fisher, AG (1999). Dynamic repositioning of genes in the nucleus of lymphocytes preparing for cell division. *Mol Cell*, 3:207–217.

Buschdorf, JP and Stratling, WH (2004). A WW domain binding region in methyl-CpG-binding protein MeCP2: impact on Rett syndrome. *J Mol Med*, 82:135–143.

Caballero, IM and Hendrich, B (2005). MeCP2 in neurons: closing in on the causes of Rett syndrome. *Hum Mol Genet*, 14 (Spec. No. 1) :R19–R26.

Campuzano, V, Montermini, L, Lutz, Y, Cova, L, Hindelang, C, Jiralerspong, S, et al. (1997). Frataxin is reduced in Friedreich ataxia patients and is associated with mitochondrial membranes. *Hum Mol Genet*, 6:1771–1780.

Campuzano, V, Montermini, L, Molto, MD, Pianese, L, Cossee, M, Cavalcanti, F, et al. (1996). Friedreich's ataxia: autosomal recessive disease caused by an intronic GAA triplet repeat expansion. *Science*, 271:1423–1427.

Cao, R, Wang, L, Wang, H, Xia, L, Erdjument-Bromage, H, Tempst, P, et al. (2002). Role of histone H3 lysine 27 methylation in Polycomb-group silencing. *Science*, 298:1039–1043.

Cardoso, C, Lutz, Y, Mignon, C, Compe, E, Depetris, D, Mattei, MG, et al. (2000). ATR-X mutations cause impaired nuclear location and altered DNA binding properties of the XNP/ATR-X protein. *J Med Genet*, 37:746–751.

Cardoso, C, Timsit, S, Villard, L, Khrestchatisky, M, Fontes, M, and Colleaux, L (1998). Specific interaction between the XNP/ATR-X gene product and the SET domain of the human EZH2 protein. *Hum Mol Genet*, 7:679–684.

Cattanach, BM (1974). Position effect variegation in the mouse. *Genet Res* 23:291–306.

Chen, RZ, Akbarian, S, Tudor, M, and Jaenisch, R (2001). Deficiency of methyl-CpG binding protein-2 in CNS neurons results in a Rett-like phenotype in mice. *Nat Genet* 27:327–331.

Chen, T, Tsujimoto, N, and Li, E (2004). The PWWP domain of Dnmt3a and Dnmt3b is required for directing DNA methylation to the major satellite repeats at pericentric heterochromatin. *Mol Cell Biol*, 24:9048–9058.

Chen, WG, Chang, Q, Lin, Y, Meissner, A, West, AE, Griffith, EC, et al. (2003). Derepression of BDNF transcription involves calcium-dependent phosphorylation of MeCP2. *Science*, 302:885–889.

Cheutin, T, McNairn, AJ, Jenuwein, T, Gilbert, DM, Singh, PB, and Misteli, T (2003). Maintenance of stable heterochromatin domains by dynamic HP1 binding. *Science*, 299:721–725.

Chiurazzi, P, Pomponi, MG, Pietrobono, R, Bakker, CE, Neri, G, and Oostra, BA (1999). Synergistic effect of histone hyperacetylation and DNA demethylation in the reactivation of the FMR1 gene. *Hum Mol Genet*, 8:2317–2323.

Coffee, B, Zhang, F, Ceman, S, Warren, ST, and Reines, D (2002). Histone modifications depict an aberrantly heterochromatinized FMR1 gene in fragile x syndrome. *Am J Hum Genet*, 71:923–932.

Coffee, B, Zhang, F, Warren, ST, and Reines, D (1999). Acetylated histones are associated with FMR1 in normal but not fragile X-syndrome cells. *Nat Genet*, 22:98–101.

Cooper, DN and Youssoufian, H (1988). The CpG dinucleotide and human genetic disease. *Hum Genet*, 78:151–155.

Czermin, B, Melfi, R, McCabe, D, Seitz, V, Imhof, A, and Pirrotta, V (2002). Drosophila enhancer of Zeste/ESC complexes have a histone H3 methyltransferase activity that marks chromosomal Polycomb sites. *Cell*, 111:185–196.

Daujat, S, Zeissler, U, Waldmann, T, Happel, N, and Schneider, R (2005). HP1 binds specifically to Lys26-methylated histone H1.4, whereas simultaneous Ser27 phosphorylation blocks HP1 binding. *J Biol Chem*, 280:38090–38095.

de la Cruz, X, Lois, S, Sanchez-Molina, S, and Martinez-Balbas, MA (2005). Do protein motifs read the histone code? *Bioessays*, 27:164–175.

de Vries, BB, Jansen, CC, Duits, AA, Verheij, C, Willemsen, R, van Hemel, JO, et al. (1996). Variable FMR1 gene methylation of large expansions leads to variable phenotype in three males from one fragile X family. *J Med Genet*, 33:1007–1010.

Dellino, GI, Schwartz, YB, Farkas, G, McCabe, D, Elgin, SC, and Pirrotta, V (2004). Polycomb silencing blocks transcription initiation. *Mol Cell*, 13:887–893.

Diaz-Meyer, N, Day, CD, Khatod, K, Maher, ER, Cooper, W, Reik, W, et al. (2003). Silencing of CDKN1C (p57KIP2) is associated with hypomethylation at KvDMR1 in Beckwith–Wiedemann syndrome. *J Med Genet*, 40:797–801.

Dillon, N and Festenstein, R (2002). Unravelling heterochromatin: competition between positive and negative factors regulates accessibility. *Trends Genet*, 18:252–258.

Durr, A, Cossee, M, Agid, Y, Campuzano, V, Mignard, C, Penet, C, et al. (1996). Clinical and genetic abnormalities in patients with Friedreich's ataxia [see comments]. *N Engl J Med*, 335:1169–1175.

Ehrlich, M, Buchanan, KL, Tsien, F, Jiang, G, Sun, B, Uicker, W, et al. (2001). DNA methyltransferase 3B mutations linked to the ICF syndrome cause dysregulation of lymphogenesis genes. *Hum Mol Genet*, 10:2917–2931.

Eissenberg, JC and Elgin, SC (2000). The HP1 protein family: getting a grip on chromatin. *Curr Opin Genet Dev*, 10:204–210.

Everett, CM and Wood, NW (2004). Trinucleotide repeats and neurodegenerative disease. *Brain*, 127:2385–2405.

Fedoriw, AM, Stein, P, Svoboda, P, Schultz, RM, and Bartolomei, MS (2004). Transgenic RNAi reveals essential function for CTCF in H19 gene imprinting. *Science*, 303:238–240.

Festenstein, R. (2006). Breaking the silence in Friedreich's ataxia. *Nat Chem Biol*, 2:512–513.

Festenstein, R and Kioussis, D (2000). Locus control regions and epigenetic chromatin modifiers. *Curr Opin Genet Dev*, 10:199–203.

Festenstein, R, Pagakis, SN, Hiragami, K, Lyon, D, Verreault, A, Sekkali, B, et al. (2003). Modulation of heterochromatin protein 1 dynamics in primary Mammalian cells. *Science*, 299:719–721.

Festenstein, R, Tolaini, M, Corbella, P, Mamalaki, C, Parrington, J, Fox, M, et al. (1996). Locus control region function and heterochromatin-induced position effect variegation. *Science*, 271:1123–1125.Filla, A, De Michele, G, Cavalcanti, F, Pianese, L, Monticelli, A, Campanella, G, et al. (1996). The relationship between trinucleotide (GAA) repeat length and clinical features in Friedreich ataxia. *Am J Hum Genet*, 59:554–560.

Fischle, W, Wang, Y, and Allis, CD (2003). Histone and chromatin cross-talk. *Curr Opin Cell Biol*, 15:172–183.

Fischle, W, Wang, Y, Jacobs, SA, Kim, Y, Allis, CD, and Khorasanizadeh, S (2003). Molecular basis for the discrimination of repressive methyl-lysine marks in histone H3 by Polycomb and HP1 chromodomains. *Genes Dev*, 17:1870–1881.

Fisher, J and Upadhyaya, M (1997). Molecular genetics of facioscapulohumeral muscular dystrophy (FSHD). *Neuromuscul Disord*, 7:55–62.

Fitzpatrick, GV, Soloway, PD, and Higgins, MJ (2002). Regional loss of imprinting and growth deficiency in mice with a targeted deletion of KvDMR1. *Nat Genet*, 32:426–431.

Fu, YH, Pizzuti, A, Fenwick, RG Jr., King, J, Rajnarayan, S, Dunne, PW, et al. (1992). An unstable triplet repeat in a gene related to myotonic muscular dystrophy. *Science*, 255:1256–1258.

Fujita, N, Watanabe, S, Ichimura, T, Tsuruzoe, S, Shinkai, Y, Tachibana, M, et al. (2003). Methyl-CpG binding domain 1 (MBD1) interacts with the Suv39h1-HP1 heterochromatic complex for DNA methylation-based transcriptional repression. *J Biol Chem*, 278:24132–24138.

Fukagawa, T, Nogami, M, Yoshikawa, M, Ikeno, M, Okazaki, T, Takami, Y, et al. (2004). Dicer is essential for formation of the heterochromatin structure in vertebrate cells. *Nat Cell Biol*, 6:784–791.

Fuks, F, Hurd, PJ, Deplus, R, and Kouzarides, T (2003). The DNA methyltransferases associate with HP1 and the SUV39H1 histone methyltransferase. *Nucleic Acids Res*, 31:2305–2312.

Fuks, F, Hurd, PJ, Wolf, D, Nan, X, Bird, AP, and Kouzarides, T (2003). The methyl-CpG-binding protein MeCP2 links DNA methylation to histone methylation. *J Biol Chem*, 278:4035–4040.

Gabellini, D, Green, MR, and Tupler, R (2002). Inappropriate gene activation in FSHD: a repressor complex binds a chromosomal repeat deleted in dystrophic muscle. *Cell*, 110:339–348.

Gabellini, D, D'Antona, G, Moggio, M, Prelle, A, Zecca, C, Adami, R, et al. (2006). Facioscapulohumeral muscular dystrophy in mice overexpressing FRG1. *Nature*, 439, 973–977.

Gabriels, J, Beckers, MC, Ding, H, De Vriese, A, Plaisance, S, van der Maarel, SM, et al. (1999). Nucleotide sequence of the partially deleted D4Z4 locus in a patient with FSHD identifies a putative gene within each 3.3 kb element. *Gene*, 236:25–32.

Gatchel, JR and Zoghbi, HY (2005). Diseases of unstable repeat expansion: mechanisms and common principles. *Nat Rev Genet*, 6:743–755.

Gellera, C, Pareyson, D, Castellotti, B, Mazzucchelli, F, Zappacosta, B, Pandolfo, M, et al. (1997). Very late onset Friedreich's ataxia without cardiomyopathy is associated with limited GAA expansion in the X25 gene. *Neurology*, 49:1153–1155.

Geschwind, DH, Perlman, S, Grody, WW, Telatar, M, Montermini, L, Pandolfo, M, et al. (1997). Friedreich's ataxia GAA repeat expansion in patients with recessive or sporadic ataxia. *Neurology*, 49:1004–1009.

Gibbons, RJ, Bachoo, S, Picketts, DJ, Aftimos, S, Asenbauer, B, Bergoffen, J, et al. (1997). Mutations in transcriptional regulator ATRX establish the functional significance of a PHD-like domain. *Nat Genet*, 17:146–148.

Gibbons, RJ, McDowell, TL, Raman, S, O'Rourke, DM, Garrick, D, Ayyub, H, et al. (2000). Mutations in ATRX, encoding a SWI/SNF-like protein, cause diverse changes in the pattern of DNA methylation. *Nat Genet*, 24:368–371.

Grabczyk, E and Usdin, K (2000). The GAA*TTC triplet repeat expanded in Friedreich's ataxia impedes transcription elongation by T7 RNA polymerase in a length and supercoil dependent manner. *Nucleic Acids Res*, 28:2815–2822.

Greco, CM, Berman, RF, Martin, RM, Tassone, F, Schwartz, PH, Chang, A, et al. (2005). Neuropathology of fragile X-associated tremor/ataxia syndrome (FXTAS). *Brain*, 129:243–255.

Grosveld, F (1999). Activation by locus control regions? *Curr Opin Genet Dev*, 9:152–157.

Guenatri, M, Bailly, D, Maison, C, and Almouzni, G (2004). Mouse centric and pericentric satellite repeats form distinct functional heterochromatin. *J Cell Biol*, 166:493–505.

Guy, J, Hendrich, B, Holmes, M, Martin, JE, and Bird, A (2001). A mouse Mecp2-null mutation causes neurological symptoms that mimic Rett syndrome. *Nat Genet*, 27:322–326.

Hagerman, RJ, Leehey, M, Heinrichs, W, Tassone, F, Wilson, R, Hills, J, et al. (2001). Intention tremor, parkinsonism, and generalized brain atrophy in male carriers of fragile X. *Neurology*, 57:127–130.

Hammer, S, Dorrani, N, Dragich, J, Kudo, S, and Schanen, C (2002). The phenotypic consequences of MECP2 mutations extend beyond Rett syndrome. *Ment Retard Dev Disabil Res Rev*, 8:94–98.

Harding, AE (1981). Friedreich's ataxia: a clinical and genetic study of 90 families with an analysis of early diagnostic criteria and intrafamilial clustering of clinical features. *Brain*, 104:589–620.

Hendrich, B and Bickmore, W (2001). Human diseases with underlying defects in chromatin structure and modification. *Hum Mol Genet*, 10:2233–2242.

Henikoff, S (1996). Dosage-dependent modification of position-effect variegation in Drosophila. *Bioessays*, 18:401–409.

Henry, I, Bonaiti-Pellie, C, Chehensse, V, Beldjord, C, Schwartz, C, et al. (1991). Uniparental paternal disomy in a genetic cancer-predisposing syndrome. *Nature*, 351:665–667.

Hermann, A, Gowher, H, and Jeltsch, A (2004). Biochemistry and biology of mammalian DNA methyltransferases. *Cell Mol Life Sci*, 61:2571–2587.

Herman, D, Jenssen, K, Burnett, R, Soragni, E, Perlman, SL, and Gottesfeld, JM (2006). Histone deacetylase inhibitors reverse gene silencing in Friedreich's ataxia. *Nat Chem Biol*, 2:551–558.

Higashimoto, K, Urano, T, Sugiura, K, Yatsuki, H, Joh, K, Zhao, W, et al. (2003). Loss of CpG methylation is strongly correlated with loss of histone H3 lysine 9 methylation at DMR-LIT1 in patients with Beckwith--Wiedemann syndrome. *Am J Hum Genet*, 73:948–956.

Hiragami, K and Festenstein, R (2005). Heterochromatin protein 1: a pervasive controlling influence. *Cell Mol Life Sci*, 62:2711–2726.

Holliday, R (2005). DNA methylation and epigenotypes. *Biochemistry (Mosc)*, 70:500–504.

Horike, S, Cai, S, Miyano, M, Cheng, JF, and Kohwi-Shigematsu, T (2005). Loss of silent-chromatin looping and impaired imprinting of DLX5 in Rett syndrome. *Nat Genet*, 37:31–40.

Horike, S, Mitsuya, K, Meguro, M, Kotobuki, N, Kashiwagi, A, Notsu, T, et al. (2000). Targeted disruption of the human LIT1 locus defines a putative imprinting control element playing an essential role in Beckwith--Wiedemann syndrome. *Hum Mol Genet*, 9:2075–2083.

Hornstra, IK, Nelson, DL, Warren, ST, and Yang, TP (1993). High resolution methylation analysis of the FMR1 gene trinucleotide repeat region in fragile X syndrome. *Hum Mol Genet*, 2:1659–1665.

Hsu, TC, Cooper, JE, Mace, ML Jr., and Brinkley, BR (1971). Arrangement of centromeres in mouse cells. *Chromosoma*, 34:73–87.

Inukai, A, Doyu, M, Kato, T, Liang, Y, Kuru, S, Yamamoto, M, et al. (2000). Reduced expression of DMAHP/SIX5 gene in myotonic dystrophy muscle. *Muscle Nerve*, 23:1421–1426.

Jeanpierre, M, Turleau, C, Aurias, A, Prieur, M, Ledeist, F, Fischer, A, et al. (1993). An embryonic-like methylation pattern of classical satellite DNA is observed in ICF syndrome. *Hum Mol Genet*, 2:731–735.

Jeppesen, P, Mitchell, A, Turner, B, and Perry, P (1992). Antibodies to defined histone epitopes reveal variations in chromatin conformation and underacetylation of centric heterochromatin in human metaphase chromosomes. *Chromosoma*, 101:322–332.

Jiang, G, Yang, F, van Overveld, PG, Vedanarayanan, V, van der Maarel, S, and Ehrlich, M (2003). Testing the position-effect variegation hypothesis for facioscapulohumeral muscular dystrophy by analysis of histone modification and gene expression in subtelomeric 4q. *Hum Mol Genet*, 12:2909–2921.

Jiang, H, Mankodi, A, Swanson, MS, Moxley, RT, and Thornton, CA (2004). Myotonic dystrophy type 1 is associated with nuclear foci of mutant RNA, sequestration of muscleblind proteins and deregulated alternative splicing in neurons. *Hum Mol Genet*, 13:3079–3088.

Jiang, YH, Bressler, J, and Beaudet, AL (2004). Epigenetics and human disease. *Annu Rev Genomics Hum Genet*, 5:479–510.

Jorgensen, HF, Ben-Porath, I, and Bird, AP (2004). Mbd1 is recruited to both methylated and nonmethylated CpGs via distinct DNA binding domains. *Mol Cell Biol*, 24:3387–3395.

Joseph, A, Mitchell, AR, and Miller, OJ (1989). The organization of the mouse satellite DNA at centromeres. *Exp Cell Res*, 183:494–500.

Kass, SU, Pruss, D, and Wolffe, AP (1997). How does DNA methylation repress transcription? *Trends Genet*, 13:444–449.

Kaukonen, J, Juselius, JK, Tiranti, V, Kyttala, A, Zeviani, M, Comi, GP, et al. (2000). Role of adenine nucleotide translocator 1 in mtDNA maintenance. *Science*, 289:782–785.

Klesert, TR, Cho, DH, Clark, JI, Maylie, J, Adelman, J, Snider, L, et al. (2000). Mice deficient in Six5 develop cataracts: implications for myotonic dystrophy. *Nat Genet*, 25:105–109.

Klesert, TR, Otten, AD, Bird, TD, and Tapscott, SJ (1997). Trinucleotide repeat expansion at the myotonic dystrophy locus reduces expression of DMAHP. *Nat Genet*, 16:402–406.

Klose, RJ, Sarraf, SA, Schmiedeberg, L, McDermott, SM, Stancheva, I, and Bird, AP (2005). DNA binding selectivity of MeCP2 due to a requirement for A/T sequences adjacent to methyl-CpG. *Mol Cell*, 19:667–678.

Kondo, T, Bobek, MP, Kuick, R, Lamb, B., Zhu, X, Narayan, A, et al. (2000). Whole-genome methylation scan in ICF syndrome: hypomethylation of non-satellite DNA repeats D4Z4 and NBL2. *Hum Mol Genet*, 9:597–604.

Kornberg, RD and Lorch, Y (1999). Twenty-five years of the nucleosome, fundamental particle of the eukaryote chromosome. *Cell*, 98:285–294.

Kourmouli, N, Sun, YM, van der Sar, S, Singh, PB, and Brown, JP (2005). Epigenetic regulation of mammalian pericentric heterochromatin in vivo by HP1. *Biochem Biophys Res Commun*, 337:901–907.

Krasilnikova, MM and Mirkin, SM (2004). Replication stalling at Friedreich's ataxia (GAA)n repeats in vivo. *Mol Cell Biol*, 24:2286–2295.

Kriaucionis, S and Bird, A (2003). DNA methylation and Rett syndrome. *Hum Mol Genet*, 12 (Spec. No. 2) :R221–R227.

Kriaucionis, S and Bird, A (2004). The major form of MeCP2 has a novel N-terminus generated by alternative splicing. *Nucleic Acids Res*, 32:1818–1823.

Kuzmichev, A, Jenuwein, T, Tempst, P, and Reinberg, D (2004). Different EZH2-containing complexes target methylation of histone H1 or nucleosomal histone H3. *Mol Cell*, 14:183–193.

Kuzmichev, A, Margueron, R, Vaquero, A, Preissner, TS, Scher, M, Kirmizis, A, et al. (2005). Composition and histone substrates of polycomb repressive group complexes change during cellular differentiation. *Proc Natl Acad Sci USA*, 102:1859–1864.

Kuzmichev, A, Nishioka, K, Erdjument-Bromage, H, Tempst, P, and Reinberg, D (2002). Histone methyltransferase activity associated with a human multiprotein complex containing the Enhancer of Zeste protein. *Genes Dev*, 16:2893–2905.

Lachner, M and Jenuwein, T (2002). The many faces of histone lysine methylation. *Curr Opin Cell Biol*, 14:286–298.

Lachner, M, O'Carroll, D, Rea, S, Mechtler, K, and Jenuwein, T (2001). Methylation of histone H3 lysine 9 creates a binding site for HP1 proteins. *Nature*, 410:116–120.

Lamont, PJ, Davis, MB, and Wood, NW (1997). Identification and sizing of the GAA trinucleotide repeat expansion of Friedreich's ataxia in 56 patients. Clinical and genetic correlates. *Brain*, 120:673–680.

Le Douarin, B, Nielsen, AL, Garnier, JM, Ichinose, H, Jeanmougin, F, Losson, R, et al. (1996). A possible involvement of TIF1 alpha and TIF1 beta in the epigenetic control of transcription by nuclear receptors. *EMBO J*, 15:6701–6715.

Lechner, MS, Schultz, DC, Negorev, D, Maul, GG, and Rauscher, FJ 3rd (2005). The mammalian heterochromatin protein 1 binds diverse nuclear proteins through a common motif that targets the chromoshadow domain. *Biochem Biophys Res Commun*, 331:929–937.

Lee, MP, DeBaun, MR, Mitsuya, K, Galonek, HL, Brandenburg, S, Oshimura, M, et al. (1999). Loss of imprinting of a paternally expressed transcript, with antisense orientation to KVLQT1, occurs frequently in Beckwith–Wiedemann syndrome and is independent of insulin-like growth factor II imprinting. *Proc Natl Acad Sci USA*, 96:5203–5208.

Lehnertz, B, Ueda, Y, Derijck, AA, Braunschweig, U, Perez-Burgos, L, Kubicek, S, et al. (2003). Suv39h-mediated histone H3 lysine 9 methylation directs DNA methylation to major satellite repeats at pericentric heterochromatin. *Curr Biol*, 13:1192–1200.

Leighton, PA, Saam, JR, Ingram, RS, Stewart, CL, and Tilghman, SM (1995). An enhancer deletion affects both H19 and Igf2 expression. *Genes Dev*, 9:2079–2089.

Lemmers, RJ, de Kievit, P, Sandkuijl, L, Padberg, GW, van Ommen, GJ, Frants, RR, et al. (2002). Facioscapulohumeral muscular dystrophy is uniquely associated with one of the two variants of the 4q subtelomere. *Nat Genet*, 32:235–236.

Lemmers, RJ, Osborn, M, Haaf, T, Rogers, M, Frants, RR, Padberg, GW, et al. (2003). D4F104S1 deletion in facioscapulohumeral muscular dystrophy: phenotype, size, and detection. *Neurology*, 61:178–183.

Lewis, A, Mitsuya, K, Umlauf, D, Smith, P, Dean, W, Walter, J, et al. (2004). Imprinting on distal chromosome 7 in the placenta involves repressive histone methylation independent of DNA methylation. *Nat Genet*, 36:1291–1295.

Lewis, A and Murrell, A (2004). Genomic imprinting: CTCF protects the boundaries. *Curr Biol*, 14:R284–R286.

Liquori, CL, Ricker, K, Moseley, ML, Jacobsen, JF, Kress, W, Naylor, SL, et al. (2001). Myotonic dystrophy type 2 caused by a CCTG expansion in intron 1 of ZNF9. *Science*, 293:864–867.

Locke, J, Kotarski, MA, and Tartof, KD (1988). Dosage-dependent modifiers of position effect variegation in Drosophila and a mass action model that explains their effect. *Genetics*, 120:181–198.

Luciani, JJ, Depetris, D, Missirian, C, Mignon-Ravix, C, Metzler-Guillemain, C, Megarbane, A, et al. (2005). Subcellular distribution of HP1 proteins is altered in ICF syndrome. *Eur J Hum Genet*, 13:41–51.

Lunt, PW (1998). 44th ENMC International workshop: facioscapulohumeral muscular dystrophy: molecular studies 19–21 July 1996, Naarden, The Netherlands. *Neuromuscul Disord*, 8:126–130.

Mager, J, Montgomery, ND, de Villena, FP, and Magnuson, T (2003). Genome imprinting regulated by the mouse Polycomb group protein Eed. *Nat Genet*, 33:502–507.

Mahadevan, M, Tsilfidis, C, Sabourin, L, Shutler, G, Amemiya, C, Jansen, G, et al. (1992). Myotonic dystrophy mutation: an unstable CTG repeat in the 3' untranslated region of the gene. *Science*, 255:1253–1255.

Maison, C, Bailly, D, Peters, AH, Quivy, JP, Roche, D, Taddei, A, et al. (2002). Higher-order structure in pericentric heterochromatin involves a distinct pattern of histone modification and an RNA component. *Nat Genet*, 30:329–334.

Mariappan, SV, Catasti, P, Silks, LA 3rd, Bradbury, EM, and Gupta, G (1999). The high-resolution structure of the triplex formed by the GAA/TTC triplet repeat associated with Friedreich's ataxia. *J Mol Biol*, 285:2035–2052.

Martinowich, K, Hattori, D, Wu, H, Fouse, S, He, F, Hu, Y, et al. (2003). DNA methylation-related chromatin remodeling in activity-dependent BDNF gene regulation. *Science*, 302:890–893.

Masny, PS, Bengtsson, U, Chung, SA, Martin, JH, van Engelen, B, van der Maarel, SM, et al. (2004). Localization of 4q35.2 to the nuclear periphery: is FSHD a nuclear envelope disease? *Hum Mol Genet*, 13:1857–1871.

Matsuura, T, Yamagata, T, Burgess, DL, Rásmussen, A, Grewal, RP, Watase, K, et al. (2000). Large expansion of the ATTCT pentanucleotide repeat in spinocerebellar ataxia type 10. *Nat Genet*, 26:191–194.

Maurer-Stroh, S, Dickens, NJ, Hughes-Davies, L, Kouzarides, T, Eisenhaber, F, and Ponting, CP (2003). The Tudor domain "Royal Family": Tudor, plant Agenet, Chromo, PWWP and MBT domains. *Trends Biochem Sci*, 28:69–74.

McDowell, TL, Gibbons, RJ, Sutherland, H, O'Rourke, DM, Bickmore, WA, Pombo, A, et al. (1999). Localization of a putative transcriptional regulator (ATRX) at pericentromeric heterochromatin and the short arms of acrocentric chromosomes. *Proc Natl Acad Sci USA*, 96:13983–13988.

Miranda, CJ, Santos, MM, Ohshima, H, Smith, J, Liangtao, L, Bunting, M, et al. (2002). Frataxin knockin mouse. *FEBS Lett*, 512:291–297.

Mitsuya, K, Meguro, M, Lee, MP, Katoh, M, Schulz, TC, Kugoh, H, et al. (1999). LIT1, an imprinted antisense RNA in the human KvLQT1 locus identified by screening for differentially expressed transcripts using monochromosomal hybrids. *Hum Mol Genet*, 8:1209–1217.

Mnatzakanian, GN, Lohi, H, Munteanu, I, Alfred, SE, Yamada, T, MacLeod, PJ, et al. (2004). A previously unidentified MECP2 open reading frame defines a new protein isoform relevant to Rett syndrome. *Nat Genet*, 36:339–341.

Monros, E, Molto, MD, Martinez, F, Canizares, J, Blanca, J, Vilchez, JJ, et al. (1997). Phenotype correlation and intergenerational dynamics of the Friedreich ataxia GAA trinucleotide repeat. *Am J Hum Genet*, 61:101–110.

Montermini, L, Richter, A, Morgan, K, Justice, CM, Julien, D, Castellotti, B, et al. (1997). Phenotypic variability in Friedreich ataxia: role of the associated GAA triplet repeat expansion. *Ann Neurol*, 41:675–682.

Moutou, C, Junien, C, Henry, I, and Bonaiti-Pellie, C (1992). Beckwith–Wiedemann syndrome: a demonstration of the mechanisms responsible for the excess of transmitting females. *J Med Genet*, 29:217–220.

Murrell, A, Heeson, S, Cooper, WN, Douglas, E, Apostolidou, S, Moore, GE, et al. (2004). An association between variants in the IGF2 gene and

Beckwith–Wiedemann syndrome: interaction between genotype and epigenotype. *Hum Mol Genet*, 13:247–255.

Nan, X, Ng, HH, Johnson, CA, Laherty, CD, Turner, BM, Eisenman, RN, et al. (1998). Transcriptional repression by the methyl-CpG-binding protein MeCP2 involves a histone deacetylase complex. *Nature*, 393:386–389.

Narlikar, GJ, Fan, HY, and Kingston, RE (2002). Cooperation between complexes that regulate chromatin structure and transcription. *Cell*, 108:475–487.

Ng, HH, Jeppesen, P, and Bird, A (2000). Active repression of methylated genes by the chromosomal protein MBD1. *Mol Cell Biol*, 20:1394–1406.

Ng, HH, Zhang, Y, Hendrich, B, Johnson, CA, Turner, BM, Erdjument-Bromage, H, et al. (1999). MBD2 is a transcriptional repressor belonging to the MeCP1 histone deacetylase complex. *Nat Genet*, 23:58–61.

Niemitz, EL, DeBaun, MR, Fallon, J, Murakami, K, Kugoh, H, Oshimura, M, et al. (2004). Microdeletion of LIT1 in familial Beckwith–Wiedemann syndrome. *Am J Hum Genet*, 75:844–849.

Ohshima, K, Kang, S, Larson, JE, and Wells, RD (1996). Cloning, characterization, and properties of seven triplet repeat DNA sequences. *J Biol Chem*, 271:16773–16783.

Ohshima, K, Montermini, L, Wells, RD, and Pandolfo, M (1998). Inhibitory effects of expanded GAA.TTC triplet repeats from intron I of the Friedreich ataxia gene on transcription and replication in vivo. *J Biol Chem*, 273:14588–14595.

Ohshima, K, Sakamoto, N, Labuda, M, Poirier, J, Moseley, ML, Montermini, L, et al. (1999). A nonpathogenic GAAGGA repeat in the Friedreich gene: implications for pathogenesis. *Neurology*, 53:1854–1857.

Okano, M, Bell, DW, Haber, DA, and Li, E (1999). DNA methyltransferases Dnmt3a and Dnmt3b are essential for de novo methylation and mammalian development. *Cell*, 99:247–257.

Osborne, CS, Chakalova, L, Brown, KE, Carter, D, Horton, A, Debrand, E, et al. (2004). Active genes dynamically colocalize to shared sites of ongoing transcription. *Nat Genet*, 36:1065–1071.

Otten, AD and Tapscott, SJ (1995). Triplet repeat expansion in myotonic dystrophy alters the adjacent chromatin structure. *Proc Natl Acad Sci USA*, 92:5465–5469.

Pal-Bhadra, M, Leibovitch, BA, Gandhi, SG, Rao, M, Bhadra, U, Birchler, JA, et al. (2004). Heterochromatic silencing and HP1 localization in Drosophila are dependent on the RNAi machinery. *Science*, 303:669–672.

Peters, AH, O'Carroll, D, Scherthan, H, Mechtler, K, Sauer, S, Schofer, C, et al. (2001). Loss of the Suv39h histone methyltransferases impairs mammalian heterochromatin and genome stability. *Cell*, 107:323–337.

Peters, AH and Schubeler, D (2005). Methylation of histones: playing memory with DNA. *Curr Opin Cell Biol*, 17:230–238.

Pieretti, M, Zhang, FP, Fu, YH, Warren, ST, Oostra, BA, Caskey, CT, et al. (1991). Absence of expression of the FMR-1 gene in fragile X syndrome. *Cell*, 66:817–822.

Pietrobono, R, Tabolacci, E, Zalfa, F, Zito, I, Terracciano, A, Moscato, U, et al. (2005). Molecular dissection of the events leading to inactivation of the FMR1 gene. *Hum Mol Genet*, 14:267–277.

Prawitt, D, Enklaar, T, Gartner-Rupprecht, B, Spangenberg, C, Oswald, M, Lausch, E, et al. (2005). Microdeletion of target sites for insulator protein CTCF in a chromosome 11p15 imprinting center in Beckwith–Wiedemann syndrome and Wilms' tumor. *Proc Natl Acad Sci, USA* 102:4085–4090.

Primerano, B, Tassone, F, Hagerman, RJ, Hagerman, P, Amaldi, F, and Bagni, C (2002). Reduced FMR1 mRNA translation efficiency in fragile X patients with premutations. *RNA*, 8:1482–1488.

Qiu, C, Sawada, K, Zhang, X, and Cheng, X (2002). The PWWP domain of mammalian DNA methyltransferase Dnmt3b defines a new family of DNA-binding folds. *Nat Struct Biol*, 9:217–224.

Rattner, JB and Hamkalo, BA (1979). Nucleosome packing in interphase chromatin. *J Cell Biol*, 81:453–457.

Ravn, K, Nielsen, JB, Uldall, P, Hansen, FJ, and Schwartz, M (2003). No correlation between phenotype and genotype in boys with a truncating MECP2 mutation. *J Med Genet*, 40:e5.

Reik, W and Walter, J (2001). Genomic imprinting: parental influence on the genome. *Nat Rev Genet*, 2:21–32.

Rijkers, T, Deidda, G, van Koningsbruggen, S, van Geel, M, Lemmers, RJ, van Deutekom, JC, et al. (2004). FRG2, an FSHD candidate gene, is transcriptionally upregulated in differentiating primary myoblast cultures of FSHD patients. *J Med Genet*, 41:826–836.

Robertson, KD (2005). DNA methylation and human disease. *Nat Rev Genet* 6:597–610.

Saito, Y, Kanai, Y, Sakamoto, M, Saito, H, Ishii, H, and Hirohashi, S (2002). Overexpression of a splice variant of DNA methyltransferase 3b, DNMT3b4, associated with DNA hypomethylation on pericentromeric satellite regions during human hepatocarcinogenesis. *Proc Natl Acad Sci USA*, 99:10060–10065.

Sakamoto, N, Chastain, PD, Parniewski, P, Ohshima, K, Pandolfo, M, Griffith, JD, et al. (1999). Sticky DNA: self-association properties of long GAA.TTC repeats in R.R.Y triplex structures from Friedreich's ataxia. *Mol Cell*, 3:465–475.

Sakamoto, N, Ohshima, H, Montermini, L, Pandolfo, M, and Wells, RD (2001). Sticky DNA, a self associated complex formed at long GAA*TTC repeats in intron 1 of the frataxin gene, inhibits transcription. *J Biol Chem*, 276:27171–27177.

Samaco, RC, Hogart, A, and LaSalle, JM (2005). Epigenetic overlap in autism-spectrum neurodevelopmental disorders: MECP2 deficiency causes reduced expression of UBE3A and GABRB3. *Hum Mol Genet*, 14:483–492.

Sarkar, PS, Appukuttan, B, Han, J, Ito, Y, Ai, C, Tsai, W, et al. (2000). Heterozygous loss of Six5 in mice is sufficient to cause ocular cataracts. *Nat Genet*, 25:110–114.

Sarsero, JP, Li, L, Wardan, H, Sitte, K, Williamson, R, and Ioannou, PA (2003). Upregulation of expression from the FRDA genomic locus for the therapy of Friedreich ataxia. *J Gene Med*, 5:72–81.

Saveliev, A, Everett, C, Sharpe, T, Webster, Z, and Festenstein, R (2003). DNA triplet repeats mediate heterochromatin-protein-1-sensitive variegated gene silencing. *Nature*, 422:909–913.

Schmiedeberg, L, Weisshart, K, Diekmann, S, Meyer Zu Hoerste, G, and Hemmerich, P (2004). High- and low-mobility populations of HP1 in heterochromatin of mammalian cells. *Mol Biol Cell*, 15:2819–2833.

Schramke, V, Sheedy, DM, Denli, AM, Bonila, C, Ekwall, K, Hannon, GJ, et al. (2005). RNA-interference-directed chromatin modification coupled to RNA polymerase II transcription. *Nature*, 435:1275–1279.

Schubeler, D, MacAlpine, DM, Scalzo, D, Wirbelauer, C, Kooperberg, C, van Leeuwen, F, et al. (2004). The histone modification pattern of active genes revealed through genome-wide chromatin analysis of a higher eukaryote. *Genes Dev*, 18:1263–1271.

Shestakova, EA, Mansuroglu, Z, Mokrani, H, Ghinea, N, and Bonnefoy, E (2004). Transcription factor YY1 associates with pericentromeric gamma-satellite DNA in cycling but not in quiescent (G0) cells. *Nucleic Acids Res*, 32:4390–4399.

Shirohzu, H, Kubota, T, Kumazawa, A, Sado, T, Chijiwa, T, Inagaki, K, et al. (2002). Three novel DNMT3B mutations in Japanese patients with ICF syndrome. *Am J Med Genet*, 112:31–37.

Smeets, DF, Moog, U, Weemaes, CM, Vaes-Peeters, G, Merkx, GF, Niehof, JP, et al. (1994). ICF syndrome: a new case and review of the literature. *Hum Genet*, 94:240–246.

Smeets, HJ, Smits, AP, Verheij, CE, Theelen, JP, Willemsen, R, van de Burgt, I, et al. (1995). Normal phenotype in two brothers with a full FMR1 mutation. *Hum Mol Genet*, 4:2103–2108.

Smilinich, NJ, Day, CD, Fitzpatrick, GV, Caldwell, GM, Lossie, AC, Cooper, PR, et al. (1999). A maternally methylated CpG island in KvLQT1 is associated with an antisense paternal transcript and loss of imprinting in Beckwith-Wiedemann syndrome. *Proc Natl Acad Sci USA*, 96:8064–8069.

Soejima, H and Wagstaff, J (2005). Imprinting centers, chromatin structure, and disease. *J Cell Biochem*, 95:226–233.

Spencer, CM, Alekseyenko, O, Serysheva, E, Yuva-Paylor, LA, and Paylor, R (2005). Altered anxiety-related and social behaviors in the Fmr1 knockout mouse model of fragile X syndrome. *Genes Brain Behav*, 4:420–430.

Stancheva, I, Collins, AL, Van den Veyver, IB, Zoghbi, H, and Meehan, RR (2003). A mutant form of MeCP2 protein associated with human Rett syndrome cannot be displaced from methylated DNA by notch in Xenopus embryos. *Mol Cell*, 12:425–435.

Strahl, BD and Allis, CD (2000). The language of covalent histone modifications. *Nature*, 403:41–45.

Stuhmer, T, Anderson, SA, Ekker, M, and Rubenstein, JL (2002). Ectopic expression of the Dlx genes induces glutamic acid decarboxylase and Dlx expression. *Development*, 129:245–252.

Suetake, I, Shinozaki, F, Miyagawa, J, Takeshima, H, and Tajima, S (2004). DNMT3L stimulates the DNA methylation activity of Dnmt3a and Dnmt3b through a direct interaction. *J Biol Chem*, 279:27816–27823.

Sutcliffe, JS, Nelson, DL, Zhang, F, Pieretti, M, Caskey, CT, Saxe, D, et al. (1992). DNA methylation represses FMR-1 transcription in fragile X syndrome. *Hum Mol Genet*, 1:397–400.

Tabolacci, E, Pietrobono, R, Moscato, U, Oostra, BA, Chiurazzi, P, and Neri, G (2005). Differential epigenetic modifications in the FMR1 gene of the fragile X syndrome after reactivating pharmacological treatments. *Eur J Hum Genet*, 13:641–648.

Tam, R, Smith, KP, and Lawrence, JB (2004). The 4q subtelomere harboring the FSHD locus is specifically anchored with peripheral heterochromatin unlike most human telomeres. *J Cell Biol*, 167:269–279.

Tang, J, Wu, S, Liu, H, Stratt, R, Barak, OG, Shiekhattar, R, et al. (2004). A novel transcription regulatory complex containing death domain-associated protein and the ATR-X syndrome protein. *J Biol Chem*, 279:20369–20377.

Tassone, F, Hagerman, RJ, Chamberlain, WD, and Hagerman, PJ (2000). Transcription of the FMR1 gene in individuals with fragile X syndrome. *Am J Med Genet*, 97:195–203.

Tassone, F, Hagerman, RJ, Taylor, AK, Gane, LW, Godfrey, TE, and Hagerman, PJ (2000). Elevated levels of FMR1 mRNA in carrier males: a new mechanism of involvement in the fragile-X syndrome. *Am J Hum Genet*, 66:6–15.

Tawil, R (2004). Facioscapulohumeral muscular dystrophy. *Curr Neurol Neurosci Rep*, 4:51–54.

Tawil, R, Figlewicz, DA, Griggs, RC, and Weiffenbach, B (1998). Facioscapulohumeral dystrophy: a distinct regional myopathy with a novel molecular pathogenesis. FSH Consortium. *Ann Neurol*, 43:279–282.

Tawil, R, Forrester, J, Griggs, RC, Mendell, J, Kissel, J, McDermott, M, et al. (1996). Evidence for anticipation and association of deletion size with severity in facioscapulohumeral muscular dystrophy. The FSH-DY Group. *Ann Neurol*, 39:744–748.

Thoma, F, Koller, T, and Klug, A (1979). Involvement of histone H1 in the organization of the nucleosome and of the salt-dependent superstructures of chromatin. *J Cell Biol*, 83:403–427.

Topcu, M, Akyerli, C, Sayi, A, Toruner, GA, Kocoglu, SR, Cimbis, M, and Ozcelik, T (2002). Somatic mosaicism for a MECP2 mutation associated with classic Rett syndrome in a boy. *Eur J Hum Genet*, 10:77–81.

Tudor, M, Akbarian, S, Chen, RZ, and Jaenisch, R (2002). Transcriptional profiling of a mouse model for Rett syndrome reveals subtle transcriptional changes in the brain. *Proc Natl Acad Sci, USA* 99:15536–15541.

Tupler, R, Berardinelli, A, Barbierato, L, Frants, R, Hewitt, JE, Lanzi, G, et al. (1996). Monosomy of distal 4q does not cause facioscapulohumeral muscular dystrophy. *J Med Genet*, 33:366–370.

Turner, BM (2000). Histone acetylation and an epigenetic code. *Bioessays*, 22:836–845.

Umlauf, D, Goto, Y, Cao, R, Cerqueira, F, Wagschal, A, Zhang, Y, et al. (2004). Imprinting along the Kcnq1 domain on mouse chromosome 7 involves repressive histone methylation and recruitment of Polycomb group complexes. *Nat Genet*, 36:1296–1300.

van Deutekom, JC, Wijmenga, C, van Tienhoven, EA, Gruter, AM, Hewitt, JE, Padberg, GW, et al. (1993). FSHD associated DNA rearrangements are due to deletions of integral copies of a 3.2 kb tandemly repeated unit. *Hum Mol Genet*, 2:2037–2042.

van Overveld, PG, Lemmers, RJ, Sandkuijl, LA, Enthoven, L, Winokur, ST, Bakels, F, et al. (2003). Hypomethylation of D4Z4 in 4q-linked and non-4q-linked facioscapulohumeral muscular dystrophy. *Nat Genet*, 35:315–317.

Verkerk, AJ, Pieretti, M, Sutcliffe, JS, Fu, YH, Kuhl, DP, Pizzuti, A, et al. (1991). Identification of a gene (FMR-1) containing a CGG repeat coincident with a breakpoint cluster region exhibiting length variation in fragile X syndrome. *Cell*, 65:905–914.

Villard, L, Kpebe, A, Cardoso, C, Chelly, PJ, Tardieu, PM, and Fontes, M (2000). Two affected boys in a Rett syndrome family: clinical and molecular findings. *Neurology*, 55:1188–1193.

Wang, J, Shiels, C, Sasieni, P, Wu, PJ, Islam, SA, Freemont, PS, et al. (2004). Promyelocytic leukemia nuclear bodies associate with transcriptionally active genomic regions. *J Cell Biol*, 164:515–526.

Wang, YH, Amirhaeri, S, Kang, S, Wells, RD, and Griffith, JD (1994). Preferential nucleosome assembly at DNA triplet repeats from the myotonic dystrophy gene. *Science*, 265:669–671.

Wang, YH and Griffith, J (1995). Expanded CTG triplet blocks from the myotonic dystrophy gene create the strongest known natural nucleosome positioning elements. *Genomics*, 25:570–573.

Wang, Z, Taylor, AK, and Bridge, JA. (1996). FMR1 fully expanded mutation with minimal methylation in a high functioning fragile X male. *J Med Genet*, 33:376–378.

Weaving, LS, Ellaway, CJ, Gecz, J, and Christodoulou, J (2005). Rett syndrome: clinical review and genetic update. *J Med Genet*, 42:1–7.

Weksberg, R, Nishikawa, J, Caluseriu, O, Fei, YL, Shuman, C, Wei, C, et al. (2001). Tumor development in the Beckwith–Wiedemann syndrome is associated with a variety of constitutional molecular 11p15 alterations including imprinting defects of KCNQ1OT1. *Hum Mol Genet*, 10:2989–3000.

Weksberg, R, Shen, DR, Fei, YL, Song, QL, and Squire, J (1993). Disruption of insulin-like growth factor 2 imprinting in Beckwith-Wiedemann syndrome. *Nat Genet*, 5:143–150.

Weksberg, R, Smith, AC, Squire, J, and Sadowski, P (2003). Beckwith–Wiedemann syndrome demonstrates a role for epigenetic control of normal development. *Hum Mol Genet*, 12 (Spec. No. 1):R61–R68.

Wells, RD, Collier, DA, Hanvey, JC, Shimizu, M, and Wohlrab, F (1988). The chemistry and biology of unusual DNA structures adopted by oligopurine. oligopyrimidine sequences. *FASEB J*, 2:2939–2949.

West, AG and Fraser, P (2005). Remote control of gene transcription. *Hum Mol Genet*, 14 (Spec. No. 1) :R101–R111.

Wijmenga, C, Hansen, RS, Gimelli, G, Bjorck, EJ, Davies, EG, Valentine, D, et al. (2000). Genetic variation in ICF syndrome: evidence for genetic heterogeneity. *Hum Mutat*, 16:509–517.

Wijmenga, C, Hewitt, JE, Sandkuijl, LA, Clark, LN, Wright, TJ, Dauwerse, HG, et al. (1992). Chromosome 4q DNA rearrangements associated with facioscapulohumeral muscular dystrophy. *Nat Genet*, 2:26–30.

Wysocka, J, Swigut, T, Milne, TA, Dou, Y, Zhang, X, Burlingame, AL, et al. (2005). WDR5 associates with histone H3 methylated at K4 and is essential for H3 K4 methylation and vertebrate development. *Cell*, 121:859–872.

Xue, Y, Gibbons, R, Yan, Z, Yang, D, McDowell, TL, Sechi, S, et al. (2003). The ATRX syndrome protein forms a chromatin-remodeling complex with Daxx and localizes in promyelocytic leukemia nuclear bodies. *Proc Natl Acad Sci USA*, 100:10635–10640.

Yip, DJ and Picketts, DJ (2003). Increasing D4Z4 repeat copy number compromises C2C12 myoblast differentiation. *FEBS Lett*, 537:133–138.

5

Genomic Perspectives of Human Development

Dhavendra Kumar

Disorders that affect tissue differentiation, organogenesis, and morphogenesis constitute a significant proportion of human genetic disease. These disorders may result from any of the known genetic mechanisms and occur singly, in combination with other malformations, or complicated with systemic manifestations. Clinical and laboratory analyses of these disorders have led to the emergence of the clinical field of dysmorphology. Clinical dysmorphology is now an integral part of medical genetics. All practicing clinical geneticists and genetic counselors are required to have the basic skills in order to deal with patients who present with a developmental malformation that may occur either singly or as part of a recognizable dysmorphic syndrome (Jones, 1997).

During the past two decades, dysmorphologists have delineated a large number of malformation syndromes that comprise multiple malformations affecting unrelated organs and tissues. Researchers and clinicians continue to generate more information, and the number of related syndromes continues to increase. Dedicated dysmorphology databases [London dysmorphology database (LDDB), GeneEye, Pictures of standardized syndromes and undiagnosed malformations (POSSUM)] are now available to help clinical geneticists and related professionals carry out searches. Although considerable progress has been made in understanding the developmental pathology and possible causes of some of these syndromes, there remains a huge gap in our understanding of abnormal human development.

Recent advances in genetics and genomics have made it possible to understand the molecular basis of developmental malformations. These fall broadly into two groups. The first group that includes advances in gene mapping, cloning, and identification of genes has received a tremendous boost from the Human Genome Project (HGP) (see Chapter 1), which has helped in identifying mutations and polymorphisms in specific human genes associated with particular syndromes. The HGP has also permitted comparisons between the human genome and the genomes of other organisms such as mouse and *Drosophila*, which has helped in the identification of new genes and the subsequent study of their role in abnormal development. The second group includes a number of highly interactive molecular pathways that govern the developmental process. Each of these molecular pathways is regulated by specific genes that belong to a *gene family*. These advances have made it possible to develop a better understanding of development and morphogenesis and to work out a plausible *genotype–phenotype* correlation.

This chapter reviews some of the basic concepts in developmental biology, which are intricately related to genes and genomes. This chapter is not intended to provide the reader with a comprehensive account of normal and abnormal human development, as this field is enormous and beyond the scope of this review. There are excellent textbooks written on this subject (see Further Reading). The main aim of this chapter is to introduce the subject from the perspective of genomics. It is anticipated that the reader may find the contents of this chapter helpful in understanding aspects of advances in genomic-related science and technology that are relevant to understanding developmental malformations which may in turn influence their practice of clinical dysmorphology.

Genome Projects

Since the 1990s, the efforts to sequence a number of eukaryotic genomes have been successful, resulting in the availability of complete genome sequences. In addition, several prokaryotic organisms have been sequenced (http://www.tigr.org/). Apart from completion of the human genome sequence (Lander et al., 2001; Venter et al., 2001), genomes that have been sequenced in their entirety include mouse, *Drosophila melanogaster* (Rubin et al., 2000), worm (*Caenorhabditis elegans*) (Chervitz et al., 1998), and budding yeast (*Saccharomyces cerevisiae*). Despite limited information on the precise gene content and gene order in the human genome, most of the approximately 35,000 human genes are comparable to the corresponding mouse genome. This subject has been covered in detail in the first section of the book. It is widely agreed that the availability of genome sequences is of paramount importance to our understanding of fundamental structural and functional units across a broad range of eukaryotic organisms, from the fruit fly to higher primates like *Homo sapiens*. Such understanding has already led to the discovery of several gene families and gene-sequence polymorphisms that regulate embryonic development, and are involved in the pathogenesis of several human diseases.

Genome Sequence Polymorphisms

Human beings do not all possess the same genomic sequence. Every person, except probably monozygotic twins, has a unique sequence that is different from any other person's by one change in every 500–1000 base pair (bp). This variation or polymorphism presumably does not carry any significance in terms of structural or functional phenotype. Nevertheless, the process of genome sequencing needs to take into account the existence of this variation between individuals, populations, and ethnic/racial groups and its possible influence on phenotype. Although yet unclear, these polymorphisms could affect the level or activity of certain classes of proteins. In humans, some of these differences may be disease-causing or result in developmental malformations. Polymorphisms could be located anywhere within the exons of genes, outside of or close to the promoter region of such variant genes. In addition, these sequence polymorphisms could be extremely important in the identification of both single loci and sets of loci that produce disease. The most common type of polymorphic variation is the single nucleotide polymorphism (SNP). Several research institutions and organizations are engaged in collecting data on SNPs that are correlated with disease states (http://snp.cshl.org/). The analysis of such data must take into account the population structure, since the expression of a particular allele may depend on the presence or absence of other genetic variability in the genome. This is a mammoth and challenging task, but it is likely to identify which of these polymorphisms are associated with abnormal human development and disease.

Comparative Genomics and Human Development

The sequencing of genomes from different organisms now provides a splendid opportunity to compare the functional significance of known or predicted genes in one species with another species. A systematic comparison of closely related organisms, such as mouse and human, in which most genes are evolutionarily conserved has revealed the location of homologous genes. This information is made available and regularly updated on the Internet (www.informatics.jax.org/; www.geneontology.org).

Similarly, comparisons between the genomes of distantly related species, such as humans and fruit fly, that share common developmental processes, a vast amount of information about the genetic basis of development, has been accumulated. These comparisons have helped in identifying genes that occur in gene families which are derived from an ancestral gene in a common branch of the evolutionary tree. The genes in such a family are called *paralogous*, and originate from repeated gene duplication events followed by evolutionary divergence of the duplicated genes (Fig. 5–1). The proteins

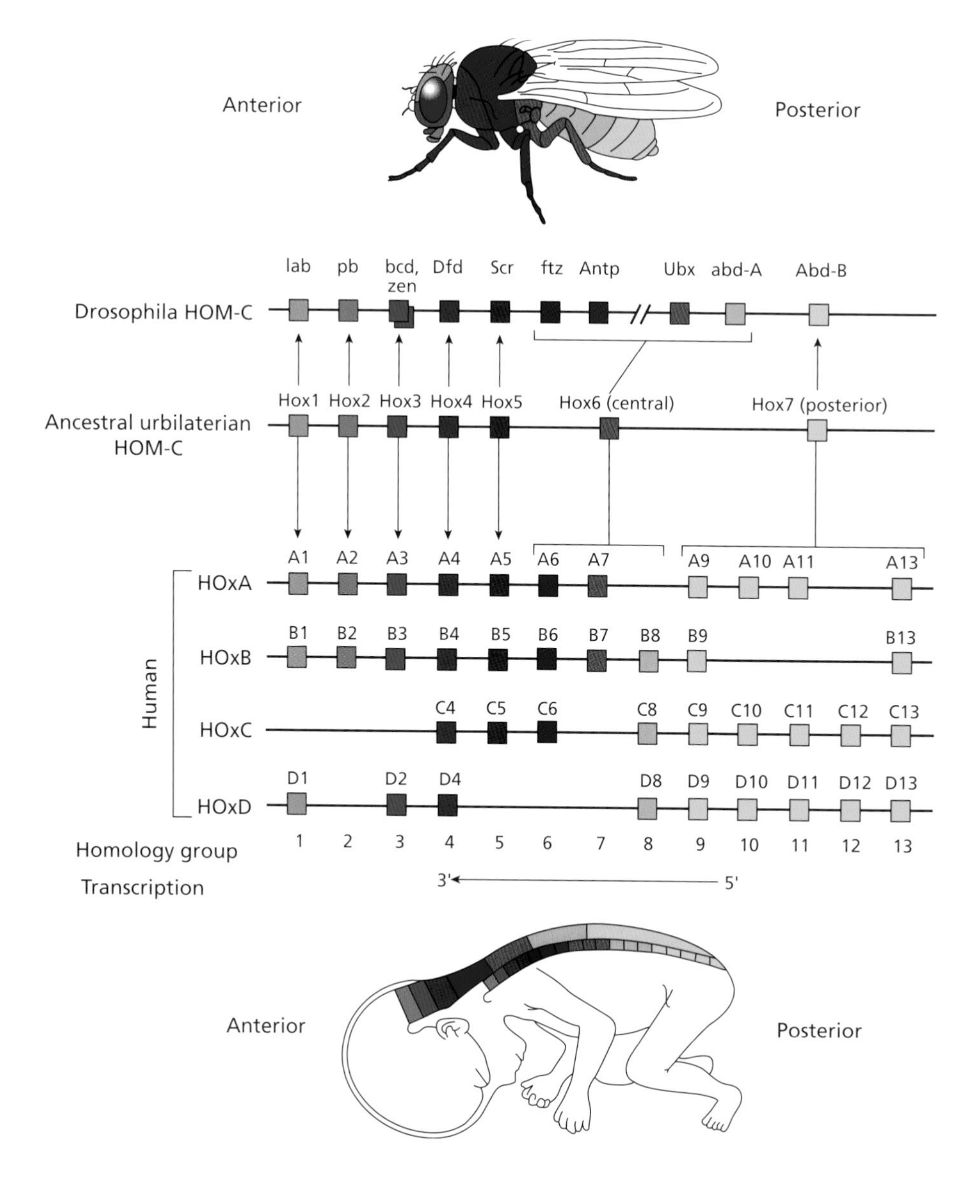

Figure 5–1 Evolutionary conservation of genomic organization and expression patterns of *Drosophila* fruit fly and mammal *Hox* genes. The human embryo shows anterior–posterior cluster of four *Hox* genes based on mouse expression studies. The fruit fly shows *Drosophila Hox* genes aligned with their mammalian orthologs and corresponding expression patterns mapped onto the body plan. (Reproduced from Bier and McGinis, 2004 with permission from Oxford University Press.)

encoded by such genes share common sequence characteristics and are referred to as *gene family*. The activities of such proteins may differ slightly, or the proteins perform a different function as a result of mutation or natural selection and they are also likely to have been produced at different stages of development. The function of the proteins may also vary in different tissues due to alternative splicing and mRNA editing causing modification of the mRNA sequence and the subsequent protein sequence. It is known that the same gene family identified by the same function can occur in two different species. Such genes are called *orthologous*. Using a cluster analysis of such proteins, it is possible to determine which of the synthesized proteins are alike and therefore likely to have similar function. This type of analysis has helped in understanding the evolution of gene families that influence development (Fig. 5–2) (Mount, 2001).

Evolutionary Conserved Regulation of Development

In order to identify genes that share sequences with similar biological function developmental, biologists search the genomes of different organisms. This is important because such genes in the two different organisms are likely to have originated from a common ancestral gene in the evolutionary past. It is likely that if mutations in one of the genes result in a developmental malformation, then the other gene presumably has a similar role in the development. This inference holds true even for strikingly dissimilar organisms, such as fruit fly and humans.

Biologists agree that evolutionarily divergent organisms use similar fundamental gene systems and associated encoded proteins during developmental processes. Developmental variation is not solely dependent on the structural variation in these genes but also on the gene function and regulation during the evolutionary cycles. This is referred to as the *evo-devo* concept of development (Tautz, 2002) and focuses on identifying basic sets of genes in organisms and studying how they are regulated in developing cells and tissues. One approach to identify such gene sets is a detailed functional identification of specific transcription factors and cis-acting binding sites for these factors (Davidson et al., 2002). Another approach is the search for common biological function in distantly related organisms, which may help in tracing the evolutionary and developmental origins of genes. For example, Gehring (2002) proposed that the genes regulating light perception in a microbe are similar to those that regulate the function of chloroplasts and, eventually, the evolution of complex function of the eye. Other *evo-devo* studies support

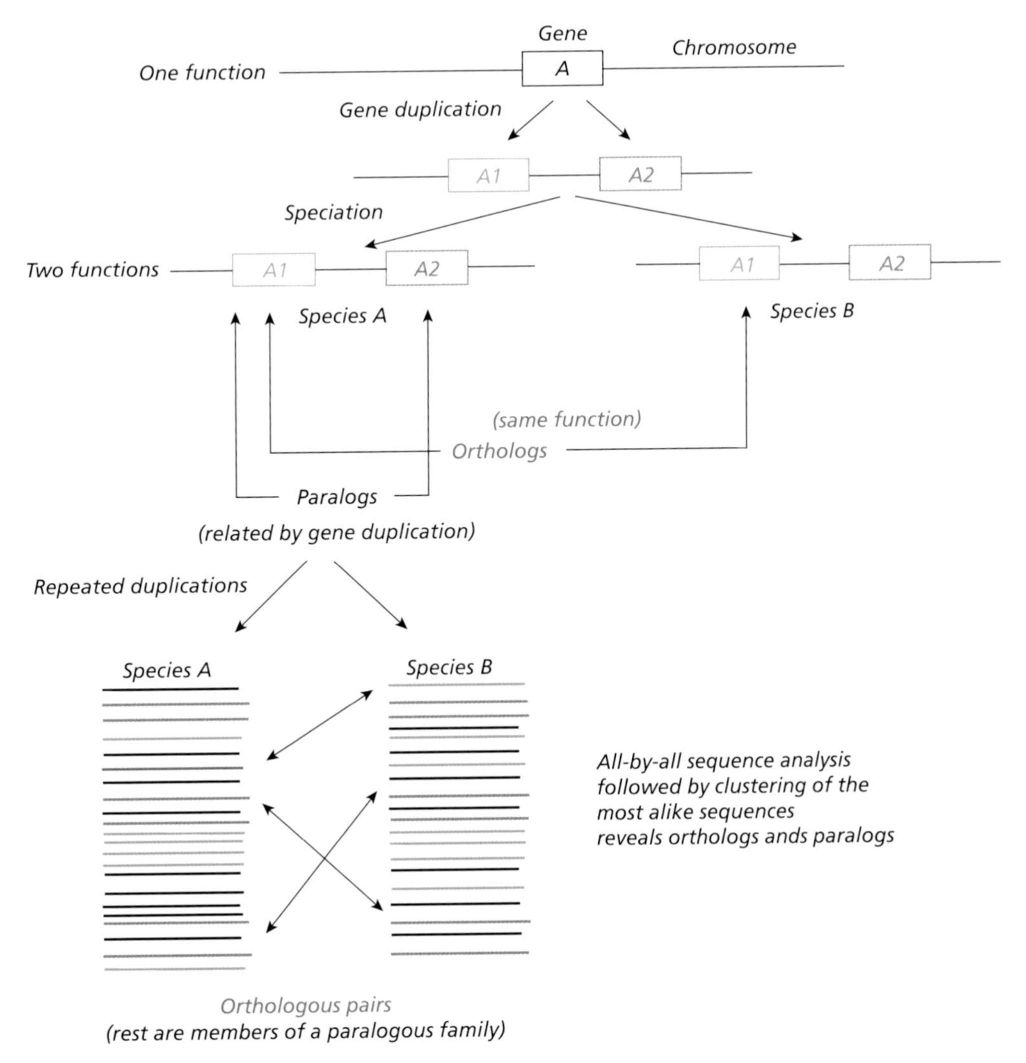

Figure 5–2 Orthologous and paralogous origin of gene families through successive gene duplication events. (Reproduced from Mount, 2004 with permission from Oxford University Press, New York.)

the premise that variations in the genes themselves are important for evolution and development.

An observation that stems from studying the effects of environmental factors such as a dietary supplement has significant relevance with respect to evolution and development. For example, a particular strain of inbred mice can have either a normal or highly defective skeleton due to the presence of immature cartilage, simply by altering the diet (Pennisi, 2002). This is probably related to the suppression of mutant *Hox* gene expression by a dietary substance, supporting the argument that expression of accumulated mutations in an important developmental gene may not occur until accompanied by an environmental change (Pennisi, 2002). It is possible that some of these evolutionary events may be neutral or deleterious. However, as with evolution in general, some of these events may be beneficial, leading to a new developmental scheme under the influence of natural selection. For example, the great diversity in butterfly wings is linked to genetic variation, and the resulting phenotypes are attributed to the cumulative action of natural selection on developmental processes (Beldade and Brakefield, 2002).

In brief, there are many developmental pathways common to invertebrates and vertebrates, including humans. These have originated from a sophisticated, bilaterally symmetric ancestral creature regulated by many architectural and organ-specific genetic systems (Fig. 5–3). These genetic systems have been conserved through successive evolutionary cycles.

Regulation of Developmental Genes

Previous sections have elucidated some basic aspects of genetics and genomics that are intricately connected with biological development across a wide range of organisms. This is referred to as developmental biology, which is a science concerned with the genetic basis of anatomy. During the past few years, we have developed a far better understanding of the basic mechanisms of developmental biology, because it is beyond the scope of this chapter (see Table 5–1).

It is important to appreciate that for each of these fundamental processes a number of genes are involved which relate to each other through developmental pathways. These genes function in a coordinated manner and encode proteins that share common biological properties. For this reason, these are also referred to as *gene family*. Several such gene families include genes that encode for a special class of proteins which bind to enhancer or promoter regions and interact to activate or repress the transcription of a particular gene. These are called *transcription factors*. Most transcription factors can bind to specific DNA sequences. These proteins can be grouped together into protein families, based on structural similarities (Table 5–2). DNA-binding sites in a particular transcription factor family are similar. Any alteration in the amino acids at the binding sites can alter the DNA sequence to which the factor binds.

Transcription Factors

Transcription factors have three major domains: firstly, a DNA-binding domain, which recognizes a particular DNA sequence; secondly, a trans-activating domain, which activates or suppresses the transcription of the gene whose enhancer or promoter it has found; finally, there may be a protein–protein interaction domain, which allows the transcription factor's activity to be modulated by transcription binding protein (TBP), transcription-associated factors (TAFs), or other transcription factors.

There are a number of diseases that result from a deficiency of transcription factors, often termed "*transcription factoropathy*." The first such human disease was androgen insensitivity syndrome (AIS) in which, despite normal testosterone production, the affected male externally develops as a phenotypic female and fails to develop secondary sexual characteristics. In AIS, the testosterone receptor is either absent or deficient, the DNA of which fails to bind the DNA of male-specific genes (Meyer et al., 1975). Conversely, the binding of DNA and consequent activation of the receptor site can lead to Waardenburg syndrome type II. In this disorder, the affected heterozygotes have a white forelock, and are deaf with multicolored irides. They possess a wild-type allele of the Microphthalmia (*MITF*) gene. Activation of this transcription factor through the protein tyrosine kinase cascade enables it to dimerize and bind to the regulatory regions of particular genes that open a region of DNA for transcription.

Transcription factors work in conjunction with other transcription factors to activate particular genes. However, the binding of a specific transcription factor to the enhancer or promoter of a gene does not always cause transcription of the gene. Some of these transcription factors, called "master regulatory genes," are important and take the lead in the transcription process. For example, *PAX6* (eye) and *MYOD* (muscles) work in concert to initiate cellular differentiation. The use of *PAX6* by different organs illustrates the

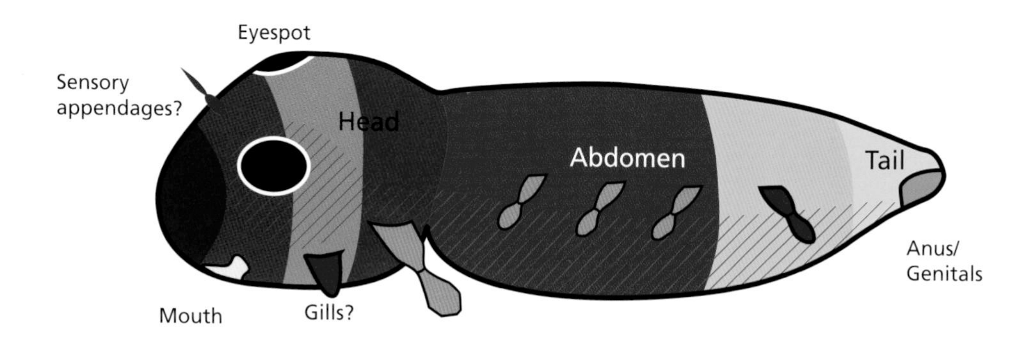

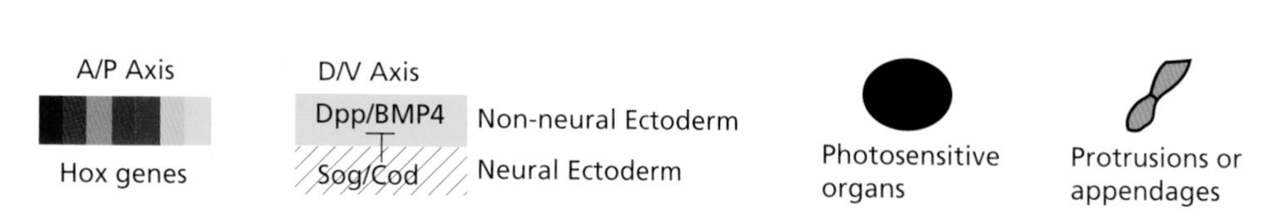

Figure 5–3 Conserved developmental patterning systems in a hypothetical ancestral creature; A/P axis segmental identity by *Hox* genes; A/P axis patterning by Hedgehog genes and through suppression of BMP signaling; D/V division by Notch signaling and promotion of appendage outgrowth by Distalless; and formation of light-sensitive organs by *Eyeless/Pax6*. (Reproduced from Bier and McGinnis, 2004 with permission from Oxford University Press.)

Table 5–1 Mechanisms in Developmental Biology (Gilbert, 2004)

1. Mechanisms of differential gene expression
2. Role of enhancers and promoters
3. Signal-transduction pathways linking cell membrane and nucleus
4. Mechanisms of origin of syndromes
5. Mechanisms producing dominant and recessive traits
6. Molecular mechanisms for morphogenetic interactions
7. Role of genome variation in morphogenesis

modular nature of transcriptional regulatory units. *PAX6* is needed for mammalian eye, nervous system, and pancreatic development. In addition to ocular abnormalities, mutations in the *PAX6* gene can cause severe nervous system and pancreatic optic abnormalities (Glaser et al., 1994). PAX6-binding sequences have been found in the enhancers of vertebrate lens crystalline genes and in the genes for insulin, glucagon, and somatostatin, which are expressed in the endothelial cells of the pancreas.

There are some basic principles that govern the role played by transcription factors. Firstly, transcription factors function in combination with other transcription factors. Secondly, transcription factors find their way to the nucleus either through cell lineage or induction. Thirdly, transcription factors can continue to be synthesized after the original signal has ceased. Finally, posttranslational modification is often necessary to ensure adequate functioning of the transcriptional factors. This mechanism is probably of major importance in differentiation and morphogenesis.

Other Mechanisms

Differential transcription of DNA is not solely dependent on transcriptional factors. There are other mechanisms that influence developmental gene regulation. Even though a particular RNA transcript is synthesized, it is not always possible to generate a functional protein. In order for the mRNA to become an active protein, a number of other steps are important including processing of mRNA by removal of introns, translocation from the nucleus to the cytoplasm, translation by the protein-synthesizing apparatus, and posttranslational modification to make an active protein molecule. Regulation can occur at any of these steps during development.

Table 5–2 Transcription Factor Families and Functions (Gilbert, 2004)

Family	Representative Transcription Factors	Key Functions
Homeodomain		
HOX	*HOXA-1, HOXB-2, HOXC, HOXD-1* etc.	Axis formation, patterning
POU	*PIT1, UNC-86, Oct-2*	Pituitary development, neural fate
LIM	*LIM-1*, Forkhead	Head development
PAX	*PAX-1*, -2, -3 etc.	Neural specification, eye development
Basic helix–loop–helix	*MYOD*, achaete	Muscle and nerve specification
Basic leucine zipper	*C/EBP, AP1*	Liver differentiation, fat cell specification
Zinc finger		
Standard	WT1, Krüppel	Kidney, gonad development, hormone receptors, estrogen receptor, secondary sex determination
Sry–Sox	*Sry, SOXD, Sox2*	Bone, primary sex determination

Regulation of Embryogenesis and Organogenesis

Soon after fertilization, the rapid cell division in the resulting embryo paves the way for the formation of ectoderm, mesoderm, and endoderm, the three primordial germ layers. Early embryogenesis is regulated by a complex system of proteins that regulate complex processes of differentiation and morphogenesis (Table 5–1; Fig. 5–4). Toward the end of embryonic development and differentiation, specific organ formation is initiated. Organogenesis is highly complex as different organs are composed of tissues derived from different primordial layers. Disruption in one of these tissues will lead to either a structural malformation or a functional abnormality in the organ. For example, in the eye, a precise arrangement of tissues forming transparent cornea, lens, vitreous, choroid, and neural retina is necessary for normal shape and function. Thus, construction of organs is accomplished by a group of cells changing the behavior of an adjacent set of cells causing them to change their shape, mitotic rate, and eventual fate. This is called *interaction*. The interaction between closely located cells or tissues is referred to as *proximate interaction*. This process is continued throughout organogenesis. Proximate interaction consists of two components—*inducers*, the tissue producing the signal, and *responders*, the tissue being induced. The ability of the tissue to respond to the induction signal is called *competence*. It is an active process, as the responding tissue undergoes several changes and interaction with other factors to ensure formation of the intended organ. For example, in the formation of the lens in the eye, *Pax6* acts as the inducer for the ectoderm in the optic vesicle, which in turn acts with fibroblast growth factor-8 (FGF8) and other factors (Sox2, Sox3, and L-Maf) to ensure the production of the lens (Vogel-Höpker et al., 2000). Further induction is called *instructive*, when the responding tissue depends on a specific tissue to begin the process. In general, there are three broad principles of *instructive induction* (Wessels, 1977): (1) tissue A is necessary for tissue B to respond in a desired manner; (2) tissue B does not respond in the desired manner in the absence of tissue A; (3) tissue B may not respond in the desired manner in the presence of another tissue, but in the absence of tissue A. An example of this form of instruction is in the optic vesicle, which when placed in another part of the developing head ectoderm can form an ectopic lens. Induction also depends on environmental factors, which is called *permissive induction*.

Regional Specificity of Induction

Induction is a dynamic process that governs the early cell and tissue differentiation leading to specific organ formation. It particularly involves an interaction between various tissues, particularly those that lie adjacent to each other (Gilbert, 2004). For example, the interaction between *epithelial cells and mesenchymal cells* is probably the most important in the development of several organs (Table 5–3). The best example of which is the skin. Skin comprises epidermis (epithelial) and dermis (mesodermal), developed from

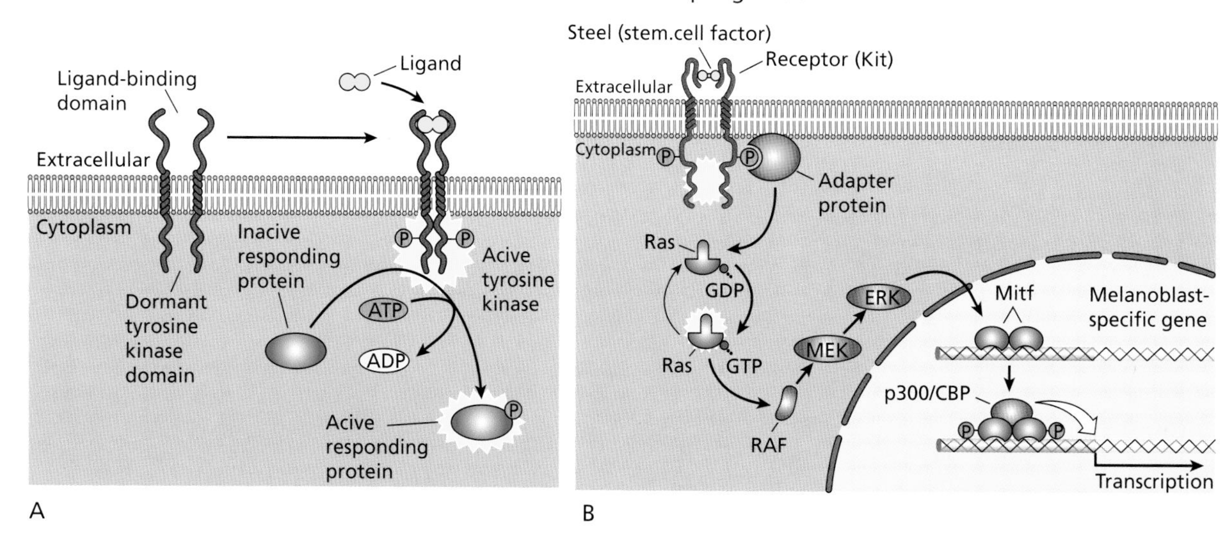

Figure 5–4 Differentiation and morphogenesis.

the interactions of sheets of epithelial cells and mesenchymal cells derived from the mesoderm (see also Chapter 34). This phenomenon is regionally specific. For example, the developing epidermis signals the underlying dermis, probably through Sonic hedgehog and transforming growth factor β (TGF-β) proteins (see later section), and the condensed dermal mesenchyme responds by secreting factors that cause the epidermis to form regionally specific cutaneous structures (Ting-Berreth and Chuong, 1996). These structures could be any of the skin appendages such as hair, nails, to sweat glands (see Chapter 34). Several other organs develop from such interactions where the mesenchymal cells take the lead in instructing different sets of genes in the responding epithelial cells.

Paracrine Factors

Transmission of signals from the inducer to the responder is a complex process and depends on several factors. The interaction is *juxtacrine* when cell membrane proteins of the responding cell are physically in close contact with the cell membrane proteins of the inducing cell. In contrast, *paracrine interaction* depends on the diffusion of proteins synthesized by the inducing cell over to the cell membrane of the responding cell. This process involves several special kinds of protein families, collectively referred to as *paracrine factors.* Essentially, these are *growth and differentiation* factors (GDFs). Paracrine factors differ from endocrine factors (hormones), as they do not travel through the blood but are secreted into spaces surrounding the target cells. These proteins are inducers and are biologically similar throughout the animal kingdom, from the fruit fly *Drosophila* to humans. There are four major classes of protein families that comprise the majority of the paracrine factors.

Table 5–3 Organs Derived from Epithelial–Mesenchymal Interactions (Gilbert, 2004)

Organ	Epithelial Component	Mesodermal Component
Skin appendages	Epidermis (ectoderm)	Dermis (mesoderm)
Tooth	Jaw epithelium (ectoderm)	Neural crest (ectodermal) mesenchyme
Gut organs	Endodermal epithelium	Mesodermal mesenchyme
Respiratory organs	Endodermal epithelium	Mesodermal mesenchyme
Kidney	Ureteric bud epithelium (mesoderm)	Metanephrogenic (mesodermal) mesenchyme

Fibroblast Growth Factors

Several FGFs genes are important for mammalian development. The FGFs code for specific proteins of which there are a number of isoforms produced by alternate RNA splicing or varying initiation codons in different tissues. These factors activate a set of receptor tyrosine kinases, namely the FGF receptors (FGFRs). Mutations in some of the FGFRs result in certain skeletal disorders. For example, mutations in *FGFR3* result in sporadic lethal thanotophoric dysplasia and autosomal dominant achondroplasia. Receptor tyrosine kinases are proteins that extend from the cell surface to the nucleus. The extracellular part binds with FGFs, and the intracellular component activates dormant tyrosine kinase. FGFs are associated with a number of developmental functions including angiogenesis (blood vessel formation), mesoderm formation, and axon extension. FGF2 is particularly important for angiogenesis, and FGF8 is important for the development of the mid-brain, eyes, and limbs (Gilbert, 2004).

Hedgehog Proteins

The Hedgehog proteins are a family of important paracrine factors that induce particular cell types and create boundaries between tissues. There are at least three homologs known for the Drosophila hedgehog gene—sonic hedgehog *(shh),* desert hedgehog *(dhh),* and Indian hedgehog *(ihh).* Desert hedgehog is expressed in the Sertoli cells of the testes and mice homozygous for a null allele of *dhh* exhibit abnormal spermatogenesis. Indian hedgehog is expressed in the gut and cartilage, and is important for postnatal skeletal growth.

Sonic hedgehog is perhaps the most widely used hedgehog protein. It is expressed in the developing notochord, and is responsible for the patterning of the neural tube in such a manner that the ventral neurons develop motor neurons and the sensory neurons are formed from the dorsal neurons (Yamada et al., 1993). It is also responsible for patterning the somites. Sonic hedgehog is crucial for the formation of left–right axis in many vertebrates. It initiates the anterior–posterior axis in limbs, induces regional specificity in the gut, and induces hair formation (Gilbert, 2004). Sonic hedgehog works in conjunction with other paracrine factors, for example, Wnt (wingless integrated) and FGF proteins.

Wnt Family Proteins

The Wnt family proteins are cysteine-rich glycoproteins, and comprise at least 15 members that are important for skeletal and muscle development in vertebrates. Wnt1 appears to be active in inducing the dorsal cells of the somites to become muscles, while sonic hedgehog proteins are important in patterning of the ventral portion of somites (Stern et al., 1995). In addition, Wnt proteins are also important in establishing the polarity of insect and vertebrate limbs, and are used in several steps of urogenital system development.

TGF-β Super Family Proteins

The TGF-β superfamily comprises 30 structurally related proteins that regulate some of the most important interactions in development. The important members are the TGF-β family, the activin family, the bone morphogenic proteins (BMPs), the Vg1 family, glia-derived neurotrophic factor (necessary for kidney and enteric neuron differentiation), and müllerian inhibitory factor (important for mammalian sex differentiation). Some of the TGF-β proteins (TGF-β-1, -2, -3, and -5) are important in regulating the formation of the extracellular matrix between cells and for regulating cell division. The members of the BMP family are named because of their ability to induce bone formation. However, bone formation is just one of the many functions including regulation of cell division, programmed cell death (apoptosis), cell migration, and differentiation (Hogan, 1996).

Other Paracrine Factors

Apart from the above family of proteins, there are several other paracrine factors that are important for vertebrate development. These include epidermal growth factor, hepatocyte growth factor, neurotrophins, and stem cell factor. Several factors are exclusively associated with the development and maturation of erythrocytes and including erythropoietin, the cytokines, and the interleukins.

Signal-transduction Pathways

As described earlier, the paracrine factors include several protein families that act as inducers. The next major class of molecules is involved in the cellular response. These act on the cell membrane by binding to different receptor molecules and trigger a cascade of interacting proteins that transmit a signal through a pathway from the bound receptor to the nucleus. These pathways between the cell membrane and the genome are called *signal-transduction pathways*. Each receptor spans the cell membrane and consists of an extracellular region, a transmembrane region, and a cytoplasmic region. When one of the paracrine factor protein molecules binds to the extracellular region, it induces a conformational change in the receptor. This is transmitted across the cell membrane and changes the shape of the cytoplasmic domains. This physical change induces an enzymatic activity, usually one of the kinases that relies on ATP for phosphorylation. This further triggers phosphorylation of other kinases and activates a dormant transcription factor, which activates or represses a particular set of genes. There are several *signal-transduction pathways* (Fig. 5–5A–E). A detailed discussion on the molecular biology of these pathways is beyond the scope of this chapter.

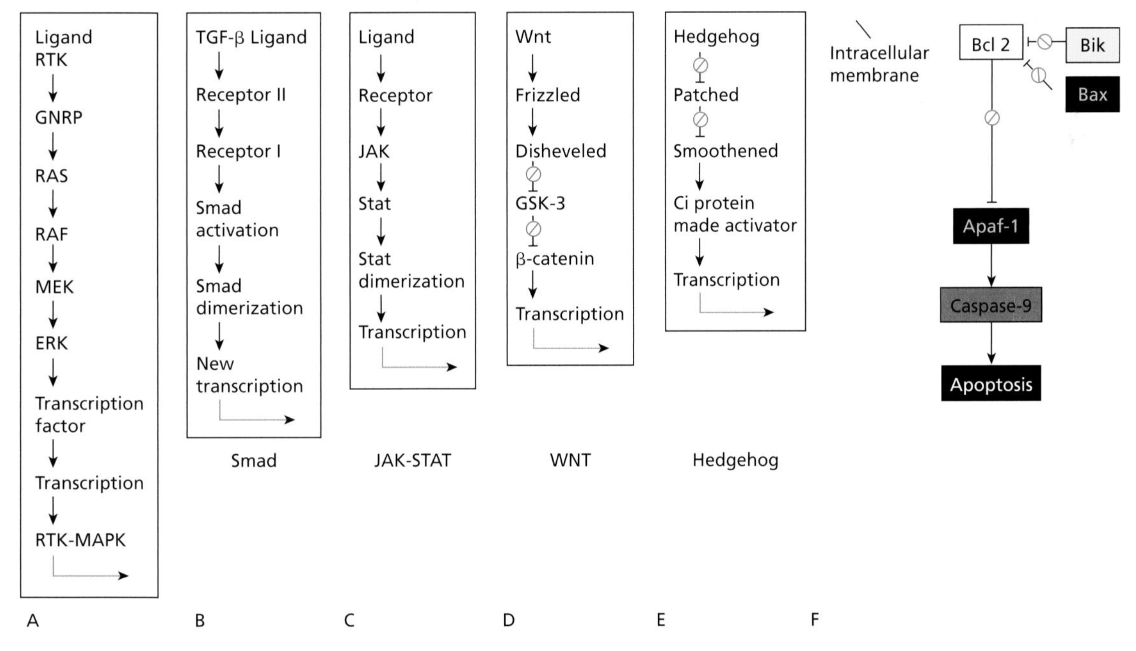

Figure 5–5 Five of the major signal-transduction pathways through which signals from the cell surface are sent into the nucleus. (*A*) The receptor tyrosine kinase-mitogen-activated protein kinase (RTK-MAPK pathway); (*B*) the Smad pathway used by transforming growth factor β (TGF-β) superfamily proteins; (*C*) the JAK-STAT pathway; (*D*) the Wnt pathway; (*E*) the Hedgehog pathway; and (*F*) one of the apoptosis pathway used by mammalian neurons. Abbreviations: ERK, extracellular signal-regulated kinase; GNRP, guanine nucleotide-releasing protein; GSK, glycogen synthase kinase; JAK, Janus kinase; MAPK, mitogen-activated protein kinase pathway; MEK, MAPK/ERK-kinases; STAT, signal transduction and activator of transcription.

The interested reader is advised to refer to one of the major texts on developmental biology (see Further Reading). However, a brief outline is summarized in Table 5–4.

The Role of the Extracellular Matrix in Development

Cells secrete several large molecules into their immediate vicinity. These form the extracellular matrix, which is largely noncellular material and resides in the interstices between the cells. The extracellular matrix has an important role in vertebrate development. A number of cellular functions including cell migration, cell adhesion, and the formation of epithelial sheets and tubes depend on the ability of cells to form attachments with extracellular matrices. Depending upon the requirements of specific cells and tissues, these attachments are of variable strength and rely largely upon the properties of the extracellular matrix.

The extracellular matrix consists of collagen, proteoglycans, and several specialized glycoproteins, including fibronectin and laminin. These large glycoproteins are responsible for organizing the matrix and cells into an ordered structure. Fibronectin is important in cell migration, especially the mesenchymal cells. These large molecules form a kind of "road" on which specific types of mesenchymal cells travel to the desired organ, for example, germ cells reach the gonads and heart cells move to the midline of the embryo, which is the mesodermal region for the development of the heart.

In contrast to fibronectin, the laminin helps to keep cells firmly in place. Laminin has a greater affinity for epithelial cells, and fibronectin for mesenchymal cells. Laminin, along with type IV collagen, is an important component of the basal lamina. The basal lamina ensures adhesion of epithelial cells in sheets. Like fibronectin, the laminin also plays a role in the assembly of the extracellular matrix promoting cell adhesion and growth, changing cell shape, and permitting cell migration (Hakamori et al., 1984).

In addition to the above-mentioned mechanical roles, the extracellular matrix also plays a role in regulating gene function. It helps in inducing specific gene expression in developing tissues, for instance, liver, testis, and mammary glands, through binding the cell substrate that is important for specific transcription factors.

Table 5–4 The Signal-transduction Pathways in Mammalian Development

Pathway	Ligands	Intermediary Proteins
The receptor tyrosine kinase-protein mitogen-activated protein kinase pathway (RTK-MAPK)	FGFs, EGFs, PDGFs, STF	GTPase-activating (GAP)
The Smad Pathway	TGF-β superfamily	Sma and Mad proteins (1–5)
The JAK-STAT Pathway	FGFRs	JAK-STAT tyrosine kinases
The Wnt-β-Catenin Pathway	Frizzled proteins	Disheveled, GSK-3, β-catenin
The Hedgehog Pathway	Patched proteins	Smoothened, Ci protein
The Notch Pathway	Delta, Jagged, Serrate	CBF-1, Suppressor of Hairless, Lag-1

Abbreviations: Ci, Cubitus interruptus; EGFs, epidermal growth factors; FGFs, fibroblast growth factors; JAK, Janus-activating kinase; PDGFs, platelet-derived growth factors; STAT, signal transducers and activators of transcription; STF, stem cell trigger factor.

The extracellular matrix also has a role in inhibiting apoptosis through integrin, which is the cell membrane receptor for fibronectin and other extracellular matrix molecules. Another major cellular function of the extracellular matrix is the ability to regulate differentiation of the chondrocytes to produce the cartilage for developing vertebrae and limbs, which is also achieved by binding to the integrin. In the absence of integrin or experimentally blocking the binding of integrin, the developing chondrocytes fail to differentiate into cartilage and bone (Hirsch et al., 1997).

Lastly, it has also been shown that branching of some parenchymal organs, such as kidney and lungs, is dependent on the extracellular matrix (Lin et al., 2001).

The Apoptosis and Development

Apoptosis or programmed cell death is a normal part of development. Cells in all animals are programmed to die every day, and approximately equal numbers of cells are replaced. For example, adult humans lose as many as 10^{11} cells every day and these are regularly replaced by other cells. It is estimated that the total weight lost every year through programmed cell death could be equal to the adult body weight. Apoptosis begins immediately after birth. It is estimated that the total number of neurons accumulated throughout the gestation period of 9 months is approximately three times that in an adult of average intelligence.

Programmed cell death is a continuous process. It is essential for creating proper spaces within an organ, as well as between organs or body parts. Examples include the middle-ear space, separation of digits to give proper shape and size of fingers and toes, and the lower vaginal space and opening (Rodrigez et al., 1997; Roberts and Miller, 1998). Clearly, through apoptosis, redundant tissues and structures are pruned away. Different tissues use varying signals for apoptosis. Among vertebrates, the BMP4-mediated signals are important. For example, the connective tissues respond to BMP4 to differentiate into bone. Similarly, the surface ectoderm responds to this by differentiating into skin. Another good example is the development of tooth enamel. After the tooth cusp has grown, the enamel knot synthesizes BMP4, which, through apoptosis, stops further enamel differentiation (Vaahtokari et al., 1996). As previously described, the erythropoietin induced red blood cell population is programmed for apoptosis. In its absence, the red cells will undergo apoptosis. This works through the JAK-STAT signal-transduction pathway (see previous section).

Apoptosis works through several pathways. One of the pathways is regulated by genes that were discovered from studies on *C. elegans*, appropriately designated as *ced-3* and *ced-4* genes. The gene product of these two genes initiates apoptosis. However, the protein product of another gene (*ced-9*) is shown to inhibit the programmed cell death. Mutations in this gene will accentuate apoptosis by withdrawing the control. This has been confirmed experimentally when inactivated CED-9 protein led to the death of an entire embryo. On the contrary, gain-of-function *ced-9* mutations can help cells to survive, which would have otherwise died. In other words, wild *ced-9* acts as a binary switch between life and death at the cellular level. In mammals, members of the *Bcl-2* gene family are the CED-9 protein homologs. This gene-family is important for red blood cell development and differentiation.

It is now confirmed that CED-3 and CED-4 proteins act at the center of the apoptosis pathway. These regulate initiation of other genes in the pathway, such as *BMP-4*. Homologs for these proteins are important for auto-cell digestion. The CED-4 protein homolog is called *Apaf-1* (apoptotic protease activating factor-1), which participates in the cytochrome-dependent activation of the mammalian

CED-3 homologs caspase-9 and caspase-3 (Cecconi et al., 1998; Yoshida et al., 1998). Activation of these caspases causes cell autodigestion leading to cell death. Mice homozygous for Apaf-1 deletions have severe craniofacial abnormalities, brain overgrowth, and syndactyly (webbing between fingers/toes).

It is important to appreciate that apoptosis can follow more than one pathway. For instance, the "death-domain," containing receptors of the tumor necrosis factor (TNF) family, can induce apoptosis in several cell systems that can also be triggered by other apoptosis-inducing factors. This is accomplished by blocking the anti-apoptosis signals sent by other factors. One of the developmentally important TNF receptors with a death domain is Edar, a protein required for the development of hair, teeth, and other cutaneous appendages. Mutations of this gene or its ligand, *Eda*, cause X-linked hypohidrotic ectodermal hypoplasia (OMIM 305100), a syndrome characterized by lack of sweat glands, sparse hairs, and poorly formed teeth. An identical syndrome results from deficiency of the adapter protein that binds the death domain of this receptor (Headon et al., 2001). Instead of resulting in cell death, the activation of the receptor enables continued development of skin appendages.

Influence of Environmental and Opportunistic Factors on Development

Developmental biologists agree that the environment plays a significant part in producing a phenotype. Nutritional factors undoubtedly result in a number of disease phenotypes, such as marasmus, Kwashiorkar, rickets, diabetes mellitus, and coronary heart disease. There are several genetic factors that are equally important for creating a pathophysiological state, which predispose to morbid effects of dietary factors (see Chapter 8). Dietary supplementation and modification is known to significantly alter the phenotype. For example, a normal daily dietary intake of vitamin C prevents the development of the clinical effects of vitamin C deficiency, as human beings lack naturally occurring vitamin C (Hypoascorbemia, OMIM 240400) due to deficient gulonolactone oxidase as a result of a mutation in the gulonolactone oxidase gene on the short arm of chromosome 8. This can result in severe childhood connective tissue disease resulting in death. Gulonic acid oxidase enzyme is the final enzyme leading to the synthesis of ascorbic acid (vitamin C). In contrast to humans, several other mammals have normal gulonolactone oxidase enzyme activity offering natural protection from the clinical effects of vitamin C deficiency.

Periconceptional folic acid supplementation is now commonplace for the prevention of recurrence of neural tube defects, and probably even for the primary prevention of some other congenital anomalies. Fetuses with mutations in genes associated with folate metabolism are at an increased risk for neural tube defects (De Marco et al., 2000). One such gene is methylene tetrahydrofolate reductase (MTHFR), which incorporates folic acid in the methylation of homocysteine to methionine. Mutations or polymorphisms in this gene result in increased homocysteine levels resulting in peripheral vascular disease, and are associated with myocardial infarction. However, the mechanisms by which folate deficiency or lack of bio-availability results in neural tube defect are not known.

Several developmental malformation syndromes are associated with defects in cholesterol biosynthesis. One of the enzymes in this pathway is 7-dehydoxycholesterol reductase. Mutations in the gene for this enzyme (*7DHCR*) result in lack of downregulation of Sonic hedgehog (*SHh*), resulting in a number of abnormal phenotypes including Smith–Lemli–Opitz syndrome (Fig. 5–6). It is possible to ameliorate some of the deleterious effects of this downregulation by dietary cholesterol supplementation. Similarly, dietary restriction of the excess metabolite in a number of inherited metabolic conditions can alter the phenotype. An excellent example is that of phenylketonuria in which the behavioral and cognitive effects of excessive accumulation of phenylalanine can be significantly reduced by dietary restriction of phenylalanine.

The genomes of primate mammals determine their final physical shape, which is also under the direct influence of the environment. For example, the facial phenotype depends on firm and regular chewing, which stimulates the facial muscle and bone (maxilla and mandible) development (Corruccini, 1984). The increased prevalence of orthodontic problems in young children and adults in modern times is attributed to a soft or mid-textured diet. This has been shown in experimental primates who were fed soft diet and developed lower jaw malocclusion similar to children requiring orthodontic treatment (Corruccini and Beecher, 1982).

In brief, the production of a phenotype such as a developmental malformation depends on the genotype, which is regulated at numerous levels. The cellular phenotype is the direct consequence of the genome within the cell and the fate of the community of cells in which it resides. Barton Childs (1999) believes that even the environment can alter gene expression!

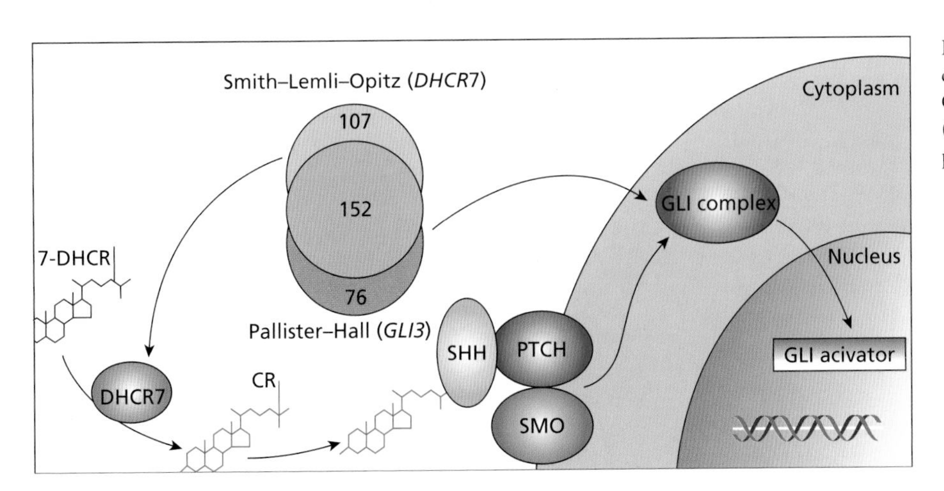

Figure 5–6 The gene pathway of 7-dehydroxy cholesterol reductase, Sonic Hedgehog, and GLI3 complex (Brunner and van Driel, 2004) (Adopted from *Nature Reviews Genetics* with permission.)

Model Organisms for Understanding Development

For several years, geneticists have used a number of model organisms to dissect how genes control metabolism, reproduction, and development. Similarly, using the same model organisms, biochemists and molecular biologists have discovered mechanisms whereby the protein products of these genes regulate biological processes. These model organisms include different microbes, the budding yeast (*S. cerevisiae*), the worm (nematode) *C. elegans*, the plant *Arabidopsis thaliana*, and the fruit fly *D. melanogaster*. For humans, the mouse model, *Mus musculus*, is ideal since the two species are closely related through evolution. With the rapid progress in DNA sequencing, sequences of the best understood genes were deposited into sequence databases such as GeneBank, maintained at the National Center of Biotechnology Information (NCBI; http://www.ncbi.nlm.nih.gov). This information is used to predict an mRNA sequence, which can be readily used to predict the amino acid sequence of the corresponding protein. These protein sequences are known as the *sequence signature* for that particular protein in other organisms. Thus, homologous genes in other organisms can be found through searching for nucleic acid sequences that, when translated, produce a similar protein sequence. This newly identified gene is then predicted to have similar biological function in the second organism.

Complete genome sequences of model organisms and the human genome sequence are available on GeneBank, which is now used in searching for new genes and the corresponding proteins. A newly identified gene sequence can be compared to the existing database of sequences by aligning the predicted protein sequences. If a gene of unknown function is similar to another of a known function, the newly discovered gene can be predicted to have similar function. However, not all gene functions can be predicted in this manner. In some situations, new genes of unknown function are found in two organisms. This problem can be alleviated by building a protein domain knowledge base, which can be searched for amino acid signatures indicative of structure or biochemical activity. If present, these signatures provide a clue for the gene function. If two predicted protein sequences from two genomes are similar, then their amino acid alignments would be the same. In such a situation, the genes can be predicted to have similar function.

Biological make up of the model organism determines the extent of information applicable to human development. Unicellular organisms are an excellent model for studying eukaryotic cell function. For example, genetic studies in the yeast (*S. cerevisiae*) and slime molds (*Dictyostelium discoideum*) can be carried out and repeated several times as a billion progeny can be produced in a relatively short time. These studies provide vital information about intragenic primary and secondary suppressor loci. Thus, unicellular organisms are immensely useful in establishing the networks of gene action involved in basic cell biological processes. However, such studies have limited applications in studying complex cellular functions such as that of the nervous system, which depend on the interaction between cells. Nevertheless, studies on unicellular organisms have helped in understanding enzymes that are involved in complex energy metabolism defects resulting in neuromuscular disorders (Darras and Friedman, 2000; Palau, 2001).

Studies on invertebrate model organisms, for example, *Drosophila* fruit fly and nematode (*C. elegans*), have contributed to the understanding of several basic biological mechanisms, including the organization of genes into independently segregating linear chromosomes, the creation of the first chromosome maps, the one gene-one protein hypothesis, radiation-induced mutagenesis, the principles of pattern formation, and the identification of genetic pathways implicated in human disease. As both fruit fly and nematodes have closely related counterparts to many human disease genes, the identification of new genes in these invertebrates will help define new candidate disease genes that are likely to be involved in the same disease process. A useful example is the Notch signaling pathway that has been shown to be important in invertebrate development. Mutations in the component genes of the Notch signaling pathway have been shown to result in notching of the wing margin in fruit flies (Wu and Rao, 1999) and defects in vulval development in worms (Wang and Sternberg, 2001). Notched wings were also observed in mutations in the ligand Delta, the Notch receptor, or the signal transducer suppressor (Hairless). Vertebrates have several common paralogs of the Notch signaling pathway components. Reduced function of the related ligand, Delta3, or the Notch homolog (Notch1) itself results in axial skeletal malformations. A good example is spondylocostal dysostosis, an autosomal recessive heterogenous condition (OMIM 277300, 608681, 609813). The characteristic feature of this disease is the spinal segmental anomaly, which is also associated with anal and urogenital anomalies (OMIM 271520).

One of the major contributions of studies in invertebrate model organisms has been the discovery of homeotic selector genes, now referred to as *Hox* genes. The term *homeostasis* was coined by William Bateson (1894) as the phenomenon in which one segment of an organism is "trans*formed in whole or part to another.*" The genetic basis of these transformations is explained by mutations in *Hox* genes. Systematic analyses of *Hox* gene mutations in the *Drosophila* fruit fly revealed an extra pair of wings due to mutations in the *Ultrabithorax* (*Ubx*) gene and an extra thoracic leg attached to the head resulting from dominant mutations in the *Antennapedia (Antp)* gene (Lewis et al., 1978; Kaufman et al., 1980). Molecular analysis of the genomes has revealed that humans and other bilateral animals have multiple *Hox* genes (Fig. 5–1), which carry a common DNA sequence motif called the homeobox. The homeobox motif encodes a similar 60-amino acid motif in Hox proteins, termed the *homeodomain*. The Hox proteins belong to the transcription factor family (Table 5–1). These exert their function through activation and repression of multiple target genes. Arrangement of these genes is strikingly similar in the fruit fly and humans. These genes are arranged in clusters. There is evidence that this clustered arrangement of *Hox* genes has been maintained for more than 500 million years, because different genes in the clusters are controlled by the same cis-acting DNA regulatory regions. In general, there are four clusters—*HOXA, HOXB, HOXC,* and *HOXD*. Each gene has a role in the anterior–posterior axis patterning of various organs and body parts. This is evident as specific human malformation syndromes are now recognized to be associated with *HOXA* (OMIM 609296; 601536), *HOXB* (OMIM 249000, Meckel syndrome, MKS1), and *HOXD* (OMIM 606708, Split Hand–Foot–Absent Uterus syndrome; OMIM 127300, Leri-Well dyschondrosteosis; OMIM 113200, Brachydactyly type D; OMIM 112500, Brachydactyly type A1). In addition to these malformations, the *HOX* genes interact with several other patterning genes with a crucial role in development.

Molecular Basis of Malformation Syndromes

Developmental biologists and dysmorphologists have been intrigued for a considerable time about the molecular basis of phenotypic variability in malformation syndromes. Frequently, questions have been asked to explain (1) why mutations at a single genetic locus are associated with multitude of phenotypic features (*pleiotropy*); (2) what is the mechanism for the same phenotype being caused by mutations in several different genes (*genetic heterogeneity*); and (3) what is the mechanism of a dominant phenotype.

Pleiotropy

Observations on the expression patterns of transcription factors and paracrine factors have revealed mechanisms that lead to different malformations caused by mutations at a single locus. This is called *pleiotropy*. There are several examples of this phenomenon. For instance, it is known that heterozygosity for MITF (in humans and mice) causes multiple malformations that include iris defects, pigmentation abnormalities, deafness, and the inability to produce a normal number of mast cells. Moreover, these abnormalities are not related to each other and can occur independently. This occurs because all body parts can use the MITF protein as a transcription factor. This type of pleiotropy is called *mosaic pleiotropy*, as the relevant organ or body part is separately affected by the mutant gene. In contrast, some malformations in the related part do not result directly from the abnormal gene function as the mutant protein is not expressed. For example, the failure of MITF expression results in the pigmented retina not being fully differentiated. This in turn causes a malformation involving the choroidal tissue, which results in drainage of the vitreous humor fluid. This further leads to failure of ocular development, leading to microphthalmia (small eye). This phenomenon, in which several developing tissues or organs might be sequentially affected even though they do not express the mutant gene, is called *relational pleiotropy*. This concept is important in dealing with complex clinical genetic situations, particularly in prenatal diagnosis where the prediction of phenotype is important in making informed choices.

Genetic Heterogeneity

An important aspect of dysmorphology is the recognition of the phenotype and the search for a syndrome of which that particular phenotype is known to be a major component. However, it is made difficult due to the fact that many syndromes can feature the same phenotype. At the molecular level, this refers to mutations in different genes being responsible for these syndromes. This is possible if the genes are part of the same signal-transduction pathway. This phenomenon is referred to as *genetic heterogeneity*. There are several examples. Genetic heterogeneity could be either within the gene (*allelic heterogeneity*) or due to genes at different loci (*locus heterogeneity*). A good example is cyclopia, which is caused by both mutations in the *sonic hedgehog* gene or the genes regulating cholesterol biosynthesis. Since they are in the same pathway, mutations in one gene generate a phenotype similar or identical to mutations in other genes. Another good example is hypohidrotic ectodermal dysplasia, which can result from mutations in *EDA* (receptor), *EDAR* (ligand), or *EDARADD* (adapter protein) (Headon et al., 2001).

Mechanisms of Dominance

Molecular analysis can help in delineating whether a particular syndrome is dominant or recessive. The true dominant phenotype only occurs in the heterozygous state. In other words, the homozygous state never exists, probably because it is lethal to the embryo and the fetus never survives. There are several possibilities that can result in a dominant phenotype.

Haploinsufficiency

This refers to the situation when one copy of the gene (haploid) is not enough to produce the normal amount of wild protein required for normal development. In other words, an abnormal phenotype can result if one of the two copies of the gene is either absent or nonfunctional. For example, in type 2 Waardenburg's syndrome the variable phenotype is associated with only about half the amount of wild-type MITF protein being present. This is not enough for full pigment cell proliferation, mast cell differentiation, or inner ear development, which manifest clinically in the variable phenotype of this autosomal dominant condition. The phenotype in several micro-deletion multiple malformation syndromes is also due to haploinsufficiency when one or more genes can be lost, which lie within the deleted chromosomal band. Thus it is likely that mutations in some of these genes could behave in a dominant manner.

Gain-of-function Mutations

An abnormal phenotype can also result from mutations in a gene causing a gain in additional or acquiring new function. A good example is the *FGFR* gene where different mutations can result in a constitutively active gene product, which can result in the potentially lethal thanatophoric dysplasia as well as the milder form of related skeletal dysplasia (achondroplasia). Other examples of *gain-in-function* type dominant phenotypes include late-onset Huntington's disease and other neurodegenerative disorders associated with expanded triplet-repeats (see Chapter 6).

Dominant-negative Allele

Another mechanism of dominance is a dominant-negative allele. This situation can occur when the active protein product is a multitimer. All constituent proteins of the multitimer have to be wild type in order for the multitimer to be active. Mutations in the gene for any of the member proteins can render that unit becoming inactive, resulting in a nonfunctional or relatively ineffective multitimer. A good example is autosomal dominant Marfan syndrome, the variable phenotype of which is due to mutations in the fibrillin-1 gene. The wild-type product of this gene is a glycoprotein that forms multitimer microfibrils in elastic connective tissue. The presence of even minute amounts of mutant fibrillin prohibits the association of wild-type fibrillin into microfibrils (Eldadah et al., 1995).

Allelic Interactions

Another mechanism of dominance is related to qualitative differences in the product made from the interactions of different alleles. Such interactions can result in a superior protein dimer made from two alleles compared to an inferior or less active product made by one allele alone (Trehan and Gill, 2002).

Functional Genomics in Syndrome-families

Delineation of a dysmorphic syndrome is complex, and involves continuous refining of the phenotype by the identification of new cases and revisiting those previously described. There are several syndromes that are apparently identical or nearly identical to the original description, and this might create a false impression of homogeneity for that syndrome (Cohen, 1989). However, this is often not true as close inspection of the phenotype can reveal distinctive features. Reports on patients with partly overlapping phenotypes lead to frequent debates between "*lumpers and splitters*" (McKusick, 1969). Apart from a few quantitative analyses, syndrome definition is largely a matter of comparing the phenotype of a suspected patient with the "best-fit" case (Verloes, 1995). Identification of pathogenic mutations in a gene might help in defining the core phenotype and its variants. However, resolving the molecular genetics does not always solve the problem of understanding the phenotypic variability, because allelic heterogeneity and the action of modifier genes can influence the phenotype (Romeo and McKusick, 1994). This is further complicated as mutations in different genes can manifest with the same or related phenotypes. Thus, a strict molecular classification of syndromes is clinically not relevant as it would obscure the relationship between molecular pathology and the phenotype (Biesecker, 1998).

It has been recognized for some time that several syndromes share overlapping dysmorphic features. It is argued that these syndromes might have a common biological relationship (Pinsky, 1977), and may be referred to as "syndrome community," "phenotype communities," and "syndrome family." The concept of a "syndrome-family" is based on the observations that several dysmorphic syndromes have common phenotypic features. The syndrome-family approach was first systematically applied to skeletal dysplasias; for example, the family of *chondrodysplasias* includes several distinct skeletal dysplasias such as achondrodysplasia, hypochondrodysplasia, and so on (Spranger, 1985). The advances in molecular genetics have vindicated this concept. For example, mutations in FGFR3 result in three distinct members of the chondrodysplasia family. The converse is true for the phenotype of the Stickler–Kniest family, which is related to mutations in three different collagen genes (*COL2A1*, *COL11A2*, and *COL11A1*) (Annunen et al., 1999). Thus, rapid advances in molecular genetics and genomics could see merger of syndromes, more splitting of syndrome families, and even the complete disappearance of syndromes. However, description and definition of syndromes and syndrome families is important. Several genetic databases [McKusick's Online Mendelian Inheriance in Man (OMIM), the LDDB and POSSUM] continue to record these syndromes as independent entities.

There are approximately 200 Mendelian syndromes in man, which appear in OMIM and other databases. Each syndrome is recognized by a specific phenotypic feature or pattern. Some syndromes differ only by a few features. It is acceptable logic that there could be some biological relationship in syndrome families that share the same phenotype. Brunner and van Driel (2004) successfully argue that a systematic analysis of such phenotype relationships could be applied in the identification of new genes, providing clues to gene interactions, molecular pathways, and functions. This has been applied in large-scale mutagenesis programs that aim to define the function of genes in a genome in relation to mutant phenotypes. This has been completed in yeast, and work is underway for other model systems such as *C. elegans*, the mouse, and the zebrafish. Although this strategy has been successful, it is not clear how many different mutants will be required for such screens to be comprehensive. It is accepted that creating a single knockout mouse model might not be sufficient to probe specific gene function in development and homeostasis. A comprehensive analysis of the mutant phenotype would require studying the functional effects of several mutations. Starting from interesting phenotypic differences and then comparing the underlying mutations might be more productive (Brunner and van Driel, 2004). Spontaneous mutations are frequent, and studying the phenotypic effect can contribute to our understanding of gene function. Thus, it is important to adopt a "*phenotype-driven*" approach that saturates the genome with mutations, either experimentally shown in animal models or in observed in human disease states.

Phenotypic similarities in different syndromes belonging to a so-called syndrome family could be a reliable indicator of shared biological mechanisms. Apparently, single-gene human genetic disease could in fact be associated with mutations in different genes, probably contributing to a common molecular pathway. This was first demonstrated by Morton (1956) in familial elliptocytosis, because linkage to Rhesus (*Rh*) blood group was not seen in all families. It is now clear that nonallelic heterogeneity is extensive in human disease. Although genetic heterogeneity could be problematic in conducting familial genetic studies, this can be viewed positively as it might reflect interactions at the protein level, for example, ligand–receptor interactions, different subunits of a multiprotein complex, or proteins that function at different steps of a metabolic pathway. Using this strategy, the other genes could be found once the first gene is discovered. It is also possible that unrelated genes that result in the same phenotype could also be found and will ultimately be shown to have a functional relationship. It is important that clinical classification is paramount followed by molecular verification. In other words, defining the phenotype and syndrome identification could become a functional genomics tool (Brunner and van Driel, 2004).

Translational Research in Dysmorphology

Clinical studies on human multiple malformation syndromes are not only helpful in medical management but are also a useful resource for research in understanding the basic mechanism of human development and eventually mammalian development in general. This approach allows researchers to gain insight into basic mechanisms of development, and how genes program organisms to achieve permanent or adult morphological shapes (Biesecker, 2002). A number of malformation syndromes have overlapping manifestations, despite being phenotypically and genetically dissimilar. This can be used in basic research, for example, developing animal model systems, such as fruit flies, mice, worms, and other simple organisms. The data thus generated can be applied in the clinical setting to understand both the problems patients suffer and the mechanisms of development. This is referred to as *translational research*. However, there are limitations to using human subjects in translational research. In humans, it is obviously not possible or ethically appropriate to experimentally manipulate genes to test hypotheses. In addition, there is inevitable difficulty in obtaining tissues or organs for study due to the lack of consent and/or availability. The final limitation or disadvantage in carrying out studies on humans is the enormous cost and length of time it can take to complete the study. Moreover, as a genetic model system, the human has a long generation time and hence individuals and families must be followed over long periods of time.

The typical example of human translational research is to creatively apply a basic laboratory discovery to the clinical management of a patient with the clinical phenotype of a particular dysmorphic syndrome. This approach would allow testing of the hypothesis as well as providing an opportunity to validate the basic laboratory finding. Any further improvement or information can be fed back to the patient, and the system thus works bidirectionally. The end result of this system would allow extracting the maximum amount of basic information from the patient. This could then be applied back in the clinical setting to explain the cause and the likely outcome of the diagnosis. Thus, researchers can develop novel hypotheses about the mechanisms of development, modifying variables, or other disorders with overlapping manifestations. The aim is to keep the cycle in motion, continually collecting and examining the new information, and constantly applying back in the improved clinical care.

The above concept has been applied in a number of multiple malformation syndromes with overlapping phenotypic manifestations, often referred to as "*family of syndromes*" (see previous section). One example is the Pallister–Hall syndrome (PHS), a typical "*syndrome family*" disorder in which affected individuals can present with one or more of a range of malformations that include hypothalamic hamartoma, imperforate anus, laryngeal anomalies, and central polydactyly, with shortened terminal digits (Biesecker and Graham, 1996). The disorder is inherited in an autosomal dominant manner with significant inter/intra-familial variability. Clinical and genetic studies in families affected with PHS revealed mutations in the zinc finger transcription factor gene, *GLI3* (Kang et al., 1997). This was an interesting finding, as previously this gene was incidentally found to be causally related to the Greig cephalopolysyndactyly syndrome (GCPS) (Vortkamp et al., 1991). Patients with GCPS had balanced chromosomal rearrangements that had apparently disrupted the *GLI3* gene. It is now accepted that GCPS is a distinct developmental syndrome caused by haploinsufficiency of *GLI3*.

This was a challenging observation as mutations in the same gene were found to be causally linked to two distinct genetic syndromes. The research group led by Dr. Leslie Biesecker at the NIH discovered that mutations in *Cubitus interruptus* or *ci*, the *Drosophila* homolog of *GLI3*, were linked to a wide range of phenotypes in fruit flies including abnormal wing vein patterning and lethal malformations manifesting with larval death (Kinzler et al., 1988). It transpired that *ci* belonged to the genetic pathway downstream of the hedgehog signaling molecule (*Hh*). It was shown that the ci protein negatively controlled downstream genes, and the cleaved ci protein turned off the expression of downstream genes.

When the same information was applied to humans, it soon became apparent that the truncated *GLI3* protein had a different clinical outcome (PHS) compared to that resulting from haploinsufficiency of the gene (GCPS). In other words, PHS phenotype was due to the qualitative change, and the quantitative change led to the GCPS phenotype. It is now shown that GLI3 protein, like ci protein, is proteolytically processed (Wang et al., 2000). This example illustrates that molecular studies in rare developmental disorders like PHS and GCPS can illustrate basic mechanisms of mammalian development. The phenotype related to *GLI3* mutations also includes polydactyly, imperforate anus and vertebral anomalies (OMIM 174100).

The genetic pathway involving *GLI3* also includes the *DCHR7* gene (Fig. 5–6), mutations in which cause Smith–Lemli–Opitz syndrome (OMIM 270400), one of the malformations syndromes related to disruption in the cholesterol biosynthesis. The *DCHR7* gene codes for an enzyme called 7-dehydroxycholesterol reductase, which regulates sonic hedgehog (*SHH)* gene function. One of the downstream effectors of *SHH* is *GLI3*. Other SHH protein homologs include patched protein homolog (PTCH) and smoothed homolog precursor (SMO), both of which play key roles in regulating *GLI3* gene function. It is likely that the phenotypic similarity in other malformation syndromes like Optiz G (OMIM 145410) and Mohr syndrome (OMIM 252100) could be due to mutations in other genes in this gene family (Brunner and van Driel, 2004).

Another illustrative example is McKusick–Kauffman syndrome (MKS), which was first described among the Old Order Amish of Lancaster County, Pennsylvania (Kaufman et al., 1972). This disorder is inherited in an autosomal recessive manner, and the phenotype includes polydactyly (central and post-axial), congenital heart disease, hydrometrocolpos due to congenital uterine outflow obstruction. This disorder is rare with fewer than 100 cases described in the literature. It is likely that the disorder, or a phenotypically similar disorder, probably occurs in other inbred populations groups (Kumar et al., 1998). Increased incidence of autosomal recessive disorders, including multiple malformation syndromes, is recognized among the Amish and other highly inbred ethnic population groups. These populations groups are ideal for conducting homozygosity mapping studies.

Using the whole genome-wide scan with 385 markers, the gene for MKS was mapped to chromosome 20. Further molecular studies identified two substitution mutations in a single mutant chromosome in one of the candidate genes. However, this was not associated with a known function. Since the Amish are closely related, it was not possible to prove that the sequence variants were pathogenic. A search for non-Amish cases was then made. This proved difficult as the reported cases were either deceased or no longer available. In some cases the diagnosis was changed. Eventually, a new-born girl with features of MKS was recruited to the study, and was found to have a 2-bp deletion on the same allele in the same gene that was identified in the Amish (Stone et al., 2000). This confirmed that mutations in the novel Amish gene (*MKKS*) caused MKS.

Since that time many patients who were originally diagnosed with MKS have developed other clinical features and have been diagnosed with Bardet–Biedl syndrome (BBS) (Beales et al., 1999). Affected individuals with BBS have post-axial polydactyly, mental retardation, progressive pigmentary retinopathy, and obesity complicated with diabetes mellitus. It soon became obvious that both MKS and BBS share common phenotypic features. BBS is caused by mutations in different genes. It is an illustrative example of a digenic or multigenic disorder, where a combination of mutations in different genes results in the BBS phenotype. Some patients who were not known to have mutations in one of the BBS genes have now been reported to have mutations in the *MKKS* gene (Katsanis et al., 2000; Slovtnik et al., 2002). These mutations are different to that found in the Amish. These findings show that it is important to review all patients with a clinical diagnosis of BBS or MKS, and apply the molecular techniques to establish diagnosis. This is important in the clinical care and offering accurate genetic counseling to the family. This is another example where the basic scientific data could be taken back to the clinical setting for the benefit of patients, parents, and other family members.

Conclusion

In conclusion, understanding normal human development is a prerequisite for analyzing the pathophysiology of developmental malformations occurring either singly or as part of a multiple malformation syndrome. Development in humans is strikingly similar to that in small creature, as well as in several other mammals including primates. The mechanisms are similar by which the individual genome specifies the physical and functional state of the human body. This chapter, in brief, summarizes a number of gene families and numerous related protein families that regulate the complex process of development. The availability of human and mouse genomes and that of several other small organisms and animals have provided developmental biologists with powerful new tools for identifying genes, and their mutations, that control the development. The interested reader is recommended to explore other literature resources on this complex and stimulating field.

References

Annunen, S, Korkko, J, Czarny, M, Warman, ML, Brunner, HG, Kaariainen, H, et al. (1999). Splicing mutations of 54-bp exons in the *COL11A1* gene cause Marshall syndrome, but other mutations cause overlapping Marshall/Stickler phenotypes. *Am J Hum Genet*, 65:974-983.

Bateson, W (1894). *Materials for the Study of Variation*. London: MacMillan.

Beales, P, Elioglu, N, Woolf, A, Parker, D, and Flinter, F (1999). New criteria for improved diagnosis of Bardet-Biedl syndrome: results of a population survey. *J Med Genet*, 36:437–446.

Beldade, P and Brakefield, PM (2002). The genetics and evo devo of butterfly wing patterns. *Nat Rev Genet*, 3:442–452.

Bier, E and McGinnis, W (2004). Model organisms in the study of development and disease. In Epstein, CJ, Erickson, RP, and Wynshaw-Boris A (eds.), *Inborn Errors of Development* (pp. 25–45). New York: Oxford University Press.

Biesecker, LG (1998). Lumping and splitting: molecular biology in the genetics clinic. *Clin Genet*, 53:3–7.

Biesecker, LG (2002). Coupling genomics and human genetics to delineate basic mechanisms of development. *Genet Med*, 4(6) (Suppl.):39S–42S.

Biesecker, LG and Graham, JM Jr. (1996). Syndrome of the month: Pallister-Hall syndrome. *J Med Genet*, 33:585–589.

Brunner, HG and van Driel MA (2004). From syndrome families to functional genomics. *Nature Rev Genet*, 5:545–551.

Cecconi, F, Alverez-Bolado, G, Meyer, BI, Roth, KA, and Gruss, P (1988). Apaf-1 (CED-4 homologue) regulates programmed cell death in mammalian development. *Cell*, 94:727–737.

Chervitz, SA, Aravind, K, Sherlock, G, Ball, CA, Koonin, EV, Dwight, SS, et al. (1998). Comparison of the complete protein sets of worm and yeast: orthology and divergence. *Science*, 282:2022–2028.

Childs, B (1999). *Genetic Medicine*. Baltimore: Johns Hopkins University Press.

Cohen, MM Jr. (1989). Syndromology: an updated conceptual overview. III. Syndrome delineation. *Int J Oral Maxillofac Surg*, 18:281–285.

Corruccini, RS (1984). An epidemiologic transition in dental occlusion in world populations. *Am J Orthod*, 86:419–426.

Corruccini, RS and Beecher, CL (1982). Occlusal variation related to soft diet in a nonhuman primate. *Science*, 218:74–76.

Darras, BT and Friedman, NR (2000). Metabolic myopathies: a clinical approach: part 1. *Pediatr Neurol*, 22:87–97.

Davidson, EH, Rast, JP, Oliveri, P, Ransick, A, Calestani, C, Yuh, C, et al. (2002). A genomic regulatory network for development. *Science*, 295: 1669–1678.

De Marco, P, Moroni, A, Merello, E, de Franchis, R, Andreussi, L, Finnell, RH, et al. (2000). Folate pathway gene alterations in patients with neural tube defects. *Am J Med Genet*, 95:216–223.

Eldadah, ZA, Brenn, T, Furthmayer, H, and Dietz, HC (1995). Expression of a mutant human fibrillin allele upon a normal human or murine genetic background recapitulates a Marfan cellular phenotype. *J Clin Invest*, 95:874–880.

Gehring, WJ (2002). The genetic control of eye development and its implications for the evolution of the various eye-types. *Int J Dev Biol*, 46:65–73.

Gilbert, SF (2004). General principles of differentiation and morphogenesis. In Epstein, CJ, Erickson, RP, and Wynshaw-Boris, A (eds.), *Inborn Errors of Development* (pp. 10–24). New York: Oxford University Press.

Glaser, T, Jepeal, L, Edwards, JG, Young, SR, Favor, J, and Maas, RL. (1994). *PAX6* gene dosage effect in a family with congenital cataracts, aniridia, anophthalmia, and central nervous system defects. *Nat Genet*, 7:463–471

Hakamori, S, Fukuda, M, Sekiguchi, K, and Carter, WG (1984). Fibronectin, laminin, and other extracellular glycoproteins. In Picz, KA and Reddi, AH (eds), *Extracellular Matrix Biochemistry* (pp. 229–276). New York: Elsevier.

Headon, DJ, Emmal, SA, Ferguson, BM, Tucker, AS, Justice, MJ, Sharpe, PT, et al. (2001). Gene defect in ectodermal dysplasia implicates a death domain adapter in development. *Nature*, 414:913–916.

Hirsch, MS, Lunsford, LE, Trinkaus-Randall, V, and Svoboda, KK (1997). Chondrocyte survival and differentiation in situ are integrin mediated. *Dev Dyn*, 210:249–263.

Hogan BLM (1996). Bone morphogenesis proteins: multifunctional regulators of vertebrate development. *Genes Dev*, 10:1580–1594.

Jones, K (1997). *Smith's Recognizable Patterns of Human Malformations* (5th edn). Philadelphia, PA: W.B. Saunders Co

Kang, S, Graham, JM, Jr, Olney, AH and Biesecker, LG (1997). GLI3 frameshift mutations cause autosomal dominant Pallister-Hall syndrome. *Nat Genet*, 15:266–268.

Katsanis, N, Beales, PL, Woods, MO, Lewis, RA, Green, JS, Parfery, PS, et al. (2000). Mutations in *MKKS* cause obesity, retinal dystrophy and renal malformations associated with Bardet-Biedl syndrome. *Nat Genet*, 26:67–70.

Kaufman, RL, Hartmann, AF, and McAlister, WH (1972). Family studies in congenital heart disease, II: a syndrome of hydrometrocolpos, postaxial polydactyly and congenital heart disease. *Birth Defects Orig Artic Ser*, 8:85–87.

Kaufman, RL, Lewis, R, and Wakimoto, B (1980). Cytogenetic analysis of chromosome 3 in *Drosophila melanogaster*: the homeotic gene complex in polytene chromosome interval 84A–B. *Genetics*, 94:115–133.

Kinzler, KW, Ruppert, JM, Bigner, SH, and Vogelstein, B (1988). The GLI gene is a member of the Kruppel family of zinc finger proteins. *Nature*, 332:371–374.

Kumar, D, Primhak, RA, and Kumar, A (1998). Variable phenotype in Kaufman–McKusick syndrome: Report of inbred Muslim family and review of literature. *Clin Dysmorphol*, 7:163–170.

Lander, ES, Linton, LM, Birren, B, Nusbaum, C, Zody, MC, Baldwon, K, et al. (2001). Initial sequencing and analysis of the human genome. *Nature*, 409:860–921.

Lewis, EB (1978). A gene complex controlling segmentation in *Drosophila*. *Nature*, 276:565–570.

Lin, Y, Shang, S, Rehn, M, Itäranta, P, Tuukkanen, J, Heljasvaara, R, et al. (2001). Induced repatterning of type XVIII collagen expression in ureter bud from kidney to lung type: association with sonic hedgehog and ectopic surfactant protein C. *Development*, 128:1573–1585.

McKusick, VA (1969). On lumpers and splitters, or the nosology of genetic disease. *Perspect Biol Med*, 12:298–312.

Meyer, WJ III, Migeon, BR, and Migeon, CJ (1975). Locus on human X chromosome for dihydrotestosterone receptor and androgen insensitivity. *Proc Natl Acad Sci USA*, 91:8856–8860.

Morton, NE (1956). The detection and estimation of linkage between the genes for elliptocytosis and the Rh blood type. *Am J Hum Genet*, 8:80–96.

Mount, DW (2001). *Bioinformatics: Sequence and Genome Analysis*. Cold Spring Harbor, NY: Cold Spring Harbor Laboratory Press.

Mount, DW (2004). Consequences of the genome project for understanding development. In Epstein, CJ, Erickson, RP, and Wynshaw-Boris, A (eds.). *Inborn Errors of Development* (pp. 46–50). New York: Oxford University Press.

Palau, F (2001). Friedreich's ataxia and frataxin: molecular genetics, evolution and pathogenesis. *Int J Mol Med*, 7:581–589.

Pennisi, E (2002). Evo-devo devotees eye ocular origins and more. *Science*, 296:1010–1011.

Pinsky, L (1977). The polythetic (phenotype community) system of classifying human malformation syndromes. *Birth Defects Orig Artic Ser*, 13:13–30.

Roberts, DS and Miller, SA (1998). Apoptosis in caviation of middle ear space. *Anat Rec*, 251:286–289.

Rodrigez, I, Araki, K, Khatib, K, Martinou, J-C, and Vassalli, P (1997). Mouse vaginal opening is an apoptosis-dependent process which can be prevented by the overexpression of Bcl2. *Dev Biol*, 184:115–121.

Romeo, G. and McKusick, VA (1994). Phenotypic diversity, allelic series and modifier genes. *Nature Genet*, 7:451–453.

Rubin, GM, Yandell, MD, Wortman, JR, Gabor Miklos, GL, Nelson, CR, Hariharan, IK, et al. (2000). The genome sequence of *Drosophila melanogaster*. *Science*, 287:2204–2215.

Slovtinek, AM, Searby, C, Al-Gazali, L, Hennekam, RC, Schrander-Stumpel, C, Orcana-Losa, M, et al. (2002). Mutation analysis of the MKKS gene in Mckusick-Kaufman syndrome and selected Bardet-Biedl syndrome patients. *Hum Genet*, 110(6):561–567.

Spranger, J (1985). Pattern recognition in bone dysplasias. *Prog Clin Biol Res*, 200:315–342.

Stern, HM, Brown, AMC, and Hauschka, SD (1995). Myogenesis in paraxial mesoderm: preferential induction by dorsal neural tube and by cells expressing Wnt-1. *Development*, 121:3675–3686.

Stone, D, Slavotinek, A, Bouffard, G, Banerjee-Basu, S, Baxevanis, A, Barr, M, et al. (2000). Mutations of a gene encoding a putative chaperonin causes McKusick-Kaufman syndrome. *Nat Genet*, 25:79–82.

Tautz, D (2002). Evo-devo graduates to new levels. *Trends Genet*, 18:66–67.

Ting-Berreth, SA and Chuong, C-M (1996). Local delivery of TGFβ-2 can substitute for placode epithelium to induce mesenchymal condensation during skin morphogenesis. *Dev Biol*, 179:347–359.

Trehan, KP and Gill, KP (2002). Epigenetics of dominance for enzyme activity. *J Biosci*, 27:127–134.

Vaahtokari, A, Aberg, T, and Thesleff, I (1996). Apoptosis in the developing tooth: association with an embryonic signalling center and suppression by EGF and FGF-4. *Development*, 122:121–129.

Venter, JC, Adams, MD, Myers, EW, Li, PW, Mural, RJ, Sutton, GG et al. (2001). The sequence of the human genome. *Science*, 291:1304–1351.

Verloes, A. (1995). Numerical syndromology: a mathematical approach to the nosology of complex phenotypes. *Am J Med Genet*, 55:433–443.

Vogel-Höpker, A, Momose, T, Rohrer, H, Yasua, K, and Rappaport, DH (2000). Multiple functions of fibroblast growth factor-8 (FGF-8) in chick eye development. *Mech Dev*, 94:25–36.

Vortkamp, A, Gessler, M, and Grzeschik, K-H (1991). GLI3 zinc finger gene interrupted by translocations in Greig syndrome families. *Nature*, 352:539–540.

Wang, B, Fallon, J, and Beachy, P (2000). Hedgehog-regulated processing of Gli3 produces an anterior/posterior repressor gradient in the developing vertebrate limb. *Cell*, 100:423–434.

Wang, B and Sternberg, PW (2001). Pattern formation during *C. elegans* vulval induction. *Curr Top Dev Biol*, 51:189–220.

Wessels, NK (1977). *Tissue Interaction and Development*. Menlo Park, CA: Benjamin Cummings.

Wu, JY and Rao, Y (1999). Fringe: defining borders by regulating the notch pathway. *Curr Opin Neurobiol*, 9:537–543.

Yamada, T, Pfaff, SL, Edlund, T, and Jessel, TM (1993). Control of cell pattern in the neural tube: motor neuron induction by diffusible factors from notochord and floor plate. *Cell*, 73:673–686.

Yoshida, H, Kong, YY, Yoshida, R, Elia, AJ, Hakem, A, Penninger, JM, et al. (1998). Apaf1 is required for mitochondrial pathways of apoptosis and brain development. *Cell*, 94:739–750.

Further Reading

Epstein, CJ, Erickson, RP, and Wynshaw-Boris, A (eds.) (2004). *Inborn Errors of Development*. New York: Oxford University Press.

Human Malformations and Related Anomalies (2nd edn). New York: Oxford University Press.

6

Genetic and Genomic Approaches to Taxonomy of Human Disease

Dhavendra Kumar

The philosophy of disease is complex and reflects the way an abnormal body state is perceived and understood on the background of sociocultural, psychological, and biomedical interpretation. The sociocultural factors influence the outcome of disease by modulating the environment. The psychological and biomedical factors provide the innate framework for generating and developing symptoms and signs of the disease, dependent on the pathology. On this basis, a disease could be best defined as an "overall perception of an abnormal body state and appreciation of the ensuing psychological and physical impact."

Historically, clinical practice has always faced its inability to differentiate events that mediate the disease process from the resulting clinical, biochemical, and pathological changes (Bell, 2003). Despite having made tremendous advances, clinicians continue to rely on phenotypic manifestations of the disease process to define most diseases. Inevitably, this approach often obscures the underlying mechanisms and thus a clinician may fail to identify significant heterogeneity. Concerns have been raised that most human disease provides only "insecure and temporary conceptions" (Lewis, 1944). Apart from infectious diseases, alarmingly, there are few diseases that have a truly mechanism-based nomenclature.

The classification of human disease is dependent on several factors ranging from perceptions and analysis of symptoms, sociocultural interpretation of varied manifestations of the disease, biological considerations, and therapeutic interventions. Conventionally, a disease refers to a particular organ or system dysfunction resulting from one or more causative factors such as physical trauma, infection, exposure to a toxic substance, or malnutrition. A large number of diseases remain unaccounted for due to lack of a clear explanation for underlying mechanisms. Terms like degenerative or autoimmune disorders are not uncommonly used to describe an organ-system dysfunction. However, a "cause and effect" relationship remains unclear. Nevertheless, associated pathological changes often provide a firm and precise basis to categorize a disease or disorder. This has helped in delineating distinct categories of diseases such as immunological and metabolic diseases.

Developments in genetics and molecular biology have provided a vast amount of data and information to support the view that most human diseases have a significant genetic component. Characterization of the genetic determinants of disease provides remarkable opportunities for clinical medicine through an improved understanding of pathogenesis, diagnosis, and therapeutic options. An understanding of the genetic basis of human disease has opened the way forward for a new taxonomy of human disease that will be free from limitations and bias in developing diagnostic criteria related to events that are often secondary and peripheral to its cause (Bell, 2003). For instance, genetic information has allowed us to identify distinct forms of diabetes mellitus, defining an autoimmune form associated with highly diverse and complex human leukocyte antigens (HLAs) and other factors that affect both expression and modification of gene products in mediating the adult form of the disease (Cardon and Bell, 2001). Similarly, a number of genetically determined molecules and pathways have been characterized that are crucial in the pathogenesis of bronchial asthma (Cookson et al., 1989). It is now widely believed that a clearer understanding of the mechanisms and pathways of a disease will assist us in delineating distinct disease subtypes and may resolve many questions relating to variable disease symptoms, progression, and response to therapy. This might help in revising the current diagnostic criteria. Eventually, genetics may contribute a new taxonomy of human disease within clinical practice.

Although genetics is acknowledged to be an important aspect in understanding the pathogenesis of disease, genetic classification of human disease has not yet received full recognition. There is ample evidence in support of the argument that genetic factors are probably associated with all human diseases except for trauma (Fig. 6–1). However, the outcome of trauma could be influenced by underlying genetic factors such as genetically determined host-response to infection and tissue damage. Various categories of genetic disorders are considered rare with a tendency to be included under the broad title of organ-system diseases. Often, these are listed as simply etiological factors rather than a distinct disease category. This concept and approach is now rapidly being outdated and replaced with new classes of diseases. This progress is seriously hampered by the lack of formal education at all levels and integration of appropriate technologies into the modern medical diagnostic and therapeutic infrastructure.

Traditionally, genetic diseases are classified as chromosomal (numerical or structural), Mendelian or single-gene disorders, multifactorial/polygenic complex diseases or congenital anomalies, and diseases associated with specific mitochondrial gene mutations (Table 6–1). Apart from chromosomal disorders, essentially all genetic disorders result from some form of alteration or mutation occurring in a specific gene (single gene diseases) or involving

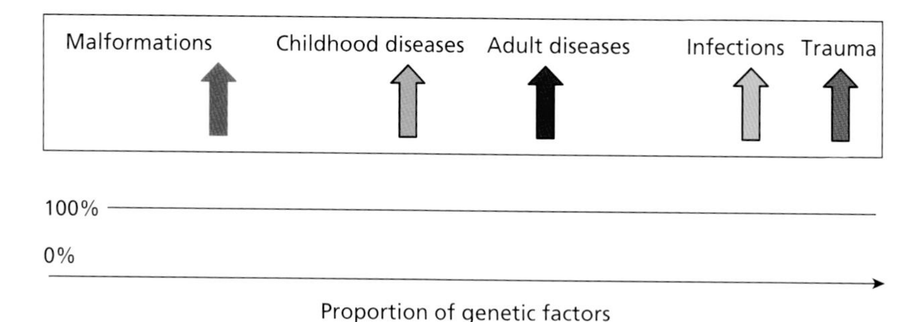

Figure 6–1 Genetic factors in human disease.

multiple loci spread across the human genome (polygenic disorders). The major impact of chromosomal disorders occurs before birth and carries a serious health burden throughout childhood and during the early years of life (Fig. 6–2). On the other hand, single gene diseases can pose a real medical and health burden from the perinatal period to adulthood with a peak around mid-childhood. In contrast, the polygenic/multifactorial diseases tend to present late, except for developmental anomalies, which will require active multidisciplinary care during early life. A brief description of the major types of genetic diseases is included here. Any leading medical genetics textbook will contain a detailed description of all these group of genetic disorders (see Further Reading).

Chromosomal Disorders

The entire human genome is spread around 23 pairs of chromosomes including one pair specifically assigned to male (XY) and female (XX) gender, designated the sex-chromosome pair. The chromosomal constitution of man is complex and comprises variable amounts of euchromatin and heterochromatin that exhibits with a characteristic "banding-pattern" and is essential for the physical and distinct appearance of a particular chromosome. Typically, a chromosome pair includes two homologues each comprising a short arm (p) and a long arm (q) separated by the central heterochromatin-G-C rich region designated the centromere. A detailed account of the chromosome structure and fundamental changes that occur during meiosis and mitosis can be found in any leading textbook on basic genetics (see Further Reading).

Table 6–1 The Classification of Genetic Disorders

Chromosomal	Numerical—aneuploidies
	Structural—deletion, duplication, inversion, isochromosome, ring chromosome, reciprocal, or Robertsonian translocation
	Mosaicism- aneuploidy or structural abnormality
Mendelian	Autosomal recessive
	Autosomal dominant
	X-linked recessive
	X-linked dominant
Multifactorial/ polygenic	Multiple gene-environment interactions; single nucleotide polymorphisms and copy number variations
Mitochondrial	Mitochondrial genome rearrangements, deletions, duplications and point mutations

Chromosomal disorders are essentially disorders of the genome resulting from either loss or addition of a whole chromosome (aneuploidy) or parts of chromosomes (structural). A chromosome abnormality results in major disturbance in the genomic arrangement since each chromosome or part thereof consists of thousands of genes and several noncoding polymorphic DNA sequences. The physical manifestations of chromosome disorders are often quite striking, characterized by growth retardation, developmental delay, and a variety of somatic abnormalities. A number of chromosomal syndromes are now recognizable. The diagnosis and genetic management of these disorders fall within the scope of the subspecialty "clinical cytogenetics."

The management of chromosomal disorders requires a coordinated and dedicated team approach involving a wide range of clinicians and health professionals. A typical example is Down syndrome resulting from either three copies of chromosome 21 (trisomy) (Fig. 6–3) or an addition to the long arm of chromosome 21 usually resulting from an unbalanced meiotic rearrangement of a parental chromosomal translocation between chromosomes 21 and one of the other acrocentric (centromere located at the end) chromosomes (Robertsonian translocation). Down syndrome occurs in approximately 1 in 800 live births and increases in frequency with advancing maternal age. It is characterized by growth and developmental delay, often severe mental retardation, and the characteristic facial appearance recognized with upward slanting eyes. A major cause of death in these individuals is associated congenital heart defects that can complicate the clinical management in a significant proportion of Down syndrome cases. Prenatal diagnosis and antenatal assessment of the maternal risk for Down syndrome employing a variety of imaging and biochemical markers is now the established clinical and public health practice in most countries.

Clinically significant chromosome abnormalities occur in nearly 1% of live-born births and account for approximately 1% of pediatric hospital admissions and 2.5% of childhood mortality (Hall et al., 1978). The loss or gain of whole chromosomes is often incompatible with survival, and such abnormalities are a major cause of spontaneous abortions or miscarriages. Almost half of the spontaneous abortuses are associated with a major chromosomal abnormality. It is estimated that about a quarter of all conceptions may suffer from major chromosome problems, because approximately 50% of all conceptions may not be recognized as established pregnancies, and 15% of these end in miscarriage. Essentially the major impact of chromosomal disorders occurs before birth or during early life (Fig. 6–2).

The delineation of rare and uncommon chromosomal disorders has been crucial in the gene mapping of several Mendelian (single gene) disorders such as the X-linked Duchenne muscular

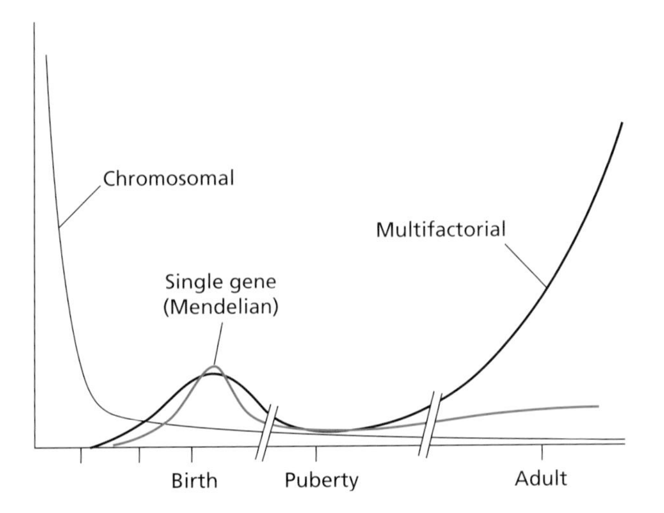

Figure 6–2 Distribution of different genetic disorders in various age groups. (Adapted from Gelehrter et al., 1998 with permission from Williams and Wilkins.)

dystrophy and type 1 neurofibromatosis. The chromosomal regions involved in deletion, duplication, inversion, and break points involved in a complex chromosomal rearrangement provide an important clue and assist the keen researcher in focusing on genes located within the chromosomal segment.

Mendelian (Single Gene) Disorders

About 4000 human diseases are caused by mutations in single genes and constitute a major health burden. Single-gene disorders account for approximately 5–10% of pediatric hospital admissions and childhood mortality. The major impact of these disorders occurs in the newborn period and early childhood. However, these also constitute a significant proportion of adulthood diseases, notably late-onset neurodegenerative diseases and various forms of familial cancer. Although the majority of single-gene diseases are rare, some are relatively common and pose a major health problem. For example, familial hypercholesterolemia, a major predisposing factor in premature coronary artery disease, occurs in 1 in 500 people. Other good examples would be familial breast and colorectal cancers (CRCs), which affect approximately 1 in 300. Some single-gene disorders are specific for certain populations, for example, Tay-Sachs disease among Ashkenazi Jews, cystic fibrosis in white Caucasians, thalassemias among people from South-East Asia and the Mediterranean countries, and sickle-cell disease in people of west African origin. Techniques in molecular biology have enabled characterization of a number of mutated genes. Sickle-cell disease was the first single-gene disorder to be defined at the molecular level. This has revolutionized the diagnosis and management of these disorders. The single-gene disorders are inherited in a simple Mendelian manner and hence justifiably called "*Mendelian disorders.*" The genetic transmission of altered genes or traits follow principles as set out by the Austrian Monk Gregor Mendel in 1865 based on his seminal work on garden pea plants. Mendel inferred that, "those characteristics that are transmitted entire, or almost unchanged by hybridization, and therefore constitute the characters of the hybrid, are termed dominant, and those that become latent in the process, recessive." The nomenclature of these disorders reflects the gender-specific transmission and is supported by localization of an altered gene on either an autosome (1–22) or the X chromosome. Mendelian disorders are described as autosomal dominant, autosomal recessive, X-linked recessive (Fig. 6–4), or X-linked dominant (Fig. 6–5). The latter pattern differs from the X-linked recessive by having an excess of affected females in a family, because the

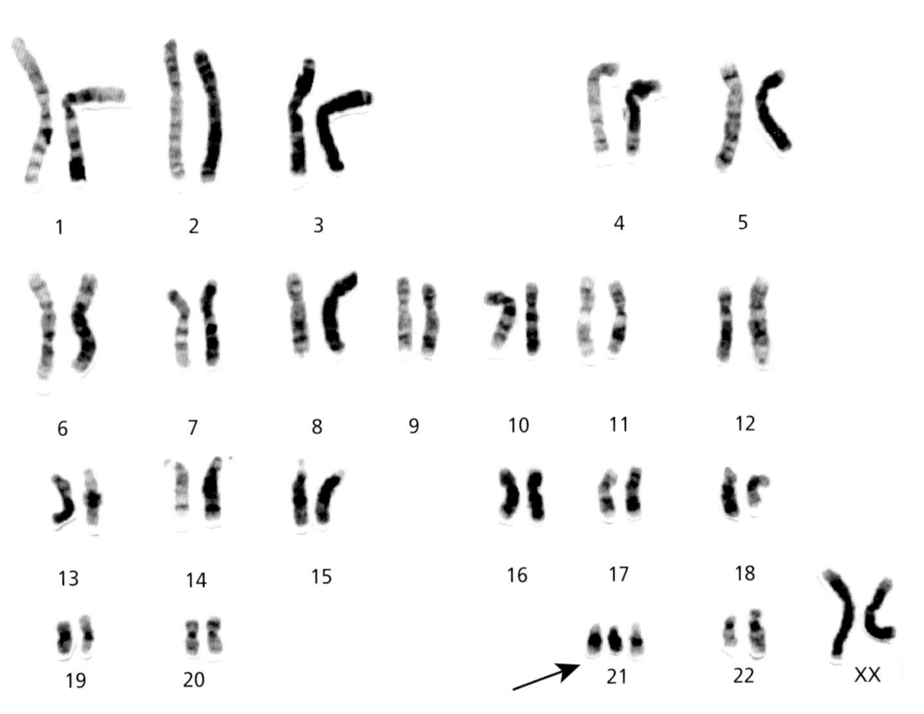

Figure 6–3 Karyotype of a female (XX) with Down syndrome—note trisomy 21.

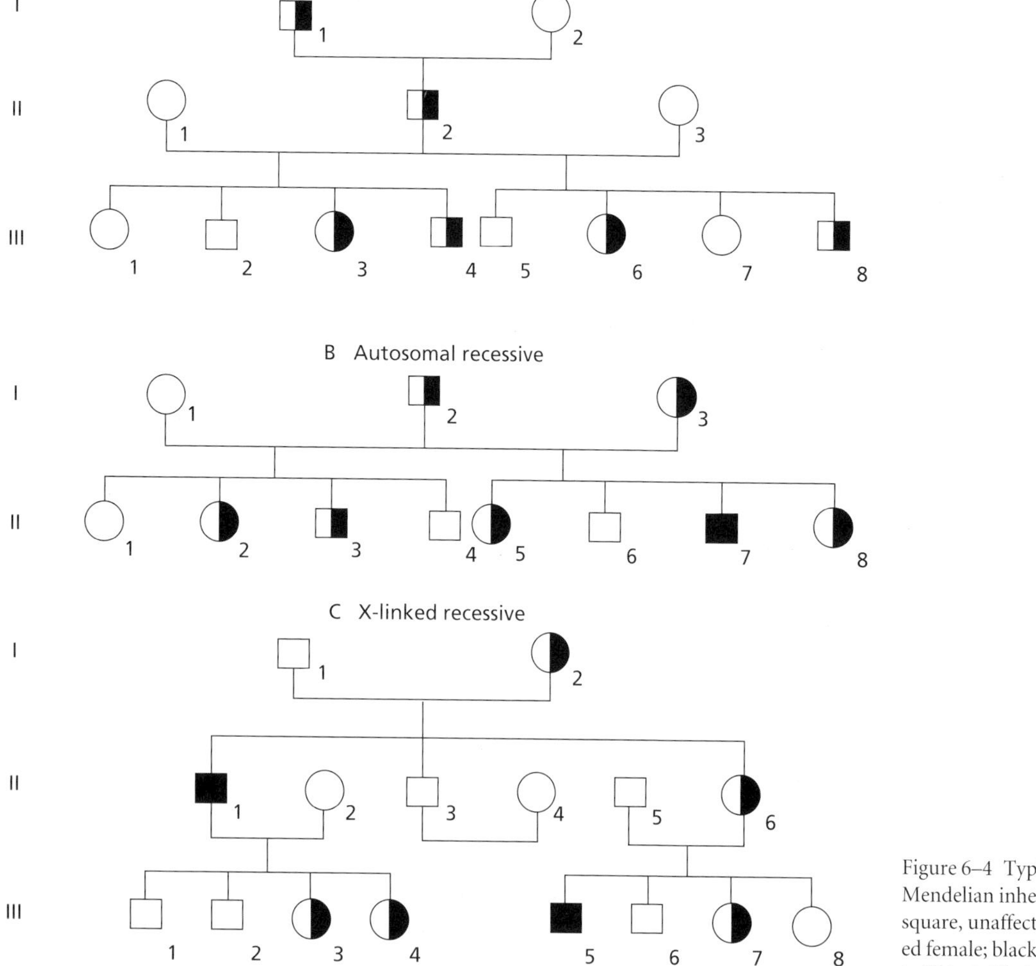

Figure 6–4 Typical pedigree appearances in Mendelian inheritance. Key to symbols: blank square, unaffected male; open circle, unaffected female; black filled, affected (homozygous); half black filled, carrier (heterozygous).

heterozygous mutation on the X chromosome can be transmitted to the daughter from an affected mother as well as the affected father. Sporadic X-linked dominant diseases are encountered in a female, as these are lethal in a male. A detailed family history and careful interpretation of the pedigree are essential prerequisites in the diagnosis of a Mendelian disease. Accurate risk-estimates, for use in genetic counseling, are not possible in the absence of a reliable and comprehensive pedigree. The major features of the individual inheritance manner are described in leading genetics textbooks (see Further Reading). All human disorders and traits that follow the Mendelian principles are listed in a major resource—"McKusick's catalogue of Mendelian inheritance of Man." An online version (OMIM) is available, which is regularly updated.

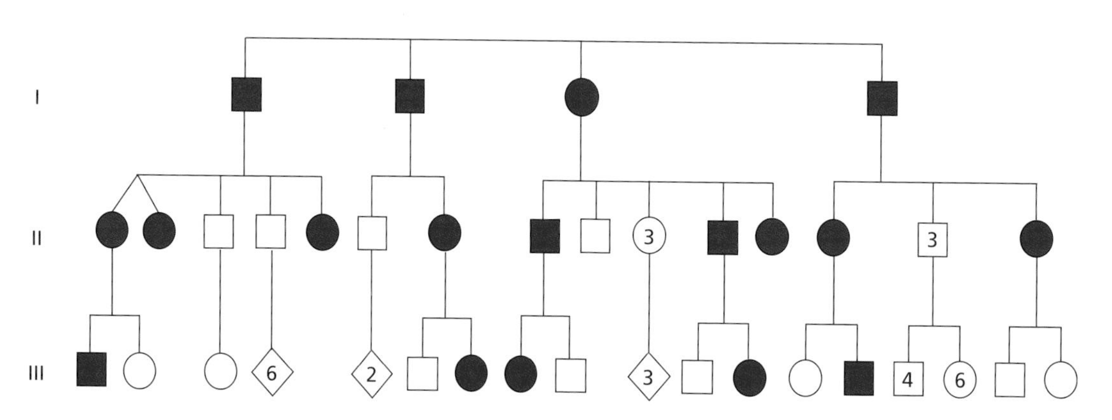

Figure 6–5 A pedigree with an X-linked dominant disorder—note absence of "male–male" transmission; all daughters of an affected male would be heterozygous and thus could be symptomatic. (Adapted from Gelehrter et al., 1998 with permission from Williams and Wilkins.)

Polygenic or Multifactorial Disorders

This group of disorders includes the most common and least understood human genetic diseases. These diseases result from interaction of certain environmental factors with multiple genes, some of which may have a major effect, but the majority carries only a relatively minor effect. The minor additive effect of these multiple loci lowers the "threshold" of an organ or body system to withstand environmental pressures resulting in either a developmental anomaly or an abnormal disease state. Examples include common congenital anomalies such as cleft lip, cleft palate, neural tube defects, and most congenital heart diseases. The common chronic medical diseases fall within this category of genetic disorders including diabetes mellitus, coronary heart disease, hypertension, arthritis, and schizophrenia. Understanding the genetic basis of common diseases remains the major challenge facing modern genetics and genomics.

The clinical impact of multifactorial diseases is significant in both the neonatal period as well as in adult life. It is estimated that approximately 25%–50% pediatric hospital admissions are related to these groups of disorders and associated with 25%–35% of childhood mortality. There is an even greater medical and health burden of these disorders during adult life due to a chronic natural history of resulting medical diseases. For instance, diabetes mellitus and obesity account for approximately 40% of the adult medical problem in the developed and developing countries.

Identification of any such disorder or condition is important in assessing risks to close relatives. A comparison of general population and multiple cases in a family would indicate a shift of the bell-shaped Gaussian curve to the right, reflecting a lowered threshold with an increased incidence (Fig. 6–6). The precise additional risk would be dependent upon the degree of relationship with the index case in the family. In addition, the gender of the index case is also important in assessing the liability. The genetic liability is estimated to be greater if the index case is of the gender with lowest incidence. For example, in the case of pyloric stenosis greater risk would be applicable if the index case were a female, which carries the lowest birth prevalence. Finally, recurrence risks for a given population group is estimated to equal the square root of the birth incidence. For instance birth incidence of ventricular septal defect is approximately 3 per 1000, the recurrence risk to first degree relative, such as the next child, would be the square root of 0.003 or 3%. These figures are useful in genetic-counseling a family after the birth of a child with a congenital anomaly.

This group of diseases poses the challenge of working out the mechanisms that determine the additive or interactive effects of many genes creating predisposition to diseases, which in turn manifest only in the presence of certain environmental factors. It is hoped that a combination of molecular genetic approaches, gene mapping, and functional genomics will allow a clearer definition of these genetic diseases. Several sections in this book will address this issue at length and focus on specific disease groups and systems.

Mitochondrial Genetic Disorders

Apart from nuclear DNA (nDNA), a small proportion of DNA is also found in mitochondria in the cytoplasm of cells (mtDNA). Each cell contains 2–100 mitochondria, each of which contains 5–10 circular chromosomes. The 16.5 kb mtDNA molecule is free from any non-coding intronic regions and encodes two ribosomal RNA (rRNA) genes, 22 transfer RNAs (tRNA), and 13 polypeptides that are parts of multisubunit enzymes involved in oxidative phosphorylation (Fig. 6–7). In comparison to the nuclear DNA, the mtDNA is 20 times more prone to recurrent mutations resulting in generation of mutagenic oxygen radicals in the mitochondria. The inheritance of mtDNA is exclusively maternal due to its cytoplasmic location. The mature sperm cytoplasm contains very little mitochondria as it is almost completely lost during the fertilization process, apparently with the loss of the tail that carries the bulk of the cytoplasm. Due to the wholly maternal cytoplasmic location, only females can transmit mitochondrial diseases to their offspring of either gender (Fig. 6–8).

Since mtDNA replicates separately from the nuclear DNA, and mitochondria segregate in daughter cells independently of the nuclear chromosomes (replicative segregation), the proportion of mitochondria carrying an mtDNA mutation can differ among somatic cells. This mitochondrial heterogeneity is also called heteroplasmy and plays an important part in the variable and tissue-specific phenotype of mitochondrial disease. Since different tissues have varying degrees of dependence on oxidative phosphorylation, with heart, muscle, and central nervous system being the most dependent, the common manifestations of mitochondrial disease

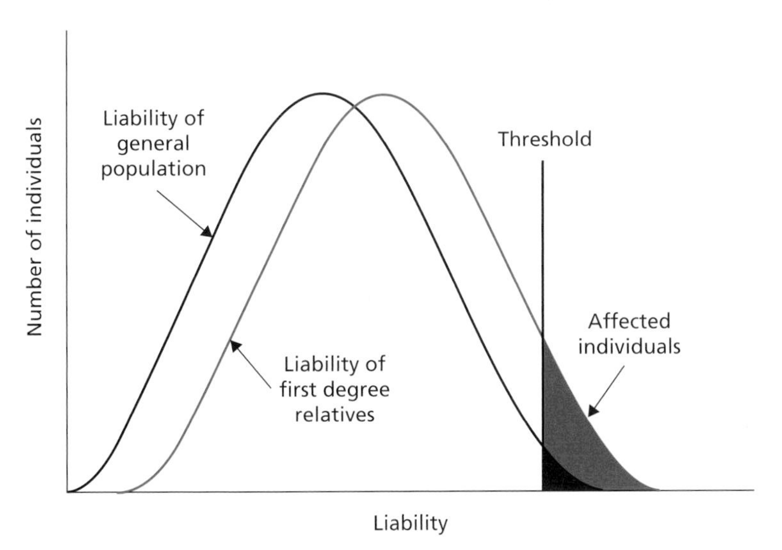

Figure 6–6 The "Gaussian" bell-shaped curve to illustrate "genetic threshold," indicated by liability in the general population (shown in black). A shift to right (in red) indicates increased liability in first-degree relatives with an increased risk of recurrence. (Reproduced from Weatherall, 1991 with permission from Oxford University Press.)

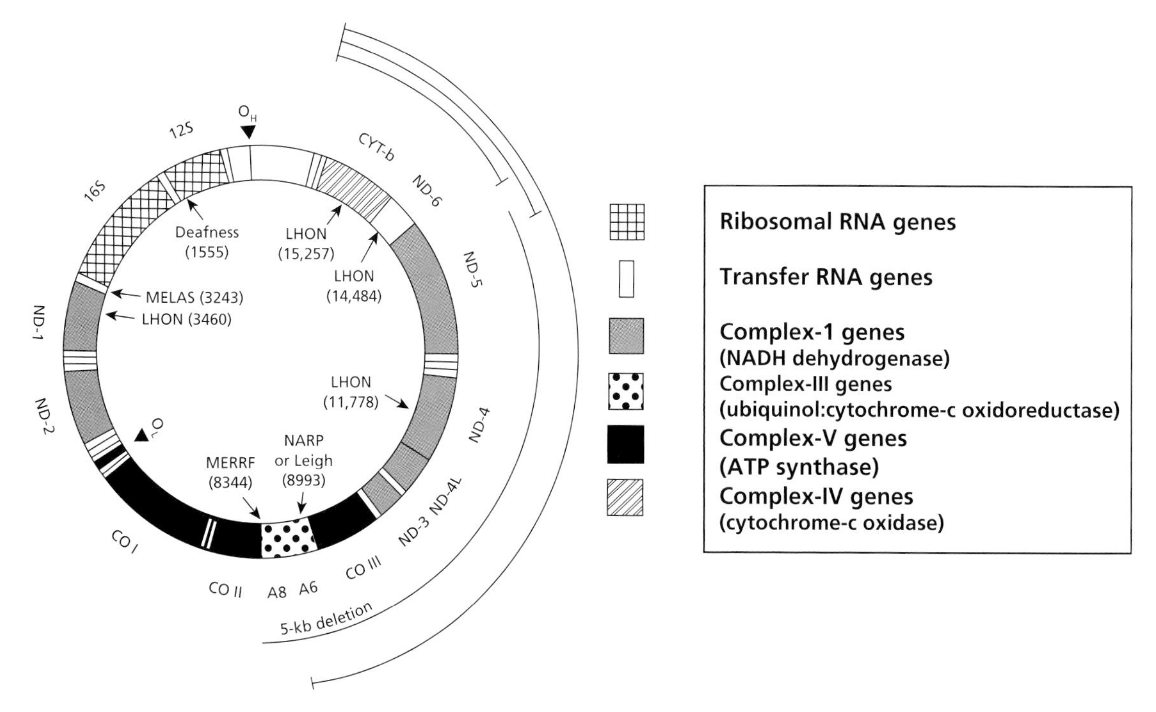

Figure 6–7 The human mitochondrial DNA molecule with examples of point mutations with their associated clinical phenotypes. (Adapted from Pulst, 2000 with permission from Oxford University Press.)

include cardiomyopathy, myopathy, and encephalopathy (Fig. 6–9). Furthermore, oxidative phosphorylation declines with age, probably related to the accumulation of successive mtDNA mutations. Thus, the clinical phenotype in a mitochondrial disease is not simply or directly related to mtDNA genotype, but reflects several factors, including the overall capacity for oxidative phosphorylation determined by both mtDNA and nuclear DNA genes, the accumulation of somatic mtDNA mutations and degree of heteroplasmy, tissue-specific requirements of oxidative phosphorylation, and age.

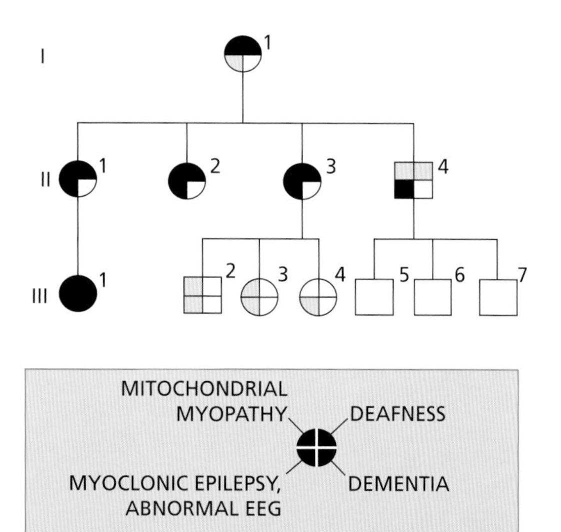

Figure 6–8 Pedigree of a family with mitochondrial encephalopathy with ragged-red muscle fibers (MERRF)—note segregation of different features with variable severity in the affected family members. (Adapted from Gelehrter, 1998 with permission from Williams and Wilkins.)

Several mitochondrial diseases have now been characterized (Table 6–2). One of the best-characterized is Leber's hereditary optic neuropathy (LHON), which exclusively affects males. There is loss of central vision secondary to optic nerve degeneration. The vision loss usually occurs in the 20s and can progress rapidly in some men. Eleven different missense mtDNA mutations in three different mitochondrial genes encoding respiratory chain enzyme subunits have been described. The phenotype in other mitochondrial diseases tends to include a combination of heart, muscle, and central nervous system manifestations with considerable intra-/inter-familial variability for the same mtDNA mutation. In addition, mitochondrial dysfunction can be part of the phenotype in some Mendelian diseases where the mutant gene-product presumably has pathogenic influence on the mitochondrial mediated metabolic pathway. Examples include autosomal recessive respiratory enzyme disorders. Genetic counseling and decision for prenatal diagnosis can be difficult in mitochondrial disorders due to difficulty in predicting the phenotype in the affected pregnancy.

Finally, a high degree of sequence variation (polymorphism) is known to occur in the noncoding region of the mitochondrial chromosome (the D-loop). This polymorphism has been used in anthropologic and evolutionary studies to trace back the origins and links of human populations. In addition, this information has been applied in forensic analysis as well to match maternal grandparent's mtDNA with an orphaned child whose parents have "disappeared" during war, a natural disaster, or in mysterious circumstances.

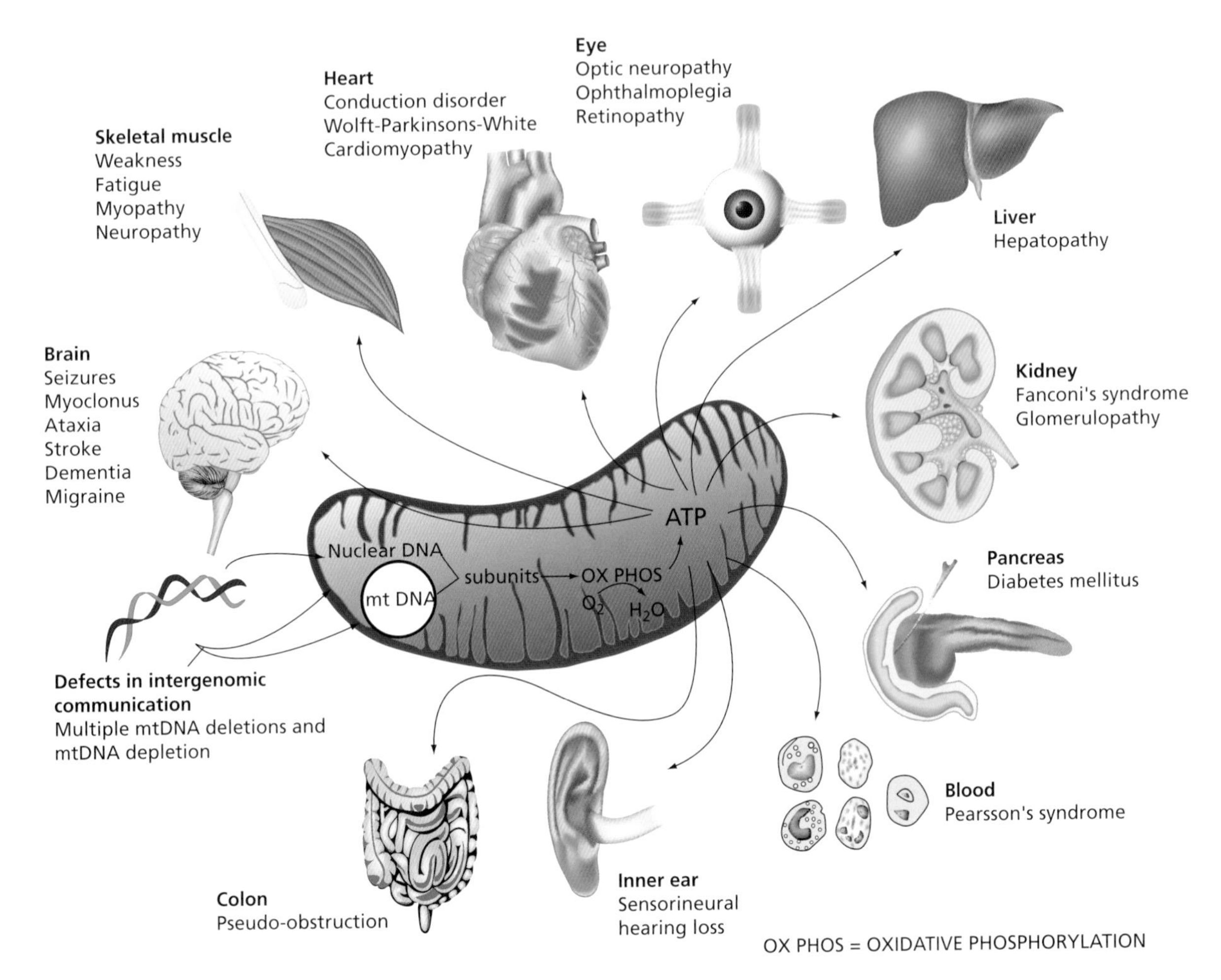

Figure 6–9 Multisystem involvement in mitochondrial genetic disorders. (Adapted from Pulst, 2000 with permission from Oxford University Press.)

Genomic Disorders

Recent advances in molecular genetics have enabled us to identify specific groups of disorders, which result from characteristic mechanisms involving specific areas of the human genome. Often, these do not conform to the standard basic principles of genetics. A broad term "genomic disorders" has been coined to describe these conditions (Table 6–3).

A number of hereditary disorders are present with complex genetic pathology that does not follow the conventional principles of inheritance as outlined in the previous sections. There is now overwhelming evidence within these disorders that indicates unusual mechanisms suggesting "nontraditional inheritance." The mechanisms involve certain genomic regions that directly or indirectly influence regulation and expression of one or more genes manifesting in complex phenotypes. Currently, some of these disorders are either listed as chromosomal or single-gene disorders.

Disorders of Genomic Imprinting: Epigenetic Diseases

The term "epigenetics" refers to heritable factors that affect gene expression without any change in the gene coding-sequence. These factors could be operational either during meiosis or mitosis and are often selective and preferential on the basis of "*parent of origin.*" The term "*imprinting*" is commonly used to describe this important biological mechanism that is recognized to influence wide ranging physical and molecular phenotypes. A number of human diseases have now been confirmed to result from epigenetic changes in various parts of the genome. The term "epigenetic diseases" or "genomic imprinting disorders" refers to this group of diseases. Basic mechanisms related to the phenomenon of epigenetics or epigenomics are reviewed separately (see also Chapter 4).

Epigenetic initiation and silencing is regulated by the complex interaction of three systems, including DNA methylation, RNA-associated silencing, and histone modification (Egger et al., 2004). The relationship between these three components is vital for expression or silencing of genes (Fig. 6–10). Disruption of one or other of these interacting systems can lead to inappropriate expression or silencing of genes leading to "epigenetic diseases." Methylation of the C^5 position of cytosine residues in DNA has long been recognized as an epigenetic silencing mechanism of fundamental importance (Holliday and Pugh, 1975). The methylation of CpG sites within the human genome is maintained by a number of DNA methyltransferases (DNMTs) and has multifaceted roles for the silencing of transportable elements, for defense against viral sequences and for transcriptional repression of certain genes. A strong suppression of the CpG methyl-acceptor site in human DNA results from mutagenic changes in 5-methylcytosine, causing C:G to T:A transitions. Normally CpG islands, which are GC rich evolutionary-conserved regions of more than 500 base-pairs, are kept free of methylation. These stretches of DNA are located within the promoter region of approximately 40% of mammalian genes and, when methylated,

Table 6–2 Genetic Classification of Mitochondrial Disorders

Disorder	Major Clinical Features	Type of Gene	Mitochondrial DNA Mutation
Chronic progressive external ophthalmoplegia (CPEO)	External ophthalmoplegia, bilateral ptosis, mild proximal myopathy	tRNA	A3243G, T8356C Rearrangement (deletion/ duplication)
Kearns–Sayre syndrome (KSS)	PEO onset <20 years, pigmentary retinopathy, cerebellar ataxia, heart block, CSF protein >1 g/l		Rearrangement (deletion/ duplication)
Pearson syndrome	Siderobalstic anemia of childhood, pancytopenia, renal tubular defects, exocrine pancreatic deficiency		Rearrangement (deletion/ duplication)
Diabetes and deafness	Diabetes mellitus, sensorineural hearing loss	tRNA	A3243G, C12258A Rearrangement (deletion/ duplication)
Leber's hereditary optic neuropathy (LHON)	Subacute painless bilateral visual loss, age of onset 24 years, males > females (~4:1), dystonia, cardiac preexcitation syndromes	Protein encoding	G11778A, T14484C, G3460A
Neurogenic ataxia with retinitis pigmentosa (NARP)	Late-childhood or adult onset peripheral neuropathy, ataxia, pigmentary retinopathy	Protein encoding	T8993G/C
Leigh syndrome (LS)	Subacute relapsing encephalopathy, cerebellar and brain-stem signs, infantile onset	Protein encoding	T8993G/C
Exercise intolerance and myoglobulinuria	Exercise induced myoglobulinuria	Protein encoding	Cyt B mutations
Mitochondrial encephalomyopathy with lactic acidosis and stroke-like episodes (MELAS)	Stroke-like episodes before 40 years, seizures and/or dementia, ragged-red fibers and/or lactic acidosis, diabetes mellitus, cardiomyopathy (HCM/ DCM), deafness, cerebellar ataxia	tRNA	A32343G, T3271C, A3251G
Myoclonic epilepsy with ragged-red fibers (MERRF)	Myoclonus, seizures, cerebellar ataxia, myopathy, dementia, optic atrophy, bilateral deafness, peripheral neuropathy, spasticity, multiple lipomata	tRNA	A8344G, T8356C
Cardiomyopathy	Hypertrophic cardiomyopathy (HCM) progressing to dilated cardiomyopathy (DCM)	tRNA	A3243G, A4269G
Infantile myopathy/encephalopathy	Early-onset progressive muscle weakness with developmental delay	tRNA	T14709C, A12320G, G1606A, T10010C
Nonsyndromic sensorineural deafness	Early-onset progressive bilateral moderate to severe sensorineural hearing loss	rRNA	A7445G
Aminoglycoside-induced nonsyndromic deafness	Early-onset nonprogressive sensorineural deafness secondary to aminoglycoside administration	rRNA	A1555G

cause stable heritable transcriptional silencing. Aberrant *de novo* methylation of CpG islands is a hallmark of human cancers and is found early during carcinogenesis (Jones and Baylin, 2002).

In addition to DNA methylation, histone modifications have also been found to have epigenetic effects. Acetylation and methylation of conserved lysine residues of the amino-terminal tail domains are the key elements in histone modification. Generally, the acetylation of histones marks active, transcriptionally competent regions, whereas hypoacetylation histones are found in transcriptionally inactive euchromatic and heterochromatic regions. On the other hand, histone methylation can be a marker for both active and inactive regions of chromatin. Methylation of lysine residue 9 on the N terminus of histone 3 (H3-K9) is a hallmark of silent DNA and is evenly distributed throughout the heterochromatic regions such as centromeres and telomeres, including the inactive X-chromosome. In contrast, methylation of lysine 4 of histone 3 (H3-K4) denotes activity and is predominantly found at promoter regions of active genes (Lachner and Jenuwein, 2002). This constitutes a "histone code," which can be read and interpreted by different cellular factors. There is evidence that DNA methylation depends on methylation of

Table 6–3 Classification of Genomic Disorders

Disorders of genomic imprinting (Epigenetic diseases)
Disorders of genome architecture
Trinucleotide repeats disorders
Complex genomic diseases

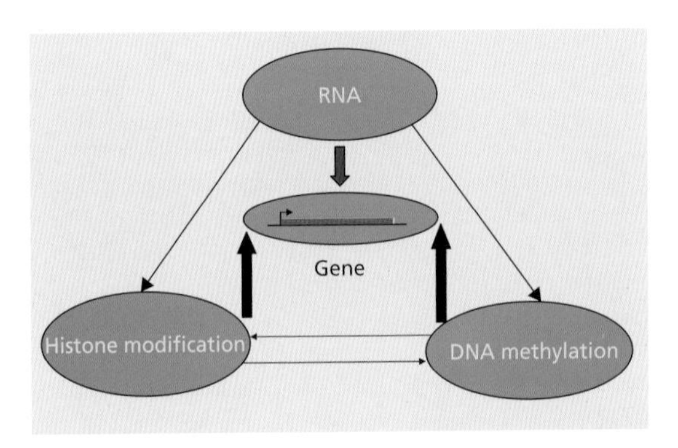

Figure 6–10 Interaction between RNA, histone modification, and DNA methylation in gene expression and silencing (Egger et al., 2004).

H3-K9 and can also be a trigger for its methylation. Recently, evidence has been accumulated on the role of RNA in posttranscriptional silencing. In addition, RNA in the form of antisense transcripts (*Xist* or RNAi) can also lead to mitotically heritable transcriptional silencing by the formation of heterochromatin. For example, transcription of antisense RNA led to gene silencing and to the methylation of the structurally normal α-globin gene in patients with alpha thalassemia. This could be one of the many human diseases resulting from epigenetic silencing due to antisense RNA transcripts (Tufarelli et al., 2003).

Mutations in genes that affect genomic epigenetic profiles can give rise to human diseases that can be inherited or somatically acquired (Table 6–4). These epigenetic mutations can be either due to hypermethylation (silencing) of a regulating gene or loss of methylation (LOM) (activation) of another gene that has a positively modifying effect on the phenotype. The parental imprinting effect can be inferred by demonstrating the parental origin of the mutant allele. Similarly, either a loss or gain of a chromosomal segment can result in the same situation. Confirmation of a specific chromosomal deletion or duplication is usually possible by using the fluorescence in situ hybridization (FISH) method. The paternal imprinting in this situation is commonly demonstrated by genotyping a set of polymorphic markers located within the chromosomal segment. Inheritance of the whole chromosomal homologue from one parent effectively confirms imprinting phenomenon since the regulatory gene sequences for the pathogenic gene would be missing from the other parent. This characteristic abnormality is commonly referred to as "uniparental disomy" or UPD. This could be either isodisomy (similar parental homologues) or heterodisomy (parental and grandparental homologues) (Fig. 6–11). The origin of UPD is believed to result through loss of the additional chromosomal homologue, failing which the conceptus would be trisomic. This mechanism is also called "trisomic rescue."

For a maternally imprinted disorder, paternal UPD would be confirmatory and maternal UPD diagnostic for the paternally imprinted condition. For example, maternal UPD is diagnostic for Prader–Willi syndrome, and paternal UPD for Angelman's syndrome, both conditions being associated with a microdeletion of 15q11 region. The parental origin of the 15q micro deletion follows the expected epigenetic pattern and is in keeping with the clinical diagnosis. Recurrence risk-estimates vary dependent on the specific epigenetic pattern. This information is crucial to offer accurate genetic counseling in any genomic imprinting disorder.

Many epigenetic diseases are associated with chromosomal alterations and manifest with physical and learning difficulties. For example, mutations in X-linked mental retardation with the alpha thallassemia phenotype (*ATRX*) result in consistent changes in the methylation pattern of ribosomal DNA, Y-specific repeats, and subtelomeric repeats. Another X-linked recessive mental retardation syndrome, associated with a visible "fragile site" on the terminal part of the long arm of the X chromosome (Fragile X syndrome), results from *de novo* silencing of the pathogenic gene *FMR1*. The syndrome is characteristically associated with an abnormal expansion of CGG triplet repeats in the *FMR1* 5′ untranslated terminal region. Methylation of the expansion leads to silencing of the *FMR1* gene and under certain cultural conditions creates the visible "*fragile site*" on the X chromosome.

Epigenetic silencing is probably also significant in other neurodevelopmental disorders. For example, in Rett syndrome, a common cause of intellectual disability in young girls, mutations of the *MeCP2* gene are seen in approximately 80% cases. The MeCP

Table 6–4 Recognizable Epigenetic Dysmorphic Syndromes (Egger et al., 2004)

Disease	Main Features	Epigenetic Mechanism
ATR-X syndrome	α-thalassemia, facial dysmorphic features, neurodevelopmental disabilities	Mutations in *ATRX* gene; hypomethylation of repeat, and satellite sequences
Fragile X syndrome	Chromosome instability, physical, and learning/behavioral difficulties	Expansion and methylation of CGG repeat in *FMR1* 5′ UTR, promoter methylation
ICF syndrome	Chromosome instability, immunodeficiency	*DNMT3* mutations; DNA hypomethylation
Angelman's syndrome	Seizures and intellectual disabilities	Deregulation of one or more imprinted genes at 15q11–13 (maternal)
Prader–Willi syndrome	Obesity, intellectual disabilities	Deregulation of one or more imprinted genes at 15q11–13 (paternal)
Beckwith–Wiedemann syndrome (BWS)	Organ overgrowth, childhood tumors	Deregulation of one or more syndrome imprinted genes at 11p15.5 (*IGF2, CDKN1C, KvDMR1* etc.)
Russell–Silver syndrome	Growth delay, body asymmetry	Deregulation of one or more imprinted genes at 7p (maternal)
Rett syndrome	Seizures, intellectual disabilities	*MeCP2* mutations
Rubinstein–Taybi syndrome	Facial dysmorphism, intellectual disabilities	Mutation in CREB-binding protein (histone acetylation)
Coffin–Lowry syndrome	Facial dysmorphism, developmental delay	Mutation in *RSk-2* (histone phosphorylation)

Abbreviations: ATR-X, α-Thalassemia, X-linked mental retardation; CREB-cAMP, response-element-binding protein; ICF, immunodeficiency, chromosome instability, facial anomalies; UTR, untranslated region.

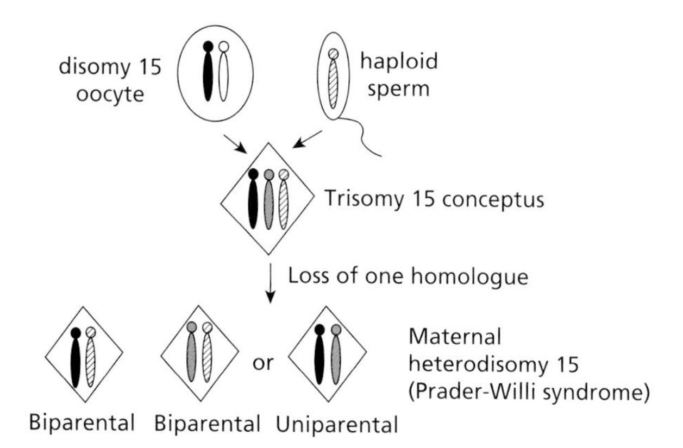

Figure 6–11 The origin of uniparental disomy 15 in Prader–Willi syndrome through trisomic rescue during early embryogenesis—note different homologues (maternal heterodisomy).

protein binds to methylcytosine residues and causes derepression of genes normally suppressed by DNA methylation. Despite lack of firm evidence, it is likely that MeCP2 might have a key role in the control of neuronal gene activity resulting in the pathology of Rett syndrome (Chen et al., 2003). Interaction with another pathogenic gene (*CTKL5* or *STK9*) in Rett syndrome is likely to be important in the pathogenesis of this neurodevelopmental disorder (Slager et al., 2003). On a wider genomic level, mutations in the *DNMT3b* gene, causing the ICF (immunodeficiency, centromeric region instability, and facial anomalies) syndrome, result in deregulation of DNA methylation patterns. A notable example is that of Beckwith–Wiedemann syndrome, an overgrowth syndrome predisposing to Wilms' tumor and other childhood tumors, which is associated with duplications and rearrangements of a small chromosomal region on the short arm of chromosome (11p15.5). This region contains a cluster of genes, which is susceptible to a number of epigenetic alterations, manifesting with the BWS phenotype and tumorigenesis, particularly Wilms' tumor and other childhood embryonal tumors (Fig. 6–12). LOM in imprinting control regions (such as *KvDMR1*) can cause deregulation of imprinting and either biallelic expression (*IGF2* and *H19*) or silencing (such as *CDKN1C*) of imprinted genes, which is seen in most sporadic cases.

The epigenetic phenomenon is probably significant for the phenotypic manifestations in some other hereditary tumors. For example, transmission of autosomal dominant familial chemodectomas (nonchromaffin paragangliomas or glomus tumors) is exclusively via the paternal line (Fig. 6–13) (van der Mey et al., 1989). The maternally derived gene is inactivated during oogenesis and can be reactivated only during spermatogenesis (Heutnik et al., 1992). This genetically heterogeneous cancer family syndrome is associated with germline mutations in succinate dehydrogenase subunits B (SDHB) and D (SDHD) (Neumann et al., 2004).

Thus, epigenetic changes are probably significant in a number of other complex phenotypes, particularly those associated with cancer and a number of degenerative diseases (see "*complex genomic diseases*").

Disorders of Genome Architecture

Recent completion of the human genome project and sequencing of the total genomes of yeast and other bacterial species have enabled investigators to view genetic information in the context of the entire genome. As a result, it is now possible to recognize mechanisms of some genetic diseases at the genomic level. The evolution of the mammalian genome has resulted in the duplication of genes, gene segments, and repeat gene clusters (Lupski, 1998). This aspect of genome architecture provides recombination hot spots between nonsyntenic regions of chromosomes that are distributed across the whole genome. These genomic regions become susceptible to further DNA rearrangements that may be associated with an abnormal phenotype. Such disorders are collectively grouped under the broad category of "*genome architecture disorders.*"

The term "genome architecture disorder" refers to a disease that is caused by an alteration of the genome that results in complete loss, gain, or disruption of the structural integrity of a dosage sensitive gene(s) (Fig. 6–14) (Shaw and Lupski, 2004; Lupski and Stankiewicz, 2005). Notable examples include a number of chromosome deletion/duplication syndromes (Table 6–5). In these conditions, there is a critical rearranged genomic segment flanked by large (usually more than 10 kb), highly homologous low-copy repeat (LCR) structures that can act as recombination substrates. Meiotic

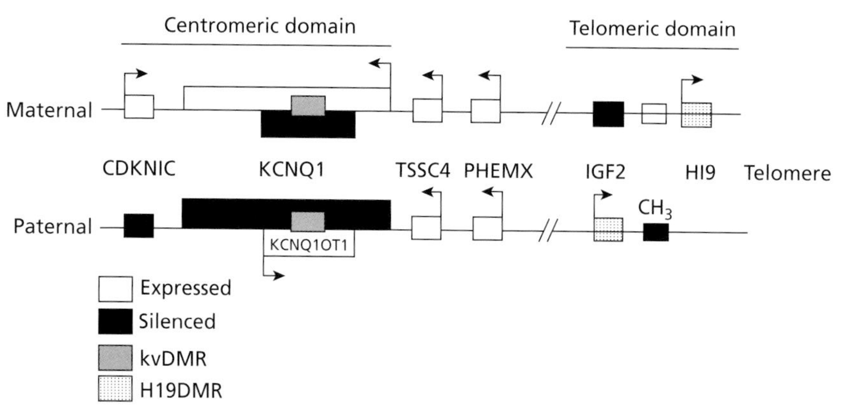

Figure 6–12 The cluster of genes on 11p15.5 associated with the phenotype of Beckwith–Wiedemann syndrome. The methylated region KvDMR1 is indicated by the grey box within the gene *KCNQ1OT1* and marked CH_3 on the maternal homologue. The methylated region between *IGF2* and *H19* genes is indicated by the hatched box and marked CH_3 on the paternal homologue. [Reproduced from Weksberg, R et al. (2001). Tumor development in the Beckwith–Wiedemann syndrome is associated with a variety of constitutional molecular 11p15 alterations including imprinting defects of KCNQ1OT1. *Hum Mol Genet* 10(26):2989–3000 with permission from Oxford University Press.]

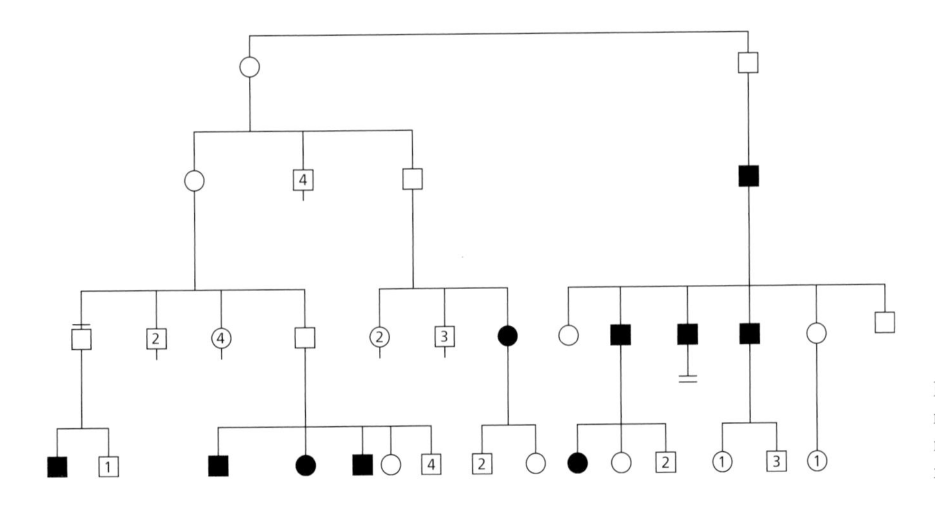

Figure 6–13 Pedigree showing paternal transmission of paraganglioma in a family: note no maternal transmission among "at-risk" family members.

recombination between nonallelic LCR copies, also known as nonallelic homologous recombination (NAHR), can result in deletion or duplication of the intervening segment.

Similarly, other chromosomal rearrangements (Table 6–6), including reciprocal, Robertsonian and jumping translocations, inversions, isochromosomes, and small marker chromosomes, may also involve susceptibility to rearrangement related to genome structure or architecture. In several cases, LCRs, A–T rich palindromes, and pericentromeric repeats are located at such rearrangement breakpoints. This susceptibility to genomic rearrangements is not only implicated in disease etiology, but also in primate genome evolution (Shaw and Lupski, 2004).

An increasing number of Mendelian diseases (Table 6–7) are recognized to result from recurrent inter and intra chromosomal rearrangements involving unstable genomic regions facilitated by LCRs. These genomic regions are predisposed to NAHR between paralogous genomic segments. LCRs usually span approximately 10–400 kb of genomic DNA, share 97% or more sequence identity, and provide the substrates for NAHR, thus predisposing to rearrangements. LCRs have been shown to facilitate meiotic DNA rearrangements associated with several multiple malformation syndromes and some disease traits (Table 6–6). Seminal examples include microdeletion syndromes (Williams–Beuren syndrome (7q11del), DiGeorge syndrome (22q11del), autosomal dominant Charcot-Marie-Tooth disease (CMTD) type 1A (PMP22 gene duplication), Hereditary Neuropathy of Pressure Palsy (HNPP: PMP22 gene deletion) mapped to 17p11.2, and Smith–Magenis, a contiguous gene syndrome (CGS) with del (17) (p11.2p11.2). Dominantly inherited male infertility related to AZF gene deletion follows a similar mechanism. In addition, this LCR-based complex genome architecture appears to play a major role in the primate karyotype evolution, pathogenesis of complex traits and human carcinogenesis.

A notable example includes genetically heterogeneous CMTD. The disorder is also known as hereditary motor and sensory neuropathy (HMSN) by virtue of being a peripheral neuropathy due to either involvement of the axonal or myelinated segments of the peripheral nerve. Genetically autosomal dominant, autosomal recessive, and X-linked dominant types are recognized. The disorder is not uncommon affecting approximately 1 in 2500 of the adult

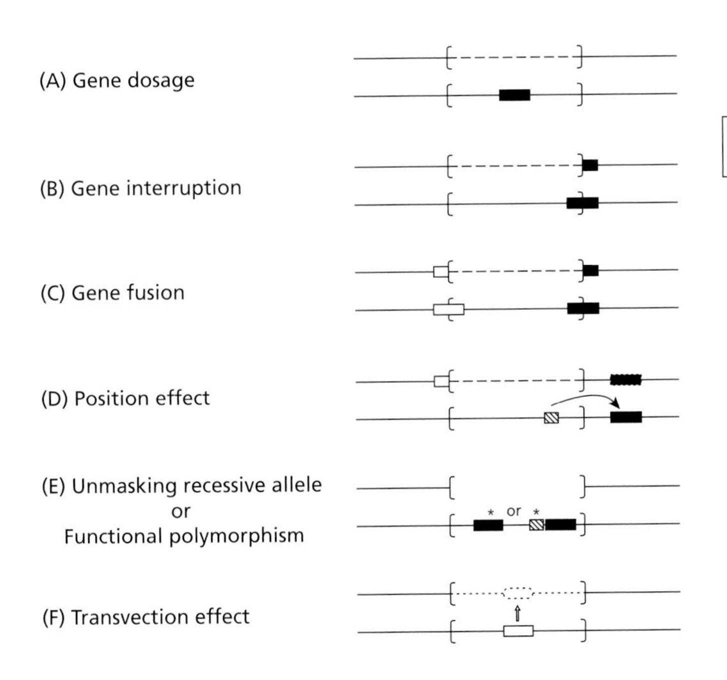

Figure 6–14 Molecular mechanisms for genomic disorders (Lupski and Stankiewicz, 2005)—dashed lines indicate either deleted or duplicated region; the rearranged genomic interval is shown in brackets; gene is depicted by filled horizontal rectangle; regulatory gene is shown as horizontal hatch-marked rectangle; asterisks denote point mutations.

Table 6–5 Contiguous Gene Syndromes as Genomic Disorders

Disorder (OMIM)	Inheritance Pattern	Locus	Gene	Rearrangement Type	Recombination Substrates Size (kb)	Recombination Substrates Repeat	% Identity	Orientation
Williams–Beuren syndrome (194050)	AD	7q11.23	*ELN*	del; inv	1600	>320	98	C
Prader–Willi syndrome (176270)	AD	5q11.2q13	?	del	3500	>500		C
Angelman's syndrome (105830)	AD	15q11.2q13	*UBE3A*	del	3500	>500		C
Dup(15)(q11.2q13)		15q11.2q13	?	dup	3500	>500		C
Triplication 15q11.2q13		15q11.2q13	?	trip		>500		C
Smith–Magenis syndrome (18290)		17p11.2	*RA13*	del	4000	~250	98	C
Dup(17(p11.2p11.2)	AD	17p11.2	*PMP22*	dup	4000	~250	98	C
VCFS(192430/ DiGeorge (188400)	AD	22q11.2	*TBX1*	del	3000/ 1500	~225–400	97–98	C
Male infertility (415000)	YL	Yq11.2	*DBY*	del	800	~10		D
AZFa microdeletion			*USP9Y*					
AZFc microdeletion (400024)	YL	Yq11.2	*RBMY*	del	3500	~220	99.9	C
			DAZ?					

Abbreviations: C, complex; D, direct; del, deletion; dup, duplication; inv, inversion.

population. This could be an underestimate, since medically the condition is benign often not requiring any medical and/or surgical intervention. However, some affected individuals experience increasingly progressive neuromuscular weakness of distal muscles of lower legs, feet, distal forearms, and hands with onset in early teens and causing severe locomotor restrictions.

An affected person usually presents late with relative hypertrophy of the upper calf muscles described as "*inverted Champagne bottle*" (Fig. 6–15) associated with pes cavus due to wasting of the small muscles of the feet. Similarly, wasting of the small muscles of hands leads to "*claw-hands*." Neurophysiological studies remain an essential method of differentiating the two major types of CMTD. A reduced motor nerve conduction velocity of less than 35 m/s helps in differentiating type 1 CMTD from type 2 CMTD, in which the motor nerve conduction velocity is usually normal, but the sensory nerve conduction is often slow. While this distinction is undoubtedly helpful in the clinical management, application for genetic counseling is limited, because both types are genetically heterogeneous. For instance, molecular characterization and gene mapping have confirmed the existence of at least four types of type 1 CMTD, autosomal dominant types 1a, 1b, and 1c and the X-linked CMT (XCMT) type. Similarly, there are distinct genetic types within the type 2 CMTD group.

Approximately two-thirds of cases of CMT1 have a detectable 1.5 Mb duplication within a proximal chromosomal segment of the short arm of chromosome 17 (17p12) (Lupski et al., 1991). This duplicated chromosomal segment contains a gene for peripheral myelin protein called *PMP22*. This duplication results in disruption of the gene leading to abnormal myelination of the peripheral nerves, an essential molecular pathological step resulting in the CMT1 phenotype designated as CMT1A. The CMT1A duplication was visualized by multiple molecular methods (Patel and Lupski, 1994), including fluorescence in situ hybridization (FISH), pulsed-field gel electrophoresis (PFGE), and dosage differences of heterozygous alleles by restriction fragment length polymorphisms (RFLPs) (Fig. 6–16). This finding led to further molecular studies on the origin of the 1.5 Mb duplicated 17p12 segment (Lupski, 2003).

Studies by several investigators have revealed a significant variation in the size of marker alleles flanking the duplicated 17p12 region. It soon became apparent that a 500 kb allele cosegregated

Table 6–6 Genomic Diseases Resulting from Recurrent Chromosomal Rearrangements

Rearrangement	Type	Recombination Substrates Repeat Size	% Identity	Orientation	Type
Inv dup(15)(q11q13)	Inverted dup	>500		C	
Inv dup(22)(q11.2)	Inverted dup	~225–400	97–98	C	
Idic(X)(p11.2)	Isodicentric			I?	
Inv dup(8)(pterp23.1::p23.2pter); Del(8)(p23.1p23.2)	inv/dup/del	~400	95–97	I	Olfactory receptor gene cluster
dup(15)(q24q26)	dup	~13–60		?	

Abbreviations: C, complex; D, direct; del, deletion; dup, duplication; I, inverted; inv, inversion.

Table 6.7 Mendelian Genomic Disorders

				Rearrangement		Recombination Substrates		
Disorders	Inheritance	Chromosome Location	Gene(s)	Type	Size (kb)	Repeat Size (kb)	% Identity	Orientation
Barter syndrome type III	AD	1p36	*CLCNKA/8*	del	11		91	D
Gaucher disease	AR	1q21	*GBA*	del	16	14		D
Familial juvenile nephronophthisis	AR	2q13	*NPHP1*	del	290	45	>97	D
Facioscapulohumeral muscular dystrophy	AD	4q35	*FRG1?*	del	25–222	3.3		D
Spinal muscular dystrophy	AR	5q13.2	*SMN*	inv/dup	500			I
Congenital adrenal hyperplasia: 21 hydroxylase deficiency	AR	6p21.3	*CYP21*	del	30		96–98	D
Glucocorticoid remediable aldosteronism	AD	8q21	*CYP11B1/2*	dup	45	10	95	D
β-thalassemia	AR	11p15.5	*β-globin*	del	4,(7)?			D
α-thalassemia	AR	16p13.3	*α-globin*	del	3,7,4.2?	4		D
Polycystic kidney disease type 1	AD	16p13.3	*PKD1*			50	95	
Charcot–Marie–Tooth (CMT1A)	AD	17p12	*PMP22*	dup	1400	24	98.7	D
Hereditary neuropathy with liability to pressure palsy (HNPP)	AD	17p12	PMP22	del	1400	24	98.7	D
Neurofibromatosis type 1 (NF1)	AD	17q11.2	*NF1*	del	1500	85		D
Pituitary dwarfism	AR	17q23.3	*GH1*	del	6.7	2.24	99	D
CYP2D6-phramcogenetic trait	AR	22q13.1	*CYP2D6*	del/dup	9.3	2.8		
Ichthyosis	XL	Xq28	*STS*	del	1900	20		D
Red-green color blindness	XL	Xq28	*RCP/GCP*	del	0	39	98	D
Incontinentia pigmenti	XL	Xq28	*NEMO*	del	10	0.870		D
Hemophilia A	XL	Xq28	*F8*	inv	300–500	9.5	99.9	I
Emery-Dreifuss muscular Dystrophy (EMD)	XL	Xq28	Emerin/ FLN1	del/ dup/ inv	48	11.3	99.2	
Hunter syndrome	XL	Xq28	*IDS*	inv/ del	20	3	>88	

Abbreviations: C, complex; D, direct; del, deletion; dup, duplication; I, inverted, inv, inversion.

with 17p duplication in all affected individuals. This suggested a stable mutation and followed a precise recombination mechanism. However, in *de novo* duplication, the presence of repeated flanking marker alleles indicated the mechanism of unequal crossing-over leading to duplication. Indeed, this was confirmed when a highly homologous >20 kb size repeat sequence was confirmed flanking the 17 p duplication. It was appropriately termed "CMT1A-REP." As predicted by the unequal crossing-over model, CMT1A-REP was found to be present in three copies on the CMT1A duplication-bearing chromosome (Pentao et al., 1992). Interestingly, the presence of only one copy was soon demonstrated in another peripheral nervous system disorder known as HNPP (Chance et al., 1994; Reiter et al., 1996). The affected individuals with this disorder present with mild to moderate episodic weakness of the lower limbs and occasionally of upper limbs when subjected to prolonged pressure, such as sitting and sleeping. The disorder is dominantly inherited in an autosomal dominant manner. This is generally a clinically mild and benign hereditary neuropathy. The presence of only one copy results from a reciprocal deletion following unequal crossing-over involving the CMT1A-REP repeat (Fig. 6–17).

Similar observations were also made in relation to Smith–Magenis syndrome (SMS), a CGS associated with a microdeletion of 17p11.2 segment (Greenberg et al., 1991). Affected children present with facial dysmorphic features, severe speech delay, and behavioral problems with signs of self-harm. A specific junction fragment was detected by PFGE (SMS-REP) that was involved in recurrent rearrangement resulting in either SMS or reciprocal 17p11.2 duplication. Pathogenic mutations in *RAI1* gene, mapped to the 17p11.2 chromosomal region, are now shown to be etiologically linked with SMS (Slager et al., 2003). It is also possible to have both duplication and deletion at the same time, resulting from DNA rearrangements on both homologues of chromosome 17. This was demonstrated in a

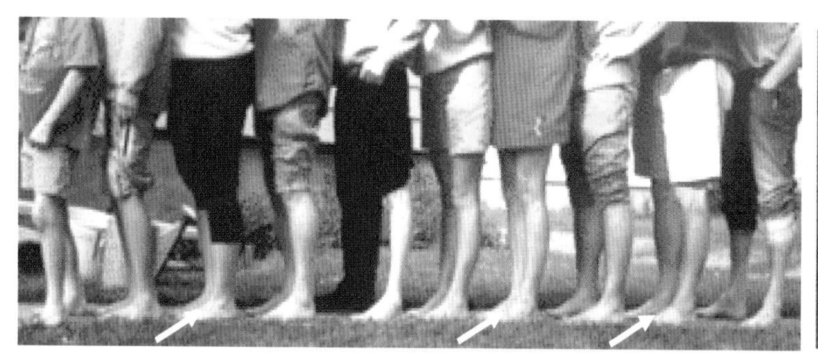
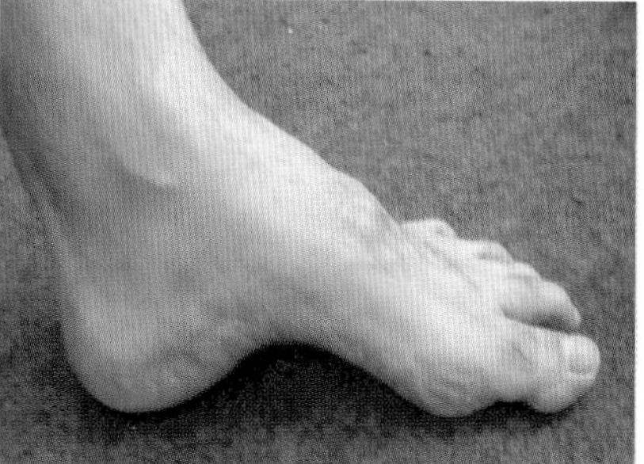

Figure 6–15 Lower legs and feet in Charcot–Marie–Tooth disease—note characteristic lower leg appearance and pes cavus.

patient with mild delay and a family history of autosomal dominant carpel-tunnel syndrome (Potocki et al., 1999). The occurrence of both the 17p11.2 duplication and HNPP deletion in this patient reflects the relatively high frequency at which these abnormalities arise and the underlying molecular characteristics of the genome in this region.

It is perfectly reasonable to accept the argument that similar molecular mechanisms apply in causing other disorders (Table 6–7). The human genome has evolved an architecture that may make us as a species more susceptible to rearrangements causing genomic disorders (Lupski, 2003).

Disorders with Trinucleotide (Triplet) Repeats

Several disorders are recognized to have a phenomenon of earlier age at onset of disease in successive generations. This is known as "*anticipation*." This observation failed to secure a valid biological explanation and had been put aside simply on the basis of biased ascertainment of probands or random variations in the age of onset. With the identification of unstable DNA repeats distributed across the genome, a molecular basis has been found for the phenomenon of anticipation. These unstable DNA repeats tend to increase in size during meiosis over successive generations. The abnormal expansion is correlated with reducing age of onset and increased severity with further expansion of DNA repeats. The characteristic pattern of the DNA repeat involving a set of three nucleotides is commonly referred to "*tri-nucleotide*" or "*triplet*" repeats. This soon became established as a novel class of mutation and offered a plausible explanation for the phenomenon of anticipation and variable clinical severity in a number of neurodegenerative diseases (Table 6–8).

The X-linked recessive spinal bulbar atrophy (SBA) was one of the first hereditary neurological disorders recognized to be associated with CAG triplet repeats (Warren, 1996).

The expanded region can occur anywhere in the gene and thus can disrupt the expression of the gene. In the case of X-linked fragile X syndrome (FRAXA), the CGG repeats are found in the 5′-untranslated region of the first exon of *FMR1*, the pathogenic gene for FRAXA (Fig. 6–18). However, in the case of Friedreich's ataxia (FA), an autosomal recessive form of spinocerebellar ataxia (SCA), the expanded triplet repeat allele (GAA) occurs in the first intron of *X25*, the gene encoding frataxin. In Huntington disease (HD) and other inherited neurodegenerative disorders, the CAG triplet repeats occur within exons and encode an elongated polyglutamine tract (Perutz, 1995) (Fig. 6–19). However, the expanded CTG triplet repeats of myotonic dystrophy (DM) are found in the 3′-untranslated region of the last exon of the DM protein kinase (myotonin) gene (*DM*).

Each class of trinucleotide repeats exists in normal individuals. A pathogenic expansion is the one that is seen in clinically symptomatic individuals. Carriers for an X-linked disease also have an expanded allele (premutation), which does not usually result in abnormal phenotype. However, it is likely that some carrier females

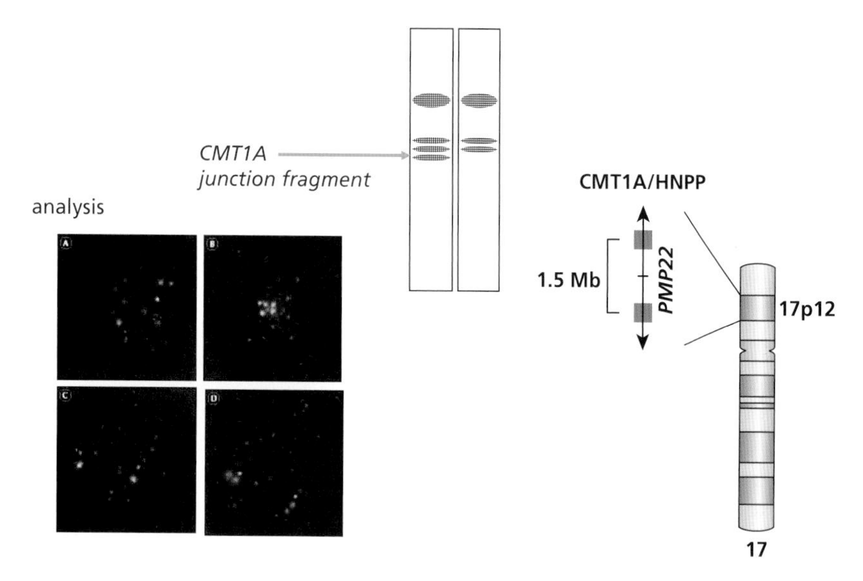

Figure 6–16 The 1.5 Mb duplicated chromosomal region of 17p12 including the *PMP22* gene—note 500 Kb junction fragment allele flanking the *CMT1A* gene detected by PFGE and Southern analysis. Note additional 17p segment (red color) by metaphase (top two pictures) and interphase (lower two pictures) FISH.

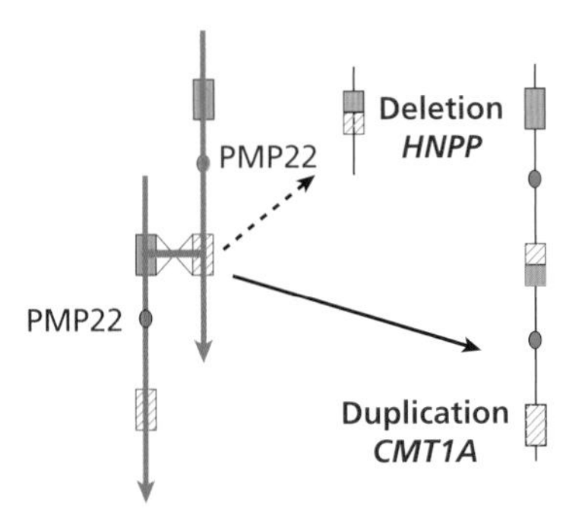

Figure 6–17 The unequal meiotic recombination (crossing-over) resulting in duplication (*CMT1A*) and deletion (*HNPP*).

might exhibit some manifestations as in Fragile X syndrome. An expanded allele in the premutation range in a male would not be associated with any clinical manifestations (normal transmitting male or NTM), but this could further expand resulting in all his daughters being carriers. However, recent studies have provided data on the existence of late-onset gait ataxia in NTMs (Greco et al., 2006). On the other hand, a normal size CGG repeat in a normal male could undergo further expansion during meiosis leading to a carrier daughter. This usually becomes known when a symptomatic grandson is confirmed to have pathogenic FRAXA expansion. Prior to availability of the molecular testing in FRAXA, this kind of unusual pedigree pattern in fragile X syndrome was called the "*Sherman paradox*" (Fig. 6–20). Detailed molecular studies in the family are often necessary to offer accurate genetic counseling to "*at-risk*" carrier females. Carrier females are at an additional risk for developing premature ovarian failure, usually diagnosed when investigated for secondary infertility (Ennis et al., 2006).

Genetic counseling in other neurodegenerative disorders with triplet repeats is often complicated. In particular, the clinical prediction in "border-line" expanded triplet repeats (intermediate reduced penetrance allele) in HD is extremely difficult due to lack of reliable data. However, recent studies have produced some data, which is likely to be helpful in genetic counseling (Quarrell et al., 2007).

Table 6–8 Disorders with Trinucleotide (triplet) Repeat Expansion

Disorder	Triplet	Location	Normal #	Mutation #
Fragile X syndrome	CGG	5′ UTR	10–50	200–2000
Friedreich's ataxia	GAA	Intronic	17–22	200–900
Kennedy disease (SBMA)	CAG	Coding	17–24	40–55
Spinocerebellar ataxia 1 (SCA1)	CAG	Coding	19–36	43–81
Huntington disease	CAG	Coding	9–35	37–100
Dentatorubral-Pallidoluysian atrophy (DRPLA)	CAG	Coding	7–23	49–(>75)
Machado–Joseph disease (SCA3)	CAG	Coding	12–36	67–(>79)
Spinocerebellar ataxia 2 (SCA2)	CAG	Coding	15–24	35–39
Spinocerebellar ataxia 6 (SCA6)	CAG	Coding	4–16	21–27
Spinocerebellar ataxia 7 (SCA7)	CAG	Coding	7–35	37–200
Spinocerebellar ataxia 8 (SCA8)	TG	UTR	16–37	100–(>500)
Myotonic dystrophy	CTG	3′ UTR	5–35	50–4000
Fragile site E (FRAXE)	CCG	Promoter	6–25	>200
Fragile site F (FRAXF)	GCC	?	6–29	>500
Fragile site 16 A (FRA16A)	CCG	?	16–49	1000–2000

Abbreviation: UTR—untranslated region.

Complex Genomic Diseases

All inherited disorders have a genetic abnormality present in the DNA of all cells in the body including germ cells (sperm and egg) and can be transmitted to subsequent generations. In contrast, a genetic abnormality present only in specific somatic cells cannot be transmitted. The genetic abnormality in a somatic cell can occur any time from the postconception stage to late adult life. The paradigm of somatic cell genetic disorder is cancer, where the development of malignancy is often the consequence of mutations in genes that control cellular growth. There are several such genes, and these are designated oncogenes. It is now accepted that all human cancer results from mutations in the nuclear DNA of a specific somatic cell, making it the most common genetic disease. The various genetic mechanisms that can result in cancer are discussed in the chapter on cancer genomics (see Chapter 24).

The clinical course and outcome of treatment in a number of acute and chronic medical conditions depend on various factors. For instance, there is overwhelming evidence that highly polymorphic cytokine, interferon, and interleukin families of complex proteins influence the host response to acute infection and physical injury. All these proteins are encoded by several genes. Similarly, association of HLAs in the pathogenesis of a number of acute and chronic medical disorders is well known (see Chapter 27). In addition, interaction of mutations within these genes and with several other genomic polymorphisms, such as single nucleotide polymorphisms (SNPs) is probably important in several acute medical conditions including trauma. This will have a major impact in critical care and acute medicine (see Chapter 17). The role of SNPs in modulating complex medical disorders, such as diabetes mellitus, coronary heart disease, hypertension, and various forms of cancer is unclear. However, the complexity of interaction of SNPs with other genetic traits and loci is probably important in the prognosis of these disorders, in particular the outcome of therapeutic interventions. This argument probably justifies separating some of these disorders under the title of "complex genomic diseases."

Various cancers and degenerative diseases occur with increasing frequency in old age. However, these may also present at a younger age, such as childhood leukemias. The molecular mechanisms in these diseases are not entirely clear, but probably include defects in DNA repair mechanisms, accelerated apoptosis, deregulation of imprinted genomic regions, and *de novo* chromosome rearrangements involving specific genomic regions. Although, these disorders can be arguably included under the broad category of multifactorial/polygenic diseases, the pattern of distribution and recurrence does not follow the agreed principles of multifactorial/polygenic inheritance as discussed elsewhere in this chapter.

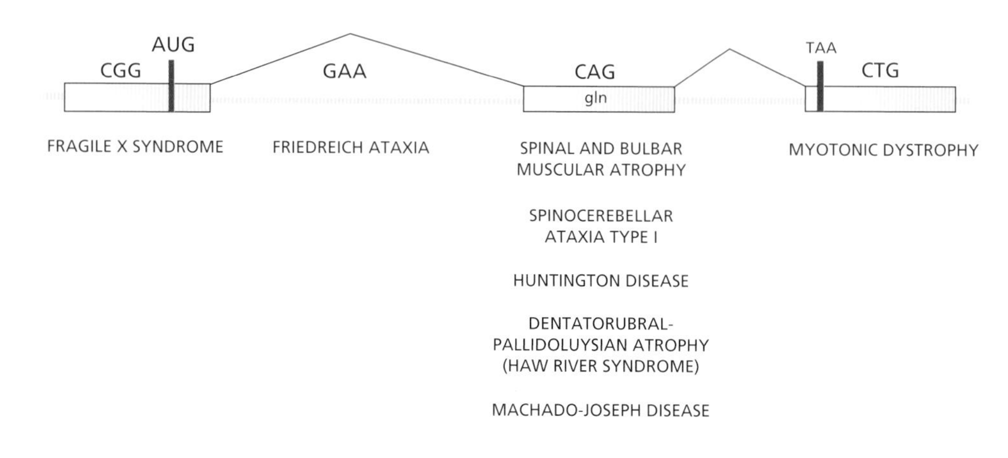

Figure 6–18 Location of four classes of triplet repeats in human diseases. Exons are shown in light pink with intervening introns as a pink solid line. The translation site AUG and termination signal TAA are indicated by red vertical bars. (Adapted from Gelehrter et al., 1998 with permission from Williams and Wilkins.)

As described in the previous section on "*epigenetics,*" epigenetic changes play a major role in the development of human cancer (Egger et al., 2004). A high percentage of patients with sporadic CRC possess microsatellite instability and show methylation and silencing of the gene encoding MLH1 (Kane et al., 1997). It is thus likely that epigenetic changes also predispose to genetic instability. In some cases, promoter-associated methylation of *MLH1* is found not only in the tumor but also in normal somatic tissues, including spermatozoa. These germline "epimutations" predispose individuals carrying abnormal methylation patterns to multiple cancers. Indeed, disruption of pathways that lead to cancer is often caused by the *de novo* methylation of the relevant gene's promoters (Jones and Baylin, 2002). Epigenetic silencing has been recognized as a third pathway satisfying Knudson's "two-hit" hypothesis for the silencing of tumor suppressor genes (Jones and Laird, 1999).

Chromosomal rearrangements have long been associated with human leukemias. These result in formation of fusion proteins including histone acetyltransferases and histone methyltransferases that influence upregulation of target genes. In acute promyelocytic leukemia, the oncogenic fusion protein PML-RARα (promyelocytic leukemia-retinoic acid receptor-α) causes repression of genes that are essential for differentiation of hematopoietic cells. Similarly, in acute myeloid leukemia, AML-ETO fusions recruit the repressive N-COR-Sin3-HDAC1 complex and inhibit myeloid development (Jones and Saha, 2002). There are further examples of complex genomic arrangements that result in other cancers and which can

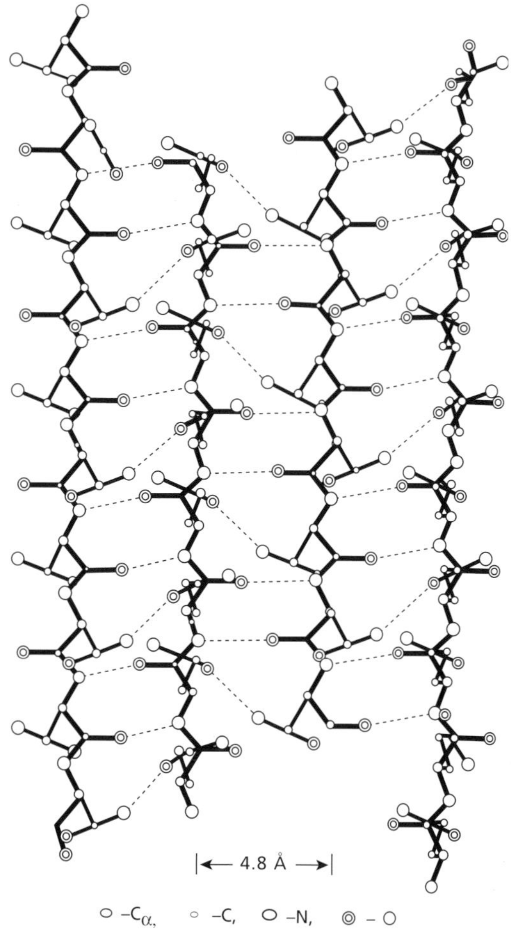

|← 4.8 Å →|

○ –Cα, ○ –C, ○ –N, ◎ – O

Figure 6–19 Schematic diagram of the polyglutamine tract resulting from abnormal expansion of CAG trinucleotide repeats. (Reproduced with permission from Perutz et al. 1994.)

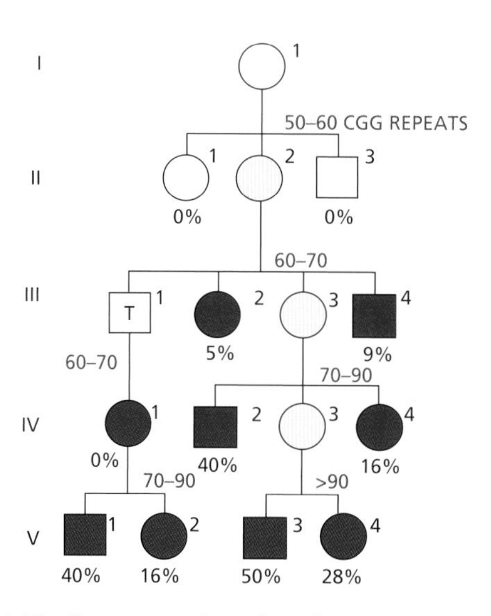

Figure 6–20 The Sherman paradox: a hypothetical pedigree showing affected members (red) and carrier females (pink); individual III.1 is a normal transmitting male; the percentage risk for mental retardation is given for respective size of the triplet (CGG) repeats. (Adapted from Gelehrter et al., 1998 with permission from Williams and Wilkins.)

modify the therapeutic response. For example, mutations in genes for ATPase complex are associated with poorer prognosis in patients with non-small-cell lung cancer (Roberts and Orkin, 2004).

Summary

Developments in genetics and subsequently sequencing of the human genome have provided us with an opportunity to review the taxonomy of human disease. Conventionally, the causation of human disease includes malformations, trauma, infection, immune dysfunction, metabolic abnormality, malignancy, and degenerative conditions associated with aging. Genetic factors have long been recognized in all of these disease groups. The traditional genetic categories of diseases include chromosomal disorders, single-gene or Mendelian diseases, and several forms of multifactorial/polygenic conditions. In addition, somatic genetic changes and mutations of the mitochondrial genome probably account for a small, albeit important, number of diseases. These groups of disorders are well recognized and have an established place in the classification of human disease.

Recent developments in genome research have provided enormous data indicating different genomic mechanisms to explain complex pathogenesis in some disorders. The spectrum of these disorders is wide and includes both acute and chronic medical and surgical diseases. Perhaps it is reasonable to identify these disorders on the basis of underlying molecular pathology including genomic imprinting, genomic rearrangements, and gene-environment interactions involving multiple genes and genomic polymorphisms. This chapter reviews the genetic and genomic approaches in the classification of human disease. The new taxonomy of human disease is likely to have a major impact on the practice of clinical medicine in future.

References

Bell, JI (2003) The double helix in clinical practice. *Nature*, 421(6921):414–416.

Cardon, IR and Bell, JI (2001). Association study designs for complex diseases. *Nature, Rev Genet*, 2:91–99.

Chance, PF, Abbas, N, Lensch, MW, Pentao, L, Roa, BB, Patel, PI, et al. (1994). Two autosomal dominant neuropathies result from reciprocal DNA duplication/deletion of a region on chromosome 17. *Hum Mol Genet*, 3:223–228.

Chen, WG, Chang, Q, Lin, Y, Meissner, A, West, AE, Griffith, EC, et al. (2003). Derepression of BDNF transcription involves calcium-dependent phosphorylation of MeCP2. *Science*, 302:885–889.

Cookson, WO, Sharp, PA, Faux, JA and Hopikin, JM (1989). Linkage between immunoglobulin E responses underlying asthma and rhinitis and chromosome 11q. *Lancet*, 1:1292–1295.

Egger, G, Liang, G, Aparicio, A and Jones, P (2004). Epigenetics in human disease and prospects of epigenetic therapy. *Nature*, 429:457–463.

Ennis, S, Ward, D, Murray, A (2006). Nonlinear association between CGG repeat number and age of menopause in FMR1 premutation carriers. *Eur J Hum Genet*, 14:253–255

Greco, CM, Berman, RF, Martin, RM, Tassone, F, Schwartz, PH, Chang, A, et al. (2006). Neuropathology of fragile X-associated tremor/ataxia syndrome (FXTAS). *Brain*, 129(Pt 1):243–255.

Greenberg, F, Guzzeta, V, Montes de Oca-Luna, R, Magenis, RE, Smith, AC, Richter, SF, et al. (1991). Molecular analysis of the Smith–Magenis syndrome: a possible contiguous gene syndrome associated with del(17)(p11.2). *Am J Hum Genet*, 49:1207–1218.

Hall, JG, Powers, EK, Mclivaine, RT and Ean, VH (1978). The frequency and financial burden of genetic disease in a pediatric hospital. *Am J Med Genet*, 1:417–436.

Heutnik, P, van der Mey, AG, Sandkuiji, LA, van Gils, AP, Bardoel, A, et al. (1992). A gene subject to genomic imprinting and responsible for hereditary paragangliomas maps to chromosome 11q23-qter. *Hum Mol Genet*, 1:7–10.

Holliday, R and Pugh, JE (1975). DNA modification mechanisms and gene activity during development. *Science*, 187:226–232.

Jones, LK and Saha, V (2002). Chromatin modification, leukemia and implications for therapy. *Br J Haematol*, 118:714–727.

Jones, P and Baylin, SB (2002). The fundamental role of epigenetic events in cancer. *Nat Rev Genet*, 3:415–428.

Jones, PA and Laird, PW (1999). Cancer epigenetics comes of age. *Nat Genet*, 21:163–167.

Kane, MF, Loda, M, Gaida, GM, Lipman, J, Mishra, R, Goldman, H, et al. (1997). Methylation of hMLH1 promoter correlates with lack of expression in sporadic colon tumors and mismatch repair defective human cancer cell lines. *Cancer Res*, 57:808–811.

Lachner, M and Jenuwin, T (2002). The many facets of histone lysine methylation. *Curr Opin Cell Biol*, 14:286–298.

Lewis, T (1944). Reflections upon medical education. *Lancet*, i:619–621.

Lupski, JR (1998). Genomic disorders: structural features of the genome can lead to DNA rearrangements and human disease traits. *Trends Genet*, 14:417–420.

Lupski, JR (2003). Genomic disorders: recombination-based disease resulting from genome architecture. *Am J Hum Genet*, 72:246–252.

Lupski, JR, Montes de Oca-Luna, R, Slaugenhaupt, S, Pentao, L, Guzzetta, V, Trask, BJ, et al. (1991). DNA duplication associated with Charcot-Marie-Tooth disease type 1A. *Cell*, 66:219–232.

Lupski JR, Stankiewicz P.(2005). Genomic disorders: molecular mechanisms for rearrangements and conveyed phenotypes. *PLoS Genet*. 2005 Dec;1(6): e49.

Neumann, HP, Pawlu, C, Peczkowska, M, Bausch, B, McWhinney, SR, Muresan, M, et al. (2004). Distinct clinical features of paraganglioma syndromes associated with SDHB and SDHD gene mutations. *JAMA*, 292(8):943–951.

Patel, P and Lupski, JR (1994). Charcot-Marie-Tooth disease: a new paradigm for the mechanism of inherited disease. *Trends Genet*, 10:128–133.

Pentao, L, Wise, CA, Chinault, AC, Patel, PI and Lupski, JR (1992). Charcot-Marie-Tooth type 1A duplication appears to arise from recombination at repeat sequences flanking the 1.5 Mb monomer unit. *Nat Genet*, 2:292–300.

Perutz, MF (1995). Glutamine repeats as polar zippers: their role in inherited neurodegenerative disease. *Mol Med*, 1:718–721.

Perutz, MF, Johnson, T, Suzuki, M and Finch, JT (1994). Glutamine repeats as polar zippers: their role in inherited neurodegenerative diseases. *Proc Natl Acad Sci USA*, 91:5355–5358.

Potocki, L, Chen, K-S, Koeuth, T, Killian, J, Iannaccone, ST, Shapira, SK, et al. (1999). DNA rearrangements on both homologues of chromosome 17 in a mildly delayed individual with a family history of autosomal dominant carpal tunnel syndrome. *Am J Hum Genet*, 64:471–478.

Quarrell OW, Rigby AS, Barron L, Crow Y, Dalton A, Dennis N, Fryer AE, Heydon F, Kinning E, Lashwood A, Losekoot M, Margerison L, McDonnell S, Morrison PJ, Norman A, Peterson M, Raymond FL, Simpson S, Thompson E, Warner J. (2007). Reduced penetrance alleles for Huntington's disease: a multi-centre direct observational study. *J Med Genet*. 2007 Mar;44(3):e68.

Reiter, LT, Murakami, T, Koeuth, T, Pentao L, Muzny, DM, et al. (1996). A recombination hotspot responsible for two inherited peripheral neuropathies is located near a *mariner* transposon-like element. *Nat Genet*, 12:288–297.

Roberts, CW and Orkin, SH (2004). The SW1/SNF complex-chromatin and cancer. *Nat Rev Cancer*, 4:133–142.

Shaw, CJ and Lupski, JR (2004). Implications of human genome architecture for rearrangement-based disorders: the genomic basis of disease. *Hum Mol Genet*, 13(1):R57–R64.

Slager, RE, Newton, TL, Vlangos, CN, Finucane, B and Elsea, SH (2003). Mutations in *RAI1* associated with Smith–Magenis syndrome. *Nat Genet*, 33:466–468.

Tufarelli, C, Stanley, JA, Garrick, D, Sharpe, JA, Ayyub, H, Wood, WG, et al. (2003). Transcription of antisense RNA leading to gene silencing and methylation as a novel cause of human genetic disease. *Nat Genet*, 203:157–165.

Van der Mey, AG, Maaswinkel-Mooy, PD, Cornelisse, CJ, Schmidt, PH, van de Kemp, JJ (1989). Genomic imprinting in hereditary glomus tumors: Evidence for new genetic theory. *Lancet*, 2:1291–1294.

Warren, ST (1996). The expanding world of trinucleotide repeats. *Science*, 271:1374–1375.

Further Reading

Emery, AEH and Rimoin, DL (2000). *Principles and Practice of Medical Genetics* (4th edn). Edinburgh: Churchill Livingstone.

Gelehrter, TD, Collins, FS, and Ginsburg, D (1998). *Principles of Medical Genetics* (2nd edn). Baltimore: Williams and Wilkins.

Harper, PS (2004). *Practical Genetic Counselling* (6th edn). London: Arnold publishers.

McKusick, VA (1994). *Mendelian Inheritance in Man* (11th edn). Baltimore: Johns Hopkins University Press. Regular updates available online (internet) at http://www.ncbi.nlm.nih.gov/OMIM/

Mueller, RF and Young, ID (2004). *Emery's Elements of Medical Genetics*. Edinburgh: Churchill Livingstone.

Pulst, S-M (2000). *Neurogenetics*. New York: Oxford University Press.

Weatherall, DJ (1991). *The New Genetics and Clinical Practice*. New York: Oxford University Press.

7

Genomic Technologies

Ian M Frayling

Understanding the structure of the genome is an essential step in relating this to its function, both in health and disease. Advances in methods for studying the structure of the genome have gone hand-in-hand with developments in the technology of the platforms to carry them out, with concomitant reductions in time and expense. This was a major factor in the sequencing of the majority of the human genome, ahead of schedule.

There are ever increasing numbers of laboratory techniques available for genetic analysis. While this can be bewildering to those first approaching the subject, the actual number of distinct technologies involved is relatively small, and most techniques are related to or derived from pre-existing methods. Techniques for genetic analysis can be broadly classified into those, which are directed at small-scale structure and variation of the genome, say one to a few hundred base-pairs of deoxyribonucleic acid (DNA), and those which are directed at larger and higher levels, from a few hundred to a hundred mega base pairs . Given the inherent overlap in coverage of some techniques, this is necessarily a somewhat artificial distinction, but it helps to distinguish genomic technologies as those used to study larger and higher order structure in the genome. Whole books could and indeed have been written about genetic laboratory techniques, so this chapter will not attempt to reiterate these works and will only be a brief overview of genomic technologies. Likewise, there is no shortage of up-to-date information available on the internet, or in the published literature, on techniques directed at studying small-scale structure and variation of the genome [Sambrook and Russell, 2000 (also at www.molecularcloning.com); Coleman and Tsongalis, 2006].

In any discussion of genomic techniques, it is useful to think of the scale over which a particular method works or is applicable, and this is illustrated in Fig. 7–1. At the highest level are techniques to study the arrangement of chromosomes in 3D and even 4D, if spatiotemporal studies are included. Conventional cytogenetics covers chromosome structure largely in 2D, in terms of metaphase spreads reducing the natural 3D arrangements of chromosomes in a nucleus to a 2D arrangement on a glass slide, though it is to be noted that even these fixed and stained chromosomes have a 3D structure visible under the microscope. Chromosome painting and comparative genomic hybridization are able to look at chromosomes at a high order and allow estimations of genome dosage in difficult circumstances, for example, tumor cells. Fluorescence in situ hybridization (FISH) uses labeled probes of the order of 100 kb in size to demonstrate both location and dosage of a region of interest in the genome. Array comparative genome hybridization (CGH) is effectively a few to several thousand FISH experiments all carried out simultaneously, using probes of the order of 100 kb in size, but only dosage and not positional information is gained. Targeted, or locus specific, arrays study a particular small region, one gene perhaps, but in great detail, using probe sizes of the order of 1 kb. Again, only information on dosage is obtained. Southern analysis ("Southern blotting") is able to give data on position, orientation, dosage, and even methylation of sections of the genome from approximately 100 bp and up (Southern, 1975, 2000). As such, it holds a unique and special position in the armamentarium of the laboratory geneticist. More recent techniques, such as multiplex ligation-dependent probe amplification (MLPA) and multiplex amplifiable probe hybridization (MAPH), may have replaced Southern analysis for routine dosage measurement of small sections of the genome, for example, exons, but they are unable in themselves to give positional or orientation information. For an excellent overview of the history of genomic techniques, the recent article by Southern cannot be bettered (Southern, 2005).

3D Mapping

The earliest observation of the nuclear localization of an interphase chromosome was probably the Barr body, also known as the sex chromatin body, which is a projection of the nuclear envelope containing an inactivated X chromosome (Barr and Bertram, 1949). The term "Barr body" itself was coined by Mary Lyon (Lyon, 1961, 1963, 1999). Since then, study of the nuclear organization of chromosomes has required the development of in situ hybridization techniques, in particular FISH (Pinkel et al., 1986; Trask et al., 1988; Trask, 1991; van Ommen et al., 1995). This method depends on the specific hybridization of probe DNA sequences to the complementary sequence in the genome. As the probe is labeled with a fluorescent dye (although radioisotopic labeling is also possible) its location is revealed by laser excitation microscopy, having counterstained with a dye such as 4′,6-diamidino-2-phenylindole (DAPI) (Q-banding, see below). Competitive hybridization by including DNA enriched for repetitive sequences, that is, C_0t-1 DNA, suppresses unwanted signals. (It is also possible to detect amplification of particular

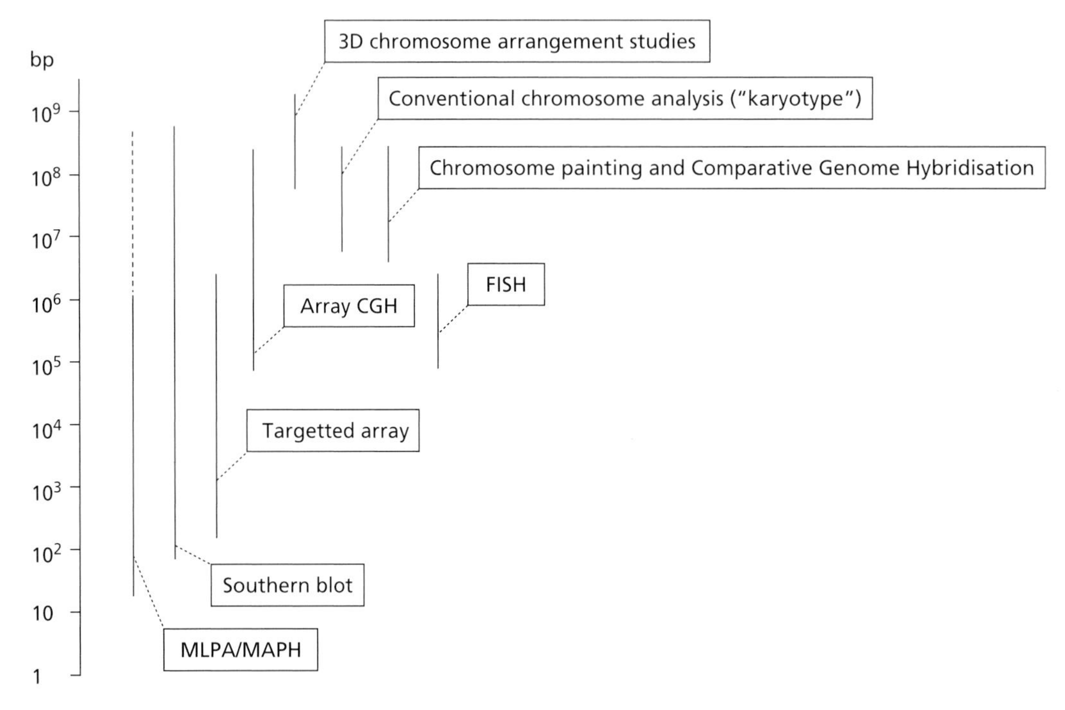

Figure 7–1 Genomic techniques: differences in genomic structure can vary by many orders of magnitude. A range of techniques is therefore necessary to cover this variety, and this diagram indicates the approximate ranges over which any one technique will give information.

DNA sequences in cells in tissue sections by in situ hybridization, for example, Her2/Neu (Wang et al., 2000).) The organization of chromosomes in interphase nuclei can be revealed by careful multicolor FISH (M-FISH; see section that follows) experiments. Considerable work has thus been carried out on chromosome organization (for reviews see Kosak and Groudine, 2004; Parada et al., 2004; Pederson, 2004) and it is revealing that it is not random, with the implication that chromosome colocation is a factor in normal gene expression and karyotype evolution, as well as in pathological states such as cancer and muscular dystrophy (Masny et al., 2004; Rijkers et al., 2004; Bolzer et al., 2005).

Cytogenetics

Classical cytogenetics is the method par excellence for rapidly carrying out a whole genome scan relatively cheaply. At the simplest level, it allows enumeration of chromosomes, but a number of different techniques for revealing structure within chromosomes, otherwise known as banding techniques, have been developed (Craig and Bickmore, 1993). Giemsa banding (G-banding) is the most widely used method for routine analysis of human chromosomes. Metaphase spreads of chromosomes are subjected to treatment with a protease, such as trypsin, after first some sort of ageing, for example, 3–5 days at room temperature, or approximately 56°C overnight (Seabright, 1972). Giemsa stain contains eosin and a mixture of thiazine dyes, and the latter bind to DNA, resulting in differential staining. Late-replicating, transcriptionally quiet, and A/T-rich DNA stains darkly (G-positive), while early-replicating, transcriptionally active, and relatively G/C-rich DNA stains light (G-negative). The highest quality G-banded preparations are able to give 850 bands/points of comparison across the genome, to which if one adds the total number of different chromosomes, totals approximately 1000 (Mitelman, 1995) (Fig. 7–2). Thus, a G-banded chromosome preparation can be thought of as a 1K whole genome array, arranged in a linear anatomical fashion. Such analysis can reveal loss or gain of material down to only a few Mb in size, though exactly how small depends on the location of the defect. It can also reveal balanced events such as translocations and inversions, where no net gain or loss of material has occurred. Thus, chromosomal analysis is capable of giving not just dosage, but also location and orientational data.

Other banding methods include Q-banding, where a fluorescent dye such as quinacrine, DAPI, or Hoechst 33258, binds preferentially to A/T-rich sequences; R-banding, which is the reverse of G-banding, uses high-molarity low pH phosphate buffer heat-denaturation of chromosomes and/or acridine orange staining, to produce a banding pattern complementary to that produced by G-banding; T-banding, which is directed at R-bands toward the telomeres by especially strong heat treatment before Giemsa staining; and C-banding, where denaturation with for example, barium hydroxide, before Giemsa staining, brings out the so-called constitutive heterochromatin at centromeres. All these techniques can reveal complementary information about the structure and organization of the genome at the chromosomal level. Q-banding is commonly used in combination with FISH to precisely locate hybridization signals.

In situ Hybridization, Including FISH

As already mentioned above, in situ hybridization techniques, in particular FISH, have allowed study of the structure of the genome at a level of detail greater than that seen by conventional banding techniques (Pinkel et al., 1986; Trask et al., 1988; Trask, 1991; van Ommen et al., 1995). The method depends on the specific hybridization of a probe DNA sequence to its complementary sequence in

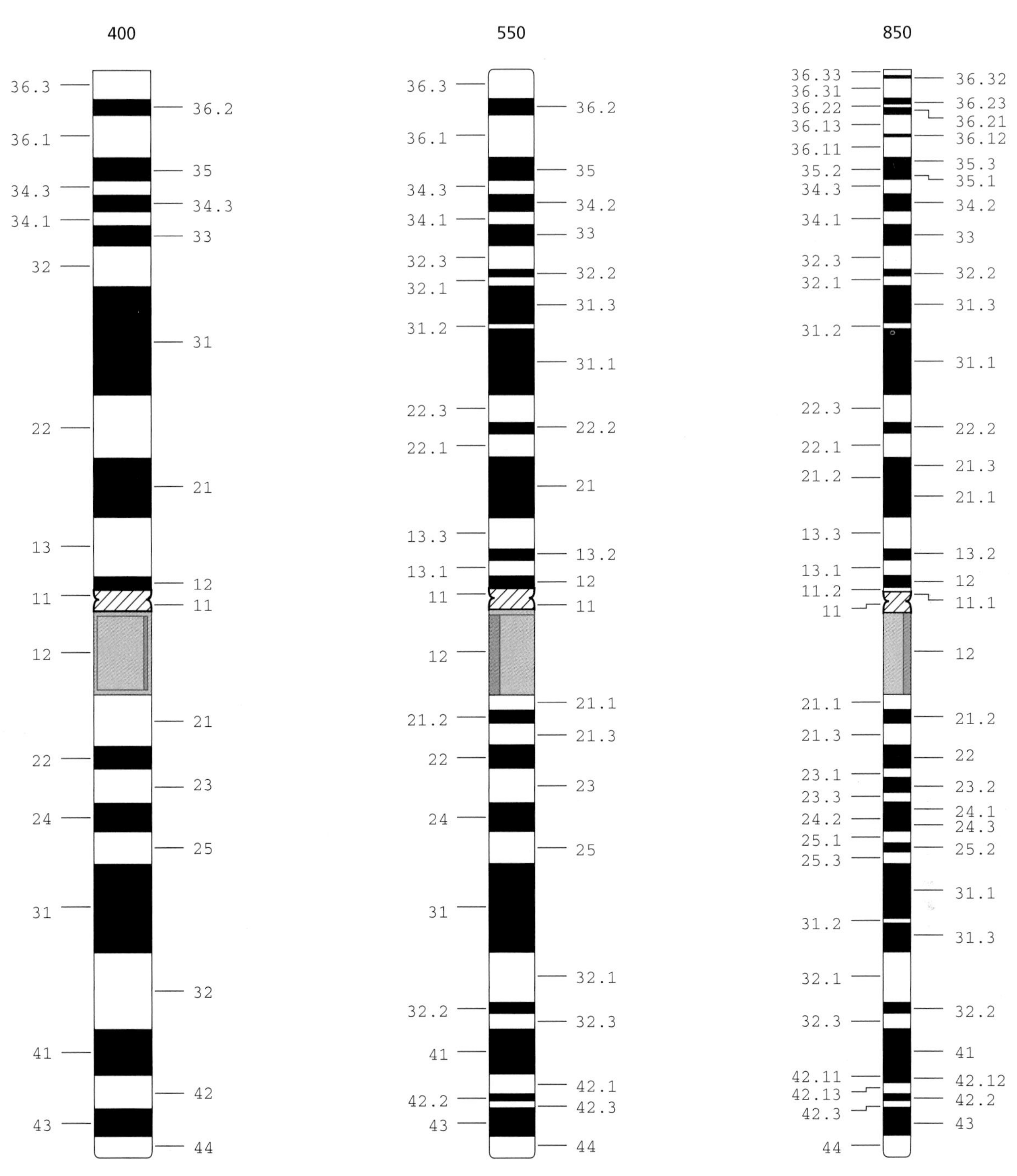

Figure 7–2 Conventional cytogenetic analysis by G-banding. Analysis of chromosome number and structure is most commonly achieved by the technique of G-banding. To achieve this, cells are grown in culture and arrested at metaphase by treatment with colchicine. They are then fixed in acetic acid/methanol and dropped onto glass slides. The nuclei burst and the chromosomes spill out as individual "spreads." The chromosomes are then treated with trypsin to partially digest their protein, and stained with Giemsa. The dark G-bands, so called heterochromatin, are relatively A + T rich, have a low gene density, and are less transcriptionally active: pale G-bands, euchromatin, have many genes, which tend to be active. The number of bands visible under the microscope depends on the cell type (for example, lymphocytes give more than amniocytes) and the precise stage of cell division that the cell was arrested. Preparations are classed according to the total number of bands visible: 400, 550, or up to 850. Ideograms show the theoretical G-banding patterns visible at different levels. (*A*) Ideograms of chromosome 1, the largest chromosome, showing expected G-banding patterns. (*B*) Ideograms of the smallest chromosome, number 21. (*C*) A normal male karyotype produced by G-banding: 23 pairs of chromosomes, consisting of 22 pairs of autosomes, plus the two sex chromosomes, in this instance X and Y. Somewhat more than 550 bands are visible in this example. Software has been used to arrange the chromosome images neatly in pairs. (Courtesy of Dr. Peter Thompson, Cytogenetics Laboratory, Institute of Medical Genetics, Cardiff, UK.)

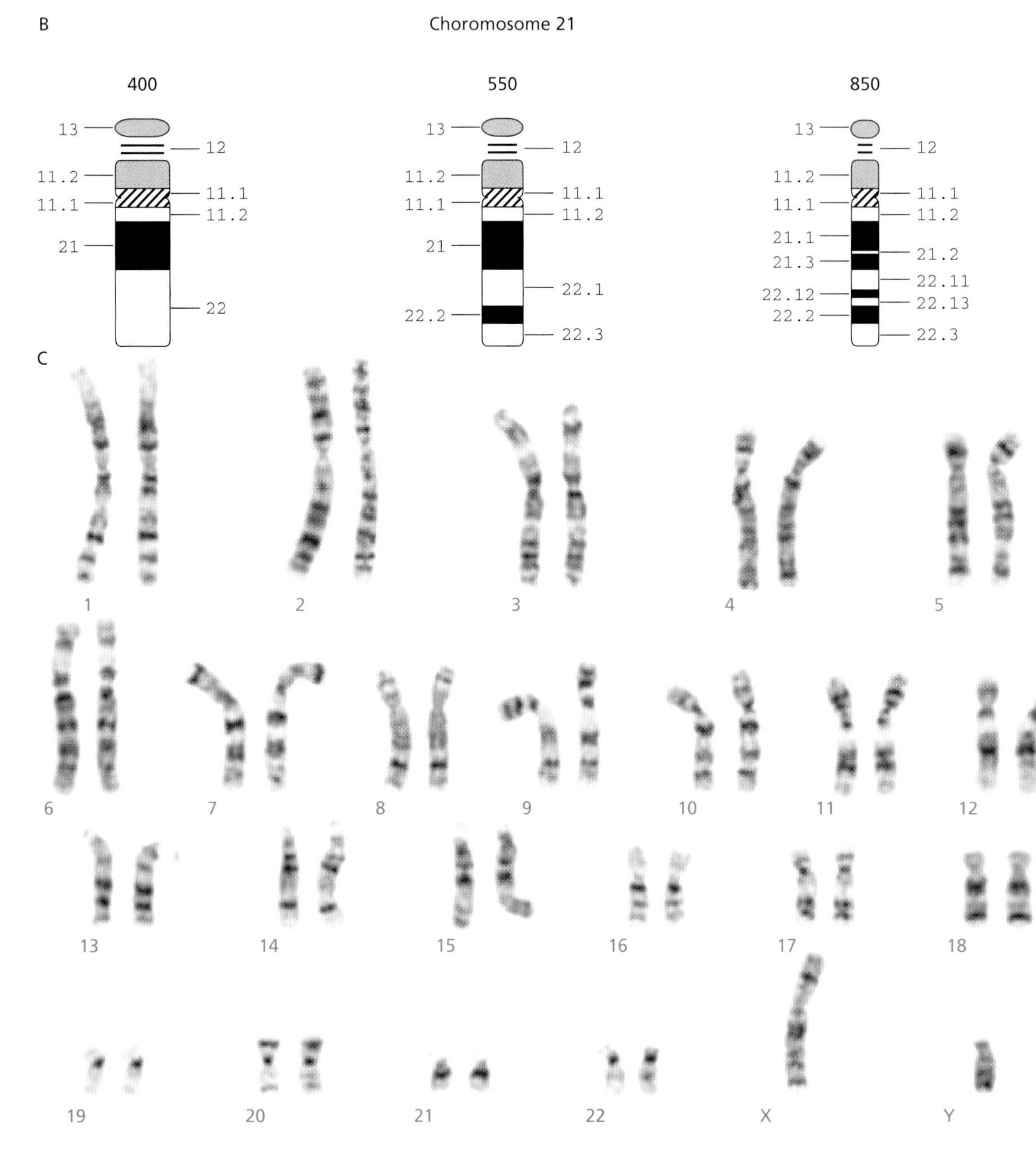

Figure 7–2 *(continued)*

the genome, and an early description shows how it was used to study satellite DNA location within cells (Jones, 1970). Labeling of the probe with a fluorescent dye allows its location to be revealed by laser excitation microscopy, usually in combination with some form of Q-banding (Fig. 7–3). C_0t-1 DNA competitive hybridization suppresses unwanted signals. The probes used in FISH experiments are typically derived from cloning human sequences into bacterial artificial chromosomes (BACs), with sizes of approximately 100 kb. For a practical consideration, see Mundle and Koska, 2006.

The human genome project has resulted in an almost complete tiling path consisting of approximately 32,000 BACs, and hence, there are only very few regions of the genome not amenable to FISH experiments. The human genome project required individual FISH experiments to ensure that any given BAC clone's sequence was defined and unique, and not subject to duplication or rearrangement. FISH can be performed on both metaphase spreads and interphase nuclei. Simple and rapid enumeration of chromosomes can be achieved on interphase nuclei, given that no culturing of cells is necessarily required (Fig. 7–4). On metaphase spreads with contrasting Q-banding, highly specific positional and dosage information can be obtained, and FISH is an excellent tool in gene mapping and diagnosis (Fig. 7–5).

One application of FISH technology that gives highly specific and accurate positional and dosage information is fiber-FISH. In this, individual strands of DNA are stretched out on a glass slide and the FISH probes hybridized. The probes can be as small as individual exons, that is, only a few hundred base pairs, far smaller than conventional FISH probes. Fiber-FISH can thus reveal deletion, duplication, or rearrangement of exons in specific genes, for example, DMD (Florijn et al., 1995).

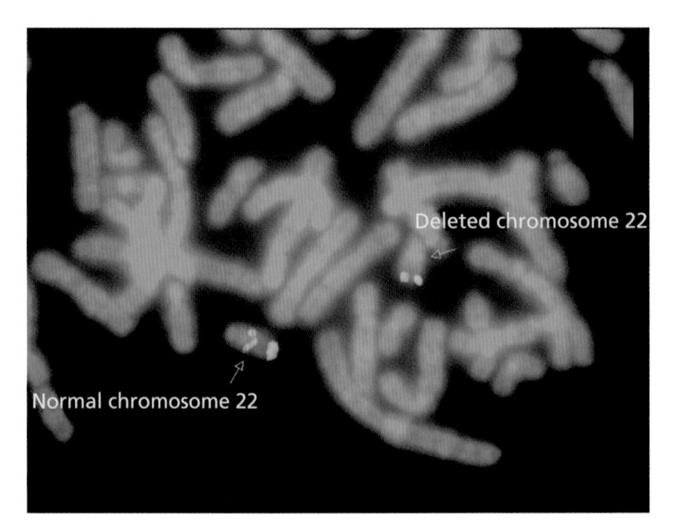

Figure 7–3 Fluorescence in situ hybridization (FISH). Specific target sequences in the genome can be analyzed by the use of DNA probes (typically 100–150 kb in size) labeled with a fluorescent dye. These can be hybridized to a metaphase chromosome spread, together with a contrasting dye that stains all chromosomes, to give information on both position within the genome and copy number. The figure shows an experiment to test a patient for the common chromosomal microdeletion on the long arm of chromosome 22 associated with Di-George syndrome: del22q11.2. A blue counter stain is used to show the chromosomes, while a control FISH probe to the end of the long arm (22q) is labeled green and is present on both copies of chromosome 22. However, only one copy of chromosome 22 is labeled (red) by the test probe to the Di-George region on 22q11.2, showing clearly that the patient has a deletion and thus confirming a diagnosis of Di-George syndrome. (Courtesy of Dr. Peter Thompson, Cytogenetics Laboratory, Institute of Medical Genetics, Cardiff, UK.)

Developments in FISH have recently been reviewed (Murthy and Demetrick, 2006).

CGH, M-FISH, and Chromosome Painting

If the probe DNA used in a FISH experiment is derived from a whole chromosome, rather than a smaller specific sequence cloned in a vector, then it will reveal a whole chromosome. The ability to flow sort chromosomes and then carry out some form of in vitro amplification before labeling with a mixture of dyes to give a specific color signal, has enabled probes to be produced for all 24 human chromosomes. If the desired experiment is simply to investigate one or two different chromosomes, that is, chromosome painting, then conventional FISH image analysis will suffice. This is good for detecting abnormal arrangements of chromosomes, such as translocations, and identification of the origin of material in for example, marker chromosomes, which may otherwise have no identifying features. If all 24 different chromosome paints are applied, the technique becomes even more sophisticated and is known by the terms multicolor FISH (M-FISH), further developed as spectral karyotyping (SKY) (Schröck et al., 1996, 1997). This is especially good at determining the many and complicated rearrangements that occur in cancer cells (Fig. 7–6). By means of specialized image analysis, dosage across the genome can be measured in an M-FISH/SKY experiment, chromosome by chromosome, and this is the technique of CGH (du Manoir et al., 1993; reviewed by Pinkel and Albertson, 2005). Again, it has found its application in delineating complex rearrangements, particularly in the field of cancer (Karpova et al., 2006; Veldman et al., 1997). In this way, careful study of regions of loss in tumors from individuals predisposed to cancer has been able to reveal the location of the gene responsible (Hemminki et al., 1997). The amassing of ever more data by techniques such as CGH is made ever more useful by the development of public databases such as the joint NCBI/NCI SKY/M-FISH and CGH Database. (http://www.ncbi.nlm.nih.gov/sky/).

Array CGH

While FISH experiments using individual probes can give useful information about the dosage and location of a particular sequence in the genome, it is simply unfeasible to carry out 32,000 different FISH experiments to cover the whole of the human genome in cases where this might be appropriate. However, a number of technological developments have made it possible to put small and uniform amounts of many different DNA probes onto a suitable substrate, for example, a $1'' \times 3''$ glass microscope slide, in the form of an array, often referred to by the colloquial name of DNA or gene "chip." If then DNA from a test subject or patient is labeled with one fluorochrome (e.g., red), and DNA from a pool of control individuals is labeled with another color (e.g., green), and equal amounts are mixed and then hybridized to the probes on the glass slide, the relative amounts of red versus green signal from each probe will indicate relative dosage, that is, loss or gain of that probe sequence in the test subject. This is array CGH (Pinkel et al., 1998; Pollack et al, 1999; Veltman et al., 2002; Mantripragada, Buckley, et al., 2004) (Fig. 7–7).

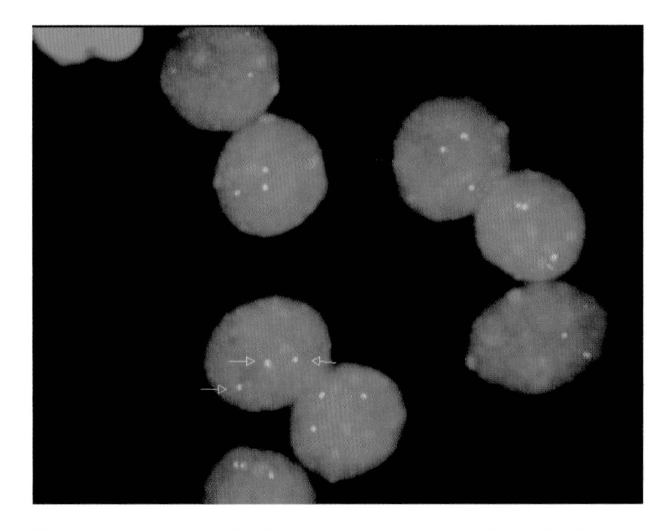

Figure 7–4 Interphase FISH. For the rapid prenatal diagnosis of common aneuploidies, FISH can be performed on cells in interphase obtained at amniocentesis, rather than having to spend time culturing them with arrest at metaphase. A probe to the chromosome of interest is hybridized to interphase nuclei, together with a counterstain, blue in this instance. In this example, the probe is to chromosome 21, and all the nuclei contain three red signals, establishing a diagnosis of trisomy 21 (Down syndrome). It is necessary to count a sufficient number of nuclei to make the analysis statistically valid. Unlike conventional cytogenetic analysis, this technique only indicates the number of copies of the probe region, which does not necessarily equate to the number of copies of whole chromosomes. (Courtesy of Dr. Peter Thompson, Cytogenetics Laboratory, Institute of Medical Genetics, Cardiff, UK.)

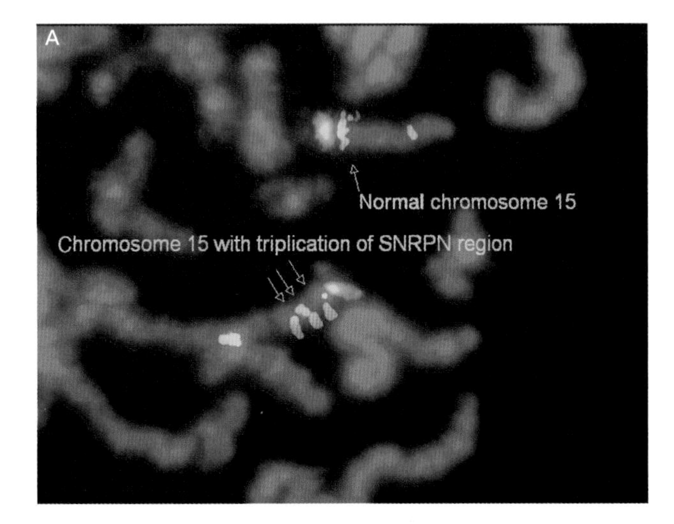

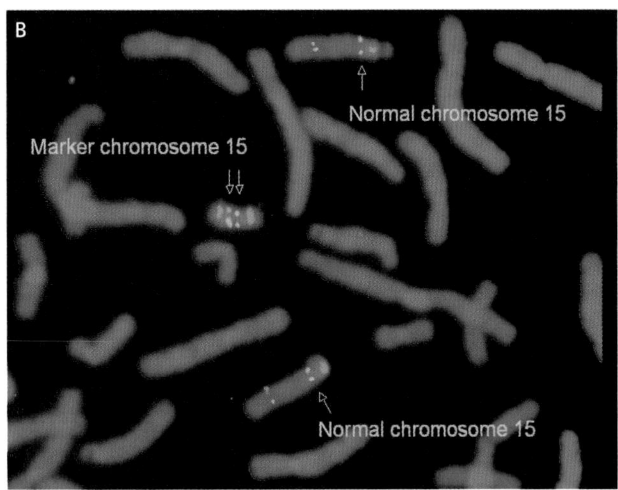

Figure 7–5 Metaphase FISH: position and dosage. (*A*) A red control probe toward the distal end of the long arm of chromosome 15 has been used in conjunction with a green control probe to the centromere, and a red probe to the *SNRPN* region (15q11-q13), which is located nearer to the centromere. On the normal chromosome 15, there is only a single copy of each probe, but on the abnormal chromosome there are three copies of the *SNRPN* probe. Marker chromosomes are small extra chromosomes that pose a particular diagnostic problem. If they are composed of heterochromatin, they are unlikely to be deleterious, but if they contain euchromatin with important transcriptionally active genes, they may well be pathogenic. Marker chromosomes are usually maternal in origin and often derived from chromosome 15. In this case, therefore, a red control probe to the distal end of the long arm of chromosome 15 (15q) has again been used in conjunction with a green control probe to the centromere and a red probe to the *SNRPN* region (15q11-q13) (*B*). The two normal chromosomes 15 are clearly shown, as well as the presence on the marker chromosome of two copies each of the green chromosome 15 centromeric probe and the red *SNRPN* probe, confirming its origin. Thus, this marker can be diagnosed as containing definitely significant euchromatin and is therefore highly likely to be pathogenic. Such a result helps considerably in genetic counseling. Prader–Willi and Angelman syndromes are classically associated with deletion of the *SNRPN* region on the paternally or maternally inherited chromosome 15 respectively, but quadruplication of *SNRPN*, as in these two cases, is associated with a phenotype of growth retardation including microcephaly, profound learning difficulties with behavioral problems, severe epilepsy refractory to treatment, plus tremor and ataxia. (Courtesy of Dr. Peter Thompson, Cytogenetics Laboratory, Institute of Medical Genetics, Cardiff, UK.)

To start with, up to a few thousand probes have been used in array CGH. Choosing every tenth BAC clone covering the genome, such that every 1 Mb or so a region of approximately 100 kb is probed requires a set of approximately 3000 BACs, that is, a 3K CGH array. More recently, it has become possible to produce arrays of 32,000 clones almost completely covering the whole genome: 32K arrays. While these clearly have the ability to detect smaller changes, it remains to be seen whether this is, in fact, useful in the clinical setting. Having obtained a result from an array CGH experiment that suggests gain or loss of a particular probe region, it is then necessary to carry out further experiments to determine if this is the case and whether there has been some other complication, such as a rearrangement. These extra confirmatory experiments can take a number of forms, from for example, one or more FISH experiments on metaphase spreads, to a Southern blot (see below), or some other molecular dosage technique. Indeed, array CGH is revealing that some apparently balanced translocations do, in fact, harbor gains or losses of material.

In the clinical setting, array CGH is looking useful in those circumstances where significant abnormalities are present in the patient, such as learning difficulties and dysmorphic feature, but no one feature, or combination of features, is diagnostic of a particular syndrome. Thus, the clinician cannot suggest the laboratory to carry out a specific FISH test, for example, a particular microdeletion syndrome, and the laboratory can do no more specific test than a karyotype analysis by G-banding, plus perhaps a screen of subtelomeric regions for deletions. However, armed with an array of 3000 or even 32,000 FISH probes in one array, it does become feasible to carry out in effect a genome-wide FISH experiment (Shaw-Smith et al., 2004). Many of those working in the field are pooling their data in the DatabasE of Chromosomal Imbalance and Phenotype in Humans using Ensembl Resources (DECIPHER; at http://www.sanger.ac.uk/PostGenomics/decipher/). There are two main reasons for this: array CGH studies are revealing the complexity of variation within the normal human genome, and the subtleties and complexity of clinical features constituting the phenotype of individual patients necessitates the most careful data collection if comparisons are to be meaningful (Buckley et al., 2005; de Bustos et al., 2006). There are understandable concerns about interpreting possible abnormality in the face of so much normal variation, but in principle the problem is not different from the interpretation of point mutations in genes, something, which molecular geneticists do not underestimate, but have been getting to grips with for some time. This author, at least, is encouraged that a problem involving a similar large amount of data processing resulted over 60 years ago in the development of the machine he is using to produce this manuscript, the electronic digital computer (Copeland, 2006). It seems unlikely to be beyond the wit of man to solve the problem of array CGH data interpretation.

Powerful approaches using a combination of techniques, such as DNA microarrays and chromosome sorting, are showing that positional data can be obtained about aberrant chromosomes, as well as dosage (Gribble et al., 2004). Cross species experiments are revealing details of primate genome evolution (Locke et al., 2003). Given the delays inherent in producing a karyotype analysis, because of the cell culturing required, prenatal testing has always been problematic, however, array CGH opens up the possibility that rapid testing, currently carried out by FISH or polymerase chain reaction (PCR) for only the most common of defects, may soon be carried out by array CGH, which if it was would have the additional benefit of the comprehensive detection of microdeletions (Le Caignec et al., 2005; reviewed by Rickman et al., 2005). One of the drawbacks of array CGH up to now has been that it requires large amounts of DNA, however, in an exciting recent advance, it has been shown that robust results can be obtained from single cells, with the use of whole genome amplification by PCR using degenerate oligonucleotide primers (Le Caignec et al., 2006).

Targeted Arrays

There are a number of forms of targeted arrays. One is simply an array of BAC clone probes, of the type used in whole genome arrays, but restricted to a particular region of interest in the genome (Bruder et al., 2001; Buckley et al., 2002; Locke et al., 2004; Sharp et al., 2005). To achieve higher density coverage in more detail, cosmid probes have been used (Mantripragada et al., 2003). One potential clinical application of this principle is in an array designed for the purposes of prenatal diagnosis of common chromosomal abnormalities, from trisomy 21 to various microdeletion syndromes and requiring only 600 BAC clones, rather than 3000 or more (Rickman et al., 2006). This would have the effect of reducing cost, because less clones are required, but also because several replicates of the array can be printed on one slide.

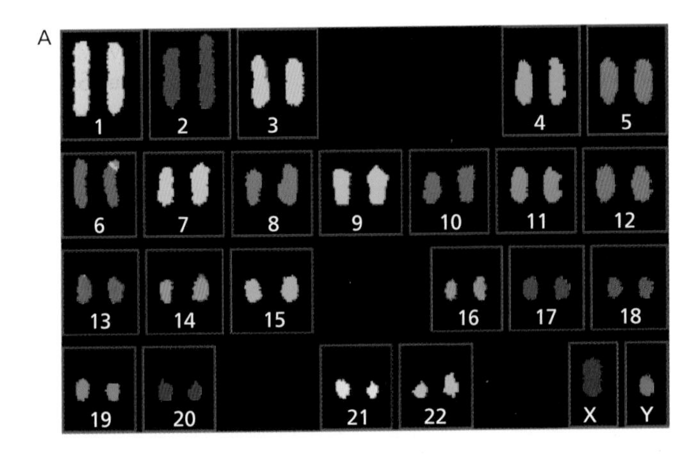

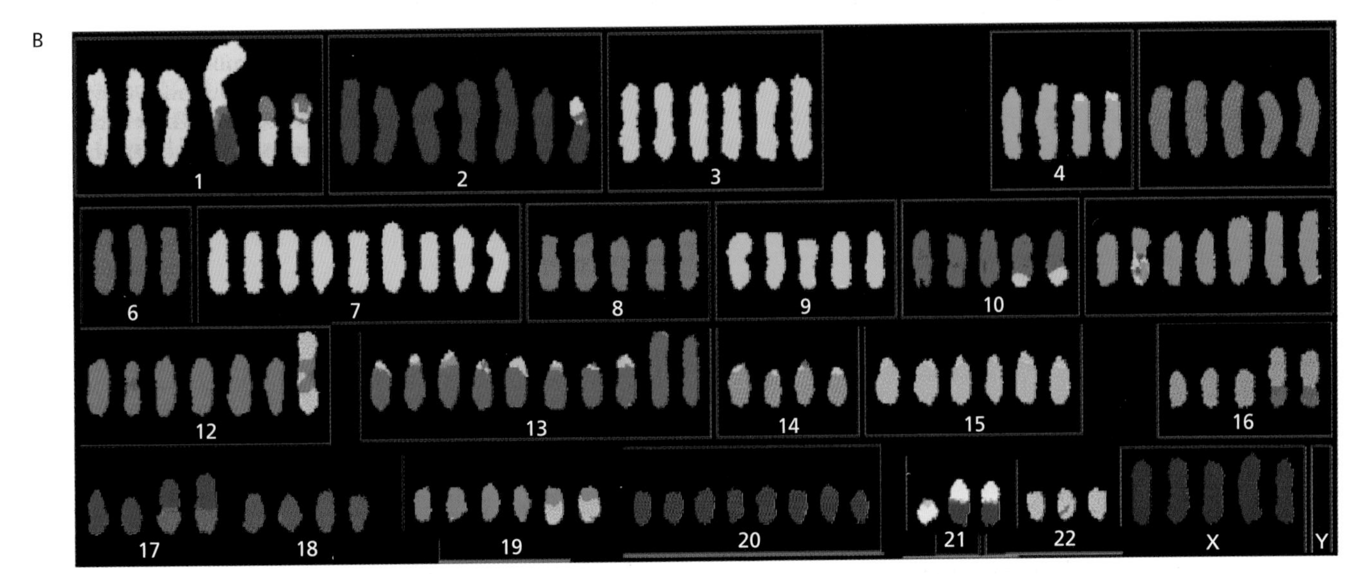

Figure 7–6 Spectral karyotyping (SKY) is a way of analyzing complex and various cytogenetic abnormalities, such as seen in malignant tumors. Chromosome-specific probes are dyed with specific colors, and a mix of probes covering all chromosomes is then hybridized with a sample. (*A*) This SKY picture of a colorectal cancer cell line (DLD-1) shows that the karyotype is essentially diploid, with a minimum of visible changes, for example, a 6:11 translocation. This cell line, however, exhibits microsatellite instability, a hallmark of loss of DNA mismatch repair, and tumors that arise along this pathway characteristically have minimal chromosomal abnormalities and are typically "near diploid." Although they do not have gross chromosomal abnormalities, such tumors have numerous small and point mutations all over the genome, invisible at this scale. (Courtesy of George Poulogiannis, Department of Pathology, University of Cambridge, England.) (*B*) This SKY picture of colorectal cancer cell line C70 shows its underlying karyotype to be triploid, but with numerous gains, losses, and translocations. Determining this by conventional cytogenetic analysis would be extremely difficult, if not impossible. The variation seen here in relative numbers of chromosomes shows that the concept of "loss of heterozygosity" is not necessarily simple. (Courtesy of George Poulogiannis, Department of Pathology, University of Cambridge, England.) (*C*) SKY analysis of two clones derived from the colorectal cancer cell line SW480 reveals differences in both relative copy number of chromosomes, as well as some structural abnormalities, for example chromosomes 2, 8, and 9. Again, this highlights potential issues with assumptions of clonality and techniques that average out relative copy number of chromosomes. (Courtesy of George Poulogiannis, Department of Pathology, University of Cambridge, England.)

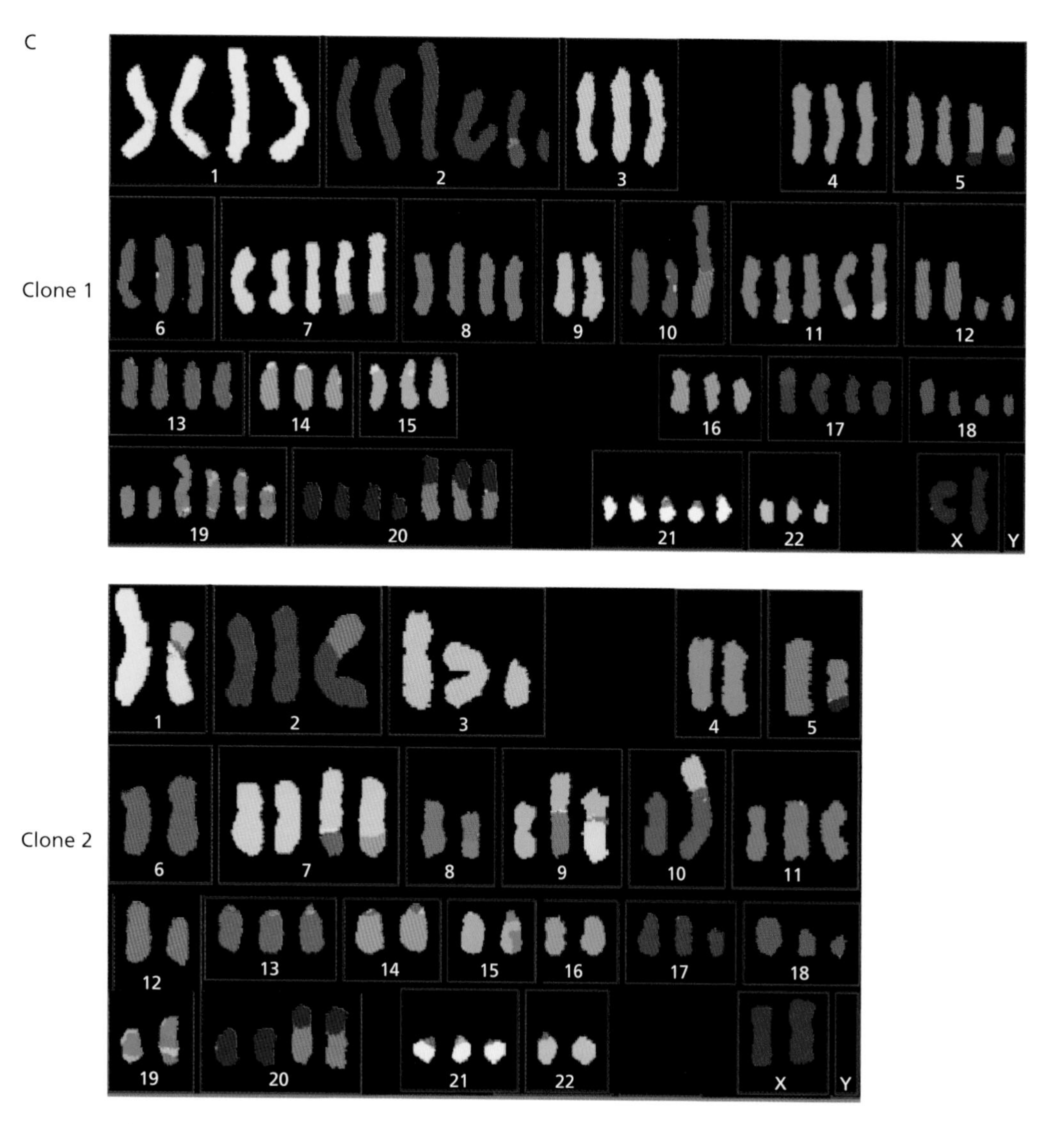

Figure 7–6 *(continued)*

Another form of targeted array involves the use of smaller probes of either PCR product size (approximately 200–2000 bp), or even oligonucleotide arrays, allowing resolution down to the level of individual exons (Mantripragada, Tapia-Paez, et al., 2004; Ren et al., 2005; Selzer et al., 2005; Mantripragada et al., 2006; Urban et al., 2006) (Fig. 7–8). Careful attention has to be taken that the PCR probes used in such arrays are derived from nonrepetitive regions of the genome, but the resolution offered by such arrays is impressive (Fig. 7–9).

Oligonucleotide arrays for the determination of single nucleotide polymorphisms (SNPs) can be used to determine copy number changes in the genome in, for example, cancer cells. Where a series of contiguous SNPs become apparently homozygous, that is, there is loss of heterozygosity, then it can be assumed that a deletion of one allele has occurred, and if no signal should be obtained then a homozygous deletion can be assumed (Lindblad-Toh et al., 2000; Wang et al., 2006; reviewed by Zhou et al., 2005).

Southern Analysis

The gel transfer method, known the world over as "Southern analysis," or more colloquially "Southern blotting," is with PCR one of the best known techniques in molecular biology. It has the unique ability to give data simultaneously on position, orientation, dosage, and even methylation of sections of the genome from approximately 100 bp up to several Mb (Southern, 1975, 2000, 2005).

The method depends on the combination of a number of techniques. Firstly, DNA is fragmented in a controlled fashion using a restriction endonuclease, which will cut the DNA at defined sites. Next, the digested DNA fragments are separated by size using gel electrophoresis, and the DNA then transferred to membrane capable of binding the DNA and immobilizing it, usually nitrocellulose or nylon, to produce a "blot." The size fractionated DNA on the membrane is then hybridized with a labeled probe DNA (usually radiolabeled), which is then put up against radiosensitive photographic film, revealing where on the membrane there are target sequences, which match the probe sequence. Hence, the actual and relative sizes of the fragments can be determined. The principle is sound and robust, but the beauty lies in the variations that can be used to address specific questions.

As there are hundreds of different restriction endonucleases ("restriction enzymes"), recognizing distinct, typically 6–8 bp

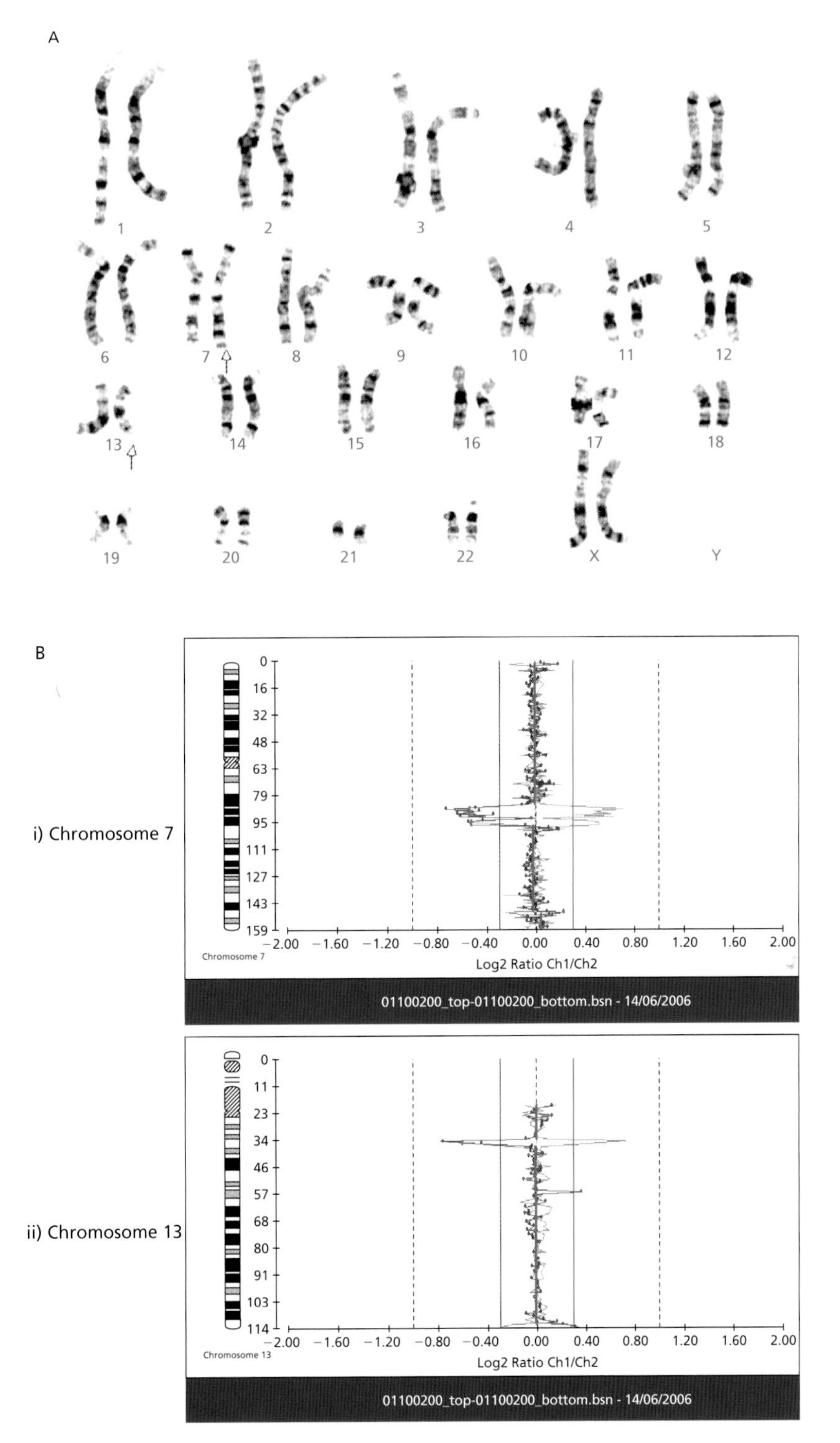

Figure 7–7 **Case A**: A 2-year old female with developmental delay, short stature, and microcephaly, otherwise not dysmorphic. Normal MRI brain scan. Found to have an apparently balanced translocation between chromosomes 7 and 13 (46,XX,t)(7;13)(q21.2;q12.3)de novo), as shown by the two arrows on the karyotype (*A*). Subsequent array CGH analysis, using a whole genome array of 0.1 Mb BAC clones spaced approximately 1 Mb apart, has shown that in the regions of the breakpoints on chromosomes 7 and 13 that there has actually been loss of DNA [indicated in red on the diagram of the chromosomes; (*B*)] and the translocation is therefore not balanced. Loss of 0.2 Mb of chromosome 13 and 8.42 Mb of chromosome 7 is consistent with the phenotype and its severity. (Courtesy of Siàn Morgan, Cytogenetics Laboratory, Institute of Medical Genetics, Cardiff, UK.)
Case B: A 3-year-old male, with developmental delay, poor speech, positional talipes, a left squint, atrial septal defect, and short stature (2nd percentile), but head circumference on the 50th percentile. Found by conventional cytogenetic analysis to have an apparently balanced translocation between chromosomes 6 and 9 (46,XY,t)(6;9)(q16.2;p22-24) de novo), as shown by the arrows. (*C*). Subsequent array CGH analysis, again using a whole genome array of 0.1 Mb BAC clones spaced approximately 1 Mb apart, has shown that there is no loss of DNA in the regions of the breakpoints on chromosomes 6 and 9. However, it reveals that there has been loss of 8.68 Mb of DNA from chromosome 4 [indicated in red on the diagram of the chromosomes; (*D*)]. The breakpoints of the chromosome 4 deletion were determined to be at 4q32.2 and 4q34.2 and in retrospect are visible (*E*). The loss of 8.68 Mb of DNA from chromosome 4 is highly likely to be the cause of the patient's phenotype, and the translocation is therefore merely an incidental finding. (Courtesy of Siàn Morgan, Cytogenetics Laboratory, Institute of Medical Genetics, Cardiff, UK.)

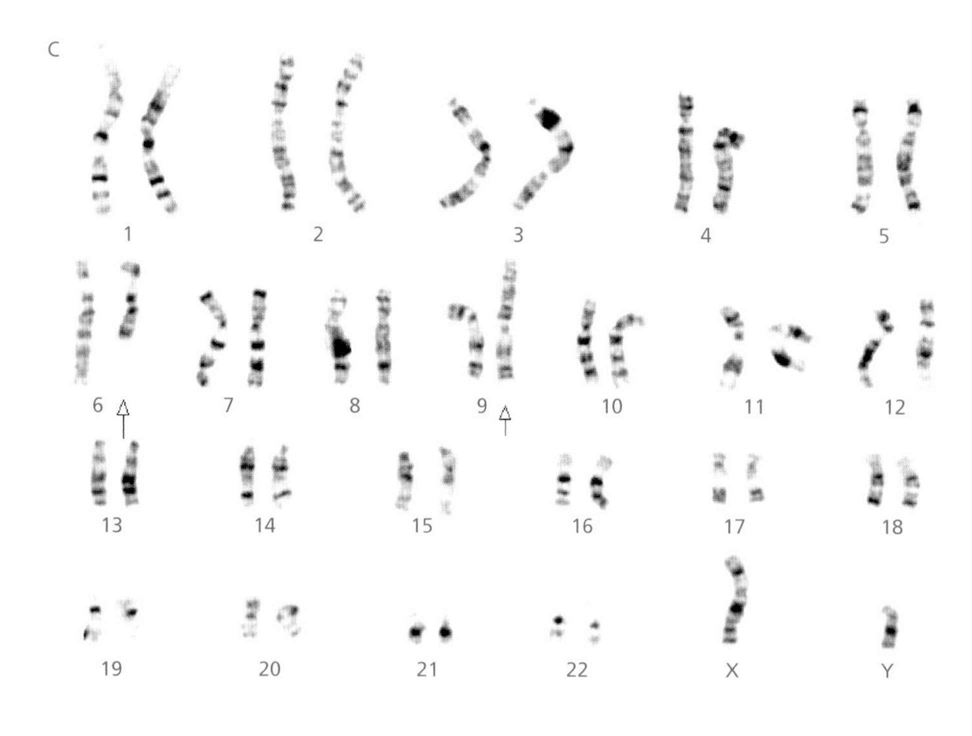

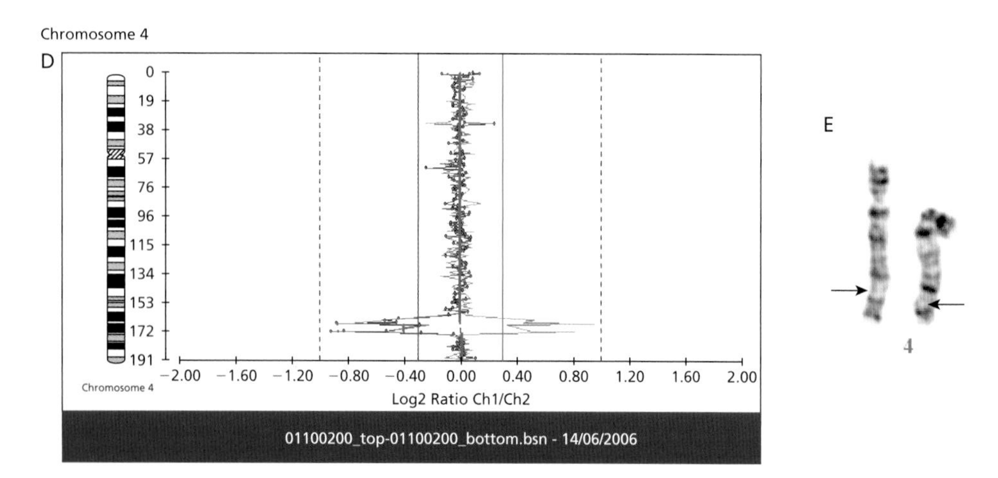

Figure 7–7 *(continued)*

sequence motifs, so information can be gleaned about hundreds of different sites in the DNA. Combinations of restriction endonucleases can be used, and the information on fragment sizes can be used to construct a genomic map of a region. Restriction endonucleases that cut at commonly occurring sites will tend to produce small fragments, while "rare cutters" will produce large fragments useful for longer-range mapping. As Southern found in his original studies, if a restriction endonuclease cutting within a repetitive sequence should be used then information about such regions is gained, such as the size of the repeats. (Southern, 2005). Some restriction endonucleases are specific for methylated or unmethylated motifs, so the methylation status of a stretch of DNA can be determined, potentially down to the level of individual restriction sites (Marcaud et al., 1981; Wolf and Migeon, 1982; Gronostajski et al., 1985). Use of a combination of methylation sensitive and insensitive restriction endonucleases is employed in the diagnosis of Fragile X(A) syndrome, where the presence is sought of an expansion at the *FMR1* locus, which is also methylated (reviewed by Tsongalis and Silverman, 1993; Oostra and Willemsen, 2001). The Southern analysis also gives dosage information (one allele in males but two in females), and often shows somatic heterogeneity/mosaicism in the sizes of pathological expansions, which is manifest as a smear on the blot (Fig. 7–10A).

If a particular restriction site should be polymorphic, then Southern analysis can be used to determine the frequency of the polymorphism, but perhaps more importantly these restriction fragment length polymorphisms (RFLPs) can be used in linkage experiments to locate genes (Goodfellow et al., 1975; Solomon et al., 1976; Jeffreys, 1979). Loss of heterozygosity for an RFLP can and has been used to locate tumor suppressor genes (Solomon et al., 1987; Ward et al., 1993).

It is possible to use different probes, and this can even be done on the same blot as probes can be stripped off under appropriate conditions, which denature the target DNA/probe hybrid. This will give information, for example, on probe sites being located together, or not, on the same restricted fragment of DNA. Cross-species

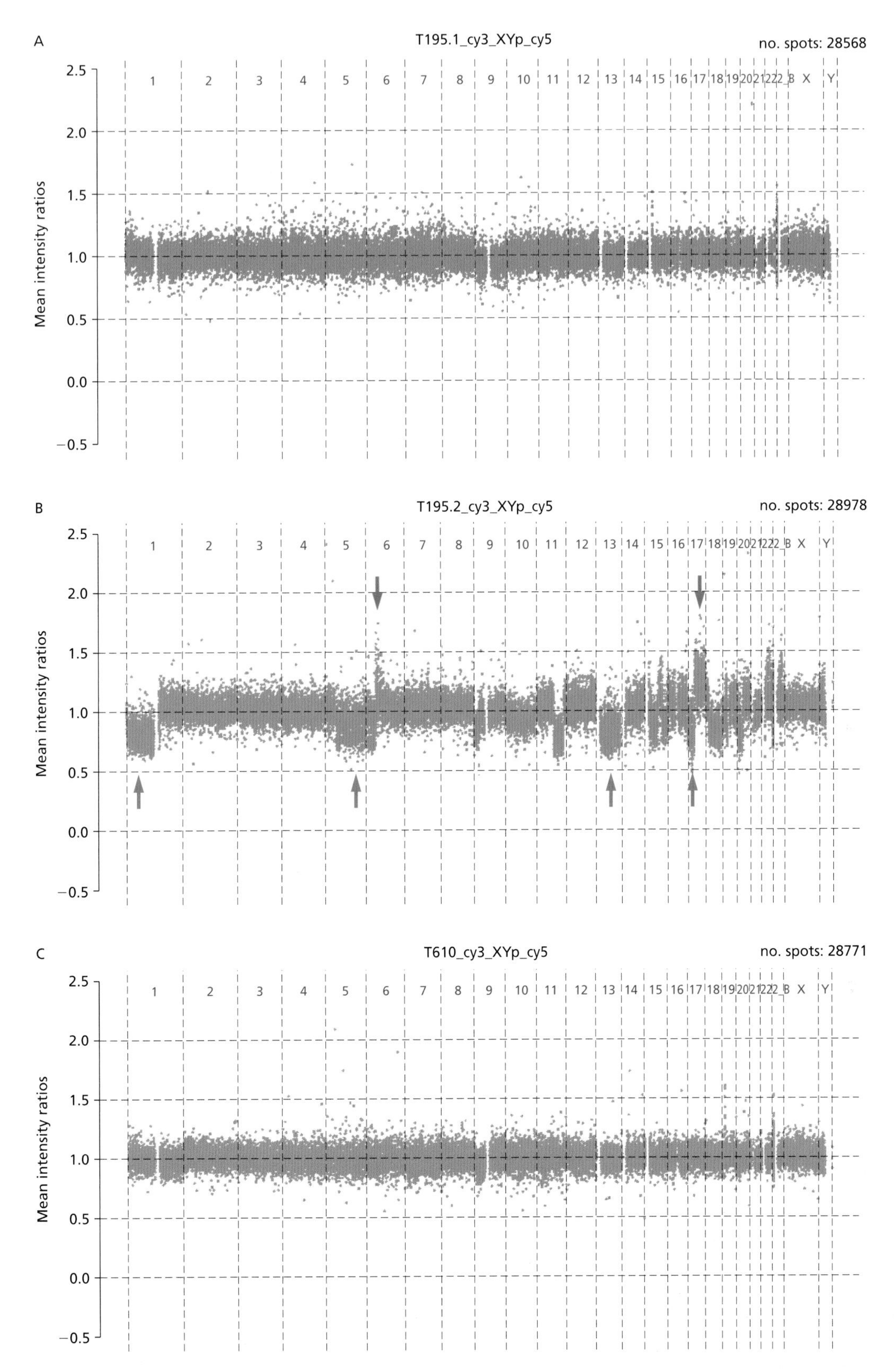

Figure 7–8 Arrays of 32,000 probes, effectively covering almost all of the genome, can be used to study the complex abnormalities that occur in the genome of tumors. (*A*) 32K array CGH analysis of a neurofibroma. (*B*) 32K array CGH analysis of a MPNST. (*C*) 32K array CGH analysis of a neurofibroma misclassified as a MPNST. Patients with neurofibromatosis type 1 typically develop numerous benign tumors called neurofibromas. Occasionally, these can undergo transformation into malignant peripheral nerve sheath tumors (MPNSTs), but the underlying genetic changes responsible for this are unknown. Analysis of a neurofibroma from a particular patient (*A*), shows that there are no obvious regions of consistent gain (an intensity ratio >1.0) or loss (an intensity ratio <1.0) in the genome. In contrast, analysis of a MPNST from the same patient (*B*) reveals a quite distinct pattern with numerous regions of gain, such as on chromosomes 6 and 17 (green arrows), and loss, such as on chromosomes 1, 5, 13, and 17 (red arrows). The result from analysis of a MPNST from another patient, however, looked more like that from a neurofibroma (*C*). On histological review, the tumor was reclassified as a benign neurofibroma. (Courtesy of Dr. Kiran Mantripragada, National Health Service R&D Laboratory, Institute of Medical Genetics, Cardiff, UK.)

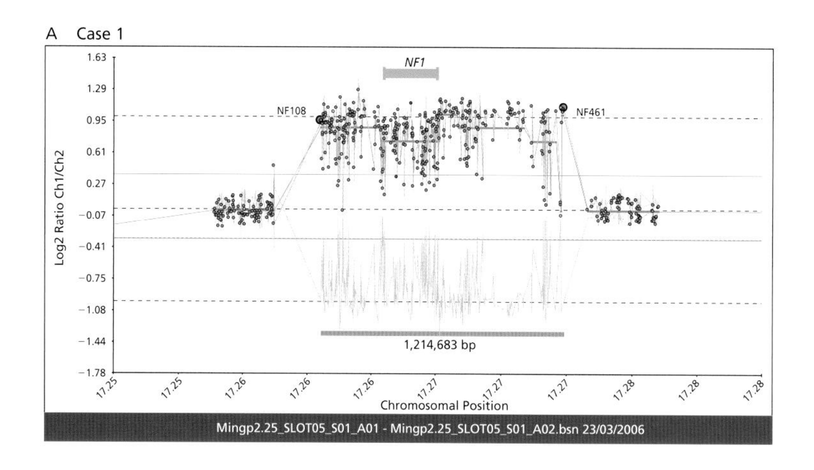

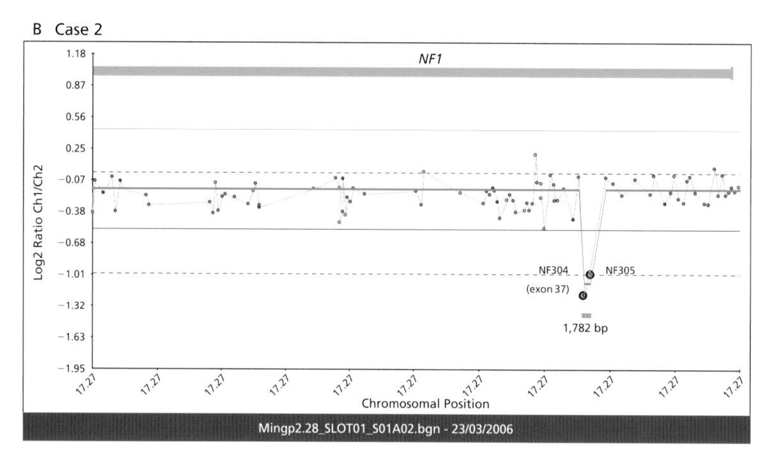

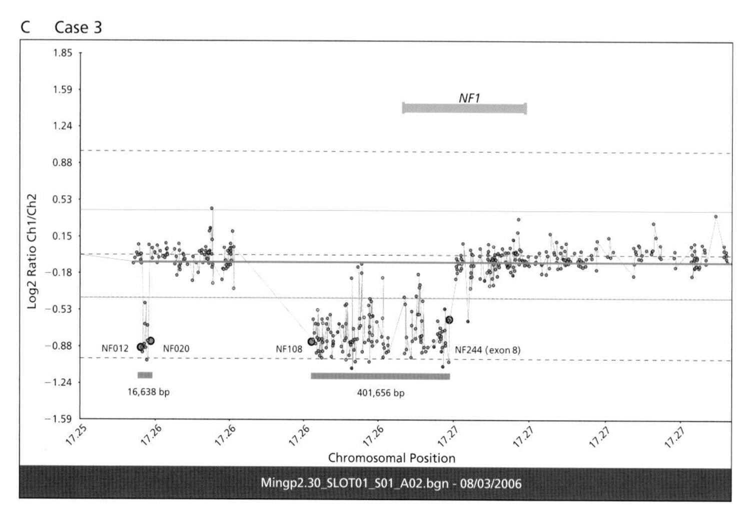

Figure 7–9 Targeted arrays can be used to analyse specific regions of the genome in great detail. Mutations in the *NF1* gene on chromosome 17 can lead to neurofibromatosis type 1. These mutations can take many forms and range in size from a single bp (point mutations) up to 1–2 Mb, or larger. Detecting and defining the larger mutations can be difficult as the *NF1* gene lies in a region of the genome containing many repeated and repetitive sequences. By constructing a targeted array of 515 probes, between 200 and 999 bp in size and generated by PCR, the whole 2 Mb region around *NF1* can be studied. Test (patient) DNA and normal DNA are labeled with different fluorescent dyes, mixed together, and then hybridized to the PCR probes immobilized as spots on the array. The positions of the probes along the chromosome are shown on the *x*-axis, with the position of the coding region of the *NF1* gene itself shown by the light blue bar (probes 209–328). The relative amounts of fluorescence at each probe are measured and hence the relative copy number of the probes determined in the test sample (*y*-axis). In Case 1 (*A*), probes 108 and 461 (shown by large red dots), and all those in between, have been deleted, encompassing a region of at least 1,214,683 bp (shown by the thick red line). The green line shows the results when test DNA is labeled green and normal DNA red, the thin red line shows the results of a "dye swop experiment" when the test and normal DNAs are labeled the other way round, this provides extra confidence in the result. In Case 2 (*B*), only two probes have been deleted, 304 (in exon 37) and 305, representing a mere 1782 bp. In Case 3 (*C*), a more complicated picture is revealed. The patient has two regions deleted, between probes 12–20 and 108–244 (in exon 8). What an array cannot determine is the relative orientation of the remaining parts of the genome, it may well be that the retained region between probes 20 and 108 has become inverted. (Courtesy of Dr. Ming Hong Shen, National Health Service R&D Laboratory, Institute of Medical Genetics, Cardiff, UK.)

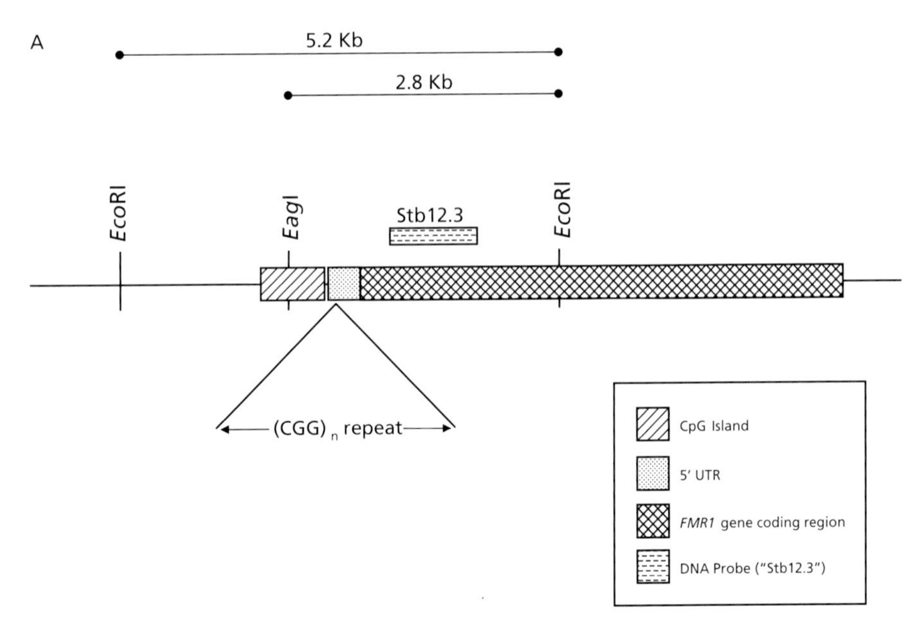

Figure 7–10 Southern analysis. *Fragile X testing.* Fragile X syndrome is caused by pathogenic expansions of a CGG repeat in the 5′ untranslated region (5′ UTR) of the *FMR1* gene, located on the X chromosome. It is so named because such expansions are liable to cause chromosome breakage at the resultant fragile site in cultured cells used for karyotype determination. Large expansions of the CGG repeat lead to methylation of it and the *FMR1* gene's promoter in the adjoining CpG island, causing down-regulation of the gene. Small expansions which do not cause methylation, but which may expand in subsequent generations are known as premutations. Methylation, and hence inactivation of the single copy of *FMR1* in males causes the syndrome, which includes mental retardation. In normal females, one copy of the X chromosome in each cell is anyway randomly and naturally inactivated by being methylated, so the condition is typically less manifest in them. Southern analysis is very useful in molecular diagnosis of the condition. A map of the region surrounding the *FMR1* locus is shown in *A*. Sites which can be cut by the restriction enzymes *Eco*RI and *Eag*I are shown, and it should be noted that the *Eag*I site lies within the CpG island. Because *Eag*I does not cut DNA that is methylated, then it is able to discriminate between methylated (inactive) and unmethylated (active) alleles, in both males and females. The approximate distances between the restriction sites is shown, as is the region of DNA which corresponds to the probe sequence, called Stb12.3. DNA digested by a combination of *Eco*RI and *Eag*I is separated by electrophoresis and then transferred to a suitable membrane that binds the DNA fragments which have been sorted by size by the gel. The probe is then radioactively labeled and hybridized with the membrane, in the process binding to unmethylated fragments (2.8 kb; *Eag*I–*Eco*RI), or methylated fragments (5.2 kb: *Eco*RI–*Eco*RI) and exposed to X-ray film. *B* shows the results obtained using a variety of controls and test samples. Lanes 1 and 14–16 show normal females, 50% of whose X chromosomes are randomly inactivated by methylation, hence two bands are seen, one of 2.8 kb plus another of 5.2 kb. Lanes 6, 9, 10, and 17 show normal males: a single unmethylated allele of 2.8 kb in size. Lane 2 shows a male with a pathological methylated expansion of more than 5.2 kb (somatic mosaicism of the unstable expanded region causes the smeared indistinct band). Lane 12 shows a male with an unmethylated, but expanded, allele, that is, a premutation. Lane 5 shows a female with a premutation: as the natural X chromosome inactivation methylates both the normal and premutation-carrying chromosomes, there are two bands seen at 2.8 and 5.2 kb, the slightly larger one in each case being derived from the premutation chromosome. (Courtesy of Dr Moira MacDonald, National Health Service Molecular Genetics Laboratory, Institute of Medical Genetics, Cardiff, UK.) *Testing for facio-scapulo-humeral muscular dystrophy (FSHD).* FSHD is a form of muscular dystrophy that affects muscles of the face and upper limb girdle. The precise underlying genetic cause remains obscure, but most cases are associated with the presence of a shortened array of repetitive DNA sequences, called *D4Z4*, toward the end of the long arm of chromosome 4. Because of the large sizes of the regions of DNA involved, sometimes more than 100 kb, a special technique called pulsed-field gel electrophoresis is necessary to give good separation of the fragments. A complication is that similar repeats are found on chromosome 10, but have nothing to do with FSHD. A map of the FSHD-associated region of chromosome 4 is shown in *C*, showing the probe region (p13-E11) between two *Eco*RI sites and next to the region of *D4Z4* repeats. Each *D4Z4* repeat is 3.3 kb in size. Normal individuals usually have at least 11 and sometimes up to approximately 90 *D4Z4* repeats on each chromosome 4, so the distance between the two *Eco*RI sites can vary between 34 and approximately 300 kb. Individuals with FSHD commonly have less than 11 repeats, thus having an *Eco*RI fragment less than 34 kb. Although there is a similar region on chromosome 10, fortunately type 10 repeats contain within them a site for the restriction enzyme *Bln*I, which distinguishes them from the true *D4Z4* repeats on chromosome 4: thus digestion with *Bln*I reduces the probe binding fragment to such a small size it does not appear. Note, however, that there is a *Bln*I site 3 kb inside the left hand of the chromosome 4 *Eco*RI fragment. Hence, in *D* in the left hand lane, which is genomic DNA digested with *Eco*RI, there are four bands at 120, 50, 29, and 20 kb, while in the right hand lane (DNA digested with both *Eco*RI and *Bln*I) there are only two bands at approximately 115 and 26 kb, the difference in size being the distance between the *Eco*RI and *Bln*I sites of approximately 3 kb. Thus, *D4Z4*-containing fragments can be identified (solid arrows), distinct from the chromosome 10-derived ones (hollow arrows). The presence in this individual of an *Eco*RI fragment derived from chromosome 4 and of more than 34 kb is consistent with their clinical diagnosis of FSHD. (Courtesy of Dr Gill Spurlock and Prof. Meena Upadhyaya, National Health Service R&D Laboratory, Institute of Medical Genetics, Cardiff, UK.)

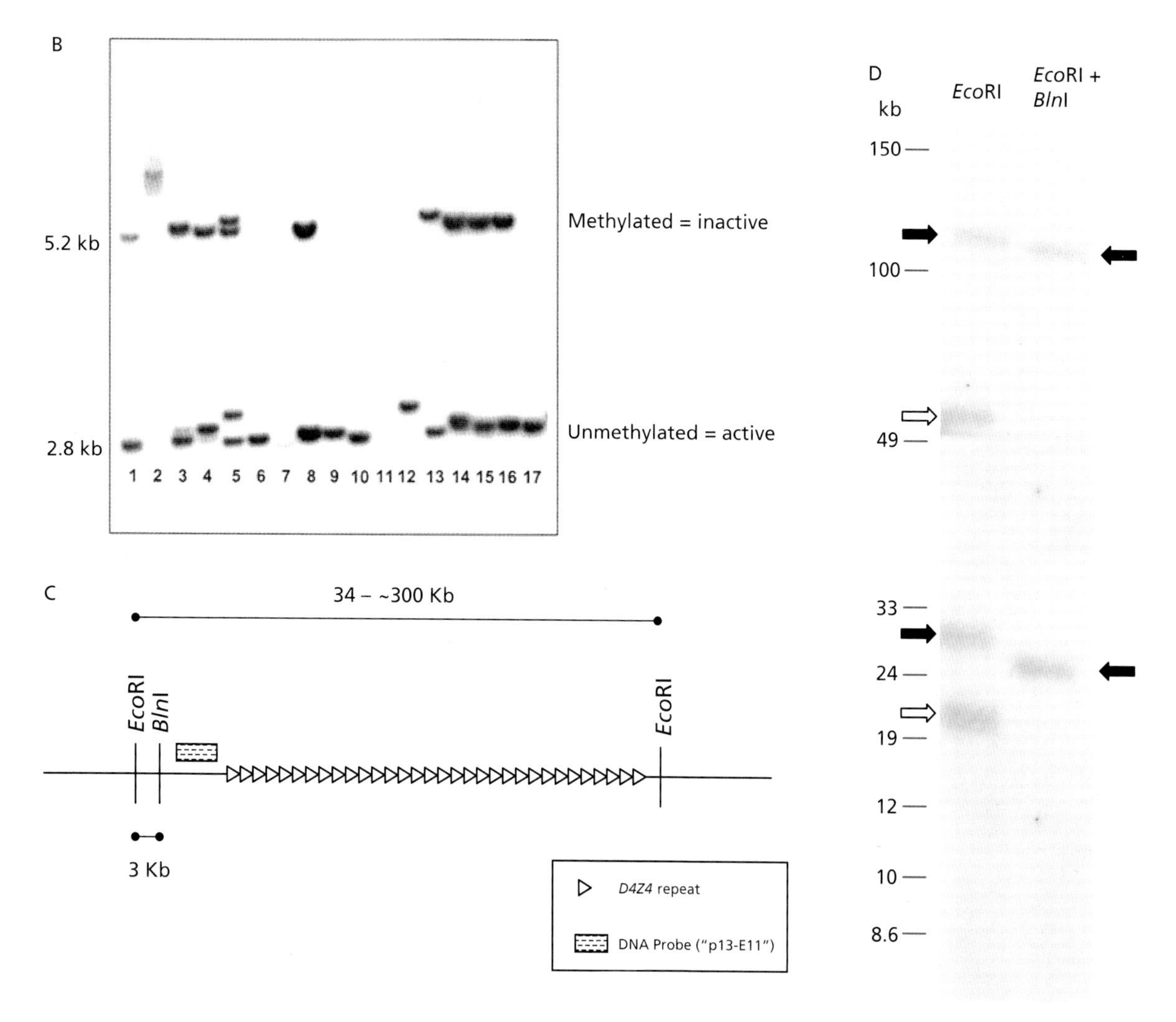

Figure 7–10 *(continued)*

hybridization experiments can also be carried out to find homologous or related genes.

Densitometric measurement of band intensities on the X-ray film, or actual counting of radioactivity on the membrane itself using, for example, a phosphorimager, can give dosage information: useful in studies of gene amplification, for example (Cowan et al., 1982; reviewed by Stark and Wahl, 1984). Given the wealth of information obtainable from a Southern blot, at one and the same time data can be obtained on copy number and rearrangements (Sabbir et al., 2006).

For a long time, one of the limitations of Southern analysis was that it could not be used to study DNA fragments any larger than approximately 50 kb in size, because that was the maximum size resolvable on conventional agarose gel electrophoresis. However, this has been overcome by an understanding of the physics of DNA migration within gels leading to the development of pulsed field gel electrophoresis, able to resolve DNA fragments of up to several Mb in size (Schwartz and Cantor, 1984; Southern et al., 1987). An ability to analyse such large fragments has been critical in helping to understand a number of conditions, for example facioscapulohumeral muscular dystrophy, by first mapping the region of interest and then facilitating laboratory diagnosis, and also fusion genes in malignancy (Min et al., 1990; Wijmenga et al., 1993; Galluzzi et al., 1999) (Fig. 7–10B).

Other Techniques: MLPA, MAPH, and Long-range PCR

As well as the techniques described above, there are a number of methods that can give genomic data. Advances in PCR now enable amplicons of up to 20 kb to be routinely generated, and with care even up to 30 kb. This has facilitated the study and identification of, for example, chromosomal breakpoints and fusion genes, as well as haplotype mapping (Waggott et al., 1995; Klockars et al., 1996; Michalatos-Beloin et al., 1996; Wu et al., 2005; reviewed by Eisenstein, 2006).

Notwithstanding the power of Southern blotting to detect deletions, duplications, and rearrangements, diagnostic laboratories have long sought a simpler and easier alternative to such mutations affecting exons. Quantitative PCR is one approach (Jung et al., 2000), but the technique of MLPA has really taken off as a robust and now increasingly routine option. It relies on first the ligation of oligonucleotides that hybridize specifically to the region/exon of interest: ligation is only possible if there is no mismatches at the adjoining ends of the primers. The primers have generic sequences at their opposing ends, which then enables a PCR amplification to produce an amplicon corresponding to the ligation product, and most importantly quantitatively. By varying the length of intervening sequences between the specific and generic parts of the primers, a whole series of reactions can be amplified in the same tube, thus allowing quantitative analysis of all exons in the *BRCA1* gene (Schouten et al., 2002; Hogervorst et al., 2003). While MLPA can confidently give information on exon dosage, it does not give data on how those exons are arranged. Another technique that gives similar data to MLPA is MAPH (Akrami et al., 2005; Patsalis et al, 2005).

References

Akrami, SM, Dunlop, MG, Farrington, SM, Frayling, IM, MacDonald, F, Harvey, JF, et al. (2005). Screening for exonic copy number mutations at MSH2 and MLH1 by MAPH. *Fam Cancer,* 4:145–149.

Barr, ML and Bertram, EG (1949). A morphological distinction between neurons of the male and female and the behavior of the nucleolar satellite during accelerated nucleoprotein synthesis. *Nature,* 163:676–677.

Bolzer, A, Kreth, G, Solovei, I, Koehler, D, Saracoglu, K, Fauth, C, et al. (2005). Three-dimensional maps of all chromosomes in human male fibroblast nuclei and prometaphase rosettes. *PLoS Biol,* 3:e157.

Bruder, CE, Hirvela, C, Tapia-Paez, I, Fransson, I, Segraves, R, Hamilton, G, et al. (2001). High resolution deletion analysis of constitutional DNA from neurofibromatosis type 2 (NF2) patients using microarray-CGH. *Hum Mol Genet,* 10:271–282.

Buckley, PG, Mantripragada, KK, Benetkiewicz, M, Tapia-Paez, I, Diaz De Stahl, T, Rosenquist, M, et al. (2002). A full-coverage, high-resolution human chromosome 22 genomic microarray for clinical and research applications. *Hum Mol Genet,* 11:3221–3229.

Buckley, PG, Mantripragada, KK, Piotrowski, A, Diaz de Stahl, T, and Dumanski, JP (2005). Copy-number polymorphisms: mining the tip of an iceberg. *Trends Genet,* 21:315–317.

Coleman, WB and Tsongalis, GJ (eds.) (2006). *Molecular Diagnostics for the Clinical Laboratarian,* 2nd edn. Totowa, New Jersey: Humana Press Inc.

Copeland, J (ed.) (2006). *Colossus: The Secrets of Bletchley Park's Code-breaking Computers.* UK: Oxford University Press. ISBN-10: 0-19-284055-X; ISBN-13: 978-0-19-284055-4.

Cowan, KH, Goldsmith, ME, Levine, RM, Aitken, SC, Douglass, E, Clendeninn, N, et al. (1982). Dihydrofolate reductase gene amplification and possible rearrangement in estrogen-responsive methotrexate-resistant human breast cancer cells. *J Biol Chem,* 257:15079–15086.

Craig, JM and Bickmore, WA (1993). Chromosome bands—flavours to savour. *Bioessays,* 15:349–354.

de Bustos, C, Diaz de Stahl, T, Piotrowski, A, Mantripragada, KK, Buckley, PG, Darai, E, et al. (2006). Analysis of copy number variation in the normal human population within a region containing complex segmental duplications on 22q11 using high-resolution array-CGH. *Genomics,* 88(2):152–162.

du Manoir, S, Speicher, MR, Joos, S, Schröck, E, Popp, S, Dohner, H, et al. (1993). Detection of complete and partial chromosome gains and losses by comparative genomic in situ hybridization. *Hum Genet,* 90:590–610.

Eisenstein, M (2006). Putting long-range mapping in reach. *Nat Methods,* 3:239.

Florijn, RJ, Bonden, LA, Vrolijk, H, Wiegant, J, Vaandrager, JW, Baas, F, et al. (1995). High-resolution DNA Fiber-FISH for genomic DNA mapping and colour bar-coding of large genes. *Hum Mol Genet,* 4:831–836.

Galluzzi, G, Deidda, G, Cacurri, S, Colantoni, L, Piazzo, N, Vigneti, E, et al. (1999). Molecular analysis of 4q35 rearrangements in fascioscapulohumeral muscular dystrophy (FSHD): application to family studies for a correct genetic advice and a reliable prenatal diagnosis of the disease. *Neuromuscul Disord,* 9:190–198.

Goodfellow, P, Jones, E, van Heyningen, V, Solomon, E, Kennett, R, Bobrow, M, et al. (1975). Linkage relationships of the HL-A system and beta2 microglobulin. *Cytogenet Cell Genet,* 14:332–337.

Gribble, SM, Fiegler, H, Burford, DC, Prigmore, E, Yang, F, Carr, P, et al. (2004). Applications of combined DNA microarray and chromosome sorting technologies. *Chromosome Res,* 12:35–43.

Gronostajski, RM, Adhya, S, Nagata, K, Guggenheimer, RA, and Hurwitz, J (1985). Site-specific DNA binding of nuclear factor I: analyses of cellular binding sites. *Mol Cell Biol,* 5:964–971.

Hemminki, A, Tomlinson, I, Markie, D, Jarvinen, H, Sistonen, P, Bjorkqvist, AM, et al. (1997). Localization of a susceptibility locus for Peutz-Jeghers syndrome to 19p using comparative genomic hybridization and targeted linkage analysis. *Nat Genet,* 15:87–90.

Hogervorst, FB, Nederlof, PM, Gille, JJ, McElgunn, CJ, Grippeling, M, Pruntel, R, et al. (2003). Large genomic deletions and duplications in the BRCA1 gene identified by a novel quantitative method. *Cancer Res,* 63:1449–1453.

Jeffreys, AJ (1979). DNA sequence variants in the G gamma-, A gamma-, delta- and beta-globin genes of man. *Cell,* 18:1–10.

Jones, KW (1970). Chromosomal and nuclear location of mouse satellite DNA in individual cells. *Nature,* 225:912–915.

Jung, R, Soondrum, K, and Neumaier, M (2000). Quantitative PCR. *Clin Chem Lab Med,* 38:833–836.

Karpova, MB, Schoumans, J, Blennow, E, Ernberg, I, Henter, JI, Smirnov, AF, et al. (2006). Combined spectral karyotyping, comparative genomic hybridization, and in vitro apoptyping of a panel of Burkitt's lymphoma-derived B cell lines reveals an unexpected complexity of chromosomal aberrations and a recurrence of specific abnormalities in chemoresistant cell lines. *Int J Oncol,* 28:605–617.

Klockars, T, Savukoski, M, Isosomppi, J, Laan, M, Jarvela, I, Petrukhin, K, et al. (1996). Efficient construction of a physical map by fiber-FISH of the CLN5 region: refined assignment and long-range contig covering the critical region on 13q22. *Genomics,* 35:71–78.

Kosak, ST and Groudine, M (2004). Form follows function: the genomic organization of cellular differentiation. *Genes Dev,* 18:1371–1384.

Le Caignec, C, Boceno, M, Saugier-Veber, P, Jacquemont, S, Joubert, M, David, A, et al. (2005). Detection of genomic imbalances by array based comparative genomic hybridisation in fetuses with multiple malformations. *J Med Genet,* 42:121–128.

Le Caignec, C, Spits, C, Sermon, K, De Rycke, M, Thienpont, B, Debrock, S, et al. (2006). Single-cell chromosomal imbalances detection by array CGH. *Nucleic Acids Res,* 34:e68.

Lindblad-Toh, K, Tanenbaum, DM, Daly, MJ, Winchester, E, Lui, WO, Villapakkam, A, et al. (2000). Loss-of-heterozygosity analysis of small-cell lung carcinomas using single-nucleotide polymorphism arrays. *Nat Biotechnol,* 18:1001–1005.

Locke, DP, Segraves, R, Carbone, L, Archidiacono, N, Albertson, DG, Pinkel, D, et al. (2003). Large-scale variation among human and great ape genomes determined by array comparative genomic hybridization. *Genome Res,* 13:347–357.

Locke, DP, Segraves, R, Nicholls, RD, Schwartz, S, Pinkel, D, Albertson, DG, et al. (2004). BAC microarray analysis of 15q11-q13 rearrangements and the impact of segmental duplications. *J Med Genet,* 41:175–182.

Lyon, MF (1961). Gene action in the X-chromosome of the mouse (*Mus musculus*). *Nature,* 190:372–373.

Lyon, MF (1963). Lyonisation of the X-chromosome. *Lancet,* 2:1120–1121.

Lyon, MF (1999). X-chromosome inactivation. *Curr Biol,* 9:R235–R237.

Mantripragada, KK, Buckley, PG, Benetkiewicz, M, De Bustos, C, Hirvela, C, Jarbo, C, et al. (2003). High-resolution profiling of an 11 Mb segment of human chromosome 22 in sporadic schwannoma using array-CGH. *Int J Oncol,* 22:615–622.

Mantripragada, KK, Buckley, PG, de Stahl, TD, and Dumanski, JP (2004). Genomic microarrays in the spotlight. *Trends Genet,* 20:87–94.

Mantripragada, KK, Tapia-Paez, I, Blennow, E, Nilsson, P, Wedell, A, and Dumanski, JP (2004). DNA copy-number analysis of the 22q11 deletion-syndrome region using array-CGH with genomic and PCR-based targets. *Int J Mol Med,* 13:273–279.

Mantripragada, KK, Thuresson, AC, Piotrowski, A, Diaz de Stahl, T, Menzel, U, Grigelionis, G, et al. (2006). Identification of novel deletion breakpoints bordered by segmental duplications in the NF1 locus using high resolution array-CGH. *J Med Genet,* 43:28–38.

Marcaud, L, Reynaud, CA, Therwath, A, and Scherrer, K (1981). Modification of the methylation pattern in the vicinity of the chicken globin genes in avian erythroblastosis virus transformed cells. *Nucleic Acids Res,* 9:1841–1851.

Masny, PS, Bengtsson, U, Chung, SA, Martin, JH, van Engelen, B, van der Maarel, SM, et al. (2004). Localization of 4q35.2 to the nuclear periphery: is FSHD a nuclear envelope disease? *Hum Mol Genet,* 13:1857–1871.

Michalatos-Beloin, S, Tishkoff, SA, Bentley, KL, Kidd, KK, and Ruano, G (1996). Molecular haplotyping of genetic markers 10 kb apart by allele-specific long-range PCR. *Nucleic Acids Res,* 24:4841–4843.

Min, GL, Martiat, P, Pu, GA, and Goldman, J (1990). Use of pulsed field gel electrophoresis to characterize BCR gene involvement in CML patients lacking M-BCR rearrangement. *Leukemia,* 4:650–656.

Mitelman, F (ed.) (1995). *An International System for Human Cytogenetic Nomenclature (ISCN).* Basel: Karger.

Mundle, SD and Koska, RJ (2006). Fluorescence in situ hybridization: A major milestone in luminous cytogenetics. In WB Coleman and GJ Tsongalis (eds.), *Molecular Diagnostics for the Clinical Laboratarian,* 2nd edn. (pp. 189–202). Totowa, New Jersey: Humana Press Inc.

Murthy, SK and Demetrick, DJ (2006). New approaches to fluorescence in situ hybridization. *Methods Mol Biol,* 319:237–259.

Oostra, BA and Willemsen, R (2001). Diagnostic tests for fragile X syndrome. *Expert Rev Mol Diagn,* 1:226–232.

Parada, LA, McQueen, PG, and Misteli, T (2004). Tissue-specific spatial organization of genomes. *Genome Biol,* 5:R44.

Patsalis, PC, Kousoulidou, L, Sismani, C, Mannik, K, and Kurg, A (2005). MAPH: from gels to microarrays. *Eur J Med Genet,* 48:241–249.

Pederson, T (2004). The spatial organization of the genome in mammalian cells. *Curr Opin Genet Dev,* 14:203–209.

Pinkel, D and Albertson, DG (2005). Comparative genomic hybridization. *Annu Rev Genomics Hum Genet,* 6:331–354.

Pinkel, D, Segraves, R, Sudar, D, Clark, S, Poole, I, Kowbel, D, et al. (1998). High resolution analysis of DNA copy number variation using comparative genomic hybridization to microarrays. *Nat Genet,* 20:207–211.

Pinkel, D, Straume, T, and Gray, JW (1986). Cytogenetic analysis using quantitative, high-sensitivity, fluorescence hybridization. *Proc Natl Acad Sci USA,* 83:2934–2938.

Pollack, JR, Perou, CM, Alizadeh, AA, Eisen, MB, Pergamenschikov, A, Williams, CF, et al. (1999). Genome-wide analysis of DNA copy-number changes using cDNA microarrays. *Nat Genet,* 23:41–46.

Ren, H, Francis, W, Boys, A, Chueh, AC, Wong, N, La, P, et al. (2005). BAC-based PCR fragment microarray: high-resolution detection of chromosomal deletion and duplication breakpoints. *Hum Mutat,* 25:476–482.

Rickman, L, Fiegler, H, Carter, NP, and Bobrow, M (2005). Prenatal diagnosis by array-CGH. *Eur J Med Genet,* 48:232–240.

Rickman, L, Fiegler, H, Shaw-Smith, C, Nash, R, Cirigliano, V, Voglino, G, et al. (2006). Prenatal detection of unbalanced chromosomal rearrangements by array CGH. *J Med Genet,* 43:353–361.

Rijkers, T, Deidda, G, van Koningsbruggen, S, van Geel, M, Lemmers, RJ, van Deutekom, JC, et al. (2004). FRG2, an FSHD candidate gene, is transcriptionally upregulated in differentiating primary myoblast cultures of FSHD patients. *J Med Genet,* 41:826–836.

Sabbir, MG, Dasgupta, S, Roy, A, Bhoumik, A, Dam, A, Roychoudhury, S, et al. (2006). Genetic alterations (amplification and rearrangement) of D-type cyclins loci in head and neck squamous cell carcinoma of Indian patients: prognostic significance and clinical implications. *Diagn Mol Pathol,* 15:7–16.

Sambrook, JF and Russell, DW (eds.) (2000). *Molecular Cloning: A Laboratory Manual,* 3rd edn. Cold Spring Harbor, NY: Cold Spring Harbor Laboratory Press (and at www.MolecularCloning.com).

Schouten, JP, McElgunn, CJ, Waaijer, R, Zwijnenburg, D, Diepvens, F, and Pals, G (2002). Relative quantification of 40 nucleic acid sequences by multiplex ligation-dependent probe amplification. *Nucleic Acids Res,* 30: e57.

Schröck, E, du Manoir, S, Veldman, T, Schoell, B, Wienberg, J, Ferguson-Smith, MA, et al. (1996). Multicolor spectral karyotyping of human chromosomes. *Science,* 273:494–497.

Schröck, E, Veldman, T, Padilla-Nash, H, Ning, Y, Spurbeck, J, Jalal, S, et al. (1997). Spectral karyotyping refines cytogenetic diagnostics of constitutional chromosomal abnormalities. *Hum Genet,* 101:255–262.

Schwartz, DC and Cantor, CR (1984). Separation of yeast chromosome-sized DNAs by pulsed field gradient gel electrophoresis. *Cell,* 37:67–75.

Seabright, M (1972). The use of proteolytic enzymes for the mapping of structural rearrangements in the chromosomes of man. *Chromosoma,* 36:204–210.

Selzer, RR, Richmond, TA, Pofahl, NJ, Green, RD, Eis, PS, Nair, P, et al. (2005). Analysis of chromosome breakpoints in neuroblastoma at sub-kilobase resolution using fine-tiling oligonucleotide array CGH. *Genes Chromosomes Cancer,* 44:305–319.

Sharp, AJ, Locke, DP, McGrath, SD, Cheng, Z, Bailey, JA, Vallente, RU, et al. (2005). Segmental duplications and copy-number variation in the human genome. *Am J Hum Genet,* 77:78–88.

Shaw-Smith, C, Redon, R, Rickman, L, Rio, M, Willatt, L, Fiegler, H, et al. (2004). Microarray based comparative genomic hybridisation (array-CGH). detects submicroscopic chromosomal deletions and duplications in patients with learning disability/mental retardation and dysmorphic features. *J Med Genet,* 41:241–248.

Solomon, E, Bobrow, M, Goodfellow, PN, Bodmer, WF, Swallow, DM, Povey, S, et al. (1976). Human gene mapping using an X/autosome translocation. *Somatic Cell Genet,* 2:125–140.

Solomon, E, Voss, R, Hall, V, Bodmer, WF, Jass, JR, Jeffreys, AJ, et al. (1987). Chromosome 5 allele loss in human colorectal carcinomas. *Nature,* 328:616–619.

Southern, E (2005). Tools for genomics. *Nat Med,* 11:1029–1034.

Southern, EM (1975). Detection of specific sequences among DNA fragments separated by gel electrophoresis. *J Mol Biol,* 98:503–517.

Southern, EM (2000). Blotting at 25. *Trends Biochem Sci,* 25:585–588.

Southern, EM, Anand, R, Brown, WR, and Fletcher, DS (1987). A model for the separation of large DNA molecules by crossed field gel electrophoresis. *Nucleic Acids Res,* 15:5925–5943.

Stark, GR and Wahl, GM (1984). Gene amplification. *Annu Rev Biochem,* 53:447–491.

Trask, B, van den Engh, G, Pinkel, D, Mullikin, J, Waldman, F, van Dekken, H, et al. (1988). Fluorescence in situ hybridization to interphase cell nuclei in suspension allows flow cytometric analysis of chromosome content and microscopic analysis of nuclear organization. *Hum Genet,* 78:251–259.

Trask, BJ (1991). Fluorescence in situ hybridisation: applications in cytogenetics and gene mapping. *Trends Genet,* 7:149–154.

Tsongalis, GJ and Silverman, LM (1993). Molecular pathology of the fragile X syndrome. *Arch Pathol Lab Med,* 117:1121–1125.

Urban, AE, Korbel, JO, Selzer, R, Richmond, T, Hacker, A, Popescu, GV, et al. (2006). High-resolution mapping of DNA copy alterations in human chromosome 22 using high-density tiling oligonucleotide arrays. *Proc Natl Acad Sci USA,* 103:4534–4539.

Van Ommen, G-JB, Breuning, MH, and Raap, AK (1995). FISH in genome research and molecular diagnostics. *Curr Opin Genet Dev,* 5:304–308.

Veldman, T, Vignon, C, Schrock, E, Rowley, JD, and Ried, T (1997). Hidden chromosome abnormalities in haematological malignancies detected by multicolour spectral karyotyping. *Nat Genet,* 15:406–410.

Veltman, JA, Schoenmakers, EF, Eussen, BH, Janssen, I, Merkx, G, van Cleef, B, et al. (2002). High-throughput analysis of subtelomeric chromosome rearrangements by use of array-based comparative genomic hybridization. *Am J Hum Genet,* 70:1269–1276.

Waggott, W, Lo, YM, Bastard, C, Gatter, KC, Leroux, D, Mason, DY, et al. (1995). Detection of NPM-ALK DNA rearrangement in CD30 positive anaplastic large cell lymphoma. *Br J Haematol,* 89:905–907.

Wang, S, Saboorian, MH, Frenkel, E, Hynan, L, Gokaslan, ST, and Ashfaq, R (2000). Laboratory assessment of the status of Her2/neu protein and oncogene in breast cancer specimens: comparison of immunohistochemistry assay with fluorescence in situ hybridisation assays. *J Clin Pathol,* 53:374–381.

Wang, Y, Makedon, F, and Pearlman, J (2006). Tumor classification based on DNA copy number aberrations determined using SNP arrays. *Oncol Rep,* 15 (Spec. no.):1057–1059.

Ward, JR, Cottrell, S, Thomas, HJ, Jones, TA, Howe, CM, Hampton, GM, et al. (1993). A long-range restriction map of human chromosome 5q21-q23. *Genomics,* 17:15–24.

Wijmenga, C, Wright, TJ, Baan, MJ, Padberg, GW, Williamson, R, van Ommen, GJ, et al. (1993). Physical mapping and YAC-cloning connects four genetically distinct 4qter loci (D4S163, D4S139, D4F35S1 and D4F104S1) in the FSHD gene-region. *Hum Mol Genet,* 2:1667–1672.

Wolf, SF and Migeon, BR (1982). Studies of X chromosome DNA methylation in normal human cells. *Nature,* 295:667–671.

Wu, WM, Tsai, HJ, Pang, JH, Wang, TH, Wang, HS, Hong, HS, et al. (2005). Linear allele-specific long-range amplification: a novel method of long-range molecular haplotyping. *Hum Mutat,* 26:393–394.

Zhou, X, Rao, NP, Cole, SW, Mok, SC, Chen, Z, and Wong, DT (2005). Progress in concurrent analysis of loss of heterozygosity and comparative genomic hybridization utilizing high density single nucleotide polymorphism arrays. *Cancer Genet Cytogenet,* 159:53–57.

8

Nutritional Genomics

Patrick J Stover

Nutrients and other food components are amongst the most persistent, variable (both in terms of the nature and abundance of the food supply), and essential environmental exposures for all life forms. The need for organisms to balance constant nutrient "need" with intermittent nutrient "availability" has driven the evolution of sophisticated yet distinct strategies to sense, store, and utilize individual nutrients to achieve internal homeostasis. Indeed, epidemiological, whole animal, and tissue- and cell-culture studies validate nutrition's pivotal role as an exogenous determinant of health. Most common human chronic diseases, including diabetes (type 2), metabolic syndrome, cardiovascular and neurological disease, and many cancers are initiated and/or accelerated by nutrient/food exposures. In the absence of adaptation, nutrient deficiencies can impair the function of transcriptional and metabolic networks whereas nutrient excesses can exceed their capacity and/or overwhelm the buffering capability of the associated signaling pathways that maintain homeostasis.

Nutritional genomics is a field that has emerged at the interface of nutrition and genomics. Genomics is defined as the "study of the functions and interactions of all the genes in the genome, including their interaction with environmental factors" (Guttmacher and Collins, 2002). Nutrients are essential environmental factors for organismal survival; the term nutrient was recently defined as a "fully characterized (physical, chemical, physiological) constituent of a diet, natural or designed, that serves as a significant energy yielding substrate or a precursor for the synthesis of macromolecules or of other components needed for normal cell differentiation, growth, renewal, repair, defense, and/or maintenance or a required signaling molecule, cofactor, or determinant of normal molecular structure/function and/or promoter of cell and organ integrity" (Young, 2002). The interactions among nutrients and the genome are seamless and essential features of organismal evolution. They are fundamental for virtually all life processes; these interactions affect both the primary sequence of deoxyribonucleic acid (DNA, the genetic code), as well as the expression of the code (Fig. 8–1). Individual nutrients influence DNA mutation rates in somatic cells, and more recently have been shown to influence both the generation and propagation of DNA mutations in the germ line and thereby facilitate the generation of human genetic variation. Genomic "signatures" can be found within DNA primary sequences that validate the role of dietary components as selective pressures throughout human evolution. The influence of nutrients on the genome is not limited to DNA primary sequence. Nutrients and metabolites function as signaling molecules that enable networks to sense and respond to their internal and external environments. In this regard, nutrients can elicit transient alterations in gene expression and/or influence more permanent and potentially heritable whole genome reprogramming events. The genome, in turn, influences diet (Fig. 8–1). Human genetic variation, including variations in primary sequence and variations in epigenetic programming, affects nutrient absorption and utilization and thereby confers differences in food tolerances and potentially nutrient requirements among human individuals and human populations.

Nutritional genomics is a multidisciplinary field that draws upon an extensive and rich foundation of knowledge in nutritional anthropology, population genetics, nutritional biochemistry, human clinical nutrition and metabolism, human genetics and development, and nutritional toxicology among other disciplines. This new field emerged as the sequence of human genomes, and the genomes of other organisms, became available. Metabolic- and nutrition-related disorders are complex traits with multiple interacting environmental and genetic determinants. Their onset and severity are modified by both diet and the genetic background of the individual. Therefore, nutritional genomics, like most "omics" fields, is focused on the biology of the *individual*, but is distinguished by its unique potential to advance our understanding of disease *prevention* and healthy aging through manipulation of gene–diet interactions. In addition, nutritional genomics has therapeutic applications through the rational design of dietary interventions to manage chronic disease.

Advances in nutritional genomics research are anticipated to illuminate the mechanisms underlying the acute and long lasting diet/nutrition–genome interactions that promote health and revolutionize both clinical and public health nutrition practice and culminate in (1) genetically informed nutrient and food-based dietary guidelines for disease prevention and healthy aging, (2) improved and/or individualized nutritional therapeutic regimes for disease management, and (3) better targeted public health nutrition interventions (e.g., micronutrient fortification and supplementation) that maximize benefit and minimize adverse outcomes within human populations. These objectives will be met only with further development of the basic science that underpins nutritional

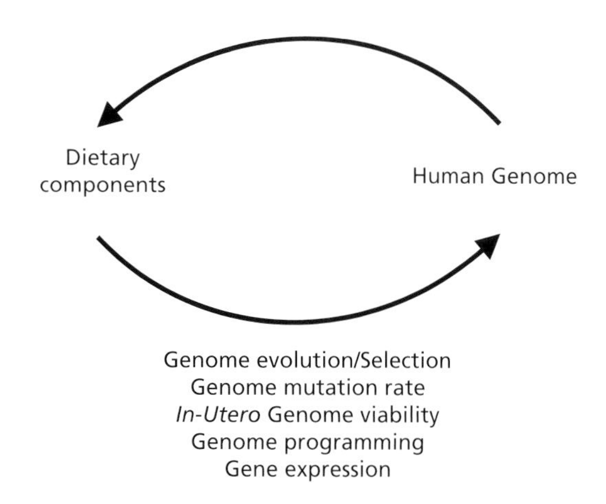

Figure 8–1 Genome–nutrient interactions.

genomics, and effective translation of this knowledge into nutrition practice in the areas of:

1. *Nutritional genetics*, the identification, classification, and characterization of human genetic variation that modifies nutritional requirements and food tolerances;
2. *Nutritional epigenetics*, the modification of chromatin structure (and hence gene expression) by diet, through postreplication and posttranslational modification of DNA and protein, respectively, that serves to program or reprogram biological networks with multigenerational consequences;
3. *Nutritional and metabolic engineering*, the targeted manipulation of biological networks (e.g., transcriptional, signaling, and metabolic) for benefit through nutrition, which includes the use of diet to affect genome expression, stability, and/or direct dietary compensation for network outputs, including the use of dietary restriction [e.g., phenylalanine restriction to prevent phenylketonuria (PKU)] or nutritional supplements (e.g., vitamin B_{12} to prevent anemia in the elderly).

This chapter summarizes the underlying scientific principles and preliminary advances in these three primary focus areas of nutritional genomics.

Nutritional Genetics

Human populations and/or individuals within populations differ in their sensitivities to nutrient deficiencies and excesses. Human genetic variation contributes to differences in physiological responses to diet. The United Nations Educational, Scientific, and Cultural Organization (UNESCO) recognized both the influence of human genetic variation on human nutrition and the concept that nutrient utilization and efficacy are unique characteristics of individuals in the Universal Declaration on The Human Genome and Human Rights, Section A, Article 3 which states:

> The human genome, which by its nature evolves, is subject to mutations. It contains potentialities that are expressed differently according to each individual's natural and social environment including the individual's state of health, living conditions, nutrition and education. (UNESCO Document 27 V/45 Adopted by the Thirty First General Assembly of UNESCO, Paris, November 11, 1997)

Indeed, human evolution is a continuous albeit irregular process made manifest through the generation and expansion of DNA mutations that permit survival amidst erratic and unpredictable environmental exposures. Changes in DNA primary sequence enable human evolution through the generation of adaptive genes and gene variants that alter an organism's response to environmental challenges and hence to its fitness. Human DNA primary sequence differs by approximately 0.2%–0.4% among humans (Venter et al., 2001; Tishkoff and Kidd, 2004), and this variation is in part a product of historical interactions among humans and their environment including dietary patterns.

Types of Genetic Variation

Differences in DNA primary sequence constitute a primary molecular basis for human phenotypic variation including metabolic efficiency and disease susceptibility. Genomic polymorphisms emerge through the sequential processes of DNA mutation and expansion of the mutation within a population. Environmental exposures can accelerate both processes. Polymorphisms are classified according to the origin and nature of the genomic mutation, and include single nucleotide polymorphisms (SNPs), microsatellite and macrosatellite repeat sequences, and viral insertions. SNPs are common nucleotide base pair differences in the primary sequence of DNA and are the most common variation in human DNA. There are more than 10 million SNPs in the human genome with over 4.5 million SNPs validated as of 2004 (Tishkoff and Kidd, 2004). SNPs can be single base pair insertions, deletions, or substitutions of one base pair for another. Nucleotide substitutions are the most common polymorphism whereas insertion/deletion mutations occur at one-tenth the frequency (Nachman and Crowell, 2000). SNPs differ from DNA mutations in two regards: they are present in the germ line and therefore are heritable and must have a prevalence of at least 1% in human populations.

Genetic variation can also result from the integration and/or transposition of retroviral DNA (Tishkoff and Verrelli, 2003). Approximately 50% of noncoding human DNA originates from transposable elements that are highly mobile and contain repetitive sequences (Batzer and Deininger, 2002; Brookfield, 2003; Van Blerkom, 2003). Retrotransposons are classified by size and include long interspersed nuclear elements (LINEs) and short interspersed nuclear elements (SINEs). Approximately 10% of human genomic sequence consists of 280 bp Alu SINE. Over 1200 Alu elements integrated into the human genome following early human migrations out of Africa. Today, there are an estimated 1.4 million in the human genome with a new Alu insertion event occurring every 200 births (Batzer and Deininger, 2002). Alu elements are believed to be catalysts for organismal evolution (Batzer and Deininger, 2002; Bamshad and Wooding, 2003). Alu nucleotide sequence does not encode function. Rather, they display promoter activity, but their transcripts lack an open reading frame and therefore are not translated into protein. Alu insertions can alter genome stability and/or the function of a single gene function near their site of integration. Alu elements contain splice acceptor sites and therefore their integration within an intron can lead to the generation of new proteins through alternative splicing of gene transcripts. Because of their repetitive sequence and high degree of sequence similarities among family members, Alu insertions can serve as nucleation sites for unequal intrachromosomal and interchromosomal homologous recombination events that result in chromosomal aberrations including deletion and translocation events.

Viral elements can confer new types of regulation to existing genes, including regulation by essential nutrients, but also interrupt genes or create new genes. Alu elements can function as transcriptional silencers or activators that are responsive to nutrient status.

Some Alu elements have retinoic acid (vitamin A) response elements that bind nuclear receptor transcription factors and therefore confer retinoic acid responsiveness to genes that neighbor the insertion site. Alu elements also contain multiple CpG "islands," a nucleotide sequence that attracts the DNA methylation of the cytosine (C) base. DNA methylation typically serves to suppress transcription at that locus, and the degree of methylation can be sensitive to dietary intake of vitamins involved in one-carbon metabolism including folic acid (Waterland and Jirtle, 2003).

Alu insertions are also associated with metabolic disease. New Alu insertions may account for up to 0.1% of human genetic disorders including Apert's syndrome, cholinesterase deficiency, and breast cancer. Alu-mediated unequal homologous recombination events are responsible for 0.3 % of inherited human genetic diseases including type 2 insulin-resistant diabetes and familial hypercholesterolemia (Batzer and Deininger, 2002; Tishkoff and Verrelli, 2003). Such recombination events are rare, because Alu-mediated unequal homologous recombination events are usually inhibited by CpG methylation of the element.

DNA Mutation Rates and Diet

The human genome is assembled from deoxynucleotide monomers that are synthesized de novo from metabolic precursors derived from energy, amino acid, and one-carbon metabolism. DNA mutations arise as a consequence of the inherent chemical instability of DNA bases, error associated with the fidelity DNA replication and recombination, exposure to chemical oxygen radicals that are generated during oxidative metabolism as well as by numerous genotoxic xenobiotics that are present in the food supply. Therefore, a portion of DNA mutations are unavoidable, although their impact is minimized by DNA repair systems that detect and correct most mutation events.

Environmental exposures including nutrient deficiencies or excesses, factors that increase cellular oxidative stress, or genetic variations that modify the metabolism of dietary components can accelerate DNA mutation rates by inducing DNA modification reactions and/or by accelerating DNA polymerase error rates. Many of the essential B vitamins (e.g., folate, niacin, flavin, vitamin B_6, vitamin B_{12}) and minerals (e.g., zinc, iron) are required for nucleotide biosynthesis; nutritional deficiencies can impair the synthesis of nucleotide precursors and lower DNA synthesis rates (rates of mitosis) and/or the fidelity of DNA replication (DNA mutation rates). For example, folate deficiency inhibits deoxythymidine monophosphate (dTMP) synthesis which increases incorporation of deoxyuridine triphosphate (dUTP) into DNA, resulting in increased frequencies of DNA strand breaks (Blount et al., 1997; Fenech, 2001, 2002a,b, 2004). Furthermore, deficiency of dietary antioxidants that scavenge chemical radicals, or excesses of prooxidant nutrients including iron, may increase mutation rates (Ames, 2001; Goswami et al., 2002; Hediger et al., 2002). Other dietary components affect DNA mutation rates by altering cellular redox states or functioning as genotoxic radicals that chemically modify purine and pyrimidine bases. Certain aflatoxins, a common class of natural xenobiotics found in soil molds that contaminate certain foods, increase DNA mutation rates leading to the transformation of somatic cells and localized cancer epidemics (Mishra and Das, 2003). However, only mutations that occur in the germ line contribute to a species' heritable genetic variation. Mutations that have no functional consequences are anticipated to be phenotypically silent and therefore selectively neutral. The frequency of silent mutations is exclusively a function of the DNA mutation rate (estimated to be 2.5×10^{-8} on average for autosomes in regions of the genome presumed to be nonfunctional including intronic and intergenic regions) (Nachman and Crowell, 2000; Tishkoff and Verrelli, 2003).

Diet and Mutation Expansions in Human Populations

There is increasing evidence that dietary challenges have increased human genetic variation by driving the expansion of rare gene variants in isolated human populations and/or isolated geographic regions. Mutations that promote survival in challenging dietary environments expand within populations through the generation of environmentally adaptive gene alleles. Mutations that expand and become fixed within a population contribute to genetic variation as polymorphisms, and this expansion within human populations is the molecular basis for the evolution of genomes. Germ line mutation alone is necessary but not sufficient for establishing genetic variation.

Not all genes "evolve" or change at the same rate. The neutral theory of evolution (DNA mutation in the absence of selection) does not account for the extent of amino acid substitutions observed in mammalian genomes (Kimura, 1968; King and Jukes, 1969; Clark et al., 2003; Wolfe and Li, 2003). Natural selection, which is the differential contribution of genetic variants to future generations, is the only evolutionary force that has adaptive consequences (Akey et al., 2002). Darwinian selection favors the conservation and expansion of favorable mutations (by positive or balancing selection), and the elimination of mutations that are deleterious (referred to as negative or purifying selection). Not all genes evolve at the same rate, because positive selection only accelerates mutation fixation at defined loci within the genome. Mutations that confer reproductive and/or survival advantage within a single environmental context expand within populations at higher rates relative to neutral mutations and replace a population's preexisting variation.

Mutations that alter physiological processes are under constraint and subject to positive, balancing, or negative selection. Patterns of genetic variation across the human genome are affected by demographic histories of populations, whereas variation at particular genetic loci is influenced by the effects of natural selection, mutation, and recombination (Tishkoff and Verrelli, 2003). Although protein-coding sequences are conserved among mammals in general, rates of amino acid substitution vary markedly among proteins compared to rates of synonymous substitution among genes (changes in the coding region of genes that do not affect protein sequence) (Wolfe and Li, 2003). The proportion of amino acid substitutions that result from positive selection is estimated to be 35%–45% (Wolfe and Li, 2003). Mutations that alter amino acid sequence, which affect protein structure and function can have physiological consequences that may be beneficial, deleterious, and/or neutral and thereby influence an organism's fitness in specific environmental/dietary contexts. Likewise, mutations that affect protein expression can alter biological network outputs leading to altered physiology.

Mutations can expand in the absence of selection and contribute to metabolic disease. The rate of mutation fixation is a function of the effective population size, population demographic history and the effect of the mutation on an organism's fitness (Tishkoff and Verrelli, 2003). Polymorphisms can expand and become fixed within a population through the processes of genetic drift or natural selection. Drift is a stochastic process resulting from the random assortment of chromosomes at meiosis. Because only a fraction of all possible zygotes are generated and/or survive to reproduce, mutations can expand through many generations by the random sampling of gametes in the absence of selection (Tishkoff and Verrelli,

2003). Drift is expected to have a greater influence on genetic variation in small populations expanding rapidly; drift in large static populations is not usually as significant. Genetic drift becomes relevant in large populations undergoing bottlenecks (massive reductions in population) or in founding events that have occurred during human migrations, e.g., population groups that include the Old Order Amish, Hutterite, and Ashkenazi Jewish (Tishkoff and Verrelli, 2003). In such populations, rare disease alleles can expand rapidly and increase the incidence of diseases including breast cancer, Tay–Sachs, Gaucher, Niemann–Pick, and familial hypercholesterolemia (Tishkoff and Verrelli, 2003). Although most human genetic variation arose as a result of the neutral processes of mutation and genetic drift, variation resulting from drift rarely has physiological consequences in static environments. However, environmental shifts, including alterations in the food supply, can challenge biological systems and convert otherwise physiological "silent" genetic variation into functional gene variants. Relevant examples are discussed below.

Identification of Nutritionally and Environmentally Sensitive Human Alleles

Candidate Gene Approach

The vast majority of known functional polymorphisms that contribute to food intolerances and metabolic disorders were first identified as highly penetrant disease alleles from epidemiological or clinical studies (Table 8–1). Candidate genes were analyzed for genetic variation; their selection as candidate genes was based on knowledge of metabolic pathways and inference that their impairment could result in metabolic phenotypes associated with a particular disease state or affect biomarkers associated with the disease. Model organisms, including yeast, drosophila, *Caenorhabditis elegans*, and mice have been excellent resources to identify potential candidate genes and to confirm their contribution to a metabolic phenotype. Other advancements, including the availability of high-density SNP maps of the human genome have accelerated the identification of human disease alleles, including low-penetrant alleles that may make relatively small contributions to the initiation and/or progression of complex disease (Tishkoff and Verrelli, 2003). Furthermore, haplotype maps of human genetic variation offer advantages for disease associational studies because of their reduced complexity compared to SNP maps (Gabriel et al., 2002), but their utility may be limited because of the variability in haplotype diversity across candidate genes (Crawford and Nickerson, 2005). The candidate gene approach, although successful in identifying alleles underlying monogenetic traits, (Ames et al., 2002; Young, 2002) is limited by incomplete knowledge of gene function, incomplete knowledge of transcriptional and metabolic networks that suggest candidate genes for analyses, and the multifactorial nature of most complex human metabolic chronic disease (most are polygenic traits with multiple environmental components which in isolation make relatively minor contributions to the disease phenotype). This is witnessed by the many inconsistent findings that have emerged within the

Table 8–1 Candidate Human Disease Alleles That Affect the Uptake or Metabolism of Dietary Components

Food Component	Gene	Polymorphic Allele	References
Vitamins			
Folate	*MTHFR* *CBS* *GCPII*	A222V 844ins68 H475Y	(Bailey, 2003) (Jacques et al., 1996) (Tsai et al., 1999) (Afman et al., 2003) (Devlin et al., 2000)
Vitamin B_{12}	*MTR* *MTRR*	N919G I22M	(Jacques et al., 1996) (Jacques et al., 1996)
Vitamin D	*VDR*	Many	(Uitterlinden et al., 2002)
Minerals			
Iron	*HFE*	C282Y	(Griffiths and Cox, 2000; Zhao et al., 2001)
Sodium	*CIC-Kb*	T481S	(Geller, 2004; Jeck et al., 2004)
Lipids			
	APOB	Many	(Bentzen et al., 2002; Hubacek et al., 2001)
	APOC3	Many	(Brown et al., 2003)
	APOE	Many	(Fullerton et al., 2000)
Alcohol			
	ADH/ *ALDH2*	Many	(Bosron and Li, 1986; Loew et al., 2003)
Carbohydrate			
Lactose	*LCT*	Promoter	(Poulter et al., 2003)
Fructose	*Aldolase B*	Many	(Esposito et al., 2002)
Detoxification/oxidative stress			
	NAT1/ *NAT2*	Many	(Hein et al., 2000, 2002)
	PON1	Q192R;L55M	(Ferre et al., 2003)
	Mn-SOD	Ala(-9)Val	(Chistyakov et al., 2001; Van Landeghem et al., 1999)

nutritional and genetic epidemiological literature, especially for the involvement of low-penetrant genetic alleles in chronic metabolic disease (Bracken, 2005).

Once candidate genes are identified, establishing alleles as disease causing is equally challenging. Haplotype associations do not identify disease-causing mutations due to genetic hitchhiking (Tishkoff and Verrelli, 2003) [polymorphisms that are in linkage disequilibrium (LD)] with a mutation that is under selection will change in frequency along with the site undergoing selection]. Because otherwise rare disease alleles can be enriched in geographically or culturally isolated populations with distinct dietary practices, full characterization of SNP diversity and haplotype structure from ethnically diverse populations is critical for the identification of metabolic or nutritional risk alleles that may be specific to small but identifiable subpopulations.

Computational Approaches

Genes that have undergone accelerated change or evolution display genomic signatures that can be identified computationally. The identifiable genomic signatures include the presence of an excess of rare variants within a population (which can be indicative of a selective sweep), large allele frequency differences among populations and a common haplotype that remains intact over long distances (Akey et al., 2002, 2004; Bamshad and Wooding, 2003; Wooding, 2004). The identification of polymorphic alleles that have arisen as a result of historical selection resulting from nutritional challenges offers the opportunity to identify genes that contribute to monogenetic metabolic disorders as well as low-penetrant alleles that contribute to complex metabolic disease (Tishkoff and Verrelli, 2003; Akey et al., 2004). The common disease–common variant hypothesis states that disease-susceptibility alleles arose before human migrations out of Africa and therefore exist at high frequency across all human populations (Chakravarti, 2001; Reich and Lander, 2001). However, both "single-gene" disorders, including cystic fibrosis and hemochromatosis, as well as complex diseases can be associated with geographically-restricted populations because the alleles arose after migrations out of Africa (Jorde et al., 2001; Pritchard, 2001; Pritchard and Cox, 2002; Bamshad and Wooding, 2003; Akey et al., 2004; Tishkoff and Kidd, 2004). Therefore, although 85%–90% of all human genetic variation is found within populations and presumably arose prior to human migrations, some of the 10%–15% of variation among populations likely arose from recent selective pressures that contribute to both simple and complex disease (Jorde et al., 2000; Jorde and Wooding, 2004).

Comparison of genomic sequence divergence among mammalian species enables the identification of ancient selection throughout the process of speciation and genetic divergence (Table 8–2). These approaches can identify single genes or pathways within biological networks that enabled adaptation. Similarly, analyses of human genomic diversity among human populations can identify genetic selection within the human species that occurred prior to and following human migrations out of Africa. These complementary approaches have permitted the identification of genes that have undergone accelerated and/or adaptive evolution (Table 8–2) (Clark et al., 2003; Akey et al., 2004). Rapidly evolving genes are inferred to have enabled adaptation and thus became fixed in populations by positive or balancing selection. Adaptive genes originate from region-specific selective factors, and therefore are expected to concentrate in specific geographic regions where the selection occurred (Tishkoff and Verrelli, 2003). The geographic origins of an individual can be predicted from genomic signatures of positive selection to the degree that different selective pressures are operative across populations, but do not always correspond to specific ethnic or racial groups because races are not homogenous (Collins, 2004; Tishkoff and Kidd, 2004). Many of the human alleles that are known to affect metabolism, food tolerances, or optimal nutrient intakes display signatures of positive selection (Table 8–2). Below is a summary of gene variants that are known to affect diet, many of which display signatures for positive selection.

Table 8–2 Diet-related Genes That Display Genomic Signatures of Adaptive Evolution

Gene	Species/Function	References
Lysozyme	Langur monkey	(Messier and Stewart, 1997) (Wolfe and Li, 2003) (Zhang et al., 2002)
Ribonuclease	Langur monkey	(Wolfe and Li, 2003) (Zhang et al., 2002)
Cox4	Primates	(Wu et al., 1997)
LCT	Human lactose metabolism	(Bersaglieri et al., 2004)
ADH1B	Human ethanol metabolism	(Osier et al., 2002)
ALDH2	Human ethanol metabolism	(Oota et al., 2004)
HFE	Human iron homeostasis	(Toomajian et al., 2003)
PPARγ	Human nuclear receptor	(Akey et al., 2002)
PTC	Human bitter-taste receptor	(Wooding et al., 2004)
TAS2R16	Human bitter taste receptor	(Bufe et al., 2002)
KEL	Human protein metabolism	(Akey et al., 2004)
TRPV5	Human calcium transport	(Akey et al., 2004)
TRPV6	Human calcium transport	(Akey et al., 2004)
ABO	Human protein metabolism	(Akey et al., 2004)
ACE2	Human protein metabolism	(Akey et al., 2004)
CYP1A2	Human arylamine metabolism	(Wooding et al., 2002)
G6PD	Human NADP metabolism	(Verrelli et al., 2002)
AGXT	Human Glyoxylate metabolism	(Danpure, 2005)
SLC23A1	Human Vitamin C transport	(Eck et al., 2004)

Bitter Taste Receptor

Recognition of bitter taste may have conferred a selective advantage by deterring the consumption of plant toxins that often elicit the sensation of bitterness (Bufe et al., 2002). The *TAS2R16* gene encodes a G protein-coupled receptor that is activated by salicin in fruit, amygdalin in almonds, and many common β-glucopyranosides that elicit cyanogenic toxicity. The K172V polymorphism increases the receptor's sensitivity to cyanogenic glycosides and displays signatures of positive selection. The adaptive allele arose in the Middle Pleistocene prior to migrations out of Africa (Bufe et al., 2002).

Lactose and Calcium Metabolism

Lactose metabolism requires the expression of lactase-phlorizin hydrolase, an enzyme encoded by the *LCT* gene. In most humans and other mammals, *LCT* expression declines after weaning resulting in primary lactose intolerance. Some human populations, including those of northwest European descent and nomads of the Afro-Arabian desert region, *LCT* expression persists into adulthood and confers the ability to effectively digest dairy products. An SNP was identified 14 kb upstream of the *LCT* transcriptional initiation site in a cis-acting transcriptional element. This SNP is enriched in individuals of northern European descent and displays genomic signatures of positive selection (Enattah et al., 2002; Poulter et al., 2003; Bersaglieri et al., 2004). Its prevalence correlates with but does not account fully for the persistence of *LCT* expression and resistance to primary lactose intolerance throughout adulthood (Poulter et al., 2003). The fixation of this polymorphism in these populations may have been driven by the benefits of milk consumption in cattle-herding populations, both as a source of liquid in arid regions and prevention of rickets and osteomalacia in regions of low solar irradiation (Wooding, 2003, 2004; Bersaglieri et al., 2004). The requirement for efficient calcium absorption may also have driven alleles for TRPV5 and TRPV6 to fixation in these same populations (Table 8–2) (Akey et al., 2004).

Fructose Metabolism

Hereditary fructose intolerance (HFI) is an autosomal recessive disorder of fructose metabolism resulting from low fructose-1, 6-aldolase activity, resulting in an accumulation of metabolic intermediate fructose-1-phosphate, which deregulates glycolysis. Twenty-five allelic variants of aldolase B, the human liver isozyme, have been identified that impair enzyme activity by altering the catalytic properties of the enzyme and/or protein stability (Esposito et al., 2002). The accumulation of fructose-1-phosphate inhibits glycogen breakdown and glucose synthesis, resulting in severe hypoglycemia following ingestion of fructose. Individuals carrying polymorphic variants of aldolase B are asymptomatic in the absence of fructose or sucrose consumption and can avoid the recurrence of symptoms by remaining on a fructose- and sucrose-free diet. Chronic fructose ingestion in infants leads ultimately to hepatic and/or renal failure and death. The prevalence of these variants differs throughout Europe; the L288delta C frameshift mutation is restricted to Sicilian subjects. These aldolase B variants likely emerged in populations through random drift; fructolysis is not an essential metabolic pathway for humans, and fructose has not been an abundant dietary component throughout most of human history. However, the incidence of HFI intolerance has increased since the widespread use of sucrose and fructose as nutrients and sweeteners, providing an excellent example whereby an environmental shift resulted in the apparent conversion of normally nonpenetrant "silent" aldolase B alleles into HFI disease-alleles (Cox, 2002).

Lipid Metabolism

Apolipoprotein E (apoE) is a polymorphic protein that functions in lipid metabolism and cholesterol transport (Fullerton et al., 2000). All human populations display apoE polymorphism. There are three common allelic variants, ε2, ε3, and ε4 whose relative distribution varies among populations; the frequency of the ε4allele declines from northern to southern Europe. These variant alleles encode proteins that differ in their affinity for both lipoprotein particles and for low-density lipoprotein receptors. The ε4 allele increases risk for late-onset Alzheimer's disease and arteriosclerosis with low penetrance. Carriers of the ε2 allele tend to display lower levels of plasma total cholesterol, whereas carriers of the ε4 allele, which may be ancestral, display higher cholesterol levels. Therefore, serum cholesterol levels are likely to be more responsive to low-fat and low-cholesterol diets in carriers of the ε4 allele (Ordovas, 2004; Ordovas and Corella, 2004).

One-carbon Metabolism

Folate-mediated one-carbon metabolism is required for purine, thymidylate and methionine biosynthesis and affects genome synthesis, stability and gene expression (Stover, 2004b). Several polymorphic alleles have been identified to be associated with metabolic perturbations that can confer both protection and risk for specific pathologies and developmental anomalies (DAs) (Stover, 2004b). SNPs in methylenetetrahydrofolate reductase (MTHFR) (A222V) and methylenetetrahydrofolate dehydrogenase (MTHFD1) (R653Q) (Brody et al., 2002), which encode folate-dependent enzymes, are associated with increased risk for neural tube defects (NTDs); the MTHFR (A222V) is protective against colon cancer in folate replete subjects (Ma et al., 1997). The MTHFR A222V variant protein has reduced affinity for riboflavin cofactors and is thermolabile, resulting in reduced cellular MTHFR activity; its stability is increased when folate is bound (Guenther et al., 1999). The prevalence of the MTHFR allelic variant varies markedly among human population and occurs with an allelic frequency of approximately 40% in some Hispanic populations, but is mostly absent in African populations (Bailey, 2003). However, it has not been reported to display signatures of positive selection. Although the biochemical role of these polymorphisms in the etiology of NTDs and cancer is unknown, it is demonstrated that some carriers of MTHFR variants require higher folate intakes than others to (1) stabilize the MTHFR protein, (2) lower the concentration the metabolic intermediate homocysteine, and (3) decrease a women's risk of bearing children with DAs including NTDs (Bailey, 2003). The fortification of the food supply with folic acid that occurs in many countries targets women of childbearing age for birth defect prevention, with genetically identifiable subgroups receiving the most benefit.

Iron Metabolism

Hereditary hemochromatosis is a recessive iron storage disease that is prevalent in populations of European descent with an incidence of 1 in 300 persons. The *HFE* gene is polymorphic and encodes a protein that regulates iron homeostasis. A common polymorphism, HFE C282Y, emerged approximately 138 generations ago (Toomajian and Kreitman, 2002; Bamshad and Wooding, 2003; Toomajian et al., 2003). This SNP is associated with the disease phenotype in 60%–100% of Europeans. The HFE C282Y allele is not present in Asian and African populations despite the presence of iron storage diseases in those populations indicating that other genes are associated with hereditary hemochromatosis. Furthermore, the penetrance of the C282Y HFE allele for the iron overload phenotype varies widely among homozygotes, with some individuals being asymptomatic, indicating the presence of modifying alleles. The recent expansion of this polymorphism may have conferred selective advantages in iron-poor environments (Toomajian and Kreitman, 2002; Toomajian et al., 2003) or resistance to microbial infection (Beutler, 2004).

Alcohol Metabolism

Ethanol metabolism efficiency varies considerably among human populations (Bosron and Li, 1986). Ethanol is oxidized to acetaldehyde by the enzyme alcohol dehydrogenase, which is encoded by the *ADH* genes. Acetaldehyde, a toxic metabolite, is subsequently oxidized to acetic acid by the enzyme aldehyde dehydrogenase, which is encoded by *ALDH2*. Seven *ADH* genes cluster on chromosome 4, which encode proteins with distinct catalytic properties and tissue-specific expression patterns. Two of the genes encoding class I enzymes (*ADH1B* and *ADH1C*) are expressed in liver, function in systemic ethanol clearance and display functional polymorphism. A variant ADH1B* 47His allele predominates in Japanese and Chinese populations but is rare in European and northern African populations (Osier et al., 2002). The variant allele encodes an enzyme with elevated enzyme activity leading to more rapid formation of acetaldehyde. The ADH1C*349Ile variant is found in Europeans whereas the ADH1B*369Arg variant is mostly restricted to individuals of African descent. ALDH2 is also highly polymorphic and Asian populations carry a common dominant null allelic variant (E487K) and develop a characteristic "flush" reaction when consuming alcohol resulting from acetaldehyde accumulation (Loew et al., 2003). ADH and ALDH alleles that predominate in East Asian populations display signatures of positive selection, and the expression of these variant alleles results in elevated acetaldehyde concentrations following alcohol consumption, which may have conferred advantage by protecting against parasite infection (Oota et al., 2004).

Glyoxylate Metabolism and Kidney Stone Disease

Kidney stone disease resulting from calcium oxalate (CaOx) formation is common in Western populations and results from multiple etiologies (Danpure, 2005). Polymorphism in the *AGXT* gene, which encodes the enzyme alanine:glyoxylate aminotransferase (AGT) results in an accumulation of glyoxylate, a toxic intermediary metabolite that is converted to oxalate. Oxalate does not undergo further metabolism and accumulates as insoluble CaOx precipitates that accumulate in the kidney and urinary tract. There are two major precursors of glyoxylate: glycolate, which is present in plant-based foods and hydroxyproline, which is present in meat collagen. Glycolate metabolism occurs in the peroxisomes whereas hydroxyproline is metabolized in mitochondria. Among mammals, the intracellular localization varies among carnivores and omnivores; it is primarily peroxisomal in herbivores, mitochondrial in carnivores and both mitochondrial and peroxisomal in omnivores. In humans, AGT is usually localized to peroxisomes. A common AGT variant, Pro11Leu, decreases enzyme activity by approximately 70% but also results in the formation of a mitochondrial leader sequence and 5% of AGT protein localization in mitochondria. This variant has been proposed to confer advantage to meat-eating populations but detrimental to vegetarians. It displays evidence for positive selection (Danpure, 2005). The allelic frequency varies among human populations and correlates with historical dietary patterns; it is present at a frequency of 28% in Saami and 2.3% and 3% in Chinese and Indian Hindus respectively.

Vitamin C Transport

The individuality of vitamin C needs was first recognized in 1967 (Williams and Deason, 1967). Vitamin C is required for the function of at least eight mammalian enzymes and can scavenge reactive oxygen species (Eck et al., 2004). There are two genes that encode sodium-dependent vitamin C transporters, *SLC23A1* and *SLC23A2*. These genes resulted from an early gene duplication event but appear to have acquired distinct functions. *SLC23A1* is responsible for intestinal and renal absorption of vitamin C and therefore has the potential to affect whole-body vitamin C accumulation (Eck et al., 2004). The overall mutation rate of these genes is similar as evidenced by the similarity in their nonsynonymous substitution rates. However, four population-specific nonsynonymous substitutions are seen in *SLC23A1* that arose after the migrations out of Africa indicating a potential role for selection in the expansion of these variant alleles (Eck et al., 2004). Only synonymous substitutions are observed in *SLC23A2* and deletion of the orthologous slc23a2 in mice is lethal indicating the critical, nonredundant function of this transporter that seems to be under selective constraint. The effects of nonsynonymous *SLC23A1* on vitamin C transport or physiology have not been investigated (Eck et al., 2004).

Energy Metabolism

The "thrifty gene" hypothesis was first proposed over 40 years ago to account for the epidemic of type 2 diabetes observed in nonwestern cultures that adopt western style diets and lifestyles (Neel, 1962; Diamond, 2003). The hypothesis states that exposure to frequent famine selected for gene variants, which enabled the more efficient conversion of food into energy and fat disposition during unpredictable and sometimes scant food supplies. The putative adaptations also may have resulted in more efficient adaptations to fasting conditions (e.g., more rapid decreases in basal metabolism) and/or physiologic responses that facilitate excessive intakes in times of plenty. Conclusive genomic data has not yet supported this hypothesis (Rockman and Wray, 2002; Diamond, 2003).

Oxidative Metabolism

Variations that impact human nutrition and metabolism may have arisen independent of direct nutritional challenges. The enzyme glucose-6-phosphate dehydrogenase is solely responsible for the generation of reduced nicotinamide adenine dinucleotide phosphate (NADPH) in red blood cells and therefore is required to prevent oxidative damage. Variants with low activity resulting from amino acid substitutions, including the G6PD-202A allele, are enriched in sub-Saharan African populations and arose 2500–6500 years ago (Verrelli et al., 2002). Presumably, this allelic variant became enriched in populations as a result of balancing selection, because it conferred resistance to malarial disease in heterozygous females and hemizygous males (Tishkoff et al., 2001; Watkins et al., 2003).

These examples illustrate the role of environmental exposures, including pathogens and dietary components, as selective forces that facilitated the fixation of alleles that alter the utilization and metabolism of dietary components. Adaptive alleles may become recessive disease alleles, or disease alleles even in heterozygote individuals, when the environmental conditions change profoundly, such as those brought about by the advent of civilization and agriculture, including alterations in the nature and abundance of the food supply (Inoue et al., 1997; Wright et al., 1999; Baier et al., 2000; Akey et al., 2002; Clark et al., 2003; Wolfe and Li, 2003; Wray et al., 2003; Wooding, 2004; Wooding et al., 2004). Adaptive alleles may be responsible for the generation of metabolic disease alleles both within and across ethnically diverse human populations, and therefore are strong, nonbiased candidate genes for disease association studies; the interacting and modifying environmental factors can be inferred from the nutrients and/or metabolites that are known to interact with the gene product (Tishkoff and Verrelli, 2003).

Nutritional Epigenetics

Traits can be inherited from one generation to the next through either genetic or epigenetic mechanisms. Classical genetic inheritance refers to the transmission of DNA primary sequence from one generation to the next. Concordance among monozygotic twins illustrates the predominant yet nonexclusive contribution of DNA primary sequence to human phenotypes; other modes of inheritance must also be operative (Dennis, 2003). Epigenetics refers to the inheritance of traits through mechanisms that are independent of DNA primary sequence. There are now many examples demonstrating the inheritance of gene expression patterns and/or levels independent of DNA primary sequence, and such differences can elicit phenotypic differences among individuals including monozygotic twins through multiple generations (Dennis, 2003).

Interest in the relationships among epigenetic events and human nutrition was ignited by the fetal origins of adult disease hypothesis, originally put forward by David Barker and colleagues (Barker, 1997). Barker proposed that nutrition acts very early in life to *program* risk for adverse outcomes in adult life (Fig. 8–2). The notion that phenotypic plasticity was associated with in utero environmental exposures had been validated in the toxicology literature, but not well considered in the nutrition literature (Blake et al., 2005). Until recently, Barker's hypothesis was supported only by epidemiological associations among early nutritional exposures and increased risk in adulthood for obesity, hypertension, and insulin resistance, which are the antecedents of adult chronic disease, diseases that include cardiovascular disease (CVD), diabetes, and metabolic syndrome (Barker, 1997; Rasmussen, 2001). But now, there is an emerging, basic science literature that supports the concept that fetal environment can, in fact, program or reprogram the fetal genome with life-long consequences. The human genome has evolved in the context of a nutrient environment that was often scant and always unpredictable. Hence, organisms developed the capacity to "sense" and "adapt" to the food supply. These genomics adaptations have been referred to as "metabolic imprinting" or "metabolic programming." Such adaptations occur within critical windows in development, are seemingly irreversible and permit in utero survival in context of a suboptimal nutrient environment, but predispose the affected individual to metabolic disease in adulthood (Waterland and Garza, 1999).

Waterland and Garza described specific criteria to differentiate adaptive metabolic imprinting phenomenon from toxicological responses that resulted in permanent genomic alterations during the affected individual's lifetime. These criteria include (1) a susceptibility window limited to a critical ontogenic window in development, (2) a persistent effect lasting through adulthood, (3) a specific and measurable outcome, and (4) a dose–response or threshold relation between a specific exposure and outcome (Waterland and Garza, 1999). Mechanisms for metabolic imprinting are beginning to be understood and modeled in animal systems, some examples are illustrated in the following section. Metabolic programming mechanisms are more complex than those associated with toxic or deficiency states. Whereas many teratogens are exogenous agents that disrupt biological processes, metabolic imprinting is a conserved and adaptive response that optimizes biological function in one life stage through genomic mechanisms result in permanent functional characteristics. These imprints, however, may prevent or limit the range of other adaptive mechanisms that are protective in subsequent life stages when environments change, e.g., in a transition from "want" to "surplus." Thus, once the program is established, a system's buffering capacity is limited. The novelty of sustained surplus food present in most western cultures may explain some of the limited response capability apparently at the core of the present obesity/type 2 diabetes epidemic.

Genome programming is now known to result from chemical modifications of chromatin, either DNA or histone proteins, at a specific loci that leads to programming of transcriptional and metabolic networks. DNA is modified by methylation of cytosine deoxyribonucleotides present in the sequence CpG. Cytosine methylation is usually associated with gene silencing; methylcytosine is bound by methylcytosine-binding proteins that heterochromatize DNA and hence silence the gene. Histone proteins are essential components of chromatin and are subject to modification by methylation, phosphorylation, ubiquitination, biotinylation, ribosylation, and acetylation all of which modify gene transcription efficiency and can influence DNA stability (Henikoff et al., 2004; Gravel and Narang, 2005; Kimmins and Sassone-Corsi, 2005). Alterations in DNA and histone methylation constitute the most likely epigenetic signatures that enable genome programming because of their potential connections to metabolic networks, chromatin structure, and transcriptional networks. Furthermore, DNA and histone methylations are metastable, heritable, and alter genome expression and stability. Methylation is a higher order genomic signal that can override transient metabolic or hormonal signals such as the regulation of transcriptional networks through nuclear receptors (e.g., vitamin A, vitamin D, and steroid hormones).

The molecular mechanisms that describe the interactions among nutrients, metabolism, and gene expression/genome programming are mostly unknown. The following section outlines two of the best-characterized examples of genome programming by nutrition.

Example 1: Maternal Folate, One-carbon Metabolism, Fetal Genome Programming, and In Utero Survival

Folate is a B vitamin and a family of metabolic cofactors that carry and chemically activate one-carbon units for the de novo synthesis of purine nucleotides and thymidylate (dTMP), and for the remethylation of homocysteine to methionine, a metabolic network known as folate-mediated one-carbon metabolism (Fig. 8–3). Methionine can in turn be adenylated to form *S*-adenosylmethionine (AdoMet), which is a cofactor for numerous cellular methylation reactions including histone and DNA methylation (Stover, 2004a). Impairments in this metabolic network by nutritional deficiencies or highly penetrant SNPs increase risk for pathologies that include cancer and CVD, and DAs such as spontaneous abortion (SA) and NTDs (Stover, 2004a). Folate supplementation can reduce risk for these disorders; maximal benefit is achieved in genetically susceptible individuals/populations.

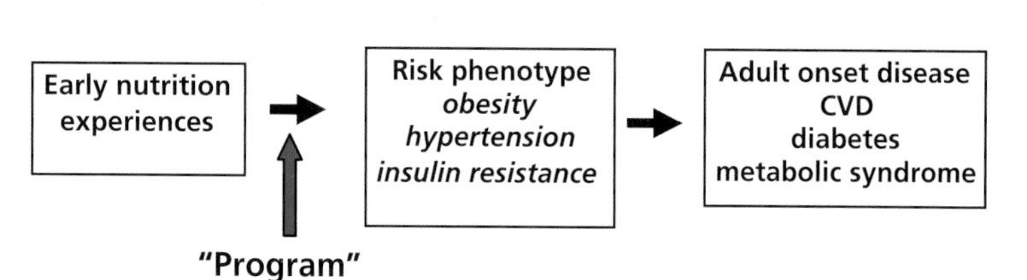

Figure 8–2 The fetal origins of disease hypothesis. Fetal environmental exposures, especially nutrition, act in early life to program risk for adult health outcomes.

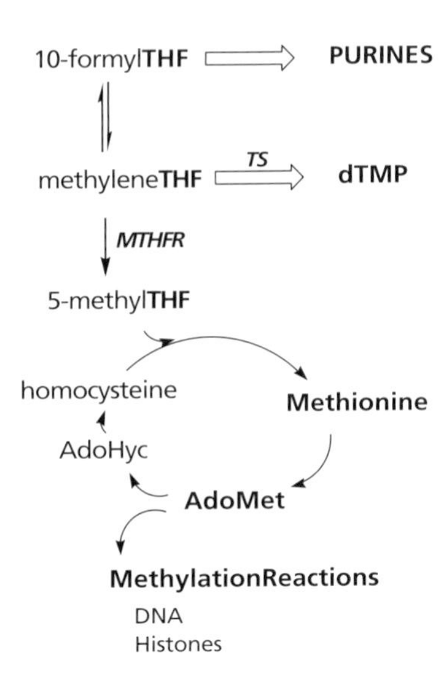

Figure 8–3 Folate-mediated one-carbon metabolism. Tetrahydrofolate (THF)-mediated one-carbon metabolism is required for the synthesis of purines, thymidylate and methionine. MTHFR, methylenetetrahydrofolate reductase; TS, thymidylate synthase; AdoMet, *S*-adenosylmethionine; AdoHcy, *S*-adenosylhomocysteine.

Methionine and dTMP synthesis are the most vulnerable pathways within the network; their impairments compromise the fidelity of DNA synthesis and cellular methylation reactions (Blount et al., 1997; Suh et al., 2001; Stover, 2004a). AdoMet-dependent methyltransferases, including histone and DNA methyltransferases, are subject to product inhibition by *S*-adenosylhomocysteine (AdoHcy), which accumulates during folate deficiency (Fig. 8–3) (Finkelstein, 2000a,b; Clarke and Banfield, 2001). Hence, methylation of chromatin is sensitive to the efficiency the one-carbon metabolic network; methyltransferases sense the efficiency of the folate metabolic network, because their activity is determined by the cellular AdoMet/AdoHcy ratio, otherwise known as the "methylation potential" of the cell (Finkelstein, 2000a,b). Global genomic methylcytosine content is highly sensitive to the AdoMet/AdoHcy ratio, which can affect both gene expression and DNA stability (Choi and Mason, 2000).

The mechanisms underlying folate-associated pathologies, including NTDs, SA, and cancers are assumed to be the result of insufficient flux through the dTMP and/or AdoMet synthesis pathways. Therefore, the etiology involves either impairments in genome synthesis (mitotic) rates, genome stability, and/or methylation-sensitive gene expression (Stover, 2004b). Numerous studies have indicated that folate-dependent dTMP synthesis, catalyzed by thymidylate synthase (TS) and 5-methylTHF synthesis (leading to AdoMet synthesis) catalyzed by MTHFR, are competitive pathways within the one-carbon network. They compete for a limiting pool of the cofactor methylenetetrahydrofolate (methyleneTHF) (Fig. 8–3) (Finkelstein, 2000a,b; Stover, 2004b). This metabolic competition is unbalanced by the previously described MTHFR A222V polymorphism. This functional SNP reduces MTHFR activity and has two effects on the network. It impairs the remethylation of homocysteine to methionine, thereby reducing global DNA methylation and thus also influences gene expression (Oyama et al., 2004; Shelnutt et al., 2004). It also increases the conversion of deoxyuridine monophosphate (dUMP) to dTMP (Quinlivan et al., 2005). These changes in the network are associated with increased risk for SA and NTDs, but decreased risk for adult colon cancer (Bailey, 2003), illustrating that optimal network function or outputs differs between the fetal and adult environments. Identifying the precise mechanism for folate-related pathologies in experimental systems is challenging, because any factor, genetic or environmental, that influences the metabolic competition for methyleneTHF may simultaneously alter the efficiency of both dTMP and AdoMet synthesis (Fig. 8–3).

Alterations in one-carbon metabolism, and the AdoMet cycle in particular, can have dramatic effects on genome methylation. Both genome-wide and allele-specific DNA methylation is influenced by alterations in folate metabolism (Zingg and Jones, 1997). DNA hypomethylation induced by folate deficiency alters transcription of genes regulated by promoter methylation, including tumor suppressor genes (Zingg and Jones, 1997; Mason and Kim, 1999), and enables interchromosomal recombination events through common retroviral repeat sequences whose activity normally is silenced by methylation (Kim et al., 2004). Patients with hyperhomocysteinemia, a clinical state that results from the inability to effectively metabolize homocysteine, accumulate cellular AdoHcy and exhibit alterations in gene expression. Patients exhibit DNA hypomethylation and a homocysteine-dependent shift from monoallelic to biallelic expression of genetically imprinted genes including H19. Folate supplementation in these patients restores homocysteine levels to baseline, reverses global DNA hypomethylation, and restores monoallelic expression of imprinted genes (Ingrosso et al., 2003). Interestingly, SNPs in the H19 gene are associated with cord blood IGF-II levels and birth size (Petry et al., 2005).

Folate-mediated alterations in genome methylation can be set irreversibly or "imprinted" during early development. In the viable yellow agouti (A^{vy}) mouse model, maternal diet determines the coat color of offspring (Waterland and Jirtle, 2003). This mouse strain contains an intracisternal A-particle (IAP) element, which is a retrotransposon that integrated into a 5′ exon of the agouti gene resulting in cryptic and constitutive expression of the agouti gene. This aberrant expression of the agouti results in a "yellow" coat color and an obesity phenotype (Morgan et al., 1999). The IAP retroviral element also attracts DNA methylation to that locus, and the degree of methylation determines the expression level of the agouti gene.

The A^{vy} mouse is sensitive to maternal folate and one-carbon status during gestation. Within a critical window in development, maternal diet determines the density of cytosine methylation at the agouti locus and hence the level of agouti gene transcription, coat color and propensity for obesity (Waterland and Jirtle, 2004). The methylation patterns and subsequent effects on coat color and, presumably, associated metabolic characteristics are maintained throughout the lifetime of experimental animals (Waterland and Jirtle, 2003) and are heritable (Morgan et al., 1999). The identification of other genes that are influenced by alterations in the AdoMet/AdoHcy ratio through chromatin modifications, and the critical developmental windows that enable genome programming, are essential to elucidate the mechanisms of folate-related pathologies and DAs. Equally important, this mouse model illustrates the concept that epigenetic modifications in the developing embryo induced by maternal diet can "rescue" deleterious genetic insults, such as retroviral insertions in gene promoters and restore the "normal" phenotype.

The ability of maternal folate and one-carbon sources to compensate or "rescue" genetic deficiencies may not be limited to the A^{vy} mouse model, and therefore may have implications for women taking nutritional supplements during pregnancy at intake levels that exceed dietary recommendations. Fetal genotypes that cannot support basic biological processes in the embryonic and fetal stages usually are eliminated. In primates, this is achieved by SA. Humans may be unique compared to other mammalian species in their high rates of fetal loss (Delhanty, 2001); high SA rates may be a selective pressure that accelerates the expansion of polymorphic alleles within human populations. Approximately 75% of human conceptions are lost spontaneously before term, 80% of all SA occur within the first trimester (Edmonds et al., 1982; Brent and Beckman, 1994; Edwards, 1997). It is estimated that half of SA occur before the first 3 weeks of gestation and generally are unnoticed; many embryos fail to implant (Wilcox et al., 1988).

Risk for SA increases and fertility decreases in women over the age of 30 years (Brock and Holloway, 1990; Bulletti et al., 1996). Although the etiologies of SA are generally not established, the etiologies of most SA are likely multifactorial. Many SA fetuses have structural and/or genetic anomalies; SA may have evolved to limit defective offspring in human populations (Cowchock et al., 1993; Brent and Beckman, 1994). Potential inducers of SA include maternal immune responses, fetal genotypes, maternal and/or fetal endocrine, nutritional or hormonal imbalances, maternal and/or fetal infection, and/or endometriosis (Table 8–3) (Bulletti et al., 1996). Few specific environmental risk factors for SA have been identified, but include low maternal folate status, diabetes (type 1), and elevated homocysteine (which most often results secondarily to primary or conditioned folate deficiency) (Nelen, Blom, Steegers, den Heijer, and Eskes, 2000; Nelen, Blom, Steegers, den Heijer, Thomas, et al., 2000; Nelen, 2001; Zetterberg et al., 2002, 2003, 2004; Gris et al., 2003). For example, variant alleles of MTHFR (A222V) (Isotalo et al., 2000; Zetterberg, Regland, Palmer, Ricksten, et al., 2002; Zetterberg, Regland, Palmer, Rymo, et al., 2002) and TCII (776G) (Zetterberg, Regland, Palmer, Rymo, et al., 2002) impair the metabolism of folic acid and homocysteine and are independent risk factors for SA.

DAs and SA have both independent and shared etiologies (Table 8–3) (Brent and Beckman, 1994). The underlying mechanisms for more than 75% of DA are unknown and assumed to be multifactorial; only 15% of those whose etiologies have been identified are solely genetic, including chromosomal abnormalities or autosomal/sex-linked genetic disease. Ten percent of DA are attributed to environmental factors, 4% of which are attributed to disruptions in maternal/fetal metabolism or suboptimal nutrition that includes micronutrient under nutrition, starvation, PKU, diabetes, alcoholism, etc. Infectious agents account for another 4% of DA, mechanical disruption accounts for 2% and known chemical/prescription toxins account for less than 1% of DA (Brent and Beckman, 1994). Polymorphic variants of two genes that encode folate-dependent enzymes, MTHFS A222V and MTHFD1 R653Q, are associated with risk for DAs, including NTDs (Brody et al., 2002). Interestingly, human alleles associated with DA that encode folate-dependent metabolic enzymes are not in Hardy–Weinberg equilibrium in some studies (alleles are not inherited at the expected frequency), consistent with evidence that elevated homocysteine are risk factors for spontaneous miscarriage and decreased fetal viability (Nelen et al.,1998; Nelen, Blom, Steegers, den Heijer, and Eskes, 2000; Nelen, Blom, Steegers, den Heijer, Thomas, et al., 2000; Brody et al., 2002; Stover, 2004b).

The concept that embryos can be rescued by maternal nutritional status, or "good diet hides genetic mutations" (Pennisi, 2002) is suggested by numerous examples of nutritional rescue or compensation (viability or phenotype) of gene disruptions through diet in mice and yeast (Pasqualetti et al., 2001; Zhao et al., 2001; Finnell et al., 2002; Stover and Garza, 2002; Pal et al., 2003; Stover, 2004). Individual nutrients can rescue severe genetic lesions in mice when administered in supra-physiological levels during critical developmental windows. Maternal retinoic acid administration between 7.5 and 9.5 days postconception rescued deafness and inner ear development in Hoxa1$^{-/-}$mice (Pasqualetti et al., 2001), and folic acid can rescue skeletal defects associated with deletion of a *Hox* gene, as well as NTDs in mice that have no evidence of disrupted folate metabolism (Pennisi, 2002). The most comprehensive studies have been performed in yeast. Gene deletion studies indicate that 80% of yeast genes are nonessential for survival under laboratory conditions.

Table 8–3 Maternal Risk Genotypes and Reproductive Outcomes

Gene Variant	Pathway	Fetal Risk	References
MTHFR V222A	One-carbon metabolism	SA	(Bailey and Gregory, 1999)
		NTD	
		Down syndrome	
		Adult CVD	
MTFD1	One-carbon metabolism	NTD	(Brody et al., 2002)
TC(transcobalamin)	Vitamin B_{12}/one-carbon metabolism	SA	(Zetterberg, 2004)
IL6 (-174 G→C)	Cytokine	SA	(von Linsingen et al., 2005)
IFN-gamma 874 A→T	Cytokine	SA	(Daher et al., 2003)(Prigoshin et al., 2004)
IL1RN*1	Cytokine	SA	(Perni et al., 2004)
IL1RN*2	Cytokine	Preterm birth	(Perni et al., 2004)
CYP17A2	Steroid biosynthesis	SA	(Sata, F, Yamada, H, Kondo, T, et al., 2003)
CYP1A1*2A	Phase 1detox	SA	(Suryanarayana et al., 2004)
PR*2	Progesterone receptor	SA	(Schweikert et al., 2004)
GSTM1	Phase 2 detox	SA	(Sata, F, Yamada, H, Yamada, A, et al., 2003)
Prothrombin G20210A	Clotting	SA	(Finan et al., 2002)
Factor V G1691A	Clotting	SA	(Finan et al., 2002)
Nos3B	Vascular function	SA	(Tempfer et al., 2001)
PGM1*2	Phosphoglucomutase	SA	(Gloria-Bottini et al., 2001)

A recent examination of yeast metabolic networks using an in-silico model revealed that culture conditions, especially the use of nutrient-rich culture media, can compensate for the disruption of 37%–68% of the organism's genes. In microbial systems, the maintenance of enzymatic flux under highly diverse environmental conditions appears to be a primary selective pressure that maintains gene sequence, e.g., starvation being among the most common environmental stresses (Gasch and Werner-Washburne, 2002). The concept of nutritional rescue of genetic mutations is exceptionally salient to human modernity because of the unprecedented degree to which we can manipulate our nutritional environments.

There have been some studies that have addressed the concept of "nutritional rescue" of human embryos. The risk for both DA and SA conferred by MTHFR and TCII polymorphic alleles, and experience that folic acid lowers DA risk associated with MTHFR SNPs, has led to the suggestion that women predisposed to hyperhomocysteinemia might benefit from folate and vitamin B_{12} supplementation to reduce their risk of SA (Zetterberg, Regland, Palmer, Ricksten, et al., 2002; Zetterberg, Regland, Palmer, Rymo, et al., 2002; Zetterberg et al., 2003, 2004). The only human study to validate the concept of embryo rescue through nutrition in humans was performed in southern Spain. This, and a subsequent follow-up study, indicated that maternal folate supplementation increased the prevalence of the MTHFR A222V allele by rescuing fetuses at risk of SA in this population (Munoz-Moran et al., 1998; Reyes-Engel et al., 2002). This observation has yet to be confirmed in other populations and has been criticized for its experimental design (Whitehead, 1998). Nonetheless, the long-term consequences of fortification and supplementation interventions that result in maternal nutrient intakes above what can be achieved with healthy, unfortified food-based diets, and their potential rescue of fetuses at risk of SA, potentially present public health and ethical challenges to communities and health professionals that serve them.

Clearly, maternal folate and other methyl donor supplementation alter the methylation status of targeted alleles in the mouse embryo, and methylation patterns and subsequent effects on gene expression persist throughout adulthood (Waterland and Jirtle, 2004). Epigenetic phenomenon may provide mechanistic insight into the many observational studies that associate risk for adult chronic disease with maternal nutrition and embryonic nutrient exposures as proposed by Barker and colleagues (Waterland and Garza, 1999).

Example 2: Glucocorticoids and Metabolic Disease: Modification of the Placental Barrier by Maternal Nutrition

The consequences of fetal glucocorticoid (GC) exposure on adult chronic disease provide some of the best supporting evidence for the fetal origins of disease hypothesis (Seckl and Walker, 2001; Seckl, 2004; Seckl and Meaney, 2004; McMillen and Robinson, 2005). Complementary human clinical and animal studies have revealed the long-term consequences and associated mechanisms of fetal GC exposure (Fig. 8–4) (Reinisch et al., 1978; French et al., 1999; Newnham et al., 1999). Fetal GC levels are maintained at low concentrations relative to maternal concentrations, primarily through the action of placental 11β-hydroxysteroid dehydrogenase type 2 (11β-HSD2) that catalyzes the oxidative inactivation of cortisol and corticosterone (White et al., 1997). Elevated fetal exposures to GC during late gestation (Nyirenda et al., 1998) (which can result from 11β-HSD2 inhibitors, rare mutations in the human *11β-HSD2* gene,

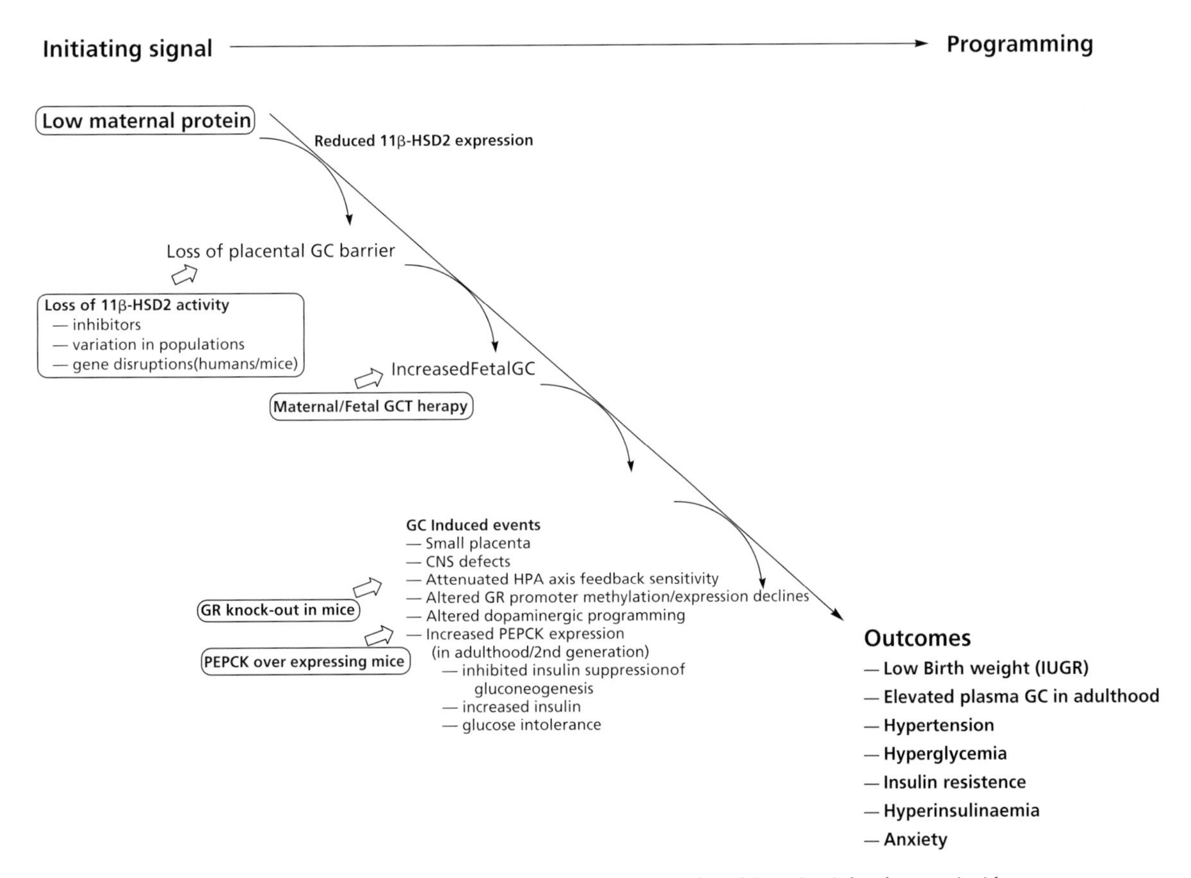

Figure 8–4 Programming of the hypothalamic-pituitary-adrenal (HPA) axis by glucocorticoids.

or large existing variation in placental 11β-HSD2 activity among humans) have life-long consequences for the fetus including low birthweight (intrauterine growth restriction; IUGR), elevated plasma GC, hypertension, hyperglycemia, insulin resistance, hyperinsulinemia, and anxiety (Seckl, 2004). IUGR and preeclampsia also are associated with elevated cortisol and low birthweight (Goland et al., 1993, 1995). Furthermore, elevated GC and the metabolic syndrome (i.e., the combination of type 2 diabetes/insulin resistance, dyslipidemia, and hypertension) are also characteristic of Cushing's syndrome (Seckl and Walker, 2001).

Interestingly, low maternal dietary protein intake during gestation causes a specific loss of placental 11β-HSD2 expression and similar fetal outcomes as elevated fetal GC exposure (Bertram et al., 2001). 11β-HSD2 activity in placenta correlates with birthweight (Stewart et al., 1994) and disruption of the murine gene encoding 11β-HSD2 reduces birthweight (Seckl, 2004). Similarly, obstetric GC therapy to accelerate lung development prior to anticipated preterm deliveries increases risk for reduced fetal birthweight, and long-term susceptibility to hypertension, hyperglycemia, CVD, and increased hypothalamic–pituitaryadrenal axis (HPA) activity. These disorders not only persist into adulthood, but also into the next generation (Drake et al., 2005).

GCs are steroid hormones that serve as ligands for the glucocorticoid receptor (GR), a member of the nuclear receptor super family (Hebbar and Archer, 2003). GRs are complexed with heat shock proteins and do not affect transcription in the absence of bound ligand. With GC bound, GR remodels chromatin through transient interactions that recruit remodeling proteins to a defined locus and "open" chromatin thereby enabling transcription factor binding (Hebbar and Archer, 2003). GR also may recruit transcription factors directly to the transcriptional preinitiation complex (Hebbar and Archer, 2003), and may target promoter demethylation (Thomassin et al., 2001). GCs are required for normal central nervous system (CNS) development, apoptosis, and synapse formation (Weinstock, 2001). Both GR and 11β-HSD2 are present in the developing brain; 11β-HSD2 expression begins to dissipate and brain GC accumulates from week 19 to 26 during end stages of neurogenesis (Stewart et al., 1994, 1995). Life-long consequences associated with fetal GC exposure may result from premature GR-mediated chromatin remodeling in the hippocampus (Seckl, 2004). Interestingly, GR programming resulting from fetal GC exposure, and its deleterious repercussions, can be erased in the adult animal by treatment with histone deacetylase inhibitors (Seckl, 2004).

GC homeostasis is maintained by the HPA axis, which is imprinted or programmed by fetal GC exposures during gestation. Plasma GC concentrations are regulated normally by a feedback loop that involves GR in the hippocampus. Persistent prenatal GC exposure decreases fetal GR expression, and these expression levels can be set or memorized with life-long consequences. Low hippocampal GR levels increase adult corticosterone plasma levels and thereby may reinforce the decreased GR expression levels. Maternal undernutrition can elicit the same effect, presumably by decreasing placental 11β-HSD2 and thereby program hypertension and hyperglycemia in the fetus. Fetal GC exposure also affects the dopaminergic system (Diaz et al., 1997) and alters amygdale function, including the regulation of fear and anxiety (Dunn and Berridge, 1990). In addition, maternal GC exposure affects glucose and insulin homeostasis (Gatford et al., 2000) by programming hepatic phosphoenolpyruvate carboxykinase (PEPCK) levels with effects that persist into adulthood (Rosella et al., 1993; Valera et al., 1994) and are carried to the next generation (Drake et al., 2005).

Interestingly, GR levels also can be imprinted epigenetically and postnatally in rodents through maternal behavior including handling, licking, and grooming (Seckl and Meaney, 2004). At day 20 postconception, the GR exon 17 promoter is unmethylated. One week after birth, offspring from low licking/grooming-arched-back nursing mothers uniformly exhibited methylated CpG islands in the NGFI-A cis-element located within the exon 17 promoter and elevated GR expression in the hippocampus, whereas offspring from high licking/grooming-arched-back nursing mothers rarely methylated this sequence (Weaver, Cervoni, et al., 2004; Weaver, Diorio, et al., 2004). The established DNA methylation pattern is specific to that sequence, and the imprinting window is limited to the first week of life. These elevations in GR expression lower plasma GC levels and the stress response through the HPA axis throughout the animal's lifetime.

Nutritional and Metabolic Engineering: Nutritional Genomics Application and Public Health

The use of genetic information in current nutrition practice and policy is very limited, but such consideration has the potential to "personalize" nutrition and thereby prevent chronic disease and promote healthy aging through diet by targeting the molecular antecedents of disease. The successes of the human genome sequencing efforts are enabling the comprehensive identification of all proteins and functional RNA molecules, and variants thereof, and expanding knowledge of individual gene function(s). These accomplishments are facilitating the assembly of gene products into biological networks and enabling the study of the complex and iterative interactions that occur among metabolic, signaling and gene expression networks in the context of human health and disease. Cellular networks are sensitive to internal perturbations that result from genetic variation and gene mutations; identification of the gene variant and/or metabolites that induce network dysfunction can be informative in the diagnoses or prediction of various health and disease outcomes. For example, classic monogenic disorders, including the inborn errors of metabolism, PKU, and galactosemia, illustrate the severe consequences that can result from catastrophic metabolic network disruptions. Perhaps more importantly, these early clinical studies also demonstrated that single-gene disorders can be managed and/or alleviated through dietary interventions (e.g., the use of phenylalanine-restricted diets to prevent or mitigate the cognitive impairments resulting from mutations in the phenylalanine hydroxylase gene) and thereby established the principles of metabolic engineering. Nutrients, like pharmaceuticals, are powerful modifiers of genome and network function and stability, and gene–nutrient interactions can be optimized for disease prevention and management (Fig. 8–1).

Personalized Nutrition

Conrad Waddington (1905–1975) was among the first to recognize the dynamic behavior of biological systems that permits survival in the midst of a myriad of environmental challenges. He proposed the concept of canalization, a now classic idea. Canalization is a measure of an organism's ability to buffer against the penetrance of genetic polymorphism and environmental challenges to preserve developmental pathways (Ruden et al., 2003, 2005). It may have evolved through stabilizing natural selection to generate robust developmental processes and optimal phenotypes that are resilient to nutritional challenges (Gibson and Wagner, 2000; Siegal and Bergman,

2002; Milton et al., 2003). Canalization is an inherent property of biological networks, (metabolic, transcriptional, and signaling) (Siegal and Bergman, 2002) and in this context describes the buffering capacity that can be achieved by altering a network's dynamics. The prevalence, limits, and functional capacity of network canalization are unknown, and the underlying mechanisms that conceal phenotypic variation and enable environmental tolerance are subjects of investigation (Ruden et al., 2003). Buffering environmental and genetic perturbations through dynamics requires that the networks have the capacity to sense the nutritional and metabolic milieu, and to allow for network adaptations that preserve the functional output. Discovering and elucidating the mechanisms for nutrient sensing and nutrient-mediated network adaptation are critical for the rational design of nutrigenomic approaches and interventions that seek to manipulate genome function with a high predictive health benefit and minimized risk of unintended and deleterious consequences.

Salient examples of nutritional engineering are provided by past experiences that demonstrate maternal and perinatal nutrition can improve birth outcomes including cognitive development (e.g., ensuring iodine sufficiency), life-long chronic disease resistance (e.g., preventing small for gestational age births), and increased longevity (e.g., optimizing dietary fat consumption and immune function) (Kornman et al., 2004). It also is apparent that inappropriate uses of nutrition to maximize reproductive outcomes will present new potential risks (Stover and Garza, 2002; Stover, 2004a). High-dose vitamin therapy has been advocated to rescue impaired metabolic reactions that result from mutations and polymorphisms that decrease the affinity of substrates and cofactors for the encoded enzyme (Ames et al., 2002). ω-3 fatty acids and tocopherols may promote healthy aging and longevity by modulating the inflammatory response by altering gene transcription (Kornman et al., 2004). Genes and their allelic variants that influence longevity (a trait that is nonadaptive) are being identified at accelerated rates (Vijg and Suh, 2005), and their penetrance can be modified by the rational design of nutrition-based interventions and therapies (Kornman et al., 2004). Repression of energy metabolism, through caloric restriction or transcriptional regulation of metabolic enzymes, reduces oxidative stress and promotes longevity in many experimental model systems. Manipulation of these transcriptional and/or metabolic networks by designer vitamin supplements may promote healthy aging. However, caution is warranted. Genes encoding virtually all physiological process are not adapted to excessive nutrient intake exposures that exceed what has been achieved in historical and healthy food-based diets. Therefore, new risks and toxicities should be anticipated in human populations or population subgroups when nutrients are administered at pharmacological levels, as illustrated by the introduction of high levels of fructose into the food supply (Cox, 2002). Some of the unintended consequences may involve genome programming, including permanent alterations in genome-wide methylation patterns in stem cell populations as observed in mouse embryos whose mothers received elevated dose of folic acid and one-carbon donors during gestation (Waterland and Jirtle, 2003). Methylation patterns that are established in utero, and perhaps in adult stem cell populations, can be metastable and influence gene expression and potentially mutation rates throughout the lifespan (Waterland and Jirtle, 2004). Furthermore, although antioxidants can decrease mutation rates, they can also function as prooxidants in vivo (Seifried et al., 2004) and may be cancer promoting when consumed at elevated intakes by inhibiting cellular death programs in transformed cells (Zeisel, 2004). In conclusion, elucidation of robust gene-by-nutrient interactions will inform dietary approaches for individuals and for populations that aim to prevent and/or manage complex metabolic disease, as has been accomplished for rare inborn errors of metabolism. Equally important, these and other examples indicate that rigorous hazard identification is essential prior to the establishment of policies that result in pharmacological intakes of nutrients and other dietary components.

Assisted Reproduction and Optimal Culture Medium

Elevated SA rates have been observed in some but not all studies of human in vitro fertilization (IVF) pregnancies compared to natural conceptions. These findings that may be the result of early harvest and early manipulations of eggs and embryos in culture medium (Stillman et al., 1986; Bulletti et al., 1996). Furthermore, other studies have found that human IVF procedures result in higher than expected incidences of IUGR (De Rycke et al., 2002). Numerous studies have shown that the composition of embryo culture medium affects the expression and methylation status of imprinted genes including *H19*, *Igf2*, and *Igf2r* in ovine and other mammalian embryos, resulting in large offspring syndrome (Sinclair et al., 1998; Khosla et al., 2001; Young et al., 2001; De Rycke et al., 2002; Gao and Latham, 2004). There appear to be many critical windows associated with the establishment of environmentally sensitive methylation patterns from early embryogenesis through the suckling period, all of which are cell-type and/or allele specific (Weaver, Cervoni, et al., 2004; Weaver, Diorio, et al., 2004). Some of these networks are sensitive to folate-mediated one-carbon metabolism, as illustrated by the impact of maternal nutrition on genomic methylation in the viable A^{vy} mouse (Waterland and Jirtle, 2003); other networks may respond to the allelic- or locus- specific targeting of methylase/demethylase activity as seen with GC receptor programming described earlier (Thomassin et al., 2001). The increasing evidence that early nutritional exposures can increase risk for late-onset metabolic diseases through epigenetic mechanisms illuminate the major challenges that are made immediate by the increased demand for assisted reproduction.

Dietary Recommendations for Populations

Food- and nutrient-based dietary guidelines were established to assist individuals and populations achieve adequate dietary patterns to maintain health. The derivation and goals of these guidelines evolve as new knowledge becomes available (Schneeman and Mendelson, 2002; Schneeman, 2003). Guidelines for single nutrients and other food components are scientifically and quantitatively derived, and usually based on the level of nutrient intake that prevents a clinical and/or biochemical outcome that is associated with a particular nutrient deficiency. Numeric standards for nutrients are essential to validate the efficacy of food-based guidelines (Schneeman, 2003). Nutrient requirements vary among individuals within all human populations, and can be modified by age, gender, and life stage, among other variables. Therefore, recommended nutrient intakes are often derived separately for population subgroups. Although genetic variation can modify the efficacy, dosage, and safety of pharmaceutical agents (Weinshilboum, 2003) and tolerance/intolerance for certain foods (Enattah et al., 2002), the contribution of genetics to nutrient requirements within and among human populations remains to be evaluated. However, the characterization of gene variants that modify optimal nutrient requirements will enable the classification of genetic subgroups for whom generalized nutritional requirements may be valid.

Recommended Daily Allowance and Upper Limit

The "recommended dietary allowance" (RDA) for each nutrient is defined as the level of dietary intake that is sufficient to meet the requirement of 97% of healthy individuals in a particular life stage and gender group (Fig. 8–5). When there is insufficient data to calculate an RDA for a nutrient, an "adequate intake" (AI), which is an estimated recommended intake value, is established. Some nutrients demonstrate toxicities at elevated intake levels, and therefore a "tolerable upper intake level" (UL), which represents the highest level of nutrient intake that can be achieved without incurring risk for adverse health effects for the vast majority of individuals in the general population, is established.

Human genetic variation is not anticipated to confer extreme variations in optimal nutrient requirement among individuals and populations. Nutrition, unlike pharmaceuticals, is an in utero and life-long exposure that can serve as a selective pressure to eliminate genomes that are not compatible with the nutrient environment. Therefore, human genotypes that do not support basic physiological processes will be selected against in large part because of fetal loss and/or the failure to survive to reproduce. Allelic variants that confer atypical nutrient requirements that cannot be met by the mother or that result in severe metabolic disruptions are expected to be embryonic lethal. Alleles that confer more subtle differences in nutrient requirements or food tolerances are expected to be enriched in subgroups or populations, and contribute to disease in certain environmental contexts (Tables 8–1 and 8–2).

The concept of generalized nutrient requirements within and among human populations is nullified only when a level of nutrient intake that represents minimal nutrient adequacy for one genetic subgroup exceeds a safe intake level for another group, assuming that nutrient deficiency and harm/toxicity avoidance are the primary criteria for requirement. For example, it is widely recognized that optimal folate intakes may differ among identifiable genetic subgroups. However, it is not at all certain that the magnitude of the genetic contribution to variations in adequate dietary folate intake warrants genotype-specific recommendations, especially considering that folic acid intakes up to 1 mg/day are not associated with known toxicities (Shelnutt et al., 2004). Iron is another candidate nutrient for genotype-specific nutrient requirement

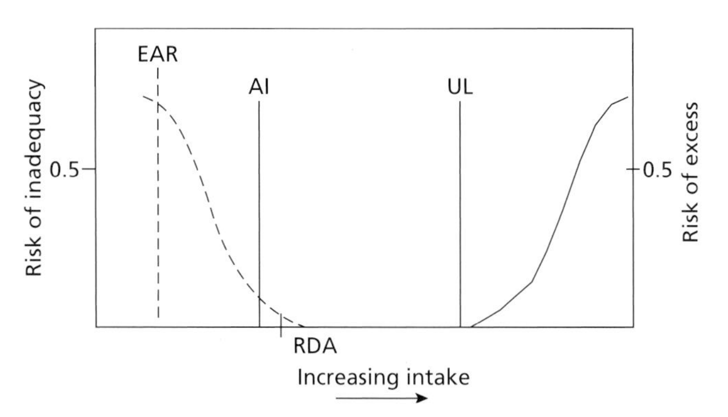

Figure 8–5 Dietary requirements. Estimated average requirement (EAR) represents the intake at which the risk of inadequacy is 0.5 (50%) to an individual. Recommended dietary allowance (RDA) represents a level of nutrient intake at which the risk of inadequacy is less than 3%. Adequate intake (AI) is not based on an EAR but is assumed to be near or above the RDA if one could be calculated. The tolerable upper intake level (UL) represents risk of excess that is set close to 0.

recommendations (Moirand et al., 2002; Swanson, 2003; Beutler, 2004; Hunt and Zeng, 2004). For these and other cases, the penetrance (contribution of the individual allele to variation in nutrient requirements) and the prevalence of these functional gene variants must be elucidated both within and among human populations to validate the concept of generalized nutrient requirements for all human populations. Unlike the effects of gender and life cycle, no common allelic variant has been shown to be sufficiently penetrant and to warrant genotype-specific numeric standards for nutrient adequacy or upper levels of intake associated with harm or toxicity. At this time, genetic variation is known only to influence nutrient and food intolerance (Tables 8–1 and 8–2). However, genetic variation has not been characterized in many geographically and culturally isolated populations that have existed in nutrient-poor or otherwise unique nutritional environments for many generations, and therefore the presence of adaptive alleles should be expected in such populations. Recent experiences have demonstrated the severe adverse health consequences that can result from rapid alterations in dietary patterns, especially energy intake, among certain populations (Baier and Hanson, 2004).

Finally, to achieve dietary guidelines that optimize health for all individuals and populations, the many functions of individual genes, and their regulation within metabolic and transcriptional networks, must be understood comprehensively. Recently, it was shown that the *LCT* gene also encodes the enzyme pyridoxine-5′-β-D-glucoside as a result of differential processing of the LCT transcript (Tseung et al., 2004). Pyridoxine-5′-β-D-glucoside activity is necessary for the bioavailability of pyridoxine-5′-β-D-glucoside, the major form of vitamin B_6 in plant-derived foods (Mackey et al., 2004). Therefore, LCT variation predicts both lactose tolerance as well as preferred dietary sources of vitamin B_6 in adulthood.

Genomic Criteria for Setting Requirements and Toxicities

Genomic technologies may provide new criteria for establishing numeric standards for adequate levels of nutrient intake by targeting the molecular antecedents of disease. DNA mutations increase risk for degenerative diseases and cancers, and can be quantified in controlled experimental settings, indicating that the effect of mineral and vitamin intake level on somatic cell DNA mutation rates should be considered when establishing RDAs (Fenech, 2004). Marginal deficiencies in folate, vitamin B_{12}, niacin, and zinc can influence genome stability, and antioxidants including carotenoids, vitamin C, and vitamin E may prevent damage resulting from oxidative stress. Validation of these protective effects on DNA mutation rates in controlled human trials may indicate benefits and lead to increased recommended intake levels, perhaps at levels not normally achievable from a natural food-based diet. Similarly, the use of functional genomic approaches, including expression profiling and proteomics to quantify gene expression and metabolomics to quantify metabolic pathway flux, may provide a comprehensive set of quantitative and physiologically relevant "biomarkers" to model and assess nutrient efficacy in the context of optimal network function (Csete and Doyle, 2004).

Other genomic outcomes are emerging as criteria for hazard identification and may influence the establishment of tolerable upper levels of nutrient intake during pregnancy. Studies of animal models are revealing that nutrients can rescue deleterious genetic mutations, leading to the concept that "good diet hides genetic mutations" (Pennisi, 2002) as previously described. This rescue phenomenon is not established in humans, but animal studies indicate that nutrients can modify the viability of genomes, including genomes that confer atypical nutrient requirements to the surviving fetus (Finnell et al., 2002).

Nutrition Interventions

Public health interventions, including micronutrient food fortifications, often have been successful in preventing adverse outcomes associated with maternal nutrient deficiencies or excesses. Successful initiatives include the fortification of grain products with folate to prevent NTDs (Honein et al., 2001) and of salt with iodine to prevent cretinism (Delange, 2003), discouraging alcohol consumption during pregnancy to prevent fetal alcohol syndrome (CDC, 2004) and the fortification of various vehicles to prevent iron deficiency (Swanson, 2003). These successes encourage further improvements of maternal diets and infant formulas to optimize short- and long-term outcomes, and predict the likely accelerated development of dietary approaches that aim to optimize functional abilities rather than the more traditional goals of preventing deficiency and deficiency-related disease. For example, there is evidence that docosahexaenoic acid (DHA) (Mellott et al., 2004) or choline (Heird and Lapillonne, 2005) supplementation during prenatal and/or early postnatal life improves CNS function and cognitive performance throughout life. Continued progress in these areas requires the identification and understanding of nutritionally modifiable networks that enable successful adaptations in successive life stages, their associated critical windows, and validation of the efficacy, effectiveness, and safety of the nutritional interventions.

Summary

Advancements in nutritional genomics promise to reveal the role that diet has played in sculpting the human genome through the processes of mutation and selection. These fundamental processes of genome evolution led to the development of sophisticated nutrient sensing and regulatory mechanisms that permit homeostasis in diverse and ever-changing environmental settings. Human genetic variation that emerged as a consequence of adaptation becomes a present day determinant of nutrient efficacy, food and nutrient tolerances/intolerances and risk for metabolic disease, and may influence human nutritional requirements for disease prevention. Truly, we are what our ancestors ate! Genetic variation that affects the utilization of lactose, iron, and alcohol, and the associated food intolerances, display genomic signatures of positive selection, indicating that these variants offered survival advantage in specific geographic regions. Environmental shifts continue to challenge the human genome. Rapid and severe alterations in the nature and abundance of the food supply can unmask previously silent genetic variation and create new or more prevalent food intolerances, as occurred with the infusion of large quantities of fructose in the food supply. Currently, no single common gene variant has been demonstrated to be sufficiently penetrant to warrant genotype-specific nutrient intake recommendations, although the affect of the MTHFR A222V variant on folate requirements has been considered. However, many human populations have existed for generations in exotic, isolated, and challenging nutrient environments and such populations may carry relatively rare gene variants that confer unique dietary requirements that may be highly prevalent in historically isolated, stable human populations. In the near future, all human genetic variation is expected to be identified and catalogued. Linking specific polymorphisms to known nutrient-sensitive human subpopulations, such as salt sensitivity in African Americans, may enable population-specific recommendations for genetic subgroups. However, caution is warranted. Polymorphisms can confer both health benefits and risks depending on the outcome of interest, and these outcomes may respond differentially to nutrient intake levels. Therefore, it is essential that the effects of genotype-specific recommendations on all known health outcomes be evaluated prior to establishing such recommendations.

The use of specific nutrients to modify genome expression, stability, and viability for disease management and prevention will expand markedly in the coming years. The promise of metabolic engineering for optimal health through diet remains a research challenge and public expectation. Billions of dollars are spent annually for dietary supplements by a general public that demands medicinal qualities from food, believing that the food supply should promote wellness throughout the life cycle in a manner that transcends mere disease prevention. The use of "food as drug" to modify or overcome inherited genetic risks has been demonstrated to be successful in the treatment of several monogenetic disorders. However, expanding this paradigm to polygenic disorders and exposing entire populations or subgroups to quantities of nutrients that are not obtainable in a healthy, food-based diet remains unproven and presents risks. Known and unknown toxicities associated with elevated nutrient intakes, and their effects on stem cell programming, provide ample evidence for caution as we proceed forward in engineering genome function for optimal health.

References

Afman, LA, Trijbels, FJ, and Blom, HJ (2003). The H475Y polymorphism in the glutamate carboxypeptidase II gene increases plasma folate without affecting the risk for neural tube defects in humans. *J Nutr*, 133:75–77.

Akey, JM, Eberle, MA, Rieder, MJ, Carlson, CS, Shriver, MD, et al. (2004). Population history and natural selection shape patterns of genetic variation in 132 genes. *PLoS Biol*, 2:e286.

Akey, JM, Zhang, G, Zhang, K, Jin, L, and Shriver, MD (2002). Interrogating a high-density SNP map for signatures of natural selection. *Genome Res*, 12:1805–1814.

Ames, BN (2001). DNA damage from micronutrient deficiencies is likely to be a major cause of cancer. *Mutat Res*, 475:7–20.

Ames, BN, Elson-Schwab, I, and Silver, EA (2002). High-dose vitamin therapy stimulates variant enzymes with decreased coenzyme binding affinity (increased K(m)): relevance to genetic disease and polymorphisms. *Am J Clin Nutr*, 75:616–658.

Baier, LJ and Hanson, RL (2004). Genetic studies of the etiology of type 2 diabetes in Pima Indians: hunting for pieces to a complicated puzzle. *Diabetes*, 53:1181–1186.

Baier, LJ, Permana, PA, Yang, X, Pratley, RE, Hanson, RL, et al. (2000). A calpain-10 gene polymorphism is associated with reduced muscle mRNA levels and insulin resistance. *J Clin Invest*, 106:R69–R73.

Bailey, LB (2003). Folate, methyl-related nutrients, alcohol, and the MTHFR 677C→T polymorphism affect cancer risk: intake recommendations. *J Nutr*, 133:3748S–3753S.

Bailey, LB and Gregory, JF 3rd (1999). Folate metabolism and requirements. *J Nutr*, 129:779–782.

Bamshad, M and Wooding, SP (2003). Signatures of natural selection in the human genome. *Nat Rev Genet*, 4:99–111.

Barker, DJ (1997). Intrauterine programming of coronary heart disease and stroke. *Acta Paediatr Suppl*, 423:178–182; discussion 183.

Batzer, MA and Deininger, PL (2002). Alu repeats and human genomic diversity. *Nat Rev Genet*, 3:370–379.

Bentzen, J, Jorgensen, T, and Fenger, M (2002). The effect of six polymorphisms in the Apolipoprotein B gene on parameters of lipid metabolism in a Danish population. *Clin Genet*, 61:126–134.

Bersaglieri, T, Sabeti, PC, Patterson, N, Vanderploeg, T, Schaffner, SF, et al. (2004). Genetic signatures of strong recent positive selection at the lactase gene. *Am J Hum Genet*, 74:1111–1120.

Bertram, C, Trowern, AR, Copin, N, Jackson, AA, and Whorwood, CB (2001). The maternal diet during pregnancy programs altered expression of the glucocorticoid receptor and type 2 11beta-hydroxysteroid dehydrogenase: potential molecular mechanisms underlying the programming of hypertension in utero. *Endocrinology*, 142:2841–2853.

Beutler, E (2004). Iron absorption in carriers of the C282Y hemochromatosis mutation. *Am J Clin Nutr*, 80:799–800.

Blake, MJ, Castro, L, Leeder, JS, and Kearns, GL (2005). Ontogeny of drug metabolizing enzymes in the neonate. *Semin Fetal Neonatal Med*, 10:123–138.

Blount, BC, Mack, MM, Wehr, CM, MacGregor, JT, Hiatt, RA, et al. (1997). Folate deficiency causes uracil misincorporation into human DNA and chromosome breakage: implications for cancer and neuronal damage. *Proc Natl Acad Sci USA*, 94:3290–3295.

Bosron, WF and Li, TK (1986). Genetic polymorphism of human liver alcohol and aldehyde dehydrogenases, and their relationship to alcohol metabolism and alcoholism. *Hepatology*, 6:502–510.

Bracken, MB (2005). Genomic epidemiology of complex disease: the need for an electronic evidence-based approach to research synthesis. *Am J Epidemiol*, 162:297–301.

Brent, RL and Beckman, DA (1994). The contribution of environmental teratogens to embryonic and fetal loss. *Clin Obstet Gynecol*, 37:646–670.

Brock, DJ and Holloway, S (1990). Fertility of older women. *Lancet*, 335:1470.

Brody, LC, Conley, M, Cox, C, Kirke, PN, McKeever, MP, et al. (2002). A polymorphism, R653Q, in the trifunctional enzyme methylenetetrahydrofolate dehydrogenase/methenyltetrahydrofolate cyclohydrolase/formyltetrahydrofolate synthetase is a maternal genetic risk factor for neural tube defects: report of the birth defects research group. *Am J Hum Genet*, 71:1207–1215.

Brookfield, JF (2003). Mobile DNAs: the poacher turned gamekeeper. *Curr Biol*, 13:R846–R847.

Brown, S, Ordovas, JM, and Campos, H (2003). Interaction between the APOC3 gene promoter polymorphisms, saturated fat intake and plasma lipoproteins *Atherosclerosis*, 170:307–313.

Bufe, B, Hofmann, T, Krautwurst, D, Raguse, JD, and Meyerhof, W (2002). The human TAS2R16 receptor mediates bitter taste in response to beta-glucopyranosides. *Nat Genet*, 32:397–401.

Bulletti, C, Flamigni, C, and Giacomucci, E (1996). Reproductive failure due to spontaneous abortion and recurrent miscarriage. *Hum Reprod Update*, 2:118–136.

Centers for Disease Control and Prevention (CDC). (2004). Alcohol consumption among women who are pregnant or who might become pregnant—United States, 2002. *Morb Mortal Wkly Rep*, 53:1178–1181.

Chakravarti, A (2001). To a future of genetic medicine. *Nature*, 409:822–823.

Chistyakov, DA, Savost'anov, KV, Zotova, EV, and Nosikov, VV (2001). Polymorphisms in the Mn-SOD and EC-SOD genes and their relationship to diabetic neuropathy in type 1 diabetes mellitus. *BMC Med Genet*, 2:4.

Choi, SW and Mason, JB (2000). Folate and carcinogenesis: an integrated scheme. *J Nutr*, 130:129–132.

Clark, AG, Glanowski, S, Nielsen, R, Thomas, PD, Kejariwal, A, et al. (2003). Inferring nonneutral evolution from human-chimp-mouse orthologous gene trios. *Science*, 302:1960–1963.

Clarke, S and Banfield, K (2001). *Homocysteine in Health and Disease* In R Carmel and DW Jacobson (Eds.), Cambridge: Cambridge Press.

Collins, FS (2004). What we do and don't know about "race," "ethnicity," genetics and health at the dawn of the genome era. *Nat Genet*, 36: S13–S15.

Cowchock, FS, Gibas, Z, and Jackson, LG (1993). Chromosome errors as a cause of spontaneous abortion: the relative importance of maternal age and obstetric history. *Fertil Steril*, 59:1011–1014.

Cox, TM (2002). The genetic consequences of our sweet tooth. *Nat Rev Genet*, 3:481–487.

Crawford, DC and Nickerson, DA (2005). Definition and clinical importance of haplotypes. *Annu Rev Med*, 56:303–320.

Csete, M and Doyle, J (2004). Bow ties, metabolism and disease. *Trends Biotechnol*, 22:446–450.

Daher, S, Shulzhenko, N, Morgun, A, Mattar, R, Rampim, GF, et al. (2003). Associations between cytokine gene polymorphisms and recurrent pregnancy loss. *J Reprod Immunol*, 58:69–77.

Danpure, CJ (2005). Molecular etiology of primary hyperoxaluria type 1: new directions for treatment. *Am J Nephrol*, 25:303–310.

Delange, FM (2003). Control of iodine deficiency in Western and Central Europe. *Cent Eur J Public Health*, 11:120–123.

Delhanty, JD (2001). Preimplantation genetics: an explanation for poor human fertility? *Ann Hum Genet*, 65:331–338.

Dennis, C (2003). Epigenetics and disease: altered states. *Nature*, 421:686–688.

De Rycke, M, Liebaers, I, and Van Steirteghem, A (2002). Epigenetic risks related to assisted reproductive technologies: risk analysis and epigenetic inheritance. *Hum Reprod*, 17:2487–2494.

Devlin, AM, Ling, EH, Peerson, JM, Fernando, S, Clarke, R, et al. (2000). Glutamate carboxypeptidase II: a polymorphism associated with lower levels of serum folate and hyperhomocysteinemia. *Hum Mol Genet*, 9:2837–2844.

Diamond, J (2003). The double puzzle of diabetes. *Nature*, 423: 599–602.

Diaz, R, Fuxe, K, and Ogren, SO (1997). Prenatal corticosterone treatment induces long-term changes in spontaneous and apomorphine-mediated motor activity in male and female rats. *Neuroscience*, 81:129–140.

Drake, AJ, Walker, BR, and Seckl, JR (2005). Intergenerational consequences of fetal programming by in utero exposure to glucocorticoids in rats. *Am J Physiol Regul Integr Comp Physiol*, 288:R34–R38.

Dunn, AJ and Berridge, CW (1990). Physiological and behavioral responses to corticotropin-releasing factor administration: is CRF a mediator of anxiety or stress responses? *Brain Res Brain Res Rev*, 15:71–100.

Eck, P, Erichsen, HC, Taylor, JG, Yeager, M, Hughes, AL, et al. (2004). Comparison of the genomic structure and variation in the two human sodium-dependent vitamin C transporters, SLC23A1 and SLC23A2. *Hum Genet*, 115:285–294.

Edmonds, DK, Lindsay, KS, Miller, JF, Williamson, E, and Wood, PJ (1982). Early embryonic mortality in women. *Fertil Steril*, 38:447–453.

Edwards, RG (1997). Recent scientific and medical advances in assisted human conception. *Int J Dev Biol*, 41:255–262.

Enattah, NS, Sahi, T, Savilahti, E, Terwilliger, JD, Peltonen, L, et al. (2002). Identification of a variant associated with adult-type hypolactasia. *Nat Genet*, 30:233–327.

Esposito, G, Vitagliano, L, Santamaria, R, Viola, A, Zagari, A, et al. (2002). Structural and functional analysis of aldolase B mutants related to hereditary fructose intolerance. *FEBS Lett*, 531:152–156.

Fenech, M (2001). Recommended dietary allowances (RDAs) for genomic stability. *Mutat Res*, 480–481:51–54.

Fenech, M (2002a). Chromosomal biomarkers of genomic instability relevant to cancer. *Drug Discov Today*, 7:1128–1137.

Fenech, M (2002b). Micronutrients and genomic stability: a new paradigm for recommended dietary allowances (RDAs). *Food Chem Toxicol*, 40:1113–1117.

Fenech, M (2004). Genome health nutrigenomics: nutrition and the science of optimal genome maintenance. *Asia Pac J Clin Nutr*, 13:S15.

Ferre, N, Camps, J, Fernandez-Ballart, J, Arija, V, Murphy, MM, et al. (2003). Regulation of serum paraoxonase activity by genetic, nutritional, and lifestyle factors in the general population. *Clin Chem*, 49:1491–1497.

Finan, RR, Tamim, H, Ameen, G, Sharida, HE, Rashid, M, et al. (2002). Prevalence of factor V G1691A (factor V-Leiden) and prothrombin G20210A gene mutations in a recurrent miscarriage population. *Am J Hematol*, 71:300–305.

Finkelstein, JD (2000a). Homocysteine: a history in progress. *Nutr Rev*, 58:193–204.

Finkelstein, JD (2000b). Pathways and regulation of homocysteine metabolism in mammals. *Semin Thromb Hemost*, 26:219–225.

Finnell, RH, Spiegelstein, O, Wlodarczyk, B, Triplett, A, Pogribny, IP, et al. (2002). DNA methylation in Folbp1 knockout mice supplemented with folic acid during gestation. *J Nutr*, 132:2457S–2461S.

French, NP, Hagan, R, Evans, SF, Godfrey, M, and Newnham, JP (1999). Repeated antenatal corticosteroids: size at birth and subsequent development. *Am J Obstet Gynecol*, 180:114–121.

Fullerton, SM, Clark, AG, Weiss, KM, Nickerson, DA, Taylor, SL, et al. (2000). Apolipoprotein E variation at the sequence haplotype level: implications for the origin and maintenance of a major human polymorphism. *Am J Hum Genet*, 67:881–900.

Gabriel, SB, Schaffner, SF, Nguyen, H, Moore, JM, Roy, J, et al. (2002). The structure of haplotype blocks in the human genome. *Science*, 296:2225–2229.

Gao, S and Latham, KE (2004). Maternal and environmental factors in early cloned embryo development. *Cytogenet Genome Res*, 105:279–284.

Gasch, AP and Werner-Washburne, M (2002). The genomics of yeast responses to environmental stress and starvation. *Funct Integr Genomics*, 2:181–192.

Gatford, KL, Wintour, EM, De Blasio, MJ, Owens, JA, and Dodic, M (2000). Differential timing for programming of glucose homoeostasis, sensitivity to insulin and blood pressure by in utero exposure to dexamethasone in sheep. *Clin Sci (Lond)*, 98:553–560.

Geller, DS (2004). A genetic predisposition to hypertension? *Hypertension*, 44:27–28. Gibson, G and Wagner, G (2000). Canalization in evolutionary genetics: a stabilizing theory? *Bioessays*, 22:372–380.

Gloria-Bottini, F, Lucarini, N, Palmarino, R, La Torre, M, Nicotra, M, et al. (2001). Phosphoglucomutase genetic polymorphism of newborns. *Am J Hum Biol*, 13:9–14.

Goland, RS, Jozak, S, Warren, WB, Conwell, IM, Stark, RI, et al. (1993). Elevated levels of umbilical cord plasma corticotropin-releasing hormone in growth-retarded fetuses. *J Clin Endocrinol Metab*, 77:1174–1179.

Goland, RS, Tropper, PJ, Warren, WB, Stark, RI, Jozak, SM, et al. (1995). Concentrations of corticotrophin-releasing hormone in the umbilical-cord blood of pregnancies complicated by pre-eclampsia. *Reprod Fertil Dev*, 7:1227–1230.

Goswami, T, Rolfs, A, and Hediger, MA (2002). Iron transport: emerging roles in health and disease. *Biochem Cell Biol*, 80:679–689.

Gravel, RA and Narang, MA (2005). Molecular genetics of biotin metabolism: old vitamin, new science. *J Nutr Biochem*, 16:428–431.

Griffiths, W and Cox, T (2000). Haemochromatosis: novel gene discovery and the molecular pathophysiology of iron metabolism. *Hum Mol Genet*, 9:2377–2382.

Gris, JC, Perneger, TV, Quere, I, Mercier, E, Fabbro-Peray, P, et al. (2003). Antiphospholipid/antiprotein antibodies, hemostasis-related autoantibodies, and plasma homocysteine as risk factors for a first early pregnancy loss: a matched case-control study. *Blood*, 102:3504–3513.

Guenther, BD, Sheppard, CA, Tran, P, Rozen, R, Matthews, RG, et al. (1999). The structure and properties of methylenetetrahydrofolate reductase from *Escherichia coli* suggest how folate ameliorates human hyperhomocysteinemia. *Nat Struct Biol*, 6:359–365.

Guttmacher, AE and Collins, FS (2002). Genomic medicine—a primer. *N Engl J Med*, 347:1512–1520.

Hebbar, PB and Archer, TK (2003). Chromatin remodeling by nuclear receptors. *Chromosoma*, 111:495–504.

Hediger, MA, Rolfs, A, and Goswami, T (2002). Iron transport and hemochromatosis. *J Investig Med*, 50:239S–246S.

Hein, DW (2002). Molecular genetics and function of NAT1 and NAT2: role in aromatic amine metabolism and carcinogenesis. *Mutat Res*, 506–507:65–77.

Hein, DW, Doll, MA, Fretland, AJ, Leff, MA, Webb, SJ, et al. (2000). Molecular genetics and epidemiology of the NAT1 and NAT2 acetylation polymorphisms. *Cancer Epidemiol Biomarkers Prev*, 9:29–42.

Heird, WC and Lapillonne, A (2005). The role of essential fatty acids in development. *Annu Rev Nutr*, 25:549–571.

Henikoff, S, McKittrick, E, and Ahmad, K (2004). Epigenetics, histone H3 variants, and the inheritance of chromatin states. *Cold Spring Harb Symp Quant Biol*, 69:235–243.

Honein, MA, Paulozzi, LJ, Mathews, TJ, Erickson, JD, and Wong, LY (2001). Impact of folic acid fortification of the US food supply on the occurrence of neural tube defects. *JAMA*, 285:2981–2986.

Hubacek, JA, Pistulkova, H, Skodova, Z, Berg, K, and Poledne, R (2001). Association between apolipoprotein B promotor haplotypes and cholesterol status. *Ann Clin Biochem*, 38:399–400.

Hunt, JR and Zeng, H (2004). Iron absorption by heterozygous carriers of the HFE C282Y mutation associated with hemochromatosis. *Am J Clin Nutr*, 80:924–931.

Ingrosso, D, Cimmino, A, Perna, AF, Masella, L, De Santo, NG, et al. (2003). Folate treatment and unbalanced methylation and changes of allelic expression induced by hyperhomocysteinaemia in patients with uraemia. *Lancet*, 361:1693–1699.

Inoue, I, Nakajima, T, Williams, CS, Quackenbush, J, Puryear, R, et al. (1997). A nucleotide substitution in the promoter of human angiotensinogen is associated with essential hypertension and affects basal transcription in vitro. *J Clin Invest*, 99:1786–1797.

Isotalo, PA, Wells, GA, and Donnelly, JG (2000). Neonatal and fetal methylenetetrahydrofolate reductase genetic polymorphisms: an examination of C677T and A1298C mutations. *Am J Hum Genet*, 67:986–990.

Jacques, PF, Bostom, AG, Williams, RR, Ellison, RC, Eckfeldt, JH, et al. (1996). Relation between folate status, a common mutation in methylenetetrahydrofolate reductase, and plasma homocysteine concentrations. *Circulation*, 93:7–9.

Jeck, N, Waldegger, S, Lampert, A, Boehmer, C, Waldegger, P, et al. (2004). Activating mutation of the renal epithelial chloride channel ClC-Kb predisposing to hypertension. *Hypertension*, 43:1175–1181.

Jorde, LB, Watkins, WS, and Bamshad, MJ (2001). Population genomics: a bridge from evolutionary history to genetic medicine. *Hum Mol Genet*, 10:2199–2207.

Jorde, LB, Watkins, WS, Bamshad, MJ, Dixon, ME, Ricker, CE, et al. (2000). The distribution of human genetic diversity: a comparison of mitochondrial, autosomal, and Y-chromosome data. *Am J Hum Genet*, 66:979–988.

Jorde, LB and Wooding, SP (2004). Genetic variation, classification and "race." *Nat Genet*, 36:S28–S33.

Khosla, S, Dean, W, Reik, W, and Feil, R (2001). Culture of preimplantation embryos and its long-term effects on gene expression and phenotype. *Hum Reprod Update*, 7:419–427.

Kim, M, Trinh, BN, Long, TI, Oghamian, S, and Laird, PW (2004). Dnmt1 deficiency leads to enhanced microsatellite instability in mouse embryonic stem cells. *Nucleic Acids Res*, 32:5742–5749.

Kimmins, S and Sassone-Corsi, P (2005). Chromatin remodelling and epigenetic features of germ cells. *Nature*, 434:583–589.

Kimura, M (1968). Evolutionary rate at the molecular level. *Nature*, 217:624–626.

King, JL and Jukes, TH (1969). Non-Darwinian evolution. *Science*, 164:788–798.

Kornman, KS, Martha, PM, and Duff, GW (2004). Genetic variations and inflammation: a practical nutrigenomics opportunity. *Nutrition*, 20:44–49.

Loew, M, Boeing, H, Sturmer, T, and Brenner, H (2003). Relation among alcohol dehydrogenase 2 polymorphism, alcohol consumption, and levels of gamma-glutamyltransferase. *Alcohol*, 29:131–135.

Ma, J, Stampfer, MJ, Giovannucci, E, Artigas, C, Hunter, DJ, et al. (1997). Methylenetetrahydrofolate reductase polymorphism, dietary interactions, and risk of colorectal cancer. *Cancer Res*, 57:1098–1102.

Mackey, AD, McMahon, RJ, Townsend, JH, and Gregory, JF 3rd (2004). Uptake, hydrolysis, and metabolism of pyridoxine-5′-beta-D-glucoside in Caco-2 cells. *J Nutr*, 134:842–846.

Mason, JB and Kim, Y (1999). Nutritional strategies in the prevention of colorectal cancer. *Curr Gastroenterol Rep*, 1:341–353.

McMillen, IC and Robinson, JS (2005). Developmental origins of the metabolic syndrome: prediction, plasticity, and programming. *Physiol Rev*, 85:571–633.

Mellott, TJ, Williams, CL, Meck, WH, and Blusztajn, JK (2004). Prenatal choline supplementation advances hippocampal development and enhances MAPK and CREB activation. *FASEB J*, 18:545–547.

Messier, W and Stewart, CB (1997). Episodic adaptive evolution of primate lysozymes. *Nature*, 385:151–154.

Milton, CC, Huynh, B, Batterham, P, Rutherford, SL, and Hoffmann, AA (2003). Quantitative trait symmetry independent of Hsp90 buffering: distinct modes of genetic canalization and developmental stability. *Proc Natl Acad Sci USA*, 100:13396–13401.

Mishra, HN and Das, C (2003). A review on biological control and metabolism of aflatoxin. *Crit Rev Food Sci Nutr*, 43:245–264.

Moirand, R, Guyader, D, Mendler, MH, Jouanolle, AM, Le Gall, JY, et al. (2002). HFE based re-evaluation of heterozygous hemochromatosis. *Am J Med Genet*, 111:356–361.

Morgan, HD, Sutherland, HG, Martin, DI, and Whitelaw, E (1999). Epigenetic inheritance at the agouti locus in the mouse. *Nat Genet*, 23:314–318.

Munoz-Moran, E, Dieguez-Lucena, JL, Fernandez-Arcas, N, Peran-Mesa, S, and Reyes-Engel, A (1998). Genetic selection and folate intake during pregnancy. *Lancet*, 352:1120–1121.

Nachman, MW and Crowell, SL (2000). Estimate of the mutation rate per nucleotide in humans. *Genetics*, 156:297–304.

Neel, JV (1962). Diabetes mellitus: a "thrifty" genotype rendered detrimental by "progress?" *Am J Hum Genet*, 14:353–362.

Nelen, WL (2001). Hyperhomocysteinaemia and human reproduction. *Clin Chem Lab Med*, 39:758–763.

Nelen, WL, Blom, HJ, Steegers, EA, den Heijer, M, and Eskes, TK (2000). Hyperhomocysteinemia and recurrent early pregnancy loss: a meta-analysis. *Fertil Steril*, 74:1196–1199.

Nelen, WL, Blom, HJ, Steegers, EA, den Heijer, M, Thomas, CM, et al. (2000). Homocysteine and folate levels as risk factors for recurrent early pregnancy loss. *Obstet Gynecol*, 95:519–524.

Nelen, WL, Blom, HJ, Thomas, CM, Steegers, EA, Boers, GH, et al. (1998). Methylenetetrahydrofolate reductase polymorphism affects the change in homocysteine and folate concentrations resulting from low dose folic acid supplementation in women with unexplained recurrent miscarriages. *J Nutr*, 128:1336–1341.

Newnham, JP, Evans, SF, Godfrey, M, Huang, W, Ikegami, M, et al. (1999). Maternal, but not fetal, administration of corticosteroids restricts fetal growth. *J Matern Fetal Med*, 8:81–87.

Nyirenda, MJ, Lindsay, RS, Kenyon, CJ, Burchell, A, and Seckl, JR (1998). Glucocorticoid exposure in late gestation permanently programs rat hepatic phosphoenolpyruvate carboxykinase and glucocorticoid receptor expression and causes glucose intolerance in adult offspring. *J Clin Invest*, 101:2174–2181.

Oota, H, Pakstis, AJ, Bonne-Tamir, B, Goldman, D, Grigorenko, E, et al. (2004). The evolution and population genetics of the ALDH2 locus: random genetic drift, selection, and low levels of recombination. *Ann Hum Genet*, 68:93–109.

Ordovas, JM (2004). The quest for cardiovascular health in the genomic era: nutrigenetics and plasma lipoproteins. *Proc Nutr Soc*, 63:145–152.

Ordovas, JM and Corella, D (2004). Nutritional genomics. *Annu Rev Genomics Hum Genet*, 5:71–118.

Osier, MV, Pakstis, AJ, Soodyall, H, Comas, D, Goldman, D, et al. (2002). A global perspective on genetic variation at the ADH genes reveals unusual patterns of linkage disequilibrium and diversity. *Am J Hum Genet*, 71:84–99.

Oyama, K, Kawakami, K, Maeda, K, Ishiguro, K, and Watanabe, G (2004). The association between methylenetetrahydrofolate reductase polymorphism and promoter methylation in proximal colon cancer. *Anticancer Res*, 24:649–654.

Pal, C, Papp, B, and Hurst, LD (2003). Genomic function: rate of evolution and gene dispensability. *Nature*, 421:496–497; discussion 497–498.

Pasqualetti, M, Neun, R, Davenne, M, and Rijli, FM (2001). Retinoic acid rescues inner ear defects in Hoxa1 deficient mice. *Nat Genet*, 29:34–39.

Pennisi, E (2002). Evolution of developmental diversity. Evo-devo devotees eye ocular origins and more. *Science*, 296:1010–1011.

Perni, SC, Vardhana, S, Tuttle, SL, Kalish, RB, Chasen, ST, et al. (2004). Fetal interleukin-1 receptor antagonist gene polymorphism, intra-amniotic interleukin-1beta levels, and history of spontaneous abortion. *Am J Obstet Gynecol*, 191:1318–1323.

Petry, CJ, Ong, KK, Barratt, BJ, Wingate, D, Cordell, HJ, et al. (2005). Common polymorphism in H19 associated with birthweight and cord blood IGF-II levels in humans. *BMC Genet*, 6:22.

Poulter, M, Hollox, E, Harvey, CB, Mulcare, C, Peuhkuri, K, et al. (2003). The causal element for the lactase persistence/non-persistence polymorphism is located in a 1 Mb region of linkage disequilibrium in Europeans. *Ann Hum Genet*, 67:298–311.

Prigoshin, N, Tambutti, M, Larriba, J, Gogorza, S, and Testa, R (2004). Cytokine gene polymorphisms in recurrent pregnancy loss of unknown cause. *Am J Reprod Immunol*, 52:36–41.

Pritchard, JK (2001). Are rare variants responsible for susceptibility to complex diseases? *Am J Hum Genet*, 69:124–137.

Pritchard, JK and Cox, NJ (2002). The allelic architecture of human disease genes: common disease–common variant...or not? *Hum Mol Genet*, 11:2417–2423.

Quinlivan, EP, Davis, SR, Shelnutt, KP, Henderson, GN, Ghandour, H, et al. (2005). Methylenetetrahydrofolate reductase 677C→T polymorphism and folate status affect one-carbon incorporation into human DNA deoxynucleosides. *J Nutr*, 135:389–396.

Rasmussen, KM (2001). The "fetal origins" hypothesis: challenges and opportunities for maternal and child nutrition. *Annu Rev Nutr*, 21:73–95.

Reich, DE and Lander, ES (2001). On the allelic spectrum of human disease. *Trends Genet*, 17:502–510.

Reinisch, JM, Simon, NG, Karow, WG, and Gandelman, R (1978). Prenatal exposure to prednisone in humans and animals retards intrauterine growth. *Science*, 202:436–438.

Reyes-Engel, A, Munoz, E, Gaitan, MJ, Fabre, E, Gallo, M, et al. (2002). Implications on human fertility of the 677C→T and 1298A→C polymorphisms of the MTHFR gene: consequences of a possible genetic selection. *Mol Hum Reprod*, 8:952–957.

Rockman, MV and Wray, GA (2002). Abundant raw material for cis-regulatory evolution in humans. *Mol Biol Evol*, 19:1991–2004.

Rosella, G, Zajac, JD, Kaczmarczyk, SJ, Andrikopoulos, S, and Proietto, J (1993). Impaired suppression of gluconeogenesis induced by overexpression of a noninsulin-responsive phosphoenolpyruvate carboxykinase gene. *Mol Endocrinol*, 7:1456–1462.

Ruden, DM, Garfinkel, MD, Sollars, VE, and Lu, X (2003). Waddington's widget: Hsp90 and the inheritance of acquired characters. *Semin Cell Dev Biol*, 14:301–310.

Ruden, DM, Xiao, L, Garfinkel, MD, and Lu, X (2005). Hsp90 and environmental impacts on epigenetic states: a model for the trans-generational effects of diethylstibesterol on uterine development and cancer. *Hum Mol Genet*, 14(Spec. No. 1):R149–R155.

Sata, F, Yamada, H, Kondo, T, Gong, Y, Tozaki, S, et al. (2003) Glutathione S-transferase M1 and T1 polymorphisms and the risk of recurrent pregnancy loss *Mol Hum Reprod* 9:165–169.

Sata, F, Yamada, H, Yamada, A, Kato, EH, Kataoka, S, et al. (2003). A polymorphism in the CYP17 gene relates to the risk of recurrent pregnancy loss. *Mol Hum Reprod*, 9:725–728.

Schneeman, BO (2003). Evolution of dietary guidelines. *J Am Diet Assoc*, 103: S5–S9.

Schneeman, BO and Mendelson, R (2002). Dietary guidelines: past experience and new approaches. *J Am Diet Assoc*, 102:1498–1500.

Schweikert, A, Rau, T, Berkholz, A, Allera, A, Daufeldt, S, et al. (2004). Association of progesterone receptor polymorphism with recurrent abortions. *Eur J Obstet Gynecol Reprod Biol*, 113:67–72.

Seckl, JR (2004). Prenatal glucocorticoids and long-term programming. *Eur J Endocrinol*, 151(Suppl. 3):U49–U62.

Seckl, JR and Meaney, MJ (2004). Glucocorticoid programming. *Ann N Y Acad Sci*, 1032:63–84.

Seckl, JR and Walker, BR (2001). Minireview: 11beta-hydroxysteroid dehydrogenase type 1- a tissue-specific amplifier of glucocorticoid action. *Endocrinology*, 142:1371–1376.

Seifried, HE, Anderson, DE, Sorkin, BC, and Costello, RB (2004). Free radicals: the pros and cons of antioxidants. Executive summary report. *J Nutr*, 134:3143S–3163S.

Shelnutt, KP, Kauwell, GP, Gregory, JF 3rd, Maneval, DR, Quinlivan, EP, et al. (2004). Methylenetetrahydrofolate reductase 677C→T polymorphism affects DNA methylation in response to controlled folate intake in young women. *J Nutr Biochem*, 15:554–560.

Siegal, ML and Bergman, A (2002). Waddington's canalization revisited: developmental stability and evolution. *Proc Natl Acad Sci USA*, 99:10528–10532.

Sinclair, KD, Dunne, LD, Maxfield, EK, Maltin, CA, Young, LE, et al. (1998). Fetal growth and development following temporary exposure of day 3 ovine embryos to an advanced uterine environment. *Reprod Fertil Dev*, 10:263–269.

Stewart, PM, Murry, BA, and Mason, JI (1994). Type 2 11 beta-hydroxysteroid dehydrogenase in human fetal tissues. *J Clin Endocrinol Metab*, 78:1529–1532.

Stewart, PM, Rogerson, FM, and Mason, JI (1995). Type 2 11 beta-hydroxysteroid dehydrogenase messenger ribonucleic acid and activity in human placenta and fetal membranes: its relationship to birth weight and putative role in fetal adrenal steroidogenesis. *J Clin Endocrinol Metab*, 80:885–890.

Stillman, RJ, Rosenberg, MJ, and Sachs, BP (1986). Smoking and reproduction. *Fertil Steril*, 46:545–566.

Stover, PJ (2004a). Nutritional genomics. *Physiol Genomics*, 16:161–165.

Stover, PJ (2004b). Physiology of folate and vitamin B12 in health and disease. *Nutr Rev*, 62:S3–S12; discussion S13.

Stover, PJ and Garza, C (2002). Bringing individuality to public health recommendations. *J Nutr*, 132:2476S–2480S.

Suh, JR, Herbig, AK, and Stover, PJ (2001). New perspectives on folate catabolism. *Annu Rev Nutr*, 21:255–282.

Suryanarayana, V, Deenadayal, M, and Singh, L (2004). Association of CYP1A1 gene polymorphism with recurrent pregnancy loss in the South Indian population. *Hum Reprod*, 19:2648–2652.

Swanson, CA (2003). Iron intake and regulation: implications for iron deficiency and iron overload. *Alcohol*, 30:99–102.

Tempfer, C, Unfried, G, Zeillinger, R, Hefler, L, Nagele, F, et al. (2001). Endothelial nitric oxide synthase gene polymorphism in women with idiopathic recurrent miscarriage. *Hum Reprod*, 16:1644–1647.

Thomassin, H, Flavin, M, Espinas, ML, and Grange, T (2001). Glucocorticoid-induced DNA demethylation and gene memory during development. *EMBO J*, 20:1974–1983.

Tishkoff, SA and Kidd, KK (2004). Implications of biogeography of human populations for "race" and medicine. *Nat Genet*, 36:S21–S27.

Tishkoff, SA, Varkonyi, R, Cahinhinan, N, Abbes, S, Argyropoulos, G, et al. (2001). Haplotype diversity and linkage disequilibrium at human G6PD: recent origin of alleles that confer malarial resistance. *Science*, 293:455–462.

Tishkoff, SA and Verrelli, BC (2003). Patterns of human genetic diversity: implications for human evolutionary history and disease. *Annu Rev Genomics Hum Genet*, 4:293–340.

Toomajian, C, Ajioka, RS, Jorde, LB, Kushner, JP, and Kreitman, M (2003). A method for detecting recent selection in the human genome from allele age estimates. *Genetics*, 165:287–297.

Toomajian, C and Kreitman, M (2002). Sequence variation and haplotype structure at the human HFE locus. *Genetics*, 161:1609–1623.

Tsai, MY, Yang, F, Bignell, M, Aras, O, and Hanson, NQ (1999). Relation between plasma homocysteine concentration, the 844ins68 variant of the cystathionine beta-synthase gene, and pyridoxal-5′-phosphate concentration. *Mol Genet Metab*, 67:352–356.

Tseung, CW, McMahon, LG, Vazquez, J, Pohl, J, and Gregory, JF 3rd (2004). Partial amino acid sequence and mRNA analysis of cytosolic pyridoxine-beta-D-glucoside hydrolase from porcine intestinal mucosa: proposed derivation from the lactase-phlorizin hydrolase gene. *Biochem J*, 380:211–218.

Uitterlinden, AG, Fang, Y, Bergink, AP, van Meurs, JB, van Leeuwen, HP, et al. (2002). The role of vitamin D receptor gene polymorphisms in bone biology. *Mol Cell Endocrinol*, 197:15–21.

Valera, A, Pujol, A, Pelegrin, M, and Bosch, F (1994). Transgenic mice overexpressing phosphoenolpyruvate carboxykinase develop non-insulin-dependent diabetes mellitus. *Proc Natl Acad Sci USA*, 91:9151–9154.

Van Blerkom, LM (2003). Role of viruses in human evolution. *Am J Phys Anthropol*, 37(Suppl.):14–46.

Van Landeghem, GF, Tabatabaie, P, Kucinskas, V, Saha, N, and Beckman, G (1999). Ethnic variation in the mitochondrial targeting sequence polymorphism of MnSOD. *Hum Hered*, 49:190–193.

Venter, JC, Adams, MD, Myers, EW, Li, PW, Mural, RJ, et al. (2001). The sequence of the human genome. *Science*, 291:1304–1351.

Verrelli, BC, McDonald, JH, Argyropoulos, G, Destro-Bisol, G, Froment, A, et al. (2002). Evidence for balancing selection from nucleotide sequence analyses of human G6PD. *Am J Hum Genet*, 71:1112–1128.

Vijg, J and Suh, Y (2005). Genetics of longevity and aging. *Annu Rev Med*, 56:193–212.

von Linsingen, R, Bompeixe, EP, and Bicalho Mda, G (2005). A case-control study in IL6 and TGFB1 gene polymorphisms and recurrent spontaneous abortion in southern Brazilian patients. *Am J Reprod Immunol*, 53: 94–99.

Waterland, RA and Garza, C (1999). Potential mechanisms of metabolic imprinting that lead to chronic disease. *Am J Clin Nutr* 69:179–197.

Waterland, RA and Jirtle, RL (2003). Transposable elements: targets for early nutritional effects on epigenetic gene regulation. *Mol Cell Biol*, 23:5293–5300.

Waterland, RA and Jirtle, RL (2004). Early nutrition, epigenetic changes at transposons and imprinted genes, and enhanced susceptibility to adult chronic diseases. *Nutrition*, 20:63–68.

Watkins, WS, Rogers, AR, Ostler, CT, Wooding, S, Bamshad, MJ, et al. (2003). Genetic variation among world populations: inferences from 100 Alu insertion polymorphisms. *Genome Res*, 13:1607–1618.

Weaver, IC, Cervoni, N, Champagne, FA, D'Alessio, AC, Sharma, S, et al. (2004). Epigenetic programming by maternal behavior. *Nat Neurosci*, 7:847–854.

Weaver, IC, Diorio, J, Seckl, JR, Szyf, M, and Meaney, MJ (2004). Early environmental regulation of hippocampal glucocorticoid receptor gene expression: characterization of intracellular mediators and potential genomic target sites. *Ann N Y Acad Sci*, 1024:182–212.

Weinshilboum, R (2003). Inheritance and drug response. *N Engl J Med*, 348:529–537.

Weinstock, M (2001). Alterations induced by gestational stress in brain morphology and behaviour of the offspring. *Prog Neurobiol*, 65:427–451.

White, PC, Mune, T, and Agarwal, AK (1997). 11 beta-Hydroxysteroid dehydrogenase and the syndrome of apparent mineralocorticoid excess. *Endocr Rev*, 18:135–156.

Whitehead, AS (1998). Changes in MTHFR genotype frequencies over time. *Lancet*, 352:1784–1785.

Wilcox, AJ, Weinberg, CR, O'Connor, JF, Baird, DD, Schlatterer, JP, et al. (1988). Incidence of early loss of pregnancy. *N Engl J Med*, 319:189–194.

Williams, RJ and Deason, G (1967). Individuality in vitamin C needs. *Proc Natl Acad Sci USA*, 57:1638–1641.

Wolfe, KH and Li, WH (2003). Molecular evolution meets the genomics revolution. *Nat Genet*, 33(Suppl.):255–265.

Wooding, S (2003). PopHist: inferring population history from the spectrum of allele frequencies. *Bioinformatics*, 19:539–540.

Wooding, S (2004). Natural selection: sign, sign, everywhere a sign. *Curr Biol*, 14:R700–R701.

Wooding, S, Kim, UK, Bamshad, MJ, Larsen, J, Jorde, LB, et al. (2004). Natural selection and molecular evolution in PTC, a bitter-taste receptor gene. *Am J Hum Genet*, 74:637–646.

Wooding, SP, Watkins, WS, Bamshad, MJ, Dunn, DM, Weiss, RB, et al. (2002). DNA sequence variation in a 3.7-kb noncoding sequence 5′ of the CYP1A2 gene: implications for human population history and natural selection. *Am J Hum Genet*, 71:528–542.

Wray, GA, Hahn, MW, Abouheif, E, Balhoff, JP, Pizer, M, et al. (2003). The evolution of transcriptional regulation in eukaryotes. *Mol Biol Evol*, 20:1377–1419.

Wright, AF, Carothers, AD, and Pirastu, M (1999). Population choice in mapping genes for complex diseases. *Nat Genet*, 23:397–404.

Wu, W, Goodman, M, Lomax, MI, and Grossman, LI (1997). Molecular evolution of cytochrome c oxidase subunit IV: evidence for positive selection in simian primates. *J Mol Evol*, 44:477–491.

Young, LE, Fernandes, K, McEvoy, TG, Butterwith, SC, Gutierrez, CG, et al. (2001). Epigenetic change in IGF2R is associated with fetal overgrowth after sheep embryo culture. *Nat Genet*, 27:153–154.

Young, VR (2002). 2001 W.O. Atwater Memorial Lecture and the 2001 ASNS President's Lecture: human nutrient requirements: the challenge of the post-genome era. *J Nutr*, 132:621–629.

Zeisel, SH (2004). Antioxidants suppress apoptosis. *J Nutr*, 134:3179S–3180S.

Zetterberg, H (2004). Methylenetetrahydrofolate reductase and transcobalamin genetic polymorphisms in human spontaneous abortion: biological and clinical implications. *Reprod Biol Endocrinol*, 2:7.

Zetterberg, H, Regland, B, Palmer, M, Ricksten, A, Palmqvist, L, et al. (2002). Increased frequency of combined methylenetetrahydrofolate reductase

C677T and A1298C mutated alleles in spontaneously aborted embryos. *Eur J Hum Genet*, 10:113–118.

Zetterberg, H, Regland, B, Palmer, M, Rymo, L, Zafiropoulos, A, et al. (2002). The transcobalamin codon 259 polymorphism influences the risk of human spontaneous abortion. *Hum Reprod*, 17:3033–3036.

Zetterberg, H, Zafiropoulos, A, Spandidos, DA, Rymo, L, and Blennow, K (2003). Gene–gene interaction between fetal MTHFR 677C>T and transcobalamin 776C>G polymorphisms in human spontaneous abortion. *Hum Reprod*, 18:1948–1950.

Zhang, J, Zhang, YP, and Rosenberg, HF (2002). Adaptive evolution of a duplicated pancreatic ribonuclease gene in a leaf-eating monkey. *Nat Genet*, 30:411–415.

Zhao, R, Russell, RG, Wang, Y, Liu, L, Gao, F, et al. (2001). Rescue of embryonic lethality in reduced folate carrier-deficient mice by maternal folic acid supplementation reveals early neonatal failure of hematopoietic organs. *J Biol Chem*, 276:10224–10228.

Zingg, JM and Jones, PA (1997). Genetic and epigenetic aspects of DNA methylation on genome expression, evolution, mutation and carcinogenesis. *Carcinogenesis*, 18:869–882.

9

Pharmacogenomics: Drug Development, Drug Response, and Precision Medicines

Michelle Penny and Duncan McHale

In its broadest sense, pharmacogenomics can be defined as the investigation of variations of DNA and RNA characteristics as related to drug response. The last decade has seen a large increase in the amount of genomics data generated and, with it, increased the expectations of how improved understanding of disease will lead to the development of more effective therapies and personalized medicines. Despite the regular media reports of novel genes being identified in a range of disorders, by 2007 the much-heralded promise of the human genome project has yet to materialize. However, it is important to realize that this does not represent a failure of the science to deliver but is a reflection of the length of time it takes to develop new drugs. The 1980s and 1990s saw a boom time for the pharmaceuticals industry producing many highly effective new classes of drugs from statins to proton-pump inhibitors and quinolone antibiotics. All were novel therapeutic approaches offering significant benefit to individuals and society. The science that drove many of these advances was based on the greater understanding of biochemistry and pharmacology that emerged during the 1970s and early 1980s. This 10–15-year time-lag from gaining scientific knowledge to developing therapies is typical for the pharmaceutical industry, and reflects the complexity of drug discovery and the time required for preclinical and clinical testing to ensure safety and efficacy. This chapter will introduce the major concepts of drug discovery and development and give a broad overview of how genetics and genomics is used across the whole drug discovery and development pipeline, from pretarget identification to postmarketing surveillance to help discover and develop improved medicines.

The Drug Discovery and Development Process

The generation of an idea that a particular protein might be a suitable therapeutic target for the treatment of a disease sets in motion what is often depicted as a linear process known as the drug discovery and development pipeline, in which new medicines follow a set route from early discovery and preclinical stages through a set of clinical development processes to the marketplace (Fig. 9–1).

Candidate Seeking

The ultimate aim of the drug discovery process is to find a chemical (e.g. small molecule) or biological reagent, such as an antibody, which has the potential to be a drug that can be moved into preclinical and then clinical testing. In order to start the process of identifying a potential drug, a biological assay testing interactions with the drug target must be developed. This assay is generally based on a cloned and expressed form of the drug target and will be converted into a format that will allow high-throughput testing, as millions of chemicals may need to be screened in the assay. The need to screen millions of chemicals means that it is usually only feasible to screen one protein variant of the target in the high-throughput screen. It is therefore vital to screen the "right" variant. In the situation where there may be more than one form of the protein that can be included in the screen, it is important to know that the most biologically relevant and/or the most common variant is screened, and it may be necessary to screen the chemical matter against more than one form of the protein. The high-throughput screens generally identify several potential "hits," which need to be tested in more rigorous biological assays to determine the type of interaction and the effects. Promising "leads" are then developed by creating chemical series based on the original lead, and the final candidate is chosen based on the selectivity and potency criteria. This candidate is then taken forward into preclinical testing.

Preclinical Testing

Once a drug candidate has been identified, it goes into a range of preclinical toxicology testing that includes in vitro screening tests to identify potential pharmacological effects at other receptors that could lead to adverse events, and genetic toxicology testing, which evaluates mutagenicity and clastogenicity. Only if these are satisfactory does animal testing begin. The animal testing is done in two species and is staged to ensure that as few animals as possible are used and that major problems are picked up early. Toxicology studies to evaluate long-term exposure, reproductive toxicological effects, and carcinogenicity are generally only performed once the data has been obtained from human studies that support safety and efficacy. Genomic data collected in preclinical testing can be used to validate the animal models in a comparative pharmacogenomic approach to ensure that the most appropriate animal models are used. To date, toxicology induced by new chemicals are identified and classified by standard phenotypic and histological changes. While this picks up the majority of potential toxic effects, it can be insensitive to subtle changes and can identify species-specific effects that are difficult to interpret. A greater understanding of the molecular changes

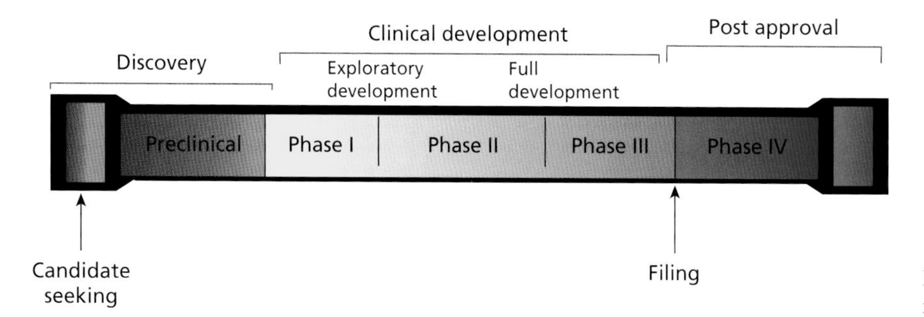

Figure 9–1 The drug discovery and development pipeline.

following drug administration could identify more subtle effects and species-specific effects. Similarly, the applicability of animal models of disease could be assessed by evaluating molecular changes rather than phenotypic similarities that can be misleading.

Clinical Development

Once the initial in vitro testing and acute animal toxicology studies (generally 14 days) have been performed then the candidate can start to be tested in humans. The human studies have traditionally been split into four phases (phase I–IV), each with specific aims:

Box 9–1 The Human Studies Have Traditionally Been Split into Four Phases (Phase I–IV)

- Phase I—Pharmacokinetic and safety profiles in healthy volunteers
- Phase II—Safety and efficacy in patients and the establishment of the dose response
- Phase III—Safety and efficacy at the chosen dosage
- Phase IV—Post approval studies to answer specific safety or efficacy questions and to support commercial strategies

Phase I

The first time a novel compound (or biological therapy) is tested in humans, it is given as a single dose in healthy volunteers. The dose is then escalated over several weeks starting at between 10- and 100-fold below the expected pharmacological exposure levels to a maximum tolerable level or several-fold beyond the expected maximum clinical dose (whichever is reached sooner). The aim is to identify common adverse events and the relationship with plasma exposure as well as to establish the basic pharmacokinetic (PK) parameters. As drug development continues, more healthy volunteer studies are performed to understand the effects of multiple dosing, specific drug–drug interactions, and food effects. The aim of these studies is to provide a more comprehensive understanding of the PKs and significant causes of variability in PK profiles. Collections of pharmacogenomic samples in phase-I clinical protocols allow the extension of these studies to start the assessment of the impact of genetic variation on drug metabolism and transport.

Phase II

Phase II is the first time that patients will receive the potential therapy and is traditionally divided into phase IIa, where the aim is to demonstrate the safety and PK parameters in patients, and IIb, where the aim is to establish efficacy and delineate the dose–response curve. However, most companies now endeavor to generate some efficacy data in the phase-IIa studies to provide confidence to progress into the more expensive and larger phase-IIb dose ranging study. This is a critical time, as up to 50% of all drug candidates will fail in phase II. Samples collected in phase II studies for pharmacogenomic analysis are useful for testing specific hypotheses on the impact of genetic variation with respect to drug response, particularly for genes with large effects, as these studies are limited in that they comprise relatively small numbers of patients (50–100). Samples for these pharmacogenomic studies may be collected with specific consent for genotyping of named genes within the protocol, which can be correlated with clinical data collected in the trial.

Phase III

Phase III trials form the basis of the regulatory approval, as they are large studies evaluating the safety and efficacy of the candidate at the clinical dose and in the population where the drug will ultimately be used. The cost of this phase of development is significantly more than the others and so failure at this point has a major impact on the company. Larger numbers of patients included in these studies provide more power for pharmacogenomic analysis. In addition, these samples also provide a useful resource for more disease-focused phenotype–genotype correlations, and often samples are collected with broad consent for genotyping that allows the investigation of many candidate genes. Additional data security measures may be used when samples are collected with broad consent for exploratory research e.g., anonymization. The anonymization process ensures that the genetic data cannot be linked back to the patient who donated the sample (Fig. 9–2). A copy of the clinical data is made, a new link to the DNA sample created, and all links back to the original identifiers, and hence the subject, are deleted before any genotyping is performed.

Phase IV

Drug testing does not stop with regulatory approval, and phase-IV studies are run after the drug has been approved. Sometimes they are a post-approval commitment required by the regulators. These generally test a specific question around safety and efficacy or are used to generate data to support commercial strategies. Studies conducted postapproval, represent an excellent prospect for the implementation of a pharmacogenomics strategy because of the availability of larger sample sets. The potential to collect genomic samples from thousands of individuals recruited into large phase-IV clinical studies presents the opportunity to link genomic data to good quality clinical data, biomarker data, and, in many cases, long-term follow up. An area where postmarket pharmacogenomic surveillance can have a great impact is in addressing safety issues. The availability of large numbers of patients on active treatments not only provides the material to look for pharmacogenomic effects but is also a valuable resource for understanding the molecular basis for

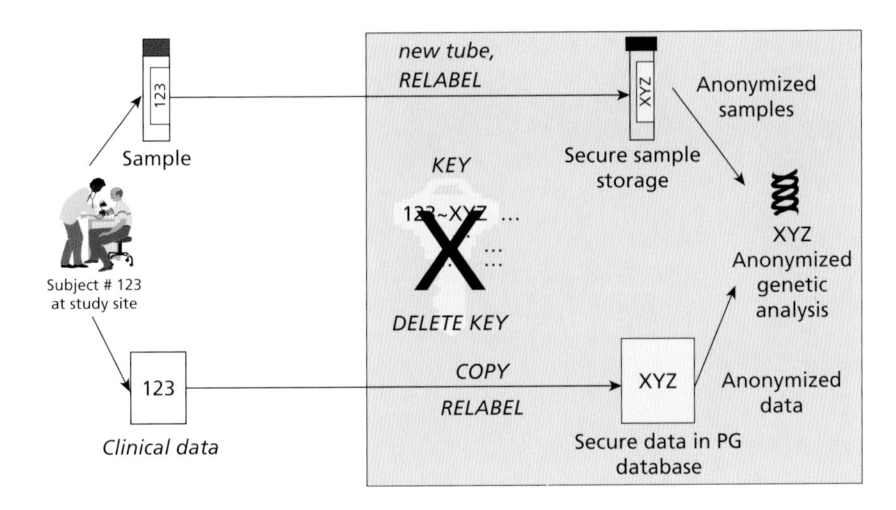

Figure 9–2 The anonymization process.

disease, which in turn feeds back into the idea generation in the early discovery section of the pipeline.

The studies performed within drug development programs are still classified according to this system, but, increasingly, companies are looking to generate potential signals of efficacy data in the early phase-I and IIa studies to provide confidence that the compound will work, before investing in the more expensive phase-IIb and III studies.

Application of Genomics to Drug Discovery and Development

Pharmaceutical companies have tended to focus their drug discovery and development programs on a few key areas to provide the bulk of their revenue and drive their future research and development strategies, the so-called "blockbuster business model." A blockbuster drug is usually defined as one with peak annual sales of greater than $1 billion and is generally developed for long-term use to treat common complex chronic disorders in the general population. A survey of the drug development costs of 68 new compounds from 10 pharmaceutical companies estimated that the cost to develop a new drug in 2000 was $802 million (DiMasi et al., 2003). The high costs of developing drugs can be attributed to two main factors, the large size and duration of the clinical trials required to provide the data to show safety and efficacy of the compound and a high rate of attrition of compounds in clinical development; less than 20% of compounds entering phase I clinical development reach the market, the majority failing in clinical development due to safety or efficacy issues. As the cost of sustaining the blockbuster model for drug development continues to increase, new business models will be required to make drug discovery and development more efficient.

The advances in genetic and genomic technologies and the completion of human genome sequencing effort offer an opportunity to extend the application of genetics to drug development and exploit the information provided by the genome to increase efficiency in drug development. Pharmacogenomics, the investigation of variations of DNA and RNA characteristics as related to drug response in individual patients or groups of patients, will be one of a number of initiatives employed by the pharmaceutical industry to provide more effective medicines in a cost-effective manner in the future. Pharmacogenomics has been heralded as a discipline that promises to revolutionize medicine by providing tailored drug therapy, allowing the use of medications that would otherwise be rejected because of side effects and by encouraging development of new medications for specific disease subtypes. In January 2001, an article in *Time* magazine suggested that because of pharmacogenomics "Doctors will treat diseases like cancer and diabetes before the symptoms even begin, using medications that boost or counteract the effect of individual proteins . . . and they will know right from the start how to select the best medicine to suit each patient." Whether or not the science lives up to its promise remains to be seen, but there are areas in pharmaceutical development where genetics and genomics is already beginning to make an impact on the drug discovery process from target selection to postapproval phase-IV studies.

Choosing the Best Drug Targets

One key area where pharmacogenomics can impact the process is in target selection. Drug response is likely to be influenced by genetic polymorphisms related to these genes (Table 9–1). A significant number of compounds fail in development, because the mechanism of action and biological impact of the target was not fully elucidated and the underlying molecular basis of the disease to which they are directed was poorly understood. Taking the view that the more you know about a drug target earlier in the discovery process the less likely it is to fail in development due to lack of confidence in rationale (CIR), many companies are now investing up front in understanding the molecular genetics of the complex diseases we treat, and using genetics to identify novel targets and prioritize target selection from candidate gene lists for drug development programs.

Before 1990, pharmaceutical companies had worked on approximately 500 potential drug targets with around 100 of these mechanisms having produced marketed drugs (Hopkins and Groom, 2002). Initial analysis of the final draft of the human genome project suggested that the total number of targets druggable with small chemicals might increase to 5000 (Drews, 2000). However, not all of these targets will be relevant to disease and therefore current estimates are that there are 600–1500 drug targets in the human genome (Hopkins and Groom, 2002). This expansion of potential targets in concert with the rising costs of drug development means that the choice of targets is increasingly important.

Given the length of time it takes to get from an idea to a compound to the market, there are few prospective examples of marketed compounds where genomics has provided a new drug target or supported its initial CIR, and thus there is insufficient

Table 9–1 Some examples of drug response related to genetic polymorphisms of drug-target genes (Evans and McLeod, 2003)

Gene or gene product	Drug	Drug response associated with genetic polymorphisms
ACE	ACE inhibitors	Renoprotective effects, blood-pressure reduction, reduction in left ventricular mass, endothelial function
	Fluvastatin	Lipid changes—reduction in low density lipoprotein cholesterol and apolipoprotein B; regression coronary atherosclerosis
Arachidonate 5-lipoxygenase	Leukotriene inhibitors	Improvement in FEV_1
β_2-Adrenergic receptor	β_2- Agonists	Bronchodilatation, susceptibility to agonist induced desensitisation, cardiovascular effects
Bradykinin B2 receptor	ACE inhibitors	ACE-inhibiotr-induced cough
Dopamine receptors	Antipsychotics	Antipsychotic response, antipsychotic-induced tardive dyskinesia and acute akathsia (D3)
Estrogen receptor-α	Conjugated estrogens	Increase in bone mineral density
	Hormone-replacement	Increase in high-density lipoprotein cholesterol
Glycoprotein IIIa subunit of	Aspirin or glycoprotein	Antiplatelet effect
Glycoprotein IIb/IIIa	IIb/IIIa inhibitors	
Serotonin (5-HT) transporter	Antidepressants	5-HT neurotransmission, antidepressant response

data to show that having genetic or genomic CIR has significantly increased candidate survival in the drug development pipeline. The genetic association study is a simple and effective way of beginning to assess the molecular evidence and provide the CIR for establishing a drug development program for a particular target. It is possible to retrospectively identify positive genetic associations between drug target and incidence or severity of disease for drugs that are currently widely prescribed, for example, angiotensin-converting enzyme inhibitors and hypertension (Zee et al., 1992; Province et al., 2003), β-agonists and asthma (Turki et al., 1995; Santillan et al., 2003), and serotonin reuptake inhibitors and depression (Ogilvie et al., 1996; Golimbet et al., 2004). Although this is not always the case, as the proton-pump inhibitors, used to treat gastroesophageal reflux disease (GERD) are one of the most frequently prescribed class of drugs worldwide, but currently very little is known about the molecular genetics of GERD and no reported association between the genes encoding the α and β subunits of the drug target hydrogen/potassium adenosine triphosphatase (ATPase) and the disease (Post et al., 2005). Knockout mouse data also provides evidence relevant to the function of target on the phenotype (Zambrowicz and Sands, 2003). The CIR for the statins, one of the most successful drug classes to be developed for the lowering of LDL cholesterol, was derived from biochemistry; the HMG-CoA reductase knockout mouse is lethal, there are very few published genetic association studies on HMG-CoA reductase (Tong et al., 2004).

There are now some compounds currently in late stage clinical development and the regulatory approval process where genetics has comprised part of the primary evidence for the establishment of the drug discovery program. The identification of CCR5 as a potential therapeutic target for HIV infection came from the identification of CCR5 and the coreceptor for HIV and a genetic study of individuals, who, despite multiple high-risk exposures, did not become infected with the virus. The association between a common mutation in the gene encoding CCR5 that resulted in a nonfunctional protein and resistance to HIV infection, identified the CCR5 receptor as a coreceptor used by HIV to infect cells in the majority of primary infections. Individuals who were homozygous for this mutation (CCR5 Δ32) and therefore had no functional CCR5 protein were apparently healthy and resistant to infection by HIV (Samson et al., 1996). Subsequent candidate gene studies have shown that the CCR5Δ32 mutation is associated with slower progression to AIDS (Michael et al., 1997). Recent data have shown that a genetic polymorphism in the promoter of the *CCR5* gene, resulting in increased CCR5 expression is more common in individuals rapidly progressing to AIDS (Salkowitz et al., 2003). Thus, within 7 years of the publication of genetic evidence that CCR5 would be a valid target in HIV therapy, clinical validation of this drug target was achieved with both Pfizer Inc. and Schering-Plough publishing data showing significant viral load drops in patients with HIV infection treated with the potent CCR5 antagonists Maraviroc and Schering C, respectively (Feinberg, 2003).

The discovery of Janus kinase (Jak) and the identification of causative mutations in the *Jak3* gene and severe combined immunodeficiency (SCID) highlighted the key role of this target in cytokine signaling and lymphocyte development and function and provided CIR for the development of a selective Jak3 antagonist for the treatment of rejection in renal transplantation and rheumatoid arthritis. As with CCR5 above, the fact that individuals with the mutations only have the very specific effects of immunodeficiency and no other apparent deleterious phenotype means that these genetic data also provide confidence in safety (CIS) for the therapeutic approach (O'Shea et al., 2004).

The ability to carry out large-scale whole genome studies in well-characterized populations extends the candidate gene approach, and has increased the potential to identify novel targets and new pathways that are relevant to disease. Linkage studies have had some success in identifying genetic variants associated with complex diseases; examples include phosphodiesterase 4D and stroke (Gretarsdottir et al., 2003), osmoprotectants taurine cyanate and nitrate (OTCN) cation transporter and DLG5 [discs large (drosophila) homologue 5] genes with inflammatory bowel disease (Peltekova et al., 2004; Stoll et al., 2004) and 5-lipoxygenase-activating protein (FLAP) and myocardial infarction and stroke (Helgadottir et al., 2004).

Family-based microsatellite linkage studies, however, do have some limitations as the ultimate resolution of the association depends on the number of informative meioses, locus heterogeneity is difficult to account for within the analysis and there is a susceptibility to errors through false recombination assignments. The single nucleotide polymorphism (SNP) and linkage disequilibrium (LD) data provided from the human genome sequencing effort and the International HapMap Project have prompted a complementary and powerful approach to finding genetic determinants of complex disease that confer a modest increased risk of disease, known as LD-based whole genome association (LD-WGA). LD-WGA provides an opportunity to generate a vast amount of information on genetic associations with complex disease without the prerequisite of knowing where to look. LD-WGA has the potential to generate a wealth of novel targets, in addition to the identification of relevant pathways, all leading to a further characterization of the genetic heterogeneity of the disease. Studies carried out to date are promising. They have confirmed associations with known candidate genes for the disease under investigation as well as identifying novel associations with other regions of the genome, which might provide new targets for treatment and launch new drug discovery programs with candidates that may not have been considered as a priority (Frazer et al., 2004; John et al., 2004, The Wellcome Trust Case Control Consortium 2007).

An alternative strategy to the candidate gene and whole genome approaches is to combine the two approaches and carry out association studies in a large subset of druggable target genes. Several companies have taken this approach to explore genetic association with as many targets as possible in many indications. Oxagen is a biopharmaceutical company specializing in understanding the genetic basis of common human diseases. One of the main areas of interest for the company is in G-protein coupled receptors (GPCRs); 20%–30% of marketed drugs are targeted to the products of this class of genes. There are over 750 *GPCR* genes, thus Oxagen applied a filtering process to select the best targets for further analysis based on expression profiling, known biology, whether they have a known drug targeted to them or are likely to be chemically tractable before high-throughput genetic analysis (Allen and Carey, 2004). GlaxoSmithKline also have a high-throughput human disease specific target program called HiTDIP that focuses on the association of all classes of tractable genes within the human genome, GPCRs, ion channels, nuclear hormone receptors, and their cofactors as well as kinases and proteases. Several thousand polymorphisms in approximately 1800 genes, including disease candidate genes, genes encoding targets of marketed products as well as the tractable drug targets have been screened in a variety of specific patient populations with common complex diseases such as asthma, obesity, metabolic syndrome, Alzheimer's disease, and diabetes. Early data from a type 2 diabetes study has been reported; 256/1405 genes were significantly associated with type 2 diabetes in 400 cases and controls, 53 of these genes were also significantly associated with disease in a secondary screen of more than 1100 cases and controls, of which 21 remained significantly associated after a correction factor was applied. Ten of the twenty-one genes encoded proteins of biological relevance to type 2 diabetes; four had already been chemically screened (Roses et al., 2005). In a disease area, where much is already known about disease pathogenesis, these data provide valuable information regarding prioritization of candidates. In disease areas, where less is known regarding etiology and pathogenesis of disease such as schizophrenia, epilepsy, and neuropathic pain, for example, this approach can provide insight into the underlying molecular basis of disease and generate new targets.

With the implementation of pharmacogenomics to drive target identification in well-defined patient populations comes the dilemma of knowing which of all the targets identified is the best to take forward. Many of the positive genetic associations with disease from LD-WGA are likely to occur in noncoding regions of the genome and the basis for a strong association, if replicated, will be unknown. Recent data investigating non-coding parts of the genome have revealed the importance of these regions in regulating gene expression (The ENCODE Project Consortium 2007).

In addition, many of the positively associated novel druggable targets will be unprecedented. As the application of whole genome technologies to understanding genetic disease increases the number of potential targets, the debate moves to whether pharmacogenomic resources should provide the basic science knowledge to support research on each potential target or whether high-throughput chemistry screens should drive target selection. Undertaking a comprehensive analysis of the genetic variation that exists in putative drug targets, whether chemistry or genomics is the primary driver for target selection, will provide information that has the potential to impact on drug discovery processes downstream.

In an internal study within Pfizer Inc. comparing coding SNP (cSNP) frequency, a selection of 111 genes encoding potential druggable targets and 160 genes considered as "non-druggable" targets identified that 15% (26/111) of the putative targets were not polymorphic at the amino acid level while 40% (45/111) had one or two cSNPs. There are also well-documented differences in the frequencies of specific polymorphisms between ethnic groups. Prior knowledge of any polymorphisms in a target can be incorporated into target validation, lead optimization, and inform preclinical projects supporting the development of the compound. The effect of genetic variation can be assessed through in vitro assays that incorporate a comparison of polymorphic targets either by using cells or biological reagents obtained from donors of known genotypes where available, or by site-directed mutagenesis. This will facilitate early assessment of the potential impact of genetic variation on the activity of compounds and offer the potential to choose candidates that are least likely to be influenced by the target polymorphism (Penny and McHale, 2005).

Gaining an early understanding of the impact of genetic variation can increase confidence in chemistry (CIC). For example, cholesterol esterase transfer protein (CETP) is a potential therapeutic target in high-density lipoprotein (HDL) modulation and cardiovascular disease. A low level of HDL cholesterol (HDL-C) is known to be a strong independent risk factor for coronary artery disease (CAD). However, there remains uncertainty as to how HDL protects from CAD and as CETP plays a central role in HDL-C metabolism it was considered a potential drug target. Much of the work undertaken to investigate the role of CETP in disease has focused on human genetic variation because of a lack of suitable animal models to study the mechanism. Associations between a mutation in the human *CETP* gene that results in reduced activity of the protein and thus high HDL levels highlighted the importance of CETP in modulating HDL levels, and supports the CIR for development of CETP inhibitors in the cardiovascular therapeutic area (Thompson et al., 2003; Thompson et al., 2005). The *CETP* gene is highly polymorphic; the majority of research is focused on the more common polymorphisms including those in the promoter and 5' region of the gene that influence gene expression. An assessment of nine less common missense SNPs showed that while they were associated with differential stability of the protein, they did not impact inhibition of Torcetrapib in vitro, providing confidence that genetic variation in the target is unlikely to cause variability in drug response due to differential compound binding (Lloyd et al., 2005).

The pharmacogenomic studies included in the preclinical phase of drug discovery that provide CIR and CIC and support nomination of a candidate for development are not intended to replace any of the clinical studies required for exploratory drug development or predict response in patient populations. The preclinical strategy will produce data to inform the pharmacogenomic plan for compounds in exploratory and full development. The challenge facing pharmacogenomics specialists in the pharmaceutical industry is to use the available genomic data to improve the efficiency of clinical trials.

Improving Disease Classification: Targeted Therapy

The need to accurately and precisely characterize the disease under investigation has important implications in drug development. The knowledge from the outset of a drug discovery program that there are two distinct molecular subtypes of disease means that appropriate preclinical experiments can be developed early to predict the likelihood of a pharmacogenomic effect, and this information can be used advantageously in the drug development program. Combining genotype data with other genomic data provides valuable information relating to disease subtype. Integration of genotyping data with gene expression has identified subtypes of obesity phenotypes in a mouse model (Schadt et al., 2005). Using similar approaches and including proteomic and metabonomic analyses in well-defined patient cohorts will provide a powerful tool to aid the dissection of the phenotype of disease in humans in order to drive the development of targeted therapies based on molecular subclassification.

One therapeutic area where using genetic and genomic technologies has undoubtedly had a major and measurable impact on understanding the molecular subtypes of disease is oncology. The advances in understanding the molecular mechanisms predisposing to cancer have seen the number of oncology compounds in clinical development rise from 10 to over 400 in a 10-year period. The majority of the new compounds now being tested are classed as "targeted biotech medicines." Imatinib mesylate (Gleevec) and trastuzumab (Herceptin) were the first two such targeted compounds approved. Herceptin is a therapy targeting the HER2/neu receptor in breast cancer. The rationale for this therapy was based on a sound understanding of the underlying molecular pathology. It was known that only 20%–30% of breast tumors overexpress this protein and it was demonstrated in the drug development program that response to Herceptin was limited to subjects whose tumors overexpressed the target (Vogel et al., 2002). Similarly, Gleevec is a therapy targeting the fusion protein product resulting from the Philadelphia chromosomal translocation observed in most cases of chronic myeloid leukemia (CML) (Deininger et al., 1997). This therapy provided dramatic efficacy in cases of CML with the chromosomal translocation and was rapidly approved by the Food and Drug Administration (FDA).

Following the rapid approval and success of Gleevec and Herceptin, many other targeted cancer therapies have entered clinical trials thus highlighting the absolute requirement to continue to investigate and understand the underlying molecular mechanisms that are associated with disease. Gefitinib (Iressa) was the first in class selective epidermal growth factor receptor (EGFR) inhibitor to receive accelerated approval based on preliminary data from phase II studies in non-small-cell lung carcinoma (NSCLC) patients. Activating mutations and overexpression of EGFR were known to occur in many cancers, providing CIR for development of an EGFR inhibitor for cancer treatment. Inactivation of the *EFGR* gene in mice did not cause any major phenotypic effects, which, in turn provided CIS with respect to pharmacological inhibition of this target (Wong, 2003). However, tumor response to treatment in the clinical trials was only observed in 9%–19% of patients. Subsequent analysis to predict factors that would indicate good response to Iressa identified that female gender, nonsmoking status and specific histological subtype of tumor was associated with better response to therapy. Investigation of biological and markers of response failed to show an association with EGFR expression levels. However, somatic mutations in the ATP-binding site of the tyrosine kinase domain of EGFR were observed more frequently in the tumors of patients who responded to Iressa. The EGFR mutations are located close to the putative binding site for compounds like Iressa and lead to increased signaling in the growth factor pathway, and thus tumors harboring these mutations are more susceptible to treatment with an EGFR inhibitor (Lynch et al., 2004). This highlights the importance of defining the molecular subtypes of disease and understanding the impact on response to therapy. Had the molecular profile of NSCLC been identified before testing in humans, it may have been possible to design preclinical cell-based assays to determine whether the genetic profile of the tumor would influence response to therapy and then inform clinical trial design.

Predicting Drug Response

There are several definitions of pharmacogenetics in the literature, but the term was originally used in 1959 by Vogel to describe the inter-individual difference in drug response due to variation in DNA (Vogel, 1959). Although this is the origin of the term, the concept of inherited differences in biochemical attributes dates back much further with Garrod describing the inheritance of alcaptonuria and phenylketonuria in 1902 and Synder in 1932 describing the inherited ability to taste (or not) of phenylthiocarbamide (Garrod, 1902; Snyder, 1932). The article by Motulsky in 1957 was the first serious attempt to understand the basis of inherited inter-individual response to drug therapies with descriptions of the effects of glucose-6-phosphate dehydrogenase (G6PD) deficiency and primaquine in African–American soldiers (Motulsky, 1957). In the Pacific theater before and during World War II, scientists from the University of Chicago observed that approximately 10% of black American soldiers and, rarely, some of the white soldiers, developed hemolytic anemia of varying severity when given conventional doses of a then-new antimalarial drug, primaquine. Further investigation revealed that this was due to the lack of the G6PD enzyme in red cells, which was the same genetic defect that had been shown to be responsible for the development of hemolytic anemia in susceptible individuals following the ingestion of fava beans. This was one of the first descriptions of a Mendelian (X-linked) pharmacogenetic trait. Also, in 1957 Kalow and Genest described a Mendelian pharmacogenetic trait although this was autosomal recessive (Kalow and Genest, 1957). Approximately 1 in 2000 subjects undergoing anesthesia develop a prolonged effect of succinyl choline due to a deficiency in the enzyme pseudocholinesterase. This autosomal recessive trait has since been recognized in a wide variety of ethnic populations and although the enzyme deficiency was identified in 1957, it was a further 30 years, before the causative genetic mutations responsible for these reactions were identified (McGuire et al., 1989).

Pharmacogenetics remained a relatively small field for the next 40 years due to the fact that although it was well recognized that all drugs exhibited significant inter-individual variability in response, the genetic tools to examine this variability were not available. Apart from a few standard approaches, for example, renal impairment studies and gender differences, there was limited investigation of this phenomenon during drug development. The approach of the drug companies and regulators alike was to ensure that all compounds had a sufficiently good therapeutic index that the average benefit significantly outweighed the potential risk. This has led to the withdrawal or termination of development of a number of compounds with good efficacy but an insufficient population based safety profile, which can often be driven by a small number of potentially serious adverse events. These events can be categorized into those which are expected based on an understanding of the pharmacological action of the drug (type A) and which correlate with plasma exposure levels or idiosyncratic (type B) (Rawlins and Thompson, 1991). The mechanism of idiosyncratic reactions are generally unknown and do not have a clear dose response relationship.

The importance of being able to predict drug response is highlighted by the fact that it has been estimated that approximately 30% of prescriptions written do not benefit the patient and even in highly controlled environments, such as clinical trials, it is rare to get response rates significantly above 70% (Silber, 2000). If we assume that subjects take the medication in the prescribed manner, then lack of efficacy may result from inadequate exposure to the drug (PK variability), an inability to respond to the therapy due to genetic variation in the target and/or downstream effectors [pharmacodynamic (PD) variability], or because the pharmacological intervention does not alter the underlying pathophysiological process (disease heterogeneity). While some commentators have suggested that differences in disease genetics (disease heterogeneity) should be considered as separate from pharmacogenetics, at a practical level, understanding this genetic variation will result in the same outcome, for example, understanding increased or decreased likelihood of response to therapy. Therefore, this group will be included in the PD variability subgroup.

PK Variability

Inter-individual variation in drug metabolism is now a well-documented phenomenon, but it was not until Mahgoub et al. (1977) described the polymorphic metabolism of debrisoquin that significant interest grew in the genetic contribution. The cytochrome P450 (CYP) enzyme family protects the body from xenobiotic agents and is the major route of metabolism of many drugs (Danielson, 2002). Several of these enzymes, for example, 2D6, 2C9, and 2C19, are known to have functional genetic polymorphisms that result in significant reductions or increases in function (Lee et al., 2002; Shimizu et al., 2003). Genetic variation in cytochrome P450 2D6 (CYP2D6) is well characterized and approximately 10% of Caucasians make no 2D6 enzyme. Experiments with the antihypertensive agent, debrisoquin, were the first proven examples of a pharmacogenetic effect. Debrisoquin is metabolized by the CYP2D6 enzyme. An individual who makes no 2D6 and takes a standard dose of debrisoquin will suffer a profound hypotensive event resulting from high plasma exposure levels due to an inability to metabolize the drug (Idle et al., 1978). Approximately 20% of all drugs are metabolized by 2D6 and subjects who are unable to make this enzyme are at increased risk of developing adverse events when taking one of these compounds (Cascorbi, 2003) (Fig. 9–3).

Inter-individual variation in drug metabolism is well documented. Approximately 20% of drugs are metabolized by the CYP2D6 enzyme (Cascorbi, 2003). The incorporation of genetic testing for CYP2D6 or related enzymes in clinical trials has the potential to identify, prospectively, subjects that are likely to have adverse events due to poor metabolism or those that may have limited response through inadequate exposure because of ultra rapid metabolism.

Many drug metabolizing enzymes have genetic variants leading to reduced or increased function with consequent impact on the PK variability. Despite this knowledge, there are no drugs for which pharmacogenetic tests are routinely applied, and only recently has it become accepted best practice to test for the presence of variation in the gene encoding the thiopurine methyltransferase (TPMT) enzyme before prescription of azathioprin and 6-mercaptopurine (4,5). Approximately 1 in 300 individuals are homozygous for mutations in the gene encoding the TPMT (Evans, 2004). If treated with a standard dose of azathioprin (6-mercaptopurine), these individuals have a substantially increased risk of developing the potentially fatal complication of red cell aplasia (Evans, 2004). Suitable dose reduction decreases this risk. The recent decision by the clinical pharmacology division of the FDA to recommend that subjects be tested for TPMT enzyme status (either phenotypically or genotypically) before

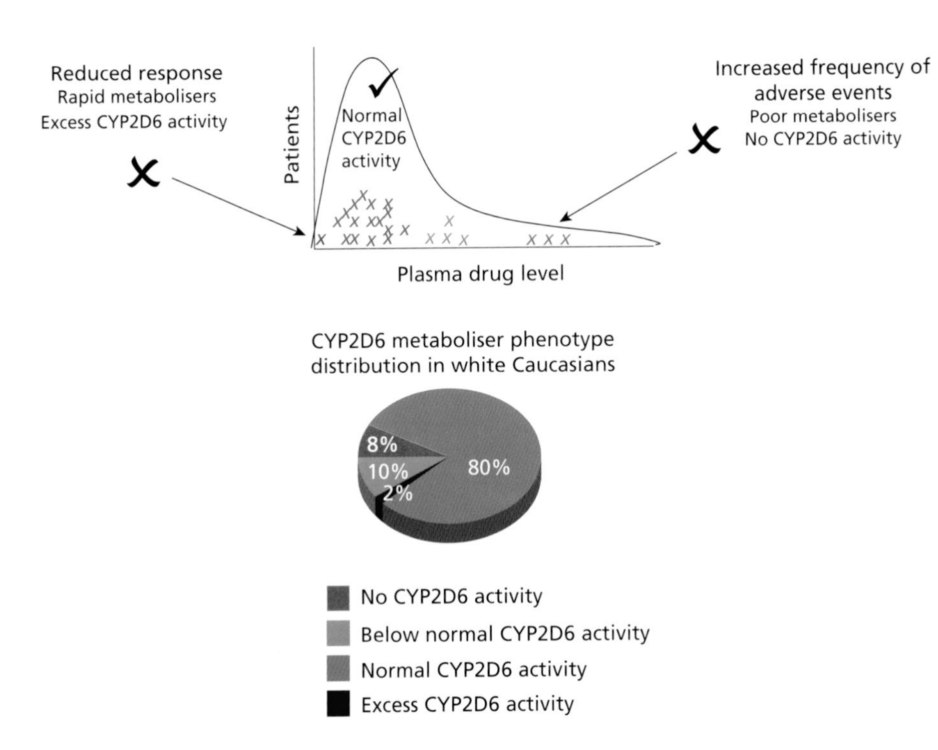

Figure 9–3 Individual variation in drug metabolism.

dosing with 6-mercaptopurine is evidence of the increasing awareness of the value of understanding inter-individual variation in drug metabolism. Similarly, the recently approved drug Strattera from Eli Lilly provides safety data for CYP2D6 poor and extensive metabolizers, and the availability of a suitable test to distinguish these two groups is also included in the label, although there is currently no recommendation about using the test and adjusting the dose according to genotype.

As the clinical value of these tests becomes established and is translated into practice so will the acceptability of requiring a metabolizing enzyme diagnostic before dispensing of the drug. Clear demonstration of the advantages of prospectively using a diagnostic test versus clinical management of drug dosing will also be vital if these tests are to be used in clinical practice. This will also allow the development of chemicals with narrow therapeutic windows and predominantly metabolized by a polymorphic enzyme. Many of these compounds have historically been terminated, as the risk of adverse events due to high plasma exposures outweighed the potential benefit. A clinically acceptable way of managing this risk would make safe use of these compounds possible.

PD Variability

Although it is accepted that genetic variation in drug targets and/or downstream effectors can lead to variability in response to therapies there are only a few good examples of this in the literature. One of the first described was a genetic polymorphism in the promoter of the gene encoding 5-lipoxygenase and response to the 5-lipoxygenase inhibitor, ABT-761 (used in the treatment of asthma) (Drazen et al., 1999). Approximately 5% of individuals are homozygous for a rare promoter polymorphism, which results in no enzyme being produced. Although these individuals do not suffer any overt symptoms from the lack of this enzyme, treating these subjects with a 5-lipoxygenase inhibitor is of little value as there is no enzyme to inhibit. This example, and the oncology examples described above, are in the minority to date, with most other reported effects reported in the literature being less dramatic e.g., response to clozapine (Lerer et al., 2002) or β-agonists (Liggett, 2000).

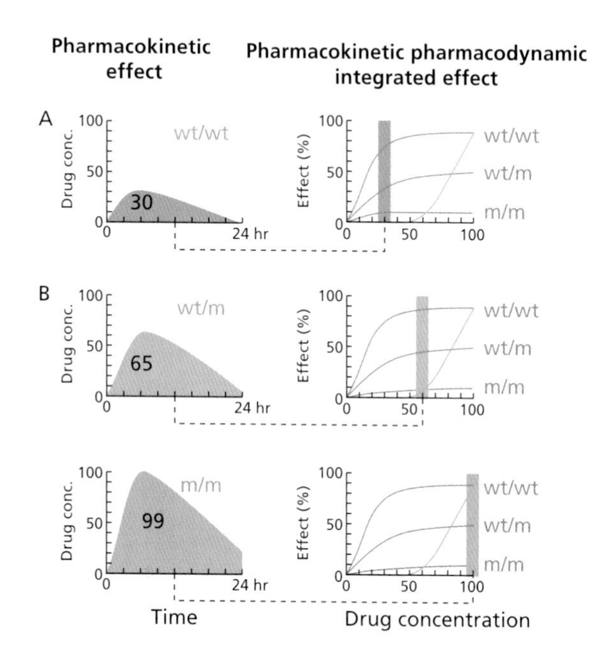

Figure 9–4 Drug response due to pharmacokinetic (PK) and pharmacodynamic (PD) interactions—The impact of genetic variation leading to altered plasma exposures is dependent upon the variation in the genes leading to the effector mechanisms of the drug.

Individualized Therapy—An Integrated Response

In real life, the response of an individual is based on both the plasma exposure and how that affects the various physiological processes in the target organs. Evans and Relling (1999) generated a hypothetical graph representing the PK and PD variation in concert.

Variation in drug metabolizing enzymes can dramatically impact plasma exposure levels (left hand column in Fig. 9–4). However, it is not until we integrate this with variation in genes affecting PD response in the right hand column that we start to get a real understanding of the impact on response for the individual. It is important to realize that dose related adverse events are observed in extensive metabolizers as well as poor metabolizers, but the incidence is dependent upon the frequency of variation in the genes affecting PD response. As the frequency of variation in genes affecting PD response approaches 0.5, the predictive power of a test solely looking at drug metabolism decreases. Similarly, the predictive power of a test evaluating variation in genes impacting PD response will vary depending upon PK variability. Most pharmacogenetic studies, which have been published to date, concentrate on single genes or small numbers of candidate genes, which are likely to impact either PK or PD variability. It is unsurprising that these studies fail to demonstrate high positive or negative predictive information for drug response that is in general due to a combination of both of these factors. As we move forward, a more holistic approach to the examination of genetic factors impacting drug response should lead to the identification of sets of SNPs with higher predictive values, leading to improved prescribing (Table 9–2).

Response to Atypical Antipsychotics—A Case Study

Schizophrenia is a major cause of psychiatric morbidity and affects up to 1% of the population. Treatment regimens depend upon a trial-and-error approach to finding the right dose of the right drug for each subject. Clozapine is an atypical antipsychotic agent and affects multiple central nervous system (CNS) receptors. It is a dopamine antagonist binding to the dopamine D1, 2, 3, 4, and 5 receptor subtypes although the affinity for subtype 4 is greater than for the other subtypes. Additionally, clozapine acts as an antagonist at adrenergic, cholinergic, histaminergic, and serotonergic receptors. It is considered to be one of the most effective therapies, but the response is heterogeneous and response rates vary from 30% to 60%. Tardive dyskinesia is a common adverse event associated with administration of antipsychotic agents and a major cause of discontinuation of therapy. It occurs in 20%–30% of all subjects taking long-term antipsychotic therapy. Variants in the drug target's dopamine D3 receptor, the drug metabolizing agents CYP2D6, and CYP1A2 have been associated with increased risk of developing tardive dyskinesia, and Basile et al. demonstrated that, by combining genetic data from the *DRD3* gene and the *CYP1A2* gene, more than 50% of cases could be predicted (Basile et al., 2002). In addition to the genetic contribution to development of tardive dyskinesia, the genetic contribution to the efficacy response has also been studied although, in general, the studies have examined a small number of candidate genes. Pharmacogenetic effects associating polymorphisms in a range of genes including *5-HT2A*, *5-HT2C*, *5-HT6*, *5-HTT*, *DRD2*, *DRD3*, *DRD4*, and *ADRA2A* and efficacy rates have been reported in the literature. However, the clinical significance of each has remained uncertain and the predictive power of any one is low. Only the serine 9 glycine polymorphism of the dopamine receptor D3 and the C102T polymorphism in the 5HT2C receptor have

Table 9–2 Examples of Drug Response Modification Associated with Genetic Polymorphisms in "Disease-modifying" or "Treatment-modifying" Genes (Evans and McLeod, 2003)

Gene or Gene Product	Disease or Drug Effect	Medication	Influence of Polymorphism
Adducin	Hypertension	Diuretics	Myocardial infarction or stroke
Apolipoprotein E (APOE)	Atherosclerosis, ischemic cardio vascular events	Statins (simvastatin)	Enhanced survival
Apolipoprotein E (APOE)	Alzheimer's disease	Tacrine	Clinical improvement
HLA	Toxicity	Abacavir	Hypersensitivity reaction
Cholesterol ester transfer protein (CETP)	Progression of atherosclerosis	Statins (e.g., pravastatin)	Slowing of atherosclerosis
Ion channels (HERG, KvLQT1, Mink MiRP1	Congenital long-QT syndrome	Erythromycin, cisapride, terfenadine, clarithromycin, quinidine)	Increased risk of drug, induced torsade de pointes
Methylguanine methyltransferase (MGMT)	Glioma	Carmustine	Response of glioma to carmustine
Parkin	Parkinson's disease	Levodopa	Clinical improvement and levodopa-induced dyskinesias
Prothrombin and factor V	Deep-vein thrombosis and cerebral vein thrombosis	Oral contraceptives	Increased risk of deep-vein and cerebral-vein thrombosis with oral contraceptives
Stromelysin-I	Atherosclerosis progression	Statins (pravastatin)	Reduction in cardio-vascular events-death, myocardial infarction, stroke, angina; reduction in risk of angioplasty

multiple positive associations reported in the literature, but even here, there are reports of studies where no association has been observed. The predictive value of any single marker is small but Arranz et al. (2000) have developed a model combining information from six polymorphisms in the 5HT2A, 5HT2C, Histamine 2 receptor, and the serotonin transporter, which has more than 75% specificity and more than 95% sensitivity for response to clozapine (Arranz et al., 2000). In their population, the serine 9 glycine polymorphism of the dopamine D3 receptor was not associated with response.

By combining genetic data from a range of SNPs tested, it is becoming possible to predict the likely response more accurately, both efficacy and safety, to clozapine and so inform the choice of therapy. The demonstration that clinically useful tests can be created from a range of SNPs reported in the literature to individually have small genetic effects, will stimulate further pharmacogenetic research and has resulted in the generation of test kits targeted at influencing prescribing in this area.

Pharmacogenomics and Marketed Drugs

Pharmacovigilance

In a recent study of adverse drug reactions (ADRs), 5% of hospital admissions in the United Kingdom were identified as being due to ADRs. Over 70% were considered avoidable, and while drug interactions accounted for the majority of the ADRs, and older drugs were implicated in the hospital admission, there is still a need to understand the underlying causes of all ADRs (Pirmohamed et al., 2004). It is difficult to detect rare adverse events in the confines of a clinical trial and the current system for monitoring ADRs has been suggested to be "too disparate." A move to a more comprehensive epidemiological approach to monitoring drug safety has been proposed. The inclusion of pharmacogenomic analyses within this approach would allow the systematic assessment of the contribution of genetic determinants to ADRs. Pharmacogenomic surveillance in large phase-IV trials of approved compounds will have a great impact in addressing safety issues.

One therapeutic area, where detailed pharmacosurveillance including pharmacogenomic analyses, post approval, is not new, is in the antiretroviral treatment of HIV infection. Viral resistance and drug toxicity are common and often lead to treatment failure. Determination of HIV genetic sequences and viral load are constantly monitored to assess viral resistance to highly active antiretroviral therapy (HAART). Polymorphisms in drug transporters and drug metabolizing enzymes have also been monitored in HIV therapy. Two retrospective studies have identified the HLA-B*5701 allele of the major histocompatibility complex (MHC) class I B gene as a genetic determinant of hypersensitivity to abacavir (Ziagen) (Hetherington et al., 2002; Mallal et al., 2002). The availability of a relatively large patient population led to the identification of the HLA-B*5701-Hsp70-Hom variant haplotype in 94.4% of cases compared to only 0.4% of controls. Analysis in different ethnic groups, however, showed that HLA-B*5701 alone would not be sufficiently predictive of hypersensitivity in diverse patient populations, suggesting that other genetic determinants of hypersensitivity remain to be identified. Implementation of pharmacogenetics post approval will have a role in increasing the CIS of new products.

Pharmacogenomics in Clinical Practice

Despite extensive knowledge of the genetics of CYP2D6 and related enzymes, and their involvement in the metabolism of many commonly used drugs, PG has made little impact in the clinic. Multiple case control studies have implicated the role of genetic variants in these enzyme systems and risk of adverse events (Brockmoller et al., 2002; Rau et al., 2004; Steimer et al., 2005). These studies have investigated both specific compounds and adverse event rates in drugs metabolized by polymorphic enzymes compared with non-polymorphic pathways. The failure to implement genetic testing for these variants in the clinic and appropriate adjustments in dosing are due to a number of factors. The lack of appropriately designed prospective trials demonstrating the benefits of this approach, and inconsistency of results in some of the retrospective case control studies, are often cited as reasons for the lack of clinical usage. Additional factors include the need for rapid easy testing and increased genetic education for many healthcare groups e.g., physicians, nurses, and pharmacists.

The degree to which pharmacogenetics is incorporated into mainstream clinical practice depends not only upon the science but also the regulatory and societal environment. To date, there has been little impact in the clinic, but the available tests have had limited predictive value, and there have been few good prospective studies performed. As the science progresses, the regulatory and societal factors will become more important. The country's regulatory authorities are responsible for ensuring that all drugs licensed for use have an appropriate risk benefit ratio. However, this ratio is an average based on efficacy across the total treated population and the adverse event rate across the same population. Approval can, and has been, refused for drugs that offer significant benefit but have serious adverse effects in a few subjects. The ability to detect the subjects at increased risk of these adverse events would allow these drugs to be used safely. The number of drugs withdrawn in the last 5–10 years reflects the increasingly risk averse regulatory environment. The potential of preventing these withdrawals in the future by identifying at risk subjects has stimulated significant interest from the regulatory authorities. It is likely that in the future, the identification or confirmation of these adverse events following a drugs launch will stimulate research into the precise mechanism of the event and strategies to identify subjects at risk, rather than an immediate withdrawal. While understanding the mechanisms of ADRs have always been attempted during drug development, pharmacogenetics offers the potential to not only understand why a reaction has occurred but also to identify who is at risk of it before administration of the drug. Initially, it is likely that the regulatory authorities will be a key driver of the use of pharmacogenetics to improve the safety profiles of drugs.

An improved efficacy profile for a compound is also important in the context of gaining drug approval but can be vital when drug reimbursement and use is considered. The use of drugs is primarily driven by the physicians who prescribe them and the healthcare infrastructures that reimburse their costs e.g., the National Health Service (NHS) in the UK or Health Maintenance Organizations (HMOs) in the US. In order for the use of newer drugs to be justified there needs to be significant benefit over existing therapies, which may be generic and have proven safety profiles. It is possible to use pharmacogenetics to improve efficacy profiles by identifying subjects likely to respond well (or those likely to get minimal benefit) and targeting the therapies accordingly. This use of pharmacogenetics is likely to be driven by the payers for the therapies as the increasing pressure on healthcare budgets means that paying for more expensive branded therapies can only be justified for subjects likely to gain significant benefit.

While the role of the regulators and healthcare payers in driving forward the use of pharmacogenetics can be postulated, it is unclear what the role of the patient will be, although it is clear that this could be significant. The risks of taking a medication must always be placed in context with the potential benefits and not treated in isolation. It may be perfectly acceptable to license and use a drug with significant risks if the potential benefits are substantial e.g., cancer therapies. Meanwhile, in other situations, very little risk of adverse events can be tolerated e.g., erectile dysfunction. The indication being treated and the current available therapies are the key determinants of the level of adverse events that would be tolerable for the efficacy observed. As the science becomes more sophisticated and the prediction gets better, it will then be possible to provide more refined risk benefit ratios for each individual. However, it is unclear as to how this range of risk benefits will be managed. Traditionally the regulatory authorities have, in conjunction with independent experts, determined what is an acceptable average risk benefit ratio? This average risk benefit ratio may soon become a range of risk and benefits and the acceptability of an individual risk benefit ratio will become a question for the patient and his or her physician rather than the regulators. As the patients' role in drug selection becomes more central to the prescribing process so does their influence on drug licensing and the use of pharmacogenetics.

Summary

Pharmacogenomics offers great promise to all stakeholders in the healthcare community. To industry, it offers the potential of improving the efficiency of drug development by reducing the current high failure rate through better choice of targets and improved understanding of drug response early in development. To the healthcare providers, it offers the potential to reduce the burden of adverse events by identifying those subjects at increased risk and offering them alternative therapies, as well as targeting its resources to use newer more expensive treatments on subjects who will derive most benefit. Finally, and most importantly, it offers to the patient the opportunity with their physician to identify from the range of available therapeutic options the one most suited to them. While pharmacogenetic testing is unlikely to be able to guarantee that the therapy will work and will not cause an adverse event, it will increase the probability that a drug will work and reduce uncertainty around adverse events and provide a rational way of choosing between therapies.

As our understanding of genomics improves so will our ability to determine key factors involved in variability of drug response. The quest for precision medicines will start at the beginning of the drug discovery process with more comprehensive understanding of the molecular basis of the disease, molecular stratification, and the role of the drug target in the pathological process. Significant variability in PKs will be explained by systematic evaluation of all the relevant metabolizing enzymes and transport proteins. The drug candidates will only be tested in patients with suitable variants of the drug target. Drugs will be approved with variable dosage levels dependent upon underlying genotypes affecting PKs and variation at the drug target. Finally, pharmacogenetics will not stop with approval, but postmarketing research will endeavor to identify the causes of rarer adverse events leading to continuous refinement of how we use drugs throughout their lifecycle.

References

Allen, MJ and Carey, AH (2004). Target identification and validation through genetics. *Drug Discov Today*, 3(5):183–191.

Arranz, MJJ, Munro, et al. (2000). Pharmacogenetic prediction of clozapine response. *Lancet*, 355(9215):1615–1616.

Basile, VSM, Masellis, et al. (2002). Pharmacogenomics in schizophrenia: the quest for individualized therapy. *Hum Mol Genet*, 11(20):2517–2530.

Brockmoller, JJ, Kirchheiner, et al. (2002). The impact of the CYP2D6 polymorphism on haloperidol pharmacokinetics and on the outcome of haloperidol treatment. *Clin Pharmacol Ther*, 72(4):438–452.

Cascorbi, I (2003). Pharmacogenetics of cytochrome p4502D6: genetic background and clinical implication. *Eur J Clin Invest*, 33(Suppl. 2):17–22.

Danielson, PB (2002). The cytochrome P450 superfamily: biochemistry, evolution, and drug metabolism in humans. *Curr Drug Metab*, 3:561–597.

Deininger, MW, Goldman, JM, et al. (1997). The tyrosine kinase inhibitor CGP57148B selectively inhibits the growth of BCR-ABL-positive cells. *Blood*, 90(9):3691–3698.

DiMasi, JA, Hansen, RW, et al. (2003). The price of innovation: new estimates of drug development costs. *J Health Econ*, 22(2):151–185.

Drazen, JM, Yandava, CN, et al. (1999). Pharmacogenetic association between ALOX5 promoter genotype and the response to anti-asthma treatment. *Nat Genet*, 22(2):168–170.

Drews, J (2000). Drug discovery: a historical perspective. *Science*, 287(5460):1960–1964.

The ENCODE Project Consortium (2007). Identification and analysis of functional elements in 1% of the human genome by the ENCODE pilot project. *Nature* 447:799–816.

Evans, WE (2004). Pharmacogenetics of thiopurine S-methyltransferase and thiopurine therapy. *Ther Drug Monit*, 26(2):186–191.

Evans, WE and Relling, MV (1999). Pharmacogenomics: translating functional genomics into rational therapeutics. *Science*, 286(5439):487–491.

Feinberg, J (2003). Meeting notes from the 43rd Interscience Conference on Antimicrobial Agents and Chemotherapy (ICAAC). New CCR5 antagonist shows antiretroviral effect. *AIDS Clin Care*, 15(11):94–95.

Frazer, K, Seymour, AB, et al. (2004). Identification of the genetic basis of individual differences in human serum high-density lipoprotein cholesterol (HDL-C) concentration. American Society of Human Genetics Los Angeles, California, USA.

Garrod, A (1902). The incidence of alcaptonuria: a study in chemical individuality. *Lancet*, 2:1616–1620.

Golimbet, VE, Alfimova, MV, et al. (2004). Serotonin transporter polymorphism and depressive-related symptoms in schizophrenia. *Am J Med Genet*, 126B(1):1–7.

Gretarsdottir, S, Thorleifsson, G, et al. (2003). The gene encoding phosphodiesterase 4D confers risk of ischemic stroke. *Nat Genet*, 35 (2):131–138.

Helgadottir, A, Manolescu, A, et al. (2004). The gene encoding 5-lipoxygenase activating protein confers risk of myocardial infarction and stroke. *Nat Genet*, 36(3):233–239.

Hetherington, S, Hughes, AR, et al. (2002). Genetic variations in HLA-B region and hypersensitivity reactions to abacavir. *Lancet*, 359(9312):1121–1122.

Hopkins, AL and Groom, CR (2002). The druggable genome. *Nat Rev Drug Discov*, 1(9):727–730.

Idle, JR, Mahgoub, A, et al. (1978). Hypotensive response to debrisoquine and hydroxylation phenotype. *Life Sci*, 22(11): 979–983.

John, S, Shephard, N, et al. (2004). Whole-genome scan, in a complex disease, using 11,245 single-nucleotide polymorphisms: comparison with microsatellites. *Am J Hum Genet*, 75(1):54–64.

Kalow, W and Genest, K (1957). A method for the detection of atypical forms of human serum cholinesterase; determination of dibucaine numbers. *Can J Biochem Physiol*, 35(6):339–346.

Lee, CR, Goldstein, JA, et al. (2002). Cytochrome P450 2C9 polymorphisms: a comprehensive review of the in-vitro and human data. *Pharmacogenetics*, 12(3):251–263.

Lerer, B, Segman, RH, et al. (2002). Pharmacogenetics of tardive dyskinesia: combined analysis of 780 patients supports association with dopamine D3 receptor gene Ser9Gly polymorphism. *Neuropsychopharmacology*, 27(1):105–119.

Liggett, SB (2000). The pharmacogenetics of beta2-adrenergic receptors: relevance to asthma. *J Allergy Clin Immunol*, 105(2 Pt 2): S487–S492.

Lloyd, DB, Lira, ME, et al. (2005). Cholesteryl ester transfer protein variants have differential stability but uniform inhibition by torcetrapib. *J Biol Chem*, 280(15):14918–14922.

Lynch, TJ., Bell, DW, et al. (2004). Activating mutations in the epidermal growth factor receptor underlying responsiveness of non-small-cell lung cancer to gefitinib. *N Engl J Med*, 350(21):2129–2139.

Mahgoub, A, Idle, JR, et al. (1977). Polymorphic hydroxylation of Debrisoquine in man. *Lancet*, 2(8038):584–586.

Mallal, S, Nolan, D, et al. (2002). Association between presence of HLA-B*5701, HLA-DR7, and HLA-DQ3 and hypersensitivity to HIV-1 reverse-transcriptase inhibitor abacavir. *Lancet*, 359(9308):727–732.

McGuire, MC, Nogueira, CP, et al. (1989). Identification of the structural mutation responsible for the dibucaine-resistant (atypical) variant form of human serum cholinesterase. *Proc Natl Acad Sci USA*, 86 (3):953–957.

Michael, NL, Louie, LG, et al. (1997). The role of CCR5 and CCR2 polymorphisms in HIV-1 transmission and disease progression. *Nat Med*, 3(10):1160–1162.

Motulsky, AG (1957). Drug reactions enzymes, and biochemical genetics. *J Am Med Assoc*, 165(7):835–837.

Ogilvie, AD, Battersby, S, et al. (1996). Polymorphism in serotonin transporter gene associated with susceptibility to major depression. *Lancet*, 347(9003):731–733.

O'Shea, JJ, Husa, M, et al. (2004). Jak3 and the pathogenesis of severe combined immunodeficiency. *Mol Immunol*, 41(6–7):727–737.

Peltekova, VD, Wintle, RF, et al. (2004). Functional variants of OCTN cation transporter genes are associated with Crohn disease. *Nat Genet*, 36(5):471–475.

Penny, MA and McHale, D (2005). Pharmacogenomics and the drug discovery pipeline: when should it be implemented? *Am J Pharmacogenomics*, 5(1):53–62.

Pirmohamed, M, James, S, et al. (2004). Adverse drug reactions as cause of admission to hospital: prospective analysis of 18 820 patients. *BMJ*, 329(7456):15–19.

Post, JC, Ze, F, et al. (2005). Genetics of pediatric gastroesophageal reflux. *Curr Opin Allergy Clin Immunol*, 5(1):5–9.

Province, MA, Kardia, SL, et al. (2003). A meta-analysis of genome-wide linkage scans for hypertension: the National Heart, Lung and Blood Institute Family Blood Pressure Program. *Am J Hypertens*, 16(2):144–147.

Rau, T, Wohlleben, G, et al. (2004). CYP2D6 genotype: impact on adverse effects and nonresponse during treatment with antidepressants-a pilot study. *Clin Pharmacol Ther*, 75(5):386–393.

Rawlins, M and Thompson, J (1991). Mechanisms of adverse drug reactions. In D Davies (ed.), *Textbook of adverse drug reactions*, (pp. 18–45). Oxford: Oxford University Press.

Roses, AD, Burns, DK, et al. (2005). Disease-specific target selection; a critical first step down the right road. *Drug Discov Today*, 10(3):177–191.

Salkowitz, JR, Bruse, SE, et al. (2003). CCR5 promoter polymorphism determines macrophage CCR5 density and magnitude of HIV-1 propagation in vitro. *Clin Immunol*, 108(3):234–240.

Samson, M, Libert, F, et al. (1996). Resistance to HIV-1 infection in caucasian individuals bearing mutant alleles of the CCR-5 chemokine receptor gene. *Nature*, 382(6593):722–725.

Santillan, AA, Camargo, CA Jr, et al. (2003). Association between beta2-adrenoceptor polymorphisms and asthma diagnosis among Mexican adults. *J Allergy Clin Immunol*, 112(6):1095–1100.

Schadt, EE, Lamb, J, et al. (2005). An integrative genomics approach to infer causal associations between gene expression and disease. *Nat Genet*, 37(7):710–717.

Shimizu, T, Ochiai, H, et al. (2003). Bioinformatics research on inter-racial difference in drug metabolism I. Analysis on frequencies of mutant alleles and poor metabolizers on CYP2D6 and CYP2C19. *Drug Metab Pharmacokinet*, 18(1):48–70.

Silber, BM (ed.) (2000). Pharmacogenomics, biomarkers and the promise of personalised medicine. *Pharmacogenetics—Pharmacogenomics*.

Snyder, L (1932). Studies in human inheritance. IX. The inheritance of taste deficiency in man. *Ohio J Sci*, 32:436–468.

Steimer, W, Zopf, K, et al. (2005). Amitriptyline or not, that is the question: pharmacogenetic testing of CYP2D6 and CYP2C19 identifies patients with low or high risk for side effects in amitriptyline therapy. *Clin Chem*, 51(2):376–385.

Stoll, M, Corneliussen, B, et al. (2004). Genetic variation in DLG5 is associated with inflammatory bowel disease. *Nat Genet*, 36(5):476–480.

Thompson, JF, Durham, LK, et al. (2005). CETP polymorphisms associated with HDL cholesterol may differ from those associated with cardiovascular disease. *Atherosclerosis*, 181(1):45–53.

Thompson, JF, Lira, ME, et al. (2003). Polymorphisms in the CETP gene and association with CETP mass and HDL levels. *Atherosclerosis*, 167(2):195–204.

Tong, Y, Zhang, S, et al. (2004). 8302A/C and (TTA)n polymorphisms in the HMG-CoA reductase gene may be associated with some plasma lipid metabolic phenotypes in patients with coronary heart disease. *Lipids*, 39(3):239–241.

Turki, J, Pak, J, et al. (1995). Genetic polymorphisms of the beta 2-adrenergic receptor in nocturnal and nonnocturnal asthma. Evidence that Gly16 correlates with the nocturnal phenotype. *J Clin Invest*, 95(4):1635–1641.

Vogel, CL, Cobleigh, MA, et al. (2002). Efficacy and safety of trastuzumab as a single agent in first-line treatment of HER2-overexpressing metastatic breast cancer. *J Clin Oncol*, 20(3):719–726.

Vogel, F (1959). Moderne probleme der humangenetik. *Ergeb Inn Med Kinderheilkd*, 12:52–125.

The Wellcome Trust Case Control Consortium (2007). Genome-wide association study of 14,000 cases of seven common diseases and 3,000 shared controls. *Nature* 447:661–678.

Wong, RW (2003). Transgenic and knock-out mice for deciphering the roles of EGFR ligands. *Cell Mol Life Sci*, 60(1):113–118.

Zambrowicz, BP and Sands, AT (2003). Knockouts model the 100 best-selling drugs–will they model the next 100? *Nat Rev Drug Discov*, 2(1):38–51.

Zee, RY, Lou, YK, et al. (1992). Association of a polymorphism of the angiotensin I-converting enzyme gene with essential hypertension. *Biochem Biophys Res Commun*, 184(1):9–15.

Part II

Clinical Genomics

10

Clinical Medicine in the Genome Era: An Introduction

Dhavendra Kumar

It was in Padua that medicine, long degraded and disguised, was now to prove her lineage as the mother of natural science, and the truth of the saying of Hippocrates, that to know the nature of man one must know the nature of all things.

—*Harveian Oration (1900), Clifford Allbutt, Regius Professor of Physic.*

Ancient civilizations in India, China, Greece, and Egypt had recognized that heredity played a significant part in human health. Writings, paintings, and sculptures belonging to people from these regions, and dating back several thousand years, provide evidence that they knew of a close association between heredity and health. Hippocrates in his medical texts noted that "heredity affects health." It is likely that this concept and its practical applications were recognized by other societies but lacked documentary evidence. But, there are some examples, such as the reference in ancient Jewish texts to a young boy given exemption from circumcision because he had a male relative who died from bleeding after ritual circumcision. This reflects the practice of heredity medicine. However, scientific support for heredity and its biological relevance was not clearly understood until the mid-nineteenth century, when the "evolution of species" theory proposed by Charles Darwin and Gregor Mendel's seminal observations on crossbreeding various forms of "green pea" plants laid the foundations of the Science of Genetics. Darwin's theory elucidated the biological importance of genetic variation, and Mendel put in place the fundamental principles of heredity based on individual hereditary factors that were eventually to be identified as genes. Unfortunately, the medical applications of these major discoveries remained unrecognized for almost half a century.

Archibald Garrod, one of the pioneers of "*Genetics in Medicine*" first began applying this knowledge to human health at the start of the twentieth century. In his Harveian Oration of 1924, he quoted from a letter written by William Harvey, the Founder of Modern Medicine:

> Nature is nowhere accustomed more openly to display her secret mysteries than in cases where she shows traces of her workings apart from the beaten path; nor is there any better way to advance the proper practice of medicine than to give our minds to the discovery of the usual law of Nature, by careful investigation of cases of rarer forms of disease. For it has been found, in almost all things, that when they contain of useful or applicable is hardly perceived unless we are deprived of them, or they become deranged in some way.

Garrod, in his seminal paper on the patient with "*alkaptonuria*," provided the necessary evidence to support the "one gene–one enzyme" theory. It not only laid the foundation for molecular medicine but also opened the way forward for genetic medicine. Surprisingly, all this happened well before the elucidation of the chromosomal and nucleic acid basis of inheritance. However, it was widely agreed that the individual genetic complement was locked in the nucleus.

The science of human genetics probably began with the discovery of the full chromosomal complement of man by Tjio and Evans in 1956 (see Chapter 1). This led to the rapid expansion of the whole field of human genetics, and divergence into various specialist fields including medical and clinical genetics. Developments in medical genetics and subsequent applications in clinical practice provided a strong base for wider applications of genetics in medicine. Genetic medicine gained wide recognition over a period of 40 years, and led to the historical decision of 1996 to sequence the whole human genome. This mammoth task, called "The Human Genome Project," was immediately acknowledged as similar to, or even larger than, the "Manhattan Project" to build the atomic bomb, or the "Apollo mission" for the safe return trip to the moon. Some enthusiasts regard this as the "genomic period" followed by the "postgenomic" phase. This is probably incorrect. The "genomic era" has only just begun. The first glimpse of the human genome was possible with the publication of the full sequence of the human mitochondrial genome (Anderson et al., 1981). The technology to sequence the human genome was provided by the publication of the genome of the bacterium, *Haemophilus influenzae.* The editorial in the science journal *Nature Genetics* (November 1999) heralded this achievement as "A point of entry into genomics" (Bork and Huynen, 1999). Perhaps, it is reasonable to identify the period before the completion of the human genome project as the "*pregenome*" era.

Dr. Francis Collins, Director of the Human Genome Project, addressed the use of "*postgenomic era*" while discussing the scope of proteomics after the completion of sequencing the human genome. He queries whether this means that from the beginning of the universe until 2001 we were in the "*pregenome era*," and then suddenly "bang" we moved into the postgenome era, leading one to wonder what happened to the genome era? He suggested that it was presumptuous to say that the Human Genome Project is already behind us. He pointed out that proteomics is a subset of genomics, and genomics is more than sequencing genomes, which will be ongoing for decades to come. The most appropriate term would be "*posthuman genome project era,*" which can be referred to as the "*postHGP era.*" This is truly the "*genomic era.*" Perhaps, "*postMendelian*" seems more appropriate as we move from an era in which genetics has been rooted in *monogenic diseases* with high *penetrance* to a greater awareness (but limited understanding) of *polygenic* diseases (and traits), often with relatively low penetrance.

The "*genomic era*" holds phenomenal promise for identifying the mechanistic basis of anatomic development, metabolic processes, and disease. This is supported by bioinformatics research, which will have a dramatic impact on improving our understanding of such diverse areas as the regulation of gene expression, protein structure determination, comparative evolution, and drug discovery. The availability of virtually complete data sets also makes negative data informative: by mapping entire pathways, for example, it becomes interesting to ask not only what is present but also what is absent (Roos, 2001). This phase can also refer to the increasing emphasis on functional genomics. With an increasing number of organisms for which we have (more or less) complete genomes, we are just beginning to glimpse the potential power of fully mapped sequences.

Perhaps, the various phases of genomics could be delineated more easily by examining the evolution of the scientific methods that led to the development of different analytic methods. In the past (pregenomic era), both experiments and analytic methods were based on hypotheses. As we enter into the genomic phase (present), the majority of experiments are still hypothesis based, but analysis is now more systematic. It is envisaged that in the future (postgenomic era) experiments will be more systematic leading to automatic analytic methods.

Functional Genomics and Molecular Medicine

Understanding the molecular basis of human disease provides insight into pathogenesis and helps in designing therapeutic interventions. The rapid developments in genomics have strengthened the field of molecular medicine. High-throughput genome sequencing and systematic experimental approaches have helped in developing strategic programs to investigate gene function at the cellular, biochemical, and organism levels. Comprehensive functional analysis of all genes and genome sequences falls within the remit of "*functional genomics.*" This field holds great promise, and heralds the beginning of the genomic era.

Understanding the functional significance of genes and genomic variants would require full knowledge of the existing human gene mutations as well as all variants, particularly single nucleotide polymorphisms (SNPs) (see also Chapter 2). Efforts are being made to catalog all known human gene mutations (http://www.hgmd.cf.ac.uk) and haplotype analyses of all known SNPs. The recent completion of the human haplotype map (HapMap) has provided a valuable resource in the study design for disease association studies in common complex diseases, such as cancer, coronary heart disease, diabetes, schizophrenia (Altshuler et al., 2005). A natural successor to the HapMap Project is likely to be the "functional-variant database," which is likely to include all SNPs that alter amino acids in proteins, and possibly gene splicing or transcription (Gibbs, 2005). This functional-variant database would be made available to all researchers, and would probably be an essential resource in all future disease–gene studies.

Functional genomics (see Chapter 3) is a systematic effort to understand the function of genes and gene products by high-throughput analysis of gene products (transcripts, proteins) and biological systems (cell, tissue, or organism) using automated procedures that allow scaling up of experiments classically performed for single genes, such as generation of mutants, analysis of transcript, protein expression, protein structure, and protein–protein interactions on a genome-wide basis (Yaspo, 2001). Functional genomics is based on two approaches: gene-driven and phenotype-driven (Fig. 10–1).

The gene-driven approach uses genomic techniques for identifying, cloning, expressing, and characterizing genes at the molecular level. On the other hand, the phenotype-driven approach depends upon analyzing phenotypes from random mutation screens or naturally occurring variants, such as mouse mutants or human disease, to identify and clone the gene(s) responsible for the phenotype, without having prior knowledge of the underlying molecular mechanisms. Both approaches are highly complementary at virtually all levels of analysis and assist in understanding genotype–phenotype correlations. An important component of functional genomics is comparative genomics, which allows in vivo understanding of the molecular mechanisms of various cellular processes.

The chapter on functional genomics (Chapter 3) reviews basic tools and routes of investigations commonly employed by researchers in this major field. Collectively, functional genomics approaches provide a matrix of information on gene products and their functional attributes. Integration with SNP (not defined in this chapter) information and allelic variants in patient populations will be particularly useful. Research in this field will lead to novel findings that will remove the current bias toward "*interesting genes*," and thus will be of particular importance in novel "*genomics-based*" therapeutic approaches.

Bioinformatics and Genomic Medicine

Bioinformatics is a rapidly emerging field of biomedical research (Kim, 2002). This relatively new discipline develops and applies informatics to the field of molecular biology. The field is broad and includes scientific tools and methods for sequence analysis (nucleotide and protein sequences), rendering of secondary and tertiary structures for these molecules, and protein-fold prediction that is crucial to targeted drug design and development (Elkin, 2003). Bioinformatics opens the way for a new approach in molecular medicine referred to as "phenomics." Clinical (medical) informatics has long been recognized as an important methodology in biomedical research and clinical care by integrating experimental and clinical information systems. Both clinical/medical informatics and bioinformatics will eventually change the current practice of medicine, including diagnostics, therapeutics, and prognostics (Maojo and Kulikowski, 2003). In some ways, this is similar to the clinical applications of biochemistry that happened about half a century ago. Postgenome informatics, equipped with high-throughput technologies and genomic-based databases, is likely to transform biomedical understanding. Some of the key applications of genome-based bioinformatics include multivariate data projection, gene-metabolic pathway mapping, automated biomolecular annotation, text mining of factual and literature databases, and the integrated management of biomolecular databases.

A New Taxonomy for Human Disease

Realizing the full beneficial potential of the human genome sequence will ultimately depend on its applications in clinical medicine. Many aspects of modern-day clinical practice will change with technological advances and the understanding of disease mechanisms, developments in diagnosis, and new drugs and therapeutic procedures.

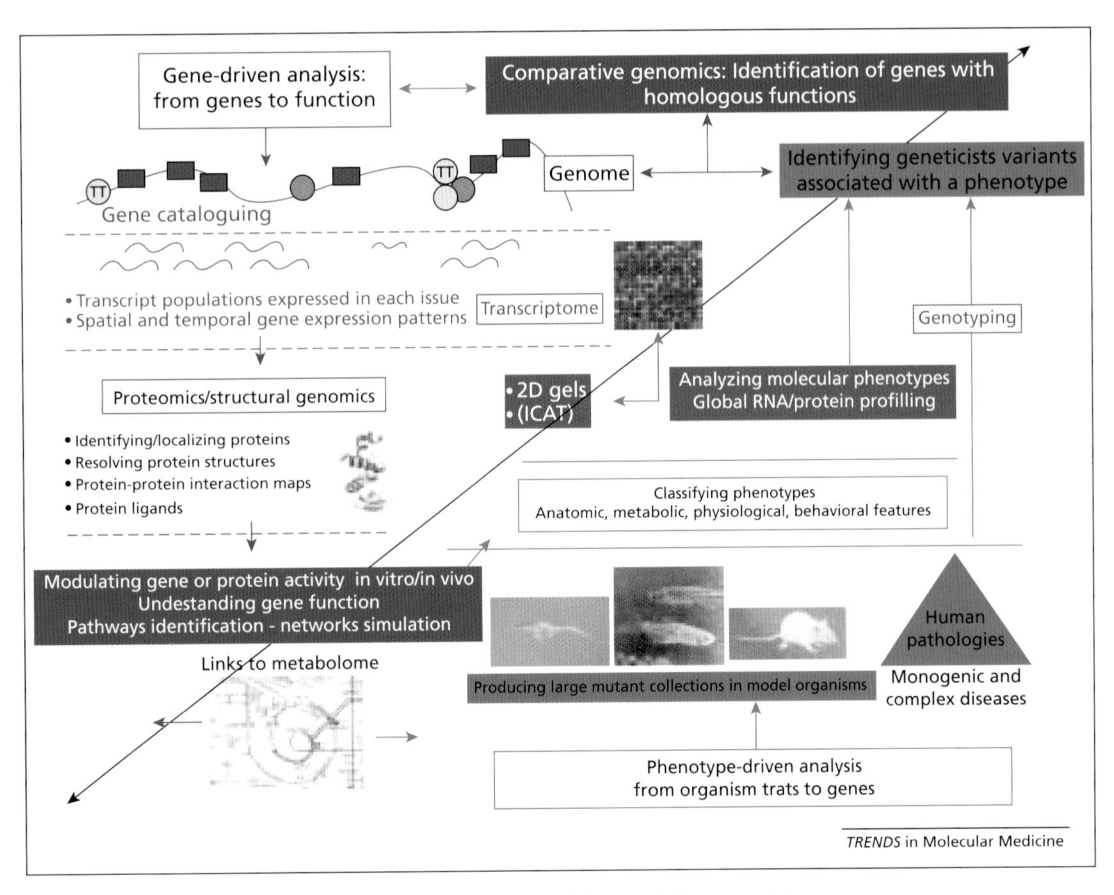

Figure 10–1 Two complementary approaches in functional genomics: gene-driven and phenotype-driven, artificially separated along the diagonal axis. Different levels of information are interconnected with the main tools of analysis and routes of investigations. [Adapted with permission from Yaspo (2001).]

These advances have largely influenced our approach toward human disease, and have contributed to developing a new taxonomy for human disease (Chapter 6).

Genetic etiology in human disease is recognized and includes diseases resulting from chromosomal abnormalities, mutations in nuclear and mitochondrial DNA (mtDNA) gene sequences, and interaction of environmental factors with several low-penetrance genes carrying a small additive effect. Advances in genomic technology have now made it possible to unravel pathogenic mechanisms in certain genetic disorders, which result from unusual genetic factors that do not comply with the traditional concept of genetic etiology. The pathogenic mechanisms include a number of complex alterations distributed across various genomic regions. These are now being termed as "genomic disorders." Distinct categories of genomic disorders include epigenomic diseases, disorders of genome architecture, tri-nucleotide repeat disorders, and diseases associated with complex genomic polymorphisms. It is likely that in the future more groups of genomic diseases will come to light. It is essential that the underlying pathogenic mechanisms in these diseases are understood to facilitate diagnosis and development of targeted specific therapy.

Toxicogenomics

Toxicogenomics is a scientific field that aims to study the complex interaction between the whole genome, chemicals in the environment, and disease. When the cells are exposed to a stress, drug, or toxicant, they respond by altering the pattern of gene expression within their chromosomes. Genes are transcribed into messenger RNA (mRNA), the chemical message by which information encoded in genes is translated into proteins that serve a variety of cellular functions in response to the exposure. The production of protein encoded by a given gene may be increased, decreased, or remain unchanged, depending upon the type of exposure and the cellular requirements. A technology that is central to the field of toxicogenomics is known as microarrays, which enable scientists to simultaneously monitor the interactions among thousands of genes within the genome. This technology (see Chapter 7) will help define the complex regulatory circuitry within a cell, tissue, or organ, and give scientists a global perspective on how an organism responds to a stress, drug, or toxicant. The data generated will provide information about the cellular networks of responding genes, define important target molecules associated with the toxicity mechanism, and provide biomarkers for epidemiological studies. Ultimately, this information may allow us to identify ways to reduce or prevent disease by pinpointing the biochemical and molecular functions that have been perturbed by environmental chemicals. DNA microarray technology will undoubtedly become a major tool in environmental medicine, because it will improve our diagnostic and prognostic capabilities for specific diseases, as well as our ability to examine drug interactions, sensitivities, and effectiveness. This technology will also aid the research on alternative model testing procedures, and support the development of new toxicity screening processes.

It is envisioned that DNA microarray technologies will permit the design of experiments in the occupational and environmental sciences that will clarify whether:

- specific toxicants have unique gene expression–profiling signatures;
- different cells in different tissues have profoundly different response signatures for a given toxicant;
- different species show similar, overlapping, or distinct patterns of gene responses to a toxicant;
- a specific toxicant signature is altered depending upon the stage in the developmental process or defined health condition;
- responses to complex chemical mixtures can be more easily elucidated and defined by their gene expression–profiling signatures;
- responses to chronic low doses of toxicants or environmental pollutants can be defined by gene expression profiling;
- specific gene polymorphisms can be defined, which are characteristic of an increased susceptibility to the pathology of environmental health diseases.

Toxicologists and environmental health scientists have studied the effects of the environment on human health for several decades. Many adverse environmental effects have been identified, and important progress has been made in preventing exposure to harmful agents such as γ-radiation, ultraviolet (UV)-light, lead, pesticides, and dioxins. Toxicological research has attempted to develop an efficient, cost-effective, comprehensive strategy for predicting and preventing toxic responses in humans. However, progress toward this goal has been proportionate to the existing technologies and level of scientific knowledge. The field of toxicology could not have risen to this challenge if it only had access to the less efficient technologies of the past several decades.

One challenge is to use the human genome sequence as a first step to understand the genetic and biological basis of complex biological traits and diseases such as cancer, diabetes, Alzheimer disease, and Parkinson's disease. Another challenge is to use the increased volume of toxicological data to construct genetic and biochemical pathways that will explain the mechanism of toxic responses. Advances in combinatorial chemistry and molecular biology have dramatically accelerated the rate of drug discovery and availability, and the rate at which populations are exposed to new drugs. Such advances intensify the burden of exposure in the population, making it critical that we rapidly increase our understanding of the consequences of such exposure.

The National Center for Toxicogenomics (NCT) of the NIH in the USA is leading the development of a unified strategy for toxicogenomics studies and a public knowledge-base. This will have an informatics infrastructure that will allow all partners in this unprecedented enterprise to share equally in its benefits and products. By providing a focus for technological coordination and basic research, a centralized public knowledge-base, and a center for coordination for all the partners in the pharmaceutical and chemical industries, the NCT will facilitate this diverse national effort. The NCT will not only achieve the economies of time, cost, and effort but also help to ensure the successful development of a broad scientific consensus on the application of toxicogenomics to the improvement of human health.

In brief, toxicogenomics combines the conventional tools of toxicology (such as enzyme assays, clinical chemistry, pathology, and histopathology) with the new approaches of transcriptomics, proteomics, metabolomics, and bioinformatics. This marriage of toxicology and genomics has created not only opportunities but also new informatics challenges. This field is likely to be of major importance in genomic medicine.

Metabonomics and Metabolomics

Genomics measures the entire genetic makeup of an organism, while proteomics measures all the proteins expressed under given conditions. Metabonomics, as the name implies, is defined as measurement of the complete metabolic response of an organism to an environmental stimulus or genetic modification. Some people use the term *metabolomics* to refer to *metabonomics* at the level of a single cell.

The *omics* can provide information for basic biological research, and for pharmaceutical and clinical applications. One of the challenges is integrating the information from the various omics and, in the process, yet another term is coined, "*systeomics*," which refers to the integration of genomics, proteomics, and metabonomics.

Metabonomics may be the most recently named of the *omics*, but it is one of the oldest. In fact, metabonomics dates back to old-fashioned biochemistry, with its emphasis on metabolism, the sum of the processes to acquire and use energy in an organism, to biosynthesize cellular components, and to catabolize waste. A number of toxicological and disease diagnostics are based on metabolic profiling. This methodology has been in existence for more than 20 years, well before the advent of genomics or proteomics. It is probably true that metabonomics is "more closely related to things in the clinical world" than either genomics or proteomics, owing to the fact that metabonomic signatures reflect both genetic information and environmental influences.

Nutrigenomics

In the past decade, nutrition research has undergone an important shift in focus from epidemiology and physiology to molecular biology and genetics. Nutrigenomics is the application of transcriptomics, proteomics, metabolomics/metabonomics, and bioinformatics in nutrition research (see Chapter 8). A European Nutrigenomics Organisation (NuGO) has recently taken over the ambitious challenge to translate the nutrigenomics data into an accurate prediction of the beneficial or adverse health effects of dietary components. This organization and associated agencies have set out to address important issues including nutrigenomics technology standardization and innovation, bioinformatics environment harmonization, and integrated information system development.

The integration of genomics and nutritional sciences has led to the field of "*nutritional genomics*." This provides a very important base from which we can study the complexity of genome responses to nutritional exposure while offering opportunities to enhance our understanding of the effectiveness of dietary interventions, both at individual and population levels. Nutrients influence multiple physiological responses that affect genome stability, imprinting, expression, and viability. Nutritional genomics challenges us to understand the complex interactions between the human genome and dietary components in normal physiology and pathophysiology. An understanding of these interactions will enable us to assess the benefits and risks of various dietary recommendations, minimizing the risk of unintended consequences. Furthermore, nutritional genomics will enable the design of effective dietary regimens for the prevention and management of complex chronic disease. Chapter 8 reviews the new perspectives in the nutritional sciences in the light of advances in genomics.

Pharmacogenomics

The study of the role of genetic inheritance in individual variation in drug response and toxicity is referred to as *pharmacogenetics*. Convergence during the past decade of advances in pharmacogenetics and human genomics has led to the emergence of the field of pharmacogenomics. Pharmacogenomics refers to the study of the relationship between the specific drug, DNA-sequence variation, and drug response (see Chapter 9).

Pharmacogenomics holds great promise for the future of medicine, and is the basis of "*personalized medicine*." This relies on genomic biomarkers for disease susceptibility including both Mendelian and complex diseases. Applications of human genomics will result in improved understanding of the pathophysiology of disease, identification of new therapeutic targets, and improved molecular classification of disease. The promise of individualized therapeutic interventions largely depends on the identification of drug toxicity genomic biomarkers that will enable differentiation of the individuals likely to show both positive and negative therapeutic response. This would also help in assessing the efficacy of new drugs, their side effects, and toxicity. Chapter 9 provides a comprehensive review of the principles and applications of pharmacogenomics.

Personalized Medicine

Although 99.9% of the human genome sequence is similar in any two individuals, sequence variation in the remainder of the genome could be linked to the functional information relevant to an individual's unique genetic constitution. This "personalized sequence variation" has major applications in genomic medicine. Functional annotation for individual sequence variation, when complete, will be of fundamental importance in diagnosing and selecting appropriate therapeutic agents. This is likely to be vastly improved with the availability of targeted sequencing of selected genes, exons, or promoters. There are fewer variants in protein coding than noncoding sequences (Bentley, 2004).

Variants that cause amino acid changes, and thus altered protein product, are in general dissimilar (non-synonymous) compared with those that lack such an association (synonymous). If an excess of nonsynonymous substitution is observed for one particular coding region, then this can be taken as an indicator of diversifying (positive) selection. With the help of newer technologies, more and more variants are being characterized and sequence annotation made available. Ultimately, a fuller picture will emerge of the variants that alter genome function, and will enable selection of those that contribute to health and disease.

The success of genomic medicine will depend upon the ability to sequence an individual's full genome. With the benefit of new technologies, it is possible to generate gigabases of data as short-sequence reads and to assemble the data accurately using the finished sequence as a template. This will provide the essential database of human genome variation for a given population. Comparison of these data sets will provide a full profile of common genome variation along each chromosome. Detection of each variant will help in estimating the recombination rates and correlation along each chromosome. This approach could give important baseline information on healthy tissue compared with pathological tissue. For example, a comparison of the cancer genome sequences could allow monitoring the DNA changes on a genome-wide basis for cancer development. A similar approach could also be applied in other diseases. This genomic information on both healthy and diseased tissue could be used in screening an individual's disease risk and devising appropriate therapy and medical advice, paving the way forward for "*personalized medicine*."

As the human genome functional annotation becomes available, the prospects of "personalized medicine" will improve. A hypothetical scenario is described by Bentley (2004), where variation in the *PPAR-γ* gene, one of the susceptibility genes in type 2 diabetes mellitus (T2DM), is employed in selection of the most appropriate oral hypoglycemic agent (Fig. 10–2).

The chromosome 3 region ("a" in Fig. 10–2) (12,300–12,450 kb), numbering as in build 34 (http://www.ensembl.org) contains the *PPAR-γ* gene structure (dark blue) with an alternative promoter (light blue), hypothetical noncoding functional variants (green-shaded boxes), and functional variants (red). Magnification of the variant segments ("b" in Fig. 10–2) shows the translated sequence with nucleotide changes (functional variants highlighted in blue) and amino acid changes (pink) ("c" in Fig. 10–2). The amino acid variant results in variation in the protein molecule as confirmed from a linked database. This variant protein molecule contains a specific drug-binding site (blue) for antidiabetic thiazolidinedione, an oral hypoglycemic agent. A number of biological consequences, such as biochemical, medical, and pharmacological, can be predicted using linked database information ("d" in Fig. 10–2). This information can be regularly updated and curated, allowing a detailed listing of the likely consequences. A small subset of this information would define the disease or drug outcome or side-effect associated with each variant, and would enable the clinician to provide specific risk information in clinical consultation. This information could be made available on the public domain subject to stringent review, and including only those data for which medical relevance were established.

The use of personal genetic information in a clinical setting could be requested and consented by the individual concerned. The individual sequence acquired could be restricted to one or two genotypes, or as much as a complete genome sequence. The information thus acquired would be exclusive and private and wholly owned by the individual. It could be stored electronically, protected by a high-security code requiring unique personal identifiers, such as multiple fingerprint or iris-pattern, for access only with the consent of the individual ("e" in Fig. 10–2). The information might be taken either before consultation or afterward, and in either case would be subject to counseling by the medical practitioner and consent by the individual.

The clinical consultation could initiate a specific investigation ("f" in Fig. 10–2). The personal genetic information would then be supplied by the individual, for interpretation with respect to an agreed set of variants and/or a specific phenotype. The clinician would use the available risk information concerning each variant to provide a genetic assessment for the individual ("g" in Fig. 10–2). In the case illustrated, the individual has the heterozygous genotype TC at position 3:12,450,610. This corresponds to having both Pro 495 and Ala 495 forms of the protein PPAR-γ. This genotype confers an increased risk of insulin-resistant DM on the individual, and also resistance to the thiazolidinedione class of antidiabetic drugs. Combining this with the risk information for other genotypes would help to make informed subsequent clinical decisions ("h" in Fig. 10–2).

Thus, with easy access to a well-annotated human genome and availability of cheap, accurate whole-genome sequencing technology, an individual could acquire either a specific or complete genetic health profile, including risk and resistance factors. The information could then be used to improve and guide important medical

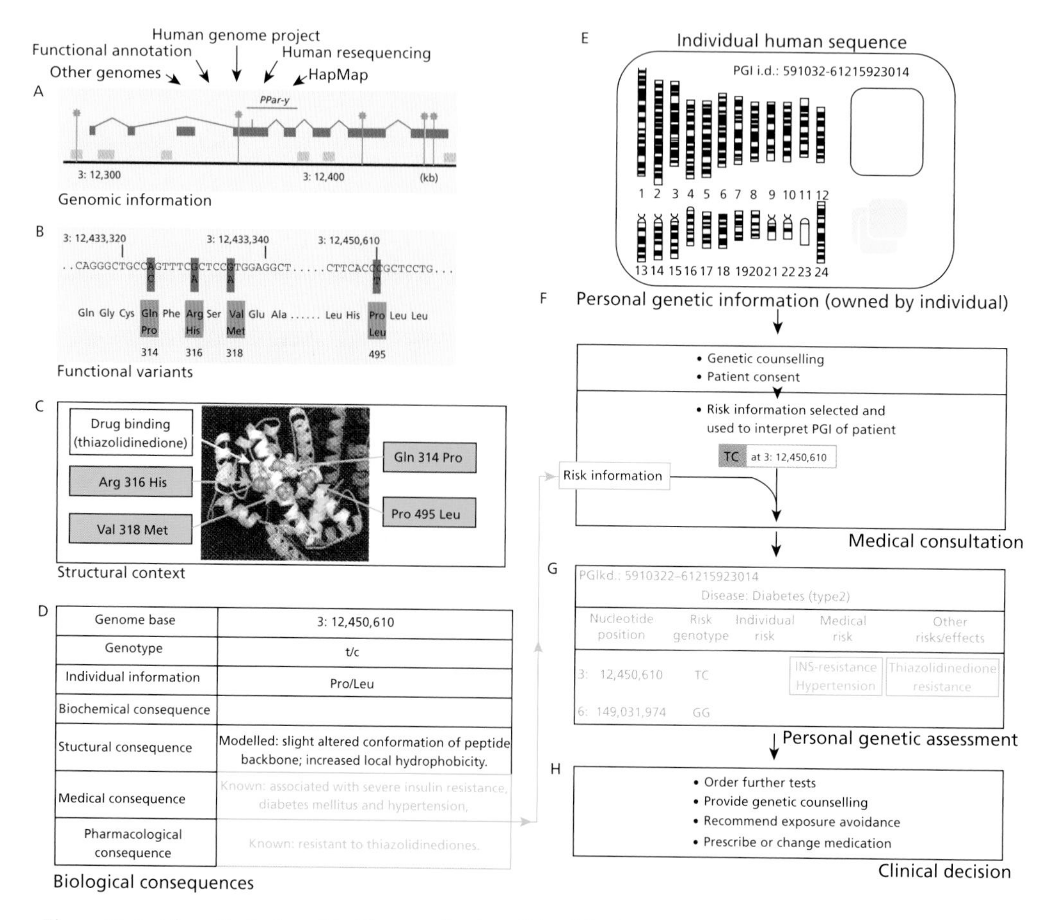

Figure 10–2 A hypothetical model of "personalized medicine." [Adapted with permission from Bentley (2004).]

decisions, to assess the risk of possible future exposures, and to select preventive treatments for improved health (Bentley, 2004).

Genomics and Infectious Diseases

Genetic factors have been recognized to influence infectious disease susceptibility, resistance, and response to antimicrobial therapy. A number of studies, including twin studies, support the importance of host–genotype interactions in wide-ranging infectious disease clinical phenotypes, population differences in susceptibility or resistance, and favorable or adverse reactions to antimicrobial therapy. Population studies have also contributed the evidence supporting the selective advantage to human evolution and population genetic structure. For instance, a high frequency of heterozygotes for sickle cell anemia and thalassaemias in some populations confer a selective advantage for malaria in the face of deleterious effects in homozygotes (Clegg and Weatherall, 1999). Interplay between the Duffy locus mutations and sickle cell anemia in certain populations is recognized to confer malaria resistance. The molecular basis of this phenomenon lies in the erythrocyte chemokine receptor that also binds the malarial parasite (Horuk et al., 1993

Similar observations have been made for another chemokine receptor, CCR5 (CKR5), which offers resistance to HIV/AIDS (human immunodeficiency virus/acquired immunodeficiency syndrome) (Dean et al., 1996). This has led to an interesting speculation that this mutation emerged as a selective force from the plague which affected the northeastern European populations (Stephens et al., 1998).

Similar observations have been made by studying host–pathogen interactions in determining infectious disease susceptibility.

A number of genome projects of both human and pathogen genomes have provided an insight into microbial ecogenetic relationships (Blackwell, 2001).

Several prominent examples of host–pathogen interactions include malaria (*Plasmodium falciparum*, *P. vivax*), tuberculosis (*Mycobacterium tuberculosis*), AIDS (HIV), cholera (*Vibrio cholerae*), and meningitis–otitis (*H. influenzae*). Genomic studies have helped to accumulate a wealth of data in this field. For example, toward the end of 2001, prokaryotic sequences were complete for 39 bacteria and 9 archaebacteria with at least 100 more species sequenced complete with annotations (Subramanian et al., 2001).

Genomics has accelerated insights in microorganisms including genome architecture, sequence similarities, mobile genetic elements, and large numbers of genes of previously unknown function. This information is vital in developing antimicrobial therapy and development of a new class of "*DNA-based*" vaccines. An illustrative example is *H. influenzae*, which was the first organism to be fully sequenced (Fleischman et al., 1995). *H. influenzae* has been immensely useful in microbial genomics research due to its small size (1830 kb), its importance as a major human pathogen, its capacity for DNA transformation as seen in the mouse model, and the rapid advance in knowledge of its genome. There are 1703 proposed genes of which 736 lack proposed functions; of these, 347 are conserved across species while 389 are unique to *H. influenzae*. These unique genes are now the target for developing selective therapeutic agents and vaccines.

Similar approaches are currently being applied to many pathogens including the flu virus and HIV. Other notable organisms targeted for this work include *M. tuberculosis*, *Escherichia coli* strain O157:H7, *V. cholerae*, *Helicobacter pylori*, and *Yersinia pestis*. The whole field of microbial genomics is promising and lies in the core sphere of genomic medicine (see Chapter 25).

Genomics in Critical Care Medicine

Advances in genomics increasingly affect all areas of clinical medicine, including critical care medicine. Survival after acute trauma and sepsis is now common with the development of improved trauma systems, advanced resuscitation methods, and organ support systems. However, this is often complicated by nosocomial infection and organ failure. Technological advances in genomics and proteomics, together with the techniques of bioinformatics, provide an opportunity to characterize the determinants of, and the responses to, injury and sepsis on a genome-wide scale. This includes large-scale collaborative efforts aimed to investigate genomic variation (polymorphisms) and characterize multiple levels of biological response (transcriptome and proteome) to injury and infection, and relate these to clinical situations.

Applications of in-depth genome-wide analysis can allow a thorough understanding of disease processes that are relevant in intensive care, such as acute trauma, sepsis, acute respiratory distress syndrome, and multiple-organ dysfunction syndrome. Understanding critical illness at the genomic level may allow for more effective stratification of patient subclasses and targeted patient-specific therapy. The related fields of pharmacogenomics and pharmacogenetics hold the promise of improved drug development and the tailoring of drug therapy based on an individual's drug metabolism profile. The "genotyping" of critically ill patients will allow us to ascertain individual cytoprotective mechanisms that are crucial to organ and tissue protection in these patients. It is important that in future all physicians caring for critically ill patients are familiar with the advances in genomic technologies and applications in clinical medicine. Chapter 16 reviews the developments in genomics and related fields relevant to the care of critically ill patients.

Genomics and Cardiovascular Diseases Including Sudden Death

Sudden death is a major public health concern. It has inevitable social, personal, and economic consequences. The extent of personal grief and long-term psychological effects associated with sudden death are impossible to assess. This is particularly true in relation to the unexplained death of an infant (cot death) or a young person, which could be due to an as yet unknown genetic, metabolic, or cardiac disease. Sudden death in an adult is often due to an underlying cardiac disease, referred to as "sudden cardiac death or SCD."

Apart from established Mendelian disorders (Marfan's syndrome, hypertrophic cardiomyopathy, long QT syndrome, etc.), SCD is commonly used to refer to coronary heart disease and heart failure. Advances in genomic science applicable to sudden death, with particular reference to SCD, are reviewed in Chapter 11. The review discusses novel bioinformatics approaches in identifying candidate genes/pathways and their functional significance. This chapter also discusses the possibility of applying high-density genome-wide SNP analysis in organizing community-based screening for genetic susceptibility to common heart diseases.

Obesity and Related Metabolic Diseases

Obesity is endemic in the developed world, and rapidly reaching epidemic proportions in the developing world. Obesity is associated with hypertension, coronary heart disease, and diabetes mellitus [DM.]. Only a small proportion of obesity is related to genetically determined causes, when it is often accompanied by involvement of other body systems. In the majority of cases, it is related to environmental factors. However, the severity and clinical outcome is modulated by multiple genetic factors consistent with polygenic/multifactorial inheritance. An association of obesity with type 2 diabetes mellitus [T2DM] is well recognized. In addition, obesity increases the risk of hypertension and coronary heart disease.

In contrast to the immunologically determined insulin-dependent type 1 diabetes mellitus (IDDM or T1DM), non-insulin-dependent T2DM (NIDDM or T2DM) is genetically heterogeneous. Several candidate genes are implicated in the pathogenesis of T2DM. A genomic approach is essential in analyzing an individual's risk for T2DM.

Obesity, hypertension, and T2DM are good examples where specifically designed microarrays could eventually be very effective in the screening of "at-risk" individuals, and in the identification of those who might benefit from appropriate lifestyle changes and prophylactic pharmacological interventions. Chapter 12 reviews the genetic and genomic aspects of DM and related metabolic diseases.

Bronchial Asthma and Chronic Obstructive Lung Diseases

Asthma is a complex genetic disorder with a heterogeneous phenotype resulting from the interactions between many genes and the environment. Numerous loci and candidate genes have been reported to show linkage and association with asthma, the asthma-associated phenotypes. These include microsatellite markers and SNPs associated with specific cytokine/chemokine and IgE regulating genes. Although significant progress has been made in the field of asthma genetics in the past decade, the clinical implications of the genetic variations within the numerous candidate asthma genes remain largely undetermined. However, new information has recently emerged from postgenomic research with the cloning of new asthma genes, such as *ADAM33* and *PHF11*. Chapter 21 highlights recent developments in genetic, genomic, and proteomic research in asthma and related respiratory diseases.

Autoimmune Disorders: Inflammatory Bowel Disease and Rheumatic Diseases

The complexity of the immune system is related to gene expression in the tissues, cells, and biologic systems. An analysis of selected gene systems, using the high-throughput whole-genome screening approach, has gained us insight into these complex systems. Development of sophisticated methodologies, such as microarray technology, allows an open-ended survey to identify comprehensively the fraction of genes that are differentially expressed between different biological samples. New developments in genomics have helped us in improving our understanding of basic and applied aspects of immunologically determined disorders. For instance, improved understanding of the molecular basis of inflammatory bowel disease (IBD) has enabled the development of new therapeutic agents (Chapter 22).

Rheumatic disorders comprise several heterogeneous diseases that impose a heavy burden on healthcare services because of the significant long-term morbidity and disability. A small number of these conditions result from single-gene connective tissue diseases, for example, Marfan's syndrome, Ehlers–Danlos syndrome, and uncommon inherited metabolic diseases. However, the etiology and pathogenesis in a large number of these diseases is poorly understood, except for some evidence of autoimmune pathogenesis supporting multifactorial/polygenic inheritance. The postgenomic advances are helping to enhance our understanding of the pathogenesis in several rheumatic diseases, for example, rheumatoid arthritis and osteoarthritis. Chapter 26 covers all these developments and highlights the future methods for diagnosing rheumatic diseases and possible therapeutic interventions.

Genomics and Neuropsychiatric Diseases

Although rapid progress has been made in mapping and characterizing the genes for several monogenic neurological diseases, the pathogenesis in a large number of neuropsychiatric disorders remains unexplained. However, there is evidence, albeit limited, that genetic factors play a significant role in the causation of these disorders, interacting with environmental factors. The list of these conditions is long and includes multiple sclerosis, Parkinson's disease, Alzheimer dementia, autism, schizophrenia, and affective disorders. Chapters 19 and 20 provide information on the progress made and developments in understanding these disorders in the genomics context.

Cancer Genomics

Cancer genetics is a relatively new, but rapidly developing, field that has acquired a prominent place in clinical genetics. However, it has largely focused on uncommon developmental malformation syndromes with malignancy, familial breast, ovarian and colorectal cancers, and some other uncommon Mendelian familial cancer syndromes. Major clinical genetic centers are now equipped with laboratory facilities offering diagnostic and presymptomatic genetic testing in selected conditions and situations. There are well-established clinical protocols dealing with clinical referrals, risk assessment, genetic counseling, genetic testing, early detection of cancer (screening) in family members at increased lifetime risk, and including provision of follow-up and long-term support. However, there are no clinically validated protocols dealing with isolated common cancers, such as lung, skin, breast, bowel, and prostate. The etiology in these malignancies is not clearly known, probably following a multifactorial pattern with multiple genes or polymorphisms conferring a genetic predisposition.

Completion of the human genome sequence and dissemination of high-throughput technology will provide opportunities for systematic analysis of cancer cells. Genome-wide mutation screens, high-resolution analysis of chromosomal aberrations, and expression profiling all give comprehensive views of genetic alterations in cancer cells. These analyses will facilitate compilation of a complete list of the genetic changes causing malignant transformation and of the therapeutic targets that may be exploited for clinical benefit. It has been suggested that using SNPs will aid in identifying individuals at high risk of developing certain cancers, and will also help develop tailored medication or identify genetic profiles of specific drug action and toxicity. This is facilitated by the introduction of new concepts, such as epigenomics, in developing targeted therapeutic tools (Chapter 4). The significance and challenges of genomics-based technologies in the diagnosis and treatment of cancer are reviewed in Chapter 23.

Genomics and Hematology

Genetic blood diseases comprise a significant part of the workload of a busy clinical hematology service. These mainly include single-gene diseases, such as hemophilia A, Christmas disease, von Willibrand's disease, sickle cell disease, and other hemoglobinopathies. Genetic factors play a significant role in the causation of several other hematological disorders including autoimmune hemolytic anemias, platelet disorders, and complex thrombosis and bleeding disorders. Genetic factors play a crucial role in both etiology and therapeutic outcome of various kinds of hematological malignancies. Advances in genomics and applications of genomics-based technology have made a promising contribution to the development of powerful diagnostic and therapeutic tools for dealing with complex hematological diseases, for example, deep vein thrombosis (DVT), disseminated intravascular coagulation (DIC), autoimmune thrombocytopenias, autoimmune hemolytic anemias, and hematological malignancies.

Some advances in genetics are currently in use in clinical hematology. For example, screening for factor V Leiden heterozygous status can help in identifying persons at risk for thrombophilia, which can clinically manifest with potentially life-threatening complications of DVT and pulmonary embolism. Approximately 4% of the population could be heterozygous (carrier) for this mutation. Similarly, confirmation of the homozygous status for the gene encoding thiopurine-S methyltransferase, an enzyme that inactivates the chemotherapeutic drug mercaptopurine, can help in selecting an alternative therapy or reducing the maintenance dose for children suffering from acute lymphoblastic leukemia. It is well known that approximately 1 in 300 children develop a serious, sometimes lethal, adverse reaction to mercaptopurine therapy.

Chapters on thrombosis (Chapter 14), platelet disorders (Chapter 15), and hematological malignancies (Chapter 24) provide comprehensive reviews on genomics-associated developments and clinical applications.

Genomics and Clinical Pediatrics

Advances in microarray technology have made a significant contribution in the care of critically sick children (Chapter 28). Examples of potential uses of this technology in clinical pediatrics include disease classification, risk stratification, pathogen detection, pathogen subtyping, antibiotic resistance analysis, newborn screening, and prediction of drug responses and adverse reactions (Bates, 2003).

The technique of comparative genomic hybridization (CGH) has now made it possible to carry out an in-depth genomic analysis in a child with unexplained developmental and physical disability, often called "unknown dysmorphic syndrome." Applications of microarrays in pediatric oncology are being used for specific tumor subtyping, which yields prognostic information and helps to plan therapy. For instance, gene expression analysis in medulloblastoma can help to distinguish between various histologically distinct tumor subtypes. This can be helpful in advising prognosis and the outcome of therapy. The identification of genes differentially expressed in patients with poor prognosis might help in developing more effective therapies for these children.

Microarrays are being used for the sensitive detection of pathogens without the need for culture. Such a method is of particular use in rapid and reliable diagnosis of *M. tuberculosis*. This can allow institution of prompt antituberculosis therapy. Various subtypes of *Mycobacterium* species can be reliably detected, which can be useful in selecting appropriate choice of antimicrobial and chemotherapy. This approach has been used in detecting rifampicin resistance, a major problem in mycobacterial therapy. In addition, the outcome of therapy can be accurately assessed in critically ill children with tuberculous meningitis. A 32-fold amplification of pathogen sequences of *E. coli* O157:H7 strain, using polymerase chain reaction (PCR), could allow the rapid and sensitive identification of pathogens and antibiotic resistance genes so that appropriate therapy can be more quickly instituted.

Genetic analyses are an obvious application of microarray technology. The simultaneous identification of specific mutations in multiple genes is now possible. This technique has tremendous potential in the care of a sick newborn. Such microarrays may be useful for newborn screening, covering several genes. For instance, a single blood sample from a baby with neonatal cholestasis might be used for the detection of mutations present in tyrosinemia, galactosemia, the various forms of familial intrahepatic cholestasis, a-1 antitrypsin deficiency, cystic fibrosis, etc. A similar technique can be used in determining the risk of polygenic diseases. This employs preparing the whole-genome profile of SNPs. This could also be used in assessing an individual's response to medication, and in the selection of the most efficacious drugs with the least risk of adverse reaction for a given patient and disorder. The application of genomic techniques in clinical pediatrics has begun, and is now poised for expansion beyond the research setting into clinical care.

Genomics and Obstetrics and Gynecology

Developments in genomic medicine have far reaching implications in all aspects of clinical medicine. A number of genomic studies concerned with clinical obstetrics and gynecology has expanded the profile of genomic medicine (Weston et al., 2003). The impact of genomics-related research is evident in the way the practice of reproductive medicine has rapidly changed during the past few years. Genomic microarray technology is being used in improving our understanding of the different aspects of reproductive medicine including physiological processes, diagnosing diseases, and drug development (Chapter 36).

Gene expression microarray studies on endometrial receptivity and implantation in both mouse and human have increased our understanding of the physiology of implantation (Kao et al., 2002; Martin et al., 2002). Such studies have been helpful in determining the causes of, and treating, implantation failure. Similar techniques have been applied in studying endometrial decidualization, ovarian follicle development, labor and normal placentation (Weston et al., 2003).

Genomic microarrays can be useful in studying specific gynecological diseases. For example, the role of microarrays in ovarian cancer has been the subject of a number of reports describing therapeutic targets and diagnostic markers (Haviv and Campbell, 2002). Similarly, both endometrial and cervical cancers have been studied using microarrays. A specific subtype of cervical cancer with resistance to radiotherapy is attributed to the genes conferring such resistance. Genomic studies in other obstetric and gynecological conditions include preeclampsia, trisomy-21 pregnancies, and endometriosis. Gene expression studies in endometriosis sufferers have been helpful in understanding the causes of implantation failure, using eutopic and ectopic endometrium (Giudice et al., 2002). Similar approaches have helped in understanding the pathophysiology of fibroid growth by studying differential gene expression in fibroid and adjacent normal myometrium (Tsibris et al., 2002). Finally, the availability of microarray chips for infectious diseases will enable prompt detection of pathogenic organisms in the reproductive tract that account for a significant proportion of secondary reproductive failure.

In future, the practice of obstetrics and gynecology could dramatically change with the availability of genomic profiling using SNPs. This would enable specific drug development targeted at an individual's genomic profile, thus avoiding the risk of serious iatrogenic effect, for example, ovarian hyperstimulation syndrome. Each patient attending an antenatal clinic in future may be offered DNA testing using the buccal swab. This might enable the attending clinician to predict the likelihood of a variety of pregnancy-related disorders, including preeclampsia and preterm premature rupture of membranes (PPROM). The DNA sample could be analyzed on a microarray chip, designed to identify a number of genetic polymorphisms known to confer an increased risk of these disease (Weston et al., 2003). For example, recently an SNP in the promoter region of the gene for matrix-metalloproteinase-9 was found to be associated with PPROM (Ferrand et al., 2002). It is likely that similar developments will transform the practice of obstetrics and gynecology. This would make it necessary for obstetricians and gynecologists to be educated and trained in the application and delivery of genomic-related methods in the diagnosis and management of various obstetric and gynecological conditions.

Gene Therapy and Stem-cell Genomics

Many people falsely assume that germline gene therapy is already taking place with regularity. News reports of parents selecting a genetically tested egg for implantation or for choosing the sex of their unborn child can mislead the public into believing that this is gene therapy. Actually, in these cases, genetic information is being used for selection. A recent review provides a detailed account of the recent developments in somatic gene transfer and the associated risks and ethical issues (Kimmelman, 2005). Regular online updates are also available (http://www.ornl.gov/hgmis/medicine/genetherapy.html).

The term "gene therapy" encompasses at least four types of application of genetic engineering for the insertion of genes into humans. The scientific requirements and the ethical issues associated with each type are discussed. *Somatic cell gene therapy* is technically the simplest and ethically the least controversial. The first clinical trials were undertaken in 1986–1987. *Germline gene therapy* will require major advances in our present knowledge, and it raises ethical issues that are now being debated. In order to provide guidelines for determining when germline gene therapy would be ethical, the author presents three criteria that should be satisfied before a human clinical trial.

Enhancement genetic engineering presents significant, and troubling, ethical concerns. Except where this type of therapy can be justified on the grounds of preventive medicine, enhancement engineering should not be performed. The fourth type, *eugenic genetic engineering*, is impossible at present and will probably remain so for the foreseeable future, despite the widespread media attention it has received (Anderson, 1985). The whole field of stem-cell genomics carries great potential in developing powerful therapeutic tools, and increasing the availability of a wide range of transplantable tissues. There is also the perceived danger of human cloning and making so-called designer babies. Various theoretical and technical aspects of stem-cell genomics are covered in Chapter 37, including ethical and legal issues related to stem-cell genomics and cell-based therapy.

Predictive Genomic Medicine

The rapid progress of genetic profiling technologies, in a whole-genome context, is creating some of the fundamental prerequisites for a new, much-heralded era of predictive genomic medicine. Nevertheless, formidable conceptual and practical obstacles must still be addressed by pharmaceutical, biotech, academic, and government research organizations before clinical applications of these technologies become commonplace. Many of these effectively translate into information technology challenges, including privacy issues related to the use of genetic data and new data-analysis approaches in dealing with complex, heterogeneous phenotypes.

An integral part of the practice of clinical genetics includes genetic testing that involves chromosomal, molecular, or biochemical testing to establish the disease-status suspected on the basis of clinical phenotype. An important aspect of clinical genetics involves assessing and discussing with an individual, usually a family member, presumed to be "at-risk" on the basis of family history or following investigations. Predictive or presymptomatic genetic testing is carried out in the absence of any symptoms, solely with the aim of verifying the genetic risk.

Genetic or genomic screening (Chapter 39) is aimed at selecting individuals at "*high-risk*" for developing a specific genetic disease from a selected population subgroup. These individuals are then referred to clinical genetics service or to an appropriate clinical service for confirmatory genetic testing and genetic counseling. Examples include antenatal screening for neural tube defects and Down syndrome. Neonatal genetic screening programs include phenylketonuria, galactosemia, cystic fibrosis, and Duchenne type muscular dystrophy. Developments in genomics now opens for large-scale population screening for complex traits in selected "*high-risk*" subgroups, such as coronary artery disease, hyperlipidemia, bronchial asthma, bipolar depression, schizophrenia, and susceptibility for infectious diseases.

The prediction of an individual's genetic risk in 1 of the 1500 so-called Mendelian disorders (online Mendelian inheritance in Man, OMIM) is possible on finding a mutation in a single gene (Yoon et al., 2000). But genetic testing for these single-gene disorders, of which autosomal dominant Huntington's disease (HD) is a prime example, has had only limited benefit in the overall healthcare. This is due to several factors including the nonavailability of effective therapeutic or preventive interventions, except for contraception or termination of a "high-risk" pregnancy. These conditions only account for approximately 5% of the total disease burden (Khoury et al., 2004). The vast majority of illnesses are common multifactorial disorders, such as coronary heart disease, DM, bronchial asthma, arthritis, and major depression (Murray and Lopez, 1997). Several studies support the "gene–environment" interaction as the conceptual basis for the etiology of these disorders. Such studies, known as genetic association, have been replicated on several occasions by substantial genetic contribution and interaction with environmental factors.

Genetic association studies help in estimating the genetic contribution in the etiology of polygenic/multifactorial disorders. This is commonly referred to as "heritability." Heritability is either expressed as a fraction of 1 or as a percentage figure. Since, currently, heritability is measured on the basis of the phenotype, this represents the phenotypic variance attributed to the genetic factors. Heritability estimates for common multifactorial disorders range from 39% to 80% (Table 10–1). Identification of genetic factors in these disorders is a prerequisite for developing predictive tests (Evans et al., 2003).

Identification of genes for a few single-gene dominantly inherited disorders, such as familial polyposis coli (*FAP*) for colorectal cancer and *BRCA1* and *BRCA2* for breast/ovarian cancers, which are strongly associated with disease risk, is a good model for developing new genetic tests for reliable prediction of the risk in common diseases (Schork, 1997). The main strategy used so far has been to look for associations between a disorder and either genetic markers or specific candidate alleles. Typically, in both approaches, the aim is to compare a case–control design with a set of specific alleles, genotypes, or genetic markers (e.g., SNPs) in individuals who have the disease (a defined phenotype) with matched controls (population, age, and gender) (Hirschhorn et al., 2002; Zondervan and Cardon, 2004). Unfortunately, most of the results have been disappointing. Meta-analyses of association studies to identify susceptibility alleles for heart disease, cancers, depression, asthma, and diabetes have shown that many initially positive findings have not been replicated in later studies (Hirschhorn et al., 2002). However, modest replications of such studies indicate that people who share these susceptibility alleles are 1.2–1.5 times more likely to develop these disorders (Ioannidis, 2003). A unsuccessful outcome of association studies depends upon several factors, including a small sample, study

Table 10–1 Heritability Estimates in Common Polygenic/Multifactorial Diseases

Disorder	Frequency (%)	Heritability
Schizophrenia	1	85
Bronchial asthma	4	80
Cleft lip ± palate	0.1	76
Pyloric stenosis	0.3	75
Ankylosing spondylitis	0.2	70
Talipes (clubfoot)	0.1	68
Coronary artery disease	3	65
Essential hypertension	5	62
Congenital dislocation hip	0.1	60
Neural tube defect (spina bifida)	0.3	60
Type 2 diabetes mellitus (T2DM)	5	50
Peptic ulcer	4	37
Congenital heart disease	0.5	35
Type 1 diabetes mellitus (T1DM)	0.4	15

design, inadequate or inappropriate selection of genetic markers, and a publication bias against the reporting of negative results (Colhoun et al., 2003). Thus, it is likely that a few rare alleles confer susceptibility in conjunction with several other genes with only modest susceptibility. For example, genetic susceptibility for colorectal cancer includes several genes triggered by mutations in the familial adenomatous polyposis gene (*APC*) (Fig. 10–3). A conservative estimate of the number of susceptibility alleles for major cancers range between tens and hundreds, all of which increase disease risk only modestly dependent on the interactions with other genes and environmental factors (Peto, 2001). The prediction of disease risk will depend on various factors: the number of genes influencing each condition, the frequency of susceptibility alleles in the population, the penetrance of these alleles, the predictive power of these alleles, how these alleles interact with each other, and under different genetic backgrounds, and, finally, interactions between these alleles and other risk factors (Hall et al., 2004). Pessimists argue that predictive genomic medicine does not hold any future as the prediction of risk for most polygenic disorders is not feasible. However, optimists point to the "common disease–common variant" (CDCV) hypothesis, which states that susceptibility alleles for common diseases reflect mutations that occurred in the human population 100,000 years ago (Balmain et al., 2003), and which can therefore be identified in large association studies with 1000–5000 cases and controls (Zondervan and Cardon, 2004). Thus, prospects for predictive genomic medicine depend heavily on validation of the CDCV hypothesis. It is true that association studies would need to employ carefully matched patients and controls and candidate alleles or genetic markers with reasonable population frequency (Hall et al., 2004).

There is continued anxiety about the feasibility of population-based predictive genomic screening for common diseases. Fewer than 5% of the available predictive genetic tests are applicable to common diseases (Yoon et al., 2000). Most of these tests employ alleles that carry a high predictive power. Thus, the use of single or few alleles will not offer good prediction, unless the lifetime risk of the disease is 5% or more and the genotype is either rare or increases disease risk 20 times or more (Holtzman and Marteau, 2000; Khoury et al., 2004). Others argue that it will simply be economically unviable for a country's healthcare system to screen the whole population for susceptibility alleles to prevent only a small number of these disorders (Vines et al., 2001). Nevertheless, a better prediction of

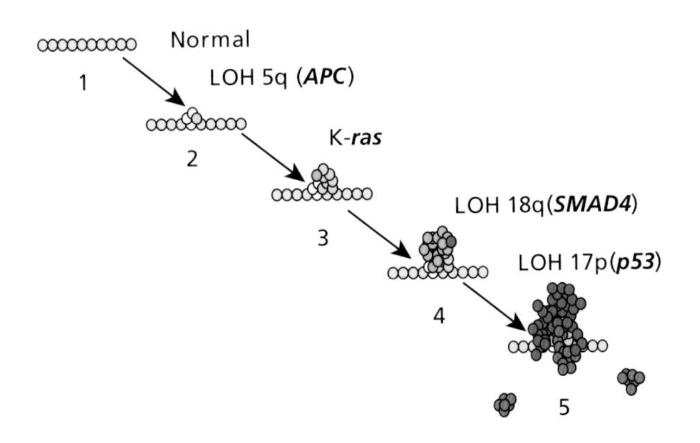

Figure 10–3 Multiple susceptibility genes in the pathogenesis of colorectal cancer. (Courtesy of Dr. Ian Frayling, Institute of Medical Genetics, University Hospital of Wales, Cardiff, Wales, UK.)

future disease risk will be possible by employing multiple genetic variants (Khoury, 2003). Results from several studies could be combined to devise a risk-scoring system for use in population-based genomic screening (Pharoah et al., 2002).

The efficiency of genomic screening could be further enhanced if the decision to test for multiple susceptibility alleles was based on a person's family history of the disease (Khoury et al., 2003). Family histories routinely carried out in any clinical setting help in categorizing the risk group; average risk, the same as the general population; moderate risk, has one or two affected close relatives; and high risk, has two or three affected family members, either first-degree relatives, or an earlier onset of the disorder. For example, most of the family history clinics for cancer employ this strategy in triaging the family history for further action. It is estimated 30%–50% of the family history would fall into the *moderate-risk* group, 10% into the *high-risk*, and approximately 40%–60% in the *average-risk* group (Scheuner et al., 1997). This approach would eliminate approximately half the people originally selected for genetic screening, thus improving the efficiency of available testing. Apart from the family history, other epidemiological factors could be relevant in selecting people for genetic testing. Such studies and surveys are currently in process and likely to add valuable information.

The success of predictive genomic medicine will depend on effective communication, provision of evidence-based effective intervention for the prevention of disease, and safeguarding that screening does not cause any kind of social, economic, or psychological disadvantage to the person concerned. Some researchers have expressed concerns that inappropriate communication of risks may instead result in demoralization and reduce a person's self-confidence, compromising their ability to change their lifestyle and make effective use of the available prevention methods or treatment (Wright et al., 2003). The outcome of disease-risk reduction will be dependent not only on the availability of treatment methods and medications but also on the person's own perception of the risk, acceptance of the risk, and motivation to make lifestyle changes and modify health behavior. This would largely be dependent on the manner information was delivered to the person concerned. The aim and method of genetic counseling in this situation will be different to that employed with a specific Mendelian disease or a chromosomal abnormality. It remains to be established whether the traditional genetic counseling approach should be adapted when used in predictive genomic medicine.

Advocates of predictive genomic medicine have expressed concerns that widespread genomics-based screening might adversely influence public health policies aimed at reducing the overall health burden on the population. The related public health strategies would include recommendations to promote a reduction in smoking and a per capita reduction in alcohol consumption, to promote healthy eating and exercising regularly to reduce the risks for high blood pressure, diabetes, and heart and lung diseases (Rose, 1992). Genomic screening would enable selection of the "*high-risk*" group and targeting them with appropriate health advice and interventions (Khoury et al., 2004). Public health policies would need to address these issues and ensure effective and efficient implementation. This will be a prerequisite before any form of predictive genomic screening is offered to a population.

Concerns about the social, ethical, and moral implications in predictive genomic medicine are largely based on experiences and issues surrounding predictive genetic testing in late-onset Mendelian disorders such as HD. Mutations that cause HD and many other serious late-onset disorders carry a strong predictive power. Thus, predictive genetic testing in closely related family members poses a potential threat that the information could be misused, leading to discrimination in careers and affecting employment prospects and financial planning. In addition, they might also experience personal and social difficulties (Marteau and Richards, 1996). By contrast, genomics-based screening will not have the benefit of a strong predictive power for common polygenic diseases. It is not clear how the outcome of genomic screening will impact upon an individual's prospects for career choice, employment, life insurance, and other personal interests.

Another important factor in the successful implementation of the predictive genomic medicine program would be public awareness and education. This is fast changing with the help of wide media coverage and "*online*" availability of a massive amount of data and information. However, it is feared that the current level of understanding of genetics and genomics is not high among the general public (Robins and Metcalfe, 2004). However, it would be wrong to assume that the public will only be passive consumers of information and genetics services (Dietrich and Schibeci, 2003). Khoury et al. (2000) are optimistic that the public will have clear perceptions and be able to distinguish between "*gene-X*," which is likely to be associated with the disease. and "*gene-Y*," which is less likely to be associated with developing disease symptoms. Nevertheless, it is essential that educational campaigns to improve public awareness on genetics and genomics be launched to provide an opportunity for understanding and appreciating the realistic applications of predictive genomic medicine.

Forensic Medicine and Genomics

DNA fingerprinting is regarded as one of the major discoveries of genetics. It is now widely used in forensic science, and is admissible as an evidence in criminal and civil legal cases all over the world. This was fully developed before sequencing of the human genome. DNA fingerprinting these days is a PCR-based technique that uses highly variable regions known as short tandem repeats (STRs) to construct a profile of an individual's DNA. This is occasionally supported by carrying out mtDNA typing, particularly when the nuclear DNA is degraded. mtDNA is inherited only from the mother, and there may be several thousand copies in a single cell. There is now a range of commercially available kits for forensic use.

New methods are on the horizon, a product of the flood of new information and new technology arising from genomics research, such as the use of SNPs, microarrays, and robotics. The new genomic technology can be used to build up the physical profile of an individual, so-called "phenotyping." Phenotyping is a hot topic in forensic science, for example, an individual's DNA left at the scene of crime can be used not only for matching with the DNA database but also in describing various physical characteristics such as skin, hair and eye color, build, and ethnic origin. However, it will require considerable time before this is accepted as a reliable evidence.

Genomics of other species, such as flies, slugs, algae, and small plants, can also be used for forensic purposes and could help in locating the scene of a crime or in collating the non-human evidence. Thus, the impact of genomics on forensic medicine is likely to be considerable. This is not addressed in any more detail in this book as forensic genomics and its scope are outside the remit of this book.

Ethical, Societal, and Legal Issues

Dr. Francis Collins, Director of the National Institute for Human Genome Research, spoke at an American Association for the Advancement of Science event on the day Bill Clinton signed an executive order prohibiting federal government agencies from obtaining genetic information from employees or job applicants or from using genetic information in hiring and job promotion decisions. He noted, "But genetic information and genetic technology can be used in ways that are fundamentally unjust. Genetic information can be used as the basis for insidious discrimination. Already, with but a handful of genetic tests in common use, people have lost their jobs, lost their health insurance, and lost their economic well-being because of the misuse of genetic information. It is estimated that all of us carry dozens of glitches in our DNA—so establishing principles of fair use of this information is important for all of us."

The practice of modern clinical genetics is inseparable from ethical, legal, social issues (ELSI) (see Chapter 40). Ethics in the new genomics era will be even more complex than at present, when it often arouses passion and confusion. Several challenges have surfaced from the scientific developments in genomics, including professional responsibility, liability, and issues regarding processing and management of genetic information, as they relate to the core principles of modern ethics, such as autonomy, beneficence, nonmalfeasance and justice. However, it is anticipated that genomic medicine will diminish rather than enhance existing sex, race, and socioeconomic-based inequalities in healthcare access and delivery. This chapter describes some aspects of the social, ethical, moral, and legal implications of genomics in clinical practice.

Implications on Community and Global Health

The innovations that have dramatically improved health in the developed world over the past century have largely not reached the developing world (see Chapter 38). In many cases, the technologies that underpin the medical tools available in the developed world will never appropriately support introduction in the developing world due to their dependence on constant energy sources, clean water and environmental conditions, cold storage of reagents, and training requirements. The opportunity exists to reduce these inequities through innovation in expanding access to existing appropriate technologies, accelerating adaptation of existing technologies to meet developing world needs, and developing entirely new technologies that will enable long-term solutions to health issues in low-resource settings. The Bill and Melinda Gates Foundation is committed to reducing global health inequities through support of such innovation as it relates to the development of effective vaccines and other prevention strategies, treatments, and diagnostics. There is an increasing awareness of the critical role of molecular diagnostics in improving global health and the lack of high performance technologies to support introduction in low-resource settings. H. Cindy Pellegrini, Legislative Director for Rep. Louise Slaughter (New York), as part of a roundtable discussion on genetic discrimination said, "It's Rep. Slaughter's view that all of us are ultimately uninsurable. The more that we learn about our genes, everyone has enough genetic flaws that we're anywhere from 5 to 30 or 50 genes, depending who you are listening to, that predispose you to major, severe illnesses at some point in your lifetime. And so really what we are doing right now by allowing discrimination to happen is punishing the people with the bad luck to have the genes we have discovered first."

Although the main focus of genomic medicine is to develop the practice of "personalized medicine," its main impact will be on community and public health. A significant proportion of the time and resources of any clinical genetics service is spent dealing with members of the extended family, who are inevitably influenced by the decisions made by one member of the family after genetic counseling. This is a good example of community genetics in practice. This will change in the postgenomic era with advances and availability of powerful and sophisticated diagnostic tools. Future developments in genomics will cause the style and practice of clinical medicine to change and, with this the expectations and aspirations of the community will rise. Genomic medicine will have a major impact on the future of preventive medicine, an essential prerequisite for optimal community health. This would have a direct effect on the way public health policies are decided and implemented. These aspects of genomic medicine are reviewed in this chapter.

Genomics has considerable potential for improving the health of the developing countries. Currently, many developing countries, such as India, have established centers and departments co-coordinating research in genomics. Recently, a joint Canadian and Indian group has made several recommendations to harness genomics to improve health in India (Acharya et al., 2004). The group has made seven major recommendations: increased funding for healthcare research with appropriate emphasis on genomics; leverage India's assets such as traditional knowledge and genomic diversity in consultation with knowledge-holders; prioritizing strategic entry points for India; improving industry–academic interface with appropriate incentives to improve public health and the nation's wealth; develop independent, accountable, transparent regulatory systems to ensure the major ethical, legal, and social issues; engage the public and ensure broad-based input into policy-setting; ensure equitable access of the poor to genomics products and services; and deliver knowledge, products, and services for public health. The group has made an Internet-based network, the Indian Genomics Policy Forum, a multi-stakeholder forum to foster further discussion on genomics-related health policy.

Another area useful for developing countries would be pathogen genomics research. This could be of enormous value in identifying a particular strain of pathogen and development of specific antimicrobial drugs and vaccines. Genomics can play a major role in specific therapeutic intervention in managing common communicable diseases including tuberculosis, HIV, and malaria. Other applications of the posthuman genome advances in healthcare in the developing countries would include high-throughput genetic screening for common single-gene diseases, such as thalassemia, and complex disorders, such as T2DM. However, applications of the development in genomics in developing countries require a cautious approach to maintain a balance with the conventional methods of public health and clinical practice. Leading international agencies, such as World Health Organization (WHO), can play a significant role in introducing new genomics-related technology in developing countries as it becomes available in future years. Implications on global health are discussed in Chapter 38, including the current state of genomics research as applied to clinical medicine and population health in both developed and developing countries.

Genomics and Medical School Curriculum

The curriculum for both undergraduate and postgraduate medical courses should reflect the current state of the science and art of clinical practice, and include prospective developments that are likely to shape future medical practice. Medical genetics has had a significant impact on the way we now consider disease causation and strive for new therapeutic avenues. It has turned a corner and has moved from the study of rare conditions to the study of common diseases that impact on the entire spectrum of medical practice and is likely to change as we rapidly move into the "*Genomics era.*"

The future generation of medical practitioners will be expected to be equipped with a new set of "*genetics or genomics*" related basic principles and clinical skills. Currently, genetics is commonly taught as a basic science, sometimes as a free-standing course, but often as part of a larger course, such as biochemistry or cell biology (Korf, 2002). Unfortunately, genetics is typically a small component of clinical teaching, leading the medical students to believe that genetics is irrelevant in medical practice. However, there has been a shift from this view, and some medical school curricula incorporate genetics in the final stages of the undergraduate medical course, such as during the pediatric and obstetrics rotations. There is a huge disparity in the curriculum across medical schools. The gap is particularly obvious and alarming in the developing or underdeveloped countries (Ghosh and Mohanty, 2002).

Even before the completion of the Human Genome Project, ambitious predictions were made of wide-ranging changes in medical practice (Collins, 1999). It has been proposed that physicians will use genetic testing routinely to determine the disorders that their patients will some day develop, in order to prescribe medications or recommend changes of lifestyle that prevent these conditions. Understandably, some clinicians have expressed concerns that this approach will create a new class of underprivileged individuals, who will be deprived life or health insurance cover, or be restricted in making appropriate career choices. The power of assisted reproduction techniques and prenatal testing has been extended to the selection of gender, physical, and mental traits in the unborn child. To the other extreme lies the perceived danger of designing children with desired physical or mental traits using "human cloning" techniques. Society will need appropriately trained and skilled physicians to guide and supervise these new developments for the benefit of patients, families, and the community at large.

It is not essential that all medical practitioners should master all of the basic principles of genetics to be able to apply them in their clinical practice. For example, any physician can request blood gases or electrolyte levels without understanding the chemistry and methodology that is used in the assays. However, the clinician is expected to interpret the results and apply these to clinical practice. Similarly, an understanding of the common genetic tests should be acquired by all clinicians. These are a set of concepts and skills that will be necessary if genetics tools are to be used wisely and efficiently.

Korf (2002) has recommended that the future class of medical practitioners should be equipped with the following set of skills:

1. the ability to obtain a family history and to recognize the major patterns of genetic transmission;
2. awareness of the indications for chromosome analysis and an ability to interpret the reports of chromosomal abnormalities;
3. an understanding of molecular genetic testing and an ability to interpret test results;
4. appreciation of the major approaches to prenatal diagnosis, including indications and limitations;
5. knowledge of the role of genetics in the pathogenesis of cancer, and opportunities for genetic testing to refine estimation of risk based on family history;
6. awareness of population screening programs, including both newborn screening for metabolic disorders and carrier screening programs;

7. understanding the role of clinical geneticists and genetic counselors;
8. sensitivity to the issues of privacy, discrimination, and family issues, which arise in dealing with genetic conditions;
9. awareness of where to go or whom to approach for up-to-date information on a genetic disease and genetic testing that can be applied at the point of clinical care.

Summary

An important milestone in the history of medical science is the recent completion of the human genome sequence. The progress on the identification of approximately 35,000 genes and their regulatory regions provides the framework for understanding the molecular basis of disease. This advance has also laid the foundation for a broad range of genomic tools that can be applied to medical science. These developments in gene and gene product analysis across the whole genome have opened the way for targeted molecular genetic testing in a number of medical disorders. This is destined to change the practice of medicine: future medical practice will be more focused and individualized, often referred to as "personalized medicine." However, despite these exciting advances, many practicing clinicians perceive the role of molecular genetics, in particular that of medical genomics, as confined to the research arena with limited clinical applications. Genomic medicine applies the knowledge and understanding of all genes and genetic variation in human disease. This chapter introduces genomics-based advances in disease susceptibility screening, diagnosis, prognostication, therapeutics, and prediction of treatment outcome in various areas of medicine.

References

Acharya, T, Kumar, NK, Muthuswamy, V, et al. (2004). "Harnessing genomics to improve health in India"—an executive course to support genomics policy. *Health Res Policy Syst*, 2(1):1.

Altshuler, D, Brooks, LD, Chakravarti, A, et al. (2005). The International HapMap Consortium—a haplotype map of the human genome. *Nature*, 437:1299–1320.

Anderson, S, Bankier, AT, Barrell, BG, et al. (1981). Sequence and organization of the human mitochondrial genome. *Nature*, 290(5806):457–465.

Anderson, WF (1985). "Human gene therapy: scientific and ethical considerations." *J Med Philos*, 10(3):275–291.

Balmain, A, Gray, J, and Ponder, B (2003). The genetics and genomics of cancer. *Nat Genet*, 33:238–244.

Bates, MD (2003). The potential of DNA microarrays for the care of children. *J Pediatr*, 142;235–239.

Bentley, DR (2004). Genomes for medicine. *Nature*, 429:440–445.

Blackwell, JM (2001). Genetics and genomics in infectious disease susceptibility. *Trends Mol Med*, 7(11):521–526.

Bork, P and Huynen, M (1999). A point of entry into genomics. Editorial. *Nat Genet*, 23:273.

Clegg, JB and Weatherall, DJ (1999). Thalassemia and malaria: new insights into an old problem. *Proc Assoc Am Physicians*, 111(4):278–282.

Collins, FS (1999). Shattuck lecture—medical and societal consequences of the Human Genome Project. *N Engl J Med*, 341:28–37.

Colhoun, HM, McKeigue, PM, and Daveysmith, G (2003). Problems of reporting genetic associations with complex outcomes. *Lancet*, 361:865–872.

Dean, M, Carrington, M, Winkler, C, et al. (1996). Genetic restriction of HIV-1 infection and progression to AIDS by a deletion allele of the CKR5 structural gene. Hemophilia Growth and Development Study, Multicenter AIDS Cohort Study, Multicenter Hemophilia Cohort Study, San Francisco City Cohort, ALIVE Study. *Science*, 273(5283):1856–1862.

Dietrich, H and Schibeci, R (2003). Beyond public perceptions of gene technology: community participation in public policy in Australia. *Public Understand Sci*, 12:381–401.

Elkin, PL (2003). Primer on medical genomics part V: bioinformatics. *Mayo Clin Proc*, 78(1):57–64.

Evans, A et al. (2003) The genetics of coronary heart disease: the contribution of twin studies. *Twin Res*, 6:432–441.

Ferrand, PE, Parry, S, Samuel, M, et al. (2002). A polymorphism in the matrix metalloproteinase-9 promoter is associated with increased risk of preterm premature rupture of membranes in African Americans. *Mol Hum Reprod*, 8:494–501.

Fleischmann, RD, Adams, MD, White, O, et al. (1995). Whole-genome random sequencing and assembly of Haemophilus influenzae Rd. *Science*, 269(5223):496–512.

Ghosh, K and Mohanty, D (2002). Teaching of medical genetics in the medical colleges of India—way ahead. *Indian J Hum Genet*, 8(2):43–44.

Gibbs, R (2005). Deeper into the genome. *Nature*, 437:1233–1234.

Giudice, LC, Telles, TL, Lobo, et al. (2002). The molecular basis for implantation failure in endometriosis—on the road to discovery. *Ann N Y Acad Sci*, 955:252–264.

Hall, WD, Morley, KI, and Lucke, JC (2004). The prediction of disease risk in genomic medicine. *EMBO Reports*, 5(Suppl.):S22–S26.

Haviv, I and Campbell, IG (2002). DNA microarrays for assessing ovarian cancer gene expression. *Mol Cell Endocrinol*, 191:121–126.

Hirschhorn, JN, Lohmueller, KE, Byrne, E, et al. (2002). A comprehensive review of genetic association studies. *Genet Med*, 4:45–61.

Holtzman, NA and Marteau, TM (2000). Will genetics revolutionize medicine? *N Engl J Med*, 343:141–144.

Horuk, R, Chitnis, CE, Darbonne, WC, et al. (1993). A receptor for the malarial parasite *Plasmodium vivax*: the erythrocyte chemokine receptor. *Science*, 261(5125):1182–1184.

Ioannidis, JP (2003). Genetic associations: false or true? *Trends Mol Med*, 9:135–138.

Kao, LC, Tulac, S, Lobo, S, et al. (2002). Global gene profiling in human endometrium during the window of implantation. *Endocrinology*, 143:2119–2138.

Khoury, MJ, Thrasher, JF, Burke, W, et al. (2000). Challenges in communicating genetics: a public health approach. *Genet Med*, 2:198–202.

Khoury, MJ (2003). Genetics and genomics in practice: the continuum from genetic disease to genetic information in health and disease. *Genet Med*, 5:261–268

Khoury, MJ, McCabe, LL, McCabe, ER (2003). Population screening in the age of genomic medicine. *N Engl J Med*, 348:50–58.

Khoury, MJ, Yang, Q, Gwinn, M, et al.(2004). An epidemiological assessment of genomic profiling for measuring susceptibility to common diseases and targeting interventions. *Genet Med*, 6:38–47.

Kim, JH (2002). Bioinformatics and genomic medicine. *Genet Med*, 4 (Suppl. 6):62S–65S.

Kimmelman, J (2005). Recent developments in gene transfer: risk and ethics. *BMJ*, 330:79–82.

Korf, BR (2002). Integration of genetics into clinical teaching in medical school education. *Genet Med*, 4(Suppl. 6):33S–38S.

Maojo, V and Kulikowski, CA (2003). Bioinformatics and medical informatics: collaboration on the road to genomic medicine? *J Am Med Inform Assoc*, 10 (6):515–522.

Marteau, TM and Richards, D (eds.) (1996). *The Troubled Helix: Social and Psychological Implications of the New Genetics*. Cambridge University Press, Cambridge, UK.

Martin, J, Dominguex, F, Avila, S, et al. (2002). Human endometrial receptivity: gene regulation. *J Repro Immunol*, 55:131–139.

Murray, CJL and Lopez, AD (1997). Mortality by cause for eight regions of the world: Global Burden of Disease Study. *Lancet*, 349:1269–1276.

Peto, J (2001). Cancer epidemiology in the last century and the next decade. *Nature*, 411:390–395.

Pharoah, PD, Antoniou, A, Bobrow, M, et al. (2002). Polygenic susceptibility to breast cancer and implications for prevention. *Nat Genet*, 31:33–36.

Robins, R and Metcalfe, S (2004). Integrating genetics as practices of primary care. *Soc Sci Med*, 59:223–233.

Roos, DS (2001). "Computational biology. Bioinformatics—trying to swim in a sea of data." *Science*, 291:1260–1261.

Rose, G (1992). *The Strategy of Preventive Medicine.* Oxford University Press, Oxford, UK.

Scheuner, MT, Wang, SJ, Raffel, LJ, et al. (1997). Family history: a comprehensive genetic risk assessment method for the chronic conditions of adulthood. *Am J Med Genet*, 71:315–324.

Schork, NJ (1997). Genetic of complex disease: approaches, problems, and solutions. *Am J Respir Crit Care Med*, 156:S103–S109.

Stephens, JC, Reich, DE, Goldstein, DB, et al. (1998). Dating the origin of the CCR5-Delta32 AIDS-resistance allele by the coalescence of haplotypes. *Am J Hum Genet*, 62(6):1507–1515.

Subramanian, G, Adams, MD, Venter, JC, et al. (2001). Implications of the human genome for understanding human biology and medicine. *JAMA*, 286(18):2296–2307.

Tsibris, JC, Segars, J, Coppola, D, et al. (2002). Insights from gene arrays on the development and growth regulation of uterine leiomyomata. *Fertil Steril*, 78:S74–S75.

Vineis, P, Schulte, P, and McMichael, AJ (2001). Misconceptions about the use of genetic tests in populations. *Lancet*, 357:709–712.

Weston, GC, Ponnampalam, A, Vollenhoven, BJ, et al. (2003). Genomics in obstetrics and gynecology. *Aust NZ J Obstet Gynecol*, 43:264–272.

Wright, AJ, Weinman, J, and Marteau, TM (2003). The impact of learning of a genetic predisposition to nicotine dependence: an analogue study. *Tob Control*, 12:227–230.

Yaspo, ML (2001). Taking a functional genomics approach in molecular medicine. *Trends Mol Med*, 7(11):494–502.

Yoon, PW, Chen, B, Faucett, AQ, et al. (2000). Public health impact of genetic tests at the end of the 20th century. *Genet Med*, 3:405–410.

Zondervan, KT and Cardon, LR (2004). The complex interplay among factors that influence allelic association. *Nat Rev Genet*, 5:89–99.

11

Complex Cardiovascular Disorders

Dan E Arking

Introduction

The genomics revolution, including mapping of the complete sequence of the human genome, has greatly advanced our understanding of the molecular basis of Mendelian diseases, which are largely the result of necessary and sufficient mutations that are individually rare. However, despite all our advances, we know almost nothing about the genetic architecture of traits with complex inheritance. These traits are governed by the interplay between genes, epigenetic factors, and the environment. Under this scenario, complex traits are likely modulated by multiple genes (and possibly multiple genetic variants within a gene), which are individually neither necessary nor sufficient. Thus, reproducibly identifying genetic variants that impact these traits has been a daunting task. In cardiovascular genetics, this has been further complicated by the multitude of different outcomes assessed, often using slightly different definitions of the phenotype in question. Genetic studies of cardiovascular disease (CVD) are largely dichotomized into studies of genes that influence CVD risk factors, such as lipids and blood pressure and studies that focus on clinical outcomes. This chapter focuses on three specific CVD events [myocardial infarction (MI), stroke, and sudden cardiac death (SCD)], with the idea that genetic variants that directly impact clinical outcomes would not necessarily be identified through studies of CVD risk factors and are likely to have the most direct clinical relevance, allowing for more accurate risk stratification. This is particularly important in SCD, where two-third of victims do not have CVD risk profiles that would warrant intervention under current guidelines. These results are summarized in Table 11–1.

MI/Ischemic Heart Disease

MI is an often fatal manifestation of coronary artery disease (CAD), with an annual incidence of approximately 865,000 in the United States, of which approximately 180,000 will die from the disease (American Heart Association 2003). The proximal cause of MI is believed to be a thrombosis event triggered by atherosclerotic plaque rupture, which occludes a coronary artery and leads to necrosis of the myocardium. In a long-term follow-up study of approximately 21,000 monozygous (MZ) and dizygous (DZ) twins, heritability of fatal coronary events was 57% and 38%, respectively, for men and women (Zdravkovic et al., 2002), underlining the importance of genetic factors in susceptibility to MI. A number of clinically relevant risk factors have been identified, many of which display significant heritability and thus are, at least in part, under genetic influence (Table 11–2). Indeed, rare Mendelian mutations that affect these risk factors can lead to premature CAD (Table 11–3), and have identified important pathways that have directly led to successful drug therapies, including the cholesterol metabolism pathway that emerged from studies of familial hypercholesterolemia (FH).

Numerous studies have attempted to identify genes that modulate the common variation in these risk factors and a number of such genes that have been observed in multiple studies are listed in Table 11–4. Aside from identifying novel molecular pathways, the importance of these genetic variants is unclear from a clinical standpoint, given that one can directly measure/observe the risk factor of interest. The exception to this is where particular variants may play a role in drug response and thus would inform as to which therapy to implement, or when the variant increases the risk for MI independently of its effect on the risk factor. Although still requiring further investigation, one particularly exciting example of this is with the *Taq1B* polymorphism in cholesteryl ester transfer protein (*CETP*).

Cholesteryl Ester Transfer Protein

CETP modifies the lipid composition of the plasma by transferring triglycerides and esters between lipoproteins, decreasing plasma high-density lipoproteins cholesterol (HDL-C) concentrations (Yen et al., 1989; Hannuksela et al., 1994). Kondo et al. (1989) demonstrated that a common variant in intron 1, denoted *Taq1B*, was associated with plasma HDL-C levels. The B2 allele (frequency, ~40% in Caucasian populations) is associated with lower levels of CETP, and hence, increased HDL-C levels. Kuivenhoven et al. (1998) were the first to link this variant to progression of CAD, and also to implicate a pharmacogenetic interaction between *Taq1B* genotype and pravastatin treatment. Since this initial study, numerous population-based studies have been performed, virtually all of which have replicated the association with plasma HDL-C, but not necessarily the association with CAD or statin therapy. It has been suggested that the *Taq1B* variant acts as a marker for a functional variant in the promoter, which has been identified 629 bp upstream of the transcription start site (−629C→A), and has been shown to directly affect *CETP* promoter activity (Dachet et al., 2000; Klerkx et al., 2003), however, not all studies are consistent with this

Table 11–1 Common Genetic Variants That Influence Susceptibility to Cardiovascular Events[a]

Event	Gene Name (Symbol)	Allele(s)	Comment
MI/IHD	Cholesteryl ester transfer protein (*CETP*)	Taq1B, −629A→C	Associated with HDL-C levels, possible interaction with statins
	Apolipoprotein E (*APOE*)	E2, E3, E4, −219G→T	Associated with LDL-C levels, gene–environment interactions
	Lipoprotein lipase (LPL)	Ser447Ter, Asp9Asn	Possible interaction with APOE genotype
	Endothelial nitric oxide synthase (*ENOS*)	Intron-4, Glu298Asp	
	Matrix metalloproteinase 3 (*MMP3*)	5A/6A	Strong interaction with smoking
	Lmyphotoxin-α (*LTA*)	252A→G, Thr26Asn	Identified through a gene-based genome-wide association study
Stroke	Factor V	Leiden	
	Methylenetetrahydrofolate reductasae (*MTHFR*)	C677T	Associated with elevated homocysteine levels
	Prothrombin	G20210A	
	Angiotensin-converting enzyme (*ACE*)	I/D	
	Phosphodiesterase 4D (*PDE4D*)	AX, G0, GX	Identified by linkage and association
	Arachidonate 5-lipoxygenase-activating protein (*ALOX5AP*)	HapA, HapB	Also associated with MI
	Cyclooxygenase 2 (*COX-2*)	−765G→C	Also associated with MI, may be relevant to ischemic events with COX-2 inhibitors
SCD/QT interval	α Nav1.5 subunit (*SCN5A*)	S1102Y, H558R	Only S1102Y directly associated with SCD
QT interval	HERG (*KCNH2*)	K897T, rs3815459	Gender-dependent associations
	α KvLQT1 (*KCNQ1*)	rs757092	Gender-dependent association
	C-terminal PDZ domain ligand of neuronal nitric oxide synthase adapter protein (*NOS1AP*)	rs10494366	Identified through genome-wide association study

Abbreviations: HDL-C, high-density lipoproteins cholesterol; IHD, ischemic heart disease; LDL-C, low-density lipoprotein cholesterol; MI, myocardial infarction; SCD, sudden cardiac death.

[a] QT interval is included as a subclinical phenotype for SCD, due to the limited direct studies of SCD.

Table 11–2 Risk Factors for Myocardial Infarction (MI)

Risk factor
Risk factors with a significant genetic component (heritability)
Total cholesterol (40%–60%)
High-density lipoprotein(HDL)-cholesterol (45%–75%)
Total triglycerides (40%–80%)
Body mass index (25%–60%)
Blood pressure (50%–70%)
Apolipoprotein(a) levels (90%)
Homocysteine levels (45%)
Type 2 diabetes (40%–80%)
Fibrinogen (20%–50%)
C-reactive protein
Gender
Age
Environmental risk factors
Smoking
Diet
Exercise
Infection

Source: Adapted from Lusis et al. (2004).

observation (Eiriksdottir et al., 2001). Indeed, it is possible that multiple variants, some of which remain to be identified, can combine to influence both HDL-C levels and the risk for CAD.

Two recent studies examined the relationship between *CETP* and cardiovascular mortality. Blankenberg et al. (2003), in a study of 1303 patients who underwent coronary angiography and had measurable stenosis (>30% in a major artery), found that the –629A allele had a strong protective effect for cardiovascular mortality, independent of its role in HDL-C and CETP activity levels. They presented additional evidence that -629 genotype was able to predict which individuals would benefit from statin therapy, with only individuals homozygous for the common allele (−629C) showing a benefit. In contrast, Carlquist et al. (2003) examined 2531 individuals with significant CAD (⩾70% in ⩾1 artery) and found that the *Taq1B* variant was associated with cardiovascular mortality only in conjunction with statin therapy. Moreover, the interaction with genotype was with homozygosity for the minor allele (B2), in which a strong reduction in mortality was observed. Aside from the possibility that the two variants are having opposing effects on mortality and drug interaction, it is important to note that the two studies had very different distributions of statin prescriptions. In the study by Carlquist and colleagues, 61% of prescriptions were for simvastatin, whereas in the study by Blankenberg and colleagues, only 17% were taking simvastatin, and the most prescribed statin was atorvastatin (25%). Interestingly, a study by van Venrooij (van Venrooij et al., 2003) found that atorvastatin therapy interacted with *Taq1B* genotype, showing a greater increase of HDL-C levels

Table 11–3 Mendelian Diseases That Exhibit Premature Coronary Artery Disease

Disease	Genes	Effect of Mutations
Familial hypercholesterolemia	*LDLR*	Defective binding of LDL by receptor
Familial defective APOB	*APOB*	Reduced binding affinity of APOB to LDLR
Sitosterolemia	*ABCG5, ABCG8*	Increased absorption of plant sterols
Autosomal recessive hypercholesterolemia	*ARH*	Defective endocytosis of LDLR
APOA1 deficiency	*APOA1*	Deletion or loss-of-function mutations that leads to very low HDL
Tangier disease	*ABCA1*	Impaired cholesterol efflux in macrophages
Homocystinuria	*CBS*	Leads to increased thrombotic tendency

Abbreviations: APOB, apolipoprotein B; HDL, high-density lipoprotein; LDLR, low-density lipoprotein receptor.
Source: Adapted from Watkins et al. (2006).

observed in the B1 homozygotes, which would be consistent with the findings of Blankenberg and colleagues.

In a meta-analysis performed by Boekholdt and colleagues (Boekholdt et al., 2005), in which they used individual data from seven large population-based studies comprised of 13,677 subjects, *Taq1B* genotype was significantly associated with HDL-C levels ($p < 0.0001$) even after adjusting for traditional HDL-C confounders [(age, sex, BMI, smoking, diabetes, and low-density lipoproteins cholesterol (LDL-C)]. Individuals homozygous for the B2 allele had significantly reduced risk of CAD [odds ratio (OR) = 0.77, $p = 0.001$], and adjustment for traditional risk factors (excluding HDL-C levels) did not change the size of the effect. However, when adjusted for HDL-C levels, the effect was attenuated and no longer significant (OR = 0.92, $p = 0.4$), but did not completely return to unity. This result is consistent with several previous studies that have noted a residual association between *Taq1B* genotype and CAD risk even after adjusting for HDL-C (Kuivenhoven et al., 1998; Kakko et al., 2001; Blankenberg et al., 2003). No interaction was observed between pravastatin therapy and *Taq1B* genotype analyzing the 5691 subjects who had been enrolled in three randomized, double-blind, placebo-controlled trials of pravastatin treatment. The authors rule out the possibility of an interaction between *Taq1B* genotype and pravastatin therapy, but in light of the findings of Carlquist and colleagues and Blankenberg and colleagues, both of which had less than 18% pravastatin use in their populations, it is quite possible that *CETP* variants interact with other statin treatments, and this should be examined in a much larger clinical trial.

Table 11–4 Genes With Common Genetic Variants Modulating Risk Factors for MI/IHD

Trait	Gene Name	Gene Symbol
LDL/VLDL	LDL receptor	*LDLR*
	Apolipoprotein E	*APOE*
HDL levels	Hepatic lipase	*LIPC*
Lp(a)	Apolipoprotein(a)	*LPA*
Homocysteine	Methylenetetrahydrofolate reductase	*MTHFR*
Blood pressure	Angiotensinogen	*AGT*
	β2-adrenergic receptor	*ADRB2*
	α-Adducin	*ADD1*
Diabetes, obesity, and insulin resistance	Peroxisome proliferator-activated receptor-γ	*PPARG*
	Calpain 10	*CAPN10*

Abbreviations: HDL, high-density lipoprotein; LDL, low-density lipoprotein; VLDL, very low-density lipoprotein.
Source: Adapted from Lusis et al. (2004).

In addition to *CETP*, numerous candidate genes have been investigated for association with CAD/MI, often with variable results (a classic example being the *ACE* I/D polymorphism, discussed in the following section in relation to stroke). Thus, the following discussion of candidate genes is limited to those that have been subject to a meta-analysis providing strong evidence for association with MI/IHD.

Apolipoprotein E

Apolipoprotein E (apoE), an LDL receptor ligand, is an important player in the metabolism of cholesterol and triglycerides, where it mediates the clearance of chylomicrons and very low-density lipoprotein (VLDL) from plasma. Utermann and colleagues (Utermann et al., 1977) first described the effects of three common allelic variants of *APOE* (termed E2, E3, and E4, and with frequencies of 8%, 77%, and 15%, respectively, in Caucasian populations) on type III familial hyperlipoproteinemia, in which more than 95% of affected individuals where homozygous for the rare allele. The common genotype, E3/E3, is used as the reference group in most studies, and individuals who carry the E2 allele have approximately 14 mg/dl lower LDL levels, and E4 carriers have approximately 7 mg/dl higher LDL levels (Motulsky and Brunzell, 2002). Numerous studies have examined the association between the E2/E3/E4 variants and coronary heart disease (CHD), and Song and colleagues (Song et al., 2004) have recently performed a meta-analysis incorporating 48 studies comprised of 15,492 case–patients and 32,965 controls. A consistent increased risk [OR 1.42, 95% confidence interval (CI) 1.26–1.61] was found for E4 carriers, whereas the E2 allele had no significant effect on risk for CHD. The exact mechanism by which *APOE* variants influence risk for CHD is not entirely clear. In addition to its effects on lipids, there is evidence that apoE may affect CHD risk through antioxidative, inflammatory, and immune activities (Davignon et al., 1999; Moghadasian et al., 2001; Stannard et al., 2001). Indeed, although most of the studies in the meta-analysis did not report ORs adjusted for lipids, and thus the meta-analysis did not determine whether the *APOE* variants influence risk for CHD independent of lipids, several studies have shown an association of the E4 allele with increased risk for CHD even after adjusting for lipid levels (Humphries et al., 2001; Lahoz et al., 2001). Adding to the confusion, Keavney and colleagues (Keavney et al., 2004) examined 4685 cases of MI and 3460 controls, and found that after adjusting for apoB/apoA(1) ratios the effect of the E4 allele went from significantly increasing risk for CHD to a *protective* effect! Thus, the clinical implementation of genotyping these variants still needs to be resolved, especially in conjunction with other known CHD risk factors.

Numerous studies have examined gene–environment interaction, reporting conflicting results for interaction with diet, sex, cholesterol-lowering agents, and smoking (Schaefer et al., 1994; Stengard et al., 1995; Ordovas 1999; Ballantyne et al., 2000; Hagberg et al., 2000; Ordovas and Schaefer, 2000; Stankovic et al., 2000; Humphries et al., 2001; Djousse et al., 2002; Eichner et al., 2002; Schaefer 2002; Keavney et al., 2003; Liu, Chen, et al., 2003). A particularly interesting finding that may help to explain the heterogeneity observed across these studies, is the effect of a common promoter polymorphism (−219G→T, frequency ~40% in Caucasians populations) on *APOE* transcript levels. The T allele has lower transcriptional activity (Artiga et al., 1998; Bullido et al., 1998; Lambert et al., 1998; Lambert et al., 1998), and has been shown to independently influence risk for CHD (Lambert et al., 2000), with a possible additive interaction with E2/E3/E4 genotype (Ye et al., 2003). These findings highlight the difficulty in identifying the "underlying functional variant" associated with a common disease, as the assumption of a single functional variant may not be correct and may not even be the norm with common variants of small effect.

Lipoprotein Lipase

Lipoprotein lipase (LPL) degrades triglyceride-rich particles, resulting in the generation of free fatty acids and glycerol and the production of HDL-C, and also plays a role in "bridging" between lipoproteins and the cell surface in receptor-mediated uptake of lipoproteins (Beisiegel et al., 1991; Mulder et al., 1993). A number of missense mutations have been identified that significantly alter triglycerides and HDL-C levels across multiple studies, however, association of these variants with CHD has been contradictory. In a meta-analysis by Wittrup and colleagues (Wittrup et al., 1999) who examined the Gly188Glu, Asp9Asn, Asn291Ser, and Ser447Ter alleles, suggestive results were observed only for Ser447Ter (allele frequency ~20% in Caucasians populations). A lack of association for the other alleles may have been due to limited power, as two subsequent prospective studies involving over 1000 events among 10,000 subjects showed an association between 9Asn genotype (frequency ~2–3% in Caucasian populations) and increased risk for events (Talmud et al., 2000; Wittrup et al., 2006). Indeed the ORs for both these studies were entirely consistent with the earlier meta-analysis (OR = 1.5, 95% CI 1.2–2.1 and OR = 1.6, 95% CI 1.1–2.4 versus OR = 1.4, 95% CI 0.8–2.4). In both these studies, adjustment for triglycerides and HDL-C levels did not significantly alter the results. Perhaps more intriguing, Wittrup and colleagues reported an interaction with *APOE* genotype, while Talmud and colleagues observed a strong interaction with smoking, suggesting the possibility of both gene–gene and gene–environment interactions.

Endothelial Nitric Oxide Synthase

Nitric oxide (NO) plays a key role in vascular biology, and studies have shown that the impairment of endothelium-dependent vasodilation in atherosclerotic vessels is due, at least in part, to reduced levels of endothelial nitric oxide synthase (*eNOS*) (Wilcox et al., 1997). NO also helps prevent atherosclerosis through inhibition of platelet adherence and aggregation, suppression of smooth muscle proliferation, and reduction of leucocytes adherence to the endothelium (for review see, Moncada and Higgs, 1993; Radomski and Salas, 1995; Cooke and Dzau, 1997). A recent meta-analysis by Casas and colleagues (Casas, Bautista, et al., 2004) including 26 studies involving 23,028 subjects examined the association between ischemic heart disease (IHD) and the three most widely studied *eNOS* polymorphisms: Glu298Asp, intron-4, and −786T→C. Homozygosity for the Asp298 was associated with an increased risk of IHD (OR 1.31, 95% CI 1.13–1.51) as was homozygosity for intron-4a (OR 1.34, 95% CI 1.03–1.75). Individuals homozygous for the intron-4a allele are relatively rare (1–2% in Caucasian populations) compared to the 298Asp allele (~10% in Caucasian populations) and thus do not contribute a great deal to the population attributable risk. As the Glu298Asp variant alters the protein structure, the effect of this variant has been extensively characterized. A number of studies have shown reduced NO synthesis (Tesauro et al., 2000; Persu et al., 2002) or endothelial function (Savvidou et al., 2001; Leeson et al., 2002) for the Asp298 allele, which is subject to selective proteolytic cleavage in endothelial cells and vascular tissue. No consistent functional associations between intron-4a and the NO pathway activity have been shown, though some studies have shown lower NO plasma levels and decreased protein expression (Tsukada et al., 1998; Wang et al., 2000).

One limitation of the meta-analysis is that it only included cross-sectional case–control studies. These types of studies are prone to population stratification. One attempt to address this issue was undertaken by Spence and colleagues (Spence et al., 2004), who examined 1023 individuals from 388 families (discordant sibships and parent–offspring trios) for association between the Glu298Asp variant and IHD. Using robust family-based association tests, they found no evidence for association between this variant and IHD. Given the small expected effect size, these studies are not conclusive, but certainly indicate that larger family-based studies are warranted to rule out false-positive findings due to population stratification.

Matrix Metalloproteinase 3

Matrix metalloproteinase 3 (MMP3) is a proteoglycanase that is largely secreted by connective tissue cells and is important in regulating the accumulation of extracellular matrix (ECM) during tissue injury (Saarialho-Kere et al., 1994). The extent of vascular remodeling is an important feature in atherosclerotic plaque formation (Newby et al., 1994), and thus MMP3 influences atherosclerosis through its ability to degrade components of the ECM, including proteoglycan, fibronectin, laminin, and type IV collagen (Wilhelm et al., 1987), during tissue injury. Localization of *MMP3* gene expression to regions particular prone to plaque rupture (Henney et al., 1991) led Ye and colleagues to examine the role of *MMP3* genetic variants in the progression of CAD (Ye et al., 1995). They identified a common variant in the promoter region, in which one allele has five adenosine nucleotides (5A) and the other, six adenosine nucleotides (6A) and demonstrated that individuals homozygous for the 6A allele show a more rapid progression of atherosclerotic disease. In a subsequent study (Ye et al., 1996), they used in vitro assays to demonstrate that the 6A allele has lower gene expression, leading to decreased MMP3 activity. This result may be somewhat surprising, as *increased* MMP activity is traditionally associated with *increased* CVD (Shah et al., 1995). Indeed, while coronary stenosis was associated with the 6A allele in three independent studies (Gnasso et al., 2000; Rauramaa et al., 2000; Rundek et al., 2002), the 5A allele has been associated with MI in multiple studies (Terashima et al., 1999; Beyzade et al., 2003; Liu, Chen, et al., 2003; Nojiri et al., 2003; Zhou et al., 2004) and ischemic stroke (Flex et al., 2004). It has been suggested (Ye 2006) that the apparent contradiction between the association of the MMP3 5A/6A variant with these two related outcomes is due to the nature of the atherosclerotic plaque formation, with individuals homozygous for the 6A allele predisposed to developing plaques that are rich in matrix proteins (and more stable) as opposed to lipid rich plaques, which are more vulnerable to rupture. In the absence of multiple studies showing both the increase in IMT with the 6A allele and increased risk for MI with the 5A allele, given the apparent contradiction, one might well have considered these results to be false positive. Instead, these findings highlight the need for caution when drawing conclusions from a subclinical phenotype and emphasize the requirement for testing genetic variants directly for clinical outcomes.

Although identifying a gene associated with MI and stroke is important for furthering the understanding of the underlying etiology of disease, the 5A/6A variant has additional clinical implications. Two studies have demonstrated a synergistic effect between the 5A allele and smoking and the risk of MI (Humphries et al., 2002; Liu, Chen, et al., 2003), and a subsequent prospective study of individuals with premature MI has additionally demonstrated that smoking cessation is particular important in individuals who carry the 5A allele (Liu et al., 2005). Carriers of the 5A allele who continued to smoke had more than four-fold increased risk of adverse event relative to those who quit, whereas individuals homozygous for the 6A allele had only a two-fold increased risk. These results highlight the importance of interactions between genetic variants and the environment.

The association between *MMP3* and CVD has led to a number of additional association studies with the other MMPs. Some suggestive results have been observed and are reviewed in Ye (2006).

Coagulation and Fibrinolytic Genes

The most common pathogenetic pathway of acute MI is through thrombosis, generally triggered by atherosclerotic plaque rupture. Thus, a great number of candidate gene studies have involved the examination of genetic variants in genes involved in coagulation and fibrinolytic pathways. Boekholdt and colleagues (Boekholdt et al., 2001) performed a meta-analysis of the association between MI and the four most studied variants: factor V Leiden, prothrombin G20210A, fibrinogen B-chain G-455A, and plasminogen activator inhibitor-1 (*PAI-1*) 4G/5G. The authors conclude that associations for these genetic variants were weak (*PAI-1*, fibrinogen) or absent (factor V, prothrombin). Despite this apparent lack of association with MI, stronger associations are seen with stroke, and these variants are discussed in greater detail later in the chapter in that context.

Given the difficulty in identifying genetic variants that play a significant role in susceptibility to MI/IHD through a candidate gene approach, which is a problem observed with for most complex diseases, family-based strategies have also been implemented. Although a number of regions have been implicated through linkage analysis, these studies often result in large candidate regions, often comprised of hundreds of genes. Thus, their utility in identifying specific gene variants associated with disease, is limited, though some successes have been reported [see section on Stroke, arachidonate 5-lipooxygenase-activating protein (*ALOX5AP*), and phosphodiesterase 4D (*PDE4D*)]. With both a rapid reduction in genotyping costs and vast increase in throughput, a new focus has been on genome-wide association studies.

Lymphotoxin-α

Ozaki and colleagues (Ozaki et al., 2002) reported the first genome-wide study, in which they successfully genotyped 65,671 gene-based single nucleotide polymorphism (SNP) in 94 individuals of Japanese descent with MI and compared allele frequencies with 658 individuals from the general Japanese population. SNPs that were significant at $p < 0.01$ in the initial screen were then examined in 1133 individuals with MI and 1006 control individuals from the general population. A SNP in intron 1 of lymphotoxin-α (*LTA*) (252A←G) was associated with MI. *LTA* encodes a cytokine with broad inflammatory activity (for review, Bazzoni and Beutler, 1996), and has been implicated in the development of autoimmunity and inflammation pathways, which are likely important in development and progression of CVD. To exclude the possibility of a false-positive finding due to differences in age between the case and controls, they genotyped a second, age-matched, control population comprising 872 individuals and similar results were observed. Subsequent fine-mapping and functional assays identified two SNPs in high linkage disequilibrium that had a strong association with MI (OR ~1.8) and affected the expression level and biological function of LTA. Using luciferase assays, they demonstrated that the 252A←G variant influenced transcriptional activity, and gel shift assays demonstrated altered transcription factor binding. A second mutation was found in the coding region, Thr26Asn, which induced greater expression of vascular cell adhesion molecule 1 (*VCAM1*) and selectin E (*SELE*) messenger RNA (mRNA) in vascular smooth muscle cells. Increased expression of these adhesion molecules has been previously implicated in CVD (O'Brien et al., 1996; Hwang et al., 1997).

As with many initial findings from association studies, attempts to replicate the results in other populations have met with mixed success. Iwagana and colleagues (Iwanaga et al., 2004) reported similar findings in a Japanese cohort comprised of 477 individuals with MI and 372 controls, whereas Yamada and colleagues (Yamada et al., 2004) found no association in a larger Japanese cohort comprised of 993 controls and 1493 cases with MI. It is somewhat difficult to reconcile these findings, but it is interesting to note that the individuals with MI have had the same genotype distributions across all three studies, and it is the controls from each study, which look different, and thus give conflicting results. In studies reported for two Caucasian populations, Koch and colleagues (Koch et al., 2001) did not find any evidence linking *LTA* genetic variants to MI, whereas a family-based study from the PROCARDIS Consortium (2004) demonstrated increased transmission of the 26Asn allele to individuals with MI. The family-based test has the advantage of being robust to population stratification, and thus is strong evidence that the 26Asn variant is indeed the risk factor for MI. However, large prospective studies will be required to confirm this finding.

In addition to the initial finding of an association between *LTA* and MI, Ozaki and colleagues in a subsequent report (Ozaki et al., 2004) used a yeast two-hybrid screen to identify proteins that physically interact with lymphotoxin-α. They identified galectin-2 (*LGALS2*) as a direct interactor and demonstrated that an intron 1 variant (3279C→T, frequency 36% in Japanese populations) was associated with risk for MI. Homozygosity for the less common 3279T allele was associated with a 1.57-fold (95% CI 1.3–1.9) reduced risk for MI in a cohort of 2302 patients with MI and 2038 controls. They further demonstrated that the 3239T allele was associated with a 50% reduction in transcriptional activity and that *LTA* and *LGALS2* were coexpressed in smooth muscle cells and macrophages in the intima of human atherosclerotic plaques but were absent in normal medial smooth muscle cells. Taken together, these data strongly implicate a role for these genes in the pathogenesis of MI, and illustrate the power of a genome-wide scan to identify genes and generate hypotheses for future study.

Stroke

Stroke is one of the leading causes of death and disability in the developed world, with annual incidence of more than 700,000 strokes in the United States alone (American Heart Association

2003). With limited treatment options available, focus has been on primary prevention largely through modification of acquired risk factors (diabetes mellitus, smoking, high blood pressure, and atrial fibrillation) (Goldstein et al., 2001). However, as with many common diseases, rare monogenic conditions that cause stroke have been identified, and a great deal of progress has been made identifying the underlying genetic defects. Indeed, the gene for cerebral autosomal dominant arteriopathy with subcortical infarcts and leukoencephalopathy (CADASIL), which has served as a model for inherited ischemic stroke, was identified as *NOTCH3* in 1996 (Joutel et al., 1996). More recently, substantial progress has been made in cerebral cavernous malformations (CCMs), with genes for all three types of CCM now identified. CCM1 is caused by *KRIT1* (Laberge-le Couteulx et al., 1999), CCM2 is caused by *MGC4607* (malcavernin; Denier et al., 2004), and CCM3 is caused by programmed cell death protein 10 (Bergametti et al., 2005). Although these genetic defects are rare in the general population, they have high penetrance and thus being able to identify carriers has a significant clinical impact. Whether any of these genes have a role in common forms of stroke remains to be determined, though a recent study by Dong and colleagues, in which they screened individuals with lacunar stroke for coding mutations in *NOTCH3* did not find any association (Dong et al., 2003). This does not, however, rule out the involvement of common variants, more likely to be found in noncoding conserved regions and involved in gene regulation.

Common forms of stroke can be divided into two major varieties: ischemic and hemorrhagic. The majority of strokes are ischemic (80–90%) and can be further subdivided into: (1) large-vessel occlusive disease, usually due to atherosclerosis and plaque formation; (2) small-vessel occlusive disease, which have involvement of small perforating end-arteries in the brain; and (3) cardiogenic stroke, which is secondary to blood clots from a diseased heart. Traditionally, occlusive disease has been considered due primarily to atherosclerosis and cardiogenic stroke due to atrial fibrillation secondary to mitral-valve stenosis, however, arguments have been put forward that all three forms likely have a significant atherosclerotic component (Gulcher et al., 2005). Given the heterogenous nature of the stroke phenotype, assumptions about the underlying etiology of disease can have a significant impact upon the ability to identify genetic determinants of stroke susceptibility.

The leap from the existence of genes for monogenic forms of stroke to the likely existence of genes contributing to risk for common forms of stroke is bolstered by numerous studies that have shown that a genetic component to susceptibility to common forms of stroke likely exists. A comprehensive analysis of these studies, in which all genetic epidemiology studies of ischemic stroke from 1966 to May 2003 were systematically reviewed, was recently conducted by Flossman and colleagues (Flossmann et al., 2004). Based on twin (OR = 1.65, 95% CI 1.2–2.3), case–control (OR = 1.76, 95% CI 1.7–1.9), and cohort (OR = 1.30, 95% CI 1.2–1.5) studies, they conclude that there is a modest but significant genetic component to risk for ischemic stroke in the general population. Most genetic studies have focused on candidate genes under case–control designs, and Casas and colleagues (Casas, Hingorani, et al., 2004) have performed a meta-analysis of these studies on ischemic stroke, incorporating 32 genes studies across approximately 18,000 cases and approximately 58,000 controls (Table 11–5). They identified significant associations for factor V Leiden (OR 1.33, 95% CI 1.12–1.58), methylenetetrahydrofolate reductase (*MTHFR*) C677T (OR = 1.24, 95% CI 1.08–1.42), prothrombin G20210A (OR = 1.44, 95% CI 1.11–1.86), and angiotensin-converting enzyme (*ACE*) insertion/deletion (OR = 1.21, 95% CI 1.08–1.35) (these variants are discussed in detail in the section that follows). They did not find any association with the next three most investigated genes, factor XIII, *APOE*, and glycoprotein IIIa. Associations were also seen for human platelet antigen type 1 (*PAI-1*) and for two different mutations in *GPIBA* (HPA2 polymorphism and Kozak sequence), however, the association studies for these three variants each had less than 1000 cases and therefore require further validation. Thus, they conclude that there are common variants in several genes involved in common forms of stroke, each with modest effect.

Factor V Leiden

Bertina and colleagues (Bertina et al., 1994) identified an arginine to glycine (R506Q) mutation (termed Leiden allele) in factor V in a family with activated protein C (*APC*) resistance and prone to thrombosis. APC limits clot formation by inactivation of factors Va and VIIIa, and the Leiden mutation is predicted to alter the amino acid at the APC cleavage site in factor V, causing factor V to be less efficiently degraded. Thus, individuals who carry factor V Leiden have increased thrombin generation and a hypercoagulable state, which could explain the increased risk for stroke associated with this allele (Dahlback 1995). Despite the small increased risk associated with this variant, the relative high frequency of carriers (~6.5% in 13,798 controls from the studies included in the meta-analysis by Casas and colleagues) underlines the importance of this variant in risk stratification.

Methylenetetrahyddrofolate Reductase

MTHFR encodes a protein that catalyzes the conversion of 5,10-methylenetetrahydrofolate to 5-methyltetrahydrofolate, a cosubstrate for homocysteine remethylation to methionine. Frosst and colleagues (Frosst et al., 1995) identified a cytosine to thymine single-base pair substitution at position 677 (C677T) that converts an alanine to a valine residue, and produces a thermolabile form of the protein. They demonstrated that this variant was associated with reduced enzyme activity and increased levels of homocysteine. Elevated levels of homocysteine are associated with increased risk for stroke (Wald et al., 2002), and thus, the C677T variant likely contributes to increased risk of stroke directly due to its reduced ability to metabolize homocysteine.

Prothrombin

Poort and colleagues (Poort et al., 1996) first identified the G20210A single-base pair substitution in the 3′ untranslated region of prothrombin and demonstrated its association with elevated plasma prothrombin levels and increased risk for venous thrombosis. Subsequent studies have demonstrated that prothrombin levels were likely due to increased thrombin generation (Franco et al., 1999). More recently, the mechanism by which G20210A alters prothrombin levels has been established. Ceelie and colleagues (Ceelie et al., 2004) have demonstrated that the G20210A mutation results in a more effective poly(A) site [a poly(A) tail is required for mRNA to be efficiently exported from the nucleus and translated into protein in the cytoplasm], leading to elevated mRNA levels, resulting in increased prothrombin production and thrombin formation. Thus, similar to factor V Leiden, the G20210A mutation likely leads to a procoagulant state, thereby increasing risk of stroke.

Angiotensin-converting Enzyme

ACE plays in important role in blood pressure regulation and electrolyte balance, and ACE inhibitors have been at the forefront of therapy for treating hypertension and reducing risk for CVD.

Table 11–5 Meta-analysis of Candidate Genes in Stroke

Gene (No. of Studies)	Polymorphism	Genetic Model	Frequency of Variant at Risk (%)*	Total No. of Cases	Total No. of Controls	OR (95% CI)	P_{Het} Value
Factor V Leiden (26)[16–41]	Arg506Gln	Dominant	6.5	4588	13,798	1.33 (1.12–1.58)	0.03
MTHFR(22)[20,28–30,34,38,40,42–56]	C677T	Recessive	13.7	3387	4597	1.24 (1.08–1.42)	0.22
Prothrombin (19)[17,20,22,23,26,28–30,32,34,38,40,57–63]	G20210A	Dominant	2.9	3028	7131	1.44 (1.11–1.86)	0.91
ACE (11)[35,64–73]	I/D	Recessive	26.4	2990	11,305	1.21 (1.08–1.35)	0.47
Factor XIII (6)[74–79]	Val→Leu	Recessive	6.6	2166	1950	0.97 (0.75–1.25)	0.08
Apolipoprotein E (10)[50,56,80–87]	ε4, ε3, ε2	Allele ε4 vs. others	29.3	1805	10,921	0.96 (0.84–1.11)	0.02
Glycoprotein IIIa (9)[23,88–95]	Leu33Pro	Dominant	27.3	1467	2537	1.11 (0.95–1.28)	0.76
eNOS (3)[96–98]	Glu298Asp	Recessive	12.5	1086	1089	0.98 (0.76–1.26)	0.4
PAI1 (4)[74,99–101]	*4G/5G*	Recessive	18.9	842	1189	1.47 (1.13–1.92)	0.75
GPIBA (3)[102–104]	VNTR	*D/D* vs. others	2.08	816	719	0.81 (0.39–1.70)	0.78
Glycoprotein IIb (3)[88,89,91]	Ile→Ser	Recessive	13.9	770	1090	0.99 (0.74–1.32)	0.42
GPIBA (4)[88,91,102,104]	Thr→Met†	Dominant	13.4	564	962	1.55 (1.14–2.11)	0.68
Factor VII (3)[23,105,108]	*A1/A2*	Dominant	24.8	545	504	1.11 (0.83–1.48)	0.2
GPIBA (3)[102,107,108]	Kozak sequence	*T/T* vs. *C/C* + *C/T*	71.7	350	549	1.88 (1.28–2.76)	<0.001
LPL (3)[108–111]	Asn291Ser	Dominant	4.78	452	8879	1.27 (0.80–2.01)	0.73

* Derived from control subjects.
† Also known as the *HPA2* polymorphism.
Abbreviations: *ACE*, gene encoding angiotensin-converting enzyme; CI, confidence interval; *D/D*, deletion/deletion; *eNOS*, endothelial nitric oxide synthase; *GPIBA*, gene encoding glycoprotein Ib-α; I/D, insertion/deletion; *LPL*, gene encoding lipoprotein lipase; *MTHFR*, gene encoding methylenetetrahydrofolate reductase; OR, odds ratio; PAI1, gene encoding plasminogen activator inhibitor 1; P_{Het}, P for heterogeneity; VNTR, variable number tandem repeat.

Indeed, ACE is an important regulator of the renin–angiotensin–aldosterone system through both its ability to hydrolyze angiotensin I into angiotensin II, a potent vasopressor, and its ability to inactivate bradykinin, a potent vasodilator that may stimulate NO production (Kim and Iwao, 2000). ACE is found on the surface of vascular endothelial cells, as well as circulating in the plasma, and animal studies have shown the importance of ACE in regulating blood pressure (Krege et al., 1995; Esther et al., 1997). In 1990, Rigat and colleagues (Rigat et al., 1990) identified an insertion/deletion polymorphism (I/D) that was responsible for up to 50% of the variation in circulating levels of ACE. Although the molecular basis of how the I/D polymorphism affects circulating ACE levels is not entirely clear, a recent study using nearby polymorphisms to perform allele specific expression indicates that the D allele leads to higher expression of *ACE* mRNA (Suehiro et al., 2004).

Given the widespread use of ACE inhibitors in clinical treatment, and high frequency of the I/D polymorphism in the general population (~30% in Caucasians populations), the *ACE* I/D polymorphism provides a prime target for testing the potential impact of a genetic variant on choice of drug therapy (pharmacogenetics). Arnett and colleagues (Arnett et al., 2005) have recently reported the results of a double-blind, active-controlled randomized trial of antihypertensive treatment in which they examined the impact of the *ACE* I/D polymorphism on response to four different medications (chlorthalidone, amlodipine, lisinopril, doxazosin). The study included 37,939 participants ⩾55 years of age with ⩾1 risk factor for CVD. These individuals were followed up for 4–8 years, with primary outcomes of fatal CHD and/or nonfatal MI, and secondary outcomes of stroke, all-cause mortality, combined CHD, and combined CVD. *ACE* I/D genotype was not predictive for CHD, though the risk for stroke was consistent with the meta-analysis of Casas and colleagues (Casas, Hingorani, et al., 2004), nor did it modify the response to treatment with the different antihypertensive medications. These results were surprising, as one would predict that those with the DD genotype and therefore higher ACE levels, would be more responsive to the ACE inhibitor therapy (lisinopril). Indeed, these results provide a warning for making the leap between genetics and treatment. Despite a functional variant in a gene whose product is a direct target of one of the therapies, choice of therapy did not affect outcome. There were however, some differences in outcome according to gender and diabetes status, but given the number of hypotheses tested, further follow-up is needed to verify any of these observations. The authors also point out a significant limitation, which is often applicable to many genetic studies: only a single variant in *ACE* was examined, and other variants, or the interaction among them, may in fact impact the effectiveness of drug therapy.

Phosphodiesterase 4D

Traditional family-based linkage studies, which have been at the forefront of identifying genes for Mendelian diseases, are often difficult to implement for phenotypes that occur late in life, such as stroke, due to the difficulty in obtaining informative pedigrees. However, deCODE Genetics has leveraged the combination of extensive genealogical records and medical records in Iceland to enable performing these types of studies.

Gretarsdottir and colleagues (Gretarsdottir et al., 2002) initially performed a genome-wide linkage scan in 476 patients with stroke within 179 extended pedigrees from Iceland and identified a locus on 5q12 [log of odds (LOD) = 4.40], designated "*STRK1*." They employed a broad definition of stroke, including individuals with either ischemic or hemorrhagic stroke, as well as transient ischemic attack (TIA), which they considered an ischemic event, arguing that the same pathophysiological mechanisms are responsible for both. Stratified analysis indicated that the susceptibility gene in this locus contributes to common forms of stroke, with the possible exception of hemorrhagic stroke, and more significantly, did not seem to act through traditional risk factors such as hypertension, diabetes, and hypercholesterolemia. In a subsequent study, the same group fine-mapped this locus using a population-based case–control study composed of 864 affected Icelandic individuals and 908 controls, implicating *PDE4D*, which encodes a regulator of intracellular levels of cyclic adenosine monophosphate (cAMP) (Gretarsdottir et al., 2003). *PDE4D* is expressed in cardiac myocytes, and may be involved in excitation–contraction coupling (Lehnart et al., 2005), though the exact nature by which variation in *PDE4D* may contribute to susceptibility to stroke remains elusive (Greenberg and Rosand, 2005). However, this association was limited to ischemic stroke, specifically to the combined cardiogenic and carotid forms [using Trial of Org 10172 in Acute Stroke Treatment (TOAST) subcategories]. Sequencing of the exons in 188 affected individuals did not reveal any coding variants that associated with disease, suggesting that the functional variant(s) was likely to be regulatory, and indeed, they demonstrated altered expression of *PDE4D* between affected and unaffected individuals in EBV-transformed B-cell lines. In the absence of identifying the functional variant(s), Gretarsdottir and colleagues (Gretarsdottir et al., 2003) used haplotype analysis to identify three distinct haplotypes that carry significantly different risk for the combined carotid and cardiogenic stroke phenotype, designated AX, G0, and GX. Haplotype GX was considered to be the wild type, as it is most common in their population (53.4% in controls) and had an intermediate phenotype with regards to risk; haplotype G0 carriers exhibited a higher risk (OR = 1.46, $p < 0.0002$) and AX carriers, a lower risk (OR = 0.70, $p < 0.006$). Relative to the protective group, the general population-attributed risk of the at-risk and wild type groups combined is estimated to be 55%. Given the nature of these findings, specifically the convergence of linkage, association, and functional data (altered gene expression) in a plausible gene, a great deal of interest has been generated. Indeed multiple groups have attempted to replicate these findings with varying success and often confusing results.

Since the initial association study identifying *PDE4D* published in 2003, eight studies including more than 2800 affected and more than 3200 unaffected individuals using both family-based and case–control designs have been published examining the connection of this gene and stroke in various populations. Five of the studies found a positive association with *PDE4D* and ischemic stroke (Meschia et al., 2005; Saleheen et al., 2005; van Rijn et al., 2005; Nakayama et al., 2006; Woo et al., 2006), though not necessarily with the same set of SNPs as Gretarsdottir and colleagues identified (Gretarsdottir et al., 2003). Additionally, although van Rijn and colleagues (van Rijn et al., 2005) found an association with ischemic stroke, in contrast to Gettarsdottir and colleagues (Gretarsdottir et al., 2003), they observed an association only with small-vessel infarction and not large-vessel stroke. Of the three negative associations (Bevan et al., 2005; Lohmussaar et al., 2005; Nilsson-Ardnor et al., 2005), one suggested an association with cardioembolic stroke (Bevan et al., 2005) and one showed linkage to the *STRK1* locus (Nilsson-Ardnor et al., 2005). Interestingly, although Nakayama and colleagues (Nakayama et al., 2006) confirmed the *PDE4D* association in their population, they also identified a second region within the *STRK1* region associated with cerebral infarction.

Taken together, the weight of these findings clearly suggests that the *STRK1* locus is associated with risk for stroke, but that the details of that association remain to be elucidated. The nature of whether the association is only with ischemic stroke, and more specifically for the forms of stroke related to atherosclerosis, is still open to debate. Indeed, Bevan and colleagues (Bevan et al., 2005) found that none of the *PDE4D* polymorphisms that they studied was associated with carotid artery intima–media thickness or plaque. However, a recent study in mice that indicates that *PDE4D* deficiency promotes heart failure and arrhythmias (Lehnart et al., 2005); phenotypes associated with increased risk for cardiogenic stroke that often occurs in the setting of atrial fibrillation or ischemic heart disease (Adams et al., 1993) may support this hypothesis. Many of the ambiguities raised in the studies could be answered by identifying the underlying functional variant and specifically testing that variant in all the population studies to date. In the absence of the functional variant, a comprehensive study using a large sample size with extensive phenotype information and a high-density set of SNPs spanning the entire region is required to implicate *PDE4D* definitively in susceptibility to common forms of stroke.

Arachidonate 5-Lipooxygenase-Activating Protein

Helgadottir and colleagues (Helgadottir et al., 2004) reported a finding of linkage and association with *ALOX5AP* and both stroke and MI in an Icelandic population. They identified a specific haplotype (HapA), which is relatively common and is carried in 27% of patients with stroke [relative risk (RR) = 1.7, $p < 0.0001$]. The association with both MI and stroke was particularly intriguing, and in the same article, Helgadottir and colleagues identified another haplotype (HapB) that was associated with MI in an independent British cohort. They went on to demonstrate that the synthesis of leukotriene B4 (LTB4) in ionomycin-stimulated neutrophils from patients with a history of MI is greater than from controls without MI and that this difference is largely accounted for by carriers of HapA haplotype. LTB4, which is a biolipid inflammatory and vasoactive mediator, is produced from LTA4, which is produced from arachidonic acid by ALOX5AP. Thus, the association study is backed by functional evidence that points to a specific mechanism by which variants in *ALOX5AP* may increase risk for both stroke and MI: elevated levels of LTB4 might contribute to increased inflammation, a known risk factor for CVD events through the development and atherosclerosis and/or plaque instability.

A number of follow-up studies have now been reported, three of which have specifically examined the association with stroke, and one that examined patients with MI. Helgadottir and colleagues (Helgadottir et al., 2005) report a replication of their initial finding in a cohort comprised of 450 patients with ischemic stroke and 710 controls from Scottish population. The HapA haplotype was significantly associated with stroke ($p = 0.007$), but with a smaller RR (1.36), and the HapB haplotype was not associated, though the authors note, "HapB was overrepresented in male patients." One concern about this finding is that the Scottish *controls* had the same frequency of HapA as the Icelandic *cases*. The interpretation of this result is unclear, though it could suggest a potential for substructure in the populations. In subsequent studies in central European (Lohmussaar et al., 2005) and North American (Meschia et al., 2005) populations, HapA was not associated with stroke, though Lohmussar and colleagues reported several SNPs showing marginal significance. Kajimoto and colleagues (Kajimoto, Shioji, Ishida, et al., 2005) examined the association of *ALOX5AP* and MI in a Japanese cohort composed of 1875 controls and 353 cases. Due to differences in allele frequencies from the Caucasian populations, HapA and HapB could not be assessed, however, they reported a strong association for a haplotype that reduced the risk for MI. As none of these studies claim to have tested the putative functional variant, the range of results may be due to differing correlation between the tested SNPs/haplotypes and the underlying functional variant, and thus, identifying this variant and directly testing is of critical importance before concluding that *ALOX5AP* variants contribute to the risk for stroke and/or MI.

Cyclooxygenase 2

Although the path to identifying genes associated with stroke are littered with reports that have not, or cannot, be replicated and are thus not discussed in this chapter, one preliminary result bears mention due to it's potential application to pharmacogenomics. In 2004, Cipollone and colleagues (Cipollone et al., 2004) reported that the –765G→C polymorphism of the cyclooxygenase 2 (*COX-2*) gene was associated with reduced risk for MI and stroke. They studied 864 patients with first MI or ischemic stroke and 864 hospitalized controls that were matched for known CVD risk factors. They went on to demonstrate an association of *COX-2*, *MMP-2*, and *MMP-9* expression levels in atherosclerotic plaques with –765G→C genotype. Given the recent reports of increased risk for ischemic events caused by widely prescribed COX-2 inhibitors, these findings may have direct clinical relevance and should be validated in a large prospective study.

Genetic Interactions

Despite our limited ability to identify even single genes that impact risk for stroke, a great deal of interest has been generated recently in analyzing gene–gene interactions or gene networks. A traditional approach was applied by Szolnoki and colleagues (Szolnoki et al., 2002), who examined some of the well-established risk variants (Factor V Leiden, *MTHFR* C677T, *ACE* I/D, *APOE*, and prothrombin G20210A) in 689 ischemic stroke patients and 653 controls from a Hungarian population. Using multiple logistic regression analysis, they observed that specific combinations of the mutations that by themselves were minor or nonsignificant, could have a highly significant, moderate genetic risk of specific stroke subtypes. More recently, Sebastiani and colleagues (Sebastiani et al., 2005) used Bayesian networks to analyze 108 SNPs in 39 candidate genes in 1398 individuals with sickle cell anemia (SCA) for prognostic modeling of stroke, a severe complication of SCA. They identified 12 genes that interact with fetal hemoglobin to modulate risk of stroke in this high-risk population. Although the specific implication of these findings on risk for stroke in the general population has yet to be assessed, this approach has huge advantages over traditional methods, including the ability to include both genotype and phenotype data, allowing for both gene–gene and gene–phenotype interactions to be examined. Given the sparse samples relative to the number of variables tested, traditional methodologies could not be employed even for simple pair–wise interactions, let alone higher order interactions identified in this study. Thus, employing Bayesian network analysis to larger population-based cohorts with robust phenotype and genotype data may prove particularly illuminating in identifying genes and genetic pathways that underlie common forms of stroke.

Sudden Cardiac Death

SCD continues to be one of the leading causes of death in the United States. According to the US Centers for Disease Control and Prevention, approximately 462,000 of the 24,00,000 (19.3%) US deaths in 1999 were classified as "SCDs" using their definition of SCD as including all deaths, ". . .due to cardiac disease that occurred out of hospital (~341,000) or in an emergency department, or one in which the decedent was reported 'dead on arrival'" (Zheng et al., 2002). From the standpoint of preventive care, SCD poses a huge burden since less than 10% of SCD victims survive, and approximately one-third of all victims manifest SCD as their first clinical event. Approximately two-thirds of SCD victims do not have clinical symptoms that would warrant preventive intervention. Thus, the ability to identify individuals who are at high risk for SCD is crucial, and advances in genetics may fill this gap.

As for stroke and MI, a great deal of progress has been made identifying the genes involved in Mendelian forms of disease that contribute to increased risk of SCD. Mutations in coding sequences in at least seven cardiac sarcolemmal, Na, K, and Ca ion channel subunit genes (i.e., *KCNQ1*, *KCNH2*, *SCN5A*, *KCNE1*, *RYR2*, *KCNE2*, and *KCNJ2*), result in increased susceptibility to SCD (Fig. 11–1) (Priori and Napolitano, 2004). Electrophysiological dysfunction, manifest as delayed myocardial cell depolarization and repolarization, is caused by mutations in the proteins encoded by these genes, as originally discovered by Keating, Schwartz, Moss, Priori, and others during the 1990s and is now known to underlie a whole family of related proarrhythmic conditions, exemplified by the Long QT and Brugada Syndromes (Splawski et al., 2000; Keating and Sanguinetti, 2001). More recent evidence suggests that mutations in these same genes may result in converse disorders, such as "Short QT Syndrome," which also enhance SCD risk (Gaita et al., 2003; Brugada et al., 2004).

Multiple etiologies likely contribute to increased SCD risk, including susceptibilities that arise from genetic variations in sarcomeric proteins, such as β-myosin heavy chain (MyHC), cardiac troponin T (cTnT), and myosin binding protein-C (MyBP-C), which underlie SCDs that occur in patients with inherited hypertrophic cardiomyopathies (Marian and Roberts, 2001). Another important factor is genetic changes that impact on early patterning events during embryogenesis and subsequently cause disturbances in cardiac electrical function from development through maturation. Chien and collaborators were among the first to report that genetic alterations impacting early transcription factor expression may lead to enhanced arrhythmia susceptibility (Nguyen-Tran et al., 2000). Similar alterations in developmental factors were also recently implicated in rare familial vascular defects that appear to result in enhanced susceptibility to MI (Wang et al., 2003).

The existence of rare inherited monogenic disorders, such as the LQTS and Brugada syndrome, demonstrate that mutations in ion channel genes, structural proteins, and calcium-handling genes can increase susceptibility to lethal arrhythmias (Splawski et al., 2000; Keating et al., 2001). It is thus a small leap to propose that more subtle variations in these same genes may predispose to the more common forms of SCD. Two recent population-based studies have demonstrated that family history of SCD increases risk for SCD approximately 1.8-fold, independent of other traditional CVD risk factors (Friedlander et al., 1998; Jouven et al., 1999). The first study, conducted in the United States by Friedlander and colleagues (Friedlander et al., 1998), analyzed associations with "primary

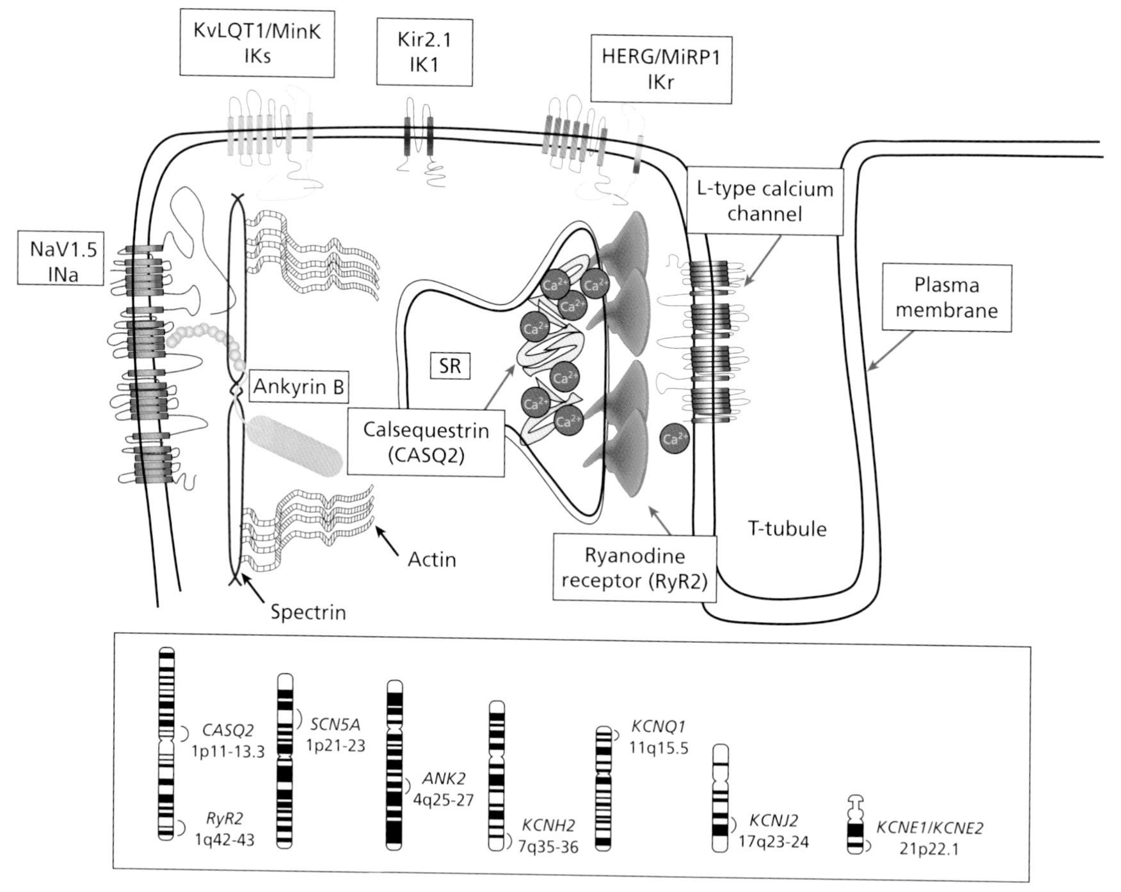

Figure 11–1 Genes involved in long QT syndrome (LQTS) and short QT syndrome (SQTS). (From Priori and Napolitano, 2004).

cardiac events" in a cohort of men and women attended by first responders in King County, Washington (235 cases and 374 controls). The second study conducted in Paris by Jouven and colleagues (Jouven et al., 1999) analyzed deaths in a cohort of 7746 asymptomatic middle aged males, using retrospective autopsy and clinical data analyses to ascribe cardiac deaths to either SCD or MI. Multifactorial statistical analyses indicated that the occurrence of SCD in a parent results in a 1.6- to 1.8-fold increase in SCD susceptibility despite controlling for conventional risk factors indicative of coronary disease (e.g., cholesterol subfractions, blood pressure, obesity, tobacco use, etc.). In a very limited number of cases in the Parisian study, where there was a history of both maternal and paternal SCD events ($n = 19$), the RR in offspring was approximately 9 ($p = 0.01$), indicating an additive genetic model. Elevated incidence of SCD in the Paris study segregated independently of elevated familial incidence of MI, suggesting genetic factors that specifically associate with risk for SCD, rather than for factors that may underlie overall CVD risk (e.g., atherosclerosis).

α Nav1.5 Subunit

The idea that ion channel sequence variations that alter cardiac de- or repolarizing currents in patients with rare inherited syndromes, like LQTS, may also contribute to enhanced SCD susceptibility seen with more common forms of cardiac disease represents an attractive hypothesis. Recent work by Splawski and colleagues (Splawski et al., 2002) has given some insight into this paradigm. In this study, a single nucleotide sequence variant in the *SCN5A* Na channel gene found in African Americans, S1102Y, was associated with a modest enhancement of arrhythmia risk. The aberrant allele was estimated to be present in up to 4.6 million African Americans: a level of prevalence far beyond all previously established SCD susceptibility alleles combined and was not identified in other ethnic populations sampled. In in vitro transfection experiments, the Y1102 allele accelerated channel activation, providing a plausible mechanism by which this variant may increase the likelihood of altered cardiac repolarization and arrhythmia. This finding has now been followed up in a series of 289 sudden deaths in blacks by Burke and colleagues (Burke et al., 2005). Individuals were classified into four categories: (1) controls, (2) cardiac deaths with clear anatomic substrate, (3) cardiac deaths with no anatomic substrate except mild to moderate cardiac hypertrophy, and (4) unexplained cardiac arrhythmias. The frequency of the Y1102 allele was significantly higher in those with SCD in the absence of a clear morphological abnormality (categories 3 and 4, combined $n = 65$). These findings strongly suggest that this allele is a risk factor for SCD in African Americans, but require confirmation in a larger cohort.

QT Interval

Although several marginal associations have been identified through case–control studies (*GPIIIa* A2 allele, alpha$_{2B}$-adenoreceptor I/D polymorphism, and *PAI-1* 4G allele), none of these studies have been replicated in larger prospective cohorts and thus remain unproven. SCD is likely the result of multiple pathways that contribute to increased susceptibility to arrhythmias, including atherosclerosis and thrombosis, electrogenesis and propagation, and initiating influences and triggers (Fig. 11–2) (Spooner et al., 2001). These pathways are likely to involve different genes, and thus extensive phenotyping of sample becomes important. In the absence of being able to distinguish SCD due to different underlying etiologies (e.g., structural defects versus ion channel defects), all samples will fall under the rubric of SCD, and the power to find genetic determinants is greatly reduced (Arking et al., 2004). Several studies are attempting to address this issue, including the Oregon Sudden Unexpected Death Study, which recruits samples through the emergency medical system and attempts to get electrocardiogram data on all SCDs and autopsy data when available (Chugh et al., 2003). There is also the need to obtain prospective data in order to assess attributable risk for SCD susceptibility variants and is ongoing in the Atherosclerosis Risk in Communities (ARIC) and CHS cohorts (ARIC-Investigators 1989; Fried et al., 1991). In the absence of large well-phenotyped SCD cohorts, a great deal of focus has been on subclinical phenotypes, including the QT interval.

The QT interval is a measure of cardiac repolarization and is under the joint control of the depolarizing Na^+ (I_{Na}), the repolarizing slow (I_{Ks}), and rapid (I_{Kr}) K^+-delayed rectifier currents. A number of studies have shown QT interval (corrected for heart rate) to be a moderately heritable trait (25%–52% heritability), indicating the likely involvement of genetic factors (Busjahn et al., 1999; Carter et al., 2000; Newton-Cheh et al., 2005). Moreover, extremes of QT interval have been associated with increased risk for SCD in both Mendelian forms (LQTS and SQTS), as well as in population-based settings (Schouten et al., 1991; Dekker et al., 1994; Elming et al., 1998; Sharp et al., 1998; de Bruyne et al., 1999; Okin et al., 2000; Dekker et al., 2004). As described in the preceding section, rare variants in a number of ion channel genes have been shown to cause LQTS and SQTS, and these findings have lead to a number of studies to determine whether more common variants in these genes influence QT interval in the general population.

Ion Channel Genes

Initial candidate gene studies have focused on five genes which encode subunits of the Na^+ and K^+ channel proteins involved in cardiac current regulation: I_{Na} current—α Nav1.5 subunit (*SCN5A*); I_{Ks} current—α KvLQT1 (*KCNQ1*) and β minK (*KCNE1*) subunits; and I_{Kr} current—α HERG (*KCNH2*) and β Mirp1 (*KCNE2*) subunits. Several studies have focused on nonsynonymous or coding region SNPs. Pietila and colleagues (Pietila et al., 2002) examined the K897T variant in exon 11 of *KCNH2* in a study of 226 males and 187 females of Finnish descent. They found that the 897T allele prolonged the QT interval, but only in females. Functional studies of this allele have noted that the I_{Kr}-897T channel exhibited a decreased current density (Paavonen et al., 2003; Anson et al., 2004). In a subsequent study of 1316 Europeans, the 897-T allele was found to be associated with shortened QT interval in both males and females, though a larger effect was observed in females (Bezzina et al., 2003). They further demonstrated that the I_{Kr}-897T channel showed a decrease in steady state activation potential, which by simulation studies predicts a shortening of action potential duration due to an increase in I_{Kr} current. Two subsequent studies in European populations have confirmed that the 897T allele is associated with shorter QT interval (Gouas et al., 2005; Pfeufer et al., 2005). Whether the initial report by Pietial and colleagues is a false positive, or reflects the existence of population-specific functional polymorphisms that influence the effect of the K897T polymorphism, remains to be determined. Several studies have also looked at the G38S variant in *KCNE1*, however, only one group reported a positive association, observed in 437 Israelis and with opposite effects in males and females (Friedlander et al., 2005), whereas three other studies, with over 1200 combined samples, did not find an association. Two studies have reported associations with SNPs in *SCN5A*, but only a nonsynonymous SNP, H558R, was in common between the two studies, and both reported an association of 558R allele with longer QT interval (Aydin et al., 2005; Gouas et al., 2005).

In the most comprehensive study published to date, Pfeufer and colleagues examined *KCNQ1*, *KCNE1*, *KCNH2*, and *KCNE2* using 174 SNPs chosen at a spacing of 1 every 5 kb, and including all known SNPs in exons and intron/exon boundaries (Pfeufer et al., 2005). These SNPs were genotyped in a screening sample of 689 population-based individuals of German ancestry. Significant findings were then confirmed in an additional 3277 individuals recruited from the same population. Using haplotype analyses of the *KCNH2* gene, they were not only able to confirm the K897T association, but also identified an additional variant, rs3815459, that marks a common haplotype (frequency ~19.5% in Caucasian populations) that does not harbor the 897T allele, and is associated with prolongation of the QT interval. They also identified a common variant in *KCNQ1*, rs757092, in which the minor allele is associated with prolongation of QT interval and had suggestive findings for a variant in *KCNE1*. The combination of the three confirmed variants (*KCNQ1*-rs757092, *KCNH2*-K897T, and *KCNH2*-rs3815459) explained 0.95% of the QT interval trait variance, acting in an additive fashion. All three SNPs showed gender-dependent associations, highlighting the importance of considering gender as a potent confounder when designing complex genotype–phenotype association studies.

Neuronal Nitric Oxide Synthase Adapter Protein (*NOS1AP*)

Although the candidate gene approach, largely based on lessons learned from Mendelian models of disease, has been successful, such an approach relies too heavily upon the current understanding of biology of cardiac repolarization and will therefore miss many important genes. Thus, many have suggested an unbiased genome-wide approach to identify genetic determinants of complex traits. One of the first such studies has been successfully employed to identify a common variant in a gene previously unsuspected to play a role in cardiac repolarization and modulates QT interval.

To identify genetic determinants of the QT interval, we chose a multistage approach, in which we performed genome-wide scans in samples from the extremes of the QT interval from a population-based German cohort (KORA S4), with follow-up of significant SNPs in the remainder of the cohort (Arking et al., 2006). For a complex phenotype, likely involving multiple genetic factors, choosing samples from the extremes of the trait distribution maximizes the genetic effect by enriching for individuals who are likely to be homozygous for the functional variants contributing to the trait. This approach therefore maximizes detectability while minimizing the amount of genotyping required. In stage I, 100 women from each extreme of the QT interval (corrected for heart rate, age, and gender, termed QTc_RAS) distribution were selected, and these samples were successfully genotyped for approximately 88,500 SNPs. Females were chosen to minimize the known gender heterogeneity of QT interval. Although no single SNP in stage I reached genome-wide significance at the $\alpha = 5.6 \times 10^{-7}$ level, the 10 most significant loci, which exhibited nominal p values $<10^{-4}$, were carried forward to stage II. Given that the underlying functional variant is unlikely to have been genotyped in stage I, each SNP selected for follow-up was enhanced with a flanking SNP to increase the likelihood of genotyping a SNP in high correlation with the hidden functional variant. In stage II, an additional 200 females each with QTc_RAS <15th or

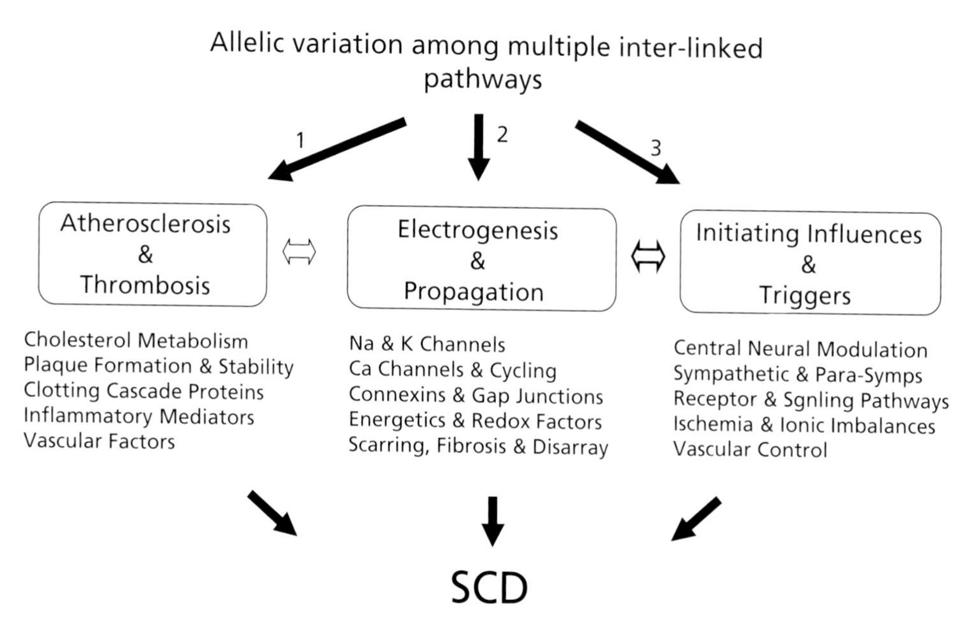

Figure 11–2 Potential genetic contributors to sudden cardiac death (SCD). Potential and documented elements of susceptibility are suggested in three broad pathways: (1) those that lead to progressive atherosclerosis and frank coronary disease and the likelihood of an occlusive infarction and ischemic arrhythmias; (2) those involved in electrogenesis and intromyocardial conduction pathways; and (3) those that may influence initiation of aberrant triggering events and the perpetuation of an arrhythmia. [Adapted from (Spooner et al., 2001) and (Arking et al., 2004)].

>85th percentiles were added and all 600 individuals were genotyped for the 10 SNPs from stage I and their flanking SNPs. In stage III, 4 SNPs significant at $p < 0.005$ in stage II, were genotyped in the remaining 3366 male and female individuals. After correcting for the number of genetic models and SNPs tested, one SNP was still highly associated with QT interval, rs10494366, in the gene *NOS1AP* (chromosome 1: $p < 10^{-7}$). *NOS1AP*, also known as *CAPON*, is the C-terminal PDZ domain ligand to neuronal nitric oxide synthase (*nNOS*) (Jaffrey et al., 1998) and affects N-methyl-D-aspartic acid (NMDA) receptor-gated calcium influx. It has not been previously suspected to play a role in cardiac repolarization but is expressed in human heart left-ventricular tissue.

As a further validation, a separate sample of 2646 individuals from the KORA F3 survey, who were collected using the same criteria as for the KORA S4 survey, were genotyped for the 4 SNPs identified through genome-wide analyses of the KORA S4 data. Again, rs10494366 in *NOS1AP* was highly significant even after adjusting for the number of SNPs tested ($p < 10^{-10}$) but none of the other SNPs was significant. The average genetic effect, measured as the difference in means of QTc_RAS between the two homozygotes, is 4.9 ms for *NOS1AP* in the total S4 sample and 7.9 ms in the F3 sample, with *NOS1AP* accounting for 1.2% and 1.9% of the variance, respectively. The minor allele frequency was the same in both populations (36%) and genetic effects were observed in both males and females. These results emphasize that a genome-wide association study, which is not limited by our current understanding of cardiac repolarization biology, can be employed to identify novel genes. Indeed, the involvement of the *NOS1AP* gene in QT interval regulation was previously unsuspected, yet it explains 1.5% of the variance in the combined sample of approximately 6600 adults. Consistent findings were also observed in approximately 1800 individuals from the Framingham Heart Study.

Cautionary Tales

MADS Box Transcription Enhancer Factor 2, Polypeptide A

One approach to identify genetic determinants of complex traits has been to identify families which exhibit monogenic forms of the phenotype of interest, with the idea any genes identified are likely to be involved in complex forms of the phenotype as well. This approach has several merits, including less phenotypic heterogeneity, since presumably all affected individuals in the pedigree are manifesting the same phenotype. Additionally, one can use traditional linkage analysis that does not require recruiting a control population, which, if not done carefully, can lead to both false positives and false negative results. This approach was adopted by Wang and colleagues (Wang et al., 2003), who studied a large family that displayed an autosomal dominant form of CAD. They performed genome-wide linkage analysis and identified a significant association on chromosome 15q26. With 93 genes in the associated region, they focused on the transcription factor *MEF2A* (MADS box transcription enhancer factor 2, polypeptide A), largely due to the overall expression pattern; *MEF2A* was expressed in blood vessels during mouse early embryogenesis (Edmondson et al., 1994), and expression was similar to vascular endothelial growth factor receptor 2 (*VEGFR2*) and the von Willebrand factor (Subramanian and Nadal-Ginard, 1996). These data led Wang and colleagues to speculate that *MEF2A* can be an early marker for vasculogenesis. They sequenced the gene in the affected individuals and identified a 21-bp deletion in exon 11 (termed Δ7aa) that resulted in the loss of seven conserved amino acids and was present in all affected members in the family. They further went on to demonstrate altered cellular trafficking for the mutant protein and that its ability to activate atrial natriuretic factor is reduced. In a subsequent study, the same group examined 207 CAD/MI patients for mutations in MEF2A, identifying three novel mutations in exon 7 in four patients and no mutations in 191 controls (Bhagavatula et al., 2004). They demonstrated that these mutations significantly reduce the transcriptional activity of *MEF2A* and suggest that, "a significant percent of the CAD/MI population may carry mutations in *MEF2A*."

The combination of both family- and population-based evidence would seem to present strong evidence for the involvement of *MEF2A* variants with risk of CAD/MI. However, a recent study by Weng and colleagues (Weng et al., 2005) has raised some doubts about the strength of the effect of *MEF2A* variants on susceptibility to CAD/MI. They sequenced *MEF2A* exons in 300 white individuals with documented CAD with onset before age 55 years (men) or 65 years (women) and in 300 elderly controls (men > 60 years, women >70 years), who did not have signs or symptoms of CAD. Of the five missense mutations identified, one was unique to the CAD individuals, one unique to the controls, and three common to both groups. They further observed the 21-bp deletion in three unrelated unaffected individuals and demonstrated that the deletion does not segregate with early CAD. They conclude that, "these studies support that *MEF2A* mutations are not a common cause of CAD in white people and argue strongly against a role for the *MEF2A* 21-bp deletion in autosomal dominant CAD." A similar negative association was reported by Kajimoto and colleagues in a Japanese population (Kajimoto, Shioji, Tago, et al., 2005) and Gonzalez and colleagues (Gonzalez et al., 2006) in a Spanish cohort for the Δ7aa allele, though Gonzalez and colleagues reported a positive association for a rare Pro279Leu mutation (OR = 3.06, 95% CI 1.17–8.06).

In an accompanying commentary to the Weng and colleagues paper, Altshuler and Hirschhorn (Altshuler et al., 2005) conclude that the role of *MEF2A* variants in CAD has not been established and use the *MEF2A* studies to illustrate a number of criteria that should be imposed when performing similar studies. They note that in a complex disease like CAD, it may not be particularly uncommon to find large pedigrees that appear to display monogenic forms of the disease. Thus, they propose looking for an unusual phenotype shared by affected individuals, such as early-onset or a syndromic form of disease. The phenotype also needs to be consistent across family members, which can be difficult with CAD, in which some individuals are "affected" by virtue of having had prior coronary events, while others may have angiographically defined disease in the absence of events. Given these concerns, linkage across multiple families with identical ascertainment criteria should be observed.

In the event a linkage signal in a region is detected, similar concerns arise when trying to identify the underlying functional variant. The observation of rare, potentially functional variants segregating within a family is not uncommon, and any such variation observed in the linkage region will, by virtue of being in a disease-linked region, segregate with the disease. Thus, Altshuler and Hirschorn propose that for a specific gene in a linked region to be associated with disease, it must meet a one or more of the following criteria: (1) multiple different mutations exist, each of which cosegregates with disease; (2) confirmatory evidence for a particular allele in a case–control study; (3) multiple rare variants that have been well ascertained in controls; (4) observation of a *do novo* mutation in an affected child (but not in his or her parent); (5) strong evidence of the effects of a human mutation in a model organism that recapitulates the human disease phenotype. With the cost of genotyping decreasing and sample collection increasing, the number of such studies being performed is on the rise, and thus, adopting rigorous criteria, as outlined in the preceding section, for deeming a gene to be involved in complex disease is warranted.

Conclusions

Although only a few of the genes studied to date for association with CVD, and in particular, clinical outcomes, have reproducibly been shown to modulate risk, a number of lessons can be learned. First, common genetic variants are of modest effect (OR < 2), which has been a significant factor in the failure to identify and reproducibly demonstrate an effect for a given variant. Most studies have simply been underpowered, greatly contributing to the confusion in the literature. Second, our ability to pick out genes involved in disease based on our limited understanding of biology is insufficient, especially given that fully half the genes in the genome are of largely unknown function. Third, most genes that have been researched have not been studied in a comprehensive way, limiting the ability to compare across studies and to truly exclude genes (as opposed to a specific variant) from association with disease. Fourth, functional variants are likely to be influenced by sex and/or environment (e.g., smoking, and diet), and thus comprehensively collecting data on study populations is essential. With whole-genome association studies rapidly becoming feasible, genotyping will no longer be an impediment, and the focus going forward will be on collecting and assessing much larger populations to both identify additional genetic variants and to establish further the true impact of these variants on the general population risk for CVD.

References

Adams, HP, Jr, Bendixen, BH, Kappelle, LJ, Biller, Love, J, Gordon, DL, and Marsh, EE III (1993). Classification of subtype of acute ischemic stroke. Definitions for use in a multicenter clinical trial. TOAST. Trial of Org 10172 in Acute Stroke Treatment. *Stroke*, 24:35–41.

Altshuler, D and Hirschhorn, JN (2005). MEF2A sequence variants and coronary artery disease: A change of heart? *J Clin Invest*, 115:831–833.

American Heart Association (2003). Heart Diseases and Stroke Statistics 2003 Update. Dallas, TX: American Heart Association.

Anson, BD, Ackerman, MJ, Tester, DJ, Will, ML, Delisle, BP, Anderson, CL, and January, CT (2004). Molecular and functional characterization of common polymo in HERG (KCNH2) potassium channels. *Am J Physiol Heart Circ Physiol*, 286:H2434–H2441.

ARIC-Investigators (1989). The Atherosclerosis Risk in Communities (ARIC) Study: Design and objectives. *Am J Epidemiol*, 129:687–702.

Arking, DE, Chugh, SS, Chakravarti, A, and Spooner, PM (2004). Genomics in sudden cardiac death. *Circ Res*, 94:712–723.

Arking, DE, Pfeufer, A, Post, W, Kao, WHL, Newton-Cheh, C, Ikeda, M et al. (2006). A common genetic variant in the nNOS regulator CAPON modulates cardiac repolarization (QT interval). *Nat Genet*, 38:644–651.

Arnett, DK, Davis, BR, Ford, CE, Boerwinkle, E, Leiendecker-Foster, C, Miller, MB, Black, H, and Eckfeldt, JH (2005). Pharmacogenetic association of the angiotensin-converting enzyme insertion/deletion polymorphism on blood pressure and cardiovascular risk in relation to antihypertensive treatment: The Genetics of Hypertension-Associated Treatment (GenHAT) study. *Circulation*, 111:3374–3383.

Artiga, MJ, Bullido, MJ, Sastre, I, Recuero, M, Garcia, MA, Aldudo, J, Vazquez, J, and Valdivieso, F (1998). Allelic polymorphisms in the transcriptional regulatory region of apolipoprotein E gene. *FEBS Lett*, 421:105–108.

Aydin, A, Bahring, S, Dahm, S, Guenther, UP, Uhlmann, R, Busjahn, A, and Luft, FC (2005). Single nucleotide polymorphism map of five long-QT genes. *J Mol Med*, 83:159–165.

Ballantyne, CM, Herd, JA, Stein, EA, Ferlic, LL, Dunn, JK, Gotto, AM, Jr, and Marian, AJ (2000). Apolipoprotein E genotypes and response of plasma lipids and progression-regression of coronary atherosclerosis to lipid-lowering drug therapy. *J Am Coll Cardiol*, 36:1572–1578.

Bazzoni, F and Beutler, B (1996). The tumor necrosis factor ligand and receptor families. *N Engl J Med*, 334:1717–1725.

Beisiegel, U, Weber, W, and Bengtsson-Olivecrona, G (1991). Lipoprotein lipase enhances the binding of chylomicrons to low density lipoprotein receptor-related protein. *Proc Natl Acad Sci USA*, 88:8342–8346.

Bergametti, F, Denier, C, Labauge, P, Arnoult, M, Boetto, S, Clanet, M, et al. (2005). Mutations within the programmed cell death 10 gene cause cerebral cavernous malformations. *Am J Hum Genet*, 76:42–51.

Bertina, RM, Koeleman, BP, Koster, T, Rosendaal, FR, Dirven, RJ, de Ronde, H, van der Velden, PA, and Reitsma, PH (1994). Mutation in blood coagulation factor V associated with resistance to activated protein C. *Nature*, 369:64–67.

Bevan, S, Porteous, L, Sitzer, M, and Markus, HS (2005). Phosphodiesterase 4D gene, ischemic stroke, and asymptomatic carotid atherosclerosis. *Stroke*, 36:949–953.

Beyzade, S, Zhang, S, Wong, YK, Day, IN, Eriksson, P, and Ye, S (2003). Influences of matrix metalloproteinase-3 gene variation on extent of coronary atherosclerosis and risk of myocardial infarction. *J Am Coll Cardiol*, 41:2130–2137.

Bezzina, CR, Verkerk, AO, Busjahn, A, Jeron, A, Erdmann, J, Koopmann, TT, et al. (2003). A common polymorphism in KCNH2 (HERG) hastens cardiac repolarization. *Cardiovasc Res*, 59:27–36.

Bhagavatula, MR, Fan, C, Shen, GQ, Cassano, J, Plow, EF, Topol, EJ, and Wang, Q (2004). Transcription factor MEF2A mutations in patients with coronary artery disease. *Hum Mol Genet*, 13:3181–3188.

Blankenberg, S, Rupprecht, HJ, Bickel, C, Jiang, XC, Poirier, O, Lackner, KJ, Meyer, J, Cambien, F, and Tiret, L (2003). Common genetic variation of the cholesteryl ester transfer protein gene strongly predicts future cardiovascular death in patients with coronary artery disease. *J Am Coll Cardiol*, 41:1983–1989.

Boekholdt, SM, Bijsterveld, NR, Moons, AH, Levi, M, Buller, HR, and Peters, RJ (2001). Genetic variation in coagulation and fibrinolytic proteins and their relation with acute myocardial infarction: A systematic review. *Circulation*, 104:3063–3068.

Boekholdt, SM, Sacks, FM, Jukema, JW, Shepherd, J, Freeman, DJ, McMahon, AD, et al. (2005). Cholesteryl ester transfer protein TaqIB variant, high-density lipoprotein cholesterol levels, cardiovascular risk, and efficacy of pravastatin treatment: Individual patient meta-analysis of 13,677 subjects. *Circulation*, 111:278–287.

Brugada, R, Hong, K, Dumaine, R, Cordeiro, J, Gaita, F, Borggrefe, M, et al. (2004). Sudden death associated with short-QT syndrome linked to mutations in HERG. *Circulation*, 109:30–35.

Bullido, MJ, Artiga, MJ, Recuero, M, Sastre, I, Garcia, MA, Aldudo, J, et al. (1998). A polymorphism in the regulatory region of APOE associated with risk for Alzheimer's dementia. *Nat Genet*, 18:69–71.

Burke, A, Creighton, W, Mont, E, Li, L, Hogan, S, Kutys, R, Fowler, D, and Virmani, R (2005). Role of SCN5A Y1102 polymorphism in sudden cardiac death in blacks. *Circulation*, 112:798–802.

Busjahn, A, Knoblauch, H, Faulhaber, HD, Boeckel, T, Rosenthal, M, Uhlmann, R, Hoehe, M, Schuster, H, and Luft, FC (1999). QT interval is linked to 2 long-QT syndrome loci in normal subjects. *Circulation*, 99:3161–3164.

Carlquist, JF, Muhlestein, JB, Horne, BD, Hart, NI, Bair, TL, Molhuizen, HO, and Anderson, JL (2003). The cholesteryl ester transfer protein Taq1B gene polymorphism predicts clinical benefit of statin therapy in patients with significant coronary artery disease. *Am Heart J*, 146:1007–1014.

Carter, N, Snieder, H, Jeffery, S, Saumarez, R, Varma, C, Antoniades, L, and Spector, TD (2000). QT interval in twins. *J Hum Hypertens*, 14:389–390.

Casas, JP, Bautista, LE, Humphries, SE, and Hingorani, AD (2004). Endothelial nitric oxide synthase genotype and ischemic heart disease: Meta-analysis of 26 studies involving 23028 subjects. *Circulation* 109:1359–1365.

Casas, JP, Hingorani, AD, Bautista, LE, and Sharma, P (2004). Meta-analysis of genetic studies in ischemic stroke: Thirty-two genes involving approximately 18,000 cases and 58,000 controls. *Arch Neurol*, 61:1652–1661.

Ceelie, H, Spaargaren-van Riel, CC, Bertina, RM, and Vos, HL (2004). G20210A is a functional mutation in the prothrombin gene; effect on protein levels and 3′-end formation. *J Thromb Haemost*, 2:119–127.

Chugh, SS, Dogra, V, Thompson, B, Ilias, N, Danna, D, John, B, et al. (2003). Annual incidence of sudden cardiac death based on operative definition: A prospective, population-based, countywide study using multiple sources. *Circulation*, 108:IV–1040.

Cipollone, F, Toniato, E, Martinotti, S, Fazia, M, Iezzi, A, Cuccurullo, C, et al. (2004). A polymorphism in the cyclooxygenase 2 gene as an inherited protective factor against myocardial infarction and stroke. *JAMA*, 291:2221–2228.

Cooke, JP and Dzau, VJ (1997). Nitric oxide synthase: Role in the genesis of vascular disease. *Annu Rev Med*, 48:489–509.

Dachet, C, Poirier, O, Cambien, F, Chapman, J, and Rouis, M (2000). New functional promoter polymorphism, CETP/–629, in cholesteryl ester transfer protein (CETP) gene related to CETP mass and high density lipoprotein cholesterol levels: Role of Sp1/Sp3 in transcriptional regulation. *Arterioscler Thromb Vasc Biol*, 20:507–515.

Dahlback, B (1995). New molecular insights into the genetics of thrombophilia. Resistance to activated protein C caused by Arg506 to Gln mutation in factor V as a pathogenic risk factor for venous thrombosis. *Thromb Haemost*, 74:139–148.

Davignon, J, Cohn, JS, Mabile, L, and Bernier, L (1999). Apolipoprotein E and atherosclerosis: Insight from animal and human studies. *Clin Chim Acta*, 286:115–143.

de Bruyne, MC, Hoes, AW, Kors, JA, Hofman, A, van Bemmel, JH, and Grobbee, DE (1999). Prolonged QT interval predicts cardiac and all-cause mortality in the elderly. The Rotterdam Study. *Eur Heart J*, 20:278–284.

Dekker, JM, Crow, RS, Hannan, PJ, Schouten, EG, and Folsom, AR (2004). Heart rate-corrected QT interval prolongation predicts risk of coronary heart disease in black and white middle-aged men and women: The ARIC study. *J Am Coll Cardiol*, 43:565–571.

Dekker, JM, Schouten, EG, Klootwijk, P, Pool, J, and Kromhout, D (1994). Association between QT interval and coronary heart disease in middle-aged and elderly men. The Zutphen Study. *Circulation*, 90:779–785.

Denier, C, Goutagny, S, Labauge, P, Krivosic, V, Arnoult, M, Cousin, A, et al. (2004). Mutations within the MGC4607 gene cause cerebral cavernous malformations. *Am J Hum Genet*, 74:326–337.

Djousse, L, Myers, RH, Province, MA, Hunt, SC, Eckfeldt, JH, Evans, G, Peacock, JM, and Ellison, RC (2002). Influence of apolipoprotein E, smoking, and alcohol intake on carotid atherosclerosis: National Heart, Lung, and Blood Institute Family Heart Study. *Stroke*, 33:1357–1361.

Dong, Y, Hassan, A, Zhang, Z, Huber, D, Dalageorgou, C, and Markus, HS (2003). Yield of screening for CADASIL mutations in lacunar stroke and leukoaraiosis. *Stroke*, 34:203–205.

Edmondson, DG, Lyons, GE, Martin, JF, and Olson, EN (1994). Mef2 gene expression marks the cardiac and skeletal muscle lineages during mouse embryogenesis. *Development*, 120:1251–1263.

Eichner, JE, Dunn, ST, Perveen, G, Thompson, DM, Stewart, KE, and Stroehla, BC (2002). Apolipoprotein E polymorphism and cardiovascular disease: A HuGE review. *Am J Epidemiol*, 155:487–495.

Eiriksdottir, G, Bolla, MK, Thorsson, B, Sigurdsson, G, Humphries, SE, and Gudnason, V (2001). The –629C>A polymorphism in the CETP gene does not explain the association of TaqIB polymorphism with risk and age of myocardial infarction in Icelandic men. *Atherosclerosis*, 159:187–192.

Elming, H, Holm, E, Jun, L, Torp-Pedersen, C, Kober, L, Kircshoff, M, Malik, M, and Camm, J (1998). The prognostic value of the QT interval and QT interval dispersion in all-cause and cardiac mortality and morbidity in a population of Danish citizens. *Eur Heart J*, 19:1391–1400.

Esther, CR, Marino, EM, Howard, TE, Machaud, A, Corvol, P, Capecchi, MR, and Bernstein, KE (1997). The critical role of tissue angiotensin-converting enzyme as revealed by gene targeting in mice. *J Clin Invest*, 99:2375–2385.

Flex, A, Gaetani, E, Papaleo, P, Straface, G, Proia, AS, Pecorini, G, Tondi, P, Pola, P, and Pola, R (2004). Proinflammatory genetic profiles in subjects with history of ischemic stroke. *Stroke*, 35:2270–2275.

Flossmann, E, Schulz, UG, and Rothwell, PM (2004). Systematic review of methods and results of studies of the genetic epidemiology of ischemic stroke. *Stroke*, 35:212–227.

Franco, RF, Trip, MD, ten Cate, H, van den Ende, A, Prins, MH, Kastelein, JJ, and Reitsma, PH (1999). The 20210 G→A mutation in the 3′-untranslated region of the prothrombin gene and the risk for arterial thrombotic disease. *Br J Haematol*, 104:50–54.

Fried, LP, NO Borhani, Enright, P, Furberg, CD, Gardin, JM, Kronmal, RA, et al. (1991). The Cardiovascular Health Study: Design and rationale. *Ann Epidemiol*, 1:263–276.

Friedlander, Y, Siscovick, DS, Weinmann, S, Austin, MA, Psaty, BM, Lemaitre, RN, Arbogast, P, Raghunathan, TE, and Cobb, LA (1998). Family history as a risk factor for primary cardiac arrest. *Circulation*, 97:155–160.

Friedlander, Y, Vatta, M, Sotoodehnia, N, Sinnreich, R, Li, H, Manor, O, Towbin, JA, Siscovick, DS, and Kark, JD (2005). Possible association of the human KCNE1 (minK) gene and QT interval in healthy subjects: Evidence from association and linkage analyses in Israeli families. *Ann Hum Genet*, 69:645–656.

Frosst, P, Blom, HJ, Milos, R, Goyette, P, Sheppard, CA, Matthews, RG, et al. (1995). A candidate genetic risk factor for vascular disease: A common mutation in methylenetetrahydrofolate reductase. *Nat Genet*, 10:111–113.

Gaita, F, Giustetto, C, Bianchi, F, Wolpert, C, Schimpf, R, Riccardi, R, Grossi, S, Richiardi, E, and Borggrefe, M (2003). Short QT syndrome: A familial cause of sudden death. *Circulation*, 108:965–970.

Gnasso, A, Motti, C, Irace, C, Carallo, C, Liberatoscioli, L, Bernardini, S, Massoud, R, Mattioli, PL, Federici, G, and Cortese, C (2000). Genetic variation in human stromelysin gene promoter and common carotid geometry in healthy male subjects. *Arterioscler Thromb Vasc Biol*, 20:1600–1605.

Goldstein, LB, Adams, R, Becker, K, Furberg, CD, Gorelick, PB, Hademenos, G, et al. (2001). Primary prevention of ischemic stroke: A statement for healthcare professionals from the Stroke Council of the American Heart Association. *Circulation*, 103:163–182.

Gonzalez, P, Garcia-Castro, M, Reguero, JR, Batalla, A, Ordonez, AG, Palop, RL, Lozano, I, Montes, M, Alvarez, V, and Coto, E (2006). The Pro279Leu variant in the transcription factor MEF2A is associated with myocardial infarction. *J Med Genet*, 43:167–169.

Gouas, L, Nicaud, V, Berthet, M, Forhan, A, Tiret, L, Balkau, B, and Guicheney, P (2005). Association of KCNQ1, KCNE1, KCNH2 and SCN5A polymorphisms with QTc interval length in a healthy population. *Eur J Hum Genet*, 13:1213–1222.

Greenberg, SM and Rosand, J (2005). The phosphodiesterase puzzle box: PDE4D and stroke. *Ann Neurol*, 58:345–346.

Gretarsdottir, S, Sveinbjornsdottir, S, Jonsson, HH, Jakobsson, F, Einarsdottir, E, Agnarsson, U, et al. (2002). Localization of a susceptibility gene for common forms of stroke to 5q12. *Am J Hum Genet*, 70:593–603.

Gretarsdottir, S, Thorleifsson, G, Reynisdottir, ST, Manolescu, A, Jonsdottir, S, Jonsdottir, T, et al. (2003). The gene encoding phosphodiesterase 4D confers risk of ischemic stroke. *Nat Genet*, 35:131–138.

Gulcher, JR, Gretarsdottir, S, Helgadottir, A, and Stefansson, K (2005). Genes contributing to risk for common forms of stroke. *Trends Mol Med*, 11:217–224.

Hagberg, JM, Wilund, KR, and Ferrell, RE (2000). APO E gene and gene–environment effects on plasma lipoprotein-lipid levels. *Physiol Genomics*, 4:101–108.

Hannuksela, ML, Liinamaa, MJ, Kesaniemi, YA, and Savolainen, MJ (1994). Relation of polymorphisms in the cholesteryl ester transfer protein gene to transfer protein activity and plasma lipoprotein levels in alcohol drinkers. *Atherosclerosis*, 110:35–44.

Helgadottir, A, Gretarsdottir, S, St Clair, D, Manolescu, A, Cheung, J, Thorleifsson, G, et al. (2005). Association between the gene encoding 5-lipoxygenase-activating protein and stroke replicated in a Scottish population. *Am J Hum Genet*, 76:505–509.

Helgadottir, A, Manolescu, A, Thorleifsson, G, Gretarsdottir, S, Jonsdottir, H, Thorsteinsdottir, U, et al. (2004). The gene encoding 5-lipoxygenase activating protein confers risk of myocardial infarction and stroke. *Nat Genet*, 36:233–239.

Henney, AM, Wakeley, PR, Davies, MJ, Foster, K, Hembry, R, Murphy, G, and Humphries, S (1991). Localization of stromelysin gene expression in atherosclerotic plaques by in situ hybridization. *Proc Natl Acad Sci USA*, 88:8154–8158.

Humphries, SE, Martin, S, Cooper, J, and Miller, G (2002). Interaction between smoking and the stromelysin-1 (MMP3) gene 5A/6A promoter polymorphism and risk of coronary heart disease in healthy men. *Ann Hum Genet*, 66:343–352.

Humphries, SE, Talmud, PJ, Hawe, E, Bolla, M, Day, IN, and Miller, GJ (2001). Apolipoprotein E4 and coronary heart disease in middle-aged men who smoke: A prospective study. *Lancet*, 358:115–119.

Hwang, SJ, Ballantyne, CM, Sharrett, AR, Smith, LC, Davis, CE, Gotto, AM, Jr, and Boerwinkle, E (1997). Circulating adhesion molecules VCAM-1, ICAM-1, and E-selectin in carotid atherosclerosis and incident coronary heart disease cases: The Atherosclerosis Risk In Communities (ARIC) study. *Circulation*, 96:4219–4225.

Iwanaga, Y, Ono, K, Takagi, S, Terashima, M, Tsutsumi, Y, Mannami, T, Yasui, N, Goto, Y, Nonogi, H, and Iwai, N (2004). Association analysis between polymorphisms of the lymphotoxin-alpha gene and myocardial infarction in a Japanese population. *Atherosclerosis*, 172:197–198.

Jaffrey, SR, Snowman, AM, Eliasson, MJ, Cohen, NA, and Snyder, SH (1998). CAPON: A protein associated with neuronal nitric oxide synthase that regulates its interactions with PSD95. *Neuron*, 20:115–124.

Joutel, A, Corpechot, C, Ducros, A, Vahedi, K, Chabriat, H, Mouton, P, et al. (1996). Notch3 mutations in CADASIL, a hereditary adult-onset condition causing stroke and dementia. *Nature*, 383:707–710.

Jouven, X, Desnos, M, Guerot, C, and Ducimetiere, P (1999). Predicting sudden death in the population: The Paris Prospective Study I. *Circulation*, 99:1978–1983.

Kajimoto, K, Shioji, K, Ishida, C, Iwanaga, Y, Kokubo, Y, Tomoike, H, Miyazaki, S, Nonogi, H, Goto, Y, and Iwai, N (2005). Validation of the association between the gene encoding 5-lipoxygenase-activating protein and myocardial infarction in a Japanese population. *Circ J*, 69:1029–1034.

Kajimoto, K, Shioji, K, Tago, N, Tomoike, H, Nonogi, H, Goto, Y, and Iwai, N (2005). Assessment of MEF2A mutations in myocardial infarction in Japanese patients. *Circ J*, 69:1192–1195.

Kakko, S, Tamminen, M, Paivansalo, M, Kauma, H, Rantala, AO, Lilja, M, Reunanen, A, Kesaniemi, YA, and Savolainen, MJ (2001). Variation at the cholesteryl ester transfer protein gene in relation to plasma high density lipoproteins cholesterol levels and carotid intima-media thickness. *Eur J Clin Invest*, 31:593–602.

Keating, MT and Sanguinetti, MC (2001). Molecular and cellular mechanisms of cardiac arrhythmias. *Cell*, 104:569–580.

Keavney, B, Palmer, A, Parish, S, Clark, S, Youngman, L, Danesh, J, et al. (2004). Lipid-related genes and myocardial infarction in 4685 cases and 3460 controls: Discrepancies between genotype, blood lipid concentrations, and coronary disease risk. *Int J Epidemiol*, 33:1002–1013.

Keavney, B, Parish, S, Palmer, A, Clark, S, Youngman, L, Danesh, J, et al. (2003). Large-scale evidence that the cardiotoxicity of smoking is not significantly modified by the apolipoprotein E epsilon2/epsilon3/epsilon4 genotype. *Lancet*, 361:396–398.

Kim, S and Iwao, H (2000). Molecular and cellular mechanisms of angiotensin II-mediated cardiovascular and renal diseases. *Pharmacol Rev*, 52:11–34.

Klerkx, AH, Tanck, MW, Kastelein, JJ, Molhuizen, HO, Jukema, JW, Zwinderman, AH, and Kuivenhoven, JA (2003). Haplotype analysis of the CETP gene: Not TaqIB, but the closely linked –629C→A polymorphism and a novel promoter variant are independently associated with CETP concentration. *Hum Mol Genet*, 12:111–123.

Koch, W, Kastrati, A, Bottiger, C, Mehilli, J, von Beckerath, N, and Schomig, A (2001). Interleukin-10 and tumor necrosis factor gene polymorphisms and risk of coronary artery disease and myocardial infarction. *Atherosclerosis*, 159:137–144.

Kondo, I, Berg, K, Drayna, D, and Lawn, R (1989). DNA polymorphism at the locus for human cholesteryl ester transfer protein (CETP) is associated with high density lipoprotein cholesterol and apolipoprotein levels. *Clin Genet*, 35:49–56.

Krege, JH, John, SW, Langenbach, LL, Hodgin, JB, Hagaman, JR, Bachman, ES, Jennette, JC, O'Brien, DA, and Smithies, O (1995). Male-female differences in fertility and blood pressure in ACE-deficient mice. *Nature*, 375:146–148.

Kuivenhoven, JA, Jukema, JW, Zwinderman, AH, de Knijff, P, McPherson, R, Bruschke, AV, Lie, KI, and Kastelein, JJ (1998). The role of a common variant of the cholesteryl ester transfer protein gene in the progression of coronary atherosclerosis. The Regression Growth Evaluation Statin Study Group. *N Engl J Med*, 338:86–93.

Laberge-le Couteulx, S., Jung, HH, Labauge, P, Houtteville, JP, Lescoat, C, Cecillon, M, Marechal, E, Joutel, A, Bach, JF, and Tournier-Lasserve, E, (1999). Truncating mutations in CCM1, encoding KRIT1, cause hereditary cavernous angiomas. *Nat Genet*, 23:189–193.

Lahoz, C, Schaefer, EJ, Cupples, LA, Wilson, PW, Levy, D, Osgood, D, Parpos, S, Pedro-Botet, J, Daly, JA, and Ordovas, JM (2001). Apolipoprotein E genotype and cardiovascular disease in the Framingham Heart Study. *Atherosclerosis*, 154:529–537.

Lambert, JC, Berr, C, Pasquier, F, Delacourte, A, Frigard, B, Cottel, D, et al. (1998). Pronounced impact of Th1/E47cs mutation compared with –491 AT mutation on neural APOE gene expression and risk of developing Alzheimer's disease. *Hum Mol Genet*, 7:1511–1516.

Lambert, JC, Brousseau, T, Defosse, V, Evans, A, Arveiler, D, Ruidavets, JB, et al. (2000). Independent association of an APOE gene promoter polymorphism with increased risk of myocardial infarction and decreased APOE plasma concentrations-the ECTIM study. *Hum Mol Genet*, 9:57–61.

Lambert, JC, Pasquier, F, Cottel, D, Frigard, B, Amouyel, P, and Chartier-Harlin, MC (1998). A new polymorphism in the APOE promoter associated with risk of developing Alzheimer's disease. *Hum Mol Genet*, 7:533–540.

Leeson, CP, Hingorani, AD, Mullen, MJ, Jeerooburkhan, N, Kattenhorn, M, Cole, TJ, Muller, DP, Lucas, A, Humphries, SE, and Deanfield, JE (2002). Glu298Asp endothelial nitric oxide synthase gene polymorphism interacts with environmental and dietary factors to influence endothelial function. *Circ Res*, 90:1153–1158.

Lehnart, SE, Wehrens, XH, Reiken, S, Warrier, S, Belevych, AE, Harvey, RD, Richter, W, Jin, SL, Conti, M, and Marks, AR (2005). Phosphodiesterase 4D deficiency in the ryanodine-receptor complex promotes heart failure and arrhythmias. *Cell*, 123:25–35.

Liu, PY, Chen, JH, Li, YH, Wu, HL, and Shi, GY (2003). Synergistic effect of stromelysin-1 (matrix metallo-proteinase-3) promoter 5A/6A polymorphism with smoking on the onset of young acute myocardial infarction. *Thromb Haemost*, 90:132–139.

Liu, PY, Li, YH, Tsai, WC, Tsai, LM, Chao, TH, Wu, HL, and Chen, JH (2005). Stromelysin-1 promoter 5A/6A polymorphism is an independent genetic prognostic risk factor and interacts with smoking cessation after index premature myocardial infarction. *J Thromb Haemost*, 3:1998–2005.

Liu, S, Ma, J, Ridker, PM, Breslow, JL, and Stampfer, MJ (2003). A prospective study of the association between APOE genotype and the risk of myocardial infarction among apparently healthy men. *Atherosclerosis*, 166:323–329.

Lohmussaar, E, Gschwendtner, A, Mueller, JC, Org, T, Wichmann, E, Hamann, G, Meitinger, T, and Dichgans, M (2005). ALOX5AP gene and the PDE4D gene in a central European population of stroke patients. *Stroke*, 36:731–736.

Lusis, AJ, Mar, R, and Pajukanta, P (2004). Genetics of atherosclerosis. *Annu Rev Genomics Hum Genet*, 5:189–218.

Marian, AJ and Roberts, R (2001). The molecular genetic basis for hypertrophic cardiomyopathy. *J Mol Cell Cardiol*, 33:655–670.

Meschia, JF, Brott, TG, Brown, RD, Jr, Crook, R, Worrall, BB, Kissela, B, et al. (2005). Phosphodiesterase 4D and 5-lipoxygenase activating protein in ischemic stroke. *Ann Neurol*, 58:351–361.

Moghadasian, MH, McManus, BM, Nguyen, LB, Shefer, S, Nadji, M, Godin, DV, et al. (2001). Pathophysiology of apolipoprotein E deficiency in mice: Relevance to apo E-related disorders in humans. *FASEB J*, 15:2623–2630.

Moncada, S and Higgs, A (1993). The L-arginine-nitric oxide pathway. *N Engl J Med*, 329:2002–2012.

Motulsky, AG and Brunzell, JD (2002). Genetics of coronary atherosclerosis. The genetic basis of common diseases. In RA King, JI Rotter and AG Motulsky (eds.). Oxford, UK, Oxford University Press..

Mulder, M, Lombardi, P, Jansen, H, van Berkel, TJ, Frants, RR, and Havekes, LM (1993). Low density lipoprotein receptor internalizes low density and very low density lipoproteins that are bound to heparan sulfate proteoglycans via lipoprotein lipase. *J Biol Chem*, 268:9369–9375.

Nakayama, T, Asai, S, Sato, N, and Soma, M (2006). Genotype and haplotype association study of the STRK1 region on 5q12 among Japanese: A case–control study. *Stroke*, 37:69–76.

Newby, AC, Southgate, KM, and Davies, M (1994). Extracellular matrix degrading metalloproteinases in the pathogenesis of arteriosclerosis. *Basic Res Cardiol*, 89(Suppl 1):59–70.

Newton-Cheh, C, Larson, MG, Corey, DC, Benjamin, EJ, Herbert, AG, Levy, D, D'Agostino, RB, and O'Donnell, CJ (2005). QT interval is a heri—quantitative trait with evidence of linkage to chromosome 3 in a genome-wide linkage analysis: The Framingham Heart Study. *Heart Rhythm*, 2:277–284.

Nguyen-Tran, VT, Kubalak, SW, Minamisawa, S, Fiset, C, Wollert, KC, Brown, AB, et al. (2000). A novel genetic pathway for sudden cardiac death via defects in the transition between ventricular and conduction system cell lineages. *Cell*, 102:671–682.

Nilsson-Ardnor, S, Wiklund, PG, Lindgren, P, Nilsson, AK, Janunger, T, Escher, SA, Hallbeck, B, Stegmayr, B, Asplund, B, and Holmberg, D (2005). Linkage of ischemic stroke to the PDE4D region on 5q in a Swedish population. *Stroke*, 36:1666–1671.

Nojiri, T, Morita, H, Imai, Y, Maemura, K, Ohno, M, Ogasawara, K, et al. (2003). Genetic variations of matrix metalloproteinase-1 and -3 promoter regions and their associations with susceptibility to myocardial infarction in Japanese. *Int J Cardiol*, 92:181–186.

O'Brien, KD, McDonald, TO, Chait, A, Allen, MD, and Alpers, CE (1996). Neovascular expression of E-selectin, intercellular adhesion molecule-1, and vascular cell adhesion molecule-1 in human atherosclerosis and their relation to intimal leukocyte content. *Circulation*, 93:672–682.

Okin, PM, Devereux, RB, Howard, BV, Fabsitz, RR, Lee, ET, and Welty, TK (2000). Assessment of QT interval and QT dispersion for prediction of all-cause and cardiovascular mortality in American Indians: The Strong Heart Study. *Circulation*, 101:61–66.

Ordovas, JM (1999). The genetics of serum lipid responsiveness to dietary interventions. *Proc Nutr Soc*, 58:171–187.

Ordovas, JM and Schaefer, EJ (2000). Genetic determinants of plasma lipid response to dietary intervention: The role of the APOA1/C3/A4 gene cluster and the APOE gene. *Br J Nutr*, 83(Suppl 1):S127–S136.

Ozaki, K, Inoue, K, Sato, H, Iida, A, Ohnishi, Y, Sekine, A, et al. (2004). Functional variation in LGALS2 confers risk of myocardial infarction and regulates lymphotoxin-alpha secretion in vitro. *Nature*, 429:72–5.

Ozaki, K, Ohnishi, Y, Iida, A, Sekine, A, Yamada, R, Tsunoda, T, Sato, H, Hori, M, Nakamura, Y, and Tanaka, T (2002). Functional SNPs in the lymphotoxin-alpha gene that are associated with susceptibility to myocardial infarction. *Nat Genet*, 32:650–654.

Paavonen, KJ, Chapman, H, Laitinen, PJ, Fodstad, H, Piippo, K, Swan, H, Toivonen, L, Viitasalo, M, Kontula, K, and Pasternack, M (2003). Functional characterization of the common amino acid 897 polymorphism of the cardiac potassium channel KCNH2 (HERG). *Cardiovasc Res*, 59:603–611.

Persu, A, Stoenoiu, MS, Messiaen, T, Davila, S, Robino, C, El-Khattabi, O, et al. (2002). Modifier effect of ENOS in autosomal dominant polycystic kidney disease. *Hum Mol Genet*, 11:229–241.

Pfeufer, A, Jalilzadeh, S, Perz, S, Mueller, JC, Hinterseer, M, Illig, T, et al. (2005). Common variants in myocardial ion channel genes modify the QT interval in the general population: Results from the KORA study. *Circ Res*, 96:693–701.

Pietila, E, Fodstad, H, Niskasaari, E, Laitinen, PP, Swan, H, Savolainen, M, Kesaniemi, YA, Kontula, K, and Huikuri, HV (2002). Association between HERG K897T polymorphism and QT interval in middle-aged Finnish women. *J Am Coll Cardiol*, 40:511–514.

Poort, SR, Rosendaal, FR, Reitsma, PH, and Bertina, RM (1996). A common genetic variation in the 3′-untranslated region of the prothrombin gene is associated with elevated plasma prothrombin levels and an increase in venous thrombosis. *Blood*, 88:3698–3703.

Priori, SG and Napolitano, C. (2004). Genetics of cardiac arrhythmias and sudden cardiac death. *Ann N Y Acad Sci*, 1015:96–110.

PROCARDIS-Consortium (2004). A trio family study showing association of the lymphotoxin-alpha N26 (804A) allele with coronary artery disease. *Eur J Hum Genet*, 12:770–774.

Radomski, MW and Salas, E (1995). Nitric oxide—biological mediator, modulator and factor of injury: Its role in the pathogenesis of atherosclerosis. *Atherosclerosis*, 118(Suppl):S69–S80.

Rauramaa, R, Vaisanen, SB, Luong, LA, Schmidt-Trucksass, A, Penttila, IM, Bouchard, C, Toyry, J, and Humphries, SE (2000). Stromelysin-1 and interleukin-6 gene promoter polymorphisms are determinants of asymptomatic carotid artery atherosclerosis. *Arterioscler Thromb Vasc Biol*, 20:2657–2662.

Rigat, B, Hubert, C, Alhenc-Gelas, F, Cambien, F, Corvol, P, and Soubrier, F (1990). An insertion/deletion polymorphism in the angiotensin I-converting enzyme gene accounting for half the variance of serum enzyme levels. *J Clin Invest*, 86:1343–1346.

Rundek, T, Elkind, MS, Pittman, J, Boden-Albala, B, Martin, S, Humphries, SE, Juo, SH, and Sacco, RL (2002). Carotid intima-media thickness is associated with allelic variants of stromelysin-1, interleukin-6, and hepatic lipase genes: The Northern Manhattan Prospective Cohort Study. *Stroke*, 33:1420–1423.

Saarialho-Kere, UK, Pentland, AP, Birkedal-Hansen, H, Parks, WC, and Welgus, HG (1994). Distinct populations of basal keratinocytes express stromelysin-1 and stromelysin-2 in chronic wounds. *J Clin Invest*, 94:79–88.

Saleheen, D, Bukhari, S, Haider, SR, Nazir, A, Khanum, S, Shafqat, S, Anis, MK, and Frossard, P (2005). Association of phosphodiesterase 4D gene with ischemic stroke in a Pakistani population. *Stroke*, 36:2275–2277.

Savvidou, MD, Vallance, PJ, Nicolaides, KH, and Hingorani, AD (2001). Endothelial nitric oxide synthase gene polymorphism and maternal vascular adaptation to pregnancy. *Hypertension*, 38:1289–1293.

Schaefer, EJ (2002). Lipoproteins, nutrition, and heart disease. *Am J Clin Nutr*, 75:191–212.

Schaefer, EJ, Lamon-Fava, S, Johnson, S, Ordovas, JM, Schaefer, MM, Castelli, WP, and Wilson, PW (1994). Effects of gender and menopausal status on the association of apolipoprotein E phenotype with plasma lipoprotein levels. Results from the Framingham Offspring Study. *Arterioscler Thromb*, 14:1105–1113.

Schouten, EG, Dekker, JM, Meppelink, P, Kok, FJ, Vandenbroucke, JP, and Pool, J (1991). QT interval prolongation predicts cardiovascular mortality in an apparently healthy population. *Circulation*, 84:1516–1523.

Sebastiani, P, Ramoni, MF, Nolan, V, Baldwin, CT, and Steinberg, MH (2005). Genetic dissection and prognostic modeling of overt stroke in sickle cell anemia. *Nature Genet*, 37:435–440.

Shah, PK, Falk, E, Badimon, JJ, Fernandez-Ortiz, A, Mailhac, A, Villareal-Levy, G, Fallon, JT, Regnstrom, J, and Fuster, V (1995). Human monocyte-derived macrophages induce collagen breakdown in fibrous caps of atherosclerotic plaques. Potential role of matrix-degrading metalloproteinases and implications for plaque rupture. *Circulation*, 92:1565–1569.

Sharp, DS, Masaki, K, Burchfiel, CM, Yano, K, and Schatz, IJ (1998). Prolonged QTc interval, impaired pulmonary function, and a very lean body mass jointly predict all-cause mortality in elderly men. *Ann Epidemiol*, 8:99–106.

Song, Y, Stampfer, MJ, and Liu, S (2004). Meta-analysis: Apolipoprotein E genotypes and risk for coronary heart disease. *Ann Intern Med*, 141:137–147.

Spence, MS, McGlinchey, PG,. Patterson, CC, Allen, AR, Murphy, G, Bayraktutan, U, Fogarty, DG, Evans, AE, and McKeown, PP (2004). Endothelial nitric oxide synthase gene polymorphism and ischemic heart disease. *Am Heart J*, 148:847–851.

Splawski, I, Shen, J, Timothy, KW, Lehmann, MH, Priori, S, Robinson, JL, et al. (2000). Spectrum of mutations in long-QT syndrome genes. KVLQT1, HERG, SCN5A, KCNE1, and KCNE2. *Circulation*, 102:1178–1185.

Splawski, I, Timothy, KW, Tateyama, M, Clancy, CE, Malhotra, A, Beggs, AH, Cappuccio, FP, Sagnella, GA, Kass, RS, and Keating MT (2002). Variant of SCN5A sodium channel implicated in risk of cardiac arrhythmia. *Science*, 297:1333–1336.

Spooner, PM, Albert, C, Benjamin, EJ, Boineau, R, Elston, RC, George, AL, Jr, et al. (2001). Sudden cardiac death, genes, and arrhythmogenesis: Consideration of new population and mechanistic approaches from a National Heart, Lung, and Blood Institute workshop, Part II. *Circulation*, 103:2447–2452.

Stankovic, S, Glisic, S, and Alavanatic, D (2000). The effect of a gender difference in the apolipoprotein E gene DNA polymorphism on serum lipid levels in a Serbian healthy population. *Clin Chem Lab Med*, 38:539–544.

Stannard, AK, Riddell, DR, Sacre, SM, Tagalakis, AD, Langer, C, von Eckardstein, A, Cullen, P, Athanasopoulos, T, Dickson, G, and Owen, JS (2001). Cell-derived apolipoprotein E (ApoE) particles inhibit vascular cell adhesion molecule-1 (VCAM-1) expression in human endothelial cells. *J Biol Chem*, 276:46011–46016.

Stengard, JH, Zerba, KE, Pekkanen, J, Ehnholm, C, Nissinen, A, and Sing, CF (1995). Apolipoprotein E polymorphism predicts death from coronary heart disease in a longitudinal study of elderly Finnish men. *Circulation*, 91:265–269.

Subramanian, SV and Nadal-Ginard, B (1996). Early expression of the different isoforms of the myocyte enhancer factor-2 (MEF2) protein in myogenic as well as non-myogenic cell lineages during mouse embryogenesis. *Mech Dev*, 57:103–112.

Suehiro, T, Morita, T, Inoue, M, Kumon, Y, Ikeda, Y, and Hashimoto, K (2004). Increased amount of the angiotensin-converting enzyme (ACE) mRNA originating from the ACE allele with deletion. *Hum Genet*, 115:91–96.

Szolnoki, Z, Somogyvari, F, Kondacs, A, Szabo, M, and Fodor, L (2002). Evaluation of the interactions of common genetic mutations in stroke subtypes. *J Neurol*, 249:1391–1397.

Talmud, PJ, Bujac, SR, Hall, S, Miller, GJ, and Humphries, SE (2000). Substitution of asparagine for aspartic acid at residue 9 (D9N) of lipoprotein lipase markedly augments risk of ischaemic heart disease in male smokers. *Atherosclerosis*, 149:75–81.

Terashima, M, Akita, H, Kanazawa, K, Inoue, N, Yamada, S, Ito, K, et al. (1999). Stromelysin promoter 5A/6A polymorphism is associated with acute myocardial infarction. *Circulation*, 99:2717–2719.

Tesauro, M, Thompson, WC, Rogliani, P, Qi, L, Chaudhary, PP, and Moss, J (2000). Intracellular processing of endothelial nitric oxide synthase isoforms associated with differences in severity of cardiopulmonary diseases: Cleavage of proteins with aspartate vs. glutamate at position 298. *Proc Natl Acad Sci USA*, 97:2832–2835.

Tsukada, T, Yokoyama, K, Arai, T, Takemoto, F, Hara, S, Yamada, A, Kawaguchi, Y, Hosoya, T, and Igari, J (1998). Evidence of association of the ecNOS gene polymorphism with plasma NO metabolite levels in humans. *Biochem Biophys Res Commun*, 245:190–193.

Utermann, G, Hees, M, and Steinmetz, A (1977). Polymorphism of apolipoprotein E and occurrence of dysbetalipoproteinaemia in man. *Nature*, 269:604–607.

van Rijn, MJ, Slooter, AJ, Schut, AF, Isaacs, A, Aulchenko, YS, Snijders, PJ, Kappelle, LJ, van Swieten, JC, Oostra, BA, and van Duijn, CM (2005). Familial aggregation, the PDE4D gene, and ischemic stroke in a genetically isolated population. *Neurology*, 65:1203–1209.

van Venrooij, FV, Stolk, RP, Banga, JD, Sijmonsma, TP, van Tol, A, Erkelens, DW, and Dallinga-Thie, GM (2003). Common cholesteryl ester transfer protein gene polymorphisms and the effect of atorvastatin therapy in type 2 diabetes. *Diabetes Care*, 26:1216–1223.

Wald, DS, Law, M, and Morris, JK (2002). Homocysteine and cardiovascular disease: Evidence on causality from a meta-analysis. *BMJ*, 325:1202.

Wang, L, Fan, C, Topol, SE, Topol, EJ, and Wang, Q (2003). Mutation of MEF2A in an inherited disorder with features of coronary artery disease. *Science*, 302:1578–1581.

Wang, XL, Sim, AS, Wang, MX, Murrell, GA, Trudinger, B, and Wang, J (2000). Genotype dependent and cigarette specific effects on endothelial nitric oxide synthase gene expression and enzyme activity. *FEBS Lett*, 471:45–50.

Watkins, H and Farrall, M (2006). Genetic susceptibility to coronary artery disease: From promise to progress. *Nat Rev Genet*, 7:163–173.

Weng, L, Kavaslar, N, Ustaszewska, A, Doelle, H, Schackwitz, W, Hebert, S, Cohen, JC, McPherson, R, and Pennacchio, LA (2005). Lack of MEF2A mutations in coronary artery disease. *J Clin Invest*, 115:1016–1020.

Wilcox, JN, Subramanian, RR, Sundell, CL, Tracey, WR, Pollock, JS, Harrison, DG, and Marsden, PA (1997). Expression of multiple isoforms of nitric oxide synthase in normal and atherosclerotic vessels. *Arterioscler Thromb Vasc Biol*, 17:2479–2488.

Wilhelm, SM, Collier, IE, Kronberger, A, Eisen, AZ, Marmer, BL, Grant, GA, Bauer, EA, and Goldberg, GI (1987). Human skin fibroblast stromelysin: Structure, glycosylation, substrate specificity, and differential expression in normal and tumorigenic cells. *Proc Natl Acad Sci USA*, 84:6725–6729.

Wittrup, HH, Andersen, RV, Tybjaerg-Hansen, A, Jensen, GB, and Nordestgaard, BG (2006). Combined analysis of six lipoprotein lipase genetic variants on triglycerides, high-density lipoprotein and ischemic heart disease: Cross-sectional, prospective and case–control studies from The Copenhagen City Heart Study. *J Clin Endocrinol Metab*, 91:1438–1445.

Wittrup, HH, Tybjaerg-Hansen, A, and Nordestgaard, BG (1999). Lipoprotein lipase mutations, plasma lipids and lipoproteins, and risk of ischemic heart disease. A meta-analysis. *Circulation*, 99:2901–2907.

Woo, D, Kaushal, R, Kissela, B, Sekar, P, Wolujewicz, M, Pal, P, et al. (2006). Association of Phosphodiesterase 4D with ischemic stroke: A population-based case-control study. *Stroke*, 37:371–376.

Yamada, A, Ichihara, S, Murase, Y, Kato, T, Izawa, H, Nagata, K, Murohara, T, Yamada, Y, and Yokota, M (2004). Lack of association of polymorphisms of the lymphotoxin alpha gene with myocardial infarction in Japanese. *J Mol Med*, 82:477–483.

Ye, S (2006). Influence of matrix metalloproteinase genotype on cardiovascular disease susceptibility and outcome. *Cardiovasc Res*, 69:636–645.

Ye, S, Dunleavey, L, Bannister, W, Day, LB, Tapper, W, Collins, AR, Day, IN, and Simpson, I (2003). Independent effects of the −219 G>T and epsilon 2/ epsilon 3/ epsilon 4 polymorphisms in the apolipoprotein E gene on coronary artery disease: The Southampton Atherosclerosis Study. *Eur J Hum Genet*, 11:437–443.

Ye, S, Eriksson, P, Hamsten, A, Kurkinen, M, Humphries, SE, and Henney, AM (1996). Progression of coronary atherosclerosis is associated with a common genetic variant of the human stromelysin-1 promoter, which results in reduced gene expression. *J Biol Chem*, 271:13055–13060.

Ye, S, Watts, GF, Mandalia, S, Humphries, SE, and Henney, AM (1995). Preliminary report: Genetic variation in the human stromelysin promoter is associated with progression of coronary atherosclerosis. *Br Heart J*, 73:209–215.

Yen, FT, Deckelbaum, RJ, Mann, CJ, Marcel, YL, Milne, RW, and Tall, AR (1989). Inhibition of cholesteryl ester transfer protein activity by monoclonal antibody. Effects on cholesteryl ester formation and neutral lipid mass transfer in human plasma. *J Clin Invest*, 83:2018–2024.

Zdravkovic, S, Wienke, A, Pedersen, NL, Marenberg, ME, Yashin, AI, and De Faire, U (2002). Heritability of death from coronary heart disease: A 36-year follow-up of 20 966 Swedish twins. *J Intern Med*, 252:247–254.

Zheng, Z, Croft, JB, Giles, WH, Ayala, CI, Greenlund, KJ, Keenan, NL, Neff, L, Wattigney, WA, Mensah, GA (2002). State-specific mortality from sudden cardiac death—United States, 1999. *Mortal Morb Wkly Rep*, 51:123–126.

Zhou, X, Huang, J, Chen, J, Su, S, Chen, R, and Gu, D (2004). Haplotype analysis of the matrix metalloproteinase 3 gene and myocardial infarction in a Chinese Han population. The Beijing atherosclerosis study. *Thromb Haemost*, 92:867–873.

12

Diabetes Mellitus and Obesity

Mark I McCarthy

An Epidemiological Perspective on Diabetes Mellitus and Obesity

Increasingly, the twin threats of diabetes and obesity are perceived as amongst the most important global contributors to human disease and death (Kopelman, 2000; Zimmet et al., 2001). The rise in the prevalence of both conditions has been staggering. Over the course of less than a generation—from 1995 to 2010—the total number of people with diabetes is projected to double (from approximately 110 to 220 million) (Fig. 12–1).

This burden of increased prevalence will fall disproportionately in the societies least able to deal with the social and economic costs of this new "epidemic," since prevalence rates are climbing fastest in Asia, Africa, and South America (Zimmet et al., 2001). These increases in diabetes prevalence are following similar trends in rates of obesity. According to the latest figures from the National Health and Nutrition Examination Survey in the USA, 65% of adults are overweight [body mass index (BMI) exceeding 25 kg m – 2], 30% obese (BMI > 30), and 5% extremely obese (BMI > 40) (Hedley et al., 2004).

Etiological Understanding as a Prelude to Better Management

As with many other common non-infectious causes of human morbidity and mortality, the etiological pathways responsible for the development of diabetes and obesity are, in most cases, obscure. At the same time, the growing prevalence of these conditions, and the burden of morbidity associated with them, provides the clearest indication possible that current therapeutic and preventative options remain woefully inadequate. Although much can be done to reduce the risk of serious complications, treatment for diabetes is almost never curative, and rates of diabetes-related complications continue to rise in many countries. The challenges associated with obtaining sustained weight reduction in those who have become overweight is, of course, a matter of common discourse, and provides the drive for a billion dollar industry based on the promise of ever more ingenious methods for achieving and maintaining weight reduction.

There is a clear need for novel therapeutic and preventative approaches for both conditions, as well as more rationale and effective use of the agents already available. An improved understanding of the etiological pathways involved in disease development and progression will provide an essential underpinning to such therapeutic developments. Advances in genetics and genomics offer much promise in this regard, not only in identifying the etiological pathways that are common to most people with diabetes and obesity (and which will form the substrate for public health measures and for the development of novel therapeutic agents for widespread use) but also in moving toward an appreciation of etiology at the individual level. The latter, of course, is an essential prerequisite for future successes in the application of personalized "genomic" medicine in this disease area (Bell, 2003).

Different Patterns of Disease

When considering "diabetes," it is important to recognize that there are, broadly speaking, two main subtypes of disease (see Table 12–1) (The Expert Committee on the Diagnosis and Classification of Diabetes Mellitus, 2003). A small, but significant, proportion of diabetes (so-called type 1) is caused by a permanent and irrevocable loss of insulin production from the β-cells of the pancreatic islets. Most type 1 diabetes is due to the development of islet autoimmunity, leading to a gradual but inexorable immune-mediated loss of β-cell capacity. Much is known about the molecular pathways responsible, and several of the major genes involved have been identified (*HLA, INS, PTPN22,* and *CTLA4*). This subtype of diabetes will not be discussed further in this chapter, and the interested reader is referred to a number of excellent reviews and papers (Bennett and Todd, 1996; Pociot and McDermott, 2002; Ueda et al., 2003; Smyth et al., 2004).

This chapter will focus on the intimately related conditions of type 2 diabetes and obesity and, to some extent, on other "partners-in-crime" within the so-called metabolic syndrome (alternatively termed the insulin resistance syndrome, Reaven syndrome, and Syndrome X). Obesity is a major risk factor for the development of type 2 diabetes, and for other metabolic syndrome components including hypertension, dyslipidemias, and cardiovascular disease (Eckel et al., 2005). Although the precise composition of the metabolic syndrome has become controversial (Gale, 2005), the concept retains the merit of emphasizing that many of these conditions coexist more frequently than one would expect by chance, and that there are likely to be

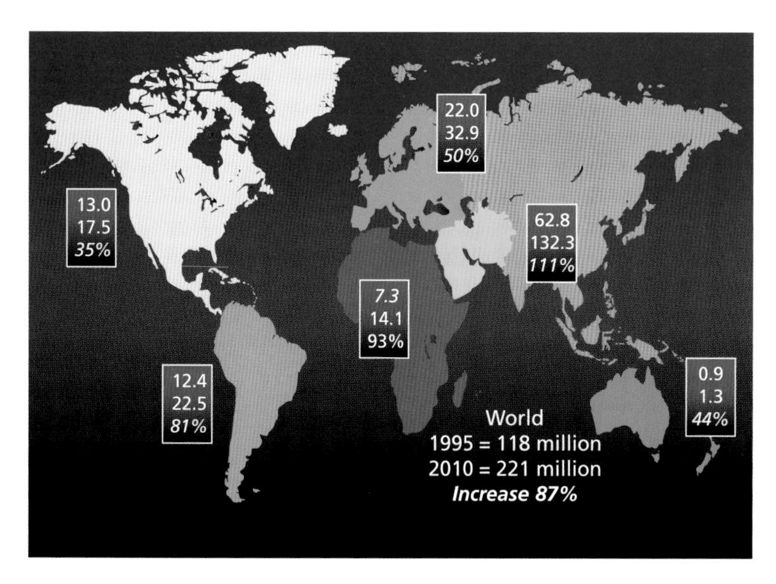

Figure 12–1 The projected global diabetic epidemic, 1995–2010.

common antecedents (including insulin resistance, see next section) (Ferranini et al., 1991). We will also discuss some of the known genetic causes of diabetes such as maturity-onset diabetes of the young (MODY). Though these are, for the purposes of disease classification (Table 12–1), regarded as distinct from type 2 diabetes, they have substantial clinical and genetic overlap with this condition.

Pathogenesis of Diabetes and Obesity

The Physiological Basis for Type 2 Diabetes

Globally, more than 95% of diabetes is type 2. Whilst this remains, to some degree, a diagnosis of exclusion (since there are no specific diagnostic markers for this subtype), the cardinal features are similar across different ethnic groups (The Expert Committee on the Diagnosis and Classification of Diabetes Mellitus, 2003). Those key features include: typical onset in middle age or later in life [though cases in children are increasingly reported (Ehtisham et al., 2001)]; a strong association with obesity (though it can certainly occur in the lean); and a physiological picture that can be described as an inadequate insulin secretory response to prevailing insulin resistance (Stumvoll et al., 2005). Indeed, whilst there has been a longstanding debate over the relative etiological importance of insulin resistance (i.e., a reduced capacity of target tissues—such as liver, fat, and muscle—to produce an adequate metabolic response to circulating insulin) and insulin secretion (i.e., the capacity of the pancreatic β-cells to release insulin in response to glycemic and other stimuli) in the pathogenesis of type 2 diabetes, it is clear that in most patients both are involved (Reaven et al., 1993). Furthermore, when one looks in groups at high future risk of diabetes (such as first-degree relatives of those with diabetes, or women with gestational diabetes) whilst they are still normoglycemic, abnormalities of both insulin secretion and action can be detected (Warram et al., 1990; Ryan et al., 1995; van Haeften et al., 1998).

Using data from both longitudinal and cross-sectional studies of the progression from normal glucose metabolism to overt

Table 12–1 Current Classification of Diabetes Subtypes

	Subtypes	Examples/Comments
I: Type 1 diabetes (β-cell destruction, usually leading to absolute insulin deficiency)		
IA	Immune-mediated	Most "classical" type 1 diabetes
IB	Idiopathic	
II: Type 2 diabetes (ranging from predominant defects of insulin action to one of predominant insulin secretion)		
III: Other specific types		
III A	Genetic defects of β-cell function	MODY, mitochondrial diabetes
III B	Genetic defects of insulin action	Leprechaunism
III C	Diseases of exocrine pancreas	Pancreatitis
III D	Endocrinopathies	Cushing's syndrome, acromegaly
III E	Drug- or chemical-induced infection	
III F		
III G	Uncommon forms of immune-mediated diabetes	
III H	Other genetic syndromes sometimes associated with diabetes	Down syndrome, Wolfram syndrome

Source: Summarized from The Expert Committee on the Diagnosis and Classification of Diabetes Mellitus (2003.)

Abbreviation: MODY, maturity-onset diabetes of the young.

diabetes, it is possible to define the natural history of the condition (Saad et al., 1988; Martin et al., 1992). Early stages of glucose intolerance are associated with an enhanced (but by definition inadequate) attempt by the β-cells to secrete additional insulin to maintain normal metabolic control in the face of prevailing insulin resistance (itself, typically the result of a combination of ageing and obesity and genetic predisposition—see below). Subsequent deterioration in this β-cell response (often glibly referred to as β-cell "exhaustion" though the specific mechanisms are now increasingly understood) leads to progressively diminished insulin production and spiraling hyperglycemia. Whilst such a description would appear to suggest that the insulin resistance component is primary, this in part reflects the relative ease with which insulin sensitivity can be measured as compared to β-cell mass and other pertinent parameters of insulin secretory capacity. Indeed, a wide range of genetic evidence suggests that defects in β-cell functional capacity (too subtle to be detected until exposed by the growing demands imposed by insulin resistance) may account for the majority of the inherited component to diabetes susceptibility (Stumvoll et al., 2005).

The Physiological Basis for Obesity

All too often, obesity is still seen as reflecting a failure of individual willpower to control dietary intake (Friedman, 2004). However, such a view is hard to square with evidence of the strong genetic basis of obesity (see later) as well as metabolic differences between obese and non-obese individuals, which generate a positive energy balance independent of the differences in food intake (Brolin, 2002). Evidence from mouse models of obesity, as well as human monogenic conditions (see below), is generating powerful insights into the specific mechanisms responsible for individual differences in the response to the increasingly obesogenic environment that prevails in many parts of the world, an environment characterized by caloric (over) sufficiency, a shift to energy-dense foods, and reduced volitional energy expenditure (Zimmet et al., 2001).

Central to this improved understanding of the pathogenesis of obesity has been a clearer view of the role of the hypothalamus. Whilst the importance of this area of the brain for control of appetite and body weight homeostasis has long been appreciated, it is only in recent years that the neuroendocrine pathways involved have been defined (Bell et al., 2005). The hypothalamus integrates a number of afferent outputs emanating from the periphery. Some of these are orexigenic, such as ghrelin and cannabinoids, others anorexigenic, including leptin, insulin, and peptide YY (PYY). Collectively, these provide the hypothalamus with a series of complementary (acute and chronic) measures of food intake, fat mass, and the like. Orexigenic signals are conveyed through neurones releasing neuropeptide Y (NPY), agouti gene-related peptide (AGRP), and melanin-concentrating hormone (MCH): the complementary network of α-melanocyte stimulating hormone (α-MSH) and cocaine- and amphetamine-regulated transcript (CART) neurones conveys the anorexigenic signals. The results of this integration generate signals sent out through a range of effector mechanisms that either stimulate or inhibit feeding, and/or modify energy expenditure. Genetic (or other) manipulations of this complex mechanism are known to modify food intake (see below).

This system is generally very efficient in maintaining weight around an individual set-point. Even in the obese, the system remains highly active, with volitional weight reduction (i.e., dieting) leading to stimulation of orexigenic pathways (leading to increased hunger) and reduced energy expenditure. These responses seek to restore weight to the previous (high) levels, explaining the enormous difficulties with long-term weight loss reduction in the previously obese (Schwartz et al., 2003). Viewed in this way, obesity is the result of the exposure of individuals predisposed to have a higher body mass set-point to the environmental conditions that permit sustained energy accumulation (Friedman, 2004). The mechanisms underlying that higher set-point remain unclear, though the fact that most obese individuals have elevated levels of the hormone leptin (released from adipose tissue in proportion to fat mass and usually a powerful anorexigenic stimulus) (Maffei et al., 1995), and display partial resistance to its effects on weight loss (Halaas et al., 1997), may provide some clues.

In understanding the genesis of obesity, it is also important to consider the expenditure part of the energy balance equation. In any given person, energy balance is usually regulated extremely tightly. Even the positive (im)balance represented by an annual weight gain of 1 kg, more than adequate to foster obesity if sustained over decades, represents only a small proportion of overall energy flux. There are, however, substantial individual differences in the extent of energy intake and expenditure, and those with lower energy expenditure are more likely to develop obesity in prospective studies (Ravussin et al., 1998). Again, through studies in man and rodents, the major players in the regulation of energy expenditure are coming to light, with an interesting focus on the molecules involved in the control of fatty acid oxidation, including acetyl coA decarboxylase, diacylglycerol acetyl transferase, and stearoyl-coA desaturase (Ntambi and Miyazaki, 2004). These enzymes are, in various ways, involved in pathways that regulate fatty acid entry into the mitochondria where they are oxidized, thereby translating fat mass into energy.

Obesity and Type 2 Diabetes: How are These Connected?

Whilst obesity and type 2 diabetes are intimately connected, the mechanisms that mediate this relationship remain controversial. Fat accumulation (particularly intra-abdominal fat) is associated with a substantial increase in insulin resistance, which, as discussed earlier, is one of the key features of type 2 diabetes. However, this insulin resistance is predominantly expressed in liver and muscle, which begs the question: how does the muscle and liver "know" about the excess fat deposition? Several complementary hypotheses are in play. First, when fat cells are overloaded, they tend to release more fatty acids into the circulation. Not only may those fatty acids provide an alternative source of fuel (reducing the requirement for the uptake and use of glucose) (Randle et al., 1965), it is now clear that one consequence is ectopic deposition of fat in tissues such as liver, muscle, and even the pancreatic islets (Yki-Jarvinen, 2002). Such local deposition of fat seems to be capable of generating local insulin resistance. Second, it has become clear over the past decade that fat tissue is far from an inert site of long-term fuel storage: instead, adipose tissue is a highly active endocrine organ, releasing a growing list of agents into the circulation (including leptin and adiponectin), which are capable of influencing the metabolic behavior of remote tissues (such as liver and muscle) (Caro et al., 1996; Arita et al., 1999).

Type 2 Diabetes and Obesity as Complex Traits: The Role of Environment

As with most common diseases, type 2 diabetes and obesity are best described as complex multifactorial traits. In other words, individual susceptibility to these conditions reflects an integration of multiple genetic and environmental factors, each of which has only a modest effect on risk.

Evidence for the environmental component of susceptibility is strong. The rapid secular changes in prevalence rates provide the most compelling argument, and also identify the likely culprits: increasing availability of cheap sources of energy-dense foods, and a shift to more sedentary lifestyles (Zimmet et al., 2001). These factors can be shown to be predictive of future diabetes and obesity (Manson et al., 1991), and potential avenues for successful intervention to reduce risk (Tuomilehto et al., 2001). It may be tempting to assume that the risk of obesity and diabetes is related not only to the volume of food consumed but also to its specific composition, but the evidence that this is so is, as yet, unconvincing (Hollis and Mattes, 2005).

Studies over the past decade have suggested an important role for early environment in the genesis of type 2 diabetes (and other components of the metabolic syndrome) in later life. The so-called fetal origins hypothesis arose out of observations that the prevalence of adult metabolic and cardiovascular disease was far greater in those with low birth weight due (presumably) to an adverse intrauterine environment. More specifically, disease rates were highest in those who were born small, but became obese as adults (Hales et al., 1991). These observations, confirmed in many populations worldwide, have important public health implications, including a potential explanation for the explosion in diabetes prevalence in many emerging societies (where large swathes of the population would have undergone the transition from early deprivation to a more secure and prosperous adulthood). The mechanism proposed to explain these life-course correlations is "programming": that is, the concept that early life deprivation leads to permanent changes in the metabolic repertoire of the individual concerned, and that it is these which predispose to future disease (Hales and Ozanne, 2003). There is abundant evidence to support this concept from animal models: human data are more limited but clearly indicate that such mechanisms also operate in man (Poulsen et al., 1997; Ravelli et al., 1998). However, it is worth pointing out that a complementary explanation for such associations between poor early growth and subsequent adult disease could be the pleiotropic effects of genes that influence both phenotypes (Hattersley and Tooke, 1999). Associations between paternal metabolic/cardiovascular disease and low offspring birth weight are consistent with the notion of transmission of genetic variants able to influence both processes (Hypponen et al., 2003). Obvious candidates for involvement in this process include genes involved in insulin secretion and action (insulin is the major growth factor in early life), since abnormalities in either could result in poor early growth and subsequent diabetes. Evidence from some monogenic forms of diabetes due to known mutations in the insulin secretion pathway provides proof of principle (Hattersley et al., 1998).

Type 2 Diabetes and Obesity as Complex Traits: The Role of Genes

Notwithstanding the compelling evidence for the contribution of environment, it is absolutely clear that genetic factors play a substantial role in both phenotypes. Indeed, obesity seems to be one of the most heritable traits known with a series of twin, adoption, and familial studies pointing to heritability estimates as high as 0.8 (Stunkard, Foch, et al., 1986; Stunkard, Sorensen, et al., 1986; Stunkard et al., 1990; Allison, Kaprio, et al., 1996). Heritability estimates for diabetes and diabetes-related phenotypes are more variable (partly because of the greater difficulties in assigning diabetes status and in trait measurement) (Newman et al., 1987; Kaprio et al., 1992), but derived measures of β-cell function in particular show a consistently high heritability (0.50–0.80) across family and twin studies (Stumvoll et al., 2005). Robust data on familial aggregation of type 2 diabetes are limited, but the oft-quoted figure for the sibling relative risk (RR) (or λs, the ratio of risk in a sibling of a diabetic individual, compared to that of the baseline population) in European populations is between 3 and 4 (Kobberling and Tillil, 1982). The equivalent figure for severe obesity is on a similar scale (Allison, Faith, et al., 1996). Other indirect evidence for the role of genes in type 2 diabetes comes from admixture studies (for example, in Native American tribes where the extent of European admixture is negatively correlated with diabetes prevalence) (Brosseau et al., 1979; Williams et al., 2000), and from migration studies [notably, the high rates of diabetes in the South Asian diaspora (Simmons and Powell, 1993)].

Of course, the strongest evidence for a role of genes in a trait is the identification of the specific genes responsible: recent progress in this respect is described below.

Embracing Complexity: Genes, Genomes, and Environments

At this stage, we have little conception of the specific ways in which genes will interact with each other and with environmental exposures. Answers to this question will have to await a more complete assessment of the genetic architecture of these traits. However, in the meantime, it is worth emphasizing the potential complexity of the system. Whilst we speak of "environment," we should (Fig. 12–2) be thinking of environments: each of us charts a specific trajectory through a series of pertinent exposures at each stage of life, and "environment" represents some sort of integration of these different factors. And on the genetic side, where we are becoming accustomed to thinking about the effects of multiple variants in the index genome, we also need to be thinking about other genomes that can modify risk, particularly that of the mother. The maternal genome (which of course contributes to 50% of the offspring genome) also has the capacity, through the effects on the maternal phenotype, to influence fetal and childhood environment and thereby exert indirect effects on the development of the offspring (Weinberg et al., 1998). Disentangling these various components will not be a trivial matter.

Current Understanding of the Genetics of Type 2 Diabetes and Obesity

The Value of Studying Monogenic and Syndromic Forms of Disease

Progress in the identification of genes involved in the development of type 2 diabetes and obesity has, not surprisingly, been swiftest for the relatively small proportion of disease attributable to monogenic and/or syndromic forms. For such individuals, disease susceptibility is determined by the segregation of mutations at a single gene, and the approaches that have proven so effective in the identification of more than 1000 Mendelian diseases can be brought to bear (Peltonen and McKusick, 1991). Whether or not the same genes are also implicated in the more common, multifactorial forms of the disease (and in the case of diabetes there is increasing evidence that this is so), studies of such rare monogenic and syndromic subtypes provide powerful insights into the pathways and networks that are critical for normal homeostasis. For tissues for which access to human tissues for direct study is impossible or difficult (such as the hypothalamus and pancreatic islets), such "accidents of nature" provide one of the few feasible avenues for the convincing attribution of function in man.

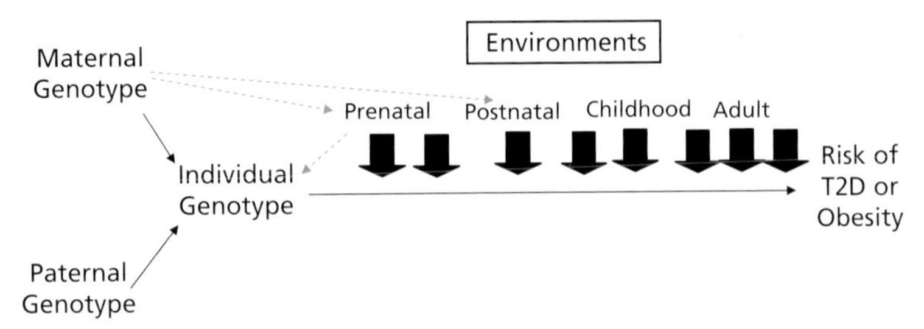

Figure 12–2 Genome–environmental interactions from a life-course perspective. Individual risk of disease represents a summation of predisposition encoded in an individual's genotype, combined with the integrated effect of pertinent environmental exposures at various stages of life. Maternal genotype contributes not only to individual genotype but also has substantial effects on early environment that may modify for example through changes in methylation patterns, the expression of the individual's genetic predisposition.

Monogenic and Syndromic Forms of Diabetes

Several different monogenic forms of diabetes are recognized, but most work has focused on maturity-onset diabetes of the young (MODY). The name harks back to former times when the two major subtypes of diabetes were described as "juvenile onset" (more or less equivalent to type 1) and "maturity onset" (type 2). It was recognized (Tattersall and Fajans, 1975) that there were certain pedigrees in which there was autosomal dominant segregation of a distinct form of diabetes characterized by early onset (typically below the age of 25) but insulin-independence (indicating that this was not type 1). It was also noted by astute clinicians at the time (Fajans, 1990) that there was considerable heterogeneity between MODY families with respect to the age at presentation, risk of complications, and requirement for therapy.

Positional cloning and candidate-based approaches to causal gene identification have, at the latest count, identified seven different genes responsible for MODY (see Table 12–2): collectively, these account for up to 90% of the families meeting strict criteria for MODY. Several important insights have flowed from these observations. First, six of the genes concerned (all except *CEL*) are now known to play important roles in the differentiation and/or function of pancreatic β-cells. This fits neatly with the clinical phenotype of MODY subjects who, unlike multifactorial type 2 diabetes, are characterized by relative insulin deficiency without insulin resistance. The recent identification of *CEL* as a seventh MODY gene is an intriguing development, since this gene is not known to be expressed in the β-cell: current indications are that this form of diabetes is initiated in the exocrine pancreas (Hattersley, 2006; Raeder et al., 2006). Second, clinical and phenotypic heterogeneity between MODY families can largely be explained by the differing molecular etiologies. Specifically, MODY due to glucokinase deficiency is associated with only modest hyperglycemia, detectable at birth, little deterioration in control over time, low risk of complications, and a high likelihood of successful control on diet alone (Owen and Hattersley, 2001). In contrast, MODY due to mutations in *TCF1* (hepatocyte-nuclear factor 1-α) is associated with onset in adolescence or early adulthood, more severe hyperglycemia, a higher rate of complications, a relatively poor response to diet but an exaggerated response to sulfonylureas as compared to type 2 diabetes (Pearson et al., 2003; Song et al., 2005). Since defining the precise molecular defect in an individual presenting with MODY conveys significant prognostic and therapeutic information for the patient (and their relatives), there is increasing use of molecular diagnostics in this situation (McCarthy and Hattersley, 2001). Third, several of these genes have now been implicated in susceptibility to multifactorial type 2 diabetes (see below).

A second, rather rarer, form of monogenic diabetes is neonatal diabetes (i.e., diabetes diagnosed before 6 months of age), historically divided into transient (TNDM or transient neonatal diabetes mellitus) and permanent (PNDM or permanent neonatal diabetes mellitus) forms. Approximately 80% of cases of TNDM are attributable to loss of imprinting in a region of chromosome 6q: the most likely candidate gene in this region (based on animal model studies) (Ma et al., 2004) is *ZAC* (aka *PLAGL1*: pleomorphic adenoma gene-like 1) (Arima et al., 2001). The clinical course of TNDM, as the name suggests, is for the diabetes to resolve during the early years of life, though such patients have a substantially increased risk of type 2 diabetes later. Until recently, the molecular basis of PNDM was not well understood, save for a few cases attributable to homozygous (or compound heterozygous) mutations in *GCK* (Gloyn, 2003). However, recent work has established that mutations in the *KCNJ11* gene (encoding a subunit of the β-cell KATP channel crucial for glucose-stimulated insulin secretion) are an important cause of both PNDM and TNDM (Gloyn et al., 2004; Gloyn et al., 2005). There is a strong relationship between the degree of impairment of channel function due to *KCNJ11* sequence variation (as measured in vitro) and the clinical phenotype. The mutations leading to TNDM are associated with a less extreme abrogation of channel function than those that lead to PNDM (Gloyn et al., 2005). In the case of the mutations with the most severe functional effects, the phenotypic consequences extend beyond those attributable to the disturbance of KATP channel activity in the islets, to include other syndromic manifestations (developmental delay, epilepsy, hypotonia) that reflect *KCNJ11* expression in other tissues. This has been termed the developmental delay, epilepsy, and neonatal diabetes (DEND) syndrome (Gloyn et al., 2004). At the other extreme, a common variant in *KCNJ11*, which results in a codon substitution (E23K), has been consistently shown to be a risk factor for common, multifactorial diabetes (Gloyn et al., 2003) (see below).

From the clinical point of view, the most dramatic consequence of this discovery has been the capacity to convert a large proportion of children suffering from PNDM from insulin therapy (most of these children have been misdiagnosed as having type 1 diabetes for which this is the standard treatment) to oral agents (such as sulfonylureas), a maneuver that has substantial clinical benefits for the patients and their families. This switch is possible since the mutated channels (which have lost the capacity to close in response to the adenosine triphosphate (ATP) signal normally generated by glucose metabolism) generally retain responsiveness to sulfonylureas (Zung et al., 2004). As with sulfonylurea sensitivity in *TCF1*-deficient MODY (Pearson et al., 2003), these findings in neonatal diabetes represent early examples of the value of genetic diagnostics to provide valuable therapeutic insights in diabetes (albeit in rare subtypes).

Table 12–2 Genes Implicated in Monogenic Forms of Diabetes (MODY and neonatal diabetes)

Gene ID	Gene Name	Condition	Notes	Reference
HNF4A	Hepatocyte nuclear factor 4-α	MODY (*MODY1*)	Rare cause of MODY	Yamagata, Oda, et al., 1996
GCK	Glucokinase	MODY (*MODY2*)	~20% of MODY (heterozygous)	Froguel et al., 1992; Hattersley et al., 1992
		PNDM	PNDM if homozygous	Gloyn, 2003
TCF1	Hepatocyte nuclear factor 1-α	MODY (*MODY3*)	Responsible for ~60% of MODY	Yamagata, Furuta, et al., 1996
IPF1	Insulin promoter factor 1	MODY (*MODY4*)	MODY when heterozygous	Stoffers et al., 1997
		Pancreatic agenesis	Agenesis when homozygous	
TCF2	Hepatocyte nuclear factor 1-β	MODY (*MODY5*)	RCAD (renal cysts and diabetes) (=MODY + genitourinary abnormalities)	Horikawa et al., 1997
NEUROD1	Neurogenic differentiation1; B-cell E-box Transactivator 2	MODY (*MODY6*)	Rare cause of MODY	Malecki et al., 1999
CEL	Carboxyl ester lipase	MODY (*MODY7*)	Rare cause of MODY	Raeder et al., 2006
KCNJ11		TNDM		Gloyn et al., 2005
		PNDM	~40% of TDNM	Gloyn et al., 2004
		DEND	Developmental delay, epilepsy and neonatal diabetes	Gloyn et al., 2004
ZAC region (*PLAGL1* and/or *HYMAI*)	Pleomorphic adenoma gene-like 1; ZAC tumor suppressor gene; hydatidiform mole associated and imprinted	TNDM	Loss of imprinting control	Arima et al., 2001

Abbreviations: DEND, developmental delay, epilepsy, and neonatal diabetes; MODY, maturity-onset diabetes of the young; TNDM, transient neonatal diabetes mellitus; PNDM, permanent neonatal diabetes mellitus.

There are several other monogenic and syndromic forms of diabetes for which the genetic basis is now clear (see McCarthy, 2004 for a review of these). Each of these provides an additional example of the mechanisms whereby glucose homeostasis can be disturbed and further insights into disease pathogenesis. Numerically, the most important of these other syndromes is that associated with variation at position 3243 in the mitochondrial genome, which may account for up to 3% of diabetes in some ethnic groups (Kadowaki et al., 1994). The clinical presentation of variation at this site ranges from diabetes alone (which may present clinically as type 1 or type 2 diabetes), via diabetes plus high-tone deafness [the so-called maternally inherited diabetes with deafness (MIDD) syndrome], to the full-blown mitochondrial syndrome known as mitochondrial encephalomyopathy, lactic acidosis, and stroke-like episodes (MELAS) (Velho et al., 1996; Suzuki et al., 2003). Since this variant tends to exhibit heteroplasmy (i.e., tissues will contain a mix of both normal and variant mitochondria), the precise pattern of disease in a given patient will depend principally on the chance segregation of abnormal mitochondria into different tissues during development (Maassen et al., 2004). The clinical relevance of a precise molecular diagnosis of mitochondrial diabetes includes prognostic information for the patient (an early need for insulin therapy and an increased risk of cardiomyopathy and neurological abnormalities) and their relatives (transmission is exclusively maternal; approximately 1 in 30 relatives will have the full MELAS syndrome) (Suzuki et al., 2003). The critical role of the mitochondria in maintaining normal glucose homeostasis, which is revealed by the 3243 variant, is consistent with other evidence implicating inherited and acquired deficiencies in oxidative phosphorylation in the development of diabetes (Mootha et al., 2003; Schrauwen and Hesselink, 2004).

Monogenic and Syndromic Forms of Obesity

There is a similar story to tell for obesity, with the genetic basis of several rare monogenic and syndromic forms now established. However, unlike diabetes, the relative paucity of effective treatments has (apart from the very few cases with congenital leptin deficiency) blunted the clinical use of molecular diagnostics. Nevertheless, these genetic discoveries have contributed substantially to our understanding of hypothalamic and neuroendocrine mechanisms involved in appetite control and weight maintenance (see above).

Nowhere is this more evident than with respect to mutations in the leptin gene and its receptor. The landmark discovery that mutations in the genes for leptin and its receptor are responsible for the obesity of the obese (ob/ob) and diabetic (db/db) mice, respectively (Zhang et al., 1994, Tartaglia et al., 1995), provided novel pathways for study in man. However, definitive proof of their contribution to human physiology was only provided with the identification of mutations in the human homologs in obese children (Montague et al., 1997; Clement et al., 1998). Though extremely rare, identification of the molecular basis of diabetes in children with homozygous leptin gene mutations (who have absolute leptin deficiency), has made it possible for them to obtain dramatic and sustained clinical benefits (reduced hunger, substantial weight loss, restoration of normal endocrine function) through administration of exogenous leptin (Farooqi et al., 1999). Heterozygous relatives of these children have low leptin levels and an increased prevalence of obesity (Farooqi et al., 2001). Individuals with mutations in the leptin receptor show similar obesity (combined with profound growth retardation), and do not of course respond to exogenous leptin (Clement et al., 1998).

Unraveling of the hypothalamic neuronal network (see above) and work in animal models have highlighted additional candidates since proven to be responsible for isolated cases of profound early-onset obesity. Alpha-MSH [derived by processing of the product of the proopiomelanocorticotropin (*POMC*) gene] is known to be a potent anorexigenic signal through its actions on melanocortin 4 receptor (MC4R)-neurones in the hypothalamus. A few cases of *POMC* mutations have been described (Krude et al., 1998): as well as obesity and hyperphagia, such individuals display other expected consequences of loss of all POMC-derived peptides (including adrenal insufficiency and red hair pigmentation). Mutations in the proprotein convertase subtilisin kexin-1 (*PCSK1*) gene, encoding a prohormone convertase responsible for the normal processing of various peptide hormones, including POMC, are also capable of producing marked early obesity (Jackson et al., 1997).

All the mutations described earlier are vanishingly rare. By contrast, mutations within the *MC4R* gene (encoding the target of α-MSH) seem to account for a significant proportion (up to 5%) of severe obesity (Vaisse et al., 1998; Yeo et al., 1998; Larsen et al., 2005). The majority of these cases are heterozygotes, with occasional homozygotes showing more marked obesity (Farooqi et al., 2003). Interestingly, the phenotypic effect of *MC4R* mutations seems most pronounced in childhood, when there is a fairly clear relationship between the degree of obesity and the in vitro properties of the mutation concerned (Farooqi et al., 2003). As carriers move into adulthood, the phenotype becomes less dramatic and the clinical overlap with multifactorial obesity becomes greater.

Finally, there are more than 20 rare syndromes for which obesity is a characteristic component of the phenotype. Usually, the constellation of ancillary syndromic components provides the specific diagnosis (Loos and Bouchard, 2003; Bell et al., 2005). Some of the more well-known examples include Prader–Willi syndrome and the various forms of the Bardet–Biedl syndrome (Loos and Bouchard, 2003; Bell et al., 2005). The genetic basis of most cases of both conditions is established (paternal deletion and/or maternal disomy at chromosome 15q11.2–12 in the case of Prader–Willi; mutations at any one of several loci in the cases of Bardet–Biedl), but the mechanisms whereby these lead to obesity have until recently remained obscure. Demonstration that Bardet–Biedl syndrome is in fact a disease of ciliary and basal body function offers intriguing new opportunities in obesity research (Beales, 2005).

Multifactorial Forms of Diabetes

As with other complex multifactorial traits, progress in identifying etiological genes influencing the development of type 2 diabetes and related traits has been slow when compared with the success in defining the causation for monogenic and syndromic forms of the disease (McCarthy, 2004). The reasons for this have been amply covered in several recent reviews (Ioannidis et al., 2001; Hirschhorn et al., 2002; Colhoun et al., 2003; Ioannidis et al., 2003; Lohmueller et al., 2003; Zondervan and Cardon 2004; Hattersley and McCarthy, 2005). The main difficulties encountered include those related to the disease itself—the fact that the effect sizes associated with any single susceptibility variant will generally be modest, with the potential for considerable conditionality due to gene–gene and gene–environment interaction—and those which can be described as methodological. Chief amongst the latter have been the use of woefully inadequate sample sizes, overenthusiastic interpretation of findings, failure to tackle variation within a gene in any comprehensive fashion, and (likely far more important than generally accepted) poor-quality genotyping. Nevertheless, several widely replicated positive associations have now been reported for type 2 diabetes: some of these—the P12A variant in peroxisomal proliferators activated receptor gamma (*PPARG*) and E23K in *KCNJ11* (Altshuler et al., 2000; Gloyn et al., 2003)—have gained general acceptance as "proven" diabetes-susceptibility genes on the basis of strong prior functional claims and replicated associations that, in combination, attain genome-wide significance (McCarthy, 2004; Parikh and Groop, 2004). The state-of-play for the long list of genes surveyed in the search for additional diabetes susceptibility variants has been the subject of several recent reviews (McCarthy, 2004; Parikh and Groop, 2004; O'Rahilly et al., 2005). Rather than repeat this exercise, the following paragraphs summarize the main conclusions that can be drawn (with illustrative examples):

1. Both candidate gene and positional-cloning based approaches have been successful in the identification of diabetes-susceptibility loci. Given that several hundred candidate genes have been examined for a role in diabetes-susceptibility (albeit that the vast majority of those studies have been methodologically inadequate), the rate of return in terms of convincingly replicated findings has been rather poor. Easily, the best two examples are those referred to above. The product of the *PPARG* gene is a key mediator of adipocyte differentiation and function, and a target of the thiazolidinedione class of drugs (which are effective in reducing insulin resistance and used in the treatment of diabetes). A common variant in the 5′ portion of this gene [which has been shown to influence capacity to transactivate responsive promoters (Deeb et al., 1998)] is associated with a 1.25-fold RR for diabetes. Since it is the common Pro allele (frequency ~85%) that is associated with increased risk of diabetes, the population attributable risk of this variant may be as high as 25% (Altshuler et al., 2000; Douglas et al., 2001; Florez, 2004). The product of the *KCNJ11* gene is an integral part of the KATP channel in the pancreatic β-cell, which (see above) plays a crucial role in normal glucose-stimulated insulin secretion. The K allele (~40% frequency) is associated with an approximately 15% increase in disease risk (there is some evidence that there is a dosage effect with an even greater risk associated with the KK genotype) (Gloyn et al., 2003; Nielsen et al., 2003; Schwanstecher and Schwanstecher, 2002). Some groups have been able to show a functional effect of this variant on channel function in vitro (Schwanstecher, Meyer, et al., 2002; Schwanstecher, Neugebauer, et al., 2002).

The complementary approach—that of genome-wide linkage followed by positional cloning—has also met with increasing success. The first example for diabetes (indeed for any complex trait, arguably) was the identification of variants within the *CAPN10* gene (encoding the widely expressed protease, calpain 10) as the likely cause of the diabetes linkage signal mapped to chromosome 2q in Mexican Americans (Hanis et al., 1996; Horikawa et al., 2000). Though these data remain compelling, follow-up of this initial finding has been somewhat disappointing. Replication studies have shown variable results, though two large meta-analyses did come out in favor of a modest association effect for certain *CAPN10* variants (Weedon et al., 2003; Song et al., 2004). Outside Mexican American populations, the effect on risk does seem modest. Efforts to understand how variation in calpain function might lead to diabetes have not reached a convincing conclusion (Sreenan et al., 2001).

There are several other linkage signals that have led to promising association results, though in several of these cases it is by no means clear that the associations found actually explain the linkage (suggesting that there are other etiological variants within the regions concerned). These include variants in the genes encoding adiponectin (like leptin, another hormone secreted by fat cells and thought to be involved in energy balance) (Vasseur et al., 2002) and *PSARL* (a mitochondrial rhomboid protease) (Walder et al., 2005), which map to a linkage signal (predominantly one for metabolic syndrome) on chromosome 3q (Kissebah et al., 2000; Vionnet et al., 2000), and common promoter variants in the *HNF4A* gene (a gene known to be causal for MODY) (Love-Gregory et al., 2004; Silander et al., 2004; Weedon et al., 2004) that may be relevant to replicated linkage signals on chromosome 20 (Ghosh et al., 2000). Most recently, researchers in Iceland have provided very strong evidence that variants within the *TCF7L2* gene (encoding a transcription factor involved in Wnt-signaling, and mapping to a signal on chromosome 10) have a powerful effect (RR ~1.5) on diabetes (Grant et al., 2006). Several other replicated linkage signals (such as those on chromosomes 1q and 10) are the subject of intensive positional cloning efforts (McCarthy, 2003).

2. Individual effect sizes are modest. Failure to find universally replicated linkage signals in genome scans for type 2 diabetes already made it plain that it was highly unlikely that there were any major gene effects in this condition (i.e., nothing akin to the effect of HLA in type 1 diabetes). This view is certainly reinforced by the susceptibility variants so far identified, which have RRs of approximately 1.2. (Only the effect size at *TCF7L2* may well turn out to be larger than this, but further replication is required before a robust estimate of effect size can be obtained). Detection and replication of signals of such modest magnitude provides a number of challenges and requires careful attention to methodological detail along with large sample sizes (in the thousands) (Hattersley and McCarthy, 2005). Studies of the role of variants in the insulin-receptor substrate-1 (*IRS1*) gene illustrate these issues well. Following an initial report that variants (notably the G972R coding polymorphism) were associated with diabetes risk (Almind et al., 1993), there were a series of conflicting replication studies. However, a meta-analysis (involving 27 studies and almost 9000 subjects) did indicate that the combined data pointed to a modest association (RR = 1.25) (Jellema et al., 2003). This conclusion is probably no longer sustainable since the two largest single studies (Florez et al., 2004; Zeggini et al., 2004), published subsequently, found no evidence of association. These (and other) examples, illustrate that, when dealing with such modest effect sizes, it is highly unlikely that any single study will provide a definitive result. Not only is replication essential to obtain a robust estimate of true effect size but it also represents an important defense against the wide range of errors, biases, and chance effects that can afflict any individual study (Page et al., 2003). Before we get too pessimistic, however (and too negative about the potential to use susceptibility variants as clinical tools—see below), it is worth remembering that, until recently, only an infinitesimal proportion of genomic variation has been sampled for its relationship to diabetes, and any view of the genetic architecture of diabetes that we base on current knowledge is unlikely to be representative. In addition, there have been only limited efforts to date to undertake systematic large-scale surveys of all variation within those genes—such as *PPARG* and *KCNJ11*—for which susceptibility effects are considered proven, to seek out other association signals indicating additional potential susceptibility variants. It is entirely possible—even probable given the proof that functional variation within these genes is capable of modifying disease risk—that these genes contain additional variants that contribute to susceptibility, or which explain the associations so far found. If so, the variants so far studied may significantly underestimate the total effect attributable to variation in these genes (Doney et al., 2004).

3. Evidence that altered gene product function influences the risk of diabetes is a strong predictor of involvement in multifactorial disease. Two of the most successful strategies to disease gene identification have been to select: (1) genes whose products are known drug targets (as with *PPARG, KCNJ11*) and (2) genes that have been implicated as causative for monogenic and syndromic forms of disease [this is true of both *KCNJ11* and *HNF4A* as described earlier; but also seems to hold for glucokinase, *TCF1* (aka *HNF1A*), *PPARG*, and possibly, *IPF1*] (Barroso et al., 1999; Frayling et al., 2005; Hegele et al., 1999; Owen et al., 2004; Weedon, Weedon, Owen, et al., 2005). Of course, this is very much to be expected. What both of these testaments to candidacy provide is the evidence that disruption/dysfunction of the gene (or its product) has some material effect on glucose homeostasis: little wonder therefore that common variants within the same genes (provided they also modify function or regulation) are capable of subtle effects that influence the risk of multifactorial disease.

Multifactorial Forms of Obesity

The story of susceptibility gene discovery in obesity recapitulates much of that of type 2 diabetes, though, arguably, there are fewer concrete examples of success. Until recently, the same strategies—candidate gene analyses on the one hand, genome-wide linkage followed by positional cloning—have predominated.

As with diabetes, the linkage approach has generated a bewildering mosaic of signals derived from the studies conducted in diverse populations, using a range of different approaches. A summary of these findings, maintained at the Obesity Gene Map database (http://obesitygene.pbrc.edu) and periodically appearing in print (Perusse et al., 2005), reveals that there are few points in the genome that have not appeared in at least one region of potential linkage. It is useful, in this situation, to look for the evidence of replication and identify regions that appear in multiple scans (reviewed in Loos and Bouchard, 2003; Bell et al., 2005). However, in the absence of any systematic method for defining replication in this situation, the selection of such regions remains rather subjective. Certainly, the two reviews quoted generate nonidentical but overlapping lists of these regions, with chromosomes 2p, 3q, 7q, 10p, 11q, and 20q being highlighted in both. As with diabetes, several are the subject of positional cloning efforts. As yet, there has been no conclusive evidence for an etiological locus in any of these signals though adiponectin (see above) may be contributing to the 3q signal, and variation within *GAD2* (see below) has been implicated in the 10q finding.

In reviewing the genetic studies of obesity, it is important to recognize that there are additional aspects of study design (above and beyond those described earlier) that may be very relevant to the question of gene identification and replication. Foremost amongst these (Bell et al., 2005) is the division of studies into those that use a case-control design [typically comparing subjects with severe, or morbid obesity (BMI > 40 kg m – 2) with controls: these will be examining the effects of variants on the risk of the extreme phenotype] and those that use a population-based approach (which will be concentrating on the effects of genomic variation on phenotypic variation across the normal range). There is every reason to suspect that the spectrum of variants detected with each design will differ substantially (*MC4R* associations, for example, are, in children at least, more likely to be detected in the case-control design—see above), even if there is some overlap in the susceptibility genes themselves. Bundling both types of study together for combined analyses may not be a particularly useful exercise.

In terms of candidate gene studies, as with diabetes, a great many genes (particularly candidates from the energy balance pathway including the uncoupling proteins and β-adrenergic receptors) have been examined (e.g., Clement et al., 1995; Large et al., 1997; Walder et al., 1998), but there has been little in the way of confirmed replication. Examination of genes implicated in monogenic and syndromic forms of obesity (such as those encoding leptin and its receptor) has generated some positive findings, but these have not been widely confirmed (Karvonen et al., 1998; Li et al., 1999; Heo et al., 2002; Jiang et al., 2004). There are interesting data emerging for common variants in *POMC* (Challis et al., 2002; Lee et al., 2006). Somewhat surprisingly, despite the known biological functions of *PPARG* (adipocyte differentiation and function), the marked effects on body weight of agonists acting on its gene product and the established effects on diabetes susceptibility evidence that the P12A variant influences BMI is modest at best (Masud et al., 2003).

Recent studies have raised interesting questions about the possible role of variants in the *GAD2* and *ENPP1* genes (Boutin et al., 2003; Meyre, Boutin, et al., 2005; Meyre, Bouatia-Naji, et al., 2005). Both represent positional candidates lying within linked regions on chromosome 10p (Hager et al., 1998) and 6q (Meyre et al., 2004), respectively. *GAD2* encodes glutamate decarboxylase (GAD65), which catalyses the formation of the neurotransmitter gamma aminobutyric acid (GABA) and is synthesized by GABAergic neurones within nuclei of the hypothalamus known to be involved in feeding behavior. Associations between *GAD2* variants and extreme obesity (and supportive functional data) were reported in two studies (Boutin et al., 2003; Meyre, Boutin, et al., 2005), but have not been sustained in subsequent publications (Swarbrick et al., 2005). The product of ectonucleotide pyrophosphatase phosphodiesterase (*ENPP1*) gene (aka PC1) directly inhibits insulin receptor activation and downstream signaling. A three-allele risk haplotype within *ENPP1* showed powerful associations with childhood obesity [odds ratio (OR)1.69], adult obesity (1.50), and type 2 diabetes (combined OR 1.56) (Meyre, Bouatia-Naji, et al., 2005). Again, replication of these findings is vital to obtain robust estimates of the true effect size.

Ongoing and Future Research into Genetics of Multifactorial Diabetes and Obesity

We are already starting to see the benefits from complex trait mapping efforts that increasingly take into account the harsh realities of multifactorial disease genetic architecture to design better studies (Clark et al., 2005; Hattersley and McCarthy, 2005). This means sample sizes in the thousands (at least), comprehensive assessment of variation within genes of interest (making use of linkage disequilibrium information to maximize the efficiency with which variation is surveyed), high-quality genotyping, and sober interpretation of the findings. It is increasingly the case that studies which fail to meet such criteria offer little in the way of useful knowledge.

At the same time, efforts to map susceptibility using candidate gene-based approaches are benefiting from the wealth of data emerging from genomic studies in rodents and man (Mootha et al., 2003), including the use of expression QTL approaches to define the variants that appear to modify key expression phenotypes as well as being associated with the disease (Mehrabian et al., 2005; Schadt et al., 2005). By increasing the prior odds that the candidates selected play important roles in the regulation of normal homeostasis, the prospects for identification of genuine susceptibility effects are enhanced.

However, at the time of writing, most excitement centers around the new insights that are emerging from genome-wide association studies (Carlson et al., 2004). Arising out of the advances in genome informatics (principally the International HapMap Consortium) (Altshuler et al., 2005), combined with progress in high-throughput genotyping platforms, it has become feasible to survey a large proportion of common variation within the genome (at least 80% in principle) in a single assay, for relatively low cost per genotype. Several large-scale studies in type 2 diabetes (cumulatively approximately 5000 case–control pairs) and obesity are underway or have recently been completed (see, e.g., the Wellcome Trust Case Control Consortium: www.wtccc.org.uk). Amongst the new insights emerging from these studies are: (a) identification of the *FTO* (fat mass and obesity associated) gene as the first widely-replicated susceptibility-locus for obesity (Frayling et al., 2007) and (b) identification of at least five novel T2D-susceptibility genes (Saxena et al., 2007; Scott et al., 2007; Sladek et al., 2007; Steinthorsdottir et al., 2007; Zeggini et al., 2007).

It is important to recognize that, as with any other novel genomic technology, genome-wide association analysis brings with it a number of challenges and limitations (Clark et al., 2005; Hirschhorn and Daly, 2005; Wang et al., 2005). First, the sheer cost of the studies (cost per genotype may be low, but costs per sample are high given the need to type several hundreds of thousands of markers) means that the sample sizes included in each study are currently relatively low (i.e., a few thousand subjects). This means that successive rounds of confirmation, extension, and replication will be required to distinguish the few real associations from the many spurious signals that attain nominal significance in the first round of genome-wide analysis (Skol et al., 2006). Second, none of the fixed-content products currently available captures all common variation: for strong functional candidates, additional genotyping is required for exclusion. Third, the necessary focus on common variants means that the contribution of rare variants to common phenotypes (which may be substantial) (Cohen et al., 2004) will not be well assayed using such approaches (Zeggini et al., 2005). Fourth, the informatics and analytical challenges associated with the processing, storage, and examination of such data are nontrivial (Clayton et al., 2005). Fifth, substantial advances are required in the capacity to integrate biological information into the analysis of genome-wide association data, so that attention can be focused on those signals with the strongest posterior probability of being true signals (Wacholder et al., 2004). Finally, it is unclear how effective single-nucleotide polymorphism (SNP)-based screening will be in detecting the functional effects due to structural variation within the genome (Feuk et al., 2006).

Notwithstanding the above, genome-wide association analysis is providing a wonderful opportunity to address the fundamental questions relevant to complex trait etiology. In the near future, we can look forward to more systematic surveys of the importance of gene–gene and gene–environment interactions. Looking further into the future, the logical conclusion of these efforts will be the ability to capture all sources of variation through complete genome resequencing, a goal that is certainly within reach (Bennett et al., 2005).

Other Genomic Approaches

To date, we have considered only the clues to etiology, which are flowing from an understanding of human genome sequence variation. Though space does not permit a complete survey of what is a massive field in its own right (especially with respect to studies in animal models), it is important to point out that, from the point-of-view of future genomic medicine, clinical applications are also likely to emerge from the consideration of other genomic modalities (including transcript profiling, proteomics, metabonomics, and epigenomics). Each of these has the potential to offer novel biomarkers that can be used for prognostic, diagnostic, and therapeutic purposes (Zhang et al., 2004; Schulte et al., 2005). In this context, genetics and genomics are likely to be complementary. Whilst information from genome sequence variation can help to define predisposition, variation at the transcriptional or proteomic (or metabonomic) level allows the pathogenetic process to be monitored in real time, and provides readouts more proximal to the disease state. As such, they may offer the capacity to refine the risks that arise from genetic predisposition, and/or to provide biomarkers that can be used to predict disease and the risk of complications. It is worth remembering that the use of biomarkers is well established in diabetes care and research (Table 12–3).

Table 12–3 Selected Biomarkers Already Used in Clinical Care of Type 2 Diabetes

Area	Biomarker	Comments
Diagnostic	Glucose	Used to define diabetes and prediabetes
	HbA1c	Epidemiological value for prevalence estimates
Differential diagnosis	C-peptide	Absent levels imply type 1 diabetes
	Insulin	Absent levels imply type 1 diabetes
	GAD antibodies	Marker of islet autoimmunity
Monitoring progression	Glucose	Monitors disease progression and treatment effectiveness
	HbA1c	Monitors disease progression and treatment effectiveness
	Fructosamine	Monitors disease progression and treatment effectiveness
Assessment of β-cell function	Insulin	Can be used (with glucose) to derive measure of islet function
	Proinsulin	Raised proinsulin/insulin ratio a sign of β-cell failure
	C-peptide	Can be used (with glucose) to derive measure of islet function
Assessment of insulin sensitivity	Insulin	Fasting insulin, a measure of insulin action
Predict complication risk	HDL-cholesterol	Helps define risk of cardiovascular disease
	LDL-cholesterol	Helps define risk of cardiovascular disease
	Triglycerides	Helps define risk of cardiovascular disease
	Urinary albumin	Widely used to identify risk of diabetic nephropathy
	CRP	Helps define risk of cardiovascular disease

Notwithstanding this promise, the use of genomic approaches to study diabetes and obesity faces clear challenges. Compared to the cancer field, the tissues involved in the pathogenesis of these conditions (particularly liver, brain, pancreas) are either impossible or difficult to access for either research or clinical purposes. This, of course, explains why so much of the research in this area has been performed in animal models. More accessible samples (plasma, serum, urine) are available for studies to define biomarkers through proteomic and metabonomic approaches, but the extent to which circulating leucocytes can offer pathogenetic insights from the application of transcriptional and epigenomic analyses remains unclear. In addition, as in other diseases, there remain substantial informatic and analytical challenges associated with the application of many (still novel) genomics technologies in the research and clinical setting (Allison et al., 2006; Jarvis and Centola, 2005). Some of the initial data on clinical applications of genomics in other disease areas have proven difficult to reproduce (Ludwig and Weinstein, 2005).

Despite these obstacles, there remains substantial optimism that advances in genomics methods, better informatic and analytical approaches, combined with more rigorous standardization (Brazma et al., 2001), will in time deliver novel biomarkers that offer genuine clinical benefits in terms of disease classification, risk stratification, monitoring of disease progression, early detection of complications, and individualized assessment of therapeutic and interventional opportunities (www.systemsx.ch; www.molpage.org).

Genomic Medicine in Diabetes and Obesity: An Appraisal

Where do we stand with respect to the application of these genetic and genomic findings into clinical practice? As we have seen in monogenic and syndromic forms of diabetes in particular (e.g., MODY, PNDM), molecular medicine is already playing a role in clinical management (www.diabetesgenes.org). In many countries, including the UK, diagnostic testing services are in place which allow suitable patients to be screened to establish (or refute) a diagnosis of MODY or neonatal diabetes (which may allow those on insulin due to a misdiagnosis of type 1 diabetes to move onto less demanding therapeutic options), and/or to identify the precise molecular cause. This information carries important diagnostic, prognostic, and therapeutic implications for patients and their relatives (McCarthy and Hattersley, 2001; Owen and Hattersley, 2001; Pearson et al., 2003; Gloyn et al., 2004). There are increasing grounds for contemplating a wider role for molecular diagnostics in the differential diagnosis of diabetes arising in early adulthood: in this case, the range of potential diagnoses is greatest (type 1, type 2, MODY, mitochondrial diabetes, etc.), and clinical features are not always reliable guides (Owen et al., 2003). The main impediment to more extensive use of diagnostics in this setting is the current cost of genetic screening. The high proportion of private mutations in MODY and neonatal diabetes usually mandates complete resequencing of target genes: ongoing developments in resequencing technologies should allow reduced costs and more widespread use of genetic diagnostics in these and other situations (Song et al., 2004). In contrast to the situation in diabetes, the application of diagnostic genetics in obesity is currently reserved for those with extreme phenotypes (e.g., severe childhood obesity).

The application of genomic medicine in multifactorial forms of diabetes and obesity is, as yet, more theoretical than practical. However, there are reasons to be optimistic that current efforts

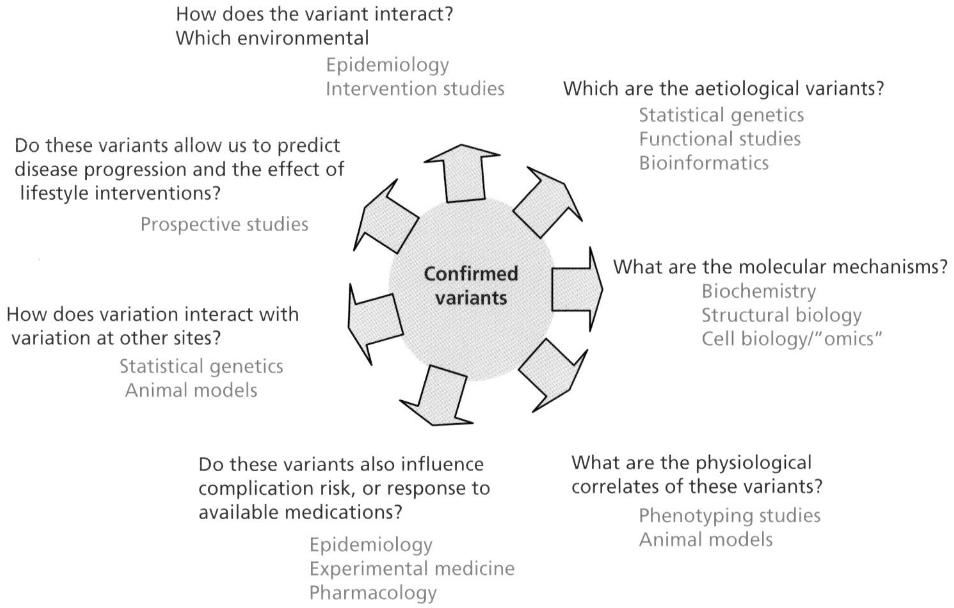

How does the variant interact?
Which environmental
Epidemiology
Intervention studies

Which are the aetiological variants?
Statistical genetics
Functional studies
Bioinformatics

Do these variants allow us to predict disease progression and the effect of lifestyle interventions?
Prospective studies

Confirmed variants

What are the molecular mechanisms?
Biochemistry
Structural biology
Cell biology/"omics"

How does variation interact with variation at other sites?
Statistical genetics
Animal models

Do these variants also influence complication risk, or response to available medications?
Epidemiology
Experimental medicine
Pharmacology

What are the physiological correlates of these variants?
Phenotyping studies
Animal models

Figure 12–3 From bench to bedside in genomic medicine. Identification of a proven susceptibility variant is just the start of the research that will allow translation of that finding into clinical practice. Some of the questions that need to be asked, and some of the methods that can be used to address them are shown.

(e.g., whole genome association studies) will deliver a more complete enumeration of the susceptibility landscape of both conditions. Given linkage data revealing no loci of overweaning importance, it seems probable that much of this susceptibility will reflect variants of modest effect, in which case the clinical information available from any single variant will be strictly limited (Sotos et al., 2000). However, combined information from multiple variants may provide the basis for useful diagnostic testing (Janssens et al., 2004; Yang et al., 2005). Efforts to obtain gene-wide estimates of effect size (i.e., the composite effect of all relevant variation within susceptibility genes) should enhance power in this regard, as will a more sophisticated understanding of the importance of gene–gene and gene–environment interactions.

Armed with a list of susceptibility variants, researchers will be able to embark on the wide range of basic and clinical studies that are essential prerequisites for rational application in patient care (see Fig. 12–3). Large-scale prospective studies with robust record linkage will be essential to such efforts. One productive area is likely to be in the analysis of pharmacogenetic effects: there have been rather few studies of the role of genetic variation in response to agents used in the treatment of diabetes and obesity (e.g., Wolford et al., 2005), and there remain challenges in obtaining appropriately powered and designed studies. Clinical trials are often too small to deliver convincing results, and observational studies within unselected populations may be confounded by channeling bias (i.e., treatment selection by medical practitioners may be influenced by the estimation of individual risk of disease progression).

Progress in identifying the genetic and genomic architecture of trait susceptibility will undoubtedly deliver an improved understanding of disease pathogenesis, which will be of value to genomic medicine. For instance, appreciation of fundamental etiological pathways will highlight potential biomarkers that can be used to track the early stages of disease development and/or enhance disease classification. In parallel, we can also hope to emerge with a set of genetic tools that can be applied to assist with the evaluation of individual diagnosis and prognosis, enhance risk stratification, and improve treatment selection.

How might the management of diabetes and obesity be influenced by such developments? It is certainly reasonable to envisage that, within one or two decades, it will be possible to obtain a "broad-brush" view of individual diatheses and predispositions through the analysis of patterns of genome sequence variation (Almond, 2006). Thereafter, one should be able to make use of an array of (existing and novel) clinical, biochemical (biomarkers), genomic, and imaging tests to monitor those liabilities and serially update individual risks of major causes of morbidity. Such "rolling" assessments of risk would then trigger interventions (or further rounds of more intensive diagnostic investigations) when the levels of risk cause concern: or equally, downgrade risk when it appears that the predisposition calculated on the basis of genome sequence variation has overestimated true risk. Such a model is, in many ways, not that dissimilar to current medical practice, though the evolution in care that will result from wider use of genomic medicine should provide greater precision in the assignment of risk and definition of pathology, and offer a more rational framework for the deployment of available therapeutic and preventative options.

References

Allison, DB, Cui, X, Page, GP, and Sabripour, M (2006). Microarray data analysis: from disarray to consolidation and consensus. *Nat Rev Genet,* 7:55–65.

Allison, DB, Faith, J, and Nathan, JS (1996). Risch's lambda values for human obesity. *Int J Obes Relat Metab Disord,* 20:990–999.

Allison, DB, Kaprio, J, Korkeila, M, Koskenvuo, M, Neale, MC, and Hayakawa, K (1996). The heritability of body mass index among an international sample of monozygotic twins reared apart. *Int J Obes Relat Metab Disord,* 20:501–506.

Almind, K, Bjørbaek, C, Vestergaard, H, Hansen, T, Echwald, S, and Pedersen, O (1993). Aminoacid polymorphisms of insulin receptor substrate-1 in non-insulin dependent diabetes mellitus. *Lancet,* 342:828–832.

Almond, B (2006). Genetic profiling of newborns: ethical and social issues. *Nat Rev Genet,* 7:67–71.

Altshuler, D, Brooks, LD, Chakravarti, A, Collins, FS, Daly, MJ, Donnelly, P, International HapMap Consortium (2005) A haplotype map of the human genome. *Nature,* 437:1299–1320.

Altshuler, D, Hirschhorn, JN, Klannemark, M, Lindgren, CM, Vohl, MC, Nemesh, J, et al. (2000). The common PPARgamma Pro12Ala polymorphism is associated with decreased risk of type 2 diabetes. *Nat Genet,* 26:76–80.

Arima, T, Drewell, RA, Arney, KL, Inoue, J, Makita, Y, Hata, A, et al. (2001). A conserved imprinting control region at the HYMAI/ZAC domain is implicated in transient neonatal diabetes mellitus. *Hum Mol Genet,* 10:1475–1483.

Arita, Y, Kihara, S, Ouchi, N, Takahashi, M, Maeda, K, Miyagawa, J, et al. (1999). Paradoxical decrease of an adipose-specific protein, adiponectin, in obesity. *Biochem Biophys Res Commun,* 257:79–83.

Barroso, I, Gurnell, M, Crowley, VEF, Agostini, M, Schwabe, JW, Soos, MA, et al. (1999). Dominant negative mutations in human PPARγ associated with severe insulin resistance, diabetes mellitus, and hypertension. *Nature,* 402:880–883.

Beales, PL (2005). Lifting the lid on Pandora's box: the Bardet-Biedl syndrome. *Curr Opin Genet Dev,* 15:315–323.

Bell, CG, Walley, AJ, and Froguel, P (2005). The genetics of human obesity. *Nat Rev Genet,* 6:221–234.

Bell, JI (2003). The double helix in clinical practice. *Nature,* 421:414–416.

Bennett, ST, Barnes, C, Cox, A, Davies, L, and Brown, C (2005). Toward the $1000 human genome. *Pharmacogenomics,* 6:373–382.

Bennett, ST and Todd, JA (1996). Human type 1 diabetes and the insulin gene: principles of mapping polygenes. *Annu Rev Genet,* 30:343–370.

Boutin, P, Dina, C, Vasseur, F, Dubois, S, Corset, L, Seron, K, et al. (2003). *GAD2* on chromosome 10p12 is a candidate gene for human obesity. *PLoS Biol,* 1:1–11.

Brazma, A, Hingamp, P, Quackenbush, J, Sherlock, G, Spellman, P, Stoeckert, C, et al. (2001). Minimum information about a microarray experiment (MIAME)-toward standards for microarray data. *Nat Genet,* 29:365–371.

Brolin, SE (2002). Bariatric surgery and long-term control of morbid obesity. *JAMA,* 288:2793–2796.

Brosseau, JD, Eelkema, RC, Crawford, AC, and Abe, TA (1979). Diabetes among the three affiliated tribes: correlation with degree of Indian inheritance. *Am J Public Health,* 69:1277–1278.

Carlson, CS, Eberle, MA, Kruglyak, L, and Nickerson, DA (2004). Mapping complex disease loci in whole-genome association studies. *Nature,* 429:446–452.

Caro, JF, Sinha, MK, Kolaczynski, JW, Zhang, PL, and Considine, RV (1996). Leptin: the tale of an obesity gene. *Diabetes,* 45:1455–1462.

Challis, BG, Pritchard, LE, Creemers, JW, Delplanque, J, Keogh, JM, Luan, J, et al. (2002). A missense mutation disrupting a dibasic prohormone processing site in pro-opiomelanocortin (*POMC*) increases susceptibility to early-onset obesity through a novel molecular mechanism. *Hum Mol Genet,* 11:1997–2004.

Clark, AG, Boerwinkle, E, Hixson, J, and Sing, CF (2005). Determinants of the success of whole-genome association testing. *Genome Res,* 15:1463–1467.

Clayton, DG, Walker, NM, Smyth, DJ, Pask, R, Cooper, JD, Maier, LM, et al. (2005). Population structure, differential bias and genomic control in a large-scale, case-control association study. *Nat Genet,* 37:1243–1246.

Clement, K, Vaisse, C, Lahlou, N, Cabrol, S, Pelloux, V, Cassuto, D, et al. (1998). A mutation in the human leptin receptor gene causes obesity and pituitary dysfunction. *Nature,* 392:398–401.

Clement, K, Vaisse, C, Manning, BS, Basdevant, A, Guy-Grand, B, Ruiz, J, et al. (1995). Genetic variation in the β3- adrenergic receptor and an increased capacity to gain weight in patients with morbid obesity. *N Engl J Med,* 333:352–354.

Cohen, JC, Kiss, RS, Pertsemlidis, A, Marcel, YL, McPherson, R, and Hobbs, HH (2004). Multiple rare alleles contribute to low plasma levels of HDL cholesterol. *Science,* 305:869–872.

Colhoun, HM, McKeigue, PM, and Smith, GD (2003). Problems of reporting genetic associations with complex outcomes. *Lancet,* 361:865–872.

Deeb, SS, Fajas, L, Nemoto, M, Pihlajamaki, J, Mykkanen, L, Kuusisto, J, et al. (1998). A Pro12Ala substitution in PPARgamma2 associated with decreased receptor activity, lower body mass index and improved insulin sensitivity. *Nat Genet,* 20:284–287.

Doney, AS, Fischer, B, Cecil, JE, Boylan, K, McGuigan, FE, Ralston, SH, et al. (2004). Association of the Pro12Ala and C1431T variants of *PPARG* and their haplotypes with susceptibility to type 2 diabetes. *Diabetologia,* 47:555–558.

Douglas, JA, Erdos, MR, Watanabe, RM, Braun, A, Johnston, CL, Oeth, P, et al. (2001). The peroxisome proliferator-activated receptor-gamma2 Pro12Ala variant: association with type 2 diabetes and trait differences. *Diabetes,* 50:886–890.

Eckel, RH, Grundy, SM, and Zimmet, PZ (2005). The metabolic syndrome. *Lancet,* 365:1415–1428.

Ehtisham, S, Kirk, J, McEvilly, A, Shaw, N, Jones, S, Rose, S, et al. (2001). Prevalence of type 2 diabetes in children in Birmingham. *Br Med J,* 322:1428.

Fajans, S (1990). Scope and heterogeneous nature of MODY. *Diabetes Care,* 13:49–64.

Farooqi, IS, Jebb, SA, Langmack, G, Lawrence, E, Cheetham, CH, Prentice, AM, et al. (1999). Effects of recombinant leptin therapy in a child with congenital leptin deficiency. *N Engl J Med,* 341:879–884.

Farooqi, IS, Keogh, JM, Kamath, S, Jones, S, Gibson, WT, Trussell, R, et al. (2001). Partial leptin deficiency and human adiposity. *Nature,* 414:34–35.

Farooqi, IS, Keogh, JM, Yeo, GS, Lank, EJ, Cheetham, T, and O'Rahilly, S (2003). Clinical spectrum of obesity and mutations in the melanocortin 4 receptor gene. *N Engl J Med,* 348:1085–1095.

Ferranini, E, Haffner, SM, Mitchell, BD, and Stern, MP (1991). Hyperinsulinaemia: the key feature of a cardiovascular and metabolic syndrome. *Diabetologia,* 34:416–422.

Feuk, L, Carson, AR, and Scherer, SW (2006). Structural variation in the human genome. *Nat Rev Genet,* 7:85–97.

Florez, JC (2004). Phenotypic consequences of the peroxisome proliferator-activated receptor-γ Pro12Ala polymorphism: the weight of the evidence in genetic association studies. *J Clin Endocrinol Metab,* 89:4234–4237.

Florez, JC, Sjogren, M, Burtt, N, Orho-Melander, M, Schayer, S, Sun, M, et al. (2004). Association testing in 9,000 people fails to confirm the association of the insulin receptor substrate-1 G972R polymorphism with type 2 diabetes. *Diabetes,* 53:3313–3318.

Frayling, TM, Timpson, NJ, Weedon, MN, Zeggine, E, Freathy, RM, Lindgren, CM, et al. (2007). A common variant in the FTO gene is associated with body mass index and predisposes to childhood and adult obesity. *Science,* 316(5826):889–894.

Friedman, JM (2004). Modern science versus the stigma of obesity. *Nat Med,* 10:563–569.

Froguel, P, Vaxillaire, M, Sun, F, Velho, G, Zouali, H, Butel, MO, et al. (1992). Close linkage of glucokinase locus on chromosome 7p to early-onset non-insulin-dependent diabetes mellitus. *Nature,* 356:162–165.

Gale, EA (2005). The myth of the metabolic syndrome. *Diabetologia,* 48:1679–1683.

Ghosh, S, Watanabe, RM, Valle, TT, Hauser, ER, Magnuson, VL, Langefeld, CD, et al. (2000). The Finland-United States Investigation of non-insulin-dependent diabetes mellitus genetics (FUSION) study. I. An autosomal genome scan for genes that predispose to type 2 diabetes. *Am J Hum Genet,* 67:1174–1185.

Gloyn, AL (2003). Glucokinase (*GCK*) mutations in hyper- and hypoglycemia: maturity-onset diabetes of the young, permanent neonatal diabetes, and hyperinsulinemia of infancy. *Hum Mutat,* 22:353–362.

Gloyn, AL, Pearson, ER, Antcliff, JF, Proks, P, Bruining, GJ, Slingerland, AS, et al. (2004). Activating mutations in the gene encoding the ATP-sensitive potassium-channel subunit Kir6.2 and permanent neonatal diabetes. *N Engl J Med,* 350:1838–1849.

Gloyn, AL, Reimann, F, Girard, C, Edghill, EL, Proks, P, Pearson, ER, et al. (2005). Relapsing diabetes can result from moderately activating mutations in *KCNJ11*. *Hum Mol Genet,* 14:925–934.

Gloyn, AL, Weedon, MN, Owen, KR, Turner, MJ, Knight, BA, Hitman, GA, et al. (2003). Large scale association studies of variants in genes encoding the pancreatic beta-cell K-ATP channel subunits Kir6.2 (KCNJ11) and SUR1 (ABCC8) confirm that the *KCNJ11* E23K variant is associated with increased risk of Type 2 Diabetes. *Diabetes,* 52:568–572.

Grant, SF, Thorleifsson, G, Reynisdottir, I, Benediktsson, R, Manolescu, A, Sainz, J, et al. (2006). Variant of transcription factor 7-like 2 (*TCF7L2*) gene confers risk of type 2 diabetes. *Nat Genet,* 38:320–323.

Hager, J, Dina, C, Francke, S, Dubois, S, Houari, M, Vatin, V, et al. (1998). A genome-wide scan for human obesity genes reveals a major susceptibility locus on chromosome 10. *Nat Genet,* 20:304–308.

Halaas, JL, Boozer, C, Blair-West, J, Fidahusein, N, Denton, DA, and Friedman, JM (1997). Physiological response to long-term peripheral and central leptin infusion in lean and obese mice. *Proc Natl Acad Sci USA,* 94:8878–8883.

Hales, CN, Barker, DJP, Clark, PMS, Cox, LJ, Fall, C, Osmond, C, et al. (1991). Fetal and infant growth and impaired glucose tolerance at age 64. *Br Med J,* 303:1019–1022.

Hales, CN and Ozanne, SE (2003). For debate: fetal and early postnatal growth restriction lead to diabetes, the metabolic syndrome and renal failure. *Diabetologia,* 46:1013–1019.

Hanis, CL, Boerwinkle, E, Chakraborty, R, Ellsworth, DL, Concannon, P, Stirling, B, et al. (1996). A genome-wide search for human non-insulin-dependent (type 2) diabetes genes reveals a major susceptibility locus on chromosome 2. *Nat Genet,* 13:161–171.

Hattersley, AT (2006). Beyond the beta cell in diabetes. *Nat Genet,* 38:12–13.

Hattersley, AT, Beards, F, Ballantyne, E, Appleton, M, Harvey, R, and Ellard, S (1998). Mutations in the glucokinase gene of the foetus result in reduced birth weight. *Nat Genet,* 19:268–270.

Hattersley, AT and McCarthy, MI (2005). A question of standards: what makes a good genetic association study? *Lancet,* 366:1315–1323.

Hattersley, AT and Tooke, JE (1999). The fetal insulin hypothesis an alternative explanation of the association of low birthweight with diabetes and vascular disease. *Lancet,* 353:1789–1792.

Hattersley, AT, Turner, RC, Permutt, MA, Patel, P, Tanizawa, Y, Chiu, KC et al. (1992). Linkage of type 2 diabetes to the glucokinase gene. *Lancet,* 339:1307–1310.

Hedley, AA, Ogden, CL, Johnson, CL, Carroll, MD, Curtin, LR, and Flegal, KM (2004). Prevalence of overweight and obesity among US children, adolescents, and adults, 1999-2002. *JAMA,* 291:2847–2850.

Hegele, RA, Cao, H, Harris, SB, Hanley, AJG, and Zinman, B (1999). The hepatic nuclear factor-1α G319S variant is associated with early-onset type 2 diabetes in Canadian Oji-Cree. *J Clin Endocrinol Metab,* 84:1077–1082.

Heo, M, Leibel, RL, Fontaine, KR, Boyer, BB, Chung, WK, Koulu, M, et al. (2002). A meta-analytic investigation of linkage and association of common leptin receptor (*LEPR*) polymorphisms with body mass index and waist circumference. *Int J Obes Relat Metab Disord,* 26:640–646.

Hirschhorn, JN and Daly, MJ (2005). Genome-wide association studies for common diseases and complex traits. *Nat Rev Genet,* 6:95–108.

Hirschhorn, JN, Lohmueller, K, Byrne, E, and Hirschhorn, K (2002). A comprehensive review of genetic association studies. *Genet Med,* 4:45–61.

Hollis, JH and Mattes, RD (2005). Are all calories created equal? Emerging issues in weight management. *Curr Diab Rep,* 5:374–378.

Horikawa, Y, Iwasaki, N, Hara, M, Furuta, H, Hinokio, Y, Cockburn, BN, et al. (1997). Mutation in hepatocyte nuclear factor-1α gene (*TCF2*) associated with MODY. *Nat Genet,* 17:384–385.

Horikawa, Y, Oda, N, Cox, NJ, Li, X, Orho-Melander, M, Hara, M, et al. (2000). Genetic variation in the gene encoding calpain-10 is associated with type 2 diabetes mellitus. *Nat Genet,* 26:163–175.

Hyponnen, E, Smith, GD, and Power, C (2003). Parental diabetes and birth weight of offspring: intergenerational cohort study. *Br Med J,* 326:19–20.

Ioannidis, JP, Ntzani, EE, Trikalinos, TA, and Contopoulos-Ioannidis, DG (2001). Replication validity of genetic association studies. *Nat Genet,* 29:306–309.

Ioannidis, JP, Trikalinos, TA, Ntzani, EE, and Contopoulos-Ioannidis, DG (2003). Genetic associations in large versus small studies: an empirical assessment. *Lancet,* 361:567–571.

Jackson, RS, Creemers, JW, Ohagi, S, Raffin-Sanson, ML, Sanders, L, Montague, CT, et al. (1997). Obesity and impaired prohormone processing associated with mutations in the human prohormone convertase 1 gene. *Nat Genet,* 16:303–306.

Janssens, ACJW, Pardo, MC, Steyerberg, EW, and van Duijn, CM (2004). Revisiting the clinical validity of multiplex genetic testing in complex diseases. *Am J Hum Genet,* 74:585–588.

Jarvis, JN and Centola, M (2005). Gene-expression profiling: time for clinical application. *Br Med J,* 365:199–200.

Jellema, A, Zeegers, MP, Feskens, EJ, Dagnelie, PC, and Mensinkm, RP (2003). Gly972Arg variant in the insulin receptor substrate-1 gene and association with type 2 diabetes: a metaanalysis of 27 studies. *Diabetologia,* 46:990–995.

Jiang, Y, Wilk, JB, Borecki, I, Williamson, S, DeStefano, AL, Xu, G, et al. (2004). Common variants in the 5′ region of the leptin gene are associated with body mass index in men from the National Heart, Lung, and Blood Institute Family Heart Study. *Am J Hum Genet,* 75:220–230.

Kadowaki, T, Kadowaki, H, Mori, Y, Tobe, K, Sakuta, R, Suzuki, Y, et al. (1994). A subtype of diabetes mellitus associated with a mutation of mitochondrial DNA. *N Engl J Med,* 330:962–968.

Kaprio, J, Tuomilehto, J, Koskenvuo, M, Romanov, K, Reunanen, A, Eriksson, J, et al. (1992). Concordance for type 1 (insulin-dependent) and type 2 (non-insulin-dependent) diabetes mellitus in a population-based cohort of twins in Finland. *Diabetologia,* 35:1060–1067.

Karvonen, MK, Pesonen, U, Heinonen, P, Laakso, M, Rissanen, A, Naukkarinen, H, et al. (1998). Identification of new sequence variants in the leptin gene. *J Clin Endocrinol Metab,* 83:3239–3242.

Kissebah, AH, Sonnenberg, GE, Myklebust, J, Goldstein, M, Broman, K, James, RG, et al. (2000). Quantitative trait loci on chromosomes 3 and 17 influence phenotypes of the metabolic syndrome. *Proc Natl Acad Sci USA,* 97:14478–14483.

Kobberling, J and Tillil, H (1982). Empirical risk figures for first degree relatives of non-insulin-dependent diabetics. In *The Genetics of Diabetes Mellitus.* London: Academic Press.

Kopelman, PG (2000). Obesity as a medical problem. *Nature,* 404:635–643.

Krude, H, Biebermann, H, Luck, W, Horn, R, Brabant, G, and Gruters, A (1998). Severe early-onset obesity, adrenal insufficiency and red hair pigmentation caused by *POMC* mutations in humans. *Nat Genet,* 19:155–157.

Large, V, Hellström, L, Reynisdottir, S, Lönnqvist, F, Eriksson, P, Lannfelt, L, et al. (1997). Human beta-2 adrenoceptor gene polymorphisms are highly frequent in obesity and associate with altered adipocyte beta-2 adrenoceptor function. *J Clin Invest,* 100:3005–3013.

Larsen, LH, Echwald, SM, Sørensen, TIA, Andersen, T, Wulff, BS, and Pedersen, O (2005). Prevalence of mutations and functional analyses of melanocortin 4 receptor variants identified among 750 men with juvenile-onset obesity. *J Clin Endocrinol Metab,* 90:219–224.

Lee, YS, Challis, BG, Thompson, DA, Yeo, GS, Keogh, JM, Madonna, ME, et al. (2006). A *POMC* variant implicates beta-melanocyte-stimulating hormone in the control of human energy balance. *Cell Metab,* 3:135–140.

Li, WD, Reed, DR, Lee, JH, Xu, W, Kilker, RL, Sodam, BR, et al. (1999). Sequence variants in the 5′ flanking region of the leptin gene are associated with obesity in women. *Ann Hum Genet,* 63:227–234.

Lohmueller, KE, Pearce, CL, Pike, M, Lander, ES, and Hirschhorn, JN (2003). Meta-analysis of genetic association studies supports a contribution of common variants to susceptibility to common disease. *Nat Genet,* 33:177–182.

Loos, RJF and Bouchard, C (2003). Obesity—is it a genetic disorder? *J Intern Med,* 254:401–425.

Love-Gregory, L, Wasson, J, Ma, J, Hin, CH, Glaser, B, Suarez, BK, et al. (2004). A common polymorphism in the upstream promoter region of the Hepatocyte Nuclear Factor-4 Gene on chromosome 20q is associated with type 2 diabetes and appears to contribute to the evidence for linkage in an Ashkenazi Jewish population. *Diabetes,* 53:1134–1140.

Ludwig, JA and Weinstein, JN (2005). Biomarkers in cancer staging, prognosis and treatment selection. *Nat Rev Cancer,* 5:845–856.

Ma, D, Shield, JP, Dean, W, Leclerc, I, Knauf, C, Burcelin, RR et al. (2004). Impaired glucose homeostasis in transgenic mice expressing the human transient neonatal diabetes mellitus locus, TNDM. *J Clin Invest,* 114:339–348.

Maassen, JA, 'T Hart, LM, Van Essen, E, Heine, RJ, Nijpels, G, Jahangir Tafrechi, RS, et al. (2004). Mitochondrial diabetes: molecular mechanisms and clinical presentation. *Diabetes,* 53(Suppl. 1):S103–S109.

Maffei, M, Halaas, J, Ravussin, E, Pratley, RE, Lee, GH, Zhang, Y, et al. (1995). Leptin levels in human and rodent: measurement of plasma leptin and *ob* RNA in obese and weight-reduced subjects. *Nat Med,* 1:1155–1161.

Malecki, MT, Jhala, US, Antonellis, A, Fields, L, Doria, A, Orban, T, et al. (1999). Mutations in *NEUROD1* are associated with the development of type 2 diabetes mellitus. *Nat Genet,* 23:323–328.

Manson, JE, Rimm, EB, Stampfer, MJ, Colditz, GA, Willett, WC, Krolewski, AS, et al. (1991). Physical activity and incidence of non-insulin-dependent diabetes mellitus in women. *Lancet,* 338:774–778.

Martin, BC, Warram, JH, Krolewski, AS, Bergman, RN, Soeldner, JS, and Kahn, CR (1992). Role of glucose and insulin resistance in development of type 2 diabetes mellitus: results of a 25-year follow-up study. *Lancet*, 340:925–929.

Masud, S, Ye, S, SAS group (2003). Effect of the peroxisome proliferator activated receptor-gamma gene Pro12Ala variant on body mass index: a metaanalysis. *J Med Genet*, 40:773–780.

McCarthy, MI (2003). Growing evidence for diabetes susceptibility genes from genome scan data. *Curr Diab Rep*, 3:159–167.

McCarthy, MI (2004). Progress in defining the molecular basis of type 2 diabetes through susceptibility gene identification. *Hum Mol Genet*, 13 (Suppl. 1):R33–R41.

McCarthy, MI and Hattersley, AT (2001). Molecular diagnostics in monogenic and multifactorial forms of type 2 diabetes. *Expert Rev Mol Diagn*, 1:403–412.

Mehrabian, M, Allayee, H, Stockton, J, Lum, PY, Drake, TA, Castellani, LW, et al. (2005). Integrating genotypic and expression data in a segregating mouse population to identify 5-lipoxygenase as a susceptibility gene for obesity and bone traits. *Nat Genet*, 37:1224–1233.

Meyre, D, Bouatia-Naji, N, Tounian, A, Samson, C, Lecoeur, C, Vatin, V, et al. (2005). Variants of *ENPP1* are associated with childhood and adult obesity and increase the risk of glucose intolerance and type 2 diabetes. *Nat Genet*, 37: 863–867.

Meyre, D, Boutin, P, Tounian, A, Deweirder, M, Aout, M, Jouret, B, et al. (2005). Is glutamate decarboxylase 2 (*GAD2*) a genetic link between low birth weight and subsequent development of obesity in children? *J Clin Endocrinol Metab*, 90:2384–2390.

Meyre, D, Lecoeur, C, Delplanque, J, Francke, S, Vatin, V, Durand, E, et al. (2004). A genome-wide scan for childhood obesity-associated traits in French families shows significant linkage on chromosome 6q22.31–q23.2. *Diabetes*, 53:803–811.

Montague, CT, Farooqi, IS, Whitehead, JP, Soos, MA, Rau, H, Wareham, NJ, et al. (1997). Congenital leptin deficiency is associated with severe early-onset obesity in humans. *Nature*, 387:903–908.

Mootha, VK, Lindgren, CM, Eriksson, K-F, Subramanian, A, Sihag, S, Lehar, J, et al. (2003). PGC-1α-responsive genes involved in oxidative phosphorylation are coordinately downregulated in human diabetes. *Nat Genet*, 34:267–273.

Newman, B, Selby, J, King, M-C, Slemenda, C, Fabsitz, R, and Friedman, GD (1987). Concordance for type 2 (non-insulin-dependent) diabetes mellitus in male twins. *Diabetologia*, 30:763–768.

Nielsen, ED, Hansen, L, Carstensen, B, Echwald, SM, Drivsholm, T, Glümer C, et al. (2003). The E23K variant of Kir6.2 associates with impaired post-OGTT serum insulin response and increased risk of type 2 diabetes. *Diabetes*, 52:573–577.

Ntambi, JM and Miyazaki, M (2004). Regulation of stearoyl-CoA desaturases and role in metabolism. *Prog Lipid Res*, 43:91–104.

O'Rahilly, S, Barroso, I, and Wareham, NJ (2005). Genetic factors in type 2 diabetes: the end of the beginning? *Science*, 307:370–373.

Owen, K and Hattersley, AT (2001). Maturity-onset diabetes of the young: from clinical description to molecular genetic characterization. *Best Pract Res Clin Endocrinol Metab*, 15:309–323.

Owen, KR, Evans, JC, McCarthy, MI, Walker, M, Hitman, G, Frayling, TM, et al. (2004). Role of the insulin promoter factor-1 D76N polymorphism in predisposing to type 2 diabetes. *Diabetologia*, 47:957–958.

Owen, KR, Stride, A, Ellard, S, and Hattersley, AT (2003). Etiological investigation of diabetes in young adults presenting with apparent type 2 diabetes. *Diabetes Care*, 26:2088–2093.

Page, GP, George, V, Page, PZ, and Allison, DB (2003). "Are we there yet?": deciding when one has demonstrated specific genetic causation in complex diseases and quantitative traits. *Am J Hum Genet*, 73:711–719.

Parikh, H and Groop, L (2004). Candidate genes for type 2 diabetes. *Rev Endocr Metab Disord*, 5:151–176.

Pearson, ER, Starkey, BJ, Powell, RJ, Gribble, FM, Clark, PM, and Hattersley, AT (2003). Genetic cause of hyperglycaemia and response to treatment in diabetes. *Lancet*, 362:1275–1281.

Peltonen, L and McKusick, V (2001). Genomics and medicine. Dissecting human disease in the postgenomic era. *Science*, 291:1224–1229.

Perusse, L, Rankinen, T, Zuberi, A, Chagnon, YC, Weisnagel, SJ, Argyropoulos, G, et al. (2005). The human obesity gene map: the 2004 update. *Obes Res*, 13:381–490.

Pociot, F and McDermott, MF (2002). Genetics of type 1 diabetes mellitus. *Genes Immunity*, 3:235–249.

Poulsen, P, Vaag, AA, Kyvik, KO, Møller Jensen, D, and Beck-Nielsen, H (1997). Low birth weight is associated with NIDDM in discordant monozygotic and dizygotic twin pairs. *Diabetologia*, 40:439–446.

Raeder, H, Johansson, S, Holm, PI, Haldorsen, IS, Mas, E, Sbarra, V, et al. (2006). Mutations in the *CEL* VNTR cause a syndrome of diabetes and pancreatic exocrine dysfunction. *Nat Genet*, 38:54–62.

Randle, PJ, Garland, PB, Hales, CN, and Newshome, EA (1965). The glucose fatty-acid cycle. Its role in insulin sensitivity and the metabolic disturbances of diabetes mellitus. *Lancet*, I:785–789.

Ravelli, ACJ, van der Meulen, JHP, Michels, RPJ, Osmond, C, Barker, DJP, Hales, CN, et al. (1998). Glucose tolerance in adults after prenatal exposure to famine. *Lancet*, 351:173–177.

Ravussin, E, Lillioja, S, Knowler, WC, Christin, L, Freymond, D, Abbott, WG, et al. (1988). Reduced rate of energy expenditure as a risk factor for body-weight gain. *N Engl J Med*, 318:467–472.

Reaven, GM, Brand, RJ, Chen, Y-D, Mathur, AK, and Goldfine, I (1993). Insulin resistance and insulin secretion are determinants of oral glucose tolerance in normal individuals. *Diabetes*, 42:1324–1332.

Ryan, EA, Imes, S, Liu, D, McManus, R, Finegood, DT, Polonsky, KS, et al. (1995). Defects in insulin secretion and action in women with a history of gestational diabetes. *Diabetes*, 44:506–512.

Saad, MF, Knowler, WC, Pettitt, DJ, Nelson, RG, Mott, DM, and Bennett, PH (1988). The natural history of impaired glucose tolerance in the Pima Indians. *N Engl J Med*, 319:1500–1506.

Saxena, R, Voight, BF, Lyssenko, V, Burtt, NP, de Bakker, PI, Chen, H, et al. (2007). Genome-wide association analysis identifies loci for type 2 diabetes and triglyceride levels. *Science*, 316(5829):1331–1336.

Schadt, EE, Lamb, J, Yang, X, Zhu, J, Edwards, S, Guhathakurta, D, et al. (2005). An integrative genomics approach to infer causal associations between gene expression and disease. *Nat Genet*, 37:710–717.

Schrauwen, P and Hesselink, MKC (2004). Oxidative capacity, lipotoxicity, and mitochondrial damage in type 2 diabetes. *Diabetes*, 53:1412–1417.

Schulte, I, Tammen, H, Selle, H, and Schulz-Knappe, P (2005). Peptides in body fluids and tissues as markers of disease. *Expert Rev Mol Diagn*, 5:145–157.

Schwanstecher, C, Meyer, U, and Schwanstecher, M (2002). KIR6.2 Polymorphism predisposes to type 2 diabetes by inducing overactivity of pancreatic β-cell ATP-sensitive K+ channels. *Diabetes*, 51:875–879.

Schwanstecher, C, Neugebauer, B, Schulz, M, and Schwanstecher, M (2002). The common single nucleotide polymorphism E23K in KIR6.2 sensitizes pancreatic β-cell ATP-sensitive potassium channels toward activation through nucleoside diphosphonates. *Diabetes*, 51(Suppl. 3):S363–S367.

Schwanstecher, C and Schwanstecher, M (2002). Nucleotide sensitivity of pancreatic ATP-sensitive potassium channels and type 2 diabetes. *Diabetes*, 51(Suppl. 3):S358–S362.

Schwartz, MW, Woods, SC, Seeley, RJ, Barsh, GS, Baskin, DG, and Leibel, RL (2003). Is the energy homeostasis system inherently biased toward weight gain? *Diabetes*, 52:232–238.

Scott, LJ, Mohlke, KL, Bonnycastle, LL, Willer, CJ, Li, Y, Duren, WL, et al. (2007). A genome-wide association study of type 2 diabetes in Finns detects multiple susceptibility variants. *Science*, 316(5829):1341–1345.

Silander, K, Mohlke, KL, Scott, LJ, Peck, EC, Hollstein, P, Skol, AD, et al. (2004). Genetic variation near the hepatocyte nuclear factor-4 gene predicts susceptibility to type 2 diabetes. *Diabetes*, 53:1141–1149.

Simmons, D and Powell, MJ (1993). Metabolic and clinical characteristics of south Asians and Europeans in Coventry. *Diabet Med*, 10:751–758.

Skol, AD, Scott, LJ, Abecasis, GR, and Boehnke, M (2006). Joint analysis is more efficient than replication-based analysis for two-stage genome-wide association studies. Nat Genet 38:209:213.

Sladek, R, Rocheleau, G, Rung, J, Dina, C, Shen, L, Serre, D, et al. (2007). A genome-wide association study identifies novel risk loci for type 2 diabetes. *Nature*, 445(7130):881–885.

Smyth, D, Cooper, JD, Collins, JE, Heward, JM, Franklyn, JA, Howson, JM, et al. (2004). Replication of an association between the lymphoid tyrosine phosphatase locus (LYP/PTPN22) with type 1 diabetes, and evidence for its role as a general autoimmunity locus. *Diabetes,* 53:3020–3023.

Song, JY, Park, HG, Jung, SO, and Park, J (2005). Diagnosis of HNF-1alpha mutations on a PNA zip-code microarray by single base extension. *Nucleic Acids Res,* 33:e19.

Song, Y, Niu, T, Manson, JE, Kwiatkowski, DJ, and Liu, S (2004). Are variants in the *CAPN10* gene related to risk of type 2 diabetes? A quantitative assessment of population and family-based association studies. *Am J Hum Genet,* 74:208–222.

Sotos, JG, Reinhoff, HY, Block, GD, Aulisio, MP, Khoury, MJ, Ference, BA, et al. (2000). Will genetics revolutionize medicine? *N Engl J Med,* 343:1496–1498.

Sreenan, SK, Zhou, YP, Otani, K, Hansen, PA, Currie, KP, Pan, CY, et al. (2001). Calpains play a role in insulin secretion and action. *Diabetes,* 50:2013–2020.

Steinthorsdottir, V, Thorleifsson, G, Reynisdottir, I, Benediktsson, R, Jonsdottir, T, Walters, GB, et al. (2007). A variant in CDKAL1 influences insulin response and risk of type 2 diabetes. *Nat Genet,* 39(6):770–775.

Stoffers, DA, Ferrer, J, Clarke, WL, and Habener, JF (1997). Early-onset type-II diabetes mellitus (MODY4) linked to *IPF1*. *Nat Genet,* 17:138–139.

Stumvoll, M, Goldstein, BJ, and van Haeften, TW (2005). Type 2 diabetes: principles of pathogenesis and therapy. *Lancet,* 365:1333–1346.

Stunkard, AJ, Foch, TT, and Hrubec, Z (1986). A twin study of human obesity. *JAMA,* 256:51–54.

Stunkard, AJ, Harris, JR, Pedersen, NL, and McClearn, GE (1990). The body-mass index of twins who have been reared apart. *N Engl J Med,* 322:1483–1487.

Stunkard, AJ, Sorensen, TI, Hanis, C, Teasdale, TW, Chakraborty, R, Schull, WJ, et al. (1986). An adoption study of human obesity. *N Engl J Med,* 314:193–198.

Suzuki, S, Oka, Y, Kadowaki, T, Kanatsuka, A, Kuzuya, T, Kobayashi, M, et al. (2003). The research committee for specific types of diabetes mellitus with gene mutations of the Japan Diabetes Society clinical features of diabetes mellitus with the mitochondrial DNA 3243 (A-G) mutation in Japanese: maternal inheritance and mitochondria-related complications. *Diabetes Res Clin Pract,* 59:207–217.

Swarbrick, MM, Waldenmaier, B, Pennacchio, LA, Lind, DL, Cavazos, MM, Geller, F, et al. (2005). Lack of support for the association between *GAD2* polymorphisms and severe human obesity. *PLOS Biol,* 3:e315.

Tartaglia, LA, Dembski, M, Weng, X, Deng, N, Culpepper, J, Devos, R, et al. (1995). Identification and expression cloning of a leptin receptor, OB-R. *Cell,* 83:1263–1271.

Tattersall, R and Fajans, S (1975). A difference between the inheritance of classic juvenile-onset and maturity-onset type diabetes of young people. *Diabetes,* 24:44–53.

The Expert Committee on the Diagnosis and Classification of Diabetes Mellitus (2003). Report of the expert committee on the diagnosis and classification of diabetes mellitus. *Diabetes Care,* 26:S5–S20.

Tuomilehto, J, Lindstrom, J, Eriksson, JG, Valle, TT, Hamalainen, H, Ilanne-Parikka, P, et al., Finnish Diabetes Prevention Study Group (2001). Prevention of type 2 diabetes mellitus by changes in lifestyle among subjects with impaired glucose tolerance. *N Engl J Med,* 344:1343–1350.

Ueda, H, Howson, JMM, Esposito, L, Heward, J, Snook, H, Chamberlain, G, et al. (2003). Association of the T-cell regulatory gene CTLA4 with susceptibility to autoimmune disease. *Nature,* 423:506–511.

Vaisse, C, Clement, K, Guy-Grand, B, and Froguel, P (1998). A frameshift mutation in human *MC4R* is associated with a dominant form of obesity. *Nat Genet,* 20:113–114.

van Haeften, TW, Dubbeldam, S, Zonderland, ML, and Erkelens, DW (1998). Insulin secretion in normal glucose-tolerance relatives of type 2 diabetic subjects. *Diabetes Care,* 21:278–282.

Vasseur, F, Helbecque, N, Dina, C, Lobbens, S, Delannoy, V, Gaget, S, et al. (2002). Single nucleotide polymorphism haplotypes in the both proximal promoter and exon 3 of the *APM1* gene modulate adipocyte-secreted adiponectin hormone levels and contribute to the genetic risk for type 2 diabetes in French Caucasians. *Hum Mol Genet,* 11:2607–2614.

Velho, G, Byrne, MM, Clément, K, Sturis, J, Pueyo, ME, Blanché, H, et al. (1996). Clinical phenotypes, insulin secretion and insulin sensitivity in kindreds with maternally inherited diabetes and deafness due to mitochondrial tRNALeu(UUR) gene mutation. *Diabetes,* 45:478–487.

Vionnet, N, Hani, EH, Dupont, S, Gallina, S, Francke, S, Dotte, S, et al. (2000). Genomewide search for type 2 diabetes-susceptibility genes in French Whites: evidence for a novel susceptibility locus for early-onset diabetes on chromosome 3q27-qter and independent replication of a type 2-diabetes locus on chromosome 1q21-q24. *Am J Hum Genet,* 67:1470–1480.

Wacholder, S, Chanock, S, Garcia-Closas, M, El Ghormli, L, and Rothman, N (2004). Assessing the probability that a positive report is false: an approach for molecular epidemiology studies. *J Natl Cancer Inst,* 96:434–442.

Walder, K, Kerr-Bayles, L, Civitarese, A, Jowett, J, Curran, J, Elliott, K, et al. (2005). The mitochondrial rhomboid protease *PSARL* is a new candidate gene for type 2 diabetes. *Diabetologia,* 48(3):459–468.

Walder, K, Norman, RA, Hanson, RL, Schrauwen, P, Neverova, M, Jenkinson, CP, et al. (1998). Association between uncoupling protein polymorphisms (*UCP2–UCP3*) and energy metabolism/obesity in Pima Indians. *Hum Mol Genet,* 7:1431–1435.

Wang, WYS, Barratt, BJ, Clayton, DG, and Todd, JA (2005). Genome-wide association studies: theoretical and practical concerns. *Nat Rev Genet,* 6:109–118.

Warram, JH, Martin, BC, Krolewski, AS, Soeldner, JS, and Kahn, CR (1990). Slow glucose removal rate and hyperinsulinaemia precede the development of type II diabetes in the offspring of diabetic parents. *Ann Intern Med,* 113:909–915.

Weedon, MN, Frayling, TM, Shields, B, Knight, B, Turner, T, Metcalf, BS, et al. (2005). Genetic regulation of birth weight and fasting glucose by a common polymorphism in the islet cell promoter of the glucokinase gene. *Diabetes,* 54:576–581.

Weedon, MN, Owen, KR, Shields, B, Hitman, G, Walker, M, McCarthy, MI, et al. (2004). Common variants of the HNF4alpha P2 promoter are associated with type 2 diabetes in the UK population. *Diabetes,* 53:3002–3006.

Weedon, MN, Owen, KR, Shields, B, Hitman, G, Walker, M, McCarthy, MI, et al. (2005). A large-scale association analysis of common variation in the HNF1alpha gene in the UK Caucasian population. *Diabetes,* 54:2487–2491.

Weedon, MN, Schwarz, PEH, Horikawa, Y, Iwasaki, N, Illig, T, Holle, R, et al. (2003). Meta-analysis confirms a role for calpain-10 variation in type 2 diabetes susceptibility. *Am J Hum Genet,* 73:1208–1212.

Weinberg, CR, Wilcox, AJ, and Lie, RT (1998). A log-linear approach to case-parent-triad data: assessing effects of disease genes that act either directly or through maternal effects and that may be subject to parental imprinting. *Am J Hum Genet,* 62:969–978.

Williams, RC, Long, JC, Hanson, RL, Sievers, ML, and Knowler, WC (2000). Individual estimates of European genetic admixture associated with lower body-mass index, plasma glucose, and prevalence of type 2 diabetes in Pima Indians. *Am J Hum Genet,* 66:527–538.

Wolford, JK, Yeatts, KA, Dhanjal, SK, Black, MH, Xiang, AH, Buchanan, TA, et al. (2005) Sequence variation in PPARG may underlie differential response to troglitazone. *Diabetes,* 54:3319–3325.

Yamagata, K, Furuta, H, Oda, N, Kasiaki, PJ, Menzel, S, Cox, NJ, et al. (1996). Mutations in the hepatocyte nuclear factor-4α gene in maturity-onset diabetes of the young (MODY1). *Nature,* 384:458–460.

Yamagata, K, Oda, N, Kaisaki, PJ, Menzel, S, Furuta, H, Vaxillaire, M, et al. (1996). Mutations in the hepatocyte nuclear factor-1α gene in maturity-onset diabetes of the young (MODY3). *Nature,* 384:455–458.

Yang, Q, Khoury, MJ, Friedman, JM, Little, J, and Flanders, WD (2005). How many genes underlie the occurrence of common complex diseases in the population? *Int J Epidemiol,* 34:1129–1137.

Yeo, GS, Farooqi, IS, Aminian, S, Halsall, DJ, Stanhope, RG, and O'Rahilly, S (1998). A frameshift mutation in MC4R associated with dominantly inherited human obesity. *Nat Genet,* 20:111–112.

Yki-Jarvinen, H (2002). Ectopic fat accumulation: an important cause of insulin resistance in humans. *J R Soc Med,* 95(Suppl. 42):39–45.

Zeggini, E, Parkinson, JR, Halford, S, Owen, KR, Frayling, TM, Walker, M, et al. (2004). Association studies of insulin receptor substrate 1 gene (IRS1) variants in type 2 diabetes samples enriched for family history and early age of onset. *Diabetes,* 53:3319–3322.

Zeggini, E, Rayner, W, Morris, AP, Hattersley, AT, Walker, M, Hitman, GA, et al. (2005). An evaluation of HapMap sample size and tagging SNP performance in large-scale empirical and simulated data sets. *Nat Genet,* 37:1320–1322.

Zeggini, E, Weedon, MN, Lindgren, CM, Frayling, TM, Elliott, KS, Lango, H, et al. (2007). Replication of genome-wide association signals in UK samples reveals risk locifor type 2 diabetes. *Science,* 316(5826):1336–1341.

Zhang, R, Barker, L, Pinchev, D, Marshall, J, Rasamoelisolo, M, Smith, C, et al. (2004). Mining biomarkers in human sera using proteomic tools. *Proteomics,* 4:244–256.

Zhang, Y, Proenca, R, Maffei, M, Barone, M, Leopold, L, and Friedman, JM (1994). Positional cloning of the mouse obese gene and its human homologue. *Nature,* 372:425–432.

Zimmet, P, Alberti, KGMM, and Shaw, J (2001). Global and societal implications of the diabetes epidemic. *Nature,* 414:782–787.

Zondervan, KT and Cardon, LR (2004). The complex interplay among factors that influence allelic association. *Nat Rev Genet,* 5:89–100.

Zung, A, Glaser, B, Nimri, R, and Zadik, Z (2004). Glibenclamide treatment in permanent neonatal diabetes mellitus due to an activating mutation in Kir6.2. *J Clin Endocrinol Metab,* 89:5504–550.

13

Chronic Renal Disease

Albert CM Ong and A Peter Maxwell

The prevalence of end-stage renal disease (ESRD) is steadily increasing, contributing to growth in the provision of renal replacement therapy (RRT) (dialysis and transplantation) for over 1 million individuals worldwide (Collins et al., 2003). Living with ESRD is challenging for individuals and the economic costs of RRT for society are enormous (El Nahas and Bello, 2005).

Figure 13–1 shows the relative prevalence of different primary renal diseases among the England and Wales population on RRT in 2003. Diabetic nephropathy is the most common cause of ESRD and is increasing in incidence worldwide. There is increasing evidence for genetic (polygenic) susceptibility for diabetic nephropathy. Autosomal dominant polycystic kidney disease (ADPKD) accounts for approximately 10% of treated ESRD and is the single most common monogenic cause of ESRD worldwide. In this chapter, we review the evidence of how genetic factors can modify the phenotypic expression of these two common diseases and illustrate the different experimental approaches that are being used to identify modifying genes, important in chronic renal failure.

ADPKD as a Complex Trait Disease

ADPKD is one of the most common human monogenic diseases with an incidence of 1:400–1:1000. It is characterized by the progressive development and enlargement of focal cysts in both kidneys, typically resulting in ESRD by the fifth decade.

Two Genes Cause ADPKD

Mutation to *PKD1* (chromosome region 16p13.3) is the most common cause of ADPKD (~86% cases) with most of the remainder due to changes to *PKD2* (4q22). PKD1 and PKD2 patients have indistinguishable renal and extrarenal phenotypes: they were only recognized as distinct diseases by genetic linkage analysis in the late 1980s (Parfrey et al., 1990). *PKD1* is a complex gene with 46 exons that generates a large transcript (~14 kb) containing a long open-reading frame predicted to encode a 4302 amino acid protein named polycystin-1. *PKD2* has 15 exons, generates approximately 5 kb transcript, and encodes polycystin-2, a protein of 968 amino acids (Mochizuki et al., 1996) (Fig. 13–2).

Gene Locus Effect

PKD2 is a significantly milder disease in terms of the mean age at diagnosis, a lower prevalence of hypertension and a later age at onset of ESRD (PKD1, 54.3 years; PKD2, 74.0 years) (Hateboer et al., 1999) Consistent with this milder phenotype, there appears to be an enrichment of PKD2 patients among ADPKD patients who present late (after 63 years) with ESRD (Torra et al., 2000). Conversely, all cases of very-early onset (VEO) ADPKD manifesting in utero or infancy are related to mutations in *PKD1* (Peral et al., 1996).

The Effect of Gender

While there is no clear gender difference in PKD1, PKD2 females have a significantly better prognosis (age at onset of ESRD: males, 68.1 years; females, 76.0 years): the reason for this is unclear (Rossetti, Burton, et al., 2002; Magistroni et al., 2003).

Allelic Heterogeneity

Both *PKD1* and *PKD2* exhibit marked allelic heterogeneity, with approximately 200 different *PKD1* and over 50 different *PKD2* mutations described (Rossetti et al., 2001; Magistroni et al., 2003; Rossetti et al., 2003). Unlike *PKD2*, *PKD1* is also highly polymorphic with at least 71 changes described so far (Rossetti, Chauveau, et al., 2002). The majority of mutations are private (that is., unique to a single family) and indicate that a significant level of new mutation is occurring (Rossetti et al., 2001). For *PKD1*, mutations 5′ to the median nucleotide (7812) are associated with more severe disease (average age at onset of ESRD: 5′, 53 years; 3′, 56 years) and a significantly greater risk of developing intracranial aneurysms (Rossetti, Burton, et al., 2002; Rossetti et al., 2003). This association is not related to mutation type and may be due to the proposed cleavage of polycystin-1 via a G-protein coupled receptor proteolytic site (GPS) into two different proteins (Fig. 13–2) with mutations to each half having potentially different phenotypic consequences (Qian et al., 2002; Rossetti et al., 2003). As yet, no clear phenotype/genotype correlations have been reported for *PKD2* (Magistroni et al., 2003).

Gene Syndromes

Patients with a contiguous gene deletion of both *PKD1* and the neighboring tuberous sclerosis gene (*TSC2*) allele have more severe early-onset PKD than those with *TSC2* mutations alone (Brook-Carter et al., 1994). Most probably, this indicates a synergistic role between polycystin-1 and tuberin (the TSC2 protein) in cyst development: tuberin may play a role in trafficking polycystin-1 to the lateral cell membrane in kidney epithelial cells (Ong et al., 1999; Kleymenova et al., 2001).

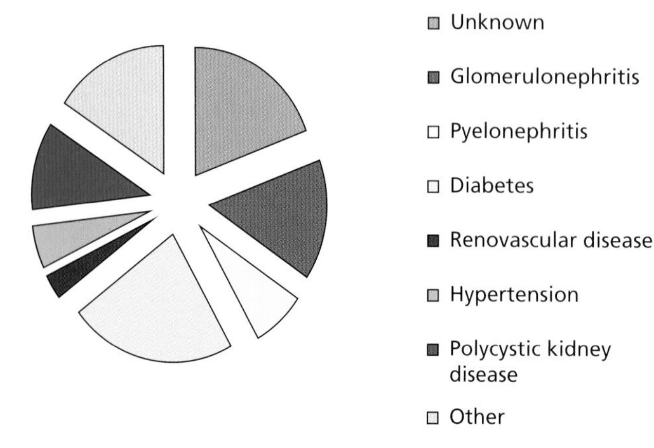

Figure 13–1 Primary diagnosis in patients of renal replacement therapy (RRT) under the age of 65 years in England and Wales (E&W) in 2003. Unknown diagnosis (17.5%), glomerulonephritis (13.5%), pyelonephritis (6.7%), diabetic nephropathy (20.4%), renovascular disease (2.8%), hypertension (5.1%), polycystic kidney disease (PKD) (9.9%), other diagnoses (14.4%), nonreturned (9.6%). The major change in the over 65-age group was an increase in renovascular disease (11.2%). The prevalence of diabetic nephropathy and PKD in this age group was lower, possibly reflecting an increase in mortality at any early age. (Data from UK Renal Registry Data, 2003.)

A family with bilineal inheritance of germ line *PKD1* and *PKD2* mutations has been described (Pei et al., 2001). Although these patients have more severe disease than cases with either mutation alone, the difference is not dramatic (that is, not every renal tubular cell gives rise to a cyst). The effect of a trans-heterozygous mutation in either gene appears to act as a modifying factor for the other in terms of the risk of cyst development or hastening its progression rather than an effect on cyst initiation itself (Wu et al., 2002).

Mutational Mechanism in ADPKD

A "two-hit" mechanism of cyst formation has been proposed for ADPKD (consisting of a germ line mutation to one allele and a somatic mutation to the other) (Fig. 13–3). There is evidence for epithelial cell clonality and a high rate of somatic mutations in cells isolated from individual kidney and liver cysts (Qian et al., 1996; Brasier and Henske, 1997; Watnick et al., 1998). Experimentally, a unique *Pkd2* knockout mouse (*Pkd2*^{WS25} mutant), which has an unstable allele (prone to inactivation by recombination) develops progressive cystic disease in a manner consistent with a two-hit model (Lu et al., 1997; Wu, Agati, et al., 1998). Certainly, a two-hit model could help explain both the focal nature of cyst development and the striking phenotypic variability seen in most families. However, there remain questions as to whether a two-hit mechanism is the only means to generate a cyst and indeed, whether these somatic events may be later events more important for cyst expansion and progression rather than initiation (Ong and Harris, 1997).

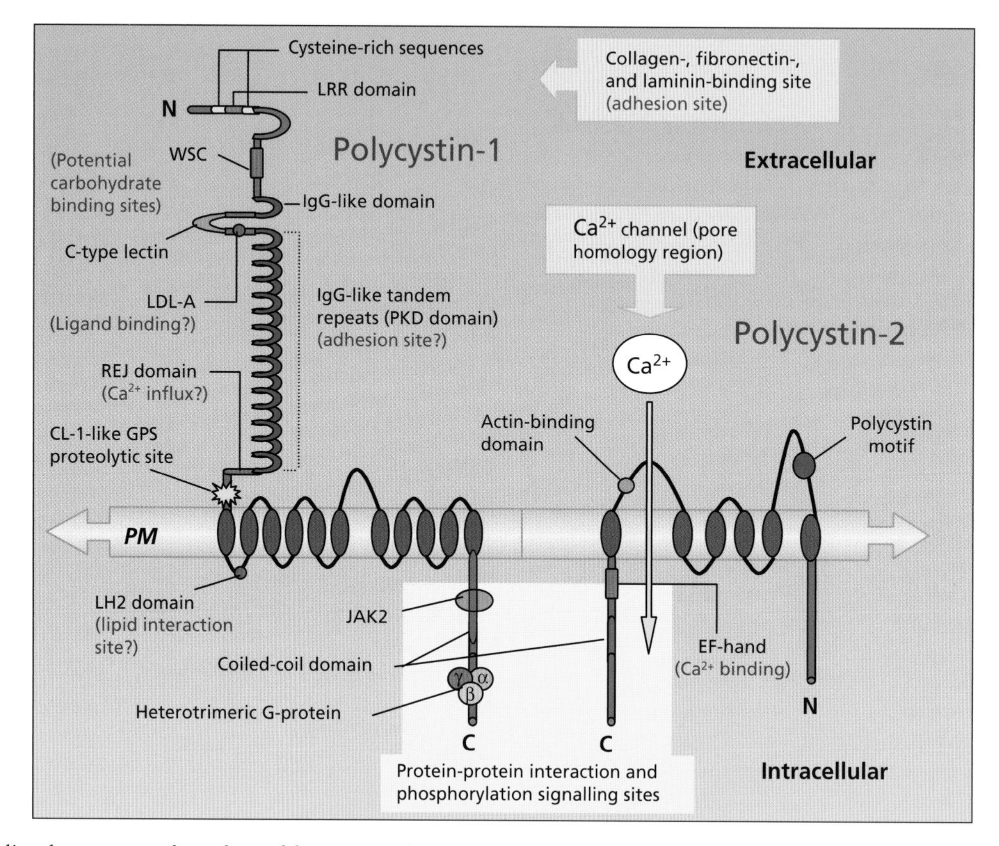

Figure 13–2 Predicted structure and topology of the autosomal dominant polycystic kidney disease (ADPKD) proteins, polycystin-1, and polycystin-2. A number of functional protein, carbohydrate, and lipid-binding functional domains of both proteins are depicted. Both proteins interact via a coiled coil domain in their C- termini. A G-protein coupled receptor proteolytic site (GPS) may allow polycystin-1 to be cleaved into two halves, which remain tethered by noncovalent bonds.

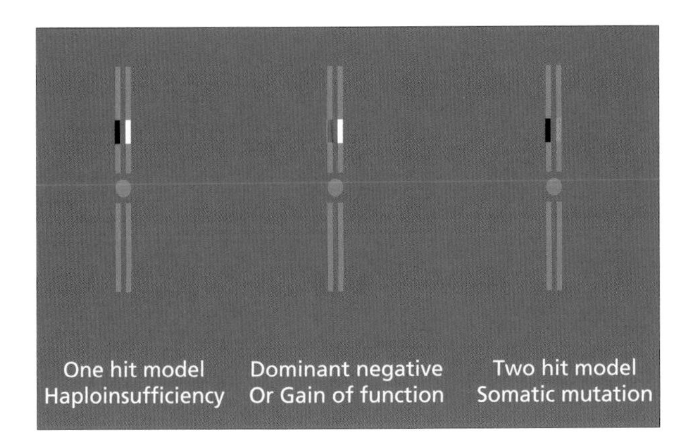

Figure 13–3 Mutational mechanisms underlying cyst initiation in autosomal dominant polycystic kidney disease (ADPKD). The presence of a germ line null mutation (black) in either PKD1 or PKD2 is necessary but not sufficient for cysts to arise from every cell. A one-hit model is consistent with the presence of a normal allele (white) whereas a two-hit model requires a somatic null mutation (grey) to occur to the normal allele. Alternatively, a germ line mutant allele (red) could exert a phenotypic change in the presence of a normal allele (white) via a dominant negative or gain of function mechanism.

A decrease (loss of a single allele or haploinsufficiency) or increase (transgenic expression) in polycystin-1 dosage may itself result in cyst formation (Pritchard et al., 2000; Lantinga-van Leeuwen et al., 2004). In addition, haploinsufficient *Pkd2* mice have reduced survival (not due to renal failure) indicating that a dosage reduction of polycystin-2 itself can exert a phenotypic load (Wu et al., 2000). Recent studies have shown that $Pkd2^{+/-}$ vascular smooth muscle cells (which express a lower level of polycystin-2) have altered intracellular Ca^{2+} homeostasis and are hyperproliferative (Qian et al., 2003). Similar findings have been shown for $Pkd1^{+/-}$ cyst epithelial cells (Yamaguchi et al., 2004).

In summary, there appear to be multiple mechanisms that could underlie cyst formation. Cyst development requires a germ line mutation but beyond this, the likelihood of cyst formation is influenced by a number of different factors which could include somatic genetic events at the other (normal) allele, mutations at the other ADPKD gene, and possibly a wide array of other genetic loci (e.g., TSC2). In effect, these loci act as modifiers of disease presentation in ADPKD. Environmental or genetic factors that modulate the rate of somatic mutation or DNA repair could modify the disease phenotype (Peters and Breuning, 2001). Beyond the genetic events, stochastic factors probably also influence whether a cell, haploinsufficient for an ADPKD mutation, is diverted into a cystogenic pathway. Another factor that could modify the cystic phenotype is the presence of partially functional (mutant) polycystin-1 protein, for example, in *Pkd1* del34 mice (Lu et al., 2001); alternatively, nonfunctional mutant protein could act in a dominant negative manner.

Intrafamilial Variability

The wide range of intrafamilial phenotypic variability in large ADPKD families and more recently for PKD1 and PKD2 has been described (Milutinovic et al., 1992; Magistroni et al., 2003; Rossetti et al., 2003). The phenomenon of genetic anticipation in ADPKD families was first postulated in 1925 with the description of VEO cases often born to relatively asymptomatic parents (Cairns, 1925). However, no evidence for dynamic unstable mutations in PKD1 or PKD2 has been found. Nevertheless, there is a high recurrence risk (45%) of a severely affected sibling in subsequent pregnancies (Zerres et al., 1993).

The causes of this variability are largely nonallelic and could include modifying genes, environmental factors, or stochastic events. For VEO cases, it is difficult to envisage (though impossible to exclude) that somatic mutations to PKD1 could be the underlying reason for increased disease severity in utero (Ong et al., 1999).

A number of studies have attempted to define the possible role of modifying genes (heritability) in determining renal outcome. An informative study of 74 parent–offspring pairs, where the age of renal death of both was known, showed a wide Gaussian distribution of values around zero (Geberth et al., 1995). Furthermore, there was no difference in the median age of renal death between two decades (1950–1971 and 1975–1985) thus excluding the effect of secular trend as a major confounding factor. Another study using sib-pairs compared the age of ESRD between 56 sibships and 9 pairs of monozygotic twins and calculated the intraclass correlation coefficient (ICC) within both groups (Persu et al., 2004). The ICC, which compares the similarity within sibships to that between sibships, was significantly higher in the twins (0.92) than that in sibs (0.49). Discordance in the renal phenotype between a pair of dizygotic twins carrying the same *PKD1* mutation has been reported in one family (Peral et al., 1996). This could imply that compared with genetic factors, fetal environment does not play a major role in determining the cystic phenotype.

The possible influence of modifying genes (heritability) in PKD1 families has also been addressed recently in two large studies. Heritability (h^2) for the age of ESRD was estimated at between 43% and 78% (Fain et al., 2005; Paterson et al., 2005). However, the number of potential modifiers, their allele frequency, and their relative effect in relation to determining loss of renal function is not clear. Paterson et al. calculate that a genome-wide scan to look for a single modifier locus will require very large cohorts to be powered adequately and thus necessitate international collaboration (Paterson et al., 2005). These studies also point to the likely significance of (unknown) environmental factors as contributing to the ESRD phenotype. These factors could include urinary tract infections in men and the number of pregnancies in women (Gabow et al., 1992), but require verification in larger scale populations.

Coexistent Diseases—Cystic Fibrosis

A significant body of evidence suggests that ADPKD cyst epithelial cells undergo a phenotypic switch to cyclic adenosine monophosphate (cAMP) responsiveness as measured by cell proliferation and fluid secretion (Grantham, 2003). The cystic fibrosis transmembrane conductance regulator (CFTR) is a cAMP-dependent chloride channel protein, which is mutated in patients with the autosomal recessive disease, cystic fibrosis (CF). CFTR is expressed in the kidney and in cyst lining cells. A family with two individuals with coexistent CF and ADPKD has been reported with normal kidney function suggesting a possible protective effect of the absence of CFTR (O'Sullivan et al., 1998). However, this was not substantiated in another family where neither heterozygous nor homozygous CFTR mutations (DeltaF508) had a protective effect on renal outcome (Persu et al., 2000).

Association Studies of Candidate Gene Polymorphisms

A number of studies examining the association between candidate gene polymorphisms [e.g., angiotensin converting enzyme (ACE) and endothelial nitric oxide synthase (eNOS)] and kidney function have been performed. Overall, these finding have proved to be inconclusive and contradictory (Persu et al., 2002; Walker et al., 2003). As discussed in the preceding section, similar future studies will need to be adequately powered to reach any meaningful conclusions.

Modifiers in Rodent PKD Models

The apparent similarities in the process of cyst formation between different rodent models and human disease have led some investigators to exploit the approach of crossbreeding a PKD allele onto different genetic backgrounds to map modifier loci. In theory, this approach allows environmental factors to be controlled for, thus permitting the contribution of genetic factors to be paramount. In one of the most studied murine PKD models, the *cpk* mouse, it is apparent that disease expression and the severity of both renal and extrarenal phenotypes can be modified by genetic background. For instance, the biliary abnormality in *cpk* is not penetrant in the original C57BL/6 (B6) background but is variably penetrant when expressed in other strain backgrounds, for example, CAST/Ei, DBA/2J, BALB/c, or CD1 (Guay-Woodford, 2003).

PKD due to the *pcy* allele is more severe in a DBA/2 than a C57BL/6 background. A good example of the approach of whole genome quantitative trait loci (QTL) mapping led to the identification of two major modifiers (MOP1 and MOP2) regulating renal disease progression in the *pcy* mouse (Woo et al., 1997). The *pcy* gene was subsequently identified as the murine homologue of the nephronophthisis. Type 3 (*NPHP3*) gene, which gives rise to adolescent-onset nephronophthisis, a phenotype, which can include cysts (Olbrich et al., 2003). It would therefore be of interest to investigate whether MOP1 and MOP2 can also act as modifiers in the human disease. Similar approaches have been taken in other rodent strains including the recessive cpk mouse and the dominant Han:SPRD (cy/+) rat, genes which have not yet been associated with human PKD (Bihoreau et al., 2002; Mrug et al., 2005). Table 13–1 summarizes the most common rodent PKD models that have been characterized and those in which QTL have been mapped. It is hoped that a comparative genome analysis will enable any modifying loci identified in a rodent model to be mapped back to a human susceptibility locus in ADPKD.

Phenotypic Differences between *Pkd1* Knockout Mice

A number of *Pkd1* deficient mice have been reported, all of which are associated with a cystic kidney phenotype (Table 13–2). Almost all are embryonic lethal in the recessive state apart from a hypomorphic allele ($Pkd1^{nl}$) (Lantinga-van Leeuwen et al., 2004). Of interest, some but not all mutants develop skeletal or cardiovascular defects, unrelated to the site of the gene disrupted. There is a high prevalence of aortic aneurysms in homozygous $Pkd1^{nl}$ mice, which has not been described in other models (Lantinga-van Leeuwen et al., 2004). The expression of these phenotypes could therefore be related to genetic background, that is, other modifying genes. This may reflect the situation in human ADPKD where not all patients (that is, ~8–10%) develop vascular complications (aneurysms) and more rarely, patients associated with a Marfanoid phenotype have been reported (Somlo et al., 1993; Rossetti et al., 2003).

Polycystin Homologues

A number of polycystin-like homologues has been identified from whole genomic and EST databases. In essence, they divide into "PC1-like" and "PC2-like" proteins (Table 13–3). Polycystin-L, a PC2-like homologue, has been shown to function as a cation channel in experimental systems (Chen et al., 1999) and polycystin-REJ, a PC1-like protein, is the homologue of a sea urchin protein, suREJ, which is implicated in the acrosome reaction (Hughes et al., 1999). In theory, these homologues could modulate the PKD phenotype caused by mutations in *PKD1* or *PKD2*. However, the restricted tissue distribution of each homologue (most are not expressed in the kidney) suggests that they are more likely to play a role in expression of the extrarenal phenotype.

Possible Pathways Determining Disease Progression

The discussion so far has focused largely on factors that could modulate the rate of cyst initiation. However, factors that could regulate the rate of cyst growth (proliferation, apoptosis, and fluid secretion) may be more appropriate therapeutic targets than

Table 13–1 Quantitative Trait Loci (QTL) Mapping in Rodent Models of Polycystic Kidney Disease (PKD)

Model	Transmission	Gene	Protein	Human Disease	Original Strain	QTL
Mouse						
bpk	AR	BiccI	Bicaudal C		BALB/c	+
cpk	AR	CysI	Cystin		C57BL/6J (B6)	+
inv	AR	Invs	Inversin	Yes	OVE210	
jck	AR	*Nek8*	Nek8		MMTV/c-myc	+
jcpk	AR	*BiccI*	Bicaudal C			
kat	AR	*Nek1*	Nek1		RBF/Dn	+
kat^{2J}					C57BL/6J	
orpk	AR	*TgN737Rpw*	Polaris		FBV/N	+
pcy	AR	*NPHP3*	Nephrocystin-3	Yes	KK DBA/2J	+
Rat						
Han:SPRD-cy	AD	*Pkdr1*	SamCystin		SPRD	+
pck	AR	*Pkhd1*	Fibrocystin	Yes	Crj:CD/SD	
wpk	AR	?	?		Wistar	

Table 13–2 Renal and Extrarenal Phenotypes of *Pkd1* and *Pkd2* Mouse Mutants

Mutant	Exons Disrupted	Embryonic Lethality	Vascular Defect	Cardiac Defect	Skeletal Defect	Heterozygous Phenotype (Cysts; Age)	References
Pkd1							
Del 34	34	+ (E18.5)	—	—	+	Renal >9 m Liver >9 m Pancreatic >22 m	Lu et al., 1997; Lu et al., 1999; Lu et al., 2001
Pkd1L	43–45	+ (E14.5–15.5)	+ (heme; edemas)	—	—	?	Kim et al., 2000
Del 17–21 βgeo	17–21	+ (E13.5–14.5)	+ (heme)	+	+	Renal >3 m Liver >19 m	Boulter et al., 2001
Null	4	+ (E13.5–16.5)	+ (edema)		+	Renal > 2.5 m Liver > 11 m Pancreatic >12 m	Lu et al., 2001
−/−	2–6	+ (E14.5)	+ (heme)	+	—	?	Muto et al., 2002
Null	1	+ (E12.5–birth)	+ (heme; edema)	—	—	Renal >3 m No gross liver cysts	Wu et al., 2002
Pkd1 (nl)	IVS1—aberrant splicing	live born	+ (aortic aneurysms)	—	—	No renal cysts >13 m	Lantinga-van Leeuwen et al., 2004
Pkd2							
WS25 (unstable)	Disrupted exon 1 in IVS1	live born	—	—	—	WS25 +/− Renal >2.5 m (many) Liver >2.5 m	Wu, Agati, 1998; Wu et al., 2000
WS183 (null)	1	+ (E13.5–18.5)	+ (heme; edema)	+	—	+/− Renal < 3 m (rare) Liver < 9 m Pancreatic none	Wu, Agati, 1998; Wu et al., 2000
−/LacZ	1	+ (E12.5–18.5)	?	+	—	?	Pennekamp et al., 2002

attempts to replace the gene (Qian et al., 2001) (Fig. 13–4). Also, the diseased ADPKD kidney is characterized by noncystic features such as interstitial fibrosis, inflammation, and vessel wall thickening, probably contributing to ischemia (Ong and Harris, 2005). Although it is unclear whether these are the direct downstream consequences of cystic transformation, alternative pharmacological strategies targeting these pathways might be effective to prevent or delay progression to ESRD.

A large number of studies have been performed to test potential drugs, treatments, or diets in rodent PKD models especially in the Han:SPRD rat (Guay-Woodford, 2003). However, it should be noted that this model displays some unusual features, for example, gender dimorphism, which could limit its applicability to human ADPKD. Moreover, it is clear that what works in one model may not always be reproduced in another (Guay-Woodford, 2003). Thus, future studies should concentrate on PKD models that are orthologous to PKD1 or PKD2. Such an approach has been successfully employed with a new class of compounds targeting the vasopressin type 2 receptor (VP2R), which are now entering clinical trials (Torres et al., 2004).

Diabetic Nephropathy as an Example of a Complex Multifactorial Renal Disease with a Genetic Predisposition

Introduction

Diabetic nephropathy is now the commonest identified cause of chronic kidney disease in Europe and the USA (UK Renal Registry, 2004; US Renal Data System, 2004). The prevalence of diabetic kidney disease continues to rise reflecting the emerging epidemic of type 2 diabetes, the improving survival of persons with diabetes and the fact that an increasing number of diabetic patients with ESRD are being accepted for RRT (Jones et al., 2005).

The clinical diagnosis of diabetic nephropathy is based on the presence of persistent proteinuria (albuminuria) in an individual with diabetes in the absence of other renal diseases, urinary tract infection, or heart failure. It is usually associated with rising blood pressure and eventual decline in glomerular filtration rate (GFR). The fall in GFR is normally identified, once the serum creatinine rises outside the normal range.

Table 13–3 Polycystin-1 and Polycystin-2 Human Homologues

Protein	Gene	Chromosome	Function	Human Disease	Tissue Distribution	References
Polycystin-1	*PKD1*	16p13.3	Adhesion Channel regulator Mechanosensor	Autosomal dominant polycystic kidney disease (ADPKD)	Widespread, higher in fetal tissue	Hughes et al., 1995
Polycystin-REJ	*PKDREJ*	22q13	Acrosome reaction	Unknown	Testis	Hughes et al., 1999
Polycystin-1L1	*PKD1L1*	7p12–13	?	Unknown	Heart, brain, placenta, testis, mammary tissue	Yuasa et al., 2002
Polycystin-1L2	*PKD1L2*	16q 22–23	?	Unknown	Heart, brain, placenta, testis, lung, pancreas, liver, skeletal muscle	Li et al., 2003
Polycystin-1L3	*PKD1L3*	16q 22–23	?	Unknown	Placenta, liver, testis	Li et al., 2003
Polycystin-2	*PKD2*	4q21–23	Ion channel LR determination ?Acrosome reaction	ADPKD	Widespread	Mochizuki et al., 1996; McGrath et al., 2003; Neill et al., 2004
Polycystin-L (Polycystin-2L; Polycystin-2L1)	*PKDL* (*PKD2L*; *PKD2L1*)	10q24–25	Ion channel	Unknown	Heart, brain, testis, retina, spleen, kidney, skeletal muscle	Wu, Hayashi, 1998; Chen et al., 1999
Polycystin-2L2	*PKD2L2*	5q31	?	Unknown	Testis	Guo et al., 2000

The early changes in kidney function and structure include development of glomerular capillary hypertension, glomerular basement membrane thickening, and mesangial expansion (Fioretto et al., 1992; Adler, 2004). These changes are associated with loss of small quantities of albumin from the glomerular capillaries through "leaky" glomerular basement membranes into the urine (microalbuminuria) (Gross et al., 2005a). Persistent glomerular injury is associated with renal tubular dysfunction, clinical proteinuria, and progressive renal failure (Figs. 13–5A–C).

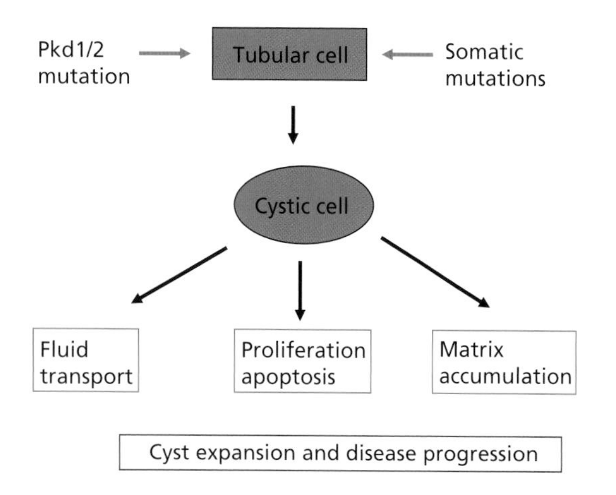

Figure 13–4 Pathophysiology of disease progression in the cystic kidney. The rate of cyst formation as well as downstream phenotypic changes in the cystic cell (cell turnover, fluid secretion, and matrix accumulation) could be modified by nonallelic and environmental factors. This may account for the phenotypic variability arising between individuals carrying the same germ line mutation.

Diabetic nephropathy is arguably the most important microvascular complication of diabetes. Individuals with diabetes have an increased mortality rate due to macrovascular disease, such as coronary heart disease and stroke (Morgan et al., 2000; Williams and Airey, 2002). The development of diabetic nephropathy further amplifies this risk of macrovascular disease, which is increased by a factor of two to four in early renal disease (microalbuminuria), nine times with established nephropathy (proteinuria) and up to 50-fold in persons with ESRD (Deckert et al., 1996; Dinneen and Gerstein, 1997; Tuomilehto et al., 1998; Fuller et al., 2001). Diabetic nephropathy is associated with prolonged duration of diabetes, poor glycemic control, and raised blood pressure and is more common in diabetics of non-Caucasian origin (The Diabetes Control and Complications Trial Research Group, 1993; Nelson et al., 1997; UKPDS, 1998). Primary prevention of diabetic nephropathy requires excellent glycemic control that is often difficult to achieve in practice (The Diabetes Control and Complications Trial Research Group, 1993; UKPDS, 1998; Barnett, 2004). Progression of renal disease can be modified by inhibition of the renin-angiotensin system (RAS) with ACE inhibitors and/or angiotensin II receptor blockers (Lewis et al., 1993; Brenner et al., 2001; Lewis et al., 2001). These agents have beneficial effects beyond blood pressure lowering, including lowering glomerular capillary pressure and reduction of proteinuria (Chiurchiu et al., 2005). Smoking cessation, aspirin therapy, and lipid lowering drugs are also part of the multiple risk factor management to reduce cardiovascular events in persons with nephropathy (Gaede et al., 2003; Fioretto and Solini, 2005). Dietary protein restriction (to 1 g/kg body weight) can also slow progression in proteinuric diabetic patients (Pedrini et al., 1996).

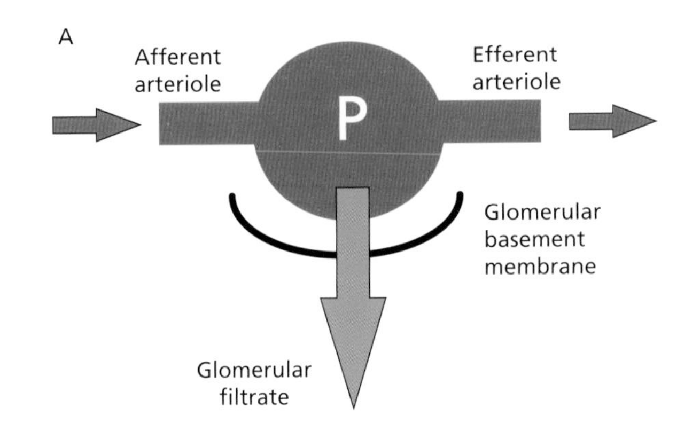

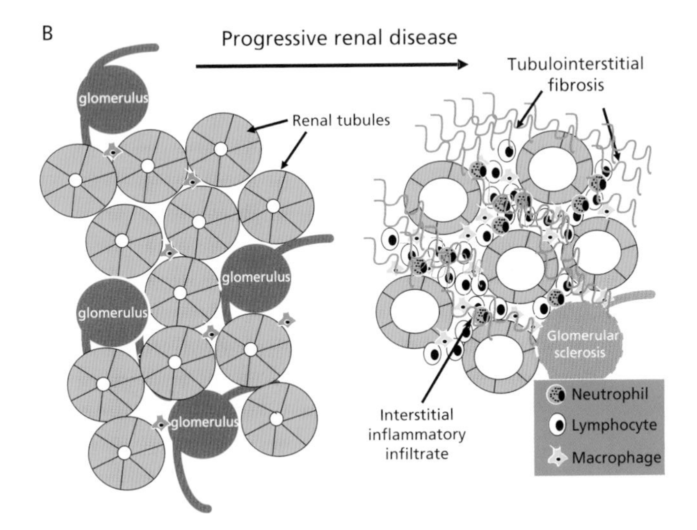

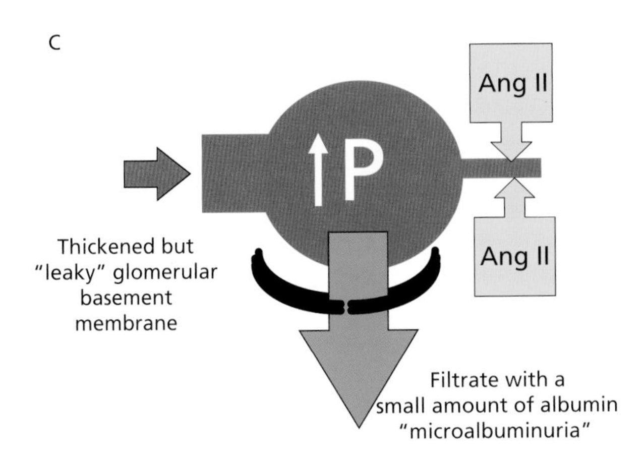

Figure 13–5 (*A*) Normal glomerular filtration. This depends on glomerular capillary pressure (P) providing a hydraulic force for filtration across the semipermeable glomerular basement membrane. Constant glomerular capillary pressure is maintained by autoregulation of glomerular blood flow by adjustment of the vascular resistance provided by the afferent (preglomerular) and efferent (postglomerular) arterioles. (*B*) Early structural and functional changes in the pathogenesis of diabetic nephropathy. Glomerular capillary pressure (P) is elevated due to vasoconstriction of the efferent arteriole mediated by higher renal tissue levels of angiotensin II. The glomerular basement membrane is widened but has decreased charge and size selectivity permitting an increased flux of albumin from blood to urinary space. (*C*) Glomerular hypertension in diabetes. Progressive glomerular sclerosis and tubulointerstitial fibrosis are associated with loss of renal function. An inflammatory interstitial infiltrate is seen in association with tubular atrophy.

Unfortunately, despite these interventions nephropathy still develops and progresses in approximately 30% of diabetics (Krolewski et al., 1996). Additional novel strategies are therefore urgently required to impact on the epidemic of diabetic renal disease (Wolf and Ritz, 2003). Of interest, diabetic nephropathy is not an inevitable complication of a prolonged duration of diabetes. Also, while the cumulative incidence of nephropathy is greatest in those persons with the worst glycemic control the majority of persons with poorly controlled diabetes still do not develop renal disease (Krolewski et al., 1996). This illustrates that hyperglycemia is necessary but in itself not sufficient for the development of renal disease. These clinical paradoxes have lead to renewed interest in the genetic susceptibility to diabetic nephropathy.

Pathophysiology

Mesangial expansion and tubulointerstitial fibrosis are two important renal structural changes associated with the progression of diabetic nephropathy. The degree of tubulointerstitial fibrosis also correlates with prognosis (Gilbert and Cooper, 1999). There is clinical and experimental evidence for a wide range of mechanisms mediating the effects of hyperglycemia and hypertension on the diabetic kidney.

Upregulation of the RAS plays a prominent role in the pathophysiology of nephropathy and clear experimental evidence indicates that treatment with ACE inhibitors can reduce proteinuria, tubulointerstitial injury, and glomerulosclerosis (Zatz et al., 1986; Gilbert and Cooper, 1999). Local activation of intrarenal RAS leads to high levels of angiotensin II in diabetic kidneys (Anderson et al., 1993). While angiotensin II mediates its effects in part through alteration in systemic blood pressure and glomerular hemodynamics, angiotensin II is also recognized to be a potent inducer of transforming growth factor-β (TGF-β). Overexpression of TGF-β is associated with tissue fibrosis (Clouthier et al., 1997).

Accumulation of extracellular matrix is a prominent histological feature of diabetic nephropathy. This extracellular matrix expansion is associated with elevated TGF-β mRNA and protein levels in both renal glomeruli and the tubulointestitium in diabetic patients and animal models of diabetes (Rocco et al., 1992; Yamamoto et al., 1993; Yamamoto et al., 1996). ACE inhibitors can reduce renal TGF-β production in diabetic rodent models.

Persistent hyperglycemia can induce formation of polyol compounds as glucose is reduced to sorbitol by the reduced form of nicotinamide adenine dinucleotide phosphate (NADPH)-dependent enzyme, aldose reductase (Dunlop, 2000). The role of this pathway in development of microvascular diabetic complications is supported by the efficacy of aldose reductase inhibitors such as tolrestat and sorbinol in animal models (Greene et al, 1999; Oates and Mylari, 1999) Toxicity of these agents has largely prevented their therapeutic use in diabetic nephropathy.

Advanced glycosylation end products (AGEs) are nonenzymatic modifications of amino acids and proteins formed in states of chronic hyperglycemia (Brownlee, 2001). Serum levels of AGEs are increased in diabetic nephropathy and are also found in glomeruli and tubules in diabetic animals (Chen et al., 2000). Receptors for AGEs (RAGE) are upregulated in diabetic nephropathy and may play a role in the epithelial to mesenchymal transdifferentiation of renal tubular cells in diabetes. Infusion of AGEs causes albuminuria and glomerulosclerosis (Vlassara et al., 1994).

Hyperglycemia may also play a direct role in upregulating the protein kinase C (PKC) family of serine threonine kinases involved in multiple cell signaling pathways that regulate gene transcription

(Murphy et al., 1998). The PKC enzymes are implicated in regulation of vascular permeability, blood flow, smooth muscle contractility, angiogenesis, and fibrinolysis (Koya et al., 1997). Thus hyperglycemia, by stimulation of PKC regulated pathways, can induce increased production of cytokines and growth factors critical to the progression of nephropathy. Experimental models of diabetes have shown that pharmacological inhibitors of PKC can reduce severity of renal injury (Koya et al., 2000).

Predisposition to Diabetic Nephropathy

A number of environmental risk factors are described that contribute to the pathogenesis of diabetic nephropathy.

Hypertension

Raised blood pressure is arguably the most important modifiable risk factor affecting progression of nephropathy (Cooper, 1998; Ritz et al., 2001; Marshall, 2004). Patients with type 1 diabetes usually develop a rise in blood pressure after the onset of microalbuminuria whereas hypertension often predates the onset of clinical renal disease in persons with type 2 diabetes (Remuzzi et al., 2002). Familial clustering of hypertension has been reported where there are diabetic offspring affected by nephropathy (De Cosmo et al., 1997; Roglic et al., 1998).

Hyperglycemia

The risk of developing diabetic nephropathy is increased by poor glycemic control (Chase et al., 1989; Molitch et al., 1993). The Diabetes Control and Complications Trial (DCCT) conclusively demonstrated that intensive glycemic control in patients with type-1 diabetes reduced the risk of developing nephropathy (microalbuminuria) and the progression to early nephropathy (proteinuria) (The Diabetes Control and Complications Trial Research Group, 1993). The effect of improved glycemic control remained after 4 years of follow up despite the fact that the difference in glycemic control between the intensive and conventional groups had begun to converge (DCCT/EDIC Research Group, 2000, 2003). The evidence for a beneficial effect of intensive glycemic control in retarding progression in patients who already have established nephropathy is less convincing. One interesting study has reported the reversal of diabetic nephropathy pathology following successful pancreatic transplantation for type 1 diabetes (Fioretto et al., 1998).

Multifactorial intervention, including improved glycemic control, can also achieve impressive reductions in the risk of nephropathy in patients with type 2 diabetes. In the Steno study, the intensively treated group had a target HbA1c of less than 6.5% and a target blood pressure of less than 130/80 mmHg (Gaede et al., 1999). In addition, the intensively treated group received low dose aspirin, ACE inhibition, and lipid lowering therapy. During 8 years of follow up, the intensive treatment achieved a 61% reduction in nephropathy and 58% reduction in retinopathy compared to the conventionally treated group (Gaede et al., 2003).

Birth Weight

Low birth weight is associated with an increased risk of cardiovascular disease and type 2 diabetes (Barker et al., 1993; Phillips et al., 1994; Hales and Barker, 2001; Barker and Bagby, 2005). It has been suggested that intra-uterine growth retardation is associated with a reduction in nephron number in man, a hypothesis supported by animal models (Brenner and Chertow, 1994; Luyckx and Brenner, 2005; Schreuder et al., 2005). Some clinical studies have challenged this concept reporting no association between birth weight and progression of diabetic nephropathy (Jacobsen et al., 2003). Careful autopsy studies have correlated the reduced nephron number with the presence of essential hypertension prior to death (Keller et al., 2003; Gross et al., 2005b).

Diet

Modification of dietary protein intake has been studied in an effort to retard progression of renal failure (Klahr et al., 1994; Levey et al., 1999). Animal studies have shown dietary protein restriction reduces glomerular hyperfiltration, intraglomerular capillary hypertension, and progression of renal disease (Molitch et al., 2003). In clinical studies, protein restricted diets have been associated with a fall in proteinuria and reduction in the rate of decline of GFR (Pedrini et al., 1996; Hansen et al., 2002). In routine practice, however, the effects of dietary protein restriction have been difficult to replicate. Interest has also been focused on the influence of dietary fat intake on the development of nephropathy and atherosclerosis. Hypercholesterolemia has been reported to be associated with more rapid fall in GFR (Breyer et al., 1996). Treatment with statin therapy reduced cardiovascular events in type 2 diabetics with microalbuminuria and proteinuria (Colhoun et al., 2004), but paradoxically statin therapy did not reduce cardiovascular mortality in dialysis-dependent diabetic patients (Wanner et al., 2005).

Smoking

Cigarette smoking is associated with development of nephropathy in patients with either type 1 or type 2 diabetes (Telmer et al., 1984; Olivarius Nde et al., 1993; Nilsson et al., 2004). Smoking increases the risk of cardiovascular death, particularly in those diabetic patients with proteinuria (Borch-Johnsen et al., 1987; Moy et al., 1990). There is also evidence that smoking is associated with more rapid progression of renal disease (Orth et al., 1997; Ritz et al., 2000). This may be induced by smoking-related changes to glomerular structure and function (Baggio et al., 2002).

The Impact of Diabetes Duration on Risk of Nephropathy

If renal disease were directly due to persistent hyperglycemia, then a linear relationship between cumulative incidence and duration of diabetes would be expected. Background diabetic retinopathy is present in almost all diabetic individuals after 50 years duration but in contrast, the cumulative incidence of nephropathy in Caucasians plateaus at 30% after approximately 25 years duration (Doria et al., 1995).

Ethnicity

The prevalence of diabetic nephropathy varies among ethnic groups, being highest among Asian, African American, and Native American populations (Cowie et al., 1989; Satko et al., 2002). After 25 years diabetes duration, the cumulative risk for diabetic nephropathy in Caucasians is approximately 30% compared with 80% in Pima Indians (Ballard et al., 1988; Nelson et al., 1993).

Familial Clustering of Renal Disease

Familial aggregation of renal disease is not explained by the known environmental risk factors, and this finding provides strong support for an inherited genetic susceptibility to nephropathy. The study of familial clustering can be achieved by comparing the incidence of renal disease in families where a proband has both diabetes and renal disease compared with the incidence of renal disease in families with diabetic probands without obvious renal disease. Such families, with multiple diabetic offspring, are more difficult to recruit than individual diabetic patients with or without nephropathy.

Familial clustering of nephropathy in Caucasian type 1 diabetics was first reported by Seaquist and colleagues (Seaquist et al., 1989). They found that renal disease was present in 83% of the diabetic siblings of probands who had received kidney transplants. In contrast, renal disease affected only 17% of diabetic siblings of probands without nephropathy ($p < 0.001$). Both groups had similar glycemic control. Confirmation of this familial clustering phenomenon in type 1 diabetes mellitus in further studies demonstrates that despite similar glycemic control, the prevalence of nephropathy was greater in diabetic siblings of a proband with diabetic renal disease (Borch-Johnsen et al., 1992; Quinn et al., 1996) (Fig. 13–6).

These observations on familial clustering of diabetic nephropathy have been extended to type 2 diabetes mellitus and are evident among different ethnic groups including Native Americans, African Americans, and Asians (Pettitt et al., 1990; Satko and Freedman, 2005). Of interest, the risk of developing renal failure is also increased in first-degree relatives of probands with ESRD due to etiologies other than diabetes suggesting common genetic risk factors for progressive kidney disease (O'Dea et al., 1998; Freedman et al., 2005).

Careful analysis of renal biopsy material has demonstrated apparent inherited differences in diabetes-induced glomerular pathology (Fioretto et al., 1999). These insights into inherited differences in glomerular structure are reinforced by assessments of the heritability of albuminuria and determinants of renal function such as GFR (Forsblom et al., 1999; Hunter et al., 2002; Langefeld et al., 2004). An inherited predisposition to progressive renal disease may exist, for example, in individuals with reduced nephron number or subtle defects in glomerular function resulting in albuminuria. The phenotype of progressive renal failure only becomes apparent if the susceptible individual is exposed to additional injury from hypertension or hyperglycemia.

Finding the Genes Responsible for Diabetic Nephropathy

The ultimate goal of genetic studies of diabetic nephropathy is the identification and characterization of the gene variants conferring susceptibility to progressive renal failure. Investigators need to bear in mind the multifactorial etiology of nephropathy and the likelihood of multiple common gene variants, each conferring a small individual relative risk of this complication, are responsible for this susceptibility.

The molecular methods for detecting variants associated with a complex disease are candidate gene analyses and whole genome scans, and both approaches have been employed for diabetic nephropathy. The studies are designed to test either association between the frequency of a specific genetic marker in different populations or linkage to assess the inheritance of a particular genetic locus within families.

Candidate Gene Analysis for Diabetic Nephropathy

In this approach, candidate genes for diabetic nephropathy are selected on the basis of an *a priori* hypothesis that the protein products of such candidate genes are involved in the pathogenesis of the disease. It is possible to prioritize the search for candidate genes in a rational manner based on knowledge of the physiological and biochemical pathways implicated in diabetic nephropathy. For instance, genes involved in the RAS regulating blood pressure or enzymes regulating glucose metabolism are studied. Problems with this approach include the relatively the small number of genes studied to date (<0.5% of the known genome) and the fact that candidates can only be selected based on current, and therefore incomplete, understanding of disease pathogenesis.

Rational selection of candidate genes can also be augmented by data derived from subtractive hybridization experiments and DNA microarrays that may detect upregulated or downregulated genes that had not been previously implicated in disease pathogenesis (Holmes et al., 1997; Murphy et al., 1999; Clarkson et al., 2002; Connolly et al., 2003). These gene expression profiles may be derived from direct analysis of mRNA derived from microdissected renal tissue from patients with diabetic nephropathy (Baelde et al., 2004). Once a candidate gene has been selected, the sequence variation within the gene can be tested for association with the disease using case–control or family-based study designs.

Case–control Studies

Case–control studies are arguably the simplest design and a powerful method to detect association between a gene variant and disease. The frequency of a marker (sequence variant) in a population with a disease like nephropathy (cases) is compared with that of a matched population without disease (controls). A 2×2 contingency table and X^2 test is used to determine if there is a statistical difference between the observed and expected marker frequency.

Case–control studies are relatively easy to undertake but are bedeviled by numerous potential sources of errors. Replication of studies where a positive association has been found has been notoriously difficult. One review of 166 candidate gene association studies

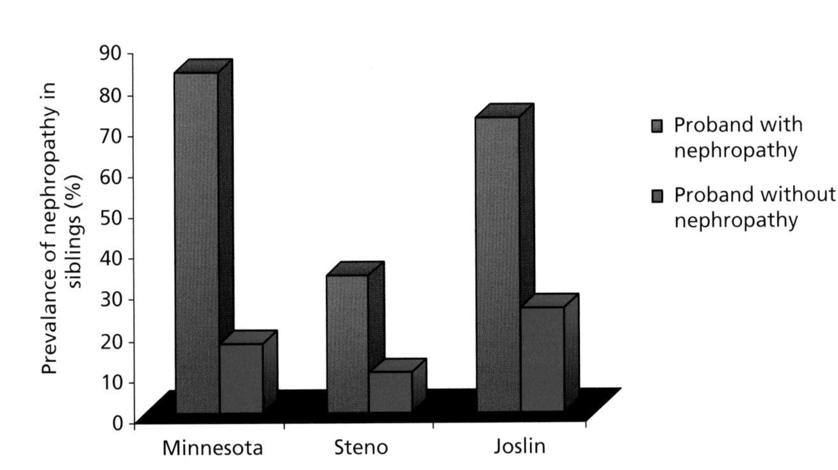

Figure 13–6 Familial clustering of diabetic nephropathy in type 1 diabetes (adopted from Minnesota, USA (Seaquist et al, 1989), Steno Clinic, Copenhagen, Denmark (Borch-Johnsen et al, 1992) and Joslin Diabetes Center, Boston, USA (Quinn et al, 1996).

that had been replicated at least three times found only six associations that remained significant (Hirschhorn et al., 2002). Nevertheless, those that did remain positive after multiple replication studies were clinically relevant and included the associations of variants of the *ApoE* gene with Alzheimer's disease, CTLA4 with Grave's disease, and Factor V Leiden with deep venous thrombosis (Hirschhorn et al., 2002). Further examination of genetic association studies in a series of meta-analyses have provided grounds for optimism that positive associations can be replicated if efforts are made to avoid smaller underpowered studies (Ioannidis et al., 2003; Lohmueller et al., 2003). It should prove possible to detect the effect of common gene variants on common disease risk provided large-scale studies are performed.

Potential pitfalls in the interpretation of case–control studies performed to date include incorrect phenotype definition, small sample size, population stratification from ethnic admixture, over interpretation of results including subgroup analysis, multiple testing, the effect of duration of exposure to risk factors, and publication bias. Sufficient cases and controls must be recruited in order to minimize the risk of identifying false associations that are due to chance alone (type 1 error) or conversely, of failing to detect a true association between a variant and a disease (type 2 error).

Family-based Investigations of Genetic Susceptibility

Family-based studies, where family members are used as controls, have been employed to avoid the problem of population stratification in case–control studies. The relatives used may be siblings or parents and are ethnically matched, thereby eliminating the risk of stratification. In the transmission disequilibrium test (TDT), first pioneered by Spielman and colleagues to study the genetics of type 1 diabetes, affected offspring together with their parents are genotyped (Spielman et al., 1993). The test compares the actual number of alleles transmitted from heterozygous parents to offspring compared to the number expected (using the MacNemar X^2 test). If the allele is transmitted more frequently than the expected 50:50 ratio, then it is likely to be associated with development of disease. The power of these studies is reduced, however, since parents may be homozygous for the marker allele and in that case, no information regarding transmission can be determined.

Unfortunately, family-based studies, using a TDT analysis, are of very limited value in later-onset disease such as nephropathy in type 2 diabetics (where parents are likely to be deceased), and this design has also been difficult to use for nephropathy in type 1 diabetics as there is familial clustering of cardiovascular disease (and premature death) in the parents of type 1 diabetic offspring with nephropathy. One further option, avoiding the requirement to recruit parents, is the use of sibling pairs in a modification known as the sib TDT analysis (Spielman and Ewens, 1998).

The criteria for ideal genetic association studies include large sample sizes, small p values, associations that make biological sense, alleles (sequence variants) that alter the gene product in a physiological way, an initial study that is independently replicated in a separate population, and where possible the association is seen in both case–control and family-based studies (Annotation, 1999). Examples of conflicting results from separate case–control studies testing association between gene variants and diabetic nephropathy are shown in Table 13–4.

Genome-wide Scans for Diabetic Nephropathy

A complimentary approach to identify candidate genes for complex disease such as diabetic nephropathy is the use of genome-wide scans. A major advantage of this approach is that no detailed understanding of the pathophysiology of the disease of interest is required, and there is no *a priori* hypothesis as to which gene or genes are implicated in causation. Previously, these screens were based on the principle of linkage disequilibrium, that is, where a genetic marker and a disease susceptibility gene are in close proximity, the marker will segregate with the disease more often than expected by chance alone. Further, detailed mapping of the chromosomal segment adjacent to the linked marker should, in theory, reveal the identity of a disease susceptibility gene. The linkage method employs several hundred genetic markers (microsatellites consisting of variable sequence lengths of repeated CA base pairs). In practice, this approach has been much more successful in identifying monogenic renal diseases such as X-linked Alport's syndrome (Barker et al., 1990) or ADPKD, where the pattern of inheritance is known.

There have been very few linkage studies in diabetic nephropathy, reflecting the practical difficulty in recruiting large families with multiple affected individual members. For instance, in a study of familial clustering of nephropathy in siblings with type 1 diabetes, Seaquist and colleagues identified a cohort of 696 patients with diabetic nephropathy who had received a renal transplant. Of these 696 patients, only 113 had a type 1 diabetic sibling, and only 26 of these siblings could be enrolled in the study (Seaquist et al., 1989).

Genome-wide linkage studies have been performed for nephropathy in persons with type 2 diabetes. The earliest study, performed in Pima Indians, employed 98 sib-pairs concordant for nephropathy and reported four chromosomal regions linked to nephropathy: 3q26.9, 7q35, 9q, and 20q (Imperatore et al., 1998). The strongest linkage was with chromosome 7q35. One issue in using sib-pairs concordant for both type 2 diabetes and nephropathy is unraveling whether the linkage is with diabetes or renal disease. To resolve this problem, a further genome-wide scan in Pima Indians showed no evidence of linkage to type 2 diabetes on chromosome 7q35 suggesting that this region may indeed harbor a susceptibility gene for nephropathy (Hanson et al., 1998).

An alternative approach to determining the genetic susceptibility to diabetic nephropathy is to study an intermediate trait such as urine protein excretion. Segregation analysis has been employed in Caucasian families to model inheritance patterns of proteinuria (urinary albumin/creatinine ratio or ACR) in type 2 diabetes (Fogarty, Rich, et al., 2000). Sib-pairs, discordant for type 2 diabetes, had urinary ACR tests performed, and this quantitative trait was modeled with age, gender, and duration of diabetes as covariables. Urinary ACR was heritable and genetically correlated to blood pressure (Fogarty, Hanna, et al., 2000). A similar segregation analysis for urine protein excretion has been performed in Pima Indians with type 2 diabetes (Fogarty, Hanna, et al., 2000). These studies are both consistent with a Mendelian model of inheritance albeit that the heritability was 27%, and proteinuria will be influenced by multiple gene variants and environmental factors (Fogarty, Rich, et al., 2000).

Until recently, investigators had only these two broad options (candidate gene studies or linkage analysis) to determine genetic susceptibility to common disease such as diabetic nephropathy. Direct association studies testing candidate gene variants for association with disease were undertaken based on an *a priori* hypothesis. The limitations of this approach included the limited number of candidate genes that have been systematically examined for any disease. The alternative approach of genome scans to identify chromosomal regions linked to common disease has proved less successful than for monogenic disorders with some notable exceptions such as type 1 diabetes (Nistico et al., 1996) and inflammatory bowel disease (Hugot et al., 2001).

Table 13–4 Candidate Genes for Diabetic Nephropathy: Conflicting Results

Gene	Symbol	Chromosome	Association
Angiotensinogen	AGT	1q41–qter	√ (Chang et al., 2003) X (Buraczynska et al., 2002)
Angiotensin II receptor, type 1	AGTR1	3q21–q25	√ (Buraczynska et al., 2002) X (Chang et al., 2003)
Angiotensin I converting enzyme 1	ACE	17q23	√ (Viswanathan et al., 2001) X (Chang et al., 2003)
Apolipoprotein E	APOE	19q13.2	√ (Chowdhury et al., 1998) X (Tarnow et al., 2000)
5,10-methylenetetrahydrofolate reductase	MTHFR	1p36.3	√ (Shcherbak et al., 1999) X (Makita Y et al., 2003)
Serine proteinase inhibitor, clade E, member 1	SERPINE1	7q21.3–q22	√ (Wong et al., 2000) X (Tarnow et al., 2000)
Natriuretic peptide precursor A	NPPA	1p36	√ (Nannipieri et al., 2001) X (Schmidt et al., 1998)
Adrenergic, beta-3-, receptor	ADRB3	8p12–8p11.1	√ (Sakane et al., 1998) X (Grzeszczak et al., 1999)
Nitric oxide synthase	NOS	Multiple members investigated	√ (Zanchi et al., 2000) X (Rippin et al., 2003)
Advanced glycosylation end product-specific receptor	AGER	6p21.3	√ (Pettersson-Fernholm et al., 2003) X (Poirier et al., 2001)
Solute carrier family 2 (facilitated glucose transporter), member 1	SLC2A1	1q35–p31.3	√ (Hodgkinson et al., 2001) X (Gutierrez et al., 1998)
Aldose reductase family 1, member B1	AKR1B1	7q35	√ (Yamamoto et al., 2003) X (Fanelli et al., 2002)
Interleukin 1	IL1	2q12–q21	√ (Loughrey et al., 1998) X (Tarnow et al., 1997)
Interleukin 1 receptor antagonist	IL1RN	2q14.2	√ (Blakemore et al., 1995) X (Tarnow et al., 1997)

Examples of investigations showing positive association with diabetic nephropathy are marked "√" and negative association "X."

Several additional strategies are now planned or are in progress to identify the genetic susceptibility to diabetic nephropathy. These are indirect genome screens for association with nephropathy (using an extended panel of genetic markers) and direct genome screens employing panels of single nucleotide polymorphisms (SNPs) to assess directly thousands of candidate genes simultaneously (Fig. 13–7).

Indirect Genome-wide Association Screens to Identify Chromosomal Regions Harboring Candidate Genes for Nephropathy

Microsatellite-based Indirect Genome-wide Association Screen

To identify chromosomal regions that harbor possible candidate genes for diabetic nephropathy, it is possible to perform a low-resolution genome-wide microsatellite association screen. This method employs as many as 6000 fluorescently labeled microsatellite markers (compared to the usual 300–400 microsatellite markers for a linkage screen) and a pooling strategy to minimize the costs of typing large numbers of samples with 6000 individual genetic markers. Careful analysis of separate pools of DNA from cases and controls generates a series of allele frequencies for the microsatellite markers. Following statistical ranking of markers associated with the pooled cases versus pooled controls, a second round of individual genotyping is performed to validate any associations found. This indirect genome screen strategy was originally employed in the Genetic Analysis of Multiple sclerosis in EuropeanS (GAMES) collaborative project (Goedde et al., 2002; Sawcer et al., 2002; Setakis, 2003). Importantly, the GAMES project succeeded in identifying the well-established association of multiple sclerosis with the major histocompatibility complex (MHC) region, confirming that the approach is able to identify microsatellites in linkage disequilibrium with genuine but modest susceptibility factors.

The same strategy (using the GAMES marker set) has also been successful in identifying regions of chromosome 10 significantly associated with diabetic nephropathy in Caucasians with type 1 diabetes (McKnight et al., 2006). The data from these indirect screens allow investigators to select candidate genes, identified close to the associated microsatellite markers, for further study.

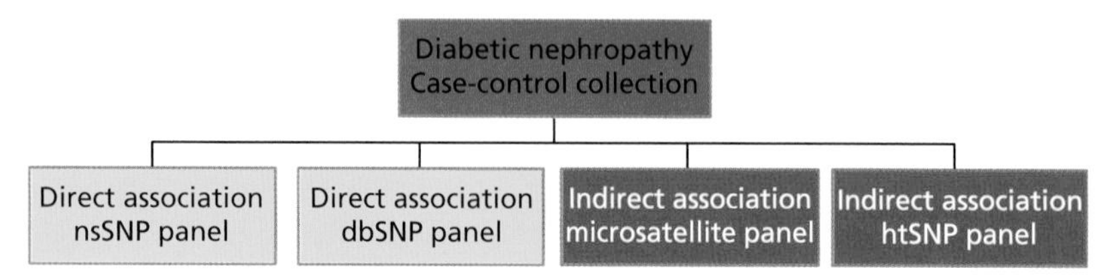

Figure 13–7 Genome-wide association screens to identify genetic susceptibility to diabetic nephropathy. nsSNP, nonsynonymous single nucleotide polymorphism (SNP); dbSNP, random SNP which may be intragenic or intergenic; microsatellite, repeated sequence e.g., CACACA. . . with variable length allele; htSNP, haplotype tagged SNP defining a chromosomal segment.

Haplotype Tagged SNP Indirect Genome-wide Association Screen

A significantly larger marker set of SNPs could be used, as an alternative to a limited microsatellite panel, in an indirect genome-wide association screen. An ideal marker density of one SNP for every 1–2 kb of the genome would provide excellent resolution to identify regions of the genome associated with complex diseases such as diabetic nephropathy. Fortunately, there is considerable SNP redundancy due to linkage disequilibrium between individual SNPs. Using the information from the International HapMap project (Altshuler et al., 2005; Phimister, 2005), it should be possible to reduce the size of the marker set required by employing haplotype tagged SNPs (htSNPs). Proof of principle for this strategy has been reported for type 1 diabetes (Johnson et al., 2001; Lowe et al., 2004). For instance, a panel of 250,000 htSNPs should provide comprehensive genome-wide coverage for an association screen (Chapman et al., 2003). Nevertheless, the current expense of an htSNP genome scan is considerable; for example, genotyping 2000 DNA samples at 5 cents per individual genotype would still be prohibitively expensive at a cost of $25 million.

Clearly, the utility of this strategy is dependent on parallel technical advances in genotyping to reduce the unit cost per individual genotype. Recent advances in novel low-cost high-throughput genotyping technologies (Oliphant et al., 2002; Hardenbol et al., 2003), and improvements in statistical methods for data analysis (Lowe et al., 2004) indicate that with continued progress, genome-wide association screens will be technically and financially viable in the future. An htSNP genome-wide association screen will detect the common variation in the genome associated with common disease. Modelling of empirical datasets indicates that this approach will, however, fail to capture efficiently the association of rare variants (minor allele frequency $<5\%$) with disease (Zeggini et al., 2005).

Direct Genome-wide Association Screens

Direct genome-wide association screens employ either a large panel of unselected SNPs randomly distributed throughout the genome or a more focused set of markers using nonsynonymous SNPs (nsSNP), which may potentially alter gene expression or protein structure.

For any screen, a large number of SNPs will be genotyped, and it is therefore likely that many putative disease-susceptibility variants would be detected. Most of the direct associations between SNPs and disease will be by chance alone (for example, one would expect 5000 SNPs in a screen employing 100,000 SNPs to be associated by chance alone at the 5% confidence level). Therefore, a major challenge will be to rationalize the output from genome screens. One possibility is to apply a correction factor for the number of SNPs genotyped. For example, in a genome screen with 100,000 markers, adoption of a significance threshold of $p = 5 \times 10^{-7}$ for individual SNPs would result in a genome-wide false positive rate of 1 in 20, the accepted level of confidence for an individual genotype. However, this correction is arguably too rigorous and may discount many true associations.

An alternative approach is to adopt a two-stage screening strategy. Markers showing association in an initial screen could be reassessed in an independent, larger sample. The threshold of significance for associated markers in the first screen would purposely be set low so that it remained sufficiently powerful to detect loci with a relatively modest effect on disease despite employing only a fraction of the total samples available. The potential for detecting susceptibility variants in this initial screen may be enhanced by including only extreme phenotypes: cases with onset of nephropathy after a relatively short duration of diabetes or despite good glycemic control and conversely controls that have not developed nephropathy despite a long duration of diabetes or poor glycemic control. The majority of associations detected in the first screen will be false positives; therefore, a much more stringent significance threshold is required for the replication study. Hence, a subset of patient samples would be screened using the complete marker set, with only a fraction of these markers being genotyped in a larger sample, thereby improving efficiency without a reduction in power while reducing the risk of detecting false positive associations.

A screen that employed a two-stage strategy similar to that described has been undertaken in a Japanese population with type 2 diabetes (Shimazaki et al., 2005). Over 80,000 SNP markers were genotyped and SNPs in two candidate genes were associated with diabetic nephropathy. The genes identified were a sodium chloride cotransporter known to be mutated in Gitelman's syndrome (Tanaka et al., 2003) and the engulfment and cell motility (*ELMO 1*) gene, which is upregulated in a high extracellular glucose and promotes accumulation of matrix proteins (Shimazaki et al., 2005). These interesting findings have not yet been replicated in other populations.

An nsSNP scan is an alternative method of directly detecting gene variants associated with disease. An nsSNP may influence phenotype by altering gene expression or protein function. Computer algorithms have been devised to select nsSNPs that may alter phenotype (Bao and Cui, 2005; Bao et al., 2005; Reumers et al., 2005). Detailed mapping of allele-specific in vivo expression will provide a rich source of additional regulatory SNPs that can be afforded higher priority in association studies (Pastinen et al., 2005). Current estimates of the number of nsSNPs in the genome are in the range 30,000–60,000 (Karchin et al., 2005). It is arguably more efficient to focus on nsSNP genotyping for a direct association screen, as for example, genotyping 2000 DNA samples at 5 cents per individual genotype would cost $2.5 million.

Conclusion

The human genome project and the HapMap initiative have accelerated the technological advances necessary to allow dissection of the molecular basis of disease. Genome-wide screens with high-density

SNP markers are now feasible and an affordable investment. To capitalize fully on these advances will require continued multicenter research collaborations to acquire large and precisely phenotypical patient collections. By combining the expertise of geneticists and clinicians, there is optimism that the genetic basis of complex renal diseases like diabetic nephropathy will be revealed. These investigations should yield new diagnostic approaches and further insights into mechanisms of disease progression. Ultimately, the research goals must include development of rational therapies to reduce the future burden of renal failure.

Acknowledgments

We thank Carol Fidler for excellent secretarial assistance and Linda Newby for Figure 13–2.

References

Adler, S (2004). Diabetic nephropathy: Linking histology, cell biology, and genetics. *Kidney Int,* 66:2095–2106.

Altshuler, D, Brooks, LD, Chakravarti, A, Collins, FS, Daly, MJ, and Donnelly, P (2005). A haplotype map of the human genome. *Nature,* 437:1299–1320.

Anderson, S, Jung, FF, and Ingelfinger, JR (1993). Renal renin-angiotensin system in diabetes: functional, immunohistochemical, and molecular biological correlations. *Am J Physiol,* 265:477–486.

Annotation (1999). Freely associating. *Nat Genet,* 22:1–2.

Baelde, HJ, Eikmans, M, Doran, PP, Lappin, DW, de Heer, E, and Bruijn, JA (2004). Gene expression profiling in glomeruli from human kidneys with diabetic nephropathy. *Am J Kidney Dis,* 43:636–650.

Baggio, B, Budakovic, A, Dalla Vestra, M, Saller, A, Bruseghin, M, and Fioretto, P (2002). Effects of cigarette smoking on glomerular structure and function in type 2 diabetic patients. *J Am Soc Nephrol,* 13:2730–2736.

Ballard, DJ, Humphrey, LL, Melton, LJ, 3rd, Frohnert, PP, Chu, PC, O'Fallon, WM, and Palumbo, PJ (1988). Epidemiology of persistent proteinuria in type II diabetes mellitus. Population-based study in Rochester, Minnesota. *Diabetes,* 37:405–412.

Bao, L, and Cui, Y (2005). Prediction of the phenotypic effects of non-synonymous single nucleotide polymorphisms using structural and evolutionary information. *Bioinformatics,* 21:2185–2190.

Bao, L, Zhou, M, and Cui, Y (2005). nsSNP Analyzer: identifying disease-associated nonsynonymous single nucleotide polymorphisms. *Nucleic Acids Res,* 33: W480–W482.

Barker, DF, Hostikka, SL, Zhou, J, Chow, LT, Oliphant, AR, Gerken, SC, Gregory, MC, Skolnick, MH, Atkin, CL, and Tryggvason, K (1990). Identification of mutations in the COL4A5 collagen gene in Alport syndrome. *Science,* 248:1224–1227.

Barker, DJ, and Bagby, SP (2005). Developmental antecedents of cardiovascular disease: a historical perspective. *J Am Soc Nephrol,* 16:2537–2544.

Barker, DJ, Gluckman, PD, Godfrey, KM, Harding, JE, Owens, JA, and Robinson, JS (1993). Fetal nutrition and cardiovascular disease in adult life. *Lancet,* 341:938–941.

Barnett, AH (2004). Treating to goal: challenges of current management. *Eur J Endocrinol,* 151 (Suppl. 2):T3–T7; discussion T29–T30.

Bihoreau, MT, Megel, N, Brown, JH, Kranzlin, B, Crombez, L, Tychinskaya, Y, Broxholme, J, Kratz, S, Bergmann, V, Hoffman, S, et al. (2002). Characterization of a major modifier locus for polycystic kidney disease (Modpkdr1) in the Han:SPRD (cy/ +) rat in a region conserved with a mouse modifier locus for Alport syndrome. *Hum Mol Genet,* 11:2165–2173.

Blakemore, AI, Watson, PF, Weetman, AP, and Duff, GW (1995). Association of Graves' disease with an allele of the interleukin-1 receptor antagonist gene. *J Clin Endocrinol Metab,* 80:111–115.

Borch-Johnsen, K, Nissen, H, Henriksen, E, Kreiner, S, Salling, N, Deckert, T, and Nerup, J (1987). The natural history of insulin-dependent diabetes mellitus in Denmark: 1. Long-term survival with and without late diabetic complications. *Diabet Med,* 4:201–210.

Borch-Johnsen, K, Norgaard, K, Hommel, E, Mathiesen, ER, Jensen, JS, Deckert, T, and Parving, HH (1992). Is diabetic nephropathy an inherited complication? *Kidney Int,* 41:719–722.

Boulter, C, Mulroy, S, Webb, S, Fleming, S, Brindle, K, and Sandford, R (2001). Cardiovascular, skeletal, and renal defects in mice with a targeted disruption of the Pkd1 gene. *Proc Natl Acad Sci USA,* 98:12174–12179.

Brasier, JL and Henske, EP (1997). Loss of the polycystic kidney disease (PKD1) region of chromosome 16p13 in renal cyst cells supports a loss-of-function model for cyst pathogenesis. *J Clin Invest,* 99:1–6.

Brenner, BM and Chertow, GM (1994). Congenital oligonephropathy and the etiology of adult hypertension and progressive renal injury. *Am J Kidney Dis,* 23:171–175.

Brenner, BM, Cooper, ME, de Zeeuw, D, Keane, WF, Mitch, WE, Parving, HH, Remuzzi, G, Snapinn, SM, Zhang, Z, and Shahinfar, S (2001). Effects of losartan on renal and cardiovascular outcomes in patients with type 2 diabetes and nephropathy. *N Engl J Med,* 345:861–869.

Breyer, JA, Bain, RP, Evans, JK, Nahman, NS, Jr., Lewis, EJ, Cooper, M, McGill, J, and Berl, T (1996). Predictors of the progression of renal insufficiency in patients with insulin-dependent diabetes and overt diabetic nephropathy. The Collaborative Study Group. *Kidney Int,* 50:1651–1658.

Brook-Carter, PT, Peral, B, Ward, CJ, Thompson, P, Hughes, J, Maheshwar, MM, Nellist, M, Gamble, V, Harris, PC, and Sampson, JR (1994). Deletion of the TSC2 and PKD1 genes associated with severe infantile polycystic kidney disease—a contiguous gene syndrome. *Nat Genet,* 8:328–332.

Brownlee, M (2001). Biochemistry and molecular cell biology of diabetic complications. *Nature,* 414:813–820.

Buraczynska, M, Ksiazek, P, Lopatynski, J, Spasiewicz, D, Nowicka, T, and Ksiazek, A (2002). Association of the renin-angiotensin system gene polymorphism with nephropathy in type II diabetes. *Pol Arch Med Wewn,* 108:725–730.

Cairns, HWB (1925). Heredity in polycystic disease of the kidneys. *Q J Med,* 18:359–392.

Chang, HR, Cheng, CH, Shu, KH, Chen, CH, Lian, JD, and Wu, MY (2003). Study of the polymorphism of angiotensinogen, anigiotensin-converting enzyme and angiotensin receptor in type II diabetes with end-stage renal disease in Taiwan. *J Chin Med Assoc,* 66:51–56.

Chapman, JM, Cooper, JD, Todd, JA, and Clayton, DG (2003). Detecting disease associations due to linkage disequilibrium using haplotype tags: a class of tests and the determinants of statistical power. *Hum Hered,* 56:18–31.

Chase, HP, Jackson, WE, Hoops, SL, Cockerham, RS, Archer, PG, and O'Brien, D (1989). Glucose control and the renal and retinal complications of insulin-dependent diabetes. *JAMA,* 261:1155–1160.

Chen, S, Cohen, MP, and Ziyadeh, FN (2000). Amadori-glycated albumin in diabetic nephropathy: pathophysiologic connections. *Kidney Int Suppl,* 77, S40–S44.

Chen, XZ, Vassilev, PM, Basora, N, Peng, JB, Nomura, H, Segal, Y, Brown, EM, Reeders, ST, Hediger, MA, and Zhou, J (1999). Polycystin-L is a calcium-regulated cation channel permeable to calcium ions. *Nature,* 401:383–386.

Chiurchiu, C, Remuzzi, G, and Ruggenenti, P (2005). Angiotensin-converting enzyme inhibition and renal protection in nondiabetic patients: the data of the meta-analyses. *J Am Soc Nephrol,* 16, (Suppl. 1):S58–S63.

Chowdhury, TA, Dyer, PH, Kumar, S, Gibson, SP, Rowe, BR, Davies, SJ, Marshall, SM, Morris, PJ, Gill, GV, Feeney, S, et al. (1998). Association of apolipoprotein epsilon2 allele with diabetic nephropathy in Caucasian subjects with IDDM. *Diabetes,* 47:278–280.

Clarkson, MR, Murphy, M, Gupta, S, Lambe, T, Mackenzie, HS, Godson, C, Martin, F, and Brady, HR (2002). High glucose-altered gene expression in mesangial cells. Actin-regulatory protein gene expression is triggered by oxidative stress and cytoskeletal disassembly. *J Biol Chem,* 277:9707–9712.

Clouthier, DE, Comerford, SA, and Hammer, RE (1997). Hepatic fibrosis, glomerulosclerosis, and a lipodystrophy-like syndrome in PEPCK-TGF-beta1 transgenic mice. *J Clin Invest,* 100:2697–2713.

Colhoun, HM, Betteridge, DJ, Durrington, PN, Hitman, GA, Neil, HA, Livingstone, SJ, Thomason, MJ, Mackness, MI, Charlton-Menys, V, and Fuller, JH (2004). Primary prevention of cardiovascular disease with atorvastatin in type 2 diabetes in the Collaborative Atorvastatin Diabetes Study

(CARDS): multicentre randomised placebo-controlled trial. *Lancet,* 364:685–696.
Collins, AJ, Kasiske, B, Herzog, C, Chen, SC, Everson, S, Constantini, E, Grimm, R, McBean, M, Xue, J, Chavers, B, et al. (2003). Excerpts from the United States Renal Data System 2003 Annual Data Report: atlas of end-stage renal disease in the United States. *Am J Kidney Dis,* 42: A5–A7.
Connolly, SB, Sadlier, D, Kieran, NE, Doran, P, and Brady, HR (2003). Transcriptome profiling and the pathogenesis of diabetic complications. *J Am Soc Nephrol,* 14:S279–S283.
Cooper, ME (1998). Pathogenesis, prevention, and treatment of diabetic nephropathy. *Lancet,* 352:213–219.
Cowie, CC, Port, FK, Wolfe, RA, Savage, PJ, Moll, PP, and Hawthorne, VM (1989). Disparities in incidence of diabetic end-stage renal disease according to race and type of diabetes. *N Engl J Med,* 321:1074–1079.
DCCT/EDIC Research Group. (2000). Retinopathy and nephropathy in patients with type 1 diabetes four years after a trial of intensive therapy. The Diabetes Control and Complications Trial/Epidemiology of Diabetes Interventions and Complications Research Group. *N Engl J Med,* 342:381–389.
DCCT/EDIC Research Group. (2003). Sustained effect of intensive treatment of type 1 diabetes mellitus on development and progression of diabetic nephropathy: the Epidemiology of Diabetes Interventions and Complications (EDIC) study. *JAMA,* 290:2159–2167.
De Cosmo, S, Bacci, S, Piras, GP, Cignarelli, M, Placentino, G, Margaglione, M, Colaizzo, D, Di Minno, G, Giorgino, R, Liuzzi, A, and Viberti, GC (1997). High prevalence of risk factors for cardiovascular disease in parents of IDDM patients with albuminuria. *Diabetologia,* 40:1191–1196.
Deckert, T, Yokoyama, H, Mathiesen, E, Ronn, B, Jensen, T, Feldt-Rasmussen, B, Borch-Johnsen, K, and Jensen, JS (1996). Cohort study of predictive value of urinary albumin excretion for atherosclerotic vascular disease in patients with insulin dependent diabetes. *BMJ,* 312:871–874.
Dinneen, SF and Gerstein, HC (1997). The association of microalbuminuria and mortality in non-insulin-dependent diabetes mellitus. A systematic overview of the literature. *Arch Intern Med,* 157:1413–1418.
Doria, A, Warram, JH, and Krolewski, AS (1995). Genetic susceptibility to nephropathy in insulin-dependent diabetes: from epidemiology to molecular genetics. *Diabetes Metab Rev,* 11:287–314.
Dunlop, M (2000). Aldose reductase and the role of the polyol pathway in diabetic nephropathy. *Kidney Int Suppl,* 77:S3–S12.
Fain, PR, McFann, KK, Taylor, MR, Tison, M, Johnson, AM, Reed, B, and Schrier, RW (2005). Modifier genes play a significant role in the phenotypic expression of PKD1. *Kidney Int,* 67:1256–1267.
Fanelli, A, Hadjadj, S, Gallois, Y, Fumeron, F, Betoule, D, Grandchamp, B, and Marre, M (2002). Polymorphism of aldose reductase gene and susceptibility to retinopathy and nephropathy in Caucasians with type 1 diabetes. *Arch Mal Coeur Vaiss,* 95:701–708.
Fioretto, P, and Solini, A (2005). Antihypertensive treatment and multifactorial approach for renal protection in diabetes. *J Am Soc Nephrol,* 16 (Suppl. 1): S18–S21.
Fioretto, P, Steffes, MW, Barbosa, J, Rich, SS, Miller, ME, and Mauer, M (1999). Is diabetic nephropathy inherited? Studies of glomerular structure in type 1 diabetic sibling pairs. *Diabetes,* 48:865–869.
Fioretto, P, Steffes, MW, Brown, DM, and Mauer, SM (1992). An overview of renal pathology in insulin-dependent diabetes mellitus in relationship to altered glomerular hemodynamics. *Am J Kidney Dis,* 20:549–558.
Fioretto, P, Steffes, MW, Sutherland, DE, Goetz, FC, and Mauer, M (1998). Reversal of lesions of diabetic nephropathy after pancreas transplantation. *N Engl J Med,* 339:69–75.
Fogarty, DG, Hanna, LS, Wantman, M, Warram, JH, Krolewski, AS, and Rich, SS (2000a). Segregation analysis of urinary albumin excretion in families with type 2 diabetes. *Diabetes,* 49:1057–1063.
Fogarty, DG, Rich, SS, Hanna, L, Warram, JH, and Krolewski, AS (2000). Urinary albumin excretion in families with type 2 diabetes is heritable and genetically correlated to blood pressure. *Kidney Int,* 57:250–257.
Forsblom, CM, Kanninen, T, Lehtovirta, M, Saloranta, C, and Groop, LC (1999). Heritability of albumin excretion rate in families of patients with Type II diabetes. *Diabetologia,* 42:1359–1366.
Freedman, BI, Volkova, NV, Satko, SG, Krisher, J, Jurkovitz, C, Soucie, JM, and McClellan, WM (2005). Population-based screening for family history of end-stage renal disease among incident dialysis patients. *Am J Nephrol,* 25:529–535.
Fuller, JH, Stevens, LK, and Wang, SL (2001). Risk factors for cardiovascular mortality and morbidity: the WHO Multinational Study of Vascular Disease in Diabetes. *Diabetologia,* 44, (Suppl. 2):S54–S64.
Gabow, PA, Johnson, AM, Kaehny, WD, Kimberling, WJ, Lezotte, DC, Duley, IT, and Jones, RH (1992). Factors affecting the progression of renal disease in autosomal-dominant polycystic kidney disease. *Kidney Int,* 41:1311–1319.
Gaede, P, Vedel, P, Larsen, N, Jensen, GV, Parving, HH, and Pedersen, O (2003). Multifactorial intervention and cardiovascular disease in patients with type 2 diabetes. *N Engl J Med,* 348:383–393.
Gaede, P, Vedel, P, Parving, HH, and Pedersen, O (1999). Intensified multifactorial intervention in patients with type 2 diabetes mellitus and microalbuminuria: the Steno type 2 randomised study. *Lancet,* 353:617–622.
Geberth, S, Ritz, E, Zeier, M, and Stier, E (1995). Anticipation of age at renal death in autosomal dominant polycystic kidney disease (ADPKD)? *Nephrol Dial Transplant,* 10:1603–1606.
Gilbert, RE and Cooper, ME (1999). The tubulointerstitium in progressive diabetic kidney disease: more than an aftermath of glomerular injury? *Kidney Int,* 56:1627–1637.
Goedde, R, Sawcer, S, Boehringer, S, Miterski, B, Sindern, E, Haupts, M, Schimrigk, S, Compston, A, and Epplen, JT (2002). A genome screen for linkage disequilibrium in HLA-DRB1*15-positive Germans with multiple sclerosis based on 4666 microsatellite markers. *Hum Genet,* 111:270–277.
Grantham, JJ (2003). Lillian Jean Kaplan International Prize for advancement in the understanding of polycystic kidney disease. Understanding polycystic kidney disease: a systems biology approach. *Kidney Int,* 64:1157–1162.
Greene, DA, Arezzo, JC, and Brown, MB (1999). Effect of aldose reductase inhibition on nerve conduction and morphometry in diabetic neuropathy. Zenarestat Study Group. *Neurology,* 53:580–591.
Gross, JL, de Azevedo, MJ, Silveiro, SP, Canani, LH, Caramori, ML, and Zelmanovitz, T (2005a). Diabetic nephropathy: diagnosis, prevention, and treatment. *Diabetes Care,* 28:164–176.
Gross, ML, Amann, K, and Ritz, E (2005b). Nephron number and renal risk in hypertension and diabetes. *J Am Soc Nephrol,* 16, (Suppl. 1):S27–S29.
Grzeszczak, W, Saucha, W, Zychma, MJ, Zukowska-Szczechowska, E, Labuz, B, Lacka, B, and Szydlowska, I (1999). Is Trp64Arg polymorphism of beta3-adrenergic receptor a clinically useful marker for the predisposition to diabetic nephropathy in Type II diabetic patients? *Diabetologia,* 42:632–633.
Guay-Woodford, LM (2003). Murine models of polycystic kidney disease: molecular and therapeutic insights. *Am J Physiol Renal Physiol,* 285: F1034–F1049.
Guo, L, Schreiber, TH, Weremowicz, S, Morton, CC, Lee, C, and Zhou, J (2000). Identification and characterization of a novel polycystin family member, polycystin-L2, in mouse and human: sequence, expression, alternative splicing, and chromosomal localization. *Genomics,* 64:241–251.
Gutierrez, C, Vendrell, J, Pastor, R, Broch, M, Aguilar, C, Llor, C, Simon, I, and Richart, C (1998). GLUT1 gene polymorphism in non-insulin-dependent diabetes mellitus: genetic susceptibility relationship with cardiovascular risk factors and microangiopathic complications in a Mediterranean population. *Diabetes Res Clin Pract,* 41:113–120.
Hales, CN and Barker, DJ (2001). The thrifty phenotype hypothesis. *Br Med Bull,* 60:5–20.
Hansen, HP, Tauber-Lassen, E, Jensen, BR, and Parving, HH (2002). Effect of dietary protein restriction on prognosis in patients with diabetic nephropathy. *Kidney Int,* 62:220–228.
Hanson, RL, Ehm, MG, Pettitt, DJ, Prochazka, M, Thompson, DB, Timberlake, D, Foroud, T, Kobes, S, Baier, L, Burns, DK, et al. (1998). An autosomal genomic scan for loci linked to type II diabetes mellitus and body-mass index in Pima Indians. *Am J Hum Genet,* 63:1130–1138.
Hardenbol, P, Baner, J, Jain, M, Nilsson, M, Namsaraev, EA, Karlin-Neumann, GA, Fakhrai-Rad, H, Ronaghi, M, Willis, TD, Landegren, U, and Davis, RW (2003). Multiplexed genotyping with sequence-tagged molecular inversion probes. *Nat Biotechnol,* 21:673–678.
Hateboer, N, v Dijk, MA, Bogdanova, N, Coto, E, Saggar-Malik, AK, San Millan, JL, Torra, R, Breuning, M, and Ravine, D (1999). Comparison of

phenotypes of polycystic kidney disease types 1 and 2. European PKD1-PKD2 Study Group. *Lancet,* 353:103–107.

Hirschhorn, JN, Lohmueller, K, Byrne, E, and Hirschhorn, K (2002). A comprehensive review of genetic association studies. *Genet Med,* 4:45–61.

Hodgkinson, AD, Millward, BA, and Demaine, AG (2001). Polymorphisms of the glucose transporter (GLUT1) gene are associated with diabetic nephropathy. *Kidney Int,* 59:985–989.

Holmes, DI, Abdel Wahab, N, and Mason, RM (1997). Identification of glucose-regulated genes in human mesangial cells by mRNA differential display. *Biochem Biophys Res Commun,* 238:179–184.

Hughes, J, Ward, CJ, Aspinwall, R, Butler, R, and Harris, PC (1999). Identification of a human homologue of the sea urchin receptor for egg jelly: a polycystic kidney disease-like protein. *Hum Mol Genet,* 8:543–549.

Hughes, J, Ward, CJ, Peral, B, Aspinwall, R, Clark, K, San Millan, JL, Gamble, V, and Harris, PC (1995). The polycystic kidney disease 1 (PKD1) gene encodes a novel protein with multiple cell recognition domains. *Nat Genet,* 10:151–160.

Hugot, JP, Chamaillard, M, Zouali, H, Lesage, S, Cezard, JP, Belaiche, J, Almer, S, Tysk, C, O'Morain, CA, Gassull, M, et al. (2001). Association of NOD2 leucine-rich repeat variants with susceptibility to Crohn's disease. *Nature,* 411:599–603.

Hunter, DJ, Lange, M, Snieder, H, MacGregor, AJ, Swaminathan, R, Thakker, RV, and Spector, TD (2002). Genetic contribution to renal function and electrolyte balance: a twin study. *Clin Sci (Lond),* 103:259–265.

Imperatore, G, Hanson, RL, Pettitt, DJ, Kobes, S, Bennett, PH, and Knowler, WC (1998). Sib-pair linkage analysis for susceptibility genes for microvascular complications among Pima Indians with type 2 diabetes. Pima Diabetes Genes Group. *Diabetes,* 47:821–830.

Imperatore, G, Knowler, WC, Pettitt, DJ, Kobes, S, Bennett, PH, and Hanson, RL (2000). Segregation analysis of diabetic nephropathy in Pima Indians. *Diabetes,* 49:1049–1056.

Ioannidis, JP, Trikalinos, TA, Ntzani, EE, and Contopoulos-Ioannidis, DG (2003). Genetic associations in large versus small studies: an empirical assessment. *Lancet,* 361:567–571.

Jacobsen, P, Rossing, P, Tarnow, L, Hovind, P, and Parving, HH (2003). Birth weight—a risk factor for progression in diabetic nephropathy? *J Intern Med,* 253:343–350.

Johnson, GC, Esposito, L, Barratt, BJ, Smith, AN, Heward, J, Di Genova, G, Ueda, H, Cordell, HJ, Eaves, IA, Dudbridge, F, et al. (2001). Haplotype tagging for the identification of common disease genes. *Nat Genet,* 29:233–237.

Jones, CA, Krolewski, AS, Rogus, J, Xue, JL, Collins, A, and Warram, JH (2005). Epidemic of end-stage renal disease in people with diabetes in the United States population: do we know the cause? *Kidney Int,* 67:1684–1691.

Karchin, R, Diekhans, M, Kelly, L, Thomas, DJ, Pieper, U, Eswar, N, Haussler, D, and Sali, A (2005). LS-SNP: large-scale annotation of coding non-synonymous SNPs based on multiple information sources. *Bioinformatics,* 21:2814–2820.

Keller, G, Zimmer, G, Mall, G, Ritz, E, and Amann, K (2003). Nephron number in patients with primary hypertension. *N Engl J Med,* 348:101–108.

Kim, K, Drummond, I, Ibraghimov-Beskrovnaya, O, Klinger, K, and Arnaout, MA (2000). Polycystin 1 is required for the structural integrity of blood vessels. *Proc Natl Acad Sci USA,* 97:1731–1736.

Klahr, S, Levey, AS, Beck, GJ, Caggiula, AW, Hunsicker, L, Kusek, JW, and Striker, G (1994). The effects of dietary protein restriction and blood-pressure control on the progression of chronic renal disease. Modification of Diet in Renal Disease Study Group. *N Engl J Med,* 330:877–884.

Kleymenova, E, Ibraghimov-Beskrovnaya, O, Kugoh, H, Everitt, J, Xu, H, Kiguchi, K, Landes, G, Harris, P, and Walker, C (2001). Tuberin-dependent membrane localization of polycystin-1: a functional link between polycystic kidney disease and the TSC2 tumor suppressor gene. *Mol Cell,* 7:823–832.

Koya, D, Haneda, M, Nakagawa, H, Isshiki, K, Sato, H, Maeda, S, Sugimoto, T, Yasuda, H, Kashiwagi, A, Ways, DK, et al. (2000). Amelioration of accelerated diabetic mesangial expansion by treatment with a PKC beta inhibitor in diabetic db/db mice, a rodent model for type 2 diabetes. *FASEB J,* 14:439–447.

Koya, D, Jirousek, MR, Lin, YW, Ishii, H, Kuboki, K, and King, GL (1997). Characterization of protein kinase C beta isoform activation on the gene expression of transforming growth factor-beta, extracellular matrix components, and prostanoids in the glomeruli of diabetic rats. *J Clin Invest,* 100:115–126.

Krolewski, M, Eggers, PW, and Warram, JH (1996). Magnitude of end-stage renal disease in IDDM: a 35 year follow-up study. *Kidney Int,* 50:2041–2046.

Langefeld, CD, Beck, SR, Bowden, DW, Rich, SS, Wagenknecht, LE, and Freedman, BI (2004). Heritability of GFR and albuminuria in Caucasians with type 2 diabetes mellitus. *Am J Kidney Dis,* 43:796–800.

Lantinga-van Leeuwen, IS, Dauwerse, JG, Baelde, HJ, Leonhard, WN, van de Wal, A, Ward, CJ, Verbeek, S, Deruiter, MC, Breuning, MH, de Heer, E, and Peters, DJ (2004). Lowering of Pkd1 expression is sufficient to cause polycystic kidney disease. *Hum Mol Genet,* 13:3069–3077.

Levey, AS, Greene, T, Beck, GJ, Caggiula, AW, Kusek, JW, Hunsicker, LG, and Klahr, S (1999). Dietary protein restriction and the progression of chronic renal disease: what have all of the results of the MDRD study shown? Modification of Diet in Renal Disease Study group. *J Am Soc Nephrol,* 10:2426–2439.

Lewis, EJ, Hunsicker, LG, Bain, RP, and Rohde, RD (1993). The effect of angiotensin-converting-enzyme inhibition on diabetic nephropathy. The Collaborative Study Group. *N Engl J Med,* 329:1456–1462.

Lewis, EJ, Hunsicker, LG, Clarke, WR, Berl, T, Pohl, MA, Lewis, JB, Ritz, E, Atkins, RC, Rohde, R, and Raz, I (2001). Renoprotective effect of the angiotensin-receptor antagonist irbesartan in patients with nephropathy due to type 2 diabetes. *N Engl J Med,* 345:851–860.

Li, A, Tian, X, Sung, SW, and Somlo, S (2003). Identification of two novel polycystic kidney disease-1-like genes in human and mouse genomes. *Genomics,* 81:596–608.

Lohmueller, KE, Pearce, CL, Pike, M, Lander, ES, and Hirschhorn, JN (2003). Meta-analysis of genetic association studies supports a contribution of common variants to susceptibility to common disease. *Nat Genet,* 33:177–182.

Loughrey, BV, Maxwell, AP, Fogarty, DG, Middleton, D, Harron, JC, Patterson, CC, Darke, C, and Savage, DA (1998). An interluekin 1B allele, which correlates with a high secretor phenotype, is associated with diabetic nephropathy. *Cytokine,* 10:984–988.

Lowe, CE, Cooper, JD, Chapman, JM, Barratt, BJ, Twells, RC, Green, EA, Savage, DA, Guja, C, Ionescu-Tirgoviste, C, Tuomilehto-Wolf, E, et al. (2004). Cost-effective analysis of candidate genes using htSNPs: a staged approach. *Genes Immun,* 5:301–305.

Lu, W, Fan, X, Basora, N, Babakhanlou, H, Law, T, Rifai, N, Harris, PC, Perez-Atayde, AR, Rennke, HG, and Zhou, J (1999). Late onset of renal and hepatic cysts in Pkd1-targeted heterozygotes [letter]. *Nat Genet,* 21:160–161.

Lu, W, Peissel, B, Babakhanlou, H, Pavlova, A, Geng, L, Fan, X, Larson, C, Brent, G, and Zhou, J (1997). Perinatal lethality with kidney and pancreas defects in mice with a targeted Pkd1 mutation. *Nat Genet,* 17:179–181.

Lu, W, Shen, X, Pavlova, A, Lakkis, M, Ward, CJ, Pritchard, L, Harris, PC, Genest, DR, Perez-Atayde, AR, and Zhou, J (2001). Comparison of Pkd1-targeted mutants reveals that loss of polycystin-1 causes cystogenesis and bone defects. *Hum Mol Genet,* 10:2385–2396.

Luyckx, VA, and Brenner, BM (2005). Low birth weight, nephron number, and kidney disease. *Kidney Int Suppl,* S68–S77.

Magistroni, R, He, N, Wang, K, Andrew, R, Johnson, A, Gabow, P, Dicks, E, Parfrey, P, Torra, R, San-Millan, JL, et al. (2003). Genotype-renal function correlation in type 2 autosomal dominant polycystic kidney disease. *J Am Soc Nephrol,* 14:1164–1174.

Makita, Y, Moczulski, DK, Bochenski, J, Smiles, AM, Warram, JH, and Krolewski, AS (2003). Methylenetetrahydrofolate reductase gene polymorphism and susceptibility to diabetic nephropathy in type 1 diabetes. *Am J Kidney Dis,* 41:1189–1194.

Marshall, SM (2004). Recent advances in diabetic nephropathy. *Clin Med,* 4:277–282.

McGrath, J, Somlo, S, Makova, S, Tian, X, and Brueckner, M (2003). Two populations of node monocilia initiate left-right asymmetry in the mouse. *Cell,* 114:61–73.

McKnight, AJ, Maxwell, AP, Sawcer, S, Compston, A, Setakis, A, Patterson, CC, Brady, HR, and Savage, DA (2006). A genome-wide DNA microsatellite

association screen to identify chromosomal regions harbouring candidate genes in diabetic nephropathy. *J Am Soc Nephrol,* 17:831–836.
Meguid El Nahas, A and Bello, AK (2005). Chronic kidney disease: the global challenge. *Lancet,* 365:331–340.
Milutinovic, J, Rust, PF, Fialkow, PJ, Agodoa, LY, Phillips, LA, Rudd, TG, and Sutherland, S (1992). Intrafamilial phenotypic expression of autosomal dominant polycystic kidney disease. *Am J Kidney Dis,* 19:465–472.
Mochizuki, T, Wu, G, Hayashi, T, Xenophontos, SL, Veldhuisen, B, Saris, JJ, Reynolds, DM, Cai, Y, Gabow, PA, Pierides, A, et al. (1996). PKD2, a gene for polycystic kidney disease that encodes an integral membrane protein. *Science,* 272:1339–1342.
Molitch, ME, DeFronzo, RA, Franz, MJ, Keane, WF, Mogensen, CE, and Parving, HH (2003). Diabetic nephropathy. *Diabetes Care,* 26: (Suppl. 1): S94–S98.
Molitch, ME, Steffes, MW, Cleary, PA, and Nathan, DM (1993). Baseline analysis of renal function in the Diabetes Control and Complications Trial. The Diabetes Control and Complications Trial Research Group [corrected]. *Kidney Int,* 43:668–674.
Morgan, CL, Currie, CJ, Stott, NC, Smithers, M, Butler, CC, and Peters, JR (2000). The prevalence of multiple diabetes-related complications. *Diabet Med,* 17:146–151.
Moy, CS, LaPorte, RE, Dorman, JS, Songer, TJ, Orchard, TJ, Kuller, LH, Becker, DJ, and Drash, AL (1990). Insulin-dependent diabetes mellitus mortality. The risk of cigarette smoking. *Circulation,* 82:37–43.
Mrug, M, Li, R, Cui, X, Schoeb, TR, Churchill, GA, and Guay-Woodford, LM (2005). Kinesin family member 12 is a candidate polycystic kidney disease modifier in the cpk mouse. *J Am Soc Nephrol,* 16:905–916.
Murphy, M, Godson, C, Cannon, S, Kato, S, Mackenzie, HS, Martin, F, and Brady, HR (1999). Suppression subtractive hybridization identifies high glucose levels as a stimulus for expression of connective tissue growth factor and other genes in human mesangial cells. *J Biol Chem,* 274:5830–5834.
Murphy, M, McGinty, A, and Godson, C (1998). Protein kinases C: potential targets for intervention in diabetic nephropathy. *Curr Opin Nephrol Hypertens,* 7:563–570.
Muto, S, Aiba, A, Saito, Y, Nakao, K, Nakamura, K, Tomita, K, Kitamura, T, Kurabayashi, M, Nagai, R, Higashihara, E, et al. (2002). Pioglitazone improves the phenotype and molecular defects of a targeted Pkd1 mutant. *Hum Mol Genet,* 11:1731–1742.
Nannipieri, M, Manganiello, M, Pezzatini, A, De Bellis, A, Seghieri, G, and Ferrannini, E (2001). Polymorphisms in the hANP (human atrial natriuretic peptide) gene, albuminuria, and hypertension. *Hypertension,* 37:1416–1422.
Neill, AT, Moy, GW, and Vacquier, VD (2004). Polycystin-2 associates with the polycystin-1 homolog, suREJ3, and localizes to the acrosomal region of sea urchin spermatozoa. *Mol Reprod Dev,* 67:472–477.
Nelson, RG, Knowler, WC, Pettitt, DJ, Saad, MF, and Bennett, PH (1993). Diabetic kidney disease in Pima Indians. *Diabetes Care,* 16:335–341.
Nelson, RG, Meyer, TW, Myers, BD, and Bennett, PH (1997). Clinical and pathological course of renal disease in non-insulin-dependent diabetes mellitus: the Pima Indian experience. *Semin Nephrol,* 17:124–131.
Nilsson, PM, Gudbjornsdottir, S, Eliasson, B, and Cederholm, J (2004). Smoking is associated with increased HbA1c values and microalbuminuria in patients with diabetes—data from the National Diabetes Register in Sweden. *Diabetes Metab,* 30:261–268.
Nistico, L, Buzzetti, R, Pritchard, LE, Van der Auwera, B, Giovannini, C, Bosi, E, Larrad, MT, Rios, M S, Chow, C C, Cockram, C S, et al. (1996). The CTLA-4 gene region of chromosome 2q33 is linked to, and associated with, type 1 diabetes. Belgian Diabetes Registry. *Hum Mol Genet,* 5:1075–1080.
Oates, PJ, and Mylari, BL (1999). Aldose reductase inhibitors: therapeutic implications for diabetic complications. *Expert Opin Investig Drugs,* 8:2095–2119.
O'Dea, DF, Murphy, SW, Hefferton, D, and Parfrey, PS (1998). Higher risk for renal failure in first-degree relatives of white patients with end-stage renal disease: a population-based study. *Am J Kidney Dis,* 32:794–801.
Olbrich, H, Fliegauf, M, Hoefele, J, Kispert, A, Otto, E, Volz, A, Wolf, MT, Sasmaz, G, Trauer, U, Reinhardt, R, et al. (2003). Mutations in a novel gene, NPHP3, cause adolescent nephronophthisis, tapeto-retinal degeneration and hepatic fibrosis. *Nat Genet,* 34:455–459.
Oliphant, A, Barker, DL, Stuelpnagel, JR, and Chee, MS (2002). BeadArray technology: enabling an accurate, cost-effective approach to high-throughput genotyping. *Biotechniques,* (Suppl. 56–58):60–61.
Olivarius Nde, F, Andreasen, AH, Keiding, N, and Mogensen, CE (1993). Epidemiology of renal involvement in newly-diagnosed middle-aged and elderly diabetic patients. Cross-sectional data from the population-based study "Diabetes Care in General Practice," Denmark. *Diabetologia,* 36:1007–1016.
Ong, AC and Harris, PC (1997). Molecular basis of renal cyst formation—one hit or two? *Lancet,* 349:1039–1040.
Ong, AC and Harris, PC (2005). Molecular pathogenesis of ADPKD: The polycystin complex gets complex. *Kidney Int,* 67:1234–1247.
Ong, AC, Harris, PC, Davies, DR, Pritchard, L, Rossetti, S, Biddolph, S, Vaux, DJ, Migone, N, and Ward, CJ (1999). Polycystin-1 expression in PKD1, early-onset PKD1, and TSC2/PKD1 cystic tissue. *Kidney Int,* 56:1324–1333.
Orth, SR, Ritz, E, and Schrier, RW (1997). The renal risks of smoking. *Kidney Int,* 51:1669–1677.
O'Sullivan, DA, Torres, VE, Gabow, PA, Thibodeau, SN, King, BF, and Bergstralh, EJ (1998). Cystic fibrosis and the phenotypic expression of autosomal dominant polycystic kidney disease. *Am J Kidney Dis,* 32:976–983.
Parfrey, PS, Bear, JC, Morgan, J, Cramer, BC, McManamon, PJ, Gault, MH, Churchill, DN, Singh, M, Hewitt, R, Somlo, S, et al. (1990). The diagnosis and prognosis of autosomal dominant polycystic kidney disease. *N Engl J Med,* 323:1085–1090.
Pastinen, T, Ge, B, Gurd, S, Gaudin, T, Dore, C, Lemire, M, Lepage, P, Harmsen, E, and Hudson, TJ (2005). Mapping common regulatory variants to human haplotypes. *Hum Mol Genet,* 14:3963–3971.
Paterson, AD, Magistroni, R, He, N, Wang, K, Johnson, A, Fain, PR, Dicks, E, Parfrey, P, St George-Hyslop, P, and Pei, Y (2005). Progressive loss of renal function is an age-dependent heritable trait in type 1 autosomal dominant polycystic kidney disease. *J Am Soc Nephrol,* 16:755–762.
Pedrini, MT, Levey, AS, Lau, J, Chalmers, TC, and Wang, PH (1996). The effect of dietary protein restriction on the progression of diabetic and nondiabetic renal diseases: a meta-analysis. *Ann Intern Med,* 124:627–632.
Pei, Y, Paterson, AD, Wang, KR, He, N, Hefferton, D, Watnick, T, Germino, GG, Parfrey, P, Somlo, S, and St George-Hyslop, P (2001). Bilineal disease and trans-heterozygotes in autosomal dominant polycystic kidney disease. *Am J Hum Genet,* 68:355–363.
Pennekamp, P, Karcher, C, Fischer, A, Schweickert, A, Skryabin, B, Horst, J, Blum, M, and Dworniczak, B (2002). The ion channel polycystin-2 is required for left-right axis determination in mice. *Curr Biol,* 12:938–943.
Peral, B, Ong, AC, San Millan, JL, Gamble, V, Rees, L, and Harris, PC (1996). A stable, nonsense mutation associated with a case of infantile onset polycystic kidney disease 1 (PKD1). *Hum Mol Genet,* 5:539–542.
Persu, A, Devuyst, O, Lannoy, N, Materne, R, Brosnahan, G, Gabow, PA, Pirson, Y, and Verellen-Dumoulin, C (2000). CF gene and cystic fibrosis transmembrane conductance regulator expression in autosomal dominant polycystic kidney disease. *J Am Soc Nephrol,* 11:2285–2296.
Persu, A, Duyme, M, Pirson, Y, Lens, XM, Messiaen, T, Breuning, MH, Chauveau, D, Levy, M, Grunfeld, JP, and Devuyst, O (2004). Comparison between siblings and twins supports a role for modifier genes in ADPKD. *Kidney Int,* 66:2132–2136.
Persu, A, Stoenoiu, MS, Messiaen, T, Davila, S, Robino, C, El-Khattabi, O, Mourad, M, Horie, S, Feron, O, Balligand, JL, et al. (2002). Modifier effect of ENOS in autosomal dominant polycystic kidney disease. *Hum Mol Genet,* 11:229–241.
Peters, DJ and Breuning, MH (2001). Autosomal dominant polycystic kidney disease: modification of disease progression. *Lancet,* 358:1439–1444.
Pettersson-Fernholm, K, Forsblom, C, Hudson, BI, Perola, M, Grant, PJ, and Groop, PH (2003). The functional -374 T/A RAGE gene polymorphism is associated with proteinuria and cardiovascular disease in type 1 diabetic patients. *Diabetes,* 52:891–894.
Pettitt, DJ, Saad, MF, Bennett, PH, Nelson, RG, and Knowler, WC (1990). Familial predisposition to renal disease in two generations of Pima Indians with type 2 (non-insulin-dependent) diabetes mellitus. *Diabetologia,* 33:438–443.

Phillips, DI, Barker, DJ, Hales, CN, Hirst, S, and Osmond, C (1994). Thinness at birth and insulin resistance in adult life. *Diabetologia,* 37:150–154.

Phimister, EG (2005). Genomic cartography—presenting the HapMap. *N Engl J Med,* 353:1766–1768.

Poirier, O, Nicaud, V, Vionnet, N, Raoux, S, Tarnow, L, Vlassara, H, Parving, HH, and Cambien, F (2001). Polymorphism screening of four genes encoding advanced glycation end-product putative receptors. Association study with nephropathy in type 1 diabetic patients. *Diabetes,* 50:1214–1218.

Pritchard, L, Sloane-Stanley, JA, Sharpe, JA, Aspinwall, R, Lu, W, Buckle, V, Strmecki, L, Walker, D, Ward, CJ, Alpers, CE, et al. (2000). A human PKD1 transgene generates functional polycystin-1 in mice and is associated with a cystic phenotype. *Hum Mol Genet,* 9:2617–2627.

Qian, F, Boletta, A, Bhunia, AK, Xu, H, Liu, L, Ahrabi, AK, Watnick, TJ, Zhou, F, and Germino, GG (2002). Cleavage of polycystin-1 requires the receptor for egg jelly domain and is disrupted by human autosomal-dominant polycystic kidney disease 1-associated mutations. *Proc Natl Acad Sci USA,* 99:16981–16986.

Qian, F, Watnick, TJ, Onuchic, LF, and Germino, GG (1996). The molecular basis of focal cyst formation in human autosomal dominant polycystic kidney disease type I *Cell,* 87:979–987.

Qian, Q, Harris, PC, and Torres, VE (2001). Treatment prospects for autosomal-dominant polycystic kidney disease. *Kidney Int,* 59:2005–2022.

Qian, Q, Hunter, LW, Li, M, Marin-Padilla, M, Prakash, YS, Somlo, S, Harris, PC, Torres, VE, and Sieck, GC (2003). Pkd2 haploinsufficiency alters intracellular calcium regulation in vascular smooth muscle cells. *Hum Mol Genet,* 12:1875–1880.

Quinn, M, Angelico, MC, Warram, JH, and Krolewski, AS (1996). Familial factors determine the development of diabetic nephropathy in patients with IDDM. *Diabetologia,* 39:940–945.

Remuzzi, G, Schieppati, A, and Ruggenenti, P (2002). Clinical practice. Nephropathy in patients with type 2 diabetes. *N Engl J Med,* 346:1145–1151.

Reumers, J, Schymkowitz, J, Ferkinghoff-Borg, J, Stricher, F, Serrano, L, and Rousseau, F (2005). SNPeffect: a database mapping molecular phenotypic effects of human non-synonymous coding SNPs. *Nucleic Acids Res,* 33: D527–D532.

Rippin, JD, Patel, A, Belyaev, ND, Gill, GV, Barnett, AH, and Bain, SC (2003). Nitric oxide synthase gene polymorphisms and diabetic nephropathy. *Diabetologia,* 46:426–428.

Ritz, E, Ogata, H, and Orth, SR (2000). Smoking: a factor promoting onset and progression of diabetic nephropathy. *Diabetes Metab,* 26 (Suppl. 4):54–63.

Ritz, E, Rychlik, I, Schomig, M, and Wagner, J (2001). Blood pressure in diabetic nephropathy—current controversies. *J Intern Med,* 249:215–223.

Rocco, MV, Chen, Y, Goldfarb, S, and Ziyadeh, FN (1992). Elevated glucose stimulates TGF-beta gene expression and bioactivity in proximal tubule. *Kidney Int,* 41:107–114.

Roglic, G, Colhoun, HM, Stevens, LK, Lemkes, HH, Manes, C, and Fuller, JH (1998). Parental history of hypertension and parental history of diabetes and microvascular complications in insulin-dependent diabetes mellitus: the EURODIAB IDDM Complications Study. *Diabet Med,* 15:418–426.

Rossetti, S, Burton, S, Strmecki, L, Pond, GR, San Millan, JL, Zerres, K, Barratt, TM, Ozen, S, Torres, VE, Bergstralh, EJ, et al. (2002). The Position of the polycystic kidney disease 1 (PKD1) gene mutation correlates with the severity of renal disease. *J Am Soc Nephrol,* 13:1230–1237.

Rossetti, S, Chauveau, D, Kubly, V, Slezak, JM, Saggar-Malik, AK, Pei, Y, Ong, AC, Stewart, F, Watson, ML, Bergstralh, EJ, et al. (2003). Association of mutation position in polycystic kidney disease 1 (PKD1) gene and development of a vascular phenotype. *Lancet,* 361:2196–2201.

Rossetti, S, Chauveau, D, Walker, D, Saggar-Malik, A, Winearls, CG, Torres, VE, and Harris, PC (2002). A complete mutation screen of the ADPKD genes by DHPLC. *Kidney Int,* 61:1588–1599.

Rossetti, S, Strmecki, L, Gamble, V, Burton, S, Sneddon, V, Peral, B, Roy, S, Bakkaloglu, A, Komel, R, Winearls, CG, and Harris, PC (2001). Mutation analysis of the entire PKD1 gene: genetic and diagnostic implications. *Am J Hum Genet,* 68:46–63.

Sakane, N, Yoshida, T, Yoshioka, K, Nakamura, Y, Umekawa, T, Kogure, A, Takakura, Y, and Kondo, M (1998). Trp64Arg mutation of beta3-adrenoceptor gene is associated with diabetic nephropathy in Type II diabetes mellitus. *Diabetologia,* 41:1533–1534.

Satko, SG, and Freedman, BI (2005). The familial clustering of renal disease and related phenotypes. *Med Clin North Am,* 89:447–456.

Satko, SG, Langefeld, CD, Daeihagh, P, Bowden, DW, Rich, SS, and Freedman, BI (2002). Nephropathy in siblings of African Americans with overt type 2 diabetic nephropathy. *Am J Kidney Dis,* 40:489–494.

Sawcer, S, Maranian, M, Setakis, E, Curwen, V, Akesson, E, Hensiek, A, Coraddu, F, Roxburgh, R, Sawcer, D, Gray, J, et al. (2002). A whole genome screen for linkage disequilibrium in multiple sclerosis confirms disease associations with regions previously linked to susceptibility. *Brain,* 125:1337–1347.

Schmidt, S, Bluthner, M, Giessel, R, Strojek, K, Bergis, KH, Grzeszczak, W, and Ritz, E (1998). A polymorphism in the gene for the atrial natriuretic peptide and diabetic nephropathy. Diabetic Nephropathy Study Group. *Nephrol Dial Transplant,* 13:1807–1810.

Schreuder, MF, Nyengaard, JR, Fodor, M, van Wijk, JA, and Delemarre-van de Waal, HA (2005). Glomerular number and function are influenced by spontaneous and induced low birth weight in rats. *J Am Soc Nephrol,* 16:2913–2919.

Seaquist, ER, Goetz, FC, Rich, S, and Barbosa, J (1989). Familial clustering of diabetic kidney disease. Evidence for genetic susceptibility to diabetic nephropathy. *N Engl J Med,* 320:1161–1165.

Setakis, E (2003). Statistical analysis of the GAMES studies. *J Neuroimmunol,* 143:47–52.

Shcherbak, NS, Shutskaya, ZV, Sheidina, AM, Larionova, VI, and Schwartz, EI (1999). Methylenetetrahydrofolate reductase gene polymorphism as a risk factor for diabetic nephropathy in IDDM patients. *Mol Genet Metab,* 68:375–378.

Shimazaki, A, Kawamura, Y Kanazawa, A, Sekine, A, Saito, S, Tsunoda, T, Koya, D, Babazono, T, Tanaka, Y, Matsuda, M, et al. (2005). Genetic variations in the gene encoding ELMO1 are associated with susceptibility to diabetic nephropathy. *Diabetes,* 54:1171–1178.

Somlo, S, Rutecki, G, Giuffra, LA, Reeders, ST, Cugino, A, and Whittier, FC (1993). A kindred exhibiting cosegregation of an overlap connective tissue disorder and the chromosome 16 linked form of autosomal dominant polycystic kidney disease. *J Am Soc Nephrol,* 4:1371–1378.

Spielman, RS, and Ewens, WJ (1998). A sibship test for linkage in the presence of association: the sib transmission/disequilibrium test. *Am J Hum Genet,* 62:450–458.

Spielman, RS, McGinnis, RE, and Ewens, WJ (1993). Transmission test for linkage disequilibrium: the insulin gene region and insulin-dependent diabetes mellitus (IDDM). *Am J Hum Genet,* 52:506–516.

Tanaka, N, Babazono, T, Saito, S, Sekine, A, Tsunoda, T, Haneda, M, Tanaka, Y, Fujioka, T, Kaku, K, Kawamori, R, et al. (2003). Association of solute carrier family 12 (sodium/chloride) member 3 with diabetic nephropathy, identified by genome-wide analyses of single nucleotide polymorphisms. *Diabetes,* 52:2848–2853.

Tarnow, L, Pociot, F, Hansen, PM, Rossing, P, Nielsen, FS, Hansen, BV, and Parving, HH (1997). Polymorphisms in the interleukin-1 gene cluster do not contribute to the genetic susceptibility of diabetic nephropathy in Caucasian patients with IDDM. *Diabetes,* 46:1075–1076.

Tarnow, L, Stehouwer, CD, Emeis, JJ, Poirier, O, Cambien, F, Hansen, BV, and Parving, HH (2000). Plasminogen activator inhibitor-1 and apolipoprotein E gene polymorphisms and diabetic angiopathy. *Nephrol Dial Transplant,* 15:625–630.

Telmer, S, Christiansen, JS, Andersen, AR, Nerup, J, and Deckert, T (1984). Smoking habits and prevalence of clinical diabetic microangiopathy in insulin-dependent diabetics. *Acta Med Scand,* 215:63–68.

The Diabetes Control and Complications Trial Research Group (1993). The effect of intensive treatment of diabetes on the development and progression of long-term complications in insulin-dependent diabetes mellitus. *N Engl J Med,* 329:977–986.

Torra, R, Badenas, C, Perez-Oller, L, Luis, J, Millan, S, Nicolau, C, Oppenheimer, F, Mila, M, and Darnell, A (2000). Increased prevalence of polycystic kidney disease type 2 among elderly polycystic patients. *Am J Kidney Dis,* 36:728–734.

Torres, VE, Wang, X, Qian, Q, Somlo, S, Harris, PC, and Gattone, VH, 2nd (2004). Effective treatment of an orthologous model of autosomal dominant polycystic kidney disease. *Nat Med,* 10:363–364.

Tuomilehto, J, Borch-Johnsen, K, Molarius, A, Forsen, T, Rastenyte, D, Sarti, C, and Reunanen, A (1998). Incidence of cardiovascular disease in Type 1 (insulin-dependent) diabetic subjects with and without diabetic nephropathy in Finland. *Diabetologia*, 41:784–790.

UK Prospective Diabetes Study (UKPDS) Group (1998). Intensive blood-glucose control with sulphonylureas or insulin compared with conventional treatment and risk of complications in patients with type 2 diabetes (UKPDS 33). *Lancet*, 352:837–853.

UK Renal Registry 2004 http://www.renalreg.com (accessed 15 December 2005

US Renal Data System Annual Data Report 2004 http://www.usrds.org/adr_2004.htm (accessed 15 December 2005)

Viswanathan, V, Zhu, Y, Bala, K, Dunn, S, Snehalatha, C, Ramachandran, A, Jayaraman, M, and Sharma, K (2001). Association between ACE gene polymorphism and diabetic nephropathy in South Indian patients. *JAPI*, 2:83–87.

Vlassara, H, Striker, LJ, Teichberg, S, Fuh, H, Li, YM, and Steffes, M (1994). Advanced glycation end products induce glomerular sclerosis and albuminuria in normal rats. *Proc Natl Acad Sci USA*, 91:11704–11708.

Walker, D, Consugar, M, Slezak, J, Rossetti, S, Torres, VE, Winearls, CG, and Harris, PC (2003). The ENOS polymorphism is not associated with severity of renal disease in polycystic kidney disease 1. *Am J Kidney Dis*, 41:90–94.

Wanner, C, Krane, V, Marz, W, Olschewski, M, Mann, JF, Ruf, G, and Ritz, E (2005). Atorvastatin in patients with type 2 diabetes mellitus undergoing hemodialysis. *N Engl J Med*, 353:238–248.

Watnick, TJ, Torres, VE, Gandolph, MA, Qian, F, Onuchic, LF, Klinger, KW, Landes, G, and Germino, GG (1998). Somatic mutation in individual liver cysts supports a two-hit model of cystogenesis in autosomal dominant polycystic kidney disease. *Mol Cell*, 2:247–251.

Williams, R, and Airey, M (2002). Epidemiology and public health consequences of diabetes. *Curr Med Res Opin*, 18 (Suppl. 1):S1–S12.

Wolf, G, and Ritz, E (2003). Diabetic nephropathy in type 2 diabetes prevention and patient management. *J Am Soc Nephrol*, 14:1396–1405.

Wong, TY, Poon, P, Szeto, CC, Chan, JC, and Li, PK (2000). Association of plasminogen activator inhibitor-1 4G/4G genotype and type 2 diabetic nephropathy in Chinese patients. *Kidney Int*, 57:632–638.

Woo, DD, Nguyen, DK, Khatibi, N, and Olsen, P (1997). Genetic identification of two major modifier loci of polycystic kidney disease progression in pcy mice. *J Clin Invest*, 100:1934–1940.

Wu, G, D'Agati, V, Cai, Y, Markowitz, G, Park, JH, Reynolds, DM, Maeda, Y, Le, TC, Hou, H Jr., Kucherlapati, R, et al. (1998). Somatic inactivation of Pkd2 results in polycystic kidney disease. *Cell*, 93:177–188.

Wu, G, Hayashi, T, Park, JH, Dixit, M, Reynolds, DM, Li, L, Maeda, Y, Cai, Y, Coca-Prados, M, and Somlo, S (1998). Identification of PKD2L, a human PKD2-related gene: tissue-specific expression and mapping to chromosome 10q25. *Genomics*, 54:564–568.

Wu, G, Markowitz, GS, Li, L, D'Agati, VD, Factor, SM, Geng, L, Tibara, S, Tuchman, J, Cai, Y, Park, JH, et al. (2000). Cardiac defects and renal failure in mice with targeted mutations in Pkd2. *Nat Genet*, 24:75–78.

Wu, G, Tian, X, Nishimura, S, Markowitz, GS, D'Agati, V, Hoon Park, J, Yao, L, Li, L, Geng, L, Zhao, H, et al. (2002). Trans-heterozygous Pkd1 and Pkd2 mutations modify expression of polycystic kidney disease. *Hum Mol Genet*, 11:1845–1854.

Yamaguchi, T, Wallace, DP, Magenheimer, BS, Hempson, SJ, Grantham, JJ, and Calvet, JP (2004). Calcium restriction allows cAMP activation of the B-Raf/ERK pathway, switching cells to a cAMP-dependent growth-stimulated phenotype. *J Biol Chem*, 279:40419–40430.

Yamamoto, T, Nakamura, T, Noble, NA, Ruoslahti, E, and Border, WA (1993). Expression of transforming growth factor beta is elevated in human and experimental diabetic nephropathy. *Proc Natl Acad Sci USA*, 90:1814–1818.

Yamamoto, T, Noble, NA, Cohen, AH, Nast, CC, Hishida, A, Gold, LI, and Border, WA (1996). Expression of transforming growth factor-beta isoforms in human glomerular diseases. *Kidney Int*, 49:461–469.

Yamamoto, T, Sato, T, Hosoi, M, Yoshioka, K, Tanaka, S, Tahara, H, Nishizawa, Y, and Fujii, S (2003). Aldose reductase gene polymorphism is associated with progression of diabetic nephropathy in Japanese patients with type 1 diabetes mellitus. *Diabetes Obes Metab*, 5:51–57.

Yuasa, T, Venugopal, B, Weremowicz, S, Morton, CC, Guo, L, and Zhou, J (2002). The sequence, expression, and chromosomal localization of a novel polycystic kidney disease 1-like gene, PKD1L1, in human. *Genomics*, 79:376–386.

Zanchi, A, Moczulski, DK, Hanna, LS, Wantman, M, Warram, JH, and Krolewski, AS (2000). Risk of advanced diabetic nephropathy in type 1 diabetes is associated with endothelial nitric oxide synthase gene polymorphism. *Kidney Int*, 57:405–413.

Zatz, R, Dunn, BR, Meyer, TW, Anderson, S, Rennke, HG, and Brenner, BM (1986). Prevention of diabetic glomerulopathy by pharmacological amelioration of glomerular capillary hypertension. *J Clin Invest*, 77:1925–1930.

Zeggini, E, Rayner, W, Morris, AP, Hattersley, AT, Walker, M, Hitman, GA, Deloukas, P, Cardon, LR, and McCarthy, MI (2005). An evaluation of HapMap sample size and tagging SNP performance in large-scale empirical and simulated data sets. *Nat Genet*, 37:1320–1322.

Zerres, K, Rudnik-Schoneborn, S, and Deget, F (1993). Childhood onset autosomal dominant polycystic kidney disease in sibs: clinical picture and recurrence risk. German Working Group on Paediatric Nephrology (Arbeitsgemeinschaft fur Padiatrische Nephrologie). *J Med Genet*, 30:583–588.

14

Hemostasis and Thrombosis

John H McVey and Edward GD Tuddenham

Vertebrates have evolved a complex system to prevent blood loss that involves coordinate vascular wall muscle contraction, cell aggregation (platelets), and the deposition of a clottable protein (fibrin). To achieve this, a complex network of positive and negative regulated reactions have evolved that result in controlled fibrin deposition and platelet activation only at the site of vascular injury without compromising blood flow through either the uninjured or damaged blood vessels.

The importance of such a system to multicellular organisms that possess a high pressure vascular system is evidenced by the recent demonstration that the coagulation network is present in its entirety in teleosts [bony fish such as zebra fish (*Danio rerio*) and Japanese puffer fish (*Fugu rubripes*)] and must therefore have evolved before the divergence of teleosts and tetrapods 430 million years ago (Hanumanthaiah et al., 2002; Davidson, Hirt, et al., 2003; Davidson, Tuddenham, et al., 2003)

The Coagulation Network

The Procoagulant Response

Blood coagulation in vivo is initiated by the exposure of factor (F) VII to cells that express the integral membrane protein tissue factor (TF). The primary control of hemostasis is the segregation of cells that express functional TF from other components of the coagulation network present in blood. TF is constitutively expressed at biological boundaries such as skin, organ surfaces, vascular adventitia, and epithelial-mesenchymal surfaces. The TF expression pattern has been described as forming a "hemostatic envelope" (Drake et al., 1989), which ensures that following disruption of vascular integrity FVII/FVIIa in blood is exposed to cells that express TF, leading to the initiation of blood coagulation (Fig. 14–1).

Conversely, it also ensures that inappropriate initiation of intravascular coagulation does not occur within intact vasculature. The formation of the TF-FVII complex promotes the activation of FVII. The TF-FVIIa complex catalyses the activation of FIX and FX.

In the absence of its cofactor FVa, FXa generates only trace amounts of thrombin. Although insufficient to initiate significant fibrin polymerization, trace amounts of thrombin formed in the "initiation" stage of coagulation are able to activate FV and FVIII by limited proteolysis in a positive feedback loop. In the "amplification" phase of coagulation, FVIIIa forms a complex with FIXa (the tenase complex) and activates sufficient FXa, which in complex with FVa (the prothrombinase complex) leads to the explosive generation of thrombin that immediately leads to the formation of a fibrin clot. Thrombin also activates FXI to FXIa in a further positive feedback loop, resulting in further generation of FIXa independent of the TF-FVIIa complex.

A key feature of these processes is the assembly of multiprotein complexes on a negatively charged phospholipid surface. Each of these complexes consists of a cofactor (TF, FVa, FVIIIa), an enzyme (FVIIa, FIXa, FXa) and a substrate that is a zymogen (FIX, FX, and prothrombin) of a serine protease. The product of one reaction becomes the enzyme in the next complex. Platelets activated at sites of vascular injury play key roles in normal hemostasis. By adhering to the exposed subendothelium and aggregating, they create a physical barrier that limits blood loss. In addition, platelets accelerate thrombin generation by providing a surface that promotes the activation of FX and prothrombin. Furthermore, they release procoagulant factors that contribute to the local coagulation response.

The importance of providing a negatively charged phospholipid surface for the assembly of the procoagulant response is seen in the extremely rare bleeding disorder, Scott syndrome [Online Mendelian Inheritance in Man (OMIM): 262890], which is characterized by a failure to expose phosphatidylserine on the outer leaflet of the plasma membrane and is associated with a moderate bleeding tendency.

The Anticoagulant Response

Following the initiation of coagulation, various inhibitory mechanisms prevent extension of the coagulation process beyond the site of vascular injury that might otherwise result in unnecessary occlusion of the blood vessel. Tissue factor pathway inhibitor (TFPI), associated with glycosaminoglycans or glycosylphosphatidylinositol (GPI) anchored proteins on the endothelial cell surface, rapidly inactivates the initiation complex by forming a quaternary inhibited complex (TF-FVIIa-FXa-TFPI). Thrombin stimulates both endothelial cells and platelets to release further TFPI.

Thrombin generated at the endothelial surface binds thrombomodulin (TM) and activates protein C (PC). The activation of PC is promoted by endothelial cell protein C receptor (EPCR), which provides a direct binding site for PC on endothelial cells and

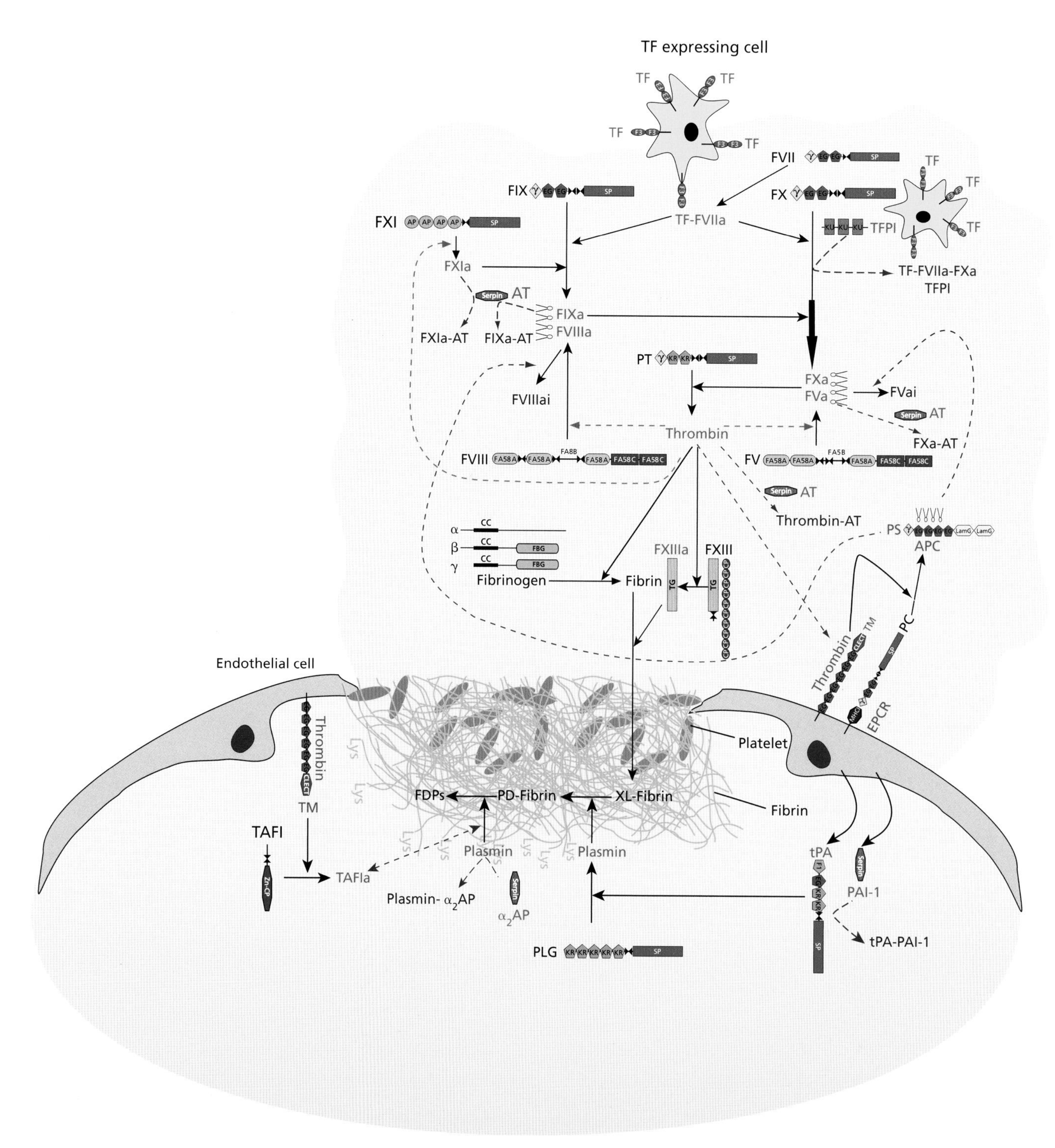

Figure 14–1 The hemostasis network. The process of blood coagulation is initiated by the exposure of cells expressing TF to flowing blood. Thrombin generation is propagated by a series of positive feedback loops, leading to fibrin deposition. This process is controlled by a series of negative feedback steps: the initiation complex is inhibited by the formation of the quaternary complex TF-FVIIa-FXa-TFPI and the active proteases FIXa, FXa, and thrombin are inactivated by the serpin AT. In addition, thrombin initiates a negative feedback pathway by activating PC, leading to the inhibition of FVa and FVIIIa. The fibrin clot is degraded by plasmin. Module organization of proteins is indicated as proposed by Bork and Bairoch; (http://smart.embl.de/browse.shtml) and all abbreviations are given in Table 14–1: Protein names are colored thus: functionally active proteins, red; inactivated or inhibited proteins, blue; precursors and fibrinogen-derived components, black. Lines and arrows are colored thus: dashed red lines, positive procoagulant feedback loops; dashed blue lines, negative feedback and inhibitory loops; solid black lines, interactions or processes.

increases the affinity of the thrombin-TM complex for PC. A PC in complex with its cofactor protein S (PS) rapidly inactivates the procoagulant cofactors FVa and FVIIIa by specific proteolysis, forming a negative feedback loop. The activated coagulation proteases FIXa, FXa, FXIa, and thrombin are all inhibited by antithrombin (AT). The rate of inhibition by AT is substantially increased by binding glycosaminoglycans on the surface of endothelial cells.

The Fibrinolytic Response

Formation of fibrin triggers activation of the fibrinolytic system and generation of the active fibrinolytic enzyme plasmin that degrades fibrin into soluble fragments disintegrating the clot. Plasmin is formed from plasminogen by limited proteolysis by the action of tissue-type plasminogen activator (tPA) or urokinase-type plasminogen activator (uPA). tPA is the most important activator in the circulation. Plasminogen activator inhibitor (PAI)-1 and PAI-2 inhibit the activities of these activators, while plasmin is primarily inhibited by plasmin inhibitor (á2-antiplasmin). tPA is a serine protease synthesized, stored, and secreted by endothelial cells. In the absence of fibrin, tPA has low activity toward plasminogen, but this activity is increased up to three-fold in the presence of fibrin. The rate enhancing effect of fibrin ensures that the formation of plasmin and the generation of fibrinolytic activity are restricted to the location of a fibrin clot.

Thrombin-activatable fibrinolysis inhibitor (TAFI) is a carboxypeptidase, synthesized in the liver, that circulates in blood as a zymogen and is activated by a single proteolytic cleavage mediated by thrombin in complex with TM. TAFIa inhibits fibrinolysis by removing the C-terminal lysine residues formed by limited plasmin proteolysis of fibrin, thus removing binding sites for plasminogen and tPA. The rate enhancement effect of fibrin is reduced and fibrinolysis is downregulated.

Molecular Genetics

The blood coagulation network is a finely balanced system that ensures that clot formation occurs rapidly and efficiently at the site of vascular injury but localizes the procoagulant response in order to minimize blood loss without compromising blood flow generally. Anticoagulant and fibrinolytic components prevent inappropriate clot formation or extension of the procoagulant response beyond the site of vascular injury. Deficiency of components of the coagulation network can lead to either a bleeding phenotype or a thrombotic phenotype depending on the role of the deficient factor within the network.

The bleeding disorders are typically single gene disorders with a defect in the gene encoding the coagulation factor. The exceptions being combined deficiency of FV and FVIII where the defect lies in two genes involved in intracellular trafficking (LMAN1 and MCFD2) (Neerman-Arbez et al., 1999; Nichols et al., 1999; Zhang et al., 2003; Zhang et al., 2006), multiple Vitamin K dependent factor deficiencies type I and II where the defect is in gamma carboxylase (Wu et al., 1997; Brenner et al., 1998; Spronk et al., 2000; Mousallem et al., 2001), or VKORC1 (Li et al., 2004; Rost et al., 2004), respectively and Scott syndrome, an extremely rare moderate bleeding disorder characterized by a failure to expose phosphatidylserine on the outer leaflet of the plasma membrane where the defect lies in the *ABCA1* gene (Albrecht et al., 2005).

The genes encoding the human coagulation/fibrinolytic proteins were identified and characterized (often completely sequenced) in the 1980s. The completion of the human genome project in 2003 therefore had less impact than in many other disciplines. Many mutations have been identified and characterized and locus-specific mutation databases have been established for many of the disorders (Table 14–1). All types of mutations have been identified: missense, nonsense, splice site, and promoter mutations as well as insertions, deletions, and rearrangements. The spectrum of mutations represented in the FVII mutation database (http://europium.csc.mrc.ac.uk/) is typical of those found in other coagulation factor gene disorders (Fig. 14–2). Recurrent mutations in unrelated individuals arise at functionally important residues and in CpG dinucleotides, which are known hot spots for mutation due to deamination of methylated cytosine residues. A mutation hot spot, specific to hemophilia A, is the recurrent rearrangement responsible for 50% of all severe cases in which homologous recombination of a sequence within intron 22 of the *F8* gene and two extragenic and telomeric copies of the sequence results in an inversion that disrupts the *F8* gene.

The Bleeding Disorders

The most common inherited bleeding disorder is von Willebrand's disease (VWD).von Willebrand factor (VWF), which is the protein deficient or defective in patients with VWD, has two distinct roles in hemostasis. First, it is responsible for enabling platelets to adhere to collagen exposed at wound sites under conditions of shear. Second, VWF is the plasma carrier for FVIII, thus deficiency of VWF leads to deficiency of FVIII. VWD is an autosomal disease with variable expressivity and reduced penetrance. It has been classified into six distinct types: types 1 and 3 result from quantitative deficiency whereas four type 2 variants (2A, 2B, 2M, and 2N) are characterized by qualitative abnormalities of VWF. Types 1, 2A, 2B, and 2M VWD are inherited as autosomal dominant whereas types 2N and 3 are inherited in autosomal recessive manner. The mutations causing type 1 VWD are poorly defined with only 14 mutations identified to date. Common mutations causing type 2A, 2B, and 2N have been reported, in contrast no common mutations have been identified in type 2M or type 3 (http://www.vwf.group.shef.ac.uk/).

FVII deficiency is an autosomal recessive disorder with a highly variable phenotype. In contrast to the other bleeding disorders, the residual clotting factor activity does not accurately predict the clinical phenotype. Complete absence of FVII activity is usually incompatible with life, and individuals die shortly after birth due to severe hemorrhage. The majority of individuals with mutations in their *f7* gene(s) are asymptomatic. A severe bleeding phenotype is only observed in individuals with residual FVII activity below 2%; however, a considerable proportion of individuals with a mild to moderate bleeding phenotype have similar FVII activity levels. The severity of the bleeding in individuals with no residual FVII activity is consistent with the key role of the TF-FVIIa complex in initiating blood coagulation in vivo. Surprisingly, no congenital abnormalities in the gene encoding TF have been described to date. Mutations in the TF gene might have been predicted to lead to either a bleeding (loss of function) or prothrombotic (gain of function) phenotype. Targeted disruption of the mouse TF gene (*f3*) results in embryonic lethality of $f3^{-/-}$ embryos at embryonic days 9.5–10.5 (Bugge et al., 1996; Carmeliet et al., 1996; Toomey et al., 1996). Hence, loss in early pregnancy most probably accounts for the lack of TF null individuals in clinical practice.

Deficiency of FVIII (hemophilia A) and FIX (hemophilia B) share an indistinguishable clinical phenotype characterized by bleeding into muscles, joints, and other organs with consequent damage. They are both X-linked recessive disorders affecting 1 in 5000 and 1 in 30,000 males, respectively. Clinically, hemophilia is

Table 14–1 Proteins in the Hemostatic Network Associated with Bleeding or Thrombosis

Common Name	Abbreviation	Sub-unit	Gene Symbol	Gene Location	Gene Size (Kbp)	No of Exons	mRNA Size (bp)	Amino acids (mature)	Mr of Monomer (kDa)	OMIM	Main Action	LMD
Tissue Factor	TF		F3	1p13	12.6	6	1852	263	44	134390	Cofactor for FVII/FVIIa	
Prothrombin	FII		F2	11p11.1	20.3	14	2028	579	72	176930	Clots FBG, activates PC, FXI, TAFI	http://www.med.unc.edu/isth/mutations-databases/prothrombin.htm
Factor V	FV		F5	1q23	72.4	25	7009	2196	330	227400	Cofactor for FXa	http://www.lumc.nl/4010/research/Factor_V_gene.html
Factor VII	FVII		F7	13q34	15.1	8	2880	416	50	227500	Activates FIX & FX	http://europium.csc.mrc.ac.uk/
Factor VIII	FVIII		F8	Xq28	187.1	26	8957	2332	330	306700	Cofactor for FIXa	http://europium.csc.mrc.ac.uk/
Factor IX	FIX		F9	Xq27	32.8	8	2831	415	56	306900	Activates FX	http://www.kcl.ac.uk/ip/petergreen/haemBdatabase.html
Factor X	FX		F10	13q34	26.8	8	1884	445	59	227600	Activates prothrombin	http://www.med.unc.edu/isth/mutations-databases/Factor_X.htm
Factor XI	FXI		F11	4q35	25.9	15	2979	607	80*	264900	Activates FIX	http://www.factorxi.com/
Factor XIII** (A chain)	FXIII	A	F13A1	6p25	177.8	15	3834	731	75**	134570	Crosslinks fibrin	http://www.med.unc.edu/isth/mutations-databases/Factor_-XIIIA.htm
Factor XIII** (B chain)	FXIII	B	F13B	1q31	28	12	2190	641	80**	134580	Stabilises FXIII A chain	http://www.med.unc.edu/isth/mutations-databases/Factor_-XIIIA.htm
Fibrinogen (α chain)***	FGN	α	FGA	4q32	7.8	6	3828	866	68***	134820	Mechanical stabilisation of clot	http://www.geht.org/databaseang/fibrinogen/
Fibrinogen (β chain)***	FGN	β	FGB	4q32	9.8	8	3693	491	52***	134830	Mechanical stabilisation of clot	http://www.geht.org/databaseang/fibrinogen/
Fibrinogen (γ chain)***	FGN	γ	FGG	4q32	23.6	10	1753	453	49***	134850	Mechanical stabilisation of clot	http://www.geht.org/databaseang/fibrinogen/
von Willebrand factor	VWF		VWF	12p13	176	52	9028	2050	255	193400	Cell adhesion & FVIII carrier	http://www.vwf.group.shef.ac.uk/
Protein C	PC		PROC	2q14.2	10.8	9	1759	419	62	176860	Inactivation of FVa and FVIIIa	http://www.xs4all.nl/%7Ereitsma/Prot_C_home.htm
Protein S	PS		PROS1	3q11.2	101.9	15	3477	676	69	176880	Inactivation of FVa and FVIIIa	
Antithrombin	AT		SERPINC1	1q23	21	9	1684	464	58	107300	Inhibits thrombin, FIX, FX, FXI	http://wwwfom.sk.med.ic.ac.uk/medicine/about/divisions/is/haemo/coag/antithrombin/default.html

(Continued)

Table 14–1 Proteins in the Hemostatic Network Associated with Bleeding or Thrombosis *(Continued)*

Common Name	Abbreviation	Sub-unit	Gene Symbol	Gene Location	Gene Size (Kbp)	No of Exons	mRNA Size (bp)	Amino acids (mature)	Mr of Monomer (kDa)	OMIM	Main Action	LMD
Plasminogen	PLG		PLG	6q27	51.1	14	2905	791	92	173350	dissolution of clot in wound repair	
Tissue plasminogen activator	TPA		PLAT	8p11.1	32.7	14	2933	562	69	173370	Plasma activator of plasminogen	
Plasminogen activator inhibitor 1	PAI-1		SERPINE1	7q22	12.3	9	3320	379	52	173360	Inhibition of TPA and UPA	
Alpha2-antiplasmin	α2-AP		SERPINF2	17p13	13.3	9	2210	452	67	262850	Inhibition of plasmin	
Thrombin-activatable fib. inhibitor	TAFI		CPB2	13q14	52.4	11	1984	401	60	603101	Inhibition of fibrinolysis	

noted to be severe, moderate, or mild and this correlates with the residual level of activity of the affected factor, which in turn relates to the precise gene defect. Thus, severely affected patients bleed spontaneously and have less than 2% of the normal level of factor in their blood; moderately affected patients bleed after minor trauma and have 2–5% of the normal level; and mildly affected individuals have more than 5% of the normal level and bleed only after severe trauma.

The severity of the bleeding in hemophilia A and B supports a key role for the FIXa-FVIIIa (tenase) complex in the amplification phase of blood coagulation. Half of all severe hemophilia A is caused by spontaneous homologous recombination resulting in inversion of the *F8* gene (Fig. 14–3).

Two types of inversion involving a sequence in intron 22 and either the proximal (int22h-2) or the distal (int22h-3) of two extragenic copies of this sequence were thought to be responsible for the two forms of the inversion observed in hemophilia A patients (Naylor et al., 1995). However, recent genomic sequence data following the completion of the DNA sequence of the human X chromosome (Ross et al., 2005) indicates that recombination leading to inversion can only occur between the intron 22 and distal copies of the sequence (Ross and Bentley, 2005). Recombination between intron 22 and the proximal copies of the sequence would be predicted to lead to duplication and deletion, since they are in the same orientation to each other. The new sequence information also revealed that int22h-2 and int22h-3 form the arms of a large palindrome. It has therefore been suggested that recombination between the arms of the palindrome may occur creating alleles where either the int22h-2 or the int22h-3 occupies the proximal position, thus explaining the observed rearrangements (Bagnall et al., 2005).

FXI deficiency (hemophilia C) is an autosomal bleeding disorder that is not completely recessive since heterozygotes have a mild but definite bleeding tendency. FXI circulates in plasma as a homodimer. The majority of identified mutations in the *f11* gene associated with deficiency either prevents or greatly reduces protein

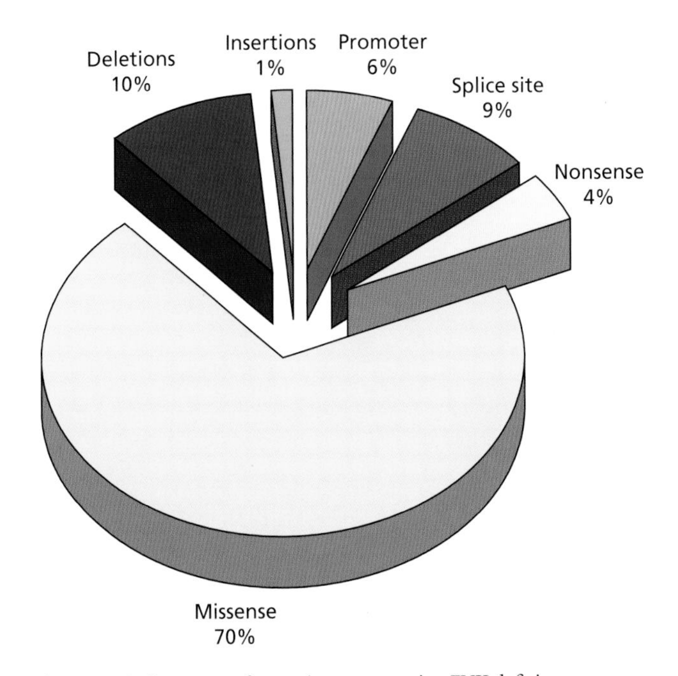

Figure 14–2 Spectrum of mutation type causing FVII deficiency. Data obtained from the FVII mutation database (http://europium.csc.mrc.ac.uk/).

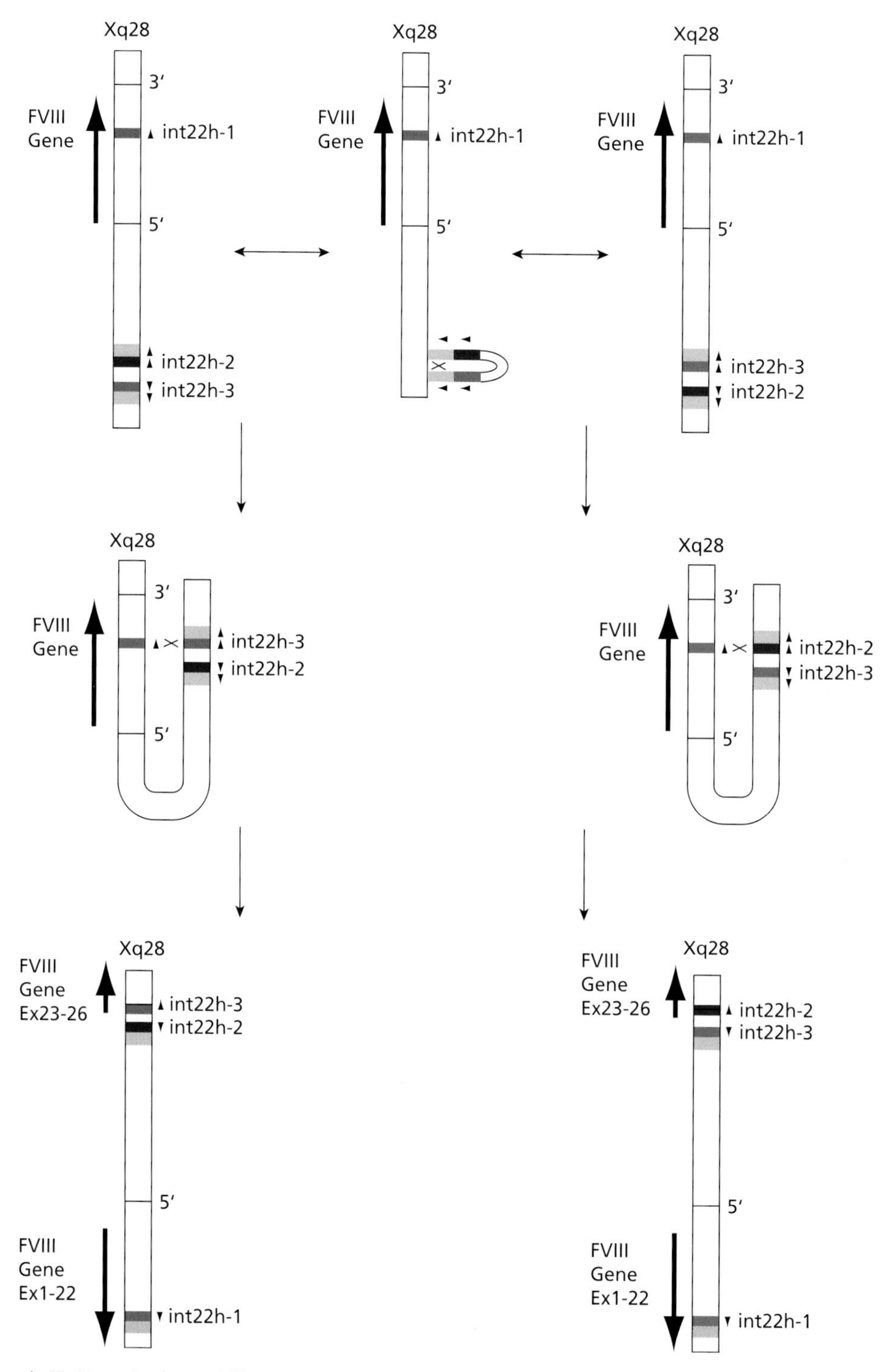

Figure 14–3 Inversions in Xq28 causing hemophilia A. Proposed mechanism causing polymorphic inversion and recombination between int22h repeats leading to inversion of the *F8* gene. The *F8* gene is indicated by the large arrow. Inversions disrupting the *F8* gene result from recombination between int22h-1 region (in intron 22 of the *F8* gene) and either int22h-2 or int22h-3, which lie 400 kb distal to *F8*. The three copies of the int22h sequence are indicated by the red, blue, and green colored boxes and their orientation relative to the intron 22 copy is indicated by the arrowhead. The distal copies (int22h-2 and int22h-3) are in opposite orientation to each other and are flanked by a large imperfect palindrome, indicated by blue boxes. Proposed recombination between the arms of the palindrome would generate alleles where either the in22h-2 or the int22h-3 is most telomeric and in the opposite orientation to int22h-1.

expression. The dominant negative effect of mutations may therefore be due to a failure to secrete heterodimers of mutant and wild-type FXI or the secretion of functionally compromised heterodimers. The severity of bleeding in this disorder is relatively mild, suggesting that the back activation of FXI by thrombin may only be required in severe trauma. In the Ashkenazi Jewish population, where the gene frequency for FXI mutations is as high as 8%, four different mutations have been identified: type I mutation occurs in the 5′ splice donor site of intron N; type II mutation in exon 5 is a nonsense mutation (FXI E117X); type III mutation is a missense mutation (FXI F283L) in exon 9; type IV mutation is a 14 base pair deletion at the junction of exon 14 and intron N. The prevalence of these mutations is due to a founder effect within this population, and these mutations account for only a small percentage of disease alleles in non-Jewish patients (reviewed in O'Connell, 2003).

Deficiency of either FV, FX, or PT is exceedingly rare, presumably because total deficiency is incompatible with life, since complete gene knockout of these genes in mice is embryonic lethal (Cui et al., 1996; Sun et al., 1998; Dewerchin et al., 2000). Fibrinogen deficiency is also rare; however, total deficiency of fibrinogen in mice is compatible with normal in utero development, although they do have a bleeding phenotype (Suh et al., 1995). Of note, some types of dysfibrinogenemia are associated with a thrombotic rather than a bleeding tendency. Deficiency of PAI-1 results in a failure to control plasmin generation adequately. Excess plasmin increases fibrin degradation and clot disintegration leading to a bleeding phenotype. In contrast, mice deficient for PAI-1 (*serpine1*$^{-/-}$) do not show spontaneous bleeding or delayed bleeding even after amputation of the tail (Carmeliet et al., 1993). Deficiency of PAI-1 is extremely rare and once again, it should be noted that raised levels of PAI-1 are associated with a thrombotic tendency.

The Thrombotic Disorders

The failure to control the generation of thrombin or impaired neutralization of thrombin causes most inherited thrombophilia. AT when bound to glycosaminoglycans on the surface of endothelial cells neutralizes the procoagulants thrombin, FIXa, FXa, and FXIa. PC controls the generation of thrombin by inactivating the procoagulant cofactors FVa and FVIIIa when in complex with its cofactor PS. Deficiency of AT, PC, or PS are all associated with a thrombotic phenotype but are relatively rare. Homozygous AT deficiency is probably incompatible with life. Individuals with homozygous PC or PS deficiency present soon after birth with purpura fulminans or severe venous thrombosis, however, these are exceedingly rare. Individuals heterozygous for deficiency of AT, PC, or PS are all susceptible to venous thrombosis.

The most frequently occurring inherited defects associated with a thrombotic tendency are resistance to activated PC caused by a point mutation in the gene for FV (FV R506Q; FV Leiden) and a mutation in the prothrombin gene (e.g., F2 G20210A). The population frequency of FV Leiden is 5% in ethnic Europeans. FVa is inactivated by activated PC through proteolytic cleavage at R506–S507 and R1765–L1766. The cleavage at R506 is rate limiting and the mutation FV R506Q is resistant to cleavage by activated PC. The mutation in the prothrombin gene occurs in the 3′ untranslated region of the gene and is associated with raised prothrombin levels. Its frequency in Europeans is approximately 2%.

Elevated levels of FVIII have also been associated with an increased risk of venous thromboembolism, however, the cause is unknown (reviewed in Seligsohn and Lubetsky, 2001; Feinbloom and Bauer, 2005). In contrast, a failure to appropriately regulate fibrinolysis and thus clot disintegration may also be associated with a thrombotic tendency, however, these conditions (deficiency of tPA and plasminogen; increased PAI-1 activity) are very rare.

Comparative Sequence Analysis

The impact of the completion sequence of the human genome on our understanding of hemostasis has been much less than for other inherited disorders, since most of the relevant genes were cloned before the genomic era. However, identification and characterization of the genes responsible for combined FV and FVIII deficiency and of Type II multiple coagulation factor deficiency were aided by this genomic information. The completion of the sequence of other vertebrate genomes has, however, provided a fascinating insight into the evolution of the coagulation network. In addition, the comparative sequence information has been invaluable in interpreting naturally occurring mutations in blood coagulation deficiencies (Gomez et al., 2004). Comparison of the amino acid sequence alignments has also provided important insights into structure–function relationships of the coagulation factors. This information may aid the design of novel recombinant coagulation factors for replacement therapy, in particular FVIII molecules with modified domains. Comparative sequence analysis of noncoding sequences may also further our understanding of regulatory gene sequences.

References

Albrecht, C, McVey, JH, Elliott, JI, Sardini, A, Kasza, I, Mumford, AD, et al. (2005). A novel missense mutation in ABCA1 results in altered protein trafficking and reduced phosphatidylserine translocation in a patient with Scott syndrome. *Blood* 106:542–549.

Bagnall, RD, Giannelli, F, and Green, PM (2005). Polymorphism and hemophilia A causing inversions in distal Xq28: a complex picture. *J Thromb Haemost*, 3:2598–2599.

Brenner, B, Sanchez-Vega, B, Wu, SM, Lanir, N, Stafford, DW, and Solera, J (1998). A missense mutation in gamma-glutamyl carboxylase gene causes combined deficiency of all vitamin K-dependent blood coagulation factors. *Blood*, 92:4554–4559.

Bugge, TH, Xiao, Q, Kombrinck, KW, Flick, MJ, Holmback, K, Danton, MJ, et al. (1996). Fatal embryonic bleeding events in mice lacking tissue factor, the cellassociated initiator of blood coagulation. *Proc Natl Acad Sci USA*, 93:6258–6263.

Carmeliet, P, Mackman, N, Moons, L, Luther, T, Gressens, P, Van, V, et al. (1996). Role of tissue factor in embryonic blood vessel development. *Nature*, 383:73–75.

Carmeliet, P, Stassen, JM, Schoonjans, L, Ream, B, van den Oord, JJ, De, MM, et al. (1993). Plasminogen activator inhibitor-1 gene-deficient mice. II. Effects on hemostasis, thrombosis, and thrombolysis. *J Clin Invest*, 92:2756–2760.

Cui, J, O'Shea, KS, Purkayastha, A, Saunders, TL, and Ginsburg, D (1996). Fatal haemorrhage and incomplete block to embryogenesis in mice lacking coagulation factor V. *Nature*, 384:66–68.

Davidson, CJ, Hirt, RP, Lal, K, Snell, P, Elgar, G, Tuddenham, EG, et al. (2003). Molecular evolution of the vertebrate blood coagulation network. *Thromb Haemost*, 89:420–428.

Davidson, CJ, Tuddenham, EG, and McVey, JH (2003). 450 million years of hemostasis. *Thromb Haemost*, 1:1487–1494.

Dewerchin, M, Liang, Z, Moons, L, Carmeliet, P, Castellino, FJ, Collen, D, et al. (2000). Blood coagulation factor X deficiency causes partial embryonic lethality and fatal neonatal bleeding in mice. *Thromb Haemost*, 83:185–190.

Drake, TA, Morrissey, JH, and Edgington, TS (1989). Selective cellular expression of tissue factor in human tissues. Implications for disorders of hemostasis and thrombosis. *Am J Pathol*, 134:1087–1097.

Feinbloom, D and Bauer, KA (2005). Assessment of hemostatic risk factors in predicting arterial thrombotic events. *Arterioscler Thromb Vasc Biol*, 25:2043–2053.

Gomez, K, Laffan, MA, Kemball-Cook, G, Pasi, J, Layton, M, Singer, JD, et al. (2004). Two novel mutations in severe factor VII deficiency. *Br J Haematol*, 126:105–110.

Hanumanthaiah, R, Day, K, and Jagadeeswaran, P (2002). Comprehensive analysis of blood coagulation pathways in teleostei: evolution of coagulation factor genes and identification of zebra fish factor VIII. *Blood Cells Mol Dis*, 29:57–68.

Li, T, Chang, CY, Jin, DY, Lin, PJ, Khvorova, A, and Stafford, DW (2004). Identification of the gene for vitamin K epoxide reductase. *Nature*, 427:541–544.

Mousallem, M, Spronk, HM, Sacy, R, Hakime, N, and Soute, BA (2001). Congenital combined deficiencies of all vitamin K-dependent coagulation factors. *Thromb Haemost*, 86:1334–1336.

Naylor, JA, Buck, D, Green, P, Williamson, H, Bentley, D, and Giannelli, F (1995). Investigation of the factor VIII intron 22 repeated region (int22h) and the associated inversion junctions. *Hum Mol Genet*, 4:1217–1224.

Neerman-Arbez, M, Johnson, KM, Morris, MA, McVey, JH, Peyvandi, F, Nichols, WC, et al. (1999). Molecular analysis of the ERGIC-53 gene in 35 families with combined factor V-factor VIII deficiency. *Blood*, 93:2253–2260.

Nichols, WC, Terry, VH, Wheatley, MA, Yang, A, Zivelin, A, Ciavarella, N, et al. (1999). ERGIC-53 gene structure and mutation analysis in 19 combined factors V and VIII deficiency families. *Blood*, 93:2261–2266.

O'Connell, NM (2003). Factor XI deficiency—from molecular genetics to clinical management. *Blood Coagul Fibrinolysis*, 14 (Suppl. 1):S59–S64.

Ross, MT and Bentley, DR (2005). More on: polymorphism and hemophilia A causing inversions in distal Xq28: a complex picture. *J Thromb Haemost*, 3:2600–2601.

Ross, MT, Grafham, DV, Coffey, AJ, Scherer, S, McLay, K, Muzny, D, et al. (2005). The DNA sequence of the human X chromosome. *Nature*, 434:325–337.

Rost, S, Fregin, A, Ivaskevicius, V, Conzelmann, E, Hortnagel, K, Pelz, HJ, et al. (2004). Mutations in VKORC1 cause warfarin resistance and multiple coagulation factor deficiency type 2. *Nature*, 427:537–541.

Seligsohn, U and Lubetsky, A (2001). Genetic susceptibility to venous thrombosis. *N Engl J Med*, 344:1222–1231.

Spronk, HM, Farah, RA, Buchanan, GR, Vermeer, C, and Soute, BA (2000). Novel mutation in the gamma-glutamyl carboxylase gene resulting in congenital combined deficiency of all vitamin K-dependent blood coagulation factors. *Blood*, 96:3650–3652.

Suh, TT, Holmback, K, Jensen, NJ, Daugherty, CC, Small, K, Simon, DI, et al. (1995). Resolution of spontaneous bleeding events but failure of pregnancy in fibrinogen-deficient mice. *Genes Dev*, 9:2020–2033.

Sun, WY, Witte, DP, Degen, JL, Colbert, MC, Burkart, MC, Holmback, K, et al. (1998). Prothrombin deficiency results in embryonic and neonatal lethality in mice. *Proc Natl Acad Sci USA*, 95:7597–7602.

Toomey, JR, Kratzer, KE, Lasky, NM, Stanton, JJ, and Broze, GJ, Jr. (1996). Targeted disruption of the murine tissue factor gene results in embryonic lethality. *Blood*, 88:1583–1587.

Wu, SM, Stafford, DW, Frazier, LD, Fu, YY, High, KA, Chu, K, et al. (1997). Genomic sequence and transcription start site for the human gamma-glutamyl carboxylase. *Blood*, 89:4058–4062.

Zhang, B, Cunningham, MA, Nichols, WC, Bernat, JA, Seligsohn, U, Pipe, SW, et al. (2003). Bleeding due to disruption of a cargo-specific ER-to-Golgi transport complex. *Nat Genet*, 34:220–225.

Zhang, B, McGee, B, Yamaoka, JS, Guglielmone, H, Downes, KA, Minoldo, S, et al. (2006). Combined deficiency of factor V and factor VIII is due to mutations in either LMAN1 or MCFD2. *Blood*, 107:1903–1907.

15

Disorders of Platelets

Wadie F Bahou

Overview of Platelet Biology and Function

Blood coagulation is controlled by a tightly regulated cascade of proteases and their cofactors that sequentially lead to the generation of a gelatinous meshwork of protein called fibrin. This coagulation mechanism ensures the normal cessation of blood flow that occurs during physiological processes such as wounding or menstruation, and is termed hemostasis. Human blood platelets play a critical function in the maintenance of primary hemostasis, not only via adhesive interactions with the subendothelial matrix but also by providing the negatively charged phospholipid required for coagulation factor assembly, amplification, and thrombin generation (Bahou, 2002).

Platelets are small, discoid cells that are 2–3 μm in diameter with an approximate cell volume of 10 fl. Generated as cytoplasmic buds from precursor bone marrow-derived megakaryocytes, the normal human platelet count is 2,50,000 ± 1,00,000/μl of whole blood. Although platelets are required for normal hemostasis, longstanding observations have established that their primary relevance to human diseases are their roles in situations of pathological thrombosis, that is, vascular disorders such as stroke or coronary occlusion with myocardial infarction (MI) (Wardle, 1973). Thus, although well-characterized (but generally rare) molecular defects involving platelet receptors and/or their signaling networks may be associated with human bleeding disorders, defects causally implicated in the more frequent "platelet prothrombotic state" remain incompletely identified. This chapter will review the principles of platelet function, molecular defects, and/or polymorphisms causally implicated in hemostatic and thrombotic defects, and the current attempts to apply whole-genome (and proteome) principles to identify the genes causally implicated in platelet-associated diseases.

Megakaryocyte Development and Proplatelet Formation

The megakaryocyte (Mk) is the largest and least frequent hematopoietic cell accounting for 0.02%–0.05% of total bone marrow nucleated cells. The endpoint of proliferation is marked by a switch to endomitosis in which deoxyribonucleic acid (DNA) duplication occurs in the absence of cellular division (Fig. 15–1). Megakaryocytes are polyploid with a nuclear modal ploidy of 16N, although the ploidy typically ranges from 2N to 64N (Paulus, 1968; Corash et al., 1989). This process allows the polyploid Mk to support a greatly enhanced cytoplasmic volume, providing for the production of thousands of platelets from individual cells. Thrombopoietin (TPO) stimulates cell growth and Mk maturation (Kaushansky, 1995), but is dispensable for terminal differentiation and may inhibit proplatelet formation (Choi et al., 1996). GATA-1 is a predominant transcriptional modulator of lineage commitment and virtually all subsequent steps involving Mk maturation occur *via* the interactions with FOG-1, Runx1, and Ets-family proteins, and the apparent activation of the p45 *NF-E2* gene (Tsang et al., 1998; Vyas et al., 1999). Roles for B- and D-type cyclins have been proposed in the regulation of Mk endomitosis (Wang et al., 1995; Matsumura et al., 2000), although more recent studies in knockout mice implicate both cyclins E1 and E2 in the endoreplicative mechanism(s) (Geng et al., 2003).

An extensive proplatelet network is required en route to Mk-derived platelet formation. This process requires an elaborate reservoir of membranes, granules, and cytoskeletal proteins, which is partially provided by the demarcation membrane system. Assembled platelet granules, ribosomes, mitochondria, and other oganelles are transported from the Mk cellular body along the proplatelet cytoskeletal tracks into the tips of these structures (Italiano et al., 1999). Actin reorganization may occur before microtubule accumulation at the cell periphery; upstream proplatelet formation depends on the activation of phosphatidyl inositol 3-kinase and protein kinase isoforms (Rojnuckarin and Kaushansky, 2001; Rojnuckarin et al., 2001). The role of β1 tubulin is demonstrated in the knockout mice that have thrombocytopenia and platelet spherocytosis, presumably because the marginal microtubule band contains—one to three coils in place of the usual eight to ten loops (Schwer et al., 2001). The membrane asymmetry of proplatelets must be maintained to minimize the exposure of negatively charged phospholipids that trigger apoptotic clearance mechanisms involving macrophages, or the assembly of the coagulation cascade. Paradoxically, transgenic mice overexpressing the anti-apoptotic proteins Bcl-2 and Bcl-xL demonstrate thrombocytopenia (Ogilvy et al., 1999), and mice deficient in the proapoptotic protein Bim have reduced platelet counts (Bouillet et al., 1999), suggesting a poorly defined mechanism whereby proplatelet production from Mks is regulated by the Bcl-2 family of proteins (Ogilvy et al., 1999). Terminally differentiated Mks cleave procaspase 3 and 9 (and the caspase gelsolin substrate),

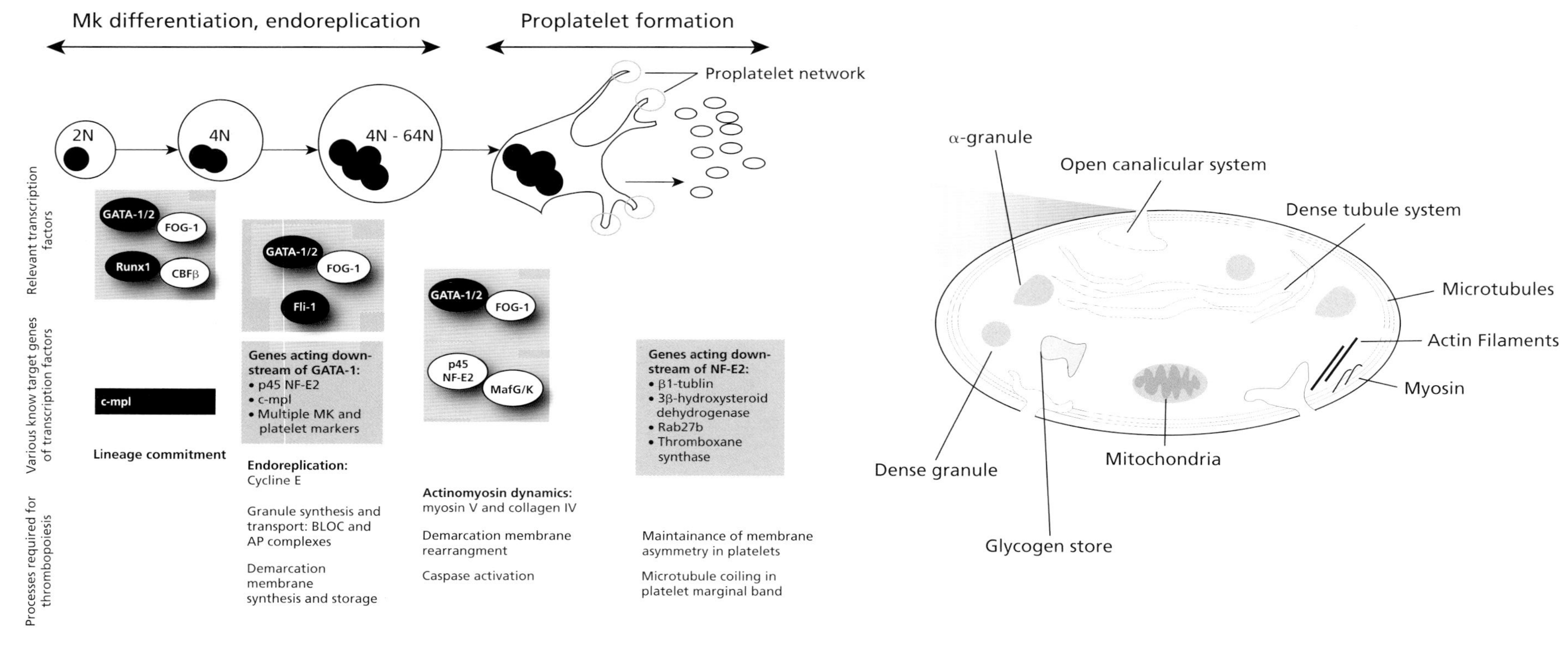

Figure 15–1 Schema outlining the stages of megakaryocyte differentiation leading to platelet formation. Shown are the important transcription factors, known target genes, and cellular processes critical to the process of thrombopoiesis, together with an ultrastructural schema of a mature platelet. Genes with molecular defects implicated in clinical thrombocytopenic syndromes are delineated by solid black backgrounds. BLOC, biogenesis of lysosome-related organelle complex; AP, adapter-related protein complex; GATA, globin transcription factor; Fli-1, friend leukemia virus integration-1; FOG-1, friend of GATA; NF-E2, nuclear factor, erythroid-derived 2. [Schema adapted from Schulze and Shivdasani (2004).]

with the release of cytochrome *c*, thereby triggering the apoptotic pathway (De Botton et al., 2002). In contrast, platelets lack caspase 9 and rely on caspase 9-independent cell-death pathways (Clarke et al., 2003). Finally, platelets maintain their inner mitochondrial membrane potential ($\Delta\varphi M$) and do not release mitochondrial cytochrome *c* until they have aged in the circulation (De Botton et al., 2002; Clarke et al., 2003).

Platelet Structural Anatomy

Platelets contain a plasma membrane, cytoskeleton, connected canalicular and dense tubular systems, and a heterogeneous repertoire of granules that coordinately function to effect hemostatic endpoints. The *plasma membrane* is a trilaminar membrane containing glycoproteins, glycolipids, and cholesterol embedded in a phospholipid bilayer composed of phosphatidylserine (PS) and phosphatidylethanolamine (PE). The plasma membrane and the *surface-connected open canalicular system* (OCS) are considered together as a single unit, since the channels of the OCS represent continuous invaginations of the cell wall. The OCS provides a route of entry and egress for molecules, an internal reservoir of membrane to facilitate platelet spreading and filopodia formation after adhesion, and a storage reservoir for membrane glycoproteins that increase on the surface after activation (Suzuki et al., 1992). Within the platelet exterior membrane are several major classes of glycoproteins that serve as receptors for the adhesive macromolecules involved in critical platelet–surface interactions, including platelet attachment to the injured vessel wall, and platelet–platelet interactions that serve to amplify and stabilize platelet plug formation. Another important property of the platelet membrane is the maintenance of transmembrane ionic gradients (Horne et al., 1981). The *dense tubular system* (DTS) is a closed-channel system consisting of narrow membrane-delimited tubules approximately 400–600 Å in diameter. The DTS is the residual endoplasmic reticulum that contains peroxidases, glucose-6-phosphatase, adenylate cyclase, and $Ca^{[2+]}$- and $Mg^{[2+]}$-activated adenosine triphosphatases (ATPases) (Daimon and Gotoh, 1982). The channel system is involved in the regulation of intracellular calcium transport and release (Cutler et al., 1978), and is also the site of platelet prostaglandin synthesis (Gerrard et al., 1978).

The platelet cytoskeleton contains 30%–50% of the total platelet protein, and is composed of three major structural components: an actin microfilament network present throughout the cytoplasm, a microtubule coil located at the platelet periphery, and a membrane skeleton comprising a network of short actin filaments that underlies the inner surface of the plasma membrane (Fox et al., 1988). Twenty to thirty percent of the total platelet protein is made up of actin; other cytoskeletal proteins such as myosin, talin, and actin-binding proteins make up 2%–5%. Platelet stimulation results in profound alteration in cytoskeletal actin reorganization that occurs concomitantly with platelet spreading, filopodia formation, and generation of high concentrations of filamentous (F)-actin. Phosphorylation of myosin light chain results in the binding of myosin to actin, which additionally provides the tension for granule centralization (Nachmias et al., 1985). A circumferential microtubule band that supports the discoid platelet is composed of two nonidentical α and β tubulin subunits associated with microtubule-associated proteins. The 25-nm diameter microtubule coil is contiguous with (but does not touch) the plasma membrane (Behnke, 1967). Microtubules may be disorganized in giant platelet syndromes (*vide infra*).

In addition to functionally important glycoprotein receptors, platelets contain four distinct populations of granules, although only the dense- and α-granules are distinguished morphologically by electron microscopy. Neither the acid hydrolase-containing lysosomes nor the peroxidase (catalase)-containing microperoxisomes are visualized without cytochemical stains. Upon platelet stimulation, granules fuse with channels of the OCS and extrude their contents. α-Granules are the predominant granules found in platelets, and store a large panoply of proteins known to regulate the hemostatic response (Table 15–1). In contrast, the principal constituents of the dense granules are the metabolic pool of adenine nucleotides [i.e., adenosine triphosphate (ATP) and adenosine diphosphate (ADP)], calcium, magnesium, and serotonin (5-hydroxytryptamine). Adenine nucleotides are synthesized and segregated by Mks, whereas serotonin is incorporated into dense granules from the plasmatic compartment. Finally, platelets contain approximately seven mitochondria apiece, and with the exception of their smaller size (0.1 μm^3), platelet mitochondria retain functions comparable to those from other cell sources.

Functional Properties and Signal Transduction

Platelet Adhesion

The vascular endothelial cells that line the intimal surfaces of all blood vessels normally function as a highly effective thrombo-resistant surface to maintain blood fluidity. In response to vessel wall damage, platelets initially adhere and spread in a monolayer to replace and cover the exposed subendothelium (platelet adhesion). The extracellular components that mediate these adhesive interactions are complex, but clearly include types I and III collagen and noncollagenous microfibrils (Baumgartner et al., 1980; Turitto et al., 1980; Fauvel et al., 1983) (Fig. 15–2). At high shear rates (>800 s^{-1}) corresponding to those typically found in the microvasculature, the large von Willebrand's factor (vWF) multimer is required for platelet attachment to the extracellular matrix (Sakariassen et al., 1979). At lower shear rates such as those found in the aorta and large vessels, platelet adhesion to collagen is predominant (Savage et al., 1998, 1999). Exposed collagen initiates two platelet functions essential for hemostasis: adhesion to sites of injury and activation of platelet signals required for thrombus growth (Kahn, 2004). Platelet adhesion to collagen occurs by two fundamental mechanisms: (1) indirectly *via* a bridging interaction of vWF to subendothelial collagen and the platelet GPIb-IX-V complex, and (2) directly *via* platelet collagen receptors. Although a number of receptors have been proposed as platelet collagen receptors (Tandon et al., 1989; Monnet and Fauvel-Lafeve, 2000; Moog et al., 2001), GPIa/IIa ($\alpha_2\beta_1$) (Santoro, 1986) and the GPVI/FcRγ complex (Clemetson et al., 1999; Jandrot-Perrus et al., 2000) appear to be the most important and best characterized. GPVI is an immunoglobulin-domain-containing protein that is found in association with the promiscuous signal transducer adapter FcRγCollagen binding to GPVI induces platelet activation through a pathway that involves phosphorylation of the FcRγ chain, followed by the binding of Syk and phosphorylation-dependent activation of phospholipase C (PLCγ2) (Asselin et al., 1997). The integrin $\alpha_2\beta_1$ is not required for platelet responses to collagen under many experimental conditions; indeed, recent data would suggest that GPVI/FcRγ intracellular signals may activate $\alpha_2\beta_{1,}$ thereby integrating reciprocal responses upon collagen activation (Chen and Kahn, 2003).

Signal Transduction and Secretory Mechanisms

The primary stimulatory pathway in platelets uses two distinct intracellular second messengers: inositol 1,4,5-triphosphate (IP_3) and 1,2-diacylglycerol (DAG), both of which are simultaneously generated from PLC-stimulated hydrolysis of membrane inositol

Table 15–1 Characteristics of Predominant Disease-associated Platelet Glycoprotein Receptors and Storage Bodies

Glycoprotein Receptors	Major Ligands[a]	Receptors/Platelet	Subunit(s)	Gene Symbol
• GPIIb/IIIa ($\alpha^{IIb}\beta^3$)	Fibrinogen, vWF	~50,000	GPIIb (α_{IIb}) GPIIa (β_3)	*ITGA2B* *ITGB3*
• GPIb-IX-V	vWF, α-thrombin	~25,000[b]	GPIbα	*GPIBA*
			GPIbβ	*GPIBB*
			GPIX	*GP9*
			GPV	*GP5*
• GPIa/IIa ($\alpha_2\beta_1$)	Collagen	~10,000	GPIa (α_2)	*ITGA2*
			GPIIa (β_1)	*ITGB1*
• GPVI/FcRγ	Collagen	~1200–1500	Glycoprotein VI	*GP6*
			Fc receptor subunit, γ polypeptide	*FCER1G*
Granules	Functional class	Components		
• α-Granules	Adhesion molecules	vWF, P-selectin, thrombospondin, GPIIb/IIIa, $\alpha_v\beta_3$, fibronectin		
	Chemokines	Platelet basic protein, β-thromboglobulin, CCL3 (MIP-1α), CCL5 (RANTES), CCL7 (MCP-3), CCL17, CXCL1 (growth-regulated oncogene-α), CXCL5 (ENA-78), CXCL8 (IL8; interleukin-8)		
	Coagulation	Factor V, multimerin, fibrinogen		
	Fibrinolysis	α2-Macroglobulin, plasminogen, PAI-1, bFGF		
	Growth and angiogenesis	Hepatocyte growth factor, insulin-like growth factor 1, TGF-β, EGF, VEGF-A, VEGF-C, PDGF		
	Miscellaneous	Histidine-rich glycoprotein, albumin, α1-antitrypsin, Gas6, HMWK, osteonectin, protease nexin-II (amyloid β precursor protein)		
• Dense granules	Ions	$Ca^{[2+]}$, $Mg^{[2+]}$, pyrophosphate		
	Nucleotides	ATP, GTP, ADP, GDP		
	Membrane proteins	CD63 (granulophysin), LAMP-2		
	Transmitters	Serotonin		
Lysosomes	Matrix-localized acid hydrolases	β-Glucuronidase, β-galactosidase, *N*-acetylglucosaminidase, cathepsins, aryl sulfatase, β-hexosaminidase, β-glycerophosphatase, elastase, collagenase, endoglucosidase		
	Membrane-localized	LAMP-1, LAMP-2, CD36/LAMP-3		

[a]*Abbreviations*: vWF, von Willebrand factor; HMWK, high molecular weight kininogen, VEGF, vascular endothelial cell growth factor; TGFβ, transforming growth factor-β; PDGF, platelet-derived growth factor; EGF, epidermal growth factor; LAMP, lysosomal-associated membrane protein; CCL, chemokine (C-C motif) ligand; CXCL, chemokine (C-X-C motif) ligand; PAI-1, plasminogen activator inhibitor-1; bFGF, basic fibroblast growth factor.

[b]Stoichiometric studies suggest that the polypeptides comprising GPIb-IX-V are in the ratio 2:2:2:1, for GPIbα, GPIbβ, GPIX, and GPV, respectively.

Source: Data are compiled from references (Handin, 1995; Stenberg and Hill, 1999; Reed, 2004).

phospholipids (Fig. 15–3). The major inhibitory pathway in platelets is coupled to adenylate cyclase with the generation of cyclic adenosine monophosphate (cAMP). Second messengers of both excitatory and inhibitory pathways are generated intracytoplasmically after receptor/ligand activation of signal-transducing guanosine triphosphate (GTP) regulatory proteins (G proteins). Platelet IP_3 induces the release of intracytosolic calcium ($Ca^{[2+]}$) from the dense tubular system, which activates calmodulin and calmodulin-dependent protein kinases (in addition to other calcium-dependent effectors) (Majerus et al., 1986); diacylglycerol kinase activates protein kinase C (Berridge, 1987); and cAMP activates a cAMP-dependent protein kinase. cAMP has various inhibitory effects in platelets, blocking

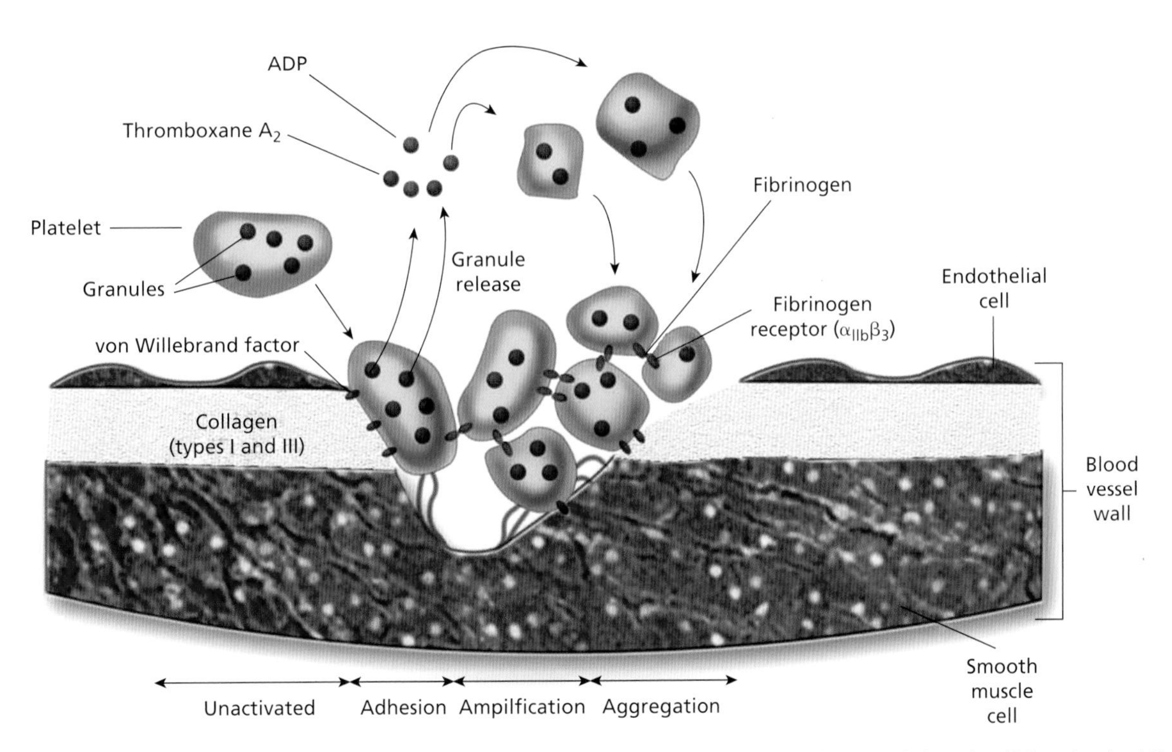

Figure 15–2 Stages of platelet activation. Focal vascular damage exposes the subendothelial matrix, followed by platelet adhesion mediated by platelet–collagen interactions. At low shear rates, the GPIa/IIa ($\alpha_2\beta_1$) receptor is important, whereas at higher shear rates, von Willebrand factor functions as a bridging link between exposed collagen and the platelet GPIb-IX-V receptor complex. Adherent platelets release prepackaged granule constituents such as ADP (adenosine diphosphate), while the hydrolysis of arachidonic acid results in generation of biologically active eicosanoids such as thromboxane A_2; ADP and thromboxane A_2 bind to their cognate receptors to amplify the activation process. The final process of platelet aggregation is mediated by fibrinogen bridging conformationally activated GPIIb/IIIa ($\alpha_2\beta_1$) receptors on opposing membranes.

both the initiation and the maintenance of stimulatory responses. Vascular prostacyclin (PGI_2) functions as a potent platelet inhibitor by raising intracellular cAMP concentrations after receptor binding, acting through the G_s inhibitory pathway.

Stimulus–response coupling in platelets results in two additional effects: (1) secretion and/or degranulation of stored granule constituents (*vide supra*) and (2) generation of arachidonic acid (AA). The latter is mobilized from the activation of phospholipase A_2 (PLA_2), which directly cleaves the free fatty acid from the C2 position of phospholipids. Although the preferred phospholipid substrate for PLC is PIP_2 (phosphatidylinositol 4,5-bisphosphate), PLA_2 acts primarily on phosphatidylcholine to liberate free AA. After deacylation, free AA is rapidly metabolized to various biologically active products (eicosanoids) through the actions of cyclooxygenase and lipoxygenase (Needleman et al., 1986). Cyclooxygenase is irreversibly acetylated by aspirin and reversibly acetylated by nonsteroidal anti-inflammatory agents, and converts AA to the labile prostaglandin endoperoxides PGG_2 and PGH_2; the latter are further metabolized by thromboxane synthase to thromboxane A_2 (TXA_2). PGG_2, PGH_2, and TXA_2 are potent platelet agonists that activate platelets by direct receptor binding, G-protein signaling, and activation of PLC.

Platelet Aggregation

Platelet aggregation (platelet–platelet interaction) requires the binding of plasma adhesive proteins (typically fibrinogen) to the activated form of the GPIIb/IIIa integrin [$\alpha_2\beta_1$; (cluster designation) CD41a]. Although platelet aggregation is induced by a variety of stimuli, physiological aggregation is essentially dependent on the $\alpha_{IIb}\beta_3$ integrin. In contrast to other integrin/ligand interactions, the binding of adhesive ligands to $\alpha_2\beta_1$ requires prior platelet activation, resulting in a conformationally active binding pocket that can interact with macromolecular adhesive proteins such as fibrinogen, vWF, fibronectin, and vitronectin (Shattil et al., 1985; Shattil, 1999). In resting platelets $\alpha_{IIb}\beta_3$ has a low activity for ligand binding, although its ability to bind ligands increases rapidly after platelet exposure to soluble agonists or subendothelial matrices (inside-out signaling). In contrast, ligand binding to the extracellular portion of $\alpha_{IIb}\beta_3$ causes integrin clustering and conformational changes in the integrin cytoplasmic domains to promote outside-in-signaling (Miyamoto et al., 1995). Soluble agonists such as thrombin, ADP, epinephrine, and TXA_2 bind G-protein-coupled receptors to generate inside-out signaling (Shattil, 1999). Although the signaling pathway to the $\alpha_{IIb}\beta_3$ cytoplasmic domains remains incompletely characterized, various reports implicate tyrosine kinases such as src and syk. Considerable attention has focused on the intracellular proteins bound directly to the $\alpha_{IIb}\beta_3$ cytoplasmic domains, with the confirmation that at least two proteins (β_3-endonexin and talin) are able to modulate inside-out-signaling (Kashiwagi et al., 1997; Calderwood et al., 1999).

Whole Genome and Proteome Approaches to Dissect Platelet Structure and Functional Mechanisms

Platelet Transcript Profiling

Platelets contain as little as 2×10^{-3} fg messenger RNA (mRNA)/cell, approximately 3–4 logs less ribonucleic acid (RNA) than a typical nucleated cell (Fink et al., 2003). The limited yield of RNA

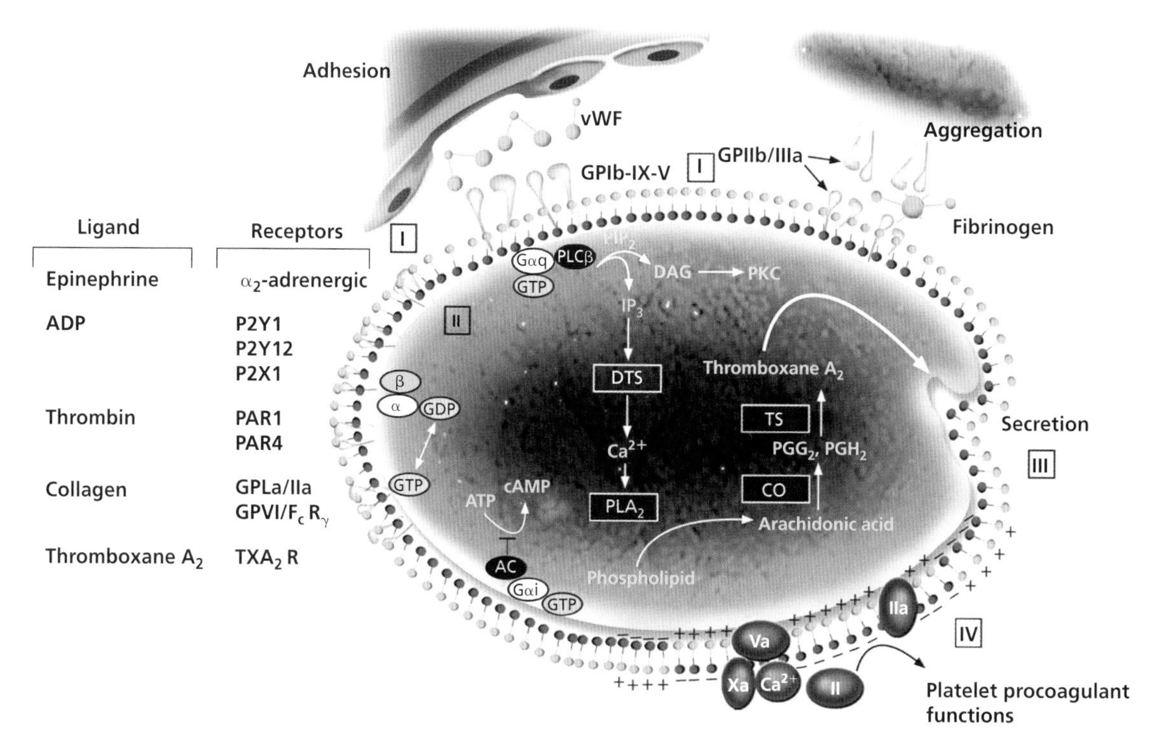

Figure 15–3 Mechanisms of platelet intracellular signaling relevant to human molecular defects. Ligand–receptor interactions are functionally coupled to discrete intracellular pathways by the heterotrimeric family of $\alpha\beta\gamma$ G proteins. The α-subunit is maintained in the inactive [guanosine diphosphate (GDP)-bound] state, but is replaced by cytosolic guanosine triphosphate (GTP) upon stimulus–response coupling. GTP-charged $G_{\alpha i}$ inhibits adenylyl cyclase (AC), resulting in diminished levels of cyclic adenosine monophosphate (cAMP) and enhanced platelet responsiveness. In contrast, the $G_{\alpha q}$-subunit activates phospholipase Cβ (PLCβ), resulting in generation of 1,4,5-inositol trisphosphates (IP_3) from phosphatidylinositol 4,5-bisphosphate (PIP_2), and release of intracytosolic calcium (Ca^{2+}) from the platelet dense tubular system (DTS). Concomitant activation of protein kinase C (PKC) from diacylglycerol (DAG) provides for a pathway linked to activation of calcium-regulated kinases, receptor tyrosine kinases (RTK), and integrins, among others. The activation of phospholipase A_2 (PLA_2) mobilizes free arachidonic acid (AA) from phospholipids, thereby providing substrates for generation of labile prostaglandin endoperoxides (PGG_2 and PGH_2) by the enzyme cyclooxygenase (CO); subsequent hydrolysis of PGG_2 and PGH_2 by thromboxane synthase (TS) generates thromboxane A_2 (TXA_2), whose release amplifies platelet activation via binding to its cognate receptor. Platelet activation results in the exposure of negatively-charged (−) phospholipid (primarily phosphatidylserine), which serves as a surface for assembly of the prothrombinase complex with subsequent thrombin generation (Bahou et al., 2004). Not shown is the GTP-coupling to $G\alpha_{12/13}$ which links extracellular signals to actin filament uncapping and cytoskeletal actin reorganization (Bahou, 2003). Qualitative defects of platelet function are broadly classified into those involving receptor/ligand interactions (I), intracellular signaling networks (II), secretory (i.e., granule) release (III), or defective prothrombinase assembly (IV). The latter defect (i.e., Scott's syndrome) is extremely rare with no clear molecular genetic etiology (Sims et al., 1989). Key enzymes are depicted by solid black shapes. ADP, adenosine diphosphate; P2Y1, purinergic receptor P2Y, G-protein coupled 1; P2Y12, purinergic receptor P2Y12, G-protein coupled 12; P2X1, purinergic receptor P2X, ligand-gated ion channel 1; PAR1, protease activated receptor 1; PAR4, protease activated receptor 4.

means that potential interference by contaminating leukocyte-derived RNA may bias transcript profiling approaches. Despite this technical concern, a limited number of published microarray studies using Mk- or platelet-derived mRNAs have been described, with generally concordant agreements on transcript quantitation and gene expression patterns (Table 15–2). Not surprisingly, platelets have less transcripts than those found in nucleated cells, ranging from approximately 1600 to 3000 mRNAs (Bugert et al., 2003; Gnatenko et al., 2003; McRedmond et al., 2004). The small number of platelet-expressed transcripts represents the lack of ongoing transcription in the anucleate platelet. The quantitative differences between studies presumably reflect, in part, the methods of isolation.

Several well-characterized platelet transcripts are amongst the most abundant genes identified in the platelet transcriptome (i.e., platelet factor 4, glycoprotein 1bβ), while numerous transcripts encoding proteins involved in cytoskeletal organization (actin, thymosin β4, cofilin) are also readily identified (Bahou and Gnatenko, 2004) (Table 15–3). A large percentage of genes with miscellaneous (25%) or unknown function (32%) as annotated by Affymetrix and RefSeq databases are also evident (Pruitt and Maglott, 2001). Thus, nearly one-half of platelet genes and gene products presumably have important, but poorly characterized, functions in platelet and/or Mk biology. The mRNAs encoding the actin-related machinery are overrepresented in these analyses (Gnatenko et al., 2003), not unexpected given prior estimates that 20%–30% of the total platelet proteome is comprised of actin or actin-related proteins such as mysosin and talin (Fox et al., 1988). In unstimulated platelets, 30%–40% of actin is polymerized as F-actin, while the balance of G-actin monomers are polymerization inhibited by sequestering proteins such as profilin and thymosin β4 (Safer et al., 1991). The high thymosin β4 transcript expression not only correlates with its known abundance in platelets but also supports the importance of actin inhibitory proteins in maintaining the nonstimulated state of circulating platelets. Genes involved in the immune response [chemokines, growth factors, and human leukocyte antigen (HLA) proteins] are also evident, consistent with the role of platelets in inflammation and immunity (Bugert et al., 2003). Finally, unexpected transcripts encoding erythroid genes (i.e., globin, ferritin) have

Table 15–2 Megakaryocyte and Platelet Microarray Studies

Genes Studied[a]	12,599	22,200	9850	12,599	12,599	22,200
P/M[b] (number)	1500–2147	NR[c]	NR	NR	NR	~10,000
P (number)	NR	~1668	~1526	2928	NR	4637–4655
Arrays (number)	3	5	6	1 in duplicate from 23 pooled donors	3	4 Pooled (six samples/array)
Cells studied	Platelets	Platelets	Platelets	Platelets	Megakaryocytes	Megakaryocytes
Cellular source	Apheresis	Apheresis	Concentrates	Blood (50 mL)	CD34 + ex vivo expanded	CD34 + ex vivo expanded
Reference	Gnatenko et al., 2003	Gnatenko et al., 2005	Bugert et al., 2003	McRedmond et al., 2004	Shim et al., 2004	Tenedini et al., 2004

[a] Number of probes (genes) represented on individual microrarray slides.

[b] P, present; M, marginal.

[c] NR, not reported.

been consistently identified in the reported platelet microarray studies. While this may suggest reticulocyte contamination during platelet isolation, residual expression from precursor cells cannot be excluded. Similarly, histone transcripts (notably histone H2A, but also H3.3, H2B.1, and H2A.2) have been observed in transcriptome analysis of highly purified platelets, the significance of which remains unestablished (Gnatenko et al., 2003).

Serial Analysis of Gene Expression

DNA microarray technology represents a closed profiling strategy limited by the target genes imprinted upon the chips. In contrast, serial analysis of gene expression (SAGE) is an open architectural system that can be used to identify novel genes, and to quantify differentially expressed mRNAs (Velculescu et al., 1995; Zhang et al., 1997). SAGE relies on the observation that short (<10 bp) sequences within 3′-mRNAs can stringently discriminate among the approximately 30,000 genes comprising the human genome, assuming a random nucleotide distribution along a 9-bp stretch (4^9 bp = 262,244 random nucleotide combinations). SAGE is completed by (1) preparation of a 3′ end-specific complementary DNA (cDNA) library, (2) generation and concatenation of "tags" from the cDNA library using an anchoring enzyme such as *Nla*III and a tagging enzyme such as *Bsm*FI, and (3) cloning of tag concatemers into a sequencing vector and performing automated sequencing. The sequence of each tag along with its positional location uniquely identifies the gene from which it is derived, and differentially expressed genes can be identified in a quantitative manner since the frequency of tag detection reflects the steady-state mRNA level of the cellular transcriptome (Velculescu et al., 1995; Zhang et al., 1997). With relatively deep sampling (>50,000 tags), genes expressed at low levels (<0.01% of total mRNA) can be readily identified.

Platelet SAGE has been completed, initial studies cataloguing a total of 2033 tags (Gnatenko et al., 2003). Interestingly, 1800 tags (89% of total) corresponded to mitochondrial-derived genes. The unusually high preponderance of mitochondrial-derived genes is not inconsistent with the known enrichment of these genomes in human platelets (Wallace, 2001), and presumably reflects persistent transcription from the mitochondrial genome in the absence of nuclear-derived transcripts.

The mitochondrial genome is a compact sequence of approximately 16.6 kb, encoding 13 genes and 2 ribosomal RNAs; the overall distribution of platelet-derived mitochondrial SAGE tags is quite similar to that found in muscle (Welle et al., 1999). Like muscle, platelets are metabolically adapted to rapidly expend large amounts of energy required for aggregation, granule release, and clot retraction. Platelet mitochondria represent the primary source of ATP, and mitochondria are also responsible for most of the toxic reactive oxygen species generated as by-products of oxidative phosphorylation. The mitochondrial DNA (mtDNA) encodes polypeptides found within four of the five multifunctional complexes that regulate oxidative phosphorylation within the platelet mitochondria (Raha and Robinson, 2001). Whether the continued generation of these polypeptides has a role in platelet energy metabolism and/or the apoptotic mechanisms regulating platelet survival remains speculative, although not inconsistent with prior evidence that platelets retain a fully functional translational apparatus (Weyrich et al., 1998).

Platelet Proteomic Studies

Current approaches to cellular protein analysis are generally accomplished using two methodologies: (1) high-resolution two-dimensional polyacrylamide gel electrophoresis (2-DE) with mass spectrometric sequence identification, or (2) microcapillary liquid chromotography with tandem mass spectrometry (μLC-MS/MS) (Link et al, 1999). While the latter approach can be automated for cellular analyses, neither "first generation" approach is quantitative nor limited by the extreme dynamic range of cellular proteins [~5-log differences between most (1×10^6 copies/cell) and least (1×10^1 copies/cell) abundant proteins]. Additional technical limitations of these methodologies include their inability to detect low-abundance proteins (Corthals et al., 2000), proteins with extremes in pI and molecular weight (Gygi, Rist, et al., 1999), and membrane-associated or bound proteins (Molloy, 2000). Recent modifications to both methods have been devised for direct comparative studies between cellular proteomes. The application of isotope-coded affinity tags (ICATs) to μLC-MS/MS represents a novel means of *quantitative* analyses (Gygi, Rochon, et al., 1999), although its applicability in whole-proteome studies remains tantalizing, but unestablished.

Initial attempts to adapt these technologies to platelet analyses have been described, initially with the characterization of a cytosolic fraction by 2-DE (Marcus et al., 2000), and subsequently with the identification of more than 500 proteins by high-throughput tandem MS/MS (O'Neill et al., 2002). More focused studies have used thrombin-stimulated platelets to identify the subsets of proteins that are secreted or tyrosine phosphorylated during platelet activation (Maguire, 2003). Detailed analysis of the secretome from thrombin-activated platelets has identified more than 300 proteins, including novel proteins such as the monocyte chemoattractant precursor, secretogranin III (Coppinger et al., 2004). In addition, a number of secreted proteins have been identified in atherosclerotic lesions,

Table 15–3 Most Highly Expressed Platelet Transcripts ($N = 50$)

Platelet Transcript[a]	Gene Symbol	*Unigene Cluster*
Platelet factor 4 (chemokine (C-X-C motif) ligand 4	*PF4*	Hs.81564
H3 histone, family 3A	*H3F3A*	Hs.53362
Rho guanine nucleotide exchange factor (GEF) 10	*ARHGEF10*	Hs.98594
β-Actin mRNA	*ACTB*	Hs.520640
H3 histone, family 3B	*H3FB*	Hs.180877
β2-Microglobulin	*B2M*	Hs.534255
Neurogranin (protein kinase C substrate)	*NRGN*	Hs.524116
Proplatelet basic protein (chemokine (C-X-C motif) ligand 7	*PPBP*	Hs.2164
Guanine nucleotide-binding protein G(s), γ-subunit	*GNAS*	Hs.125898
Thymosin β4	*TMSB4X*	Hs.522584
Ferritin, light polypeptide	*FTL*	Hs.433670
Ferritin, heavy polypeptide 1	*FTH1*	Hs.446345
Glutathione peroxidase 1	*GPX1*	Hs.76686
Vinculin	*VCL*	Hs.500101
Major histocompatibility complex, class I, B	*HLA-B*	Hs.77961
Glycoprotein Ibβ	*GP1BB*	Hs.517410
Major histocompatibility complex, class IC	*HLA-C*	Hs.534125
Nuclear receptor coactivator 4	*NCOA4*	Hs.522932
Ornithine decarboxylase antizyme 1	*OAZ1*	Hs.446427
Protein tyrosine phosphatase, receptor-type O	*PTPRO*	Hs.160871
Cofilin 1 (non-muscle)	*CFL1*	Hs.170622
Chemokine (C-C motif) ligand 5	*CCL5*	Hs.514821
Transgelin 2	*TAGLN2*	Hs.517168
Coagulation factor XIII, A1 polypeptide	*F13A1*	Hs.335513
Ribosomal protein L41	*RPL41*	Hs.381172
SH3 domain binding glutamic acid–rich protein like 3	*SH3BGRL3*	Hs.109051
Clusterin (complement lysis inhibitor)	*CLU*	Hs.436657
Nerve growth factor receptor-associated protein 1	*NGFRAP1*	Hs.448588
Major histocompatibility complex, class I, E	*HLA-E*	Hs.381008
Major histocompatibility complex, class I, A	*HLA-A*	Hs.549038
Tumor protein, translationally controlled 1	*TPT1*	Hs.374596
Regulator of G-protein signaling 10	*RGS10*	Hs.501200
SPARC/osteonectin mRNA	*SPARC*	Hs.111779
Homo sapiens nuclear protein SDK3 mRNA	*PNN*	Hs.409965
Monocyte to macrophage differentiation-associated	*MMD*	Hs.463483
Hemoglobin β-gene	*HBB*	Hs.523443
MutL homolog 3 (*E. coli*)	*MLH3*	Hs.279843
Ubiquitin C	*UBC*	Hs.520348
Chromosome 21 open reading frame 7	*C21orf7*	Hs.222802
Ubiquitin B	*UBB*	Hs.356190
CD99 antigen	*CD99*	Hs.495605
Spermidine/spermine *N*1-acetyltransferase	*SAT*	Hs.28491
Glyceraldehyde-3-phosphate dehydrogenase	*GAPD*	Hs.479728
Progesterone receptor membrane component 1	*PGRMC1*	Hs.90061
Guanine nucleotide binding protein (G protein), γ 11	*GNG11*	Hs.83381
Myosin, light polypeptide 6, smooth muscle and non-muscle	*MYL6*	Hs.505705
Nucleosome assembly protein 1-like	*NAP1L1*	Hs.524599
Integral membrane protein 2B	*ITM2B*	Hs.446450
Hemoglobin, α1-gene	*HBA1*	Hs.449630
Actin related protein 2/3 complex, subunit 1B	ARPC1B	Hs.489284

[a] Transcripts are rank-ordered using log_2-intensities of average-difference values from five normal donors, normalized across individual arrays; data and methods updated from Gnatenko et al. (2005, 2003).

suggesting a potential role in the development of atherothrombosis and/or coronary artery disease.

Integration of Proteomic and Genomic Technologies for Platelet Studies

Proteomic and genomic technologies are being coupled to identify new drug targets and therapies for common diseases in which platelets play a role such as MI and stroke. Integration of genomic and proteomic datasets is helpful for among other reasons: (1) protein identification is facilitated by the knowledge on mRNA transcripts and (2) the role of poorly characterized genes can often be surmised based on protein expression. A large percentage of platelet mRNAs have corresponding proteins (McRedmond et al., 2004). In addition, the most abundant mRNAs correlate well with constitutively expressed proteins. A one-to-one correlation between mRNA and protein products is not expected because platelets adsorb proteins on their membranes, many of which enter the cytoplasm. Finally, there is clear evidence that low-abundance transcripts are translated into proteins upon platelet activation, which suggests that the platelet proteome is dynamic (Weyrich et al., 2003).

Genetic Disorders Causing Thrombocytopenia

The inherited platelet disorders are generally categorized into those causing quantitative or qualitative defects in platelet number or function. In some syndromes, both quantitative and qualitative platelet phenotypes are evident [i.e., Bernard–Soulier Syndrome (BSS), Grey Platelet syndrome]. Many of the patients with hereditary thrombocytopenia have large platelets (with elevated mean platelet volumes) and an autosomal dominant mode of inheritance (Guerois et al., 1988; Najean and Lecompte, 1990). Platelet survival is reduced in some situations, but usually is normal. This subset of disorders is either due to the defects involving Mk commitment and differentiation or caused by defects of platelet formation (Table 15–4; refer to Fig. 15–1). In unusual situations, the thrombocytopenia occurs because of accelerated platelet clearance, or due to inherent platelet receptor defects (platelet-type von Willebrand's disease).

Inherited Thrombocytopenias Caused by Hypo- or Dysmegakaryocytopoiesis

Mediterranean Macrothrombocytopenia

This autosomal dominant disorder is characterized by a variable proportion of large platelets (macrothrombocytes) in peripheral blood and generally normal *in vitro* platelet function. Mediterranean macrothrombocytopenia is probably one of the most common forms of congenital thrombocytopenia, and is likely heterogeneous in nature. Apart from the presence of macrothrombocytes, platelet ultrastructure is normal, and *in vivo* survival of transfused platelets is also unaffected (Najean and Lecompte, 1990; Fabris et al., 1997). Although bone marrow megakaryocytopoiesis appears normal, detailed investigations suggest excessive proliferation of immature Mks that display premature apoptosis (Fabris et al., 1997). The molecular basis of this disorder is not well-characterized, although (in Italy) many patients have heterozygous BSS as the molecular etiology (Savoia et al., 2001) (*vide infra*).

Congenital Amegakaryocytic Thrombocytopenia

Congenital amegakaryocytic thrombocytopenia (CAMT) is a rare disorder manifest in infancy and characterized by isolated thrombocytopenia with megakaryocytopenia and no physical anomalies. Studies in patients with CAMT demonstrate a defective response to TPO, decreased numbers of erythroid and myeloid progenitors in hematopoietic cultures, and a lack of the TPO receptor (c-mpl) mRNA in bone marrow mononuclear cells (Muraoka et al., 1997). Studies in a 10-year old girl born to nonconsanguineous Japanese parents detected compound heterozygosity for two mutations of the c-*mpl* gene: a Gln^{186}-to-STOP substitution in exon 4 and a single nucleotide deletion in exon 10 (Ihara et al., 1999). Subsequent studies in nine additional patients demonstrated high TPO levels in the sera of all patients; however, platelets and hematopoietic progenitor cells demonstrated no TPO reactivity by Mk colony assays. Flow cytometry revealed absent surface expression of the c-mpl receptor in three of three patients analyzed; sequence analysis of the c-*mpl* gene in eight of the patients revealed point mutations in all (Ballmaier et al., 2001). Retrospective comparison of 18 CAMT patients' clinical data from different German clinics resulted in a division into two distinct groups: Group I patients (~60%) presented with a more severe form of CAMT and early progression from isolated thrombocytopenia into pancytopenia. Group II patients demonstrated a transient increase of platelet counts during the first year of life followed by later development of pancytopenia (Ballmaier et al., 2001).

Thrombocytopenia with Absent Radii

Since the initial description in 1956 (Shaw and Oliver, 1959), there has been little progress in delineating the molecular basis of this syndrome. Thrombocytopenia with absent radii (TAR) is distinct from Fanconi's anemia, since there is no erythroid involvement. Thrombocytopenia usually is evident early in life but is transient; thus, the process is more benign than Fanconi's anemia in which leukemia is a further complication. Parental consanguinity seems to be rare. A clinical study of 34 patients with TAR syndrome, all of whom had documented thrombocytopenia and bilateral radial aplasia, demonstrated the following clinical features: 47% had lower limb anomalies, 47% intolerance to cow's milk, 23% renal anomalies, and 15% cardiac anomalies. Congenital anomalies also included facial capillary hemangiomata, intracranial vascular malformation, sensorineural hearing loss, and scoliosis (Greenhalgh et al., 2002). Megakaryocyte colony growth *in vitro* was virtually absent in optimally stimulated cultures of a patient's bone marrow progenitors, whereas erythroid and myeloid colony growth were preserved (Homans et al., 1988). High levels of Mk colony stimulating activity comparable to the levels found in adults with aplastic anemia have been detected in the serum from an infant with TAR. The elevated serum activity decreased by 6 months of age at which time partial platelet recovery had occurred (Homans et al., 1988). Screening of the coding and promoter regions of the gene encoding the TPO receptor in four unrelated patients affected by TAR syndrome failed to identify any mutations, and the molecular basis of this syndrome remains unestablished.

Amegakaryocytic Thrombocytopenia with Radioulnar Synostosis

Evidence exists that this syndrome is caused by a mutation in the *HOXA11* gene which maps to 7p15. Full clinical details on the two families with autosomal dominant amegakaryocytic thrombocytopenia with radioulnar synostosis have been provided (Thompson et al., 2001). In one family, two affected sisters with thrombocytopenia were successfully treated with bone marrow transplantation. The father had proximal radioulnar synostosis, bilateral clinodactyly of the fifth digits (present also on the right hand of the older sister), and

Table 15–4 Inherited Thrombocytopenias

Disorder	OMIM[a]	Inheritance[b]	Gene	Key Features
Defects of megakaryocytic commitment and/or differentiation				
• Mediterranean macrothrombocytopenia	153670	D	—	Macrothromobocytes, dysmegakaryocytopoiesis
• Congenital amegakaryocytic thrombocytopenia (CAMT)	604498	R	*MPL*	Decreased megakaryocytes; normal platelet volumes; progression to pancytopenia
• Thrombocytopenia with absent radii	274000	R	—	Decreased megakaryocytes; bilateral radial aplasia, typically with other anomalies (skeletal and non-skeletal); normal platelet volume
• Amegakaryocytic thrombocytopenia with radioulnar synostosis	605432	D	*HOXA11*	Decreased megakaryocytes; radioulnar synostosis, typically with other malformations; normal platelet volume
• X-linked thrombocytopenia with anemia	314050	X	*GATA1*	Macrothrombocytes with dysmegakaryocytopoiesis; anemia with dyserythropoietic features
• Thrombocytopenia 2	313900	D	*FLJ14813*	??Mile dysmegakaryocytopoiesis; normal platelet size
• Paris-Trousseau syndrome	188025	D	11q23	Dysmegakaryocytopoiesis with giant platelet α-granules; mental retardation with facial and cardiac anomalies
	600588		Deletion	
• Jacobsen's syndrome	147791		FLI1	
• Familial thrombocytopenia with associated myeloid leukemia	601399	D	*CBFA2*	Decreased megakaryocytes with normal platelet size; tendency to develop myelodysplastic syndrome or acute leukemia
Defects of platelet production				
• MYH9-type disorders		D	*MYH9*	Continuous clinical spectrum characterized by macrothrombocytes; neutrophil cytoplasmic inclusions; varying degrees of sensorineural deafness, cataracts, and nephritis
— May-Hegglin anomaly	155100			
— Sebastian syndrome	605249			
— Fechtner syndrome	153640			
— Epstein syndrome	153650			
• Bernard–Soulier syndrome	321200	R	*GPIBA* *GPIBB* *GP9*	Macrothrombocytes with mild thrombocytopenia; defective ristocetin-induced platelet aggregation essentially diagnostic
Shortened platelet survival				
• Wiskott–Aldrich syndrome (X-linked thrombocytopenia)	301000	X	*WAS*	Small platelets; severe immunodeficiency; eczema; defective WAS protein
• Platelet-type or pseudo-von Willebrand disease	177820	D	*GPIBA*	Mild thrombocytopenia with variable macrothrombocytes; enhanced platelet aggregation to low-dose ristocetin

[a] Online Mendelian Inheritance in Man.
[b] D, autosomal dominant; R, autosomal recessive; X, X-linked.

webbed fingers, but no hematologic abnormality; both sisters had hip dysplasia with dislocation. The second pedigree included an affected sister and brother; the father had radioulnar synostosis and bilateral fifth digit clinodactyly. Subsequent studies demonstrated that a mutation in the homeo box *HOXA11* gene was responsible for the phenotype.

Macrothrombocytopenia, X-Linked (with Anemia)

This syndrome (and the closely related dyserythropoietic anemia with thrombocytopenia) has been found to be caused by mutations in the *GATA1* gene. The original description involved a woman with mild chronic thrombocytopenia who had two pregnancies complicated by severe fetal anemia requiring *in utero* red blood cell transfusions. The offspring were male half-sibs who were anemic and severely thrombocytopenic at birth. Peripheral blood smear demonstrated thrombocytopenia, and erythrocytes were abnormal in size and shape (demonstrating anisopoikilocytosis). There were three asymptomatic female sibs; genetic studies identified a hemizygous mutation in the *GATA1* gene, and the mother was heterozygous for the mutation (Nichols et al., 2000). Subsequent studies described a family with isolated X-linked macrothrombocytopenia without anemia (but with dyserythropoiesis) in 13 males in 9 sibships of 3 generations connected through carrier females. Electron microscopy of the patients' platelets showed giant platelets with cytoplasmic clusters consisting of smooth endoplasmic reticulum and abnormal membrane complexes. A two-generation family with macrothrombocytopenia and marked anemia was also identified, containing an Asp^{218} Tyr substitution in *GATA1* (Freson et al., 2002). Compared to individuals with the Asp^{218}Gly mutation (Freson, Devriendt, et al., 2001), platelet and erythrocyte morphology as well as expression levels of the platelet GATA1-target gene products were more profoundly affected. Since the mutation was not expressed in the platelets of the female carrier (while her leukocytes showed a skewed X-inactivation pattern), the nature of the amino acid substitution at *GATA1* position 218 may be of crucial importance in determining the severity of the phenotype in X-linked macrothrombocytopenia patients (Freson, Devriendt, et al., 2001).

Thrombocytopenia 2

This autosomal dominant, nonsyndromic thrombocytopenia maps to chromosome 10 and has been associated with a mutation in *FLJ14813*, a gene with no clear function. Platelet counts usually fluctuate between 30,000 and 80,000/μl. The number and appearance of Mks are generally normal and the platelets, although somewhat larger than normal, are otherwise morphologically normal (Stavem et al., 1986). Although the disease is clearly heterogeneous in nature, suggestive clinical findings include a familial inheritance pattern, normal autologous and homologous platelet life span, increased mean platelet volume without Dohle bodies, absence of any functional platelet abnormalities, and a normal Mk count (Najean and Lecompte, 1990). Treatment with corticosteroids, immunoglobulins, androgens, and immunosuppressive agents are generally ineffective. Inheritance is typically autosomal dominant with many instances of male-to-male transmission (Najean and Lecompte, 1990). Early onset thrombocytopenia in other members of the family and failure of treatment with corticosteroids suggests this entity. Splenectomy may or may not be helpful in ameliorating the thrombocytopenia (Najean and Lecompte, 1990; Majado et al., 1992). A genome-wide search in a large Italian family affected by autosomal dominant thrombocytopenia demonstrated that the disease locus (THC2) was linked to 10p12–p11.1 (Savoia et al., 1999). A similar locus-susceptibility region was established in a second large family with autosomal-dominant, lifelong thrombocytopenia, in this case electron microscopic studies indicating that Mks had markedly delayed nuclear and cytoplasmic differentiation (Drachman et al., 2000). Subsequent molecular studies identified a missense mutation in the *FLJ14813* gene that segregated perfectly with thrombocytopenia in their kindred of 51 family members. The mutation was neither detected in 94 random unrelated and unaffected individuals nor reported in a single-nucleotide polymorphism (SNP) database (Gandhi et al., 2003).

Paris-Trousseau Thrombocytopenia and Jacobsen's Syndrome

These related entities are contiguous gene syndromes, initially described in patients with 11q23 deletions (Wardinsky et al., 1990; Breton-Gorius et al., 1995). The syndrome is typically associated with congenital dysmegakaryopoietic thrombocytopenia associated with giant platelet α-granules (Breton-Gorius et al., 1995). Platelet electron microscopy of an infant with an 11q23.3→qter deletion and clinical features of Jacobsen syndrome identified giant α-granules identical to those described in Paris-Trousseau syndrome, suggesting that the latter syndrome may be a variant of Jacobsen syndrome and that the thrombocytopenia in all cases of 11q23.3 deletion is due to dysmegakaryopoiesis with the formation of giant α-granules (Krishnamurti et al., 2001). Lentivirus-mediated overexpression of Fli-1 in $CD34^{+}$-positive cells of patients with Paris-Trousseau syndrome restored megakaryopoiesis in vitro, suggesting that the Fli-1 hemizygous deletion contributes to the hematopoietic defects in the disorder (Raslova et al., 2004). Furthermore, these authors proposed that the absence of Fli-1 expression might prevent their differentiation, resulting in the segregation of the Paris-Trousseau Mks into two subpopulations: one normal and the other composed of small immature Mks undergoing spontaneous lysis (Raslova et al., 2004). The putative involvement of Fli-1 in this syndrome is supported by recent evidence that Fli-1 interacts with GATA-1 (see Fig. 15–1), thereby synergistically activating the transcription of genes involved in Mk differentiation and/or maturation (Eisbacher et al., 2003).

Familial Thrombocytopenia with Associated Myeloid Leukemia

The transcription factor CBF (core binding factor) consists of two subunits, CBFα (also known as Runx1) and CBFβRunx1 and CBFβ levels increase during Mk differentiation and are downregulated during erythroid differentiation, imputing a role in hematopoietic cell divergence. Runx1 binds DNA as a monomer through the Runt homology domain, although DNA binding is increased by heterodimerization that stabilizes the interaction of the complex with DNA. Molecular mutations or intragenic deletions of the *CBFA2* gene (i.e., *RUNX1*) have been detected in familial platelet disorders with predisposition to acute myelogenous leukemia (AML) (Song et al., 1999). This autosomal disorder is characterized by thrombocytopenia, abnormal platelet aggregation responses, and development of leukemia. Analysis of bone marrow or peripheral blood cells from affected individuals showed a decrease in Mk colony formation, suggesting that *CBFA2* dosage affects megakaryopoiesis. The involvement of *CBFA2* suggested a model for this familial platelet disorder in which haploinsufficiency of *CBFA2* causes an autosomal dominant congenital platelet defect and predisposes to an acquisition of additional mutations that cause leukemia (Song et al., 1999). Genetic studies of an additional 12 families with this syndrome failed to identify *CBFA2/RUNX1* gene mutations, providing evidence for genetic heterogeneity of the syndrome (Minelli et al., 2004). Finally, murine models do not support gene dosage as an important modulator of *CBFA2* function. Heterozygous *CBFA2* mutants demonstrated minimal phenotypic changes, although homozygosity resulted in embryonic death with the involvement of all hematopoietic lineages (Wang et al., 1996).

Inherited Thrombocytopenias with Defects of Platelet Production (Thrombopoiesis)

MYH9-type Disorders Class II myosins comprise filament-forming subtypes found in muscle and non-muscle cells. These hexameric proteins composed of two heavy and two pairs of light chains interact with actin to produce movement, and hydrolyze ATP. Neutrophils and platelets (including Mks) exclusively express nonmuscle myosin-IIA, which has been postulated to function in proplatelet formation (Hartwig and Italiano, 2003). Heterozygous mutations in the *MYH9* gene encoding for nonmuscle myosin heavy chain-IIA (NMMHC-IIA) result in a heterogeneous complex of disorders collectively termed MYH9-related disorders (Seri et al., 2003). Affected individuals demonstrate enlarged thrombocytes, thrombocytopenia, and leukocyte inclusions containing ribosomes and thin filaments containing NMMHC-IIA (Pecci et al., 2002). Patients with isolated macrothrombocytopenia and leukocyte inclusions had previously been classified as having *May–Hegglin anomaly* or *Sebastian syndrome*, whereas the presence of co-morbid conditions such as renal failure, deafness, or cataracts were used as the basis for the diagnosis of *Epstein Syndrome* or *Fechtner syndrome*. More detailed re-evaluation of these syndromes demonstrated clear clinical and immunocytochemical overlap (as demonstrated using anti-NMMHC-IIA antibodies), suggesting that these syndromes represent a single disorder with a continuous clinical spectrum. Most of the *MYH9* defects are missense mutations that affect the motor or coiled-coil domain of NMMHC-IIA, although the various nonsense or frameshift mutations all involve the last exon (Hu et al., 2002). A specific $Asp^{142}Asn$ mutation appears to be responsible for an unstable protein, whereas various deletion mutants appear to exert dominant negative effects on the wild-type allele (Deutsch et al., 2003; Kunishima et al., 2003). These molecular studies support a current model invoking the role of Mk microtubules in the actomysosin ring assemblies required during proplatelet formation (Hartwig and Italiano, 2003).

Bernard–Soulier Syndrome Intensively studied because of its associated hemostatic defect, BSS is a heterogenous disorder caused by molecular defects involving the GPIb-IX-V complex (Fig. 15–4). Although the extracellular domain of GPIbα mediates the initial interaction with vWF, the GPIbα cytoplasmic domain serves as a cytoskeletal anchor through filamin, while also interacting with several platelet activation molecules such as 14-3-3ζ, calmodulin, and possibly phosphoinositide 3-kinase (Lopez et al., 1998).

Mutations involving *GPIbα*, *GPIbβ*, or *GPIX* are collectively implicated in the development of the BSS, an autosomal recessive disorder manifest by macrothrombocytopenia, normal platelet survival, and platelet dysfunction. The majority of the more than 30 mutations prevent the cell-surface association and/or expression of the glycoprotein complex. A smaller subset of patient mutations demonstrate normal platelet expression with disordered vWF binding due to mutations in the GPIbα leucine-rich repeats (variant BSS). In all phenotypes, platelets are considerably larger than normal and fail to bind vWF as demonstrated using ristocetin-induced platelet aggregometry. Heterozygote mutations cause a milder or non-existent phenotype with normal or slightly reduced platelet counts, and normal ristocetin-induced aggregations (Savoia et al., 2001). While the clinical (hemostatic) phenotype is clearly explained by the quantitative or qualitative abnormalities of the GPIb-IX-V complex, the molecular mechanism(s) causally implicated in the development of macrothrombocytopenia remain unestablished. Nonetheless, targeted disruption of the *GPIbα* gene in mice phenocopies the human disorder, clearly implicating this gene in the development of both the thrombocytopenia and macrothrombocytes (Ware et al., 2000).

Inherited Thrombocytopenias with Shortened Platelet Survival

Wiskott–Aldrich Syndrome and X-Linked Thrombocytopenia Patients with Wiskott–Aldrich syndrome (WAS) have an immunodeficiency syndrome involving both cellular and humoral immunity, with associated thrombocytopenia and small platelets. X-linked thrombocytopenia (XLT) represents a milder form of WAS without associated immunodeficiency (Sullivan et al., 1994). The WAS protein (WASp) is expressed exclusively in hematopoietic-derived cells, and functions in the reorganization of the actin cytoskeleton, regulating shape change, cell division, locomotion, and chemotaxis (Snapper and Rosen, 2003). More specifically, WASp interacts with the actin-related Arp 2/3 complex to initiate nucleation of actin filaments (Welch et al., 1997). Both WAS and XLT are caused by mutations of the *WAS* gene that is located on Xp11.23. More than 150 mutations have been identified, although more than 50% of all

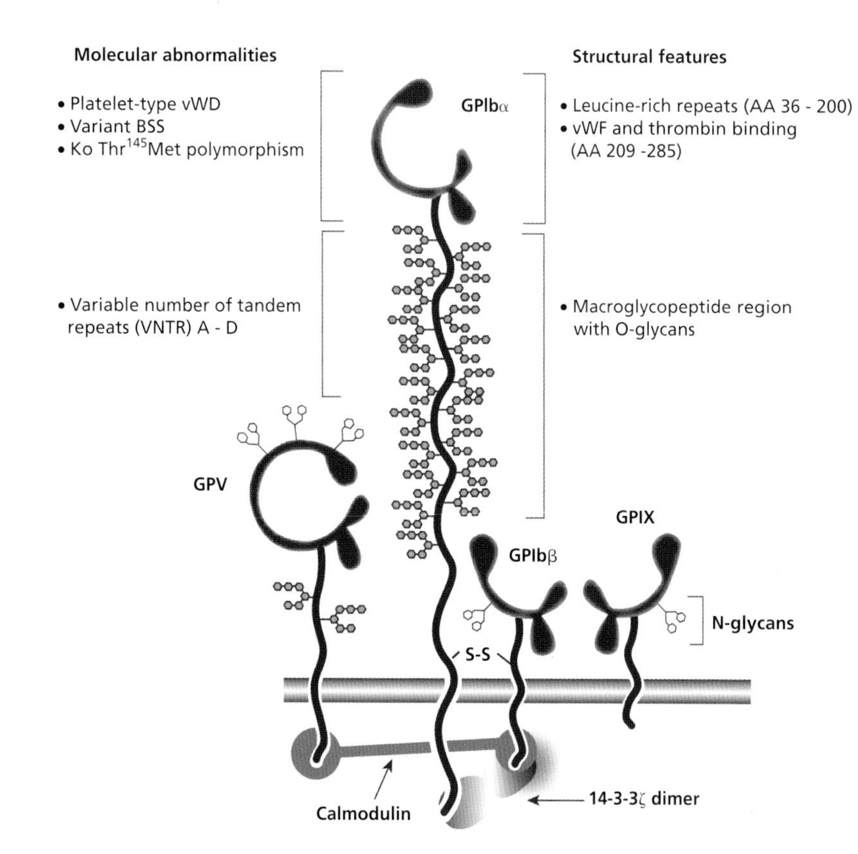

Figure 15–4 Hypothetical schematic representation of the GPIb-IX-V receptor complex. Key structural features and sites of relevant molecular phenotypes are delineated (see text for details). The polypeptide cytoplasmic domains associate with several proteins, including 14-3-3-ζ, calmodulin, and actin-binding protein; the latter links the complex to the actin microfilaments comprising the membrane cytoskeleton (not shown). vWD, von Willebrand's disease; BSS, Bernard–Soulier syndrome; vWF, von Willebrand factor.

mutations are clustered within the first four exons. Amino acid substitutions located at the N-terminus result in reduced levels of WASp, and are typically associated with mild XLT. Bleeding manifestations in WAS range from minor petechiae to serious gastrointestinal or intracranial hemorrhage (Sullivan et al., 1994). Serious bleeding events are comparable among platelet count subgroups, suggesting the presence of a concomitant qualitative platelet disorder. Indeed, several WAS patients demonstrated a marked reduction in dense granules, adenine nucleotide storage pools, and platelet aggregation (Grottum et al., 1969). Enhanced bleeding does not appear to be present in WAS-deficient mice; furthermore, their platelets are of normal size, and the mechanism of the small platelets in the human disease remains unestablished (Grottum et al., 1969; Snapper and Rosen, 2003).

Platelet-type von Willebrand's Disease Mutations in the carboxy-terminal flanking sequence of the GPIbα leucine-rich repeats (Gly[233]Val and Met[239]Val) increase the affinity of the receptor for vWF and cause platelet-type vWD (Miller et al., 1991). This autosomal dominant disorder is usually characterized by mild thrombocytopenia, variable degree(s) of platelet macrocytosis, and a prolonged bleeding time. Its clinical and laboratory features are similar to those of type 2B von Willebrand's disease, a disorder caused by enhanced affinity of vWF for GPIbα, deriving from mutations within the *vWF* gene (Cooney et al., 1991). In both situations, spontaneous association of vWF to GPIbα is evident, resulting in *in vivo* platelet binding and clumping, with shortened platelet survival and thrombocytopenia.

Genetic Defects Causing Qualitative Disorders of Platelet Function

Molecular Defects Involving Cell-surface Receptors

Glanzmann's Thrombasthenia

Glanzmann's thrombasthenia (GT) is characterized by a normal platelet count, normal platelet morphology, and failure to aggregate when exposed to physiological platelet agonists. A lifelong bleeding diathesis, coupled with the diagnostic platelet aggregometry studies, is characteristic of the disorder. GT is inherited as an autosomal recessive trait, with either complete absence or marked reduction in the GPIIb/IIIa ($\alpha_{IIb}\beta b_3$) heterodimeric complex (Newman et al., 1991). The genes encoding GPIIb (*ITGA2B*) and GPIIIa (*ITGB3*) are closely linked and reside on the long arm of chromosome 17q21–q23. This close physical linkage raises the possibility the GPIIb/IIIa cell-surface expression are coordinately regulated, not dissimilar to that of the globin genes. Patients were initially classified into two categories (type I and type II), based on the severity of their defects. Type I patients have less than 5% of the normal GPIIb/IIIa, have absent clot retraction and an absence of fibrinogen in the α granules (Nurden et al., 1985). Type II thrombasthenic platelets have 10%–20% of the normal GPIIb/IIIa cell-surface expression, maintaining varying degrees of functional competency in fibrinogen binding. Subsequently, variant GT patients have been described in which cell-surface GPIIb/IIIa expression is near-normal, but dysfunctional as demonstrated by the inability of the receptor to display activation-dependent fibrinogen binding. These defects are especially interesting since they have identified cytoplasmic and metal-ion-dependent domains that are necessary for optimal receptor activation and binding (Nurden, 1999). Not surprisingly, a wide spectrum of defects involving both *ITGA2B* and *ITGB3* have been described, and are largely due to splice and nonsense mutations (Newman et al., 1991; Djaffar and Rosa, 1993; Djaffar et al., 1993; Schlegel et al., 1995) (Fig. 15–5). As expected, patients with *GPIIIa* (β_3) mutations have a parallel defect in vitronectin receptor ($\alpha_v\beta_3$) expression, since the vitronectin receptor (VnR) shares the GPIIIa polypeptide (Coller et al., 1991). An updated database cataloging the mutations identified in patients with GT is available (http://sinaicentral.mssm.edu/intranet/research/glanzmann/menu).

Collagen Receptor Defects

Although platelet glycoproteins GPIa/IIa ($\alpha_2\beta_1$) and GPVI/FcRγ represent the predominant collagen receptors on human platelets, molecular defects involving cognate receptors associated with clinically relevant hemostatic disorders have not been described. Earlier studies had described a patient displaying approximately 15%–25% of normal GPIa (α_2) expression, with concomitant defects in collagen-induced aggregation and adhesion (Nieuwenhuis et al., 1985). Isolated impairment in collagen responses with an associated mild bleeding diathesis have also been described in patients with GPVI/FcRγ deficiency (Moroi et al., 1989; Arai et al., 1995). Despite these initial observations, no associated molecular defects involving these or other associated receptors have been established. In contrast, molecular defects involving another putative collagen receptor, GPIV (CD36), have been described, although they do not have a concomitant bleeding disorder (Yamamoto et al., 1990; Diaz-Ricart et al., 1993). Individuals deficient in GPIV are found in approximately 3% of the Japanese and approximately 0.3% of the US population, and platelets deficient in GPIV demonstrate reduced adhesion to collagen in flowing blood chambers (Diaz-Ricart et al., 1993), although they maintain normal collagen-induced platelet aggregation and intracellular calcium mobilization. Subsets of GPIV-deficient patients have a Pro[90]Ser mutation with persistent mRNA transcripts, which may be causally implicated in the functional defect (Kashiwagi, Honda, Take, et al., 1993; Kashiwagi, Honda, Tomiyama, et al., 1993) (see CD36, below).

Defects in Platelet ADP Receptors

The purinergic receptors P2Y1, P2Y12, and P2X1 mediate platelet/ADP interactions. ADP-induced aggregation requires coactivation of both P2Y1 and P2Y12, whereas P2X1 functions as a cation channel (Dorsam and Kunapuli, 2004). The P2Y1 receptor is coupled to PLC activation and intracellular calcium release, whereas P2Y12 inhibits cAMP formation by adenylyl cyclase. Defective ADP-induced signaling and/or binding have been described in some patients (Cattaneo et al., 1992; Nurden et al., 1995; Levy-Toledano et al., 1998), and the molecular defects elucidated. A single-allele two-nucleotide deletional mutant has been described that unexpectantly is associated with near-complete loss of P2Y12 receptors, suggesting that receptor expression is solely derived from the mutant allele (Nurden et al., 1995; Hollopeter et al., 2001). Subsequent mutations involving the *P2Y12* gene have subsequently been described. Compound heterozygote mutations that do not affect P2Y12 expression or ADP binding—but do alter signal transduction pathways—have been associated with a hemostatic bleeding disorder (Cattaneo et al., 2003). Finally, homozygous deletions resulting in premature termination and lack of P2Y12 protein expression have also been described (Conley and Delaney, 2003). While the majority of defects involve the *P2Y12* gene, a recent mutation involving the P2X1 receptor has been reported in a patient with selective impairment of ADP-induced aggregation (Oury et al., 2000). This mutant lacks one leucine within a stretch of four leucine residues in its second transmembrane domain (amino acids 351–354). When co-expressed with the wild-type receptor, the mutated protein exhibited a dose-dependent dominant negative effect on ADP-induced P2X1

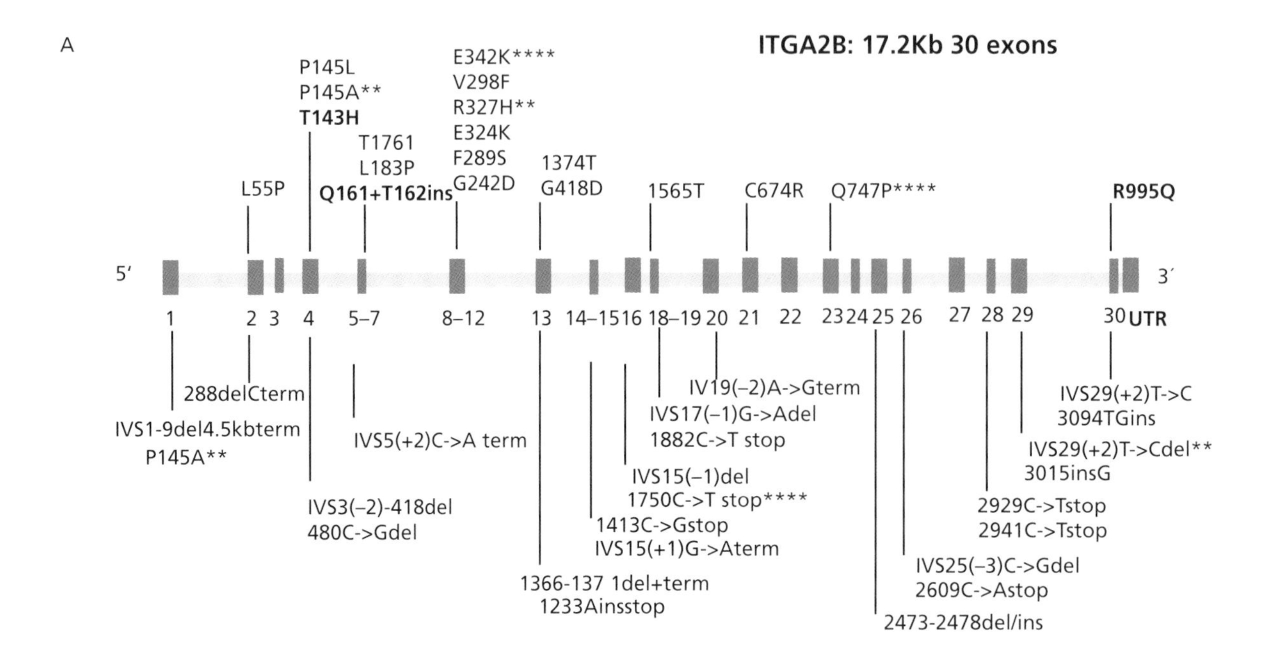

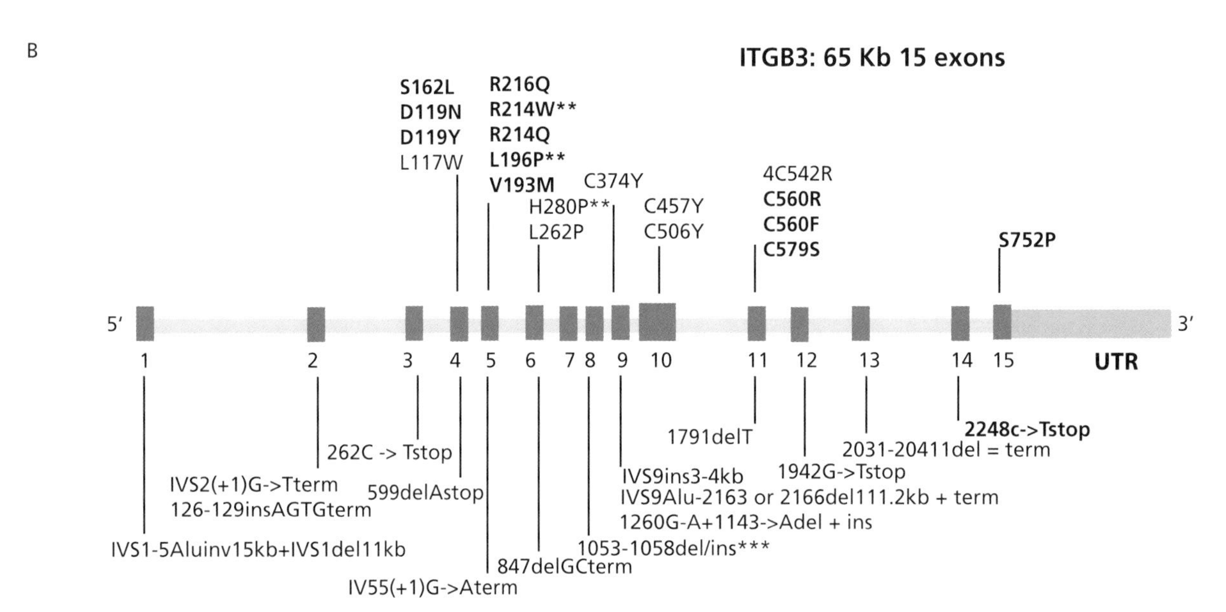

Figure 15–5 Schema of genetic defects causally implicated in development of Glanzmann's thrombasthenia. Boxes delineate individual exons; asterisks (*) refer to the number of times the identical genetic defect has been identified in unrelated pedigrees; mutations in bold refer to those causing variant Glanzmann's thrombasthenia. (*A*) ITGA2B: 17.2 kb 30 exons. (*B*) ITGB3: 65 kb 15 exons. ITGA2B, gene encoding integrin α_{IIb} subunit; ITGB3, gene encoding integrin β_3 subunit; UTR: 3′-untranslated region; del, deletion; ins, insertion; inv, inversion; term, premature termination. [Adapted from Nurden and Nurden (2002).]

channel activity. The association of this defect with a hemostatic bleeding disorder implies a previously unknown function of P2X1 in normal hemostasis (Oury et al., 2000).

Thromboxane A_2 Receptor Defects

The arachidonate metabolite thromboxane A_2 (TXA_2) is a potent stimulator of platelet aggregation and a constrictor of vascular and respiratory smooth muscles. Two isoforms of the human TXA_2 receptor have been cloned, both of which demonstrate similar ligand-binding characteristics and PLC activation, but opposite-regulated adenylyl cyclase activity (Hirata et al., 1996). In two unrelated patients with a mild bleeding disorder whose platelets demonstrated impaired aggregation responses to TXA_2 and its analogs (despite a normal response to thrombin), an $Arg^{60}Leu$ mutation involving the *TBXA2R* has been described (Hirata et al., 1996). Although the patients' platelets exhibited normal binding activities to TXA_2 analogs (Ushikubi et al., 1987; Fuse et al., 1993), they showed decreased guanosine triphosphatase (GTPase) activity and second messenger formation. The proband in one family was homozygous for the mutation, and the heterozygotes demonstrated a similar but slightly attenuated phenotype of impaired platelet aggregation.

Molecular Defects Involving Intracellular Constituents

Intracellular Signaling Defects

Rare patients with congenital enzymatic deficiencies have been described, involving any of a large number of intracellular platelet enzymes, such as cyclooxygenase, prostaglandin H synthetase-1, thromboxane synthetase (Rao et al., 2004). In general, molecular defects causally implicated in the development of these functionally relevant disorders have been incompletely elucidated. Interestingly, two unrelated families have been described that have a bleeding diathesis and inducible hyperactivity involving the $G\alpha_s$ pathway (Freson et al., 2001b). The $G\alpha_s$ gene (*GNAS1*) contains three cryptic alternative promoters and exons, thus encoding an elongated $G\alpha_s$ gene termed *XLαs* (Rickard and Wilson, 2003). A paternally inherited functional polymorphism in *XLαs* exon 1 consisting of a 36-bp duplication and two nucleotide substitutions resulted in an $Ala^{138}D$ and $Pro^{161}R$ mutations associated with $G\alpha_s$ platelet hyperfunction and enhanced trauma-related bleeding tendency (Freson et al., 2001b). An additional eight patients with brachydactyly who inherited the same *XLαs* defect also demonstrated platelet hyper-reactivity, and had an elongated $G\alpha_s$ protein as a consequence of the paternally inherited insertion (Freson et al., 2003). Although the mechanism leading to the functional defect remains unestablished, defective association(s) between wild-type and mutant proteins may result in unimpeded receptor-stimulated activation (Freson et al., 2003). Another patient with a mild bleeding diathesis had selective decrease in platelet $G\alpha_q$, with platelet aggregometry demonstrating less robust responses to various agonists (Gabbeta et al., 1997). Although the coding region was normal, specific reduction of platelet-derived $G\alpha_q$ mRNA was evident to the exclusion of that found in neutrophils. Correlative functional studies in $G\alpha_q$-deficient mice have identified a comparable hemostatic defect and diffuse agonist loss in platelet activation (Offermanns et al., 1997).

Defects of Secretion (Platelet Granule Defects)

α-Granule Disorders

Isolated deficiencies of α-granule-stored constituent proteins can occur in inherited disorders of the corresponding plasma protein (i. e., factor V in congenital factor V deficiency, fibrinogen in Glanzmann's thrombasthenia, vWF in von Willebrand's disease). In contrast, α-granule deficiency is found in patients with *gray platelet syndrome* (GPS, OMIM %139090), a well-characterized hemostatic disorder manifest by macrothrombocytopenia and the presence of pale platelets on Wright–Giemsa stain. The endothelial cell storage organelles (Weibel–Palade bodies), like the α-granules of Mks and platelets, contain vWF and P-selectin, although the defect in GPS does not extend to these structures (Gebrane-Younes et al., 1993). The molecular defect in GPS remains unknown although it presumably interferes with the ability of Mks to package secretory proteins into the developing granules. SNARE (soluble N-ethylmaleimide sensitive factor-associated protein receptors) proteins constitute a superfamily of proteins that regulate membrane fusion events involved in platelet exocytosis and vesicle trafficking (Polgar and Reed, 1999), and are likely candidate genes that may be implicated in the development of this disorder (Reed et al., 2000). Alternatively, the molecular basis of GPS has been hypothesized to involve a cytoskeletal protein, based on the studies using Wistar Furth rats (Pestina et al., 1995).

In contrast to the global defect involving α-granules found in patients with GPS, platelets from patients with the *Quebec platelet syndrome* (OMIM %601709) demonstrate a restricted defect involving proteolytic degradation of various α-granule proteins such as factor V, vWF, fibrinogen, thrombospondin, and osteonectin (Hayward et al., 1997). In this hemostatic syndrome with an autosomal dominant mode of inheritance, the pathologic proteolysis of α-granular contents—rather than a defect in targeting proteins to α-granules—represents the etiology of the disorder. Unlike normal platelets, Quebec platelets contain large amounts of urokinase-type plasminogen activators (Kahr et al., 2001), partially (if not fully) providing an explanation for the bleeding disorder, and likely explaining the efficacy of fibrinolytic inhibitors in treating the hemostatic defect (McKay et al., 2004).

Disorders of Platelet Dense (δ) Granules

Platelet activation is followed by membrane fusion of the dense granule with the platelet membrane, followed by exocytosis of granular constituents, the most important being ADP, ATP, and serotonin. ADP activates and recruits other platelets, and serotonin causes vasoconstriction and facilitates binding of adhesive proteins to their endogenous platelet receptors (McNicol and Israels, 1999). *Storage pool disease* (OMIM %185050) refers to a heterogeneous group of bleeding disorders, collectively grouped together because of defective dense granule release identifiable by platelet aggregometry (Nieuwenhuis et al., 1987). Characteristics of defective dense granule release are (1) absent ADP-induced second wave in citrated platelet-rich plasma, (2) attenuated and delayed collagen response, and (3) absent or reduced response to epinephrine. Rare hybrid disorders that include defective dense granule release with variable degrees of α-granule deficiency are referred to as αδ-SPD; to date, the molecular etiolog(ies) of this heterogeneous group of disorders are unknown.

In contrast, the genetic defects causing *Hermansky–Pudlak syndrome* (HPS) and *Chediak–Higashi syndrome* (CHS) have been recently elucidated. HPS and CHS represent distinctive, autosomal recessive diseases associated with the absence of platelet dense granules, deficient skin and hair pigmentation (affecting melanosomes), and defective lysosomal function in other cell types. All HPS patients have oculocutaneous albinism and variable hypopigmentation of hair, skin, and irides; the diagnosis relies on the clinical manifestations with confirmatory absence of dense granules by platelet electron microscopy (Gunay-Aygun et al., 2004).

A number of distinct HPS subtypes have been described, although the most common type (HPS-1) is found in northwest Puerto Ricans (Toro et al., 1999). Despite the rarity of HPS, the gene (s) causally implicated have been well-characterized, and all encode novel proteins with no recognizable homology to other proteins (Table 15–5). Many of the human disorders have murine models with comparable phenotypic findings (Reed, 2003; Gunay-Aygun et al., 2004). Some HPS proteins interact with each other in oligomeric complexes referred to as biogenesis of lysosome-related organelles complex (BLOCs), which explains the comparable phenotypes among the subsets (Nazarian et al., 2003) (Fig. 15–6). In addition to oculocutaneous albinism and defective dense granules, patients with CHS have progressive neurological dysfunction and an immune deficiency that does not infrequently lead to a lymphoproliferative disorder (Karim et al., 2002). Ultrastructural features of CHS are the giant inclusion bodies identifiable in

Table 15–5 Genetic Etiology of Major Platelet Dense Granule Disorders and Murine Phenotypic Homologues

Hermansky–Pudlak Subtype	OMIM	Gene	Protein (M_r)	Murine Phenotype	Reference
• HPS1[a]	604982	*HPS1*	79.3	Pale ear	(Oh et al., 1996; Shotelersuk et al., 1998; Horikawa et al., 2000)
• HPS2	608233	*AP3B1*	140.0	Pearl	(Dell'Angelica et al., 1999)
• HPS3	606118	*HPS3*	113.7	Cocoa	(Anikster et al., 2001; Huizing et al., 2001)
• HPS4	606682	*HPS4*	76.9	Light ear	(Suzuki et al., 2002; Anderson et al., 2003; Bachli et al., 2004)
• HPS5	607521	*HPS5*	127.4	Ruby eye 2	(Zhang et al., 2003)
• HPS6	607522	*HPS6*	83.0	Ruby eye	(Zhang et al., 2003)
• HPS7	607145	*DTNBP1*	39.5	Sandy	(Li et al., 2003)
Chediak–Higashi Syndrome	214500	*LYST*	425.0	—	(Barbosa et al., 1997; Karim et al., 1997; Karim et al., 2002)

[a] Most common subtype.

a variety of granule-containing cells including platelets. CHS results from mutations in the lysosomal trafficking regulator (*LYST*) gene, which encodes a large 425-kDa cytoplasmic protein with unknown function. Alternatively, LYST may act as an adaptor protein that brings proteins into close proximity during intracellular membrane fusion events during vesicle transport and/or fusion (Nagle et al., 1996). The lack of identifiable *LYST* mutations in some CHS patients suggests an alternative disease-susceptibility locus (Karim et al., 2002). Finally, patients with *WAS* and *TAR syndrome* have variable reductions in dense granules and platelet aggregometry; these disorders have been discussed in greater detail earlier.

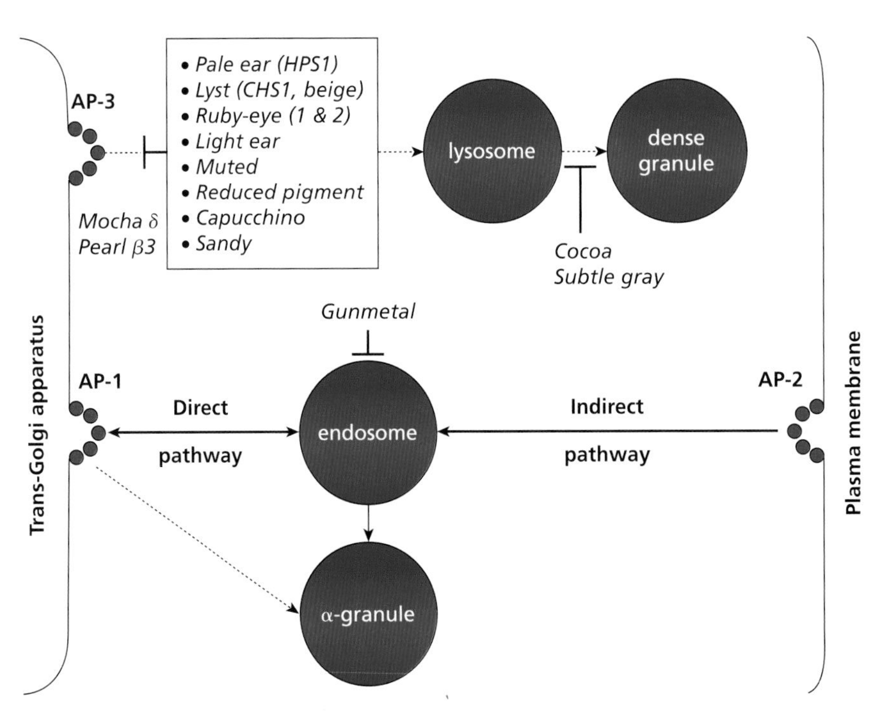

Figure 15–6 Molecular mechanisms of granule development relevant to human and murine disease phenotypes. Based on murine and human models, current evidence suggests that three distinct adapter protein (AP) complexes regulate the development and packaging of megakaryocyte secretory granules: AP-1, AP-2, and AP-3. Proteins may be sorted from the trans-Golgi apparatus to the α-granules by vesicle trafficking through the endosome in an AP-1-mediated direct pathway, or an AP-2-restricted indirect (i.e., endocytic) pathway; direct trafficking to the α-granule is also plausible (dotted arrow). Human and murine phenotypes displaying Hermansky–Pudlak (HPS) and Chediak–Higashi syndromes (CHS) have abnormal dense granules and lysosomes, suggesting that these organelles develop through an AP-3-dependent pathway. Similar studies in *mocha* and *pearl* mice with defective dense granules and lysosomes also suggest that these organelles develop through an AP-3-dependent pathway. Numerous other mice with abnormal dense granules and lysosomes have been described (boxed), presumably linked to the AP-3-dependent pathway. Since *gunmetal* mice have both α- and dense granule defects, it is plausible that a common (presumably endosomal) vesicular intermediary may regulate organelle development. [Schematic model adapted and updated from Reed (2002).]

Platelet Polymorphisms in Human Diseases

Given their importance in mediating platelet functions, it is not unexpected that receptor polymorphisms may be causally implicated in various platelet disorders. While platelets clearly regulate arterial thrombotic events, polymorphisms predisposing to cardiovascular or cerebrovascular stroke are not as well-characterized as those causing neonatal alloimmune thrombocytopenia or posttransfusion purpura.

Polymorphisms Associated with Acquired Thrombocytopenias

Neonatal Alloimmune Thrombocytopenia

Neonatal alloimmune thrombocytopenia (NAIT) refers to the immune-mediated destruction of fetal platelets by maternally transferred alloantibodies, and is pathophysiologically comparable to hemolytic disease of the newborn (Rh-incompatability). Specifically, maternal antibodies directed against fetal platelet antigens inherited from the father, but absent in the mother, cross the placenta and induce severe thrombocytopenia. The frequency of NAIT is 0.1%, although, unlike hemolytic disease of the newborn, it may affect the first pregnancy (Bussel et al., 1997). The thrombocytopenia is present at birth, and may be associated with intracranial hemorrhage in 10%–20% of newborns; spontaneous platelet recovery typically occurs in 2–3 weeks. The antigenic sequences causally implicated in this disorder are well known, although the nomenclature for the distinct polymorphisms is confusing since the antigens were initially named according to affected patients (i.e., PlA, Bak, Pen). A universal antisera-dependent Human Platelet Alloantigen (HPA) system has been developed to bypass this confusion (von dem Borne and Decary, 1990) although its genetic limitations have been outlined; thus, distinct alloantigens categorized in the HPA system are identical at the molecular level (Newman, 1994). Table 15–6 categorizes the important polymorphisms using both nomenclatures, along with detailed nucleotide, protein, and molecular information.

Incompatibility for the Pl^{A} ($Leu^{33}Pro$) alloantigen is the most common etiology of NAIT in Caucasians, accounting for nearly 80% of all cases; antibodies directed against HPA-5 (Br^{b}/Br^{a}) and HPA-8b (Gov^{a}/Gov^{b}) alloantigens are also not infrequently seen (Schuh et al., 2002). The allelic frequency of the Pl^{A1} Leu^{33} polymorphism (~0.85) makes this the common allele found worldwide, compared with the less frequent Pro^{33} allele. Typically, the mother is Pl^{A2} (Pro^{33}/Pro^{33}) homozygous, and develops alloantibodies against the paternally inherited PL^{A1} (Leu^{33}) antigen present on fetal platelets. Given the known allelic frequencies in Caucasians, there is a 98% probability that the father will be Pl^{A1}-positive (i.e., express at least one of the immunodominant Pl^{A1} (Leu^{33}) alleles), based on Hardy–Weinberg predictions:

$p^2 + 2pq + q^2 = 1$, where $p\{(Leu^{33}) = 0.85$; and $q(Pro^{33}) = 0.15$

Probability of Pl^{A1}-positivity (i.e., presence of at least one Leu^{33} allele) $= p^2 + 2pq\ p^2\{(0.85)(0.15) = 0.98$

Clearly, such alloantigen incompatibility is necessary but not sufficient for the generation of the alloimmune response, since the observed incidence of this syndrome (~1/1000–1/1500 live births) is considerably less than that predicted based on prevalence (i.e., 2% or 1/50 pregnancies). It is now known that the risk of developing alloantibodies is determined by the HLA DRB3*0101 genotype. The chance of antibody formation in a HPA-1a negative (PL^{A2}), HLA DRB3*0101-positive mother is 1 in 3; in contrast, if the mother is HLA DRB3*0101-negative, the risk is reduced to 1 in 300–400 (L'Abbe et al., 1992). While genotyping can clearly identify the polymorphic sites, this approach should be viewed as complementary to the serologic studies that focus on detection of the platelet-specific Pl^{A1} antibody, since this is the requisite "effector" for the development of thrombocytopenia. Management of NAIT is complex, and should be handled by physicians with a complete understanding of the pathophysiology, risks, and natural history of the disorder (Blanchette et al., 2000). In general, compatible platelets can be transfused to prevent and/or treat postnatal bleeding complications. Since the mother's platelets are phenotypically distinct from those of the fetus, washed maternal platelets (to remove residual anti-Pl^{A1} antibody) are an excellent source for compatible platelets.

Genetic counseling for subsequent pregnancies should focus on the father's genotype, since documentation of paternal homozygosity coupled with the presence of maternal anti-Pl^{A2} antibodies represents a strong, predictive risk for the development of NAIT homozygosity (Bray et al., 1994). In situations of paternal heterozygosity, fetal amniocyte genotyping may be useful (Khouzami et al., 1995). Documentation of maternal–fetal incompatability would necessitate close observation with early intervention, given the relatively high incidence of prenatal intracranial hemorrhage.

Posttransfusion Purpura

Posttransfusion purpura (PTP) typically occurs 5–10 days after a blood transfusion, presenting with profound, sudden-onset thrombocytopenia. In the absence of confounding clinical manifestations, the disorder is self-limiting with complete recovery 1–4 weeks after initial development. Comparable to the situation in NAIT, the recipient is typically Pl^{A2}-positive, and develops an alloantibody directed against a donor platelet Pl^{A1} antigen. Involvement of other polymorphic epitopes involving Pen, Bak, and Br has also been described. PTP likely represents an anamnestic response to a previously exposed (foreign) antigen in the host, although it remains unestablished why development of an anti-donor Pl^{A1} antibody should lead to the destruction of endogenous ($Pl^{A2/A2}$) platelets. A variety of explanations have been proposed, including the development of immune complexes or cross-reactive antibodies with affinity for autologous platelets (Lau et al., 1980). Platelets expressing the offending antigen (usually Pl^{A1}) are contraindicated, and, when appropriate, $Pl^{A2/A2}$-homozygous platelets should be used for transfusion support.

Platelet Polymorphisms Associated with Coronary Thrombosis

Given the importance of platelets in arterial thromboses, it is not surprising that considerable attention has focused on platelet polymorphisms as the risk factors for coronary thrombotic syndromes. Estimates of the effects of inheritance on MI range from 20% to 80%. Twin studies have identified a genetic component on MI risk that is most significant at younger ages (Pastinen et al., 1998); furthermore, the genetic component may be more pronounced in women than in men (Marenberg et al., 1994). While a large number of polymorphisms have been studied as cardiovascular risk factors, only those polymorphisms that have been characterized in most detail are outlined below.

Table 15–6 Common Platelet Polymorphisms Implicated in Human Disease

Polymorphisms with Altered Antigenic Determinants[1]						Genotype Frequency				
HPA	Polymorphism	Original	Polymorphism	Allele	Polymorphism	Caucasian	African/American	East Asian	Syndrome[2]	Comment
HPA-1	HPA-1a	PL^{A}	PL^{A1}	ITGB3	GPIIIa (β3) Leu^{33}	0.84–0.89	0.92	0.98–0.99	NAIT	Accounts for the majority of alloimmune syndromes worldwide
	HPA-1b		PL^{A2}	1565T/C	GPIIIa (β3) Pro^{33}	0.11–0.16	0.08	0.01–0.02	PTP, CAD	
HPA-2	HPA-2a	Ko	Ko^{b}	GPIBA	GP1bα (Thr^{145})	0.91–0.93	0.82–0.88	0.87–0.93	NAIT	Located within the vWF and thrombin-binding sites; alloimmune syndromes usually IgM antibodies
	HPA-2b		Ko^{a}	542G/A	GP1bα (Met^{145})	0.07–0.09	0.12–0.18	0.07–0.13	PTP, CAD	
HPA-3	HPA-3a	Bak	Bak^{a}	ITGA2B	GPIIb (αIIb) Ile^{843}	0.61	—	0.54	NAIT	Bak^{a} appears to be in linkage disequilibrium with Pl^{A1}
	HPA-3b		Bak^{b}	2622T/G	GPIIb (αIIb) Ser^{843}	0.39		0.46	PTP	
HPA-4	HPA-4a	Pen	Pen^{a}	ITGB3	GPIIIa (β3) Arg^{143}	1.00	1.00	0.99	NAIT	More frequent cause of NAIT in Japanese (compared to Caucasians) because of prevalence of HPA-4b phenotype
	HPA-4b		Pen^{b}	526G/A	GPIIIa (β3) Gln^{143}	0.00	0.00	0.01	PTP	
HPA-5	HPA-5a	Br	Br^{b}	ITGA2	GPIa (α_2) Glu^{505}	0.87–0.95	0.79	0.95–0.97	NAIT	Most common etiology for NAIT in Japan; second most common in Caucasians
	HPA-5b		Br^{a}	1648G/A	GPIa (α_2) Lys^{505}	0.05–0.13	0.21	0.03–0.05	PTP	

(Continued)

Table 15–6 Common Platelet Polymorphisms Implicated in Human Disease *(Continued)*

Polymorphisms with Altered Antigenic Determinants[1]						Genotype Frequency				
HPA	Polymorphism	Original	Polymorphism	Allele	Polymorphism	Caucasian	African/American	East Asian	Syndrome[2]	Comment
HPA-8	HPA-8b	Gov	Gov^a	CD109	CD109 Tyr^{703}	0.53	—	—	NAIT	CD109 is a glycosylphosphatidyl inositol (GPI)-anchored protein; third most common cause of NAIT
			Gov^b	2108A/C	CD109 Ser^{703}	0.47			PTP	
HPA-8	HPA-8b	Nak	Nak^a	CD36	CD36 Pro^{90}	0.94	0.70	0.70	NAIT	CD36 deficiency is prevalent in Japan (~4% of population)
			Nak^{a-}	478C/T	CD36 Ser^{90}	0.06	0.30	0.30	PTP, CAD	
Polymorphisms without altered antigenic determinants										
Glycoprotein Ia [α_2 polypeptide])	ITGA2	807C	0.54–0.65	—	0.61	CAD	807T allele has also been identified as a risk factor for stroke (Carlsson et al., 1999; Reiner et al., 2000).			
	807C/T	807T	0.35–0.46		0.39					
Glycoprotein Ibα	GPIBA	A	0.001	—	0.06	CAD	VNTR B allele appears to confer a significant risk for the development of anterior ischemic optic neuropathy (Salomon et al., 2004)			
	VNTR	B	0.07–0.1	0.19–0.2	0.006–0.01					
		C	0.80–0.84	0.70–0.75	0.47–0.60					
		D	0.08–0.11	0.03–0.10	0.33–0.41					

[1] The human platelet antigen (HPA) system was developed to standardize platelet alloantigen nomenclature, although it is evident that different alloantigens named in the HPA system (i.e., HPA-1a, HPA-4a) are identical at the molecular level. Since many of the alloantigens continue to be delineated by their original designation, both nomenclatures are outlined.

[2] NAIT, neonatal alloimmune thrombocytopenic purpura; PTP, posttransfusion purpura; CAD, coronary artery disease.

Note: Data compiled and updated from (Roychoudhury and Nei, 1988; Warkentin and Kelton, 1995; Afshar-Kharghan and Bray, 2002). Refer to text for details.

GPIIIa (β_3) Leu^{33}/Pro^{33} Dimorphism

The first suggestion that a gain-of-function mutation could lead to a prothrombotic phenotype was an association study between the GPIIIa Pro^{33} (Pl^{A2}) polymorphism and the development of acute coronary syndrome (Weiss et al., 1996). The genetic risk appeared to be more significant at a younger age. These initial conclusions have been supported by comparable results from other investigators (Mikkelsson, Perola, Kauppila, et al., 1999; Mikkelsson, Perola, Laippala, et al. 1999), although the exact risks appeared variable and associated with other genetic risk markers such as the plasminogen activator inhibitor-1 (PAI-1) 4G allele (Pastinen et al., 1998). Subsequent meta-analyses have concluded that Pro^{33} is a modest risk factor for coronary thrombosis (but not cerebrovascular stroke), with higher odds ratios for younger subjects and those undergoing revascularization procedures (Di Castelnuovo et al., 2001; Wu and Tsongalis, 2001). Furthermore, the polymorphism may be associated with enhanced thrombin generation and impaired antithrombotic action of aspirin (Undas et al., 2001). More recent data have demonstrated that GPIIIa Pro^{33} carriers have the most significant risks when associated with high serum fibrinogen levels (Boekholdt et al., 2004). Thus, although the initial events of platelet activation may not be influenced by the presence of the Pl^{A2} polymorphism, an integrated, hypothetical model would suggest that excessive fibrinogen binding to the conformationally active GPIIb/IIIa receptor would result in enhanced "outside-in" signaling (Shattil, 1999), and consolidation of platelet responses that would favor stable thrombus formation.

GPIa C807T Allele

The GPIa/IIa ($\alpha_2\beta_1$) receptor is converted to a high-affinity collagen receptor upon platelet activation. The silent C/T polymorphism in exon 7 of *ITGA2* was first implicated as a risk factor for MI in homozygous carriers of the 807T genotype (Moshfegh et al., 1999). Since the initial report, the results have been confirmed by other groups (Roest et al., 2000; Casorelli et al., 2001), with additional evidence that the 807T allele is associated with risks for cerebrovascular stroke (Carlsson et al., 1999; Reiner et al., 2000). Comparable to the data found in patients with MI, the risk appears to be strongest for young patients. Platelets from individuals with the 807T allele have greater $\alpha_2\beta_1$ receptor density. Nonetheless, the 807T allele neitheralters an amino acid nor is there any evidence that it is located within a transcriptionally relevant region of the gene (Kunicki et al., 1997), suggesting its unlikely involvement in modulating $\alpha_2\beta_1$ receptor density. As suggested, the 807T allele is linked to a polymorphism of the *ITGA2* gene (−52C/T), which may regulate transcription (Jacquelin et al., 2001). The quantitative differences in $\alpha_2\beta_1$ receptor density associated with the 807T allele have been functionally characterized. When compared with platelets homozygous for the 807C allele, 807T/T platelets demonstrate enhanced attachment to collagen type I-coated surfaces (Kritzik et al., 1998), thereby providing a putative mechanism for the enhanced prothrombotic risk associated with the 807T allele.

Finally, a number of additional polymorphisms within *ITGA2* have been identified, some of which are closely linked to the C807T allele (Fig. 15–7). Among the best characterized, evidence exists that the haplotype that encompasses the Br^a (Glu^{505}Lys) variant (refer to Table 15–6) is linked to 807C (Kritzik et al., 1998). This G → A substitution at nucleotide 1648 results in the amino acid 505 polymorphism, resulting in a three-allele combination within *ITGA2*. The 1648A (Lys^{505}) allele appears to be in linkage disequilibrium with 807C, with evidence that the 807C/1648A allele may be ancestral (Reiner et al., 1998). Further studies will be needed to assess

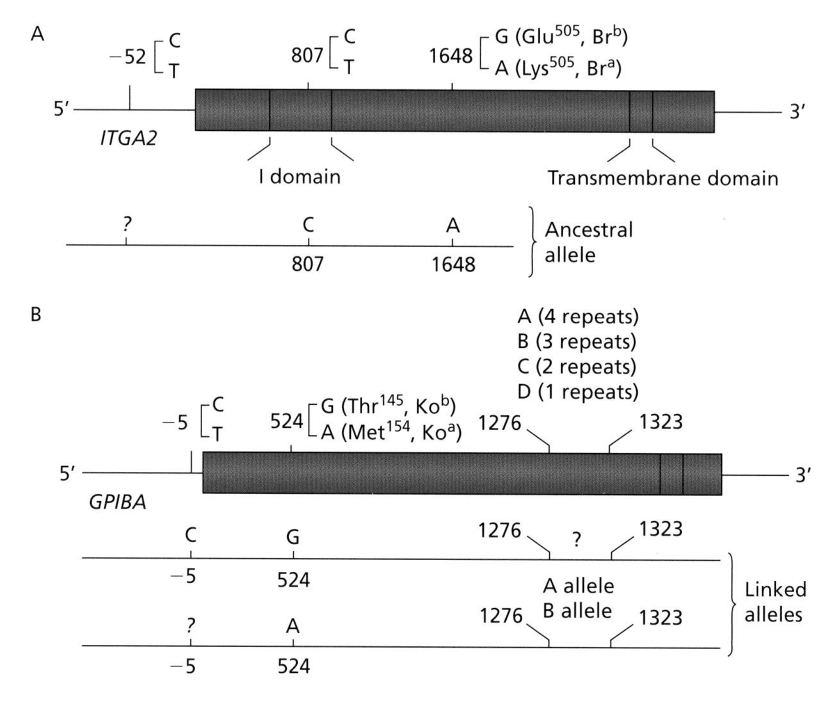

Figure 15–7 Schematic structure of ITGA2 and GPIBA cDNAs. (*A*) The three most intensively studied *ITGA2* polymorphisms are delineated, along with the ancestral allele encompassing the 807C/1648A allele; the position of the functionally important I domain that mediates the interaction with collagen is also shown (Bahou et al., 1994). (*B*) The key polymorphisms within *GPIBA* are depicted, inclusive of the VNTRs (variable number of tandem repeats) embedded within the GPIbα macroglycopeptide region (refer to Fig. 15–4). Alleles demonstrating linkage disequilibrium are depicted (refer to text for details). ITGA2, gene encoding glycoprotein Ia (α_2) subunit; GPIBA, gene encoding glycoprotein Ibα subunit.

linkage and haplotypes as allelic risks for MI, both within *ITGA2* and other candidate genes.

GPIbα Polymorphisms

The GPIbα polypeptide of GPIb-IX-V is the largest subunit (~135 kDa) of the multi-subunit receptor complex. Containing binding sites for vWF and α-thrombin (Mazzucato et al., 1998), GPIbα plays a crucial role in mediating platelet adhesion in situations of high-shear, and in modulating platelet procoagulant activity (Dormann et al., 2000). Three polymorphic sites within *GPIBA* have been extensively studied as risk factors for thromboses: a Kozak dimorphism (−5T/C), the Ko (HPA-2) $Thr^{145}Met$ dimorphism, and the VNTRs (variable number of tandem repeats) allelic polymorphisms (see Figs. 15–4 and 15–7).

Located five nucleotides upstream from the translation initiation site, the *Kozak–5T/C polymorphism* occurs within an area known to be important for translational regulation of eukaryotic genes (Kozak, 1989). The −5C variant is only found on the Thr^{145} allele, whereas the −5T variant can be located on either the Thr^{145} (Ko^b) or Met^{145} (Ko^a) alleles (Frank et al., 2001). The − 5C allele is associated with enhanced surface expression of GPIbαon platelets and heterologous cell lines (Afshar-Kharghan and Li, 1999), although this observation has not been confirmed by others (Corral et al., 2000). Similarly, its role as a risk factor for coronary thrombosis appears unestablished (Corral et al., 2000).

A G→A substitution at position 524 of *GPIBA* is responsible for the *Thr^{145}Met polymorphism*. The notion that these two polymorphisms might be associated with distinct functions was suggested by their location in the vWF- and thrombin-binding N-terminal domain of GPIbα, and by the function of some anti-Ko antibodies in preventing ristocetin-induced platelet aggregation (Kuijpers et al., 1992). Recent data have confirmed that vWF bound with higher affinity to the Thr^{145} GPIbα isoform, using both platelets and recombinant fragments of expressed GPIbα (Ulrichts et al., 2003). Similarly, carriers of the Thr^{145} polymorphism demonstrated increased platelet agglutination, increased P-selectin expression and aggregation under high-shear, and shortened closure teams using *in vitro* methods of platelet function (Yee and Bray, 2004). Nonetheless, clinical evidence consistently linking this polymorphism to cardiovascular risk is equivocal, and will require more studies (Yee and Bray, 2004).

The *GPIBA VNTR alleles* represent a series of identical 39-nucleotide repeats encoding 13 amino acids in the GPIbα mucin-like domain (Lopez et al., 1992). The A, B, C, and D alleles contain four, three, two, and one repeats, respectively (Ishida et al., 1995), encoding protein isoforms of molecular mass 168 (A allele), 162 (B allele), 159 (C allele), and 153 (D allele) kDa (Moroi et al., 1984). Although the A allele was initially considered to be restricted to East Asians, low genotypic frequencies have been described in American Indians and Caucasians from Southern Spain (Corral et al., 1998; Aramaki and Reiner, 1999). Several reports have outlined linkage disequilibrium between the Ko polymorphism and VNTR alleles, which are separated by less than 1 kb. For most populations, the Ko^a (Met^{145}) polymorphism has been found on the VNTR A and B alleles, although cross-overs have occurred in African–Americans, and Ko^b was found to be allelic with VNTR A and B alleles in populations in Southern Spain (Ishida et al., 1995; Corral et al., 1998). By studying well-matched controls in patients with various *GPIBA* polymorphisms, strong associations between Met^{145} and VNTR B and C genotypes have been described in patients with acute coronary syndrome and stroke (Gonzalez-Conejero et al., 1998), with evidence that the CC genotype (homozygous with two repeats) may be protective in African–Americans (Afshar-Kharghan et al., 2004). Detailed studies outlining the role of the VNTR alleles on GPIbα function remain incomplete, although one could speculate that the resulting GPIbαisoforms would have distinct endpoints in stabilizing platelet adhesion to subendothelial matrices in situations of high shear. In summary, though various studies implicate the Kozak − 5C, Met^{145} and VNTR (B) genotypes as the risk factors for arterial vascular diseases, the findings remain supportive but generally inconsistent (Yee and Bray, 2004).

GPVI/FcRγ

This complex serves as the predominant collagen signaling receptor, with the identification of at least five dimorphisms resulting in amino acid substitutions (Croft et al., 2001). A case–control study of patients with MI demonstrated an association between the Ser^{219}-Pro dimorphism and cardiovascular thrombotic risk, most prominent in patients older than 60 years of age (Croft et al., 2001). A subsequent study from Japan reported an association between MI risk and a $Thr^{249}Ala$ dimorphism (Takagi et al., 2002). Variability of GPVI expression among individuals appears broad, ranging between 1.4- and 5-fold (Chen et al., 2002; Best et al., 2003). Correlative functional studies demonstrate that the low-frequency Pro^{219} polymorphism is associated with fewer GPVI receptors (Best et al., 2003; Joutsi-Korhonen et al., 2003), and diminished functional responses as quantified using platelet aggregometry, fibrinogen binding, and downstream signaling events (Joutsi-Korhonen et al., 2003). The mechansism(s) by which reduced GPVI/FcRγ cell-surface expression and functional sequelae apparently enhance cardiovascular risk remains unestablished.

CD36 CD36 is an integral membrane cell-adhesion receptor known to have numerous potential physiological functions, including binding to type I collagen, thrombospondin, anionic phospholipids, and oxidized low-density lipoprotein (LDL); further evidence suggests that it binds and may transport long-chain fatty acids (Abumrad et al., 1993). Defects in CD36 are the cause of platelet GPIV deficiency (also known as CD36 deficiency). The CD36 antigenic epitope is referred to as Nak (refer to Table 15–6), and platelets deficient in CD36 are Nak^{a-} [note that the Nak^a alloantigen system predisposes individuals to the development of NAIT and platelet refractoriness when $Naka^-$ individuals are transfused with Nak^a (CD36-positive) platelets; see Polymorphisms Associated with Acquired Thrombocytopenias, earlier]. Platelet CD36 deficiency is evident in approximately 2% − 4% of Japanese (and other East Asian populations) and African–American cohorts, but is seen in less than 0.3% of American (Caucasian) cohorts (Tomiyama et al., 1990). CD36 deficiency can be classified into two subgroups: the type I phenotype is characterized by complete absence in platelets and monocytes/macrophages, while the type II phenotype demonstrates absent platelet CD36 with near-normal expression in monocytes/macrophages (Kashiwagi et al., 1995); although a number of defects have been described, the original $Pro^{90}Ser$ mutation appears to be relatively common (Kashiwagi et al., 2001). An increased incidence of CD36-deficiency has been identified in Japanese individuals with coronary heart disease, typically associated with type II diabetes and hypertriglyceridemia (Tanaka et al., 1997; Miyaoka et al., 2001). Nak^{a-} platelets display abnormal aggregation to fibrillar collagen with preserved aggregatory responses to other agonists (McKeown et al., 1994), and CD36 expression is important in collagen adhesive interactions in flowing blood (Diaz-Ricart et al., 1993). Despite both molecular and epidemiological evidence implicating CD36 deficiency in the development of coronary heart disease, there is no evidence for an associated, clinically relevant hemostatic defect (Yamamoto et al., 1990).

Indications for Polymorphism Screening in Clinical Diagnostics

Given the potentially serious consequences of known polymorphisms associated with the development of NAIT and PTP, pursuit of diagnostics both at the molecular level and antibody screening are appropriate. This is especially relevant for NAIT where confirmation of paternal incompatibility at the PL^A locus has considerable implications for the management of subsequent pregnancies. As implied earlier, the utility of generalized screening for the various platelet polymorphisms known to be associated with enhanced arteriothromobotic risk is more limited. To date, the data demonstrate moderate risk and inconsistent hazard ratios, with variability among ethnic groups and within clinically diverse substratification endpoints. Future association studies using larger numbers of genes, and/or focusing on informative haplotypes may be more useful. Such studies are more likely to be informative if they incorporate gene–gene (i.e., ethnicity) and gene–environment (i.e., plasma fibrinogen concentrations) interactions.

References

Abumrad, NA, el-Maghrabi, MR, Amri, EZ, Lopez, E, and Grimaldi, PA (1993). Cloning of a rat adipocyte membrane protein implicated in binding or transport of long-chain fatty acids that is induced during preadipocyte differentiation: Homology with human CD36. *J Biol Chem*, 268:17665–17668.

Afshar-Kharghan, V and Bray, PF (2002). Platelet polymorphisms. In AD Michelson (ed.), *Platelets* (pp. 157–169). London: Academic Press.

Afshar-Kharghan, V and Li, C (1999). Kozak sequence polymorphism of glycoprotein (GP) Ibalpha gene is a major determinant of the plasma membrane levels of the platelet GP Ib-IX-V complex. *Blood*, 94:186–191.

Afshar-Kharghan, V, Matijevic-Aleksic, N, Ahn, C, Boerwinkle, E, Wu, KK, and Lopez, JA (2004). The variable number of tandem repeat polymorphism of platelet glycoprotein Ibalpha and risk of coronary heart disease. *Blood*, 103:963–965.

Anderson, PD, Huizing, M, Claassen, DA, White, J, and Gahl, WA (2003). Hermansky-Pudlak syndrome type 4 (HPS-4): Clinical and molecular characteristics. *Hum Genet*, 113:10–17.

Anikster, Y, Huizing, M, White, J, Shevchenko, YO, Fitzpatrick, DL, Touchman, JW, Compton, JG, Bale, SJ, Swank, RT, Gahl, WA, et al. (2001). Mutation of a new gene causes a unique form of Hermansky-Pudlak syndrome in a genetic isolate of central Puerto Rica. *Nat Genet*, 28:376–380.

Arai, M, Yamamoto, N, Moroi, M, Akamatsu, N, Fukutake, K, and Tanoue, K (1995). Platelets with 10% of the normal amount of glycoprotein VI have an impaired response to collagen that results in a mild bleeding tendency. *Br J Haematol*, 89:124–130.

Aramaki, KM and Reiner, AP (1999). A novel isoform of platelet glycoprotein Ib alpha is prevalent in African Americans. *Am J Hematol*, 60:77–79.

Asselin, J, Gibbins, JM, Achison, M, Lee, YH, Morton, LF, Farndale, RW, Barnes, MJ, and Watson, SP (1997). A collagen-like peptide stimulates tyrosine phosphorylation of syk and phospholipase C gamma2 in platelets independent of the integrin alpha2beta1. *Blood*, 89:1235–1242.

Bachli, EB, Brack, T, Eppler, E, Stallmach, T, Trueb, RM, Huizing, M, and Gahl, WA (2004). Hermansky-Pudlak syndrome type 4 in a patient from Sri Lanka with pulmonary fibrosis. *Am J Med Genet A*, 127:201–207.

Bahou, WF (2002). Attacked from within, blood thins. *Nat Med*, 8:1082–1083.

Bahou, WF (2003). Protease-activated receptors. *Curr Top Dev Biol*, 54:343–369.

Bahou, WF and Gnatenko, DV (2004). Platelet transcriptome: The application of microarray analysis to platelets. *Semin Thromb Hemost*, 30:473–484.

Bahou, WF, Potter, CL, and Mirza, H (1994). The VLA-2 (alpha 2 beta 1) I domain functions as a ligand-specific recognition sequence for endothelial cell attachment and spreading: Molecular and functional characterization. *Blood*, 84:3734–3741.

Bahou, WF, Scudder, L, Rubenstein, D, and Jesty, J (2004). A shear-restricted pathway of platelet procoagulant activity is regulated by IQGAP1. *J Biol Chem*, 279:22571–22577.

Ballmaier, M, Germeshausen, M, Schulze, H, Cherkaoui, K, Lang, S, Gaudig, A, Krukemeier, S, Eilers, M, Strauss, G, and Welte, K (2001). c-mpl mutations are the cause of congenital amegakaryocytic thrombocytopenia. *Blood*, 97:139–146.

Barbosa, MD, Barrat, FJ, Tchernev, VT, Nguyen, QA, Mishra, VS, Colman, SD, Pastural, E, Dufourcq-Lagelouse, R, Fischer, A, Holcombe, RF, et al. (1997). Identification of mutations in two major mRNA isoforms of the Chediak-Higashi syndrome gene in human and mouse. *Hum Mol Genet*, 6:1091–1098.

Baumgartner, HR, Turitto, V, and Weiss, HJ (1980). Effect of shear rate on platelet interaction with subendothelium in citrated and native blood. II. Relationships among platelet adhesion, thrombus dimensions, and fibrin formation. *J Lab Clin Med*, 95:208–221.

Behnke, O (1967). Incomplete microtubules observed in mammalian blood platelets during microtubule polymerization. *J Cell Biol*, 34:697–701.

Berridge, MJ (1987). Inositol trisphosphate and diacylglycerol: Two interacting second messengers. *Annu Rev Biochem*, 56:159–193.

Best, D, Senis, YA, Jarvis, GE, Eagleton, HJ, Roberts, DJ, Saito, T, Jung, SM, Moroi, M, Harrison, P, Green, FR, et al. (2003). GPVI levels in platelets: Relationship to platelet function at high shear. *Blood*, 102:2811–2818.

Blanchette, VS, Johnson, J, and Rand, M (2000). The management of alloimmune neonatal thrombocytopenia. *Baillieres Best Pract Res Clin Haematol*, 13:365–390.

Boekholdt, SM, Peters, RJ, de Maat, MP, Zwinderman, AH, van Der Wall, EE, Reitsma, PH, Jukema, JW, and Kastelein, JJ (2004). Interaction between a genetic variant of the platelet fibrinogen receptor and fibrinogen levels in determining the risk of cardiovascular events. *Am Heart J*, 147:181–186.

Bouillet, P, Metcalf, D, Huang, DC, Tarlinton, DM, Kay, TW, Kontgen, F, Adams, JM, and Strasser, A (1999). Proapoptotic Bcl-2 relative Bim required for certain apoptotic responses, leukocyte homeostasis, and to preclude autoimmunity. *Science*, 286:1735–1738.

Bray, PF, Jin, Y, and Kickler, T (1994). Rapid genotyping of the five major platelet alloantigens by reverse dot-blot hybridization. *Blood*, 84:4361–4367.

Breton-Gorius, J, Favier, R, Guichard, J, Cherif, D, Berger, R, Debili, N, Vainchenker, W, and Douay, L (1995). A new congenital dysmegakaryopoietic thrombocytopenia (Paris-Trousseau) associated with giant platelet alpha-granules and chromosome 11 deletion at 11q23. *Blood*, 85:1805–1814.

Bugert, P, Dugrillon, A, Gunaydin, A, Eichler, H, and Kluter, H (2003). Messenger RNA profiling of human platelets by microarray hybridization. *Thromb Haemost*, 90:738–748.

Bussel, JB, Zabusky, MR, Berkowitz, RL, and McFarland, JG (1997). Fetal alloimmune thrombocytopenia. *N Engl J Med*, 337:22–26.

Calderwood, DA, Zent, R, Grant, R, Rees, DJ, Hynes, RO, and Ginsberg, MH (1999). The Talin head domain binds to integrin beta subunit cytoplasmic tails and regulates integrin activation. *J Biol Chem*, 274:28071–28074.

Carlsson, L, Santoso, S, Spitzer, C, Kessler, C, and Greinacher, A (1999). The alpha2 gene coding sequence T807/A873 of the platelet collagen receptor integrin alpha2 beta1 might be a genetic risk factor for the development of stroke in younger patients. *Blood*, 93:3583–3586.

Casorelli, I, De Stefano, V, Leone, AM, Chiusolo, P, Burzotta, F, Paciaroni, K, Rossi, E, Andreotti, F, Leone, G, and Maseri, A (2001). The C807T/G873A polymorphism in the platelet glycoprotein Ia gene and the risk of acute coronary syndrome in the Italian population. *Br J Haematol*, 114:150–154.

Cattaneo, M, Lecchi, A, Randi, AM, McGregor, JL, and Mannucci, PM (1992). Identification of a new congenital defect of platelet function characterized by severe impairment of platelet responses to adenosine diphosphate. *Blood*, 80:2787–2796.

Cattaneo, M, Zighetti, ML, Lombardi, R, Martinez, C, Lecchi, A, Conley, PB, Ware, J, and Ruggeri, ZM (2003). Molecular bases of defective signal transduction in the platelet P2Y12 receptor of a patient with congenital bleeding. *Proc Natl Acad Sci USA*, 100:1978–1983.

Chen, H and Kahn, ML (2003). Reciprocal signaling by integrin and nonintegrin receptors during collagen activation of platelets. *Mol Cell Biol*, 23:4764–4777.

Chen, H, Locke, D, Liu, Y, Liu, C, and Kahn, ML (2002). The platelet receptor GPVI mediates both adhesion and signaling responses to collagen in a receptor density-dependent fashion. *J Biol Chem*, 277:3011–3019.

Choi, ES, Hokom, MM, Chen, JL, Skrine, J, Faust, J, Nichol, J, and Hunt, P (1996). The role of megakaryocyte growth and development factor in terminal stages of thrombopoiesis. *Br J Haematol*, 95:227–233.

Clarke, MC, Savill, J, Jones, DB, Noble, BS, and Brown, SB (2003). Compartmentalized megakaryocyte death generates functional platelets committed to caspase-independent death. *J Cell Biol*, 160:577–587.

Clemetson, JM, Polgar, J, Magnenat, E, Wells, TN, and Clemetson, KJ (1999). The platelet collagen receptor glycoprotein VI is a member of the immunoglobulin superfamily closely related to FcalphaR and the natural killer receptors. *J Biol Chem*, 274:29019–29024.

Coller, BS, Cheresh, DA, Asch, E, and Seligsohn, U (1991). Platelet vitronectin receptor expression differentiates Iraqi-Jewish from Arab patients with Glanzmann thrombasthenia in Israel. *Blood*, 77:75–83.

Conley, PB and Delaney, SM (2003). Scientific and therapeutic insights into the role of the platelet P2Y12 receptor in thrombosis. *Curr Opin Hematol*, 10:333–338.

Cooney, KA, Nichols, WC, Bruck, E, Bahou, WF, Shapiro, AD, Bowie, EJW, Gralnick, HR, and Ginsburg, D (1991). The molecular defect in type IIB von Willebrand disease. Identification of four potential missense mutations within the putative GpIb binding domain. *J Clin Invest*, 87:1227–1233.

Coppinger, JA, Cagney, G, Toomey, S, Kislinger, T, Belton, O, McRedmond, JP, Cahill, DJ, Emili, A, Fitzgerald, DJ, and Maguire, PB (2004). Characterization of the proteins released from activated platelets leads to localization of novel platelet proteins in human atherosclerotic lesions. *Blood*, 103:2096–2104.

Corash, L, Levin, J, Mok, Y, Baker, G, and McDowell, J (1989). Measurement of megakaryocyte frequency and ploidy distribution in unfractionated murine bone marrow. *Exp Hematol*, 17:278–286.

Corral, J, Gonzalez-Conejero, R, Lozano, ML, Rivera, J, and Vicente, V (1998). New alleles of the platelet glycoprotein Ibalpha gene. *Br J Haematol*, 103:997–1003.

Corral, J, Lozano, ML, Gonzalez-Conejero, R, Martinez, C, Iniesta, JA, Rivera, J, and Vicente, V (2000). A common polymorphism flanking the ATG initiator codon of GPIb alpha does not affect expression and is not a major risk factor for arterial thrombosis. *Thromb Haemost*, 83:23–28.

Corthals, G, Wasinger, V, Hochstrasser, D, and Sanchez, J (2000). The dynamic range of protein expression: A challenge for proteomic research. *Electrophoresis*, 21:1104–1115.

Croft, SA, Samani, NJ, Teare, MD, Hampton, KK, Steeds, RP, Channer, KS, and Daly, ME (2001). Novel platelet membrane glycoprotein VI dimorphism is a risk factor for myocardial infarction. *Circulation*, 104:1459–1463.

Cutler, L, Rodan, G, and Feinstein, MB (1978). Cytochemical localization of adenylate cyclase and of calcium ion, magnesium ion-activated ATPases in the dense tubular system of human blood platelets. *Biochim Biophys Acta*, 542:357–371.

Daimon, T and Gotoh, Y (1982). Cytochemical evidence of the origin of the dense tubular system in the mouse platelet. *Histochemistry*, 76:189–196.

De Botton, S, Sabri, S, Daugas, E, Zermati, Y, Guidotti, JE, Hermine, O, Kroemer, G, Vainchenker, W, and Debili, N (2002). Platelet formation is the consequence of caspase activation within megakaryocytes. *Blood*, 100:1310–1317.

Dell'Angelica, EC, Shotelersuk, V, Aguilar, RC, Gahl, WA, and Bonifacino, JS (1999). Altered trafficking of lysosomal proteins in Hermansky-Pudlak syndrome due to mutations in the beta 3A subunit of the AP-3 adaptor. *Mol Cell*, 3:11–21.

Deutsch, S, Rideau, A, Bochaton-Piallat, ML, Merla, G, Geinoz, A, Gabbiani, G, Schwede, T, Matthes, T, Antonarakis, SE, and Beris, P (2003). Asp1424Asn MYH9 mutation results in an unstable protein responsible for the phenotypes in May-Hegglin anomaly/Fechtner syndrome. *Blood*, 102:529–534.

Di Castelnuovo, A, de Gaetano, G, Donati, MB, and Iacoviello, L (2001). Platelet glycoprotein receptor IIIa polymorphism PLA1/PLA2 and coronary risk: A meta-analysis. *Thromb Haemost*, 85:626–633.

Diaz-Ricart, M, Tandon, NN, Carretero, M, Ordinas, A, Bastida, E, and Jamieson, GA (1993). Platelets lacking functional CD36 (glycoprotein IV) show reduced adhesion to collagen in flowing whole blood. *Blood*, 82:491–496.

Djaffar, I, Caen, JP, and Rosa, JP (1993). A large alteration in the human platelet glycoprotein IIIa (integrin beta 3) gene associated with Glanzmann's thrombasthenia. *Hum Mol Genet*, 2:2183–2185.

Djaffar, I and Rosa, JP (1993). A second case of variant of Glanzmann's thrombasthenia due to substitution of platelet GPIIIa (integrin beta 3) Arg214 by Trp. *Hum Mol Genet*, 2:2179–2180.

Dormann, D, Clemetson, KJ, and Kehrel, BE (2000). The GPIb thrombin-binding site is essential for thrombin-induced platelet procoagulant activity. *Blood*, 96:2469–2478.

Dorsam, RT and Kunapuli, SP (2004). Central role of the P2Y12 receptor in platelet activation. *J Clin Invest*, 113:340–345.

Drachman, JG, Jarvik, GP, and Mehaffey, MG (2000). Autosomal dominant thrombocytopenia: Incomplete megakaryocyte differentiation and linkage to human chromosome 10. *Blood*, 96:118–125.

Eisbacher, M, Holmes, ML, Newton, A, Hogg, PJ, Khachigian, LM, Crossley, M, and Chong, BH (2003). Protein-protein interaction between Fli-1 and GATA-1 mediates synergistic expression of megakaryocyte-specific genes through cooperative DNA binding. *Mol Cell Biol*, 23:3427–3441.

Fabris, F, Cordiano, I, Salvan, F, Ramon, R, Valente, M, Luzzatto, G, and Girolami, A (1997). Chronic isolated macrothrombocytopenia with autosomal dominant transmission: A morphological and qualitative platelet disorder. *Eur J Haematol*, 58:40–45.

Fauvel, F, Grant, ME, Legrand, YJ, Souchon, H, Tobelem, G, Jackson, DS, and Caen, JP (1983). Interaction of blood platelets with a microfibrillar extract from adult bovine aorta: Requirement for von Willebrand factor. *Proc Natl Acad Sci USA*, 80:551–554.

Fink, L, Holschermann, H, Kwapiszewska, G, Muyal, JP, Lengemann, B, Bohle, RM, and Santoso, S (2003). Characterization of platelet-specific mRNA by real-time PCR after laser-assisted microdissection. *Thromb Haemost*, 90:749–756.

Fox, JE, Boyles, JK, Berndt, MC, Steffen, PK, and Anderson, LK (1988). Identification of a membrane skeleton in platelets. *J Cell Biol*, 106:1525–1538.

Frank, MB, Reiner, AP, Schwartz, SM, Kumar, PN, Pearce, RM, Arbogast, PG, Longstreth, WT, Jr, Rosendaal, FR, Psaty, BM, and Siscovick, DS (2001). The Kozak sequence polymorphism of platelet glycoprotein Ibalpha and risk of nonfatal myocardial infarction and nonfatal stroke in young women. *Blood*, 97:875–879.

Freson, K, Devriendt, K, Matthijs, G, Van Hoof, A, De Vos, R, Thys, C, Minner, K, Hoylaerts, MF, Vermylen, J, and Van Geet, C (2001). Platelet characteristics in patients with X-linked macrothrombocytopenia because of a novel GATA1 mutation. *Blood*, 98:85–92.

Freson, K, Hoylaerts, MF, Jaeken, J, Eyssen, M, Arnout, J, Vermylen, J, and Van Geet, C (2001). Genetic variation of the extra-large stimulatory G protein alpha-subunit leads to Gs hyperfunction in platelets and is a risk factor for bleeding. *Thromb Haemost*, 86:733–738.

Freson, K, Jaeken, J, Van Helvoirt, M, de Zegher, F, Wittevrongel, C, Thys, C, Hoylaerts, MF, Vermylen, J, and Van Geet, C (2003). Functional polymorphisms in the paternally expressed XLalphas and its cofactor ALEX decrease their mutual interaction and enhance receptor-mediated cAMP formation. *Hum Mol Genet*, 12:1121–1130.

Freson, K, Matthijs, G, Thys, C, Marien, P, Hoylaerts, MF, Vermylen, J, and Van Geet, C (2002). Different substitutions at residue D218 of the X-linked transcription factor GATA1 lead to altered clinical severity of macrothrombocytopenia and anemia and are associated with variable skewed X inactivation. *Hum Mol Genet*, 11:147–152.

Fuse, I, Mito, M, Hattori, A, Higuchi, W, Shibata, A, Ushikubi, F, Okuma, M, and Yahata, K (1993). Defective signal transduction induced by thromboxane A2 in a patient with a mild bleeding disorder: Impaired phospholipase C activation despite normal phospholipase A2 activation. *Blood*, 81:994–1000.

Gabbeta, J, Yang, X, Kowalska, MA, Sun, L, Dhanasekaran, N, and Rao, AK (1997). Platelet signal transduction defect with Galpha subunit dysfunction and diminished Galphaq in a patient with abnormal platelet responses. *Proc Natl Acad Sci USA*, 94:8750–8755.

Gandhi, MJ, Cummings, CL, and Drachman, JG (2003). FLJ14813 missense mutation: A candidate for autosomal dominant thrombocytopenia on human chromosome 10. *Hum Hered*, 55:66–70.

Gebrane-Younes, J, Cramer, EM, Orcel, L, and Caen, JP (1993). Gray platelet syndrome: Dissociation between abnormal sorting in megakaryocyte alpha-granules and normal sorting in Weibel-Palade bodies of endothelial cells. *J Clin Invest*, 92:3023–3028.

Geng, Y, Yu, Q, Sicinska, E, Das, M, Schneider, JE, Bhattacharya, S, Rideout, WM, Bronson, RT, Gardner, H, and Sicinski, P (2003). Cyclin E ablation in the mouse. *Cell*, 114:431–443.

Gerrard, JM, White, JG, and Peterson, DA (1978). The platelet dense tubular system: Its relationship to prostaglandin synthesis and calcium flux. *Thromb Haemost*, 40:224–231.

Gnatenko, DV, Cupit, LD, Huang, EC, Dhundale, A, Perrotta, PL, and Bahou, WF (2005). Platelets express steroidogenic 17beta-hydroxysteroid dehydrogenases. Distinct profiles predict the essential thrombocythemic phenotype. *Thromb Haemost*, 94:412–421.

Gnatenko, DV, Dunn, JJ, McCorkle, SR, Weissmann, D, Perrotta, PL, and Bahou, WF (2003). Transcript profiling of human platelets using microarray and serial analysis of gene expression. *Blood*, 101:2285–2293.

Gonzalez-Conejero, R, Lozano, ML, Rivera, J, Corral, J, Iniesta, JA, Moraleda, JM, and Vicente, V (1998). Polymorphisms of platelet membrane glycoprotein Ib associated with arterial thrombotic disease. *Blood*, 92: 2771–2776.

Greenhalgh, KL, Howell, RT, Bottani, A, Ancliff, PJ, Brunner, HG, Verschuuren-Bemelmans, CC, Vernon, E, Brown, KW, and Newbury-Ecob, RA (2002). Thrombocytopenia-absent radius syndrome: A clinical genetic study. *J Med Genet*, 39:876–881.

Grottum, KA, Hovig, T, Holmsen, H, Abrahamsen, AF, Jeremic, M, and Seip, M (1969). Wiskott-Aldrich syndrome: Qualitative platelet defects and short platelet survival. *Br J Haematol*, 17:373–388.

Guerois, G, Gruel, Y, Petit, A, Kaplan, C, Champeix, P, Delahousse, B, and Leroy, J (1988). Familial macrothrombocytopenia. Clinical, ultrastructural and biochemical study. *Curr Stud Hematol Blood Transfus*, 55:153–161.

Gunay-Aygun, M, Huizing, M, and Gahl, WA (2004). Molecular defects that affect platelet dense granules. *Semin Thromb Hemost*, 30:537–547.

Gygi, S, Rist, B, Gerber, SA, Turecek, F, Gelb, MH, and Aebersold, R (1999). Quantative analysis of complex protein mixtures using isotope-coded affinity tags. *Nat Biotechnol*, 17:994–999.

Gygi, S, Rochon, Y, Franza, B, and Aebersold, R (1999). Correlation between protein and mRNA abundance in yeast. *Mol Cell Biol*, 19:1720–1730.

Handin, R (1995). Platelet membrane proteins and their disorders. In R Handin, SE Lux, and T Stossel (eds), *Blood: Principles and Practice of Hematology* (pp. 1049–1068). Philadelphia: JB Lippincott.

Hartwig, J and Italiano, J, Jr (2003). The birth of the platelet. *J Thromb Haemost*, 1:1580–1586.

Hayward, CP, Cramer, EM, Kane, WH, Zheng, S, Bouchard, M, Masse, JM, and Rivard, GE (1997). Studies of a second family with the Quebec platelet disorder: Evidence that the degradation of the alpha-granule membrane and its soluble contents are not secondary to a defect in targeting proteins to alpha-granules. *Blood*, 89:1243–1253.

Hirata, T, Ushikubi, F, Kakizuka, A, Okuma, M, and Narumiya, S (1996). Two thromboxane A2 receptor isoforms in human platelets: Opposite coupling to adenylyl cyclase with different sensitivity to Arg60 to Leu mutation. *J Clin Invest*, 97:949–956.

Hollopeter, G, Jantzen, HM, Vincent, D, Li, G, England, L, Ramakrishnan, V, Yang, RB, Nurden, P, Nurden, A, Julius, D, et al. (2001). Identification of the platelet ADP receptor targeted by antithrombotic drugs. *Nature*, 409:202–207.

Homans, AC, Cohen, JL, and Mazur, EM (1988). Defective megakaryocytopoiesis in the syndrome of thrombocytopenia with absent radii. *Br J Haematol*, 70:205–210.

Horikawa, T, Araki, K, Fukai, K, Ueda, M, Ueda, T, Ito, S, and Ichihashi, M (2000). Heterozygous HPS1 mutations in a case of Hermansky-Pudlak syndrome with giant melanosomes. *Br J Dermatol*, 143:635–640.

Horne, WC, Norman, NE, Schwartz, DB, and Simons, ER (1981). Changes in cytoplasmic pH and in membrane potential in thrombin-stimulated human platelets. *Eur J Biochem*, 120:295–302.

Hu, A, Wang, F, and Sellers, JR (2002). Mutations in human nonmuscle myosin IIA found in patients with May-Hegglin anomaly and Fechtner syndrome result in impaired enzymatic function. *J Biol Chem*, 277:46512–46517.

Huizing, M, Anikster, Y, Fitzpatrick, DL, Jeong, AB, D'Souza, M, Rausche, M, Toro, JR, Kaiser-Kupfer, MI, White, JG, and Gahl, WA (2001). Hermansky-Pudlak syndrome type 3 in Ashkenazi Jews and other non-Puerto Rican patients with hypopigmentation and platelet storage-pool deficiency. *Am J Hum Genet*, 69:1022–1032.

Ihara, K, Ishii, E, Eguchi, M, Takada, H, Suminoe, A, Good, RA, and Hara, T (1999). Identification of mutations in the c-mpl gene in congenital amegakaryocytic thrombocytopenia. *Proc Natl Acad Sci USA*, 96:3132–3136.

Ishida, F, Furihata, K, Ishida, K, Yan, J, Kitano, K, Kiyosawa, K, and Furuta, S (1995). The largest variant of platelet glycoprotein Ib alpha has four tandem repeats of 13 amino acids in the macroglycopeptide region and a genetic linkage with methionine145. *Blood*, 86:1357–1360.

Italiano, JE, Jr, Lecine, P, Shivdasani, RA, and Hartwig, JH (1999). Blood platelets are assembled principally at the ends of proplatelet processes produced by differentiated megakaryocytes. *J Cell Biol*, 147:1299–1312.

Jacquelin, B, Tarantino, MD, Kritzik, M, Rozenshteyn, D, Koziol, JA, Nurden, AT, and Kunicki, TJ (2001). Allele-dependent transcriptional regulation of the human integrin alpha2 gene. *Blood*, 97:1721–1726.

Jandrot-Perrus, M, Busfield, S, Lagrue, AH, Xiong, X, Debili, N, Chickering, T, Le Couedic, JP, Goodearl, A, Dussault, B, Fraser, C, et al. (2000). Cloning, characterization, and functional studies of human and mouse glycoprotein VI: A platelet-specific collagen receptor from the immunoglobulin superfamily. *Blood*, 96:1798–1807.

Joutsi-Korhonen, L, Smethurst, PA, Rankin, A, Gray, E, IJsseldijk, M, Onley, CM, Watkins, NA, Williamson, LM, Goodall, AH, de Groot, PG, et al. (2003). The low-frequency allele of the platelet collagen signaling receptor glycoprotein VI is associated with reduced functional responses and expression. *Blood*, 101:4372–4379.

Kahn, ML (2004). Platelet-collagen responses: Molecular basis and therapeutic promise. *Semin Thromb Hemost*, 30:419–425.

Kahr, WH, Zheng, S, Sheth, PM, Pai, M, Cowie, A, Bouchard, M, Podor, TJ, Rivard, GE, and Hayward, CP (2001). Platelets from patients with the Quebec platelet disorder contain and secrete abnormal amounts of urokinase-type plasminogen activator. *Blood*, 98:257–265.

Karim, MA, Nagle, DL, Kandil, HH, Burger, J, Moore, KJ, and Spritz, RA (1997). Mutations in the Chediak-Higashi syndrome gene (CHS1) indicate requirement for the complete 3801 amino acid CHS protein. *Hum Mol Genet*, 6:1087–1089.

Karim, MA, Suzuki, K, Fukai, K, Oh, J, Nagle, DL, Moore, KJ, Barbosa, E, Falik-Borenstein, T, Filipovich, A, Ishida, Y, et al. (2002). Apparent genotype-phenotype correlation in childhood, adolescent, and adult Chediak-Higashi syndrome. *Am J Med Genet*, 108:16–22.

Kashiwagi, H, Honda, S, Take, H, Mizutani, H, Imai, Y, Furubayashi, T, Tomiyama, Y, Kurata, Y, and Yonezawa, T (1993). Presence of the entire coding region of GP IV mRNA in Nak(a)-negative platelets. *Int J Hematol*, 57:153–161.

Kashiwagi, H, Honda, S, Tomiyama, Y, Mizutani, H, Take, H, Honda, Y, Kosugi, S, Kanayama, Y, Kurata, Y, and Matsuzawa, Y (1993). A novel polymorphism in glycoprotein IV (replacement of proline-90 by serine) predominates in subjects with platelet GPIV deficiency. *Thromb Haemost*, 69:481–484.

Kashiwagi, H, Schwartz, MA, Eigenthaler, M, Davis, KA, Ginsberg, MH, and Shattil, SJ (1997). Affinity modulation of platelet integrin alphaIIbbeta3 by beta3-endonexin, a selective binding partner of the beta3 integrin cytoplasmic tail. *J Cell Biol*, 137:1433–1443.

Kashiwagi, H, Tomiyama, Y, Honda, S, Kosugi, S, Shiraga, M, Nagao, N, Sekiguchi, S, Kanayama, Y, Kurata, Y, and Matsuzawa, Y (1995). Molecular basis of CD36 deficiency. Evidence that a 478 C→T substitution (proline90→serine) in CD36 cDNA accounts for CD36 deficiency. *J Clin Invest*, 95:1040–1046.

Kashiwagi, H, Tomiyama, Y, Nozaki, S, Kiyoi, T, Tadokoro, S, Matsumoto, K, Honda, S, Kosugi, S, Kurata, Y, and Matsuzawa, Y (2001). Analyses of genetic abnormalities in type I CD36 deficiency in Japan: Identification and cell biological characterization of two novel mutations that cause CD36 deficiency in man. *Hum Genet*, 108:459–466.

Kaushansky, K (1995). Thrombopoietin: The primary regulator of platelet production. *Blood*, 86:419.

Khouzami, AN, Kickler, TS, Bray, PF, Callan, NA, Sciscione, AC, Shumway, JB, and Blakemore, KJ (1995). Molecular genotyping of fetal platelet antigens with uncultured amniocytes. *Am J Obstet Gynecol*, 173:1202–1206.

Kozak, M (1989). An analysis of 5′-noncoding sequences from 699 vertegrate messenger RNA's. *Nucleic Acids Res*, 15:8125–8148.

Krishnamurti, L, Neglia, JP, Nagarajan, R, Berry, SA, Lohr, J, Hirsch, B, and White, JG (2001). Paris-Trousseau syndrome platelets in a child with Jacobsen's syndrome. *Am J Hematol*, 66:295–299.

Kritzik, M, Savage, B, Nugent, DJ, Santoso, S, Ruggeri, ZM, and Kunicki, TJ (1998). Nucleotide polymorphisms in the alpha2 gene define multiple alleles that are associated with differences in platelet alpha2 beta1 density. *Blood*, 92:2382–2388.

Kuijpers, RW, Ouwehand, WH, Bleeker, PM, Christie, D, and von dem Borne, AE (1992). Localization of the platelet-specific HPA-2 (Ko) alloantigens on the N-terminal globular fragment of platelet glycoprotein Ib alpha. *Blood*, 79:283–288.

Kunicki, TJ, Kritzik, M, Annis, DS, and Nugent, DJ (1997). Hereditary variation in platelet integrin alpha 2 beta 1 density is associated with two silent polymorphisms in the alpha 2 gene coding sequence. *Blood*, 89:1939–1943.

Kunishima, S, Matsushita, T, Kojima, T, Sako, M, Kimura, F, Jo, EK, Inoue, C, Kamiya, T, and Saito, H (2003). Immunofluorescence analysis of neutrophil nonmuscle myosin heavy chain-A in MYH9 disorders: Association of subcellular localization with MYH9 mutations. *Lab Invest*, 83:115–122.

L'Abbe, D, Tremblay, L, Filion, M, Busque, L, Goldman, M, Decary, F, and Chartrand, P (1992). Alloimmunization to platelet antigen HPA-1a (PIA1) is strongly associated with both HLA-DRB3*0101 and HLA-DQB1*0201. *Hum Immunol*, 34:107–114.

Lau, P, Sholtis, CM, and Aster, RH (1980). Post-transfusion purpura: An enigma of alloimmunization. *Am J Hematol*, 9:331–336.

Levy-Toledano, S, Maclouf, J, Rosa, JP, Gallet, C, Valles, G, Nurden, P, and Nurden, AT (1998). Abnormal tyrosine phosphorylation linked to a defective interaction between ADP and its receptor on platelets. *Thromb Haemost*, 80:463–468.

Li, W, Zhang, Q, Oiso, N, Novak, EK, Gautam, R, O'Brien, EP, Tinsley, CL, Blake, DJ, Spritz, RA, Copeland, NG, et al. (2003). Hermansky-Pudlak syndrome type 7 (HPS-7) results from mutant dysbindin, a member of the biogenesis of lysosome-related organelles complex 1 (BLOC-1). *Nat Genet*, 35:84–89.

Link, AJ, Eng, J, Schieltz, DM, Carmack, E, Mize, GJ, Morris, DR, Garvik, BM, Yates, JR. (1999). Direct analysis of protein complexes using mass spectrometry. *Nat Biotechnol*, 17:676–682.

Lopez, JA, Andrews, RK, Afshar-Kharghan, V, and Berndt, MC (1998). Bernard-Soulier syndrome. *Blood*, 91:4397–4418.

Lopez, JA, Ludwig, EH, and McCarthy, BJ (1992). Polymorphism of human glycoprotein Ib alpha results from a variable number of tandem repeats of a 13-amino acid sequence in the mucin-like macroglycopeptide region. Structure/function implications. *J Biol Chem*, 267:10055–10061.

Maguire, PB (2003). Platelet proteomics: Identification of potential therapeutic targets. *Pathophysiol Haemost Thromb*, 33:481–486.

Majado, MJ, Gonzalez, C, Tamayo, M, Sanchez, A, and Moreno, M (1992). Effective splenectomy in familial isolated thrombocytopenia. *Am J Hematol*, 39:70.

Majerus, PW, Connolly, TM, Deckmyn, H, Ross, TS, Bross, TE, Ishii, H, Bansal, VS, and Wilson, DB (1986). The metabolism of phosphoinositide-derived messenger molecules. *Science*, 234:1519–1526.

Marcus, K, Immler, D, Sternberger, J, and Meyer, H (2000). Identification of platelet proteins separated by two-dimensional gel electrophoresis and analyzed by matrix assisted laser desorption/ionization-time of flight-mass spectrometry and detection of tyrosine-phosphorylated proteins. *Electrophoresis*, 21:2622–2636.

Marenberg, ME, Risch, N, Berkman, LF, Floderus, B, and de Faire, U (1994). Genetic susceptibility to death from coronary heart disease in a study of twins. *N Engl J Med*, 330:1041–1046.

Matsumura, I, Tanaka, H, Kawasaki, A, Odajima, J, Daino, H, Hashimoto, K, Wakao, H, Nakajima, K, Kato, T, Miyazaki, H, et al. (2000). Increased D-type cyclin expression together with decreased cdc2 activity confers megakaryocytic differentiation of a human thrombopoietin-dependent hematopoietic cell line. *J Biol Chem*, 275:5553–5559.

Mazzucato, M, DeMarco, L, Masotti, A, Pradella, P, Bahou, WF, and Ruggeri, ZM (1998). Characterization of the initial α-thrombin interaction with glycoprotein Ibα in relation to platelet activation. *J BiolChem*, 273:1880–1887.

McKay, H, Derome, F, Haq, MA, Whittaker, S, Arnold, E, Adam, F, Heddle, NM, Rivard, GE, and Hayward, CP (2004). Bleeding risks associated with inheritance of the Quebec platelet disorder. *Blood*, 104:159–165.

McKeown, L, Vail, M, Williams, S, Kramer, W, Hansmann, K, and Gralnick, H (1994). Platelet adhesion to collagen in individuals lacking glycoprotein IV. *Blood*, 83:2866–2871.

McNicol, A and Israels, SJ (1999). Platelet dense granules: Structure, function and implications for haemostasis. *Thromb Res*, 95:1–18.

McRedmond, JP, Park, SD, Reilly, DF, Coppinger, JA, Maguire, PB, Shields, DC, and Fitzgerald, DJ (2004). Integration of proteomics and genomics in platelets: A profile of platelet proteins and platelet-specific genes. *Mol Cell Proteomics*, 3:133–144.

Mikkelsson, J, Perola, M, Kauppila, LI, Laippala, P, Savolainen, V, Pajarinen, J, Penttila, A, and Karhunen, PJ (1999). The GPIIIa Pl(A) polymorphism in the progression of abdominal aortic atherosclerosis. *Atherosclerosis*, 147:55–60.

Mikkelsson, J, Perola, M, Laippala, P, Savolainen, V, Pajarinen, J, Lalu, K, Penttila, A, and Karhunen, PJ (1999). Glycoprotein IIIa Pl(A) polymorphism associates with progression of coronary artery disease and with myocardial infarction in an autopsy series of middle-aged men who died suddenly. *Arterioscler Thromb Vasc Biol*, 19:2573–2578.

Miller, JL, Cunningham, D, Lyle, VA, and Finch, CN (1991). Mutation in the gene encoding the alpha chain of platelet glycoprotein Ib in platelet-type von Willebrand disease. *Proc Natl Acad Sci USA*, 88:4761–4765.

Minelli, A, Maserati, E, Rossi, G, Bernardo, ME, De Stefano, P, Cecchini, MP, Valli, R, Albano, V, Pierani, P, Leszl, A, et al. (2004). Familial platelet disorder with propensity to acute myelogenous leukemia: Genetic heterogeneity and progression to leukemia via acquisition of clonal chromosome anomalies. *Genes Chromosomes Cancer*, 40:165–171.

Miyamoto, S, Teramoto, H, Coso, OA, Gutkind, JS, Burbelo, PD, Akiyama, SK, and Yamada, KM (1995). Integrin function: Molecular hierarchies of cytoskeletal and signaling molecules. *J Cell Biol*, 131:791–805.

Miyaoka, K, Kuwasako, T, Hirano, K, Nozaki, S, Yamashita, S, and Matsuzawa, Y (2001). CD36 deficiency associated with insulin resistance. *Lancet*, 357:686–687.

Molloy, M (2000). Two-dimensional electrophoresis of membrane proteins using immobilized pH gradients. *Anal Biochem*, 280:1–10.

Monnet, E and Fauvel-Lafeve, F (2000). A new platelet receptor specific to type III collagen. Type III collagen-binding protein. *J Biol Chem*, 275:10912–10917.

Moog, S, Mangin, P, Lenain, N, Strassel, C, Ravanat, C, Schuhler, S, Freund, M, Santer, M, Kahn, M, Nieswandt, B, et al. (2001). Platelet glycoprotein V binds to collagen and participates in platelet adhesion and aggregation. *Blood*, 98:1038–1046.

Moroi, M, Jung, SM, Okuma, M, and Shinmyozu, K (1989). A patient with platelets deficient in glycoprotein VI that lack both collagen-induced aggregation and adhesion. *J Clin Invest*, 84:1440–1445.

Moroi, M, Jung, SM, and Yoshida, N (1984). Genetic polymorphism of platelet glycoprotein Ib. *Blood*, 64:622–629.

Moshfegh, K, Wiullemin, W, Redondo, M, Lammle, B, Beer, J, Liechti-Gallati, S, and Meyer, B (1999). Association of two silent polymorphisms of platelet glycoprotein Ia/IIa receptor with risk of myocardial infarction: A case-control study. *Lancet*, 353:351–354.

Muraoka, K, Ishii, E, Tsuji, K, Yamamoto, S, Yamaguchi, H, Hara, T, Koga, H, Nakahata, T, and Miyazaki, S (1997). Defective response to thrombopoietin and impaired expression of c-mpl mRNA of bone marrow cells in congenital amegakaryocytic thrombocytopenia. *Br J Haematol*, 96:287–292.

Nachmias, VT, Kavaler, J, and Jacubowitz, S (1985). Reversible association of myosin with the platelet cytoskeleton. *Nature*, 313:70–72.

Nagle, DL, Karim, MA, Woolf, EA, Holmgren, L, Bork, P, Misumi, DJ, McGrail, SH, Dussault, BJ, Jr, Perou, CM, Boissy, RE, et al. (1996). Identification and mutation analysis of the complete gene for Chediak-Higashi syndrome. *Nat Genet*, 14:307–311.

Najean, Y and Lecompte, T (1990). Genetic thrombocytopenia with autosomal dominant transmission: A review of 54 cases. *Br J Haematol*, 74:203–208.

Nazarian, R, Falcon-Perez, JM, and Dell'Angelica, EC (2003). Biogenesis of lysosome-related organelles complex 3 (BLOC-3): A complex containing the Hermansky-Pudlak syndrome (HPS) proteins HPS1 and HPS4. *Proc Natl Acad Sci USA*, 100:8770–8775.

Needleman, P, Turk, J, Jakschik, BA, Morrison, AR, and Lefkowith, JB (1986). Arachidonic acid metabolism. *Annu Rev Biochem*, 55:69–102.

Newman, PJ (1994). Nomenclature of human platelet alloantigens: A problem with the HPA system? *Blood*, 83:1447–1451.

Newman, PJ, Seligsohn, U, Lyman, S, and Coller, BS (1991). The molecular genetic basis of Glanzmann thrombasthenia in the Iraqi-Jewish and Arab populations in Israel. *Proc Natl Acad Sci USA*, 88:3160–3164.

Nichols, KE, Crispino, JD, Poncz, M, White, JG, Orkin, SH, Maris, JM, and Weiss, MJ (2000). Familial dyserythropoietic anaemia and thrombocytopenia due to an inherited mutation in GATA1. *Nat Genet*, 24:266–270.

Nieuwenhuis, HK, Akkerman, J, Houdijk, W, and Sixma, J (1985). Human blood platelets showing no response to collagen failed to express surface glycoprotein Ia. *Nature*, 318:470–470.

Nieuwenhuis, HK, Akkerman, JW, and Sixma, JJ (1987). Patients with a prolonged bleeding time and normal aggregation tests may have storage pool deficiency: Studies on one hundred six patients. *Blood*, 70:620–623.

Nurden, A and Nurden, P (2002). Inherited disorders of platelet function. In AD Michelson (ed.), *Platelets* (pp. 681–700). London: Academic Press.

Nurden, AT (1999). Inherited abnormalities of platelets. *Thromb Haemost*, 82:468–480.

Nurden, AT, Didry, D, Kieffer, N, and McEver, RP (1985). Residual amounts of glycoproteins IIb and IIIa may be present in the platelets of most patients with Glanzmann's thrombasthenia. *Blood*, 65:1021–1024.

Nurden, P, Savi, P, Heilmann, E, Bihour, C, Herbert, JM, Maffrand, JP, and Nurden, A (1995). An inherited bleeding disorder linked to a defective interaction between ADP and its receptor on platelets. Its influence on glycoprotein IIb-IIIa complex function. *J Clin Invest*, 95:1612–1622.

O'Neill, EE, Brock, CJ, von Kriegsheim, AF, Pearce, AC, Dwek, RA, Watson, SP, and Hebestreit, HF (2002). Towards complete analysis of the platelet proteome. *Proteomics*, 2:288–305.

Offermanns, S, Toombs, CF, Hu, YH, and Simon, MI (1997). Defective platelet activation in G alpha(q)-deficient mice. *Nature*, 389:183–186.

Ogilvy, S, Metcalf, D, Print, CG, Bath, ML, Harris, AW, and Adams, JM (1999). Constitutive Bcl-2 expression throughout the hematopoietic compartment affects multiple lineages and enhances progenitor cell survival. *Proc Natl Acad Sci USA*, 96:14943–14948.

Oh, J, Bailin, T, Fukai, K, Feng, GH, Ho, L, Mao, JI, Frenk, E, Tamura, N, and Spritz, RA (1996). Positional cloning of a gene for Hermansky-Pudlak syndrome, a disorder of cytoplasmic organelles. *Nat Genet*, 14:300–306.

Oury, C, Toth-Zsamboki, E, Van Geet, C, Thys, C, Wei, L, Nilius, B, Vermylen, J, and Hoylaerts, MF (2000). A natural dominant negative P2X1 receptor due to deletion of a single amino acid residue. *J Biol Chem*, 275:22611–22614.

Pastinen, T, Perola, M, Niini, P, Terwilliger, J, Salomaa, V, Vartiainen, E, Peltonen, L, and Syvanen, A (1998). Array-based multiplex analysis of candidate genes reveals two independent and additive genetic risk factors for myocardial infarction in the Finnish population. *Hum Mol Genet*, 7:1453–1462.

Paulus, JM (1968). Cytophotometric measurements of DNA in thrombopoietic megakaryocytes. *Exp Cell Res*, 53:310–313.

Pecci, A, Noris, P, Invernizzi, R, Savoia, A, Seri, M, Ghiggeri, GM, Sartore, S, Gangarossa, S, Bizzaro, N, and Balduini, CL (2002). Immunocytochemistry for the heavy chain of the non-muscle myosin IIA as a diagnostic tool for MYH9-related disorders. *Br J Haematol*, 117:164–167.

Pestina, TI, Jackson, CW, and Stenberg, PE (1995). Abnormal subcellular distribution of myosin and talin in Wistar Furth rat platelets. *Blood*, 85:2436–2446.

Polgar, J and Reed, GL (1999). A critical role for N-ethylmaleimide-sensitive fusion protein (NSF) in platelet granule secretion. *Blood*, 94:1313–1318.

Pruitt, K and Maglott, DR (2001). RefSeq and LocusLink: NCBI gene-centered resources. *Nucleic Acids Res*, 29:137–140.

Raha, S and Robinson, BH (2001). Mitochondria, oxygen free radicals, and apoptosis. *Am J Med Genet*, 106:62–70.

Rao, AK, Jalagadugula, G, and Sun, L (2004). Inherited defects in platelet signaling mechanisms. *Semin Thromb Hemost*, 30:525–535.

Raslova, H, Komura, E, Le Couedic, JP, Larbret, F, Debili, N, Feunteun, J, Danos, O, Albagli, O, Vainchenker, W, and Favier, R (2004). FLI1 monoallelic expression combined with its hemizygous loss underlies Paris-Trousseau/Jacobsen thrombopenia. *J Clin Invest*, 114:77–84.

Reed, GL (2002). Platelet secretion. In AD. Michelson (ed.), *Platelets* (pp. 181–196). London: Academic Press.

Reed, GL (2003). Good breeding matters: In-bred rodents provide genetic insights into platelet secretion. *Thromb Haemost*, 89:951–952.

Reed, GL (2004). Platelet secretory mechanisms. *Semin Thromb Hemost*, 30:441–450.

Reed, GL, Fitzgerald, ML, and Polgar, J (2000). Molecular mechanisms of platelet exocytosis: Insights into the "secrete" life of thrombocytes. *Blood*, 96:3334–3342.

Reiner, AP, Aramaki, KM, Teramura, G, and Gaur, L (1998). Analysis of platelet glycoprotein Ia (alpha2 integrin) allele frequencies in three North American populations reveals genetic association between nucleotide 807C/T and amino acid 505 Glu/Lys (HPA-5) dimorphisms. *Thromb Haemost*, 80:449–456.

Reiner, AP, Kumar, PN, Schwartz, SM, Longstreth, WT, Jr, Pearce, RM, Rosendaal, FR, Psaty, BM, and Siscovick, DS (2000). Genetic variants of platelet glycoprotein receptors and risk of stroke in young women. *Stroke*, 31:1628–1633.

Rickard, SJ and Wilson, LC (2003). Analysis of GNAS1 and overlapping transcripts identifies the parental origin of mutations in patients with sporadic Albright hereditary osteodystrophy and reveals a model system in which to observe the effects of splicing mutations on translated and untranslated messenger RNA. *Am J Hum Genet*, 72:961–974.

Roest, M, Banga, JD, Grobbee, DE, de Groot, PG, Sixma, JJ, Tempelman, MJ, and van der Schouw, YT (2000). Homozygosity for 807 T polymorphism in alpha(2) subunit of platelet alpha(2)beta(1) is associated with increased risk of cardiovascular mortality in high-risk women. *Circulation*, 102:1645–1650.

Rojnuckarin, P and Kaushansky, K (2001). Actin reorganization and proplatelet formation in murine megakaryocytes: The role of protein kinase calpha. *Blood*, 97:154–161.

Rojnuckarin, P, Miyakawa, Y, Fox, NE, Deou, J, Daum, G, and Kaushansky, K (2001). The roles of phosphatidylinositol 3-kinase and protein kinase Czeta for thrombopoietin-induced mitogen-activated protein kinase activation in primary murine megakaryocytes. *J Biol Chem*, 276:41014–41022.

Roychoudhury, AK and Nei, M (1988). *Human Polymorphic Genes: World Distribution*. New York: Oxford University Press.

Safer, D, Elzinga, M, and Nachmias, VT (1991). Thymosin beta 4 and Fx, an actin-sequestering peptide, are indistinguishable. *J Biol Chem*, 266:4029–4032.

Sakariassen, KS, Bolhuis, PA, and Sixma, JJ (1979). Human blood platelet adhesion to artery subendothelium is mediated by factor VIII-Von Willebrand factor bound to the subendothelium. *Nature*, 279:636–638.

Salomon, O, Rosenberg, N, Steinberg, DM, Huna-Baron, R, Moisseiev, J, Dardik, R, Goldan, O, Kurtz, S, Ifrah, A, and Seligsohn, U (2004). Non-arteritic anterior ischemic optic neuropathy is associated with a specific platelet polymorphism located on the glycoprotein Ibalpha gene. *Ophthalmology*, 111:184–188.

Santoro, SA (1986). Identification of a 160,000 dalton platelet membrane protein that mediates the initial divalent cation-dependent adhesion of platelets to collagen. *Cell*, 46:913–920.

Savage, B, Almus-Jacobs, F, and Ruggeri, ZM (1998). Specific synergy of multiple substrate-receptor interactions in platelet thrombus formation under flow. *Cell*, 94:657–666.

Savage, B, Ginsberg, MH, and Ruggeri, ZM (1999). Influence of fibrillar collagen structure on the mechanisms of platelet thrombus formation under flow. *Blood*, 94:2704–2715.

Savoia, A, Balduini, CL, Savino, M, Noris, P, Del Vecchio, M, Perrotta, S, Belletti, S, Poggi, and Iolascon, A (2001). Autosomal dominant macrothrombocytopenia in Italy is most frequently a type of heterozygous Bernard-Soulier syndrome. *Blood*, 97:1330–1335.

Savoia, A, Del Vecchio, M, Totaro, A, Perrotta, S, Amendola, G, Moretti, A, Zelante, L, and Iolascon, A (1999). An autosomal dominant thrombocytopenia gene maps to chromosomal region 10p. *Am J Hum Genet*, 65:1401–1405.

Schlegel, N, Gayet, O, Morel-Kopp, MC, Wyler, B, Hurtaud-Roux, MF, Kaplan, C, and Mc Gregor, J (1995). The molecular genetic basis of Glanzmann's thrombasthenia in a gypsy population in France: Identification of a new mutation on the alpha IIb gene. *Blood*, 86:977–982.

Schuh, AC, Watkins, NA, Nguyen, Q, Harmer, NJ, Lin, M, Prosper, JY, Campbell, K, Sutherland, DR, Metcalfe, P, Horsfall, W, et al. (2002). A tyrosine703serine polymorphism of CD109 defines the Gov platelet alloantigens. *Blood*, 99:1692–1698.

Schulze, H and Shivdasani, RA (2004). Molecular mechanisms of megakaryocyte differentiation. *Semin Thromb Hemost*, 30:389–398.

Schwer, HD, Lecine, P, Tiwari, S, Italiano, JE, Jr, Hartwig, JH, and Shivdasani, RA (2001). A lineage-restricted and divergent beta-tubulin isoform is essential for the biogenesis, structure and function of blood platelets. *Curr Biol*, 11:579–586.

Seri, M, Pecci, A, Di Bari, F, Cusano, R, Savino, M, Panza, E, Nigro, A, Noris, P, Gangarossa, S, Rocca, B, et al. (2003). MYH9-related disease: May-Hegglin anomaly, Sebastian syndrome, Fechtner syndrome, and Epstein syndrome are not distinct entities but represent a variable expression of a single illness. *Medicine (Baltimore)*, 82:203–215.

Shattil, S (1999). Signaling through platelet integrin $\alpha IIb\beta 3$: Inside-out, outside-in, and sideways. *Thromb Haemost*, 82:318–325.

Shattil, SJ, Hoxie, JA, Cunningham, M, and Brass, LF (1985). Changes in the platelet membrane glycoprotein IIb.IIIa complex during platelet activation. *J Biol Chem*, 81:533–543.

Shaw, S and Oliver, RA (1959). Congenital hypoplastic thrombocytopenia with skeletal deformaties in siblings. *Blood*, 14:374–377.

Shim, MH, Hoover, A, Blake, N, Drachman, JG, and Reems, JA (2004). Gene expression profile of primary human CD34 + CD38lo cells differentiating along the megakaryocyte lineage. *Exp Hematol*, 32:638–648.

Shotelersuk, V, Hazelwood, S, Larson, D, Iwata, F, Kaiser-Kupfer, MI, Kuehl, E, Bernardini, I, and Gahl, WA (1998). Three new mutations in a gene causing Hermansky-Pudlak syndrome: Clinical correlations. *Mol Genet Metab*, 64:99–107.

Sims, PJ, Wiedmer, T, Esmon, CT, Weiss, HJ, and Shattil, SJ (1989). Assembly of the platelet prothrombinase complex is linked to vesiculation of the platelet plasma membrane. Studies in Scott syndrome: An isolated defect in platelet procoagulant activity. *J Biol Chem*, 264:17049–17057.

Snapper, SB and Rosen, FS (2003). A family of WASPs. *N Engl J Med*, 348:350–351.

Song, WJ, Sullivan, MG, Legare, RD, Hutchings, S, Tan, X, Kufrin, D, Ratajczak, J, Resende, IC, Haworth, C, Hock, R, et al. (1999). Haploinsufficiency of CBFA2 causes familial thrombocytopenia with propensity to develop acute myelogenous leukaemia. *Nat Genet*, 23:166–175.

Stavem, P, Abrahamsen, AF, Vartdal, F, Nordhagen, R, and Rootwelt, K (1986). Hereditary thrombocytopenia with excessively prolonged bleeding time, corrected by infusions of platelet poor plasma. *Scand J Haematol*, 37:210–214.

Stenberg, PE and Hill, RL (1999). Platelets and megakaryocytes. In G Lee, J Foerster, J Lukens, F Paraskevas, J Greer, and G Rodgers (eds), *Wintrobe's Clinical Hematology* (pp. 615–660). Philadelphia: Lippincott Williams & Wilkins.

Sullivan, KE, Mullen, CA, Blaese, RM, and Winkelstein, JA (1994). A multiinstitutional survey of the Wiskott-Aldrich syndrome. *J Pediatr*, 125:876–885.

Suzuki, H, Nakamura, S, Itoh, Y, Tanaka, T, Yamazaki, H, and Tanoue, K (1992). Immunocytochemical evidence for the translocation of alpha-granule membrane glycoprotein IIb/IIIa (integrin alpha IIb beta 3) of human platelets to the surface membrane during the release reaction. *Histochemistry*, 97:381–388.

Suzuki, T, Li, W, Zhang, Q, Karim, A, Novak, EK, Sviderskaya, EV, Hill, SP, Bennett, DC, Levin, AV, Nieuwenhuis, HK, et al. (2002). Hermansky-Pudlak syndrome is caused by mutations in HPS4, the human homolog of the mouse light-ear gene. *Nat Genet*, 30:321–324.

Takagi, S, Iwai, N, Baba, S, Mannami, T, Ono, K, Tanaka, C, Miyata, T, Miyazaki, S, Nonogi, H, and Goto, Y (2002). A GPVI polymorphism is a risk factor for myocardial infarction in Japanese. *Atherosclerosis*, 165: 397–398.

Tanaka, T, Sohmiya, K, and Kawamura, K (1997). Is CD36 deficiency an etiology of hereditary hypertrophic cardiomyopathy? *J Mol Cell Cardiol*, 29:121–127.

Tandon, NN, Kralisz, U, and Jamieson, GA (1989). Identification of glycoprotein IV (CD36) as a primary receptor for platelet-collagen adhesion. *J Biol Chem*, 264:7576–7583.

Tenedini, E, Fagioli, ME, Vianelli, N, Tazzari, PL, Ricci, F, Tagliafico, E, Ricci, P, Gugliotta, L, Martinelli, G, Tura, S, et al. (2004). Gene expression profiling of normal and malignant CD34-derived megakaryocytic cells. *Blood*, 104:3126–3135.

Thompson, AA, Woodruff, K, Feig, SA, Nguyen, LT, and Schanen, NC (2001). Congenital thrombocytopenia and radio-ulnar synostosis: A new familial syndrome. *Br J Haematol*, 113:866–870.

Tomiyama, Y, Take, H, Ikeda, H, Mitani, T, Furubayashi, T, Mizutani, H, Yamamoto, N, Tandon, NN, Sekiguchi, S, Jamieson, GA, et al. (1990). Identification of the platelet-specific alloantigen, Naka, on platelet membrane glycoprotein IV. *Blood*, 75:684–687.

Toro, J, Turner, M, and Gahl, WA (1999). Dermatologic manifestations of Hermansky-Pudlak syndrome in patients with and without a16-base pair duplication in the HPS1 gene. *Arch Dermatol*, 135:774–780.

Tsang, AP, Fujiwara, Y, Hom, DB, and Orkin, SH (1998). Failure of megakaryopoiesis and arrested erythropoiesis in mice lacking the GATA-1 transcriptional cofactor FOG. *Genes Dev*, 12:1176.

Turitto, VT, Weiss, HJ, and Baumgartner, HR (1980). The effect of shear rate on platelet interaction with subendothelium exposed to citrated human blood. *Microvasc Res*, 19:352–365.

Ulrichts, H, Vanhoorelbeke, K, Cauwenberghs, S, Vauterin, S, Kroll, H, Santoso, S, and Deckmyn, H (2003). von Willebrand factor but not alpha-thrombin binding to platelet glycoprotein Ibalpha is influenced by the HPA-2 polymorphism. *Arterioscler Thromb Vasc Biol*, 23:1302–1307.

Undas, A, Brummel, K, Musial, J, Mann, KG, and Szczeklik, A (2001). Blood coagulation at the site of microvascular injury: Effects of low-dose aspirin. *Blood*, 98:2423–2431.

Ushikubi, F, Okuma, M, Kanaji, K, Sugiyama, T, Ogorochi, T, Narumiya, S, and Uchino, H (1987). Hemorrhagic thrombocytopathy with platelet thromboxane A2 receptor abnormality: Defective signal transduction with normal binding activity. *Thromb Haemost*, 57:158–164.

Velculescu, V, Zhang, L, Vogelstein, B, and Kinzler, K (1995). Serial analysis of gene expression. *Science*, 270:484–487.

von dem Borne, AE and Decary, F (1990). ICSH/ISBT Working Party on platelet serology: Nomenclature of platelet-specific antigens. *Vox Sang*, 58:176.

Vyas, P, Ault, K, Jackson, CW, Orkin, SH, and Shivdasani, RA (1999). Consequences of GATA-1 deficiency in megakaryocytes and platelets. *Blood*, 93:2867–2875.

Wallace, DC (2001). Mouse models for mitochondrial disease. *Am J Med Genet*, 106:71–93.

Wang, Q, Stacy, T, Binder, M, Marin-Padilla, M, Sharpe, AH, and Speck, NA (1996). Disruption of the Cbfa2 gene causes necrosis and hemorrhaging in the central nervous system and blocks definitive hematopoiesis. *Proc Natl Acad Sci USA*, 93:3444–3449.

Wang, Z, Zhang, Y, Kamen, D, Lees, E, and Ravid, K (1995). Cyclin D3 is essential for megakaryocytopoiesis. *Blood*, 86:3783–3788.

Wardinsky, TD, Weinberger, E, Pagon, RA, Clarren, SK, and Thuline, HC (1990). Partial deletion of the long arm of chromosome 11 [del(11)(q23.3 → qter)] with abnormal white matter. *Am J Med Genet*, 35:60–63.

Wardle, EN (1973). Long-term blood abnormalities in thrombosis. *Lancet*, 2:133–134.

Ware, J, Russell, S, and Ruggeri, ZM (2000). Generation and rescue of a murine model of platelet dysfunction: The Bernard-Soulier syndrome. *Proc Natl Acad Sci USA*, 97:2803–2808.

Warkentin, TE and Kelton, JG (1995). The platelet life cycle: quantitative disorders. In R Handin, SE Lux, and T Stossel (eds), *Blood: Principles and Practice of Medicine* (pp. 973–1048). Philadelphia: JB Lippincott.

Weiss, EJ, Bray, PF, Tayback, M, Schulman, SP, Kickler, TS, Becker, LC, Weiss, JL, Gerstenblith, G, and Goldschmidt-Clermont, PJ (1996). A polymorphism of a platelet glycoprotein receptor as an inherited risk factor for coronary thrombosis. *N Engl J Med*, 334:1090–1094.

Welch, M, Iwamatsu, A, and Mitchison, T (1997). Actin polymerization is induced by Arp2/3 protein complex at the surface of Listeria monocytogenes. *Nature*, 385:265–269.

Welle, S, Bhatt, K, and Thornton, C (1999). Inventory of high-abundance mRNAs in skeletal muscle of normal men. *Genome Res*, 9:506–513.

Weyrich, A, Dixon, D, Pabla, R, Elstad, M, McIntyre, T, Prescott, S, and Zimmerman, G (1998). Signal-dependent translation of a regulatory protein, Bcl-2, in activated human platelets. *Proc Nat Acad Sci USA*, 95:5556–5561.

Weyrich, AS, Lindemann, S, and Zimmerman, GA (2003). The evolving role of platelets in inflammation. *J Thromb Haemost*, 1:1897–1905.

Wu, AH and Tsongalis, GJ (2001). Correlation of polymorphisms to coagulation and biochemical risk factors for cardiovascular diseases. *Am J Cardiol*, 87:1361–1366.

Yamamoto, N, Ikeda, H, Tandon, NN, Herman, J, Tomiyama, Y, Mitani, T, Sekiguchi, S, Lipsky, R, Kralisz, U, and Jamieson, GA (1990). A platelet membrane glycoprotein (GP) deficiency in healthy blood donors: Nak^{a-} platelets lack detectable GPIV (CD36). *Blood*, 76:1698–1703.

Yee, DL and Bray, PF (2004). Clinical and functional consequences of platelet membrane glycoprotein polymorphisms. *Semin Thromb Hemost*, 30:591–600.

Zhang, L, Zhou, W, Velculescu, V, Kern, S, Hruban, R, Hamilton, S, Vogelstein, B, and Kinzler, K (1997). Gene expression profiles in normal and cancer cells. *Science*, 276:1268–1272.

Zhang, Q, Zhao, B, Li, W, Oiso, N, Novak, EK, Rusiniak, ME, Gautam, R, Chintala, S, O'Brien, EP, Zhang, Y, et al. (2003). Ru2 and Ru encode mouse orthologs of the genes mutated in human Hermansky-Pudlak syndrome types 5 and 6. *Nat Genet*, 33:145–153.

16

Applications in Critical Care Medicine

Christopher S Garrard, Charles Hinds, and Julian Knight

The Critical Care Patient Population

The development of intensive care units (ICUs) has arisen from the need to support and effectively treat the most severely ill patients within the hospital population, in particular those with impairment or failure of vital organs and systems. In some respects, such supportive care follows the ancient dictum "The physician should not treat the disease but the patient who is suffering from it" (Moses Maimonides, 1135–1204). It is also humbling to acknowledge that there are precious few specific and effective therapies for such patients, and that recovery depends primarily on supporting vital organ systems until physiological and biochemical equilibrium are restored. As Samuel Pepys (1666) observed, "Ye function of ye physician is to entertain ye patient until nature takes its course or ye patient dies."

There is little doubt that our skills in supporting failing organ systems have advanced significantly over the past two or three decades. The ability to safely apply positive pressure ventilation for prolonged period of time, to perform renal replacement therapy and support the circulation with combinations of pharmacological agents and mechanical devices should theoretically ensure that no patient should die. It has been a source of considerable frustration for the clinician that this is not, of course, the case and that mortality rates, particularly from severe life-threatening infection, remain stubbornly high, ranging from 15% to more than 75% depending on the severity of the physiological insult and the number of organs that have failed.

It is now well understood that the progression and outcome of critical illness can be influenced by a wide variety of factors including age, comorbidities, and the nature and the severity of the insult, as well as being improved by the prompt institution of effective therapy, including resuscitation and measures to control infection. Nevertheless, it is common experience that the pattern of the host response, disease progression, and outcome can vary dramatically, even between ostensibly very similar patients. For example, two similar-aged patients both with fecal peritonitis may be admitted to the same ICU, having undergone the same surgical procedure. Both will have received early and appropriate antibiotic therapy, and will have been well resuscitated with intravenous fluid. During the following 24–48 hours, however, one patient may be successfully weaned from mechanical ventilation, require only a brief period of vasopressor therapy, and maintain excellent renal function, whereas the other may remain hypotensive, despite high doses of vasopressor agents with deteriorating lung and renal function. Circulatory failure persists and progresses with increasing resistance to pressor agents, and this patient finally dies in ICU, 5 days after admission. Such examples are not rare and are not confined to peritonitis, but may be encountered with many other infectious (e.g., pneumonia, urosepsis, meningitis) or non-infectious processes (e.g., trauma, anaphylaxis, or acute pancreatitis). As a consequence of such experiences, clinicians involved in the care of the critically ill have long suspected that inherited factors could influence susceptibility and host response to severe life-threatening illness.

Models of the Systemic Response to Infection

A general, adult, non-specialized ICU will admit a mixture of patients, including those with surgical problems such as trauma, infection, and a range of non-infective medical conditions. Despite this diagnostic heterogeneity, the greatest single cause of death is infection (either primary or acquired), and its systemic manifestations, sepsis, septic shock, and multiple-organ failure.

Sepsis presents a difficult challenge for the innate immune system. On the one hand, after microbial invasion, an immunocompetent host must initiate an immediate aggressive response to limit and eradicate the infecting organism. On the other hand, the resultant systemic immune response may itself cause detrimental complications such as circulatory failure, vital organ dysfunction, and immune paresis.

Early concepts were based upon the notion that sepsis represented an uncontrolled, exaggerated inflammatory response to infection. Twenty-five years ago, Lewis Thomas wrote, "Our arsenals for fighting off bacteria are so powerful, and involve so many different defense mechanisms, that we are more in danger from them than from the invaders. We live in the midst of explosive devices; we are mined" (Thomas, 1972). This disseminated inflammatory reaction was later defined clinically by consensus as the systemic inflammatory response syndrome (SIRS). SIRS can be associated with a range of insults other than infection (e.g., trauma, acute pancreatitis, anaphylaxis) although the prognosis from these is better than that of sepsis. In addition, there may be important pathophysiological differences between sepsis and non-infection-related SIRS, such as the suppression of important immune mediators (follicular dendritic cell, B cells, CD4 T cells) (Hotchkiss et al., 1999, 2001, 2002).

The features of SIRS include simple clinical and laboratory findings, such as fever, tachypnoea, tachycardia, and leukocytosis, that still serve as an invaluable clinical tool for the early recognition of sepsis. The SIRS model has been refined and extended with the recognition that there are both pro- and anti-inflammatory processes involved, and that other clinical and laboratory measures may improve the diagnostic precision [compensatory antiinflammatory response syndrome (CARS), predisposition, infection/insult, response, organ dysfunction (PIRO)] (Bone, 1996). This pro- and anti-inflammatory concept was supported by the apparently pivotal role played by mediators such as tumor necrosis factor (TNF), interleukin (IL)-1, IL-6, and IL-10 in experimental animal and human models. It was also appreciated that the initial proinflammatory period may be followed by, or even accompanied by a counter-regulatory, anti-inflammatory process. This model can be further developed by the consideration of a wider range of cellular processes that include immune paresis, coagulation abnormalities, altered cellular energetics, and apoptosis. The extreme complexity of the systemic response to infection largely underlies the failure of "anti-inflammatory" clinical strategies in the treatment of sepsis. These have included monoclonal antibodies targeting endotoxin and TNF, high-dose systemic steroids, IL-1 receptor antagonists, and many others.

Viewed from the clinician's perspective, the greatest threat posed by sepsis is the development of multi-organ failure. The combination of shock (circulatory failure), acute respiratory distress syndrome (ARDS) (respiratory failure), renal failure, disseminated intravascular coagulation (DIC), liver failure, and neurological impairment may all contribute to the patient's death have become less clear. As techniques for individual organ support have improved, however, the exact processes leading to death. As techniques for individual organ support have improved, however the exact processes leading to death have become less clear. In patients in whom there is rapidly progressive circulatory shock that is uninfluenced by fluid resuscitation, inotropes, or pressors, death is neither unexpected nor difficult to understand. In other cases there is a longer, drawn-out process of persistent organ failure with gradually increasing resistance to inotrope/pressor therapy and secondary infection, often culminating in the withdrawal of active therapy on the basis of futility. Here the process leading to death is not so clear. Two broad pathophysiological mechanisms have been proposed as contributing to the development of organ failures. One favors a micocirculatory process characterized by intravascular fibrin deposition, white cell margination, and vascular shunting. The other favors a process of direct tissue cytotoxicity or dysoxia. However, it is likely that both processes can be implicated in the resulting tissue hypoxia and hypoperfusion. It has also been proposed that the temporary "shutting down" or "hibernation" of vital organs may represent a protective mechanism. Furthermore, the varied patterns of organ dysfunction in individual patients suggest that injury may preferentially affect different organs. Thus, some patients may exhibit shock and renal failure while the lungs remain unimpaired. Conversely, acute lung injury may occur in a patient in whom renal function remains intact.

Sepsis: A Clinical Challenge

Sepsis is the commonest single cause of death in adult ICUs. There are up to 750,000 cases of severe sepsis a year in the USA (Angus and Wax, 2001), and approximately 21,000 cases per annum in the UK, while the incidence of severe sepsis in hospitals has been conservatively estimated at 2 per 100 admissions (Sands et al., 1997) and the incidence amongst patients in ICUs at approximately 6–10 per 100 admissions (Brun-Buisson et al., 1995; Angus and Wax, 2001). In the UK, severe sepsis accounts for approximately 46% of all bed days in ICUs. Mortality rates, which are closely related to the severity of illness and the number of organs that fail, are high (20–60%) and there may be more than 200,000 deaths from severe sepsis every year in the USA, and more than 1400 deaths per day worldwide (Angus and Wax, 2001). The impact on health care expenditure and resource utilization has been considerable (annual total hospital costs for these patients in the USA have been estimated at approximately $17 billion and in Europe at €7.6 billion).

Early attempts to combat the high mortality associated with sepsis concentrated on cardiovascular and respiratory support. Despite some success, mortality rates remained unacceptably high, and often death was merely postponed until they were overwhelmed by a dysfunctional host response (characterized by persistent or recurrent sepsis, intractable hypotension, and failure of vital organs). Current evidence suggests that this response is largely independent of the site of infection and the responsible organism. Efforts to further reduce mortality by manipulating hemodynamics or modulating the host response have generally proved disappointing, perhaps in part because of our limited understanding of the complex mechanisms that regulate innate immunity and the inflammatory response (Ziegler et al., 1991; Hayes et al., 1994; Fisher et al., 1996). It has also been naively assumed that outcome could be improved by inhibiting just one component of the innate immune response. Importantly, such interventions have usually been applied unselectively to heterogeneous groups of patients, without considering the potential influence of their genetic diversity on the response to treatment.

It is, however, worth acknowledging that as experience has been gained in the design of multicenter interventional studies and the deficiencies of earlier sepsis studies has been recognized, there have been some encouraging developments for the clinician. These have included interventions such as early goal directed therapy, administration of activated Protein C (aPC), the use of "replacement doses" of steroids (Annane et al., 2002), and the tight control of blood sugar levels with insulin infusion (Bernard et al., 2001; Rivers et al., 2001; van den Berghe et al., 2001).

Many now believe that genomics research has the potential to substantially reduce the mortality, morbidity and costs associated with overwhelming infection by yielding new insights into the pathogenesis of sepsis and organ failure, revealing new therapeutic targets, and identifying high-risk individuals who might benefit from early "tailored" treatments or preventative measures.

Application of Genomics to Critical Care

Genomics involves the study of genes and their function in the context of all the genetic information found in a given organism. This approach, of considering the whole genome or total set of genes carried by an individual or cell, reflects the dramatic recent advances made in the field of genetic research, which has provided the entire DNA sequence for a rapidly growing number of species including man (Lander et al., 2001; Venter et al., 2001). This allows us not only to resolve individual genes and their encoded proteins but also to discover how gene expression is regulated in order to coordinate responses to environmental stimuli such as infectious organisms. Study of the interacting networks coordinating genome-wide gene expression will allow us to understand the complex biological processes occurring within the cell and individual critically ill patient. An array of new terminology (Box 16–1) has grown up to reflect the different layers of complexity in this process of regulation from the physical study of the make up of DNA in the genome (structural genomics) and its function (functional genomics), of the encoded RNA (transcriptome) and proteins (proteome), to the level of networks and pathways (physiome) and cellular phenotype (biome). For the critical care physician, the implications of genomic research in the future are likely to be profound. This field of research has the potential to radically alter our understanding of disease processes and our ability to diagnose, treat, and screen specific conditions in individual patients. The principles of applying genomic approaches to critically ill patients will be outlined in this section before more detailed examples are reviewed in the remainder of the chapter.

Box 16–1 A Classification of Regulatory Processes

Structural genomics	• Generation, assembly, and analysis of DNA sequence
Functional genomics	• Analysis of the biological function of genes and their resulting proteins
Pharmacogenomics	• Study of inherited differences in drug disposition and metabolism that influence inter-individual variability in drug response
Transcriptomics	• Study of expression of mRNA molecules (transcripts)
Proteomics	• Study of all proteins defined by a given genome, their structure, function, and interactions
Physiome	• Networks and pathways
Biome	• Cellular phenotype

Genetic Variation and Disease Susceptibility

To date, genomic research involving the critically ill patient has focused primarily on single genes and analysis of disease susceptibility or outcome based on the association with genetic polymorphisms of those genes. These studies have analyzed genetic association through case control studies to determine the relationship between possessing a given genetic variant and disease susceptibility or outcome. Gene selection for such studies has been largely based on the existing knowledge of disease pathogenesis (the so-called candidate gene approach), and has met with some success in the field of critical illness, notably for sepsis. Here, polymorphisms of a number of cytokine genes including *TNF* (Mira et al., 1999; Appoloni et al., 2001; Waterer et al., 2001), *LTA* (Stuber et al., 1996), IL-1 receptor anatagonist (Fang et al., 1999), *IL-6* (Sutherland et al., 2005b; Watanabe et al., 2005), and *IL-10* (Stuber, 2003) have been associated with disease severity and outcome. Polymorphisms in heat-shock genes (Waterer et al., 2003), Toll-like receptors (TLRs) (Lorenz et al., 2002; Barber et al., 2004), CD14 (Gibot et al., 2002; Sutherland et al., 2005a), lipopolysaccharide (LPS)-binding protein (LBP) (Hubacek et al., 2001), and *NOD2* (Sgambato, 2005) have also been implicated. These studies and others will be reviewed in detail later in this chapter. For many of these associations, the results have often proved difficult to replicate, and in very few cases have specific functional polymorphisms been convincingly shown to influence susceptibility, disease progression, or outcome.

A major problem has been the heterogeneity of the patient population, which makes it very difficult to define and study a clear disease phenotype. This is likely to confound analysis, and accounts in part for the often conflicting and inconclusive literature that has arisen in this field. The difficulty of analyzing the genetic component of complex disease traits is not unique to critical care (Risch and Merikangas, 1996; Glazier, et al., 2002). Investigators are typically studying diseases in which there will be significant and multifactorial environmental effects, with any genetic component likely to be modest and involve many genes. This limits the use of conventional approaches such as linkage analysis, and in an often elderly critical care population it is frequently difficult to identify and locate the surviving family members for such an approach.

There are a number of prerequisites for the successful application of population genetic approaches such as association studies. These include the need to recruit sufficient numbers of patients to achieve statistical power, and the need to take into account and avoid potential confounding due to population stratification in terms of the ethnic mix. Apparent associations should be replicated, ideally in an independent population. The analysis of such studies is also reliant on understanding the coinheritance of genetic variants or linkage disequilibrium between genetic markers. This means that although a genetic association may be observed with a given genetic marker, it does not necessarily implicate that variant as functionally important as it may only be serving as a marker for a polymorphism present and coinherited on the same region of DNA. The resolution of the underlying haplotypic structure is therefore critical to understanding and interpreting any genetic associations, as we will see in the examples discussed later in the chapter. Furthermore, the availability of experimental methods to define the consequences of the genetic variation for the function of the encoded protein or control of gene expression becomes essential. Finally, the advent of genome-wide association studies will dramatically increase the number of genetic markers that can be analyzed, and will allow the field to progress from analyzing small numbers of single nucleotide polymorphisms (SNPs) in candidate genes to a whole genome approach (Carlson et al., 2004) (Box 16–2).

Box 16–2 The Genome-wide Approach to Association Analysis

Genome-wide association analysis: Recent studies have validated the approach of genome-wide association analysis to define genetic susceptibility loci. Within the British population, the Wellcome Trust Case Control Consortium has analysed 2000 patients for each of seven common diseases (type 1 diabetes, type 2 diabetes, coronary heart disease, hypertension, bipolar disorder, rheumatoid arthritis and Crohn's disease). In total, 17,000 samples were genotyped for 500,000 single nucleotide polymorphisms (Wellcome Trust Case Control Consortium 2007).

Gene Expression Profiling

The application of genomics to critical illness, beyond investigating the relationship between genetic diversity and disease susceptibility, is still in its early stages. Functional genomic analysis of patterns of gene expression through microarray platforms has been undertaken for a number of specific diseases encountered in critical care, most notably sepsis (Chung et al., 2002). Such studies allow us to gain insights into disease pathogenesis by understanding how sets of genes are coordinately expressed. This approach also has the potential to redefine how we categorize and diagnose specific disease states, for example, based on a characteristic gene expression profile. This may improve diagnostic power and remove some of the apparent heterogeneity between patients with a given disease encountered in clinical practice. For the critically ill patient, use of microarray or other high throughput platforms to analyze changes in differential gene expression may in the future provide a real-time monitor of disease process and repair, thus superseding the use of multiple single laboratory measurements of discrete biochemical markers (Sander, 2000; Hopf, 2003). Prediction and monitoring of sepsis-induced immune suppression, for example, may be possible through microarray analysis, as recently demonstrated with major histocompatability complex (MHC) class II genes in whole blood from patients with septic shock (Pachot et al., 2005b), or by quantitative reverse transcription polymerase chain reaction (RT-PCR) for specific immunological mediators (Pachot et al., 2005a). Expression of TLR proteins has been associated with survival and mortality in animal models (Williams et al., 2003), and with sepsis in monocytes from ICU patients (Armstrong et al., 2004). As an example of the use of biomarkers to predict disease onset, the levels of immunoglobulin sTREM-1 in bronchoalveolar lavage have been associated with the onset of ventilator-associated pneumonia (Determann et al., 2005).

Pharmacogenomics

The critical care physician is typically required to manage patients using a cocktail of different drugs, many of which are of unpredictable efficacy and potential toxicity, and require careful therapeutic monitoring of the individual patient. The ability of genomics to assist the clinician by prospectively identifying the patients at risk and those patients most likely to benefit from drug treatments has the potential to dramatically improve patient care. Most pharmacogenomics data have, however, been derived outside of the critical care environment where polypharmacy, co-morbidity, and other confounding factors make genetic studies fraught with difficulty. It is, however, within the ICU that pharmacogenomics has the potential to have greatest benefit both in terms of clinical outcome for the individual patient and health economics.

There is a growing list of clinical situations where pharmacogenomics has been shown to play an important role (reviewed in Freeman and McLeod, 2004; McLeod, 2005). Genetic variation between individuals has been shown to be a powerful predictor of warfarin dose requirements and bleeding risk. These include polymorphism of the cytochrome P450 2C9 system and of the vitamin K epoxide reductase complex subunit 1 gene (*VKORC1*) (King et al., 2004; D'Andrea et al., 2005). Clinical studies have shown that genotype can be usefully incorporated into warfarin dosage algorithms to create individualized regimens (Sconce et al., 2005). Another example is seen with genetic heterogeneity in the DNA sequence of CYP2C19, which has been shown to reduce the catalytic activity of the encoded enzyme, reducing systemic clearance of proton pump inhibitors (PPIs) and peptic ulcer cure rates (Furuta et al., 1998). In high allele frequency populations such as Japan, genotyping of this gene is now entering clinical practice to aid drug selection and dose. As a further example, for azathioprine the risk of severe or fatal hematological toxicity has been associated with coding sequence polymorphism of the thiopurine methyltransferase (*TPMT*) gene, and now is being prospectively tested in the clinic in some centers (Evans, 2004).

The *TNF* gene, encoding tumor necrosis factor, illustrates both the potential and the difficulties of applying advances in genomics to critically ill patients. TNF fulfills Koch's postulates for having a causative role in septic shock as it is produced during the septic shock syndrome; it causes the development of the syndrome when given to uninfected animals; and neutralizing TNF in infected septic animals prevents the development of shock (Tracey and Cerami, 1993). Genetic variation at the *TNF* locus has been associated with many infectious, autoimmune, and inflammatory diseases including septic shock, although many studies appear conflicting and specific functional variants remain hard to localize due to linkage disequilibrium between SNPs across the locus (Knight, 2005). In the context of sepsis, genetic variation at the neighboring gene *LTA*, encoding lymphotoxin α, may be more significant with functional mechanisms postulated through changes in amino acid composition and gene expression. Anti-TNF therapies are of significant clinical benefit in rheumatoid arthritis and inflammatory bowel disease, while clinical trials in sepsis have failed to show a significant benefit (Reinhart and Karzai, 2001). There is some evidence from patients with rheumatoid arthritis (Ranganathan, 2005) and Crohn's disease (Taylor et al., 2001) that use of anti-TNF therapies can be targeted to patients based on their individual genetic profile, for example, by genotyping specific SNPs or haplotypes spanning the *TNF* locus. This pharmacogenomics approach also has the potential to address some of the heterogeneity between patients included in clinical trials. The question arises as to whether clinical trials of anti-TNF therapies in sepsis would have shown a significant benefit if patients had been stratified for genotype. Pharmacogenomics is likely to have broad implications for how clinical trials are conducted in a critically ill population with our growing ability to acquire genomic data (Cariou et al., 2002).

The potential benefits of pharmacogenomics to tailor drug use to the individual patient for maximum benefit and reduced risk of adverse drug reactions are apparent. The ongoing advances in genomic medicine will, however, raise many ethical questions for the critical care physician in the future. With limited health care resources, should genomic profiling be a criterion for targeting of treatment or even admission to ICU? How can genomic information be stored and to whom should it be available? What are the cost benefits of this form of analysis? How can we validate genomic approaches and establish a clinical level of quality control in the laboratory? In a private health care setting, should information be available to health insurance companies? For health care professionals and society as a whole, the field of genomics is raising profound ethical issues that will take many years to resolve but need to be given a high priority if the research is to advance into clinical practice.

Genomic Research in Critical Care: Sepsis and the Systemic Inflammatory Response

Sepsis is the epitome of a complex polygenic disorder with many interactions between genes and between genes and the environment. Nevertheless, there is a substantial body of evidence to support a significant heritable component to the variation observed between individuals in response to infection. In a landmark study of adoptees, early death of a biological parent from infection increased the risk of death of the child from an infectious disease nearly six-fold (Sorensen et al., 1988). Twin studies show significant concordance for many infectious diseases, including TB, leprosy, polio, and hepatitis B (Cooke and Hill, 2001). At a population level, differences in disease susceptibility have been reported. For example, in West Africa the Fulani are more resistant to severe malaria than other sympatric ethnic groups (Modiano et al., 1996).

The remainder of this chapter will examine a number of the putative molecular pathways that underlie the development of sepsis and multi-organ failure, and present evidence that, taken as a whole, supports the hypothesis that outcome from sepsis depends to a significant degree on genetically determined elements of the host response.

Organism/Antigen Recognition Pathways

Patients may be admitted to ICU with an infection acquired in the community or may develop secondary infection after several days of ICU admission. The latter, "nosocomial" infections, are common and are associated with significant morbidity and mortality (Vincent et al., 1995). The risk of ICU-acquired infection is related to the length of time the patient remains in the ICU, the level and number of invasive procedures (e.g., mechanical ventilation), and the state of the patient's immune defenses. Resident tissue macrophages, as guardians of the host immune response, are responsible for the recognition of infection through pattern recognition molecules and triggering of a complex cytokine and chemokine cascade. The innate immune system recognizes limited, but highly conserved, structures on pathogens called pathogen-associated molecular patterns (PAMPs) using germ-line-encoded pattern recognition receptors (PRRs) (Janeway, 1989). Organism recognition pathways appear to be multiple and exhibit a significant degree of redundancy. Examples of PRRs are TLR-4 and CD14, which are expressed on the surface of monocytes and macrophages and are important for LPS recognition (Opal and Huber, 2002). Other cellular receptors that recognize microorganisms include cytoplasmic surveillance proteins such as nucleotide-binding oligomerization domain (NOD) receptors (Matzinger, 2002) (Box 16–3).

Box 16–3 Examples of Organism and Antigen Recognition Pathways

- TLRs
- CD14
- LBP/BPI
- MD2
- NODs
- TREM-1
- DAP12
- MBL/MBP

Toll-like Receptors

TLRs are expressed in the cell types that are involved in first-line host defense such as macrophages, neutrophils, epithelial cells derived from gut or lungs, and dendritic cells (Becker et al., 2000; Cario et al., 2000; Muzio et al., 2000; Faure et al., 2001; Hertz et al., 2001; Visintin et al., 2001). Importantly, TLRs are also expressed on B and T lymphocytes (Mokuno et al., 2000; Ogata et al., 2000). So far, 10 TLR homologs have been identified, and a number of PAMPs recognized by individual or combinations of TLRs have been defined (Rock et al., 1998; Takeuchi et al., 1999; Chuang and Ulevitch, 2000; Du et al., 2000; Chuang and Ulevitch, 2001; Takeda et al., 2003). TLRs share a common intracellular Toll-IL-1 receptor (TIR) domain, which interacts with a group of TIR domain containing adaptor proteins, myeloid differentiation protein 88 (MyD88), TIR-domain-containing adapter-inducing interferon-β (TRIF), TIR domain-containing adapter protein (TIRAP), and TRIF-related adaptor molecules (TRAM) (Li and Qin, 2005). This in turn triggers a cascade of kinases that result in the activation of the transcription factor NF-κB and subsequent cytokine production. Combinations of TLRs also appear in different cell types acting in pairs, which bind a wide range of microbial ligands.

TLR2 has been shown to signal the presence of bacterial lipoproteins and lipoteichoic acids (Aliprantis et al., 1999; Aderem and Ulevitch, 2000), whereas TLR4 is now established as a signal-transducing receptor for LPS. As such, TLR4-mutant mice are hyporesponsive to LPS (Poltorak et al., 1998; Rhee and Hwang, 2000), as are humans with a polymorphic TLR4 allele (Arbour et al., 2000). TLR4 also has been shown to form a complex with MD2, a cell surface protein that is required for surface expression and LPS-regulated activation of TLR4 (Shimazu et al., 1999; Yang et al., 2000). In a caecal ligation and puncture (CLP) mouse sepsis model, both liver and lung TLR2 and TLR4 gene expression are increased, as are TLR4 protein concentrations when compared with controls. Furthermore, the increased *TLR2/4* gene and TLR4 protein expression correlated with mortality, whereas blunting of *TLR* gene and protein expression using glucan improved long-term survival (Williams et al., 2003).

Several functional, human TLR-4 polymorphisms have been identified, which affect the response to Gram-negative infection. Lorenz and colleagues studied the TLR-4 Asp299Gly and TLR-4 Thr399Ile mutations in patients with septic shock and healthy controls (Lorenz et al., 2002) The TLR-4 Asp299Gly mutation was found only in septic shock patients, suggesting that the TLR-4 Asp299Gly mutation interrupted LPS signaling. Furthermore, patients with septic shock with the TLR4 Asp299Gly/Thr399Ile alleles had a higher prevalence of Gram-negative infections. Interestingly, there were two patients from the sepsis group with TLR-2 Arg753Gln mutations, both of whom had staphylococcal sepsis (Lorenz et al., 2002). A recent study of severe burns patients showed that carriage of the TLR4 + 896 G-allele imparted a 1.8-fold increased risk of developing severe sepsis compared to AA homozygotes. In the same group of patients, carriage of the TNF-308 A-allele also imparted a similar increased risk, relative to GG homozygotes, although neither TLR nor TNF SNPs appeared to influence the risk of death (Barber et al., 2004).

The expression of TLRs may themselves be altered by the inflammatory mediators present in patients with sepsis (Armstrong et al., 2004). For example, monocyte TLR-2 mRNA is upregulated in ICU patients with Gram-positive and Gram-negative sepsis, and this is associated with increased monocyte cell surface TLR-2 protein compared with non-septic controls. TLR-4 mRNA is also increased in Gram-positive infections, but without the corresponding increase in TLR-4 protein (Armstrong et al., 2004). Although TLR-4 mRNA expression in healthy control monocytes can be modulated in vitro by culture with LPS or IL-10, this appears not to occur in the monocytes obtained from sepsis and ITU control subjects, suggesting that the in vivo conditions of sepsis or critical illness may alter the regulation of TLR-4 mRNA expression. Differential growth of subpopulations of monocytes in response to TNF and IL-6 may also influence the level of expression of TLR4 (Skinner et al., 2005).

CD14

CD14 is glycosylphosphatidylinositol membrane-anchored protein found on monocytes, macrophages, and neutrophils, which promotes the TLR-4 response to LPS and peptidoglycan. Interaction of these bacterial products with CD14 and TLR-4 leads to increased synthesis and secretion of IL-1, IL-6, IL-8, TNF, and platelet-activating factor (PAF). These cytokines bind to cytokine receptors on target cells and initiate inflammation, as well as activating both the complement and the coagulation pathways. Interacting with MD-2 and TLR-4, CD14 has been identified as a key component of the LPS recognition pathway (Triantafilou and Triantafilou, 2002). CD14 lacks transmembrane and intracellular domains and cannot itself initiate signal transduction, but the formation of the CD14/LPS complex considerably reduces (by a factor of 100–1000-fold) the concentration of LPS required for the activation of macrophages as compared with LPS alone. Thus, CD14 knockout mice are highly resistant to LPS-induced shock, and monocytes derived from these mice are insensitive to LPS (Haziot et al., 1996). CD14 seems to possess an ability to discriminate between bacterial products and sort their signals to different TLRs (Muroi et al., 2002). CD14 monoclonal antibodies administered to rabbits with *Escherichia coli* pneumonia improved blood pressure and reduced fluid requirements. Despite the improved hemodynamics, lung function and the bacterial burden in the lung fluid was worse (Frevert et al., 2000).

A polymorphism at −159 involving a cytosine to thymidine transition has been identified in the *CD14* gene with those who carry the T allele having greater circulating levels of soluble CD14 (Baldini et al., 1999). Transgenic mice that overexpress CD14 are highly susceptible to septic shock, hence increased expression of CD14 may be an important risk factor for septic shock (Ferrero et al., 1993). Gibot and colleagues (Gibot et al., 2002) found that carriage of the CD14 −159 TT genotype was more common in septic shock patients than in healthy controls.

LBP and Bactericidal/Permeability Increasing Protein

LBP and bactericidal/permeability increasing protein (BPI) are structurally and functionally related lipid transfer proteins that recognize the lipid A component of LPS proteins with high affinity for LPS (Weiss and Olsson, 1987; Schumann et al., 1990; Schumann et al., 1994). The LBP and BPI genes are localized to human chromosome 20 (Gray et al., 1993), and their genomic structures are almost identical (Hubacek et al., 1997; Kirschning et al., 1997).

LBP is secreted by hepatocytes. Activation of inflammatory cells by bacterial pathogens or their products requires the interaction between CD14 and LBP. LBP facilitates the binding of LPS to soluble CD14 or membrane CD14, and initiates the CD14-dependent clearance of bacteria from the site of infection (Schumann et al., 1990). BPI is produced only by the precursors of neutrophils, and is stored in the primary granules of these cells. The potent cytotoxicity of BPI is limited to Gram-negative bacteria, reflecting the high affinity of BPI for LPS. Binding of BPI to live bacteria via LPS causes immediate cessation of bacterial growth. Bacterial killing follows later with damage to the inner baterial cell membrane.

In patients with sepsis, the plasma concentration of BPI and LBP are increased (Froon et al., 1995), while administration of the N-terminal fragment of BPI appears to increase survival in animal models of *E. coli* bacteraemia (Ammons et al., 1994). The same fragment of BPI given, in an open label study, to children with severe meningococcal sepsis reduced mortality (Giroir et al., 1997). However, a subsequent, possibly underpowered randomized study using recombinant BPI failed to confirm a survival benefit, although complications were reduced (Levin et al., 2000). Anti-neutrophil cytoplasmatic antibodies against BPI (BPI-ANCA) have been found in patients with cystic fibrosis, suggesting a mechanism of chronic inflammation or infection based on reduced BPI activity (Carlsson et al., 2003; Dorlochter et al., 2004).

Common polymorphisms have been identified in both BPI and LBP. Polymorphisms in the gene for LBP in male patients were found to be associated with an increased risk for the development of sepsis, and and were associated with an unfavorable outcome (Hubacek et al., 2001). The same study found no associations between the polymorphisms of BPI and neither the incidence nor the outcome of sepsis (Hubacek et al., 2001).

Myeloid Differentiation Protein-2

Myeloid differentiation protein-2 (MD-2) is a 20- to 25-kDa glycoprotein that is produced by several cell types, binds to LPS, and confers LPS responsiveness to TLR4-expressing cells (Shimazu et al., 1999; Yang et al., 2000; Akashi et al., 2001; Miyake, 2003). The LPS receptor complex may, however, vary according to cell type (Aderem and Ulevitch, 2000); in myeloid cells it comprises TLR4, CD14, and MD-2, whereas in endothelial cells only TLR4 and MD-2 are expressed (Aderem and Ulevitch, 2000). LPS responsiveness in endothelial cells requires the presentation of LPS to the TLR4/MD-2 receptor by LBP and soluble CD14 (Henneke and Golenbock, 2002). Epithelial cells may only express TLR4, and whether a cofactor is required for these cells to respond to LPS remains uncertain (Abreu et al., 2001; Abreu et al., 2002; Zarember and Godowski, 2002). In intestinal epithelial cells, TLR4 is located in the Golgi apparatus with no detectable surface expression, although it colocalizes with internalized LPS (Hornef et al., 2002). Thus, antibody directed against TLR4-MD-2 does not block TLR4-mediated responses in these cells, but does so in peritoneal macrophages (Akashi et al., 2000). MD-2 expression may modulate lung (airway) epithelial responses to endotoxin, as might be encountered in Gram-negative pneumonia. Under basal, unstimulated conditions, MD-2 expression in airway epithelial cells is low and endotoxin responsiveness is limited. Upregulation of MD-2 by a variety of stimuli such as inflammatory mediators can greatly enhance cellular responses to endotoxin (Jia et al., 2004). Stimulated pulmonary macrophages might also generate sufficient MD-2 to enhance TLR-4 signaling in the lung.

In addition to being involved in the recognition of LPS, MD-2 plays a role in the regulation of the cellular distribution of TLR4 (Fig. 16–1). Thus, in mouse cells MD-2 seems to be essential to bring TLR4 to the cell surface, and in MD-2-deficient embryonic fibroblasts TLR4 is not transported to the cell surface but accumulates in the Golgi apparatus, whereas in the wild-type the surface expression of TLR4 is highly concentrated at the leading edge surface of embryonic fibroblasts (Nagai et al., 2002). When this process is deficient, for example, in MD-2 knockout mice, the animal is protected against the lethal effects of LPS (Nagai et al., 2002) although these mice are more susceptible to *Salmonella typhimurium* infection. Such phenotypes are identical to those of TLR4 knockout mice, indicating that MD-2 is essential for TLR4-dependent LPS responses in vivo (Nagai et al., 2002). A proportion of MD-2 is detached from the TLR4 complex, and is secreted in the extracellular tissues and into the plasma (Henneke and Golenbock, 2002; Schromm et al., 2001). This "soluble" MD-2 in the plasma and urine of patients with sepsis enables LPS activation of TLR4-expressing epithelial cells (Pugin et al., 2004). This effect can be mimicked by recombinant soluble MD-2 or blocked by anti-MD-2 monoclonal antibody (Pugin et al., 2004). A comprehensive review of the roles of the TLR4/MD-2 complex in macrophage activation can be found elsewhere (Fujihara et al., 2003).

A rare MD-2 mutation with A to G substitution at position 103 resulted in reduced LPS-induced signaling in one of the 20 patients screened in a recent report (Hamann et al., 2004). This patient had

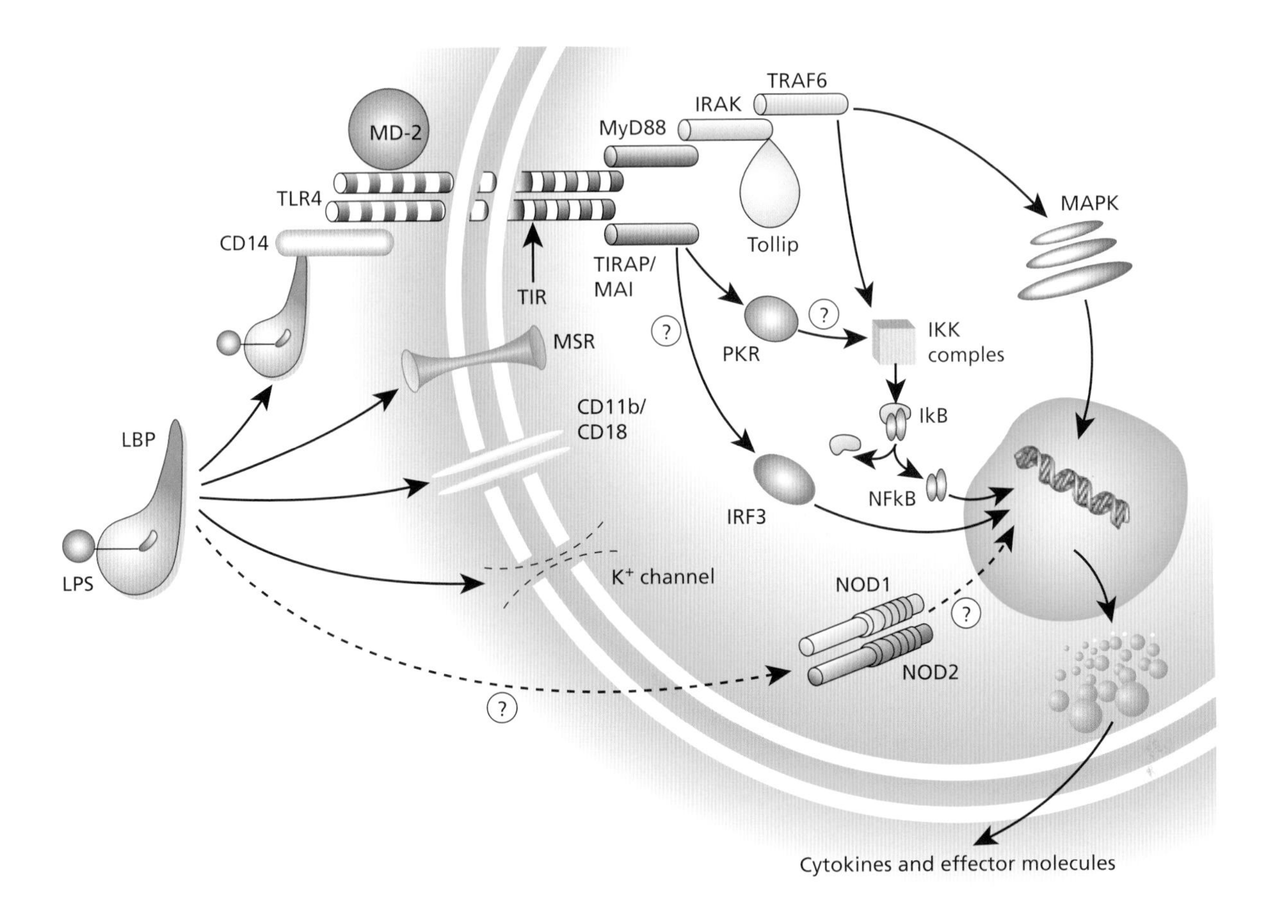

Figure 16–1 Schematic of lipopolysaccharide (LPS) sensing via LPS-binding protein (LBP)–LPS complex and then signaling through the Toll-like receptor 4 (TLR-4)–MD-2 complex. Other cell surface molecules include macrophage scavenger receptor (MSR), CD11b/CD18, and ion channels. Intracellular signaling depends on binding of the intracellular TLR domain, TIR (Toll-IL-1 receptor homology domain), to IRAK (IL-1 receptor-associated kinase), facilitated by two adapter proteins, MyD88 (myeloid differentiation protein 88) and TIRAP (TIR domain-containing adapter protein), and inhibited by a third protein, Tollip (Toll-interacting protein). (Reproduced from Cohen, 2002 with permission.)

slightly decreased TNF-α levels after in vitro stimulation with LPS as compared to wild-type patients. The possible effects of MD-2 polymorphism in human sepsis have not yet been reported.

NOD Proteins

NODs are cytosolic proteins that contain a NOD and act as important regulators of apoptosis and pathogen resistance (Fritz and Girardin, 2005). Both NOD1 and NOD2 recognize peptidoglycan (PGN), which is the major constituent of the cell wall of Gram-positive bacteria (Philpott and Girardin, 2004). NOD2 can detect PGN from both Gram-positive and Gram-negative organisms. The reason for this may be that NOD2 can detect the minimal bioactive structure of PGN, muramyl dipeptide (Girardin and Philpott, 2004). In Gram-negative bacteria, PGN is found in a thin layer in the periplasmic space between the outer and cytoplasmic membranes. As macrophages phagocytose bacteria or their peptidoglycan fragments, the PGN is broken down into muramyl dipeptides. Binding of the muramyl dipetides to NOD1 or NOD2 leads to the activation of genes coding for proinflammatory cytokines in a manner similar to the cell surface TLRs. A number of NODs contain leucine-rich repeats (LRRs), and are therefore referred to as NOD-LRR proteins.

Genetic variations in several NOD-LRR proteins are associated with inflammatory disease or increased susceptibility to microbial infections (Inohara et al., 2005). Ileal Crohn's disease, a chronic inflammatory disease of the distal small bowel, is commonly associated with the mutations of the gene encoding NOD2 (Wehkamp and Stange, 2005), while NOD2 mutations occur approximately three times more frequently amongst septic as compared with non-septic patients (Sgambato, 2005).

TREM-1

The triggering receptor expressed on myeloid cells (TREM-1) is a member of the immunoglobulin superfamily. After bacterial invasion, tissues are infiltrated with polymorphonuclear leukocytes and monocytes that express TREM-1. Under resting conditions TREM-1 has a short intracellular domain but, when bound to its ligand, is associated with a signal transduction molecule called DAP12 that triggers the secretion of proinflammatory cytokines (Lanier et al., 1998). When TREM-1 is bound to its ligand, it acts synergistically with LPS to induce cytokines, amplifying the host response to bacterial infection. Blockade of TREM-1 using a fusion protein of murine TREM-1 extracellular domain and human immunoglobulin-G (IgG1) Fc portion (mTREM-1/IgG1) protects mice against LPS-induced shock and death, as well as microbial sepsis caused by live *E. coli* or CLP (Bouchon et al., 2001).

TREM-1 is upregulated by extracellular bacteria such as *Staphylococcus aureus* or *Pseudomonas aeruginosa*, in cell culture, in peritoneal-lavage fluid and tissue samples from patients with infection

(Bouchon et al., 2001). In contrast, TREM-1 is only weakly expressed in patients with non-infectious inflammatory disorders (Bouchon et al., 2001).

A soluble form of TREM-1 (sTREM-1) can be measured in various body fluids, possibly reflecting TREM-1 shed from the membranes of activated phagocytes (Bouchon et al., 2001; Gibot et al., 2004). Indeed, increased levels of sTREM-1 in bronchoalveolar fluid may serve as a biomarker for pulmonary infection (Gibot et al., 2004; Determann et al., 2005). In ventilator-associated pneumonia, increased sTREM-1 levels in bronchoalveolar fluid appear to precede the development of clinical signs of infection (Determann et al., 2005). This response seems to be limited to the lung compartment, since plasma levels remain at basal levels (Determann et al., 2005).

Cells may also be able to respond to LPS through intracellular NOD proteins. Expression of NOD1 and NOD2 confer responsiveness to Gram-negative LPS but not to lipoteichoic acid, which is found in Gram-positive bacteria.

Using the RT-PCR, Wang et al. examined the expression of TREM-1 mRNA in mild and severe acute pancreatitis compared with healthy controls. They found that the expression of TREM-1 mRNA correlated with the severity of acute pancreatitis, suggesting a causative role (Wang et al., 2004). In a separate study, surface expression of TREM-1 was analyzed in a range of cell types in a murine sepsis model, and clearly demonstrated that in sepsis TREM-1 gene expression was strongly upregulated (Gibot et al., 2005). No equivalent human data are available.

Mannose-Binding Lectin/Protein

Mannose binding lectin (MBL) or mannan-binding protein (MBP) is a liver derived, "acute phase" circulating plasma protein that plays a key role in innate immunity. MBL functions as a pathogen recognition molecule, opsonizing organisms and initiating the complement cascade. MBL binds to the mannose groups of many microbial carbohydrates, and is equivalent to C1q in the classical complement pathway. When MBL binds to the mannose groups of microbial carbohydrates, two other lectin pathway proteins called MASP1 and MASP2 also bind to the MBL. This forms an enzyme similar to C1 of the classical complement pathway, which is able to cleave C4 and C2 to form C4bC2a, the C3 convertase capable of enzymatically splitting C3 into C3a and C3b. As a consequence of triggering the complement cascade, phagocytes are chemotactically attracted to the infection site, opsonization is enhanced, and harmful immune complexes are removed. The regulation of MBL has not been fully elucidated. It appears to be regulated independently of other acute phase proteins (Thiel et al., 1992; Liu et al., 2001; Hansen et al., 2003), but may be controlled in part by IL-6 and IL-1.

MBL deficiency arising from mutations and promoter polymorphisms in the *MBL2* gene is common, and has been associated with risk, severity, and frequency of infection in a number of clinical settings. The MBL gene, *MBL2*, is located on chromosome 10q11.2–q21 (Sastry et al., 1989). Within the MBL gene, three point mutations have been identified in the collagen-encoding region of exon 1 (Lipscombe et al., 1992a,b). These mutations and a number of polymorphisms in the promoter region of the *MBL2* gene markedly affect the circulating levels of functional MBL (Lipscombe et al., 1992a; Hibberd et al., 1999). The combination of promoter haplotypes and structural mutations results in a common deficiency state, with approximately 25% of individuals producing very little MBL (Eisen et al., 2004). Variant MBL alleles (*MBL-2* gene exon 1 and promoter polymorphisms) were found to be overrepresented in children who developed SIRS and were admitted to a pediatric ICU. Critically ill children with variant alleles for the MBL gene had a seven-fold increased risk of developing SIRS. In the children with infection, variant MBL alleles were associated with an increased systemic response, including the development of shock (Fidler et al., 2004). MBL serum levels correlated with the genotype and indicated that MBL levels less than 1000 ng/ml are associated with a greatly increased risk of SIRS (Fidler et al., 2004). A similar increased overrepresentation of variant MBL alleles with increased severity of sepsis has also been shown in adult patients (Garred et al., 2003). The reason for the observed effect of MBL on the development of SIRS remains unclear. High levels of MBL have been shown to inhibit TNF, IL-6, and IL-1 release from monocytes in a human whole blood model (Jack et al., 2001). Intensive insulin therapy appears to suppress the observed MBL acute phase response in sepsis patients with prolonged stay in ICU (Hansen et al., 2003). Genetic polymorphisms resulting in mannose-binding lectin deficiency appear to be associated with increased susceptibility to sepsis. The relationship between polymorphic variants and plasma MBL concentration persists during sepsis, but individual levels vary widely. Lower circulating MBL levels are associated with a poor outcome (Gordon et al., 2006).

MBL exhibits a variable acute phase response in the clinical setting of sepsis and septic shock. In one report, approximately one-third of patients maintained consistent MBL levels throughout hospital stay, one-third demonstrated a positive acute phase response, and a negative acute phase response was observed in the remaining third (Dean et al., 2005). The positive response was generally observed in individuals with wild-type *MBL2* genes. When a positive acute phase response was observed in patients with coding mutation, these individuals had a normal MBL level on admission to hospital. No patient, regardless of genotype, who was MBL deficient at admission was able to demonstrate a positive acute phase response thereafter (Dean et al., 2005).

Nuclear Factor κB

There are several members of the nuclear factor κB (NF-κB) transcription factor family in mammals, namely p50/p105, p65/RelA, c-Rel, RelB, and p52/p100. Although a number of dimeric forms of NF-κB have been described, the classic form of NF-κB appears to be the heterodimer of the p65/RelA and p50 subunits. NF-κB is activated by many bacteria, bacterial toxins, and proinflammatory mediators, and plays a central role in the pathophysiology of sepsis, severe sepsis, and septic shock (reviewed in Liu and Malik, 2006). NF-κB triggers the transcription of a large number of genes, whose products are important mediators in sepsis. Thus, inhibition of NF-κB or its activation reverses systemic hypotension, reduces myocardial dysfunction (Liu et al., 1997; Sheehan et al., 2003), inhibits proinflammatory gene expression, reduces intravascular coagulation, (Levi and van der Poll, 2004) reduces tissue neutrophil influx, and limits endothelial leakage. These widespread effects are summarized in Fig. 16–2 (Liu and Malik, 2006).

The inflammatory role of NF-κB is exemplified by the effect of anti-inflammatory compounds such as aspirin, non-steroidal anti-inflammatories, and corticosteroids, which act through that pathway (Barnes and Karin, 1997; Yin et al., 1998; Yamamoto et al., 1999; De Bosscher et al., 2000).

Indirect evidence for a pivotal role for NF-κB in sepsis can be inferred from the knowlegdge that mice deficient in a wide range of NF-κB-dependent genes are resistant to the development of septic shock and are less likely to die (Xu et al., 1994; Blackwell and Christman, 1997; Li et al., 1997; Kristof et al., 1998; Hollenberg et al., 2000; Raeburn et al., 2002; Chauhan et al., 2003). Blocking the NF-κB pathway in LPS animal models of septic shock improves outcome (Shanley et al., 1995; Altavilla et al., 2002; Sheehan et al., 2002),

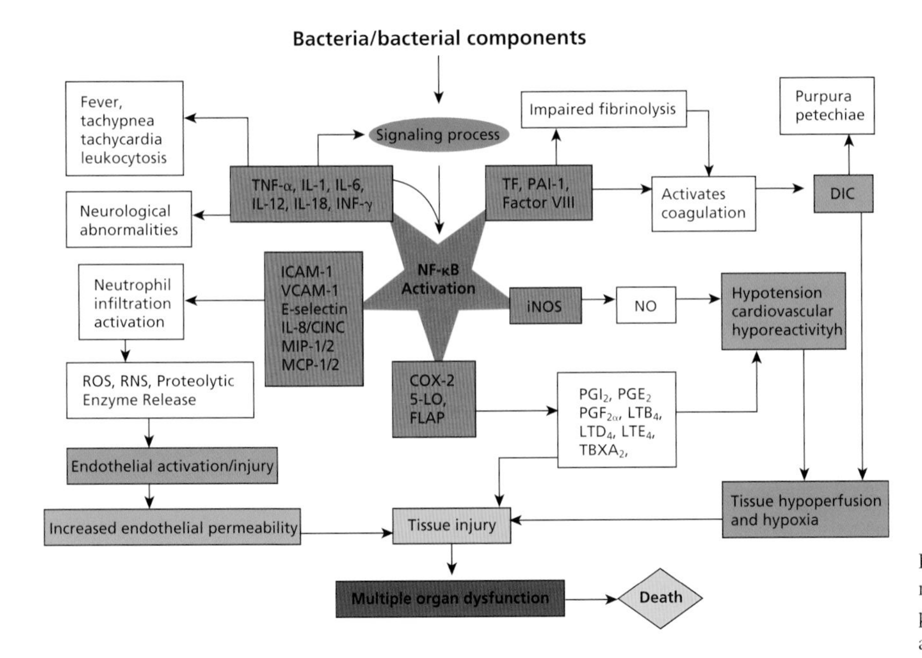

Figure 16–2 Schematic of signaling pathways for nuclear factor κB (NF-κB) and its role in the pathophysiology of sepsis. (Reproduced from Liu and Malik, 2006 with permission.)

although experience in more relevant bacteria-induced sepsis models has been less convincing (Sha et al., 1995; Zingarelli et al., 2003; Wang et al., 2005b). However, increased plasma levels of NF-κB activity are associated with a higher mortality rate and worse clinical outcome in human sepsis (Schwartz et al., 1996; Bohrer et al., 1997; Arnalich et al., 2000). IL-10, a potent inhibitor of NF-κB activation, inhibits downstream inflammatory cytokine production (Grunig et al., 1997), reduces bacteria-induced lung injury (Sawa et al., 1997; Fan et al., 2001), and improves survival in LPS challenged animals (Howard et al., 1993).

The gene encoding NF-κB1 is located on chromosome 4q23–q24 and is composed of 24 exons. NF-κB1 polymorphism may play a role in the susceptibility to type 1 diabetes mellitus and inflammatory bowel disease (Hegazy et al., 2001; Karban et al., 2004), but no associations have been reported for sepsis. Members of the NF-κB family are fundamental mediators of the transcriptional regulation of genes in the immune and inflammatory response, which are likely to be under strong evolutionary pressure to remain highly conserved. As NF-κB is activated through diverse signaling cascades, it is within these that genetic diversity may be more likely to act.

Acute Phase Protein Responses in Sepsis

Short Pentraxins—C-reactive Protein and Serum Amyloid P

Pentraxins include C-reactive protein (CRP) and serum amyloid P (SAP), together known as the short pentraxins, and are produced in the liver in response to inflammatory stimuli, principally IL-6 (Pepys and Hirschfield, 2003). CRP is an acute phase protein, which is a sensitive systemic marker of inflammation, cardiovascular disease, and tissue damage (Hirschfield et al., 2003; Pepys and Hirschfield, 2003), and plays a role in eliminating bacteria. Human CRP binds with highest affinity to phosphocholine residues, but it also binds to a variety of other autologous and extrinsic ligands. Extrinsic ligands include the constituents of microorganisms, such as capsular and somatic components of bacteria, and fungi. When aggregated or bound to macromolecular ligands, human CRP is recognized by C1q and potently activates the classical complement pathway, engaging C3, the main adhesion molecule of the complement system, and the terminal membrane attack complex, C5–C9 (Volanakis, 1982; Mold et al., 1999) (Box 16–4).

Box 16–4 Examples of Acute Phase Protein Responses Implicated in Sepsis

A1-acid glycoprotein
Angiotensin
C-reactive protein (CRP)
Ferritin
Fibronectin
Haptoglobin
Phospholipase a2
Pertussis toxin-3
Serum amyloid A

Plasma CRP is predominantly produced by hepatocytes under the transcriptional control of cytokines such as interleukin IL-6, TNF, and IL-1β (Gabay and Kushner, 1999). The plasma CRP level is less than 10 mg/L in healthy adults and rises in response to an inflammatory stimulus. Most infections and inflammations result in CRP levels above 100 mg/L. In critically ill patients, CRP levels are increased by a range of disorders other than sepsis, such as myocardial infarction, surgery, trauma, burns, acute pancreatitis, inflammatory disease, and malignancy (Pepys and Hirschfield, 2003; Povoa et al., 2005). CRP levels can be suppressed by immunosuppression either by corticosteroid or cyclosporine (Selberg et al., 2000; Luzzani et al., 2003; Mitaka, 2005).

CRP transgenic mice exhibit increased resistance to lethal infection against *S. typhimurium* (Szalai et al., 2000). Several CRP promoter polymorphisms have been found to influence CRP levels in a large cohort study of cardiovascular risk in European and African-American adults (Carlson et al., 2005), and a +1444C→T polymorphism in the CRP gene has been associated with altered CRP concentrations in patients with periodontitis (D'Aiuto et al., 2005).

SAP, named for its universal presence in amyloid deposits, is a constitutive, non-acute-phase plasma glycoprotein in humans and other species, except the mouse, in which it is the major acute-phase protein. A SAP-derived peptide, consisting of amino acids 27–39, inhibits LPS-mediated effects in the presence of whole human blood (de Haas et al., 1999).

Serum Amyloid A

Serum amyloid A (SAA) is an acute-phase protein that increases in response to tissue injury and inflammation. It is so-named as it is related to the A proteins of secondary amyloidosis. It may also act as a cytokine, influencing cell adhesion, migration, proliferation, and aggregation.

SAA seems to be a reliable early marker for the diagnosis of late-onset sepsis in preterm infants (Enguix et al., 2001; Arnon et al., 2002), and in children with suspected pneumonia, meningitis, or sepsis (Huttunen et al., 2003). High levels of SAA occur at sepsis onset, and gradually decline thereafter. Levels of SAA returned to baseline faster than CRP levels. Receiver-operating characteristic analysis values revealed that SAA has a high sensitivity of 95%–100%, and a negative predictive value 97%–100% (Arnon et al., 2005). Although homozygosity for the SAA1.3 allele is a predictor of survival in Japanese rheumatoid arthritis patients, there are no published reports of genomic influences on SAA expression in sepsis.

Long Pentraxins—PTX3

Pentraxin 3 (PTX3) is a long pentraxin produced in response to microbial infections, and increased levels are reported in patients with sepsis (al-Ramadi et al., 2004). Unlike the short pentraxin CRP, which is made in the liver and induced by IL-6, PTX3 is induced by inflammatory cytokines and microbial products in macrophages and endothelial cells (Alles et al., 1994). PTX3 is elevated in ICU patients with SIRS, sepsis, and septic shock. The greatest increases in PTX3 are observed in patients with septic shock and with high disease severity scores (APACHE II) (Muller et al., 2001a).

Hormokine Responses in Sepsis

Adrenomedullin

Adrenomedullin (ADM) is a 52-amino acid peptide with immune, metabolic, and vasodilator activity (Nishio et al., 1997). Its widespread vasodilator action helps to maintain blood supply to vital organs (Eto, 2001; Marutsuka et al., 2001). In addition, it has bactericidal effects and modulates complement activity (Marutsuka et al., 2001; Christ-Crain et al., 2005) (Box 16–5).

Box 16–5 Examples of Hormokines Implicated in Sepsis

Adrenomedullin
CGRP
Calcitonin precursors
IL-6
Leptin
MIF

In a rat CLP model of sepsis, plasma levels of ADM increase significantly 2 hours after CLP and coincide with the onset of a hyperdynamic (vasodilated, high cardiac output) circulation. Levels of ADM continued to increase progressively over a 5–20 hour-time-period (Wang, 1998). These increased levels of plasma ADM correlated with the upregulation of ADM mRNA in the small intestine, left ventricle, and thoracic aorta, but not in renal or hepatic tissues (Wang, 1998). As sepsis progresses, however, the vascular response to ADM is blunted by a sepsis-induced decrease in the binding protein for ADM (i.e., ADM binding protein-1) rather than a change in the gene expression of the components of ADM receptors (Fowler and Wang, 2002). This may explain in part the "late" hypodynamic stage of sepsis. Treatment of septic animals with ADM, together with its binding protein, prevents this late hypodynamic phase of sepsis and improves survival (Fowler and Wang, 2002).

The measurement of ADM is technically difficult because of its rapid clearance and the presence of a binding protein (complement factor H) (Eto, 2001), rendering it less accessible to immunoassay. By measuring a more stable midregional fragment of pro-ADM (MR-proADM), large increases in ADM can be inferred in human sepis (Struck et al., 2004; Christ-Crain et al., 2005). Furthermore, MR-proADM plasma levels are significantly higher in non-survivors than in survivors. Although it is known that ADM is elevated in sepsis, reported concentrations are much lower than the observed increases in MR-proADM levels (Nishio et al., 1997; Ueda et al., 1999).

ADM is widely expressed and extensively synthesized during sepsis, similar to other calcitonin peptides including procalcitonin (PCT) and calcitonin-gene-related peptides (Becker et al., 2004). The term "hormokine" was proposed to encompass the cytokine-like behavior of these hormones, specifically calcitonin peptides, in inflammation and infections (Muller et al., 2001b). LPS and pro-inflammatory cytokines upregulate ADM gene expression in many tissues in animals and humans (Shoji et al., 1995; Linscheid et al., 2005). A microsatellite DNA polymorphism of ADM gene may be associated with a genetic predisposition to essential hypertension, (Ishimitsu et al., 2001), but no equivalent sepsis association has yet been established.

Leptin

Leptin is a hormone that is involved in the regulation of body weight. Leptin, produced by fat cells, the gut, and placenta, circulates in the blood and acts on the hypothalamus, skeletal muscle, and liver. It is a protein that consists of 167 amino acids and has a molecular weight of approximately 16 kDa. The DNA sequence for the protein was determined in 1994 and was found on the obese (ob) gene (Zhang et al., 1994; Halaas et al., 1995).

In addition to weight regulation, leptin appears to have a role in immune modulation. Serum leptin levels are increased in rodents by the administration of endotoxin or inflammatory cytokines (Grunfeld et al., 1996; Sarraf et al., 1997). Increased plasma concentrations of IL-6 and leptin have been reported in adult patients with acute sepsis, and the highest leptin levels on admission were found in survivors (Bornstein et al., 1998). In another study of patients with severe sepsis and septic shock, a two- to four-fold increase in leptin levels was demonstrated (Arnalich et al., 1999). Leptin levels were positively correlated with IL-1 receptor antagonist (IL-1ra) and IL-10, but not with body mass index. Again the highest leptin levels predicted survival (Arnalich et al., 1999). Such responses have not, however, been confirmed in children with sepsis (mostly meningococcal disease) (Blanco-Quiros et al., 2002). In 20 subjects with intra-abdominal sepsis, Carlson et al. (1999) showed serum leptin concentrations to be similar in septic and control subjects. In these patients, serum leptin concentrations correlated significantly with percent body fat in both septic patients and healthy subjects. Multiple regression showed that percent body fat, fasting plasma insulin, and plasma cortisol, but not sepsis, were independent determinants of serum leptin levels (Carlson et al., 1999). It is possible that sepsis in this small number of patients may have been less severe than in the patients studied by Arnalich et al. (1999).

Various functional SNPs have been reported to be associated with obesity, cardiovascular disease risk, pre-eclampsia, insulin resistance, prostate cancer, and type II diabetes mellitus, but not as yet in sepsis (Oksanen et al., 1998; Stefan et al., 2002; Chiu et al., 2004; Ribeiro et al., 2004; Muy-Rivera et al., 2005; Porreca Febbo et al., 2005).

Macrophage Migration Inhibitory Factor

Migration inhibitory factor (MIF) was originally described as a soluble factor released by activated lymphocytes that inhibited the migration of macrophages in peritoneal fluid. It is now recognized as a powerful proinflammatory cytokine with an important role in sepsis and septic shock. Monocytes and macrophages constitutively express large amounts of MIF, which is released in response to bacterial toxins and other cytokines. MIF acts by modulating the expression of TLR-4, and thus its downstream effects (see Fig. 16–3).

Neutralization of MIF activity with anti-MIF antibodies attenuates TNF production and protects mice from endotoxin-induced shock and also toxic shock from Gram-positive organisms (Calandra et al., 1998). The same beneficial effects are seen in the mouse MIF deletion model (Calandra et al., 1998; Bozza et al., 1999).

Raised MIF levels are seen in patients with severe sepsis and septic shock due to both Gram-positive and Gram-negative bacteria (Calandra et al., 2000). The most severely ill patients have the greater elevations in circulating levels of MIF, and non-survivors the largest elevations of all (Calandra et al., 2000; Beishuizen et al., 2001).

MIF is encoded by a functionally polymorphic gene [−794 CATT(5–8) microsatelite repeat and −173 G/C SNP] that has been linked to several inflammatory conditions including inflammatory arthritis and malaria (Calandra and Roger, 2003; Gregersen and Bucala, 2003; Zhong et al., 2005). The haplotype containing both these linked polymorphisms was associated with a three-fold increased risk of developing inflammatory polyarthritis (Barton et al., 2003).

Intracellular Mediators/Receptor Responses in Sepsis

Heat-shock Proteins

Heat-shock proteins (HSPs) are highly conserved proteins ranging from 8 to 110 kDa, which are classified into families based roughly on the molecular mass of a typical member (Lindquist and Craig, 1988). The best characterized are the HSP-70 family, which are found in inducible and constitutive forms in many cells. The family of HSP-70 contains HSP-72, HSP-73, HSP-75, and HSP-78. HSP-72 is an inducible form, while the others are constitutive (Kiang and Tsokos, 1998). The inducible HSP-72 is synthesized after exposure to various harmful stimuli such as heat, cytokines, hypoxia, endotoxin challenge, various chemicals, and oxygen-free radicals (De Maio, 1999).

HSP exhibits protective effects in animal sepsis models. (Box 16–6) This protective effect is probably achieved by the heat-shock proteins recognizing and forming complexes with denatured proteins, thus inducing correct protein folding and proteolytic degradation where necessary. They also appear to protect normal functional proteins against degradation and inhibit apoptosis (Ellis and van der Vies, 1991; Welch, 1992; Georgopoulos and Welch, 1993; Parsell and Lindquist, 1993). In recognition of these effects, HSPs have often been referred to as "molecular chaperones" (Ellis and van der Vies, 1991; Georgopoulos and Welch, 1993). HSPs also play a role in antigen presentation to T lymphocytes (Pockley, 2003).

Box 16–6 Examples of Intracellular Mediators or Receptor Responses in Sepsis

Heat-shock proteins
High mobility group-1
NO

Several studies demonstrate that a preceding heat-shock reaction reduces both mortality and organ dysfunction in experimentally induced endotoxaemia or severe sepsis. HSP-70, expressed in animal models subjected to heat stress or sodium arsenite, reduces lethality and organ injury after LPS challenge or experimental

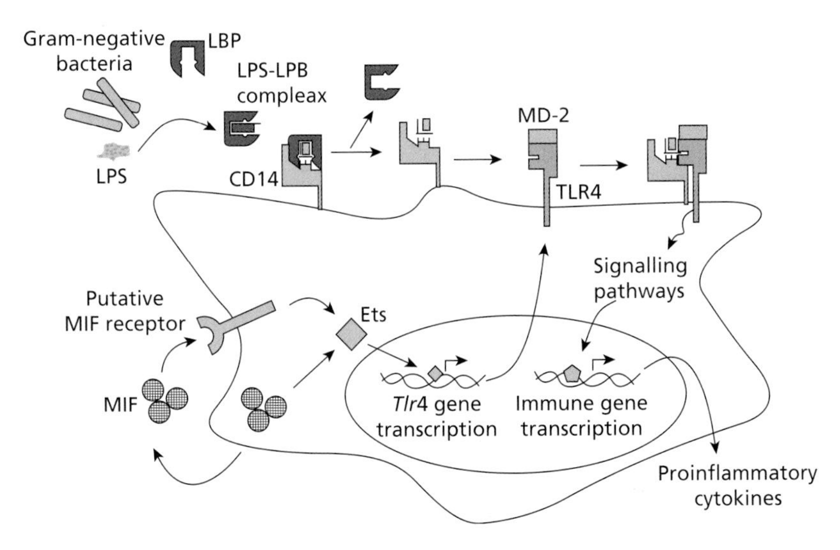

Figure 16–3 Role of the cytokine migration inhibitory factor (MIF) in recognition of endotoxin and Gram-negative bacteria by macrophages. MIF promotes the recognition of endotoxin-containing elements by upregulating the basal expression of Toll-like receptor (TLR) 4 mRNA and protein. MIF is constitutively expressed by quiescent macrophages, and may exert its effects via a direct intracellular pathway or by being first released from macrophages to then act in an autocrine fashion by binding to an unidentified putative MIF receptor. MIF effects on Tlr4 gene transcription are mediated by transcription factors of the Ets family. CD14, surface antigen expressed on myeloid cells that binds LPS–LBP complexes; LBP, LPS-binding protein; MD-2, protein associated with TLR4 extracellular domain that amplifies responses to LPS. (Reproduced from Calandra and Roger, 2003 with permission.)

peritonitis (Hotchkiss et al., 1993; Ribeiro et al., 1994; Villar et al., 1994). In a porcine model, the expression of HSP-70 reduced proinflammatory cytokines, shock, and the rate of apoptosis of lung, liver, and kidney cells (Klosterhalfen et al., 1997). HSP-70 expression in the rat lung can be increased using an adenovirus (AdHSP) carrying the gene for HSP-70 (Weiss et al., 2002). AdHSP has been shown to attenuate neutrophil accumulation, septal thickening, interstitial fluid accumulation, and alveolar protein exudate, features of sepsis-induced ARDS, 48 hours after caecal ligation and double-puncture (Weiss et al., 2002). Furthermore, AdHSP administration significantly decreased 48-hour mortality. Caution should be exercised in extrapolating these findings to human sepsis. In human sepsis-induced ARDS, death is rarely a direct consequence of the lung injury. Nevertheless, reduction in multi-organ failure in human sepsis would be expected to impinge on survival.

Heat shock-induced HSP-70 confers protection against endotoxin-induced circulatory depression by inhibiting inducible nitric oxide synthase (iNOS) mRNA and HSP-70 upregulation in the rostral ventrolateral medulla (RVLM), suggesting a key role for HSP-70 in the autonomic control of the circulation in sepsis (Chan et al., 2004).

Heat shock pretreatment in rats also prevents the decline in myocardial adenosine triphosphate (ATP) seen during late sepsis (Chen et al., 2003), protects against sepsis-induced apoptosis (Chen et al., 2000), and attenuates endotoxemia-induced histological changes in the liver (Chen et al., 2005).

Heat shock inhibits proinflammatory cytokine production induced by LPS in vitro, and also attenuates tissue injury induced by various insults including sepsis, radiation, and ischemia/reperfusion (Villar et al., 1994; Buzzard et al., 1998; Uchinami et al., 2002). This protective effect may be, in part, attributed to the attenuation of endotoxin-induced hepatic NF-κB activation (Chen et al., 2005).

The ex vivo expression of HSP-70 induced by LPS in peripheral blood monocytes is inhibited to a greater extent in patients with severe sepsis than healthy controls; an effect that might be attributable to the high levels of circulating proinflammatory cytokines in sepsis (Schroeder et al., 1999a). Expression of HSP-27, HSP-60, and HSP-70 in polymorphonuclear leukocytes (PMNLs) from burns patients has been shown to be significantly elevated for up to 28 days after injury. During this time, oxidative activity in these patient's PMNLs is significantly enhanced and PMNL apoptosis is inhibited (Ogura et al., 2002).

Intravenous glutamine administration increases serum and tissue HSP-70 expression in animal models and critically ill patients with sepsis (Wischmeyer et al., 2001; Singleton et al., 2005; Ziegler et al., 2005). Glutamine, given after initiation of sepsis in a rat CLP model, enhanced pulmonary heat-shock factor-1 phosphorylation, HSP-70, HSP-25, and attenuated lung injury after sepsis (Singleton et al., 2005). An almost four-fold increase in serum HSP-70 concentrations was observed in one study in which critically ill patients received glutamine supplemented parenteral nutrition (Ziegler et al., 2005). Patients receiving glutamine supplementation had a decreased ICU length of stay.

Three human HSP-70 genes are located in the gene cluster of the human leukocyte antigen (HLA) class III genes on the short arm of chromosome 6, and are denoted as HSP-70-1, HSP-70-2, and HSP-70-HOM genes (Milner and Campbell, 1990). Two biallelic polymorphisms within the coding region of the constitutively expressed HSP70-HOM C/T and the stress-inducible HSP70-2 G/A have been studied in patients with severe sepsis (Schroeder et al., 1999b). Neither polymorphism was found to be associated with susceptibility to nor outcome from severe sepsis. In contrast, a relationship between the risk of septic shock and HSP70-2 + 1267 AA genotype was later demonstrated in another group of patients with community-acquired pneumonia (relative risk, 3.5) (Waterer et al., 2003). In addition to the HSP70-2 + 1267 genotype, an LTA + 252 genotype was also measured. Distributions of both HSP70-2 + 1267 and LTA + 252 were consistent with Hardy–Weinberg equilibrium, and linkage disequilibrium was present between the two loci such that the HSP70-2 + 1267 G allele was invariably associated with the LTA + 252 allele. Neither allele predicted mortality (Waterer et al., 2003).

High-mobility Group Box 1 Protein

High-mobility group box 1 (HMGB1) protein, also known as amphoterin, is a nuclear factor and a secreted protein. It acts in the cell nucleus as an architectural chromatin-binding factor that bends DNA and promotes protein assembly on specific DNA loci. Outside the cell, it activates inflammatory responses in immune cells and endothelial cells, and transduces cellular signals through RAGE (the receptor for advanced glycation end products) and other receptors, such as TLR2 and TLR4 (Lotze and Tracey, 2005). HMGB1 is secreted by activated monocytes and macrophages, and is passively released by necrotic or damaged cells (Scaffidi et al., 2002). HMGB1 facilitates activation of the transcription factor NF-κB and mitogen-activated protein kinase (MAPK), leading to downstream expression of proinflammatory mediators in monocytes (Andersson et al., 2000). In endothelial cells, HMGB1 induces intercellular adhesion molecules-1 (ICAM-1), vascular cell adhesion molecule-1 (VCAM-1), and RAGE expression, and secretion of monocyte chemotactic protein-1 (MCP-1), plasminogen activator inhibitor-1 (PAI-1), TNF, IL-8, and tissue plasminogen activator (Fiuza et al., 2003). A summary of intra- and extracellular roles of HMGB1 is shown in Fig. 16–4 (Lotze and Tracey, 2005).

HMGB1 is released in the liver after exposure to toxins and thermal injury, resulting in increased transaminase levels and the production of acute-phase reactants (Fang et al., 2002). Hemorrhagic shock–induced liver damage is associated with increased serum levels of liver enzymes and HMGB1. HMGB1 is passively released by necrotic or damaged cells, but not from cells undergoing apoptotic cell death even after they undergo secondary necrosis or autolysis. HMGB1 is bound to chromatin as a result of under-acetylation of histone, and is released extracellularly if chromatin deacetylation is prevented, thus promoting inflammation. Accordingly, cells undergoing apoptosis are programmed to suppress the signal that is normally broadcast by cells which have been damaged or killed (Fang et al., 2002; Scaffidi et al., 2002; Tsung et al., 2005).

HMGB1 acts as a late mediator in animal models of LPS-induced toxicity, as well as in human sepsis, and late administration of antibodies to HMGB1 reduces LPS lethality in mice (Box 16–6) (Wang et al., 1999; Sunden-Cullberg et al., 2005). HMGB1 is released by cultured macrophages more than 8 hours after stimulation with LPS, TNF, or IL-1 (Wang et al., 1999). In most patients with septic shock, HMGB1 levels remain high for up to 1 week after admission to ICU, whereas other inflammatory mediators have usually returned to baseline levels (Sunden-Cullberg et al., 2005). No associations between sepsis susceptibility or outcome and functional SNPs have yet been reported.

Nitric Oxide, Nitric Oxide Synthase

Nitric oxide (NO) is generated from L-arginine in various tissues by nitric oxide synthase (NOS). There are three isoforms of NOS labeled in relation to their activity or the tissue type of origin. These include neuronal NOS (nNOS), iNOS, and endothelial NOS

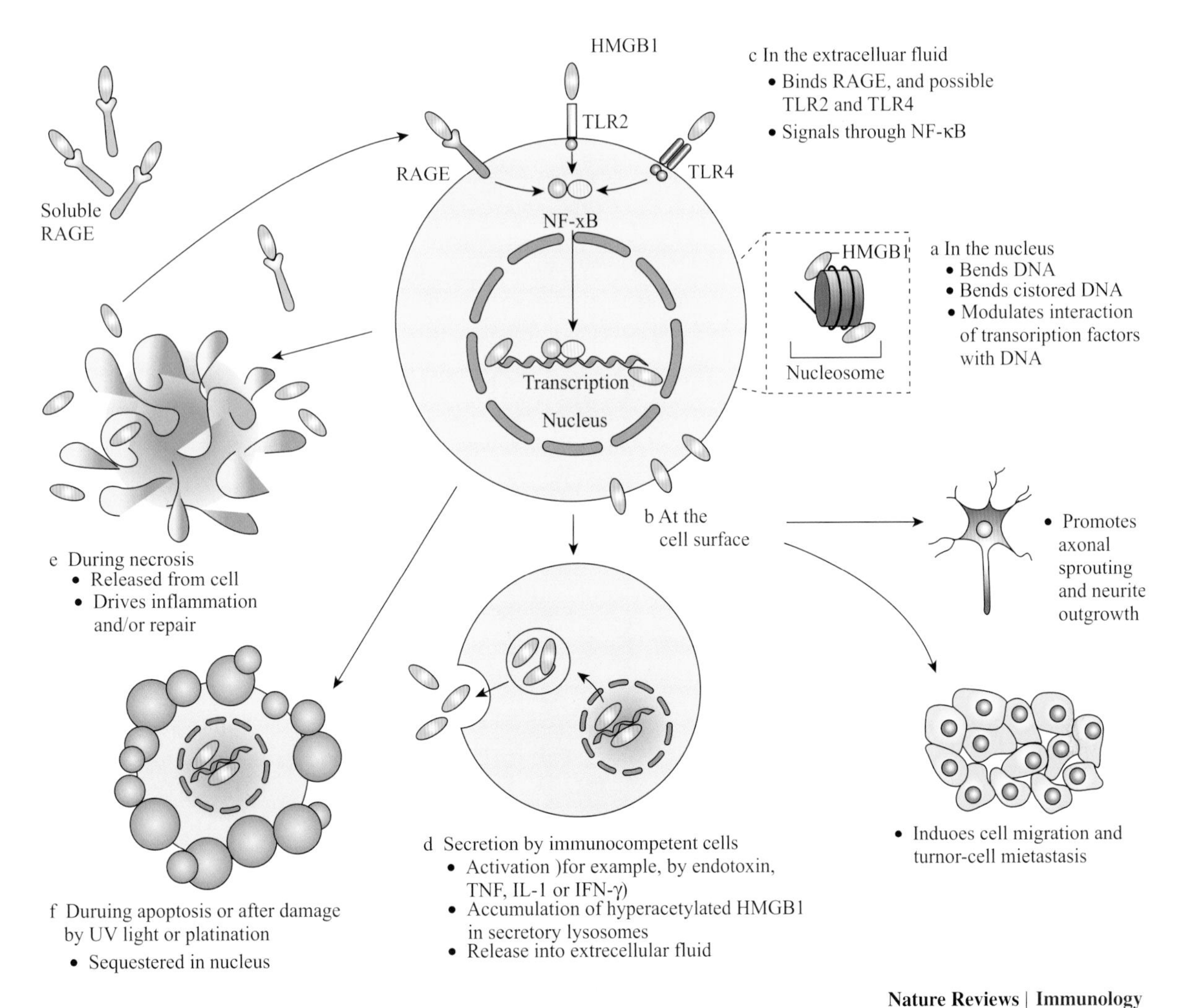

Figure 16–4 Intranuclear and extranuclear roles of high-mobility group box 1 (HMGB1) protein. (*A*) In the nucleus, HMGB1 protein is associated with nucleosomes. HMGB1 is important for spatial segregation and controlling transcription of DNA, and for interaction with the nuclear matrix. (*B*) In the developing nervous system and after axonal damage, HMGB1 present at the cell surface promotes axonal sprouting and neurite outgrowth. Membrane-bound HMGB1 also promotes cell migration and tumor metastasis. (*C*) In the extracellular fluid, HMGB1 signals through RAGE (receptor for advanced glycation end products), TLR2 (Toll-like receptor), and TLR4, activating pathways that involve NF-κB and the mitogen-activated protein kinase p38. (*D*) HMGB1 can be secreted by immunocompetent cells, such as macrophages and dendritic cells, after activation by bacterial products, such as endotoxin, or pro-inflammatory cytokines tumor necrosis factor (TNF), interleukin-1 (IL-1), and interferon-γ (IFN-γ. (*E*) During necrotic cell death, HMGB1 is released from the cell and drives inflammation and repair processes. (*F*) During apoptosis, HMGB1 is sequestered in the nucleus and prevented from release. (Reproduced from Lotze and Tracey, 2005 with permission.)

(eNOS), which are also referred to as NOS1, NOS2, and NOS3, respectively. Despite these labels, all three isoforms can be found in a variety of cell types. iNOS is present in several organ tissues such as the small intestine, lung, and platelets. Although not normally active. iNOS becomes activated in response to proinflammatory factors such as TNF, IL-1, interferon (IFN)-γ, and PAF. In the presence of cofactors such as calmodulin and the substrate L-arginine, NO is produced. NO binds to guanylate cyclase, releasing cyclic guanosine monophosphate (cGMP) (Ravichandran and Johns, 1995). cGMP-mediated effects include relaxation of smooth and myocardial muscle and inhibition of platelet aggregation. NO also reacts with superoxide anion to form peroxynitrite, which has oxidant toxic effects on nucleic acids, lipids, and proteins. Notably, peroxynitrite may impair glycolysis and ATP production by its effects upon mitochondria (Szabo et al., 1997). NO produces a negative feedback on both constitutive and inducible forms of NOS (Ravichandran et al., 1995).

In addition to its vasoactive role, NO is intimately involved in the immunological responses to infection, malignant disease, chronic degenerative disease, and autoimmune processes (Box 16–6). These aspects are well reviewed elsewhere (Bogdan, 2001). The role of NO in sepsis is complex and not entirely resolved. This is in part due to inter-species differences and the use of different models of sepsis. The characteristic vasodilatation seen in human septic shock is attributed to increased NO production (Kilbourn and Griffith, 1992) although other mediators, such as carbon monoxide (CO), also contribute (Yet et al., 1997; Kirkeboen and Strand, 1999; Kaide et al., 2001; Zhang et al., 2001). Pharmacological blockade of NO production in sepsis re-establishes the pressor response to catecholamines (Hollenberg et al., 1993), while iNOS gene deficiency is associated with the loss of NO-induced vasodilation (Gunnett et al., 1998). In addition to the generalized vasodilatory effects, NO influences coronary perfusion, myocardial contractility (Panas et al., 1998), and moderates the vasoconstriction seen in the pulmonary circulation during sepsis (Petros et al., 1994; Cobb et al., 1995).

Indirect evidence of excess NO synthesis in sepsis can be found in the studies that report significant increases in nitrite and nitrate plasma levels during endotoxin shock in animal models and human sepsis (Nava et al., 1991; Evans et al., 1994; MacKenzie et al., 2001). iNOS knockout mice appear, in one study, to be more susceptible to certain types of infection and yet exhibit resistance to LPS-induced shock (Wei et al., 1995). Survival of iNOS knockout mice from sepsis following CLP was reduced in one study (Cobb et al., 1999) and increased in another (Hollenberg et al., 2000). Endotoxin shock may be ameliorated by inhibitors of NO synthesis (Lorente et al., 1993; Kilbourn et al., 1997), and much lower nitrite and nitrate levels are induced by inhibition of NO synthesis inhibition (Evans et al., 1994). In other clinical and animal models of sepsis pharmacological inhibition of NO synthesis may reverse some of the hemodynamic disturbances, but this may not be translated into a survival benefit (Fukatsu et al., 1995; Park et al., 1996; Hobbs et al., 1999).

Various polymorphisms of the eNOS gene have been reported as susceptibility genes in a number of cardiovascular diseases (Veldman et al., 2002; Wang et al., 2002; Boutlis et al., 2003), while interactive effects of the angiotensin-converting enzyme (ACE) DD polymorphism with the NOS3 homozygous G849T variant has been shown to alter endothelial function in coronary artery disease (Schmidt et al., 2003). Polymorphisms in the inducible NOS gene (*NOS2*) promoter region have also been associated with differences in clinical outcome from malaria (Hobbs et al., 2002; Boutlis et al., 2003). Large, well-powered genetic association studies of NO-related polymorphisms in human sepsis have not yet been reported.

Complement System

Complement C3, C5

The complement system is an integral component of the innate immune response and plays an important role in controlling bacterial infections, particularly those caused by Gram-negative organisms (Box 16–6). Complement pathways are activated on bacterial surfaces by LPS, acute-phase protein immune complexes, and other factors. The cleaved complement proteins then assemble membrane pores that lead to lysis of the bacterial cell wall. Not unexpectedly, activation of the complement system and the generation of anaphylatoxins, especially C3a and C5a (Goya et al., 1994), are key features of sepsis and are associated with significant neutrophil dysfunction (Solomkin et al., 1981a,b; Goya et al., 1994; Botha et al., 1995a,b). Generation of C5a during sepsis has been described as "too much of a good thing," and may be associated with a worse outcome (Gerard, 2003). Furthermore, blockade of either C5a or its receptor (C5aR) improves survival of animals with sepsis (Czermak et al., 1999; Huber-Lang et al., 2002a). During experimental sepsis, circulating polymorphonuclear neutrophils (PMNs) have impaired C5a binding (Huber-Lang et al., 2001) due to receptor downregulation (Guo et al., 2003; Huber-Lang et al., 2005) while C5aR is upregulated in an IL-6-dependent manner in the heart, lungs, kidneys, liver, and vasculature (Riedemann et al., 2003b). In an LPS-induced mouse sepsis model, C3aR expression appears to be increased on both epithelial and smooth muscle cells, supporting the role of complement receptors in the systemic inflammatory response of sepsis (Drouin et al., 2001).

In severe human sepsis, C5a is elevated and is associated with reduced survival and increased multiorgan failure, compared to survivors and the less severely septic patients (Nakae et al., 1994; Nakae et al., 1996). The mechanisms by which C5a exerts its harmful effects during sepsis are not fully resolved, but the generation of C5a during sepsis impairs the innate immune functions of PMNs, their ability to generate a respiratory burst and express cytokines, as well as promoting intravascular coagulation (Huber-Lang et al., 2002b; Laudes et al., 2002; Guo et al., 2003; Riedemann et al., 2003a). In sepsis, apoptosis occurs in many organs and cell types through a process linked, at least in part, to C5a and the C5a receptor (C5aR). In the early stages of sepsis, increased expression of C5aR occurs in thymocytes, which increases their susceptibility to C5a-induced apoptosis (Riedemann et al., 2002).

The C5 gene is located in chromosome 9q34.1 and spans 75 kb, interrupted by 41 exon–intron boundaries. C5 complement deficiency, a rare autosomal recessive inherited disease, is associated with recurrent infectious episodes, particularly meningitis (Peter et al., 1981). Sequences in the 5′-flanking region may play a role in the regulation of C5 expression in inflammation. Several polymorphisms are recognized, but no associations with sepsis have been reported.

Pro-inflammatory Mediator Responses in Sepsis

Tumor Necrosis Factor

TNF-α is a key mediator of the inflammatory response to infection (Beutler, 1993). The systemic administration of TNF mimics the symptoms and signs of sepsis and LPS infusion increases serum TNF levels (Eskandari et al., 1992). The increases in TNF levels after infusion of LPS in mice are, however, 300-fold greater than those achieved by CLP, leading many to question whether LPS infusion is a valid model for either animal or human sepsis. Although clinical studies have shown that high TNF levels are associated with a poor outcome from meningococcal infection (Girardin et al., 1988; Tracey and Lowry, 1990), trials of TNF blockade have failed to demonstrate benefit in CLP mice (Eskandari et al., 1992; Newcomb et al., 1998) or in human sepsis (Reinhart et al., 2001) (Box 16–7).

Box 16–7 Examples of Pro-inflammatory Mediators Implicated in Sepsis

Adhesion molecules	IL-1 decoy factors
Chemokines	Leukotrienes
Elastase	MIF
Growth factors	NO
Endothelin	Prostaglandins
IL-1β	PAF
IL-6	Proteases
IL-8	RANTES
IL-12	Reactive oxygen intermediates
IL-15	TNF
IL-18	Thromboxanes

Specific HLA markers have been shown to be associated with variations in TNF expression (Price et al., 1999). The highly polymorphic nature of the TNF locus has been recognized, with many SNPs and microsatellites being identified. Several SNPs within the TNF/LTA locus are thought to influence TNF production, and have therefore been identified as candidate genes that might influence susceptibility to infectious disease and its outcome (Box 16–7). Interest has particularly focused on the promoter TNF-308 G/A SNP (Price et al., 1999) and the LTA −252 G/A SNP, which are now known to be in linkage disequalibrium. Several studies have found an association between the A allele at TNF-308 and a predisposition to septic shock and a worse outcome from sepsis, but findings have been contradictory (Stuber et al., 1995; Mira et al., 1999; Tang et al., 2000; Appoloni et al., 2001; Waterer et al., 2001; O'Keefe et al., 2002; Gordon et al., 2004). Moreover, although the A allele of the −308 SNP has been shown to increase TNF transcription in some in vitro studies, in others this has not been the case, leading some to question the functional importance of this SNP (Brinkman et al., 1995; Wilson et al., 1997). It is possible that linkage disequilibrium (LD), which is known to exist between this SNP and other functionally important polymorphisms such as LTA − 252 in the region, in part explains these contradictions. Thus, for example, in a study of 280 patients with community-acquired pneumonia the risk of developing septic shock was associated with the ATA-252AA genotype (high secretors of TNF), whereas no significant association was found between the TNF-308A allele and susceptibility to sepsis, septic shock, or death. Interestingly, however, there was an association between a specific LTA + 252:TNF-308 haplotype and the risk of septic shock (Waterer et al., 2001). In contrast, in a recent study of more than 200 Caucasian patients with severe sepsis there were no significant associations between TNF and its receptor polymorphisms, or their haplotypes and susceptibility, disease progression or outcome, despite plasma levels of TNF and its receptors being significantly higher in those patients who died (Gordon et al., 2004).

IL-1 and IL-1ra

IL-1 is a proinflammatory cytokine produced by monocytes, macrophages, dendritic cells, and a variety of other cells. It has effects similar to TNF, and acts synergistically with TNF to enhance inflammation (Dinarello and Wolff, 1993). In human sepsis, increased levels of IL-1 are detectable in plasma in many but not all cases, compared with control, non-septic patients (Casey et al., 1993; Samson et al., 1997) (Box 16–7). There is a parallel increase in the levels of IL-1ra, which is several orders of magnitude (1000-fold) greater than for the levels of IL-1β. Cells from women with sepsis produce significantly less IL-1β in response to stimulation with LPS, possibly because of alterations in MAPK phosphorylation (Imahara et al., 2005).

There are three IL-1 genes in a cluster on human chromosome 2q13. Two of these genes encode proinflammatory proteins, IL-1A produces IL-1α and IL-1B produces IL-1β The third gene, *IL-1RN*, produces IL-1ra. IL-1ra is a naturally occurring inhibitor of IL-1 activity, which competes with IL-1 for cell-surface receptors, but without agonist activity (Vannier et al., 1992). A polymorphic region within intron 2 of the *IL-1ra* gene contains a variable number of 86 bp tandem repeats, and six alleles at this polymorphic site have been identified (Ku et al., 1992). The allele frequency in healthy control subjects is 54% for IL-lraA1 (four repeats) and 34% for IL-1raA2 (two repeats), with the other alleles being much less common (Haskill et al., 1991). The IL-1raA2 allele occurs with increased frequency in patients with severe sepsis compared with normal individuals (Vannier et al., 1992), but does not appear to be associated with a worse outcome (Haskill et al., 1991). However, coincidence of homozygous TNF-A2 allele and IL-1raA2 genotypes identified a group with a 100% mortality rate from sepsis, suggesting a potentially important interaction of TNF and IL-1ra (Haskill et al., 1991). A later study of the same SNP revealed a six-fold increase in the risk of death associated with lower plasma levels of IL-1ra in critically ill patients with sepsis (Arnalich et al., 2002).

Interleukin-6

IL-6 is a cytokine crucial in both adaptive and innate immunity with both pro- and anti-inflammatory effects, which is produced by many cells including T-lymphocytes, macrophages, monocytes, endothelial cells, and fibroblasts. IL-6 stimulates the liver to produce acute phase proteins, stimulates the proliferation of B-lymphocytes, and increases neutrophil production. Serum concentrations of IL-6 increase during sepsis, higher levels being associated with an increased incidence of shock and a poor outcome (Reinhart et al., 2001; Wunder et al., 2004; Oda et al., 2005) (Box 16–7). IL-6 has a longer half-life than either TNF-a or IL-1B, and the measurement of IL-6 blood levels may predict severity and outcome not only in sepsis (Presterl et al., 1997; van der Poll and van Deventer, 1999) but also in trauma (Martin et al., 1997), severe acute pancreatitis (Berney et al., 1999), and cardiogenic shock (Geppert et al., 2002). IL-6 is an important inducer of the C5aR in sepsis (Riedemann et al., 2004). C5a exerts proinflammatory effects such as chemotactic responses of neutrophils, release of granular enzymes from phagocytes, vasodilatation, increased vascular permeability, and induction of thymocyte apoptosis during sepsis. In humans, C5a has been described as a serum marker that correlates with the severity of sepsis (Nakae et al., 1994).

In longitudinal studies of human sepsis, the most frequently detectable plasma cytokines were TNF-α and IL 6, with higher levels of IL 6 being found in non-surviving patients (Munoz et al., 1991). This study also found no correlation between plasma cytokine levels and cell-associated cytokines, suggesting that activation of monocytes may still be present in the absence of detectable plasma levels (Munoz et al., 1991). IL-6 knockout mice appear to have elevated levels of proinflammatory cytokines in response to pulmonary infection with *Streptococcus pneumoniae* and increased mortality, compared with wild-type animals (van der Poll et al., 1997).

Recently, a G to C polymorphism at −174 in the promoter region of the *IL-6* gene, found on chromosome 7, was reported (Fishman et al., 1998). This polymorphism has been associated with an adverse outcome in a number of inflammatory diseases, although reports including those relating to sepsis have been conflicting (Fishman et al., 1998; Burzotta et al., 2001; Heesen et al., 2002a; Heesen et al., 2002b; Marshall et al., 2002b). In a recent study in critically ill SIRS patients, there was no association between the −174 G or C alleles and outcome. However, this SNP contributed to the IL-6 C/C/G, G/G/G, and G/C/C haplotype clades, which were associated with a poor outcome and more organ failure (Sutherland et al., 2005b). Thus, a more complex regulatory haplotype that extends upstream of the promoter region may provide a better measure of IL-6 regulation in several disease phenotypes including SIRS and sepsis (Fife et al., 2005; Sutherland and Russell, 2005; Sutherland et al., 2005).

Chemokines

Chemokines comprise a superfamily of small cytokine-like molecules that regulate white cell transport by promoting the adhesion of leukocytes to endothelial cells and transendothelial leukocyte migration, and tissue invasion. Chemokines, produced by leukocytes, endothelial cells, epithelial cells, fibroblasts, facilitate the migration of leukocytes from the intravascular space to the extravascular tissues at local and systemic sites of inflammation. Chemokines also trigger the release of cytotoxic cytokines from leukocytes and promote phagocytosis of damaged cells. In excess, chemokines can lead to tissue damage as seen in pneumonia (Standiford et al., 1996), ARDS (Donnelly et al., 1993; Miller et al., 1996; Wiedermann et al., 2004), and sepsis (Miller et al., 1996). Examples include IL-8 (CXCL8), MIP-1α (CCL3), MIP-1β (CCL4), MCP-1 (CCL2), MCP-2 (CCL8), MCP-3 (CCL7), GRO-α (CXCL1), GRO-β (CXCL2), GRO-γ (CXCL3), and RANTES (regulated on activation normal T cell expressed and secreted) (CCL5). Approximately, 50 chemokines have been identified that fall into four subclasses (CXC, CC, C, and CX3C). These classes form the basis of a new nomenclature derived from established chemokine gene classification.

For a comprehensive review of the role of chemokines in sepsis and septic shock, the reader is referred to the excellent review by Murdoch and Finn (Murdoch and Finn, 2003) (Box 16–7). Several human studies show raised chemokine levels in sepsis and septic shock. Bossink et al. measured circulating MCP-1 (CCL2) and MCP-2 (CCL8) levels in septic and bacteraemic patients. CCL2 and CCL8 levels were elevated in more than half of the septic patients, compared with healthy volunteers. Levels of CCL2 were highest in patients with septic shock or fatal outcome, and serial observations showed that chemokine plasma levels remained elevated for 48 hours (Bossink et al., 1995).

Vermont et al. measured the chemokines MCP-1 (CCL2), MIP 1α, (CCL3), GRO-α (CXCL1), and IL-8 (CXCL8) in children with meningococcal sepsis or septic shock on admission to ICU and at 24 hours. High levels of all chemokines were measured in the childrens' acute-phase sera, and were significantly higher in non-survivors and in patients with septic shock. Twenty-four hours after admission, chemokine levels were much reduced (Vermont et al., 2006).

Other CXC chemokines have been associated with the development and persistence of sepsis-related ARDS (Goodman et al., 1996; Kiehl et al., 1998; Adams et al., 2001). Logically, one might predict that the chemokine receptor that binds these ligands would be downregulated in sepsis. This does appear to be the case for CXCR2 (Cummings et al., 1999), and in separate work α-chemokine receptor blockade has been found to reduce HMGB1-induced inflammatory injury in mice (Feterowski et al., 2004; Lin et al., 2005).

IL-8 (CXCL8) polymorphism has been shown to be associated with *Helicobacter pylori*-induced duodenal ulcer disease (Gyulai et al., 2004), and with susceptibilty to *Clostridium difficile* diarrhea (Jiang et al., 2006). Diffuse panbronchiolitis (DPB), a distinctive chronic inflammatory lung disease of Asian populations, has an association with an IL-8 (CXCL8) microsatelite polymorphism (Emi et al., 1999). AA genotype at the −251 position of the IL-8 (*CXCL8*) gene was associated with the occurrence of enteroaggregative *E. coli*-associated diarrhea and increased levels of fecal IL-8 (Jiang et al., 2003).

RANTES (CCL5) gene polymorphism (position-403) is associated with rheumatoid arthritis and polymyalgia rheumatica (Makki, et al., 2000; Wang et al., 2005a), while a −2518 promotor polymorphism in the *MCP-1* gene has been shown to have an association with systemic sclerosis (Karrer et al., 2005). Polymorphism of MCP-1 (CCL2) and its promoter may determine the disease behavior in Crohn's disease (Herfarth et al., 2003) and the development of lupus nephritis (Tucci et al., 2004).

To date, there have been no reported associations between chemokine gene polymorphism and the susceptibility to and severity of sepsis in humans.

Interferons

IFNs are cytokines produced by the cells of the immune system in response to challenges by viruses, bacteria, parasites, and tumor cells. There are three major classes of IFNs in humans: type I consisting of 13 different α-isoforms (including IFN-α and IFN-β; type II consisting of IFN-γ, its sole member; and a recently discovered third class, IFN-λ with three isoforms.

IFN-γ is a 34-kDa cytokine secreted by activated monocytes, T-lymphocytes, and natural killer (NK) cells. IFN-γ promotes T-cell differentiation to the Th1 phenotype, increases MHC class I and II expression on various cell types, activates mononuclear phagocytes, and increases production of IL-12, TNF, and positively feeds back to induce further IFN-γ expression (Lammas et al., 2000). IFN-γ intron 1 polymorphisms have been shown to play a role in hypersensitivity diseases, the severity of rheumatoid arthritis and the development of pulmonary fibrosis after lung transplantation (Awad et al., 1999; Khani-Hanjani et al., 2000; Winning et al., 2006), while homozygotes for the D allele of the intronic IFN-γ gene have an increased risk of sepsis after traumatic injury (Stassen et al., 2002).

Antiinflammatory Cytokine Responses in Sepsis

Interleukin-10

IL-10, an anti-inflammatory cytokine, is an inhibitor of activated macrophages and dendritic cells, and as such regulates innate immunity and cell-mediated immunity. T-helper lymphocytes and macrophages produce IL-10 that suppresses the production of IFN-γ IL-1, IL-6, and TNF-a (Terry et al., 2000). In critically ill patients with SIRS, a change in the ratio of IL-6 to IL-10 predicts a poor outcome (Taniguchi et al., 1999a,b). Polymorphisms of the IL-10 promoter region on chromosome 1, IL-10–592A and IL-10–819T, are associated with lower IL-10 production and a better response to IFN-γ therapy in hepatitis C than the other haplotypes (Edwards-Smith et al., 1999). Lowe et al. have shown that the C allele of the SNP at position "592 bp" in the promoter region of the *IL-10* gene is associated with the survival of critically ill patients and linked to increased IL-10 release (Lowe et al., 2003). In a later study, a polymorphism at position −1082 in the promoter region of the *IL-10* gene was also found to be associated with susceptibility to severe sepsis(Box 16–8). However, in this later report the −592 SNP was not associated with the incidence or the outcome of severe sepsis (Shu et al., 2003).

Box 16–8 Examples of Antiinflammatory Cytokines Implicated in Sepsis

IL-4
IL-6
IL-10
IL-11
IL-13
IL-ra
Leptin
NO
TGF-B
TNFr

Adhesion Molecules

Adhesion molecules play a pivotal role in cellular interactions and the response to perceived pathogen invasion or attack (Box 16–9). Leukocyte adhesion facilitates T lymphocyte and NK cell cytotoxicity, B cell maturation to antibody-secreting plasma cells, leukocyte migration, and myeloid cell phagocytosis and chemotaxis. Adhesion molecules can be classified within several molecular families that include integrins, cadherins, the immunoglobulin superfamily, and selectins. Leukocyte-specific integrins, β_2 integrins, (CD11:CD18) are among the most important. The β_2 integrins bind to ICAMs and to other soluble proteins, many of which are involved in inflammation. Leukocyte integrins are not constitutively active, but require activation to exhibit adhesive properties. Leukocyte (L)-selectin, and endothelial (E)-selectin and platelet (P)-selectin, mediate the initial step of leukocyte adhesion and rolling on activated endothelium. ICAM and VCAM encourage firm adhesion between leukocytes and endothelial cells

Box 16–9 Examples of Adhesion Molecules Implicated in Sepsis

- Integrins
- Cadherins
- Immunoglobulin superfamily
- Selectins

Adhesion molecules appear to perform an important pathophysiological role in response to sepsis, contributing to tissue injury and organ failure by facilitating leukocyte recruitment and activation. A significant reduction in mortality rate (from 45% to 12.5%) was observed in ICAM-1$^{-/-}$ mice versus WT in a CLP model. Furthermore, no significant histological changes were apparent in pulmonary or hepatic tissue, and pro- and anti-inflammatory cytokine synthesis was significantly attenuated in ICAM-1$^{-/-}$ animals (Hildebrand et al., 2005). However, in a baboon sepsis model the administration of ICAM-1 antibodies not only increased IL-1, IL-6, IL-8, and TNFR-1 but was also associated with a worse outcome (Welty-Wolf et al., 2000). Elevated plasma levels of E-selectin, ICAM-1, and VCAM-1 have been measured in children with sepsis, compared with non-septic controls. Persistent elevation of ICAM-1 and VCAM-1 was characteristic of those children with multi-organ failure (Whalen et al., 2000).

ICAM-1 has been described as being functionally promiscuous, since it is a ligand for leukocyte integrins and fibrinogen and also a receptor for a number of organisms such as *Plasmodium falciparum* and some viruses (Jenkins et al., 2005). Surprisingly, an ICAM-1 polymorphism (ICAM-1^{Kilifi}) has been shown to be associated with an increased risk of cerebral complication and death from malaria, compared with the wild type. However, this polymorphism, which is common in sub-Saharan Africa, may be protective in non-malarial febrile illnesses, reflecting a more general effect on the inflammatory response to bacterial infections (Jenkins et al., 2005). Other ICAM-1 polymorphisms may also be protective against severe malaria (Amodu et al., 2005).

It has been postulated that suppression of cell adhesion within the lung might prevent the development of ARDS, but this has not been confirmed in studies using antibody directed against ICAM-1 (Welty-Wolf et al., 2001). In a study of pneumonia patients, an ICAM Gly241Arg polymorphism (associated with polymyalgia rheumatica, ulcerative colitis, and giant-cell arteritis) was unrelated to the subsequent development of either sepsis or ARDS (Quasney et al., 2002). Decreased plasma levels of ICAM-1 have been demonstrated in schizophrenics, which appear not to be associated with the ICAM-1 G241A SNP (unlike non-schizophrenic controls) (Schwarz et al., 2000).

Cell Energetics/Cell Survival

Mitochondrial DNA

Mitochondrial dysfunction has been invoked as a pathophysiological mechanism in animal models of endotoxic shock, hemorrhagic shock, and human sepsis (Box 16–10) (Singer and Brealey, 1999; Fink, 2001; Sayeed, 2002). Other reports have also investigated the modulation of mitochondrial function by pharmacologic agents and mediators such as NO (Brealey et al., 2002; Crouser et al., 2002; Davies et al., 2003).

Box 16–10 Examples of Pathways Involving Cell Energetics or Survival That May Be Implicated in Sepsis

- Mitochondrial DNA
- Angiotensin converting enzyme
- Cell apoptosis

Several essential proteins in the mitochondrial respiratory cascade are encoded by human mitochondrial DNA (mtDNA), which is maternally transmitted (DiMauro and Schon, 2003). Within the mitochondrial genome, there are a number of mtDNA haplogroups that incorporate specific SNPs (Ruiz-Pesini et al., 2004; Baudouin et al., 2005). Recent evidence suggests that the most common European mtDNA haplogroup, haplogroup H, is associated with enhanced respiratory-chain activity and may be responsible for increased survival in sepsis (Ruiz-Pesini et al., 2000; Baudouin et al., 2005).

Angiotensin-converting Enzyme

Large inter-subject variations in plasma ACE concentrations have been demonstrated in humans, but these are less marked within families, consistent with a genetic effect. There is evidence to suggest that activation of a pulmonary renin–angiotensin system might influence the pathogenesis of ARDS via such mechanisms as vasomotor tone, vascular permeability, fibroblast activity, and increased alveolar epithelial cell apoptosis. It has been shown recently that recombinant ACE2 can protect mice from severe acute lung injury, possibly by inactivating angiotensin II (Imai et al., 2005).

The human ACE gene (*DCP1*), located on chromosome 17q23, contains a restriction fragment length polymorphism as an insertion, (I) or deletion, (D) of a 287-bp alu repeat sequence in intron 16. The DD genotype frequency is increased in patients with ARDS compared with the ICU and population-based controls, and is associated with increased mortality in the patients with ARDS (Marshall et al., 2002; Quasney, 2006). Extending this observation to sepsis has produced inconsistent findings (Box 16–10). In a group of ventilated, very low birth weight infants, no effect of the I/D polymorphism on survival from sepsis could be demonstrated

(Baier et al., 2005), while in children with meningococcal disease the DD allele appears to be associated with ncreased illness severity (Harding et al., 2002).

Cell Apoptosis

Cells can die either as a result of damage to their cell membrane followed by necrosis or by a process of programmed cell death known as apoptosis. Apoptosis is a normal cellular process that facilitates tissue remodeling and development, and allows a variety of tissue cells to be selectively removed (Hengartner, 2000). The regulation of apoptosis can be broadly considered as a process mediated by membrane signaling through a family of "death-receptors," such as Fas, TNF-type I receptor (TNF-R1/R2), TRAIL-receptors (DR4/5), or mediated via mitochondrial function in association with a variety of mediators such as corticosteroids, reactive oxygen species, NO, chemokines, and cytokines, or due to a loss of essential growth factors such as the interleukins (IL-2, IL-4) or granulocyte–macrophage-colony stimulating factor (GM-CSF) (Chung et al., 2000; Hengartner, 2000; Ayala et al., 2003).

Sepsis patients exhibit alterations in apoptosis in a variety of tissues including immune cells, and extensive lymphocyte apoptosis is seen in experimental and human sepsis (Box 16–10) (Hotchkiss et al., 1999). Septic patients are often lymphopenic, and analysis of autopsy tissue samples has shown selective depletion of B and CD4+ lymphocytes (Hotchkiss et al., 2001). The consequences of lymphocyte depletion, immune hyporesponsiveness, and anergy are believed to contribute to the state of relative immunosuppression (CARS) (Bone, 1996), which contrasts with the marked inflammatory response seen in earlier stages of sepsis. Recent evidence suggests that apoptosis, particularly of monocytes, might be an early feature of recovery from sepsis (Giamarellos-Bourboulis et al., 2006).

Although there is general acceptance that apoptosis is increased in lymphoid cells during sepsis, evidence for parenchymal cell apoptosis in other organs during sepsis is less convincing. In an animal endotoxemia model, hepatocyte apoptosis is only seen with the simultaneous presence of transcriptional inhibition by either D-galactosamine or actinomycin-D or the presence of caspase-3 activity (Leist et al., 1995; Mignon et al., 1999). Remarkably, hepatocyte apoptosis appears not to be common in other models of sepsis or endotoxaemic shock. Hepatocyte apoptosis is not seen in animals dying of high-dose endotoxin shock, although hepatocyte necrosis may be widespread (Bohlinger et al., 1996). Autopsy evidence from patients dying with sepsis reveals increased apoptosis in gut lymphoid tissues, spleen, and thymus. In these autopsy studies, increased apoptosis is evident in epithelial cells of the gut, but not in parenchymal cells of the heart, lungs, kidney, or liver (Hotchkiss et al., 1999). Marked increases in activated caspase-3 and reduced Bcl-2 expression were seen only in tissues where apoptosis was evident (Fig. 16–5).

TNF and Fas ligand (FasL), glucocorticoids, and granzymes induce apoptosis, while other cytokines such as IL-1, IL-6, and G-CSF are able to inhibit the process. Increased levels of TNF, IL-1β, IL-6, and G-CSF have been described in sepsis, trauma and endotoxaemia (Casey et al., 1993; Martin et al., 1997). In addition, increased FasL expression has been reported in rodent models of endotoxemia and in generalized peritonitis (Tannahill et al., 1999); elevated soluble Fas ligand has also been detected in cerebrospinal and bronchial fluid of critically ill patients with ARDS and head trauma (Ertel et al., 1997; Matute-Bello et al., 1999; Albertine et al., 2002). For a comprehensive discussion of the role of apoptosis in sepsis and an exploration of the possible therapeutic opportunities, the reader is directed to the excellent review by Oberholzer et al. (2001).

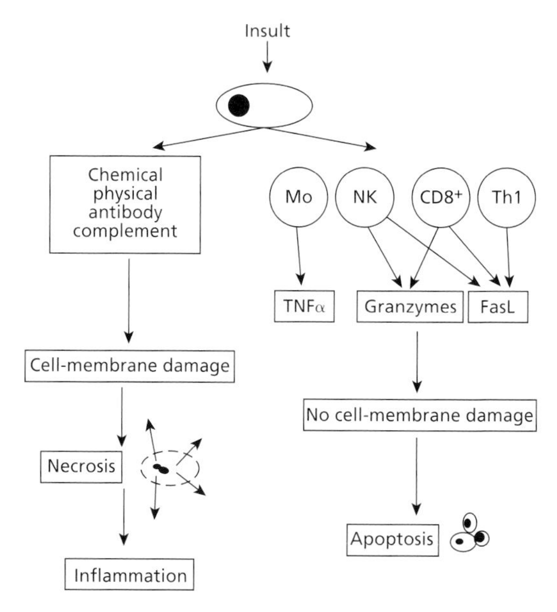

Figure 16–5 Insult-dependent pathways leading to necrosis or apoptosis. Cells can die in two ways: either by cell membrane–damaging mechanisms like chemical, physical, complement, or antibody injury leading to cellular necrosis, or by programmed cell death (apoptosis) induced by soluble or membrane-bound tumor necrosis factor (TNF) or Fas ligand (FasL) produced by macrophages (Mo), cytotoxic-T cells (CD8+), or T helper cells (TH1). Moreover, cytotoxic T cells (CD8+) or natural killer (NK) cells can induce apoptosis in the target cells through the release of granzymes. (Reproduced from Oberholzer et al., 2001 with permission.)

Fas signaling plays an important role in lymphocyte homeostasis. Repeated activation of antigen receptors on T cells induces FasL expression, leading to Fas-transduced apoptosis. Fas and FasL exist in both membrane-bound and soluble forms. Fas (APO-1/CD95) is a type I transmembrane protein and a member of the TNF-R family (Ashkenazi and Dixit, 1998). The ligand of membrane-bound Fas, FasL (CD95L or APO-1L), is synthesized as a transmembrane molecule and belongs to the family of TNF-related cytokines.

The alveolar epithelial injury in humans with ALI or ARDS has been shown in part to be associated with the local upregulation of the Fas/FasL system and activation of the apoptotic cascade in the alveolar epithelial cells (Albertine et al., 2002).

Caspases mediate proteolytic events in inflammatory cascades and in the apoptotic cell death pathway. Human caspases form into two distinct subgroups: caspase-1, -4, and -5 involved in cytokine maturation and caspase-2, -3, -6, -7, -8, -9, and -10 that are involved in cellular apoptosis. Although caspase-12 is related to the former group, it has been proposed as a mediator of apoptosis induced by endoplasmic reticulum stress. Csp12-L, a mediator of apoptosis, attenuates the inflammatory and innate immune response after LPS challenge and may constitute a risk factor for developing sepsis (Winoto, 2004). In a preliminary study, Saleh et al. demonstrated that the frequency of the Csp12-L allele was increased in African-American patients with severe sepsis (Saleh et al., 2004).

The apoptosis gene, *Bcl-2*, was initially implicated in a common form of leukaemia, but later found to play a role in the inhibition of apoptosis. There are likely to be a large number of key genes involved in apoptosis, which can be grouped into categories according to their functional and structural features. They include, but are not limited to, the TNF ligands (including Fas), TNF receptors, Bcl-2, caspases, IAP (inhibitor of apoptosis), TRAF (TNF receptor-associated factor), CARD (caspase recruitment domain), and death domain family members. Evidence has also suggested a role for the anti-apoptotic protein Bcl-2 in the development of autoimmune diseases, including type 1 diabetes mellitus. A Bcl-2 polymorphism (Ala43Thr) appears to be associated with type 1 diabetes mellitus in a Japanese population, but not in Danish, Finnish, and Basque Type 1 diabetes families (Heding et al., 2001).

Due to the number of effective pro- and anti-apoptotic factors, the role of apoptosis in the genesis and outcome from sepsis is likely to be complex, particularly in regard to the effect on the cells of the immune system. There are currently no comprehensive or systematic studies of genetic polymorphism in apoptosis during sepsis in humans although this remains a potentially fruitful area for investigation.

Coagulation Cascade

Abnormal coagulation is common in sepsis, with almost half of such patients manifesting the most severe clinical variant, DIC (Box 16–11) (Levi and ten Cate, 1999). The procoagulant state is induced primarily by proinflammatory cytokine mediators such as IL-1 and IL-6, in contrast to anti-inflammatory cytokines like IL-10 that prevent coagulation by inhibiting the expression of tissue factor (TF) (van der Poll et al., 1999). Animal models show that lack of TF can reduce mortality associated with septic shock, and that deficiency of thrombomodulin and components of the PC pathway can increase endotoxin-induced mortality (Weiler et al., 2001; Levi et al., 2003; Pawlinski et al., 2004). Downregulation of anticoagulant proteins (antithrombin, PC, and TF pathway inhibitor) accompanies the procoagulant state of sepsis. These natural anticoagulants have anti-inflammatory properties, by inhibiting activation of NF-κB and activator protein-1 (Okajima, 2001) (Fig. 16–6).

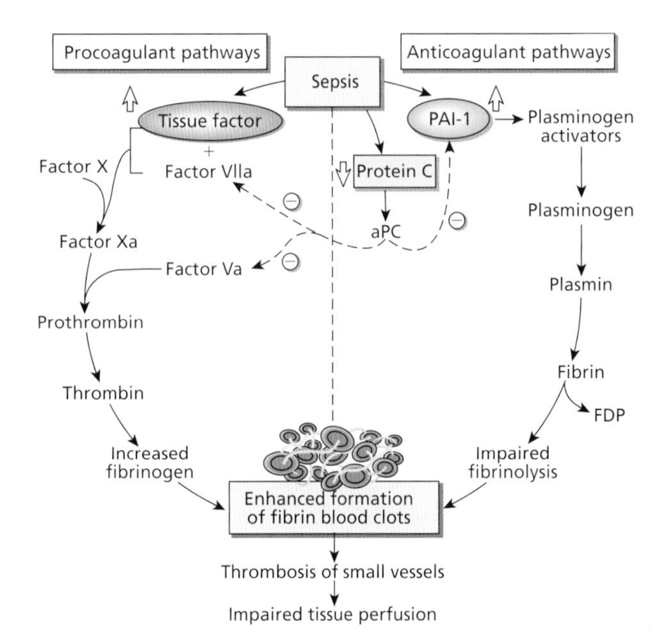

Figure 16–6 Coagulant pathways in sepsis. Tissue factor expression is enhanced leading to increased production of prothrombin that is converted to thrombin, and that in turn generates fibrin from fibrinogen. Simultaneously, levels of the plasminogen-activator inhibitor-1 (PAI-1) are increased, resulting in impaired production of plasmin and thus failure of normal fibrinolytic mechanisms by which fibrin is converted to degradation products (FDP). Sepsis also causes a fall in the levels of the natural anticoagulant protein C. The activated form of protein C, aPC, dissociates from the endothelial protein C receptor to inactivate factors Va and VIIa and inhibit PAI-1 activity; hence reduced levels of protein C result in further procoagulant effect. The net result is enhanced formation of fibrin clots in the microvasculature, leading to impaired tissue oxygenation and cell damage. (Reproduced from Cohen, 2002 with permission.)

Box 16–11 Examples of Coagulation Factors Implicated in Sepsis

- Protein C
- Endothelial cell protein-C receptor
- Factor V leiden
- Plasminogen activator inhibitor-1
- Antithrombin III
- Tissue factor
- Thrombomodulin

Activated Protein C

PC is a vitamin K-dependent zymogen of a serine protease. The active form, aPC, regulates the blood coagulation cascade by the inactivation of activated factors, Va and VIIIa. The normal expression of aPC on endothelial cells requires the presence and interaction of thrombin, thrombomodulin, PC, and endothelial cell PC receptor (EPCR), and failure of any one of these components will result in deficient aPC generation and a procoagulant, pro-thombotic state (Esmon et al., 1997).

aPC exerts anti-inflammatory effects by downregulating thrombin generation and by its actions on factors Va and VIIIa, by interfering with thrombin-induced proinflammatory activities, and by upregulation of leukocyte adhesion molecules. aPC also directly suppresses inflammation by inhibiting macrophage expression of TF, TNF, and NF-κB translocation (de Wouwer et al., 2004).

Acquired PC deficiency in severe sepsis patients is associated with lower survival rates and higher incidence of shock (Yan et al., 2001). The important role of PC was emphasized by a phase III, randomized, controlled trial of recombinant human aPC in severe sepsis patients (PROWESS), in which administration of aPC was responsible for a 19% relative reduction in mortality (Bernard et al., 2001).

The gene for PC, located on chromosome 2q13–q14, is 11 kb in length and contains nine exons (Miao et al., 1996). The gene shares significant organizational similarity with the genes coding for the other vitamin K-dependent proteins such as factors VII, IX, and X, prothrombin, protein S, and protein Z. PC polymorphism has been described with respect to thrombotic risk, but not so far with regard to the evolution of sepsis (Aiach et al., 1999).

Endothelial Cell PC Receptor

EPCR is expressed either as a membrane-bound form at the surface of large vessels or as a plasma-soluble form (sEPCR) as a result of metalloprotease-mediated shedding of membrane-bound EPCR (Esmon, 2003). The membrane-bound EPCR increases PC activation, while sEPCR binds PC and aPC with the same affinity but has an inverse effect on PC activity. Thus, PC binding to sEPCR inhibits the anticoagulant activity of a APC by blocking the inactivation of factors Va and VIIIa. The human EPCR gene is located on chromosome 20, spans 6 kb, and encodes 238 amino acids. Several polymorphisms that may have pro-thrombotic activity have been described, but functional associations in sepsis have not been reported (Simmonds and Lane, 1999; Biguzzi et al., 2001; Espana et al., 2002).

Factor V Leiden

Factor V Leiden mutation (R506Q) is a prothrombotic gene polymorphism that renders factor Va partially resistant to inactivation by aPC, and might be predicted to modify the beneficial effects of aPC. However, a systematic examination of almost 4000 severe sepsis patients (combined PROWESS and ENHANCE studies) failed to reveal any reduction in the therapeutic benefit of aPC in Factor V heterozygous carriers (no homozygotes were available to study) (Yan and Nelson, 2004). In meningococcal sepsis, patients heterozygous for factor V Leiden experienced more complications of purpura fulminans in the form of more skin grafting and more amputations of ischemic limbs. Mortality, however, was not altered (Kondaveeti et al., 1999). In contrast, heterozygosity for factor V Leiden appears to carry a survival benefit in a mouse LPS model (Kerlin et al., 2003) and in the PROWESS study on patients not receiving a PC.

Plasminogen Activator Inhibitor-1

Increased PAI-1 plasma levels are associated with a worse outcome in patients with sepsis and septic shock (Kornelisse et al., 1996; Mesters et al., 1996a,b). In meningococcal sepsis, PAI-1 levels are directly related to the severity and outcome of the disease, circulating cytokine levels and coagulation factors (Hermans et al., 1999). A well-studied polymorphism within the promoter region of the *PAI-1* gene is the biallelic single base-pair insertion/deletion polymorphism at position −675. The homozygous 4G/4G genotype leads to higher PAI-1 concentrations on stimulation with IL-1β, and is associated with non-survival in meningococcal septic shock (Dawson et al., 1993; Kornelisse et al., 1996). A study by Hermans et al. (Hermans et al., 1999) showed that the 4G/4G genotype was associated with a two-fold greater risk of death from meningococcal disease compared with those with the 4G/5G or 5G/5G genotype.

Antithrombin III

The serpin antithrombin III (AT III), is a natural inhibitor of thrombin activity with anti-inflammatory effects that are at least in part mediated by blockade of the activation of NF-κB (Oelschlager et al., 2002). Although AT III replacement has decreased mortality in experimental models of sepsis, its use in the early stages of adult, human sepsis appears to offer no survival benefit (Warren et al., 2001).

The AT III gene (*SERPINC1*) is located at 1q23–q25.1. A number of polymorphisms are described that result in amino acid change, but no association with sepsis has as yet been explored.

Tissue Factor

TF is a 47-kDa membrane-bound protein expressed in many tissues including hematopoietic cells. TF is not normally expressed by circulating blood cells or by endothelial cells, but during sepsis a variety of stimuli such as proinflammatory cytokines increase its expression on macrophages and initiate the coagulation cascade. Two TF gene-promoter haplotypes have been described—1208 D (deletion) and 1208 I (insertion). The wild type, 1208 D haplotype, is associated with reduced plasma TF concentration and a lower risk of venous thrombotic episodes (Arnaud et al., 2000). In a mouse endotoxemia model, mice bred to express low levels of tissue TF have reduced coagulation, inflammation, and mortality compared with control mice (Pawlinski et al., 2004). No associations in human sepsis have yet been explored.

Thrombomodulin

A point mutation into the mouse TM (THBD) gene (Glu404Pro) disrupts the thrombin–TM complex and impairs the activation of PC. Mice with this mutation have increased mortality after intraperitoneal injection of endotoxin (Weiler et al., 2001). In humans, functional polymorphisms in the general population are rare and have been described primarily in families with venous thromboembolic disease, but not in association with sepsis (Kottke-Marchant, 2002).

Conclusion

Genomic influences on disease susceptibility, progression, and outcome in critical illness, and particularly sepsis, are complex and remain poorly understood. This chapter has illustrated this complexity and highlighted the growing number of novel pathways that may be involved. Advances in this area of research have considerable potential to enhance our knowledge of disease pathogenesis and to improve patient management. An expanding body of evidence indicates that an individual's risk of sepsis and his or her response to treatment is significantly modulated by genetic factors. To date, most research has focused on susceptibility to sepsis and outcome through a candidate gene approach. Although this approach has identified some notable disease associations, it has also been plagued by difficulties, in common with other complex disease traits, relating to phenotypic and alleleic heterogeneity, linkage disequilibrium, and study design. Further large, well-powered studies, which also incorporate functional genomic approaches, are required if we are to begin understanding the effects of genetic variation. Adopting genome-wide association approaches offers significant potential for improved detection of modest multiple gene effects, as is likely to be the case in complex traits such as sepsis.

References

Abreu, MT, Arnold, ET, et al. (2002). TLR4 and MD-2 expression is regulated by immune-mediated signals in human intestinal epithelial cells. *J Biol Chem*, 277(23):20431–20437.

Abreu, MT, Vora, P, et al. (2001). Decreased expression of Toll-like receptor-4 and MD-2 correlates with intestinal epithelial cell protection against dysregulated proinflammatory gene expression in response to bacterial lipopolysaccharide. *J Immunol*, 167(3):1609–1616.

Adams, JM, Hauser, CJ, et al. (2001). Early trauma polymorphonuclear neutrophil responses to chemokines are associated with development of sepsis, pneumonia, and organ failure. *J Trauma*, 51(3):452–456; discussion 456–457.

Aderem, A and Ulevitch, RJ (2000). Toll-like receptors in the induction of the innate immune response. *Nature*, 406(6797):782–787.

Aiach, M, Nicaud, V, et al. (1999). Complex association of protein C gene promoter polymorphism with circulating protein C levels and thrombotic risk. *Arterioscler Thromb Vasc Biol*, 19(6):1573–1576.

Akashi, S, Nagai, Y, et al. (2001). Human MD-2 confers on mouse Toll-like receptor 4 species-specific lipopolysaccharide recognition. *Int Immunol*, 13 (12):1595–1599.

Akashi, S, Shimazu, R, et al. (2000). Cutting edge: cell surface expression and lipopolysaccharide signaling via the toll-like receptor 4-MD-2 complex on mouse peritoneal macrophages. *J Immunol*, 164(7):3471–3475.

al-Ramadi, BK, Ellis, M, et al. (2004). Selective induction of pentraxin 3, a soluble innate immune pattern recognition receptor, in infectious episodes in patients with haematological malignancy. *Clin Immunol*, 112 (3):221–224.

Albertine, KH, Soulier, MF, et al. (2002). Fas and Fas ligand are up-regulated in pulmonary edema fluid and lung tissue of patients with acute lung injury and the acute respiratory distress syndrome. *Am J Pathol*, 161 (5):1783–1796.

Aliprantis, AO, Yang, RB, et al. (1999). Cell activation and apoptosis by bacterial lipoproteins through toll-like receptor-2. *Science*, 285 (5428):736–739.

Alles, VV, Bottazzi, B, et al. (1994). Inducible expression of PTX3, a new member of the pentraxin family, in human mononuclear phagocytes. *Blood*, 84(10):3483–3493.

Altavilla, D, Squadrito, G, et al. (2002). Inhibition of nuclear factor-κB activation by IRFI 042, protects against endotoxin-induced shock. *Cardiovasc Res*, 54(3):684–693.

Ammons, WS, Kohn, FR, et al. (1994). Protective effects of an N-terminal fragment of bactericidal/permeability-increasing protein in rodent models of gram-negative sepsis: role of bactericidal properties. *J Infect Dis*, 170 (6):1473–1482.

Amodu, OK, Gbadegesin, RA, et al. (2005). *Plasmodium falciparum*, malaria in south-west Nigerian children: is the polymorphism of ICAM-1 and E-selectin genes contributing to the clinical severity of malaria? *Acta Trop*, 95(3):248–255.

Andersson, U, Wang, H, et al. (2000). High mobility group 1 protein (HMG-1) stimulates proinflammatory cytokine synthesis in human monocytes. *J Exp Med*, 192(4):565–570.

Angus, DC and Wax, RS (2001). Epidemiology of sepsis: an update. *Crit Care Med*, 29(Suppl. 7): S109–S116.

Annane, D, Sebille, V, et al. (2002). Effect of treatment with low doses of hydrocortisone and fludrocortisone on mortality in patients with septic shock. *JAMA*, 288(7): 862–871.

Appoloni, O, Dupont, E, et al. (2001). Association of tumor necrosis factor-2 allele with plasma tumor necrosis factor-α levels and mortality from septic shock. *Am J Med*, 110(6):486–488.

Arbour, NC, Lorenz, E, et al. (2000). TLR4 mutations are associated with endotoxin hyporesponsiveness in humans. *Nat Genet*, 25 (2):187–191.

Armstrong, L, Medford, AR, et al. (2004). Differential expression of Toll-like receptor (TLR)-2 and TLR-4 on monocytes in human sepsis. *Clin Exp Immunol*, 136(2):312–319.

Arnalich, F, Garcia-Palomero, E, et al. (2000). Predictive value of nuclear factor κB activity and plasma cytokine levels in patients with sepsis. *Infect Immun*, 68(4):1942–1945.

Arnalich, F, Lopez, J, et al. (1999). Relationship of plasma leptin to plasma cytokines and human survival in sepsis and septic shock. *J Infect Dis*, 180 (3):908–911.

Arnalich, F, Lopez-Maderuelo, D, et al. (2002). Interleukin-1 receptor antagonist gene polymorphism and mortality in patients with severe sepsis. *Clin Exp Immunol*, 127(2):331–336.

Arnaud, E, Barbalat, V, et al. (2000). Polymorphisms in the 5′ regulatory region of the tissue factor gene and the risk of myocardial infarction and venous thromboembolism: the ECTIM and PATHROS studies. Etude Cas-Temoins de l'Infarctus du Myocarde. Paris Thrombosis case-control Study. *Arterioscler Thromb Vasc Biol*, 20(3):892–898.

Arnon, S, Litmanovitz, I, et al. (2002). Serum amyloid A protein in the early detection of late-onset bacterial sepsis in preterm infants. *J Perinat Med*, 30 (4):329–332.

Arnon, S, Litmanovitz, I, et al. (2005). Serum amyloid A protein is a useful inflammatory marker during late-onset sepsis in preterm infants. *Biol Neonate*, 87(2):105–110.

Ashkenazi, A and Dixit, VM (1998). Death receptors: signaling and modulation. *Science*, 281(5381):1305–1308.

Awad, M, Pravica, V, et al. (1999). CA repeat allele polymorphism in the first intron of the human interferon-γ gene is associated with lung allograft fibrosis. *Hum Immunol*, 60(4):343–346.

Ayala, A, Lomas, JL, et al. (2003). Pathological aspects of apoptosis in severe sepsis and shock? *Int J Biochem Cell Biol*, 35(1):7–15.

Baier, JR, Loggins, J, et al. (2005). Angiotensin converting enzyme insertion/deletion polymorphism does not alter sepsis outcome in ventilated very low birth weight infants. *J Perinatol*, 25(3):205–209.

Baldini, M, Lohman, IC, et al. (1999). A Polymorphism* in the 5′ flanking region of the CD14 gene is associated with circulating soluble CD14 levels and with total serum immunoglobulin E. *Am J Respir Cell Mol Biol*, 20 (5):976–983.

Barber, RC, Aragaki, CC, et al. (2004). TLR4 and TNF-α polymorphisms are associated with an increased risk for severe sepsis following burn injury. *J Med Genet*, 41(11):808–813.

Barnes, PJ and Karin, M (1997). Nuclear factor-κB—a pivotal transcription factor in chronic inflammatory diseases. *N Engl J Med*, 336(15):1066–1071.

Barton, A, Lamb, R, et al. (2003). Macrophage migration inhibitory factor (MIF) gene polymorphism is associated with susceptibility to but not severity of inflammatory polyarthritis. *Genes Immun*, 4(7):487–491.

Baudouin, SV, Saunders, D, et al. (2005). Mitochondrial DNA and survival after sepsis: a prospective study. *Lancet*, 366(9503):2118–2121.

Becker, KL, Nylen, ES, et al. (2004). Clinical review 167: procalcitonin and the calcitonin gene family of peptides in inflammation, infection, and sepsis: a journey from calcitonin back to its precursors. *J Clin Endocrinol Metab*, 89 (4):1512–1525.

Becker, MN, Diamond, G, et al. (2000). CD14-dependent lipopolysaccharide-induced β-defensin-2 expression in human tracheobronchial epithelium. *J Biol Chem*, 275(38):29731–29736.

Beishuizen, A, Thijs, LG, et al. (2001). Macrophage migration inhibitory factor and hypothalamo-pituitary-adrenal function during critical illness. *J Clin Endocrinol Metab*, 86(6):2811–2816.

Bernard, GR, Vincent, JL, et al. (2001). Efficacy and safety of recombinant human activated protein C for severe sepsis. *N Engl J Med*, 344 (10):699–709.

Berney, T, Gasche, Y, et al. (1999). Serum profiles of interleukin-6, interleukin-8, and interleukin-10 in patients with severe and mild acute pancreatitis. *Pancreas*, 18(4):371–377.

Beutler, B (1993). Endotoxin, tumor necrosis factor, and related mediators: new approaches to shock. *New Horiz*, 1(1):3–12.

Biguzzi, E, Merati, G, et al. (2001). A 23 bp insertion in the endothelial protein C receptor (EPCR) gene impairs EPCR function. *Thromb Haemost*, 86 (4):945–948.

Blackwell, TS and JW Christman (1997). The role of nuclear factor-κB in cytokine gene regulation. *Am J Respir Cell Mol Biol*, 17(1):3–9.

Blanco-Quiros, A, Casado-Flores, J, et al. (2002). Influence of leptin levels and body weight in survival of children with sepsis. *Acta Paediatr*, 91(6):626–631.

Bogdan, C (2001). Nitric oxide and the immune response. *Nat Immunol*, 2(10):907–916.

Bohlinger, I, Leist, M, et al. (1996). DNA fragmentation in mouse organs during endotoxic shock. *Am J Pathol*, 149(4):1381–1393.

Bohrer, H, Qiu, F, et al. (1997). Role of NFκB in the mortality of sepsis. *J Clin Invest*, 100(5):972–985.

Bone, RC (1996). Sir Isaac Newton, sepsis, SIRS, and CARS. *Crit Care Med*, 24 (7):1125–1128.

Bornstein, SR, Licinio, J, et al. (1998). Plasma leptin levels are increased in survivors of acute sepsis: associated loss of diurnal rhythm, in cortisol and leptin secretion. *J Clin Endocrinol Metab*, 83(1):280–283.

Bossink, AW, Paemen, L, et al. (1995). Plasma levels of the chemokines monocyte chemotactic proteins-1 and -2 are elevated in human sepsis. *Blood*, 86 (10):3841–3847.

Botha, AJ, Moore, FA, et al. (1995a). Postinjury neutrophil priming and activation states: therapeutic challenges. *Shock*, 3(3):157–166.

Botha, AJ, Moore, FA, et al. (1995b). Postinjury neutrophil priming and activation: an early vulnerable window. *Surgery*, 118(2):358–364; discussion 364–365.

Bouchon, A, Facchetti, F, et al. (2001). TREM-1 amplifies inflammation and is a crucial mediator of septic shock. *Nature*, 410(6832):1103–1107.

Boutlis, CS, Hobbs, MR, et al. (2003). Inducible nitric oxide synthase (NOS2) promoter CCTTT repeat polymorphism: relationship to in vivo nitric oxide production/NOS activity in an asymptomatic malaria-endemic population. *Am J Trop Med Hyg*, 69(6):569–573.

Bozza, M, Satoskar, AR, et al. (1999). Targeted disruption of migration inhibitory factor gene reveals its critical role in sepsis. *J Exp Med*, 189(2):341–346.

Brealey, D, Brand, M, et al. (2002). Association between mitochondrial dysfunction and severity and outcome of septic shock. *Lancet*, 360 (9328):219–223.

Brinkman, BM, Zuijdeest, D, et al. (1995). Relevance of the tumor necrosis factor α (TNF-α) − 308 promoter polymorphism in TNF α gene regulation. *J Inflamm*, 46(1):32–41.
Brun-Buisson, C, Doyon, F, et al. (1995). Incidence, risk factors, and outcome of severe sepsis and septic shock in adults. A multicenter prospective study in intensive care units. French ICU Group for Severe Sepsis. *JAMA*, 274 (12):968–974.
Burzotta, F, Iacoviello, L, et al. (2001). Relation of the − 174 G/C polymorphism of interleukin-6 to interleukin-6 plasma levels and to length of hospitalization after surgical coronary revascularization. *Am J Cardiol*, 88 (10):1125–1128.
Buzzard, KA, Giaccia, AJ, et al. (1998). Heat shock protein 72 modulates pathways of stress-induced apoptosis. *J Biol Chem*, 273(27):17147–17153.
Calandra, T, Echtenacher, B, et al. (2000). Protection from septic shock by neutralization of macrophage migration inhibitory factor. *Nat Med*, 6 (2):164–170.
Calandra, T and Roger, T (2003). Macrophage migration inhibitory factor: a regulator of innate immunity. *Nat Rev Immunol*, 3(10):791–800.
Calandra, T, Spiegel, LA, et al. (1998). Macrophage migration inhibitory factor is a critical mediator of the activation of immune cells by exotoxins of Gram-positive bacteria. *Proc Natl Acad Sci USA*, 95(19):11383–11388.
Cario, E, Rosenberg, IM, et al. (2000). Lipopolysaccharide activates distinct signaling pathways in intestinal epithelial cell lines expressing Toll-like receptors. *J Immunol*, 164(2):966–972.
Cariou, A, Chiche, JD, et al. (2002). The era of genomics: impact on sepsis clinical trial design. *Crit Care Med*, 30(Suppl. 5): S341–S348.
Carlson, CS, Aldred, SF, et al. (2005). Polymorphisms within the C-reactive protein (CRP) promoter region are associated with plasma CRP levels. *Am J Hum Genet*, 77(1):64–77.
Carlson, CS, Eberle, MA, et al. (2004). Mapping complex disease loci in whole-genome association studies. *Nature*, 429(6990):446–452.
Carlson, GL, Saeed, M, et al. (1999). Serum leptin concentrations and their relation to metabolic abnormalities in human sepsis. *Am J Physiol*, 276(4 Pt 1): E658–E662.
Carlsson, M, Eriksson, L, et al. (2003). *Pseudomonas*-induced lung damage in cystic fibrosis correlates to bactericidal-permeability increasing protein (BPI)-autoantibodies. *Clin Exp Rheumatol*, 21(6 Suppl. 32): S95–S100.
Casey, LC, Balk, RA, et al. (1993). Plasma cytokine and endotoxin levels correlate with survival in patients with the sepsis syndrome. *Ann Intern Med*, 119(8):771–778.
Chan, JY, Ou, CC, et al. (2004). Heat shock protein 70 confers cardiovascular protection during endotoxemia via inhibition of nuclear factor-κB activation and inducible nitric oxide synthase expression in the rostral ventrolateral medulla. *Circulation*, 110(23):3560–3566.
Chauhan, SD, Seggara, G, et al. (2003). Protection against lipopolysaccharide-induced endothelial dysfunction in resistance and conduit vasculature of iNOS knockout mice. *FASEB J*, 17(6):773–775.
Chen, D, Pan, J, et al. (2005). Induction of the heat shock response in vivo inhibits NF-κB activity and protects murine liver from endotoxemia-induced injury. *J Clin Immunol*, 25(5):452–461.
Chen, HW, Hsu, C, et al. (2000). Attenuation of sepsis-induced apoptosis by heat shock pretreatment in rats. *Cell Stress Chaperones*, 5(3):188–195.
Chen, HW, Hsu, C, et al. (2003). Heat shock pretreatment prevents cardiac mitochondrial dysfunction during sepsis. *Shock*, 20(3):274–279.
Chen, HW, Kuo, HT, et al. (2005). In vivo heat shock protein assembles with septic liver NF-κB/I-κB complex regulating NF-κB activity. *Shock*, 24(3):232–238.
Chiu, KC, Chu, A, et al. (2004). Association of leptin receptor polymorphism with insulin resistance. *Eur J Endocrinol*, 150(5):725–729.
Christ-Crain, M, Morgenthaler, NG, et al. (2005). Mid-regional pro-adrenomedullin as a prognostic marker in sepsis: an observational study. *Crit Care*, 9(6): R816–R824.
Chuang, T and Ulevitch, RJ (2001). Identification of hTLR10: a novel human Toll-like receptor preferentially expressed in immune cells. *Biochim Biophys Acta*, 1518(1–2):157–161.
Chuang, TH and Ulevitch, RJ (2000). Cloning and characterization of a sub-family of human toll-like receptors: hTLR7, hTLR8 and hTLR9. *Eur Cytokine Netw*, 11(3):372–378.
Chung, CS, Song, GY, et al. (2000). Septic mucosal intraepithelial lymphoid immune suppression: role for nitric oxide not interleukin-10 or transforming growth factor-β. *J Trauma*, 48(5):807–812; discussion 812–813.
Chung, TP, Laramie, JM, et al. (2002). Functional genomics of critical illness and injury. *Crit Care Med*, 30(Suppl. 1): S51–S57.
Cobb, JP, Hotchkiss, RS, et al. (1999). Inducible nitric oxide synthase (iNOS) gene deficiency increases the mortality of sepsis in mice. *Surgery*, 126 (2):438–442.
Cobb, JP, Natanson, C, et al. (1995). Differential hemodynamic effects of L-NMMA in endotoxemic and normal dogs. *Am J Physiol*, 268(4 Pt 2): H1634–H1642.
Cohen, J (2002). The immunopathogenesis of sepsis. *Nature*, 420 (6917):885–891.
Cooke, GS and Hill, AV (2001). Genetics of susceptibility to human infectious disease. *Nat Rev Genet*, 2(12):967–977.
Crouser, ED, Julian, MW, et al. (2002). Cyclosporin A ameliorates mitochondrial ultrastructural injury in the ileum during acute endotoxemia. *Crit Care Med*, 30(12):2722–2728.
Cummings, CJ, Martin, TR, et al. (1999). Expression and function of the chemokine receptors CXCR1 and CXCR2 in sepsis. *J Immunol*, 162 (4):2341–2346.
Czermak, BJ, Sarma, V, et al. (1999). Protective effects of C5a blockade in sepsis. *Nat Med*, 5(7):788–792.
D'Aiuto, F, Casas, JP, et al. (2005). C-reactive protein (+ 1444C>T) polymorphism influences CRP response following a moderate inflammatory stimulus. *Atherosclerosis*, 179(2):413–417.
D'Andrea, G, D'Ambrosio, RL, et al. (2005). A polymorphism in the VKORC1 gene is associated with an interindividual variability in the dose-anticoagulant effect of warfarin. *Blood*, 105(2):645–649.
Davies, NA, Cooper, CE, et al. (2003). Inhibition of mitochondrial respiration during early stage sepsis. *Adv Exp Med Biol*, 530:725–736.
Dawson, SJ, Wiman, B, et al. (1993). The two allele sequences of a common polymorphism in the promoter of the plasminogen activator inhibitor-1 (PAI-1) gene respond differently to interleukin-1 in HepG2 cells. *J Biol Chem*, 268(15):10739–10745.
De Bosscher, K, Vanden Berghe, W, et al. (2000). Glucocorticoids repress NF-κ B-driven genes by disturbing the interaction of p65 with the basal transcription machinery, irrespective of coactivator levels in the cell. *Proc Natl Acad Sci USA*, 97(8):3919–3924.
de Haas, CJ, van der Zee, R, et al. (1999). Lipopolysaccharide (LPS)-binding synthetic peptides derived from serum amyloid P component neutralize LPS. *Infect Immun*, 67(6):2790–2796.
De Maio, A (1999). Heat shock proteins: facts, thoughts, and dreams. *Shock*, 11 (1):1–12.
de Wouwer, MV, Collen, D, et al. (2004). Thrombomodulin-protein C-EPCR system: integrated to regulate coagulation and inflammation. *Arterioscler Thromb Vasc Biol*, 24(8):1374–1383.
Dean, MM, Minchinton, RM, et al. (2005). Mannose binding lectin acute phase activity in patients with severe infection. *J Clin Immunol*, 25(4):346–352.
Determann, RM, Millo, JL, et al. (2005). Serial changes in soluble triggering receptor expressed on myeloid cells in the lung during development of ventilator-associated pneumonia. *Intensive Care Med*, 31(11):1495–1500.
DiMauro, S and EA Schon (2003). Mitochondrial respiratory-chain diseases. *N Engl J Med*, 348(26):2656–2668.
Dinarello, CA and Wolff, SM (1993). The role of interleukin-1 in disease. *N Engl J Med*, 328(2):106–113.
Donnelly, SC, Strieter, RM, et al. (1993). Interleukin-8 and development of adult respiratory distress syndrome in at-risk patient groups. *Lancet*, 341 (8846):643–647.
Dorlochter, L, Carlsson, M, et al. (2004). Anti-neutrophil cytoplasmatic antibodies and lung disease in cystic fibrosis. *J Cyst Fibros*, 3(3):179–183.
Drouin, SM, Kildsgaard, J, et al. (2001). Expression of the complement anaphylatoxin C3a and C5a receptors on bronchial epithelial and smooth muscle cells in models of sepsis and asthma. *J Immunol*, 166(3):2025–2032.
Du, X, Poltorak, A, et al. (2000). Three novel mammalian toll-like receptors: gene structure, expression, and evolution. *Eur Cytokine Netw*, 11 (3):362–371.

Edwards-Smith, CJ, Jonsson, JR, et al. (1999). Interleukin-10 promoter polymorphism predicts initial response of chronic hepatitis C to interferon alfa. *Hepatology*, 30(2):526–530.

Eisen, DP, Liley, HG, et al. (2004). Alternatives to conventional vaccines—mediators of innate immunity. *Curr Drug Targets*, 5(1):89–105.

Ellis, RJ and van der Vies, SM (1991). Molecular chaperones. *Annu Rev Biochem*, 60:321–347.

Emi, M, Keicho, N, et al. (1999). Association of diffuse panbronchiolitis with microsatellite polymorphism of the human interleukin 8 (IL-8) gene. *J Hum Genet*, 44(3):169–172.

Enguix, A, Rey, C, et al. (2001). Comparison of procalcitonin with C-reactive protein and serum amyloid for the early diagnosis of bacterial sepsis in critically ill neonates and children. *Intensive Care Med*, 27 (1):211–215.

Ertel, W, Keel, M, et al. (1997). Detectable concentrations of Fas ligand in cerebrospinal fluid after severe head injury. *J Neuroimmunol*, 80(1–2):93–96.

Eskandari, MK, Bolgos, G, et al. (1992). Anti-tumor necrosis factor antibody therapy fails to prevent lethality after cecal ligation and puncture or endotoxemia. *J Immunol*, 148(9):2724–2730.

Esmon, CT (2003). The protein C pathway. *Chest*, 124(Suppl. 3):26S–32S.

Esmon, CT, Ding, W, et al. (1997). The protein C pathway: new insights. *Thromb Haemost*, 78(1):70–74.

Espana, F, Medina, P, et al. (2002). Inherited abnormalities in the protein C activation pathway. *Pathophysiol Haemost Thromb*, 32(5–6):241–244.

Eto, T (2001). A review of the biological properties and clinical implications of adrenomedullin and proadrenomedullin N-terminal 20 peptide (PAMP), hypotensive and vasodilating peptides. *Peptides*, 22(11):1693–1711.

Evans, T, Carpenter, A, et al. (1994). Inhibition of nitric oxide synthase in experimental gram-negative sepsis. *J Infect Dis*, 169(2):343–349.

Evans, WE (2004). Pharmacogenetics of thiopurine S-methyltransferase and thiopurine therapy. *Ther Drug Monit*, 26(2):186–191.

Fan, J, Ye, RD, et al. (2001). Transcriptional mechanisms of acute lung injury. *Am J Physiol Lung Cell Mol Physiol*, 281(5): L1037–L1050.

Fang, WH, Yao, YM, et al. (2002). The significance of changes in high mobility group-1 protein mRNA expression in rats after thermal injury. *Shock*, 17 (4):329–333.

Fang, XM, Schroder, S, et al. (1999). Comparison of two polymorphisms of the interleukin-1 gene family: interleukin-1 receptor antagonist polymorphism contributes to susceptibility to severe sepsis. *Crit Care Med*, 27 (7):1330–1334.

Faure, E, Thomas, L, et al. (2001). Bacterial lipopolysaccharide and IFN-γ induce Toll-like receptor 2 and Toll-like receptor 4 expression in human endothelial cells: role of NF-κB activation. *J Immunol*, 166(3):2018–2024.

Ferrero, E, Jiao, D, et al. (1993). Transgenic mice expressing human CD14 are hypersensitive to lipopolysaccharide. *Proc Natl Acad Sci USA*, 90 (6):2380–2384.

Feterowski, C, Mack, M, et al. (2004). CC chemokine receptor 2 regulates leukocyte recruitment and IL-10 production during acute polymicrobial sepsis. *Eur J Immunol*, 34(12):3664–3673.

Fidler, KJ, Wilson, P, et al. (2004). Increased incidence and severity of the systemic inflammatory response syndrome in patients deficient in mannose-binding lectin. *Intensive Care Med*, 30(7):1438–1445.

Fife, MS, Ogilvie, EM, et al. (2005). Novel IL-6 haplotypes and disease association. *Genes Immun*, 6(4):367–370.

Fink, MP (2001). Cytopathic hypoxia. Mitochondrial dysfunction as mechanism contributing to organ dysfunction in sepsis. *Crit Care Clin*, 17(1):219–237.

Fisher, CJ Jr., Agosti, JM, et al. (1996). Treatment of septic shock with the tumor necrosis factor receptor:Fc fusion protein. The Soluble TNF Receptor Sepsis Study Group. *N Engl J Med*, 334(26):1697–1702.

Fishman, D, Faulds, G, et al. (1998). The effect of novel polymorphisms in the interleukin-6 (IL-6) gene on IL-6 transcription and plasma IL-6 levels, and an association with systemic-onset juvenile chronic arthritis. *J Clin Invest*, 102(7):1369–1376.

Fiuza, C, Bustin, M, et al. (2003). Inflammation-promoting activity of HMGB1 on human microvascular endothelial cells. *Blood*, 101(7):2652–2660.

Fowler, DE and Wang, P (2002). The cardiovascular response in sepsis: proposed mechanisms of the beneficial effect of adrenomedullin and its binding protein (review). *Int J Mol Med*, 9(5):443–449.

Freeman, BD and McLeod, HL (2004). Challenges of implementing pharmacogenetics in the critical care environment. *Nat Rev Drug Discov*, 3(1):88–93.

Frevert, CW, Matute-Bello, G et al. (2000). Effect of CD14 blockade in rabbits with *Escherichia coli* pneumonia and sepsis. *J Immunol*, 164 (10):5439–5445.

Fritz, JH and Girardin, SE (2005). How toll-like receptors and nod-like receptors contribute to innate immunity in mammals. *J Endotoxin Res*, 11(6):390–394.

Froon, AH, Dentener, MA, et al. (1995). Lipopolysaccharide toxicity-regulating proteins in bacteremia. *J Infect Dis*, 171(5):1250–1257.

Fujihara, M, Muroi, M, et al. (2003). Molecular mechanisms of macrophage activation and deactivation by lipopolysaccharide: roles of the receptor complex. *Pharmacol Ther*, 100(2):171–194.

Fukatsu, K, Saito, H, et al. (1995). Detrimental effects of a nitric oxide synthase inhibitor (N-omega-nitro-L-arginine-methyl-ester) in a murine sepsis model. *Arch Surg*, 130(4):410–414.

Furuta, T, Ohashi, K, et al. (1998). Effect of genetic differences in omeprazole metabolism on cure rates for *Helicobacter pylori* infection and peptic ulcer. *Ann Intern Med*, 129(12):1027–1030.

Gabay, C and Kushner, I (1999). Acute-phase proteins and other systemic responses to inflammation. *N Engl J Med*, 340(6):448–454.

Garred, P, J Strøm J, et al. (2003). Association of mannose-binding lectin polymorphisms with sepsis and fatal outcome, in patients with systemic inflammatory response syndrome. *J Infect Dis*, 188(9):1394–1403.

Georgopoulos, C and Welch, WJ (1993). Role of the major heat shock proteins as molecular chaperones. *Annu Rev Cell Biol*, 9:601–634.

Geppert, A, Steiner, A, et al. (2002). Multiple organ failure in patients with cardiogenic shock is associated with high plasma levels of interleukin-6. *Crit Care Med*, 30(9):1987–1994.

Gerard, C (2003). Complement C5a in the sepsis syndrome—too much of a good thing? *N Engl J Med*, 348(2):167–169.

Giamarellos-Bourboulis, E, Routsi, C, et al. (2006). Early apoptosis of blood monocytes in the septic host: is it a mechanism of protection in the event of septic shock? *Crit Care*, 10(3):R76.

Gibot, S, Cariou, A, et al. (2002). Association between a genomic polymorphism within the CD14 locus and septic shock susceptibility and mortality rate. *Crit Care Med*, 30(5):969–973.

Gibot, S, Cravoisy, A, et al. (2004). Soluble triggering receptor expressed on myeloid cells and the diagnosis of pneumonia. *N Engl J Med*, 350 (5):451–458.

Gibot, S, Massin, F, et al. (2005). Surface and soluble triggering receptor expressed on myeloid cells-1: expression patterns in murine sepsis. *Crit Care Med*, 33(8):1787–1793.

Girardin, E, Grau, GE, et al. (1988). Tumor necrosis factor and interleukin-1 in the serum of children with severe infectious purpura. *N Engl J Med*, 319 (7):397–400.

Girardin, SE and Philpott, DJ (2004). Mini-review: the role of peptidoglycan recognition in innate immunity. *Eur J Immunol*, 34(7):1777–1782.

Giroir, BP, Quint, PA, et al. (1997). Preliminary evaluation of recombinant amino-terminal fragment of human bactericidal/permeability increasing protein in children with severe meningococcal sepsis. *Lancet*, 350 (9089):1439–1443.

Glazier, AM, Nadeau, JH, et al. (2002). Finding genes that underlie complex traits. *Science*, 298(5602):2345–2349.

Goodman, RB, Strieter, RM, et al. (1996). Inflammatory cytokines in patients with persistence of the acute respiratory distress syndrome. *Am J Respir Crit Care Med*, 154(3 Pt 1):602–611.

Gordon, AC, Lagan, AL, et al. (2004). TNF and TNFR polymorphisms in severe sepsis and septic shock: a prospective multicentre study. *Genes Immun*, 5(8):631–640.

Gordon, AC, Waheed, U, et al. (2006). Mannose-binding lectin polymorphisms in severe sepsis: Relationship to levels, incidence, and outcome. *Shock*, 25(1):88–93.

Goya, T, Morisaki, T, et al. (1994). Immunologic assessment of host defense impairment in patients with septic multiple organ failure: relationship between complement activation and changes in neutrophil function. *Surgery*, 115(2):145–155.

Gray, PW, Corcorran, AE, et al. (1993). The genes for the lipopolysaccharide binding protein (LBP) and the bactericidal permeability increasing protein (BPI) are encoded in the same region of human chromosome 20. *Genomics*, 15(1):188–190.

Gregersen, PK and Bucala, R (2003). Macrophage migration inhibitory factor, MIF alleles, and the genetics of inflammatory disorders: incorporating disease outcome into the definition of phenotype. *Arthritis Rheum*, 48 (5):1171–1176.

Grunfeld, C, Zhao, C, et al. (1996). Endotoxin and cytokines induce expression of leptin, the ob gene product, in hamsters. *J Clin Invest*, 97(9):2152–2157.

Grunig, G, Corry, DB, et al. (1997). Interleukin-10 is a natural suppressor of cytokine production and inflammation in a murine model of allergic bronchopulmonary aspergillosis. *J Exp Med*, 185(6):1089–1100.

Gunnett, CA, Chu, Y, et al. (1998). Vascular effects of LPS in mice deficient in expression of the gene for inducible nitric oxide synthase. *Am J Physiol*, 275 (2 Pt 2): H416–H421.

Guo, RF, Riedemann, NC, et al. (2003). Neutrophil C5a receptor and the outcome in a rat model of sepsis. *FASEB J*, 17(13):1889–1891.

Gyulai, Z, Klausz, G, et al. (2004). Genetic polymorphism of interleukin-8 (IL-8) is associated with *Helicobacter pylori*-induced duodenal ulcer. *Eur Cytokine Netw*, 15(4):353–358.

Halaas, JL, Gajiwala, KS, et al. (1995). Weight-reducing effects of the plasma protein encoded by the obese gene. *Science*, 269(5223):543–546.

Hamann, L, Kumpf, O, et al. (2004). A coding mutation within the first exon of the human MD-2 gene results in decreased lipopolysaccharide-induced signaling. *Genes Immun*, 5(4):283–288.

Hansen, TK, Thiel, S, et al. (2003). Intensive insulin therapy exerts antiinflammatory effects in critically ill patients and counteracts the adverse effect of low mannose-binding lectin levels. *J Clin Endocrinol Metab*, 88(3):1082–1088.

Harding, D, Baines, PB, et al. (2002). Severity of meningococcal disease in children and the angiotensin-converting enzyme insertion/deletion polymorphism. *Am J Respir Crit Care Med*, 165(8):1103–1106.

Haskill, S, Martin, G, et al. (1991). cDNA cloning of an intracellular form of the human interleukin 1 receptor antagonist associated with epithelium. *Proc Natl Acad Sci USA*, 88(9):3681–3685.

Hayes, MA, Timmins, AC, et al. (1994). Elevation of systemic oxygen delivery in the treatment of critically ill patients. *N Engl J Med*, 330(24):1717–1722.

Haziot, A, Ferrero, E, et al. (1996). Resistance to endotoxin shock and reduced dissemination of gram-negative bacteria in CD14-deficient mice. *Immunity*, 4(4):407–414.

Heding, PE, Karlsen, AE, et al. (2001). No evidence of a functionally significant polymorphism of the BCL2 gene in Danish, Finnish and Basque type 1 diabetes families. *Genes Immun*, 2(7):398–400.

Heesen, M, Bloemeke, B, et al. (2002a). Can the interleukin-6 response to endotoxin be predicted? Studies of the influence of a promoter polymorphism of the interleukin-6 gene, gender, the density of the endotoxin receptor CD14, and inflammatory cytokines. *Crit Care Med*, 30(3):664–669.

Heesen, M, Obertacke, U, et al. (2002b). The interleukin-6 G(-174)C polymorphism and the ex vivo interleukin-6 response to endotoxin in severely injured blunt trauma patients. *Eur Cytokine Netw*, 13(1):72–77.

Hegazy, DM, O'Reilly, DA, et al. (2001). NFκB polymorphisms and susceptibility to type 1 diabetes. *Genes Immun*, 2(6):304–308.

Hengartner, MO (2000). The biochemistry of apoptosis. *Nature*, 407 (6805):770.

Henneke, P and DT Golenbock (2002). Innate immune recognition of lipopolysaccharide by endothelial cells. *Crit Care Med*, 30(Suppl. 5): S207–S213.

Herfarth, H, Goke, M, et al. (2003). Polymorphism of monocyte chemoattractant protein 1 in Crohn's disease. *Int J Colorectal Dis*, 18(5):401–415.

Hertz, CJ, Kiertscher, SM, et al. (2001). Microbial lipopeptides stimulate dendritic cell maturation via Toll-like receptor 2. *J Immunol*, 166 (4):2444–2450.

Hibberd, ML, Sumiya, M, et al. (1999). Association of variants of the gene for mannose-binding lectin with susceptibility to meningococcal disease. Meningococcal Research Group. *Lancet*, 353(9158):1049–1053.

Hildebrand, F, Pape, HC, et al. (2005). Role of adhesion molecule ICAM in the pathogenesis of polymicrobial sepsis. *Exp Toxicol Pathol*, 56(4–5):281–290.

Hirschfield, GM, Herbert, J, et al. (2003). Human C-reactive protein does not protect against acute lipopolysaccharide challenge in mice. *J Immunol*, 171 (11):6046–6051.

Hobbs, AJ, Higgs, A, et al. (1999). Inhibition of nitric oxide synthase as a potential therapeutic target. *Annu Rev Pharmacol Toxicol*, 39:191–220.

Hobbs, MR, Udhayakumar, V, et al. (2002). A new NOS2 promoter polymorphism associated with increased nitric oxide production and protection from severe malaria in Tanzanian and Kenyan children. *Lancet*, 360 (9344):1468–1475.

Hollenberg, SM, Broussard, M, et al. (2000). Increased microvascular reactivity and improved mortality in septic mice lacking inducible nitric oxide synthase. *Circ Res*, 86(7):774–778.

Hollenberg, SM, Cunnion, RE, et al. (1993). Nitric oxide synthase inhibition reverses arteriolar hyporesponsiveness to catecholamines in septic rats. *Am J Physiol*, 264(2 Pt 2): H660–H663.

Hopf, HW (2003). Molecular diagnostics of injury and repair responses in critical illness: what is the future of monitoring in the intensive care unit? *Crit Care Med*, 31(Suppl. 8): S518–S523.

Hornef, MW, Frisan, T, et al. (2002). Toll-like receptor 4 resides in the Golgi apparatus and colocalizes with internalized lipopolysaccharide in intestinal epithelial cells. *J Exp Med*, 195(5):559–570.

Hotchkiss, R, Nunnally, I, et al. (1993). Hyperthermia protects mice against the lethal effects of endotoxin. *Am J Physiol*, 265(6 Pt 2): R1447–R1457.

Hotchkiss, RS, Swanson, PE, et al. (1999). Apoptotic cell death in patients with sepsis, shock, and multiple organ dysfunction. *Crit Care Med*, 27 (7):1230–1251.

Hotchkiss, RS, Tinsley, KW, et al. (2001). Sepsis-induced apoptosis causes progressive profound depletion of B and CD4 + T lymphocytes in humans. *J Immunol*, 166(11):6952–6963.

Hotchkiss, RS, Tinsley, KW, et al. (2002). Depletion of dendritic cells, but not macrophages, in patients with sepsis. *J Immunol*, 168(5):2493–2500.

Howard, M, Muchamuel, T, et al. (1993). Interleukin 10 protects mice from lethal endotoxemia. *J Exp Med*, 177(4):1205–1208.

Hubacek, JA, Buchler, C, et al. (1997). The genomic organization of the genes for human lipopolysaccharide binding protein (LBP) and bactericidal permeability increasing protein (BPI) is highly conserved. *Biochem Biophys Res Commun*, 236(2):427–430.

Hubacek, JA, Stuber, F, et al. (2001). Gene variants of the bactericidal/permeability increasing protein and lipopolysaccharide binding protein in sepsis patients: gender-specific genetic predisposition to sepsis. *Crit Care Med*, 29 (3):557–561.

Huber-Lang, M, Sarma, JV, et al. (2005). Changes in the novel orphan, C5a receptor (C5L2), during experimental sepsis and sepsis in humans. *J Immunol*, 174(2):1104–1110.

Huber-Lang, M, Sarma, VJ, et al. (2001). Role of C5a in multiorgan failure during sepsis. *J Immunol*, 166(2):1193–1199.

Huber-Lang, MS, Riedeman, NC, et al. (2002a). Protection of innate immunity by C5aR antagonist in septic mice. *FASEB J*, 16(12):1567–1574.

Huber-Lang, MS, Younkin, EM, et al. (2002b). Complement-induced impairment of innate immunity during sepsis. *J Immunol*, 169(6):3223–3231.

Huttunen, T, Teppo, AM, et al. (2003). Correlation between the severity of infectious diseases in children and the ratio of serum amyloid A protein and C-reactive protein. *Scand J Infect Dis*, 35(8):488–490.

Imahara, SD, Jelacic, S, et al. (2005). The influence of gender on human innate immunity. *Surgery*, 138(2):275–282.

Imai, Y, Kuba, K, et al. (2005). Angiotensin-converting enzyme 2 protects from severe acute lung failure. *Nature*, 436(7047):112–116.

Inohara, Chamaillard, et al. (2005). NOD-LRR proteins: role in host-microbial interactions and inflammatory disease. *Annu Rev Biochem*, 74:355–383.

Ishimitsu, T, Hosoya, K, et al. (2001). Microsatellite DNA polymorphism of human adrenomedullin gene in normotensive subjects and patients with essential hypertension. *Hypertension*, 38(1):9–12.

Jack, DL, Read, RC, et al. (2001). Mannose-binding lectin regulates the inflammatory response of human professional phagocytes to Neisseria meningitidis serogroup B. *J Infect Dis*, 184(9):1152–1162.

Janeway, CA Jr. (1989). Approaching the asymptote? Evolution and revolution in immunology. *Cold Spring Harb Symp Quant Biol*, 54 Pt 1:1–13.

Jenkins, NE, Mwangi, TW, et al. (2005). A polymorphism of intercellular adhesion molecule-1 is associated with a reduced incidence of nonmalarial febrile illness in Kenyan children. *Clin Infect Dis*, 41(12):1817–1819.

Jia, HP, Kline, JN, et al. (2004). Endotoxin responsiveness of human airway epithelia is limited by low expression of MD-2. *Am J Physiol Lung Cell Mol Physiol*, 287(2): L428–L437.

Jiang, ZD, Dupont, HL, et al. (2006). A common polymorphism in the interleukin 8 gene promoter is associated with *Clostridium difficile* diarrhea. *Am J Gastroenterol*, 101: 1112–1116.

Jiang, ZD, Okhuysen, PC, et al. (2003). Genetic susceptibility to enteroaggregative *Escherichia coli* diarrhea: polymorphism in the interleukin-8 promotor region. *J Infect Dis*, 188(4):506–511.

Kaide, JI, Zhang, F, et al. (2001). Carbon monoxide of vascular origin attenuates the sensitivity of renal arterial vessels to vasoconstrictors. *J Clin Invest*, 107 (9):1163–1171.

Karban, AS, Okazaki, T, et al. (2004). Functional annotation of a novel NFKB1 promoter polymorphism that increases risk for ulcerative colitis. *Hum Mol Genet*, 13(1):35–45.

Karrer, S, Bosserhoff, AK, et al. (2005). The −2518 promotor polymorphism in the MCP-1 gene is associated with systemic sclerosis. *J Invest Dermatol*, 124(1):92–98.

Kerlin, BA, Yan, SB, et al. (2003). Survival advantage associated with heterozygous factor V Leiden mutation in patients with severe sepsis and in mouse endotoxemia. *Blood*, 102(9):3085–3092.

Khani-Hanjani, A, Lacaille, D, et al. (2000). Association between dinucleotide repeat in non-coding region of interferon-γ gene and susceptibility to, and severity of, rheumatoid arthritis. *Lancet*, 356(9232):820–825.

Kiang, JG and Tsokos, GC (1998). Heat shock protein 70 kDa: molecular biology, biochemistry, and physiology. *Pharmacol Ther*, 80(2):183–201.

Kiehl, MG, Ostermann, H, et al. (1998). Inflammatory mediators in bronchoalveolar lavage fluid and plasma in leukocytopenic patients with septic shock-induced acute respiratory distress syndrome. *Crit Care Med*, 26 (7):1194–1199.

Kilbourn, RG and Griffith, OW (1992). Overproduction of nitric oxide in cytokine-mediated and septic shock (editorial). *J Natl Cancer Inst*, 84 (11):827–831.

Kilbourn, RG, Szabo, C, et al. (1997). Beneficial versus detrimental effects of nitric oxide synthase inhibitors in circulatory shock: lessons learned from experimental and clinical studies. *Shock*, 7(4):235–246.

King, BP, Khan, TI, et al. (2004). Upstream and coding region CYP2C9 polymorphisms: correlation with warfarin dose and metabolism. *Pharmacogenetics*, 14(12):813–822.

Kirkeboen, KA and Strand, OA (1999). The role of nitric oxide in sepsis—an overview. *Acta Anaesthesiol Scand*, 43(3):275–288.

Kirschning, CJ, Au-Young, J, et al. (1997). Similar organization of the lipopolysaccharide-binding protein (LBP) and phospholipid transfer protein (PLTP) genes suggests a common gene family of lipid-binding proteins. *Genomics*, 46(3):416–425.

Klosterhalfen, B, Hauptmann, S, et al. (1997). The influence of heat shock protein 70 induction on hemodynamic variables in a porcine model of recurrent endotoxemia. *Shock*, 7(5):358–363.

Knight, JC (2005). Regulatory polymorphisms underlying complex disease traits. *J Mol Med*, 83(2):97–109.

Kondaveeti, S, Hibberd, ML, et al. (1999). Effect of the factor V Leiden mutation on the severity of meningococcal disease. *Pediatr Infect Dis J*, 18 (10):893–896.

Kottke-Marchant, K (2002). Genetic polymorphisms associated with venous and arterial thrombosis: an overview. *Arch Pathol Lab Med*, 126(3):295–304.

Kristof, AS, Goldberg, P, et al. (1998). Role of inducible nitric oxide synthase in endotoxin-induced acute lung injury. *Am J Respir Crit Care Med*, 158 (6):1883–1889.

Ku, G, Thomas, CE, et al. (1992). Induction of interleukin 1 β expression from human peripheral blood monocyte-derived macrophages by 9-hydroxyoctadecadienoic acid. *J Biol Chem*, 267(20):14183–14188.

Lammas, DA, Casanova, JL, et al. (2000). Clinical consequences of defects in the IL-12-dependent interferon-γ (IFN-γ) pathway. *Clin Exp Immunol*, 121 (3):417–425.

Lander, ES, Linton, LM, et al. (2001). Initial sequencing and analysis of the human genome. *Nature*, 409(6822):860–921.

Lanier, LL, Corliss, BC, et al. (1998). Immunoreceptor DAP12 bearing a tyrosine-based activation motif is involved in activating NK cells. *Nature*, 391 (6668):703–707.

Laudes, IJ, Chu, JC, et al. (2002). Anti-c5a ameliorates coagulation/fibrinolytic protein changes in a rat model of sepsis. *Am J Pathol*, 160(5):1867–1875.

Leist, M, Gantner, F, et al. (1995). Tumor necrosis factor-induced hepatocyte apoptosis precedes liver failure in experimental murine shock models. *Am J Pathol*, 146(5):1220–1234.

Levi, M, Dorffler-Melly, J, et al. (2003). Aggravation of endotoxin-induced disseminated intravascular coagulation and cytokine activation in heterozygous protein-C-deficient mice. *Blood*, 101(12):4823–4827.

Levi, M and ten Cate, H (1999). Disseminated intravascular coagulation. *N Engl J Med*, 341(8):586–592.

Levi, M and van der Poll, T (2004). Coagulation in sepsis: all bugs bite equally. *Crit Care*, 8(2):99–100.

Levin, M, Quint, PA, et al. (2000). Recombinant bactericidal/permeability-increasing protein (rBPI21) as adjunctive treatment for children with severe meningococcal sepsis: a randomised trial. rBPI21 Meningococcal Sepsis Study Group. *Lancet*, 356(9234):961–967.

Li, P, Allen, H, et al. (1997). Characterization of mice deficient in interleukin-1 β converting enzyme. *J Cell Biochem*, 64(1):27–32.

Li, X and Qin, J (2005). Modulation of Toll-interleukin 1 receptor mediated signaling. *J Mol Med*, 83(4):258–266.

Lin, X, Yang, H, et al. (2005). α-chemokine receptor blockade reduces high mobility group box 1 protein-induced lung inflammation and injury and improves survival in sepsis. *Am J Physiol Lung Cell Mol Physiol*, 289(4): L583–L590.

Lindquist, S and Craig, EA (1988). The heat-shock proteins. *Annu Rev Genet*, 22:631–677.

Linscheid, P, Seboek, D, et al. (2005). Autocrine/paracrine role of inflammation-mediated calcitonin gene-related peptide and adrenomedullin expression in human adipose tissue. *Endocrinology*, 146(6):2699–2708.

Lipscombe, RJ, Lau, YL, et al. (1992a). Identical point mutation leading to low levels of mannose binding protein and poor C3b mediated opsonisation in Chinese and Caucasian populations. *Immunol Lett*, 32(3):253–257.

Lipscombe, RJ, Sumiya, M, et al. (1992b). High frequencies in African and non-African populations of independent mutations in the mannose binding protein gene. *Hum Mol Genet*, 1(9):709–715.

Liu, H, Jensen, L, et al. (2001). Characterization and quantification of mouse mannan-binding lectins (MBL-A and MBL-C) and study of acute phase responses. *Scand J Immunol*, 53(5):489–497.

Liu, SF and Malik, AB (2006). NF-κB activation as a pathological mechanism of septic shock and inflammation. *Am J Physiol Lung Cell Mol Physiol*, 290(4): L622–L645.

Liu, SF, Ye, X, et al. (1997). In vivo inhibition of nuclear factor-κB activation prevents inducible nitric oxide synthase expression and systemic hypotension in a rat model of septic shock. *J Immunol*, 159(8):3976–3983.

Lorente, JA, Landin, L, et al. (1993). Role of nitric oxide in the hemodynamic changes of sepsis. *Crit Care Med*, 21(5):759–767.

Lorenz, E, Mira, JP, et al. (2002). Relevance of mutations in the TLR4 receptor in patients with gram-negative septic shock. *Arch Intern Med*, 162(9):1028–1032.

Lotze, MT and Tracey, KJ (2005). High-mobility group box 1 protein (HMGB1): nuclear weapon in the immune arsenal. *Nat Rev Immunol*, 5(4):331–342.

Lowe, PR, Galley, HF, et al. (2003). Influence of interleukin-10 polymorphisms on interleukin-10 expression and survival in critically ill patients. *Crit Care Med*, 31(1):34–38.

Luzzani, A, Polati, E, et al. (2003). Comparison of procalcitonin and C-reactive protein as markers of sepsis. *Crit Care Med*, 31(6):1737–1741.

MacKenzie, IM, Garrard, CS, et al. (2001). Indices of nitric oxide synthesis and outcome in critically ill patients. *Anaesthesia*, 56(4):326–330.

Maimonides, M (1970). *The Medical Aphorisms of Moses Maimonides.* New York: Yeshiva University Press.

Makki, RF, al Sharif, F, et al. (2000). RANTES gene polymorphism in polymyalgia rheumatica, giant cell arteritis and rheumatoid arthritis. *Clin Exp Rheumatol*, 18(3):391–393.

Marshall, RP, Webb, S, et al. (2002a). Angiotensin converting enzyme insertion/deletion polymorphism is associated with susceptibility and outcome in acute respiratory distress syndrome. *Am J Respir Crit Care Med*, 166(5):646–650.

Marshall, RP, Webb, S, et al. (2002b). Genetic polymorphisms associated with susceptibility and outcome in ARDS. *Chest*, 121(Suppl. 3):68S–69S.

Martin, C, Boisson, C, et al. (1997). Patterns of cytokine evolution (tumor necrosis factor-α and interleukin-6) after septic shock, hemorrhagic shock, and severe trauma. *Crit Care Med*, 25(11):1813–1819.

Marutsuka, K, Nawa, Y, et al. (2001). Adrenomedullin and proadrenomudullin N-terminal 20 peptide (PAMP) are present in human colonic epithelia and exert an antimicrobial effect. *Exp Physiol*, 86(5):543–545.

Matute-Bello, G, Liles, WC, et al. (1999). Soluble Fas ligand induces epithelial cell apoptosis in humans with acute lung injury (ARDS). *J Immunol*, 163(4):2217–2225.

Matzinger, P (2002). The danger model: a renewed sense of self. *Science*, 296(5566):301–305.

McLeod, HL (2005). Pharmacogenetic analysis of clinically relevant genetic polymorphisms. *Clin Infect Dis*, 41(Suppl. 7): S449–S452.

Mesters, RM, Florke, N, et al. (1996a). Increase of plasminogen activator inhibitor levels predicts outcome of leukocytopenic patients with sepsis. *Thromb Haemost*, 75(6):902–907.

Mesters, RM, Mannucci, PM, et al. (1996b). Factor VIIa and antithrombin III activity during severe sepsis and septic shock in neutropenic patients. *Blood*, 88(3):881–886.

Miao, CH, Ho, W-T., et al. (1996). Transcriptional regulation of the gene coding for human protein C. *J Biol Chem*, 271(16):9587–9594.

Mignon, A, Rouquet, N, et al. (1999). LPS Challenge in D-galactosamine-sensitized mice accounts for caspase-dependent fulminant hepatitis, not for septic shock. *Am J Respir Crit Care Med*, 159(4):1308–1315.

Miller, EJ, Cohen, AB, et al. (1996). Increased interleukin-8 concentrations in the pulmonary edema fluid of patients with acute respiratory distress syndrome from sepsis. *Crit Care Med*, 24(9):1448–1454.

Milner, CM and Campbell, RD (1990). Structure and expression of the three MHC-linked HSP70 genes. *Immunogenetics*, 32(4):242–251.

Mira, JP, Cariou, A, et al. (1999). Association of TNF2, a TNF-α promoter polymorphism, with septic shock susceptibility and mortality: a multicenter study. *JAMA*, 282(6):561–568.

Mitaka, C (2005). Clinical laboratory differentiation of infectious versus non-infectious systemic inflammatory response syndrome. *Clin Chim Acta*, 351(1–2):17–29.

Miyake, K (2003). Innate recognition of lipopolysaccharide by CD14 and toll-like receptor 4-MD-2: unique roles for MD-2. *Int Immunopharmacol*, 3(1):119–128.

Modiano, D, Petrarca, V, et al. (1996). Different response to *Plasmodium falciparum* malaria in west African sympatric ethnic groups. *Proc Natl Acad Sci USA*, 93(23):13206–13211.

Mokuno, Y, Matsuguchi, T, et al. (2000). Expression of toll-like receptor 2 on γ delta T cells bearing invariant V γ 6/V delta 1 induced by *Escherichia coli* infection in mice. *J Immunol*, 165(2):931–940.

Mold, C, Gewurz, H, et al. (1999). Regulation of complement activation by C-reactive protein. *Immunopharmacology*, 42(1–3):23–30.

Muller, B, Peri, G, et al. (2001a). Circulating levels of the long pentraxin PTX3 correlate with severity of infection in critically ill patients. *Crit Care Med*, 29(7):1404–1407.

Muller, B, White, JC, et al. (2001b). Ubiquitous expression of the calcitonin-i gene in multiple tissues in response to sepsis. *J Clin Endocrinol Metab*, 86(1):396–404.

Munoz, C, Misset, B, et al. (1991). Dissociation between plasma and monocyte-associated cytokines during sepsis. *Eur J Immunol*, 21(9):2177–2184.

Murdoch, C and Finn, A (2003). The role of chemokines in sepsis and septic shock. *Contrib Microbiol*, 10:38–57.

Muroi, M, Ohnishi, T, et al. (2002). Regions of the mouse CD14 molecule required for toll-like receptor 2- and 4-mediated activation of NF-κB. *J Biol Chem*, 277(44):42372–42379.

Muy-Rivera, M, Ning, Y, et al. (2005). Leptin, soluble leptin receptor and leptin gene polymorphism in relation to preeclampsia risk. *Physiol Res*, 54(2):167–174.

Muzio, M, Bosisio, D, et al. (2000). Differential expression and regulation of toll-like receptors (TLR) in human leukocytes: selective expression of TLR3 in dendritic cells. *J Immunol*, 164(11):5998–6004.

Nagai, Y, Akashi, S, et al. (2002). Essential role of MD-2 in LPS responsiveness and TLR4 distribution. *Nat Immunol*, 3(7):667–672.

Nakae, H, Endo, S, et al. (1994). Serum complement levels and severity of sepsis. *Res Commun Chem Pathol Pharmacol*, 84(2):189–195.

Nakae, H, Endo, S, et al. (1996). Chronological changes in the complement system in sepsis. *Surg Today*, 26(4):225–229.

Nava, E, Palmer, RM, et al. (1991). Inhibition of nitric oxide synthesis in septic shock: how much is beneficial? *Lancet*, 338(8782–8783):1555–1557.

Newcomb, D, Bolgos, G, et al. (1998). Antibiotic treatment influences outcome in murine sepsis: mediators of increased morbidity. *Shock*, 10(2):110–117.

Nishio, K, Akai, Y, et al. (1997). Increased plasma concentrations of adrenomedullin correlate with relaxation of vascular tone in patients with septic shock. *Crit Care Med*, 25(6):953–957.

O'Keefe, GE, Hybki, DL, et al. (2002). The G→A single nucleotide polymorphism at the −308 position in the tumor necrosis factor-α promoter increases the risk for severe sepsis after trauma. *J Trauma*, 52(5):817–825; discussion 825–826.

Oberholzer, C, Oberholzer, A, et al. (2001). Apoptosis in sepsis: a new target for therapeutic exploration. *FASEB J*, 15(6):879–892.

Oda, S, Hirasawa, H, et al. (2005). Sequential measurement of IL-6 blood levels in patients with systemic inflammatory response syndrome (SIRS)/sepsis. *Cytokine*, 29(4):169–175.

Oelschlager, C, Romisch, J, et al. (2002). Antithrombin III inhibits nuclear factor κB activation in human monocytes and vascular endothelial cells. *Blood*, 99(11):4015–4020.

Ogata, H, Su, I, et al. (2000). The toll-like receptor protein RP105 regulates lipopolysaccharide signaling in B cells. *J Exp Med*, 192(1):23–29.

Ogura, H, Hashiguchi, N, et al. (2002). Long-term enhanced expression of heat shock proteins and decelerated apoptosis in polymorphonuclear leukocytes from major burn patients. *J Burn Care Rehabil*, 23(2):103–109.

Okajima, K (2001). Regulation of inflammatory responses by natural anticoagulants. *Immunological Reviews*, 184(1):258–274.

Oksanen, L, Palvimo, JJ, et al. (1998). Functional analysis of the C(−188)A polymorphism of the human leptin promoter. *Hum Genet*, 103(4):527–528.

Opal, SM and Huber, CE (2002). Bench-to-bedside review: Toll-like receptors and their role in septic shock. *Crit Care*, 6(2):125–136.

Pachot, A, Monneret, G, et al. (2005a). Longitudinal study of cytokine and immune transcription factor mRNA expression in septic shock. *Clin Immunol*, 114(1):61–69.

Pachot, A, Monneret, G, et al. (2005b). Messenger RNA expression of major histocompatibility complex class II genes in whole blood from septic shock patients. *Crit Care Med*, 33(1):31–38; discussion 236–237.

Panas, D, Khadour, FH, et al. (1998). Proinflammatory cytokines depress cardiac efficiency by a nitric oxide-dependent mechanism. *Am J Physiol*, 275(3 Pt 2): H1016– H1023.

Park, JH, Chang, SH, et al. (1996). Protective effect of nitric oxide in an endotoxin-induced septic shock. *Am J Surg*, 171(3):340–345.

Parsell, DA and Lindquist, S (1993). The function of heat-shock proteins in stress tolerance: degradation and reactivation of damaged proteins. *Annu Rev Genet*, 27:437–496.

Pawlinski, R, Pedersen, B, et al. (2004). Role of tissue factor and protease-activated receptors in a mouse model of endotoxemia. *Blood*, 103(4):1342–1347.

Pepys, MB and Hirschfield, GM (2003). C-reactive protein: a critical update. *J Clin Invest*, 111(12):1805–1812.

Pepys, S (1666). In *Diary of Samuel Pepys* (Robert Latham and William Matthews (eds.), vol. 7, University of California press 2000, ISBN: 0520226984.

Peter, G, Weigert, MB, et al. (1981). Meningococcal meningitis in familial deficiency of the fifth component of complement. *Pediatrics*, 67(6):882.

Petros, A, Lamb, G, et al. (1994). Effects of a nitric oxide synthase inhibitor in humans with septic shock. *Cardiovasc Res*, 28(1):34–39.

Philpott, DJ and Girardin, SE (2004). The role of toll-like receptors and nod proteins in bacterial infection. *Mol Immunol*, 41(11):1099–1108.

Pockley, AG (2003). Heat shock proteins as regulators of the immune response. *Lancet*, 362(9382):469–76.

Poltorak, A, He, X, et al. (1998). Defective LPS signaling in C3H/HeJ and C57BL/10ScCr mice: mutations in Tlr4 gene. *Science*, 282 (5396):2085–2088.

Porreca, E, Di Febbo, C, et al. (2005). Microsatellite polymorphism of the human leptin gene (LEP) and risk of cardiovascular disease. *Int J Obes (Lond)*, 30(2):209–213.

Povoa, P, Coelho, L, et al. (2005). C-reactive protein as a marker of infection in critically ill patients. *Clin Microbiol Infect*, 11(2):101–108.

Presterl, E, Staudinger, T, et al. (1997). Cytokine profile and correlation to the APACHE III and MPM II scores in patients with sepsis. *Am J Respir Crit Care Med*, 156(3):825–832.

Price, P, Witt, C, et al. (1999). The genetic basis for the association of the 8.1 ancestral haplotype (A1, B8, DR3) with multiple immunopathological diseases. *Immunol Rev*, 167:257–274.

Pugin, J, Stern-Voeffray, S, et al. (2004). Soluble MD-2 activity in plasma from patients with severe sepsis and septic shock. *Blood*, 104(13):4071–4079.

Quasney, MW (2006). Angiotensin-converting enzyme genetic variation and lung injury: are we genetically predisposed to develop acute respiratory distress syndrome? *Crit Care Med*, 34(4):1261–1262.

Quasney, MW, Waterer, GW, et al. (2002). Intracellular adhesion molecule Gly241Arg polymorphism has no impact on ARDS or septic shock in community-acquired pneumonia. *Chest*, 121(Suppl. 3):85S–86S.

Raeburn, CD, Calkins, CM, et al. (2002). ICAM-1 and VCAM-1 mediate endotoxemic myocardial dysfunction independent of neutrophil accumulation. *Am J Physiol Regul Integr Comp Physiol*, 283(2): R477–R486.

Ranganathan, P (2005). Pharmacogenomics of tumor necrosis factor antagonists in rheumatoid arthritis. *Pharmacogenomics*, 6(5):481–490.

Ravichandran, LV and Johns, RA (1995). Up-regulation of endothelial nitric oxide synthase expression by cyclic guanosine 3′,5′-monophosphate. *FEBS Lett*, 374(2):295–298.

Ravichandran, LV, Johns, RA, et al. (1995). Direct and reversible inhibition of endothelial nitric oxide synthase by nitric oxide. *Am J Physiol*, 268(6 Pt 2): H2216–H2223.

Reinhart, K and Karzai, W (2001). Anti-tumor necrosis factor therapy in sepsis: update on clinical trials and lessons learned. *Crit Care Med*, 29(Suppl. 7): S121–S125.

Reinhart, K, Menges, T, et al. (2001). Randomized, placebo-controlled trial of the anti-tumor necrosis factor antibody fragment afelimomab in hyperinflammatory response during severe sepsis: the RAMSES Study. *Crit Care Med*, 29(4):765–769.

Rhee, SH and Hwang, D (2000). Murine toll-like receptor 4 confers lipopolysaccharide responsiveness as determined by activation of NF κB and expression of the inducible cyclooxygenase. *J Biol Chem*, 275(44):34035–34040.

Ribeiro, R, Vasconcelos, A, et al. (2004). Overexpressing leptin genetic polymorphism (-2548 G/A) is associated with susceptibility to prostate cancer and risk of advanced disease. *Prostate*, 59(3):268–274.

Ribeiro, SP, Villar, J, et al. (1994). Sodium arsenite induces heat shock protein-72 kilodalton expression in the lungs and protects rats against sepsis. *Crit Care Med*, 22(6):922–929.

Riedemann, NC, Guo, RF, et al. (2002). C5a receptor and thymocyte apoptosis in sepsis. *FASEB J*, 16(8):887–888.

Riedemann, NC, Guo, RF, et al. (2003a). Regulation by C5a of neutrophil activation during sepsis. *Immunity*, 19(2):193–202.

Riedemann, NC, Guo, RF, et al. (2004). Regulatory role of C5a in LPS-induced IL-6 production by neutrophils during sepsis. *FASEB J*, 18(2):370–372.

Riedemann, NC, Neff, TA, et al. (2003b). Protective effects of IL-6 blockade in sepsis are linked to reduced C5a receptor expression. *J Immunol*, 170(1):503–507.

Risch, N and Merikangas, K (1996). The future of genetic studies of complex human diseases. *Science*, 273(5281):1516–1517.

Rivers, E, Nguyen, B, et al. (2001). Early goal-directed therapy in the treatment of severe sepsis and septic shock. *N Engl J Med*, 345(19):1368–1377.

Rock, FL, Hardiman, G, et al. (1998). A family of human receptors structurally related to Drosophila Toll. *Proc Natl Acad Sci USA*, 95(2):588–593.

Ruiz-Pesini, E, Lapena, AC, et al. (2000). Human mtDNA haplogroups associated with high or reduced spermatozoa motility. *Am J Hum Genet*, 67(3):682–696.

Ruiz-Pesini, E, Mishmar, D, et al. (2004). Effects of purifying and adaptive selection on regional variation in human mtDNA. *Science*, 303(5655):223–226.

Saleh, M, Vaillancourt, JP, et al. (2004). Differential modulation of endotoxin responsiveness by human caspase-12 polymorphisms. *Nature*, 429(6987):75.

Samson, LM, Allen, UD, et al. (1997). Elevated interleukin-1 receptor antagonist levels in pediatric sepsis syndrome. *J Pediatr*, 131(4):587–591.

Sander, C (2000). Genomic medicine and the future of health care. *Science*, 287(5460):1977–1978.

Sands, KE, Bates, DW, et al. (1997). Epidemiology of sepsis syndrome in 8 academic medical centers. Academic Medical Center Consortium Sepsis Project Working Group. *JAMA*, 278(3):234–240.

Sarraf, P, Frederich, RC, et al. (1997). Multiple cytokines and acute inflammation raise mouse leptin levels: potential role in inflammatory anorexia. *J Exp Med*, 185(1):171–175.

Sastry, K, Herman, GA, et al. (1989). The human mannose-binding protein gene. Exon structure reveals its evolutionary relationship to a human pulmonary surfactant gene and localization to chromosome 10. *J Exp Med*, 170(4):1175–1189.

Sawa, T, Corry, DB, et al. (1997). IL-10 improves lung injury and survival in *Pseudomonas aeruginosa* pneumonia. *J Immunol*, 159(6):2858–2866.

Sayeed, MM (2002). Mitochondrial dysfunction in sepsis: a familiar song with new lyrics. *Crit Care Med*, 30(12):2780–2781.

Scaffidi, P, Misteli, T, et al. (2002). Release of chromatin protein HMGB1 by necrotic cells triggers inflammation. *Nature*, 418(6894):191–195.

Schmidt, MA, Chakrabarti, AK, et al. (2003). Interactive effects of the ACE DD polymorphism with the NOS III homozygous G849T (Glu298→Asp) variant in determining endothelial function in coronary artery disease. *Vasc Med*, 8(3):177–183.

Schroeder, S, Bischoff, J, et al. (1999a). Endotoxin inhibits heat shock protein 70 (HSP70) expression in peripheral blood mononuclear cells of patients with severe sepsis. *Intensive Care Med*, 25(1):52–57.

Schroeder, S, Reck, M, et al. (1999b). Analysis of two human leukocyte antigen-linked polymorphic heat shock protein 70 genes in patients with severe sepsis. *Crit Care Med*, 27(7):1265–1270.

Schromm, AB, Lien, E, et al. (2001). Molecular genetic analysis of an endotoxin nonresponder mutant cell line: a point mutation in a conserved region of MD-2 abolishes endotoxin-induced signaling. *J Exp Med*, 194(1):79–88.

Schumann, RR, Leong, SR, et al. (1990). Structure and function of lipopolysaccharide binding protein. *Science*, 249(4975):1429–1431.

Schumann, RR, Rietschel, ET, et al. (1994). The role of CD14 and lipopolysaccharide-binding protein (LBP) in the activation of different cell types by endotoxin. *Med Microbiol Immunol (Berl)*, 183(6):279–297.

Schwartz, MD, Moore, EE, et al. (1996). Nuclear factor-κB is activated in alveolar macrophages from patients with acute respiratory distress syndrome. *Crit Care Med*, 24(8):1285–1292.

Schwarz, MJ, Riedel, M, et al. (2000). Decreased levels of soluble intercellular adhesion molecule-1 (sICAM-1) in unmedicated and medicated schizophrenic patients. *Biol Psychiatry*, 47(1):29–33.

Sconce, EA, Khan, TI, et al. (2005). The impact of CYP2C9 and VKORC1 genetic polymorphism and patient characteristics upon warfarin dose requirements: proposal for a new dosing regimen. *Blood*, 106(7):2329–2333.

Selberg, O, Hecker, H, et al. (2000). Discrimination of sepsis and systemic inflammatory response syndrome by determination of circulating plasma

concentrations of procalcitonin, protein complement 3a, and interleukin-6. *Crit Care Med*, 28(8):2793–2798.
Sgambato, E (2005). [Nosocomial sepsis and polymorphism of NOD-2 gene]. *Clin Ter*, 156(1–2):7–10.
Sha, WC, Liou, HC, et al. (1995). Targeted disruption of the P50 subunit of NF-κB leads to multifocal defects in immune-responses. *Cell*, 80(2):321–330.
Shanley, TP, Warner, RL, et al. (1995). The role of cytokines and adhesion molecules in the development of inflammatory injury. *Mol Med Today*, 1 (1):40–45.
Sheehan, M, Wong, HR, et al. (2002). Protective effects of isohelenin, an inhibitor of nuclear factor κB, in endotoxic shock in rats. *J Endotoxin Res*, 8(2):99–107.
Sheehan, M, Wong, HR, et al. (2003). Parthenolide improves systemic hemodynamics and decreases tissue leukosequestration in rats with polymicrobial sepsis. *Crit Care Med*, 31(9):2263–2270.
Shimazu, R, Akashi, S, et al. (1999). MD-2, a molecule that confers lipopolysaccharide responsiveness on Toll-like receptor 4. *J Exp Med*, 189(11):1777–1782.
Shoji, H, Minamino, N, et al. (1995). Endotoxin markedly elevates plasma concentration and gene transcription of adrenomedullin in rat. *Biochem Biophys Res Commun*, 215(2):531–537.
Shu, Q, Fang, X, et al. (2003). IL-10 polymorphism is associated with increased incidence of severe sepsis. *Chin Med J (Engl)*, 116(11):1756–1759.
Simmonds, RE and Lane, DA (1999). Structural and functional implications of the intron/exon organization of the human endothelial cell protein C/activated protein C receptor (EPCR) gene: comparison with the structure of CD1/major histocompatibility complex α1 and α2 domains. *Blood*, 94(2):632–641.
Singer, M and Brealey, D (1999). Mitochondrial dysfunction in sepsis. *Biochem Soc Symp*, 66:149–166.
Singleton, KD, Serkova, N, et al. (2005). Glutamine attenuates lung injury and improves survival after sepsis: role of enhanced heat shock protein expression. *Crit Care Med*, 33(6):1206–1213.
Skinner, NA, MacIsaac, CM, et al. (2005). Regulation of Toll-like receptor (TLR)2 and TLR4 on CD14dimCD16+ monocytes in response to sepsis-related antigens. *Clin Exp Immunol*, 141(2):270–278.
Solomkin, JS, Bauman, MP, et al. (1981a). Neutrophils dysfunction during the course of intra-abdominal infection. *Ann Surg*, 194(1):9–17.
Solomkin, JS, Jenkins, MK, et al. (1981b). Neutrophil dysfunction in sepsis. II. Evidence for the role of complement activation products in cellular deactivation. *Surgery*, 90(2):319–327.
Sorensen, TIA, Nielsen, GG, et al. (1988). Genetic and environmental influences on premature death in adult adoptees. *N Engl J Med*, 318:727–732.
Standiford, TJ, Kunkel, SL, et al. (1996). Expression and regulation of chemokines in bacterial pneumonia. *J Leukoc Biol*, 59(1):24–28.
Stassen, NA, Leslie-Norfleet, LA et al. (2002). Interferon-γ gene polymorphisms and the development of sepsis in patients with trauma. *Surgery*, 132(2):289–292.
Stefan, N, Vozarova, B, et al. (2002). The Gln223Arg polymorphism of the leptin receptor in Pima Indians: influence on energy expenditure, physical activity and lipid metabolism. *Int J Obes Relat Metab Disord*, 26(12):1629–1632.
Struck, J, Tao, C, et al. (2004). Identification of an Adrenomedullin precursor fragment in plasma of sepsis patients. *Peptides*, 25(8):1369–1372.
Stuber, F (2003). Another definite candidate gene for genetic predisposition of sepsis: interleukin-10. *Crit Care Med*, 31(1):314–315.
Stuber, F, Petersen, M, et al. (1996). A genomic polymorphism within the tumor necrosis factor locus influences plasma tumor necrosis factor-α concentrations and outcome of patients with severe sepsis. *Crit Care Med*, 24(3):381–384.
Stuber, F, Udalova, IA, et al. (1995). −308 tumor necrosis factor (TNF) polymorphism is not associated with survival in severe sepsis and is unrelated to lipopolysaccharide inducibility of the human TNF promoter. *J Inflamm*, 46(1):42–50.
Sunden-Cullberg, J, Norrby-Teglund, A, et al. (2005). Persistent elevation of high mobility group box-1 protein (HMGB1) in patients with severe sepsis and septic shock. *Crit Care Med*, 33(3):564–573.
Sutherland, AM and Russell, JA (2005). Issues with polymorphism analysis in sepsis. *Clin Infect Dis*, 41(Suppl. 7): S396–S402.
Sutherland, AM, Walley, KR, et al. (2005a). Polymorphisms in CD14, mannose-binding lectin, and Toll-like receptor-2 are associated with increased prevalence of infection in critically ill adults. *Crit Care Med*, 33(3):638–644.
Sutherland, AM, Walley, KR, et al. (2005b). The association of interleukin 6 haplotype clades with mortality in critically ill adults. *Arch Intern Med*, 165(1):75–82.
Szabo, C, Cuzzocrea, S, et al. (1997). Endothelial dysfunction in a rat model of endotoxic shock. Importance of the activation of poly (ADP-ribose) synthetase by peroxynitrite. *J Clin Invest*, 100(3):723–735.
Szalai, AJ, VanCott, JL, et al. (2000). Human C-reactive protein is protective against fatal *Salmonella enterica* serovar typhimurium infection in transgenic mice. *Infect Immun*, 68(10):5652–5656.
Takeda, K, Kaisho, T, et al. (2003). Toll-like receptors. *Annu Rev Immunol*, 21:335–376.
Takeuchi, O, Kawai, T, et al. (1999). TLR6: a novel member of an expanding toll-like receptor family. *Gene*, 231(1–2):59–65.
Tang, GJ, Huang, SL, et al. (2000). Tumor necrosis factor gene polymorphism and septic shock in surgical infection. *Crit Care Med*, 28(8):2733–2736.
Taniguchi, T, Koido, Y, et al. (1999a). Change in the ratio of interleukin-6 to interleukin-10 predicts a poor outcome in patients with systemic inflammatory response syndrome. *Crit Care Med*, 27(7):1262–1264.
Taniguchi, T, Koido, Y, et al. (1999b). The ratio of interleukin-6 to interleukin-10 correlates with severity in patients with chest and abdominal trauma. *Am J Emerg Med*, 17(6):548–551.
Tannahill, CL, Fukuzuka, K, et al. (1999). Discordant tumor necrosis factor-α superfamily gene expression in bacterial peritonitis and endotoxemic shock. *Surgery*, 126(2):349–357.
Taylor, KD, Plevy, SE, et al. (2001). ANCA pattern and LTA haplotype relationship to clinical responses to anti-TNF antibody treatment in Crohn's disease. *Gastroenterology*, 120(6):1347–1355.
Terry, CF, Loukaci, V, et al. (2000). Cooperative influence of genetic polymorphisms on interleukin 6 transcriptional regulation. *J Biol Chem*, 275(24):18138–18144.
Thiel, S, Holmskov, U, et al. (1992). The concentration of the C-type lectin, mannan-binding protein, in human plasma increases during an acute phase response. *Clin Exp Immunol*, 90(1):31–35.
Thomas, L (1972). Germs. *N Engl J Med*, 287(11):553–555.
Tracey, KJ and Cerami, A (1993). Tumor necrosis factor, other cytokines and disease. *Annu Rev Cell Biol*, 9:317–343.
Tracey, KJ and Lowry, SF (1990). The role of cytokine mediators in septic shock. *Adv Surg*, 23:21–56.
Triantafilou, M and Triantafilou, K (2002). Lipopolysaccharide recognition: CD14, TLRs and the LPS-activation cluster. *Trends Immunol*, 23(6):301–304.
Tsung, A, Sahai, R, et al. (2005). The nuclear factor HMGB1 mediates hepatic injury after murine liver ischemia-reperfusion. *J Exp Med*, 201(7):1135–1143.
Tucci, M, Barnes, EV, et al. (2004). Strong association of a functional polymorphism in the monocyte chemoattractant protein 1 promoter gene with lupus nephritis. *Arthritis Rheum*, 50(6):1842–1849.
Uchinami, H, Yamamoto, Y, et al. (2002). Effect of heat shock preconditioning on NF-κB/I-κB pathway during I/R injury of the rat liver. *Am J Physiol Gastrointest Liver Physiol*, 282(6):G962–G971.
Ueda, S, Nishio, K, et al. (1999). Increased plasma levels of adrenomedullin in patients with systemic inflammatory response syndrome. *Am J Respir Crit Care Med*, 160(1):132–136.
van den Berghe, G, Wouters, P, et al. (2001). Intensive insulin therapy in the critically ill patients. *N Engl J Med*, 345(19):1359–1367.
van der Poll, T, de Jonge, E, et al. (1999). Pathogenesis of DIC in sepsis. *Sepsis*, 3(2):103.
van der Poll, T, Keogh, CV, et al. (1997). Interleukin-6 gene-deficient mice show impaired defense against pneumococcal pneumonia. *J Infect Dis*, 176(2):439–444.
van der Poll, T and van Deventer, SJ (1999). Cytokines and anticytokines in the pathogenesis of sepsis. *Infect Dis Clin North Am*, 13(2):413–426, ix.

Vannier, E, Miller, LC, et al. (1992). Coordinated antiinflammatory effects of interleukin 4: interleukin 4 suppresses interleukin 1 production but up-regulates gene expression and synthesis of interleukin 1 receptor antagonist. *Proc Natl Acad Sci USA*, 89(9):4076–4080.

Veldman, BA, Spiering, W, et al. (2002). The Glu298Asp polymorphism of the NOS 3 gene as a determinant of the baseline production of nitric oxide. *J Hypertens*, 20(10):2023–2027.

Venter, JC, Adams, MD, et al. (2001). The sequence of the human genome. *Science*, 291(5507):1304–1351.

Vermont, C, Hazelzet, J, et al. (2006). CC and CXC chemokine levels in children with meningococcal sepsis accurately predict mortality and disease severity. *Crit Care*, 10(1):R33.

Villar, J, Ribeiro, SP, et al. (1994). Induction of the heat shock response reduces mortality rate and organ damage in a sepsis-induced acute lung injury model. *Crit Care Med*, 22(6):914–921.

Vincent, JL, Bihari, DJ, et al. (1995). The prevalence of nosocomial infection in intensive care units in Europe. Results of the European prevalence of infection in intensive care (EPIC) study. EPIC International Advisory Committee. *JAMA*, 274(8):639–644.

Visintin, A, Mazzoni, A, et al. (2001). Regulation of Toll-like receptors in human monocytes and dendritic cells. *J Immunol*, 166(1):249–255.

Volanakis, JE (1982). Complement activation by C-reactive protein complexes. *Ann NY Acad Sci*, 389:235–250.

Wang, CR, Guo, HR, et al. (2005a). RANTES promoter polymorphism as a genetic risk factor for rheumatoid arthritis in the Chinese. *Clin Exp Rheumatol*, 23(3):379–384.

Wang, DY, Qin, RY, et al. (2004). Expression of TREM-1 mRNA in acute pancreatitis. *World J Gastroenterol*, 10(18):2744–2746.

Wang, H, Bloom, O, et al. (1999). HMG-1 as a late mediator of endotoxin lethality in mice. *Science*, 285(5425):248–251.

Wang, J, Barke, RA, et al. (2005b). Morphine impairs host innate immune response and increases susceptibility to *Streptococcus pneumoniae* lung infection. *J Immunol*, 174(1):426–434.

Wang, J, Dudley, D, et al. (2002). Haplotype-specific effects on endothelial NO synthase promoter efficiency: modifiable by cigarette smoking. *Arterioscler Thromb Vasc Biol*, 22(5):e1–e4.

Wang, P (1998). Adrenomedullin in sepsis and septic shock. *Shock*, 10(5):383–384.

Warren, BL, Eid, A, et al. (2001). High-dose antithrombin III in severe sepsis: a randomized controlled trial. *JAMA*, 286(15):1869–1878.

Watanabe, E, Hirasawa, H, et al. (2005). Cytokine-related genotypic differences in peak interleukin-6 blood levels of patients with SIRS and septic complications. *J Trauma*, 59(5):1181–1189; discussion 1189–1190.

Waterer, GW, El Bahlawan, L, et al. (2003). Heat shock protein 70–2 + 1267 AA homozygotes have an increased risk of septic shock in adults with community-acquired pneumonia. *Crit Care Med*, 31(5):1367–1372.

Waterer, GW, Quasney, MW, et al. (2001). Septic shock and respiratory failure in community-acquired pneumonia have different TNF polymorphism associations. *Am J Respir Crit Care Med*, 163(7):1599–1604.

Wehkamp, J and Stange, EF (2005). NOD2 mutation and mice: no Crohn's disease but many lessons to learn. *Trends Mol Med*, 11(7):307–309.

Wei, XQ, Charles, IG, et al. (1995). Altered immune responses in mice lacking inducible nitric oxide synthase. *Nature*, 375(6530):408–411.

Weiler, H, Lindner, V, et al. (2001). Characterization of a mouse model for thrombomodulin deficiency. *Arterioscler Thromb Vasc Biol*, 21(9):1531–1537.

Weiss, J and Olsson, I (1987). Cellular and subcellular localization of the bactericidal/permeability-increasing protein of neutrophils. *Blood*, 69(2):652–659.

Weiss, YG, Maloyan, A, et al. (2002). Adenoviral transfer of HSP-70 into pulmonary epithelium ameliorates experimental acute respiratory distress syndrome. *J Clin Invest*, 110(6):801–806.

Welch, WJ (1992). Mammalian stress response: cell physiology, structure/function of stress proteins, and implications for medicine and disease. *Physiol Rev*, 72(4):1063–1081.

Wellcome Trust Case Control Consortium 2007. Genome-wide association study of 14,000 cases of seven common diseases and 3,000 shared controls. *Nature*, 447:661–678.

Welty-Wolf, KE, Carraway, MS, et al. (2000). Proinflammatory cytokines increase in sepsis after anti-adhesion molecule therapy. *Shock*, 13(5):404–409.

Welty-Wolf, KE, Carraway, MS, et al. (2001). Antibody to intercellular adhesion molecule 1 (CD54) decreases survival and not lung injury in baboons with sepsis. *Am J Respir Crit Care Med*, 163(3 Pt 1):665–673.

Whalen, MJ, Doughty, LA, et al. (2000). Intercellular adhesion molecule-1 and vascular cell adhesion molecule-1 are increased in the plasma of children with sepsis-induced multiple organ failure. *Crit Care Med*, 28(7):2600–2607.

Wiedermann, FJ, Mayr, AJ, et al. (2004). Alveolar granulocyte colony-stimulating factor and α-chemokines in relation to serum levels, pulmonary neutrophilia, and severity of lung injury in ARDS. *Chest*, 125(1):212–219.

Williams, DL, Ha, T, et al. (2003). Modulation of tissue Toll-like receptor 2 and 4 during the early phases of polymicrobial sepsis correlates with mortality. *Crit Care Med*, 31(6):1808–1818.

Wilson, AG, Symons, JA, et al. (1997). Effects of a polymorphism in the human tumor necrosis factor α promoter on transcriptional activation. *Proc Natl Acad Sci USA*, 94(7):3195–3199.

Winning, J, Claus, RA, et al. (2006). Molecular biology on the ICU. From understanding to treating sepsis. *Minerva Anestesiol*, 72(5):255–267.

Winoto, A (2004). Cutting into innate immunity. *Nat Immunol*, 5(6):563–564.

Wischmeyer, PE, Kahana, M, et al. (2001). Glutamine induces heat shock protein and protects against endotoxin shock in the rat. *J Appl Physiol*, 90(6):2403–2410.

Wunder, C, Eichelbronner, O, et al. (2004). Are IL-6, IL-10 and PCT plasma concentrations reliable for outcome prediction in severe sepsis? A comparison with APACHE III and SAPS II. *Inflamm Res*, 53(4):158–163.

Xu, H, Gonzalo, JA, et al. (1994). Leukocytosis and resistance to septic shock in intercellular adhesion molecule 1-deficient mice. *J Exp Med*, 180(1):95–109.

Yamamoto, Y, Yin, M-J. et al. (1999). Sulindac Inhibits Activation of the NF-κB Pathway. *J Biol Chem*, 274(38):27307–27314.

Yan, SB, Helterbrand, JD, et al. (2001). Low levels of protein c are associated with poor outcome in severe sepsis. *Chest*, 120(3):915–922.

Yan, SB and Nelson, DR (2004). Effect of factor V Leiden polymorphism in severe sepsis and on treatment with recombinant human activated protein C. *Crit Care Med*, 32(Suppl. 5): S239–S246.

Yang, H, Young, DW, et al. (2000). Cellular events mediated by lipopolysaccharide-stimulated toll-like receptor 4. MD-2 is required for activation of mitogen-activated protein kinases and Elk-1. *J Biol Chem*, 275(27):20861–20866.

Yet, SF, Pellacani, A, et al. (1997). Induction of heme oxygenase-1 expression in vascular smooth muscle cells. A link to endotoxic shock. *J Biol Chem*, 272(7):4295–4301.

Yin, M-J., Yamamoto, Y, et al. (1998). The anti-inflammatory agents aspirin and salicylate inhibit the activity of IκB kinase-β. *Nature*, 396(6706):77.

Zarember, KA and Godowski, PJ (2002). Tissue expression of human Toll-like receptors and differential regulation of Toll-like receptor mRNAs in leukocytes in response to microbes, their products, and cytokines. *J Immunol*, 168(2):554–561.

Zhang, F, Kaide, JI, et al. (2001). Vasoregulatory function of the heme-heme oxygenase-carbon monoxide system. *Am J Hypertens*, 14(6 Pt 2):62S–67S.

Zhang, Y, Proenca, R, et al. (1994). Positional cloning of the mouse obese gene and its human homologue. *Nature*, 372(6505):425–432.

Zhong, XB, Leng, L, et al. (2005). Simultaneous detection of microsatellite repeats and SNPs in the macrophage migration inhibitory factor (MIF) gene by thin-film biosensor chips and application to rural field studies. *Nucleic Acids Res*, 33(13):e121.

Ziegler, EJ, Fisher, CJ Jr., et al. (1991). Treatment of gram-negative bacteremia and septic shock with HA-1A human monoclonal antibody against endotoxin. A randomized, double-blind, placebo-controlled trial. The HA-1A Sepsis Study Group. *N Engl J Med*, 324(7):429–436.

Ziegler, TR, Ogden, LG, et al. (2005). Parenteral glutamine increases serum heat shock protein 70 in critically ill patients. *Intensive Care Med*, 31(8):1079–1086.

Zingarelli, B, Sheehan, M, et al. (2003). Peroxisome proliferator activator receptor-γ ligands, 15-deoxy-delta(12,14)-prostaglandin J2 and ciglitazone, reduce systemic inflammation in polymicrobial sepsis by modulation of signal transduction pathways. *J Immunol*, 171(12):6827–6837.

17

The Epilepsies

Mark Gardiner

Epilepsy is one of the most common neurological problems, affecting an estimated 50 million people worldwide. It is defined as a predisposition to recurrent unprovoked epileptic seizures, which are the clinical manifestations of transient abnormalities of neuronal activity in the cerebral cortex.

Epileptic seizures represent a transient malfunction that can occur in any mammalian brain and may be provoked by a wide variety of factors from mechanical trauma to hypoglycemia. As might be anticipated in such a complex organ, their manifestations are highly variable, from the well-known, generalized tonic–clonic seizure or "grand-mal" convulsion in which consciousness is lost and the limbs shake, to brief sensory or cognitive disturbances only apparent to the patient. Epileptic seizures fall into two main categories: generalized, which begin simultaneously in both cerebral hemispheres, and partial (or focal), which originate in one or more localized foci in the brain.

Epilepsy is an end point, rather like anemia, of a very diverse set of biological processes, and the epilepsies are a heterogeneous group of disorders with many causes. A genetic component is estimated to be present in up to 40% of patients. Epilepsies and epilepsy syndromes are classified according to whether the predominant seizure type is generalized or partial, and then by etiology into idiopathic epilepsies in which recurrent seizures occur in isolation and symptomatic epilepsies in which some obvious anatomic or biochemical cause is present such as a brain tumor, malformation or metabolic derangement.

The human genetic epilepsies can be further categorized according to the genetic mechanism involved. There are more than 200 monogenic disorders that include recurrent seizures as a component of the phenotype. Advances in genomics over the last decade have facilitated identification of more than 70 genes underlying a host of monogenic epilepsies in man and other species, especially rodents. Of particular interest is the emergence of most autosomal dominant idiopathic epilepsies as ion channelopathies. Progress in the investigation of epilepsies with "complex" inheritance has been slower, but advances in genomics including completion of the HapMap Project should facilitate their dissection at a molecular level.

In this chapter, human genetic epilepsies are considered with a particular focus on the molecular genetic basis of monogenic epilepsies. The contribution of comparative genomics in relation to animal models of epilepsy are reviewed, and potential applications of genomics to future understanding of epilepsies with "complex" inheritance and the neurobiology of seizure generation are evaluated. Finally, the potential implications for translational clinical applications in diagnosis and therapy are described. Several recent reviews are available (Scheffer and Berkovic, 2003; Noebels, 2003a; George, 2004; Steinlein, 2004a; Callenbach et al., 2005).

Human Genetic Epilepsies

Approximately 40% of patients with epilepsy are estimated to have a genetic component to their etiology.

There are several hundred chromosomal imbalances in which seizures or electroencephalogram (EEG) abnormalities occur (Singh et al., 2002) but few chromosomal syndromes specifically associated with epilepsy. In patients with epilepsy and learning difficulties plus multiple congenital abnormalities, up to 50% have a chromosomal abnormality. Those with a high association with epilepsy include ring chromosomes 20 and 14, terminal deletions of chromosome 1, the inversion duplication 15 syndrome, and three eponymous syndromes: Angelman (del 15q11-q13), Wolf–Hirschhorn (4p-), and Miller–Dieker syndrome (del 17p13-3) in which cortical malformations occur.

There are more than 200 monogenic disorders that include epilepsy as a component of the phenotype. Most idiopathic monogenic epilepsies have emerged as ion channelopathies, but symptomatic monogenic epilepsies are caused by genes with a diverse range of functions, the disruption of which may affect brain development, neuronal survival, or energy metabolism. However, monogenic epilepsies are individually rare, and in the majority of patients with familial epilepsy, the mode of inheritance is "complex."

Genetic epilepsies with "complex" inheritance are predominantly idiopathic generalized epilepsies (IGEs). These include several forms such as childhood absence epilepsy (CAE), juvenile absence epilepsy (JAE), and juvenile myoclonic epilepsy (JME). On occasion, similar phenotypes may display Mendelian inheritance, a not uncommon feature of human genetic diseases.

Idiopathic Epilepsies: Ion Channelopathies

In the last decade, it has emerged that mutations in at least 10 ion channel genes underlie at least six idiopathic epilepsy phenotypes, with a striking degree of locus heterogeneity (same phenotype caused by different genes) and variable expression (mutations in same gene causing different phenotypes). The genes involved encode both ligand-gated and voltage-gated ion channels, and the phenotypes encompass a wide range of severity and age at onset (Table 17–1). Epilepsy thus joins the other paroxysmal disorders of excitable tissues such as skeletal muscle and the heart as a channelopathy. The disorders caused by mutations in each channel type are considered in turn.

Potassium Channel Epilepsies

Mutations in the voltage-gated potassium channel genes *KCNQ2* (chromosome 20q) and *KCNQ3* (chromosome 8q) cause benign familial neonatal convulsions (BFNC), a rare autosomal dominant syndrome characterized by onset of seizures soon after birth, which usually resolve within the first few months of life. The genes were identified by linkage analysis followed by positional cloning, a strategy made possible by advances in genomics (Charlier et al., 1998; Singh et al., 1998).

Potassium channels are probably the most diverse group of ion channels encoded by more than 100 genes in the human genome (Miller, 2000). Mutations in approximately 10 potassium channel genes cause human disease, and 4 of these belong to the small KCNQ family: *KCNQ1* mutations underlie long QT syndrome (LQTS) and *KCNQ4* mutations account for one form of dominant deafness (DFNA2). KCNQ proteins have six transmembrane domains (TMDs) with a pore-forming A loop, and four subunits are believed to combine to form a functional channel.

KCNQ2 and *KCNQ3* are expressed predominantly in brain and appear to contribute to the so-called M-current that regulates excitability of many neurons. Inhibition of M-currents leads to enhanced neuronal excitability, and the mutations in these genes are loss of function, associated with quite small reductions in currents mediated by heteromeric mutant channels in vitro (Jentsch, 2000). Approximately 30 mutations in *KCNQ2*, but only 3 in *KCNQ3* have now been identified. Most mutations (60%) in *KCNQ2* are found in the C-terminus and are predicted to cause truncation of the C-terminus with haploinsufficiency (Singh et al., 2003). The reasons for the age-specific penetrance of this phenotype remain unclear.

Mutations in two additional potassium channel genes have been described in patients with epilepsy and movement disorders. In two families with episodic ataxia type 1 (EA1) due to mutations in *KCNA1*, missense mutations were associated with focal-onset epilepsy. Most recently, a missense mutation in *KCNMA1*, encoding a big K^+ (BK) channel, has been described in a family segregating generalized epilepsy and paroxysmal dyskinesia (GEPD) (Du et al., 2005).

Sodium Channel Epilepsies

Voltage-gated sodium channels (Na_vChs) are important for the initiation and propogation of action potentials in electrically excitable tissues such as skeletal and cardiac muscle and nerve. Activation of these channels underlies the initial upstroke of the action potential triggering other events such as muscle contraction or neuronal firing.

Table 17–1 Idiopathic Epilepsy: Ion Channelopathies

Gene	Locus	Protein	Phenotype
Potassium channels			
KCNQ2	20q/EBN1	$K_V7.2$	Benign familial neonatal convulsions (BFNCs)
KCNQ3	8q/EBN2	$K_V7.3$	BFNCs
Sodium channels			
SCN1A	2q/GEFS2	$Na_V1.1$	Generalized epilepsy with febrile seizures (GEFS+)
			Severe myoclonic epilepsy of infancy (SMEI)
			Intractable childhood epilepsy with generalized tonic–clonic seizures (ICEGTCS)
			Infantile spasms (ISs)
SCN2A	2q/BFNIS	$Na_V1.2$	Benign familial neonatal–infantile seizures (BFNISs)
SCN1B	19q/ GEFS1	Na_V∃1	GEFS plus (GEFS+)
Chloride channels			
CLCN2	3q/ECA3	CLC-2	Idiopathic generalized epilepsy (IGE) subtypes including juvenile myoclonic epilepsy (JME), childhood absence epilepsy (CAE), juvenile absence epilepsy (JAE), and epilepsy with grand-mal seizures on awakening (EGMA)
GABRA1	5q/EJM1	$\alpha1$	JME
GABRG2	5q/ECA2	$\gamma2$	CAE
	/GEFS3		GEFS+
Calcium channels			
CACNA1H		$Ca_V3.2$	CAEs
Neuronal nicotonic receptors			
CHRNA4	20q/ ADNFLE1	$\alpha4$	Autosomal dominant nocturnal frontal lobe epilepsy (ADNFLE)
CHRNB2	1q/ ADNFLE3	$\beta2$	ADNFLE

The pore-forming α subunits are composed of four homologous domains, each with six transmembrane segments and well-characterized voltage sensor and pore regions. Each sodium channel α subunit is associated with an assembly of signaling and cytoskeletal molecules, including one or more β subunits (β1–4). In contrast to the large number of genes encoding potassium channels, there are just nine genes encoding human sodium channel α subunits that vary in both tissue-specific expression and biophysical properties.

Approximately 20 human diseases arising from mutations in sodium channel genes have been described (George, 2005). These include seven skeletal muscle disorders caused by mutations in *SCN4A* such as paramyotonia congenita and hyperkalemic periodic paralysis, and seven cardiac phenotypes associated with mutations in *SCN5A* including LQTS and Brugada syndrome of idiopathic ventricular fibrillation. In the nervous system, disorders of the peripheral nerve are associated with mutations in *SCN9A* and at least four epilepsy phenotypes are associated with mutations in *SCN1A*, *SCN2A*, and *SCN1B* (Meisler and Kearney, 2005).

Generalized epilepsy with febrile seizures plus (GEFS+) is an autosomal dominant inherited epilepsy characterized by febrile seizures in childhood that persist beyond age 6 years and epilepsy later in life characterized by a multitude of different generalized seizure types: absence, myoclonic, atonic, and myoclonic–astatic. GEFS + is caused most commonly by mutations in *SCN1A* (Escayg, MacDonald, et al., 2000; Wallace, Scheffer, et al., 2001), but also independently by mutations in *SNCN1B* (Wallace et al., 1998), *SCN2A* and the gene for a γ-aminobutyric acid ($GABA_A$) receptor subunit, *GABRG2*.

However, mutations in *SCN1A* are also associated with severe myoclonic epilepsy of infancy (SMEI or Dravet syndrome) in which more than 150 mutations have been described (Claes et al., 2001). Approximately 90% of these are de novo, and 50% are predicted to result in protein truncation with haploinsufficiency, the abnormal function resulting from quantitative reduction in gene expression to 50% of normal levels (Claes et al., 2003). Rarely, *SCN1A* mutations have also been described in other phenotypes, including Intractable childhood epilepsy with generalized tonic–clonic seizures (ICEGTCSs) (Fujiwara et al., 2003) and infantile spasms (ISs) (Wallace et al., 2003). Missense mutations in *SCN2A* have been found in benign familial neonatal–infantile seizures (BFNISs), an early onset syndrome that does not progress to adult epilepsy (Berkovic et al., 2004).

A number of *SCN1A* missense mutations have been evaluated using in vitro functional assays (Kanai et al., 2004; Rhodes et al., 2004; Yamakawa, 2005). The most common abnormality is impaired channel inactivation leading to a persistent inward sodium current, a gain of function. This would be predicted to lead to hyperexcitability of individual neurons, but it is less easy to see how haploinsufficiency increases neuronal hyperexcitability at the level of an individual cell. A striking feature of sodium channelopathy epilepsies is the variable expression: individuals with the same mutation show striking differences in clinical severity. This may arise from differences in genetic background or modifier genes segregating in the families as well as environmental and stochastic influences. The recent description of a mutation in *SCN1A* in a pedigree with familial hemiplegic migraine 1 (FHM1) reinforces the long-suspected links between epilepsy and migraine at a molecular level (Dichgans et al., 2005).

Chloride Channel Epilepsies

Chloride channels, specific for anions rather than cations, are involved in a broad range of functions including stabilization of membrane potential in excitable cells, synaptic inhibition, cell volume regulation, transepithelial transport, extracellular and vesicular acidification, and endocytic trafficking (Jentsch et al., 2002). Three gene families of Cl^- channels are recognized: the CLC channels that are often voltage-gated, ligand-gated channels such as $GABA_A$ and glycine receptors, and the cystic fibrosis transmembrane conductance regulator (CFTR).

The CLC gene family encodes nine different chloride channels in mammals, and human mutations in CLC channels cause several diverse diseases reflecting their variety of functions (Jentsch et al., 2005). These include myotonia (*CLCN1*), Bartter syndrome of renal salt loss (*CLCNKB*), X-linked Dent disease of nephrolithiasis (*CLCN5*), osteopetrosis (*CLCN7*), and IGE (*CLCN2*).

A genome-wide search in a large number of families with common IGE subtypes identified a locus on 3q2b (Sander et al., 2000). A functional positional candidate in the critical region was *CLCN2*. Mutational analysis of this gene in affected individuals from 46 pedigrees led to identification of three different heterozygous mutations (Haug et al., 2003) in which affected individuals had any of the four most common IGE subtypes: CAE, JAE, JME, and epilepsy with grand-mal seizures on awakening (EGMA). The mutations resulted in a premature stop codon (M200fs x231), an atypical splicing event (del 74–117), and a non-synonymous amino acid change (G715E). The gene was subsequently screened in 112 unrelated patients with idiopathic epilepsy (D'Agostino et al., 2004). Numerous common single-nucleotide polymorphisms (SNPs) were identified together with three variants of possible functional significance found only in patients. These included a missense mutation E718D affecting the carboxy-terminal regulatory region close to G718E. The role of *CLCN2* in IGE has been reviewed (Heils, 2005).

CLC-2 is an 898-amino-acid protein with 18 membrane-crossing α helices expressed in brain, especially in neurons inhibited by GABA but also in glia. The inhibitory effects of GABA depend on a low intracellular chloride concentration, and CLC-2 may have a role in establishing this, although it is known that the principle mediator of this is the K-Cl cotransporter, KCC2. Functional studies of the three mutant channels originally described have given conflicting results. The truncation mutation was reported to have a dominant-negative effect, and G715E was reported to shift voltage dependence in a $[Cl^-]$-dependent manner to more positive voltages, but these results have not been replicated. Even without a dominant-negative effect, the truncation mutation should create haploinsufficiency and loss of function of CLC-2 with a resulting loss of sustained GABA inhibition.

$GABA_A$ receptors are ligand-gated chloride channels and the major sites of fast synaptic inhibition in the brain. They are members of the cys-loop superfamily, with a heteropentameric structure assembled from at least seven different subunit classes: α1–6, β1–4, γ1–3, δ, ε, π, and ρ. Random co-assembly of these subunits could generate more than 1,50,000 receptor subtypes, but the majority of $GABA_A$ receptors in mammalian brain appear to be composed of α, β, and δ subunits. So far, epilepsy-carrying mutations have been described in *GABRA1*, *GABRG2*, and *GABRD* genes encoding the α1, γ2, and δ subunits, respectively.

A mutation of *GABRA1* was identified in a large French–Canadian pedigree with an autosomal dominant form of a phenotype consistent with JME (Cossette et al., 2002). A genome screen provided evidence of linkage to 5q34 including the site of a cluster of $GABA_A$ receptor subunit genes. A missense mutation, Ala322Asp, segregated in a heterozygous state with the affected individuals.

In *GABRG2*, mutations are associated with the GEFS + phenotype and with a phenotype of CAE and febrile seizures (Baulac et al.,

2001; Wallace, Marini, et al., 2001; Harkin et al., 2002; Kananura et al., 2002), and in *GABRD*, two missense mutations are associated with GEFS+ (Dibbens et al., 2004). The majority of mutations found are loss of function, with a reduction of GABA-mediated chloride currents.

Calcium Channel Epilepsies

Calcium ions are distinct in both carrying charge and functioning as the most ubiquitous of diffusible second messengers. Voltage-dependent calcium channels (VDCCs), therefore, allow cells to couple electrical activity to intracellular calcium signaling, which controls processes such as gene transcription and neurotransmitter release. VDCCs are heteromultimers with an α-pore-forming subunit that assembles with β (four subtypes 1–4) and α2δ2 (subtypes 1–4), and γ (subtypes 1–8) subunits.

Although no VDCC gene is associated with any form of human idiopathic monogenic epilepsy, they are mentioned here because there is strong evidence for their role in seizure generation as discussed elsewhere. Spike-wave seizures in mice are associated with all four calcium channel subunit types (see Comparative Genomics: Animal Models of Epilepsy) and mutations in *CACNA1A* cause EA1 in man in which some patients are found to have epilepsy (Jouvenceau et al., 2001; Imbrici et al., 2004). A truncating mutation in *CACNB4* was found in a small family segregating JME (Escayg, De, et al., 2000). Most recently, missense mutations in *CACNA1H* were found in a small number of Chinese patients with sporadic CAE (Chen et al., 2003). Although these findings have not been replicated in other patient groups (Heron et al., 2004), the amino acid changes do appear to have functional consequences (Khosravani et al., 2004; Vitko et al., 2005).

Neuronal Nicotinic Acetylcholine Receptor Epilepsy

The existence of neuronal nicotinic acetylcholine receptors (nAChRs) in the mammalian brain has long been established, but their exact function remains uncertain. Available evidence indicates that the majority are presynaptic and modulate the release of neurotransmitters such as glutamate. They are heteropentameric channels containing usually a combination of three α and two β subunits. In humans, genes for eight α and three β subunits are known, and the α4, β2, or α7 subtypes are most highly expressed in the mammalian central nervous system (CNS).

Mutations in either *CHRNA4*, encoding the α4 subunit, or *CHRNB2*, encoding the β2 subunit, are associated with autosomal dominant nocturnal frontal lobe epilepsy (ADNFLE). ADNFLE is a rare idiopathic partial epilepsy with onset typically in childhood. Clusters of frequent brief partial seizures occur mainly during sleep. There is incomplete penetrance of approximately 70%. The first *CHRNA4* mutation—an S248F amino acid substitution in the second TMD—was identified in 1995 (Steinlein et al., 1995), and several mutations were subsequently described in *CHRNA4* and *CHRNB2* on chromosome 1 (Phillips et al., 2001). All of the mutations described so far are located within or close to the second or third TMD, both of which contribute to the channel pore.

In vitro experiments have explored the functional effects of these mutations (Steinlein, 2004b). All appear to increase the acetylcholine sensitivity of nAChRs, indicating that a gain of function underlies this form of epilepsy. In addition, some mutations reduce calcium permeability (e.g., S248F and 4776 in S3 in α4). Why these changes in nAChRs should cause an age-dependent partial epilepsy remains unknown.

Another locus for ADNFLE has been mapped to chromosome 15q24 close to a cluster of nAChRs subunit genes (*CHRNA3*, *CHRNA5*, and *CHRNB4*), but these genes map outside the region, and the responsible gene remains unidentified (Phillips et al., 1998). It is noteworthy that most ADNFLE families map to none of these loci, so the additional genes causing this syndrome are yet to be discovered.

Idiopathic Monogenic Epilepsies: Non-ion Channelopathies

Autosomal dominant lateral temporal lobe epilepsy (ADLTLE), also known as autosomal dominant partial epilepsy with auditory features (ADPEAF), is an idiopathic epilepsy characterized by simple partial seizures with acoustic or visual hallucinations, mostly having a focal onset in or near the auditory center in the temporal lobe of the brain. A locus was mapped to chromosome 10q22-24 (Ottman et al., 1995), and mutations were identified in *LGI1* (leucine-rich, glioma-inactivated 1) in five families (Kalachikov et al., 2002). Additional mutations were subsequently identified (Gu et al., 2002; Morante-Redolat et al., 2002; Fertig et al., 2003; Michelucci et al., 2003; Hedera et al., 2004; Flex et al., 2005), most of which are predicted to cause protein truncation but some of which are missense.

The human *LGI1* gene, also called epitempin, was originally identified in a glioblastoma cell line as a possible tumor suppressor but its function is unknown. The encoded protein has several remarkable features. There are three leucine-rich repeats (LRRs) in the N-terminal region. This domain usually acts as a framework for protein–protein interactions and is present in numerous proteins with diverse functions including signal transduction and regulation of cell growth. The second hallmark is a 50-residue tandem repeat located in the C-terminus, variously called the epilepsy-associated repeat (EAR) or the epitempin repeat (EPTP) and predicted to form a seven-bladed β-propeller fold. The function of this repeat is unknown, but it is also found in the *MASS1/VLGR1* gene, which has been implicated in epilepsy in mouse and humans. The homologous mouse gene *MASS1* is mutated in the Frings mouse, a model of audiogenic seizures, and mutations in *MASS1/VLGR1* are a rare cause of febrile seizures in humans.

It is possible that the EPTP is involved in some presently unknown mechanism of epileptogensis such as synaptogenesis or axon guidance.

Symptomatic Monogenic Epilepsies

There are more than 200 Mendelian disorders in which epilepsy is present but is accompanied by other neurological problems such as cognitive deficiency and an obvious disorder of metabolism or structural abnormality in the brain. In these symptomatic inherited epilepsies, the epileptic seizures are the end point of the defective function of genes involved in a wide range of functions, from energy metabolism to cortical development.

Advances in genomics have facilitated major advances in understanding the molecular basis of many of these disorders. This chapter reviews advances in three areas: the progressive myoclonus epilepsies (PMEs), developmental malformations of the cerebral cortex, and new metabolic disorders.

PMEs

The PMEs are a heterogeneous group of inherited disorders characterized by epilepsy, including myoclonic seizures and progressive neurological deterioration often with cerebellar signs and cognitive

Table 17–2 Progressive Myoclonus Epilepsies

Disease/Phenotype	Locus	Gene	Protein
Unverricht–Lundborg disease	EPM1A/21q	*CSTB*	Cystatin B
	EPM1B/12	—	—
Lafora's disease	EPM2A/6q	*EPM2A*	Laforin
	EPM2B/6q	*NHLRC1*	Malin
Neuronal ceroid lipofuscinosis			
Infantile	CLN1/1p	*PPT1*	Palmitoyl-protein
Late-infantile, classical	CLN2/11p	*TPP1*	Thioesterase 1
			Tripeptidyl peptidase 1
Finnish variant late-infantile	CLN5/13q	*CLN5*	Novel membrane protein
Variant late infantile	CLN6/15q	*CLN6*	Novel membrane protein
Juvenile	CLN3/16p	*CLN3*	Novel membrane protein
Northern epilepsy	CLN8/8p	*CLN8*	Novel membrane protein
Sialidosis			
Type 1 and 2	NEU1/6p	*NEU1*	Lysosomal sialidase 1
Dentatorubral-pallidoluysian atrophy (DRPLA)	ATN1/12p	*ATN1*	Phosphoprotein: atrophin 1
Myoclonic epilepsy with ragged red fibers (MERRF)	MTTK/mtDNA	*MTTK*	$tRNA^{Lys}$

decline (Lehesjoki, 2003; Chan et al., 2005; Shahwan et al., 2005). They encompass five main disease groups, which differ in clinical features and pathogenesis (see Table 17–2), together with some additional rather rare disorders. Genomic advances have allowed their molecular genetic basis to be elucidated.

Unverricht–Lundborg Disease: EPM1

This is an autosomal recessive disorder characterized by onset at 6–15 years of severe stimulus sensitive progressive myoclonus, generalized tonic–clonic seizures, and later development of cerebellar signs and mild dementia. It was initially recognized in the isolated Finnish population but an identical disorder occurs in Mediterranean populations, and also worldwide. Histopathology of the brain reveals widespread non-specific degenerative changes including loss of Purkinje cells but not intracellular storage material.

The disease gene, *EPM1A*, was mapped to chromosome 21q22.3 and identified by positional cloning (Lehesjoki et al., 1991, 1992, 1993; Pennacchio et al., 1996). It encodes cystatin B (CSTB), a cysteine proteinase inhibitor. The most common mutation, accounting for approximately 90% of disease alleles, is an unstable expansion of a dodecamer (126p) repeat located upstream of the promoter region (Virtaneva et al., 1997). This minisatellite repeat is polymorphic with normal alleles of two to three copies, but EPM1-disease-associated alleles contain at least 30 repeat copies. Six further mutations have been described, most of which create splicing defects or a truncated protein.

The molecular pathology of cell loss and seizure generation in EPM1 remains uncertain. CSTB is a small, 98-amino-acid, ubiquitously expressed protein that binds tightly in vitro to cathepsins H, L, and S. Loss of CSTB inhibition of cysteine proteases may dysregulate apoptosis and cause neurodegeneration. There is evidence that CSTB participates in the formation of a multiprotein complex with a specific cerebral function (Lehesjoki, 2003). A mouse model for Unverricht–Lundborg disease (ULD) has been generated by targeted disruption of the homologous mouse *Cstb* gene (Pennacchio et al., 1998). The mice show apoptotic death of cerebellar granule cells and hippocampal cells and a phenotype with progressive myoclonic seizures and ataxia, but no tonic–clonic seizures or photosensitivity (Shannon et al., 2002).

A second locus for ULD has recently been mapped to chromosome 12 in an inbred Arab family. The locus is designated *EPM1B*. The chromosomal region does not contain any genes related to CSTB (Berkovic et al., 2005).

Lafora's Disease

Lafora's disease (LD) is an autosomal recessive inherited disorder characterized by adolescent onset of progressive epilepsy with several seizure types including myoclonic, cognitive decline, ataxia, and death within 10 years of onset. Lafora bodies, periodic acid-Schiff-positive intracellular polyglucosan inclusion bodies found in neurons, heart, skeletal muscle, liver, and sweat gland duct cells, are pathognomonic.

Up to 80% of patients have a mutation in *EPM2A* on chromosome 6q, which encodes laforin, a dual-specificity protein tyrosine phosphatase (Minassian et al., 1998). The function of laforin is unknown, but mutations found appear to interfere with its phosphatase activity and to disrupt its capacity to bind with glycogen and polyglycosans. A second gene, *EPM2B* or *NHLRC1* on chromosome 6p, was recently identified (Chan et al., 2003) in a French–Canadian isolate. This encodes a 395-amino acid protein called malin ("mal" for seizure in French), which contains a zinc finger of the RING-HC type and six NHL-repeat protein–protein interaction domains. This predicts an E3 ubiquitin ligase function and suggests a role in the pathway specifying substrates to be removed by the proteasome system. Recent evidence suggests that malin interacts with and polyubiquinylates laforin, leading to its degradation (Gentry et al., 2005). Recent evidence suggests the existence of a third LD locus (Chan et al., 2004).

An Epm2a-knock-out mice model has been created, and a canine epilepsy model has been described in miniature wirehaired dachshunds (MWHDs) with a mutation in the homologous *EPM2B* gene (Lohi et al., 2005)—see the following.

Neuronal Ceroid Lipofuscinosis

The neuronal ceroid lipofuscinosis (NCL) are the most common form of inherited neurodegenerative disease of childhood onset and are characterized by the intracellular accumulation of autofluorescent lipopigment storage material and progressive neuronal cell loss. They were originally classified according to age at onset and ultrastructural appearances of the storage material, but the identification of six human genes (Mole, 2004; Mole et al., 2005) underlying these disorders in the last decade has allowed a new molecular genetic dimension to their nosology. Common clinical features include retinopathy with visual impairment, epilepsy including myoclonic seizures, and progressive dementia and ataxia. The different varieties are considered in turn:

CLN1: Palmitoyl Protein Thioesterase 1 (PPT1) *CLN1* encodes a ubiquitously expressed lysosomal enzyme that removes palmitate residues from proteins. In neurons, it is preferentially targeted to axons and presynaptic terminals. More than 40 mutations have now been described in *CLN1*, approximately half of which are missense mutations altering highly conserved amino acids. All cases of NCL caused by mutations in *CLN1/PPT1* are characterized by granular storage material described as granular osmiophilic deposits (GRODs). There is a reliable enzyme diagnostic test in which PPT activity is measured in peripheral white cells, fibroblasts, or saliva.

Mutations in *CLN1* classically cause infantile-onset NCL (Vesa et al., 1995), first described in Finland and known as Haltia–Santavuori disease, but the clinical spectrum is now known to be extremely wide with age at onset varying through to adulthood (Mitchison et al., 1998; van Diggelen et al., 2001). Infantile NCL presents at approximately 1 year with slowed development progressing to seizures, visual loss, and increasing spasticity with death between 8 and 13 years. Cranial magnetic resonance imaging (MRI) reveals progressive cerebral atrophy (Vanhanen et al., 2004). CLN1 disease may also have late-infantile, juvenile, or adult onset, and there is an evident genotype–phenotype correlation: mutations predicted to result in severely truncated protein or decreased enzyme stability tend to have earlier onset and more rapid progression. The FIN-major mutation is a missense, R122W, associated with severely reduced enzyme activity.

CLN2: Tripeptidyl Peptidase 1 (TPP1) *CLN2* encodes a lysosomal enzyme, TPP1, that removes tripeptides from the N-terminus of small proteins and is implicated in the degradation of certain neurotransmitters and hormones (Sleat et al., 1997). More than 50 mutations have been described in *CLN2*, which usually cause classical late-infantile NCL with onset between 2 and 4 years of seizures, including myoclonic jerks followed by developmental regression and visual failure. The basic ultrastructural defect is "curvilinear" bodies found in neurons and extraneural tissues including lymphocytes.

Classical late-infantile NCL is one of the most common inherited neurodegenerative disorders in Europe with an incidence of 0.46 in 100,000 live births. The two most common mutations are 1V5-1G>C and R208X. A few patients show delayed disease onset and slower progression. Enzyme assays are available for diagnosis of *CLN1* and *CLN2*.

CLN3 Mutations in *CLN3* cause most cases of juvenile onset NCL, also called Batten or Spielmeyer–Vogt disease. The gene on chromosome 16p encodes a 438-amino acid glycosylated membrane protein of unknown but fundamental cellular function: orthologous genes are present in eukaryotes including yeasts (The International Batten Disease Consortium, 1995). The most common mutation is a 1-kb deletion present on approximately 85% of disease chromosomes and predicted to encode a truncated protein that is retained in the endoplasmic reticulum (ER).

Classical juvenile onset NCL presents with progressive visual failure in the early school years with progression to epilepsy and deterioration of cognitive function, speech, and mobility. On examination, there is pigmentary retinopathy. Vacuolated lymphocytes are seen in blood, and the ultrastructural hallmark is the fingerprint profiles that are found in many tissues. The EEG is abnormal, but has no specific features, and the erythrocyte glutathione reductase (EGR)is reduced from an early stage.

There is some correlation between *CLN3* genotype and disease phenotype. Compound heterozygotes for the 1-kb deletion and a second allele exhibit a spectrum of disease severity depending on the nature of the second allele (Savukoski et al., 1998). Missense mutations affecting less highly conserved amino acids are associated with milder disease with later onset and slower progression.

CLN5, CLN6, and CLN8 These three genes underlie "variant" forms of NCL. Mutations in *CLN5* underlie a Finnish variant of late-infantile NCL first described in 1991. Onset is later than in the classical form with clumsiness and hypotonia at approximately 5 years followed by visual impairment and then ataxia and seizures at 7–10 years. *CLN5* encodes a 407-amino acid putative membrane protein of unknown function. The most common mutation accounting for 95% of Finnish patients is a 2-bp deletion in exon 4 predicted to cause a truncated protein, Y392X. The ultrastructural findings encompass classical fingerprint and curvilinear profiles and sometimes lamellar inclusions.

Mutations in *CLN6* cause a variant form of late-infantile NCL with a later onset of symptoms, slower rate of disease progression, and a mixed pattern of curvilinear fingerprints and rectilinear patterns on ultrastructual analysis. This form is found in a number of different geographical populations including patients from Venezuela, Costa Rica, Portugal, Italy, Pakistan, and the Czech Republic. *CLN6* encodes a 311-amino-acid membrane protein located in the ER (Wheeler et al., 2002). There is a naturally occurring mouse model, nclf. Eighteen mutations in *CLN6* have been reported, mostly associated with a similar disease course.

Mutations in *CLN8* cause two distinct diseases: progressive epilepsy with mental retardation (EPMR) or Northern epilepsy in Finland and a form of variant late-infantile NCL in families of Turkish origin (Ranta et al., 1999). *CLN8* encodes a predicted membrane protein of 286 amino acids, which shuffles between the ER and ER–Golgi intermediate complex (ERGIC). The exact function is unknown. Five mutations have been identified. Ultrastructural analysis of Turkish patients shows fingerprint profiles or a mixed pattern of fingerprint profiles and curvilinear bodies. A naturally occurring mouse model exists, the motor neurodegeneration (*mnd*) mouse.

CLN7 and CLN4 The locus *CLN7* was originally assigned to a group of consanguineous families from Turkey with variant LINCL. A subset of these had mutations in *CLN8*, and it is not yet known whether the remaining families represent a novel locus (*CLN7*) or are allelic variants of other forms.

The gene symbol *CLN4* is currently assigned to a group of adult onset forms of NCL (ANCL) not yet linked to a specific locus. ANCL, sometimes called Kufs' disease, may display autosomal recessive or autosomal dominant inheritance. The extent to which cases of adult onset are caused by mutations in novel genes or represent milder mutations in known genes remains uncertain. Two adult siblings with mutations in *CLN1* have been identified.

Rare Causes of PME

Other less common causes of PME include mitochondrial disorders such as myoclonic epilepsy with ragged red fibers (MERRFs), the sialidosis, and dentatorubral-pallidoluysian atrophy (DRPLA). MERRF is considered in the following under mitochondrial disorders.

Sialidosis is characterized by the lysosomal storage of sialidated glycopeptides and oligosaccharides with the accumulation or secretion of sialic acid (N-acetylneuraminic acid) due to a deficiency of neuraminidase activity. This occurs in three inherited disorders: sialidosis type I and II, which have isolated neuraminidase deficiency; I-cell disease, in which neuraminidase is one of the many deficient lysosomal hydrolases; and a third entity in which there is a combined deficiency of neuraminidase and β-galactosidase due to deficiency of cathepsin A (galactosialidosis). Sialidosis type I is the milder form, also known as the "normosomatic" type, or cherry-red spot myoclonus syndrome. Type II is more severe with dysmorphic features including coarse facies, corneal clouding, and dementia, in addition to the myoclonus. More than 39 mutations have been described in the disease gene, *NEU1*, on chromosome 6p. Enzyme deficiency can be confirmed in white cells or cultured fibroblasts.

DRPLA is a rare autosomal dominant neurodegenerative disorder first described in Japan. The phenotype includes various combinations of myoclonus, dementia, cerebellar signs, and psychiatric features, with patients presenting before age 20 years often manifesting as PME. It is caused by expansion of a trinucleotide (CAG) repeat of a gene on chromosome 12p13.31 designated Atrophia 1 (*ATN1*), which encodes a phosphoprotein. Pedigrees display anticipation (accelerated age at onset in succeeding generations), and there is an inverse correlation between age at onset and the number of repeats.

Cortical Malformations

The introduction of noninvasive MRI of the brain has led to increasing recognition of disorders of cortical development as a cause of epilepsy in conjunction with developmental delay and cognitive defects (Guerrini and Filippi, 2005). Although often referred to as neuronal migration disorders, advances in understanding their molecular basis have shown that the problem may lie in several different steps of cortical development: proliferation of neuronal progenitor cells, migration of neurons, and maintenance of the integrity of the pial surface (Mochida and Walsh, 2004; Barkovich et al., 2005).

Normal development of the human cerebral cortex starts with proliferation of neurons in the zone surrounding the cerebral ventricles. Postmitotic neurons migrate toward the surface, guided by a scaffold of specialized cells called radial glia. A striking feature is the formation of the six-layered cortex in an inside-out fashion: newly arrived neurons pass the older neurons, so that the outer layers are formed last. Most neurons have reached their destination in the cortex by the 24th week of gestation. The integrity of the pial surface prevents overmigration of neurons onto the outside of the brain.

Disorders of Neuronal Proliferation or Survival: Microcephaly These include the inherited microcephalies in which epilepsy is uncommon. Inheritance is autosomal recessive, and at least seven loci are recognized: *MCPH1–MCPH7* (Woods et al., 2005). Four genes have been identified: *MCPH1* encodes microcephalin, which may be involved in chromosome condensation and DNA repair. *MCPH3* encodes CDK5RAP2. The *MCPH5* gene, *ASPM*, appears to have a role in the proliferation of neuronal progenitor cells. ASPM and its homologues have a motif rich in isoleucine and glutamine, the IQ domain, and the number of IQ domain appears to be increased in organisms with larger brain size. More IQ domains do not however mean a higher IQ, as less intelligent mammals have the same number of domains as humans. *MCPH6* encodes CENPJ.

Disorders of Neuronal Migration: Lissencephaly and Heterotopia At least four malformations of cortical development can be ascribed to faults in neuronal migration: lissencephaly (literally "smooth brain"), lissencephaly with cerebellar hypoplasia, X-linked lissencephaly with abnormal genitalia, and periventricular nodular heterotopia.

In classical lissencephaly, normal gyration of the cortex is absent or reduced, and the cortex is thickened with four layers instead of six. It is caused by abnormalities in two genes: *LIS1* on chromosome 17p and *DCX* (or doublecortin) on chromosome Xq. Chromosomal deletions including *LIS1* cause Miller–Dieker syndrome with unique facial features, whereas point mutations cause "isolated lissencephaly sequence" (Hattori et al., 1994). Females who are heterozygous for *DCX* mutations have a "double cortex" or subcortical band heterotopia associated with epilepsy and mild-to-moderate learning difficulties (des et al., 1998; Gleeson et al., 1998). In contrast, males who are hemizygous have lissencephaly very similar to that found with mutations of *LIS1*. The protein products of both genes appear to regulate microtubules, components of the cytoskeleton important in regulatory mobility of cells (Francis et al., 1999; Gleeson et al., 1999).

Mutations in the *RELN* (reelin) gene have been identified in some patients with the autosomal recessive condition lissencephaly with cerebellar hypoplasia (Hong et al., 2000). Mutations in the mouse homologue, *Reln*, cause the "reeler" mouse, which has a similar CNS phenotype. The reelin protein is secreted by Cajal–Retzius cells and may have a role in halting neuronal migration at the correct position. Patients with X-linked lissencephaly and abnormal genitalia plus agenesis of the corpus callosum have mutations in *ARX*, the Aristaless-related homeobox transcription factor gene (Kitamura et al., 2002).

Periventricular nodular heterotopia is a heterogeneous condition caused in some patients by mutations in *FLNA* (filamin A), which encodes a large cytoplasmic actin-binding protein (Fox et al., 1998). Heterozygous females have epilepsy, often with normal cognitive function, but hemizygous males usually die in utero. This condition appears to represent a defect in the initiation of neuronal migration, reflecting the importance of the actin cytoskeleton in cell mobility.

Disorders of the Integrity of the Pial Surface: Cobblestone Dysplasia Three autosomal recessive disorders occur that are characterized by cobblestone dysplasia (also called type II lissencephaly), in which neurons migrate thorough the pial surface onto the outside of the brain, creating a nodular surface abnormality. In addition, there is muscular dystrophy, a variety of eye abnormalities, and cerebellar polymicrogyria.

Genes for all three conditions have been identified. Fukuyama-type congenital muscular dystrophy is caused by mutations in the *FCMD* (fukutin) gene, which has a probable role in glycosylation of dystroglycan (Kobayashi et al., 1998). Genes encoding glycosyltransferases are mutated in the other two conditions: the protein o-mannose β-1, 2-N-acetyglucosaminyltransferase (*POMGNT1*) gene in muscle–eye–brain-disease (MEB) and the protein o-mannosyltransferase (*POMT1*) gene in some cases of Walker–Warburg syndrome (Yoshida et al., 2001; Beltran-Valero de et al., 2002; Diesen et al., 2004). It is proposed that failure of glycosylation of α-dystroglycan leading to disruption of the pial surface is central to the pathogenesis of these disorders. Further support for this comes from the identification of mutations in the *LARGE* gene in a patient with congenital muscular dystrophy and structural brain abnormalities on cranial MRI.

Mitochondrial Disorders

The mitochondrial encephalomyopathies are a heterogeneous group of disorders primarily affecting organs and tissues with a high energy demand, such as brain or muscle (DiMauro, 2004).

The proteins of the respiratory complex are predominantly nuclear encoded, but a significant minority are of course encoded by the maternally inherited mitochondrial genome. In the latter, heteroplasmy of mitochondrial DNA (mtDNA; different proportions of normal and mutant mtDNA molecules in different tissues) accounts for the highly variable clinical phenotype. A well-known mitochondrial syndrome in which epilepsy is a dominant feature is MERRF.

Patients with MERRF manifest a wide spectrum of clinical features including myoclonic seizures, ataxia, muscle weakness, and hearing loss. Muscle biopsy shows "ragged red fibers," which reflect proliferation of mitochondria and plasma lactate and pyruvate levels are raised. A missense mutation in the gene for transfer RNA for lysine, *MTTK*, an A-to-G mutation at nucleotide 8344, accounts for up to 90% of cases (Shoffner et al., 1990). This mutation results in multiple deficiencies in the enzyme complexes of the respiratory chain. Maternal inheritance is found in pedigrees.

Patients occur whose phenotypes overlap the syndrome of mitochondrial encephalomyopathy with lactic acidosis and stroke-like episodes (MELAS). Mutations in these cases may involve genes encoding specific subunits of the respiratory complex, such as ND5 (Naini et al., 2005)

Metabolic Disorders

Epilepsy is a component of the phenotype in a host of inherited metabolic disorders. The biochemical basis of most disorders of intermediary metabolism involving amino or organic acids, carbohydrate metabolism, storage of sphingolipids or mucopolysaccharides, or defective function of organelles such as the peroxisome were established before the genomic era, although genomic advances have facilitated elucidation of their molecular genetic basis. Some are included among the PMEs (see preceding text), and disorders of the mitochondrion can be classed as "metabolic."

An important disorder recognized in 1991 is glucose transporter type 1 (GLUT-1) deficiency syndrome (De et al., 1991). The mammalian brain is highly dependent on glucose transported across the blood–brain barrier by a stereo-specific, high-capacity carrier. In vertebrates, there is a family of such transporters including *GLUT1* expressed in the brain endothelial cells and erythrocytes. The gene *GLUT1* was positionally cloned and mapped to chromosome 1p35-p31.3. Since 1991, more than 80 patients with this syndrome have been identified. Clinical features are variable, including seizures (generalized tonic–clonic, absence, partial, myoclonic, or astatic), developmental delay, acquired microcephaly, ataxia, and dystonia (Klepper and Voit, 2002; Leary et al., 2003). The hallmark of the disease is a low cerebrospinal fluid (CSF) glucose concentration (hypoglycorrhachia) in the presence of normoglycemia (CSF plasma/glucose ratio <0.4). Truncation mutations cause haploinsufficiency, and autosomal dominant inheritance has been observed (Seidner et al., 1998). Early diagnosis is important, as some antiepilepsy drugs (AEDs) (such as phenobarbitone) are ineffective or detrimental, and patients may respond to a ketogenic diet as ketones may serve as an alternative fuel. Increased awareness will probably lead to higher prevalence estimates as many cases are probably undiagnosed.

A disruption of ganglioside biosynthesis has recently been shown to underlie an autosomal recessive infantile onset symptomatic epilepsy syndrome (Simpson et al., 2004). A nonsense mutation in *SIAT9*, predicted to result in premature termination of the GM3 synthase enzyme (also called lactosylceramide α-2,3 sialyltransferase), was identified in a large Old Order Amish Pedigree segregating this severe syndrome characterized by multiple seizure types and profound developmental delay.

Epilepsies Inherited As "Complex" Traits

Monogenic epilepsies are rare, and such patients compromise only a small fraction of those with a genetic component to the etiology of their epilepsy. In the majority, epilepsy arises from the combined effect of several genes, sometimes in combination with environmental factors. The mechanism of inheritance of such traits is variously described as "complex," polygenic, non-Mendelian, or multifactorial. These terms are not entirely synonymous, and the terms "complex" causes particular difficulty as it is also applied in the common usage of indicating "complexity," for example, in describing the lack of correlation between genotype and phenotype in monogenic epilepsies. It is worth clarifying these terms.

Monogenic, synonymous with Mendelian, implies that mutation at a single locus is necessary and (usually) sufficient for a trait to be manifest. However, other loci, known as modifier genes, may influence the penetrance and expression. These are the equivalent of genetic background for inbred rodent models. Non-Mendelian implies that familial segregation is not accounted for by a single locus. Usually, this is accounted for by polygenic inheritance, but of course, mitochondrial inheritance does not display a Mendelian pattern. Polygenic implies several contributing loci (>5), all of small effect, whereas oligogenic implies fewer (<5). Multifactorial implies an environmental influence such as head trauma in the case of epilepsy, in addition to genetic factors.

The IGEs encompass the largest group of patients in whom familial clustering indicates inheritance as a "complex trait." The IGEs affect approximately 0.2% of the population and account for 25% of all epilepsy patients. The recurrence risk is 70%–95% in monozygous twins and 5%–8% in first-degree relatives, giving a ratio of sibling risk to population prevalence, λ, of approximately 10. The IGEs can be subdivided into more narrowly defined phenotypes, including JME, CAE, JAE, and EGMA. It is also possible to identify endophenotypes, such as photosensitivity or spike-wave seizures, that cross these categories.

In common with many other "complex" traits, progress in elucidating the molecular genetic basis of these epilepsies has been slow compared to the striking advances in understanding Mendelian epilepsies, some of which are of course IGEs. Strategies for positional cloning by linkage analysis are of course difficult to apply, and a large number of association studies have predictably failed to give replicable or convincing results (Tan, Mulley, et al. 2004). Advances have been made, but it is in the analysis of this group of epilepsies that the most recent advances in genomics promise new hope of further progress.

Different IGE syndromes frequently cluster within a single pedigree, but two recent studies have clarified the genetic architecture of IGE and indicate that CAE and JAE share a close genetic relationship, and that JME is a more distinct entity (Winawer et al., 2003; Marini et al., 2004). Recent advances in the IGEs, JME, and the absence epilepsies are considered in turn.

IGES

A locus has been identified on chromosome 18 that appears to be common to several adolescent-onset IGE syndromes, including JME, JAE, and EGTCs (Durner et al., 2001). Analysis using both case–control and family-based association methods provided evidence for the *ME2* gene encoding malic-enzyme 2 as a susceptibility locus. A 9-SNP haplotype in *ME2* in the homozygous state increased the risk for IGE six-fold (Greenberg et al., 2005). *ME2* is an enzyme that converts malate to pyruvate and is involved in the neuronal synthesis of GABA, suggesting the plausible hypothesis that disruption of GABA synthesis predisposes IGE.

JME (Impulsive Petit Mal, Janz Syndrome)

JME is a common subtype of IGE affecting up to 26% of all patients with IGE and representing 5%–10% of epilepsy as a whole. Onset is usually in adolescence with early morning myoclonic jerks typically of the upper limbs. Nearly all patients develop generalized tonic–clonic seizures, often precipitated by alcohol or sleep deprivation, and approximately one-third have absence seizures. The interictal EEG may be normal or may show bilateral, symmetrical 4-6 H_2 polyspike, and wave.

A genetic contribution to the etiology of JME is well established, and it appears that several modes of inheritance may underlie the phenotype, including rare pedigrees with autosomal recessive and autosomal dominant inheritance. As discussed earlier, mutations in *GABRA1* were identified in a large French–Canadian pedigree, and mutations in *CLCN2* were found in a family with JME and other IGE phenotypes. A mutation in *CACNB4* has been found in a female with JME. However, "complex" inheritance is apparent in the majority of cases.

Four studies have provided evidence for the existence of a locus predisposing to JME on chromosome 6p (Greenberg et al., 1988; Durner et al., 1991; Weissbecker et al., 1991; Liu et al., 1995), although two further studies failed to replicate these results (Whitehouse et al., 1993; Elmslie et al., 1996). Initial linkage was to the human leukocyte antigen (HLA) region at 6p21.2-p11, and this locus was designated *EJM1*. Subsequently, *EJM1* has been assigned to a different JME phenotype linked to 6p12-p11 and corresponding to the *EFHC1* gene in which mutations have been identified in some families. A separate locus at 6p21 is designated *EJM3*, for which the candidate causative gene is *BRD2*.

In exploring linkage at 6p21, a highly significant linkage disequilibrium was identified between JME and a core haplotype of five SNPs and microsatellite markers in the region with LD peaking in the *BRD2* gene (Pal et al., 2003). Sequencing of 20 probands from families consistent with linkage revealed two JME-associated SNPs in the BRD2 promoter region. *BRD2* encodes a putative nuclear transcriptional regulator expressed during brain development, and disruption could cause the cortical microdysgenesis sometimes reported in JME.

Analysis of 31 new JME families from Mexico generated significant log of odds (LOD) scores in an interval at 6p12-p11, 30 cm centromeric to the HLA region. Several heterozygous mutations were subsequently identified in the *EFHC1* gene in affected members of six unrelated families (Suzuki et al., 2004). *EFHC1* encodes a protein, designated myoclonin 1, which localizes in the soma and dendrites of neurons in many brain regions, and is proapoptotic. It is hypothesized that loss of function in this gene may result in reduced apoptotic shedding of superfluous neurons during brain development, resulting in an overpopulated, epileptogenic cerebral cortex.

Evidence for a further locus, designated *EJM2*, on chromosome 15q14 harboring the *CHRNA7* gene has been obtained (Elmslie et al., 1997) but was not replicated in a subsequent study (Sander et al., 1999).

Absence Epilepsies

There are at least three distinct epilepsy subtypes in which absence seizures are the predominant seizure type: CAE, JAE, and eyelid myoclonia with absences (EMAs). Many early genetic studies of individuals with absence seizures used definitions of the trait under study that do not correspond with current terminology. Although there is a clear genetic predisposition, familial clustering is consistent with inheritance of these syndromes as "complex" traits rather than as monogenic disorders, although occasional extended pedigrees do occur.

Absence seizures are unusual in having a rather distinct electrophysiological hallmark, the 3-Hz spike wave, and their molecular pathophysiology has been extensively studied (Crunelli and Leresche, 2002). Several excellent naturally occurring mouse and rat models exist (see the following text), for which the underlying genes have been identified.

At present, knowledge of the molecular genetic basis of absence epilepsies in humans remains limited. Absence seizures were a component of the phenotype in a family segregating a mutation in *GABRG2* and in the families with IGE in which mutations in *CLCN2* were identified. These loci are designated *ECA2* and *ECA3*. A locus for a form of childhood onset absence epilepsy persisting into adulthood was mapped to a 3.2-cm interval of chromosome 8q24 in an extended pedigree from Mumbai and five smaller families (Fong et al., 1998). A causative gene is yet to be identified at this locus, designated *ECA1*.

Suggestive evidence of linkage to *CACNG3* and the GABA receptor gene cluster on 15q has been obtained (Robinson et al., 2002). Most recently, the *CACNA1H* gene was screened for mutations in 118 sporadic CAE cases of Han ethnicity. Twenty-nine variants were discovered in the coding regions that were not present in 203 unrelated controls, of which 12 were missense and predicted to alter highly conserved amino acids (Escayg, De, et al., 2000). Subsequent functional studies on five of the missense variants revealed changes to channel properties that could predispose to epileptogenesis in three (Khosravani et al., 2004; Vitko et al., 2005). The exons in which most variants were found were subsequently screened in another resource of 192 patients with IGE (including 34 CAE and 15 JAE patients). None of the nine Chinese missense variants were found, and four novel variants that were identified did not co-segregate with a specific phenotype in families (Heron et al., 2004). These observations are consistent with a contribution of rare, population-specific alleles to a "complex trait."

Comparative Genomics: Animal Models of Epilepsy

Epileptic seizures can be regarded as an emergent potential property of complex nervous systems (Noebels, 2003b). Moreover, they represent a detectable and non-lethal phenotype, so it is not surprising that animal models of epilepsy have been documented over the last few decades in a wide range of organisms, including flies, mice, rats, and dogs, in particular. Advances in techniques for genetic manipulation, particularly in mice, together with the sequencing of the genomes of these species, have enormously enhanced the value of these animal models for exploring the molecular pathophysiology of epileptic seizures.

There are several quite distinct ways in which genomic analysis has been applied to these animal models (Meisler et al., 2001; Noebels, 2001). The genes for monogenic naturally occurring animal epilepsy syndromes have been identified by "positional cloning," on occasion exploiting comparative genomics to evaluate the homologues of human genes known to cause similar phenotypes. Alternatively, strategies exist for mutating single genes in mice and determining whether epilepsy develops. Genes may be chosen on biological grounds or because they are known to be mutated in monogenic epilepsy syndromes in man. In the latter case, an "animal model" of the human disease is created.
A selection of examples is considered below, but many more exist.

Mouse Models of Epilepsy

The analysis of several monogenic spike-wave epilepsies in mice identified genes encoding voltage-gated calcium channel subunits (VGCCS) as the underlying sites of mutations. Naturally occurring mouse models of NCL proved to be caused by mutations in homologues of human *NCL* genes, and gene targeting has successfully created mouse models of NCL and LD and identified novel and unsuspected epilepsy genes.

Spontaneous Mouse Mutants With Epilepsy

The most important are a subgroup with generalized spike-wave EEG discharges associated with behavioral arrest and representing a model for human absence seizures. The underlying genes encode three distinct protein types: calcium channel subunits, the subunit AP3δ of a complex involved in sorting molecules for incorporation into synaptic vesicles, and one form of sodium–hydrogen exchanger.

VGCCS VGCCs are composed of a large pore-forming α subunit and three auxiliary subunits, β, γ, and $\alpha 2\delta$. Multiple isoforms and genes exist for each subunit. The tottering, lethargic, stargazer, and ducky mutants bear mutations in the genes *Cacna1a*, *Cacnb4*, *Cacng2*, and *Cacnα2δ*, respectively (Fletcher et al., 1996; Burgess et al., 1997; Letts et al., 1998; Barclay et al., 2001). These mutants display cerebellar dysfunction (ataxia) in addition to spike-wave seizures.

The experimental analyses possible in such models have illuminated several possible mechanisms of epileptogenesis (Kim et al., 2001; Song et al., 2004; Letts et al., 2005; Pietrobon, 2005). The *Cacna1a* gene encodes $Ca_v2.1\alpha$, subunits of the P/Q type. Available evidence suggests that these channels have a specific role in fast synaptic transmission at central excitatory synapses by controlling neurotransmitter release, suggesting that $Ca_v2.1$ channelopathies are primarily synaptic diseases. Spontaneous mutations in the corresponding human *CACNA1A* gene cause several autosomal dominant neurological diseases, including FHM1, episodic ataxia type 2 (EA2), spinocerebellar ataxia type 6 (SCA6), and rarely, absence seizures. The different phenotypes presumably arise from disturbed neurotransmission, predominantly in the cortex (migraine), thalamus (spike-wave seizures), or cerebellum (ataxia). A survey of human epilepsy patients identified two potential mutations in the human ortholog of the lethargic gene, *CACNB4*. Recent work suggests that the stargazin protein also functions in the synaptic localization of α-amino-3-hydroxy-5-methylisoxazole-4-propionic acid (AMPA) receptors (Chen et al., 2000; Letts, 2005), and there is uncertainty about the extent to which the family of γ subunit proteins function primarily as calcium channel components.

The mocha2j (*mh2j*) mouse, a mouse model for the Hermansky–Pudlak syndrome, shows frequent 6-Hz spike-wave discharges (SWDs) and tonic–clonic seizures. The gene encodes an adaptin, the AP3δ subunit of a complex involved in sorting molecules destined for synaptic vesicles, lack of which leads to absence of vesicular zinc sequestration in the mouse brain (Kantheti et al., 2003).

The slow-wave epilepsy (SWE) mouse is caused by a mutation in the *Slc9a1* gene encoding NHE1, one of five genes encoding sodium–hydrogen exchange in neurons (Cox et al., 1997). The involvement of this gene in pH regulation is intriguing in relation to the provocation of spike-wave seizures in humans by hyperventilation.

Mouse Models of NCL Both spontaneous and targeted (see the following) mouse models exist for several of the NCLs. The naturally occurring mnd mouse is caused by a mutation in *Cln8*, the mouse homologue of *CLN8* (the gene for so-called Northern epilepsy). A homozygous 1-bp insertion is predicted to cause a frameshift and protein truncation (Ranta et al., 1999). Although seizures are not a prominent feature in the *mnd* mouse, *Cln8* expression is rapidly up-regulated in hippocampal pyramidal and granular neurons in the hippocampal electrical kindling mouse model of epilepsy (Lonka et al., 2005).

A second mouse mutant, *nclf*, develops retinal atrophy and motor degeneration with accumulation of storage material with a fingerprint pattern (Bronson et al., 1998). This was mapped to mouse chromosome 9, a region of conserved synteny with human chromosome 15q21, to which the *CLN6* gene maps. Cloning of the *CLN6* gene was accompanied by identification of a 1-bp insertion in the orthologous *cln6* gene in the nclf mouse (Wheeler et al., 2002). Ovine ceroid lipofuscinosis (OCL) in Merino and South Hampshire sheep maps to the chromosomal region OAR 7q13-15, also a region of conserved synteny with HSA 15q21-23, indicating that this arises from mutations in the ovine ortholog of *CLN6* (Broom and Zhou, 2001). Indeed, a mutation in exon 2 has been identified in Merino sheep (Cook et al., 2002).

Mouse Epilepsies Created by Gene-targeting

These models fall into two categories. Firstly, a gene known to cause epilepsy in humans is targeted. Secondly, those in which targeted mutagenesis of a particular gene results, sometimes unexpectedly, in an epilepsy phenotype, are targeted.

A number of genes known to cause human epilepsy have been modeled in mice, including those for human PMEs and others. Techniques allow the orthologous gene to be "knocked-out" or specific mutations to be introduced. A CSTB knockout mouse, a model for EPM1/ULD, displayed a partially overlapping neurological phenotype with ataxia, accompanied by apoptotic cell death of cerebellar granule cells and seizures and myoclonus during sleep (Pennacchio et al., 1998).

In addition to the two naturally occurring mouse models for *CLN6* (*nclf*) and *CLN8* (*mnd*), models have been created using gene-targeting for *CLN1*, *CLN2*, *CLN3*, and *CLN5*. These are providing major new insights, particularly into the mechanisms of neuronal cell death in NCL and other aspects of pathophysiology (Mitchison et al., 2004; Pears et al., 2005).

A number of ion channel genes have been targeted in mice. The transgenic mouse model *Scn2a* has an epilepsy phenotype as a result of a mutation that slows channel inactivation (Kearney et al., 2001). Mutations in human *SCN2A* are associated with BFNIS. Of particular interest is the influence of genetic background on phenotypic severity in these mice, an observation that has allowed modifier loci to be mapped (Bergren et al., 2005). Mice lacking sodium channel β-2 subunits display spontaneous seizures, reductions of sodium channel expression, and alterations in nodal architecture in the optic nerve (Chen et al., 2004). Targeted deletion of *Kcna1* (encoding Kv1.1) results in severe seizures and cold-induced neuromyotonia. In humans, mutations in *KCNA1* are found in patients with EA1 and myokymia and occasional patients with partial epilepsy. Analysis of double knockout, $Ca_v2.1^{-/-}/Ca_v3.1^{-/-}$ and double mutant tg/tg/$Ca_v3.1^{-/-}$ mice, has provided in vivo evidence that $Ca_v3.1$ channels have a crucial role in the generation of spike-wave seizures arising from loss of function of $Ca_v2.1$ channels.

The number of knockout mice with epilepsy phenotypes has steadily increased with the widespread application of targeted mutagenesis. A very wide range of genes has been implicated, including those for neurotransmitters (NPY, 5HT2c), receptors (GluRB, GABRβ3), transporters (Glyt-1), and a miscellany of others such as cathepsin D and GHLH transcription factor (NeuroD). These models provide new insights into the pathogenesis of epilepsy, and of course, many represent novel candidate genes for human epilepsies.

Rat Models of Epilepsy

Creation of the Rat Genome Database is greatly enhancing the value of well-characterized rat models of epilepsy. Two in particular have been investigated for many years: genetic absence epilepsy rats from Strasbourg (GAERS) and the Wistar Albino Glaxo rat (WAG/Rij). Like several mouse models, these animals display SWDs and are therefore models for absence epilepsy. In contrast to the mouse models, however, their inheritance is "complex" rather than monogenic. A methodology for introducing mutations into the rat genome comparable to that available for the mouse does not exist, of course, but the rat is a more tractable organism for physiological investigation.

In the GAERS strain, SWDs are concomitant with immobility and unresponsiveness, and are suppressed by the main AEDs effective against absence seizures in humans. They are generated in a neuronal network involving cortical and thalamic regions. Genetic analysis combined with electrophysiological measurements experimentally cross derived from GAERS and Brown Norway (BN) rats has allowed identification of three quantitative trait loci (QTLs) in rat chromosomes 4, 7, and 8, influencing specific SWD variables, designated Swd/gaers1, Swd/gaers2, and Swd/gaers3. Genomic data allowed the position of these loci in the human and mouse genomes to be inferred and a search for candidate genes to be made (Rudolf et al., 2004). The region of mouse chromosome 9 conserved with Swd/gaers3 contains the QTL E14 associated with seizures in a cross derived from the EL/Suz mouse. Several ion channels map to these GAERS QTL, including *Cacng2* (stargazin), *Kcnj4*, and *Scn2b*. In parallel, sequencing of the rat homologue of KCNK9, which encodes the twik-like acid-sensitive potassium channel *TASK3* in GAERS rats, identified an additional alanine residue in a polyalanine tract within the C-terminal intracellular domain (Holter et al., 2005). In humans, this gene maps to the *ECA1* locus on chromosome 8q24.

The WAG/Rij rat exhibits two types of SWD. Type 1 are generalized, bilateral, and symmetrical, whereas type 2 are localized to the occipitoparietal brain region (Coenen and Van Luijtelaar, 2003). A genome-wide scan using 145 microsatellite loci together with quantification of type 1 and type 2 SWD phenotypes in an intercross derived from WAG/Rij and AC1 inbred strains demonstrated independent control of type 1 and type 2 SWDs by QTL on chromosomes 5 and 9. Again, comparative genome mapping data allows identification of candidate genes in these regions. These include several functional candidates including *Kcnab2* and *Slc9a1*, the gene mutated in the SWE mouse (Gauguier et al., 2004). Most recently, a morphometric study of the so-called cortical focal zone in WAG/Rij rats showed alterations in dendritic arborization supporting a local, morphological basis for absence seizures in these animals (Karpova et al., 2005).

Canine Models of Epilepsy

Online Mendelian Inheritance in Animals (OMIA) documents 479 inherited traits in the domestic dog species, *Canis familiaris*, of which 116 are monogenic and 47 have had a causative mutation identified at the DNA level. Idiopathic epilepsy is the most common canine brain disorder, and is associated with significant morbidity and mortality. It affects nearly 35 of the (Chen et al., 2004) 150 different purebred dog breeds and is an area of active research. Sequencing of the dog genome has facilitated recent identification of three canine epilepsy genes that are homologues of human PMEs: *EPM2B*, *CLN8*, and *CLN5*.

More than 5% of purebred MWHDs are affected by an autosomal recessive PME, the locus for which was mapped to canine chromosome 35, which displays conserved synteny in its entirety to human 6p21-25, the location of human *EPM2B*. Canine Epm2b was cloned, and a sequence identified in the 5′ half of the gene's single exon containing two consecutive identical dodecamers (D), and a third copy differing by a single nucleotide (T). The affected dogs were homozygous for expansion of the dodecamer repeat with 19–26 copies of the D sequence (Lohi et al., 2005). This was associated with a dramatic reduction in messenger RNA (mRNA) levels. The corresponding region in other species is not repetitive and shorter by one D repeat. However, most other dog breeds have three repeats (two D's and one T), suggesting that its origin predates dogs, and indeed it was found in other *Canidae* species, wolves, foxes, jackals, etc. Of course, it is noteworthy that expansion of a dodecamer repeat in the CSTB promoter causes most human ULD (EPM1).

In the 1950s, the Norwegian veterinarian Nils Koppang discovered an NCL-like disease in English setters (Koppang, 1973). Affected dogs are normal at birth but develop seizures, cognitive decline, and visual impairment from 1 to 2 years of age. Mutations in canine orthologs of *CLN2* and *CLN3* were excluded,. the locus mapped to canine chromosome 37 (CFA37). Megablast searches of the first build of the canine genome for candidate genes located the homologue of *CLN8* to this region and sequence analysis revealed a T–C transition in the canine gene predicting an L164P missense mutation (Katz et al., 2005). This mutation was not found in more than 200 control dogs, and the leucine at position 164 is conserved in other mammalian species, confirming *CLN8* as the causative gene. This large-animal model may prove particularly useful for the development of therapeutic interventions.

NCL was also described in the Border Collie breed in Australia in 1991 (Studdert and Mitten, 1991). Pedigree data suggested autosomal recessive inheritance. Linkage analysis excluded regions corresponding to the human *CLN1*, *CLN2*, *CLN3*, and *CLN6* genes, but positive linkage was found to the region on *CFA22* in a region of conserved synteny with HSA13q containing CLN5. Sequencing of canine *CLN5* revealed a nonsense mutation (Q206X) within exon 4, predicting synthesis of a truncated protein and correlating with the disease (Melville et al., 2005). This observation will, of course, allow elimination of the disorder from Border Collies by selective breeding, and also provides another large-animal model for the human NCL variant.

Future of Epilepsy Genomics

It can be anticipated that genomics will continue to facilitate our understanding of the molecular basis of both inherited and acquired epilepsies. In particular, the combination of association studies using common variants based on the HapMap, together with deep resequencing of candidate genes, should allow the identification of susceptibility loci and their alleles that underlie those common familial IGEs inherited as "complex" traits. The use of microarrays for transcriptional profiling of brain tissue excised from patients with focal epilepsy should provide new insights into alterations in gene expression associated with seizure activity. Finally, the application of pharmacogenomics should allow the genetic contribution to the variation in response to antiepilepsy medication to be identified. Ultimately, this new knowledge will be translated into new diagnostic and therapeutic strategies for the benefit of patients.

Molecular Genetic Basis of "Complex" Trait IGEs

As discussed earlier, progress in understanding the molecular genetics of common familial epilepsies inherited as "complex" traits has been slow despite much effort expanded on linkage and association analysis in the last decade. Of course, this is a common pattern for most complex traits. Advances in genomics have provided new and more powerful strategies for investigating these "complex" traits, the success of which depends to a considerable extent on their underlying genetic architecture. This is of course unknown at present. The relevant parameters include the number and effect size of susceptibility loci and their mode of interaction, and the frequency and diversity of causal alleles. These issues have been widely discussed.

Two strategies are now available. Completion of the HapMap Project, a documentation of common variation across the human genome together with the pattern of linkage disequilibrium, provides the basis for efficient genome-wide association analysis of common SNPs. Technology for typing in the order of 500K SNPs is now available, and it is estimated that analysis of between 500 and 2000 patients should provide sufficient power to detect susceptibility loci of moderate effect size. However, this "indirect" approach is dependent on causal alleles that are in linkage disequilibrium and occurring at a similar frequency to the surrogate SNPs analyzed. It is only likely to succeed in those traits for which the "common disease–common variant" hypothesis is true.

Availability of the complete transcript map and gene complement of the human genome allows the alternative strategy of "deep" resequencing of candidate genes in a large number of patients. This would allow identification of low-frequency, diverse causal alleles difficult to detect by the "indirect" association approach, but does depend on a correct identification of candidate genes. This is possibly easier to do in the case of epilepsy than it is in those neurological disorders such as autism or the major psychosis in which the underlying molecular pathophysiology is entirely obscure. It is not unreasonable to imagine that genes encoding ion channels or related molecules may underlie complex idiopathic epilepsies, in so far as these clearly modulate neuronal excitability and several monogenic idiopathic epilepsies are channelopathies. This reduces the target gene population by a factor of approximately 100 to approximately 250–300 candidate genes. Direct sequencing of exons and 5′ regions of approximately 250 ion channel genes is being undertaken in 500 patients with epilepsy at Baylor College of Medicine. If this set includes the true susceptibility loci and the relevant causal alleles are rare, this approach should identify them. However, the problem of demonstrating causality may remain challenging, especially if epistasis is significant and the sequence changes are in as yet ill-defined regulatory regions.

Pharmacogenomics of Epilepsy

Most patients with epilepsy are treated with one or even several of a wide variety of the so-called AEDs, which include such compounds as phenytoin, carbamazepine, and sodium valproate. A common mode of action of AEDs is sodium channel inhibition, which is assumed to act through a non-specific reduction in neuronal hyperexcitability. However, individual patient response to these drugs is quite variable, and there is considerable potential scope for the kind of personalized medicine based on DNA profiling, which has driven a proportion of the investment in genomics by the pharmaceutical industry. Some preliminary results are available in a field that should see significant advances.

In an initial study, it was hypothesized that resistance to AED therapy might be influenced by variation in the *ABCB1* gene (also known as MDR1 and P-glycoprotein 170), which encodes a transporter involved in drug efflux (Siddiqui et al., 2003). Most AEDs are planar lipophilic agents and are therefore theoretical substrates for this transporter. A known synonymous polymorphism, 3435C>T in exon 26, was genotyped in 315 patients with epilepsy classified as drug-resistant in 200 and drug-responsive in 115, and 200 control subjects without epilepsy. Polymorphic Alu insertion markers were used to control for population stratification, and the pattern of linkage disequilibrium across the gene was characterized. Compared with patients with drug-responsive epilepsy, patients with drug-resistant epilepsy were more likely to have the CC genotype than the TT genotype: odds ratio (OR), 2.66; 95% confidence interval, 1.32–5.38, $p = 0.006$. This polymorphism resides within an extensive block of LD, although the extent of the associated interval is difficult to define. Subsequent resequencing of derived and ancestral chromosomes at 3435C>T identified additional variants with a claim for being "causal," including IVS 26 + 80 T>C (Soranzo et al., 2004).

Subsequent replication studies have given conflicting results. Analysis of an *ABCB1* haplotype with three SNPs, including 3435C>T in 210 patients with temporal lobe epilepsy (TLE) stratified according to their degree of drug resistance, identified a common haplotype that increased the risk for pharmacoresistance in the homozygote state (Zimprich et al., 2004). The associated CGC haplotype includes the C allele of 3435C>T. However, a replication study in 401 drug-resistant and 208 drug-responsive subjects with epilepsy showed no significant association between the CC genotype and drug-resistant epilepsy (Tan, Heron, et al., 2004). The reasons for these diverse results are many and have been discussed at length (Ott, 2004). Interpretation hinges to some considerable extent on the extent to which biological plausibility influences the prior probability of a true association.

In a further study, variation in *ABCB1*, *SCN1A*, and *CYP2C9* was analyzed in relation to the clinical usage of the AEDs carbamazepine and phenytoin in cohorts of 425 and 281 patients, respectively (Tate et al., 2005). A known functional polymorphism in *CYP2C9* was associated with the maximal dose of phenytoin used, and an intronic polymorphism in *SCN1A* showed significant association with the maximum doses in regular use of both carbamazepine and phenytoin. These interesting observations require confirmation by independent replication, but suggest that pharmacogenomics may improve dosing decisions in the use of AEDs in the future.

Microarrays and Epilepsy Research: Functional Genomics

The combination of fully sequenced genomes and advances in microarray technology is providing new opportunities for exploring the neurobiology of seizures at the cellular and tissue levels. In the last decade, microarray platforms have advanced to allow a higher feature density and increased flexibility of content and design. It is no longer sufficient to focus entirely on gene-expression data. New and emerging areas include the analysis of splice-variants and microRNAs, and the development of tiling arrays allows the genome-wide study of gene regulation by mapping sites of transcription factor binding and chromatin modification.

In the epilepsy field, initial work has focused on the partial epilepsies, in particular TLE, in which human tissue may be available from surgical intervention (Majores et al., 2004). In TLE, seizure origin typically involves the hippocampal formation, but early stages of epilepsy development are unfortunately not available for analysis in humans. However, animal models of TLE allow the study of molecular mechanisms of dynamic processes such as development of hyperexcitability and pharmacoresistance. The use of laser-microdissection techniques and quantitative reverse transcription-polymerase chain reaction (RT-PCR) methodologies allows observations to be focused on well-defined anatomical regions or cellular subpopulations (Becker et al., 2002).

References

Barclay, J, Balaguero, N, Mione, M, Ackerman, SL, Letts, VA, Brodbeck, J, et al. (2001). Ducky mouse phenotype of epilepsy and ataxia is associated with mutations in the Cacna2d2 gene and decreased calcium channel current in cerebellar Purkinje cells. *J Neurosci*, 21(16): 6095–6104.

Barkovich, AJ, Kuzniecky, RI, Jackson, GD, Guerrini, R, and Dobyns, WB (2005). A developmental and genetic classification for malformations of cortical development. *Neurology*, 65(12):1873–1887.

Baulac, S, Huberfeld, G, Gourfinkel-An, I, Mitropoulou, G, Beranger, A, Prud'homme, J, et al. (2001). First genetic evidence of GABA(A) receptor dysfunction in epilepsy: a mutation in the gamma2-subunit gene. *Nat Genet*, 28(1):46–48.

Becker, AJ, Wiestler, OD, and Blumcke, I (2002). Functional genomics in experimental and human temporal lobe epilepsy: powerful new tools to identify molecular disease mechanisms of hippocampal damage. *ProgBrain Res*, 135:161–173.

Beltran-Valero de, BD, Currier, S, Steinbrecher, A, Celli, J, van, BE, van der, ZB, et al. (2002). Mutations in the O-mannosyltransferase gene POMT1 give rise to the severe neuronal migration disorder Walker–Warburg syndrome. *Am J Hum Genet*, 71(5):1033–1043.

Bergren, SK, Chen, S, Galecki, A, and Kearney, JA (2005). Genetic modifiers affecting severity of epilepsy caused by mutation of sodium channel Scn2a. *Mamm Genome*, 16(9):683–690.

Berkovic, SF, Heron, SE, Giordano, L, Marini, C, Guerrini, R, Kaplan, RE, et al. (2004). Benign familial neonatal–infantile seizures: characterization of a new sodium channelopathy. *Ann Neurol*, 55(4):550–557.

Berkovic, SF, Mazarib, A, Walid, S, Neufeld, MY, Manelis, J, Nevo, Y, et al. (2005). A new clinical and molecular form of Unverricht–Lundborg disease localized by homozygosity mapping. *Brain*, 128(Pt 3): 652–658.

Bronson, RT, Donahue, LR, Johnson, KR, Tanner, A, Lane, PW, and Faust, JR (1998). Neuronal ceroid lipofuscinosis (nclf), a new disorder of the mouse linked to chromosome 9. *Am J Med Genet*, 77(4):289–297.

Broom, MF and Zhou, C (2001). Fine mapping of ovine ceroid lipofuscinosis confirms orthology with CLN6. *Eur J Paediatr Neurol*, 5(Suppl. A):33–35.

Burgess, DL, Jones, JM, Meisler, MH, and Noebels, JL (1997). Mutation of the Ca2 + channel b subunit gene Cchb4 is associated with ataxia and seizures in the lethargic (lh) mouse. *Cell*, 88(3):385–392.

Callenbach, PM, van den Maagdenberg, AM, Frants, RR, and Brouwer, OF (2005). Clinical and genetic aspects of idiopathic epilepsies in childhood. *Eur J Paediatr Neurol*, 9(2):91–103.

Chan, EM, Andrade, DM, Franceschetti, S, and Minassian, B (2005). Progressive myoclonus epilepsies: EPM1, EPM2A, EPM2B. *Adv Neurol*, 95:47–57.

Chan, EM, Omer, S, Ahmed, M, Bridges, LR, Bennett, C, Scherer, SW, et al. (2004). Progressive myoclonus epilepsy with polyglucosans (Lafora disease): evidence for a third locus. *Neurology*, 63(3):565–567.

Chan, EM, Young, EJ, Ianzano, L, Munteanu, I, Zhao, X, Christopoulos, CC, et al. (2003). Mutations in NHLRC1 cause progressive myoclonus epilepsy. *Nat Genet*, 35(2):125–127.

Charlier, C, Singh, NA, Ryan, SG, Lewis, TB, Reus, BE, Leach, RJ, et al. (1998). A pore mutation in a novel KQT-like potassium channel gene in an idiopathic epilepsy family. *Nat Genet*, 18:53–55.

Chen, C, Westenbroek, RE, Xu, X, Edwards, CA, Sorenson, DR, Chen, Y, et al. (2004). Mice lacking sodium channel beta1 subunits display defects in neuronal excitability, sodium channel expression, and nodal architecture. *J Neurosci*, 24(16):4030–4042.

Chen, L, Chetkovich, DM, Petralia, RS, Sweeney, NT, Kawasaki, Y, Wenthold, RJ, et al. (2000). Stargazin regulates synaptic targeting of AMPA receptors by two distinct mechanisms. *Nature*, 408(6815):936–943.

Chen, Y, Lu, J, Pan, H, Zhang, Y, Wu, H, Xu, K, et al. (2003). Association between genetic variation of CACNA1H and childhood absence epilepsy. *Ann Neurol*, 54(2):239–243.

Claes, L, Ceulemans, B, Audenaert, D, Smets, K, Lofgren, A, Del-Favero, J, et al. (2003). De novo SCN1A mutations are a major cause of severe myoclonic epilepsy of infancy. *Hum Mutat*, 21(6):615–621.

Claes, L, Del-Favero, J, Ceulemans, B, Lagae, L, Van, BC, and De, JP (2001). De novo mutations in the sodium-channel gene SCN1A cause severe myoclonic epilepsy of infancy. *Am J Hum Genet*, 68(6):1327–1332.

Coenen, AM and Van Luijtelaar, EL (2003). Genetic animal models for absence epilepsy: a review of the WAG/Rij strain of rats. *Behav Genet*, 33 (6):635–655.

Cook, RW, Jolly, RD, Palmer, DN, Tammen, I, Broom, MF, and McKinnon, R (2002). Neuronal ceroid lipofuscinosis in Merino sheep. *Aust Vet J*, 80(5):292–297.

Cossette, P, Liu, L, Brisebois, K, Dong, H, Lortie, A, Vanasse, M, et al. (2002). Mutation of GABRA1 in an autosomal dominant form of juvenile myoclonic epilepsy. *Nat Genet*, 31(2):184–189.

Cox, GA, Lutz, CM, Yang, CL, Biemesderfer, D, Bronson, RT, Fu, A, et al. (1997). Sodium/hydrogen exchanger gene defect in slow-wave epilepsy mutant mice. *Cell*, 91(1):139–148.

Crunelli, V and Leresche, N (2002). Childhood absence epilepsy: genes, channels, neurons and networks. *Nat Rev Neurosci*, 3(5):371–382.

D'Agostino, D, Bertelli, M, Gallo, S, Cecchin, S, Albiero, E, Garofalo, PG, et al. (2004). Mutations and polymorphisms of the CLCN2 gene in idiopathic epilepsy. *Neurology*, 63(8):1500–1502.

De, V, Trifiletti, RR, Jacobson, RI, Ronen, GM, Behmand, RA, and Harik, SI (1991). Defective glucose transport across the blood–brain barrier as a cause of persistent hypoglycorrhachia, seizures, and developmental delay. *N Engl J Med*, 325(10):703–709.

des, PV, Pinard, JM, Billuart, P, Vinet, MC, Koulakoff, A, Carrie, A, et al. (1998). A novel CNS gene required for neuronal migration and involved in X-linked subcortical laminar heterotopia and lissencephaly syndrome. *Cell*, 92(1):51–61.

Dibbens, LM, Feng, HJ, Richards, MC, Harkin, LA, Hodgson, BL, Scott, D, et al. (2004). GABRD encoding a protein for extra- or peri-synaptic GABAA receptors is a susceptibility locus for generalized epilepsies. *Hum Mol Genet*, 13(13):1315–1319.

Dichgans, M, Freilinger, T, Eckstein, G, Babini, E, Lorenz-Depiereux, B, Biskup, S, et al. (2005). Mutation in the neuronal voltage-gated sodium channel SCN1A in familial hemiplegic migraine. *Lancet*, 366 (9483):371–377.

Diesen, C, Saarinen, A, Pihko, H, Rosenlew, C, Cormand, B, Dobyns, WB, et al. (2004). POMGnT1 mutation and phenotypic spectrum in muscle-eye-brain disease. *J Med Genet*, 41(10):e115.

DiMauro, S (2004). Mitochondrial diseases. *Biochim Biophys Acta*, 1658 (1–2):80–88.

Du, W, Bautista, JF, Yang, H, ez-Sampedro, A, You, SA, Wang, L, et al. (2005). Calcium-sensitive potassium channelopathy in human epilepsy and paroxysmal movement disorder. *Nat Genet*, 37(7):733–738.

Durner, M, Keddache, MA, Tomasini, L, Shinnar, S, Resor, SR, Cohen, J, et al. (2001). Genome scan of idiopathic generalized epilepsy: evidence for major susceptibility gene and modifying genes influencing the seizure type. *Ann Neurol*, 49(3):328–335.

Durner, M, Sander, T, Greenberg, DA, Johnson, K, Beck-Mannagetta, G, and Janz, D (1991). Localisation of idiopathic generalised epilepsy on chromosome 6p in families of juvenile myoclonic epilepsy patients. *Neurology*, 41(10):1651–1655.

Elmslie, FV, Rees, M, Williamson, MP, Kerr, M, Juel Kjeldsen, M, Pang, KA, et al. (1997). Genetic mapping of a major susceptibility locus for juvenile myoclonic epilepsy on chromosome 15q. *Hum Mol Genet*, 6(8):1329–1334.

Elmslie, FV, Williamson, MP, Rees, M, Kerr, M, Juel Kjeldsen, M, Pang, KA, et al. (1996). Linkage analysis of juvenile myoclonic epilepsy and microsatellite loci spanning 61 cM of human chromosome 6p in 19 nuclear pedigrees provides no evidence for a susceptibility locus in this region. *Am J Hum Genet*, 59:653–663.

Escayg, A, De, WM, Lee, DD, Bichet, D, Wolf, P, Mayer, T, et al. (2000). Coding and noncoding variation of the human calcium-channel beta4-subunit gene CACNB4 in patients with idiopathic generalized epilepsy and episodic ataxia. *Am J Hum Genet*, 66(5):1531–1539.

Escayg, A, MacDonald, BT, Meisler, MH, Baulac, S, Huberfeld, G, An-Gourfinkel, I, et al. (2000). Mutations of SCN1A, encoding a neuronal sodium channel, in two families with GEFS + 2. *Nat Genet*, 24(4):343–345.

Fertig, E, Lincoln, A, Martinuzzi, A, Mattson, RH, and Hisama, FM (2003). Novel LGI1 mutation in a family with autosomal dominant partial epilepsy with auditory features. *Neurology*, 60(10):1687–1690.

Fletcher, CF, Lutz, CM, O"Sullivan, TN, Shaughnessy, JD, Hawkes, R, Frankel, WN, et al. (1996). Absence epilepsy in tottering mutant mice is associated with calcium channel defects. *Cell*, 87:607–617.

Flex, E, Pizzuti, A, Di, BC, Douzgou, S, Egeo, G, Fattouch, J, et al. (2005). LGI1 gene mutation screening in sporadic partial epilepsy with auditory features. *J Neurol*, 252(1):62–66.

Fong, GC, Shah, PU, Gee, MN, Serratosa, JM, Castroviejo, IP, Khan, S, et al. (1998). Childhood absence epilepsy with tonic–clonic seizures and electroencephalogram 3-4-Hz spike and multispike-slow wave complexes: linkage to chromosome 8q24. *Am J Hum Genet*, 63(4):1117–1129.

Fox, JW, Lamperti, ED, Eksioglu, YZ, Hong, SE, Feng, Y, Graham, DA, et al. (1998). Mutations in filamin 1 prevent migration of cerebral cortical neurons in human periventricular heterotopia. *Neuron*, 21(6):1315–1325.

Francis, F, Koulakoff, A, Boucher, D, Chafey, P, Schaar, B, Vinet, MC, et al. (1999). Doublecortin is a developmentally regulated, microtubule-associated protein expressed in migrating and differentiating neurons. *Neuron*, 23(2):247–256.

Fujiwara, T, Sugawara, T, Mazaki-Miyazaki, E, Takahashi, Y, Fukushima, K, Watanabe, M, et al. (2003). Mutations of sodium channel alpha subunit type 1 (SCN1A) in intractable childhood epilepsies with frequent generalized tonic–clonic seizures. *Brain*, 126(Pt 3):531–546.

Gauguier, D, van, LG, Bihoreau, MT, Wilder, SP, Godfrey, RF, Vossen, J, et al. (2004). Chromosomal mapping of genetic loci controlling absence epilepsy phenotypes in the WAG/Rij rat. *Epilepsia*, 45(8):908–915.

Gentry, MS, Worby, CA, and Dixon, JE (2005). Insights into Lafora disease: malin is an E3 ubiquitin ligase that ubiquitinates and promotes the degradation of laforin. *Proc Natl Acad Sci USA*, 102(24):8501–8506.

George, AL Jr (2004). Molecular basis of inherited epilepsy. *Arch Neurol*, 61 (4):473–478.

George, AL Jr (2005). Inherited disorders of voltage-gated sodium channels. *J Clin Invest*, 115(8):1990–1999.

Gleeson, JG, Allen, KM, Fox, JW, Lamperti, ED, Berkovic, S, Scheffer, I, et al. (1998). Doublecortin, a brain-specific gene mutated in human X-linked lissencephaly and double cortex syndrome, encodes a putative signaling protein. *Cell*, 92(1):63–72.

Gleeson, JG, Lin, PT, Flanagan, LA, and Walsh, CA (1999). Doublecortin is a microtubule-associated protein and is expressed widely by migrating neurons. *Neuron*, 23(2):257–271.

Greenberg, DA, Cayanis, E, Strug, L, Marathe, S, Durner, M, Pal, DK, et al. (2005). Malic enzyme 2 may underlie susceptibility to adolescent-onset idiopathic generalized epilepsy. *Am J Hum Genet*, 76(1):139–146.

Greenberg, DA, Delgado-Escueta, AV, Widelitz, H, Sparkes, RS, Treiman, L, Maldonado, HM, et al. (1988). Juvenile myoclonic epilepsy may be linked to the BF and HLA loci on human chromosome 6. *Am J Med Genet*, 31(1):185–192.

Gu, W, Brodtkorb, E, and Steinlein, OK (2002). LGI1 is mutated in familial temporal lobe epilepsy characterized by aphasic seizures. *Ann Neurol*, 52(3):364–367.

Guerrini, R and Filippi, T (2005). Neuronal migration disorders, genetics, and epileptogenesis. *J Child Neurol*, 20(4):287–299.

Harkin, LA, Bowser, DN, Dibbens, LM, Singh, R, Phillips, F, Wallace, RH, et al. (2002). Truncation of the GABA(A)-receptor gamma2 subunit in a family with generalized epilepsy with febrile seizures plus. *Am J Hum Genet*, 70(2):530–536.

Hattori, M, Adachl, H, Tsujlmoto, M, Aral, H, and Inoue, K (1994). Miller-Dieker lissencephaly gene encodes a subunit of brain platelet-activating factor. *Nature*, 370:216–218.

Haug, K, Warnstedt, M, Alekov, AK, Sander, T, Ramirez, A, Poser, B, et al. (2003). Mutations in CLCN2 encoding a voltage-gated chloride channel are associated with idiopathic generalized epilepsies. *Nat Genet*, 33(4):527–532.

Hedera, P, Abou-Khalil, B, Crunk, AE, Taylor, KA, Haines, JL, and Sutcliffe, JS (2004). Autosomal dominant lateral temporal epilepsy: two families with novel mutations in the LGI1 gene. *Epilepsia*, 45(3):218–222.

Heils, A (2005). CLCN2 and idiopathic generalized epilepsy. *Adv Neurol*, 95:265–271.

Heron, SE, Phillips, HA, Mulley, JC, Mazarib, A, Neufeld, MY, Berkovic, SF, et al. (2004). Genetic variation of CACNA1H in idiopathic generalized epilepsy. *Ann Neurol*, 55(4):595–596.

Holter, J, Carter, D, Leresche, N, Crunelli, V, and Vincent, P (2005). A TASK3 channel (KCNK9) mutation in a genetic model of absence epilepsy. *J Mol Neurosci*, 25(1):37–51.

Hong, SE, Shugart, YY, Huang, DT, Shahwan, SA, Grant, PE, Hourihane, JO, et al. (2000). Autosomal recessive lissencephaly with cerebellar hypoplasia is associated with human RELN mutations. *Nat Genet*, 26(1):93–96.

Imbrici, P, Jaffe, SL, Eunson, LH, Davies, NP, Herd, C, Robertson, R, et al. (2004). Dysfunction of the brain calcium channel CaV2.1 in absence epilepsy and episodic ataxia. *Brain*, 127(Pt 12):2682–2692.

Jentsch, TJ (2000). Neuronal KCNQ potassium channels: physiology and role in disease. *Nat Rev Neurosci*, 1(1):21–30.

Jentsch, TJ, Poet, M, Fuhrmann, JC, and Zdebik, AA (2005). Physiological functions of CLC Cl- channels gleaned from human genetic disease and mouse models. *Annu Rev Physiol*, 67:779–807.

Jentsch, TJ, Stein, V, Weinreich, F, and Zdebik, AA (2002). Molecular structure and physiological function of chloride channels. *Physiol Rev*, 82 (2):503–568.

Jouvenceau, A, Eunson, LH, Spauschus, A, Ramesh, V, Zuberi, SM, Kullmann, DM, et al. (2001). Human epilepsy associated with dysfunction of the brain P/Q-type calcium channel. *Lancet*, 358(9284):801–807.

Kalachikov, S, Evgrafov, O, Ross, B, Winawer, M, Barker-Cummings, C, Martinelli, BF, et al. (2002). Mutations in LGI1 cause autosomal-dominant partial epilepsy with auditory features. *Nat Genet*, 30(3):335–341.

Kanai, K, Hirose, S, Oguni, H, Fukuma, G, Shirasaka, Y, Miyajima, T, et al. (2004). Effect of localization of missense mutations in SCN1A on epilepsy phenotype severity. *Neurology*, 63(2):329–334.

Kananura, C, Haug, K, Sander, T, Runge, U, Gu, W, Hallmann, K, et al. (2002). A splice-site mutation in GABRG2 associated with childhood absence epilepsy and febrile convulsions. *Arch Neurol*, 59(7):1137–1141.

Kantheti, P, Diaz, ME, Peden, AE, Seong, EE, Dolan, DF, Robinson, MS, et al. (2003). Genetic and phenotypic analysis of the mouse mutant mh2J, an Ap3d allele caused by IAP element insertion. *Mamm Genome*, 14 (3):157–167.

Karpova, AV, Bikbaev, AF, Coenen, AM, and van, LG (2005). Morphometric Golgi study of cortical locations in WAG/Rij rats: the cortical focus theory. *Neurosci Res*, 51(2):119–128.

Katz, ML, Khan, S, Awano, T, Shahid, SA, Siakotos, AN, and Johnson, GS (2005). A mutation in the CLN8 gene in English setter dogs with neuronal ceroid-lipofuscinosis. *Biochem Biophys Res Commun*, 327(2):541–547.

Kearney, JA, Plummer, NW, Smith, MR, Kapur, J, Cummins, TR, Waxman, SG, et al. (2001). A gain-of-function mutation in the sodium channel gene Scn2a results in seizures and behavioral abnormalities. *Neuroscience*, 102(2):307–317.

Khosravani, H, Altier, C, Simms, B, Hamming, KS, Snutch, TP, Mezeyova, J, et al. (2004). Gating effects of mutations in the Cav3.2 T-type calcium channel associated with childhood absence epilepsy. *J Biol Chem*, 279(11):9681–9684.

Kim, D, Song, I, Keum, S, Lee, T, Jeong, MJ, Kim, SS, et al. (2001). Lack of the burst firing of thalamocortical relay neurons and resistance to absence seizures in mice lacking alpha(1G) T-type Ca(2 +) channels. *Neuron*, 31(1):35–45.

Kitamura, K, Yanazawa, M, Sugiyama, N, Miura, H, Iizuka-Kogo, A, Kusaka, M, et al. (2002). Mutation of ARX causes abnormal development of forebrain and testes in mice and X-linked lissencephaly with abnormal genitalia in humans. *Nat Genet*, 32(3):359–369.

Klepper, J and Voit, T (2002). Facilitated glucose transporter protein type 1 (GLUT1) deficiency syndrome: impaired glucose transport into brain—a review. *Eur J Pediatr*, 161(6):295–304.

Kobayashi, K, Nakahori, Y, Miyake, M, Matsumura, K, Kondo-Iida, E, Nomura, Y, et al. (1998). An ancient retrotransposal insertion causes Fukuyama-type congenital muscular dystrophy. *Nature*, 394 (6691):388–392.

Koppang, N (1973). Canine ceroid-lipofuscinosis—a model for human neuronal ceroid-lipofuscinosis and aging. *Mech Ageing Dev*, 2(6):421–445.

Leary, LD, Wang, D, Nordli, DR Jr, Engelstad, K, and De, V (2003). Seizure characterization and electroencephalographic features in Glut-1 deficiency syndrome. *Epilepsia*, 44(5):701–707.

Lehesjoki, AE (2003). Molecular background of progressive myoclonus epilepsy. *EMBO J*, 22(14):3473–3478.

Lehesjoki, AE, Koskiniemi, M, Norio, R, Tirrito, S, Sistonen, P, Lander, E, et al. (1993). Localization of the EPM1 gene for progressive myoclonus epilepsy on chromosome 21: linkage disequilibrium allows high resolution mapping. *Hum Mol Genet*, 3(8):1229–1234.

Lehesjoki, AE, Koskiniemi, M, Sistonen, P, Miao, J, Hastbacka, J, Norio, R, et al. (1991). Localisation of a gene for progressive myoclonus epilepsy to chromosome 21q22. *Proc Natl Acad Sci USA*, 88(9):3696–3699.

Lehesjoki, AE, Koskiniemi, M, Sistonen, P, Pandolfo, M, Antonelli, A, Kyllerman, M, et al. (1992). Linkage studies in progressive myoclonic epilepsy: Unverricht–Lundborg and Lafora disease. *Neurology*, 42(8):1545–1550.

Letts, VA (2005). Stargazer-a mouse to seize! *Epilepsy Curr*, 5(5):161–165.

Letts, VA, Felix, R, Biddlecome, GH, Arikkath, J, Mahaffey, CL, Valenzuela, A, et al. (1998). The mouse stargazer gene encodes a neuronal Ca2 + -channel gamma subunit. *Nat Genet*, 19(4):340–347.

Letts, VA, Mahaffey, CL, Beyer, B, and Frankel, WN (2005). A targeted mutation in Cacng4 exacerbates spike-wave seizures in stargazer (Cacng2) mice. *Proc Natl Acad Sci USA*, 102(6):2123–2128.

Liu, AW, Delgado-Escueta, AV, Serratosa, JM, Alonso, ME, Medina, MT, Gee, MN, et al. (1995). Juvenile myoclonic epilepsy locus in chromosome 6p21.2-p11: linkage to convulsions and electroencephalography trait. *Am J Hum Genet*, 57:368–381.

Lohi, H, Young, EJ, Fitzmaurice, SN, Rusbridge, C, Chan, EM, Vervoort, M, et al. (2005). Expanded repeat in canine epilepsy. *Science*, 307(5706):81.

Lonka, L, Aalto, A, Kopra, O, Kuronen, M, Kokaia, Z, Saarma, M, et al. (2005). The neuronal ceroid lipofuscinosis Cln8 gene expression is developmentally regulated in mouse brain and up-regulated in the hippocampal kindling model of epilepsy. *BMC Neurosci*, 6(1):27.

Majores, M, Eils, J, Wiestler, OD, and Becker, AJ (2004). Molecular profiling of temporal lobe epilepsy: comparison of data from human tissue samples and animal models. *Epilepsy Res*, 60(2–3):173–178.

Marini, C, Scheffer, IE, Crossland, KM, Grinton, BE, Phillips, FL, McMahon, JM, et al. (2004). Genetic architecture of idiopathic generalized epilepsy: clinical genetic analysis of 55 multiplex families. *Epilepsia*, 45(5):467–478.

Meisler, MH and Kearney, JA (2005). Sodium channel mutations in epilepsy and other neurological disorders. *J Clin Invest*, 115(8):2010–2017.

Meisler, MH, Kearney, J, Ottman, R, and Escayg, A (2001). Identification of epilepsy genes in human and mouse. *Annu Rev Genet*, 35:567–588.

Melville, SA, Wilson, CL, Chiang, CS, Studdert, VP, Lingaas, F, and Wilton, AN (2005). A mutation in canine CLN5 causes neuronal ceroid lipofuscinosis in Border collie dogs. *Genomics*, 86(3):287–294.

Michelucci, R, Poza, JJ, Sofia, V, de Feo, MR, Binelli, S, Bisulli, F, et al. (2003). Autosomal dominant lateral temporal epilepsy: clinical spectrum, new epitempin mutations, and genetic heterogeneity in seven European families. *Epilepsia*, 44(10):1289–1297.

Miller, C (2000). An overview of the potassium channel family. *Genome Biol*, 1(4):REVIEWS0004.

Minassian, BA, Lee, JR, Herbrick, JA, Huizenga, J, Soder, S, Mungall, AJ, et al. (1998). Mutations in a gene encoding a novel protein tyrosine phosphatase cause progressive myoclonus epilepsy. *Nat Genet*, 20(2):171–174.

Mitchison, HM, Hofmann, SL, Becerra, CH, Munroe, PB, Lake, BD, Crow, YJ, et al. (1998). Mutations in the palmitoyl-protein thioesterase gene (PPT; CLN1) causing juvenile neuronal ceroid lipofuscinosis with granular osmiophilic deposits. *Hum Mol Genet*, 7(2):291–297.

Mitchison, HM, Lim, MJ, and Cooper, JD (2004). Selectivity and types of cell death in the neuronal ceroid lipofuscinoses. *Brain Pathol*, 14(1):86–96.

Mochida, GH and Walsh, CA (2004). Genetic basis of developmental malformations of the cerebral cortex. *Arch Neurol*, 61(5):637–640.

Mole, SE (2004). The genetic spectrum of human neuronal ceroid-lipofuscinoses. *Brain Pathol*, 14(1):70–76.

Mole, SE, Williams, RE, and Goebel, HH (2005). Correlations between genotype, ultrastructural morphology and clinical phenotype in the neuronal ceroid lipofuscinoses. *Neurogenetics*, 6(3):107–126.

Morante-Redolat, JM, Gorostidi-Pagola, A, Piquer-Sirerol, S, Saenz, A, Poza, JJ, Galan, J, et al. (2002). Mutations in the LGI1/Epitempin gene on 10q24 cause autosomal dominant lateral temporal epilepsy. *Hum Mol Genet*, 11(9):1119–1128.

Naini, AB, Lu, J, Kaufmann, P, Bernstein, RA, Mancuso, M, Bonilla, E, et al. (2005). Novel mitochondrial DNA ND5 mutation in a patient with clinical features of MELAS and MERRF. *Arch Neurol*, 62(3):473–476.

Noebels, JL (2001). Modeling human epilepsies in mice. *Epilepsia*, 42 (Suppl. 5):11–15.

Noebels, JL (2003a). Exploring new gene discoveries in idiopathic generalized epilepsy. *Epilepsia*, 44(Suppl. 2):16–21.

Noebels, JL (2003b). The biology of epilepsy genes. *Annu Rev Neurosci*, 26:599–625.

Ott, J (2004). Association of genetic loci: replication or not, that is the question. *Neurology*, 63(6):955–958.

Ottman, R, Risch, N, Hauser, WA, Pedley, TA, Lee, JH, Barker-Cummings, C, et al. (1995). Localization of a gene for partial epilepsy to chromosome 10q. *Nat Genet*, 10:56–60.

Pal, DK, Evgrafov, OV, Tabares, P, Zhang, F, Durner, M, and Greenberg, DA (2003). BRD2 (RING3) is a probable major susceptibility gene for common juvenile myoclonic epilepsy. *Am J Hum Genet*, 73(2):261–270.

Pears, MR, Cooper, JD, Mitchison, HM, Mortishire-Smith, RJ, Pearce, DA, and Griffin, JL (2005). High resolution 1H NMR based metabolomics indicates a neurotransmitter cycling deficit in cerebral tissue from a mouse model of batten disease. *J Biol Chem*, 280(52):42508–42514.

Pennacchio, LA, Bouley, DM, Higgins, KM, Scott, MP, Noebels, JL, and Myers, RM (1998). Progressive ataxia, myoclonic epilepsy and cerebellar apoptosis in cystatin B-deficient mice. *Nat Genet*, 20(3):251–258.

Pennacchio, LA, Lehesjoki, AE, Stone, NE, Willour, VL, Virtaneva, K, Miao, J, et al. (1996). Mutations in the gene encoding cystatin B in progressive myoclonus epilepsy (EPM1). *Science*, 271:1731–1734.

Phillips, HA, Favre, I, Kirkpatrick, M, Zuberi, SM, Goudie, D, Heron, SE, et al. (2001). CHRNB2 is the second acetylcholine receptor subunit associated with autosomal dominant nocturnal frontal lobe epilepsy. *Am J Hum Genet*, 68(1):225–231.

Phillips, HA, Scheffer, IE, Crossland, KM, Bhatia, KP, Fish, DR, Marsden, CD, et al. (1998). Autosomal dominant nocturnal frontal-lobe epilepsy: genetic heterogeneity and evidence for a second locus at 15q24. *Am J Hum Genet*, 63(4):1108–1116.

Pietrobon, D (2005). Function and dysfunction of synaptic calcium channels: insights from mouse models. *Curr Opin Neurobiol*, 15(3):257–265.

Ranta, S, Zhang, Y, Ross, B, Lonka, L, Takkunen, E, Messer, A, et al. (1999). The neuronal ceroid lipofuscinoses in human EPMR and mnd mutant mice are associated with mutations in CLN8. *Nat Genet*, 23(2):233–236.

Rhodes, TH, Lossin, C, Vanoye, CG, Wang, DW, and George, AL Jr (2004). Noninactivating voltage-gated sodium channels in severe myoclonic epilepsy of infancy. *Proc Natl Acad Sci USA*, 101(30):11147–11152.

Robinson, R, Taske, N, Sander, T, Heils, A, Whitehouse, W, Goutieres, F, et al. (2002). Linkage analysis between childhood absence epilepsy and genes encoding GABAA and GABAB receptors, voltage-dependent calcium channels, and the ECA1 region on chromosome 8q. *Epilepsy Res*, 48(3):169–179.

Rudolf, G, Bihoreau, MT, Godfrey, RF, Wilder, SP, Cox, RD, Lathrop, M, et al. (2004). Polygenic control of idiopathic generalized epilepsy phenotypes in the genetic absence rats from Strasbourg (GAERS). *Epilepsia*, 45 (4):301–308.

Sander, T, Schulz, H, Saar, K, Gennaro, E, Riggio, MC, Bianchi, A, et al. (2000). Genome search for susceptibility loci of common idiopathic generalised epilepsies. *Hum Mol Genet*, 9(10):1465–1472.

Sander, T, Schulz, H, Vieira-Saeker, AM, Bianchi, A, Sailer, U, Bauer, G, et al. (1999). Evaluation of a putative major susceptibility locus for juvenile myoclonic epilepsy on chromosome 15q14. *Am J Med Genet*, 88 (2):182–187.

Savukoski, M, Klockars, T, Holmberg, V, Santavuori, P, Lander, ES, and Peltonen, L (1998). CLN5, a novel gene encoding a putative transmembrane protein mutated in Finnish variant late infantile neuronal ceroid lipofuscinosis. *Nat Genet*, 19(3):286–288.

Scheffer, IE and Berkovic, SF (2003). The genetics of human epilepsy. *Trends Pharmacol Sci*, 24(8):428–433.

Seidner, G, Alvarez, MG, Yeh, JI, O'Driscoll, KR, Klepper, J, Stump, TS, et al. (1998). GLUT-1 deficiency syndrome caused by haploinsufficiency of the blood–brain barrier hexose carrier. *Nat Genet*, 18(2):188–191.

Shahwan, A, Farrell, M, and Delanty, N (2005). Progressive myoclonic epilepsies: a review of genetic and therapeutic aspects. *Lancet Neurol*, 4(4):239–248.

Shannon, P, Pennacchio, LA, Houseweart, MK, Minassian, BA, and Myers, RM (2002). Neuropathological changes in a mouse model of progressive myoclonus epilepsy: cystatin B deficiency and Unverricht–Lundborg disease. *J Neuropathol Exp Neurol*, 61(12):1085–1091.

Shoffner, JM, Lott, MT, Lezza, AMS, Seibel, P, Ballinger, SW, and Wallace, DC (1990). Myoclonic epilepsy and ragged-red fiber disease (MERRF) is associated with a mitochondrial DNA tRNALys mutation. *Cell*, 61:931–937.

Siddiqui, A, Kerb, R, Weale, ME, Brinkmann, U, Smith, A, Goldstein, DB, et al. (2003). Association of multidrug resistance in epilepsy with a polymorphism in the drug-transporter gene ABCB1. *N Engl J Med*, 348(15):1442–1448.

Simpson, MA, Cross, H, Proukakis, C, Priestman, DA, Neville, DC, Reinkensmeier, G, et al. (2004). Infantile-onset symptomatic epilepsy syndrome caused by a homozygous loss-of-function mutation of GM3 synthase. *Nat Genet*, 36(11):1225–1229.

Singh, NA, Charlier, C, Stauffer, D, DuPont, BR, Leach, RJ, Melis, R, et al. (1998). A novel potassium channel gene, KCNQ2, is mutated in an inhertied epilepsy of newborns. *Nat Genet*, 18:25–29.

Singh, NA, Westenskow, P, Charlier, C, Pappas, C, Leslie, J, Dillon, J, et al. (2003). KCNQ2 and KCNQ3 potassium channel genes in benign familial neonatal convulsions: expansion of the functional and mutation spectrum. *Brain*, 126(Pt 12):2726–2737.

Singh, R, Gardner, RJ, Crossland, KM, Scheffer, IE, and Berkovic, SF (2002). Chromosomal abnormalities and epilepsy: a review for clinicians and gene hunters. *Epilepsia*, 43(2):127–140.

Sleat, DE, Donnelly, RJ, Lackland, H, Liu, CG, Sohar, I, Pullarkat, RK, et al. (1997). Association of mutations in a lysosomal protein with classical late-infantile neuronal ceroid lipofuscinosis. *Science*, 277(5333):1802–1805.

Song, I, Kim, D, Choi, S, Sun, M, Kim, Y, and Shin, HS (2004). Role of the alpha1G T-type calcium channel in spontaneous absence seizures in mutant mice. *J Neurosci*, 24(22):5249–5257.

Soranzo, N, Cavalleri, GL, Weale, ME, Wood, NW, Depondt, C, Marguerie, R, et al. (2004). Identifying candidate causal variants responsible for altered activity of the ABCB1 multidrug resistance gene. *Genome Res*, 14(7):1333–1344.

Steinlein, OK (2004a). Genes and mutations in human idiopathic epilepsy. *Brain Dev*, 26(4):213–218.

Steinlein, OK (2004b). Nicotinic receptor mutations in human epilepsy. *Prog Brain Res*, 145:275–285.

Steinlein, OK, Mulley, JC, Propping, P, Wallace, RH, Phillips, HA, Sutherland, GR, et al. (1995). A missense mutation in the neuronal nicotinic receptor a4 subunit is associated with autosomal dominant nocturnal frontal lobe epilepsy. *Nat Genet*, 11:201–203.

Studdert, VP and Mitten, RW (1991). Clinical features of ceroid lipofuscinosis in border collie dogs. *Aust Vet J*, 68(4):137–140.

Suzuki, T, Delgado-Escueta, AV, Aguan, K, Alonso, ME, Shi, J, Hara, Y, et al. (2004). Mutations in EFHC1 cause juvenile myoclonic epilepsy. *Nat Genet*, 36(8):842–849.

Tan, NC, Heron, SE, Scheffer, IE, Pelekanos, JT, McMahon, JM, Vears, DF, et al. (2004). Failure to confirm association of a polymorphism in ABCB1 with multidrug-resistant epilepsy. *Neurology*, 63(6):1090–1092.

Tan, NC, Mulley, JC, and Berkovic, SF (2004). Genetic association studies in epilepsy: the truth is out there. *Epilepsia*, 45(11):1429–1442.

Tate, SK, Depondt, C, Sisodiya, SM, Cavalleri, GL, Schorge, S, Soranzo, N, et al. (2005). Genetic predictors of the maximum doses patients receive during clinical use of the anti-epileptic drugs carbamazepine and phenytoin. *Proc Natl Acad Sci USA*, 102(15):5507–5512.

The International Batten Disease Consortium (1995). Isolation of a novel gene underlying Batten disease, CLN3. *Cell*, 82: 949–957.

van Diggelen, OP, Thobois, S, Tilikete, C, Zabot, MT, Keulemans, JL, van Bunderen, PA, et al. (2001). Adult neuronal ceroid lipofuscinosis with palmitoyl-protein thioesterase deficiency: first adult-onset patients of a childhood disease. *Ann Neurol*, 50(2):269–272.

Vanhanen, SL, Puranen, J, Autti, T, Raininko, R, Liewendahl, K, Nikkinen, P, et al. (2004). Neuroradiological findings (MRS, MRI, SPECT) in infantile neuronal ceroid-lipofuscinosis (infantile CLN1) at different stages of the disease. *Neuropediatrics*, 35(1):27–35.

Vesa, J, Hellsten, E, Verkruyse, LA, Camp, LA, Rapola, J, Santavuori, P, et al. (1995). Mutations in the palmitoyl protein thioesterase gene causing infantile neuronal ceroid lipofuscinosis. *Nature*, 376:584–587.

Virtaneva, K, D'Amato, E, Miao, J, Koskiniemi, M, Norio, R, Avanzini, G, et al. (1997). Unstable minisatellite expansion causing recessively inherited myoclonus epilepsy, EPM1. *Nat Genet*, 15:393–396.

Vitko, I, Chen, Y, Arias, JM, Shen, Y, Wu, XR, and Perez-Reyes, E (2005). Functional characterization and neuronal modeling of the effects of childhood absence epilepsy variants of CACNA1H, a T-type calcium channel. *J Neurosci*, 25(19):4844–4855.

Wallace, RH, Hodgson, BL, Grinton, BE, Gardiner, RM, Robinson, R, Rodriguez-Casero, V, et al. (2003). Sodium channel alpha1-subunit mutations in severe myoclonic epilepsy of infancy and infantile spasms. *Neurology*, 61(6):765–769.

Wallace, RH, Marini, C, Petrou, S, Harkin, LA, Bowser, DN, Panchal, RG, et al. (2001). Mutant GABA(A) receptor gamma2-subunit in childhood absence epilepsy and febrile seizures. *Nat Genet*, 28(1):49–52.

Wallace, RH, Scheffer, IE, Barnett, S, Richards, M, Dibbens, L, Desai, RR, et al. (2001). Neuronal sodium-channel alpha1-subunit mutations in generalized epilepsy with febrile seizures plus. *Am J Hum Genet*, 68(4):859–865.

Wallace, RH, Wang, DW, Singh, R, Scheffer, IE, George, AL Jr, Phillips, HA, et al. (1998). Febrile seizures and generalized epilepsy associated with a mutation in the Na+-channel beta1 subunit gene SCN1B. *Nat Genet*, 19 (4):366–370.

Weissbecker, KA, Durner, M, Janz, D, Scaramelli, A, Sparkes, RS, and Spence, MA (1991). Confirmation of linkage between juvenile myoclonic epilepsy locus and the HLA region on chromosome 6. *Am J Med Genet*, 38(1):32–36.

Wheeler, RB, Sharp, JD, Schultz, RA, Joslin, JM, Williams, RE, and Mole, SE (2002). The gene mutated in variant late-infantile neuronal ceroid lipofuscinosis (CLN6) and in nclf mutant mice encodes a novel predicted transmembrane protein. *Am J Hum Genet*, 70(2):537–542.

Whitehouse, WP, Rees, M, Curtis, D, Sundqvist, A, Parker, K, Chung, E, et al. (1993). Linkage analysis of idiopathic generalised epilepsy (IGE) and marker loci on chromosome 6p in families of patients with juvenile myoclonic epilepsy: no evidence for an epilepsy locus in the HLA region. *Am J Hum Genet*, 53(3):652–662.

Winawer, MR, Rabinowitz, D, Pedley, TA, Hauser, WA, and Ottman, R (2003). Genetic influences on myoclonic and absence seizures. *Neurology*, 61 (11):1576–1581.

Woods, CG, Bond, J, and Enard, W (2005). Autosomal recessive primary microcephaly (MCPH): a review of clinical, molecular, and evolutionary findings. *Am J Hum Genet*, 76(5):717–728.

Yamakawa, K (2005). Epilepsy and sodium channel gene mutations: gain or loss of function? *Neuroreport*, 16(1):1–3.

Yoshida, A, Kobayashi, K, Manya, H, Taniguchi, K, Kano, H, Mizuno, M, et al. (2001). Muscular dystrophy and neuronal migration disorder caused by mutations in a glycosyltransferase, POMGnT1. *Dev Cell*, 1(5):717–724.

Zimprich, F, Sunder-Plassmann, R, Stogmann, E, Gleiss, A, Dal-Bianco, A, Zimprich, A, et al. (2004). Association of an ABCB1 gene haplotype with pharmacoresistance in temporal lobe epilepsy. *Neurology*, 63 (6):1087–1089.

18

Neurodegenerative Disorders: Tauopathies and Synucleinopathies

Huw R Morris and Andrew Singleton

Neurodegenerative disorders are common and usually affect the elderly. This is set to become the most pressing health concern in the developed world, because of the projected rapid expansion in the aged population. The prevalence of Alzheimer disease (AD) is predicted to quadruple in the United States over the next 50 years (Kawas and Brookmeyer, 2001). Unlike cancer and vascular disease, there are no disease-modifying therapies for patients with neurodegenerative diseases. There is an immediate need to develop new treatments, based on an understanding of the underlying disease biology. The genomic analysis of neurodegenerative disease is progressing rapidly, and unlike other common sporadic conditions, this has largely been guided by the study of rare Mendelian families with inherited late onset neurodegenerative disease. Although, genome-wide screens have been performed in series of affected sibling pairs (ASPs), small multiplex families and more recently in case–control series, more work has been done on systematic selective candidate gene analysis. Despite the inherent difficulties in studying Mendelian families with late onset disease, a series of genes have been identified which can be responsible, in mutated forms, for AD, amyotrophic lateral sclerosis (ALS), Parkinson's disease (PD), and other conditions (Hardy and Gwinn-Hardy, 1998). The identification of these genes has led to intensive work on basic protein biology, with the development of in vitro and transgenic in vivo disease models, and exploration of the role of these Mendelian disease proteins in sporadic human disease. Some investigators initially felt that Mendelian disease genes were unlikely to be relevant to sporadic disease. However, molecular neuropathology, and more recently, genomic association analysis have confirmed the validity of a Mendelian disease-driven approach.

The convergence of molecular pathology and genetics has led to apparently disparate diseases being grouped by the protein which seems to be primarily important in the disease. Often molecular genetics has led to the identification of an abnormal gene/protein which can be primarily responsible for the disease, and immunocytochemistry has confirmed the importance of the protein in the disease pathology. For example, spinocerebellar ataxia type 1 (SCA1) and Huntington's disease are both classed as polyglutamine diseases based on the occurrence of a pathogenic translated CAG repeat expansion, in different genes, and of intraneuronal polyglutamine inclusions (Hardy and Gwinn-Hardy, 1998). Intranuclear inclusions can rarely occur without a defined genetic/familial defect. Similarly, frontotemporal dementia with parkinsonism linked to chromosome 17 (FTDP-17) and corticobasal degeneration (CBD) are both classed as tau-related disorders, with tau deposition as neurofibrillary tangles (NFTs) and/or globose intracellular tau inclusions. However, in this case FTDP-17 is invariably due to autosomal dominant mutations in the tau gene (*MAPT*), and CBD is nearly always a sporadic condition. This chapter deals with the genomic analysis of movement disorders and dementia related to the deposition of tau and α-synuclein and will emphasize the impact of Mendelian genetics on the genomic investigation of sporadic disease.

Disorders with Tau Protein Deposition

Tau Protein

Tau is one of a family of microtubule-associated proteins (MAPs), which bind to microtubules and modulate microtubule function. MAPT is encoded on chromosome 17q21 (Neve et al., 1986). It consists of 16 exons, 3 of which (exons 4a, 6, and 8) are not expressed in the human brain, 1 of which (exon 1) is part of the promoter and transcribed but not translated, and 3 of which (exons 2, 3, and 10) are alternatively spliced to produce six different tau isoforms in the adult human brain (Andreadis et al., 1992). These isoforms range in length from 352 to 441 amino acids. The interaction between tau and the microtubules is mediated by a repeated 18 amino acid sequence encoded by exons 9, 10, 11, and 12, separated by an imperfectly repeating 13 or 14 amino acid sequence, via a flexible array of distributed weak sites. Tau isoforms containing exon 10 are referred to as four-repeat (4R) tau with those lacking exon 10 designated three-repeat (3R) tau. The ratio between 4R and 3R tau is normally tightly regulated, and in the normal human adult brain the ratio is usually estimated to be approximately 0.8, that is, 4R tau is produced at slightly lower levels than 3R tau (Hutton et al., 1998). The alternative splicing of tau is developmentally regulated, the fetal brain containing only the smallest 352 amino acid 3R tau without an amino terminus (exon 2/3), or carboxy terminus (exon 10) insert. Microtubules are polymers made from monomeric tubulin whose functions include maintaining cellular stability, axoplasmic transport, stabilizing mitotic processes, and establishing neurite outgrowth and cellular polarity (Ebneth et al., 1998). These functions are central to maintaining neuronal function, and disturbance of the microtubule function is likely to have detrimental effects. In cell-free systems, tau promotes the polymerization of tubulin into microtubules and stabilizes microtubules against de-polymerization by agents such as nocadazole (Drubin and Kirschner, 1986). In neural cells, the importance of tau is supported by the observations that (1) the levels of tau protein correlate with the extent of microtubule formation and neurite outgrowth (Drubin et al., 1985), (2) induction of the expression of tau messenger ribonucleic acid (mRNA) in differentiating cells occurs at the onset of axonal outgrowth, (Drubin et al., 1985), and (3) suppression of normal tau expression by the administration of tau antisense oligonucleotides inhibits normal axonal outgrowth and the establishment of neural polarity (Caceres et al., 1991).

Molecular Pathology of Tauopathies

A number of neurodegenerative diseases are characterized by the deposition of hyperphosphorylated tau as abnormal aggregates ("tauopathies"), most commonly NFTs (Morris, Lees, et al., 1999; Goedert, 2004). AD is the commonest tauopathy, and initially interest in the role of tau in neurodegeneration centered on AD. The influence of Mendelian genetics on sporadic neurodegenerative disease is exemplified by the interpretation of the importance of genes involved in autosomal dominant AD. AD is characterized by the formation of extracellular amyloid plaques and predominantly intracellular NFTs. The description of amyloid processing (presenilin-1 and 2) and amyloid precursor protein (APP) mutations in familial AD focused attention on amyloid dysfunction as a primary cause of AD and suggested that in AD tau NFT deposition was a secondary "downstream" phenomenon (Hutton and Hardy, 1997). However, tau NFTs are deposited in a wide range of other neurodegenerative conditions. The APP and presenilin mutations both favor the production of Aβ1–42, and in families with these mutations, tau NFT deposition follows on from a primary abnormality in Aβ production. However, two lines of evidence suggest that tau may have a more central role in the neurodegenerative process: (1) the pattern of tau deposition closely corresponds to the pattern of cell loss in AD, whereas amyloid deposition can occur in many cortical areas without cell loss (Arriagada et al., 1992; Goedert, 1996; Delacourte et al., 1999) and (2) the extent of tau deposition more closely follows the clinical disease course than does amyloid deposition.

In AD, tau is deposited predominantly as intracellular NFTs, which are visualized as paired helical filaments (PHFs) through the electron microscope. The PHFs of AD consist of intertwined fibers with a regular repeating diameter variation from 8 to 20 nm (Goedert, 1993). Biochemical analysis of AD tau indicates that PHF-tau is deposited in a sarkosyl insoluble form and that the sarkosyl insoluble tau fraction runs as a major triplet of tau bands at 55, 64, and 68 kDa on protein immunoelectrophoresis (Western blotting), together with a minor band at 72 kDa. (Goedert et al., 1992; Goedert, 1996). These tau protein bands are hyperphosphorylated, since, following alkaline phosphatase treatment the abnormal tau bands resolve to form the six tau isoforms that can be identified in the normal brain (Goedert et al., 1992; Goedert, 1996). In addition, PHF-tau is recognized by a range of phosphorylation-dependent antibodies which do not recognize soluble, dephosphorylated tau. The phosphorylation of tau occurs at the Ser/Thr-Pro sites clustered around the microtubule binding domains (Goedert, 1993). This phosphorylation is mediated by a number of protein kinases, most importantly mitogen-activated protein kinase and glycogen synthase kinase-3 (Lovestone and Reynolds, 1997). A number of other conditions are characterized by the deposition of tau-containing NFTs: progressive supranuclear palsy (PSP; Steele–Richardson–Olszewski disease), CBD, Pick's disease (PiD), FTDP-17, postencephalitic parkinsonism (PEP), the parkinsonism-dementia complex of Guam (PDC; bodig), Niemann–Pick disease type C, subacute sclerosing panencephalitis (SSPE) and posttraumatic parkinsonism (PTP; "dementia pugilistica") (Dickson, 1997). In addition, mutations in the parkin and LRRK-2 genes, responsible for Mendelian parkinsonism, are also sometimes associated with NFT pathology in a PD-like distribution (Hattori et al., 1998; Kitada et al., 1998; Wszolek et al., 2004).

In contrast to AD, in PSP the NFTs are straight filaments, which are composed of only two of the major triplet of bands at 64 and 68 kDa (Roy et al., 1974; Mailliot et al., 1998). Dephosphorylation analysis indicates that this PSP tau protein is made up mostly of 4R tau (Mailliot et al., 1998). In PiD, the tau protein is made up of only the smallest two bands at 55 and 64 kDa, which are composed of a separate tau isoform subpopulation of 3R tau (Delacourte et al., 1996; Mailliot et al., 1998). The classification of tauopathies has been helped by the development of 3R- and 4R-specific antibodies which confirm the findings of dephosphorylation analysis (de Silva, Lashley, et al., 2003). Thus, tauopathies can be classified according to the tau isoforms deposited: PSP, CBD, and argyrophillic grain disease (AGD) are all 4R tauopathies, PiD, and myotonic dystrophy are 3R tauopathies, and other disease including AD and PDC involve the deposition of all six isoforms of tau.

PSP is the most extensively studied of the nonamyloid tauopathies, and causes a balance and eye movement disorder with parkinsonism, later followed by progressive dysphagia, dysarthria, and immobility. PSP has a distinctive topographical and molecular pathology. The main lesions are in the substantia nigra pars compacta and reticulata, the globus pallidus, the subthalamic nucleus, and the midbrain and pontine reticular formation. This destruction of midbrain and pontine structures leads to identifiable changes of brainstem atrophy on neuroimaging with dilatation of the cerebral aqueduct, thinning of the midbrain tegmentum, and dilation of the fourth ventricle. The brainstem reticular pathology in PSP includes the midbrain and pontine nuclei involved in the supranuclear control of gaze: the rostral interstitial nucleus of the medial longitudinal fasciculus, the interstitial nucleus of Cajal, the nucleus of Darkschewitsch, and the raphé nucleus interpositus (Juncos et al., 1991; Revesz et al., 1996).

CBD, PEP, PDC, and to a lesser extent PiD all share a similar propensity to damage the globus pallidus and substantia nigra (Gibb et al., 1989; Geddes et al., 1993; Feany and Dickson, 1996). The clinical observation that many of these conditions affect the supranuclear control of gaze further suggests that these diseases, all of which involve tau protein deposition, share similarities in their topographic pathology. Cortical damage and glial pathology are more variable. CBD and PiD lead to asymmetric frontoparietal and frontotemporal atrophy, respectively. Amnesia may be a prominent feature in PTP and PDC with mesial temporal pathology involving the hippocampus, and entorhinal and transentorhinal cortex, whereas these areas are relatively spared in PSP and CBD (Corsellis et al., 1973; Dickson, 1998). Distinctive glial inclusions have been reported to be relatively specific for some diseases, and these seem to relate to compartmentalization of tau within glial cells. In PSP, tau accumulates in the cell body producing the tufted astrocyte appearance; in CBD, tau accumulates at the distal astrocytic processes producing astrocytic plaques; and in PiD tau is distributed more diffusely through the cell soma (Chin and Goldman, 1996). Although distinctions can be made between these diseases, pathological and clinical similarities suggest that cell groups in the basal ganglia and brainstem share a common mode of cell damage, with abnormal tau accumulation.

The importance of tau was confirmed in 1998 with the identification of mutations in MAPT in families with FTDP-17, and the inference that abnormalities in tau can be primarily pathogenic in familial and sporadic diseases.

Clinical Features of Tauopathies

FTD and FTDP-17

FTD is a common but probably under-recognized cause of dementia and is equally as frequent as AD in patients who develop symptoms before the age of 65 (Ratnavalli et al., 2002). The differences between AD and FTD reflect the pathological topography of the diseases: AD involves the mesial temporal cortex early in the disease process and usually presents with amnesia, whereas FTD affects the frontal and temporal cortices and presents with behavioral or language impairment. FTD has a number of different pathological substrates defined by intracellular inclusions. These pathologies include PiD, motor neuron disease (MND)-inclusion dementia, neurofilament inclusion dementia, and dementia lacking distinctive histopathology (Kertesz et al., 2005). In addition, some patients with FTD develop swollen achromatic neurons which stain positively with antibodies to α-B crystallin ("ballooned neurons"). Approximately one-third of patients with FTD have an autosomal dominant family history. The identification of linkage to chromosome 17q21 in the dementia-disinhibition-parkinsonism-amyotrophy complex family (DDPAC, family Mo) led to a series of reports of further families with linkage to this region, summarized in a 1996 workshop report on FTDP-17 (Wilhelmsen et al., 1994; Foster et al., 1997).

The original report of the DDPAC family described a clinical syndrome in which insidious personality change with disinhibition began at an average age of 45. Subsequent reports of the clinical features of chromosome 17-linked families described broadly similar syndromes although each was given a distinct name. These included pallido-pontine-nigral degeneration (PPND) (Wijker et al., 1996), progressive subcortical gliosis (PSG) (Petersen et al., 1995), hereditary frontotemporal dementia (HFTD), and familial multiple system tauopathy with presenile dementia (FMSTD) (Murrell et al., 1997). These families all had similar clinical features with disease onset in the 40–50s, early personality change with disinhibition, and later withdrawal and apathy, with extrapyramidal features. The pathological topography of these families is similar. Frontal and temporal atrophy was a consistent finding, together with involvement of the substantia nigra, amygdala, and caudate/putamen/globus pallidus (Foster et al., 1997). There was confusion in the description of the immunocytochemical pathology. Two Dutch families linked to 17q21, HFTD-1, and HFTD-2, were initially reported to have no tau deposition after immunohistochemistry with both tau and PHF specific antibodies (Heutink et al., 1997) and yet were subsequently shown to involve both mutations in MAPT and extensive tau deposition (Spillantini, Crowther, Kamphorst, et al., 1998). Similarly the report of the PSG-1 family's linkage to chromosome 17 described prion protein deposition without tau deposition. (Peterson et al., 1995; Goedert et al., 1999) The report of the 1996 workshop concluded that the FTDP-17 families could be pathologically divided into four groups: (1) tau pathology with ballooned neurons, (2) tau pathology without ballooned neurons, (3) ballooned neurons alone, and (4) dementia without specific pathological features (Foster et al., 1997). Although many of these families have tau pathology and MAPT mutations, some of these families have now been shown to have mutations in the progranulin gene with ubquitin inclusion pathology.

MAPT Mutations in FTDP-17

MAPT was the primary candidate gene for FTDP-17, and three reports in 1998 defined pathogenic mutations. The first report that MAPT mutations were responsible for FTDP-17 was made by Poorkaj and co-workers who described a coding change, V337M in exon 12 of MAPT in Seattle family A (Poorkaj et al., 1998). This sequence change was reported, together with a number of polymorphic variants within MAPT. However, this report was qualified by the absence of MAPT mutations in other FTDP-17 families. In June 1998, Hutton and co-workers were able to conclusively demonstrate that MAPT was the causative gene for FTDP-17 by describing six separate coding and intronic mutations in MAPT and providing an explanation for the mechanism of action of the intronic mutations (Hutton et al., 1998). The coding mutations described were G272V and P301L, which disrupted a conserved PGGG motif in the microtubule binding domains of exons 9 and 10, respectively, and the exon 13 mutation R406W which occurred adjacent to two serine residues, Ser396 and Ser404, which may act as sites for phosphorylation by Ser/Thr-Pro-directed protein kinases (Grover et al., 1999). The splicing mutations exon 10 +13, +14, and +16 were thought to alter the splicing of exon 10 and both in vitro evidence, using an exon trapping system, and in vivo evidence, based on RNA analysis from postmortem tissue, were presented which supported the role of MAPT splicing mutations in FTDP-17 (Grover et al., 1999). The original chromosome 17-linked family DDPAC was shown to have an MAPT exon 10 + 14 mutation. The description of these mutations was accompanied by the description of a further MAPT exon 10 splice site mutation, exon 10 +3, in the best characterized of the FTDP-17 families, FMSTD, by Spillantini and co-workers (Spillantini, Murell, et al., 1998).

Since the original description of MAPT mutations in 1998, 35 MAPT mutations have been described, including codon deletions and one family with apparent autosomal recessive disease (Table 18–1) (Rademakers et al., 2004). The mutations are clustered around the microtubule-binding domains of tau, exons 9–12, although two mutations have been described in exon 1 (R5H and R5L) and one in exon 13 (G389R and R406W). Most mutations are protein coding, although some affect the alternative splicing of exon 10. In vitro, the protein coding mutations have been shown to have two effects: (1) impairment of the microtubule stabilizing properties of tau and (2) acceleration of the rate of filament formation by tau. The relative importance of these effects is not known. Most MAPT coding mutations have a decreased ability to stabilize microtubules, although two mutations, Q336R and S305N, enhance microtubule assembly. Similarly, most MAPT coding mutations lead to an enhanced ability to form filamentous aggregates. The P301S mutation is one of the most malignant MAPT mutations leading to disease onset during the 20s. The P301S mutation has a marked effect on filament formation but a small effect on microtubule destabilization, providing some clinical-genotype evidence suggesting that accelerated filament formation may be the most important pathogenic mechanism. Comparison with Mendelian AD, PD, and MND suggests that accelerated aggregate/filament formation may be a common pathogenic theme in late-onset neurodegenerative diseases. A third potential mechanism is the alteration of the phosphorylation state of tau via alteration of Ser/Thr-Pro phosphorylation sites. This has not been demonstrated in postmortem tissue, but at least two mutations (K257T, P301S) could potentially alter the tau phosphorylation state.

Interestingly, there are a number of noncoding mutations within MAPT, which alter the alternative splicing of exon 10 (Table 18–1). These include splice site mutations (e.g., MAPT exon 10 + 16) which are predicted to disrupt a stem–loop structure at the exon–intron junction, and exon 10 mutations (e.g., L284L) which affect exonic splice enhancer/promoter sequences. Nearly all these

Table 18–1 Brief Overview of MAPT Mutations and their Functional Effects

No.	Exon	Tau Mutation	Phenotype	Age at Onset	Microtubule Interaction (Binding, Stabilization or Assembly)[a]	Filament Formation	4R:3R Tau RNA or Protein Isoform Ratio	References[b]
1	1	R5H	FTD Park	76	↓	↑	N	Hayashi et al., 2002 11921059
2	1	R5L	PSP	62	↑	NA	N	Poorkaj et al., 2002 12325083
3	9	K257T	FTD	47	↓	↑	3R ↑	Rizzini et al., 2000 11089577
4	9	I260V	FTD	68	↓	↑	N	Grover et al., 2003 14637086
5	9	L266V	FTD	33	↓	↑	4R ↑	Hogg et al., 2003 12883828
6	9	G272V	FTD	46	↓	↑	NA	Hutton et al., 1998 9641683
7	10	N279K	FTD + PSP	43	↔	↑	↑	Clark et al., 1998 9789048
8	10	DK280	FTD	53	↓	–	↓	Rizzu et al., 1999 9973279
9	10	L284L	FTD	52	↔	↔	↑	D'Souza et al., 1999 10318930
10	10	DN296[c]	PSP	38	↓	↑	N	Pastor et al., 2001 11220749
11	10	N296H	FTD	57	↓	↑	4R ↑	Iseki et al., 2001 11585254
12	10	N296N	FTD + Park + SNGP	56	↔	↔	4R ↑	Spillantini et al., 2000 11117553
13	10	P301L	FTD	50	↓	↑	↔	Hutton et al., 1998 9641683
14	10	P301S	FTD + CBD	34	↓	↑	NA	Bugiani et al., 1999 10374757
15	10	S305N	FTD	35	↑	↔	4R↑	Kobayashi et al., 2003 12928922
16	10	S305S	FTD + Aphasia	36	↔	↔	4R↑	Stanford et al., 2000 10775534
17	Intronic	Exon 10 +3	FTD + PSP	49	↔	↔	4R↑	Spillantini, Murell, et al., 1998 9636220
18	Intronic	Exon 10 +11	FTD + Park + PSP	48	↔	↔	4R ↑	Miyamoto et al., 2001 11456301
19	Intronic	Exon 10 +12	FTD	53	↔	↔	4R ↑	Yasuda et al., 2000 10762152
20	Intronic	Exon 10 +13	FTD	65	↔	↔	4R ↑	Hutton et al., 1998 9641683
21	Intronic	Exon 10 +14	FTD + Park	45	↔	↔	4R ↑	Hutton et al., 1998 9641683
22	Intronic	Exon 10 +16	FTD + Park	53	↔	↔	4R ↑	Hutton et al., 1998 9641683
23	Intronic	+19	FTD	52	↔	↔	3R ↑	Stanford et al., 2003 12615641
24	Intronic	+29	FTD	52	↔	↔	3R ↑	Stanford et al., 2003 12615641
25	11	L315R	FTD + Semantic impairment	51[d]	↓	↔	NA	van Herpen et al., 2003 14595646
26	11	K317M	FTD + MND	48	NA	NA	NA	Zarranz et al., 2005 15883319

(Continued)

Table 18–1 Brief Overview of MAPT Mutations and their Functional Effects *(Continued)*

No.	Exon	Tau Mutation	Phenotype	Age at Onset	Microtubule Interaction (Binding, Stabilization or Assembly)[a]	Filament Formation	4R:3R Tau RNA or Protein Isoform Ratio	References[b]
27	11	S320F	FTD	38	↓	NA	NA	Rosso et al., 2002 11891833
28	12	G335V	FTD	25	↓	↑	NA	Neumann et al., 2005 15765246
29	12	Q336R	FTD	58	↑	↑	NA	Pickering-Brown, Baker, Nonaka, et al., 2004 15047590
30	12	V337M	Psychosis + FTD	53	↓	↑	NA	Poorkaj et al., 1998 9629852
31	12	E342V	FTD	48	NA	NA	4R ↑	Lippa et al., 2000 11117541
32	12	S352L[c]	Respiratory failure	29	↓	↑	NA	Nicholl et al., 2003 14595660
33	12	K369I	FTD	52	↓	↓	NA	Neumann et al., 2001 11601501
34	13	G389R	FTD	38	↓	NA	NA	Murrell et al., 1999 10604746
35	13	R406W	FTD	59	↓	↔	↑	Hutton et al., 1998 9641683

[a] Some mutations have different effects on rate of microtubule assembly, amount of microtubule synthesis etc. Many mutations have clinical and pathological heterogeneity between and within families, and in different assays and parameters of microtubule interaction and filament formation. Please see original references for further details.

[b] References—number under the author's name indicates PubMed/Medline entry.

[c] Mutations described as homozygous MAPT mutations.

[d] Age at onset in individuals harboring L315R mutations is highly variable.

Abbreviations: CBD, corticobasal degeneration; FTD, frontotemporal dementia; N, normal; NA, not available; Park, parkinsonism; PSP, progressive supranuclear palsy; SNGP, supranuclear gaze palsy.

mutations are predicted or have been shown to increase the ratio of 4R:3R MAPT RNA, although the +19 and +29 mutations are predicted to lead to a decrease in 3R MAPT RNA. Using reverse transcriptase polymerase chain reaction (rtPCR) and the postmortem brain it has been confirmed that these mutations lead to an increase in the ratio of 4R:3R RNA and 4R:3R tau protein. It is not immediately clear why an alteration in the tau isoform ratio should lead to neurodegeneration. It is thought that there are separate binding sites for 4R and 3R tau on microtubules, which could lead to saturation of 4R tau binding sites and deposition of excess tau when the 4R:3R ratio is changed (Goode et al., 2000). The existence of sporadic 4R tauopathies such as PSP immediately suggests a link between familial and sporadic tauopathy.

On a population basis, the most common MAPT mutations are exon 10 +16, P301L, P301S, and N279K, all of which have been described in more than two apparently separate families worldwide. The N279K mutation appears to have arisen independently in Japanese and French populations, whereas haplotype analysis of the exon 10 +16 mutation families suggests that this mutation, found around the world, has originated from a common founder in the North Wales region of the UK (Pickering-Brown, Baker, Bird, et al., 2004). In general, exon 10 mutations are particularly likely to lead to parkinsonism, and mutations outside exon 10 are more likely to lead to a pure cognitive presentation, although there are a number of exceptions to this pattern. Some families show phenotypic heterogeneity with a CBD-like picture, and an FTD-like picture, occurring in the same family as manifestations of the same mutation. Usually mutations in and around exon 10 lead to neuronal and glial tau deposition, whereas nonexon 10 mutations lead to neuronal tau deposition in Figure 18–1 (Ingram and Spillantini, 2002). K257T and G389R lead to a PiD type pathology with small round tau immunoreactive inclusions within the dentate fascia of the hippocampus. K257T causes predominant 3R tau deposition, similar to sporadic PiD. The ultrastructural morphology of the tau deposits varies with the MAPT mutation (Ingram and Spillantini, 2002). Intronic MAPT splicing mutations lead to extensive neuronal and glial tau deposits, which appear as twisted ribbon filaments under the electron microscope. Protein-coding mutations in exon 10 lead to narrow twisted ribbon filaments. Protein-coding mutations outside exon 10 lead to a variety of different ultrastructural appearances, including PHFs, straight filaments, and twisted filaments. Exceptions to the general pattern include R5H and R5L which lead to significant glial tau deposition, and in the case of R5L, a PSP-like movement disorder. For some of these mutations, there is insufficient pathological material from single families to be certain that these pathological hallmark features relate directly to the MAPT mutation, rather than interindividual variation. However, the variation in tau pathology ultrastructure and clinical features across the different MAPT mutation families mirrors the variation seen in the sporadic tauopathies. It seems a reasonable inference that this

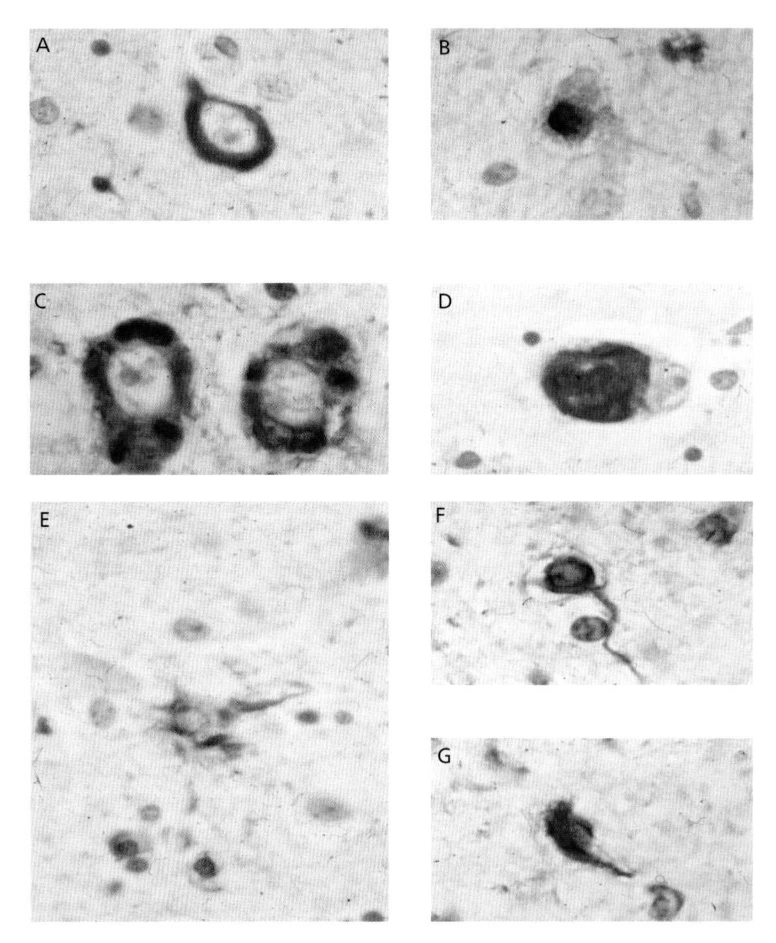

Figure 18–1 Neuronal and glial tau deposition patterns in FTDP-17T. High power views stained with tau antibody AT-8 in a patients with a MAPT exon10 + 16 mutation: A–D neuronal tau deposition showing: (*A*) coiled (*B*) dense Pick body-like (*C*) diffuse granular with dense deposits and (*D*) globose neuronal inclusions; and E–G glial tau deposition showing (*E*) tufted astrocyte and (*F*) (*G*) coiled oligodendroglial inclusions. Courtesy of Prof. Tamas Revesz, Institute of Neurology, Queen Square, London UK.

variation relates mostly to the specific mutation and that similar mechanisms govern phenotypic variation in sporadic tauopathies.

MAPT and Genetic Susceptibility to PSP

In 1997, Conrad and colleagues described the first evidence of a genetic link between MAPT and sporadic neurodegeneration (Conrad et al., 1997). They identified a TG repeat within the MAPT cosmid clone k6, lying approximately 7 kb downstream of exon 9 and 5 kb upstream of exon 10. Variability at this polymorphism was analyzed in normal controls, AD patients, and 22 patients with pathologically confirmed PSP. They reported five alleles: A0 with 11 TG repeats, and A1, A2, A3, and A4 defined by the presence of 12, 13, 14, or 15 TG repeats. Both the A0 allele and the A0/A0 genotype were positively associated with PSP, in comparison with normal controls and patients with AD (Conrad et al., 1997). In control subjects, 57.4% of genotypes were A0/A0 and 74.6% of alleles were A0, compared to 95.5% of PSP genotypes and 97.7% of PSP alleles, giving an odds ratio (OR) of 15.6 for A0A0 genotype status. After this publication, this association was replicated in a series of geographically separate studies (Conrad et al., 1997; Oliva et al., 1998; Baker et al., 1999; Hoenicka et al., 1999; Morris, Janssen, et al., 1999). The replication of the MAPT-PSP association across these different studies confirms that this is a robust and accurate finding. Ethnic differences in variability in MAPT were noted in these early studies, with two groups reporting no difference between the PSP and control A0 frequency in the Japanese population (Conrad et al., 1998; Okuizumi et al., 1998). This lack of association was interpreted as suggesting either that there is a significant linkage disequilibrium effect in the Caucasian population, not present in the Japanese, between the A0 allele and an adjacent functionally important variable site, or that there is a difference in the role of MAPT in the development of PSP in Caucasian and Japanese populations. These data highlight the importance of the background allele frequency in genetic association studies. The A0 allele is virtually ubiquitous within the Japanese population, and therefore genetic association studies would not be expected to establish allelic association even if there was a significant biological effect. MAPT is an excellent candidate gene for determining susceptibility to PSP as tau is the major constituent of NFTs in PSP, and the identification of MAPT mutations in FTDP-17 reinforces the likely primary role of MAPT in neurodegeneration. The more plausible interpretation of the Japanese studies is that the biological effect of MAPT predisposition to PSP is identical but that this effect is not detectable against the background Japanese allele distribution. A better test of a possible biological role for the A0 allele in the Japanese and Caucasian populations would be an epidemiological study. The working hypothesis would be that the population genetic differences in MAPT genotypes would lead to a doubling of the prevalence of PSP in Japan. Clearly, epidemiological work of this type presents formidable methodological problems, and to date the population prevalence of PSP in Japan has not been reported.

MAPT and Other Diseases

Corticobasal Degeneration

Like PSP, CBD involves the deposition of 4R tau protein. An initial study did not identify an association between CBD and the A0 MAPT allele, using clinically defined cases (Morris, Janssen, et al., 1999). After the early work on the MAPT intronic dinucleotide polymorphism, the A0 allele was identified to lie on an "H1" MAPT haplotype and most subsequent evaluated association with MAPT H1. Two further studies recruited patients with pathologically diagnosed CBD and identified a positive association between MAPT H1 and CBD, similar to PSP (Di Maria et al., 2000; Houlden et al., 2001).

Argyrophillic Grain Disease

Argyrophillic grains are comma- or spindle-shaped tau immunoreactive inclusions. AGD is a neurodegenerative disorder presenting with predominant involvement of the mesial temporal lobes and an amnestic syndrome and has been thought to be analogous to AD (Togo et al., 2002). However, argyrophillic grains are also commonly seen in PSP and CBD, and recent work has shown that AGD involves the deposition of 4R tau, and like CBD and PSP is a 4R tauopathy. AGD is also associated with H1, meaning that all three known sporadic 4R tauopathies are associated with the same MAPT haplotype.

Sporadic PiD

PiD is a well-defined pathological substrate of FTD involving the formation of small, round tau immunoreactive inclusions, particularly in the frontotemporal cortex and in the dentate fascia of the hippocampus. Unlike PSP and CBD, PiD preferentially involves the deposition of 3R tau, and inclusions do not stain with tau monoclonal antibody 12-E8 which recognizes phosphorylated codon 262 and 356. A number of FTDP-17 mutations have been described, which lead to a pathological phenotype reminiscent of PiD, but to date a change in RNA splicing leading to overexpression of 3R tau RNA has not been described. A case–control study looking at pathologically confirmed PiD did not identify an association between the A0 allele/ H1 haplotype and Pick's (Morris et al., 2002).

Sporadic FTD

Clinically diagnosed FTD is a pathologically heterogeneous disorder (Kertesz et al., 2005). A number of association studies have been done which have not shown any consistent association between MAPT and FTD, although some studies report a possible interaction between apolipoprotein E (ApoE) and MAPT. Given the different pathological phenotypes underlying FTD, it is likely that this relationship between FTD and MAPT will only be resolved with large-scale clinico-pathological series.

Alzheimer Disease

The association between AD and the ApoE ε4 allele is well established, yet only a small proportion of patients with sporadic late-onset AD carry an ε4 allele. However, the analysis of a large number of alternative candidate polymorphisms has not identified any reproducible associations. Although the evidence from Mendelian AD supports a primary role for Aβ in AD, cell biology suggests that ApoE may be important in Aβ (and/or tau deposition. A meta-analysis of association studies did not identify any association between AD and MAPT, even when stratifying for ApoE e4 status, using 1603 samples (Russ et al., 2001). However, in a small series there is a suggestion of an association between AD and a subhaplotype of H1, designated H1c (see below).

Further Analysis of the Tau Region

After the identification of the first autosomal dominant mutations in MAPT, Baker and colleagues identified a series of single nucleotide polymorphisms (SNPs), which were in complete linkage disequilibrium in Caucasian subjects, spanning approximately 62 kb across MAPT and forming two haplotypes (Baker et al., 1999). The haplotypes were designated H1, the common haplotype with a population frequency of 78% in Caucasians, and H2. The A0 allele and less common A1 and A2 alleles lie on the H1 haplotype background, and the H1 haplotype is associated with PSP (Baker et al., 1999). This was one of the earliest reports of a genomic haplotype block and its use in allelic association. Before the identification of the MAPT H1 haplotype, statistical analysis incorrectly clumped non-A0 versus A0 alleles to achieve statistical power. Assuming that the functional polymorphism is bi-allelic and lies within the H1 haplotype block, the use of microsatellite markers will have led to a loss of power in the MAPT/PSP association analysis. The initial studies of variability in the region suggested microsatellite slippage on the H1/H2 haplotype background and the occurrence of rarer alleles on the H1 background. It was inferred that the MAPT H1/H2 division was ancient and that there was recombination suppression, or selection against recombinant alleles in the MAPT region. Further analysis of the MAPT region indicated that the H1/H2 haplotype split included the promoter region as well as a nested gene, saitohin, lying within intron 9 of MAPT. The involvement of this wider area implied that the area of linkage disequilibrium spanned at least 120 kb, around the average extent of a haplotype block of linkage disequilibrium in the human genome (de Silva et al., 2001; Pastor et al., 2002). The publication of the draft sequence of the human genome enabled a search for the maximum extent of the LD block surrounding MAPT. Surprisingly, this area is massive, initially estimated by Pittman and colleagues to extend up to 2 Mb, and as expected the whole of this extended H1 haplotype is associated with PSP (Pastor et al., 2002; Pittman et al., 2004; Pastor et al., 2004). This wide area of allelic association potentially implicates a large number of genes and regulatory regions as important in the pathogenesis of PSP. Having established this association, two main approaches were used to further explore the relationship between MAPT and PSP. First, of searching for H1/H2 subhaplotypes which might highlight the functionally important variation in the MAPT region and second, looking at molecular pathology and in vivo and in vitro models which might further explain the functionally relevant genomic variability. Further analysis by Pittman and colleagues, and Skipper and colleagues using tagging SNPs have identified a number of H1 subhaplotypes, which may inform the MAPT-PSP association (Skipper and Farrer, 2002; Pittman et al., 2005). An MAPT H1 subhaplotype, H1c, has been identified which is strongly associated with PSP. There is potential difficulty in separating the protective effect of the H2 haplotype from the risk effect of H1. However, this H1c association seems to occur independently of the protective association with MAPT H2. In other words, in both independent US and UK samples, the majority of the loss of H2 alleles in the PSP population is accounted for by H1c MAPT alleles, rather than other H1 subhaplotypes. This suggests that the 56 kb region, defined by H1c, spanning the region before MAPT exon1 to intron 9, may harbor the pathogenic variant responsible for PSP.

Aside from the detailed analysis of the allelic association between MAPT and PSP, a great deal of interest is focused on the mechanism and history of the MAPT H1/H2 split. A recent study has reported that the split seems to have occurred on the basis of a 1.4 Mb inversion in the chromosome 17q21/MAPT region (Stefansson et al., 2005). As outlined above, the H2 haplotype was initially identified to be very rare in the Japanese population. In fact, the H2 haplotype is very strongly associated with Caucasian population ancestry, with the highest H2 halotype frequencies seen in northern European populations, and a decrease in H2 frequency in mixed nonCaucasian populations (Evans et al., 2004; Stefansson et al., 2005). Recently, Hardy and colleagues have suggested that the MAPT H2 allele may have been introduced into the Caucasian population by mating between *Homo neanderthalis* and *Homo sapiens* (Hardy et al., 2005). This focus on ethnic differences in the H2 frequency has underlined the likelihood of racial variation in neurodegenerative disease susceptibility—the Caucasian population is relatively protected from PSP by virtue of the H2 haplotype.

MAPT is one of the few genes in neurogenetics which harbor both rare mutations and disease-causing common variation, which can lead to inherited and sporadic disease, respectively. The identification of a number of different possible mechanisms for the toxicity of MAPT mutations in FTDP-17 suggested that it should be possible to identify possible pathogenic mechanisms for the MAPT H1 association in PSP, relevant to sporadic disease.

Biological Effects of the Tau H1/H2 Haplotypes

Although the full H1 haplotype includes a number of genes and regulatory regions MAPT remains the strongest functionally important candidate gene. The identification of the H1c haplotype, which narrows the association down to the promoter region and amino terminus of MAPT, seems to confirm its importance. A number of investigators have sought to establish the functional role of the H1 haplotype. The two main possibilities are that the susceptibility allele alters the overall level of expression of MAPT or, analogous to FTDP-17, alters the alternative splicing of exon 2/3 and/or exon 10. PSP involves a predominant deposition of 4R tau protein so that a primary alteration in exon 10 RNA splicing may explain the pathogenesis of PSP (Liu et al., 2001). Two studies have reported an increase in 4R:3R RNA levels in PSP, but this has not been directly related to the MAPT H1 haplotype (Chambers et al., 1999; Takanashi et al., 2002). Chambers and colleagues reported that there is a change in the 4R:3R MAPT mRNA ratio in PSP brain but only in the brainstem and not in other areas known to be affected by the disease process (Chambers et al., 1999). Takanashi and colleagues similarly report an increase in 4R:3R RNA ratios in Japanese patients compared to controls, who were all H1H1 homozygotes. In contrast, Hoenicka and colleagues report no difference between 4R:3R ratios in PSP and control cases (Hoenicka et al., 1999). An overall change in tau mRNA expression levels has been suggested by Kwok and colleagues on the basis of in vitro luciferase reported assays comparing the activity of the H1 and H2 promoter haplotypes (Kwok et al., 2004). Postmortem analysis of RNA, important to either a splicing based or total transcription theory of the pathogenesis of PSP, is confounded by changes in RNA produced by the agonal and disease state. Although changes in RNA splicing have been convincingly demonstrated in FTDP-17, it is worthwhile noting that this is on the basis of a systemic mutation. Within the sporadic diseases there may be cellular and regional variation in alternative splicing, and neurons with the most abnormal levels of RNA alternative splicing may be absent late in the disease course.

From a therapeutic perspective, the data from splice mutation FTDP-17 clearly point toward the possible usefulness of agents that might restore the normal 4R:3R RNA or protein ratios. Beyond this a variety of agents may be of use, including agents that reduce the total amount of tau expression, agents that prevent or enhance the degradation of tau pre-tangles or NFTs, or agents that alter the tau phosphorylation state.

Other Genetic Risk Factors in Tauopathies

To date there have been no reported genome-wide searches either in ASPs or in case–control series in the common non-AD tauopathies. A case–control genome-wide search using microsatellite markers has been undertaken in PDC, under the assumption that affected individuals on Guam are related to a common founder. This study identified areas of interest on chromosomes 14 and 20, but these areas were not replicated in a family-based linkage study. A number of other candidate genes have been evaluated in PSP including α-synuclein and ApoE, but they have not identified an association (Morris et al., 2000). This is in keeping with a recent report that ApoE status does not modify the age at onset within FTD families with known tau mutations and adds to the data suggesting that, among neurodegenerative conditions, ApoE specifically modulates only amyloid deposition disorders (Houlden et al., 1999).

It is likely that the coming decade will see the start of systematic genome-wide approaches to susceptibility gene for tauopathies with the use of genome-wide SNPs in case–control series.

Synucleinopathies

α-Synuclein Protein

The protein α-synuclein was initially identified as a constituent of extracellular senile plaques in AD (Ueda et al., 1993). Ueda and colleagues raised antibodies against synthetic peptides, based on an unknown protein within an amyloid preparation, and showed that these antibodies labeled neuritic plaques in autopsy material from AD patients. The peptide was initially named NAC (non-Aβ component of amyloid) and its precursor, cloned from a complementary DNA (cDNA) library, NACP. The protein NACP was later identified as the α member of the synuclein family, which includes α-, β-, and γ-synuclein. A role for α-synuclein in neurodegenerative diseases was further suggested by the finding that mutations in the gene-encoding α-synuclein can cause Parkinson's/Lewy body disease, and this protein is a major component of the neuropathological hallmark of these disorders—the Lewy body (discussed in detail below).

In 1995, the gene-encoding NACP/α-synuclein was mapped to human chromosome 4q21.3–22 by several groups using a battery of techniques (Campion et al., 1995; Chen et al., 1995; Spillantini et al., 1995). *SNCA*-encoding α-synuclein is a six-exon gene spanning approximately 110 kb. Two primary isoforms have been identified, both expressed in adult and fetal brain (Ueda et al., 1994). These isoforms are 140 and 112 amino acids in length, the latter produced by in-frame splicing out of the fourth exon (Ueda et al., 1994). α-synuclein, originally named because of its apparent localization to the synapse and the nucleus, contains six imperfect repeats including a KTKEGV consensus motif, which stretch through the N-terminal amphipathic region and into the central NAC domain. The C-terminal of α-synuclein consists of an acidic region. α-synuclein is a natively unfolded protein that may also exist as an oligomeric species. α-synuclein undergoes posttranslational modification, including phosphorylation, nitration, and ubiquitination

(Souza et al., 2000; Fujiwara et al., 2002; Nonaka et al., 2005). Polyubiquitination of α-synuclein is an event that has received much attention, in part because mutations in *PRKN*, an E3 ubiquitin-ligase, cause a young-onset parkinsonism but also because dysfunction or alteration of a system designed to mark proteins for destruction is a pathway of interest in a disease marked by abnormal protein accumulation. In dementia with Lewy bodies (DLB) ubiquitinated α-synuclein species are present, and this process appears to be modulated by the phosphorylation of α-synuclein (Hasegawa et al., 2002), although the exact significance of phosphorylation in this process is controversial (Nonaka et al., 2005). α-Synuclein is basally phosphorylated at Ser and tyr residues. Phosphorylation of Ser-129 in vitro promotes fibril formation (Fujiwara et al., 2002). Using a *Drosophila* model of PD in vivo suggests that phosphorylation of Ser-129 by the G protein coupled receptor kinase-2 (Gprk2) enhances α-synuclein toxicity and that in contrast to in vitro data, blocking of Ser-129 phosphorylation increases fibril formation (Chen and Feany, 2005). These data suggest that sequestration of α-synuclein into insoluble aggregates may be protective against neuronal death and that this event may be mediated by phosphorylation.

The normal function of α-synuclein is unknown; however, the localization of α-synuclein within presynaptic nerve terminals has led to the supposition that it may play a role in synaptic plasticity or vesicle trafficking. Early support for the role of α-synuclein in synaptic plasticity came from the observation that the ortholog of this protein in the Zebra Finch, synelfin, is highly regulated during song learning (George et al., 1995). Despite the enrichment of α-synuclein within presynaptic terminals, it is clear that knockout of this protein (and α-synuclein) does not disrupt basic synaptic function and that α-synuclein may contribute to the long-term regulation and/or maintenance of presynaptic function (Chandra et al., 2004). α-synuclein has also been implicated in the recycling of synaptic vesicles (Abeliovich et al., 2000).

Molecular Pathology of Synucleinopathies

Classically, neurodegenerative diseases have been described according to the clinical phenotype and the regional neuropathology. However, as is the case for tauopathies, the identification of α-synuclein has led to a new grouping of neurodegenerative diseases based on a protein of known primary pathogenic importance. The term synucleinopathy is used to describe diseases where the primary deposited protein species is α-synuclein, and these diseases are thought to have a shared molecular etiology. Several neurodegenerative disorders are characterized by the deposition and accumulation of insoluble α-synuclein in the brain. The most notable α-synuclein positive pathology is the Lewy body, which is the pathognomic hallmark lesion of both PD and DLB. In PD this intraneuronal inclusion body exists predominantly within the neuromelanin containing neurons of the substantia nigra pars compacta, and it was this structure that was initially described by Frederich Lewy in the early 1900s, hence called the classical Lewy body. The classical Lewy body is a spherical intracytoplasmic neuronal inclusion, with a dense core, surrounded by a more diffuse halo (Figure 18–2). Cortical Lewy bodies, first described in 1961 (Okazaki et al., 1961), are most notable within DLB; although similar to classical Lewy bodies, these structures lack the dense core associated with classical Lewy bodies and may appear a little more diffuse and disorganized in structure. In addition to α-synuclein, Lewy bodies are highly immunoreactive to antibodies raised against numerous proteins, most notably ubiquitin and the neurofilaments. Until the advent of ubiquitin immunocytochemisty Lewy pathology was considered a relatively rare phenomenon, traditionally seen through tinctorial methods such as haemotxylin and eosin staining. However, the application of first anti-ubiquitin and then anti-α-synuclein immunostaining showed that classical, and particularly, cortical Lewy bodies are more common than previously thought (Lennox et al., 1989; Spillantini et al., 1997; Spillantini, Crowther, Jakes, Hasegawa, et al., 1998). An additional α-synuclein positive inclusion revealed using immunocytochemistry is the Lewy neurite, a thread-like intraneuronal inclusion (Figure 18–2). Lewy neurites occur both in PD and DLB, and there has been some suggestion that this deposit may presage the formation of a full Lewy body. Although this is a difficult hypothesis to prove, the staging data presented by Braak and colleagues are consistent with this notion, Lewy neurites being the first observable change in early stages of PD (Braak et al., 2004).

In addition to the primary neuronal α-synuclein positive lesions, this protein is also found as a predominant component of an oligodendroglial lesion, the glial cytoplasmic inclusion (GCI) of multiple system atrophy (MSA) (Spillantini, Crowther, Jakes, Cairns, et al., 1998). MSA is a severe progressive neurodegenerative disorder characterized by degeneration of the substantia nigra, putamen, olivary nucleus, pontine nuclei, and cerebellum (Dickson et al., 1999). Although neuronal inclusions are found in MSA, the predominant pathology is the GCI, which exhibits strong immunoreactivity to α-synuclein and ubiquitin (Figure 18–2).

Soluble α-synuclein exists with little secondary structure, primarily as a monomer. However α-synuclein can and does aggregate, a process thought to be mediated in part by a tendency of the central hydrophobic region to self-associate. It has been suggested that this self-association leads initially to oligomeric, although still soluble, forms of α-synuclein, which in turn may give rise to highly insoluble polymers or fibrils. The role of α-synuclein fibrils as a major component of Lewy pathology is supported by evidence that shows α-synuclein can form fibrils in vitro and that α-synuclein immunochemistry labels similar fibrils extracted from Lewy bodies

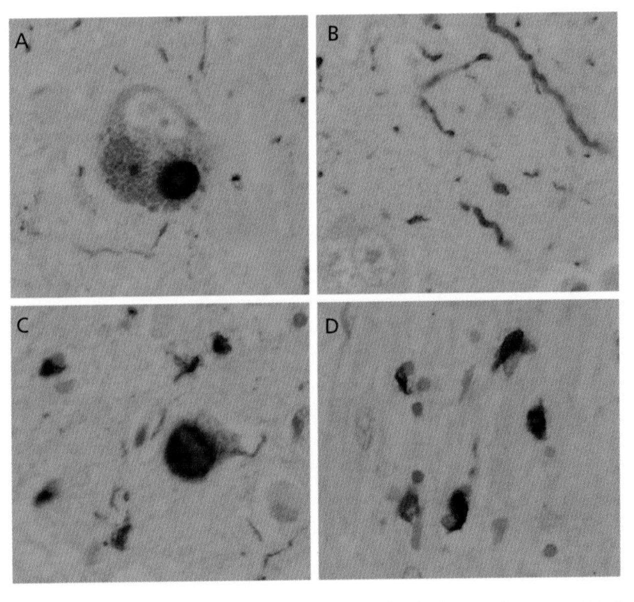

Figure 18–2 α-synuclein deposition in Lewy body dementia and multiple system tropy. Anti-synuclein immunocytocheimstry at × 400 magnification. (*A*). Lewy bodies in the substantia nigra in Lewy body dementia (*B*). Lewy neurites in CA2 region of the hippocampus in Lewy body dementia (*C*). Neuronal cytoplasmic inclusions in multiple system atrophy (*D*). Glial cytoplasmic inclusions in the pons in multiple system atrophy. Courtesy of Dr. Dennis Dickson, Neuropathology Laboratory, Mayo Clinic Jacksonville, FL, USA.

(Crowther et al., 2000). As with the tauopathies, compelling evidence from molecular genetics and molecular pathology places α-synuclein center stage in the pathogenesis of the diseases defined by α-synuclein deposition. What remains unclear and contentious is which form(s) of α-synuclein is/are toxic. The initial focus was on the aggregated fibrillar variety of this protein; however in recent years the focus has shifted toward the prefibrillar oligomeric species as the pathogenic form. This shift has resulted in part from the observation that although the pathogenic A53T mutation promotes the formation of fibrils, the A30P mutation does not, but instead this mutation pushes the balance toward formation of oligomeric species (Conway et al., 2000). In addition, aggregated α-synuclein positive Lewy bodies exist within neurons, which have survived the disease process, suggesting that the aggregates in Lewy bodies may be protective for these neurons.

α-Synuclein and PD

The deposition of insoluble α-synuclein, in the form of Lewy bodies, and relatively selective degeneration of nigrostriatal dopaminergic neurons are the hallmark features of PD. However, it is clear that the disease extends beyond these features. A landmark publication by Braak and colleagues describes the thorough analysis of brain pathology in sporadic PD at different disease stages and proposes a PD pathology-staging scheme (Braak et al., 2003). The authors suggest that the progression of PD neuropathology occurs in a predictable manner that may be separated into six individual stages [for comprehensive review see (Braak et al., 2004)]. Briefly, this staging occurs as (1) lesions of the dorsal motor nucleus and/or intermediate reticular zone; (2) lesions in the caudal raphe nucleus, gigantocellular reticular nucleus and coeruleus-subcoeruleus complex; (3) midbrain pathology, notably within the substantia nigra pars compacta; (4) prosencephalic lesions; (5) neocortical lesion, particularly high-order sensory association areas; (6) more extensive neocortical involvement, including first-order sensory association areas or the neocortex and premotor areas. It is noteworthy that the early, preclinical stages (1) and (2) occur before significant degeneration of the substantia nigra has occurred.

α-Synuclein and Lewy-body Dementia

DLB is the second most common form of progressive dementia in the elderly, after AD. There are considerable overlaps between DLB and PD, particularly PD where dementia occurs as a later feature (PD dementia; PDD). The clinical distinction between PDD and DLB is one that is made primarily based on the timeline of symptoms. If dementia presents before or no later than a year after the onset of motor symptoms a diagnosis of DLB is made; if dementia is a later feature, the disease is categorized as PDD (McKeith et al., 2005). Neuropathologically, DLB involves nigrostriatal dopaminergic cell loss as occurs in PD, in addition to a more widespread cortical atrophy. Lewy body pathology is found in the brainstem and neocortex, and indeed the presence of Lewy body pathology in these two areas are required neuropathological features of the disease. In addition, DLB cases often exhibit widespread senile plaques and infrequent NFTs (McKeith et al., 2005). The distinction between DLB and PDD is, both clinically and neuropathologically, a difficult one to make. It has been argued that these disorders are distinct entities; however, it appears reasonable to suppose that these two diseases simply represent different presentations of the same underlying molecular disease process and that the presentation is dependent on which cellular systems are more vulnerable in each patient. This viewpoint is supported by the occurrence of either PDD or DLB in families with Mendelian abnormalities of α-synuclein.

α-Synuclein and AD

Given the strong genetic and neuropathological association of α-synuclein with PD and DLB, it is often forgotten that this protein was originally discovered as a constituent of senile plaques. It is clear that in addition to extracellular deposition, α-synuclein is also found as intracellular Lewy bodies in AD, notably patients with presenilin-1 (PS-1), presenilin-2 (PS-2), or APP mutations, which cause a young-onset autosomal dominant form of AD, often exhibiting Lewy body pathology (Lantos et al., 1992; Lantos et al., 1994; Lippa et al., 1998). Lewy body pathology in familial AD cases is prominent in the amygdala but also found in cortex and brainstem. Furthermore neuropathological examination of patients with Down syndrome, who develop Alzheimer-like pathology and age- associated dementia, also demonstrates the presence of α-synuclein positive Lewy bodies, again most notably within the amygdala (Lippa et al., 1999). The common link between these forms of AD is that they are all believed to be driven by the overproduction of amyloid, particularly the amyloidogenic Aβ-42 form, which is processed from the APP. In the case of APP and PS mutations, this is thought to be caused by a shift in the processing of APP, either to produce more Aβ, or to produce proportionately more Aβ-42. In Down syndrome, the increase in Aβ is simply thought to be a function of an increased genetic load of APP, which is encoded by the APP gene on chromosome 21. This is believed to be a mechanism that is common to all forms of AD, familial and sporadic (the Amyloid cascade hypothesis (Hardy and Higgins, 1992). The advent of α-synuclein immunohistochemistry has revealed α-synuclein positive neuropathology as a common feature of sporadic typical AD, present in 30–50% of cases examined (Parkkinen et al., 2003; Mikolaenko et al., 2005). These data collectively suggest that in AD, α-synuclein accumulation and deposition is an event downstream of aberrant APP production.

α-Synuclein Mutation and Disease

The majority of PD cases are apparently sporadic, with little family history of the disease. However, traditional molecular genetic techniques take advantage of rare familial forms of disease in order to identify genes and proteins central to the disease process. The basic tenet of this approach is that understanding the etiology will reveal molecular pathways amenable to therapeutic intervention. The primary underlying assumption of this work is that familial and sporadic diseases, which overlap clinically and/or pathologically, also share a molecular etiology. Although this is a reasonable assumption for which AD research provides proof of principle, it should still be recognized as a potential flaw when interpreting experiments based on these findings.

In 1996, the first genetic linkage for PD was described in a large Italian–American family with autosomal dominant disease. The disease-causing lesion in this family was assigned to the long arm of chromosome 4 (Polymeropoulas et al., 1996). A year later, Polymeropolous and colleagues reported the identification of a disease segregating mutation in the same kindred and in three additional families originating from Greece (Table 18–2) (Polymeropoulas et al., 1997). The mutation identified altered amino acid 53 (A53T) within the protein α-synuclein. Two additional point mutations were subsequently identified in this gene, both causing an autosomal dominant Lewy body disorders. The first, A30P, was identified in a German family in 1998 (Kruger et al., 1998), the second identified in 2004 described an E46K mutation segregating with disease in a Spanish family (Zarranz et al., 2004).

Table 18–2 Overview of α-Synuclein Mutations

No.	α-Synuclein Mutation	Phenotype	Age at Onset	References
1	A30P	Parkinson disease (PD) with impairment in neuro-psychological tests	60	Kruger et al., 1998 9462735
2	E46K	PD and Lewy body dementia	60	Zarranz et al., 2004 14755719
3	A53T	PD	46	Polymeropoulos et al., 1997 9197268
4	Gene duplication	PD	48	Chartier-Harlin et al., 2004 15451224 Ibanez et al., 2004 15451225
5	Gene triplication	PD and PD dementia	34	Singleton et al., 2003 14593171

In 2003, a gene dosage mutation altering the genetic load of *SNCA* was described in a large North American family with autosomal dominant Lewy body disease (known as the Iowa kindred, or Spellman–Muenter, Waters–Miller kindred and Muenter et al, 1998) (Singleton et al., 2003). The pathogenic lesion in this family was a genomic triplication of a region on the long arm of chromosome 4, approximately 1.6 million bp in size. The region includes the α-synuclein gene, in addition to 17 annotated or putative genes. In this kindred, the cause of disease is quite simple—the affected members carry four copies of *SNCA* rather than the normal two. When the blood of mutation-carrying patients is assessed, they show a doubling in the level of α-synuclein protein compared to mutation-free relatives (Miller et al., 2004). Subsequent to the discovery of the α-synuclein triplication mutation, additional families have been reported with similar multiplication mutations, these include families with disease caused both by genomic triplication and by genomic duplication of the *SNCA* locus (Chartier-Halin et al., 2004; Farrer et al., 2004; Ibanez et al., 2004; Nishioka et al., 2006).

The vague clinical and neuropathological boundary between PD (particularly PDD) and DLB is seldom more apparent than when looking at genetic forms of synucleinopathies. This is particularly highlighted by the paper describing the discovery of the E46K mutation of α-synuclein, entitled "The new mutation, E46K, of α-synuclein causes Parkinson and Lewy body dementia" (Zarranz et al., 2004), which clearly highlights the observation that a single disease causing mutation, and presumably a single predominant etiology may cause disparate clinical and pathological endpoints, albeit with a common protein deposit. Examination of the clinical and pathological phenotype of disease associated with other missense and dose mutations of the α-synuclein gene further defines this overlap. Disease caused by missense mutation of the α-synuclein gene results in a young-onset disease that usually presents with parkinsonism, often progressing to include dementia. Neuropathologically, the disease in these cases is reminiscent of DLB. The clinical phenotype noted in the Iowan kindred, harboring the α-synuclein triplication, is of a disease with an average age at onset in the mid-30s with a disease course of approximately 10 years. The patients typically present with PD progressing to include dementia as a later feature (Berg et al., 2005). Neuropathological examination of these patients' brains reveals widespread pathology, which is above and beyond the typical neuropathological changes observed in PD. The pathology includes extensive deposition of α-synuclein, both as typical Lewy body and Lewy neurite pathology in the brainstem and neocortex, and also as glial pathology in the white matter reminiscent of GCIs observed in MSA (Berg et al., 2005). This phenotype contrasts with a milder presentation and course observed in patients who carry the duplication mutation of α-synuclein. *SNCA* duplication patients present in the mid-40s and generally do not exhibit demonstrable cognitive decline or dementia (Chartier-Harlin et al., 2004; Ibanez et al., 2004), although a family from Japan has recently been described where dementia was a later feature (Nishioka et al., 2006). The discovery that α-synuclein is a major component of Lewy bodies in sporadic disease came within a month of the publication of mutations within *SNCA* as a cause of PD (Spillantini et al., 1997). Not only did this finding show that identification of genes that cause rare familial forms of disease is relevant to the more common sporadic disorder, but it also placed α-synuclein as the first unequivocal and by primacy a central player in the molecular etiology of Lewy body diseases. Thus despite the relative rarity of α-synuclein mutations as a cause of Lewy body disease (El-Agnaf et al., 1998; Ho et al., 1998; Vaughan, Durr, et al., 1998; Vaughan, Farrer, et al., 1998; Wang et al., 1998; Hope et al., 2004; Johnson et al., 2004; Berg et al., 2005), these discoveries have been incredibly informative within the field of PD research and have driven much of the molecular genetic investigation of this disease. Given that the mutations originally identified were missense, this work initially focused on how qualitative changes in α-synuclein altered the proteins effects on various disease-associated pathways (for example cell death, toxicity, aggregation, protein clearance). The discovery of the *SNCA* triplication mutation and description of additional multiplication mutations expanded this research to include investigation of how quantitative changes in α-synuclein may affect these processes.

α-Synuclein and Genetic Susceptibility to PD

As noted above with *MAPT*, common genetic variability within a monogenic disease gene is an excellent candidate mechanism for predisposition to sporadic disease. Given that α-synuclein is the major deposited species in Lewy body diseases and that mutations in *SNCA* cause familial PD/DLB, polymorphism of this gene is a strong candidate as a modulator of nonfamilial disease. In this context, the most commonly assessed variant within *SNCA* is an imperfect dinucleotide repeat within the promoter region, approximately 10,000 bp upstream of the translation start site. Variability within this repeat was first implicated as a risk factor for PD, in combination with certain *APOE* genotypes, in 1999 (Kruger et al., 1999). The variant assessed here, called REP1, has to date been analyzed in several populations with conflicting results (Parsian et al., 1998; Kruger et al., 1999; Farrer, Maraganore, et al., 2001;

Khan et al., 2001; Spadafora et al., 2003; Tan et al., 2004; Mellick et al., 2005; Hadjigeorgiou et al., 2006). A meta-analysis of the majority of these studies suggests that the 0 allele of REP1 is protective against PD, exerting an effect with an OR of approximately 0.8 (Mellick et al., 2005), and this is wholly consistent with subsequent association data (Hadjigeorgiou et al., 2006). The functional consequence of this polymorphism has been studied in vitro, and these data suggest that the shorter 0 allele drives a lower level of α-synuclein expression than the longer alleles (Chiba-Falek and Nussbaum, 2001), and this effect appears to be driven primarily by the length of the repeat and not by the exact nucleotide sequence in each REP allele (Chiba-Falek et al., 2003). Although these data are not unequivocal, they are consistent, and the notion that decreased α-synuclein expression is protective against disease is biologically plausible in a disorder of protein deposition.

More detailed analyses of variability in *SNCA* have been performed. These typically involve assay of several polymorphisms across the 140 kb gene, in an attempt to define a risk haplotype that harbors a susceptibility allele (Farrer, Maraganore, et al., 2001; Mueller et al., 2005). These data suggest that although variability of the *SNCA* REP1 allele may modulate disease, non-coding polymorphisms in other areas of the gene also confer risk for PD, independent of this effect. It is unclear what the pathogenic role of this variability is, but it presumably affects splicing and/or expression.

α-Synuclein, Dose, and Disease

The discovery of α-synuclein triplication mutations showed that doubling the amount of genomic α-synuclein can cause a severe early onset Lewy body disease. The observation of a similar but less severe and later onset disease in patients with α-synuclein duplications suggests that the relationship between α-synuclein levels and clinical disease is dose dependent (Table 18–2) (Singleton and GwinnHardy, 2004). The *SNCA* REP1 genetic association data are consistent with the idea that increased α-synuclein expression may contribute to a lifetime risk of the disease; one can certainly imagine that if a doubling of α-synuclein leads to disease in the thirties and a 50% increase causes disease in the 40s, and relatively modest increases in α-synuclein expression would result in disease in later life. As previously discussed, the idea that increased expression of a neurodegenerative disease gene modulates the disease process makes biologic sense (Singleton, Myers, et al., 2004). Parallel mechanisms exist in AD, where mutations that cause disease result in increased processing and production of the deposited species, and in the situation that closely resembles *SNCA* multiplication mutation, patients with Down syndrome, who possess three copies of the APP gene, exhibit Alzheimer pathology. Recently families have been identified with APP multiplications with early onset AD which exactly parallels the situation in PD (Rovelet-Lecrux et al., 2006). These data also suggest that not only will expression of α-synuclein modulate the onset and progression of Lewy body disease but also that clearance of this protein may attenuate the disease process; furthermore, these data suggest that modulation of these processes is a viable therapeutic goal. A major goal in the investigation of sporadic synucleinopathies is to evaluate the role of increased disease gene expression in nonfamilial disease.

Other Gene Mutations that Cause Synucleinopathies

Mutation of genes encoding α-synuclein, APP, and the presenilins may lead to deposition of α-synuclein in the form of Lewy body pathology. In the previous 6 years, mutations in four additional genes encoding parkin (*PRKN*), DJ-1 (*DJ1*), Pten-induced kinase (*PINK1*), and dardarin (*LRRK2*) have been linked to familial forms of parkinsonism. Little is currently known about the neuropathology of *PINK1* and *DJ1*-linked disease; however there are several reports describing neuronal degeneration and protein deposition in the brains of patients with parkinsonism caused by *PRKN* or *LRRK2* mutation. *PRKN* mutations were the first identified cause of autosomal recessive young-onset parkinsonism (Kitada et al., 1998). Parkin is an E3-ubiquitin-ligase, involved in the ubiquitin pathway which marks proteins for destruction in the proteasome (Shimura et al., 2000). *PRKN* mutations thus far linked with disease include loss of function mutations, suggesting that a failure to clear proteins may play a role in the disease process. Initially the neuropathology described in parkin-linked cases of juvenile parkinsonism were notable for the lack of α-synuclein immunoreactive pathology in the presence of nigrostriatal dopaminergic neuronal loss (Kitada et al., 1998; Mori et al., 1998; Hayashi et al., 2000; van de Warrenburg et al., 2001). This observation furthered the argument that familial forms of a disease may be etiologically unrelated to sporadic equivalents. This of course is a key issue, and millions of dollars and research resources are spent creating molecular models of parkinsonism using mutations in genes such as *PRKN* with the assumption that these mechanisms will overlap with typical PD. Researchers using these models have argued that at the very least the two diseases share the selective (or preferential) vulnerability of dopaminergic neurons of the substantia nigra. Confirmation of the link between parkin disease and typical PD has been brought by reports from several groups indicating that α-synuclein positive pathology, including Lewy body pathology, may indeed be observed in some *PRKN*-linked cases (Farrer, Chan, et al., 2001; Sasaki et al., 2004; Pramstaller et al., 2005). A similar situation exists with disease caused by mutation of *LRRK2* (leucine-rich repeat kinase 2), the gene most recently identified to contain mutations that cause PD (Paisan-Ruiz et al., 2004; Zimprich et al., 2004). *LRRK2* encodes the protein dardarin, a 2527 amino acid protein of unknown function. Mutation of *LRRK2* is the most common genetic cause of PD identified to date. In the Caucasian population, a single mutation within the kinase domain of dardarin accounts for 1–2% of all apparently sporadic cases of PD and approximately 5% of all familial cases of PD (Di Fonzo et al., 2005; Gilks et al., 2005; Nichols et al., 2005).

The occurrence of the mutation in PD patients without a family history further validates the Mendelian gene approach to neurodegeneration. Although mutations in this gene have only recently been identified, there are a considerable number of reports describing neuropathological changes associated with this form of disease. The original linkage between disease and the genomic region containing *LRRK2* was described in a Japanese family called the Sagamihara kindred. The disease in this family, which was inherited in an autosomal dominant mode, had an onset in the mid-50s. The neuropathology was notable for the absence of α-synuclein positive inclusions, the affected family members presenting with a pure nigral degeneration (Funayama et al., 2002). Subsequent reports have described a range of pathologies associated with *LRRK2* mutation, all including death and degeneration of the nigrostriatal dopaminergic neurons and also including an array of protein inclusions and aggregates, such as Lewy body pathology typical of PD, Lewy body pathology similar to diffuse Lewy body disease, tau reactive pathology reminiscent of that seen in PSP, and nigral degeneration without distinctive pathology (the latter two in the absence of notable α-synuclein pathology) (Zimprich et al., 2004; Khan et al., 2005). This example of divergent pathologies caused by a single pathogenesis (*LRRK2* mutation) provides a cautionary tale with regard to compartmentalizing diseases etiologically based on the observed pathological endpoint. It is clear that the main driving force behind the onset and progression of *LRRK2* disease is mutation of *LRRK2*; parsimony suggests that the same or an extensively overlapping molecular aetiology occurs in these patients, so why do they show such a strikingly varied pathological endpoint? To simplify this problem, let us just consider that patients with these mutations show pathologies with and without α-synuclein deposition. It is clear that we should not assume that the absence of aggregated α-synuclein indicates that α-synuclein does not play a role in the *LRRK2* disease process. As outlined above, the issue over whether α-synuclein aggregates are causal or protective against disease has not yet been settled, so one can imagine that in the α-synuclein negative cases the vulnerable neurons were unable to sequester α-synuclein into aggregates. Alternatively, it is possible that α-synuclein involvement in *LRRK2* disease, and by extrapolation in the typical disease process, occurs late in the etiology and is not required for, but is often involved in, the degeneration of dopaminergic nigrostriatal neurons. A third option is that α-synuclein is a bystander in the disease process, a protein that happens to be expressed in the substantia nigra and that readily aggregates in neurons under stress. Although this is certainly a possibility, the overwhelming evidence implicating normal and mutant α-synuclein in the disease process makes this unlikely. Regardless, it is hoped that the continued investigation of the role that the protein α-synuclein, dardarin, and parkin play in neurodegenerative disease and particularly the relationship of these proteins with each other will reveal the molecular cascade of events occurring in PD and related disorders.

Conclusions

In the past decade, molecular genetic analysis of neurodegenerative disorders has become a mainstay in the investigation of the underlying etiology of these diseases. With respect to tau and α-synuclein, in both instances these proteins were originally brought to the attention of scientists because they were identified as components of protein deposits in hallmark pathological lesions of neurodegenerative diseases. Mutations in the genes encoding tau and α-synuclein have been identified to be sufficient in themselves to cause progressive degenerative disease, and this has brought the analysis of MAPT and SNCA to the forefront of molecular genetic investigation of a family of neurodegenerative diseases involving parkinsonism and dementia. Creation of antibodies to α-synuclein and tau has revealed that these proteins are constituent parts of protein lesions in an array of neurological disorders, and these data suggest that both proteins are not only involved in familial neurodegeneration when mutated but also in sporadic neurodegenerative diseases. These discoveries have led the way to more accurate clinical diagnosis and genetic counselling in members of affected families. The insights provided by the molecular pathology has led to a new classification of neurodegenerative disease, presumed to be related more closely to the underlying aetiology. Over the next 10 years, the focus will shift to investigation of primary pathogenic mechanisms in patients with sporadic neurodegeneration. The insights into disease mechanism provided by the characterization of disease gene mutations and risk factors will lead to potentially new treatments for patients with neurodegeneration. The next period will see further molecular characterization of these disease pathways and identification of other key proteins involved in this process, with the aim ultimately of identifying a pathogenic pathway or interaction that is amenable to therapeutic intervention.

References

Abeliovich, A, Schmitz, Y, Farinas, I, Choi-Lundberg, D, Ho, WH, Castillo, PE, et al. (2000). Mice lacking α-synuclein display functional deficits in the nigrostriatal dopamine system. *Neuron*, 25(1):239–252.

Andreadis, A, Brown, WM, and Kosik, KS (1992). Structure and novel exons of the human tau gene. *Biochemistry*, 31(43):10626–10633.

Arriagada, PV, Growdon, JH, Hedley-Whyte, ET, and Hyman, BT (1992). Neurofibrillary tangles but not senile plaques parallel duration and severity of Alzheimer's disease. *Neurology*, 42(3 Pt 1):631–639.

Baker, M, Litvan, I, Houlden, H, Adamson, J, Dickson, D, Perez-Tur, J, et al. (1999). Association of an extended haplotype in the tau gene with progressive supranuclear palsy. *Hum Mol Genet*, 8(4):711–715.

Berg, D, Niwar, M, Maass, S, Zimprich, A, Moller, JC, Wuellner, U, et al. (2005). α-Synuclein and Parkinson's disease: implications from the screening of more than 1900 patients. *Mov Disord*, 20(9):1191–1194.

Braak, H, Del Tredici, K, Rub, U, de Vos, RA, Jansen Steur, EN, and Braak, E (2003). Staging of brain pathology related to sporadic Parkinson's disease. *Neurobiol Aging*, 24(2):197–211.

Braak, H, Ghebremedhin, E, Rub, U, Bratzke, H, and Del Tredici, K (2004). Stages in the development of Parkinson's disease-related pathology. *Cell Tissue Res*, 318(1):121–134.

Bugiani, O, Murrell, JR, Giaccone, G, Hasegawa, M, Ghigo, G, Tabaton, M, et al. (1999). Frontotemporal dementia and corticobasal degeneration in a family with a P301S mutation in tau. *J Neuropathol Exp Neurol*, 58 (6):667–677.

Caceres, A, Potrebic, S, and Kosik, KS (1991). The effect of tau antisense oligonucleotides on neurite formation of cultured cerebellar macroneurons. *J Neurosci*, 11(6):1515–1523.

Campion, D, Martin, C, Heilig, R, Charbonnier, F, Moreau, V, Flaman, JM, et al. (1995). The NACP/synuclein gene: chromosomal assignment and screening for alterations in Alzheimer disease. *Genomics*, 26(2):254–257.

Chambers, CB, Lee, JM, Troncoso, JC, Reich, S, and Muma, NA (1999). Overexpression of four-repeat tau mRNA isoforms in progressive supranuclear palsy but not in Alzheimer's disease. *Ann Neurol*, 46(3):325–332.

Chandra, S, Fornai, F, Kwon, HB, Yazdani, U, Atasoy, D, Liu, X, et al. (2004). Double-knockout mice for α- and β-synucleins: effect on synaptic functions. *Proc Natl Acad Sci USA*, 101(41):14966–14971.

Chartier-Harlin, MC, Kachergus, J, Roumier, C, Mouroux, V, Douay, X, Lincoln, S, et al. (2004). α-Synuclein locus duplication as a cause of familial Parkinson's disease. *Lancet*, 364(9440):1167–1169.

Chen, L and Feany, MB (2005). α-Synuclein phosphorylation controls neurotoxicity and inclusion formation in a Drosophila model of Parkinson disease. *Nat Neurosci*, 8(5):657–663.

Chen, X, de Silva, HA, Pettenati, MJ, Rao, PN, St George-Hyslop, P, Roses, AD, et al. (1995). The human NACP/α-synuclein gene: chromosome assignment to 4q21.3-q22 and TaqI RFLP analysis. *Genomics*, 26(2):425–427.

Chiba-Falek, O and Nussbaum, RL (2001). Effect of allelic variation at the NACP-Rep1 repeat upstream of the α-synuclein gene (SNCA) on transcription in a cell culture luciferase reporter system. *Hum Mol Genet*, 10 (26):3101–3109.

Chiba-Falek, O, Touchman, JW, and Nussbaum, RL (2003). Functional analysis of intra-allelic variation at NACP-Rep1 in the α-synuclein gene. *Hum Genet*, 113(5):426–431.

Chin, SS and Goldman, JE (1996). Glial inclusions in CNS degenerative diseases. *J Neuropathol Exp Neurol*, 55(5):499–508.

Clark, LN, Poorkaj, P, Wszolek, Z, Geschwind, DH, Nasreddine, ZS, Miller, B, et al. (1998). Pathogenic implications of mutations in the tau gene in pallido-ponto-nigral degeneration and related eurodegenerative disorders linked to chromosome 17. *Proc Natl Acad Sci USA*, 95(22):13103–13107.

Conrad, C, Amano, N, Andreadis, A, Xia, Y, Namekataf, K, Oyama, F, et al. (1998). Differences in a dinucleotide repeat polymorphism in the tau gene between Caucasian and Japanese populations: implication for progressive supranuclear palsy. *Neurosci Lett*, 250(2):135–137.

Conrad, C, Andreadis, A, Trojanowski, JQ, Dickson, DW, Kang, D, Chen, X, et al. (1997). Genetic evidence for the involvement of tau in progressive supranuclear palsy. *Ann Neurol*, 41(2):277–281.

Conway, KA, Lee, SJ, Rochet, JC, Ding, TT, Harper, JD, Williamson, RE, et al. (2000). Accelerated oligomerization by Parkinson's disease linked α-synuclein mutants. *Ann NY Acad Sci*, 920:42–45.

Corsellis, JA, Bruton, CJ, and Freeman-Browne, D (1973). The aftermath of boxing. *Psychol Med*, 3(3):270–303.

Crowther, RA, Daniel, SE, and Goedert, M (2000). Characterisation of isolated α-synuclein filaments from substantia nigra of Parkinson's disease brain. *Neurosci Lett*, 292(2):128–130.

Delacourte, A, David, JP, Sergeant, N, Buée, L, Wattez, A, Vermersch, P, et al. (1999). The biochemical pathway of neurofibrillary degeneration in aging and Alzheimer's disease. *Neurology*, 52(6):1158–1165.

Delacourte, A, Robitaille, Y, Sergeant, N, Buee, L, Hof, PR, Wattez, A, et al. (1996). Specific pathological Tau protein variants characterize Pick's disease. *J Neuropathol Exp Neurol*, 55(2):159–168.

de Silva, R, Hope, A, Pittman, A, Weale, ME, Morris, HR, Wood, NW, et al. (2003). Strong association of the Saitohin gene Q7 variant with progressive supranuclear palsy. *Neurology*, 61(3):407–409.

de Silva, R, Lashley, T, Gibb, G, Hanger, D, Hope, A, Reid, A, et al. (2003). Pathological inclusion bodies in tauopathies contain distinct complements of tau with three or four microtubule-binding repeat domains as demonstrated by new specific monoclonal antibodies. *Neuropathol Appl Neurobiol*, 29(3):288–302.

de Silva, R, Weiler, M, Morris, HR, Martin, ER, Wood, NW, and Lees, AJ (2001). Strong association of a novel Tau promoter haplotype in progressive supranuclear palsy. *Neurosci Lett*, 311(3):145–148.

Dickson, DW (1997). Neurodegenerative diseases with cytoskeletal pathology: a biochemical classification. *Ann Neurol*, 42(4):541–544.

Dickson, DW (1998). Pick's disease: a modern approach. *Brain Pathol*, 8 (2):339–354.

Dickson, DW, Lin, W, Liu, WK, and Yen, SH (1999). Multiple system atrophy: a sporadic synucleinopathy. *Brain Pathol*, 9(4):721–732.

Di Fonzo, A, Rohe, CF, Ferreira, J, Chien, HF, Vacca, L, Stocchi, F, et al. (2005). A frequent LRRK2 gene mutation associated with autosomal dominant Parkinson's disease. *Lancet*, 365(9457):412–415.

Di Maria, E, Tabaton, M, Vigo, T, Abbruzzese, G, Bellone, E, Donati, C, et al. (2000). Corticobasal degeneration shares a common genetic background with progressive supranuclear palsy. *Ann Neurol*, 47(3):374–377.

Drubin, DG, Feinstein, SC, Shooter, EM, and Kirschner, MW (1985). Nerve growth factor-induced neurite outgrowth in PC12 cells involves the coordinate induction of microtubule assembly and assembly-promoting factors. *J Cell Biol*, 101(5 Pt 1):1799–1807.

Drubin, DG, and Kirschner, MW (1986). Tau protein function in living cells. *J Cell Biol*, 103(6 Pt 2):2739–2746.

D'Souza, I, Poorkaj, P, Hong, M, Nochlin, D, Lee, VM, Bird, TD, et al. (1999). Missense and silent tau gene mutations cause frontotemporal dementia with parkinsonism-chromosome 17 type, by affecting multiple alternative RNA splicing regulatory elements. *Proc Natl Acad Sci USA*, 96 (10):5598–5603.

Ebneth, A, Godemann, R, Stamer, K, Illenberger, S, Trinczek, B, and Mandelkow, E (1998). Overexpression of tau protein inhibits kinesin-dependent trafficking of vesicles, mitochondria, and endoplasmic reticulum: implications for Alzheimer's disease. *J Cell Biol*, 143(3):777–794.

El-Agnaf, OM, Curran, MD, Wallace, A, Middleton, D, Murgatroyd, C, Curtis, A, et al. (1998). Mutation screening in exons 3 and 4 of α-synuclein in sporadic Parkinson's and sporadic and familial dementia with Lewy bodies cases. *Neuroreport*, 9(17):3925–3927.

Evans, W, Fung, HC, Steele, J, Eerola, J, Tienari, P, Pittman, A, et al. (2004). The tau H2 haplotype is almost exclusively Caucasian in origin. *Neurosci Lett*, 369(3):183–185.

Farrer, M, Chan, P, Chen, R, Tan, L, Lincoln, S, Hernandez, D, et al. (2001). Lewy bodies and parkinsonism in families with parkin mutations. *Ann Neurol*, 50(3):293–300.

Farrer, M, Kachergus, J, Forno, L, Lincoln, S, Wang, DS, Hulihan, M, et al. (2004). Comparison of kindreds with parkinsonism and α-synuclein genomic multiplications. *Ann Neurol*, 55(2):174–179.

Farrer, M, Maraganore, DM, Lockhart, P, Singleton, A, Lesnick, TG, de Andrade, M, et al. (2001) α-Synuclein gene haplotypes are associated with Parkinson's disease. *Hum Mol Genet*, 10(17):1847–1851.

Feany, MB and Dickson, DW (1996). Neurodegenerative disorders with extensive tau pathology: a comparative study and review. *Ann Neurol*, 40 (2):139–148.

Foster, NL, Wilhelmsen, K, Sima, AA, Jones, MZ, D'Amato, CJ, and Gilman, S (1997). Frontotemporal dementia and parkinsonism linked to chromosome 17: a consensus conference. Conference Participants. *Ann Neurol*, 41 (6):706–715.

Fujiwara, H, Hasegawa, M, Dohmae, N, Kawashima, A, Masliah, E, Goldberg, MS, et al. (2002). α-Synuclein is phosphorylated in synucleinopathy lesions. *Nat Cell Biol*, 4(2):160–164.

Funayama, M, Hasegawa, K, Kowa, H, Saito, M, Tsuji, S, and Obata, F (2002). A new locus for Parkinson's disease (PARK8) maps to chromosome 12p11.2-q13.1. *Ann Neurol*, 51(3):296–301.

Geddes, JF, Hughes, AJ, Lees, AJ, and Daniel, SE (1993). Pathological overlap in cases of parkinsonism associated with neurofibrillary tangles. A study of recent cases of postencephalitic parkinsonism and comparison with progressive supranuclear palsy and Guamanian parkinsonism-dementia complex. *Brain*, 116(Pt 1):281–302.

George, JM, Jin, H, Woods, WS, and Clayton, DF (1995). Characterization of a novel protein regulated during the critical period for song learning in the zebra finch. *Neuron*, 15(2):361–372.

Gibb, WR, Luthert, PJ, and Marsden, CD (1989). Corticobasal degeneration. *Brain*, 112(Pt 5):1171–1192.

Gilks, WP, Abou-Sleiman, PM, Gandhi, S, Jain, S, Singleton, A, Lees, AJ, et al. (2005). A common LRRK2 mutation in idiopathic Parkinson's disease. *Lancet*, 365(9457):415–416.

Goedert, M (1993). Tau protein and the neurofibrillary pathology of Alzheimer's disease. *Trends Neurosci*, 16(11):460–465.

Goedert, M (1996). Tau Protein and the neurofibrillary pathology of Alzheimer's disease. *Ann NY Acad Sci*, 777:121–131.

Goedert, M (2004). Tau protein and neurodegeneration. *Semin Cell Dev Biol*, 15(1):45–49.

Goedert, M, Spillantini, MG, Cairns, NJ and Crowther, RA (1992). Tau proteins of Alzheimer paired helical filaments: abnormal phosphorylation of all six brain isoforms. *Neuron*, 8(1):159–168.

Goedert, M, Spillantini, MG, Crowther, RA, Chen, SG, Parchi, P, Tabaton, M, et al. (1999). Tau gene mutation in familial progressive subcortical gliosis. *Nat Med*, 5(4):454–457.

Goode, BL, Chau, M, Denis, PE, and Feinstein, SC (2000). Structural and functional differences between 3-repeat and 4-repeat tau isoforms. Implications for normal tau function and the onset of neurodegenetative disease. *J Biol Chem*, 275(49):38182–38189.

Grover, A, England, E, Baker, M, Sahara, N, Adamson, J, Granger, B, et al. (2003). A novel tau mutation in exon 9 (1260V) causes a four-repeat tauopathy. *Exp Neurol*, 184(1):131–140.

Grover, A, Houlden, H, Baker, M, Adamson, J, Lewis, J, Prihar, G, et al. (1999). 5′ splice site mutations in tau associated with the inherited dementia FTDP-17 affect a stem–loop structure that regulates alternative splicing of exon 10. *J Biol Chem*, 274(21):15134–15143.

Hadjigeorgiou, GM, Xiromerisiou, G, Gourbali, V, Aggelakis, K, Scarmeas, N, Papadimitriou, A, et al. (2006). Association of α-synuclein Rep1 polymorphism and Parkinson's disease: Influence of Rep1 on age at onset. *Mov Disord*, 21(4):534–539.

Hardy, J and Gwinn-Hardy, K (1998). Genetic classification of primary neurodegenerative disease. *Science*, 282(5391):1075–1079.

Hardy, JA and Higgins, GA (1992). Alzheimer's disease: the amyloid cascade hypothesis. *Science*, 256(5054):184–185.

Hardy, J, Pittman, A, Myers, A, Gwinn-Hardy, K, Fung, HC, de Silva, R, et al. (2005). Evidence suggesting that Homo neanderthalensis contributed the H2 MAPT haplotype to Homo sapiens. *Biochem Soc Trans*, 33(Pt 4):582–585.

Hasegawa, M, Fujiwara, H, Nonaka, T, Wakabayashi, K, Takahashi, H, Lee, VM, et al. (2002). Phosphorylated α-synuclein is ubiquitinated in α-synucleinopathy lesions. *J Biol Chem*, 277(50):49071–49076.

Hattori, N, Kitada, T, Matsumine, H, Asakawa, S, Yamamura, Y, Yoshino, H, et al. (1998). Molecular genetic analysis of a novel Parkin gene in Japanese families with autosomal recessive juvenile parkinsonism: evidence for variable homozygous deletions in the Parkin gene in affected individuals. *Ann Neurol*, 44(6):935–941.

Hayashi, S, Toyoshima, Y, Hasegawa, M, Umeda, Y, Wakabayashi, K, Tokiguchi, S, et al. (2002). Late-onset frontotemporal dementia with a novel exon 1 (Arg5His) tau gene mutation. *Ann Neurol*, 51(4):525–530.

Hayashi, S, Wakabayashi, K, Ishikawa, A, Nagai, H, Saito, M, Maruyama, M, et al. (2000). An autopsy case of autosomal-recessive juvenile parkinsonism with a homozygous exon 4 deletion in the parkin gene. *Mov Disord*, 15 (5):884–888.

Heutink, P, Stevens, M, Rizzu, P, Bakker, E, Kros, JM, Tibben, A, et al. (1997). Hereditary frontotemporal dementia is linked to chromosome 17q21–q22: a genetic and clinicopathological study of three Dutch families. *Ann Neurol*, 41(2):150–159.

Ho, SL and Kung, MH (1998). G209A mutation in the α-synuclein gene is rare and not associated with sporadic Parkinson's disease. *Mov Disord*, 13 (6):970–971.

Hoenicka, J, Perez, M, Perez-Tur, J, Barabash, A, Godoy, M, Vidal, L, et al. (1999). The tau gene A0 allele and progressive supranuclear palsy. *Neurology*, 53(6):1219–1225.

Hogg, M, Grujic, ZM, Baker, M, Demirci, S, Guillozet, AL, Sweet, AP, et al. (2003). The L266V tau mutation is associated with frontotemporal dementia and Pick-like 3R and 4R tauopathy. *Acta Neuropathol (Berl)*, 106 (4):323–336.

Hope, AD, Myhre, R, Kachergus, J, Lincoln, S, Bisceglio, G, Hulihan, M, et al. (2004). α-Synuclein missense and multiplication mutations in autosomal dominant Parkinson's disease. *Neurosci Lett*, 367(1):97–100.

Houlden, H, Baker, M, Morris, HR, MacDonald, N, Pickering-Brown, S, Adamson, J, et al. (2001). Corticobasal degeneration and progressive supranuclear palsy share a common tau haplotype. *Neurology*, 56 (12):1702–1706.

Houlden, H, Rizzu, P, Stevens, M, de Knijff, P, van Duijn, CM, van Swieten, JC, et al. (1999). Apolipoprotein E genotype does not affect the age of onset of dementia in families with defined tau mutations. *Neurosci Lett*, 260 (3):193–195.

Hutton, M and Hardy, J (1997). The presenilins and Alzheimer's disease. *Hum Mol Genet*, 6(10):1639–1646.

Hutton, M, Lendon, CL, Rizzu, P, Baker, M, Froelich, S, Houlden, H, et al. (1998). Association of missense and 5′-splice-site mutations in tau with the inherited dementia FTDP-17. *Nature*, 393(6686):702–705.

Ibanez, P, Bonnet, AM, Debarges, B, Lohmann, E, Tison, F, Pollak, P, et al. (2004). Causal relation between α-synuclein gene duplication and familial Parkinson's disease. *Lancet*, 364(9440):1169–1171.

Ingram, EM and Spillantini, MG (2002). Tau gene mutations: dissecting the pathogenesis of FTDP-17. *Trends Mol Med*, 8(12):555–562.

Iseki, E, Matsumura, T, Marui, W, Hino, H, Odawara, T, Sugiyama, N, et al. (2001). Familial frontotemporal dementia and parkinsonism with a novel N296H mutation in exon 10 of the tau gene and a widespread tau accumulation in the glial cells. *Acta Neuropathol (Berl)*, 102(3):285–292.

Johnson, J, Hague, SM, Hanson, M, Gibson, A, Wilson, KE, Evans, EW, et al. (2004). SNCA multiplication is not a common cause of Parkinson disease or dementia with Lewy bodies. *Neurology*, 63(3):554–556.

Juncos, JL, Hirsch, EC, Malessa, S, Duyckaerts, C, Hersh, LB, and Agid, Y (1991). Mesencephalic cholinergic nuclei in progressive supranuclear palsy. *Neurology*, 41(1):25–30.

Kawas, CH and Brookmeyer, R (2001). Aging and the public health effects of dementia. *N Engl J Med*, 344(15):1160–1161.

Kertesz, A, McMonagle, P, Blair, M, Davidson, W, and Munoz, DG (2005). The evolution and pathology of frontotemporal dementia. *Brain*, 128(Pt 9):1996–2005.

Khan, N, Graham, E, Dixon, P, Morris, C, Mander, A, Clayton, D, et al. (2001). Parkinson's disease is not associated with the combined α-synuclein/apolipoprotein E susceptibility genotype. *Ann Neurol*, 49(5):665–668.

Khan, NL, Jain, S, Lynch, JM, Pavese, N, Abou-Sleiman, P, Holton, JL, et al. (2005). Mutations in the gene LRRK2 encoding dardarin (PARK8) cause familial Parkinson's disease: clinical, pathological, olfactory and functional imaging and genetic data. *Brain*, 128(Pt 12):2786–2796.

Kitada, T, Asakawa, S, Hattori, N, Matsumine, H, Yamamura, Y, Minoshima, S, et al. (1998). Mutations in the parkin gene cause autosomal recessive juvenile parkinsonism. *Nature*, 392(6676):605–608.

Kobayashi, K, Kidani, T, Ujike, H, Hayashi, M, Ishihara, T, Miyazu, K, et al. (2003). Another phenotype of frontotemporal dementia and parkinsonism linked to chromosome-17 (FTDP-17) with a missense mutation of S305N closely resembling Pick's disease. *J Neurol*, 250 (8):990–992.

Kruger, R, Kuhn, W, Muller, T, Woitalla, D, Graeber, M, Kosel, S, et al. (1998). Ala30Pro mutation in the gene encoding α-synuclein in Parkinson's disease. *Nat Genet*, 8(2):106–108.

Kruger, R, Vieira-Saecker, AM, Kuhn, W, Berg, D, Muller, T, Kuhnl, N, et al. (1999). Increased susceptibility to sporadic Parkinson's disease by a certain combined α-synuclein/apolipoprotein E genotype. *Ann Neurol*, 45 (5):611–617.

Kwok, JB, Teber, ET, Loy, C, Hallupp, M, Nicholson, G, Mellick, GD, et al. (2004). Tau haplotypes regulate transcription and are associated with Parkinson's disease. *Ann Neurol*, 55(3):329–334.

Lantos, PL, Luthert, PJ, Hanger, D, Anderton, BH, Mullan, M, and Rossor, M (1992). Familial Alzheimer's disease with the amyloid precursor protein position 717 mutation and sporadic Alzheimer's disease have the same cytoskeletal pathology. *Neurosci Lett*, 137(2):221–224.

Lantos, PL, Ovenstone, IM, Johnson, J, Clelland, CA, Roques, P, and Rossor, MN (1974). Lewy bodies in the brain of two members of a family with the 717 (Val to Ile) mutation of the amyloid precursor protein gene. *Neurosci Lett*, 172(1–2):77–79.

Lennox, G, Lowe, J, Morrell, K, Landon, M, and Mayer, RJ (1989). Anti-ubiquitin immunocytochemistry is more sensitive than conventional techniques in the detection of diffuse Lewy body disease. *J Neurol Neurosurg Psychiatry*, 52(1):67–71.

Lippa, CF, Fujiwara, H, Mann, DM, Giasson, B, Baba, M, Schmidt, ML, et al. (1998). Lewy bodies contain altered α-synuclein in brains of many familial Alzheimer's disease patients with mutations in presenilin and amyloid precursor protein genes. *Am J Pathol*, 153(5):1365–1370.

Lippa, CF, Schmidt, ML, Lee, VM, and Trojanowski, JQ (1999). Antibodies to α-synuclein detect Lewy bodies in many Down's syndrome brains with Alzheimer's disease. *Ann Neurol*, 45(3):353–357.

Lippa, CF, Zhukareva, V, Kawarai, T, Uryu, K, Shafiq, M, Nee, LE, et al. (2000). Frontotemporal dementia with novel tau pathology and a Glu342Val tau mutation. *Ann Neurol*, 48(6):850–858.

Liu, WK, Le, TV, Adamson, J, Baker, M, Cookson, N, Hardy, J, et al. (2001). Relationship of the extended tau haplotype to tau biochemistry and neuropathology in progressive supranuclear palsy. *Ann Neurol*, 50(4):494–502.

Lovestone, S and Reynolds, CH (1997). The phosphorylation of tau: a critical stage in neurodevelopment and neurodegenerative processes. *Neuroscience*, 78(2):309–324.

Mailliot, C, Sergeant, N, Bussiere, T, Caillet-Boudin, ML, Delacourte, A, and Buee, L (1998). Phosphorylation of specific sets of tau isoforms reflects different neurofibrillary degeneration processes. *FEBS Lett*, 433 (3):201–204.

McKeith, IG, Dickson, DW, Lowe, J, Emre, M, O'Brien, JT, Feldman, H, et al. (2005). Diagnosis and management of dementia with Lewy bodies: third report of the DLB Consortium. *Neurology*, 65(12):1863–1872.

Mellick, GD, Maraganore, DM, and Silburn, PA (2005). Australian data and meta-analysis lend support for α-synuclein (NACP-Rep1) as a risk factor for Parkinson's disease. *Neurosci Lett*, 375(2):112–116.

Mikolaenko, I, Pletnikova, O, Kawas, CH, O'Brien, R, Resnick, SM, Crain, B, et al. (2005). α-synuclein lesions in normal aging, Parkinson disease, and Alzheimer disease: evidence from the Baltimore Longitudinal Study of Aging (BLSA). *J Neuropathol Exp Neurol*, 64(2):156–162.

Miller, DW, Hague, SM, Clarimon, J, Baptista, M, Gwinn-Hardy, K, Cookson, MR, et al. (2004). α-Synuclein in blood and brain from familial Parkinson disease with SNCA locus triplication. *Neurology*, 62(10):1835–1838.

Miyamoto, K, Kowalska, A, Hasegawa, M, Tabira, T, Takahashi, K, Araki, W, et al. (2001). Familial frontotemporal dementia and parkinsonism with a novel mutation at an intron 10 + 11-splice site in the tau gene. *Ann Neurol*, 50(1):117–120.

Mori, H, Kondo, T, Yokochi, M, Matsumine, H, Nakagawa-Hattori, Y, Miyake, T, et al. (1998). Pathologic and biochemical studies of juvenile parkinsonism linked to chromosome 6q. *Neurology*, 51(3):890–892.

Morris, HR, Baker, M, Yasojima, K, Houlden, H, Khan, MN, Wood, NW, et al. (2002). Analysis of tau haplotypes in Pick's disease. *Neurology*, 59 (3):443–445.

Morris, HR, Janssen, JC, Bandmann, O, Daniel, SE, Rossor, MN, Lees, AJ, et al. (1999). The tau gene A0 polymorphism in progressive supranuclear palsy and related neurodegenerative diseases. *J Neurol Neurosurg Psychiatry*, 66 (5):665–667.

Morris, HR, Lees, AJ, and Wood, NW (1999). Neurofibrillary tangle parkinsonian disorders—tau pathology and tau genetics. *Mov Disord*, 14 (5):731–736.

Morris, HR, Vaughan, JR, Datta, SR, Bandopadhyay, R, Rohan De Silva, HA, Schrag, A, et al. (2000). Multiple system atrophy/progressive supranuclear palsy: α-synuclein, synphilin, tau, and APOE. *Neurology*, 55 (12):1918–1920.

Mueller, JC, Fuchs, J, Hofer, A, Zimprich, A, Lichtner, P, Illig, T, et al. (2005). Multiple regions of α-synuclein are associated with Parkinson's disease. *Ann Neurol*, 57(4):535–541.

Muenter, MD, Forno, LS, Hornykiewicz, O, Kish, SJ, Maraganore, DM, Caselli, RJ, et al. (1998). Hereditary form of parkinsonism—dementia. *Ann Neurol*, 43(6):768–781.

Murrell, JR, Koller, D, Foroud, T, Goedert, M, Spillantini, MG, Edenberg, HJ, et al. (1997). Familial multiple-system tauopathy with presenile dementia is localized to chromosome 17. *Am J Hum Genet*, 61(5):1131–1138.

Murrell, JR, Spillantini, MG, Zolo, P, Guazzelli, M, Smith, MJ, Hasegawa, M, et al. (1999). Tau gene mutation G389R causes a tauopathy with abundant pick body-like inclusions and axonal deposits. *J Neuropathol Exp Neurol*, 58 (12):1207–1226.

Neumann, M, Diekmann, S, Bertsch, U, Vanmassenhove, B, Bogerts, B, and Kretzschmar, HA (2005). Novel G335V mutation in the tau gene associated with early onset familial frontotemporal dementia. *Neurogenetics*, 6 (2):91–95.

Neumann, M, Schulz-Schaeffer, W, Crowther, RA, Smith, MJ, Spillantini, MG, Goedert, M, et al. (2001). Pick's disease associated with the novel Tau gene mutation K369I. *Ann Neurol*, 50(4):503–513.

Neve, RL, Harris, P, Kosik, KS, Kurnit, DM, and Donlon, TA (1986). Identification of cDNA clones for the human microtubule-associated protein tau and chromosomal localization of the genes for tau and microtubule-associated protein 2. *Brain Res*, 387(3):271–280.

Nicholl, DJ, Greenstone, MA, Clarke, CE, Rizzu, P, Crooks, D, Crowe, A, et al. (2003). An English kindred with a novel recessive tauopathy and respiratory failure. *Ann Neurol*, 54(5):682–686.

Nichols, WC, Pankratz, N, Hernandez, D, Paisan-Ruiz, C, Jain, S, Halter, CA, et al. (2005). Genetic screening for a single common LRRK2 mutation in familial Parkinson's disease. *Lancet*, 365(9457):410–412.

Nishioka, K, Hayashi, S, Farrer, MJ, Singleton, AB, Yoshino, H, Imai, H, et al. (2006). Clinical heterogeneity of α-synuclein gene duplication in Parkinson's disease. *Ann Neurol*, 59(2):298–309.

Nonaka, T, Iwatsubo, T, and Hasegawa, M (2005). Ubiquitination of α-synuclein. *Biochemistry*, 44(1):361–368.

Okazaki, H, Lipkin, LE, and Aronson, SM (1961). Diffuse intracytoplasmic ganglionic inclusions (Lewy type) associated with progressive dementia and quadriparesis in flexion. *J Neuropathol Exp Neurol*, 20:237–244.

Okuizumi, K, Takano, H, Seki, K, Onishi, Y, Horikawa, Y, and Tsuji, S (1998). Genetic polymorphism of the tau gene and neurodegenerative diseases with tau pathology among Japanese. *Ann Neurol*, 44(4):707–708.

Oliva, R, Tolosa, E, Ezquerra, M, Molinuevo, JL, Valldeoriola, F, Burguera, J, et al. (1998). Significant changes in the tau A0 and A3 alleles in progressive supranuclear palsy and improved genotyping by silver detection. *Arch Neurol*, 55(8):1122–1124.

Paisan-Ruiz, C, Jain, S, Evans, EW, Gilks, WP, Simon, J, van der Brug, M, et al. (2004). Cloning of the gene containing mutations that cause PARK8-linked Parkinson's disease. *Neuron*, 44(4):595–600.

Parkkinen, L, Soininen, H, and Alafuzoff, I (2003). Regional distribution of α-synuclein pathology in unimpaired aging and Alzheimer disease. *J Neuropathol Exp Neurol*, 62(4):363–367.

Parsian, A, Racette, B, Zhang, ZH, Chakraverty, S, Rundle, M, Goate, A, et al. (1998). Mutation, sequence analysis, and association studies of α-synuclein in Parkinson's disease. *Neurology*, 51(6):1757–1759.

Pastor, P, Ezquerra, M, Perez, JC, Chakraverty, S, Norton, J, Racette, BA, et al. (2004). Novel haplotypes in 17q21 are associated with progressive supranuclear palsy. *Ann Neurol*, 56(2):249–258.

Pastor, P, Ezquerra, M, Tolosa, E, Munoz, E, Marti, MJ, Valldeoriola, F, et al. (2002). Further extension of the H1 haplotype associated with progressive supranuclear palsy. *Mov Disord*, 17(3):550–556.

Pastor, P, Pastor, E, Carnero, C, Vela, R, Garcia, T, Amer, G, et al. (2001). Familial atypical progressive supranuclear palsy associated with homozigosity for the delN296 mutation in the tau gene. *Ann Neurol*, 49 (2):263–267.

Petersen, RB, Tabaton, M, Chen, SG, Monari, L, Richardson, SL, Lynch, T, et al. (1995). Familial progressive subcortical gliosis: presence of prions and linkage to chromosome 17. *Neurology*, 45(6):1062–1067.

Pickering-Brown, S, Baker, M, Bird, T, Trojanowski, J, Lee, V, Morris, H, et al. (2004). Evidence of a founder effect in families with frontotemporal dementia that harbor the tau +16 splice mutation. *Am J Med Genet B Neuropsychiatr Genet*, 125(1):79–82.

Pickering-Brown, SM, Baker, M, Nonaka, T, Ikeda, K, Sharma, S, Mackenzie, J, et al. (2004). Frontotemporal dementia with Pick-type histology associated with Q336R mutation in the tau gene. *Brain*, 127(Pt 6):1415–1426.

Pittman, AM, Myers, AJ, Abou-Sleiman, P, Fung, HC, Kaleem, M, Marlowe, L, et al. (2005). Linkage disequilibrium fine mapping and haplotype association analysis of the tau gene in progressive supranuclear palsy and corticobasal degeneration. *J Med Genet*, 42(11):837–846.

Pittman, AM, Myers, AJ, Duckworth, J, Bryden, L, Hanson, M, Abou-Sleiman, P, et al. (2004). The structure of the tau haplotype in controls and in progressive supranuclear palsy. *Hum Mol Genet*, 13(12):1267–1274.

Polymeropoulos, MH, Higgins, JJ, Golbe, LI, Johnson, WG, Ide, SE, Di Iorio, G, et al. (1996). Mapping of a gene for Parkinson's disease to chromosome 4q21–q23. *Science*, 274(5290):1197–1199.

Polymeropoulos, MH, Lavedan, C, Leroy, E, Ide, SE, Dehejia, A, Dutra, A, et al. (1997). Mutation in the α-synuclein gene identified in families with Parkinson's disease. *Science*, 276(5321):2045–2047.

Poorkaj, P, Bird, TD, Wijsman, E, Nemens, E, Garruto, RM, Anderson, L, et al. (1998). Tau is a candidate gene for chromosome 17 frontotemporal dementia. *Ann Neurol*, 43(6):815–825.

Poorkaj, P, Muma, NA, Zhukareva, V, Cochran, EJ, Shannon, KM, Hurtig, H, et al. (2002). An R5L tau mutation in a subject with a progressive supranuclear palsy phenotype. *Ann Neurol*, 52(4):511–516.

Pramstaller, PP, Schlossmacher, MG, Jacques, TS, Scaravilli, F, Eskelson, C, Pepivani, I, et al. (2005). Lewy body Parkinson's disease in a large pedigree with 77 Parkin mutation carriers. *Ann Neurol*, 58(3):411–422.

Rademakers, R, Cruts, M, and van Broeckhoven, C (2004). The role of tau (MAPT) in frontotemporal dementia and related tauopathies. *Hum Mutat*, 24(4):277–295.

Ratnavalli, E, Brayne, C, Dawson, K, and Hodges JR (2002). The prevalence of frontotemporal dementia. *Neurology*, 58(11):1615–1621.

Revesz, T, Sangha, H, and Daniel, SE (1996). The nucleus raphe interpositus in the Steele-Richardson-Olszewski syndrome (progressive supranuclear palsy). *Brain*, 119(Pt 4):1137–1143.

Rizzini, C, Goedert, M, Hodges, JR, Smith, MJ, Jakes, R, Hills, R, et al. (2000). Tau gene mutation K257T causes a tauopathy similar to Pick's disease. *J Neuropathol Exp Neurol*, 59(11):990–1001.

Rizzu, P, Van Swieten, JC, Joosse, M, Hasegawa, M, Stevens, M, Tibben, A, et al. (1999). High prevalence of mutations in the microtubule-associated protein tau in a population study of frontotemporal dementia in the Netherlands. *Am J Hum Genet*, 64(2):414–421.

Rosso, SM, van Herpen, E, Deelen, W, Kamphorst, W, Severijnen, LA, Willemsen, R, et al. (2002) A novel tau mutation, S320F, causes a tauopathy with inclusions similar to those in Pick's disease. *Ann Neurol*, 51 (3):373–376.

Rovelet-Lecrux, A, Hannequin, D, Raux, G, Meur, NL, Laquerriere, A, Vital, A, et al. (2006). APP locus duplication causes autosomal dominant early-onset Alzheimer disease with cerebral amyloid angiopathy. *Nat Genet*, 38 (1):24–26.

Roy, S, Datta, CK, Hirano, A, Ghatak, NR, and Zimmerman, HM (1974). Electron microscopic study of neurofibrillary tangles in Steele-Richardson-Olszewski syndrome. *Acta Neuropathol (Berl)*, 29(2):175–179.

Russ, C, Powell, JF, Zhao, J, Baker, M, Hutton, M, Crawford, F, et al. (2001). The microtubule associated protein Tau gene and Alzheimer's disease—an association study and meta-analysis. *Neurosci Lett*, 314(1–2):92–96.

Sasaki, S, Shirata, A, Yamane, K, and Iwata, M (2004). Parkin-positive autosomal recessive juvenile Parkinsonism with α-synuclein-positive inclusions. *Neurology*, 63(4):678–682.

Shimura, H, Hattori, N, Kubo, S, Mizuno, Y, Asakawa, S, Minoshima, S, et al. (2000). Familial Parkinson disease gene product, parkin, is a ubiquitin-protein ligase. *Nat Genet*, 25(3):302–305.

Singleton, AB, Farrer, M, Johnson, J, Singleton, A, Hague, S, Kachergus, J, et al. (2003). α-Synuclein locus triplication causes Parkinson's disease. *Science*, 302(5646):841.

Singleton, A and Gwinn-Hardy, K (2004). Parkinson's disease and dementia with Lewy bodies: a difference in dose? *Lancet*, 364(9440):1105–1107.

Singleton, A, Myers, A, and Hardy, J (2004). The law of mass action applied to neurodegenerative disease: a hypothesis concerning the etiology and pathogenesis of complex diseases. *Hum Mol Genet*, 13(Spec. No 1):R123–R126.

Skipper, L and Farrer, M (2002). Parkinson's genetics: molecular insights for the new millennium. *Neurotoxicology*, 23(4–5):503–514.

Souza, JM, Giasson, BI, Chen, Q, Lee, VM, and Ischiropoulos, H (2000). Dityrosine cross-linking promotes formation of stable α-synuclein polymers. Implication of nitrative and oxidative stress in the pathogenesis of neurodegenerative synucleinopathies. *J Biol Chem*, 275(24):18344–18349.

Spadafora, P, Annesi, G, Pasqua, AA, Serra, P, Ciro Candiano, IC, Carrideo, S, et al. (2003). NACP-REP1 polymorphism is not involved in Parkinson's disease: a case–control study in a population sample from southern Italy. *Neurosci Lett*, 351(2):75–78.

Spillantini, MG, Crowther, RA, Jakes, R, Cairns, NJ, Lantos, PL, and Goedert, M (1998). Filamentous α-synuclein inclusions link multiple system atrophy with Parkinson's disease and dementia with Lewy bodies. *Neurosci Lett*, 251(3):205–208.

Spillantini, MG, Crowther, RA, Jakes, R, Hasegawa, M, and Goedert, M (1998). α-Synuclein in filamentous inclusions of Lewy bodies from Parkinson's disease and dementia with Lewy bodies. *Proc Natl Acad Sci USA*, 95 (11):6469–6473.

Spillantini, MG, Crowther, RA, Kamphorst, W, Heutink, P, and van Swieten, JC (1998). Tau pathology in two Dutch families with mutations in the microtubule-binding region of tau. *Am J Pathol*, 153(5):1359–1363.

Spillantini, MG, Divane, A, and Goedert, M (1995). Assignment of human α-synuclein (SNCA) and α-synuclein (SNCB) genes to chromosomes 4q21 and 5q35. *Genomics*, 27(2):379–381.

Spillantini, MG, Murrell, JR, Goedert, M, Farlow, MR, Klug, A, and Ghetti, B (1998). Mutation in the tau gene in familial multiple system tauopathy with presenile dementia. *Proc Natl Acad Sci USA*, 95(13):7737–7741.

Spillantini, MG, Schmidt, ML, Lee, VM, Trojanowski, JQ, Jakes, R, and Goedert, M (1997) α-Synuclein in Lewy bodies. *Nature*, 388(6645):839–840.

Spillantini, MG, Yoshida, H, Rizzini, C, Lantos, PL, Khan, N, Rossor, MN, et al. (2000). A novel tau mutation (N296N) in familial dementia with swollen achromatic neurons and corticobasal inclusion bodies. *Ann Neurol*, 48 (6):939–943.

Stanford, PM, Halliday, GM, Brooks, WS, Kwok, JB, Storey, CE, Creasey, H, et al. (2000). Progressive supranuclear palsy pathology caused by a novel silent mutation in exon 10 of the tau gene: expansion of the disease phenotype caused by tau gene mutations. *Brain*, 123(Pt 5):880–893.

Stanford, PM, Shepherd, CE, Halliday, GM, Brooks, WS, Schofield, PW, Brodaty, H, et al. (2003). Mutations in the tau gene that cause an increase in three repeat tau and frontotemporal dementia. *Brain*, 126(Pt 4):814–826.

Stefansson, H, Helgason, A, Thorleifsson, G, Steinthorsdottir, V, Masson, G, Barnard, J, et al. (2005). A common inversion under selection in Europeans. *Nat Genet*, 37(2):129–137.

Takanashi, M, Mori, H, Arima, K, Mizuno, Y, and Hattori, N (2002). Expression patterns of tau mRNA isoforms correlate with susceptible lesions in progressive supranuclear palsy and corticobasal degeneration. *Brain Res Mol Brain Res*, 104(2):210–219.

Tan, EK, Chai, A, Teo, YY, Zhao, Y, Tan, C, Shen, H, et al. (2004). α-synuclein haplotypes implicated in risk of Parkinson's disease. *Neurology*, 62 (1):128–131.

Togo, T, Sahara, N, Yen, SH, Cookson, N, Ishizawa, T, Hutton, M, et al. (2002). Argyrophilic grain disease is a sporadic 4-repeat tauopathy. *J Neuropathol Exp Neurol*, 61(6):547–556.

Ueda, K, Fukushima, H, Masliah, E, Xia, Y, Iwai, A, Yoshimoto, M, et al. (1993). Molecular cloning of cDNA encoding an unrecognized component of amyloid in Alzheimer disease. *Proc Natl Acad Sci USA*, 90 (23):11282–11286.

Ueda, K, Saitoh, T, and Mori, H (1994). Tissue-dependent alternative splicing of mRNA for NACP, the precursor of non-A (component of Alzheimer's disease amyloid. *Biochem Biophys Res Commun*, 205 (2):1366–1372.

van de Warrenburg, BP, Lammens, M, Lucking, CB, Denefle, P, Wesseling, P, Booij, J, et al. (2001). Clinical and pathologic abnormalities in a family with parkinsonism and parkin gene mutations. *Neurology*, 56 (4):555–557.

van Herpen, E, Rosso, SM, Serverijnen, LA, Yoshida, H, Breedveld, G, van de Graaf, R, et al. (2003). Variable phenotypic expression and extensive tau pathology in two families with the novel tau mutation L315R. *Ann Neurol*, 54(5):573–581.

Vaughan, J, Durr, A, Tassin, J, Bereznai, B, Gasser, T, Bonifati, V, et al. (1998). The α-synuclein Ala53Thr mutation is not a common cause of familial Parkinson's disease: a study of 230 European cases. European Consortium on genetic susceptibility in Parkinson's disease. *Ann Neurol*, 44 (2):270–273.

Vaughan, JR, Farrer, MJ, Wszolek, ZK, Gasser, T, Durr, A, Agid, Y, et al. (1998). Sequencing of the α-synuclein gene in a large series of cases of familial Parkinson's disease fails to reveal any further mutations. The European Consortium on Genetic Susceptibility in Parkinson's Disease (GSPD). *Hum Mol Genet*, 7(4):751–753.

Wang, WW, Khajavi, M, Patel, BJ, Beach, J, Jankovic, J, and Ashizawa, T (1998). The G209A mutation in the α-synuclein gene is not detected in familial cases of Parkinson disease in non-Greek and/or Italian populations. *Arch Neurol*, 55(12):1521–1523.

Wijker, M, Wszolek, ZK, Wolters, EC, Rooimans, MA, Pals, G, Pfeiffer, RF, et al. (1996). Localization of the gene for rapidly progressive autosomal dominant parkinsonism and dementia with pallido-ponto-nigral degeneration to chromosome 17q21. *Hum Mol Genet*, 5(1):151–154.

Wilhelmsen, KC, Lynch, T, Pavlou, E, Higgins, M, and Nygaard, TG (1994). Localization of disinhibition-dementia-parkinsonism-amyotrophy complex to 17q21–22. *Am J Hum Genet*, 55(6):1159–1165.

Wszolek, ZK, Pfeiffer, RF, Tsuboi, Y, Uitti, RJ, McComb, RD, Stoessl, AJ, et al. (2004). Autosomal dominant parkinsonism associated with variable synuclein and tau pathology. *Neurology*, 62(9):1619–1622.

Yasuda, M, Takamatsu, J, D'Souza, I, Crowther, RA, Kawamata, T, Hasegawa, M, et al. (2000). A novel mutation at position +12 in the intron following exon 10 of the tau gene in familial frontotemporal dementia (FTD-Kumamoto). *Ann Neurol*, 47(4):422–429.

Zarranz, JJ, Alegre, J, Gomez-Esteban, JC, Lezcano, E, Ros, R, Ampuero, I, et al. (2004). The new mutation, E46K, of alpha-synuclein causes Parkinson and Lewy body dementia. *Ann Neurol*, 55(2):164–173.

Zarranz, JJ, Ferrer, I, Lezcano, E, Forcadas, MI, Eizaguirre, B, Atares, B, et al. (2005). A novel mutation (K317M) in the MAPT gene causes FTDP and motor neuron disease. *Neurology*, 64(9):1578–1585.

Zimprich, A, Biskup, S, Leitner, P, Lichtner, P, Farrer, M, Lincoln, S, et al. (2004). Mutations in LRRK2 cause autosomal-dominant parkinsonism with pleomorphic pathology. *Neuron*, 44(4):601–607.

19

Neuropsychiatric Diseases I: Schizophrenia

Patrick F Sullivan

Neuropsychiatric disorders are not uncommon and amount to a major global health burden. The field is broad and covers a wide range of distinct clinical entities with overlapping features. Psychiatric disorders are commonly components of the phenotype of several seizure disorders (Chapter 18), progressive neurodegenerative disorders (Chapter 19), and disorders of learning and behavior (Chapter 29). This chapter reviews the present understanding of genetic and genomic basis of schizophrenia (Part I) and mood disorders (Part II), the two most common psychiatric diseases that affect populations across all continents. There have been impressive developments on several fronts, particularly regarding the molecular genetics of these two complex disorders of mind and brain. At the same time, a number of critically important and unresolved issues remain, which qualify the ultimate clinical and scientific validity of the results. Progress in genetic and genomic research is discussed with its limitations and the implications for clinical research and clinical practice.

The public health importance of schizophrenia is clear (Sullivan, 2005). The median lifetime morbid risk of schizophrenia is 0.7% (Saha et al., 2005), with the onset typically ranging from adolescence to early adulthood and a course of illness typified by exacerbations, remissions, and substantial residual symptoms and functional impairment (McGlashan, 1988). Morbidity is substantial, and schizophrenia ranks ninth in the global burden of illness (Murray and Lozpe, 1996). In addition, schizophrenia is often complicated with drug dependence (principally alcohol, nicotine, cannabis, and cocaine) and important medical conditions (obesity, type 2 diabetes mellitus) (Jeste et al., 1996). Mortality due to natural and unnatural causes is considerable, and the projected lifespan for individuals with schizophrenia is approximately 15 years less than the general population (Harris and Barraclough, 1998). The personal, familial, and societal costs of schizophrenia are enormous.

Etiological Clues

A substantial body of epidemiological research has established a set of risk factors for schizophrenia. A subset of this work is summarized in Figure 19–1. Of a large set of pre- and antenatal risk factors (Murray et al., 2003), having a first-degree relative with schizophrenia is associated with an odds ratio of almost 10. The general impact of some of the risk factors in Figure 19–1 remains uncertain, and, in addition, migrant status, urban residence, cannabis use, and sex are supported as risk factors for schizophrenia. Although the attributable risk of some of these risk factors may be greater (e.g., place and season of birth) (Mortensen et al., 1999), the size of the odds ratio for family history suggests that searching for the familial determinants of schizophrenia is rational for etiological research.

Unpacking the Family History Risk Factor

Studies of families, adoptees, and twins have been widely used to attempt understanding the relative contributions of genetic and environmental effects upon risk for schizophrenia. These "old genetics" approaches use phenotypic resemblance of relatives as an indirect means by which to infer the roles of genes and environment. There are many important assumptions and methodological issues with these studies (Plomin et al., 2003); however, genetic epidemiological studies of schizophrenia have yielded a remarkably consistent set of findings, as summarized in Table 19–1 (Korstanje and Paigen, 2002; Sullivan et al., 2003).

To summarize this literature in brief, schizophrenia is familial or "runs" in families. Both adoption and twin studies indicate that the familiality of schizophrenia is due mainly to genetic effects. Twin studies suggest the relevance of small but significant shared environmental influences that are likely prenatal in origin. Thus, schizophrenia is best viewed as a complex trait resulting from both genetic and environmental etiological influences. These results are only broadly informative, as they provide no information about the location of the genes or the identity of the environmental factors that predispose or protect against schizophrenia. Searching for genetic influences that mediate vulnerability to schizophrenia is rational, given the larger overall effect size and lesser error of measurement in comparison with typical assessments of environmental effects. It can be noted that high heritability is no guarantee of success in the efforts to identify candidate genes.

A review of the literature indicates many family studies of schizophrenia involving more than 40,000 family members of affected individuals. These empirical studies showed that approximately 5%–6% of the parents of affected persons with a form of schizophrenia were also affected (Gottesman and Shields, 1982). Brothers and sisters of probands with schizophrenia have approximately a 10% lifetime risk for being affected, which is approximately 10 times

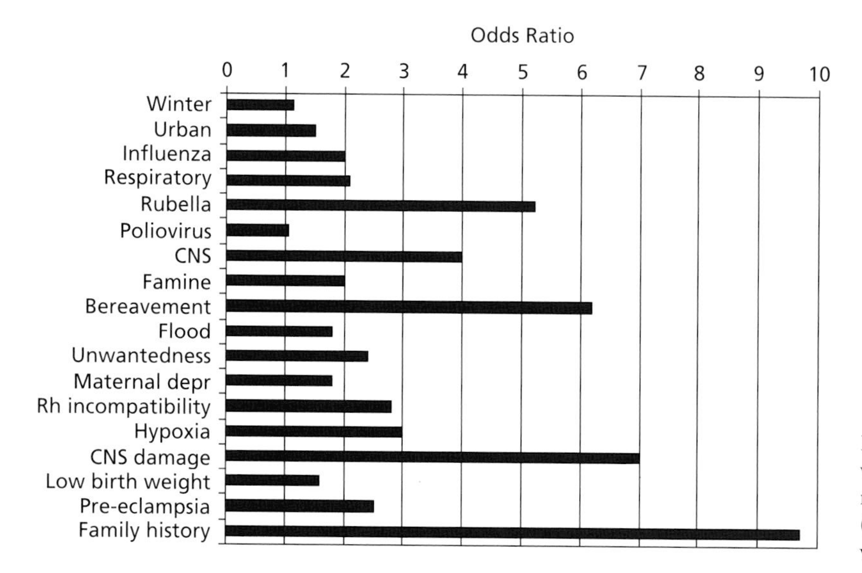

Figure 19–1 Comparison of a selected set of relatively well-established risk factors for schizophrenia, focusing mainly on preconceptual and antenatal factors (Adapted from Murray et al., 2003). CNS, central nervous system; depr, depression; Rh, Rhesus.

higher than the risk to the general population. The risk to children if one parent has schizophrenia is approximately 10–15%. Limited data exist on the risk to offspring if both parents have schizophrenia, but the risk might approach 50%. If a parent and child are affected, the risk to other siblings is approximately 15–20%. Similarly, twin studies indicate high concordance rate of between 60% and 65% for identical (monozygotic, MZ) twins compared with approximately 10–15% for non-identical (dizygotic, DZ) twins. Studies on adopted children of schizophrenic parents report recurrence figures similar to those who were cared for by the biological parents. One study compared children of schizophrenic mothers who were separated by day 3 of life with a control group of children separated from healthy mothers. None of the children in the later group was diagnosed with schizophrenia, while 17% of the former were schizophrenic, and many more had other psychiatric problems. The risk to nephews and nieces of persons with schizophrenia is approximately 2%, about twice the lifetime risk of the general population. On a practical note, the risk is age-dependent, for example, few "at-risk" persons actually manifest after age 40 and almost none after 50.

Genome-Wide Linkage Studies of Schizophrenia

Modern genotyping technologies and statistical analyses have enabled the discovery of genetic loci related to the etiology of many complex traits (Korstanje and Paigen, 2002), such as type 2 diabetes mellitus, obesity, and Alzheimer's disease. These "discovery science" approaches have been applied to schizophrenia, and are summarized in Figure 19–2. The 27 samples shown here included from 1 to 294 multiplex pedigrees (median 34) containing 32–669 (median 101) individuals affected with a narrow definition of schizophrenia. There were 310–950 (median 392) genetic markers in the first-stage genome scans.

"Hard" replication—implication of the same markers, alleles, and haplotypes in the majority of samples—is elusive. It is evident from Figure 19–2 that these studies are inconsistent, and no genomic region was implicated in more than four of the 27 samples. The Lewis et al. (2003) meta-analysis included most of the studies in Figure 19–2, and found that one region on chromosome 2 was

Table 19–1 Summary of Studies on the Genetic Epidemiology of Schizophrenia

Study Type	Conceptual Basis	Studies	Findings
Family	Risk of schizophrenia in first-degree relatives of cases with schizophrenia vs. controls	11	10/11 studies show familiality of schizophrenia
			Significant familial aggregation of schizophrenia:odds ratio: 9.8 (95% CI 6.2–15.5)
Adoption	Risk of schizophrenia in adoption cluster (offspring of one set of parents raised from early in life by unrelated strangers)	5	Effect of postnatal environment negligible
			Adoptees with schizophrenia: increased risk in biological vs. adoptive parents (OR = 5.0 (95% C I 2.4–10.4)
			Parents with schizophrenia:increased risk in biological vs. control offspring 3.5 (95% CI 1.9–6.4)
Twin	Risk of schizophrenia in monozygotic vs. dizygotic twins	12	Heritability in liability of schizophrenia:81% (95% CI 73–90%)
			Environmental effects shared by members of a twin pair: 11% (95% CI 3–19)

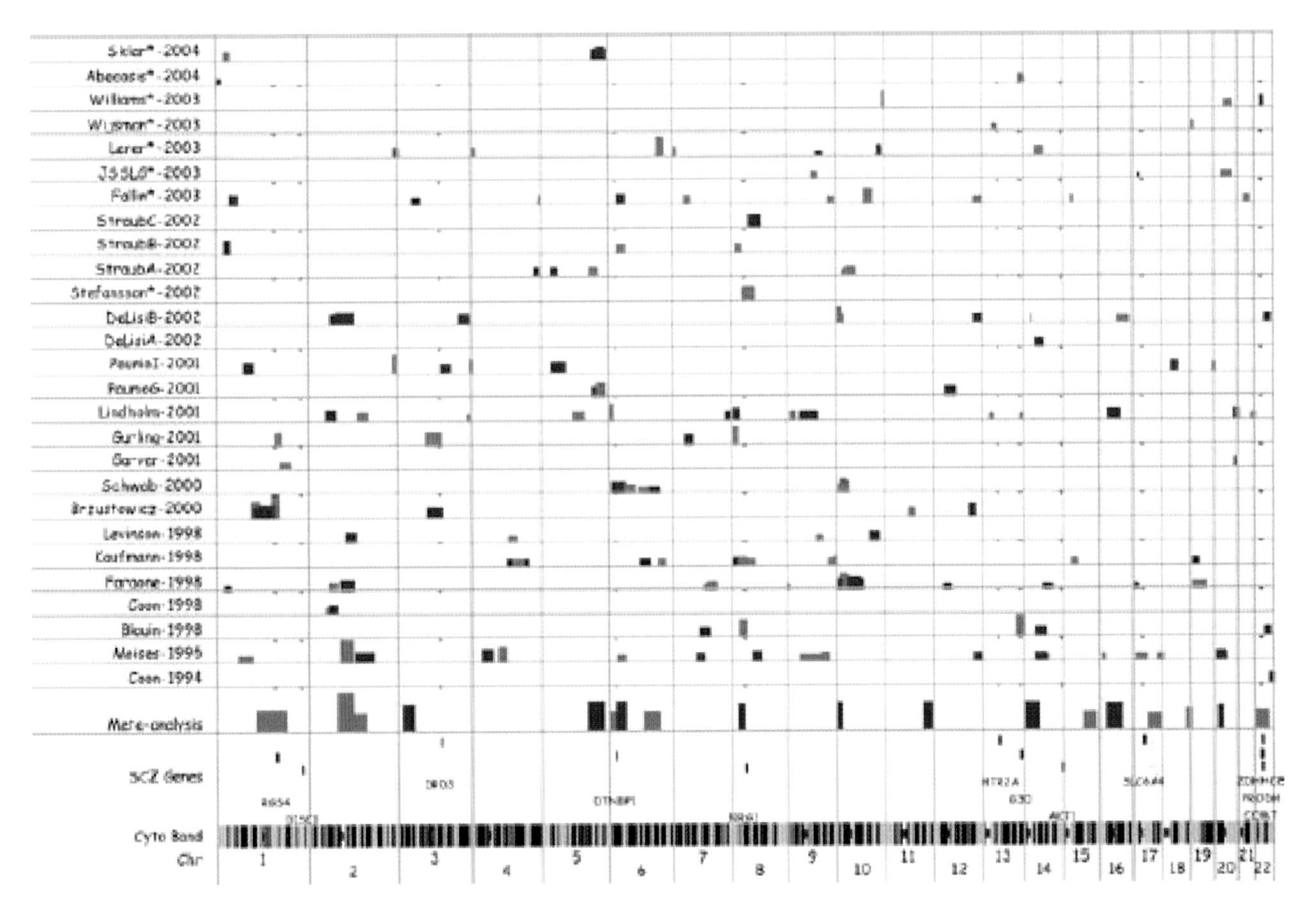

Figure 19–2 Summary of genome-wide linkage studies of schizophrenia. The *x*-axis shows the location on the genome, from the telomere of the short arm of chromosome 1 to the telomere of the long arm of chromosome 22 (bottom row), along with 303 band chromosomal staining on the second-to-bottom row. The *y*-axis shows the 27 primary samples that reported rst-stage genome scans for schizophrenia (i.e., excluding. ne-mapping or partial reports) along with the results of a meta-analysis including most of the primary samples (12) (studies not included are shown with asterisks). Within each row, the height and color of the bars are proportional to the ($\log_{10}$(p-value), and the width of the bar shows the genomic location implicated by a particular sample. A selected set of candidate genes for schizophrenia are also shown. All genomic locations are per the hg16 build (http://genome.ucsc.edu). The physical positions of an inclusive set of the markers showing the best findings in the primary samples were plotted (assuming a confidence interval of ± 10 cM or, if mapping was uncertain, ± 10 megabases; seven markers from the primary samples did not map). [Adapted with permission from Sullivan (2005).]

stringently significant and several additional regions neared significance. Our focus on first-stage genome scans does not adequately capture the evidence supporting replication for certain regions (e.g., 6p) (Moises et al., 1995; Straub et al., 1995; Schizophrenia Linkage Collaborative Group, 1996; Schwab et al., 2000; Lindholm et al., 2001; Maziade et al., 2001;). However, there appears to be "soft" replication across studies.

It is unlikely that all of these linkage findings are true. The regions suggested by the Lewis et al. (2003) meta-analysis implicate more than 3000 genes (18% of all known genes). For the 27 samples in Figure 19–2, the percentages of all known genes implicated by 0, 1, 2, 3, and 4 linkage studies were 42, 35, 14, 6, and 3%, respectively. This crude summation suggests that linkage analysis is an imprecise tool—implausibly large numbers of genes are implicated and few genes are consistently identified in more than a small subset of studies.

There are several potential reasons why clear-cut or "hard" replication was not found. With respect to the teams that conducted these enormously effortful studies, it is possible that no study possessed sufficient statistical power to detect the subtle genetic effects suspected for schizophrenia. For example, it would require 4900 pedigrees to have 80% power to detect a locus accounting for 5% of variance in liability to schizophrenia at $\alpha = 0.001$. These calculations make highly optimistic assumptions, and less favorable assumptions can lead to sample size requirements above 50,000 sibling pairs. For comparison, the total number of pedigrees in Figure 19–2 is less than 2000.

In addition, it is possible that etiological heterogeneity (different combinations of genetic and environmental causes between samples) and technical differences (different ascertainment, assessment, genotyping, and statistical analysis between samples) contributed; however, their impact is uncertain, whereas insufficient power is clear. If correct, the implication is that Figure 19–2 contains a mix of true and false positive findings.

Association Studies of Schizophrenia

Schizophrenia, like most other complex traits in biomedicine, has had a large number of genetic case-control association studies

Table 19–2 Evidence Supporting 12 Potential Candidate Genes for Schizophrenia

Gene[a]	Description	OMIM[b]	Cytogenetic Band	Cytogenetic Abnomalities	Genome Scan Meta-Analysis[c]	Linkage Evidence[d]	Association Study Support[e]	Expression in PFC[f]	Functional Studies: Plausibility?
AKT1	V-AKT murline thymorma viral oncogene homolog 1	164730	14q32.33	No	No	No	2 + and 1 − studies	+ +	Yes
COMT	Catechol-*O*-methyltransferase	116790	22q11.21	Yes	Yes	Yes	Some studies +	+ +	Yes
DISC1	Disrupted in schizophrenia 1	605210	1q42.2	Yes	No	Yes	Multiple studies +	+	Yes
DRD3	Dopamine receptor D3	126451	3q13.31	No	No	Inconsistent	Meta-analysis +	−	Yes
DTNBP1	Dystrobrevin binding protein 1	607415	6p22.3	No	Yes	Yes	Multiple studies +	+ +	Yes
G30/G72	Putative proteins LG30 and G72	607415	13q14.2	No	No	Inconsistent data	Multiple studies +		Insufficient data
HTR2A	Serotonin receptor 2A	182135	13q33.2	No	No	Inconsistent	Meta-analysis +	+ +	Yes
NRG1	Neuregulln 1	142445	8p12	No	Nearby	Yes	Multiple studies +	+	Yes
PRODH	Proline dehydrogenase 1	606810	22q11.21	Yes	Yes	Yes	−	+ +	Yes
RGS4	Regulator of G-protein signaling 4	602516	1q23.3	No	Yes	Yes	Multiple studies +	+ +	Yes
SLC6A4	Serotonin transporter	182138	17q11.2	No	Nearby	Inconsistent	Meta-analysis +	+ +	Yes
ZDHHC8	Zinc finger/DHHC domain protein 8	608784	22q11.21	Yes	Yes	Yes	2 + and 1 − studies	+ +	Yes

[Adapted with permission from Sullivan (2005).]

[a]Standard gene name (http://www.gene.ucl.ac.uk/nomenclature).

[b]Online Mendelian Inheritance in Man (http://www.ncbi.nlm.nih.gov/entrez/query.fcgi?db=OMIM).

[c]Gene lies in a genomic region ("bin") implicated at a suggestive level in the Lewis et al. (2003) meta-analysis.

[d]Evidence here includes studies not found in Figure 19–2 (e.g., fine-mapping studies or studies targeted to a particular region).

[e] +, positive study; −, negative study.

[f]From the Novartis Research Foundation (http://symatlas.gnf.org). +, expression above median over all tissues; + +, expression above 75th percentile.

(Sullivan et al., 2001). Although research practice is changing, interpretation of many studies is hindered by small sample sizes and a tendency to genotype a single genetic marker of the hundreds that might be available in a gene. For example, a widely studied functional genetic marker in *COMT* (rs4680) is probably not associated with schizophrenia (Fan et al., 2005), but nearby genetic markers assessed in a minority of studies may be (Shifman et al., 2002). However, as discussed in the next section, a number of methodologically adequate association studies of schizophrenia appear to support the role of several candidate genes in the etiology of schizophrenia. Similar to the linkage study data, "hard" replication remains elusive.

Synthesis

Despite the limitations of the accumulated linkage and association studies, there are suggestions that these studies have identified plausible candidate genes for schizophrenia. Table 19–2 summarizes the evidence in support of a set of possible candidate genes for schizophrenia. Reports supporting the role of many of these genes have appeared in top-tier international journals known for rigorous peer review. The evidence for several genes is encouraging, but currently insufficient to declare any a clear-cut cause of schizophrenia.

The accumulated data provide particular support for *DISC1*, *DTNBP1*, *NRG1*, and *RGS4*. Each of these genes has received support from multiple lines of evidence with imperfect consistency: (1) the case for each of these as a candidate gene for schizophrenia is supported by linkage studies; (2) the preponderance of association study findings provides further support; (3) mRNA from each gene is expressed in the prefrontal cortex as well as in other areas of the brain; and (4) additional neurobiological data link the functions of these genes to biological processes thought to be related to schizophrenia. For example, *DISC1* modulates neurite outgrowth, there is an extensive literature on the involvement of *NRG1* in the development of the CNS, and *RGS4* may modulate intracellular signaling for many G-protein-coupled receptors. Moreover, *DTNBP1* and *RGS4* have been reported to be differentially expressed in the postmortem brain samples of individuals with schizophrenia.

This encouraging summation of work in progress masks a critical issue—the lack or consistent replication for the same markers and haplotypes across studies. The literature supports the contention that genetic variation in these genes is associated with schizophrenia, but it lacks impressive consistency in the precise genetic regions and alleles implicated. In contrast, association studies of other complex human genetic diseases have produced unambiguous, consistent, and clear-cut ("hard") replication.

For example, (1) in type 1 diabetes mellitus, the bulk of both the linkage and association data implicate the *HLA* region and *INS* (Florez et al., 2003); (2) for type 2 diabetes mellitus, there are a number of findings in the literature where the association evidence appears to be consistent and compelling (*CAPN10*, *KCNJ11*, and *PPARG*)—the data indicate that the same marker allele is significantly associated and has an effect size of similar direction and magnitude (Florez et al., 2003) (the linkage data are less congruent, probably due to power considerations); and (3) for age-related macular degeneration, at least five studies show highly significant association with the same *CFH* Y402H polymorphism (Edwards et al., 2005; Hageman et al., 2005; Haines et al., 2005; Klein et al., 2005; Zareparsi et al., 2005) in a region strongly implicated by multiple linkage studies. For these findings, the data are highly compelling and consistent, and provide a solid foundation for the next generation of studies to investigate the mechanisms of the gene--phenotype connection. Power/type 2 error appears to be a major factor—if the genetic effect is sufficiently large (*HLA* in type 1 diabetes mellitus or *CFH* in age-related macular degeneration)—or, if the sample size is large, then there appears to be a greater chance of "hard" replication.

At present, the data for schizophrenia are confusing, and there are two broad possibilities. The first possibility is that the current findings for some of the best current genes are true. This implies that the genetics of schizophrenia are different from other complex traits in the existence of very high degrees of etiological heterogeneity: schizophrenia is hypercomplex, and we need to invoke more complicated genetic models than other biomedical disorders. The alternative possibility is that the current findings are clouded by type 1 and type 2 error. Schizophrenia is similar to other complex traits: it is possible that there are kernels of wheat, but it is highly likely that there is a lot of chaff. At present, the second and more parsimonious possibility has not been rigorously excluded. The impact of type 1/type 2 error is likely, and it is not clear why schizophrenia should be inherently more complex. At present, we cannot resolve these possibilities.

Public Health Implications

The public health importance of schizophrenia is clear, and the rationale for the search for genetic causes is strong. Schizophrenia research has never been easy: the current epoch of investigation into the genetics of schizophrenia provides a set of tantalizing clues, but definitive answers are not yet fully established. Findings from the accumulated literature appear to be more than chance, yet sufficiently variable as to render "hard" replication elusive. The currently murky view of this literature may result from the competing filters of type 1 and type 2 error. The current literature could be a mix of true and false positive findings; however, it would be a momentous advance for the field if even one of the genes in Table 19–2 were a true positive result.

This body of work is not yet ready for wholesale translation into clinical practice. However, it is not premature to inform patients that this work is advancing and that it holds promise for new insights into etiology, pathophysiology, and treatment on the 5- to 10-year horizon. On a larger scale, the treatment of the mentally ill mirrors the humanity of a society; in many societies, the return image is not flattering. If a specific genetic variation were proven to be causal to schizophrenia, this poor reflection might improve (Braslow, 1995).

Acknowledgements

The section on schizophrenia in this chapter is reproduced with permission from a review article [Sullivan, PF (2005). The genetics of schizophrenia. *PLoS Med* 2(7):e212.] originally published in *PLoS Medicine* (www.plosmedicine.org). This is an open-access article distributed under the terms of the Creative Commons Attribution License, which permits unrestricted use, distribution, and reproduction in any medium, provided the original work is properly cited.

References

Braslow, JT (1995). Effect of therapeutic innovation on perception of disease and the doctor-patient relationship: a history of general paralysis of the insane and malaria fever therapy, 1910–1950. *Am J Psychiatry*, 152:660–665.

Edwards, AO, Ritter, R, III, Abel, KJ, Manning, A, Panhuysen, C, et al. (2005). Complement factor H polymorphism and age-related macular degeneration. *Science,* 308:421–424.

Fan, JB, Zhang, CS, Gu, NF, Li, XW, Sun, WW, et al. (2005) Catechol-O-methyltransferase gene fever therapy, 1910–1950. *Am J Psychiatry,* 152:660–665.

Florez, JC, Hirschhorn, J, Altshuler, D (2003). The inherited basis of diabetes mellitus: Implications for the genetic analysis of complex traits. *Annu Rev Genomics Hum Genet,* 4:257–291.

Gottesman, II and Shields, J (1982). *Schizophrenia: The Epigenetic Puzzle.* Cambridge, UK: Cambridge University Press.

Hageman, GS, Anderson, DH, Johnson, LV, Hancox, LS, Taiber, AJ, et al. (2005). From the cover: A common haplotype in the complement regulatory gene factor H (HF1/CFH) predisposes individuals to age-related macular degeneration. *Proc Natl Acad Sci USA,* 102:7227–7232.

Haines, JL, Hauser, MA, Schmidt, S, Scott, WK, Olson, LM, et al. (2005). Complement factor H variant increases the risk of age-related macular degeneration. *Science,* 308:419–421.

Harris, EC and Barraclough, BB (1998). Excess mortality of mental disorder. *Br J Psychiatry,* 173:11–53.

Jeste, DV, Gladsjo, JA, Lindamer, LA, Lacro, JP (1996). Medical comorbidity in schizophrenia. *Scizophrenia Bull,* 22:413–430.

Klein, RJ, Zeiss, C, Chew, EY, Tsai, JY, Sackler, RS, et al. (2005). Complement factor H polymorphism in age-related macular degeneration. *Science,* 308:385–389.

Korstanje, R and Paigen, B (2002) From QTL to gene: The harvest begins. *Nat Genet,* 31:235–236.

Lewis, CM, Levinson, DF, Wise, LH, DeLisi, LE, Straub, RE, et al. (2003). Genome scan metaanalysis of schizophrenia and bipolar disorder, part II: Schizophrenia. *Am J Hum Genet,* 73:34–48.

Lindholm, E, Ekholm, B, Shaw, S, Jalonen, P, Johansson, G, et al. (2001). A schizophrenia susceptibility locus at 6q25, in one of the world's largest reported pedigrees. *Am J Hum Genet,* 69:96–105.

Maziade, M, Roy, MA, Rouillard, E, Bissonnette, L, Fournier, JP, et al. (2001). A search for specific and common susceptibility loci for schizophrenia and bipolar disorder: A linkage study in 13 target chromosomes. *Mol Psychiatry,* 6:684–693.

McGlashan, TH (1988). A selective review of recent North American long-term follow up studies of schizophrenia. *Schizophr Bull,* 14:515–542.

Moises, HW, Yang, L, Kristbjarnarson, H, Wiese, C, Byerley, W, et al. (1995). An international two-stage genome-wide search for schizophrenia susceptibility genes. *Nat Genet,* 11:321–324.

Mortensen, PB, Pedersen, CB, Westergaard, T, Wohlfahrt, J, Ewald, H, et al. (1999). Effects of family history and place and season of birth on the risk of schizophrenia. *N Eng J Med,* 340:603–608.

Murray, CJL and Lozpe, AD (1996). *The Global Burden of Disease: A Comprehensive Assessment of Mortality and Disability From Diseases, Injuries, and Risk Factors in 1990 and Projected to 2020.* Boston: Harvard University Press. 900p.

Murray, RM, Jones, PB, Susser, E, van Os, J, Cannon, M (2003). *The Epidemiology of Schizophrenia.* Cambridge: Cambridge University Press. 470p.

Plomin, R, DeFries, JC, Craig, IW, McGuffin, P (2003). *Behavioral Genetics in the Postgenomic era,* 3rd Edn. Washington DC: APA Books. 414p.

Saha, S, Welham, J, Chant, D, McGrath, J (2005). The epidemiology of schizophrenia. *PLoS Med* 2:e141.

Schizophrenia Linkage Collaborative Group (1996). Additional support for schizophrenia linkage on chromosomes 6 and 8: A multicentre study. Schizophrenia Linkage Collaborative Group for Chromosomes 3, 6 and 8. *Am J Med Genet,* 67:580–594.

Schwab, SG, Hallmayer, J, Albus, M, Lerer, B, Eckstein, GN, et al. (2000). A genome-wide autosomal screen for schizophrenia susceptibility loci in 71 families with affected siblings: Support for loci on chromosome 10p and 6. *Mol Psychiatry,* 5:638–649.

Shifman, S, Bronstein, M, Sternfeld, M, Pisante-Shalom, A, Lev-Lehman, E, et al. (2002). A siblings: support for loci on chromosome 10p and 6. *Mol Psychiatry,* 5:638–649.

Straub, RE, MacLean, CJ, ONeill, FA, Burke, J, Murphy, B, et al. (1995). A potential vulnerability locus for schizophrenia on chromosome 6p24–22: Evidence for genetic heterogeneity. *Nat Genet* 11:287–293.

Sullivan, PF (2005). The genetics of schizophrenia. *PLoS Med,* 2 (7):0614–0618.

Sullivan, PF, Eaves, LJ, Kendler, KS, Neale, MC (2001). Genetic case-control association studies in neuro-psychiatry. *Arch Gen Psychiatry,* 58:1015–1024.

Sullivan, PF, Kendler, KS, Neale, MC (2003). Schizophrenia as a complex trait: Evidence from a meta-analysis of twin studies. *Arch Gen Psychciatry,* 60:1187–1192.

Sullivan, PF, Owen, MJ, O'Donovan, MC, and Freedman, RR (2005). *Textbook of Schizophrenia.* Lieberman, J, Stroup, T, and Perkins, D (eds), (pp. 39–53). Washington, DC: American Psychiatric Publishing, Inc.

Zareparsi, S, Branham, KE, Li, M, Shah, S, Klein, RJ, et al. (2005). Strong association of the Y402H variant in complement factor H at 1q32 with susceptibility to age-related macular degeneration. *Am J Hum Genet,* 173:11–53.

20

Neuropsychiatric Diseases II: Mood Disorders

Dhavendra Kumar

Disorders of mood affect all people irrespective of age, gender, culture, religion, and social and working background. Although, women are thought to have two to three times higher prevalence of mood disorders, a number of studies have failed to provide a proof for this. It is probably due to the fact that women tend to seek medical help more frequently compared with men. Depression is a term used to classify persons suffering from a prolonged disturbance of mood (Reilly, 2004). It is not uncommon and the prevalence is rising at an alarmingly rapid rate. Some health economists are convinced that by 2020 depression will rank second only to coronary heart disease in its economic impact on society.

There are different forms of depression. The standard diagnostic criteria are set forth in the *Diagnostic and Statistical Manual IV* (DSM IV), a reference work recognized by most psychiatrists as containing the most current and widely held rules for making a diagnosis. There is no biochemical or any other test applicable to making a diagnosis of any form of depression. There are two recognizable clinical forms of depression:

1) Unipolar depression (UPD), which is often described as a major depressive disease (MDD) requiring medical intervention (clinical depression). Mild depression is probably common with a life time risk of between 5% and 15%. This may remain undiagnosed for long time, which is brought about by not uncommon challenging situations, for example, examination failure, relationship breakdown, loss of a close relative or a friend, and unpleasant job-related situations. The diagnosis of UPD depends upon recognizing at least five of the nine symptoms more than a period of 2 weeks as set out in DSM IV. For example, feelings of worthlessness, fatigue, sleep disturbance, unintended weight loss, recurrent thoughts of death, anhedonia, and depression. The symptoms should not be secondary to an underlying disease, for example, hypothyroidism, and must impair important areas of the patient's daily living.
2) Bipolar depression (BPD), which includes mood swings with variable periods of depression and mania. The manic phase is at times associated with psychotic symptoms, and BPD can be appropriately termed as "manic-depressive psychosis." It is estimated that approximately 1% of the population probably suffers from BPD. To satisfy the DSM IV criteria for BPD, the patient must have periods of depression (as in UPD) with episodic manic behavior. The patient must experience a period of elevated, expansive, or irritable mood for at least 1 week. During the episode, the patient must have experienced at least three of the seven symptoms—a sense of grandiosity, a reduced need for sleep, a marked increase in talkativeness, a sense that his or her thoughts are racing, easy distractibility, a high increase in goal-oriented activity, or an unusual involvement in pleasure-seeking activities. These symptoms must impair his or her normal routine and there should not be an underlying cause, such as use of alcohol or illicit drugs. Patients with BPD go through a cycle of symptoms from one end of the spectrum (depression) to the other extreme (mania).

Genetic Basis of Depression

Family and Twin Studies

Clues for the genetic causation of depression are from a number of family studies. These studies demonstrate that the first-degree relatives of persons with depression have approximately 3–5 times higher risk for being diagnosed with depression compared with the relatives of control group of individuals who are asymptomatic (Table 20–1). Recurrence risk figures from several studies demonstrate that there are probably higher underlying genetic factors manifesting with BPD compared with UPD.

Adoption studies strongly support a major genetic influence on risk for depression. One large study (Reilly, 2004) found that 30% of the biological parents of children who had been given for adoption at birth and who were later diagnosed with bipolar disorder had either BPD or UPD. Several twin studies have also found that the concordance rate for both these disorders is much higher (60%) for identical (monozygotic, MZ) twins compared with non-identical (dizygotic, DZ) twins (20%).

Empiric risk figures to close relatives of persons diagnosed with either UPD or BPD are based from several studies (Table 20–1). Lifetime risk for any form of depression to offspring, irrespective of the gender, is estimated to be approximately 20%. In a large study, for a couple in which one parent was diagnosed with depression, the lifetime risk for major depressive illness was 8% in the son compared with approximately 15% in the daughter. The lifetime risk to children (regardless of their gender) of developing bipolar disorder was 7%–14%. Within this group, the lifetime risk was higher for a child

Table 20–1 Risk of Depression [Unipolar Depression (UPD) or Dipolar Depression (BPD)] Among First-degree Relatives with UPD and BPD

Diagnosis in First-degree Relatives	Range of Risk (%)	Average
Unipolar depression		
Unipolar depression	6–28	14.0
Bipolar depression	0–5	1.4
Bipolar depression		
Unipolar depression	6–23	16.0
Bipolar depression	2–16	6

Based on several studies: Reilly, 2004.

who was of the same sex as the affected parent. If both parents have been diagnosed with major depression, and there are no other affected family members, then the lifetime risk of major depression is approximately 15% for the son and approximately 25% for the daughter. Where both parents and sibs are affected, the lifetime risk to other children might approach 50%, with an average figure of 30%–40%.

Genetic Linkage Studies

During the past two decades, several large-scale studies have undertaken genetic linkage studies in the search for possible gene loci in all forms of depression. These studies have largely concentrated on bipolar disorder (BPD), as the case identification and verification of the family history is probably more reliable. Several different genomic regions were implicated harboring susceptibility loci. Early reports suggested chromosome 11 (Egeland et al., 1987), the X chromosome (Baron et al., 1987), and chromosome 18 (Berrettini et al., 1994). Several other studies are published, but none of these agree on a major gene locus to support the genetic etiology of depression. This lack of agreement among such studies could be false-negative result due to inadequate power of the selected cohort, genetic heterogeneity, and inconsistent and weak evidence of linkage with variable peak locations (Roberts et al., 1999). On the other hand, several studies might have produced false-positive results simply by chance, particularly when multiple models were tested. However, it is also true that some reports could be true positives that are difficult to replicate, because of substantial locus heterogeneity for BPD susceptibility across samples and across families within samples. Meta-analysis has been widely used to resolve disputes between several studies. These can provide valuable statistical evidence either in support or against genetic linkage.

The genome scan meta-analysis (GSMA) approach (Wise et al., 1999) was applied to 18 BPD genome scan studies with more than 20 affected cases to determine whether statistical support might be achieved for any chromosomal regions (Segurado et al., 2003). Meta-analysis of genome scans presents with numerous methodological difficulties, because of the use of diverse phenotypic and transmission models, linkage analysis methods, marker maps and map densities, samples sizes, pedigree structures, and ethnic backgrounds. Levinson et al. (2003) examined the above issues encountered in GSMA in relation to schizophrenia and BPD.

The outcome of GSMA in depressive illnesses (Segurado et al., 2003) has been inconclusive. No genomic region achieved statistical significance using several simulation-based criteria. The two primary analyses used available linkage date for "very narrow" (BP-I and schizoaffective disorder-BP), "narrow" (adding BP-II disorder), and "broad" (adding recurrent major depression) models. The most significant *p*-values (<0.1) were observed on chromosome 9p22.3-21.1 (very narrow) and 14q24.1-32.12 (narrow). In addition, nominal significant values were observed for all models on adjacent bins involving 9p, 14q, and 18p-q. It is important to note that very few BPD studies have been undertaken for the "narrow" disease model compared to schizophrenia, where more significant results were produced. The present results for the "very narrow" model are promising, but suggest that more and larger data sets are needed. Alternatively, genetic linkage might be detected in certain populations or subsets of pedigrees. Although narrow and broad data sets had more power, they did not produce more significant evidence for linkage. This large GSMA was inconclusive and did not provide support for linkage, but could not disprove linkage to any candidate region either.

For complex disorders like UPD or BPD, genome-wide association or linkage scans are essential. Association scans on a large population sample can help in selecting possible chromosomal regions that could be used for genome scans for linkage using carefully selected design models and patient and age–sex matched cohort drawn from the same population or ethnic background. Future studies could be designed using various pedigrees as part of a large collaborative effort, or just on one or more extended families. Both strategies are important and should be pursued. The meta-analyses of such studies should be able to draw reasonable conclusion using agreed disease models, removing possible selection bias, correction of any methodological errors, and working out the statistical significance.

The Future of Neuropsychiatry in the Genomics Era

The future of neuropsychiatry in the genomics era looks promising (Insel and Collins, 2003). However, the task of finding and confirming susceptibility genes for schizophrenia (Saha et al., 2005 and see Chapter 19), mood disorders, and a range of other psychiatric illnesses is daunting. More than 99% of what has been published about genes and the brain structure and function has focussed on approximately 300 genes only, which is less than 1% of the genome. It is estimated that approximately half of the mouse genome ($\sim$16,500 genes) is expressed in the brain (Sandberg et al., 2000). This being applied to humans would be overwhelming as several new genes remain to be explored, including many more that might indirectly influence the few genes connected with several genes for proteins regulating neurotransmitters and molecular signaling pathways. Undoubtedly, the strategy would be to search for mutations and polymorphisms in multiple genes that are more likely to yield positive results rather than rare mutations in few selected genes, each of which will have a weak effect. In addition, non-genetic factors are equally important in the pathogenesis of mental disorders, which cumulatively make an individual susceptible or vulnerable. However, these are difficult to quantify or validate as compared with other multifactorial disorders such as diabetes mellitus, coronary heart disease, and hypertension, where simple factors like blood glucose and blood pressure are likely to be helpful in selecting vulnerable individuals. Carefully collected family history is probably the most important tool in identifying an individual likely to manifest with schizophrenia or depressive illness. Other phenotypic measures, such as eye tracking, sensory-motor gating, and working memory measures in schizophrenia, might be useful in assessing the transmission of mental disorders (Gottesmann and Gould, 2003).

In this chapter and the previous chapter, several candidate genes and chromosomal regions are listed as possible candidate genes for schizophrenia and mood disorders. Among the candidate genes for schizophrenia that deserve to be high on the list include neuregulin-1 (Stefansson et al., 2002, 2003), catechol *O*-methyltransferase (Egan et al., 2001), dysbindin (Straub et al., 2002; Schwab et al., 2003), G72 (Chumakov et al., 2002), and 2′,3′-cyclic nucleotide 3′-phosphodiestrase, *CNP* (Peirce et al., 2006). In this context, research based on the animal model has been rewarding. Niculescu and colleagues (2000) initiated such quest by employing GeneChip microarrays in analyzing specific brain regions of rats treated with methamphentamine as an animal model for psychotic mania. The data was crossmatched against the human genomic loci associated with either schizophrenia or bipolar disorder. The group identified several novel candidate genes involved with signal transduction molecules, transcription factors, and metabolic enzymes, which are probably implicated in the pathogenesis of either schizophrenia or mood disorders. One of these genes, G protein-coupled receptor kinase 3 (GRK3), was found to be correlated with disease severity. This report and other reports acknowledge the application of microarray genomic technology as a major strategy in the identification of susceptibility candidate genes in psychiatric disorders (Bunney et al., 2003). This strategy would also be useful in pharmacogenomic research to design and develop new psychopharmacological agents. Successful application of this technology could also open the way forward for constructing diagnostic microarrays that could be used in the future for selecting vulnerable individuals who could be targeted for appropriate personalized psychopharmacotherapy.

Conclusion

Depression is a major health burden that is likely to rapidly rise to epidemic proportions worldwide. Several studies, both at family and population levels, support the multifactorial etiology related to several genes with small additive effect. In this context, data based on adoption and twin studies are strong. The past two decades have seen several impressive reports published, each making a case in support of a candidate gene pointing to a specific chromosomal region. Despite these reports, however, no single or a selection of candidate genes has emerged as a favourite contender. This is not surprising and favors the universally accepted argument for multifactorial-polygenic model for depression. This is also shown to be the case for autism, panic disorders, and a range of behavioral disorders.

The future of neuropsychiatric genetics is promising with the increasing application of genomic microarrays technology. This technology is likely to provide good results both for identifying candidate genes and for establishing their functional significance. This is important for developing reliable diagnostic tools and in pharmacogenomic research for the design and development of new psychopharmacological drugs.

Acknowledgments

Both chapters in the neuropsychiatric section were gratefully reviewed by Dr. Ravi Mehrotra, Consultant Psychiatrist, West Middlesex Hospital, and Hon. Senior Lecturer in Psychiatry, Imperial College of Medicine, University of London, England, UK.

References

Baron, M, Risch, N, Hamburger, R, Mandel, B, et al. (1987). Genetic linkage between X-chromosome markers and bipolar affective illness. *Nature,* 326:289–292.

Berrettini, WH, Ferraro, TN, Goldin, LR, Detera-Wadleigh, SD, et al. (1994). A linkage study of bipolar illness. *Arch Gen Psychiatry,* 54:27–35.

Bunney, WE, Bunney, BG, Vawter, MP, Tomita, H, et al. (2003). Microrray technology: a review of new strategies to discover candidate vulnerability genes in psychiatric disorders. *Am J Psychiatry,* 160(4):657–666.

Chumakov, I, Blumenfeld, M, Guerassimenko, O, Cavarec, L, et al. (2002). Genetic and phsyiological data implicating the new human gene G72 and the gene for D-amino acid oxidase in schizophrenia. *Proc Natl Acad Sci USA,* 99:13765–13680.

Egan, MF, Gldberg, TE, Kolachana, BS, Callicott, JH, et al. (2001). Effect of COMT val108/158 Met genotype on frontal lobe function and risk for schizophrenia. *Proc Natl Acad Sci USA,* 98:6971–6922.

Egeland, JA, Gerhard, DS, Pauls, DL, Sussex, JN, et al. (1987). Bipolar affective disorders linked to DNA markers on chromosome 11. *Nature,* 325:238–246.

Gottesmann, II and Gould, TD (2003). The endophenotype concept in psychiatry: etymology and strategic intentions. *Am J Psychiatry,* 160:636–645.

Insel, TR and Collins, FS (2003). Psychiatry in genomics era. *Am J Psychiatry* 160:616–620.

Levinson, DF, Levinson, MD, Segurado, R, and Lewis, CM (2003). Genome scan meta-analysis of schizophrenia and bipolar disorder, part I: methods and power analysis. *Am J Hum Genet,* 73:17–33.

Niculescu, AB, III, Segal, DS, Kuczenski, R, Barrett, T, et al. (2000). Identifying a series of candidate genes for mania and psychosis: a convergent functional genomics approach. *Physiol Genomics,* 4:83–91.

Peirce, TR, Bray, NJ, Williams, NM, Norton, N, et al. (2006). Convergent evidence for 2′,3′-cyclic nucleotide 3′-phosphodiestrase as a possible susceptibility gene for schizophrenia. *Arch Gen Psychiatry,* 63:18–24.

Reilly, PR (2004). *Is It in Your Genes? The Influence of Genes on Common Disorders and Diseases that Affect You and Your Family.* New York: Cold Spring Harbor Laboratory, pp. 219–233.

Roberts, SB, McLean, CJ, Neale, MC, Eaves, LJ, et al. (1999). Replication of linkage studies of complex traits: an examination of variation in location estimates. *Am J Hum Genet,* 65:876–884.

Saha, S, Welham, J, Chant, D, and McGrath, J (2005). The epidemiology of schizophrenia. *PLoS Med,* 2:e141

Sandberg, R, Yasuda, R, Pankratz, DG, Carter, TA, et al. (2000). Regional and strain-specific gene expression mapping in the adult mouse brain. *Proc Natl Acad Sci USA,* 97:11038–11043.

Schwabb, SG, Knapp, M, Mondabon, S, Hallmayer, J, et al. (2003). Support for association of schizophrenia with genetic variation in the 6p22.3 gene, dysbindin, in sib-pair families with linkage and in an additional sample of triad families. *Am J Hum Genet,* 72:185–190.

Segurado, R, Detera-Wadleigh, SD, Levinson, DF, Lewis, CM, et al. (2003). Genome can meta-analysis of schizophrenia and bipolar disorder, part III: bipolar disorder. *Am J Hum Genet,* 73:49–62.

Stefansson, H, Sigurdsson, E, Steinthorsdottir, V, Bjornsdottir, S, et al. (2002). Neuroregulin 1 and susceptibility to schizophrenia. *Am J Hum Genet,* 71:877–892.

Stefansson, H, Sarginson, J, Kong, A, Yates, P, et al. (2003). Association of neuregulin 1 and susceptibility to schizophrenia confirmed in a Scottish population. *Am J Hum Genet,* 72:83–87.

Straub, RE, Jiang, Y, MacLean, CJ, Ma, Y, et al. (2002). Genetic variation in the 6p22.3 gene DTNBP1, the human ortholog of the mouse dysbindin gene, is associated with schizophrenia. *Am J Hum Genet,* 71:337–348.

Wise, LH, Lanchbury, JS, and Lewis, CM (1999). Meta-analysis of genome scans. *Ann Hum Genet,* 63:63–72.

21

Asthma and Chronic Obstructive Pulmonary Disease

William OC Cookson

Asthma is a disease of the airways of the lung. Inflammation and intermittent constriction of these airways give rise to symptoms of wheeze, cough, chest tightness, and shortness of breath. Chronic disease may lead to airway scarring and irreversible airflow limitation. Asthma affects 155 million individuals in the world.

Asthma is most prevalent in children and young adults and in 90% of cases is associated with immunoglobulin E (IgE)-mediated reactions to common inhaled proteins, which are called allergens. Typical allergen sources include house dust mite (HDM), grass pollens, and animal danders (sheddings from skin and fur). The word "atopy" (originally meaning "strange disease") is now commonly used to indicate the presence of positive prick skin tests to allergens, or elevations of the total or specific serum IgE.

Atopic asthma runs strongly in families. Twin studies show a heritability of approximately 60% (Duffy et al., 1990), and segregation analyses suggest that this is due to a few genes of moderate effect (oligogenes) rather than many genes of small effect (polygenes) (Lawrence et al., 1994; Holberg et al., 1996; Jenkins et al., 1997). It is estimated that genes and environment contribute approximately equally to the disease (Palmer et al., 2000). A substantial international effort is now under way to identify these genes.

Chronic obstructive pulmonary disease (COPD) is an adult disease that is characterized by shortness of breath, which is due to a severe limitation in lung airflow that is not fully reversible. The airflow limitation is usually progressive and is often associated with an abnormal inflammatory response of the lungs to toxic particles or gases. COPD is usually a result of cumulative exposure to tobacco smoke, occupational dusts or vapors, and indoor or outdoor air pollution. COPD ranks as the fourth leading cause of death worldwide, trailing only cardiovascular disease, pneumonia, and HIV/AIDS (human immunodeficiency virus/acquired immunodeficiency syndrome) (Calverley and Walker, 2003).

A genetic component to COPD is suggested by the observation that most smokers do not develop chronic irreversible airflow limitation. In addition, familial clustering of COPD is well recognized and well documented (Kueppers et al., 1977; Silverman et al., 1998), and the rate of decline in lung function in the general population has a heritability of approximately 50% (Gottlieb et al., 2001).

Asthma and COPD both affect the airway and share several characteristic features. Both diseases have been associated with an accelerated decline in lung function during adulthood, increased airways responsiveness, and the development of chronic airflow obstruction (CAO). Some patients with asthma develop irreversible airway obstruction and thus may be classified as having COPD (ATS, 1987). In addition, there is extensive evidence that an important subset of people with COPD have features that are usually associated with asthma such as variable airflow limitation, peripheral eosinophilia, and steroid responsiveness (O'Byrne and Postma, 1999).

In 1961, it was proposed that asthma, chronic bronchitis, and emphysema are expressions of one disease entity, and may stem from a common genetic root (Orie et al., 1961). This concept, later termed "The Dutch hypothesis" (Fletcher et al., 1976), suggested that asthma and CAO share common pathogenic mechanisms, and that asthmatic patients and many smokers who develop CAO share a common underlying predisposition to atopy and increased airways responsiveness (Pride, 1986). If true, the Dutch hypothesis is likely to have important implications for our understanding of the pathogenesis of both diseases and for clinical practice.

COPD is unique amongst complex genetic diseases in that the environmental inducer of the disease is usually obvious and that the level of exposure can often be documented with precision. The high mortality and morbidity associated with COPD and its chronic and progressive nature has prompted the use of molecular genetic studies in an attempt to identify susceptibility factors for the disease. The eventual aim of such studies is to develop effective therapy. In addition, the early identification of genetic susceptibility to COPD amongst cigarette smokers may be an essential element in prevention of disease.

Candidate Genes

The science of genetics depends on the study of variation (polymorphism), and the molecular geneticist is trying to identify polymorphism in deoxyribonucleic acid (DNA) that results in an altered phenotype (the presence of disease) in particular environmental circumstances. Polymorphisms may be identified in genes of interest and then are tested for differences in frequency in cases and controls with disease. Association studies may also be carried out in families. Family-based tests of association such as the transmission disequilibrium test (TDT) may be of particular value because they control for the effects of genetic admixture in the population to be studied (Spielman and Ewens, 1996).

Candidate gene studies address only a narrow range of hypotheses, which revolve around the known function of the gene selected for study. The biomedical literature is now filling rapidly with such investigations. Unfortunately, many reports of positive results fail to be reproduced. In interpreting such studies, it is important to realize that any gene may contain polymorphisms (usually single nucleotide polymorphisms or SNPs), which occur approximately every 500 bp. The great majority of these SNPs will not affect coding sequences. Although susceptibility to common complex diseases may often be mediated through regulatory sequences of genes, most SNPs do not have any impact on gene function.

The large number of SNPs in any given gene makes it facile for inexperienced investigators to carry out multiple comparisons which are often uncorrected and which easily lead to spurious claims of significance. These difficulties have been reviewed by Ott (Ott, 2004). He suggests that the prior probability that any SNP will indeed impact on disease susceptibility is very small and that this prior probability should be taken into account in estimating whether a novel association is real. He indicates that the minimum criterion for acceptance of a report of association in a biomedical journal should be a *p* value less than 0.005, after correction for multiple comparisons (Ott, 2004). In order to take a candidate gene study seriously, it is also desirable, if not mandatory, for some evidence to be provided that the associated SNPs impact on function.

Candidate Genes and Asthma

Candidate gene studies have identified numerous genes that may be involved in asthma susceptibility, many of which may exert their effects within the mucosa. For example, *IL-13* polymorphism influences mucus production as well as serum IgE levels (Kuperman et al., 2002) through a receptor encoded by the polymorphic *IL4-R*α (Ober et al., 2000). *Fc*ε*RI*-β variants modify the activity of the high-affinity receptor for IgE on mast cells, possibly through modulation of the level of expression of the receptor on the surface of the cell (Donnadieu et al., 2003; Traherne et al., 2003), and a receptor expressed by T cells for the key mast-cell signaling factor prostanoid DP has also been reported to be associated with asthma (Oguma et al., 2004). These findings indicate that the role of mast cells in epithelial inflammation may also be a potential target in asthma therapy.

Other asthma susceptibility genes include the pattern-recognition receptors (PRRs) of the innate immune system, which are expressed by dendritic cells and other cells, and recognize specific microbial components and activate innate immune responses. Polymorphism in *CD14* (Baldini et al., 1999), *TLR-2* (Toll-like receptor-2) (Eder et al., 2004), *NOD2* (nucleotide-binding oligomerization domain-2) (Kabesch et al., 2003), NOD1 (Hysi et al., 2005), and *TIM-1* (T-cell immunoglobulin domain and mucin domain-1) (McIntire et al., 2003) have all been shown to influence asthma susceptibility, indicating that these genes may be important in providing the link between microbial exposure and reduced susceptibility to asthma. While *CD14* polymorphisms have been associated with total serum IgE levels (Baldini et al., 1999), *TLR-4* does not seem to be associated with asthma (Raby et al., 2002; Noguchi et al., 2004), and although *TLR-2* polymorphisms may show association in children raised on farms (Eder et al., 2004), they do not seem to be associated with asthma in the general population (Noguchi et al., 2004). *TLR-10*, which responds to an unknown ligand, has recently been associated with asthma (Lazarus et al., 2004). However, none of these studies has tested for IgE responses to particular allergens, so systematic studies of PRR activation in asthma and atopic dermatitis (AD) are now desirable.

Other recognized effects are from tumor necrosis factor (*TNF*), (Moffatt and Cookson, 1997) which encodes a potent proinflammatory cytokine released by many cells, including airway epithelial cells, and tumor growth factor-β (*TGF*-β) (Pulleyn et al., 2001), which is an important local regulator of epithelial inflammation.

Several other recent observations have indicated the importance of proteins that are expressed by epithelial cells in conferring susceptibility to (or protection against) other diseases of epithelial surfaces.

Susceptibility to inflammatory bowel disease (IBD) is conferred by *NOD2* on chromosome 16p (Hugot et al., 2001; Ogura et al., 2001), *DLG5* on chromosome 10q23 (Stoll et al., 2004), and *OCTN* cation transporter genes (*SLC22A4* and *SLC22A5*) on chromosome 5q31 (Peltekova et al., 2004). *NOD2* mutants interfere with the function of Paneth cells, which are most numerous in the terminal ileum and are critically important in enteric antibacterial defenses (Hisamatsu et al., 2003; Lala et al., 2003). *DLG5* and *OCTN* are most highly expressed in terminally differentiating epithelial cells. *DLG5* encodes a scaffolding protein involved in the maintenance of epithelial integrity (Stoll et al., 2004).

The OCTN genes (*SLC22A4* and *SLC22A5*) are predicted to be cation transporters on the basis of their sequence homologies, but they do not have a known substrate or function. Other equally mysterious cation transporters involved in the genetic predisposition to epithelial disease include *SLC12A8* and psoriasis from chromosome 3q21 (Hewett et al., 2002), and *CLCA1* which has recently been implicated in asthma (Kamada et al., 2004) and COPD (Hegab et al., 2004) and is found at high levels in the mucus.

The list of susceptibility genes so far discovered for each of these diseases contains many surprises and many unanswered questions. However, the concentration of the expression of many of these genes in the skin or mucosa suggests that their function lies in these tissues. Although previous understanding of asthma, AD, psoriasis, and IBD has centered on mechanisms in the adaptive immune system, often with an emphasis on the T_H1–T_H2 paradigm, these results from genetic studies indicate that further understanding of innate mechanisms of epithelial defense is essential to the treatment and prevention of these disorders.

Candidate Genes in COPD

There have been several studies of various candidate genes in COPD [reviewed in Sandford and Silverman (2002)]. The archetypical COPD susceptibility gene is α_1-antitrypsin. α_1-Antitrypsin is a serine proteinase inhibitor (SERPIN), which has a highly selective action against neutrophil elastase.

Given the specificity of the action of α_1-antitrypsin for a specific substrate, it may be naïve to carry out a blanket search for the effects of other proteinase inhibitors on COPD susceptibility. The suggestion that α_1-antichymotrypsin may be a susceptibility factor has not been replicated (Sandford and Silverman, 2002). Matrix metalloproteinases have been implicated in murine models of COPD (Shapiro, 2002), but polymorphisms in these genes have not yet found a clear role in human COPD. However, other SERPINS may have effects on COPD, in particular those which may be discovered to be expressed in the airway epithelium. SERPINE2 is an inhibitor of plasminogen activator and has been implicated by genetic mapping studies on chromosome 2q33-35 (DeMeo et al., 2004).

Other studies have examined xenobiotic metabolizing enzymes [such as microsomal epoxide hydrolase 1 (EPHX1and glutathione *S*-transferases (GSTM1, GSTT1, and GSTP1) (Sandford and

Silverman, 2002)]. Antioxidants such as heme oxygenase 1 (HMOX1) have also been tested. The results from all these studies have not in general been replicated (Sandford and Silverman, 2002; Hersh et al., 2005).

TGF-β1 is an important antiinflammatory and profibrotic chemokines, and polymorphisms in its gene have been consistently associated with COPD in different studies (Celedon et al., 2004; Wu et al., 2004). The understanding of the mechanisms through which this gene influences COPD is rudimentary. TNF-α is a proinflammatory cytokine with known polymorphisms that influence many inflammatory disorders, but the evidence of its involvement in COPD has been equivocal (Sandford and Silverman, 2002).

The results of several studies have indicated the potential importance of apoptosis in generating emphysema and COPD (Segura-Valdez et al., 2000; Taraseviciene-Stewart et al., 2005), and it is quite possible that polymorphisms in genes upstream of apoptosis effectors may be of relevance to the disease process. Histone deacetylases (HDACs) show disordered regulation in COPD (Ito et al., 2005), and these genes might also be considered novel candidates.

Behavioral Candidate Genes for COPD

Patients with COPD exhibit exceptional patterns of behavior, in that they continue to smoke cigarettes despite feeling the dramatic effects of progressive ill-health over a prolonged time. Addictive behavior is among the most heritable psychiatric traits (Goldman et al., 2005), and the public health impact of gene discovery for addiction vulnerability is potentially very large (Merikangas and Risch, 2003). Intriguingly, variants in the SLC6A3 dopamine transporter and the dopamine receptor D2 (DRD2) have been suggested to modify smoking behavior (Sabol et al., 1999; Erblich et al., 2004). It might therefore be worthwhile to focus some attention on the type of study design that could systematically identify genes for cigarette addiction in patients with COPD.

Positional Cloning

Positional cloning is much more likely to identify novel genetic effects than candidate gene studies. Positional cloning begins with the finding of chromosomal regions, which are consistently inherited with the disease in families (genetic linkage). Genetic linkage studies are very powerful for the study of single-gene disorders but have limited power in complex genetic disorders when many genes are likely to be acting, and there is no established model for the inheritance of a given disease. Genetic linkage studies in a complex disorder such as asthma typically give an imprecise signal for the localization of the disease gene that may extend more than tens of megabases of DNA. This is because only a proportion of families will actually be linked to the locus, while others will randomly appear linked and nonlinked. In addition, the proportion of individuals with susceptibility alleles who develop the disease will vary between families (i.e., the alleles will have variable penetrance). For this reason, even highly ambitious genetic linkage studies involving several hundreds of families have often failed to deliver conclusive results.

Linkage Studies in Asthma and COPD

At least 11 full-genome screens have been reported for asthma and its associated phenotypes [reviewed in Cookson (2003) and Wills-Karp and Ewart (2004)]. These have identified 10 regions of linkage that were reproducible between screens and four regions that were statistically significant but not replicated by other groups (Cookson, 2003). Those regions that were consistently identified are likely to contain the genes with the strongest effect on disease.

A linkage study for COPD in families with severe early-onset COPD identified several regions of weak linkage, but no region that showed genome-wide significance or unequivocal evidence for a susceptibility gene on a particular chromosomal segment Silverman, Mosley, et al., 2002). Subsequent recruitment of further families and the study of additional phenotypes such as the FEV_1/FVC ratio (Silverman, Palmer, et al., 2002) have increased the evidence of linkage on chromosome 2. An independent genome-wide scan of pulmonary function measures in the National Heart, Lung, and Blood Institute Family Heart Study found that the FEV_1/FVC ratio was significantly linked to the short arm of chromosome 4, with no obvious overlap of linkages with the COPD studies described above (Wilk et al., 2003).

Genome screens have also been carried out for many other immune diseases that have a genetic basis and have identified regions of linkage that are shared (Becker et al., 1998). Asthma consistently shows linkage to the major histocompatibility complex (MHC) on the short arm of chromosome 6 (Cookson, 2002), and linkage regions for asthma also overlap with loci for ankylosing spondylitis (on chromosomes 1p31-36, 7p13, and 16q23); type 1 diabetes (on chromosomes 1p32-34, 11q13, and 16q22-24); IBD (on chromosomes 7p13 and 12q12-14); and multiple sclerosis and rheumatoid arthritis (on chromosome 17q22-24) (Cookson, 2002).

These shared loci are of particular interest, because they may lead to the elucidation of as yet unknown immune processes with important general effects on disease susceptibility.

Single-gene Disorders

The positional cloning of novel genes from regions that are linked to complex diseases is a long and difficult business. By contrast, the identification of Mendelian (single-gene) disorders is much more straightforward and can sometimes give insight into complex diseases. Several Mendelian diseases show strong features of atopy.

Hyper IgE

The hyper-IgE syndrome (HIES) is a rare primary immunodeficiency characterized by recurrent skin abscesses, pneumonia, and highly elevated levels of serum IgE. It can be transmitted as an autosomal dominant trait with variable expressivity. Linkage analyses in extended families with multiple cases of HIES have identified genetic linkage to chromosome 4q12, near to the marker D4S428 (Grimbacher et al., 1999). It is of interest that linkage to the same region has been identified in two genome screens for asthma (Daniels et al., 1996; Laitinen et al., 2001). The gene has not yet been identified.

Wiskott–Aldrich Syndrome

Wiscott–Aldrich syndrome (WAS) is a rare X-linked disorder of T and B cell function, which is typified by recurrent infections and thrombocytopenia. Many boys with the disease also develop a rash, which is indistinguishable from AD. A study of the WAS gene region has been carried out in Swedish families with AD (Bradley et al., 2001). One marker monoamine oxidase B (MAOB) showed linkage ($p < 0.05$) to the severity score of AD, but association to AD was not seen. These results should provoke further study of the gene in AD.

Familial Eosinophilia

Familial eosinophilia (FE) is an autosomal dominant disorder characterized by peripheral hypereosinophilia of unidentifiable cause with or without other organ involvement (Rioux et al., 1998). It has been localized on chromosome 5q34, near the *IL-4* cytokine cluster and serine protease inhibitor Kazal-type 5 precursor (*SPINK5*). Its gene has not yet been identified.

Netherton's Disease

Netherton's disease is a rare recessive disorder characterized by generalized erythroderma, symptoms of atopic disease (hay fever, food allergy, urticaria, and asthma) and very high levels of IgE (Chavanas, Garner, et al., 2000). Mutations in the gene encoding a serine protease inhibitor known as *SPINK5* or lymphoepithelial Kazal-type-related inhibitor (LEKTI) were shown to be causing disease in these patients (Mägert et al., 1999; Chavanas, Bodemer, et al., 2000). Subsequent work has shown that a common polymorphism in *SPINK5* (particularly Glu420-Lys) modifies the risk of developing AD, asthma, and elevated serum IgE levels (Walley et al., 2001; Kato et al., 2003; Nishio et al., 2003; Kabesch et al., 2004), indicating that SPINK5 might be involved in an unexpected pathway in the development of atopic disease.

The SPINK5 protein contains 13 active protease inhibitory domains, which are joined together by linking domains. The sequence of each of the SPINK5 protease inhibitory domains are slightly different (Mägert et al., 1999), suggesting a polyvalent action against multiple substrates. SPINK5 is expressed in the skin in the outer epidermis, the sebaceous glands and around the shafts of hair follicles (Komatsu et al., 2002), and is expressed in the airway epithelium, indicating that it might be important for the inhibition of environmental proteases. These might arise from bacteria (Miedzobrodzki et al., 2002) or from the HDM, *Dermatophagoides pteronyssinus* (Winton et al., 1998).

COPD

Several single-gene disorders show some elements of COPD. Bare lymphocyte syndrome type 1 can be caused by mutation in the *TAP2*, *TAP1*, or *TAPBP* genes (de la Salle et al., 1994) and is characterized by human leukocyte antigen (HLA) class I deficiency with normal expression of class II molecules. It results in chronic bacterial infections that are restricted to the respiratory tract and extend from the upper to the lower airway. The syndrome may cause bronchiectasis, emphysema, panbronchiolitis, and bronchial obstruction. Marfan syndrome is due to mutations in the structural fibrillin-1 gene, and patients with this disorder and other collagenoses develop emphysema due to loss of lung elastic tissue.

Genes Positionally Cloned for Asthma

The positional cloning of susceptibility genes for asthma has been successful, with the identification of four novel genes: *DPP10* (dipeptidyl peptidase 10; chromosome 2q14) (Allen et al., 2003), *GPRA* (G-protein-coupled receptor for asthma susceptibility; chromosome 7p14) (Laitinen et al., 2004), *PHF11* (plant homeodomain finger protein-11; chromosome 13q12) (Zhang et al., 2003), and *ADAM33* (a disintegrin and metalloproteinase 33; chromosome 20p) (Van Eerdewegh et al., 2002).

The functions and activities of these genes are as yet poorly understood, but they certainly do not fit into classical pathways of asthma pathogenesis. *ADAM33* is expressed in bronchial smooth muscle and is thought to alter the hypertrophic response of bronchial smooth muscle to inflammation (a component of airway remodeling) (Van Eerdewegh et al., 2002). *PHF11* encodes a nuclear receptor that is part of a complex containing a histone methyl transferase (*SETDB1*), a regulator of HDAC (*RCBTB1*), and a nuclear transport molecule (*karyopherin α3*). Their function in asthma is unknown. *DPP10* encodes a prolyl dipeptidase, which may remove the terminal two peptides from certain proinflammatory chemokines. It is uncertain whether this would activate or deactivate them, but if the substrate for DPP10 is as predicted, and if chemokines are activated, then DPP10 might be the target for a new asthma therapy (Allen et al., 2003). *GPRA* encodes an orphan G-protein-coupled receptor with isoforms that show distinct patterns of expression in bronchial epithelial cells and smooth muscle cells in asthmatic versus healthy individuals (Laitinen et al., 2004). G-protein-coupled receptors are in general good targets for pharmaceutical therapy, but more needs to be known about GPRA before it can become a focus for treating or preventing asthma.

Even though their functions are still largely unknown, it is of interest that the expression of both *DPP10* and *GPRA* is concentrated in the terminally differentiating bronchial epithelium, the layer in the epidermis corresponding to the site of highest expression of *SPINK5* and genes of the epidermal differentiation complex (EDC). This suggests that all these genes play a role in the maintenance of the epithelial barrier or that they are involved in the first lines of response once the barrier has been breached.

Genes Positionally Cloned for COPD

The strongest public linkage of COPD is to chromosome 2q33-35, in a region that contains the *SERPINE2* gene (Demeo et al., 2006). SERPINE2 has been shown to be expressed in normal lung epithelium, and SNPs in this gene show associations to COPD severe, early-onset COPD pedigrees, and case–control analyses. These results indicate that *SERPINE2* is a COPD susceptibility gene that results from a gene-by-smoking interaction (Demeo et al., 2006).

Post-genome Genetics

Although positional cloning can discover more that is new and unexpected than candidate gene studies, it is now recognized that genetic linkage statistics are far less powerful than those which detect association (Risch and Merikangas, 1996). The challenge has then become the development of systematic association studies of all human genes, or "whole-genome association." The estimates of the number of polymorphisms necessary to cover all human genes vary between 1,00,000 and 5,00,000.

After the completion of the human genome project, the entire human genetic sequence is publicly available, and a systematic search for all common human genetic variation (the Hapmap Project: http://www.hapmap.org/) is well advanced. These advances in knowledge have been matched with advances in technology, which allow high-throughput genotyping of thousands of polymorphisms on hundreds or thousands of subjects, together with the examination of the expression of all human genes in samples of cells or tissues. As a consequence, the molecular geneticist now has a remarkable set of tools to aid in gene discovery.

Conclusions

Genetic studies of the pathogenesis of asthma are showing increasing consistency and power, whereas studies of COPD are at an earlier stage. However, awareness of the need for large samples of carefully phenotyped patients, together with the ready availability of technology which allows large-scale genotyping of candidate genes, as well as whole-genome association testing, will have a significant impact on asthma and on COPD within the short-to-medium term.

References

Allen, M, Heinzmann, A, Noguchi, E, Abecasis, G, Broxholme, J, Ponting, CP, Bhattacharyya, S, et al. (2003). Positional cloning of a novel gene influencing asthma from chromosome 2q14. *Nat Genet*, 35:258–263.

ATS (1987). Standards for the diagnosis and care of patients with chronic obstructive pulmonary disease (COPD) and asthma. This official statement of the American Thoracic Society was adopted by the ATS Board of Directors, November 1986. *Am Rev Respir Dis*, 136:225–244.

Baldini, M, Lohman, I, Halonen, M, Erickson, R, Holt, P, and Martinez, F (1999). A polymorphism* in the 5′ flanking region of the CD14 gene is associated with circulating soluble CD14 levels and with total serum immunoglobulin E. *Am J Respir Cell Mol Biol*, 20:976–983.

Becker, K, Simon, R, Bailey-Wilson, J, Freidlin, B, Biddison, W, McFarland, H, and Trent, J (1998). Clustering of non-major histocompatibility complex susceptibility candidate loci in human autoimmune diseases. *Proc Natl Acad Sci USA*, 95:9979–9984.

Bradley, M, Soderhall, C, Wahlgren, CF, Luthman, H, Nordenskjold, M, and Kockum, I (2001). The Wiskott-Aldrich syndrome gene as a candidate gene for atopic dermatitis. *Acta Derm Venereol*, 81:340–342.

Calverley, PM and Walker, P (2003). Chronic obstructive pulmonary disease. *Lancet*, 362:1053–1061.

Celedon, JC, Lange, C, Raby, BA, Litonjua, AA, Palmer, LJ, DeMeo, DL, Reilly, JJ, et al. (2004). The transforming growth factor-beta1 (TGFB1) gene is associated with chronic obstructive pulmonary disease (COPD). *Hum Mol Genet*, 13:1649–1656.

Chavanas, S, Bodemer, C, Rochat, A, Hamel-Teillac, D, Ali, M, Irvine, AD, Bonafe, JL, et al. (2000). Mutations in SPINK5, encoding a serine protease inhibitor, cause Netherton syndrome. *Nat Genet*, 25:141–142.

Chavanas, S, Garner, C, Bodemer, C, Ali, M, Hamel-Teillac, D, Wilkinson, J, Bonafe, JL, et al. (2000). Localization of the Netherton syndrome gene to chromosome 5q32, by linkage analysis and homozygosity mapping. *Am J Hum Genet*, 66:914–921.

Cookson, W (2002). Genetics and genomics of asthma and allergic diseases. *Immunol Rev*, 190:195–206.

Cookson, W (2003). A new gene for asthma: would you ADAM and Eve it? *Trends Genet*, 19:169–172.

Daniels, SE, Bhattacharrya, S, James, A, Leaves, NI, Young, A, Hill, MR, Faux, JA, et al. (1996). A genome-wide search for quantitative trait loci underlying asthma. *Nature*, 383:247–250.

de la Salle, H, Hanau, D, Fricker, D, Urlacher, A, Kelly, A, Salamero, J, Powis, SH, et al. (1994). Homozygous human TAP peptide transporter mutation in HLA class I deficiency. *Science*, 265:237–241.

DeMeo, DL, Celedon, JC, Lange, C, Reilly, JJ, Chapman, HA, Sylvia, JS, Speizer, FE, et al. (2004). Genome-wide linkage of forced mid-expiratory flow in chronic obstructive pulmonary disease. *Am J Respir Crit Care Med*, 170:1294–1301.

Demeo, DL, Mariani, TJ, Lange, C, Srisuma, S, Litonjua, AA, Celedon, JC, Lake, SL, et al. (2006). The SERPINE2 gene is associated with chronic obstructive pulmonary disease. *Am J Hum Genet*, 78:253–264.

Donnadieu, E, Jouvin, MH, Rana, S, Moffatt, MF, Mockford, EH, Cookson, WO, and Kinet, JP (2003). Competing functions encoded in the allergy-associated F(c)epsilonRIbeta gene. *Immunity*, 18:665–674.

Duffy, DL, Martin, NG, Battistutta, D, Hopper, JL, and Mathews, JD (1990). Genetics of asthma and hay fever in Australian twins. *Am Rev Respir Dis*, 142:1351–1358.

Eder, W, Klimecki, W, Yu, L, von Mutius, E, Riedler, J, Braun-Fahrlander, C, Nowak, D, et al. (2004). Toll-like receptor 2 as a major gene for asthma in children of European farmers. *J Allergy Clin Immunol*, 113:482–488.

Erblich, J, Lerman, C, Self, DW, Diaz, GA, and Bovbjerg, DH (2004). Stress-induced cigarette craving: effects of the DRD2 TaqI RFLP and SLC6A3 VNTR polymorphisms. *Pharmacogenomics J*, 4:102–109.

Fletcher, C, Peto, R, Tinker, C, and Speizer, FE (1976). *The Natural History of Chronic Bronchitis and Emphysema*. Oxford: Oxford University Press.

Goldman, D, Oroszi, G, and Ducci, F (2005). The genetics of addictions: uncovering the genes. *Nat Rev Genet*, 6:521–532.

Gottlieb, DJ, Wilk, JB, Harmon, M, Evans, JC, Joost, O, Levy, D, O'Connor, GT, et al. (2001). Heritability of longitudinal change in lung function. The Framingham study. *Am J Respir Crit Care Med*, 164:1655–1659.

Grimbacher, B, Schaffer, AA, Holland, SM, Davis, J, Gallin, JI, Malech, HL, Atkinson, TP, et al. (1999). Genetic linkage of hyper-IgE syndrome to chromosome 4. *Am J Hum Genet*, 65:735–744.

Hegab, AE, Sakamoto, T, Uchida, Y, Nomura, A, Ishii, Y, Morishima, Y, Mochizuki, M, et al. (2004). CLCA1 gene polymorphisms in chronic obstructive pulmonary disease. *J Med Genet*, 41:e27.

Hersh, CP, Demeo, DL, Lange, C, Litonjua, AA, Reilly, JJ, Kwiatkowski, D, Laird, N, et al. (2005). Attempted replication of reported chronic obstructive pulmonary disease candidate gene associations. *Am J Respir Cell Mol Biol*, 33:71–78.

Hewett, D, Samuelsson, L, Polding, J, Enlund, F, Smart, D, Cantone, K, See, CG, et al. (2002). Identification of a psoriasis susceptibility candidate gene by linkage disequilibrium mapping with a localized single nucleotide polymorphism map. *Genomics*, 79:305–314.

Hisamatsu, T, Suzuki, M, Reinecker, HC, Nadeau, WJ, McCormick, BA, and Podolsky, DK (2003). CARD15/NOD2 functions as an antibacterial factor in human intestinal epithelial cells. *Gastroenterology*, 124:993–1000.

Holberg, CJ, Elston, RC, Halonen, M, Wright, AL, Taussig, LM, Morgan, WJ, and Martinez, FD (1996). Segregation analysis of physician-diagnosed asthma in Hispanic and non-Hispanic white families. A recessive component? *Am J Respir Crit Care Med*, 154:144–150.

Hugot, JP, Chamaillard, M, Zouali, H, Lesage, S, Cezard, JP, Belaiche, J, Almer, S, et al. (2001). Association of NOD2 leucine-rich repeat variants with susceptibility to Crohn's disease. *Nature*, 411:599–603.

Hysi, P, Kabesch, M, Moffatt, MF, Schedel, M, Carr, D, Zhang, Y, Boardman, B, et al. (2005). NOD1 variation, immunoglobulin E and asthma. *Hum Mol Genet*, 14:935–941.

Ito, K, Ito, M, Elliott, WM, Cosio, B, Caramori, G, Kon, OM, Barczyk, A, et al. (2005). Decreased histone deacetylase activity in chronic obstructive pulmonary disease. *N Engl J Med*, 352:1967–1976.

Jenkins, MA, Hopper, JL, and Giles, GG (1997). Regressive logistic modeling of familial aggregation for asthma in 7394 population-based nuclear families. *Genet Epidemiol*, 14:317–332.

Kabesch, M, Carr, D, Weiland, SK, and von Mutius, E (2004). Association between polymorphisms in serine protease inhibitor, kazal type 5 and asthma phenotypes in a large German population sample. *Clin Exp Allergy*, 34:340–345.

Kabesch, M, Peters, W, Carr, D, Leupold, W, Weiland, SK, and von Mutius, E (2003). Association between polymorphisms in caspase recruitment domain containing protein 15 and allergy in two German populations. *J Allergy Clin Immunol*, 111:813–817.

Kamada, F, Suzuki, Y, Shao, C, Tamari, M, Hasegawa, K, Hirota, T, Shimizu, M, et al. (2004). Association of the hCLCA1 gene with childhood and adult asthma. *Genes Immun*, 5(7):540–547.

Kato, A, Fukai, K, Oiso, N, Hosomi, N, Murakami, T, and Ishii, M (2003). Association of SPINK5 gene polymorphisms with atopic dermatitis in the Japanese population. *Br J Dermatol*, 148:665–669.

Komatsu, N, Takata, M, Otsuki, N, Ohka, R, Amano, O, Takehara, K, and Saijoh, K (2002). Elevated stratum corneum hydrolytic activity in Netherton syndrome suggests an inhibitory regulation of desquamation by SPINK5-derived peptides. *J Invest Dermatol*, 118:436–443.

Kueppers, F, Miller, RD, Gordon, H, Hepper, NG, and Offord, K (1977). Familial prevalence of chronic obstructive pulmonary disease in a matched pair study. *Am J Med*, 63:336–342.

Kuperman, DA, Huang, X, Koth, LL, Chang, GH, Dolganov, GM, Zhu, Z, Elias, JA, et al. (2002). Direct effects of interleukin-13 on epithelial cells cause airway hyperreactivity and mucus overproduction in asthma. *Nat Med*, 8:885–889.

Laitinen, T, Daly, MJ, Rioux, JD, Kauppi, P, Laprise, C, Petays, T, Green, T, et al. (2001). A susceptibility locus for asthma-related traits on chromosome 7

revealed by genome-wide scan in a founder population. *Nat Genet*, 28:87–91.
Laitinen, T, Polvi, A, Rydman, P, Vendelin, J, Pulkkinen, V, Salmikangas, P, Makela, S, et al. (2004). Characterization of a common susceptibility locus for asthma-related traits. *Science*, 304:300–304.
Lala, S, Ogura, Y, Osborne, C, Hor, SY, Bromfield, A, Davies, S, Ogunbiyi, O, et al. (2003). Crohn's disease and the NOD2 gene: a role for paneth cells. *Gastroenterology*, 125:47–57.
Lawrence, S, Beasley, R, Doull, I, Begishvili, B, Lampe, F, Holgate, ST, and Morton, NE (1994). Genetic analysis of atopy and asthma as quantitative traits and ordered polychotomies. *Ann Hum Genet*, 58:359–368.
Lazarus, R, Raby, BA, Lange, C, Silverman, EK, Kwiatkowski, DJ, Vercelli, D, Klimecki, WJ, et al. (2004). TOLL-like receptor 10 genetic variation is associated with asthma in two independent samples. *Am J Respir Crit Care Med*, 170:594–600.
Mägert, HJ, Standker, L, Kreutzmann, P, Zucht, HD, Reinecke, M, Sommerhoff, CP, Fritz, H, et al. (1999). LEKTI, a novel 15-domain type of human serine proteinase inhibitor. *J Biol Chem*, 274:21499–21502.
McIntire, JJ, Umetsu, SE, Macaubas, C, Hoyte, EG, Cinnioglu, C, Cavalli-Sforza, LL, Barsh, GS, et al. (2003). Immunology: hepatitis A virus link to atopic disease. *Nature*, 425:576.
Merikangas, KR and Risch, N (2003). Genomic priorities and public health. *Science*, 302:599–601.
Miedzobrodzki, J, Kaszycki, P, Bialecka, A, and Kasprowicz, A (2002). Proteolytic activity of *Staphylococcus aureus* strains isolated from the colonized skin of patients with acute-phase atopic dermatitis. *Eur J Clin Microbiol Infect Dis*, 21:269–276.
Moffatt, MF and Cookson, WO (1997). Tumour necrosis factor haplotypes and asthma. *Hum Mol Genet*, 6:551–554.
Nishio, Y, Noguchi, E, Shibasaki, M, Kamioka, M, Ichikawa, E, Ichikawa, K, Umebayashi, Y, et al. (2003). Association between polymorphisms in the SPINK5 gene and atopic dermatitis in the Japanese. *Genes Immun*, 4:515–517.
Noguchi, E, Nishimura, F, Fukai, H, Kim, J, Ichikawa, K, Shibasaki, M, and Arinami, T (2004). An association study of asthma and total serum immunoglobin E levels for Toll-like receptor polymorphisms in a Japanese population. *Clin Exp Allergy*, 34:177–183.
Ober, C, Leavitt, SA, Tsalenko, A, Howard, TD, Hoki, DM, Daniel, R, Newman, DL, et al. (2000). Variation in the interleukin 4-receptor alpha gene confers susceptibility to asthma and atopy in ethnically diverse populations. *Am J Hum Genet*, 66:517–526.
O'Byrne, PM and Postma, DS (1999). The many faces of airway inflammation. Asthma and chronic obstructive pulmonary disease. Asthma Research Group. *Am J Respir Crit Care Med*, 159:S41–S63.
Oguma, T, Palmer, LJ, Birben, E, Sonna, LA, Asano, K, and Lilly, CM (2004). Role of prostanoid DP receptor variants in susceptibility to asthma. *N Engl J Med*, 351:1752–1763.
Ogura, Y, Bonen, DK, Inohara, N, Nicolae, DL, Chen, FF, Ramos, R, Britton, H, et al. (2001). A frameshift mutation in NOD2 associated with susceptibility to Crohn's disease. *Nature*, 411:603–606.
Orie, NGM, Sluiter, HJ, de Vries, K, Tammeling, GJ, and Witkop, J (1961). The host factor in bronchitis. In NGM Orie and HJ Sluiter (eds.), *Bronchitis* (pp. 43–59). Assen, Netherlands: Royal Vangorcum.
Ott, J (2004). Association of genetic loci: replication or not, that is the question. *Neurology*, 63:955–958.
Palmer, LJ, Burton, PR, Faux, JA, James, AL, Musk, AW, and Cookson, WO (2000). Independent inheritance of serum immunoglobulin E concentrations and airway responsiveness. *Am J Respir Crit Care Med*, 161:1836–1843.
Peltekova, VD, Wintle, RF, Rubin, LA, Amos, CI, Huang, Q, Gu, X, Newman, B, et al. (2004). Functional variants of OCTN cation transporter genes are associated with Crohn disease. *Nat Genet*, 36:471–475.
Pride, N (1986). Smoking, allergy and airways obstruction: revival of the Dutch hypothesis. *Clin Allergy*, 16:3–6.
Pulleyn, LJ, Newton, R, Adcock, IM, and Barnes, PJ (2001). TGFbeta1 allele association with asthma severity. *Hum Genet*, 109:623–627.
Raby, BA, Klimecki, WT, Laprise, C, Renaud, Y, Faith, J, Lemire, M, Greenwood, C, et al. (2002). Polymorphisms in toll-like receptor 4 are not associated with asthma or atopy-related phenotypes. *Am J Respir Crit Care Med*, 166:1449–1456.
Rioux, J, Stone, V, Daly, M, Cargill, M, Green, T, Nguyen, H, Nutman, T, et al. (1998). Familial eosinophilia maps to the cytokine gene cluster on human chromosomal region 5q31-q33. *Am J Hum Genet*, 63:1086–1094.
Risch, N and Merikangas, K (1996). The future of genetic studies of complex human diseases. *Science*, 273:1516–1517.
Sabol, SZ, Nelson, ML, Fisher, C, Gunzerath, L, Brody, CL, Hu, S, Sirota, LA, et al. (1999). A genetic association for cigarette smoking behavior. *Health Psychol*, 18:7–13.
Sandford, AJ and Silverman, EK (2002). Chronic obstructive pulmonary disease. 1: Susceptibility factors for COPD the genotype–environment interaction. *Thorax*, 57:736–741.
Segura-Valdez, L, Pardo, A, Gaxiola, M, Uhal, BD, Becerril, C, and Selman, M (2000). Upregulation of gelatinases A and B, collagenases 1 and 2, and increased parenchymal cell death in COPD. *Chest*, 117:684–694.
Shapiro, SD (2002). Proteinases in chronic obstructive pulmonary disease. *Biochem Soc Trans*, 30:98–102.
Silverman, EK, Chapman, HA, Drazen, JM, Weiss, ST, Rosner, B, Campbell, EJ, O'Donnell, WJ, et al. (1998). Genetic epidemiology of severe, early-onset chronic obstructive pulmonary disease. Risk to relatives for airflow obstruction and chronic bronchitis. *Am J Respir Crit Care Med*, 157:1770–1778.
Silverman, EK, Mosley, JD, Palmer, LJ, Barth, M, Senter, JM, Brown, A, Drazen, JM, et al. (2002). Genome-wide linkage analysis of severe, early-onset chronic obstructive pulmonary disease: airflow obstruction and chronic bronchitis phenotypes. *Hum Mol Genet*, 11:623–632.
Silverman, EK, Palmer, LJ, Mosley, JD, Barth, M, Senter, JM, Brown, A, Drazen, JM, et al. (2002). Genomewide linkage analysis of quantitative spirometric phenotypes in severe early-onset chronic obstructive pulmonary disease. *Am J Hum Genet*, 70:1229–1239.
Spielman, RS and Ewens, WJ (1996). The TDT and other family-based tests for linkage disequilibrium and association. *Am J Hum Genet*, 59:983–989.
Stoll, M, Corneliussen, B, Costello, CM, Waetzig, GH, Mellgard, B, Koch, WA, Rosenstiel, P, et al. (2004). Genetic variation in DLG5 is associated with inflammatory bowel disease. *Nat Genet*, 36:476–480.
Taraseviciene-Stewart, L, Scerbavicius, R, Choe, KH, Moore, M, Sullivan, A, Nicolls, MR, Fontenot, AP, et al. (2005). An animal model of autoimmune emphysema. *Am J Respir Crit Care Med*, 171:734–742.
Traherne, JA, Hill, MR, Hysi, P, D'Amato, M, Broxholme, J, Mott, R, Moffatt, MF, et al. (2003). LD mapping of maternally and non-maternally derived alleles and atopy in Fc{varepsilon}RI-{beta}. *Hum Mol Genet*, 12:2577–2585.
Van Eerdewegh, P, Little, RD, Dupuis, J, Del Mastro, RG, Falls, K, Simon, J, Torrey, D, et al. (2002). Association of the ADAM33 gene with asthma and bronchial hyperresponsiveness. *Nature*, 418:426–430.
Walley, AJ, Chavanas, S, Moffatt, MF, Esnouf, RM, Ubhi, B, Lawrence, R, Wong, K, et al. (2001). Gene polymorphism in Netherton and common atopic disease. *Nat Genet*, 29:175–178.
Wilk, JB, DeStefano, AL, Arnett, DK, Rich, SS, Djousse, L, Crapo, RO, Leppert, MF, et al. (2003). A genome-wide scan of pulmonary function measures in the National Heart, Lung, and Blood Institute Family Heart Study. *Am J Respir Crit Care Med*, 167:1528–1533.
Wills-Karp, M and Ewart, SL (2004). Time to draw breath: asthma-susceptibility genes are identified. *Nat Rev Genet*, 5:376–387.
Winton, HL, Wan, H, Cannell, MB, Thompson, PJ, Garrod, DR, Stewart, GA, and Robinson, C (1998). Class specific inhibition of house dust mite proteinases which cleave cell adhesion, induce cell death and which increase the permeability of lung epithelium. *Br J Pharmacol*, 124:1048–1059.
Wu, L, Chau, J, Young, RP, Pokorny, V, Mills, GD, Hopkins, R, McLean, L, et al. (2004). Transforming growth factor-beta1 genotype and susceptibility to chronic obstructive pulmonary disease. *Thorax*, 59:126–129.
Zhang, Y, Leaves, NI, Anderson, GG, Ponting, CP, Broxholme, J, Holt, R, Edser, P, et al. (2003). Positional cloning of a quantitative trait locus on chromosome 13q14 that influences immunoglobulin E levels and asthma. *Nat Genet*, 34:181–186.

22

Inflammatory Bowel Disease

Saad Pathan and Derek Jewell

Ulcerative colitis (UC) and Crohn's disease (CD) are chronic inflammatory diseases of the intestine that can begin at any age, even in children. Both diseases are commonly known as inflammatory bowel disease (IBD). They represent a major challenge, because their causation is unclear, medical treatment is far from satisfactory, and many patients require major surgery. Over the last 10 years, the application of genome-wide markers has reduced the components of familial aggregation to novel locations within the human genome. Subsequently, a variety of genetic mutations have been discovered in some of the novel loci that could render individuals susceptible to these diseases and might also influence the behavior of the disease once it has occurred. More recently, a new quantum wave of genome-wide case-control revaluation has followed the advent of a higher density of markers and high-resolution linkage disequilibrium (LD) maps. These investigations have rapidly implicated novel disease influencing genes and regions as well as replicated previously associated regions. The narrowing of the loci will require complementary investigations in gene function and epidemiology. Nevertheless, there is accumulating evidence that an unregulated triggering of the mucosal immune system toward normal intestinal flora influences the development of chronic inflammatory disease. Furthermore, inflammatory processes are also likely to be provoked by genetic variants that adversely affect innate immunity including mucosal barrier integrity and autophagy. The combination of genetics and functional characterization should provide a framework to dissect the molecular mechanisms of disease.

A Complex Clinical Predisposition with Complex Complications

In 1859, Samuel Wilks, at Guy's Hospital in London, differentiated UC from bacterial dysentery. By 1931, Sir Arthur Hurst described clinical and sigmoidoscopic features of the disease and showed that the rectum was always affected with extension proximally to involve a variable length of the colon. The disease is most commonly seen in individuals between 20 and 40 years, but can be present in the first few months of life or in adults in their 80s. The disease is characterized by a relapsing and remitting course in the majority of patients.

The formal recognition of CD as distinct from UC had to await the finding in 1932 by Crohn, Ginzburg, and Oppenheimer. The granulomatous nature of the disease was recognized, but only in approximately 65% of patients, and it soon became clear that the disease could affect any part of the gastrointestinal tract. Nevertheless, it is by no means clear that UC and CD are two homogeneous disease entities, and the terms may well encompass a variety of disorders with shared clinical characteristics. When CD is confined to the colon, it may well be difficult to distinguish it from UC in 5%–10%—these patients are currently labeled as "colitis not yet classified."

Epidemiology

Over the last century, CD and UC have manifested in increasing incidence in various global populations with a current estimated prevalence of 0.15% in northwest Europe and North America (Binder, 2004). The environmental influences associated with modern urban developments have coincided with increasing incidence and prevalence of IBD in industrialized countries (see Fig. 22–1). The clinical manifestation of IBDs is likely to be due to a provocation of the mucosal immune system by commensal intestinal flora in susceptible individuals—in most individuals, these would be nonpathogenic (Podolsky, 2002). Accumulating evidence from genetic epidemiology implies that the host genome predisposes to autoimmune and/or inflammatory diseases (Ahmad et al., 2004; Schreiber et al., 2005). However, genetics alone are not adequate in explaining the disease, since animal models that are genetically altered for predisposition to colitis do not develop the phenotype when kept in a germ-free environment (Strober, 1985; Blumberg et al., 1999).

Nature versus Nurture

The manifestation of familial clustering as observed by Kirsner in 1963 set the foundation for subsequent genetic investigations. Furthermore, within families, there was a striking concordance of clinical characteristics (Bayless et al., 1996; Satsangi et al., 1996). A more rigorous validation was provided from twin studies that compared disease concordance in monozygotic (MZ) and dizygotic (DZ) twins, because familial clustering does not entirely exclude environmental influences (Tysk et al., 1988; Thompson et al., 1996; Orholm et al., 2000; Halfvarson et al., 2003). The twin studies provide an indication of the maximum risk of recurrence. The respective concordance rates among MZ twins with CD and UC

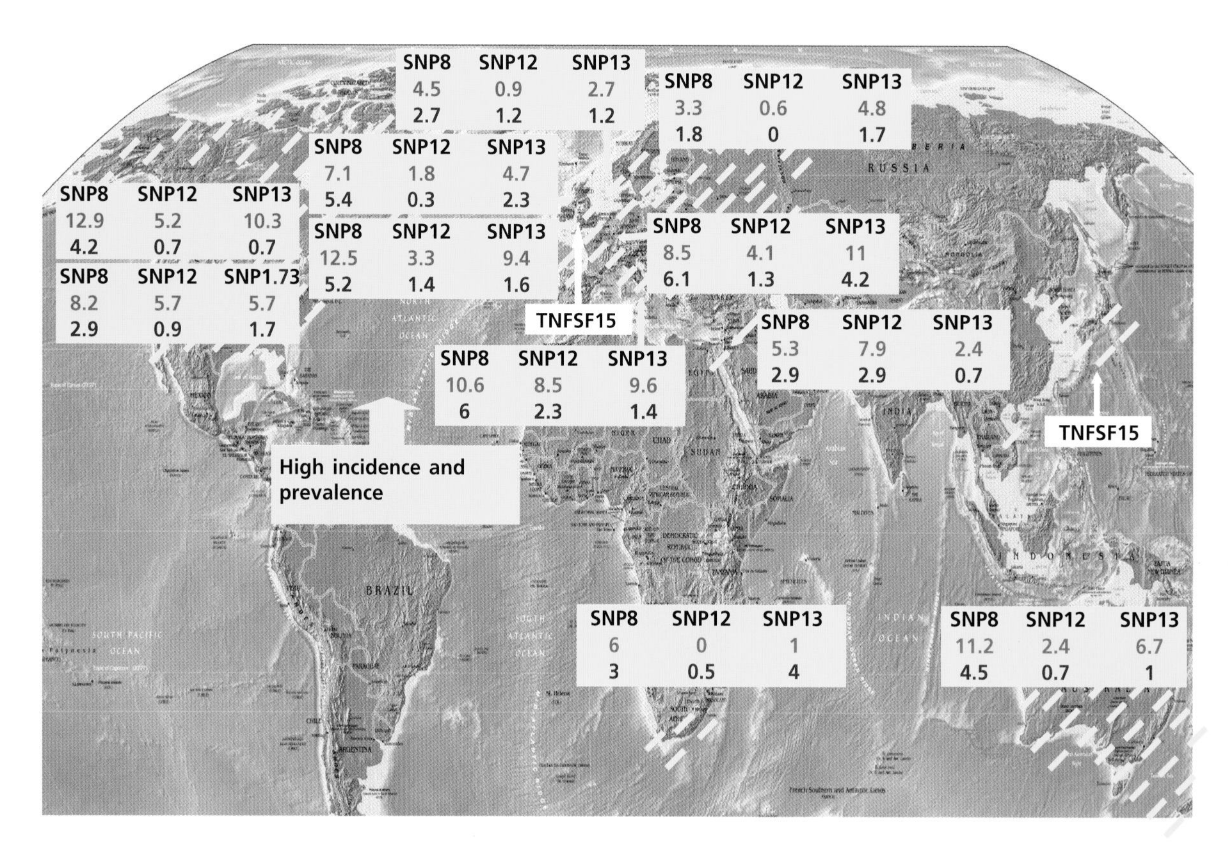

Figure 22–1 The regions known for high incidence and prevalence of inflammatory bowel diseases are highlighted on the global atlas. The allele frequencies (shown in red with controls in blue) indicate association of the heterozygote caspase recruitment domain family, member 15 (CARD15) mutations Arg702Trp (SNP8), Gly908Arg (SNP12), and the frameshift 3020 insC (SNP13) among populations of European descent and European admixture populations affected with Crohn's disease (see section Epidemiology of CARD15; Hampe et al., 2002; Cavanaugh et al., 2003; Fidder et al., 2003; Palmieri et al., 2003; Zaahl et al., 2005). In contrast, a core TNFSF15 haplotype is associated with CD in both Caucasians and Japanese populations (see section TNFSF15).

range from 37% to 10%, while the corresponding concordance rates among DZ twins is only 10% and 3%. Since twins in general share similar environments during childhood, the higher disease concordance rates in MZ twins are a very strong argument for genetic susceptibility.

An alternative to a comparison of the concordance rate between MZ and DZ twins affected with IBD is the measurement of risk-to-relative ratio for a relative of type R, known as λ_R; this method also yields an assessment of the genetic component (Risch, 1990). The λ_R quantifies the familial component of a discrete trait that describes the relationship: first-degree sibs, closely related family members, and/or relatives. Therefore, if the risk of a relative or affected sib is 3%–3.5% for CD, and the population prevalence is 0.1%–0.2%, then the λ_S value is 15–35 (Satsangi et al., 1994). The λ_S value between 6 and 9 for UC (Orholm et al., 1991) implies that the genetic influence is not as powerful as it is for CD. This measurement could include discrepancies due to the introduction of a bias if the family members are exposed to the same environmental factors (Guo, 1998). The λ_S value may fluctuate among different populations.

Markers for Mapping Genetic Susceptibility

In some of the earlier studies, there was the suggestion that CD could segregate as an autosomal recessive phenotype and UC in an autosomal dominant mode (Orholm et al., 1993). However, more recent formal segregation analysis in families with IBD fails to provide consistent estimates of modes of inheritance, mainly due to the involvement of a number of genes as well as interaction with environment (Koutroubakis et al., 1996). These IBD susceptibility genes tend to aggregate in almost all families rather than segregate within families.

The genotyping tools of human molecular genetics have been instrumental in substantiating the genetic component of IBD (Table 22–1). Linkage analysis with evenly spaced microsatellite markers throughout the different chromosomes of the genome in a collection of families, then denser linkage mapping and, further fine mapping by association with single nucleotide polymorphisms (SNPs) and other microsatellite markers are steps that have led to the identification of caspase recruitment domain family, member 15 (CARD15 formerly known as NOD2) at 16q12 [see section "IBD1 at chromosome 16q (CARD15)"], an unsuspected gene (Hugot et al., 2001). Ideally, genome-wide linkage studies are undertaken to map genes that would otherwise not be considered, as well as for enabling the coverage of the exhaustive list of candidate genes and/or loci that require systematic analysis. Advances in high-throughput genotyping to efficiently search for susceptibility loci were pioneered with the implementation of fluorescently labeled genotyping microsatellite markers in a genome-wide linkage to search in families with type 1 diabetes (Davies et al., 1994). Initially, for IBD, there was a general lack of replication for the numerous loci of linkage that were reported, and this added to the burden of drawing conclusions from genome search results, as was the case for other complex

Table 22–1 Comparison of Crohn's disease with ulcerative colitis (Adapted from the Oxford Textbook of Medicine)

	Crohn's Disease	Ulcerative Colitis
Genetic Epidemiology		
Prevalence	27–106 per 1,00,000	80–150 per 1,00,000
Incidence	4–10 per 1,00,000	6–15 per 1,00,000
Sibling affected risk ratio	30–42	4
Concordance between identical twins	37%	10%
Concordance between nonidentical twins	10%	3%
Mode of inheritance	Complex	Complex
Loci predominantly linked with either Crohn's or ulcerative colitis	IBD1, IBD5	IBD2
Associated genes	*CARD15* (ileal disease), *TNFSF15, IL23R, ATG16L1, IRGM, OCTN, PTPN2*	*IL23R*
Major histocompatibility complex (MHC)	HLA DR7, DRB3*0301, DQ4, and DRB1*0103 (associated with colonic Crohn's)	HLA class II DR2, and DRB1*0103
Predisposing environmental factor	Smoking	
Protective environmental factor		Smoking
Clinical Features		
Bloody diarrhea	Less common	Common
Abdominal mass	Common	Rare
Perianal disease	Common	Less common
Malabsorption	Frequent (ileal disease)	Never
Radiological/Endoscopic features		
Rectal involvement	Frequently spared	Invariable
Distribution	Segmental discontinuous	Continuous
Mucosa	Cobblestones, fissure ulcers	Fine ulceration, "double contour"
Strictures (narrowing)	Common	Rare
Fistulas (opening)	Frequent	Rare
Histological Features		
Distribution	Transmural	Mucosal
Cellular infiltrate	Lymphocytes, plasma cells, macrophages	Lymphocytes, polymorphs, plasma cells, eosinophils
Glands	Gland preservation	Mucus depletion, gland destruction, crypt abscess
Special features	Aphthoid ulcers, granulomas, histiocyte-lined fissures	None

diseases (Altmuller et al., 2001). Nevertheless, there was rapid replication of the novel highly significant 16q12 locus, designated as IBD1 (Hugot et al., 1996; Cavanaugh, 2001). To date, there have been up to a dozen independent genome search results that have been reported from multiply affected IBD families of European ancestry and an emerging pattern of replication now reaffirms that the following loci are significantly linked with IBD (Fig. 22–2): 16q, 12q, 6p, 14q, 5q, 19p, 1p, 16p, and 3p (Ahmad et al., 2004; Gaya et al., 2006). Furthermore, both of the recent meta-analysis by pooling the results from the different genome search have shown that IBD3 is significantly linked with disease, and that IBD1 and IBD6 show suggestive linkage (Williams et al., 2002; van Heel et al., 2004). The reasons for these loci not being consistently replicated in all of the independent investigations could be due to one or more of the following: a false positive; a true locus could be population specific; low power; variable number of CD and UC patients; different criteria for diagnosis; different markers; and variation in genotyping quality. Similarly, pooling the data for meta-analysis can be hindered by these differences between the different investigations.

The identification of the tumor necrosis factor ligand superfamily member 15 *(TNFSF15)* locus at 9q32 (see section on *TNFSF15*) marked a new era for genotyping across the whole genome for case–control investigations with SNP markers in CD, as a successor to genome-wide search with dinucleotide repeat sizing for linkage analysis in families. For fine mapping and genome-wide association (GWA), the availability of a higher density of SNPs has offered a higher resolution and is better at detecting association in a region of short-range LD (Kruglyak, 1999). Directly genotyping variants with potential for function, such as non-synonymous genome-wide SNPs has successfully implicated the autophagy related gene, the autophagy 16-like gene *(ATG16L1)* at chromosome 2q37 in a German population case-control investigation (Hampe et al., 2007). Microarray platforms that utilize genome-wide haplotype-tagged SNPs derived from the HapMap project have also efficiently led to the identification of novel genetic variants that are associated with IBD. This is exemplified by the implication of a subunit of the receptor for the proinflammatory cytokine interleukin-23 *(IL23R)* gene at chromosome 1p31 with non-coding variants that increase

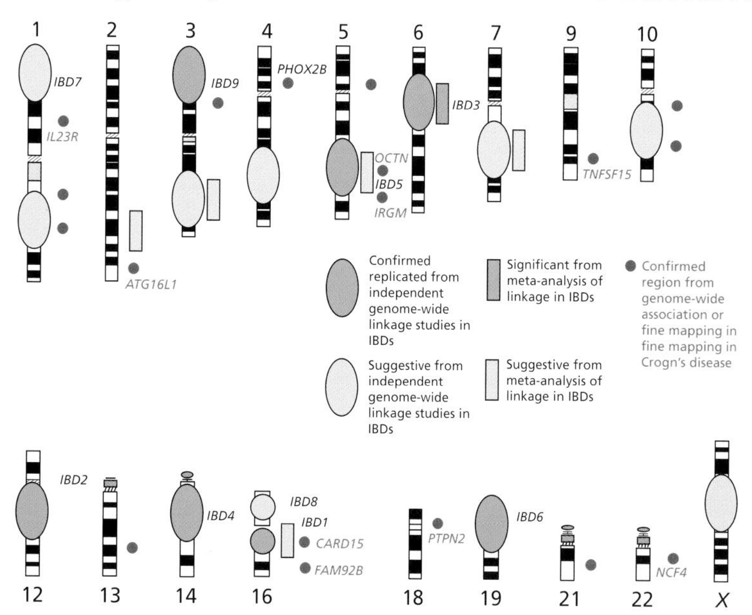

Figure 22–2 Loci that have been confirmed and replicated in independent genome searches according to the Lander and Kruglyak criteria (Adapted from Ahmad et al., 2001). Meta-analysis has confirmed linkage to inflammatory bowel disease-3 (IBD3) and suggestive linkage to chromosomes 2, 3, 5, and 7 (van Heel et al., 2004). The regions confirmed from genome-wide association (GWA) and fine-mapping approaches have been marked in red.

the risk of disease and a rare coding variant that protects from disease in North American populations of European ancestry (Duerr et al, 2006). The other novel genome-wide significant associated regions from an extension to this investigation included paired-like homeobox 2B (*PHOX2B*) 4p13, neutrophil cytosolic factor 4 (*NCF4*) at 22q12, a predicted gene (*FAM92B*) at 16q24, an intergenic locus at 10q21, as well as additional evidence for *ATG16L1* (Rioux et al., 2007). A similar approach for GWA in a Belgian investigation discovered a novel association of CD with a gene desert region with no recognized genes other than CpG islands on chromosome 5p13 (Libioulle et al., 2007). A larger scale GWA with a high-density randomly genotyped SNPs in the UK has independently unraveled another novel autophagy-inducing gene, the interferon-inducible protein 1 (*IRGM*) gene on chromosome 5q33 (The Wellcome Trust Case Control Consortium 2007; Parkes et al., 2007). This investigation also revealed novel association with gene deserts on chromosome 1q. The other novel loci included 3p21, 5q33, 10q24 and 21q22. Also, the analysis of the GWA results among several diseases showed a novel association with the T-cell protein tyrosine phosphatase (*PTPN2*) gene locus at chromosome 18p in both CD and type 1 diabetes, and subsequently replicated in other cases of both the diseases (Parkes et al., 2007; Todd et al., 2007).

There is a consistent pattern of replication between the different GWAs but the intersection between genome wide linkage and association is only visible at the IBD1 and IBD5 locus (see Fig. 22–2). This correlation might also occur at IBD3 and IBD9—these loci are gene dense and show higher levels of LD that makes it difficult to parse out the causative variants. However, the other loci show no correlation between the previously undertaken genome-wide linkage scans and the recently undertaken GWA. The meticulous genotyping of microsatellite markers can yield equivalent information in comparison to that of a higher density of SNP markers for linkage analysis and thus their use might not have been a major reason for the lack of correlation between genome-wide linkage and association studies. In retrospect, an over optimistic estimation of weak replication between genome-wide linkage scans to reaffirm loci might be one of the contributing factors that has lead to a lack of correlation between the more robust GWA. In addition, the regions of aggregation in linkage are broader and their boundaries are not precisely demarcated. For the GWAs, however, the first generation of commercially available microarray platforms can yield a similar overall coverage in comparison to each other, but is not complete in different regions of the genome (Barret and Cardon, 2006). Therefore, a systematic meta-analysis that combines the results from the recently undertaken GWAs with different platforms might refine the locations of overlap between the IBD loci.

Unlike the common SNP markers of the *CARD15* gene, associated with CD (Vermeire et al., 2002) the rarer disease variants that could also be associated are not necessarily identical by descent between different families within a closely related population—even if the defective gene is the same. Therefore, family studies using linkage analysis, rather than assessing association with LD mapping, might be more appropriate in assessing contribution to disease, because the very rare disease allele that is associated with the disease would be more likely to occur in certain families rather than that at the level of a population, and it would not be sufficient to lead to association. Thus, although linkage is robust for a rare disease causing variant, a population case–control study for association would not be appropriate for some of the very rare putative CD susceptibility variants, for example, those described within the *CARD15* gene (Lesage et al., 2002). The rare variants are, however, more likely to be of a greater functional significance, but whether some of these very rare alleles that have been identified within the *CARD15* could segregate with the disease within a family is unknown.

Other markers, such as trinucleotide repeats might be of direct relevance in IBD susceptibility, if the expansion of their repeat size is associated with disease (Polito et al., 1996), similar to "anticipation" that has previously been described for Huntington's disease (Vegter-van der Vlis et al., 1976; Ridley et al., 1988, 1991). There is however no conclusive evidence regarding the role of anticipation in IBD, since this observation could merely be an earlier diagnosis in the later generation members of a family that are under clinical investigation, rather than genuine earlier onset of disease than in the preceding generation (Faybush et al., 2002).

Variation in the copy number of genes, such as beta-defensins, may also predispose to disease (Feller et al., 2006). Further insights of the contribution of copy number variants (CNV) in IBDs may be gained from the emerging technologies for typing SNPs as markers of copy number polymorphisms (CNP) across the genome (Komura et al., 2006).

Genome-wide Association in the Identification of TNFSF15

A prelude to the prolific nature of genome-wide SNP genotyping has been ascertained by defining the *TNFSF15* gene locus in the predisposition to CD in a Japanese population study with 94 CD patients and 752 common disease control subjects (Yamazaki et al., 2005). The technology included the application of multiplex polymerase chain reaction (PCR) and invader assay (Ohnishi et al., 2001) for the inclusion of up to 72,738 genome-wide SNPs, and 1888 of these SNPs displayed p-values <0.01. The second phase of genotyping from the genome-wide association investigation included 484 CD and 345 unaffected controls, and from this 22 SNPs showed p-values $<1 \times 10^{-4}$ of which several of these were within 280 kb of 9q32. Intriguingly, this region had previously shown a nominal multipoint locus near D9S2157 [MLOD (multipoint logarithm of odds score) = 1.41, p = 0.0054] (Cho et al., 1998), but was neglected until this re-evaluation from genome-wide association due to a lack of replication from genome-wide linkage studies. The detailed follow-up of this region proceeded with resequencing that revealed 143 SNPs including 28 novel SNPs. The region that spanned the *TNFSF15* gene displayed the most marked association, and the marker *TNFSF15_28* (14, 340T→C in intron 3 of *TNFSF15*), with $\chi^2 = 58.8$, $p = 1.71 \times 10^{-14}$, and odds ratio (OR) of 2.17 [95% (confidence interval) CI, 1.78–2.66]. Several other SNPs showed association within this *TNFSF15* genetic region, but none outside. In addition, a high-resolution LD map within this region showed 100 SNPs with a minor allele frequency of more than 15%. The LD block containing the *TNFSF15* gene and exon 1 of the *LOC389786* gene was more strongly associated in comparison to the LD blocks that did not span this *TNFSF15* gene. Therefore, logistic regression was applied to distinguish between the associations of the two different genes within this haplotype. The marker *TNFSF15_28*, however, remained associated with disease, even after adjusting for the other markers that were in the same LD block.

A replication study in the UK was then performed in 347 IBD trio families (161 CD, 175 UC, and 11 with indeterminate colitis), 233 IBD multiplex families (containing 263 individuals affected with CD and 196 individuals affected with UC), and an independent sample of 363 CD patients and 372 healthy controls. It was found that among the 10 selected markers, 5 were not informative. Nevertheless, the other five polymorphic markers were associated with the IBD phenotype. Although not all of the polymorphic markers in the Japanese were heterozygous in the UK, the variant markers did show similar trends in allelic and haplotypic association with disease. Furthermore, the recent genome-wide association from the UK has replicated this locus (The Wellcome Trust Case Control Consortium 2007). Thus, in contrast to CARD15 mutations and the IBD5 haplotype, a core *TNFSF15* haplotype is found to be associated with CD in both Caucasian and Japanese populations (see Fig. 22–1).

Structure and Function of TNFSF15

TNFSF15, otherwise known as *TL1A* or vascular endothelial growth factor 1, is expressed on endothelial cells (Yue et al., 1999), and is an endogenous inhibitor of angiogenesis. It mediates signaling via its cognate receptor DR3, a death domain receptor (Migone et al., 2002), and it is overexpressed in the mucosa of patients with active IBD (Bamias et al., 2003). Preliminary data suggest that lippopolysaccharides (LPS) can induce transcription of *TNFSF15* in leukocytes (Yamazaki et al., 2005).

The IBD Loci

IBD1 at Chromosome 16q (CARD15)

The initial genome-wide linkage analysis search for IBD susceptibility made a great leap forward with the semiautomated genotyping of 270 fluorescently labeled dinucleotide repeat microsatellite markers in 78 multiple sibs affected with CD (Hugot et al., 1996). This model-free approach identified a striking novel MLOD score of 3.17 between markers D16S408 and D16S409 spanning a 40 cM region that is pericentromeric at chromosome 16q (IBD1). Subsequent to the completion of other genome searches, it was found that the IBD1 locus was much more consistently replicated in comparison to other regions that were reported to confer susceptibility to IBD (Cho et al., 1998; Hampe et al., 1999; Williams et al., 2002). However, even this relatively well-replicated locus did not yield linkage in all the different genome searches, possibly reflecting lack of power to detect linkage and relative proportions and total numbers of CD and UC patients, as well as genetic heterogeneity (Satsangi, Parkes et al., 1996; Ma et al., 1999; Duerr et al., 2000; Paavola-Sakki et al., 2003; Barmada et al., 2004; Vermeire et al., 2004). Furthermore, the locus specific sibling λ_{locus} was estimated at 1.3 as predicted by the proportion of expected allele sharing by affected sibs pairs identical by descent (i.e., 25%) to the observed allele sharing (19.2%) (Hugot et al., 1996). From this λ_{locus} value it can be calculated (Risch, 1987) that the 16q (IBD1) might contribute less than 20% of the genetic influence in CD (Hugot, 2003). Linkage of the disease to 16q has been confirmed by two meta-analyses of published genome-wide screens (Williams et al., 2002; van Heel et al., 2004).

The fine mapping with association did not, however, await extensive replication from other centers, and was initiated as soon as multipoint linkage was ascertained between the microsatellite markers D16S409 and D16S419. The trios from the linked families were included for investigating association using the transmission disequilibrium test (TDT). The 16 microsatellite markers were spaced up to 1 cM apart across the region of linkage (up to 1 cM density). A positive TDT test is often regarded as a gold standard for association, since it can overcome the adverse effects of population stratification. However, the borderline association of $p < 0.05$ with an allelic association of marker D16S3136 in 108 families did not meet the strict criteria for TDT. The result was further put into question when 76 additional families displayed weak association with a different allele of marker D16S3136. The authors of this investigation hypothesized the role of allelic heterogeneity for the association with different alleles of the same marker. The other concern was that the sample size was not adequate to confirm or refute association. Furthermore, in other similarly sized collections, a negative association was detected with the same allele of this marker (van Heel, McGovern, et al., 2002). Nevertheless, fine mapping identified a 260 kb interval, which was then searched by association and fluorescent sequencing technology to detect mutations associated with CD. The expressed sequence tag (EST) showed homology to a novel gene with 11 SNPs and 2 microsatellite markers. The 11 exons of this putative gene revealed 2 missense SNPs (Arg702Trp), SNP12 (Gly908Arg), and a frameshift mutation SNP13 (3020 insC). The positional cloning result was confirmed by undertaking a case–control investigation that yielded much more significant association than with alleles of other microsatellite markers within the region of LD. This seminal discovery is regarded as the first novel unequivocal identification of a susceptibility gene using nonparametric approaches (Hugot et al., 2001). At the same time, equally remarkable were the results from the positional candidate gene approach that was undertaken simultaneously by an independent research group (Ogura, Bonen, et al., 2001), and subsequently replicated (Hampe et al., 2001). The rationale for selecting CARD15 as a positional candidate gene related to functional investigations that were emerging for the role of this gene in innate immunity (Ogura, Inohara et al., 2001). The successful genetic incrimination of *CARD15* gene in CD has opened new questions of how the mutations cause disease and their implications for the epidemiology of disease.

CARD15 Gene Structure and Function

The *CARD15* gene belongs to a family of cytosolic pattern-recognition receptors (Inohara and Nunez, 2003). Its role in innate immunity is based on its ability to recognize conserved structures within the gut flora (Inohara et al., 2003). There are now more than 20 different NOD-like human proteins, and they show varying degrees of homology to plant cytosolic R proteins. An acronym put forward to describe NOD-like human proteins is CATERPILLAR because leucine-rich repeats are found in the carboxy-termini [CATERPILLAR—CARD, transcription enhancer, R (purine)-binding, pyrin, and many leucine repeats (Harton et al., 2002)].

The *CARD15* gene consists of a central nucleotide-binding domain (NBD), a leucine-rich repeat, and two N-terminal CARD domains. The three common variants of CARD15 (Fig. 22–3) are located within the leucine-rich repeat; this includes the Arg702Trp (tryptophan substituted for arginine at codon 702), the Gly908Arg (arginine substituted for glycine at codon 908), and Leu1007finsC (frameshift mutation truncates ending 3% of protein). Subsequently, many other private mutations in the *CARD15* gene have been found in patients with CD (Lesage et al., 2002). Other novel rare mutations in the NBD were found to segregate with Blau syndrome: an autosomal granulomatous disease (Kanazawa et al., 2005). These mutations in the *CARD15* gene have been shown to provide a gain of function and increase nuclear factor-κB (NF-κB) activation in Blau syndrome (Miceli-Richard et al., 2001). However, not all granulomatous diseases are associated with CARD15 mutations as they have not been found in sarcoidosis (Schurmann et al., 2003; Ho et al., 2005), but association is possible in the early onset of disease (Kanazawa et al., 2005).

CARD15 is an intracellular pattern-recognition protein for bacterial detection (Chamaillard et al., 2003). The recognition of the peptidoglycan component of bacterial cell walls is dependent on the detection of muramyl dipeptide (MDP), a minimal motif that is an almost universal constituent of peptoglycans found in gram-negative and gram-positive bacteria (Girardin et al., 2003; Inohara et al., 2003). Specifically, it is the leucine-rich repeat region of CARD15 that has a role in binding to MDP, and consequently activating NF-κB through a number of other intracellular molecules (Chamaillard et al., 2003; Girardin et al., 2003; Inohara et al., 2003). Paradoxically, in vitro experiments have shown that common variants of CARD15 associated with CD decrease NF-κB activation (Bonen et al., 2003; Chamaillard et al., 2003; Chamaillard et al., 2003). It was hypothesized that this contradiction between in vitro and in vivo finding might be explained by *CARD15*-independent bacterial activation of NF-κB mediated by Toll-like receptors on macrophages (Inohara et al., 2002). Experimental results from in vivo experiments have been consistent with the hypothesis (Watanabe et al., 2004). Further insights have come from the transfection of epithelial cell lines with the CD-associated variants of *CARD15*. The transfected cells failed to kill intracellular *Salmonella typhimurium* compared to cells transfected with wild-type *CARD15* (Hisamatsu et al., 2003). A very novel mechanism whereby *CARD15* mutations may put individuals at risk by disease has recently been proposed. The expression of *CARD15* has been shown in human Paneth cells (Lala et al., 2003; Ogura et al., 2003). Subsequently, CD patients with *CARD15* mutations have been shown to have a reduced release of the antimicrobial peptides, α-defensins 4 and 5, from Paneth cells (Wehkamp et al., 2004, 2005a, 2005b; Wehkamp, Salzman, et al., 2005; Wehkamp, Schmid, et al., 2005; Wehkamp and Stange, 2005). Since Paneth cells are predominantly found in the

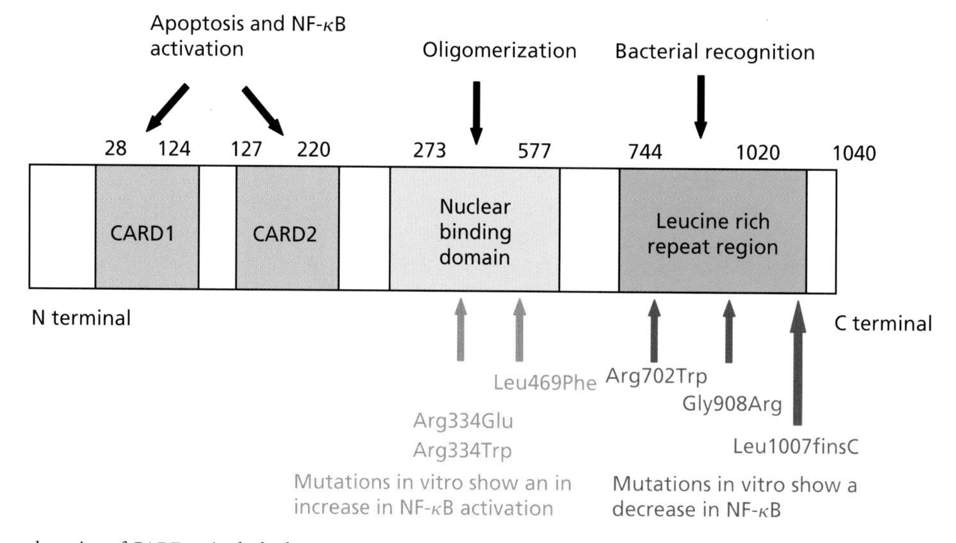

Figure 22–3 The three domains of CARD15 include the : caspase recruitment, nuclear binding and leucine rich domain. The three known variants that are associated with Crohn's disease display a loss of function, but the mutations that are associated with Blau syndrome reveal a gain in function in vitro.

terminal ileum and right colon, deficiency of antimicrobial peptides in the intestinal lumen may allow bacterial-induced inflammation to occur.

CARD15 knockout murine models have also been used to address the inconsistency of MDP activation between in vitro and in vivo (Kobayashi et al., 2005). It was found that the detection of bacterial MDP was abolished in the absence of *CARD15* in mice. Furthermore, the CARD15-deficient mice were more susceptible than wild-type mice to infection with bacteria via oral route but not with intravenous (IV) or intraperitoneal (IP) routes of infection. However, the *CARD15*-deficient mice show no evidence of intestinal inflammation. These mice also showed a reduced release of cryptdins from intestinal Paneth cells—cryptdins being the murine equivalent of human defensins.

In contrast to the loss of function in cell lines transfected with *CARD15* variants or the *CARD15* knockout models, knock in of the *CARD15* frameshift mutation has shown a much more efficient induction of cytokine interleukin-1β and NF-κB in response to MDP (Maeda et al., 2005). Perhaps, the presence of an alternative pathway for positive regulation of NF-κB activation in response to MDP could offer a possible reconciliation between elevated NF-κB activation in the experimental models using CARD15 variants.

Epidemiology of CARD15 in CD

The functional insights have been further supported by the compelling weight of replication of epidemiological disease association with populations of European descent (Hampe et al., 2001; Hugot et al., 2001; Ogura et al., 2001; Abreu et al., 2002; Ahmad et al., 2002; Cuthbert et al., 2002). These investigations have indicated that 10%–30% of CD patients are heterozygotes for one of the three common mutations (see Figs. 22–1 and 22–3). Another 3%–15% of patients are either homozygotes or compound heterozygotes. Although this relative risk of 20–40 in homozygotes (with a 1/25 risk of developing disease) is likely to be detected within a population case–control investigation, it would have been unlikely to have influenced the high-linkage score that was obtained from the genome search, because homozygotes tend to reduce the LOD score (Brant and Shugart, 2004). Furthermore, families unaffected by disease, yet possessing the high-risk alleles, have also been reported (Ahmad et al., 2002; Radlmayr et al., 2002; Linde et al., 2003). Moreover, within different populations of European descent, there is a variable association of these alleles with disease (see Fig. 22–1), even though the control allele frequencies within these populations are similar (Economou et al., 2004). These differences in attributable risk of the CARD15 disease susceptibility variants could in some cases correlate with the apparent North, South gradient (Vind et al., 2005). However, there is a reduced association of these susceptibility alleles with CD in Scotland (Arnott et al., 2004; Russell et al., 2005), Ireland (Bairead et al., 2003), and Finland (Helio et al., 2003). Moreover, in northern Europe the higher incidence and prevalence of disease is likely to be due to other genetic and environmental influences (Halfvarson et al., 2005). There is a notably reduced allele frequency of the disease susceptibility alleles within Afro-American communities (Kugathasan et al., 2005).There is also an absence or much rarer occurrence of CARD15 disease susceptibility alleles within a number of Asian populations, such as the Koreans (Croucher et al., 2003; Lee et al., 2005), Japanese (Inoue et al., 2002; Yamazaki et al., 2002) Chinese (Leong et al., 2003; Guo et al., 2004). In other communities, such as the Ashkenazi Jews, the much higher incidence and prevalence of CD could be irrespective of geographic and environmental influences in comparison to other groups (Zhou et al., 2002). In general, however, these differences of disease incidence and prevalence are usually thought to be due in part to migration and changes in the environment (Lanzarotto et al., 2005). Although the much higher Gly908Arg allele frequencies within the Askenazi appear to correlate with disease (Bonen and Cho, 2003), this particular genetic influence does not alone explain the much higher prevalence of CD within this community.

CARD15 Mutations and Phenotype

The association between the three common variants of *CARD15* has been consistently replicated with CD and only occasional weak associations with UC have been noted. Within CD (Ahmad et al., 2002), it is reported that the association was particularly strong for ileal CD. This has now been widely replicated and confirmed within a meta-analysis with populations of European descent (Economou et al., 2004). How *CARD15* variants influence the anatomical site of disease is not clear, but the possibility that Paneth cells are involved, as discussed in the preceding section, is intriguing. Paneth cells are predominantly located in the distal ileum, and it is possible that the lack of defensin production in individuals with CARD15 mutations may impair ileal antimicrobial defense mechanisms. CARD15 variants have also been reported to predispose to early onset of disease and to stricturing disease. However, these associations are not completely established, although they receive some support from the meta-analysis (Economou et al., 2004).

IBD2 Chromosome 12q24

The UK genome search of IBD reported linkage to a 41-cM region of chromosome 12q24 with a peak at marker D12S83 for both UC and CD (Satsangi et al., 1996). This was possibly influenced by the inclusion of a number of sibs affected with UC in comparison with the preceding genome search. This IBD2 locus was examined by the International IBD Consortium with the exclusion of the 68 families that were originally used in the UK (Cavanaugh, 2001). The linkage results did not add the same level of support for chromosome 12: the multipoint analysis LOD score was 1.2 at marker D12S368, and there was linkage specifically with UC at marker D12S85. Another study with model-free linkage analyses with microsatellite markers was undertaken in 122 North American Caucasian families of IBD-affected relative pairs (Duerr et al., 1998). This confirmed linkage to chromosome 12 [LOD* score 2.76 (p = 0.00016) between D12S1724 and D12S90]. The linkage results were further confirmed to have a greater influence in UC-affected relative pairs rather than CD (Parkes et al., 2000). A further study of North American families has replicated the linkage between UC and IBD2 (Achkar et al., 2006). In addition, from North America, TDTs for association revealed that there could be two distinct regions that are responsible for CD as well as UC (Duerr et al., 1998). The TDT was positive at the D12S83 locus with a global $\chi^2 = 16.41$, 6 df, and $p = 0.012$. There is also the possibility that LD could make it difficult to distinguish between two distinct regions that are within close physical proximity and share a similar population history of the immediate chromosomal environment. Furthermore, chromosome 12 has manifested a high degree of LD, as well as one of the largest haplotype blocks in comparison to that of the other chromosomes of the human genome (Scherer et al., 2006). In regions of long LD, positional candidates could be derived from expression profile results, in particular those that are differentially expressed in disease (Lawrance et al., 2001).

IBD3 (6p21)

At human chromosome 6, the genetic region of the major histocompatibility complex (MHC) is known to confer crucial immunological function and is linked with susceptibility to IBD (Yap et al., 2004;

Ahmad et al., 2006), and many other autoimmune diseases. This locus has been the most investigated over a period of 30 years, and more recently researched at a higher resolution with the tools of molecular genetics. However, a good number of investigations have yielded inconsistent results (Stokkers et al., 1999). Unlike most other suspected autoimmune diseases, IBD has not shown the same level of consistent linkage to the MHC from independent genome-wide searches. In fact, it was only when a denser set of markers were applied in the MHC locus that it became possible to detect linkage (Hampe et al., 1999). The most consistent data within the human leukocyte antigen (HLA) class II region have concerned HLA-DR2 and HLA-DRB1*0103. The DRB1*1502 allele of DR2 has been found predominantly in Japanese populations of UC patients (OR 3.74, CI 2.20–6.38), but has also been reported from North America and the UK (Stokkers et al. 1999; Futami et al., 1995; Yoshitake et al., 1999; Trachtenberg et al., 2000; Ahmad et al., 2003). The higher frequency of this allele in the Japanese general population, compared with its low frequency in Caucasians, probably explains the more consistent data from Japan. The association between UC and HLA-DRB1*0103 has been convincingly replicated in several large studies, which have also shown that patients possessing this allele are at risk of having severe colitis with a risk of colectomy (Bouma et al., 1997; Roussomoustakaki et al., 1997; Ahmad et al., 2003). More recently, it has been shown that this allele also influences the natural history of CD of the colon, that is, it is associated with severe disease and risk of colectomy early in the course of disease (Silverberg et al., 2003; Hancock et al., 2006). Whether this association with DRB1*0103 is causative or whether this allele is in tight LD with a more relevant allele is uncertain. However, DRB1*0103 appears to be present on only two small haplotypes, and thus a genuine association seems likely. The meta-analysis in 1999 also showed that there was a negative association between HLA-DR4 and UC (OR 0.54, CI 0.43–0.68). Subsequently, this protective effect was shown to be due to the DRB1*0401 molecular subtype but only when it is present on the DRB1*0401-DQB1*0301 haplotype (Ahmad et al., 2003).

For CD, meta-analysis has confirmed significant positive associations with DR7 (OR 1.42, CI 1.16–1.74), DRB3*0301 (OR 2.18, CI 1.25–3.80), and DQ4 (OR 1.88, CI 1.16–3.05) but, in contrast to UC, a negative association with DR2. The associations with DR7 and DR2 have been confirmed in a subsequent study (Akolkar et al., 2001).

Adjacent to the boundary between the HLA class I and class III regions, there are members of the nonclassical MHC class-I-related chain (MIC) gene family. *MICA* and *MICB* are polymorphic and are the only two members of the family to encode functional transcripts. They are predominantly expressed on the basal-lateral surface of epithelial cells and interact with NKG2D and a variety of natural killer (NK) cells, T cells, and macrophages, but particularly the CD8α T cells and the γT cells found in the intraepithelial compartment. So far, no consistent associations between UC or CD and polymorphisms in the *MICA* or *MICB* genes have been reported, and the most comprehensive study was negative (Ahmad et al., 2002).

Between the MHC class I and class II region is the highly gene-dense class III region spanning approximately 900 kb. The genes for TNFα, lymphotoxin α, and heat-shock proteins are found within class III, and polymorphisms within these genes have been associated with IBD. For TNFα, a number of promoter polymorphisms have been studied (at positions −1031, −308, and −857). In the UK, TNF-857C was associated with CD patients not possessing CARD15 mutations (van Heel et al., 2002) but, in Australia, TNF-857C was present only in patients with *CARD15* variants (O'Callaghan et al., 2003). Published data on the other promoter polymorphisms have been equally inconsistent (Ahmad et al., 2006). The common promoter haplotype (TNF −1031T, −863C, −857C, −308G, −380G, and −238G) has been shown to be associated with distal UC, which is stable over time and, at least for a minimum of a 10-year follow-up, does not extend (Ahmad et al., 2003). The mechanisms are unknown but patients homozygous for this haplotype might well have impaired TNF production, which could influence disease activity.

Polymorphisms in the lymphotoxin α gene and within a number of heat-shock protein genes are not consistently associated with either UC or CD (reviewed in Yap et al., 2004).

Genes within the HLA region on Chr 6p may also determine whether patients with either UC or CD are at risk of developing extraintestinal manifestations. Thus, the arthropathies, uveitis, erythema nodosum, recurrent mouth ulcers, and primary sclerosing cholangitis have been reported in association with class I and class II alleles (Ahmad et al., 2006). For example, the reactive large joint arthropathy seen in some patients with active colitis is strongly associated with HLA-B*27, HLA-B*35, and HLA DRB1*0103, whereas the symmetrical small-joint arthropathy is associated with HLA-B*44 (Orchard et al., 2000). Uveitis has been associated with HLA-B*27 and DRB1*0103, and erythema nodosum with TNF-1031C (Orchard et al., 2002). These associations have yet to be replicated and the data are inevitably based on small numbers. Nevertheless, it seems likely that the phenotypic heterogeneity seen in the clinic may partly be explained by genetic polymorphisms within the HLA region.

IBD4 (14q11-12)

This locus with suggestive linkage (MLOD > 2) was first shown on chromosome 14q11.2 in 222 individuals from 46 families (20 Jewish and 26 non-Jewish), with a total of 65 sib pairs diagnosed with CD (Ma et al., 1999). Another investigations was undertaken with a higher density of 751 microsatellite markers in 127 CD-affected relative pairs from 62 families (Duerr et al., 2000). Single-point nonparametric linkage (NPL) analysis revealed evidence for linkage to the adjacent D14S261 and D14S283 loci on chromosome 14q11-12 (LOD = 3.00 and 1.70, respectively), and the maximal multipoint significant linkage at D14S261 (LOD = 3.6). The finding of suggestive and significant linkage at the same locus from both of these independent studies satisfied the criteria for confirmed linkage. In addition, genome scanning in a Flemish IBD population found nominal evidence for linkage 14q11, overlapping with the previous genome scan results, with linkage on 14q11-12, and could support the IBD4 locus with a variable interaction with other loci (epistasis), as well as the environment (Vermeire et al., 2004). There has also been further evidence for replication and the interaction of the IBD locus with environmental factors such as smoking (Pierik et al., 2005). This was undertaken as part of an ongoing international collaborative effort with a total of 733 IBD families that consisted of 892 affected sibling pairs that were genotyped for microsatellite markers within the IBD4 region. The results showed that the predisposition of this locus was increased in smokers, but did not address the mechanism by which smoking influences IBD4.

The *interleukin-25* (*IL-25*) gene was investigated as a positional candidate in the 14q11.2 region, because it is a proinflammatory cytokine (Buning et al., 2003). Furthermore, Th2 responses can be induced with cytokines such as *IL-4*, *IL-5*, and *IL-13*, and therefore it was hypothesized that an imbalance of the Th1/Th2 might be due to variation in the *IL-25* gene. However, typing the 424C/A coding polymorphism in the *IL-25* gene in patients with CD or UC did not reveal any significant difference from controls. The genotype–phenotype relations in patients with CD showed a comparable distribution of the 424C/A polymorphism in all subgroups of the Vienna classification, and this led the authors to conclude that genetic alterations in the coding regions of the *IL-25* gene are not likely to have a major influence in IBD.

IBD5 5q31 Cytokine Gene Cluster (250 kb Cytokine Gene Cluster)

As in the case of CARD15, the narrowing of the 250 kb cytokine gene cluster has made it to the Hall of Fame for the mapping of a complex diseases. The genome-wide significance with a higher density of microsatellite markers at 5q31 locus was reported in 158 sib-pair families from Canada (Rioux et al., 2000). In fact, only suggestive linkage was shown by two previous studies (Cho et al., 1998; Ma et al., 1999), but it was convincingly replicated by the pooling data from various genome searches (van Heel et al., 2004). The fine mapping with 56 microsatellite markers spanning the 18 cm linked region in 256 mother–father–affected offspring trios yielded a significant association in two markers with a ($p < 0.0005$) (Rioux et al., 2001). This result was followed up by typing additional markers within the 1 Mb region that encompassed the two markers that were significantly associated. In general, the other markers between the two associated markers sustained the significant association ($p < 0.0005$). A multimarker haplotype analysis further narrowed this region to 500 kb. However, there was no observation of causal sequence variants among eight individuals that were screened. Nevertheless, a total of 651 SNPs were identified within the 285 kb most associated region and a 185 kb region that was sequenced either side. This was followed by selecting 301 SNPs for genotyping 139 trios. The novel finding of this ultra-high-resolution genotyping was that 2–4 haplotypes represented 90%–98% of that in the population within discrete blocks (up to 100 kb), and recombination was interpreted as separating these regions of high LD, but there was some haplotype correlation between different blocks (Daly et al., 2001). The capture of SNPs in a region of high LD with a minimum subset of common SNPs is known as haplotype tagging (Johnson et al., 2001) and these investigations have offered a paradigm for using this approach in investigating association in other complex diseases. The multilocus analyses showed a haplotype of 250 kb with maximum transmission with a ratio of 2:5:1 and a p-value of 0.002 and $p < 0.05$. This association was also replicated in four different collections of trios (Armuzzi et al., 2003; Giallourakis et al., 2003; Mirza et al., 2003). The overall accumulated association from the pooling of these independent investigations yielded a p-value of 10^{-10} and a modest odds ratio between 1.3 and 1.4 and therefore replicated the smaller studies. The replication of this association by independent groups in European populations has validated that association investigations can yield more power than linkage (Giallourakis et al., 2003; Mirza et al., 2003). The further fine mapping of this region by association has been hampered by LD. Nevertheless, some initial functional investigations have suggested that two putative functional SNPs in the carnitine/organic cations transporter (*OCTN*) cluster encoded by the adjacent genes *SLC22A4* and *SLC22A5* that are in LD could be responsible for the IBD5 at risk haplotype (Peltekova et al., 2004). The two different SNPs in the respective genes could collectively impair *OCTN*. In this population, the two mutations were shown to be in LD with each other, and the association of this *OCTN-TC* haplotype was found to be independent of the extended 250 kb IBD5 haplotype. The *OCTN-TC* haplotype was also replicated in a German population, but was not observed to be independent of the extended IBD5 haplotype (Torok et al., 2005).

Experimental results from *OCTN* have shown gene expression in various tissues such as the brain, intestine, skeletal muscle, heart, kidney, and intestines, suggesting that its function could influence disease (Tamai et al., 1997; Kekuda et al., 1998; Wu et al., 1998; Lahjouji et al., 2001). However, other putative functional SNPs within the associated haplotype block have not yet been observed. Furthermore, a number of population case–control investigations with specific phenotypes have not been consistent, such as ileal, colonic, and perianal disease (Rioux et al., 2001; Armuzzi et al., 2003; Giallourakis et al., 2003; Mirza et al., 2003; Negoro et al., 2003). However, the genome-wide search from the UK has shown association to this locus with CD (The Wellcome Trust Case Control Consortium 2007).

IBD6 (Chromosome 19)

The IBD6 locus was first shown in families that were predominantly affected with CD (Rioux et al., 2000). The IBD6 locus shows a similar degree of linkage replication as IBD1 and IBD3, and in the meta-analysis, it shows linkage for families that were affected with CD (van Heel et al., 2004). The most recent replication with CD was from Oxford with suggestive linkage on 19p13 to CD (LOD 1.59), but not with UC (Low et al., 2004). This locus has been replicated in various other inflammatory and autoimmune diseases, and there has been a good deal of overlap between the investigated candidate genes, such as intracellular adhesion molecule-1 (*ICAM-1*) (Nejentsev et al., 2003), suggesting that it could serve as a common pathway for disease. Experimental investigations suggested that one of the possible roles of *ICAM1* is in the recruitment of lymphocytes (Miner et al., 2004).

As was the case for the 16q locus, where the subsequent associations of the known *CARD15* disease susceptibility alleles did not entirely explain the residual linkage at this locus, it is possible that the identification of the common disease associated variants at chromosome 19 may also not entirely explain this linkage. In addition, it could be that the variants are phenotype influencing rather than influencing the overall disease susceptibility. There are a number of positional candidates that could influence the phenotype of CD by disrupting barrier function and/or contribute to the inflammatory process, such as *EMR1, 2, 3,* and *CD97* (Hooper et al., 2001). The *EMR1* is a member of a class B seven-span transmembrane receptor family. The expression is most apparent on leukocytes, and with the potential to modulate myeloid–myeloid interactions. An Oxford case–control investigation has surveyed *EMR1, 2, 3,* and *CD97* with 379 Caucasian patients with CD and 377 with UC (Unpublished). Among these four selected candidate genes, the EMR3 shows the most support for influencing CD. In particular, Q127 associated with ileal disease ($p = 0.005$, OR 2.7) and fibrostenosing disease ($p < 0.0001$, OR 3.8). These investigations of disease subphenotypes are impeded by small sample size, which makes it more difficult to detect or refute association. The functional significance of these results has yet to be explained. There are also a number of other candidate genes from at least 160 known genes that could influence CD within the 19p locus. These candidate genes include CARD8 (formerly known as TUCAN) (Arnott et al., 2004). However 60 positional candidates genes on chromosome 19p have not shown significant association (Tello-Ruiz et al., 2006), therefore shifting the spotlight to those regions of 19p and 19q that have not yet been examined. However, if significant association is not detected with the other genes it could be that the linkage shown in affected sibs was due to the collective contribution of a number of genes with CD and/or disease subtypes of CD. There is also the possibility that association within this locus can be shown in UC.

1p36 (IBD7)

The novel chromosome 1p36 locus was initially shown in a genome-wide search from North America with maximum LOD score of 2.65, $p = 2.4 \times 10^{-4}$ (Cho et al., 1998). Linkage was also shown in a Chaldean population—the susceptibility variant is likely to be relatively recent within this population, which would explain the longer-range LD (Cho et al., 1998). Among Chaldeans, a dominant pattern of inheritance can be seen but also some recessive inheritance as well (Brant and Shugart, 2004). The recombination breakpoint mapping between different families could assist localization of the susceptibility gene. In addition to the 1p36 locus, further suggestive evidence for linkage was observed on chromosome 1q with a maximum LOD score of 2.08 in a large European cohort (Hampe, Schreiber, et al., 1999).

Chromosome 10 and Drosophila Discs Large Homologue 5 (DLG5)

Linkage to the chromosome 10 locus was revealed by a genome-wide search in families from Europe (Hampe, Schreiber, et al., 1999). Subsequently, the undertaking of fine mapping with association narrowed the region of interest to highlight the drosophila discs large homologue 5 (*DLG5*) gene (Stoll et al., 2004). This included two different haplotypes in the region that were associated with IBD and CD: the at-risk "haplotype D" consisted of a nonsynonymous SNP 113G->A coding for an amino acid substitution at R30Q. The association displayed was $\chi^2 = 8.1$ ($p = 0.004 \times 2 = 4.2$, $p = 0.04$) with independent replication in their case–control cohort ($p = 0.0001$, OR = 1.6). Furthermore, there was a protective haplotype that was identified by eight haplotype-tagging SNPs. This was under-transmitted in IBD trios but confirmed in case–control sample. The evidence for epistasis, that is nonlinear interaction, was between *DLG5* and *CARD15*. However, association with *DLG5* has not been widely replicated (Torok et al., 2005), and no locus–locus interactions with *CARD15* were detected. Furthermore, no association with any haplotype and IBD has been found (Noble et al., 2005; Tenesa et al., 2006). In addition, no genetic epistasis with *DLG5* could be confirmed, nor could association be detected with subphenotypes. However, studies of three populations of European descent replicated the association of IBD with R30Q in two of the populations (Daly et al., 2005).

The conflicting results of association with *DLG5* in case–control studies could in part be due to the inherent problems of a lack of standardization for defining the clinical subphenotypes, adjusting for others confounders, as well as a lack of power to detect association. In addition, it is possible that the application of experimental SNP genotyping, different genotyping platforms, and different quality control procedures applied by different centers could also in part contribute towards the conflicting results of association. Therefore, guidelines for high stringency quality control procedures have been suggested to address this concern when investigating the genetic association of complex diseases (Hattersley and McCarthy, 2005).

CARD4

The discovery of modest linkage at the 7p14 locus was first reported from the British genome-wide search (Satsangi, Parkes et al., 1996). Two other genome searches also showed some evidence of linkage to this locus (Cho et al., 1998; Rioux et al., 2000), and this evidence was reinforced with the meta-analysis (van Heel et al., 2004). The interest in this locus was reasserted following the realization that the *CARD4* (formerly known as *NOD1*) gene was a positional candidate. In addition, the multiple-drug-resistant gene 1 (MDR1) is also a positional candidate gene within this locus (Brant et al., 2003). The *NOD1* gene has structural similarity to *CARD15* gene that includes the leucine-rich repeat region (Fig. 22–3), but it has a single CARD domain rather than a double domain as in the case of CARD15. This CARD domain activates NF-κB and assists with apoptosis. A unique tripeptide motif (diaminopimelic acid) found in gram-negative bacterial peptidoglycan is recognized by CARD4 (Chamaillard, Hashimoto et al., 2003). The first case–control investigation with genotyping SNPs that were derived from sequencing CD patients was negative (Zouali et al., 2003). Additional variants for case–control investigations were derived from resequencing patients with asthma (Hysi et al., 2005). Twelve common SNPs in the *CARD4* gene that were identified were investigated in a panel of 556 IBD trios (McGovern et al., 2005). A significant association was indicated with a common deletion allele of a complex functional CARD4 indel polymorphism ($ND_1 + 32656^{*}1$) with IBD $p < 0.02$, CD $p < 0.05$, and UC $p < 0.07$. Furthermore, a significant association was also obtained with the haplotype in the terminal exonic region of *CARD4* gene and IBD ($p = 0.0000003$). The results from this investigation suggest that CARD4 could influence IBD to a similar degree as IBD5, but further replications are required.

Chromosome X

A broad coverage of linkage analysis of chromosome X was ascertained from 145 affected relative pairs (Vermeire et al., 2001). The first line of investigation was linkage analysis of 79 (68 CD and 11 mixed) IBD patients with 12 microsatellite markers. This was followed up with another 12 microsatellite markers spanning 146 cM of chromosome X. The results from the analysis with IBD yielded an LOD score of 1.52 with a $p = 0.0041$. For CD, there was an LOD score of 1.40, $p = 0.006$ and an LOD of 1.14 was associated with UC. Subsequently, 10 additional markers in X-pericentromeric region were genotyped with 62 new families. This yielded evidence of linkage over a 30-cM pericentromeric region with the following markers: DX991, DX990, and DXS8096. In the CD subgroup, the multipoint maximum LOD score was 2.5 and $p = 0.0003$. A higher density of markers shifted the evidence of linkage to Xq21.3. The DX1203 marker yielded an NPL of 2.90 ($p = 0.0017$). The positional candidate genes of interest for IBD include *IL2*γ, *IL13 receptor* α*2*.

Turner's syndrome, a recognizable multiple anomaly syndrome due to a number of constitutional X chromosome abnormalities, is associated with IBD. More than 20 reported cases have shown this association (Hayward et al., 1996). The reported frequency of IBD has been found to be much higher in patients affected with Turner's syndrome than in the general population. There has also been the report of a monogenic form of immunodeficiency and IBD being associated with a splice site mutation in the *NEMO* gene on chromosome X (Orstavik et al., 2006). Mutations in the *NEMO* gene result in the variable phenotype of incontinentia pigmenti (IP), an X-linked dominant disorder, predominantly seen in females, as it is lethal in males.

The Ancestor's Tale of Mutations That Predispose to IBD

The CARD15 mutations that are associated with CD are likely to post date the "out-of-Africa migration" since this has so far not been detected in a number of populations outside of European ancestry (Inoue et al., 2002; Stockton et al., 2004), and there is reduced allele frequency within the African–American communities (Kugathasan et al., 2005). The proline to serine amino acid substitution at position

268 of the *CARD15* gene has also not been detected in various global populations other than in those of European descent(Inoue et al., 2002; Stockton et al., 2004). The substitution at position 268 has, however, been found in LD with all of the three common variants that predispose to CD in European populations (Lesage et al., 2002). Although the evolutionary significance of this association is not understood (Cho, 2001), the absence of the variant allele at position 268 in various global communities other than in those of European descent suggests that it is not likely to precede the "out-of-Africa migration." In addition, the LD between the variant allele at position 268 and the three other known disease susceptibility alleles could have occurred by chance alone, since there is limited haplotype diversity within such a narrow region. The age of the mutation and geographical distribution of allele frequencies could provide clues for the survival advantage of this mutation.

Conclusion

Considerable progress has been made in understanding the role of genetics in both CD and UC. This has been achieved by assembling large cohorts of patients who have been meticulously documented in terms of clinical phenotype and by the application of new technologies developed for studying polygenic disorders. The genome-wide association studies have recently unraveled a number of unsuspected genes as well as novel regions of interest that should also lead to gene identification. The genetic knowledge so far has highlighted the important role of the host's innate immunity and its interaction with commensal bacteria including autophagy. Further functional insights are likely to come from expression microarrays and proteomic technology.

References

Abreu, MT, Taylor, KD, et al. (2002). Mutations in NOD2 are associated with fibrostenosing disease in patients with Crohn's disease. *Gastroenterology*, 123(3):679–688.

Achkar, JP, Dassopoulos, T, et al. (2006). Phenotype-stratified genetic linkage study demonstrates that IBD2 is an extensive ulcerative colitis locus. *Am J Gastroenterol*, 101(3):572–580.

Ahmad, T, Armuzzi, A, et al. (2002). The molecular classification of the clinical manifestations of Crohn's disease. *Gastroenterology*, 122(4):854–866.

Ahmad, T, Armuzzi, A, et al. (2003). The contribution of human leucocyte antigen complex genes to disease phenotype in ulcerative colitis. *Tissue Antigens*, 62(6):527–535.

Ahmad, T, Marshall, SE, et al. (2002). High resolution MIC genotyping: design and application to the investigation of inflammatory bowel disease susceptibility. *Tissue Antigens*, 60(2):164–179.

Ahmad, T, Marshall, SE, et al. (2006). Genetics of IBD: Topic highlights for the role of the HLA complex. *World J Gastroenterol*, 12(23):3628–3635.

Ahmad, T, Tamboli, CP, et al. (2004). Clinical relevance of advances in genetics and pharmacogenetics of IBD. *Gastroenterology*, 126(6):1533–1549.

Ahmad, T, Satsangi, J, et al. (2001). Review article: the genetics of inflammatory bowel disease. *Aliment Pharmacol Ther* 15(6):731–48.

Akolkar, PN, Gulwani-Akolkar, B, et al. (2001). The IBD1 locus for susceptibility to Crohn's disease has a greater impact in Ashkenazi Jews with early onset disease. *Am J Gastroenterol*, 96(4):1127–1132.

Altmuller, J, Palmer, LJ, et al. (2001). Genomewide scans of complex human diseases: true linkage is hard to find. *Am J Hum Genet*, 69(5):936–950.

Armuzzi, A, Ahmad, T, et al. (2003). Genotype–phenotype analysis of the Crohn's disease susceptibility haplotype on chromosome 5q31. *Gut*, 52(8):1133–1139.

Arnott, ID, Nimmo, ER, et al. (2004). NOD2/CARD15, TLR4 and CD14 mutations in Scottish and Irish Crohn's disease patients: evidence for genetic heterogeneity within Europe? *Genes Immun*, 5(5):417–425.

Bairead, E, Harmon, DL, et al. (2003). Association of NOD2 with Crohn's disease in a homogenous Irish population. *Eur J Hum Genet*, 11(3):237–244.

Bamias, G, Martin, C, III, et al. (2003). Expression, localization, and functional activity of TL1A, a novel Th1-polarizing cytokine in inflammatory bowel disease. *J Immunol*, 171(9):4868–4874.

Barmada, MM, Brant, SR, et al. (2004). A genome scan in 260 inflammatory bowel disease-affected relative pairs. *Inflamm Bowel Dis*, 10(1):15–22.

Barrett, JC and Cardon, LR (2006). Evaluating coverage of genome-wide association studies. *Nat Genet*, 38(6):659–662.

Bayless, TM, Tokayer, AZ, et al. (1996). Crohn's disease: concordance for site and clinical type in affected family members—potential hereditary influences. *Gastroenterology*, 111(3):573–579.

Binder, V (2004). Epidemiology of IBD during the twentieth century: an integrated view. *Best Pract Res Clin Gastroenterol*, 18(3):463–479.

Blumberg, RS, Saubermann, LJ, et al. (1999). Animal models of mucosal inflammation and their relation to human inflammatory bowel disease. *Curr Opin Immunol*, 11(6):648–656.

Bonen, DK and Cho, JH (2003). The genetics of inflammatory bowel disease. *Gastroenterology*, 124(2):521–536.

Bonen, DK, Ogura, Y, et al. (2003). Crohn's disease-associated NOD2 variants share a signaling defect in response to lipopolysaccharide and peptidoglycan. *Gastroenterology*, 124(1):140–146.

Bouma, G, Oudkerk Pool, M, et al. (1997). Evidence for genetic heterogeneity in inflammatory bowel disease (IBD); HLA genes in the predisposition to suffer from ulcerative colitis (UC) and Crohn's disease (CD). *Clin Exp Immunol*, 109(1):175–179.

Brant, SR and Shugart, YY (2004). Inflammatory bowel disease gene hunting by linkage analysis: rationale, methodology, and present status of the field. *Inflamm Bowel Dis*, 10(3):300–311.

Brant, SR, Panhuysen, CI, et al. (2003). MDR1 Ala893 polymorphism is associated with inflammatory bowel disease. *Am J Hum Genet*, 73 (6):1282–1292.

Buning, C, Genschel, J, et al. (2003). The interleukin-25 gene located in the inflammatory bowel disease (IBD) 4 region: no association with inflammatory bowel disease. *Eur J Immunogenet*, 30(5):329–333.

Cavanaugh, J (2001). International collaboration provides convincing linkage replication in complex disease through analysis of a large pooled data set: Crohn disease and chromosome 16. *Am J Hum Genet*, 68(5):1165–1171.

Cavanaugh, JA, Adams, KE, et al. (2003). CARD15/NOD2 risk alleles in the development of Crohn's disease in the Australian population. *Ann Hum Genet*, 67(Pt 1):35–41.

Chamaillard, M, Girardin, SE, et al. (2003). Nods, Nalps and Naip: intracellular regulators of bacterial-induced inflammation. *Cell Microbiol*, 5(9):581–592.

Chamaillard, M, Hashimoto, M, et al. (2003). An essential role for NOD1 in host recognition of bacterial peptidoglycan containing diaminopimelic acid. *Nat Immunol*, 4(7):702–707.

Chamaillard, M, Philpott, D, et al. (2003). Gene-environment interaction modulated by allelic heterogeneity in inflammatory diseases. *Proc Natl Acad Sci USA*, 100(6):3455–3460.

Cho, JH (2001). The Nod2 gene in Crohn's disease: implications for future research into the genetics and immunology of Crohn's disease. *Inflamm Bowel Dis*, 7(3):271–275.

Cho, JH, Nicolae, DL, et al. (1998). Identification of novel susceptibility loci for inflammatory bowel disease on chromosomes 1p, 3q, and 4q: evidence for epistasis between 1p and IBD1. *Proc Natl Acad Sci USA*, 95(13):7502–7507.

Croucher, PJ, Mascheretti, S, et al. (2003). Haplotype structure and association to Crohn's disease of CARD15 mutations in two ethnically divergent populations. *Eur J Hum Genet*, 11(1):6–16.

Cuthbert, AP, Fisher, SA, et al. (2002). The contribution of NOD2 gene mutations to the risk and site of disease in inflammatory bowel disease. *Gastroenterology*, 122(4):867–874.

Daly, MJ, Pearce, AV, et al. (2005). Association of DLG5 R30Q variant with inflammatory bowel disease. *Eur J Hum Genet*, 13(7):835–839.

Daly, MJ, Rioux, JD, et al. (2001). High-resolution haplotype structure in the human genome. *Nat Genet*, 29(2):229–232.

Davies, JL, Kawaguchi, Y, et al. (1994). A genome-wide search for human type 1 diabetes susceptibility genes. *Nature*, 371(6493):130–136.

Duerr, RH, Barmada, MM, et al. (1998). Linkage and association between inflammatory bowel disease and a locus on chromosome 12. *Am J Hum Genet*, 63(1):95–100.

Duerr, RH, Barmada, MM, et al. (2000). High-density genome scan in Crohn disease shows confirmed linkage to chromosome 14q11-12. *Am J Hum Genet*, 66(6):1857–1862.

Duerr, R. H., K. D. Taylor, et al. (2006). A genome-wide association study identifies IL23R as an inflammatory bowel disease gene. *Science*, 314 (5804):1461–1463.

Economou, M, Trikalinos, TA, et al. (2004). Differential effects of NOD2 variants on Crohn's disease risk and phenotype in diverse populations: a metaanalysis. *Am J Gastroenterol*, 99(12):2393–2404.

Faybush, EM, Blanchard, JF, et al. (2002). Generational differences in the age at diagnosis with Ibd: genetic anticipation, bias, or temporal effects. *Am J Gastroenterol*, 97(3):636–640.

Fidder, HH, Olschwang, S, et al. (2003). Association between mutations in the CARD15 (NOD2) gene and Crohn's disease in Israeli Jewish patients. *Am J Med Genet A*, 121(3):240–244.

Futami, S, Aoyama, N, et al. (1995). HLA-DRB1*1502 allele, subtype of DR15, is associated with susceptibility to ulcerative colitis and its progression. *Dig Dis Sci*, 40(4):814–818.

Gaya, DR, Russell, RK, et al. (2006). New genes in inflammatory bowel disease: lessons for complex diseases? *Lancet*, 367(9518):1271–1284.

Giallourakis, C, Stoll, M, et al. (2003). IBD5 is a general risk factor for inflammatory bowel disease: replication of association with Crohn disease and identification of a novel association with ulcerative colitis. *Am J Hum Genet*, 73(1):205–211.

Girardin, SE, Boneca, IG, et al. (2003). Nod2 is a general sensor of peptidoglycan through muramyl dipeptide (MDP) detection. *J Biol Chem*, 278 (11):8869–8872.

Guo, QS, Xia, B, et al. (2004). NOD2 3020insC frameshift mutation is not associated with inflammatory bowel disease in Chinese patients of Han nationality. *World J Gastroenterol*, 10(7):1069–1071.

Guo, SW (1998). Inflation of sibling recurrence-risk ratio, due to ascertainment bias and/or overreporting. *Am J Hum Genet*, 63(1):252–258.

Halfvarson, J, Bodin, L, et al. (2003). Inflammatory bowel disease in a Swedish twin cohort: a long-term follow-up of concordance and clinical characteristics. *Gastroenterology*, 124(7):1767–1773.

Halfvarson, J, Bresso, F, et al. (2005). CARD15/NOD2 polymorphisms do not explain concordance of Crohn's disease in Swedish monozygotic twins. *Dig Liver Dis*, 37(10):768–772.

Hampe, J, Cuthbert, A, et al. (2001). Association between insertion mutation in NOD2 gene and Crohn's disease in German and British populations. *Lancet*, 357(9272):1925–1928.

Hampe, J., A. Franke, et al. (2007). A genome-wide association scan of nonsynonymous SNPs identifies a susceptibility variant for Crohn disease in ATG16L1. *Nat Genet*, 39(2):207–211.

Hampe, J, Schreiber, S, et al. (1999). A genomewide analysis provides evidence for novel linkages in inflammatory bowel disease in a large European cohort. *Am J Hum Genet*, 64(3):808–816.

Hampe, J, Shaw, SH, et al. (1999). Linkage of inflammatory bowel disease to human chromosome 6p. *Am J Hum Genet*, 65(6):1647–1655.

Hampe, J, Grebe, J, et al. (2002). Association of NOD2 (CARD 15) genotype with clinical course of Crohn's disease: a cohort study. *Lancet*, 359 (9318):1661–1665.

Hancock, L, Ahmad, T, et al. (2006). Clinical and molecular characteristics of isolated colonic Crohn's disease. *Gut*, 55(Suppl. 2):A1.

Harton, JA, Linhoff, MW, et al. (2002). Cutting edge: CATERPILLER: a large family of mammalian genes containing CARD, pyrin, nucleotide-binding, and leucine-rich repeat domains. *J Immunol*, 169(8):4088–4093.

Hattersley, AT and McCarthy, MI, (2005). What makes a good genetic association study? *Lancet*, 366(9493):1315–1323.

Hayward, PA, Satsangi, J, et al. (1996). Inflammatory bowel disease and the X chromosome. *QJM*, 89(9):713–718.

Helio, T, Halme, L, et al. (2003). CARD15/NOD2 gene variants are associated with familially occurring and complicated forms of Crohn's disease. *Gut*, 52 (4):558–562.

Hisamatsu, T, Suzuki, M, et al. (2003). CARD15/NOD2 functions as an antibacterial factor in human intestinal epithelial cells. *Gastroenterology*, 124 (4):993–1000.

Ho, LP, Merlin, F, et al. (2005). CARD 15 gene mutations in sarcoidosis. *Thorax*, 60(4):354–355.

Hooper, LV, Wong, MH, et al. (2001). Molecular analysis of commensal host-microbial relationships in the intestine. *Science*, 291(5505):881–884.

Hugot, JP, Chamaillard, M, et al. (2001). Association of NOD2 leucine-rich repeat variants with susceptibility to Crohn's disease. *Nature*, 411 (6837):599–603.

Hugot, JP, Laurent-Puig, P, et al. (1996). Mapping of a susceptibility locus for Crohn's disease on chromosome 16. *Nature*, 379(6568):821–823.

Hugot, JP, Zouali, H, et al. (2003). Lessons to be learned from the *NOD2* gene in Crohn's disease. *Eur J Gastroenterol Hepatol*, 15(6):593–597.

Hysi, P, Kabesch, M, et al. (2005). NOD1 variation, immunoglobulin E and asthma. *Hum Mol Genet*, 14(7):935–941.

Inohara, N and Nunez, G (2003). NODs: intracellular proteins involved in inflammation and apoptosis. *Nat Rev Immunol*, 3(5):371–382.

Inohara, N, Ogura, Y, et al. (2002). Nods: a family of cytosolic proteins that regulate the host response to pathogens. *Curr Opin Microbiol*, 5(1):76–80.

Inohara, N, Ogura, Y, et al. (2003). Host recognition of bacterial muramyl dipeptide mediated through NOD2. Implications for Crohn's disease. *J Biol Chem*, 278(8):5509–5512.

Inoue, N, Tamura, K, et al. (2002). Lack of common NOD2 variants in Japanese patients with Crohn's disease. *Gastroenterology*, 123(1):86–91.

Johnson, GC, Esposito, L, et al. (2001). Haplotype tagging for the identification of common disease genes. *Nat Genet*, 29(2):233–237.

Kanazawa, N, Okafuji, I, et al. (2005). Early-onset sarcoidosis and CARD15 mutations with constitutive nuclear factor-kappaB activation: common genetic etiology with Blau syndrome. *Blood*, 105(3):1195–1197.

Kekuda, R, Prasad, PD, et al. (1998). Cloning and functional characterization of a potential-sensitive, polyspecific organic cation transporter (OCT3) most abundantly expressed in placenta. *J Biol Chem*, 273(26):15971–15979.

Kobayashi, KS, Chamaillard, M, et al. (2005). Nod2-dependent regulation of innate and adaptive immunity in the intestinal tract. *Science*, 307 (5710):731–734.

Koutroubakis, I, Manousos, ON, et al. (1996). Environmental risk factors in inflammatory bowel disease. *Hepatogastroenterology*, 43(8):381–393.

Kruglyak, L (1999). Prospects for whole-genome linkage disequilibrium mapping of common disease genes. *Nat Genet*, 22(2):139–144.

Kugathasan, S, Loizides, A, et al. (2005). Comparative phenotypic and CARD15 mutational analysis among African American, Hispanic, and White children with Crohn's disease. *Inflamm Bowel Dis*, 11(7):631–638.

Lahjouji, K, Mitchell, GA, et al. (2001). Carnitine transport by organic cation transporters and systemic carnitine deficiency. *Mol Genet Metab*, 73 (4):287–297.

Lala, S, Ogura, Y, et al. (2003). Crohn's disease and the NOD2 gene: a role for paneth cells. *Gastroenterology*, 125(1):47–57.

Lanzarotto, F, Akbar, A, et al. (2005). Does innate immune response defect underlie inflammatory bowel disease in the Asian population? *Postgrad Med J*, 81(958):483–485.

Lawrance, IC, Fiocchi, C, et al. (2001). Ulcerative colitis and Crohn's disease: distinctive gene expression profiles and novel susceptibility candidate genes. *Hum Mol Genet*, 10(5):445–456.

Lee, GH, Kim, CG, et al. (2005). Frequency analysis of NOD2 gene mutations in Korean patients with Crohn's disease. *Korean J Gastroenterol*, 45 (3):162–168.

Leong, RW, Armuzzi, A, et al. (2003). NOD2/CARD15 gene polymorphisms and Crohn's disease in the Chinese population. *Aliment Pharmacol Ther*, 17 (12):1465–1470.

Lesage, S, Zouali, H, et al. (2002). CARD15/NOD2 mutational analysis and genotype–phenotype correlation in 612 patients with inflammatory bowel disease. *Am J Hum Genet*, 70(4):845–857.

Libioulle, C., E. Louis, et al. (2007). Novel Crohn disease locus identified by genome-wide association maps to a gene desert on 5p13.1 and modulates expression of PTGER4. *PLoS Genet 3*(4):e58.

Linde, K, Boor, PP, et al. (2003). Card15 and Crohn's disease: healthy homozygous carriers of the 3020insC frameshift mutation. *Am J Gastroenterol*, 98 (3):613–617.

Low, JH, Williams, FA, et al. (2004). Inflammatory bowel disease is linked to 19p13 and associated with ICAM-1. *Inflamm Bowel Dis*, 10(3):173–181.

Ma, Y, Ohmen, JD, et al. (1999). A genome-wide search identifies potential new susceptibility loci for Crohn's disease. *Inflamm Bowel Dis*, 5 (4):271–278.

Maeda, S, Hsu, LC, et al. (2005). Nod2 mutation in Crohn's disease potentiates NF-kappaB activity and IL-1beta processing. *Science*, 307(5710):734–738.

McGovern, DP, Hysi, P, et al. (2005). Association between a complex insertion/deletion polymorphism in NOD1 (CARD4) and susceptibility to inflammatory bowel disease. *Hum Mol Genet*, 14(10):1245–1250.

Miceli-Richard, C, Lesage, S, et al. (2001). CARD15 mutations in Blau syndrome. *Nat Genet*, 29(1):19–20.

Migone, TS, Zhang, J, et al. (2002). TL1A is a TNF-like ligand for DR3 and TR6/DcR3 and functions as a T cell costimulator. *Immunity*, 16(3):479–492.

Miner, P, Wedel, M, et al. (2004). An enema formulation of alicaforsen, an antisense inhibitor of intercellular adhesion molecule-1, in the treatment of chronic, unremitting pouchitis. *Aliment Pharmacol Ther*, 19(3):281–286.

Mirza, MM, Fisher, SA, et al. (2003). Genetic evidence for interaction of the 5q31 cytokine locus and the CARD15 gene in Crohn disease. *Am J Hum Genet*, 72(4):1018–1022.

Negoro, K, McGovern, DP, et al. (2003). Analysis of the IBD5 locus and potential gene–gene interactions in Crohn's disease. *Gut*, 52(4):541–546.

Nejentsev, S, Guja, C, et al. (2003). Association of intercellular adhesion molecule-1 gene with type 1 diabetes. *Lancet*, 362(9397):1723–1724.

Noble, CL, Nimmo, ER, et al. (2005). DLG5 variants do not influence susceptibility to inflammatory bowel disease in the Scottish population. *Gut*, 54(10):1416–1420.

O'Callaghan, NJ, Adams, KE, et al. (2003). Association of TNF-alpha-857C with inflammatory bowel disease in the Australian population. *Scand J Gastroenterol*, 38(5):533–534.

Ogura, Y, Bonen, DK, et al. (2001). A frameshift mutation in NOD2 associated with susceptibility to Crohn's disease. *Nature*, 411(6837):603–606.

Ogura, Y, Inohara, V, et al. (2001). Nod2, a Nod1/Apaf-1 family member that is restricted to monocytes and activates NF-kappaB. *J Biol Chem*, 276(7):4812–4818.

Ogura, Y, Lala, S, et al. (2003). Expression of NOD2 in Paneth cells: a possible link to Crohn's ileitis. *Gut*, 52(11):1591–1597.

Ohnishi, Y, Tanaka, T, et al. (2001). A high-throughput SNP typing system for genome-wide association studies. *J Hum Genet*, 46(8):471–477.

Orchard, TR, Chua, CN, et al. (2002). Uveitis and erythema nodosum in inflammatory bowel disease: clinical features and the role of HLA genes. *Gastroenterology*, 123(3):714–718.

Orchard, TR, Thiyagaraja, S, et al. (2000). Clinical phenotype is related to HLA genotype in the peripheral arthropathies of inflammatory bowel disease. *Gastroenterology*, 118(2):274–278.

Orholm, M, Binder, V, et al. (2000). Concordance of inflammatory bowel disease among Danish twins. Results of a nationwide study. *Scand J Gastroenterol*, 35(10):1075–1081.

Orholm, M, Iselius, L, et al. (1993). Investigation of inheritance of chronic inflammatory bowel diseases by complex segregation analysis. *BMJ*, 306(6869):20–24.

Orholm, M, Munkholm, P, et al. (1991). Familial occurrence of inflammatory bowel disease. *N Engl J Med*, 324(2):84–88.

Orstavik, KH, Kristiansen, M, et al. (2006). Novel splicing mutation in the NEMO (IKK-gamma) gene with severe immunodeficiency and heterogeneity of X-chromosome inactivation. *Am J Med Genet A*, 140(1):31–39.

Paavola-Sakki, P, Ollikainen, V, et al. (2003). Genome-wide search in Finnish families with inflammatory bowel disease provides evidence for novel susceptibility loci. *Eur J Hum Genet*, 11(2):112–120.

Palmieri, O, Toth, S, et al. (2003). CARD15 genotyping in inflammatory bowel disease patients by multiplex pyrosequencing. *Clin Chem*, 49 (10):1675–1679.

Parkes, M, Barmada, MM, et al. (2000). The IBD2 locus shows linkage heterogeneity between ulcerative colitis and Crohn disease. *Am J Hum Genet*, 67(6):1605–1610.

Parkes, M., J. C. Barrett, et al. (2007). Sequence variants in the autophagy gene IRGM and multiple other replicating loci contribute to Crohn's disease susceptibility. *Nat Genet*, 39(7):830–832.

Peltekova, VD, Wintle, RF, et al. (2004). Functional variants of OCTN cation transporter genes are associated with Crohn disease. *Nat Genet*, 36(5):471–475.

Pierk M, Yang H, et al. (2005) IBD International Genetics Consortium. The IBD international consortium provides further evidence for linkage to IBD4 and shows gene-environment interaction. *Inflamm Bowel Dis*, 11(1):1–7.

Podolsky, DK (2002). The current future understanding of inflammatory bowel disease. *Best Pract Res Clin Gastroenterol*, 16(6):933–943.

Polito, JM, II, Rees, RC, et al. (1996). Preliminary evidence for genetic anticipation in Crohn's disease. *Lancet*, 347(9004):798–800.

Radlmayr, M, Torok, HP, et al. (2002). The c-insertion mutation of the NOD2 gene is associated with fistulizing and fibrostenotic phenotypes in Crohn's disease. *Gastroenterology*, 122(7):2091–2092.

Ridley, RM, Frith, CD, et al. (1988). Anticipation in Huntington's disease is inherited through the male line but may originate in the female. *J Med Genet*, 25(9):589–595.

Ridley, RM, Frith, CD, et al. (1991). Patterns of inheritance of the symptoms of Huntington's disease suggestive of an effect of genomic imprinting. *J Med Genet*, 28(4):224–231.

Rioux, JD, Daly, MJ, et al. (2001). Genetic variation in the 5q31 cytokine gene cluster confers susceptibility to Crohn disease. *Nat Genet*, 29 (2):223–228.

Rioux, JD, Silverberg, MS, et al. (2000). Genomewide search in Canadian families with inflammatory bowel disease reveals two novel susceptibility loci. *Am J Hum Genet*, 66(6):1863–1870.

Rioux, JD, Xavier, RJ, et al. (2007). Genome-wide association study identifies new susceptibility loci for Crohn disease and implicates autophagy in disease pathogenesis. *Nat Genet*, 39(5):596–604.

Risch, N (1990). Linkage strategies for genetically complex traits. II. The power of affected relative pairs. *Am J Hum Genet*, 46(2):229–241.

Risch, N (1987). Assessing the role of HLA-linked and unlinked determinants of disease. *Am J Hum Genet*, 40(1):1–14.

Roussomoustakaki, M, Satsangi, J, et al. (1997). Genetic markers may predict disease behavior in patients with ulcerative colitis. *Gastroenterology*, 112(6):1845–1853.

Russell, RK, Drummond, HE, et al. (2005). Genotype–phenotype analysis in childhood-onset Crohn's disease: NOD2/CARD15 variants consistently predict phenotypic characteristics of severe disease. *Inflamm Bowel Dis*, 11(11):955–964.

Satsangi, J, Grootscholten, C, et al. (1996). Clinical patterns of familial inflammatory bowel disease. *Gut*, 38(5):738–41.

Satsangi, J, Jewell, DP, et al. (1994). Genetics of inflammatory bowel disease. *Gut*, 35(5):696–700.

Satsangi, J, Parkes, M, et al. (1996). Two stage genome-wide search in inflammatory bowel disease provides evidence for susceptibility loci on chromosomes 3, 7 and 12. *Nat Genet*, 14(2):199–202.

Scherer, SE, Muzny, DM, et al. (2006). The finished DNA sequence of human chromosome 12. *Nature*, 440(7082):346–351.

Schreiber, S, Rosenstiel, P, et al. (2005). Genetics of Crohn disease, an archetypal inflammatory barrier disease. *Nat Rev Genet*, 6(5):376–388.

Schurmann, M, Valentonyte, R, et al. (2003). CARD15 gene mutations in sarcoidosis. *Eur Respir J*, 22(5):748–754.

Silverberg, MS, Mirea, L, et al. (2003). A population- and family-based study of Canadian families reveals association of HLA DRB1*0103 with colonic involvement in inflammatory bowel disease. *Inflamm Bowel Dis*, 9(1):1–9.

Stockton, JC, Howson, JM, et al. (2004). Polymorphism in NOD2, Crohn's disease, and susceptibility to pulmonary tuberculosis. *FEMS Immunol Med Microbiol*, 41(2):157–160.

Stokkers, PC, Reitsma, PH, et al. (1999). HLA-DR and -DQ phenotypes in inflammatory bowel disease: a meta-analysis. *Gut*, 45(3):395–401.

Stoll, M, Corneliussen, B, et al. (2004). Genetic variation in DLG5 is associated with inflammatory bowel disease. *Nat Genet*, 36(5):476–480.

Strober, W (1985). Animal models of inflammatory bowel disease—an overview. *Dig Dis Sci*, 30(Suppl.12):3S–10S.

Tamai, I, Yabuuchi, H, et al. (1997). Cloning and characterization of a novel human pH-dependent organic cation transporter, OCTN1. *FEBS Lett*, 419(1):107–111.

Tello-Ruiz, MK, Curley, C, et al. (2006). Haplotype-based association analysis of 56 functional candidate genes in the IBD6 locus on chromosome 19. *Eur J Hum Genet*, 14:780–790.

Tenesa, A, Noble, C, et al. (2006). Association of DLG5 and inflammatory bowel disease across populations. *Eur J Hum Genet*, 14(3):259–260; author reply 260–261.

The Wellcome Trust Case Control Consortium (2007). Genome-wide association study of 14,000 cases of seven common diseases and 3,000 shared controls. *Nature*, 447(7145):661–678.

Thompson, NP, Driscoll, R, et al. (1996). Genetics versus environment in inflammatory bowel disease: results of a British twin study. *BMJ*, 312(7023):95–96.

Todd, JA, Walker, NM, et al. (2007). Robust associations of four new chromosome regions from genome-wide analyses of type 1 diabetes. *Nat Genet*, 39(7):857–64.

Torok, HP, Glas, J, et al. (2005). Polymorphisms in the DLG5 and OCTN cation transporter genes in Crohn's disease. *Gut*, 54(10):1421–1427.

Trachtenberg, EA, Yang, H, et al. (2000). HLA class II haplotype associations with inflammatory bowel disease in Jewish (Ashkenazi) and non-Jewish caucasian populations. *Hum Immunol*, 61(3):326–333.

Tysk, C, Lindberg, E, et al. (1988). Ulcerative colitis and Crohn's disease in an unselected population of monozygotic and dizygotic twins. A study of heritability and the influence of smoking. *Gut*, 29(7):990–996.

van Heel, DA, Fisher, SA, et al. (2004). Inflammatory bowel disease susceptibility loci defined by genome scan meta-analysis of 1952 affected relative pairs. *Hum Mol Genet*, 13(7):763–770.

van Heel, DA, McGovern, DP, et al. (2002). Fine mapping of the IBD1 locus did not identify Crohn disease-associated NOD2 variants: implications for complex disease genetics. *Am J Med Genet*, 111(3):253–259.

van Heel, DA, Udalova, IA, et al. (2002). Inflammatory bowel disease is associated with a TNF polymorphism that affects an interaction between the OCT1 and NF(-kappa)B transcription factors. *Hum Mol Genet*, 11(11):1281–1289.

Vegter-van der Vlis, M, Volkers, WS, et al. (1976). Ages of death of children with Huntington's chorea and of their affected parents. *Ann Hum Genet*, 39(3):329–334.

Vermeire, S, Rutgeerts, P, et al. (2004). Genome wide scan in a Flemish inflammatory bowel disease population: support for the IBD4 locus, population heterogeneity, and epistasis. *Gut*, 53(7):980–986.

Vermeire, S, Satsangi, J, et al. (2001). Evidence for inflammatory bowel disease of a susceptibility locus on the X chromosome. *Gastroenterology*, 120(4):834–840.

Vermeire, S, Wild, G, et al. (2002). CARD15 genetic variation in a Quebec population: prevalence, genotype–phenotype relationship, and haplotype structure. *Am J Hum Genet*, 71(1):74–83.

Vind, I, Vieira, A, et al. (2005). NOD2/CARD15 gene polymorphisms in Crohn's disease: a genotype-phenotype analysis in Danish and Portuguese patients and controls. *Digestion*, 72(2–3):156–163.

Watanabe, T, Kitani, A, et al. (2004). NOD2 is a negative regulator of Toll-like receptor 2-mediated T helper type 1 responses. *Nat Immunol*, 5 (8):800–808.

Wehkamp, J, Fellermann, K, et al. (2005a). Mechanisms of disease: defensins in gastrointestinal diseases. *Nat Clin Pract Gastroenterol Hepatol*, 2 (9):406–415.

Wehkamp, J, Fellermann, K, et al. (2005b). Human defensins in Crohn's disease. *Chem Immunol Allergy*, 86:42–54.

Wehkamp, J, Harder, J, et al. (2004). NOD2 (CARD15) mutations in Crohn's disease are associated with diminished mucosal alpha-defensin expression. *Gut*, 53(11):1658–1664.

Wehkamp, J, Salzman, NH, et al. (2005). Reduced Paneth cell alpha-defensins in ileal Crohn's disease. *Proc Natl Acad Sci USA*, 102(50):18129–18134.

Wehkamp, J, Schmid, M, et al. (2005). Defensin deficiency, intestinal microbes, and the clinical phenotypes of Crohn's disease. *J Leukoc Biol*, 77 (4):460–465.

Wehkamp, J and Stange, EF (2005). NOD2 mutation and mice: no Crohn's disease but many lessons to learn. *Trends Mol Med*, 11(7):307–309.

Williams, CN, Kocher, K, et al. (2002). Using a genome-wide scan and meta-analysis to identify a novel IBD locus and confirm previously identified IBD loci. *Inflamm Bowel Dis*, 8(6):375–381.

Wu, X, Kekuda, R, et al. (1998). Identity of the organic cation transporter OCT3 as the extraneuronal monoamine transporter (uptake2) and evidence for the expression of the transporter in the brain. *J Biol Chem*, 273 (49):32776–32786.

Yamazaki, K, McGovern, D, et al. (2005). Single nucleotide polymorphisms in TNFSF15 confer susceptibility to Crohn's disease. *Hum Mol Genet*, 14(22):3499–3506.

Yamazaki, K, Takazoe, M, et al. (2002). Absence of mutation in the NOD2/CARD15 gene among 483 Japanese patients with Crohn's disease. *J Hum Genet*, 47(9):469–472.

Yap, LM, Ahmad, T, et al. (2004). The contribution of HLA genes to IBD susceptibility and phenotype. *Best Pract Res Clin Gastroenterol*, 18(3):577–596.

Yoshitake, S, Kimura, A, et al. (1999). HLA class II alleles in Japanese patients with inflammatory bowel disease. *Tissue Antigens*, 53(4 Pt 1):350–358.

Yue, TL, Ni, J, et al. (1999). TL1, a novel tumor necrosis factor-like cytokine, induces apoptosis in endothelial cells. Involvement of activation of stress protein kinases (stress-activated protein kinase and p38 mitogen-activated protein kinase) and caspase-3-like protease. *J Biol Chem*, 274 (3):1479–1486.

Zaahl, MG, Winter, T, et al. (2005). Analysis of the three common mutations in the CARD15 gene (R702W, G908R and 1007fs) in South African colored patients with inflammatory bowel disease. *Mol Cell Probes*, 19(4):278–281.

Zhou, Z, Lin, XY, et al. (2002). Variation at NOD2/CARD15 in familial and sporadic cases of Crohn's disease in the Ashkenazi Jewish population. *Am J Gastroenterol*, 97(12):3095–3101.

Zouali, H, Lesage, S, et al. (2003). CARD4/NOD1 is not involved in inflammatory bowel disease. *Gut*, 52(1):71–74.

23

Genomics and Cancer: Mechanisms and Applications

Mark Davies and Julian Sampson

Cancer is diagnosed in one in three of those living in the "developed world" and causes the death of one in four. Longevity is a major factor in the health impact of cancer since the incidence of most common cancers shows a log-linear relationship with age (Hanson et al., 1980). Environmental and social factors including smoking, diet and the trend to a delay, and reduction in numbers of pregnancies are also of striking importance in determining the pattern of cancers seen in contemporary society. Nonetheless, cancer is fundamentally a genetic disease. Its development is determined by the accumulation of mutations in genes that normally coordinate the birth and death of cells to ensure that they are produced only as and when they are required and their regulate behavior such that they remain under control at the level of tissue, organ, and organism. Over the last three decades, our understanding of cancer has been transformed by the identification of many genes that are mutated, or whose expression is altered in cancers, and by the characterization of their functions that are lost or changed as a cancer develops. Key mechanisms of tumorigenesis have been established, and many of the signaling and regulatory pathways that are hijacked by cancers have been elucidated, often through genetic approaches. Many advances that have impacted on knowledge of cancer have come initially from genetic studies of simple organisms, notably bacteria and yeast where comprehensive experimental strategies could be adopted. The more recent, availability of the human and other genome sequences, together with the development of technologies for assessment of entire genomes or transcriptomes have started to build a comprehensive picture of the changes that lead to cancer. The application of genome-wide approaches to complex biological systems has also brought about considerable challenges in distinguishing the changes that drive cancer development from others that are unimportant consequences of the genomic instability that characterizes cancer. Despite these difficulties, new knowledge gained through genomics is beginning to translate into diagnostic and prognostic approaches and opportunities for better targeted, more individualized, and sometimes entirely novel therapeutics. Cancer research has been and continues to be the greatest focus for innovation and application in genomics. This chapter reviews briefly the foundations of cancer genetics and assesses the progress made and promise genomics holds in relation to the understanding and clinical management of cancer.

Foundations of Cancer Genetics

Mutation, Epimutation, and Genome Instability in Cancer

The progressive changes in cell behavior that lead to cancer reflect an accumulation of genetic changes and their consequences for the transcriptome and proteome. Cancers are characterized by losses, gains, and rearrangements in their genomes, by deoxyribonucleic acid (DNA) sequence changes that arise through a variety of mechanisms and by physiologically inappropriate gene silencing or activation through chemical changes to DNA and chromatin (epimutation). Genetic instability is a hallmark of cancer, and in recent years it has been recognized that rather than being a mere consequence of cancer, it is the driver of cancer progression. Genetic instability in cancer cells reflects a breakdown in the all-important mechanisms that normally govern genome integrity. These are chiefly the pathways involved in DNA repair and chromatin modification and the maintenance of chromosome number and structure. Although DNA repair pathways have been subject to long-standing and intensive study our knowledge of their involvement in cancer is far from complete. The dynamics of chromatin regulation and the pathways that govern chromosome structure and function have only been studied more recently in relation to cancer and knowledge remains patchy. It is clear, however, that genes whose products mediate genome integrity are themselves targets for somatic mutation, loss or epigenetic silencing in sporadic cancers and that some of these genes are also mutated in the germline in inherited cancer syndromes. Once these protective pathways are compromised, the stage is set for an acceleration of the rounds of mutation and selection through which cancer cells acquire autonomous growth and proliferation, promote angiogenesis, evade host immune mechanisms, invade surrounding tissues, and metastasize.

Tumor Suppressor Genes and Oncogenes

Genes involved in cancer are often classified into those that contribute to tumorigenesis through loss of function and are termed tumor suppressor genes, and those that play a role through a gain of function and are termed oncogenes. Traditionally, tumor suppressor genes are considered to be recessive at a cellular level, that is, both alleles (of autosomal genes) must be inactivated before significant functional consequences occur. This concept, the "two-hit hypothesis," was articulated particularly by Knudson, based initially upon his classical statistical study of inherited and sporadic forms of retinoblastoma (Knudson, 1971). Later, following identification of the *Rb* gene, the hypothesis was supported by direct molecular genetic analysis of the genes concerned. In sporadic cancers, both alleles of a tumor suppressor gene are inactivated by intragenic somatic mutation, loss of heterozygosity (LOH), or epigenetic silencing. LOH is the conversion at a particular locus of two heterozygous alleles to either a hemizygous state by loss of one allele or a homozygous state by loss of one allele and its replacement with a copy of the other. LOH can occur through a number of mechanisms such as deletion or mitotic recombination (Fig. 23–1). An inherited predisposition to cancer (a "family cancer syndrome") may result when a tumor suppressor gene is inactivated in the germline. In this situation, the inherited mutation constitutes the "first-hit," and only a single somatic ("second-hit") mutation is required for initiation of tumorigenesis in a susceptible cell.

Individuals carrying inherited mutations in tumor suppressors may have a very high risk of cancer. Their tumors may be multiple and tend to occur at an earlier age than in sporadic cases. Although there is much experimental data to support the "two-hit hypothesis" of tumor suppressor gene action, recent studies suggest that, for a number of tumor suppressors, inactivation of only one allele may have functional and phenotypic consequences (Di Cristofano et al., 1998; Alberici et al., 2006). In other situations, a "third-hit" affecting an allele whose function is already compromised but not extinguished may be required for tumorigenesis. Thus, while the "two-hit" tumor suppressor gene concept remains a useful model, we must acknowledge the existence of a diversity of related mechanisms through which loss of gene function may impact on cancer development. Tumor suppressor genes serve a multitude of roles including

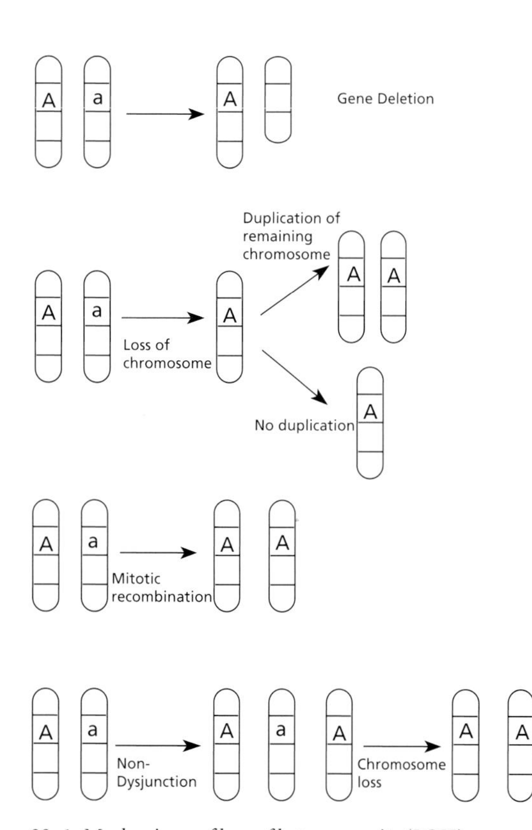

Figure 23–1 Mechanisms of loss of heterozygosity (LOH).

DNA transcription and repair and apoptosis (*TP53, hMLH1*, and *hMSH2*), cell cycle regulation (*Rb, p16*) and cell growth control (*TSC1* and *TSC2, LKB1*), and angiogenesis (*VHL*). Many encode multifunctional proteins that serve roles in several cancer-related pathways. For example, p53 signals to DNA repair, apoptosis and cell cycle arrest pathways in response to DNA damage, and many other tumor suppressors exhibit pleiotropic effects on cell growth, proliferation, and death as a result of the signaling links and cross talk between pathways.

In contrast to tumor suppressors, oncogenes encode proteins that promote cell growth and proliferation through a wide variety of mechanisms. Their products include secreted growth factors [platelet-derived growth factor (PDGF)], cell surface receptors for growth factors [colony-stimulating factor 1 receptor (CSF1R), epidermal growth factor receptor (EGFR)], signaling pathway components (Ras and Raf) and genes involved in transcription control (MYC, JUN). Activation or overexpression of oncogenes in tumors occurs through many mechanisms. Gene amplification can lead to either a modest or massive increase in their copy number. A translocation or other structural chromosomal rearrangement may result in the production of an oncogenic fusion protein or lead to overexpression of a structurally normal product by perturbing transcriptional control. Constitutive activation of oncoproteins may result from missense or more rarely truncating mutations in oncogenes that alter the conformation of the encoded protein. Inherited mutations or epimutations affecting some oncogenes can give rise to family cancer syndromes and to congenital dysmorphic syndromes that may be associated with increased cancer risk (Arighi et al., 2005; Kerr et al., 2006). In some cases, exogenous oncogenes can be introduced into cells by viruses. Their expression plays a critical role in the development of cervical cancer, Burkitt's lymphoma, and some T cell leukemias.

Functions of Cancer-associated Genes

Although mutations or altered expression of a huge number of genes have been implicated in cancer, some processes are consistently involved. The genes involved are particularly important to our understanding of cancer and to the identification of opportunities for intervention.

Maintenance of Genome Integrity

Cancers display a variety of types of genetic instability. Kinzler and Vogelstein (1996) suggest that the genes whose mutation or altered expression lead to increased mutation in cancer cells can be considered as "caretaker genes" whose normal functions are concerned with maintenance of genome integrity.The identities and roles of some, but certainly not all, of these genes have been established. Chromosomal instability (CIN) was the first form of genetic instability to be recognized in cancer and it is the most frequent. Indeed, the majority of epithelial cancers have been found to display numeric karyotypic abnormalities. In some cases, CIN results from inactivation of the genes regulating the mitotic spindle checkpoint, the centrosome or chromosome dynamics but usually the cause is unknown (Cahill et al., 1998; Gisselsson, 2005). Inherited mutations of the mitotic checkpoint gene, *BUB1B*, have been demonstrated recently in mosaic variegated aneuploidy, a very rare recessive condition characterized by growth failure, childhood cancer, and constitutional mosaicism for chromosomal gains and losses. This finding provides evidence for a causal relationship between CIN and cancer (Hanks et al., 2004). Microsatellite instability (MIN or MSI) is a less common and subtler form of genetic instability. MIN describes the mitotic instability of repetitive DNA sequences (microsatellites). In MIN tumors, there are widespread changes in the lengths of microsatellite sequences compared to the host patient's constitutive DNA. MIN has been studied most extensively in colorectal cancers, where it is observed in approximately 15% of cases. MIN results from inactivation of genes encoding proteins involved in the DNA mismatch repair (MMR) system, usually due to methylation of the *MLH1* promoter in sporadic colorectal cancers (Herman et al., 1998; Veigl et al., 1998). A small proportion of colorectal cancers (perhaps ~2%–3%) occur in the context of hereditary nonpolyposis colorectal cancer (HNPCC), a family cancer syndrome associated with a very high risk of colorectal, endometrial, and other cancers. HNPCC results from inherited mutations in genes of the MMR system, notably *MLH1*, *MSH2*, and *MSH6* (Kinzler and Vogelstein, 1996). Mutation rates are greatly elevated in MIN cells, leading to accelerated carcinogenesis through the generation of frequent frameshift mutations in tumor suppressor genes (Parsons et al., 1993).

The other major DNA repair pathways, nucleotide excision repair (NER), base excision repair (BER), and double-strand break repair (DSBR) (mediated via homologous recombination or nonhomologous end joining) are all implicated in cancer. Defective NER leads to xeroderma pigmentosum, a rare recessive susceptibility to skin cancers that result from ultraviolet (UV) light exposure. Inherited mutations in the *MUTYH* gene that plays a central role in BER through its glycosylase activity for adenines mis-incorporated opposite 8-oxoguanine are associated with predisposition to colorectal tumors (Cheadle and Sampson, 2003). Polymorphisms in other BER genes have also been linked with cancer susceptibility in genetic association studies and reduced activity of 8-oxoguanine DNA-glycosylase 1 (OGG1), a glycosylase that removes 8-oxoguanine, has been reported in patients with lung cancer, but consistent, conclusive, and replicated evidence is lacking. Evidence for somatic inactivation of BER genes in sporadic, colorectal, and other cancers has not been reported (Halford et al., 2003). DSBR is mediated by homologous recombination in cells that have replicated their DNA and by nonhomologous end joining in nondividing cells. The latter process is error prone due to the lack of a template for repair. *BRCA1*, *BRCA2*, and *ATM* are among the many genes involved in DSBR.

Cell Cycle Control

A cancer can only develop if there is failure to regulate cell division. In the majority of cancers, this occurs through a combination of mutations affecting genes whose products play a role in the retinoblastoma (Rb) pathway and altered expression, degradation, or phosphorylation of cyclin-dependent kinases (CDKs) and their activators (cyclins) and inhibitors (CDKIs). Together these molecules normally regulate progression through the cell cycle and past checkpoints that ensure genomic integrity prior to DNA replication (in S phase) and cell division (in M phase) (Sherr, 1996; Sherr, 2000). p53 is the key protein linking detection of DNA damage to the cell cycle checkpoint at the critical G_1 to S phase transition. Remarkably, the *TP53* gene is mutated in 80% of cancer, reflecting the requirement for cancer cells to propagate mutations to their progeny as well as their ability to survive despite the presence of DNA damage through abrogation of p53 dependent apoptosis.

Cell Growth Control

Many cancers achieve autonomy of growth control by synthesizing their own growth factors (autocrine), overexpressing cell surface receptors, stimulating neighboring cells to secrete growth factors (paracrine), and through inactivation of pathways that respond to growth inhibitory signals (Fig. 23–2). Growth regulating pathways including the mammalian target of rapamycin (mTOR) pathway (Shaw and Cantley, 2006) serve to regulate cell growth in response to growth factors and nutrient, energy, and oxygen availability. Components of the mTOR pathway, including phosphatase and tensin homolog deleted on chromosome 10 (PTEN), are inactivated in a significant proportion of cancers, and inherited mutations in the genes encoding pathway components are mutated in a number of hamartoma and cancer predisposition syndromes, including Cowden disease (*PTEN*), Peutz–Jegers syndrome (*LKB1/STK11*), and tuberous sclerosis (*TSC1/2*).

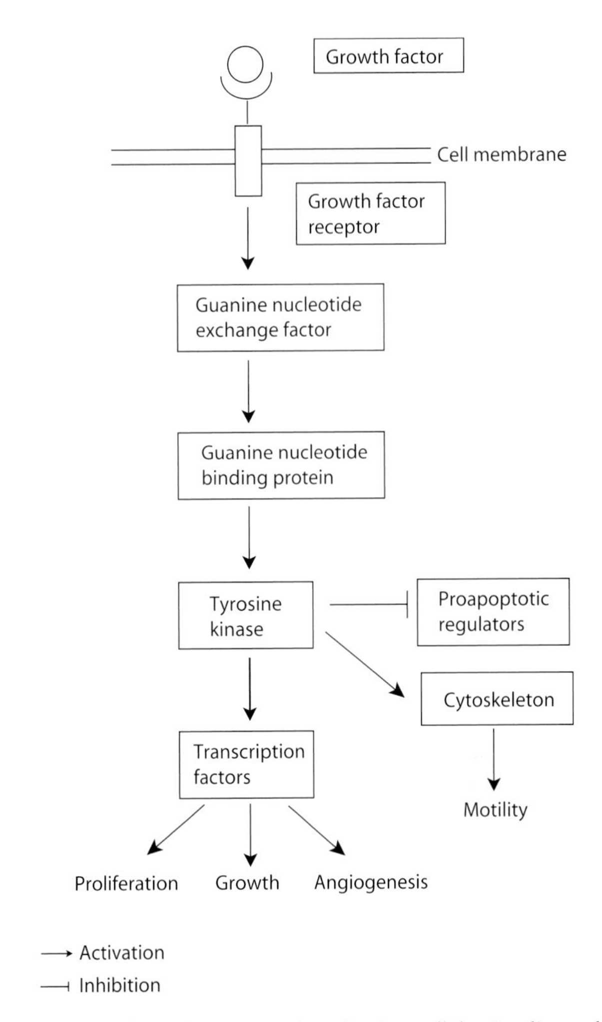

Figure 23–2 Schematic representation of an intracellular signaling pathway regulating cell growth and proliferation.

Senescence

Most somatic cells have a finite capacity for proliferation, typically approximately 70 doublings, after which they senesce, that is, they cease to proliferate even in the presence of appropriate stimuli and nutrients. The major factor determining senescence appears to be the shortening to a critical length of the TTAGGG repeat sequences at telomeres, the structures that form chromosome ends. Normally, these telomeric repeats are shortened at each cell division, since the replication machinery cannot synthesize the ends of DNA molecules. Cancer cells can overcome senescence through expression of telomerase reverse transcriptase encoded by *TERT* and telomerase ribonucleic acid (RNA) encoded by *TERC*. The telomerase ribonucleoprotein DNA polymerase is normally only expressed in germ cells but is frequently expressed in cancers. A proportion of cancers do not, however, express telomerase, but the mechanisms through which they maintain telomeres is not understood (Artandi and Attardi, 2005; Shin et al., 2006).

Apoptosis (Programmed Cell Death)

Apoptosis serves essential roles during development through the remodeling or removal of developing or temporary body structures, maintenance of cell number in the developed organism, and defense against viral infection. It also enables cells to self-destruct when cellular or genetic damage is not or cannot be repaired. The widespread genomic abnormalities, activation of oncogenes, and inactivation of tumor suppressors that characterize cancer cells would normally provoke their suicide. However, cancer cells circumvent this protective mechanism through silencing of proapoptotic genes (e.g., Bax, Fas) and by activating and overexpressing their antiapoptotic counterparts (e.g., Bcl2) (Gerl and Vaux, 2005; Kim et al., 2006).

Angiogenesis

In order to sustain their blood supply, cancers invoke angiogenesis (the development of new blood vessels) through the expression of many proangiogenic factors, most notably vascular endothelial growth factor (VEGF) (Tonini et al., 2003; Shibuya and Claesson-Welsh, 2006).

Metastasis

Metastasis, the spread of a tumor from its primary site to other parts of the body, requires a sequential series of steps including the separation of the cancer cells from the primary tumor, invasion through the basement membrane and into surrounding tissue and into the vascular, lymphatic, or peritoneal compartments, invasion of a target organ, and proliferation at this site. The traditional view was that metastatic capability developed as a result of acquired somatic mutations in genes involved in the metastatic process within a small subpopulation of cancer cells. For example, e-cadherin, which encodes a cell–cell adhesion molecule is commonly lost in metastatic cells (Cavallaro and Christofori, 2004). However, recent studies suggest that metastatic capability may arise in substantial proportion of tumor cells early in tumor progression. In addition, there is increasing evidence that the host genetic background may substantially alter metastatic efficiency, giving rise to the possibility that some individuals have a predisposition to metastases if they develop cancer (Hunter, 2006).

Mitochondria and the Mitochondrial Genome

Mitochondria are organelles that play a key role in cellular energy production and the control of apoptosis. The mitochondrial genome consists of a 16.5-kb circular DNA molecule encoding a number of genes involved in oxidative phosphorylation and protein synthesis. Nuclear genes encode the remaining proteins involved in mitochondrial functioning. Many sporadic tumors contain mutations in mitochondrial DNA (Czarnecka et al., 2006), and they are likely to contribute to the mitochondrial dysfunction often seen in cancer cells and may promote tumor growth (Petros et al., 2005; Shidara et al., 2005). A number of mitochondrial proteins encoded by nuclear genes can act as tumor suppressors, and inherited mutations in these genes result in family cancer syndromes. Fumarate hydratase is a component of the tricarboxylic acid (TCA) cycle, and germline mutations in the gene encoding this enzyme predispose to leiomyoma, leiomyosarcoma, or renal cell carcinoma. Germline mutations in genes encoding components of succinate dehydrogenase, a TCA cycle enzyme and member of the electron transport chain, result in the development of paraganglioma or pheochromocytoma (Gottlieb and Tomlinson, 2005).

Genetic Testing for Familial Cancer

Genetic testing for the family cancer syndromes is now central to their clinical management. Identification of individuals carrying inherited cancer-predisposing mutations provides opportunities for surveillance and early intervention. Such individuals may come to clinical attention through their family history of cancer or because they manifest characteristic combinations of precursor lesions or cancers. In some scenarios, there is little doubt that a specific inherited mutation will be present, for example, in an individual with multifocal retinoblastoma or when two close relatives have unifocal retinoblastoma. In these settings, identification of the causative mutation will enable predictive testing of at-risk offspring, confirmation of genetic status in adult family members, and prenatal or preimplantation diagnosis. Comparable strategies are employed for the management of extended families with familial adenomatous polyposis, multiple endocrine neoplasias, and von Hippel–Lindau disease. Management is often implemented through genetic registers that coordinate gene testing with surveillance for carriers. However, in many settings it is initially unclear whether a germline mutation is segregating in a family or whether there is a chance clustering of cancers. Algorithms to determine the likelihood of detecting germline mutations in families where, for example, *BRCA1* or *BRCA2* involvement is suspected have been developed and are used widely in clinical cancer genetics to guide selection of families for testing.

Cancer Genomics

The most comprehensive approach to the cancer genome would be to characterize all genomic alterations associated with the disease. This is essentially the goal of The Cancer Genome Atlas (TCGA), a collaborative effort by the National Cancer Institute and the National Human Genome Research Institute (http://cancergenome.nih.gov/index.asp). A TCGA Pilot Project will initially focus on a small number of types of cancers, to address the many technical and logistical challenges that must be overcome in order to undertake a large-scale human cancer genome project. This ambitious project is driven by the rationale that increasing our understanding of genomic alterations in cancer will lead to improved clinical management, suggest novel therapeutic targets and more accurately identify those at high risk of the disease, thus, directing screening or other interventions. To achieve a truly comprehensive catalog of the cancer genome the use of a range of approaches such as high-throughput resequencing and genome-wide analysis of RNA expression, of DNA copy number changes and of epigenetic regulation will be required. These approaches have already yielded insights into the biology of cancer and impacted on clinical practice.

High Throughput Resequencing

The Cancer Genome Project aims to systematically screen the coding exons and splice site junctions of all known genes for somatically acquired small intragenic mutations, utilizing DNA from primary tumors and matched normal genomic DNA and also DNA from cancer cell lines. The results of the study are freely accessible on the COSMIC (Catalog Of Small Mutations In Cancer) database (http://www.sanger.ac.uk/genetics/CGP/cosmic).

This project has initially screened 518 members of the protein kinase family in a variety of cancer types. In a series of 16 breast cancers, no mutations in protein kinases were detected in 12 of the samples, 2 had a single mutation each, 1 had two mutations, and 1 had 52 mutations (suggestive of a mutator phenotype because of a DNA repair defect or prior exposure to a mutagenic agent). This contrasted with a series of breast cancer cell lines where eight out of nine samples contained at least one mutation. In total, 92 mutations were observed, 90 base substitution, and two in-frame deletions. Of the base substitutions, 58 caused missense mutations, 12 nonsense mutations, 6 were splice site mutations, and 14 were apparently silent (synonymous) changes (Stephens et al., 2005). In a series of non-small-cell lung cancers (NSCLCs), a substantial number of mutations which were likely to contribute to the oncogenic phenotype were identified. These mutations were distributed across a range of kinases and mutations in any particular kinase were relatively infrequent (Davies, Hunter, et al., 2005). This project has also demonstrated that an activating mutation in the BRAF kinase is present in approximately two-thirds of melanomas (Davies et al., 2002), leading to constitutive activation of the mitogen-activated protein kinase (MAPK) pathway. A single V599E mutation accounts for more than 80% of the mutations described to date (Brose et al., 2002). These findings have led to interest in the use of inhibitors of BRAF and other components of the MAPK pathway in the treatment of melanoma.

Genome-wide RNA Expression Analysis

Gene expression microarray enables the levels of expression of thousands of genes, at the level of messenger RNA (mRNA), to be measured simultaneously. The resulting gene expression data are highly complex—the composite of multiple interactions between the cancer genome, the normal cell genome, and the environment. At it simplest level the data consists of the expression levels of multiple genes. Higher dimensional analysis allows the assessment of the behavior of clusters of genes, identifying "signatures" of potential clinical or biological importance. The main objectives of most gene expression microarray studies using clinical samples have been to define new discrete subsets of a cancer based upon gene expression profiles (class discovery), to compare gene expression between predefined classes (class comparison), or to develop a gene expression-based function which predicts the membership of new samples to predefined classes (class prediction) (Golub et al., 1999; Simon et al., 2003). Irrespective of the question being addressed, gene expression microarray experiments have a number of inherent problems. Variability can arise as a result of different methods of sample handling and the platforms used, making comparison between studies difficult. Most studies published to date have been retrospective and of relatively small sample sizes. The number of variables tested, that is, the expression levels of thousands of genes usually far exceed the number of samples used, leading to false positive results and over-fitting of statistical models.

Perou et al. (2000) were the first to demonstrate that subtypes of breast cancer could be defined on the basis of differences in gene expression profiles. Using a panel of breast cancers, including paired samples before and after doxorubicin chemotherapy, they selected an "intrinsic gene set" of 496 genes, consisting of genes with greater variation between different tumors than between paired samples from the same tumor. This intrinsic gene set was then used to divide the tumors into a number of subgroups. A fundamental division was into two groups which largely corresponded with whether the tumors where pathologically classified as estrogen receptor (ER) positive or ER negative. The ER positive profile was characterized by many genes expressed by breast luminal cells. The ER negative group could be divided into three subtypes: "basal like" expressing genes characteristic of breast basal epithelial cells, the ERBB2+ subtype, which overexpressed the Erb-B2 oncogene, and the "normal breast" subtype which expressed genes characteristic of adipose tissue and basal epithelial cells.

Subsequent studies demonstrated that the ER positive/luminal group could be divided into at least two subgroups termed luminal A and B (Sorlie et al., 2001) and that the five subtypes could be repeatedly demonstrated in independent sample sets (Sorlie et al., 2003) and were associated with differences in clinical outcomes. Patients with basal like tumors had shortest overall and disease-free survival while patients with luminal A type tumors had the best prognosis (Sorlie et al., 2001).

Van't Veer et al. identified a 70-gene signature, which could be used to classify young (<55 years) node-negative patients into two groups: those with a good prognosis (no recurrence in 5 years of follow-up) and with poor prognosis (recurrence within 5 years). They developed this signature in a training set of 78 breast cancer patients, all of whom where under 55, had tumors under 5 cm, had no lymph node involvement, and who could be classified into the good or poor prognosis group based on clinical follow-up data. They then validated the signature in a test set of 19 similar patients. The genes included in the signature had roles in cell cycle, invasion, metastasis, angiogenesis, and signal transduction. Some were almost exclusively expressed by the stromal cell, suggestive of the importance of the local tumor environment on metastasis (van't Veer et al., 2002).

In a follow-up study using the 70-gene signature in a cohort of 295 patients (both lymph node negative and lymph node positive) the mean overall 10-year survival rate ($\pm$standard error) of the 115 patients with a good prognostic signature was 94.5 $\pm$ 2.6% compared to 54.6 $\pm$ 4.4% for the 180 patients with a poor signature. At 10-year follow-up 85.2 $\pm$ 4.3% were metastasis free in the good prognostic signature group and 50.6 $\pm$ 4.5% in the poor prognostic signature group. In the 151 patients with lymph node negative disease, of the 60 with a good prognostic signature, 86.8 $\pm$ 4.8% were free of metastasis at 10 years compared to 44.1 $\pm$ 6.3% of the 91 with the poor prognostic signature. Multivariate analysis showed that the gene expression signature was a strong independent factor in predicting outcome (van de Vijver et al., 2002).

The 70-gene signature will be evaluated in the Microarray In Node-negative Disease may Avoid Chemotherapy (MINDACT) prospective trial. This trial will recruit women with early stage, node-negative breast cancer. Those considered at high risk for metastasis by both gene signature and clinical–pathological criteria will receive adjuvant chemotherapy. Those patients who are low risk by both criteria will not receive chemotherapy. Patients who are considered to be high risk by one criterion and low risk by the other will be randomized to either a group where the clinical–pathological features will be used to guide treatment decisions or a group where the gene signature alone determines whether chemotherapy is given. As a first phase of this trial, the 70-gene signature was evaluated in a series of 302 node-negative, systemically untreated early breast cancer patients from a number of institutions. The signature was found to add independent prognostic information to clinicopathologic risk factors, but its predictive value was far lower than in the original study (Loi et al., 2006).

These studies make no use of a priori knowledge of gene's function. An alternative approach is to generate gene expression signatures from cells manipulated in vitro to simulate a biological process important in cancer progression. Wound healing involves many of the processes implicated in cancer progression such as cell motility, angiogenesis, and matrix remodeling. Using an in vitro model of wound healing Chang et al. identified a set of "core serum response (CSR) genes" involved in this process and applied this to the expression data of the 295 patients used to validate the 70-gene signature. Both the overall survival and the metastasis-free survival were reduced in patients whose tumors expressed the CSR signature (Chang et al., 2004; Chang, Nuyten, et al., 2005). Chi et al. obtained a gene expression signature of the cellular response to hypoxia from cells cultured in hypoxic conditions and by applying this to the 70-gene signature validation set showed that the hypoxic response signature was a predictor of worse survival and increased risk of metastasis (Chi et al., 2006).

There is very little overlap between the expression signatures that successfully predict prognosis in breast cancer. For example, no genes are present in all three of the CSR signature, the 70-gene signature, and the original intrinsic gene list, and only 22 are shared by two signatures. Ein-Dor et al. demonstrated that multiple gene sets with equal prognostic value could be obtained from a single data set. They took the van't Veer data set and showed that the set of genes most strongly correlated with survival in a given analysis is heavily influenced by which patients are included in the training set as opposed to the test set. By varying the patients included in the training and test sets, they showed that many more genes than the 70 included in the original signature are related to survival and a number of 70-gene signatures with equivalent prognostic value can be generated (Ein-Dor et al., 2005). Focusing on a specific signature may have clinical utility but may mean that genes of biological significance or those that could be targeted therapeutically are overlooked.

A comparison by Chang et al. of the CSR signature, the 70-gene signature, and the intrinsic gene set showed that they gave generally consistent predictions of outcome but there were some discrepancies. For example, a small group of patients with a poor prognosis was not identified by the 70-gene signature but was highlighted by the CSR signature. In a multivariate analysis incorporating established clinical risk factors, only the 70-gene and the CSR signatures provided independent and significant prognostic information. By developing a decision tree incorporating the 70-gene signature and the CSR signature, the authors were able to refine the prognostic stratification (Chang, Nuyten, et al., 2005). Pitman et al. demonstrated the value of combining genomic and clinical data by developing a "clinicogenomic model" that evaluated the contribution of multiple gene expression profiles and clinical risk factors (ER and lymph node status) to optimize prognostic information (Pittman et al., 2004).

A number of studies have developed signatures, which predict response to therapy. Chang et al. identified gene signature which can be used to predict response to neoadjuvant docetaxel in breast cancer. Samples were taken from pretreatment biopsies in 24 patients. Response was based on clinical measurement, and tumors were defined as sensitive (25% or less residual disease) or resistant (>25% residual disease) after treatment with the taxane. A 92-gene signature was constructed from the genes differentially expressed in the sensitive and resistant tumors. In leave-one-out cross-validation analysis 10 out of 11 responders and 11 out of 13 nonresponders were correctly classified for an overall accuracy of 83%. The positive and negative predictive values of the signature were 92% and 83%, respectively. The signature correctly predicted response in a small validation set of six patients (Chang et al., 2003). In a follow-up study, the authors attempted to examine gene expression patterns associated with incomplete response to neoadjuvant docetaxel, using surgical specimens from 13 of the same patients in the original study, resected after completion of neoadjuvant docetaxel therapy. No statistically significant differences were found between the gene expression profiles of the residual tumors that had been classified as sensitive or resistant. This similarity may result from the clonal selection of a resistant phenotype. Elevated expression of genes encoding components of the mTOR pathway was seen in the profiles obtained from the tumors originally classified as sensitive, suggesting that mTOR inhibition may increase the efficacy of docetaxel (Chang, Wooten, et al., 2005). Iwao-Koizumi et al. developed an algorithm that could predict neoadjuvant response to docetaxel in breast cancer based on gene expression profiling by high-throughput reverse transcriptase polymerase chain reaction (PCR). In a small validation set of 26 patients, the algorithm had an accuracy of approximately 80%. Elevated expression of genes controlling the cellular redox environment was seen in non-responders, and over-expression of redox genes in a breast cancer cell line induced resistance to docetaxel (Iwao-Koizumi et al., 2005).

Gene expression profiling has also been used to predict response to tamoxifen. Paik et al. used the expression levels of 21 genes in an algorithm that classified node-negative, tamoxifen-treated breast cancer patients into a low, intermediate, or high-risk group. This algorithm was validated prospectively in patients enrolled in a large multicenter clinical trial. Kaplan–Meier analysis of the rates of recurrence in the low, intermediate, and high-risk groups were 6.8%, 14.3%, and 30.5%, respectively (Paik et al., 2004). Ma et al. used expression profiling of the tumors of 60 patients with hormone receptor positive primary breast cancer, who were treated with adjuvant tamoxifen to identify a two-gene expression ratio [of homeobox 13 (HOXB13) to interleukin-17B receptor (IL17BR)] that was predictive of recurrence. This finding was validated by real-time quantitative PCR analysis of RNA extracted from paraffin embedded tumor for a cohort of 20 patients (Ma et al., 2004).

There is immediate clinical utility in the application of gene expression signatures as prognostic and predictive tools, but these profiles also contain higher order structure that delineates the biological processes, which confer these differences in risk or response and which may be amenable to therapeutic intervention. Gene set enrichment analysis (GSEA) is one approach that can identify coordinate changes in the expression of a priori specified gene sets, for example, biologically coherent groups obtained from pathway databases such as KEGG (Kyoto Encyclopedia of Genes and Genomes) (Mootha et al., 2003). Majumder et al. used GSEA to identify loss of expression of Hif-1α targets as the dominant transcriptional response to mTOR inhibition in a mouse prostate intraepithelial neoplasia model (Majumder et al., 2004). Several studies have examined biological themes by integrating multiple independent data sets. Ramaswamy et al. by comparing the gene expression profiles of adenocarcinoma metastases of multiple tumor types to unmatched primary adenocarcinomas, identified a 17-gene expression signature, which distinguishes primary from metastatic adenocarcinomas. Applying this to data on 279 primary tumors, they found that the signature was associated with metastasis and poor outcome, suggesting that metastatic potential may be present in a subset of tumor cells early in tumor progression (Ramaswamy et al., 2003). Segal et al. developed a module map of cancer. They analyzed a "cancer compendium" of expression profiles from 26 studies and collected a large number of biologically meaningful gene sets, including groups of genes co-expressed in specific tumor types or belonging to the same functional category or pathway. Applying these modules they demonstrated that activation or repression of some modules (e.g., those involved in cell cycle control) are shared across multiple tumor types while others were more tumor-type-specific, for example, a growth inhibitory module was repressed in acute lymphoblastic leukemia (Segal et al., 2004).

A number of tumor suppressor or oncogene specific expression signatures have been constructed from both human tumors and animal and cell culture models. Miller et al. described a 32-gene expression profile associated with p53 mutation in human breast cancer, which is also associated with p53 mutation in hepatocellular carcinoma and which predicts survival in breast cancer patients in independent data sets of patients treated with different therapeutic modalities (Miller et al., 2005). Sweet-Cordero identified a KRAS oncogene expression signature by cross species gene expression analysis, comparing the gene expression profiles derived from a mouse model K-ras-mediated lung cancer model and human lung cancers known to carry K-ras mutations (Sweet-Cordero et al., 2005). Bild et al. derived a number of oncogenic pathway signatures by using recombinant adenoviruses to express oncogenic activities in human primary epithelial cell cultures. Integrated analysis of the multiple pathway activation signatures was able to predict survival in cohorts of patients with breast and ovarian cancer. In addition, the signatures could be used to predict the sensitivity of breast cancer cell lines to drugs that targeted specific components of the pathways (Sweet-Cordero et al., 2005).

Genome-wide Loss and Amplification Analysis

A number of high-throughput techniques have been developed that allow the entire genome to be surveyed for regions that are recurrently lost or amplified in different types of cancers. Such regions would be expected to contain genes, which play important roles in the disease. Gene amplification is an important mechanism of oncogene activation, and amplification of individual genes is already used clinically as markers of prognosis or as predictors of response, for example, amplification of N-MYC is associated with adverse outcome in neuroblastoma (Seeger et al., 1985), while the amplification and overexpression of ERBB2 is associated with an adverse outcome in breast cancer and is a predictor of response to the monoclonal antibody, trastuzumab (Nahta and Esteva, 2006). Tumor suppressor genes are often inactivated by mechanisms, which lead to LOH, and many, including *Rb* and *PTEN*, were originally identified by localizing regions of homozygous deletion.

Comparative genomic hybridization (CGH) is a technique that detects changes in DNA copy number (amplification and deletions). In this method, test (tumor) and reference (normal genomic) DNA samples are differentially labeled with fluorescent dyes and the resulting fluorescent intensities compared when the labeled samples are co-hybridized to a representation of the genome. Use of array-based formats allows genome-wide coverage with high resolution (Davies, Wilson, et al., 2005). For example, in tiling path arrays, where samples are hybridized to overlapping bacterial artificial chromosome (BAC) clones that contiguously cover the human genome, regions of gain and loss as small as 40–80 kb are detectable (Ishkanian et al., 2004).

Array CGH has been used to demonstrate associations between specific copy number aberrations and prognosis in a variety of tumor types. Callagy et al. used a CGH array containing clones from 57 oncogenes to identify amplified genes in tumors from a cohort of breast cancer patients and tested their prognostic significance both as single markers and as part of a panel. Four genes, considered to be most significant, were validated in a larger series using immunohistochemistry and fluorescent in situ hybridization. This approach demonstrated that amplification of TOP2A and of ERBB2 was associated with adverse disease-related outcome (Callagy et al., 2005). Using a BAC clone array to give genome-wise coverage, Paris found that gain at 11q13.1 was predictive of postoperative recurrence in prostate cancer (Paris et al., 2004) and Weiss, using a similar genome-wide array, identified a number of copy number aberrations associated with poor survival in gastric adenocarcinoma (Weiss et al., 2004).

Adler et al. (2006) devised a method to identify regulator genes for which a change in copy number can induce a cancer-associated gene-expression signature. The method involves examining data from gene expression profiling and CGH to establish linkage between a chromosomal locus and a signature. Candidate regulators at that locus can then be filtered by establishing concordance between their mRNA expression and the signature. Applying this approach, Adler and colleagues were able to identify two genes, *MYC* and *CSN5*, whose amplification was associated with the wound signature. They validated their findings in a breast cancer cell line. Overexpression of these genes simultaneously in a breast cancer cell line led to induction of the majority of the genes comprising the wound signature (Adler et al., 2006).

A single nucleotide polymorphism (SNP) is a single nucleotide position at which two or more alternative nucleotides are present at appreciable frequency (traditionally, at least 1%) in a population. It is estimated that there are more than 10 million SNPs distributed through the human genome. Studies focusing on single or a few SNPs have been used to associate SNPs with the risk of developing cancer, drug toxicity, or response to treatment. Developments in technology are making high-throughput SNP genotyping possible, facilitating genome-wide association studies. SNP genotyping, but not CGH, can detect LOH events which do not alter copy number (e.g., gene conversion), and SNP microarrays have been used to generate global LOH patterns (Lindblad-Toh et al., 2000; Mei et al., 2000; Zhou et al., 2004). LOH profiling by genotyping cannot detect genomic amplification, but a recent approach that uses the hybridization intensity on SNP arrays to generate copy number data allows both LOH and copy number aberrations to be analyzed simultaneously on a single platform (Zhao et al., 2004).

Epigenomics

Epigenetics (see also Chapter 4) refers to changes of gene expression and the resultant phenotypic effects that are not mediated by nucleotide sequence variation, and that are reversible but also potentially stable during cell division. Epigenetic changes are mediated by chromatin remodeling, histone modification, and DNA methylation that collectively result in alterations of transcriptional activity (Laird, 2005).

Inroads have been made into elucidating the contribution of epigenetics to cancer at a genome-wide level. Analysis of the "epigenome" requires the use and integration of high-throughput techniques of DNA, RNA, and protein analysis. Genome-wide methylation patterns can be detected by high-throughput pyrosequencing, matrix-assisted laser desorption/ionization mass spectrometry (MALDI-MS), methylation specific arrays or restriction landmark genomic scanning, a type of two-dimensional electrophoresis using methylation sensitive restriction enzymes. These techniques have been used to demonstrate that different tumor types exhibit specific patterns of methylation (Costello et al., 2000; Adorjan et al., 2002). Cluster analysis of neuroblastomas based on methylation of 10 genes could separate the tumors into several prognostic groups (Alaminos et al., 2004). A global analysis of DNA methylation in 10 patients with stage III and IV ovarian carcinomas identified two groups of patients with distinct methylation profile and significantly different progression-free survival (Wei et al., 2002). Epigenetic silencing has also been shown to have predictive implications. Promoter methylation of the O6-methylguanine-DNA methyltransferase (*MGMT*) DNA-repair gene has been associated with longer survival in patients with glioblastoma who receive alkylating agents (Hegi et al., 2005).

Comparing gene expression patterns in model systems before and after treatment with agents that remove epigenetic silencing has the potential to identify genes involved in the oncogenic process (Suzuki et al., 2002; Ibanez de Caceres et al., 2006). Such agents are also being introduced into clinical practice, where their mechanism of action may involve reactivation of silenced tumor suppressor genes (Yoo and Jones, 2006).

Targeted Therapy for Cancer

Clinical trials are currently evaluating many novel therapeutic agents directed against targets identified by studying the molecular lesions underlying cancer or the cellular processes subverted in the disease. Examples include BRAF inhibitors, mTOR inhibitors, and antiangiogenic agents (Bianco et al., 2006; Eckhardt, 2006). Imatinib treatment for gastrointestinal stromal tumors (GISTs) is a model of the development of molecularly targeted therapy. GISTs are mesenchymal tumors of the human gastrointestinal tract and are relatively resistant to conventional chemotherapy agents. Nearly all GISTs express the c-kit tyrosine receptor kinase (KIT), and most have gain-of-function mutations in *KIT*. Many of the GISTs without *KIT* mutations have gain-of-function mutations in the PDGF receptor gene (Hirota and Isozaki, 2006). Imatinib was initially identified in a screen for PDGF receptor inhibitors but was found to also be a potent inhibitor of ABL tyrosine kinases, including the BCR–ABL fusion protein associated with chronic myeloid leukemia (CML) and of KIT. Imatinib was first evaluated as a treatment for CML, where dramatic response rates were seen and subsequently in patients with GISTs, where it was associated with complete or partial responses in a large percentage of patients (Sawyers, 2003). Some tumors, however, show primary resistance to imatinib treatment, and most others become resistant during treatment. There is a correlation between the type of mutation in *KIT* or *PDGFR* and the initial response to imatinib. The most common site of mutation in *KIT* is in exon 11, which encodes the juxtamembrane domain. Patients with mutations in exon 11 are more likely to achieve a response, than those mutations in other exons or in whom no mutation in *KIT* or *PDGFR* is detectable (Heinrich et al., 2003; Debiec-Rychter et al., 2004). The most common mechanism for acquired resistance is the acquisition of second mutations within the KIT kinase domain leading to reactivation of the kinase (Antonescu et al., 2005; Wardelmann et al., 2006). This is analogous to CML where second site mutation in BCR–ABL is the predominant form of acquired resistance (Branford et al., 2003).

Gefitinib is a tyrosine kinase inhibitor of EGFR-mediated signaling (Wakeling et al., 2002), but unlike imatinib, its initial clinical development was not guided primarily by knowledge of the mutations in its target in human cancers. In preclinical studies, gefitinib showed activity against a variety of cancer types. Gefitinib was evaluated both as a single agent and in combination with chemotherapy in a series phase II and III trials in NSCLC. The phase II studies showed that gefitinib produced a dramatic response in as small proportion of patients (Fukuoka et al., 2003; Kris et al., 2003). A placebo-controlled phase III trial [Iressa Survival Evaluation in Lung cancer (ISEL)] showed that treatment with single-agent gefitinib was not associated with a statistically significant improvement in survival in the patient population as a whole, although there was marked heterogeneity in survival outcomes between patient groups (Thatcher et al., 2005). Trials of platinum-based chemotherapy in combination with gefitinib failed to show any benefit over chemotherapy alone. However, in these studies there was a marked heterogeneity of outcome between patient groups within the study population (Giaccone et al., 2004; Herbst et al., 2004).

A great deal of effort has gone into establishing factors that predict gefitinib sensitivity. A number of studies have shown a correlation between mutations in *EDGRFR* and sensitivity to gefitinib. However, this correlation is not perfect; some NSCLs carrying mutations are initially insensitive to the drug, while some NSCLs without apparent mutations do respond (Lynch et al., 2004; Paez et al., 2004; Cappuzzo et al., 2005; Tsao et al., 2005). Selecting patients who are most likely to respond in to clinical trials to improve trial efficiency is likely to be a key challenge in the development of targeted therapy.

DNA repair is an example of a biological process that can be exploited therapeutically in cancer. There is increasing evidence that impairment of DNA repair pathways, often by epigenetic modification, is important in the development of some sporadic cancers (Turner et al., 2004; Lyakhovich and Surralles, 2006). Tumor cells with impaired ability to repair particular types of DNA damage may be far more sensitive than normal cells to drugs that cause such damage. [MD join this and next paragraph to make one single paragraph.]

In vitro evidence suggests that cells deficient in BRCA1 or BRCA2 are particularly sensitive to drugs, such as cisplatin and carboplatin, which generate interstrand cross links, a type of DNA damage believed to be repaired by BRCA-mediated homologous recombination (Bhattacharyya et al., 2000; Tassone et al., 2003). Whether this in vitro sensitivity is of clinical relevance is being investigated in a multinational trial, in which *BRCA1* or *BRCA2* mutation carriers with metastatic breast cancer will be randomized to receive either docetaxel or carboplatin chemotherapy.

BRCA deficient cells are also particularly sensitive to poly (ADP-ribose) polymerase (PARP) inhibitors. This enzyme is involved in the repair of DNA single-stranded breaks. Inhibition of PARP leads to persistence of single-strand breaks which in BRCA competent cells would be repaired by BRCA-mediated homologous recombination but persist in BRCA deficient cells, leading to CIN and apoptosis (Paik et al., 2004; Farmer et al., 2005). PARP inhibitors are in early stage clinical trials, and it will be important to evaluate their efficacy in BRCA deficient tumors.

While mutations in *BRCA1* or *BRCA2* are rare in sporadic tumors, it is becoming apparent that the BRCA pathway may be defective in a number of sporadic tumors by events such *BRCA1* promoter methylation, giving rise to "BRCAness," the occurrence of traits in sporadic cancers usually seen in those occurring in *BRCA1* or *BRCA2* mutation carriers. Tumors which exhibit "BRCAness" may be susceptible to the same therapies being developed for familial BRCA deficient tumors (Turner et al., 2004).

MMR deficiency is linked to resistance to a wide range of DNA damaging drugs (Fedier and Fink, 2004). This may reflect a role of the MMR system in triggering apoptosis following recognition of DNA damage (Jiricny, 2006). Cisplatin resistance has been associated with loss of MMR proteins both in vitro and in vivo (Brown et al., 1997; Watanabe et al., 2001). MMR deficient cells appear to be resistant to 5-fluorouracil (Meyers et al., 2001). However, conflicting data exist about the relevance of MIN in predicting prognosis and benefit of 5-fluorouracil-based chemotherapy in patients with colorectal cancer (Benatti et al., 2005; Westra et al., 2005; Jover et al., 2006), and these discrepancies reflect the difficulties of translating experimental observations into clinical practice.

Conclusion

The integration of multiple levels of genomic information will increasingly drive improvements in the diagnosis and treatment of cancer. Limited genetic information is already utilized in clinical oncology, but a more comprehensive genomic approach has the potential to transform practice. To enable this genomic-based prognostic and predictive tools will have to be assessed in prospective trials and experimental platforms and analytical procedures will need to be refined. The development of molecularly targeted therapy will require a number of challenges to be overcome. Not all potential targets will be readily "tractable" and lend themselves to therapeutic intervention; associated toxicity is likely to be unacceptable for some targets serving essential cellular function. Targeting multiple pathways or processes may be necessary to overcome compensatory mechanisms and resistance may arise quickly in the context of genomic instability. Nevertheless, the expectation that therapeutic gains will be made by exploiting our increasing knowledge of the cancer genome will continue to drive research in this area, in both the academic and pharmaceutical sectors.

References

Adler, AS, Lin, M, Horlings, H, Nuyten, DS, van de Vijver, MJ, and Chang, HY (2006). Genetic regulators of large-scale transcriptional signatures in cancer. *Nat Genet*, 38:421–430.

Adorjan, P, Distler, J, Lipscher, E, Model, F, Muller, J, Pelet, C, et al. (2002). Tumour class prediction and discovery by microarray-based DNA methylation analysis. *Nucleic Acids Res*, 30:e21.

Alaminos, M, Davalos, V, Cheung, NK, Gerald, WL, and Esteller, M (2004). Clustering of gene hypermethylation associated with clinical risk groups in neuroblastoma. *J Natl Cancer Inst*, 96:1208–1219.

Alberici, P, Jagmohan-Changur, S, De Pater, E, Van Der Valk, M, Smits, R, Hohenstein, P, et al. (2006). Smad4 haploinsufficiency in mouse models for intestinal cancer. *Oncogene*, 25:1841–1851.

Antonescu, CR, Besmer, P, Guo, T, Arkun, K, Hom, G, Koryotowski, B, et al. (2005). Acquired resistance to imatinib in gastrointestinal stromal tumor occurs through secondary gene mutation. *Clin Cancer Res*, 11:4182–4190.

Arighi, E, Borrello, MG, and Sariola, H (2005). RET tyrosine kinase signaling in development and cancer. *Cytokine Growth Factor Rev*, 16:441–467.

Artandi, SE and Attardi, LD (2005). Pathways connecting telomeres and p53 in senescence, apoptosis, and cancer. *Biochem Biophys Res Commun*, 331:881–890.

Benatti, P, Gafa, R, Barana, D, Marino, M, Scarselli, A, Pedroni, M, et al. (2005). Microsatellite instability and colorectal cancer prognosis. *Clin Cancer Res*, 11:8332–8340.

Bhattacharyya, A, Ear, US, Koller, BH, Weichselbaum, RR, and Bishop, DK (2000). The breast cancer susceptibility gene BRCA1 is required for subnuclear assembly of Rad51 and survival following treatment with the DNA cross-linking agent cisplatin. *J Biol Chem*, 275:23899–23903.

Bianco, R, Melisi, D, Ciardiello, F, and Tortora, G (2006). Key cancer cell signal transduction pathways as therapeutic targets. *Eur J Cancer*, 42:290–294.

Branford, S, Rudzki, Z, Walsh, S, Parkinson, I, Grigg, A, Szer, J, et al. (2003). Detection of BCR-ABL mutations in patients with CML treated with imatinib is virtually always accompanied by clinical resistance, and mutations in the ATP phosphate-binding loop (P-loop) are associated with a poor prognosis. *Blood*, 102:276–283.

Brose, MS, Volpe, P, Feldman, M, Kumar, M, Rishi, I, Gerrero, R, et al. (2002). BRAF and RAS mutations in human lung cancer and melanoma. *Cancer Res*, 62:6997–7000.

Brown, R, Hirst, GL, Gallagher, WM, McIlwrath, AJ, Margison, GP, van der Zee, AG, et al. (1997). hMLH1 expression and cellular responses of ovarian tumour cells to treatment with cytotoxic anticancer agents. *Oncogene*, 15:45–52.

Cahill, DP, Lengauer, C, Yu, J, Riggins, GJ, Willson, JK, Markowitz, SD, et al. (1998). Mutations of mitotic checkpoint genes in human cancers. *Nature*, 392:300–303.

Callagy, G, Pharoah, P, Chin, SF, Sangan, T, Daigo, Y, Jackson, L, et al. (2005). Identification and validation of prognostic markers in breast cancer with the complementary use of array-CGH and tissue microarrays. *J Pathol*, 205: 388–396.

Cappuzzo, F, Hirsch, FR, Rossi, E, Bartolini, S, Ceresoli, GL, Bemis, L, et al. (2005). Epidermal growth factor receptor gene and protein and gefitinib sensitivity in non-small-cell lung cancer. *J Natl Cancer Inst*, 97:643–655.

Cavallaro, U and Christofori, G (2004). Cell adhesion and signalling by cadherins and Ig-CAMs in cancer. *Nat Rev Cancer*, 4:118–132.

Chang, HY, Nuyten, DS, Sneddon, JB, Hastie, T, Tibshirani, R, Sorlie, T, et al. (2005). Robustness, scalability, and integration of a wound-response gene expression signature in predicting breast cancer survival. *Proc Natl Acad Sci USA*, 102:3738–3743.

Chang, HY, Sneddon, JB, Alizadeh, AA, Sood, R, West, RB, Montgomery, K, et al. (2004). Gene expression signature of fibroblast serum response predicts human cancer progression: similarities between tumors and wounds. *PLoS Biol*, 2:E7.

Chang, JC, Wooten, EC, Tsimelzon, A, Hilsenbeck, SG, Gutierrez, MC, Elledge, R, et al. (2003). Gene expression profiling for the prediction of therapeutic response to docetaxel in patients with breast cancer. *Lancet*, 362:362–369.

Chang, JC, Wooten, EC, Tsimelzon, A, Hilsenbeck, SG, Gutierrez, MC, Tham, YL, et al. (2005). Patterns of resistance and incomplete response to docetaxel by gene expression profiling in breast cancer patients. *J Clin Oncol*, 23:1169–1177.

Cheadle, JP and Sampson, JR (2003). Exposing the MYtH about base excision repair and human inherited disease. *Hum Mol Genet*, 12(Spec. No 2): R159–R165.

Chi, JT, Wang, Z, Nuyten, DS, Rodriguez, EH, Schaner, ME, Salim, A, et al. (2006). Gene expression programs in response to hypoxia: cell type specificity and prognostic significance in human cancers. *PLoS Med*, 3:e47.

Costello, JF, Fruhwald, MC, Smiraglia, DJ, Rush, LJ, Robertson, GP, Gao, X, et al. (2000). Aberrant CpG-island methylation has non-random and tumour-type-specific patterns. *Nat Genet*, 24:132–138.

Czarnecka, AM, Golik, P, and Bartnik, E (2006). Mitochondrial DNA mutations in human neoplasia. *J Appl Genet*, 47:67–78.

Davies, H, Bignell, GR, Cox, C, Stephens, P, Edkins, S, Clegg, S, et al. (2002). Mutations of the BRAF gene in human cancer. *Nature*, 417:949–954.

Davies, H, Hunter, C, Smith, R, Stephens, P, Greenman, C, Bignell, J, et al. (2005). Somatic mutations of the protein kinase gene family in human lung cancer. *Cancer Res*, 65:7591–7595.

Davies, JJ, Wilson, IM, and Lam, WL (2005). Array CGH technologies and their applications to cancer genomes. *Chromosome Res*, 13:237–248.

Debiec-Rychter, M, Dumez, H, Judson, I, Wasag, B, Verweij, J, Brown, M, et al. (2004). Use of c-KIT/PDGFRA mutational analysis to predict the clinical response to imatinib in patients with advanced gastrointestinal stromal tumours entered on phase I and II studies of the EORTC Soft Tissue and Bone Sarcoma Group. *Eur J Cancer*, 40:689–695.

Di Cristofano, A, Pesce, B, Cordon-Cardo, C, and Pandolfi, PP (1998). Pten is essential for embryonic development and tumour suppression. *Nat Genet*, 19:348–355.

Eckhardt, S (2006). Molecular targeted therapy: a strategy of disillusions or optimism? *J Lab Clin Med*, 147:108–113.

Ein-Dor, L, Kela, I, Getz, G, Givol, D, and Domany, E (2005). Outcome signature genes in breast cancer: is there a unique set? *Bioinformatics*, 21:171–178.

Farmer, H, McCabe, N, Lord, CJ, Tutt, AN, Johnson, DA, Richardson, TB, et al. (2005). Targeting the DNA repair defect in BRCA mutant cells as a therapeutic strategy. *Nature*, 434:917–921.

Fedier, A and Fink, D (2004). Mutations in DNA mismatch repair genes: implications for DNA damage signaling and drug sensitivity (review). *Int J Oncol*, 24:1039–1047.

Fukuoka, M, Yano, S, Giaccone, G, Tamura, T, Nakagawa, K, Douillard, JY, et al. (2003). Multi-institutional randomized phase II trial of gefitinib for previously treated patients with advanced non-small-cell lung cancer (The IDEAL 1 Trial) [corrected]. *J Clin Oncol*, 21:2237–2246.

Gerl, R and Vaux, DL (2005). Apoptosis in the development and treatment of cancer. *Carcinogenesis*, 26:263–270.

Giaccone, G, Herbst, RS, Manegold, C, Scagliotti, G, Rosell, R, Miller, V, et al. (2004). Gefitinib in combination with gemcitabine and cisplatin in advanced non-small-cell lung cancer: a phase III trial–INTACT 1. *J Clin Oncol*, 22:777–784.

Gisselsson, D (2005). Mitotic instability in cancer: is there method in the madness? *Cell Cycle*, 4:1007–1010.

Golub, TR, Slonim, DK, Tamayo, P, Huard, C, Gaasenbeek, M, Mesirov, JP, et al. (1999). Molecular classification of cancer: class discovery and class prediction by gene expression monitoring. *Science*, 286:531–537.

Gottlieb, E and Tomlinson, IP (2005). Mitochondrial tumour suppressors: a genetic and biochemical update. *Nat Rev Cancer*, 5:857–866.

Halford, SE, Rowan, AJ, Lipton, L, Sieber, OM, Pack, K, Thomas, HJ, et al. (2003). Germline mutations but not somatic changes at the MYH locus contribute to the pathogenesis of unselected colorectal cancers. *Am J Pathol*, 162:1545–1548.

Hanks, S, Coleman, K, Reid, S, Plaja, A, Firth, H, Fitzpatrick, D, et al. (2004). Constitutional aneuploidy and cancer predisposition caused by biallelic mutations in BUB1B. *Nat Genet*, 36:1159–1161.

Hanson, MR, McKay, FW, and Miller, RW (1980). Three-dimensional perspective of U.S. cancer mortality. *Lancet*, 2:246–247.

Hegi, ME, Diserens, AC, Gorlia, T, Hamou, MF, de Tribolet, N, Weller, M, et al. (2005). MGMT gene silencing and benefit from temozolomide in glioblastoma. *N Engl J Med*, 352:997–1003.

Heinrich, MC, Corless, CL, Demetri, GD, Blanke, CD, von Mehren, M, Joensuu, H, et al. (2003). Kinase mutations and imatinib response in patients with metastatic gastrointestinal stromal tumor. *J Clin Oncol*, 21:4342–4349.

Herbst, RS, Giaccone, G, Schiller, JH, Natale, RB, Miller, V, Manegold, C, et al. (2004). Gefitinib in combination with paclitaxel and carboplatin in advanced non-small-cell lung cancer: a phase III trial–INTACT 2. *J Clin Oncol*, 22:785–794.

Herman, JG, Umar, A, Polyak, K, Graff, JR, Ahuja, N, Issa, JP, et al. (1998). Incidence and functional consequences of hMLH1 promoter hypermethylation in colorectal carcinoma. *Proc Natl Acad Sci USA*, 95:6870–6875.

Hirota, S and Isozaki, K (2006). Pathology of gastrointestinal stromal tumors. *Pathol Int*, 56:1–9.

Hunter, K (2006). Host genetics influence tumour metastasis. *Nat Rev Cancer*, 6:141–146.

Ibanez de Caceres, I, Dulaimi, E, Hoffman, AM, Al-Saleem, T, Uzzo, RG, and Cairns, P (2006). Identification of novel target genes by an epigenetic reactivation screen of renal cancer. *Cancer Res*, 66:5021–5028.

Ishkanian, AS, Malloff, CA, Watson, SK, DeLeeuw, RJ, Chi, B, Coe, BP, et al. (2004). A tiling resolution DNA microarray with complete coverage of the human genome. *Nat Genet*, 36:299–303.

Iwao-Koizumi, K, Matoba, R, Ueno, N, Kim, SJ, Ando, A, Miyoshi, Y, et al. (2005). Prediction of docetaxel response in human breast cancer by gene expression profiling. *J Clin Oncol*, 23:422–431.

Jiricny, J (2006). The multifaceted mismatch-repair system. *Nat Rev Mol Cell Biol*, 7:335–346.

Jover, R, Zapater, P, Castells, A, Llor, X, Andreu, M, Cubiella, J, et al. (2006). Mismatch repair status in the prediction of benefit from adjuvant fluorouracil chemotherapy in colorectal cancer. *Gut*, 55:848–855.

Kerr, B, Delrue, MA, Sigaudy, S, Perveen, R, Marche, M, Burgelin, I, et al. (2006). Genotype-phenotype correlation in Costello syndrome: HRAS mutation analysis in 43 cases. *J Med Genet*, 43:401–405.

Kim, R, Emi, M, Tanabe, K, Uchida, Y, and Arihiro, K (2006). The role of apoptotic or nonapoptotic cell death in determining cellular response to anticancer treatment. *Eur J Surg Oncol*, 32:69–277.

Kinzler, KW and Vogelstein, B (1996). Lessons from hereditary colorectal cancer. *Cell*, 87:159–170.

Knudson, AG Jr, (1971). Mutation and cancer: statistical study of retinoblastoma. *Proc Natl Acad Sci USA*, 68:820–823.

Kris, MG, Natale, RB, Herbst, RS, Lynch Jr TJ, Prager, D, Belani, CP, et al. (2003). Efficacy of gefitinib, an inhibitor of the epidermal growth factor receptor tyrosine kinase, in symptomatic patients with non-small cell lung cancer: a randomized trial. *JAMA*, 290:2149–2158.

Laird, PW (2005). Cancer epigenetics. *Hum Mol Genet*, 14(Spec. No 1): R65–R76.

Lindblad-Toh, K, Tanenbaum, DM, Daly, MJ, Winchester, E, Lui, WO, Villapakkam, A, et al. (2000). Loss-of-heterozygosity analysis of small-cell lung carcinomas using single-nucleotide polymorphism arrays. *Nat Biotechnol*, 18:1001–1005.

Loi, S, Sotiriou, C, Buyse, M, Rutgers, E, Van't Veer, L, Piccart, M, et al. (2006). Molecular forecasting of breast cancer: time to move forward with clinical testing. *J Clin Oncol*, 24:721–722; author reply 722–723.

Lyakhovich, A and Surralles, J (2006). Disruption of the Fanconi anemia/BRCA pathway in sporadic cancer. *Cancer Lett*, 232:99–106.

Lynch, TJ, Bell, DW, Sordella, R, Gurubhagavatula, S, Okimoto, RA, Brannigan, BW, et al. (2004). Activating mutations in the epidermal growth factor receptor underlying responsiveness of non-small-cell lung cancer to gefitinib. *N Engl J Med*, 350:2129–2139.

Ma, XJ, Wang, Z, Ryan, PD, Isakoff, SJ, Barmettler, A, Fuller, A, et al. (2004). A two-gene expression ratio predicts clinical outcome in breast cancer patients treated with tamoxifen. *Cancer Cell*, 5:607–616.

Majumder, PK, Febbo, PG, Bikoff, R, Berger, R, Xue, Q, McMahon, LM, et al. (2004). mTOR inhibition reverses Akt-dependent prostate intraepithelial neoplasia through regulation of apoptotic and HIF-1-dependent pathways. *Nat Med*, 10:594–601.

Mei, R, Galipeau, PC, Prass, C, Berno, A, Ghandour, G, Patil, N, et al. (2000). Genome-wide detection of allelic imbalance using human SNPs and high-density DNA arrays. *Genome Res*, 10:1126–1137.

Meyers, M, Wagner, MW, Hwang, HS, Kinsella, TJ, and Boothman, DA (2001). Role of the hMLH1 DNA mismatch repair protein in fluoropyrimidine-mediated cell death and cell cycle responses. *Cancer Res*, 61:5193–5201.

Miller, LD, Smeds, J, George, J, Vega, VB, Vergara, L, Ploner, A, et al. (2005). An expression signature for p53 status in human breast cancer predicts mutation status, transcriptional effects, and patient survival. *Proc Natl Acad Sci USA*, 102:13550–13555.

Mootha, VK, Lindgren, CM, Eriksson, KF, Subramanian, A, Sihag, S, Lehar, JP, et al. (2003). PGC-1alpha-responsive genes involved in oxidative phosphorylation are coordinately downregulated in human diabetes. *Nat Genet*, 34:267–273.

Nahta, R and Esteva, FJ (2006). Herceptin: mechanisms of action and resistance. *Cancer Lett*, 232:123–138.

Paez, JG, Janne, PA, Lee, JC, Tracy, S, Greulich, H, Gabriel, S, et al. (2004). EGFR mutations in lung cancer: correlation with clinical response to gefitinib therapy. *Science*, 304:1497–1500.

Paik, S, Shak, S, Tang, G, Kim, C, Baker, J, Cronin, M, et al. (2004). A multigene assay to predict recurrence of tamoxifen-treated, node-negative breast cancer. *N Engl J Med*, 351:2817–2826.

Paris, PL, Andaya, A, Fridlyand, J, Jain, AN, Weinberg, V, Kowbel, D, et al. (2004). Whole genome scanning identifies genotypes associated with recurrence and metastasis in prostate tumors. *Hum Mol Genet*, 13:1303–1313.

Parsons, R, Li, GM, Longley, MJ, Fang, WH, Papadopoulos, N, Jen, J, et al. (1993). Hypermutability and mismatch repair deficiency in RER+ tumor cells. *Cell*, 75:1227–1236.

Perou, CM, Sorlie, T, Eisen, MB, van de Rijn, M, Jeffrey, SS, Rees, CA, et al. (2000). Molecular portraits of human breast tumours. *Nature*, 406:747–752.

Petros, JA, Baumann, AK, Ruiz-Pesini, E, Amin, MB, Sun, CQ, Hall, J, et al. (2005). mtDNA mutations increase tumorigenicity in prostate cancer. *Proc Natl Acad Sci USA*, 102:719–724.

Pittman, J, Huang, E, Dressman, H, Horng, CF, Cheng, SH, Tsou, MH, et al. (2004). Integrated modeling of clinical and gene expression information for

personalized prediction of disease outcomes. *Proc Natl Acad Sci USA*, 101:8431–8436.
Ramaswamy, S, Ross, KN, Lander, ES, and Golub, TR (2003). A molecular signature of metastasis in primary solid tumors. *Nat Genet*, 33: 49–54.
Sawyers, CL (2003). Opportunities and challenges in the development of kinase inhibitor therapy for cancer. *Genes Dev*, 17:2998–3010.
Seeger, RC, Brodeur, GM, Sather, H, Dalton, A, Siegel, SE, Wong, KY, et al. (1985). Association of multiple copies of the N-myc oncogene with rapid progression of neuroblastomas. *N Engl J Med*, 313:1111–1116.
Segal, E, Friedman, N, Koller, D, and Regev, A (2004). A module map showing conditional activity of expression modules in cancer. *Nat Genet*, 36:1090–1098.
Shaw, RJ and Cantley, LC (2006). Ras, PI(3)K and mTOR signalling controls tumour cell growth. *Nature*, 441:424–430.
Sherr, CJ (1996). Cancer cell cycles. *Science*, 274:1672–1677.
Sherr, CJ (2000). The Pezcoller lecture: cancer cell cycles revisited. *Cancer Res*, 60:3689–3695.
Shibuya, M and Claesson-Welsh, L (2006). Signal transduction by VEGF receptors in regulation of angiogenesis and lymphangiogenesis. *Exp Cell Res*, 312:549–560.
Shidara, Y, Yamagata, K, Kanamori, T, Nakano, K, Kwong, JQ, Manfredi, G, et al. (2005). Positive contribution of pathogenic mutations in the mitochondrial genome to the promotion of cancer by prevention from apoptosis. *Cancer Res*, 65:1655–1663.
Shin, JS, Hong, A, Solomon, MJ, and Lee, CS (2006). The role of telomeres and telomerase in the pathology of human cancer and aging. *Pathology*, 38:103–113.
Simon, R, Radmacher, MD, Dobbin, K, and McShane, LM (2003). Pitfalls in the use of DNA microarray data for diagnostic and prognostic classification. *J Natl Cancer Inst*, 95:14–18.
Sorlie, T, Perou, CM, Tibshirani, R, Aas, T, Geisler, S, Johnsen, H, et al. (2001). Gene expression patterns of breast carcinomas distinguish tumor subclasses with clinical implications. *Proc Natl Acad Sci USA*, 98:10869–10874.
Sorlie, T, Tibshirani, R, Parker, J, Hastie, T, Marron, JS, Nobel, A, et al. (2003). Repeated observation of breast tumor subtypes in independent gene expression data sets. *Proc Natl Acad Sci USA*, 100:8418–8423.
Stephens, P, Edkins, S, Davies, H, Greenman, C, Cox, C, Hunter, C, et al. (2005). A screen of the complete protein kinase gene family identifies diverse patterns of somatic mutations in human breast cancer. *Nat Genet*, 37:590–592.
Suzuki, H, Gabrielson, E, Chen, W, Anbazhagan, R, van Engeland, M, Weijenberg, MP, et al. (2002). A genomic screen for genes upregulated by demethylation and histone deacetylase inhibition in human colorectal cancer. *Nat Genet*, 31:141–149.
Sweet-Cordero, A, Mukherjee, S, Subramanian, A, You, H, Roix, JJ, Ladd-Acosta, C, et al. (2005). An oncogenic KRAS2 expression signature identified by cross-species gene-expression analysis. *Nat Genet*, 37:48–55.
Tassone, P, Tagliaferri, P, Perricelli, A, Blotta, S, Quaresima, B, Martelli, ML, et al. (2003). BRCA1 expression modulates chemosensitivity of BRCA1-defective HCC1937 human breast cancer cells. *Br J Cancer*, 88:1285–1291.
Thatcher, N, Chang, A, Parikh, P, Rodrigues Pereira, J, Ciuleanu, T, von Pawel, J, et al. (2005). Gefitinib plus best supportive care in previously treated patients with refractory advanced non-small-cell lung cancer: results from a randomised, placebo-controlled, multicentre study (Iressa Survival Evaluation in Lung Cancer). *Lancet*, 366:1527–1537.
Tonini, T, Rossi, F, and Claudio, PP (2003). Molecular basis of angiogenesis and cancer. *Oncogene*, 22:6549–6556.
Tsao, MS, Sakurada, A, Cutz, JC, Zhu, CQ, Kamel-Reid, S, Squire, J, et al. (2005). Erlotinib in lung cancer—molecular and clinical predictors of outcome. *N Engl J Med*, 353:133–144.
Turner, N, Tutt, A, and Ashworth, A (2004). Hallmarks of "BRCAness" in sporadic cancers. *Nat Rev Cancer*, 4:814–819.
van de Vijver, MJ, He, YD, van't Veer, LJ, Dai, H, Hart, AA, Voskuil, DW, et al. (2002). A gene-expression signature as a predictor of survival in breast cancer. *N Engl J Med*, 347:1999–2009.
van't Veer, LJ, Dai, H, van de Vijver, MJ, He, YD, Hart, AA, Mao, M, et al. (2002). Gene expression profiling predicts clinical outcome of breast cancer. *Nature*, 415:530–536.
Veigl, ML, Kasturi, L, Olechnowicz, J, Ma, AH, Lutterbaugh, JD, Periyasamy, S, et al. (1998). Biallelic inactivation of hMLH1 by epigenetic gene silencing, a novel mechanism causing human MSI cancers. *Proc Natl Acad Sci USA*, 95:8698–8702.
Wakeling, AE, Guy, SP, Woodburn, JR, Ashton, SE, Curry, BJ, Barker, AJ, et al. (2002). ZD1839 (Iressa): an orally active inhibitor of epidermal growth factor signaling with potential for cancer therapy. *Cancer Res*, 62:5749–5754.
Wardelmann, E, Merkelbach-Bruse, S, Pauls, K, Thomas, N, Schildhaus, HU, Heinicke, T, et al. (2006). Polyclonal evolution of multiple secondary KIT mutations in gastrointestinal stromal tumors under treatment with imatinib mesylate. *Clin Cancer Res*, 12:1743–1749.
Watanabe, Y, Koi, M, Hemmi, H, Hoshai, H, and Noda, K (2001). A change in microsatellite instability caused by cisplatin-based chemotherapy of ovarian cancer. *Br J Cancer*, 85:1064–1069.
Wei, SH, Chen, CM, Strathdee, G, Harnsomburana, J, Shyu, CR, Rahmatpanah, F, et al. (2002). Methylation microarray analysis of late-stage ovarian carcinomas distinguishes progression-free survival in patients and identifies candidate epigenetic markers. *Clin Cancer Res*, 8:2246–2252.
Weiss, MM, Kuipers, EJ, Postma, C, Snijders, AM, Pinkel, DS, Meuwissen, G, et al. (2004). Genomic alterations in primary gastric adenocarcinomas correlate with clinicopathological characteristics and survival. *Cell Oncol*, 26:307–317.
Westra, JL, Schaapveld, M, Hollema, H, de Boer, JP, Kraak, MM, de Jong, D, et al. (2005). Determination of TP53 mutation is more relevant than microsatellite instability status for the prediction of disease-free survival in adjuvant-treated stage III colon cancer patients. *J Clin Oncol*, 23:5635–5643.
Yoo, CB and Jones, PA (2006). Epigenetic therapy of cancer: past, present and future. *Nat Rev Drug Discov*, 5:37–50.
Zhao, X, Li, C, Paez, JG, Chin, K, Janne, PA, Chen, TH, et al. (2004). An integrated view of copy number and allelic alterations in the cancer genome using single nucleotide polymorphism arrays. *Cancer Res*, 64: 3060–3071.
Zhou, X, Mok, SC, Chen, Z, Li, Y, and Wong, DT (2004). Concurrent analysis of loss of heterozygosity (LOH) and copy number abnormality (CNA) for oral premalignancy progression using the Affymetrix 10K SNP mapping array. *Hum Genet*, 115:327–330.

24

Hematological Malignancies: The Paradigm of Acute Myeloid Leukemia

Kenneth I Mills and Alan Burnett

Acute myeloid leukemia (AML) is a clinically and molecular heterogeneous set of hematopoietic disorders (Lowenberg et al., 1999) which account for 77% of acute leukemia (Walker et al., 1994). It is the most common form of acute leukemia in adults but forms only a minor fraction (10%–15%) of childhood leukemia (Lightfoot, 2005). The overall incidence of AML is approximately 4 in 1,00,000 in the United States and United Kingdom, and increases with age from 1 in 1,00,000 in people younger than 35 years old at diagnosis to 15 in 1,00,000 in those older than 75 years. The incidence rates are slightly higher in males and in Whites.

AML is characterized by the infiltration of bone marrow by abnormal hematopoietic progenitors that disrupts normal production of erythroid, myeloid, and/or megakaryocytic cells. The deregulated proliferation and impaired differentiation result in an increased survival advantage for the leukemia cell (Bishop, 1997; Lowenberg et al., 2003; Stone et al., 2004). AML arises from a clone of poorly differentiated hematopoietic stem cells that contain mutated genes which confer an advantage for growth or survival. The clonal leukemia cells develop, expand in number, and eventually dominate normal hematopoiesis.

Most patients have a presentation that arises from the disruption of normal blood component production. These include fatigue and shortness of breath (anemia), bleeding and bruising (thrombocytopenia), and fever with or without infection (neutropenia). The most common infections are upper respiratory infections, though the occasional patient will present with pneumonia. Bone pain and skin infiltrates (leukemia cutis) or gingival often occur with monocytic variants. Less often, patients may have collections of myeloblasts/chloromas or granulocytic sarcomas in any soft tissue. Laboratory studies usually reveal pancytopenia, although a combination of anemia, thrombocytopenia, and leukopenia or leukocytosis may exist. Blasts and other immature white blood cells may be present in the peripheral blood, and because of rapid cell turnover, lactic dehydrogenase and uric acid levels may be elevated. In acute promyelocytic leukemia (APL), coagulation profiles should be evaluated carefully as disseminated intravascular coagulation (DIC) may occur when these cells lyse and release factors to promote coagulation. Leukocytosis may also be present and can cause headaches, altered mental status, dyspnea, and ultimately fatal pulmonary or intracranial hemorrhages. AML can usually be diagnosed from the peripheral smear, but bone marrow aspirate and biopsy should always be performed to determine proper classification. The bone marrow aspirate has more than 30% blasts (<5% blasts is normal) of which more than 3% are positive for myeloperoxidase (MPO) stain. Special stains help determine specific subtypes. Patients with M4 or M5 morphology are positive for monocytic stains (peroxidase and esterase), while patients with M6 disease have diffuse cytoplasmic positivity for periodic acid Schiff (PAS) stain. Histochemical stains on bone marrow blasts are crucial for diagnosis, but cell immunophenotype and electron microscopy are helpful in difficult cases. Patients negative by all stains (AUL, acute undifferentiated leukemia) are usually treated as AML if immunophenotyping demonstrates myeloid markers. Immunophenotyping of the AML blood or marrow will show myeloid markers such as CD13, CD14, CD15, and CD33, while Auer rods and cytoplasmic inclusion bodies found in myeloblasts and monoblasts are indicative of AML. Additional studies (e.g., CD41, electron microscopy) may help in diagnosing subtype M7. In APL, M3 subtype, the bone marrow may have less than 30% blasts but usually more than 70% pro-granulocytes. Cytogenetics is usually done on bone marrow, and is abnormal in 70% of patients.

Development of AML

The cause of AML is unknown in most cases, but the development has been strongly linked to radiation, chemical exposure (benzene), or prior use of alkylating agents. For example, there is a 20-fold increase of AML in Japanese atomic bomb survivors, while the incidence of AML is 10 times higher in manufacturing workers exposed to benzene (Deininger et al., 1998; Andersen et al., 2001). Several congenital disorders terminate in AML—Down syndrome (20 times the risk), Fanconi's anemia, Klinefelter's syndrome, Turner's syndrome, and Wiskott-Aldrich syndrome (Perry et al., 1980).

In a small group of patients, AML evolves from a prior hematological disorder, e.g., severe congenital neutropenia, chronic myeloid leukemia (CML) or Myelodysplastic syndrome (MDS) (Levine and Bloomfield, 1992), a congenital or inherited disorder such as Diamond Blackfan anemia, Fanconi's anemia (Wasser et al., 1978); (Auerbach and Allen, 1991) or as a result of exposure to alkylating agents or topoisomerase II inhibitors used in the treatment of other neoplasias (Stanulla, Wang, Chervinsky, Thandla, et al., 1997; Stanulla, Wang, Chervinsky, and Aplan, 1997); (Blanco et al., 2001). Secondary AML may occur up to 7 years after treatment with alkylating agents or other chemotherapy drugs and has a poorer prognosis than de novo leukemia if, as is usually the case, cytogenetics are abnormal. In addition, patients with MDS may develop AML; 25% of AML patients have a history of MDS. These leukemia are generally designated as secondary leukemia and are of interest due to the probable multiple genetic aberrations required for the evolution into AML.

Disease Classification

AML has been classified using the French–American–British (FAB) system, which is based on morphological grounds such as degree of lineage, commitment, and differentiation.(Bennett et al., 1976). AML FAB subtypes (M0–M7) were determined by cell morphology with particular subtypes such as M3 (APL). Recently, the World Health Organization (WHO) introduced a new classification of hematopoietic and lymphoid neoplasm by combining morphology, genetic, immunophenotype, and biologic and clinical information (Vardiman et al., 2002). The WHO classification places the most common and well-defined recurring cytogenetic abnormalities into separate major groups with further defined classified subgroups. Cytogenetic analysis of leukemia samples has identified various non-random chromosomal aberrations in AML. Many types of AML are associated with balanced reciprocal translocations involving two transcription factors. These transcription factors are frequently conserved in evolution and are important both in mammalian development as well as in normal hematopoiesis. Transcription factors involved in human leukemia include core binding factor (CBF), retinoic acid receptor α (RARα), homeobox (HOX) family members, and members of the ETS family of transcription factors. Some of these chromosomal abnormalities correlate with specific FAB subtypes, for instance, the translocation t(8;21) is present in approximately 40% of the FAB-M2 leukemia (Miyoshi et al., 1991). Translocations involving the retinoic receptor α (RARα) on chromosome 17 in APL (APML) FAB-M3 t(15;17) is present in 98% of cases (Warrell, et al., 1993) and inv(16) or t(16;16) involved in 80% of the FAB-M4-Eo cases (Le Beau et al., 1983; Larson et al., 1986).

Based on the molecular and cytogenetic aspects, patients have been classified into three distinct prognostic subgroups or risk groups: those with favorable or good risk, intermediate or standard risk, and unfavorable, adverse, or poor risk disease (Fig. 24–1). The favorable risk group includes approximately 15% of patients with an

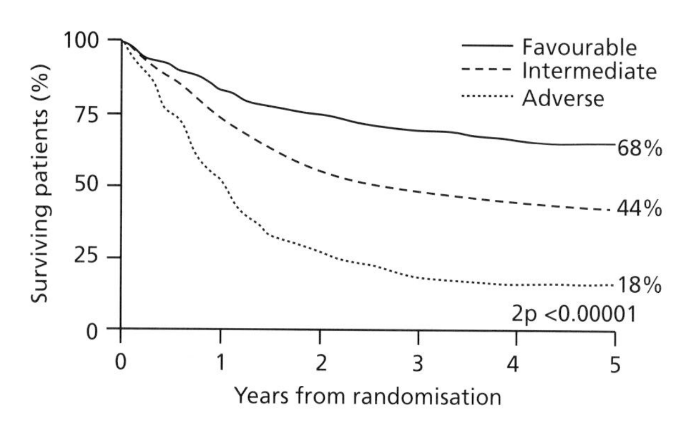

Figure 24–1 Survival rates of acute myeloid leukemia (AML) patients with different cytogenetic abnormalities. Favorable or good risk patients have either t(8;21), inv(16) or t(15;17) cytogenetic translocations. The intermediate or standard risk groups mainly consist of patients with apparent normal cytogenetics or those who do not fall into the other two groups. Patients within the unfavorable, adverse or poor risk group have complex cytogenetic changes, del(5q), −5, −7 or abn(3q) abnormalities.

age below 60 years and is defined by the presence of either t(15;17), t(8;21) and inv(16), or t(16;16) cytogenetic translocations. Patients presenting with these mutations usually have an increased rate of complete remission and are at relatively low risk of relapse (Zwaan and Kaspers, 2004). The poor prognosis group of AML patients have cytogenetic abnormalities such as deletion of chromosomes 5 or 7 (or the short-arms of the chromosomes) abnormalities in the long-arm of chromosome 3 and AML with complex karyotype (three or more cytogenetic aberrations) (Kalwinsky et al., 1990; Grimwade et al., 1998). Cytogenetic mutations involving the 11q23 region are also considered a marker for poor prognosis. Overall, patients with adverse risk-type mutations generally have a probability of 5-year survival of approximately 20%. Patients presenting with AML with other type of cytogenetic abnormalities or without any apparent cytogenetic abnormalities (i.e., an apparently normal karyotype) are considered to have an intermediate risk (Mrozek et al., 1997) (Grimwade et al., 1998; Lowenberg et al., 1999).

Cytogenetics and Mutations

Cytogenetic Rearrangements

Cytogenetics, including fluorescence in situ hybridization (FISH), can provide an insight into the heterogeneity of the disease by identifying two major groups of AML: in one group, chromosomal aberrations occur in approximately 52% of de novo AML, and in the second group no karyotype abnormalities can be detected. Within the first group two further subtypes can be further distinguished. One subgroup has balanced aberrations mainly consisting of t(8;21), t(15;17), and inv(16). The second subgroup has cases with unbalanced aberrations such as 5q−, 7q−, −5, −7, or complex cytogenetic abnormalities.

Balanced translocations result in the expression of fusion proteins, which frequently contain a truncated transcription factor joined with another unrelated protein. The balanced translocations result in a deregulation of transcription factors and can be seen as a main pathogenic event (Rowley, 2001). The simultaneous occurrence of more than one of these recurrent, balanced translocations is rare. The detection of subtype specific abnormalities such as the t(15;17) or t(8;21) translocations has resulted in the characterization of the involved genes, the identification of abnormal fusion proteins, provided new diagnostic targets, but also identified distinct AML subtypes.

t(15;17)—PML-RARα

APL (FAB subtype M3) accounts for approximately 10% of all AML cases (Grimwade et al., 2001) and is characterized by translocations involving the *RARα* gene located at 17q12. In the majority of the cases, the partner chromosome in the translocation is chromosome 15 resulting in the fusion of large parts of RARα to most of PML coding sequence, generating the PML-RARα fusion protein (Pandolfi, 2001a, b). Rarer cases of APL show translocations involving t(11;17)(q23;q12), t(5;17)(q35q12), t(17;17)(q11;q12), or t(11;17)(q13;q12). These result in fusion proteins which also contain RARα but with different partner genes: PLZF (promyelocytic leukemia zinc finger), NPM (nucleophosmin), Stat5b or NuMA (Chen, SJ et al., 1993; Chen, Z et al., 1993; Wells et al., 1996, 1997; Arnould et al., 1999; Redner et al., 2000; Redner et al., 2006).

RARα is a nuclear hormone receptor and in the absence of the ligand retinoic acid transcription is repressed through the recruitment of co-repressors and HDACs. When retinoic acid binds to

RARα, the heterodimeric complex of RARα and RXRα releases the co-repressors and HDACs, converting the complex into a transcriptional activator. However, PML-RARα disrupts this response through the retention of the DNA binding specificity; physiological concentration of retinoic acid cannot induce the conformational change necessary to release co-repressors. Therefore any RARα target genes remain repressed (Grignani et al., 1998; He et al., 1998; Lin et al., 1998). Treatment with pharmacological levels of all-*trans*-retinol acid (ATRA) reverses this process and restores the transcriptional activity of the retinoic acid promoter resulting in the differentiation of the leukemia cells along the granulocytic lineage (Fenaux et al., 2001). APL is a classic example of how basic research can influence clinical therapy directly as a result of identifying the disease-specific molecular defect and its functional repair by a targeted therapeutic approach.

The over-expression of PML-RARα alone is insufficient to induce an undifferentiated phenotype in mouse bone marrow: only 15–20% of the transplanted mice develop a disease with undifferentiated characteristics and only after a long latency. This would suggest that a second event is required for the development of APL (Grignani et al., 2000). Co-expression of PML-RARα with Flt3-ITD mutations enhances the frequency of disease in recipient mice; mice reconstituted with Flt3-ITD do not develop an acute leukemia-like disease again indicating mutational cooperation (Le Beau et al., 2002, 2003).

In the translocation t(15;17), the reciprocal RARα-PML transcript (Alcalay et al., 2001) is also expressed in some patients and may contribute to leukemia development as transgenic mice expressing both PML-RARα and RARα-PML have an increased frequency of APL (Pollock et al., 2001). In contrast to PML-RARα, PLZF-RARα, the product of t(11;17), does not respond to high doses of retinoic acid. Similar to RARα, PLZF can bind co-repressors by itself, a mechanism that cannot be overcome by ATRA (Grignani et al., 1998). PLZF contains an oligomerization domain available for PLZF or PLZF-RARα homodimers as well as for heterodimers with regulatory proteins. Therefore, PLZF-RARα forms high molecular weight nuclear complexes in vivo, displacing the normal PLZF from its correct location in the cell. This would suggest that the fusion protein disturbs both RARα and PLZF function (Dong et al., 1996)

An important target gene of both PML-RARα and PLZF-RARα is cyclin A1, a homolog of cyclin A2 that is specifically expressed in hematopoietic cells (Yasmeen et al., 2003). ATRA exposure decreases cyclin A1 expression in the PML-RARα positive NB4 cell line, Cyclin A1 is induced by PML-RARα and PLZF-RARα (Muller, Readhead, et al., 2000; Muller, Yang, et al., 2000), while over-expression in a transgenic mouse model induces a myeloproliferative disease with long latency in a number of transgenic mice (Liao et al., 2001). These findings provide direct evidence for a role of PML-RARα in accelerating cell cycle progression.

CBF Translocations

The other balanced abnormalities are the t(8;21) and inv 16 translocations which disrupt the "CBF complex," and this type of translocation is observed in approximately 20% of AML cases, particularly in those patients less than 60 years of age at diagnosis. (Downing, 1999). The CBF consists of two subunits: AML1 (also referred to as RUNX1, CBPα2, and PEBP2αB) and is located on chromosome 21q22 and CBFβ, which is encoded by a gene on 16q22. The two subunits form a heterodimer transcription activator, although only the AML1 subunit contains DNA binding ability.

The most frequent translocation is the t(8;21) translocation, found in 10%–15% of adult patients with this disease (Downing, 1999) and results in the fusion protein AML1-ETO. The C-terminus of the transcriptional activator AML1 is replaced by the transcriptional repressor ETO, resulting in the fusion protein AML1-ETO (CBFA2-CBFA2T1) (Meyers et al., 1995). The ETO protein binds to co-repressors (e.g., NcoR, SMRT, and mSin3) and histone deacetylases (HDACs) (Gelmetti et al., 1998; Amann et al., 2001) (Fig. 24–2). The recruitment of HDACs to the complex changes the histone acetylation status and as a result alters the DNA conformation, making it less accessible for the basal transcription machinery and results in a repression of AML1 target genes. Target gene include granulocyte–macrophage colony-stimulating factor (GM-CSF) (Frank et al., 1995), for interleukin-3 (IL-3 and NP3 (Westendorf et al., 1998).

AML1-ETO influences various other transcription factors including MEF, c/EBPα, AP-1, and Pu.1 (Muller-Tidow et al., 2001; Vangala et al., 2003; Mueller and Pabst, 2006). c/EBPα is a major regulator of early granulocytic differentiation and AML-ETO physically interacts with c/EBPα and downregulates it at the transcriptional level (Mueller and Pabst, 2006), which has an important influence on granulocytic differentiation. An additional function of AML1-ETO in leukemia transformation is through its influence on cellular proliferation and survival. It has recently been shown that AML1-ETO directly inhibits the expression of $p14^{ARF}$ (Linggi et al., 2002), and this has a role in inducing p53-dependent proliferation arrest and apoptosis (Hiebert et al., 2003). Low levels of p14ARF are predictors of poorer outcome for AML patients (Muller-Tidow, Metzelder, et al., 2004). AML1-ETO also has a further role in apoptosis as it up-regulates Bcl2 (Klampfer et al., 1996), although contrasting results have been observed in an established monoblastic leukemia cell line apoptosis (Burel et al., 2001). The

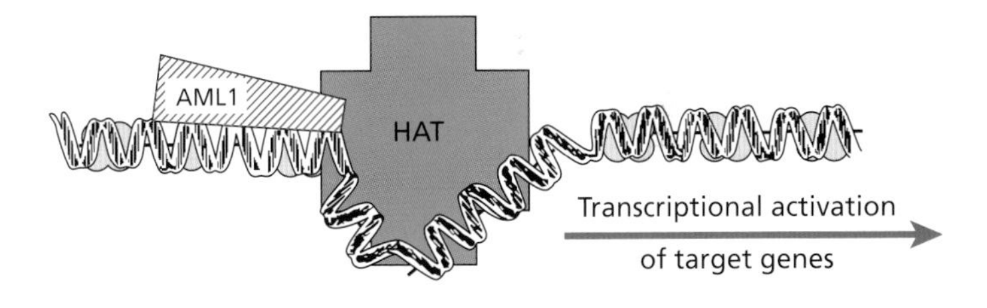

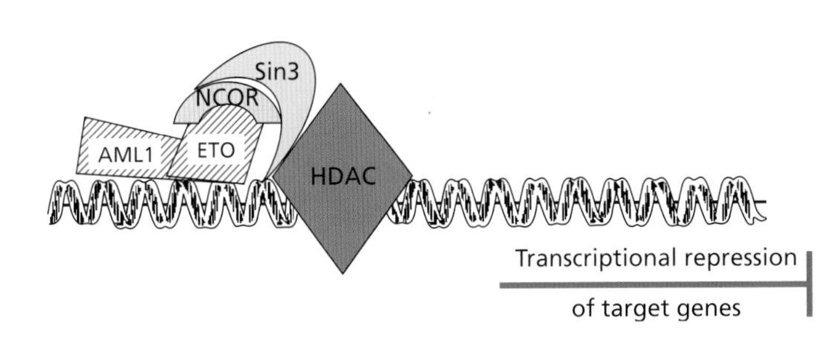

Figure 24–2 In normal cells, the acute myeloid leukemia 1 (AML1) protein binds to its target DNA sequence and recruits several co-activitators including histone acetyl transferase (HAT), acetylation of the histone tails by HAT causes the DNA to unwind allowing the transcription machinery access resulting in activation. In leukemia cells have the t(8;21) translocation and express the AML1-ETO fusion protein, the ETO portion of the protein recruits a series of co-repressors including histone deacetylase (HDAC). The deacetylated histone tails bind the DNA in a closed configuration resulting in transcriptional repression.

down-regulation of c/EBPα by AML1-ETO resulting in the release of c/EBPα-mediated cell cycle control leads to enhanced cellular proliferation.

Other CBF leukemia associated translocations in AML are inv (16) and t(16;16) in which CBFβ is fused to MYH11, also referred to as the smooth muscle myosin heavy chain (*SMMHC*) gene (Liu et al., 1993), generating a CBFβ/SMMHC fusion protein. The fusion protein binds to AML1, and through DNA contact acts as a repressor of AML1 function (Lutterbach et al., 1999). In addition, the SMMHC part of the fusion protein contains a repressor domain that directly represses AML1 activity when bound to transcriptional active sites (Lutterbach et al., 1999).

Recently, studies on shared target genes of the fusion proteins AML1-ETO, PML-RARα, and PLZF-RARα in U937 cells reported that several of the genes regulated by all three fusion proteins are associated with Wnt-signaling (Muller-Tidow, Steffen, et al., 2004). The Wnt-signaling pathway involves both known tumor suppressors or proto-oncogenes in the pathogenesis of epithelial cancers and have been implicated in the self-renewal and proliferation of hematopoietic progenitor cells (Polakis, 2000). A positive transcriptional regulator induced by this pathway is γ-catenin (plakoglobin), which was strongly induced by all three fusion proteins. The enhanced transcriptional activity of γ-catenin leads to induction of Wnt target genes resulting in enhanced self-renewal and proliferative capacity of myeloid progenitor cells.

Transcription Factor Mutations

The transcription factors Pu.1, CEBPα, and GATA-1 are important in myeloid differentiation and the observed functional inactivation by AML-associated fusion proteins, and although they have been examined, no involvement in chromosomal translocations has been observed, but mutations have been identified in subgroups of AML patients.

Mutations of Pu.1 mutations were observed in approximately 7% of AML patients (Mueller et al., 2002) disrupting the function of one allele without generating a dominant negative protein. The multi-lineage importance of Pu.1 was confirmed by the finding of mutations in patients with an immature phenotype (FAB M0), myelomonocytic or monocytic subtypes (FAB M4 or M5), or with erythroleukemia (FAB M6).

Two types of somatic c/EBPα mutations have been reported in approximately 10% of AML cases with similar frequencies (Preudhomme et al., 2002), predominantly FAB M1 or M2 subtypes. One type of mutation is found within the basic zipper region and disrupts DNA-binding as well as dimerization of the protein with other c/EBP family members. Patients with mutations in the DNA binding domain mutations also have mutations in the second allele, leading to a complete loss of c/EBPα function (Preudhomme et al., 2002). The second type of mutations causes a frameshift within the area encoding the N-terminal region of the protein, resulting in the translation of a shorter form of the c/EBPα protein (30 kDa), only containing the N-terminal portion (Preudhomme et al., 2002; Mueller and Pabst, 2006). This protein inhibits the normal c/EBPα in a dominant negative fashion (Mueller and Pabst, 2006). The presence of c/EBPα mutations seems to be a relatively good prognostic indicator (Barjesteh van Waalwijk van Doorn-Khosrovani et al., 2003).

The inhibition of the function of Pu.1 and c/EBPα defines a common theme for several primary genetic defects in AML. C/EBPα is inactivated or has transcription repressed by Flt3-ITD and AML1-ETO, whereas Pu.1 can also be transcriptionally repressed by Flt3-ITD, although the physiological Flt3 signal induces Pu.1 function (Mizuki et al., 2003).

While the importance of the disruption of Pu.1 function is still not understood, c/EBPα is increasingly recognized to be a tumor suppressor protein beyond its function as a differentiation inducer. For example, it inhibits cell proliferation by increasing p21 gene expression and protein stability (Timchenko et al., 1996). C/EBPα also directly binds to CDK2 and CDK4, inhibits their function, and thereby cell cycle (G1-S) progression. In addition, c/EBPα can enhance the degradation of CDK2 (Wang et al., 2002). Interestingly, the two functions of c/EBPα: cell cycle inhibition and differentiation can be structurally separated within the protein. Mutations of c/EBPα lacking DNA binding capacity slow down G1-S transition but are insufficient to induce granulocytic differentiation. In contrast, deletion mutations of c/EBPα lacking regions that interact with CDK2 and CDK4 inhibit cell cycle progression in myeloid cell lines (Wang et al., 2002). Furthermore, c/EBPα inhibits E2F pathways and thereby downregulates C-MYC and cell proliferation (Porse et al., 2005). Interruption of c/EBPα function thus not only inhibits granulocytic differentiation but also releases proliferation of myeloid cells.

MLL gene rearrangements and mutations

A subgroup of AML patients have translocations involving the MLL gene located on chromosome 11q23. The MLL protein, the human homolog of *Drosophila* trithorax, consists of 3969 amino acids and the gene spans 90 kb. MLL has histone methyltransferase activity and assembles in very large multi-protein complexes composed of more than 29 proteins (Nakamura et al., 2002). The DNA binding domain of MLL contains three AT-hooks that change DNA conformation and thereby allow regulatory transcription factors to bind to DNA (Aravind and Landsman, 1998). In addition to the chromosomal translocations involving numerous partner chromosomes and genes, partial tandem duplications (PTDs) also occur. Translocations involving the *MLL* gene are found in up to 80% of leukemia diagnosed in children younger than 24 months (Satake et al., 1999; Schoch et al., 2003). Other common translocations involving chromosome 11q23 result from a fusion chromosomes 4, 9, or 19 and only occur in 2%–5% of all adult AML (Schoch et al., 2003). More than 60 cytogenetic translocations involving the region 11q23 have been described, with at least 30 of the fusion partners of MLL defined (Huret et al., 2004). All fusion proteins contain the N-terminal part including the DNA binding domain and the repressor domain of MLL fused to the C-terminal part of the translocation partner. Most of the known MLL fusion partners, for example, ENL, AF9, and AF4 contain transcriptional activation domains. MLL fusion proteins do regulate the expression of *HOX* genes, and this is partially responsible for their ability to immortalize myeloid progenitor cells. It has been shown that MLL-ENL fails to immortalize myeloid progenitors from HOXA9- or HOXA7-deficient bone marrow (Ayton and Cleary, 2003). MLL regulates HOX genes expression through direct binding to promoter sequences, and HOXA9 is highly expressed in MLL leukemia (Armstrong et al., 2002).

In addition to chromosomal translocations, a different type of rearrangement of the *MLL* gene has been described in AML. PTDs result in a duplication of a portion of the gene and have been observed in 3%–10% of adult AML, often associated with trisomy 11 (Steudel et al., 2003; Schnittger et al., 2005), but more frequently in 10% of cases without normal cytogenetics. The incidence of MLL PTD in Flt3-ITD positive patients is significantly higher than in Flt3-ITD negative patients (Steudel et al., 2003). This may be due to the fact that Flt3 and MLL loci are susceptible to similar agents and mechanisms of DNA damage and links a mechanism leading to enhanced proliferation with a mechanism resulting in blocked differentiation.

HOX Gene Translocations

HOX gene-family members are also involved in AML-associated chromosomal translocations. Examples include the translocations t(7;11) and t(2;11) generating the fusion proteins NUP98/HoxA9 and NUP98/HOXD13. NUP98 encodes for a component of the nuclear core complex. The fusion protein consists of the homeodomain of HOXA9 and the phenylalanine–glycine repeats from NUP98 (Kawakami et al., 2002; Taketani et al., 2002; Ghannam et al., 2003), which contains an interaction domain with the transcriptional activator CBP/p300. Over-expression of the fusion protein induces a myeloproliferative disease, followed by the occurrence of AML in transplanted mice (Kasper et al., 1999).

Receptor Tyrosine Kinase Mutations

FLT Gene Mutations

Activating mutations within the Flt3 gene was first described in 1996, and the protein has been identified as both a possible progression factor and a drug target in AML. Flt3 (Fig. 24–3) (Fms-like tyrosine kinase 3, also designated Flk-2 or STK-1) is a member of the class III RTKs and shares high homology with the other members of this family, the PDGF (platelet-derived growth factor) receptors, c-FMS/M-CSF (macrophage colony stimulating factor) receptor, and Kit (receptor for SCF, the stem cell factor). Common features of class III RTK include an extra-cellular domain consisting of five to seven immunoglobulin-like subdomains and a split kinase domain (Rosnet et al., 1993). The gene encoding Flt3 is located on chromosome 13q12 (Carow et al., 1995). The gene contained 24 exons, with exons 14 and 15 encoding the juxta-membrane region of the receptor and exon 20 encoding the tyrosine kinase domain (TKD) (Abu-Duhier et al., 2001a).

Flt3 is predominantly, but not exclusively, expressed in hematopoietic tissues but can be detected in placenta, brain, testis, lymph nodes, thymus and in the liver of 6–8-week-old mice (Maroc et al., 1993). In hematopoiesis, Flt3 is expressed in early multi-potent progenitor cells but not in primitive stem cells. Flt3 expression seems to be accompanied by a loss of self-renewal capacity in the hematopoietic stem cell compartment and remains detectable during monocytic differentiation but is lost during B-lymphoid and granulocytic differentiation (Rappold et al., 1997; Adolfsson et al., 2001; Christensen and Weissman, 2001).

After ligand binding, Flt3 activates the signal pathway via receptor auto-phosphorylation and through subsequent binding of proteins, which specifically recognize phosphorylated tyrosines on the receptor. These include the recruitment and activation of the Src kinase family and the activation of the Ras/MAPK signaling cascade via direct interaction with Grb2 (Marchetto et al., 1999). Flt3 induces the phosphorylation of other substrates such as Gab (Zhang and Broxmeyer, 2000) and Cbl proteins that are involved in receptor internalization and degradation (Lavagna-Sevenier et al., 1998). A weak but consistent phosphorylation of STAT5a, but not STAT5b, can be observed after ligand-induced activation of Flt3, but it is not associated with STAT5 DNA-binding or expression of STAT5 target genes (Mizuki et al., 2000).

ITD mutations of FLT3 In AML, Flt3 is highly expressed in 60%–92% of the cases (Stacchini et al., 1996), and this is at a higher level compared to normal hematopoietic progenitors. Flt3 has been shown to have in-frame insertions of several nucleotides in the mRNA sequence encoding the juxtamembrane domain of the receptor in AML patients(Nakao et al., 1996). These insertions are internal tandem duplications (ITDs) of varying lengths and result in the repetition of between 3 and 50 amino acids in the juxtamembrane region of the Flt3 receptor. Most patients also express the wild-type allele. The prevalence of Flt3-ITD in AML patients have indicated an incidence of between 20% and 30% (Abu-Duhier et al., 2000; Kottaridis et al., 2001; Schnittger et al., 2002; Mills et al., 2005). The frequency of in FAB subgroups of AML differs with a relatively high frequency in the APL (Schnittger et al., 2002). Highest prevalence of Flt3-ITD mutations was seen in cases with normal karyotype but occur rarely in CBF [inv(16) or t(8;21)] leukemia (Schnittger et al., 2002), where conversely Kit mutations are seen with high frequency.

In APL, Flt3-ITD mutations are associated with high white blood counts, a high percentage of bone marrow blasts, and high levels of LDH at diagnosis (Kiyoi et al., 1999; Kottaridis et al., 2001; Kottaridis et al., 2003; Gale et al., 2005). The presence of Flt3-ITD has been associated with an unfavorable prognosis in all FAB subgroups as a result of reduced disease free and overall survival (Kiyoi et al., 1999; Abu-Duhier et al., 2000; Kottaridis et al., 2003). ITD mutations have a higher incidence with increasing age (Stirewalt et al., 2001).

The ITD-mutated alleles and the wild-type allele are usually transcribed so that both are co-expressed in AML blasts. However, the wild-type allele is frequently lost in some leukemia clones, resulting in loss of expression of the wild-type mRNA which has been shown to be due to homologous recombination and subsequent loss of heterozygosity as a form of uniparental disomy (Kottaridis et al., 2003; Fitzgibbon et al., 2005). In some patients, Flt3-ITD mutations

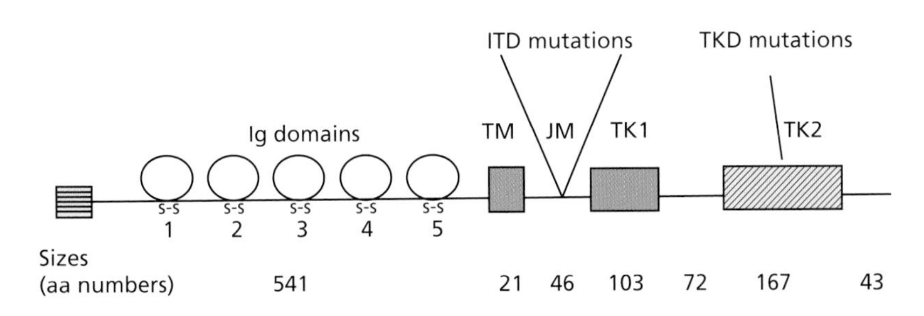

Figure 24–3 Schematic of the structure of the receptor tyrosine kinase FLT3. Internal tandem duplications (ITDs) occur in the juxtamembrance region in approximately 25% of AML patients. Point mutations occur within the tyrosine kinase domain (TKD) in a further 5%–10% of patients.

were not detected at initial diagnosis but were found to be present at relapse, while in others they were present at first diagnosis and undetectable at relapse (Kottaridis et al., 2001; Shih et al., 2004). Therefore the use of Flt3-mutations for minimal residual disease detected may be limited.

Activating Mutations of FLT3 A different type of Flt3 mutation has also been identified within the activation loop, or the TKD, in approximately 7% of AML patients (Abu-Duhier et al., 2001b; Mills et al., 2005). The activation loop is a highly conserved protein structure in the membrane-distal portion of the kinase domain of several receptor tyrosine kinases. This type of mutation usually affects the codon 835 (D835) where the point mutations alter the amino acid at that codon. Few studies have correlated TKD mutations with prognosis (Frohling et al., 2002). Numerous other point mutations within the TKD have been observed near codon 835, for example, at codons 836, 839, or 840 (Spiekermann et al., 2002; Mills et al., 2005).

Analysis of the signal transduction and biological consequences of activating Flt3 mutations (Hayakawa et al., 2000) showed that activation of STAT5a is weak, whereas STAT5a is strongly activated. Flt3-ITD activates different Bcl2 family members, and consequently different anti-apoptotic pathways, from those induced by wild-type Flt3 (Minami et al., 2003).

KIT Mutations

Activating mutations of Kit, also a class III RTK, occur in subgroups of AML affecting two regions of the receptor: the juxta-membrane region and also in residues within in the activation loop (Nagata et al., 1995). These mutations are very frequent in the AML patients with a translocation affecting the CBF: t(8;21) and inv (16) chromosomal translocations. In one study, 24% of patients with inv(16) had an exon 8 mutation, but in only 2% in patients with t(8;21). However, mutations of residue 816 in exon 17 occurred with similar frequencies (7.9% and 10.6%, respectively) (Care et al., 2003). This indicates that different mutations in tyrosine kinase receptors may occur in the different types of AML, and they are probably dependent on, and highly specificfor, accompanying molecular alterations in transcription factors.

RAS Gene Mutations

Activating Ras mutations have been described in many reports. These are usually at codon 12 and 13 of N-Ras (Bos et al., 1985), also codon 61 of N-Ras but rarely in K-Ras, and very rarely in H-Ras. Overall, the incidence of N-Ras mutations is between 10% and 27% (Bowen et al., 2005). The presence of a Ras mutations has been reported to correlate with low blast counts and an improved survival of patients, but the contrast has also been reported in patients with Ras mutations having a worse outcome when compared to patients without tyrosine kinase or Ras mutations (Bowen et al., 2005).

Ras mutations MDSs patients correlated with an increased risk of progression to AML (Paquette et al., 1993; Bowen et al., 2005), but the Ras mutation status in patients at first diagnosis and at relapse revealed a marked instability of these mutations, with loss of mutations in some cases, sometimes replaced by mutations in different codons (Farr et al., 1988). This would suggest that Ras mutations are not an initiating event but may be important for disease progression.

Mutations within the Flt3, Kit, and Ras genes have indicated a strong involvement in the development of AML; but it is also likely that the activation of other, as yet unidentified, signaling pathways occur whether as a result of, or independent of, mutation will contribute to AML development.

The over-expression of non-mutated receptor tyrosine kinases may result in ligand-independent activation, such as might occur in cases of Flt3 over-expression (Armstrong et al., 2003). High expression of wild-type Flt3 in AML blasts can induce spontaneous Flt3 auto-phosphorylation (Armstrong et al., 2003), which may contribute to leukemia transformation.

Pathway analysis via gene expression profiling of AML samples might reveal activated signaling pathways that could be traced back to identify candidates that provide key signals important for leukemia transformation. Since more and more microarray data of AML bone marrow are being added to the databases, such approaches become increasingly feasible.

Microarray Genomics

Disease Classification in AML

The first approach to disease classification, based on microarrays, was described in 1999 (Golub et al., 1999) and used an unsupervised, class discovery approach, identifying previously unrecognized subtypes, to uncover the distinction between AML and acute lymphoblastic leukemia (ALL). The authors speculated that (Golub et al., 1999) a larger sample would enable finer subclassifications to be identified that would correspond to existing subclassifications for AML or define new groupings. This concept sparked numerous publications trying to identify gene expression profiles or signatures that could distinguish between the major subclasses within AML.

Specific Cytogenetic Subgroups

Most of the early microarray studies were done on small numbers (>60 patients) with cytogenetically defined groups of AML samples; these groupings included t(8;21); t(15;17), inv(16), 11q23 abnormalities, or normal cytogenetics (Virtaneva et al., 2001; Schoch et al., 2002; Debernardi et al., 2003; Morikawa et al., 2003; Vey et al., 2004; Gutierrez et al., 2005; Haferlach et al., 2005b). For example, oligonucleotide arrays were used to examine 37 samples from distinct cytogenetic subtypes of AML: t(8;21), inv(16), and t(15;17), which are associated with four morphological subgroups (M2, M4Eo, M3, and M3v). Significance of microarray analysis (SAM) (Tusher et al., 2001), and other analyses were used to identify genes whose profiles accurately discriminated between the cytogenetic classes. Furthermore, they showed that only 13 genes were needed to separate the AML samples into their distinctive subgroups, indicating that AML-specific cytogenetic aberrations can be correlated with corresponding gene expression profiles.

A similar study on 28 AML patients used statistical group comparison to identify 145 genes and then hierarchical clustering to separate major cytogenetic classes: t(8;21), t(15;17), inv(16), 11q23, and normal karyotype (Debernardi et al., 2003). The study identified that members of the class I HOX A and B gene families showed a distinct up-regulation within the normal karyotype group, which may imply a common genetic lesion within these pseudo-normal cytogenetics.

Patients with normal cytogenetics constitute the single largest group in AML; however, trisomy 8 is the most common numerical aberration observed as either a sole abnormality or part of more complex cytogenetics. Patients with normal cytogenetic versus patients with trisomy 8 as the sole abnormality (Virtaneva et al., 2001) could be divided into two genetically very similar populations with 29 genes up-regulated and located on chromosome 8.

A more focused approach compared patients with APL with the t(15;17) translocation and patients with AML FAB group M1 (AML without maturation characterized by morphologically and phenotypically immature AML blasts and no recurrent chromosomal abnormalities) (Morikawa et al., 2003). A multivariate sigma-classifier algorithm could distinguish FAB-M3 with t(15;17) from the AML M1 subgroup, although unsurprisingly, this study showed that two morphologically defined subgroups (FAB-M1 and FAB-M3) had distinct gene expression signatures. Patients with variants of APL [M3 (abnormal pro-myelocytes with heavy granulation and bundles of Auer rods) or M3v (non- or hypogranular cytoplasm and a bi-lobed nucleus)] could be separated from other AML with defined cytogenetic abnormalities or those with a normal cytogenetic (Haferlach et al., 2005b). Further supervised pair-wise comparison showed discrimination between M3 and M3v, based on gene signatures, with a median classification accuracy of 90%.

Disease Classification

A more recent study (Bullinger et al., 2004) used cDNA arrays on 116 adults with AML (including 45 with a normal cytogenetic) and identified 6283 genes that had variable expression. They found that, in common with previous studies, samples with t(15;17) had a highly correlated pattern of expression. However, they also identified gene signatures for patients with t(8;21), inv(16) or normal cytogenetics, although each group was further subdivided into separate clusters. Samples with a normal cytogenetic segregated into two distinct groups, which also included AML samples from other classes with M1 or M2, was significantly more common in one group, whereas M4 or M5 were more common in the second group. A combination of unsupervised and supervised analysis methods identified a list of genes to predict clinical outcome. The subgroup of samples predicted to have a poor outcome correlated with significantly shorter survival than those samples predicted to have a good outcome. However, it should be noted that when a different analysis algorithm was used [(PAM) (Tibshirani et al., 2002)] no difference in outcome could be determined, and this may be due a relatively small sample or an inherently poorer performance of patients with a normal cytogenetic.

A second study was published in 2004 (Valk et al., 2004) on 285 AML patients fully characterized for the presence of mutations within the FLT3 ITD or TKD region, NRAS, KRAS, or CEBPα and the over-expression of EVI1. Data were analyzed by SAM (Tusher et al., 2001) and PAM (Tibshirani et al., 2002), in addition to using the Pearson correlation algorithms within the OmniViz software package. Their analysis produced 16 distinct groups of patients on the basis of strong similarities in gene-expression profiles. Distinct clusters of t(8;21), inv(16), and t(15;17) were readily identified, which emphasized the strong effect of these distinctive and recurrent translocations on gene expression. In addition, separate clusters were identified containing samples with monosomy 7, mutations of FLT3, or over-expression of EVI1. In addition, clusters were defined for 11q23 abnormalities or CEPBα mutations. Patients with t(15;17) also clustered into one main group although two subgroups could be identified separating patients with high or low white blood cell count, which also correlated with the presence of FLT3 mutations. Several clusters were not associated with any defined morphological, cytogenetic or mutation status. The best survival rates were for those clusters containing the recurrent translocations [t(8;21), t(15;17) and inv(16)]; these cytogenetic abnormalities have previously been associated with a favorable prognosis (Grimwade et al., 1998). Signatures associated with FLT3 ITD were not distinctive; and the authors suggested that this type of mutation, which occurred across the clusters, reflect the heterogeneity of AML.

The molecular characterization of different ALL or AML were analyzed (Kohlmann et al., 2003) and only a small set of differentially expressed genes was needed to accurately discriminate eight clinically relevant acute leukemia subgroups, including those with t (8;21), t(15;17), t(11q23)/MLL, or inv(16), as well as precursor B-ALL with t(9;22), t(8;14), or t(11q23)/MLL and precursor T-ALL. This approach was expanded in a study (Haferlach et al., 2005a) of 892 patients with mainly AML or ALL, which also included CLL, CML, and some non-leukemia samples (Fig. 24–4). The

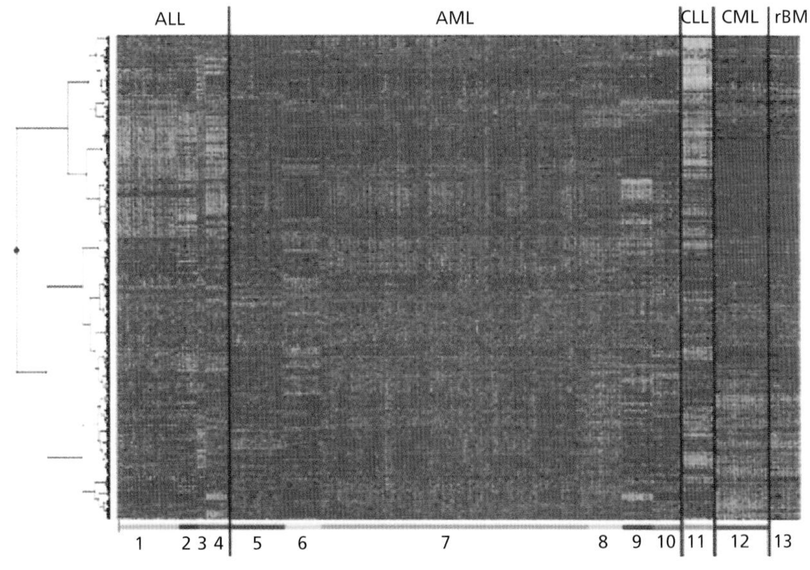

Figure 24–4 Hierarchical cluster analysis of 937 samples using 1019 differentially expressed genes. Red color indicates high expression, and green indicates low expression. Major leukemia types could be distinguished from each other using 100 differentially expressed genes for each of the 13 classes. Of these 281 were repeatedly identified as important diagnostic markers and overlapped among the lists resulting in 1019 non-overlapping genes. Reproduced from Haferlach et al. (2005a).

authors used a combination of unsupervised and supervised statistical methods to identify gene lists that reproduced 12 predefined cytogenetic or morphological leukemia classification groupings, with a 95.1% accuracy, distinct from each other and the non-leukemia samples. Within AML, specificities of 100% were obtained for t(15;17), t(8;21), and inv(16); 97.7% for patients with 11q23 abnormalities and those with complex cytogenetic abnormalities; and 93.7% for the larger and diverse subgroup of "normal cytogenetics and other abnormalities."

Other Mutational Subgroups

As discussed above, several other mutations can occur in AML (FLT3, MLL, CEBPα, or RAS) and are usually observed across FAB, cytogenetic, or WHO subgroups. The heterogeneity of AML masks the identification of a FLT3 specific gene expression profile (Bullinger et al., 2004; Valk et al., 2004). If the heterogeneity is removed by studying only patients with APL [M3 containing the t (15;17) translocation], then those patients with FLT3 mutations, either ITD or TKD, clearly clustered differently from patients with wild-type FLT3 (Gale et al., 2005) reinforcing the idea that FLT3 mutations are secondary and may be "over-shadowed" in a heterogeneous AML population.

Some of the larger studies (Schoch et al., 2002; Bullinger et al., 2004; Valk et al., 2004) showed that gene signatures from patients with FLT3, and in particular those for MLL and RAS mutations, were more difficult to identify than for recurrent translocations. Patients with normal cytogenetics, characterized for FLT3-ITD, FLT3-TKD, and NRAS-PM at diagnosis were used to identify a 10-gene signature by neural network analysis that allowed for the correct classification of FLT3-ITD, FLT3-TKD, NRAS-PM, MLL-PTD, and wild-type samples with an accuracy of 83.7% (Neben et al., 2005). Their data again showed that FLT3-ITD and FLT3-TKD mutations had a distinct clinical phenotype.

Neither did any of the larger studies (Bullinger et al., 2004; Valk et al., 2004) identify any specific cluster associated with MLL PTD nor did a study of pediatric AML, despite clearly identifying the other major cytogenetic types including a subgroup containing patients with rearrangements of the MLL gene (Ross et al., 2004).

A high proportion of AML with normal cytogenetics have a mutation in the NPM gene and unsupervised clustering which clearly distinguished NPM mutants from NPM wild-type AML, with the expression profile dominated by a stem cell molecular signature (Alcalay et al., 2005).

Diagnostic potential

All the studies discussed above are diverse in sample size, patient characteristics, cytogenetic background, and mutational status, but it is reassuring to note that each study has attempted, and in most cases succeeded, to obtain a molecular-based classification. But how relevant are the classifications? For each individual study, they are very relevant, but comparing across studies is perhaps less relevant due to potential differences in extraction procedures, labeling protocols, array platforms, and statistical analyses. Ideally, the data should be harmonized or combined to produce a clear molecular classification of AML based initially on established cytogenetic groupings. The use of microarrays for a global approach to leukemia classification was suggested by Haferlach et al. (Haferlach et al., 2005a) in 2005, in which he stated that a large multi-center comparison assessing microarray diagnosis with standard diagnostics was required. A major step toward this type of diagnostic classification is the launch of an international multi-center clinical research program (MILE Study) to assess the application of a microarray test and its potential for use in the diagnosis and subclassification of hematologic malignancies. The MILE study research program was launched in 10 centers: 7 European centers in association with the European Leukemia Network (ELN)http://www.leukemia-net.org/index.htm and 3 centers from the United States. This study will include 4000 patients (from all types of leukemia) and assess the clinical accuracy of gene expression profiles of 16 acute and chronic leukemia subclasses including MDS with current gold-standard routine diagnostic work-up (Haferlach, Kohlmann, Basso et al., 2005). The advantage of this study is that each participating center will use an identical microarray protocol, laboratory equipment, kits, and reagents for target preparation, hybridization, washing, scanning, and statistical interpretation. Furthermore, the analysis of the data may lead to the further definition and refinement of leukemia classification.

Potential for prognosis prediction

Current classifications of AML based on cytogenetics and morphology are not robust enough to predict the prognosis for each patient. Broad cytogenetic-based risk groups (Grimwade et al., 1998) can be used for stratification, but within these groups are patients with better than expected outcome and, of more concern, those with worse clinical outcome. It was, and still is, the expectation that new prognostic stratification schemes will be formulated based on the global profiling of gene expression identifying genes whose expression is associated with clinical outcome.

The approximately 50% of adult AML patients who have normal cytogenetics (NC) and hierarchical clustering confirmed this diversity as a supervised analysis produced two subgroups: one subgroup which was called a homogeneous or "pure" NC cluster; the other NC-AML had gene signatures that were "translocation like" (Vey et al., 2004). More significantly, clinical outcome data could discriminate between a poor prognosis group (containing a majority of "pure" NC) and the "translocation-like" AML, which

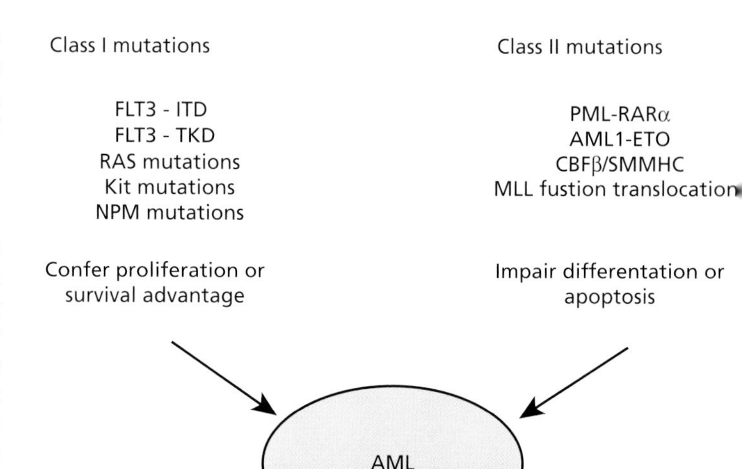

Figure 24–5 The development of AML is dependent the acquisition of genetic abnormalities in two groups. One group that confers a proliferative or survival advantage and the second group that is blocks differentiation or impair apoptosis. The mix of these abnormalities will result in the spectrum of disease types seen in AML.

were in the good prognosis class. Another study identified two groups from non-monocytic leukemia, of which one cluster contained all the eight patients that were either refractory to treatment or that relapsed after treatment (Gutierrez et al., 2005).

A limited number of studies have examined the possible predictive ability of array technologies. One series of childhood AML patients with either FLT3 ITD or TKD mutations identified gene signatures associated with a high risk of treatment failure (Lacayo et al., 2004) However, the analysis revealed that the expression levels of two genes (elevated expression of RUNX3 and decreased expression of ATRX) correlated with higher risk of treatment failure. Another study, again of pediatric AML patients, identified a prognostic set of genes, most of which had not previously been linked to prognosis or correlated with FAB or cytogenetics (Yagi et al., 2003).

Conclusion

The genomics of AML is one of the fastest moving areas of cancer research. It is probably one of the best-understood malignant diseases, and its dissection and understanding of the molecular abnormalities can be used as a paradigm for numerous other diseases. Mutations that block differentiation combine with those that promote proliferation or survival to form the heterogeneous spectrum of phenotypes seen in AML (Gilliland, 2002) (Fig. 24–5). These mechanisms are now being dissected from all angles: expression profiling, mutational screening, epigenetics, in-vivo, and transgeneic models through to clinical trials and therapeutic interventions.

REFERENCES

Abu-Duhier, FM, Goodeve, AC, Wilson, GA, Care, RS, Peake, IR, and Reilly, JT (2001a). Genomic structure of human FLT3: implications for mutational analysis. *Br J Haematol*, 113:1076–1077.

Abu-Duhier, FM, Goodeve, AC, Wilson,GA, Care, RS, Peake, IR, and Reilly, JT (2001b). Identification of novel FLT-3 Asp835 mutations in adult acute myeloid leukaemia. *Br J Haematol*, 113:983–988.

Abu-Duhier, FM, Goodeve, AC, Wilson, GA, Gari, MA, Peake, IR, Rees, DC, et al. (2000). FLT3 internal tandem duplication mutations in adult acute myeloid leukaemia define a high-risk group. *Br J Haematol*, 111:190–195.

Adolfsson, J, Borge, OJ, Bryder, D, Theilgaard-Monch, K, Astrand-Grundstrom, I, Sitnicka, E, et al. (2001). Upregulation of Flt3 expression within the bone marrow Lin(−)Sca1(+)c-kit(+) stem cell compartment is accompanied by loss of self-renewal capacity. *Immunity*, 15:659–669.

Alcalay, M, Orleth, A, Sebastiani, C, Meani, N, Chiaradonna, F, Casciari, C, et al. (2001). Common themes in the pathogenesis of acute myeloid leukemia. *Oncogene*, 20:5680–5694.

Alcalay, M, Tiacci, E, Bergomas, R, Bigerna, B, Venturini, E, Minardi, SP, et al. (2005).Acute myeloid leukemia bearing cytoplasmic nucleophosmin (NPMc + AML) shows a distinct gene expression profile characterized by up-regulation of genes involved in stem cell maintenance. *Blood*, 106:899–902.

Amann, JM, Nip, J, Strom, DK, Lutterbach, B, Harada, H, Lenny, N, et al. (2001). ETO, a target of t(8;21) in acute leukemia, makes distinct contacts with multiple histone deacetylases and binds mSin3A through its oligomerization domain. *Mol Cell Biol*, 21:6470–6483.

Andersen, MK, Christiansen, DH, Jensen, BA, Ernst, P, Hauge, G, and Pedersen-Bjergaard, J (2001). Therapy-related acute lymphoblastic leukaemia with MLL rearrangements following DNA topoisomerase II inhibitors, an increasing problem: report on two new cases and review of the literature since 1992. *Br J Haematol*, 114:539–543.

Aravind, L and Landsman, D (1998). AT-hook motifs identified in a wide variety of DNA-binding proteins. *Nucleic Acids Res*, 26: 4413–4421.

Armstrong, SA, Golub, TR, and Korsmeyer, SJ (2003). MLL-rearranged leukemias: Insights from gene expression profiling. *Semin Hematol*, 40:268–273.

Armstrong, SA, Staunton, JE, Silverman, LB, Pieters, R, Den Boer, ML, Minden, MD, et al. (2002). MLL translocations specify a distinct gene expression profile that distinguishes a unique leukemia. *Nat Genet*, 30:41–47.

Arnould, C, Philippe, C, Bourdon, V, Gr goire, MJ, Berger, R, and Jonveaux, P (1999). The signal transducer and activator of transcription STAT5b gene is a new partner of retinoic acid receptor alpha in acute promyelocytic-like leukaemia. *Hum Mol Genet*, 8:1741–1749.

Auerbach, AD and Allen, RG (1991). Leukemia and preleukemia in Fanconi anemia patients. A review of the literature and report of the International Fanconi Anemia Registry. *Cancer Genet Cytogenet*, 51:1–12.

Ayton, PM and Cleary, ML (2003). Transformation of myeloid progenitors by MLL oncoproteins is dependent on Hoxa7 and Hoxa9. *Genes Dev*, 17:2298–2307.

Barjesteh van Waalwijk van Doorn-Khosrovani, S Erpelinck, C, Meijer, J, van Oosterhoud, S, van Putten, WL, Valk, PJ, et al. (2003). Biallelic mutations in the CEBPA gene and low CEBPA expression levels as prognostic markers in intermediate-risk AML. *Hematol J*, 4:31–40.

Bennett, JM, Catovsky, D, Daniel, MT, Flandrin, G, Galton, DA, Gralnick, HR, et al. (1976). Proposals for the classification of the acute leukaemias. French-American-British (FAB) co-operative group. *Br J Haematol*, 33:451–458.

Bishop, JF (1997). The treatment of adult acute myeloid leukemia. *Semin Oncol*, 24:57–69.

Blanco, JG, Dervieux, T, Edick, MJ, Mehta, PK, Rubnitz, JE, Shurtleff, S, et al. (2001). Molecular emergence of acute myeloid leukemia during treatment for acute lymphoblastic leukemia. *Proc Natl Acad Sci USA*, 98:10338–10343.

Bos, JL, Toksoz, D, Marshall, CJ, Verlaan de Vries, M, Veeneman, GH, van der Eb, AJ, et al. (1985). Amino-acid substitutions at codon 13 of the N-ras oncogene in human acute myeloid leukaemia. *Nature*, 315:726–730.

Bowen, DT, Frew, ME, Hills, R, Gale, RE, Wheatley, K, Groves, MJ, et al. (2005). RAS mutation in acute myeloid leukemia is associated with distinct cytogenetic subgroups but does not influence outcome in patients <60 yrs. *Blood*, 106:2113–2119.

Bullinger, L, Dohner, K, Bair, E, Frohling, S, Schlenk, RF, Tibshirani, et al. (2004). Use of gene-expression profiling to identify prognostic subclasses in adult acute myeloid leukemia. *N Engl J Med*, 350:1605–1616.

Burel, SA, Harakawa, N, Zhou, L, Pabst, T, Tenen, DG, and Zhang, DE (2001). Dichotomy of AML1-ETO functions: growth arrest versus block of differentiation. *Mol Cell Biol*, 21:5577–5590.

Care, RS, Valk, PJ, Goodeve, AC, Abu-Duhier, FM, Geertsma-Kleinekoort, WM, Wilson, GA, et al. (2003). Incidence and prognosis of c-KIT and FLT3 mutations in core binding factor (CBF) acute myeloid leukaemias. *Br J Haematol*, 121:775–777.

Carow, CE, Kim, E, Hawkins, AL, Webb, HD, Griffin, CA, Jabs,EW, et al. (1995). Localization of the human stem cell tyrosine kinase-1 gene (FLT3) to 13q12–>q13. *Cytogenet Cell Genet*, 70:255–257.

Chen, SJ, Zelent, A, Tong, JH, Yu, HQ, Wang, ZY, Derre, J, et al. (1993). Rearrangements of the retinoic acid receptor alpha and promyelocytic leukemia zinc finger genes resulting from t(11; 17)(q23;q21) in a patient with acute promyelocytic leukemia. *J Clin Invest*, 91:2260–2267.

Chen, Z, Brand, NJ, Chen, A, Chen, SJ, Tong, JH, Wang, ZY, et al. (1993). Fusion between a novel Kruppel-like zinc finger gene and the retinoic acid receptor-alpha locus due to a variant t(11;17) translocation associated with acute promyelocytic leukaemia. *Embo J*, 12:1161–1167.

Christensen, JL and Weissman, IL (2001). Flk-2 is a marker in hematopoietic stem cell differentiation: a simple method to isolate long-term stem cells. *Proc Natl Acad Sci USA*, 98:14541–14546.

Debernardi, S, Lillington, DM, Chaplin, T, Tomlinson, S, Amess, J, Rohatiner, A, et al. (2003). Genome-wide analysis of acute myeloid leukemia with normal karyotype reveals a unique pattern of homeobox gene expression distinct from those with translocation-mediated fusion events. *Genes Chromosomes Cancer*, 37:149–158.

Deininger, MWN, Bose, S, Gora-Tybor, J, Yan, XH, Goldman, JM, and Melo, JV (1998). Selective induction of leukemia-associated fusion genes by high-dose ionizing radiation. *Cancer Res*, 58:421–425.

Dong, S, Zhu, J, Reid, A, Strutt, P, Guidez, F, Zhong, HJ, et al. (1996). Amino-terminal protein-protein interaction motif (POZ-domain) is responsible for activities of the promyelocytic leukemia zinc finger- retinoic acid receptor-alpha fusion protein. *Proc Natl Acad Sci USA*, 93:3624–3629.

Downing, JR (1999). The AML1-ETO chimaeric transcription factor in acute myeloid leukaemia: biology and clinical significance [In Process Citation]. *Br J Haematol*, 106:296–308.

Farr, CJ, Saiki, RK, Erlich, HA, McCormick, F, and Marshall, CJ (1988). Analysis of RAS gene mutations in acute myeloid leukemia by polymerase chain reaction and oligonucleotide probes. *Proc Natl Acad Sci USA*, 85:1629–1633.

Fenaux, P, Chomienne, C, and Degos, L (2001). Treatment of acute promyelocytic leukaemia. *Best Pract Res Clin Haematol*, 14:153–174.

Fitzgibbon, J, Smith, LL, Raghavan, M, Smith, ML, Debernardi, S, Skoulakis, S, et al. (2005). Association between acquired uniparental disomy and homozygous gene mutation in acute myeloid leukemias. *Cancer Res*, 65:9152–9154.

Frank, R, Zhang, J, Uchida, H, Meyers, S, Hiebert, SW, and Nimer, SD (1995). The AML1/ETO fusion protein blocks transactivation of the GM-CSF promoter by AML1B. *Oncogene*, 11:2667–2674.

Frohling, S, Schlenk, RF, Breitruck, J, Benner, A, Kreitmeier, S, Tobis, K, et al. (2002). Prognostic significance of activating FLT3 mutations in younger adults (16 to 60 years) with acute myeloid leukemia and normal cytogenetics: a study of the AML Study Group Ulm. *Blood*, 100:4372–4380.

Gale, RE, Hills, R, Pizzey, AR, Kottaridis, PD, Swirsky, D, Gilkes, AF, et al. (2005). The relationship between FLT3 mutation status, biological characteristics and response to targeted therapy in acute promyelocytic leukemia. *Blood*, 106(12):3768–3776.

Gelmetti, V, Zhang, J, Fanelli, M, Minucci, S, Pelicci, PG, and Lazar, MA (1998). Aberrant recruitment of the nuclear receptor corepressor-histone deacetylase complex by the acute myeloid leukemia fusion partner ETO. *Mol Cell Biol*, 18:7185–7191.

Ghannam, G, Takeda, A, Camarata, T, Moore, MA, Viale, A, and Yaseen, NR (2003). The oncogene Nup98-HOXA9 induces gene transcription in myeloid cells. *J Biol Chem*, 279:866–875.

Gilliland, DG (2002). Molecular genetics of human leukemias: new insights into therapy. *Semin.Hematol*, 39:6–11.

Golub, TR, Slonim, DK, Tamayo, P, Huard, C, Gaasenbeek, M, Mesirov, JP, et al. (1999). Molecular classification of cancer: class discovery and class prediction by gene expression monitoring. *Science*, 286:531–537.

Grignani, F, De Matteis, S, Nervi, C, Tomassoni, L, Gelmetti, V, Cioce, M, et al. (1998). Fusion proteins of the retinoic acid receptor-alpha recruit histone deacetylase in promyelocytic leukaemia. *Nature*, 391:815–818.

Grignani, F, Valtieri, M, Gabbianelli, M, Gelmetti, V, Botta, R, Luchetti, L, et al. (2000). PML/RAR alpha fusion protein expression in normal human hematopoietic progenitors dictates myeloid commitment and the promyelocytic phenotype. *Blood*, 96:1531–1537.

Grimwade, D, Walker, H, Harrison, G, Oliver, F, Chatters, S, Harrison, CJ, et al. (2001). The predictive value of hierarchical cytogenetic classification in older adults with acute myeloid leukemia (AML): analysis of 1065 patients entered into the United Kingdom Medical Research Council AML11 trial. *Blood*, 98:1312–1320.

Grimwade, D, Walker, H, Oliver, F, Wheatley, K, Harrison, C, Harrison, G, et al. (1998). The importance of diagnostic cytogenetics on outcome in AML: analysis of 1,612 patients entered into the MRC AML 10 trial. The Medical Research Council Adult and Children's Leukaemia Working Parties. *Blood*, 92:2322–2333.

Gutierrez, NC, Lopez-Perez, R, Hernandez, JM, Isidro, I, Gonzalez, B, Delgado, M, et al. (2005). Gene expression profile reveals deregulation of genes with relevant functions in the different subclasses of acute myeloid leukemia. *Leukemia*, 3:402–409.

Haferlach, T, Kohlmann, A, Basso, G, Bene, MC, Downing, JR, Hernandez, JM, et al. (2005). A Multi-Center and Multi-National Program to Assess the Clinical Accuracy of the Molecular Subclassification of Leukemia by Gene Expression Profiling. *Blood*, 106:757.

Haferlach, T, Kohlmann, A, Schnittger, S, Dugas, M, Hiddemann, W, Kern, W, et al. (2005a). A global approach to the diagnosis of leukemia using gene expression profiling. *Blood*, 106:1189–1198.

Haferlach, T, Kohlmann, A, Schnittger, S, Dugas, M, Hiddemann, W, Kern, W, et al. (2005b). AML M3 and AML M3 variant each have a distinct gene expression signature but also share patterns different from other genetically defined AML subtypes. *Genes Chromosomes Cancer*, 43:113–127.

Hayakawa, F, Towatari, M, Kiyoi, H, Tanimoto, M, Kitamura, T, Saito, H, et al. (2000). Tandem-duplicated Flt3 constitutively activates STAT5 and MAP kinase and introduces autonomous cell growth in IL-3-dependent cell lines. *Oncogene*, 19:624–631.

He, LZ, Guidez, F, Tribioli, C, Peruzzi, D, Ruthardt, M, Zelent, A, et al. (1998). Distinct interactions of PML-RARalpha and PLZF-RARalpha with co-repressors determine differential responses to RA in APL. *Nat Genet*, 18:126–135.

Hiebert, SW, Reed-Inderbitzin, EF, Amann, J, Irvin, B, Durst, K, and Linggi, B (2003). The t(8;21) fusion protein contacts co-repressors and histone deacetylases to repress the transcription of the p14(ARF) tumor suppressor. *Blood Cells Mol Dis*, 30:177–183.

Huret, JL, Senon, S, Bernheim, A, and Dessen, P (2004). An Atlas on genes and chromosomes in oncology and haematology. *Cell Mol.Biol.(Noisy-le-grand)*, 50:805–807.

Kalwinsky, DK, Raimondi, SC, Schell, MJ, Mirro, J Jr, Santana, VM, Behm, F, et al. (1990). Prognostic importance of cytogenetic subgroups in de novo pediatric acute nonlymphocytic leukemia. *J Clin Oncol*, 8:75–83.

Kasper, LH, Brindle, PK, Schnabel, CA, Pritchard, CE, Cleary, ML, and van Deursen, JM (1999). CREB binding protein interacts with nucleoporin-specific FG repeats that activate transcription and mediate NUP98-HOXA9 oncogenicity. *Mol Cell Biol*, 19:764–776.

Kawakami, K, Miyanishi, S, Nishii, K, Usui, E, Murata, T, Shinsato, I, (2002). A case of acute myeloid leukemia with t(7;11)(p15;p15) mimicking myeloid crisis of chronic myelogenous leukemia. *Int J Hematol*, 76:80–83.

Kiyoi, H, Naoe, T, Nakano, Y, Yokota, S, Minami, S, Miyawaki, S, et al. (1999). Prognostic implication of FLT3 and N-RAS gene mutations in acute myeloid leukemia. *Blood*, 93:3074–3080.

Klampfer, L, Zhang, J, Zelenetz, AO, Uchida, H, and Nimer, SD (1996). The AML1/ETO fusion protein activates transcription of BCL-2. *Proc Natl Acad Sci USA*, 93:14059–14064.

Kohlmann, A, Schoch, C, Schnittger, S, Dugas, M, Hiddemann, W, Kern, W, et al. (2003). Molecular characterization of acute leukemias by use of microarray technology. *Genes Chromosomes Cancer*, 37:396–405.

Kottaridis, PD, Gale, RE, Frew, ME, Harrison, G, Langabeer, SE, Belton, AA, et al. (2001). The presence of a FLT3 internal tandem duplication in patients with acute myeloid leukemia (AML) adds important prognostic information to cytogenetic risk group and response to the first cycle of chemotherapy: analysis of 854 patients from the United Kingdom Medical Research Council AML 10 and 12 trials. *Blood*, 98:1752–1759.

Kottaridis, PD, Gale, RE, and Linch, DC (2003) Prognostic implications of the presence of FLT3 mutations in patients with acute myeloid leukemia. *Leuk Lymphoma*, 44:905–913.

Lacayo, NJ, Meshinchi, S, Kinnunen, P, Yu, R, Wang, Y, Stuber, CM, et al. (2004). Gene expression profiles at diagnosis in de novo childhood AML patients identify FLT3 mutations with good clinical outcomes. *Blood*, 104:2646–2654.

Larson, RA, Williams, SF, Le Beau, MM, Bitter, MA, Vardiman, JW, and Rowley, JD (1986). Acute myelomonocytic leukemia with abnormal eosinophils and inv(16) or t(16;16) has a favorable prognosis. *Blood*, 68:1242–1249.

Lavagna-Sevenier, C, Marchetto, S, Birnbaum, D, and Rosnet, O (1998). The CBL-related protein CBLB participates in FLT3 and interleukin-7 receptor signal transduction in pro-B cells. *J Biol.Chem.*, 273:14962–14967.

Le Beau, MM, Bitts, S, Davis, EM, and Kogan, SC (2002). Recurring chromosomal abnormalities in leukemia in PML-RARA transgenic mice parallel human acute promyelocytic leukemia. *Blood*, 99:2985–2991.

Le Beau, MM, Davis, EM, Patel, B, Phan, VT, Sohal, J, and Kogan, SC (2003). Recurring chromosomal abnormalities in leukemia in PML-RARA transgenic mice identify cooperating events and genetic pathways to acute promyelocytic leukemia. *Blood*, 102:1072–1074.

Le Beau, MM, Larson, RA, Bitter, MA, Vardiman, JW, Golomb, HM, and Rowley, JD (1983). Association of an inversion of chromosome 16 with abnormal marrow eosinophils in acute myelomonocytic leukemia. A unique cytogenetic-clinicopathological association. *N Engl J Med*, 309:630–636.

Levine, EG and Bloomfield, CD (1992). Leukemias and myelodysplastic syndromes secondary to drug, radiation, and environmental exposure. *Semin Oncol*, 19:47–84.

Liao, C, Wang, XY, Wei, HQ, Li, SQ, Merghoub, T, Pandolfi, PP, (2001). Altered myelopoiesis and the development of acute myeloid leukemia in transgenic mice overexpressing cyclin A1. *Proc Natl Acad Sci USA*, 98:6853–6858.

Lightfoot, T. (2005) Aetiology of childhood leukemia. *Bioelectromagnetics*, (Suppl. 7):S5–S11.

Lin, RJ, Nagy, L, Inoue, S, Shao,W, Miller, WH Jr, and Evans, RM (1998). Role of the histone deacetylase complex in acute promyelocytic leukaemia. *Nature*, 391:811–814.

Linggi, B, Muller-Tidow, C, van de, LL, Hu, M, Nip, J, Serve, H, et al. (2002). The t(8;21) fusion protein, AML1 ETO, specifically represses the transcription of the p14(ARF) tumor suppressor in acute myeloid leukemia. *Nat Med*, 8:743–750.

Liu, P, Tarle, SA, Hajra, A, Claxton, DF, Marlton, P, Freedman, M, et al. (1993). Fusion between transcription factor CBF beta/PEBP2 beta and a myosin heavy chain in acute myeloid leukemia. *Science*, 261:1041–1044.

Lowenberg, B, Downing, JR, and Burnett, A (1999). Acute myeloid leukemia. *N Engl J Med*, 341:1051–1062.

Lowenberg, B, Griffin, JD, and Tallman, MS (2003). Acute myeloid leukemia and acute promyelocytic leukemia. *Hematology (Am Soc Hematol Educ Program)*, 82–101.

Lutterbach, B, Hou, Y, Durst, KL, and Hiebert, SW (1999). The inv(16) encodes an acute myeloid leukemia 1 transcriptional corepressor. *Proc Natl Acad Sci USA*, 96:12822–12827.

Marchetto, S, Fournier, E, Beslu, N, Aurran-Schleinitz, T, Dubreuil, P, Borg, JP, et al. (1999). SHC and SHIP phosphorylation and interaction in response to activation of the FLT3 receptor. *Leukemia*, 13:1374–1382.

Maroc, N, Rottapel, R, Rosnet, O, Marchetto, S, Lavezzi, C, Mannoni, P, et al. (1993). Biochemical characterization and analysis of the transforming potential of the FLT3/FLK2 receptor tyrosine kinase. *Oncogene*, 8:909–918.

Meyers, S, Lenny, N, and Hiebert, SW (1995). The t(8;21) fusion protein interferes with AML-1B-dependent transcriptional activation. *Mol Cell Biol*, 15:1974–1982.

Mills, KI, Gilkes, AF, Walsh, V, Sweeney, M, and Gale, R (2005). Rapid and sensitive detection of internal tandem duplication and activating loop mutations of FLT3. *Br J Haematol*, 130:203–208.

Minami, Y, Yamamoto, K, Kiyoi, H, Ueda, R, Saito, H, and Naoe, T (2003). Different antiapoptotic pathways between wild-type and mutated FLT3: insights into therapeutic targets in leukemia. *Blood*, 102:2969–2975.

Miyoshi, H, Shimizu, K, Kozu, T, Maseki, N, Kaneko, Y, and Ohki, M (1991). t(8;21) breakpoints on chromosome 21 in acute myeloid leukemia are clustered within a limited region of a single gene, AML1. *Proc Natl Acad Sci USA*, 88:10431–10434.

Mizuki, M, Fenski, R, Halfter, H, Matsumura, I, Schmidt, R, Muller, C, et al. (2000). Flt3 mutations from patients with acute myeloid leukemia induce transformation of 32D cells mediated by the Ras and STAT5 pathways. *Blood*, 96:3907–3914.

Mizuki, M, Schwable, J, Steur, C, Choudhary, C, Agrawal, S, Sargin, B, et al. (2003). Suppression of myeloid transcription factors and induction of STAT response genes by AML-specific Flt3 mutations. *Blood*, 101:3164–3173.

Morikawa, J, Li, H, Kim, S, Nishi, K, Ueno, S, Suh, E, (2003). Identification of signature genes by microarray for acute myeloid leukemia without maturation and acute promyelocytic leukemia with t(15;17)(q22;q12)(PML/RARalpha). *Int J Oncol*, 23, 617–625.

Mrozek, K, Heinonen, K, de la,CA, and Bloomfield, CD (1997). Clinical significance of cytogenetics in acute myeloid leukemia. *Semin Oncol*, 24:17–31.

Mueller, BU and Pabst,T (2006) C/EBPalpha and the pathophysiology of acute myeloid leukemia. *Curr Opin Hematol*, 13:7–14.

Mueller, BU, Pabst, T, Osato, M, Asou, N, Johansen, LM, Minden, MD, et al. (2002). Heterozygous PU.1 mutations are associated with acute myeloid leukemia. *Blood*, 100:998–1007.

Muller, C, Readhead, C, Diederichs, S, Idos, G, Yang, R, Tidow, N, et al. (2000). Methylation of the cyclin A1 promoter correlates with gene silencing in somatic cell lines, while tissue-specific expression of cyclin A1 is methylation independent. *Mol Cell Biol*, 20:3316–3329.

Muller, C, Yang, R, Park, DJ, Serve, H, Berdel, WE, and Koeffler, HP (2000). The aberrant fusion proteins PML-RAR alpha and PLZF-RAR alpha contribute to the overexpression of cyclin A1 in acute promyelocytic leukemia. *Blood*, 96:3894–3899.

Muller-Tidow, C, Metzelder, SK, Buerger, H, Packeisen, J, Ganser, A, Heil, G, et al. (2004a). Expression of the p14ARF tumor suppressor predicts survival in acute myeloid leukemia. *Leukemia*, 18:720–726.

Muller-Tidow, C, Steffen, B, Cauvet, T, Tickenbrock, L, Ji, P, Diederichs, S, et al. (2004b). Translocation products in acute myeloid leukemia activate the Wnt signaling pathway in hematopoietic cells. *Mol Cell Biol*, 24:2890–2904.

Muller-Tidow, C, Wang, W, Idos, GE, Diederichs, S, Yang, R, Readhead, C, et al. (2001). Cyclin A1 directly interacts with B-myb and cyclin A1/cdk2 phosphorylate B-myb at functionally important serine and threonine residues: tissue-specific regulation of B-myb function. *Blood*, 97:2091–2097.

Nagata, H, Worobec, AS, Oh, CK, Chowdhury, BA, Tannenbaum, S, Suzuki, Y, et al. (1995). Identification of a point mutation in the catalytic domain of the protooncogene c-kit in peripheral blood mononuclear cells of patients who have mastocytosis with an associated hematologic disorder. *Proc Natl Acad Sci USA*, 92:10560–10564.

Nakamura, T, Mori, T, Tada, S, Krajewski, W, Rozovskaia, T, Wassell, R, et al. (2002). ALL-1 is a histone methyltransferase that assembles a supercomplex of proteins involved in transcriptional regulation. *Mol Cell*, 10:1119–1128.

Nakao, M, Yokota, S, Iwai, T, Kaneko, H, Horiike, S, Kashima, K, et al. (1996). Internal tandem duplication of the flt3 gene found in acute myeloid leukemia. *Leukemia*, 10:1911–1918.

Neben, K, Schnittger, S, Brors, B, Tews, B, Kokocinski, F, Haferlach, T, et al. (2005). Distinct gene expression patterns associated with FLT3- and NRAS-activating mutations in acute myeloid leukemia with normal karyotype. *Oncogene*, 24:1580–1588.

Pandolfi, PP (2001a). Oncogenes and tumor suppressors in the molecular pathogenesis of acute promyelocytic leukemia. *Hum Mol Genet*, 10:769–775.

Pandolfi, PP (2001b). In vivo analysis of the molecular genetics of acute promyelocytic leukemia. *Oncogene*, 20:5726–5735.

Paquette, RL, Landaw, EM, Pierre, RV, Kahan, J, Lubbert, M, Lazcano, O, et al. (1993). N-ras mutations are associated with poor prognosis and increased risk of leukemia in myelodysplastic syndrome. *Blood*, 82:590–599.

Perry, GS, III, Spector, BD, Schuman, LM, Mandel, JS, Anderson, VE, McHugh, RB, et al. (1980). The Wiskott-Aldrich syndrome in the United States and Canada (1892–1979). *J Pediatr*, 97:72–78.

Polakis, P (2000). Wnt signaling and cancer. *Genes Dev*, 14:1837–1851.

Pollock, JL, Westervelt, P, Walter, MJ, Lane, AA, and Ley, TJ (2001). Mouse models of acute promyelocytic leukemia. *Curr Opin Hematol*, 8:206–211.

Porse, BT, Bryder, D, Theilgaard-Monch, K, Hasemann, MS, Anderson, K, Damgaard, I, et al. (2005). Loss of C/EBP alpha cell cycle control increases myeloid progenitor proliferation and transforms the neutrophil granulocyte lineage. *J Exp Med*, 202:85–96.

Preudhomme, C, Sagot, C, Boissel, N, Cayuela, JM, Tigaud, I, De Botton, S, et al. (2002). Favorable prognostic significance of CEBPA mutations in patients with de novo acute myeloid leukemia: a study from the Acute Leukemia French Association (ALFA). *Blood*, 100:2717–2723.

Rappold, I, Ziegler, BL, Kohler, I, Marchetto, S, Rosnet, O, Birnbaum, D, (1997). Functional and phenotypic characterization of cord blood and bone marrow subsets expressing FLT3 (CD135) receptor tyrosine kinase. *Blood*, 90:111–125.

Redner, RL, Chen, JD, Rush, EA, Li, H, and Pollock, SL (2000). The t(5;17) acute promyelocytic leukemia fusion protein NPM-RAR interacts with co-repressor and co-activator proteins and exhibits both positive and negative transcriptional properties. *Blood*, 95:2683–2690.

Redner, RL, Contis, LC, Craig, F, Evans, C, Sherer, ME, and Shekhter-Levin, S (2006). A novel t(3;17)(p25;q21) variant translocation of acute promyelocytic leukemia with rearrangement of the RARA locus. *Leukemia*, 20:376–379.

Rosnet, O, Schiff, C, Pebusque, MJ, Marchetto, S, Tonnelle, C, Toiron, Y, et al. (1993). Human FLT3/FLK2 gene: cDNA cloning and expression in hematopoietic cells. *Blood*, 82:1110–1119.

Ross, ME, Mahfouz, R, Onciu, M, Liu, HC, Zhou, X, Song, G, et al. (2004). Gene expression profiling of pediatric acute myelogenous leukemia. *Blood*, 104:3679–3687.

Rowley, JD (2001). Chromosome translocations: dangerous liaisons revisited. *Nat Rev Cancer*, 1:245–250.

Satake, N, Maseki, N, Nishiyama, M, Kobayashi, H, Sakurai, M, Inaba, H, et al. (1999). Chromosome abnormalities and MLL rearrangements in acute myeloid leukemia of infants. *Leukemia*, 13:1013–1017.

Schnittger, S, Schoch, C, Dugas, M, Kern, W, Staib, P, Wuchter, C, et al. (2002). Analysis of FLT3 length mutations in 1003 patients with acute myeloid leukemia: correlation to cytogenetics, FAB subtype, and prognosis in the AMLCG study and usefulness as a marker for the detection of minimal residual disease. *Blood*, 100:59–66.

Schnittger, S, Schoch, C, Kern, W, Mecucci, C, Tschulik, C, Martelli, MF, et al. (2005). Nucleophosmin gene mutations are predictors of favourable prognosis in acute myelogenous leukemia with a normal kayotype. *Blood*, 106:3733–3739.

Schoch, C, Kohlmann, A, Schnittger, S, Brors, B, Dugas, M, Mergenthaler, S, et al. (2002). Acute myeloid leukemias with reciprocal rearrangements can be distinguished by specific gene expression profiles. *Proc Natl Acad Sci USA*, 99:10008–10013.

Schoch, C, Schnittger, S, Klaus, M, Kern, W, Hiddemann, W, and Haferlach, T (2003). AML with 11q23/MLL abnormalities as defined by the WHO classification: incidence, partner chromosomes, FAB subtype, age distribution, and prognostic impact in an unselected series of 1897 cytogenetically analyzed AML cases. *Blood*, 102:2395–2402.

Shih, LY, Huang, CF, Wu, JH, Wang, PN, Lin, TL, Dunn, P, et al. (2004). Heterogeneous patterns of FLT3 Asp(835) mutations in relapsed de novo acute myeloid leukemia: a comparative analysis of 120 paired diagnostic and relapse bone marrow samples. *Clin Cancer Res*, 10:1326–1332.

Spiekermann, K, Bagrintseva, K, Schoch, C, Haferlach, T, Hiddemann, W, and Schnittger, S (2002). A new and recurrent activating length mutation in exon 20 of the FLT3 gene in acute myeloid leukemia. *Blood*, 100:3423–3425.

Stacchini, A, Fubini, L, Severino, A, Sanavio, F, Aglietta, M, and Piacibello, W (1996). Expression of type III receptor tyrosine kinases FLT3 and KIT and responses to their ligands by acute myeloid leukemia blasts. *Leukemia*, 10:1584–1591.

Stanulla, M, Wang, J, Chervinsky, DS, and Aplan, PD (1997). Topoisomerase II inhibitors induce DNA double-strand breaks at a specific site within the AML1 locus. *Leukemia*, 11:490–496.

Stanulla, M, Wang, J, Chervinsky, DS, Thandla, S, and Aplan, PD (1997b). DNA cleavage within the MLL breakpoint cluster region is a specific event which occurs as part of higher-order chromatin fragmentation during the initial stages of apoptosis. *Mol Cell Biol*, 17:4070–4079.

Steudel, C, Wermke, M, Schaich, M, Schakel, U, Illmer, T, Ehninger, G, et al. (2003). Comparative analysis of MLL partial tandem duplication and FLT3 internal tandem duplication mutations in 956 adult patients with acute myeloid leukemia. *Genes Chromosomes Cancer*, 37:237–251.

Stirewalt, DL, Kopecky, KJ, Meshinchi, S, Appelbaum, FR, Slovak, ML, Willman, CL, et al. (2001). FLT3, RAS, and TP53 mutations in elderly patients with acute myeloid leukemia. *Blood*, 97:3589–3595.

Stone, RM, O'Donnell, MR, and Sekeres, MA (2004). Acute myeloid leukemia. *Hematology (Am Soc Hematol Educ Program)*, 98–117.

Taketani, T, Taki, T, Ono, R, Kobayashi, Y, Ida, K, and Hayashi, Y (2002). The chromosome translocation t(7;11)(p15;p15) in acute myeloid leukemia results in fusion of the NUP98 gene with a HOXA cluster gene, HOXA13, but not HOXA9. *Genes Chromosomes Cancer*, 34:437–443.

Tibshirani, R, Hastie, T, Narasimhan, B, and Chu, G (2002). Diagnosis of multiple cancer types by shrunken centroids of gene expression. *Proc Natl Acad Sci USA*, 99:6567–6572.

Timchenko, NA, Wilde, M, Nakanishi, M, Smith, JR, and Darlington, GJ (1996). CCAAT/enhancer-binding protein alpha (C/EBP alpha) inhibits cell proliferation through the p21 (WAF-1/CIP-1/SDI-1) protein. *Genes Dev*, 10:804–815.

Tusher, VG, Tibshirani, R, and Chu, G (2001). Significance analysis of microarrays applied to the ionizing radiation response. *Proc Natl Acad Sci USA*, 98:5116–5121.

Valk, PJ, Verhaak, RG, Beijen, MA, Erpelinck, CA, Barjesteh van Waalwijk van Doorn-Khosrovani, Boer, JM, et al (2004). Prognostically useful gene-expression profiles in acute myeloid leukemia. *N Engl J Med*, 350:1617–1628.

Vangala, RK, Heiss-Neumann, MS, Rangatia, JS, Singh, SM, Schoch, C, Tenen, DG, et al. (2003). The myeloid master regulator transcription factor PU.1 is inactivated by AML1-ETO in t(8;21) myeloid leukemia. *Blood*, 101:270–277.

Vardiman, JW, Harris, NL, and Brunning, RD. (2002). The World Health Organization (WHO) classification of the myeloid neoplasms. *Blood*, 100:2292–2302.

Vey, N, Mozziconacci, MJ, Groulet-Martinec, A, Debono, S, Finetti, P, Carbuccia, N, et al. (2004). Identification of new classes among acute myelogenous leukaemias with normal karyotype using gene expression profiling. *Oncogene*, 23:9381–9391.

Virtaneva, K, Wright, FA, Tanner, SM, Yuan, B, Lemon, WJ, Caligiuri, MA, et al. (2001). Expression profiling reveals fundamental biological differences in acute myeloid leukemia with isolated trisomy 8 and normal cytogenetics. *Proc Natl Acad Sci USA*, 98:1124–1129.

Walker, H, Smith, FJ, and Betts, DR (1994). Cytogenetics in acute myeloid leukaemia. *Blood Rev*, 8:30–36.

Wang, H, Goode, T, Iakova, P, Albrecht, JH, and Timchenko, NA (2002). C/EBPalpha triggers proteasome-dependent degradation of cdk4 during growth arrest. *EMBO J*, 21:930–941.

Warrell, RP Jr, De The, H, Wang, ZY, and Degos, L (1993). Acute promyelocytic leukemia. *N Engl J Med*, 329:177–189.

Wasser, JS, Yolken, R, Miller, DR, and Diamond, L (1978). Congenital hypoplastic anemia (Diamond-Blackfan syndrome) terminating in acute myelogenous leukemia. *Blood*, 51:991–995.

Wells, RA, Catzavelos, C, and Kamel-Reid, S (1997). Fusion of retinoic acid receptor alpha to NuMA, the nuclear mitotic apparatus protein, by a variant translocation in acute promyelocytic leukaemia. *Nat Genet*, 17:109–113.

Wells, RA, Hummel, JL, De, KA, Zipursky, A, Kirby, M, Dube, I, et al. (1996). A new variant translocation in acute promyelocytic leukaemia: molecular characterization and clinical correlation. *Leukemia*, 10:735–740.

Westendorf, JJ, Yamamoto, CM, Lenny, N, Downing, JR, Selsted, ME, and Hiebert, SW (1998) The t(8;21) fusion product, AML-1-ETO, associates with C/EBP- alpha, inhibits C/EBP-alpha-dependent transcription, and blocks granulocytic differentiation. *Mol Cell Biol*, 18:322–333.

Yagi, T, Morimoto, A, Eguchi, M, Hibi, S, Sako, M, Ishii, E, et al. (2003). Identification of a gene expression signature associated with pediatric AML prognosis. *Blood*, 102:1849–1856.

Yasmeen, A, Berdel, WE, Serve, H, and Muller-Tidow, C (2003). E- and A-type cyclins as markers for cancer diagnosis and prognosis. *Expert Rev Mol Diagn*, 3:617–633.

Zhang, S and Broxmeyer, HE (2000). Flt3 ligand induces tyrosine phosphorylation of gab1 and gab2 and their association with shp-2, grb2, and PI3 kinase. *Biochem Biophys Res Commun*, 277:195–199.

Zwaan, C and Kaspers, GJ (2004). Possibilities for tailored and targeted therapy in paediatric acute myeloid leukaemia. *Br J Haematol*, 127:264–279.

25

Genomics and Infectious Diseases: Susceptibility, Resistance, Response, and Antimicrobial Therapy

Sarra E Jamieson and Christopher S Peacock

Mankind has been afflicted with infectious diseases throughout history and, at the start of the third millennium, the global burden of infectious disease is still extremely high, with World Health Organization (WHO) figures indicating that infectious diseases are the second leading cause of death after cardiovascular diseases (WHO, 2004) (see Fig. 25–1).

Of the 57 million deaths worldwide in 2002, an estimated 14.9 million (~26%) were due to infectious or parasitic disease, with the majority of this burden falling on children in the African and South East Asian regions (Black et al., 2003; Bryce et al., 2005). Of those 14.9 million deaths, large proportions were due to well-known infectious diseases, namely HIV/AIDS (human immunodeficiency virus/acquired immunodeficiency syndrome) (2.77 million deaths in 2002), tuberculosis (TB) (1.57 million deaths), and malaria (1.27 million deaths) as illustrated in Figure 25–2. As well as mortality figures, the overall burden placed on a population by a particular disease can also be measured in disability-adjusted life years (DALYs) (Murray and Acharya, 1997). This measurement takes into account not only the number of years of healthy life lost due to early mortality, but also the number of years of healthy life lost due to disability or poor health, which are combined and used to quantify the overall disease burden. Using this measure, the global burden of infectious and parasitic diseases was 445 million DALYs in 2002, as compared to cardiovascular diseases with a burden of approximately 148 million DALYs in 2002 (WHO, 2004).

In addition to the well-known existing infectious and parasitic diseases, a number of newly emerging or re-emerging infectious diseases have also been noted throughout the world in recent times. While humans and pathogens have evolved together throughout history, the ever-changing pattern of land usage, farming practices, and human demographics coupled with increased global travel means that humans have ever-increasing contact with novel infectious agents of indigenous animal populations resulting in zoonotic, or cross-species, transmissions. Of all newly emerging infectious diseases, approximately three quarters are estimated to be of a zoonotic origin (Taylor et al., 2001). In the last century, the HIV-1 (group M) virus is believed to have jumped from chimpanzees to humans sometime close to 1931 (reviewed by Sharp et al., 2001), although it was not identified as a human pathogen until the early 1980s, by which time the infection was already widespread in many African countries (Weiss, 2001). In the last two decades, HIV/AIDS has gone on to reach pandemic proportions, resulting in more than 20 million deaths globally, with a further 40.3 million believed to be infected with the virus as of 2005 (www.unaids.org).

There are also relatively recent examples of emerging zoonotic infections, including the prion disease variant Creutzfeldt–Jakob disease (vCJD) in the United Kingdom, believed to be linked to the human consumption of bovine spongiform encephalopathy (BSE) contaminated beef products (Beisel and Morens, 2004), and viral infections such as severe acute respiratory syndrome (SARS) (Wong and Yuen, 2005) and the H5N1 avian influenza (Fleming, 2005). All of these infectious diseases have resulted in human deaths and, in the case of SARS and avian influenza, to speculation of a new pandemic (Lee and Krilov, 2005). Many of the same biological, environmental, social and economic factors known to play a role in emergence of new infectious diseases also play a role in the re-emergence or reappearance of infectious diseases known to afflict man in the past. This category currently includes diseases such as multidrug resistant (MDR) TB, drug-resistant malaria, plague, cholera, dengue, and West Nile virus (Morens et al., 2004).

While there have been many advances in the last few decades in sanitation and hygiene, and in the development of antimicrobials, vaccines and novel therapeutics that have undoubtedly represented huge advances in mankind's fight against infection, it is clear that infectious diseases still represent a significant concern in the modern world. Research is ongoing in the field of infectious diseases and is making full use of up-to-date scientific advances in both genetics and genomics to better understand human susceptibility to infection, the complex mechanisms of the host's immune response and the rapidly evolving evasion mechanisms of the pathogen in order to develop and continually improve diagnostic protocols, treatment regimens and prevention strategies. This chapter aims to highlight some of this ongoing research.

Human Genetic Susceptibility to Infectious Disease

Human susceptibility to infectious disease is multifactorial, being the result of interactions between environmental factors, host-related factors and the infecting pathogen itself. While infection with the pathogen is required for disease it is not the sole requirement as not

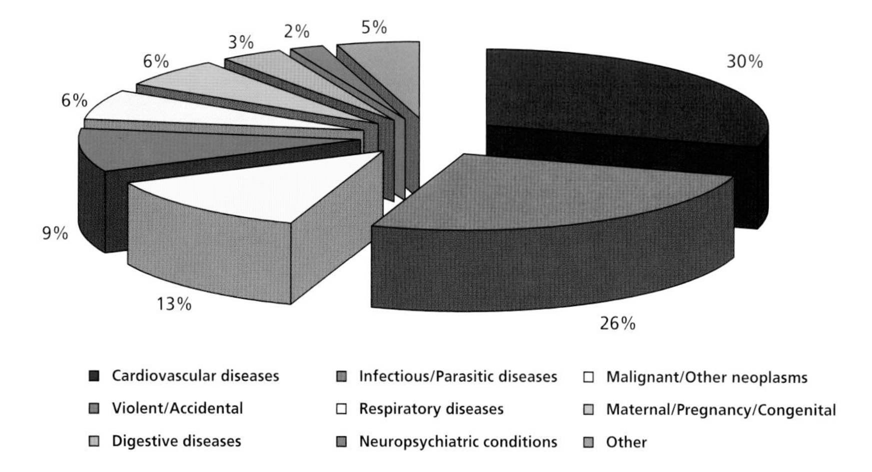

Figure 25–1 World Health Organization figures for worldwide causes of death in 2002.

all exposed individuals will go on to develop clinical disease. Accordingly, there is now a large body of evidence demonstrating that host genetics plays an important role in the susceptibility, or resistance, of individuals to infectious disease. This includes epidemiological evidence from ethnic differences in infectious disease susceptibility coupled with familial studies of rare families who exhibit Mendelian susceptibility to weakly pathogenic infectious agents, familial aggregation studies, complex segregation analyses, as well as twin and adoptee studies, all of which confirm the role played by host genetics in infectious disease susceptibility.

Ethnic Differences in Susceptibility to Infectious Disease

A number of racial or ethnic differences in infectious disease susceptibility have been noted. For instance, it was noted many years ago that sub-Saharan African populations were seemingly resistant to infection by the normally pathogenic malarial parasite *Plasmodium vivax* (Miller et al., 1975). The basis of this resistance has subsequently been shown to be a polymorphism in the *FY* gene encoding the Duffy antigen, resulting in a Duffy-negative allele (*FY* 0*) that completely inhibits *P. vivax* invasion of erythrocytes (Kwiatkowski, 2005).

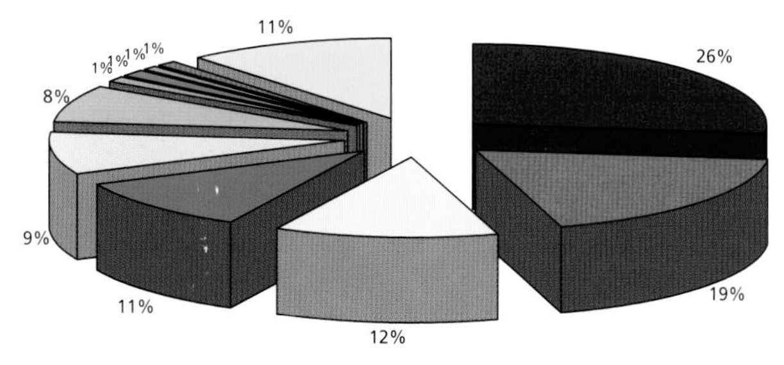

Figure 25–2 Breakdown of the infectious/parasitic diseases category to show specific causes of death. WHO figures, 2002 (Tropical Diseases*—Trypanosomiasis, Chagas Disease, Schistosomiasis, Leishmaniasis, Leprosy, Ascariasis, Lymphatic Filariasis, Onchocerciasis, Hookworm and other intestinal infections. Childhood Diseases‡—Pertussis, Poliomyelitis, Diphtheria, Measles and Tetanus.)

It has also been observed that Caucasian populations appear to have a greater resistance to *Mycobacterium tuberculosis* (*M. tb*) than that seen in African populations, possibly due to the relatively recent exposure of African populations to the causative agent as compared to European populations. In agreement with this, an investigation of infection rates in 165 racially integrated nursing homes in the United States, including more than 25,000 residents, indicated that African–American residents were almost twice as likely (relative risk (RR) = 1.9) to become infected, as compared to Caucasian American residents (Stead et al., 1990). A similar investigation of a TB outbreak in an ethnically mixed elementary school, while not detecting any racial difference in susceptibility to disease, did detect a difference in disease progression rates between African-American and Caucasian American children (Hoge et al., 1994). Furthermore, investigations of the geographic distribution of a gene shown to be significantly associated with TB resistance, namely the *CD209* gene encoding the C-type lectin DC-SIGN, demonstrated that resistance associated alleles were confined mainly to Eurasian populations (Barreiro et al., 2006), potentially as a result of the stronger selective pressure exerted by *M. tb* on non-African populations.

Similar ethnic differences have also been reported in studies of hepatitis B viral infection in Zimbabwe, where white hospital workers were shown to have significantly reduced markers of infection (Emmanuel et al., 1988), in studies of malaria, with similarly exposed tribal groups showing differing levels of parasitism (Modiano et al., 1996) and in studies of *Helicobacter pylori* infection with Mexican–Americans and African–Americans demonstrating increased rates of seropositivity as compared to non-Hispanic Caucasian Americans (Everhart et al., 2000). While these studies give an indication that host genetics is important in susceptibility to infection and also concomitantly indicate the potential heterogeneity of susceptibility and resistance loci in different ethnic groups, it is important to remember that these results could potentially be influenced by differing sociodemographic characteristics between the study groups.

Mendelian Susceptibility to Infectious Disease

A number of families demonstrating Mendelian susceptibility to infectious diseases have been described (reviewed by Casanova and Abel, 2005; Casanova et al., 2005) and, in some cases, the single gene defects responsible have been identified. Briefly, these include mutations in *IFNγR1* (Jouanguy et al., 1996; Newport et al., 1996; Jouanguy et al., 1997), *IFNγR2* (Dorman and Holland, 1998), *IL12Rβ1* (Altare, Durandy, et al., 1998; de Jong et al., 1998; Altare et al., 2001), *IL12B* (Altare, Lammas, et al., 1998; Picard et al., 2002) and *STAT1* (Dupuis et al., 2003) genes that all result in disseminated infections with weakly pathogenic atypical mycobacterial or salmonella species. In addition, mutations in *EVER1* and *EVER2*, encoding transmembrane channel-like proteins, cause a rare autosomal recessive susceptibility to specific human papillomaviruses, resulting in Epidermodysplasia verruciformis (Ramoz et al., 2002); mutations in *SH2D1A*, encoding a signaling lymphocyte-activation molecule (SLAM)-associated protein, cause X-linked susceptibility to Epstein–Barr virus (EBV) infection (Purtilo et al., 1977; Dutz et al., 2001); while mutations in a number of components of the complement pathway, in particular C7, lead to recurrent infections with *Neisseria meningitidis* (Nurnberger et al., 1989; Wurzner et al., 1992).

Furthermore, there are a number of immunodeficiencies in which no single gene defect has, as yet, been described. These include both autosomal dominant and autosomal recessive susceptibility to a fungal infection resulting in chronic mucocutaneous candidiasis (Lilic, 2002) and an autosomal recessive, as opposed to the previously described X-linked, susceptibility to EBV (Fleisher et al., 1982). Mutations such as these are, however, rare, and more common variants with more subtle effects would be expected to influence susceptibility to infectious disease at a population level.

Familial Studies of Infectious Disease Susceptibility

The suggestion that human genetics plays a part in susceptibility to infection is not a new one. The occurrence of familial clustering in diseases such as leprosy and TB led to the common belief that these diseases were an inherited defect (discussed in Levin and Newport, 2000) until the discovery of the causative mycobacterial agents by Gerhard Hansen in 1873 (Hansen and Freney, 2002) and Robert Koch in 1882 (Gradmann, 2006), respectively. Such familial clustering is the result not only of shared genetic factors but also of shared familial environment and shared exposure to the infectious agent. The relative contribution made by host genetics, as opposed to other non-genetic factors, has been investigated in a number of infectious diseases, including bacterial, parasitic and helminthic infections. For example, studies of familial clustering in visceral leishmaniasis due to *Leishmania chagasi* (Cabello et al., 1995), of clustering in both cutaneous leishmaniasis (CL) and mucocutaneous leishmaniasis (ML) due to *Leishmania braziliensis* (Castellucci et al., 2005), of the intestinal helminth *Ascaris lumbricoides* (Williams-Blangero et al., 1999), of lymphatic filariasis caused by *Brugia malayi* (Terhell et al., 2000) and of clustering in malaria due to *Plasmodium falciparum* infection (Mackinnon et al., 2000; Domarle et al., 2002) have all concluded a genetic component to the observed familial clustering.

More conclusive evidence of the role played by host genetics in familial clustering is provided by studies looking at the recurrence risk ratio for siblings or other relatives of affected individuals compared to that of the general population. Familial risk ratios indicative of a genetic component to infectious disease susceptibility have been reported in studies of meningococcal disease due to *N. meningitidis* ($\lambda_s = 8.2–11.9$) (Haralambous et al., 2003), of leprosy per se due to *Mycobacterium leprae* ($\lambda_R > 1$) (Wallace et al., 2003) and in cerebral malaria and severe malarial anemia [odds ratio (OR) = 6.15] (Ranque et al., 2005). In addition, studies of heritability estimates, based on the correlation between all genetically related individuals, including those who may not share the same household, also indicate a role for host genetics. Heritability estimates of 57% for leprosy per se (Bakker et al., 2005), of up to 25% in mild or hospital-admitted malaria cases (Mackinnon et al., 2005) and of 58% for schistosomiasis due to infection with *Schistosoma japonicum* (Ellis et al., 2006) have been reported.

Segregation analysis has also been used in the familial investigations of infectious disease susceptibility. This is a statistical method that allows the model of disease transmission in multicase families to be examined and the likelihood of various models to be ascertained in order to determine the most appropriate model to explain the variation in the phenotype under investigation. Detection of a single major gene does not necessarily indicate that this is the only locus involved in the variability of the phenotype in question, only that this gene has a genetic effect large enough to be detected in a statistically significant manner. While not exhaustive, Table 25–1 illustrates a large number of those segregation analyses undertaken in infectious diseases, including viral, bacterial, parasitic and helminthic diseases, the result of which seem to confirm a role for host genetics.

In addition to evidence derived from familial clustering, a number of studies have also employed twins with an observation of increased concordance rates for the disease of interest in genetically identical monozygous (MZ) as compared to dizygous (DZ) twins, which share only half their genes on average, indicating a genetic component to disease susceptibility. In infectious disease research, a number of twin studies have demonstrated that MZ twins do indeed show increased concordance rates in comparison to DZ twins. For example, twin studies investigating TB susceptibility (Kallmann and Reisner, 1943; Comstock, 1978), susceptibility to leprosy per se, but not to leprosy type (reviewed by Fine, 1988), the clinical presentation, but not infection per se, of malaria (Jepson et al., 1995), as well as the acquisition of *H. pylori* infection (Malaty et al., 1994) have all indicated higher concordance in MZ twins. In addition, a number of twin and triplet studies of otitis media, a common condition of childhood involving bacterial or viral infection of the middle ear, have documented a significant heritable component in the occurrence of both acute (Kvaerner et al., 1997; Rovers et al., 2002) and chronic episodes of otitis media (Casselbrant et al., 1999). While some care may be necessary in the interpretation of the more historical twin studies due to potential inaccuracies in the determination of zygosity, it is clear that the results of twin studies such as these are consistent with a genetic component in susceptibility to infection.

Twin studies have also been used to investigate the heritability of cellular and humoral immune responses to a variety of immunomodulatory antigens. Studies looking at antibody responses to intact or immunomodulatory epitopes of malarial antigens (Sjoberg et al., 1992) or to purified protein derivative (PPD) of *M. tb* (Jepson et al., 1997; Jepson et al., 2001) as well as looking at cytokine production in meningococcal disease (Westendorp et al., 1997) have all concluded a heritable component to these responses, results that have implications in vaccine design. Indeed, investigation of variation in immune response to vaccination in infant twins demonstrated significant heritability in antibody and/or cytokine responses to hepatitis B, oral polio, tetanus, diphtheria and pertussis vaccines as well as to some Bacille Calmette-Guérin (BCG) vaccine subunits, indicating that genetic factors play an essential role in the regulation of the immune response to vaccine (Newport et al., 2004).

While twin studies indicate a genetic component to infectious disease susceptibility, twins are, in general, raised together, meaning they share not only genetics but also their environment. Non-familial adoptee studies have been used in order to try and assess the relative contribution of genetics versus environment by investigating causes of death in both the biological and adoptive parents and assessing the RR posed to the adoptee as a result. Such studies demonstrate that the early death of a biological parent from infectious disease results in a nearly six-fold increase in risk (RR = 5.81) of death from the same cause in the adoptee. This is in comparison to

Table 25–1 Summary of Segregation Analyses Carried in Infectious Diseases

Disease	Population	Conclusion	Reference
Lepromatous leprosy	Philippines	Autosomal recessive inheritance	Smith, 1979
Tuberculoid leprosy	Indian	Recessive model	Haile et al., 1985
Lepromatous leprosy	Australian Aboriginal	Autosomal recessive inheritance	Rawlinson et al., 1988
Leprosy per se	Caribbean	Recessive major gene	Abel and Demenais, 1988
Leprosy per se	Thai	Generalized major gene	Wagener et al., 1988
Tuberculoid leprosy	Thai	Recessive model	Wagener et al., 1988
HBV[a]	Chinese	Recessive model	Shen et al., 1991
Schistosomiasis	Brazilian	Codominant major gene	Abel et al., 1991
Malaria (Parasite levels)	Cameroon	Recessive major gene	Abel et al., 1992
Leprosy per se	Vietnamese	Codominant major gene	Abel et al., 1995
Leprosy per se	Brazilian	Recessive major gene	Feitosa, et al., 1995
Cutaneous leishmaniasis	Peruvian	Recessive gene plus modifier	Shaw et al., 1995
Leprosy (Mitsuda reaction)	Brazilian	Segregation of a major gene	Feitosa, et al., 1996
Schistosomiasis (IL5 levels)	Brazilian	Codominant major gene	Rodrigues et al., 1996
Tuberculosis	Brazilian	General two-locus model	Shaw et al., 1997
Mucocutaneous leishmaniasis	Bolivian	Recessive major gene	Alcais et al., 1997
Malaria (Parasite levels)	Burkino Faso	Complex inheritance, major effect	Rihet et al., 1998
Malaria (Parasite levels)	Cameroon	Complex mode of inheritance	Garcia et al., 1998
Schistosomiasis (Severity[b])	Sudanese	Codominant major gene	Dessein et al., 1999
Filarial Loa Loa	Cameroon	Dominant major gene	Garcia et al., 1999
HTLV-1[c]	French Guiana	Dominant major gene	Plancoulaine et al., 2000
Visceral leishmaniasis	Brazilian	General single locus model	Peacock et al., 2001
HHV-8[d]	French Guiana	Recessive major gene	Plancoulaine et al., 2003
Leprosy (Mitsuda reaction)	Vietnamese	Major gene model	Ranque et al., 2005

[a] Hepatitis B virus resulting in primary hepatocellular carcinoma.
[b] Measured by hepatic fibrosis.
[c] Human T Lymphotrophic virus.
[d] Human Herpesvirus 8.

four-fold increase (RR = 4.52) for cardiovascular or cerebrovascular causes. In contrast, the early death of the adoptive parent from infectious disease resulted in no increased risk of early death from the same cause in the adoptee (Sorensen et al., 1988), indicating that genetic rather than shared environment is important.

Identification of Candidate Genes

Overall epidemiological and familial studies provide convincing evidence that genetic variation in human populations contributes to infectious disease susceptibility to a diverse range of pathogens; however, they provide no information as to the actual genes involved in that susceptibility. In order to identify those causative loci, two basic study designs are employed—the whole-genome linkage scan and the candidate gene association study, both of which have been used in the investigation of infectious disease susceptibility.

The Genome-wide Linkage Approach

Linkage scans of the entire human genome are a useful tool in the field of human genetics, as they assume no prior knowledge of the causative gene or its location in the human genome. Such genome-wide linkage scans are performed by typing a large number (usually ~400) of evenly spaced [every 5–10 centiMorgans (cM)] microsatellite markers (DNA regions containing a variable number of short tandem repeats) in a number of pedigrees containing multiple affected individuals. Linkage analysis methods are then used to determine whether any of the typed markers demonstrate co-segregation with the causative disease gene by virtue of their close proximity to that disease gene on the same chromosome. Classic parametric, or model-based, linkage analyses then uses the LOD ($\log_{10}$ of the odds) score to test the hypothesis of linkage versus no linkage at various genetic distances (or recombination fractions—θ). Traditionally, a LOD score of + 3 and above (i.e., 1000:1 odds in favor of linkage) is taken as an indication of significant linkage between marker and disease locus (Nyholt, 2000). However, Lander and Kruglyak (1995) recommended that, for acceptance of genome-wide significance, a LOD score of + 3.3 is required. In the absence of a specific model of disease inheritance, a nonparametric method of linkage analysis may be employed. These methods use the extent of allele sharing identical by descent (IBD) between affected relative pairs at concordant intervals to narrow the region of interest, providing a defined candidate region containing a putative susceptibility locus.

While linkage analysis is a powerful tool in the investigation of simple Mendelian traits where a single major locus is expected, it is important to remember that human infectious disease is a complex trait and as such, multiple genetic and environmental factors will contribute to an individual's risk of being affected. Unlike the majority of Mendelian disorders, where a simple pattern of inheritance can be observed, in complex traits, there is no simple pattern of inheritance and the presence or absence of a risk allele may have only a slight effect on the individual's risk of disease. Given the difficulties encountered in defining an accurate model of disease inheritance for complex traits, linkage analysis of such traits more commonly uses the nonparametric linkage approach. This scenario is complicated even further by the potential occurrence of complex interactions between susceptibility alleles of multiple genes, which may individually confer little risk to disease development, as well as by the existence of phenotypic and genotypic heterogeneity. Nevertheless, genome-wide linkage studies have been successfully employed in the investigation of human susceptibility to a number of infectious diseases. Some of these studies are illustrated in Table 25–2, a number of which will be considered in more detail.

Schistosomiasis

The first attempt to map a human gene contributing to infectious disease susceptibility using the whole-genome scan approach was done using a number of Brazilian pedigrees with multiple individuals affected with the helminth infection schistosomiasis (Marquet et al., 1996). Previous segregation analysis using the same Brazilian families had provided evidence for a codominant major gene accounting for 66% of the variance in infection intensity with *Schistosoma mansoni* (Abel et al., 1992; Marquet et al., 1996) and provided an accurate model of disease inheritance in this endemic population. Using individual *S. mansoni* egg output as a measure of infection intensity, parametric linkage analysis of whole-genome scan data identified a single major peak with a LOD score of + 4.74. This peak of linkage falls in the 5q31-q33 region of the genome, a region known to encompass a number of interesting candidate genes in terms of the immune response to schistosomiasis. These include the genes of the T_H2 cytokine cluster, containing the interleukin 13 *(IL13)*, *IL4*, *IL5*, *IL3*, *IL9* and *IL12B* genes, as well as the interferon regulatory factor 1 (*IRF1*), granulocyte–macrophage colony stimulating factor 2 (*CSF2*) and the colony stimulating factor-1 receptor (*CSF1R*).

While not at a genome-wide level of significance (LOD = + 3.3) (Lander and Kruglyak, 1995), three additional regions of the genome (1p22.2, 21q22-qter and 7q36) also provided some evidence of linkage (LOD > + 0.83; $p > 0.025$) to *S. mansoni* infection intensity (Marquet et al., 1999). However, only one of these regions (7q35-q36) was subsequently confirmed using a more appropriate nonparametric analysis method as opposed to parametric linkage analysis using the single major gene model (Zinn-Justin et al., 2001). Further evidence of linkage to the 5q31-q33 region was additionally provided by a second study carried out in extended multicase pedigrees from a recently exposed population in northern Senegal (Muller-Myhsok et al., 1997). The causal locus in this region has still to be identified.

Tuberculosis

A number of whole-genome scans have also been performed in multicase TB families from various endemic regions. The first of these was a genome scan performed using families from Gambia and South Africa (Bellamy et al., 2000). Combined nonparametric analysis provided suggestive evidence of linkage to TB susceptibility at 15q11.2 [maximum LOD score (MLS) = + 2.00] and Xq26 (MLS = +1.77), although neither region reaches genome-wide significance (Lander and Kruglyak, 1995).

Table 25–2 Summary of Genome-wide Linkage Studies Carried Out in Infectious Diseases

Disease Phenotype	Population	Linked Regions	MLS	Reference
Schistosomiasis	Brazilian	5q31-q33	4.74	Marquet et al., 1996
		7q35-q36		Zinn-Justin et al., 2001
	Senegalese	5q31-q33	2.01	Muller-Myhsok et al., 1997
Tuberculosis	Gambian/S. African	15q11.2	2.00	Bellamy et al., 2000
		Xq26	1.77	Bellamy et al., 2000
	Brazilian	11q13	1.84	Miller et al., 2004
		20p12.3	1.78	Miller et al., 2004
		10q25-q26	1.31	Miller et al., 2004
Leprosy	Indian	10p13 (PB)	4.09	Siddiqui et al., 2001
		20p12 (PB)	3.48	Tosh et al., 2002
	Vietnamese	6q25-q26	4.31	Mira et al., 2003
		10p13 (PB)	1.74	Mira et al., 2003
	Brazilian	6p21.32	3.23	Miller et al., 2004
		17q22	2.38	Miller et al., 2004
		20p13	1.51	Miller et al., 2004
Visceral leishmaniasis	Sudanese	22q12	3.90	Bucheton et al., 2003
		2q34-q24	2.29	Bucheton et al., 2003
		2q35	1.00	Bucheton et al., 2003
Roundworm	Nepalese	13q32-q34	4.43	Williams-Blangero et al., 2002
		1p32	3.01	Williams-Blangero et al., 2002
		8q12-q14	2.09	Williams-Blangero et al., 2002
Helicobacter pylori	Senegalese	6q23.3	3.10	Thye et al., 2003
Otitis media	USA	19q13.43	2.61	Daly et al., 2004
		10q26.3	1.68	Daly et al., 2004

Abbreviations: MLS, maximum LOD score reported; PB, paucibacillary form of leprosy

Fine-mapping of the 15q11.2 region, using an additional 13 polymorphic microsatellite, insertion/deletion (in/del), and single nucleotide polymorphisms (SNPs), was subsequently undertaken in the same Gambian families used in the genome screen, plus additional families from Guinea-Conakry (Cervino et al., 2002). Allelic association analysis using the transmission disequilibrium test (TDT) (Spielman et al., 1993) demonstrated that a 7-bp deletion polymorphism in the *UBE3A* gene, encoding an ubiquitin ligase expressed in cells of the macrophage lineage, was significantly associated with TB, identifying this as the putative susceptibility locus in this region.

A second genome screen has also been performed using a cohort of extended multicase pulmonary TB families from the northeast of Brazil (Miller et al., 2004). Nonparametric linkage analysis demonstrated three regions, 10q25-q26 (MLS = +1.31), 11q13 (MLS = +1.84) and 20p12.1 (MLS = +1.78), with suggestive evidence of linkage, although again these do not reach genome-wide significance levels (Lander and Kruglyak, 1995). It is of interest to note that no overlap is observed between the regions detected in the Brazilian genome scan and those previously reported in the African genome scan, providing an indication of the potential genetic heterogeneity that underlies TB, and indeed all infectious disease, susceptibility in distinct populations.

Leprosy

As with TB, a number of genome scans for leprosy susceptibility loci have been undertaken in a number of diverse populations. The first of these was carried out using affected sib pairs (ASPs) from the Tamil Nadu region of India (Siddiqui et al., 2001). Nonparametric analysis provided significant evidence of linkage at one region of the human genome in this population, specifically 10p13 (MLS = + 4.09). Furthermore, estimation of the locus-specific sibling recurrence risk (λ_s), that is, the ratio of risk for siblings of affected patients compared to the risk of the general population, for the peak of linkage at 10p13 indicates that this locus may make an appreciable contribution to the overall genetic component of leprosy susceptibility, indicative of a major susceptibility locus in this region in this particular population.

An additional region on 20p12 showing suggestive evidence of linkage was also followed up using further families from the Tamil Nadu region plus a number of families from Andhra Pradesh region of India (Tosh et al., 2002). Analysis stratified by region and disease type demonstrated significant evidence of linkage at 20p12 (MLS = + 3.48) in the paucibacillary form of leprosy in the Tamil Nadu region and indicated that there was no evidence of interaction with the previously described 10p13 region.

A second independent genome scan undertaken using Vietnamese families affected with leprosy per se identified a further major peak on 6q25-q26 [maximum likelihood binomial (MLB) LOD = + 4.31] and provided confirmation of a gene in the 10p13 region identified in India as contributing to paucibacillary leprosy susceptibility (Mira et al., 2003). Fine mapping of the 6q25-q26 region was subsequently undertaken (as discussed in detail in the next section) and the *PARK/PACRG* genes identified as the causal loci associated with leprosy (Mira et al., 2004).

In contrast, linkage to the 10p13 and 6q25-q26 regions was not observed in a third independent genome scan performed using multicase leprosy per se families from the Northeast of Brazil (Miller et al., 2004). The results of this genome scan did, however, provide some overlap with previously reported results with a region of linkage identified at 20p13 (MLS = + 1.51), approximately 3.5 mega bases (Mb) distal to that reported in the Indian genome scan (Tosh et al., 2002). The Brazilian genome screen also provided evidence for two further peaks of linkage, one at the human leukocyte antigen (HLA) complex on 6p21.32 (MLS = + 3.23) and a second at 17q22 (MLS = + 2.38).

As seen in TB, these leprosy genome scans illustrate the potential presence of genetic heterogeneity and population-specific susceptibility loci, with unique regions of linkage being identified in all three populations investigated to date. However, these studies also revealed a number of linkage peaks in common between a number of the populations, namely 10p13 in India and Vietnam and 20p13 in India and Brazil, raising the possibility that some of these susceptibility loci may be more general and are shared across populations. Interestingly, these two regions have also been investigated in leprosy families from Northern Malawi in a linkage study adopting a candidate region approach to identify regions of the genome linked to leprosy type (Wallace et al., 2004). Results from this study demonstrated no evidence for linkage in the 20p13 region, only nominal evidence of linkage to the paucibacillary leprosy type at 10p13 (MLS = + 1.0), and identified a new peak of linkage at 21q22 (MLS = + 1.4) confirming the idea that numerous loci are likely to be involved in the genetic susceptibility to an infectious disease such as leprosy.

As illustrated in Table 25–2, a number of other genome-wide scans have been performed in other infectious diseases or diseases with an infectious etiology but, to date, these have not been replicated in other populations. Of the eleven genome scans listed in Table 25–2, it is of interest to note that only one region identified as being linked to disease susceptibility has subsequently been fine mapped sufficiently to identify the putative susceptibility locus, that being the *PARK/PACRG* genes in leprosy (Mira et al., 2004). This is predominantly due to the fact that most linkage peaks cover a relatively broad region of the genome, usually approximately 10–20 cM, encompassing numerous genes, all of which could potentially be the disease locus. Furthermore, many of the cohorts used in genome-wide studies to date actually have limited power to fine map using allelic association, necessitating the need for further collections of more appropriate cohorts.

The Candidate Gene Association Approach

The identification of a candidate gene or gene region can be done in a number of ways. The results of genome-wide linkage studies will highlight regions of the genome that are linked to disease susceptibility and therefore contain a disease susceptibility locus; however, as mentioned in the preceding section, these regions tend to be large and may contain many genes that could be the potential disease locus. Alternatively, interesting candidate genes can be identified on the basis of homology with previously identified murine susceptibility genes or on the basis of plausible biological function. All of these methods have been successfully employed in recent years in the study of human susceptibility to infectious disease, and brief examples of these will be discussed in further detail.

Having identified a candidate gene, or gene region, these can be investigated using population-based case–control or family-based allelic association methods in order to try to pinpoint the disease-causing variant. While linkage and association methods rely on the same basic genetic principle to identify disease genes, namely regions of the genome transmitted together more frequently than would be expected given independent segregation, there are subtle differences in how this is achieved. While linkage methods rely on the observed pattern of recombination events that occur in the relatively small number of generations found within an extended family, allelic association methods rely on the recombination-driven breakdown of the physical association that originally exists between two ancestral alleles spanning many generations in an unrelated population.

The choice of whether a population-based case–control study or the alternative family-based association study is more appropriate depends on a number of factors. The case–control method has the advantage of requiring smaller numbers of individuals, generally using one (matched) control per case to achieve equivalent power to detect a specific genetic effect size (odds ratio) in comparison to the trio-based family study. For example, a study consisting of 1500 case–control pairs requires genotyping of 3000 individuals, whereas the equivalent family-based study requires 4500 individuals (i.e., 1500 cases plus their parents) to be typed, significantly increasing the cost. However, the case–control approach is susceptible to the presence of population stratification, the existence of subpopulations within the study group, which will potentially increase the type I error rate and lead to spurious associations. Family-based methods circumvent this issue through the use of internal controls, generated by considering the untransmitted allele as the control for the transmitted allele observed in the case, and are therefore robust to population stratification.

The Candidate Gene Region Approach—*PARK/PACRG*

Having identified a region of the genome linked to disease susceptibility, the causal variant can potentially be identified using allelic association to fine map the region. Such an approach was adopted in the subsequent analysis of the 6q25-q26 region shown to be linked to leprosy per se in the previously discussed genome-wide screen of multicase Vietnamese families (Mira et al., 2003). Fine mapping of this 6q25-q26 region was undertaken using a panel of 64 SNPs located within or close to all 31 genes falling within the 90% confidence intervals (CIs) of the linkage peak in 197 Vietnamese trios (Mira et al., 2004). Results demonstrated significant associations at a number of SNPs clustered within the shared promoter region of the Parkinson's disease gene, *PARK2*, and the coregulated *PACRG* gene. A combination of comparative sequencing and database searches identified a further 81 polymorphic SNPs in the region of the *PARK2/PACRG* genes that were subsequently genotyped in the same trios to confirm the association observed in the 5′ region of the *PARK2/PACRG* genes. Linkage disequilibrium (LD) analysis identified a specific approximately 80 kilobase (kb) block containing 17 leprosy associated markers, of which two were identified as being haplotype tagging SNPs (htSNPs), that is, SNPs shown to capture all the diversity in the haplotype block. Independent verification of this association between the *PARK2/PACRG* genes and leprosy susceptibility was provided by typing 13 of the 17 associated SNPs in an independent case–control cohort from Brazil, the results of which confirmed the significant association previously observed. Furthermore, functional studies confirmed the expression of these genes in biologically relevant tissues in the context of *M. leprae* infection, namely macrophages and Schwann cells.

This example clearly highlights that, not only is it possible to identify the causative loci initially identified through the use of linkage based methods, but that genome-wide linkage studies can also be a very powerful tool for identifying disease genes when no prior assumptions as to the location or functional candidacy of those genes are made. The surprising identification of a leprosy susceptibility locus in the regulatory region of the Parkinson's disease gene, *PARK2*, and the coregulated *PACRG* gene provides a clear example of this with neither gene having previously been associated with an infectious disease of any kind.

The Mouse to Man Approach—*SLC11A1*

The identification of a disease susceptibility gene in the mouse leads to the possibility that the homologous human gene also plays a role in human disease. A prime example of this can be taken from the solute carrier family 11a member 1 (*SLC11A1*) story. The discovery in the mouse of a functional null mutation in *Slc11a1* that results in the loss of resistance to a wide range of intracellular pathogens, including *Leishmania donovani, Salmonella typhimurium* and *Mycobacterium bovis* (Vidal et al., 1995), led to speculation that the human homologue may be involved in human susceptibility to similar infectious diseases. Subsequent studies confirmed that, while the same functional null mutation is not observed, a number of other polymorphic variants (reviewed by Buu et al., 2000), of which some are known to be functional (Searle and Blackwell, 1999), are indeed associated with a variety of infectious diseases, including TB (Bellamy et al., 1998), leprosy (Meisner et al., 2001), visceral leishmaniasis (Bucheton et al., 2003) and HIV (Donninger et al., 2004); further details of these studies plus many more in the area of infectious disease are given in the Candidate Gene Tables available at http://blackwell.cimr.cam.ac.uk. Interestingly, a number of functional variants in the *SLC11A1* gene, including the $GT_{(3)}$ promoter polymorphism that results in increased expression of SLC11A1, are also associated with a number of autoimmune diseases, including rheumatoid arthritis, type 1 diabetes and sarcoidosis, and Crohn's disease (CD) (Blackwell et al., 2001) reflecting the multiple pleiotropic effects of SLC11A1 on macrophage activation.

The Biological Candidate Approach—*IL6*

Knowledge of the mechanisms underlying an infectious disease can also point to plausible biological candidates that may be associated with disease susceptibility and a prime example of such an approach is provided by a study investigating the functional –174G/C SNP located in the *IL6* promoter and its potential involvement in susceptibility to ML (Castellucci et al., 2006). ML develops in a fraction of CL patients after infection with *L. braziliensis.* Metastatic spread from the original CL lesion to the naso-pharyngeal mucus membranes results in the progressive inflammation and destruction of the nasal mucosa. While the immune response to CL is characterized by a T_H1 polarized immune response with interferon-γ (IFN-γ) and tumor necrosis factor-α (TNF-α) required for control of the parasite, in ML an exacerbated T_H1-type immune response is observed with poor modulation by regulatory cytokines such as transforming growth factor-β (TGF-β) and IL-10 resulting in immune-mediated pathology (Bacellar et al., 2002). In addition, the cytokine IL-6, secreted by numerous cell types including macrophages, dendritic cells, T cells and B cells, has been shown to play a dual role in the balance between the T_H1 and T_H2 type immune response via independent mechanisms (Diehl and Rincon, 2002), thus making it an interesting and plausible biological candidate in the context of CL and ML disease.

In view of this, the functional –174G/C SNP located in the *IL6* promoter has been investigated using case–control and family-based cohorts identified from an area of Brazil endemic for *L. braziliensis* (Castellucci et al., 2006). Initial population-based analysis performed in the case–control cohort indicated an association between ML disease specifically and the variant C allele of the –174G/C polymorphism. Family-based analysis subsequently confirmed this association and functional analysis carried out in vitro demonstrated that macrophages, being the primary site of infection, from CC individuals released significantly lower levels of IL-6 after stimulation with soluble leishmania antigen (SLA) compared to macrophages of GG individuals. Thus, it would seem that low levels of IL-6 production in individuals who carry the variant C allele at the functional –174G/C polymorphism in the *IL6* promoter results in an increased risk of progressing to ML disease. This may potentially be due to the reduced capacity of IL-6 to regulate the balance between the T_H1 and T_H2 type immune response leading to the characteristic exacerbated T_H1 response observed in ML patients.

Power Considerations in Association Studies

As mentioned previously, infectious disease is a complex genetic trait and, as such, numerous genetic and environmental factors will influence an individual's susceptibility to infection. In addition, the genetic factors involved in that susceptibility to infection, or to any complex disease, are expected to have relatively small effect sizes (i.e., an odds ratio <1.5). This raises implications in terms of the power required to detect such a small genetic effect with credible statistical support.

While there are currently numerous association studies in the field of susceptibility to infectious disease reported in the literature, (for a comprehensive review of those in the areas of TB, leprosy, leishmania, malaria and HIV/AIDS please refer to the Candidate Gene Tables available at http://blackwell.cimr.cam.ac.uk), it is important to note that many of these studies are generally underpowered with only small numbers of cases and controls or families used (usually far less than 500 cases), mainly due to limitations of budget and problems of case recruitment. As illustrated in Figure 25–3, the percentage power available in a cohort of 50 or 100 cases and controls, a sample size commonly observed in the literature, to detect an odds ratio of 1.5 at a significance level of $p = 0.001$ is very limited, even when the frequency of the risk allele is relatively high. Indeed, even a case–control cohort of 500 cases and controls has limited power to detect the same genetic effect if the risk allele is at a low frequency in the population (<60% power with risk allele frequency <15%). In reality, a cohort of at least 1500 cases and 1500 controls is required in order to detect an effect size of 1.5 at a significance level of $p = 0.001$. However it is likely that the majority of genetic effects influencing susceptibility to such complex genetic traits will be somewhat smaller than an odds ratio of 1.5, meaning that the numbers of cases and controls recruited will have to be increased accordingly in order to avoid spurious association (Lohmueller et al., 2003; Wang et al., 2005).

It is also of importance to note that, while a number of candidate genes have been investigated in independent studies in numerous populations both within and between infectious diseases, replication of the original report of association is often not observed. This may be due to real differences in susceptibility to infection between the populations under study; however, it may also be an artifact of the spurious associations that undoubtedly appear in the literature as a result of underpowered studies. Attempts to address this problem are being made with the formation of organizations such as HuGENet (www.cdc.gov/genomics/hugenet/). This is an international collaboration of investigators applying genetic epidemiological methods, including the use of meta-analyses, to the large volumes of data available in the published literature, with the aim of generating reliable evidence of genetic risk in complex disease (Ioannidis et al., 2006).

Genome-wide Association Studies

The completion of the human genome project (The International Human Genome Sequencing Consortium, 2004) provided extensive information on the large number of SNPs distributed throughout the human genome. As of September 2005, the dbSNP database (available at www.ncbi.nlm.nih.gov/projects/SNP/) contained 10.4 million reference SNPs, of which nearly 5 million are validated and nearly 3 million have associated frequency data in at least one human population. In addition, following on from the completion of the human genome project, it has become apparent that far more variation exists in the genome than originally believed with large numbers of polymorphic copy-number variants (CNVs), including insertions, deletions and duplications, and other structural variants being reported (Feuk et al., 2006), of which some have been shown to be involved in human susceptibility to infectious disease. For example, the C-C chemokine, *CCL3L1*, gene is present in the human genome in variable copy number (Modi, 2004) and, in individuals with a lower than average copy number, is associated with an increased risk of HIV-1 infection (Gonzalez et al., 2005).

The availability of information on the large numbers of SNPs in the human genome has also led to the formation of large international collaborations such as the International HapMap Project (Altshuler et al., 2005), which ultimately aims to genotype approximately 1 SNP every 600 bp in DNA samples from 30 Nigerian trios, 45 unrelated individuals from Japan, 45 unrelated individuals

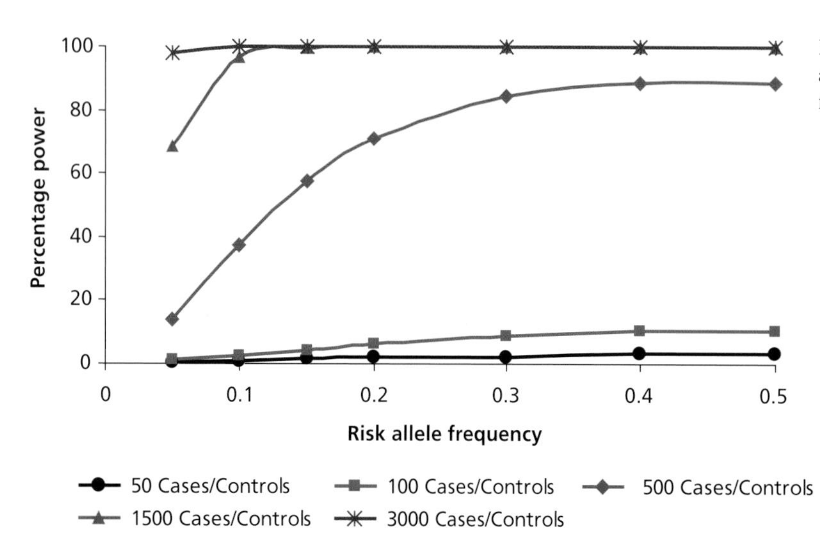

Figure 25–3 Graph to illustrate the percentage power available to detect an odds ratio of 1.5 at various risk allele frequencies at a significance level of $p = 0.001$.

from China and 30 US trios with northern and western European ancestry. It is hoped that this large-scale genotyping project will provide information on the patterns of LD that exist in different human populations and, in so doing, will provide a resource of useful htSNPs distributed throughout the genome that will assist researchers in the future identification of genes involved in health and disease (www.hapmap.org/index.html). Assessment of the transferability of such htSNPs identified by the HapMap project indicates that, while they have been determined in only four populations, they do demonstrate good transferability to other populations (Bonnen et al., 2006; Montpetit et al., 2006; Stankovich et al., 2006). This makes the data generated by the HapMap project an extremely useful tool for the design of future association studies allowing common variation in the human genome to be covered with a minimal set of SNPs.

With the increasing availability of polymorphic SNPs (i.e., those having a minor allele frequency of more than 5%) combined with improved association study design (Skol et al., 2006), advances in highly multiplexed genotyping technologies (Syvanen, 2005), analytical considerations (Clayton et al., 2005), and rapidly decreasing costs per genotype mean, it is now feasible to perform genome-wide allelic association studies using dense SNP array platforms. While no genome-wide association studies have been reported to date for an infectious disease, a number of such studies have been successfully undertaken for other complex genetic traits, including age-related macular degeneration (Klein et al., 2005), CD (Yamazaki et al., 2005), alcoholism (Namkung et al., 2005), and schizophrenia (Mah et al., 2006).

Such genome-wide association studies in the field of infectious disease are, however, currently underway. The Wellcome Trust Case–Control Consortium (WTCCC) aims to genotype approximately 675,000 SNPs using the commercially available Affymetrix 500K SNP chip alongside a custom array generated by Perlegen Sciences specifically designed to harness information from Phase II data of the International HapMap Project, thus maximizing the diversity captured across the genome by this project (www.wtccc.org.uk/info/overview.shtml). Among the complex diseases to be investigated by the WTCCC are two infectious disease cohorts consisting of 2000 African malaria cases, 2000 African TB cases, and 2000 shared African controls, the results of which will hopefully provide convincing statistical evidence for association with disease susceptibility at a number of loci across the genome.

Shaping of the Human Genome by Infectious Disease

J.B.S. Haldane originally made the suggestion that infectious disease has played a role in the shaping of the human genome, with the selection pressure exerted by pathogen exposure during human history resulting in the increased frequency of the certain immune-related genes. The observation that approximately 7% of the mammalian genome is devoted to immune-related genes and that many of these are highly polymorphic and frequently clustered in the genome, the major histocompatibility complex (MHC) located on chromosome 6 being the archetypal example of both features, lends weight to the theory that infectious disease represents a major selective force in shaping the human genome (Trowsdale and Parham, 2004).

The generation of a database containing all currently known human immune-related genes, the Immunogenetic Related Information Source (IRIS; www.immunegene.org/) has allowed a formal analysis of how immune genes are clustered in the human genome (Kelley et al., 2005). Results of this analysis reveal that a number of chromosomes contain higher than expected densities of immune genes, namely chromosomes 6, 11, 17 and 19, while others contain fewer than expected, specifically chromosomes 13, 14, 18, and X. Such clustering can be explained by the repeated duplication of genes, such as that observed in the *KIR* cluster on chromosome 19 (Martin et al., 2000). However, it has also been shown that genes of related function but of unrelated sequence, such as seen in the MHC complex (Trowsdale and Parham, 2004), are clustered in the genome (Lee and Sonnhammer, 2003), suggesting an advantage in this proximity, possibly in coordinate expression.

In addition, the selection pressure exerted by pathogens on individual genes leaves a number of potentially detectable signals of positive evolution that are not observed for genes undergoing neutral evolution. Possibly the most extensively studied of these selection signatures are those seen at a number of genes responsible for certain erythrocyte defects, some of which are responsible for the most common Mendelian diseases in man but that concomitantly confer resistance to malaria (Kwiatkowski, 2005). In particular, the *HbS* allele of the β-globin locus, which, in the homozygous form, results in sickle cell anemia and is a cause of significant morbidity and mortality. Nonetheless, this allele has been maintained at a high frequency (~10%) in certain populations as a result of the approximately 10-fold reduced risk of severe malaria it confers to heterozygote carriers, a so-called heterozygote advantage or balanced polymorphism (Clegg and Weatherall, 1999; Allison, 2004). In addition, as mentioned earlier in this chapter, the Duffy-negative allele (*FY* 0*) is found at high frequency in certain African populations with homozygosity for this allele conferring total resistance to *P. vivax* infection. Comparison of DNA sequence variation at this locus with that of a 1 kb region located 5–6 kb away demonstrated a significantly reduced level of variation thus indicating a departure from neutral selection at the FY*0 allele in African populations (Hamblin and Di Rienzo, 2000).

In the last decade, attention has also focused on the C-C chemokine receptor 5 (*CCR5*) gene and, specifically, a 32-bp deletion in this gene (*CCR5-Δ32*) that confers almost complete protection against HIV-1 infection in homozygous carriers and protection against progression to AIDS in heterozygous carriers (Blanpain et al., 2002; Dean et al., 2002). The observation that this apparently recent mutation (estimated age ~700 years) occurs at high frequency (5–14%) and in a region of extended LD in the European population, with an apparent north–south and east–west cline, but is absent in African and Asian populations led to speculation that this gene has been subject to strong selective pressure, possibly from an epidemic such as the bubonic plague (Stephens et al., 1998). However, taking advantage of the denser maps of polymorphic markers and increased information on the extent of LD in the human genome that are now available, re-evaluation of this locus has been undertaken (Sabeti et al., 2005). The results of this re-analysis suggest that this mutation is actually much older than originally believed (more than 5000 years) and that the pattern of variation observed at this locus is actually more consistent with neutral evolution as opposed to positive selection, illustrating the value of genome-wide data in deciphering the mechanisms that shaped our genome and, in particular, the genomic signatures left by host–pathogen coevolution.

There is no doubt that recent and ongoing advances in genomic technology will have a dramatic effect on the ability of researchers to pinpoint genes involved in complex human disease, such as susceptibility to infection, with increasing confidence. The availability of the complete human genome sequence has provided a wealth of information regarding the types and numbers of informative genetic markers that appear throughout our genome. This information, coupled with the ongoing large-scale collaborative investigations of population-specific variation, is providing denser and more informative maps of markers that will allow more efficient association studies to be undertaken. Furthermore, this invaluable resource is allowing novel investigations of our own genome, providing insight into how it has been shaped by a history of host–pathogen coevolution and illustrating how our immune system has interacted with and adapted to a broad variety of viral, microbial, and parasitic pathogens.

However, this advanced genomic technology is not just being applied to the investigation of our own genome; it is also being employed to scrutinize the genomes and transcriptomes of the pathogens themselves. Such investigations are allowing an advanced understanding of how pathogens interact with our immune system and, indeed, how pathogens adapt and evolve as a result of such an interaction, allowing the development of novel diagnostic technologies, therapeutic interventions and preventative strategies.

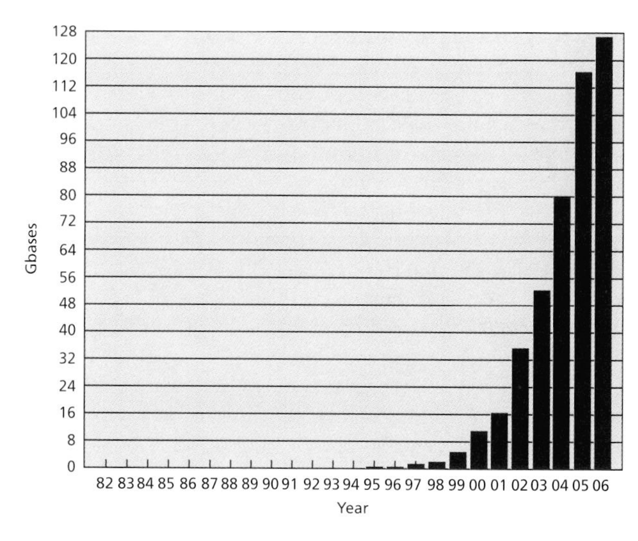

Figure 25–4 The amount of sequence data submitted to EMBL, the European sequence data repository. The rate of data generation is increasing at an incredible rate. This represents 69 million entries with sequence from approximately 2,00,000 species across all life forms. Further statistics on the EMBL nucleotide database can be found at www3.ebi.ac.uk/Services/DBStats/.

Microbial Genomics

The dramatic impact of the first antimicrobial drugs, derived predominantly from natural sources, led to a wave of optimism regarding treatment of infectious diseases. Whether viral, bacterial or parasitic diseases, most were on the decrease, thanks to the combined onslaught of antimicrobial therapy, vaccination, and environmental control measures. The effective elimination of childhood diseases like polio and measles in the industrialized nations and the total global eradication of smallpox are measures of this success. The subsequent decline in development of measures for antimicrobial control has been reversed in recent years in response to outbreaks and increases in new and re-emerging infectious diseases (Cohen, 2000).

Arguably, one of the most significant and innovative areas of pathogen research in the last 10 years has been the advent of DNA/RNA sequence technology and its high-throughput application. Unsurprisingly, the first complete genome, published in 1977, was that of a virus (Sanger et al., 1977), sequenced using the revolutionary dideoxy method of sequencing based on chain terminating inhibitors. This phiX174 virus was only 5375 bp in size and took 4 years to complete. Since this first breakthrough in sequencing technology, the availability of genome sequence has increased dramatically (see Fig. 25–4). Advances in technology during the 1980s and 1990s with the introduction of fluorescent dye technology (Smith et al., 1986), random shotgun sequencing (Sanger et al., 1980), polymerase chain reaction (PCR) (Saiki et al., 1985), improvements in DNA cloning (Burke et al., 1987), and the production of dedicated sequencing machines together with development of computer hardware has revolutionized sequencing. Now, almost 30 years later, it is possible to sequence a viral genome of this size in an afternoon, and the next generation of sequencing machines are expected to achieve outputs of 20 million bases in 4 hours!

In 1995, the first bacterial genome, that of *Haemophilus influenzae* consisting of 1,830,137 bp, was sequenced using whole-genome shotgun sequence methodology (Fleischmann et al., 1995). A mere 10 years later, European Molecular Biology Laboratory (EMBL)reached the milestone of having 100 gigabases (Gb) (100 billion bases) of sequence submitted, representing 200,000 organisms (Kanz et al., 2005). In the last 2 years, the amount of data has risen exponentially, such that, despite being in existence for 27 years, the EMBL database has doubled in size in the last 2 years alone (Fig. 25–4). This massive generation of data is driven both by technological advances and increased numbers of researchers generating sequence. The improvement in technology, in particular the cost and speed at which sequence can be generated, means genome sequencing is no longer restricted to large labs with sizable budgets; it can now routinely be employed for practical medical applications such as diagnostics, epidemiological surveillance, and identification of drug and vaccine targets.

There are approximately 2500 complete genomes publicly available (www.ebi.ac.uk/genomes/), ranging from organelles, viruses and bacteria up to various mammalian species including the human genome. Given the relatively small genome size of most microbial organisms, it is not surprising that they make up a majority of these complete datasets. Medical relevance has dictated that many of the fully sequenced smaller genomes are human pathogens and, as technology and reliability have improved, the use of genome analysis has extended from defining the basic biological processes in these organisms to more targeted questions relating to drug efficacy and resistance, pathogenicity, virulence and immune avoidance mechanisms. The fact that infectious disease is a major cause of morbidity and mortality (see Fig. 25–2), not just in the developing world but also increasingly in developed nations, has given this type of approach increased impetuous. The dramatic increase in global travel allows unprecedented opportunity for novel and virulent human pathogens to race across the globe far quicker than preventative measures can be put in place to prevent their spread. The prolonged use of frontline antimicrobials combined with their inappropriate application has led to drug resistance becoming increasingly common and, in the worst cases, has resulted in the emergence of multidrug resistant pathogens that are effectively resistant to almost all currently licensed drugs. In addition, cost efficiencies in hospitals, high turnover of patients and lack of aseptic technique has led to a rise in transmission of resistant strains within high-risk groups and subsequent higher mortality rates. This situation is further compounded by the emergence of HIV/AIDS, where the standard immunosuppressive treatments leave patients more susceptible to normally avirulent microbes.

The availability of genome sequence has had a remarkable impact on research into microbial pathogens. The accumulation of sequence data coupled with impressive advances made in bioinformatics and computer orientated analysis has culminated in the ability to sequence and compare the complete genetic complement of numerous species and strains to establish which genes determine disease phenotype. Differences between organisms can vary between whole chromosomal re-arrangements, sequence deletions or acquisitions down to single base changes and the ability to identify such genetic differences between virulent, drug resistant and attenuated strains of microbes has subsequently driven an increased use of comparative genomics. Comparative sequencing of microbial pathogens has revealed that many have a highly conserved core genome that codes for genes essential to existence, while the variable or accessory portion of the genome is more usually associated with genes involved in host interaction, immune avoidance, virulence factors and drug resistance.

Genome Variation

With the numerous comparative genome analyses of bacteria, it has become apparent that their genomes are very dynamic, with constant change in response to environmental pressure. Bacteria and viruses, and to a lesser extent eukaryotic and fungal microbes, have a great capacity for change. Equally significant is the fact that this constant change is coupled with a high rate of asexual replication such that beneficial variants can quickly become established within a clonal population. Microbes alter their genetic complement through a whole raft of mechanisms and comparative genomics of related species allows the manner in which genome variation occurs to be identified. Perhaps the simplest form of change is the SNP and, while those occurring in noncoding regions or those that do not result in an amino acid change (synonymous substitution) will often have no visible effect, those that do cause an amino acid change in a coding region, especially in an active site, can result in changes to protein function. A single base change can have a major effect on gene function, even to the point that important virulence or immune evasion genes can become totally ineffective. Such minor changes can even introduce stop codons or frameshifts such that the coding region becomes a pseudogene with the subsequent total loss of the functional protein.

Another relatively subtle change occurs in repetitive regions, both within coding regions and in intergenic sequences. Hypervariable short sequences can vary in size during replication or undergo rearrangement such that the genes encoded are activated or inactivated. Variable repeat sequences are often found in surface expressed genes that interact with the host, and are commonly used as a mechanism to alter cell surface antigens, thus avoiding immune recognition by the host. Phase variation in the duplication of coding regions can also lead to functional variation. In most organisms, duplication can lead to speciation, whereby one copy evolves separately, leading to expansion of functionality. In pathogens belonging to the protozoal group Kinetoplastida, where almost all protein expression is controlled, not at the level of transcription, but by posttranscriptional modification, gene duplication is commonplace as a means of enhancing protein expression. So much so that the recently published genome of the Kinetoplastid pathogen *Trypanosoma cruzi* demonstrated that as much as 50% of the genome is composed of duplicated DNA (El-Sayed et al., 2005).

Furthermore, it is obvious from comparisons of some related species that much larger changes can also occur. For example, two strains of *Escherichia coli* have been shown to differ in genome size by almost 30%, with the larger *E. coli* strain *0157:H7* containing more than 1300 strain specific genes (Perna et al., 2001). Even the homologous chromosomes in the protozoan pathogen *Trypanosoma brucei* differ by as much as 20%, with the subtelomeric regions of the two copies often appearing totally different in sequence (Melville et al., 1999). In addition, genomes can undergo gene loss leading to a reduction in genome size; an extreme example of this is *M. leprae*, which has effectively lost half of its genes during evolution from its ancestral mycobacterial species (Cole et al., 2001). While gene acquisition or gene duplication might be seen as the most beneficial scenario, examination of some virulent pathogens has shown that gene loss can also enhance virulence and survival. *Bordettella pertussis* is thought to have gained virulence by the loss of genes coding for some immunogenic cell surface antigens, thereby enhancing its ability to evade the hosts immune surveillance system (Parkhill et al., 2003).

Comparison of virulent and nonvirulent bacteria have shown that the acquisition of genes is the most rapid mechanism by which bacteria change their genetic makeup in response to adverse environmental stresses. While most microbes have a stable conserved core component of their genome, encoding proteins essential for survival, other areas of the genome, containing genes for non-essential functions, are often prone to large-scale variation, including that occurring through acquisition of new genetic material. Bacteria in particular, have a great capacity for exchanging genetic material by horizontal or lateral transfer and the variable portion of their genome usually contains structural signatures allowing for the insertion and excision of mobile genetic elements. Such acquisition of new genetic material can occur in three different ways, through natural transformation (the uptake of free DNA), through transduction (the bacteriophage mediated transfer of DNA) and through conjugation (transfer of mobile genetic elements between two bacteria temporarily joined by a physical bridge or pili) (Thomas and Nielsen, 2005). It is this inherent ability of pathogenic bacteria to rapidly acquire new genes that make them so adaptable against therapeutic treatment, as in many microbes the genes for virulence and drug resistance are contained within these highly variable regions.

Microbial Genome Sequencing Projects

The practical benefits of microbial genome projects are extensive across various disciplines in microbial research and can broadly be divided into three distinct areas; providing insight into novel aspects of their biology, advancing discovery and development of new therapeutic treatments and vaccines, and developing better means of diagnosis and epidemiological surveillance.

New Insights into Microbial Biology

Prior to the advent of high throughput sequencing machines, sequence data were still generated and submitted to the public databases. However, although useful, especially when combined with functional information, the interpretation and subsequent implementation of such data was limited in practical terms. With the advent of whole genome sequencing, the benefits are obvious; the full complement of genes are available, such that complete metabolic and physiological pathways can be mapped in full with no missing components. The usefulness of this increases with time as the functional characteristics of more and more genes are added to the public databases.

As well as increasing the volume and quality of the sequence data, two other advances have dramatically improved genome annotation. The initial emergence of numerous fractured databases that were incompatible with each other has made way for incredibly large public data repositories that not only standardize data storage formats but also routinely exchange data, such that any sequence submitted to one is quickly visible in the others. There are three such public data repositories, EMBL in Europe (www.ebi.ac.uk/), GenBank in the United States (www.ncbi.nlm.nih.gov) and the DNA Databank of Japan (DDBJ; http://www.ddbj.nig.ac.jp/). Advances in computing power and computer programming have provided researchers with the analysis programs, search facilities and web-based access to specialized databases required to characterize genomes, from making simple comparisons, such as comparing two orthologous genes, through to defining complete pathways and genomes. This information, when fed back to the public database repositories, adds value to future genome annotation.

This process started as soon as the first two genomes were sequenced in 1995. *H. influenzae* had 1703 protein coding genes and *Mycoplasma genitalium* only 468. Despite this difference in size, genome content and physiology, the two genomes were compared to try and find the minimum gene content necessary for life (Mushegian and Koonin, 1996). Despite some gaps in the metabolic pathways and with almost half of the predicted genes having no known function, a gene set of 256 was identified as being required for survival as a free-living cell. However, this type of comparative analysis is speculative, as even the absence of so-called, "non-essential" genes will result in a steady decline of the viability of microbial cultures. What it does do, however, is steadily increase our understanding of the physiology of all living organisms, with the more data added leading to increased benefits in comparative sequence analysis.

As more information is added to the public domain, more accurate and specialized databases are generated to aid in the characterization of new genomes. The current Molecular Biology Database Collection has 858 databases listed (Galperin, 2006). Some databases provide searches based on specific criteria, including specific pathways, protein domains, families or motifs, and metabolic pathways. There are also now an increasing number of databases that cater specifically to certain organisms, the main advantage being that these databases curate additional information and often combine post-genomic and genome data to make a centralized repository that can be extensively queried (reviewed by Chaudhuri and Pallen, 2006).

Knowledge of metabolic pathways, signaling and cell communication networks, cytoskeletal and structural proteins, immune evasion, molecular transport, transcriptional regulation, and cell motility have all been acquired in this way, and, with regards to pathogens, can be applied to the fight against infectious disease. The complete genome sequence also underpins the development of whole-genome and high-throughput application of post genomic analyses, such as global analysis of gene transcription, whole genome protein expression analysis, large-scale mutagenesis studies to identify gene function, screens and assays for drug design, protein structural predictions, metabolic pathway reconstruction and additional comparative genomics.

Development of New Drugs and Vaccines

The science of genomics has advanced sufficiently to be applied to specifically answer more practical issues that can be used to fight infectious diseases. Almost half of the sequenced microbes are pathogens and most of those sequenced early on represented the main causes of human infectious diseases. Every major human microbial pathogen has a representative genome in the database, with many having several additional strains available. Initially, the main driving force behind these projects was to learn more about the pathogens and to catalogue their complete gene repertoire, however, many of the more recent additions to the genome databases have been sequenced specifically to help identify microbial genes that could be targeted for therapeutic treatment or as vaccine candidates.

The urgency of this application has increased dramatically as antibiotics, originally credited with the massive reduction in morbidity and mortality due to infectious diseases, are now proving to have a finite lifespan with more microbes developing drug resistance. Traditional drug screening, used for almost all of the currently available drugs, relies on the screening of natural product compound libraries to find molecules that inhibit pathogen cell growth either in vitro or in vivo. Subsequently, nearly all new antimicrobial drugs are structural and biochemically related to those already in use (Rogers, 2004). This leads to an increased likelihood of pathogens quickly developing drug resistance. Subsequently, identifying novel pathways and enzymes that are critical for survival and/or persistence in humans is crucial for providing drug development programs with a short list of alternative drug targets.

Prior to comparative genomics, identification of essential genes relied on the use of random gene finding techniques such as transposon mutagenesis and conditional lethal mutations. With the availability of whole genome sequence and by applying the usual criteria as to what makes a viable drug target, the complete genome can be scanned for potentially interesting drug targets. Due to the vast amount of information this approach can generate, the list of potential candidates can be prioritized so that only the most promising go through to further development. For such proteins to make effective drug targets they must be essential to either the survival or virulence of the pathogen in the life stages present in the human host, and ideally, to avoid cytotoxicity issues, should be involved in non-redundant pathways and have no significant similarity to any human genes.

Genome data also serves as a starting point that leads to a whole range of post-genomic experimentation, such as comparisons of the differential transcription of genes or expression of proteins, much of which can be carried out on whole genome datasets. For example, the transcriptome of an organism can be analyzed by using DNA/RNA microarrays where gene specific oligonucleotide or PCR products are immobilized on glass or silicone slides and serve as hybridiation templates for messenger RNA (mRNA) molecules. Micorarrays can be used to compare gene expression patterns between species, strains or life cycle stages or used in a more targeted manner to compare different strains exhibiting diverse drug resistance or virulence profiles. By comparing normal cell cultures with those exposed to drugs, toxins or growth inhibitors, the genes responsible for drug resistance can be identified.

Since not all metabolic processes are under transcriptional control, the complete set of proteins, or the proteome, can be similarly studied. Differential post-translational modifications to proteins such as phosphorylation, methylation, acylation and prenylation also determine protein regulation. By combining two-dimensional (2D) gel electrophoresis and mass spectrometry, many of the expressed proteins can be separated and identified from genome databases. One drawback of this methodology is that many proteins, particularly membrane bound moieties, cannot be separated by gel electrophoresis. However, for those that can be separated, it does show a more accurate representation of protein activity.

The same comparisons between different organisms described for microarrays can also be applied to proteomic analyses and is of particular use in those organisms for which post-transcriptional control is the main means of regulation. For example, the recently published genome sequences from the trio of human pathogens that cause leishmaniasis, sleeping sickness and Chagas disease, all of which belong to the Kinetoplastid order, demonstrated that these organisms have a very unusual means of gene transcription (Ivens et al., 2005). The protein-coding genes of these species are organized in long, polycistronic clusters, each containing tens to hundreds of seemingly cotranscribed genes of unrelated function, indicating that post-transcriptional control mechanisms are essential in these organisms.

Functional characterization of genes identified as "interesting" using techniques such as whole genome microarray, proteomic screening or gel electrophoresis can then be validated using molecular genetic tools that target specific genes. Thus, the more labor intensive methodologies of transposon mutagenesis, targeted gene knockout or RNA interference (RNAi) "knockdown" can be systematically applied to the gene list that has been whittled down from the complete genome using in silico predictions.

Diagnosis and Epidemiological Surveillance

The specificity and sensitivity of using amplified DNA/RNA sequence for diagnostic applications has been utilised extensively in the field of microbial diseases and the availability of genome data has revolutionized DNA amplification methods as most microbial resistance and virulence variants are now well characterized. Currently, there are numerous ways of identifying a specific locus following PCR amplification. Those methods that are rapid, portable, relatively affordable, practical for use with clinical specimens and adaptable for use in a high-throughput setting have gained favor, many having been commercialized, leading to methods that are specific, sensitive, reproducible and affordable.

Monitoring the incidence of microbial infections is critical for managing outbreaks, the emergence of drug resistant strains or the occurrence of new infectious agents. Methods used for strain surveillance must undergo rigorous quality control and be capable of discriminating between numerous virulent and resistant strains. One recent example of the use of large-scale sequencing projects for immune surveillance is the development of a diagnostic test to monitor variation in the capsule polysaccharides of *Streptococcus pneumoniae*. Like many bacteria, the *S. pneumoniae* capsule has a polysaccharide covering, variation in which helps avoid immune recognition. While a vaccine has been developed for the commonest of the 90 known variants, sequencing of the capsular biosynthetic genes that determine all 90 serotypes has been undertaken with the aim of generating a diagnostic test to monitor the response of the vaccination program (Bentley et al., 2006).

Application to Viral Disease

The phiX174 virus was the first genome to be completed almost 30 years ago (Sanger et al., 1977). There are now more than 1000 completed viral genomes published and publicly available (www.ebi.ac.uk/genomes/virus/) and many more that are not yet submitted to the public databases. The relatively small size of the virus genome makes them very good candidates for whole genome comparisons since most comparative software and computers can easily handle this amount of data. Nevertheless, compared to the number of available antibiotics, antiviral drugs are quite rare and to date there are fewer than 50 antiviral drugs, most of which have been developed in response to the high profile HIV pandemic. There have, however, been two recent outbreaks of human infective viral disease in which the benefits of genome sequencing can clearly be seen.

Severe Acute Respiratory Syndrome

In 2002, an outbreak of atypical pneumonia in Guangdong province of China showed just how efficiently a new emerging pathogen could spread. The disease, termed severe acute respiratory syndrome (SARS) was caused by a coronavirus (SARS-CoV) that was only poorly related to other known members of this group and was estimated to have mortality rates as high as 55% in susceptible patients. Despite massive efforts to contain the outbreak, infections quickly spread to 29 other countries, leading to cases as far away as Canada. By the time the epidemic was under control more than 8000 patients were infected with approximately a 10% mortality rate (Poon et al., 2004). Identification of the causative agent was done using a DNA-based microarray chip, developed specifically for the purpose of rapidly characterizing viruses, including novel members of existing families. The microarray chip contained numerous highly conserved DNA sequences from every fully sequenced reference viral genome in the public databases (Wang et al., 2003), allowing sequences derived from viral culture to be identified as belonging to the Coronaviridae group. Within weeks of isolating the causative agent, the viral genome had been completely sequenced and made publicly available (Marra et al., 2003), a scenario that resulted in the novel situation where the availability of the genome sequence preceded any knowledge of the biology of the pathogen. It was now possible for the first time to see that the virus was dissimilar in sequence across its whole length to other recognized Coronaviridae viruses and was not derived by recombination between any known human or animal viruses. The sequencing of the genome paved the way for the development of diagnostic tests that would help rapidly identify the virus directly from clinical samples, a critical element in trying to contain an epidemic.

The earliest specific real-time PCR (RT-PCR) tests were based on the coronavirus replicase gene, the first part of the genome to be sequenced (Drosten et al., 2003) but these first tests were found to be relatively insensitive in the first few days of the infection, indicating low viral rates. Quantitative RT-PCR, performed at various time points, indicated that, unlike other respiratory viral infections, viral loads in the respiratory tract increased up to day 10 after infection, the point at which it is most contagious (Wu et al., 2004). RNA detection methods also showed that the virus could be detected in the feces and urine, revealing that the infection was systemic and was not restricted to aerosol transmission (Chan et al., 2003). The rapid development of more sensitive detection techniques using quantitative nested RT-PCR combined with automated specimen extraction not only allowed a rapid high-throughput system to be implemented, but also improved detection in the early stages of the disease and indicated that presence of high viral load in the nasopharyngeal tract was correlated with poor prognosis, providing important information for patient management (Jiang et al., 2004). Greater amplification of viral RNA, targeting of several viral sequences in the RT-PCR and testing of several specimens from different sites have improved the successful diagnosis in the first few days after infection (Poon et al., 2004); serodiagnosis, on the other hand, is generally only useful 2–3 weeks after infection.

Examination of the SARS-CoV genome predicted 14 genes, 8 of which had no detectable homology to other genes in the public databases. Genome analysis identified three genes that appeared to be both essential for maintaining the viral life cycle and sufficiently different to any present in the human host, identifying these as potential drug targets. Of these three genes, one encodes a helicase protein responsible for RNA synthesis and cap formation and high-throughput screening of a compound library identified two inhibitors of this enzyme confirming it as a potential drug target using in vitro assays (Kao et al., 2004). The second encodes an RNA-dependent RNA polymerase (RdRp), which has been shown to be unaffected by non-nucleoside analogue inhibitors, possibly due to the absence of a hydrophobic pocket near the polymerase active site, and resistant to the nucleoside analogue inhibitor ribavirin, probably due to alteration of a normally conserved motif required for control of rNTP binding (Xu et al., 2003). Development of a novel inhibitor is, however, underway including aurintricarboxylic acid (ATA), which has been shown to inhibit viral replication and appears to bind to conserved residues on the RdRp protein (Yap et al., 2005). The third enzyme is the main protease, subsequently identified as a chymotrypsin-like cysteine proteinase (CCP), required for processing polypeptides for viral replication and transcription. Using a novel virtual screening program against a computer predicted model of the structure of the SARS cysteine proteinase, a CCP inhibitor was identified as a potential therapeutic treatment (Dooley et al., 2006).

The small number of predicted genes in the SARS virus means that there are a limited number of epitopes to test as vaccine candidates. The likelihood that a protective vaccine can be generated is high, since re-infection with the virus only causes mild illness and most deaths have been recorded in older people with diminished humoral and cellular immune responses. Peptide and DNA vaccines against the spike protein of various SARS-CoV strains, required for virus binding, fusion and entry in host cells thus making it a good vaccine candidate, have been tested in mice, hamsters and primates (reviewed by Zhang et al., 2005). It has been shown that these vaccines are capable of generating neutralizing and protective antibodies, such that the animals were protected from further challenge. The nuclear capsid protein, which binds to the viral RNA genome to form a helical core, is also likely to be a good vaccine candidate. Accordingly, an experimental DNA vaccine has been shown to induce specific nuclear capsid protein antibodies and cytotoxic T-cell activity (Zhu et al., 2004).

Avian Bird Flu

A more recent outbreak, and a cause of considerable concern, is the spread of an avian influenza strain that demonstrates virulence in humans and, since the first case in South East Asia in 2003, has resulted in 173 human cases, leading to 93 deaths (www.who.int/csr/disease/avian_influenza/country/). Previous influenza pandemics, including the devastating Spanish influenza of 1918 that caused 40 million deaths, appear to have had avian influenza genetic elements on a human influenza strain background (Kawaoka et al., 1989), with the expression of avian influenza strain derived surface antigens being responsible for the ability of the virus to avoid immune recognition in the human host (Reid et al., 2004).

Complete genome sequencing of the H5N1 strain, the causative agent of the current bird flu epidemic, has shown that it consists of eight RNA segments that encode 11 proteins, all of which are of avian origin. Continued genomic surveillance of isolates from both bird and human cases is important to try and identify genetic variation that can account for the virulence of this H5N1 strain in humans. Such analyses need to answer two questions: firstly, what genetic variants determine virulence and secondly, bearing in mind that this virus is not a chimera of a human and avian virus, what mutations in which genes have allowed the virus to cross the species barrier and to proliferate in humans.

Analysis of avian influenza strains adapted to a mammalian host, in this instance the mouse, identified mutations in the viral polymerase gene, neuraminidase gene and hemagglutinin gene that contributed to virulence in the mouse adapted strain and therefore appear associated with the adaptation of avian influenza strains to a mammalian host (Gabriel et al., 2005). In addition, meta analysis of sequence data from 336 avian influenza viruses collected over a period of 30 years combined with genome data for avian and human influenza strains deposited in the public databases has provided the most detailed information on variation in avian influenza viruses to date. In addition to significant variability in the surface expressed hemagglutinin and neuraminidase genes, significant variation in the NS gene, encoding two non-structural proteins NS1 and NS2, also apparently influences the course of human infection (Obenauer et al., 2006). The NS1 protein is only expressed when the virus infects the host cell and appears to impede intracellular signaling pathways, specifically IFN-dependent and TNF-dependent pathways (Hayman et al., 2006), apparently via a specific binding motif present on most avian influenza virus NS1 proteins and those human influenza viruses of avian origin, suggesting that this motif may be involved in human pathogenicity (Obenauer et al., 2006).

The rapid mutation rate observed in such surface expressed genes, such as hemagglutinin or neuramidase, poses significant problems in the development of influenza vaccines, which are typically aimed at molecules such as these (Gillis, 2006). In addition, with regard to treatment, drug resistance to commonly used antivirals has recently been documented. Resistance to oseltamivir, a neuraminidase inhibitor and one of the main antiviral treatments in use for avian flu, was identified in two Vietnamese patients, who both subsequently died despite prompt treatment. Sequencing of the viral isolates and comparison to those from patients who responded to treatment identified a single mutation in the neuraminidase gene that led to an amino acid substitution of a tyrosine residue with a histidine (de Jong et al., 2005).

Application to Bacterial Disease

The current number of publicly available whole genome bacterial datasets is close to 300 but there are at least twice as many in progress (Fraser-Liggett, 2005). Genetic exchange by horizontal transfer, rapid proliferation and global mobility of infected hosts, be it humans or livestock, make bacteria very successful pathogens. Knowledge of the complete genome content of bacteria is even more advantageous, because bacteria are haploid, as such the genotype should be reflected in the phenotype. Transfer of a resistance gene will therefore lead to acquisition of resistance and so, unlike diploid organisms, direct assumptions can be derived from the genetic complement.

The small size of bacterial genomes, combined with the large number of submitted sequences, has led to extensive use of genome sequencing. After the initial sequencing efforts to provide the whole genome complement of reference strains for the major human bacterial pathogens, it soon became apparent that this would be insufficient to answer the most important questions, those of virulence and drug resistance. One obvious method to determine which genes are necessary for survival in the host is to compare free-living microbes with their pathogenic relatives, a process referred to as "differential genome display" (Huynen et al., 1998). Similarly, attenuated or avirulent strains can be compared to clinical isolates or pathogenic reference strains to identify genes, operons or pathogenicity islands that confer virulence in the human host.

Staphylococcus aureus

S. aureus is a ubiquitous bacterium intermittently carried by the majority of the population as a commensal organism. Although invasive infections were a major cause of death in the past, the advent of antibiotics in the 1940s effectively controlled such disease. However, in the last 15 years, it has re-emerged as a serious problem among patients in the hospital environment and is now the most common cause of hospital-acquired infection (Jones, 2003). The spread of *S. aureus* infections has been accompanied by the emergence of virulent and drug resistant strains, resulting in thousands of deaths. In 2004, there were 7684 cases of methicillin resistant *S. aureus* (MRSA) bacteraemia in the UK, leading to more than 1000 deaths (http://www.dh.gov.uk/en/Publicationsandstatistics/Publications/PublicationsStatistics/DH_4085951—MRSA surveillance system: Department of Health). The ability of the bacteria to rapidly acquire new sequence has led to the successive emergence of resistance to most antibiotics, a process that is leading inexorably to multidrug resistant strains unresponsive to all available therapies. MRSA is now endemic in many hospital environments (Enright et al., 2002) and recently, resistance has been seen to the drug vancomycin (VRSA).

The interest in the developing resistance of this pathogen has led to the sequencing of seven strains of *S. aureus*, all but one being clinical isolates from virulent or drug resistant cases (Lindsay and Holden, 2006). Although there are single nucleotide differences (SNPs) along the genome, the main mechanism by which *S. aureus* adapts so rapidly is the mobility and horizontal transfer of regions of the genome between bacteria (Lindsay and Holden, 2004). Alignment of chromosomes from these seven species, together with the attenuated laboratory strain has identified variable regions that contain several types of mobile genetic elements (see Fig. 25–5), and groups of genes involved in pathogenicity and resistance are coded on these elements.

Most horizontal transfer in *S. aureus* is mediated by bacteriophages, viruses that infect bacteria, integrating their own DNA into the bacterial chromosome. Such phage genomes are known to contain numerous virulence genes, the functions of which are passed onto the bacterium. While most *S. aureus* strains contain integrated phages that are dormant and replicated with the rest of the bacterial DNA, stress induced by DNA damaging chemicals, antibiotics or other mechanisms can induce the phage to replicate independently of the bacterium. This results in production of numerous infectious phage particles that eventually cause the bacterium to lyse and release the phage particles that allow them to bind to and re-infect new bacterial cells, thereby transferring any virulence or resistance genes they carry.

Staphylococcal cassette chromosomes are additional mobile elements and can be transferred between bacteria, probably by transduction. They have been shown to encode a range of antibiotic resistance genes, including those for both methicillin (Ito et al., 2001) and fusidic acid (Holden et al., 2004) resistance. However, these cassettes may also contain integrated plasmids that themselves

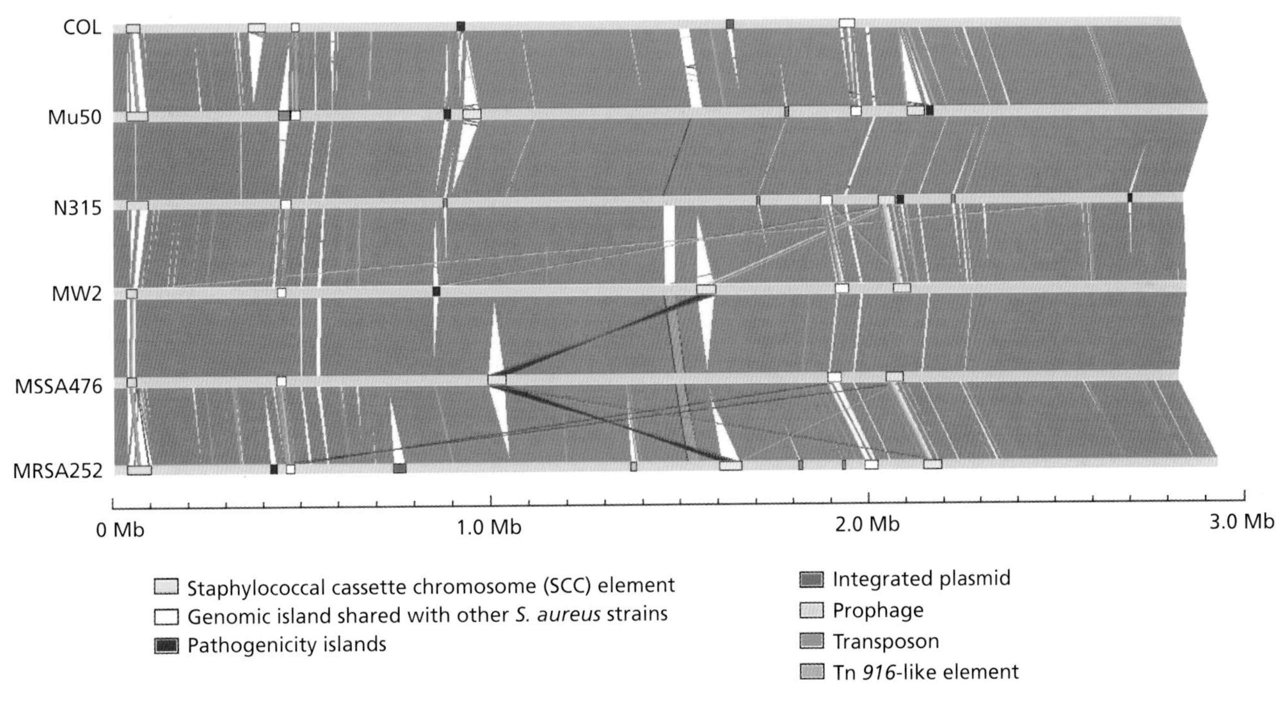

Figure 25–5 Comparative sequencing of six strains of *Staphylococcus aureus* showing locations of mobile genetic elements. Graphical represents pairwise comparisons of the *S. aureus* strains COL, Mu50, N315, MW2, MSSA476 and MRSA252 chromosomes displayed in the Artemis Comparison Tool (ACT). The sequences are aligned with the predicted replication origins on the right and the termination of replication in the centre. The colored bars separating each chromosome represent orthologous matches identified by reciprocal FASTA analysis and represent the conserved regions. The variable regions described in the text are shown in colored boxes and represent the mobile genetic elements that contain resistance and virulence factors (Reproduced with permission of M. Holden and Springer Science and Business Media from the review paper by Lindsay JA and Holden MT; Understanding the rise of the superbug: investigation of the evolution and genomic variation of *Staphlococcus aureus.* Funct Integr Genomics. 2006 Feb 2:1–16).

contain further resistance genes (Hiramatsu et al., 2001) and frequently contain smaller mobile elements called transposons that are inserted and rearranged, leading to greater variability in pathogenicity between related strains. While smaller plasmids appear to move by generalized transduction, some larger plasmids can also be transmitted between bacteria by conjugation and, although usually remaining independent of the bacterial chromosome, smaller plasmids can integrate into the host DNA, usually in staphylococcal cassette chromosomes (Holden et al., 2004).

Plasmids have been shown to confer resistance to most types of antimicrobial drugs including β-lactams (such as methicillin), heavy metals, antiseptics and aminoglycosides. Transposons that encode a transposase gene have the ability to excise, insert and replicate at various positions in the chromosome. If they insert into a plasmid or other mobile genetic elements, they can then be transferred between bacteria. Several resistance genes have been associated with transposons, including erythromycin (Phillips and Novick, 1979) and tetracycline resistance (Kuroda et al., 2001).

Given the range of drug resistance in *S. aureus* populations and implications for patient management in hospitals, it is imperative to monitor clinical isolates and environmental samples for these known variants and there are currently numerous methods in use to type strains for epidemiological surveillance. Ribotyping of the 16S and 23S ribosomal genes (Kostman et al., 1995), pulse-field gel electrophoresis (Bannerman et al., 1995) and multilocus typing (Enright et al., 2000) are commonly used for strain identification and, while they often provide a limited repertoire of markers, such methods are quick and, for identification of a particular resistance gene, they are ideal for most laboratories. For example, the identification of MRSA strains by typing the methicillin resistance gene *mecA* using the commercially available PCR based TaqMan™ assay takes as little as 6 h (Francois et al., 2003). A more comprehensive typing method uses an Affymetrix gene chip (Dunman et al., 2004), PCR-based microarray (Witney et al., 2005) or oligonucleotide microarrays (Koessler et al., 2006). These systems have the capacity to type genes and variable sequences from all sequenced strains and additional novel genes submitted to the public databases, thereby covering as many of the variants as possible for maximum discrimination.

Mycobacteria

Of all the bacterial diseases, *M. tb* poses the greatest threat to human health. Once considered to be in rapid decline, it is now second only to HIV as a cause of death (see Fig 25–2) and, unlike many pathogens, it is indiscriminate in the selection of victims. The current preventative measure, a live vaccine made of *M. bovis* Bacillus Calmette-Guerin (BCG), be Bacillus Calm is the commonest vaccination in the world, with more than 3 billion recipients; however its efficacy at protecting against TB ranges from 0% up to 80% (Brewer, 2000). Two factors are driving the increase in the incidence of TB. Firstly, the association of TB and HIV infections where, in Africa, it is the main contributing cause of death in people with HIV. Secondly, the dramatic increase in MDR-TB has led to the threat of strains that are resistant to all available drugs. As it is, current chemotherapy relies on a combination of drugs taken for a minimum period of 6 months. This extensive period of treatment, much longer than the usual period for conventional antibiotic therapy, is due to the ability of these bacteria to enter a dormant or persistent state that avoids immune surveillance and the actions of antimycobacterial drugs. Extended treatment regimens create problems in compliance, particularly in areas of poor medical infrastructure, leading to the possible selection of resistant strains. Even after prolonged treatment with good compliance, cure is generally not sterile, and latent bacteria can be reactivated by immunosuppression in later years. Given that TB poses a huge global threat, it is noticeable that there have been no new drugs for TB in the last 30 years.

The complete genome sequence of the laboratory reference strain H37Rv (Cole et al., 1998) has been followed by five more mycobacterial genomes (www.ebi.ac.uk/genomes/bacteria), including a clinical isolate of *M. tb* (Fleischmann et al., 2002) plus a number of other members of the genus mycobacterium, including a number of predominantly non-human pathogens. These other members comprise *M. leprae*, the causative agent of leprosy (Cole et al., 2001), *Mycobacterium avium* (Li et al., 2005), the causative agent of chronic granulomatous enteritis in cattle and implicated as an etiologic agent in CD in man (Greenstein, 2003), as well as *M. bovis* and *Mycobacterium marinum* (www.sanger.ac.uk/Projects/M_marinum), predominantly pathogens of cattle or fish and frogs, respectively, but both known to infect humans (Katoch, 2004; Esteban et al., 2005). The comparative genome projects have contributed to knowledge of genes that control persistence or dormancy, energy metabolism, cell wall synthesis, transcription regulation and cell signaling.

Finding and developing drugs to kill persistent mycobacteria is a high priority, as this would significantly shorten treatment times and reduce diseases due to reactivation, a serious concern considering the HIV/AIDS pandemic. Although the nature of how mycobacteria maintain dormancy is not well understood, there are several genes known to be involved in the mechanisms of persistence. These include a novel methyl transferase involved in modification of mycolic acid (Glickman et al., 2000), an isocitrate lyase that is part of the glyoxylate shunt (McKinney et al., 2000), and a hyperphosphorylated guanine (ppGpp) synthase involved in the coordinated shutdown of active metabolism required for dormancy (Dahl et al., 2003). A search of the human genome reveals that neither methyl transferase nor isocitrate lyase has a human homologue and as such would make good drug targets.

Specialized databases can be used to compare pathogen and host systems in the search for drug targets. Most of the current drugs are enzyme inhibitors and, given the evolutionary gulf between pathogen and host and the dramatic difference in physiology and environment, it is likely that *M. tb* will have a number of metabolic pathways that are absent in humans that can be targeted in the search for new drugs. A comprehensive in silico comparison of the complete metabolic pathways in the KEGG metabolic pathway database for *M. tb* against *Homo sapiens* specific proteins revealed six mycobacterial pathways absent in humans and a total of 185 enzymes that could be considered as potential drug targets (Anishetty et al., 2005). Although many of the 185 targets have been previously described, 67 had not and, of these targets, some have orthologs in other bacteria, including the related *M. leprae*, and as such, could be considered for more broad-spectrum drug development.

Comparative genomics of the *M. tb* complex has also been used to identify genes and pathways that appear to determine pathogenicity and drug resistance and have led to a list of potential drug targets that can now be explored in a systematic and rational manner. The members of the *M. tb* complex so far sequenced are predicted to have in the region of 4000 genes and all share a high degree of similarity at the DNA level. Using whole genome DNA microarrays, the attenuated BCG strains of *M. bovis*, currently used as vaccines, show sequence deletions compared to other mycobacteria and comparison to *M. tb* reveals 18 regions, containing approximately 120 genes, that have been deleted (RD1-18) compared to pathogenic members of the *M. tb* complex, although only one region is consistently absent in all BCG strains (RD1) (Behr et al., 1999). The RD1 region contains nine genes and has been extensively studied. Functional analysis indicates that these genes are functionally related and have a role in bacterial virulence and dissemination (Dietrich et al., 2006). At least three of the proteins encoded by this region are strongly immunogenic and act as potent T-cell antigens, and as such would potentially make good vaccine candidates.

Having identified differences at the genome level, it is necessary to determine if the selected genes are essential components of the life cycle and to ensure they are expressed during infection. Numerous downstream post genomic methodologies have been applied to the whole genome dataset to verify potential drug or vaccine targets derived from bioinformatic screening. Differential expression of customized amplification libraries (DECAL) (Alland et al., 1998) and selective capture of transcribed sequences (SCOTS) (Graham and Clark-Curtiss, 1999), both use subtractive hybridization to identify genes that are differentially expressed. DECAL was used to identify three genes whose expression was up-regulated in response to isoniazid treatment, and SCOTS was used to identify genes up-regulated upon mycobacterial entry into the human macrophage. Of the genes identified, isocitrate lyase, previously implicated in the initiation of persistence and without any human homologue, appears to be the best candidate. Mycobacterial gene expression in human macrophages has also been examined using microarrays, comparing the expression profiles seen when in human macrophages compared with synthetic medium (Cappelli et al., 2006). This method allows genes that are down-regulated as well as those that that are up-regulated to be identified. Of an almost complete genome complement, 214 genes, representing approximately 5% of the genome, were significantly differentially expressed, most being up-regulated. A third of these genes had no functional characterization, illustrating how much remains unknown about mycobacterial survival in human cells. Classification of the characterized genes into functional categories demonstrated that genes involved in replication, repair and nucleic acid transcription were up-regulated in the macrophage, while most other categories of genes remained relatively static. This suggests that the genes involved in replication, repair and those that help prevent oxidative and nitrosative damage are very important in promoting survival in the hostile environment of the macrophage and as such would make attractive drug targets.

Comparative analysis of the sequenced members of the *M. tb* complex reveals the high degree of similarity between them and as such highlights the few differences that might explain the different host specificities. Another approach to shortlist new drug targets is to compare the genome of the two major human pathogens in the mycobacterial group. *M. leprae*, the causative agent of leprosy, could not be further removed from *M. tb* in terms of genome architecture. The leprosy bacillus occupies the same cellular niche as the tubercle bacillus, in that it resides in host macrophages; however, the host response, tissue tropism and disease phenotypes are very different. The fully sequenced genome revealed that *M. leprae* has undergone a massive reduction in the coding genome, such that it has less than half the number of genes of *M. tb* (Cole et al., 2001), a phenomenon that may potentially explain the extraordinarily slow growth of leprosy. The *M. leprae* genome contains a huge number of pseudogenes that represent genes present in *M. tb* but that have degenerated in *M. leprae*. While reductive evolution is present in other specialized obligate intracellular pathogens, the leprosy bacillus appears to have taken this to an extreme and it is predicted that *M. leprae* has lost more than 2000 genes during its evolution from an ancestral mycobacterial organism (Cole et al., 2001), such that the remaining genes may represent the minimal set of genes required for pathogenic existence. Comparison of the two genomes identifies 1433 genes that are functional in both organisms. Further database searches revealed that 219 of these genes appear to be restricted solely to mycobacteria (Cole, 2002). Again, these criteria make them suitable for progressing further as potential drug targets, as they are likely to be essential for survival or human pathogenicity and lack orthologs in eukaryotes.

The definitive diagnostic test for *M. tb* is either the identification of acid-fast bacilli in a sputum smear stained with Ziehl-Nielsen or auramine (Fite Faraco staining method for *M. leprae*) or by culturing in media, but both methods have restrictions. While staining of smears is relatively rapid and cheap, it is positive in only approximately 50% of cases, as many individuals have low numbers of bacilli. A second disadvantage is that, although it is likely to be *M. tb* in a sputum or tissue aspirate, this test will not discriminate between species and will not identify any potential non-tuberculous mycobacterial infections. Culturing of mycobacteria, on the other hand is remarkably slow such that, although culture methods have improved, a result can take days or weeks to obtain. However, there are now numerous DNA sequence tests that use PCR amplification of conserved species specific sequences before identification by restriction enzyme cleavage, gel electrophoresis, DNA probe hybridization, or fluorescently labeled RT-PCR (reviewed by Glennon and Cormican, 2001). Depending on the target sequence, such methods will discriminate between not only species but also different isolates and, where the genome sequences have provided information on variable regions involved in drug resistance and virulence, to discriminate between strains, allowing for targeted treatment of drug resistant cases and surveillance of outbreaks.

The search for immunodominant antigens that could replace or complement the BCG vaccine has been given a significant boost by the publication of the genome sequences. By starting with the complete gene set for both *M. tb* and *M. leprae*, a reduced list of potential candidate proteins likely to be specifically immunostimulatory can be deduced. Motif searches for transmembrane domains and signal peptides will identify proteins that are localized at the cell surface or secreted from the mycobacterial cell. One such group has been classed the early secreted antigenic target family ([ESAT). Short-term culture filtrates from mycobacterial cultures are enriched in proteins secreted from the mycobacteria; fractionation of the filtrate followed by separation using 2D gel electrophoresis and protein identification by mass spectrometry revealed a number of antigens that stimulated infected human cells to produce protective cell mediated cytokines (Dietrich et al., 2006). One group of these, the ESAT-6 family, was subsequently shown to be a potent T-cell antigen that stimulates a protective T_H1 cytokine response (Andersen et al., 1995). Interestingly, a comparison of the *M. tb* and BCG genomes revealed that ESAT-6 is present on the RD1 region of virulent and clinical isolates of mycobacteria but absent from all attenuated BCG strains. This comparison also identified further potent immunodominant antigens in the RD1 region that are recognized by T cells from patients with both early and latent infection (Dietrich et al., 2006). A recombinant BCG expressing this RD1 region was subsequently shown to give an increased protective response when compared to vaccination with BCG alone (Pym et al., 2003).

The availability of the genome sequences have given some insight into the immunodominant antigens found via other means and have allowed a more systematic approach to generating attenuated *M. tb* and BCG strains that can be tested as vaccines. Since the likelihood that human variability in the innate and adaptive immune response will lead to variable efficacy to any one target or epitope, potential vaccine candidates are being fused either as recombinant fusion proteins or DNA molecules for use as vaccines that will hopefully stimulate a range of host protective responses. To this end, a fusion of ESAT-6 with another potent T-cell antigen, Ag85B, was shown to give greater protection than either of these given individually and to be comparable with BCG (Doherty et al., 2004).

Application to Protozoan Pathogens

The larger genome sizes of the unicellular eukaryotic pathogens have taken longer to sequence. The recent publication of the *Leishmania major* and *T. brucei* genomes (Berriman et al., 2005; Ivens et al., 2005) was the culmination of 10 years of work, with 7 years of actual genome sequencing. In comparison, the second species of *Leishmania*, started in 2004, took only a matter of months to complete a whole genome shotgun. However, annotation of these eukaryotic pathogens has taken much longer and remains an area of development. The more complex genome structure, presence of gene splicing and paucity of sequences available for similarity searching has meant that the bottleneck is not so much in the sequencing, but more in the understanding of the genome. Since 2002, the complete genomes of representative members of most of the human protozoan pathogens have been published. Projects to sequence the much larger and more complex helminth pathogen genomes are underway and represent the last major area of infectious disease that have yet to profit from this methodology.

Malaria

One of the first protozoan infectious diseases to benefit from whole-genome sequencing was malaria. Of all of the eukaryotic infectious diseases, malaria has by far the highest impact on mortality and morbidity and as such is the most researched. Estimates of the number of people killed annually from malaria range from the conservative figure of approximately 1.3 million per year reported by the WHO (see Fig. 25–2) to as much as 2.5 million, with as many as 660 million infections (Snow et al., 2005). Consequently, *P. falciparum*, the causative agent of almost all fatal malarial infections, was one of the first eukaryotic pathogens to be sequenced and published (Gardner et al., 2002; Hall et al., 2002). The completion of the rodent malaria *Plasmodium yoelii yoelii* at the same time and the subsequent sequencing of other non-human infective species have led to new leads for drug and vaccine discovery programs. Apart from the original reference strain of *P. falciparum*, there are now a further eight species that have undergone whole genome shotgun sequencing including five non-human species and *P. vivax*, a human pathogen that normally causes a non fatal infection (www.sanger.ac.uk/Projects/Protozoa/).

In addition to defining many of the complex metabolic pathways and other features of the parasite's life cycle, comparative analysis has defined both the core proteome shared by all malaria species and the species-specific genes that define host tropism and pathogenicity. The different species of *Plasmodium* to be sequenced are extremely host specific and as such, those genes that differ between species are likely to be those that determine this specificity. Of a total gene complement of about 5500 genes, approximately 4500 make up the core proteome. The remainder are now under close scrutiny to help understand the complexities of the parasite-host interaction. In silico predictions of protein motifs and domains have helped identify the most promising proteins to pursue for vaccine and drug development.

Given the acute nature of the disease, the most effective means of controlling malaria would be with vaccination. Repeated infection with malaria does provide some degree of protection to future episodes, at least to severe life threatening diseases. This immunity is non-sterile and relies on the presence of live parasites and re-infection. As far back as the 1960s, this was used to generate lasting immunity in humans by using the infective stage parasites, or sporozoites, that were weakened by irradiation such that these attenuated sporozoites could infect the liver cells but were unable to proliferate (Clyde et al., 1973). Comparative analysis of several malaria genomes coupled with expression analysis in the different life cycle stages has revealed conserved genes that are only expressed by the parasite in the sporozoite stage (Mueller et al., 2005; van Dijk et al., 2005). Using the rodent pathogen *Plasmodium berghei* and the mouse model of infection, parasites with mutated forms of some of these conserved stage specific genes have been tested. A number of them survive sufficiently long enough to generate infections that were sufficient to promote immunity but unable to develop to the next stage in the hepatocyte, thereby promoting an immune response that was protective without progression of infection.

The sequenced genome provides the complete set of potential vaccine candidates. In silico analysis of these sequences will identify a small subset of these that can be taken through the process of validation. For instance, proteins expressed on the cell surface will be exposed to the host immune system. Predicted signal peptide motifs identify those genes that are transported to the cell surface or presented on the surface of the host erythrocyte. Of the 500 or so proteins that have predicted signal peptides, less than 200 are expressed in stages identified as immune stimulating (Waters, 2006). Genes coding for proteins known to interact with and challenge the immune system are often under positive selective pressure to change residues after immune recognition. Analysis of the rates of synonymous versus non-synonymous nucleotides changes between different species will identify those genes that have undergone accelerated evolution. Approximately half of those surface expressed genes show evidence of increased non-synonymous change, presumably in response to host immune pressures. Thus, from a list of more than 5000 genes, comparative analysis and routine motif searching has generated a list of approximately 100 potential vaccine candidates.

One of the complicating factors with the blood stage (merozoite) of *Plasmodium* infection is the presence of a parasite protein on the red blood cell surface that undergoes significant antigenic variation, thereby preventing sustained recognition and subsequent effective antibody response. The *P. falciparum* genome project has revealed that there are three large families of highly polymorphic genes that are responsible for this antigenic variation. Most of these genes reside in the subtelomeric regions of the chromosomes. Comparative genome analysis has revealed that these regions are very variable in size between different strains of *P. falciparum* and may help to explain the variation in strain virulence.

One of these families consists of 59 *var* genes that are expressed clonally at any given time. The *var* genes code for *P. falciparum* erythrocyte membrane protein 1 (PfEMP-1), a multimodular adhesin protein that determines the ability of the infected red blood cells to bind specifically to other cells. The adherence and sequestration of erythrocytes leads to accumulation and microhemorrhaging of blood vessels in the brain, resulting in life threatening conditions and determining the severity of cerebral malaria. Identifying how these and other virulence genes interact with the host is critical to future development of effective therapies that will inhibit parasite development and survival.

The absence of any vaccine has meant that prophylaxis and therapeutic treatment are imperative for controlling malaria infections. The acute nature of the cerebral malaria has meant that even in treated patients, the effect of drug resistance has devastating effects on mortality rates. In silico analysis of the data from the genome projects has provided the research community with a wealth of information that will greatly speed up drug research. It has provided the data for large-scale whole genome proteomic work, which will help identify proteins that are expressed in the human infective life stages. Proteomic analysis and microarrays have also been used to answer specific questions related to drug resistance and virulence genes, with comparison of 2D gels from drug resistant or virulent strains compared with less virulent ones, identifying which proteins are up-regulated and potentially responsible for conferring drug resistance.

One area that has attracted considerable interest recently is the existence of the apicoplast, an organelle found in all species of the phylum Apicomplexa and related to the chloroplast found in plants. Although originally acquired from an endosymbiotic cyanobacterium, it is essential for parasite survival (He et al., 2001). The bacterial origin of the apicoplast means that most of its enzymes are fundamentally different from human orthologs and as such make good potential drug targets. Prior to the completion of the genome project, the problem had been in identifying these proteins. Although it has its own self replicating DNA, most of the proteins in the apicoplast are encoded in the nucleus and then transferred to the organelle. By looking for a series of specific sequences, such as secretory motifs and transit peptides, which allow this transfer to happen, these proteins can be identified (Foth et al., 2003). The bacterial and algal origins of this organelle raises the prospect that the malaria parasite, and indeed other intracellular parasites of the Ampicomplexan phylum such as *Toxoplasma gondii*, will be susceptible to antibiotics and herbicidal drugs already licensed for use (Wiesner and Seeber, 2005).

Unlike the other infections described in this review, the advent of genome technology has had little impact on the field of malaria diagnosis. Poor laboratory infrastructure, economics and the rapid onset of life threatening illness often precludes the use of more sophisticated equipment needed for DNA typing. Currently, a Giemsa stained blood smear is still the cheapest, easiest and fastest way to definitively identify a life threatening infection with *P. falciparum*, where the usual high parasitemia make this an effective means of diagnosis.

Ongoing Advances—Metagenomics and Metaproteomics

With improvements in the rate of sequencing, reduction in costs, and improvements in assembly and analysis software, we are now entering an era of metagenomics, entailing the sampling of clinical or environmental specimens rather than isolated or cultured organisms (Schloss and Handelsman, 2005; Tringe and Rubin, 2005). Such an approach has already been applied to identifying the causative organism of viral meningitis in clinical specimens. A high-resolution diagnostic microarray with conserved sequences from the 13 known causes of meningitis identified nine viral causes in 41 different samples of cerebral spinal fluid (Boriskin et al., 2004). A more extensive system, using more than 10,000 small subunit ribosomal DNA sequences from diverse microbial populations has also been developed to allow the characterization of complex microbial communities such as those taken from clinical samples (Palmer et al., 2006). In addition, metaproteomic techniques are being developed that will allow investigation of functional gene expression patterns in similar complex microbial communities (Wilmes and Bond, 2006) and may, hopefully, be applied in the clinical setting.

Conclusions

Overall, the genomic era is revolutionizing the manner in which infectious disease research is being undertaken, both from the point of view of the human host and the infecting pathogen, allowing the generation of new and exciting discoveries at a phenomenal rate. The ongoing development of technologies that are amenable to large-scale, high-throughput projects is allowing both human and pathogen genomes and transcriptomes to be considered in their entirety. With respect to human susceptibility to infectious disease, the availability of the human genome sequence coupled with dense SNP array technology is now allowing researchers to interrogate the whole genome simultaneously. It is hoped that such an approach will identify the many genes involved in susceptibility to certain infections and ultimately allow the complex genetic interactions that exist between these genes and the environment to be untangled. Not only will this provide valuable information about the pathways involved in the innate and adaptive immune responses, but may also eventually allow the more targeted application of appropriate drug therapies.

With respect to pathogens, the advent of the genomic era, coupled with general use of powerful bioinformatics capabilities and global access to data via the Internet, pathogen research has become information driven. There is not a species of bacterial or viral human pathogen that does not have a related genome represented in the public databases. In many cases, there are numerous strains, including attenuated laboratory variants, that make comparative analysis an important part of microbiological research. This has resulted in the identification of genes involved in virulence and drug resistance as well as the identification of novel drug targets in numerous infectious diseases, and hopefully, will continue to be a source of such valuable information in the future.

References

Abel, L, Cot, M, Mulder, L, Carnevale, P, and Feingold, J (1992). Segregation analysis detects a major gene controlling blood infection levels in human malaria. *Am J Hum Genet*, 50:1308–1317.

Abel, L and Demenais, F (1988). Detection of major genes for susceptibility to leprosy and its subtypes in a Caribbean island: Desirade island. *Am J Hum Genet*, 42:256–266.

Abel, L, Demenais, F, Prata, A, Souza, AE, and Dessein, A (1991). Evidence for the segregation of a major gene in human susceptibility/resistance to infection by Schistosoma mansoni. *Am J Hum Genet*, 48:959–970.

Abel, L, Vu, DL, Oberti, J, Nguyen, VT, Van, VC, Guilloud-Bataille, M, et al. (1995). Complex segregation analysis of leprosy in southern Vietnam. *Genet Epidemiol*, 12:63–82.

Alcais, A, Abel, L, David, C, Torrez, ME, Flandre, P, and Dedet JP (1997). Evidence for a major gene controlling susceptibility to tegumentary leishmaniasis in a recently exposed Bolivian population. *Am J Hum Genet*, 61:968–979.

Alland, D, Kramnik, I, Weisbrod, TR, Otsubo, L, Cerny, R, Miller, LP, et al. (1998). Identification of differentially expressed mRNA in prokaryotic organisms by customized amplification libraries (DECAL): the effect of isoniazid on gene expression in *Mycobacterium tuberculosis*. *Proc Natl Acad Sci USA*, 95:13227–13232.

Allison, AC (2004). Two lessons from the interface of genetics and medicine. *Genetics*, 166:1591–1599.

Altare, F, Durandy, A, Lammas, D, Emile, JF, Lamhamedi, S, Le Deist, F, et al. (1998). Impairment of mycobacterial immunity in human interleukin-12 receptor deficiency. *Science*, 280:1432–1435.

Altare, F, Ensser, A, Breiman, A, Reichenbach, J, Baghdadi, JE, Fischer, A, et al. (2001). Interleukin-12 receptor beta1 deficiency in a patient with abdominal tuberculosis. *J Infect Dis*, 184:231–236.

Altare, F, Lammas, D, Revy, P, Jouanguy, E, Doffinger, R, Lamhamedi, S, et al. (1998). Inherited interleukin 12 deficiency in a child with bacille Calmette-Guerin and *Salmonella enteritidis* disseminated infection. *J Clin Invest*, 102:2035–2040.

Altshuler, D, Brooks, LD, Chakravarti, A, Collins, FS, Daly, MJ, and Donnelly, P (2005). A haplotype map of the human genome. *Nature*, 437:1299–1320.

Andersen, P, Andersen, AB, Sorensen, AL, and Nagai, S (1995). Recall of long-lived immunity to *Mycobacterium tuberculosis* infection in mice. *J Immunol*, 154:3359–3372.

Anishetty, S, Pulimi, M, and Pennathur, G (2005). Potential drug targets in *Mycobacterium tuberculosis* through metabolic pathway analysis. *Comput Biol Chem*, 29:368–378.

Bacellar, O, Lessa, H, Schriefer, A, Machado, P, Ribeiro de Jesus, A, Dutra, WO, et al. (2002). Up-regulation of Th1-type responses in mucosal leishmaniasis patients. *Infect Immun*, 70:6734–6740.

Bakker, MI, May, L, Hatta, M, Kwenang, A, Klatser, PR, Oskam, L, et al. (2005). Genetic, household and spatial clustering of leprosy on an island in Indonesia: a population-based study. *BMC Med Genet*, 6:40.

Bannerman, TL, Hancock, GA, Tenover, FC, and Miller, JM (1995). Pulsed-field gel electrophoresis as a replacement for bacteriophage typing of *Staphylococcus aureus*. *J Clin Microbiol*, 33:551–555.

Barreiro, LB, Neyrolles, O, Babb, CL, Tailleux, L, Quach, H, McElreavey, K, et al. (2006). Promoter variation in the DC-SIGN-encoding gene CD209 is associated with tuberculosis. *PLoS Med*, 3:e20.

Behr, MA, Wilson, MA, Gill, WP, Salamon, H, Schoolnik, GK, Rane, S, et al. (1999). Comparative genomics of BCG vaccines by whole-genome DNA microarray. *Science*, 284:1520–1523.

Beisel, CE and Morens, DM (2004). Variant Creutzfeldt-Jakob disease and the acquired and transmissible spongiform encephalopathies. *Clin Infect Dis*, 38:697–704.

Bellamy, R, Beyers, N, McAdam, KP, Ruwende, C, Gie, R, Samaai, P, et al. (2000). Genetic susceptibility to tuberculosis in Africans: a genome-wide scan. *Proc Natl Acad Sci USA*, 97:8005–8009.

Bellamy, R, Ruwende, C, Corrah, T, McAdam, KP, Whittle, HC, and Hill, AV (1998). Variations in the NRAMP1 gene and susceptibility to tuberculosis in West Africans. *N Engl J Med*, 338:640–644.

Bentley, SD, Aanensen, DM, Mavroidi, A, Saunders, D, Rabbinowitsch, E, Collins, M, et al. (2006). Genetic analysis of the capsular biosynthetic locus from all 90 pneumococcal serotypes. *PLoS Genetics*, 2(3):e31.

Berriman, M, Ghedin, E, Hertz-Fowler, C, Blandin, G, Renauld, H, Bartholomeu, DC, et al. (2005). The genome of the African trypanosome *Trypanosoma brucei*. *Science*, 309:416–422.

Black, RE, Morris, SS, and Bryce, J (2003). Where and why are 10 million children dying every year? *Lancet*, 361:2226–2234.

Blackwell, JM, Goswami, T, Evans, CA, Sibthorpe, D, Papo, N, White, JK, et al. (2001). SLC11A1 (formerly NRAMP1) and disease resistance. *Cell Microbiol*, 3:773–784.

Blanpain, C, Libert, F, Vassart, G, and Parmentier, M (2002). CCR5 and HIV infection. *Receptors Channels*, 8:19–31.

Bonnen, PE, Pe'er, I, Plenge, RM, Salit, J, Lowe, JK, Shapero, MH, et al. (2006). Evaluating potential for whole-genome studies in Kosrae, an isolated population in Micronesia. *Nat Genet*, 38:214–217.

Boriskin, YS, Rice, PS, Stabler, RA, Hinds, J, Al-Ghusein, H, Vass, K, et al. (2004). DNA microarrays for virus detection in cases of central nervous system infection. *J Clin Microbiol*, 42:5811–5818.

Brewer, TF (2000). Preventing tuberculosis with bacillus Calmette-Guerin vaccine: a meta-analysis of the literature. *Clin Infect Dis*, 31 (Suppl. 3): S64–S67.

Bryce, J, Boschi-Pinto, C, Shibuya, K, and Black, RE (2005). WHO estimates of the causes of death in children. *Lancet*, 365:1147–1152.

Bucheton, B, Abel, L, Kheir, MM, Mirgani, A, El-Safi, SH, Chevillard, C, et al. (2003). Genetic control of visceral leishmaniasis in a Sudanese population: candidate gene testing indicates a linkage to the NRAMP1 region. *Genes Immun*, 4:104–109.

Burke, DT, Carle, GF, and Olson, MV (1987). Cloning of large segments of exogenous DNA into yeast by means of artificial chromosome vectors. *Science*, 236:806–812.

Buu, N, Sanchez, F, and Schurr, E (2000). The Bcg host-resistance gene. *Clin Infect Dis*, 31 (Suppl. 3):S81–S85.

Cabello, PH, Lima, AM, Azevedo, ES, and Krieger, H (1995). Familial aggregation of *Leishmania chagasi* infection in northeastern Brazil. *Am J Trop Med Hyg*, 52:364–365.

Cappelli, G, Volpe, E, Grassi, M, Liseo, B, Colizzi, V, and Mariani, F (2006). Profiling of *Mycobacterium tuberculosis* gene expression during human macrophage infection: upregulation of the alternative sigma factor G, a group of transcriptional regulators, and proteins with unknown function. *Res Microbiol*, 157(5):445–455.

Casanova, JL and Abel, L (2005). Inborn errors of immunity to infection: the rule rather than the exception. *J Exp Med*, 202:197–201.

Casanova, JL, Fieschi, C, Bustamante, J, Reichenbach, J, Remus, N, von Bernuth, H, et al. (2005). From idiopathic infectious diseases to novel primary immunodeficiencies. *J Allergy Clin Immunol*, 116:426–430.

Casselbrant, ML, Mandel, EM, Fall, PA, Rockette, HE, Kurs-Lasky, M, Bluestone, CD, et al. (1999). The heritability of otitis media: a twin and triplet study. *JAMA*, 282:2125–2130.

Castellucci, L, Cheng, LH, Araujo, C, Guimaraes, LH, Lessa, H, Machado, P, et al. (2005). Familial aggregation of mucosal leishmaniasis in northeast Brazil. *Am J Trop Med Hyg*, 73:69–73.

Castellucci, L, Menezes, E, Oliveira, J, Magalhaes, A, Guimaraes, LH, Lessa, M, et al. (2006). Interleukin6-174G/C promoter polymorphism influences susceptibility to mucosal but not localised cutaneous leishmaniasis in Brazil. *J Infect Dis*, 194(4):519–527.

Cervino, AC, Lakiss, S, Sow, O, Bellamy, R, Beyers, N, Hoal-van Helden, E, et al. (2002). Fine mapping of a putative tuberculosis-susceptibility locus on chromosome 15q11-13 in African families. *Hum Mol Genet*, 11:1599–1603.

Chan, PK, Tam, JS, Lam, CW, Chan, E, Wu, A, Li, CK, et al. (2003). Human metapneumovirus detection in patients with severe acute respiratory syndrome. *Emerg Infect Dis*, 9:1058–1063.

Chaudhuri, RR and Pallen, MJ (2006). xBASE, a collection of online databases for bacterial comparative genomics. *Nucleic Acids Res*, 34: D335–D337.

Clayton, DG, Walker, NM, Smyth, DJ, Pask, R, Cooper, JD, Maier, LM, et al. (2005). Population structure, differential bias and genomic control in a large-scale, case-control association study. *Nat Genet*, 37:1243–1246.

Clegg, JB and Weatherall, DJ (1999). Thalassemia and malaria: new insights into an old problem. *Proc Assoc Am Physicians*, 111:278–282.

Clyde, DF, Most, H, McCarthy, VC, and Vanderberg, JP (1973). Immunization of man against sporozite-induced falciparum malaria. *Am J Med Sci*, 266:169–177.

Cohen, ML (2000). Changing patterns of infectious disease. *Nature*, 406:762–767.

Cole, ST (2002). Comparative and functional genomics of the *Mycobacterium tuberculosis* complex. *Microbiology*, 148:2919–2928.

Cole, ST, Brosch, R, Parkhill, J, Garnier, T, Churcher, C, Harris, D, et al. (1998). Deciphering the biology of *Mycobacterium tuberculosis* from the complete genome sequence. *Nature*, 393:537–544.

Cole, ST, Eiglmeier, K, Parkhill, J, James, KD, Thomson, NR, Wheeler, PR, et al. (2001). Massive gene decay in the leprosy bacillus. *Nature*, 409:1007–1011.

Comstock, GW (1978). Tuberculosis in twins: a re-analysis of the Prophit survey. *Am Rev Respir Dis*, 117:621–624.

Dahl, JL, Kraus, CN, Boshoff, HI, Doan, B, Foley, K, Avarbock, D, et al. (2003). The role of RelMtb-mediated adaptation to stationary phase in long-term persistence of *Mycobacterium tuberculosis* in mice. *Proc Natl Acad Sci USA*, 100:10026–10031.

Daly, K.A. et al (2004). Chronic and recurrent otitis media: a genome scan for susceptibility loci. *Am J Hum Genet* 75, 988–97.

de Jong, MD, Bach, VC, Phan, TQ, Vo, MH, Tran, TT, Nguyen, BH, et al. (2005). Fatal avian influenza, A (H5N1) in a child presenting with diarrhea followed by coma. *N Engl J Med*, 352:686–691.

de Jong, R, Altare, F, Haagen, IA, Elferink, DG, Boer, T, van Breda Vriesman, PJ, et al. (1998). Severe mycobacterial and Salmonella infections in interleukin-12 receptor-deficient patients. *Science*, 280:1435–1438.

Dean, M, Carrington, M, and O'Brien, SJ (2002). Balanced polymorphism selected by genetic versus infectious human disease. *Annu Rev Genomics Hum Genet*, 3:263–292.

Dessein, AJ, Marquet, S, Henri, S, El Wali, NE, Hillaire, D, Rodrigues, V, et al. (1999). Infection and disease in human schistosomiasis mansoni are under distinct major gene control. *Microbes Infect*, 1:561–567.

Diehl, S and Rincon, M (2002). The two faces of IL-6 on Th1/Th2 differentiation. *Mol Immunol*, 39:531–536.

Dietrich, J, Weldingh, K, and Andersen, P (2006). Prospects for a novel vaccine against tuberculosis. *Vet Microbiol*, 112:163–169.

Doherty, TM, Olsen, AW, Weischenfeldt, J, Huygen, K, D'Souza, S, Kondratieva, TK, et al. (2004). Comparative analysis of different vaccine constructs expressing defined antigens from *Mycobacterium tuberculosis*. *J Infect Dis*, 190:2146–2153.

Domarle, O, Migot-Nabias, F, Pilkington, H, Elissa, N, Toure, FS, Mayombo, J, et al. (2002). Family analysis of malaria infection in Dienga, Gabon. *Am J Trop Med Hyg*, 66:124–129.

Donninger, H, Cashmore, TJ, Scriba, T, Petersen, DC, Janse van Rensburg, E, and Hayes, VM (2004). Functional analysis of novel SLC11A1 (NRAMP1) promoter variants in susceptibility to HIV-1. *J Med Genet*, 41:e49.

Dooley, AJ, Shindo, N, Taggart, B, Park, JG, and Pang, YP (2006). From genome to drug lead: identification of a small-molecule inhibitor of the SARS virus. *Bioorg Med Chem Lett*, 16:830–833.

Dorman, SE and Holland, SM (1998). Mutation in the signal-transducing chain of the interferon-gamma receptor and susceptibility to mycobacterial infection. *J Clin Invest*, 101:2364–2369.

Drosten, C, Gunther, S, Preiser, W, van der Werf, S, Brodt, HR, Becker, S, et al. (2003). Identification of a novel coronavirus in patients with severe acute respiratory syndrome. *N Engl J Med*, 348:1967–1976.

Dunman, PM, Mounts, W, McAleese, F, Immermann, F, Macapagal, D, Marsilio, E, et al. (2004). Uses of *Staphylococcus aureus* GeneChips in genotyping and genetic composition analysis. *J Clin Microbiol*, 42:4275–4283.

Dupuis, S, Jouanguy, E, Al-Hajjar, S, Fieschi, C, Al-Mohsen, IZ, Al-Jumaah, S, et al. (2003). Impaired response to interferon-alpha/beta and lethal viral disease in human STAT1 deficiency. *Nat Genet*, 33:388–391.

Dutz, JP, Benoit, L, Wang, X, Demetrick, DJ, Junker, A, de Sa, D, et al. (2001). Lymphocytic vasculitis in X-linked lymphoproliferative disease. *Blood*, 97:95–100.

Ellis, MK, Li, Y, Rong, Z, Chen, H, and McManus, DP (2006). Familial aggregation of human infection with *Schistosoma japonicum* in the Poyang Lake region, China. *Int J Parasitol*, 36:71–77.

El-Sayed, NM, Myler, PJ, Bartholomeu, DC, Nilsson, D, Aggarwal, G, Tran, AN, et al. (2005). The genome sequence of *Trypanosoma cruzi*, etiologic agent of Chagas disease. *Science*, 309:409–415.

Emmanuel, JC, Bassett, MT, and Smith, HJ (1988). Risk of hepatitis B infection among medical and paramedical workers in a general hospital in Zimbabwe. *J Clin Pathol*, 41:334–336.

Enright, MC, Day, NP, Davies, CE, Peacock, SJ, and Spratt, BG (2000). Multilocus sequence typing for characterization of methicillin-resistant and methicillin-susceptible clones of *Staphylococcus aureus*. *J Clin Microbiol*, 38:1008–1015.

Enright, MC, Robinson, DA, Randle, G, Feil, EJ, Grundmann, H, and Spratt, BG (2002). The evolutionary history of methicillin-resistant *Staphylococcus aureus* (MRSA). *Proc Natl Acad Sci USA*, 99:7687–7692.

Esteban, J, Robles, P, Soledad Jimenez, M, and Fernandez Guerrero, ML (2005). Pleuropulmonary infections caused by *Mycobacterium bovis*: a re-emerging disease. *Clin Microbiol Infect*, 11:840–843.

Everhart, JE, Kruszon-Moran, D, Perez-Perez, GI, Tralka, TS, and McQuillan, G (2000). Seroprevalence and ethnic differences in *Helicobacter pylori* infection among adults in the United States. *J Infect Dis*, 181:1359–1363.

Feitosa, MF, Borecki, I, Krieger, H, Beiguelman, B, and Rao, DC (1995). The genetic epidemiology of leprosy in a Brazilian population. *Am J Hum Genet*, 56:1179–1185.

Feitosa, M, Krieger, H, Borecki, I, Beiguelman, B, and Rao, DC (1996). Genetic epidemiology of the Mitsuda reaction in leprosy. *Hum Hered*, 46:32–35.

Feuk, L, Carson, AR, and Scherer, SW (2006). Structural variation in the human genome. *Nat Rev Genet*, 7:85–97.

Fine, PE (1988). Implications of genetics for the epidemiology and control of leprosy. *Philos Trans R Soc Lond B Biol Sci*, 321:365–376.

Fleischmann, RD, Adams, MD, White, O, Clayton, RA, Kirkness, EF, Kerlavage, AR, et al. (1995). Whole-genome random sequencing and assembly of *Haemophilus influenzae* Rd. *Science*, 269:496–512.

Fleischmann, RD, Alland, D, Eisen, JA, Carpenter, L, White, O, Peterson, J, et al. (2002). Whole-genome comparison of *Mycobacterium tuberculosis* clinical and laboratory strains. *J Bacteriol*, 184:5479–5490.

Fleisher, G, Starr, S, Koven, N, Kamiya, H, Douglas, SD, and Henle, W (1982). A non-x-linked syndrome with susceptibility to severe Epstein-Barr virus infections. *J Pediatr*, 100:727–730.

Fleming, D (2005). Influenza pandemics and avian flu. *BMJ*, 331:1066–1069.

Foth, BJ, Ralph, SA, Tonkin, CJ, Struck, NS, Fraunholz, M, Roos, DS, et al. (2003). Dissecting apicoplast targeting in the malaria parasite *Plasmodium falciparum*. *Science*, 299:705–708.

Francois, P, Pittet, D, Bento, M, Pepey, B, Vaudaux, P, Lew, D, et al. (2003). Rapid detection of methicillin-resistant *Staphylococcus aureus* directly from sterile or nonsterile clinical samples by a new molecular assay. *J Clin Microbiol*, 41:254–260.

Fraser-Liggett, CM (2005). Insights on biology and evolution from microbial genome sequencing. *Genome Res*, 15:1603–1610.

Gabriel, G, Dauber, B, Wolff, T, Planz, O, Klenk, HD, and Stech, J (2005). The viral polymerase mediates adaptation of an avian influenza virus to a mammalian host. *Proc Natl Acad Sci USA*, 102:18590–18595.

Galperin, MY (2006). The Molecular Biology Database Collection: 2006 update. *Nucleic Acids Res*, 34:D3–D5.

Garcia, A, Cot, M, Chippaux, JP, Ranque, S, Feingold, J, Demenais, F, et al. (1998). Genetic control of blood infection levels in human malaria: evidence for a complex genetic model. *Am J Trop Med Hyg*, 58:480–488.

Garcia, A, Abel, L, Cot, M, Richard, P, Ranque, S, Feingold, J, et al. (1999). Genetic epidemiology of host predisposition microfilaraemia in human loiasis. *Trop Med Int Health*, 4:565–574.

Gardner, MJ, Shallom, SJ, Carlton, JM, Salzberg, SL, Nene, V, Shoaibi, A, et al. (2002). Sequence of *Plasmodium falciparum* chromosomes 2, 10, 11 and 14. *Nature*, 419:531–534.

Gillis, JS (2006). An avian influenza vaccine for humans targeting the polymerase B2 protein inside the capsid instead of hemagglutinin or neuramidase on the virus surface. *Med Hypotheses*, 66:975–977.

Glennon, M and Cormican, M (2001). Detection and diagnosis of mycobacterial pathogens using PCR. *Expert Rev Mol Diagn*, 1:163–174.

Glickman, MS, Cox, JS, and Jacobs, WR Jr. (2000). A novel mycolic acid cyclopropane synthetase is required for cording, persistence, and virulence of *Mycobacterium tuberculosis*. *Mol Cell*, 5:717–727.

Gonzalez, E, Kulkarni, H, Bolivar, H, Mangano, A, Sanchez, R, Catano, G, et al. (2005). The influence of CCL3L1 gene-containing segmental duplications on HIV-1/AIDS susceptibility. *Science*, 307:1434–1440.

Gradmann, C (2006). Robert Koch and the white death: from tuberculosis to tuberculin. *Microbes Infect*, 8:294–301.

Graham, JE and Clark-Curtiss, JE (1999). Identification of *Mycobacterium tuberculosis* RNAs synthesized in response to phagocytosis by human macrophages by selective capture of transcribed sequences (SCOTS). *Proc Natl Acad Sci USA*, 96:11554–11559.

Greenstein, RJ (2003). Is Crohn's disease caused by a mycobacterium? Comparisons with leprosy, tuberculosis, and Johne's disease. *Lancet Infect Dis*, 3:507–514.

Haile, RW, Iselius, L, Fine, PE, and Morton, NE (1985). Segregation and linkage analyses of 72 leprosy pedigrees. *Hum Hered*, 35:43–52.

Hall, N, Pain, A, Berriman, M, Churcher, C, Harris, B, Harris, D, et al. (2002). Sequence of *Plasmodium falciparum* chromosomes 1, 3–9 and 13. *Nature*, 419:527–531.

Hamblin, MT and Di Rienzo, A (2000). Detection of the signature of natural selection in humans: evidence from the Duffy blood group locus. *Am J Hum Genet*, 66:1669–1679.

Hansen, W and Freney, J (2002). [Armauer Hansen (1841–1912), portrait of a Nordic pioneer]. *Hist Sci Med*, 36:75–81.

Haralambous, E, Weiss, HA, Radalowicz, A, Hibberd, ML, Booy, R, and Levin, M (2003). Sibling familial risk ratio of meningococcal disease in UK Caucasians. *Epidemiol Infect*, 130:413–418.

Hayman, A, Comely, S, Lackenby, A, Murphy, S, McCauley, J, Goodbourn, S, et al. (2006). Variation in the ability of human influenza A viruses to induce and inhibit the IFN-β pathway. *Virology*, 347(1):52–64.

He, CY, Shaw, MK, Pletcher, CH, Striepen, B, Tilney, LG, and Roos, DS (2001). A plastid segregation defect in the protozoan parasite *Toxoplasma gondii*. *EMBO J*, 20:330–339.

Hiramatsu, K, Cui, L, Kuroda, M, and Ito, T (2001). The emergence and evolution of methicillin-resistant *Staphylococcus aureus*. *Trends Microbiol*, 9:486–493.

Hoge, CW, Fisher, L, Donnell, HD, Jr., Dodson, DR, Tomlinson, GV, Jr., Breiman, RF, et al. (1994). Risk factors for transmission of *Mycobacterium tuberculosis* in a primary school outbreak: lack of racial difference in susceptibility to infection. *Am J Epidemiol*, 139:520–530.

Holden, MT, Feil, EJ, Lindsay, JA, Peacock, SJ, Day, NP, Enright, MC, et al. (2004). Complete genomes of two clinical *Staphylococcus aureus* strains: evidence for the rapid evolution of virulence and drug resistance. *Proc Natl Acad Sci USA*, 101:9786–9791.

Huynen, M, Dandekar, T, and Bork, P (1998). Differential genome analysis applied to the species-specific features of *Helicobacter pylori*. *FEBS Lett*, 426:1–5.

Ioannidis, JP, Gwinn, M, Little, J, Higgins, JP, Bernstein, JL, Boffetta, P, et al. (2006). A road map for efficient and reliable human genome epidemiology. *Nat Genet*, 38:3–5.

Ito, T, Katayama, Y, Asada, K, Mori, N, Tsutsumimoto, K, Tiensasitorn, C, et al. (2001). Structural comparison of three types of staphylococcal cassette chromosome mec integrated in the chromosome in methicillin-resistant *Staphylococcus aureus*. *Antimicrob Agents Chemother*, 45:1323–1336.

Ivens, AC, Peacock, CS, Worthey, EA, Murphy, L, Aggarwal, G, Berriman, M, et al. (2005). The genome of the kinetoplastid parasite, *Leishmania major*. *Science*, 309:436–442.

Jepson, AP, Banya, WA, Sisay-Joof, F, Hassan-King, M, Bennett, S, and Whittle, HC (1995). Genetic regulation of fever in *Plasmodium falciparum* malaria in Gambian twin children. *J Infect Dis*, 172:316–319.

Jepson, A, Banya, W, Sisay-Joof, F, Hassan-King, M, Nunes, C, Bennett, S, et al. (1997). Quantification of the relative contribution of major histocompatibility complex (MHC) and non-MHC genes to human immune responses to foreign antigens. *Infect Immun*, 65:872–876.

Jepson, A, Fowler, A, Banya, W, Singh, M, Bennett, S, Whittle, H, et al. (2001). Genetic regulation of acquired immune responses to antigens of *Mycobacterium tuberculosis*: a study of twins in West Africa. *Infect Immun*, 69:3989–3994.

Jiang, SS, Chen, TC, Yang, JY, Hsiung, CA, Su, IJ, Liu, YL, et al. (2004). Sensitive and quantitative detection of severe acute respiratory syndrome coronavirus infection by real-time nested polymerase chain reaction. *Clin Infect Dis*, 38:293–296.

Jones, RN (2003). Global epidemiology of antimicrobial resistance among community-acquired and nosocomial pathogens: a five-year summary from the SENTRY Antimicrobial Surveillance Program (1997–2001). *Semin Respir Crit Care Med*, 24:121–134.

Jouanguy, E, Altare, F, Lamhamedi, S, Revy, P, Emile, JF, Newport, M, et al. (1996). Interferon-gamma-receptor deficiency in an infant with fatal bacille Calmette-Guerin infection. *N Engl J Med*, 335:1956–1961.

Jouanguy, E, Lamhamedi-Cherradi, S, Altare, F, Fondaneche, MC, Tuerlinckx, D, Blanche, S, et al. (1997). Partial interferon-gamma receptor 1 deficiency in a child with tuberculoid bacillus Calmette-Guerin infection and a sibling with clinical tuberculosis. *J Clin Invest*, 100:2658–2664.

Kallmann FJ and Reisner, D (1943). Twin studies on the significance of genetic factors in tuberculosis. *Am Rev Tuber*, 47:549–574.

Kanz, C, Aldebert, P, Althorpe, N, Baker, W, Baldwin, A, Bates, K, et al. (2005). The EMBL Nucleotide Sequence Database. *Nucleic Acids Res*, 33: D29–D33.

Kao, RY, Tsui, WH, Lee, TS, Tanner, JA, Watt, RM, Huang, JD, et al. (2004). Identification of novel small-molecule inhibitors of severe acute respiratory syndrome-associated coronavirus by chemical genetics. *Chem Biol*, 11:1293–1299.

Katoch, VM (2004). Infections due to non-tuberculous mycobacteria (NTM). *Indian J Med Res*, 120:290–304.

Kawaoka, Y, Krauss, S, and Webster, RG (1989). Avian-to-human transmission of the PB1 gene of influenza A viruses in the 1957 and 1968 pandemics. *J Virol*, 63:4603–4608.

Kelley, J, de Bono, B, and Trowsdale, J (2005). IRIS: a database surveying known human immune system genes. *Genomics*, 85:503–511.

Klein, RJ, Zeiss, C, Chew, EY, Tsai, JY, Sackler, RS, Haynes, C, et al. (2005). Complement factor H polymorphism in age-related macular degeneration. *Science*, 308:385–389.

Koessler, T, Francois, P, Charbonnier, Y, Huyghe, A, Bento, M, Dharan, S, et al. (2006). Use of oligoarrays for characterization of community-onset methicillin-resistant *Staphylococcus aureus*. *J Clin Microbiol*, 44:1040–1048.

Kostman, JR, Alden, MB, Mair, M, Edlind, TD, LiPuma, JJ, and Stull, TL (1995). A universal approach to bacterial molecular epidemiology by polymerase chain reaction ribotyping. *J Infect Dis*, 171:204–208.

Kuroda, M, Ohta, T, Uchiyama, I, Baba, T, Yuzawa, H, Kobayashi, I, Cui, L, et al. (2001). Whole genome sequencing of meticillin-resistant *Staphylococcus aureus*. *Lancet*, 357:1225–1240.

Kvaerner, KJ, Tambs, K, Harris, JR, and Magnus, P (1997). Distribution and heritability of recurrent ear infections. *Ann Otol Rhinol Laryngol*, 106:624–632.

Kwiatkowski, DP (2005). How malaria has affected the human genome and what human genetics can teach us about malaria. *Am J Hum Genet*, 77:171–192.

Lander, E and Kruglyak, L (1995). Genetic dissection of complex traits: guidelines for interpreting and reporting linkage results. *Nat Genet*, 11:241–247.

Lee, JM and Sonnhammer, EL (2003). Genomic gene clustering analysis of pathways in eukaryotes. *Genome Res*, 13:875–882.

Lee, PJ and Krilov, LR (2005). When animal viruses attack: SARS and avian influenza. *Pediatr Ann*, 34:42–52.

Levin, M and Newport, M (2000). Inherited predisposition to mycobacterial infection: historical considerations. *Microbes Infect*, 2:1549–1552.

Li, L, Bannantine, JP, Zhang, Q, Amonsin, A, May, BJ, Alt, D, et al. (2005). The complete genome sequence of *Mycobacterium avium* subspecies paratuberculosis. *Proc Natl Acad Sci USA*, 102:12344–12349.

Lilic, D (2002). New perspectives on the immunology of chronic mucocutaneous candidiasis. *Curr Opin Infect Dis*, 15:143–147.

Lindsay, JA and Holden, MT (2004). *Staphylococcus aureus*: superbug, super genome? *Trends Microbiol*, 12:378–385.

Lindsay, JA and Holden, MT (2006). Understanding the rise of the superbug: investigation of the evolution and genomic variation of *Staphylococcus aureus*. *Funct Integr Genomics*, 6(3): 186–201.

Lohmueller, KE, Pearce, CL, Pike, M, Lander ES, and Hirschhorn, JN (2003). Meta-analysis of genetic association studies supports a contribution of common variants to susceptibility to common disease. *Nat Genet*, 33:177–182.

Mackinnon, MJ, Gunawardena, DM, Rajakaruna, J, Weerasingha, S, Mendis, KN, and Carter, R (2000). Quantifying genetic and nongenetic contributions to malarial infection in a Sri Lankan population. *Proc Natl Acad Sci USA*, 97:12661–12666.

Mackinnon, MJ, Mwangi, TW, Snow, RW, Marsh, K, and Williams, TN (2005). Heritability of Malaria in Africa. *PLoS Med*, 2:e340.

Mah, S, Nelson, MR, Delisi, LE, Reneland, RH, Markward, N, James, MR, et al. (2006). Identification of the semaphorin receptor PLXNA2 as a candidate for susceptibility to schizophrenia. *Mol Psychiatry*, 11(5):471–478.

Malaty, HM, Engstrand, L, Pedersen, NL, and Graham, DY (1994). *Helicobacter pylori* infection: genetic and environmental influences. A study of twins. *Ann Intern Med*, 120:982–986.

Marquet, S, Abel, L, Hillaire, D, and Dessein, A (1999). Full results of the genome-wide scan which localises a locus controlling the intensity of infection by *Schistosoma mansoni* on chromosome 5q31-q33. *Eur J Hum Genet*, 7:88–97.

Marquet, S, Abel, L, Hillaire, D, Dessein, H, Kalil, J, Feingold, J, et al. (1996). Genetic localization of a locus controlling the intensity of infection by *Schistosoma mansoni* on chromosome 5q31-q33. *Nat Genet*, 14:181–184.

Marra, MA, Jones, SJ, Astell, CR, Holt, RA, Brooks-Wilson, A, Butterfield, YS, et al. (2003). The Genome sequence of the SARS-associated coronavirus. *Science*, 300:1399–1404.

Martin, AM, Freitas, EM, Witt, CS, and Christiansen, FT (2000). The genomic organization and evolution of the natural killer immunoglobulin-like receptor (KIR) gene cluster. *Immunogenetics*, 51:268–280.

McKinney, JD, Honer zu Bentrup, K, Munoz-Elias, EJ, Miczak, A, Chen, B, Chan, WT, et al. (2000). Persistence of *Mycobacterium tuberculosis* in macrophages and mice requires the glyoxylate shunt enzyme isocitrate lyase. *Nature*, 406:735–738.

Meisner, SJ, Mucklow, S, Warner, G, Sow, SO, Lienhardt, C, and Hill, AV (2001). Association of NRAMP1 polymorphism with leprosy type but not susceptibility to leprosy per se in West Africans. *Am J Trop Med Hyg*, 65:733–735.

Melville, SE, Gerrard, CS, and Blackwell, JM (1999). Multiple causes of size variation in the diploid megabase chromosomes of African tyrpanosomes. *Chromosome Res*, 7:191–203.

Miller, EN, Jamieson, SE, Joberty, C, Fakiola, M, Hudson, D, Peacock, CS, et al. (2004). Genome-wide scans for leprosy and tuberculosis susceptibility genes in Brazilians. *Genes Immun*, 5:63–67.

Miller, LH, Mason, SJ, Dvorak, JA, McGinniss, MH, and Rothman, IK (1975). Erythrocyte receptors for (*Plasmodium knowlesi*) malaria: Duffy blood group determinants. *Science*, 189:561–563.

Mira, MT, Alcais, A, Nguyen, VT, Moraes, MO, Di Flumeri, C, Vu, HT, et al. (2004). Susceptibility to leprosy is associated with PARK2 and PACRG. *Nature*, 427:636–640.

Mira, MT, Alcais, A, Van Thuc, N, Thai, VH, Huong, NT, Ba, NN, et al. (2003). Chromosome 6q25 is linked to susceptibility to leprosy in a Vietnamese population. *Nat Genet*, 33:412–415.

Modi, WS (2004). CCL3L1 and CCL4L1 chemokine genes are located in a segmental duplication at chromosome 17q12. *Genomics*, 83:735–738.

Modiano, D, Petrarca, V, Sirima, BS, Nebie, I, Diallo, D, Esposito, F, et al. (1996). Different response to *Plasmodium falciparum* malaria in west African sympatric ethnic groups. *Proc Natl Acad Sci USA*, 93:13206–13211.

Montpetit, A, Nelis, M, Laflamme, P, Magi, R, Ke, X, Remm, M, et al. (2006). An Evaluation of the performance of tag SNPs derived from HapMap in a Caucasian population. *PLoS Genet*, 2:e27.

Morens, DM, Folkers, GK, and Fauci, AS (2004). The challenge of emerging and re-emerging infectious diseases. *Nature*, 430:242–249.

Mueller, AK, Camargo, N, Kaiser, K, Andorfer, C, Frevert, U, Matuschewski, K, et al. (2005). Plasmodium liver stage developmental arrest by depletion of a protein at the parasite-host interface. *Proc Natl Acad Sci USA*, 102:3022–3027.

Muller-Myhsok, B, Stelma, FF, Guisse-Sow, F, Muntau, B, Thye, T, Burchard, GD, et al. (1997). Further evidence suggesting the presence of a locus, on human chromosome 5q31-q33, influencing the intensity of infection with *Schistosoma mansoni*. *Am J Hum Genet*, 61:452–454.

Murray, CJ and Acharya, AK (1997). Understanding DALYs (disability-adjusted life years). *J Health Econ*, 16:703–730.

Mushegian, AR and Koonin, EV (1996). A minimal gene set for cellular life derived by comparison of complete bacterial genomes. *Proc Natl Acad Sci USA*, 93:10268–10273.

Namkung, J, Kim, Y, and Park, T (2005). Whole-genome association studies of alcoholism with loci linked to schizophrenia susceptibility. *BMC Genet*, 6 (Suppl. 1):S9.

Newport, MJ, Goetghebuer, T, Weiss, HA, Whittle, H, Siegrist, CA, and Marchant, A (2004). Genetic regulation of immune responses to vaccines in early life. *Genes Immun*, 5:122–129.

Newport, MJ, Huxley, CM, Huston, S, Hawrylowicz, CM, Oostra, BA, Williamson, R, et al. (1996). A mutation in the interferon-gamma-receptor gene and susceptibility to mycobacterial infection. *N Engl J Med*, 335:1941–1949.

Nurnberger, W, Pietsch, H, Seger, R, Bufon, T, and Wahn, V (1989). Familial deficiency of the seventh component of complement associated with recurrent meningococcal infections. *Eur J Pediatr*, 148:758–760.

Nyholt, DR (2000). All LODs are not created equal. *Am J Hum Genet*, 67:282–288.

Obenauer, JC, Denson, J, Mehta, PK, Su, X, Mukatira, S, Finkelstein, DB, et al. (2006). Large-Scale Sequence Analysis of Avian Influenza Isolates. *Science*, 311:1576–1580.

Palmer, C, Bik, EM, Eisen, MB, Eckburg, PB, Sana, TR, Wolber, PK, et al. (2006). Rapid quantitative profiling of complex microbial populations. *Nucleic Acids Res*, 34:e5.

Parkhill, J, Sebaihia, M, Preston, A, Murphy, LD, Thomson, N, Harris, DE, et al. (2003). Comparative analysis of the genome sequences of *Bordetella pertussis*, *Bordetella parapertussis* and *Bordetella bronchiseptica*. *Nat Genet*, 35:32–40.

Peacock, C.S. et al. (2001). Genetic epidemiology of visceral leishmaniasis in northeastern Brazil. *Genet Epidemiol*, 20: 383–396.

Perna, NT, Plunkett, G, 3rd, Burland, V, Mau, B, Glasner, JD, et al. (2001). Genome sequence of enterohaemorrhagic *Escherichia coli* O157:H7. *Nature*, 409:529–533.

Phillips, S and Novick, RP (1979). Tn554—a site-specific repressor-controlled transposon in *Staphylococcus aureus*. *Nature*, 278:476–478.

Picard, C, Fieschi, C, Altare, F, Al-Jumaah, S, Al-Hajjar, S, Feinberg, J, et al. (2002). Inherited interleukin-12 deficiency: IL12B genotype and clinical phenotype of 13 patients from six kindreds. *Am J Hum Genet*, 70:336–348.

Plancoulaine, S. et al. (2000). Detection of a major gene predisposing to human T lymphotropic virus type I infection in children among an endemic population of African origin. *J Infect Dis*, 182: 405–512.

Plancoulaine, S, Gessain, A, van Beveren, M, Tortevoye, P, Abel, L. (2003). Evidence for a recessive major gene predisposing to human herpesvirus 8 (HHV-8) infection in a population in which HHV-8 is endemic. *J Infect Dis* 187: 1944–1950.

Poon, LL, Guan, Y, Nicholls, JM, Yuen, KY, and Peiris, JS (2004). The aetiology, origins, and diagnosis of severe acute respiratory syndrome. *Lancet Infect Dis*, 4:663–671.

Purtilo, DT, DeFlorio, D, Jr., Hutt, LM, Bhawan, J, Yang, JP, Otto, R, et al. (1977). Variable phenotypic expression of an X-linked recessive lymphoproliferative syndrome. *N Engl J Med*, 297:1077–1080.

Pym, AS, Brodin, P, Majlessi, L, Brosch, R, Demangel, C, Williams, A, et al. (2003). Recombinant BCG exporting ESAT-6 confers enhanced protection against tuberculosis. *Nat Med*, 9:533–539.

Ramoz, N, Rueda, LA, Bouadjar, B, Montoya, LS, Orth, G, and Favre, M (2002). Mutations in two adjacent novel genes are associated with epidermodysplasia verruciformis. *Nat Genet*, 32:579–581.

Ranque, S, Safeukui, I, Poudiougou, B, Traore, A, Keita, M, Traore, D, et al. (2005). Familial aggregation of cerebral malaria and severe malarial anemia. *J Infect Dis*, 191:799–804.

Rawlinson, WD, Basten, A, Britton, WJ, and Serjeantson, SW (1988). Leprosy and immunity: genetics and immune function in multiple case families. *Immunol Cell Biol*, 66(Pt 1):9–21.

Reid, AH, Taubenberger, JK, and Fanning, TG (2004). Evidence of an absence: the genetic origins of the 1918 pandemic influenza virus. *Nat Rev Microbiol*, 2:909–914.

Rihet, P, Abel, L, Traoré, Y, Traoré-Leroux, T, Aucan C, and Fumoux, F (1998). Human malaria: segregation analysis of blood infection levels in a suburban area and a rural area in Burkina Faso. *Genet Epidemiol*, 15:435–450.

Rodrigues, V, Jr, Abel, L, Piper, K, and Dessein, AJ (1996). Segregation analysis indicates a major gene in the control of interleukine-5 production in humans infected with Schistosoma mansoni. *Am J Hum Genet*, 59:453–461.

Rogers, BL (2004). Bacterial targets to antimicrobial leads and development candidates. *Curr Opin Drug Discov Devel*, 7:211–222.

Rovers, M, Haggard, M, Gannon, M, Koeppen-Schomerus, G, and Plomin, R (2002). Heritability of symptom domains in otitis media: a longitudinal study of 1,373 twin pairs. *Am J Epidemiol*, 155:958–964.

Sabeti, PC, Walsh, E, Schaffner, SF, Varilly, P, Fry, B, Hutcheson, HB, et al. (2005). The case for selection at CCR5-Delta32. *PLoS Biol*, 3:e378.

Saiki, RK, Scharf, S, Faloona, F, Mullis, KB, Horn, GT, Erlich, HA, et al. (1985). Enzymatic amplification of beta-globin genomic sequences and restriction site analysis for diagnosis of sickle cell anemia. *Science*, 230:1350–1354.

Sanger, F, Air, GM, Barrell, BG, Brown, NL, Coulson, AR, Fiddes, CA, et al. (1977). Nucleotide sequence of bacteriophage phi X174 DNA. *Nature*, 265:687–695.

Sanger, F, Coulson, AR, Barrell, BG, Smith, AJ, and Roe, BA (1980). Cloning in single-stranded bacteriophage as an aid to rapid DNA sequencing. *J Mol Biol*, 143:161–178.

Schloss, PD and Handelsman, J (2005). Metagenomics for studying unculturable microorganisms: cutting the Gordian knot. *Genome Biol*, 6:229.

Searle, S and Blackwell, JM (1999). Evidence for a functional repeat polymorphism in the promoter of the human NRAMP1 gene that correlates with autoimmune versus infectious disease susceptibility. *J Med Genet*, 36:295–299.

Sharp, PM, Bailes, E, Chaudhuri, RR, Rodenburg, CM, Santiago, MO, and Hahn, BH (2001). The origins of acquired immune deficiency syndrome viruses: where and when? *Philos Trans R Soc Lond B Biol Sci*, 356:867–876.

Shaw, MA, Collins, A, Peacock, CS, Miller, EN, Black, GF, Sibthorpe, D, et al. (1997). Evidence that genetic susceptibility to *Mycobacterium tuberculosis* in a Brazilian population is under oligogenic control: linkage study of the candidate genes NRAMP1 and TNFA. *Tuber Lung Dis*, 78:35–45.

Shaw, MA, Davies, CR, Llanos-Cuentas, EA, and Collins, A (1995). Human genetic susceptibility and infection with Leishmania peruviana. *Am J Hum Genet*, 57:1159–1168.

Shen, FM, Lee, MK, Gong, HM, Cai, XQ, and King, MC (1991). Complex segregation analysis of primary hepatocellular carcinoma in Chinese families: interaction of inherited susceptibility and hepatitis B viral infection. *Am J Hum Genet*, 49:88–93.

Siddiqui, MR, Meisner, S, Tosh, K, Balakrishnan, K, Ghei, S, Fisher, SE, et al. (2001). A major susceptibility locus for leprosy in India maps to chromosome 10p13. *Nat Genet*, 27:439–441.

Sjoberg, K, Lepers, JP, Raharimalala, L, Larsson, A, Olerup, O, Marbiah, NT, et al. (1992). Genetic regulation of human anti-malarial antibodies in twins. *Proc Natl Acad Sci USA*, 89:2101–2104.

Skol, AD, Scott, LJ, Abecasis, GR, and Boehnke, M (2006). Joint analysis is more efficient than replication-based analysis for two-stage genome-wide association studies. *Nat Genet*, 38:209–213.

Smith, DG (1979). The genetic hypothesis for susceptibility to lepromatous leprosy. *Hum Genet*, 50:163–177.

Smith, LM, Sanders, JZ, Kaiser, RJ, Hughes, P, Dodd, C, Connell, CR, et al. (1986). Fluorescence detection in automated DNA sequence analysis. *Nature*, 321:674–679.

Snow, RW, Guerra, CA, Noor, AM, Myint, HY, and Hay, SI (2005). The global distribution of clinical episodes of *Plasmodium falciparum* malaria. *Nature*, 434:214–217.

Sorensen, TI, Nielsen, GG, Andersen, PK, and Teasdale, TW (1988). Genetic and environmental influences on premature death in adult adoptees. *N Engl J Med*, 318:727–732.

Spielman, RS, McGinnis, RE, and Ewens, WJ (1993). Transmission test for linkage disequilibrium: the insulin gene region and insulin-dependent diabetes mellitus (IDDM). *Am J Hum Genet*, 52:506–516.

Stankovich, J, Cox, CJ, Tan, RB, Montgomery, DS, Huxtable, SJ, Rubio, JP, et al. (2006). On the utility of data from the International HapMap Project for Australian association studies. *Hum Genet*, 119:220–222.

Stead, WW, Senner, JW, Reddick, WT, and Lofgren, JP (1990). Racial differences in susceptibility to infection by *Mycobacterium tuberculosis*. *N Engl J Med*, 322:422–427.

Stephens, JC, Reich, DE, Goldstein, DB, Shin, HD, Smith, MW, Carrington, M, et al. (1998). Dating the origin of the CCR5-Delta32 AIDS-resistance allele by the coalescence of haplotypes. *Am J Hum Genet*, 62:1507–1515.

Syvanen, AC (2005). Toward genome-wide SNP genotyping. *Nat Genet* 37 (Suppl.):S5–S10.

Taylor, LH, Latham, SM, and Woolhouse, ME (2001). Risk factors for human disease emergence. *Philos Trans R Soc Lond B Biol Sci*, 356:983–989.

Terhell, AJ, Houwing-Duistermaat, JJ, Ruiterman, Y, Haarbrink, M, Abadi, K, and Yazdanbakhsh, M (2000). Clustering of *Brugia malayi* infection in a community in South-Sulawesi, Indonesia. *Parasitology*, 120(Pt 1):23–29.

The International Human Genome Sequencing Consortium (2004). Finishing the euchromatic sequence of the human genome. *Nature*, 431:931–945.

Thomas, CM and Nielsen, KM (2005). Mechanisms of, and barriers to, horizontal gene transfer between bacteria. *Nat Rev Microbiol*, 3:711–721.

Tosh, K, Meisner, S, Siddiqui, MR, Balakrishnan, K, Ghei, S, Golding, M, et al. (2002). A region of chromosome 20 is linked to leprosy susceptibility in a South Indian population. *J Infect Dis*, 186:1190–1193.

Tringe, SG and Rubin, EM (2005). Metagenomics: DNA sequencing of environmental samples. *Nat Rev Genet*, 6:805–814.

Trowsdale, J and Parham, P (2004). Mini-review: defense strategies and immunity-related genes. *Eur J Immunol*, 34:7–17.

van Dijk, MR, Douradinha, B, Franke-Fayard, B, Heussler, V, van Dooren, MW, van Schaijk, B, et al. (2005). Genetically attenuated, P36p-deficient malarial sporozoites induce protective immunity and apoptosis of infected liver cells. *Proc Natl Acad Sci USA*, 102:12194–12199.

Vidal, S, Tremblay, ML, Govoni, G, Gauthier, S, Sebastiani, G, Malo, D, et al. (1995). The Ity/Lsh/Bcg locus: natural resistance to infection with intracellular parasites is abrogated by disruption of the Nramp1 gene. *J Exp Med*, 182:655–666.

Wagener, DK, Schauf, V, Nelson, KE, Scollard D, Brown, A, and Smith, T (1988). Segregation analysis of leprosy in families of northern Thailand. *Genet Epidemiol*, 5:95–105.

Wallace, C, Clayton, D, and Fine, P (2003). Estimating the relative recurrence risk ratio for leprosy in Karonga District, Malawi. *Lepr Rev*, 74:133–140.

Wallace, C, Fitness, J, Hennig, B, Sichali, L, Mwaungulu, L, Ponnighaus, JM, et al. (2004). Linkage analysis of susceptibility to leprosy type using an IBD regression method. *Genes Immun*, 5:221–225.

Wang, D, Urisman, A, Liu, YT, Springer, M, Ksiazek, TG, Erdman, DD, et al. (2003). Viral discovery and sequence recovery using DNA microarrays. *PLoS Biol*, 1:E2.

Wang, WY, Barratt, BJ, Clayton, DG, and Todd, JA (2005). Genome-wide association studies: theoretical and practical concerns. *Nat Rev Genet*, 6:109–118.

Waters, A (2006). Malaria: new vaccines for old? *Cell*, 124:689–693.

Weiss, RA (2001). The Leeuwenhoek Lecture 2001. Animal origins of human infectious disease. *Philos Trans R Soc Lond B Biol Sci*, 356:957–977.

Westendorp, RG, Langermans, JA, Huizinga, TW, Elouali, AH, Verweij, CL, Boomsma, DI, et al. (1997). Genetic influence on cytokine production and fatal meningococcal disease. *Lancet*, 349:170–173.

WHO (2004). *Changing History*. WHO, Geneva.

Wiesner, J and Seeber, F (2005). The plastid-derived organelle of protozoan human parasites asa target of established and emerging drugs. *Expert Opin Ther Targets*, 9:23–44.

Williams-Blangero, S, Subedi, J, Upadhayay, RP, Manral, DB, Rai, DR, Jha, B, et al. (1999). Genetic analysis of susceptibility to infection with *Ascaris lumbricoides*. *Am J Trop Med Hyg*, 60:921–926.

Williams-Blangero, S. et al. (2002). Genes on chromosomes 1 and 13 have significant effects on Ascaris infection. *Proc Natl Acad Sci USA*, 99: 5533–5588.

Wilmes, P and Bond, PL (2006). Metaproteomics: studying functional gene expression in microbial ecosystems. *Trends Microbiol*, 14:92–97.

Witney, AA, Marsden, GL, Holden, MT, Stabler, RA, Husain, SE, Vass, JK, et al. (2005). Design, validation, and application of a seven-strain *Staphylococcus aureus* PCR product microarray for comparative genomics. *Appl Environ Microbiol*, 71:7504–7514.

Wong, SS and Yuen, KY (2005). The severe acute respiratory syndrome (SARS). *J Neurovirol*, 11:455–468.

Wu, HS, Chiu, SC, Tseng, TC, Lin, SF, Lin, JH, Hsu, YH, et al. (2004). Serologic and molecular biologic methods for SARS-associated coronavirus infection, Taiwan. *Emerg Infect Dis*, 10:304–310.

Wurzner, R, Orren, A, and Lachmann, PJ (1992). Inherited deficiencies of the terminal components of human complement. *Immunodefic Rev*, 3:123–147.

Xu, X, Liu, Y, Weiss, S, Arnold, E, Sarafianos, SG, and Ding, J (2003). Molecular model of SARS coronavirus polymerase: implications for biochemical functions and drug design. *Nucleic Acids Res*, 31:7117–7130.

Yamazaki, K, McGovern, D, Ragoussis, J, Paolucci, M, Butler, H, Jewell, D, et al. (2005). Single nucleotide polymorphisms in TNFSF15 confer susceptibility to Crohn's disease. *Hum Mol Genet*, 14:3499–3506.

Yap, Y, Zhang, X, Andonov, A, and He, R (2005). Structural analysis of inhibition mechanisms of aurintricarboxylic acid on SARS-CoV polymerase and other proteins. *Comput Biol Chem*, 29:212–219.

Zhang, DM, Wang, GL, and Lu, JH (2005). Severe acute respiratory syndrome: vaccine on the way. *Chin Med J (Engl)*, 118:1468–1476.

Zhu, MS, Pan, Y, Chen, HQ, Shen, Y, Wang, XC, Sun, YJ, et al. (2004). Induction of SARS-nucleoprotein-specific immune response by use of DNA vaccine. *Immunol Lett*, 92:237–243.

Zinn-Justin, A, Marquet, S, Hillaire, D, Dessein, A, and Abel, L (2001). Genome search for additional human loci controlling infection levels by *Schistosoma mansoni*. *Am J Trop Med Hyg*, 65:754–758.

26

Rheumatoid Arthritis and Related Arthropathies

Pille Harrison and Paul Wordsworth

Chronic inflammatory joint diseases of noninfective origin are common and cause significant long-term morbidity. Their etiology is not entirely clear but genetic factors have been suspected for a long time. Rheumatoid arthritis (RA) and most of the related forms of inflammatory arthritis are multifactorial diseases. Although they are often referred to as "*autoimmune disorders*," it is far from clear what triggers the abnormal inflammatory processes that characterize these conditions. In contrast, there has been substantial progress in determining the nature of genetic susceptibility to these potentially crippling disorders. After 30 years in which the human leukocyte antigen (*HLA*) genes have stood out as the only oasis in a genetic desert the application of genome-wide linkage and association techniques is beginning to bear fruit. The striking recent example of the association of the protein tyrosine phosphatase, *PTPN22*, with RA and several related autoimmune diseases seems likely to herald a wealth of new genetic information to inform our knowledge of the etiology and pathogenesis of these conditions.

The seronegative spondyloarthropathies exhibit many similarities clinically and pathologically with RA. They also have a major genetic basis, the full extent of which is currently being unravelled.

Rheumatoid Arthritis

RA is a chronic immune-mediated inflammatory disorder predominantly, but not exclusively, affecting synovial joints (Fig. 26–1). It is clinically heterogeneous in its pattern of joint inflammation, disease progression, the presence of extraarticular features, and response to therapy. Systemic features, including weight loss, malaise, low-grade fever, and vasculitis are also common but variable accompaniments. Some commentators have suggested that it is better to regard RA as a syndrome rather than a single specific disease entity although for disease classification purposes the 1987 criteria developed by the American College of Rheumatology (Arnett et al., 1988) have proved relatively robust (Table 26–1).

Rheumatoid synovitis typically affects the diarthrodial joints of the hands and feet first, causing progressive cartilage and bone destruction in the affected joints (Fig. 26–2). Subsequently, the more proximal joints, including hip and knee, may become involved leading to a severe functional disability. In the course of the disease, any synovial joint can be affected, including the temporomandibular, and even the cricoarytenoid joints.

RA is one of the most common autoimmune disorders affecting approximately 1% of most populations worldwide. However, it is relatively uncommon in much of sub-Saharan Africa and is less common in many rural communities than their urban equivalents; for example, rural South Africa versus townships. The peak age at onset is in the fifth decade, although it can start at virtually any age. Overall, the disease is three times more common in women, but this varies with the age at onset; at 30 years of age, the incidence is 10 times higher in women than men but at the age of 65 there is no gender difference. RA usually has a subacute (20%) or insidious (60%) onset with inflammatory symptoms of pain, swelling, and stiffness typically in the extremities, gradually involving more joints over weeks or months. Approximately 5% of patients have an acute severe polyarticular onset, and a few start with palindromic rheumatism characterized by short-lived episodes (<48 hours) of synovitis, recurring at variable intervals. The course of the disease is variable; mild self-limiting disease is common in the community while the development of bony erosions early in the course of the disease indicates a poor prognosis. The presence of certain types of autoantibody in the blood is not only highly predictive for the development of RA but also often predates the onset of clinical manifestations by months or years. These antibodies include classic immunoglobulin M (IgM) rheumatoid factor (RF) found in approximately 85% of the patients and the more specific antibodies to citrullinated cyclic peptides (anti-CCP antibodies); the latter have a greater positive predictive value and are gradually replacing RF as a diagnostic test in many centers.

Bone loss in RA is both focal and more generalized. Juxtaarticular osteopenia adjacent to inflamed joints and the presence of focal erosions of subchondral bone at the joint margins occur in the areas of direct invasion by synovial pannus. This is mediated through the action of osteoclast-like cells believed to be derived from macrophage/monocyte precursors under the influence of growth factors such as RANKL [receptor activator of (nuclear factor κB) NFκB ligand] and proinflammatory cytokines, such as tumor necrosis factor (TNF) produced in large quantities locally in the inflamed synovium by activated macrophages. The inflammatory process can also lead to a more generalized bone loss causing as osteopenia and osteoporosis of the axial and appendicular skeleton (Goldring and Gravallese, 2000).

There are conflicting reports about age-related differences in the clinical features and outcome of RA (van Schaardenburg et al.,

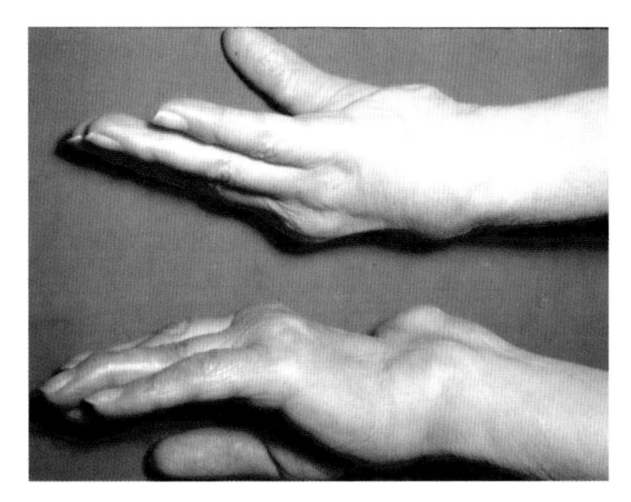

Figure 26–1 Rheumatoid arthritis of the hands showing volar subluxation at the wrist and metacarpophalangeal joints with associated soft tissue swelling.

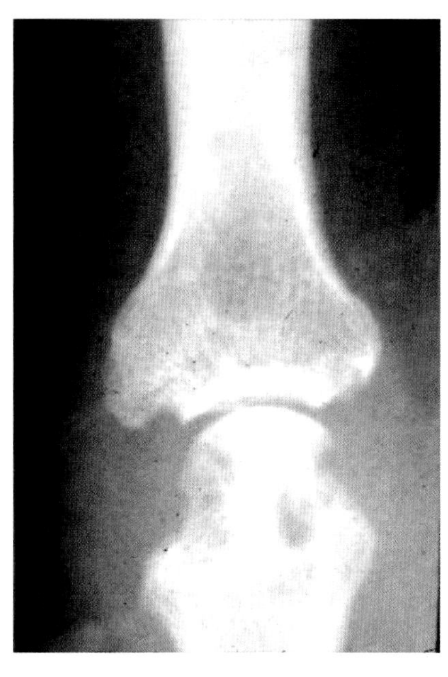

Figure 26–2 Classic juxtaarticular erosions in rheumatoid arthritis occurring at the point of synovial insertion (arrows).

1993; Peltomaa et al., 2000). However, some of the apparently greater damage in those older than 55 years of age at first presentation can probably be attributed to the preexisting osteoarthritic joint damage (Khanna et al., 2005).

Extraarticular manifestations of RA are common and accentuate the systemic nature of the disease. They can occur at any time in the course of the disease and often occur quite independently of the current severity of the joint disease (Turesson et al., 1999). In population-based studies, 56% of male patients and 42% of female patients have evidence of nonarticular rheumatoid organ disease (Weyand et al., 1998). Smoking and early progression to disability seem to increase the risk of severe extraarticular manifestations (Turesson, C et al., 2003).

Extraskeletal manifestations of RA can affect almost any organ and include rheumatoid nodules, vasculitis, serositis, pulmonary fibrosis and hematological and neurological abnormalities. These extraarticular manifestations have a defining impact on disease outcome, including increased mortality not only in relation to the general population but also in comparison to RA in general (Turesson, I et al., 2003). Standardized mortality rates are increased two or three times, largely due to increased cardiovascular disease and infections. Some differences in extraarticular manifestations have been observed between men and women; episodes of major vasculitis affect one in ten men with RA but are approximately three times less common in women. Major systemic upset, including weight loss and malaise, is also a component of RA presenting in a subset of middle-aged males. In contrast, women are more likely to describe symptoms related to secondary Sjogren syndrome than men (Thomas and Hay, 1998). Lung disease occurs more commonly in men with long-standing illness, positive RF, and subcutaneous nodules. Pleural involvement, often asymptomatic, is the most common manifestation of lung disease in RA and may occur concurrently with pulmonary nodulosis or interstitial disease (Anaya et al., 1995). The classic cardiovascular manifestations of RA include pericarditis, nodular valve disease, and major vessel vasculitis. It is also increasingly apparent that the risk of myocardial infarction or congestive heart failure in RA patients is significantly higher than in age and sex-matched individuals but that no more than half of this excess can be attributed to classic risk factors, such as hypertension, smoking, and dyslipidemia (Gabriel et al., 1999a; Nicola and Marandit-Kremers, 2005). Cardiovascular events are four-fold more common than in the general population (del Rincon et al., 2001). The reasons for increased cardiovascular morbidity and mortality are likely to be multifactorial, but it is likely that the chronic systemic inflammation found in rheumatoid disease promotes endothelial dysfunction and induces premature atherosclerosis (Ridker et al., 2002). Clearly, aggressive control of classic risk factors for cardiovascular disease is indicated in RA along similar lines to those recommended for diabetes mellitus but there is also emerging evidence that the use of disease-modifying therapies with potent antiinflammatory properties can also improve endothelial dysfunction and may reduce cardiovascular events in RA (Gonzalez-Juanatey et al., 2004).

Felty syndrome describes the combination of chronic polyarthritis, neutropenia, and splenomegaly which is often associated with high titres of RF and other extraarticular manifestations (Campion et al., 1990). It affects approximately 1% of those with RA, typically with long-standing disease, and may sometimes be associated with the large granular lymphocyte syndrome. It is

Table 26–1 Classification Criteria for Rheumatoid Arthritis developed by the American College of Rheumatology

1. Morning stiffness for at least 1 hour and present for at least 6 weeks
2. Arthritis in three or more joint areas for at least 6 weeks (involvement of PIP, MCP, wrist, elbow, knee, ankle, and MTP joints)
3. Arthritis of the hand joints (swelling of PIP, MCP, wrist joints)
4. Symmetric arthritis
5. Rheumatoid nodules
6. Presence of rheumatoid factor
7. Hand radiographic changes typical of RA (must include erosions or unequivocal bony decalcification localized or most marked adjacent to the involved joints)

Abbreviations: MCP, metacarpophalangeal; MTP, metatarsophalangeal; PIP, proximal interphalangeal; RA, rheumatoid arthritis.

particularly strongly associated with the *HLA-DRB1*0401* allele (Lanchbury et al., 1991). Vasculitic manifestations in RA are highly variable, ranging from benign nailfold lesions through isolated digital infarcts to systemic involvement with major organ damage, such as mononeuritis multiplex, massive skin ulcers, myocardial infarction and bowel necrosis. Peripheral neuropathy in RA is usually mild but has been associated with excess mortality in RA (Turesson et al., 2002). It is suggested that major vasculitic features are more common in those who are homozygous for *HLA-DRB1*04* alleles (Weyand et al.,1994).

Diagnosis

For epidemiologic classification purposes, RA is defined by relatively nonspecific criteria developed by the American College of Rheumatology (Table 26–1) (Arnett et al., 1988). The presence of four or more of these seven criteria is considered to define RA but they should be applied with caution to the diagnosis of individual patients in the clinic, particularly in the early stages of the disease.

RF constitutes the classic immunologic marker for RA but lately the anti-CCP autoantibodies directed against proteins containing the nonstandard amino acid citrulline have provoked great interest (Schellekens et al., 1998; Girbal-Neuhauser et al., 1999). Besides their high specificity (~98% for RA), anti-CCP antibodies are present early in the disease, often before clinical onset (Kurki et al., 1992; Rantapaa-Dahlqvist et al., 2003). They appear to be produced locally in the inflamed synovium, since they constitute a higher proportion (7.5-fold) of immunoglobulin G (IgG) in the synovial compartment than in paired serum samples (Masson-Bessiere et al., 2000). In addition, the existence of citrulline-reactive antibodies has been associated with more active and severe forms of the disease (Bas et al., 2000; Paimela et al., 2001). The extent to which either RF or anti-CCP antibodies are involved in the pathogenesis of RA is unclear but demands further exploration. A strong association of anti CCP antibodies with major histocompatibility complex (MHC) shared epitope (SE) haplotypes (see below), has also been reported (van Gaalen et al., 2004). The potential association of RA with the gene for peptidylarginine deiminase *PADI4* (see below) is also of great interest because of its role in the alteration of arginine to citrulline residues in situ in polypeptides.

Early RA can be difficult to differentiate from self-limiting or other forms of inflammatory arthritis. Radiographic changes are often absent at presentation (van der Heijde and van Leeuwen, 1995) and conventional radiographic methods do not provide much information about the soft tissues and synovium. Ultrasonography and magnetic resonance imaging (MRI) have shown promising results for detecting synovitis and early erosive change but their utility is currently limited by cost and availability (Hoving and Buchbinder, 2004; Lopez-Ben and Bernreuter, 2004; Szkudlarek and Narvestad, 2004). Many attempts to utilize genetic profiling in the diagnosis and prognosis of early inflammatory arthritis have been described but to date have met with limited success.

Genetics of RA

RA is a classic example of a complex disease with a genetic component characterized by incomplete penetrance, genetic heterogeneity and a role for several or many disease genes. Concordance rates in monozygotic (MZ) twins vary widely between studies although there is broad agreement that the MZ concordance rate is approximately four times that of dizygotic (DZ) twins. A survey of twins in the UK at the end of 1960s reported a concordance rate for MZ twins of 30% and 5% for DZ for seropositive RA (Lawrence et al., 1970). This was accepted as standard for many years, until in 1986 a Finnish survey, based on nationwide twin registry that probably included less severe index cases, reported much lower rates for both MZ and DZ (12% and 4%, respectively). Similar results were obtained more recently in a further UK study by (Silman et al., 1993) (15.4% MZ; 3.6% DZ). Notably, the concordance rates were highest in those twin pairs with the well-established *HLA-DRB1*04* RA susceptibility allele (Jawaheer et al., 1994). Overall, the excess of MZ over DZ concordance rates in twins suggest an important genetic contribution with heritability estimates generally around 50% (MacGregor et al., 2000) although there have been exceptions with much lower estimates (Svendsen et al., 2002).

There is a significant excess risk of RA to the siblings of index cases compared to the general population ($\lambda_s \sim 6$) but this is dependent on the criteria used to define the disease. Thus, in the UK in the 1960s, λ_s was estimated at only 1.1 where the index case had mild, seronegative disease, in contrast to λ_s of approximately 6–10 for seropositive, erosive RA (Lawrence, 1970). The sibling recurrence risk ratio also tends to be greater with younger age at onset of the proband, with more severe disease, and in multiplex families. The siblings of those with RA who share both HLA haplotypes identical by descent to the proband are at particular risk (Deighton et al., 1993). Observations from several population studies suggest λ_s is approximately 8 (Wordsworth and Bell, 1992). This is substantially less than some other inflammatory rheumatic diseases, such as ankylosing spondylitis (AS) ($\lambda s \sim 60$) and systemic lupus erythematosus ($\lambda s \sim 20$), (Hochberg, 1987; Vyse and Todd, 1996; Brown et al., 1998a) but substantially greater than for large-joint osteoarthritis ($\lambda_s \sim 2.5$) or bronchial asthma ($\lambda_s \sim 3$).

Population Prevalence

The prevalence of RA is relatively constant at approximately 0.5%–1% in most populations of European origin. However, there are some important differences in the prevalence of RA in other ethnic groups that illustrate the potential importance of environmental and/or genetic factors in its etiology. RA is exceptionally rare in most of sub-Saharan Africa and also has a relatively low prevalence in Southeast Asia, China, and Japan compared to Europe and North America. A significantly lower frequency of disease has also been found in individuals of Afro-Caribbean origin living in Manchester, UK compared to the local white Caucasian population (MacGregor et al., 1994). In contrast, RA is more common than expected (~7%) in some native Amerindian tribes. Whether environmental factors contribute to this is not known but the high frequency of certain RA-associated HLA class II alleles could explain this in the Chippewa (*HLA-DRB1*04*) and Yakima (*HLA-DRB1*1402*). Studies in many populations worldwide have shown an association (odds ratio ~4) between RA and various *HLA-DRB1*04* alleles. However, this association is subject to ethnic variability and not all *HLA-DRB1*04* alleles contribute equally to RA susceptibility. The *HLA DRB1*0401* and **0404* alleles have been most commonly associated with RA in Europe and North America (Nepom et al., 1986; Wordsworth et al., 1989). In contrast, in East Asian populations, *DRB1*0401* is relatively rare and *DRB1*0405* is the most frequent RA-associated *DRB1*04* allele (Kim et al., 1995). *HLA-DRB1*0101* is also associated with RA in many populations although the attributable risk is relatively weak compared to *DRB1*0401* or **0404*, particularly in northern Europe,. *HLA-DRB1*01* appears to be more strongly associated with RA in more southerly parts of Europe, such as Spain and the former Yugoslavia than in northern Europe (Wordsworth and Bell, 1992). Independent association of *HLA-DRB1*01* with RA has been identified in other populations as diverse as Israeli Jews in whom the HLA-DRB1*0102 allele predominates (de Vries et al., 1993) and northern Indians living in the UK (Ollier et al., 1991). It may be more associated with seronegative and mild RA in some populations (Ploski and Mellbye, 1994). Although *HLA-DRB1*1001* is uncommon in most European populations, it is relatively common among Spanish patients with RA (Sanchez and Moreno, 1990) and has also been associated with RA in Greece, India, and Singapore populations (Carthy et al., 1993; Chan et al., 1994; Taneja et al., 1996). Although the effect is weak in British patients with RA, an association with *HLA-DRB1*10* can be detected if the analysis is stratified to take account of the stronger influences from *DRB1*04* and *DRB1*01* (Wordsworth et al., 1991). Finally, *HLA-DRB1*1402*, a common allele among Yakima American Indians, is strongly associated with RA in this nation (Willkens et al., 1991).

Table 26–2 HLA-DRB1 genotype relative risks for rheumatoid arthritis

DRB1 Genotype	Relative Risk	*p*-value
DRX/DRX	1.0	
0101/DRX	2.3	10^{-3}
0401/DRX	4.7	10^{-12}
0404/DRX	5.0	10^{-9}
0101/0401	6.4	10^{-4}
0401/0401	18.9	10^{-19}
0401/0404	31.3	10^{-33}

[a]Data from Hall et al., 1996.
DRX represents any non-DRB1*04 or DRB1*01 allele

HLA and RA Susceptibility

The association of RA with *HLA-DRB1* was first discovered by Stastny (1976) using a combination of cellular and serological reagents that identified various specificities of the HLA class II transplant antigen series encoded at the *HLA-DRB1* locus among others. The original observations by Stastny and others have been amply confirmed in numerous studies and using DNA typing methods have been extended into many diverse populations worldwide. Attempts to reconcile the apparently disparate associations with diverse HLA-DRB1 alleles, including *0101, *0401, *0404, *0405, *1001, and *1402, have invoked several possible explanations. These include: the possibility of linked genes being the primary source of susceptibility; that there are combinations of genes on particular haplotypes acting synergistically to contribute susceptibility; that there are conserved functional elements shared between the various RA-associated HLA-DRB1 alleles. In 1976, Stastny demonstrated that, in allogeneic mixed lymphocyte reactions, lymphocytes from patients with RA tended to be mutually nonstimulatory compared to controls. This observation could be explained on the basis that the Dw4 specificity (encoded by *DRB1*0401*) was markedly overrepresented in the RA population. These experiments were subsequently extended to large populations using serologic reagents including HLA-DR4, corresponding to *HLA-DRB1*04* (Stastny, 1978). The relationship between HLA and RA remains the strongest genetic link with RA and has been confirmed by both association and linkage studies. MHC class II molecules are heterodimeric (α and β chain) transmembrane glycoproteins, typically but not exclusively, displayed at the surface of specialized antigen-presenting cells of the immune system. The α and β chain of each MHC II molecule is encoded by separate genes in the MHC class II regions. Three MHC class II isotypes exist in humans, designated HLA-DR, HLA-DQ, and HLA-DP. All three isotypes serve similar functions and are critically involved in regulating the response to antigenic peptides presented to T cells. The genetic contribution to RA attributable to *HLA* genes is estimated to be approximately 30% (Deighton et al., 1989). To date, the strongest association with RA is from *HLA-DRB1* alleles that encode a conserved sequence of amino acids (shared epitope - SE) comprising residues 70–74 of the third hypervariable region of the DRβ1 chain that contributes one side of the antigen binding-site. The putative susceptibility epitope is glutamine–leucine–arginine–alanine, (QKRAA) or the QRRAA, or RRRAA variants of this. This sequence is encoded by several *HLA-DRB1* genes, associated with RA, including DRB1*0401 (^{69}QKRAA^{23}), DRB1*0101, DRB1*0404, DRB1*B405, DRB1*0408, DRB1*1402 (QRRAA), and DRB1*10 (RRRAA). There is a hierarchy of strength of the association of the different SE-positive HLA-DRB1 alleles in European populations (Table 26–2). The precise role of HLA-DRB1 is uncertain; it could be critical in influencing the binding of potentially arthritogenic peptides, in shaping the T cell repertoire necessary for the development of autoimmunity or by acting itself as an autoantigen. However, whatever mechanism is invoked, the SE model has some inconsistencies. For example, it does not explain the effect of HLA-DRB1*0901 alleles lacking the SE motif which has been positively associated with RA in Chile and UK (Gonzalez et al., 1992; Milicic et al., 2002); nor does it account for the fact that at least 15% of patients with RA lack the SE.

It has become apparent that not all HLA-DR4 haplotypes are equally associated with RA, and some may have a protective effect. Thus, HLA-DRB1*0402 is the most frequent DRB1*04 allele among Israeli Jews (Gao et al., 1991) but it is not associated with RA; this accounts for the lack of HLA-DRB1*04 association overall in this population. HLA-DRB1*0402 shares a similar amino acid motif, 69IDEAA73, with other HLA-DRB1 alleles, such as DRB1*0103, 1102, 1301, 1302, 1304 which are also negatively associated with RA (Hall et al., 1996; Barrera et al., 2000).

HLA and RA Severity

It now appears that rather than contributing merely to susceptibility to RA, *HLA-DRB1* genes probably play a major role in determining severity of the disease. *HLA DRB1*0401* is found commonly in RF-positive patients and appears to confer risk for the formation of a more rapidly progressive synovitis. In community studies of RA which include milder cases, the association with *HLA-DRB1*04* is weak or even absent, in contrast to hospital-based studies of more severe disease (Westedt et al., 1986; Thomson et al., 1999). In predicting progression, *HLA-DRB1*04* positive individuals are more likely to develop early erosive disease (Young et al., 1980; Goronzy et al., 2004). However, a systematic study of an inception cohort of 524 patients with RA showed only a very weak effect on progression of erosions after more than 2 years if all DRB1 SE alleles were considered rather than just DRB1*04 (Harrison et al., 1998). The exact mechanisms by which *HLA* genes influence the inflammatory synovitis are unknown. It is possible that HLA polymorphism could regulate the T-cell selection in thymus, resulting in smaller T-cell diversity and playing an important role in the formation of different lymphoid microstructures (Weyand and Goronzy, 1994). On the other hand, carrier status of DRB1*1301 and DRB1*1302 has been reported to protect against severe erosions among patients who do not have SE-positive alleles (Mattey et al., 1999).

A number of studies have indicated a role for HLA-DR4-related genes in extraarticular RA (Turessonet et al., 2004). Up to 95% of Felty syndrome patients are HLA-DRB1*04 positive, most commonly carrying the DRB1*0401 allele (Westedt et al., 1986; Lanchbury et al., 1991). Recently, a study looking into the impact of HLA on extraarticular manifestations in RA confirmed that patients with severe extraarticular RA were more likely to carry *HLA-DRB1*04* alleles. Genotypes featuring two DRB1*04 alleles are associated with rheumatoid vasculitis, Felty syndrome, and all extraarticular RA combined (Westedt et al., 1986; Turesson et al., 2005). In addition, severe extraarticular manifestations tend to cluster in individual patients with RA. This may in part be due to shared genetic factors such as *HLA-DRB1*04* alleles. We and others have also suggested that compound heterozygosity for individual *HLA-DRB1*04* alleles (e.g., **0401/*0404*) is particularly associated with susceptibility to RA (Nepom et al., 1986; Wordsworth et al., 1992) Thus, the *HLA-DRB1*0401/*0404* genotype may be associated with an odds ratio of 49 compared with only 6 for either allele alone.

Other HLA Genes

Candidate genes, such as *HLA-DQ*, *TNF*, *HSP-70*, complement components and the HLA class I loci do not appear to be associated with RA independently of HLA-DRB1. However, there is emerging evidence to suggest that there may be extended haplotypic effects in RA (Mulcahy et al., 1996). In particular, an extended MHC class III haplotype has been identified which is positively associated with RA in the context of *HLA-DRB1*0404* but not **0401* (Newton et al., 2003). The interval implicated in susceptibility has subsequently been narrowed by high density single nucleotide polymorphism (SNP) mapping to 126 kb segment at the junction of the MHC class III and class I regions (Kilding et al., 2004; Newton et al., 2004).This includes several poorly characterized immunomodulatory genes, including allograft inflammatory factor 1 (AIF1) which is expressed in rheumatoid synovium (Kimura et al., 2007), lymphocyte antigen complex, and also HLA-B-associated transcript 1 (BAT1) which may have a role in down-regulating cytokine production.

Noninherited Maternal HLA Antigens

Even in hospital-based studies of relatively severe RA, no more than 85% of the patients carry HLA-DRB1 SE alleles. It has, therefore, been suggested that exposure to SE-bearing noninherited maternal HLA antigens (NIMAs) could be important, particularly since this type of exposure is thought to be relevant to tolerance in transplantation biology. Initial studies in the Netherlands suggested an excess of HLA-DRB1*04 NIMA in RA patients who lacked the SE alleles (van der Horst-Bruinsma and Hazes, 1998). Harney and co-workers (2003) replicated the initial observation in the UK population where there was a small excess of DRB1*04 and also SE-positive NIMA in *HLA-DRB1*04* negative RA.

Genome-wide Linkage Studies

Genetic linkage studies of RA in multiplex families unsurprisingly indicate strong linkage to the HLA region but suggest that this constitutes no more than 40% of the genetic component of susceptibility (Deighton et al., 1989; Wordsworth and Bell, 1991). For this reason during the past decade, several systematic efforts to identify genomic regions that contain disease predisposing genes have been undertaken using genome-wide linkage approaches. The first of these studies, involving French, Italian, Spanish, and Belgian patient populations was carried out by Cornélis et al., in 1998. This reported significant linkage only for HLA, although nominal linkage was reported for 19 markers in 14 other regions, among them insulin-dependent diabetes mellitus 6 (IDDM6) and IDDM9 which had already been implicated in type 1 DM. Several subsequent major genome-wide linkage studies for RA have been carried out in the UK (MacKay et al., 2002; Eyre et al., 2004), Japan (Shiozawa et al., 1998), and North America (Jawaheer et al., 2001; Jawaheer et al., 2003) using microsatellite markers in affected sib-pair or multicase RA families. These studies have demonstrated a number of regions of the genome likely to be linked with RA. However, apart from the well-established HLA susceptibility locus, replication studies confirming the linkage of any other reported significant regions, have been relatively unsuccessful. A combined analysis of the US and UK families suggests an interaction between HLA-DRB1 and loci on chromosomes 6q and 16p (John et al., 2006). This is particularly interesting in the light of subsequent whole genome association study implicating a locus on 6q (see below, WTCCC). The limited success of linkage analysis for RA is partly due to the studies being too small to detect genes of modest effect reliably. For example, we have previously reported that 200 families are required to detect an effect equivalent to HLA in RA with 80% power (Brown and Wordsworth, 1998). Simulation modeling shows that increasing the sample size will improve the likelihood of detecting the genes with weaker effect, but not indefinitely (Eaves and Meyer, 1994; Risch and Zhang, 1995). Meta-analysis is a useful approach to measuring the information available from linkage studies. In RA, although consistency was found using such an approach in four chromosomal regions: 6p (containing HLA), 16p, 6q, and 8p, there were also slight differences in the reported results. This may be explained in part by differences in the studies included for analysis and also the meta-analysis methods. Fisher et al., (2003) analyzed genome data obtained from linkage graphs, whereas Choi et al., (2006) evaluated linkage scores with $p < 0.05$. Although increasing the sample size in meta-analysis may introduce heterogeneity arising from phenotypic differences between different populations, it is a powerful tool providing a basis for further targeted linkage and candidate gene studies.

Linkage analysis can only hint at the approximate chromosomal localizations of genetic effects. However, important information has arisen from such linkage studies; for example, the region on 6q replicated in female/female sib pairs in UK (Eyre et al., 2004) was subsequently confirmed in the combined UK/US data set (John et al., 2006) and probably contains the locus identifyied by genome wide association (see below Wellcome Trust Case Control Consortium, 2007). Likewise, the region on 1p13 identified in the US data set (Jawaheer et al., 2003) contains the major susceptibility gene *PTPN22*. Much interest now focuses on the possibility of using whole genome association data to increase the power to detect relevant genes and to refine the primary associations. Association studies using high-density SNP should generally be more effective in precise gene identification. Until recently, the technology to undertake such studies was not available but several new developments now make this feasible. One of the first approaches in RA involved a three-stage approach using 27,000 microsatellites spanning the genome (Tamiya et al., 2005). This approach used pooled DNA samples from 300 patients and 300 controls to screen for association with all 27,000 microsatellites initially, thereafter concentrating only on those showing positive association in two further rounds of analysis with similar number of patients and controls. In addition to HLA, a region on chromosome 11q13.4 containing eight candidate genes was identified; other potential loci include *PADI4* on chromosome 1p36, 10q13, and 14q23.1. In striking contrast, the gene *PTPN22* on chromosome 1 which has been confirmed from numerous studies as a susceptibility locus for RA was not identified in this study. This probably reflects a lack of power to detect relatively weak genetic effects using this type of sequential study design (with a first phase of only 300 patients and controls) although ethnic differences in susceptibility between Japanese and Europeans has not been formally addressed.

A study in the UK instigated by the Wellcome Trust has now yielded an important result using genome-wide association. A panel of 2,000 patients with RA was compared with 3,000 geographically matched controls as part of the Wellcome Trust Case Control Consortium (http://www.wtccc.org.uk) investigating the genetic component of common diseases. As well as confirming the known associations with *HLA-DRB1* alleles and *PTPN22*, a strong sex differentiated association was identified in chromosome 7. Furthermore, 9 SNPs with nominal significance were found that all map to genes with plausible biological relevance. Independent replication study of these 9 SNPs incorporating a large RA cohort collected from multiple centres all over the UK is currently underway. There was nominal significance with 2 markers in CTLA4 and association at 6q23 which does not fit with any currently identified gene.

Other Genes

Susceptibility to RA is likely to involve several genes of weak effect. Many candidate genes have been reported to be associated with susceptibility to RA: *TNFR2* on 1p36-p36.2, *PTPN22* on 1p13, *PADI4* on 1p36.13, *FcGR3A* on 1q23, *IL10* on 1q32, *PARP* on 1q41-1q42, *CTLA-4* on 2q33, *SLC22A4* on 5q31, *IFNG* on 12q21.1, *MHC2TA* on 16p13, and *MIF* on 22q11.23. Several candidate genes have been replicated in many studies on different populations. However, some candidate gene polymorphisms are more variable between different ethnic groups. Two large and comprehensive association studies on Japanese RA patients have identified two new genes to be significantly associated with RA: *PADI4* on chromosome 1 and *SLC22A4* on chromosome 5. *PADI4* is a particularly interesting candidate gene as it is responsible for the in situ conversion of positively charged arginine to polar but uncharged citrulline by deamination and the presence of anticitrulline antibodies is highly specific for RA. This association was initially described and replicated in Japanese (Suzuki et al., 2003). Although association with *PADI4* genes has been replicated in a Korean cohort of RA patient (Kang et al., 2006), studies in Caucasian population have been inconsistent, showing weak association at best (Barton et al., 2005; Harney et al., 2005; Martinez et al., 2005). This may be explicable by ethnic differences in allele frequencies although a recent large analysis of US and Swedish patients suggests that the association is real in Caucasians as well as Japanese with an odds ration of 1.24 (Plenge et al., 2005).

A missense SNP in PTPN22 encoding a protein tyrosine phosphatase, *PTPN22* (1p13), has been found in association with multiple autoimmune diseases, including RA, systemic lupus erythematosus, autoimmune thyroid disease, and type 1 DM (Begovich et al., 2004; Criswell et al., 2005). The *PTPN22* gene encodes the protein Lyp, an important negative regulator of T-cell activation. The presence of the *PTPN22* 1858 T/T genotype, encoding an R620W amino acid substitution, increases the risk for RA more than two-fold. The risk allele *PTPN22* C1858T is present in up to 17% of Caucasian patients disrupts the interaction of Lyp with the cytoplasmic tyrosine kinase (CSK) altering its natural function as a negative modulator of T-cell activation (Online Mendelian Inheritance in Man or OMIM 124095). This discovery demonstrates effectively that autoimmune diseases can have a common pathway and in part may account for some of the familial aggregation of these conditions.

Many of the observed genetic associations in RA (including HLA-DRB1) place the immunological synapse between antigen presenting cells and T cells right at the heart of the immunopathology of RA. Interaction between HLA molecules, the immunogenic peptides that bind to them and the T-cell antigen receptor lie at the centre of the synapse but for T-cell activation to occur a variety of costimulatory molecules are also involved. These include interactions between CD28 and CTLA-4 (cytotoxic T lymphocyte-associated-4) expressed differentially on CD4-positive T cells with their ligands, CD80 and CD86, on antigen presenting cells. Polymorphisms of these essential components of costimulation of T cells are potentially important in the pathogenesis of RA as are more peripheral components, such as programmed cell death 1 (PD-1) and its ligands (Grakoui et al., 1999).

CTLA-4 (2q34) is a costimulatory factor that acts as a negative regulator of T-cell activation. It balances the activating T-cell signals derived from CD28 and inducible co stimulatory (ICOS) molecule. CTLA-4 is related to CD28, but in contrast to CD28, it is not present on resting T cells, but is expressed several days after T-cell activation. Polymorphisms in the *CTLA-4* gene have been associated with multiple autoimmune diseases (Nistico et al., 1996; Donner et al., 1997; Heward et al., 1999). Studies in RA have been controversial, with positive association found in some populations, but not replicated in others. Interestingly, in the combined analysis of the US and UK linkage data, 2q34 locus was identified in the early onset RA subgroup. Combined analysis of European and US data confirm a significant association of *CTLA4* with RA with an odds ratio of 1.2 (Plenge et al., 2005). This is of particular interest in view of the highly beneficial effects of the human recombinant CTLA4 1g fusion protein in treatment of RA (Westhovens et al., 2006).

The PD-1 molecule is another costimulatory factor belonging to the immunoglobulin receptor super family and is a negative regulator of T cells (Nishimura et al., 2001). PD-$1^{-/-}$ mice on the C57BL/6 background show spontaneous development of lupus-like glomerulonephritis and destructive arthritis (Nishimura et al., 1999). The human gene for PD-1, programmed death cell domain 1 (*PDCD1*) is localized on 2q37.3, telomeric to *CD28/CTLA-4/ICOS*. Initially, four SNPs in *PDCD1* have been associated with lupus. Recent studies have found independent SNPs in association with RA (Kong et al., 2005; James et al., 2005).

Pathology

The normal synovial membrane which is one to two cell layers thick is transformed in the rheumatoid joint into a proliferating cell mass, pannus, which destroys surrounding tissue and bone. A cellular composition of rheumatoid synovitis is relatively simple and consists of resident cells and inflammatory infiltrate. Synovial membrane includes resident fibroblast-like (type A) and macrophage-like (type B) synoviocytes and endothelial cells. A complex cascade of cytokine-mediated events alter the phenotype of synoviocytes (especially type B cells) and they show increased expression of oncogenes and resist apoptosis (Yamanishi et al., 2005). As a result, the synoviocytes become highly proliferative, resulting in widespread joint inflammation and matrix degradation (Ritchlin, 2000). A broad display of macrophage and fibroblast cytokines like interleukin-1 (IL-1), IL-6, IL-12, IL-15, IL17, IL-18, TNF-, granulocyte–macrophage colony-stimulating factor (GM-CSF) and many more are produced by RA synovium that play a central role in maintaining synovial inflammation (Bucala et al., 1991). The inflammatory infiltrate in RA synovitis consists of T cells, B cells, dentritic cells, macrophages, plasma cells, mast cells, and granulocytes. Histology has shown that synovial infiltrates can form different architectural patterns. The most common form of rheumatoid synovitis is a diffuse infiltrate, where inflammatory cells are scattered among synovial cells without forming any higher organized structures. In approximately 40% of patients, inflammatory cells are organized into follicular structures and can form germinal centers. In a minority of patients, formation of granulomas, resembling rheumatoid nodules, have been observed. These different patterns are important, as they may influence therapeutic response.

T lymphocytes have a central role in formation and maintenance of chronic inflammatory process in RA. So, it came as a surprise, that RA synovial fluid and tissue consist of relatively low concentrations of T cell cytokines such as IL-2 and interferon-γ (IFN-γ), in contrast to the macrophage and fibroblast products (Firestein et al., 1990). Besides the accumulation of activated T cells in the RA synovium, the entire T-cell pool phenotype and homeostasis is also abnormal. Significant telomere shortening found in the $CD4^+$ T cells of RA patients indicates that they have undergone intensive replication (Wagner et al., 1998). $CD4^+$ T cells express CCR5 and CXCR3 chemokine receptors. These chemoattractants mark the subsets of cells with capacity for migration to inflammatory sites (Qin and Rottman, 1998). In addition to the aberrations in the global and in the synovial T cell receptor repertoire, clonal proliferation of T cell population in RA patients has been observed. Proliferating CD4 cells are functionally different from classic T-helper cells and are uniformly present in the circulation and infiltrate into synovial lesions (Goronzy et al., 1994; Waase et al., 1996). A set of clonally expanded $CD4^+$ T cells lack surface expression of the major costimulatory molecule CD28. This is caused by a transcriptional block of the *CD28* gene, due to the lack of two gene-specific nuclear proteins, necessary for the initiation of CD28 transcription (Vallejo et al., 1998). The $CD4^+$CD28- T cells are potentially cytotoxic, as they produce inflammatory cytokines (IFN-γ) and contain intracellular cytolytic substances (perforin and granzyme B) necessary to lyse target cells. In the absence of CD28 molecule, these unusual cells use alternative costimulatory pathways and de novo express killer cell immunoglobulin-like receptors (KIR2DS2, NKG2D). These cells may be related to natural killer (NK) cells as they express MHC class I-recognizing receptors allowing them to participate in innate and adaptive immune processes. Furthermore, the increased frequency of $CD4^+CD28^-$ cells is a predictor of progressive erosive disease in early RA (Goronzy et al., 2004).

TNF and IL-1 are the two main cytokines contributing to the joint inflammation in RA. TNF is a proinflammatory cytokine, produced by monocytes, T lymphocytes, macrophages, and fibroblasts. TNF stimulates prostaglandin E_2 and collagenase release from human synovial cells and potentiates osteoclastogenesis. IL-2 is typically synthesized and released simultaneously with TNF and the two cytokines share similar biologic properties. Elevated concentrations of proinflammatory cytokines have been found in peripheral blood and synovial fluid of RA patients and the concentrations correlate with disease activity and bone resorption (Eastgate et al., 1988; Neidel et al., 1995). Both IL-1 and TNF, as well as GM-CSF, stimulate fibroblast growth and pannus formation. These inflammatory cytokines also interact with each other and promote inflammatory process. Normal bone remodeling depends on the coordinated activity of the osteoblasts and osteoclasts. Under normal physiological conditions, osteoclasts are primarily responsible for bone resorption. These multinucleated cells, derived from the monocyte/macrophage line, also mediate focal bone resorption in RA. Osteoclast precursors have been identified in the bone resorption lacuna adjacent to the invading pannus (Bromley and Woolley, 1984). The differentiation of osteoclast precursors into mature bone resorbing cells depends on cytokine network including TNF, IL-1, and IL-6. Osteoblasts and bone stromal cells produce a cytokine RANKL, which in conjugation with M-CSF is the essential factor for osteoclast differentiation. RANKL binds to its cell-surface receptor RANK located on dentritic, T-lymphocyte, osteoclast precursors, and osteoclasts. RANK mediates its signal transduction via TNF receptor-associated factor (TRAF) proteins leading to differentiation of precursors, fusion to form multicellular osteoclasts and the activation of the osteoclasts to resorb bone (Gravallese et al., 2001). Osteoblasts also secrete osteoprotegerin (OPG), a soluble decoy receptor for RANKL. By binding to RANKL with high affinity, OPG prevents RANKL from interacting with its cognate receptor (RANK). It is the relative amounts of RANKL and OPG that regulate osteoclast activity (Lacey et al., 1998). Confirming evidence for the role of RANKL in bone resorption comes from studies of rat adjuvant arthritis (Kong et al., 1999). Arthritic rats, treated with OPG at the onset of disease had a minimal loss of cortical and trabecular bone in contrast to the untreated control animals, where bone loss was severe. OPG treatment also prevented osteoclast accumulation. However, the treatment did not decrease joint inflammation, indicating that the effect was specific only for osteoclast-mediated bone loss. RANKL-knockout mice, in whom inflammatory arthritis was induced, demonstrated a reduced degree of bone erosions compared to the wild-type animals (Pettit, 2001). In addition, RA synovial tissue produces other cytokines like IL-1α, IL-1β, and IL-11 that enhances osteoclast formation, activity, and survival (Deleuran et al., 1992a; Deleuran et al., 1992b).

Treatment

RA treatment aims to relieve symptoms, improve overall function, and prevent disease progression. There is little evidence that standard pre-biologic treatments directed toward symptom relief significantly lessen joint destruction and disease progression (Callahan et al., 1997). Previously, well-accepted step-up or pyramidal approach to treating RA underwent a significant reconsideration when observational and MRI studies revealed that joint erosions and more importantly functional decline emerge early in the disease (McQueen et al., 1998). As a result, standard disease-modifying antirheumatic drugs (DMARD) were increasingly introduced early in the disease course, and combinations of different DMARDs became common practice (Calguneri et al., 1999; Hochberg, 1999). Treatment with methotrexate (MTX), leflunomide, and sulfasalazine (SSZ) has been shown to reduce swollen/tender joint counts as well as decrease radiographic progression (Rich et al., 1999; Smolen et al., 1999; Sharp et al., 2000). In recent years, RA treatment has been revolutionized by the development and introduction of TNF-targetted biologic agents. The most widely used are etanercept and infliximab. Etanercept is a human recombinant fusion protein combining two soluble p75 TNF receptors linked to the Fc part of human IgG1 (Immunex Corporation: Seattle 2002). This compound blocks the activity of TNF (and also the related cytokine lymphotoxin α) by competitively blocking the association of TNF with its cell-surface receptors (Moreland et al., 1997). Infliximab is a mouse/human chimeric monoclonal antibody that binds with high affinity and selectivity to TNF (Remicade. Malvern PA: Centocor, Inc. 2002). Like etanercept, infliximab prevents the association of TNF with its cell-surface receptor and blocks the further signaling activity. However, infliximab does not neutralize the activity of lymphotoxin-α, which also exerts its biologic actions via the TNF receptor (Mikuls et al., 2001). Infliximab may also cause complement-mediated lysis of TNF-expressing cells (Lorenz et al., 1996). Whether this is relevant to the relative efficacy of infliximab compared to etanercept in inflammatory bowel disease is a moot point.

It has been suggested that polymorphisms of the *TNF* gene associated with differences in expression of TNF may influence the efficacy of antiTNF therapy in RA, but this requires further confirmation. Anakinra is another class of biologic therapy that has been approved for the treatment of RA. Anakinra is a human recombinant IL-1 receptor antagonist and closely resembles naturally occurring IL-receptor antagonist. Anakinra competitively inhibits binding of IL-1 to its cell-surface receptor thus blocking the biologic activity of IL-1 (Hannum et al., 1990).

Promising results in treating refractory RA have also been reported with rituximab (Edwards et al., 2001; Leandro et al., 2002). Rituximab is a chimeric monoclonal antibody directed against CD20. The CD20 antigen is present on the cell surface of differentiating B cells, though its expression ceases when B cells terminally differentiate into plasma cells. Rituximab causes prolonged and sustained improvement in disease activity and is useful in patients who have failed anti TNF treatment (Edwards et al., 2004).

Other biologic agents modulating the inflammatory response which have proved beneficial in rheumatoid arthritis include antibodies targeting the soluble IL-6 receptor (Maini et al., 2006) and CTLA4 1g fusion protein blocking interactions between the costimulatory molecules CD28 and CD80/CD86 (Westhovens et al., 2006; Sacre et al., 2005).

Seronegative Spondyloarthropathies

In contrast to RA, this group of inflammatory arthropathies is characterized by the absence of RF from the serum, a predilection for axial skeletal involvement and variable genetic association with HLA-B27. The occurrence of synovitis is common but, in contrast to RA, inflammation of the entheses is characteristic. Enthesitis constitutes a distinct anatomical form of inflammation particularly affecting sites of high mechanical stress, such as the origin and insertion of tendons and ligaments in addition to fibrocartilage joints such as the sacroiliac (SI) joints and costochondral junctions.

SI and spinal joint arthritis are commonly associated with asymmetric large peripheral joint arthritis. Extraskeletal ossification and new bone growth around peripheral joints with a tendency to ankylosis commonly occur. There are also extraarticular features such as various skin manifestions, uveitis and bowel inflammation that vary with the individual spondyloarthropathies. The constituent members of this group include:

1. Reactive arthritis,
2. Ankylosing spondylitis (AS)
3. Psoriatic arthritis (PsA),
4. Enteropathic arthritis,
5. Undifferentiated spondyloarthropathy (SpA).

Reactive Arthritis

This type of inflammatory arthritis predominantly affects the large lower limb joints asymmetrically and is triggered by antecedent infection, usually of the gastrointestinal or genitourinary tract. The classic triad of arthritis, conjunctivitis, and urethritis (Reiter syndrome) is relatively uncommon and the term reactive arthritis is, therefore, to be preferred. Diagnostic criteria have been proposed (Willkens et al., 1981).

Clinical Features

Joint inflammation typically affects the lower limbs in an oligoarticular asymmetric pattern. Prior infection with a variety of organisms may not be immediately obvious. The most common triggering organisms for enteric reactive arthritis are *Campylobacter sp*, *Salmonella sp*, *Shigella flexneri*, and *Yersinia enterocolitica*. *Chlamydia trachomatis* is usually responsible for the genital form. Rarely, the condition has also been described after immunization. Dactylitis ("sausage digits") is quite typical of both reactive and PsA (Fig. 26–3). Sequential involvement of a small number of joints is typical. In most cases, this will settle spontaneously over the course of several weeks to months but recurrent or persistent disease affects approximately 40% of the patients (Leirisalo-Repo, 1998).

Extraarticular features are common: mucocutaneous involvement includes mouth ulcers (typically shallow painless ulceration of the palate), urethritis, circinate balanitis, and keratoderma blennorrhagica; conjunctivitis commonly precedes the arthritis; systemic features may be severe with fever, malaise, and weight loss even prompting the search for underlying deep-seated malignancy or infection.

Epidemiolgy and Etiology

Approximately three-quarters of patients with reactive arthritis are positive for HLA-B27 and in epidemics of enteritis associated with potentially arthrogenic bacteria approximately 20% of HLA-B27-positive individuals develop reactive arthritis. Typically, the triggering bacteria cannot be cultured from joint fluid aspirated from the affected joints but variable success in isolating bacterial proteins and nucleic acid is described (Inman et al., 2000). It has been proposed that host reactivity to foreign antigens presented to the immune system in the context of HLA-B27 trigger a cross-reactive immune response to a similar self-peptide but direct evidence for this is limited. Others have proposed that the association of reactive arthritis with HLA-B27 reflects unusual biochemical characteristics of the HLA-B27 molecule which has a capacity to form homodimers which may be capable of interacting with other elements of the immune system such as KIR (killer immunoglobulin-like receptors)on NK cells (Chan et al., 2005).

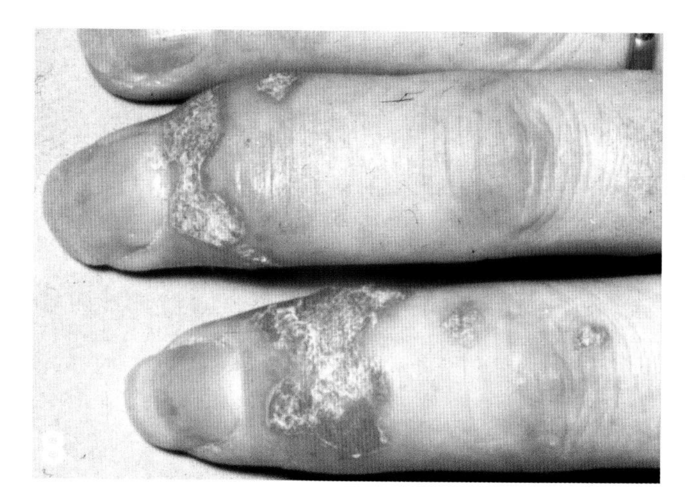

Figure 26–3 Psoriatic dactylitis of the second and third fingers. Skin lesions and nail dystrophy are both evident as well as swelling of the distal and proximal interphalangeal joints.

Management

Many milder cases of reactive arthritis can be managed with appropriate rest and nonsteroidal antiinflammatory drugs (NSAIDs). Local steroid injections of the affected joints with rest or splintage are valuable in selected cases and some individuals with more persistent or recurrent disease require the use of slow-acting antirheumatic agents, such as MTX, SSZ, or in particularly severe cases antiTNF biologics. There is debate about the utility of antibiotics although prolonged courses of antibiotics may have some influence on the late outcome of the arthritis (Yli-Kerttula et al., 2003). Testing HLA-B27 in individuals with suspected reactive arthritis may be of some help diagnostically and prognostically although it is not a substitute for full evaluation of the clinical presentation including a search for potential triggering organisms. If, present HLA-B27 is a significant adverse prognostic indicator since chronic forms of reactive arthritis are much more likely in those who are HLA-B27 positive, 20% of whom may ultimately proceed to develop AS (Leirisalo-Repo, 1998).

Ankylosing Spondylitis

The relationship between AS and reactive arthritis is tantalizing. Both are strongly associated with HLA-B27 and a significant proportion of B27-positive individuals with reactive arthritis develop the axial skeletal involvement characteristic of AS over a period of time. Perhaps for this reason, there has historically been substantial interest in the possibility of identifying bacterial triggers for AS analogous to the situation in reactive arthritis. In contrast to reactive arthritis, AS is characterized particularly by involvement of the axial skeleton and, in particular, the SI joints. For classification purposes, the New York Diagnostic Criteria have been developed (Table 26–3).

These are heavily dependent on the presence of radiographic evidence of sacroiliitis coupled with at least one clinical criterion indicative of either axial skeletal stiffness or inflammatory back pain. Further, these criteria are not particularly sensitive, not least, since it may take years for radiographic evidence of sacroiliitis to be apparent. MRI [particularly "short tau inversion recovery (STIR)" sequences] is very effective at detecting bone marrow edema, a characteristic of enthesitis (Fig. 26–4). These changes can be seen much earlier than changes on plain radiography and are currently being evaluated for inclusion in new diagnostic criteria.

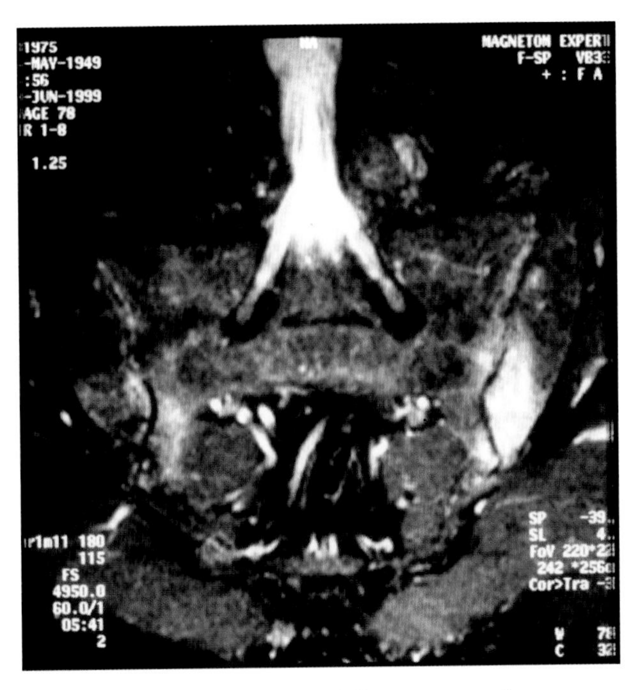

Figure 26–4 Bilateral sacroiliitis revealed by STIR magnetic resonance imaging. High signal is apparent on both sides of the sacroiliac joints bilaterally indicating bone marrow edema due to inflammation.

Epidemiology

The prevalence of AS varies worldwide and strongly reflects the allele frequency of the main genetic susceptibility factor, HLA-B27, in different populations (Khan, 1997). The apparent prevalence also varies with the method of ascertainment of cases. From a study of hospital records Gran and Husby (1993) found a prevalence of 0.5–2.3 per 1000 and a male-to-female ratio of 3 to 8:1. General population surveys have revealed significantly higher prevalence of the disease ranging between 2.4 and 14 per 1000 (van der Linden et al., 1984b; Gran et al., 1985). Early studies hinted at a heavy male preponderance of cases but later studies suggest that the ratio is approximately 2.5 to 1 and this may be age dependent. More than 90% of individuals with AS occur before the age of 40 (van der Linden et al., 1998) and after the age of 40 the gender ratio falls to only 1.8 to 1 (Kennedy et al., 1993).

Table 26–3 Revised New York Criteria for Ankylosing Spondylitis (Modified From van der Linden et al., 1984a).

Low back pain ⩾ 3 months duration (improved by exercise and not relieved by rest)
Limited back movement (sagittal and coronal)
Reduced chest expansion (compared to age and sex-matched values)
Bilateral sacroiliitis (grade ⩾ 2)
Unilateral sacroiliitis (grade ⩾ 3)

HLA-B27 and AS

The prevalence of AS strongly reflects the allele frequency of HLA-B27 in the UK where the prevalence of HLA-B27 is 8%. AS affects approximately two per thousand of the population; there is a much prevalence of AS in some populations with higher frequency of HLA-B27, including Haida Indians (B27 18%–50%), Lapps (B27 25%–30%), and Alaskan Eskimos (B27 25%–40%). In contrast, the low frequency of HLA-B27 in other ethnic groups such as South African Blacks (0.5%), Australian Aboriginals (absent), and South American Indians corresponds with the absence of AS or only extremely low prevalence.

HLA-B27 initially defined by serological reagents is now recognized to have approximately 30 different allelic variants, some of which are very rare. There has been considerable interest to see if all of these are equally associated with AS. Systematic studies for each of the variants are impossible due to their rarity but in most cases the individual HLA-B27 variants are positively associated with the disease although there are some important exceptions.

The initial association studies using serological reagents in the early 1970s showed strong relative risk in excess of 100 associated with carriage of HLA-B27 (Brewerton et al., 1973; Caffrey and James, 1973; Schlosstein et al., 1973). Numerous studies worldwide have confirmed this strength of association and rigorous systematic studies have confirmed the association with the main allelic subtypes, *HLA-B27* 2702, * 2704, * 2705*. With many of the rarer subtypes such as *HLA-B* 2701, * B2707, * B2708* systematic association studies are impossible because of their rarity but individual cases have been described. In contrast, there has been a substantial interest in the three alleles (*B*2703*, **2706*, and **2709*) which appeared to have negative association with AS. Hill et al. (1991) described the prevalence of *B*2703* as 6% in the Gambia where AS is extremely rare. Subsequently, our field studies in the Gambia (Brown, Jepson, et al., 1997) confirmed the rarity of AS even in populations chosen specifically for are relatively high prevalence of HLA-B27. However, the *HLA-B2705* allele was equally prevalent with *B*2703*, indicating that in this population the standard *HLA-B*27* associations do not seem to apply. Whether this reflects an environmental effect or the influence of other protective genes in this population is unclear. Although *HLA-B27* may be less associated with AS than other B27 alleles there are reports of the disease occurring in American Blacks and among West Africans in Senegal. *HLA-B*2706* appears to be negatively associated with AS from systematic studies in Southeast Asia (Lopez-Larrea et al., 1995) and from Indonesia (Nasution et al.,1997). Finally *HLA-B*2709*, a major B27 subtype in Sardinia, has yet to be reported in association with AS. There is evidence that these subtypes which vary relatively little from the more widespread *B*2705*, apart from single amino acid substitutions in the floor of the antigen-binding site, are either less strongly associated or even negatively associated with susceptibility to AS.

HLA-B27 in Diagnosis

The positive predictive value of routine HLA-B27 testing in the diagnosis of AS is dependent on the pretest probability of disease. In the community setting using HLA-B27 testing is of very limited value when applied to a subject in the community developing back pain unless there is strongly supportive clinical evidence. In the clinical situation, a patient with a history suspicious of inflammatory back pain with some limitation of spinal movement but normal spinal radiographs, positive B27 testing increases the probability of AS from approximately 0.12 to 0.62 which is only just greater than random and, therefore, adds relatively little to the clinical decision-making process (Baron and Zendele, 1989). In contrast, a negative B27 would make the diagnosis unlikely. Khan (1980) estimates that with a pretest probability for AS at 0.5, the presence of B27 increases the likelihood to 0.92, whereas a negative result reduces it to 0.08.

The sibling recurrence risk for AS is approximately 6% (Carter et al., 2000). Sibling recurrence risk to siblings carrying HLA-B27 is 12%. The likelihood of the offspring of patients with AS also developing the disease is 10% increasing to 20% in those who are HLA-B27 positive. In certain circumstances, screening of offspring and siblings may be justified particularly if there is great anxiety to the first-degree relatives of patients with particularly severe disease.

Genetics of AS

The importance of genetic factors in susceptibility to AS is clearly evident not only from the association with HLA-B27 but it is also clearly evident from studies of familial aggregation. Twin studies conducted from the early 1960s demonstrated much higher concordance rates for MZ twins compared to DZ twin pairs. From a recent large twin study in the UK broad sense, heritability of AS has been estimated to be in excess of 92%. The concordance rate for DZ twins is approximately 13% compared to at least 64% in MZ twins (Brown, Kennedy, et al., 1997). Some of the genetic component is clearly related to HLA-B27 but even among B27-positive DZ twins concordance is only 23%, clearly indicating a role for nonHLA-linked genes.

Familial aggregation of AS in nontwin relatives can be used to assess disease heritability. Several studies have assessed the frequency of AS in first-degree relatives and found that the recurrence of disease is almost entirely restricted to HLA-B27-positive individuals (van der Linden and Khan, 1984; Said-Nahal et al., 2002). HLA-B27-positive individuals with a first-degree relative with AS have a 6–16 times higher risk of developing the disease than B-27-positive individuals without the family background (Calin et al., 1983; van der Linden et al., 1983).

Recurrence risk in families also differs according to the gender of the proband, with an increased risk to the relatives of young female probands. (Calin et al., 1999; Miceli-Richard et al., 2000). An increased risk of AS in first-born children and in the children of younger mothers has also been reported in one study (Baudoin et al., 2000) although this was not replicated in multicase SpA families (Said-Nahal et al., 2001). Susceptibility to the disease is probably oligogenic (~3–9genes) although the presence of HLA-B27 is apparently almost a prerequisite (Brown et al., 2000). The HLA-linked contribution to developing AS ranges from 15% to 53% of the genetic component, indicating that a significant effect arises from outside the MHC.

More than 95% of the patients are HLA-B27 positive, yet only a small proportion of subjects (2%–8%) carrying B27, develop the disease (Braun et al., 1998). A possible explanation may lie in the variations within the *B27* gene itself. To date, over 31 B27 subtypes have been identified. Most of the defining polymorphisms occur in exons 2 and 3, encoding the $\alpha1$ and $\alpha2$ domains of the B27 peptide-binding domain. The B27 subtypes have a varied racial and ethnic prevalence throughout the world. However, HLA-B2705 is thought to be the common ancestor of the other subtypes and is the most common in European Caucasian populations (Blanco-Gelaz et al., 2001). Whether all the subtypes are equally associated with disease remains unclear (see above). In some communities, certain subtypes of HLA-B27 may have a protective role. HLA B-2709 is commonly present in Sardinia and not associated with the disease. A similar pattern of negative disease association has been observed with HLA-B2706 in Thai and Indonesian populations. The only difference between the AS-associated HLA-B2705 and protective HLA-B2709 is the exchange of position 116 from aspartate to histidine, located in the peptide-binding groove and has an effect on anchoring the C-terminal peptide residues. (Boyle and Gaston, 2003; Montserrat et al., 2003).

NonHLA-B27 Genetic Effects in AS

As outlined above, HLA-B27 alone is unlikely to account for all the genetic basis of susceptibility to AS. There are in excess of 160 genes within the MHC alone, many of which have important functions in immunity and inflammation, which play a major role in the immune response. Several other *MHC* genes in addition to HLA-B27 have been implicated in AS. HLA-B60 was first reported to have an effect in B27-positive individuals, increasing their chances of being affected by AS up to three-fold (Robinson et al., 1989). Subsequently, this observation has been confirmed in other populations and also extended to nonHLA-B27 individuals with AS (Rubin et al., 1994; Brown et al., 1996; Wei et al., 2004; Madhavan et al., 2002; Said-Nahal et al., 2002). Some early studies have demonstrated an association of *HLA-DRB1*01* with AS in the UK (odds ratio 1.5) with a suggestion that an extended *HLA-B*27/DRB1*01* haplotype was overrepresented in patients (Brown et al., 1998). Subsequent work in the same population suggests that the excess of *HLA-DRB1*01* actually arises from an excess of transmission of the non-HLA-B27 haplotype. It remains to be shown whether *DRB1*01* or a linked gene is primarily involved. There is no shortage of potential candidate genes in the linkage region, including some which are involved in the processing and transport of antigens for presentation to the immune system in the context of HLA class I molecules, such as HLA-B27. These include the low molecular weight components of the proteasome (LMP) and the transporters associated with antigen processing (TAP).

Despite initial suggestions that certain TAP polymorphisms increase susceptibility to AS and reactive arthritis, this has not been confirmed. LMP2 polymorphisms have been associated with juvenile AS in Mexico and Norway, and also in Canadian AS patients with uveitis but have not been reliably replicated. LMP-7 alleles have been associated with AS in Spain. LMP-2 and 7 genes encode two subunits of proteasome complex that degrades cytosolic proteins into peptides, that are subsequently bound by MHC class I molecules. These findings need to be further replicated to confirm the association.

Polymorphisms in TNF-α and heat-shock protein 70 (HSP-70) have been found in certain AS patient populations, although these observations have not yet been widely replicated. Exhaustive mapping of the MHC has now revealed a 270 kb region, containing 23 genes, likely to be involved in AS (Sims et al., 2007).

NonMHC Genes in AS

Family studies have provided strong evidence that a limited number (probably ~–3–9) of nonMHC genes, each of modest effect are involved in susceptibility to AS. Four genome-wide linkage studies and a subsequent meta-analysis of these scans have identified several overlapping and distinct chromosomal regions linked with susceptibility to AS and/or undifferentiated SpA (Brown et al., 1998a; Laval et al., 2001; Miceli-Richard et al., 2004; Carter et al., 2007; Zhang et al., 2004). The individual linkage intervals are generally broad, but by combining both linkage and association approaches to positional mapping and taking into account our knowledge of the pathophysiology of these conditions there is a realistic expectation that at least some of the genes responsible will be identified. Recently, an international consortium drawing particularly on British, North American, and French families has collaborated on completing the meta-analysis of genome-wide association in study of 500,000 SNPs in more than 2000 patients with AS. A number of interesting potential candidate genes or linkage intervals have already been identified.

IL-1 Gene Cluster

Two genome-wide scans have demonstrated linkage to chromosome 2q. This area harbours the IL-1 gene cluster, encoding IL-1α, IL-1β, IL-1 receptor antagonist, and six other homologous genes *IL-IF5-10*, some of which have agonist and others antagonist actions. Initial case–control studies on Caucasian AS patients indicated an association of allele 2 of the *IL-1RN* variable number of tandem repeats (VNTR) with the disease (McGarry et al., 2001; van der Paardt et al., 2002). These findings have not been subsequently replicated. Studies on a Taiwanese AS population revealed associations with other IL-1 family genes, indicating that possible linkage disequilibrium exists with the true disease-causing variant (Chou et al., 2006). Extensive linkage disequilibrium mapping study in British AS cases and families has demonstrated that the most likely candidate genes are *IL-1A* and *IL-1F5-10* (Timms et al., 2004). Furthermore, Spondyloarthritis Research Consortium of Canada (SPARCC) and the North American Spondyloarthritis Consortium (NASC) have independently demonstrated association with *IL1-1A*.

Cytochrome P450 2D6

Genetic polymorphisms leading to loss of function of the cytochrome P450 gene debrisoquine 4-hydroxylase (*CYP2D6*) are associated with AS (Beyeler et al., 1996; Brown et al., 2000). This enzyme deficiency affects approximately 6% of UK Caucasians leading to poor metabolism of drugs, such as codeine and tricyclic antidepressants. However, the mechanism(s) by which *CYP2D6* might influence susceptibility to AS are unknown. Possible explanations include failure to metabolize a natural toxin or influences the production of peptide antigens presented by HLA-B27.

Caspase Recruitment Domain Family Member 15

Caspase recruitment domain family member 15 (*CARD15*) [also known as nucleotide oligomerization domain 2 (NOD2)] is expressed in monocytes and mediates the activation of NFκB, a key transcription factor in the activation of numerous inflammatory genes. Mutations in *CARD15* were first described in Crohn disease (Hugot et al., 2001; Ogura et al., 2001) and subsequently in a chronic granulomatous inflammatory disorder, Blau syndrome, associated with arthritis. Since AS is strongly associated with inflammatory bowel disease (IBD) and inflammatory changes are well described in the terminal ileum in AS (Mielants et al., 1991) *CARD15* is clearly an important candidate gene in AS. Several studies have been conducted to assess its potential involvement, but the results have been generally negative (Crane et al., 2002; D'Amato, 2002; Miceli-Richard et al., 2002). Furthermore, although genetic linkage studies in AS initially suggested that there might be overlap with the *CARD15* interval on chromosome 16, subsequent finer mapping appears to have excluded the *CARD15* region in AS (Crane et al., 2002).

Transforming Growth Factor β-1 (TGF-β1)

The *TGF-β1* locus is on chromosome 19q13.1, previously found to be associated with AS in linkage studies. However, subsequent studies in different populations have yielded variable results. TGFB has pleiomorphic actions on inflammation fibrosis and ossification making it an interesting candidate for AS but on current evidence it plays little part in the aetiology of AS (Jaakkola et al., 2004).

The Ankylosis Progressive Homolog Gene

The role of ankylosis progressive homolog (*ANKH*) gene in AS susceptibility has been implicated though the studies initially performed in the *ank/ank* mouse in which there is a loss-of-function mutation in the corresponding *ank* gene. This mouse develops widespread joint ankylosis in the spine and peripheral joints due to ectopic calcium hydroxyapatite crystal deposition (Sampson, 1988). Early studies suggested that some *ANKH* variants may be associated with susceptibility to AS in a gender specific manner (Tsui et al., 2003). However, it is now clear that AS is not associated with variants of this gene (Timms et al., 2003). The function of ANKH as a transmembrane transporter of inorganic pyrophosphate plays a critical role in bone and soft tissue mineralization, but this seems more relevant to pyrophosphate arthritis than AS (Williams et al., 2002; Zhang et al., 2005).

Many other candidate genes have been studied: matrix metalloproteinase 3 (*MMP3*) (Jin et al., 2005), nucleotide pyrophosphatase (*NPPS*) (Mori et al., 2003), androgenic receptor (Mori et al., 2000), T-cell receptor α and β genes (Brown et al., 1998b); reporting either weak evidence or no association with AS at all.

Genetic Factors and Severity of AS

AS is a chronic disease characterized by relapsing-remitting course and variable prognosis. Several outcome measures have been designed to define disease activity and severity. These include the Bath AS Metrology Index (BASMI), the Bath AS Functional Index (BASFI), the Bath AS Disease Activity Index (BASDAI) and the Bath AS Patient Global Score (BAS-G). The BASFI, BASDAI, and BAS-G are all questionnaires utilizing a 100 mm visual analogue scale. BASMI aims to summarize the patient mobility by applying a simple scoring system to five clinical measurements of axial skeletal movement (Jenkinson et al., 1994).

There is some evidence that severity of AS is under genetic control since markers of disease severity are more similar in MZ than DZ twins (Brown, Kennedy, et al., 1997). Segregation analysis reveals that AS disease severity is largely genetically determined. Heritability of BASDAI was estimated at 51% and heritability of BASFI at 68% (Hamersma et al., 2001). High heritability values ($h^2 = 0.62$) have also been demonstrated for the degree of radiographic SI joint involvement (Brophy et al., 2004). Homozygosity for HLA-B27 is somewhat more common than expected in AS. Anterior uveitis is more common in B27+ (50%) than B27− (16%) cases. The age of onset of AS is also younger in B27+ disease (Jaakkola et al., 2006).

Pathology

AS and related spondyloarthropathies have a unique pathology, initially described by Ball (1971) further followed by Bywaters (1983). In contrast to RA, the primary pathological site is the enthesis rather than the synovium. The enthesis is a metabolically active site, perhaps explaining the initiation of pathological changes occurring commonly during growth in the adolescent years. The evolving enthesopathy is characterized by fibrosis and ossification rather than the joint destruction and instability. In the spine, enthesopathic changes at the site of insertion of the outer fibers of the annulus fibrosus resulting in characteristic pattern, ultimately leading to ankylosis and "bamboo spine" formation. The SI joints are almost universally involved in the disease. Osteoporosis is an early progressive feature and may add to poor posture and disability. Synovitis itself does occur in peripheral joints and can mimic RA, although the process is typically less severe. Villous synovial hyperplasia is present with diffuse inflammatory infiltrate that has the tendency to perivascular aggregation. Compared to RA, synovial fluid in AS contains less polymorphs and has a higher lymphocyte count (Revell and Mayston, 1982). Immune histology on SI biopsy material has demonstrated predominance of macrophages and $CD8^+$ cytotoxic T cells (Braun et al., 1995). Presence of TNF, TGF-β, and messenger ribonucleic acid (mRNA) in SI joints has also been noted. Inflammatory infiltrate involving entheses is also rich with $CD8^+$ cytotoxic cells (Laloux et al., 2001). However, these results were obtained in long-standing SpA and may not apply to early enthesitis.

Diagnosis

The first criteria set for the diagnosis of AS was defined in Rome in 1961 (Kellgren et al., 1963). This has now been modified twice and the one currently used for AS diagnosis is the modified New York Criteria (Table 26–3) (van der Linden et al., 1984). These criteria have low sensitivity, especially toward diagnosing early AS and limited specificity to grade 2 sacroiliitis (van Tubergen et al., 2003). However, the radiographic diagnosis of sacroiliitis is complicated by the slow evolution of changes in many patients. Thus the SI radiographs may be normal in the early phase of disease (Mau et al., 1988). Radiological changes reflect the consequence of inflammation leading to bone damage, but not inflammation itself. MRI studies have demonstrated widespread inflammation in SI joints without chronic changes (Braun et al., 1994). Two sets of classification criteria have been developed that allow earlier diagnosis in patients with predominant axial and peripheral joint involvement: the European Spondyloarthropathy Study Group (ESSG) criteria (Dougados et al., 2000) and the Amor criteria (Amor et al., 1995). In both sets of criteria SI joint involvement is included as a component, but not as a precondition for diagnosis. Their sensitivity and specificity are approximately 85% (Boyer et al., 1993; Collantes-Estevez et al., 1995).

There are many radiographic changes described in AS and they reflect the pathological process. In the lumbosacral spine, these features include erosions, squaring, sclerosis, syndesmophytes, bridging syndesmophytes, and spondylodiscitis. This structural damage reflects the consequence of inflammation. Much earlier changes have been described using fat-suppressed MRI (STIR) revealing bone marrow edema. Osteoporosis is a common feature and often present from an early stage of the illness (Will et al., 1990; Donnelly et al., 1994; O'Neill et al., 1994). Spinal fractures have often been observed in longstanding AS due to increased bone fragility and reduced function (Ralston et al., 1990). The exact mechanisms of this advanced osteoporosis are yet unknown, but reduced mobility due to ankylosis and increased bone turnover due to inflammation may play a central role. SI inflammation is commonly present in AS and in the majority of cases the disease usually starts in this region. The earliest inflammatory changes observed using MRI typically affects dorsocaudal parts of the synovial joint and the bone marrow. Initial inflammation leads to the formation of erosions that are seen as irregular variations in the width of the SI joint space and loss of clear definition of the joint line. Overall no major differences using MRI have been documented in the early and later disease stages; this is consistent with the view that AS does not burn out.

Management AS is associated with significant pain and functional disability affecting quality of life. Pain relief and structured exercise to reduce disability have been the major facets of treatment. Benefit from intensive exercise regimens in a hospital setting and under physiotherapist supervision has been well documented. For pain relief, non-steroidal anti-inflammatory drugs have been used with some evidence that continuous treatment might reduce paraspinal ossification (Wanders et al., 2005). Benefit from SSZ and low-dose MTX is limited to patients with peripheral joint arthritis, as these drugs have shown no effect on spinal disease. Bisphosphonates are believed to have anti-inflammatory properties and initial trials using intravenous pamidronate infusion for spinal symptoms in AS looked promising (Maksymowych et al., 2002). However, more recent studies on NSAID-resistant AS patients have not demonstrated significant effect (Grover et al., 2005).

Anti-TNF agents have revolutionized the management of AS. Several large multicenter randomized controlled trials have shown that all three antiTNF agents: Infliximab, etanercept, and adalimumab, lead to impressive improvements in spinal pain, functional disability, and inflammatory markers. This effect seems to persist with ongoing treatment and discontinuation can cause AS relapse (Baraliakos et al., 2005). Hopefully ongoing studies will help to clarify the optimal therapeutic strategies for antiTNF in AS.

Anakinra, a recombinant IL-1 receptor antagonist, has shown controversial results in the treatment of AS. Trial results have been conflicting and there is to date insufficient evidence to support its use in AS (Tan et al., 2004; Haibel et al., 2005).

Psoriatic Arthritis

PsA is defined as an inflammatory arthritis associated with psoriasis (see also Chapter 32). Although the first classical description of arthritis associated with psoriasis comes from Aliberti in 1818, it has only recently been recognized as a distinct clinical entity. PsA occurs approximately in 7%, but in some series up to 42% of the patients with cutaneous disease (Green et al., 1981; Stern, 1985). PsA is clinically a heterogeneous disease, comprising of a five subgroups with distinct clinical forms, varying from RA to seronegative SpA (Moll and Wright, 1973). The majority of patients with psoriasis who develop arthritis clinically similar to RA are RF negative, but it is unclear whether psoriasis itself can inhibit RF production (Wright, 1959). The prevalence of RF positivity and CCP antibodies is low in psoriatic arthritis. Enthesitis is more common in patients with PsA and is thought by some researchers to be the original site of inflammation in PsA (McGonagle et al., 1998; McGonagle et al., 2002). In some patients, the symptoms related to enthesitis can be more debilitating than those from synovitis. A typical feature of PsA is dactylitis (Fig. 26–3) resulting from tenosynovitis and synovitis involving two to three consecutive joints of the same finger, giving an appearance of "a sausage finger." Dactylitis can occur in over one-third of the patients with PsA (Gladman, 1987). Dystrophic nail lesions of psoriatic origin, manifesting in ridging, pitting, and onycholyis have been found more commonly in patients with PsA but are often overlooked (Gladman, 1986; Williamson et al., 2004a). One study found the psoriatic nail lesions to be particularly common in patients with distal interphalangeal joint disease (Cohen et al., 1999). Spinal involvement tends to be more advanced among men, but overall the clinical picture is less severe compared to classical AS (Gladman et al., 1992).

Genetic Factors

There is overwhelming evidence that psoriasis is an inherited disease (discussed in detail in Chaper 32). Family studies have shown up to 23% of first-degree relatives are affected, and twin studies reveal 70% of MZ twins are concordant (Brandrup, et al., 1978). The sibling concordance rate for PsA has been estimated to be 5.5%, although a frequency of up to 14% has been reported in a more recent study. However, the individual arthritis subtypes have not shown distinct inheritance patterns (Moll and Wright, 1973; Myers et al., 2005). Several large population studies have also suggested a parental sex effect as children of affected fathers are more likely to develop psoriasis or PsA than children of affected mothers (16.2% versus 8.3%, respectively) (Burden et al., 1998). A variety of genetic factors have been associated with PsA, with the strongest effect residing within the MHC. This locus has been designated as "*PSORS1*" (OMIM 177900) and initially located in the class I or class III regions of the MHC with a strong association to HLA-Cw6. Uncomplicated psoriasis has long been associated with HLA-Cw6, in addition to the alleles in the linkage disequilibrium (HLA-B13, -B37, -B57) (Tiilikainen et al., 1980). However, HLA-Cw6 is unlikely to be the PSORS1-susceptibility allele as many psoriatic patients do not carry HLA-Cw6 (Jenisch et al., 1998; Leder et al., 1998; Jenisch et al., 1999). Furthermore, the most recent linkage disequilibrium mapping and haplotypic studies of the HLA class I region localized the *PSORS1* gene to a 60 kb interval, excluding HLA-C locus and corneodesmosin from involvement (Nair et al., 2000).

Sacroiliitis is common in patients with psoriatic arthritis but is often asymptomatic (Williamson et al., 2004b). HLA-B27 is variably associated with sacroiliitis and to a lesser extent spondyloarthritis (Gladman et al., 1986; Queiro et al., 2002; Williamson et al., 2004b). *HLA-DRB1* 04* has been associated with peripheral joint arthritis in some studies (McHugh et al., 1987).

In addition, associations with TNF promoter, *TAP*, and *MICA* gene variants have been demonstrated in PsA, suggesting that other *MHC* genes may have a role in susceptibility to PsA. However, the presence of linkage disequilibrium with HLA-B27 and Cw6 may have influenced the results. So far only one genome-wide scan has been completed in PsA, localizing a candidate region on chromosome 16q (Karason et al., 2003). Polymorphisms in *CARD15* have been found to be associated with PsA in one patient population (Rahman et al., 2003). Further studies have not replicated these findings (Giardina et al., 2004; Lascorz et al., 2005; Jenisch et al., 2006).

Pathogenesis

T lymphocytes are thought to play a central role in the pathogenesis of psoriasis and PsA. Interestingly, unlike the skin in psoriasis, synovial lymphocyte population has not demonstrated up-regulation of cutaneous lymphocyte-associated (CLA) antigen, indicating that different lymphocyte populations are involved in skin as opposed to synovial tissues (Pitzalis et al., 1996). $CD4^+$ cells are the most common cells in the psoriatic tissues, whereas, $CD8^+$ cells dominate in the synovial fluid and at the enthesis (Costello et al., 1999; Laloux et al., 2001). In contrast to RA, synovial tissue in PsA contains reduced numbers of macrophages (Veale et al., 1993).

Prominent angiogenesis, affecting the skin, nailfolds, and synovial membrane, is an early feature in psoriasis and PsA, linked to specific up-regulation of signaling and growth factors including TGFβ, platelet-derived growth factor, vascular endothelial growth factor, and angiopoietins (Creamer et al., 1997). It is likely that cytokines, in particular TNF, may be critically involved in driving the inflammatory process leading to the production of MMPs, and cartilage and bone degradation.

Treatment

NSAIDs and local intraarticular glucocorticoid injections have shown good effect on peripheral arthritis. Oral glucocorticoids are not favored in PsA due to observed cutaneous flares after withdrawal.

The action of commonly used DMARDs including leflunomide SSZ, MTX, and cyclosporin A appears to be confined to skin manifestations and peripheral arthritis without any compelling evidence of their benefit in axial disease. Etanercept, infliximab, and adalimumab have been shown to provide a significant improvement in disease activity, physical function, and quality of life, either as monotherapy or as add-on therapy to other DMARDs in both psoriasis and PsA. They seem to have an excellent effect on both axial and peripheral arthritis. Furthermore, all three agents have demonstrated to retard radiological progression in PsA. Etanercept has been reported to be the least effective at treating the skin manifestations (Gottlieb and Antoni, 2004).

Enteropathic ArthritisEntheropathic arthritis is a form of seronegative SpA variably associated with two different types of IBD - ulcerative colitis (UC) and Crohn disease. Inflammatory arthropathy is the most common extraintestinal manifestation of IBD affecting between 2% and 26% of patients (see also Chapter 22) (Greenstein et al., 1976; Palm et al., 2001). Peripheral arthritis, enthesopathy, and involvement of other periarticular structures have been described in 10%–20% of patients but objective signs of joint inflammation probably affect 5%–10% of patients only. Patients with IBD are more likely to develop peripheral arthritis if the colon is extensively involved (Palumbo et al., 1973; Greenstein et al., 1976). The arthritis may precede the onset of IBD and is of variable character. Two distinct forms of peripheral arthritis have been described based on their clinical manifestations and immunogenetic associations (Orchard et al., 1998; Orchard et al., 2000).

Neither form is usually erosive, although exceptions do occur. The *type 1 arthropathy* is self-limiting, pauciarticular, affecting lower limb joints, especially knees and ankles, and is strongly associated with other extraintestinal manifestations, such as erythema nodosum and uveitis. Flares of type 1 arthropathy generally follow increased intestinal disease activity and are typically self-limiting in weeks or months. This type is associated with *HLA-B*27, HLA-B*35*, and *HLA-DRB1*0103*. The *Type 2 arthropathy* is symmetric and polyarticular, commonly affecting hands and wrists, and follows a more prolonged course over months or even years. Arthritis flares do not often mirror the intestinal disease. In this type, significant association with *HLA-B*44* has been identified. Both arthritis types occur about twice as commonly in Crohn disease and UC (Orchard et al.,1998). The true prevalence of axial skeletal involvement is less well established although it seems to be more common in Crohn disease (5%–22%) than UC (2%–6%) (Gravallese and Kantrowitz, 1988). Sacroiliitis accompanying IBD is often under-reported due to its insidious onset and asymptomatic nature. Radiographic evidence of sacroiliitis has been reported in 20%–25% of patients with IBD but the prevalence of changes may be higher using MRI. AS is present in 3%–10% of IBD patients but there are some differences in the genetic predisposition between "enteropathic" AS and the isolated "pure" form of the disease (Schorr-Lesnick and Brandt, 1988). The association between *HLA-B27* and AS associated with IBD is less strong than with classic AS; typically approximately 60% – 80% are B27 positive compared with more than 90% with classic AS. Isolated sacroiliitis without obvious spinal ankylosis does not seem be associated with *HLA-B27* (Orchard et al., 1998). In contrast, patients with Crohn disease who are HLA-B27 positive will almost invariably develop AS. Evidence supporting the contribution of genetic factors in the susceptibility to IBD is overwhelming and is described in more detail in Chapter 23. Based on data from three large twin cohorts from Sweden, Denmark, and the UK, the combined concordance rate was 36% in MZ and only 4% in DZ twin pairs (Tysk et al., 1988; Thompson et al., 1996; Orholm et al., 2000). The Swedish study also provided evidence that disease phenotype in Crohn disease is genetically determined (Halfvarson et al., 2003). The genetic effect on susceptibility to UC seems to be weaker as illustrated by lower combined concordance rates (16%), indicating the presence of a stronger environmental component. The twin data is supported by various family aggregation studies showing increased sibling recurrence risk of both forms of IBD and higher familial prevalence of early onset disease (Polito et al., 1996; Satsangi et al., 1996; Freeman, 2001). Based on whole genome linkage studies many susceptibility loci have been implicated in IBD (Zheng et al., 2003). Only seven such loci, named IBD1-7 have been replicated in independent studies and meet the strict criteria for significant linkage (Ahmad et al., 2004). These loci position are in chromosomes 16q (IBD1), 12q (IBD2), 6p (IBD3), 14q (IBD4), 5q (IBD5), 19p (IBD6), and 1p (IBD7) (Zheng et al., 2003). Some of these loci are specific to either Crohn disease (IBD1) or UC (IBD2), whereas others seem to confer overall susceptibility to IBD (IBD3) (Hampe et al., 1999; Parkes et al., 2000; Cavanaugh, 2001). Polymorphisms in *CARD15* gene within IBD1 locus have been found to be strongly associated with Crohn disease. One frameshift and two missense mutations increase the risk of Crohn disease by three-fold in the heterozygous and between 38-fold and 44-fold, in simple homozygous individuals (Hugot et al., 1996). A meta-analysis of 16 published case–control studies revealed that the presence of *CARD15* mutation is associated more with familial disease, stenosing clinical features, and ileal location (Economou et al., 2004). Phenotype–genotype studies showing *CARD15* polymorphism associations with extraintestinal manifestations have been controversial (Miceli-Richards, 2002; Peeters et al., 2004). No associations between *CARD15* and susceptibility to UC have been found.

Very recent emerging data have helped to underline the aetiological links between the various spondyloarthropaties, including ankylosing spondylitis, psoriatic arthritis and enteropathic arthritis. IL-23R was identified as a susceptibility factor for inflammatory bowel disease by Duerr et al. (2006) and subsequently in psoriasis (Cargill et al., 2007). Similar associations between AS and IL-23R have been identified through international collaborations, independent of whether the probands have isolated AS, psoriatic spondylitis or enteropathic spondylitis. These findings clearly highlight the importance of certain proinflammatory pathways, including those mediated through the Th17 subset of T-helper $CD4^+$ lymphocytes in the pathogenesis of these disorders. Translation of these results to effective therapies should be practicable in the near future.

References

Ahmad, T, Tamboli, CP, Jewell, D, and Colombel, JF (2004). Clinical relevance of advances in genetics and pharmacogenetics of IBD. *Gastroenterology*, 126 (6):1533–1549.

Amor, B, Dougados, M, and Khan, MA (1995). Management of refractory ankylosing spondylitis and related spondyloarthropathies. *Rheum Dis Clin North Am*, 21(1):117–128.

Anaya, JM, Diethelm, L, Ortiz, LA, Gutierrez, M, Citera, G, Welsh, RA, et al. (1995). Pulmonary involvement in rheumatoid arthritis. *Semin Arthritis Rheum*, 24(4):242–254.

Arnett, FC, Edworthy, SM, Bloch, DA, McShane, DJ, Fries, JF, Cooper, NS, et al. (1988). The American Rheumatism Association 1987 revised criteria for the classification of rheumatoid arthritis. *Arthritis Rheum*, 31(3):315–324.

Ball, J (1971). Enthesopathy of rheumatoid and ankylosing spondylitis. *Ann Rheum Dis*, 30(3):213–223.

Baraliakos, X, Brandt, J, Listing, J, Haibel, H, Sorensen, H, Rudwaleit, M, et al. (2005). Outcome of patients with active ankylosing spondylitis after two years of therapy with etanercept: clinical and magnetic resonance imaging data. *Arthritis Rheum*, 53(6):856–863.

Baron, M and Zendel, I (1989). HLA-B27 testing in ankylosing spondylitis: an analysis of the pretesting assumptions. *J Rheumatol*, 16(5):631–634; discussion 634–636.

Barrera, P, Balsa, A, Alves, H, Westhovens, R, Maenaut, K, Cornelis, F, et al. (2000). Noninherited maternal antigens do not play a role in rheumatoid arthritis susceptibility in Europe. European Consortium on Rheumatoid Arthritis Families. *Arthritis Rheum*, 43(4):758–764.

Barton, A, Eyre, S, Bowes, J, Ho, P, John, S, and Worthington, J (2005). Investigation of the SLC22A4 gene (associated with rheumatoid arthritis in a Japanese population) in a United Kingdom population of rheumatoid arthritis patients. *Arthritis Rheum*, 52(3):752–758.

Bas, S, Perneger, TV, Mikhnevitch, E, Seitz, M, Tiercy, JM, Roux-Lombard, P, et al. (2000). Association of rheumatoid factors and anti-filaggrin antibodies with severity of erosions in rheumatoid arthritis. *Rheumatology (Oxford)*, 39(10):1082–1088.

Baudoin, P, van der Horst-Bruinsma, IE, Dekker-Saeys, AJ, Weinreich, S, Bezemer, PD, and Dijkmans, BA (2000). Increased risk of developing ankylosing spondylitis among first-born children. *Arthritis Rheum*, 43 (12):2818–2822.

Begovich, AB, Carlton, VE, Honigberg, LA, Schrodi, SJ, Chokkalingam, AP, Alexander, HC, et al. (2004). A missense single-nucleotide polymorphism in a gene encoding a protein tyrosine phosphatase (PTPN22) is associated with rheumatoid arthritis. *Am J Hum Genet*, 75(2):330–337.

Beyeler, C, Armstrong, M, Bird, HA, Idle, JR, and Daly, AK (1996). Relationship between genotype for the cytochrome P450 CYP2D6 and susceptibility to ankylosing spondylitis and rheumatoid arthritis. *Ann Rheum Dis*, 55 (1):66–68.

Blanco-Gelaz, MA, Lopez-Vazquez, A, Garcia-Fernandez, S, Martinez-Borra, J, Gonzalez, S, and Lopez-Larrea, C (2001). Genetic variability, molecular evolution, and geographic diversity of HLA-B27. *Hum Immunol*, 62 (9):1042–1050.

Boyer, GS, Templin, DW, and Goring, WP (1993). Evaluation of the European Spondylarthropathy Study Group preliminary classification criteria in Alaskan Eskimo populations. *Arthritis Rheum*, 36(4):534–538.

Boyle, LH and Hill Gaston, JS (2003). Breaking the rules: the unconventional recognition of HLA-B27 by CD4+ T lymphocytes as an insight into the pathogenesis of the spondyloarthropathies. Rheumatology (Oxford), 42 (3):404–412.

Brandrup, F, Hauge, M, Henningsen, K, and Eriksen, B (1978). Psoriasis in an unselected series of twins. *Arch Dermatol*, 114(6):874–878.

Braun, J, Bollow, M, Eggens, U, Konig, H, Distler, A, and Sieper, J (1994). Use of dynamic magnetic resonance imaging with fast imaging in the detection of early and advanced sacroiliitis in spondylarthropathy patients. *Arthritis Rheum*, 37(7):1039–1045.

Braun, J, Bollow, M, Neure, L, Seipelt, E, Seyrekbasan, F, Herbst, H, et al. (1995). Use of immunohistologic and in situ hybridization techniques in the examination of sacroiliac joint biopsy specimens from patients with ankylosing spondylitis. *Arthritis Rheum*, 38(4):499–505.

Braun, J, Bollow, M, Remlinger, G, Eggens, U, Rudwaleit, M, Distler, A, et al. (1998). Prevalence of spondylarthropathies in HLA-B27 positive and negative blood donors. *Arthritis Rheum*, 41(1):58–67.

Brewerton, DA, Hart, FD, Nicholls, A, Caffrey, M, James, DC, and Sturrock, RD (1973). Ankylosing spondylitis and HL-A 27. *Lancet*, 1(7809):904–907.

Bromley, M and Woolley, DE (1984). Histopathology of the rheumatoid lesion. Identification of cell types at sites of cartilage erosion. *Arthritis Rheum*, 27 (8):857–863.

Brophy, S, Hickey, S, Menon, A, Taylor, G, Bradbury, L, Hamersma, J, et al. (2004). Concordance of disease severity among family members with ankylosing spondylitis? *J Rheumatol*, 31(9):1775–1778.

Brown, M, Bunce, M, Calin, A, Darke, C, and Wordsworth, P (1996). HLA-B associations of HLA-B27 negative ankylosing spondylitis: comment on the article by Yamaguchi et al. *Arthritis Rheum*, 39(10):1768–1769.

Brown, MA, Edwards, S, Hoyle, E, Campbell, S, Laval, S, Daly, AK, et al. (2000). Polymorphisms of the CYP2D6 gene increase susceptibility to ankylosing spondylitis. *Hum Mol Genet*, 9(11):1563–1566.

Brown, MA, Jepson, A, Young, A, Whittle, HC, Greenwood, BM, and Wordsworth, BP (1997a). Ankylosing spondylitis in West Africans—evidence for a non-HLA-B27 protective effect. *Ann Rheum Dis*, 56(1):68–70.

Brown, MA, Kennedy, LG, MacGregor, AJ, Darke, C, Duncan, E, Shatford, JL, et al. (1997b). Susceptibility to ankylosing spondylitis in twins: the role of genes, HLA, and the environment. *Arthritis Rheum*, 40(10):1823–1828.

Brown, MA, Laval, SH, Brophy, S, and Calin, A (2000). Recurrence risk modelling of the genetic susceptibility to ankylosing spondylitis. *Ann Rheum Dis*, 59(11):883–886.

Brown MA, Pile KD, Kennedy LG, Andrew L, Butcher S, Gibson K, et al. (1998a). Genome wide screen in ankylosing spondylitis. *Arthritis Rheum* 41:588–595.

Brown, MA, Rudwaleit, M, Pile, KD, Kennedy, LG, Shatford, J, Amos, CI, et al. (1998b). The role of germline polymorphisms in the T-cell receptor in susceptibility to ankylosing spondylitis. *Br J Rheumatol*, 37(4):454–458.

Brown, MA and Wordsworth, BP (1998). Genetic studies of common rheumatological diseases. *Br J Rheumatol*, 37(8):818–823.

Bucala, R, Ritchlin, C, Winchester, R, and Cerami, A (1991). Constitutive. production of inflammatory and mitogenic cytokines by rheumatoid synovial fibroblasts. *J Exp Med*, 173(3):569–574.

Burden, AD, Javed, S, Bailey, M, Hodgins, M, Connor, M, and Tillman, D (1998). Genetics of psoriasis: paternal inheritance and a locus on chromosome 6p. *J Invest Dermatol*, 110(6):958–960.

Bywaters, EG (1983). Historical perspectives in the aetiology of ankylosing spondylitis. *Br J Rheumatol*, 22(4 Suppl. 2):1–4.

Caffrey, MF and James, DC (1973). Human lymphocyte antigen association in ankylosing spondylitis. *Nature*, 242(5393):121.

Calguneri, M, Pay, S, Caliskaner, Z, Apras, S, Kiraz, S, Ertenli, I, et al. (1999). Combination therapy versus monotherapy for the treatment of patients with rheumatoid arthritis. *Clin Exp Rheumatol*, 17(6):699–704.

Calin A, Brophy S, and Blake D (1999). Impact of sex on inheritance of ankylosing spondylitis: a cohort study. *Lancet*, 354(9191):1687–1690.

Calin, A, Marder, A, Becks, E, and Burns, T (1983). Genetic differences between B27 positive patients with ankylosing spondylitis and B27 positive healthy controls. *Arthritis Rheum*, 26(12):1460–1464.

Callahan, LF, Pincus, T, Huston JW 3rd, Brooks, RH, Nance EP Jr, and Kaye, JJ (1997). Measures of activity and damage in rheumatoid arthritis: depiction of changes and prediction of mortality over five years. *Arthritis Care Res*, 10 (6):381–394.

Campion, G, Maddison, PJ, Goulding, N, James, I, Ahern, MJ, Watt, I, et al. (1990). The Felty syndrome: a case-matched study of clinical manifestations and outcome, serologic features, and immunogenetic associations. *Medicine (Baltimore)*, 69(2):69–80.

Cargill M, Schrodi SJ, Chang M, Garcia VE, Brandon R, Callis KP, et al. (2007). A large-scale genetic association study confirms IL12B and leads to the identification of IL23R as psoriasis-risk genes. *Am J Hum Genet* 80 (2):273–290.

Carter, N, Williamson, L, Kennedy, LG, Brown, MA, and Wordsworth, BP (2000). Susceptibility to ankylosing spondylitis. *Rheumatology (Oxford)*, 39 (4):445

Carter KW, Pluzhnikov A, Timms AE, Miceli-Richard C, Bourgain C, Wordsworth BP, et al. (2007). Combined analysis of three whole genome linkage scans for ankylosing spondylitis. *Rheumatol* 46:763–771.

Carthy, D, Ollier, W, Papasteriades, C, Pappas, H, and Thomson, W (1993). A shared HLA-DRB1 sequence confers RA susceptibility in Greeks. *Eur J Immunogenet*, 20(5):391–398.

Cavanaugh, J and IBD International Genetics Consortium (2001). International collaboration provides convincing linkage replication in complex disease through analysis of a large pooled data set: Crohn disease and chromosome 16. *Am J Hum Genet*, 68(5):1165–1171.

Chan AT, Kollnberger SD, Wedderburn LR, and Bowness P (2005). Expansion and enhanced survival of natural killer cells expressing the killer immunoglobulin-like receptor KIR3DL2 in spondylarthritis. *Arthritis Rheum* 52 (11):3586–3595.

Chan, SH, Lin, YN, Wee, GB, Koh, WH, and Boey, ML (1994). HLA class 2 genes in Singaporean Chinese rheumatoid arthritis. *Br J Rheumatol*, 33 (8):713–717.

Choi, SJ, Rho, YH, Ji, JD, Song, GG, and Lee, YH (2006). Genome scan meta-analysis of rheumatoid arthritis. *Rheumatology (Oxford)*, 45 (2):166–170.

Chou, CT, Timms, AE, Wei, JC, Tsai, WC, Wordsworth, BP, and Brown MA (2006). Replication of association of IL1 gene complex members with ankylosing spondylitis in Taiwanese Chinese. *Ann Rheum Dis*, 65 (8):1106–1109.

Cohen, MR, Reda, DJ, and Clegg, DO (1999). Baseline relationships between psoriasis and psoriatic arthritis: analysis of 221 patients with active psoriatic arthritis. Department of Veterans Affairs Cooperative Study Group on Seronegative Spondyloarthropathies. *J Rheumatol*, 26(8):1752–1756.

Collantes-Estevez, E, Cisnal del Mazo, A, and Munoz-Gomariz, E (1995). Assessment of 2 systems of spondyloarthropathy diagnostic and classification criteria (Amor and ESSG) by a Spanish multicenter study. European Spondyloarthropathy Study Group. *J Rheumatol*, 22(2):246–251.

Cornélis, F, Faure, S, Martinez, M, Prud'homme, JF, Fritz, P, Dib, C, et al. (1998). New susceptibility locus for rheumatoid arthritis suggested by a genome-wide linkage study. *Proc Natl Acad Sci USA*, 95(18):10746–10750.

Costello, P, Bresnihan, B, O'Farrelly, C, and FitzGerald, O (1999). Predominance of CD8$^+$ T lymphocytes in psoriatic arthritis. *J Rheumatol*, 26 (5):1117–1124.

Crane, AM, Bradbury, L, van Heel, DA, McGovern, DP, Brophy, S, Rubin, L, et al. (2002). Role of NOD2 variants in spondylarthritis. *Arthritis Rheum*, 46 (6):1629–1633.

Creamer, D, Jaggar, R, Allen, M, Bicknell, R, and Barker, J (1997). Overexpression of the angiogenic factor platelet-derived endothelial cell growth factor/thymidine phosphorylase in psoriatic epidermis. *Br J Dermatol*, 137 (6):851–855.

Criswell, LA, Pfeiffer, KA, Lum, RF, Gonzales, B, Novitzke, J, Kern, M, et al. (2005). Analysis of families in the multiple autoimmune disease genetics consortium (MADGC) collection: the PTPN22 620W allele associates with multiple autoimmune phenotypes. *Am J Hum Genet*, 76 (4):561–571.

D'Amato, M (2002). The Crohn's associated NOD2 3020InsC frameshift mutation does not confer susceptibility to ankylosing spondylitis. *J Rheumatol*, 29(11):2470–2471.

Deighton, CM, Cavanagh, G, Rigby, AS, Lloyd, HL, and Walker, DJ (1993). Both inherited HLA-haplotypes are important in the predisposition to rheumatoid arthritis. *Br J Rheumatol*, 32(10):893–898.

Deighton, CM, Walker, DJ, Griffiths, ID, and Roberts, DF (1989). The contribution of HLA to rheumatoid arthritis. *Clin Genet*, 36(3):178–182.

Deleuran, BW, Chu, CQ, Field, M, Brennan, FM, Katsikis, P, Feldmann, M, et al. (1992a). Localization of interleukin-1 alpha, type 1 interleukin-1 receptor and interleukin-1 receptor antagonist in the synovial membrane and cartilage/pannus junction in rheumatoid arthritis. *Br J Rheumatol*, 31 (12):801–809.

Deleuran, BW, Chu, CQ, Field, M, Brennan, FM, Mitchell, T, Feldmann, M, et al. (1992b). Localization of tumor necrosis factor receptors in the synovial tissue and cartilage-pannus junction in patients with rheumatoid arthritis. Implications for local actions of tumor necrosis factor alpha. *Arthritis Rheum*, 35(10):1170–1178.

del Rincon, ID, Williams, K, Stern, MP, Freeman, GL, and Escalante, A (2001). High incidence of cardiovascular events in a rheumatoid arthritis cohort not explained by traditional cardiac risk factors. *Arthritis Rheum*, 4 (12):2737–2745.

De Vries N, Ronningen KS, Tilanus MG, Bouwens-Rombouts A, Segal R, Egeland T, et al. (1993). HLA-DR1 and rheumatoid arthritis in Israeli Jews: sequencing reveals that DRB1*0102 is the predominant HLA-DR1 subtype. *Tissue Antigens* 41:26–30.

Donnelly, S, Doyle, DV, Denton, A, Rolfe, I, McCloskey, EV, and Spector, TD (1994). Bone mineral density and vertebral compression fracture rates in ankylosing spondylitis. *Ann Rheum Dis*, 53(2):117–121.

Donner, H, Rau, H, Walfish, PG, Braun, J, Siegmund, T, Finke, R, et al. (1997). CTLA4 alanine-17 confers genetic susceptibility to Graves' disease and to type 1 diabetes mellitus. *J Clin Endocrinol Metab*, 82(1):143–146.

Dougados, M, Leclaire, P, van der Heijde, D, Bloch, DA, Bellamy, N, and Altman, RD (2000). Response criteria for clinical trials on osteoarthritis of the knee and hip: a report of the Osteoarthritis Research Society International Standing Committee for Clinical Trials response criteria initiative. *Osteoarthr Cartil*, 8(6):395–403.

Duerr RH, Taylor KD, Brandt SR, Rioux JD, Silverberg MS, Daly MJ, et al. (2006). A genome-wide association study identifies IL23R as an inflammatory bowel disease gene. *Science* 314(5804):1461–1463.

Eastgate, JA, Symons, JA, Wood, NC, Grinlinton, FM, di Giovine, FS, and Duff, GW (1988). Correlation of plasma interleukin 1 levels with disease activity in rheumatoid arthritis. *Lancet*, 2(8613):706–709.

Eaves, L and Meyer, J (1994). Locating human quantitative trait loci: guidelines for the selection of sibling pairs for genotyping. *Behav Genet*, 24 (5):443–455.

Economou, M, Trikalinos, TA, Loizou, KT, Tsianos, EV, and Ioannidis, JP (2004). Differential effects of NOD2 variants on Crohn's disease risk and phenotype in diverse populations: a metaanalysis. *Am J Gastroenterol*, 99 (12):2393–2404.

Edwards, JC and Cambridge, G (2001). Sustained improvement in rheumatoid arthritis following a protocol designed to deplete B lymphocytes. *Rheumatology (Oxford)*, 40(2):205–211.

Edwards, JC, Szczepanski, L, Szechinski, J, Filipowicz-Sosnowska, A, Emery, P, Close, DR, et al. (2004). Efficacy of B-cell-targeted therapy with rituximab in patients with rheumatoid arthritis. *N Engl J Med*, 350(25):2572–2581.

Eyre S, Barton A, Shepherd N, Hinks A, Brintnell W, MacKay K, et al. (2004). Investigation of susceptibility loci identified in the UK rheumatoid arthritis whole-genome scan in a further series of 217 UK affected sibling pairs. *Arthritis Rheum* 50(3):729–735.

Firestein, GS, Alvaro-Gracia, JM, and Maki, R (1990). Quantitative analysis of cytokine gene expression in rheumatoid arthritis. *J Immunol*, 144 (9):3347–3353.

Fisher, SA, Lanchbury, JS, and Lewis, CM (2003). Meta-analysis of four rheumatoid arthritis genome-wide linkage studies: confirmation of a susceptibility locus on chromosome 16.*Arthritis Rheum*, 48(5):1200–1206.

Freeman, HJ (2001). Familial occurrence of lymphocytic colitis. *Can J Gastroenterol*, 15(11):757–760.

Gabriel, SE, Crowson, CS, and O'Fallon, WM (1999). Comorbidity in arthritis. *J Rheumatol*, 26(11):2475–2479.

Gao, XJ, Brautbar, C, Gazit, E, Segal, R, Naparstek, Y, Livneh, A, et al. (1991). A variant of HLA-DR4 determines susceptibility to rheumatoid arthritis in a subset of Israeli Jews. *Arthritis Rheum*, 34(5):547–551.

Giardina, E, Novelli, G, Costanzo, A, Nistico, S, Bulli, C, Sinibaldi, C, et al. (2004). Psoriatic arthritis and CARD15 gene polymorphisms: no evidence for association in the Italian population. *J Invest Dermatol*, 122 (5):1106–1107.

Girbal-Neuhauser, E, Durieux, JJ, Arnaud, M, Dalbon, P, Sebbag, M, Vincent, C, et al. (1999). The epitopes targeted by the rheumatoid arthritis-associated antifilaggrin autoantibodies are posttranslationally generated on various sites of (pro)filaggrin by deimination of arginine residues. *J Immunol*, 162 (1):585–594.

Gladman, DD, Anhorn, KA, Schachter, RK, and Mervart, H (1986). HLA antigens in psoriatic arthritis. *J Rheumatol*, 13(3):586–592.

Gladman, DD, Brubacher, B, Buskila, D, Langevitz, P, and Farewell, VT (1992). Psoriatic spondyloarthropathy in men and women: a clinical, radiographic, and HLA study. *Clin Invest Med*, 15(4):371–375.

Gladman, DD, Shuckett, R, Russell, ML, Thorne, JC, and Schachter, RK (1987). Psoriatic arthritis (PSA)—an analysis of 220 patients. *QJM*, 62 (238):127–141.

Goldring, SR and Gravallese, EM (2000). Mechanisms of bone loss in inflammatory arthritis: diagnosis and therapeutic implications. *Arthritis Res*, 2 (1):33–37.

Gonzalez A, Nicovani S, Massardo L, Bull P, Rodriguez L and Jacobelli S (1992). Novel genetic markers of rheumatoid arthritis in Chilean patients, by DR serotyping and restriction fragment length polymorphism analysis. *Arthritis Rheum* 35(3):282–289.

Gonzalez-Juanatey, C, Testa, A, Garcia-Castelo, A, Garcia-Porrua, C, Llorca, J, and Gonzalez-Gay, MA (2004). Active but transient improvement of endothelial function in rheumatoid arthritis patients undergoing long-term treatment with anti-tumor necrosis factor alpha antibody. *Arthritis Rheum*, 51(3):447–450.

Goronzy, JJ, Bartz-Bazzanella, P, Hu, W, Jendro, MC, Walser-Kuntz, DR, and Weyand, CM. (1994) Dominant clonotypes in the repertoire of peripheral $CD4^+$ T cells in rheumatoid arthritis. *J Clin Invest*, 94(5):2068–2076.

Goronzy, JJ, Matteson, EL, Fulbright, JW, Warrington, KJ, Chang-Miller, A, Hunder, GG, et al. (2004). Prognostic markers of radiographic progression in early rheumatoid arthritis. *Arthritis Rheum*, 50(1):43–54.

Gottlieb, AB and Antoni, CE (2004). Treating psoriatic arthritis: how effective are TNF antagonists? *Arthritis Res Ther*, 6(Suppl. 2):S31–S35.

Grakoui, A, Bromley, SK, Sumen, C, Davis, MM, Shaw, AS, Allen PM, et al. (1999). The immunological synapse: a molecular machine controlling T cell activation. *Science*, 285(5425):221–227.

Gran, JT and Husby, G (1993). The epidemiology of ankylosing spondylitis. *Semin Arthritis Rheum*, 22(5):319–334.

Gran, JT, Husby, G, and Hordvik, M (1985). Prevalence of ankylosing spondylitis in males and females in a young middle-aged population of Tromso, northern Norway. *Ann Rheum Dis*, 44(6):359–367.

Gravallese, EM, Galson, DL, Goldring, SR, and Auron, PE (2001). The role of TNF-receptor family members and other TRAF-dependent receptors in bone resorption. *Arthritis Res*, 3(1):6–12.

Gravallese, EM and Kantrowitz, FG (1988). Arthritic manifestations of inflammatory bowel disease. *Am J Gastroenterol*, 83(7):703–709.

Green L, Meyers OL, Gordon W, and Briggs B. (1981). Arthritis in psoriasis. *Ann Rheum Dis*, Aug;40(4):366–369.

Greenstein, AJ, Janowitz, HD, and Sachar, DB (1976). The extra-intestinal complications of Crohn's disease and ulcerative colitis: a study of 700 patients. *Medicine (Baltimore)*, 55(5):401–412.

Grover, SA, Coupal, L, and Zowall, H (2005). Treating osteoarthritis with cyclooxygenase-2-specific inhibitors: what are the benefits of avoiding blood pressure destabilization? *Hypertension*, 45(1):92–97.

Haibel, H, Rudwaleit, M, Braun, J, and Sieper, J (2005). Six months open label trial of leflunomide in active ankylosing spondylitis. *Ann Rheum Dis*, 64 (1):124–126.

Halfvarson, J, Bodin, L, Tysk, C, Lindberg, E, and Jarnerot, G (2003). Inflammatory bowel disease in a Swedish twin cohort: a long-term follow-up of concordance and clinical characteristics. *Gastroenterology,* 124 (7):1767–1773.

Hall, FC, Weeks, DE, Camilleri, JP, Williams, LA, Amos, N, Darke, C, et al. (1996). Influence of the HLA-DRB1 locus on susceptibility and severity in rheumatoid arthritis. *QJM,* 89(11):821–829.

Hamersma, J, Cardon, LR, Bradbury, L, Brophy, S, van der Horst-Bruinsma, I, Calin, A, et al. (2001). Is disease severity in ankylosing spondylitis genetically determined? *Arthritis Rheum,* 44(6):1396–1400.

Hampe, J, Shaw, SH, Saiz, R, Leysens, N, Lantermann, A, Mascheretti, S, et al. (1999). Linkage of inflammatory bowel disease to human chromosome 6p. *Am J Hum Genet,* 65(6):1647–1655.

Hannum, CH, Wilcox, CJ, Arend, WP, Joslin, FG, Dripps, DJ, Heimdal, PL, et al. (1990). Interleukin-1 receptor antagonist activity of a human interleukin-1 inhibitor. Nature, 343(6256):336–340.

Harney, SM, Meisel, C, Sims, AM, Woon, PY, Wordsworth, BP, and Brown, MA (2005). Genetic and genomic studies of PADI4 in rheumatoid arthritis. *Rheumatology (Oxford),* 44(7):869–872.

Harney S, Newton J, Milicic A, Brown MA, and Wordsworth BP (2003). Non-inherited maternal HLA alleles are associated with rheumatoid arthritis. *Rheumatology* 42:171–174.

Harrison, BJ, Symmons, DP, Barrett, EM, and Silman, AJ (1998). The performance of the 1987 ARA classification criteria for rheumatoid arthritis in a population based cohort of patients with early inflammatory polyarthritis. *J Rheumatol,* 25(12):2324–2330.

Heward, JM, Allahabadia, A, Armitage, M, Hattersley, A, Dodson, PM, Macleod, K, et al. (1999). The development of Graves' disease and the CTLA-4 gene on chromosome 2q33. *J Clin Endocrinol Metab,* 84 (7):2398–2401.

Hill, AV, Allsopp, CE, Kwiatkowski, D, Anstey, NM, Greenwood, BM, and McMichael, AJ (1991). HLA class I typing by PCR: HLA-B27 and an African B27 subtype. *Lancet,* 337(8742):640–642.

Hochberg, MC (1987). Mortality from systemic lupus erythematosus in England and Wales, 1974–1983. *Br J Rheumatol,* 26(6):437–441.

Hochberg, MC (1999). Early aggressive DMARD therapy: the key to slowing disease progression in rheumatoid arthritis. *Scand J Rheumatol Suppl,* 112:3–7.

Hoving, JL, Buchbinder, R, Hall, S, Lawler, G, Coombs, P, McNealy, S, et al. (2004). A comparison of magnetic resonance imaging, sonography, and radiography of the hand in patients with early rheumatoid arthritis. *J Rheumatol,* 31(4):663–675.

Hugot, JP, Chamaillard, M, Zouali, H, Lesage, S, Cezard, JP, Belaiche, J, et al. (2001). Association of NOD2 leucine-rich repeat variants with susceptibility to Crohn's disease. *Nature,* 411(6837):599–603.

Hugot, JP, Laurent-Puig, P, Gower-Rousseau, C, Olson, JM, Lee, JC, Beaugerie, L, et al. (1996). Mapping of a susceptibility locus for Crohn's disease on chromosome 16. *Nature,* 379(6568):821–823.

Inman, RD, Whittum-Hudson, JA, Schumacher, HR, and Hudson, AP (2000). Chlamydia and associated arthritis. *Curr Opin Rheumatol,* 12(4):254–262.

Jaakkola E, Crane AM, Laiho K, Herzberg I, Sims AM, Bradbury L, et al. (2004). The effect of transforming growth factor beta 1 gene polymorphisms in ankylosing spondylitis. *Rheumatology (Oxford)* 43:32–38.

Jaakkola E, Herzberg I, Laiho K, Barnardo M, Pointon J, Kauppi M, et al. (2006). Finnish HLA studies confirm the increased risk conferred by HLA-B27 homozygosity in ankylosing spondylitis. *Ann Rehum Dis* 65:775–780.

James, ES, Harney, S, Wordsworth, BP, Cookson, WO, Davis, SJ, and Moffatt, MF (2005). PDCD1: a tissue-specific susceptibility locus for inherited inflammatory disorders. *Genes Immun,* 6(5):430–437.

Jawaheer D, Thomson W, MacGregor AJ, Carthy D, Davidson J, Dyer PA, et al. (1994). 'Homozygosity' for the HLA-DR shared epitope contributes the highest risk for rheumatoid arthritis concordance in identical twins. *Arthritis Rheum* 37(5):681–686.

Jawaheer D, Seldin MF, Amos CI, Chen WV, Shigeta R, Etzel C, et al. (2003). Screening the genome for rheumatoid arthritis susceptibility genes: a repliation study and combied analysis of 512 multicase families. *Arthritis Rheum* 48(4):906–916.

Jawaheer D, Seldin MF, Amos CI, Chen WV, Shigeta R, Monteiro J, et al. (2001). A genomewide screen in multiplex rheumatoid arthritis families suggests genetic overlap with other autoimmune diseases. *Am J Hum Genet* 68 (4):927–936.

Jenisch, S, Hampe, J, Elder, JT, Nair, R, Stuart, P, Voorhees, JJ, et al. (2006). CARD15 mutations in patients with plaque-type psoriasis and psoriatic arthritis: lack of association. *Arch Dermatol Res,* 297(9):4.

Jenisch, S, Henseler, T, Nair, RP, Guo, SW, Westphal, E, Stuart, P, et al. (1998). Linkage analysis of human leukocyte antigen (HLA) markers in familial psoriasis: strong disequilibrium effects provide evidence for a major determinant in the HLA-B/-C region. *Am J Hum Genet,* 63(1):191–199.

Jenkinson, TR, Mallorie, PA, Whitelock, HC, Kennedy, LG, Garrett, SL, and Calin, A (1994). Defining spinal mobility in ankylosing spondylitis (AS). The Bath AS Metrology Index. *J Rheumatol,* 21(9):1694–1698.

Jin, L, Weisman, M, Zhang, G, Ward, M, Luo, J, Bruckel, J, et al. (2005). Lack of association of matrix metalloproteinase 3 (MMP3) genotypes with ankylosing spondylitis susceptibility and severity. *Rheumatology (Oxford),* 44 (1):55–60.

John S, Amos C, Shepherd N, Chen W, Butterworth A, Etzel C, et al. (2006). Linkage analysis of rheumatoid arthritis in US and UK families reveals interactions between HLA-DRB1 and loci on chromosomes 6q and 16p. *Arthritis Rheum* 54(5):1482–1490.

Kang, CP, Lee, HS, Ju, H, Cho, H, Kang, C, and Bae, SC (2006). A functional haplotype of the PADI4 gene associated with increased rheumatoid arthritis susceptibility in Koreans. *Arthritis Rheum,* 54(1):90–96.

Karason, A, Gudjonsson, JE, Upmanyu, R, Antonsdottir, AA, Hauksson, VB, Runasdottir, EH, et al. (2003). A susceptibility gene for psoriatic arthritis maps to chromosome 16q: evidence for imprinting. *Am J Hum Genet,* 72 (1):125–131

Kellgren, JH, Lawrence, JS, and Bier, F (1963). Genetic factors in generalized osteo-arthrosis. *Ann Rheum Dis,* 22:237–255

Kennedy, LG, Will, R, and Calin, A (1993a). Sex ratio in the spondyloarthropathies and its relationship to phenotypic expression, mode of inheritance and age at onset. *J Rheumatol,* 20(11):1900–1904.

Khan, MA (1980). Clinical application of the HLA-B27 test in rheumatic diseases. A current perspective. *Arch Intern Med,* 140(2):177–180.

Khan, MA (1997) A worldwide overview: the epidemiology of HLA-B27 and associated spondyloarthritides. In A Calin, and J Taurog (eds), *The Spondyloarthritides* (pp. 17–26). Oxford: Oxford University Press.

Khanna, D, Ranganath, VK, Fitzgerald, J, Park, GS, Altman, RD, Elashoff, D, et al. (2005). Increased radiographic damage scores at the onset of seropositive rheumatoid arthritis in older patients are associated with osteoarthritis of the hands, but not with more rapid progression of adamage. *Arthritis Rheum,*52(8):2284–2292.

Kilding R, Iles MM, Timms JM, Worthington J, Wilson AG (2004). Additional genetic susceptibility for rheumatoid arthritis telomeric of the DRB1 locus. *Arthritis Rheum* 50(3):763–769.

Kim, HY, Kim, TG, Park, SH, Lee, SH, Cho, CS, and Han, H (1995). Predominance of HLA-DRB1*0405 in Korean patients with rheumatoid arthritis. *Ann Rheum Dis,* 54(12):988–990.

Kimura M, Kiwahito Y, Obayashi H, Ohta M, Hara H, Adachi T, et al. (2007). A critical role for allograft inflammatory factor-1 in the pathogenesis of rheumatoid arthritis. *J Immunol* 178:336–3322.

Kong, YY, Feige, U, Sarosi, I, Bolon, B, Tafuri, A, Morony, S, et al. (1999). Activated T cells regulate bone loss and joint destruction in adjuvant arthritis through osteoprotegerin ligand. *Nature,* 402(6759):304–309.

Kong, EK, Prokunina-Olsson, L, Wong, WH, Lau, CS, Chan, TM, Alarcon-Riquelme, M, et al. (2005). A new haplotype of PDCD1 is associated with rheumatoid arthritis in Hong Kong Chinese. *Arthritis Rheum,* 52(4):1058–1062.

Kurki, P, Aho, K, Palosuo, T, and Heliovaara, M (1992). Immunopathology of rheumatoid arthritis. Antikeratin antibodies precede the clinical disease. *Arthritis Rheum,* 35(8):914–917.

Lacey, DL, Timms, E, Tan, HL, Kelley, MJ, Dunstan, CR, Burgess, T, et al. (1998).Osteoprotegerin ligand is a cytokine that regulates osteoclast differentiation and activation. *Cell,* 93(2):165–176.

Laloux, L, Voisin, MC, Allain, J, Martin, N, Kerboull, L, Chevalier, X, et al. (2001). Immunohistological study of entheses in spondyloarthropathies: comparison in rheumatoid arthritis and osteoarthritis. *Ann Rheum Dis,* 60(4):316–321.

Lanchbury, JS, Jaeger, EE, Sansom, DM, Hall, MA, Wordsworth, P, Stedeford, J, et al. (1991). Strong primary selection for the Dw4 subtype of DR4 accounts for the HLA-DQw7 association with Felty's syndrome. *Hum Immunol,* 32(1):56–64.

Lascorz, J, Burkhardt, H, Huffmeier, U, Bohm, B, Schurmeyer-Horst, F, Lohmann, J, et al. (2005). Lack of genetic association of the three more common polymorphisms of CARD15 with psoriatic arthritis and psoriasis in a German cohort. *Ann Rheum Dis,* 64(6):951–954.

Laval SH, Timms A, Edwards S, Bradbury L, Brophy S, Milicic, A et al. (2001). Whole genome screening in ankylosing spondylitis: evidence of non MHC genetic susceptibility loci. *Am J Hum Genet* 68:918–926.

Lawrence JS (1970). Rheumatoid arthritis – nature or nurture? *Ann Rheum Dis* 29:357–379.

Leandro, MJ, Edwards, JC, and Cambridge, G (2002). Clinical outcome in 22 patients with rheumatoid arthritis treated with B lymphocyte depletion. *Ann Rheum Dis,* 61(10):883–888.

Leirisalo-Repo, M (1998). Prognosis, course of disease, and treatment of the spondyloarthropathies. *Rheum Dis Clin North Am,* 24(4):737–751.

Lopez-Ben, R, Bernreuter, WK, Moreland, LW, and Alarcon, GS (2004). Ultrasound detection of bone erosions in rheumatoid arthritis: a comparison to routine radiographs of the hands and feet. *Skeletal Radiol,* 33 (2):80–84.

Lopez-Larrea, C, Sujirachato, K, Mehra, NK, Chiewsilp, P, Isarangkura, D, Kanga, U, et al. (1995). HLA-B27 subtypes in Asian patients with ankylosing spondylitis. Evidence for new associations. *Tissue Antigens,* 45 (3):169–176.

Lorenz, HM, Antoni, C, Valerius, T, Repp, R, Grunke, M, Schwerdtner, N, et al. (1996). In vivo blockade of TNF-alpha by intravenous infusion of a chimeric monoclonal TNF-alpha antibody in patients with rheumatoid arthritis. Short term cellular and molecular effects. *J Immunol,* 156 (4):1646–1653.

MacGregor, AJ, Riste, LK, Hazes, JM, and Silman, AJ (1994). Low prevalence of rheumatoid arthritis in black-Caribbeans compared with whites in inner city Manchester. *Ann Rheum Dis,* 53(5):293–297.

MacGregor, AJ, Snieder, H, Rigby, AS, Koskenvuo, M, Kaprio, J, Aho, K, et al. (2000). Characterizing the quantitative genetic contribution to rheumatoid arthritis using data from twins. *Arthritis Rheum,* 43:30–37.

MacKay K, Eyre S, Myerscough A, Milicic A, Barton A, Laval S, et al. (2002). Whoe-genome linkage analysis of rheumatoid arthritis susceptibility loci in 252 affected sibling pairs in the United Kingdom. *Arthritis Rheum* 46:632–639.

Madhavan, R, Parthiban, M, Rajendran, CP, Chandrasekaran, AN, Zake, L, and Sanjeevi, CB (2002). HLA class I and class II association with ankylosing spondylitis in a southern Indian population. *Ann N Y Acad Sci,* 958:403–407.

Maini RN, Taylor PC, Szechinski J, Pavelka K, Broll J, Balint G et al. (2006). Double blind randomized controlled clinical trial of the interleukin-6 receptor antagonist, tocilizumab, in European patients with rheumatoid arthritis who had an incomplete response to methotrexate. *Arthritis Rheum* 54:2817–2829.

Maksymowych, WP, Jhangri, GS, Fitzgerald, AA, LeClercq, S, Chiu, P, Yan, A, et al. (2002). A six-month randomized, controlled, double-blind, dose-response comparison of intravenous pamidronate (60 mg versus 10 mg) in the treatment of nonsteroidal antiinflammatory drug-refractory ankylosing spondylitis. *Arthritis Rheum,* 46(3):766–773.

Martinez, A, Valdivia, A, Pascual-Salcedo, D, Lamas, JR, Fernandez-Arquero, M, Balsa, A, et al. (2005). PADI4 polymorphisms are not associated with rheumatoid arthritis in the Spanish population. *Rheumatology (Oxford),* 44 (10):1263–1266.

Masson-Bessiere, C, Sebbag, M, Durieux, JJ, Nogueira, L, Vincent, C, Girbal-Neuhauser, E, et al. (2000). In the rheumatoid pannus, anti-filaggrin autoantibodies are produced by local plasma cells and constitute a higher proportion of IgG than in synovial fluid and serum. *Clin Exp Immunol,* 119(3):544–552.

Mattey, DL, Hassell, AB, Plant, MJ, Cheung, NT, Dawes, PT, Jones, PW, et al. (1999). The influence of HLA-DRB1 alleles encoding the DERAA amino acid motif on radiological outcome in rheumatoid arthritis. *Rheumatology (Oxford),* 38(12):1221–1227.

Mau, W, Zeidler, H, Mau, R, Majewski, A, Freyschmidt, J, Stangel, W, et al. (1988) Clinical features and prognosis of patients with possible ankylosing spondylitis. Results of a 10-year followup. *J Rheumatol,* 15(7):1109–1114.

McGarry, F, Neilly, J, Anderson, N, Sturrock, R, and Field, M (2001). A polymorphism within the interleukin 1 receptor antagonist (IL-1Ra) gene is associated with ankylosing spondylitis. *Rheumatology (Oxford),* 40 (12):1359–1364.

McGonagle, D, Benjamin, M, Marzo-Ortega, H, and Emery, P (2002). Advances in the understanding of entheseal inflammation. *Curr Rheumatol Rep,* 4(6):500–506.

McGonagle, D, Gibbon, W, and Emery, P (1998). Classification of inflammatory arthritis by enthesitis. *Lancet,* 352(9134):1137–1140.

McHugh, NJ, Laurent, MR, Treadwell, BL, Tweed, JM, and Dagger, J (1987). Psoriatic arthritis: clinical subgroups and histocompatibility antigens. *Ann Rheum Dis,* 46(3):184–188.

McQueen, FM, Stewart, N, Crabbe, J, Robinson, E, Yeoman, S, Tan, PL, et al. (1998). Magnetic resonance imaging of the wrist in early rheumatoid arthritis reveals a high prevalence of erosions at four months after symptom onset. *Ann Rheum Dis,* 57(6):350–356.

Miceli-Richard, C, Said-Nahal, R, and Breban, M (2000). Impact of sex on inheritance of ankylosing spondylitis. *Lancet,* 355(9209):1097–1098

Miceli-Richard, C, Zouali, H, Lesage, S, Thomas, G, Hugot, JP, and Said-Nahal, R, et al. (2002). CARD15/NOD2 analyses in spondylarthropathy. *Arthritis Rheum,* 46(5):1405–1406.

Miceli-Richard C, Zouali H, Said-Nahal R, Lesage S, Merlin F, De Toma C, et al. (2004). Significant linkage to spondyloarthropathy on 9q31–34. *Hum Mol Genet* 13(15):1641–1648.

Mielants, H, Veys, EM, Goemaere, S, Goethals, K, Cuvelier, C, and De Vos, M (1991). Gut inflammation in the spondyloarthropathies: clinical, radiologic, biologic and genetic features in relation to the type of histology. A prospective study. *J Rheumatol,* 18(10):1542–1551.

Mikuls, TR and Moreland, LW (2001). TNF blockade in the treatment of rheumatoid arthritis: infliximab versus etanercept. *Expert Opin Pharmacother,* 2(1):75–84.

Milicic A, Lee D, Brown MA, Darke C and Wordsworth BP (2002). HLA-DR/DQ haplotype in rheumatoid arthritis: novel allelic associations in UK Caucasians. *J Rheumatol* 29(9):1821–1826.

Moll, JM and Wright, V (1973). Familial occurrence of psoriatic arthritis. *Ann Rheum Dis,* 32(3):181–201.

Moreland, LW, Baumgartner, SW, Schiff, MH, Tindall, EA, Fleischmann, RM, Weaver, AL, et al. (1997). Treatment of rheumatoid arthritis with a recombinant human tumor necrosis factor receptor (p75)-Fc fusion protein. *N Engl J Med,* 337(3):141–147.

Mori, K, Ushiyama, T, Inoue, K, and Hukuda, S (2000). Polymorphic CAG repeats of the androgen receptor gene in Japanese male patients with ankylosing spondylitis. *Rheumatology (Oxford),* 39(5):530–532.

Mori, K, Chano, T, Ikeda, T, Ikegawa, S, Matsusue, Y, Okabe, H, et al. (2003). Decrease in serum nucleotide pyrophosphatase activity in ankylosing spondylitis. *Rheumatology (Oxford),* 42(1):62–65.

Montserrat, V, Marti, M, and Lopez de Castro, JA (2003) Allospecific T cell epitope sharing reveals extensive conservation of the antigenic features of peptide ligands among HLA-B27 subtypes differentially associated with spondyloarthritis. *J Immunol,* 1:170(11):5778–5785.

Mulcahy, B, Waldron-Lynch, F, McDermott, MF, Adams, C, Amos, CI, Zhu, DK, et al. (1996). Genetic variability in the tumor necrosis factor-lymphotoxin region influences susceptibility to rheumatoid arthritis. *Am J Hum Genet,* 59(3):676–683.

Myers, A, Kay, LJ, Lynch, SA, and Walker, DJ (2005). Recurrence risk for psoriasis and psoriatic arthritis within sibships. *Rheumatology (Oxford),* 44(6):773–776.

Nair, RP, Stuart, P, Henseler, T, Jenisch, S, Chia, NV, Westphal, E, et al. (2000). Localization of psoriasis-susceptibility locus PSORS1 to a 60-kb interval telomeric to HLA-C. *Am J Hum Genet,* 66(6):1833–1844.

Nasution AR, Mardjuadi A, Kunmartini S, Suryadhana NG, Setyohadi B, Sudarsono D, et al. (1997). HLA-B27 subtypes positively and negatively associate with spondyloarthropathy. *J Rheumatol* 24(6):1111–1114.

Neidel, J, Schulze, M, and Lindschau, J (1995). Association between degree of bone-erosion and synovial fluid-levels of tumor necrosis factor alpha in the knee-joints of patients with rheumatoid arthritis.*Inflamm Res,* 44 (5):217–221.

Nepom, GT, Seyfried, CE, Holbeck, SL, Wilske, KR, and Nepom, BS (1986). Identification of HLA-Dw14 genes in DR4^{+} rheumatoid arthritis. *Lancet,* 2 (8514):1002–1005.

Newton, J, Brown, MA, Milicic, A, Ackerman, H, Darke, C, Wilson, JN, et al. (2003). The effect of HLA-DR on susceptibility to rheumatoid arthritis is influenced by the associated lymphotoxin alpha-tumor necrosis factor haplotype. *Arthritis Rheum,* 48(1):90–96.

Newton, JL, Harney, SM, Timms, AE, Sims, AM, Rockett, K, Darke, C, et al. (2004). Dissection of class III major histocompatibility complex haplotypes associated with rheumatoid arthritis. *Arthritis Rheum,* 50(7):2122–2129.

Nicola, PJ, Maradit-Kremers, H, Roger, VL, Jacobsen, SJ, Crowson, CS, Ballman, KV, et al. (2005). The risk of congestive heart failure in rheumatoid arthritis: a population-based study over 46 years. *Arthritis Rheum,* 52 (2):412–420.

Nishimura, H, Nose, M, Hiai, H, Minato, N, and Honjo, T (1999). Development of lupus-like autoimmune diseases by disruption of the PD-1 gene encoding an ITIM motif-carrying immunoreceptor. *Immunity,* 11 (2):141–151.

Nishimura, H, Okazaki, T, Tanaka, Y, Nakatani, K, Hara, M, Matsumori, A, et al. (2001). Autoimmune dilated cardiomyopathy in PD-1 receptor-deficient mice. *Science,* 291(5502):319–322.

Nistico, L, Buzzetti, R, Pritchard, LE, Van der Auwera, B, Giovannini, C, Bosi, E, et al. (1996). The CTLA-4 gene region of chromosome 2q33 is linked to, and associated with, type 1 diabetes. Belgian Diabetes Registry. *Hum Mol Genet,* 5(7):1075–1080.

Ogura, Y, Bonen, DK, Inohara, N, Nicolae, DL, Chen, FF, Ramos, R, et al. (2001). A frameshift mutation in NOD2 associated with susceptibility to Crohn's disease. *Nature,* 411(6837):603–606.

Ollier WE, Stephens C, Awad J, Carthy D, Gupta A, Perry D, Jawad A and Festenstein H (1991). Is rheumatoid arthritis in Indians associated with HLA antigens sharing a DR beta 1 epitope? *Ann Rheum Dis* 50(5):295–297.

O'Neill, TW, Cooper, C, Cannata, JB, Diaz Lopez, JB, Hoszowski, K, Johnell, O, et al. (1994). Reproducibility of a questionnaire on risk factors for osteoporosis in a multicentre prevalence survey: the European Vertebral Osteoporosis Study. *Int J Epidemiol,* 23(3):559–565.

Orchard, TR, Satsangi, J, Van Heel, D, and Jewell, DP (2000). Genetics of inflammatory bowel disease: a reappraisal. *Scand J Immunol,* 51(1):10–17.

Orchard, TR, Wordsworth, BP, and Jewell, DP (1998). Peripheral arthropathies in inflammatory bowel disease: their articular distribution and natural history. *Gut,* 42(3):387–391.

Orholm, M, Binder, V, Sorensen, TI, Rasmussen, LP, and Kyvik, KO (2000). Concordance of inflammatory bowel disease among Danish twins. Results of a nationwide study. *Scand J Gastroenterol,* 35(10):1075–1081.

Paimela, L, Palosuo, T, Aho, K, Lukka, M, Kurki, P, Leirisalo-Repo, M, et al. (2001). Association of autoantibodies to filaggrin with an active disease in early rheumatoid arthritis. *Ann Rheum Dis,* 60(1):32–35.

Palm, O, Moum, B, Jahnsen, J, and Gran, JT (2001). The prevalence and incidence of peripheral arthritis in patients with inflammatory bowel disease, a prospective population-based study (the IBSEN study). *Rheumatology (Oxford),* 40(11):1256–1261.

Palumbo, PJ, Ward, LE, Sauer, WG, and Scudamore, HH (1973). Musculoskeletal manifestations of inflammatory bowel disease. Ulcerative and granulomatous colitis and ulcerative proctitis. *Mayo Clin Proc,* 48(6):411–416.

Parkes, M, Barmada, MM, Satsangi, J, Weeks, DE, Jewell, DP, and Duerr, RH (2000). The IBD2 locus shows linkage heterogeneity between ulcerative colitis and Crohn disease. *Am J Hum Genet,* 67(6):1605–1610.

Peeters, H, Vander Cruyssen, B, Laukens, D, Coucke, P, Marichal, D, Van Den Berghe, M, et al. (2004). Radiological sacroiliitis, a hallmark of spondylitis, is linked with CARD15 gene polymorphisms in patients with Crohn's disease. *Ann Rheum Dis,* 63(9):1131–1134.

Peltomaa, R, Leirisalo-Repo, M, Helve, T, and Paimela, L (2000). Effect of age on 3 year outcome in early rheumatoid arthritis. *J Rheumatol,* 27 (3):638–643.

Pettit, AR, Ji, H, von Stechow, D, Muller, R, Goldring, SR, Choi, Y, et al. (2001). TRANCE/RANKL knockout mice are protected from bone erosion in a serum transfer model of arthritis. *Am J Pathol,* 159(5):1689–1699.

Pitzalis, C, Cauli, A, Pipitone, N, Smith, C, Barker, J, Marchesoni, A, et al. (1996). Cutaneous lymphocyte antigen-positive T lymphocytes preferentially migrate to the skin but not to the joint in psoriatic arthritis. *Arthritis Rheum,* 39(1):137–145.

Plenge RM, Padukov L, Remmers EF, Purcell S, Lee AT, Karlson EW, Wolfe F, Kastner DL, Alfredsson L, Altshuler D, Gregersen PK, Klareskog L and Rioux JD (2005). Replication of putative candidate-gene associations with rheumatoid arthritis in >4,000 samples from North America and Sweden: association of susceptibility with PTPN22, CTLA4, and PADI4. *Am J Hum Genet,* 77(6):1044–1060.

Polito, JM 2nd, Rees, RC, Childs, B, Mendeloff, AI, Harris, ML, and Bayless, TM (1996). Preliminary evidence for genetic anticipation in Crohn's disease. *Lancet,* 347(9004):798–800.

Ploski, R, Mellbye, OJ, Ronningen, KS, Forre, O, and Thorsby, E (1994). Seronegative and weakly seropositive rheumatoid arthritis differ from clearly seropositive rheumatoid arthritis in HLA class II associations. *J Rheumatol,* 21(8):1397–1402.

Queiro, R, Sarasqueta, C, Belzunegui, J, Gonzalez, C, Figueroa, M, and Torre-Alonso, JC (2002). Psoriatic spondyloarthropathy: a comparative study between HLA-B27 positive and HLA-B27 negative disease. *Semin Arthritis Rheum,* 31(6):413–418.

Rahman, P, Bartlett, S, Siannis, F, Pellett, FJ, Farewell, VT, Peddle, L, et al. (2003). CARD15: a pleiotropic autoimmune gene that confers susceptibility to psoriatic arthritis. *Am J Hum Genet,* 73(3):677–681.

Qin, S, Rottman, JB, Myers, P, Kassam, N, Weinblatt, M, Loetscher, M, et al. (1998). The chemokine receptors CXCR3 and CCR5 mark subsets of T cells associated with certain inflammatory reactions. *J Clin Invest,* 101 (4):746–754.

Ralston, SH, Urquhart, GD, Brzeski, M, and Sturrock, RD (1990). Prevalence of vertebral compression fractures due to osteoporosis in ankylosing. *BMJ,* 300(6724):563–565.

Rantapaa-Dahlqvist, S, de Jong, BA, Berglin, E, Hallmans, G, Wadell, G, Stenlund, H, et al. (2003). Antibodies against cyclic citrullinated peptide and IgA rheumatoid factor predict the development of rheumatoid arthritis. *Arthritis Rheum,* 48(10):2741–2749.

Revell, PA and Mayston, V (1982). Histopathology of the synovial membrane of peripheral joints in ankylosing spondylitis. *Ann Rheum Dis,* 41 (6):579–586.

Rich, E, Moreland, LW, and Alarcon, GS (1999). Paucity of radiographic progression in rheumatoid arthritis treated with methotrexate as the first disease modifying antirheumatic drug. *J Rheumatol,* 26(2):259–261.

Ridker, PM, Rifai, N, Rose, L, Buring, JE, and Cook, NR (2002).Comparison of C-reactive protein and low-density lipoprotein cholesterol levels in the prediction of first cardiovascular events. *N Engl J Med,* 347(20):1557–1565.

Risch, N and Zhang, H (1995). Extreme discordant sib pairs for mapping quantitative trait loci in humans. *Science,* 268(5217):1584–1589.

Ritchlin, C (2000). Fibroblast biology. Effector signals released by the synovial fibroblast in arthritis. *Arthritis Res,* 2(5):356–360.

Robinson, WP, van der Linden, SM, Khan, MA, Rentsch, HU, Cats, A, Russell, A, et al. (1989). HLA-Bw60 increases susceptibility to ankylosing spondylitis in HLA-B27^{+} patients. *Arthritis Rheum,* 32(9):1135–1141.

Rubin, LA, Amos, CI, Wade, JA, Martin, JR, Bale, SJ, Little, AH, et al. (1994). Investigating the genetic basis for ankylosing spondylitis. Linkage studies with the major histocompatibility complex region. *Arthritis Rheum,* 37 (8):1212–1220.

Sacre, SM, Andreakos, E, Taylor, P, Feldmann, M, and Foxwell, BM (2005). Molecular therapeutic targets in rheumatoid arthritis. *Expert Rev Mol Med,* 7(16):1–20.

Sanchez, B, Moreno, I, Magarino, R, Garzon, M, Gonzalez, MF, Garcia, A, et al.. (1990). HLA-DRw10 confers the highest susceptibility to rheumatoid arthritis in a Spanish population. *Tissue Antigens,* 36(4):174–176.

Said-Nahal R, Miceli-Richard C, Dougados M, and Breban M (2001). Increased risk of ankylosing spondylitis among first-born children: comment on article by Baudoin et al. *Arthritis Rheum* 44(8):1964–1965.

Said-Nahal, R, Miceli-Richard, C, Gautreau, C, Tamouza, R, Borot, N, Porcher, R, et al. (2002). The role of HLA genes in familial spondyloarthropathy: a comprehensive study of 70 multiplex families. *Ann Rheum Dis,* 61 (3):201–206.

Sampson, HW (1988). Spondyloarthropathy in progressive ankylosis (ank/ank) mice: morphological features. *Spine*, 13(6):645–649.

Satsangi, J, Grootscholten, C, Holt, H, and Jewell, DP (1996). Clinical patterns of familial inflammatory bowel disease. *Gut*, 38(5):738–741.

Schellekens, GA, de Jong, BA, van den Hoogen, FH, van de Putte, LB, and van Venrooij, WJ (1998). Citrulline is an essential constituent of antigenic determinants recognized by rheumatoid arthritis-specific autoantibodies. *J Clin Invest*, 101(1):273–281.

Schlosstein, L, Terasaki, PI, Bluestone, R, and Pearson, CM (1973). High association of an HL-A antigen, W27, with ankylosing spondylitis.*N Engl J Med*, 288(14):704–706.

Schorr-Lesnick, B and Brandt, LJ (1988). Selected rheumatologic and dermatologic manifestations of inflammatory bowel disease. *Am J Gastroenterol*, 83 (3):216–223.

Sharp, JT, Strand, V, Leung, H, Hurley, F, and Loew-Friedrich, I (2000). Treatment with leflunomide slows radiographic progression of rheumatoid arthritis: results from three randomized controlled trials of leflunomide in patients with active rheumatoid arthritis. Leflunomide Rheumatoid Arthritis Investigators Group. *Arthritis Rheum*, 43 (3):495–505.

Shiozawa S, Hayashi S, Tsukamoto Y, Goko H, Kawasaki H, Wade T, Shimizu K, et al. (1998). Identification of the gene loci that predispose to rheumatoid arthritis. *Int Immunol* 10(12):1891–1895.

Silman, AJ, MacGregor, AJ, Thomson, W, Holligan, S, Carthy, D, Farhan, A, et al. (1993). Twin concordance rates for rheumatoid arthritis: results from a nationwide study. *Br J Rheumatol*, 32(10):903–907.

Sims AM, Barnardo M, Herzberg I, Bradbury L, Calin A, Wordsworth BP, et al. (2007). Non-B27 MHC associations of ankylosing spondylitis. *Genesand Immunity* 8:115–123.

Smolen, JS, Kalden, JR, Scott, DL, Rozman, B, Kvien, TK, Larsen, A, et al. (1999). Efficacy and safety of leflunomide compared with placebo and sulphasalazine in active rheumatoid arthritis: a double-blind, randomised, multicentre trial. European Leflunomide Study Group. *Lancet*, 353 (9149):259–266.

Stastny, P (1976). Mixed lymphocyte cultures in rheumatoid arthritis. *J Clin Invest*, 57(5):1148–1157.

Stastny P (1978). Association of the B-cell alloantigen DRw4 with rheumatoid arthritis. *N Engl J Med* 298(16):869–871.

Stern, RS (1985). The epidemiology of joint complaints in patients with psoriasis. *J Rheumatol*, 12(2):315–320.

Suzuki, K, Sawada, T, Murakami, A, Matsui, T, Tohma, S, Nakazono, K, et al. (2003). High diagnostic performance of ELISA detection of antibodies to citrullinated antigens in rheumatoid arthritis. *Scand J Rheumatol*, 32 (4):197–204.

Svendsen, AJ, Holm, NV, Kyvik, K, Petersen, PH, and Junker, P (2002). Relative importance of genetic effects in rheumatoid arthritis: historical cohort study of Danish nationwide twin population. *BMJ*, 324(7332):264–266.

Szkudlarek, M, Narvestad, E, Klarlund, M, Court-Payen, M, Thomsen, HS, and Ostergaard, M (2004). Ultrasonography of the metatarsophalangeal joints in rheumatoid arthritis: comparison with magnetic resonance imaging, conventional radiography, and clinical examination. *Arthritis Rheum*, 50 (7):2103–2112.

Tamiya, G, Shinya, M, Imanishi, T, Ikuta, T, Makino, S, Okamoto, K, et al. (2005). Whole genome association study of rheumatoid arthritis using 27 039 microsatellites. *Hum Mol Genet*, 14(16):2305–2321.

Tan, AL, Marzo-Ortega, H, O'Connor, P, Fraser, A, Emery, P, and McGonagle, D (2004). Efficacy of anakinra in active ankylosing spondylitis: a clinical and magnetic resonance imaging study. *Ann Rheum Dis*, 63(9):1041–1045.

Taneja, V, Giphart, MJ, Verduijn, W, Naipal, A, Malaviya, AN, and Mehra, NK (1996). Polymorphism of HLA-DRB, -DQA1, and -DQB1 in rheumatoid arthritis in Asian Indians: association with DRB1*0405 and DRB1*1001. *Hum Immunol*, 46(1):35–41.

Thomas, E, Hay, EM, Hajeer, A, and Silman, AJ (1998). Sjogren's syndrome: a community-based study of prevalence and impact. *Br J Rheumatol*, 37 (10):1069–1076.

Thompson, NP, Driscoll, R, Pounder, RE, and Wakefield, AJ (1996). Genetics versus environment in inflammatory bowel disease: results of a British twin study. *BMJ*, 312(7023):95–96.

Thomson, W, Harrison, B, Ollier, B, Wiles, N, Payton, T, Barrett, J, et al. (1999). Quantifying the exact role of HLA-DRB1 alleles in susceptibility to inflammatory polyarthritis: results from a large, population-based study. *Arthritis Rheum*, 42(4):757–762.

Tiilikainen, A, Lassus, A, Karvonen, J, Vartiainen, P, and Julin, M (1980). Psoriasis and HLA-Cw6. *Br J Dermatol*, 102(2):179–184.

Timms A, Crane AM, Sims A-M,Cordell H, Bradbury L, Abbott A, et al. (2004). The interleukin 1 gene cluster contains a major susceptibility locus for ankylosing spondylitis. *Am J Hum Genet* 75:587–595.

Timms, AE, Zhang, Y, Bradbury, L, Wordsworth, BP, and Brown, MA (2003). Investigation of the role of ANKH in ankylosing spondylitis. *Arthritis Rheum*, 48(10):2898–2902.

Tsui, FW, Tsui, HW, Cheng, EY, Stone, M, Payne, U, Reveille, JD, et al. (2003). Novel genetic markers in the 5'-flanking region of ANKH are associated with ankylosing spondylitis. *Arthritis Rheum*, 48(3):791–797.

Turesson, C, Jacobsson, L, and Bergstrom, U (1999). Extra-articular rheumatoid arthritis: prevalence and mortality. *Rheumatology (Oxford)*, 38 (7):668–674.

Turesson, C, O'Fallon, WM, Crowson, CS, Gabriel, SE, and Matteson, EL (2002). Occurrence of extraarticular disease manifestations is associated with excess mortality in a community based cohort of patients with rheumatoid arthritis. *J Rheumatol*, 29(1):62–67.

Turesson, C, Schaid, DJ, Weyand, CM, Jacobsson, LT, Goronzy, JJ, Petersson, IF, et al. (2005). The impact of HLA-DRB1 genes on extra-articular disease manifestations in rheumatoid arthritis. *Arthritis Res Ther*, 7(6): R1386–R1393.

Turesson, C, Weyand, CM, and Matteson, EL (2004). Genetics of rheumatoid arthritis: is there a pattern predicting extraarticular manifestations? *Arthritis Rheum*, 51(5):853–863.

Turesson, I, Carlsson, J, Brahme, A, Glimelius, B, Zackrisson, B, Stenerlow, B, et al. (2003). Biological response to radiation therapy. *Acta Oncol*, 42 (2):92–106.

Tysk, C, Lindberg, E, Jarnerot, G, and Floderus-Myrhed, B (1988). Ulcerative colitis and Crohn's disease in an unselected population of monozygotic and dizygotic twins. A study of heritability and the influence of smoking. *Gut*, 29 (7):990–996.

Vallejo, AN, Nestel, AR, Schirmer, M, Weyand, CM, and Goronzy, JJ (1998). Aging-related deficiency of CD28 expression in $CD4^+$ T cells is associated with the loss of gene-specific nuclear factor binding activity. *J Biol Chem*, 273(14):8119–8129.

van der Heijde, DM, van Leeuwen, MA, van Riel, PL, and van de Putte, LB (1995). Radiographic progression on radiographs of hands and feet during the first 3 years of rheumatoid arthritis measured according to Sharp's method (van der Heijde modification). *J Rheumatol*, 22(9):1792–1796.

van der Horst-Bruinsma, IE, Hazes, JM, Schreuder, GM, Radstake, TR, Barrera, P, van de Putte, LB, et al. (1998). Influence of non-inherited maternal HLA-DR antigens on susceptibility to rheumatoid arthritis. *Ann Rheum Dis*, 57 (11):672–675.

van der Linden, S, Valkenburg, HA, and Cats, A (1984a). Evaluation of diagnostic criteria for ankylosing spondylitis. A proposal for modification of the New York criteria. *Arthritis Rheum*, 27(4):361–368.

van der Linden, SM, Valkenburg, HA, de Jongh, BM, and Cats, A (1984b). The risk of developing ankylosing spondylitis in HLA-B27 positive individuals. A comparison of relatives of spondylitis patients with the general population. *Arthritis Rheum*, 27(3):241–249.

van der Linden, S and van der Heijde, D (1998). Ankylosing spondylitis. Clinical features. *Rheum Dis Clin North Am*, 24(4):663–676.

van der Linden, SM and Khan, MA (1984). The risk of ankylosing spondylitis in HLA-B27 positive individuals: a reappraisal. *J Rheumatol*, 11 (6):727–728.

van der Paardt, M, Crusius, JB, Garcia-Gonzalez, MA, Baudoin, P, Kostense, PJ, Alizadeh, BZ, et al. (2002). Interleukin-1beta and interleukin-1 receptor antagonist gene polymorphisms in ankylosing spondylitis. *Rheumatology (Oxford)*, 41(12):1419–1423.

van Gaalen, FA, van Aken, J, Huizinga, TW, Schreuder, GM, Breedveld, FC, Zanelli, E, et al. (2004). Association between HLA class II genes and autoantibodies to cyclic citrullinated peptides (CCPs) influences the severity of rheumatoid arthritis. .*Arthritis Rheum*, 50(7):2113–2121.

van Schaardenburg, D, Hazes, JM, de Boer, A, Zwinderman, AH, Meijers, KA, and Breedveld, FC (1993). Outcome of rheumatoid arthritis in relation to age and rheumatoid factor at diagnosis. *J Rheumatol,* 20(1):45–52.

van Tubergen, A, Heuft-Dorenbosch, L, Schulpen, G, Landewe, R, Wijers, R, van der Heijde, D, et al. (2003), Radiographic assessment of sacroiliitis by radiologists and rheumatologists: does training improve quality? *Ann Rheum Dis,* 62(6):519–525.

Veale, D, Yanni, G, Rogers, S, Barnes, L, Bresnihan, B, and Fitzgerald, O (1993). Reduced synovial membrane macrophage numbers, ELAM-1 expression, and lining layer hyperplasia in psoriatic arthritis as compared with rheumatoid. *Arthritis Arthritis Rheum,* 36(7):893–900.

Vyse, TJ and Todd, JA (1996). Genetic analysis of autoimmune disease. *Cell,* 85(3):311–318.

Waase, I, Kayser, C, Carlson, PJ, Goronzy, JJ, and Weyand, CM (1996). Oligoclonal T cell proliferation in patients with rheumatoid arthritis and their unaffected siblings. *Arthritis Rheum,* 39(6):904–913.

Wagner, UG, Koetz, K, Weyand, CM, and Goronzy, JJ (1998). Perturbation of the T cell repertoire in rheumatoid arthritis. *Proc Natl Acad Sci USA,*95 (24):14447–14452.

Wanders, A, Landewe, R, Dougados, M, Mielants, H, van der Linden, S, and van der Heijde, D (2005). Association between radiographic damage of the spine and spinal mobility for individual patients with ankylosing spondylitis: can assessment of spinal mobility be a proxy for radiographic evaluation? *Ann Rheum Dis,* 64(7):988–994.

Wei, JC, Tsai, WC, Lin, HS, Tsai, CY, and Chou, CT (2004). HLA-B60 and B61 are strongly associated with ankylosing spondylitis in HLA-B27-negative Taiwan Chinese patients. *Rheumatology (Oxford),* 43(7):839–842.

Wellcome Trust Case Control Consortium (2007). Genome-wide association study of 14,000 cases of seven common diseases and 3,000 shared controls. *Nature* 447(7145):661–678.

Westedt, ML, Breedveld, FC, Schreuder, GM, D'Amaro, J, Cats, A, and de Vries, RR (1986). Immunogenetic heterogeneity of rheumatoid arthritis. *Ann Rheum Dis,* 45(7):534–538.

Westhovens R, Cole JC, Li T, Martin M, Maclean R, Lin P, et al. (2006). Improved health-related quality of life for rheumatoid arthritis patients treated with abatacept who have inadequate response to anti-TNF therapy in a double-blind, placebo-controlled, multicentre randomized clinical trial. *Rheumatology (Oxford)* 45(19):1238–1246.

Weyand, CM and Goronzy, JJ (1994). Disease mechanisms in rheumatoid arthritis: gene dosage effect of HLA-DR haplotypes. *J Lab Clin Med,* 124(3):335–338.

Weyand, CM, Hicok, KC, Conn, DL, and Goronzy JJ (1992). The influence of HLA-DRB1 genes on disease severity in rheumatoid arthritis. *Ann Intern Med,*17(10):801–806.

Weyand, CM, Schmidt, D, Wagner, U, and Goronzy, JJ (1998). The influence of sex on the phenotype of rheumatoid arthritis. *Arthritis Rheum,* 41 (5):817–822.

Will, R, Palmer, R, Bhalla, AK, Ring, F, and Calin, A (1990). Bone loss as well as bone formation is a feature of progressive ankylosing spondylitis. *Br J Rheumatol,* 29(6):498–499.

Williams, CJ, Zhang, Y, Timms, A, Bonavita, G, Caeiro, F, Broxholme, J, et al. (2002). Autosomal dominant familial calcium pyrophosphate dihydrate deposition disease is caused by mutation in the transmembrane protein ANKH. *Am J Hum Genet,* 71:985–991.

Williamson, L, Dalbeth, N, Doherty, J, Gee, BC, Weatherall, R, and Wordsworth, BP (2004a). Nail disease in psoriatc arthritis. *Rheumatolology,* 43:790–794.

Williamson, L, Doherty, JL, Dalbeth, N, McNally, E, Ostlere, S, and Wordsworth, BP (2004b). Clinical assessment of sacroiliitis and HLA-B27 are poor predictors of sacroiliitis diagnosed by magnetic resonance imaging in psoriatic arthritis. *Rheumatology,* 43:85–88.

Willkens, RF, Arnett, FC, Bitter, T, Calin, A, Fisher, L, Ford, DK, et al. (1981). Reiter's syndrome. Evaluation of preliminary criteria for definite disease. *Arthritis Rheum,* 24(6): 844–849.

Willkens, RF, Nepom, GT, Marks, CR, Nettles, JW, and Nepom, BS (1991). Association of HLA-Dw16 with rheumatoid arthritis in Yakima Indians. Further evidence for the "shared epitope" hypothesis. *Arthritis Rheum,* 34(1):43–47.

Wordsworth, BP and Bell, JI (1992). The immunogenetics of rheumatoid arthritis. *Springer Semin Immunopathol,* 14(1):59–78.

Wordsworth, BP, Lanchbury, JS, Sakkas, LI, Welsh, KI, Panayi, GS, and Bell, JI (1989). HLA-DR4 subtype frequencies in rheumatoid arthritis indicate that DRB1 is the major susceptibility locus within the HLA class II region. *Proc Natl Acad Sci USA,* 86(24):10049–10053.

Wordsworth, BP, Stedeford, J, Rosenberg, WM, and Bell, JI (1991). Limited heterogeneity of the HLA class II contribution to susceptibility to rheumatoid arthritis is suggested by positive associations with HLA-DR4, DR1 and DRw10. *Br J Rheumatol,* 30(3):178–180.

Wordsworth, P and Bell, J (1991). Polygenic susceptibility in rheumatoid arthritis. *Ann Rheum Dis,* 50(6):343–346.

Wordsworth, P, Pile, KD, Buckely, JD, Lanchbury, JS, Ollier, B, Lathrop, M, et al. (1992). HLA heterozygosity contributes to susceptibility to rheumatoid arthritis. *Am J Hum Genet,* 51(3):585–591.

Wright, V (1959). Psoriatic arthritis; a comparative study of rheumatoid arthritis, psoriasis, and arthritis associated with psoriasis. *AMA Arch Derm,* 80(1):27–35.

Yamanishi, Y, Boyle, DL, Green, DR, Keystone, EC, Connor, A, and Zollman S (2005). p53 tumor suppressor gene mutations in fibroblast-like synoviocytes from erosion synovium and non-erosion synovium in rheumatoid arthritis. *Arthritis Res Ther,* 7(1):R12–8.

Yli-Kerttula, T, Luukkainen, R, Yli-Kerttula, U, Mottonen, T, Hakola, M, Korpela, M, et al. (2003). Effect of a three month course of ciprofloxacin on the late prognosis of reactive arthritis. *Ann Rheum Dis,* 62(9):880–884.

Young, A, Corbett, M, and Brook, A (1980). The clinical assessment of joint inflammatory activity in rheumatoid arthritis related to radiological progression. *Rheumatol Rehabil,* 19(1):14–19.

Zhang G, Luo J, Bruckel J, Weisman MA, Schumacher HR, Khan MA, et al. (2004). Genetic studies in familial ankylosing spondylitis suceptibility. *Arthritis Rheum* 50:2246–2254.

Zhang Y, Johnson K, Russell RGG, Wordsworth BP, Carr AJ, Terkeltaub RA, and Brown MA (2005). Association of sporadic chondrocalcinosis with a -4-basepair G-to-A transition in the 5'- untranslated region of ANKH that promotes enhanced expression of ANKH protein and excess generation of extracellular inorganic pyrophosphate. *Arthritis Rheum* 52:1110–1117.

Zheng, CQ and Hu, GZ (2003). Genome research on susceptibility loci in inflammatory bowel disease. *Zhonghua Nei Ke Za Zhi,* 42(9):660–662.

27

Immunological Disorders

Tineke CMT van der Pouw Kraan and Cornelis L Verweij

The analysis of gene expression in tissues, cells, and biological systems has evolved in the past decade from the analysis of a selected set of genes to an efficient high-throughput whole-genome screening approach of potentially all genes expressed in a tissue or cell sample. This development allows an open-ended survey to identify comprehensively the fraction of genes that define the sample's unique biology. This discovery-based research provides the opportunity to characterize either new genes with unknown function or genes not previously known to be involved in a biological process. The latter category may hold surprises that sometimes urge us to redirect our thinking. Besides a more detailed understanding in basic processes in immunology, microarray studies have increased our insight in the immunological mechanisms that contribute to chronic immune-related diseases. Of particular interest is the overlap of genes identified in these diseases with the genes that are induced in response to a microbial stimulus. Moreover, genomics studies have provided evidence for the existence of genetically determined differences in immune activity that might have consequences for susceptibility and/or outcome of immune-related diseases.

Large-scale gene-expression analysis is of great relevance in the field of immunology to generate a global view of how the immune systems attacks invading microorganisms, maintains tolerance, or creates a memory for past infections. The exciting part of microarray studies is that the many data points that are generated give rise to unpredictable and unexpected results, which may lead to new insights in immunology.

The present chapter reviews microarray research to characterize gene expression in immunology to broaden our understanding of the biology of immunological processes and immune-related diseases.

Innate Immunity

The innate immune system is central to the type of the immune response that is mounted in a given situation. This system acts effectively to sense altered signals without previous exposure to a pathogen such as a bacterium or virus. The innate immune system controls and assists the acquired antigen-specific immune response represented by T- and B-lymphocytes, which both carry an infinite number of receptor specificities that need to be exploited when necessary. The influence of the innate immune system on the adaptive immune response has long been recognized and used for its adjuvant effect of (dead) microbes in antibody responses. Since this system is also intimately linked to inflammatory and modulatory processes, many human diseases result from an aberrant primary or secondary innate immunity. The application of genomics technology provides answers to basic questions related to the regulation of innate and adaptive immunity. The extensive experience that immunologists have with cellular aspects of immunology is highly beneficial in the generation of homogeneous cell populations for genomics analyses of specific cell subsets. Within such a framework, we may gain an understanding of immune deregulation in autoimmune diseases, inflammatory responses, and certain tumors.

Response Programs of the Innate Immune System to Microorganisms

The innate immune system distinguishes between self and non-self by the creation of a variety of pattern-recognition receptors, called Toll-like receptors (TLRs) that recognize conserved motifs on pathogens which are not found in higher eukaryotes. Each TLR member recognizes and responds to different microbial components. At present, 10 different mammalian TLRs have been recognized, and their number is likely to increase. For example, lipopolysaccharides (LPS) from gram-negative bacteria signals through TLR4, while the constituents of the cell wall of gram-positive bacteria, lipoteichoic acid (LTA) and muramyl dipeptide (MDP) both activate TLR2. Unmethylated CpG-containing bacterial deoxyribonucleic acid (DNA) sequences are able to activate TLR9. Double-stranded ribonucleic acid (dsRNA) molecules that are produced during the replication process of many RNA and DNA viruses mediate an activation signal through TLR3. Signaling through different TLRs (TLR9, 2, and 4) leads to activation of c-Jun N-terminal kinase (JNK), p38, and nuclear factor-κB (NF-κB), which induces the production of cytokines such as tumor necrosis factora (TNFα), which after binding to its receptor in turn induces more NF-κB activation and induction of many other genes (Beutler, 2004).

To gain insight into the host response against different pathogens and their products, a meta-analysis of 32 studies has been performed by Jenner and Young (2005), in which stimulation of macrophages, dendritic cells (DCs), whole blood, peripheral blood mononuclear cells (PBMCs), T and B cells, and other nonimmune cell types were compared. Several different activators such as cytokines, parasites, bacteria, viruses, yeast and microbial components that bind to specific TLRs were used to stimulate the different cell types. Based on these comparisons, a common host-transcriptional response was defined. Several different pathogens induced a group of genes, with a correlated expression profile, involved in inflammation including TNFα, interleukin-1β (IL-1β), IL-6, IL-8, NF-κB family members, and several chemokines (CCL3/MIP1α, CCL4/MIP1β, CCL20/MIP3α, CXCL1/GRO1, CXCL2/MIP2α/GRO2, and CXCL3/MIP2β/GRO3). A comparison of the different components known to activate distinct TLRs, such as TLR2 by LTA and MDP, TLR3 by Poly I:C, and TLR4 by LPS, revealed that all three TLRs activated this set of inflammatory/chemotactic genes in macrophages (Nau et al., 2002; Jenner and Young, 2005). These genes are expressed not only in macrophages but also in DCs, leukocytes, PBMCs, and whole blood during acute infection with several bacteria. In the meta-analysis, it was shown that these genes are induced not only by bacteria but also by the protozoan, *Leishmania chagasi*, in macrophages, and by several viruses in peripheral blood cells and other nonimmune cell types such as endothelial and epithelial cells.

This group of co-regulated genes was previously recognized in PBMC stimulated with several bacteria and LPS (Boldrick et al., 2002). Because many of these genes are known to be induced by TNFα, which increases NF-κB activation, it has been referred to as the TNF/NF-κB response set. The TNF/NF-κB regulon thus resembles part of the common innate response to infection (Fig. 27–1). Interferon (IFN)-stimulated genes comprise another set of co-regulated genes included in the common innate response program. These genes include *OAS1* (2′,5′-oligo-adenylate synthetase 1), *OAS2, OASL, MX1, MX2, G1P2/ISG15, IFI16, ISG20/HEM45, IFIM1,* IFN-inducible chemokines *CCL8/MCP2, CXCL9/MIG, CXCL10/IP10,* and *CXCL11/I-TAC,* and metallothioneins. Most of the genes induced by viruses appear in this cluster. Many cell types are able to induce this IFN-stimulated gene program, including macrophages, fibroblasts, PBMCs, DCs, B cells, as well as endothelial and epithelial cells. However, it has now become clear that other organisms such as bacteria and fungi are also able to induce an IFN-response program, through activation of the TLRs and induction of IFNβ, a concept that has been highlighted by Hertzog et al. (2003). However, unlike stimulation of TLR3 and TLR4 by poly I:C and LPS, respectively, the bacterial products, LTA and MDP, from gram-positive bacteria were not able to induce an IFN-response program (Nau et al., 2002; Jenner and Young, 2005). Thus type 1 IFNs, IFNα, and IFNβ, are involved in the expression program of many cells types, in response to parasites, fungi, bacterial, and viral stimuli that activate different

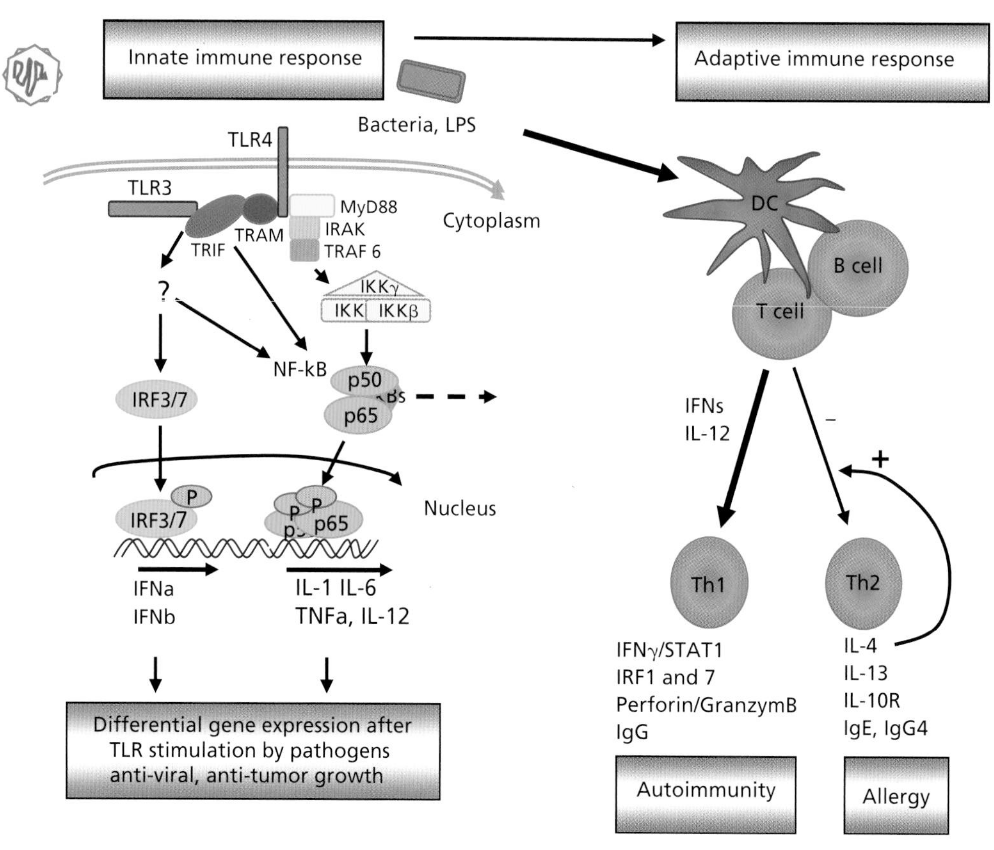

Figure 27–1 Ligation of Toll-like receptor 3 (TLR3) and TLR4 activate the (IFN)-and nuclear factor-κB (NF-κB) pathway. TLR3 stimulation predominantly induces phosphorylation of IFN regulatory factor 3 (IRF3) and IRF7, which activate type 1 IFN genes. TLR4 stimulation predominantly induces the release of inhibitory κB (IκBs) from NF-κB allowing phosphorylation and translocation of NF-κB to the nucleus where target genes are activated. TLR3 and TLR4 both activate the IFN- and NF-κB pathway; however, TLR4 stimulation leads to a more profound activation of the NF-kB pathway with the production of proinflammatory cytokines (Jenner and Young, 2005). The innate immune response profoundly affects the adaptive/specific immune response. In particular, the production of IFNs and IL-12 lead to differentiation of T-helper cells to Th1 cells, while IL-4 leads to Th2 differentiation. Gene expression profiling has identified specific gene expression profiles of these differentiated T cells that explain their function in autoimmunity and allergy (Rogge et al., 2000; Chtanova et al., 2001; Hamalainen et al., 2001).

TLRs. Type 2 IFN, IFNγ, is produced by T cells, natural killer (NK) cells, and NKT cells in response to mitogens and cytokines. IFNs regulate many important activities of macrophages and DCs, and are required for an effective innate immune response, which in turn directs the adaptive immune response.

Adaptive Immunity

Both B- and T-lymphocytes can be recognized as *the* key players in the adaptive immune response as a consequence of their molecular and functional adaptation to invading pathogens. Within both types of lymphocytes, subtypes and distinct stages of differentiation exist, which are a consequence of their functional adaptation to different antigens.

Differences between T and B cells are reflected by the differential expression of many genes. Each lymphocyte type has its specific gene expression signature. The T cell signature includes CD2, T-cell receptor (TCR), and genes encoding TCR signaling proteins [linker for activation of T cells (LAT) and fyn]. The B-cell signature contains genes like *CD20*, immunoglobulin genes, and genes encoding B-cell receptor (BCR) components such as CD79B. Although these two cell types are clearly distinct, they also share remarkable similarities in their activation program. The "activation gene expression signature" of T- and B-lymphocytes includes many genes that are either induced or repressed upon activation (Teague et al., 1999a; Alizadeh et al., 2000; Glynne et al., 2000). Central to the activation status is the expression of the transcription factor, NF-κB. NF-κB factors play a central role in immunity, and members of this family (NF-κB1, NF-κB2, RelB, c-Rel, and IkBα) are included in the activation gene expression signature together with a prominent group of target genes, consisting of cytokines, chemokines, and antiapoptotic genes like A1 (an antiapoptotic bcl-2 family member) and c-IAP2. Differences that exist are at the level of, for example, CCR7, CXCR5, and BCL-2 expression, which are expressed at much higher levels in activated B cells. The elevated expression of chemokine receptors is likely to be involved in lymphocyte homing of activated B cells. One of the features of activated T cells is the repressed expression of the genes for IFN receptors and for STAT-1, which is involved in IFN signal transduction (Teague et al., 1999b).

T lymphocytes

A nice example of the way microarray studies may help to address basic questions in immunology is revealed by a study aimed to unravel the pathway of CD28 co-stimulation in T cells. Co-stimulation of T cells via both the TCR and CD28 allows for sustained activation and cell cycle entry. Diehn et al. (2002) have used human peripheral blood T cells that were stimulated with antibodies to CD3 and/or CD28. The data showed that signals induced by anti-CD3 were amplified by co-stimulation with anti-CD28, with the most striking enhancement of transcriptional activity of genes that are responsive to the transcription factors, NFAT (nuclear factor of activated T cells). This effect could be blocked by FK506, an inhibitor of the phosphatase calcineurin, thereby interfering with the nuclear translocation of NFAT proteins. Conventional biochemical techniques were then used to demonstrate that stimulation of CD28 leads to reduced nuclear export of NFATc protein caused by increased phosphorylation (and thus inactivation) of the nuclear export kinase, glycogen synthase kinase-3 (GSK3). Even in the absence of CD3 stimulation, engagement of CD28 stimulated the phosphorylation of GSK3, which becomes more effective when NFAT is activated by CD3 stimulation. Without the use of microarrays, the NFAT-dependent pathway induced by CD28 co-stimulation would not have been recognized easily.

Among effector/memory T-helper (Th) cells, a distinction can be made based on the pattern of secreted proteins, dividing them into Th1 cells (expressing IFNγ and TNFα) involved in phagocyte-dependent immune response and Th2 cells (expressing IL-4, IL-5, and IL-13) that promote immunoglobulin E (IgE) production and eosinophil function. A dysbalance in Th differentiation may lead to pathology. Th1 cells may be held responsible for tissue damage in inflammatory diseases such as rheumatoid arthritis (RA) and inflammatory bowel disease. Th2 cells play a role in the pathogenesis of allergic reactions and asthma (Fig. 27–1). Studies to gain insight in the program that affect the creation and interconversion of Th1 and Th2 cells are of importance to define the molecular markers to distinguish these subsets in a clinical setting and design strategies to selectively silence or reverse a detrimental Th cell subset. Many genes were discovered that were differentially expressed in Th1 and Th2 cells including transcription factors, cytokines and receptors, costimulatory molecules, apoptosis-related genes, genes involved in migration and adhesion, and genes with unknown function (Rogge et al., 2000; Chtanova et al., 2001; Hamalainen et al., 2001). Studies performed on fully polarized human Th1 and Th2 cells at different time points of stimulation identified a set of approximately 100 differentially expressed genes that encode for lytic proteins, co-stimulatory membrane molecules, cytokines, chemokines, and genes indicating differential signaling, transcriptional control, and cell differentation. The identification of these differentially expressed genes provides a starting point for many future studies (van der Pouw Kraan TCTM, van Baarsen L, and Verweij CL, unpublished observations).

B lymphocytes

Gene expression profiling studies have also provided a molecular definition of different phases in B cell differentiation. The characteristic signature of B cells in the anergic state contained genes that have a negative regulatory function (Glynne et al., 2000). On the other hand, B-cell activation-specific genes, including the proto-oncogenes LSIRF/IRF4 and c-myc, the antiapoptotic molecule A1, and the inhibitory transcription factor LKLF (lung Krüppel-like factor), implicated in lymphocyte mitogenesis were blocked in anergic cells. This gene profile may explain how tolerance is maintained in B cells.

Germinal center (GC) B cells differentiate into plasma cells or memory B cells, while undergoing affinity maturation and somatic mutation. GC B cells and plasma cells show distinct expression signatures (Shaffer et al., 2001). Plasma cells show a higher expression of plasmacytic differentiation regulators blimp-1 and XBP-1. GC B cells show an expression signature that interestingly is retained in lymphoma cell lines (Alizadeh et al., 2000), (see below). This latter profile can be explained by the high expression of BCL-6, because the dominant negative form of BCL-6 introduced into GC-derived B cell lines is able to block terminal differentiation into plasma cells (Shaffer et al., 2001).

Microarray Studies in Immunological Diseases

Information about the differences in gene expression between cells or tissues of healthy and diseased individuals provides markers to differentiate between health and disease, and disease states. In combination with biochemical research, the expression profiles may provide clear insights into the disease mechanisms.

Infectious Diseases

Bacteria and Viruses

In vivo analysis of gene expression changes during viral infections have been mainly addressed in animal studies. The response of nonhuman primates to viruses such as smallpox and simian and human immunodeficiency chimeric virus (SHIV) are quite informative. Rubins et al. (2004) found a prominent upregulation of genes that represent an IFN response, cell cycle/cell proliferation, and genes that encode for immunoglobulins. Strikingly, a TNFα/NF-κB-induced response was lacking, while these genes were induced during infection with another fatal systemic virus, Ebola. Possibly, this response was modulated by the smallpox virus, as they are known to encode TNFα receptor homologs such as CrmB.

Bosinger et al. (2004) used SHIV, which is persistent and causes a permanent decrease in $CD4^+$ cell counts in macaques. The SHIV challenge induced a suppression of genes regulating innate immunity including the LPS receptors, CD14 and TLR4. This may impair the host response to opportunistic infections such as gram-negative bacteria, cytomegalovirus (CMV) and *Candida albicans.* Genes that were persistently and most strongly up-regulated during infection were the IFN-response genes, such as *MX1, MX2, OAS1, OAS2, OAS3, ISG20, ISG15, STAT1, IFIT1, IFITM1,* and *CXCL10.* Like the smallpox virus, SHIV also failed to induce a TNFα/NF-κB response. Because in addition to the IFN response, TNFα importantly contributes to the anti-viral response (Guidotti and Chisari, 2001), these viruses may evade the host immune response allowing their persistent presence.

During HIV infection, B cells are indirectly affected as well, indicated by hypergammaglobulinemia, increased expression of activation markers and terminal differentiation, and increased susceptibility to apoptosis. Interestingly, in HIV-viremic patients, B cells show an increased expression of IFN-responsive genes such as *MX1, ISG15,* and *OAS1* (Moir et al., 2004). These genes partly overlapped with IFN-response genes that are overexpressed in approximately 50% of the systemic lupus erythematosus (SLE) patients (Bennett et al., 2003), a disease in which B-cell perturbations are prominent as well. In addition, increased expression of CD95, the Fas receptor (known to be up-regulated by IFN), was associated with increased susceptibility to CD95-mediated apoptosis. In line with this observation, an inverse correlation of total numbers of B cells and HIV plasma viremia was found.

In peripheral blood cells from chronic hepatitis C patients, an enhanced expression of IFNα and IFNβ (messenger RNA) mRNA has been found and induction of the IFN-response gene *MX1* (Mihm et al., 2004). Unfortunately, microarray studies on chronic in vivo bacterial infections are lacking. It would be interesting to verify whether the in vitro response to bacteria reflects the in vivo response. For HIV, it is clear that the in vitro response (Izmailova et al., 2003) activates the same pathways as the in vivo response measured in peripheral blood (Bosinger et al., 2004).

Parasites

Usage of a combination of murine complementary DNA (cDNA) microarrays, cytokine-deficient mice, and confirmatory biological assays led to the identification of gene expression profiles that associate with lethal type 1 (IL-12/IFN-γ) and type 2 (IL-4) polarized immune responses during infection with the parasitic trematode, *Schistosoma mansomi.* These results uncovered the contributions of previously unappreciated disease mechanisms that contribute to parasite-induced pathogenesis. The granuloma formation around the schistosomal eggs in the livers differed not in size but in qualitative aspects revealed by gene expression levels of infected livers. In the type 2 polarized animals, increased expression of genes involved in collagen deposition, wound repair, and matrix remodeling genes, and fibrosis associated with increased morbidity were observed. The type 1 polarized animals showed increased expression of proinflammatory cytokine genes and genes involved in apoptosis in association with increased mortality.

The reciprocal approach to host–pathogen interactions is to gain a better understanding of pathogen responses to host contact upon infection. These experiments require the genome sequence of the pathogen and microarrays containing gene sequences of the respective pathogenic microorganisms. The results of this pathogen genome mining approach may yield the identification of potential vaccine candidates, as was exemplified by studies using *Neisseria meningitidis* group B bacteria (Pizza et al., 2000).

Autoimmune Diseases

Systemic Lupus Erythematosus

Another study revealed remarkable results upon gene expression profiling within SLE. A characteristic feature of the inflammatory autoimmune disease SLE is the heterogeneity in the clinical presentation, therapy responsiveness, and severity of disease. Baechler et al. (2003) determined the mRNA expression profiles of unstimulated PBMC from 48 SLE patients and controls. Distinct gene expression patterns, consisting of 161 genes, were identified for patients and controls. Strikingly, in approximately 50% of the patients an up-regulated expression of IFN-response genes, including enhanced expression of mRNA for STAT-1, MX1, and ISG15, was observed. In addition, this pathway was associated with more severe disease involving the kidneys, hematopoetic cells, and/or the central nervous system. However, mRNA levels of the IFNs itself were not different between patients and controls, which probably indicates that the IFN-response genes are a more sensitive readout for the activation of this pathway, or alternatively, another cytokine or mediator induces the expression of these genes.

The overexpression of IFN-response genes in SLE patients has been confirmed by others (Bennett et al., 2003). In addition, a granulocyte-specific gene response associated with increased numbers of immature granulocytes in peripheral blood of SLE patients was found.

The IFN-response genes most likely reflect the activity of type I IFNs, (possibly IFNα) after comparison of the altered gene expression of PBMC in response to in vitro IFNγ or, IFNα treatment (Kirou et al., 2004). Analysis of a selection of three IFNα-induced genes by real-time polymerase chain reaction (real-time PCR) revealed a higher expression in a panel of SLE patients, compared with healthy controls and RA patients. (Kirou et al., 2005). Within the SLE patients, a higher expression of IFNα-induced genes was associated with a higher prevalence of renal disease and disease activity scores, lower C3 levels, hemoglobin, and albumin levels, and higher anti-double-stranded DNA titers and erythrocyte sedimentation rate.

Rheumatoid Arthritis

To generate a molecular description of synovial tissue from RA patients that would allow us to unravel novel aspects of pathogenesis and identification of different forms of disease, we profiled gene expression in synovial tissue from the affected joint tissues of patients with RA ($n = 21$) using the "Lymphochip" and microarrays containing approximately 24,000 cDNA spots, manufactured at Stanford University (van der Pouw Kraan, Van Gaalen, Huizinga, et al., 2003, van der Pouw Kraan, Van Gaalen, Kasperkovitz, et al., 2003). Unsupervised hierarchical clustering of the RA gene expression signatures revealed considerable variability within the RA tissues resulting in the identification of at least two molecular distinct forms of RA tissues. The first group revealed abundant expression of clusters of genes indicative of an involvement of the adaptive immune response with evidence for a prominent role of an activated STAT-1 pathway, including STAT-1 signaling receptors and STAT-1 target genes, among these STAT-1 itself (Lehtonen et al., 1997), MMPs, GBP1, ICSBP, IP-10, caspase-1, TAP-1, and IRF-1. Among the genes overexpressed in this group of patients, a relative majority of nine genes is located on chromosome 6p21.3, which harbors the human leukocyte antigen (HLA) genes and is responsible for one-third of the genetic influence on the development of RA. The expression profiles of the second group of RA tissues revealed an increased tissue remodeling activity and a low inflammatory gene expression signature, which is associated with fibroblast dedifferentiation.

In addition, fibroblast-like synoviocytes (FLS) were cultured from high and low inflammatory RA synovial tissues, and their expression profile was examined on Stanford microarrays with a complexity of 24,000 cDNA spots. A most intriguing finding was that the expression profiles of FLS show the imprint of the tissues they were derived from. FLS that are related to tissues with a low inflammation profile are characterized by increased expression of IGF2 and IGFBP5, indicative for an autocrine signaling of the IGF receptor, which supports FLS proliferation and survival. FLS from high-inflammatory tissues exhibit a gene expression program that is indicative for an activated TGFβ/activin-induced pathway. A hallmark of this pathway is the induction of smooth muscle actin, a myofibroblast marker. The presence of myofibroblasts in RA synovial tissues was confirmed by immunohistochemical staining of smooth muscle actin. Myofibroblasts are involved in tissue repair/wound healing, but also contribute to the inflammatory process by the secretion of cytokines and chemokines, e.g., IL-8, IL-6, and CXCL12.

In conclusion, these results confirm the heterogeneous nature of RA and suggest the existence of distinct pathogenic mechanisms that contribute to RA, as was suggested by Firestein and Zvaifler (2002).

Gene expression profiling in whole blood cells from RA patients revealed significant variation between RA patients, which allows stratification of patients on the basis of distinct molecular characteristics (Tineke van der Pouw Kraan, Carla Wijbrandts, Paul-Peter Tak en Cornelis Verweij, manuscript in preparation). Interestingly, in a subset of patients an expression profile was induced that shows similarities to a profile observed after microbial infections, suggesting that an infectious agent may be involved in the pathogenesis of RA.

Differences in gene expression were also observed between early arthritis (disease duration less than 2 years) and established RA patients (with an average disease duration of 10 years) (Olsen et al., 2004). Analysis of the expression of 4300 genes in PBMC revealed that in early RA genes such as colony-stimulating factor 3 receptor, cleavage stimulation factor, and TGFβ receptor II, which affect B-cell function, are expressed at higher levels than during established disease. Genes involved in immune/inflammatory processes and genes related to cell proliferation and neoplasia were expressed at late stages of disease. Comparison between the genes that were expressed at higher levels in early RA and influenza-induced genes indicated that about a quarter of the early arthritis genes overlapped with the influenza-induced genes. This finding led the authors to suggest that the early arthritis signature may partly reflect the response to an unknown infectious agent.

The differences in expression profiles between subgroups of RA patients confirm the heterogeneous character of the disease and provide opportunities to stratify patients based on molecular criteria for clinical studies and evaluation of targeted therapies.

Multiple Sclerosis

Only a limited number of studies are reported on microarray analysis of multiple sclerosis (MS) lesions obtained at autopsy. Analysis of affected brain tissues from four MS patients compared with two controls revealed increased expression of transcripts encoding inflammatory cytokines such as IL-6, IL-17, and IFN-γ, cytokine receptors IL-1R, IL-18Rtype2, IL-11Ra, and TNFRp75, and downstream signaling pathways (Lock et al., 2002). Comparison of the gene expression profiles of active or acute lesions with silent lesions revealed a number of genes whose transcripts were at least two-fold differentially expressed. Among these genes were included those encoding FGF-12, IL-17, and the IFN-induced 17-kDa/15-kDa protein mRNA.

Achiron et.al. (2004) identified a unique gene expression signature of 1109 genes in PBMC from 26 MS patients in comparison with 18 healthy subjects. The signature contains genes that implicate processes including T-cell activation, inflammation, and apoptosis. Among the up-regulated genes were lymphotoxin β(LTB), several integrins, which may play a role in adhesion to vascular endothelial adhesion structures, permitting their penetrance through the blood–brain barrier, cathepsin K and -B, and antiapoptotic genes. Interestingly, genes involved in the TNFα/ NF-κB pathway, including TNFα itself, TNFα-induced protein 6, TNF receptor-associated factor 6 (TRAF6), NF-κB1, NF-κB2, and NF-κBIA were reduced in MS patients.

Combined Analysis

Maas et al. (2002) compared the expression of 4300 genes in PBMC of control individuals ($n = 9$) before and after immunization with influenza vaccine to those of individuals with four distinct autoimmune diseases including RA ($n = 20$), SLE ($n = 24$), type I diabetes and MS ($n = 5$) (Maas et al., 2002). The induced gene expression profile of the normal immune response exhibited coordinate changes in the expression of genes with related functions over time. Three days after immunization, genes encoding proteins involved in key signal transduction pathways (e.g., PKC, phospholipase C, 1,2-DAG kinase, MAPK, STATs and STAT inhibitors, AP-1, IFN regulatory proteins, and cell cycle proteins) were expressed. After 3 weeks, the program had shifted toward functional activity (chemokines, complement components, and IFN-inducible proteins and leukocyte homing/adhesion molecules). During the entire course of the immune response to influenza, genes encoding ribosomal proteins were suppressed, a process that has been linked to differentiation of eukaryotic cells. The autoimmune patients displayed a remarkably similar gene expression profile, which was unrelated to the pattern of the immunized group. Two gene clusters were differentially expressed between patients with an autoimmune disease and controls. Surprisingly, genes with a distinct expression pattern in autoimmunity were not necessarily "immune response" genes, but were down-regulated genes that encode proteins involved in, e.g., apoptosis (such as TRADD, TRAP1, TRIP, TRAF2, CASP6, and CASP8), ubiquitin/protesome function (UBE2M, UBE2G2, and POH1), inhibitors of cell cycle progression (CDKN1B, CDKN2A, and BRCA1), and inducers of cell differentiation (LIF and CD24). No genes could be selected that distinguished among the different autoimmune diseases.

Cancer

Microarrays have been particularly useful in the classification of cancer. Alizadeh et al. (2000) performed a systematic survey to define the gene expression programs of normal lymphocytes in different stages of differentiation and determined the extent to which these programs are inherited by human lymphoid malignancies. The idea was to extend markers for determining the cellular origins of lymphomas that so far relied on the analysis of somatic hypermutation in the variable regions. Such a reference database provides a rich framework allowing interpretation of the gene expression signature in lymphoid malignancies. The knowledge gained from the assembly of a reference database was applied to classify diffuse large B-cell lymphoma (DLBCL), a clinically heterozygous disease in terms of treatment response and subsequent survival, for which no good diagnostic parameters are available. Unsupervised hierarchical clustering performed on expression data of lymph node biopsies from DLBCL patients revealed two groups of patients, one group showed characteristics of the gene expression profile of GC B cells, while the other showed expression profiles characteristic of activated peripheral blood B cells. This distinction turned out to be a good predictor for survival. The 5-year survival probability for the GC B-like group was 70%, whereas only 20% of the activated B-like DLBCL patients survived for 5 years.

Golub et al. (1999) clustered 38 tumor samples by creating self-organizing maps (SOM) with a preset number of two or four clusters (groups of patients) for new class discovery. Subsequently, class predictors were identified from these SOMs. For testing the class predictors consisting of various numbers of genes (10–100), the one with the highest cross validation accuracy rate was used. A set of known ALL and AML tumor samples was used to test how well this class prediction method performed (Golub et al., 1999). With two exceptions, all 38 samples were correctly grouped into AML and ALL, while the distinction between B-cell and T-cell ALL was made as well.

Others have identified sets of genes that predict the clinical outcome of breast cancer patients. van 't Veer et al. (2002) have used another strategy to classify breast tumors according to their clinical behavior. A total of 78 sporadic lymph-node negative patients were selected to identify a prognostic gene expression signature for clinical outcome: metastasis or disease-free status within 5 years. From the 25,000 genes on the arrays, a selection was made of 5000 genes that showed a set threshold of variation between the samples. A supervised method was then applied to these genes for the identification of the genes that are the best predictors for clinical outcome. A list of 70 genes turned out to be the best predictors for disease development with 83% accuracy. So only a small group of genes within a large pool that showed inter-individual variation in expression determines one important aspect of disease, namely metastasis. In a follow-up study (van de Vijver et al., 2002), expression analysis of these 70 genes in biopsies from 295 patients again proved to harbor prognostic value. From the patients with a poor-prognosis profile only 50.6% remained free of over a period of 10 years, while 85.2% of the patients with a good-prognosis profile remained free of metastasis.

Genetic Variation in Immune Activity

Many immune-related diseases have a genetic component that contributes to susceptibility and/or disease severity. Part of the genetic basis of disease may be explained by genetically determined alterations in the immune response between individuals. In particular, the often observed disease association with HLA alleles supports this concept. Besides HLA, there might be other partly subtle genetic variation that could dramatically influence the outcome of an immune response.

Since the innate immune system initiates the host defense against pathogens and is required for an effective adaptive immune response, genetically determined differences in host effector programs against pathogens are likely to affect the control of infectious diseases, autoimmune disease, and cancer. Accordingly, Westendorp et al. have provided proof for genetically determined differences in IL-10 and TNF production capacity, which could have dramatic clinical consequences (Westendorp et al., 1997). The assembly of a database containing expression signatures of various immune cells in different phases of activation and differentiation would provide a valuable resource to understand inter-individual differences in gene expression.

A detailed study on inter-individual differences in unstimulated blood cell gene expression profiles has been reported by Whitney et al. (2003). The genes that show intrinsic differences in gene expression between 75 individuals were male- and female-specific genes, genes involved in antigen presentation such as major histocompatibility complex (MHC) class II genes and *TAP-2*, and IFN-responsive genes. The most consistent donor-dependent gene of all was *DDX17*, a RNA helicase, which is suggested to function as a pre-mRNA splicing regulator (Lee, 2002). Possibly, differential expression of *DDX17* gives rise to inter-individual differences in splice variants. The inter-individual variation in IFN-induced genes was also observed by Radich et al. (2004). In this study, gene expression profiling of peripheral blood lymphocytes from unrelated individuals over a period of several weeks to months also revealed a high level of variation between different individuals, in particular in IFN-type 1 (IFNα and β) response genes, including CIG5, IFIT1, IFIT4, MX1, and USP18). Approximately 10%–20% of the individuals show a higher baseline level. Interestingly, the inducibility of these genes to in vitro IFN stimulation was lower in those individuals.

These observations are in line with the inter-individual differences in gene expression reported by Cobb et al. (2005). Gene expression levels of unstimulated peripheral blood cells show a remarkable constant profile within the same individual over a 24-hour period. The average difference from the mean expression values is less than 10%. In contrast, expression values between different donors (inter-individual variation) shows a greater variation than the variation within the same individual. As expected, traumatic injury induced major changes in gene expression with a magnitude greater than the inter-individual variance.

Accordingly, Boldrick et al. observed inter-individual differences in response to some bacteria and their products (Boldrick et al., 2002). Part of the inductive response was repeatedly reduced in one donor. Most of the genes that showed a poor induction in this donor fit in a group of IFN/STAT-1-responsive genes, including some genes that are important for an anti-viral responses, such as *OAS1* and *MX1* (Fig. 27–1). Taking into account that donors with an increased baseline IFN-response gene program show a reduced induction of these genes upon in vitro stimulation with IFN (Radich et al., 2004), the poor response by the individual may reflect high baseline levels of IFN-response genes.

The results from these studies indicate that the individual gene expression levels in peripheral blood cells is a stable feature over time, while differences between individuals may reflect genetic variation although subclinical infections cannot be ruled out. In addition, during clinically manifest disease, a disruption of immunological homeostasis is detected in the gene expression program in peripheral blood cells with a variation that is larger than the inter-individual differences, and indicates processes induced by infection, malignancies, or trauma.

The assumption that gene expression levels are (in part) genetically determined has been verified by Cheung and Morley (Cheung et al., 2003; Morley et al., 2004). They established that the inter-individual differences in gene expression levels indeed carries a genetic component (Cheung et al., 2003). This conclusion was based on the analysis of immortalized human B cells from unrelated individuals, siblings from the same family and between monozygotic twins. In a follow-up study, this genetic component was analyzed in more detail (Morley et al., 2004). Baseline gene expression levels in immortalized B cell lines from 94 unrelated individuals from 14 families were studied from which genotypes for genetic markers (SNP genotypes) were available. Linkage analysis was performed on genes that varied more between individuals than between replicates from the same individual (3554 genes out of a total of 8500 genes). This analysis resulted in the discovery of genes that are linked to markers (potential regulators of transcription) within 5 Mb of the gene (cis-acting transcriptional regulators) and markers more distant than 5 Mb from the target gene or markers on a different chromosome (trans-acting transcriptional regulators). The majority of the genes were regulated by trans-acting elements. The *DDX17* gene, identified as individual specific by Whitney et.al. (2003), was also identified in this study and shown to be linked to a cis-acting transcriptional regulator on chromosome 22q13. In addition, master regulators of transcription, or hotspots of genetic determinants, were identified, which regulate transcription of many genes. These master regulators are located within a 5 Mb region on chromosome 14q32 and on chromosome 20q13. For the genes with cis-acting linkages, individuals with allelic marker differences indeed showed different expression levels. These results were confirmed by allele-specific real-time PCR in heterozygous individuals, in which one allele is expressed at higher levels than the other allele.

Individual-specific gene expression signatures are thus, at least in part, genetically determined. This knowledge provides new opportunities to explore gene expression profiles in pathological conditions. Expression profiles from patients, but also from unrelated individuals and family members, can be used to identify polymorphic genes that influence the immune response and susceptibility to infectious and (auto)immune diseases (Fig. 27–2).

Conclusions—The Future

Gene expression profiling will lead to a molecular definition of disease. Consequently, autoimmune, immune deficiency diseases, and malignancies that are currently defined by their clinical phenotypes will soon be classified into molecular subtypes based on functional genomics experiments. Since most infectious processes result in an ordered series of host regulatory responses, gene expression profiles of host cells could also be used to define the stage of the host response in individual patients. In the ideal situation, pathogen-specific profiles may be recognized, which might be particularly useful for rapid diagnosis, especially in those cases where it is

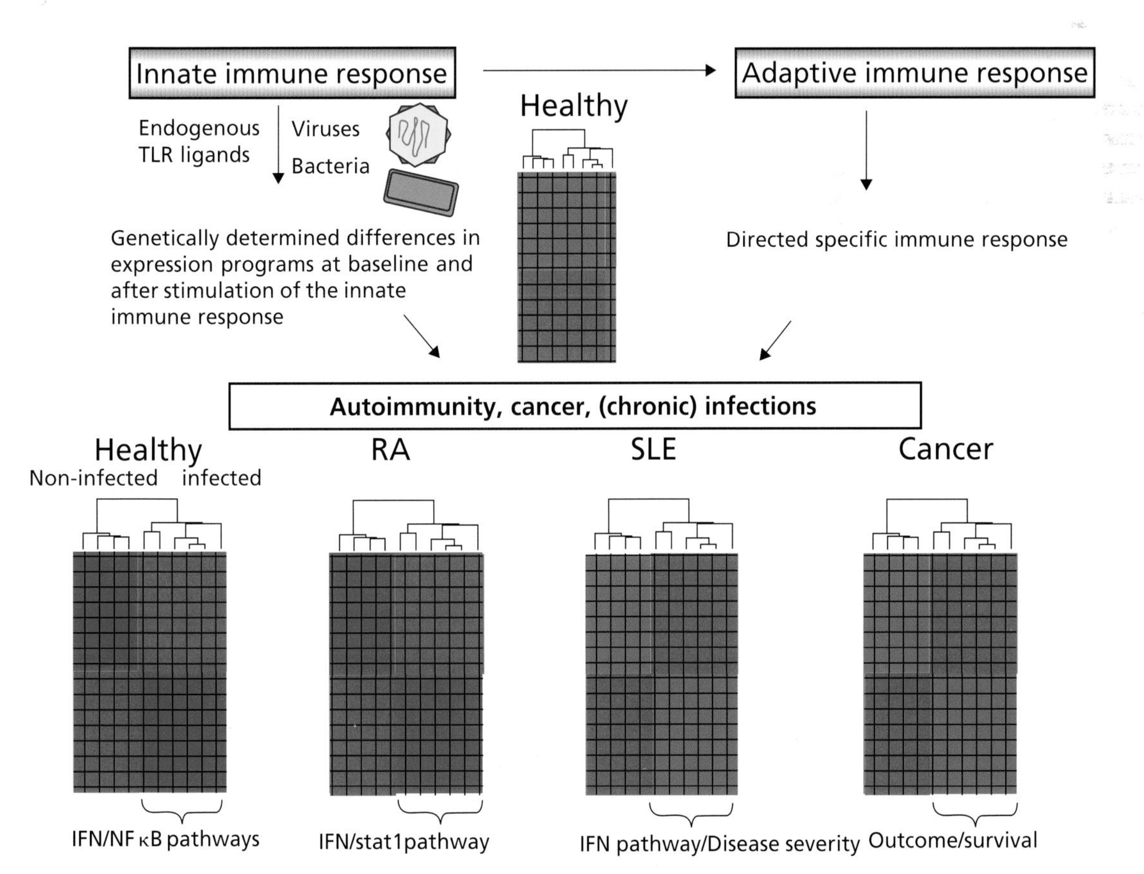

Figure 27–2 Immune response in chronic diseases and infections. Gene expression profiling of immune cells from healthy individuals at baseline and challenged with different microorganisms identified, at least in part, genetically determined expression programs that influence the adaptive immune response. The genetically determined differences in the innate immune response play a key role in the capacity to clear infections. Disease profiling identifies patient-specific and disease-specific expression programs that partly overlap with the expression programs induced by

technically difficult or impossible to culture the pathogen. In addition to diagnostic purposes, this will also have considerable impact on therapy of infectious diseases.

In addition, the assembly of comprehensive gene expression databases that contain molecular signatures of specific cell subsets and biological networks will be useful to understand the biological basis of dysregulation in disease. Knowledge that will be gained by identifying the genes that are expressed in a cell or tissue will make the biology of the system more comprehensible. While this is in part true, it is clear that transcriptomics information does not address the regulation that occurs downstream of transcription, involving mRNA splicing, protein expression and post-translational modification. A systems approach requires integration of as many of different data types as possible including, besides gene expression data, information of the proteome (the global inventory of all proteins in an organ, tissue of individual cell type) and the metabolome (global inventory of metabolites). Together the data on the different levels of information transfer from genome to phenotype defines a new approach to biology, which is termed "systems biology" (Kitano, 2002; Hood, 2003): a computational modeling approach aimed to understand the structure and dynamics of cellular and organismal functions. Systems biology will certainly become a mainstay of drug development, allowing identification of novel drug target followed by target-directed drug effects combined with tailored therapy. It can further be anticipated that this information will allow us to determine how the disease pathway can be manipulated pharmacologically with greater specificity than achieved by current drugs.

Furthermore, differences in responses between individuals might provide a lead to explain how genetic variation affects the immune system. Evidence exists that these inter-individual differences are genetically determined. One of the distinctive gene expression programs between individuals involves the IFN/STAT signaling pathway (Boldrick et al., 2002; Whitney et al., 2003; Radich et al., 2004). These findings make it tempting to speculate that inter-individual differences that exist in healthy controls might explain heterogeneity which is observed in immune-related diseases such as SLE, RA, and MS (Fig. 27–1). Because baseline and microbial-induced alterations of transcript levels of immune-related genes differs between individuals, these expression profiles may become informative for the prediction of an individual's response to infectious agents and susceptibility to cancer and autoimmune diseases. The activation of the IFN pathway in SLE and RA may therefore reflect an inherited genetic capacity for an altered activity of the IFN-response program. Analogous to these findings, the most consistent intrinsic variable gene in healthy donors, *DDX17* (Whitney et al., 2003), is selectively expressed in a subgroup of RA patients that showed a high degree of inflammation and high expression of IFN-responsive genes (van der Pouw Kraan, Van Gaalen, Huizinga, et al., 2003).

In the near future, new analysis methods will evolve to further improve the identification of relationships between the expression levels of many genes, to identify new signaling pathways, determine the genetic differences, and further improve the prediction of disease susceptibility and subclassification of patients within a disease, finally leading to a more effective and personalized treatment.

Acknowledgments

Financial support was provided by the Netherlands Organization of Pure Scientific Research, the Dutch Arthritis Association, the European Union FP6 STREP program; "AUTOROME" (www.AUTOROME.de), and the Center for Medical Systems Biology (CMSB), a center of excellence financed by the Netherlands Genomics Initiative.

References

Achiron, A, Gurevich, M, Friedman, N, Kaminski, N, and Mandel, M (2004). Blood transcriptional signatures of multiple sclerosis: unique gene expression of disease activity. *Ann Neurol*, 55(3):410–417.

Alizadeh, AA, Eisen, MB, Davis, RE, Ma, C, Lossos, IS, Rosenwald, A, Boldrick, JC, Sabet, H, Tran T, Yu X, Powell JI, Yang, L, et al. (2000). Distinct types of diffuse large B-cell lymphoma identified by gene expression profiling. *Nature*, 403(6769):503–511.

Baechler, EC, Batliwalla, FM, Karypis, G, Gaffney, PM, Ortmann, WA, Espe, KJ, Shark, KB, Grande, WJ, Hughes, KM, Kapur, V, Gregersen, PK, and Behrens, TW (2003). Interferon-inducible gene expression signature in peripheral blood cells of patients with severe lupus. *Proc Natl Acad Sci USA*, 100(5):2610–2615.

Bennett, L, Palucka, AK, Arce, E, Cantrell, V, Borvak, J, Banchereau, J, and Pascual, V (2003). Interferon and granulopoiesis signatures in systemic lupus erythematosus blood. *J Exp Med*, 197(6):711–723.

Beutler, B (2004). Inferences, questions and possibilities in Toll-like receptor signalling. *Nature*, 430(6996):257–263.

Boldrick, , Alizadeh, AA, Diehn, M, Dudoit, S, Liu, CL, Belcher, CE, Botstein, D, Staudt, LM, Brown, PO, and Relman, DA (2002). Stereotyped and specific gene expression programs in human innate immune responses to bacteria. *Proc Natl Acad Sci USA*, 99(2):972–977.

Bosinger, SE, Hosiawa, KA, Cameron, MJ, Persad, D, Ran, L, Xu, L, Boulassel, MR, Parenteau, M, Fournier, J, Rud, EW, and Kelvin, DJ (2004). Gene expression profiling of host response in models of acute HIV infection. *J Immunol*, 173(11):6858–6863.

Cheung, VG, Conlin, LK, Weber, TM, Arcaro, M, Jen, KY, Morley, M, and Spielman, RS (2003). Natural variation in human gene expression assessed in lymphoblastoid cells. *Nat Genet*, 33(3):422–425.

Chtanova, T, Kemp, RA, Sutherland, AP, Ronchese, F, and Mackay, CR (2001). Gene microarrays reveal extensive differential gene expression in both CD4 (+) and CD8(+) type 1 and type 2 T cells. *J Immunol*, 167(6):3057–3063.

Cobb, JP, Mindrinos, MN, Miller-Graziano, C, Calvano, SE, Baker, HV, Xiao, W, Laudanski, K, Brownstein, BH, Elson, CM, Hayden, DL, Herndon, DN, Lowry, SF, et al. (2005). Application of genome-wide expression analysis to human health and disease. *Proc Natl Acad Sci USA*, 102(13):4801–4806.

Diehn, M, Alizadeh, AA, Rando, OJ, Liu, CL, Stankunas, K, Botstein, D, Crabtree, GR, and Brown, PO (2002). Genomic expression programs and the integration of the CD28 costimulatory signal in T cell activation. *Proc Natl Acad Sci USA*, 99(18):11796–11801.

Firestein, GS and Zvaifler, NJ (2002). How important are T cells in chronic rheumatoid synovitis?: II. T cell-independent mechanisms from beginning to end. *Arthritis Rheum*, 46(2):298–308.

Glynne, R, Akkaraju, S, Healy, JI, Rayner, J, Goodnow, CC, and Mack, DH (2000). How self-tolerance and the immunosuppressive drug FK506 prevent B-cell mitogenesis. *Nature*, 403(6770):672–676.

Golub, TR, Slonim, DK, Tamayo, P, Huard, C, Gaasenbeek, M, Mesirov, JP, Coller, H, Loh ML, Downing, JR, Caligiuri, MA, Bloomfield, CD, and Lander, ES (1999). Molecular classification of cancer: class discovery and class prediction by gene expression monitoring. *Science*, 286(5439):531–537.

Guidotti, LG and Chisari, FV (2001). Noncytolytic control of viral infections by the innate and adaptive immune response. *Annu Rev Immunol*, 19:65–91.

Hamalainen, H, Zhou, H, Chou, W, Hashizume, H, Heller, R, and Lahesmaa, R (2001). Distinct gene expression profiles of human type 1 and type 2 T helper cells. *Genome Biol*, 2(7):RESEARCH0022.

Hertzog, PJ, O'Neill, LA, and Hamilton, JA (2003). The interferon in TLR signaling: more than just antiviral. *Trends Immunol*, 24(10):534–539.

Hood, L (2003). Systems biology: integrating technology, biology, and computation. *Mech Ageing Dev*, 124(1):9–16.

Izmailova, E, Bertley, FM, Huang, Q, Makori, N, Miller, CJ, Young, RA, and Aldovini, A (2003). HIV-1 Tat reprograms immature dendritic cells to express chemoattractants for activated T cells and macrophages. *Nat Med*, 9(2):191–197.

Jenner, RG and Young, RA (2005). Insights into host responses against pathogens from transcriptional profiling. *Nat Rev Microbiol*, 3(4):281–294.

Kirou, KA, Lee, C, George, S, Louca, K, Papagiannis, IG, Peterson, MG, Ly, N, Woodward, RN, Fry, KE, Lau, AY, Prentice, JG, Wohlgemuth, JG, et al. (2004). Coordinate overexpression of interferon-alpha-induced genes in systemic lupus erythematosus. *Arthritis Rheum*, 50(12):3958–3967.

Kirou, KA, Lee, C, George, S, Louca, K, Peterson, MG, and Crow, MK (2005). Activation of the interferon-alpha pathway identifies a subgroup of systemic lupus erythematosus patients with distinct serologic features and active disease. *Arthritis Rheum*, 52(5):1491–1503.

Kitano, H (2002). Systems biology: a brief overview. *Science*, 295(5560):1662–1664.

Lee, CG (2002). RH70, a bidirectional RNA helicase, co-purifies with U1snRNP. *J Biol Chem*, 277(42):39679–39683.

Lehtonen, A, Matikainen, S, and Julkunen, I (1997). Interferons up-regulate STAT1, STAT2, and IRF family transcription factor gene expression in human peripheral blood mononuclear cells and macrophages. *J Immunol*, 159(2):794–803.

Lock, C, Hermans, G, Pedotti, R, Brendolan, A, Schadt, E, Garren, H, Langer-Gould, A, Strober, S, Cannella, B, Allard, J, Klonowski, P, Austin, A, et al. (2002). Gene-microarray analysis of multiple sclerosis lesions yields new targets validated in autoimmune encephalomyelitis. *Nat Med*, 8(5):500–508.

Maas, K, Chan, S, Parker, J, Slater, A, Moore, J, Olsen, N, and Aune TM (2002). Cutting edge: molecular portrait of human autoimmune disease. *J Immunol*, 169(1):5–9.

Mihm, S, Frese, M, Meier, V, Wietzke-Braun, P, Scharf, JG, Bartenschlager, R, and Ramadori, G (2004). Interferon type I gene expression in chronic hepatitis C. *Lab Invest*, 84(9):1148–1159.

Moir, S, Malaspina, A, Pickeral, OK, Donoghue, ET, Vasquez, J, Miller, NJ, Krishnan, SR, Planta, MA, Turney, JF, Justement, JS, Kottilil, S, Dybul M, et al. (2004). Decreased survival of B cells of HIV-viremic patients mediated by altered expression of receptors of the TNF superfamily. *J Exp Med*, 200(7):587–599.

Morley, M, Molony, CM, Weber, TM, Devlin, JL, Ewens, KG, Spielman, RS, and Cheung, VG (2004). Genetic analysis of genome-wide variation in human gene expression. *Nature*, 430 (7001):743–747.

Nau, GJ, Richmond, JF, Schlesinger, A, Jennings, EG, Lander, ES, and Young, RA (2002). Human macrophage activation programs induced by bacterial pathogens. *Proc Natl Acad Sci USA*, 99(3):1503–1508.

Olsen, N, Sokka, T, Seehorn, CL, Kraft, B, Maas, K, Moore, J, and Aune, TM (2004). A gene expression signature for recent onset rheumatoid arthritis in peripheral blood mononuclear cells. *Ann Rheum Dis*, 63(11):1387–1392.

Pizza, M, Scarlato, V, Masignani, V, Giuliani, MM, Arico, B, Comanducci, M, Jennings, GT, Baldi, L, Bartolini, E, Capecchi, B, Galeotti, CL, Luzzi, E, et al. (2000). Identification of vaccine candidates against serogroup B meningococcus by whole-genome sequencing. *Science*, 287(5459):1816–1820.

Radich, JP, Mao, M, Stepaniants, S, Biery, M, Castle, J, Ward, T, Schimmack, G, Kobayashi, S, Carleton, M, Lampe, J, and Linsley, PS (2004). Individual-specific variation of gene expression in peripheral blood leukocytes. *Genomics*, 83(6):980–988.

Rogge, L, Bianchi, E, Biffi, M, Bono, E, Chang, SY, Alexander, H, Santini, C, Ferrari, G, Sinigaglia, L, Seiler, M, Neeb, M, Mous, J, et al. (2000). Transcript imaging of the development of human T helper cells using oligonucleotide arrays. *Nat Genet*, 25(1):96–101.

Rubins, KH, Hensley, LE, Jahrling, PB, Whitney, AR, Geisbert, TW, Huggins, JW, Owen, A, Leduc, JW, Brown, PO, and Relman, DA (2004). The host response to smallpox: analysis of the gene expression program in peripheral blood cells in a nonhuman primate model. *Proc Natl Acad Sci USA*, 101(42):15190–15195.

Shaffer, AL, Rosenwald, A, Hurt, EM, Giltnane, JM, Lam, LT, Pickeral, OK, and Staudt LM (2001). Signatures of the immune response. *Immunity*, 15(3):375–385.

Teague, TK, Hildeman, D, Kedl, RM, Mitchell, T, Rees, W, Schaefer, BC, Bender, J, Kappler, J, and Marrack, P (1999a). Activation changes the spectrum but not the diversity of genes expressed by T cells. *Proc Natl Acad Sci USA*, 96(22):12691–12696.

Teague, TK, Hildeman, D, Kedl, RM, Mitchell, T, Rees, W, Schaefer, BC, Bender, J, Kappler, J, and Marrack, P (1999b). Activation changes the spectrum but not the diversity of genes expressed by T cells. *Proc Natl Acad Sci USA*, 96(22):12691–12696.

van de Vijver, MJ, He, YD, van't Veer, LJ, Dai, H, Hart, AA, Voskuil, DW, Schreiber, GJ, Peterse, JL, Roberts, C, Marton, MJ, Parrish, M, Atsma, D, et al. (2002). A gene-expression signature as a predictor of survival in breast cancer. *N Engl J Med*, 347(25):1999–2009.

van der Pouw Kraan, TC, Van Gaalen, FA, Kasperkovitz, PV, Verbeet, NL, Smeets, TJ, Kraan, MC, Fero, M, Tak, PP, Huizinga, TW, Pieterman, E, Breedveld, FC, Alizadeh, AA, et al. (2003). Rheumatoid arthritis is a heterogeneous disease: evidence for differences in the activation of the STAT-1 pathway between rheumatoid tissues. *Arthritis Rheum*, 48(8):2132–2145.

van der Pouw Kraan, TC, Van Gaalen, FA, Huizinga, TW, Pieterman, E, Breedveld, FC, and Verweij, CL (2003). Discovery of distinctive gene expression profiles in rheumatoid synovium using cDNA microarray technology: evidence for the existence of multiple pathways of tissue destruction and repair. *Genes Immun*, 4(3):187–196.

van 't Veer, LJ, Dai H, van de Vijver, MJ, He, YD, Hart, AA, Mao, M, Peterse, HL, van der, KK, Marton, MJ, Witteveen, AT, Schreiber, GJ, Kerkhoven, RM, et al. (2002). Gene expression profiling predicts clinical outcome of breast cancer. *Nature*, 415(6871):530–536.

Westendorp, RG, Langermans JA, Huizinga, TW, Elouali, AH, Verweij, CL, Boomsma, DI, and Vandenbroucke, JP (1997). Genetic influence on cytokine production and fatal meningococcal disease. *Lancet*, 349(9046):170–173.

Whitney, AR, Diehn, M, Popper, SJ, Alizadeh, AA, Boldrick, JC, Relman, DA, and Brown, PO (2003). Individuality and variation in gene expression patterns in human blood. *Proc Natl Acad Sci USA*, 100(4):1896–1901.

28

Applications in Clinical Pediatrics

Michael R Konikoff and Michael D Bates

Recent advances in human genetics and genomics, such as the Human Genome Project, characterization of human genomic variations [e.g., single nucleotide polymorphisms (SNPs) in the HapMap Project], comparative genomics, and others are providing new insights and suggesting new approaches to the application of genomics to human health and disease. However, special consideration needs to be given to the application of these advances in the care of children. In this chapter, we will describe the origins of using rudimentary genetic information in early diagnosis and preventative intervention in the care of children (through neonatal screening programs), and then discuss how this model informs the application of human genomic approaches to the care of children in the post-genome era.

Neonatal Screening

Large-scale newborn screening began in September 1962 when the Diagnostic Laboratories of the Massachusetts Department of Public Health in the United States began collecting blood specimens from all newborns in the state to test for phenylketonuria (PKU) (Levy, 2003), a rare defect in phenylalanine metabolism that results in mental retardation in children if left untreated. Dr. Robert Guthrie had recently developed the bacterial inhibition assay for phenylalanine (Guthrie, 1992), which made practical the monitoring of blood levels of children with known PKU to assess the efficacy of dietary restriction therapy. Since this dietary therapy was shown to be effective in preventing mental retardation in children with PKU if instituted in the neonatal period (Armstrong et al., 1957; Horner and Streamer, 1959), Dr. Guthrie adapted his phenylalanine assay to work with blood dried on filter paper, which could be easily obtained from newborns by means of a simple heel stick. Guthrie met Dr. Robert MacCready, the Director of the Diagnostic Laboratories at the Massachusetts Department of Public Health, by chance at an annual meeting of the National Association for Retarded Children, during which MacCready learned of the promise of preventing mental retardation in PKU by the application of Guthrie's test. Upon his return to Massachusetts, MacCready established Guthrie's phenylalanine assay in the Diagnostic Laboratories and began screening all newborns for PKU using dried blood collected on filter paper, now called Guthrie cards, by the hospitals' pathologists (Fig. 28–1).

Nine cases of PKU were detected in the first 53,000 infants screened. This finding of PKU in 1 in 6000 live births was much higher than previously thought (1 in 20,000) and attracted national attention to newborn screening. Massachusetts soon passed a law requiring mandatory newborn screening for PKU, and the majority of states in the United States followed suit. Building on the success of PKU screening, laboratories adapted Guthrie's assay to detect other amino acids and sugars, which allowed screening for additional metabolic disorders whose deleterious effects could be prevented by early dietary intervention, including maple syrup urine disease (leucine) (Naylor and Guthrie, 1978), homocystinuria (methionine), and galactosemia (galactose) (Guthrie, 1968; Paigen et al., 1982). With the availability of excess blood on the Guthrie cards, other disorders for which a metabolite, hormone, or protein could be assayed for diagnosis using a screening blood test were added to the newborn screening panels in the United States. These included hemoglobinopathies (Garrick et al., 1973), congenital hypothyroidism (Dussault et al., 1975), congenital adrenal hyperplasia (CAH) (Pang et al., 1977), and biotinidase deficiency (BD) (Heard et al., 1984). Table 28–1 gives a listing of the core conditions currently included in the newborn screening programs in the United States. In each disorder, it has been determined that clinical disease can be mitigated or prevented by early dietary or medical intervention.

Since their inception in the United States, newborn screening programs have been created in many countries and regions all over the world and have become an indispensable part of the health care system and preventative medicine around the world (Loeber et al., 1999). Most programs have focused on the detection of two diseases, namely PKU and congenital hypothyroidism. Other disorders have been included depending on local prevalence, interest of professionals, technological availability, and feasibility of interventions. Table 28–2 gives a listing of the conditions included in the newborn screening programs of various countries. Not only do the number and types of conditions tested differ widely between countries but the design of these programs with respect to legislation, funding mechanisms, and follow-up also vary. Although the cost-effectiveness of newborn screening for rare conditions has been debated in the international community (Grosse, 2005), the creation of new programs and the expansion of existing programs continue.

Work in human genetics has elucidated the underlying molecular defects responsible for many inherited congenital disorders. By applying molecular techniques to newborn screening, many genetic

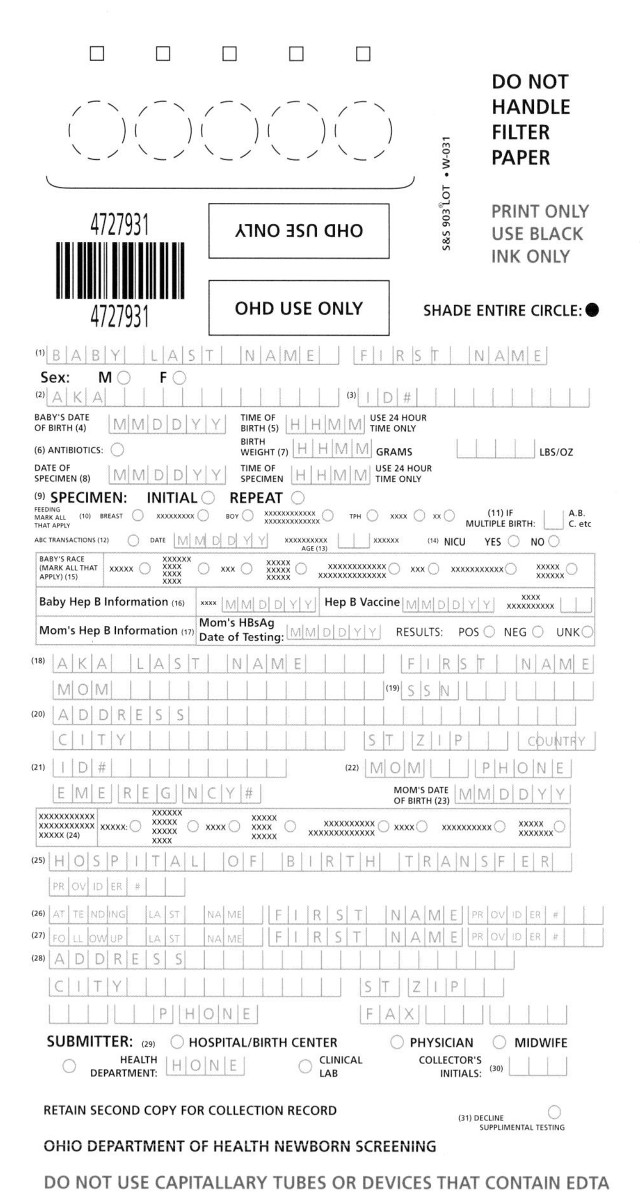

DO NOT HANDLE FILTER PAPER

S&S 903 LOT • W-031

4727931

OHD USE ONLY

PRINT ONLY USE BLACK INK ONLY

4727931

OHD USE ONLY

SHADE ENTIRE CIRCLE: ●

(1) BABY LAST NAME FIRST NAME

Sex: M ○ F ○

(2) AKA (3) ID #

BABY'S DATE OF BIRTH (4) MMDDYY TIME OF BIRTH (5) HHMM USE 24 HOUR TIME ONLY

(6) ANTIBIOTICS: ○ BIRTH WEIGHT (7) HHMM GRAMS LBS/OZ

DATE OF SPECIMEN (8) MMDDYY TIME OF SPECIMEN HHMM USE 24 HOUR TIME ONLY

(9) SPECIMEN: INITIAL ○ REPEAT ○

FEEDING MARK ALL THAT APPLY (10) BREAST ○ xxxxxxxxx ○ BOY ○ xxxxxxxxxxxx xxxxxxxxxxxxx ○ TPH ○ xxxx ○ xx ○ (11) IF MULTIPLE BIRTH: A.B. C. etc

ABC TRANSACTIONS (12) ○ DATE MMDDYY xxxxxxxxxx AGE (13) xxxxxx (14) NICU YES ○ NO ○

BABY'S RACE (MARK ALL THAT APPLY) (15) xxxxx ○ xxxx xxxx xxxx ○ xxx ○ xxxxx xxxxx xxxxx ○ xxxxxxxxxxxx xxxxxxxxxxxxxx ○ xxx ○ xxxxxxxxxxx ○ xxxxx xxxxxx ○

Baby Hep B Information (16) xxxx MMDDYY Hep B Vaccine MMDDYY xxxx xxxxxxxxxx

Mom's Hep B Information (17) Mom's HBsAg Date of Testing: MMDDYY RESULTS: POS ○ NEG ○ UNK ○

(18) AKA LAST NAME FIRST NAME

MOM (19) SSN

(20) ADDRESS

CITY ST ZIP COUNTRY

(21) ID # (22) MOM PHONE

EMER REG NCY # MOM'S DATE OF BIRTH (23) MMDDYY

xxxxxxxxxxx xxxxxxxxxxx xxxxx (24) xxxxx: ○ xxxxxx xxxxx xxxxx xxxxx xxxx ○ xxxx ○ xxxxx xxxx xxxxx ○ xxxxxxxxxxx xxxxxxxxxxxxxx ○ xxxx ○ xxxxxxxxxx ○ xxxxx xxxxxxx ○

(25) HOSPITAL OF BIRTH TRANSFER

PR OV ID ER #

(26) ATTENDING LAST NAME FIRST NAME PROVIDER #

(27) FOLLOWUP LAST NAME FIRST NAME PROVIDER #

(28) ADDRESS

CITY ST ZIP

PHONE FAX

SUBMITTER: (29) ○ HOSPITAL/BIRTH CENTER ○ PHYSICIAN ○ MIDWIFE

○ HEALTH DEPARTMENT: HONE ○ CLINICAL LAB COLLECTOR'S INITIALS: (30)

RETAIN SECOND COPY FOR COLLECTION RECORD (31) DECLINE SUPPLIMENTAL TESTING ○

OHIO DEPARTMENT OF HEALTH NEWBORN SCREENING

DO NOT USE CAPITALLARY TUBES OR DEVICES THAT CONTAIN EDTA

Figure 28–1 Example of a Guthrie card for collection of blood specimens for newborn screening (from the Ohio Department of Health, USA). Blood samples are collected within the dashed circles on filter paper attached to the top of the form.

diseases can be detected shortly after birth, before their full or irreversible phenotypic expression. Disorders with well-defined genetics and common mutations represented in a significant proportion of affected individuals are particularly well suited to this form of analysis. Genomic deoxyribonucleic acid (DNA) can be easily extracted from newborn blood specimens by inexpensive, automated methods, further enhancing the feasibility of this approach (Heath et al., 1999). Traditional molecular methodologies, however, analyze only one gene or mutation at a time, requiring separate assays for each disease. Thus, while excellent in a research setting, these procedures are not amenable to population-based high-throughput newborn screening programs. They are primarily used clinically as second-tier confirmatory assays for such disorders as cystic fibrosis, sickle cell anemia and other hemoglobinopathies, and medium-chain acyl-CoA dehydrogenase (MCAD) deficiency (Dobrowolski et al., 1999).

The recent application of tandem mass spectrometry (MS/MS) to the specimens on Guthrie cards has further expanded the ability of newborn screening programs to identify congenital disorders. MS/MS recognizes amino acids and acylcarnitines by their characteristic mass spectra, and is thus able to identify more than 20 inborn errors of metabolism not previously a part of newborn screening programs (Chace et al., 2003; Fearing and Marsden, 2003). MS/MS also has the advantage of being able to measure concentrations of multiple compounds, including amino acids and acylcarnitines, in the same sample, greatly increasing the utility of this method. MS/MS has revolutionized the field of newborn screening by allowing the detection of multiple metabolic disorders in a single step. Previously, each screened disease required a separate assay. This was true not only for older assays, like the bacterial inhibition assay for PKU, but also for newer molecular assays for genetic disorders. Thus, MS/MS is a straightforward, efficient, and cost-effective method for the detection of diseases involving altered amino acid and acylcarnitine metabolism. Additional work is expanding the range of the utility of MS/MS to include identification of hexose monophosphates in patients with galactosemia (Jensen et al., 2001) and of steroids in patients with congenital adrenal hypoplasia (Lacey et al., 2004).

The ability to extract DNA from the newborn blood specimens and recent advances in microarray technology offer the opportunity for primary screening of a vast number of genetic disorders that could not previously be identified in the newborn. DNA microarrays, which contain representations of thousands of genes, are fabricated either by robotically spotting gene fragments onto a glass slide or by synthesizing oligonucleotides onto glass by photolithographic techniques (Lipshutz et al., 1999). Microarrays can be used to measure the expression of thousands of genes simultaneously by the specific hybridization of nucleic acid strands having complementary nucleotide sequences. This stringent hybridization can also be applied to the analysis of genomic DNA to detect SNPs, point mutations, insertions, and deletions (Hacia, 1999).

Microarrays have been designed to test these concepts using well-characterized genetic disorders such as hemoglobinopathies, α_1-antitrypsin deficiency, and factor V Leiden (Dobrowolski et al., 1999). Oligonucleotides representing multiple disease alleles in the above disorders were spotted onto a microarray, and DNA with known mutations was extracted from newborn blood specimens and hybridized to the array. Hybridization signals were detected corresponding to the various known mutations, indicating that the microarray correctly identified the tested gene mutations. The entire process from DNA extraction to data analysis can be automated and, with the addition of multiplex amplification, several disorders may be assayed on a single microarray. This will allow population-based first-tier molecular screening for a large number of genetic disorders, which has not previously been possible.

Primary newborn screening using microarray technology offers several advantages over traditional methods (Table 28–3). Just as MS/MS can measure multiple analytes from a single sample, DNA microarrays can screen many genes for mutations in a single step, with little additional cost for each disorder. The major limiters at present are the feature density of the microarray and the resolution of laser scanners. The completion of the sequencing of the human genome facilitates the design of "resequencing" microarrays to screen for and identify *any* mutations of a specific gene, rather than only common or previously characterized ones. Because the entire process can be automated with high throughput, microarrays are well suited to population screening. DNA microarrays are able to screen for certain disorders that are not amenable to traditional protein-based methods, such as fragile X syndrome (through

Table 28–1 Core Conditions Currently Included in Newborn Screening Programs in the United States

Disorder	Screening Method	Treatment
Phenylketonuria	Guthrie bacterial inhibition assay	Diet restricting phenylalanine
	Fluorescence assay	
	Amino acid analyzer	
	Tandem mass spectrometry	
Congenital hypothyroidism	Measurement of thyroxine and thyrotropin	Oral levothyroxine
Hemoglobinopathies	Hemoglobin electrophoresis	Prophylactic antibiotics
	Isoelectric focusing	Immunization against
	High-performance liquid chromatography	*Streptococcus pneumoniae* and *Haemophilus influenzae*
	Follow-up DNA analysis	
Galactosemia	Beutler test	Galactose-free diet
	Paigen test	
Maple syrup urine disease	Guthrie bacterial inhibition assay	Diet restricting intake of branched-chain amino acids
	Tandem mass spectrometry	
Homocystinuria	Guthrie bacterial inhibition assay	Vitamin B_{12}
	Tandem mass spectrometry	Diet restricting methionine and supplementing cystine
Biotinidase deficiency	Colorimetric assay	Oral biotin
Congenital adrenal hyperplasia	Radioimmunoassay	Glucocorticoids
	Enzyme immunoassay	Mineralocorticoids
		Salt
Cystic fibrosis	Immunoreactive trypsinogen assay followed by DNA testing	Improved nutrition
	Sweat chloride testing	Management of pulmonary symptoms
Fatty acid disorders[a]	Tandem mass spectrometry	Various
Organic acid disorders[b]	Tandem mass spectrometry	Various
Other amino acid disorders	Tandem mass spectrometry	Various
Hearing screen	Audiometry	Hearing aids
		Cochlear implants

[a] Carnitine uptake defect, very long-chain hydroxyacyl-CoA dehydrogenase deficiency, long-chain hydroxyacyl-CoA dehydrogenase deficiency, medium-chain hydroxyacyl-CoA dehydrogenase deficiency, trifunctional protein deficiency.

[b] Glutaric acidemia type 1,3-hydroxy 3-methylglutaric aciduria, isovaleric acidemia, 3-methylcrotonyl-CoA carboxylase deficiency, methylmalonic acidemia, beta ketothiolase deficiency, propionic acidemia, multiple carboxylase deficiency, argininosuccinate acidemia, citrullinemia type 1, tyrosinemia type 2.

Source: Data from Khoury et al., 2003 and National Newborn Screening and Genetics Resource Center, 2005.

identification of untranslated, triplet-repeat DNA sequences) (Bailey, 2004), the various severe combined immunodeficiencies (through detection of excised T-cell receptor sequences rather than of various T-cell related proteins) (Kalman et al., 2004), and cystic fibrosis. For example, the DF508 mutation of the cystic fibrosis transmembrane conductance regulator (CFTR), which accounts for 80% of mutant cystic fibrosis alleles in the Caucasian population but only 30% of mutant alleles in Asians or other nonEuropean individuals (Grody et al., 2001), is the only mutation assayed in the newborn screening program of at least one state in the United States (Green et al., 2004). There are now more than 1000 known mutations in the CFTR gene, and all of these could be assayed using one microarray, which would improve detection of cystic fibrosis cases in ethnic groups where uncommon mutations are overrepresented. Microarrays also have the potential to be more specific, while maintaining sensitivity, than traditional metabolite- or protein-based assays, which commonly use threshold values that yield ratios of false to true positives ranging from 10:1 to 75:1 (Green and Pass, 2005). More specific tests would reduce the rate of false positives, thus decreasing the need for follow-up testing and its resultant costs, not to mention the increased anxiety for parents and medical caregivers. Microarrays are able to distinguish disease gene carriers (heterozygotes) by the selective detection of hybridization signal intensity and the simultaneous probing for wild-type alleles. Carriers could then be excluded from follow-up testing and laboratory reporting, which now diverts resources from detecting diseased individuals, the primary mission of newborn screening programs. Alternatively, carriers could be easily identified and reported if this was appropriate from ethical and/or legal perspectives. In addition, microarrays could be used to improve phenotype prediction of Mendelian "single-gene" disorders by genotyping known modifying or interacting genes, which would have the potential to improve the

Table 28–2 Neonatal Screening Programs in Various Countries per 2003

Nation	PKU	CH	CAH	CF	BD	Gal	MCAD	MSUD	HC	SCD	G6PD	Other
Argentina	X	X	X	X	X	X						
Australia	X	X	X*	X	X*	X*		X				
Austria	X	X	X	X	X	X	X					
Bangladesh		P										
Belarus	X	X	P	P			X					
Belgium	X	X	X	X	X	X	X	X	X			1
Bosnia-Herzegovina												
Brazil	X	X	X	X	X	X		X	X	X	X	2,3
Bulgaria	X	X										
Canada	X	X	X*		X*	X*						
China	X	X										
Croatia	X	X										
Cyprus	X	X										1P
Czech Republic	X	X	X*			X*			X			
Denmark	X	X										
Estonia	X	X										
Finland	X	X										
France	X	X	X	X						X		1P,5
Georgia												
Germany	X	X	X	X	X	X	X	X	X		X	1P
Greece	X	X				P					X	
Hungary	X	X			X	X						
Iceland	X	X										
India	P	P	P									
Ireland	X	X				X						
Israel	X	X								X		
Italy	X	X	X*	X*	X*	X	X	X			X	
Japan	X	X	X			X		X	X			
Korea	X	X				X		X	X			
Hong Kong		X									X	
Latvia	X	X										
Lichtenstein	X	X										
Lithuania	X	X										
Luxembourg	X	X	X									
Macedonia		X										
Malta												
Moldova	X											
Montenegro												
Netherlands	X	X	X				X					
New Zealand	X	X	X	X	X	X		X				
Norway	X	X										
Philippines	X	X	X			X					X	
Poland	X	X		P								
Portugal	X	X	X			X						
Romania												
Russia	X	X	X	X	X	X						
Scotland	X	X		X								6
Serbia	X	X										
Singapore		X									X	

(Continued)

Table 28–2 Neonatal Screening Programs in Various Countries per 2003 *(Continued)*

Nation	PKU	CH	CAH	CF	BD	Gal	MCAD	MSUD	HC	SCD	G6PD	Other
Slovakia												
Slovenia												
Spain	X	X	X	X	X		X			X		6
Sweden	X	X	X		X	X						2,4
Switzerland	X	X	X		X	X						
Taiwan	X	X	X*			X			X			
Thailand	X	X										
Turkey	X	X			X							
Ukraine												
United Kingdom (excl. Scotland)	X	X		X						X		
United States	X	X	X	X*	X	X*	X*	X*	X*	X*	X*	2*,4*,6*

X, regular program; P, pilot program; *, not in all parts of the country; 1, Duchenne muscular dystrophy (pilot); 2, congenital toxoplasmosis; 3, Chagas' disease; 4, tyrosinemia; 5, neuroblastoma; 6, HIV antibody.

Abbreviations: BD, biotinidase deficiency; CAH, congenital adrenal hyperplasia; CF, cystic fibrosis; CH, congenital hypothyroidism; G6PD, glucose-6-phosphate dehydrogenase deficiency; Gal, galactosemia; HC, homocystinuria; MCAD, medium-chain acyl-coA dehydrogenase deficiency; MSUD, maple syrup urine disease; PKU, phenylketonuria; SCD, sickle cell disorders.

Source: Data from Loeber et al., 1999; International Society for Neonatal Screening 2003; and National Newborn Screening and Genetics Resource Center, 2005.

prediction of disease severity that impacts treatment strategies, planning, and counseling. Finally, DNA-based analysis is not affected by clinical factors that may interfere with the analyses of blood samples for metabolites, hormones, or proteins. Factors such as prematurity, medications of the mother or child, or parenteral nutrition will not affect DNA-based analysis.

DNA microarrays do have several limitations that must be overcome before their widespread use in newborn screening programs will become practical (Table 28–3). Cost has been a major factor limiting their use; but newer manufacturing procedures have eased this constraint somewhat, and economies of scale would be achieved if standardized microarrays were developed for use in newborn screening programs. Still, the cost of single-use microarrays is much higher than standard assays used in public health laboratories, many of which cost a few US dollars at the most. The high-throughput, automated technology on which PCR-driven microarrays depend may also be another source of potential problems. Contamination of samples and resulting PCR artifacts must be minimized, and systems for organizing, storing, and accessing the vast amounts of generated data will need to be developed. In addition to cost and technical issues, the very nature of the information derived from microarrays can be a limitation in some cases. For example, if a previously uncharacterized point mutation is identified, it may not be possible to predict whether the mutation is deleterious. This approach will not detect post-transcriptional and post-translational modifications or impairments (Mendell and Dietz, 2001), as well as nonMendelian genetic events such as gene duplications, epigenetic influences, and nonrandom X chromosome inactivation, and thus may lead to false-positive or false-negative screening results. Some disorders that have a final common phenotype may also be caused by mutations in different genes, and all such genes would need to be included in the microarray. For

Table 28–3 Potential Advantages and Limitations of Using DNA Microarrays for Newborn Screening

Advantages
Screen many genes in a single assay
Identify any mutations rather than only common or previously characterized mutations
Can be automated for high-throughput
Potential for increased specificity with high sensitivity
Can detect heterozygotes (disease gene carriers)
Allows genotyping of modifying genes to improve phenotype prediction of "single-gene" disorders
Less affected by clinical factors (prematurity, diet, medications, etc.)
Limitations
Cost
Amplification of genomic DNA by polymerase chain reaction (PCR) can result in artifacts or contamination
May identify polymorphisms/mutations that have no effect on gene or protein function
Does not identify alterations in post-transcriptional or post-translational processes
Does not identify nonMendelian genetic events (e.g., gene duplications, epigenetic influences, and nonrandom X chromosome inactivation)
Not applicable to disorders with no known genetic basis (e.g., congenital hypothyroidism)

example, PKU is usually caused by a mutation in the phenylalanine hydroxylase gene, but mutations in the cofactor pathway of tetrahydrobiopterin synthesis may cause a form of atypical PKU (Green and Pass, 2005). Yet other disorders have no known genetic basis, such as congenital hypothyroidism, in which only a small percentage of cases have a readily identifiable genetic cause (Gruters et al., 2004). These disorders will not be amenable to DNA-based microarray screening unless a genetic etiology is elucidated.

With the advent of new technologies such as MS/MS and DNA microarrays, newborn screening programs are poised to significantly expand the number and types of disorders that are detectable. The principles of population screening developed in 1968 by Wilson and Jungner form a basis for applying these new technologies to newborn screening (Table 28–4) (Wilson and Jungner, 1968). In particular, the ability to screen for a disorder using these new technologies does not imply the obligation to do so. The early identification of a genetic disorder is of no benefit (1) if treatment is not available or (2) if an abnormal finding determines only susceptibility for and not certainty of disease.

An example of a genetic disorder for which effective treatment is currently not available is Huntington's disease, a devastating neurological disorder of adult onset, for which there continues to be much debate regarding the clinical and ethical merits of presymptomatic diagnosis (European Community Huntington's Disease Collaborative Study Group, 1993; Paulson and Prior, 1997; Wusthoff, 2003). Even for disorders that present in childhood, such as familial adenomatous polyposis and Duchenne muscular dystrophy, presymptomatic diagnosis is controversial because of the lack of preventative treatments (Ross, L. F., 2002).

Potential examples for susceptibility screening include predispositional screening for adult-onset, multifactorial disorders such as type II diabetes mellitus or cardiovascular disease. Screening for these disorders is often not medically beneficial and may have negative psychological consequences to both the affected individual and his or her family. Currently, pilot studies have been undertaken to detect the predisposition to type I diabetes mellitus by typing multiple HLA DQ alleles (Wion et al., 2003; Barker et al., 2004). If successful, this effort would have enormous public health implications as type I diabetes mellitus affects over 3,00,000 people in the United States alone and generates huge direct and indirect health care costs. However, no treatment is available to prevent the development of diabetes in predisposed individuals, so early identification will not necessarily improve the outcome for these children. This information may, in fact, become available to insurance providers or employers, and could lead to discrimination of identified children or their families (see section "Ethical, Legal, and Social Implications Unique to Pediatrics").

Limiting newborn screening to disorders currently assayed by screening programs, though, will not take full advantage of the public health potential of microarray technology. It will be necessary to carefully evaluate each new disorder added to the panel to determine immediate and future ramifications of screening, as well as to develop methods of obtaining informed consent from parents of newborns, which is now not required in mandatory screening programs.

Other Genetic Disorders

Genetic analyses for the identification and characterization of genetic disorders are an obvious application of genomics in pediatrics. As noted above, "resequencing" DNA microarrays can be used to identify mutations and polymorphisms in either single large genes or panels of genes whose mutation can result in clinically similar phenotypes. Such disease-specific microarrays might be developed commercially to provide rapid diagnosis. For example, we have previously suggested that such an approach could be used to rapidly differentiate among different genetic disorders that all present with neonatal cholestasis as a major feature (Bates, 2003). Such resequencing arrays could not be easily designed without the use of the human genomic sequence determined in the Human Genome Project.

Microarray analyses using genomic DNA may also be useful for testing panels of polymorphisms or mutations that act as modifiers of risk in genetic disorders or for combinations of genes that result in multigenic disorders. For such analyses, panels of genes might be resequenced or profiles of large numbers of SNPs linked to disease genes or modifier genes could be analyzed (Hacia and Collins, 1999). Such specific information could be obtained for use in the promotion of early intervention in cases of increased risk. To date, no such analyses have been reported for pediatric disorders.

Finally, the genomics-based tools that have been developed in the wake of the Human Genome Project are being applied to

Table 28–4 Criteria for a Screening Program that Benefits Society (Wilson and Jungner, 1968)

1. The condition sought should be an important health problem
2. There should be an accepted treatment for patients with recognized disease
3. Facilities for diagnosis and treatment should be available
4. There should be a recognizable latent or early symptomatic stage
5. There should be a suitable test or examination
6. The test should be acceptable to the population
7. The natural history of the condition, including development from latent to declared disease, should be adequately understood
8. There should be an agreed policy on whom to treat as patients
9. The cost of case-finding (including diagnosis and treatment of patients diagnosed) should be economically balanced in relation to possible expenditure on medical care as a whole
10. Case-finding should be a continuing process and not a "once and for all" project

Source: Wilson, JMG and Jungner, G (1968). Principles and practice of screening for disease. Public Health Papers (WHO), no. 34. World Health Organization, Geneva.

determine the genetic basis of pediatric disorders, including congenital malformations. The relatively small numbers of patients for many pediatric disorders means that multi-center recruitment of subjects or intensive study of large kindreds with a particular disorder will be necessary.

Genomic Classification of Pediatric Disorders

Genomics techniques are being taken to classify many different kinds of diseases (see Chapter 6) and to determine the risk associated with the various classifications to refine therapeutic approaches. Among the pediatric disorders reported to have been studied to date are psychiatric and developmental disorders, rheumatological and immune disorders, digestive system disorders (such as celiac disease, inflammatory bowel disease, and biliary atresia), and renal disorders (see Fig. 28–2 and Table 28–5). However, the most work has been done in oncology, in order to identify prospectively the

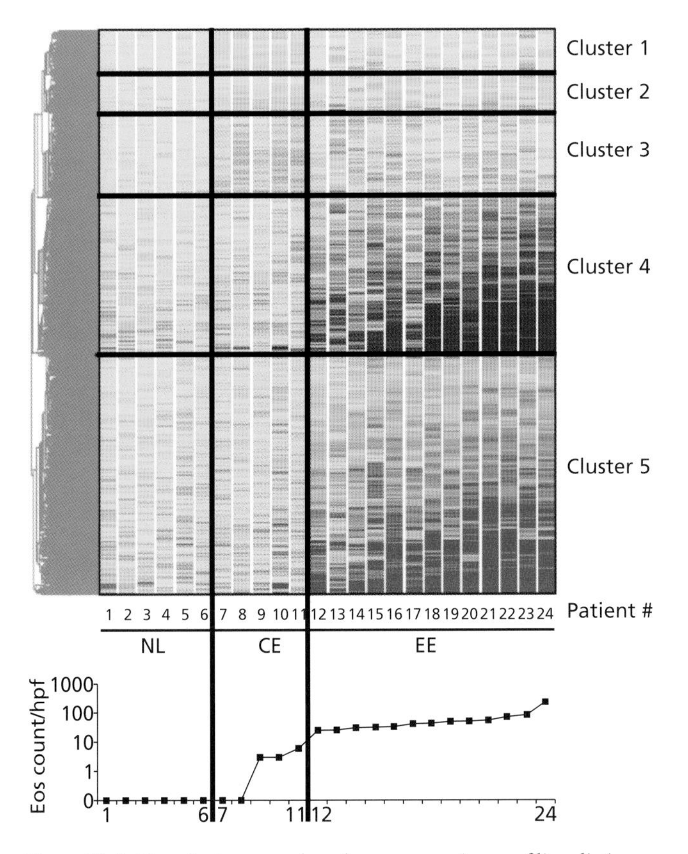

Figure 28–2 Use of microarray-based gene expression profiling distinguishes among esophageal biopsies from normal children (NL) and those with chronic esophagitis (CE, such as from gastroesophageal reflux disease) or eosinophilic esophagitis (EE) (Blanchard et al., 2006). Columns represent biopsies from individual patients, and rows represent individual genes. Clusters 1, 2, and 3 include 228 genes expressed differently in the CE group compared with normal healthy patients. Clusters 4 and 5 include 574 genes expressed significantly differently ($p < 0.01$) in the EE group compared with normal healthy patients. The eosinophil (Eos) count in each patient's esophageal biopsy is shown in the lower panel. [Reprinted from the *Journal of Clinical Investigation* by Blanchard et al., copyright 2006 by the American Society for Clinical Investigation. Reproduced with permission of the American Society for Clinical Investigation in the format Other Book via Copyright Clearance Center.]

Table 28–5 Examples of Genomics Approaches to the Study of Specific Pediatric Disorders

Disorder	Reference
Cardiology	
Kawasaki disease	Nomura et al., 2005
Kawasaki disease (response to therapy)	Abe et al., 2005
Right ventricular outflow obstructive lesions	Konstantinov et al., 2004
Dentistry	
Dental caries	Paakkonen et al., 2005
Dermatology	
Giant hairy nevi	Dasu et al., 2004
Hypertrophic and normal scars	Tsou et al., 2000
Gastroenterology/hepatology	
Biliary atresia	Bezerra et al., 2002
Celiac disease	Diosdado et al., 2004
Eosinophilic esophagitis	Blanchard et al., 2006
Inflammatory bowel disease	Dieckgraefe et al., 2000
Genetics	
Duchenne muscular dystrophy	Noguchi et al., 2003
Mitochondrial disorders	Crimi et al., 2005
Rett Syndrome	Colantuoni et al., 2001
X-Linked myotubular myopathy	Noguchi et al., 2005
Infectious diseases	
Neisseria meningitides	Sun et al., 2000
Parvovirus B19 infection	Kerr et al., 2005
Vaccine-derived poliovirus	Laassri et al., 2005
Neonatology/obstetrics	
Premature rupture of membranes	Tromp et al., 2004
Nephrology	
Acute renal allograft rejection	Sarwal et al., 2003
Focal segmental glomerulosclerosis	Schwab et al., 2004
Neurology/psychiatry	
Brain arteriovenous malformations	Hashimoto et al., 2004
Neurofibromatosis type 1	Tang et al., 2004
Neurologic diseases	Tang et al., 2005
Tardive dyskinesia in therapy of schizophrenia	Nikoloff et al., 2002
Oncology	
Acute megakaryoblastic leukemia in Down syndrome	Lightfoot et al., 2004
Acute myelogenous leukemia	Ross, M. E. et al., 2004
Acute myeloid leukemia	Yagi et al., 2003
Ependymoma	Suarez-Merino et al., 2005
Ewing's sarcoma	Ohali et al., 2004
Juvenile pilocytic astrocytomas	Wong et al., 2005
Large cell lymphomas	Thompson et al., 2005
Leukemias	Moos et al., 2002
Medulloblastoma	Pomeroy et al., 2002; Fernandez-Teijeiro et al., 2004
Neuroblastoma	Hiyama et al., 2004
Retinoblastoma	Gratias et al., 2005
Wilms' tumor	Williams et al., 2004

(Continued)

Table 28–5 Examples of Genomics Approaches to the Study of Specific Pediatric Disorders *(Continued)*

Rheumatology/immunology	
Hyper-IgE syndrome	Tanaka et al., 2005
Rheumatoid diseases	Barnes et al., 2004
Systemic lupus erythematosus	Bennett et al., 2003
Other	
Muscle wasting following burns	Barrow et al., 2003; Dasu et al., 2005
Recurrent aphthous ulceration	Borra et al., 2004
Unexplained drowning (molecular autopsy)	Tester et al., 2005

prognoses associated with histologically similar tumors based on significant differences in their gene expression signatures. The first example of this approach in pediatric oncology was reported by Pomeroy et al., who demonstrated that medulloblastoma gene expression profiles could be used not only to distinguish between different types of tumors but also could predict survival (Fig. 28–3) (Pomeroy et al., 2002). Since then, studies have been reported for a variety of leukemias, lymphomas, and solid-tumor malignancies, in particular brain tumors. From the standpoint of cancer biology and pathophysiology, the identification of genes differentially expressed in tumors with a poor prognosis might suggest approaches to the development of novel and potentially more effective therapies for patients with such tumors. Clinically, such gene expression profiling provides a potentially more objective approach to disease classification than traditional histopathological approaches.

Infectious Diseases

A key challenge in patients with infectious diseases is to detect and characterize pathogens, particularly for organisms that are difficult to culture such as *Mycobacterium tuberculosis*, so that therapy can be initiated in a timely manner. Genomics approaches can also be applied to such challenges once the relevant pathogen genome(s) have been sequenced. This sequence information is necessary for the preparation of microarrays that can detect genomic DNA from the pathogen(s) of interest. Hanson et al. have pointed out that the use of microarrays can potentially decrease the time necessary to identify a central nervous system pathogen from many days currently to just a few hours (Hanson et al., 2004), which would allow targeted therapy to begin sooner. These assays can also be used to identify pathogen subtypes and the presence of antimicrobial resistance markers. Examples of such approaches include the use of microarrays to detect, subtype, and determine rifampin resistance for various *Mycobacterium* species (Gingeras et al., 1998), and for the sensitive detection of *Escherichia coli* O157:H7 (Call et al., 2001), an important food-borne cause of hemorrhagic colitis and the hemolytic-uremic syndrome.

Pharmacogenetics and Pharmacogenomics

An exciting current frontier for genomic medicine is to use an individual's genomic information to "personalize" drug therapy through the approaches of pharmacogenetics and pharmacogenomics (see also Chapter 9). Such personalization will require attention to both the patient's age and genome, because the targets and toxicities of medications change with age (Stephenson, 2005). Thus, investigation of developmental pharmacogenetics and pharmacogenomics is needed to understand the relevant age-related changes in the expression of drug targets and in drug metabolism, so that the most efficacious drugs with the least risk for adverse reactions for a patient at a given age can be identified.

Ethical, Legal, and Social Implications Unique to Pediatrics

No discussion on the use of genomics in clinical pediatrics would be complete without a thorough consideration of the ethical, legal, and social implications of this emerging technology. Traditional medical ethics has been governed by four principles: beneficence (doing good), nonmaleficence (doing no harm), autonomy (the right to choose), and justice (being fair and equitable) (Beauchamp and Childress, 1994). These principles also apply to the use of genomic information in pediatrics, although the "usual" ethics may need to be modified to deal with the unique ethical quandaries found in genomic medicine (Anonymous Editorial, 2001).

Genomic medicine, by generating vast amounts of personal genetic data, poses significant ethical dilemmas both for individuals and for society, and the potential for abuse of this information is real. A major concern is that the existence of such detailed genetic information could result in so-called "genetic discrimination" to deny access to health and life insurance or employment (Lapham et al., 1996). In fact, public pressure in the United States has resulted in the passage of laws prohibiting genetic discrimination and regulating the use of genetic information, and US federal legislation has been debated in many recent sessions of Congress (Clayton, 2003). Although genetic discrimination may be a widespread public concern, there are few examples of such discrimination and no evidence that it is common (Greely, 2005). Most lawsuits that have been filed involve cases of actual disease identified by genetic testing, not presymptomatic or predispositional conditions. There are almost no well-documented cases of health insurers collecting or using presymptomatic genetic test results in their underwriting decisions, and little evidence that such testing impacts a person's ability to obtain health insurance (Hall and Rich, 2000). Despite this apparent lack of improper use of genetic information, as our knowledge of genomics increases, these dilemmas are likely to increase in both frequency and importance. Deciding what to do about such predispositions will depend on the likelihood of occurrence in at-risk individuals and the natural history of the disease. The challenge for clinicians will be to discuss with their patients the possible adverse social consequences of testing so that patients can make informed decisions about whether or not to proceed with the testing (Clayton, 2003).

Genetic testing of children presents unique ethical considerations. Such testing may be diagnostic or predictive. Diagnostic genetic testing refers to testing of a symptomatic child's genetic material to establish a medical diagnosis. This type of testing, with direct benefit to the child who is symptomatic, is generally noncontroversial. Predictive genetic testing, on the other hand, refers to the testing of asymptomatic children for genetic conditions that may present months to years in the future, if at all. Predictive testing can be presymptomatic (i.e., virtually all children with the specific genotype will develop the disease) or predispositional (i.e., children with the specific genotype are at risk for developing the disease, but not certain to do so). Predictive genetic testing can be performed for

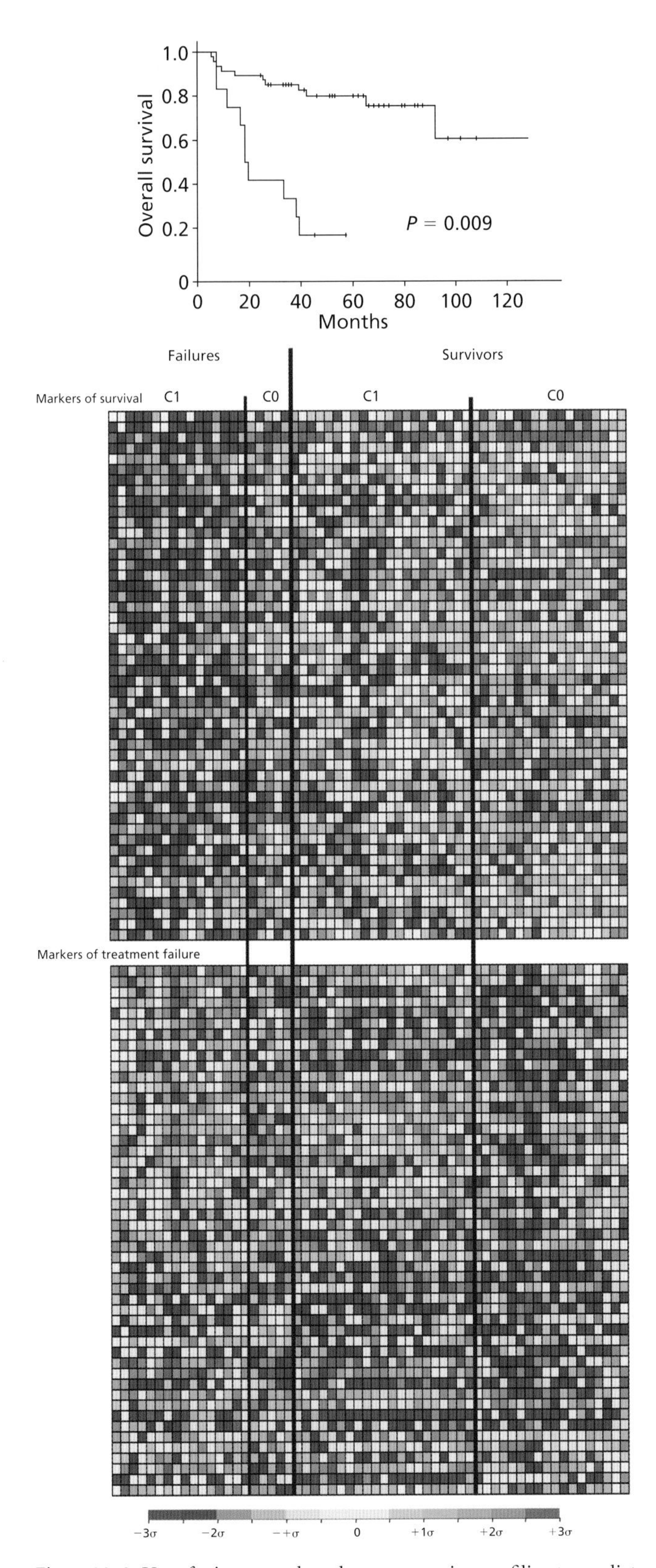

Figure 28–3 Use of microarray-based gene expression profiling to predict medulloblastoma outcome in children (Pomeroy et al., 2002). Top panel, Kaplan–Meier overall survival curves for favorable and unfavorable outcomes using an 8-gene *k*-nearest neighbor analysis. Five-year overall survivals were 80% for the favorable group, compared with 17% for the unfavorable group. Middle and bottom panels, 50 genes most highly associated with survival or with treatment failure. Columns represent individual tumor samples, and rows represent individual genes. Tumor samples are further sorted according to their membership in the two unsupervised self-organizing maps-derived clusters (C0, C1). The first four genes in each group were the ones used in the 8-gene *k*-nearest neighbor analysis. For each gene, red indicates a high level of expression relative to the mean, while blue indicates a low expression. The standard deviation (s) from the mean is indicated. [Adapted by permission from Macmillan Publishers Ltd: *Nature* 415(6870):436–442, copyright 2002.]

diseases with either childhood or adult-onset, and this testing may offer direct, indirect, or no benefit to the child. If there are treatments or preventative measures available for the condition being tested, predictive genetic testing of children should be offered, and, in some cases, strongly advised (Hook et al., 2004), similar to the considerations given to neonatal screening. If there are no treatments or preventative measures available, or the condition is adult-onset, testing of children becomes more controversial (Khoury et al., 2003), as is the testing of children for carrier status. Here the need-to-know of the child and parents must be weighed against the adverse effects of a harmful diagnosis on family relationships and the child's self-esteem and developing social identity (Wertz et al., 1994). In addition, mature adolescents who desire genetic testing for future reproductive planning or other reasons must be considered differently than younger children who have not yet reached the stage of formal operational thought and thus are unable to give assent. In any case, thorough genetic counseling is a necessity for any genetic testing of children, and should include the child, the parents, and the siblings. In most of these situations, genetic testing of children should be deferred until the child reaches adulthood or can make reproductive decisions (Hook et al., 2004).

Summary

The application of genomic medicine approaches to the care of children is not as advanced as for adults because of the relatively small numbers of affected individuals for pediatric as opposed to adult disorders, as well as because of the unique ethical and social implications for such application in the care of children. As experience with genomic medicine is gained by its application in adults, and as genomic medicine in general becomes acceptable to society, the approaches of genomic medicine would be expected to play an increasing role in the care of children, as for adults.

References

Abe, J, Jibiki, T, Noma, S, Nakajima, T, Saito, H, and Terai, M (2005). Gene expression profiling of the effect of high-dose intravenous Ig in patients with Kawasaki disease. *J Immunol*, 174(9):5837–5845.

Anonymous Editorial (2001). Defining a new bioethic. *Nat Genet*, 28 (4):297–298.

Armstrong, MD, Low, NL, and Bosma, JF (1957). Studies on phenylketonuria. IX. Further observations on the effect of phenylalanine-restricted diet on patients with phenylketonuria. *Am J Clin Nutr*, 5(5):543–554.

Bailey, DB, Jr. (2004). Newborn screening for fragile X syndrome. *Ment Retard Dev Disabil Res Rev*, 10(1):3–10.

Barker, JM, Barriga, KJ, Yu, L, Miao, D, Erlich, HA, Norris, JM, et al. (2004). Prediction of autoantibody positivity and progression to type 1 diabetes: Diabetes Autoimmunity Study in the Young (DAISY). *J Clin Endocrinol Metab*, 89(8):3896–3902.

Barnes, MG, Aronow, BJ, Luyrink, LK, Moroldo, MB, Pavlidis, P, Passo, MH, et al. (2004). Gene expression in juvenile arthritis and spondyloarthropathy: pro-angiogenic ELR$^+$ chemokine genes relate to course of arthritis. *Rheumatology (Oxford)*, 43(8):973–979.

Barrow, RE, Dasu, MR, Ferrando, AA, Spies, M, Thomas, SJ, Perez-Polo, et al. (2003). Gene expression patterns in skeletal muscle of thermally injured children treated with oxandrolone. *Ann Surg*, 237(3):422–428.

Bates, MD (2003). The potential of DNA microarrays for the care of children. *J Pediatr*, 142(3):235–239.

Beauchamp, TL and Childress, JF (1994). *Principles of Biomedical Ethics*,. New York:Oxford University Press. Inc.

Bennett, L, Palucka, AK, Arce, E, Cantrell, V, Borvak, J, Banchereau, et al. (2003). Interferon and granulopoiesis signatures in systemic lupus erythematosus blood. *J Exp Med*, 197(6):711–723.

Bezerra, JA, Tiao, G, Ryckman, FC, Alonso, M, Sabla, GE, Shneider, et al. (2002). Genetic induction of proinflammatory immunity in children with biliary atresia. *Lancet*, 360(9346):1653–1659.

Blanchard, C, Wang, N, Stringer, KF, Mishra, A, Fulkerson, PC, Abonia, JP, et al. (2006). Eotaxin-3 and a uniquely conserved gene-expression profile in eosinophilic esophagitis. *J Clin Invest*, 116(2):536–547.

Borra, RC, Andrade, PM, Silva, ID, Morgun, A, Weckx, LL, Smirnova, AS, et al. (2004). The Th1/Th2 immune-type response of the recurrent aphthous ulceration analyzed by cDNA microarray. *J Oral Pathol Med*, 33 (3):140–146.

Call, DR, Brockman, FJ, and Chandler, DP (2001). Detecting and genotyping *Escherichia coli* O157:H7 using multiplexed PCR and nucleic acid microarrays. *Int J Food Microbiol*, 67(1–2):71–80.

Chace, DH, Kalas, TA, and Naylor, EW (2003). Use of tandem mass spectrometry for multianalyte screening of dried blood specimens from newborns. *Clin Chem*, 49(11):1797–1817.

Clayton, EW (2003). Ethical, legal, and social implications of genomic medicine. *N Engl J Med*, 349(6):562–569.

Colantuoni, C, Jeon, OH Hyder, K, Chenchik, A, Khimani, AH, Narayanan, V, et al. (2001). Gene expression profiling in postmortem Rett Syndrome brain: differential gene expression and patient classification. *Neurobiol Dis*, 8(5):847–865.

Crimi, M, Bordoni, A, Menozzi, G, Riva, L, Fortunato, F, Galbiati, S, et al. (2005). Skeletal muscle gene expression profiling in mitochondrial disorders. *FASEB J*, 19(7):866–868.

Dasu, MR, Barrow, RE, Hawkins, HK, and McCauley, RL (2004). Gene expression profiles of giant hairy naevi. *J Clin Pathol*, 57(8):849–855.

Dasu, MR, Barrow, RE, and Herndon, DN (2005). Gene expression changes with time in skeletal muscle of severely burned children. *Ann Surg*, 241 (4):647–653.

Dieckgraefe, BK, Stenson, WF, Korzenik, JR, Swanson, PE, and Harrington, CA (2000). Analysis of mucosal gene expression in inflammatory bowel disease by parallel oligonucleotide arrays. *Physiol Genomics*, 4(1):1–11.

Diosdado, B, Wapenaar, MC, Franke, L, Duran, KJ, Goerres, MJ, Hadithi, M, et al. (2004). A microarray screen for novel candidate genes in coeliac disease pathogenesis. *Gut*, 53(7):944–951.

Dobrowolski, SF, Banas, RA, Naylor, EW, Powdrill, T, and Thakkar, D (1999). DNA microarray technology for neonatal screening. *Acta Paediatr Suppl*, 88 (432):61–64.

Dussault, JH, Coulombe, P, Laberge, C, Letarte, J, Guyda, H, and Khoury, K (1975). Preliminary report on a mass screening program for neonatal hypothyroidism. *J Pediatr*, 86(5):670–674.

European Community Huntington's Disease Collaborative Study Group (1993). Ethical and social issues in presymptomatic testing for Huntington's disease: a European Community collaborative study. *J Med Genet*, 30 (12):1028–1035.

Fearing, MK and Marsden, D (2003). Expanded newborn screening. *Pediatr Ann*, 32(8):509–515.

Fernandez-Teijeiro, A, Betensky, RA, Sturla, LM, Kim, JY, Tamayo, P, and Pomeroy, SL, (2004). Combining gene expression profiles and clinical parameters for risk stratification in medulloblastomas. *J Clin Oncol*, 22 (6):994–998.

Garrick, MD, Dembure, P, and Guthrie, R (1973). Sickle-cell anemia and other hemoglobinopathies. Procedures and strategy for screening employing spots of blood on filter paper as specimens. *N Engl J Med*, 288 (24):1265–1268.

Gingeras, TR, Ghandour, G, Wang, E, Berno, A, Small, PM, Drobniewski, F, et al. (1998). Simultaneous genotyping and species identification using hybridization pattern recognition analysis of generic *Mycobacterium* DNA arrays. *Genome Res*, 8(5):435–448.

Gratias, S, Schuler, A, Hitpass, LK, Stephan, H, Rieder, H, Schneider, S, et al. (2005). Genomic gains on chromosome 1q in retinoblastoma: consequences on gene expression and association with clinical manifestation. *Int J Cancer*, 116(4):555–563.

Greely, HT (2005). Banning genetic discrimination. *N Engl J Med*, 353 (9):865–867.

Green, NS, Dolan, SM, and Oinuma, M (2004). Implementation of newborn screening for cystic fibrosis varies widely between states. *Pediatrics*, 114 (2):515–516.

Green, NS and Pass, KA (2005). Neonatal screening by DNA microarray: spots and chips. *Nat Rev Genet*, 6(2):147–151.

Grody, WW, Cutting, GR, Klinger, KW, Richards, CS, Watson, MS, and Desnick, RJ (2001). Laboratory standards and guidelines for population-based cystic fibrosis carrier screening. *Genet Med*, 3(2):149–154.

Grosse, SD (2005). Does newborn screening save money? The difference between cost-effective and cost-saving interventions. *J Pediatr*, 146 (2):168–170.

Gruters, A, Krude, H, and Biebermann, H (2004). Molecular genetic defects in congenital hypothyroidism. *Eur J Endocrinol*, 151. (Suppl. 3):U39–U44.

Guthrie, R (1968). Screening for "inborn errors of metabolism" in the newborn infant—a multiple test program. *Birth Def Orig Art Ser*, 4:92–98.

Guthrie, R (1992). The origin of newborn screening. *Screening*, 1:5–15.

Hacia, JG (1999). Resequencing and mutational analysis using oligonucleotide microarrays. *Nat Genet*, 21(Suppl. 1):42–47.

Hacia, JG and Collins, FS (1999). Mutational analysis using oligonucleotide microarrays. *J Med Genet*, 36(10):730–736.

Hall, MA and Rich, SS (2000). Laws restricting health insurers' use of genetic information: impact on genetic discrimination. *Am J Hum Genet*, 66 (1):293–307.

Hanson, EH, Niemeyer, DM, Folio, L, Agan, BK, and Rowley, RK (2004). Potential use of microarray technology for rapid identification of central nervous system pathogens. *Mil Med*, 169(8):594–599.

Hashimoto, T, Lawton, MT, Wen, G, Yang, GY, Chaly, T, Jr, Stewart, CL, et al. (2004). Gene microarray analysis of human brain arteriovenous malformations. *Neurosurgery*, 54(2):410–423.

Heard, GS, Secor McVoy, JR, and Wolf, B (1984). A screening method for biotinidase deficiency in newborns. *Clin Chem*, 30(1):125–127.

Heath, EM, O'Brien, DP, Banas, R, Naylor, EW, and Dobrowolski, S (1999). Optimization of an automated DNA purification protocol for neonatal screening. *Arch Pathol Lab Med*, 123(12):1154–1160.

Hiyama, E, Hiyama, K, Yamaoka, H, Sueda, T, Reynolds, CP, and Yokoyama, T (2004). Expression profiling of favorable and unfavorable neuroblastomas. *Pediatr Surg Int*, 20(1):33–38.

Hook, CC, DiMagno, EP, and Tefferi, A (2004). Primer on medical genomics. Part XIII: ethical and regulatory issues. *Mayo Clin Proc*, 79(5):645–650.

Horner, FA and Streamer, CW (1959). Phenylketonuria treated from earliest infancy; report of three cases. *AMA J Dis Child*, 97(3):345–347.

International Society for Neonatal Screening (2003). Neonatal screening in Europe, data 2003. Available: http://www.isns-neoscreening.org/pdf/Europeanscreening2003overview2.PDF [accessed February 1, 2006].

Jensen, UG, Brandt, NJ, Christensen, E, Skovby, F, Norgaard-Pedersen, B, and Simonsen, H (2001). Neonatal screening for galactosemia by quantitative analysis of hexose monophosphates using tandem mass spectrometry: a retrospective study. *Clin Chem*, 47(8):1364–1372.

Kalman, L, Lindegren, ML, Kobrynski, L, Vogt, R, Hannon, H, Howard, JT, et al. (2004). Mutations in genes required for T-cell development: IL7R, CD45, IL2RG, JAK3, RAG1, RAG2, ARTEMIS, and ADA and severe combined immunodeficiency: HuGE review. *Genet Med*, 6(1):16–26.

Kerr, JR, Kaushik, N, Fear, D, Baldwin, DA, Nuwaysir, EF, and Adcock, IM (2005). Single-nucleotide polymorphisms associated with symptomatic infection and differential human gene expression in healthy seropositive persons each implicate the cytoskeleton, integrin signaling, and oncosuppression in the pathogenesis of human parvovirus B19 infection. *J Infect Dis*, 192(2):276–286.

Khoury, MJ, McCabe, LL, and McCabe, ER (2003). Population screening in the age of genomic medicine. *N Engl J Med*, 348(1):50–58.

Konstantinov, IE, Coles, JG, Boscarino, C, Takahashi, M, Goncalves, J, Ritter, J, et al. (2004). Gene expression profiles in children undergoing cardiac surgery for right heart obstructive lesions. *J Thorac Cardiovasc Surg*, 127(3):746–754.

Laassri, M, Dragunsky, E, Enterline, J, Eremeeva, T Ivanova, O, Lottenbach, K, et al. (2005). Genomic analysis of vaccine-derived poliovirus strains in stool specimens by combination of full-length PCR and oligonucleotide microarray hybridization. *J Clin Microbiol*, 43(6):2886–2894.

Lacey, JM, Minutti, CZ, Magera, MJ, Tauscher, AL, Casetta, B, McCann, M, et al. (2004). Improved specificity of newborn screening for congenital adrenal hyperplasia by second-tier steroid profiling using tandem mass spectrometry. *Clin Chem*, 50(3):621–625.

Lapham, EV, Kozma, C, and Weiss, JO (1996). Genetic discrimination: perspectives of consumers. *Science*, 274(5287):621–624.

Levy, HL (2003). Lessons from the past—looking to the future. Newborn screening. *Pediatr Ann*, 32(8):505–508.

Lightfoot, J, Hitzler, JK, Zipursky, A, Albert, M, and Macgregor, PF (2004). Distinct gene signatures of transient and acute megakaryoblastic leukemia in Down syndrome. *Leukemia*, 18(10):1617–1623.

Lipshutz, RJ, Fodor, SP, Gingeras, TR, and Lockhart, DJ (1999). High density synthetic oligonucleotide arrays. *Nat Genet*, 21(Suppl. 1):20–24.

Loeber, G, Webster, D, and Aznarez, A(1999). Quality evaluation of newborn screening programs. *Acta Paediatr Suppl*, 88(432):3–6.

Mendell, JT and Dietz, HC (2001). When the message goes awry: disease-producing mutations that influence mRNA content and performance. *Cell*, 107(4):411–414.

Moos, PJ, Raetz, EA, Carlson, MA, Szabo, A, Smith, FE, Willman, C, et al. 2002). Identification of gene expression profiles that segregate patients with childhood leukemia. *Clin Cancer Res*, 8(10):3118–3130.

National Newborn Screening & Genetics Resource Center (2005). National Newborn Screening Status Report: U.S. National Screening Status Report, Updated 01/10/06. Available: http://genes-r-us.uthscsa.edu/nbsdisorders.pdf [accessed February 1, 2006].

Naylor, EW and Guthrie, R (1978). Newborn screening for maple syrup urine disease (branched-chain ketoaciduria). *Pediatrics*, 61(2):262–266.

Nikoloff, D, Shim, JC, Fairchild, M, Patten, N, Fijal, BA, Koch, et al. (2002). Association between CYP2D6 genotype and tardive dyskinesia in Korean schizophrenics. *Pharmacogenomics J*, 2(6):400–407.

Noguchi, S, Fujita, M, Murayama, K, Kurokawa, R, and Nishino, I (2005). Gene expression analyses in X-linked myotubular myopathy. *Neurology*, 65 (5):732–737.

Noguchi, S, Tsukahara, T, Fujita, M, Kurokawa, R, Tachikawa, M., Toda, T, et al. (2003). cDNA microarray analysis of individual Duchenne muscular dystrophy patients. *Hum Mol Genet*, 12(6):595–600.

Nomura, I, Abe, J, Noma, S, Saito, H, Gao, B, Wheeler, G, et al. (2005). Adrenomedullin is highly expressed in blood monocytes associated with acute Kawasaki disease: a microarray gene expression study. *Pediatr Res* 57(1):49–55.

Ohali, A, Avigad, S, Zaizov, R, Ophir, R, Horn-Saban, S, Cohen, IJ, et al. (2004). Prediction of high risk Ewing's sarcoma by gene expression profiling. *Oncogene*, 23(55):8997–9006.

Paakkonen, V, Ohlmeier, S, Bergmann, U, Larmas, M, Salo, T, and Tjaderhane, L (2005). Analysis of gene and protein expression in healthy and carious tooth pulp with cDNA microarray and two-dimensional gel electrophoresis. *Eur J Oral Sci*, 113(5):369–379.

Paigen, K, Pacholec, F, and Levy, HL (1982). A new method of screening for inherited disorders of galactose metabolism. *J Lab Clin Med*, 99(6):895–907.

Pang, S, Hotchkiss, J, Drash, AL, Levine, LS, and New, MI (1977). Microfilter paper method for 17 alpha-hydroxyprogesterone radioimmunoassay: its application for rapid screening for congenital adrenal hyperplasia. *J Clin Endocrinol Metab*, 45(5):1003–1008.

Paulson, GW and Prior, TW (1997). Issues related to DNA testing for Huntington's disease in symptomatic patients. *Semin Neurol*, 17(3):235–238.

Pomeroy, SL, Tamayo, P, Gaasenbeek, M, Sturla, LM, Angelo, M, McLaughlin, et al. (2002). Prediction of central nervous system embryonal tumour outcome based on gene expression. *Nature*, 415(6870):436–442.

Ross, LF (2002). Predictive genetic testing for conditions that present in childhood. *Kennedy Inst Ethics J*, 12(3):225–244.

Ross, ME, Mahfouz, R, Onciu, M, Liu, HC, Zhou, X, Song, G, et al. (2004). Gene expression profiling of pediatric acute myelogenous leukemia. *Blood*, 104 (12):3679–3687.

Sarwal, M, Chua, MS, Kambham, N, Hsieh, SC, Satterwhite, T Masek, M, et al. (2003). Molecular heterogeneity in acute renal allograft rejection identified by DNA microarray profiling. *Engl J Med*, 349(2):125–138.

Schwab, K, Witte, DP, Aronow, BJ, Devarajan, P, Potter, SS, and Patterson, LT (2004). Microarray analysis of focal segmental glomerulosclerosis. *Am J Nephrol*, 24(4):438–447.

Stephenson, T (2005). How children's responses to drugs differ from adults. *Br J Clin Pharmacol*, 59(6):670–673.

Suarez-Merino, B, Hubank, M, Revesz, T, Harkness, W, Hayward, R, Thompson, D, et al. (2005). Microarray analysis of pediatric ependymoma identifies a cluster of 112 candidate genes including four transcripts at 22q12.1-q13.3. *Neuro-oncol*, 7(1):20–31.

Sun, YH, Bakshi, S, Chalmers, R, and Tang, CM (2000). Functional genomics of *Neisseria meningitidis* pathogenesis. *Nat Med*, 6(11):1269–1273.

Tanaka, T, Takada, H, Nomura, A, Ohga, S, Shibata, R, and Hara, T (2005). Distinct gene expression patterns of peripheral blood cells in hyper-IgE syndrome. *Clin Exp Immunol*, 140(3):524–531.

Tang, Y, Gilbert, DL, Glauser, TA, Hershey, AD, and Sharp, FR (2005). Blood gene expression profiling of neurologic diseases: a pilot microarray study. *Arch Neurol*, 62(2):210–215.

Tang, Y, Lu, A, Ran, R, Aronow, BJ, Schorry, EK, Hopkin, RJ, et al. (2004). Human blood genomics: distinct profiles for gender, age and neurofibromatosis type 1. *Brain Res Mol Brain Res*, 132(2):155–167.

Tester, DJ, Kopplin, LJ, Creighton, W, Burke, AP, and Ackerman, MJ (2005). Pathogenesis of unexplained drowning: new insights from a molecular autopsy. *Mayo Clin Proc*, 80(5):596–600.

Thompson, MA, Stumph, J, Henrickson, SE, Rosenwald, A, Wang, Q, Olson, S, et al. (2005). Differential gene expression in anaplastic lymphoma kinase-positive and anaplastic lymphoma kinase-negative anaplastic large cell lymphomas. *Hum Pathol*, 36(5):494–504.

Tromp, G, Kuivaniemi, H, Romero, R, Chaiworapongsa, T, Kim, YM, Kim, MR, et al. (2004). Genome-wide expression profiling of fetal membranes reveals a deficient expression of proteinase inhibitor 3 in premature rupture of membranes. *Am J Obstet Gynecol*, 191(4):1331–1338.

Tsou, R, Cole, JK, Nathens, AB, Isik, FF, Heimbach, DM, Engrav, LH, et al. (2000). Analysis of hypertrophic and normal scar gene expression with cDNA microarrays. *J Burn Care Rehabil*, 21(6):541–550.

Wertz, DC, Fanos, JH, and Reilly, PR (1994). Genetic testing for children and adolescents. Who decides? *JAMA*, 272(11):875–881.

Williams, RD, Hing, SN, Greer, BT, Whiteford, CC, Wei, JS, Natrajan, R, et al. (2004). Prognostic classification of relapsing favorable histology Wilms tumor using cDNA microarray expression profiling and support vector machines. *Genes Chromosomes Cancer*, 41(1):65–79.

Wilson, JMG and Jungner, G (1968). Principles and practice of screening for disease. Public Health Papers (WHO), no. 34. World Health Organization, Geneva.

Wion, E, Brantley, M, Stevens, J, Gallinger, S, Peng, H, Glass, M, et al. (2003). Population-wide infant screening for HLA-based type 1 diabetes risk via dried blood spots from the public health infrastructure. *Ann NY Acad Sci*, 1005:400–403.

Wong, KK, Chang, YM, Tsang, YT, Perlaky, L, Su, J, Adesina, A, et al. (2005). Expression analysis of juvenile pilocytic astrocytomas by oligonucleotide microarray reveals two potential subgroups. *Cancer Res*, 65 (1):76–84.

Wusthoff, C (2003). The dilemma of confidentiality in Huntington disease. *JAMA*, 290(9):1219–1220.

Yagi, T, Morimoto, A, Eguchi, M, Hibi, S, Sako, M, Ishii, E, et al. (2003). Identification of a gene expression signature associated with pediatric AML prognosis. *Blood*, 102(5):1849–1856.

29

Learning and Behavioral Disorders

Lucy Raymond and James Cox

At present, compared to many areas of medicine, relatively little is known about the underlying cellular mechanisms that lead to learning and behavioral disorders in humans. The reasons for this are many, including social, political, medical, and practical issues. The advent of the Human Genome Project and developments in molecular genetics over the last 20 years now means that, from a practical point of view, this area can be considered in detail, which was not possible previously.

Why should one even consider trying to understand the genetic basis of disorders of learning or behavior? The overwhelming reason is that families that are caring for children with severe disabilities wish to understand how and why this has happened to their children. If a mechanism of disease can be identified, there is an opportunity to understand and to clarify the cause of the problems. Knowledge of the disease can also lead to a greater understanding of the disease and provide prognostic information. In many cases, having a diagnosis means that the lengthy search for the cause of the disease can now cease, and further unnecessary investigations can be avoided. Also, the identification of a specific diagnosis may mean that other medical systems, e.g., cardiac or nutritional status, now need closer surveillance than would have been done if the diagnosis was not made. This can prevent further morbidity and avoid early mortality.

This chapter focuses on the identification of novel genes and abnormalities of these genes that cause mental retardation (MR). In this chapter, the terms MR, cognitive impairment, intellectual disability, and learning disability are used interchangeably. The audience will interpret this from their own cultural and academic background, and the language and terminology may not be acceptable to some readers. This is unavoidable, since the literature uses different words according to its origin.

The genetic basis of behavioral disorders is understood even less well than that of learning disability. However, close analysis of a few conditions with striking behavioral phenotypes give us astonishing insights into the workings of the human brain. The areas that are covered are speech and language disorders, dyslexia, autism, and the commoner microdeletion syndromes, and defects in single genes that are associated with unusual behavior.

Mental Retardation

MR is a developmental disability that manifests during childhood and is characterized by deficiencies in both cognitive abilities and adaptive behavior (such as daily living, social, and communication skills) (Stevenson et al., 2000; Chelly and Mandel, 2001; Ropers et al., 2003). In practical terms, MR is defined as an intelligence quotient (IQ) of less than 70 in standardized cognitive tests. On the basis of IQ scores, MR can be subdivided into mild MR (IQ 50–70), moderate MR (IQ 35–49), severe MR (IQ 20–34), and profound MR (IQ < 20).

Causes of MR: Epidemiology

The underlying causes of MR are extremely heterogeneous, as the developing brain is susceptible to a wide variety of environmental and genetic insults (Stevenson et al., 2000; Chelly and Mandel, 2001). In 1995, the consensus conference, held under the auspices of the American College of Medical Genetics, undertook a literature review of nine surveys of the intellectually disabled that were carried out between 1977 and 1994 (Curry et al., 1997). This review enabled a general distribution of the causes of MR to be determined. Table 29–1 summarizes their findings, together with a more recent survey of 429 individuals with MR carried out as part of the Australian Child and Adolescent (ACAD) study (Partington et al., 2000). Overall, MR (IQ < 70) is a common condition that affects approximately 1%–1.5% of the population (McLaren and Bryson, 1987). Moderate-to-profound MR (IQ < 50) is estimated to affect 0.3%–0.5% of the population. The causes of MR are divided into environmental and genetic, although both these causes frequently coexist in an individual. Environmental exposure divides into prenatal, perinatal, and postnatal exposure. The long-term consequences of extreme prematurity and the associated medical complications accounts for an increasingly significant proportion of children with MR, whereas the proportion of children suffering from postnatal infections is diminishing (Stevenson et al., 2000). Perinatal injury due to birth problems remains common, as does the prenatal exposure to teratogens during pregnancy. These include fetal exposure to sodium valproate and other anticoagulants, alcohol, and high levels of blood glucose in diabetic mothers. A small proportion of children also suffer from accidents or infections beyond the postnatal period that result in MR. The genetic contribution to MR is high, as empiric recurrence risks for siblings of severely affected individuals are 5%–8% if a single case is observed and 12%–15% if two siblings are affected (Bundey et al., 1985; Crow and Tolmie, 1998). These figures include both chromosomal abnormalities and single gene defects. Chromosomal abnormalities are collectively the single most frequent identifiable cause of MR (Table 29–1). The most common

Table 29–1 Comparison of the Causes of Mental Retardation Reported by the Consensus Conference (1997) and the Australian Child and Adolescent Development (ACAD) Study (2000)

Category/Group	Consensus Conference (%)	ACAD Study (%)
Chromosomal abnormalities	4–28	21
Down syndrome	12.9–16.1	15
Recognizable syndromes	3–	
Provisional unique syndromes	1–	
Known monogenic conditions	3–	
Structural central nervous system (CNS) abnormalities	7–17	6
Complications of prematurity	2–10	8
Environmental/teratogenic	5–13	8
Metabolic/endocrine	1–	
Unknown	30–50	46
Male/female ratio (all categories)	1.35–1.4	1.38
Male/female ratio (unknown diagnosis subgroup)	—	1.77

Source: Adapted from Partington et al., 2000.

chromosomal cause of MR is trisomy of chromosome 21 that underlies Down syndrome. Other abnormalities associated with MR include autosomal and sex chromosome aneuploidies, partial chromosomal deletions and duplications, translocations, insertions, inversions, ring chromosomes, and uniparental disomies. Single gene disorders both as recognized syndromes and known monogenic disorders collectively include a significant portion (7%–21%) of the identified causes of MR.

A striking statistic from Table 29–1 is that, in many cases (up to 40%), the underlying etiology remains unknown, but this figure is gradually reducing with improved technology and understanding.

The relative excess of males in the population with severe MR (1.3–1.7:1) suggests that X-linked disease genes are a significant contribution to the overall genetic etiology (Penrose, 1938; Stevenson et al., 2000). Recent studies using modern cytogenetic techniques, to exclude chromosome abnormalities, suggest that where two male siblings are severely retarded, in the absence of a chromosome abnormality, the likelihood of this being due to an X-linked gene abnormality is as high as 80% (Turner and Partington, 2000). Although a diagnosis (genetic or environmental) cannot be reached in approximately 40% of patients with an IQ less than 50, this figure rises steeply to approximately 76% for patients with mild MR (IQ 50–70) (Flint and Wilkie, 1996). Multifactorial causation is assumed to be responsible for most cases of mild MR (Stevenson et al., 2000).

Genetic Causes of MR

Many chromosome abnormalities are detected by the visual inspection of the whole genome at the resolution of approximately 5–10 Mb. This will detect large gain or loss of chromosome material and rearrangements that are almost always of clinical significance. Although the technique of karyotyping is ultimately limited by the resolution of the microscope, the quality of the chromosome preparations have gradually improved over the last 10 years, and reevaluating a patient's chromosomes where a chromosome abnormality is suspected is well worthwhile.

There are recurrent small microdeletions of the genome that are associated with characteristic syndromes. Routine chromosome analysis would appear normal, as these microdeletions are too small to be detected by G-banding and light microscopy, and would only be detected using specific genomic deoxyribonucleic acid (DNA) fluorescent probes that fail to bind where there is deletion. The history of detecting microdeletion syndromes started with the recognition of discrete syndromic phenotypes associated with MR, such as Prader–Willi, Miller–Dieker, Angelman's, and Williams syndromes. Patients with similar phenotypes were then found to have similar submicroscopic deletions in discrete genomic regions that then defined the condition (Ledbetter et al., 1981; Schwartz et al., 1988; Knoll et al., 1989; Ewart et al., 1993). Since then, a large number of microdeletion syndromes have been described, each with a discrete phenotype and microdeletion, e.g., Rubinstein–Taybi, Smith–Magenis, Williams, DiGeorge or 22q11 deletion syndrome. Of note is that most of the microdeletion syndromes are associated with a degree of MR. The spectrum of disability depends on the specific deletion syndrome and the extent of the deletion. The intellectual disability in 22q11 deletion syndrome is frequently mild and may be unrecognized clinically, whereas the degree of disability in Rubinstein–Taybi and Angelman's syndromes is always very significant. Larger deletions in any one area tend to be associated with more severe MR than small deletions as more genes are deleted, but the correlation entirely depends on the nature of the genes within the deleted region. In Williams syndrome, although there is a general reduction in intellectual ability, the affected individuals are frequently described as being excessively sociable and have relatively preserved expressive speech and language. In contrast, a patient with a 1.5-Mb duplication that includes the minimum Williams syndrome critic region has recently been reported where expressive speech and language is severely delayed relative to the global intellectual problems, suggesting that a locus exists within 7q11.23 that is responsible for speech and language development and is exquisitely sensitive to gene dosage (Somerville et al., 2005).

The observation that microdeletions within the genome results in MR as a near universal feature led Flint et al. in 1995 to develop a more systematic screening of the genome for microdeletions in a cohort of patients with MR. They screened for the abnormal inheritance of subtelomeric DNA polymorphisms in individuals with MR (Flint et al., 1995). They found that three out of ninety-nine patients had an abnormality at the end of one of the chromosomes. This initial experiment did not screen all ends of the chromosomes, telomeres, as not all were easily studied; however, the technology has since been developed to provide this as a clinical service for suitably selected patients, using FISH (fluorescent in situ hybridization) on a regular basis. The range of diagnostic yield is approximately 3.5%–11% (Knight et al., 1999; Slavotinek et al., 1999; Flint and Knight, 2003). The development of subtelomeric screening of patients with MR has resulted from our increased knowledge of the genome arrangement and from the direct use of the bacterial artificial chromosome (BAC) clones that were used to generate the sequencing tiling path necessary to establish the human genome sequence.

As a result of more systematic screening of the human genome in patients with MR, new syndromes have emerged in which the identification of the deletion initially defined the condition. Where many cases are reported, the clinical picture of the condition emerges and gradually assumes the status of recognizable syndromes, such as 1p36 or 2q37.3 microdeletion syndromes, as common clinical features have emerged once sufficient patients with the same deletion have been collected together (Shapira et al., 1997; Heilstedt et al., 2003; Aldred et al., 2004). However, several of the subtelomeric deletions have been reported only rarely to date, and the delineation of the associated clinical phenotype is therefore much more difficult and has only been elucidated recently such a 3q29 microdeletion syndrome (Willatt et al., 2005).

Genomic Microdeletions and Duplications

Therefore, the principle is well established that small deletions or duplications of chromosome material can lead to MR. The recent challenge has been to identify small deletions and duplications in the whole genome in a single or few procedures. The simultaneous use of multiple probes is now possible using probes of known location on the genome 1 Mb apart (Veltman et al., 2002). Recent publications have established that a further 10% of patients with MR carry deletions or duplications (Vissers et al., 2003; Shaw-Smith et al., 2004). Although this technique is not yet in routine clinical practice, it is likely to be so soon as the diagnostic yield is approaching twice that of a routine karyotype analysis (de Vries et al., 2005). Having clones located 1 Mb apart means by definition that small deletions or duplications will not be detected. The possibility remains that a much larger number of patients have MR due to these small intragenic abnormalities. Recently, a set of probes covering the complete sequence of the X chromosome has been developed, and the identification of a recurrent duplication in Xq28 has been identified as responsible for severe MR in two families (Veltman et al., 2004; Lugtenberg et al., 2006).

Not only has the Human Genome Project enabled new syndromes to be identified, the increasing knowledge of the sequence has led to a furthering of our understanding of the molecular mechanism by which duplications and deletions occur. The presence of low copy repeat sequences at and near the location of deletion breakpoints is thought to predispose the genome to acquire recurrent deletion events by nonhomologous recombination (Shaw et al., 2004; Willatt et al., 2005). This is emerging as a common feature underlying the mechanism of formation of deletion and duplication syndromes. It is also emerging that the genome contains many polymorphic copy number variants (CNVs) where deletions and duplications of regions throughout the genome are not thought to be of any pathological significance (Sebat et al., 2004; Sharp et al., 2005). This data will give further insight into human variation and help when interpreting observed copy number changes in families with MR (Conrad et al., 2006; Hinds et al., 2006). It is emerging that variation within the genome may not itself be pathological, but rearrangements may predispose to subsequent deletions in subsequent generations. In patients with Williams syndrome, a recent observation showed that the chromosome carrying the de novo deletion is frequently carrying a small genomic inversion, suggesting that the presence of one relatively benign genomic variation in a parent may predispose to an additional pathological variation in a subsequent generation (Osborne et al., 2001). It seems that natural variation in the genome may predispose to pathological changes in subsequent generations and further examples of this will emerge in the literature in the near future. (see also Chapter 7).

Identification of Single Genes that Cause MR

The Human Genome Project has had a significant impact on the rate of identification of genes that cause MR. The focus has been the identification of novel genes on the X chromosome, as this is thought to be where the bulk of the genes lie, based on the observation of male excess.

A consistent finding from surveys of the mentally retarded is an excess of males over females, in a proportion of approximately three to two (male/female ratio of 1.38) (Penrose, 1938; Penrose, 1949; Partington et al., 2000). This male excess is very apparent (male/female ratio of 1.77) in the group of patients in which a diagnosis cannot be made (Table 29–1). Historically, a variety of explanations have been proposed for the excess of males observed in the mentally retarded population, such as (1) "the greater size of the male head exposing the infant to greater difficulties and injuries during labour"; (2) bias of ascertainment, with more males likely to come to the attention of the authorities because affected males "are more difficult to manage in the house than affected females" or because parents seek assistance more frequently for boys than for girls because of different expectations for males; (3) greater mortality among females with MR; and (4) hormonal contributions to the causation of MR (Ireland, 1900; Dewey et al., 1965; Nance and Engel, 1972; Partington et al., 2000; Stevenson et al., 2000). However, none of these explanations have been substantiated.

In 1972, it was proposed by Lehrke that major genes related to intellectual function were located on the X chromosome, and that the male excess was due to differences in the sex chromosome constitution between males and females; that is, a mutant gene on the X chromosome could result in MR in a male, whereas the presence of a second normal allele on a female's other X chromosome may protect her from being affected (Lehrke, 1972). This hypothesis was initially supported by reports of large families in which MR was segregating in an undoubted X-linked fashion, and then by surveys that described a marked excess of brothers with MR compared to affected sisters (Lehrke, 1972; Turner and Turner, 1974; Herbst and Miller, 1980). The concept of X-linked MR (XLMR) has subsequently been reinforced by the mapping of several well-recognized MR syndromes to the X chromosome and by the cloning of numerous genes for syndromic and nonsyndromic XLMR (NS-XLMR). Analysis of the OMIM (online Mendelian Inheritance in Man) database (http://ncbi.gov.nih/OMIM) indicates that, compared to the autosomes, the X chromosome contains a significantly higher number of genes that, when mutated, cause mental impairment (Zechner et al., 2001). After taking into account the + bias caused by a single gene being responsible for different entries in the OMIM database due to imprecise mapping information and allelic heterogeneity, and also the ascertainment bias caused by the X chromosome being a focus for mapping and identification of genes for mental disability due to the preferential expression of X-linked disorders in hemizygous males, Zechner et al. (2001) reported that mental impairment was still 3.1 times more-frequently associated with X-chromosomal genes than with autosomes.

X-linked Mental Retardation

XLMR is currently recognized to represent up to 5% of all MR, although if XLMR was to explain the approximate 30% excess of males, then it would account for approximately 14% of all MR (Stevenson et al., 2000). The shortfall between the identifiable and the estimated prevalence of XLMR is likely to be explained by difficulties in assigning X-linkage to MR patients (e.g., the family pedigree is small, if the mutation is very recent, or if some carrier females have manifestations).

Historically, XLMR has been subdivided into syndromic and nonsyndromic forms. Syndromic XLMR (MRXS) is defined as MR associated with one or more distinguishing somatic, neurological, behavioral, or metabolic manifestations for which there is good evidence of an X-linked pattern of inheritance. Currently, 137 MRXS disorders have been described for which 30 MRXS genes have been cloned (Chelly and Mandel, 2001; Chiurazzi et al., 2001). Fragile X syndrome (birth prevalence of 1 in 4000–6000 males and 1 in 8000–10,000 females) and Duchenne muscular dystrophy (birth prevalence of 1 in 3500–4500 males) are the most common (Duchenne, 1968; Turner et al., 1996; de Vries et al., 1998). However, most XLMR syndromes are rare, and some are represented by only single families.

NS-XLMR is defined as MR caused by mutations in genes on the X chromosome, which manifests in the absence of other distinguishing somatic, neurological, behavioral, or metabolic findings. NS-XLMR is extremely heterogeneous and has an estimated incidence of 0.9–1.4 in 1000 males (Kerr et al., 1991). Comparison of this incidence to the overall incidence of XLMR of 1.8 males per 1000 shows NS-XLMR to account for approximately two-third of all XLMR (Herbst and Miller, 1980). Collectively, NS-XLMR is much more common than fragile X syndrome (estimated incidence of 0.22 in 1000 males), although each NS-XLMR condition is individually rare (Turner et al., 1996).

The distinction between syndromic and nonsyndromic forms of XLMR is not always clear-cut (Stevenson et al., 2000; Frints et al., 2002). For example, Fragile X syndrome was initially classified as nonsyndromal as no distinguishing features in addition to MR were identified in 11 affected males from a two-generation family (Martin, 1943). However, when the family was restudied several years later, the affected males were found to have prognathism, large ears, and macroorchidism, clinical features typical of Fragile X syndrome (Richards et al., 1981). Hence, once a disease-causing gene is identified, then genotype–phenotype correlations in families that initially appear to have a nonsyndromal form of XLMR may actually show features of syndromic XLMR upon clinical reevaluation. Cases of nonsyndromal XLMR can also be wrongly classified as syndromic XLMR. For example, affected males from a family may have distinctive clinical features that occur by chance or heritable family characteristics that are unrelated to the gene that produced MR.

The recent progress in the cloning of several XLMR genes has shown that mutations in the same gene may be responsible for both nonsyndromic and syndromic forms of XLMR. This further makes the distinction between MRXS and NS-XLMR less straightforward. An interesting example is the Aristaless-related paired-class homeobox (*ARX*) gene. Here, the same 24-bp insertion duplication within exon 2 is found in patients with NS-XLMR, X-linked West syndrome (characterized by infantile spasms, hypsarrhythmias, and MR), and Partington syndrome (an extrapyramidal neurological disorder characterized by dystonic movements of the hands, dysarthria, and MR) (Bienvenu et al., 2002; Stromme et al., 2002). It has been suggested that the phenotypic heterogeneity caused by the same mutation in different families could be due to differences in genetic and environmental backgrounds, which are specific to each family (Bienvenu et al., 2002). Further examples of allelism where mutations at the same locus give rise to both NS-XLMR and MRXS include mutations within *MECP2* (Rett syndrome), *RPS6KA3* (Coffin–Lowry syndrome), and *ATRX* (a-thalassaemia MR syndrome, X-linked) (Gibbons et al., 1995; Trivier et al., 1996; Amir et al., 1999; Merienne et al., 1999; Guerrini et al., 2000; Couvert et al., 2001). The mutations in these genes in NS-XLMR families are presumed to cause only a partial loss of function of the encoded proteins, which could explain the absence of syndromal features (Chelly and Mandel, 2001).

How Many NS-XLMR Genes Exist?

In general, MRXS disorders lend themselves to conventional positional cloning strategies because families that share similar clinical phenotypes can be pooled for linkage analysis to narrow down a candidate interval. In contrast, identification of NS-XLMR genes is less straightforward due to the extensive genetic heterogeneity and the lack of consistent phenotypic features in addition to MR that can be used to distinguish between affected males in different NS-XLMR families (Gedeon et al., 1996b).

NS-XLMR mapping studies using linkage have depended on the analysis of large NS-XLMR families that show clear evidence of MR segregating in an X-linked manner. Such families, which have a total logarithm of the odds (LOD) score of greater than 2, which is statistically significant for disorders with an X-linked mode of inheritance, are assigned an MRX number (MRX1, MRX2, etc.) (Mulley et al., 1992). To date, 78 MRX families have been described. However, the candidate region for each family is very large, usually in the 20–30 cM range, and may contain 100–300 genes (Chelly and Mandel, 2001). Furthermore, linkage results cannot be pooled, even from families in which NS-XLMR genes map to overlapping regions, because such families may carry mutations in different genes. This means that currently the only approach to estimate the number of genes from linkage data is by determining the number of nonoverlapping regional localizations of linkage groups. It is currently recognized that there are up to 15 linkage intervals that may correspond to NS-XLMR genes (Fryns et al., 2000). To date, 21 genes (Table 29–2) have been identified that, when mutated, result in NS-XLMR. However, in most of these genes, mutations have turned out to be very rare, with each gene identified within a linkage interval accounting for only some of the MRX families that map to that region. Although this may be due to limitations in mutation screening, it is more likely that this is due to locus heterogeneity within each region. Extrapolation of these findings suggests that close to 100 different genes might be involved in NS-XLMR, 5–10 times more than previously thought (Herbst and Miller, 1980; Gedeon et al., 1996a; Gecz and Mulley, 2000; Ropers et al., 2003).

Methods of Identifying Candidate Genes for MR

The methods by which the single genes have been identified have included characterization of balanced chromosome translocations and positional cloning using these X chromosomal rearrangements, deletion mapping of rare chromosome abnormalities, linkage analysis of informative families, candidate gene analysis, and systematic screening of genes within a chromosomal region. These will be discussed in the following with illustrations.

Table 29–2 Genes Mutated in Nonsyndromic X-linked Mental Retardation (NS-XLMR)

HUGO Gene Symbol	Gene Name	Locus	Cloning Strategy	Potential Function
NLGN4X	Neuroligin 4, X linked	Xp22.32	Candidate gene	Forms a neuronal intercellular adhesion complex with β-neurexin and is important in synaptogenesis
RPS6KA3	Ribosomal protein S6 kinase, 90 kDa, polypeptide 3	Xp22.12	Candidate gene	Serine/threonine protein kinase functioning in the ras/ERK/CREB pathway. May also play a role in chromatin remodeling
ARX	Aristaless-related homeobox	Xp22.11	Candidate gene	Member of PAX family of homeobox transcription factors. Speculated to regulate essential events during embryogenesis and head development, in particular
IL1RAPL1	Interleukin 1 receptor accessory protein-like 1	Xp21.3	Deletion mapping	Interacts with neuronal calcium sensor-1 (NCS1) and may regulate neurotransmitter release
TM4SF2	Transmembrane 4 superfamily member 2	Xp11.4	Breakpoint cloning	A tetraspanin protein that interacts with integrins and is involved in regulating actin cytoskeleton dynamicsDelete?
ZNF41	Zinc finger protein 41	Xp11.3	Breakpoint cloning	Member of Krueppel-type zinc-finger gene family. Likely to be involved in regulating gene expression
ZNF81	Zinc finger protein 81	Xp11.23	Breakpoint cloning	Member of Krueppel-type zinc-finger gene family. Likely to be involved in regulating gene expression
FTSJ1	FtsJ homolog 1 (*Escherichia coli*)	Xp11.23	Candidate gene	Homolog of *Escherichia coli* RNA methyltransferase FtsJ/RrmJ and may play a role in the regulation of translation
PQBP1	Polyglutamine-binding protein 1	Xp11.23	Candidate gene	Implicated in transcription and mRNA processing
JARID1C	Jumonji AT-rich interactive domain 1C	Xp11.23	Candidate gene	Member of highly conserved ARID protein family. Contains several DNA-binding motifs that link it to transcriptional regulation and chromatin remodeling
OPHN1	Oligophrenin 1	Xq12	Breakpoint cloning	RhoGAP involved in regulating actin cytoskeletal dynamics and dendritic spine length
DLG3	Discs, large homolog 3 (neuroendocrine-dlg, Drosophila)	Xq13.1	Candidate gene	Scaffolding protein in post-synaptic density. Binds NMDA receptors and may be involved in NMDA receptor clustering and trafficking to post-synaptic membrane
ATRX	Alpha thalassemia/mental retardation syndrome X-linked (RAD54 homolog, *S. cerevisiae*)	Xq21.1	Candidate gene	Forms an ATP-dependent chromatin-remodeling complex with a transcription cofactor called Daxx
ACSL4	Acyl-CoA synthetase long-chain family member 4	Xq23	Deletion mapping	Catalyzes the formation of acyl-CoA esters from fatty acids, ATP, and coenzyme A. Brain function/development likely to be disrupted as a result of imbalance between long-chain free fatty acids and esters
PAK3	p21 (CDKN1A)-activated kinase 3	Xq23	Candidate gene	A Rac/Cdc42 effector involved in regulating actin cytoskeleton dynamics/neuronal morphogenesis
AGTR2	Angiotensin II receptor, type 2	Xq23	Breakpoint cloning	Angiotensin II-specific receptor; precise function in brain development unclear
ARHGEF6	Rac/Cdc42 guanine nucleotide exchange factor (GEF) 6	Xq26.3	Breakpoint cloning	Activates Rac1 and Cdc42 Rho GTPases and binds PAK3. Provides a link between integrin-mediated signaling and actin cytoskeletal dynamics
FMR2	Fragile X mental retardation 2	Xq28	Fragile site and deletion	Nuclear protein and transcription activator
SLC6A8	Solute carrier family 6 (neurotransmitter transporter, creatine), member 8	Xq28	Candidate gene	Creatine transporter
MECP2	Methyl-CpG-binding protein 2 (Rett syndrome)	Xq28	Candidate gene	Methyl CpG DNA-binding protein and transcriptional repressor
GDI1	GDP dissociation inhibitor 1	Xq28	Candidate gene	Helps to maintain a soluble pool of RabGDP. May be involved in neurotransmitter release or have a general role in endo/exocytic pathways and thus neuronal morphogenesis

Candidate Gene Approach

Twelve of the twenty-one genes that cause nonsyndromic MR have been identified by the candidate gene approach. Four of these genes (*RPS6KA3*, *ATRX*, *SLC6A8*, and *MECP2*) were previously found to be involved in MRXS and hence were strong candidates for NS-XLMR in families that mapped to the corresponding regions of the X chromosome. *RPS6KA3* is located in Xp22.12 and is mutated in Coffin–Lowry syndrome, which is characterized by facial and digital dysmorphisms, progressive skeletal deformations, and generally severe MR (Trivier et al., 1996). Mutation screening in an NS-XLMR family (MRX19) that mapped to a 42-cM region in Xp22 identified a missense mutation in *RPS6KA3* (Merienne et al., 1999). This mutation reduced but did not abolish the phosphorylation activity of the encoded kinase, which may help explain the absence of Coffin–Lowry syndromal features seen in affected males with this mutation. *ATRX* is mutated in a series of MRXS conditions, which are all characterized by severe MR and facial dysmorphisms including α-thalassemia/MR syndrome, Juberg–Marsidi syndrome, Carpenter–Waziri syndrome, severe MR with spastic paraplegia, Holmes–Gang syndrome, and Smith–Fineman–Myers syndrome (Gibbons et al., 1995; Villard et al., 1996; Abidi et al., 1999; Lossi et al., 1999; Villard et al., 2000). However, *ATRX* mutations have also been identified in NS-XLMR patients with mild-to-moderate MR without the characteristic facial dysmorphisms seen in those MRXS disorders (Yntema, Poppelaars, et al., 2002). *SLC6A8*, a creatine transporter gene, was recently found to be mutated in a novel MRXS disorder resulting from creatine deficiency in the brain (Salomons et al., 2001; Hahn et al., 2002). Patients with this disorder show expressive speech and language delay and epilepsy in association with MR. A high prevalence (2.1%) of *SLC6A8* mutations have now been identified in a recent survey of 288 NS-XLMR patients collected by the European XLMR Consortium, indicating that *SLC6A8* deficiency may be a comparatively common cause of NS-XLMR, although other cohorts have not confirmed thus relatively high prevalence (Rosenberg et al., 2004). *MECP2* was originally identified as the gene mutated in Rett syndrome, but mutations have also been identified in male patients with NS-XLMR (Amir et al., 1999; Orrico et al., 2000; Couvert et al., 2001; Winnepenninckx et al., 2002). Initially, it was reported that the frequency of *MECP2* mutations in the mentally retarded population was comparable to that of CGG expansions in the Fragile X syndrome gene *FMR1*, but a more recent survey indicates that the prevalence of *MECP2* mutations in mentally retarded males was probably overestimated (Couvert et al., 2001; Yntema, Oudakker, et al., 2002).

NLGN4X was identified as a candidate NS-XLMR gene not as a result of being mutated in MRXS but through being mutated in a condition that is frequently associated with MR, namely autism. MR is present in approximately 70% of individuals with autism, and the IQ of half of these individuals is evaluated as less than 50 (Fombonne, 2003). After the identification of a nonsense mutation in *NLGN4X* in a family with two mentally-retarded brothers with autism and Asperger's syndrome, respectively, a 2-bp deletion leading to a premature stop codon was found in affected individuals from a large French family with NS-XLMR with or without autism or pervasive developmental disorder (Jamain et al., 2003; Laumonnier et al., 2004). A mutation in the highly homologous *NLGN3* gene has also been reported in two members of a family with autism, and MR and mutation screening of *NLGN3* may reveal mutations in individuals with NS-XLMR (Jamain et al., 2003).

A further seven genes (*ARX*, *FTSJ1*, *PQBP1*, *JARID1C*, *DLG3*, *PAK3*, and *GDI1*) have been identified by a candidate gene approach. The initial step in identifying these genes has typically been to perform linkage studies to localize the disease gene to a chromosomal subregion. However, the genetic heterogeneity of NS-XLMR means that linkage data from different families cannot be pooled, in turn meaning that mapping intervals have remained comparatively large and usually contain numerous candidate genes to screen. Good candidates include genes that are expressed in regions of the brain known to play a role in learning and memory, such as the hippocampus, and genes with known roles in brain development and function (Olton et al., 1986; Bliss and Collingridge, 1993). However, gene candidature is not so straightforward, because approximately one-third of all human genes are expressed in the brain, and the human genome still contains numerous uncharacterized genes (Takahashi et al., 1995). Nevertheless, *ARX*, *PAK3*, and *GDI1* were selected from linkage intervals as potential good candidates, and mutations in these genes were subsequently identified in patients with NS-XLMR (Allen et al., 1998; Bienvenu et al., 1998; D'Adamo et al., 1998; Bienvenu et al., 2002; Stromme et al., 2002).

A spin-off from the Human Genome Project has been the development of sequencing technology that enables automated high-throughput mutation screening. This technology has recently been put to good use to enable the identification of several novel NS-XLMR genes, *DLG3*, *FTSJ1*, *PQBP1*, *JARID1C*, *AP1S2*, *CUL4B*, *ZDHHC9*, and *MED12* (Kalscheuer et al., 2003; Freude et al., 2004; Ramser et al., 2004; Tarpey et al., 2004; Jensen et al., 2005; Tarpey et al., 2006; Tarpey et al., 2007; Raymond et al., 2007; Schwartz et al., 2007).

Positional Cloning Using X Chromosomal Rearrangements

Three types of X chromosome rearrangement in patients with nonsyndromic MR have proved useful for the identification of NS-XLMR candidate genes, namely cytogenetically balanced X;autosome translocations where the normal X chromosome is preferentially inactivated in affected females; submicroscopic deletions in affected males; and X chromosome inversions in affected males. X chromosome genes structurally affected by breakpoints or lost by deletion represent naturally occurring human gene knockouts probably accounting for most of the symptoms in the patients who have the chromosomal aberrations. Moreover, these same genes become instant candidates for familial NS-XLMR mapping by linkage to the same locations (Gecz and Mulley, 2000).

Characterization of breakpoints in rare patients with apparently balanced constitutional chromosome rearrangements and phenotypic abnormalities has proved an invaluable strategy for identifying not only NS-XLMR genes but also genes for other diseases (Tommerup, 1993; Collins, 1995; Bugge et al., 2000). The first gene to be cloned from an X;autosome translocation was the gene for Duchenne muscular dystrophy, and analysis of balanced translocations has subsequently helped in the cloning of other disease genes, such as in X-linked lissencephaly, autosomal dominant familial adenomatous polyposis, and in the autosomal recessive condition Alstrom syndrome (Ray et al., 1985; Boyd et al., 1987; Cockburn et al., 1992; van der Luijt et al., 1995; Matsumoto et al., 1998; Matsumoto et al., 2000; Hearn et al., 2002). Phenotypic abnormalities seen in cases with apparently balanced chromosome rearrangements are typically explained by the disruption of a gene at the breakpoint causing the loss of gene function. However, other mechanisms of disease causation have also been described where (1) a cryptic deletion or duplication is identified at the breakpoint or (2) a breakpoint disrupts or alters gene expression through a position effect (Puissant et al., 1988; Kleinjan and van Heyningen, 1998; Kumar et al., 1998; Mohrschladt et al., 1999; Wirth et al., 1999; Cox JJ et al., 2003). An advantage of the breakpoint cloning strategy over the candidate gene approach is that genes that initially seem unusual candidates for NS-XLMR are not excluded, and so genes that play a novel role in human brain development can be isolated. However, the technique is limited by the number of patients who carry these rare chance rearrangements, and so breakpoint cloning cannot be used as a systematic strategy to identify all X-linked disease-causing genes.

The Human Genome Project has significantly improved the efficiency by which chromosomal breakpoints can be mapped, because large-insert clones [(BACs and P1 artificial chromosomes (PACs)] that are anchored to the genomic sequence are now readily available (Cheung et al., 2001). After the localization of a breakpoint to a chromosomal band by high-resolution G-banding, the large-insert clones can be fluorescently labeled and hybridized to the patient's chromosomes to fine map a breakpoint.

Six NS-XLMR genes (*TM4SF2, ZNF41, ZNF81, OPHN1, AGTR2,* and *ARHGEF6*) were identified after being mapped to an X chromosome breakpoint in an individual with an X;autosome translocation and nonsyndromic MS (Billuart et al., 1998; Kutsche et al., 2000; Zemni et al., 2000; Vervoort et al., 2002; Shoichet et al., 2003; Kleefstra et al., 2004). Notably, for five out of six of these genes, the translocation patients were females (see Fig. 29–1). Males with balanced X;autosome translocations are rare because all de novo X; autosome translocations are usually paternal in origin, and so males only have female children who carry an X;autosome translocation (McKinlay Gardner and Sutherland, 2003). Furthermore, females who carry an X;autosome translocation have significantly reduced fertility and are almost always infertile if the X chromosome breakpoint involves the critical region Xq13-q26 due to gonadal dysgenesis (Therman et al., 1990). Those female X;autosome carriers that are fertile have a 50:50 chance of having a male child, but the chance of passing on the normal balanced translocation is small.

When studying females with balanced X;autosome translocations, it is important to determine their X-inactivation pattern. X-inactivation is the process by which mammalian females achieve dosage compensation by transcriptionally silencing one X

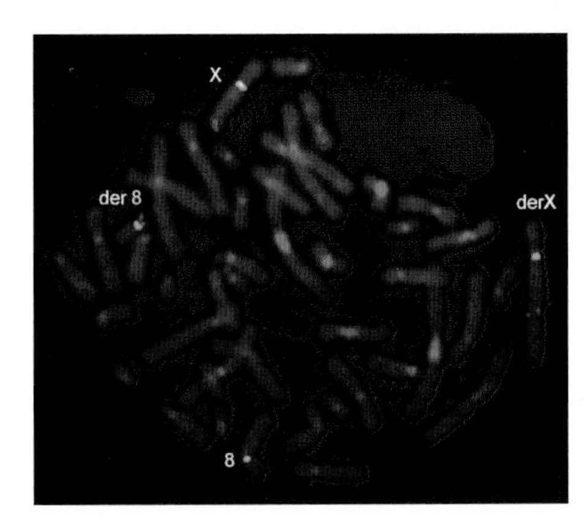

Figure 29–1 X;autosome translocation in mapping of genes for nonsyndromic X-linked mental retardation. FISH (fluorescent in situ hybridization) analysis with genomic clones from the end of a duplicated region in Xq28. The clone, RP13-156P1, spans the breakpoint of an X;autosome translocation due to a small duplication of 600 kb at the X chromosome breakpoint as seen by the red signals on the normal X, derivative X and derivative 8 chromosomes. The centromeres are labeled in green.

chromosome (Lyon, 1961). This inactivation occurs early in development and results in individuals who are essentially cellular mosaics, with either the maternal or paternal X chromosome inactivated. The choice of which X chromosome is inactivated is random, and in chromosomally normal females, approximately half their cells have the maternal X chromosome inactivated, and the other half have the paternal X chromosome inactivated. However, most females who carry a balanced X;autosome translocation have a skewed X inactivation pattern, with 100% of cells having the same X chromosome inactivated (Mattei et al., 1982; Waters et al., 2001). This is because cells in which the structurally normal X chromosome is inactivated are selected over cells in which the derivative X chromosome is inactivated, because the latter show functional autosomal monosomy (assuming that the portion of autosomal material on the derivative X chromosome is inactivated), and functional X chromosome disomy for the portion of the X chromosome translocated onto the autosome.

In the cases of *TM4SF2, ZNF41, ZNF81, OPHN1,* and *ARHGEF6,* the coding sequence of the genes were disrupted by the translocation breakpoint. However, for *AGTR2,* the breakpoint was 150 kb upstream of the gene, suggesting that the breakpoint had separated distant regulatory regions from *AGTR2.* Mutations in all these genes were subsequently identified in families with NS-XLMR, which confirmed their role in the etiology of nonsyndromic MR. At least two other X-linked genes (*DXS6673E,* and *KLF8*) have been identified from X;autosome translocations in mentally retarded patients, but have subsequently not been found to be mutated in familial NS-XLMR, and hence are not thought to be common causes of MR (van der Maarel et al., 1996; Lossi et al., 2002).

Submicroscopic deletions of the X chromosome are often identified when a male with a known X-linked disease phenotype has MR as an accompanying feature. In these contiguous gene syndromes, the gene known to cause the X-linked disease is deleted, together with genetic material that is responsible for the patient's MR. For example, *FACL4* was cloned as a result of comparing the extent of deletions in patients with Alport's syndrome alone (a hereditary glomerulonephritis) with those with Alport's syndrome and MR. A critical region for MR of approximately 380 kb was identified in Xp22.3, and mutation screening of genes within this interval in NS-XLMR families identified *FACL4* as an NS-XLMR gene (Meloni et al., 2002).

Positional cloning from within deletions has identified two further NS-XLMR genes, *FMR2* and *IL1RAPL1.* In 1992, a folate-sensitive fragile site in Xq28 was identified (FRAXE), which was distinct from the FRAXA fragile site associated with fragile X syndrome (Sutherland and Baker, 1992). The underlying molecular basis for FRAXE was found to be an unstable repeat $(CCG)_n$ identical to that of FRAXA (Knight et al., 1993). However, the gene associated with FRAXE (*FMR2*) was not identified until a submicroscopic deletion in a developmentally delayed boy was found to have the FRAXE fragile site at its 5' end (Gedeon et al., 1995; Gecz et al., 1996; Gu et al., 1996). Expansion of the unstable CCG repeat located in the 5' region of *FMR2* is associated with abnormal methylation and suppression of *FMR2* transcription, which in turn causes mild cognitive impairment. *IL1RAPL1* was identified as an NS-XLMR gene because of being interrupted by a submicroscopic deletion in family MRX34, and a nonsense mutation was subsequently identified in a different NS-XLMR family (Carrie et al., 1999).

Deletion mapping has identified at least two other candidate NS-XLMR genes, *VCX-A* and *RPS6KA6,* but mutations in these genes have not been found in NS-XLMR families without chromosomal aberrations. *VCX-A* was identified from a 15-kb critical region for MR, which was found by comparing the extent of deletions in 15 males with Xp deletions with and without MR (Fukami et al., 2000). Similarly, *RPS6KA6*, which is related to the XLMR gene *RPS6KA3*, is also a good candidate NS-XLMR gene, because its deletion correlates with cognitive impairment in patients with contiguous gene syndromes caused by large Xq21 deletions (Yntema et al., 1999).

Characterization of X chromosome inversions in mentally retarded individuals has also led to the identification of candidate XLMR genes, including *SOX3*, *NXF5*, and *RLGP*. After the identification of *SOX3* at an inversion breakpoint, a disease-causing mutation was identified in a large family with MR and growth hormone deficiency (Laumonnier et al., 2002). *NXF5* has been screened in a cohort of NS-XLMR families, but no disease-causing mutations have been identified, whereas *RLGP* has yet to be screened, and so this novel Ras-like GTPase in Xq22.2 may yet prove to be a gene that causes familial XLMR (Saito-Ohara et al., 2002; Frints et al., 2003).

Autosomal MR

Although there are likely to be a significant number of genes on autosomes that cause MR, the identification of these genes has been slow. Three genes, *PRSS12* on chromosome 4q26, *CRBN* on chromosome 3p26, and most recently *CC2D1A* on chromosome 19p13.12, have been identified using autozygosity mapping of highly consanguineous families (Molinari et al., 2002; Higgins et al., 2004; Basel-Vanagaite et al., 2006). The principle is that in consanguineous families, areas of the genome are homozygous by descent, given that they are closely related. In addition, where there are areas that are homozygous and are also shared by affected individuals who are more distantly related, the recessive gene causing the disease must lie within the area of shared homozygosity. This method is powerful for detecting autosomal recessive genes that cause disease, but require the identification of rare families with the condition. The anticipation is that many new genes will emerge using this technique in the next 5–10 years.

Autosomal-dominant genes that cause MR are going to be more difficult to find, because when the condition is associated with significant intellectual problems, this also frequently reduces reproductive fitness, and consequently, large pedigrees are not seen in the population so ascertainment of suitable families for linkage and gene mapping are not possible. Techniques such as whole-genome hybridization, deletion mapping, and breakpoint mapping of autosomal-balanced translocations are mainly being used.

The key to understanding autosomal genes may be the identification and characterization of syndromic conditions, and gradually deducing the role of single genes in intellectual development along with its other features. Recently, the single gene that is mainly responsible for the clinical features of Smith–Magenis syndrome has been identified (Slager et al., 2003). This required the delineation of patients with smaller and smaller deletions and the sequencing of the few genes within the minimum deletion interval in a series of patients who were clinically affected but did not have a deletion. Mutations in *RAI1* were identified in this syndrome. Of particular interest is that the behavioral phenotype of self-hugging, head banging, onychotillomania, and polyembolokoilamania, characteristic of Smith–Magenis syndrome, were also seen in the patients with intragenic mutations in RAI1, suggesting that single gene disorders are sufficient to alter behavior quite significantly.

Similarly, a common microdeletion is associated with Rubinstein–Taybi syndrome, but mutations in CRB are sufficient to develop the full phenotype (Petrij et al., 1995).

The identification of mutations in the autosomal homologues of the XLMR genes and the identification of disrupted genes at breakpoints of reciprocal balanced autosomal translocations will define yet further genes on autosomes that cause MR in the coming years.

Single Gene Abnormalities Associated with Behavioral Phenotype

The idea that human behavior can be determined by a few genes is somewhat surprising, and is likely not to be the case in most cases. Complex human interactions including speech, language, and nonverbal communication is likely to be ordered by the equally complex constellation of genes and their subtle interactions. A somewhat surprising observation is that there are a number of single genes that when mutated result in dramatic changes in behavior or language. The example of *RAI1* resulting in specific behavioral features of Smith–Magenis syndrome is one such example. Another is the identification of a mutation in the gene *MAOA* in a particular family where five affected individuals have a range of behaviors that include impulsive aggressive behavior, arson, attempted rape, and exhibitionism (Brunner et al., 1993). No other families have been described to have mutations in this gene, and it does not appear to be a common cause of behavioral disturbance in groups of individuals with abnormal behavior (Schuback et al., 1999). While it is reasonable to infer that normal functioning of *MAOA* is necessary for the control of impulsive behavior, it does not imply that this gene is any more important than any other in the complex process of neurotransmitter formation and release in normal individuals.

Speech and Language

A family was described in the literature in which absence of speech and language was inherited in an autosomal dominant manner. The language impairment was part of a broader syndrome that included impairment of articulation, including nonlinguistic oral and facial movements (Hurst et al.. 1990; Vargha-Khadem et al., 1995). A single base abnormality was identified in the family after the identification of a balanced translocation with a similar phenotype that disrupts the gene (Lai et al., 2001). A truncating mutation in a single unrelated individual has also been described that causes a similar defect in speech and language (MacDermot et al., 2005). How a transcription factor *FOXP2* causes this abnormality of speech and language is little understood to date, although gene expression appears to be concentrated in areas of the adult brain responsible for pathology of speech and language formation (Lai et al., 2003). Further work on the biological effect of *FOXP2* and the identification of further genes that cause abnormalities of speech need to be done before we can really have an understanding of the complex genes, their interactions and the biological mechanisms that are involved in the development and abnormal processings of speech and language. Analysis of large cohorts of children with other more common abnormalities of speech and language has failed to identify pathological sequence variants in this gene, and furthermore, common intragenic microsatellite polymorphic markers within this gene are not associated with autism or specific language impairment (SLI) (Newbury et al., 2002).

Susceptibility to Language Impairment, Dyslexia

No other single genes have been identified to date that carry mutations that can predict SLI, dyslexia, or pure autism in an individual. It is likely that, for the most part, these are complex traits in which the combination of variants within several genes simultaneously confers a susceptibility to develop the disorder.

For SLI, the condition is highly heritable as the prevalence of the condition within the family is 24%–78% (mean 46%), whereas in control groups it is 3%–46% (mean 18%) (Stromswold, 1998). Similarly, twin studies show a significant concordance rate in monozygotic twins compared to dizygotic twins (Bishop et al., 1995). For SLI, two whole-genome scans have been performed to date to identify linkage regions, and three loci have been identified on chromosome 16q (SLI1), 19q (SLI2), and 13q (SLI3) (Bartlett et al., 2002; SLI Consortium, 2002). The locus on 19q has been further strengthened by screening 184 additional families in a second independent whole-genome scan to that of the original identification of the SLI1 locus. No single gene or allele within the linked interval has been identified yet, but with the rapid annotation of the genome this is now feasible.

For dyslexia, a similar approach has identified several regions of linkage on whole-genome scans. Two areas have been refined, and the region on 6p23 (DYX1) has been narrowed down to a single gene KIAA0319, which is proposed to be a susceptibility gene for developmental dyslexia (Cardon et al., 1994; Francks et al., 2004; Cope et al., 2005). As this data is still new, further substantiation of this finding is needed, but this is encouraging.

Genetic Basis of Autism

Autism is a common childhood disorder, and the estimated prevalence in the population is approximately 4 in 10,000. The criteria for diagnosis include impairments in social interactions, communication, restricted and stereotyped patterns of interests and activities, and the presence of developmental abnormalities by 3 years of age (WHO, 1992). Males are more commonly affected, and the sibling recurrence risk is 3%–10%. Twin studies show a much higher concordance rates in monozygotic as compared dizygotic twins, suggesting a major genetic component to the underlying etiology (Bailey et al., 1995). Identifying the genetic basis of autism has been an area of sustained work for the past 5–10 years, and to date we do not have any clear answers as the genetics appears to be significantly more complex than initially thought. Although cases of autism clusters within families, the models of inheritance make it unlikely that it is frequently due to high-penetrant single gene defects as found in XLMR. Mutations in a rare X-linked gene, NLGN4, have been described that cause autism and associated severe MR, but this is very uncommon (Jamain et al., 2003; Gauthier et al., 2005).The disease models suggest a polygenic mode of inheritance to be more likely, whereby disease only occurs in an individual if alterations or variants are present in several genes within the same individual. No one variant is sufficient to cause disease as in monogenic disease. In 1998, the first genome-wide linkage was performed on 99 families with affected sib pairs with disease. This identified six susceptibility loci on chromosomes 4, 7q, 10, 16p, 19p, and 22, with a maximum LOD score of more than 1.0 (IMGSAC, 1998). The locus on 7q was the most significant and has since been replicated with additional sample sets (IMGSAC, 2001). Additional whole-genome-wide scans of different cohorts have identified numerous other loci, including 3q, 15q, 4q, 5p, 6q, 10q, 18q, and Xp (Philippe et al., 1999; Risch et al., 1999; Liu et al., 2001; Alarcon et al., 2002; Auranen et al., 2002; Yonan et al., 2003). While each study identifies different areas of the genome that carries susceptibility loci, two particular regions on chromosomes 7q22-31 and 2q32 appear to be consistently identified as regions of the genome where a common susceptibility gene lies. The remaining many susceptibility loci identified for autism either indicate that the studies have been underpowered to detect truly significant linkage or that the constellation of genes that underlie autism are very many. Furthermore, it is likely that the clinical entity autism is highly heterogeneous in etiology and reflects a final common pathway of a series of complex brain developmental abnormalities.

Table 29–3 Candidate Genes for Autism Based on Association Studies or Identification of Rare Variants in the Autistic Population

Genes where association has been identified	UBE3A, Engrailed 2, CAMP-GEFII, RELN, GABRA4, GABRB1, SCL6A4, GRN2A, ABAT, LAMB1, and NRCAM
Genes where no association has been identified	WNT2, 5HTT, ApoE, 22q11 deletion, FOXP2, DCX, SLC25A12, CMYA3, and PTPR21

Having identified susceptibility loci, the task has been to identify variants in single genes within these large chromosomal regions that confer disease susceptibility. This has been done either by using association studies of common single polymorphic variants (single nucleotide polymorphisms, SNPs) or by identifying rare variants in candidate genes within the linkage interval. Table 29–3 lists genes that have been published where variants are associated with disease compared to controls, or where there are rare variants in the candidate genes in autistic individuals but is not found in the control population. Where studies have found an association and others have failed to identify this association, they have been placed in the no-association category. The number of positive associations is large, and the next 5 years will be concentrated on verifying them and trying to understand the biological basis for these observations.

The assumption of many researchers when they began the road to identifying the genetic basis of autism was that as the condition had such a high heritability, understanding the genetic basis of the diseases was within reach. Some 15 years later, experts in the field are suggesting that it has and continues to be significantly harder to identify the genetic basis of this condition than initially thought, but it is likely that we will have a better understanding of this condition in the next 5–10 years. The hope and reality of being able to provide treatment, or reliable predictive information for families, still seems a very long way off.

Conclusion

The first clear epidemiological study which showed that intellectual disability is frequently familial was in 1938 (Penrose, 1938). However, not until 1959 was the basis for Down syndrome clearly identified as due to additional chromosome material in every cell (Lejeune et al., 1959). In the next 40 years, concentrated in the last 10 years, an astonishing identification of many single genes that cause intellectual impairment when abnormal has been made. This avalanche is likely to continue at an equally astonishing pace for the next few years when mutations in very many genes are going to be identified and found to result in intellectual disabilities in humans.

The technology that has been developed to identify more "conventional genes" is going to be applied to new areas of behavioral phenotypes, which is a field in its infancy. The identification of *FOXP2* still fascinates many in that a single base change in a single gene is sufficient to profoundly interrupt normal speech and language in a consistent manner. The suspicion is that, in the next 10 years, more genes will be identified that are vital to some small aspect of behavior.

What is emerging in the genetics of MR is that there are very many genes and that each gene abnormality is a rare cause of MR. It is likely that this pattern will be followed for other phenotypes, and that there are a number of genes which, when abnormal, give rise to unusual behavior, but each gene is a rare cause of problems. Similarly, in the MR field, the assumption that much of the male excess of affected individuals is due to X-linked single gene abnormalities is likely to be incorrect, and polygenic disease may be more important as has already be shown in SLI, dyslexia, and autism.

The power of the Human Genome Project has meant that our ability to analyze large numbers of genes in very many clinical samples has increased annually and exponentially, but the power of many studies are now limited not by the molecular genetic technology but the availability of suitable characterized samples and families. For behavioral genetics and the study of intellectual development, this is a major issue and has not yet been resolved. Currently, the focus is still on identifying the genes that underlie these disorders. The next challenge is to understand the cellular interactions and mechanisms by which the disorders are manifest and to understand the complex phenotypic variations that we see even in abnormalities within the same gene in different individuals.

Acknowledgments

JJC and FLR are supported by research grants from the Wellcome Trust, Birth Defects Foundation, and the Medical Research Council, UK.

References

Abidi, F, Schwartz, CE, Carpenter, NJ, Villard, L, Fontes, M, and Curtis, M (1999). Carpenter–Waziri syndrome results from a mutation in XNP. *Am J Med Genet*, 85:249–251.

Alarcon, M, Cantor, RM, Liu, J, Gilliam, TC, and Geschwind, DH (2002). Evidence for a language quantitative trait locus on chromosome 7q in multiplex autism families. *Am J Hum Genet*, 70:60–71.

Aldred, MA, Sanford, RO, Thomas, NS, Barrow, MA, Wilson, LC, Brueton, LA, et al. (2004). Molecular analysis of 20 patients with 2q37.3 monosomy: definition of minimum deletion intervals for key phenotypes. *J Med Genet*, 41(6):433–439.

Allen, KM, Gleeson, JG, Bagrodia, S, Partington, MW, MacMillan, JC, Cerione, RA, et al. (1998). PAK3 mutation in nonsyndromic X-linked mental retardation. *Nat Genet*, 20:25–30.

Amir, RE, Van den Veyver, IB, Wan, M, Tran, CQ, Francke, U, and Zoghbi, HY (1999). Rett syndrome is caused by mutations in X-linked MECP2, encoding methyl- CpG-binding protein 2. *Nat Genet*, 23:185–188.

Auranen, M, Vanhala, R, Varilo, T, Ayers, K, Kempas, E, Ylisaukko-Oja, T, et al. (2002). A genomewide screen for autism-spectrum disorders: evidence for a major susceptibility locus on chromosome 3q25-27. *Am J Hum Genet*, 71:777–790.

Bailey, A, Le Couteur, A, Gottesman, I, Bolton, P, Simonoff, E, Yuzda, E, et al. (1995). Autism as a strongly genetic disorder: evidence from a British twin study. *Psychol Med*, 25:63–77.

Bartlett, CW, Flax, JF, Logue, MW, Vieland, VJ, Bassett, AS, Tallal, P, et al. (2002). A major susceptibility locus for specific language impairment is located on 13q21. *Am J Hum Genet*, 71:45–55.

Basel-Vanagaite, L, Attia, R, Yahav, M, Ferland, RJ, Anteki, L, Walsh, CA, et al. (2006). The CC2D1A, a member of a new gene family with C2 domains, is involved in autosomal recessive nonsyndromic mental retardation. *J Med Genet*, 43:203–210.

Bienvenu, T, des Portes, V, Saint Martin, A, McDonell, N, Billuart, P, Carrie, A, et al. (1998). Non-specific X-linked semidominant mental retardation by mutations in a Rab GDP-dissociation inhibitor. *Hum Mol Genet*, 7:1311–1315.

Bienvenu, T, Poirier, K, Friocourt, G, Bahi, N, Beaumont, D, Fauchereau, F, et al. (2002). ARX, a novel Prd-class-homeobox gene highly expressed in the telencephalon, is mutated in X-linked mental retardation. *Hum Mol Genet*, 11:981–991.

Billuart, P, Bienvenu, T, Ronce, N, des Portes, V, Vinet, MC, Zemni, R, et al. (1998). Oligophrenin-1 encodes a rhoGAP protein involved in X-linked mental retardation. *Nature*, 392:923–926.

Bishop, DV, North, T, and Donlan, C (1995). Genetic basis of specific language impairment: evidence from a twin study. *Dev Med Child Neurol*, 37:56–71.

Bliss, TV and Collingridge, GL (1993). A synaptic model of memory: long-term potentiation in the hippocampus. *Nature*, 361:31–39.

Boyd, Y, Munro, E, Ray, P, Worton, R, Monaco, T, Kunkel, L, et al. (1987). Molecular heterogeneity of translocations associated with muscular dystrophy. *Clin Genet*, 31:265–272.

Brunner, HG, Nelen, M, Breakefield, XO, Ropers, HH, and van Oost, BA (1993). Abnormal behavior associated with a point mutation in the structural gene for monoamine oxidase A. *Science*, 262:578–580.

Bugge, M, Bruun-Petersen, G, Brondum-Nielsen, K, Friedrich, U, Hansen, J, Jensen, G, et al. (2000). Disease associated balanced chromosome rearrangements: a resource for large scale genotype–phenotype delineation in man. *J Med Genet*, 37:858–865.

Bundey, S, Webb, TP, Thake, A, and Todd, J (1985). A community study of severe mental retardation in the West Midlands and the importance of the fragile X chromosome in its aetiology. *J Med Genet*, 22:258–266.

Cardon, LR, Smith, SD, Fulker, DW, Kimberling, WJ, Pennington, BF, and DeFries, JC (1994). Quantitative trait locus for reading disability on chromosome 6. *Science*, 266:276–279.

Carrie, A, Jun, L, Bienvenu, T, Vinet, MC, McDonell, N, Couvert, P, et al. (1999). A new member of the IL-1 receptor family highly expressed in hippocampus and involved in X-linked mental retardation. *Nat Genet*, 23:25–31.

Chelly, J and Mandel, JL (2001). Monogenic causes of X-linked mental retardation. *Nat Rev Genet*, 2:669–680.

Cheung, VG, Nowak, N, Jang, W, Kirsch, IR, Zhao, S, Chen, XN, et al. (2001). Integration of cytogenetic landmarks into the draft sequence of the human genome. *Nature*, 409:953–958.

Chiurazzi, P, Hamel, BC, and Neri, G (2001). XLMR genes: update 2000. *Eur J Hum Genet*, 9:71–81.

Cockburn, DJ, Munro, EA, Craig, IW, and Boyd, Y (1992). Mapping of X chromosome translocation breakpoints in females with Duchenne muscular dystrophy with respect to exons of the dystrophin gene. *Hum Genet*, 90:407–412.

Collins, FS (1995). Positional cloning moves from perditional to traditional. *Nat Genet*, 9:347–350.

Conrad, DF, Andrews, TD, Carter, NP, Hurles, ME, and Pritchard, JK (2006). A high-resolution survey of deletion polymorphism in the human genome. *Nat Genet*, 38:75–81.

Cope, N, Harold, D, Hill, G, Moskvina, V, Stevenson, J, Holmans, P, et al. (2005). Strong evidence that KIAA0319 on chromosome 6p is a susceptibility gene for developmental dyslexia. *Am J Hum Genet*, 76:581–591.

Couvert, P, Bienvenu, T, Aquaviva, C, Poirier, K, Moraine, C, Gendrot, C, et al. (2001). MECP2 is highly mutated in X-linked mental retardation. *Hum Mol Genet*, 10:941–946.

Cox, JJ, Holden, ST, Dee, S, Burbridge, JI, and Raymond, FL (2003). Identification of a 650 kb duplication at the X chromosome breakpoint in a patient with 46,X,t(X;8)(q28;q12) and non-syndromic mental retardation. *J Med Genet*, 40:169–174.

Crow, YJ and Tolmie, JL (1998). Recurrence risks in mental retardation. *J Med Genet*, 35:177–182.

Curry, CJ, Stevenson, RE, Aughton, D, Byrne, J, Carey, JC, Cassidy, S, et al. (1997). Evaluation of mental retardation: recommendations of a Consensus Conference: American College of Medical Genetics. *Am J Med Genet*, 72:468–477.

D'Adamo, P, Menegon, A, Lo Nigro, C, Grasso, M, Gulisano, M, Tamanini, F, et al. (1998). Mutations in GDI1 are responsible for X-linked non-specific mental retardation. *Nat Genet*, 19:134–139.

de Vries, BB, Halley, DJ, Oostra, BA, and Niermeijer, MF (1998). The fragile X syndrome. *J Med Genet*, 35:579–589.

de Vries, BB, Pfundt, R, Leisink, M, Koolen, DA, Vissers, LE, Janssen, IM, et al. (2005). Diagnostic genome profiling in mental retardation. *Am J Hum Genet*, 77:606–616.

Dewey, WJ, Barrai, I, Morton, NE, and Mi, MP (1965). Recessive genes in severe mental defect. *Am J Hum Genet*, 17:237–256.

Duchenne, GB (1968). Studies on pseudohypertrophic muscular paralysis or myosclerotic paralysis. *Arch Neurol*, 19:629–636.

Ewart, AK, Morris, CA, Atkinson, D, Jin, W, Sternes, K, Spallone, P, et al. (1993). Hemizygosity at the elastin locus in a developmental disorder, Williams syndrome. *Nat Genet*, 5:11–16.

Flint, J and Knight, S (2003). The use of telomere probes to investigate submicroscopic rearrangements associated with mental retardation. *Curr Opin Genet Dev*, 13:310–316.

Flint, J and Wilkie, AO (1996). The genetics of mental retardation. *Br Med Bull*, 52:453–464.

Flint, J, Wilkie, AO, Buckle, VJ, Winter, RM, Holland, AJ, and McDermid, HE (1995). The detection of subtelomeric chromosomal rearrangements in idiopathic mental retardation. *Nat Genet*, 9:132–140.

Fombonne, E (2003). Epidemiological surveys of autism and other pervasive developmental disorders: an update. *J Autism Dev Disord*, 33:365–382.

Francks, C, Paracchini, S, Smith, SD, Richardson, AJ, Scerri, TS, Cardon, LR, et al. (2004). A 77-kilobase region of chromosome 6p22.2 is associated with dyslexia in families from the United Kingdom and from the United States. *Am J Hum Genet*, 75:1046–1058.

Freude, K, Hoffmann, K, Jensen, LR, Delatycki, MB, des Portes, V, Moser, B, et al. (2004). Mutations in the FTSJ1 gene coding for a novel S-adenosylmethionine-binding protein cause nonsyndromic X-linked mental retardation. *Am J Hum Genet*, 75:305–309.

Frints, SG, Froyen, G, Marynen, P, and Fryns, JP (2002). X-linked mental retardation: vanishing boundaries between non-specific (MRX) and syndromic (MRXS) forms. *Clin Genet*, 62:423–432.

Frints, SG, Jun, L, Fryns, JP, Devriendt, K, Teulingkx, R, Van den Berghe, L, et al. (2003). Inv(X)(p21.1;q22.1) in a man with mental retardation, short stature, general muscle wasting, and facial dysmorphism: clinical study and mutation analysis of the NXF5 gene. *Am J Med Genet*, 119A:367–374.

Fryns, JP, Borghgraef, M, Brown, TW, Chelly, J, Fisch, GS, Hamel, B, et al. (2000). 9th international workshop on fragile X syndrome and X-linked mental retardation. *Am J Med Genet*, 94:345–360.

Fukami, M, Kirsch, S, Schiller, S, Richter, A, Benes, V, Franco, B, et al. (2000). A member of a gene family on Xp22.3, VCX-A, is deleted in patients with X-linked nonspecific mental retardation. *Am J Hum Genet*, 67:563–573.

Gauthier, J, Bonnel, A, St-Onge, J, Karemera, L, Laurent, S, Mottron, L, et al. (2005). NLGN3/NLGN4 gene mutations are not responsible for autism in the Quebec population. *Am J Med Genet B Neuropsychiatr Genet*, 132:74–75.

Gecz, J, Gedeon, AK, Sutherland, GR, and Mulley, JC (1996). Identification of the gene FMR2, associated with FRAXE mental retardation. *Nat Genet*, 13:105–108.

Gecz, J and Mulley, J (2000). Genes for cognitive function: developments on the X. *Genome Res*, 10:157–163.

Gedeon, AK, Donnelly, AJ, Mulley, JC, Kerr, B, and Turner, G (1996a). How many X-linked genes for non-specific mental retardation (MRX) are there? *Am J Med Genet*, 64:158–162.

Gedeon, AK, Donnelly, AJ, Mulley, JC, Kerr, B, and Turner, G (1996b). How many X-linked genes for non-specific mental retardation (MRX) are there? [letter]. *Am J Med Genet*, 64:158–162.

Gedeon, AK, Meinanen, M, Ades, LC, Kaariainen, H, Gecz, J, Baker, E, et al. (1995). Overlapping submicroscopic deletions in Xq28 in two unrelated boys with developmental disorders: identification of a gene near FRAXE. *Am J Hum Genet*, 56:907–914.

Gibbons, RJ, Picketts, DJ, Villard, L, and Higgs, DR (1995). Mutations in a putative global transcriptional regulator cause X-linked mental retardation with alpha-thalassemia (ATR-X syndrome). *Cell*, 80:837–845.

Gu, XX, Decorte, R, Marynen, P, Fryns, JP, Cassiman, JJ, and Raeymaekers, P (1996). Localisation of a new gene for non-specific mental retardation to Xq22-q26 (MRX35). *J Med Genet*, 33:52–55.

Guerrini, R, Shanahan, JL, Carrozzo, R, Bonanni, P, Higgs, DR, and Gibbons, RJ (2000). A nonsense mutation of the ATRX gene causing mild mental retardation and epilepsy. *Ann Neurol*, 47:117–121.

Hahn, KA, Salomons, GS, Tackels-Horne, D, Wood, TC, Taylor, HA, Schroer, RJ, et al. (2002). X-linked mental retardation with seizures and carrier manifestations is caused by a mutation in the creatine-transporter gene (SLC6A8) located in Xq28. *Am J Hum Genet*, 70:1349–1356.

Hearn, T, Renforth, GL, Spalluto, C, Hanley, NA, Piper, K, Brickwood, S, et al. (2002). Mutation of ALMS1, a large gene with a tandem repeat encoding 47 amino acids, causes Alstrom syndrome. *Nat Genet*, 31:79–83.

Heilstedt, HA, Ballif, BC, Howard, LA, Lewis, RA, Stal, S, Kashork, CD, et al. (2003). Physical map of 1p36, placement of breakpoints in monosomy 1p36, and clinical characterization of the syndrome. *Am J Hum Genet*,72 (5):1200–1212.

Herbst, DS and Miller, JR (1980). Nonspecific X-linked mental retardation II: the frequency in British Columbia. *Am J Med Genet*, 7:461–469.

Higgins, JJ, Pucilowska, J, Lombardi, RQ, and Rooney, JP (2004). A mutation in a novel ATP-dependent Lon protease gene in a kindred with mild mental retardation. *Neurology*, 63:1927–1931.

Hinds, DA, Kloek, AP, Jen, M, Chen, X, and Frazer, KA (2006). Common deletions and SNPs are in linkage disequilibrium in the human genome. *Nat Genet*, 38:82–85.

Hurst, JA, Baraitser, M, Auger, E, Graham, F, and Norell, S (1990). An extended family with a dominantly inherited speech disorder. *Dev Med Child Neurol*, 32:352–355.

IMGSAC (1998). A full genome screen for autism with evidence for linkage to a region on chromosome 7q. International Molecular Genetic Study of Autism Consortium. *Hum Mol Genet*, 7:571–578.

IMGSAC (2001). A genomewide screen for autism: strong evidence for linkage to chromosomes 2q, 7q, and 16p. *Am J Hum Genet*, 69:570–581.

Ireland, WW (1900). *The Mental Affections of Children: Idiocy, Imbecility and Insanity*. London: J & A Churchill.

Jamain, S, Quach, H, Betancur, C, Rastam, M, Colineaux, C, Gillberg, IC, et al. (2003). Mutations of the X-linked genes encoding neuroligins NLGN3 and NLGN4 are associated with autism. *Nat Genet*, 34:27–29.

Jensen, LR, Amende, M, Gurok, U, Moser, B, Gimmel, V, Tzschach, A, et al. (2005). Mutations in the JARID1C gene, which is involved in transcriptional regulation and chromatin remodeling, cause X-linked mental retardation. *Am J Hum Genet*, 76:227–236.

Kalscheuer, VM, Freude, K, Musante, L, Jensen, LR, Yntema, HG, Gecz, J, et al. (2003). Mutations in the polyglutamine binding protein 1 gene cause X-linked mental retardation. *Nat Genet*, 35:313–315.

Kerr, B, Turner, G, Mulley, J, Gedeon, A, and Partington, M (1991). Non-specific X linked mental retardation. *J Med Genet*, 28:378–382.

Kleefstra, T, Yntema, HG, Oudakker, AR, Banning, MJ, Kalscheuer, VM, Chelly, J, et al. (2004). Zinc finger 81 (ZNF81) mutations associated with X-linked mental retardation. *J Med Genet*, 41:394–399.

Kleinjan, DJ and van Heyningen, V (1998). Position effect in human genetic disease. *Hum Mol Genet*, 7:1611–1618.

Knight, SJ, Flannery, AV, Hirst, MC, Campbell, L, Christodoulou, Z, Phelps, SR, et al. (1993). Trinucleotide repeat amplification and hypermethylation of a CpG island in FRAXE mental retardation. *Cell*, 74:127–134.

Knight, SJ, Regan, R, Nicod, A, Horsley, SW, Kearney, L, Homfray, T, et al. (1999). Subtle chromosomal rearrangements in children with unexplained mental retardation. *Lancet*, 354:1676–1681.

Knoll, JH, Nicholls, RD, Magenis, RE, Graham, JM Jr, Lalande, M, and Latt, SA (1989). Angelman and Prader–Willi syndromes share a common chromosome 15 deletion but differ in parental origin of the deletion. *Am J Med Genet*, 32:285–290.

Kumar, A, Becker, LA, Depinet, TW, Haren, JM, Kurtz, CL, Robin, NH, et al. (1998). Molecular characterization and delineation of subtle deletions in de novo "balanced" chromosomal rearrangements. *Hum Genet*, 103:173–178.

Kutsche, K, Yntema, H, Brandt, A, Jantke, I, Nothwang, HG, Orth, U, et al. (2000). Mutations in ARHGEF6, encoding a guanine nucleotide exchange factor for Rho GTPases, in patients with X-linked mental retardation. *Nat Genet*, 26:247–250.

Lai, CS, Fisher, SE, Hurst, JA, Vargha-Khadem, F, and Monaco, AP (2001). A forkhead-domain gene is mutated in a severe speech and language disorder. *Nature*, 413:519–523.

Lai, CS, Gerrelli, D, Monaco, AP, Fisher, SE, and Copp, AJ (2003). FOXP2 expression during brain development coincides with adult sites of pathology in a severe speech and language disorder. *Brain*, 126:2455–2462.

Laumonnier, F, Bonnet-Brilhault, F, Gomot, M, Blanc, R, David, A, Moizard, MP, et al. (2004). X-linked mental retardation and autism are associated with a mutation in the NLGN4 gene, a member of the neuroligin family. *Am J Hum Genet*, 74:552–557.

Laumonnier, F, Ronce, N, Hamel, BC, Thomas, P, Lespinasse, J, Raynaud, M, et al. (2002). Transcription factor SOX3 is involved in X-linked mental retardation with growth hormone deficiency. *Am J Hum Genet*, 71:1450–1455.

Ledbetter, DH, Riccardi, VM, Airhart, SD, Strobel, RJ, Keenan, BS, and Crawford, JD (1981). Deletions of chromosome 15 as a cause of the Prader–Willi syndrome. *N Engl J Med*, 304:325–329.

Lehrke, R (1972). Theory of X-linkage of major intellectual traits. *Am J Ment Defic*, 76:611–619.

Lejeune, J, Turpin, R, and Gautier, M (1959). Chromosomic diagnosis of mongolism. *Arch Fr Pediatr*, 16:962–963.

Liu, J, Nyholt, DR, Magnussen, P, Parano, E, Pavone, P, Geschwind, D, et al. (2001). A genomewide screen for autism susceptibility loci. *Am J Hum Genet*, 69:327–340.

Lossi, AM, Laugier-Anfossi, F, Depetris, D, Gecz, J, Gedeon, A, Kooy, F, et al. (2002). Abnormal expression of the KLF8 (ZNF741) gene in a female patient with an X;autosome translocation t(X;21)(p11.2;q22.3) and non-syndromic mental retardation. *J Med Genet*, 39:113–117.

Lossi, AM, Millan, JM, Villard, L, Orellana, C, Cardoso, C, Prieto, F, et al. (1999). Mutation of the XNP/ATR-X gene in a family with severe mental retardation, spastic paraplegia and skewed pattern of X inactivation: demonstration that the mutation is involved in the inactivation bias. *Am J Hum Genet*, 65:558–562.

Lugtenberg, D, de Brouwer, AP, Kleefstra, T, Oudakker, AR, Frints, SG, Schrander-Stumpel, CT, et al. (2006). Chromosomal copy number changes in patients with non-syndromic X-linked mental retardation detected by array CGH. *J Med Genet*, 43:362–370.

Lyon, MF (1961). Gene action in the X-chromosome of the mouse (*Mus musculus* L.). *Naturwissenschaften*, 190:372–373.

MacDermot, KD, Bonora, E, Sykes, N, Coupe, AM, Lai, CS, Vernes, SC, et al. (2005). Identification of FOXP2 truncation as a novel cause of developmental speech and language deficits. *Am J Hum Genet*, 76:1074–1080.

Martin, JP and Bell, J (1943). A pedigree of mental defect showing sex-linkage. *J Neurol Psychiatry*, 6:154–157.

Matsumoto, N, David, DE, Johnson, EW, Konecki, D, Burmester, JK, Ledbetter, DH, et al. (2000). Breakpoint sequences of an 1;8 translocation in a family with Gilles de la Tourette syndrome. *Eur J Hum Genet*, 8:875–883.

Matsumoto, N, Pilz, DT, Fantes, JA, Kittikamron, K, and Ledbetter, DH (1998). Isolation of BAC clones spanning the Xq22.3 translocation breakpoint in a lissencephaly patient with a de novo X;2 translocation. *J Med Genet*, 35:829–832.

Mattei, MG, Mattei, JF, Ayme, S, and Giraud, F (1982). X-autosome translocations: cytogenetic characteristics and their consequences. *Hum Genet*, 61:295–309.

McKinlay Gardner, RJ and Sutherland, GR (2003). *Chromosome Abnormalities and Genetic Counseling*, 3rd edn. USA: Oxford University Press.

McLaren, J and Bryson, SE (1987). Review of recent epidemiological studies of mental retardation: prevalence, associated disorders, and etiology. *Am J Ment Retard*, 92:243–254.

Meloni, I, Muscettola, M, Raynaud, M, Longo, I, Bruttini, M, Moizard, MP, et al. (2002). FACL4, encoding fatty acid-CoA ligase 4, is mutated in nonspecific X-linked mental retardation. *Nat Genet*, 30:436–440.

Merienne, K, Jacquot, S, Pannetier, S, Zeniou, M, Bankier, A, Gecz, J, et al. (1999). A missense mutation in RPS6KA3 (RSK2) responsible for non-specific mental retardation. *Nat Genet*, 22:13–14.

Mohrschladt, MF, Bijlsma, EK, Sluijter, S, De Coo, RF, Hoovers, JM, and Leschot, NJ (1999). A patient with a de novo t (6;9) and an interstitial duplication of (9)(q21.2q22.1). *Clin Dysmorphol*, 8:211–214.

Molinari, F, Rio, M, Meskenaite, V, Encha-Razavi, F, Auge, J, Bacq, D, et al. (2002). Truncating neurotrypsin mutation in autosomal recessive nonsyndromic mental retardation. *Science*, 298:1779–1781.

Mulley, JC, Kerr, B, Stevenson, R, and Lubs, H (1992). Nomenclature guidelines for X-linked mental retardation. *Am J Med Genet*, 43:383–391.

Nance, WE and Engel, E (1972). One X and four hypotheses: response to Lehrke's "A theory of X-linkage of major intellectual traits." *Am J Ment Defic*, 76:623–625.

Newbury, DF, Bonora, E, Lamb, JA, Fisher, SE, Lai, CS, Baird, G, et al. (2002). FOXP2 is not a major susceptibility gene for autism or specific language impairment. *Am J Hum Genet*, 70:1318–1327.

Olton, DS, Wible, CG, and Shapiro, ML (1986). Mnemonic theories of hippocampal function. *Behav Neurosci*, 100:852–855.

Orrico, A, Lam, C, Galli, L, Dotti, MT, Hayek, G, Tong, SF, et al. (2000). MECP2 mutation in male patients with non-specific X-linked mental retardation. *FEBS Lett*, 481:285–288.

Osborne, LR, Li, M, Pober, B, Chitayat, D, Bodurtha, J, Mandel, A, et al. (2001). A 1.5 million-base pair inversion polymorphism in families with Williams-Beuren syndrome. *Nat Genet*, 29:321–325.

Partington, M, Mowat, D, Einfeld, S, Tonge, B, and Turner, G (2000). Genes on the X chromosome are important in undiagnosed mental retardation. *Am J Med Genet* 92:57–61.

Penrose (1938). *A Clinical and Genetic Study of 1280 Cases of Mental Defect*, vol. 229. London: HMSO.

Penrose (1949). *The Biology of Mental Defect*. UK: Anchor Press Ltd.

Petrij, F, Giles, RH, Dauwerse, HG, Saris, JJ, Hennekam, RC, Masuno, M, et al. (1995). Rubinstein–Taybi syndrome caused by mutations in the transcriptional co-activator CBP. *Nature*, 376:348–351.

Philippe, A, Martinez, M, Guilloud-Bataille, M, Gillberg, C, Rastam, M, Sponheim, E, et al. (1999). Genome-wide scan for autism susceptibility genes. Paris Autism Research International Sibpair Study. *Hum Mol Genet*, 8:805–812.

Puissant, H, Azoulay, M, Serre, JL, Piet LL, and Junien, C (1988). Molecular analysis of a reciprocal translocation t(5;11) (q11;p13) in a WAGR patient. *Hum Genet*, 79:280–282.

Ramser, J, Winnepenninckx, B, Lenski, C, Errijgers, V, Platzer, M, Schwartz, CE, et al. (2004). A splice site mutation in the methyltransferase gene FTSJ1 in Xp11.23 is associated with non-syndromic mental retardation in a large Belgian family (MRX9). *J Med Genet*, 41:679–683.

Ray, PN, Belfall, B, Duff, C, Logan, C, Kean, V, Thompson, MW, et al. (1985). Cloning of the breakpoint of an X;21 translocation associated with Duchenne muscular dystrophy. *Nature*, 318:672–675.

Raymond, FL, Tarpey, PS, Edkins, S, Tofts, C, O'Meara, S, Teague, J, et al. (2007). Mutations in ZDHHC9, which encodes a palmitoyltransferase of NRAS and HRAS, cause X-linked mental retardation associated with a Marfanoid habitus. *Am J Hum Genet*, 80(5):982–987.

Richards, BW, Sylvester, PE, and Brooker, C (1981). Fragile X-linked mental retardation: the Martin-Bell syndrome. *J Ment Defic Res*, 25 (Pt 4):253–256.

Risch, N, Spiker, D, Lotspeich, L, Nouri, N, Hinds, D, Hallmayer, J, et al. (1999). A genomic screen of autism: evidence for a multilocus etiology. *Am J Hum Genet*, 65:493–507.

Ropers, HH, Hoeltzenbein, M, Kalscheuer, V, Yntema, H, Hamel, B, Fryns, JP, et al. (2003). Nonsyndromic X-linked mental retardation: where are the missing mutations? *Trends Genet*, 19:316–320.

Rosenberg, EH, Almeida, LS, Kleefstra, T, deGrauw, RS, Yntema, HG, Bahi, N, et al. (2004). High prevalence of SLC6A8 deficiency in X-linked mental retardation. *Am J Hum Genet*, 75:97–105.

Saito-Ohara, F, Fukuda, Y, Ito, M, Agarwala, KL, Hayashi, M, Matsuo, M, et al. (2002). The Xq22 inversion breakpoint interrupted a novel Ras-like GTPase gene in a patient with Duchenne muscular dystrophy and profound mental retardation. *Am J Hum Genet*, 71:637–645.

Salomons, GS, van Dooren, SJ, Verhoeven, NM, Cecil, KM, Ball, WS, Degrauw, TJ, et al. (2001). X-linked creatine-transporter gene (SLC6A8) defect: a new creatine-deficiency syndrome. *Am J Hum Genet*, 68:1497–1500.

Schuback, DE, Mulligan, EL, Sims, KB, Tivol, EA, Greenberg, BD, Chang, SF, et al. (1999). Screen for MAOA mutations in target human groups. *Am J Med Genet*, 88:25–28.

Schwartz, CE, Johnson, JP, Holycross, B, Mandeville, TM, Sears, TS, Graul, EA, et al. (1988). Detection of submicroscopic deletions in band 17p13 in patients with the Miller-Dieker syndrome. *Am J Hum Genet*, 43:597–604.

Schwartz, CE, Tarpey, PS, Lubs, HA, Verloes, A, May, MM, Risheg, H, et al. (2007). The original Lujan syndrome family has a novel missense mutation (p.N1007S) in the MED12 gene. *J Med Genet*, 44(7):472–477.

Sebat, J, Lakshmi, B, Troge, J, Alexander, J, Young, J, Lundin, P, et al. (2004). Large-scale copy number polymorphism in the human genome. *Science*, 305:525–528.

Shapira, SK, McCaskill, C, Northrup, H, Spikes, AS, Elder, FF, Sutton, VR, et al. (1997). Chromosome 1p36 deletions: the clinical phenotype and molecular characterization of a common newly delineated syndrome. *Am J Hum Genet*, 61(3):642–650.

Sharp, AJ, Locke, DP, McGrath, SD, Cheng, Z, Bailey, JA, Vallente, RU, et al. (2005). Segmental duplications and copy-number variation in the human genome. *Am J Hum Genet*, 77:78–88.

Shaw, CJ, Withers, MA, and Lupski, JR (2004). Uncommon deletions of the Smith-Magenis syndrome region can be recurrent when alternate low-copy repeats act as homologous recombination substrates. *Am J Hum Genet*, 75:75–81.

Shaw-Smith, C, Redon, R, Rickman, L, Rio, M, Willatt, L, Fiegler, H, et al. (2004). Microarray based comparative genomic hybridisation (array-CGH) detects submicroscopic chromosomal deletions and duplications in patients with learning disability/mental retardation and dysmorphic features. *J Med Genet*, 41:241–248.

Shoichet, SA, Hoffmann, K, Menzel, C, Trautmann, U, Moser, B, Hoeltzenbein, M, et al. (2003). Mutations in the ZNF41 gene are associated with cognitive deficits: identification of a new candidate for X-linked mental retardation. *Am J Hum Genet*, 73:1341–1354.

Slager, RE, Newton, TL, Vlangos, CN, Finucane, B, and Elsea, SH (2003). Mutations in RAI1 associated with Smith-Magenis syndrome. *Nat Genet*, 33:466–468.

Slavotinek, A, Rosenberg, M, Knight, S, Gaunt, L, Fergusson, W, Killoran, C, et al. (1999). Screening for submicroscopic chromosome rearrangements in children with idiopathic mental retardation using microsatellite markers for the chromosome telomeres. *J Med Genet*, 36:405–411.

SLI Consortium (2002). A genomewide scan identifies two novel loci involved in specific language impairment. *Am J Hum Genet*, 70:384–398.

Somerville, MJ, Mervis, CB, Young, EJ, Seo, EJ, del Campo, M, Bamforth, S, et al. (2005). Severe expressive-language delay related to duplication of the Williams-Beuren locus. *N Engl J Med*, 353:1694–1701.

Stevenson, RE, Schwartz, CE, and Schroer, RJ (2000). *X-linked Mental Retardation*. Oxford: Oxford University Press.

Stromme, P, Mangelsdorf, ME, Shaw, MA, Lower, KM, Lewis, SM, Bruyere, H, et al. (2002). Mutations in the human ortholog of Aristaless cause X-linked mental retardation and epilepsy. *Nat Genet*, 30:441–445.

Stromswold, K (1998). Genetics of spoken language disorders. *Hum Biol*, 70:297–324.

Sutherland, GR and Baker, E (1992). Characterisation of a new rare fragile site easily confused with the fragile X. *Hum Mol Genet*, 1:111–113.

Takahashi, N, Hashida, H, Zhao, N, Misumi, Y, and Sakaki, Y (1995). High-density cDNA filter analysis of the expression profiles of the genes preferentially expressed in human brain. *Gene*, 164:219–227.

Tarpey, P, Parnau, J, Blow, M, Woffendin, H, Bignell, G, Cox, C, et al. (2004). Mutations in the DLG3 gene cause nonsyndromic X-linked mental retardation. *Am J Hum Genet*, 75:218–324.

Tarpey, PS, Stevens, C, Teague, J, Edkins, S, O'Meara, S, Avis, T, et al. (2006). Mutations in the gene encoding the Sigma 2 subunit of the adaptor protein 1 complex, AP1S2, cause X-linked mental retardation. *Am J Hum Genet*, 79 (6):1119–1124.

Tarpey, PS, Raymond, FL, O'Meara, S, Edkins, S, Teague, J, Butler, A, et al. (2007). Mutations in CUL4B, which encodes a ubiquitin E3 ligase subunit, cause an X-linked mental retardation syndrome associated with aggressive outbursts, seizures, relative macrocephaly, central obesity, hypogonadism, pes cavus, and tremor. *Am J Hum Genet*, 80(2):345–352.

Therman, E, Laxova, R, and Susman, B (1990). The critical region on the human Xq. *Hum Genet*, 85:455–461.

Tommerup, N (1993). Mendelian cytogenetics. Chromosome rearrangements associated with mendelian disorders. *J Med Genet*, 30:713–727.

Trivier, E, De Cesare, D, Jacquot, S, Pannetier, S, Zackai, E, Young, I, et al. (1996). Mutations in the kinase Rsk-2 associated with Coffin-Lowry syndrome. *Nature*, 384:567–570.

Turner, G and Partington, M (2000). Recurrence risks in undiagnosed mental retardation. *J Med Genet*, 37:E45.

Turner, G and Turner, B (1974). X-linked mental retardation. *J Med Genet*, 11:109–113.

Turner, G, Webb, T, Wake, S, and Robinson, H (1996). Prevalence of fragile X syndrome. *Am J Med Genet*, 64:196–197.

van der Luijt, RB, Tops, CM, Khan, PM, van der Klift, HM, Breukel, C, van Leeuwen-Cornelisse, IS, et al. (1995). Molecular, cytogenetic, and phenotypic studies of a constitutional reciprocal translocation t(5;10)(q22;q25) responsible for familial adenomatous polyposis in a Dutch pedigree. *Genes Chromosomes Cancer*, 13:192–202.

van der Maarel, SM, Scholten, IH, Huber, I, Philippe, C, Suijkerbuijk, RF, Gilgenkrantz, S, et al. (1996). Cloning and characterization of DXS6673E, a candidate gene for X-linked mental retardation in Xq13.1. *Hum Mol Genet*, 5:887–897.

Vargha-Khadem, F, Watkins, K, Alcock, K, Fletcher, P, and Passingham, R (1995). Praxic and nonverbal cognitive deficits in a large family with a genetically transmitted speech and language disorder. *Proc Natl Acad Sci USA*, 92:930–933.

Veltman, JA, Schoenmakers, EF, Eussen, BH, Janssen, I, Merkx, G, van Cleef, B, et al. (2002). High-throughput analysis of subtelomeric chromosome rearrangements by use of array-based comparative genomic hybridization. *Am J Hum Genet*, 70:1269–1276.

Veltman, JA, Yntema, HG, Lugtenberg, D, Arts, H, Briault, S, Huys, EH, et al. (2004). High resolution profiling of X chromosomal aberrations by array comparative genomic hybridisation. *J Med Genet*, 41:425–432.

Vervoort, VS, Beachem, MA, Edwards, PS, Ladd, S, Miller, KE, de Mollerat, X, et al. (2002). AGTR2 mutations in X-linked mental retardation. *Science*, 296:2401–2403.

Villard, L, Gecz, J, Mattei, JF, Fontes, M, Saugier-Veber, P, Munnich, A, et al. (1996). XNP mutation in a large family with Juberg-Marsidi syndrome. *Nat Genet*, 12:359–360.

Villard, L, Kpebe, A, Cardoso, C, Chelly, PJ, Tardieu, PM, and Fontes, M (2000). Two affected boys in a Rett syndrome family: clinical and molecular findings. *Neurology*, 55:1188–1193.

Vissers, LE, de Vries, BB, Osoegawa, K, Janssen, IM, Feuth, T, Choy, CO, et al. (2003). Array-based comparative genomic hybridization for the genome-wide detection of submicroscopic chromosomal abnormalities. *Am J Hum Genet*, 73:1261–1270.

Waters, JJ, Campbell, PL, Crocker, AJ, and Campbell, CM (2001). Phenotypic effects of balanced X-autosome translocations in females: a retrospective survey of 104 cases reported from UK laboratories. *Hum Genet*, 108:318–327.

WHO (1992). The ICD-10 classification of mental and behavioural disorders. WHO.

Willatt, L, Cox J, Barber, J, Cabanas, ED, Collins, A, Donnai, D, FitzPatrick, DR, et al. (2005). 3q29 microdeletion syndrome: clinical and molecular characterization of a new syndrome. *Am J Hum Genet*, 77:154–160.

Winnepenninckx, B, Errijgers, V, Hayez-Delatte, F, Reyniers, E, and Frank Kooy, R (2002). Identification of a family with nonspecific mental retardation (MRX79) with the A140V mutation in the MECP2 gene: is there a need for routine screening? *Hum Mutat*, 20:249–252.

Wirth, J, Nothwang, HG, van der Maarel, S, Menzel, C, Borck, G, Lopez-Pajares, I, et al. (1999). Systematic characterisation of disease associated balanced chromosome rearrangements by FISH: cytogenetically and genetically anchored YACs identify microdeletions and candidate regions for mental retardation genes. *J Med Genet*, 36:271–278.

Yntema, HG, Oudakker, AR, Kleefstra, T, Hamel, BC, van Bokhoven, H, Chelly, J, et al. (2002). In-frame deletion in MECP2 causes mild nonspecific mental retardation. *Am J Med Genet*, 107:81–83.

Yntema, HG, Poppelaars, FA, Derksen, E, Oudakker, AR, van Roosmalen, T, Jacobs, A, et al. (2002). Expanding phenotype of XNP mutations: mild to moderate mental retardation. *Am J Med Genet*, 110:243–247.

Yntema, HG, van den Helm, B, Knoers, NV, Smits, AP, van Roosmalen, T, Smeets, DF, et al. (1999). X-linked mental retardation: evidence for a recent mutation in a five-generation family (MRX65) linked to the pericentromeric region. *Am J Med Genet*, 85:305–308.

Yonan, AL, Alarcon, M, Cheng, R, Magnusson, PK, Spence, SJ, Palmer, AA, et al. (2003). A genomewide screen of 345 families for autism-susceptibility loci. *Am J Hum Genet*, 73:886–897.

Zechner, U, Wilda, M, Kehrer-Sawatzki, H, Vogel, W, Fundele, R, and Hameister, H (2001). A high density of X-linked genes for general cognitive ability: a run-away process shaping human evolution? *Trends Genet*, 17:697–701.

Zemni, R, Bienvenu, T, Vinet, MC, Sefiani, A, Carrie, A, Billuart, P, et al. (2000). A new gene involved in X-linked mental retardation identified by analysis of an X;2 balanced translocation. *Nat Genet*, 24:167–170.

30

Complex Ophthalmic Disorders

Forbes DC Manson, Andrew R Webster, and Graeme CM Black

How Much of Clinical Ophthalmology is Genetic?

As in many other clinical specialties, the overall contribution of genetics to the whole of ophthalmic disease is broader than generally imagined. For some conditions, this is obvious and when faced in the clinic with a patient with retinitis pigmentosa (RP), the ophthalmologist is immediately aware of the inherited basis of a patient's condition. However less straightforward, but no less significant, are the genetic etiologies that are beginning to emerge for rare sporadic disease such as microphthalmia or optic nerve hypoplasia; for common early-onset phenotypes such as squint or myopia; and for later-onset phenotypes such as senile cataract or macular degeneration.

Monogenic or Mendelian forms of ocular diseases affect every tissue of the eye and encompass all age groups and subspecialties. As a result of their relative ease of study and the Human Genome Project, these single gene disorders have received a considerable, and perhaps, disproportionate focus in recent times. Equally, it may be argued that the intensive study of monogenic disorders is fully justified as they often represent an excellent model for complex diseases and are our best source for novel drug and genetic medicines (Antonarakis and Beckmann, 2006; Brinkman et al., 2006; O'Connor and Crystal, 2006). Many such conditions are early-onset and the pediatric ophthalmologists constitute a large and important group that underlies perhaps as many as half of those children registered blind or partial-sighted in industrialized countries (Gilbert and Foster, 2001; Alagaratnam et al., 2002; Rahi and Cable, 2003). Many later-onset disorders including macular (e.g., Sorsby fundus dystrophy and autosomal dominant drusen) and corneal dystrophies (e.g., Fuchs' dystrophy) represent important models of common phenotypes and are ideal starting point for studying the primary mechanisms which underlie the pathogenesis of the more common and earlier-onset diseases. Human Mendelian conditions associated with ophthalmic phenotypes account for perhaps a third of the single gene disorders documented in Online Mendelian Inheritance in Man (OMIM) (http://www.ncbi.nlm.nih.gov/entrez/query.fcgi?db = OMIM). Many of those that lack systemic associations have little impact upon reproductive fitness and therefore represent a powerful resource of extensively characterized conditions which have proved to be extremely valuable as a resource for gene identification programs.

However, many of the commonest groups of ophthalmic conditions (many of which contribute significantly to world blindness) that the general ophthalmologist encounters on a frequent basis, including those of early- (myopia, squint) and late-onset [age-related macular degeneration (AMD), senile cataract, and primary open-angle glaucoma (POAG)], have a strong genetic component. For age-related cataract, segregation analyses suggest the existence of a major genetic component that accounts for approximately half of cataract variability while twin studies also show significant heritability for both cortical (53%–58%) and nuclear cataracts (48%) (Hammond et al., 2000, 2001; Heiba et al., 1995). There is also a range of evidence for the existence of genetic factors contributing to AMD including concordance studies (which suggest a heritability of ~45%), an increased risk for siblings, and reports of single families in which apparently monogenic subforms of AMD segregate (Hammond et al., 2002). POAG also has a strong genetic component based on family and twin concordance studies. The influence of functional genomics continues to impact upon the clinician. Having begun slowly with large-scale positionally based or candidate gene approaches, more recent genome-wide searches have identified potential loci for POAG (9q22, 20p12), cataract (6p12-q12), and AMD (1q31, 10q26, and 17q25) (Iyengar et al., 2004; Weeks et al., 2004; Wiggs et al., 2004). The increasing impact in the study of common disease is illustrated by the recent success of a combined positional/candidate gene approach that has identified four genes as having an important contribution to AMD (see below). Thus, there may now be a realistic prospect that we may begin to better understand groups of conditions whose complex multigenic nature has previously hindered progress. By identification of the complement system in AMD pathogenesis, for example, now presents novel therapeutic targets for a disease which was previously difficult to study and treat.

The need for a working knowledge of genetics can no longer be regarded as an arcane practice for academics as the field has progressed to a point where it now actively contributes to clinical practice. For some conditions, such as retinoblastoma, X-linked RP, and Sorsby fundus dystrophy, genetic testing already has a part to play in the diagnosis, counseling and management of patients and their families. In the near future, such services will be more comprehensive and widely available. Whilst genetic testing contributes to the clinician's ability to diagnose and has a potentially profound impact upon decision-making for patients and families, seldom has the power to alter disease progression or management been so tangible. This is exemplified in the progress that has been made in the development of novel gene-based therapies. Viral-based gene delivery approaches to the retina and/or retinal pigment epithelium (RPE) have now been attempted in rodents and larger animals such as the RPE65-deficient Briard dog (Acland et al., 2001). Although in its infancy these gene therapy studies demonstrate that not only can the retinal morphology and function be improved, but that the disease progression can also be altered.

The retina's central nervous system origins bestow it with a broad range of gene expression, and the huge progress in ophthalmic genetics confirms how powerful a tool the human eye remains for the study of genes and proteins in disease. This is demonstrated by the numbers of mapped and cloned genes known to underlie retinal disease which in the 10 years to 2005 increased four- and five-fold, respectively to 110 and 158 (RetNet: http://www.sph.uth.tmc.edu/RetNet/home.htm).

Monogenic Disease and Ophthalmology

The impact and importance of classical Mendelian genetics in ophthalmology has been profound and is well illustrated using the examples of ocular patterning defects and RP.

Defects of Ocular Patterning

Defects of early ocular development are often lumped into a single heterogeneous group termed microphthalmia. This represents not one but a spectrum of clinical entities which can be referred to as MAC (microphthalmia, anophthalmia, and coloboma). These may be unilateral or bilateral and may range from complete anophthalmia, through microphthalmia (defined as an axial length of <19.5 mm at the age of 1 year) to nanophthalmia (defined as a short axial length, high hypermetropia, and shallow anterior segment but normal visual function). Such defects result from interruptions in the earliest processes of ocular development including the formation of the optic vesicle (which forms as neuroectodermal evaginations of the forebrain), the induction of the lens placode/vesicle, and the differentiation of the neural retina and RPE. Since in many cases there may be an associated failure of the optic fissure to close (resulting in the formation of a colobomatous defect), this too is seen as a key event.

Heterogeneity of Ocular Developmental Abnormalities

The underlying cause of ocular developmental abnormalities is varied and may be associated with a large number of single gene disorders as well as developmental abnormalities of a broad etiology, including chromosomal abnormalities, maternal infection or toxin ingestion (e.g., antiepileptic drugs or alcohol), environmental influences and fetal disruptions. There has been considerable progress in understanding the single gene defects underlying these conditions, both in humans and in other mammals. Two such genes are *CHX10* and *RAX* (neural retina-specific homeodomain-containing transcription factors) which have, to date, been defined in only a very small number of patients (Percin et al., 2000; Voronina et al., 2004). Although of limited clinical value, the major benefit of such observations is in understanding the interconnecting pathways and cell-specific defects that can lead to defects in organogenesis. In the case of *RAX*, the evidence from human, rodent, and other vertebrate models suggests that it acts as a transcriptional activator that determines retinal cell fate (Furukawa et al., 1997; Kimura et al., 2000). In contrast, pathogenic (usually de novo) mutations in the genes encoding the transcription factors SOX2 and OTX2 have been found in significant numbers (~15%–20%) of patients with severe, bilateral anophthalmia and microphthalmia (Ragge et al., 2005a,b). This presents a realistic prospect for the genetic screening of patients with the severest forms of bilateral anophthalmia/microphthalmia.

The RPE and Ocular Patterning

Structures such as the lens are repeatedly implicated as a cause of microphthalmia and anophthalmia (reviewed by Manson et al., 2005) and are consequently implicated as being crucial to normal ocular development. Defects in lens-derived transcription regulators (e.g., MAF, BCOR) and structural molecules (e.g., crystallins) can cause both cataract and microphthalmia (Graw et al., 2001; Jamieson et al., 2002; Ng et al., 2004). Although it has long been recognized that Mitf is a key regulator of RPE gene expression whose interruption underlies the *microphthalmia* series of murine mutants (Steingrimsson et al., 2004), the role of the RPE in early ocular patterning has come under renewed scrutiny. OTX2 also has a crucial role in regulating RPE-specific gene expression as it induces a pigmentary cell phenotype in transfected retina cells, and with MITF co-activates a number of RPE-specific genes including *Tyr* and *TRP-1*. Furthermore MITF and OTX2 co-localize in RPE cells and are able to interact through the MITF basic helix–loop–helix domain. Defects in other RPE-specific genes including *VMD2*, a chloride channel whose expression is regulated by MITF, and *MFRP*, a frizzled-related protein of unknown function, have been described in patients with milder reductions in ocular size including nanophthalmos and extreme long-sightedness (Yardley et al., 2004; Sundin et al., 2005). These data demonstrate the importance of posterior segment structures, specifically of the RPE, in both the development and signaling of the entire eye and in the patterning of anterior segment structures such as the lens (since such patients can develop premature cataract and glaucoma).

Diagnostic Impact of Gene Discovery

Whilst the continued identification of the genetic causes of developmental eye defects allows increased molecular diagnosis and the construction of developmental pathways, there remain many challenges ahead. In many familial cases there is considerable intrafamilial phenotypic variability (e.g., optic fissure closure defects) suggesting the existence of genetic and/or environmental modifiers and there is considerable potential for progress in this area, especially in the identification of digenic effects. Microphthalmia can be an isolated condition, but in approximately half of cases it also has multisystemic associations and the careful clinical phenotyping of patients who carry such mutations will enable the delineation of gene-specific structural abnormalities. For example, patients carrying *PAX6* mutations may also have impaired olfaction, cognitive dysfunction or autism (Heyman et al., 1999; Malandrini et al., 2001; Chao et al., 2003). Magnetic resonance imaging (MRI) scans of patients with an *OTX2* mutation have been described with specific hippocampal abnormalities while patients with a *SOX2* mutation also have a clinically recognizable multi-system disorder (Ragge et al., 2005a,b). By improving the targeting of genetic testing such observations can facilitate an improvement in recurrent risk estimation for these patients. Combining clinical descriptions with knowledge of the underlying genetics causes, underlines the observation that many of the transcriptional regulators of ocular development are critical for controlling the development of other tissues. Hence, the insights gained from the characterization of the molecules and pathways that are fundamental to ocular development may have a broader impact.

The Inherited Retinal Dystrophies Including RP

In the field of ophthalmology, no one condition exemplifies the progress that has been made using human molecular genetics better than RP. RP is a blanket description for a large group of inherited retinal degenerations that, despite the name, do not involve inflammation. The condition is relatively common with an incidence estimated to be between 1 in 3000–5000 (Heckenlively, 1989; Dryja and Li, 1995). Classically, RP presents with an initial and progressive degeneration of the rod photoreceptors that starts in the peripheral retina giving tunnel vision and night blindness (nyctalopia). As the disease progresses, the central retina and cone photoreceptors become involved, resulting in a concomitant reduction in central vision including loss of visual acuity and color perception. As the neural retina dies, the retinal arterioles become attenuated and the characteristic waxy optic disc pallor develops. In addition, cells from the underlying RPE migrate into the retina and may form bone spicules, one of the most characteristic features of the disease. An initial reduction of the dark-adapted rod (scotopic) electroretinogram (ERG) is followed by a reduction of the light adapted, cone-mediated (photopic) ERG. Ultimately, the ERG becomes unrecordable.

Clinically, RP is also highly heterogeneous in terms of onset and severity. This is reflected at the genetic level with autosomal dominant (ad), autosomal recessive (ar), X-linked (xl), and digenic forms all described. In addition, RP may form part of a number of syndromic disorders. Forty-three loci have been described for non-syndromic RP, of which 35 genes have been identified. A further 118 genes or loci have been described that are associated with retinal disease (RetNet).

Cloning of Single Genes Causing RP: A Powerful Approach

Rhodopsin was identified as the first RP gene (causing adRP) using a positional candidate gene approach in 1990 (Dryja et al., 1990). It was also found to be responsible for an arRP 2 years later (Rosenfeld et al., 1992) and remains the only molecule that causes both dominant and recessive forms of RP. Since that time the molecular basis of many monogenic retinal dystrophies have been identified using techniques that, although in routine use in human genetics, demonstrate the extraordinary power of genetic approaches in the dissection of heterogeneous disorders. Positional approaches have been used extremely successfully in linked families to identify mutated genes in obvious pathways (such as photoreceptor specific pathways) as well as in less obvious pathways. This latter group includes the transcription factors cone-rod homeobox (CRX) and neural retina leucine zipper (NRL), whose function was previously undefined, and the ubiquitous splicing factors PRPF3, 8, 31 and PAP-1. These molecules are just a few examples that could not have been identified using hypothesis-driven approaches and may provide the most interesting and unsuspected areas for subsequent investigation.

Pathways Commonly Disrupted in Retinal Degenerations

The proteins responsible for RP have now been successfully grouped according to their normal function (Rivolta et al., 2002; Hims et al., 2003; Kennann et al., 2005). Those responsible for adRP include structural photoreceptor proteins, ribonucleic acid (RNA) splicing regulators and transcription factors, as well as proteins involved in the photoreceptor's intracellular transport. In addition to structural proteins and transcription factors, proteins mutated in arRP also include several that are part of the phototransduction cascade and the visual cycle, but no splicing factors. Both proteins identified as causing xlRP are implicated in intracellular transport, making this function common to all three modes of inheritance and highlights the crucial and vulnerable nature of this process in photoreceptors. Although such groupings can be useful for our understanding of RP and the possible identification of further genes they are not infallible, and when proteins involved in intracellular transport are considered no pattern emerges.

As can been seen from Table 30–1 there is, given the large number of genes and proteins implicated in retinal dystrophies, a functional convergence of such abnormalities. The disruption of a surprisingly small number of key photoreceptor pathways underlie the majority of those defects that have been identified. This is illustrated below in regard of two such groups: messenger RNA (mRNA) splicing factors and proteins implicated in photoreceptor cilium function.

Table 30–1 Retinitis Pigmentosa Genes Grouped According to Function

			Structural						
	Photo-Transduction Cascade	Visual Cycle	Disc	Connect-ing Cilium	Plasma Membrane	Transcrip-tion Factor	Splicing Factor	Intracellular Transport	Misc.
ad	Rhodopsin		**Rhodopsin** peripherin **fascin** ROM1[a]	RP1		CRX, NRL	PRPF3 PRPF8 PRPF31 PAP-1	Fascin?, RP1	CA4, IMPDH1, guanylate activating protein
ar	Phospho-diesterase α and β arrestin rhodopsin	RPE65 RLBP ABCA4 LRAT RGR	Rhodopsin		Cyclic nu-cleotide gated channel α 1 and 2	NR2E3, NRL		Tubby-like protein	CERKL, MERTK, CRB1, my-osin VIIa
xl								RPGR, RP2	

[a] ROM1 causes digenic RP in conjunction with RDS.
Those in bold can be assigned to multiple roles.

Defects of RNA Splicing Factors

Four dominant forms of RP (RP13, RP11, RP18, and RP9) are caused by mutations in splicing factors (McKie et al., 2001; Vithana et al., 2001; Chakarova et al., 2002; Keen et al., 2002; Maita H et al., 2004). These proteins form part of the spliceosome, a large multiprotein complex that removes intronic sequences from the primary RNA transcript before the mRNA of a gene can be translated into protein. An explanation for the possible disease mechanism resulting from the mutations in these splicing factors has presented researchers with a problem as they are widely or ubiquitously expressed and yet result in a photoreceptor-specific phenotype. One hypothesis suggests that because the rods have such a huge demand for protein synthesis (the outer segment is turned over every 10 days), and so for splicing, any compromise of the splicing machinery will be deleterious. The retina also has one of the highest metabolic rates of any tissue in the body, which may add further demands on to an already "stressed" splicing system (Chakarova et al., 2002). It has also been suggested that the mutated splicing factors decrease the rate at which the spliceosome is activated, making this a rate-limiting step in photoreceptors (Kuhn et al., 2002; Pacione et al., 2003). Lastly, the retinal specificity may be due to a common interacting co-factor that is, spatially restricted to the retina (Chakarova et al., 2002). The only mechanism that has any experimental evidence demonstrates a specific interaction between a splicing factor and a retina-specific protein. Studies on PRPF31, which is mutated in RP11, show that the mutant factor causes cell death in primary cultured cells and significantly inhibits the pre-mRNA splicing of rhodopsin intron 3. This functionally links two major adRP genes (*RHO* and *PRPF31*) and provides an explanation for the retinal specificity (Yuan et al., 2005).

Defects of the Photoreceptor Cilium

The analysis of molecules expressed in the photoreceptor cilium has been the subject of a considerable degree of interest and several such molecules have recently been implicated in several forms of syndromic RP, nonsyndromic RP, and other retinal degenerations. The connecting cilium is the conduit between the inner and outer segments and is the key to the process of intracellular transport (Rosenbaum and Witman, 2002).

Nonsyndromic RP is an extremely well covered topic in the scientific literature with many recent reviews (Rivolta et al., 2002; Hims et al., 2003; Kalloniatis and Fletcher, 2004; Ben-Arie-Weintrob et al., 2005; Kennann et al., 2005; Wang et al., 2005). Of these *RPGR*, the gene mutated in the RP3 form of xlRP, represents the first such molecule to be localized to the cilium and illustrates the importance of its integrity in maintaining photoreceptor health. In addition to xlRP, mutations in *RPGR* are responsible for a heterogeneous collection of retinal phenotypes that include cone dystrophy (Yang et al., 2002), cone-rod dystrophy (Ayyagari et al., 2002), and atrophic macular dystrophy (Demirci et al., 2002). *RPGR* is also associated with syndromic RP and interacts with a number of other proteins that cause retinal dystrophies that include nehprocystin-5 mutated in the childhood renal abnormality nephronophthisis (NPHP) with retinal degeneration (Otto et al., 2005) and RP GTPase regulator-interacting protein (RPGRIP) which is mutated in a proportion of those with the severe recessive retinal degeneration Leber's congenital amaurosis (LCA) (Dryja et al., 2001; Gerber et al., 2001).

A possible role for RPGR in the cilium has been suggested for many years by the clinical observations of a small number of xlRP families with abnormal nasal cilia and axonomes (Arden and Fox, 1979; Fox et al., 1980; Hunter et al., 1988). A putative role for RPGR in syndromic RP was derived from the description of an xlRP family who were susceptible to respiratory infections (recurrent bronchitis and sinusitis) (van Dorp, 1992) and were later found to have an *RPGR* splice site mutation (Dry et al., 1999). The condition was indistinguishable from immotile cilia syndrome 1 (ICS1) except for the lack of deafness or sterility. Koenekoop et al. then described five members of an xlRP family with a frameshift in exon 10 of *RPGR* with hearing loss (Koenekoop et al., 2003). Since then other families have been described with *RPGR* mutations that have RP, hearing loss (sensorineural or conductive) and recurring sinorespiratory infections (Iannaccone et al., 2003; Zito et al., 2003; Iannaccone et al., 2004).

Although no function has yet been ascribed to RPGR a number of observations are beginning to suggest the functional pathways in which it may play a role. Mice with a knockout of *Rpgr* develop normal retinas which subsequently underwent a slow retinal degeneration. Abnormal distribution of rod and cone opsins suggested Rpgr may play a role in the transport of these proteins across the connecting cilium (Hong et al., 2000). Rpgr is localized to the connecting cilium by Rpgrip, and when this is knocked out the mice have a more severe phenotype compared to the Rpgr knockout. Retinal development was abnormal and then degenerated rapidly. Outer segments also generated grossly over-sized discs. As Rpgrip tethers Rpgr to the connecting cilium, an Rpgrip knockout is functionally and phenotypically indistinguishable from a double Rpgr-Rpgrip knockout (Zhao et al., 2003). In cultured cells $RPGR^{ORF15}$, the predominant isoform of RPGR in the retina interacts with nucleophosmin in the centrioles, which are the structures that give rise to the basal body from which cilia originate. However, $RPGR^{ORF15}$ and nucleophosmin could not be co-localized at the connecting cilium and the interaction between these proteins is thought to occur in the nucleus. It may be that both proteins participate in the assembly of centrosomal proteins and that mutation affecting the interaction between $RPGR^{ORF15}$ and nucleophosmin affects centrosomal duplication (Shu et al., 2005). The significance of this interaction in a postmitotic tissue such as the retina is unclear.

The importance of functional integrity for the centriole/basal body/connecting cilium is further illustrated in a syndromic RP in which another RPGRIP is mutated. Ten percent of patients with NPHP, the most common genetic cause of chronic renal failure in children, also have RP. Together, these two conditions constitute Senior-Loken syndrome (SLSN). A novel gene, *IQCB1* (*NPHP5*), found to be mutated in a consanguineous family with NPHP was also found to be mutated in 16 other NPHP patients. As three other NPHP genes were localized to the primary cilia of the renal epithelium and all 16 NPHP patients also had RP (i.e., were SLSN) the authors tested whether NPHP5 interacted with RPGR, a cilia protein that causes RP. NPHP5 also interacts with calmodulin by virtue of two IQ domains, and all three proteins can be co-immunoprecipitated from retinal extracts and localize to the connecting cilium of photoreceptors and the primary cilia of renal epithelial cells (Otto et al., 2005).

Syndromic RP and the Connecting Cilium

A search of OMIM for phenotypically described forms of RP with a known molecular basis returned 76 hits. As there are 26 known nonsyndromic RP genes, this leaves in the region of 50 syndromes that include RP as part of their phenotypic description.

Bardet–Biedl Syndrome

Bardet–Biedl syndrome (BBS) is a rare condition ($\sim$1 in 1,00,000) with a range of features including RP, obesity, polydactyly, kidney

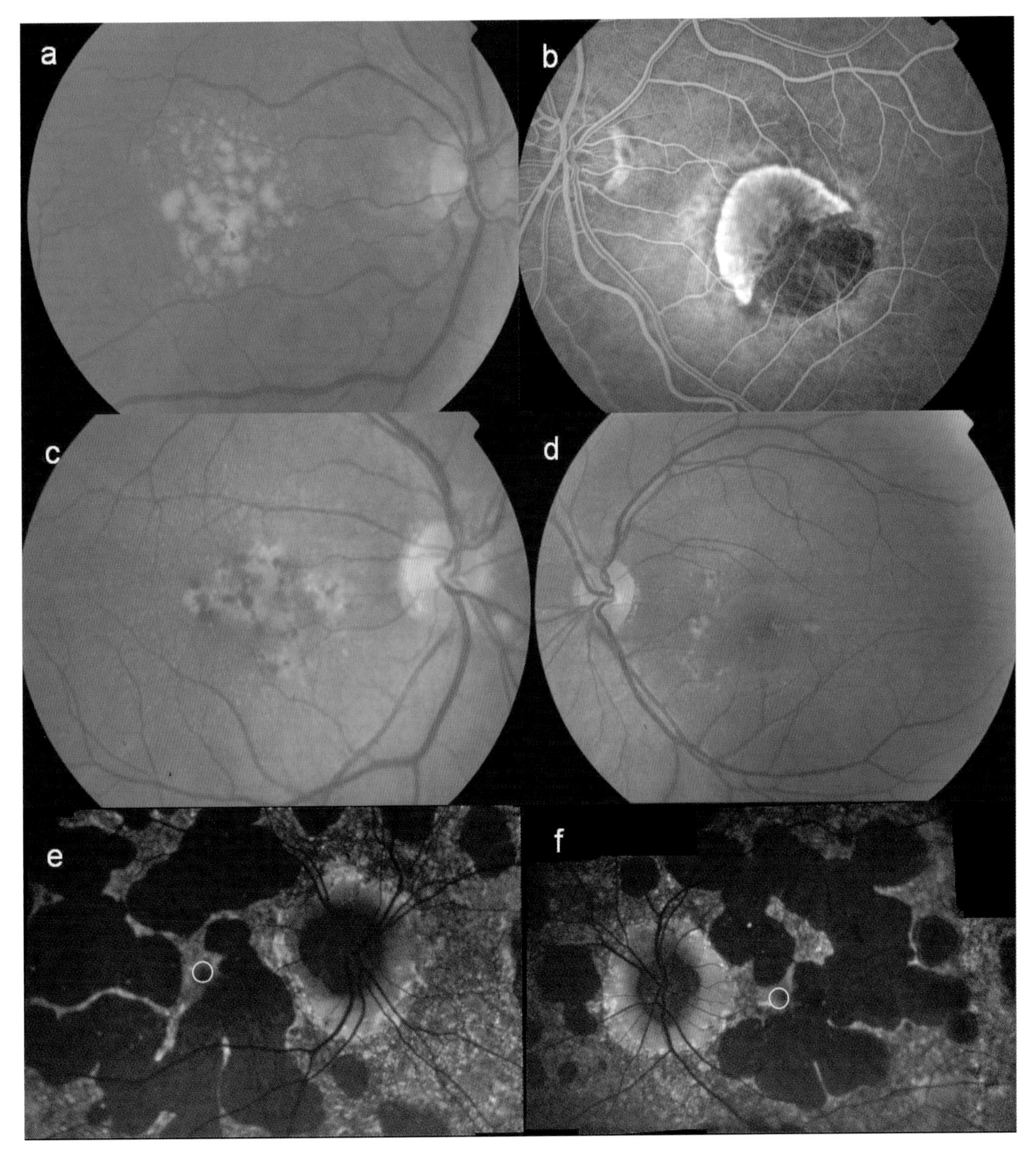

Figure 30–1 (*A*) Color image of the right retina of a patient affected with drusen deposit due to age-related macular degeneration (AMD). (*B*) Image of the left retina obtained following intravenous (IV) injection of 1 g of sodium fluorescein showing the new blood vessels occurring in AMD; this component of the disease is called choroidal neovascularization. (*C* and *D*). Right and left retina color images of a 32-year-old patient with early-onset drusen due to the single-gene disorder, Dominant Drusen, caused by an arginine to tryptophan missense mutation at codon 345 of the fibulin-3 gene. (*E* and *F*) right and left autofluorescence images of a 40-year-old patient with macular dystrophy. The dark areas indicate regions in which the outer layers of the retina have atrophied. The white circles delineate the foveae, the region of the retina where fixation and detailed vision occur. In this patient, these regions remain preserved despite surrounding atrophy and maintain normal visual acuity. Two missense variants, p.Arg24His and p.Thr1526Met were found in the *ABCA4* gene.

abnormalities, learning difficulties, heart defects, and genital anomalies (Mykytyn and Sheffield, 2004). Like RP, it is genetically heterogeneous with at least eight loci. There was a stepwise identification of BBS genes and the discovery of their possible function. The genes for BBS2, BBS4, and BBS6 were identified by positional cloning but none of the predicted protein products showed homology either with each other or with other previously known proteins, giving no clue as to their function (Katsanis et al., 2000; Slavotinek et al., 2000; Nishimura et al., 2001; Mykytyn et al., 2001). Limited homology searches allowed the identification of two further genes, *BBS1* and *BBS8* (Mykytyn et al., 2002; Ansley et al., 2003), and it was homology to the latter of these that provided a clue to a possible common cilium function for the BBS proteins. Several lines of evidence suggest that BBS8 was in some way involved in cilium function. First, it showed homology to *PilF*, a prokaryotic domain involved in pilus formation and twitching mobility. Second, BBS patients who were homozygous null for BBS8 showed anomalies in their left–right body axis symmetry, which is known to be caused by defects in the nodal cilium. Third, BBS8 localized to the centrosome and basal body of ciliated structures including the retinal connecting cilium and bronchial ciliated columnar epithelial cells, and co-localized with pericentriolar material 1 protein (PCM1), a protein that interacts with BBS4 and that probably has a role in ciliogenesis. Lastly, *Caenorhabditis elegans* homologues to BBS1, BBS2, BBS7, and BBS8 are all exclusively expressed in ciliated neurones (Ansley et al., 2003). The specific role of each BBS proteins in the cilium remains unknown, as does whether the proteins interact or provide any degree of redundancy. There are likely to be several more BBS loci and it will be interesting to see, as might be predicted, if they too locate to the basal body.

Usher's Syndrome

A second syndrome commonly associated with an RP phenotype is Usher's syndrome (USH). This recessive syndrome is associated with bilateral sensorineural hearing loss and RP. USH occurs at a frequency of 1 in 16,000–50,000 (Keats and Corey, 1999) and has been estimated to account for 18% of RP cases (Boughman et al., 1983). As with RP and BBS, USH is clinically and genetically heterogeneous with three clinical types and 11 loci recognized. USH type 1 (USH1) is characterized by severe-to-profound hearing loss, vestibular dysfunction, and RP that starts in early childhood. Of the seven USH1 loci, USH1B is the most important and accounts for approximately 75% of USH1 clinical cases. The gene (*MYO7A*) encodes unconventional myosin 7a that, in conjunction with vezatin, strengthens the adhesion between cells and between stereocilia (Petit, 2004). However not all the USH genes encode cilia-related proteins. Of the three loci responsible for USH type 2 (USH2), USH2A is the most common and accounts for 70% of all Usher patients. USH2 patients have moderate-to-severe hearing loss with normal vestibular function and a later-onset RP that typically manifests in the teenage years. *USH2A* encodes a novel glycoprotein that interacts with collagen intravenous (IV), fibronectin and integrins in retinal basement membranes (Bhattacharya and Cosgrove, 2005). The most recently described Usher gene (*VLGR1*) is that for USH2C and was identified through a candidate gene approach. The very large G-protein receptor is predicted to have both intracellular and extracellular protein–protein interaction domains that may allow a wide range of cellular and signaling roles, although no ligands have yet been identified (Weston et al., 2004). VLGR1 mutations were previously associated with a reflex-seizure phenotype in humans and mice. Only one locus has been described for USH type 3 (USH3), a largely Finnish-associated condition which is distinguished by a progressive hearing loss (Joensuu et al., 2001). The gene, *USH3A*, encodes clarin-1, one of three human paralogs which have no prokaryotic orthologs. The glycoprotein is predicted to have four transmembrane domains and shows limited homology to the synapse protein stargazing (Adato et al., 2002).

Common Disease and Ophthalmology: AMD

AMD is the most common registered cause of blindness in the Western world and accounts for more than 50% of all cases in the UK. Its prevalence is rising, a fact that cannot be fully explained by the increasing age of the population (Evans et al., 1996). In AMD, the macula can be destroyed by several mechanisms, resulting in the loss of central vision. In the so-called "wet" form of AMD, new blood vessel growth arising from the choroid (choroidal neovascularization) leads to hemorrhage and subsequent scarring and in the "dry" form there is progressive geographic atrophy of the macular retina. These processes occur as a consequence of age changes at the level of the RPE and the underlying Bruch's membrane, and are recognized clinically as focal deposits (drusen) and irregularities of RPE pigmentation. These early changes are collectively termed age-related maculopathy (ARM), although classification of the disease is not straightforward due to its variability. However, an international classification is recognized (Bird et al., 1995) and additional histopathological findings such as diffuse thickening of Bruch's membrane, an increase in lipofuscin and residual bodies in the RPE, and a narrowing of the choroidal capillaries have aided precise diagnosis (Ramrattan et al., 1994; Okubo et al., 1999).

Several treatments aimed at the ablation of new blood vessel growth (e.g., radiotherapy, surgical excision, surgical repositioning of the neurosensory macula, laser treatment, and photodynamic therapy) benefit a small minority of patients but only serve to delay or, at best reduce, the magnitude of visual loss. Moreover, these therapies are generally destructive in nature and may themselves lead to retinal damage, with consequent visual reduction. None has had a significant impact on visual loss in the community. However, the development of molecules that inhibit vascular endothelial growth factor (VEGF), which is present in the new blood vessels that cause visual loss in AMD, gives renewed hope for an effective treatment. A humanized monoclonal antibody fragment, ranibizumab (Lucentis; Genentech) has been shown in phase 3 clinical trials to improve the mean visual acuity of those treated. Furthermore, the corresponding full-size humanized monoclonal antibody, bevacizumab (Avastin; Genentech) appears to show a similar efficacy. The effect of these agents is anticipated to have a major impact on the preservation of visual acuity in AMD, and the outlook is now more optimistic.

Genetic and Environmental Susceptibility

It is generally believed that AMD is a complex disease which manifests in those with a genetic susceptibility when combined with the appropriate environmental factors. The importance of environmental factors is based on the increasing prevalence of the disease in developed nations and its emergence in the elderly in those communities with changing environment such as India and Eastern Asia (notably Japan). Despite much effort, the evidence for epidemiological factors that influence susceptibility is scant. Many factors have been proposed including smoking, hypertension, plasma and dietary anti-oxidants such as carotenoids, iris color, skin-tanning sensitivity, and wine consumption (Seddon et al., 1994, 1996; Christen et al., 1996; Smith et al., 1998). Apart from smoking, none of these factors are clearly established or appear to be clinically significant. However, the population attributable risk of smoking is significant and when analyzed in the light of UK demographic data would represent 36,100 persons with visual impairment and 17,800 persons with legal blindness having AMD attributable solely to smoking in the UK alone (Kelly et al., 2004).

By contrast, the existence of a strong genetic component in the genesis of AMD has been known for some time. Twin concordance rates are high, 90%–100% in monozygotic twins compared to 48% in dizygotic twins (Klein et al., 1994; Meyers et al., 1995; Gottfredsdottier, 1999), and sibship studies suggest a relative risk to sibs of AMD patients versus population controls to be 1.9–9.0 (Heiba et al., 1994; Silvestri et al., 1994; Seddon et al., 1997; Klaver et al., 1998a;). Studies of both early ARM and visual loss due to AMD in larger cohorts of monozygotic and dizygotic twins estimate a heritability of 40%–80% (Hammond et al., 2002; Seddon et al., 2005).

Macular Dystrophy: Identification of Candidate Genes

Many monogenic disorders manifest certain components of AMD, and for this reason the associated causative genes have been the subject of intense investigation. Examples include *EFEMP1* (mutations cause autosomal dominant drusen); *VMD2* (mutated in Best's macular dystrophy); *TIMP3* (mutated in Sorsby disease); *ABCA4* (mutated in Stargardt macular dystrophy); and *RDS* (which is mutated in a variety of autosomal dominant macular dystrophies). Mutations in *RDS* show phenotypic heterogeneity since they may also cause autosomal dominant RP, and genetic background (or other modifying factors) may modulate the phenotype caused by the same *RDS* mutation (Weleber et al., 1993). However, there is no evidence to suggest that specific coding variants or haplotypes of any of the genes that cause monogenic macular disease confer a risk to the development of AMD, despite the phenotypic similarities.

The *ABCA4* gene, originally termed *ABCR*, is deserving of particular mention. After its identification as the gene mutated in Stargardt's disease 1 (STGD1, an autosomal recessive macular disorder) (Allikmets et al., 1997b), the gene was subsequently screened in cohorts of AMD patients and controls, and it was concluded that genetic variation in *ABCA4* might be responsible for up to 16% of AMD cases (Allikmets et al., 1997a). Since publication, such findings have not been reproduced (Stone et al., 1998; Kuroiwa et al., 1999; De la Paz et al., 1999) and the original study has since been criticized as being flawed (Dryja et al., 1998; Klaver CCW, et al., 1998). The concerns included the observations that the study patients had not been clinically phenotyped accurately, had been recruited from different centers with differing referral patterns and specialist interests, and that the control cohort were not age-matched and were not matched for ethnicity with those from the disease cohort. *ABCA4* has since been found to be highly polymorphic in the general population and that the prevalence of these polymorphisms varies greatly in different populations. Subsequent larger studies on two polymorphisms (p.Gly1961Glu and p.Asp2177Asn) have produced conflicting results (Allikmets, 2000; Guymer et al., 2001).

A number of other candidate genes have been screened to assess their contribution to AMD. For example, two groups have investigated the association of apolipoprotein E (APOE protein is a constituent of retinal drusen: Klaver et al., 1998b; Mullins et al., 2000) alleles with AMD (Souied et al., 1998; Klaver et al., 1998b). The APOE haplotype is determined by two closely linked coding single nucleotide polymorphisms (SNPs) that give three possible alleles; ε2, ε3, and ε4. The least common allele, ε4, is more prevalent in those with Alzheimer disease, coronary artery disease and stroke, and it was expected that it would also be enriched in patients with AMD. Surprisingly, both groups found that ε4 was instead enriched in controls, suggesting it had a protective influence (odds ratio 0.43). The significance of this gene remains to be determined on further cohorts of patients from other centers, but its analysis is confounded by the detrimental effect of the ε4 allele on longevity.

The Identification of Novel Genetic Determinants of AMD

Complement Factor H

In order to discover novel genes that contribute to the risks of developing a complex disorder such as AMD, the standard methods for studying monogenic conditions such as candidate gene analysis and linkage analysis are not sufficiently powerful. Instead, allele sharing in affected sibs has been instituted, an approach adopted by the ambitious Biobank Project (http://www.ukbiobank.ac.uk/). In this strategy affected siblings undergo a full genome scan to determine whether each sib-pair shares 0, 1 or 2 alleles. If parental deoxyribonucleic acid (DNA) is available, the power of this strategy is increased massively, such that identity-by-descent (IBD) of chromosomal regions can be inferred when the parental genotypes are informative. When parental information is absent, as is usually the case for a late-onset disorder such as AMD, the degree of IBD has to be inferred statistically. Despite this, and the variable nature of the condition, many affected sib-pairs have been collected and analyzed using genomic markers at a density of 10 cM or more. This has pointed toward a number of chromosomal regions, found in two or more independent studies, which show promise for further investigation. A meta-analysis of six independent sib-pair studies has suggested the existence of susceptibility loci at 1q, 10q, and 16 (Fisher et al., 2005).

In 2005, four independent papers identified variations in the complement factor H (*CFH*) gene, as being the single most significant risk for developing AMD. The protein encoded by *CFH* regulates the alternative pathway of complement system. Two studies used sib-pair analyses to examine a previously identified critical region on chromosome 1q using association of SNPs in cases against controls (Edwards et al., 2005, Haines et al., 2005). A third study screened SNPs over a candidate gene in the 1q region AMD patients and controls (Hageman et al., 2005), but the fourth study employed a different methodology entirely (Klein et al., 2005). Making no prior assumptions about the possible location of AMD susceptibility genes, Klein et al. performed a full-genome association study on 96 patients and 50 controls using a high-density 100 kb SNP array. Two SNPs showed a significant skew in AMD patients compared to controls, one of which was highly noteworthy. This SNP, and the third most significant one, both resided in the intronic sequence of *CFH*, one of an array of genes within the regulation of complement activation (RCA) region of chromosome 1q. Further analysis of the gene by all four groups showed a significant association of AMD with those heterozygous and homozygous for the minor histidine allele at codon 402, (as opposed to tyrosine) (allele frequency 0.3 and 0.7, respectively). It is astonishing to many that in the latter study, an association scan of 1,00,000 SNPs (with an average spacing of 3000 bases) on such a small sample of patients and controls was so successful. If unsuccessful, the approach would have been viewed as underpowered and naive. However, the strategy suggests that as dense DNA-arrays become increasingly available, full-genome association in individual cases, instead of linkage in family members, will be an important and accessible tool in determining the genetic factors associated with complex disease. One important complicating factor is the chance association of one of many independent comparisons. Replication in independent data sets remains the key before confirming associations found in this way.

CFH is critical to the regulation of the alternative pathway of complement system. The primary site of synthesis is the liver (for export to the serum), but it is also locally expressed in the RPE (Hageman et al., 2005). It remains to be determined whether alteration of local or systemic expression, or a combination of both, affects retinal susceptibility to AMD. The protein contains 20 short consensus repeat (SCR) sequences and one, harboring residue 402, is important in the binding of heparin and C-reactive protein. It is of interest that loss of gene function, either naturally in patients or engineered in mice, causes renal failure due to membranoproliferative glomerulonephritis (Ying et al., 1999; Pickering et al., 2002) and early-onset macular drusen in patients (Duvall-Young et al., 1989). This has pointed toward complement system dysregulation as an important factor in the pathogenesis of AMD and indicates that presymptomatic genetic testing for assessing a person's risk would be beneficial (odds ratio—Y/H H/H). It also suggests other important components of the complement system may have a role in AMD etiology (see below).

LOC387715: A Novel AMD Locus

A second major AMD susceptibility locus has been identified by two groups using a combination of linkage analysis, association, and ultimately, SNP analysis. The first study examined 612 families with one or more AMD members, and 184 controls, using markers across the candidate 10q region (Jakobsdottir et al., 2005). A combination of linkage analysis and association identified two adjacent genes in the region as possible susceptibility factors: *PLEKHA1*, which encodes TAPP1, a leukocyte activating protein; and *LOC387715*, a transcript of unknown function expressed in the placenta. There was evidence of a further contribution from a nearby gene cluster but high linkage disequilibrium, particularly between *PLEKHA1* and *LOC387715*, hindered its identification. Thus, while an important locus was confirmed on 10q26 the study was unable to identify a specific gene or nucleotide variant. In contrast, a second study, performed on 1166 patients and 945 controls, using a case–control approach, confirmed that SNPs within the *PLEKHA1/LOC387715* region were significantly skewed in patients versus controls (Rivera et al., 2005). Detailed SNP analysis within both genes conclusively showed that a single nonsynonymous coding SNP in *LOC387715*, p.Ala69Ser, accounted for the association in the region. Intriguingly, *LOC387715* has no ortholog in other organisms apart from in primates, shows no protein homology, and has a restricted tissue expression pattern which includes the retina. Susceptibility to AMD of carrying the p.Ala69Ser polymorphism, like *CFH* p.Tyr402-His, behaves in a semi-dominant manner (increasing with one or two copies of the serine allele). It is a common polymorphism and, similar to the *CFH* allele, is estimated to account for approximately 50% of the population attributable risk. Furthermore, the risks determined by the *CFH* and *LOC387715* loci suggest an additive model in which the disease odds ratio of 57.6 [95% confidence interval (CI): 37.2, 89.0] is conferred by homozygosity for both the risk alleles (Rivera et al., 2005). Unlike *CFH*, more work is required to discover the role of this gene in health and disease and this may illuminate a novel biological pathway important in pathogenesis.

The link between the complement system and AMD risk has been strengthened by a study that tested the involvement of two other regulatory proteins in the system (Gold et al., 2006). The complement factor B (*BF*) gene, which acts as a complement activator and is tightly coupled with factor H in the control of complement activation, and the neighboring complement component 2 (*C2*) gene, are separated by only 500 bp and lie on chromosome 6p21 in the major histocompatibility complex (MHC) 3 region, a region not previously highlighted during sib-pair studies (Fisher et al., 2005). It is interesting to note that, in previous studies of the *CFH* knockout mouse, Pickering and colleagues (Pickering et al., 2002) were able to rescue the renal failure by breeding the $CFH^{-/-}$ into a $BF^{-/-}$ strain. Gold and colleagues showed a significant association of SNPs within both the *BF* and *C2* genes. The most common haplotype, H1, conferred a significant risk for AMD, while two other haplotypes, H7 and H10, were highly protective. Control individuals were enriched for homozygotes or mixed heterozygotes for these two protective haplotypes. The odds ratio for homozygotes was approximately twice that for heterozygotes and this was deemed consistent with a codominant model. Finally, the enrichment of protective haplotypes in controls was greatest amongst those with the high-risk CFH His/His genotype, intermediate in His/Tyr heterozygotes, and weakest in Tyr/Tyr homozygotes. Put another way, individuals at the greatest risk of developing AMD because of their *CFH* haplotype (i.e., homozygous for p.Tyr402His) but who have not developed the disease have a high frequency of the protective allele(s) at the *C2/BF* locus (Gold et al., 2006). This indicates that susceptibility conferred by *CFH* and *C2/BF* does not act independently, consistent conceptually with them acting upon the same biological pathway.

The discovery of these three AMD loci represents a dramatic increase in our understanding of genes that underlie this common and blinding disease. More progress has been made, in terms of the proportion of the identifiable population risk than with any other complex disorder. This knowledge has illuminated an important biological pathway to investigate further, namely the control of the complement cascade, and there is now hope of developing agents that may prevent the disorder in those at risk. Susceptibility to AMD can be predicted with reasonable accuracy by the genotyping of no more than four SNPs. It is likely too that other genes and other pathways will be discovered in the future (Haines et al., 2006), further increasing our understanding, and the prospects of, developing preventative strategies.

Translating the Benefits of Molecular Genetic Advances to Clinical Practice

As discussed, ophthalmology has been at the vanguard of developments in genomic medicine both for monogenic and complex disorders. As with many other branches of medicine the translation of these benefits to the clinic has lagged behind the research progress. This situation is now changing in both the provision of genetic testing for Mendelian disorders that affect vision and in the development of novel therapies for them.

Genetic Testing for Ophthalmic Disorders

It has long been accepted that once the genetic basis of an inherited disorder has been identified that genetic testing, usually performed on an affected family member, can provide valuable information regarding diagnosis, prognosis, and reproductive risks. Since such developments are not easily translated into routinely available genetic tests this technological gap has often resulted in a shortfall in our ability to meet expectations of both patients and clinicians alike.

The majority of analyses aim to test for mutations in known genes and ultimately rely upon the identification of a sequence variation which alters the expression or function of a gene or protein, and which can be shown to be pathogenic or disease-causing. In cases where there is a strong relationship between phenotype and genotype, genetic analysis is straightforward. Such a link exists amongst some of the monogenic macular dystrophies such as Sorsby dystrophy and Doyne honeycomb dystrophy (Malattia Leventinese) where the vast majority of cases result from single point mutations in the *TIMP3* and *EFEMP1* genes, respectively (Stone et al., 1999; Li et al., 2005). The same is also true for those corneal dystrophies linked to chromosome 5q31 that result from alterations in the *TGFBI* gene, where specific mutations cause granular, lattice type 1 and Bowman's layer (Thiel–Behnke and Reis-Buckler) dystrophies (Munier et al., 2002). In these circumstances, the molecular identification of pathogenic mutations is straightforward. Another area of success in the delivery of genetic testing has been in the diagnosis and counseling of families with retinoblastoma, a condition resulting from mutations in a single large gene.

Improvements in the ability to deliver high-throughput techniques, such as direct sequencing, mean that is, now realistic to offer such services for a wide range of disorders. However, this remains relatively labor-intensive and time-consuming for conditions of high heterogeneity such as the retinal dystrophies. A simple means of delivering a genotyping service remains to be developed. In most cases where a mutation is suspected the whole gene must be screened; in the case of *ABCA4* this encompasses 50 exons and more than 7000 bp of DNA, a task beyond the scope of most diagnostic laboratories. Furthermore since the pickup rate amongst those known to harbor mutations in *ABCA4* is considerably less than 100% a negative result is of limited value. In addition, for many retinal genes the pathogenic mutations are often the result of missense (amino acid) substitutions (e.g., Webster et al., 2001). Since for many of these genes, there is a significant degree of normal variation at both gene and protein level (i.e., the existence of polymorphisms), the task of defining whether a variation is pathogenic is difficult and, in the absence of a functional assay, may be impossible.

Thus, the current process of mutation screening for conditions such as RP, where mutations in several different genes can cause an identical phenotype and where there is no way to choose the correct one from a list of several possibilities, comprehensive genetic testing remains impractical. Nonetheless, selective testing through direct sequencing of genes which cause common forms of RP represents a valuable and achievable goal. Since *RPGR* mutations account for 70%–75% of xlRP and the majority (~60 %) of the mutations are present in the ORF15 exon (Vervoort et al., 2000), this represents the most common cause of RP accounting for up to approximately 17% of all cases (Vervoort and Wright, 2002). The next most common cause of RP is a mutation in rhodopsin and the p.Pro23His mutation alone accounts for 12% of RP patients in the USA (although this figure is lower in other populations) (Wang et al., 2005). Although *RPGR* and *RHO* represent major RP loci along with other genes such as *RDS*, *RP1*, *SAG*, *PDE6A*, and *PDE6B*, they probably account for less than 50% of all cases, and it is estimated that 33%–55% of RP cases remain to be genetically ascribed (Rivolta et al., 2002; Wang et al., 2005). Other estimates predict that the remaining orphan loci will increase the number of RP genes by two to three times (Wright and van Heyningen, 2001). To address the burgeoning number of genes to be screened additional methodologies need to be developed and the advent of microarray-based gene "chip" testing is now becoming a realistic prospect (Klevering et al., 2005; Zernant et al., 2005). However, in the immediate future the genetic testing of all loci is unlikely to be achieved. The demand for such services will be largely determined by the needs of those being tested and, where such provision is based on counseling needs, this is likely to be confined to conditions where diagnosis may have an impact upon family decision making (e.g., xlRP, LCA). However should gene-based therapeutics become a realistic prospect, the provision of a wider testing platform will be necessary (see below).

Ophthalmic Genetics and the Development of Novel Therapies

When the words "genetics" and "therapy" are used together in the context of treating genetic disease most people think of gene replacement strategies, or "gene therapy" where a faulty gene is replaced by a normal copy. While conceptually simple, this approach has been fraught with technical difficulties and the initial successes that sparked much of the ensuing excitement (in both the scientific and popular press) have suffered a number of well-publicized set-backs (Hacein-Bey-Abina et al., 2003; Kohn et al., 2003). In fact, the fields of genetics and molecular biology have led to several different approaches to treating genetic disease, some of which are already being trialed in humans. The eye, and in particular the retina and RPE, are uniquely suited to act as a model system in which to develop novel gene-based therapies because of their anatomically and immunologically privileged status. There has already been considerable work to develop effective delivery systems, including a range of viral-based vectors in a wide range of animal models (mammalian and nonmammalian) to allow for a broad-based assessment of novel strategies. As a result, it appears that not only is it possible to deliver biologically active molecules to both the retina and RPE, but also there is evidence to indicate that in optimal circumstances they may actually be effective. The challenge will be transferring such therapies to the human arena.

When reviewing the progress of ophthalmic gene therapy, it is useful to use retinal dystrophies as a paradigm. This is not the only area amenable to novel therapies, as the work in the fields of retinal neovascularization or uveitis amply demonstrate, but this is a field in which there has been considerable progress. It is worth emphasizing, when considering the development of novel therapies to treat inherited disease, not to overlook conventional therapies. Absent protein function that results from a genetic mutation could theoretically be nullified by dietary modification, such as the use of an arginine-restricted diet to slow the progression of visual loss in patients with gyrate atrophy (Kaiser-Kupfer et al., 2004), or modified pharmaceutically. A notable success in the latter, which also illustrates the power of molecular genetics and the use of animal models, is in the development of a potential therapy for STGD. Using a combination of linkage analysis and a candidate gene approach mutations in the *ABCA4* gene were identified as one cause of STGD, a recessive macular dystrophy (Allikmets et al., 1997b). Mutations in *ABCA4* can also cause a variety of other retinal degenerations and all involve the accumulation of lipofuscin pigment in the RPE, especially STGD (Delori et al., 1995). *ABCA4* encodes the RIM1 protein which is thought to transport N-retinylidene phosphatidylethanolamine (*N*-RPE, a product of 11-cis retinal in the phototransduction cascade) out of the disc before it is transported to the RPE as part of the light cycle to regenerate the chromophore. The mouse knockout model showed that mutations in RIM1 caused *N*-retinylidene-*N*-retinylethanolamine (A2E, which is formed from N-RPE) to accumulate in the discs (Weng et al., 1999). Thus when the discs were phagocytosed by RPE cells during the normal turnover of outer segments, they ingested high levels of A2E, which is the major constituent of lipofuscin and is toxic to them. This mouse model has now been used to test efficacy of a therapy that aims to reduce the levels of A2E. *N*-(4-hydroxyphenyl) retinamide (HPR) is a retinoic acid analogue that has been extensively tested as a cancer treatment (Naik et al., 1995). It lowers circulating levels of retinol by competing for binding on retinol-binding protein (RBP) which is subsequently excreted into the urine bound to HPR (Noy and Xu, 1990). As A2E biosynthesis is dependent upon circulating retinol levels researchers tested whether HPR would prevent lipofuscin accumulation and macular degeneration in the *Abca4* mouse knockout model. Mice treated with HPR showed an immediate and dose-dependent reduction in serum retinol and RBP. Prolonged administration also significantly reduced visual cycle retinoids and retinal A2E. The structure and function of the rod photoreceptors in the treated mice was normal compared to wild-type controls with only difference observed being a modest delay in dark adaptation. Lipofuscin accumulation in the RPE was comparable to wild-type controls, as was retinal morphology (Radu et al., 2005). Thus, this therapy seems to promise much in preventing retinal degeneration where the mechanism of photoreceptor cell death is due to the toxic effects of A2E on RPE cells. This may be (relatively) rapidly made available to clinicians as a treatment option given the well established pharmacokinetics of HPR, although these may have to be reassessed for the elder target population (Radu et al., 2005).

However, in this section, we will focus upon genetically based technologies, which can be loosely grouped into three categories:

1. Heterogeneous gene expression,
2. Modification of endogenous gene expression,
3. Cell replacement techniques.

Therapy Employing Heterogeneous Gene Expression

This group of potential therapies includes "classical" gene replacement as well as attempts to express other agents that can either promote cell growth, such as growth and survival factors, or inhibit it, such as antiangiogenic factors. The greatest successes in this area have been in the treatment of recessive retinal dystrophies where the aim is to deliver those molecules which are absent from the retina or RPE.

Gene Replacement Therapies* for *Recessive Retinal Dystrophies The work that has received the most public attention has been the successful treatments of severely visually impaired Briard dogs which lack Rpe65 and mimic the human disorder, LCA. A single subretinal injection of an adenoassociated virus (AAV) containing the normal *Rpe65* gene under the control of the chicken β-actin promoter restored photoreceptor function (as measured by ERG) and vision (as measured by visual assessment tests) (Acland et al., 2001). There have been other promising results however. For example amongst rodents, the murine model of LCA caused by a lack of Rpgrip has also been successfully treated by a subretinal injection of the normal gene in an AAV vector (Pawlyk et al., 2005). Similarly photoreceptor function (as measured by ERG) was restored in the *Rds* mouse model that lacks peripherin, a protein that forms a complex with Rom1 to stabilize the edges of discs in photoreceptors (Ali et al., 2000; Schlichtenbrede et al., 2003a), although the effect does not appear to be long-lasting. Attempts to extend this effect have been disappointing, for example, when combined with ciliary neurotrophic factor (CNTF), a factor previously associated with increased cell survival, the rescue was completely negated (Schlichtenbrede et al., 2003b).

Introduction of Growth Factors into the Eye to Prevent Retinal Degeneration There is a large body of research covering the introduction of heterogeneous molecules into the eye in an attempt to improve its health (for a review see Thanos and Emerich, 2005). Intravitreal injection of lens epithelium-derived growth factor (LEDGF) protected photoreceptor integrity in light damaged rats and in the Royal College of Surgeons (RCS rat model of RP (*Mertk*$^{-/-}$) (Machida et al., 2001). Pigment epithelium-derived factor (PEDF) delivered subretinally by simian immunodeficiency virus (SIV) also prevented photoreceptor loss in the RCS rat (Miyazaki et al., 2003). Photoreceptor apoptosis was prevented in rats with an induced retinal detachment after the subretinal injection of AAV expressing glial cell line-derived neurotrophic factor (GDNF) (Wu et al., 2002) and CNTF expressed from encapsulated human retina-derived cells preserved retinal structure in a canine model of RP (*Pde6B*$^{-/-}$) (Tao et al., 2002). The control of angiogenesis is currently receiving much attention in ophthalmology (Campochiaro, 2006) with various factors used to either promote (Otani et al., 2002) or inhibit it [e.g., PEDF, tissue inhibitor of metalloproteinase-3 (TIMP3), endostatin; Auricchio et al., 2002].

Modification of Endogenous Gene Expression

There are a range of technologies that can either enhance or prevent gene expression whose aims have important differences.

Enhancement of Gene Expression Technologies for enhancing gene expression may be useful in particular for the treatment of recessive disease. An example is the translation of nonsense-mediated decay (NMD)-targeted transcripts by aminoglycoside treatment. Aminoglycosides, such as gentomycin, act by interfering in protein translation such that stop codons are ignored. Researchers have used this observation to allow genes with premature stop codons to avoid NMD and so allow a protein product to be generated from a mutant gene that otherwise would not have been blocked (Bedwell et al., 1997; Ainsworth, 2005). To date this has not been successfully trialed as a potential therapy for retinal disease; an in vitro study treated lymphoblastoid cell lines from xlRP patients with a premature stop codon in exon 2 of the *RP2* gene with gentomycin. No full-length or truncated RP2 protein could be detected in the patient cells after gentomycin treatment (Grayson et al., 2002).

Prevention of Gene Expression: Treatments for Dominant Retinal Disease *Ribozymes.* Ribozymes are synthetic catalytic RNA molecules that can target and cleave mutated mRNA molecules. This approach may be useful for preventing the expression of mutant proteins in dominant negative diseases (but not those that occur as a result of haploinsufficiency). Although now out of vogue this technology may still offer viable therapy alternatives. Hammerhead ribozymes were used to knockdown rod (PDE in mice to generate a model of RP. The authors suggested that in addition to being a valuable research tool this approach could be used as a general therapy for dominant RP. Instead of targeting specific alleles (mutant forms) a more general ribozyme could be used to knockdown the mutant and wild-type alleles and be supplemented with a "hardened" replacement gene that is, resistant to the ribozyme (Liu et al., 2005). In two mouse models of proliferative retinopathy, aberrant retinal angiogenesis was inhibited by the expression of a ribozyme against insulin-like growth factor receptor in proliferating endothelial cells. Normal retinal vessels were preserved suggesting that this selective therapy may offer a safer treatment for the treatment of angiogenesis associated with retinopathy (Shaw et al., 2006).

RNA interference. RNA interference (RNAi) is a form of post-translational gene silencing that was originally described as an antiviral defense mechanism in plants. Short interfering RNAs (siRNAs) can either be generated naturally by the Dicer enzyme as part of this mechanism, or can be generated synthetically. The siRNA molecules pair with the target mRNAs and initiate their cleavage (and later degradation) by the RISC complex. In addition to the use of siRNA per se, similar active molecules can be produced from short hairpin RNAs (shRNAs) which can be generated from plasmids, giving the advantages of spatial and temporal control (Wiznerowicz and Trono, 2003; Chang et al., 2004), more than 100 different rhodopsin mutations cause a variety of retinal degenerations. Allele-specific therapies for such heterogeneous mutations are probably unrealistic and so a mutation-independent approach, as with ribozymes above, is desirable. shRNA against wild-type and mutant (p.Pro23His) rhodopsin silenced both alleles by 94% in human retinal embryoblasts. A codon exchanged mRNA of rhodopsin, in which the degeneracy of the genetic code was used to produce a wild-type version of rhodopsin that was resistant to the shRNA, was unaffected (Cashman et al., 2005). Data from the first clinical trial of an RNAi-based drug designed to inhibit angiogenesis in AMD has shown it to be safe and to act in a dose-dependent manner. The small study has also found that six out of seven of patients had an improved retinal thickness by optical coherence tomography and could read additional letters on a vision test chart (Whelan, 2005).

Cell Replacement Techniques

Tissues or structures can be replaced using (1) heterogeneous tissues that are then forced to transdifferentiate or (2) pluripotent stem cells that can differentiate to replace the lost cells.

Cells derived from adult iris and ciliary tissue of rats and monkeys can be induced into expressing photoreceptor-specific markers by transfection with various transcription factors (Akagi et al., 2004, 2005). In the rat, the transcription factor Crx-induced iris cells to express rhodopsin, recoverin, cyclic guanosine monophosphate (cGMP)-gated channel, arrestin, Iw, rhodopsin kinase, and NeuroD. A combination of Crx and NeuroD similarly induced monkey iris cells to express photoreceptor-specific markers. Both transduced cell types were also capable of integrating into developing retinal explant cultures (Akagi et al., 2005). This approach would be very attractive for replacing lost tissue as the autologous cells for transfection could be obtained relatively easily.

Fish and amphibia have the capacity for continual retinal growth throughout life from stem cells in the circumferential germinal zone (Otteson and Hitchcock, 2003). Potentially homologous cells in the adult mouse retina have also been described that can differentiate into retinal cell types including rod photoreceptors, bipolar cells and Müller cells, however compared to neural stem cells they proliferate less and were less plastic (Tropepe et al., 2000). Similar cells have also been identified in postnatal chickens. Although these cells expressed markers associated with embryonic retinal multipotent progenitors, their proliferation did not increase in response to acute damage (Fischer and Reh, 2000). When the sonic hedgehog (Shh) mitogenic factor was partially activated in mice by inactivating one allele of its receptor, Patched, an increased number of neural progenitors were present at every stage of retinal development. In addition, these mice crossed with mice transgenic for mutant rhodopsin (p.Pro23His, which causes adRP) had significantly more proliferating ciliary marginal zone (CMZ) cells at the retinal margin (Moshiri and Reh, 2004). This response is similar to the CMZ of lower vertebrates and demonstrates that Shh activation allows limited retinal cell regeneration.

In general, the novel therapies that have been attempted target somatic cells and as a result they will have no effect on future generations, a situation which is ethically relatively uncomplicated. More recent novel gene modification technologies carry the potential to correct mutated genomic DNA in the germ cell. While these have not yet been assayed in retinal disease they nevertheless represent a pure approach in therapeutic terms.

Zinc fingers are protein domains of approximately 30 amino acids that are held together by a zinc ion that can bind to DNA in a sequence-specific manner. There are more than 900 proteins containing this motif including many transcription factors. Zinc finger domains have been fused to other proteins to regulate gene expression and most recently have been fused to the nonspecific DNA cleavage domain of the restriction enzyme Fok I to create a zinc finger nuclease (ZFN) (High, 2005). This development allows a combination of zinc finger domains to target and cut a specific sequence in the genome. A strategy using two ZFNs to remove a piece of mutant DNA in cells for subsequent replacement by a wild-type donor piece of DNA by homologous recombination has now been developed in a human tumorigenic lymphoblastoid cell line (K562). Urnov et al. were first able to inactivate the *IL2R* (gene (mutated in severe combined immune deficiency) and then were able to correct 15%–20% of mutated chromosomes back to the wild-type sequence using the principle above. Impressively, approximately 7% of cells had both copies of the X chromosome corrected (without the need for any selection) and a similar efficiency (5%) was also achieved in primary human T cells, in which introduction efficiencies are usually much lower (Urnov et al., 2005). There are many obstacles, such as delivery and specificity, which have to be overcome before this technology can be used as a therapy. Ultimately if it can be developed for use in germ cells the ultimate goal of gene therapy will have been realized, although many ethical questions about its use may arise.

Conclusions

The management of inherited ophthalmic disease and the introduction of both testing and, potentially, therapeutics is a complex area which raises ethical questions for clinicians and scientists alike. In an area with no sanctioned guidelines care must be taken. While much has been written on the implications of our new genetic knowledge on the diagnosis of a wide spectrum of genetic conditions, little has been done to apply these to ophthalmology. In such circumstances, it is important to realize that new technologies will have to be applied and that perhaps the lessons learned from outside one specialty may need to be examined and applied to others. While it is tempting to implement new methodologies as rapidly as possible, it is perhaps wise to seek advice in guiding practice. The areas of genetic ethics, genetic testing, presymptomatic and carrier testing are, for example, routinely applicable to inherited ophthalmic conditions. However, how conditions such as Sorsby fundus dystrophy and xlRP impact upon patients, and how differently they ought to be managed (if they should) remains to be defined. Furthermore, as the number of ocular conditions for which genetic testing increases it will be possible to test more of those at risk of developing the condition, either before or after birth. Prenatal diagnosis is therefore a service whose impact upon ophthalmic genetics can only increase. This too is a subject which arouses strong emotions amongst patients and professionals. Advising individuals whether to opt for testing during pregnancy is a time-consuming complex process that will need to be approached sensitively, and without judgment, by those who have experience of such situations.

References

Acland, GM, et al. (2001). Gene therapy restores vision in a canine model of childhood blindness. *Nat Genet*, 28:92–95.

Adato, A, et al. (2002). USH3A transcripts encode clarin-1, a four-transmembrane-domain protein with a possible role in sensory synapses. *Eur J Hum Genet*, 10:339–350.

Ainsworth, C (2005). Nonsense mutations: running the red light. *Nature*, 438:726–728.

Akagi T, et al. (2004). Otx2 homeobox gene induces photoreceptor-specific phenotypes in cells derived from adult iris and ciliary tissue. *Invest Ophthalmol Vis Sci*, 45:4570–4575.

Akagi, T, et al. (2005). Iris-derived cells from adult rodents and primates adopt photoreceptor-specific phenotypes. *Invest Ophthalmol Vis Sci*, 46:3411–3419.

Alagaratnam, J, et al. (2002). A survey of visual impairment in children attending the Royal Blind School, Edinburgh using the WHO childhood visual impairment database. *Eye*, 16(5):557–561.

Ali, RR, et al. (2000). Restoration of photoreceptor ultrastructure and function in retinal degeneration slow mice by gene therapy. *Nat Genet*, 25:306–310.

Allikmets, R, et al. (1997a). Mutation of the Stargardt disease gene (ABCR) in age-related macular degeneration. *Science*, 277:1805–1807.

Allikmets, R, et al. (1997b). A photoreceptor cell-specific ATP-binding transporter gene (ABCR) is mutated in recessive Stargardt macular dystrophy. *Nat Genet*, 15:236–246.

Allikmets, R (2000). Further evidence for an association of ABCR alleles with age-related macular degeneration. The International ABCR Screening Consortium. *Am J Hum Genet*, 67(2):487–491.

Ansley, SJ, et al. (2003). Basal body dysfunction is a likely cause of pleiotropic Bardet-Biedl syndrome. *Nature*, 425:628–633.

Antonarakis, SE and Beckmann, JS (2006). Mendelian disorders deserve more attention. *Nat Rev Genet*, 7:277–282.

Arden, GB and Fox, B (1979). Increased incidence of abnormal nasal cilia in patients with retinitis pigmentosa. *Nature*, 279:534–536.

Auricchio, A, et al. (2002). Inhibition of retinal neovascularization by intraocular viral-mediated delivery of anti-angiogenic agents. *Mol Ther*, 6:490–494.

Ayyagari, R, et al. (2002). X-linked recessive atrophic macular degeneration from RPGR mutation. *Genomics*, 80:166–171.

Bedwell, DM, et al. (1997). Suppression of a CFTR premature stop mutation in a bronchial epithelial cell line. *Nat Med*, 3:1280–1284.

Ben-Arie-Weintrob Y, et al. (2005). Histopathologic-genotypic correlations in retinitis pigmentosa and allied diseases. *Ophthalmic Genet*, 26:91–100.

Bhattacharya, G and Cosgrove, D (2005). Evidence for functional importance of usherin/fibronectin interactions in retinal basement membranes. *Biochemistry*, 44:1518–1524.

Bird, AC, et al. (1995). An international classification and grading system for age-related maculopathy and age-related macular degeneration. The International ARM Epidemiological Study Group. *Surv Ophthalmol*, 39(5):367–374.

Boughman, JA, et al. (1983). Usher syndrome: definition and estimate of prevalence from two high-risk populations. *J Chronic Dis*, 36:595–603.

Brinkman, RR, et al. (2006). Human monogenic disorders—a source of novel drug targets. *Nat Rev Genet*, 7:249–260.

Campochiaro, PA (2006). Ocular versus extraocular neovascularization: mirror images or vague resemblances. *Invest Ophthalmol Vis Sci*, 47:462–474.

Cashman, SM, et al. (2005). Towards mutation-independent silencing of genes involved in retinal degeneration by RNA interference. *Gene Ther*, 12:1223–1228.

Chakarova, CF, et al. (2002). Mutations in HPRP3, a third member of pre-mRNA splicing factor genes, implicated in autosomal dominant retinitis pigmentosa. *Hum Mol Genet*, 11:87–92.

Chang, HS, et al. (2004). Using siRNA technique to generate transgenic animals with spatiotemporal and conditional gene knockdown. *Am J Pathol*, 165:1535–1541.

Chao, LY, et al. (2003). Missense mutations in the DNA-binding region and termination codon in PAX6. *Hum Mutat*, 21:138–145.

Christen, WG, et al. (1996). A prospective study of cigarette smoking and risk of age-related macular degeneration in men. *JAMA*, 276:1147–1151.

De La Paz, MA, et al. (1998). ABCR gene and age-related macular degeneration. *Science*, 279:1107.

Delori, FC, et al. (1995). In vivo measurement of lipofuscin in Stargardt's disease—fundus flavimaculatus. *Invest Ophthalmol Vis Sci*, 36:2327–2331.

Demirci, FY, et al. (2002). X-linked cone-rod dystrophy (locus COD1): identification of mutations in RPGR exon ORF15. *Am J Hum Genet*, 70:1049–1053.

Dry, KL, et al. (1999). Identification of a 5' splice site mutation in the RPGR gene in a family with X-linked retinitis pigmentosa (RP3). *Hum Mutat*, 13:141–145.

Dryja, TP, et al. (1990). A point mutation of the rhodopsin gene in one form of retinitis pigmentosa. *Nature*, 343:364–366.

Dryja, TP, et al. (1998). ABCR gene and age-related macular degeneration. *Science*, 279:1107.

Dryja, TP and Li, T (1995). Molecular genetics of retinitis pigmentosa. *Hum Mol Genet*, 4:1739–1743.

Dryja, TP, et al. (2001). Null RPGRIP1 alleles in patients with Leber congenital amaurosis. *Am J Hum Genet*, 68:1295–1298.

Duvall-Young, J, et al. (1989). Fundus changes in mesangiocapillary glomerulonephritis type II: clinical and fluorescein angiographic findings. *Br J Ophthalmol*, 73(11):900–906.

Edwards, AO, et al. (2005). Complement factor H polymorphism and age-related macular degeneration. *Science*, 308(5720):421–424.

Evans, W and Wormald, R (1996). Is the incidence of registrable age-related macular degeneration increasing? *Br J Ophthalmol*, 80:9–14.

Fischer, AJ and Reh, TA (2000). Identification of a proliferating marginal zone of retinal progenitors in postnatal chickens. *Dev Biol*, 220:197–210.

Fisher, SA, et al. (2005). Meta-analysis of genome scans of age-related macular degeneration. *Hum Mol Genet*, 14(15):2257–2264.

Fox, B, et al. (1980). Variations in the ultrastructure of human nasal cilia including abnormalities found in retinitis pigmentosa. *J Clin Pathol*, 33:327–335.

Furukawa, T, et al. (1997). Rax, a novel paired-type homeobox gene, shows expression in the anterior neural fold and developing retina. *Proc Natl Acad Sci USA*, 94:3088–3093.

Gerber S, et al. (2001). Complete exon-intron structure of the RPGR-interacting protein (RPGRIP1) gene allows the identification of mutations underlying Leber congenital amaurosis. *Eur J Hum Genet*, 9:561–571.

Gilbert, C and Foster, A (2001). Childhood blindness in the context of VISION 2020—the right to sight. *Bull World Health Organ*, 79:227–232.

Gold, B, et al. (2006). Variation in factor B (BF) and complement component 2 (C2) genes is associated with age-related macular degeneration. *Nat Genet*, 38:458–462.

Graw, J, et al. (2001). Characterization of a new, dominant V124E mutation in the mouse alphaA-crystallin-encoding gene. *Invest Ophthalmol Vis Sci*, 42:2909–2915.

Grayson, C, et al. (2002). In vitro analysis of aminoglycoside therapy for the Arg120stop nonsense mutation in RP2 patients. *J Med Genet*, 39:62–67.

Guymer, RH, et al. (2001). Variation of codons 1961 and 2177 of the Stargardt disease gene is not associated with age-related macular degeneration. *Arch Ophthalmol*, 119:745–751.

Hacein-Bey-Abina, S, et al. (2003). LMO2-associated clonal T cell proliferation in two patients after gene therapy for SCID-X1. *Science*, 302:415–419.

Hageman, GS, et al. (2005). A common haplotype in the complement regulatory gene factor H (*HF1/CFH*) predisposes individuals to age-related macular degeneration *Proc Natl Acad Sci USA*, 102:7227–7232.

Haines, JL, et al. (2005). Complement factor H variant increases the risk of age-related macular degeneration. *Science*, 308:419–421.

Haines, JL, et al. (2006). Functional candidate genes in age-related macular degeneration: significant association with VEGF, VLDLR, and LRP6. *Invest Ophthalmol Vis Sci*, 47:329–335.

Hammond, CJ, et al. (2000). Genetic and environmental factors in age-related nuclear cataracts in monozygotic and dizygotic twins. *New England J Med*, 342:1786–1790.

Hammond, CJ, et al. (2001). The heritability of age-related cortical cataract: the twin eye study. *Invest Ophthalmol Vis Sci*, 42:601–605.
Hammond, CJ, et al. (2002). Genetic influence on early age-related maculopathy: a twin study. *Ophthalmology*, 109:730–836.
Heckenlively, JR (1989). *Retinitis Pigmentosa*. Philadelphia: Lippincott Company.
Heiba, IM, et al. (1994). Sibling correlations and segregation analysis of age-related maculopathy: the Beaver Dam Eye Study. *Genet Epidemiol*, 11:51–67.
Heiba, IM, et al. (1995). Evidence for a major gene for cortical cataract. *Invest Ophthalmol Vis Sci*, 36:227–235.
Heyman, I, et al. (1999). Psychiatric disorder and cognitive function in a family with an inherited novel mutation of the developmental control gene PAX6. *Psychiatr Genet*, 9:85–90.
High, KA (2005). Gene therapy: the moving finger. *Nature*, 435:577–579.
Hims, MM, et al. (2003). Retinitis pigmentosa: genes, proteins and prospects. *Dev Ophthalmol*, 37:109–125.
Hong, DH, et al. (2000). A retinitis pigmentosa GTPase regulator (RPGR)-deficient mouse model for X-linked retinitis pigmentosa (RP3). *Proc Natl Acad Sci USA*, 97:3649–3654.
Hunter, DG, et al. (1988). Abnormal axonemes in X-linked retinitis pigmentosa. *Arch Ophthalmol*, 106:362–368.
Iannaccone, A, et al. (2003). Clinical and immunohistochemical evidence for an X linked retinitis pigmentosa syndrome with recurrent infections and hearing loss in association with an RPGR mutation. *J Med Genet*, 40: e118.
Iannaccone, A, et al. (2004). Increasing evidence for syndromic phenotypes associated with RPGR mutations. *Am J Ophthalmol*, 137:785–786.
Iyengar, SK, et al. (2004). Identification of a major locus for age-related cortical cataract on chromosome 6p12-q12 in the Beaver Dam Eye Study. *Proc Natl Acad Sci USA*, 101:14485–14490.
Jakobsdottir, J, et al (2005). Susceptibility genes for age-related maculopathy on chromosome 10q26. *Am J Hum Genet*, 77(3):389–407.
Jamieson, RV, et al. (2002). Domain disruption and mutation of the bZIP transcription factor, MAF, associated with cataract, ocular anterior segment dysgenesis and coloboma. *Hum Mol Genet*, 11:33–42.
Joensuu, T, et al. (2001). Mutations in a novel gene with transmembrane domains underlie Usher syndrome type 3. *Am J Hum Genet*, 69:673–684.
Kaiser-Kupfer, MI, et al. (2004). Use of an arginine-restricted diet to slow progression of visual loss in patients with gyrate atrophy. *Arch Ophthalmol*, 122:982–984.
Kalloniatis, M and Fletcher, EL (2004). Retinitis pigmentosa: understanding the clinical presentation, mechanisms and treatment options. *Clin Exp Optom*, 87:65–80.
Katsanis, N, et al. (2000). Mutations in MKKS cause obesity, retinal dystrophy and renal malformations associated with Bardet-Biedl syndrome. *Nat Genet*, 26:67–70.
Keats, BJ and Corey, DP (1999). The usher syndromes. *Am J Med Genet*, 89:158–166.
Keen, TJ, et al. (2002). Mutations in a protein target of the Pim-1 kinase associated with the RP9 form of autosomal dominant retinitis pigmentosa. *Eur J Hum Genet*, 10:245–249.
Kelly, SP, et al. (2004). Smoking and blindness. *BMJ*, 328(7439):537–538. No abstract available.
Kennan, A, et al. (2005). Light in retinitis pigmentosa. *Trends Genet*, 21:103–110.
Kimura, A, et al. (2000). Both PCE-1/RX and OTX/CRX interactions are necessary for photoreceptor-specific gene expression. *J Biol Chem*, 275:1152–1160.
Klaver, CC, et al. (1998a). Genetic risk of age-related maculopathy. Population-based familial aggregation study. *Arch Ophthalmol*,116:1646–1651.
Klaver, CC, et al. (1998b). Genetic association of apolipoprotein E with age-related macular degeneration. *Am J Hum Genet*, 63:200–206.
Klaver, CCW, et al. (1998). ABCR gene and age-related macular degeneration. *Science*, 279:1107.
Klein, ML, et al. (1994). Heredity and age-related macular degeneration. Observations in monozygotic twins. *Arch Ophthalmol*, 112:932–937.
Klein, RJ, et al. (2005). Complement factor H polymorphism in age-related macular degeneration. *Science*, 308:385–389.
Klevering, BJ, et al. (2005). The spectrum of retinal phenotypes caused by mutations in the ABCA4 gene. *Graefes Arch Clin Expl Ophthalmol*, 243:90–100.
Koenekoop, RK, et al. (2003). Novel RPGR mutations with distinct retinitis pigmentosa phenotypes in French-Canadian families. *Am J Ophthalmol*, 136: 678–687.
Kohn, DB, et al. (2003). Occurrence of leukaemia following gene therapy of X-linked SCID. *Nat Rev Cancer*, 3:477–488.
Kuhn, AN, et al. (2002). Distinct domains of splicing factor Prp8 mediate different aspects of spliceosome activation. *Proc Natl Acad Sci USA*, 99:9145–9149.
Kuroiwa, S, et al. (1999). ATP binding cassette transporter retina genotypes and age related macular degeneration: an analysis on exudative non-familial japanese patients. *Br J Ophthalmol*, 83(5):613–615.
Li, Z, et al. (2005). TIMP3 mutation in Sorsby's fundus dystrophy: molecular insights. *Expert Rev Mol Med*, 7:1–15.
Liu, J, et al. (2005). Ribozyme knockdown of the gamma-subunit of rod cGMP phosphodiesterase alters the ERG and retinal morphology in wild-type mice. *Invest Ophthalmol Vis Sci*, 46:3836–3844.
Machida, S, et al. (2001). Lens epithelium-derived growth factor promotes photoreceptor survival in light-damaged and RCS rats. *Invest Ophthalmol Vis Sci*, 42, 1087–1095.
Maita, H, et al. (2004). PAP-1, the mutated gene underlying the RP9 form of dominant retinitis pigmentosa, is a splicing factor. *Exp Cell Res*, 300: 283–296.
Malandrini, A, et al. (2001). PAX6 mutation in a family with aniridia, congenital ptosis, and mental retardation. *Clin Genet*, 60:151–154.
Manson, FD, et al. (2005). Inherited eye disease: cause and late effect. *Trends Mol Med*, 11:449–455.
McKie, AB, et al. (2001). Mutations in the pre-mRNA splicing factor gene PRPC8 in autosomal dominant retinitis pigmentosa (RP13). *Hum Mol Genet*, 10:1555–1562.
Meyers, SM, (1995). A twin study of age-related macular degeneration. *Am J Ophthalmol*, 120:757–766.
Miyazaki, M, et al. (2003). Simian lentiviral vector-mediated retinal gene transfer of pigment epithelium-derived factor protects retinal degeneration and electrical defect in Royal College of Surgeons rats. *Gene Ther*, 10:1503–1511.
Moshiri, A and Reh, TA (2004). Persistent progenitors at the retinal margin of ptc +/− mice. *J Neurosci*, 24:229–237.
Mullins, RF, et al. (2000). Drusen associated with aging and age-related macular degeneration contain proteins common to extracellular deposits associated with atherosclerosis, elastosis, amyloidosis, and dense deposit disease. *FASEB J*, 214:835–846.
Munier, FL, et al. (2002). BIGH3 mutation spectrum in corneal dystrophies. *Invest Ophthalmol Vis Sci*, 43:949–954.
Mykytyn, K, et al. (2001). Identification of the gene that, when mutated, causes the human obesity syndrome BBS4. *Nat Genet*, 28:188–191.
Mykytyn, K, et al. (2002). Identification of the gene (BBS1) most commonly involved in Bardet-Biedl syndrome, a complex human obesity syndrome. *Nat Genet*, 31:435–438.
Mykytyn, K and Sheffield, VC (2004). Establishing a connection between cilia and Bardet-Biedl Syndrome. *Trends Mol Med*, 10:106–109.
Naik, HR et al. (1995). 4-Hydroxyphenylretinamide in the chemoprevention of cancer. *Adv Pharmacol*, 33:315–347.
Ng, D, et al. (2004). Oculofaciocardiodental and Lenz microphthalmia syndromes result from distinct classes of mutations in BCOR. *Nat Genet*, 36:411–416.
Nishimura, DY, et al. (2001). Positional cloning of a novel gene on chromosome 16q causing Bardet-Biedl syndrome (BBS2). *Hum Mol Genet*, 10:865–874.
Noy, N and Xu, ZJ (1990). Interactions of retinol with binding proteins: implications for the mechanism of uptake by cells. *Biochemistry*, 29:3878–3883.
O'Connor TP, Crystal, RG (2006). Genetic medicines: treatment strategies for hereditary disorders. *Nat Rev Genet*, 7:61–76.

Okubo, A, et al. (1999). The relationships between age changes in retinal pigment epithelium and Bruch's membrane. *Invest Ophthalmol Vis Sci*, 40:443–449.

Otani, A, et al. (2002). Bone marrow-derived stem cells target retinal astrocytes and can promote or inhibit retinal angiogenesis. *Nat Med*, 8:1004–1010.

Otteson, DC and Hitchcock, PF (2003). Stem cells in the teleost retina: persistent neurogenesis and injury-induced regeneration. *Vision Res*, 43:927–936.

Otto, EA, et al. (2005). Nephrocystin-5, a ciliary IQ domain protein, is mutated in Senior-Loken syndrome and interacts with RPGR and calmodulin. *Nat Genet*, 37:282–288.

Pacione, LR, et al. (2003). Progress toward understanding the genetic and biochemical mechanisms of inherited photoreceptor degenerations. *Annu Rev Neurosci*, 26:657–700.

Pawlyk, BS, et al. (2005). Gene replacement therapy rescues photoreceptor degeneration in a murine model of Leber congenital amaurosis lacking RPGRIP. *Invest Ophthalmol Vis Sci*, 46:3039–3045.

Percin, EF, et al. (2000). Human microphthalmia associated with mutations in the retinal homeobox gene CHX10. *Nat Genet*, 25:397–401.

Petit, C, (2004). Memorial lecture-hereditary sensory defects: from genes to pathogenesis. *Am J Med Genet A*, 130:3–7.

Pickering, MC, et al. (2002). Uncontrolled C3 activation causes membranoproliferative glomerulonephritis in mice deficient in complement factor H. *Nat Genet*, 31:424–428.

Radu, RA, et al. (2005). Reductions in serum vitamin A arrest accumulation of toxic retinal fluorophores: a potential therapy for treatment of lipofuscin-based retinal diseases. *Invest Ophthalmol Vis Sci*, 46:4393–4401.

Ragge, NK, et al. (2005a). Heterozygous mutations of OTX2 cause severe ocular malformations. *Am J Hum Genet*, 76:1008–1022.

Ragge, NK, et al. (2005b). SOX2 anophthalmia syndrome. *Am J Med Genet*, 135:1–7.

Rahi, JS and Cable, N (2003). British Childhood Visual Impairment Study Group. Severe visual impairment and blindness in children in the UK. *Lancet*, 362:1359–1365.

Ramrattan, RS, et al. (1994). Morphometric analysis of Bruch's membrane, the choriocapillaris and the choroid in aging. *Invest Ophthalmol Vis Sci*, 35:2857–2864.

Rivera, A, et al. (2005). Hypothetical LOC387715 is a second major susceptibility gene for age-related macular degeneration, contributing independently of complement factor H to disease risk. *Hum Mol Genet*, 14:3227–3236.

Rivolta, C, et al. (2002). Retinitis pigmentosa and allied diseases: numerous diseases, genes, and inheritance patterns. *Hum Mol Genet*, 11:1219–1227.

Rosenbaum, JL and Witman, GB (2002). Intraflagellar transport. *Nat Rev Mol Cell Biol*, 3:813–825.

Rosenfeld, PJ, et al. (1992). A null mutation in the rhodopsin gene causes rod photoreceptor dysfunction and autosomal recessive retinitis pigmentosa. *Nat Genet*, 1:209–213.

Schlichtenbrede, FC, et al. (2003a). Long-term evaluation of retinal function in Prph2Rd2/Rd2 mice following AAV-mediated gene replacement therapy. *J Gene Med*, 5:757–764.

Schlichtenbrede, FC, et al. (2003b). Intraocular gene delivery of ciliary neurotrophic factor results in significant loss of retinal function in normal mice and in the Prph2Rd2/Rd2 model of retinal degeneration. *Gene Ther*, 10:523–527.

Seddon, JM, et al. (1994). Dietary carotenoids, vitamins A, C, and E, and advanced age-related macular degeneration. Eye Disease Case-Control Study Group. *JAMA*, 272:1413–1420.

Seddon, JM, et al. (1996). A prospective study of cigarette smoking and age-related macular degeneration in women. *JAMA*, 276:1141–1146.

Seddon, JM, et al. (1997). Familial aggregation of age-related maculopathy. *Am J Ophthalmol*, 123:199–206.

Seddon, JM, et al. (2005). The US twin study of age-related macular degeneration: relative roles of genetic and environmental influences. *Arch Ophthalmol*, 123:321–327.

Shaw, LC, et al. (2006). Proliferating endothelial cell-specific expression of IGF-I receptor ribozyme inhibits retinal neovascularization. *Gene Ther*, 3 (9):752–760.

Shu, X, et al. (2005). RPGR ORF15 isoform co-localizes with RPGRIP1 at centrioles and basal bodies and interacts with nucleophosmin. *Hum Mol Genet*, 14:1183–1197.

Silvestri, G, et al. (1994). Is genetic predisposition an important risk factor in age-related macular degeneration? *Eye*, 8:564–568.

Slavotinek, AM, et al. (2000). Mutations in MKKS cause Bardet-Biedl syndrome *Nat Genet*, 26:15–16.

Smith, W, et al. (1998). Plasma fibrinogen levels, other cardiovascular risk factors, and age-related maculopathy: the Blue Mountains Eye Study. *Arch Ophthalmol*, 116:583–587.

Souied, EH, et al. (1998). The epsilon4 allele of the apolipoprotein E gene as a potential protective factor for exudative age-related macular degeneration. *Am J Ophthalmol*, 125:353–359.

Steingrimsson, E, et al. (2004). Melanocytes and the microphthalmia transcription factor network. *Ann Rev Genet*, 38:365–411.

Stone, EM, et al. (1998). Allelic variation in ABCR associated with Stargardt disease but not age-related macular degeneration. *Nat Genet*, 20:328–329.

Stone, EM, et al. (1999). A single *EFEMP1* mutation associated with both Malattia Leventinese and Doyne honeycomb retinal dystrophy *Nat Genet*, 22:199–202.

Sundin, OH, et al. (2005). Extreme hyperopia is the result of null mutations in MFRP, which encodes a Frizzled-related protein. *Proc Natl Acad Sci*, 102:9553–9558.

Tao, W, et al. (2002). Encapsulated cell-based delivery of CNTF reduces photoreceptor degeneration in animal models of retinitis pigmentosa. *Invest Ophthalmol Vis Sci*, 43:3292–3298.

Thanos, C and Emerich, D (2005). Delivery of neurotrophic factors and therapeutic proteins for retinal diseases. *Expert Opin Biol Ther*, 5:1443–1452.

Tropepe, V, et al. (2000). Retinal stem cells in the adult mammalian eye. *Science*, 287:2032–2036.

Urnov, FD, et al. (2005). Highly efficient endogenous human gene correction using designed zinc-finger nucleases. *Nature*, 435:646–651.

van Dorp, DB, et al (1992). A family with RP3 type of X-linked retinitis pigmentosa: an association with ciliary abnormalities. *Hum Genet*, 88:331–334.

Vervoort, R, et al. (2000). Mutational hot spot within a new RPGR exon in X-linked retinitis pigmentosa. *Nat Genet*, 25:462–466.

Vervoort, R and Wright, AF (2002). Mutations of RPGR in X-linked retinitis pigmentosa (RP3). *Hum Mutat*, 19:486–500.

Vithana, EN, et al (2001). A human homolog of yeast pre-mRNA splicing gene, PRP31, underlies autosomal dominant retinitis pigmentosa on chromosome 19q13.4 (RP11). *Mol Cell*, 8:375–381.

Voronina, VA, et al. (2004). Mutations in the human RAX homeobox gene in a patient with anophthalmia and sclerocornea. *Hum Mol Genet*, 13(3):315–322.

Wang, DY, et al. (2005). Gene mutations in retinitis pigmentosa and their clinical implications. *Clin Chim Acta*, 351:5–16.

Webster, AR, et al. (2001). An Analysis of Allelic Variation in the ABCA4 Gene. *Invest Ophthalmol Vis Sci*, 42:1179–1189.

Weeks, DE, et al. (2004). Age-related maculopathy: a genomewide scan with continued evidence of susceptibility loci within the 1q31, 10q26, and 17q25 regions. *Am J Hum Genet*, 75:174–189.

Weleber, RG, et al. (1993). Phenotypic variation including retinitis pigmentosa, pattern dystrophy, and fundus flavimaculatus in a single family with a deletion of codon 153 or 154 of the peripherin/RDS gene. *Arch Ophthalmol*, 111:1531–1542.

Weng, J, et al. (1999). Insights into the function of Rim protein in photoreceptors and etiology of Stargardt's disease from the phenotype in abcr knockout mice. *Cell*, 98:13–23.

Weston, MD, et al. (2004). Mutations in the VLGR1 gene implicate G-protein signaling in the pathogenesis of Usher syndrome type II. *Am J Hum Genet*, 74:357–366.

Whelan, J (2005). First clinical data on RNAi. *Drug Discov Today*, 10:1014–1015.

Wiggs, JL, et al. (2004). A genomewide scan identifies novel early-onset primary open-angle glaucoma loci on 9q22 and 20p12. *Am J Hum Genet*, 74:1314–1320.

Wiznerowicz, M and Trono, D (2003). Conditional suppression of cellular genes: lentivirus vector-mediated drug-inducible RNA interference. *J Virol*, **77**:8957–8961.

Wright, AF and Van Heyningen, V (2001). Short cut to disease genes. *Nature*, 414:705–706.

Wu, WC, et al. (2002). Gene therapy for detached retina by adeno-associated virus vector expressing glial cell line-derived neurotrophic factor. *Invest Ophthalmol Vis Sci*, 43:3480–3488.

Yang, Z, et al. (2002). Mutations in the RPGR gene cause X-linked cone dystrophy. *Hum Mol Genet*, 11:605–611.

Yardley, J, et al. (2004). Mutations of VMD2 splicing regulators cause nanophthalmos and autosomal dominant vitreoretinochoroidopathy (ADVIRC). *Invest Ophthalmol Vis Sci*, 45:3683–3689.

Ying, L, et al. (1999). Complement factor H gene mutation associated with autosomal recessive atypical hemolytic uremic syndrome. *Am J Hum Genet*, 65:1538–1546.

Yuan, L, et al. (2005). Mutations in PRPF31 inhibit pre-mRNA splicing of rhodopsin gene and cause apoptosis of retinal cells. *J Neurosci*, 25:748–757.

Zernant, J, et al. (2005). Genotyping microarray (disease chip) for Leber congenital amaurosis: detection of modifier alleles. *Invest Ophthalmol Vis Sci*, 46:3052–3059.

Zhao, Y, et al. (2003). The retinitis pigmentosa GTPase regulator (RPGR)-interacting protein: subserving RPGR function and participating in disk morphogenesis. *Proc Natl Acad Sci USA*, 100:3965–3970.

Zito, I, et al. (2003). RPGR mutation associated with retinitis pigmentosa, impaired hearing, and sinorespiratory infections. *J Med Genet*, 40:609–615.

31

Applications in Audiological Medicine

Lut Van Laer and Guy Van Camp

The hearing apparatus is a complex system that requires the interaction of many different proteins. It is therefore not unexpected that defects in many genes can lead to hearing loss (HL).

HL is the most common sensory handicap. One in six hundred and fifty children is born permanently hearing impaired, or will develop severe-to-profound HL before speech is acquired (Mehl and Thomson, 2002). In addition, 1 in 1000 children will become permanently or severely hearing impaired before adulthood (Morton, 1991). In developed countries, at least 50% of permanent childhood HL is attributable to genetic causes (Marazita et al., 1993). The remaining cases are due to environmental factors that lead to HL: for example, prenatal and postnatal infectious diseases, hypoxia, ototoxic drugs, and acoustic trauma. It is estimated that approximately 30% of inherited childhood HL is syndromic, whereas the remaining 70% is nonsyndromic (Gorlin et al., 1990; Cohen and Gorlin, 1995; Gorlin et al., 1995). The most common mode of transmission of permanent childhood hereditary HL is autosomal recessive (80%). Only 15% of the cases are estimated to have an autosomal dominant mode of inheritance, whereas 2%–3% of the cases show an X-linked inheritance pattern (Newton, 1985; Morton, 1991).

In the adult population, the prevalence of HL increases. Although several autosomal dominant types of HL have an adult age at onset, HL at older ages usually is complex in etiology, involving the interaction of several genes and environmental factors. Several types of complex HL exist, including age-related hearing impairment (ARHI), noise-induced HL (NIHL), drug-induced HL (DIHL), and otosclerosis (OTSC).

Monogenic (Mendelian) HL

Nonsyndromic HL

The molecular aspects of nonsyndromic HL were almost unexplored until the early 1990s. The last decade, however, has seen rapid and constant progress. To date, more than 100 loci for nonsyndromic HL have been mapped to the human genome: 54 autosomal dominant (DFNA), 59 autosomal recessive (DFNB), and 6 X-linked (DFN) loci. In addition, two modifier loci have been reported. Thirty-eight genes have been identified (Table 31–1). These genes belong to very different gene families with various functions, including transcription factors, extracellular matrix molecules, cytoskeletal components, and ion channels and transporters (Table 31–1). A regularly updated overview of the progress being made for monogenic HL can be found on the Hereditary Hearing Loss Homepage (Van Camp, G and Smith, RJH; http://webhost.ua.ac.be/hhh/).

For three nonsyndromic HL loci (Table 31–1; DFNB1, DFNA2, and DFNA3), more than one gene has been identified, indicating the presence of clusters of HL genes in those particular chromosomal regions. In fact for the DFNA2 locus, the existence of a third gene is expected (Van Hauwe et al., 1999). On the other hand, a single nonsyndromic HL gene can be responsible for the HL associated with more than one locus name (Table 31–1; *ACTG1* for DFNA20/26, *TECTA* for DFNA8/12, *TMC1* for DFNB7/11, *TMPRSS3* for DFNB8/10, and *WFS1* for DFNA6/14/38). This is due to initial erroneous linkage intervals, or because a new locus was designated despite the fact that previous linkage reports had already mapped a nonsyndromic HL locus to exactly the same genomic region.

It is striking that mutations in one gene can cause autosomal dominant as well as autosomal recessive nonsyndromic HL (Table 31–1). For instance, mutations in *MYO7A*, an unconventional myosin, cause DFNB2 (Liu, Walsh, Mburu, et al., 1997; Weil et al., 1997) as well as DFNA11 (Liu, Walsh, Tamagawa, et al., 1997). The gap junction protein connexin 26 (*GJB2*) is involved both in DFNB1 (Kelsell et al., 1997) and DFNA3 (Denoyelle et al., 1998). Several other examples exist (Table 31–1: *TECTA*, *ESPN*, *GJB3*, *GJB6*, *MYO6*, *TMC1*). Furthermore, mutations in one gene can lead to both nonsyndromic and syndromic HL (Tables 31–1 and 32–2): for example, mutations in *MYO7A* and *SLC26A4* lead to Usher type 1B (Weil et al., 1995) and Pendred syndrome (Everett et al., 1997), respectively, but can also cause nonsyndromic HL, DFNB2 (Liu, Walsh, Mburu, et al., 1997; Weil et al., 1997) and DFNB4 (Li et al., 1998), respectively. Indeed, *SLC26A4* mutations seem to give rise to a broad spectrum of clinical symptoms: from nonsyndromic HL to nonsyndromic HL with an enlarged vestibular aqueduct (EVA; Usami et al., 1999), and to full-blown Pendred with thyroid goiter. Additional examples are given in Table 31–2.

One HL gene that has been of particular interest is *GJB2*. It is further elaborated below.

GJB2 (Connexin 26)

GJB2 was identified as a gene for autosomal recessive HL (DFNB1) in 1997 (Kelsell et al., 1997), and as a gene for autosomal dominant HL (DFNA3) in 1998 (Denoyelle et al., 1998). Although autosomal

Table 31–1 Nonsyndromic Hearing Loss Genes

Gene Symbol (Gene Product)	Gene Function	NSHL Locus (Reference)
ACTG1 (actin type G1)	Cytoskeletal component	DFNA20/26 (van Wijk et al., 2003; Zhu et al., 2003)
ATP2B2 (PMCA2)	Plasma membrane calcium pump	HL modifier gene (Schultz et al., 2005)
CDH23 (cadherin 23)	Cell adhesion protein	DFNB12 (Bork et al., 2001)
CLDN14 (claudin 14)	Tight junction protein	DFNB29 (Wilcox et al., 2001)
COCH (cochlin)	Extracellular matrix protein	DFNA9 (Robertson et al., 1998)
COL11A2 (type XI collagen α2)	Extracellular matrix protein	DFNA13 (McGuirt et al., 1999)
CRYM	NADP-regulated thyroid hormone-binding protein	Dominant HL (Abe et al., 2003)
DFNA5	Unknown function	DFNA5 (Van Laer et al., 1998)
DIAPH1 (diaphanous)	Actin polymerization	DFNA1 (Lynch et al., 1997)
ESPN (espin)	Actin-bundling in hair cell stereocilia	DFNB36 (Naz et al., 2004) Dominant HL (Donaudy et al., 2005)
EYA4	Transcription factor	DFNA10 (Wayne et al., 2001)
GJB2 (Connexin 26)	Gap junction protein, involved in ion homeostasis	DFNA3 (Denoyelle et al., 1998) DFNB1 (Kelsell et al., 1997)
GJB3 (Connexin 31)	Gap junction protein, involved in ion homeostasis	DFNA2 (Xia et al., 1998) Recessive HL (Liu et al., 2000)
GJB6 (Connexin 30)	Gap junction protein, involved in ion homeostasis	DFNA3 (Grifa et al., 1999) DFNB1 (del Castillo et al., 2002)
KCNQ4	Potassium channel, involved in K^+-recycling	DFNA2 (Kubisch et al., 1999)
MYH9 (myosin heavy chain 9)	Cytoskeletal component	DFNA17 (Lalwani et al., 2000)
MYH14 (myosin heavy chain 14)	Cytoskeletal component	DFNA4 (Donaudy et al., 2004)
MYO1A (myosin 1A)	Cytoskeletal component	DFNA48 (Donaudy et al., 2003)
MYO3A (myosin 3A)	Cytoskeletal component	DFNB30 (Walsh et al., 2002)
MYO6 (myosin 6)	Cytoskeletal component	DFNA22 (Melchionda et al., 2001) DFNB37 (Ahmed, Morell, et al., 2003)
MYO7A (myosin 7A)	Cytoskeletal component	DFNA11 (Liu, Walsh, Tamagawa, et al., 1997) DFNB2 (Liu, Walsh, Mburu, et al., 1997; Weil et al., 1997)
MYO15 (myosin 15)	Cytoskeletal component	DFNB3 (Wang et al., 1998)
OTOA (otoancorin)	Extracellular matrix protein	DFNB22 (Zwaenepoel et al., 2002)
OTOF (otoferlin)	Unknown function	DFNB9 (Yasunaga et al., 1999)
PCDH15 (protocadherin 15)	Cell adhesion protein	DFNB23 (Ahmed, Riazuddin, et al., 2003)
POU3F4	Transcription factor	DFN3 (de Kok et al., 1995)
POU4F3	Transcription factor	DFNA15 (Vahava et al., 1998)
PRES (prestin)	Outer hair cell motor protein	Recessive HL (Liu et al., 2003)
SLC26A4 (pendrin)	Ion homeostasis	DFNB4 (Li et al., 1998) NSHL with EVA (Usami et al., 1999)
STRC (stereocilin)	Cytoskeletal component	DFNB16 (Verpy et al., 2001)
TECTA (α-tectorin)	Extracellular matrix, component of the tectorial membrane	DFNA8/12 (Verhoeven et al., 1998) DFNB21 (Mustapha et al., 1999)
TFCP2L3	Transcription factor	DFNA28 (Peters et al., 2002)
TMC1	Unknown function	DFNA36 (Kurima et al., 2002) DFNB7/11 (Kurima et al., 2002)
TMIE	Unknown function	DFNB6 (Naz et al., 2002)
TMPRSS3	Transmembrane serine protease	DFNB8/10 (Scott et al., 2000, 2001)
USH1C (harmonin)	Unknown function	DFNB18 (Ahmed et al., 2002)
WFS1 (wolframin)	Unknown function	DFNA6/14/(38) (Bespalova et al., 2001; Young et al., 2001)
WHRN (whirlin)	Elongation/maintenance of hair cell stereocilia	DFNB31 (Mburu et al., 2003)

Abbreviations: EVA, enlarged vestibular aqueduct; NSHL, nonsyndromic hearing loss.

dominant HL due to *GJB2* mutations is rare, it has become clear that a large proportion of the cases of prelingual HL is due to mutations in *GJB2* (Denoyelle et al., 1997; Zelante et al., 1997; Kelley et al., 1998). To date, 92 recessive and 9 dominant disease-causing *GJB2* mutations have been identified (Gasparini P and Estivill X, The Connexin-deafness Homepage, http://davinci.crg.es/deafness/June 2005). One mutation, the deletion of one guanosine residue from a stretch of six between nucleotide positions 30 and 35 (35delG), is the most common deafness-causing variant of *GJB2* in sporadic patients and autosomal recessive families of Caucasian origin (Denoyelle et al., 1997; Zelante et al., 1997; Estivill, Fortina, et al., 1998; Kelley et al., 1998; Rabionet et al., 2000). Furthermore, the carrier frequency of 35delG in the European population ranges from 3% in southern

Table 31–2 Nonsyndromic Hearing Loss Genes Causing Syndromic Hearing Loss

Gene Symbol	Syndromic Hearing Loss	Reference
CDH23	Usher syndrome type 1D	Bolz et al., 2001; Bork et al., 2001
COl11A2	Stickler syndrome type 3	Vikkula et al., 1995
	OSMED syndrome	Pihlajamaa et al., 1998
EYA4	Dilated cardiomyopathy and HL	Schonberger et al., 2005
GJB2 (Cx26)	Palmoplantar kerato-derma and HL	Richard et al., 1998
	Vohwinkel's syndrome	Maestrini et al., 1999
	Palmoplantar hyperkera-tosis and HL	Heathcote et al., 2000
	KID	Richard et al., 2002
GJB3 (Cx31)	Peripheral neuropathy and HL	Lopez-Bigas et al., 2001
MYH9	Fechtner syndrome	May-Hegglin/Fechtner Syndrome Consortium, 2000
	Epstein syndrome	Heath et al., 2001
	Alport syndrome with macrothrombocytopenia	Heath et al., 2001
MYO7A	Usher syndrome type 1B	Weil et al., 1995
PCDH15	Usher syndrome type 1F	Ahmed et al., 2001; Alagramam et al., 2001
SLC26A4 (PDS)	Pendred syndrome	Everett et al., 1997
USH1C	Usher syndrome type 1C	Bitner-Glindzicz et al., 2000; Verpy et al., 2000
WFS1	Wolfram syndrome (DIDMOAD)	Inoue et al., 1998

Abbreviations: DIDMOAD, diabetes insipidus, diabetes mellitus, optic atrophy and deafness; HL, hearing loss; KID, keratitis-ichthyosis-deafness syndrome; OSMED, oto-spondylo-megaepiphyseal dysplasia.

Europe to 1% in central and northern Europe (Estivill, Fortina, et al., 1998; Gasparini et al., 2000). In non-Caucasian populations, the 35delG mutation is either not found or is very rare, with other "common" mutations prevailing, such as the 235delC in the Japanese and the Korean (Abe et al., 2000; Kudo et al., 2000; Park et al., 2000), the 167delT in the Ashkenazi Jewish (Morell et al., 1998; Sobe et al., 1999), and the R143W mutation in an African village (Brobby et al., 1998). It has been shown that the high frequency of three of these frequent mutations (167delT, 35delG, and 235delC) reflect a founder effect (Morell et al., 1998; Van Laer et al., 2001; Ohtsuka et al., 2003). In Spain, two deletions near *GJB2*, one 309 kb deletion called del(GJB6-D13S1830) and one 232 kb deletion called del(GJB6-D13S1854), are frequently found in compound heterozygosity with a *GJB2* mutation (del Castillo et al., 2002; del Castillo et al., 2005). These deletions leave the *GJB2* coding region intact, but delete a large region close to *GJB2* and truncate another connexin gene (*GJB6*, connexin 30) that is located within 50 kb of *GJB2*. To date, it remains unclear whether these deletions affect *GJB2* expression levels through the deletion of a putative regulatory element of *GJB2* or whether *GJB2* and *GJB6* act simultaneously to cause HL (digenic model of inheritance).

GJB2 encodes connexin 26, a gap junction protein that is a component of the potassium recycling pathway in the inner ear. Loss or malfunction of these gap junctions may disrupt electric and metabolic coupling of cellular networks in the cochlea, leading to hearing impairment. The resultant HL involves all frequencies, is mostly severe to profound (although rare, mild and moderate cases occur), rarely progresses, and is most frequently symmetrical between two ears (Cohn et al., 1999; Denoyelle et al., 1999; Murgia et al., 1999).

GJB2 is a small gene, so cost- and labor-efficient mutation screening can be performed. Therefore, and because of the high prevalence of *GJB2* mutations, the mutation analysis of *GJB2* facilitates the diagnosis of congenital permanent HL despite the unprecedented genetic heterogeneity of nonsyndromic HL. Recently, a genotype–phenotype correlation was published. This study demonstrated that it is possible to predict the HL phenotype for different *GJB2* genotypes with a certain degree of probability, which is of considerable value for genetic counseling (Cryns et al., 2004).

Syndromic HL

Several hundred genetic syndromes have been described in which HL is accompanied by various additional symptoms (Gorlin et al., 1990, 1995). The mode of transmission of these syndromic HL genes can be autosomal dominant [e.g., Alport, Waardenburg, Treacher-Collins, branchio-oto-renal syndrome (BOR)], autosomal recessive (e.g., Usher, certain types of Alport, Pendred), or X-linked (e.g., certain types of Alport). It is beyond the scope of this overview to go into detail on these and various other syndromes. However, in Table 31–3, a selection of currently identified genes responsible for syndromic HL is presented. A number of syndromes with HL, where deafness is either a minor symptom or a secondary consequence of the primary defect (e.g., neurofibromatosis type 2, Hunter, Hurler, Crouzon, Cockayne), are omitted from Table 31–3.

Otosclerosis

OTSC is a common bone abnormality of the otic capsule leading to conductive hearing impairment due to a bony fixation of the stapedial footplate in the oval window (Hueb et al., 1991). In some cases, an additional sensorineural component develops across all frequencies, leading to mixed HL (Browning and Gatehouse, 1984; Ramsay and Linthicum, 1994). The age at onset of OTSC is between 20 and 40. The disease can be progressive in some periods of life, but profound HL due to OTSC is rare. An amelioration of hearing can be realized by stapedial microsurgery for the conductive component of the HL, but the sensorineural component cannot be corrected surgically (Somers et al., 1997).

Clinical OTSC, with hearing impairment as a primary characteristic, shows a prevalence of 0.3%–0.4% in the Caucasian population (Declau et al., 2001). Women are affected twice as often as men. The majority of OTSC patients show no familial history, or patients occur in small families with an unclear inheritance pattern (complex type of OTSC, see section on complex types of HL for discussion). However, rare monogenic forms of OTSC exist, mainly characterized by an autosomal dominant inheritance pattern with reduced penetrance. At present, four loci for these rare monogenic types of OTSC have been mapped (OTSC1, Tomek et al., 1998; OTSC2, Van Den Bogaert et al., 2001; OTSC3, Chen et al., 2002; and OTSC5, Van Den Bogaert et al., 2004), but no genes have yet been identified.

Mitochondrial HL

Not all DNA of a eukaryotic cell is included in the cell nucleus; a small part is found in the mitochondrion. Each mitochondrion contains

Table 31–3 Selected Genes for Syndromic Hearing Loss

Syndrome	Most Important Additional Symptom(s)	Gene Symbol (Gene Product)	Gene Function	Reference
Alport syndrome (Autosomal recessive)	Renal disease	*COL4A3/COL4A4* (type IV collagen α3/α4)	Extracellular matrix	Mochizuki et al., 1994
Alport syndrome (X-linked)	Renal disease	*COL4A5/COL4A6* (type IV collagen α5/α6)	Extracellular matrix	Barker et al., 1990; Zhou et al., 1993
BOR (branchio-oto-renal syndrome)	Branchial anomalies;	*EYA1*	Transcription factor	Abdelhak et al., 1997
	Renal manifestations	*SIX1*	Transcription factor	Ruf et al., 2004
DFN1[a]	Blindness, dystonia, fractures, mental deficiency	*TIMM8A* (deafness dystonia peptide)	Mitochondrial import protein	Jin et al., 1996
Jervell and Lange-Nielsen	Cardiac symptoms	*KCNQ1* (KLVQT1)	Potassium channel	Neyroud et al., 1997
		KCNE1 (minK)	Potassium channel	Schulze-Bahr et al., 1997; Tyson et al., 1997
Norrie	Congenital blindness; Mental retardation	*NDP* (norrin)	Vascular growth factor	Chen et al., 1992
Pendred	Thyroid dysfunction	*SLC26A4* (pendrin)	Ion homeostasis	Everett et al., 1997
Stickler type 1	Arthro-ophthalmopathy; Connective tissue dysplasia	*COL2A1* (type II collagen α1)	Extracellular matrix	Ahmad et al., 1991
Stickler type 2	Arthro-ophthalmopathy; Connective tissue dysplasia	*COL11A1* (type XI collagen α1)	Extracellular matrix	Richards et al., 1996
Stickler type 3	Arthro-ophthalmopathy; Connective tissue dysplasia	*COL11A2* (type XI collagen α2)	Extracellular matrix	Vikkula et al., 1995
Treacher Collins syndrome	Abnormal craniofacial development	*TCOF1* (treacle)	Nucleolar phosphoprotein	The Treacher Collins Syndrome Collaborative Group, 1996
Usher syndrome type 1B	Usher 1: Retinitis pigmentosa; No vestibular response	*MYO7A* (myosin 7A)	Cytoskeletal protein	Weil et al., 1995
Usher syndrome type 1C		*USH1C* (harmonin)	Cohesion of hair cell bundle	Verpy et al., 2000; Bitner-Glindzicz et al., 2000
Usher syndrome type 1D		*CDH23* (cadherin 23)	Adhesion molecule	Bork et al., 2001; Bolz et al., 2001
Usher syndrome type 1F		*PDCH15* (protocadherin 15)	Adhesion molecule	Ahmed et al., 2001; Alagramam et al., 2001
Usher syndrome type 1G	*USH1G* (SANS)	Cohesion of hair cell bundle	Weil et al., 2003	
Usher syndrome type 2A	Usher 2: Retinitis pigmentosa; Normal vestibular response	*USH2A* (usherin)	Extracellular matrix	Eudy et al., 1998
Usher syndrome type 2C	*VLGR1* (MASS1)	G-coupled receptor	Weston et al., 2004	
Usher syndrome type 3	Usher 3: Retinitis pigmentosa; Variable vestibular responses	*USH3* (clarin 1)	Unknown function	Joensuu et al., 2001
Waardenburg type 1	Pigmentary disturbances; Distopia canthorum	*PAX3*	Transcription factor	Tassabehji et al., 1992
Waardenburg type 2	Pigmentary disturbances; No distopia canthorum	*MITF*	Transcription factor	Tassabehji et al., 1994
	SNAI2	Transcription factor	Sanchez-Martin et al., 2002	
Waardenburg type 3	Type 1 and limb abnormalities	*PAX3*	Transcription factor	Hoth et al., 1993
Waardenburg type 4	Type 2 and Hirschprung[b]	*EDNRB*	Endothelin B receptor	Attie et al., 1995
Waardenburg type 4	Type 2 and Hirschprung[b]	*EDN3* (endothelin 3)	Ligand to receptor	Edery et al., 1996

[a] DFN1 was originally described as nonsyndromic HL, however re-examination of the family revealed several additional symptoms. Therefore DFN-1 should be classified among the syndromic HL loci.

[b] Aganglionic megacolon.

2–10 mitochondrial chromosomes, and each cell contains hundreds of mitochondria. A mitochondrial DNA molecule contains the information to encode 37 mitochondrion-specific molecules. It is replicated and transcribed within the mitochondrion. Mitochondrial DNA is maternally inherited; the paternal sperm cell does not contribute to the mitochondrial DNA of the zygote.

As mitochondria have a crucial function in nearly every cell, it is not unexpected that mitochondrial DNA mutations cause multisystemic diseases. Over the past decade, it has been shown that some mutations in mitochondrial DNA are also associated with HL. Usually, patients with mitochondrial mutations exhibit generalized neuromuscular dysfunction in which HL is a minor clinical feature. Surprisingly, some nonsyndromic mitochondrial HL mutations have been described as well. Why only the auditory system is affected remains to be elucidated. An overview of syndromic and nonsyndromic mitochondrial HL mutations is given in Table 31–4.

Heteroplasmy (i.e., a mixture of mutant and normal mitochondrial DNA in the same individual) is a common feature of syndromic mitochondrial disease. The amount of heteroplasmy varies in different tissues, and among individuals. It has been suggested that heteroplasmy is responsible for the extensive variability in penetrance, leading to various phenotypes in one pedigree, a characteristic feature of mitochondrial disease. In contrast, mitochondrial mutations leading to nonsyndromic HL usually are found in a homoplasmic state. The phenotypic variation observed in families affected by nonsyndromic mitochondrial HL can therefore only be explained by the influence of environmental factors and modifier genes.

One mutation, 1555A→G, is of particular importance because it is a frequent cause of HL in several populations (Estivill, Govea, et al., 1998). Carriers of the 1555A→G mutation show an increased susceptibility to aminoglycoside antibiotics, leading to aminoglycoside-induced ototoxic HL. However, HL can also develop in the absence of aminoglycoside treatment (Prezant et al., 1993). Extensive clinical heterogeneity is observed in all pedigrees harboring this mutation. HL can occur in early childhood, later in life, or only after aminoglycoside exposure (Prezant et al., 1993; Matthijs et al., 1996; Usami et al., 1997; Estivill, Govea, et al., 1998). A first nuclear modifier gene (*MTO1*) putatively involved in the pathogenesis of 1555A→G mitochondrial HL was recently identified (Li et al., 2002). Variations in this gene might partly explain the phenotypic variability observed in 1555A→G patients. For a more thorough discussion of mitochondrial mutations in HL, see reviews by Fischel-Ghodsian (2003) and Van Camp and Smith (2000).

Polygenic (Complex) HL

Age-related Hearing Loss

The next 50 years will witness a significant increase in ageing in the European Union, the United States, and Japan, with the number of people aged 65 and above growing significantly. ARHI, also called presbyacusis, is the most frequent sensory disability of the elderly. In its most typical presentation, ARHI is mid to late adult onset, progressive, bilaterally symmetrical, sensorineural, and most pronounced in the high frequencies. Thirty-seven percent of people aged between 61 and 70 have a significant HL of at least 25 dB (Davis, 1994). This prevalence increases with older ages; 60% of 71–80-year-olds are affected by ARHI (Davis, 1994). The prevalence

Table 31–4 Mitochondrial Mutations Involved in Maternally Inherited Syndromic and Nonsyndromic Hearing Loss

	Syndromic Mitochondrial Hearing Loss		
Gene symbol	Mutation	Syndrome	Reference
$tRNA^{Leu(UUR)}$	3243A→G	MELAS MIDD	Goto et al., 1990 van den Ouweland et al., 1992
$tRNA^{Ser(UCN)}$	7512T→C	Progressive myoclonic epilepsy, ataxia and HL	Jaksch et al., 1998
$tRNA^{Lys}$	8344A→G	MERFF	Shoffner et al., 1990
	8356T→C	MERFF/MELAS	Zeviani et al., 1993
	8296A→G	MIDD	Kameoka et al., 1998
$tRNA^{Glu}$	14709T→C	MIDD	Hao et al., 1995
Several	Large deletion/duplication	MIDD	Ballinger et al., 1992
Several	Large deletions	KSS	Moraes et al., 1989
Nonsyndromic Mitochondrial Hearing Loss			
Gene Symbol	Mutation	Remark/Additional Symptoms[a]	Reference of First Report[b]
12SrRNA	961(different mutations)	Aminoglycoside induced/worsened	Bacino et al., 1995; Casano et al., 1999
	1494C→T	Aminoglycoside induced/worsened	Zhao et al., 2004
	1555A→G	Aminoglycoside induced/worsened	Prezant et al., 1993
	1556C→T	Aminoglycoside induced	Tanimoto et al., 2004
$tRNA^{ser(UCN)}$	7445A→G	Palmoplantar keratoderma	Reid et al., 1994
	7472insC	Neurological dysfunction[c]	Tiranti et al., 1995
	7510T→C	No additional symptoms reported	Hutchin et al., 2000
	7511T→C	No additional symptoms reported	Sue et al., 1999

[a] The additional symptoms reported for the mutations that are classified as nonsyndromic mitochondrial HL are often mild and can easily be overlooked. For example, palmoplantar keratoderma in case of 7445A→G was only revealed upon re-examination of the originally described families, and the neurological symptoms accompanying HL caused by 7472insC presented as a "general clumsiness" in the family described by Verhoeven et al. (Verhoeven et al., 1999). The additional symptoms are also observed only in some of the affected subjects within the respective families.

[b] Additional references can be found for most of these mutations at the Hereditary Hearing Loss Homepage (Van Camp G, and Smith RJH, http://webhost.ua.ac.be/hhh/).

[c] The main neurological characteristics were ataxia, dysarthria, and focal myoclonus.

Abbreviations: MELAS, mitochondrial encephalomyopathy, lactic acidosis, stroke-like symptoms; MERFF, myoclonus epilepsy with ragged-red fibers; MIDD, maternally inherited diabetes mellitus and deafness; KSS, Kearns-Sayre syndrome.

of ARHI is gender related; in general, men are more severely affected (Davis, 1994). On average, ARHI thresholds increase approximately 1 dB per year for individuals aged 60 and above (Lee et al., 2005). However, ARHI shows extensive variation; age at onset, progression, and severity of HL vary considerably among the elderly. The ISO 7029 standard perfectly illustrates this variation (International Organization for Standardization, 1998). The largest variation is found at the high frequencies and at older ages. For instance, at 60 years of age, the best-hearing 10% of the population display high-frequency thresholds better than 10 dB, whereas the worst-hearing 10% suffer from a HL of 55–75 dB at the high frequencies (International Organization for Standardization, 1998).

ARHI has a deleterious impact on social life. Reduced communication skills frequently result in poor psychosocial functioning and consequently in isolation of the ageing individual. Currently, hearing aids are the only possibility for therapeutic intervention of ARHI. Unfortunately, these are suitable only for a limited number of people. Although hearing aids succeed in sufficient amplification of sound, the gain in speech recognition is often experienced as poor, especially in noisy environments. In addition, many do not accept hearing aids because of social stigmatization. Future therapies for ARHI will have to rely on basic rather than symptomatic approaches. This requires a thorough knowledge of the etiological factors leading to ARHI.

Heritability of ARHI

The hypothesis that ARHI has a genetic basis has been put forward for many years in many publications, but the scientific basis for this claim has only recently been laid. Three separate studies have estimated heritability values for ARHI, and have shown that ARHI is a complex disorder with genetic and environmental etiological factors.

In one study, a Swedish male twin population comprised of monozygotic and dizygotic twins aged between 36 and 80 years was studied using a combination of audiometric and questionnaire data (Karlsson et al., 1997). This resulted in a heritability estimate of 0.47 for the age group above 65 years, which indicates that approximately half of the population variance for high frequency hearing ability above the age of 65 is caused by genetic differences, and half by environmental differences (Karlsson et al., 1997). A second study compared the auditory status in genetically unrelated (spouse pairs) and genetically related people (sibling pairs, parent–child pairs). This study showed a clear familial aggregation for age-related hearing levels, the aggregation levels being stronger in women than in men. The heritability estimates of this study suggested that 35%–55% of the variance of ARHI is attributable to the effects of genes (Gates et al., 1999). A Danish twin study evaluated the self-reported reduced hearing abilities in twins of 75 years of age and older. The heritability value was estimated at 40% (Christensen et al., 2001). Because self-assessment of HL only partly corresponds to audiometric measures of HL and frequently results in misclassification (Matthews et al., 1990), this heritability value may represent an underestimation of the involvement of genetic factors in ARHI.

Although it has been proven that ARHI is a complex disease caused by an interaction of environmental and genetic factors, so far nothing is known about the genes that contribute to ARHI in humans. In contrast, the environmental factors involved in ARHI are relatively well documented.

Environmental Risk Factors for ARHI

Several environmental risk factors have been put forward as being involved in the development of ARHI (noise, drugs, chemicals, etc). However, considerable controversy exists concerning the role of many of these. The best known and the most studied risk factor leading to HL is noise exposure. It is not always possible to make a clear distinction between ARHI and NIHL (see section on NIHL). Exposure to excessive occupational noise is seen as a cause for NIHL rather than for ARHI. On the other hand, constant low-level noise (the noise of every day life in our industrialized and urbanized environment, also called "acoustic smog") is often considered an important environmental risk factor for ARHI. From experiments in a mouse model for early-onset ARHI (C57BL/6J), it became clear that noise exposure might reveal a genetic predisposition to ARHI earlier in life. In other words, genes that are associated with ARHI might render the cochlea more susceptible to noise (Erway et al., 1996; Ohlemiller, Wright, et al., 2000). It has been a point of debate whether ageing and noise act in an additive or an interactive way to produce permanent HL. If the latter is more important, the question remains whether they amplify each other or tone each other down (Corso, 1992). The assumption of an additive effect has been most widely accepted.

Besides noise, several other environmental factors have been implicated in the etiology of ARHI, an overview of which is given in Table 31–5. Ototoxic medication can be problematic especially in

Table 31–5 Environmental Risk Factors for Age-related Hearing Loss

Factor	Effect[a]	Reference(s)
Ototoxic medication	Salicylate	Stypulkowski, 1990; Rosenhall et al., 1993
	β-Adrenergic drugs	Mills et al., 1999
	Aminoglycosides	Aran et al., 1992
	Loop diuretics	Aran et al., 1992
Chemicals	Organic solvents	Rybak, 1992; Johnson and Nylen, 1995
	Heavy metals	Rybak, 1992
Smoking	Causing hearing loss	Rosenhall et al., 1993; Cruickshanks et al., 1998; Itoh et al., 2001; Uchida et al., 2005
	Having no effect	Gates et al., 1993; Brant et al., 1996
	Passive smoking: causing hearing loss	Cruickshanks et al., 1998
Alcohol	Abuse causing hearing loss	Rosenhall et al., 1993; Popelka et al., 2000
	Abuse having no effect	Itoh et al., 2001
	Moderate use: protective effect	Popelka et al., 2000; Itoh et al., 2001
Head trauma	Whiplash	Fitzgerald, 1996
Nutrition status	Low vitamin B_{12}	Houston et al., 1999
	Low folate	Houston et al., 1999
Caloric-restricted diet	No, or a very small, protective effect	Willott et al., 1995
Socioeconomic	Low social class	Sixt and Rosenhall, 1997
	Lower level of education	Sixt and Rosenhall, 1997

[a] Unless indicated otherwise the environmental factors listed here cause hearing loss.

the elderly because they tend to take more medication and for longer periods compared to other age groups, and also because they have altered liver and renal functions which can cause medicational blood levels to rise above certain critical levels (Stypulkowski, 1990; Aran et al., 1992; Rosenhall et al., 1993; Mills et al., 1999). The detrimental effects of some chemicals on hearing levels are indisputable (Rybak, 1992; Johnson and Nylen, 1995). The effect of tobacco smoking and alcohol (ab)use on HL remains controversial (Gates et al., 1993; Rosenhall et al., 1993; Brant et al., 1996; Cruickshanks et al., 1998; Popelka et al., 2000; Itoh et al., 2001; Uchida et al., 2005). HL due to head trauma could possibly be caused by the disruption of the membranous portion of the cochlea, by disturbance in the cochlear microcirculation or by hemorrhage into the fluids of the cochlea (Fitzgerald, 1996). Nutritional status also seems to have importance (Houston et al., 1999), whereas caloric restriction does not have much effect (Willott et al., 1995). Finally, even the socioeconomic status has been implicated as a contributing factor. Interestingly, this effect remained even when noise exposure was taken into account (Sixt and Rosenhall, 1997). Clearly, it will be very difficult to assess what the contribution of all separate factors will be on the final outcome, that is, the level of HL.

Medical Risk Factors for ARHI

Several medical conditions have been postulated as risk factors for ARHI. A possible relation between ARHI and cardiovascular disease was investigated in the Framingham cohort (Gates et al., 1993). Cardiovascular disease was associated with ARHI, although predominantly in the low-frequency range and in women. Of the cardiovascular disease risk factors, hypertension and systolic blood pressure were related to hearing thresholds both in men and in women, whereas high-density lipoprotein and blood glucose levels were associated with low-frequency pure-tone averages only in women (Gates et al., 1993). The ISO 7029 norms show that ARHI mainly affects the high frequencies, indicating that other factors may be more important than cardiovascular disease in many patients. Brant et al. (1996) later confirmed the relationship between ARHI and systolic blood pressure for the speech frequencies, while Lee et al. (1998) could confirm the effect of high-density lipoproteins on ARHI in women. Classically, low-frequency ARHI has been associated with microvascular disease, leading to atrophy of the stria vascularis (Schuknecht and Gacek, 1993). Another indication that vascular abnormalities might be important in the development of ARHI was recently obtained in C57BL/6J mice, where a significant reduction of cochlear vascular endothelial growth factor (VEGF) expression was observed as a function of age (Picciotti et al., 2004).

Patients who suffer from chronic renal failure and undergo dialysis are also at risk of developing high-frequency HL. Either the disease itself (due to uraemic neuropathy, electrolyte imbalance, premature cardiovascular disease, or shared antigenicity between cochlea and kidney) or the treatment (chronic dialysis most often followed by kidney transplantation and an accompanied use of ototoxic medication) might be responsible for the increased risk in renal patients (Antonelli et al., 1990). Demineralization of the cochlear capsule in conjunction with age-related bone mass loss may lead to ARHI. This was reported by Clark et al. (1995), who could demonstrate that the femoral neck bone mass, but not the radial bone mass was associated with ARHI in a population of rural women aged 60–85 years. Another study could not demonstrate a relation between hip bone mineral density and hearing abilities (Purchase-Helzner et al., 2004). Several investigators have observed an association between diabetes mellitus and high-frequency HL. This association might be explained either by micro-angiopathic lesions in the inner ear (cochlear loss) or by primary neuropathy of the acoustic nerve (retro-cochlear loss) (Kurien et al., 1989).

Finally, a role for the immune system in the development of ARHI has been suggested. When accelerated senescence mice (SAMP1) were bred in a specific pathogen-free (SPF) environment, the age-related diseases typically observed in these mice (including age-related HL) were delayed in onset, when compared to mice bred in pathogenic environments. The involvement of autoimmune mechanisms was excluded (Iwai et al., 2003). It was argued that the stress a host experiences due to pathogen-induced infections impairs the immune functions, preceding a general decline in various physiological functions (Iwai et al., 2003).

Genetic Risk Factors for ARHI

In contrast to the vast information available regarding environmental and medical risk factors involved in the development of ARHI, only a minimal amount of information regarding genetic risk factors can be found in the literature. Most of the studies describe work on animal models; only a few studies have attempted to identify ARHI genes in human, and none have been identified so far.

Mouse strains with age-related HL (AHL) may represent valuable models and may be used for the investigation of genetic factors in human ARHI. The first major, recessive gene affecting AHL in C57BL/6J mice (designated *Ahl*) was localized to chromosome 10 (Johnson et al., 1997) and was identified in 2003. In exon 7 of *Cdh23* (cadherin 23), a hypomorphic single-nucleotide polymorphism (753A) leading to in-frame skipping of exon 7 showed significant association with *Ahl* (Noben-Trauth et al., 2003). The AHL of inbred strains homozygous for this polymorphism may be due to altered adhesion properties or reduced stability of the CDH23 protein lacking exon 7 (Noben-Trauth et al., 2003). A second locus affecting AHL (*Ahl2*) was mapped to chromosome 5 (Johnson and Zheng, 2002), while a third locus (*Ahl3*) was positioned on chromosome 17 (Nemoto et al., 2004).

Oxidative Stress

Reactive oxygen species (ROS) have been implicated in HL associated with ageing and noise exposure. ROS are a normal by-product of cellular metabolism, more in particular of the oxidative phosphorylation process. ROS are potentially toxic and can cause DNA, cellular, and tissue damage if not inactivated by cellular antioxidant protection systems (glutathione and glutathione-related enzymes, superoxide dismutases and catalase). ROS can also cause direct damage to mitochondrial DNA.

Significantly decreased glutathione levels have been observed in the auditory nerve, but not in other cochlear parts, in 24-month-old rats (Lautermann et al., 1997). SOD1-deficient mice displayed a more pronounced AHL than wild-type mice (McFadden, Ding, Burkard, et al., 1999; McFadden, Ding, Reaume, et al., 1999). However, SOD1 overexpression did not protect against AHL, indicating that the oxidative metabolism may be more complex than previously assumed (Coling et al., 2003). Antioxidants, which block and scavenge ROS, thereby reducing the deleterious impact of ROS at the molecular level, might attenuate ARHI. This has been demonstrated in rats with oversupplementation of vitamins E and C (Seidman, 2000), and with two mitochondrial metabolites (acetyl-1-carnitine and a-lipoic acid) (Seidman et al., 2000). Although in previous studies in human, no or only a very small effect had been observed (Table 31–5; Willott et al., 1995), caloric restriction, which is thought to reduce levels of oxidative stress, reduced the rate of AHL in rats (Seidman, 2000). Supplementation with lecithin, a polyunsaturated phosphatidylcholine that plays a role in SOD activation, also resulted in significant protection (Seidman et al., 2002). Hypoperfusion leads to the formation of ROS. As a reduction in blood flow to several tissues, including the cochlea, has been associated with ageing, this might mean that hypoperfusion of the cochlea may be an important causative factor for ROS formation and subsequent HL (Seidman et al., 2002).

Mitochondrial Deletions

Mitochondrial DNA has a high mutation rate. This is partly due to the fact that mitochondrial DNA is in close vicinity of the mitochondrial inner membrane, which is the major source of ROS. When sufficient mitochondrial DNA damage accumulates, the affected cell will become bioenergetically deficient. The most vulnerable cells are found in muscle and nerve tissue (including the cochlea), because these cells require high energy levels. In addition, cochlear cells are terminally differentiated and damaged cells will not be replaced. As a result, cochlear tissues are very sensitive to mitochondrial damage caused by oxidative stress.

Specific acquired mitochondrial mutations have been proposed as one of the causes of ARHI. They occur more frequently with increasing age and with the progression of ARHI. The so-called "common ageing" mitochondrial deletion involves 4977 bp in human (Seidman et al., 1996; Bai et al., 1997; Dai et al., 2004). An accumulation of many different acquired mitochondrial mutations was detected in auditory tissues of at least a proportion of ARHI patients (Fischel-Ghodsian et al., 1997).

Identification of Genes Involved in ARHI

The first requirement to tackle genetic studies for ARHI is a clear description of the phenotype. Recently, a novel *Z*-score-based method was published that allows to describe ARHI as a quantitative trait (Fransen et al., 2004). The *Z*-score, which is based on the ISO 7029 standard (International Organization for Standardization, 1998), gives an indication of the affection status of an individual independent of age and gender. *Z*-scores are calculated for each frequency as units of standard deviations from the median value for a particular age and gender. A negative *Z*-score indicates a person with better than median hearing, whereas a positive *Z*-score indicates hearing that is worse than the median value of otologically normal persons (Fransen et al., 2004).

In general, two possible study designs can be used for the identification of susceptibility genes for complex diseases: linkage and association studies. Linkage studies try to identify regions of the genome that harbor susceptibility genes on the basis of the inheritance pattern of the disease and genetic markers, whereas association studies analyze genetic variations in unrelated individuals and try to identify those variations that are more frequent in affected individuals compared to unaffected individuals. A major problem in collecting families for linkage studies of ARHI is the late onset of the disease, which means that the parents are frequently deceased. Currently, only a single whole genome linkage scan for ARHI has been published. This scan was performed using the Framingham cohort. DNA and audiological information was collected in a first phase for the parents (between 1973 and 1975), and in a second phase for the children (between 1995 and 1999). Pure-tone averages of medium and low frequencies were adjusted for cohort, sex, age, age squared, and age cubed and a genome-wide linkage scan was performed. This led to the identification of several chromosomal regions that showed suggestive evidence for linkage: 11p, 11q13.5, and 14 q (DeStefano et al., 2003).

Only a few association studies have been published for ARHI. Van Laer et al. (2002) studied the involvement of *DFNA5* in a Flemish set of samples and in a set derived from the Framingham cohort. *DFNA5* was selected as candidate ARHI susceptibility gene because mutations in *DFNA5* cause a type of HL that closely resembles the most typical type of ARHI (i.e., high-frequency, progressive, sensorineural HL). Two SNPs leading to an amino acid substitution in DFNA5 were analyzed. However, no significant association was detected in either sample collection (Van Laer et al., 2002). Recently, a second study was published. To investigate the hypothesis that variations in glutathione-related antioxidant enzyme levels are associated with the risk of ARHI, Ates et al. (2005) analyzed three glutathione *S*-transferase polymorphisms (GSTM1, GSTT1, and GSTP1) using a case–control association study. This study could not demonstrate a significant association either.

Noise-induced Hearing Loss

In the USA and Europe, 400–500 million people are at risk of developing NIHL. NIHL is also a rapidly increasing problem in countries such as India and China. This makes NIHL the leading occupational disease in industrialized countries and the second most common form of sensorineural HL, after ARHI. It is estimated that NIHL causes substantial economic losses, at a minimum level of 0.2% of the national net income. This equals approximately 400 billion Euros annually at the European Community level. This amount does not even include factors related to the reduced quality of life that affected persons experience. These numbers indicate that hearing protection for workers exposed to noise is far from being effective. Thus, NIHL has been classified as high priority for research by a number of organizations.

Heritability of NIHL

NIHL is a complex disorder, influenced by environmental and genetic factors. Firm evidence for the involvement of genetic factors in human NIHL is rare, as no formal heritability studies have been realized. In animals, however, it is well known that genetic factors can influence individual susceptibility to noise. A review has been presented by Borg et al. (1995). For example, rats with hereditary hypertension are more susceptible to noise than normotensive rats or rats with renal hypertension. Several studies have demonstrated that some mouse strains exhibiting AHL are more susceptible to noise than other strains (Erway et al., 1996; Davis et al., 2001) and in one case a responsible genetic factor has been identified (CDH23; Noben-Trauth et al., 2003) (also see the section on ARHI above). In addition, several knockout mice including PMCA2$^{-/-}$ (Kozel et al., 2002), CDH23$^{+/-}$ (Holme and Steel, 2004), SOD1$^{-/-}$ (Ohlemiller et al., 1999), and GPX1$^{-/-}$ (Ohlemiller, McFadden, et al., 2000) mice were shown to be more sensitive to NIHL than their wild-type littermates.

The ISO 1999 standard describes the average amount of NIHL as a function of noise exposure level and exposure time (International Organization for Standardization, 1990). It also indicates a large individual variability, which is one of the most remarkable features of NIHL. Approximately 10% of the population exhibits an increased susceptibility to noise. This inter-individual variability is another indication that NIHL is a complex disease that is due to an interaction of susceptibility genes and environmental factors.

Environmental and Medical Risk Factors for NIHL

Noise has a mechanical and metabolic effect on the inner ear. Although exposure to excessive occupational noise might exert the most serious insult on hearing capabilities, noise exposure due to leisure activities (rock, classical, or jazz music, personal listening devices e.g., walkmans, and "household" noise) in addition to recreational hunting or target shooting should also be taken into consideration (Clark, 1991).

Among the environmental factors known to cause a raised susceptibility to NIHL are coexposure to organic solvents and heavy metals (Sliwinska-Kowalska et al., 2004). A synergistic effect between noise exposure and exposure to chemicals has also been demonstrated in laboratory studies (Lataye et al., 2000). Additional environmental factors such as ototoxic substances, drugs, heat, nutrient disturbances and vibrations can also influence susceptibility to NIHL (for a review, see Borg et al., 1995). Furthermore, some studies have demonstrated that individual factors such as smoking, elevated blood pressure, and cholesterol levels may influence the degree of NIHL (Toppila et al., 2000, 2001).

Genetic Risk Factors for NIHL

Genetic variants within susceptibility genes may either directly lead to disease, or may render individuals more susceptible to disease under the influence of other external factors (physical, chemical, and infectious). However, at present, nothing is known about the variants that might contribute to NIHL susceptibility.

Similar to ARHI, NIHL seems to be mediated by ROS (Henderson, McFadden, et al., 1999). Antioxidant therapy can protect against NIHL, while chemicals that produce oxidative stress potentiate NIHL (Lautermann et al., 1997; Hight et al., 2003). Animal studies have shown an increase in glutathione and catalase expression following noise exposure (Jacono et al., 1998; Yamasoba et al., 1998), while two of the noise susceptible knockout mice described above involve antioxidant genes (*SOD1* and *GPX1*) (Ohlemiller et al., 1999, Ohlemiller, McFadden, 2000).

Only a few association studies on candidate genes have been performed. Rabinowitz et al. (2002) demonstrated a possible association between the absence of GSTM1 (glutathione *S*-transferase μ, an enzyme with an important anti-oxidative function) and NIHL in a study with very limited sample size. The authors pointed out that due to the low power of their study, this positive association might be due to chance. As Carlsson et al. (2005) could not confirm the observations of Rabinowitz et al. in a study with sufficient power, this association indeed seemed to be spurious. In the same study, Carlsson et al. (2005) did not find significant differences between susceptible and resistant groups for several investigated polymorphisms in genes involved in oxidative stress. In another study, using the same set of samples, Carlsson et al. (2004) concluded that the role of the Connexin 26 35delG mutation as a determining factor in noise susceptibility is negligible in the Swedish population.

Drug-induced Hearing Loss

Many clinically used substances may affect the inner ear. More than 130 drugs and chemicals have been reported to be potentially ototoxic. The best documented are anti-inflammatory agents (including salicylates), aminoglycoside antibiotics (streptomycin, kanamycin, neomycin, gentamicin and others), antineoplastic agents such as cisplatin and carboplatin, and loop diuretics such as furosemide and ethacrynic acid (Huang and Schacht, 1989; Seligmann et al., 1996). Several of these drugs are either used to fight life-threatening illnesses such as cancer and infectious diseases (cisplatin and aminoglycosides) or in situations where the risk of ototoxicity is considered to be minor compared to the beneficial therapeutic effects (loop diuretics and salicylates). The influence of some ototoxic drugs have already been described in the sections on mitochondrial HL and on environmental factors involved in ARHI.

Typically, DIHL initially results in deterioration of the high frequencies. However, lesions may progress and finally include the lower frequencies as well (Forge and Schacht, 2000). The site of action of the ototoxic substance depends on its nature; the sites most affected by different types of drugs are the cochlea, the vestibulum, and the stria vascularis. Probably the most important parameters to be considered when treatments consisting of potentially ototoxic substances are prescribed are the given doses, the dosing intervals, and the form of administration. For example, serum concentrations of 20–50 mg/dl of salicylates cause serious HL of more than 30 dB, whereas serum concentrations that remain below the border of 20 mg/dl are harmless (Brien, 1993). In general, high-dose therapies (loop diuretics and anti-inflammatory agents) are associated with acute, transient HL, whereas long-term administration of ototoxic substances (antineoplastics and aminoglycosides) leads to irreversible, permanent HL. Another important parameter is the appearance of synergistic effects when prescribing more than one ototoxic substance. Synergism has indeed been observed: a combination of different drugs always results in an increase in ototoxicity. A final consideration is whether the patient who is to receive the treatment belongs to a population at risk or not. Although age does not seem to significantly influence ototoxicity, premature infants are particularly sensitive to ototoxic substances. In addition, children are a high-risk population for cisplatin-induced ototoxicity (Weatherly et al., 1991). Subjects harboring the 1555A→G mitochondrial DNA mutations are extremely sensitive to aminoglycoside-induced HL (see also the section on mitochondrial HL).

NIHL and DIHL seem to have some common features: both have been linked to the production of ROS (Henderson, Hu, et al., 1999). Several studies have shown that antioxidant therapy protects against the adverse effect of ototoxic substances (Song et al., 1997; Schacht, 1999; Minami et al., 2004). A mimetic of superoxide dismutase (SOD), M40403, protects against gentamicin ototoxicity, but not against cisplatin otoxicity (McFadden et al., 2003). Overexpression of Cu/Zn superoxide dismutase (SOD1) protected from kanamycin-induced HL (Sha et al., 2001). Recently, it was demonstrated that antioxidant gene therapy resulting in overexpression of cochlear catalase and Mn superoxide dismutase (SOD2) could protect hair cells against ototoxicity (Kawamoto et al., 2004).

Prevention of DIHL can be partly achieved by an improved medical awareness, accomplished through an adjustment of the administered doses and the dosing interval, the monitoring of serum levels to avoid toxic drug concentrations, monitoring auditory performance before and after each treatment with sensitive audiometric methods (such as high frequency audiometry, transient evoked otoacoustic emissions, and distortion-product otoacoustic emissions), and considering certain risk factors including an individual's genetic background (for example in the presence of a 1555A→G mitochondrial DNA mutation) (Garcia et al., 2001).

Otosclerosis

OTSC is one of the most frequent causes of HL among white adults. It can be considered as a complex disease with rare monogenic forms (see section on monogenic type). It is hypothesized that, besides genetic factors, mainly immunologic factors and perhaps viral infections (measles) are involved in the etiology of OTSC. The possibility of a viral etiology for OTSC was proposed because of strong clinical similarities between OTSC and Paget's disease, a bone disease putatively caused by the consequences of viral infection (Mills et al., 1984). Several studies found evidence for an association of measles virus and OTSC (McKenna and Mills, 1989; Niedermeyer and Arnold, 1995; Arnold et al., 1996). However, the contribution of viral factors in the development of OTSC remains controversial.

The first genetic factor that was identified for OTSC also remains controversial. McKenna et al. (1998) performed an association study on three collagen candidate genes (*COL1A1*, *COL1A2* and *COL2A1*). *COL1A1* and *COL1A2* were selected because of clinical and histopathological similarities between OTSC and type I osteogenesis imperfecta, which is caused by mutations in these two genes, whereas *COL2A1* was selected because of its high expression in the otic capsule. McKenna et al. (1998) found a significant association between the presence of OTSC and certain variants of *COL1A1*. This might suggest some common features for the molecular pathology of OTSC and osteogenesis imperfecta. However, these results could not be reproduced in an independent Spanish population (Rodriguez et al., 2004). Recently, an association between osteoporosis and OTSC in women has been described (Clayton et al., 2004), which might be mediated by a Sp1 binding site polymorphism in the *COL1A1* gene (McKenna et al., 2004).

Molecular Diagnosis of Hearing Loss

As no genes for complex types of HL have yet been identified, diagnostic applications are limited to monogenic types of HL. Monogenic HL is phenotypically and genetically very heterogeneous. The fact that mutations in a single gene can cause both syndromic and nonsyndromic HL on one hand, and can give rise to both dominant and recessive forms of nonsyndromic HL on the other hand obviously complicates matters. Nevertheless, the genetic origin of HL can be determined in a number of cases.

Several types of syndromic HL are known, and in many cases the underlying gene has been identified, allowing molecular diagnostic testing. For genetically homogeneous syndromes, DNA testing is relatively simple, especially if recurrent mutations are known. In contrast, in cases of genetically heterogeneous syndromic HL (for example in Usher syndrome), several, often large genes need to be screened entirely, which is an expensive and time-consuming procedure.

Molecular diagnostics for nonsyndromic HL is even more difficult, due to its extreme heterogeneous character and the fact that the individual contribution of each gene can be very small (Fig. 31–1). Only for a few genes are DNA diagnostic procedures widely available, an overview of which is given in Table 31–6. Although recurrent mutations are known in some of these genes, one should always keep the ethnicity of the sample in mind when performing diagnostic testing as different mutations prevail in different populations. In addition, private mutations are frequent for many genes. Therefore, mutation analysis of the complete coding region can be necessary for sensitive DNA testing.

GJB2 and *GJB6*

By far the most important genetic test for nonsyndromic HL is molecular screening of *GJB2*, which is useful for children with congenital or childhood onset nonsyndromic sensorineural HL with normal hearing parents. Recurrent mutations are easy to screen using standard laboratory techniques, and can be adjusted depending on the genetic background of the individual under study. But currently, most diagnostic laboratories analyze the complete coding region in addition to a few known mutations outside the coding region. In a relatively large fraction of patients with presumed autosomal recessive HL, only one mutant *GJB2* allele is detected. In such cases, testing the del(GJB6-D13S1830) and del(GJB6-D13S1854) deletions can be considered. These deletions are particularly frequent in the Spanish population. Results from studies in other countries indicate that each country will have to design its own protocol for diagnostic testing of recessive prelingual nonsyndromic HL.

SLC26A4

The *SLC26A4* gene, encoding pendrin, was found to be mutated in patients with Pendred syndrome (Everett et al., 1997) and in patients with HL linked to the DFNB4 locus (Li et al., 1998). Most patients with Pendred syndrome are prelingually deaf and have cochlear malformations known as Mondini dysplasia. This form of dysplasia is typically associated with a EVA. EVA is a shared clinical feature between Pendred and DFNB4 patients, while the presence of thyroid abnormalities and Mondini dysplasia are unique to patients with Pendred syndrome. It seems that a variable phenotype, ranging from full-blown Pendred syndrome to nonsyndromic HL with EVA, is associated with *SLC26A4* mutations. Therefore screening of this gene should be considered both for patients with Pendred syndrome and nonsyndromic HL associated with a dilatation of the vestibular aqueduct (EVA). In families from Western Europe and North America, three missense mutations and one splice site mutation (L236P, T416P, E384G and IVS8 + 1G1A) account for 74% of all Pendred syndrome cases (Coyle et al., 1998). In German Pendred families, the V138F mutation seems to be frequent (Borck et al., 2003), whereas in Asian patients with nonsyndromic HL and EVA the H723R and IVS7-2A1G variants are most commonly reported (Usami et al., 1999; Kitamura et al., 2000; Sato et al., 2001; Park et al., 2003). The latter two mutations have also been found in patients with classical Pendred syndrome (Van Hauwe et al., 1998; Coucke et al., 1999).

COCH

A gene with important diagnostic implications in Belgium and the Netherlands is *COCH* (coagulation factor C homologue). Mutations in *COCH* cause DFNA9 (Robertson et al., 1998), the only nonsyndromic HL locus reported to involve vestibular symptoms. One particular mutation (P51S) is a frequent cause of HL and vestibular dysfunction in families originating from Belgium and the Netherlands (de Kok et al., 1999; Fransen et al., 1999). If Belgian or Dutch patients with progressive HL and accompanying vestibular symptoms are encountered in clinical practice, screening of the P51S mutation should be considered.

WFS1

Mutations in *WFS1* account for most of the cases of low-frequency autosomal dominant HL (DFNA6/14) (Bespalova et al., 2001; Young et al., 2001; Cryns et al., 2002). DNA diagnostics of *WFS1* is offered to all patients presenting with low-frequency nonsyndromic HL in several centers. No recurrent mutations have been observed. *WFS1* mutations also underlie autosomal recessive Wolfram syndrome, alias DIDMOAD (diabetes insipidus, diabetes mellitus, optic atrophy and deafness) (Inoue et al., 1998; Strom et al., 1998). Remarkably, HL in patients with Wolfram syndrome mainly affects the higher frequencies (Higashi, 1991). It is currently not known why some heterozygous *WFS1* mutations selectively affect the low frequencies (DFNA6/14), whereas homozygous *WFS1* mutations affect the higher frequencies (DIDMOAD).

OTOF

Migliosi et al. (2002) reported that one particular mutation in *OTOF* (Q829X) is the third most frequent cause of recessive nonsyndromic HL in the Spanish population. It accounts for 4.4% of recessive prelingual nonsyndromic HL negative for *GJB2* mutations. Mutations in *OTOF* may be associated with auditory neuropathy, which is characterized by an absent or abnormal brainstem response with

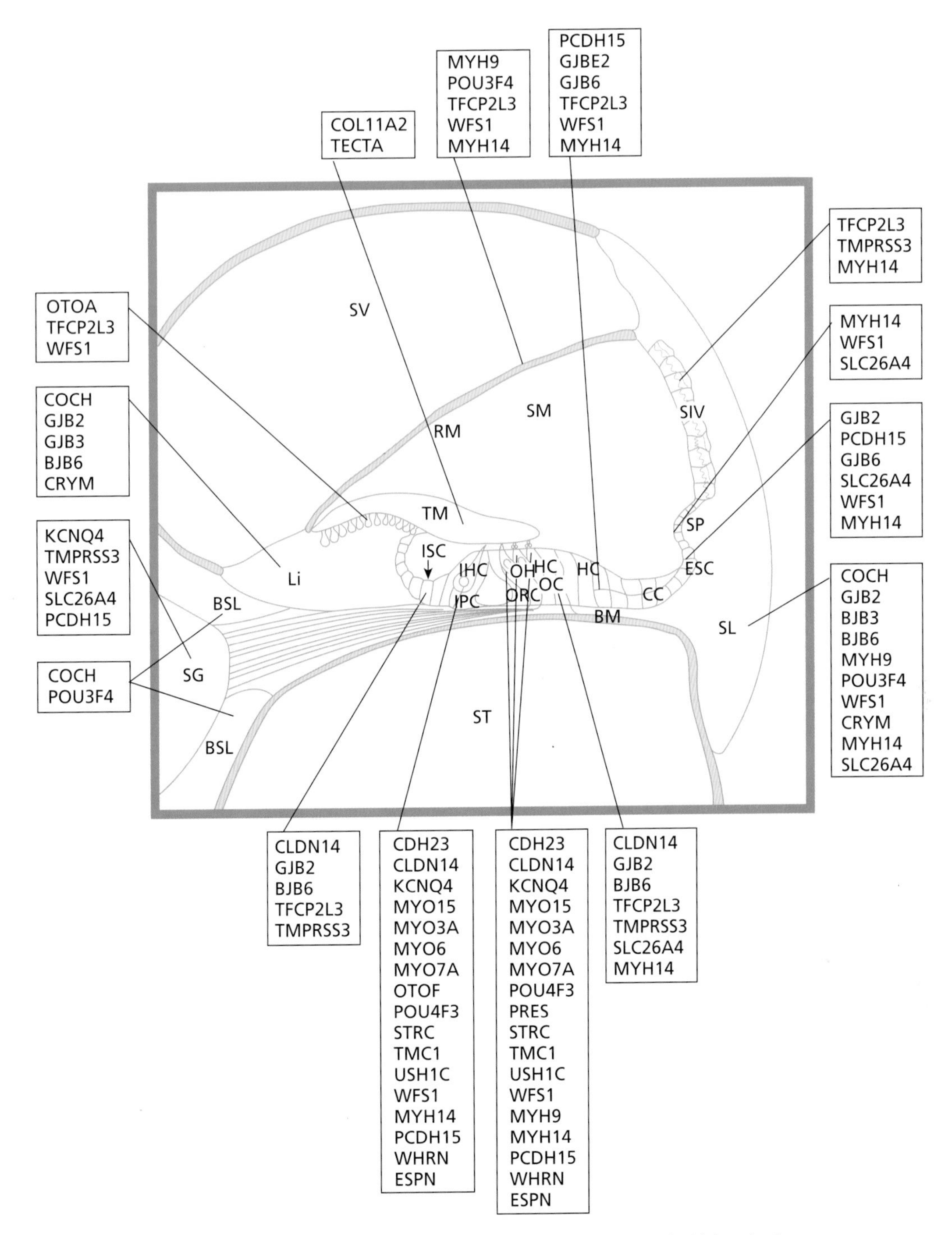

Figure 31–1 Designated expression sites for various genes associated with hearing loss.

normal outer hair cell function (Varga et al., 2003). Although this gene is not yet included in many routine molecular diagnostic laboratory examinations, it may have important future applications for nonsyndromic recessive HL patients without *GJB2* mutations, and especially for patients presenting with nonsyndromic recessive auditory neuropathy.

Mitochondrial Mutations

Molecular diagnosis is possible for syndromic and nonsyndromic maternally inherited mitochondrial HL as only a limited number of mutations are known. One mutation that is of particular importance is the 1555A→G mutation, which is a frequent cause of HL in some populations (Estivill, Govea, et al., 1998). The phenotype of 1555A→G patients ranges from profound congenital HL, through progressive moderate HL starting in adult life to complete normal hearing. In some patients HL develops only after aminoglycoside exposure, whereas in others it develops without this exposure.

Future Prospects

Although exploration of the molecular basis of monogenic types of HL has progressed at an amazing speed during the last decade, the unraveling of the genetic basis of the complex types of HL will

Table 31–6 Diagnostic Testing of Nonsyndromic Hearing Loss-Currently Available as Clinical Service

Criteria	Gene	Recurrent mutation
Prelingual autosomal recessive HL	*GJB2*	35delG, 235delC, 167delT, R143W dependent on the ethnicity
Prelingual autosomal recessive HL with only one mutant *GJB2* allele	*GJB6*	del(GJB6-D13S1830) and del (GJB6-D13S1854), especially in Spain
Progressive HL with vestibular dysfunction	*COCH*	P51S, in Belgian and Dutch patients
Autosomal dominant low-frequency HL	*WFS1*	None, but mutations cluster in C-terminal of wolframin
Nonsyndromic HL associated with EVA	*SLC26A4*	Only recurrent mutations for Pendred syndrome
Prelingual autosomal recessive HL negative for *GJB2* mutations	*OTOF*	Q829X, especially in Spain
Maternally inherited HL, aminoglycoside susceptibility	*12SrRNA*	1555A→G

Abbreviations: EVA, enlarged vestibular aqueduct; HL, hearing loss.

demand further efforts in particular. Hopefully, the first susceptibility genes for the different types of complex HL will be identified in the next few years. Ultimately, by integrating information on genetic and environmental risk factors, it may become clear how the vulnerability of a person's hearing system correlates with his genetic background. It might be that certain environmental risk factors are potentially harmful only in a limited number of individuals, depending on their genetic background. If this is the case, diagnostic tests can be developed to identify individuals who are at high risk of developing HL, personalized guidelines for HL prevention can be set up, and measures can be taken to limit HL in these individuals. This can lead to a reduction of the most severely affected HL cases, and to a reduction of the resulting economic and social costs.

Future therapies for hearing impairment will have to rely on basic rather than on symptomatic approaches. A better understanding of the basic molecular and cellular processes involved in both monogenic and complex types of HL may uncover new leads for pharmaceutical intervention, leading to the development of appropriate anti-HL drugs using pharmacogenomic methods (i.e., adapting drugs to an individual's genetic background). The personalized treatment that is aimed at in pharmacogenomics can only be achieved if the contributions of different genetic factors are resolved at the level of the individual.

In the far future, therapies based on hair cell regeneration resulting from gene delivery to the inner ear or from stem cell treatment, in combination with the correction of mutated genes through cochlear gene therapy, might be the preferred treatment for HL. Of late, a few very promising results have been presented: auditory hair cell replacement by means of *Math1* gene therapy led to improved hearing in experimental animals (Izumikawa et al., 2005); in the absence of the retinoblastoma protein, proliferation of functional hair cells was observed (Sage et al., 2005); and conversion of embryonic stem cells to hair cells has been observed (Li et al., 2003). Despite these very exciting results, there is still a long way to go and a lot of work to be done before the different types of HL can be cured.

References

Abdelhak, S, Kalatzis, V, Heilig, R, Compain, S, Samson, D, Vincent, C, et al. (1997). A human homologue of the Drosophila eyes absent gene underlies branchio-oto-renal (BOR) syndrome and identifies a novel gene family. *Nat Genet*, 15:157–164.

Abe, S, Katagiri, T, Saito-Hisaminato, A, Usami, S, Inoue, Y, Tsunoda, T, et al. (2003). Identification of CRYM as a candidate responsible for nonsyndromic deafness, through cDNA microarray analysis of human cochlear and vestibular tissues. *Am J Hum Genet*, 72:73–82.

Abe, S, Usami, S, Shinkawa, H, Kelley, PM, and Kimberling, WJ (2000). Prevalent connexin 26 gene (GJB2) mutations in Japanese. *J Med Genet*, 37:41–43.

Ahmad, NN, Ala-Kokko, L, Knowlton, RG, Jimenez, SA, Weaver, EJ, Maguire, JI, et al. (1991). Stop codon in the procollagen II gene (COL2A1) in a family with the Stickler syndrome (arthro-ophthalmopathy). *Proc Natl Acad Sci USA*, 88:6624–6627.

Ahmed, ZM, Morell, RJ, Riazuddin, S, Gropman, A, Shaukat, S, Ahmad, MM, et al. (2003). Mutations of MYO6 are associated with recessive deafness, DFNB37. *Am J Hum Genet*, 72:1315–1322.

Ahmed, ZM, Riazuddin, S, Ahmad, J, Bernstein, SL, Guo, Y, Sabar, MF, et al. (2003). PCDH15 is expressed in the neurosensory epithelium of the eye and ear and mutant alleles are responsible for both USH1F and DFNB23. *Hum Mol Genet*, 12:3215–3223.

Ahmed, ZM, Riazuddin, S, Bernstein, SL, Ahmed, Z, Khan, S, Griffith, AJ, et al. (2001). Mutations of the protocadherin gene PCDH15 cause Usher syndrome type 1F. *Am J Hum Genet*, 69:25–34.

Ahmed, ZM, Smith, TN, Riazuddin, S, Makishima, T, Ghosh, M, Bokhari, S, et al. (2002). Nonsyndromic recessive deafness DFNB18 and Usher syndrome type IC are allelic mutations of USHIC. *Hum Genet*, 110:527–531.

Alagramam, KN, Yuan, H, Kuehn, MH, Murcia, CL, Wayne, S, Srisailpathy, CRS, et al. (2001). Mutations in the novel protocadherin PCDH15 cause Usher syndrome type 1F. *Hum Mol Genet*, 10:1709–1718.

Antonelli, AR, Bonfioli, F, Garrubba, V, Ghisellini, M, Lamoretti, MP, Nicolai, P, et al. (1990). Audiological findings in elderly patients with chronic renal failure. *Acta Otolaryngol Suppl*, 476:54–68.

Aran, JM, Hiel, H, and Hayashida, T (1992). Noise, aminoglycosides, diuretics. In A Dancer, D Henderson, R Salvi, and R Hamernik (eds), *Noise Induced Hearing Loss* (pp. 175–187). St. Louis, USA: Mosby.

Arnold, W, Niedermeyer, HP, Lehn, N, Neubert, W, and Hofler, H (1996). Measles virus in otosclerosis and the specific immune response of the inner ear. *Acta Otolaryngol*, 116:705–709.

Ates, NA, Unal, M, Tamer, L, Derici, E, Karakas, S, Ercan, B, et al. (2005). Glutathione S-transferase gene polymorphisms in presbycusis. *Otol Neurotol*, 26:392–397.

Attie, T, Till, M, Pelet, A, Amiel, J, Edery, P, Boutrand, L, et al. (1995). Mutation of the endothelin-receptor B gene in Waardenburg-Hirschsprung disease. *Hum Mol Genet*, 4:2407–2409.

Bacino, C, Prezant, TR, Bu, X, Fournier, P, and Fischel-Ghodsian, N (1995). Susceptibility mutations in the mitochondrial small ribosomal RNA gene in aminoglycoside induced deafness. *Pharmacogenetics*, 5:165–172.

Bai, U, Seidman, MD, Hinojosa, R, and Quirk, WS (1997). Mitochondrial DNA deletions associated with aging and possibly presbycusis: a human archival temporal bone study. *Am J Otol*, 18:449–453.

Ballinger, SW, Shoffner, JM, Hedaya, EV, Trounce, I, Polak, MA, Koontz, DA, et al. (1992). Maternally transmitted diabetes and deafness associated with a 10.4 kb mitochondrial DNA deletion. *Nat Genet*, 1:11–15.

Barker, DF, Hostikka, SL, Zhou, J, Chow, LT, Oliphant, AR, Gerken, SC, et al. G (1990). Identification of mutations in the COL4A5 collagen gene in Alport syndrome. *Science*, 248:1224–1227.

Bespalova, IN, Van Camp, G, Bom, SJH, Brown, DJ, Cryns, K, DeWan, AT, et al. (2001). Mutations in the Wolfram syndrome 1 gene (WFS1) are a common cause of low frequency sensorineural hearing loss. *Hum Mol Genet*, 10:2501–2508.

Bitner-Glindzicz, M, Lindley, KJ, Rutland, P, Blaydon, D, Smith, VV, Milla, PJ, et al. (2000). A recessive contiguous gene deletion causing infantile hyperinsulinism, enteropathy and deafness identifies the Usher type 1C gene. *Nat Genet*, 26:56–60.

Bolz, H, von Brederlow, B, Ramirez, A, Bryda, EC, Kutsche, K, Nothwang, HG, et al. (2001). Mutation of CDH23, encoding a new member of the cadherin gene family, causes Usher syndrome type 1D. *Nat Genet*, 27:108–112.

Borck, G, Roth, C, Martine, U, Wildhardt, G, and Pohlenz, J (2003). Mutations in the PDS gene in German families with Pendred's syndrome: V138F is a founder mutation. *J Clin Endocrinol Metab*, 88:2916–2921.

Borg, E, Canlon, B, and Engstrom, B (1995). Noise-induced hearing loss. Literature review and experiments in rabbits. Morphological and electrophysiological features, exposure parameters and temporal factors, variability and interactions. *Scand Audiol Suppl*, 40:1–147.

Bork, JM, Peters, LM, Riazuddin, S, Bernstein, SL, Ahmed, ZM, Ness, SL, et al. (2001). Usher syndrome 1D and nonsyndromic autosomal recessive deafness DFNB12 are caused by allelic mutations of the novel cadherin-like gene CDH23. *Am J Hum Genet*, 68:26–37.

Brant, LJ, Gordon-Salant, S, Pearson, JD, Klein, LL, Morrell CH, Metter EJ, et al. (1996). Risk factors related to age-associated hearing loss in the speech frequencies. *J Am Acad Audiol*, 7:152–160.

Brien, JA (1993). Ototoxicity associated with salicylates. A brief review. *Drug Saf*, 9:143–148.

Brobby, GW, Müller-Myhsok, B, and Horstmann, RD (1998). Connexin 26 R143W mutation associated with recessive nonsyndromic sensorineural deafness in Africa. *N Engl J Med*, 338:548–549.

Browning, GG and Gatehouse, S (1984). Sensorineural hearing loss in stapedial otosclerosis. *Ann Otol Rhinol Laryngol*, 93:13–16.

Carlsson, PI, Borg, E, Grip, L, Dahl, N, and Bondeson, ML (2004). Variability in noise susceptibility in a Swedish population: The role of 35delG mutation in the Connexin 26 (GJB2) gene. *Audiol Med*, 2:123–130.

Carlsson, PI, Van Laer, L, Borg E, Bondeson ML, Thys M, Fransen E, et al. (2005). The influence of genetic variation in oxidative stress genes on human noise susceptibility. *Hear Res*, 202:87–96.

Casano, RA, Johnson, DF, Bykhovskaya, Y, Torricelli, F, Bigozzi, M, and Fischel-Ghodsian, N (1999). Inherited susceptibility to aminoglycoside ototoxicity: genetic heterogeneity and clinical implications. *Am J Otolaryngol*, 20:151–156.

Chen, W, Campbell, CA, Green, GE, Van Den Bogaert, K, Komodikis, C, Manolidis, LS, et al. (2002). Linkage of otosclerosis to a third locus (OTSC3) on human chromosome 6p21.3–22.3. *J Med Genet*, 39:473–477.

Chen, ZY, Hendriks, RW, Jobling, MA, Powell, JF, Breakefield, XO, Sims, KB, et al. (1992). Isolation and characterization of a candidate gene for Norrie disease. *Nat Genet*, 1:204–208.

Christensen, K, Frederiksen, H, and Hoffman, HJ (2001). Genetic and environmental influences on self-reported reduced hearing in the old and oldest old. *J Am Geriatr Soc*, 49:1512–1517.

Clark, K, Sowers, MR, Wallace, RB, Jannausch, ML, Lemke, J, and Anderson, CV (1995). Age-related hearing loss and bone mass in a population of rural women aged 60 to 85 years. *Ann Epidemiol*, 5:8–14.

Clark, WW (1991). Noise exposure from leisure activities: a review. *J Acoust Soc Am*, 90:175–181.

Clayton, AE, Mikulec, AA, Mikulec, KH, Merchant, SN, and McKenna, MJ (2004). Association between osteoporosis and otosclerosis in women. *J Laryngol Otol*, 118:617–621.

Cohen, M and Gorlin, R (1995). Epidemiology, Etiology, and Genetic Patterns. In R Gorlin, H Toriello, and M Cohen (eds), *Hereditary Hearing Loss and Its Syndromes* (pp. 9–21).Oxford, UK: Oxford University Press.

Cohn, ES, Kelley, PM, Fowler, TW, Gorga, MP, Lefkowitz, DM, Kuehn, HJ, et al. (1999). Clinical studies of families with hearing loss attributable to mutations in the connexin 26 gene (GJB2/DFNB1). *Pediatrics*, 103:546–550.

Coling, DE, Yu, KCY, Somand, D, Satar, B, Bai, U, Huang, TT, et al. (2003). Effect of SOD1 overexpression on age- and noise-related hearing loss. *Free Radic Biol Med*, 34:873–880.

Corso, JF (1992). Support for Corso's hearing loss model. Relating aging and noise exposure. *Audiology*, 31:162–167.

Coucke, PJ, Van Hauwe, P, Everett, LA, Demirhan, O, Kabakkaya, Y, Dietrich, NL, et al. (1999). Identification of two different mutations in the PDS gene in an inbred family with Pendred syndrome. *J Med Genet*, 36:475–477.

Coyle, B, Reardon, W, Herbrick, JA, Tsui, LC, Gausden, E, Lee, J, et al. (1998). Molecular analysis of the PDS gene in Pendred syndrome. *Hum Mol Genet*, 7:1105–1112.

Cruickshanks, KJ, Klein, R, Klein, BEK, Wiley, TL, Nondahl, DM, and Tweed, TS (1998). Cigarette smoking and hearing loss: the epidemiology of hearing loss study. *J Am Med Ass*, 279:1715–1719.

Cryns, K, Orzan, E, Murgia, A, Huygen, PL, Moreno, F, del Castillo, I, et al. (2004). A genotype-phenotype correlation for GJB2 (connexin 26) deafness. *J Med Genet*, 41:147–154.

Cryns, K, Pfister, M, Pennings, RJE, Bom, SJH, Flothmann, K, Caethoven, G, et al. (2002). Mutations in the WFS1 gene that cause low-frequency sensorineural hearing loss are small non-inactivating mutations. *Hum Genet*, 110:389–394.

Dai, P, Yang, W, Jiang, S, Gu, R, Yuan, H, Han, D, et al. (2004). Correlation of cochlear blood supply with mitochondrial DNA common deletion in presbyacusis. *Acta Otolaryngol*, 124:130–136.

Davis, A (1994). Prevalence of Hearing Impairment. In A Davis (ed.), *Hearing in adults* (pp. 43–321). London, UK: Whurr Publishers Ltd.

Davis, RR, Newlander, JK, Ling, X, Cortopassi, GA, Krieg, EF, and Erway, LC (2001). Genetic basis for susceptibility to noise-induced hearing loss in mice. *Hear Res*, 155:82–90.

de Kok, YJ, van der Maarel, SM, Bitner-Glindzicz, M, Huber, I, Monaco, AP, Malcolm, S, et al. (1995). Association between X-linked mixed deafness and mutations in the POU domain gene POU3F4. *Science*, 267:685–688.

de Kok, YJM, Bom, SJH, Brunt, TM, Kemperman, MH, van Beusekom, E, van der Velde-Visser, SD, et al. (1999). A Pro51Ser mutation in the COCH gene is associated with late onset autosomal dominant progressive sensorineural hearing loss with vestibular defects. *Hum Mol Genet*, 8:361–366.

Declau, F, Van Spaendonck, M, Timmermans, JP, Michaels, L, Liang, J, Qiu, JP, et al. (2001). Prevalence of otosclerosis in an unselected series of temporal bones. *Otol Neurotol*, 22:596–602.

Del Castillo, FJ, Rodriguez-Ballesteros, M, Alvarez, A, Hutchin, T, Leonardi, E, de Oliveira, CA, et al. (2005). A novel deletion involving the connexin-30 gene, del(GJB6-d13s1854), found in trans with mutations in the GJB2 gene (connexin-26) in subjects with DFNB1 non-syndromic hearing impairment. *J Med Genet*, 42:588–594.

del Castillo, I, Villamar, M, Moreno-Pelayo, MA, del Castillo, FJ, Alvarez, A, Telleria, D, et al. (2002). A deletion involving the connexin 30 gene in nonsyndromic hearing impairment. *N Engl J Med*, 346:243–249.

Denoyelle, F, Lina-Granade, G, Plauchu, H, Bruzzone, R, Chaib, H, Levi-Acobas, F, et al. (1998). Connexin 26 gene linked to a dominant deafness. *Nature*, 393:319–320.

Denoyelle, F, Marlin, S, Weil, D, Moatti, L, Chauvin, P, Garabedian, EN, et al. (1999). Clinical features of the prevalent form of childhood deafness, DFNB1, due to a connexin-26 gene defect: implications for genetic counselling. *Lancet*, 353:1298–1303.

Denoyelle, F, Weil, D, Maw, MA, Wilcox, SA, Lench, NJ, Allen-Powell, DR, et al. (1997). Prelingual deafness: high prevalence of a 30delG mutation in the connexin 26 gene. *Hum Mol Genet*, 6:2173–2177.

DeStefano, AL, Gates, GA, Heard-Costa, N, Myers, RH, and Baldwin, CT (2003). Genomewide linkage analysis to presbycusis in the Framingham Heart Study. *Arch Otolaryngol Head Neck Surg*, 129:285–289.

Donaudy, F, Ferrara, A, Esposito, L, Hertzano, R, Ben-David, O, Bell, RE, et al. (2003). Multiple mutations of MYO1A, a cochlear-expressed gene, in sensorineural hearing loss. *Am J Hum Genet*, 72:1571–1577.

Donaudy, F, Snoeckx, R, Pfister, M, Zenner, HP, Blin, N, Di Stazio, M, et al. (2004). Nonmuscle myosin heavy-chain gene MYH14 is expressed in cochlea and mutated in patients affected by autosomal dominant hearing impairment (DFNA4). *Am J Hum Genet*, 74:770–776.

Donaudy, F, Zheng, L, Ficarella, R, Ballana, E, Carella, M, Melchionda, S, et al. (2005). Espin gene (ESPN) mutations associated with autosomal dominant hearing loss cause defects in microvillar elongation or organization. *J Med Genet*, 43:157–161.

Edery, P, Attie, T, Amiel, J, Pelet, A, Eng, C, Hofstra, RM, et al. (1996). Mutation of the endothelin-3 gene in the Waardenburg-Hirschsprung disease (Shah-Waardenburg syndrome). *Nat Genet*, 12:442–444.

Erway, LC, Shiau, YW, Davis, RR, and Krieg, EF (1996). Genetics of age-related hearing loss in mice. III. Susceptibility of inbred and F1 hybrid strains to noise-induced hearing loss. *Hear Res*, 93:181–187.

Estivill, X, Fortina, P, Surrey, S, Rabionet, R, Melchionda, S, D'Agruma, L, et al. (1998). Connexin-26 mutations in sporadic and inherited sensorineural deafness. *Lancet*, 351:394–398.

Estivill, X, Govea, N, Barcelo, E, Badenas, C, Romero, E, Moral, L, et al. (1998). Familial progressive sensorineural deafness is mainly due to the mtDNA A1555G mutation and is enhanced by treatment of aminoglycosides. *Am J Hum Genet*, 62:27–35.

Eudy, JD, Weston, MD, Yao, S, Hoover, DM, Rehm, HL, Ma-Edmonds, M, et al. (1998). Mutation of a gene encoding a protein with extracellular matrix motifs in Usher syndrome type IIa. *Science*, 280:1753–1757.

Everett, LA, Glazer, B, Beck, JC, Idol, JR, Buchs, A, Heyman, M, et al. (1997). Pendred syndrome is caused by mutations in a putative sulphate transporter gene (PDS). *Nat Genet*, 17:411–422.

Fischel-Ghodsian, N (2003). Mitochondrial deafness. *Ear Hear*, 24:303–313.

Fischel-Ghodsian, N, Bykhovskaya, Y, Taylor, K, Kahen, T, Cantor, R, Ehrenman, K, et al. (1997). Temporal bone analysis of patients with presbycusis reveals high frequency of mitochondrial mutations. *Hear Res*, 110:147–154.

Fitzgerald, DC (1996). Head trauma: hearing loss and dizziness. *J Trauma*, 40:488–496.

Forge, A and Schacht, J (2000). Aminoglycoside antibiotics. *Audiol Neurootol*, 5:3–22.

Fransen, E, Van Laer, L, Lemkens, N, Caethoven, G, Flothmann, K, Govaerts, P, et al. (2004). A novel Z-score-based method to analyze candidate genes for age-related hearing impairment. *Ear Hear*, 25:133–141.

Fransen, E, Verstreken, M, Verhagen, WIM, Wuyts, FL, Huygen, PLM, D'Haese, P, et al. (1999). High prevalence of symptoms of Meniere's disease in three families with a mutation in the COCH gene. *Hum Mol Genet*, 8:1425–1429.

Garcia, VP, Martinez, AF, Agusti, EB, Mencia, LA, and Asenjo, VP (2001). Drug-induced otoxicity: Current status. *Acta Otolaryngol*, 121:569–572.

Gasparini, P, Rabionet, R, Barbujani, G, Melchionda, S, Petersen, M, Brondum-Nielsen, K, et al. (2000). High carrier frequency of the 35delG deafness mutation in European populations. The Genetic Analysis Consortium of GJB2 35delG. *Eur J Hum Genet*, 8:19–23.

Gates, GA, Cobb, JL, D'Agostino, RB, and Wolf, PA (1993). The relation of hearing in the elderly to the presence of cardiovascular disease and cardiovascular risk factors. *Arch Otolaryngol Head Neck Surg*, 119:156–161.

Gates, GA, Couropmitree, NN, and Myers, RH (1999). Genetic associations in age-related hearing thresholds. *Arch Otolaryngol Head Neck Surg*, 125:654–659.

Gorlin, RJ, Cohen, MM, and Levin, SL (1990). *Syndromes of the Head and Neck*. Oxford University Press, New York, Oxford.

Gorlin, RJ, Toriello, HV, and Cohen, MM (1995). *Hereditary hearing loss and its syndromes*. Oxford University Press, Oxford.

Goto, Y, Nonaka, I, and Horai, S (1990). A mutation in the tRNA(Leu)(UUR) gene associated with the MELAS subgroup of mitochondrial encephalomyopathies. *Nature*, 348:651–653.

Grifa, A, Wagner, CA, D'Ambrosio, L, Melchionda, S, Bernardi, F, Lopez-Bigas, N, et al. (1999). Mutations in GJB6 cause nonsyndromic autosomal dominant deafness at DFNA3 locus. *Nat Genet*, 23:16–18.

Hao, H, Bonilla, E, Manfredi, G, DiMauro, S, and Moraes, CT (1995). Segregation patterns of a novel mutation in the mitochondrial tRNA glutamic acid gene associated with myopathy and diabetes mellitus. *Am J Hum Genet*, 56:1017–1025.

Heath, KE, Campos-Barros, A, Toren, A, Rozenfeld-Granot, G, Carlsson, LE, Savige, J, et al. (2001). Nonmuscle myosin heavy chain IIA mutations define a spectrum of autosomal dominant macrothrombocytopenias: May-Hegglin anomaly and Fechtner, Sebastian, Epstein, and Alport-like syndromes. *Am J Hum Genet*, 69:1033–1045.

Heathcote, K, Syrris, P, Carter, ND, and Patton, MA (2000). A connexin 26 mutation causes a syndrome of sensorineural hearing loss and palmoplantar hyperkeratosis (MIM 148350). *J Med Genet*, 37:50–51.

Henderson, D, Hu, B, McFadden, S, and Zheng, X (1999). Evidence of a Common Pathway in Noise-Induced Hearing Loss and Carboplatin Ototoxicity. *Noise Health*, 2:53–70.

Henderson, D, McFadden, SL, Liu, CC, Hight, N, and Zheng, XY (1999). The role of antioxidants in protection from impulse noise. *Ann NY Acad Sci*, 884:368–380.

Higashi, K (1991). Otologic findings of DIDMOAD syndrome. *Am J Otol*, 12:57–60.

Hight, NG, McFadden, SL, Henderson, D, Burkard, RF, and Nicotera, T (2003). Noise-induced hearing loss in chinchillas pre-treated with glutathione monoethylester and R-PIA. *Hear Res*, 179:21–32.

Holme, RH and Steel, KP (2004). Progressive hearing loss and increased susceptibility to noise-induced hearing loss in mice carrying a Cdh23 but not a Myo7a mutation. *J Assoc Res Otolaryngol*, 5:66–79.

Hoth, CF, Milunsky, A, Lipsky, N, Sheffer, R, Clarren, SK, and Baldwin, CT (1993). Mutations in the paired domain of the human PAX3 gene cause Klein-Waardenburg syndrome (WS-III) as well as Waardenburg syndrome type I (WS-I). *Am J Hum Genet*, 52:455–462.

Houston, DK, Johnson, MA, Nozza, RJ, Gunter, EW, Shea, KJ, Cutler, GM, et al. (1999). Age-related hearing loss, vitamin B-12, and folate in elderly women. *Am J Clin Nutr*, 69:564–571.

Huang, MY and Schacht, J (1989). Drug-induced ototoxicity. Pathogenesis and prevention. *Med Toxicol Adverse Drug Exp*, 4:452–467.

Hueb, MM, Goycoolea, MV, Paparella, MM, and Oliveira, JA (1991). Otosclerosis: the University of Minnesota temporal bone collection. *Otolaryngol Head Neck Surg*, 105:396–405.

Hutchin, TP, Parker, MJ, Young, ID, Davis, AC, Pulleyn, LJ, Deeble, J, et al. (2000). A novel mutation in the mitochondrial tRNA(Ser(UCN)) gene in a family with non-syndromic sensorineural hearing impairment. *J Med Genet*, 37:692–694.

Inoue, H, Tanizawa, Y, Wasson, J, Behn, P, Kalidas, K, Bernal, M, et al. (1998). A gene encoding a transmembrane protein is mutated in patients with diabetes mellitus and optic atrophy (Wolfram syndrome). *Nat Genet*, 20:143–148.

International Organization for Standardization (1990). Acoustics: determination of occupational noise exposure and estimation of noise-induced hearing impairment: ISO1999. Geneva.

International Organization for Standardization (1998). Acoustics-threshold of hearing by air conduction as a function of age and sex for otologically normal persons: ISO 7029. Geneva.

Itoh, A, Nakashima, T, Arao, H, Wakai, K, Tamakoshi, A, Kawamura, T, et al. (2001). Smoking and drinking habits as risk factors for hearing loss in the elderly: epidemiological study of subjects undergoing routine health checks in Aichi, Japan. *Public Health*, 115:192–196.

Iwai, H, Lee, S, Inaba, M, Sugiura, K, Baba, S, Tomoda, K, et al. (2003). Correlation between accelerated presbycusis and decreased immune functions. *Exp Gerontol*, 38:319–325.

Izumikawa, M, Minoda, R, Kawamoto, K, Abrashkin, KA, Swiderski, DL, Dolan, DF, et al. (2005). Auditory hair cell replacement and hearing improvement by Atoh1 gene therapy in deaf mammals. *Nat Med*, 11:271–276.

Jacono, AA, Hu, B, Kopke, RD, Henderson, D, Van De Water, TR, and Steinman, HM (1998). Changes in cochlear antioxidant enzyme activity after sound conditioning and noise exposure in the chinchilla. *Hear Res*, 117:31–38.

Jaksch, M, Klopstock, T, Kurlemann, G, Dorner, M, Hofmann, S, Kleinle, S, et al. (1998). Progressive myoclonus epilepsy and mitochondrial myopathy associated with mutations in the tRNA(Ser(UCN)) gene. *Ann Neurol*, 44:635–640.

Jin, H, May, M, Tranebjaerg, L, Kendall, E, Fontan, G, Jackson, J, et al. (1996). A novel X-linked gene, DDP, shows mutations in families with deafness (DFN-1), dystonia, mental deficiency and blindness. *Nat Genet*, 14:177–180.

Joensuu, T, Hamalainen, R, Yuan, B, Johnson, C, Tegelberg, S, Gasparini, P, et al. (2001). Mutations in a novel gene with transmembrane domains underlie Usher syndrome type 3. *Am J Hum Genet*, 69:673–684.

Johnson, AC and Nylen, PR (1995). Effects of industrial solvents on hearing. *Occup Med*, 10:623–640.

Johnson, KR, Erway, LC, Cook, SA, Willott, JF, and Zheng, QY (1997). A major gene affecting age-related hearing loss in C57BL/6J mice. *Hear Res*, 114:83–92.

Johnson, KR and Zheng, QY (2002). Ahl2, a second locus affecting age-related hearing loss in mice. *Genomics*, 80:461–464.

Kameoka, K, Isotani, H, Tanaka, K, Azukari, K, Fujimura, Y, Shiota, Y, et al. (1998). Novel mitochondrial DNA mutation in tRNA(Lys) (8296A→G) associated with diabetes. *Biochem Biophys Res Commun*, 245:523–527.

Karlsson, KK, Harris, JR, and Svartengren, M (1997). Description and primary results from an audiometric study of male twins. *Ear Hear*, 18:114–120.

Kawamoto, K, Sha, SH, Minoda, R, Izumikawa, M, Kuriyama, H, Schacht, J, et al. (2004). Antioxidant gene therapy can protect hearing and hair cells from ototoxicity. *Mol Ther*, 9:173–181.

Kelley, PM, Harris, DJ, Comer, BC, Askew, JW, Fowler, T, Smith, SD, et al. (1998). Novel mutations in the connexin 26 gene (GJB2) that cause autosomal recessive (DFNB1) hearing loss. *Am J Hum Genet*, 62:792–799.

Kelsell, DP, Dunlop, J, Stevens, HP, Lench, NJ, Liang, JN, Parry, G, et al. (1997). Connexin 26 mutations in hereditary non-syndromic sensorineural deafness. *Nature*, 387:80–83.

Kitamura, K, Takahashi, K, Noguchi, Y, Kuroishikawa, Y, Tamagawa, Y, Ishikawa, K, et al. (2000). Mutations of the Pendred syndrome gene (PDS) in patients with large vestibular aqueduct. *Acta Otolaryngol*, 120:137–141.

Kozel, PJ, Davis, RR, Krieg, EF, Shull, GE, and Erway, LC (2002). Deficiency in plasma membrane calcium ATPase isoform 2 increases susceptibility to noise-induced hearing loss in mice. *Hear Res*, 164:231–239.

Kubisch, C, Schroeder, BC, Friedrich, T, Lütjohann, B, El-Amraoui, A, Marlin, S, et al. (1999). KCNQ4, a novel potassium channel expressed in sensory outer hair cells, is mutated in dominant deafness. *Cell*, 96:437–446.

Kudo, T, Ikeda, K, Kure, S, Matsubara, Y, Oshima, T, Watanabe, K, et al. (2000). Novel mutations in the connexin 26 gene (GJB2) responsible for childhood deafness in the Japanese population. *Am J Med Genet*, 90:141–145.

Kurien, M, Thomas, K, and Bhanu, TS (1989). Hearing threshold in patients with diabetes mellitus. *J Laryngol Otol*, 103:164–168.

Kurima, K, Peters, LM, Yang, Y, Riazuddin, S, Ahmed, ZM, Naz, S, et al. (2002). Dominant and recessive deafness caused by mutations of a novel gene, TMC1, required for cochlear hair-cell function. *Nat Genet*, 30:277–284.

Lalwani, AK, Goldstein, JA, Kelley, MJ, Luxford, WM, Castelein, CM, and Mhatre, AN (2000). Human nonsyndromic hereditary deafness DFNA17 is due to a mutation in nonmuscle myosin MYH9. *Am J Hum Genet*, 67:1121–1128.

Lataye, R, Campo, P, and Loquet, G (2000). Combined effects of noise and styrene exposure on hearing function in the rat. *Hear Res*, 139:86–96.

Lautermann, J, Crann, SA, McLaren, J, and Schacht, J (1997). Glutathione-dependent antioxidant systems in the mammalian inner ear: effects of aging, ototoxic drugs and noise. *Hear Res*, 114:75–82.

Lee, FS, Matthews, LJ, Dubno, JR, and Mills, JH (2005). Longitudinal study of pure-tone thresholds in older persons. *Ear Hear*, 26:1–11.

Lee, FS, Matthews, LJ, Mills, JH, Dubno, JR, and Adkins, WY (1998). Analysis of blood chemistry and hearing levels in a sample of older persons. *Ear Hear*, 19:180–190.

Li, H, Roblin, G, Liu, H, and Heller, S (2003). Generation of hair cells by stepwise differentiation of embryonic stem cells. *Proc Natl Acad Sci USA*, 100:13495–13500.

Li, X, Li, R, Lin, X, and Guan, MX (2002). Isolation and characterization of the putative nuclear modifier gene MTO1 involved in the pathogenesis of deafness-associated mitochondrial 12 S rRNA A1555G mutation. *J Biol Chem*, 277:27256–27264.

Li, XC, Everett, LA, Lalwani, AK, Desmukh, D, Friedman, TB, Green, ED, et al. (1998). A mutation in PDS causes non-syndromic recessive deafness. *Nat Genet*, 18:215–217.

Liu, XZ, Ouyang, XM, Xia, XJ, Zheng, J, Pandya, A, Li, F, et al. (2003). Prestin, a cochlear motor protein, is defective in non-syndromic hearing loss. *Hum Mol Genet*, 12:1155–1162.

Liu, XZ, Walsh, J, Mburu, P, Kendrick-Jones, J, Cope, MJ, Steel, KP, et al. (1997). Mutations in the myosin VIIA gene cause non-syndromic recessive deafness. *Nat Genet*, 16:188–190.

Liu, XZ, Walsh, J, Tamagawa, Y, Kitamura, K, Nishizawa, M, Steel, KP, et al. (1997). Autosomal dominant non-syndromic deafness caused by a mutation in the myosin VIIA gene. *Nat Genet*, 17:268–269.

Liu, XZ, Xia, XJ, Xu, LR, Pandya, A, Liang, CY, Blanton, SH, et al. (2000). Mutations in connexin31 underlie recessive as well as dominant non-syndromic hearing loss. *Hum Mol Genet*, 9:63–67.

Lopez-Bigas, N, Olive, M, Rabionet, R, Ben-David, O, Martinez-Matos, JA, Bravo, O, et al. (2001). Connexin 31 (GJB3) is expressed in the peripheral and auditory nerves and causes neuropathy and hearing impairment. *Hum Mol Genet*, 10:947–952.

Lynch, ED, Lee, MK, Morrow, JE, Welcsh, PL, Leon, PE, and King, MC (1997). Nonsyndromic deafness DFNA1 associated with mutation of a human homolog of the Drosophila gene diaphanous. *Science*, 278:1315–1318.

Maestrini, E, Korge, BP, Ocana-Sierra, J, Calzolari, E, Cambiaghi, S, Scudder, PM, et al. (1999). A missense mutation in connexin26, D66H, causes mutilating keratoderma with sensorineural deafness (Vohwinkel's syndrome) in three unrelated families. *Hum Mol Genet*, 8:1237–1243.

Marazita, ML, Ploughman, LM, Rawlings, B, Remington, E, Arnos, KS, and Nance, WE (1993). Genetic epidemiological studies of early-onset deafness in the U.S. school-age population. *Am J Med Genet*, 46:486–491.

Matthews, LJ, Lee, FS, Mills, JH, and Schum, DJ (1990). Audiometric and subjective assessment of hearing handicap. *Arch Otolaryngol Head Neck Surg*, 116:1325–1330.

Matthijs, G, Claes, S, Longo-Mbenza, B, and Cassiman, JJ (1996). Non-syndromic deafness associated with a mutation and a polymorphism in the mitochondrial 12S ribosomal RNA gene in a large Zairean pedigree. *Eur J Hum Genet*, 4:46–51.

May-Hegglin/Fechtner Syndrome Consortium (2000). Mutations in MYH9 result in the May-Hegglin anomaly, and Fechtner and Sebastian syndromes. *Nat Genet*, 26:103–105.

Mburu, P, Mustapha, M, Varela, A, Weil, D, El-Amraoui, A, Holme, RH, et al. (2003). Defects in whirlin, a PDZ domain molecule involved in stereocilia elongation, cause deafness in the whirler mouse and families with DFNB31. *Nat Genet*, 34:421–428.

McFadden, SL, Ding, D, Burkard, RF, Jiang, H, Reaume, AG, Flood, DG, et al. (1999). Cu/Zn SOD deficiency potentiates hearing loss and cochlear pathology in aged 129,CD-1 mice. *J Comp Neurol*, 413:101–112.

McFadden, SL, Ding, D, Reaume, AG, Flood, DG, and Salvi, RJ (1999). Age-related cochlear hair cell loss is enhanced in mice lacking copper/zinc superoxide dismutase. *Neurobiol Aging*, 20:1–8.

McFadden, SL, Ding, D, Salvemini, D, and Salvi, RJ (2003). M40403, a superoxide dismutase mimetic, protects cochlear hair cells from gentamicin, but not cisplatin toxicity. *Toxicol Appl Pharmacol*, 186:46–54.

McGuirt, WT, Prasad, SD, Griffith, AJ, Kunst, HP, Green, GE, Shpargel, KB, et al. (1999). Mutations in *COL11A2* cause non-syndromic hearing loss (DFNA13). *Nat Genet*, 23:413–419.

McKenna, MJ, Kristiansen, AG, Bartley, ML, Rogus, JJ, and Haines, JL (1998). Association of COL1A1 and otosclerosis: evidence for a shared genetic etiology with mild osteogenesis imperfecta. *Am J Otol*, 19:604–610.

McKenna, MJ and Mills, BG (1989). Immunohistochemical evidence of measles virus antigens in active otosclerosis. *Otolaryngol Head Neck Surg*, 101:415–421.

McKenna, MJ, Nguyen-Huynh, AT, and Kristiansen, AG (2004). Association of otosclerosis with Sp1 binding site polymorphism in COL1A1 gene: evidence for a shared genetic etiology with osteoporosis. *Otol Neurotol*, 25:447–450.

Mehl, AL and Thomson, V (2002). The Colorado newborn hearing screening project, 1992-1999: on the threshold of effective population-based universal newborn hearing screening. *Pediatrics*, 109:1–8.

Melchionda, S, Ahituv, N, Bisceglia, L, Sobe, T, Glaser, F, Rabionet, R, et al. (2001). MYO6, the human homologue of the gene responsible for deafness in Snell's waltzer mice, is mutated in autosomal dominant nonsyndromic hearing loss. *Am J Hum Genet*, 69:635–640.

Migliosi, V, Modamio-Hoybjor, S, Moreno-Pelayo, MA, Rodriguez-Ballesteros, M, Villamar, M, Telleria, D, et al. (2002). Q829X, a novel mutation in the gene encoding otoferlin (OTOF), is frequently found in Spanish patients with prelingual non-syndromic hearing loss. *J Med Genet*, 39:502–506.

Mills, BG, Singer, FR, Weiner, LP, Suffin, SC, Stabile, E, and Holst, P (1984). Evidence for both respiratory syncytial virus and measles virus antigens in the osteoclasts of patients with Paget's disease of bone. *Clin Orthop Relat Res*:303–311.

Mills, JH, Matthews, LJ, Lee, FS, Dubno, JR, Schulte, BA, and Weber, PC (1999). Gender-specific effects of drugs on hearing levels of older persons. *Ann NY Acad Sci*, 884:381–388.

Minami, SB, Sha, SH, and Schacht, J (2004). Antioxidant protection in a new animal model of cisplatin-induced ototoxicity. *Hear Res*, 198:137–143.

Mochizuki, T, Lemmink, HH, Mariyama, M, Antignac, C, Gubler, MC, Pirson, Y, et al. (1994). Identification of mutations in the alpha 3(IV) and alpha 4(IV) collagen genes in autosomal recessive Alport syndrome. *Nat Genet*, 8:77–81.

Moraes, CT, DiMauro, S, Zeviani, M, Lombes, A, Shanske, S, Miranda, AF, et al. (1989). Mitochondrial DNA deletions in progressive external ophthalmoplegia and Kearns-Sayre syndrome. *N Engl J Med*, 320:1293–1299.

Morell, RJ, Kim, HJ, Hood, LJ, Goforth, L, Friderici, K, Fisher, R, et al. (1998). Mutations in the connexin 26 gene (GJB2) among Ashkenazi Jews with nonsyndromic recessive deafness. *N Engl J Med*, 339:1500–1505.

Morton, NE (1991). Genetic epidemiology of hearing impairment. *Ann NY Acad Sci*, 630:16–31.

Murgia, A, Orzan, E, Polli, R, Martella, M, Vinanzi, C, Leonardi, E, et al. (1999). Cx26 deafness: mutation analysis and clinical variability. *J Med Genet*, 36:829–832.

Mustapha, M, Weil, D, Chardenoux, S, Elias, S, El-Zir, E, Beckmann, JS, et al. (1999). An alpha-tectorin gene defect causes a newly identified autosomal recessive form of sensorineural pre-lingual non-syndromic deafness, DFNB21. *Hum Mol Genet*, 8:409–412.

Naz, S, Giguere, CM, Kohrman, DC, Mitchem, KL, Riazuddin, S, Morell, RJ, et al. (2002). Mutations in a Novel Gene, TMIE, Are Associated with Hearing Loss Linked to the DFNB6 Locus. *Am J Hum Genet*, 71:632–636.

Naz, S, Griffith, AJ, Riazuddin, S, Hampton, LL, Battey, JF Jr., Khan, SN, et al. (2004). Mutations of ESPN cause autosomal recessive deafness and vestibular dysfunction. *J Med Genet*, 41:591–595.

Nemoto, M, Morita, Y, Mishima, Y, Takahashi, S, Nomura, T, Ushiki, T, et al. (2004). Ahl3, a third locus on mouse chromosome 17 affecting age-related hearing loss. *Biochem Biophys Res Commun*, 324:1283–1288.

Newton, VE (1985). Aetiology of bilateral sensorineural hearing loss in young children. *J Laryngol Otol (Suppl)*, 10:1–57.

Neyroud, N, Tesson, F, Denjoy, I, Leibovici, M, Donger, C, Barhanin, J, et al. (1997). A novel mutation in the potassium channel gene KVLQT1 causes the Jervell and Lange-Nielsen cardioauditory syndrome. *Nat Genet*, 15:186–189.

Niedermeyer, HP and Arnold, W (1995). Otosclerosis: a measles virus associated inflammatory disease. *Acta Otolaryngol*, 115:300–303.

Noben-Trauth, K, Zheng, QY, and Johnson, KR (2003). Association of cadherin 23 with polygenic inheritance and genetic modification of sensorineural hearing loss. *Nat Genet*, 35:21–23.

Ohlemiller, KK, McFadden, SL, Ding, DL, Flood, DG, Reaume, AG, Hoffman, EK, et al. (1999). Targeted deletion of the cytosolic Cu/Zn-superoxide dismutase gene (Sod1) increases susceptibility to noise-induced hearing loss. *Audiol Neurootol*, 4:237–246.

Ohlemiller, KK, McFadden, SL, Ding, DL, Lear, PM, and Ho, YS (2000). Targeted mutation of the gene for cellular glutathione peroxidase (Gpx1) increases noise-induced hearing loss in mice. *J Assoc Res Otolaryngol*, 1:243–254.

Ohlemiller, KK, Wright, JS, and Heidbreder, AF (2000). Vulnerability to noise-induced hearing loss in "middle-aged" and young adult mice: a dose-response approach in CBA, C57BL, and BALB inbred strains. *Hear Res*, 149:239–247.

Ohtsuka, A, Yuge, I, Kimura, S, Namba, A, Abe, S, Van Laer, L, et al. (2003). GJB2 deafness gene shows a specific spectrum of mutations in Japan, including a frequent founder mutation. *Hum Genet*, 112:329–333.

Park, HJ, Hahn, SH, Chun, YM, Park, K, and Kim, HN (2000). Connexin26 mutations associated with nonsyndromic hearing loss. *Laryngoscope*, 110:1535–1538.

Park, HJ, Shaukat, S, Liu, XZ, Hahn, SH, Naz, S, Ghosh, M, et al. (2003). Origins and frequencies of SLC26A4 (PDS) mutations in east and south Asians: global implications for the epidemiology of deafness. *J Med Genet*, 40:242–248.

Peters, LM, Anderson, DW, Griffith, AJ, Grundfast, KM, San Agustin, TB, Madeo, AC, et al. (2002). Mutation of a transcription factor, TFCP2L3, causes progressive autosomal dominant hearing loss, DFNA28. *Hum Mol Genet*, 11:2877–2885.

Picciotti, P, Torsello, A, Wolf, FI, Paludetti, G, Gaetani, E, and Pola, R (2004). Age-dependent modifications of expression level of VEGF and its receptors in the inner ear. *Exp Gerontol*, 39:1253–1258.

Pihlajamaa, T, Prockop, DJ, Faber, J, Winterpacht, A, Zabel, B, Giedion, A, et al. (1998). Heterozygous glycine substitution in the *COL11A2* gene in the original patient with the Weissenbacher-Zweymüller syndrome demonstrates its identity with heterozygous OSMED (nonocular Stickler syndrome). *Am J Med Genet*, 80:115–120.

Popelka, MM, Cruickshanks, KJ, Wiley, TL, Tweed, TS, Klein, BE, Klein, R, et al. (2000). Moderate alcohol consumption and hearing loss: a protective effect. *J Am Geriatr Soc*, 48:1273–1278.

Prezant, TR, Agapian, JV, Bohlman, MC, Bu, X, Oztas, S, Qiu, WQ, et al. (1993). Mitochondrial ribosomal RNA mutation associated with both antibiotic-induced and non-syndromic deafness. *Nat Genet*, 4:289–294.

Purchase-Helzner, EL, Cauley, JA, Faulkner, KA, Pratt, S, Zmuda, JM, Talbott, EO, et al. (2004). Hearing sensitivity and the risk of incident falls and fracture in older women: the study of osteoporotic fractures. *Ann Epidemiol*, 14:311–318.

Rabinowitz, PM, Pierce Wise, J Sr., Hur Mobo, B, Antonucci, PG, Powell, C, and Slade, M (2002). Antioxidant status and hearing function in noise-exposed workers. *Hear Res*, 173:164–171.

Rabionet, R, Zelante, L, Lopez-Bigas, N, D'Agruma, L, Melchionda, S, Restagno, G, et al. (2000). Molecular basis of childhood deafness resulting from mutations in the GJB2 (connexin 26) gene. *Hum Genet*, 106:40–44.

Ramsay, HAW and Linthicum, FH Jr. (1994). Mixed hearing loss in otosclerosis: indication for long-term follow-up. *Am J Otol*, 15:536–539.

Reid, FM, Vernham, GA, and Jacobs, HT (1994). A novel mitochondrial point mutation in a maternal pedigree with sensorineural deafness. *Hum Mutat*, 3:243–247.

Richard, G, Rouan, F, Willoughby, CE, Brown, N, Chung, P, Ryynanen, M, et al. (2002). Missense mutations in GJB2 encoding connexin-26 cause the ectodermal dysplasia keratitis-ichthyosis-deafness syndrome. *Am J Hum Genet*, 70:1341–1348.

Richard, G, White, TW, Smith, LE, Bailey, RA, Compton, JG, Paul, DL, et al. (1998). Functional defects of Cx26 resulting from a heterozygous missense mutation in a family with dominant deaf-mutism and palmoplantar keratoderma. *Hum Genet*, 103:393–399.

Richards, AJ, Yates, JR, Williams, R, Payne, SJ, Pope, FM, Scott, JD, et al. (1996). A family with Stickler syndrome type 2 has a mutation in the COL11A1 gene resulting in the substitution of glycine 97 by valine in alpha 1 (XI) collagen. *Hum Mol Genet*, 5:1339–1343.

Robertson, NG, Lu, L, Heller, S, Merchant, SN, Eavey, RD, McKenna, M, et al. (1998). Mutations in a novel cochlear gene cause DFNA9, a human nonsyndromic deafness with vestibular dysfunction. *Nat Genet*, 20:299–303.

Rodriguez, L, Rodriguez, S, Hermida, J, Frade, C, Sande, E, Visedo, G, et al. (2004). Proposed association between the COL1A1 and COL1A2 genes and otosclerosis is not supported by a case-control study in Spain. *Am J Med Genet A*, 128:19–22.

Rosenhall, U, Sixt, E, Sundh, V, and Svanborg, A (1993). Correlations between presbyacusis and extrinsic noxious factors. *Audiology*, 32:234–243.

Ruf, RG, Xu, PX, Silvius, D, Otto, EA, Beekmann, F, Muerb, UT, et al. (2004). SIX1 mutations cause branchio-oto-renal syndrome by disruption of EYA1-SIX1-DNA complexes. *Proc Natl Acad Sci USA*, 101:8090–8095.

Rybak, LP (1992). Hearing: the effects of chemicals. *Otolaryngol Head Neck Surg*, 106:677–686.

Sage, C, Huang, M, Karimi, K, Gutierrez, G, Vollrath, MA, Zhang, DS, et al. (2005). Proliferation of functional hair cells in vivo in the absence of the retinoblastoma protein. *Science*, 307:1114–1118.

Sanchez-Martin, M, Rodriguez-Garcia, A, Perez-Losada, J, Sagrera, A, Read, AP, and Sanchez-Garcia, I (2002). SLUG (SNAI2) deletions in patients with Waardenburg disease. *Hum Mol Genet*, 11:3231–3236.

Sato, E, Nakashima, T, Miura, Y, Furuhashi, A, Nakayama, A, Mori, N, et al. (2001). Phenotypes associated with replacement of His by Arg in the Pendred syndrome gene. *Eur J Endocrinol*, 145:697–703.

Schacht, J (1999). Antioxidant therapy attenuates aminoglycoside-induced hearing loss. *Ann N Y Acad Sci*, 884:125–130.

Schonberger, J, Wang, L, Shin, JT, Kim, SD, Depreux, FF, Zhu, H, et al. (2005). Mutation in the transcriptional coactivator EYA4 causes dilated cardiomyopathy and sensorineural hearing loss. *Nat Genet*, 37:418–422.

Schuknecht, HF and Gacek, MR (1993). Cochlear pathology in presbycusis. *Ann Otol Rhinol Laryngol*, 102:1–16.

Schultz, JM, Yang, Y, Caride, AJ, Filoteo, AG, Penheiter, AR, Lagziel, A, et al. (2005). Modification of human hearing loss by plasma-membrane calcium pump PMCA2. *N Engl J Med*, 352:1557–1564.

Schulze-Bahr, E, Wang, Q, Wedekind, H, Haverkamp, W, Chen, Q, and Sun, Y (1997). KCNE1 mutations cause Jervell and Lange-Nielsen syndrome. *Nat Genet*, 17:267–268.

Scott, HS, Kudoh, J, Wattenhofer, M, Shibuya, K, Berry, A, Chrast, R, et al. (2001). Insertion of beta-satellite repeats identifies a transmembrane protease causing both congenital and childhood onset autosomal recessive deafness. *Nat Genet*, 27:59–63.

Scott, HS, Wattenhofer, M, Shibuya, K, Berry, A, Kudoh, J, Guipponi, M, et al. (2000). A novel mutation mechanism, insertion of beta-satellite repeats, in a transmembrane protease gene causes the autosomal recessive deafness DFNB10. *Am J Hum Genet*, 67(Suppl. 2):13.

Seidman, MD (2000). Effects of dietary restriction and antioxidants on presbyacusis. *Laryngoscope*, 110:727–738.

Seidman, MD, Bai, U, Khan, MJ, Murphy, MJ, Quirk, WS, Castora, FJ, et al. (1996). Association of mitochondrial DNA deletions and cochlear pathology: a molecular biologic tool. *Laryngoscope*, 106:777–783.

Seidman, MD, Khan, MJ, Bai, U, Shirwany, N, and Quirk, WS (2000). Biologic activity of mitochondrial metabolites on aging and age-related hearing loss. *Am J Otol*, 21:161–167.

Seidman, MD, Khan, MJ, Tang, WX, and Quirk, WS (2002). Influence of lecithin on mitochondrial DNA and age-related hearing loss. *Otolaryngol Head Neck Surg*, 127:138–144.

Seligmann, H, Podoshin, L, Ben-David, J, Fradis, M, and Goldsher, M (1996). Drug-induced tinnitus and other hearing disorders. *Drug Saf*, 14:198–212.

Sha, SH, Zajic, G, Epstein, CJ, and Schacht, J (2001). Overexpression of copper/zinc-superoxide dismutase protects from kanamycin-induced hearing loss. *Audiol Neurootol*, 6:117–123.

Shoffner, JM, Lott, MT, Lezza, AM, Seibel, P, Ballinger, SW, and Wallace, DC (1990). Myoclonic epilepsy and ragged-red fiber disease (MERRF) is associated with a mitochondrial DNA tRNA(Lys) mutation. *Cell*, 61:931–937.

Sixt, E and Rosenhall, U (1997). Presbyacusis related to socioeconomic factors and state of health. *Scand Audiol*, 26:133–140.

Sliwinska-Kowalska, M, Zamyslowska-Szmytke, E, Szymczak, W, Kotylo, P, Fiszer, M, Wesolowski, W, et al. (2004). Effects of coexposure to noise and mixture of organic solvents on hearing in dockyard workers. *J Occup Environ Med*, 46:30–38.

Sobe, T, Erlich, P, Berry, A, Korostichevsky, M, Vreugde, S, Avraham, KB, et al. (1999). High frequency of the deafness-associated 167delT mutation in the connexin 26 (GJB2) gene in Israeli Ashkenazim. *Am J Med Genet*, 86:499–500.

Somers, T, Govaerts, P, Janssens de Varebeke, S, and Offeciers, E (1997). Revision stapes surgery. *J Laryngol Otol*, 111:233–239.

Song, BB, Anderson, DJ, and Schacht, J (1997). Protection from gentamicin ototoxicity by iron chelators in guinea pig in vivo. *J Pharmacol Exp Ther*, 282:369–377.

Strom, TM, Hortnagel, K, Hofmann, S, Gekeler, F, Scharfe, C, Rabl, W, et al. (1998). Diabetes insipidus, diabetes mellitus, optic atrophy and deafness (DIDMOAD) caused by mutations in a novel gene (wolframin) coding for a predicted transmembrane protein. *Hum Mol Genet*, 7:2021–2028.

Stypulkowski, PH (1990). Mechanisms of salicylate ototoxicity. *Hear Res*, 46:113–145.

Sue, CM, Tanji, K, Hadjigeorgiou, G, Andreu, AL, Nishino, I, Krishna, S, et al. (1999). Maternally inherited hearing loss in a large kindred with a novel T7511C mutation in the mitochondrial DNA tRNA(Ser(UCN)) gene. *Neurology*, 52:1905–1908.

Tanimoto, H, Nishio, H, Matsuo, M, and Nibu, K (2004). A novel mitochondrial mutation, 1556C→T, in a Japanese patient with streptomycin-induced tinnitus. *Acta Otolaryngol*, 124:258–261.

Tassabehji, M, Newton, VE, and Read, AP (1994). Waardenburg syndrome type 2 caused by mutations in the human microphthalmia (MITF) gene. *Nat Genet*, 8:251–255.

Tassabehji, M, Read, AP, Newton, VE, Harris, R, Balling, R, Gruss, P, et al. (1992). Waardenburg's syndrome patients have mutations in the human homologue of the Pax-3 paired box gene. *Nature*, 355:635–636.

Tiranti, V, Chariot, P, Carella, F, Toscano, A, Soliveri, P, Girlanda, P, et al. (1995). Maternally inherited hearing loss, ataxia and myoclonus associated with a novel point mutation in mitochondrial tRNASer(UCN) gene. *Hum Mol Genet*, 4:1421–1427.

Tomek, MS, Brown, MR, Mani, SR, Ramesh, A, Srisailapathy, CRS, Coucke, P, et al. (1998). Localization of a gene for otosclerosis to chromosome 15q25-q26. *Hum Mol Genet*, 7:285–290.

Toppila, E, Pyykko, I, and Starck, J (2001). Age and noise-induced hearing loss. *Scand Audiol*, 30:236–244.

Toppila, E, Pyykko, II, Starck, J, Kaksonen, R, and Ishizaki, H (2000). Individual risk factors in the development of noise-induced hearing loss. *Noise Health*, 2:59–70.

The Treacher Collins Syndrome Collaborative Group (1996). Positional cloning of a gene involved in the pathogenesis of Treacher Collins syndrome. *Nat Genet*, 12:130–136.

Tyson, J, Tranebjaerg, L, Bellman, S, Wren, C, Taylor, JF, Bathen, J, et al. (1997). IsK and KvLQT1: mutation in either of the two subunits of the slow component of the delayed rectifier potassium channel can cause Jervell and Lange-Nielsen syndrome. *Hum Mol Genet*, 6:2179–2185.

Uchida, Y, Nakashimat, T, Ando, F, Niino, N, and Shimokata, H (2005). Is there a relevant effect of noise and smoking on hearing? A population-based aging study. *Int J Audiol*, 44:86–91.

Usami, S, Abe, S, Kasai, M, Shinkawa, H, Moeller, B, Kenyon, JB, et al. (1997). Genetic and clinical features of sensorineural hearing loss associated with the 1555 mitochondrial mutation. *Laryngoscope*, 107:483–490.

Usami, S, Abe, S, Weston, MD, Shinkawa, H, Van Camp, G, and Kimberling, WJ (1999). Non-syndromic hearing loss associated with enlarged vestibular aqueduct is caused by PDS mutations. *Hum Genet*, 104:188–192.

Vahava, O, Morell, R, Lynch, ED, Weiss, S, Kagan, ME, Ahituv, N, et al. (1998). Mutation in transcription factor POU4F3 associated with inherited progressive hearing loss in humans. *Science*, 279:1950–1954.

Van Camp, G and Smith, RJH (2000). Maternally inherited hearing impairment. *Clin Genet*, 57:409–414.

Van Den Bogaert, K, De Leenheer, EMR, Chen, W, Lee, Y, Nurnberg, P, Pennings, RJE, et al. (2004). A fifth locus for otosclerosis, OTSC5, maps to chromosome 3q22-24. *J Med Genet*, 41:450–453.

Van Den Bogaert, K, Govaerts, PJ, Schatteman, I, Brown, MR, Caethoven, G, Offeciers, FE, et al. (2001). A second gene for otosclerosis, OTSC2, maps to chromosome 7q34-36. *Am J Hum Genet*, 68:495–500.

van den Ouweland, JM, Lemkes, HH, Ruitenbeek, W, Sandkuijl, LA, de Vijlder, MF, Struyvenberg, PA, et al. (1992). Mutation in mitochondrial tRNA (Leu)(UUR) gene in a large pedigree with maternally transmitted type II diabetes mellitus and deafness. *Nat Genet*, 1:368–371.

Van Hauwe, P, Coucke, PJ, Declau, F, Kunst, H, Ensink, RJ, Marres, HA, et al. (1999). Deafness linked to DFNA2: one locus but how many genes? *Nat Genet*, 21:263.

Van Hauwe, P, Everett, LA, Coucke, P, Scott, DA, Kraft, ML, Ris-Stalpers, C, et al. (1998). Two frequent missense mutations in Pendred syndrome. *Hum Mol Genet*, 7:1099–1104.

Van Laer, L, Coucke, P, Mueller, RF, Caethoven, G, Flothmann, K, Prasad, SD, et al. (2001). A common founder for the 35delG GJB2 gene mutation in connexin 26 hearing impairment. *J Med Genet*, 38:515–518.

Van Laer, L, DeStefano, AL, Myers, RH, Flothmann, K, Thys, S, Fransen, E, et al. (2002). Is DFNA5 a susceptibility gene for age-related hearing impairment? *Eur J Hum Genet*, 10:883–886.

Van Laer, L, Huizing, EH, Verstreken, M, van Zuijlen, D, Wauters, JG, Bossuyt, PJ, et al. (1998). Nonsyndromic hearing impairment is associated with a mutation in DFNA5. *Nat Genet*, 20:194–197.

van Wijk, E, Krieger, E, Kemperman, MH, De Leenheer, EMR, Huygen, PLM, Cremers, CW et al. (2003). A mutation in the gamma actin 1 (*ACTG1*) gene causes autosomal dominant hearing loss (DFNA20/26). *J Med Genet*, 40:879–884.

Varga, R, Kelley, PM, Keats, BJ, Starr, A, Leal, SM, Cohn, E, et al. (2003). Nonsyndromic recessive auditory neuropathy is the result of mutations in the otoferlin (OTOF) gene. *J Med Genet*, 40:45–50.

Verhoeven, K, Ensink, RJH, Tiranti, V, Huygen, PLM, Johnson, DF, Schatteman, I, et al. (1999). Hearing impairment and neurological dysfunction associated with a mutation in the mitochondrial tRNASer(UCN) gene. *Eur J Hum Genet*, 7:45–51.

Verhoeven, K, Van Laer, L, Kirschhofer, K, Legan, PK, Hughes, DC, Schatteman, I, et al. (1998). Mutations in the human alpha-tectorin gene cause autosomal dominant non-syndromic hearing impairment. *Nat Genet*, 19:60–62.

Verpy, E, Leibovici, M, Zwaenepoel, I, Liu, XZ, Gal, A, Salem, N, et al. (2000). A defect in harmonin, a PDZ domain-containing protein expressed in the inner ear sensory hair cells, underlies Usher syndrome type 1C. *Nat Genet*, 26:51–55.

Verpy, E, Masmoudi, S, Zwaenepoel, I, Leibovici, M, Hutchin, TP, del Castillo, I, et al. (2001). Mutations in a new gene encoding a protein of the hair bundle cause non-syndromic deafness at the DFNB16 locus. *Nat Genet*, 29:345–349.

Vikkula, M, Mariman, ECM, Lui, VCH, Zhidkova, NI, Tiller, GE, Goldring, MB, et al. (1995). Autosomal dominant and recessive osteochondrodysplasias associated with the *COL11A2* locus. *Cell*, 80:431–437.

Walsh, T, Walsh, V, Vreugde, S, Hertzano, R, Shahin, H, Haika, S, et al. (2002). From flies' eyes to our ears: mutations in a human class III myosin cause progressive nonsyndromic hearing loss DFNB30. *Proc Natl Acad Sci USA*, 99:7518–7523.

Wang, A, Liang, Y, Fridell, RA, Probst, FJ, Wilcox, ER, Touchman, JW, et al. (1998). Association of unconventional myosin MYO15 mutations with human nonsyndromic deafness DFNB3. *Science*, 280:1447–1451.

Wayne, S, Robertson, NG, Declau, F, Chen, N, Verhoeven, K, Prasad, S, et al. (2001). Mutations in the transcriptional activator EYA4 cause late-onset deafness at the DFNA10 locus. *Hum Mol Genet*, 10:195–200.

Weatherly, RA, Owens, JJ, Catlin, FI, and Mahoney, DH (1991). cis-platinum ototoxicity in children. *Laryngoscope*, 101:917–924.

Weil, D, Blanchard, S, Kaplan, J, Guilford, P, Gibson, F, Walsh, J, et al. (1995). Defective myosin VIIA gene responsible for Usher syndrome type 1B. *Nature*, 374:60–61.

Weil, D, El-Amraoui, A, Masmoudi, S, Mustapha, M, Kikkawa, Y, Laine, S, et al. (2003). Usher syndrome type I G (USH1G) is caused by mutations in the gene encoding SANS, a protein that associates with the USH1C protein, harmonin. *Hum Mol Genet*, 12:463–471.

Weil, D, Kussel, P, Blanchard, S, Levy, G, Levi-Acobas, F, Drira, M, et al. (1997). The autosomal recessive isolated deafness, DFNB2, and the Usher 1B syndrome are allelic defects of the myosin-VIIA gene. *Nat Genet*, 16:191–193.

Weston, MD, Luijendijk, MW, Humphrey, KD, Moller C, and Kimberling WJ (2004). Mutations in the VLGR1 gene implicate G-protein signaling in the pathogenesis of Usher syndrome type II. *Am J Hum Genet*, 74:357–366.

Wilcox, ER, Burton, QL, Naz, S, Riazuddin, S, Smith, TN, Ploplis, B, et al. (2001). Mutations in the gene encoding tight junction claudin-14 cause autosomal recessive deafness DFNB29. *Cell*, 104:165–172.

Willott, JF, Erway, LC, Archer, JR, and Harrison, DE (1995). Genetics of age-related hearing loss in mice. II. Strain differences and effects of caloric restriction on cochlear pathology and evoked response thresholds. *Hear Res*, 88:143–155.

Xia, JH, Liu, CY, Tang, BS, Pan, Q, Huang, L, Dai, HP, et al. (1998). Mutations in the gene encoding gap junction protein beta-3 associated with autosomal dominant hearing impairment. *Nat Genet*, 20:370–373.

Xiao, S, Yu, C, Chou, X, Yuan, W, Wang, Y, Bu, L, et al. (2001). Dentinogenesis imperfecta 1 with or without progressive hearing loss is associated with distinct mutations in DSPP. *Nat Genet*, 27:201–204.

Yamasoba, T, Harris, C, Shoji, F, Lee, RJ, Nuttall, AL, and Miller, JM (1998). Influence of intense sound exposure on glutathione synthesis in the cochlea. *Brain Res*, 804:72–78.

Yasunaga, S, Grati, M, Cohen-Salmon, M, El-Amraoui, A, Mustapha, M, Salem, N, et al. (1999). A mutation in OTOF, encoding otoferlin, a FER-1-like protein, causes DFNB9, a nonsyndromic form of deafness. *Nat Genet*, 21:363–369.

Young, TL, Ives, E, Lynch, E, Person, R, Snook, S, MacLaren, L, et al. (2001). Non-syndromic progressive hearing loss DFNA38 is caused by heterozygous missense mutation in the Wolfram syndrome gene WFS1. *Hum Mol Genet*, 10:2509–2514.

Zelante, L, Gasparini, P, Estivill, X, Melchionda, S, D'Agruma, L, Govea, N, et al. (1997). Connexin 26 mutations associated with the most common form of non-syndromic neurosensory autosomal recessive deafness (DFNB1) in Mediterraneans. *Hum Mol Genet*, 6:1605–1609.

Zeviani, M, Muntoni, F, Savarese, N, Serra, G, Tiranti, V, Carrara, F, et al. (1993). A MERRF/MELAS overlap syndrome associated with a new point mutation in the mitochondrial DNA tRNA(Lys) gene. *Eur J Hum Genet*, 1:80–87.

Zhao, H, Li, R, Wang, Q, Yan, Q, Deng, JH, Han, D, et al. (2004). Maternally inherited aminoglycoside-induced and nonsyndromic deafness is associated with the novel C1494T mutation in the mitochondrial 12S rRNA gene in a large Chinese family. *Am J Hum Genet*, 74:139–152.

Zhou, J, Mochizuki, T, Smeets, H, Antignac, C, Laurila, P, de Paepe, A, et al. (1993). Deletion of the paired alpha 5(IV) and alpha 6(IV) collagen genes in inherited smooth muscle tumors. *Science*, 261:1167–1169.

Zhu, M, Yang, T, Wei, S, DeWan, AT, Morell, RJ, Elfenbein, JL, et al. (2003). Mutations in the gamma-actin gene (*ACTG1*) are associated with dominant progressive deafness (DFNA20/26). *Am J Hum Genet*, 73:1082–1091.

Zwaenepoel, I, Mustapha, M, Leibovici, M, Verpy, E, Goodyear, R, Liu, XZ, et al. (2002). Otoancorin, an inner ear protein restricted to the interface between the apical surface of sensory epithelia and their overlying acellular gels, is defective in autosomal recessive deafness DFNB22. *Proc Natl Acad Sci USA*, 99:6240–6245.

32

Complex Skin Diseases I: Psoriasis

Colin Veal and David Burden

Psoriasis is one of the most common immune-mediated inflammatory diseases, affecting 2%–3% of Caucasian populations. It manifests in the skin as extremely well demarcated, red plaques covered in silver-colored scales. There is considerable variation in the psoriasis phenotype between individuals and in the same individual at different times. Chronic plaque psoriasis (psoriasis vulgaris) accounts for approximately 80% of disease and presents with plaques of variable diameter most frequently affecting the scalp, extensor aspects of elbows and knees, and the sacrum (Fig. 32–1).

Guttate psoriasis presents with the synchronous development of many small papules of psoriasis over the trunk and proximal limbs, often in childhood and following a sore throat (Fig. 32–2). When the majority of the skin surface is affected, the disease is called erythrodermic psoriasis (Fig. 32–3). In some, rather than red scaly plaques, the disease is characterized by the presence of readily visible pustules (Fig. 32–4). These either affect large areas of skin (generalized pustular psoriasis) or are restricted to smaller areas (localized pustular psoriasis), usually the palms and soles (palmoplantar pustulosis). Approximately 30% of those affected by psoriasis also develop a distinctive seronegative spondyloarthropathy. This arthritis is clinically diverse and has usually been categorized into asymmetric oligoarticular, distal interphalangeal joints only, polyarticular, arthritis mutilans, and spondylitis. Crohn's disease also occurs more frequently in those with psoriasis compared to the general population.

The disease runs an unpredictable course, varying in its severity at different times. Some of this fluctuation is due to the modifying effects of environmental factors such as ultraviolet radiation, streptococcal infection, cutaneous trauma, and stress. The etiology and pathogenesis of psoriasis are not yet fully defined. There are no naturally occurring examples of psoriasis in animals, and other than xenograft experiments, animal models of the disease are limited. Lesional skin shows increased proliferation and abnormal differentiation of keratinocytes, infiltration by activated T lymphocytes and neutrophils, and activation of the cutaneous vasculature. These changes correspond to overexpression of growth factors and their receptors, proinflammatory cytokines, and angiogenic peptides. Controversy concerning the primacy of keratinocyte, lymphocyte, and the vasculature in initiating and maintaining psoriasis dates from the late-nineteenth century and remains unresolved. Over the past 20 years, based initially on the response of the disease to immunosuppression, most attention has focussed on immunological abnormalities in psoriasis. Assuming psoriasis is induced by an immunological reaction, the antigen, if there is one, is not known. Candidates for a conventional antigen include an autoantigen, for instance, a sequestered stratum corneum antigen or a microbial antigen.

There is an excess of cardiovascular mortality in those severely affected. However, the major burden of the disease is produced by its deleterious effect on quality of life, that is profound. In addition to the soreness and irritation of the skin, the majority of patients report difficulties establishing social contacts and developing relationships. Some patients fear to have children in case they pass the disease to their offspring. At the most extreme, some patients describe feeling "untouchable" or "like a leper." There is no curative treatment, but in the majority the disease can be suppressed. For milder and particularly more localized disease, topical therapy with vitamin D analogues, corticosteroids, tars, or dithranol can be effective. For more extensive psoriasis, ultraviolet phototherapy is very effective, sometimes combined with retinoids, or photosensitizing psoralens. Severe disease is often treated by systemic immunosuppression (e.g., methotrexate, ciclosporin) or targeted immunobiological therapy (e.g., antiTNFα, antiCD11a).

Genetic Epidemiology

The reported prevalence of psoriasis in the general population ranges from 1% to 4.8% in Northern Europe (Lomholt, 1963; Hellgren, 1967) and there is a wider range in the different racial groups of Asia, Africa, and the Americas (Farber and Nall, 1991). It is difficult to assess the incidence with any accuracy, although a figure of one new case per 1000 per year has been estimated (Lomholt, 1963). Psoriasis affects both sexes equally. It may start at any age, but the peak age at onset is in adolescence and early adult life. The age at onset appears to be bimodal, leading to the concept that there are two types of psoriasis, analogous to the situation in diabetes mellitus (Henseler and Christophers, 1985). The type 1 psoriasis is of early onset <40 years), is more often familial, and more severe, whereas type 2 psoriasis is of later onset and less familial.

Many individuals with psoriasis have affected family members, but only occasional families are described in which multiple members are affected over many generations. The earliest large-scale study to determine the epidemiology of psoriasis was a census study performed by Gunnar Lomholt (1963). He personally took a detailed history and examined the skin of 10,984 individuals in the

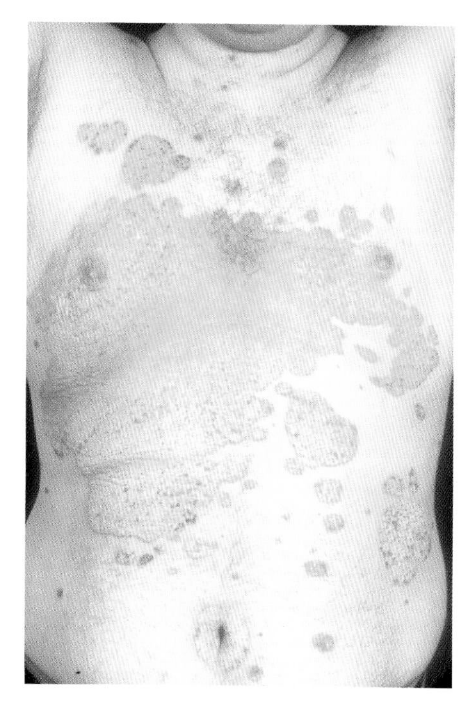

Figure 32–1 Typical appearance in chronic plaque psoriasis (psoriasis vulgaris).

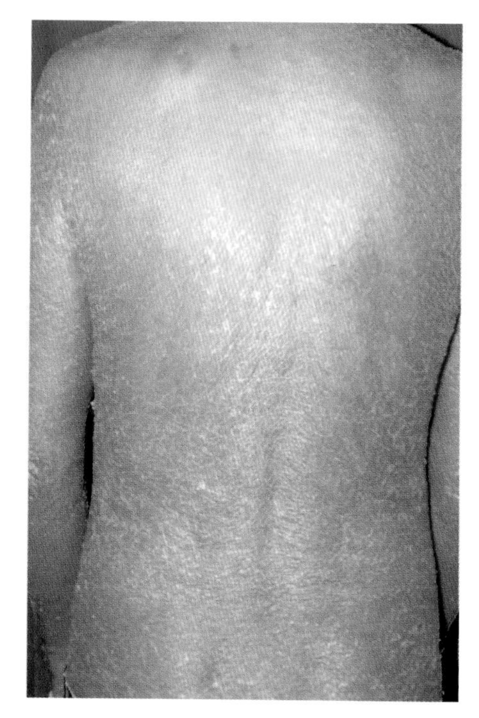

Figure 32–3 Erythrodermic psoriasis.

Faeroe Islands and identified 312 cases, giving prevalence in that population of 2.8%. He reported psoriasis in 90% of first- or second-degree relatives of psoriasis patients. The risk to siblings of psoriatics was also found to be greater in those families in which a parent was also affected. This and subsequent population-based studies point to a strong genetic component to psoriasis but with no common Mendelian mode of inheritance and it has been concluded that the disease is multifactorial. This is supported by a calculation of Risch's risk ratio, λ_R (Risch, 1990a). λ_R is defined as the risk of disease in relative of degree R in relation to the population prevalence. For a single gene disorder, $\lambda_R - 1$ will decrease by a factor of 2 with each degree of relationship. In contrast for a multilocus disorder, $\lambda_R - 1$ will decrease by more than a factor of 2 with each degree of relationship. In psoriasis, $\lambda_R - 1$ decreases by a factor of 6 from the first to the second generation, suggesting a multilocus model (Risch, 1990b; Elder et al., 1994).

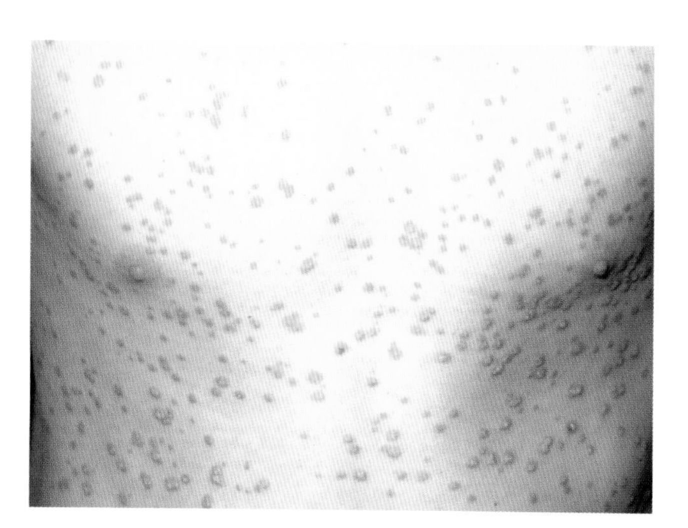

Figure 32–2 Guttate psoriasis—note small papules over trunk.

Several studies of psoriasis in twins have revealed a concordance of 50%–80% in identical twins compared with approximately 20% in monozygotic twins (Lomholt, 1963; Brandrup et al., 1978). The heritability calculated from these studies is approximately 80%. Concordant monozygotic twins have also been shown to have similar ages at onset, distribution pattern, severity, and progression, that are not seen in dizygotic twins, suggesting that genetic factors are also involved in influencing various aspects of the phenotype (Farber et al., 1974).

Mapping of Psoriasis Loci

Early Association Studies

With evidence of a genetic component to this common disorder, various regions of the genome have been examined for susceptibility loci. Before genomic maps of deoxyribonucleic acid (DNA) markers,

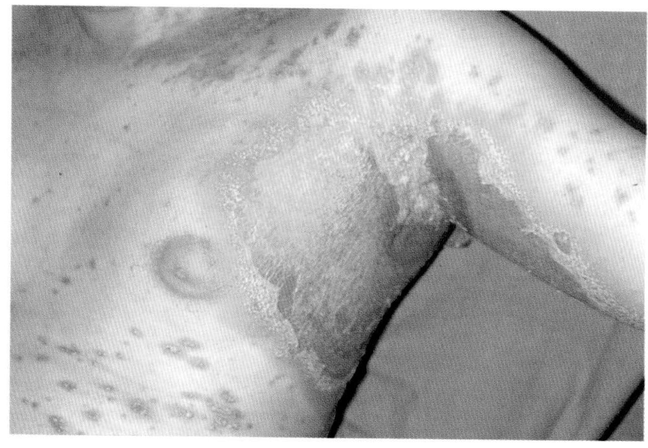

Figure 32–4 Generalized pustular psoriasis.

studies were limited to serologically typing protein polymorphisms, the most studied of which are histocompatibility antigens (human leukocyte antigen or HLA) on chromosome 6. The first associations to be reported were between psoriasis and HLA-B13 (Russell et al., 1972). Many studies followed in which several HLA alleles were shown to have strong association with psoriasis, in particular HLA-Cw6 (Elder et al., 1994). Approximately 75% of plaque psoriasis patients have at least one HLA-Cw6 allele compared to a population frequency of approximately 10%. Thus, HLA-Cw6 became the strongest candidate for a psoriasis susceptibility gene. A threonine to alanine polymorphism at position 73 from the translation start site was reported to be associated with psoriasis in a Japanese cohort (Asahina et al., 1991). However, subsequent studies have shown that, although an alanine at position 73 is present in Caucasian psoriatics, it is also present to a high degree in controls (Mallon et al., 1997). Furthermore, the same amino acid substitution is present in a number of HLA-C alleles that are not associated with psoriasis. Alanine at residue 73 is therefore unlikely to be a causative allele for psoriasis.

Early Linkage Studies

The advancement of microsatellites as informative molecular markers and the construction of genome-wide microsatellite linkage maps enabled systematic scanning of the entire genome for common disease susceptibility loci including those for psoriasis. The first study to use microsatellite markers for the identification of psoriasis susceptibility regions included eight American families, each with multiple individuals affected by psoriasis and psoriatic arthritis (PsA) (Tomfohrde et al., 1994). Of the 69 markers typed, a maximum logarithm of odds (LOD) score of 5.3 was seen in a single family for markers spanning chromosome 17q, that also generated a significant nonparametric score using allele-sharing analysis. Additional typing of markers close to the original set detected evidence of linkage in a further family. This locus is now designated *PSORS2* and, although it has been replicated in other studies, in many populations there is no evidence for linkage (Burden et al., 1998). In one early study, seven Irish families were examined with chromosome 17 markers, one of which generated a LOD score of 1.24—the maximum score the family was capable of producing (Matthews et al., 1996). The remaining families were then subjected to the first full genome-wide scan in psoriasis. Parametric LOD score analysis revealed linkage to chromosome 4 (LOD score of 3.03) in the region surrounding D4S171. This region is now designated *PSORS3*, but has not been robustly replicated in other datasets.

The first linkage evidence to give support to the association studies of HLA alleles on chromosome 6 was produced in 1997 (Trembath et al., 1997). A genome scan of a cohort of 41 English families using a nonparametric approach gave strong evidence of linkage to the major histocompatibility complex (MHC). Increased allele sharing in affected siblings was demonstrated in three further regions on chromosomes 2, 8, and 20. No support for linkage between psoriasis and previously reported regions on chromosomes 4 or 17 was seen. This region of chromosome 6 has been replicated in most subsequent genome scans of sufficient power and is now designated *PSORS1*, in acknowledgment of the initial association with HLA alleles. The locus or loci within the MHC is widespread in both sporadic and familial psoriasis and is likely to be the major psoriatic determinant, accounting for more than a third of the heritability of the disease.

MHC and Psoriasis

Linkage and association studies have confirmed that a major susceptibility locus resides within the MHC, which is an interesting and unique region of the genome. It was originally described as containing the antigens that caused the rejection of transplanted tissues, that have now been shown to include antigen presenting molecules of the adaptive immune system.

Genetic Structure of the MHC

The MHC is a region in the chromosomal band 21.3 on the short arm of chromosome 6. Analysis of the MHC sequence has identified 224 putative known and unknown gene loci, including 96 pseudogenes. An estimated 40% of the remaining genes have predicted immune system functions (1999). The MHC can be separated into three sections (Fig. 32–5): the largest region is the class I section positioned at the telomeric end of the MHC and contains genes encoding class I MHC antigen presentation molecules. The class II region is located at the centromeric end of the MHC and harbors genes for class II MHC antigen presentation molecules. This region is also unique in that nearly all the genes in this section have an immune-related function, for example, the antigen processing genes, *TAP1* and *TAP2*. The intervening class IIIregion is densely packed with genes, many of which are involved in innate immunity and inflammation although many are of unknown function. Both class I and class II regions have a large number of pseudogenes, that are absent in the intervening class III region.

Evolution of the MHC

The adaptive immune system, including the MHC, is not present in either the invertebrates or the jawless vertebrates, and appears to have evolved in the primitive jawed vertebrates, such as the cartilaginous fish (Matsunaga and Rahman, 1998). Genomic linkage of the genes is present in the amphibian *Xenopus* suggesting that the MHC locus may be 370 million years old, though multiple clusters are present in many animals, such as the chicken in which the MHC is split into two clusters (Nonaka et al., 1997).

The origin of the MHC antigen presentation genes is as yet unknown. It has been suggested that plasma membrane cell adhesion proteins may be the ultimate ancestors (Ohno, 1987). The class II antigen presentation molecules show orthologous relationship between mammals of different orders, and these genes may have arisen before the divergence of placental mammals (Hughes and Nei, 1990). However, the class I genes do not show any orthologous relationships between mammals of different orders, in particular the *HLA-C* locus has only been found in human, gorilla, and chimpanzee genomes (Boyson et al., 1996). Genes within the MHC have been subject to duplication either before or after the formation of the MHC as exhibited by evolutionary paralogous regions on chromosomes 1q, 9, and 19p (Kasahara, 1999).

Polymorphism Within the MHC

The MHC classical genes are highly polymorphic, a feature which is also seen at the DNA level across the entire MHC interval. The main source of variability comes from single nucleotide polymorphisms (SNPs) (Parham et al., 1995), that are present at a higher frequency in this region than other areas of chromosome 6 (Fig. 32–5). Interestingly, there is no evidence that the mutation rate is higher within the MHC and the variation is localized to certain genes and their close surroundings. There is strong evidence that selection is acting on the

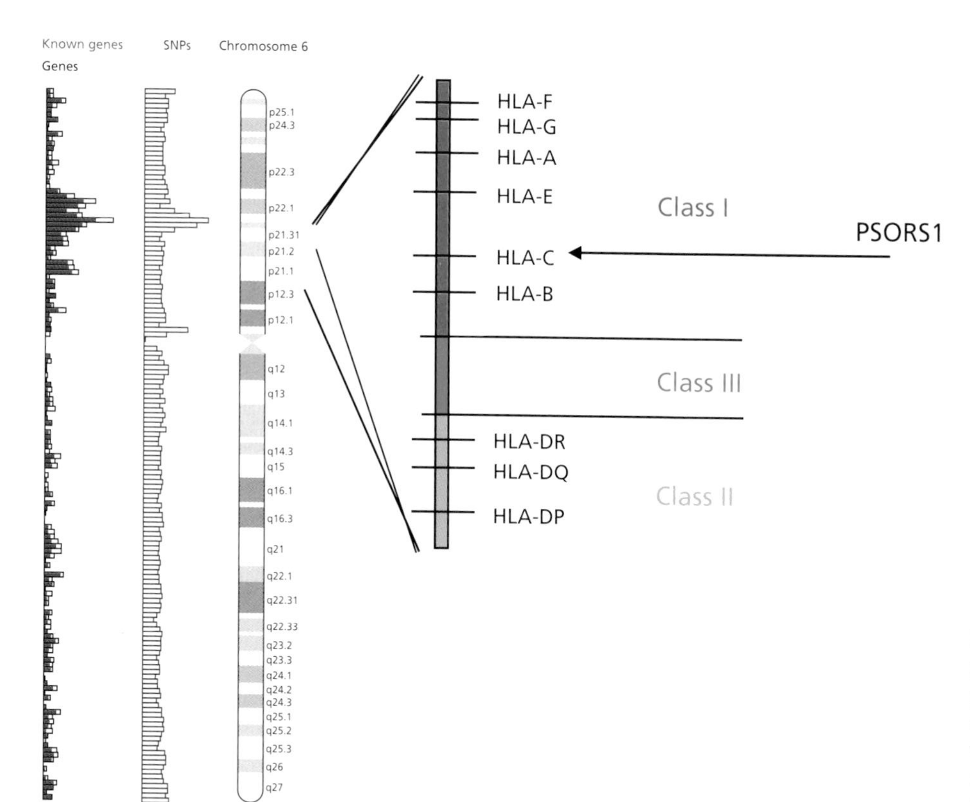

Figure 32–5 Structure and genomic location of the major histocompatibility complex (MHC). The MHC is located on the short arm of chromosome 6, coinciding with the greatest density of genes and single nucleotide polymorphisms. The structure of the MHC is displayed to the right and the approximate position of PSORS1 is noted.

maintenance of diversity within the MHC, as selection will favor rare haplotypes—pathogens are more likely to evolve mechanisms to evade the common haplotypes (Potts and Slev, 1995). Selection should also favor heterozygotes that will have greater diversity of antigen presentation molecules and are therefore more likely to be able to present foreign antigen to the immune system. This theory is supported by the high ratio of nonsynonymous to synonymous mutations within the codons for antigen recognition, which is similar to the pattern predicted for overdominant selection of heterozygote advantage (Ritte et al., 1991).

Linkage Disequilibrium Across the MHC

As the MHC has an important role in host defense, strong selective pressures from past pathogens will have shaped its present-day genomic structure. The alleles of genes that produce beneficial protein interactions will result in selective pressure for those genes to be inherited as a unit and therefore have close genetic linkage (Beck and Trowsdale, 2000). In contrast, the shuffling of MHC haplotypes will prevent fixation of haplotypes and protect against novel pathogens. Measurement of linkage disequilibrium between specific HLA alleles has revealed significant association for alleles more than 1 Mb apart (Sanchez-Mazas et al., 2000). Analysis of the recombination rate across the MHC in centre d'Etude du polymorphisme humain (CEPH) families has revealed a sex-averaged reduced rate of recombination of 0.63 cM/Mb, with a particularly low rate of 0.15 cM/Mb for the class I region (Martin et al., 1995). Estimated recombination rates from studies of human sperm are 0.49 cM/Mb for the MHC, which is lower than both the male-specific estimate for the entire genome (0.92 cM/Mb) and a chromosome 6 specific estimate (0.71 cM/Mb) (Cullen et al., 2002). Recombination can occur across the MHC but some regions have higher levels of recombination than average and a large section from *HLA-C* to *HLA-A* has reduced levels of recombination. High-resolution SNP analysis of the class II region from *HLA-DMA* to *TAP2* has revealed the existence of three extended domains of linkage disequilibrium, with abrupt breakdown at *HLA-DMA*, *HLA-DMB*, and *TAP2*. Sperm crossover analysis shows that recombination clusters into narrow hot spots (1–2 kb), of which there could be several in a region of linkage disequilibrium breakdown, including three near *HLA-DMA* and two near *HLA-DMB* (Jeffreys et al., 2001).

Refinement of the PSORS1 Interval Within the MHC

Genome-wide scans offer an efficient low-resolution method of detecting susceptibility loci. However, the region identified is often large and not well defined, and so likely to include a large number of candidate genes. Following a genome-wide scan, it is therefore necessary to further refine the regions of interest that have been detected. Both *HLA-Cw6* and *HLA-B57* are consistently associated with psoriasis in studies from different populations. The corresponding genetic alleles of these two proteins are in linkage disequilibrium with each other in normal individuals and are part of an ancestral MHC haplotype designated *EH57.1*. Carriers of the *EH57.1* haplotype have a 26 times greater risk of developing psoriasis than individuals without *EH57.1*, and both *HLA-Cw6* and *HLA-B57* are more closely associated with psoriasis than the class II alleles (Jenisch et al., 1998). Therefore, though the entire MHC demonstrates strong association with psoriasis, stronger peaks of association are present with the class I region. This still represents a large gene dense region. Three studies have used high-resolution microsatellite analyses in an attempt to disentangle the extensive linkage disequilibrium observed in the class I region of the MHC in relation to the association seen with psoriasis (Balendran et al., 1999; Oka et al., 1999; Nair et al., 2000). The results for association and linkage are plotted in 32- 6A for all studies. Family-based reconstruction of haplotypes in 78 families allowed the comparison of relative frequencies of haplotypes between affected sibs, unaffected sibs, and unaffected spouse, leading to the determination of two haplotypes showing association with psoriasis, both of which contained *HLA-Cw0602* (Balendran et al., 1999). Nair studied microsatellites spaced throughout the MHC in 478 families and reconstructed maximum likelihood haplotypes (Nair et al., 2000). Six risk haplotypes demonstrated increased transmission to affected offspring and two core regions of similarity between the risk haplotypes were isolated: *RH1* from 30 to 90 kb telomeric of *HLA-C* and *RH2* from 100 to 210 kb telomeric of *HLA-C*. One haplotype that demonstrated association with psoriasis (designated haplotype 17) contained only *RH1* and not *RH2* and therefore the 60 kb *RH1* region was considered the minimal segment to harbor the susceptibility locus. However, the haplotype is rare (<1% haplotype frequency) and transmission disequilibrium test (TDT) analysis was of only borderline significance. To assess the critical refinement of haplotype 17, an international collaboration investigated the association of marker alleles specific to haplotype 17 in a large cohort. This showed no evidence for association between haplotype 17 and psoriasis, and therefore *RH1* was withdrawn as the minimal interval (Abecasis et al., 2005). Further to this, it has been proposed that the breakdown of haplotypes between the *RH1* and *RH2* regions and between *RH1* and *HLA-C* may be due to microsatellite mutation rather than double recombination as previously suggested. Microsatellite mutation is not likely to be an isolated event and this may be present in other studies of this kind, including the other two microsatellite studies referred to. Overall examination of reported data leads the conclusion that a susceptibility locus resides in the vicinity extending from *HLA-C* to telomeric of *CDSN* (Fig. 32–6).

This interval spanning *HLA-C* to *CDSN* has been studied with a high-density panel of SNPs (Veal et al., 2002). When transmission of these SNPs to affected offspring from parents is assessed, individually and in three marker haplotypes, the entire region is highly associated with psoriasis. However, there is a distinct peak of association with psoriasis 10 kb centromeric of *HLA-C*. Construction of full-length haplotypes reveals a set of very similar haplotypes that are associated with psoriasis (Fig. 32–6C). Only one other haplotype demonstrates marginal association and this contains portions centromeric and telomeric of the strongly associated haplotypes, but *HCR*-associated SNPs are not present on this haplotype. Two SNPs within the 10 kb region centromeric of *HLA-C* are exclusive to the associated haplotypes, as is a specific *CDSN* haplotype. The presence of this rare haplotypethat appears to have occurred by recombination, has led to the examination of all the haplotypes for similar recombination events, revealing that recombination has occurred in these haplotypes and opening the possibility of further dissection of the risk haplotypes.

Candidate Genes Within *PSORS1*

In the class I interval of the MHC, besides the HLA genes, four genes have been examined for association with psoriasis. Strong association with psoriasis has been described for *HCR*, *CDSN*, and *OTF3*. Two of these genes exhibit high levels of polymorphism (*CDSN* and *HCR*), whereas the other two exhibit little polymorphism content (*OTF3* and *SC1*), and the polymorphism demonstrating association in *OTF3* is silent. The high level of polymorphism within *HCR* and *CDSN* may be attributable to a hitchhiking effect with balancing selection on the *HLA-C* gene (Guillaudeux et al., 1998). *OTF3* and *SC1* may not tolerate the high level of variation and not show the

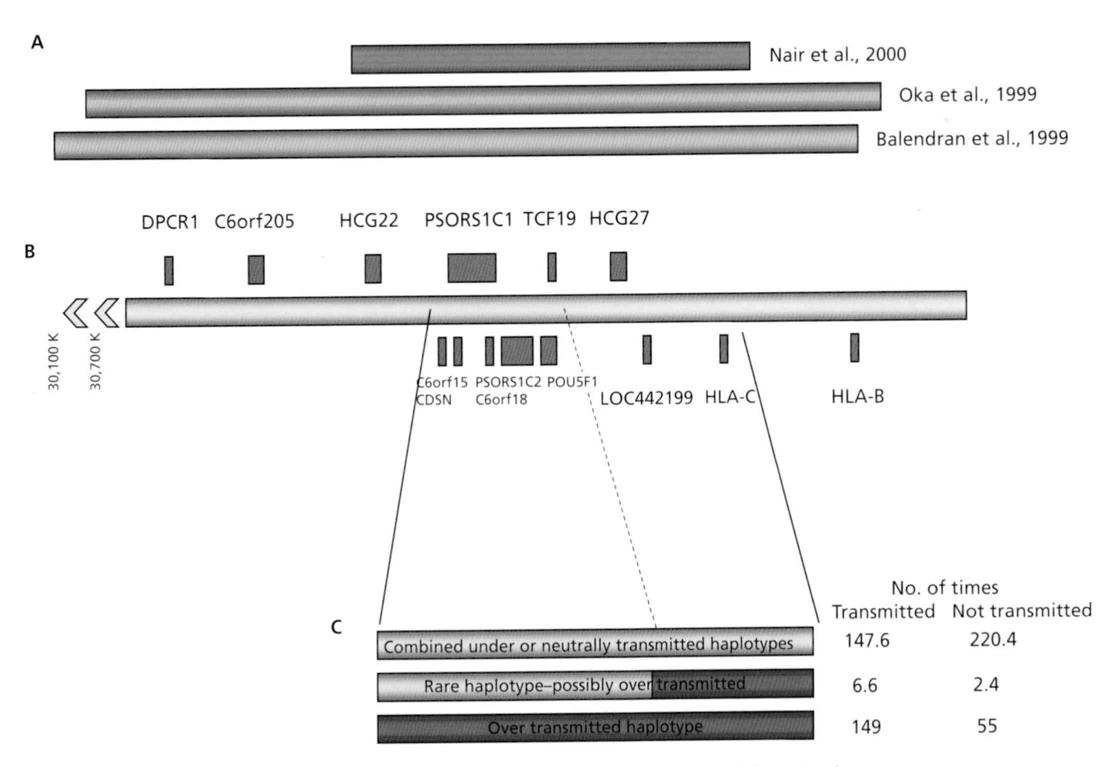

Figure 32–6 Refinement of the PSORS1 locus within the class I region of the major histocompatibility complex (MHC). (*A*) Approximate boundaries of refinement for PSORS1 from three major microsatellite studies. (*B*) Map of the class I region surrounding PSORS1. (*C*) Single nucleotide polymorphism (SNP)-based refinement of PSORS1. (Adopted from Veal et al., 2002.)

hitchhiking effect. There have been attempts to disentangle the strong associations with psoriasis for these genes, though these have led to contradictory reports of which gene demonstrates the most significant association, or is independent of the other genes. These studies have often examined individual polymorphisms against more informative haplotypes or used small sample size and there is currently an international collaborative systematic study of this interval.

Octamer-binding transcription factor 3 (*OTF3*) is expressed in pluripotent stem cells and plays a part in the early development of embryonic cells. A single *OTF3* polymorphism has strong association with psoriasis but, as the polymorphism is silent and the protein product has a low functional candidacy, the association between the polymorphism and psoriasis is probably a result of strong linkage disequilibrium between *OTF3* and a true susceptibility locus (Gonzalez et al., 2000).

CDSN is a gene whose product, corneodesmosin, is important in the tight junctions of late differentiating keratinocytes. It has a high level of polymorphism and one *CDSN* haplotype demonstrates highly significant association with psoriasis, thereby making it a positional and functional candidate in disease pathogenesis or progression. However, none of the observed polymorphisms detected so far indicate the presence of a mutation exclusive to the disease population. In fact, the associated alleles are of a high frequency in both affected and unaffected relatives. It thus remains possible that this gene is in linkage disequilibrium with the true susceptibility locus. With regard to the high level of polymorphism (>1 polymorphism per 100 bp) it is uncertain whether this is due to tolerance of polymorphism generated by mechanisms involved in HLA variation, or necessary for maintenance of function.

Strong association between alleles within *HCR* and psoriasis has been demonstrated in Europeans and Gujarati Indians (Asumalahti et al., 2000). However, the association is no more significant than that of alleles of the nearby genes, *HLA-C* and *CDSN*.

In several studies, the strongest association in this region is with *HLA*0602* (Veal et al., 2002). This gene is also a strong functional candidate as the gene product is involved in antigen presentation and also innate immunity.

The psoriasis-associated haplotype of alleles from *HLA-C*, *HCR*, and *CDSN* from Northern European patients is also present in Gujarati Indian and Japanese samples. This raises the possibility that the causal mutation/s for the major psoriasis susceptibility locus predate the divergence of the three populations.

Refinement of Susceptibility Loci Outside the MHC

PSORS2 Region on Chromosome 17q

Since the original report of linkage in a few large pedigrees 10 years ago, the *PSORS2* region has been further examined in a large number of small, nuclear families with some evidence for linkage close to a novel cluster of genes related to the immunoglobulin superfamily (Speckman et al., 2003). The strongest evidence for linkage was observed at markers *D17S1301* and *17ms39*, distinct from the linkage originally described at *D17S784*, raising the possibility that there are two independent loci on chromosome 17q. Recent SNP analysis across 200 kb from this interval has also revealed two regions of association, corresponding with the two regions demonstrating linkage (Helms et al., 2003). The first association corresponds to the original linkage peak and is located within the *RAPTOR* gene. *RAPTOR* is a ligand for mTOR (mammalian target of rapamycin), an important regulator of T-cell function, making it an interesting functional candidate gene. The association with *RAPTOR* SNPs has been replicated in an independent cohort of families from the UK (Capon et al., 2004).

The second region of association is 6 Mb telomeric to the first and contains two genes: *SLC9A3R1* and *NAPT9*. Sequence analysis of these two genes has not revealed a mutation. However, one associated SNP was observed to destroy a putative RUNX1 transcription factor binding site within region 1.2 kb 3′ of these two genes (Helms et al., 2003). There are other examples of loss of a runt related transcription factor 1 (RUNX1) binding site associated with susceptibility to autoimmune disease (systemic lupus erythematosus with *PDCD1* and rheumatoid arthritis with *SLC22A4*). *SLC9A3R1* is a scaffold protein that links plasma membrane proteins to the cytoskeleton. Its role in the T cell may influence the immunological synapse between antigen presenting cells and T cells, suggesting possible mechanisms for involvement in psoriasis. Most recently, a genome scan of patients with psoriasis from Taiwan has revealed strong evidence for a third locus on the distal end of chromosome 17q, distinct from the above two loci (Hwu et al., 2005).

PSORS3 Region on Chromosome 4q

The second locus identified by linkage analysis was the *PSORS3* locus on chromosome 4q (Matthews et al., 1996). However, this locus has not been replicated in other genome wide studies. Type I interferon has been suggested a plausible positional candidate gene at this locus but association studies with SNPs and microsatellite markers have been inconclusive.

PSORS4 Region on Chromosome 1q

The *PSORS4* region was first identified in a genome scan of large Italian families that gave no evidence of linkage to *PSORS1* and *PSORS2* (Capon et al., 1999). However, when the 1q-linked families were studied alone, there was evidence of association with *HLA-C*, suggesting epistasis between *PSORS4* and *PSORS1*. *PSORS4* contains the epidermal differentiation complex, a cluster of 20 or more genes expressed during epithelial differentiation. The critical region of *PSORS4* has subsequently been refined to a 100 kb interval containing the gene encoding loricrin, which is reduced in expression in lesional psoriatic skin in linked families (Giardina et al., 2004). Another cluster of genes in this interval are the chemoattractant S100 proteins including psoriasin (S100A7), which is markedly overexpressed in psoriatic lesions. In addition, two other members of this group (S100A8 and S100A9) have been shown as necessary to the development of the phenotype in an interesting recent mouse model in which inducible epidermal deletion of JunB and its functional companion c-Jun results in psoriasis-like skin and joint disease (Zenz et al., 2005).

PSORS5 Region on Chromosome 3q

The *PSORS5* locus was originally described in Sweden from a cohort of small nuclear families with psoriasis, in which there was also evidence of linkage to *PSORS1* and *PSORS2* (Samuelsson et al., 1999). Linkage was seen to chromosome 3q only in the group without joint symptoms, when the cohort was stratified in this way. In a subsequent study in which the families were stratified on the basis of their parental origin, linkage was strongest in a group that originated from a region in southwest Sweden, suggesting a founder effect (Enlund et al., 1999). In fine mapping studies, the linked region has been refined to a 160 kb interval that includes a solute carrier *SLC12A8* (Hewett et al., 2002). This association has recently been independently replicated in a German cohort (Huffmeier et al.,

2005). Two independent association signals are seen in extended haplotype analysis, suggesting allelic heterogeneity.

PSORS6 Region on Chromosome 19p

Linkage to markers on 19p13 has been reported in a genome scan of German extended pedigrees, which were also linked to *PSORS1* (Lee et al., 2000). This has been replicated by association analysis in small nuclear families from Northern Germany, with some evidence for both susceptibility and protective loci in this region (Hensen et al., 2003). The *PSORS6* region includes the gene encoding JunB that has been implicated in the mouse model of psoriasis and PsA referred to above.

PSORS7 Region on Chromosome 1p

Linkage to this locus was reported in a genome scan of small nuclear families from the UK (Veal et al., 2001). These families also demonstrated strong evidence of linkage to *PSORS1* with evidence of a parent of origin effect—linkage was strongest when the disease was inherited from the father. In addition, linkage to *PSORS7* was seen principally in those families that were linked to *PSORS1*, suggesting epistasis between *PSORS7* and *PSORS1*.

PSORS9 Region on Chromosome 4q

In one of the few genome scans of psoriasis to study races other than Caucasians, Han families from East and Southeast China confirmed linkage to *PSORS1* and also revealed a novel locus on chromosome 4q, proximal to and distinct from *PSORS3* (Zhang et al., 2002).

Replication of Linkage

Replication of susceptibility loci often requires substantially larger cohorts than for that in which the original locus was identified. A recent international collaboration examined 14 previously reported candidate regions for psoriasis susceptibility in a large psoriasis cohort, comprising of 710 families (Allen et al., 2003). The analysis demonstrated that the major psoriasis susceptibility locus (*PSORS1*) exhibited the most significant evidence for linkage to psoriasis. The only other loci that generated significant values in favor of linkage in the combined cohort were those on chromosome 16 and chromosome 17 (*PSORS2*). However, marked heterogeneity was observed between the independent collections comprising this cohort and parametric analysis revealed that different loci have different modes of inheritance. This suggests that the majority of reported susceptibility loci do not have a large effect in the general psoriasis population and may be restricted to specific geographic populations or particular family structures, such as unusually large psoriasis pedigrees. Only *PSORS1* demonstrates a universal presence.

Relationships Between Susceptibility Loci

Evolutionary paralogy has been described between the regions on chromosome 6p and 19p, this paralogy also extending to the *PSORS4* psoriasis susceptibility locus on chromosome 1q. Paralogy between the MHC on chromosome 6 and the other regions suggests an immune relationship between the loci and psoriasis. However, the *PSORS4* locus also harbors the epidermal differentiation complex, containing gene families with skin specific functions. Loricrin, a structural protein in the cornified cell envelope of the epidermis, encoded within the epidermal differentiation complex, is homologous to *CDSN*, in the class Iinterval of the *MHC*, a cell adhesion molecule found in the corneodesmosomes of late differentiated keratinocytes. This suggests the possibility that the link between the three paralogous regions on chromosomes 1q, 6p, and 19p may not be immune related, but related to genes with skin restricted expression.

Overlap of NonMHC Susceptibility Loci with Other Complex Disorders

Heterogeneity is considered a hallmark of diseases with complex inheritance. It clearly would be of interest if there were any overlap between putative psoriasis susceptibility loci and other immune and/or skin-related common complex disorders such as asthma and eczema, reflecting a restricted range of human common disease susceptibility alleles. Becker described clustering of loci for five immune-related diseases; multiple sclerosis, Crohn's disease, psoriasis, asthma, and type-I diabetes (IDDM; insulin-dependent diabetes mellitus) (Becker et al., 1998). Comparison of these data with the psoriasis susceptibility loci, reveals distinct overlap of putative susceptibility regions; multiple sclerosis is seen to have reported susceptibility loci overlapping with psoriasis susceptibility loci on chromosomes 1p, 4q, 7p, and 17q (Sawcer et al., 1996), susceptibility loci for Crohn's disease overlap with psoriasis regions on chromosomes 1p, 7p, and 16q (Hugot et al., 1996), asthma loci are overlapping with regions for psoriasis on chromosomes 4q, 7p, and 16q (Daniels et al., 1996), and IDDM loci overlap with psoriasis loci on chromosomes 7p, 16q, and 17q (Davies et al., 1994). Furthermore, there is distinctive overlap between loci for eczema and psoriasis on chromosomes 1q21, 17q25, 20p, and 3q21 despite the diseases being clinically and pathologically distinct (Bowcock and Cookson, 2004).

Expression Studies

Microarray expression studies provide a powerful mechanism to examine the transcriptional changes that occur during an outbreak of psoriasis. Three major microarray studies have been reported to date. Bowcock used an Affymetrix array of 12,000 genes to identify 177 genes that were differentially expressed (161 up-regulated, 16 down-regulated) (Bowcock et al., 2001). These genes reflected the proliferative and inflammatory characteristics of psoriatic skin. Samples taken from lesional skin were found to show similar patterns of changed expression, whereas nonlesional skin was similar to normal controls. It was also noted that a number of these genes were localized to regions demonstrating linkage to psoriasis. This work was expanded to 63,000 element Affymetrix arrays and a total of 1338 genes were found to be differentially expressed (Zhou et al., 2003). These genes included those involved in epidermal differentiation, immune regulation, and cell proliferation. Itoh examined 2007 genes that were expressed in epithelial tissues and noted 241 genes with altered expression in psoriasis (Itoh et al., 2005). Of further interest, seven genes that were highly expressed in uninvolved skin of patients, were related to apoptosis or were antiproliferative. These studies highlight potential pathways important in the pathogenesis of psoriasis and provide potential candidates in regions of the genome that are genetically linked to psoriasis.

Genotype–Phenotype Correlation

Nearly all the genetic analyses of psoriasis described above have been restricted to patients with chronic plaque disease although the *PSORS1* locus has been studied in several clinical variants of the disease. In an association analysis of 134 patients with guttate psoriasis, the disease was shown to be genetically indistinguishable from plaque psoriasis at *PSORS1* and the association with risk alleles at *HLA-C*, *CDSN*, and *HCR* was even stronger than for plaque psoriasis (Asumalahti et al., 2003). In contrast to guttate psoriasis, which is generally of early age at onset, plaque psoriasis that starts at the age of 50 years or above (late-onset psoriasis, or "type 2 psoriasis") has been shown to lack any significant association with risk alleles at *PSORS1* (Allen et al., 2005). Similarly palmoplantar pustulosis (also usually of late age at onset) has been shown to have no association with psoriasis risk alleles at *PSORS1* and would thus imply that susceptibility loci other than *PSORS1* may be implicated in phenotypic variation (Asumalahti et al., 2003). No genetic studies of generalized pustular psoriasis have been reported. In a genome scan of 238 Icelandic families, the only significant linkage in the entire cohort was to *PSORS1*. However when families were divided according to age at onset, the adult onset group (>17 years of age at onset) demonstrated linkage near *PSORS3*, and stratification for site of onset suggested linkage to chromosome 10q in the group with onset in the scalp (Karason et al., 2005). It is possible that whereas *PSORS1* is the major determinant of psoriasis susceptibility, other loci determine the presentation of the disease.

Studies of the genetics of PsA are comparatively sparse. What little data are available suggest a strong genetic predisposition (Moll and Wright, 1973). It is associated with class I HLA alleles (B27, B38, and B39) but the association with HLA-Cw6 is less marked than in psoriasis. In linkage studies in psoriasis, when families are segregated on the basis of additional PsA, evidence for linkage to *PSORS1* is greatly reduced (Burden et al., 1998). In addition, stratification for PsA has yielded evidence for linkage to *PSORS2* in the group with PsA, but not in the entire cohort (Enlund et al., 1999). The pedigrees in which *PSORS2* was first identified also contained PsA, raising the possibility that this is primarily a PsA locus (Tomfohrde et al., 1994).

Only one genome scan has been reported in PsA per se (Karason et al., 2003). In 39 Icelandic PsA families, linkage to a locus on chromosome 16q was observed when the analysis was conditioned on paternal transmission to affected individuals. The locus, which has previously been implicated in psoriasis, may be involved in paternal transmission of PsA. This finding is supported by the observation of an association between allelic variants of caspase recruitment domain 15 (CARD15) at this locus and PsA (Rahman et al., 2003). It is however noteworthy that no association has been identified between CARD15 variants and psoriasis (Young et al., 2003).

Pharmacogenetics

The characterization of susceptibility and severity genes, combined with the associated clinical phenotypes, will facilitate rational drug design and may enable treatment selection for individual patients. Currently, this area is poorly developed in psoriasis with few published studies. In one early report, the effect of polymorphisms in the apolipoprotein E (ApoE) gene on response of psoriasis to treatment with the retinoid acitretin was investigated with negative results (Campalani et al., 2004). However, the major driver to developments in this area is the development of targeted immunobiological treatment of psoriasis. It seems probable that cytokine gene polymorphisms are associated with disease severity and possibly also treatment responsiveness. For instance, tumour necrosis factor α (TNFα) is a key cytokine in the development of psoriasis and PsA. Functional allelic variants of this cytokine have been associated with both psoriasis and PsA, and shown to contribute to disease severity (Mossner et al., 2005). Blockade of the activity of TNFα is one of the most effective treatments for psoriasis. It remains to be seen whether individual response to treatment can be predicted on the basis of TNFα allelic variation.

Conclusions

Psoriasis and PsA are complex genetic diseases with a strong genetic component. The dominant genetic determinant of cutaneous psoriasis is *PSORS1* on chromosome 6, within the MHC region. This locus is recognized in all populations of patients with plaque psoriasis studied and is particularly implicated in disease of early onset. International collaborative efforts to identify the gene(s) at this locus are ongoing. Other loci may be implicated in other psoriasis phenotypes such as PsA, or may be rare genes of high penetrance responsible for large psoriasis pedigrees.

Over the past decade, substantial progress has been made in dissecting the molecular and cellular pathways of inflammation that mediate the psoriasis phenotype. Immunobiological T-cell-targeted and cytokine therapy have proved effective and over the past 2 years have entered routine clinical care. It is hoped that as the genetic basis of the disease is further unravelled, understanding the genetic heterogeneity of the disease will provide a more useful clinical classification that correlates with prognosis and treatment selection.

References

Abecasis, G, Allen, M, Barker, JNWN, Burden, D, Capon, F, Christophers, E, et al. (2005). Fine mapping of the psoriasis susceptibility gene PSORS1: a reassessment of risk associated with a putative risk haplotype lacking HLA Cw6. *J Invest Dermatol*, 124:921–930.

Allen, M, Barker, JNWN, Bowcock, AM, Burden, AD, Chia, N, Capon, F, et al. (2003). The international psoriasis genetics study: assessing linkage to 14 disease susceptibility loci in a 942 affected sib-pair cohort. *Am J Hum Genet*, 73:430–437.

Allen, MH, Ameen, M, Veal, C, Evans, J, Ramrakha-Jones, VS, Marsland, AM, et al. (2005). The major psoriasis susceptibility locus PSORS1 is not a risk factor for late onset psoriasis. *J Invest Dermatol*, 124:103–107.

Asahina, A, Akazaki, S, Nakagawa, H, Kuwata, S, Tokunaga, K, Ishibashi, Y, et al. (1991). Specific nucleotide sequence of HLA-C is strongly associated with psoriasis vulgaris. *J Invest Dermatol*, 97:254–258.

Asumalahti, K, Ameen, M, Suomela, S, Hagforsen, E, Michaëlsson, G, Evans, J, et al. (2003). Genetic analysis of PSORS1 distinguishes guttate psoriasis and palmoplantar pustulosis. *J Invest Dermatol*, 120(4):627–632.

Asumalahti, K, Laitinen, T, Itkonen-Vatjus, R, Lokki, ML, Suomela, S, Snellman, E, et al. (2000). A candidate gene for psoriasis near HLA-C, HCR (Pg8), is highly polymorphic with a disease associated susceptibility allele. *Hum Mol Genet*, 9:1533–1542.

Balendran, N, Clough, RL, Arguello, JR, Barber, R, Veal, C, Jones, AB, et al. (1999). Characterization of the major susceptibility region for psoriasis at chromosome 6p21.3. *J Invest Dermatol*, 113:322–328.

Beck, S and Trowsdale, J (2000). The human major histocompatability complex: lessons from the DNA sequence. *Annu Rev Genomics Hum Genet*, 1:117–137.

Becker, KG, Simon, RM, Bailey-Wilson, JE, Freidlin, B, Biddison, WE, McFarland, HF, et al. (1998). Clustering of non-major histocompatability complex susceptibility candidate loci in human autoimmune diseases. *Proc Natl Acad Sci USA*, 95:9979–9984.

Bowcock, AM and Cookson, WO (2004). The genetics of psoriasis, psoriatic arthritis and atopic dermatitis. *Hum Mol Genet*, 13(Spec. No 1):R43–R55.

Bowcock, AM, Shannon, W, Du, F, Duncan, J, Cao, K, Aftergut, K, et al. (2001). Insights into psoriasis and other inflammatory diseases from large-scale gene expression studies. *Hum Mol Genet*, 10:1793–1805.

Boyson, JE, Shufflebotham, C, Cadavid, LF, Urvater, JA, Knapp, LA, Hughes, AL, et al. (1996). The MHC class I genes of the rhesus monkey. Different evolutionary histories of MHC class I and II genes in primates. *J Immunol*, 156:4656–4665.

Brandrup, F, Hauge, M, Henningsen, K, and Eriksen, B (1978). Psoriasis in an unselected series of twins. *Arch Dermatol*, 114:874–878.

Burden, AD, Javed, S, Bailey, M, Hodgins, M, Connor, M., and Tillman, D (1998). Genetics of psoriasis: paternal inheritance and a locus on chromosome 6p. *J Invest Dermatol*, 110:958–960.

Campalani, E, Allen, M, Fairhurst, D, Young, H, Burden, D, Griffiths, C, et al. (2004). Apolipoprotein E polymorphisms and psoriasis. *Br J Dermatol,* 151:12.

Capon, F, Helms, C, Veal, CD, Tillman, D, Burden, AD, Barker, JN, et al. (2004). Genetic analysis of PSORS2 markers in a UK dataset supports the association between RAPTOR SNPs and familial psoriasis. *J Med Genet,* 41:459–460.

Capon, F, Novelli, G, Semprini, M, Clementi, M, Nudo, M, Vultaggio, P, et al. (1999). Searching for psoriasis susceptibility genes in Italy: genome scan and evidence for a new locus on chromosome 1. *J Invest Dermatol,* 112:32–35.

Cullen, M, Perfetto, SP, Klitz, W, Nelson, G, and Carrington, M (2002). High-resolution patterns of meiotic recombination across the human major histocompatibility complex. *Am J Hum Genet,* 71:759–776.

Daniels, SE, Bhattacharrya, S, James, A, Leaves, NI, Young, A, Hill, MR, et al. (1996). A genome-wide search for quantitative trait loci underlying asthma. *Nature,* 383:247–250.

Davies, JL, Kawaguchi, Y, Bennett, ST, Copeman, JB, Cordell, HJ, Pritchard, LE, et al. (1994). A genome-wide search for human type 1 diabetes susceptibility genes. *Nature,* 371:130–136.

Elder, JT, Nair, RP, Guo, SW, Henseler, T, Christophers, E, and Voorhees, JJ (1994). The genetics of psoriasis. *Arch Dermatol,* 130:216–224.

Enlund, F, Samuelsson, L, Enerbäck, C, Inerot, A, Wahlström, J, Yhr, M, et al. (1999). Psoriasis susceptibility locus in chromosome region 3q21 identified in patients from southwest Sweden. *Eur J Hum Genet,* 7:783–790.

Farber, EM and Nall, L (1991). Epidemiology: natural history and genetics. In HH Roenigk and HI Maibach (eds), *Psoriasis* (pp. 209–258). New York: Marcel Decker Inc.

Farber, EM, Nall, ML, and Watson, W (1974). Natural history of psoriasis in 61 twin pairs. *Arch Dermatol,* 109:207–211.

Giardina, E, Capon, F, de Rosa, MC, Mango, R, Zambruno, G, Orecchia, A, et al. (2004). Characterization of the loricrin (LOR) gene as a positional candidate for the PSORS4 psoriasis susceptibility locus. *Ann Hum Genet,* 68:639–645.

Gonzalez, S, Martinez-Borra, J, Del Rio, JS, Santos-Juanes, J, Lopez-Vazquez, A, Blanco-Gelaz, M, et al. (2000). The OTF3 gene polymorphism confers susceptibility to psoriasis independent of the association of HLA-Cw*0602. *J Invest Dermatol,* 115:824–828.

Guillaudeux, T, Janer, M, Wong, GKS, Spies, T, and Geraghty, DE (1998). The complete genomic sequence of 424,015 bp at the centromeric end of the HLA class I region: gene content and polymorphism. *Proc Natl Acad Sci USA,* 95:9494–9499.

Hellgren, L (1967). *Psoriasis: The Prevalence in Sex, Age and Occupational Groups in Total Populations in Sweden. Morphology, Inheritance and Association With Other Skin and Rheumatic Diseases.* (Stockholm: Almquist & Wiksell).

Helms, C, Krueger, JG, Wijsman, EM, Chamian, F, Gordon, D, Heffernan, M, et al. (2003). A putative RUNX1 binding site variant between SLC9A3R1 and NAT9 is associated with susceptibility to psoriasis vulgaris. *Nat Genet,* AOP 1–8.

Henseler, T and Christophers, E (1985). Psoriasis of early and late onset: characterization of two types of psoriasis vulgaris. *J Am Acad Dermatol,* 13:450–456.

Hensen, P, Windemuth, C, Huffmeier, U, Ruschendorf, F, Stadelmann, A, Hoppe, V, et al. (2003). Association scan of the novel psoriasis susceptibility region on chromosome 19: evidence for both susceptible and protective loci. *Exp Dermatol,* 12:490–496.

Hewett, D, Samuelsson, L, Polding, J, Enlund, F, Smart, D, Cantone, K, et al. (2002). Identification of a psoriasis susceptibility candidate gene by linkage disequilibrium mapping with a localised single nucleotide polymorphism map. *Genomics,* 79:305–314.

Huffmeier, U, Lascorz, J, Traupe, H, Bohm, B, Schurmeier-Horst, F, Stander, M, et al. (2005). Systematic linkage disequilibrium analysis of SLC12A8 to PSORS5 confirms a role in susceptibility to psoriasis vulgaris. *J Invest Dermatol,* 125:906–910.

Hughes, AL and Nei, M (1990). Evolutionary relationships of class II major-histocompatibility-complex genes in mammals. *Mol Biol Evol,* 7:491–514.

Hugot, JP, Laurent-Puig, P, Gower-Rousseau, C, Olson, JM, Lee, JC, Beaugerie, L, et al. (1996). Mapping of susceptibility locus for Crohn's disease on chromosome 16. *Nature,* 379:821–823.

Hwu, WL, Yang, CF, Fann, CSJ, Chen, CL, Tsai, TF, Chien, YH, et al. (2005). Mapping of psoriasis to 17q terminus. *J Med Genet,* 42:152–158.

Itoh, K, Kawasaki, S, Kawamoto, S, Seishima, M, Chiba, H, Michibata, H, et al. (2005). Identification of differentially expressed genes in psoriasis using expression profiling approaches. *Exp Dermatol,* 14:667–674.

Jeffreys, AJ, Kauppi, L, and Neumann, R (2001). Intensely punctate meiotic recombination in the class II region of the major histocompatibility complex. *Nat Genet,* 29:217–222.

Jenisch, S, Henseler, T, Nair, RP, Guo, SW, Westphal, E, Stuart, P, et al. (1998). Linkage analysis of human leukocyte antigen (HLA) markers in familial psoriasis: strong disequilibrium effects provide evidence for a major determinant in the HLA-B/-C region. *Am J Hum Genet,* 63:191–199.

Karason, A, Gudjonsson, JE, Johann, E, Jonsson, HH, Hauksson, VB, Runarsdottir, EH, et al. (2005). Genetics of psoriasis in Iceland: evidence for linkage of subphenotypes to distinct loci. *J Invest Dermatol,* 124:1177–1185.

Karason, A, Gudjonsson, JE, Upmanyu, R, Antonsdottir, A, Hauksson, VB, Runasdottir, EH, et al. (2003). A susceptibility gene for psoriatic arthritis maps to chromosome 16q: evidence for imprinting. *Am J Hum Genet,* 72:125–131.

Kasahara, M (1999). Genome dynamics of the major histocompatibility complex: insights from genome paralogy. *Immunogenetics,* 50:134–145.

Lee, YA, Rüschendorf, F, Windemuth, C, Schmitt-Egenolf, M, Stadelmann, A, Nürnberg, G, et al. (2000). Genomewide scan in German families reveals evidence for a novel psoriasis-susceptibility locus on chromosome 19q13. *Am J Hum Genet,* 67:1020–1024.

Lomholt, G (1963). *Psoriasis: Prevalence, Spontaneous Course and Genetics.* (Copenhagen: G.E.C GAD).

Mallon, E, Bunce, M, Wojnarowska, F, and Welsh, K (1997). HLA-CW*0602 is a susceptibility factor in type I psoriasis, and evidence Ala-73 is increased in male type I psoriatics. *J Invest Dermatol,* 109:183–186.

Martin, M, Mann, D, and Carrington, M (1995). Recombination rates across the HLA complex: use of microsatellites as a rapid screen for recombinant chromosomes. *Hum Mol Genet,* 4:423–428.

Matsunaga, T and Rahman, A (1998). What brought the adaptive immune system to vertebrates?—The jaw hypothesis and the seahorse. *Immunol Rev,* 166:177–186.

Matthews, D, Fry, L, Powles, A, Weber, J, McCarthy, M, Fisher, E, et al. (1996). Evidence that a locus for familial psoriasis maps to chromosome 4q. *Nat Genet,* 14:231–233.

Moll, JM and Wright, V (1973). Familial occurrence of psoriatic arthritis. *Ann Rheum Dis,* 32:181–201.

Mossner, R, Kingo, K, Kleensang, A, Kruger, U, Konig, IR, Silm, H, et al. (2005). Association of TNF -238 and -308 promotor polymorphisms with psoriasis vulgaris and psoriatic arthritis but not with pustulosis palmoplantaris. *J Invest Dermatol,* 124:282–284.

Nair, RP, Stuart, P, Henseler, T, Jenisch, S, Chia, NV, Westphal, E, et al. (2000). Localization of psoriasis-susceptibility locus PSORS1 to a 60-kb interval telomeric to HLA-C. *Am J Hum Genet,* 66:1833–1844.

Nonaka, M, Namikawa, C, Kato, Y, Sasaki, M, Salter-Cid, L, and Flajnik, MF (1997). Major histocompatibility complex gene mapping in the amphibian Xenopus implies a primordial organization. *Proc Natl Acad Sci USA,* 94:5789–5791.

Ohno, S (1987). The ancestor of the adaptive immune system was the CAM system for organogenesis. *Exp Clin Immunogenet,* 4:181–192.

Oka, A, Tamiya, G, Tomizawa, M, Ota, M, Katsuyama, Y, Makino, S, et al. (1999). Association analysis using refined microsatellite markers localizes a susceptibility locus for psoriasis vulgaris within a 111 kb segment telomeric to the HLA-C gene. *Hum Mol Genet,* 8:2165–2170.

Parham, P, Adams, EJ, and Arnett, KL (1995). The origins of HLA-A,B,C polymorphism. *Immunol Rev,* 143:141–180.

Potts, WK and Slev, PR (1995). Pathogen-based models favoring MHC genetic diversity. *Immunol Rev,* 143:181–197.

Rahman, P, Bartlett, S, Siannis, F, Pellett, FJ, Farewell, VT, Peddle, L, et al. (2003). CARD15: a pleiotropic autoimmune gene that confers susceptibility to psoriatic arthritis. *Am J Hum Genet,* 73:677–681.

Risch, N (1990a). Linkage strategies for genetically complex traits. I. Multilocus models. *Am J Hum Genet,* 46:222–228.

Risch, N (1990b). Linkage strategies for genetically complex traits. II. The power of affected relative pairs. *Am J Hum Genet*, 46:229–241.

Ritte, U, Neufeld, E, O'hUigin, C, Figueroa, F, and Klein, J (1991). Origins of H-2 polymorphism in the house mouse. II. Characterization of a model population and evidence for heterozygous advantage. *Immunogenetics*, 34:164–173.

Russell, TJ, Schultes, LM, and Kuban, DJ (1972). Histocompatability (HL-A) antigens associated with psoriasis. *N Engl J Med*, 287:738–740.

Samuelsson, L, Enlund, F, Torinsson, A, Yhr, M, Inerot, A, Enerback, C, et al. (1999). A genome-wide search for genes predisposing to familial psoriasis by using a stratification approach. *Hum Genet*, 105:523–529.

Sanchez-Mazas, A, Djoulah, S, Busson, M, Le Monnier de Gouville, I, Poirier, JC, Dehay, C, et al. (2000). A linkage disequilibrium map of the MHC region based on the analysis of 14 loci haplotypes in 50 French families. *Eur J Hum Genet*, 8:33–41.

Sawcer, S, Jones, HB, Feakes, R, Gray, J, Smaldon, N, Chataway, J, et al. (1996). A genome screen in multiple sclerosis reveals susceptibility loci on chromosome 6p21 and 17q22. *Nat Genet*, 13:464–468.

Speckman, RA, Wright Daw, JA, Helms, C, Duan, S, Cao, L, Taillon-Miller, P, et al. (2003). Novel immunoglobulin superfamily gene cluster, mapping to a region of human chromosome 17q25, linked to psoriasis susceptibility. *Hum Genet*, 112:34–41.

The MHC Sequencing Consortium (1999). Complete sequence and gene map of a human major histocompatibility complex. *Nature*, 401:921–923.

Tomfohrde, J, Silverman, A, Barnes, R, Fernandez, V, Young, M, Lory, D, et al. (1994). Gene for familial psoriasis susceptibility mapped to the distal end of human chromosome 17q. *Science*, 264:1141–1145.

Trembath, RC, Clough, RL, Rosbotham, JL, Jones, AB, Camp, RD, Frodsham, A, et al. (1997). Identification of a major susceptibility locus on chromosome 6p and evidence for further disease loci revealed by a two stage genome-wide search in psoriasis. *Hum Mol Genet*, 6:813–820.

Veal, CD, Capon, F, Allen, MH, Heath, EK, Evans, JC, Jones, A, et al. (2002). Family-based analysis using a dense single-nucleotide polymorphism-based map defines genetic variation at PSORS1, the major psoriasis-susceptibility locus. *Am J Hum Genet*, 71:554–564.

Veal, CD, Clough, RL, Barber, RC, Mason, S, Tillman, D, Ferry, B, et al. (2001). Identification of a novel psoriasis susceptibility locus at 1p and evidence of epistasis between PSORS1 and candidate loci. *J Med Genet*, 38(1):7–13.

Young, C, Allen, MH, Cuthbert, A, Ameen, M, Veal, C, Leman, J, et al. (2003). Crohn's disease associated insertion polymorphism (3020insC) in the NOD2 gene is not associated with psoriasis vulgaris, palmo-plantar pustulosis (PPP) or guttate psoriasis. *Exp Dermatol*, 12:506–509.

Zenz, R, Eferl, R, Kenner, L, Florin, L, Hummerich, L, Mehic, D, et al. (2005). Psoriasis-like skin disease and arthritis caused by inducible epidermal deletion of Jun proteins. *Nature*, 437:369–375.

Zhang, XJ, He, PP, Wang, ZX, Zhang, J, Li, YB, Wang, HY, et al. (2002). Evidence for a major psoriasis susceptibility locus at 6p21 (PSORS1) and a novel candidate region at 4q31 by genome-wide scan in Chinese Hans. *J Invest Dermatol*, 119:1361–1366.

Zhou, X, Krueger, JG, Kao, MC, Lee, E, Du, F, Menter, A, et al. (2003). Novel mechanisms of T-cell and dendritic cell activation revealed by profiling of psoriasis on the 63,100-element oligonucleotide array. *Physiol Genomics*, 13:69–78.

33

Complex Skin Diseases II: Atopic Dermatitis

Nilesh Morar

Atopic dermatitis (AD) or atopic eczema is a chronic, itchy, inflammatory skin condition that predominantly affects the skin flexures. Approximately 70% of cases start in children younger than 5 years of age (Williams, 2000). The prevalence has increased three-fold over the past three decades in developed countries, with a higher prevalence in urban regions and in higher social classes (Taylor et al., 1984). The lifetime prevalence in children is 10%–20% and in adults 1%–3% (Schultz-Larsen, 2002). Though 60% of children with AD are free of symptoms in adolescence (Rystedt, 1985), up to 50% may have recurrences in adulthood (Lammintausta, 1991). Asthma develops in 30% of children with AD and allergic rhinitis in 35% (Luoma et al., 1983).

Diagnostic Criteria

There has been substantial variation in disease definition and diagnosis due to the variable clinical appearance, distribution patterns, and the intermittent nature of the disease. On the basis of the original consensus criteria by Hanifin and Rajka (1980), a set of minimum and validated discriminators have been determined by the UK Working Party (Williams et al., 1994). The diagnosis requires evidence of itchy skin plus three or more of the following: history of involvement of the skin creases, history of asthma or hay fever, history of generally dry skin, onset in a child under 2 years of age, and visible flexural dermatitis. These robust criteria are crucial for effective and reproducible genetic studies.

AD and Atopy

AD, asthma, and hay fever are considered to be part of a common syndrome of atopic diseases. The word atopy means "strange disease" and was coined in 1923 by Coca (1923). The atopic state is recognized by positive skin prick tests to common allergens, by the presence of allergen-specific immunoglobulin E (IgE) in serum and by elevations of total serum IgE. Eighty percent of infants with AD have raised total serum IgE levels (Juhlin et al., 1969). Two types of AD have been described: an extrinsic type associated with IgE-mediated sensitization and an intrinsic type, which is clinically identical, with normal total serum IgE and no specific IgE responses to common allergens (Johansson et al., 2001).

The association between AD and atopy is not clear-cut as both forms of AD have the same phenotype clinically. There is some suggestion from hospital studies that higher IgE levels are correlated with severe forms of AD, worse long-term prognosis (Flohr et al., 2004), and an increased likelihood of developing asthma. There are immunological differences between these two types of AD that lie in different expression levels in skin related cytokines that determine T-cell activation patterns (Jeong et al., 2003). Inflammatory dendritic epidermal cells bear the high affinity receptor for IgE (FcεRI) on their cell surface (Novak et al., 2003). This allows allergens penetrating the impaired skin barrier in AD to be focussed and efficiently presented by FcεRI-bound IgE to T cells (Novak and Bieber, 2003; Novak et al., 2003). In intrinsic AD, there is lower surface expression of FcεRI on epidermal dendritic cells (Oppel et al., 2000). Interleukin 13 (IL13) induces the production of IgE by B-cells (Punnonen et al., 1993). IL13 is decreased in intrinsic AD (Punnonen et al., 1993; Jeong et al., 2003). Intrinsic and extrinsic forms of AD may, however, be part of a continuum in the natural history of AD (Novembre et al., 2001). The fact that phenotypically both are identical suggests that IgE sensitization is not a prerequisite for developing the skin lesions of AD. Omalizumab, a monoclonal anti-IgE antibody currently approved for the treatment of asthma, failed to improve AD in all patients that were treated (Krathen et al., 2005; Lane et al., 2006). The diagnostic criteria of AD can also be fulfilled in the absence of elevated IgE (Hanifin, 1980; Williams et al., 1994). Subgroup analysis in future genetic studies may help dissect these differences better. Therefore though atopic mechanisms dominate our understanding of the pathogenesis of AD, other mechanisms may predispose to AD independently of atopy.

Genetic Studies of AD

In genetic-epidemiologic and population based twin studies of AD, the concordance rates for monozygotic twins was 0.72–0.86 and 0.21–0.23 for dizygotic twins (Larsen and Holm, 1986; Schultz Larsen, 1993). Asthma shows similar concordance rates of 0.65 in monozygotic twins and 0.25 in dizygotic twins (Duffy et al., 1990). Total serum IgE shows heritability of approximately 47.3% (Palmer et al., 2000). Strong genetic factors therefore underlie the development of AD and atopic disorders. However, inheritance of AD is complex and does not follow Mendel's laws of inheritance.

A number of genes may be contributory. In rare Mendelian diseases, polymorphisms commonly cause mutations by altering protein coding sequences. In common diseases such as AD gene function may be altered by more subtle mechanisms. These may affect initiation of gene transcription in exons or factors that affect gene expression and splicing in introns. The genes causing complex disease are likely to show high-frequency allelic variation (frequency of >1%) that is the basis of the "common disease–common variant" hypothesis (Cargill et al., 1999; Halushka et al., 1999; Cheung and Spielman, 2002). There is complex interplay between environmental factors and the alleles of many genes and interactions between genes themselves (epistasis) that plays an important role in determining disease expression.

Two traditional approaches are used to identify genes: case–control comparisons of polymorphisms in known candidate genes or the identification of new genes by positional cloning. Recently, with the sequencing of the human genome, genome-wide gene expression profiling with deoxyribonucleic acid (DNA) microarrays allows global analyses of the expression of thousands of genes, and is a high-throughput method for gene identification.

Genomic approaches to understanding AD are new, and significant inroads have been made only in the past 5 years. To date, the following observations have been made in the genetics of AD that have given us new insights into the pathogenesis of AD.

Positional Cloning Studies

In positional cloning, families are studied to identify chromosomal regions that are co-inherited or linked with disease. A "genome screen" is a systematic search for such regions using polymorphic microsatellite markers. The markers or microsatellites typically contain repeat sequences of DNA, such as CA repeats that vary between individuals. In linkage analyses, regions of the genome are searched for a higher than expected number of shared alleles among affected individuals within a family (Carlson et al., 2004). Genetic linkage is said to exist if the marker and the phenotype are co-inherited. Genetic linkage identifies broad chromosomal regions linked to disease normally exceeding 10 cM (~10 million bases). This area has to be saturated with closely spaced markers to reduce the interval for localization to between 5 and 40 Mb. However, fortunately there is nonrandom association between alleles of individual single nucleotide polymorphisms (SNPs) and neighboring SNPs, known as linkage disequilibrium that extends for shorter distances than genetic linkage. The remaining chromosomal region of 0.5–1 Mb can then be systematically dissected. Because no assumptions are made on disease etiology positional cloning has the ability to identify novel genes and mechanisms. It has been highly successful in identifying genes underlying single gene disorders such as cystic fibrosis and Huntington's disease. Genome screens, however, are difficult to replicate because they are sensitive to population stratification, disease definition, the panels of markers chosen for genotyping, environmental factors, and the frequency of the phenotype in the population. Also, stringent statistical criteria that take into account the hundreds of markers tested and many phenotypes studied are needed to confirm whether linkage is real (Lander and Kruglyak, 1995).

There have been four genome screens for AD (Lee et al., 2000; Cookson et al., 2001; Bradley et al., 2002). The first screen carried out in German and Scandinavian families with childhood AD showed linkage to AD on chromosome 3q21 (Lee et al., 2000). There was also linkage of total serum IgE to 3q21. The second genome screen was in British families identified through a proband with childhood AD attending a tertiary referral clinic (Cookson et al., 2001). Linkage to AD or to AD and asthma combined was found to 1q21, 17q25, and 20p. These are coincident with regions linked to psoriasis (Tomfohrde et al., 1994; Trembath et al., 1997; Capon et al., 1999). Linkage of total serum IgE was also found to chromosome 5q31 and 16qtel. In both studies, evidence of linkage to total serum IgE was lower than evidence of linkage to AD. An adult Swedish study found evidence of linkage to AD to chromosome 3p24-22 (Bradley et al., 2002). For another phenotype, AD combined with raised allergen-specific IgE levels, a suggestive linkage was found to chromosome region 18q21. When a semiquantitative phenotype looking at severity scores in AD was used, suggestive linkage was found to chromosomal regions 3q14, 13q14, 15q14-15, and 17q21. It is possible that the 3q14 and 17q21 loci correspond to the previously identified AD loci in children. Chromosome 13q14 has been previously linked to children with AD (Beyer et al., 2000) and to atopy and asthma (Anderson et al., 2002). The most recent genome screen was in Danish families ascertained via affected sibpairs with both clinical AD and confirmed specific allergy. Linkage to AD was found for 3p26-24, 4p15-14, and 18q11-12 (Haagerup et al., 2004). A study of selected candidate regions found evidence of linkage of AD to the 14q11 locus in Swedish families (Soderhall et al., 2001).

Candidate Gene Studies

Candidate genes are studied based on their known biological functions in relation to disease. The genes are sequenced and known and newly identified polymorphisms are genotyped in cases and controls. A selection of candidate gene studies in AD will be discussed.

The FcεRI

FcεRI is a multimeric cell surface receptor that binds the constant region of the IgE molecule and is expressed on basophils, mast cells (MCs), monocytes, platelets, and eosinophils. Its expression is also up-regulated on both Langerhans cells and inflammatory dendritic epidermal cells (Jurgens et al., 1995; Novak et al., 2002). The β subunit of FcRI, FcεRIβ, serves as an amplifier of FcεRI biological function and stabilizes the surface expression of the receptor (Lin et al., 1996; Turner and Kinet, 1999). Cross-linking of IgE bound to its receptor on cells by multivalent allergens initiates a cascade of events resulting in allergic immune responses. MCs and basophils are involved in the early, immediate response, which is marked by cellular degranulation and the release of preformed proinflammatory mediators, including histamine, proteases, leukotrienes, and prostaglandins that cause the symptoms of allergic reactions.

The *FcεRIβ* gene on chromosome 11q 12-13 has been linked to atopy (Cookson et al., 1989; Young et al., 1992). Polymorphisms in this gene have been associated with AD in two panels of British families where the majority of probands had severe skin disease and 60% had severe asthma (Cox et al., 1998). Significant association with two intronic SNPs were seen for AD and asthma. Significant sharing of maternal alleles has also been seen for AD. As none of these variants were functional their relevance to AD is not known. Coding variants have been identified that are associated with atopy but not AD (Traherne et al., 2003).

Mast Cell Chymase

MCs are key effector cells of IgE dependent immediate reactions and also to certain IgE mediated, late phase reactions (Williams and Galli, 2000). Mast cell chymase (MCC) is a chymotryptic serine protease present in the secretory granules of MC and has various functions including promoting endothelial and epithelial permeability and recruitment of neutrophils and leukocytes in vivo (Miller and Pemberton, 2002). Its gene *CMA1* maps close to the long arm of chromosome 14q11. This region has been previously linked to AD (Soderhall et al., 2001). MCC is increased in chronic AD skin lesions (Badertscher et al., 2005) and a potential role of chymase in contributing to the skin barrier defect has been suggested (Groneberg et al., 2005). There have been reports of significant associations between a *CMA1* promotor polymorphism and AD in Japanese adults and schoolchildren (Mao et al., 1996; Mao et al., 1998; Tanaka et al., 1999). The latter two studies found that the association with AD was independent of total serum IgE levels suggesting that the mechanism by which MCC influences disease outcome is IgE independent. These findings were not confirmed by another Japanese (Kawashima et al., 1998) and Italian study (Pascale et al., 2001). Recently, this polymorphism was associated with serum total IgE levels in adult AD in a family-based association study in Caucasians (Iwanaga et al., 2004) and with the AD phenotype alone in a German case–control study (Weidinger et al., 2005). This supports the notion that *CMA1* serves as a candidate gene for AD. The therapeutic potential of chymase inhibitors in AD has been successfully explored in preliminary animal studies where symptoms and signs of AD were ameliorated with MCC inhibitors (Imada et al., 2002; Watanabe et al., 2002).

CC Chemokines: RANTES and Eotaxin

Infiltration of inflammatory cells in tissues is regulated by chemokines (Alam, 1997). *RANTES* (regulated on activation, normal T cell expressed and secreted) is a CC chemokine, and enhanced serum levels have been found in AD that correlated with total serum IgE levels and eosinophil numbers (Kaburagi et al., 2001). A functional point mutation in the proximal promoter region of the *RANTES* gene that results in a new consensus binding site for the GATA transcription factor family has been identified on chromosome 17q 11.2 (Nickel et al., 2000). The mutant allele was associated with AD but not asthma in children of the German Multicentre Allergy Study (MAS-90). Using transient transfection of human cell lines, the polymorphism was also shown to exert higher transcriptional activity. This polymorphism is associated with enhanced *RANTES* production in AD patients hence the polymorphism is involved in AD by up-regulating *RANTES* protein expression (Bai et al., 2005). These findings were not replicated in a study of 128 Hungarian children with AD (Kozma et al., 2002).

Eotaxin is a potent chemoattractant and activator of eosinophils and T helper type 2 (TH2) lymphocytes. The eotaxin gene is located at chromosome 17q21.1-q21.2 adjacent to the chromosome 17 linkage region. Eotaxin is overexpressed in AD skin (Yawalkar et al., 1999; Taha et al., 2000) and serum and plasma levels are also increased (Hossny et al., 2001). A number of polymorphisms have been identified including a coding SNP that results in a nonconservative amino acid change of alanine to threonine (Tsunemi, Saeki, Nakamura, Sekiya, Hirai, Fujita, et al., 2002). This SNP and SNPs in the promoter and exon regions are not associated with susceptibility to AD but two of them in the promoter region are associated with the total serum IgE levels in AD. Tacrolimus ointment, a macrolide lactone effective in treating AD, suppresses the expression of eotaxin and *RANTES* in AD skin (Park et al., 2005).

The Cytokine Gene Cluster

The cytokine gene cluster is on chromosome 5q31-33. Genetic linkage of genes in this region to AD (Forrest et al., 1999; Tanaka et al., 1999; Beyer et al., 2000; Soderhall et al., 2001), atopy, and the total serum IgE was found in several studies (Cookson et al., 2001). It contains several interleukins, *SPINK5* [the gene encoding the serine protease inhibitor, lymphoepithelial Kazal-type related inhibitor (LEKTI)], *GM-CSF* (granulocyte–macrophage colony-stimulating factor), *CD14* antigen, and *TIM-1* (T-cell immunoglobulin domain mucin domain protein-1). Polymorphisms in *IL13* (Liu et al., 2000; Tsunemi, Saeki, Nakamura, Sekiya, Hirai, Kakinuma, et al., 2002; He et al., 2003; Hummelshoj et al., 2003), *IL4* (Kawashima et al., 1998; He et al., 2003), *CD14* (Lange et al., 2005), *GM-CSF* (Rafatpanah et al., 2003), *TIM-1* (Chae et al., 2003), and *SPINK5* (Walley et al., 2001; Kato et al., 2003; Nishio et al., 2003) have been associated with AD.

SPINK5 has been identified as the gene defective in Netherton's syndrome (NS) (Chavanas et al., 2000). NS is a rare severe autosomal recessive genodermatosis where atopy is a universal manifestation (Judge et al., 1994). LEKTI is expressed on epithelial and mucosal surfaces as well as in the thymus (Magert et al., 1999; Chavanas et al., 2000). A number of mutations have been found in the *SPINK5* gene following sequencing of NS patients (Chavanas et al., 2000; Bitoun et al., 2002). The majority result in premature termination codons causing abolition in protein production. Further polymorphisms were found on sequencing the gene in children with severe AD and normal controls (Walley et al., 2001). Genotyping the SNPs in two independent families showed significant associations between the Glu420Lys variant and atopy and AD (Walley et al., 2001). These findings have been confirmed in two Japanese studies (Nishio et al., 2003) but not in a population in Northern Germany (Folster-Holst et al., 2005).

Significant trypsin inhibiting activity has been demonstrated for 3 of the 15 LEKTI domains (Magert et al., 2002). LEKTI is also implicated in the cornification process (Descargues et al., 2005). Many allergens are serine proteinases and their protease activity may encourage their allergenicity (Thomas et al., 2002). Proteinases are also involved in T- and B-cell maturation. Proteinase inhibitors are important negative regulators of proteinase action in vivo (Magert et al., 2002). Taken together, the findings in NS is an example of a monogenic disorder providing insights into one of the plausible mechanisms in AD where there is a barrier defect, increased skin permeability, and predisposition to recurrent infections. The house dust mite, *Dermatophagoides sp.* feces (Thomas et al., 2002) and *Staphylococcus aureus* (Miedzobrodzki et al., 2002) are potent sources of proteinases in AD skin lesions that enhance the disruption of the skin barrier. Treatment of AD with a1-protease inhibitors was effective in a pilot study in patients with recalcitrant AD (Wachter and Lezdey, 1992).

The Epidermal Differentiation Complex and AD

The peak of linkage of AD on chromosome 1q21 overlies the epidermal differentiation complex (EDC). These genes regulate epidermal differentiation. They are expressed during terminal differentiation of the epidermis.

The EDC comprises a number of gene families such as the S100 calcium-binding proteins, small proline rich proteins, and the late expressed cornified envelope proteins (Mischke et al., 1996; Marshall et al., 2001). It also includes members of the peptidoglycan recognition protein family and a number of single copy genes such as loricrin, involucrin, trichohyalin, and filaggrin. The formation of the cornified cell envelope (CE) is characteristic of terminally differentiated epidermal keratinocytes. Loricrin is the major CE protein and is cross-linked by both disulfide and N ε-(γ-glutamyl) lysine isodipeptide bonds to the CE (Hohl et al., 1991). Mutations in the loricrin gene have been seen in Vohwinkel's syndrome, a mutilating keratoderma (Maestrini et al., 1996). DNA microarray analysis has shown decreased expression of loricrin in AD skin (Sugiura et al., 2005).

S100 proteins are low molecular weight calcium-binding proteins of the elongation factor-hand (EF-hand) superfamily. They are induced during terminal differentiation of the epidermis and are involved in the regulation of a number of cellular processes such as cell cycle progression and differentiation (Schafer et al., 1995). *S100A7* (psoriasin) is highly up-regulated in psoriatic epidermis as well as in primary human keratinocytes undergoing abnormal differentiation (Hoffmann et al., 1994). It is a potent chemotaxin for neutrophils and $CD4^+$ T lymphocytes (Jinquan et al., 1996). Profilaggrin and trichohyalin are expressed later in terminally differentiating granular cells and are also members of the S100 family. Profilaggrin, by a process of dephosphorylation and proteolysis is processed to filaggrin that functions as a matrix for packing keratin filaments in differentiated cornified cells (Dale et al., 1985). Decreased expression levels of filaggrin was shown in AD skin by DNA microarrays (Sugiura et al., 2005) and immunohistochemistry (Seguchi et al., 1996). Recently, loss of function genetic variants in the gene encoding filaggrin (*FLG*) have been associated with AD in populations of European origin (Palmer et al., 2006). These variants in *FLG* are also the cause of icthyosis vulgaris that is often associated with the atopic diathesis (Smith et al., 2006). Further studies are required to identify variants in non-European populations. The EDC therefore contains several candidate genes for AD that are involved in the integrity, differentiation, and immunity of the skin barrier.

Defective Skin Barrier in AD

Several lines of evidence point toward a defective skin barrier predisposing to AD. There is a global decrease in lipids in AD skin and selective deficiency in stratum corneum ceramide content (Melnik et al., 1988) that is correlated with diminished barrier function in AD (Di Nardo et al., 1998). Ceramide dominant moisturizers are effective therapeutically as they hydrate the delipidized stratum corneum and reduce transepidermal water loss (Chamlin et al., 2002). Studies of the EDC gene products transcribed within terminally differentiating keratinocytes also reveal that the skin is not a passive barrier. The S100 calcium-binding protein family (S100 proteins, profilaggrin, filaggrin) have a wide range of immunological and structural functions as described above. Percutaneous sensitization with house dust mite antigen after barrier disruption elicits a TH2 cytokine response in mice (Kondo et al., 1998) that is relevant to the disrupted barrier in AD. The *NS* gene that encodes the serine protease inhibitor, LEKTI is found in the uppermost epidermis and pilosebaceous units has been associated with AD (Walley et al., 2001). Stratum corneum chymotryptic enzyme (SCCE), which is a serine protease, plays a central role in desquamation, and a polymorphism in *SCCE* has been associated with AD (Vasilopoulos et al., 2004). Treatment with protease inhibitors also ameliorates symptoms of AD (Imada et al., 2002).

AD and Asthma

Asthma is an inflammatory disease of the airways of the lung. It is characterized by narrowing of the airways due to inflammation and mucous secretion and bronchospasm as a result of contraction of bronchial smooth muscles to nonspecific stimuli. There is strong association between AD and asthma with 60% of children with AD also having asthma (Cox et al., 1998). Though severe AD and asthma often occur together (Williams et al., 1999), both conditions respond to different determinants in population studies (Riedler et al., 2001). Children with parental AD have a high risk for AD compared with children with parental asthma or parental allergic rhinitis (Dold et al., 1992) suggesting the presence of eczema specific genes. More than a dozen genome screens have been reported for asthma and its related phenotypes (Daniels et al., 1996; Hizawa et al., 1998; Wjst et al., 1999; Dizier et al., 2000; Ober et al., 2000; Howard et al., 2001; Laitinen et al., 2001; Mathias et al., 2001; Haagerup et al., 2002; Hakonarson et al., 2002; Koppelman et al., 2002; Bouzigon et al., 2004; Dizier et al., 2005; Meyers et al., 2005; Wang et al., 2005). Three genes underlying asthma have been identified by positional cloning. *ADAM33* on chromosome 20p13 is a membrane anchored metalloprotease with diverse functions such as the shedding of cell-surface proteins such as cytokines and cytokine receptors (Van Eerdewegh et al., 2002). *PHF11* on chromosome 13q12 contains two plant homeodomain (PHD) zinc fingers and probably regulates transcription (Zhang et al., 2003). The final gene is *DPP10* that encodes a homolog of dipeptidyl peptidases (DPPs) that cleaves terminal dipeptides from cytokines and chemokines (Allen et al., 2003). *PHF11* may have a potential role in AD as chromosome 13q12 does show linkage to AD (Beyer et al., 2000) and polymorphisms in *PHF11* are strongly associated with increased IgE levels in families comprising children with AD (Zhang et al., 2003). The protein binding zinc fingers it encodes may modify immunoglobulin production and clonal expansion of B-cells (Zhang et al., 2003).

The results from the asthma genome screens show loci that are not generally shared with regions of linkage to AD, with a few exceptions (Cookson et al., 2001), suggesting that both diseases may result from distinct mechanisms.

Defective Innate Immunity and AD

Innate immunity is a primitive and conserved response where the host's primary defenses recognize pathogen-associated molecular patterns (PAMPs) on microorganisms by germline encoded pathogen recognition receptors (PRRs) (Weidinger et al., 2005). An intact innate immune response to microbial organisms is important in skin as the skin is the initial barrier to microbes and the environment. CD14 receptor (Baldini et al., 1999; Koppelman et al., 2001), Toll-like receptors (TLRs) (Arbour et al., 2000) and CARD15 (NOD2) (Carneiro et al., 2004) are PRRs important in innate immune responses. They interact with bacterial lipopolysaccharides that are associated with interferon-γ (IFN-γ) dominated TH1 immunologic responses. Allergic responses are driven by TH2 responses (Romagnani, 2000). Variations in genes responding to microbial matter can affect the TH1/TH2 balance and predispose to allergic disorders (Kabesch et al., 2003). Recently, it was found that TH cell help and generation of T-dependent antigen-specific antibody responses requires activation of TLRs in B-cells (Pasare and Medzhitov, 2005). Hence, apart from inducing innate immune responses, TLRs can mediate control of adaptive immunity.

Keratinocytes are highly active immunologically and express PRRs such as CD14 and TLR-4 (Song et al., 2002). They can induce inflammatory responses without preinduction by other cells. Patients with AD are prone to infections for example by *S. aureus*, herpes simplex, and mollusca contagiosa viruses. Keratinocytes in

AD have also been shown to be deficient in defensins that are components of the innate immune system that normally have antimicrobial activity against bacteria, viruses, and fungi (Ong et al., 2002).

Polymorphisms in PRR such as *CARD4* (Weidinger et al., 2005), *CARD15* (Kabesch et al., 2003) and *CD14* (Lange et al., 2005) have been associated with AD. A *TLR2* polymorphism has been associated with *S. aureus* infection and has been shown to be significantly less responsive to bacterial peptides (Lorenz et al., 2000). Patients carrying this variant were found to have a distinct phenotype of severe AD suggesting that the *TLR2* polymorphism may increase the susceptibility to infections and chronic colonization of the skin in AD (Ahmad-Nejad et al., 2004).

Genetics of AD Compared to Other Inflammatory Diseases

Crohn's Disease

CARD15 polymorphisms are associated with susceptibility to Crohn's disease (Hampe et al., 2001; Hugot et al., 2001; Ogura et al., 2001) and with AD and allergic rhinitis (Kabesch et al., 2003). Recently, associations between *CARD15* variants and ulcerative colitis have also been described (McGovern et al., 2003). *CARD4* has recently been associated with inflammatory bowel disease (McGovern et al., 2005), AD (Weidinger et al., 2005), and atopy (Hysi et al., 2005; Weidinger et al., 2005). This shared genetic background between inflammatory bowel disease and atopy suggests that defective recognition of microbes at barrier defenses in skin and mucosa contributes to excessive responses that could be either TH1 or TH2 dominated (Kabesch et al., 2003).

Psoriasis

The genome screens in the UK subset do not overlap as commonly with asthma as one would expect but rather with psoriasis susceptibility loci (Cookson et al., 2001). The German genome screen locus on chromosome 3q21 also overlaps with another psoriasis locus (Enlund et al., 1999). This suggests that these genes in these chromosomal regions are polymorphic and have general effects on dermal inflammation and immunity.

Epidermodysplasia Verruciformis and Leprosy

It is notable that the linkage region to AD and psoriasis maps to the same region on chromosome 17qter that is a susceptibility locus for epidermodysplasia verruciformis (EV) (Ramoz et al., 1999). This is a lifelong disease characterized by infection with oncogenic human papillomavirus type 5 (HPV5). Patients with psoriasis also harbor a reservoir for HPV5 (Majewski et al., 1998). Genetic susceptibility to leprosy also shows linkage to chromosome 20p (Tosh et al., 2002), which is the area of linkage to the distinctive phenotype of AD and asthma combined in the UK genome screen (Cookson et al., 2001). These shared areas of linkage of AD with infectious diseases could underscore the importance of infections or dysregulation of responses to infections as a trigger factor to the sequelae that determines the AD phenotype.

Autoimmune Diseases

The 17q25 locus also shows linkage to multiple sclerosis (Sawcer et al., 1996; Kuokkanen et al., 1997) and rheumatoid arthritis (Jawaheer et al., 2001). The 20p locus has also been linked to systemic lupus erythematosus (Gaffney et al., 2000). *ADAM33* is localized to this region but does not seem a candidate for linkage (Van Eerdewegh et al., 2002).

Maternal Transmission of AD

Maternal influence of linkage to atopic disease was first described for chromosome 11q that harbors the FcεRIβ locus where maternal linkage was seen for the atopy phenotype as well as skin prick tests, radioallergosorbent tests (RASTs), and elevated total IgE (Cookson et al., 1992; Moffatt et al., 1992). Polymorphisms in FcεRIβ that are associated with AD and asthma described earlier showed significant transmission only of maternally-derived alleles for these two phenotypes (Cox et al., 1998). The reason for this maternal influence could be related to maternal–fetal interactions influencing the infants IgE responses via the placenta or breast milk (Cox et al., 1998), or genomic imprinting where the allele from one parent is differentially expressed compared to the allele from the other parent (Hall, 1990; Cookson et al., 1992). Parent of origin effects have also been noted in other immunological disease such as type 1 diabetes (Warram et al., 1984; Bennett and Todd, 1996), rheumatoid arthritis (Koumantaki et al., 1997), psoriasis (Burden et al., 1998), inflammatory bowel disease (Akolkar et al., 1997), and selective IgA deficiency (Vorechovsky et al., 1999). It has been suggested that parent of origin effects are somewhat adaptive and a general phenomenon affecting several immune related loci and diseases (Cookson, 2002).

Adult AD

Adult-onset AD is a poorly defined illness. There is limited data on adults who present for the first time with AD in adulthood (Kawashima et al., 1989; Tay et al., 1999; Bannister and Freeman, 2000; Ingordo et al., 2003; Ozkaya, 2005). Apart from typical lesions, these patients may present with atypical morphologies such as nonflexural involvement, nummular and prurigo-like lesions (Ozkaya, 2005). In a study of the genetic basis of AD persisting into adulthood, polymorphisms in candidate genes involved in asthma, atopy, and childhood AD were tested for association in an adult cohort. A polymorphism in *FcεRIβ* and an SNP in the *TLR9* was found to be associated with adult AD. Associations were, however, weaker than those seen in childhood studies suggesting the role of environmental influences or different genes involved in adult AD (Moffatt et al., 2005). Genetic studies in adults with AD may be confounded by the fact that they may not be accommodated in certain criteria such as the UK Working Party criteria (Williams et al., 1994) commonly used in epidemiological studies where age at onset and clinical distribution of lesions such as flexural involvement are important diagnostic criteria (Ozkaya, 2005).

Microarray Studies of AD

Genomic expression profiling using DNA microarrays allow for the simultaneous comparisons of thousands of messenger ribonucleic acids (mRNAs) hybridized to complementary DNA (cDNA) clones on chips, and can identify disease specific tissue responses by comparing gene expression levels. There have been a limited number of gene expression studies with small sample sizes using DNA microarrays in AD. RNA extraction from skin biopsies showed a distinct pattern of gene expression in AD compared to psoriasis (Nomura, Gao, et al., 2003). AD skin showed increased expression of CC chemokines (*CCL-13/MCP-4*, *CCL-18/PARC*, and *CCL-27/CTACK*) known to attract TH2 cells, and psoriasis skin expressed CXC chemokines that are known to elicit TH1 responses. There was also decreased expression of antimicrobial genes in AD skin lesions compared to psoriasis skin lesions (Nomura, Goleva, et al., 2003). A limiting factor in these studies is tissue heterogeneity in biopsy samples that may limit interpretation of expression results, since tissue heterogeneity is a major confounding variable in microarray experiments (Tumor Analysis Best Practices Working Group, 2004). There have been a few studies where isolated cell types were arrayed. When mRNA transcripts in peripheral blood mononuclear cells were compared in AD patients and normal controls, four transcripts, *IFN-γ*, *TRAIL* [tumor necrosis factor (TNF)-related apoptosis-inducing ligand], *ISGF-3 (STAT1)*, and *defensin-1* were found to be differentially expressed (Heishi et al., 2002). Monocytes from AD patients were also examined and various genes were up-regulated including those involved in major histocompatibility complex (MHC) class I antigen presentation, recognition of microorganisms and genes involved in apoptosis (Nagata et al., 2003). Both these studies, however, have modest sample sizes.

Recently, there have been novel studies using microarrays to study complex traits. Classical quantitative traits represent gross clinical measurements that may not be related directly to cellular or biological processes giving rise to that trait. Using transcript abundances as quantitative traits (known as expression quantitative trait loci or eQTL) is more directly related to pathophysiologic processes (Schadt et al., 2003). As these traits are directly related to the effects of genetic polymorphisms, they offer increased power over clinical phenotypes. By combining gene expression data with genetic inheritance information, it has been demonstrated that gene expression data can be used to refine the disease phenotype and implicate pathways for disease causation (Schadt et al., 2003). The heritability of gene expression traits has been demonstrated in segregating human populations (Monks et al., 2004).

For most complex diseases, there has been limited success of replication with linkage analysis that fails to explain the genetic epidemiology of the disease (Altmuller et al., 2001). Success depends on various factors that are difficult to control. Confounding factors include the low heritability of complex traits, modest effect of these traits, and different polymorphic markers used in different studies and variation in disease definition and severity. Association analysis is more powerful for the detection of common disease alleles that confer moderate disease risks (Risch and Merikangas, 1996) though candidate gene studies also have their drawbacks when the fundamental causes of the disease are not known.

With the completion of the human genome sequence, improvements in SNP genotyping technology and the initiation of the International HapMap Project (The International HapMap Project, 2003) genome, wide association studies (reviewed in Hirschhorn and Daly, 2005) are now possible where a dense set of SNPs across the whole genome is genotyped to survey the most common genetic variation for a role in disease or to identify the quantitative traits that are risks factors for disease (Hirschhorn and Daly, 2005). The genome-wide association approach is unbiased as no assumption is made on the causal gene. The genome-wide approach affords increased power and efficiency over linkage studies. This approach has been successful in identifying genes associated with susceptibility to myocardial infarction (Ozaki et al., 2002), age-related macular degeneration (Klein et al., 2005), rheumatoid arthritis (Yamamoto and Yamada, 2005), and Parkinson's disease (Maraganore et al., 2005). In order to study the genetic basis of natural variation in gene expression, genome-wide linkage analysis and the determinants of approximately 1000 expression phenotypes were mapped (Morley et al., 2004). These determinants of human gene expression were recently mapped by regional and genome-wide association (Cheung et al., 2005). The genome-wide association analysis confirmed the linkage results and pointed to the same location as the genome scans in approximately 50% of the phenotypes (Cheung et al., 2005). This suggests that large-scale association studies are feasible to identify genes for complex traits.

Conclusion

The standard model that atopic diseases are mediated by IgE responses to common allergens and that the inflammatory milieu is maintained by T cells skewed to enhance continued IgE production is now being challenged. The results from the genome screens and candidate gene studies for AD suggest that the predisposition to AD rests within the skin itself. Genetic studies have identified classes of genes by candidate gene studies that are associated with atopic mechanisms (*FcεRIβ*, *MCC*, *CC chemokines*), mediators of the TH2 dominated inflammatory milieu (*IL4*, *IL13*), barrier function and protease activity of the skin (*FLG*, *SPINK5*, *SCCE*), and innate immunity genes (*CD14*, *TLR2*, *CARD15*, *CARD4*). Genome screens have identified few regions containing candidate genes. The advent of DNA microarrays with eQTL profiling and whole-genome association studies will identify novel pathways and genes in this complex disease.

Identifying the alleles that affect the risk of developing AD will help the understanding of disease etiology and classification and their genes and proteins may serve as future therapeutic targets.

Acknowledgments

Dr. N. Morar is supported by the Wellcome Trust and has also received funding from START (Skin Treatment and Research Trust).

References

Ahmad-Nejad, P, Mrabet-Dahbi, S, Breuer, K, Klotz, M, Werfel, T, Herz, U, et al. (2004). The toll-like receptor 2 R753Q polymorphism defines a subgroup of patients with atopic dermatitis having severe phenotype. *J Allergy Clin Immunol*, 113:565–567.

Akolkar, PN, Gulwani-Akolkar, B, Heresbach, D, Lin, XY, Fisher, S, Katz, S, et al. (1997). Differences in risk of Crohn's disease in offspring of mothers and fathers with inflammatory bowel disease. *Am J Gastroenterol*, 92:2241–2244.

Alam, R (1997). Chemokines in allergic inflammation. *J Allergy Clin Immunol*, 99:273–277.

Allen, M, Heinzmann, A, Noguchi, E, Abecasis, G, Broxholme, J, Ponting, CP, et al. (2003). Positional cloning of a novel gene influencing asthma from chromosome 2q14. *Nat Genet*, 35:258–263.

Altmuller, J, Palmer, LJ, Fischer, G, Scherb, H, and Wjst, M (2001). Genome-wide scans of complex human diseases: true linkage is hard to find. *Am J Hum Genet*, 69:936–950.

Anderson, GG, Leaves, NI, Bhattacharyya, S, Zhang, Y, Walshe, V, Broxholme, J, et al. (2002). Positive association to IgE levels and a physical map of the 13q14 atopy locus. *Eur J Hum Genet*, 10:266–270.

Arbour, NC, Lorenz, E, Schutte, BC, Zabner, J, Kline, JN, Jones, M, et al. (2000). TLR4 mutations are associated with endotoxin hyporesponsiveness in humans. *Nat Genet*, 25:187–191.

Badertscher, K, Bronnimann, M, Karlen, S, Braathen, LR, and Yawalkar, N (2005). Mast cell chymase is increased in chronic atopic dermatitis but not in psoriasis. *Arch Dermatol Res,* 296:503–506.

Bai, B, Tanaka, K, Tazawa, T, Yamamoto, N, and Sugiura, H (2005). Association between RANTES promoter polymorphism -401A and enhanced RANTES production in atopic dermatitis patients. *J Dermatol Sci,* 39:189–191.

Baldini, M, Lohman, IC, Halonen, M, Erickson, RP, Holt, PG, and Martinez, FD (1999). A Polymorphism* in the 5' flanking region of the CD14 gene is associated with circulating soluble CD14 levels and with total serum immunoglobulin. *E Am J Respir Cell Mol Biol,* 20:976–983.

Bannister, MJ and Freeman, S (2000). Adult-onset atopic dermatitis. *Australas J Dermatol,* 41:225–228.

Bennett, ST and Todd, JA (1996). Human type 1 diabetes and the insulin gene: principles of mapping polygenes. *Annu Rev Genet,* 30:343–370.

Beyer, K, Nickel, R, Freidhoff, L, Bjorksten, B, Huang, SK, Barnes, KC, et al. (2000). Association and linkage of atopic dermatitis with chromosome 13q12-14 and 5q31-33 markers. *J Invest Dermatol,* 115:906–908.

Bitoun, E, Chavanas, S, Irvine, AD, Lonie, L, Bodemer, C, Paradisi, M, et al. (2002). Netherton syndrome: disease expression and spectrum of SPINK5 mutations in 21 families. *J Invest Dermatol,* 118:352–361.

Bouzigon, E, Dizier, MH, Krahenbuhl, C, Lemainque, A, Annesi-Maesano, I, Betard, C, et al. (2004). Clustering patterns of LOD scores for asthma-related phenotypes revealed by a genome-wide screen in 295 French EGEA families. *Hum Mol Genet,* 13:3103–3113.

Bradley, M, Soderhall, C, Luthman, H, Wahlgren, CF, Kockum, I, and Nordenskjold, M (2002). Susceptibility loci for atopic dermatitis on chromosomes 3, 13, 15, 17 and 18 in a Swedish population. *Hum Mol Genet,* 11:1539–1548.

Burden, AD, Javed, S, Bailey, M, Hodgins, M, Connor, M, and Tillman, D (1998). Genetics of psoriasis: paternal inheritance and a locus on chromosome 6p. *J Invest Dermatol,* 110:958–960.

Capon, F, Novelli, G, Semprini, S, Clementi, M, Nudo, M, Vultaggio, P, et al. (1999). Searching for psoriasis susceptibility genes in Italy: genome scan and evidence for a new locus on chromosome 1. *J Invest Dermatol,* 112:32–35.

Cargill, M, Altshuler, D, Ireland, J, Sklar, P, Ardlie, K, Patil, N, et al. (1999). Characterization of single-nucleotide polymorphisms in coding regions of human genes. *Nat Genet,* 22:231–238.

Carlson, CS, Eberle, MA, Kruglyak, L, and Nickerson, DA (2004). Mapping complex disease loci in whole-genome association studies. *Nature,* 429:446–452.

Carneiro, LA, Travassos, LH, and Philpott, DJ (2004). Innate immune recognition of microbes through Nod1 and Nod2: implications for disease. *Microbes Infect,* 6:609–616.

Chae, SC, Song, JH, Lee, YC, Kim, JW, and Chung, HT (2003). The association of the exon 4 variations of Tim-1 gene with allergic diseases in a Korean population. *Biochem Biophys Res Commun,* 312:346–350.

Chamlin, SL, Kao, J, Frieden, IJ, Sheu, MY, Fowler, AJ, Fluhr, JW, et al. (2002). Ceramide-dominant barrier repair lipids alleviate childhood atopic dermatitis: changes in barrier function provide a sensitive indicator of disease activity. *J Am Acad Dermatol,* 47:198–208.

Chavanas, S, Bodemer, C, Rochat, A, Hamel-Teillac, D, Ali, M, Irvine, AD, et al. (2000). Mutations in SPINK5, encoding a serine protease inhibitor, cause Netherton syndrome. *Nat Genet,* 25:141–142.

Chavanas, S, Garner, C, Bodemer, C, Ali, M, Teillac, DH, Wilkinson, J, et al. (2000). Localization of the Netherton syndrome gene to chromosome 5q32, by linkage analysis and homozygosity mapping. *Am J Hum Genet,* 66:914–921.

Cheung, VG and Spielman, RS (2002). The genetics of variation in gene expression. *Nat Genet,* 32 (Suppl.):522–525.

Cheung, VG, Spielman, RS, Ewens, KG, Weber, TM, Morley, M, and Burdick JT (2005). Mapping determinants of human gene expression by regional and genome-wide association. *Nature,* 437:1365–1369.

Coca, AF and Cooke, RA (1923). On the phenomenon of hypersensitiveness. *J Immunol,* 8:163–182.

Cookson, W (2002). Genetics and genomics of asthma and allergic diseases. *Immunol Rev,* 190:195–206.

Cookson, WO, Sharp, PA, Faux, JA, and Hopkin, JM (1989). Linkage between immunoglobulin E responses underlying asthma and rhinitis and chromosome 11q. *Lancet,* 1:1292–1295.

Cookson, WO, Ubhi, B, Lawrence, R, Abecasis, GR, Walley, AJ, Cox, HE, et al. (2001). Genetic linkage of childhood atopic dermatitis to psoriasis susceptibility loci. *Nat Genet,* 27:372–373.

Cookson, WO, Young, RP, Sandford, AJ, Moffatt, MF, Shirakawa, T, Sharp, PA, et al. (1992). Maternal inheritance of atopic IgE responsiveness on chromosome 11q. *Lancet,* 340:381–384.

Cox, HE, Moffatt, MF, Faux, JA, Walley, AJ, Coleman, R, Trembath, RC, et al. (1998). Association of atopic dermatitis to the beta subunit of the high affinity immunoglobulin E receptor. *Br J Dermatol,* 138:182–187.

Dale, BA, Resing, KA, and Lonsdale-Eccles, JD (1985). Filaggrin: a keratin filament associated protein. *Ann N Y Acad Sci,* 455:330–342.

Daniels, SE, Bhattacharrya, S, James, A, Leaves, NI, Young, A, Hill, MR, et al. (1996). A genome-wide search for quantitative trait loci underlying asthma. *Nature,* 383:247–250.

Descargues, P, Deraison, C, Bonnart, C, Kreft, M, Kishibe, M, Ishida-Yamamoto, A, et al. (2005). Spink5-deficient mice mimic Netherton syndrome through degradation of desmoglein 1 by epidermal protease hyperactivity. *Nat Genet,* 37:56–65.

Di Nardo, A, Wertz, P, Giannetti, A, and Seidenari, S (1998). Ceramide and cholesterol composition of the skin of patients with atopic dermatitis. *Acta Derm Venereol,* 78:27–30.

Dizier, MH, Besse-Schmittler, C, Guilloud-Bataille, M, Annesi-Maesano, I, Boussaha, M, Bousquet, J, et al. (2000). Genome screen for asthma and related phenotypes in the French EGEA study. *Am J Respir Crit Care Med,* 162:1812–1818.

Dizier, MH, Bouzigon, E, Guilloud-Bataille, M, Betard, C, Bousquet, J, Charpin, D, et al. (2005). Genome screen in the French EGEA study: detection of linked regions shared or not shared by allergic rhinitis and asthma. *Genes Immun,* 6:95–102.

Dold, S, Wjst, M, von Mutius, E, Reitmeir, P, and Stiepel, E (1992). Genetic risk for asthma, allergic rhinitis, and atopic dermatitis. *Arch Dis Child,* 67:1018–1022.

Duffy, DL, Martin, NG, Battistutta, D, Hopper, JL, and Mathews, JD (1990). Genetics of asthma and hay fever in Australian twins. *Am Rev Respir Dis,* 142:1351–1358.

Enlund, F, Samuelsson, L, Enerback, C, Inerot, A, Wahlstrom, J, Yhr, M, et al. (1999). Psoriasis susceptibility locus in chromosome region 3q21 identified in patients from southwest Sweden. *Eur J Hum Genet,* 7:783–790.

Flohr, C, Johansson, SG, Wahlgren, CF, and Williams, H (2004). How atopic is atopic dermatitis? *J Allergy Clin Immunol,* 114:150–158.

Folster-Holst, R, Stoll, M, Koch, WA, Hampe, J, Christophers, E, and Schreiber, S (2005). Lack of association of SPINK5 polymorphisms with nonsyndromic atopic dermatitis in the population of Northern Germany. *Br J Dermatol,* 152:1365–1367.

Forrest, S, Dunn, K, Elliott, K, Fitzpatrick, E, Fullerton, J, McCarthy, M, et al. (1999). Identifying genes predisposing to atopic eczema. *J Allergy Clin Immunol,* 104:1066–1070.

Gaffney, PM, Ortmann, WA, Selby, SA, Shark, KB, Ockenden, TC, Rohlf, KE, et al. (2000). Genome screening in human systemic lupus erythematosus: results from a second Minnesota cohort and combined analyses of 187 sib-pair families. *Am J Hum Genet,* 66:547–556.

Groneberg, DA, Bester, C, Grutzkau, A, Serowka, F, Fischer, A, Henz, BM, et al. (2005). Mast cells and vasculature in atopic dermatitis—potential stimulus of neoangiogenesis. *Allergy,* 60:90–97.

Tumor Analysis Best Practices Working Group (2004). Expression profiling-best practices for data generation and interpretation in clinical trials. *Nat Rev Genet,* 5:229–237.

Haagerup, A, Bjerke, T, Schiotz, PO, Dahl, R, Binderup, HG, Tan, Q, et al. (2004). Atopic dermatitis—a total genome-scan for susceptibility genes. *Acta Derm Venereol,* 84:346–352.

Haagerup, A, Bjerke, T, Schiotz, PO, Binderup, HG, Dahl, R, and Kruse, TA (2002). Asthma and atopy—a total genome scan for susceptibility genes. *Allergy,* 57:680–686.

Hakonarson, H, Bjornsdottir, US, Halapi, E, Palsson, S, Adalsteinsdottir, E, Gislason, D, et al. (2002). A major susceptibility gene for asthma maps to chromosome 14q24. *Am J Hum Genet,* 71:483–491.

Hall, JG (1990). Genomic imprinting. *Arch Dis Child,* 65:1013–1015.

Halushka, MK, Fan, JB, Bentley, K, Hsie, L, Shen, N, Weder, A, et al. (1999). Patterns of single-nucleotide polymorphisms in candidate genes for blood-pressure homeostasis. *Nat Genet,* 22:239–247.

Hampe, J, Cuthbert, A, Croucher, PJ, Mirza, MM, Mascheretti, S, Fisher, S, et al. (2001). Association between insertion mutation in NOD2 gene and Crohn's disease in German and British populations. *Lancet,* 357:1925–1928.

Hanifin, JM and Rajka, G (1980). Diagnostic features of atopic dermatitis. *Acta Derm Venereol Suppl (Stockh),* 92:44–47.

He, JQ, Chan-Yeung, M, Becker, AB, Dimich-Ward, H, Ferguson, AC, Manfreda, J, et al. (2003). Genetic variants of the IL13 and IL4 genes and atopic diseases in at-risk children. *Genes Immun,* 4:385–389.

Heishi, M, Kagaya, S, Katsunuma, T, Nakajima, T, Yuki, K, Akasawa, A, et al. (2002). High-density oligonucleotide array analysis of mRNA transcripts in peripheral blood cells of severe atopic dermatitis patients. *Int Arch Allergy Immunol,* 129:57–66.

Hirschhorn, JN and Daly, MJ (2005). Genome-wide association studies for common diseases and complex traits. *Nat Rev Genet,* 6:95–108.

Hizawa, N, Freidhoff, LR, Chiu, YF, Ehrlich, E, Luehr, CA, Anderson, JL, et al. (1998). Genetic regulation of *Dermatophagoides pteronyssinus*-specific IgE responsiveness: a genome-wide multipoint linkage analysis in families recruited through 2 asthmatic sibs. Collaborative Study on the Genetics of Asthma (CSGA). *J Allergy Clin Immunol,* 102:436–442.

Hoffmann, HJ, Olsen, E, Etzerodt, M, Madsen, P, Thogersen, HC, Kruse, T, et al. (1994). Psoriasin binds calcium and is upregulated by calcium to levels that resemble those observed in normal skin. *J Invest Dermatol,* 103:370–375.

Hohl, D, Mehrel, T, Lichti, U, Turner, ML, Roop, DR, and Steinert, PM (1991). Characterization of human loricrin. Structure and function of a new class of epidermal cell envelope proteins. *J Biol Chem,* 266:6626–6636.

Hossny, E, Aboul-Magd, M, and Bakr, S (2001). Increased plasma eotaxin in atopic dermatitis and acute urticaria in infants and children. *Allergy,* 56:996–1002.

Howard, TD, Whittaker, PA, Zaiman, AL, Koppelman, GH, Xu J, Hanley, MT, et al. (2001). Identification and association of polymorphisms in the interleukin-13 gene with asthma and atopy in a Dutch population. *Am J Respir Cell Mol Biol,* 25:377–384.

Hugot, JP, Chamaillard, M, Zouali, H, Lesage, S, Cézard, JP, Belaiche, J, et al. (2001). Association of NOD2 leucine-rich repeat variants with susceptibility to Crohn's disease. *Nature,* 411(6837):599–603.

Hummelshoj, T, Bodtger, U, Datta, P, Malling, HJ, Oturai, A, Poulsen, LK, et al. (2003). Association between an interleukin-13 promoter polymorphism and atopy. *Eur J Immunogenet,* 30:355–359

Hysi, P, Kabesch, M, Moffatt, MF, Schedel, M, Carr, D, Zhang, Y, et al. (2005). NOD1 variation, immunoglobulin E and asthma. *Hum Mol Genet,* 14:935–941.

Imada, T, Komorita, N, Kobayashi, F, Naito, K, Yoshikawa, T, Miyazaki, M, et al. (2002). Therapeutic potential of a specific chymase inhibitor in atopic dermatitis. *Jpn J Pharmacol,* 90:214–217.

Ingordo, V, D'Andria, G, and D'Andria, C (2003). Adult-onset atopic dermatitis in a patch test population. *Dermatology,* 206:197–203.

Iwanaga, T, McEuen, A, Walls, AF, Clough, JB, Keith, TP, Rorke, S, et al. (2004). Polymorphism of the mast cell chymase gene (CMA1) promoter region: lack of association with asthma but association with serum total immunoglobulin E levels in adult atopic dermatitis. *Clin Exp Allergy,* 34:1037–1042.

Jawaheer, D, Seldin, MF, Amos, CI, Chen, WV, Shigeta, R, Monteiro, J, et al. (2001). A genomewide screen in multiplex rheumatoid arthritis families suggests genetic overlap with other autoimmune diseases. *Am J Hum Genet,* 68:927–936.

Jeong, CW, Ahn, KS, Rho, NK, Park, YD, Lee, DY, Lee, JH, et al. (2003). Differential in vivo cytokine mRNA expression in lesional skin of intrinsic vs. extrinsic atopic dermatitis patients using semiquantitative RT-PCR. *Clin Exp Allergy,* 33:1717–1724.

Jinquan, T, Vorum, H, Larsen, CG, Madsen, P, Rasmussen, HH, Gesser B, et al. (1996). Psoriasin: a novel chemotactic protein. *J Invest Dermatol,* 107:5–10.

Johansson, SG, Hourihane, JO, Bousquet, J, Bruijnzeel-Koomen, C, Dreborg, S, Haahtela, T, et al. (2001). A revised nomenclature for allergy. An EAACI position statement from the EAACI nomenclature task force. *Allergy,* 56:813–824.

Judge, MR, Morgan, G, and Harper, JI (1994). A clinical and immunological study of Netherton's syndrome. *Br J Dermatol,* 131:615–621.

Juhlin, L, Johansson, GO, Bennich, H, Hogman, C, and Thyresson, N (1969). Immunoglobulin E in dermatoses. Levels in atopic dermatitis and urticaria. *Arch Dermatol,* 100:12–16.

Jurgens, M, Wollenberg, A, Hanau, D, de la Salle, H, and Bieber, T (1995). Activation of human epidermal Langerhans cells by engagement of the high affinity receptor for IgE, Fc epsilon RI. *J Immunol,* 155:5184–5189.

Kabesch, M, Peters, W, Carr, D, Leupold, W, Weiland, SK, and von Mutius, E (2003). Association between polymorphisms in caspase recruitment domain containing protein 15 and allergy in two German populations. *J Allergy Clin Immunol,* 111:813–817.

Kaburagi, Y, Shimada, Y, Nagaoka, T, Hasegawa, M, Takehara, K, and Sato, S (2001). Enhanced production of CC-chemokines (RANTES, MCP-1, MIP-1alpha, MIP-1beta, and eotaxin) in patients with atopic dermatitis. *Arch Dermatol Res,* 293:350–355.

Kato, A, Fukai, K, Oiso, N, Hosomi, N, Murakami, T, and Ishii, M (2003). Association of SPINK5 gene polymorphisms with atopic dermatitis in the Japanese population. *Br J Dermatol,* 148:665–669.

Kawashima, T, Kobayashi, S, Miyano, M, Ohya, N, Naruse, C, and Tokuda, Y (1989). [Senile type atopic dermatitis]. *Nippon Hifuka Gakkai Zasshi,* 99:1095–1103.

Kawashima, T, Noguchi, E, Arinami, T, Kobayashi, K, Otsuka, F, and Hamaguchi, H (1998). No evidence for an association between a variant of the mast cell chymase gene and atopic dermatitis based on case-control and haplotype-relative-risk analyses. *Hum Hered,* 48:271–274.

Kawashima, T, Noguchi, E, Arinami, T, Yamakawa-Kobayashi, K, Nakagawa, H, Otsuka, F, et al. (1998). Linkage and association of an interleukin 4 gene polymorphism with atopic dermatitis in Japanese families. *J Med Genet,* 35:502–504.

Klein, RJ, Zeiss, C, Chew, EY, Tsai, JY, Sackler, RS, Haynes, C, et al. (2005). Complement factor H polymorphism in age-related macular degeneration. *Science,* 308:385–389.

Kondo, H, Ichikawa, Y, and Imokawa, G (1998). Percutaneous sensitization with allergens through barrier-disrupted skin elicits a Th2-dominant cytokine response. *Eur J Immunol,* 28:769–779.

Koppelman, GH, Stine, OC, Xu, J, Howard, TD, Zheng, SL, Kauffman, HF, et al. (2002). Genome-wide search for atopy susceptibility genes in Dutch families with asthma. *J Allergy Clin Immunol,* 109:498–506.

Koppelman, GH, Reijmerink, NE, Colin Stine, O, Howard, TD, Whittaker, PA, Meyers, DA, et al. (2001). Association of a promoter polymorphism of the CD14 gene and atopy. *Am J Respir Crit Care Med,* 163:965–969.

Koumantaki, Y, Giziaki, E, Linos, A, Kontomerkos, A, Kaklamanis, P, Vaiopoulos, G, et al. (1997). Family history as a risk factor for rheumatoid arthritis: a case–control study. *J Rheumatol,* 24:1522–1526.

Kozma, GT, Falus, A, Bojszko, A, Krikovszky, D, Szabo, T, Nagy, A, et al. (2002). Lack of association between atopic eczema/dermatitis syndrome and polymorphisms in the promoter region of RANTES and regulatory region of MCP-1. *Allergy,* 57:160–163.

Krathen, RA and Hsu, S (2005). Failure of omalizumab for treatment of severe adult atopic dermatitis. *J Am Acad Dermatol,* 53:338–340.

Kuokkanen, S, Gschwend, M, Rioux, JD, Daly, MJ, Terwilliger, JD, Tienari, PJ, et al. (1997). Genomewide scan of multiple sclerosis in Finnish multiplex families. *Am J Hum Genet,* 61:1379–1387.

Laitinen, T, Daly, MJ, Rioux, JD, Kauppi, P, Laprise, C, Petays, T, et al. (2001). A susceptibility locus for asthma-related traits on chromosome 7 revealed by genome-wide scan in a founder population. *Nat Genet,* 28:87–91.

Lammintausta, K, Kalimo, K, Raitala, R, and Forsten, Y (1991). Prognosis of atopic dermatitis. A prospective study in early adulthood. *Int J Dermatol,* 30:563–568.

Lander, E and Kruglyak, L (1995). Genetic dissection of complex traits: guidelines for interpreting and reporting linkage results. *Nat Genet,* 11:241–247.

Lane, JE, Cheyney, JM, Lane, TN, Kent, DE, and Cohen, DJ (2006). Treatment of recalcitrant atopic dermatitis with omalizumab. *J Am Acad Dermatol,* 54:68–72.

Lange, J, Heinzmann, A, Zehle, C, and Kopp, M (2005). CT genotype of promotor polymorphism C159T in the CD14 gene is associated with lower prevalence of atopic dermatitis and lower IL-13 production. *Pediatr Allergy Immunol,* 16:456–457.

Larsen, FS, Holm, NV, and Henningsen, K (1986). Atopic dermatitis. A genetic-epidemiologic study in a population-based twin sample. *J Am Acad Dermatol,* 15:487–494.

Lee, YA, Wahn, U, Kehrt, R, Tarani, L, Businco, L, Gustafsson, D, et al. (2000). A major susceptibility locus for atopic dermatitis maps to chromosome 3q21. *Nat Genet,* 26:470–473.

Lin, S, Cicala, C, Scharenberg, AM, and Kinet, JP (1996). The Fc(epsilon)RIbeta subunit functions as an amplifier of Fc(epsilon)RIgamma-mediated cell activation signals. *Cell,* 85:985–995.

Liu, X, Nickel, R, Beyer, K, Wahn, U, Ehrlich, E, Freidhoff, LR, et al. (2000). An IL13 coding region variant is associated with a high total serum IgE level and atopic dermatitis in the German multicenter atopy study (MAS-90). *J Allergy Clin Immunol,* 106:167–170.

Lorenz, E, Mira, JP, Cornish, KL, Arbour, NC, and Schwartz, DA (2000). A novel polymorphism in the toll-like receptor 2 gene and its potential association with staphylococcal infection. *Infect Immun,* 68:6398–6401.

Luoma, R, Koivikko, A, and Viander, M (1983). Development of asthma, allergic rhinitis and atopic dermatitis by the age of five years. A prospective study of 543 newborns. *Allergy,* 38:339–346.

Maestrini, E, Monaco, AP, McGrath, JA, Ishida-Yamamoto, A, Camisa, C, Hovnanian, A, et al. (1996). A molecular defect in loricrin, the major component of the cornified cell envelope, underlies Vohwinkel's syndrome. *Nat Genet,* 13:70–77.

Magert, HJ, Kreutzmann, P, Standker, L, Walden, M, Drogemuller, K, and Forssmann, WG (2002). LEKTI: a multidomain serine proteinase inhibitor with pathophysiological relevance. *Int J Biochem Cell Biol,* 34:573–576.

Magert, HJ, Standker, L, Kreutzmann, P, Zucht, HD, Reinecke, M, Sommerhoff, CP, et al. (1999). LEKTI, a novel 15-domain type of human serine proteinase inhibitor. *J Biol Chem,* 274:21499–21502.

Majewski, S, Favre, M, Orth, G, and Jablonska, S (1998). Is human papillomavirus type 5 the putative autoantigen involved in psoriasis? *J Invest Dermatol,* 111:541–542.

Mao, XQ, Shirakawa, T, Enomoto, T, Shimazu, S, Dake, Y, Kitano, H, et al. (1998). Association between variants of mast cell chymase gene and serum IgE levels in eczema. *Hum Hered,* 48:38–41.

Mao, XQ, Shirakawa, T, Yoshikawa, T, Yoshikawa, K, Kawai, M, Sasaki, S, et al. (1996). Association between genetic variants of mast-cell chymase and eczema. *Lancet,* 348:581–583.

Maraganore, DM, de Andrade, M, Lesnick, TG, Strain, KJ, Farrer, MJ, Rocca, WA, et al. (2005). High-resolution whole-genome association study of Parkinson disease. *Am J Hum Genet,* 77:685–693.

Marshall, D, Hardman, MJ, Nield, KM, and Byrne, C (2001). Differentially expressed late constituents of the epidermal cornified envelope. *Proc Natl Acad Sci USA,* 98:13031–13036.

Mathias, RA, Freidhoff, LR, Blumenthal, MN, Meyers, DA, Lester, L, King, R, et al. (2001). Genome-wide linkage analyses of total serum IgE using variance components analysis in asthmatic families. *Genet Epidemiol,* 20:340–355.

McGovern, DP, Hysi, P, Ahmad, T, van Heel, DA, Moffatt, MF, Carey, A, et al. (2005). Association between a complex insertion/deletion polymorphism in NOD1 (CARD4) and susceptibility to inflammatory bowel disease. *Hum Mol Genet,* 14:1245–1250.

McGovern, DP, Van Heel, DA, Negoro, K, Ahmad, T, and Jewell, DP (2003). Further evidence of IBD5/CARD15 (NOD2) epistasis in the susceptibility to ulcerative colitis. *Am J Hum Genet,* 73:1465–1466.

Melnik, B, Hollmann, J, and Plewig, G (1988). Decreased stratum corneum ceramides in atopic individuals—a pathobiochemical factor in xerosis? *Br J Dermatol* 119:547–549.

Meyers, DA, Postma, DS, Stine, OC, Koppelman, GH, Ampleford, EJ, Jongepier, H, et al. (2005). Genome screen for asthma and bronchial hyperresponsiveness: interactions with passive smoke exposure. *J Allergy Clin Immunol,* 115:1169–1175.

Miedzobrodzki, J, Kaszycki, P, Bialecka, A, and Kasprowicz, A (2002). Proteolytic activity of *Staphylococcus aureus* strains isolated from the colonized skin of patients with acute-phase atopic dermatitis. *Eur J Clin Microbiol Infect Dis,* 21:269–276.

Miller, HR and Pemberton, AD (2002). Tissue-specific expression of mast cell granule serine proteinases and their role in inflammation in the lung and gut. *Immunology,* 105:375–390.

Mischke, D, Korge, BP, Marenholz, I, Volz, A, and Ziegler A (1996). Genes encoding structural proteins of epidermal cornification and S100 calcium-binding proteins form a gene complex ("epidermal differentiation complex") on human chromosome 1q21. *J Invest Dermatol,* 106:989–992.

Moffatt, MF AM, Reynolds, NJ, Meggitt, ST, Cookson, WO, and Barker, JN (2005). A Case-Control Study Investigating Candidate Genes in Adult Eczema. *35th Annual European Society of Dermatological Research Meeting.* Tubingen, Germany: Blackwell, A22.

Moffatt, MF, Sharp, PA, Faux, JA, Young, RP, Cookson, WO, and Hopkin, JM (1992). Factors confounding genetic linkage between atopy and chromosome 11q. *Clin Exp Allergy,* 22:1046–1051.

Monks, SA, Leonardson, A, Zhu, H, Cundiff, P, Pietrusiak, P, Edwards, S, et al. (2004). Genetic inheritance of gene expression in human cell lines. *Am J Hum Genet,* 75:1094–1105.

Morley, M, Molony, CM, Weber, TM, Devlin, JL, Ewens, KG, Spielman, RS, et al. (2004). Genetic analysis of genome-wide variation in human gene expression. *Nature,* 430:743–747.

Nagata, N, Oshida, T, Yoshida, NL, Yuyama, N, Sugita, Y, Tsujimoto, G, et al. (2003). Analysis of highly expressed genes in monocytes from atopic dermatitis patients. *Int Arch Allergy Immunol,* 132:156–167.

Nickel, RG, Casolaro, V, Wahn, U, Beyer, K, Barnes, KC, Plunkett, BS, et al. (2000). Atopic dermatitis is associated with a functional mutation in the promoter of the C-C chemokine RANTES. *J Immunol,* 164:1612–1616.

Nishio, Y, Noguchi, E, Shibasaki, M, Kamioka, M, Ichikawa, E, Ichikawa, K, et al. (2003). Association between polymorphisms in the SPINK5 gene and atopic dermatitis in the Japanese. *Genes Immun,* 4:515–517.

Nomura, I, Gao, B, Boguniewicz, M, Darst, MA, Travers, JB, Leung, DY (2003). Distinct patterns of gene expression in the skin lesions of atopic dermatitis and psoriasis: a gene microarray analysis. *J Allergy Clin Immunol,* 112:1195–1202.

Nomura, I, Goleva, E, Howell, MD, Hamid, QA, Ong, PY, Hall, CF, et al. (2003). Cytokine milieu of atopic dermatitis, as compared to psoriasis, skin prevents induction of innate immune response genes. *J Immunol,* 171:3262–3269.

Novak, N and Bieber, T (2003). Allergic and nonallergic forms of atopic diseases. *J Allergy Clin Immunol,* 112:252–262.

Novak, N, Allam, JP, and Bieber, T (2003). Allergic hyperreactivity to microbial components: a trigger factor of "intrinsic" atopic dermatitis? *J Allergy Clin Immunol,* 112:215–216.

Novak, N, Kraft, S, Haberstok, J, Geiger, E, Allam, P, and Bieber, T (2002). A reducing microenvironment leads to the generation of FcepsilonRIhigh inflammatory dendritic epidermal cells (IDEC). *J Invest Dermatol,* 119:842–849.

Novembre, E, Cianferoni, A, Lombardi, E, Bernardini, R, Pucci, N, and Vierucci, A (2001). Natural history of "intrinsic" atopic dermatitis. *Allergy,* 56:452–453.

Ober, C, Tsalenko, A, Parry, R, and Cox, NJ (2000). A second-generation genomewide screen for asthma-susceptibility alleles in a founder population. *Am J Hum Genet,* 67:1154–1162.

Ogura, Y, Bonen, DK, Inohara, N, Nicolae, DL, Chen, FF, Ramos, R, et al. (2001). A frameshift mutation in NOD2 associated with susceptibility to Crohn's disease. *Nature,* 411:603–606.

Ong, PY, Ohtake, T, Brandt, C, Strickland, I, Boguniewicz, M, Ganz, T, et al. (2002). Endogenous antimicrobial peptides and skin infections in atopic dermatitis. *N Engl J Med,* 347:1151–1160.

Oppel, T, Schuller, E, Gunther, S, Moderer, M, Haberstok, J, Bieber, T, et al. (2000). Phenotyping of epidermal dendritic cells allows the differentiation between extrinsic and intrinsic forms of atopic dermatitis. *Br J Dermatol,* 143:1193–1198.

Ozaki, K, Ohnishi, Y, Iida, A, Sekine, A, Yamada, R, Tsunoda, T, et al. (2002). Functional SNPs in the lymphotoxin-alpha gene that are associated with susceptibility to myocardial infarction. *Nat Genet,* 32:650–654.

Ozkaya, E (2005). Adult-onset atopic dermatitis. *J Am Acad Dermatol,* 52:579–582.

Palmer, LJ, Burton, PR, Faux, JA, James, AL, Musk, AW, and Cookson, WO (2000). Independent inheritance of serum immunoglobulin E concentrations and airway responsiveness. *Am J Respir Crit Care Med,* 161:1836–1843.

Palmer, CN, Irvine, AD, Terron-Kwiatkowski, A, Zhao, Y, Liao, H, Lee, SP, et al. (2006). Common loss-of-function variants of the epidermal barrier protein filaggrin are a major predisposing factor for atopic dermatitis. *Nat Genet,* 38 (4):441–446.

Park, CW, Lee, BH, Han, HJ, Lee, CH, and Ahn, HK (2005). Tacrolimus decreases the expression of eotaxin, CCR3, RANTES and interleukin-5 in atopic dermatitis. *Br J Dermatol,* 152:1173–1181.

Pasare, C and Medzhitov, R (2005). Control of B-cell responses by Toll-like receptors. *Nature,* 438:364–368.

Pascale, E, Tarani, L, Meglio, P, Businco, L, Battiloro, E, Cimino-Reale, G, et al. (2001). Absence of association between a variant of the mast cell chymase gene and atopic dermatitis in an Italian population. *Hum Hered,* 51:177–179.

Punnonen, J, Aversa, G, Cocks, BG, McKenzie, AN, Menon, S, Zurawski, G, et al. (1993). Interleukin 13 induces interleukin 4-independent IgG4 and IgE synthesis and CD23 expression by human B cells. *Proc Natl Acad Sci USA,* 90:3730–3734.

Rafatpanah, H, Bennett, E, Pravica, V, McCoy, MJ, David, TJ, Hutchinson, IV, et al. (2003). Association between novel GM-CSF gene polymorphisms and the frequency and severity of atopic dermatitis. *J Allergy Clin Immunol,* 112:593–598.

Ramoz, N, Rueda, LA, Bouadjar, B, Favre, M, and Orth, G (1999). A susceptibility locus for epidermodysplasia verruciformis, an abnormal predisposition to infection with the oncogenic human papillomavirus type 5, maps to chromosome 17qter in a region containing a psoriasis locus. *J Invest Dermatol,* 112:259–263.

Riedler, J, Braun-Fahrlander, C, Eder, W, Schreuer, M, Waser, M, Maisch, S, et al. (2001). Exposure to farming in early life and development of asthma and allergy: a cross-sectional survey. *Lancet,* 358:1129–1133.

Risch, N and Merikangas, K (1996). The future of genetic studies of complex human diseases. *Science,* 273:1516–1517.

Romagnani, S (2000). The role of lymphocytes in allergic disease. *J Allergy Clin Immunol,* 105:399–408.

Rystedt, I (1985). Long term follow-up in atopic dermatitis. *Acta Derm Venereol Suppl (Stockh),* 114:117–120.

Sawcer, S, Jones, HB, Feakes, R, Gray, J, Smaldon, N, Chataway, J, et al. (1996). A genome screen in multiple sclerosis reveals susceptibility loci on chromosome 6p21 and 17q22. *Nat Genet,* 13:464–468.

Schadt, EE, Monks, SA, Drake, TA, Lusis, AJ, Che, N, Colinayo, V, et al. (2003). Genetics of gene expression surveyed in maize, mouse and man. *Nature,* 422:297–302.

Schafer, BW, Wicki, R, Engelkamp, D, Mattei, MG, and Heizmann, CW (1995). Isolation of a YAC clone covering a cluster of nine S100 genes on human chromosome 1q21: rationale for a new nomenclature of the S100 calcium-binding protein family. *Genomics,* 25:638–643.

Schultz Larsen, F (1993). Atopic dermatitis: a genetic-epidemiologic study in a population-based twin sample. *J Am Acad Dermatol,* 28:719–723.

Schultz-Larsen, F and Hanifin, J (2002). Epidemiology of atopic dermatitis. *Immunol Allergy Clin North Am,* 22:1–24.

Seguchi, T, Cui, CY, Kusuda, S, Takahashi, M, Aisu, K, and Tezuka, T (1996). Decreased expression of filaggrin in atopic skin. *Arch Dermatol Res,* 288:442–446.

Smith, FJ, Irvine, AD, Terron-Kwiatkowski, A, Sandilands, A, Campbell, LE, Zhao, Y, et al. (2006). Loss-of-function mutations in the gene encoding filaggrin cause ichthyosis vulgaris. *Nat Genet,* 38:337–342.

Soderhall, C, Bradley, M, Kockum, I, Wahlgren, CF, Luthman, H, and Nordenskjold, M (2001). Linkage and association to candidate regions in Swedish atopic dermatitis families. *Hum Genet,* 109:129–135.

Song, PI, Park, YM, Abraham, T, Harten, B, Zivony, A, Neparidze ,N, et al. (2002). Human keratinocytes express functional CD14 and toll-like receptor 4. *J Invest Dermatol,* 119:424–432.

Sugiura, H, Ebise, H, Tazawa, T, Tanaka, K, Sugiura, Y, Uehara, M, et al. (2005). Large-scale DNA microarray analysis of atopic skin lesions shows overexpression of an epidermal differentiation gene cluster in the alternative pathway and lack of protective gene expression in the cornified envelope. *Br J Dermatol,* 152:146–149.

Taha RA, Minshall EM, Leung DY, Boguniewicz M, Luster A, Muro S, et al. (2000). Evidence for increased expression of eotaxin and monocyte chemotactic protein-4 in atopic dermatitis. *J Allergy Clin Immunol,* 105:1002–1007.

Tanaka, K, Sugiura, H, Uehara, M, Sato, H, Hashimoto-Tamaoki, T, and Furuyama, J (1999). Association between mast cell chymase genotype and atopic eczema: comparison between patients with atopic eczema alone and those with atopic eczema and atopic respiratory disease. *Clin Exp Allergy,* 29:800–803.

Tay, YK, Khoo, BP, and Goh, CL (1999). The profile of atopic dermatitis in a tertiary dermatology outpatient clinic in Singapore. *Int J Dermatol,* 38:689–692.

Taylor, B, Wadsworth, J, Wadsworth, M, and Peckham, C (1984). Changes in the reported prevalence of childhood eczema since the 1939–45 war. *Lancet,* 2:1255–1257.

The International HapMap Project (2003). *Nature,* 426:789–796.

Thomas, WR, Smith, WA, Hales, BJ, Mills, KL, and O'Brien, RM (2002). Characterization and immunobiology of house dust mite allergens. *Int Arch Allergy Immunol,* 129:1–18.

Tomfohrde, J, Silverman, A, Barnes, R, Fernandez-Vina, MA, Young, M, Lory, D, et al. (1994). Gene for familial psoriasis susceptibility mapped to the distal end of human chromosome 17q. *Science,* 264:1141–1145.

Tosh, K, Meisner, S, Siddiqui, MR, Balakrishnan, K, Ghei, S, Golding, M, et al. (2002). A region of chromosome 20 is linked to leprosy susceptibility in a South Indian population. *J Infect Dis,* 186:1190–1193.

Traherne, JA, Hill, MR, Hysi, P, D'Amato, M, Broxholme, J, Mott, R, et al. (2003). LD mapping of maternally and non-maternally derived alleles and atopy in FcepsilonRI-beta. *Hum Mol Genet,* 12:2577–2585.

Trembath, RC, Clough, RL, Rosbotham, JL, Jones, AB, Camp, RD, Frodsham, A, et al. (1997). Identification of a major susceptibility locus on chromosome 6p and evidence for further disease loci revealed by a two stage genome-wide search in psoriasis. *Hum Mol Genet,* 6:813–820.

Tsunemi, Y, Saeki, H, Nakamura, K, Sekiya, T, Hirai, K, Fujita, H, et al. (2002). Eotaxin gene single nucleotide polymorphisms in the promoter and exon regions are not associated with susceptibility to atopic dermatitis, but two of them in the promoter region are associated with serum IgE levels in patients with atopic dermatitis. *J Dermatol Sci,* 29:222–228.

Tsunemi, Y, Saeki, H, Nakamura, K, Sekiya, T, Hirai, K, Kakinuma, T, et al. (2002). Interleukin-13 gene polymorphism G4257A is associated with atopic dermatitis in Japanese patients. *J Dermatol Sci,* 30:100–107.

Turner, H and Kinet, JP (1999). Signalling through the high-affinity IgE receptor Fc epsilonRI. *Nature,* 402:B24–B30.

Van Eerdewegh, P, Little, RD, Dupuis, J, Del Mastro, RG, Falls, K, Simon, J, et al. (2002). Association of the ADAM33 gene with asthma and bronchial hyperresponsiveness. *Nature,* 418:426–430.

Vasilopoulos, Y, Cork, MJ, Murphy, R, Williams, HC, Robinson, DA, Duff, GW, et al. (2004). Genetic association between an AACC insertion in the 3'UTR of the stratum corneum chymotryptic enzyme gene and atopic dermatitis. *J Invest Dermatol,* 123:62–66.

Vorechovsky, I, Webster, AD, Plebani, A, and Hammarstrom, L (1999). Genetic linkage of IgA deficiency to the major histocompatibility complex: evidence for allele segregation distortion, parent-of-origin penetrance differences, and the role of anti-IgA antibodies in disease predisposition. *Am J Hum Genet,* 64:1096–1109.

Wachter, AM and Lezdey, J (1992). Treatment of atopic dermatitis with alpha 1-proteinase inhibitor. *Ann Allergy,* 69:407–414.

Walley, AJ, Chavanas, S, Moffatt, MF, Esnouf, RM, Ubhi, B, Lawrence, R, et al. (2001). Gene polymorphism in Netherton and common atopic disease. *Nat Genet,* 29:175–178.

Wang, JY, Lin, CG, Bey, MS, Wang, L, Lin, FY, Huang, L, et al. (2005). Discovery of genetic difference between asthmatic children with high IgE level and normal IgE level by whole genome linkage disequilibrium mapping using 763 autosomal STR markers. *J Hum Genet,* 50:249–258.

Warram, JH, Krolewski, AS, Gottlieb, MS, and Kahn, CR (1984). Differences in risk of insulin-dependent diabetes in offspring of diabetic mothers and diabetic fathers. *N Engl J Med,* 311:149–152.

Watanabe, N, Tomimori, Y, Saito, K, Miura, K, Wada, A, Tsudzuki, M, et al. (2002). Chymase inhibitor improves dermatitis in NC/Nga mice. *Int Arch Allergy Immunol,* 128:229–234.

Weidinger, S, Rummler, L, Klopp, N, Wagenpfeil, S, Baurecht, HJ, Fischer, G, et al. (2005). Association study of mast cell chymase polymorphisms with atopy. *Allergy,* 60:1256–1261.

Williams, HC, Burney, PG, Pembroke, AC, and Hay, RJ (1994). The U.K. Working Party's Diagnostic Criteria for Atopic Dermatitis. III. Independent hospital validation. *Br J Dermatol,* 131:406–416.

Williams, HC (2000). Epidemiology of atopic dermatitis. *Clin Exp Dermatol,* 25:522–529.

Williams, CM and Galli, SJ (2000). The diverse potential effector and immunoregulatory roles of mast cells in allergic disease. *J Allergy Clin Immunol,* 105:847–859.

Williams, H, Robertson, C, Stewart, A, Ait-Khaled, N, Anabwani, G, Anderson, R, et al. (1999). Worldwide variations in the prevalence of symptoms of atopic eczema in the International Study of Asthma and Allergies in Childhood. *J Allergy Clin Immunol,* 103:125–138.

Wjst, M, Fischer, G, Immervoll, T, Jung, M, Saar, K, Rueschendorf, F, et al. (1999). A genome-wide search for linkage to asthma. German Asthma Genetics Group. *Genomics,* 58:1–8.

Yamamoto, K and Yamada, R (2005). Genome-wide single nucleotide polymorphism analyses of rheumatoid arthritis. *J Autoimmun,* 25(Suppl.):12–5.

Yawalkar, N, Uguccioni, M, Scharer, J, Braunwalder, J, Karlen, S, Dewald, B, et al. (1999). Enhanced expression of eotaxin and CCR3 in atopic dermatitis. *J Invest Dermatol,* 113:43–48.

Young, RP, Sharp, PA, Lynch, JR, Faux, JA, Lathrop, GM, Cookson, WO, et al. (1992). Confirmation of genetic linkage between atopic IgE responses and chromosome 11q13. *J Med Genet,* 29:236–238.

Zhang, Y, Leaves, NI, Anderson, GG, Ponting, CP, Broxholme, J, Holt, R, et al. (2003). Positional cloning of a quantitative trait locus on chromosome 13q14 that influences immunoglobulin E levels and asthma. *Nat Genet,* 34:181–186.

34

Diseases of the Epidermis and Appendages, Skin Pigmentation, and Skin Cancer

Eugene Healy, Alan D Irvine, John T Lear, and Colin S Munro

Skin, the largest organ in the body, has multiple functions. It forms a physical envelope that is tough but flexible and capable of self-repair; it facilitates touch and other sensory inputs, and enables grip, friction, and dexterity; it has major roles in homeostasis of water and heat; it must defend itself and the body against physical and chemical attack and ultraviolet radiation (UVR); it functions in both innate and acquired immunity; and it manufactures vitamin D on exposure to UV light. Finally, skin, its appendages, and vasculature play important roles in nonverbal communication.

The structure of skin is correspondingly complex. Its principal structural elements are dermis and epidermis. Dermis is composed mainly of fibrillar proteins, collagen, and elastin, which are manufactured by resident fibroblasts. Within this connective tissue matrix are found blood and lymphatic vasculature, neural tissues, epidermal appendages and smooth muscle, and resident and migratory immunocytes (lymphocytes, mast cells, macrophages, and dendritic cells). In the epidermis, bound to the upper dermis by specific structures, the dominant cell type is the keratinocyte. Keratinocytes arise by division and proliferation from stem cells in the basal layer of the epidermis; differentiate to form the spinous layer, which has structural, immunological, and metabolic functions; and ultimately transform into the waterproof and wear-resistant cornified envelopes of the stratum corneum. The epidermis is host to melanocytes and Merkel cells (sensory cells of neural crest origin) in the basal layer, and resident immunocytes in the form of Langerhans' cells in the spinous layer. Appendageal structures arise from the epidermis under dermal influence, but extend into the dermis and may contain dermally derived elements and innervation. Such appendages include hair follicles with associated sebaceous glands and apocrine sweat glands, eccrine sweat glands, and nails. Skin exhibits significant topological variation in structure, reflecting specific roles, for example, in the appendages of hair bearing or flexural skin, or the thickened and ridged epidermis of palmoplantar skin.

The potential for genetic polymorphisms and disorders to affect skin is large. It is readily observed that "normal" skin varies between individuals and populations, leading to distinction by features such as pigmentation of skin and hair, the distribution of terminal hair growth or hair structure. Other less apparent variances include age or hormonally related phenomena such as "dry" or "greasy" skin, acne, predisposition to flushing, pigmentary response to radiation, number of melanocytic nevi (moles), and senescent change (e.g., baldness or gray hair). Aging changes induced by environmental factors such as UVR or tobacco toxins are likely also to be genetically modulated.

Variations in the properties of skin are common in genetic disease. Banal cutaneous manifestations [e.g., hirsutes, pigmented macules, or acanthosis nigricans (AN)] may signal the presence of a systemic disorder. However, where defects relate to functions, cell types, or pathways that are specific to skin, cutaneous features predominate. The former category is vast. This chapter will deal with specifically or importantly cutaneous conditions where there is evidence of significant genomic influence. The topics addressed are the epidermis and its appendages including hair, cutaneous pigmentation, and the major cutaneous malignancies, basal cell carcinomas (BCCs), squamous cell carcinomas (SCCs), and melanoma.

Box 34–1 Epidermis and Its Appendages (Box 34–1) Key points

- Common variances such as dry skin, acne, and pattern alopecia are due to genetic influences on epidermal and appendageal physiology .
- There are many inherited disorders of the epidermatis and appendages, but many genetically determined diseases primarily affecting other tissues produce direct or indirect effects on the epidermis.

The Epidermis

The epidermis is constantly renewed by proliferation of stem cells resident in the basal layer. Keratinocyte precursors are amplified by division in this layer before migration to the upper layers, where they differentiate sequentially to perform two major physical roles. In the spinous layer, they maintain epidermal integrity. Later, terminal differentiation in the granular layer results in the formation of a water and chemical resistant barrier in the horny layer (stratum corneum). Ultimately, keratinocytes are shed as squames.

The study of single gene disorders affecting the epidermis has provided considerable insight into the biology of the epidermis (Table 34–1). Basal and spinous layer keratinocytes contain an intracellular network of intermediate filaments, anchored to

Table 34–1 Genetic Disorders of Cornification

Ichthyoses	OMIM #	Inh	Cutaneous Findings	Extracutaneous Findings	Gene Defect(s)	Protein(s)	Class of Protein/ Function
Bullous congenital ichthyosiform erythroderma (epidermolytic hyperkeratosis)	113800	AD	Warty hyperkeratosis	None	*KRT1* and *KRT10*	Keratins 1 and 10	Cytoskeleton structural protein
Chanarin–Dorfman syndrome (Neutral lipid storage disease; also termed NCIE2)	275630	AR	Fine scales with occasional background erythema	Myopathy Hepatosplenomegaly	*CGI58*	CGI-58	Enzyme, a member of the esterase/ lipase/thioesterase subfamily
CHILD syndrome	308050	XD	Unilateral ichthyosiform erythroderma	Chondrodysplasia punctata Cataracts Limb reduction defects Asymmetric organ hypoplasia	*NSDHL*	3-β-hydroxysteroid-delta(8), delta (7)-isomerase	Enzyme involved in cholesterol biosynthesis
CHIME syndrome	280000	AR	Ichthyotic erythema Occasionally migratory plaques	Colobomas Conductive hearing loss Mental retardation	NK	NK	NK
Cyclic ichthyosis with epidermolytic hyperkeratosis	607602	AD	Cyclical occurrence of polycyclic hyperkeratotic plaques	None	*KRT1*	Keratins 1	Cytoskeleton structural protein
Ectodermal dysplasia/skin fragility syndrome	604536	AR	Skin fragility Keratotic plaques on limbs Palmoplantar keratoderma Alopecia	Diminished sweating	*PKP1*	Plakophilin-1	Desmosomal component
Familial peeling skin syndrome	270300	AR	Superficial acral peeling	None	*TGM5*	Keratinocyte transglutaminase 5	Enzyme
Gaucher syndrome, type 2	230900	AR	Collodion baby, mild scaling later	Hepatosplenomegaly retroflexion of the head, strabismus, dysphagia, choking spells, hypertonicity Death usually occurs in the first year	*GBA*	Acid β-glucosidase	Enzyme
Harlequin ichthyosis	242500	AR	Rigid plates	Ectropion, Eclabion	*ABCA12*	ATP-binding cassette (ABC), subfamily a, member 12	ABC transporter
Ichthyosis bullosa of Siemens	146800	AD	Mild flexural hyperkeratosis	None	*KRT2E*	Keratin 2 e	Cytoskeletal structural protein in suprabasal layer
Ichthyosis hystrix (Curth Macklin)	146590	AD	Spiky hyperkeratosis	None	*KRT1*	Keratin 1	Cytoskeletal structural protein
	146600						
Ichthyosis vulgaris	146700	AD Semiconductor	Fine, white scale	None	FLG	Proflagrin	Structural component of stratum corneum
IFAP syndrome (ichthyosis follicularis)	398205	XR Semiconductor			NK	NK	NK

(Continued)

			Spiny follicular ichthyosis Nail dystrophy Alopecia	Photophobia Psychomotor delay Short stature			
Keratitis ichthyosis deafness syndrome (KID; includes HID syndrome)	242150 602540	AD	Veruccous plaques Stippled pattern of keratoderma	Keratitis Sensorineural deafness	*GJB2*	Connexin 26	Gap junction protein
Lamellar ichthyosis Type-1	242300	AR	Large adherent plates	None	*TGM1*	Keratinocyte transglutaminase 1	Enzyme involved in cross linking of stratum corneum
Lamellar ichthyosis Type-2	601277	AR	Large adherent plates	None	*ABCA12*	ATP-binding cassette, subfamily a, member 12	ABC transporter
Lamellar ichthyosis Type-3	604777	AR	Large adherent plates	None	Mapped to 19p12-q12	NK	NK
Lamellar ichthyosis (autosomal dominant form)	146750	AD	Large adherent plates	None	NK	NK	NK
Multiple sulfatase deficiency	272200	AR	Mild scale	Mental retardation Mucopolysaccharidosis Metachromatic leukodystrophy	*SUMF1*	Sulfatase-modifying factor-1	Modifier of sulfatase enzyme activity
Netherton syndrome	256500	AR	Erythroderma in infancy Ichthyosis linearis circumflexa	Atopic diathesis, food allergies Structural hair defects (trichorrhexis nodosa) Growth delay	*SPINK5*	LETKI	Serine protease inhibitor
Neu–Laxova syndrome	256520	AR	Variable: mild scaling to harlequin ichthyosis appearance	Intrauterine growth retardation Microcephaly Sensorineural hearing loss Subcapsular cataracts Nystagmus, Strabismus Mental retardation	NK	NK	NK
Non-bullous congenital ichthyosiform erythroderma	242100	AR	Fine white scales, background erythema	None	*TGM1* *ALOX12B* ALOXE3 Also mapped to loci on 4q23	Keratinocyte transglutaminase 1 Arachidonate 12-lipoxygenase, r type Arachidonate lipoxygenase 3 NK	Transglutaminase, cornified envelope cross linking Lipoxygenase Epoxy alcohol synthase NK
Nonbullous/congenital ichthyosiform erythroderma	None assigned	AR	Fine white scales, background erythema Mild PPK White nails	None	*ICHTHYIN*	Ichthyin, member of DUF803 protein family	Uncertain, probably a membrane-bound receptor
Nonlamellar, non-erythrodermic ichthyosis phenotype	604781	AR	Fine, nonerythematous scaling more prominent in the knees, ankles, and ears	None	Mapped to 19p13.1-p13.2	NK	NK
Refsum's disease	266500	AR	Late onset, fine scale	Retinitis pigmentosa Cardiac failure	*PAHX* or *PHYH* *PEX7*		

(Continued)

Table 34–1 Genetic Disorders of Cornification *(Continued)*

Ichthyoses	OMIM #	Inh	Cutaneous Findings	Extracutaneous Findings	Gene Defect(s)	Protein(s)	Class of Protein/ Function
						Phytanoyl-CoA hydroxylase Peroxin-7	Enzymes involved in phytanic acid metabolism
Sjögren–Larsson syndrome	270200	AR	Fine lamellar scale	Di- or tetraplegia Retinal glistening white dots			
Trichothiodystrophy syndromes	601675	AR	May have collodion membrane Can vary from mild scaling to marked adherent plaques	Photosensitivity Brittle hair with "tiger tail" pattern Decreased fertility Short stature Susceptibility to infection	*XPD* or *ERCC2* *XPB* or *ERCC3*	Xeropigmentosum group D protein Xeropigmentosum group B protein	DNA repair enzymes also involved in regulation of transcription
X-linked Chondrodysplasia punctata (Conradi–Hünermann syndrome)	302960	XD	Striated ichthyosiform hyperkeratosis Follicular atrophoderma Alopecia	Cataracts Frontal bossing Short proximal limbs	*EBP*	Emopamil-binding protein	Enzyme involved cholesterol biosynthesis
X-linked ichthyosis	308100	XR	Large, dark scales	Prolongation of labor Cryptorchidism Corneal opacities	*STS*	Steroid sulfatase	Enzyme
Focal keratoses							
Darier–White disease (including acral hemorrhagic variant)	124200	AD	Yellow-brown, greasy papules	Neuropsychiatric features Frequent HSV infections	*ATP2A2*	SERCA2 Ca^{2+}-ATPase	Ca^{2+} ion pump protein
Keratosis pilaris	604093	AD	Follicular hyperkeratosis	None	NK	NK	NK
Keratosis pilaris spinulosa decalvans	308800	XD	Follicular hyperkeratosis	Corneal degeneration Alopecia, loss of eyebrows	Mapped to Xp22.1, some evidence that mutations in SAT may be relevant	*SAT* encodes Spermidine/spermine N(1)-acetyltransferase	Enzyme (N(1)-acetylation)
Erythrokeratodermias							
Erythrokeratodermia variabilis	133200	AD	Hyperkeratotic patches Figurate erythema	None	*GJB3* *GJB4*	Connexin 31 Connexin 30.3	Gap junction protein
Erythrokeratodermia variabilis with erythema gyratem repens	133200	AD	Hyperkeratotic patches Migratory annular morphology	None	*GJB4*	Connexin 30.3	Gap junction protein
Progressive symmetric erythrokeratodermia	602036	AD	Hyperkeratotic patches	None	*LOR*	Loricrin (1 report)	Structural protein in stratum corneum

Abbreviations: AD Autosomal dominant; AR, androgen receptor; ATP, adenosine triphosphate; CGI, comparative gene identification; CHILD, congenital hemidysplasia with ichthyosiform erythroderma and limb defects; CHIME, The colobomas of the eye, heart defects, ichthyosiform dermatosis, mental retardation, and ear defects; HID, hystrix-like ichthyosis-deafnesss; IFAP, Ichthyosis follicularis with atrichia and photophobia; KRT, keratin; NK, OMIM, Online Mendelian inheritance in man.

intercellular junctions (desmosomes). Mutations in desmosomal proteins, intermediate filament keratins, and related structural genes give rise to numerous disorders of epidermal adhesion and integrity, manifest by epidermal fragility, blistering, or hyperkeratosis (Irvine and McLean, 1999; McGrath, 1999). Other gene defects, such as those in genes encoding connexins required for gap junction communication, alter intercellular signaling and growth regulation (Richard, 2000). The epidermis responds to continued physical (and UV) stress by an increase in thickness (acanthosis). Acanthosis of the neck and flexures associated with increased pigmentation is recognized clinically as AN. This is more prevalent in some racial groups; in the presence of obesity, it reflects insulin resistance and predicts diabetes (Hud et al., 1992). It is also seen with genetically determined insulin receptor defects. The mechanism appears to be direct or indirect activation of the insulin-like growth factor receptor by high levels of circulating insulin 1. Activating mutations in related tyrosine kinase receptors, fibroblast growth factor receptor types 2 and 3, mediate AN in genetically determined disorders such as Crouzon, Beare–Stevenson, and SADDAN (severe achondroplasia with developmental delay and AN) syndromes (Torley et al., 2002). Keratinocytes are also important to innate immunity, for example, in production of cytokines or antibacterial peptides, and variation in these responses may be a factor in inflammatory skin disorders such as psoriasis (see Chapter 32) and eczema (see Chapter 33).

Increased understanding of the barrier and defensive functions of the stratum corneum has been reviewed by Elias (2005). Formation of the cornified envelope is a complex process involving specialized proteins such as filaggrin, involucrin, loricrin, envoplakin, and the numerous small proline-rich proteins. Many of these are encoded by genes present in the epidermal differentiation complex (EDC) at 1p21. Under the actions of epidermal transglutaminase, these proteins are cross-linked to form a protein envelope, which is covalently bound to an outer ceramide layer, the lipid envelope. The intercellular space of the stratum corneum contains multilayered lipid-rich lamellae, secreted from the granular layer. It is also rich in hydrolytic enzymes. Desquamation is facilitated by cleavage of cholesterol sulfatase to cholesterol by enzymes present on the cell membrane surface. As with the spinous layer, many rare single gene disorders affecting this process have been identified, producing ichthyosis and/or barrier defects (Table 34–1). The rare recessive ichthyosis, Netherton's syndrome is due to mutations in SPINK5, encoding a serine protease inhibitor, LETKI (Chavanas et al., 2000), and is of particular interest, because the ichthyosis and barrier defect is associated with severe atopic disease. However, a much commoner clinically significant form of ichthyosis, ichthyosis vulgaris (IV), has a prevalence of approximately 1:250 individuals in England (Wells and Kerr, 1966). IV has recently been shown to be due to mutations in the gene encoding profilaggrin, which prevent cleavage of the latter to filaggrin subunits (Smith, Irvine et al., 2006). Filaggrin condenses with keratin intermediate filaments during cornified envelope formation in the granular layer. Although IV has been regarded as an autosomal dominant disorder, it is in fact semidominant; heterozygotes within pedigrees of IV have mild degrees of ichthyosis (Fig. 34–1), and as 7%–9% of Caucasian populations are heterozygous for one of the common mutations; this genotype may be an important cause of "dry skin." Heterozygous loss of mutations in FLG which cause IV also predispose to atopic dermatitis (discussed in Chapter 33), and atopic dermatitis associated with asthma (Palmer et al., 2006). However, the complexity of proteins and enzymes involved in cornification and desquamation raise the possibility that polymorphisms in the other genes of the EDC may also affect cutaneous phenotype and barrier function and immune reactivity. In some types of ichthyosis, there are major systemic associations, for example, defects in lipid metabolism in Refsum, Sjögren–Larssen, and other syndromes. Other manifestations are more subtle; female carriers of X-linked ichthyosis have reduced blood levels of steroid sulfatase, but are asymptomatic. Male fetuses may present with prolonged labor due to placental sulfatase deficiency (Koppe et al., 1978).

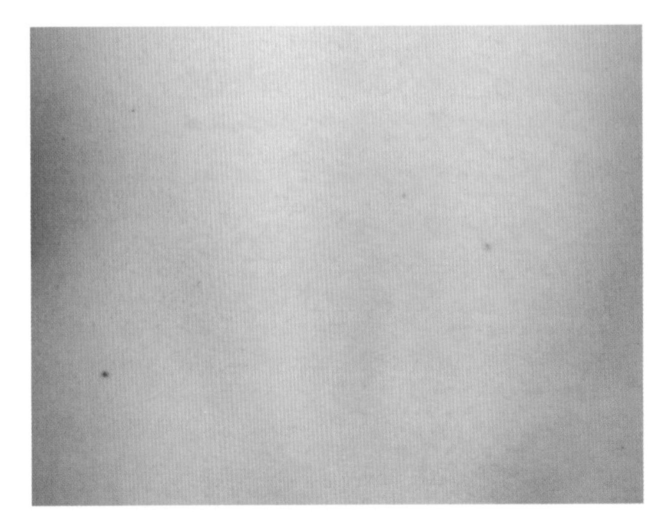

Figure 34–1 "Dry skin": fine scale characteristic of ichthyosis vulgaris in a subject heterozygous for a loss of function mutation in profilaggrin.

The specialized skin of the palms and soles is the site or focus of many disorders of keratinization (Table 34–1), but again milder defects with a presumed genetic basis, such as recurrent palmar peeling, are quite common.

Acne and Other Pilosebaceous Disorders

Acne mainly occurs in areas of epidermis rich in sebaceous glands but not terminal hair (face, chest, and upper back). It principally affects adolescents, and is characterized by occlusion of sebaceous gland orifices producing comedones and secondary inflammatory lesions, cysts, and scars. The major factors in its etiology are the increased sebum excretion rate of adolescence (which is under gonadal hormone influence), occluding hyperkeratosis of the sebaceous duct (hormones again, but also bacterial products), and proliferation of the Corynebacterium, *Propionibacterium acnes.*

Acne is so common as to be normal in adolescence (Healy and Simpson, 1994), but there is strong evidence that the tendency to acne is inherited (Goulden et al., 1999, Bataille et al., 2002). The major genetic factors in this tendency have not been identified, although sebum excretion rate in twins is correlated (Walton et al., 1988). Androgen levels are relevant; acne is a variable feature in polycystic ovarian syndrome (Fratantonio et al., 2005) and congenital adrenal hyperplasia (Trakakis et al., 2005). Severe nodulocystic acne may be a sign of the XYY karyotype (Voorhees et al., 1972). However, circulating androgen levels are not well correlated with acne in either sex, and variations in the pathways of cutaneous androgen metabolism are candidates for phenotype variation (Chen et al., 2002). Polymorphisms in cytochrome P450 (CYP) 1A1 and the polymorphic epidermal mucin gene have been associated with acne (Ando et al., 1998; Paraskevaidis et al., 1998). Acne is a component of Apert's syndrome (Solomon et al., 1970), which is due to specific activating mutations in fibroblast growth factor type 2

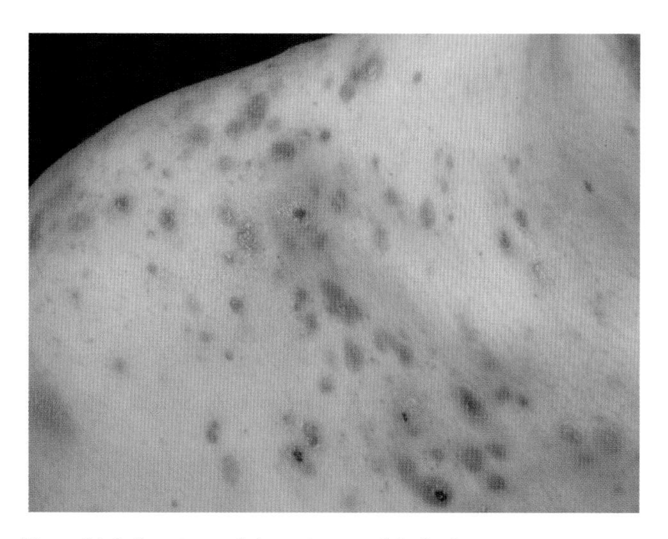

Figure 34–2 Scarring nodulocystic acne of the back.

(Wilkie et al., 1995), indicating the relevance of this signaling pathway, but other pathogenic activating mutations in this gene do not have the same effect.

The severity of acne may also be under genetic control: inflammatory cystic and scarring acne (Fig. 34–2) may in part be due to aberrant inflammatory responses. A syndrome of cystic acne, pyoderma gangrenosum, and sterile pyogenic arthritis is due to mutations in the CD2-binding protein 1, which may affect the signaling pathways necessary for the regulation of inflammatory responses (Wise et al., 2002). Severe acne is also sometimes associated with recurrent flexural abscesses (hidradenitis suppurativa), which may itself be familial (Fitzsimmons et al., 1985). Although termed hidradenitis, apocrine sweat glands are probably only incidentally affected. At least one cause of familial hidradenitis suppurativa is mutation in keratin 17 as part of the pachyonychia congenital/steatocystoma multiplex syndrome (McLean et al., 1995), presumably secondary to follicular occlusion.

Sweat Glands

Sweating is important in thermoregulation, grip (on the palms), and in emotional responses. Hyperhidrosis of palms and soles may be genetically determined (Ro et al., 2002). Apocrine axillary glands produce volatile compounds such as E-3-methyl-2-hexenoic acid, which are carried to the skin surface by binding proteins such as apolipoprotein D, and released by bacterial action (Zeng et al., 1996). Variation in pheromone production and release has obvious implications for reproductive success, but although nasal receptors may be present in man (Rodriguez et al., 2000), this possibility has been little investigated in humans.

In female carriers of X-linked hypohidrotic ectodermal dysplasia, random inactivation of the X-chromosome in epidermal precursors results in a mosaic pattern of sweat gland distribution, following the Lines of Blaschko (Happle and Frosch, 1985).

Hair

Formation of hair is a complex biological process dependent on many genes and influenced by many hundreds of others (reviewed in Irvine and Christiano, 2001). Hair follicle morphogenesis begins in the embryo with placode placement, determined by complex molecular interactions between mesenchyme and epithelium. Progressive development through well-recognized stages, again under the control of mesenchyme–epithelium interactions, leads to formation of a fully differentiated follicle. The final product of this fully differentiated follicle is the hair shaft, structurally composed of three subtypes of epithelial cell: the cuticle, cortical, and medullary cells. The shaft is supported by inner and outer root sheaths, themselves multilayered. Hair follicles remodel themselves in a predictable cyclical fashion throughout life (Cotsarelis and Millar, 2001), synchronously in many mammals but asynchronously in humans. The phases are anagen (growth phase); catagen (regression, where much of the follicle undergoes programmed cell death by apoptosis); telogen (resting phase), and finally exogen (shedding). New anagen follicles are formed from cells derived from epithelial stem cells located at the bulge region of the residual follicle (Cotsarelis et al., 1990).

Single Gene Disorders Associated with Hair Defects

In excess of 100 single gene disorders include an abnormal hair phenotype (Irvine and Christiano, 2001). Defects associated with abnormal hair phenotypes fall into several broad categories. Defects in the morphogenesis of hair, such as those associated with ectodermal dystrophies (Priolo and Lagana, 2001), may result from defective signaling pathways. These include fibroblast growth factors, bone morphogenetic proteins, Notch-1, and δ-1, β-catenin, Lef-1, sonic hedgehog (SHH), and ectodysplasin and its receptor (reviewed in Irvine and Christiano, 2001). Secondly, defects may occur in genes involved in hair cycling. The molecular signals controlling the process are beginning to be elucidated (Stenn and Paus, 2001). Important molecules have been identified as a result of studying single gene disorders. *GJB6* encodes connexin 30, a constituent of gap junctions required for intercytoplasmic cell–cell communication, and is mutated in Clouston syndrome (Lamartine et al., 2000). The human analogue of the mouse hairless gene is mutated in congenital atrichia with cysts (Ahmad et al., 1997). *Hairless* seems to have a crucial role in maintenance of the balance between cell proliferation, differentiation, and apoptosis in the follicle.

A third class of disease is due to defects in hair structure. In monilethrix mutations in hair, basic keratins cause a phenotype of beaded hairs, which show increased fragility and weathering. The severity of alopecia in monilethrix shows marked interindividual variation and often improves at puberty, indicating that other factors contribute to the phenotype. Other affected structural components of hair include desmosomal proteins plakophilin 1 (ectodermal dysplasia/skin fragility) (McGrath et al, 1997) and plakoglobin (woolly hair, striate keratoderma, and ventricular arrhythmias). In Menkes disease, kinking of the hair shaft is associated with mutation in a membrane protein that transports copper from the gut to the bloodstream. Curly, woolly, or uncombable hair is a feature of numerous ectodermal or neuroectodermal syndromes, including tricho-dento-osseous syndrome, giant axonal neuropathy, and pachyonychia congenita. A polymorphism in a gene encoding keratin 6hf, present in the companion layer of the hair follicle, has been shown to predispose to the curled, ingrowing beard hairs, which cause pseudofolliculitis (Winter et al., 2004).

Male- and Female-pattern Alopecia

Age-related pattern alopecia in both sexes is associated with progressive reduction in anagen duration, progressive miniaturization of affected hair follicles and, in men at least, a prolongation of the latent period in the hair cycle (Olsen et al., 2005). In men, many lines of evidence support the critical role of androgens, and the term androgenetic alopecia (AGA) is appropriate (Ellis and Harrap, 2001). Although AGA occurs in women in the context of polycystic ovarian syndrome and other hyperandrogenic disorders, the majority of cases of female-pattern alopecia are likely to be distinct, not being associated with increased androgens and having a different distribution (Olsen et al., 2005). In men, AGA has been regarded as autosomal dominant in males and recessive in females, but Küster and Happle (1984) pointed out that the high prevalence of the trait, the Gaussian distribution curve, and other factors suggest a polygenic trait. Candidate genes investigated include those involved in peripheral androgen metabolism such as 5-α hydroxylase (Ellis et al., 1998), with largely negative results. Ellis, Stebbing, et al. (2001) did identify polymorphisms in the androgen receptor (AR) gene, which were associated with AGA, and Hillmer et al. (2005) suggest that a GGN repeat in exon 1 of AR is a plausible functional cause. However, the AR locus is X-linked, and the strong concordance for baldness between fathers and sons identified in previous studies (e.g., Ellis and Harrap, 2001) is inconsistent with an X-linked trait.

Alopecia Areata

Alopecia areata (AA; MIM 104000) is a common disease in children and adults, characterized by intermittent patchy and circumscribed areas of loss (Fig. 34–3), with an estimated lifetime risk of 1%–2%. Recovery of hair growth is the rule, but the disease may relapse and/or progress to become confluent. In severe cases, extensive hair loss may denude the whole scalp (alopecia totalis) or even the whole body (alopecia universalis). Lymphocytic infiltrates on biopsy and response to high-dose oral steroid argue for an immunological pathogenesis, but evidence from twin and family studies suggests that AA behaves as a complex trait (McDonagh and Tazi-Ahnini, 2002; Martinez-Mir et al., 2003). There is evidence of linkage between AA and major histocompatibility complex (MHC) loci (reviewed in McDonagh and Tazi-Ahnini, 2002), in particular human leukocyte antigen (HLA) alleles DQB1*0301 and DRB1*1104 (Morling et al., 1991; Welsh et al., 1994, de Andrade et al., 1999). Polymorphisms in the interleukin-1 receptor antagonist (*IL-1RN*) gene and the IL-1RN homolog IL-1L1 (Tarlow et al., 1994; Barahmani et al., 2002; Tazi-Ahnini et al., 2002) are associated with AA, and polymorphisms in tumor necrosis factor α (TNFα) may affect its extent (Galbraith and Pandey, 1995). Patients with trisomy 21 (Down syndrome) have an increased incidence of AA. An intron 6 polymorphisms in the *MX1* gene at 21q22.3, encoding the interferon-induced p78 protein MxA, was found to be in strong linkage disequilibrium with mild AA (Tazi-Ahnini et al., 2000). AA was found in 37% of patients with autoimmune polyglandular syndrome type 1, which is caused by mutations in the *AIRE* gene at 21q22.3 (Betterle et al., 1998). A recent genome-wide study of 20 families from the United States and Israel supported the association with the HLA locus and provided evidence of new susceptibility loci on chromosomes 16 and 18 (Martinez-Mir et al., 2007)

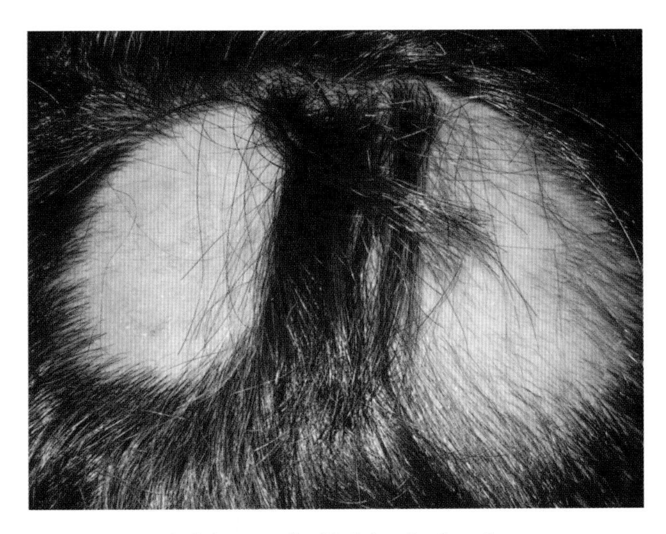

Figure 34–3 Typical circumscribed hair loss in alopecia areata.

Nails

Nails are another specialized ectodermal structure. Nail dystrophy is a feature of disorders of epidermal adhesion and keratinization and ectodermal dystrophies (van Steensel et al., 2004), and genetically influenced inflammatory disease such as psoriasis and AA. Nails may also offer a clue to systemic disorders such as nail–patella syndrome (Fistarol and Itin, 2002). However, more subtle variances in the appearance of nails (such as the visibility of lunulae) and their resistance to environmental damage may reflect genetic heterogeneity. It has been suggested (Zaias et al., 1996; Faergemann et al., 2005) that predisposition to fungal nail infection is genetically determined.

Box 34–2 Cutaneous Pigmentation (Box 34–2) Key points

- Genetic control of skin and hair pigmentation in man is complex and incompletely elucidated.
- Genes controlling skin color include the melanocortin-1 receptor and *SLC24A5*.
- Inherited disorders of pigmentation affect melanocyte migration and survival, synthesis of melanin, and its distribution to keratinocytes.
- The acquired depigmenting disorder vitiligo is strongly genetically influenced.

Skin and hair pigmentation are determined mainly by the total and relative amounts of two types of melanin, brown-black eumelanin and red-yellow pheomelanin in these tissues (Prota and Thomson, 1976; Thody et al., 1991). Melanin is synthesized in membrane-bound organelles called melanosomes by specialized pigment cells (melanocytes) in the hair follicle and in the basal layer of human epidermis, with eumelanin synthesized in ellipsoid melanosomes and pheomelanin produced in spherical melanosomes (Seiji et al., 1963; Jimbow et al., 1983). These melanin-laden melanosomes are transported in vivo along the dendrites of the melanocyte from which they are transferred to the growing hair and to the epidermal keratinocytes. Melanocytes are derived from the neural crest from where they migrate as melanoblasts to the skin during embryogenesis (Rawles, 1947). In humans, the numbers of epidermal melanocytes vary according to the skin site, with greater numbers on the face and genitals, but do not differ significantly across racial groups (Staricco and Pinkus, 1957; Glimcher et al., 1973). However, melanocytes in African-origin skin produce greater numbers of and larger melanosomes than Caucasian melanocytes, and the skin of people from darker-skinned populations generally contains more melanin than that in Caucasian skin (Szabo et al., 1969; Hunt et al., 1995).

Based on the differences in skin cancer incidence and susceptibility to sunburn between racial groups, melanin (in particular eumelanin) is considered photoprotective, preventing UVR-induced deoxyribonucleic acid (DNA) damage in skin (Crombie, 1979b; Kaidbey et al., 1979; Cress and Holly, 1997). This photoprotective role is further supported by the fact that melanin, which has

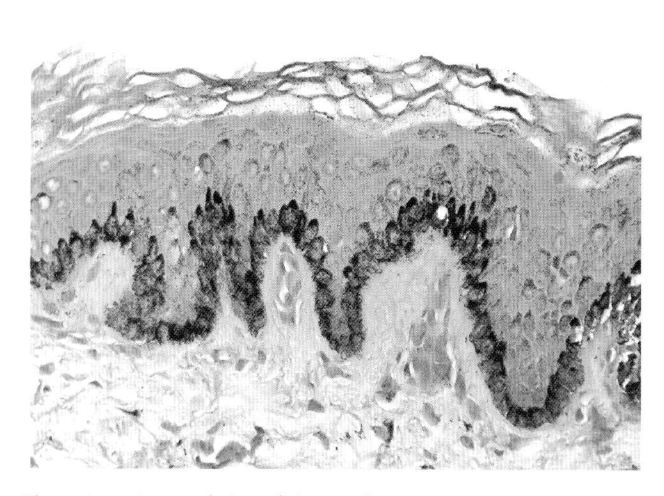

Figure 34–4 Accumulation of pigment-bearing melanosomes overlying basal cell nuclei in normal epidermis.

been transferred into neighboring keratinocytes in the basal layer of human epidermis, tends to cluster above the nucleus as a type of "biological sunhat" (Fig. 34–4). One of the most important environmental influences on human skin pigmentation is UVR from sunshine, which stimulates melanin synthesis and its redistribution within the epidermis, resulting in tanning. However, tanning ability differs between individuals, and this variation in individual tanning responses following UVR, coupled to variation in sunburning responses, have been used as a mechanism for phenotyping humans based on their "skin type."

Normal Variation in Skin and Hair Color

Human skin and hair color varies widely. Normal skin color ranges from very light to very dark, and normal hair color ranges from blond to red to black. In addition, Caucasian people with similar light constitutive skin coloring differ in their ability to tan following exposure to UVR, both after single doses and multiple repeated exposures. This variation in response to UVR was originally used to grade patients with psoriasis according to their "Fitzpatrick skin type" in order to calculate the UVR treatment doses for these patients and avoid UVR-induced burning (Fitzpatrick, 1988). However, Fitzpatrick skin type is based on a combination of erythemal and tanning responses, and many subjects do not fit neatly into the distinct subgroups (Rampen et al., 1988). Despite this, in the absence of simple alternatives, the Fitzpatrick scale or modifications of it have been employed in studies investigating the genetics of normal skin color. Previous studies had suggested that the range in skin color from fair-skinned Caucasians to dark-skinned Africans result from the additive interaction of three to six genes (Harrison and Owen, 1964; Stern, 1970). It is known that more than 120 loci affect pigmentation in mice (Bennett and Lamoureux, 2003), yet only approximately 50 of these loci affect pigmentation in humans. Although investigations have been carried out on several pigmentation genes, looking at their role in normal variation in human skin and hair color, to date genetic alterations at a single locus, the melanocortin-1 receptor (*MC1R*) gene on chromosome 16q24.3, have been comprehensively shown to be involved in this variation. Recent evidence suggests that a second gene, *SLC24A5*, on chromosome 15q21.1, is involved in determining variation in skin color between Caucasian and Asian/African races.

MC1R encodes for a seven-pass transmembrane G-protein-coupled receptor, which is expressed by melanocytes. Its ligand, α melanocyte-stimulating hormone (αMSH), stimulates pigmentation of melanocytes and melanoma cells in culture and causes skin darkening when injected into humans (Lerner and McGuire, 1961). Although the human receptor is only 317 amino acids in length, more than 30 polymorphisms have been identified in the single 951 bp exon (Rees, 2004). Alternative splicing of a second exon can lead to a larger receptor with an additional 65 amino acids (Tan et al., 1999), but no polymorphisms in this second exon have been reported to date and the single exon encoded receptor is thought to be the dominant isoform in human skin and hair. Several investigations on Caucasians worldwide have indicated that *MC1R* variants are associated with red hair and with fair skin type (Valverde et al., 1995; Box et al., 1997; Smith et al., 1998; Flanagan et al., 2000; Healy et al., 2000). Occasional frameshift alterations have been reported in subjects with red hair, with the interpretation being that these alterations lead to the red hair phenotype (Flanagan et al., 2000). Transfection studies have suggested that variant receptors bind αMSH with similar affinity as the wild-type receptor (a single report showed reduced binding with the Val92Met variant, but later studies suggest that this is not the case) (Xu et al., 1996; Frandberg et al., 1998; Schioth et al., 1999; Robinson and Healy, 2002). Several variants (Val60Leu, Val92Met, Arg142His, Arg151Cys, Arg160Trp, and Asp294His), which have been investigated in terms of their capacity to signal via cyclic-AMP, have exhibited reduced signaling capacity, with some variants more functionally compromised than others (Schioth et al., 1999; Robinson and Healy, 2002; Newton et al., 2005). In addition, a recent report suggests that certain variants (Arg151Cys, Arg160Trp) are expressed at a lower level at the cell surface, accounting for some of their compromised function (Beaumont et al., 2005). Transgenic mice have demonstrated that the Arg151Cys, Arg160Trp, and Asp294His variants fail to rescue pigmentation in a knockout mouse model to the same extent as wild-type *MC1R*, indicating that these three variants preferentially promote the synthesis of pheomelanin in vivo (Healy et al., 2001). Indeed, analysis of the odds ratios in the association studies suggests that these variants (Arg151Cys, Arg160Trp, and Asp294His) are more likely to result in red hair and fair skin than a number of other variants (Val60Leu, Val92Met, and Arg163Gln) (Flanagan et al., 2000; Duffy et al., 2004). As a general rule of thumb regarding the Arg151Cys, Arg160Trp, and Asp294His variants, individuals with two variant *MC1R* alleles tend to have red hair whereas those with a single variant *MC1R* allele tend to have fair skin. However, based on a lack of association of some other *MC1R* variants with red hair, and in the absence of signaling studies on these variants, they have been labeled as pseudovariants, with the implication that they do not differ functionally from wild-type *MC1R*. Similarly, the role of certain *MC1R* variants in regulating hair color is unclear, for example, the Val60Leu has been weakly associated with red hair in some studies, but other studies have suggested a weak association with blond/light colored hair (Box et al., 1997; Smith et al., 1998). Single nucleotide polymorphisms (SNPs) have also been reported upstream to the *MC1R* coding region, but to date no association has been found between these SNPs and variation in hair or skin color (Makova et al., 2001; Smith et al., 2001). Perhaps predictably, considering the association between red hair, fair skin, and freckling, *MC1R* variants has also been associated with freckling (Smith et al., 1998; Bastiaens, ter Huurne, Gruis, et al., 2001).

The recent report of an evolutionary conserved ancestral allele Ala111 of *SLC24A5* in the majority of African and East Asian populations but a Thr111 allele in European American populations, in addition to the relationship of these alleles to variation in skin pigmentation levels measured by reflectance in African–American and African–Caribbean admixed populations, suggest that *SLC24A5* is a determinant of human skin color (Lamason et al., 2005). In the same article, the authors showed that zebrafish golden mutants result from an alteration in the fish homolog of this gene, with rescue of darker pigmentation by wild-type *slc24a5*. Indeed, the evidence indicates that *SLC24A5* may account for the differences in melanosome size between African and Caucasian individuals.

Other genes that have been investigated for a role in variation in normal skin and hair color include tyrosinase (*TYR*), TYR-related protein-1 (TRP1), agouti-signaling peptide (*ASIP*), and the *P* gene. No sequence variation in the *TRP1* gene was noted in relation to skin type or hair color (Box et al., 1998), and limited studies of nonpathologic polymorphisms [i.e., which do not cause oculocutaneous albinism type 1 (OCA1)] in the *TYR* gene suggest no relation to normal skin and hair color (Tripathi et al., 1991; Johnston et al., 1992). Although an association between certain SNPs in the *P* gene and eye color has been identified, these SNPs do not seem to influence normal pigmentation in skin and hair, although the *P* gene remains a candidate as a gene, which is likely to influence normal variation in pigmentation at these sites (Zhu et al., 2004). Indeed, in one study in a Tibetan population, an epistatic interaction between the *P* gene and *MC1R* was reported (despite no association being identified between SNPs in *MC1R* or the *P* gene and skin pigmentation in that study) (Akey et al., 2001). An association between a g.8818A-G SNP in the 3' untranslated region of *ASIP* and dark hair and brown eyes has been reported in one study, but this was not the case in an unrelated study (Voisey et al., 2001; Kanetsky et al., 2002). In a third study, which examined skin color by reflectance in African–Americans, the analysis suggested that the g.8818A-G polymorphism leads to lighter skin pigmentation in this population (Bonilla et al., 2005).

Inherited Disorders of Pigmentation

Disease processes affecting pigmentation can be congenital or acquired, and involve many aspects of melanocyte biology including migration and survival of melanoblasts, melanocyte cell viability, melanin synthesis, and melanin distribution from melanocytes to keratinocytes. Many of these disorders of pigmentation can affect animals as well as humans, and much of the knowledge about the genetics of normal and abnormal human pigmentation is derived originally from mouse genetics. This process was chiefly due to the identification of monogenic traits in this animal as a result of inbreeding by mouse fanciers (i.e., individuals with an interest in collecting mice with various coat colors/coat color patterns) during the last century.

Failures of melanocyte development and migration may occur, such as in piebaldism, due to mutations in the *KIT* protooncogene (Fleischman et al., 1991; Giebel and Spritz, 1991, Spritz et al., 1992). KIT is a transmembrane tyrosine kinase receptor whose ligand is stem cell factor (also known as kit ligand, steel factor, and mast cell growth factor). Signaling through the receptor is necessary for melanoblast survival during embryogenesis as they migrate through the mesoderm and for subsequent proliferation of melanoblasts/melanocytes in the epidermis, in addition to its requirement for normal hematopoiesis and gametogenesis (Steel et al., 1992; Yoshida et al., 1996). The various forms of Waardenburg's syndrome (WS types 1–3) are mediated by effects on microphthalmia-associated transcription factor (MITF), which is critical for neural crest as well as melanocyte development, either by mutations of MITF itself or mutations in genes (*PAX3* and *SOX10*) whose products regulate MITF gene expression (Watanabe et al., 1998; Bondurand et al., 2000). WS type 4 (associated with aganglionic megacolon) is caused by mutations in the endothelin-B receptor (*EDNRB*) gene and endothelin-3 gene (*EDB3*, encoding the EDNRB ligand) leading to defects in signaling required for the terminal migration of melanoblasts (and enteric neuron precursor cells) (Edery et al., 1996; Lee et al., 2003).

Other pigmentary disorders are due to defects in melanin synthesis. In OCA1 missense, nonsense, and frameshift mutations in the *TYR* gene have been reported (Tomita et al., 1989; Giebel et al., 1990; King, Mentink et al., 1991). TYR catalyses the formation of various melanin intermediates during the synthesis of melanin [e.g., dihydroxyphenylalanine (DOPA)/dopaquinone from tyrosine, and indole-5,6-quinone from dihydroxyindole]. Based on the clustering of mutations in certain areas (including those encoding copper-binding sites necessary for enzymatic activity), it was considered that mutations in both copies of the *TYR* gene resulted in inactive forms of TYR in the melanosome (King, Mentink et al., 1991). However, TYR normally traffics from the endoplasmic reticulum via the trans-Golgi network to the melanosome, and subsequent investigation identified that mutant TYR may be retained by the endoplasmic reticulum and fail to reach the melanosome (Halaban et al., 2000). A temperature-sensitive variant of OCA1 has also been described, with some peripheral pigmentation on colder areas of skin, including the arms and legs of affected subjects, and is due to the missense mutations in *TYR* allowing the retention of partial TYR activity at lower temperatures (31°C) but resulting in almost complete lack of enzyme activity at 37°C (Giebel et al., 1991; King, Townsend et al., 1991).

OCA2 (TYR-positive albinism) can be phenotypically similar in Caucasians to OCA1, but is more frequent in African populations in whom it causes yellow (or yellow-red) hair and multiple pigmented freckles ranging from millimeters to centimeters in diameter. In contrast to OCA1, pigmented nevi may be present in Caucasian individuals with OCA2. Nystagmus and photophobia may also be present. Missense, nonsense, and frameshift mutations in the *P* gene, the human homolog of the mouse pink-eyed dilution locus, and deletions of this locus are responsible (chromosome 15q11.2-q12) (Rinchik et al., 1993; Durham-Pierre et al., 1994; Lee et al., 1994). The *P* gene encodes for a transmembrane protein, which functions in controlling the pH within the melanosome, and thus *P* gene defects are thought to impair TYR activity due to their failure to regulate the pH (Puri et al., 2000; Ancans et al., 2001). The phenotype in darker subjects suggests that this causes a problem with the synthesis of eumelanin, but that pheomelanin can still be produced. In addition, *MC1R* gene variants can modify the OCA2 phenotype, resulting in red rather than yellow/blond hair in affected individuals (King et al., 2003). In both Prader–Willi and Angelman's syndromes, which result from a combination of genomic imprinting and heterozygous deletion of 15q, the presence of lighter pigmentation may be due to the deletion of a single copy of the *P* gene (Gardner et al., 1992).

In other syndromic disorders, pigmentary anomalies are associated with defects affecting lysosome-related organelles, which include melanosomes. In Hermansky–Pudlak syndrome, of which there are at least seven genotypes, TYR-positive albinism is associated with a platelet defect and pigmented deposits in cells of the reticuloendothelial system (Hermansky and Pudlak, 1959). The phenotype results from mutations causing a variety of defects in the biogenesis of lysosomes and related organelles. Chediak–Higashi syndrome is a rare autosomal recessive syndrome with diluted pigmentation of the skin, hair, and eyes, and markedly increased susceptibility to bacterial and viral infections often resulting in death during the first decade of life. There are abnormal inclusions in a variety of cells, including giant granules in leukocytes. The disease is due to mutations in the lysosomal trafficking regulator (*LYST*) gene on chromosome 1q42-43 (Barbosa et al., 1996; Nagle et al., 1996). This allows the normal synthesis of secretory lysosomes, but these then fuse to form giant secretory lysosomes with defective function (and hence defective neutrophil, natural killer cell activity, etc.) (Abo et al., 1982; Stinchcombe et al., 2000). In melanosomes, the lack of ability to fuse to the cell membrane may prevent their passage to neighboring keratinocytes, and recent evidence also suggests that there may be a failure to repair plasma membrane lesions because of the inability of lysosomes to fuse with the cell membrane (Huynh et al., 2004).

Vitiligo

Vitiligo, a complex disorder in which death of melanocytes in the skin (and also in the hair bulb) results in localized hypopigmented patches, is considered an acquired disease of unknown cause, but there is strong evidence of genetic factors in its pathogenesis. It can occur at any age, but half of all cases begin within the first two decades. It affects all races, but is more obvious (and socially debilitating) in darker-skinned populations. Typically, the patches are white, but lesions in darker-skinned subjects may exhibit a tanned margin between the normal brown skin color and the central white area ["trichrome vitiligo;" (Fig. 34–5)]; in some cases, the entire patch may be this intermediate color. The margins of the lesions are often irregular, appearing to "invade" the normally pigmented skin. The number and size of the lesions and extent of skin involvement is very variable, but it commonly involves the periorificial areas (eyes, mouth, and nose), axillae, groins, genitals, and extensor bony surfaces along the elbows, knees, digits; the latter sites of predilection may be due to repeated trauma at these sites as traumatized skin may develop vitiligo (known as the Koebner phenomenon). The exact mechanism of initial melanocyte cell death is debated, and may relate to the loss of expression of c-kit by melanocytes or the generation of reactive oxygen species, but immunological factors including cytotoxic T-lymphocytes and autoantibodies against melanocyte antigens seem to be involved in the disease process in a proportion of cases (Schallreuter et al., 1991; Norris et al., 1996; Ogg et al., 1998; Kemp et al., 2002).

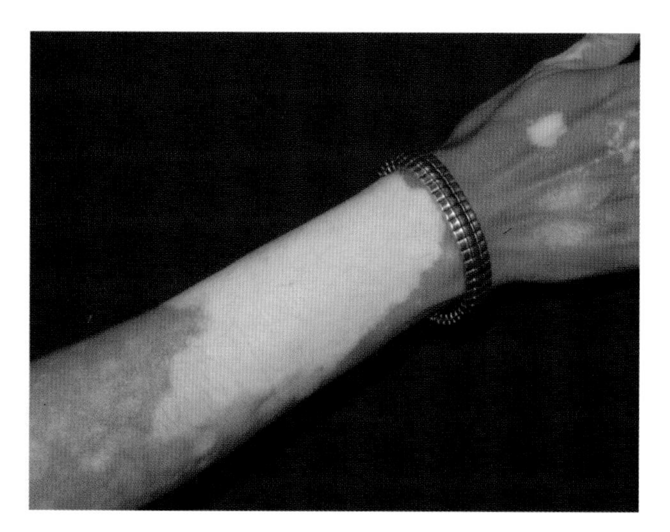

Figure 34–5 Vitiligo in a dark-skinned individual. Note the "trichrome" appearance.

A family history of vitiligo is seen in up to a third of cases (Lerner 1959; Nath et al., 1994; Boisseau-Garsaud et al., 2000). Most of the evidence indicates polygenetic inheritance, with one genetic model suggesting that recessive alleles at multiple unlinked autosomal loci interact epistatically in the pathogenesis of vitiligo (Majumder et al., 1988; Nath et al., 1994). Various HLA associations (class I and class II) have been reported, but these differ according to the individual studies (Dunston and Halder, 1990; Lorini et al., 1992; al-Fouzan et al., 1995; Buc et al., 1998; Zamani et al., 2001; de Vijlder et al., 2004; Tastan et al., 2004; Fain et al., 2006). Susceptibility to vitiligo has shown weak associations with polymorphisms in genes involved in autoimmunity and in other candidate genes, but to date many of these have either not been confirmed* by independent groups or have been refuted**. These include the gene encoding lymphoid protein tyrosine phosphatase (*PTPN22*)* (Canton et al., 2005), the cytotoxic T-lymphocyte antigen4 (*CTLA4*) gene* (Blomhoff et al., 2005), the angiotensin-converting enzyme (*ACE*) gene** (Jin et al., 2004; Akhtar et al., 2005), the transporter associated with antigen processing-1 (*TAP1*) gene* (Casp et al., 2003), and the catalase (*CAT*) gene* (Casp et al., 2002). In a study investigating linkage in subjects with vitiligo from families with systemic lupus erythematosus, evidence for linkage was detected on chromosome 17p13 SLEV1 locus (Nath et al., 2001). However, a genome-wide scan in vitiligo identified linkage to chromosome 1p31 AIS1 locus and weaker linkage was also noted at other loci on chromosomes 1, 7, 8, 11, 19, and 22 (Fain et al., 2003). This latter study failed to detect linkage to several other candidate genes, including *VIT1* (2p16), *CTLA4* (2q33), *MITF* (3p14.1-p12.3), *MHC* (6p21.3), *CAT* (11p13), *GCH1* (14q22.1-q22.2), *SLEV1* (17p13), and *AIRE* (21q22.3), although weak linkage was observed to *COMT* (22q11.21). Conversely, linkage to the MHC region at 6p21.3-21.4 was observed in another study, which specifically looked at this area of the genome (Arcos-Burgos et al., 2002). Thus, further studies are necessary to elucidate the genetics of susceptibility to this disorder and the exact pathogenetic mechanisms responsible for melanocyte cell loss from the epidermis. Treatment of vitiligo is difficult and generally unsatisfactory, with some physicians advocating potent topical steroids, UVR, and minigrafting, and others adopting a less interventional approach with cosmetic camouflage and sunscreens. Repigmentation, when it occurs, begins at the hair follicles and spreads outwards consistent with melanocyte precursors in the stem cell niche in the hair follicle not being affected by the same pathological process (Nishimura et al., 2002).

Box 34–3 Nonmelanoma Skin Cancer (Box 34–3) Key points

- Nonmelanoma skin cancer (NMSC) is the commonest human malignancy.
- Ultraviolet radiation is a major factor in NMSC, but genetic factors, including skin color and hair distribution, are important.
- Other genetic susceptibility factors implicated in NMSC include immune response genes, detoxifying enzymes, and deoxyribonucleic acid (DNA) repair mechanisms.
- Defects in *PATCH* in the sonic hedgehog signaling pathway predispose to basal cell carcinoma.
- Characteristic somatic defects in NMSC include ultraviolet (UV)-induced mutations of p53 in squamous cell carcinoma and loss of heterozygosity (LOH) for 9q and 1q in basal cell carcinoma.

Nonmelanoma skin cancer (NMSC), the great majority of which is due to basal and SCC, is common. Its high prevalence has historically been underrecognized, but its increasing incidence in aging Western populations is now recognized to constitute an important public health problem. Due to increased UVR exposure, even in temperate zones, the age-standardized incidence has dramatically increased in fair-skinned populations during the late twentieth century. A detailed population-based study in South Wales found that the crude incidence of NMSC increased from 174 to 265 per 100,000 population per annum between 1988 and 1998 (Holme et al., 2000).

Basal cell carcinomas (BCCs; Fig. 34–6) arise in hair-bearing epidermis (most areas of skin except palms and soles), and may be of follicular rather than epidermal origin (Kruger et al., 1999). The incidence of BCCs exceeds that of SCC by approximately 4 to 1, and they are often multiple. Variation of clinical morphology and behavior—for example, nodular, sclerosing, or superficial types-may reflect genetic heterogeneity, as may phenotypic variation in site and mode of presentation. Those who present with multiple tumors, and those with truncal lesions, may represent biologically distinct subgroups (Ramachandran et al., 1999, 2000). Many rarer appendageal lesions such as trichoepitheliomas and cylindromas can resemble BCCs.

Although less common than BCCs, SCCs are more invasive and have a higher metastatic rate, necessitating efficient diagnosis. In contrast to BCC, for which there is no clinically identifiable precursor lesion, premalignant skin lesions may lead to SCC. Intraepidermal squamous cell carcinoma (Bowen's disease) presents as persistent, scaling erythematous, often multiple, lesions with a risk of invasive cancer of approximately 3%. Actinic keratoses, dysplastic hyperkeratotic lesions resulting from disorganized keratinization in chronically light exposed skin, are important markers of UV damage and risk of NMSC, but the rate of progression to SCC is only approximately 0.1% (Salasche, 2000). Finally, a rapidly growing but self-healing keratinizing tumor, the keratocanthoma, may resemble SCC clinically.

UVR is the most important risk factor for both SCCs and BCCs, but there is a better relation between cumulative UVR and SCC (Vitasa et al., 1990). Though exposure to UVR is critical for BCC, the quantitative relationship with risk is unclear and intermittent rather than cumulative exposure may be important (Kricker, 1995; Rosso et al., 1996). SCCs can also arise from prolonged exposure to infrared radiation or chronic inflammation and several genodermatoses, notably dystrophic epidermolysis bullosa, and keratitis, ichthyosis, and deafness (KID) syndrome, are associated with a high risk of SCC. Arsenic and photochemotherapy (PUVA), both used to treat psoriasis, also increase the risk of BCC and SCC (Stern and Lunder, 1998). Iatrogenic immunosuppression, similarly increases the frequency of NMSC, with a 50–250-fold higher risk of SCC and a 5–10-fold greater risk of BCC in renal allograft recipients (Bouwes Bavinck, 1995; McGregor and Proby, 1995). SCCs in transplant patients develop especially on sun exposed sites and particularly in individuals with fair skin (Bavinck et al., 1993) and have an association with the presence of viral warts (Penn, 1993; Bouwes Bavinck, 1995). Indeed, human papillomavirus (HPV) DNA has been identified in more than 80% of immunosuppressed and 30% of immunocompetent SCC patients (Harwood and Proby, 2002; Pfister, 2003), and in 60% of BCCs from immunosuppressed subjects in contrast to 36% of BCCs from nonimmunosuppressed individuals (Shamanin et al., 1996).

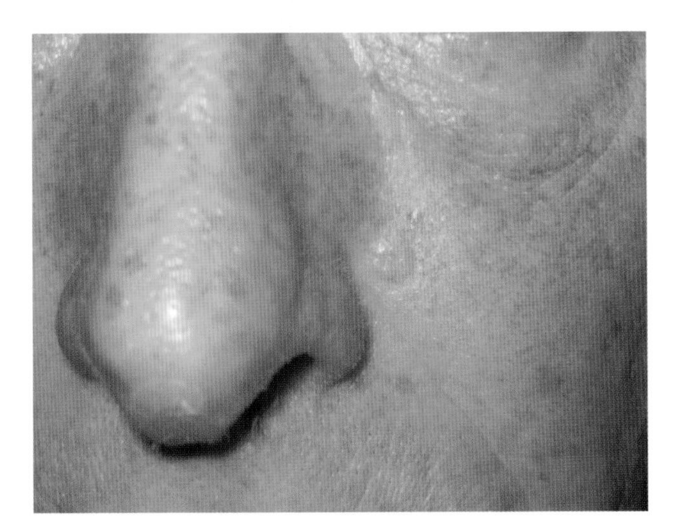

Figure 34–6 Typical basal cell carcinoma of the cheek in a subject with the red hair/fair skin phenotype associated with MC1R variants.

Germline Genetic Susceptibility to NMSC

The risk of NMSC development and/or progression is likely to result from the combined effect of many genes each with a relatively weak individual contribution. A number of syndromes predispose to NMSC, but genes may influence numbers, growth rate, and site of tumors, as well as their occurrence. Similarly, variants in several genes may alter protection against UVR and modify its effects; these include those involved in pigmentation and tanning, DNA damage repair, defense against oxidative stress, apoptosis, and immune modulation (Tilli et al., 2005). Most obviously, patients with cutaneous albinism, particularly in countries with high solar exposure, have a high prevalence of SCC at an early age (Perry and Silverberg, 2001), but in Caucasians, *MC1R* variants predisposing to red hair and fair skin/poor tanning ability determine susceptibility to NMSC (Smith et al., 1998; Bastiaens, ter Huurne, Kielich, et al., 2001; Box, Duffy, et al., 2001). Genetically determined pattern alopecia—male—pattern baldness—resulting in loss of the sun protection provided by scalp hair, is another major risk factor for actinic keratoses and SCCs on the scalp.

Genes affecting immune response play a role in NMSC development. The presence of HLA DR7 is associated with BCC (de Jong-Tieben et al.,, 2000) and HLA-DR1 is weakly associated with the development of multiple BCCs at an early age (Czarnecki et al., 1992), whereas HLA DR4 is decreased among patients with BCC, especially those with multiple BCCs located on the trunk (Rompel et al., 1995). However, varying genetic backgrounds and environmental factors may produce discrepant results—HLA A11 was associated with resistance to skin cancer in study of renal transplant recipients (Bavinck et al., 1990), but increased risk in another (Bouwes Bavnick et al., 1997). In another study of patients with multiple BCCs, BCC number was significantly influenced by the TNFα allele haplotype a2b4d5 (mean BCC number; 8.1 versus 3.7 in other allele combinations) (Hajeer et al., 2000).

Polymorphisms in genes encoding for products involved in the chemical detoxification of xenobiotics appear relevant to the growth of NMSCs. Members of the CYP supergene family catalyze the biotransformation of a range of xenobiotics, often as the first of a two-phase detoxication. The resultant potentially highly reactive intermediate is then a substrate for phase-two enzymes including members of the glutathione *S*-transferases (GSTs) supergene family. GST isozymes also catalyze detoxification of the products of oxidative stress such as lipid and DNA hydroperoxides. Polymorphisms in CYP2D6 have been associated with BCC susceptibility and especially multiple tumors (Lear et al., 1996; Ramachandran et al., 1999). Polymorphisms in GST family members (Pemble et al., 1994; Hayes and Pulford, 1995) associated with impaired detoxification seem to influence the risk for several cancers, including NMSC (Heagerty et al., 1994; Heagerty et al., 1996). *GSTM1* and *GSTT1* genotypes have been shown to influence the inflammatory response following UVR exposure, possibly reflecting the link between oxidative stress and eicosanoid mobilization (Kerb et al., 1997). The *GSTM1* null genotype may increase the risk of actinic keratoses (Carless et al., 2002) and predispose to BCC (Lear, 1996; Lear et al., 1997), whereas a polymorphism of *GSTM3* has also been shown to increase the risk for BCC (Yengi, 1996).

Impaired capacity to repair UV-induced DNA damage in the various complementation groups of the rare genodermatosis xeroderma pigmentosum (XP) reflects mutations in different genes of the DNA repair complex (Cleaver, 1968); Patients with XP develop severe photosensitivity and have a 2000-fold increased risk of SCC and melanoma. Polymorphism in these genes has been investigated for relevance to sporadic disease. The poly-AT polymorphism in intron 9 of *XPC* may be associated with increased risk of head and neck SCC (Shen et al., 2001) but is probably not a major factor in cutaneous SCC (Nelson et al., 2005). In a series of nested case–control studies, Han et al. (2004a, 2004b, 2005) identified alleles of *XRCC1*, *XRCC2*, *XRCC3*, *ligase IV*, and *XPD* associated with increased or decreased risk of SCC and/or BCC. Cockayne's syndrome resembles XP and is due to mutation in the *ERCC8* gene encoding the group 8 excision repair cross-complementing protein (Henning et al., 1995), but does not seem to carry an increased risk of skin cancer (Nance and Berry, 1992).

NMSC is not a major feature of other forms of genomic instability disorders such as Werner's, Bloom's, and Li Fraumeni syndromes or dyskeratosis congenita (Tilli et al., 2005), although it is an uncommon complication of the Rothmund–Thomson syndrome (Wang et al., 2001), which is due to defects in a DNA helicase (Kitao et al., 1999). In the autosomal dominant Muir–Torre syndrome, sebaceous carcinomas and BCC with sebaceous differentiation, with or without keratoacanthomas, are associated with a high incidence of gastric, colonic, and endometrial carcinoma (Schwartz and Torre, 1995); the syndrome is due to mutations in certain DNA mismatch repair genes (*MSH2* and *MLH1*), which also cause hereditary nonpolyposis colonic carcinoma (Kolodner et al., 1994; Bapat et al., 1996).

Nevoid basal cell carcinoma syndrome (NBCCS; Gorlin's syndrome) is an autosomal dominant disorder of developmental abnormalities and a predisposition to multiple BCC due to inactivating germline mutations in the *PTCH* gene at 9q22 (Hahn et al., 1996; Johnson et al., 1996). *PTCH* is a member of the highly conserved *hedgehog* signaling pathway, critical to determining cell fate and organogenesis in many species (Wicking et al., 1999).

Ferguson-Smith syndrome is expressed as multiple self-healing squamous epitheliomas resembling keratoacanthomas. This dominant disorder is closely linked to the *PTCH* locus at 9p21, but the suspected tumor suppressor mechanism has yet to be identified (Bose et al., 2006).

In the rare recessive disorder epidermodysplasia verruciformis (EV), HPV types 5 and 8 are associated with warty lesions and SCC in the sun exposed skin of patients (Galloway et al., 1989). EV is due to mutations in the genes *EVER1* and *EVER2* encoding related calnexin-associated transmembrane proteins (Ramoz et al., 2002).

Somatic Genetic Alterations in NMSC

Loss of heterozygosity (LOH) occurs frequently in SCC, especially on chromosomes 13q, 9p, 17p, 17q, and 3p, but the exact role of these allelic losses to the malignant phenotype in this tumor is unclear, because actinic keratoses, which are benign dysplastic lesions, have a similar degree of LOH (Quinn et al., 1994; Rehman et al., 1994). However, BCCs display a very different allelotype with LOH largely limited to 9q and 1q (Bare et al., 1992; Quinn et al., 1994), which may account for the low tendency of this cancer to metastasize. Up to 90% of cutaneous SCC lesions have UV-induced signature mutations of thymidine dimers in the p53 tumor suppressor gene (Brash et al., 1991, Ziegler, 1994). UV-induced mutations in p53 are also found in actinic keratoses, and as expanded clonal microscopic lesions in sun exposed normal human skin. (Ziegler et al., 1994; Jonason et al., 1996). p53 mutations have been detected in approximately half of all BCCs, but in contrast to SCC concomitant LOH of 17q is infrequent (van der Riet, Nawroz, et al., 1994; Demirkan et al., 2000; Auepemkiate et al., 2002). Aberrant mitosis, one of the hallmarks of p53 dysfunction, has not been observed in BCC (Pritchard, 1993) and some researchers consider that p53 mutations in BCCs may be due to coincidental sun damage (Rosenstein et al., 1999). However, aggressive BCCs are significantly associated with increased p53 expression and mutation may be a crucial but late event in BCC progression (van der Riet, Karp, et al., 1994).

The *CDKN2A* gene at 9p21, encodes p16 (INK4a) and p14 (ARF), cell cycle regulatory proteins in the p53 and retinoblastoma (RB) pathways (Brown et al., 2004). LOH of 9p21 markers has been seen in some cases of SCC; mutational analysis has confirmed point mutations or promoter methylation of p16 (INK4a) and p14(ARF) other cases. Alterations at this locus are significantly more common in tumors from immunocompetent compared with immunosuppressed individuals (Brown et al., 2004).

Somatic *PTCH* mutations are one of the most frequently observed alterations in sporadic BCC (Aszterbaum et al., 1998; Gailani, Stahle-Backdahl, et al., 1996). Loss of the wild-type allele has been demonstrated in BCC from both NBCCS and in up to 68% of sporadic patients (Gailani, Stahle-Backdahl, et al., 1996). Most mutations to *PTCH* in sporadic BCCs are not typical of UVB exposure, and in NBCCS inactivation of the second *PTCH* allele through LOH is unlikely to be due to UVB (Gailani, Leffell, et al., 1996). An association between *PTCH* and cutaneous SCC is not established, although introduction of wild-type *PTCH* into human SCC lines expressing mutant *PTCH* suppressed their oncogenic potential (Koike et al., 2002). An increased incidence of SCC has been observed in UV-irradiated heterogeneous *PTCH* knockout mice (Aszterbaum et al., 1999), and SCCs from patients with multiple BCCs may contain *PTCH* mutations (Ping et al., 2001), however, mutation and, separately, LOH for *PTCH* is less frequent in SCC than in BCC (Asplund et al., 2005).

Disrupted expression of other members of the *hedgehog* pathway have been demonstrated in BCCs, for example, activating *SHH* mutations have been described in sporadic BCC (Oro et al., 1997) and activating mutations of smoothened *SMOH*, have been described in 20% of sporadic BCCs (Dahmane et al., 1997; Kallassy et al., 1997; Xie et al., 1998; Grachtchouk et al., 2000).

Activating mutations of the Ras oncogene, usually point mutations within specific codons of the *H-Ras*, *N-Ras*, and *K-Ras* genes, are amongst the most common genetic changes in human cancer. Activating *H-Ras* mutations were observed in 35%–46% of SCCs (Pierceall, 1991; Kreimer-Erlacher, 2001) and 12% of actinic keratoses (Spencer et al., 1995).

Box 34–4 Cutaneous Melanoma (Box 34–4) Key points

- Environmental ultraviolet radiation (UVR), particularly intense exposure, plays a major role in the incidence of melanoma.
- Phenotype risk factors for melanoma include fair skin and reduced tanning ability, and increased numbers of melanocytic nevi.
- Melanoma is associated with inherited defects in the *CDKN2A* or *CDK-4* genes, but these cause a minority of familial cases.
- Other genes implicated in melanoma development and metastatic potential include *N-Ras, BRAF*, and *MITF*.

Cutaneous melanoma is a malignant neoplasm of melanocytes in the skin. The incidence of this tumor has risen steadily over the past few decades, reaching more than 50 per 100,000 in Caucasians living in sunny climates such as Australia. Melanoma is more frequent in Caucasians, especially those of Celtic ancestry, than in other racial groups (Crombie, 1979b; Cress and Holly, 1997). Phenotypic risk factors include red or light colored hair, fair skin type, a tendency to sunburn, a reduced ability to tan, freckling, light colored eyes, and increased numbers of melanocytic nevi (moles) (reviewed in Gandini et al., 2005). Melanoma occasionally occurs in sun-protected skin such as the sole of the foot (Fig. 34–7), but exposure to UVR, and especially short, intense exposure, is considered the major environmental risk factor for the development of melanoma. Some evidence indicated that exposure to UVR during childhood in amounts sufficient to cause sunburn is particularly harmful (Whiteman et al., 2001). UVR causes multiple effects in skin, including DNA damage (cyclobutane pyrimidine dimers, 6–4 photoproducts, strand breaks, etc.) in melanocytes, which if not repaired leads to permanent genetic alterations/gene mutation (Hemminki et al., 2001). Similar to other cancer types, genetic alterations, which activate protooncogenes or inactivate tumor suppressor genes involved in the control of the cell cycle lead to clonal expansion of mutated melanocytes. Further exposure to UVR causes more DNA damage leading to mutation in other genes involved in cell–cell interactions, cell-extracellular matrix interactions, evasion of immune regulation, etc., enabling the neoplastic cells to invade into the dermis and ultimately metastasize.

Multiple melanocytic nevi (Fig. 34–8) are a risk factor for melanoma development, both in sporadic cases and in families with an inherited predisposition to this tumor, but there has been considerable debate about whether most melanomas arise de novo from melanocytes or from melanocytic nevi (Clark et al., 1984; Ackerman and Mihara, 1985). In addition, cells resembling nevus cells are frequently seen on histological examination of melanomas, but the interpretation of this observation differs, with some pathologists considering it to signify the presence of an original precursor nevus whereas others think this represents nevoid differentiation of melanoma cells. Despite this uncertainty about the exact precursor lesion, much of the study on the early somatic genetic events in melanoma development has focused on alterations in melanocytic nevi rather than in melanocytes. Similar to other cancer types, the overall "picture" of the genetic events responsible for melanoma development and progression has been built up from a variety of sources, and any final interpretation needs to bear in mind the limitations of the various experimental approaches. These sources include genetic alterations in excised primary melanomas (which provide a genetic snapshot of the lesion at that stage and some indication of its history), metastatic melanomas (which may contain an amount of secondary genetic alterations not relevant to the development of the tumor but instead due to late-stage widespread genomic instability), melanoma cell lines (whose genetic mutations may or may not be present in tumors in vivo), and transgenic mouse models (which allow continuous observation of phenotype in relation to genotype, but which generally do not contain epidermal melanocytes except in the skin of their ears and tails and thus may or may not mirror human melanoma in vivo). However, despite these limitations, there is good evidence for multiple genetic alterations being important in melanoma development and progression.

The somatic genetic alterations, which have been detected within sporadic cutaneous melanomas include allelic losses of chromosomes 9p, 10q, and 6q with a lower frequency of loss of other chromosome arms (e.g., 8p, 11q, and 18q) (Healy et al., 1996; Herbst et al., 1999; Takata et al., 2000). Loss of 9p has been detected in dysplastic nevi and in early stage melanomas with a depth of invasion of less than 1 mm, whereas losses of 10q and 6q seem to occur later and are associated with a poorer clinical outcome (Healy et al., 1995; Healy et al., 1998; Tran et al., 2002). Although somatic deletions or mutations of the CDKN2A (p16INK4) tumor suppressor gene on 9p

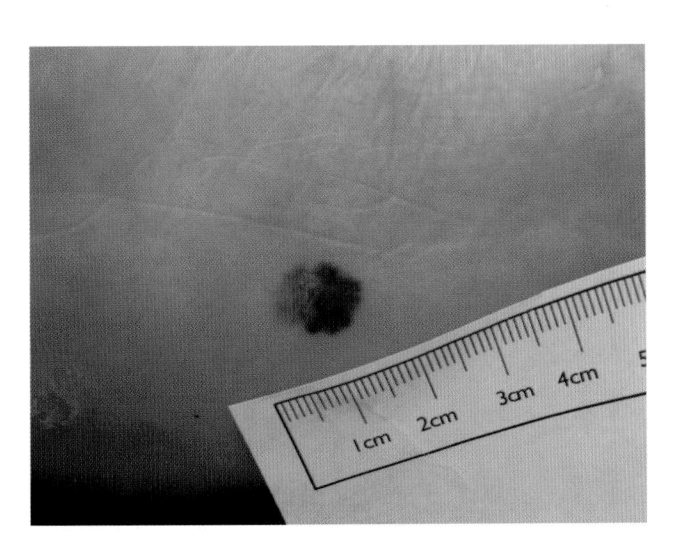

Figure 34–7 Acral lentiginous melanoma of the sole of the foot.

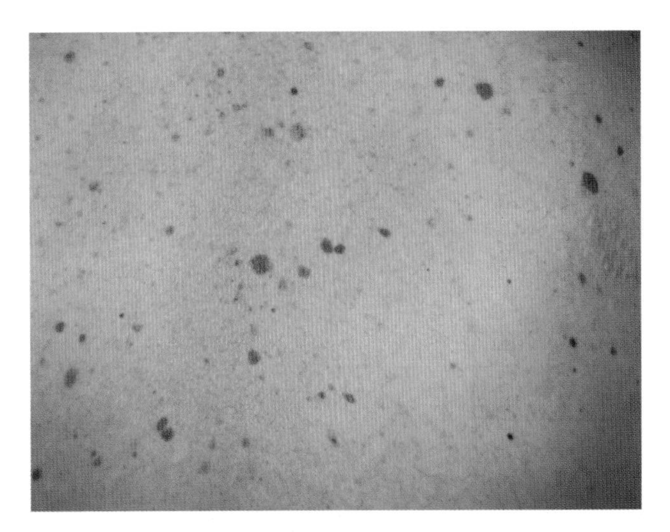

Figure 34–8 Multiple melanocytic nevi, also displaying variation in size shape and color.

are seen in patients with inherited melanoma who carry germline alterations in this gene, mutations in *CDKN2A* are uncommon in sporadic cases despite LOH at this locus and inactivation of the second allele seems to occur through alternative mechanisms including promoter CpG methylation and homozygous deletion (especially in later stage tumors where it seems to play a role in reduced patient survival (Hussussian et al., 1994; Gruis et al., 1995; Healy et al., 1996; Gonzalgo et al., 1997; Fujimoto et al., 1999; Grafstrom et al., 2005). The *CDKN2A* locus encodes two distinct gene products [p16 (INK4A) and p14(ARF)], which interact with the RB and p53 protein pathways, with the evidence stronger for a role of p16(INK4A) than p14(ARF) in melanoma development (FitzGerald et al., 1996; Fargnoli et al., 1998; Bishop et al., 2002). The exact tumor suppressor gene on 10q has been debated, with evidence from cell lines implicating *PTEN* as the most likely candidate, but mutations in this gene are infrequent in melanomas in vivo, despite frequent LOH at this locus (Deichmann et al., 2002; Pollock et al., 2002). However, allelic losses are also encountered in melanoma at other loci on 10q, including the *MXI1* gene, and the possibilities remain that haploinsufficiency of one or more genes on chromosome 10q or biallelic inactivation of the *PTEN* and/or *MXI1* genes (or another gene) through epigenetic mechanisms are responsible for tumor development and/or progression (Zhou et al., 2000; Ariyanayagam-Baksh et al., 2003).

Comparative genomic hybridization and fluorescence in situ hybridization have documented chromosomal gains most often at 1q, 6p, 7q, 8q, and 20q (Barks et al., 1997; Bastian et al., 1998). Oncogenes, which have been implicated in melanoma development include *N-Ras* (chromosome 1p) and *BRAF* (chromosome 7q), although some of the evidence indicates that these alterations are also found frequently in nevi suggesting that these genetic changes may be permissive rather than causal in melanoma development (van 't Veer et al., 1989; Davies et al., 2002; Pollock et al., 2003; Kumar et al., 2004; Curtin et al., 2005). Both the N-Ras and BRAF protein products are part of the Ras signaling pathway, leading to stimulation of cell proliferation and inhibition of apoptosis. *N-Ras* mutations seem more common in melanomas from patients with germline *CDKN2A*, and in sporadic cases have been documented more often in cutaneous melanomas on sun-exposed sites (van't Veer et al., 1989; van Elsas et al., 1996; Jiveskog et al., 1998; Eskandarpour et al., 2003). Perhaps surprisingly, the most prevalent *BRAF* mutation (V599E), which has been detected in approximately 70%–80% of melanomas does not exhibit any hallmark of UV induction (Davies et al., 2002); this does not preclude UVR as the causative factor (e.g., UVA rather than UVB) but raises questions about another possible environmental factor being responsible for melanocytic nevus and/or melanoma development.

MITF on chromosome 3p is considered a master regulator of melanocyte development during embryogenesis (Goding, 2000), and recently *MITF* amplification has been documented as being prevalent in metastatic melanoma, and is associated with reduced patient survival (Garraway et al., 2005). In addition, loss of expression of the apoptosis effector APAF1 on chromosome 12q (through allelic loss and promoter methylation) in primary and metastatic melanoma is thought to render the cancer more resistant to chemotherapeutic agents and correlates with a worse prognosis (Soengas et al., 2001; Dai et al., 2004; Fujimoto et al., 2004). The high prevalence of APAF1 alterations may account for the low frequency of *TP53* gene mutations, which are observed in approximately 0%–25% of cases (Castresana et al., 1993; Lubbe et al., 1994, Sparrow et al., 1995; Zerp et al., 1999).

An inherited predisposition to melanoma is well documented in kindreds with germline alterations in *CDKN2A* and less frequently cyclin dependent kinase-4 (*CDK-4*) genes, however, genetic susceptibility factors have also been identified for the development of melanoma in sporadic cases. These include *MC1R* variants, which may cause melanoma through nonpigmentary mechanisms (melanoma cell proliferation and extracellular matrix interactions) as well as via their effects on pigmentation (Valverde et al., 1996; Palmer et al., 2000; Kennedy et al., 2001; Robinson and Healy, 2002). Indeed, *MC1R* variants also increase the risk of melanoma development in individuals from families with *CDKN2A* mutations (Box, Duffy, Chen et al., 2001; van der Velden et al., 2001). Furthermore, null alleles at the *GST* M1 and T1 loci have been associated with melanoma in subjects with red or blond hair, but not in other hair color groups (Kanetsky et al., 2001). There is also some evidence that *XPD* gene polymorphisms predispose to melanoma development, most likely through effects on DNA repair following exposure to UVR, although in some studies this association has only been seen in subgroup analysis (e.g., older age group and people with low tanning ability) (Tomescu et al., 2001; Baccarelli et al., 2004; Millikan et al., 2005).

FUTURE PROSPECTS

The genomic mechanisms discussed in this chapter represent an important but incomplete understanding of the biology and pathology of the skin, but already indicate numerous potential targets for intervention. The opportunities to prevent and treat cutaneous neoplasis are major drivers of current research. However, the cosmetic functions of skin, obvious but sometimes underrecognized aspects of its biology in man, also encourage investigation. For example, defects in the skin barrier predispose to environment damage or dermatitis, but the vast majority of emollient use is to affect the way skin looks, feels, or smells. Modifiers of gene expression in the stratum corneum, for example, by altering the amount of ceramide or other lipids, offer the prospect of "performance" moisturizers and sun creams. Similarly, vitiligo and albinism are cosmetically distressing, and increase cancer risk, but there is a large demand for "smart tanning"—enhancing cosmetic pigmentation and providing a physiological and long-acting sunscreen. On the other hand, darker pigmentation is undesirable in some racial groups and preparations that downregulate melanogenesis are also needed. Hair is of minor biological importance in man. Even so, what hair humans do possess has massive psychological importance. Modulating the genetic mechanisms of physiological and acquired baldness (and hirsutes) also represents a large economic target.

These considerations should not detract from the potential of therapeutic intervention in disease based on better genetic understanding. Modification of barrier and defense functions will have important preventative and therapeutic roles. Vaccines which generate immune responses to viruses or to somatic mutations in p53, may prevent or treat cancers. The induction or modulation of immune or apoptotic responses offers nonsurgical strategies for the management of skin neoplasia, including melanoma, lymphoid and vascular tumors as well as nonmelanoma skin cancer. The accessibility of the skin as an organ for study and experiment means that it is likely to continue to lead in the early application of gene-based therapeutics.

References

Abo, T, Roder, JC, Abo, W, Cooper, MD, and Balch, CM (1982). Natural killer (HNK-1 +) cells in Chediak-Higashi patients are present in normal numbers but are abnormal in function and morphology. *J Clin Invest*, 70:193–197.

Ackerman, AB and Mihara, I (1985). Dysplasia, dysplastic melanocytes, dysplastic nevi, the dysplastic nevus syndrome, and the relation between dysplastic nevi and malignant melanomas. *Hum Pathol*, 16:87–91.

Ahmad, M, ul Haque, MF, Brancolini, V, Tsou, HC, ul Haque, S, Lam, H, et al. (1997). Alopecia universalis associated with a mutation in the human hairless gene. *Science*, 279:720–724.

Akey, JM, Wang, H, Xiong, M, Wu, H, Liu, W, Shriver, MD, et al. (2001). Interaction between the melanocortin-1 receptor and P genes contributes to inter-individual variation in skin pigmentation phenotypes in a Tibetan population. *Hum Genet*, 108:516–520.

Akhtar, S, Gavalas, NG, Gawkrodger, DJ, Watson, PF, Weetman, AP, and Kemp, EH (2005). An insertion/deletion polymorphism in the gene encoding angiotensin converting enzyme is not associated with generalised vitiligo in an English population. *Arch Dermatol Res*, 297:94–98.

al-Fouzan, A, al-Arbash, M, Fouad, F, Kaaba, SA, Mousa, MA, and al-Harbi, SA (1995). Study of HLA class I/IL and T lymphocyte subsets in Kuwaiti vitiligo patients. *Eur J Immunogenet*, 22:209–213.

Ancans, J, Tobin, DJ, Hoogduijn, MJ, Smit, NP, Wakamatsu, K, and Thody, AJ (2001). Melanosomal pH controls rate of melanogenesis, eumelanin/phaeomelanin ratio and melanosome maturation in melanocytes and melanoma cells. *Exp Cell Res*, 268: 26–35.

Ando, I, Kukita, A, Soma G, and Hino, H (1998). A large number of tandem repeats in the polymorphic epithelial mucin gene is associated with severe acne. *J Dermatol*, 25:150–152.

Arcos-Burgos, M, Parodi, E, Salgar, M, Bedoya, E, Builes, J, Jaramillo, D, et al. (2002). Vitiligo: complex segregation and linkage disequilibrium analyses with respect to microsatellite loci spanning the HLA. *Hum Genet*, 110:334–342.

Ariyanayagam-Baksh, SM, Baksh, FK, Swalsky, PA, and Finkelstein, SD (2003). Loss of heterozygosity in the MXI1 gene is a frequent occurrence in melanoma. *Mod Pathol*, 16:992–995.

Asplund, A, Gustafsson, AC, Wikonkal, NM, Sela, A, Leffell, DJ, Kidd, K, et al. (2005). PTCH codon 1315 polymorphism and risk for nonmelanoma skin cancer. *Br J Dermatol*, 152:868–873.

Aszterbaum, M, Rothman, A, Johnson, RL, Fisher, M, Xie, J, Bonifas, JM, et al. (1998). Identification of mutations in the human PATCHED gene in sporadic basal cell carcinomas and in patients with the basal cell nevus syndrome. *J Invest Dermatol*, 110:885–888.

Aszterbaum, M, Epstein, J, Oro, A, Douglas, V, LeBoit, PE, Scott, MP, et al. (1999). Ultraviolet and ionizing radiation enhance the growth of BCCs and trichoblastomas in patched heterozygous knockout mice. *Nat Med*, 5:1285–1291.

Auepemkiate, S, Boonyaphiphat, P, and Thongsuksai P (2002). p53 expression related to the aggressive infiltrative histopathological feature of basal cell carcinoma. *Histopathology*, 40: 568–573.

Baccarelli, A, Calista, D, Minghetti, P, Marinelli, B, Albetti, B, Tseng, T, et al. (2004). XPD gene polymorphism and host characteristics in the association with cutaneous malignant melanoma risk. *Br J Cancer*, 90:497–502.

Barahmani, N, de Andrade, M, Slusser, J, Zhang, Q, and Duvic, M (2002). Interleukin-1 receptor antagonist allele 2 and familial alopecia areata. *J Invest Dermatol*, 118: 335–337.

Bapat, B, Xia, L, Madlensky, L, Mitri, A, Tonin, P, Narod, SA, et al. (1996). The genetic basis of Muir-Torre syndrome includes the hMLH1 locus. *Am J Hum Genet*, 59:736–739.

Barbosa, MDFS, Nguyen, QA, Tchernev, VT, Ashley, JA, Detter, JC, Blaydes, SM, et al. (1996). Identification of the homologous beige and Chediak-Higashi syndrome genes. *Nature*, 382:262–265.

Bare, JW, Lebo, RV, and Epstein, EH Jr (1992). Loss of heterozygosity at chromosome 1q22 in basal cell carcinomas and exclusion of the basal cell nevus syndrome gene from this site. *Cancer Res*, 52:1494–1498.

Barks, JH, Thompson, FH, Taetle, R, Yang, JM, Stone, JF, Wymer, JA, et al. (1997). Increased chromosome 20 copy number detected by fluorescence in situ hybridization (FISH) in malignant melanoma. *Genes Chromosomes Cancer*, 19:278–285.

Bastiaens, M, ter Huurne, J, Gruis, N, Bergman, W, Westendorp, R, Vermeer, BJ, et al. (2001). The melanocortin-1-receptor gene is the major freckle gene. *Hum Mol Genet*, 10:1701–1708.

Bastiaens, MT, ter Huurne, JA, Kielich, C, Gruis, NA, Westendorp, RG, Vermeer, BJ, et al. (2001). Melanocortin-1 receptor gene variants determine the risk of nonmelanoma skin cancer independently of fair skin and red hair. *Am J Hum Genet*, 68:884–894.

Bastian, BC, LeBoit, PE, Hamm, H, Brocker EB, and Pinkel, D (1998). Chromosomal gains and losses in primary cutaneous melanomas detected by comparative genomic hybridization. *Cancer Res*, 58:2170–2175.

Bataille, V, Snieder, H, MacGregor, AJ, Sasieni, P, and Spector, TD (2002). The influence of genetics and environmental factors in the pathogenesis of acne: a twin study of acne in women. *J Invest Dermatol*, 119:1317–1322.

Bavinck, JN, De Boer, A, Vermeer, BJ, Hartevelt, MM, van der Woude, FJ, Claas, FH, et al. (1993). Sunlight, keratotic skin lesions and skin cancer in renal transplant recipients. *Br J Dermatol*, 129:242–249.

Bavinck, JN, Kootte, AM, van der Woude, F, Vandenbroucke, JP, Vermeer, BJ, and Claas, FH (1990). HLA-A11-associated resistance to skin cancer in renal-transplant patients. *N Engl J Med*, 323:1350.

Beaumont, KA, Newton, RA, Smit, DJ, Leonard, JH, Stow, JL, and Sturm, RA (2005). Altered cell surface expression of human MC1R variant receptor alleles associated with red hair and skin cancer risk. *Hum Mol Genet*, 14:2145–2154.

Bennett, C and Lamoreux, ML (2003). The color loci of mice-a genetic century. *Pigment Cell Res*, 16:333–344.

Betterle, C, Greggio, NA, and Volpato M (1998). Clinical review 93: Autoimmune polyglandular syndrome type 1. *J Clin Endocrinol Metab*, 83:1049–1055.

Bishop, DT, Demenais, F, Goldstein, AM, Bergman, W, Bishop, JN, Bressac-de Paillerets, et al. (2002). Geographical variation in the penetrance of CDKN2A mutations for melanoma. *J Natl Cancer Inst*, 94:894–903.

Blomhoff, A, Kemp, EH, Gawkrodger, DJ, Weetman, AP, Husebye, ES, Akselsen, HE, et al. (2005). CTLA4 polymorphisms are associated with vitiligo, in patients with concomitant autoimmune diseases. *Pigment Cell Res*, 18:55–58.

Boisseau-Garsaud, AM, Garsaud, P, Cales-Quist, D, Helenon, R, Queneherve, C, and Claire, RC (2000). Epidemiology of vitiligo in the French West Indies (Isle of Martinique). *Int J Dermatol*, 39:18–20.

Bondurand, N, Pingault, V, Goerich, DE, Lemort, N, Sock, E, Caignec, CL, et al. (2000). Interaction among SOX10, PAX3 and MITF, three genes altered in Waardenburg syndrome. *Hum Mol Genet*, 9:1907–17.

Bonilla, C, Boxill, LA, Donald, SA, Williams, T, Sylvester, N, Parra, EJ, et al. (2005). The 8818G allele of the agouti signaling protein (ASIP) gene is ancestral and is associated with darker skin color in African Americans. *Hum Genet*, 116:402–406.

Bose, S, Morgan, LJ, Booth, DR, Goudie, DR, Ferguson-Smith, MA, and Richards, FM (2006). The elusive multiple self-healing squamous epithelioma (MSSE) gene:further mapping, analysis of candidates, and loss of heterozygosity. *Oncogene*, 25: 806–812.

Bouwes Bavinck, JN (1995). Epidemiological aspects of immunosuppression: role of exposure to sunlight and human papillomavirus on the development of skin cancer. *Hum Exp Toxicol*, 14:98.

Bouwes Bavinck, JN, Claas, FH, Hardie, DR, Green, A, Vermeer, BJ, and Hardie, IR (1997). Relation between HLA antigens and skin cancer in renal transplant recipients in Queensland, Australia. *J Invest Dermatol*, 108:708–711.

Box, NF, Duffy, DL, Chen, W, Stark, M, Martin, NG, Sturm, RA, et al. (2001). MC1R genotype modifies risk of melanoma in families segregating CDKN2A mutations. *Am J Hum Genet*, 69:765–773.

Box, NF, Duffy, DL, Irving, RE, Russell, A, Chen, W, Griffyths, LR, et al. (2001). Melanocortin-1 receptor genotype is a risk factor for basal and squamous cell carcinoma. *J Invest Dermatol*, 116:224–229.

Box, NF, Wyeth, JR, Mayne, CJ, O'Gorman, LE, Martin, NG, and Sturm, RA (1998). Complete sequence and polymorphism study of the human TYRP1 gene encoding tyrosinase-related protein 1. *Mamm Genome*, 9:50–53.

Box, NF, Wyeth, JR, O'Gorman, LE, Martin, NG, and Sturm, RA (1997). Characterization of melanocyte stimulating hormone receptor variant alleles in twins with red hair. *Hum Mol Genet*, 6:1891–1897.

Brash, DE, Rudolph, JA, Simon, JA, Lin, A, McKenna, GJ, Baden, HP, et al. (1991). A role for sunlight in skin cancer: UV-induced p53 mutations in squamous cell carcinoma. *Proc Natl Acad Sci USA*, 88:10124–10128.

Brown, VL, Harwood, CA, Crook, T, Cronin, JG, Kelsell, DP, and Proby, CM (2004). p16INK4a and p14ARF tumor suppressor genes are commonly inactivated in cutaneous squamous cell carcinoma. *J Invest Dermatol*, 122:1284–1292.

Buc, M, Fazekasova, H, Cechova, E, Hegyi, E, Kolibasova, K, and Ferencik, S (1998). Occurrence rates of HLA-DRB1, HLA-DQB1, and HLA-DPB1 alleles in patients suffering from vitiligo. *Eur J Dermatol*, 8:13–15.

Canton, I, Akhtar, S, Gavalas, NG, Gawkrodger, DJ, Blomhoff, A, Watson, PF, et al. (2005). A single-nucleotide polymorphism in the gene encoding lymphoid protein tyrosine phosphatase (PTPN22) confers susceptibility to generalised vitiligo. *Genes Immun*, 6:584–587.

Carless, MA, Lea, RA, Curran, JE, Appleyard, B, Gaffney, P, Green A, et al. (2002). The GSTM1 null genotype confers an increased risk for solar keratosis development in an Australian Caucasian population. *J Invest Dermatol*, 119:1373–1378.

Casp, CB, She, JX, and McCormack, WT (2002). Genetic association of the catalase gene (CAT) with vitiligo susceptibility. *Pigment Cell Res*, 15:62–66.

Casp, CB, She, JX, and McCormack, WT (2003). Genes of the LMP/TAP cluster are associated with the human autoimmune disease vitiligo. *Genes Immun*, 4:492–499.

Castresana, JS, Rubio, MP, Vazquez, JJ, Idoate, M, Sober, AJ, Seizinger, BR, et al. (1993). Lack of allelic deletion and point mutation as mechanisms of p53 activation in human malignant melanoma. *Int J Cancer*, 55:562–565.

Chavanas, S, Bodemer, C, Rochat, A, Hamel-Teillac, D, Ali, M, Irvine, AD, et al. (2000). Mutations in SPINK5, encoding a serine protease inhibitor, cause Netherton syndrome. *Nat Genet*, 25:141–142.

Chen, W, Thiboutot, D, and Zouboulis, CC (2002). Cutaneous androgen metabolism: basic research and clinical perspectives. *J Invest Dermatol*, 119:992–1007.

Clark, WH Jr, Elder, DE, Guerry, D 4th, Epstein, MN, Greene, MH, and Van Horn, M (1984). A study of tumor progression: the precursor lesions of superficial spreading and nodular melanoma. *Hum Pathol*, 15:1147–1165.

Cleaver, JE (1968). Defective repair replication of DNA in xeroderma pigmentosum. *Nature*, 218:652–656.

Cotsarelis, G, and Millar, SE (2001). Towards a molecular understanding of hair loss and its treatment. *Trends Mol Med*, 7:293–301.

Cotsarelis, G, Su, TT, and Lavke, RM (1990). Label-retaining cells reside in the bulge area of pilosebaceous unit:implications for follicular stem cells, hair cycle, and skin carcinogenesis. *Cell*, 61:1329–1337.

Cress, RD and Holly, EA (1997). Incidence of cutaneous melanoma among non Hispanic whites, Hispanics, Asians, and blacks: an analysis of california cancer registry data, 1988–93. *Cancer Causes Control*, 8: 246–252.

Crombie, IK (1979a). Variation of melanoma incidence with latitude in North America and Europe. *Br J Cancer*, 40:774–781.

Crombie, IK (1979b). Racial differences in melanoma incidence. *Br J Cancer*, 40: 185–193.

Curtin, JA, Fridlyand, J, Kageshita, T, Patel, HN, Busam, KJ, Kutzner, H, et al. (2005). Distinct sets of genetic alterations in melanoma. *N Engl J Med*, 353:2135–2147.

Czarnecki, D, Lewis, A, Nicholson, I, and Tait, B (1992). HLA-DR1 is not a sign of poor prognosis for the development of multiple basal cell carcinomas. *J Am Acad Dermatol*, 26:717–719.

Dahmane, N, Lee, J, Robins, P, Heller, P, Ruiz, I, and Altaba, A (1997). Activation of the transcription factor Gli1 and the Sonic hedgehog signalling pathway in skin tumours. *Nature*, 389:876–881.

Dai, DL, Martinka, M, Bush, JA, and Li, G (2004). Reduced Apaf1 expression in human cutaneous melanomas. *Br J Cancer*, 91:1089–1095.

Davies, H, Bignell, GR, Cox, C, Stephens, P, Edkins, S, Clegg, S, et al. (2002). Mutations of the BRAF gene in human cancer. *Nature*, 417:949–954.

de Andrade, M, Jackow, CM, Dahm, N, Hordinsky, M, Reveille, JD, and Duvic, M (1999). Alopecia areata in families: association with the HLA locus. *J Investig Dermatol Symp Proc*, 4:220–223.

de Jong-Tieben, LM, Berkhout, RJ, ter Schegget, J, Vermeer, BJ, de Fijter, JW, Bruijn, JA, et al. (2000). The prevalence of human papillomavirus DNA in benign keratotic skin lesions of renal transplant recipients with and without a history of skin cancer is equally high: a clinical study to assess risk factors for keratotic skin lesions and skin cancer. *Transplantation*, 15;69:44–49.

de Vijlder, HC, Westerhof, W, Schreuder, GM, de Lange, P, and Claas, FH (2004). Difference in pathogenesis between vitiligo vulgaris and halo nevi associated with vitiligo is supported by an HLA association study. *Pigment Cell Res*, 17:270–274.

Deichmann, M, Thome, M, Benner, A, Egner, U, Hartschuh, W, and Naher, H (2002). PTEN/MMAC1 expression in melanoma resection specimens. *Br J Cancer*, 87:1431–1436.

Demirkan, NC, Colakoglu, N, and Duzcan, E (2000). Value of p53 protein in biological behavior of basal cell carcinoma and in normal epithelia adjacent to carcinomas. *Pathol Oncol Res*, 6:272–274.

Duffy, DL, Box, NF, Chen, W, Palmer, JS, Montgomery, GW, James, MR, et al. (2004). Interactive effects of MC1R and OCA2 on melanoma risk phenotypes. *Hum Mol Genet*, 13:447–461.

Dunston, GM and Halder, RM (1990). Vitiligo is associated with HLA-DR4 in black patients. A preliminary report. *Arch Dermatol*, 126:56–60.

Durham-Pierre, D, Gardner, JM, Nakatsu, Y, King, RA, Francke, U, Ching, A, et al. (1994). African origin of an intragenic deletion of the human P gene in tyrosinase positive oculocutaneous albinism. *Nat Genet*, 7:176–179.

Edery, P, Attie, T, Amiel, J, Pelet, A, Eng, C, Hofstra, RMV, et al. (1996). Mutation of the endothelin-3 gene in the Waardenburg-Hirschsprung disease (Shah-Waardenburg syndrome). *Nature Genet*, 12:442–444.

Elias, PM (2005). Stratum corneum defensive functions: an integrated view. *J Invest Dermatol*, 125:183–200.

Ellis, JA, and Harrap, SB (2001). The genetics of androgenetic alopecia. *Clin Dermatol*, 19:149–154.

Ellis, JA, Stebbing, M, and Harrap, SB (1998). Genetic analysis of male pattern baldness and the 5alpha-reductase genes. *J Invest Dermatol*, 110:849–853.

Ellis, JA, Stebbing, M, and Harrap, SB (2001). Polymorphism of the androgen receptor gene is associated with male pattern baldness. *J Invest Dermatol*, 116:452–455.

Eskandarpour, M, Hashemi, J, Kanter, L, Ringborg, U, Platz, A, and Hansson, J (2003). Frequency of UV inducible NRAS mutations in melanomas of patients with germline CDKN2A mutations. *J Natl Cancer Inst*, 95:790–798.

Faergemann, J, Correia, O, Nowicki, R, and Ro, BI (2005). Genetic predisposition—understanding underlying mechanisms of onychomycosis. *J Eur Acad Dermatol Venereol*, 19(Suppl. 1):17–19.

Fain, PR, Babu, SR, Bennett, DC, and Spritz, RA (2006). HLA class II haplotype DRB1*04-DQB1*0301 contributes to risk of familial generalized vitiligo and early disease onset. *Pigment Cell Res*, 19:51–57.

Fain, PR, Gowan, K, LaBerge, GS, Alkhateeb, A, Stetler, GL, Talbert, J, et al. (2003). A genomewide screen for generalized vitiligo: confirmation of AIS1 on chromosome 1p31 and evidence for additional susceptibility loci. *Am J Hum Genet*, 72:1560–1564.

Fargnoli, MC, Chimenti, S, Keller, G, Soyer, HP, Dal Pozzo, V, Hofler, H, et al. (1998). CDKN2a/p16INK4a mutations and lack of p19ARF involvement in familial melanoma kindreds. *J Invest Dermatol*, 111:1202–1206.

Fistarol, SK, and Itin, PH (2002). Nail changes in genodermatoses. *Eur J Dermatol* 12:119–28.

FitzGerald, MG, Harkin, DP, Silva-Arrieta, S, MacDonald, DJ, Lucchina, LC, Unsal, H, et al. (1996). Prevalence of germ-line mutations in p16, p19ARF, and CDK4 in familial melanoma: analysis of a clinic-based population. *Proc Natl Acad Sci USA*, 93:8541–8545.

Fitzpatrick, TB (1988). The validity and practicality of sun-reactive skin types I through VI. *Arch Dermatol*, 124: 869–871.

Fitzsimmons, JS, Guilbert, PR, and Fitzsimmons, EM (1985). Evidence of genetic factors in hidradenitis suppurativa. *Br J Dermatol*, 113:1–8.

Flanagan, N, Healy, E, Ray, A, Philips, S, Todd, C, Jackson, IJ, et al. (2000). Pleiotropic effects of the melanocortin 1 receptor (MC1R) gene on human pigmentation. *Hum Mol Genet*, 9: 2531–2537.

Fleischman, RA, Saltman, DL, Stastny, V, and Zneimer, S (1991). Deletion of the c-kit protooncogene in the human developmental defect piebald trait. *Proc Nat Acad Sci*, 88:10885–10889.

Frandberg, PA, Doufexis, M, Kapas, S, and Chhajlani, V (1998). Human pigmentation phenotype: a point mutation generates nonfunctional MSH receptor. *Biochem Biophys Res Commun*, 245:490–492.

Fratantonio, E, Vicari, E, Pafumi, C, and Calogero, AE (2005). Genetics of polycystic ovarian syndrome. *Reprod Biomed Online*, 10:713–720.

Fujimoto, A, Morita, R, Hatta, N, Takehara, K, and Takata, M (1999). p16INK4a inactivation is not frequent in uncultured sporadic primary cutaneous melanoma. *Oncogene*, 18:2527–2532.

Fujimoto, A, Takeuchi, H, Taback, B, Hsueh, EC, Elashoff, D, Morton DL, et al. (2004). Allelic imbalance of 12q22–23 associated with APAF-1 locus correlates with poor disease outcome in cutaneous melanoma. *Cancer Res*, 64:2245 2250.

Gailani, MR, Leffell, DJ, Ziegler, AJ, Gross, EG, Brash, DE, and Bale, AE (1996). Relationship between sunlight exposure and a key genetic alteration in basal cell carcinoma. *J Natl Cancer Inst*, 88:349–354.

Gailani, MR, Stahle-Backdahl, M, Leffell, DJ, Glynn, M, Zaphiropoulos, PG, Pressman, C, et al. (1996). The role of the human homologue of Drosophila patched in sporadic basal cell carcinomas. *Nat Genet*, 14:78–81.

Galloway, DA, and McDougall, JK (1989). Human papillomaviruses and carcinomas. *Adv Virus Res*, 37:125–171.

Galbraith, GM and Pandey, JP (1995). Tumor necrosis factor alpha (TNF-alpha) gene polymorphism in alopecia areata. *Hum Genet*, 96:433–436.

Gandini, S, Sera, F, Cattaruzza, MS, Pasquini P, Abeni D, Boyle P, et al. (2005). Meta-analysis of risk factors for cutaneous melanoma: I. Common and atypical naevi. *Eur J Cancer*, 41:28–44.

Gardner, JM, Nakatsu, Y, Gondo, Y, Lee, S, Lyon, MF, King, RA, et al. (1992). The mouse pink-eyed dilution gene: association with human Prader-Willi and Angelman syndromes. *Science*, 257:1121–1124.

Garraway, LA, Widlund, HR, Rubin, MA, Getz, G, Berger, AJ, Ramaswamy, S, et al. (2005). Integrative genomic analyses identify MITF as a lineage survival oncogene amplified in malignant melanoma. *Nature*, 436:117–122.

Giebel, LB and Spritz, RA (1991). Mutation of the KIT (mast/stem cell growth factor receptor) protooncogene in human piebaldism. *Proc Nat Acad Sci*, 88:8696–8699.

Giebel, LB, Strunk, KM, King, RA, Hanifin, JM, and Spritz, RA (1990). A frequent tyrosinase gene mutation in classic, tyrosinase-negative (type IA) oculocutaneous albinism. *Proc Nat Acad Sci*, 87:3255–3258.

Giebel, LB, Tripathi, RK, King, RA, and Spritz, RA (1991). A tyrosinase gene missense mutation in temperature-sensitive type I oculocutaneous albinism. A human homologue to the Siamese cat and the Himalayan mouse. *J Clin Invest*, 87:1119–1122.

Glimcher, ME, Kostick, RM, and Szabo, G (1973). The epidermal melanocyte system in newborn human skin. A quantitative histologic study. *J Invest Dermatol*, 61:344–347.

Goding, CR (2000). MITF from neural crest to melanoma: signal transduction and transcription in the melanocyte lineage. *Genes Dev*, 14:1712–1728.

Gonzalgo, ML, Bender, CM, You, EH, Glendening, JM, Flores, JF, Walker, GJ, et al. (1997). Low frequency of p16/CDKN2A methylation in sporadic melanoma: comparative approaches for methylation analysis of primary tumors. *Cancer Res*, 57:5336–347.

Goulden, V, McGeown, CH, and Cunliffe, WJ (1999). The familial risk of adult acne: a comparison between first-degree relatives of affected and unaffected individuals. *Br J Dermatol*, 141:297–300.

Grachtchouk, M, Mo, R, Yu, S, Zhang X, Sasaki, H, Hui, et al. (2000). Basal cell carcinomas in mice overexpressing Gli2 in skin. *Nat Genet*, 24:216–217.

Grafstrom, E, Egyhazi, S, Ringborg, U, Hansson, J, and Platz A (2005). Biallelic deletions in INK4 in cutaneous melanoma are common and associated with decreased survival. *Clin Cancer Res*, 11:2991–2997.

Gruis, NA, van der Velden, PA, Sandkuijl, LA, Prins, DE, Weaver-Feldhaus, J, Kamb, A, et al. (1995). Homozygotes for CDKN2 (p16) germline mutation in Dutch familial melanoma kindreds. *Nat Genet*, 10:351–353.

Hahn, H, Wicking, C, Gailani, MR, Shanley, S, Chidambaram, A, Vorechovsky, I, et al. (1996). Mutations of the human homolog of Drosophila patched in the nevoid basal cell carcinoma syndrome. *Cell*, 85:841–851.

Hajeer, AH, Lear, JT, Ollier, WE, Naves, M, Worthington, J, Bell, DA, et al. (2000). Preliminary evidence of an association of tumour necrosis factor microsatellites with increased risk of multiple basal cell carcinomas. *Br J Dermatol*, 142:441–445.

Halaban, R, Svedine, S, Cheng, E, Smicun, Y, Aron, R, and Hebert, DN (2000). Endoplasmic reticulum retention is a common defect associated with tyrosinase-negative albinism. *Proc Natl Acad Sci*, 97:5889–5894.

Han, J, Colditz, GA, Liu, JS, and Hunter, DJ (2005). Genetic variation in XPD, sun exposure, and risk of skin cancer. *Cancer Epidemiol Biomarkers Prev*, 14:1539–1544.

Han, J, Colditz, GA, Samson, LD, and Hunter, DJ (2004a) Polymorphisms in DNA double-strand break repair genes and skin cancer risk. *Cancer Res*, 64:3009–3013.

Han, J, Hankinson, SE, Colditz, GA, and Hunter, DJ (2004b). Genetic variation in XRCC1, sun exposure, and risk of skin cancer. *Br J Cancer*, 91:1604–1609.

Happle, R and Frosch, PJ (1985). Manifestation of the lines of Blaschko in women heterozygous for X-linked hypohidrotic ectodermal dysplasia. *Clin Genet*, 27:468–471.

Harrison, GA and Owen, JJ (1964). Studies on the inheritance of human skin colour. *Ann Hum Genet*, 28:27–37.

Harwood, CA and Proby, CM (2002). Human papillomaviruses and non-melanoma skin cancer. *Curr Opin Infect Dis*, 15:101–114.

Hayes, JD and Pulford, DJ (1995). The glutathione S-transferase supergene family: regulation of GST and the contribution of the isoenzymes to cancer chemoprotection and drug resistance. *Crit Rev Biochem Mol Biol*, 30:445–600.

Heagerty, A, Smith, A, English, J, Lear, J, Perkins, W, Bowers, B, et al. (1996). Susceptibility to multiple cutaneous basal cell carcinomas: significant interactions between glutathione S-transferase GSTM1 genotypes, skin type and male gender. *Br J Cancer*, 73:44–48.

Heagerty, AH, Fitzgerald, D, Smith, A, Bowers, B, Jones, P, Fryer, AA, et al. (1994). Glutathione S-transferase GSTM1 phenotypes and protection against cutaneous tumours. *Lancet*, 343:266–268.

Healy, E, Belgaid, C, Takata, M, Harrison, D, Zhu, NW, Burd, DA, et al. (1998). Prognostic significance of allelic losses in primary melanoma. *Oncogene*, 16:2213–2218.

Healy, E, Belgaid, CE, Takata, M, Vahlquist, A, Rehman, I, Rigby, et al. (1996). Allelotypes of primary cutaneous melanoma and benign melanocytic nevi. *Cancer Res*, 56:589–93.

Healy, E, Flanagan, N, Ray, A, Todd, C, Jackson, IJ, Matthews, JNS, et al. (2000). Melanocortin-1-receptor gene and sun sensitivity in individuals without red hair. *Lancet*, 355:1072–1073.

Healy, E, Jordan, SA, Budd, PS, Suffolk, R, Rees JL, and Jackson IJ (2001). Functional variation of MC1R alleles from red-haired individuals. *Hum Molec Genet*, 10:2397–2402.

Healy, E, Rehman, I, Angus, B, and Rees, JL (1995). Loss of heterozygosity in sporadic primary cutaneous melanoma. *Genes Chromosomes Cancer*, 12:152–156.

Healy, E, Sikkink, S, and Rees, JL (1996). Infrequent mutation of p16INK4 in sporadic melanoma. *J Invest Dermatol*, 107:318–321.

Healy, E and Simpson, N (1994). Acne vulgaris. *BMJ*, 308:831–833.

Hemminki, K, Xu, G, and Le, Curieux F (2001). Ultraviolet radiation-induced photoproducts in human skin DNA as biomarkers of damage and its repair. *IARC Sci Publ*, 154:69–79.

Henning, KA, Li, L, Iyer, N, McDaniel, LD, Reagan, MS, Legerski, R, et al. (1995). The Cockayne syndrome group A gene encodes a WD repeat protein that interacts with CSB protein and a subunit of RNA polymerase II TFIIH. *Cell*, 82:555–564.

Herbst, RA, Gutzmer, R, Matiaske, F, Mommert, S, Casper, U, Kapp, A, et al. (1999). Identification of two distinct deletion targets at 11q23 in cutaneous malignant melanoma. *Int J Cancer*, 80:205–209.

Hermansky, F and Pudlak P (1959). Albinism associated with hemorrhagic diathesis and unusual pigmented reticular cells in the bone marrow: report of two cases with histochemical studies. *Blood*, 14:162–169.

Hillmer, AM, Hanneken, S, Ritzmann, S, Becker, T, Freudenberg, J, Brockschmidt, FF, et al. (2005). Genetic variation in the human androgen receptor gene is the major determinant of common early-onset androgenetic alopecia. *Am J Hum Genet* 77:140–148.

Holme, SA, Malinovszky, K, and Roberts, DL (2000). Changing trends in non-melanoma skin cancer in South Wales, 1988–98. *Br J Dermatol*, 143:1224–1229.

Hud, JA Jr, Cohen, JB, Wagner, JM, Cruz, and PD Jr (1992). Prevalence and significance of acanthosis nigricans in an adult obese population. *Arch Dermatol*, 128:941–944.

Hunt, G, Kyne, S, Ito, S, Wakamatsu, K, Todd, C, and Thody, A (1995). Eumelanin and phaeomelanin contents of human epidermis and cultured melanocytes. *Pigment Cell Res*, 8:202–208.

Hussussian, CJ, Struewing, JP, Goldstein, AM, Higgins, PA, Ally, DS, Sheahan, MD, et al. (1994). Germline p16 mutations in familial melanoma. *Nat Genet*, 8:15–21.

Huynh, C, Roth, D, Ward, DM, Kaplan, J, and Andrews, NW (2004). Defective lysosomal exocytosis and plasma membrane repair in Chediak-Higashi/beige cells. *Proc Natl Acad Sci*, 101:16795–16800.

Irvine, AD and Christiano, AM (2001). Hair on a gene string: recent advances in understanding the molecular genetics of hair loss. *Clin Exp Dermatol*, 26:59–71.

Irvine, AD and McLean, WH (1999). Human keratin diseases: the increasing spectrum of disease and subtlety of the phenotype-genotype correlation. *Br J Dermatol*, 140:815–828.

Jimbow, K, Ishida, O, Ito, S, Hori, Y, Witkop, CJ Jr, and King, RA (1983). Combined chemical and electron microscopic studies of pheomelanosomes in human red hair. *J Invest Dermatol*, 81:506–511.

Jin, SY, Park, HH, Li, GZ, Lee, HJ, Hong, MS, Hong, SJ, et al. (2004). Association of angiotensin converting enzyme gene I/D polymorphism of vitiligo in Korean population. *Pigment Cell Res*, 17:84–86.

Jiveskog, S, Ragnarsson-Olding, B, Platz, A, and Ringborg, U (1998). N-ras mutations are common in melanomas from sun-exposed skin of humans but rare in mucosal membranes or unexposed skin. *J Invest Dermatol*, 111:757–761.

Johnson, RL, Rothman, AL, Xie, J, Goodrich, LV, Bare, JW, Bonifas, JM, et al. (1996). Human homolog of patched, a candidate gene for the basal cell nevus syndrome. *Science*, 272:1668–16671.

Johnston, JD, Winder, AF, and Breimer, LH (1992). An MboI polymorphism at codon 192 of the human tyrosinase gene is present in Asians and Afrocaribbeans. *Nucleic Acids Res*, 20:1433.

Jonason, AS, Kunala, S, Price, GJ, Restifo, RJ, Spinelli, HM, Persing, JA, et al. (1996). Frequent clones of p53-mutated keratinocytes in normal human skin. *Proc Natl Acad Sci USA*, 93:14025–14029.

Kaidbey, KH, Agin, PP, Sayre, RM, and Kligman, AM (1979). Photoprotection by melanin—a comparison of black and Caucasian skin. *J Am Acad Dermatol*, 1:249–260.

Kallassy, M, Toftgard, R, Ueda, M, Nakazawa, K, Vorechovsky, I, Yamasaki, H, et al. (1997). Patched (ptch)-associated preferential expression of smoothened (*SMOH*) in human basal cell carcinoma of the skin. *Cancer Res*, 57:4731–4735.

Kanetsky, PA, Holmes, R, Walker, A, Najarian, D, Swoyer, J, Guerry, D, et al. (2001). Interaction of glutathione S transferase M1 and T1 genotypes and malignant melanoma. *Cancer Epidemiol Biomarkers Prev*, 10:509–513.

Kanetsky, PA, Swoyer, J, Panossian, S, Holmes, R, Guerry, D, and Rebbeck, TR (2002). A polymorphism in the agouti signaling protein gene is associated with human pigmentation. *Am J Hum Genet*, 70:770–775.

Kemp, EH, Waterman, EA, Hawes, BE, O'Neill, K, Gottumukkala, RV, et al. (2002). The melanin—concentrating hormone receptor 1, a novel target of autoantibody responses in vitiligo. *J Clin Invest*, 109:923–930.

Kennedy, C, ter Huurne, J, Berkhout, M, Gruis, N, Bastiaens, M, Bergman W, et al. (2001). Melanocortin 1 receptor (MC1R) gene variants are associated with an increased risk for cutaneous melanoma which is largely independent of skin type and hair color. *J Invest Dermatol*, 117:294–300.

Kerb, R, Brockmoller, J, Reum, T, and Roots, I (1997). Deficiency of glutathione S-transferases T1 and M1 as heritable factors of increased cutaneous UV sensitivity. *J Invest Dermatol*, 108:229–232.

King, RA, Mentink, MM, and Oetting, WS (1991). Non-random distribution of missense mutations within the human tyrosinase gene in type I (tyrosinase-related) oculocutaneous albinism. *Mol Biol Med*, 8:19–29.

King, RA, Townsend, D, Oetting, W, Summers, CG, Olds, DP, White, JG, et al. (1991). Temperature-sensitive tyrosinase associated with peripheral pigmentation in oculocutaneous albinism. *J Clin Invest*, 87:1046–1053.

King, RA, Willaert, RK, Schmidt, RM, Pietsch, J, Savage, S, Brott, MJ, et al. (2003). MC1R mutations modify the classic phenotype of oculocutaneous albinism type 2 (OCA2). *Am J Hum Genet*, 73:638–645.

Kitao, S, Lindor, NM, Shiratori, M , Furuichi, Y, and Shimamoto, A (1999). Genomics. Rothmund-Thomson syndrome responsible gene, RECQL4. *Genomic Structure and Products*, 61:268–276.

Koike, C, Mizutani, T, Ito, T, and Shimizu, Y (2002). Introduction of wild-type patched gene suppresses the oncogenic potential of human squamous cell carcinoma cell lines including A431. *Oncogene*, 21:2670–2678.

Kolodner, RD, Hall, NR, Lipford, J, Kane, MF, Rao, MR, Morrison, P, et al. (1994). Structure of the human MSH2 locus and analysis of two Muir-Torre kindreds for msh2 mutations. *Genomics*, 24:516–526.

Koppe, G, Marinkovic-Ilsen, A, Rijken, Y, De Groot, WP, and Jobsis, AC (1978). X-linked ichthyosis. A sulphatase deficiency. *Arch Dis Child*, 53:803–806.

Kreimer-Erlacher, H, Seidl, H, Back, B, Kerl, H, and Wolf, P (2001). High mutation frequency at Ha-ras exons 1–4 in squamous cell carcinomas from PUVA-treated psoriasis patients. *Photochem Photobiol*, 74:323–330.

Kricker, A, Armstrong, BK, English, DR, and Heenan, PJ (1995). A dose-response curve for sun exposure and basal cell carcinoma. *Int J Cancer*, 60:482–488.

Kruger, K, Blume-Peytavi, U, and Orfanos, CE (1999). Basal cell carcinoma possibly originates from the outer root sheath and/or the bulge region of the vellus hair follicle. *Arch Dermatol Res*, 291:253–259.

Kumar, R, Angelini, S, Snellman, E, and Hemminki, K (2004). BRAF mutations are common somatic events in melanocytic nevi. *J Invest Dermatol*, 122:342–348.

Küster, W and Happle R (1984). The inheritance of common baldness: two B or not two B? *J Am Acad Dermatol*, 11:921–926.

Lamartine, J, Munhoz Essenfelder, G, Kibar, Z, Lanneluc, I, Callouet, E, Laoudj, D, et al. (2000). Mutations in GJB6 cause hidrotic ectodermal dysplasia. *Nat Genet*, 26:142–4.

Lamason, RL, Mohideen, MA, Mest, JR, Wong, AC, Norton, HL, Aros, MC, et al. (2005). SLC24A5, a putative cation exchanger, affects pigmentation in zebrafish and humans. *Science*, 310:1782–1786.

Lear, JT, Heagerty, AH, Smith, A, Bowers, B, Payne, CR, Smith, CA, et al. (1996). Multiple cutaneous basal cell carcinomas:glutathione S-transferase (GSTM1, GSTT1) and cytochrome P450 (CYP2D6, CYP1A1) polymorphisms influence tumour numbers and accrual. *Carcinogenesis*, 17:1891–1896.

Lear, JT, Smith, AG, Heagerty, AH, Bowers, B, Jones, PW, Gilford, J, et al. (1997). Truncal site and detoxifying enzyme polymorphisms significantly reduce time to presentation of further primary cutaneous basal cell carcinoma. *Carcinogenesis*, 18:1499–503.

Lee, HO, Levorse, JM, and Shin, MK (2003). The endothelin receptor-B is required for the migration of neural crest-derived melanocyte and enteric neuron precursors. *Dev Biol*, 259:162–175.

Lee, ST, Nicholls, RD, Bundey, S, Laxova, R, Musarella, M, and Spritz, RA (1994). Mutations of the P gene in oculocutaneous albinism, ocular albinism, and Prader-Willi syndrome plus albinism. *N Engl J Med*, 330: 529–534.

Lerner, AB and McGuire, JS (1961). Effect of alpha- and beta-melanocyte stimulating hormones on the skin colour of man. *Nature*, 189:176–179.

Lerner, AB (1959). Vitiligo. *J Invest Dermatol*, 32:285–310.

Lorini, R, Orecchia, G, Martinetti, M, Dugoujon, JM, and Cuccia, M (1992). Autoimmunity in vitiligo: relationship with HLA, Gm and Km polymorphisms. *Autoimmunity*, 11:255–260.

Lubbe, J, Reichel, M, Burg, G, and Kleihues, P (1994). Absence of p53 gene mutations in cutaneous melanoma. *J Invest Dermatol*, 102:819–821.

Majumder, PP, Das, SK, and Li, CC (1988). A genetical model for vitiligo. *Am J Hum Genet*, 43:119–125.

Martinez-Mir, A, Zlotogorski, A, Gordon, D, Petukhova, L, Mo, J, Gilliam, TC, et al. (2007). Genomewide scan for linkage reveals evidence of several susceptibility loci for alopecia areata. *Am J Hum Genet*, 80:316–328.

Martinez-Mir, A, Zlotogorski, A, Ott, J, Gordon, D, and Christiano, AM (2003). Genetic linkage studies in alopecia areata. *J Investig Dermatol Symp Proc*, 8:199–203.

Makova, KD, Ramsay, M, Jenkins, T, and Li, WH (2001). Human DNA sequence variation in a 6.6-kb region containing the melanocortin 1 receptor promoter. *Genetics*, 158:1253–1268.

McDonagh, AJG and Tazi-Ahnini, R (2002). Epidemiology and genetics of alopecia areata. *Clin Exp Dermatol*, 27:405–409.

McGrath, JA, McMillan, JR, Shemanko, CS, Runswick, SK, Leigh, IM, Lane, EB, et al. (1997). Mutations in the plakophilin 1 gene can result in ectodermal dysplasia/skin fragility syndrome. *Nat Genet*, 17:240–244.

McGrath, JA (1999). Hereditary diseases of desmosomes. *J Dermatol Sci*, 20:85–91

McGregor, JM and Proby, CM. (1995). Skin cancer in transplant recipients. *Lancet*, 346:964–965.

McLean, WH, Rugg, EL, Lunny, DP, Morley, SM, Lane, EB, Swensson, O, et al. (1995). Keratin 16 and keratin 17 mutations cause pachyonychia congenita. *Nat Genet*, 9:273–278.

Millikan, RC, Hummer, A, Begg, C, Player, J, de Cotret, AR, Winkel, S, et al. (2005). Polymorphisms in nucleotide excision repair genes and risk of multiple primary melanoma: the genes environment and melanoma study. *Carcinogenesis*, 27:610–618.

Morling, N, Frentz, G, Fugger, L, Georgsen, J, Jakobsen, B, Odum, N, et al. (1991). DNA polymorphism of HLA class II genes in alopecia areata. *Dis Markers*, 9:35–42.

Nagle, DL, Karim, MA, Woolf, EA, Holmgren, L, Bork, P, Misumi DJ, McGrail, SH, et al. (1996). Identification and mutation analysis of the complete gene for Chediak-Higashi syndrome. *Nature Genet*, 14: 307–311.

Nance, MA and Berry, SA (1992). Cockayne syndrome: review of 140 cases. *Am J Med Genet*, 42:68–84.

Nath, SK, Kelly, JA, Namjou, B, Lam, T, Bruner, GR, Scofield, RH, et al. (2001). Evidence for a susceptibility gene, SLEV1, on chromosome 17p13 in families with vitiligo-related systemic lupus erythematosus. *Am J Hum Genet*, 69:1401–1406.

Nath, SK, Majumder, PP, and Nordlund, JJ (1994). Genetic epidemiology of vitiligo: multilocus recessivity cross-validated. *Am J Hum Genet*, 55:981–990.

Nelson, HH, Christensen, B, and Karagas, MR (2005). The XPC poly-AT polymorphism in non-melanoma skin cancer. *Cancer Lett*, 222:205–209.

Newton, RA, Smit, SE, Barnes, CC, Pedley, J, Parsons, PG, and Sturm, RA (2005). Activation of the cAMP pathway by variant human MC1R alleles expressed in HEK and in melanoma cells. *Peptides*, 26:1818–1824.

Nishimura, EK, Jordan, SA, Oshima, H, Yoshida, H, Osawa, M, Moriyama, M, et al. (2002). Dominant role of the niche in melanocyte stem-cell fate determination. *Nature*, 416:854–860.

Norris, A, Todd, C, Graham, A, Quinn, AG, and Thody, AJ (1996). The expression of the c-kit receptor by epidermal melanocytes may be reduced in vitiligo. *Br J Dermatol*, 134:299–306.

Ogg, GS, Rod Dunbar, P, Romero, P, Chen JL, and Cerundolo, V (1998). High frequency of skin-homing melanocyte-specific cytotoxic T lymphocytes in autoimmune vitiligo. *J Exp Med*, 188:1203–1208.

Olsen, EA, Messenger, AG, Shapiro, J, Bergfeld, WF, Hordinsky, MK, Roberts, JL, et al. (2005). Evaluation and treatment of male and female pattern hair loss. *J Am Acad Dermatol*, 52:301–311.

Oro, AE, Higgins, KM, Hu, Z, Bonifas, JM, Epstein, EH Jr, and Scott, MP (1997). Basal cell carcinomas in mice overexpressing sonic hedgehog. *Science*, 276:817–821.

Palmer, CN, Irvine, AD, Terron-Kwiatkowski, A, Zhao, Y, Liao, H, Lee, SP, et al. (2006). Common loss-of-function variants of the epidermal barrier protein filaggrin are a major predisposing factor for atopic dermatitis. *Nat Genet*, 38:441–446.

Palmer, JS, Duffy, DL, Box, NF, Aitken, JF, O'Gorman, LE, Green, AC, et al. (2000). Melanocortin-1 receptor polymorphisms and risk of melanoma: is the association explained solely by pigmentation phenotype? *Am J Hum Genet*, 66:176–186.

Paraskevaidis, A, Drakoulis, N, Roots, I, Orfanos, CE, and Zouboulis, CC (1998). Polymorphisms in the human cytochrome P-450 1A1 gene (CYP1A1) as a factor for developing acne. *Dermatology*, 196:171–175.

Pemble, S, Schroeder, KR, Spencer, SR, Meyer, DJ, Hallier, E, Bolt, HM, et al. (1994). Human glutathione S-transferase theta (GSTT1): cDNA cloning and the characterization of a genetic polymorphism. *Biochem J*, 300:271–276.

Penn, I (1993). Tumors after renal and cardiac transplantation. *Hematol Oncol Clin North Am*, 7:431–445.

Perry, PK and Silverberg, NB (2001). Cutaneous malignancy in albinism. *Cutis*, 67:427–30.

Pfister, H (2003). Human papillomavirus and skin cancer. *J Natl Cancer Inst Monogr*, 31:52–56.

Pierceall, WE, Goldberg, LH, Tainsky, MA, Mukhopadhyay, T, and Ananthaswamy, HN (1991). Ras gene mutation and amplification in human nonmelanoma skin cancers. *Mol Carcino*, 4:196–202.

Ping, XL, Ratner, D, Zhang, H, Wu, XL, Zhang, MJ, Chen, FF, et al. (2001). PTCH mutations in squamous cell carcinoma of the skin. *J Invest Dermatol*, 116:614–616.

Pollock, PM, Harper, UL, Hansen, KS, Yudt, LM, Stark, M, Robbins, CM, et al. (2003). High frequency of BRAF mutations in nevi. *Nat Genet*, 33:19–20.

Pollock, PM, Walker, GJ, Glendening, JM, Que Noy, T, Bloch, NC, Fountain, JW, et al. (2002). PTEN inactivation is rare in melanoma tumours but occurs frequently in melanoma cell lines. *Melanoma Res*, 12:565–575.

Priolo, M and Lagana, C (2001). Ectodermal dysplasias: a new clinical-genetic classification. *J Med Genet*, 38:579–85.

Pritchard, BN and Youngberg, GA (1993). Atypical mitotic figures in basal cell carcinoma. A review of 208 cases. *Am J Dermatopathol*, 15:549–552.

Prota, G and Thomson, RH (1976). Melanin pigmentation in mammals. *Endeavour*, 35:32–38.

Puri, N, Gardner, JM, and Brilliant, MH (2000). Aberrant pH of melanosomes in pink-eyed dilution (p) mutant melanocytes. *J Invest Dermatol*, 115:607–613.

Quinn, AG, Sikkink, S, and Rees, JL (1994). Basal cell carcinomas and squamous cell carcinomas of human skin show distinct patterns of chromosome loss. *Cancer Res*, 54:4756–4759.

Ramachandran, S, Fryer, AA, Smith, AG, Lear, JT, Bowers, B, Griffiths, CE, et al. (2000). Basal cell carcinoma: tumor clustering is associated with increased accrual in high-risk subgroups. *Cancer*, 89: 1012–1018.

Ramachandran, S, Lear, JT, Ramsay, H, Smith, AG, Bowers, B, Hutchinson, PE, et al. (1999). Presentation with multiple cutaneous basal cell carcinomas: association of glutathione S-transferase and cytochrome P450 genotypes with clinical phenotype. *Cancer Epidemiol Biomarkers Prev*, 8:61–67.

Ramoz, N, Rueda, LA, Bouadjar, B, Montoya, LS, Orth, G, and Favre, M (2002). Mutations in two adjacent novel genes are associated with epidermodysplasia verruciformis. *Nat Genet*, 32:579–581.

Rampen, FH, Fleuren, BA, de Boo, TM, and Lemmens, WA (1988). Unreliability of self-reported burning tendency and tanning ability. *Arch Dermatol*, 124:885–888.

Rawles, ME (1947). Origin of pigment cells from the neural crest in the mouse embryo. *Physiol Zool*, 20:248–266.

Rees, JL (2004). The genetics of sun sensitivity in humans. *Am J Hum Genet*, 75:739–751.

Rehman, I, Quinn, AG, Healy, E, and Rees, JL (1994). High frequency of loss of heterozygosity in actinic keratoses, a usually benign disease. *Lancet*, 344:788–789.

Richard, G (2000). Connexins:a connection with the skin. *Exp Dermatol*, 9:77–96.

Rinchik, EM, Bultman, SJ, Horsthemke, B, Lee, ST, Strunk, KM, Spritz, RA, et al. (1993). A gene for the mouse pink-eyed dilution locus and for human type II oculocutaneous albinism. *Nature*, 361:72–76.

Ro, KM, Cantor, RM, Lange, KL, and Ahn, SS (2002). Palmar hyperhidrosis: evidence of genetic transmission. *J Vasc Surg*, 35:382–386.

Robinson, SJ and Healy, E (2002). Human melanocortin 1 receptor (MC1R) gene variants alter melanoma cell growth and adhesion to extracellular matrix. *Oncogene*, 21:8037–8046.

Rodriguez, I, Greer, CA, Mok, MY, and Mombaerts, P (2000). A putative pheromone receptor gene expressed in human olfactory mucosa. *Nat Genet*, 26:18–19.

Rompel, R, Petres, J, Kaupert, K, and Muller-Eckhardt, G (1995). Human leukocyte antigens and multiple basal cell carcinomas. *Recent Results Cancer Res*, 139:297–302.

Rosenstein, BS, Phelps, RG, Weinstock, MA, Bernstein, JL, Gordon, ML, Rudikoff, D, et al. (1999). p53 mutations in basal cell carcinomas arising in routine users of sunscreens. *Photochem Photobiol*, 70:798–806.

Rosso, S, Zanetti R, Martinez, C, Tormo, MJ, Schraub, S, Sancho-Garnier, H, et al. (1996). The multicentre south European study "Helios" II: different sun exposure patterns in the aetiology of basal cell and squamous cell carcinomas of the skin. *Br J Cancer*, 73:1447–1454.

Salasche, SJ (2000). Epidemiology of actinic keratoses and squamous cell carcinoma. *J Am Acad Dermatol*, 42:4–7.

Schallreuter, KU, Wood, JM, and Berger, J (1991). Low catalase levels in the epidermis of patients with vitiligo. *J Invest Dermatol*, 97:1081–1085.

Schioth, HB, Phillips, SR, Rudzish, R, Birch-Machin, MA, Wikberg, JES, and Rees, JL (1999). Loss of function mutations of the human melanocortin 1 receptor are common and are associated with red hair. *Biochem Biophys Res Commun*, 260:488–491.

Schwartz, RA and Torre, DP (1995). The Muir-Torre syndrome: a 25-year retrospect. *J Am Acad Dermatol*, 33:90–104.

Seiji, M, Fitzpatrick, TB, Simpson, RT, and Birbeck, MS (1963). Chemical composition and terminology of specialized organelles (melanosomes and melanin granules) in mammalian melanocytes. *Nature*, 197:1082–1084.

Shamanin, V, zur Hausen, H, Lavergne, D, Proby, CM, Leigh, IM, Neumann, C, et al. (1996). Human papillomavirus infections in nonmelanoma skin cancers from renal transplant recipients and nonimmunosuppressed patients. *J Natl Cancer Inst*,88:802–811.

Shen, H, Sturgis, EM, Khan, SG, Qiao, Y, Shahlavi, T, Eicher, SA, et al. (2001). An intronic poly (AT) polymorphism of the DNA repair gene XPC and risk of squamous cell carcinoma of the head and neck:a case-control study. *Cancer Re*, 61:3321–3325.

Smith, FJD, Irvine, AD, Terron-Kwiatkowski, A, Sandilands, A, Campbell, LE, Zhao, Y, et al. (2006). Loss-of-function mutations in the filaggrin gene cause ichthyosis vulgaris. *Nature Genet*, 38:337–342.

Smith, R, Healy, E, Siddiqui, S, Flanagan, N, Steijlen, PM, Rosdahl, I, et al. (1998). Melanocortin 1 receptor variants in an Irish population. *J Invest Dermatol*, 111:119–122.

Soengas, MS, Capodieci, P, Polsky, D, Mora, J, Esteller, M, Opitz Araya, X, et al. (2001). Inactivation of the apoptosis effector Apaf 1 in malignant melanoma. *Nature*, 409:207–211.

Solomon, LM, Fretzin, D, and Pruzansky, S (1970). Pilosebaceous abnormalities in Apert's syndrome. *Arch Dermatol*, 102:381–385.

Sparrow, LE, Soong, R, Dawkins, HJ, Iacopetta, BJ, and Heenan, PJ (1995). p53 gene mutation and expression in naevi and melanomas. *Melanoma Res*, 5:93–100.

Spencer, JM, Kahn, SM, Jiang, W, DeLeo, VA, and Weinstein, IB (1995). Activated ras genes occur in human actinic keratoses, premalignant precursors to squamous cell carcinomas. *Arch Dermatol*, 131:796–800.

Spritz, RA, Giebel, LB, and Holmes, SA (1992). Dominant negative and loss of function mutations of the c-kit (mast/stem cell growth factor receptor) proto-oncogene in human piebaldism. *Am J Hum Genet*, 50: 261–269.

Staricco, RJ and Pinkus, H (1957). Quantitative and qualitative data on the pigment cells of adult human epidermis. *J Invest Dermatol*, 28:33–45.

Steel, KP, Davidson, DR, and Jackson, IJ (1992). TRP-2/DT, a new early melanoblast marker, shows that steel growth factor (c-kit ligand) is a survival factor. *Development*, 115:1111–1119.

Stenn, KS and Paus, R (2001). Controls of hair follicle cycling. *Physiol Rev*, 81:449–494.

Stern, C (1970). Model estimates of the number of gene pairs involved in pigmentation variability of the Negro American. *Hum Hered*, 20:165–168.

Stern, RS and Lunder, EJ (1998). Risk of squamous cell carcinoma and methoxsalen (psoralen) and UV-A radiation (PUVA): a meta-analysis. *Arch Dermatol*, 134:1582–1585.

Stinchcombe, JC, Page, LJ, and Griffiths, GM (2000). Secretory lysosome biogenesis in cytotoxic T lymphocytes from normal and Chediak Higashi syndrome patients. *Traffic*, 1:435–444.

Szabo, G, Gerald, AB, Pathak, MA, and Fitzpatrick, TB (1969). Racial differences in the fate of melanosomes in human epidermis. *Nature*, 222:1081–1082.

Takata, M, Morita, R, and Takehara, K (2000). Clonal heterogeneity in sporadic melanomas as revealed by loss-of-heterozygosity analysis. *Int J Cancer*, 85:492–497.

Tan, CP, McKee, KK, Weinberg, DH, MacNeil, T, Palyha, OC, Feighner, SD, et al. (1999). Molecular analysis of a new splice variant of the human melanocortin-1 receptor. *FEBS Lett*, 451:137–141.

Tarlow, JK, Clay, FE, Cork, MJ, Blakemore, AI, McDonagh, AJ, Messenger AG, et al. (1994). Severity of alopecia areata is associated with a polymorphism in the interleukin-1 receptor antagonist gene. *J Invest Dermatol*, 103:387–390.

Tastan, HB, Akar, A, Orkunoglu, FE, Arca, E, and Inal, A (2004). Association of HLA class I antigens and HLA class II alleles with vitiligo in a Turkish population. *Pigment Cell Res*, 17:181–184.

Tazi-Ahnini, R, di Giovine, FS, McDonagh, AJ, Messenger, AG, Amadou, C, Cox, A, et al. (2000). Structure and polymorphism of the human gene for the interferon-induced p78 protein (MX1): evidence of association with alopecia areata in the Down syndrome region. *Hum Genet*, 106:639–645.

Tazi-Ahnini, R, Cox, A, McDonagh, AJ, Nicklin, MJ, di Giovine, FS, Timms, JM, et al. (2002). Genetic analysis of the interleukin-1 receptor antagonist and its homologue IL-1L1 in alopecia areata: strong severity association and possible gene interaction. *Eur J Immunogenet*, 29:25–30.

Thody, AJ, Higgins, EM, Wakamatsu, K, Ito, S, Burchill, SA, and Marks, JM (1991). Pheomelanin as well as eumelanin is present in human epidermis. *J Invest Dermatol*, 97:340–344.

Tilli, CMLJ, Van Steensel, MAM, Krekels, GAM, Neumann, HA, and Ramaekers, FC (2005). Molecular aetiology and pathogenesis of basal cell carcinoma. *Br J Dermatol*, 152:1108–1124.

Tomescu, D, Kavanagh, G, Ha, T, Campbell, H, and Melton, DW (2001). Nucleotide excision repair gene XPD polymorphisms and genetic predisposition to melanoma. *Carcinogenesis*, 22:403–408.

Tomita, Y, Takeda, A, Okinaga, S, Tagami, H, and Shibahara, S (1989). Human oculocutaneous albinism caused by single base insertion in the tyrosinase gene. *Biochem Biophys Res Commun*, 164: 990–996.

Torley, D, Bellus, GA, and Munro, CS (2002). Genes, growth factors and acanthosis nigricans. *Br J Dermatol*, 147: 1096–1101.

Trakakis, E, Laggas, D, Salamalekis, E, and Creatsas, G (2005). 21-Hydroxylase deficiency: from molecular genetics to clinical presentation. *J Endocrinol Invest*, 28:187–192.

Tran, TP, Titus-Ernstoff, L, Perry, AE, Ernstoff, MS, and Newsham, IF (2002). Alteration of chromosome 9p21 and/or p16 in benign and dysplastic nevi suggests a role in early melanoma progression. *Cancer Causes Control*, 13:675–682.

Tripathi, RK, Giebel, LB, Strunk, KM, and Spritz, RA (1991). A polymorphism of the human tyrosinase gene is associated with temperature-sensitive enzymatic activity. *Gene Expr*, 1:103–110.

Valverde, P, Healy, E, Jackson, I, Rees, JL, and Thody, AJ (1995). Variants of the melanocyte-stimulating hormone receptor gene are associated with red hair and fair skin in humans. *Nature Genet*, 11:328–330.

Valverde, P, Healy, E, Sikkink, S, Haldane, F, Thody, AJ, Carothers, A, et al. (1996). The Asp84Glu variant of the melanocortin 1 receptor (MC1R) is associated with melanoma. *Hum Mol Genet*, 5(10):1663–1666.

van der Riet, P, Karp, D, Farmer, E, Wei Q, Grossman L, Tokino K, et al. (1994). Progression of basal cell carcinoma through loss of chromosome 9q and inactivation of a single p53 allele. *Cancer Res*, 54:25–27.

van der Riet, P, Nawroz, H, Hruban, RH, Corio, R, Tokino, K, Koch, W, et al. (1994). Frequent loss of chromosome 9p21–22 early in head and neck cancer progression. *Cancer Res*, 54:1156–1158.

van der Velden, PA, Sandkuijl, LA, Bergman, W, Pavel, S, van Mourik, L, Frants, RR, et al. (2001). Melanocortin-1 receptor variant R151C modifies melanoma risk in Dutch families with melanoma. *Am J Hum Genet*, 69:774–779.

van Elsas, A, Zerp, SF, van der Flier, S, Kruse, KM, Aarnoudse, C, Hayward, NK, et al. (1996). Relevance of ultraviolet-induced N-ras oncogene point mutations in development of primary human cutaneous melanoma. *Am J Pathol*, 149:883–893.

van Steensel, MA, van Geel, M, and Steijlen PM (2004). Molecular genetics of hereditary hair and nail disease. *Am J Med Genet C Semin Med Genet*, 131C:52–60.

van 't Veer, LJ, Burgering, BM, Versteeg, R, Boot, AJ, Ruiter, DJ, Osanto, S, et al. (1989). N-ras mutations in human cutaneous melanoma from sun-exposed body sites. *Mol Cell Biol*, 9:3114–116.

Vitasa, BC, Taylor, HR, Strickland, PT, Rosenthal, FS, West, S, Abbey, H, et al. (1990). Association of nonmelanoma skin cancer and actinic keratosis with cumulative solar ultraviolet exposure in Maryland watermen. *Cancer*, 65:2811–2817.

Voisey, J, Box, NF, and van Daal, A (2001). A polymorphism study of the human Agouti gene and its association with MC1R. *Pigment Cell Res*, 14:264–267.

Voorhees, JJ, Wilkins, JW Jr, Hayes E, Harrell, ER (1972). Nodulocystic acne as a phenotypic feature of the XYY genotype. *Arch Dermatol*, 105:913–919.

Walton, S, Wyatt, EH, and Cunliffe, WJ (1988). Genetic control of sebum excretion and acne—a twin study. *Br J Dermatol*, 118:393–396.

Watanabe, A, Takeda, K, Ploplis, B, and Tachibana, M (1998). Epistatic relationship between Waardenburg syndrome genes MITF and PAX3. *Nat Genet*, 18:283–286.

Wang, LL, Levy, ML, Lewis, RA, Chintagumpala, MM, Lev, D, Rogers, M, et al. (2001). Clinical manifestations in a cohort of 41 Rothmund-Thomson syndrome patients. *Am J Med Genet*, 102:11–117.

Wells, RS and Kerr, CB (1966). Clinical features of autosomal dominant and sex-linked ichthyosis in an English population. *Br Med J*, 1:947–950.

Welsh, EA, Clark, HH, Epstein, SZ, Reveille, JD, and Duvic, M. (1994) Human leukocyte antigen-DQB1*03 alleles are associated with alopecia areata. *J Invest Dermatol*, 103: 758–63.

Whiteman, DC, Whiteman, CA, and Green, AC (2001). Childhood sun exposure as a risk factor for melanoma: a systematic review of epidemiologic studies. *Cancer Causes Control*, 12:69–82.

Wicking, C, Smyth, I, and Bale, A (1999). The hedgehog signalling pathway in tumorigenesis and development. *Oncogene*, 20;18(55):7844–7851.

Wilkie, AO, Slaney, SF, Oldridge, M, Poole, MD, Ashworth, GJ, Hockley, AD, et al. (1995). Apert syndrome results from localized mutations of FGFR2 and is allelic with Crouzon syndrome. *Nat Genet*, 9:165–172.

Winter, H, Schissel, D, Parry, DA, Smith, TA, Liovic, M, Lane, EB, et al. (2004). An unusual Ala12Thr polymorphism in the 1A alpha-helical segment of the companion layer-specific keratin K6hf:evidence for a risk factor in the etiology of the common hair disorder pseudofolliculitis barbae. *J Invest Dermatol*, 122:652–657.

Wise, CA, Gillum, JD, Seidman, CE, Lindor, NM, Veile, R, Bashiardes, S, et al. (2002). Mutations in CD2BP1 disrupt binding to PTP PEST and are responsible for PAPA syndrome, an autoinflammatory disorder. *Hum Mol Genet*, 11:961–969.

Xie, J, Murone, M, Luoh, SM, Ryan, A, Gu, Q, Zhang, C, et al. (1998). Activating Smoothened mutations in sporadic basal-cell carcinoma. *Nature*, 391:90–92.

Xu, X, Thornwall, M, Lundin, LG, and Chhajlani, V (1996). Val92Met variant of the melanocyte stimulating hormone receptor gene. *Nature Genet*, 14:384.

Yengi, L, Inskip, A, Gilford, J, Alldersea, J, Bailey, L, Smith, A, et al. (1996). Polymorphism at the glutathione S-transferase locus GSTM3:interactions with cytochrome P450 and glutathione S-transferase genotypes as risk factors for multiple cutaneous basal cell carcinoma. *Cancer Res*, 56:1974–1977.

Yoshida, H, Kunisada, T, Kusakabe, M, Nishikawa, S, and Nishikawa, SI (1996). Distinct stages of melanocyte differentiation revealed by anlaysis of non-uniform pigmentation patterns. *Development*, 122:1207–1214.

Zaias, N, Tosti, A, Rebell, G, Morelli, R, Bardazzi, F, Bieley, H, et al. (1996). Autosomal dominant pattern of distal subungual onychomycosis caused by Trichophyton rubrum. *J Am Acad Dermatol*, 34:302–304.

Zamani, M, Spaepen, M, Sghar, SS, Huang, C, Westerhof, W, Nieuweboer-Krobotova, L, et al. (2001). Linkage and association of HLA class II genes with vitiligo in a Dutch population. *Br J Dermatol*, 145:90–94.

Zeng, C, Spielman, AI, Vowels, BR, Leyden, JJ, Biemann, K, and Preti, G (1996). A human axillary odorant is carried by apolipoprotein D. *Proc Natl Acad Sci USA*, 93:6626–6630.

Zerp, SF, van Elsas, A, Peltenburg, LT, and Schrier, PI (1999). p53 mutations in human cutaneous melanoma correlate with sun exposure but are not always involved in melanomagenesis. *Br J Cancer*, 79:921–926.

Zhou, XP, Gimm, O, Hampel, H, Niemann, T, Walker, MJ, and Eng, C (2000). Epigenetic PTEN silencing in malignant melanomas without PTEN mutation. *Am J Pathol*, 157:1123–1128.

Zhu, G, Evans, DM, Duffy, DL, Montgomery, GW, Medland, SE, Gillespie, NA, et al. (2004). A genome scan for eye color in 502 twin families: most variation is due to a QTL on chromosome 15q. *Twin Res*, 7:197–210.

Ziegler, A, Jonason, AS, Leffell, DJ, Simon, JA, Sharma, HW, Kimmelman, J, et al. (1994). Sunburn and p53 in the onset of skin cancer. *Nature*,. 372:773–776.

35

Osteoporosis and Related Disorders

Yoshiji Yamada

Osteoporosis is a major health problem of the elderly, which is evenly distributed across the world. It is characterized by a reduction in bone mineral density (BMD) and deterioration in the microarchitecture of bone, both of which result in predisposition to fractures (Kanis et al., 1994). The clinical importance of osteoporosis lies in its association with fractures (Ralston, 2001). Osteoporotic fractures are common, expensive to treat, and a substantial cause of morbidity and mortality in most developed countries. The most common sites for osteoporotic fractures are the hip, spine, and distal forearm. The incidence of such fractures increases with age, and is greater for women than for men as a result of the increase in the rate of bone loss after menopause (Melton, 1995). BMD is one of the most important determinants of the risk for osteoporotic fracture (Cummings et al., 1993). Other factors include susceptibility to falling (Dargent-Molina et al., 1996), bone quality, and skeletal geometry (Faulkner et al., 1993).

Bone mass is determined by various mechanisms that are affected by environmental factors as well as by multiple genes. The environmental factors that contribute to the determination of bone mass include calcium intake (Holbrook et al., 1988), smoking (Krall and Dawson-Hughes, 1999; Hannan et al., 2000), alcohol consumption (Hannan et al., 2000), and physical activity (Cooper et al., 1988). Despite the importance of these environmental factors, genetic factors play the predominant role in the regulation of bone mass and also contribute substantially to other determinants of osteoporotic fracture risk (Ralston, 2001). Twin studies have suggested that most of the variance in BMD is genetically determined, with the actual contribution depending on the site examined. For example, the heritability of BMD for the spine and hip have been estimated at between 70% and 85%, compared with 50% and 60% for the wrist (Pocock et al., 1987; Flicker et al., 1995; Arden et al., 1996; Howard et al., 1998). Segregation studies in families also support a strong genetic contribution to the regulation of bone mass, with a pattern of inheritance that is most consistent with the additive action of several genes, each with modest effects, rather than with that of a few genes with large effects (Gueguen et al., 1995).

Other determinants of the risk for osteoporotic fracture probably have a genetic component, as reflected by maternal or grandmaternal history of hip fracture independent of the bone mass (Cummings et al., 1995; Stewart et al., 1996). This observation could be explained by the genetic effects on hip axis length, ultrasound properties of bone (Arden et al., 1996), or bone turnover (Garnero, Arden, et al., 1996). The heritability estimates for quantitative ultrasound properties of bone, hip axis length, and other aspects of femoral neck geometry range from 60% to 85% (Arden et al., 1996; Garnero, Arden, et al., 1996; Slemenda et al., 1996). Although a positive family history is a risk factor for fracture (Keen et al., 1999), twin and family studies indicate that the heritability of fracture itself is only in the order of 25%–35% (Deng et al., 2000; MacGregor et al., 2000). This heritability is much lower than that of the skeletal phenotypes that predispose to fracture, probably because of the importance of fall-related factors in determining the fracture risk (Kannus et al., 1999; Ralston, 2002).

Personalized prevention and treatment of osteoporosis and osteoporotic fractures are important public health goals. One approach is to identify disease susceptibility genes. Although genetic linkage analyses and candidate gene association studies have implicated several loci and genes in the regulation of BMD and the pathogenesis of osteoporosis or fractures, the genes that confer susceptibility to osteoporosis remain to be identified.

This chapter describes the genetic and genomic aspects of osteoporosis including an overview of related monogenic bone diseases, and the candidate genes (Table 35–1) identified by linkage analyses or association studies in BMD or predisposition to fractures. Such studies may provide insight into the function of these genes as well as into the role of genetics in the development of osteoporosis or osteoporotic fractures (Box 35–1).

Box 35–1 Key Points

- Osteoporosis is characterized by a reduction in BMD and deterioration in the microarchitecture of bone, both of which result in the predisposition to fractures.
- Although reproductive, nutritional, and lifestyle factors influence BMD, family and twin studies have suggested that BMD is largely heritable and under the control of multiple genes.
- Personalized prevention and treatment of osteoporosis and osteoporotic fractures are important public health goals.
- Several candidate genes are known, which may confer susceptibility to osteoporosis and related phenotypes.

Table 35–1 Candidate Genes for Susceptibility to Bone Mineral Density (BMD) and Osteoporosis

- Estrogen receptor α
- Interleukin-6
- Tumor necrosis factor receptor superfamily, member 11B (osteoprotegerin)
- Low-density lipoprotein receptor-related protein 5
- Vitamin D receptor
- Type I collagen α1

Genetics of Osteoporosis

Genes for Monogenic Bone Disease

In several rare inherited bone diseases, mutations in individual genes have profound effects on bone mass or the risk of fragility fractures (Ralston, 2002). The classic example of such diseases is osteogenesis imperfecta, which is characterized by severe bone fragility, reduced bone mass, and other connective tissue manifestations (Rauch and Glorieux, 2004). Most cases of osteogenesis imperfecta are caused by mutations in type I collagen genes (*COL1A1* or *COL1A2*), although clinical features of this condition also occur in individuals from families with no abnormalities of collagen genes (Glorieux et al., 2002). Osteoporosis is also a prominent manifestation in patients with inactivating mutations of the genes for aromatase (*CYP19A1*, for cytochrome P450, family 19, subfamily A, polypeptide 1), or estrogen receptor a (*ESR1*), as a result of the absence of estrogen action in bone in these individuals (Smith et al., 1994; Morishima et al., 1995).

Osteoporosis-pseudoglioma syndrome is a rare autosomal recessive disease characterized by reduced bone mass, vitreous opacities, and fragility fractures. Positional cloning identified homozygous inactivating mutations in the low-density lipoprotein (LDL) receptor-related protein 5 gene (*LRP5*) as the cause of this syndrome (Gong et al., 2001). A monogenic syndrome characterized by unusually high BMD was also mapped to the osteoporosis-pseudoglioma syndrome locus on chromosome 11q12-13, and shown to be caused by an activating mutation of *LRP5* (Boyden et al., 2002; Little et al., 2002). Mutations in the region of the transforming growth factor-β1 gene (*TGFB1*) that encodes the latency-activating peptide domain have been shown to be responsible for Camurati-Engelmann disease (Janssens et al., 2000), a condition characterized by osteosclerosis of the diaphysis of long bones, whereas mutations in the coding or regulatory regions of the sclerostin gene (*SOST*) have been identified as the cause of the sclerosing bone dysplasias known as sclerosteosis and van Buchem disease (Balemans et al., 2001, 2002).

Inactivating mutations of the T-cell immune regulator 1 gene (*TCIRG1*), which encodes a subunit of the vacuolar proton pump of osteoclasts, are responsible for autosomal recessive osteopetrosis (Frattini et al., 2000), whereas inactivating mutations in the chloride channel 7 gene (*CLCN7*), which encodes a chloride channel expressed in osteoclasts, cause severe infantile osteopetrosis. Although haploinsufficiency of *CLCN7* does not result in an obvious bone phenotype, specific heterozygous mutations in the gene give rise to autosomal dominant osteopetrosis (Albers-Schönberg disease), presumably by exerting a dominant negative effect on chloride channel function (Cleiren et al., 2001). Finally, pyknodysostosis, an autosomal recessive osteochondrodysplasia characterized by osteosclerosis and short stature, has been found to be caused by nonsense or missense mutations in the cathepsin K gene (*CTSK*) (Gelb et al., 1996), whereas a missense mutation of the carbonic anhydrase II gene (*CA2*) was shown to be responsible for an autosomal recessive syndrome of osteopetrosis with renal tubular acidosis and cerebral calcification (Sly et al., 1983; Roth et al., 1992).

Some of these genes for monogenic bone diseases may also contribute to the regulation of BMD in the general population. For example, polymorphisms of *COL1A1* (Grant et al., 1996; Uitterlinden et al., 1998), *TGFB1* (Langdahl et al., 1997; Yamada et al., 1998, 2000; Yamada, Ando, et al., 2001; Yamada, Miyauchi, et al., 2001), *LRP5* (Ferrari, Deutsch, et al., 2004), and *SOST* (Uitterlinden et al., 2004) have been shown to be associated with BMD or the prevalence of osteoporosis or fractures.

Strategies for Genetic Analysis of Osteoporosis and Related Phenotypes

There are two basic strategies for identifying genes that influence BMD or the predisposition to osteoporosis or fractures (Nguyen et al., 2000): the candidate gene approach in which individual genes are examined directly for a possible role in determination of the trait of interest, and the genome-wide approach in which all genes are examined systematically with panels of microsatellite DNA markers or single nucleotide polymorphisms (SNPs) uniformly distributed throughout the genome. In each of these approaches, susceptibility genes or loci are identified by the demonstration of a significant linkage or association. Linkage analysis involves the proposition of a model to account for the pattern of inheritance of a phenotype observed in a pedigree. It determines whether the phenotypic locus is transmitted together with the genetic markers of known chromosomal position. Association studies determine whether a certain allele occurs at a frequency higher than that expected by chance in individuals with a particular phenotype. Such an association is thus suggested by a statistically significant difference in the prevalence of alleles with respect to the phenotype (Nguyen et al., 2000). The genetic analysis of complex multifactorial diseases is described in more detail in Chapter 3.

Linkage Analysis of BMD or Osteoporosis

The published results of whole-genome and partial-genome linkage analyses for BMD or osteoporosis are summarized in Table 35–2, and those of linkage analyses of candidate genes or loci are listed in Table 35–3. Initial linkage analyses found that a locus linked to low BMD mapped to chromosomal region 1p36.3-36.2 (Devoto et al., 1998, 2001, 2005), whereas a locus linked to high BMD mapped to 11q12-13 (Johnson et al., 1997). Genes responsible for other BMD-related phenotypes such as autosomal recessive osteoporosis-pseudoglioma syndrome (Gong et al., 1996) and autosomal dominant osteopetrosis (Van Hul et al., 2002) are also located at 11q12-13. Osteoporosis-pseudoglioma syndrome and high-bone-mass syndrome are caused by loss-of-function and activating mutations, respectively, of *LRP5* at this locus (Gong et al., 2001; Boyden et al., 2002; Little et al., 2002), as described in detail in other section.

Additional studies have indicated that BMD is linked to multiple chromosomal loci and that such loci differ for BMD at different sites (Table 35–4). Osteoporosis has also been linked to

Table 35–2 Whole-genome and Partial-genome Linkage Analyses of Bone Mineral Density (BMD) or Osteoporosis

Chromosomal Locus	Marker	Phenotype	Reference
1p36.3	*D1S2694*	BMD (femoral neck)	Devoto et al., 2005
1p36.3-36.2	*D1S214*	Low BMD (femoral neck)	Devoto et al., 2001
1p36	*D1S450*	Low BMD (hip)	Devoto et al., 1998
1p36		BMD (whole body)	Wilson et al., 2003
1q21-23	*D1S484*	BMD (lumbar spine)	Koller et al., 2000
1q22-23	*D1S445*	BMD (lumbar spine)	Econs et al., 2004
2p25	*D2S1780*	BMD (femoral neck)	Kammerer et al., 2003
2p24-23	*D2S149*	Low BMD (spine)	Devoto et al., 1998
2p24-21.1	*D2S2976, D2S1400, D2S405*	BMD (radius)	Niu et al., 1999
3p26	*D3S1297*	BMD (wrist)	Deng et al., 2002
3q21	*D3S1298, D3S1285*	BMD (lumbar spine)	Wilson et al., 2003
3q25	*D3S1279, D3S1565*	BMD (lumbar spine)	Ralston et al., 2005
4p	*D4S2639*	BMD (radius)	Kammerer et al., 2003
4q25	*D4S1572, D4S406*	BMD (femoral neck)	Ralston et al., 2005
4q32	*D4S413*	BMD (spine, wrist)	Deng et al., 2002
4q32-34	*D4S1539, D4S1554*	Low BMD (hip)	Devoto et al., 1998
5q33-35	*D5S422*	BMD (femoral neck)	Koller et al., 2000
6p21.2	*D6S2427*	BMD (femoral neck, lumbar spine)	Karasik et al., 2002
6p12-11	*D6S257, D6S462*	BMD (lumbar spine)	Koller et al., 2000
6q27	*D6S446*	BMD (trochanter)	Devoto et al., 2005
7p15	*D7S493*	BMD (lumbar spine)	Devoto et al., 2005
7p14	*D7S516, D7S510*	BMD (femoral neck)	Ralston et al., 2005
7q22	*D7S531*	BMD (spine)	Deng et al., 2002
8q24.3	*D8S373*	BMD (Ward's triangle)	Karasik et al., 2002
10q21	*D10S196, D10S537*	BMD (femoral neck)	Ralston et al., 2005
10q26	*D10S1651*	BMD (hip)	Deng et al., 2002
11q12-13	*D11S987*	High BMD (lumbar spine)	Johnson et al., 1997
11q12-13	*D11S1313, D11S935*	BMD (lumbar spine, femoral neck)	Koller et al., 2000
12q24	*D12S1723*	BMD (spine)	Deng et al., 2002
12q24	*D12S2070*	BMD (radius)	Kammerer et al., 2003
12q24.2	*D12S395*	BMD (lumbar spine)	Karasik et al., 2002
13q14-22	*D13S788*	BMD (femoral neck, trochanter)	Kammerer et al., 2003
13q21-34	*D13S788, D13S800*	BMD (radius)	Niu et al., 1999
14q	*D14S592, D14S588*	BMD (trochanter)	Peacock et al., 2004
14q31	*D14S587*	BMD (lumbar spine)	Karasik et al., 2002
15q	*D15S1507*	BMD (femoral neck)	Peacock et al., 2004
16p13	*D16S3075, D16S261*	BMD (femoral neck)	Ralston et al., 2005
16q23	*D16S3091, D16S520*	BMD (lumbar spine)	Ralston et al., 2005
17p13	*D17S1852*	BMD (wrist)	Deng et al., 2002
18p11	*D18S53, D18S478*	BMD (lumbar spine)	Ralston et al., 2005
20p12.3	*D20S882, D20S900*	Osteoporosis	Styrkarsdottir et al., 2003
20q13	*D20S196, D20S173*	BMD (lumbar spine)	Ralston et al., 2005
21q22.2	*D21S2055*	BMD (trochanter)	Karasik et al., 2002
21qter	*D21S1446*	BMD (trochanter)	Karasik et al., 2002

Table 35–3 Candidate Gene or Locus Linkage Analysis of Bone Mineral Density (BMD) or Osteoporosis

Chromosomal Locus	Marker	Candidate Gene	Phenotype	Reference
1p36	*D1S507*		Low BMD (femoral neck)	Wynne et al., 2003
1q25-31	*D1S3737*	Osteocalcin (*BGLAP*)	BMD (femoral neck)	Raymond et al., 1999
2p25-22	*D2S168*		Low BMD (lumbar spine)	Wynne et al., 2003
3p22-21.2	*D3S3559, D3S1289*	PTH receptor 1 (*PTHR1*)	BMD (femoral neck, lumbar spine)	Duncan et al., 1999
6p21.3		Tumor necrosis factor (*TNF*)	Osteoporosis	Ota et al., 2000
7p21		Interleukin 6 (*IL6*)	Osteopenia (radius)	Ota et al., 1999
11q12-13	*D11S4148*	LRP5 (*LRP5*)	Low BMD (lumbar spine)	Wynne et al., 2003

Abbreviations: LRP5, LDL receptor-related protein 5; PTH, parathyroid hormone.

chromosomal region 20p12.3 (Styrkarsdottir et al., 2003). Recent family-based genome-wide scans have revealed that the loci that affect BMD are largely gender-, age-, and site-specific (Ralston et al., 2005).

Association Studies of BMD, Osteoporosis, or Fractures

Polymorphisms of a variety of candidate genes have been associated with BMD or with the genetic susceptibility to osteoporosis or osteoporotic fracture. Table 35–5 lists genes including *MTHFR, TNFRSF1B, PRDM2, ALPL,* and *BGLAP* (chromosome 1); *QPCT, IL1B, IL1RN, LCT,* and *TANK* (chromosome 2); *PPARG, CCR2,* and *CASR* (chromosome 3); *MTP* (chromosome 4); *PDE4D* and *PDLIM4* (chromosome 5); *RUNX2* and *ESR1* (chromosome 6); *IL6, CALCR, PON1, PON2,* and *COL1A2* (chromosome 7); *GNRH1* and *TNFRSF11B* (chromosome 8); *VLDLR* (chromosome 9); *CYP17A1* (chromosome 10); *CDKN1C, DRD4, PTH, CALCA, LRP5, TCIRG1,* and *MMP1* (chromosome 11); *VDR* and *IGF1* (chromosome 12); *KL* (chromosome 13); *ESR2, GC,* and *BMP4* (chromosome 14); *CYP19A1* and *CYP1A1* (chromosome 15); *SOST, COL1A1,* and *GH1* (chromosome 17); *TGFB1* and *APOE* (chromosome 19); *COMT* (chromosome 22); and *AR* (chromosome X). These genes are the candidate loci for the determination of BMD or susceptibility to osteoporosis or osteoporotic fractures. However, it is also possible that polymorphisms in these genes are in linkage disequilibrium with other polymorphisms in nearby genes that are the actual determinants of BMD or susceptibility to osteoporosis or fracture. In the following sections, six candidate genes (*ESR1, IL6, TNFRSF11B, LRP5, VDR,* and *COL1A1*) that are of particular interest in the genetics of osteoporosis are reviewed.

Estrogen Receptor α

The importance of estrogen receptor (in the regulation of bone mass was indicated by the occurrence of osteoporosis in a man with a nonsense mutation in *ESR1* (Smith et al., 1994), as well as by the observation that BMD in mice lacking a functional *ESR1* allele is 20%–25% less than that in wild-type mice (Korach, 1994). Two SNPs have been identified in the first intron of *ESR1*: a T→C polymorphism that is recognized by the restriction endonuclease *Pvu*II [*T* and *C* alleles correspond to the presence (*p* allele) and absence (*P* allele) of the restriction site, respectively], and an A→G polymorphism that is recognized by *Xba*I [*A* and *G* alleles correspond to the presence (*x* allele) and absence (*X* allele) of the restriction site, respectively]. These SNPs, alone or in combination, have been associated with BMD in postmenopausal (Kobayashi et al., 1996; Albagha et al., 2001) or premenopausal (Willing et al., 1998; Patel et al., 2000) women or with response to hormone replacement therapy (Salmén et al., 2000). However, other studies did not confirm these observations (Han et al., 1997; Gennari et al., 1998; Vandevyver et al., 1999; Becherini et al., 2000; Langdahl et al., 2000b; Brown et al., 2001). Albagha et al. (2001) found no association between BMD in women and either of these two *ESR1* polymorphisms alone, but did detect an association of BMD for the lumbar spine or femoral neck with the haplotype of the SNPs. In addition, a microsatellite (TA repeat) polymorphism in the promoter region of *ESR1* was associated with BMD and with the prevalence of fractures in studies in which such an association with the T→C or A→G SNPs in the first intron was not apparent (Becherini et al., 2000; Langdahl et al., 2000b; Albagha et al., 2001). No association was detected between estrogen responsiveness of BMD and *ESR1* genotype in postmenopausal Korean women who had undergone

Table 35–4 Genes Related to Bone Mineral Density

Anatomic Site	Chromosomal Map (References)
Radius or wrist	2p24-21.1 (Niu et al., 1999), 3p26 (Deng et al., 2002), 4p (Kammerer et al., 2003), 4q32 (Deng et al., 2002), 12q24 (Kammerer et al., 2003), 13q21-34 (Niu et al., 1999), and 17p13 (Deng et al., 2002)
Lumbar spine	1q21-23 (Econs et al., 2004; Koller et al., 2000), 2p24-23 (Devoto et al., 1998), 3q21 (Wilson et al., 2003), 3q25 (Ralston et al., 2005), 4p32 (Deng et al., 2002), 6p21.2 (Karasik et al., 2002), 6p12-11 (Koller et al., 2000), 7p15 (Devoto et al., 2005), 7q22 (Deng et al., 2002), 11q12-13 (Johnson et al., 1997; Koller et al., 2000), 12q24-24.2 (Deng et al., 2002; Karasik et al., 2002), 14q31 (Karasik et al., 2002), 16q23 (Ralston et al., 2005), and 20q13 (Ralston et al., 2005)
Hip-femoral neck, trochanter, and Ward's triangle	1p36 (Devoto et al., 1998, 2001, 2005; Wilson et al., 2003), 2p25 (Kammerer et al., 2003), 4q25 (Ralston et al., 2005), 4q32-34 (Devoto et al., 1998), 5q33-35 (Koller et al., 2000), 6p21.2 (Karasik et al., 2002), 6q27 (Devoto et al., 2005), 7p14 (Ralston et al., 2005), 8q24.3 (Karasik et al., 2002), 10q21 (Ralston et al., 2005), 10q26 (Deng et al., 2002), 11q12-13 (Koller et al., 2000), 13q14-22 (Kammerer et al., 2003), 14q (Peacock et al., 2004), 15q (Peacock et al., 2004), 16p13 (Ralston et al., 2005), 18p11 (Ralston et al., 2005), and 21q22.2-qter (Karasik et al., 2002)

Table 35–5 Candidate Gene Association Studies of Bone Mineral Density (BMD), Osteoporosis, or Fractures

Candidate Gene	Chromosomal Locus	Phenotype	Reference
Methylenetetrahydrofolate reductase (*MTHFR*)	1p36.3	BMD	Miyao, Morita, et al., 2000
Tumor necrosis factor receptor 2 (*TNFRSF1B*)	1p36.3-36.2	Low BMD (spine)	Spotila et al., 2000
Retinoblastoma protein-binding zinc finger protein RIZ (*PRDM2*)	1p36	BMD	Grundberg et al., 2004
Alkaline phosphatase, liver (*ALPL*)	1p36.1-34	BMD	Goseki-Sone et al., 2005
Gamma-carboxyglutamic acid protein, bone (*BGLAP*)	1q25-31	BMD, osteopenia	Dohi et al., 1998
Glutaminyl-peptide cyclotransferase (*QPCT*)	2p22.2	BMD	Ezura et al., 2004
Interleukin-1β (*IL1B*)	2q14	BMD	Nemetz et al., 2001
Interleukin-1 receptor antagonist (*IL1RN*)	2q14.2	Bone loss (spine)	Keen et al., 1998
		Osteoporotic fracture	Langdahl et al., 2000a
Lactase (*LCT*)	2q21	BMD	Obermayer-Pietsch et al., 2004
Tumor necrosis factor receptor-associated factor-interacting protein (*TANK*)	2q24-31	BMD	Ishida et al., 2003
Peroxisome proliferator-activated receptor-γ (*PPARG*)	3p25	BMD	Ogawa et al., 1999
CC chemokine receptor 2 (*CCR2*)	3p21	BMD	Yamada et al., 2002a
Calcium-sensing receptor (*CASR*)	3q13.3-21	BMD	Lorentzon et al., 2001
Microsomal triglyceride transfer protein (*MTP*)	4q22-24	BMD	Yamada et al., 2005
Phosphodiesterase 4D, cAMP-specific (*PDE4D*)	5q12	BMD	Reneland et al., 2005
PDZ and LIM domain protein 4 (*PDLIM4*)	5q31.1	BMD	Omasu et al., 2003
Runt-related transcription factor 2 (*RUNX2*)	6p21	BMD, osteoporotic fracture	Vaughan et al., 2002
Estrogen receptor α (*ESR1*)	6q25.1	BMD	Kobayashi et al., 1996
Interleukin-6 (*IL6*)	7p21	BMD	Lorentzon et al., 2000
		BMD	Murray et al., 1997
Calcitonin receptor (*CALCR*)	7q21.3	BMD, osteoporotic fracture	Taboulet et al., 1998
Paraoxonase 1 and 2 (*PON1, PON2*)	7q21.3	BMD	Yamada, Ando, Niino, Miki, et al., 2003
Type I collagen α2 (*COL1A2*)	7q22.1	BMD	Lei et al., 2005
Gonadotropin-releasing hormone 1 (*GNRH1*)	8p21-11.2	BMD	Iwasaki et al., 2003
Tumor necrosis factor receptor superfamily, member 11B (*TNFRSF11B*)	8q24	Vertebral fracture	Langdahl et al., 2002
		BMD	Ohmori et al., 2002
Very low-density lipoprotein receptor (*VLDLR*)	9p24	BMD	Yamada et al., 2005
Cytochrome P450, family 17, subfamily A, polypeptide 1 (*CYP17A1*)	10q24.3	BMD	Yamada et al., 2005
Cyclin-dependent kinase inhibitor 1C (*CDKN1C*)	11p15.5	BMD	Urano et al., 2000
Dopamine receptor D4 (*DRD4*)	11p15.5	BMD	Yamada et al., 2003a
Parathyroid hormone (*PTH*)	11p15.3-15.1	BMD	Hosoi et al., 1999
Calcitonin (*CALCA*)	11p15.2-15.1	BMD	Miyao, Hosoi, et al., 2000
Low-density lipoprotein receptor-related protein 5 (*LRP5*)	11q13.4	BMD	Ferrari, Deutsch, et al., 2004
T cell immune regulator 1 (*TCIRG1*)	11q13.4-13.5	BMD	Sobacchi et al., 2004
Matrix metalloproteinase 1 (*MMP1*)	11q22-23	BMD	Yamada et al., 2002b
Vitamin D receptor (*VDR*)	12q12-14	BMD	Morrison et al., 1994; Arai et al., 2001
Insulin-like growth factor-I (*IGF1*)	12q22-24.1	BMD, osteoporosis	Kim et al., 2002
Klotho (*KL*)	13q12	BMD	Kawano et al., 2002
Estrogen receptor β (*ESR2*)	14q	BMD	Shearman et al., 2004
Vitamin D-binding protein (*GC*)	14q12	BMD	Ezura et al., 2003
Bone morphogenetic protein 4 (*BMP4*)	14q22-23	BMD	Ramesh Babu et al., 2005
Aromatase (*CYP19A1*)	15q21.1	BMD, osteoporosis, spinal fracture	Masi et al., 2001

(Continued)

Table 35–5 Candidate Gene Association Studies of Bone Mineral Density (BMD), Osteoporosis, or Fractures *(Continued)*

Candidate Gene	Chromosomal Locus	Phenotype	Reference
Cytochrome P450, subfamily I, polypeptide 1 (*CYP1A1*)	15q22-24	BMD	Napoli et al., 2005
Sclerostin (*SOST*)	17q12-21	BMD	Uitterlinden et al., 2004
Type I collagen α1 (*COL1A1*)	17q21.31-22	Osteoporosis	Grant et al., 1996
		Osteoporotic fracture	Uitterlinden et al., 1998
		BMD	Garcia-Giralt et al., 2002
Growth hormone (*GH1*)	17q22-24	BMD	Dennison et al., 2004
Transforming growth factor-β1 (*TGFB1*)	19q13.1	BMD	Langdahl et al., 1997
		BMD, osteoporosis	Yamada et al., 1998
			Yamada, Miyauchi, et al., 2001
		Vertebral fracture	Yamada et al., 2000
Apolipoprotein E (*APOE*)	19q13.2	BMD	Shiraki et al., 1997
Catechol-*O*-methyltransferase (*COMT*)	22q11.2	BMD	Lorentzon et al., 2004
Androgen receptor (*AR*)	Xq11-12	BMD	Sowers et al., 1999

hormone replacement therapy (Han et al., 1997). In contrast, women with the *TT* genotype (*Pvu*II SNP) have been suggested to be relatively estrogen-insensitive; those with the *C* allele thus appeared to benefit more from the protective effect of hormone replacement therapy with regard to fracture risk than did women with the *TT* genotype (Salmén et al., 2000).

A meta-analysis of the relations of the A→G (*Xba*I) and T→C (*Pvu*II) SNPs of *ESR1* to BMD and fracture risk in 5834 women from 30 study groups showed that homozygotes for the *G* allele of the A→G SNP had a higher BMD and a lower risk of fractures compared with carriers of the *A* allele, whereas the T→C SNP was not associated with either BMD or fracture risk (Ioannidis et al., 2002). A recent meta-analysis of the relations of three polymorphisms (A→G and T→C SNPs in intron 1 and the TA repeat polymorphism in the promoter) of *ESR1* and their haplotypes to BMD and fractures in 18,917 individuals in eight European centers showed that none of the polymorphisms or haplotypes had any significant effect on BMD. However, in women homozygous for the *G* allele of the A→G SNP, the risk for all fractures was reduced by 19% (odds ratio, 0.81) and that for vertebral fractures by 35% (odds ratio, 0.65). No significant effects on fracture risk were apparent for the T→C SNP or TA repeat polymorphism (Ioannidis et al., 2004).

The molecular mechanisms that underlie the association of the T→C (*Pvu*II) or A→G (*Xba*I) SNPs or the TA repeat polymorphism of *ESR1* with BMD or with genetic susceptibility to osteoporosis or fractures remain unclear. However, genotyping of *ESR1* may prove beneficial for the assessment of genetic predisposition to osteoporosis or fracture risk.

Interleukin-6

Interleukin-6 (IL-6) is a multifunctional cytokine that is produced by several cell types in the bone microenvironment. It acts as a modulator of the differentiation and function of osteoclasts, and has an important role in the development of postmenopausal osteoporosis (Manolagas and Jilka, 1995). A variable number of tandem repeats (VNTR) polymorphism in the 3′ flanking region of *IL6* has been associated with BMD in postmenopausal white women (Murray et al., 1997), and sibling-pair analysis has provided evidence of linkage between the *IL6* locus and reduced BMD in postmenopausal Japanese women (Ota et al., 1999). A –174G→C SNP in the *IL6* promoter has been shown to affect both promoter activity and the plasma concentration of IL-6 (Fishman et al., 1998). This polymorphism was also associated with peak bone mass in healthy white men (Lorentzon et al., 2000), with bone resorption and reduced bone mass in older postmenopausal women (Ferrari et al., 2001), with hip BMD in late postmenopausal women not receiving estrogen replacement therapy and with inadequate calcium intake (Ferrari, Karasik, et al., 2004), and with bone loss and fracture risk in elderly women (Moffett et al., 2004). In each of these studies, the *G* allele was considered to be a risk factor for reduced bone mass or fracture. However, this is not supported by the studies in Japanese population (Y.Y., unpublished data).

Three polymorphisms of *IL6* have been identified in Japanese women, among which a –634C→G SNP in the promoter region has been associated with BMD for the radius in postmenopausal Japanese women, with BMD decreasing according to the rank order of genotypes *CC* > *CG* > *GG* (Ota et al., 2001). An association of the –634C→G SNP with BMD at various sites was also detected in a population-based study, with the *GG* genotype representing a risk factor for reduced BMD (Yamada et al., 2003b). The molecular mechanism responsible for the association of the –634C→G SNP of *IL6* with BMD has not been determined.

These various observations suggest that SNPs in the promoter of *IL6* are important determinants of BMD in postmenopausal women. These polymorphisms may thus prove useful in assessing the genetic risk for predisposition to osteoporosis in such individuals.

Tumor Necrosis Factor Receptor Superfamily, Member 11B

Tumor necrosis factor (TNF) receptor superfamily, member 11B (TNFRSF11B, or osteoprotegerin) is a secreted glycoprotein that was independently identified by three groups of research workers (Simonet et al., 1997; Tan et al., 1997; Tsuda et al., 1997). In vitro studies suggest that TNFRSF11B inhibits osteoclastogenesis by interrupting intercellular signaling between osteoblastic stromal cells and osteoclast progenitors (Simonet et al., 1997). TNFRSF11B-deficient mice exhibit a condition similar to juvenile Paget's disease that is characterized by a marked decrease in trabecular and cortical bone density, pronounced thinning of the parietal bone of the skull, and a high incidence of fractures (Bucay et al., 1998), whereas hepatic expression of TNFRSF11B in transgenic mice results in osteopetrosis and a coincident decrease in the proportion of osteoclasts in the later stages of differentiation (Simonet et al., 1997). The systemic administration of recombinant TNFRSF11B also results in a marked increase in BMD in normal rats as well as in the prevention of bone loss in ovariectomized rats (Simonet et al., 1997; Yasuda, Shima, Nakagawa, Mochizuki, et al., 1998). Furthermore, a single subcutaneous injection of TNFRSF11B reduced bone resorption in postmenopausal women (Bekker et al., 2001). Similar treatment with a recombinant TNFRSF11B construct suppressed bone resorption in patients with multiple myeloma or breast cancer with bone metastases (Body et al., 2003).

Osteoclastogenesis is regulated by three TNF- or TNF receptor-related proteins: TNF receptor superfamily, member 11A [TNFRSF11A, or receptor activator of nuclear factor-κB (RANK)] (Anderson et al., 1997; Hsu et al., 1999), TNF ligand superfamily, member 11 [TNFSF11, or RANK ligand (RANKL)] (Lacey et al., 1998; Yasuda, Shima, Nakagawa, Yamaguchi, et al., 1998), and TNFRSF11B (Simonet et al., 1997; Tan et al., 1997; Tsuda et al., 1997). TNFSF11 expressed on the surface of bone marrow stromal cells induces the differentiation of osteoclasts, enhances the activity of mature osteoclasts, and inhibits osteoclast apoptosis by binding to its functional receptor, TNFRSF11A, expressed on osteoclasts or their progenitors (Fuller et al., 1998; Lacey et al., 1998; Yasuda, Shima, Nakagawa, Yamaguchi, et al., 1998; Burgess et al., 1999). The interaction between TNFSF11 and TNFRSF11A is antagonized by TNFRSF11B, which acts as a decoy receptor for TNFSF11. The biological effects of TNFRSF11B include inhibition of the later stages of osteoclastogenesis (Simonet et al., 1997; Takai et al., 1998; Yasuda, Shima, Nakagawa, Mochizuki, et al., 1998), suppression of the activation of mature osteoclasts (Hakeda et al., 1998; Lacey et al., 1998), and induction of osteoclast apoptosis (Akatsu et al., 1998). The balance between TNFRSF11B and TNFSF11 may thus represent an important determinant of bone resorption (Takai et al., 1998; Hofbauer and Schoppet, 2004). The importance of TNFRSF11B in the regulation of bone remodeling in humans has been indicated by the occurrence of juvenile Paget's disease, characterized by rapid remodeling of woven bone, osteopenia, fractures, and progressive skeletal deformity, in Navajo individuals homozygous for a deletion of approximately 100 kb in *TNFRSF11B* (Whyte et al., 2002).

Several SNPs have been detected in *TNFRSF11B*, some of which have been shown to be associated with BMD or fractures. Both 209G→A and 245T→G SNPs of the *TNFRSF11B* promoter were found to be associated with BMD for the lumbar spine in postmenopausal Slovenian women, with the 209*GA*/245*TG* genotype representing a risk factor for reduced BMD (Arko et al., 2002). Both 163A→G and 245T→G SNPs were associated with vertebral fractures in Danish women and men, with the *G* allele of each polymorphism representing a risk factor for fracture (Langdahl et al., 2002). The 245T→G SNP was associated with BMD at various sites in postmenopausal Japanese women, with the *GG* genotype representing a risk factor for reduced BMD (Yamada et al., 2003c). A 950T→C SNP was also associated with BMD in postmenopausal Japanese women (Ohmori et al., 2002). In contrast, SNPs of *TNFRSF11B* were not related to BMD or the prevalence of fractures in elderly Swedish women (Brändström et al., 2004). The effects of the various SNPs in the *TNFRSF11B* promoter on transcriptional activity have not been determined.

Several lines of evidence thus indicate that TNFRSF11B plays an important role in bone remodeling. The fact that SNPs in *TNFRSF11B* have been associated with BMD or fractures in different populations and ethnic groups further suggests that the encoded protein is a prominent regulator of bone mass and determinant of the predisposition to osteoporosis or fractures.

Low-Density Lipoprotein Receptor-related Protein 5

LRP5 is expressed in osteoblasts and regulates the proliferation, survival, and activity of these cells. Population genetics and in vitro studies have shown that LRP5 is a key regulator of bone metabolism and acts via the Wnt signaling pathway (Westendorf et al., 2004; Koay and Brown, 2005). Mutations in *LRP5* cause osteoporosis-pseudoglioma syndrome (Gong et al., 2001), an autosomal recessive condition of juvenile onset that is characterized both by blindness, resulting from aberrant vitreo-retinal vascular growth, and by osteoporosis, resulting in fractures and deformation. Analysis of bone biopsy specimens from affected individuals has revealed that the volume of trabecular bone is reduced, but the number and appearance of osteoblasts and osteoclasts are normal. Six disease-causing frameshift or nonsense mutations in the regions of *LRP5* encoding the signal peptide, the epidermal growth factor (EGF)-like domain, the YWTD domains, and the LDL receptor-like domains have been identified, suggesting that osteoporosis-pseudoglioma syndrome results from loss of LRP5 function. The likely consequent down-regulation of the Wnt signaling pathway may impair bone formation and thereby lead to osteoporosis.

In contrast to osteoporosis-pseudoglioma syndrome, activating mutations in *LRP5* cause an autosomal dominant syndrome characterized by increased bone mass. Affected individuals show enhanced bone synthesis as assessed by serum markers, whereas bone resorption and bone architecture appear normal. Affected members of families that manifest this condition, but not unaffected relatives or unrelated controls, harbor a G→T transversion in exon 3 of *LRP5* (Boyden et al., 2002; Little et al., 2002). This mutation results in the substitution of a conserved glycine residue at position 171 by valine. The mutant LRP5 protein constitutively activates the Wnt signaling pathway and renders it resistant to Dkk1-mediated inhibition. The activation of LRP5 function thus results in excessive bone accumulation.

LRP5 mutations have also been detected in individuals with a heterogeneous group of conditions characterized by autosomal recessive (Heaney et al., 1998) or autosomal dominant (Van Hul et al., 2002) forms of osteopetrosis. The increased BMD apparent in affected individuals results from disruption of bone turnover and remodeling. Some forms of osteopetrosis, especially those attributable to osteoclast dysfunction, are associated with increased fracture risk. However, this is not seen in the high-bone-mass syndrome. Four mutations of *LRP5* [331G→T (Asp111Tyr), 511G→C (Gly171Arg), 724G→A (Ala242Thr), and 758C→T (Thr253Ile)] have been identified in five families affected with osteopetrosis (Van Wesenbeeck et al., 2003).

LRP5 polymorphisms have also been associated with endosteal hyperostosis and its variant, van Buchem disease. Similar to osteopetrosis, these conditions are marked by excessive accrual of bone, although, in this instance, such accrual is limited to the inner (endosteal) surface, leading to obliteration of the medullary space. Two coding polymorphisms of *LRP5* [640G→A (Ala214Thr) and 724G→A (Ala242Thr)] have been identified in individuals with these conditions (Van Wesenbeeck et al., 2003).

Several studies have implicated common polymorphisms of *LRP5* in BMD variation in the general population. A 2047G→A (Val667Met) SNP in exon 9 was thus significantly associated with bone mineral content of the lumbar spine, with bone area, and with stature in a cohort of 889 healthy white men and women, with the association being most pronounced in adult men (Ferrari, Deutsch, et al., 2004). Haplotype analyses of five SNPs at the *LRP5* locus suggested that additional genetic variation at this locus might also contribute to the determination of bone mass and size. SNPs of *LRP5* were associated with BMD for the lumbar spine, total hip, or femoral neck in a cohort of 909 British Caucasian adults (Koay et al., 2004). Family studies identified one SNP (171346C→A in intron 21) with a significant relation to BMD, with the association being stronger in men than in women. The association of *LRP5* polymorphisms with BMD was also observed as the overrepresentation of variants of three SNPs in osteoporotic probands compared with unrelated postmenopausal women with increased BMD. In addition, an association between haplotypes of SNPs and BMD was detected by comparison of the osteoporotic subjects with the individuals with a high BMD. Association of SNPs in *LRP5* with BMD has also been observed in Japanese (Mizuguchi et al., 2004) and Australian (Bollerslev et al., 2005) women. Although a significant association was observed between *LRP5* SNPs and BMD for the hip or spine in a study of healthy premenopausal white women, only a small proportion of the total variation in BMD was accounted for by these SNPs (Koller et al., 2005). The genotyped SNPs thus accounted for approximately 0.8% of the variation in BMD for the femoral neck and 1.1% of that for the lumbar spine, suggesting that natural variation in and around *LRP5* is not a major contributing factor to the observed variability in peak BMD at either of these sites in white women.

Genetic studies of families affected by extreme bone phenotypes have thus demonstrated that *LRP5* regulates bone mass, as revealed by the fact that loss-of-function and activating mutations result in osteoporosis and high bone mass, respectively. Genetic epidemiological studies have also demonstrated that polymorphisms of *LRP5* are associated with variation of BMD in the general population. *LRP5* may thus be an important regulator of bone mass and determinant of predisposition to osteoporosis.

Vitamin D Receptor

Vitamin D is a potent regulator of bone and calcium homeostasis, as well as of cellular differentiation and proliferation in many tissues. Its active form, 1,25-dihydroxyvitamin D_3 (calcitriol), interacts with the highly specific vitamin D receptor (VDR) and thereby affects the expression of target genes. A *Bsm*I restriction fragment length polymorphism (RFLP) of *VDR* was associated with BMD in Australian women (Morrison et al., 1994). Of the many studies performed subsequently, some (Tokita et al., 1996; Uitterlinden et al., 1996; Sainz et al., 1997) have supported this association whereas others (Hustmyer et al., 1994; Melhus et al., 1994; Garnero, Borel, et al., 1996) have not. These apparently contradictory results are possibly attributable to the differences in several factors among the studies, including size, age, and ethnic background of the study populations as well as environmental aspects such as diet, especially vitamin D and calcium intake. In addition, the effects of *VDR* genotype on BMD appear to be relatively small. A meta-analysis of 16 studies concluded that BMD was 2.5% lower for the spine and 2.4% lower for the femoral neck in individuals with the *BB* genotype (*B* allele, absence of the *Bsm*I restriction site) than in those with the *bb* genotype (*b* allele, presence of the restriction site) (Cooper and Umbach, 1996). Another meta-analysis of 75 studies, including those examining four polymorphisms (*Bsm*I, *Apa*I, *Taq*I, and *Fok*I), confirmed the association of *VDR* polymorphisms with BMD (Gong et al., 1999). Furthermore, a more recent meta-analysis of the relation between *VDR Bsm*I genotype and either BMD or osteoporosis in women revealed that individuals with the *BB* genotype had a lower BMD than did those with the *Bb* or *bb* genotypes at baseline, which led to a greater proportional loss in BMD in those with the *BB* genotype over time (Thakkinstian et al., 2004).

Determination of the nucleotide sequence of human *VDR* (Baker et al., 1988) revealed two potential translation initiation sites, the most 5′ of which is affected by a T→C SNP (ATG→ACG). Individuals with the *T* allele of this SNP thus have two start sites and are able to initiate translation from the first ATG codon, whereas those with the *C* allele initiate translation at the second ATG. The predicted protein produced by initiation at the first ATG is three amino acids longer than that generated by initiation at the second start site. The T→C SNP of *VDR* has been associated with BMD in postmenopausal Mexican–American women (Gross et al., 1996), premenopausal American white women (Harris et al., 1997), and Japanese women (Arai et al., 1997), with the *TT* genotype implicated as a risk factor for reduced BMD. However, no association of this SNP with BMD was detected in premenopausal American black (Harris et al., 1997) or premenopausal French (Eccleshall et al., 1998) women. The T→C SNP of *VDR* was found to affect not only the molecular mass of the encoded protein (*T* allele, 50 kDa; *C* allele, 49.5 kDa) but also the transcriptional activation of the gene by vitamin D (*T* allele < *C* allele) (Arai et al., 1997). However, this latter observation was not independently confirmed (Gross et al., 1998). The functional impact of this SNP thus remains to be determined. A −3731A→G SNP that affects the binding site of the caudal-related homeodomain protein Cdx-2 in the *VDR* promoter was associated both with transcriptional activity of the promoter and with BMD for the lumbar spine in Japanese women, with the *G* allele corresponding to reduced transcriptional activity and low BMD (Arai et al., 2001).

These various observations thus suggest that *VDR* is a susceptibility locus for reduced BMD or osteoporosis. SNPs of *VDR* may therefore prove helpful for the assessment of predisposition to osteoporosis.

Type I collagen α1

Type I collagen is the most abundant protein of bone matrix. Mutations in the coding regions of the genes for the two type I collagen chains (*COL1A1* and *COL1A2*) result in a severe autosomal dominant pediatric condition known as osteogenesis imperfecta (Sykes, 1990). A G→T SNP at the first base of a consensus binding site for the transcription factor Sp1 in the first intron of *COL1A1* was associated not only with BMD in white women (Grant et al., 1996) but also with osteoporotic fracture in postmenopausal women (Langdahl et al., 1998; Uitterlinden et al., 1998). Other studies, however, showed only a weak association of this SNP with BMD or osteoporotic fracture in premenopausal French women (Garnero et al., 1998), or a lack of association in postmenopausal Swedish women (Liden et al., 1998), in American women (Hustmyer et al., 1999), or in postmenopausal Danish women (Heegaard et al., 2000). The *T* allele of the Sp1 binding site polymorphism affects collagen gene regulation in such a manner that it increases the production of the α1(I) collagen chain relative to that of the α2(I) chain and leads to reduced bone strength by a mechanism that is partly independent of bone mass (Mann et al., 2001). A −1997G→T SNP in the *COL1A1* promoter was also associated with BMD for the lumbar spine (Garcia-Giralt et al., 2002). The −1997G→T SNP and the G→T SNP of the Sp1 binding site were shown to be in linkage disequilibrium (Garcia-Giralt et al., 2002).

A meta-analysis of the effect of the Spl binding site polymorphism of *COLIA1* on the prevalence of fractures in 3641 subjects revealed that the risk was 1.7-fold greater in *ss* (*TT*) homozygotes versus *SS* (*GG*) homozygotes, 1.4-fold greater in *ss* homozygotes versus *Ss* (*GT*) heterozygotes, and 1.3-fold greater in *Ss* heterozygotes versus *SS* homozygotes. The effects of this polymorphism were slightly increased when the analysis was limited to vertebral fractures (odds ratios, 2.1, 1.5, and 1.3, respectively) (Efstathiadou et al., 2001). Another meta-analysis of an association of this polymorphism with BMD or osteoporotic fracture in 7849 participants revealed that BMD for the lumbar spine was significantly lower in individuals with the *Ss* genotype than in those with the *SS* genotype. BMD for the femoral neck was also lower in individuals with the *Ss* genotype or those with the *ss* genotype than in those with the *SS* genotype. Analysis of the prevalence of fractures showed an increased odds ratio for any fracture in subjects with the *Ss* genotype (1.3), and an even greater increase in those with the *ss* genotype (1.8). Increased risk was largely attributable to vertebral fracture, for which the odds ratio was 1.4 for individuals with the *Ss* genotype and 2.5 for those with the *ss* genotype (Mann and Ralston, 2003).

These various observations thus suggest that genetic variants that affect type I collagen metabolism are important determinants of the development of osteoporosis or osteoporotic fractures. Such variants might thus prove useful for the assessment of individuals at risk for these conditions (Box 35–2).

Box 35–2 Molecular Genetics of Bone Mineral Density (BMD) and Osteoporosis

Key Points

- Genetic factors play an important role in the regulation of BMD and other determinants of osteoporotic fracture risk.
- These phenotypes are under polygenic control. However, certain forms of osteoporosis and abnormally high bone mass, such as osteoporosis-pseudoglioma syndrome and high-bone-mass syndrome, occur as the result of mutations in a single gene (*LRP5*).
- Linkage analyses have identified several loci that show linkage to BMD, but, in most studies, the causative genes remain to be identified.
- Completion of the Human Genome Project has reawakened interest in candidate gene association studies, and SNPs in candidate genes are being examined extensively to identify the genetic markers for osteoporosis.
- The future challenge is to identify the genes that determine BMD and to investigate the molecular mechanisms by which polymorphisms affect the function of such genes and bone remodeling.

Perspectives

Osteoporosis and osteoporotic fractures are major age-related disorders of the human skeleton. The studies described here indicate the existence of a substantial genetic component of osteoporosis. However, despite the identification of a variety of candidate genes related to osteoporosis or BMD, the replicability of such findings is relatively low, mainly because of the limited population size of the studies, the ethnic diversity of gene polymorphisms, gene–gene interactions, complicating environmental factors, gene–environment interactions, and linkage disequilibrium with nearby genes. The development of personalized prevention of osteoporosis and osteoporotic fractures will require the performance of large-scale linkage analyses and population-based association studies in various ethnic groups to identify definitively the genes that determine bone mass or susceptibility to these conditions.

Hyperlipidemia and Osteoporosis

Oxidative stress is thought to impair cellular functions in various pathological conditions, including osteoporosis (Garrett et al., 1990; Maggio et al., 2003). Several lines of evidence suggest that lipid and bone metabolism are closely related and that an atherogenic lipid profile has adverse effects on bone remodeling (Diascro et al., 1998; McFarlane et al., 2002; Orozco, 2004; Parhami et al., 1999, 2000; Yamaguchi et al., 2002; Poli et al., 2003). Bone loss is associated with an expansion of adipose tissue in bone marrow (Bergman et al., 1996), and osteoblasts and adipocytes share a common progenitor derived from stromal cells in bone marrow (Parhami et al., 1999). Products of lipoprotein oxidation and an atherogenic diet also inhibit preosteoblast differentiation and result in reduced bone mineralization (Diascro et al., 1998; Parhami et al., 1999). In addition, lipid-lowering agents (statins) stimulate bone formation and inhibit bone resorption, resulting in the prevention of both bone loss and osteoporotic fractures (McFarlane et al., 2002).

Early postmenopausal women with an atherogenic serum lipid profile [total cholesterol of 240 mg/dL or more (6.24 mmol/L), LDL cholesterol of 160 mg/dL or more (4.16 mmol/L), or lipoprotein(a) of 25 mg/dL or more (0.65 mmol/L)] had a lower BMD for the lumbar spine or femoral neck and an increased risk of osteopenia compared with those with a normal lipid profile (Orozco, 2004). Postmenopausal women with increased plasma concentrations of LDL cholesterol have also been shown to be at greater risk of developing osteopenia than those with normal concentrations, suggesting that an increased plasma level of LDL cholesterol is a risk factor for reduced BMD (Poli et al., 2003). Plasma concentrations of LDL cholesterol and high-density lipoprotein (HDL)–cholesterol were inversely and positively correlated with BMD, respectively, whereas low concentrations of plasma triglycerides were associated with an increased prevalence of vertebral fractures, in postmenopausal women (Yamaguchi et al., 2002). A relation between serum lipid profile and BMD was also observed in a population-based study (Yamada et al., 2005). For women, BMD for the radius or total body was significantly related to the serum concentrations of total cholesterol, HDL cholesterol, LDL cholesterol, and triglycerides, and BMD for the distal radius, lumbar spine, femoral neck, or trochanter was significantly related to the serum concentrations of total cholesterol, LDL cholesterol, and triglycerides. For men, BMD for the radius was significantly related to the serum concentrations of LDL cholesterol and total cholesterol, and BMD for the femoral neck or trochanter was significantly related to the serum concentration of HDL cholesterol. These observations suggest the existence of a close relation between lipid and bone metabolism, as well as demonstrating adverse effects of an atherogenic lipid profile on bone remodeling.

Although hyperlipidemia has been shown to be related to reduced BMD or osteoporosis (Poli et al., 2003; Orozco, 2004), the mechanism that underlies this relation remains unknown. A contributing factor might be the fact that estrogen deficiency results both in the impairment of lipid metabolism and in the acceleration of bone resorption (Orozco, 2004). Alternatively, bone formation might be impaired by ischemia of bone tissue caused by hyperlipidemia or dyslipidemia, given that osteoblast progenitors are located adjacent to the subendothelial matrix of bone vessels and atherosclerosis may influence the function of these bone-forming cells (Orozco, 2004). Another possibility relates to the potential interaction between cholesterol synthesis and regulation of bone metabolism, given that statins have beneficial effects on bone mass (McFarlane et al., 2002). It is also possible that some genetic variants affect both lipid metabolism and bone remodeling. Indeed, polymorphisms of genes that contribute to lipid metabolism, including *PPARG* (Ogawa et al., 1999), *MTP* (Yamada et al., 2005), *PON1* and *PON2* (Yamada, Ando, Niino, Miki, et al., 2003), *VLDLR* (Yamada et al., 2005), *LRP5* (Ferrari, Deutsch et al., 2004), and *APOE* (Shiraki et al., 1997), have been shown to be associated with BMD. The molecular mechanisms responsible for atherogenic lipid profiles and their relevance to the pathophysiology of osteoporosis thus warrant further investigation.

Clinical Diagnosis and Management

The increasing body of information garnered from studies on the genetics of osteoporosis has resulted in the emergence of a greater understanding of the biology of bone in health and disease. Such knowledge has clinical implications for the prediction, diagnosis, prognosis, and treatment of osteoporosis. The genes responsible for the regulation of BMD and bone fragility, as well as their encoded proteins, are potentially important therapeutic targets in the design of new treatments for bone disease. Genetic markers are potential diagnostic tools for the assessment of individuals at risk of developing osteoporotic fractures. Genetic markers of bone fragility or bone loss, together with measurements of BMD, might also form the basis for the instigation of preventive therapies in individuals at risk of fracture. It should be remembered, however, that gene–gene and gene–environment interactions make the interpretation of information based on genetic markers of osteoporosis more complex than that of information based on markers for monogenic bone diseases. Another use of genetic markers might be to distinguish treatment responders from nonresponders and to identify patients who are at risk of developing unfavorable side effects. It is likely that the use of gene polymorphisms to predict the response to and adverse effects of therapies for osteoporosis will increase in the future and will give rise to major advances in patient care (Hobson and Ralston, 2001; Ralston, 2002). Genetic analysis of osteoporosis or fractures is thus likely to have important direct clinical applications.

Summary

This chapter describes selected monogenic bone diseases in the context of osteoporosis and reduced BMD. Data on the candidate loci and polymorphisms in candidate genes related to BMD or predisposition to osteoporosis and/or osteoporotic fractures are provided and supported by relevant references to the literature. Although candidate genes for osteoporosis and related phenotypes have been identified, the reproducibility of such findings is poor. Further investigation based on a comprehensive approach, such as whole-genome association studies, and large-scale longitudinal population-based studies are required to identify susceptibility genes for osteoporosis or predisposition to fractures. In addition, clarification of ethnic divergence of gene polymorphisms associated with osteoporosis and related phenotypes would require international collaboration involving various ethnic groups. Clinical implication of these studies include identification of genetic markers for the assessment of susceptibility to osteoporosis and fractures, as well as discovery of novel molecular targets for the design of drugs that can be used for the treatment of osteoporosis. Understanding genetic variability in BMD will improve our understanding of the genetics of osteoporosis and fractures, and facilitate the development of the personalized prevention and treatment of these conditions.

References

Akatsu, T, Murakami, T, Nishikawa, M, Ono, K, Shinomiya, N, Tsuda, E, et al. (1998). Osteoclastogenesis inhibitory factor suppresses osteoclast survival by interfering in the interaction of stromal cells with osteoclast. *Biochem Biophys Res Commun*, 250:229–234.

Albagha, OME, McGuigan, FEA, Reid, DM, and Ralston, SH (2001). Estrogen receptor α gene polymorphisms and bone mineral density: haplotype analysis in women from the United Kingdom. *J Bone Miner Res*, 16:128–134.

Anderson, DM, Maraskovsky, E, Billingsley, WL, Dougall, WC, Tometsko, ME, Roux, ER, et al. (1997). A homologue of the TNF receptor and its ligand enhance T-cell growth and dendritic-cell function. *Nature*, 390:175–179.

Arai, H, Miyamoto, K, Taketani, Y, Yamamoto, H, Iemori, Y, Morita, K, et al. (1997). A vitamin D receptor gene polymorphism in the translation initiation codon: effect on protein activity and relation to bone mineral density in Japanese women. *J Bone Miner Res*, 12:915–921.

Arai, H, Miyamoto, K, Yoshida, M, Yamamoto, H, Taketani, Y, Morita, K, et al. (2001). The polymorphism in the caudal-related homeodomain protein Cdx-2 binding element in the human vitamin D receptor gene. *J Bone Miner Res*, 16:1256–1264.

Arden, NK, Baker, J, Hogg, C, Baan, K, and Spector, TD (1996). The heritability of bone mineral density, ultrasound of the calcaneus and hip axis length: a study of postmenopausal twins. *J Bone Miner Res*, 11:530–534.

Arko, B, Prezelj, J, Komel, R, Kocijancic, A, Hudler, P, and Marc, J (2002). Sequence variations in the osteoprotegerin gene promoter in patients with postmenopausal osteoporosis. *J Clin Endocrinol Metab*, 87:4080–4084.

Baker, AR, McDonnell, DP, Hughes, M, Crisp, TM, Mangelsdorf, DJ, Haussler, MR, et al. (1988). Cloning and expression of full-length cDNA encoding human vitamin D receptor. *Proc Natl Acad Sci USA*, 85:3294–3298.

Balemans, W, Ebeling, M, Patel, N, Van Hul, E, Olson, P, Dioszegi, M, et al. (2001). Increased bone density in sclerosteosis is due to the deficiency of a novel secreted protein (SOST). *Hum Mol Genet*, 10:537–543.

Balemans, W, Patel, N, Ebeling, M, Van Hul, E, Wuyts, W, Lacza, C, et al. (2002). Identification of a 52 kb deletion downstream of the SOST gene in patients with van Buchem disease. *J Med Genet*, 39:91–97.

Becherini, L, Gennari, L, Masi, L, Mansani, R, Massart, F, Morelli, A, et al. (2000). Evidence of a linkage disequilibrium between polymorphisms in the human estrogen receptor α gene and their relationship to bone mass variation in postmenopausal Italian women. *Hum Mol Genet*, 9:2043–2050.

Bekker, PJ, Holloway, D, Nakanishi, A, Arrighi, M, Leese, PT, and Dunstan, CR (2001). The effect of a single dose of osteoprotegerin in postmenopausal women. *J Bone Miner Res*, 16:348–360.

Bergman, RJ, Gazit, D, Kahn, AJ, Gruber, H, McDougall, S, and Hahn, TJ (1996). Age-related changes in osteogenic stem cells in mice. *J Bone Miner Res*, 11:568–577.

Body, JJ, Greipp, P, Coleman, RE, Facon, T, Geurs, F, Fermand, JP, et al. (2003). A phase I study of AMGN-0007, a recombinant osteoprotegerin construct, in patients with multiple myeloma or breast carcinoma related bone metastases. *Cancer*, 97:887–892.

Bollerslev, J, Wilson, SG, Dick, IM, Islam, FM, Ueland, T, Palmer, L, et al. (2005). LRP5 gene polymorphisms predict bone mass and incident fractures in elderly Australian women. *Bone*, 36:599–606.

Boyden, LM, Mao, J, Belsky, J, Mitzner, L, Farhi, A, Mitnick, MA, et al. (2002). High bone density due to a mutation in LDL-receptor-related protein 5. *N Engl J Med*, 346:1513–1521.

Brändström, H, Gerdhem, P, Stiger, F, Obrant, KJ, Melhus, H, Ljunggren, Ö, et al. (2004). Single nucleotide polymorphisms in the human gene for osteoprotegerin are not related to bone mineral density or fracture in elderly women. *Calcif Tissue Int*, 74:18–24.

Brown, MA, Haughton, MA, Grant, SFA, Gunnell, AS, Henderson, NK, and Eisman, JA (2001). Genetic control of bone density and turnover: role of the collagen 1(1, estrogen receptor, and vitamin D receptor genes. *J Bone Miner Res*, 16:758–764.

Bucay, N, Sarosi, I, Dunstan, CR, Morony, S, Tarpley, J, Capparelli, C, et al. (1998). Osteoprotegerin-deficient mice develop early onset osteoporosis and arterial calcification. *Genes Dev*, 12:1260–1268.

Burgess, TL, Qian, Y, Kaufman, S, Ring, BD, Van, G, Capparelli, C, et al. (1999). The ligand for osteoprotegerin (OPGL) directly activates mature osteoclasts. *J Cell Biol*, 145:527–538.

Cleiren, E, Benichou, O, Van Hul, E, Gram, J, Bollerslev, J, Singer, FR, et al. (2001). Albers-Schonberg disease (autosomal dominant osteopetrosis, type II) results from mutations in the ClCN7 chloride channel gene. *Hum Mol Genet*, 10:2861–2867.

Cooper, C, Barker, DJP, and Wickham, C (1988). Physical activity, muscle strength, and calcium intake in fracture of the proximal femur in Britain. *Br Med J*, 297:1443–1446.

Cooper, GS and Umbach, DM (1996). Are vitamin D receptor polymorphisms associated with bone mineral density? A meta-analysis. *J Bone Miner Res*, 11:1841–1849.

Cummings, SR, Black, DM, Nevitt, MC, Browner, W, Cauley, J, Ensrud, K, et al. (1993). Bone density at various sites for prediction of hip fractures. The Study of Osteoporotic Fractures Research Group. *Lancet*, 341:72–75.

Cummings, SR, Nevitt, MC, Browner, WS, Stone, K, Fox, KM, Ensrud, KE, et al. (1995). Risk factors for hip fracture in white women. Study of Osteoporotic Fractures Research Group. *N Engl J Med*, 332:767–773.

Dargent-Molina, P, Favier, F, Grandjean, H, Baudoin, C, Schott, AM, Hausherr, E, et al. (1996). Fall-related factors and risk of hip fracture: the EPIDOS prospective study. *Lancet*, 348:145–149.

Deng, HW, Chen, WM, Recker, S, Stegman, MR, Li, JL, Davies, KM, et al. (2000). Genetic determination of Colles' fracture and differential bone mass in women with and without Colles' fracture. *J Bone Miner Res*, 15:1243–1252.

Deng, HW, Xu, FH, Huang, QY, Shen, H, Deng, H, Conway, T, et al. (2002). A whole-genome linkage scan suggests several genomic regions potentially containing quantitative trait loci for osteoporosis. *J Clin Endocrinol Metab*, 87:5151–5159.

Dennison, EM, Syddall, HE, Rodriguez, S, Voropanov, A, Day, IN, Cooper C, et al. (2004). Polymorphism in the growth hormone gene, weight in infancy, and adult bone mass. *J Clin Endocrinol Metab*, 89:4898–4903.

Devoto, M, Shimoya, K, Caminis, J, Ott, J, Tenenhouse, A, Whyte, MP, et al. (1998). First-stage autosomal genome screen in extended pedigrees suggests genes predisposing to low bone mineral density on chromosomes 1p, 2p and 4q. *Eur J Hum Genet*, 6:151–157.

Devoto, M, Specchia, C, Li, HH, Caminis, J, Tenenhouse, A, Rodriguez, H, et al. (2001). Variance component linkage analysis indicates a QTL for femoral neck bone mineral density on chromosome 1p36. *Hum Mol Genet*, 10:2447–2452.

Devoto, M, Spotila, LD, Stabley, DL, Wharton, GN, Rydbeck, H, Korkko, J, et al. (2005). Univariate and bivariate variance component linkage analysis of a whole-genome scan for loci contributing to bone mineral density. *Eur J Hum Genet*, 13:781–788.

Diascro, DD Jr, Vogel, RL, Johnson, TE, Witherup, KM, Pitzenberger, SM, Rutledge, SJ, et al. (1998). High fatty acid content in rabbit serum is responsible for the differentiation of osteoblasts into adipocyte-like cells. *J Bone Miner Res*, 13:96–106.

Dohi, Y, Iki, M, Ohgushi, H, Gojo, S, Tabata, S, Kajita, E, et al. (1998). A novel polymorphism in the promoter region for the human osteocalcin gene: the possibility of a correlation with bone mineral density in postmenopausal Japanese women. *J Bone Miner Res*, 13:1633–1639.

Duncan, EL, Brown, MA, Sinsheimer, J, Bell, J, Carr, AJ, Wordsworth, BP, et al. (1999). Suggestive linkage of the parathyroid receptor type 1 to osteoporosis. *J Bone Miner Res*, 14:1993–1999.

Eccleshall, TR, Garnero, P, Gross, C, Delmas, PD, and Feldman, D (1998). Lack of correlation between start codon polymorphism of the vitamin D receptor gene and bone mineral density in premenopausal French women: The OFELY Study. *J Bone Miner Res*, 13:31–35.

Econs, MJ, Koller, DL, Hui, SL, Fishburn, T, Conneally, PM, Johnston CC Jr, et al. (2004). Confirmation of linkage to chromosome 1q for peak vertebral bone mineral density in premenopausal white women. *Am J Hum Genet*, 74:223–228.

Efstathiadou, Z, Tsatsoulis, A, and Ioannidis, JP (2001). Association of collagen Ia 1 Sp1 polymorphism with the risk of prevalent fractures: a meta-analysis. *J Bone Miner Res*, 16:1586–1592.

Ezura, Y, Kajita, M, Ishida, R, Yoshida, S, Yoshida, H, Suzuki, T, et al. (2004). Association of multiple nucleotide variations in the pituitary glutaminyl cyclase gene (QPCT) with low radial BMD in adult women. *J Bone Miner Res*, 19:1296–1301.

Ezura, Y, Nakajima, T, Kajita, M, Ishida, R, Inoue, S, Yoshida, H, et al. (2003). Association of molecular variants, haplotypes, and linkage disequilibrium within the human vitamin D-binding protein (DBP) gene with postmenopausal bone mineral density. *J Bone Miner Res*, 18:1642–1649.

Faulkner, KG, Cummings, SR, Black, D, Palermo, L, Gluer, CC, and Genant, HK (1993). Simple measurement of femoral geometry predicts hip fracture: the study of osteoporotic fractures. *J Bone Miner Res*, 8:1211–1217.

Ferrari, SL, Deutsch, S, Choudhury, U, Chevalley, T, Bonjour, JP, Dermitzakis, ET, et al. (2004). Polymorphisms in the low-density lipoprotein receptor-related protein 5 (LRP5) gene are associated with variation in vertebral bone mass, vertebral bone size, and stature in whites. *Am J Hum Genet*, 74:866–875.

Ferrari, SL, Garnero, P, Emond, S, Montgomery, H, Humphries, SE, and Greenspan, SL (2001). A functional polymorphic variant in the interleukin-6 gene promoter associated with low bone resorption in postmenopausal women. *Arthritis Rheum*, 44:196–201.

Ferrari, SL, Karasik, D, Liu, J, Karamohamed, S, Herbert, AG, Cupples, LA, et al. (2004). Interactions of interleukin-6 promoter polymorphisms with dietary and lifestyle factors and their association with bone mass in men and women from the Framingham Osteoporosis Study. *J Bone Miner Res*, 19:552–559.

Fishman, D, Faulds, G, Jeffery, R, Mohamed-Ali, V, Yudkin, JS, Humphries, S, et al. (1998). The effect of novel polymorphisms in the interleukin-6 (IL-6) gene on IL-6 transcription and plasma IL-6 levels, and an association with systemic-onset juvenile chronic arthritis. *J Clin Invest*, 102:1369–1376.

Flicker, L, Hopper, JL, Rodgers, L, Kaymakci, B, Green, RM, and Wark, JD (1995). Bone density in elderly women: a twin study. *J Bone Miner Res*, 10:1607–1613.

Frattini, A, Orchard, PJ, Sobacchi, C, Giliani, S, Abinun, M, Mattsson, JP, et al. (2000). Defects in TCIRG1 subunit of the vacuolar proton pump are responsible for a subset of human autosomal recessive osteopetrosis. *Nat Genet*, 25:343–346.

Fuller, K, Wong, B, Fox, S, Choi, Y, and Chambers, TJ (1998). TRANCE is necessary and sufficient for osteoblast-mediated activation of bone resorption in osteoclasts. *J Exp Med*, 188:997–1001.

Garcia-Giralt, N, Nogués, X, Enjuanes, A, Puig, J, Mellibovsky, L, Bay-Jensen, A, et al. (2002). Two new single-nucleotide polymorphisms in the *COL1A1* upstream regulatory region and their relationship to bone mineral density. *J Bone Miner Res*, 17:384–393.

Garnero, P, Arden, NK, Griffiths, G, Delmas, PD, and Spector, TD (1996). Genetic influence on bone turnover in postmenopausal twins. *J Clin Endocrinol Metab*, 81:140–146.

Garnero, P, Borel, O, Grant, SFA, Ralston, SH, and Delmas, PD (1998). Collagen Iα1 Sp1 polymorphism, bone mass, and bone turnover in healthy French premenopausal women: The OFELY Study. *J Bone Miner Res*, 13:813–817.

Garnero, P, Borel, O, Sornay-Rendu, E, Arlot, ME, and Delmas, PD (1996). Vitamin D receptor gene polymorphisms are not related to bone turnover, rate of bone loss, and bone mass in postmenopausal women: The OFELY Study. *J Bone Miner Res*, 11:827–834.

Garrett, IR, Boyce, BF, Oreffo, RO, Bonewald, L, Poser, J, and Mundy, GR (1990). Oxygen-derived free radicals stimulate osteoclastic bone resorption in rodent bone in vitro and in vivo. *J Clin Invest*, 85:632–639.

Gelb, BD, Shi G-P, Chapman, HA, and Desnick, RJ (1996). Pycnodysostosis, a lysosomal disease caused by cathepsin K deficiency. *Science*, 273:1236–1238.

Gennari, L, Becherini, L, Masi, L, Mansani, R, Gonnelli, S, Cepollaro, C, et al. (1998). Vitamin D and estrogen receptor allelic variants in Italian postmenopausal women: evidence of multiple gene contribution to bone mineral density. *J Clin Endocrinol Metab*, 83:939–944.

Glorieux, FH, Ward, LM, Rauch, F, Lalic, L, Roughley, PJ, and Travers, R (2002). Osteogenesis imperfecta type VI: a form of brittle bone disease with a mineralization defect. *J Bone Miner Res*, 17:30–38.

Gong, G, Stern, HS, Cheng, SC, Fong, N, Mordeson, J, Deng, HW, et al. (1999). The association of bone mineral density with vitamin D receptor gene polymorphisms. *Osteoporos Int*, 9:55–64.

Gong, Y, Slee, RB, Fukai, N, Rawadi, G, Roman-Roman, S, Reginato, AM, et al. (2001). LDL receptor-related protein 5 (LRP5) affects bone accrual and eye development. *Cell*, 107:513–523.

Gong, Y, Vikkula, M, Boon, L, Liu, J, Beighton, P, Ramescar, R, et al. (1996). Osteoporosis-pseudoglioma syndrome, a disorder affecting skeletal strength and vision, is assigned to chromosome region 11q12–13. *Am J Hum Genet*, 59:146–151.

Goseki-Sone, M, Sogabe, N, Fukushi-Irie, M, Mizoi, L, Orimo, H, Suzuki, T, et al. (2005). Functional analysis of the single nucleotide polymorphism (787T>C) in the tissue-nonspecific alkaline phosphatase gene associated with BMD. *J Bone Miner Res*, 20:773–782.

Grant, SFA, Reid, DM, Blake, G, Herd, R, Fogelman, I, and Ralston, SH (1996). Reduced bone density and osteoporosis associated with a polymorphic Sp1 binding site in the collagen type Iα1 gene. *Nat Genet*, 14:203–205.

Gross, C, Eccleshall, TR, Malloy, PJ, Villa, ML, Marcus, R, and Feldman, D (1996). The presence of a polymorphism at the translation initiation site of the vitamin D receptor gene is associated with low bone mineral density in postmenopausal Mexican-American women. *J Bone Miner Res*, 11:1850–1855.

Gross, C, Krishnan, AV, Malloy, PJ, Eccleshall, TR, Zhao X-Y, and Feldman, D (1998). The vitamin D receptor gene start codon polymorphism: a functional analysis of *Fok* I variants. *J Bone Miner Res*, 13:1691–1699.

Grundberg, E, Carling, T, Brandstrom, H, Huang, S, Ribom, EL, Ljunggren, O, et al. (2004). A deletion polymorphism in the RIZ gene, a female sex steroid hormone receptor coactivator, exhibits decreased response to estrogen in vitro and associates with low bone mineral density in young Swedish women. *J Clin Endocrinol Metab*, 89:6173–6178.

Gueguen, R, Jouanny, P, Guillemin, F, Kuntz, C, Pourel, J, and Siest, G (1995). Segregation analysis and variance components analysis of bone mineral density in healthy families. *J Bone Miner Res*, 10:2017–2022.

Hakeda, Y, Kobayashi, Y, Yamaguchi, K, Yasuda, H, Tsuda, E, Higashio, K, et al. (1998). Osteoclastogenesis inhibitory factor (OCIF) directly inhibits bone-resorbing activity of isolated mature osteoclasts. *Biochem Biophys Res Commun*, 251:796–801.

Han, KO, Moon, IG, Kang, YS, Chung, HY, Min, HK, and Han, IK (1997). Nonassociation of estrogen receptor genotypes with bone mineral density and estrogen responsiveness to hormone replacement therapy in Korean postmenopausal women. *J Clin Endocrinol Metab*, 82:991–995.

Hannan, MT, Felson, DT, Dawson-Hughes, B, Tucker, KL, Cupples, LA, Wilson, PW, et al. (2000). Risk factors for longitudinal bone loss in elderly men and women: The Framingham Osteoporosis Study. *J Bone Miner Res*, 15:710–720.

Harris, SS, Eccleshall, TR, Gross, C, Dawson-Hughes, B, and Feldman, D (1997). The vitamin D receptor start codon polymorphism (*Fok* I) and bone mineral density in premenopausal American black and white women. *J Bone Miner Res*, 12:1043–1048.

Heaney, C, Shalev, H, Elbedour, K, Carmi, R, Staack, JB, Sheffield, VC, et al. (1998). Human autosomal recessive osteopetrosis maps to 11q13, a position predicted by comparative mapping of the murine osteosclerosis (oc) mutation. *Hum Mol Genet*, 7:1407–1410.

Heegaard, A, Jorgensen, HL, Vestergaard, AW, Hassager, C, and Ralston, SH (2000). Lack of influence of collagen type Iα1 Sp1 binding site polymorphism on the rate of bone loss in a cohort of postmenopausal Danish women followed for 18 years. *Calcif Tissue Int*, 66:409–413.

Hobson, EE and Ralston, SH (2001). Role of genetic factors in the pathophysiology and management of osteoporosis. *Clin Endocrinol*, 54:1–9.

Hofbauer, LC and Schoppet, M (2004). Clinical implications of the osteoprotegerin/RANKL/RANK system for bone and vascular diseases. *JAMA*, 292:490–495.

Holbrook, TL, Barrett-Connor, E, and Wingard, DL (1988). Dietary calcium and risk of hip fracture: 14 year prospective population study. *Lancet*, ii:1046–1049.

Hosoi, T, Miyao, M, Inoue, S, Hoshino, S, Shiraki, M, Orimo, H, et al. (1999). Association study of parathyroid hormone gene polymorphism and bone mineral density in Japanese postmenopausal women. *Calcif Tissue Int*, 64:205–208.

Howard, GM, Nguyen, TV, Harris, M, Kelly, PJ, and Eisman, JA (1998). Genetic and environmental contributions to the association between quantitative ultrasound and bone mineral density measurements: a twin study. *J Bone Miner Res*, 13:1318–1327.

Hsu, H, Lacey, DL, Dunstan, CR, Solovyev, I, Colombero, A, Timms, E, et al. (1999). Tumor necrosis factor receptor family member RANK mediates osteoclast differentiation and activation induced by osteoprotegerin ligand. *Proc Natl Acad Sci USA*, 96:3540–3545.

Hustmyer, FG, Liu, G, Johnston, CC, Christian, J, and Peacock, M (1999). Polymorphism at an Sp1 binding site of COL1A1 and bone mineral density in premenopausal female twins and elderly fracture patients. *Osteoporos Int*, 9:346–350.

Hustmyer, FG, Peacock, M, Hui, S, Johnston, CC, and Christian, J (1994). Bone mineral density in relation to polymorphism at the vitamin D receptor gene locus. *J Clin Invest*, 94:2130–2134.

Ioannidis, JP, Ralston, SH, Bennett, ST, Brandi, ML, Grinberg, D, Karassa, FB, et al. (2004). Differential genetic effects of ESR1 gene polymorphisms on osteoporosis outcomes. *JAMA*, 292:2105–2114.

Ioannidis, JP, Stavrou, I, Trikalinos, TA, Zois, C, Brandi, ML, Gennari, L, et al. (2002). Association of polymorphisms of the estrogen receptor a gene with bone mineral density and fracture risk in women: a meta-analysis. *J Bone Miner Res*, 17:2048–2060.

Ishida, R, Ezura, Y, Emi, M, Kajita, M, Yoshida, H, Suzuki, T, et al. (2003). Association of a promoter haplotype (–1542G/–525C) in the tumor necrosis factor receptor associated factor-interacting protein gene with low bone mineral density in Japanese women. *Bone*, 33:237–241.

Iwasaki, H, Emi, M, Ezura, Y, Ishida, R, Kajita, M, Kodaira, M, et al. (2003). Association of a Trp16Ser variation in the gonadotropin releasing hormone signal peptide with bone mineral density, revealed by SNP-dependent PCR typing. *Bone*, 32:185–190.

Janssens, K, Gershoni-Baruch, R, Guanabens, N, Migone, N, Ralston, S, Bonduelle, M, et al. (2000). Mutations in the latency-associated peptide of TGF-β1 cause Camurati-Engelmann disease. *Nat Genet*, 26:273–275.

Johnson, ML, Gong, G, Kimberling, W, Recker, SM, Kimmel, DB, and Recker, RB (1997). Linkage of a gene causing high bone mass to human chromosome 11 (11q12–13). *Am J Hum Genet*, 60:1326–1332.

Kammerer, CM, Schneider, JL, Cole, SA, Hixson, JE, Samollow, PB, O'Connell, JR, et al. (2003). Quantitative trait loci on chromosomes 2p, 4p, and 13q influence bone mineral density of the forearm and hip in Mexican Americans. *J Bone Miner Res*, 18:2245–2252.

Kanis, JA, Melton LJ III, Christiansen, C, Johnston, CC, and Khaltaev, N (1994). The diagnosis of osteoporosis. *J Bone Miner Res*, 9:1137–1141.

Kannus, P, Palvanen, M, Kaprio, J, Parkkari, J, and Koskenvuo, M (1999). Genetic factors and osteoporotic fractures in elderly people: prospective 25 year follow up of a nationwide cohort of elderly Finnish twins. *Br Med J*, 319:1334–1337.

Karasik, D, Myers, RH, Cupples, LA, Hannan, MT, Gagnon, DR, Herbert, A, et al. (2002). Genome screen for quantitative trait loci contributing to normal variation in bone mineral density: the Framingham Study. *J Bone Miner Res*, 17:1718–1727.

Kawano, K, Ogata, N, Chiano, M, Molloy, H, Kleyn, P, Spector, TD, et al. (2002). Klotho gene polymorphisms associated with bone density of aged postmenopausal women. *J Bone Miner Res*, 17:1744–1751.

Keen, RW, Hart, DJ, Arden, NK, Doyle, DV, and Spector, TD (1999). Family history of appendicular fracture and risk of osteoporosis: a population-based study. *Osteoporos Int*, 10:161–166.

Keen, RW, Woodford-Richens, KL, Lanchbury, JS, and Spector, TD (1998). Allelic variation at the interleukin-1 receptor antagonist gene is associated with early postmenopausal bone loss at the spine. *Bone*, 23:367–371.

Kim, JG, Roh, KR, and Lee, JY (2002). The relationship among serum insulin-like growth factor-I, insulin-like growth factor-I gene polymorphism, and bone mineral density in postmenopausal women in Korea. *Am J Obstetr Gynecol*, 186:345–350.

Koay, MA and Brown, MA (2005). Genetic disorders of the LRP5-Wnt signalling pathway affecting the skeleton. *Trends Mol Med*, 11:129–137.

Koay, MA, Woon, PY, Zhang, Y, Miles, LJ, Duncan, EL, Ralston, SH, et al. (2004). Influence of LRP5 polymorphisms on normal variation in BMD. *J Bone Miner Res*, 19:1619–1627.

Kobayashi, S, Inoue, S, Hosoi, T, Ouchi, Y, Shiraki, M, and Orimo, H (1996). Association of bone mineral density with polymorphism of the estrogen receptor gene. *J Bone Miner Res*, 11:306–311.

Koller, DL, Econs, MJ, Morin, PA, Christian, JC, Hui, SL, Parry, P, et al. (2000). Genome screen for QTLs contributing to normal variation in bone mineral density and osteoporosis. *J Clin Endocrinol Metab*, 85:3116–3120.

Koller, DL, Ichikawa, S, Johnson, ML, Lai, D, Xuei, X, Edenberg, HJ, et al. (2005). Contribution of the LRP5 gene to normal variation in peak BMD in women. *J Bone Miner Res*, 20:75–80.

Korach, KS (1994). Insights from the study of animals lacking functional estrogen receptor. *Science*, 266:1524–1527.

Krall, EA and Dawson-Hughes, B (1999). Smoking increases bone loss and decreases intestinal calcium absorption. *J Bone Miner Res*, 14:215–220.

Lacey, DL, Timms, E, Tan, HL, Kelley, MJ, Dunstan, CR, Burgess, T, et al. (1998). Osteoprotegerin ligand is a cytokine that regulates osteoclast differentiation and activation. *Cell*, 93:165–176.

Langdahl, BL, Carstens, M, Stenkjaer, L, and Eriksen, EF (2002). Polymorphisms in the osteoprotegerin gene are associated with osteoporotic fractures. *J Bone Miner Res*, 17:1245–1255.

Langdahl, BL, Knudsen, JY, Jensen, HK, Gregersen, N, and Eriksen, EF (1997). A sequence variation: 713-8delC in the transforming growth factor-beta 1 gene has higher prevalence in osteoporotic women than in normal women and is associated with very low bone mass in osteoporotic women and increased bone turnover in both osteoporotic and normal women. *Bone*, 20:289–294.

Langdahl, BL, Lokke, E, Carstens, M, Stenkjaer, LL, and Eriksen, EF (2000a). Osteoporotic fractures are associated with an 86-base pair repeat polymorphism in the interleukin-1-receptor antagonist gene but not with polymorphisms in the interleukin-1β gene. *J Bone Miner Res*, 15:402–414.

Langdahl, BL, Lokke, E, Carstens, M, Stenkjaer, LL, and Eriksen, EF (2000b). A TA repeat polymorphism in the estrogen receptor gene is associated with osteoporotic fractures but polymorphisms in the first exon and intron are not. *J Bone Miner Res*, 15:2222–2230.

Langdahl, BL, Ralston, SH, Grant, SF, and Eriksen, EF (1998). An Sp1 binding site polymorphism in the COLIA1 gene predicts osteoporotic fractures in both men and women. *J Bone Miner Res*, 13:1384–1389.

Lei, SF, Deng, FY, Dvornyk, V, Liu, MY, Xiao, SM, Jiang, DK, et al. (2005). The (GT)*n* polymorphism and haplotype of the *COL1A2* gene, but not the (AAAG)*n* polymorphism of the *PTHR1* gene, are associated with bone mineral density in Chinese. *Hum Genet*, 116:200–207.

Liden, M, Wilen, B, Ljunghall, S, and Melhus, H (1998). Polymorphism at the Sp1 binding site in the collagen type Iα1 gene does not predict bone mineral density in postmenopausal women in Sweden. *Calcif Tissue Int*, 63:293–295.

Little, RD, Carulli, JP, Del Mastro, RG, Dupuis, J, Osborne, M, Folz, C, et al. (2002). A mutation in the LDL receptor-related protein 5 gene results in the autosomal dominant high-bone-mass trait. *Am J Hum Genet*, 70:11–19.

Lorentzon, M, Eriksson, AL, Mellstrom, D, and Ohlsson, C (2004). The COMT Val158Met polymorphism is associated with peak BMD in men. *J Bone Miner Res*, 19:2005–2011.

Lorentzon, M, Lorentzon, R, Lerner, UH, and Nordström, P (2001). Calcium sensing receptor gene polymorphism, circulating calcium concentrations and bone mineral density in healthy adolescent girls. *Eur J Endocrinol*, 144:257–261.

Lorentzon, M, Lorentzon, R, and Nordström, P (2000). Interleukin-6 gene polymorphism is related to bone mineral density during and after puberty in healthy white males: a cross-sectional and longitudinal study. *J Bone Miner Res*, 15:1944–1949.

MacGregor, AJ, Snieder, H, and Spector, TD (2000). Genetic factors and osteoporotic fractures in elderly people. *Br Med J*, 320:1669–1670.

Maggio, D, Barabani, M, Pierandrei, M, Polidori, MC, Catani, M, Mecocci, P, et al. (2003). Marked decrease in plasma antioxidants in aged osteoporotic women: results of a cross-sectional study. *J Clin Endocrinol Metab*, 88:1523–1527.

Mann, V, Hobson, EE, Li, B, Stewart, TL, Grant, SFA, Robins, SP, et al. (2001). A *COL1A1* Sp1 binding site polymorphism predisposes to osteoporotic fracture by affecting bone density and quality. *J Clin Invest*, 107:899–907.

Mann, V and Ralston, SH (2003). Meta-analysis of *COL1A1* Sp1 polymorphism in relation to bone mineral density and osteoporotic fracture. *Bone*, 32:711–717.

Manolagas, SC and Jilka, RL (1995). Bone marrow, cytokines, and bone remodeling. Emerging insights into the pathophysiology of osteoporosis. *N Engl J Med*, 332:305–311.

Masi, L, Becherini, L, Gennari, L, Amedei, A, Colli, E, Falchetti, A, et al. (2001). Polymorphism of the aromatase gene in postmenopausal Italian women: distribution and correlation with bone mass and fracture risk. *J Clin Endocrinol Metab*, 86:2263–2269.

McFarlane, SI, Muniyappa, R, Francisco, R, and Sowers, JR (2002). Pleiotropic effects of statins: lipid reduction and beyond. *J Clin Endocrinol Metab*, 87:1451–1458.

Melhus, H, Kindmark, A, Amér, S, Wilén, B, Lindh, E, and Ljunghall, S (1994). Vitamin D receptor genotypes in osteoporosis. *Lancet*, 344:949–950.

Melton LJ III (1995). How many women have osteoporosis now? *J Bone Miner Res*, 10:175–177.

Miyao, M, Hosoi, T, Emi, M, Nakajima, T, Inoue, S, Hoshino, S, et al. (2000). Association of bone mineral density with a dinucleotide repeat polymorphism at the calcitonin (CT) locus. *J Hum Genet*, 45:346–350.

Miyao, M, Morita, H, Hosoi, T, Kurihara, H, Inoue, S, Hoshino, S, et al. (2000). Association of methylenetetrahydrofolate reductase (MTHFR) polymorphism with bone mineral density in postmenopausal Japanese women. *Calcif Tissue Int*, 66:190–194.

Mizuguchi, T, Furuta, I, Watanabe, Y, Tsukamoto, K, Tomita, H, Tsujihata, M, et al. (2004). LRP5, low-density-lipoprotein-receptor-related protein 5, is a determinant for bone mineral density. *J Hum Genet*, 49:80–86.

Moffett, SP, Zmuda, JM, Cauley, JA, Stone, KL, Nevitt, MC, Ensrud, KE, et al. (2004). Association of the G–174C variant in the interleukin-6 promoter region with bone loss and fracture risk in older women. *J Bone Miner Res*, 19:1612–1618.

Morishima, A, Grumbach, MM, Simpson, ER, Fisher, C, and Qin, K (1995). Aromatase deficiency in male and female siblings caused by a novel mutation and the physiological role of estrogens. *J Clin Endocrinol Metab*, 80:3689–3698.

Morrison, NA, Qi, JC, Tokita, A, Kelly, PJ, Crofts, L, Nguyen, TV, et al. (1994). Prediction of bone density from vitamin D receptor alleles. *Nature*, 367:284–287.

Murray, RE, McGuigan, F, Grant, SFA, Reid, DM, and Ralston, SH (1997). Polymorphisms of the interleukin-6 gene are associated with bone mineral density. *Bone*, 21:89–92.

Napoli, N, Villareal, DT, Mumm, S, Halstead, L, Sheikh, S, Cagaanan, M, et al. (2005). Effect of CYP1A1 gene polymorphisms on estrogen metabolism and bone density. *J Bone Miner Res*, 20:232–239.

Nemetz, A, Toth, M, Garcia-Gonzalez, MA, Zagoni, T, Feher, J, Pena, AS, et al. (2001). Allelic variation at the interleukin 1β gene is associated with decreased bone mass in patients with inflammatory bowel diseases. *Gut*, 49:644–649.

Nguyen, TV, Blangero, J, and Eisman, JA (2000). Genetic epidemiological approaches to the search for osteoporosis genes. *J Bone Miner Res*, 15:392–401.

Niu, T, Chen, C, Cordell, H, Yang, J, Wang, B, Wang, Z, et al. (1999). A genome-wide scan for loci linked to forearm bone mineral density. *Hum Genet*, 104:226–233.

Obermayer-Pietsch, BM, Bonelli, CM, Walter, DE, Kuhn, RJ, Fahrleitner-Pammer, A, Berghold, A, et al. (2004). Genetic predisposition for adult lactose intolerance and relation to diet, bone density, and bone fractures. *J Bone Miner Res*, 19:42–47.

Ogawa, S, Urano, T, Hosoi, T, Miyao, M, Hoshino, S, Fujita, M, et al. (1999). Association of bone mineral density with a polymorphism of the peroxisome proliferator-activated receptor γ gene: PPARγ expression in osteoblasts. *Biochem Biophys Res Commun*, 260:122–126.

Ohmori, H, Makita, Y, Funamizu, M, Hirooka, K, Hosoi, T, Orimo, H, et al. (2002). Linkage and association analysis of the osteoprotegerin gene locus with human osteoporosis. *J Hum Genet*, 47:400–406.

Omasu, F, Ezura, Y, Kajita, M, Ishida, R, Kodaira, M, Yoshida, H, et al. (2003). Association of genetic variation of the RIL gene, encoding a PDZ-LIM domain protein and localized in 5q31.1, with low bone mineral density in adult Japanese women. *J Hum Genet*, 48:342–345.

Orozco P (2004). Atherogenic lipid profile and elevated lipoprotein (a) are associated with lower bone mineral density in early postmenopausal overweight women. *Eur J Epidemiol*, 19:1105–1112.

Ota, N, Hunt, SC, Nakajima, T, Suzuki, T, Hosoi, T, Orimo, H, et al. (1999). Linkage of interleukin 6 locus to human osteopenia by sibling pair analysis. *Hum Genet*, 105:253–257.

Ota, N, Hunt, SC, Nakajima, T, Suzuki, T, Hosoi, T, Orimo, H, et al. (2000). Linkage of human tumor necrosis factor-α to human osteoporosis by sib pair analysis. *Genes Immun*, 1:260–264.

Ota, N, Nakajima, T, Nakazawa, I, Suzuki, T, Hosoi, T, Orimo, H, et al. (2001). A nucleotide variant in the promoter region of the interleukin-6 gene associated with decreased bone mineral density. *J Hum Genet*, 46:267–272.

Parhami, F, Garfinkel, A, and Demer, LL (2000). Role of lipids in osteoporosis. *Arterioscler Thromb Vasc Biol*, 20:2346–2348.

Parhami, F, Jackson, SM, Tintut, Y, Le, V, Balucan, JP, Territo, M, et al. (1999). Atherogenic diet and minimally oxidized low density lipoprotein inhibit osteogenic and promote adipogenic differentiation of marrow stromal cells. *J Bone Miner Res*, 14:2067–2078.

Patel, MS, Cole, DEC, Smith, JD, Hawker, GA, Wong, B, Trang, H, et al. (2000). Alleles of the estrogen receptor α-gene and an estrogen receptor cotranscriptional activator gene, amplified in breast cancer-1 (*AIB11*), are associated with quantitative calcaneal ultrasound. *J Bone Miner Res*, 15:2231–2239.

Peacock, M, Koller, DL, Hui, S, Johnston, CC, Foroud, T, and Econs, MJ (2004). Peak bone mineral density at the hip is linked to chromosomes 14q and 15q. *Osteoporos Int*, 15:489–496.

Pocock, NA, Eisman, JA, Hopper, JL, Yeates, MG, Sambrook, PN, and Eberl, S (1987). Genetic determinants of bone mass in adults: a twin study. *J Clin Invest*, 80:706–710.

Poli, A, Bruschi, F, Cesana, B, Rossi, M, Paoletti, R, and Crosignani, PG (2003). Plasma low-density lipoprotein cholesterol and bone mass densitometry in postmenopausal women. *Obstetr Gynecol*, 102:922–926.

Ralston, SH (2001). Genetics of osteoporosis. *Rev Endocr Metab Disord*, 2:13–21.

Ralston, SH (2002). Genetic control of susceptibility to osteoporosis. *J Clin Endocrinol Metab*, 87:2460–2466.

Ralston, SH, Galwey, N, MacKay, I, Albagha, OM, Cardon, L, Compston, JE, et al. (2005). Loci for regulation of bone mineral density in men and women identified by genome wide linkage scan: the FAMOS study. *Hum Mol Genet*, 14:943–951.

Ramesh Babu, L, Wilson, SG, Dick, IM, Islam, FM, Devine, A, and Prince, RL (2005). Bone mass effects of a BMP4 gene polymorphism in postmenopausal women. *Bone*, 36:555–561.

Rauch F and Glorieux, FH (2004). Osteogenesis imperfecta. *Lancet*, 363:1377–1385.

Raymond, MH, Schutte, BC, Torner, JC, Burns, TL, and Willing, MC (1999). Osteocalcin: genetic and physical mapping of the human gene BGLAP and its potential role in postmenopausal osteoporosis. *Genomics*, 60:210–217.

Reneland, RH, Mah, S, Kammerer, S, Hoyal, CR, Marnellos, G, Wilson, SG, et al. (2005). Association between a variation in the phosphodiesterase 4D gene and bone mineral density. *BMC Med Genet*, 6:9.

Roth, DE, Venta, PJ, Tashian, RE, and Sly, WS (1992). Molecular basis of human carbonic anhydrase II deficiency. *Proc Natl Acad Sci USA*, 89:1804–1808.

Sainz, J, Van Tornout, JM, Loro, L, Sayre, J, Roe, TF, and Gilsanz, V (1997). Vitamin D-receptor gene polymorphisms and bone density in prepubertal American girls of Mexican descent. *N Engl J Med*, 337:77–82.

Salmén, T, Heikkinen A-M, Mahonen, A, Kröger, H, Komulainen, M, Saarikoski, S, et al. (2000). The protective effect of hormone-replacement therapy on fracture risk is modulated by estrogen receptor α genotype in early postmenopausal women. *J Bone Miner Res*, 15:2479–2486.

Shearman, AM, Karasik, D, Gruenthal, KM, Demissie, S, Cupples, LA, Housman, DE, et al. (2004). Estrogen receptor beta polymorphisms are associated with bone mass in women and men: The Framingham Study. *J Bone Miner Res*, 19:773–781.

Shiraki, M, Shiraki, Y, Aoki, C, Hosoi, T, Inoue, S, Kaneki, M, et al. (1997). Association of bone mineral density with apolipoprotein E phenotype. *J Bone Miner Res*, 12:1438–1445.

Simonet, WS, Lacey, DL, Dunstan, CR, Kelley, M, Chang, MS, Luthy, R, et al. (1997). Osteoprotegerin: a novel secreted protein involved in the regulation of bone density. *Cell*, 89:309–319.

Slemenda, CW, Turner, CH, Peacock, M, Christian, JC, Sorbel, J, Hui, SL, et al. (1996). The genetics of proximal femur geometry, distribution of bone mass and bone mineral density. *Osteoporos Int*, 6:178–182.

Sly, WS, Hewett-Emmett, D, Whyte, MP, Yu, YS, and Tashian, RE (1983). Carbonic anhydrase II deficiency identified as the primary defect in the autosomal recessive syndrome of osteopetrosis with renal tubular acidosis and cerebral calcification. *Proc Natl Acad Sci USA*, 80:2752–2756.

Smith, EP, Boyd, J, Frank, GR, Takahashi, H, Cohen, RM, Specker, B, et al. (1994). Estrogen resistance caused by a mutation in the estrogen-receptor gene in a man. *N Engl J Med*, 331:1056–1061.

Sobacchi, C, Vezzoni, P, Reid, DM, McGuigan, FE, Frattini, A, Mirolo, M, et al. (2004). Association between a polymorphism affecting an AP1 binding site in the promoter of the TCIRG1 gene and bone mass in women. *Calcif Tissue Int*, 74:35–41.

Sowers, M, Willing, M, Burns, T, Deschenes, S, Hollis, B, Crutchfield, M, et al. (1999). Genetic markers, bone mineral density, and serum osteocalcin levels. *J Bone Miner Res*, 14:1411–1419.

Spotila, LD, Rodriguez, H, Koch, M, Adams, K, Caminis, J, Tenenhouse, HS, et al. (2000). Association of a polymorphism in the TNFR2 gene with low bone mineral density. *J Bone Miner Res*, 15:1376–1383.

Stewart, A, Torgerson, DJ, and Reid, DM (1996). Prediction of fractures in perimenopausal women: a comparison of dual energy X-ray absorptiometry and broadband ultrasound attenuation. *Ann Rheum Dis*, 55:140–142.

Styrkarsdottir, U, Cazier, JB, Kong, A, Rolfsson, O, Larsen, H, Bjarnadottir, E, et al. (2003). Linkage of osteoporosis to chromosome 20p12 and association to BMP2. *PLoS Biol*, 1:351–360.

Sykes, B (1990). Human genetics. Bone disease cracks genetics. *Nature*, 348:18–20.

Taboulet, J, Frenkian, M, Frendo, JL, Feingold, N, Jullienne, A, and de Vernejoul, MC (1998). Calcitonin receptor polymorphism is associated with a decreased fracture risk in post-menopausal women. *Hum Mol Genet*, 7:2129–2133.

Takai, H, Kanematsu, M, Yano, K, Tsuda, E, Higashio, K, Ikeda, K, et al. (1998). Transforming growth factor-β stimulates the production of osteoprotegerin/osteoclastogenesis inhibitory factor by bone marrow stromal cells. *J Biol Chem*, 273:27091–27096.

Tan, KB, Harrop, J, Reddy, M, Young, P, Terrett, J, Emery, J, et al. (1997). Characterization of a novel TNF-like ligand and recently described TNF ligand and TNF receptor superfamily genes and their constitutive and inducible expression in hematopoietic and non-hematopoietic cells. *Gene*, 204:35–46.

Thakkinstian, A, D'Este, C, Eisman, J, Nguyen, T, and Attia, J (2004). Meta-analysis of molecular association studies: vitamin D receptor gene polymorphisms and BMD as a case study. *J Bone Miner Res*, 19:419–428.

Tokita, A, Matsumoto, H, Morrison, NA, Tawa, T, Miura, Y, Fukamauchi, K, et al. (1996). Vitamin D receptor alleles, bone mineral density and turnover in premenopausal Japanese women. *J Bone Miner Res*, 11:1003–1009.

Tsuda, E, Goto, M, Mochizuki, S, Yano, K, Kobayashi, F, Morinaga, T, et al. (1997). Isolation of a novel cytokine from human fibroblasts that specifically inhibits osteoclastogenesis. *Biochem Biophys Res Commun*, 234:137–142.

Uitterlinden, AG, Arp, PP, Paeper, BW, Charmley, P, Proll, S, Rivadeneira, F, et al. (2004). Polymorphisms in the sclerosteosis/van Buchem disease gene

(SOST) region are associated with bone-mineral density in elderly whites. *Am J Hum Genet*, 75:1032–1045.

Uitterlinden, AG, Burger, H, Huang, Q, Yue, F, McGuigan, FEA, Grant, SFA, et al. (1998). Relation of alleles of the collagen type Iα1 gene to bone density and the risk of osteoporotic fractures in postmenopausal women. *N Engl J Med*, 338:1016–1021.

Uitterlinden, AG, Pols, HAP, Burger, H, Huang, Q, Van Daele, PLA, Van Dujin, CM, et al. (1996). A large-scale population-based study of the association of vitamin D receptor gene polymorphisms with bone mineral density. *J Bone Miner Res*, 11:1241–1248.

Urano, T, Hosoi, T, Shiraki, M, Toyoshima, H, Ouchi, Y, and Inoue, S (2000). Possible involvement of the p57 (kip2) gene in bone metabolism. *Biochem Biophys Res Commun*, 269:422–426.

Van Hul, E, Gram, J, Bollerslev, J, Van Wesenbeeck, L, Mathysen, D, Andersen, PE, et al. (2002). Localization of the gene causing autosomal dominant osteopetrosis type I to chromosome 11q12–13. *J Bone Miner Res*, 17:1111–1117.

Van Wesenbeeck, L, Cleiren, E, Gram, J, Beals, RK, Benichou, O, Scopelliti, D, et al. (2003). Six novel missense mutations in the LDL receptor-related protein 5 (LRP5) gene in different conditions with an increased bone density. *Am J Hum Genet*, 72:763–771.

Vandevyver, C, Vanhoof, J, Declerck, K, Stinissen, P, Vandervorst, C, Michiels, L, et al. (1999). Lack of association between estrogen receptor genotypes and bone mineral density, fracture history, or muscle strength in elderly women. *J Bone Miner Res*, 14:1576–1582.

Vaughan, T, Pasco, JA, Kotowicz, MA, Nicholson, GC, and Morrison, NA (2002). Alleles of *RUNX2/CBFA1* gene are associated with differences in bone mineral density and risk of fracture. *J Bone Miner Res*, 17:1527–1534.

Westendorf, JJ, Kahler, RA, and Schroeder, TM (2004). Wnt signaling in osteoblasts and bone diseases. *Gene*, 341:19–39.

Whyte, MP, Obrecht, SE, Finnegan, PM, Jones, JL, Podgornik, MN, McAlister, WH, et al. (2002). Osteoprotegerin deficiency and juvenile Paget's disease. *N Engl J Med*, 347:175–184.

Willing, M, Sowers, M, Aron, D, Clark, MK, Burns, T, Bunten, C, et al. (1998). Bone mineral density and its change in white women: estrogen and vitamin D receptor genotypes and their interaction. *J Bone Miner Res*, 13:695–705.

Wilson, SG, Reed, PW, Bansal, A, Chiano, M, Lindersson, M, Langdown, M, et al. (2003). Comparison of genome screens for two independent cohorts provides replication of suggestive linkage of bone mineral density to 3p21 and 1p36. *Am J Hum Genet*, 72:144–155.

Wynne, F, Drummond, FJ, Daly, M, Brown, M, Shanahan, F, Molloy, MG, et al. (2003). Suggestive linkage of 2p22–25 and 11q12–13 with low bone mineral density at the lumbar spine in the Irish population. *Calcif Tissue Int*, 72:651–658.

Yamada, Y, Ando, F, Niino, N, Miki, T, and Shimokata, H (2003). Association of polymorphisms of paraoxonase 1 and 2 genes, alone or in combination, with bone mineral density in community-dwelling Japanese. *J Hum Genet*, 48:469–475.

Yamada, Y, Ando, F, Niino, N, and Shimokata, H (2001). Transforming growth factor-β1 gene polymorphism and bone mineral density. *JAMA*, 285:167–168.

Yamada, Y, Ando, F, Niino, N, and Shimokata, H (2002a). Association of a polymorphism of the CC chemokine receptor–2 gene with bone mineral density. *Genomics*, 80:8–12.

Yamada, Y, Ando, F, Niino, N, and Shimokata, H (2002b). Association of a polymorphism of the matrix metalloproteinase-1 gene with bone mineral density. *Matrix Biol*, 21:389–392.

Yamada, Y, Ando, F, Niino, N, and Shimokata, H (2003a). Association of a polymorphism of the dopamine receptor D4 gene with bone mineral density in Japanese men. *J Hum Genet*, 48:629–633.

Yamada, Y, Ando, F, Niino, N, and Shimokata, H (2003b). Association of polymorphisms of interleukin-6, osteocalcin, and vitamin D receptor genes, alone or in combination, with bone mineral density in community-dwelling Japanese women and men. *J Clin Endocrinol Metab*, 88:3372–3378.

Yamada, Y, Ando, F, Niino, N, and Shimokata, H (2003c). Association of polymorphisms of the osteoprotegerin gene with bone mineral density in Japanese women but not men. *Mol Genet Metab*, 80:344–349.

Yamada, Y, Ando, F, and Shimokata, H (2005). Association of polymorphisms in *CYP17*, *MTP*, and *VLDLR* with bone mineral density in community-dwelling Japanese women and men. *Genomics*, 86:76–85.

Yamada, Y, Miyauchi, A, Goto, J, Takagi, Y, Okuizumi, H, Kanematsu, M, et al. (1998). Association of a polymorphism of the transforming growth factor-β1 gene with genetic susceptibility to osteoporosis in Japanese women. *J Bone Miner Res*, 13:1569–1576.

Yamada, Y, Miyauchi, A, Takagi, Y, Nakauchi, K, Miki, N, Mizuno, M, et al. (2000). Association of a polymorphism of the transforming growth factor beta-1 gene with prevalent vertebral fractures in Japanese women. *Am J Med*, 109:244–247.

Yamada, Y, Miyauchi, A, Takagi, Y, Tanaka, M, Mizuno, M, Harada A. (2001) Association of the $C^{-509} \rightarrow T$ polymorphism, alone or in combination with the $T^{869} \rightarrow C$ polymorphism, of the transforming growth factor-β1 gene with bone mineral density and genetic susceptibility to osteoporosis in Japanese women. *J Mol Med*, 79:149–156.

Yamaguchi, T, Sugimoto, T, Yano, S, Yamauchi, M, Sowa, H, Chen, Q, et al. (2002). Plasma lipids and osteoporosis in postmenopausal women. *Endocr J*, 49:211–217.

Yasuda, H, Shima, N, Nakagawa, N, Mochizuki, SI, Yano, K, Fujise, N, et al. (1998). Identity of osteoclastogenesis inhibitory factor (OCIF) and osteoprotegerin (OPG): a mechanism by which OPG/OCIF inhibits osteoclastogenesis in vitro. *Endocrinology*, 139:1329–1337.

Yasuda, H, Shima, N, Nakagawa, N, Yamaguchi, K, Kinosaki, M, Mochizuki, S, et al. (1998). Osteoclast differentiation factor is a ligand for osteoprotegerin/osteoclastogenesis-inhibitory factor and is identical to TRANCE/RANKL. *Proc Natl Acad Sci USA*, 95:3597–3602.

36

Applications in Obstetrics and Gynecology and Reproductive Medicine

Gareth C Weston, Anna Ponnampalam, and Peter AW Rogers

The total number of genes in the human genome has been estimated at between 32,000 and 39,000 (Bork, and Copley, 2001). Genomics concerns the study of these genes in context within our genetic system, rather than as individuals. Large numbers of genes, or even all of the known genes in the genome, can have either their level of activity or their structure examined simultaneously.

The very existence of genomics as a field of study owes its success to two major developments. The first of these was the completion of the Human Genome Project, providing us with a complete template of the human genetic code. The second has been the development of tools enabling the interrogation of differences in structure or function of large numbers of genes in the same experiment. The most commonly used of these tools has been "microarrays."

Microarrays consist of a pattern (array) of spots on a "micro" scale designed to examine thousands of components of the genome simultaneously. The two aspects of the genome most commonly studied are variations in genetic sequence or structure (such as single nucleotide polymorphisms) and changes in gene expression or function (via measurement in differences in messenger ribonucleic acid [mRNA] abundance). The predominant use of microarrays in the study of obstetrics, gynecology, and reproductive medicine, as in other medical fields, is to analyze differences in gene expression. A detailed description of the gene expression microarray laboratory technique is beyond the scope of this chapter.

In this chapter, we will review the existing literature dealing with the application of genomics to obstetrics, gynecology, and reproductive medicine. Gene expression microarrays will be the main genomic studies that are dealt with, as they constitute the vast majority of genomic studies published in the reproductive field.

Endometrium

The endometrium is a dynamic tissue that responds to multiple stimuli, depending on physiological and environmental conditions, including steroid hormones, an implanting conceptus, withdrawal of steroid hormones, contraceptive steroids, selective steroid hormone receptor modulators, infection, transient cell populations, and metaplastic and neoplastic agents (Giudice, 2003). Microarray gene expression profiling allows large numbers of genes to be investigated simultaneously, and the resulting gene expression patterns can then be correlated with different structural, functional, or clinical parameters (Liotta, and Petricoin, 2000). A brief overview of some of the existing literature of global gene expression profiling in human endometrium is reviewed herein.

Caution should be taken when comparing data obtained using endometrial cell cultures with that from endometrial biopsies, as cells may have different responses in vitro compared with in vivo. These differences may be because of the processing of the cells, the culture medium used and loss of paracrine interactions with other cells in situ (Giudice, 2003). Especially important when dealing with endometrial tissue is careful clinical characterization of the subjects donating the tissue, their hormonal stage, age, history of endometrial pathologies, and other medical conditions which is also important to minimize intersubject variability and to maximize data reliability.

Gene Expression Profiling of the Whole Endometrial Tissue

There have been five microarray publications in recent years on human endometrium (Carson et al., 2002; Kao et al., 2002; Borthwick et al., 2003; Riesewijk et al., 2003, Mirkin et al., 2005). All five used the Affymetrix HG-U95A chip (~12,000 transcripts); Borthwick et al. (2003) also used HG-U95 B-E chips (~60,000 transcripts).

Kao et al. (2002) looked into differential expression between the average values of four late proliferative (LP) phase samples (cycle day 8–10) and samples from seven other individuals in the midsecretory (MS) phase, Luteinizing hormone (LH) LH + 8 to LH + 10). They found 114 up-regulated and 218 down-regulated genes in the MS stage. Borthwick et al. (2003) compared global gene expression profiles of pooled samples of five women in the proliferative phase (day 9–11) with pooled samples of five women in the secretory phase (LH + 6 to LH + 8) and found 89 up-regulated and 57 down-regulated genes in the secretory phase. Both experiments were identical except that Borthwick et al. (2003) pooled before hybridizing onto the microarray. There is approximately 25%–30% agreement between the studies for genes that were up-regulated and 10% agreement for genes that were down-regulated.

Carson et al. (2002) compared three pooled samples from the early secretory (ES) phase (LH day 2–4) with three pooled samples from MS phase (LH day 7–9) and found 98 up-regulated and 119 down-regulated genes in the MS phase. Riesewijk et al. (2003) compared gene expression between five pairs of ES (LH + 2) and MS (LH + 7) samples from five individual women. They found 153 up-regulated and 58 down-regulated genes in the MS phase.

These two groups have only approximately 10% agreement between their up-regulated gene set and even less agreement between their down-regulated genes. Mirkin et al. (2005) investigated the gene expression profiling of three samples from ES stage (LH + 3) versus five samples from MS stage and found 49 up-regulated and 58 down-regulated genes in MS stage compared to ES stage of the menstrual cycle. A detailed comparison of some of the findings of these five studies is summarized in Table 36–1. The lack of agreement between the different studies highlights the difficulties faced with a highly variable dynamic tissue such as the endometrium.

In an ambitious project, we have attempted to investigate the transcriptional profile of human endometrium during the menstrual cycle through a seven-phase time course analysis (Ponnampalam et al., 2004). As well as identifying 425 genes that showed at least two-fold up-regulation in at least one stage of the menstrual cycle, the study has also shown that it is possible to predict menstrual cycle stages based on the transcriptional profile of the samples and has identified different patterns of expression for groups of genes that are associated with different endometrial biological processes such as implantation and menstruation. The study also suggests that discordant patterns of gene expression may help identify endometrial samples with subtle abnormalities not readily apparent by routine histopathology.

Figures 36–1A and B show the groupings of their endometrial samples based on the global gene expression profile. The length and the subdivision of the dendrogram branches show the relatedness of the samples, with samples joined by short branches being most similar in expression profiles. The 43 samples were sorted into nine clusters based on the similarity of the gene expression profiles of the samples: midproliferatory (MP)–early-proliferatory (EP) ($n = 2$); EP–MP ($n = 5$); MP ($n = 5$); LP–ES ($n = 5$); ES–MS ($n = 6$); MS–LS (late-secretory) ($n = 7$); LS ($n = 3$); LS–M (menstrual) ($n = 2$); M ($n = 4$). Samples in these nine groupings agreed with their histopathology by either being in the same group or an adjacent group, apart from four samples (E355 [LP], EA30 [MS], EA85 [LP], and EA2 [EP]). These four outliers were removed from further analysis (Figs. 36–1A, B). Three out of the four outliers (355, A30, and A2) found by hierarchical clustering showed significant inter-observer variability between the two histopathology evaluations, while there was agreement between the two pathologists' reports for all other samples. Visual inspection of the fourth outlier's (A85) microarray image showed that it had very poor hybridization. Following the removal of outliers, group M–EP was excluded from further analysis because of lack of replicates, and MS and LS groups were merged to produce the final molecularly defined classification of the menstrual cycle. The final groupings, shown in Figure 36–1B, were

Table 36–1 Genes That are Significantly Up- and Down-regulated During the Window of Implantation: Comparison of Array Data from Five Previously Published Studies

Gene Name	Accession Number	Kao et al., 2002	Carson et al., 2002	Riesewijk et al., 2003	Borthwick et al., 2003	Mirkin et al., 2005
Up-regulated genes						
Osteopontin	AF052124	+	+	+	+	+
Apolipoprotein D	J02611	+	+	+	+	
Dickkopf/DKK1	AB020315	+	+	+	+	
Placental protein-14/Glycodelin	J04129	+		+	+	
Decay accelerating factor for complement (CD55, Cromer blood group system)	M31516	+		+	+	
Adipsin/Complement factor D	M84526	+		+	+	
Guanylate-binding protein 2, interferon-inducible	M55543		+	+	+	
Claudin 4/CEP-R	AB000712	+	+	+		
Monoamine oxidase A (MAOA)	AA420624/ M68840	+		+	+	
Growth arrest and DNA-damage-inducible protein (GADD45)	M60974	+		+	+	
NIP2	AB002365		+	+	+	
Serine (or cysteine) proteinase inhibitor, clade G (SERPINGI)	X54486			+	+	+
Interleukin 15 (IL15)	AF031167	+	+	+		+
Annexin IV (ANXA4)	M82809	+		+	+	
Mitogen-activated protein kinase kinase kinase 5 (MAP3K5)	U67156	+		+		
Inhibitor of DNA binding 4, dominant negative helix-loop-helix protein (ID4)	AL022726	+		+		
Complement component 1, r subcomponent (CIR)	M14058					
Down-regulated genes						
Olfactomedin-related ER localized protein	U79299	+	+	+	+	
Msh homeobox homolog 1 (MSX1)	M97676	+		+		
GATA binding protein 2 (GATA2)	M68891					

EP–MP ($n = 5$), MP ($n = 5$), LP–ES ($n = 5$), ES–MS ($n = 6$), MS–LS ($n = 7$), LS–M ($n = 5$), M ($n = 4$) (Ponnampalam et al., 2004).

It is hoped that the data will not only contribute to the understanding of the endometrial biology but also improve the accuracy and reproducibility of endometrial dating procedures.

Endometrial Stromal Cell Differentiation

Placental trophoblasts invade into the maternal endometrial stromal compartment during implantation in humans. Decidualization of stromal cells is essential for successful implantation and represents differentiation to an exceptional morphologic, biosynthetic, and secretory phenotype (Irwin, and Giudice 1999). It is initiated independently of conception during the LS phase of the normal menstrual cycle. The molecular pathways that are involved in the process of decidualization are very poorly understood. Decidualization in vitro can be induced by number of factors including progesterone primed with estrogen and cyclic adenosine monophosphate (cAMP). Few transcriptional profiling studies have been undertaken to shed light in the understanding of the decidualization process.

In the study by Popovici et al. (2000), human endometrial stromal cells were decidualized in vitro in response to progesterone or cAMP and the gene expression profile was compared to nondecidualized stromal cells. Of the 588 genes analyzed, significant upregulation of cytokines, growth factors, nuclear transcription factors, cyclin family members, and mediators of the cAMP signal transduction pathways were found. Many of these genes have been implicated in decidualization in vivo. The group identified new members of transforming growth factor-β (TGF-β) family

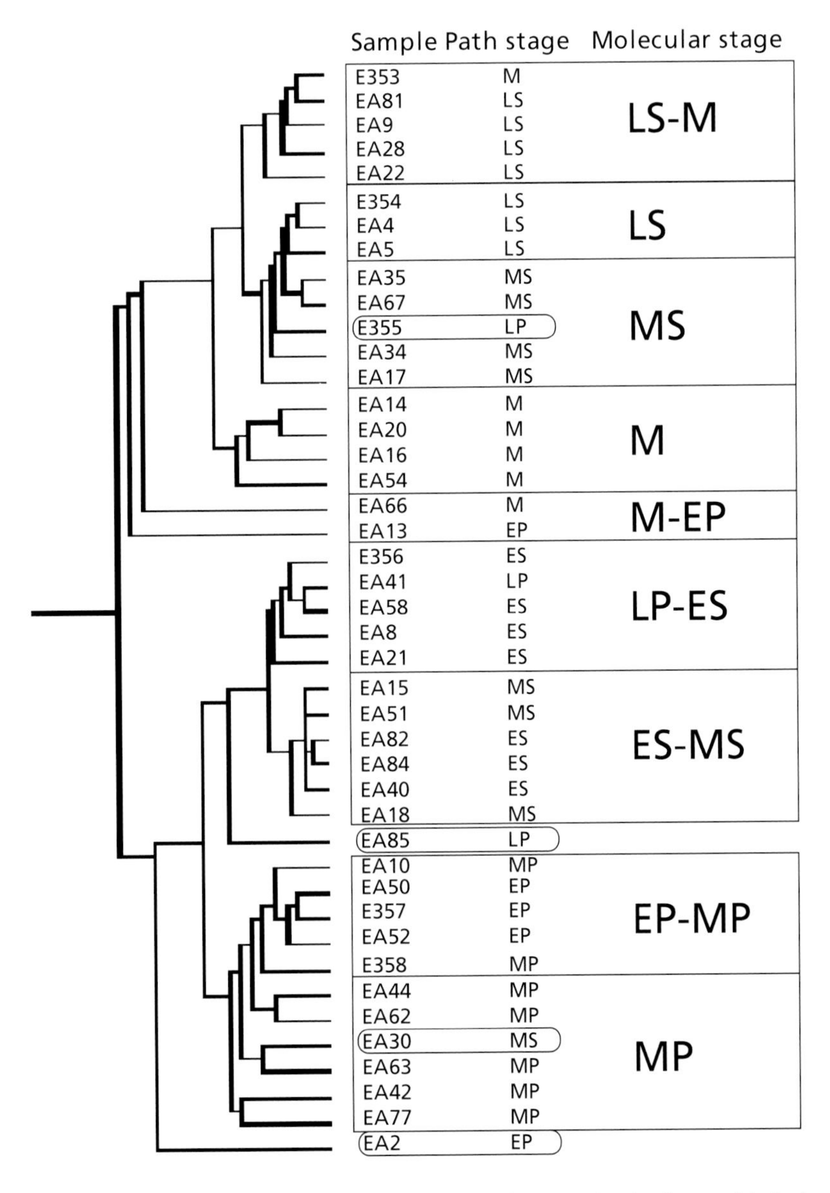

Figure 36–1 Hierarchical clustering of 43 endometrial samples. Branch length is inversely proportional to degree of similarity. (*A*) Initial nine clusters of 43 samples. Outliers are circled. (*B*) Final molecular classification of the menstrual cycle based on 37 samples divided into seven groups (outliers and excluded samples are also shown, circled for reference). EP–MP, early-proliferative–midproliferative; ES–MS, early-secretory–midsecretory; LP–ES, late-proliferative–early-secretory; LS, late-secretory; LS–M, late-secretory–menstrual; M, menstrual; M–EP, menstrual–early-proliferative; MP, midproliferative; MS, midsecretory.

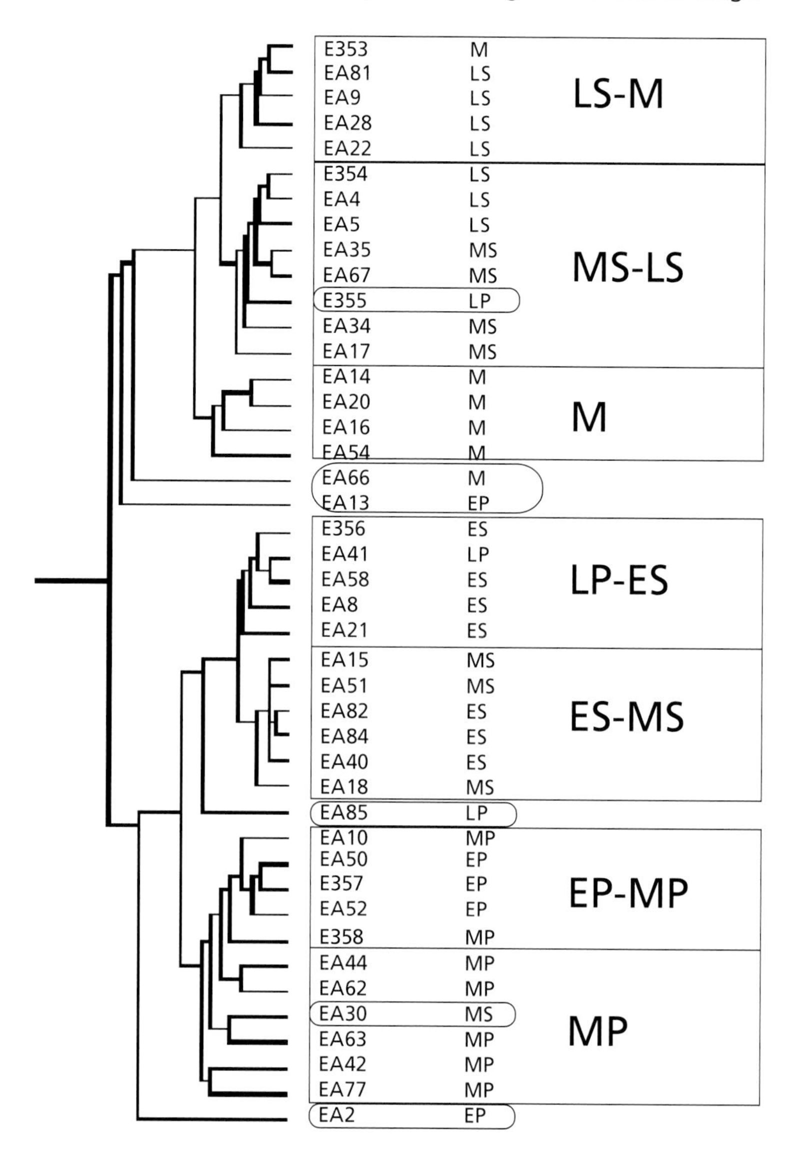

Figure 36–1 *(continued)*

up-regulated during decidualization and expression of neuromodulators and neurotransmitter receptors, many of which had not been previously documented in the differentiation process.

Temporal expression and regulation of nearly 7,000 genes during the decidualization process was investigated by two groups. Brar et al. (2001) examined the gene expression profile of human termed decidual fibroblasts in response to estrogen + progesterone + 8Br-cAMP at 0, 2, 4, 6, 9, 12, and 15 days. Of 6,918 genes analyzed, 121 genes were induced by more than two-fold, 110 were down-regulated, and 50 showed biphasic behavior. Dynamically regulated genes clustered into nine patterns of gene expression as a function of time. These genes were biologically classified into five catogories: cell and tissue function, cell and tissue structure, regulation of gene expression, expressed sequence tags, and functions unknown. Many genes were reprogrammed (i.e., changed expression) within specific functional groups. Tierney et al. (2002) investigated genes expressed by human endometrial stromal cells decidualized in vitro in response to 8Br-cAMP at 0, 2, 4, 6, 12, 24, and 48 hours. They also found reprogramming of genes and gene families during the decidualization process. Many up- and down-regulated genes were commonly found in both studies and they both demonstrated the reprogramming of genes during the temporal sequence of stromal differentiation, underscoring multiple processes occurring during differentiation of the stromal cell to decidual phenotype. They have also provided important insight into stromal functions in cycling endometrium and decidua.

Specific Disorders in the Reproductive Field

The fields of obstetrics, gynecology, and reproductive medicine cover a very broad range of disease conditions, as well as natural physiological processes such as pregnancy, parturition, and the menstrual cycle. Unique in this field of study is the need to consider the functioning of more than one organism: mother and fetus (in pregnancy) or mother and father (in fertilization and infertility).

Infertility

Fertility is affected by a wide range of conditions dealt with elsewhere in this chapter, such as endometriosis and polycystic ovarian syndrome (PCOS). Because fertility depends on function of three major components—the mother, the father, and the embryo—the genomic/microarray studies in this area of interest have adopted a wide range of experimental approaches to different aspects of the problem.

A key problem limiting success during in vitro fertilization (IVF) is diminished reserve of ovarian follicles, with increasing maternal age being the greatest factor influencing success. Unfortunately, there are no reliable predictors of ovarian reserve to inform patients of the chances of success with treatment—information which could help individuals make decisions regarding participation in often expensive treatment. Mechanisms of oocyte ageing have been investigated in a mouse model by one group (Hamatani et al., 2004), by using oligonucleotide microarrays to compare gene expression in oocytes collected from the ovaries of "young" mice (5–6-week-old) to that of oocytes from "old" mice (42–45-week-old). They found a number of genes with altered expression in ageing oocytes, including genes involved in genome stability, oxidative stress, and mitochondrial function. Luteinized granulosa cells, collected from human subjects during oocyte retrieval for IVF, have been screened by gene expression microarrays for potential markers of poor ovarian reserve (Chin et al., 2002). By comparing patterns of gene expression from women with normal and reduced ovarian reserve, the investigators found preliminary evidence of a molecular expression pattern that may predict for poor ovarian reserve.

It has been shown that up to 61% of first trimester miscarriages after IVF have chromosomal abnormalities detectable by comparative genomic hybridization (CGH) (Tan et al., 2003). CGH, and array CGH, are laboratory techniques allowing scanning of whole genomes for structural changes, such as either missing or extra pieces of chromosomes present in a cell or tissue. Genomic screening of single cells taken from embryos before transfer, with tools such as CGH, is a possible way of reducing miscarriage rates. The first live healthy birth after use of such a technique was reported in 2001 from Melbourne, Australia (Wilton et al., 2001). The same group has reported successful use of CGH in a series of 20 women with recurrent implantation failure (Wilton et al., 2003). They reported higher implantation and pregnancy rates with CGH, which examines the structure of the entire genome of the embryo, compared with conventional fluorescent in situ hybridization (FISH), which only verifies the numbers of five chromosomes. The feasibility of using array CGH on single cells taken from embryos has been proven in principle, with reported detection of trisomy 13 and 15 with amplified deoxyribonucleic acid (DNA) from a single cell (Hu et al., 2004). The cumulus cells surrounding oocytes in culture have been examined for gene expression changes between oocytes that successfully fertilize and those that do not (Zhang et al., 2005). In this study, 160 genes had altered expression in cumulus cells of oocytes that failed to fertilize, pointing to possible markers of embryo quality which could be tested objectively by sampling of cumulus cells without harming the developing embryo.

Some investigators are employing microarrays to study male factors infertility. Gene expression profiles from ejaculated spermatozoa of normal fertile men have been characterized (Ostermeier et al., 2002). They were found to correlate with gene expression profiles from pooled normal testicular tissue, showing that it may be possible to investigate genetic control of normal and abnormal spermatogenesis in the testis from analysis of readily available sperm samples. Having a "normal" gene expression profile is important as a baseline to compare with "diseased" (infertile or subfertile) samples, to identify differences in gene activity, and identify individual genes with dysregulated activity in infertile males. In an attempt to identify genes with a role in spermatogenesis, gene expression profiles of human adult and fetal testis have been compared, as well as spermatozoal mRNA with adult testis mRNA (Wang et al., 2004). Presumably, genes up-regulated in adult testis and spermatozoa in these experiments, respectively, are increasingly likely to have an important role in the spermatogenic process. Furthermore, the authors (Wang et al., 2004) compared mRNA profiles in ejaculated spermatozoa with normal and impaired mobility, finding that two motility-related genes (*TPX-1*, *LDHC*) could distinguish between the specimens. Although they used quantitative real-time-polymerase chain reaction (RT-PCR), rather than microarrays, it raises exciting possibilities of molecular profiling of ejaculates providing useful information about sperm malfunction. Testicular biopsies have been investigated by microarrays, and have been shown to be able to successfully distinguish between normal testicular tissue, and testes with absent germ cells (Sertoli cell-only syndrome) (Fox et al., 2003). Whether molecular profiling is capable of detecting more subtle testicular and spermatogenic defects remain to be seen.

Endometriosis

Endometriosis is defined as the presence of endometrial tissue in ectopic locations outside the uterine cavity, typically on the pelvic peritoneal surfaces of the uterus and its ligaments, on the ovaries, and in the rectovaginal septum (Giudice, and Kao, 2004). Between 5% and 10% of women have endometriosis, which causes a range of symptoms from pain to infertility.

Genes That are Abnormally Expressed in Eutopic Endometrium From Women with Endometriosis Compared to Women Without Endometriosis

The uterine (eutopic) endometrium from women with endometriosis may have endogenous abnormalities that promote the establishment of the disease. Matrix metalloproteinases (MMPs) such as MMP7 and MMP9 are normally expressed in endometrium during menstruation and are inhibited by progesterone during the secretory phase of the cycle. However, in women with endometriosis, these enzymes are constantly expressed during the secretory phase. This may facilitate retrograde endometrial tissue to occupy into the peritoneal surface. Other studies also found that eutopic endometrium from women with endometrium expressed higher levels of MMP2, membranous type 1 MMP, MMP3, and urokinase-type plasminogen activator (uPA) and lower levels of TIMP-2 than endometrium from normal women (D'Hooghe et al., 2004).

Up to 50% of women with infertility have endometriosis. Endometriosis-related infertility may be because of toxicity to sperm and embryos by peritoneal fluid in women with endometriosis and implantation failure because of abnormal expression of several genes (that either regulate or inhibit implantation) in women with endometriosis compared to normal women. Eutopic endometrium from women with endometriosis may be histologically normal but biochemically abnormal (Lessey et al., 1994). During the window of implantation, the following genes are aberrantly expressed in women with endometriosis compared to controls: α-v, β-3 integrin, MMPs (MMP2, MMP3, MMP7, and MMP11), transcription factors such as hepatocyte nuclear factor, endometrial bleeding factor, enzymes involved in steroid hormone metabolism (aromatase, 17b-hydroxysteroid dehydrogenase), leukemia inhibitory factor (LIF), HOX genes, and progesterone receptor isoforms (D'Hooghe et al., 2004).

Kao et al. (2003) used Affymetrix microarrays to compare gene profiling in endometrial biopsies obtained during the window of implantation (6–8 days after LH search) between women with and without endometriosis. Of the 12,686 genes analyzed, 91 were significantly increased more than two-fold in their expression and 115 were decreased more than two-fold. The data associate some genes in disease pathogenensis, such as aromatase and progesterone receptor, and the selective dysregulation of other genes involved in embryonic attachment and toxicity, immune function, and apoptotic responses in a reduced likelihood of implantation. The group also identified two groups of genes which differ with respect to the implantion window and the presence of disease.

Genes Abnormally Expressed in Ectopic Endometrium When Compared with Eutopic Endometrium from Women with Endometriosis

Ectopic endometrial tissue differs in many ways from eutopic endometrium, including clonality of origin, enzymic activity, and histologic and morphologic characteristics (Redwine, 2002). Endometrial lesions show evidence of differential expression of steroid hormone receptors, adhesion molecules and their receptors, lower apoptosis, and loss of heterozygosity. Few groups have used microarrays to compare the transcriptional profile of ectopic and eutopic endometrium.

Eyster et al. (2002) investigated the expression of 4,133 genes in ovarian endometriomas versus eutopic endometrium. They found up-regulation of β-actin, α_2-actin, vimentin, and several immune genes in ovarian endometriomas compared to eutopic endometrium. Arimoto et al. (2003) using a similar approach reported 15 up-regulated and 337 down-regulated genes in ovarian endometriomas compared to eutopic endometrium during both proliferative and secretory phases of the menstrual cycle. However, whether genes up-regulated in ovarian endometriomas are the same as for pelvic peritoneal endometriosis (the most common type of endometriosis) remains to be determined.

Two major problems in sampling ectopic endometriotic tissue from the pelvic peritoneum for microarray analysis are, first, that the tissue samples may be surrounded with contaminating "normal" tissue from the surrounding ovary or peritoneal tissue, and second, that the amounts of tissue collected will often be small. In addition, because electrodiathermy is frequently used in laparoscopic "keyhole" surgery for endometriosis, the RNA may be damaged by the excisional surgery. A solution to some of these problems may be with the use of laser capture microdissection to obtain a pure population of endometriosis cells, as has been used in one study (Matsuzaki et al., 2004). It is essential to have a good relationship between scientist and surgeon to obtain optimal samples for RNA extraction (e.g., by avoiding use of diathermy until after sample has been removed).

Knowledge of gene expression in normal endometrium and how they are changing during the menstrual cycle is a prerequisite to an understanding of gene expression data in endometrial disorders. Studying gene expression of normal endometrium via high-throughput approach of DNA microarrays also allows for the identification of new signature genes or biomarkers of normal physiological functions of endometrium such as implantation and menstruation. It is likely that some biomarker genes may be useful in molecular or immunohistochemical diagnostics or as molecular targets for drug discovery and therapeutic intervention.

One of the most promising and clinically relevant applications of global transcript analysis with microarray is the gene expression profiling of endometrial disorders to establish a genome-based diagnosis. That promise has not been delivered yet, although remarkable progress has been made. The data generated with microarrays certainly contribute to our understanding of endometrial disorders.

Polycystic Ovarian Syndrome

PCOS is the most common endocrine condition afflicting women, occurring in 5% of the population (Balen, 2003). It is characterized by enlarged ovaries with multiple small follicles, anovulation, and an androgen-secreting stroma, presumably from the theca cells of the follicles. One of the problems with studies of this condition is obtaining tissue—women with PCOS are unlikely to consent to removal of ovarian tissue, and there are no suitable animal models for the condition. However, genomic studies are beginning to be used to improve our understanding of the etiology of this condition.

Both whole ovarian tissue (Jansen et al., 2004) and isolated theca cells (Wood et al., 2003) from normal and PCOS individuals have had gene expression profiles compared. Wood et al. (2003) isolated and cultured theca cells from four normal and five PCOS women. They found that, even after four passages in culture, the PCOS theca cells showed an increase in gene transcription for aldehyde dehydrogenase 6 and retinol dehydrogenase 2. Both of these changes may contribute to the excess circulating androgens seen in PCOS. Jansen et al. (2004) found differences in gene expression between normal and PCOS whole ovarian tissue in Wnt signaling, extracellular matrix components, and immunological factors.

Uterine Fibroids

Fibroids are the most common solid tumor to afflict women in their reproductive years, derived of uterine smooth muscle (Healy et al., 2003). Both estrogen and progesterone appear to promote their growth. They are the most frequent indication for hysterectomy in Australia and the United States (Treloar et al., 1999; Farquhar et al., 2002), because of the gynecological problems of menorrhagia and abdominal discomfort.

An increasing number of gene expression microarray studies have been published concerning fibroids. The condition is common and treatment often involves myomectomy (surgical excision of the fibroid) or hysterectomy, meaning that tissue collection is simple and the tumors are usually large, yielding ample tissue for mRNA samples. Thus, it is not surprising that there are more microarray studies published on fibroids than on PCOS, or ectopic endometriosis tissue, where obtaining samples is more problematic.

There have been a number of genomic studies comparing gene expression of fibroid with adjacent normal myometrium as a control. Tsibiris et al. (2002) used nine paired hysterectomy specimens (myometrial and fibroid tissue) and identified 145 genes differentially expressed based on fold changes in gene expression. In a similar study, although using differential display rather than microarrays, it has been reported that early growth response-1 (EGR-1) is down-regulated in fibroids (Pambuccian et al., 2002). Overexpression of neuron-specific protein *PEP-19* has also been reported in fibroid in a genomic study, although only one sample was used for the microarray, and only five samples for the confirmatory Northern blots (Kanamori et al., 2003).

In addition to comparing whole tissue gene expression differences between myometrium and fibroid, Luo, Ding, Xu, Williams, and Chegini (2005) used gonadotropin-releasing hormone agonist (GnRHa) to identify estrogen-responsive genes in fibroids. They used a combination of microarrays on whole tissue as well as on cultured smooth muscle cells from the fibroid and found 48 genes differentially expressed in common between the whole tissue and in vitro systems. Only a fraction of the genes with altered expression by GnRHa were common to both fibroid tissue and fibroid cultured cells, highlighting the importance of validating in vitro systems before using them as a model for disease in vivo. Oligonucleotide microarrays were used to investigate gene expression differences between whole tissue fibroid and myometrium from the same uterus in two further studies, with seven paired specimens used in one study (Wang et al., 2003) and eight paired specimens in the other (Hoffman et al., 2004).

While there are many concordant genes identified in these studies, there are sufficient differences in results to support the view that all microarray results should be subjected to independent verification by different laboratory techniques (Weston, 2004). For example, in a recent microarray study comparing gene expression in 10 matched pairs of fibroid and myometrium (Lee et al., 2005), 67 genes were identified with differential expression, only 18 of which (27%) had been detected in previous studies. Particularly given the lack of large numbers of replicates in some of the studies above as well as the biological variability displayed by uterine fibroids, reproducibility is a key issue.

In one of the larger sets of replicates of complementary DNA (cDNA) microarrays (24 microarrays) on fibroids, our group compared gene expression between fibroids and myometrium in a crossover experiment (Weston et al., 2003). Twelve fibroids were compared with pooled myometrial mRNA, and then 12 individual myometrial samples were compared with pooled fibroid as a reference. A "dye-swap" design (Sterrenburg et al., 2002) was employed to ensure that identified genes were because of true gene expression differences, and not because of different performance of the Cy5 and Cy3 fluorescent dyes. The main aim of the experiment was to investigate differences in angiogenesis-related genes, because fibroid tissue is relatively avascular when compared with myometrium, despite having a vessel-rich capsule surrounding them (Casey et al., 2000). It is a good example of how microarrays can quickly answer very specific questions about the tissue being profiled. Given that the microarray that we used was not solely designed to examine angiogenesis genes but was a general cancer array, we listed all of the genes on the array related to angiogenesis (see Table 36–2). Although our relatively stringent statistical analysis of the microarray results revealed 25 genes with differential expression between fibroid and myometrium, only 3 of these genes were from our angiogenesis-related genes list. Genes of interest showing differential expression were confirmed with quantitative RT-PCR, an essential part of the experiment given the relatively high false-positive rate for microarray studies. CYR61 (cysteine-rich angiogenic inducer 61) (Babic et al., 1998) and CTGF (connective tissue growth factor) (Babic et al., 1999) are both angiogenesis promoters which were down-regulated in fibroids compared with myometrium in our arrays. These two important angiogenesis promoters were also shown to be markedly down-regulated in fibroids in other recent fibroid microarray studies (Hoffman et al., 2004; Lee et al., 2005). Collagen 4 $\alpha 2$ (COL4A2) is a precursor for the angiogenesis inhibitor, canstatin (Kamphaus et al., 2000), and was up-regulated in fibroid compared with adjacent myometrium. This was also confirmed in another recent array experiment (Lee et al., 2005). The differential expression of this triad of genes was confirmed by quantitative RT-PCR (see Fig. 36–2). It is possible that the antiangiogenic gene expression profiles shown by these three genes may in part explain the reduced microvessel density observed in fibroids compared with myometrium.

Studies such as those described above are often labeled "fishing expeditions" by critics of microarrays, as they lack a traditional a priori hypothesis. The key to finding clinical applications for the microarray results will lie in defining the molecular pathways involved in fibroid growth and regression. It is encouraging to see genomic studies of fibroids taking this next step. Using cultured smooth muscle cells from fibroids and myometrium, the autocrine and paracrine action of TGF-β on gene expression has been analyzed recently by microarray (Luo, Ding, Xu, and Chegini, 2005). Fibroid and myometrial smooth muscle cells were found to respond differently to TGF-β treatment, possibly explaining the altered, excessive, extracellular matrix produced in fibroids compared with myometrium. Estrogen responses in cultured fibroid smooth muscle cells have been studied using microarrays, supporting a role for IGF-1 (insulin-like growth factor-1) in mediating estrogen-induced fibroid growth (Swartz et al., 2005). It is likely that many other genomic studies on hormone and growth factor responses in fibroid smooth muscle cells will be performed in the near future. If genetic pathways can be elucidated which differ between fibroid and myometrium, it may be possible to create medical treatments which selectively inhibit fibroid growth without interfering in myometrial function. This is particularly important in premenopausal women, whose families have not yet been completed.

Uterovaginal Prolapse

Affymetrix U95A gene expression microarrays have been used to investigate genes with a possible role in the development of uterovaginal prolapse (Visco and Yuan, 2003). Five pubococcygeus muscle specimens from the pelvic floor of women with major uterovaginal prolapse were compared with five similar specimens from normal controls in a series of 10 microarrays. Of the genes with reduced expression in prolapse specimens, there were several structural genes related to actin and myosin, as well as extracellular matrix proteins, all with a biologically plausible role in the development of prolapse. Whether the gene expression changes are cause or effect of the prolapse needs to be established with further experiments.

Obstetrics

Two examples of uses of microarrays in the obstetric field are in the study of preeclampsia (hypertension of pregnancy) and screening for trisomy 21 (T21) (Down syndrome) in pregnancy.

Preeclampsia

This is a pregnancy-specific syndrome complicating approximately 5% of pregnancies and characterized by hypertension and proteinuria (Sibai et al., 2005). It is a significant obstetric problem, accounting for 15%–20% of maternal mortality in developed countries. The etiology is unknown, although placentas from women with preeclampsia display poor placentation and reduced perfusion. One of the two main theories of the etiology of preeclampsia is that ischemia because of a poorly perfused placenta leads to the release of circulating toxins to the vascular endothelium, which in turn leads to the clinical manifestations of the syndrome (Sibai et al., 2005). The other is that preeclampsia is a maternal immune reaction to the placenta.

There have been a number of microarray studies seeking to define gene expression differences in preeclamptic placentas compared with those from normal pregnancies. Reimer et al. (2002) pooled RNA from six normal and six preeclamptic placentas for comparison on a HuGeneFL microarray. They found 44 up-regulated and 15 down-regulated genes in preeclamptic placentas. In particular, they found increased expression of placental obese gene, as well as elevated plasma levels of its protein product, leptin, in preeclamptic patients. In another study employing a cDNA microarray specific for cytokines and interleukins, it was found that 162 of the

Table 36–2 Angiogenesis Genes Represented on the Microarray

Angiogenesis Promoters		Angiogenesis Inhibitors					
Accession Number	Gene Name	Accession Number	Gene Name	Accession Number	Gene Name	Accession Number	Gene Name
x89576	*MMP17*	aw044503	angiopoietin 1	r50407	GRO 2	aa460968	Interferon (IFN)-inducible protein kinase
x83535	*MMP14*	ai823651	TGF α	n81029	Collagen, type XVIII, α 1	aa459289	Laminin α 5
w51794	*MMP3*	ai762428	FGF13	m59906	TIMP1	aa457042	IFN γ-inducible protein p78
w51760	*bFGF*	ai630825	FGF14	h95960	Osteonectin	aa456022	Fibronectin transmembrane protein 3
w46900	*GRO 1*	ai439851	prostaglandin F receptor	ai803560	IL-12a	aa454813	IFN-related developmental factor 2
u14750	*CTGF*	ai292342	FGF rec 1 oncogene partner	ai391632	IFN binding protein	aa449440	IFN γ rec 2
t72581	*MMP9*	ai278780	angiopoietin-like factor	ai338894	Fibronectin transmembrane protein 2	aa448478	IFN γ-inducible protein
t49540	*PDGF-B*	ai268937	MCP2	ai278780	Angiopoietin-like factor	aa446251	Laminin β 1
t49539	*PDGF-B*	aa972293	PG E rec 1	aa969504	IFN γ	aa443090	IFN γ regulatory factor 7
r92994	*MMP12*	aa936799	MMP2	aa827287	IFN-induced protein 35	aa437226	IL-10 rec α
r56211	*PDGF rec B*	aa885619	MCP family protein	aa699697	TNF	aa432030	IFN γ-inducible protein
r54846	*FGF rec 1*	aa777187	CYR61	aa677534	Laminin γ 2	aa430540	Collagen, type IV, α 2
r52798	*HGF*	aa703094	FGF rec activating protein 1	aa676598	IFN-related developmental factor 1	aa406020	IFN-stimulated protein 15kD
r45059	*VEGF*	aa701860	Follistatin	aa630800	IFN γ-inducible protein IP-30	aa291389	IFN-stimulated tissue factor 38
r42600	*MMP17*	aa699697	TNF	aa504832	IFN γ-inducible protein 41	aa157813	IFN γ-inducible protein 27
r38539	*bFGF*	aa669222	MMP19	aa490996	IFN γ-inducible protein 16	aa156802	Laminin β 2
r36467	*TGF β*	aa630120	VEGF-B	aa490471	SPARC-like 1	aa152486	Platelet factor 4
r19956	*VEGF*	aa490300	PDGF-A associated protein 1	aa489640	IFN γ-inducible protein	aa131406	IFN γ-induced monokine
nm_006690	*MMP24*	aa487034	TGF β rec 2	aa486849	IFN γ-inducible guanylate binding protein 1	aa130238	Prolactin
nm_005941	*MMP16*	aa456160	FGF rec 2	aa486393	IL-10 rec β	aa125872	Angiopoietin 2

(Continued)

Table 36–2 Angiogenesis Genes Represented on the Microarray *(Continued)*

Angiogenesis Promoters			Angiogenesis Inhibitors				
nm_004771	*MMP20*	aa446223	TGF β 1-induced apoptotic factor 1	aa478043	IFN regulatory factor 1	aa058323	IFN-induced transmembrane protein 1
nm_002429	*MMP19*	aa436163	PG E synthase	aa464532	Thrombospondin 1	aa010079	IFN-inducible protein kinase
n69322	*MMP13*	aa431475	Transmembrane protein (EGF and follistatin-like domains)	aa464417	Fibronectin transmembrane protein 3	aa001432	Laminin α 3
n33214	*MMP14*	aa425102	MCP1				
m37484	*IGF1*	aa417274	Follistatin-like 3				
j03209	*MMP3*	aa406362	PG E rec 3				
h79306	*PDGF rec A*	aa256532	IGF1 rec				
h62473	*TGFβ rec 3*	aa256419	IGF1 rec				
h44861	*FGF12b*	aa125872	Angiopoietin 2				
h23235	*PDGF rec A*	aa102526	IL-8				
h19129	*FGF12b*	aa058828	VEGF				
h09997	*MMP16*	aa055440	FGF signalling antagonist				
h07899	*VEGF-C*	aa045731	TGF β early growth response				
d86331	*MMP15*	aa040170	MCP3				
aw072298	*MMP23b*	aa019996	PG E rec 4				

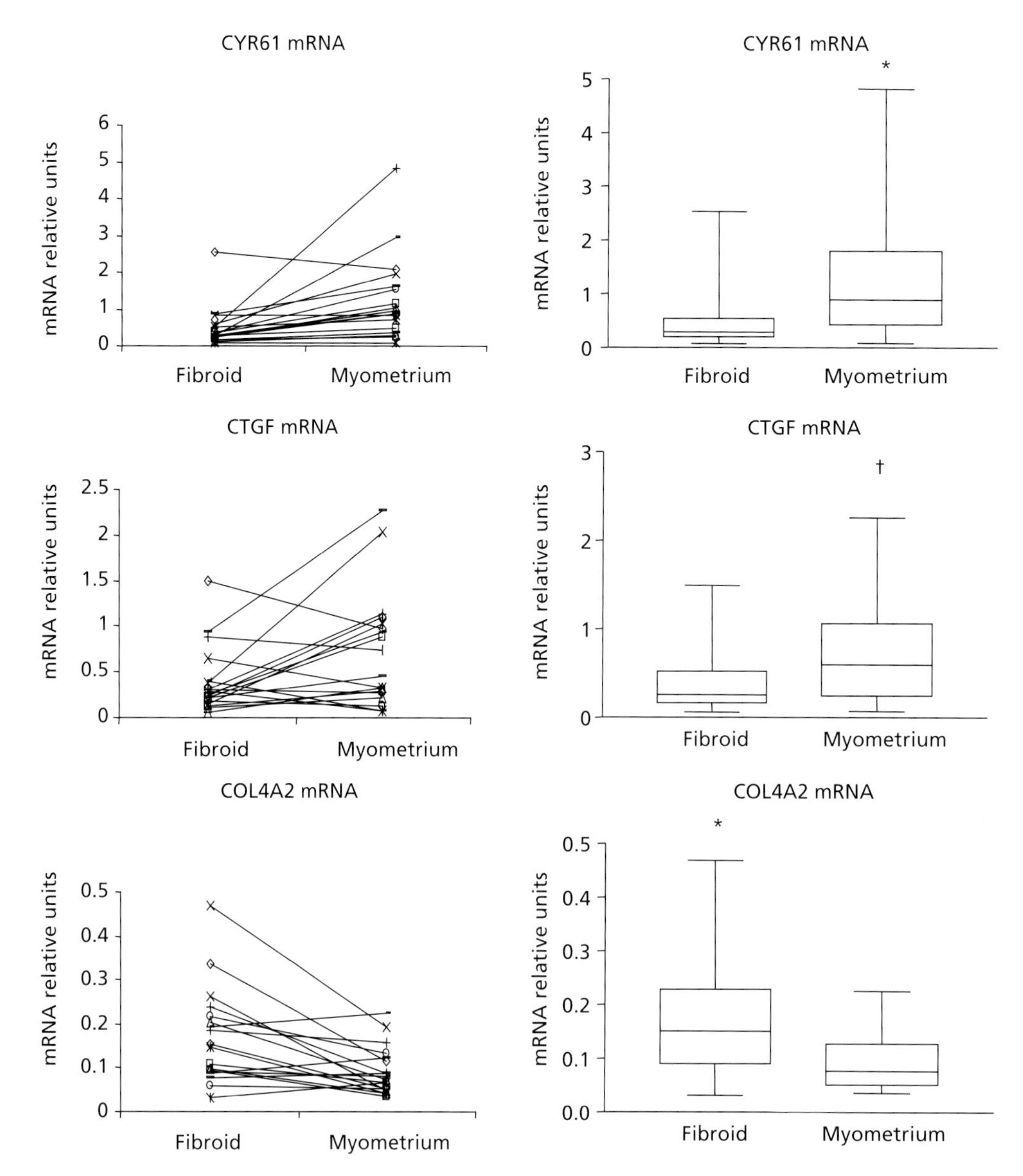

Figure 36–2 Quantitative RT-PCR results for fibroid- related angiogenesis genes identified by gene expression micro array.

221 genes on the array had different expression (defined as two-fold change) in preeclamptic compared with normal placental tissue (Pang and Xing, 2003). Comparing four normal and four preeclamptic placental specimens, using GeneFilter cDNA microarrays, Tsoi et al. (2003) found that placental glycogen phosphorylase-M mRNA was elevated in preeclamptic placentas, which may be a response to hypoxia. Reduced mRNA levels of human leukocyte antigen-G (HLA-G), which has immunomodulatory functions, has also been reported in preeclamptic placentas using gene expression microarray (Hviid et al., 2004).

Direct genomic evidence for hypoxia, presumably because of inadequate placental vascular development after implantation, was obtained in a recent microarray study (Soleymanlou et al., 2005). Gene expression profiles by microarray analysis were obtained from placentas from high-altitude pregnancies (a relatively hypoxic environment), as well as from placental tissue explants cultured with 20% (normal) and 3% (hypoxic) oxygen. These expression profiles were then compared with those from preeclamptic placentas. Similar expression profiles were obtained for high-altitude and preeclamptic placentas as for hypoxic cultured placenta, supporting the view that the preeclamptic placenta functions in a relatively hypoxic environment. In an attempt to define the systemic endothelial dysfunction hypothesized to be caused by circulating toxic factors released from the preeclamptic placenta, microarrays were performed on endothelial cell cultures incubated with serum from patients with severe preeclampsia (Donker et al., 2005). Despite examining three different time points (4, 12, and 24 hours), no significant gene expression changes were identified despite the array screening 17,000 genes. The authors concluded that endothelial dysfunction seen in preeclamptic patients may rely on mechanisms other than soluble toxic factors.

T21 Pregnancies

T21 (Down syndrome) is the most common cause of genetic mental retardation, occurring in 1 in 700 live births (Korf, 1991). Testing for Down syndrome is offered as a part of routine antenatal care for women in Australia during pregnancy. Unfortunately, noninvasive screening tests of maternal serum screening markers are insufficient to detect a Down syndrome pregnancy. The panel of serum tests used in antenatal screening for Down syndrome includes free β human chorionic gonadoptropin (βhCG), unconjugated estriol (uE3), pregnancy-associated plasma protein A (PAPP-A), and dimeric inhibin A (Neilson and Alfirevic, 2006). Recently, ADAM12, a novel serum marker with biological properties similar to PAPP-A, has been shown to improve detection rate (Laigard et al., 2003). Several studies have advocated improved performance of serum screening when combined with first trimester ultrasonographic measurement of nuchal translucency (Snijders et al., 1998). However, it is difficult to implement this as a population-based screening programme because of shortage of adequately trained sonographers and problems with quality assurance. Objective of the antenatal Down syndrome screening is selecting high-risk pregnancies among younger women under 35 years of age, who are then offered one of the two invasive tests for fetal chromosome analysis. The invasive tests required (amniocentesis or chorionic villous sampling of amniotic liquor) entail a miscarriage risk of 0.5%–1.0%.

Traditionally, fetal chromosome analysis involves cell culture and full karyotyping. This is slow and expensive, but assured a comprehensive analysis minimizing risks for failure to detect any other chromosomal abnormality. However, rapid introduction of molecular cytogenetic methodology led to increased use of commercially available fluorescent hybridization kits for selective cytogenetic analysis for chromosomal aneuploidies of 13, 18, 21 chromosomes. The technique has improved to employ PCR (quantitative fluorescent-PCR: qf-PCR). It is claimed that a selective, but fast and less expensive cytogenetic analysis using qf-PCR is more cost-effective when used in a carefully selected cohort of 17,500 women using a combined screening strategy of serum testing with increased nuchal translucency (>4 mm) (Chitty et al., 2006).

Microarray studies of T21 pregnancies include an attempt to identify new maternal serum biochemical markers for Down syndrome pregnancy by cDNA microarrays (Gross et al., 2002). Placentas from seven Down syndrome pregnancies were compared with seven normal matched controls. The study identified seven expressed sequence tags with increased gene expression in T21 placentas, two of which were confirmed by Northern blot analysis.

The possibility of faster detection of chromosomal aneuploidies has been proposed using CGH arrays. One group (Larrabee et al., 2004) has reported the use of cell-free fetal DNA from the amniotic fluid to form fluorescent probes for hybridization to CGH arrays. They found that this allowed rapid detection of abnormal chromosome number of both sex chromosomes and autosomes.

Gynecological Cancers

The three most common cancers of the female reproductive tract (ovarian, endometrial, and cervical cancer) have been extensively studied by microarrays. A brief review of the literature on each of them follows below.

Ovarian Cancer

The use of gene expression microarrays has now permeated almost all aspects of ovarian cancer research and has been the subject of recent reviews (Haviv, 2002; Bandera et al., 2003; Russo et al., 2003). With increasing access of the new technology to cancer researchers, this is not surprising. While basic research continues to unravel the molecular pathways of ovarian cancer development, there are more directly clinically applicable applications of microarray research in the field.

While ovarian cancers are uncommon, with a lifetime incidence of 1.4%, their vague symptomatology means that patients present with advanced disease in more than two-third of cases (Berek, 1994). The search for markers allowing early detection of ovarian cancer is important, as, if cancers are treated earlier, survival may be increased. Because the most common ovarian tumors originate in the ovarian epithelium, the outer layer of the ovary, a number of strategies have been employed to purify samples and obtain sufficient tissue for microarray analysis. Also, the most relevant markers are going to be obtained from early tumors, which are difficult to obtain because of the usual late clinical presentation. Options chosen by various groups have included short-term culture to expand the ovarian surface epithelium before RNA extraction (Ismail et al., 2000; Mok et al., 2001), and purification of cell populations through the use of glass or immunobeads (Ono et al., 2000; Welsh et al., 2001). Others have performed microarrays on whole tissue from normal and cancerous ovaries (Martoglio et al., 2000). Laser capture microdissection to isolate individual ovarian epithelial cells, followed by RNA amplification, is a further option, but is likely to introduce bias into the gene expression patterns (Wang et al., 2000). Although in vitro culture itself has the potential to alter gene expression and bias microarray results, one group has recently reported promising preliminary data in three serum markers obtained through arrays of cell cultures: prostatin (Mok, 2000), osteopontin (Kim et al., 2002), and creatine kinase B (Huddleston et al., 2005). Purification of cell populations with glass or immunobeads may bias results because of the lengthy process degrading the mRNA (Haviv and Campbell, 2002).

Ovarian cancer often shares a similar histopathology (adenocarcinoma, especially poorly differentiated) and pattern of spread (intraperitoneal) to a variety of other cancers, such as colon, stomach, and breast. This can make it problematic for the pathologist to establish the diagnosis of origin of the tumor. Because different cancers respond to different subsequent treatments and often require different chemotherapy regimes, cancers of unknown primary can have therapeutic implications for the patients. It has recently been shown that gene expression profiles of cancers with unknown primary, including ovarian cancers, can be successfully used to predict the correct site of origin (Tothill et al., 2005). This has exciting implications for helping to deal with a common clinical problem in the treatment of ovarian cancer.

Cervical Cancer

Radiotherapy is the primary treatment for cervical cancer once it has reached a size considered unresectable surgically, or has spread beyond the cervix, particularly to pelvic lymph nodes. Because one of the key factors in survival of these patients is the response of the tumor to radiotherapy, there have been several studies looking for genes whose expression determine radiosensitivity of the cervical tumor (Achary et al., 2000; Kitahara et al., 2002; Harima et al., 2003; Wong et al., 2003; Harima et al., 2004; Chung et al., 2005). Some studies have used cell lines (Achary et al., 2000; Chung et al., 2005), with the same potential biases of in vitro culture on mRNA expression as already outlined for ovarian cancer. In one of the cell culture studies (Chung et al., 2005), it was shown that intracellular adhesion molecule-3 (ICAM-3) expression was associated with colonies of cells displaying radiation-resistance. Other studies have used biopsies of cervical cancer tissue before radiotherapy, and then compared the gene expression profiles of the samples from patients who responded to radiotherapy to those who did not respond (Kitahara et al., 2002; Harima et al., 2003; Wong et al., 2003; Harima et al., 2004). These studies aim to find markers for response to radiotherapy, such as reduced levels of Ku80 protein, reported to indicate increased radiosensitivity (Harima et al., 2003) or to identify therapeutic targets to improve radiation response, and ultimately, patient survival.

It has been shown that genomic profiling by DNA microarray is possible with RNA obtained from both exfoliated cells from the cervix obtained at Pap (Papanicolaou) smear (Habis et al., 2004) and liquid cytology-based Pap smears (Hildebrandt et al., 2003). This opens up interesting possibilities for the use of genomic assessment of cervical cells, either precancerous dysplasia or cancerous cells. In preinvasive dysplastic lesions, this may be used to detect different serotypes of human papillomavirus (HPV), a major causative factor in the development of cervical cancer, with either a DNA microarray chip (Cho et al., 2003) or an HPV oligonucleotide microarray (Kim et al., 2003). Alternatively, it may be used to detect markers discovered in the future.

Endometrial Cancer

Endometrial carcinoma (EC) is the fourth most common cancer in women and the most common gynecologic malignancy (Jemal et al., 2004). Microarray technology is highly effective in screening for differential gene expression profiles. Hence, it has become a popular tool for the molecular profiling of cancer. The technology allows for two main types of descriptive analysis: first, the identification of genes which may be responsible for a clinicopathological feature or phenotype, and second, the genomic classification of tissue. The ultimate goal is to improve clinical outcome by adapting therapy based on the molecular characteristics of a tumor (Beer et al., 2002).

Using oligonucleotide arrays, Mutter et al. (2001) investigated genes that distinguish neoplastic transformation from normal human endometrium of both proliferative and secretory phase. They identified 50 genes which provided discrimination between normal and malignant endometria. The genes were predominantly characterized by diminished expression in cancers. One hundred genes that were hormonally regulated in normal endometrium were expressed in a disordered and heterogenous fashion in malignant endometria, with tumors resembling proliferative more than secretory endometrium in gene expression profiles. Thus, they concluded that neoplastic transformation is accompanied by predominant loss of activity of many genes constitutively expressed in normal endometrium and absence of expression profiles that characterize the antitumorigenic progestin response.

Moreno-Bueno et al. (2003) used cDNA microarrays containing 6386 different genes to investigate the differences in gene expression profiles between the two types of ECs: endometrioid endometrial carcinomas (EEC, type I), which are estrogen-related tumors, frequently euploid, and have good prognosis; and nonendometrioid endometrial carcinomas (NEEC, type 2), which are not estrogen related, are frequently aneuploid, and are clinically aggressive. Analysis of gene expression profiles in 24 EECs and 11 NEECs revealed at least two-fold difference in 66 genes between the two types. The 31 genes that were up-regulated in EECs compared to NEECs, include genes that are known to be hormonally regulated during the cycle and to be important in endometrial homeostasis. It supports the notion that EEC is a hormone-related neoplasm. Conversely, of the 35 genes overexpressed in NEECs, 3 genes, *STK15*, *BUB1*, and *CCNB2*, are involved in the regulation of the mitotic spindle checkpoint. STK15 overexpression is associated with aneuploidy and aggressive phenotype in other human tumors as well. The group independently confirmed the STK15 overexpression in ECs using FISH. Risinger et al. (2003) also investigated gene expression profiles among the two types of endometrial cancer. They identified 75 genes that were differentially expressed. RT-PCR of six genes validated the microarray findings. In addition, class prediction models reproducibly defined 30 out of 32 cancers as either type 1 or type 2.

Ferguson et al. (2004) used Human Genome U133A gene chip from Affymetrix to test the hypothesis that tamoxifen-associated ECs have a distinct gene expression profiling compared to matched cases not associated with this exposure. While their supervised class comparison revealed no statistically significant difference between the two groups, unsupervised hierarchical clustering of the entire sample of tumors revealed two groups with extremely diverse molecular profiles that were independent of tamoxifen exposure. Histological grade of the sample correlated best with these two molecular classes.

Summary

The use of microarrays has permeated most areas of study within the reproductive, obstetrics, and gynecology fields. The number of diseases and conditions within this broad field currently undergoing genomic studies is too large to be summarized in total within the context of this chapter. The rate at which microarray studies are being published makes it certain that this chapter will be out of date before it can be put into print. Nonetheless, we have attempted to give a sense of the way in which genomic studies are helping to advance knowledge in our field.

References

Achary, MP, Jaggernauth, W, Gross, E, Alfieri, AH, Klinger, HP, et al. (2000). Cell lines from the same cervical carcinoma but with different radiosensitivities exhibit different cDNA microarray patterns of gene expression. *Cytogenet Cell Genet*, 91:39–43.

Arimoto, T, Katagiri, T, Oda, K, Tsunoda, T, and Yasugi, T (2003). Genome-wide cDNA microarray analysis of gene-expression profiles involved in ovarian endometriosis. *Int J Oncol*, 22(3):551–560.

Babic, AM, Chen, CC, and Lau LF (1999). Fisp12/mouse connective tissue growth factor mediates endothelial cell adhesion and migration through integrin alphavbeta3, promotes endothelial cell survival, and induces angiogenesis in vivo. *Mol Cell Biol*, 19(4):2958–2966.

Babic, AM, Kireeva, ML, Kolesnikova, TV, and Lau LF (1998). CYR61, a product of a growth factor-inducible immediate early gene, promotes angiogenesis and tumor growth. *Proc Natl Acad Sci USA*, 95 (11):6355–6360.

Balen, A (2003). The polycystic ovarian syndrome. In RW Shaw, WP Soutter, SL Stanton (eds), *Gynaecology*, London, UK: Churchill-Livingstone, pp. 259–270.

Bandera, CA, Ye, B, and Mok, SC (2003). New technologies for the identification of markers for early detection of ovarian cancer. *Curr Opin Obstet Gynecol*, 15:51–55.

Beer, DG, Kardia, SL, Huang, CC, Giordano, TJ, Levin, AM, et al. (2002). Gene-expression profiles predict survival of patients with lung adenocarcinoma. *Nat Med*, 8(8):816–824.

Berek, JS (1994). Epithelial ovarian cancer. In JS Berek, NF Hacker (eds), *Practical gynecological oncology*, Baltimore, USA: Williams and Wilkins, pp. 327–375.

Bork, P and Copley, R (2001). Filling in the gaps. *Nature*, 409:818–820.

Borthwick, JM, Charnock-Jones, DS, Tom, BD, Hull, ML, Teirney, R, et al. (2003). Determination of the transcript profile of human endometrium. *Mol Hum Reprod*, 9:19–33.

Brar, AK, Handwerger, S, Kessler, CA, and Aronow, BJ (2001). Gene induction and categorical reprogramming during in vitro human endometrial fibroblast decidualization. *Physiol Genomics*, 7(2):135–148.

Carson, DD, Lagow, E, Thathiah, A, Al-Shami, R, Farach-Carson, MC, et al. (2002). Changes in gene expression during the early to mid-luteal (receptive phase) transition in human endometrium detected by high-density microarray screening. *Mol Hum Reprod*, 8:871–879.

Casey, R, Rogers, PAW, and Vollenhoven, BJ (2000). An immunohistochemical analysis of fibroid vasculature. *Hum Reprod*, 15:1469–1475.

Chin, KV, Seifer, DB, Feng, B, Lin, Y, and Shih, WC (2002). DNA microarray analysis of the expression profiles of luteinized granulose cells as a function of ovarian reserve. *Fertil Steril*, 77(6):1214–1218.

Chitty, LS, Kagan, KO, Molina, FS, Waters, JJ, and Nicolaides, KH (2006). Fetal nuchal translucency scan and early diagnosis of chromosomal abnormalities by rapid aneuploidy screening: an observational study. *BMJ*, 332:452–454.

Cho, NH, An, HJ, Jeong, JK, Kang, S, Kim, JW, et al. (2003). Genotyping of 22 human papillomavirus types by DNA chip in Korean women: comparison with cytological diagnosis. *Am J Obstet Gynecol*, 188:56–62.

Chung, YM, Kim, BG, Park, CS, Huh, SJ, Kim, J, et al. (2005). Increased expression of ICAM-3 is associated with radiation resistance in cervical cancer. *Int J Cancer*, 117:194–201.

D'Hooghe, TM, Kyama, C, Debrock, S, Meuleman, C, and Mwenda, JM (2004). Future directions in endometriosis research. *Ann N Y Acad Sci*, 1034:316–325.

Donker, RB, Asgeirsdottir, SA, Gerbens, F, van Pampus, MG, Kallenberg, CG, et al. (2005). Plasma factors in severe early-onset preeclampsia do not substantially alter endothelial gene expression in vitro. *J Soc Gynecol Investig*, 12:98–106.

Eyster, KM, Boles, AL, Brannian, JD, and Hansen, KA (2002). DNA microarray analysis of gene expression markers of endometriosis. *Fertil Steril*, 77 (1):38–42.

Farquhar, CM and Steiner, CA (2002). Hysterectomy rates in the United States 1990–1997. *Obstet Gynecol*, 99:229–234.

Ferguson, SE, Olshen, AB, Viale, A, Awtrey, CS, Barakat, RR, et al. (2004). Gene expression profiling of tamoxifen-associated uterine cancers: evidence for two molecular classes of endometrial carcinoma. *Gynecol Oncol*, 92 (2):719–725.

Fox, MS, Ares, VX, Turek, PJ, Haqq, C, and Pera, RAR (2003). Feasibility of global gene expression analysis in testicular biopsies from infertile men. *Mol Reprod Dev*, 66:403–421.

Giudice, LC (2003). Elucidating endometrial function in the post-genomic era. *Hum Reprod Update*, 9(3):223–235.

Giudice, LC and Kao, LC (2004). Endometriosis. *Lancet*, 364:1789–1799.

Gross, SJ, Ferreira, JC, Morrow, B, Dar, P, Funke, B, et al. (2002). Gene expression profile of trisomy 21 placentas: a potential approach for designing noninvasive techniques of prenatal diagnosis. *Am J Obstet Gynecol*, 187:457–462.

Habis, AH, Vernon, SD, Lee, DR, Verma, ME, and Unger, R (2004). Molecular quality of exfoliated cervical cells: implications for molecular epidemiology and biomarker discovery. *Cancer Epidemiol Biomarkers Prev*, 13:492–496.

Hamatani, T, Falco, G, Carter, MG, Akutsu, H, and Stagg, CA (2004). Age-associated alteration of gene expression patterns in mouse oocytes. *Hum Mol Genet*, 13(19):2263–2278.

Harima, Y, Togashi, A, Horikoshi, K, Imamura, M, Sougawa, M, et al. (2004). Prediction of outcome of advanced cervical cancer to thermoradiotherapy according to expression profiles of 35 genes selected by cDNA microarray analysis. *Int J Radiation Oncology Biol Phys*, 60:237–248.

Harima, Y, Sawada, S, Miyazaki, Y, Kin, K, Ishihara, H, et al. (2003). Expression of Ku80 in cervical cancer correlates with response to radiotherapy and survival. *Am J Clin Oncol*, 26:e80–e85.

Haviv, I and Campbell, IG (2002). DNA microarrays for assessing ovarian cancer gene expression. *Mol Cell Endocrinol*, 191:121–126.

Healy, DL, Vollenhoven, B, and Weston, G (2003). Uterine fibroids. In RW Shaw, WP Soutter, SL Stanton (eds), *Gynaecology*, London, UK: Churchill-Livingstone, pp. 477–492.

Hildebrandt, EH, Lee, JR, Crosby, JH, Ferris, DG, and Anderson, MG (2003). Liquid-based Pap smears as a source of RNA for gene expression analysis. *Appl Immunohistochem Mol Morphol*, 11:345–351.

Hoffman, PJ, Milliken, DB, Gregg, LC, Davis, RR, and Gregg, JP (2004). Molecular characterization of uterine fibroids and its implication for understanding mechanisms of pathogenesis. *Fertil Steril*, 82 (3):639–649.

Hu, DG, Webb, G, and Hussey, N (2004). Aneuploidy detection in single cells using DNA array-based comparative genomic hybridization. *Mol Hum Reprod*, 10:283–289.

Huddleston, HG, Wong, K, Welch, WR, Berkowitz, RS, and Mok, SC (2005). Clinical applications of microarray technology: creatine kinase B is an up-regulated gene in epithelial ovarian cancer and shows promise as a serum marker. *Gynecol Oncol*, 96:77–83.

Hviid, TVF, Larsen, LG, Hoegh, AM, and Bzorek, M (2004). HLA-G expression in placenta in relation to HLA-G genotype and polymorphisms. *Am J Reprod Immunol*, 52:212–217.

Irwin, JC and Giudice, LC (1999). Decidua. In E Knobil, J Niall (eds), *Encyclopedia of reproduction*. New York, USA: Academic Press.

Ismail, RS, Baldwin, RL, Fang, J, Browning, D, Karlan, BY, et al. (2000). Differential gene expression between normal and tumor-derived ovarian epithelial cells. *Cancer Res*, 60:6744–6749.

Jansen, E, Laven, JSE, Dommerholt, HBR, Polman, J, van Rijt, C, et al. (2004). Abnormal gene expression profiles in human ovaries from polycystic ovary syndrome patients. *Mol Endocrinol*, 18(12):3050–3063.

Jemal, A, Tiwari, RC, Murray, T, Ghafoor, A, Samuels, A, et al. (2004). Cancer statistics, 2004. *CA Cancer J Clin*, 54(1):8–29.

Kamphaus, GD, Colorado, PC, Panka, DJ, Hopfer, H, Ramchandran, R, et al. (2000). Canstatin, a novel matrix-derived inhibitor of angiogenesis and tumor growth. *J Biol Chem*, 275(2):1209–1215.

Kanamori, T, Takakura, K, Mandai, M, Kariya, M, Fukuhara, K, et al. (2003). PEP-19 overexpression in human uterine leiomyoma. *Mol Hum Reprod*, 9:709–717.

Kao, LC, Germeyer, A, Tulac, S, Lobo, S, Yang, JP, et al. (2003). Expression profiling of endometrium from women with endometriosis reveals candidate genes for disease-based implantation failure and infertility. *Endocrinology*, 144(7):2870–2881.

Kao, LC, Tulac, S, Lobo, S, Imani, B, Yang, JP, et al. (2002). Global gene profiling in human endometrium during the window of implantation. *Endocrinology*, 143:2119–2138.

Kim, CJ, Jeong, JK, Park, M, Park, TS, Park, TC, et al. (2003). HPV oligonucleotide microarray-based detection of HPV genotypes in cervical neoplastic lesions. *Gynecol Oncol*, 89:210–217.

Kim, J, Skates, SJ, Uede, T, Wong, K, Schorge, JO, et al (2002). Osteopontin as a potential diagnostic biomarker for ovarian cancer. *JAMA*, 287:1671–1679.

Kitahara, O, Katagiri, T, Tsunoda, T, Harima, Y, and Nakamura, Y (2002). Classification of sensitivity or resistance of cervical cancers to ionizing radiation according to expression profiles of 62 genes selected by cDNA microarray analysis. *Neoplasia*, 4:295–303.

Korf, BR (1991). Chromosomal abnormalities. In HW Taeusch, RA Ballard, ME Avery, WB Saunders (eds), *Diseases of the Newborn*. WB, Saunders: Philadelphia, USA.

Laigard, J, Sorensen, T, Frohlich, C, Pedersen, BN, Christiansen, M, et al. (2003). ADAM12: a novel first-trimester maternal serum marker for Down syndrome. *Prenat Diagn*, 23:1086–1091.

Larrabee, PB, Johnson, KL, Pestova, E, Lucas, M, Wilber, K, et al. (2004). Microarray analysis of cell-free fetal DNA in amniotic fluid: a prenatal molecular karyotype. *Am J Hum Genet*, 75:485–491.

Lee, EJ, Kong, G, Lee, SH, Rho, SB, Park, CS, et al. (2005). Profiling of differentially expressed genes in human uterine leiomyomas. *Int J Gynecol Cancer*, 15:146–154.

Lessey, BA, Castelbaum, AJ, Sawin, SW, Buck, CA, Schinnar, R, et al. (1994). Aberrant integrin expression in the endometrium of women with endometriosis. *J Clin Endocrinol Metab*, 79(2):643–649.

Liotta, L and Petricoin, E (2000). Molecular profiling of human cancer. *Nat Rev Genet*, 1(1):48–56.

Luo, X, Ding, L, Xu, J, and Chegini, N (2005). Gene expression profiling of leiomyoma and myometrial smooth muscle cells in response to transforming growth factor-beta. *Endocrinology*, 146(3):1097–1118.

Luo, X, Ding, L, Xu, J, Williams, RS, and Chegini, N (2005). Leiomyoma and myometrial gene expression profiles and their responses to gonadotropin-releasing hormone analog therapy. *Endocrinology*, 146(3):1074–1096.

Martoglio, AM, Tom, BD, Starkey, M, Corps, AN, Charnock-Jones, DS, et al. (2000). Changes in tumorigenesis- and angiogenesis-related gene transcript abundance profiles in ovarian cancer detected by tailored high density cDNA arrays. *Mol Med*, 6:750–765.

Matsuzaki, S, Canis, M, Vaurs-Barriere, C, Pouly, JL, Boespflug-Tanguy, O, et al. (2004). DNA microarray analysis of gene expression profiles in deep endometriosis using laser capture microdissection. *Mol Hum Reprod*, 10:719–728.

Mirkin, S, Arslan, M, Churikov, D, Corica, A, Diaz, JI, et al. (2005). In search of candidate genes critically expressed in the human endometrium during the window of implantation. *Hum Reprod*, 20(8):2104–2117.

Mok, SC, Chao, J, Skates, S, Wong, K, Yiu, GK, Muto, MG, Berkowitz, RS, Cramer, DW et al. (2001). Prostasin, a potential serum marker for ovarian cancer: identification through microarray technology. *J Natl Cancer Inst*, 93:1458–1464.

Moreno-Bueno, G, Sanchez-Estevez, C, Cassia, R, Rodriguez-Perales, S, Diaz-Uriarte, R, et al. (2003). Differential gene expression profile in endometrioid and nonendometrioid endometrial carcinoma: STK15 is frequently overexpressed and amplified in nonendometrioid carcinomas. *Cancer Res*, 63(18):5697–5702.

Mutter, GL, Baak, JP, Fitzgerald, JT, Gray, R, Neuberg, D, et al. (2001). Global expression changes of constitutive and hormonally regulated genes during endometrial neoplastic transformation. *Gynecol Oncol*, 83(2):177–185.

Neilson, JP and Alfirevic, Z (2006). Optimising prenatal diagnosis of Down's syndrome. *BMJ*, 332:433–434.

Ono, K, Tanaka, T, Tsunoda, T, Kitahara, O, Kihara, C, et al. (2000). Identification by cDNA microarray of genes involved in ovarian carcinogenesis. *Cancer Res*, 60:5007–5011.

Ostermeier, GC, Dix, DJ, Miller, D, Khatri, P, and Krawetz, SA (2002). Spermatozoal RNA profiles of normal fertile men. *Lancet*, 360:772–777.

Pambuccian, CA, Oprea, GM, and Lakatua, DJ (2002). Reduced expression of early growth response-1 gene in leiomyoma as identified by mRNA differential display. *Gynecol Oncol*, 84:431–436.

Pang, PZ and Xing, FQ (2003). Comparative study on the expression of cytokine-receptor genes in normal and preeclamptic human placentas using DNA microarrays. *J Perinat Med*, 31(2):153–162.

Ponnampalam, AP, Weston, GC, Trajstman, AC, Susil, B, and Rogers, PA (2004). Molecular classification of human endometrial cycle stages by transcriptional profiling. *Mol Hum Reprod*, 10(12):879–893.

Popovici, RM, Kao, LC, and Giudice, LC (2000). Discovery of new inducible genes in in vitro decidualized human endometrial stromal cells using microarray technology. *Endocrinology*, 141(9):3510–3513.

Redwine, DB (2002). Was Sampson wrong? *Fertil Steril*, 78(4):686–693.

Reimer, T, Koczan, D, Gerber, B, Richter, D, Thiesen, HJ, et al. (2002). Microarray analysis of differentially expressed genes in placental tissue of pre-eclampsia: up-regulation of obesity-related genes. *Mol Hum Reprod*, 8(7):674–680.

Riesewijk, A, Martin, J, van Os, R, Horcajadas, JA, Polman, J, et al. (2003). Gene expression profiling of human endometrial receptivity on days LH + 2 versus LH + 7 by microarray technology. *Mol Hum Reprod*, 9:253–264.

Risinger, JL, Maxwell, GL, Chandramouli, GV, Jazaeri, A, Aprelikova, O, et al. (2003). Microarray analysis reveals distinct gene expression profiles among different histologic types of endometrial cancer. *Cancer Res*, 63(1):6–11.

Russo, G, Zegar, C, and Giordano, A (2003). Advantages and limitations of microarray technology in human cancer. *Oncogene*, 22:6497–6507.

Sibai, B, Dekker, G, and Kupferminc, M (2005). Pre-eclampsia. *Lancet*, 365:785–799.

Snijders, RJM, Noble, P, Sebire, N, Souka, A, and Nicolaides, KH (1998). UK multicentre project on assessment of risk of trisomy 21 by maeranl age and fetal nuchal translucency thickness at 10-14 weeks of gestation. *Lancet*, 351:343–346.

Soleymanlou, N, Jursica, I, Nevo, O, Ietta, F, Zhang, X, et al. (2005). Molecular evidence of placental hypoxia in preeclampsia. *J Clin Endocrinol Metab*, 90 (7):4299–4308.

Sterrenburg, E, Turk, R, Boer, JM, Van Ommen, GB, and Den Dunnen, JT (2002). A common reference for cDNA microarray hybridizations. *Nucleic Acids Res*, 30:e116.

Swartz, CD, Afshari, CA, Yu, L, Hall, KE, and Dixon, D (2005). Estrogen-induced changes in IGF-I, Myb family and MAP kinase pathway genes in human uterine leiomyoma and normal uterine smooth muscle cell lines. *Mol Hum Reprod*, 11(6):441–450.

Tan, YQ, Hu, L, Lin, G, Sham, JS, Gong, F, et al. (2003). Genetic changes in human fetuses from spontaneous abortion after in vitro fertilization detected by comparative genomic hybridization. *Biol Reprod*, 70:1495–1499.

Tierney, EP, Tulac, S, Huang, STJ, and Giudice, LC (2002). Sequential of the induction of gene expression during human endometrial stromal cell decidualization using microarray expression profile analysis. *Proceedings of the 84th Annual Meeting of the Endocrine Society*, San Francisco 537:P3-192 (Abstract).

Tothill, RW, Kowalczyk, A, Rischin, D, Bousioutas, A, Haviv, I, et al. (2005). An expression-based site of origin diagnostic method designed for clinical application to cancer of unknown origin. *Cancer Res*, 65:4031–4040.

Treloar, SA, Do, KA, O'Connor, VM, O'Connor, DT, Yeo, MA, et al. (1999). Predictors of hysterectomy: an Australian study. *Am J Obstet Gynecol*, 180:945–954.

Tsibiris, JC, Segars, J, Coppola, D, Mane, S, Wilbanks, GD, et al. (2002). Insights from gene arrays on the development and growth regulation of uterine leiomyomata. *Fertil Steril*, 78:114–121.

Tsoi, SC, Cale, JM, Bird, IM, and Kay, HH (2003). cDNA microarray analysis of gene expression profiles in human placenta: up-regulation of the transcript encoding muscle subunit of glycogen phosphorylase in preeclampsia. *J Soc Gynecol Investig*, 10:496–502.

Visco, AG and Yuan, L (2003). Differential gene expression in pubococcygeus muscle from patients with pelvic organ prolapse. *Am J Obstet Gynecol*, 189:102–112.

Wang E, Miller, LD, Ohnmacht, GA, Liu, ET, and Marincola, FM (2000). High-fidelity mRNA amplification for gene profiling. *Nat Biotechnol* 18:457–459.

Wang, H, Mahadevappa, M, Yamamoto, K, Wen, Y, Chen, B, et al. (2003). Distinctive proliferative phase differences in gene expression in human myometrium and leiomyomata. *Fertil Steril*, 80(2):266–276.

Wang, H, Zhou, Z, Xu, M, Li, J, Xiao, J, et al. (2004). A spermatogenesis-related gene expression profile in human spermatozoa and its potential clinical applications. *J Mol Med*, 82:317–324.

Welsh, JB, Zarrinkar, PP, Sapinoso, LM, Kern, SG, Behling, CA, et al. (2001). Analysis of gene expression profiles in normal and neoplastic ovarian tissue samples identifies candidate molecular markers of epithelial ovarian cancer. *Proc Natl Acad Sci USA*, 98:1176–1181.

Weston, G, Trajstman, AC, Gargett, CE, Manuelpillai, U, Vollenhoven, BJ, et al. (2003). Fibroids display an anti-angiogenic gene expression profile when compared with adjacent myometrium. *Mol Hum Reprod*, 9(9):541–549.

Weston, GC (2004). An introduction to genomics in reproduction. In S Daya, RF Harrison, RD Kempers (eds), *Advances in Fertility and Reproductive Medicine*. Amsterdam, The Netherlands: Elsevier, pp. 367–374.

Wilton, L, Voullaire, L, Sargeant, P, Williamson, R, and McBain, J (2003). Preimplantation aneuploidy screening using comparative genomic hybridization or fluorescence in situ hybridization of embryos from patients with recurrent implantation failure. *Fertil Steril*, 80:860–868.

Wilton, L, Williamson, R, McBain, J, Edgar, D, and Voullaire, L (2001). Birth of a healthy infant after preimplantation confirmation of euploidy by comparative genomic hybridization. *N Engl J Med*, 345:1537–1541.

Wong, WY, Selvanayagam, ZE, Wei, N, Porter, J, Vittal, R, et al. (2003). Expression genomics of cervical cancer: molecular classification and prediction of radiotherapy response by DNA microarray. *Clin Cancer Res*, 9:5486–5492.

Wood, JR, Nelson, VL, Ho, C, Jansen, E, Wang, CY, et al. (2003). The molecular phenotype of polycystic ovary syndrome (PCOS) theca cells and new candidate PCOS genes defined by microarray analysis. *J Biol Chem*, 278 (29):26380–26390.

Zhang, X, Jafari, N, Barnes, RB, Confino, E, Milad, M, et al. (2005). Studies of gene expression in human cumulus cells indicate pentraxin 3 as a possible marker for oocyte quality. *Fertil Steril*, 83(Suppl. 1):1169–1179.

37

Stem Cell Genomics and Regenerative Medicine

Phil Gaughwin, Wai-Leong Tam and Bing Lim

The goal of this chapter is to examine the relevance of genome biology in accelerating the application of stem cells in human therapy. We will first discuss how stem cells may be defined and consider their relevance to clinical medicine as transplantable cell types for tissue repair, as well as experimental models for addressing fundamental mechanisms of disease. The therapeutic manipulation of somatic stem cells, their tissue niches, the clinical and research potential of human embryonic stem cells (ESCs) and their animal model correlate, mouse ESCs, will be highlighted.

Evaluations of the genetic and epigenetic properties of somatic and embryonic stem cells have increased fundamental understanding of the molecular basis of stemness—the core biochemistry that underlies common stem cell properties. The application of genomics technologies, in conjunction with experimental platforms for specific cell isolation and characterization, have enabled researchers to understand the endogenous environment of somatic stem cells, isolate somatic stem cells, derive differentiated lineage from human ESCs, as well as characterize their transcriptional and epigenetic properties in detail (Fig. 37–1). These high-resolution platforms of analysis continue to refine tissue-type matching for allogenic cell transplantation procedures essential for the long-term survival of transplanted tissue and avoidance of graft-versus-host disease. The over-expression of cell type-specific genes has enabled the controlled differentiation of stem cells to derive appropriately specified cell types for transplantation or research. The evaluation of stem cell differentiation after transplantation has also indicated surprising avenues for differentiation and mechanisms of repair such as trans-differentiation and cell fusion, although the genomic reorganization that occurs during these events remains unclear (Box 37–1).

Box 37–1 Key Points

- The terms "cell-based therapy" and "regenerative medicine" have been coined to describe the transplantation or stimulation of cells to promote tissue reconstitution or repair in the body.
- Tissue-resident somatic (adult) stem cells and human ESCs are emerging as cells for cell-based therapy capable of direct or indirect tissue repair, as well as important models of disease progression.
- Phenotypic similarities of different stem cell types suggest a common molecular basis for the stem cell identity – also known as "stemness", but evidence for this remains elusive.
- The transplantation of allogenic tissue carries risks of immune rejection. Molecular profiling, stem cell banking, chimeric tolerance, somatic cell nuclear transfer (SCNT) and directed reprogramming of somatic cells represent potential, stem-cell-based solutions to immunogenicity and graft rejection.
- Dysregulation of the genetic and epigenetic control of stem cell self-renewal properties in cancer tissues has suggested the existence of cancer stem cells as the driver of the disease.
- Deciphering stem cell differentiation mechanisms will enable the targeted manipulation of stem cells in culture to derive specific cell types for transplantation or research.
- Stem cells may initiate tissue repair as well as cause disease through as yet undefined mechanisms, such as trans-differentiation or cell fusion.

What is a Stem Cell?

What are the intrinsic properties of stem cells that distinguish them from other differentiated cells? A universally acceptable definition of a stem cell has proven challenging as it differs depending on the biological context in which it is considered. During the embryonic development of an organism, the term "stem cell" may describe a transitory cell that is ancestral to one or more cell types or cell lineages. In the post-natal and adult organism, "stem cell" is used to describe a self-renewing cell phenotype that persists for the lifetime of an organism, and capable of generating differentiated progeny (Weissman, 2000b; Morrison, 2001; Seaberg and van der Kooy, 2003).

It is important to consider stem cells from an environmental perspective. The scientific definition of a stem cell has been established based on the functional properties observed within their biological context. Stem cells require external stimuli to persist in an undifferentiated state and their potential for differentiation is strongly influenced by changes in this extracellular environment. Stem cell properties are therefore attributed to sub-populations of tissue-resident cells that (1) self-renew and (2) give rise to

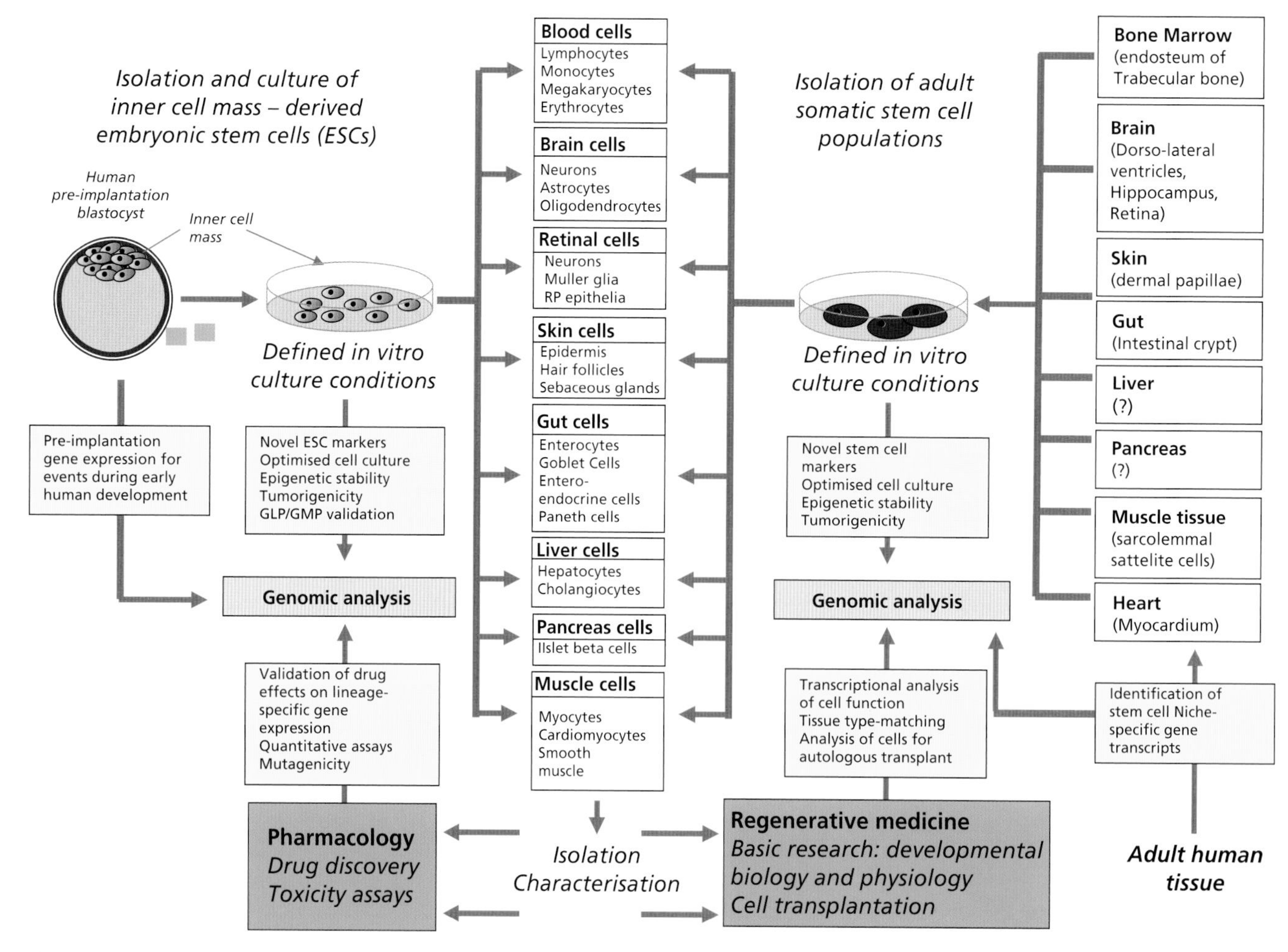

Figure 37–1 Genomic applications in the study of somatic stem cell and human ESC biology.

differentiated cells that integrate with and contribute to the tissue's normal functions (Spradling et al., 2001; Li and Xie, 2005). The identification of stem cells is sometimes reliant on patterns of gene expression that reliably correlate with a self-renewing cell population. The experimental validation of a putative stem cell phenotype is a challenging problem. A range of techniques have been developed to selectively isolate stem cell populations and examine their properties in experimental systems that test whether a single, candidate cell can regenerate itself, as well as establish a defined subset of differentiated cell types (Potten and Loeffler, 1990) (Box 37–2).

Somatic Stem Cells

The extensive degeneration and loss of certain tissues as a consequence of irradiation-induced failure, observed in the aftermath of the detonation of atomic bombs over Hiroshima and Nagasaki in 1945 highlighted the importance of radiation-sensitive, dividing cell populations in the maintenance of normal physiology. Stem cells residing in the differentiated parenchyma of organs in a post-natal animal are referred to as somatic stem cells. To distinguish them from tissue-resident stem cells present during embryonic or prenatal development, the term adult stem cell has also been applied to describe somatic stem cells. Somatic stem cells were first characterized in the attempt to understand the basis of regeneration of highly proliferative tissues affected by radiation. Tissue-specific stem cell populations have since been attributed to the bone marrow, intestinal lining, skin, muscle, gametes, and bone marrow-derived connective tissue that include bone, fat and cartilage. More surprising is the existence of neural stem cells (NSCs) and glial stem cells attributed to the brain, a largely terminally differentiated organ (Box 37–3).

Embryonic Stem Cells

The earliest antecedent stem cell population of embryonic cell lineages can be discerned in the inner cell mass (ICM) of the preimplantation blastocyst, distinguishing then from later somatic stem cell lineages (Fig. 37–2). These pluripotent cells represent a transitory cellular structure that eventually differentiates into mesodermal, endodermal, and ectodermal lineages in a regulated manner during the course of development (Pera et al., 2000; Pera and Trounson, 2004). Mouse ESCs represent the progeny of ICM cells manually isolated from preimplantation mouse embryos, and adapted to the artificial environment of in vitro culture. Several observations suggest that while considered an artifact of culture, mouse ESCs retain several properties of their ICM counterparts, and are distinct from the otherwise phenotypically similar EC cells. Mouse ESCs can form teratomas after ectopic engraftment into immune-compromised mouse tissue, indicating the potential for differentiation into the

Box 37-2 Identification and Isolation of Stem Cells

- Monoclonal antibodies to specific transmembrane and extracellular proteins, glycolipids and proteoglycans have been used to identify and characterize putative stem cells. Historically, human ESC-specific antigens SSEA3 and SSEA4 were identified through the study of human teratocarcinoma cells (Kannagi et al., 1983; Andrews et al., 1982), and mouse ESC-specific antigen SSEA1 was identified after immunizing mice with embryonal carcinoma (EC) cells (Solter and Knowles, 1978). Subsequently, these surface markers are routinely used to isolate and characterize ESC populations. SSEA1 has also been employed to characterize NSC populations in vivo (Capela and Temple, 2002). Antibody-based cell sorting was first used widely in the study of hematopoietic cell lineages, especially using the cluster of differentiation (CD) markers (Barclay et al., 1997). Examples include CD4 and CD8 commonly employed to isolate different subtypes of T lymphocytes, and CD34 and CD133, antigens associated with hematopoietic stem cells (HSCs). The panel of antibodies, CD133, CD34, CD45 and CD24 has also been employed to characterize the epitopes present or absent on non-HSCs such as NSCs obtained from human fetal brain homogenate (Tamaki et al., 2002). The advent of high-throughput gene expression detection methods such as the DNA microarray chip has enabled greater ease in the identification of cell-specific markers. Based on the gene expression profile of highly enriched mouse HSCs, the signalling lymphocytic activation molecule (SLAM) family receptors consisting of CD150, CD244 and CD48 were identified and shown to be sufficient for distinguishing HSCs from multipotent hematopoietic progenitors and restricted progenitors (Kiel et al., 2005).
- The prospective use of stem cells or progenitors in regenerative medicine is largely dependent on the ability to highly enrich stem cells from tissue samples or heterogeneous cell populations. To identify and enrich for sub-populations of cells, fluorescent- or magnetic-activated cell sorting (FACS or MACS) techniques have been developed to achieve mechanical separation. Cell sorting strategies exploit stem cell surface-associated antigens that recognized by specific antibodies conjugated to magnetic particles or fluorescent molecules, the latter isolated by virtue of an electrostatic charge that is imparted to fluorescing cells.
- The definitive test for stem cell potential of these isolated cells has been based on in vivo models. In an irradiated mouse that has its HSCs ablated, the transplant of highly enriched HSCs has been shown to repopulate the hematopoietic system. This is the most rigorous validation of whether a pool of cells contains *bona fide* stem cells. In vitro assays have also been employed as a complementary strategy to ascertain the self-renewal and differentiation potentials of putative stem cells. In vitro "neurosphere" assays are commonly used to identify NSC-containing populations. Under these conditions, a NSC would be capable of generating sequential populations of proliferating, heterogeneous aggregates (also termed *neural precursor cells*) that can undergo multi-lineage differentiation (Reynolds and Weiss, 1992; reviewed in Reynolds & Rietze, 2005). In this context, multiple sequential rounds of clonal expansion serve as evidence for the presence of self-renewing NSCs in the starting population.

Box 37-3 Somatic Stem Cells

***Intestinal Stem Cells*:** The single-layer epithelial lining of the human small intestine undergoes complete renewal on a weekly basis, an ability that can be eliminated by targeted irradiation. In mice, lineage tracing experiments used the uptake of labeled DNA nucleotides incorporated into dividing cells to show that anatomically restricted stem cells in the intestinal crypts proximal to Paneth cells served as lineage precursors to the different intestinal cell phenotypes (Potten and Loeffler, 1987; Potten et al., 2003).

***Epithelial stem cells*:** The skin exhibits a constant turnover of epithelial tissue to maintain an overlying barrier of keratinized epithelial tissue, hair and sebaceous glands. This regenerative process is maintained by stem cells localized to the basal layer of the epithelium, in the bulge region of the hair follicle between the dermal papilla and the sebaceous gland (Fuchs and Segre, 2000).

***Hematopoietic stem cells (HSCs)*:** Two types of stem cells have been associated with the human bone marrow. Hematopoietic stem cell-dependent regeneration of the myeloid and lymphoid lineages can be compromised by whole body irradiation-induced chromosome damage. Transplantation of homogenized tissue and single bone marrow-derived cells to irradiated recipients was sufficient to repopulate hematopoietic lineage cells in the recipient animal, suggestive of a bone marrow resident HSCs in the donor bone marrow (Siminovitch et al., 1963; McCulloch and Till, 1964).

***Mesenchymal stem cells*:** Mesenchymal stem cells, also present within the bone marrow, retain the potential to generate adipose, cartilage and bone, and participate in the reconstruction of an irradiated hematopoietic "niche" environment (Gronthos et al., 2003; Short et al., 2003).

***Muscle stem cells*:** Regeneration of muscle tissue after acute injury is mediated by tissue-resident muscle stem cells. Muscle-resident satellite cells, a heterogenous population of myoblasts that persist in the basal lamina of the sarcolemma, are recruited to sites of muscle damage, proliferate and form multinucleated myotubes (Chen and Goldhamer, 2003).

***Germ-line stem cells*:** Male reproductive capacity is maintained by the long-term regeneration of meiosis-competent gametes. It has been demonstrated that spermatogonia, transplanted under the skin of immuno-deficient castrated mice recipients, are capable of recapitulating the formation of gametes that (1) are competent for fertilization and (2) restore androgen levels (Honaramooz et al., 2002). Germ-line stem cells exhibit several phenotypic similarities to ESCs, and their mutation in vivo is associated with teratocarcinoma formation from which EC cells can be derived. Human germ-line stem cell is currently an area of active research (Shamblott et al., 1998, 2001; Gearhart, 2004).

***Glial stem cells*:** At a cellular level, the brain comprises multiple neuronal subtypes and two classes of glial cells that provide biochemical and structural support for neurons (astrocytes) and the synthesis of myelin sheaths necessary for propagation of action potentials (oligodendrocytes). Dividing cell populations capable of regenerating oligodendrocytes have been well-documented and are thought to mediate myelin repair in response to demyelination (Raff et al., 1983; Franklin, 2002).

***Neural stem cells (NSCs)*:** A substantial body of observations indicates that endogenous, neuron-competent stem cells (Reynolds and Weiss, 1992) contribute neurons of specific phenotype to discrete circuits, which are incorporated into the brain on a daily basis (Doetsch et al., 1999; Gage, 2000; Alvarez-Buylla and Lim, 2004). Constitutive neuronal regeneration is proscribed to the hippocampus, a structure involved in memory processing, and the olfactory bulb. The functional purpose of de novo neuronal integration at these loci remains unclear (Kempermann et al., 2004). Human NSCs have been identified by labeled nucleotide incorporation and autopsy (Eriksson et al., 1998), and have been routinely obtained from the brain of human cadavers and cultured in vitro (Palmer et al., 2001).

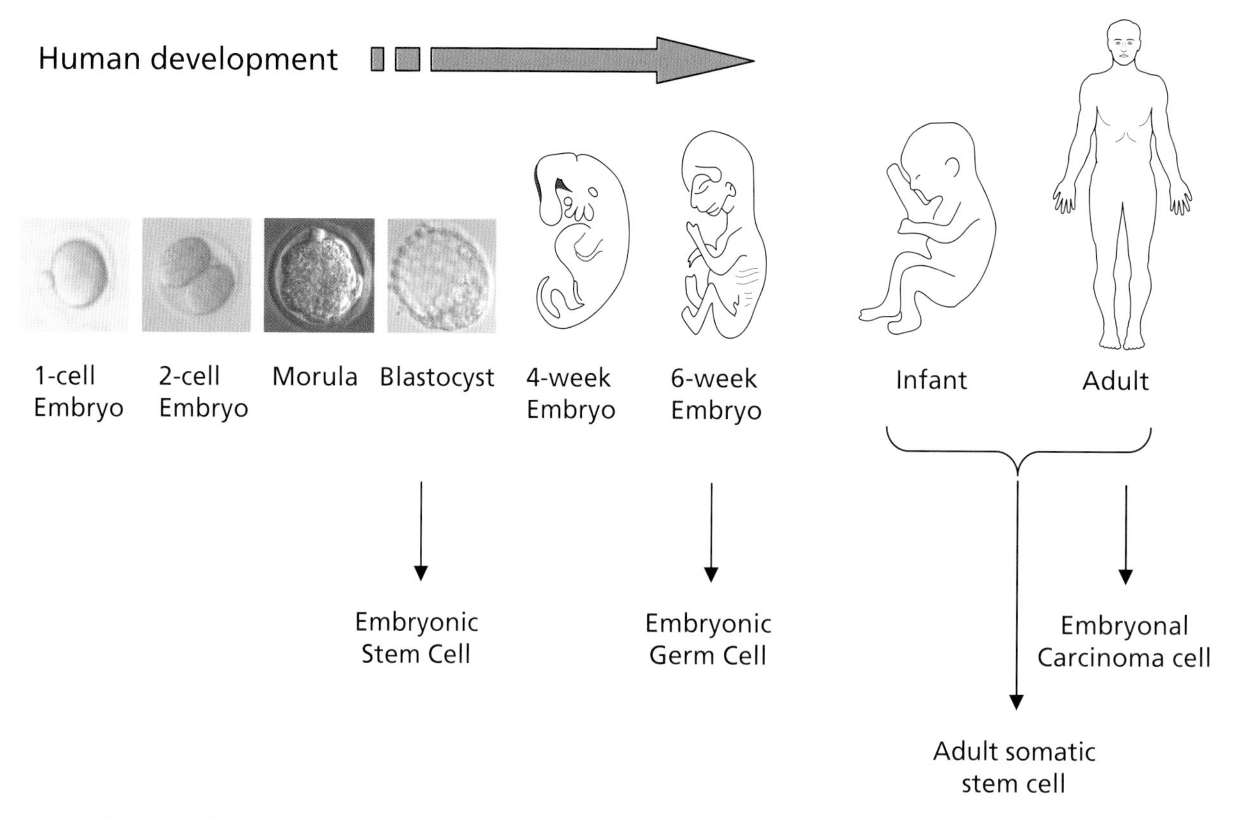

Figure 37–2 Derivation and distinction of somatic stem cells and human ESCs. Human ESCs are isolated from the inner cell mass during the blastocyst stage. Embryonic germ cells, also known as primordial germ cells, are extracted from the gonadal ridge of a 6-week embryo. During various stages of fetal development, various sources of fetal tissue containing somatic stem cell populations can be isolated. Somatic stem cells can also be derived from adult tissues. Embryonic carcinoma cells have been derived from malignant, testicular germ cell tumors.

endodermal, mesodermal, and ectodermal lineages (Evans and Kaufman, 1981; Martin, 1981). Embryonic stem cells retain the ability to self-renew for extended periods in culture, suggesting that these have become isolated from developmental events that would normally direct ICM development. Importantly, these cells contribute to the formation of chimeric mice after engraftment to a host ICM, and continue to give rise to all three germ lineages (Bradley and Robertson, 1986; Bradley, 1990). As a technology for genetic engineering, the potential for germ-line transmission in chimeras has enabled ESCs to find widespread application in basic functional genomics and applied pharmacogenetics research through the generation of transgenic mice (McNeish, 2004).

The derivation of human ESCs from donated human blastocysts provides a model to study the molecular and cellular events that regulate early human development (Thomson et al., 1998). Knowledge gained from previous investigations with mouse ESCs may become applicable and transferable to studying human ESCs. However, there are considerable differences in early embryonic development among human, mouse and non-human primate species. The human epiblast exhibits a longer developmental time course, a radically different morphology, distinct patterns of gene expression, and differences in extraembryonic endoderm composition compared to the mouse counterparts (Pera and Trounson, 2004). The successful derivation of human ESCs represents an avenue for addressing the basis of these species-dependent differences to better understanding human development. Derivation of human ESC lines at present requires the destruction of the source blastocyst and therefore has considerable ethical and legal ramifications (Meyer and Nelson, 2001; McLaren, 1998, 2002; Nelson and Meyer, 2005). In vitro and in vivo analyses indicate that, although important differences between mouse and human ESC properties exist, human ESCs undergo multi-lineage differentiation as teratomas after engraftment to immune-compromised mice, similar to mouse ESCs (Amit et al., 2000). Ethical objections preclude human–animal or donor–recipient human chimera studies, but experimental cell culture observations suggest that human ESCs may be capable of giving rise to meiotically competent gametes and therefore have the potential to contribute to the germ-line, similar to mouse ESCs (West and Daley, 2004; Kehler et al., 2005). However, in contrast to mouse ESCs, human ESCs appear genetically less stable with the propensity for aneuploidy. Cytogenetic analysis of chromosome structure indicates that at initial passages aneuploidy is not induced by the process of extraction, although chromosome duplication may occur as a consequence of long-term culture adaptation and poor culture technique (Pera, 2004; Andrews et al., 2005).

Stem Cells in Cell-based Therapy

It is important to define "cell-based therapy" and "regenerative medicine" as terms that describe the use of well-characterized transplantable cell populations to promote the repair of damaged tissues and organs (Weissman, 2000a; Lagasse et al., 2001). Any scientific consensus on "cell-based therapy" strategies requires an informed understanding of (1) the mechanism of disease, (2) adequate characterization of the donor cell population, (3) the pharmacological management of immune tolerance and rejection that may result from transplantation, (4) the development of in vitro and in vivo models of disease for characterizing the efficacy and mechanism of repair, and (5) continual refinement of tissue-matching processes. In addition to stem cells, other sources of less well-characterized transplantable biomaterials include aborted human fetuses, autologous heterotypic cell types, and tissue derived from consenting donors or cadavers. Few cell transplantation based therapeutic interventions satisfy these criteria and are considered experimental in nature. In particular, the application of human ESCs as a transplantable cell type is clearly experimental. Human ESC-based cell transplantation strategies require the use of cells that have completely defined culture conditions free of animal products (Box 37–4 and Fig. 37–3). Clinical application of human ESCs in vitro or in vivo further requires that ethical and moral concerns of clinicians and patients be addressed (Scolding, 2001a, b, 2005).

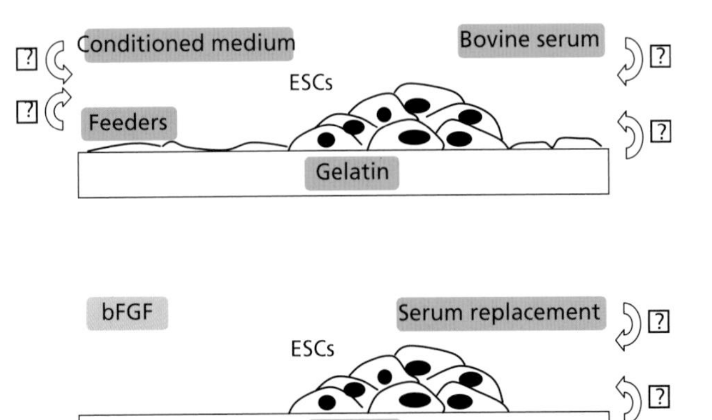

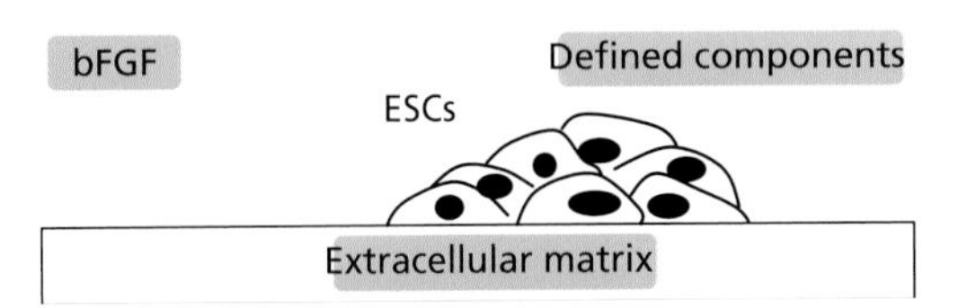

Figure 37–3 Improvization of human ESC culture techniques to achieve clinical grade cells without contamination by xeno-materials. Standard human ESC culture involves the use of either mouse or human fibroblast cells as a feeder layer that provides physical support and cell-contact dependent and soluble survival factors; conditioned medium obtained from mouse embryonic fibroblasts may also be supplemented. As the complete repertoire of human ESC survival and growth factors is poorly understood, animal serum-derived supplements are used for optimal in vitro culture. However, extracellular matrix supplements required for human ESC attachment and growth are derived from animal sources, and serum contains undefined components. On-going improvisations have resulted in the culture of feeder-free human ESCs maintained with a fully defined serum replacement supplement and recombinant growth factors (basic fibroblast growth factor, bFGF), although growth under completely defined conditions has not yet been achievable (Adapted from Pera, 2005.)

Box 37–4 Human ESC Derivatives and Cell Transplantation

The limited understanding of human ESC developmental origins, genomic stability, lineage potential and mechanisms of directed differentiation are presently hamper their potential clinical usefulness as a transplantable cell population. A defining property of undifferentiated human ESCs is the capacity to form teratomas, and it is essential that the risks of tumor formation in human ESC-derived progenies be thoroughly addressed. Cell culture conditions commonly used for the propagation of human ESCs, including cell lines that are federally approved for study in the U.S.A. as of August 2001, require the presence of undefined, soluble and contact-mediated signaling factors, adhesion substrates derived from animal proteins, and bovine serum supplementation (Hoffman and Carpenter, 2005). Concerns that animal-derived pathogenic agents may compromise transplant function have been well-documented in organ transplantations and xeno-transplantations (Dawson et al., 2003). For human ESCs that can be derived without exposure to xenobiotic agents, screening and approval processes that meet good manufacturing practice (GMP) requirements of the US Food and Drug Administration need to be satisfied prior to clinical trials (Pera, 2005; Rodriguez et al., 2006).

Stem cells are useful in cell-based therapy as these serve as sources of phenotype-specific cells for transplantation or as an experimental model for characterization of stem cells and their derivatives. The targeted manipulation of the normal environment occupied by stem cells in vivo—the stem cell niche, has emerged as an important target of therapy. To appreciate the difficulties of translating cell-based therapies to the clinic, it is helpful to highlight come examples of strategies exploiting (1) exogenous transplantation, and (2) endogenous stem cells.

Exogenous Cell Transplantation

Bone Marrow Transplantation

Bone marrow-derived hematopoietic tissue containing stem cells was successfully used for the regeneration of the hematopoietic system subsequent to irradiation in human patients (Siminovich et al., 1963), and has been reviewed extensively (Weissman et al., 2001; McCulloch and Till, 2005). The success of bone marrow transplantation for the treatment of hematopoietic malignancy has led to similar cell replacement strategies for the repair of other cell phenotypes in non-hematopoietic systems.

Neural Transplantation

Exogenous cell transplants for the brain have been performed for neurodegenerative conditions such as in Parkinson's disease where dopamine neurons are lost (Bjorklund et al., 2003; Ostenfeld and Svendsen, 2003; Lindvall and Bjorklund, 2004). To facilitate the regeneration or replacement of dopamine-releasing neurons, several tissue sources were considered, including immature dopamine-neuron containing tissue derived from the midbrain of aborted fetuses, and dopamine-expressing autologous grafts of adrenal gland tissue. However, the clinical outcomes from experimental trials have been variable. The further development of neural cell transplantation treatments requires (1) development of homogenous sources of donor tissue and (2) improved understanding of the cellular and molecular heterogeneity of the underlying pathology (Barker, 2002; Harrower and Barker, 2004).

Alternatively, transplantation of exogenously cultured somatic cell populations has been used as delivery vehicles to provide extracellular signals for promoting the survival of endogenous differentiated cells (Behrstock and Svendsen, 2004). Undifferentiated human neural cell populations manipulated to overexpress glial-derived neurotrophic factor (GDNF) and transplanted into animal models of neurodegenerative disease have shown to promote neuronal survival and functional recovery (Klein et al., 2005; Behrstock et al., 2006). This is consistent with the improvements in dopamine neuronal function observed after chronic infusion of GDNF to the putamen of Parkinson's disease patients (Gill et al., 2003).

Glial Cell Transplantation

Depletion of a type of glial cell, the oligodendrocyte, is associated with demyelination in the central nervous system. Demyelination is observed in the context of acute spinal cord injury and autoimmune and neurodegenerative disorders such as transverse myelitis and multiple sclerosis. The oligodendrocyte precursor cells which are widely distributed normally induce the regeneration of oligodendrocytes and myelin sheaths (remyelination). In chronic demyelinating disease, this process eventually fails for reasons not well-understood (Franklin, 2002b). Transplantation of exogenous myelinating cells have been considered to circumvent the failure of the endogenous cell population and to ameliorate the symptoms attributed to demyelination. Prospective sources of remyelinating tissue include cells derived from the olfactory epithelium (olfactory ensheathing cells), the peripheral nervous system (Schwann cells), and human ESC-derived oligodendrocytes (Franklin, 2002c; Jani and Raisman, 2004; Ibrahim et al., 2006). However, the diffused multifocal nature of demyelination continues to remain a major challenge for site-specific, transplantation-based remyelination strategies.

Myoblast Transplantation

The discovery of myogenic precursors resident in adult muscle is of relevance to the treatment of muscular dystrophy, which presently has no effective remedy (Charge and Rudnicki, 2004). The transplantation of isolated muscle fibers containing satellite cells into dystrophic mice was able to functionally regenerate recipient muscle tissue (Montarras et al., 2005), although this method is not in clinical development at present (Skuk, 2004). Autologous myoblast transplantation for the treatment of heart disease, a process termed cellular cardiomyoplasty, has been applied clinically. Initial observations are suggestive of some clinical benefit, but the mechanism of tissue repair remains unclear. Controlled, blinded clinical trials will be necessary before the application of exogenous transplantation of cells for heart disease can be considered for widespread use (Wollert and Drexler, 2004, 2005).

Manipulation of Endogenous Stem Cell Repair

The persistence of stem cells in vivo is dependent on the stem cell niche which is an anatomically and pharmacologically defined micro-environment of extracellular cues which are soluble and/or contact-mediated, that maintain stem cells in a state of self-renewal. The stem cell niche concept was originally proposed in mammalian systems (Spradling et al., 2001; Ohlstein et al., 2004) but anatomically defined niches were first described for germ-line stem cells in *Drosophila* (Lin, 2002). The anatomical composition of putative niches in mammalian tissue have since been described for the intestinal crypt (Potten and Loeffler, 1990; Wong, 2004), the endosteum of trabecular bone (Arai et al., 2005), bone marrow stroma (Baksh et al., 2004), the hair follicle bulge (Tumbar et al., 2004), testes (McLean, 2005), and the subventricular zone and hippocampus of the brain (Doetsch, 2003b). Studies are beginning to define the molecular and cellular interactions in some of these tissue-specific stem cell niches and how they can be manipulated for endogenous stem cell-mediated repair.

Recruitment of HSCs to the niche is actively exploited in bone marrow transplantation (Nilsson and Simmons, 2004). Hematopoietic stem cells exhibit a constitutive level of mobilization observed at low frequency under normal physiological conditions (Blau et al., 2001). Manipulation of this phenomenon may represent a clinical target for directing stem cells to repopulate the niche or mobilize to a particular tissue. Cytokine-dependent niche-derived signaling that regulate HSC mobilization, such as the cytokine granulocyte colony-stimulating factor, has been routinely used to improve the efficiency of stem cell engraftment by attracting mobile stem cells to the niche (Wright et al., 2001; Lapidot and Petit, 2002). Surprisingly, it has been proposed that central nervous system-derived, noradrenalin-dependent signals can repress the transcription of the chemokine CXCL12, leading to HSC extrusion from the niche and mobilization (Katayama et al., 2006).

By contrast, NSC-derived migratory progeny are unable to mobilize to distant sites of injury as the brain parenchyma is largely refractory to cell migration. Neural injury is associated with hypertrophy and reactivity of glial cells proximal to the lesion site which present migration-inhibiting molecules such as proteoglycans and extracellular matrix proteins. Techniques that allow for the modulation of the biochemical composition of the extracellular environment with specific enzymes may improve migration or plasticity of endogenous cells (Fawcett and Asher, 1999; Prestoz et al., 2001; Picard-Riera et al., 2004; Properzi and Fawcett, 2004). Alternatively, the introduction of a pro-migratory extracellular matrix molecule, TENASCIN-R, can improve the migration of NSCs-derived neuroblasts from the subventricular zone (Saghatelyan et al., 2004).

Stem Cell Genomics

In our attempt to understand and dissect the genetic elements and molecular interactions that underlie stem cell properties, genomics-based technologies have been ingeniously adapted. Stem cell genomics represents the interface between stem cell biology and computational science which utilizes large information sets from the transcriptome and genome to understand key features of stem cells and answer fundamental questions about developmental biology (Mikkers and Frisen, 2005). How does the cellular genome maintain the unique properties of stem cells? What are molecular mechanisms that regulate the choice between self-renewal and terminal differentiation? What are the cues in the extracellular environment critical for directing differentiation? To what extent is differentiation irreversible, and can the genome of differentiated cells be reprogrammed into stem cells?

Common stem cell-associated properties include the absence of lineage-specific markers characteristic of differentiated cells, the presence of ATP-binding cassette (ABC) transporter-mediated dye extrusion, specific cell-adhesion proteins, gap junction proteins, receptors for mitogens, survival factors and other signaling ligands, and altered levels of activity in pathways such as cell cycle progression, apoptosis, DNA repair and telomerase activity necessary for long-term self-renewal (reviewed in Cai et al., 2004).

One key application of genomic information is the search for specific genetic networks that maintain a stem cell in its undifferentiated state. Another promising application is the elucidation of molecular mechanisms directing stem cell differentiation into specialized and functional states. The appreciation of transcription networks governing stem cell pluripotency versus differentiation is fundamental to understanding mammalian embryonic development, and for realizing the therapeutic potential of these cells. Although significant progress has been made towards dissecting the molecular aspects of these properties, particularly in the area of individual gene characterization, we have yet to grasp a comprehensive picture of the global pattern of gene regulation, expression and epigenetic elements. To this end, genomics-based technologies have contributed to the acquisition of information regarding global transcriptome (total cellular transcripts) changes that occur during differentiation, as well as comparisons made with non-stem cell populations.

Interrogating the Stem Cell Transcriptome

A rationale for the genomic approach to unravel stem cell identity is the potential for answering fundamental questions about development and biology. The term "stemness" has been coined to describe the hypothesis that, within a given species, a subset of molecular mechanisms correlate with the maintenance of the stem cell state, and this can be observed across all stem cell phenotypes. The concept of stemness does not extend from the concept that all stem cells are identical, but rather that stem cells exhibit commonalities in phenotype considered to be predictive of conserved mechanisms (Cai et al., 2004; Mikkers and Frisen, 2005).

Several technologies have been applied to define the molecular signatures associated with stem cells and as they differentiate (Table 37–1). Microarray chip analysis has been extensively employed to interrogate the transcriptome to define which genes may confer stem cells their unique identity. The underlying principle is that a putative core set of genes may be required for maintaining stem cells in their proliferative and undifferentiated state. Current evidence suggests that, while common gene profiles have been described across somatic stem cell populations, no clear molecular signature associates with both somatic and embryonic stem cell types (Terskikh et al., 2001; Fortunel et al., 2003). Each stem cell population seems to be rather unique in its expression profile, which is a likely explanation for the specific functional significance associated with the terminal cell phenotypes. Based on two independent studies, the overlap of "stemness" genes between datasets from Ramlho-Santos et al. (2002) and Ivanova et al. (2002), in which both used embryonic, hematopoietic, and neural stem cells, yielded only six common genes. The observation was instructive in that it emphasized the difficulty of comparing cells on the basis of gene transcription alone, and indicated important roles of gene regulation downstream of transcription. Technical inconsistencies in microarray analysis platforms associated with differences in array construction, oligonucleotide attachment, empirical optimization of hybridization and exposure conditions often limit the value of cross-dataset comparison (Karsten et al., 2004). It is also difficult to obtain populations of purified stem cells for comparison, as stem cells are often contained within a heterogeneous mix of stem cell-derived progeny cells and other supporting cells. Stem cells isolated from their nominal context and adapted to culture (for example, ESCs or NSCs) may exhibit artifactual changes in gene expression that disguises real commonalities between stem cell populations residing in the niche (Stiles, 2003).

Other technologies, such as massively parallel signature sequencing (MPSS), serial analysis of gene expression (SAGE), and gene identification signature paired-end ditags (GIS-PET), have been utilized in examining mouse and human ESCs. Using MPSS, similarities and differences between the two types of ESCs can be examined (Wei et al., 2005). It was demonstrated that human and mouse ESCs employ divergent paths to maintain the stem cell state. In mouse ESCs, the maintenance of the undifferentiated state is dependent on the addition of a cytokine, leukemia inhibitory factor (LIF). Based on the MPSS, the genes involved in the LIF-signal transducers and activators of transcription (STAT) pathway are clearly expressed. Many of these are absent in the human ESCs, which do not require LIF for continuous culture. The transcriptome analysis has also demonstrated that the human and mouse ESCs are similar in many aspects at the molecular level. Well-known ESC-specific genes related to the self-renewal pathway were conspicuously conserved. These include OCT4, SOX2, BMPR, NODAL, LEFTY,

Table 37–1 Summary of recent transcription-level, high-throughput genomic platforms in understanding the gene expression of stem cells and regulation by transcription factors

Genomic Platform	Application in Stem Cell	Selected Reference
Microarray	Embryonic stem cell	Fortunel et al., 2003 Ramalho-Santos et al., 2002 Ivanova et al., 2002
	Embryonal carcinoma cell	Liu et al., 2006
	Mesenchyma stem cell	Wagner et al., 2006
	Hematopoietic stem cell	Kiel et al., 2005
	Neural stem cell	Cai, Wu, et al., 2006
MPSS	Embryonic stem cell	Brandenberger et al., 2004 Wei et al., 2005
	Neural stem cell	Cai, Shin, et al., 2006
SAGE	Embryonic stem cell	Richards et al., 2004 Anisimov, Tarasov, Tweedie, et al., 2002
	Embryonal carcinoma cell	Anisimov, Tarasov, Riordon, et al., 2002
	Hematopoietic stem cell	Lee et al., 2001 Zhou et al., 2001
ChIP-PET	Embryonic stem cell	Loh et al., 2006
ChIP-on-chip (whole genome)	Embryonic stem cell	Boyer et al., 2005

TERT and CRIPTO. Several additional genes not previously known to be conserved or elevated in ESCs were identified as well.

One major advantage in using high-throughput gene expression analyses for studying ESCs lies in their ability to provide a comprehensive picture to suggest which genes may be the key in regulating the undifferentiated state. Notably, the use of bioinformatics to mine microarray datasets led to the identification of NANOG as an important regulator in human and mouse ESC pluripotency (Mitsui et al., 2003; Chambers et al., 2003). Knowledge of the expression levels for each gene and its specificity to a particular cell type provides valuable preliminary information on its putative relevance. This is arguably more efficient and effective than conventional gene expression assessment methods which rely on painstaking measurements of individual gene expression. Stem cell biologists can rapidly select the top candidate genes for in-depth functional studies in a particular stem cell type.

The advances made in high-throughput technologies, along with reducing cost, has enabled the development of whole genome DNA microarrays. Using genome scale analysis which anchors on combining chromatin immunoprecipitation (ChIP) coupled with DNA microarrays (ChIP-on-chip; Fig. 37–4) or ChIP followed by high-throughput sequencing of tagged-DNAs (ChIP-PET, Table 37–1), we can potentially uncover all the target genes bound by ESC transcription regulators in a manner that is biologically and physiologically relevant. While gene expression analysis itself can provide descriptive transcriptome information pertaining to the ESC state, it does not reveal how such a state is maintained through the interaction between genes. By contrast, ChIP-on-chip technology highlights which downstream genes are being directly bound and regulated by transcription factors. Such a strategy has already been employed for transcription network analysis in the yeast model (Blais and Dynlacht, 2005; Shannon and Rao, 2002; Futcher, 2000), and recently applied to human and mouse ESCs discussed later (Boyer et al., 2005; Boyer et al., 2006). This development is in part propelled by more accurate and complete annotation of the human and mouse genomes.

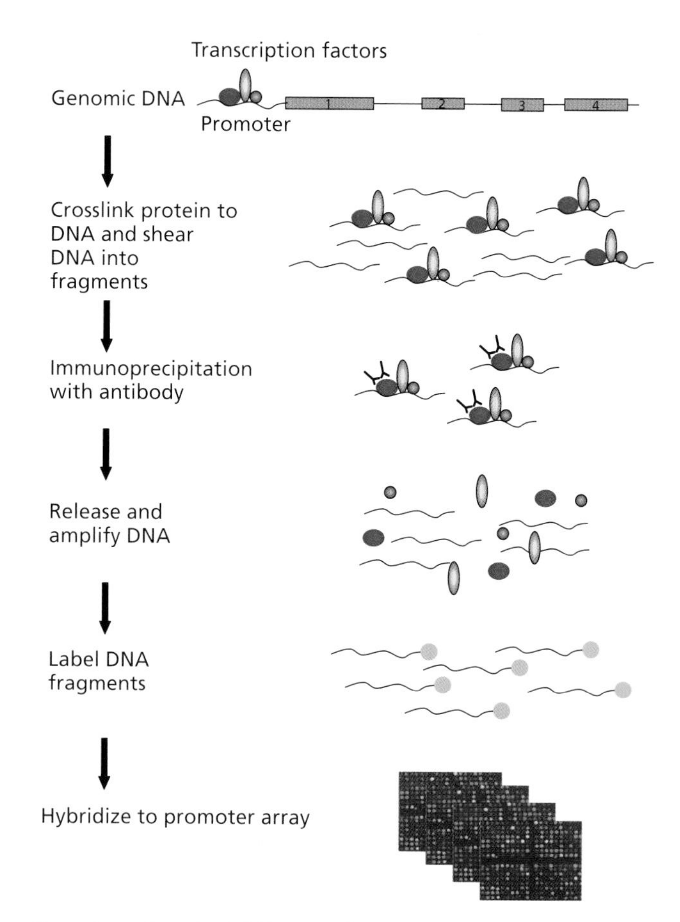

Figure 37–4 Overview of ChIP-on-chip. ChIP is performed by *in situ* cross-linking of a specific transcription factor (TF) onto its DNA binding site. The genomic DNA is sheared and the TF-bound DNA fragment is immunoprecipitated by TF-specific antibody. The immunoprecipitate is enriched for DNA fragments bound to the TF, separated from the cross-linked protein, and amplified. Fragments are labeled with a fluorescent probe and hybridized to tiled arrays (promoter or whole genome). In this way, the DNA binding sites of specific TFs can be assessed in a genome-wide manner.

Mapping Transcription Networks

The study of mouse and human ESCs provides a useful example of how such high-throughput analysis methods can define stem cell properties and key transcriptional networks (Robson, 2004). The mechanism by which OCT4, SOX2 and NANOG maintains ESC pluripotency and self-renewal is beginning to be understood. The emergent view is that these transcription factors act together in a tightly regulated transcription framework to control similar as well as divergent downstream factors. Knowledge of the global set of genes regulated by these transcription factors can therefore reveal how the phenotypic consequences of ESC differentiation are conferred when either OCT4 or NANOG is disrupted. In human ESCs, the occupancy of three key pluripotency-associated transcription factors OCT4, SOX2, and NANOG have been individually mapped to many target genes (Boyer et al., 2005). These three factors also co-occupy the promoters of a large subset of genes. Interestingly, many of these targets encode for developmentally important homeodomain transcription factors including the homeobox (HOX) and LIM homeobox (LHX) families, suggesting that these downstream regulators contribute to specialized networks in hESCs. Transcription factors binding to target genes at the promoters can represent both transcriptionally active and repressed states. The intersection of the co-occupied sets of genes by OCT4, SOX2, and NANOG with gene expression datasets from ESCs would suggest which genes are actively transcribed. Conversely, genes that are bound but not expressed in ESCs would represent gene repression during the pluripotent state.

The emergent picture of maintenance of pluripotency in ESCs appears to be dependent on the activation of pluripotency-associated genes and suppression of differentiation-promoting factors regulated, in part, by a few key genes. Among the co-occupied, transcriptionally active these genes are pluripotency associated transcription factors (OCT4, SOX2, NANOG, STAT3, ZIC3), members of the Tgfβ pathway (TDGF1, LEFTY2), and Wnt signaling pathway (DKK1, FRAT2). OCT4, SOX2 and NANOG are already very well-established members of the ESC core transcription machinery, and STAT3 is responsible for the activation of the LIF-STAT pathway unique to mouse ESCs, and ZIC3 could be important as depletion can cause mouse ESC differentiation into the endoderm lineage (Lim et al., 2007). There is cumulating evidence that Tgfβ and Wnt signaling are crucial to the maintenance of pluripotency and mESCs and human ESCs. These results indicate that there is a good degree of self-regulation, which provides positive feedback loops for maintaining the transcripts of OCT4, SOX2 and NANOG in ESCs, as well as proximal regulation of key signaling pathways. Genes repressed by OCT4, SOX2 and NANOG would be expected to regulate exit from the pluripotent state. A number of co-occupied genes encode transcription factors that specify differentiation into the endodermal, mesodermal, ectodermal, and extraembryonic lineages.

One important lesson gained from these studies has been the usefulness of genomic approaches in scanning the genome and short-listing a manageable number of putative important genes. A high-throughout approach to gene functionalization can subsequently lead to identification of factors important for ESC growth and differentiation. Such an approach, for example, led to the recent discovery that Sall4 is a critical upstream regulator of OCT4 transcription factor, and hence playing a crucial role not only during preimplantation embryo development but also during postimplantation embryogenesis (Zhang et al., 2006). The integration of novel, validated components of key regulatory transcription factors into ever-expanding but finite networks represents an emerging, systems biology conceptualization of human ESCs (Fig. 37–5).

Several studies have reported the attempt to obtain genomic information on the transcriptome of somatic stem cells. However, unlike ESCs, in which large numbers can be obtained, the difficulty in isolating pure somatic stem cells has been a limiting factor. The molecular profiling of a somatic stem cell's resident environment may enable the identification of potential factors that promote endogenous stem cell mobilization or recruitment to stem cell niches without direct transplantation. Gene expression analysis of HSCs (Kiel et al., 2005) and bone marrow and fetal liver-derived niche stromal tissue has been applied to identify potential regulatory mechanisms (Ueki et al., 2003; Iwata et al., 2004; Martin and Bhatia, 2005; Wagner et al., 2005). Microarray profiling has also been used to define gene expression changes during neural progenitor cell differentiation and within the dorsolateral sub-ventricular zone NSC niche to identifying novel factors that may regulate NSC self-renewal and differentiation (Geschwind et al., 2001; Karsten et al., 2003, 2004). However, global transcriptional assays of niche-associated transcripts have been hampered by the cytological complexity of the niche tissue. For instance, the dorsolateral sub-ventricular zone comprises multiple, differentiated and dividing cell phenotypes in close apposition (Doetsch, 2003). Careful profiling of gene expression and DNA-protein interaction of the individual components is needed to fully appreciate the complexity of gene regulation in the niche context (Liu, 2005). Quantification of the binding affinity of stem cell-specific proteins is an important area for examination. The affinity of a protein for a particular DNA binding site (motif) is regulated by protein-protein interactions with other proteins bound to adjacent repressors or promoters (module). The composition of the module is, in turn, regulated by cell signaling pathways that translate intracellular or extracellular (ligand-activated) changes into altered motif–module interactions (Liu, 2003). When examining the genomic sequences, it is considerably difficult to identify regulatory sequences or motifs compared to obvious protein-coding genes. The rules for identifying such these regulatory modules and motifs are more complex, and less obvious. Like protein-coding regions, many of these regulatory sequences have been conserved through evolution, thus allowing us a glimpse into which could be functionally important elements in the human genome. However, gene regulation is also governed by stable and heritable modifications to the DNA sequence or DNA-associated nucleosomes ("Epigenetic Regulation of Stem Cell Pluripotency", discussed below). Such modifications further complicate gene regulation by allowing even conserved motifs to be responsive in a spatial- and temporal-specific manner.

Non-coding RNA Modulation of Stem Cell Behaviour

The role of non-coding RNAs in the regulation of stem cell identity is beginning to be explored, especially in the contexts of stem cell differentiation and epigenetic regulation (Murchison and Hannon, 2004; Mattick and Makunin, 2005). It is thought that microRNAs behave as "micromanagers" of gene expression patterns established by "directors" at the transcriptional level. Each microRNA complementary site within the mRNA 3′untranslated region (UTR) may be considered analogous to a translation "resistor" (Bartel and Chen, 2004). Each target gene, with its 3′UTR, can potentially house a combination of microRNA target sites, and hence be regulated in a temporal and spatial manner by many microRNAs that exist at a specific time point. In part, this may explain the difficulties of functionalizing single microRNAs.

The restoration of a single microRNA, miR-430, in zebrafish that could no longer produce endogenous microRNA ameliorated deficits in zebrafish neuroectodermal development and neuronal differentiation (Giraldez et al., 2005). In mammals, specific microRNAs have been shown to regulate B-cell differentiation (miR-181; Chen et al., 2004), adipocyte differentiation (miR-143; Esau et al., 2004), and insulin secretion (miR-375; Poy et al., 2004). MicroRNA regulation of Hox gene expression was observed to modulate developmental patterning processes to allow the generation of asymmetric morphology (miR-196; Mansfield et al., 2004; Yekta et al., 2004). Consequently, the role of microRNAs as participants in the epigenetic mechanisms that regulate self-renewal in somatic cell populations has been investigated. MicroRNAs have thus been identified as participants in genomic imprinting (Seitz et al., 2003), a process known to require differential allele methylation (da Rocha and Ferguson-Smith, 2004), and may participate in NSC regulation (Cheng et al., 2005). Consistent with their role in regulating cell differentiation, apoptosis and epigenetic transcriptional repression, alterations of microRNA expression are consistently observed in several forms of human cancer and cancer cell lines (McManus, 2003; Takamizawa et al., 2004; Lu et al., 2005; Miska, 2005). In human and mouse ESCs, microRNAs that are highly expressed have been identified (Sun et al., 2004; Houbaviy et al., 2003; Houbaviy et al., 2005). However, none of these microRNAs have yet been shown to directly regulate ESC pluripotency. Despite this, there is emerging evidence pointing to the importance of microRNAs in ESCs as these cells lacking the machinery for processing mature microRNAs display severe defects in differentiation both in vitro and in vivo (Murchison et al., 2005; Kanellopoulou et al., 2005). Our recent data suggests that miR-134 can promote mouse ESC differentiation into neuroectodermal lineages. The number of potential microRNAs in the human and mouse genome remains controversial. Putative microRNA genes are interspersed along the genomic sequence, with many embedded within the intronic regions of classical protein-coding genes. MicroRNA genes are subject to regulation in known transcriptional networks: for example, OCT4, SOX2 and NANOG in human ESCs target 14 microRNA genes of which two (miR-137 and miR-301) are co-occupied by all three. However, the subtle effect of microRNA in stem cell function has proved intractable to conventional gene analysis strategies such as anti-sense knockdown, genetic knockout and over-expression. Technical hurdles will remain in demonstrating the protein translational inhibition and mRNA transcript cleavage by a single microRNA on the numerous predicted target genes. Furthermore, computational prediction identifies many more microRNAs than can be experimentally validated. The predictions that emerge from the few computational algorithms available appear largely unreliable

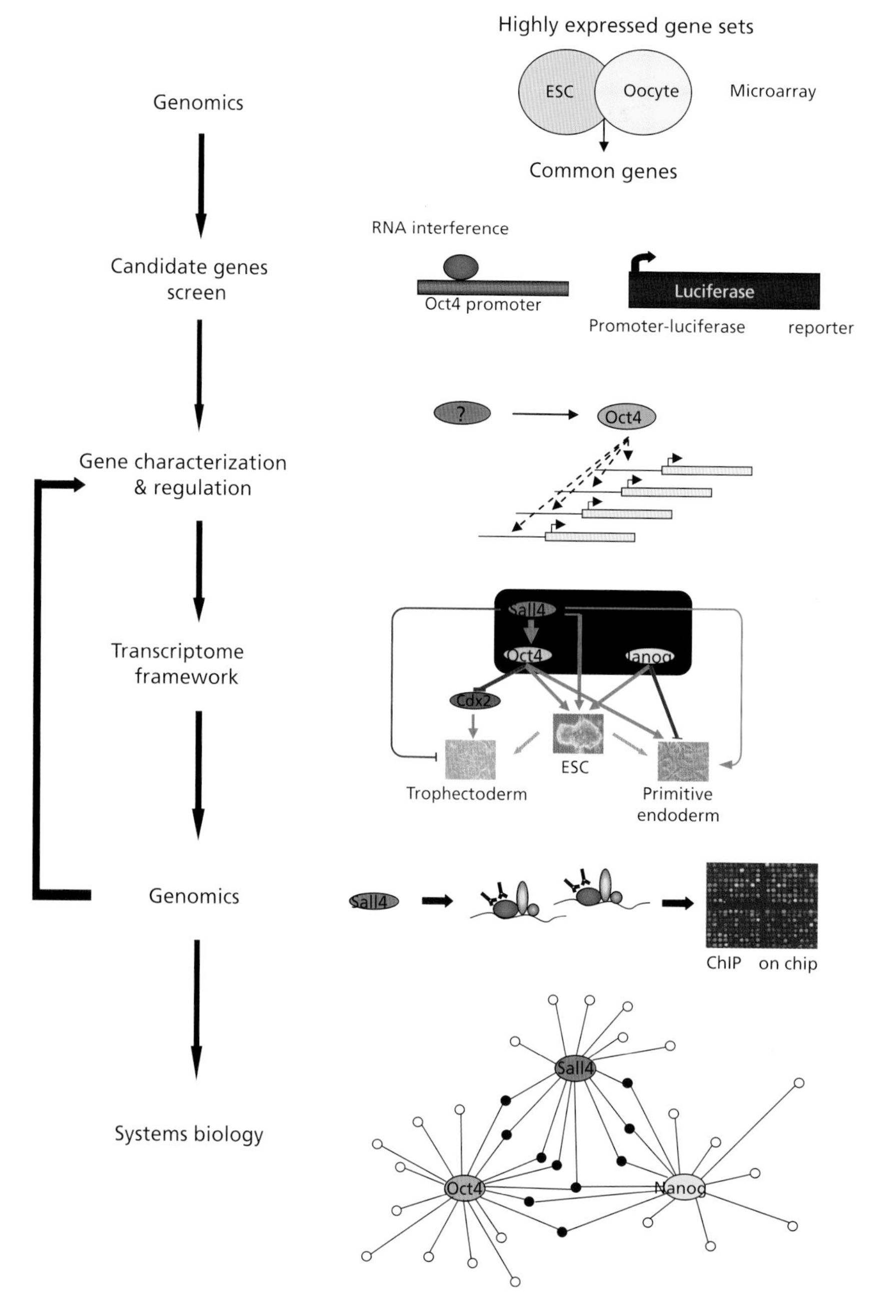

Figure 37–5 Systems biology. Genomics information is utilized for building transcription frameworks in stem cells. Both ESCs and oocytes share a common property of being pluripotent. The key regulator OCT4 is thought to maintain this unique state by controlling a whole repertoire of downstream genes; but little is known regarding the transcriptional control of OCT4. Microarray comparison of highly expressed gene sets helps to identify a list of common genes up-regulated in both cell types. Candidate genes of interest are depleted via RNA interference to assay for their effect on the OCT4 promoter activity in ESCs. Factors which down-regulate the luciferase activity when depleted are potential regulators of OCT4. Gene characterization confirms the specific regulation of a regulator through its binding and transcription regulation. A novel factor, SALL4, is identified and added to the known transcription framework helmed by OCT4 and NANOG. Each of these key factors binds a myriad of downstream target genes as determined by ChIP-on-chip. Interesting findings are re-fed into the loop, and new components are added to the framework. The amalgamation of these data, together with an understanding of the control hierarchy, adds predictive value to gene function, and is termed systems biology.

as experimental platforms. Most published computational algorithms have yet to demonstrate any large-scale experimental validation of their accuracies in a biological cell-based system. It will, nonetheless, be clear that the hunt for a mechanistic explanation of microRNA regulation of the ESC state is crucial for a more complete understanding of gene regulation.

Epigenetic Regulation of Stem Cell Pluripotency

In discrete and well-defined networks, transcription circuitry helps to generate a hierarchical order for key genes, correlate their relative positions with respect to one another, as well as implicating their limited interacting partners. However, this excludes how regulatory networks may be influenced by additional interactions associated with chromatin modification, DNA methylation and other transcription factors. The regulatory influence of such factors may in turn vary in a temporal or spatial manner, thereby modulating the activity of a core transcriptional regulatory network in different ways. ChIP-on-chip strategies centered on single protein analyses provide high micro-scale but low macro-scale 'resolution' of the protein-DNA interactions that regulate transcription.

Epigenetics refers to stable and heritable modifications made to the genome that do not alter the core information content of the DNA sequence template. Chromatin structures can be described in terms of euchromatin (non-uniform chromatin in a format "open" to transcriptional activity), facultative heterochromatin (heterochromatin that forms as a function of developmental progression), and constitutive heterochromatin (permanent chromatin structures that remain "closed" transcriptional activity). Proximity of a gene or its promoter to heterochromatin is generally associated with transcriptional silencing, a process regulated by DNA sequences and an array of nucleosomal and accessory proteins (reviewed in Dillon and Festenstein, 2002; Dillon, 2004; Orkin, 2005).

During embryonic development and stem cell differentiation, the process of cell fate determination is coupled to epigenetic modifications of the genome that include DNA methylation, histone modifications, and associated changes in the chromatin architecture (Fig. 37–6). These heritable epigenetic modifications provide a means of controlling gene expression to ensure appropriate gene activation or silencing in a cell-lineage specific manner. The process of hypermethylation of cytosine moieties in the CpG islands found in many vertebrate promoters is a conserved epigenetic modification of the somatic genome (Robertson, 2002; Laird et al., 2003).

DNA hypermethylation has been found to be associated with stable transcriptional gene silencing and stably repressed chromatin in differentiated cells (Bird, 2002; Jaenisch and Bird, 2003). During embryonic development, the promoter regions of key pluripotency-specific genes, such as OCT4 and NANOG, become highly methylated and transcriptionally silenced. These methylation patterns in differentiated cells are subject to active maintenance by methyltransferases during DNA replication, but can be erased under certain exceptional conditions. For example, during somatic cell nuclear transfer (SCNT) when a differentiated somatic cell is introduced into an enucleated oocyte, the donor nucleus becomes reverted to a more primitive state through a process termed reprogramming. This reprogrammed nucleus is subsequently capable of recapitulating development and generation of appropriate embryonic and extra-embryonic lineages. The reprogrammed status of the nucleus can be verified by the re-expression of pluripotency and early embryonic markers, and is associated with a reversion of the epigenetic state.

The observed hypermethylation of CpG islands upstream of putative tumor suppressor genes in carcinoma tissue stimulated interest in DNA methylation as a potential diagnostic tool and therapeutic target (Baylin and Herman, 2000). Alternatively, hypomethylation in constitutive heterochromatin correlates with genomic instability (Robertson and Wolffe, 2000). An increase in the incidence of imprinting disorders, such as Beckwith-Wiedemann syndrome and Algelman syndrome, in infants conceived using IVF has been demonstrated. Therefore, preimplantation genetic diagnosis of DNA methylation cis-elements in embryo-derived stem cells would be beneficial to the subsequent health and development of the implanted embryo (Allegrucci et al., 2004). Pharmacological agents that initiate hypomethylation of DNA by substituting cytosine analogs resistant to methylation (5-azacytidine) or by inhibition of methyltransferase activity (5-aza-2′-deoxycitidine) have been used to inhibit progressive hypermethylation in stem cell-associated tumorigenesis (Yoo and Jones, 2006).

Studies have further indicated the feasibility of temporal changes in DNA methylation patterns as a diagnostic evaluation of human ESC quality maintained in long-term culture. Methylation-dependent imprinting patterns on genes necessary for early developmental functions do not destabilize due to prolonged culture in some human ESC lines (Rugg-Gunn et al., 2005). However, qualitative comparison of methylation patterns across 14 methylation-sensitive promoters in other human ESC lines suggested that methylation patterns can become destabilized in culture (Maitra et al., 2005; Plaia et al., 2005). The reasons for the different observations are not clear. This may be attributed to the differences in the limited number of promoters analyzed, and hence a more global analysis may be necessary.

In addition to methylation changes, it has been demonstrated that post-translational covalent modifications on the histone moieties of nucleosomes is an important epigenetic regulatory

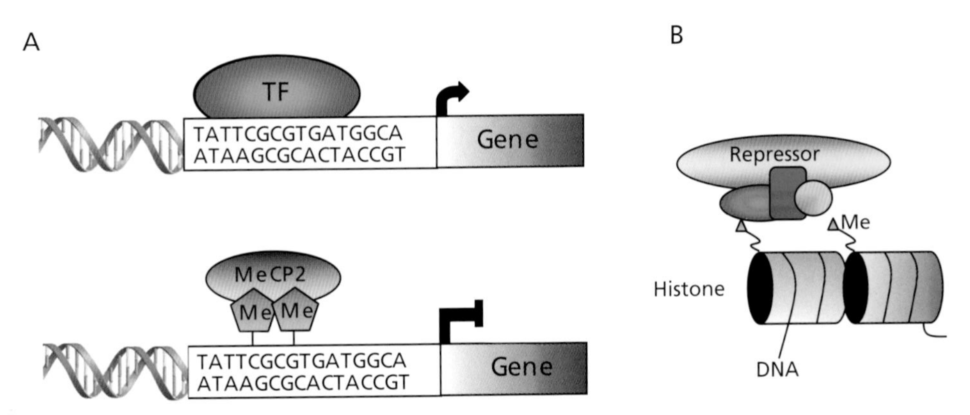

Figure 37–6 DNA methylation and chromatin-dependent mechanisms of epigenetic gene regulation. (A) Transcription factors (TFs) bind to the promoter region of a gene to activate transcription. Methylation (Me) of cytosine on the DNA by de novo DNA methyltransferase enzymes (DNMTs) recruit methylation-specific binding proteins (methyl-CpG binding protein, MeCP2), which in turn form a complex with additional co-repressors to prevent TF binding and transcript activation. (B) Histone modifications such as methylation of histone H3 at lysine (K) 27 (also H3-K9 and H4-K20) is associated with the recruitment of repressor complexes to the nucleosomes, leading to transcriptional silencing.

mechanism. Eukaryotic genomic DNA in the nucleus is packaged into a highly structured entity termed chromatin, which is composed of periodic arrays of the nucleosome core particle. The nucleosome core consists of 147 bp of DNA wrapped around a histone octamer core with two of each histone H2A, H2B, H3, and H4. Numerous studies have indicated that a variety of post-translational modifications such as acetylation, phosphorylation, methylation and ubiquitinylation, at key specific residues on the N-terminal tails of these histones represent a "histone code" that is important for the regulation of gene expression (Strahl and Allis, 2000; Jenuwein and Allis, 2001; Cosgrove et al., 2004; Buszczak and Spradling, 2006).

Although acetylation of histones has been widely associated with active gene transcription, methylation of certain residues on specific histones was found to be inhibitory. Methylation of the lysine 9 residue of histone H3, as well as methylation of lysine 20 of histone H4, has been found to be closely associated with regions of silenced genes and inactive heterochromatin (Margueron et al., 2005). Notably, Bernstein et al. (2006) identified a combination of H3 Lysine 4 and Lysine 27 methylation patterns that were proposed to regulate transcription factor expression concomitant with differentiation. Another report revealed that inactivation and transcriptional repression of the *Oct4* promoter during ESCs differentiation is coupled with an increase in tri-methylation of H3-Lysine 9 by the G9α histone methyltransferase (Feldman et al., 2006).

It is not clear whether the demethylation of histones is required for the re-activation of Oct4 during reprogramming, in which a differentiated somatic cell is transformed back to a stem cell state. Two specific histone demethylases have been identified recently (Shi et al., 2004; Tsukada et al., 2006). The precise mechanisms of epigenetic changes during reprogramming of somatic cells has recently attracted attention (Li, 2002; Simonsson and Gurdon, 2004; Cowan et al., 2005). Particularly, Lee et al. (2006) and Boyer et al. (2006) have attempted to distinguish ESCs from differentiated somatic cells by defining the chromatin organization, gene expression patterns and regulatory mechanisms.

Epigenetic regulation involves DNA methyltransferases (DNMTs) that attach methyl groups to the DNA, and polycomb group (PcG) proteins, which modify the histone proteins around which the DNA is wrapped. In human ESCs, the polycomb repressor complex (PRC2) subunit SUZ12 has been mapped to specific locations across the entire genome using ChIP-on-chip (Lee et al., 2006). PcG proteins are important developmental regulators, first described in the fruitfly *Drosophila* where they suppress homeotic genes controlling segment identity in the embryo (Kennison, 2004). The interplay between this suppression and activation by the trithorax group (trxG) proteins helps maintain epigenetic patterns of gene expression (Ringrose and Paro, 2004). The PcG proteins mammalian counterpart is highly conserved as the PRCs. In mammals, subunits EED, EZH2, and SUZ12 form the PRC2 complex that initiates gene silencing by catalyzing histone H3 methylation on lysine 27 (H3K27) at various loci (Kirmizis et al., 2004). Mediated by PRC1, transcriptional silencing is achieved either through chromatin condensation or through transcription initiation interference (Levine et al., 2004; Lund and van Lohuizen, 2004).

In human ESCs, SUZ12 was associated with the promoters of approximately 8% of annotated human protein genes (Lee et al., 2006). Many of these belong to developmental transcription factor families such as HOX protein, Forkhead box (FOX), GATA binding protein (GATA), SRY box (SOX), NK transcription factor related (NKX), LIM homeobox LHX, T-box (TBX), caudal type homeobox (CDX), distal-less homeobox (DLX), paired box and paired-like (PAX), myogenic basic domain (MYO), and NEUROD. Notably, some of the specific gene examples are key mediators of lineage specification either in stem cell differentiation or in embryo development. For example, CDX2 is required for trophectoderm formation during the blastocyst stage, GATA4 for primitive endoderm development, Brachyury for mesoderm development, NEUROD1 for neurogenesis, NEUROG2 for dopaminergic neuron formation, MYOD1 and PAX3 for myogenesis, and OLIG2 for oligodendrocyte formation. It appears likely that genes preferentially bound by the SUZ12 have functions that promote differentiation, and hence are suppressed by the PRC2 complex in pluripotent ESCs. Lee et al. (2006) further validated the gene expression of some of these genes by demonstrating that the expression levels were generally lower in ESCs compared with differentiated human cell and tissue types. To determine which genes are activated in ESCs, target loci bound by RNA polymerase II were mapped. Approximately 33% of the annotated protein-coding genes were identified, and most of these have basic cellular functions related to cell cycle, RNA and DNA metabolism, and protein biogenesis. Remarkably, a comparison of gene sets occupied by SUZ12 with those occupied by RNA polymerase II revealed that the two lists were largely mutually exclusive. This strongly suggests tight epigenetic regulation of the chromatin is one of the mechanisms that exist to prevent premature activation of differentiation-promoting genes.

It has been widely held that DNA methylation mediated by DNMTs and histones modifications by PcG proteins are discrete epigenetic pathways. In mammals, DNMT3A and -3B help to establish methylation at previously unmethylated sites, while DNMT1 serves as the major maintenance methyltransferase to recapitulate the existing methylation patterns during cell division. Apart from SUZ12, the histone methyltransferase EZH2 is part of the PRC2 complex that methylates histone H3 at the lysine 27 position. The link between DNMTs and PcG proteins was bridged when it was demonstrated that EZH2 interacts with all three DNMTs both in vitro and in vivo (Vire et al., 2006). This important observation lends weight to the suggestion that gene repression is mediated in a stepwise manner. It was known that EZH2 adds methyl groups to histone proteins. Vire et al. (2006) further showed that EZH2 interacts with DNMTs by recruiting them to regulatory regions of target genes. Subsequently, the DNMTs are able to methylate DNA either through de novo or maintenance methylation. It is also postulated that these methylation marks can aid in recruiting further repressive complexes such as PRC1 or histone deacetylase corepressors.

In summary, these studies by Lee et al. (2006) and Boyer et al. (2006) used human ESCs to examine how epigenetic regulation can interact with a well-characterized core regulatory network, defined by OCT4, SOX2 and NANOG, to maintain stem cell properties (Fig. 37–7). Comparison of the gene set bound and repressed by OCT4, SOX2, and NANOG with those occupied by PRC2 complex indicated that each of the three transcription factors occupied one-third of the PRC2-bound genes encoding differentiation-promoting transcription factors (Lee et al., 2006). Nearly all genes repressed by OCT4, SOX2 and NANOG in Boyer et al. (2005) were also occupied by PRC2. This supports the link between the repression of differentiation-promoting factors mediated by PRCs and their physical associations with OCT4, SOX2 and NANOG, which mark the target genes for either positive or negative transcription regulation. The studies in the epigenetic changes and the analysis of the mechanistic aspects of DNA and histone demethylation in ESCs will provide insights into early embryonic development as well as strategies to more easily manipulate ESCs.

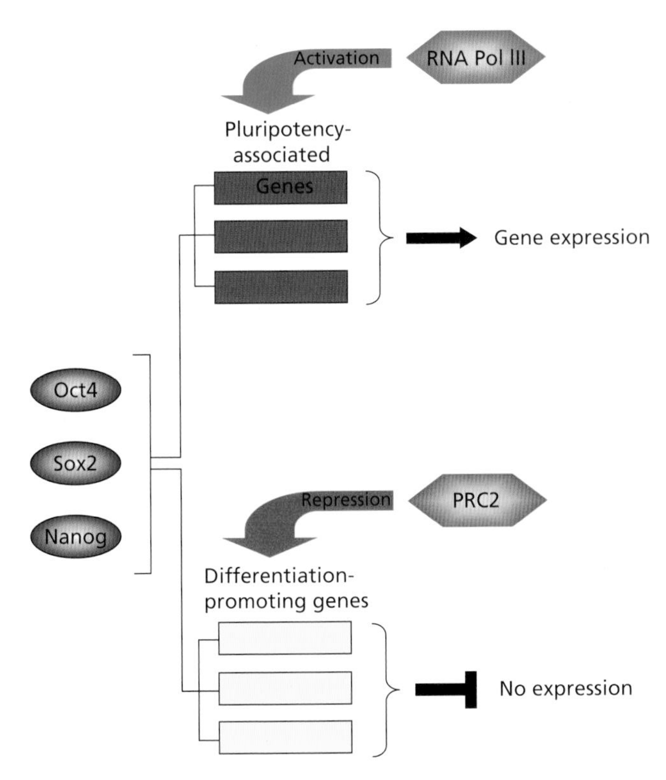

Figure 37–7 Regulation of chromatin architecture in ESCs by transcription factors. The ESC transcription factors, OCT4, SOX2 and NANOG bind to sets of genes that are also epigenetically regulated. Consistent with their positive expression, RNA polymerase II also binds to the promoters of OCT4, SOX2 and NANOG-regulated genes that are necessary for stem cell pluripotency and self-renewal. Conversely, genes repressed by the same transcription factors are associated with the PcG protein complexes. The interaction of high affinity transcription factors and specific regulators of gene transcription and chromatin modification thus establishes a functional state of transcriptional regulation of very specific genes.

Applications of Stem Cell Genomics in Cell-based Therapy

Stem Cells, Immune Rejection, and Allogenic Transplantation

The success of somatic stem cell transplantation can be influenced by variables such as the age of donor, quality of donor tissue, and heterogeneity of the underlying pathology of the recipient. Furthermore, there is little understanding about the immunogenicity of cell transplantation and how this contributes to poor clinical outcome in the long-term (Bradley et al., 2002). One exception is bone marrow transplantation which has built on a mature scientific foundation that has enabled a clinical transition from an experimental, relatively high-risk procedure to a standard medical treatment for hematopoietic malignancy (Weissman, 2000a). The success of technical and clinical advances in the engraftment procedure itself has been reliant on the parallel development of serological, transcriptional, and genomic tissue-typed stem cell donor registries. Recently, private companies are offering the cryo-preservation of cord blood cells. Fetal cord blood contains cells that can contribute to the hematopoietic and mesenchymal lineages and can be preserved for expansion and autologous cell transplantation when required for the treatment of bone marrow stem cell-related diseases (Dawson et al., 2003; Faden et al., 2003).

Organ transplantation and serology studies have identified mismatches of known, donor-associated major histocompatibility complex class I and class II alleles as primary correlates of allogenicity and graft rejection. Important roles in immune response have been attributed to natural killer (NK) cell-mediated recognition, polymorphic peptides capable of presentation as minor histocompatibility complex antigens, and polymorphisms in cytokine ligands (Petersdorf and Deeg, 1992; Mullighan and Petersdorf, 2006; Petersdorf and Malkki, 2006). In treatment of myeloma using autologous stem cell transplantation, gene expression profiling has identified aberrant gene transcription patterns in bone marrow-derived cell populations that, upon re-incorporation after high-dose therapy, associate with poor outcome and survival (Shaughnessy and Barlogie, 2003).

Even with minimal tissue mismatch and immuno-suppression therapy, immune rejection continues to present a considerable barrier to somatic or embryonic stem cell transplantation. To improve the reconstitution and engraftment potential of bone marrow-derived stem cells, the introduction of characterized, drug-resistance genes into allogenic somatic stem cells for efficient repopulation, and elimination of host immune cells have been proposed (Zaboikin et al., 2006). However, this strategy is confounded by the recent discovery of constitutively expressed drug-resistance genes in endogenous HSCs (Cai et al., 2004).

A chief advantage of exploiting human ESCs for cell-based therapy is their capacity for long-term cryo-preservation and multi-lineage differentiation. With parallel advances in improved pathogen- and xenogen-free culture conditions, the development of extensive, human tissue-typed ESC libraries would constitute a useful resource for human ESC-related transplantation strategies. Several countries have established publicly funded stem cell banks, with the short-term goal of somatic and embryonic stem cell collection, characterization, and storage, as well as the potential long-term goal of informing transplantation risk relative to genotype. The refinement of genomics platforms for assessing the immuno-haplotype of ESC lines will be an important component of cataloguing such a library, in addition to appropriate management of microbiological contamination and patient data (Cobo et al., 2005; Rodriguez et al., 2006).

Considerable interest has been raised by the alternative solution of matching recipient tissues using SCNT. Even if the considerable technical and ethical barriers to this apparent solution to immune rejection can be resolved, tissue derived using the SCNT procedure would still has to contend with the minor histocompatibility antigen recognition as a consequence of maternal oocyte-derived mitochondrial gene expression (Bradley et al., 2002).

The genetic modification of human ESCs to prevent the expression of specific antigens for creating universal donor cells, or the direct inhibition of inflammatory pathways before ESC differentiation and engraftment, have been suggested as potential methods for preventing immune rejection. However, such approaches may contribute to pathogenesis in the advent of ESC mutagenesis or viral infection. One interesting possibility is the creation of "micro-chimeras" to limit immune rejection. In this strategy, ESC-derived HSCs and ESC-derived differentiated transplant cells are co-engrafted after myeloablation. In mice, the transplantation of donor mismatched ESC-like cells into recipients served to condition the immune system before cardiac allograft (Fandrich et al., 2002). Alternatively, the repopulation of a recipient organism with

ESC-derived HSCs differentiated in vivo may also allow for long-term tolerance (Bradley et al., 2002).

Cancer and Stem Cells

It has been hypothesized that cancer may arise from a stem cell-specific alteration in genome structure, resulting in the dysregulation of intrinsic stem cell self-renewal and differentiation properties, and alteration of their interactions with the microenvironment. Therefore, understanding how genome integrity is maintained in stem cells may inform situations where this maintenance fails in cancer (Reya et al., 2001; Andrews et al., 2005). Identifying disparities between somatic stem cells and putative cancer stem cells may provide a basis for targeting genetically abnormal cells for destruction, and enable the preservation of normal somatic stem cells during chemotherapy (Raetz & Moos, 2004; Jordan, 2004; Bjerkvig et al., 2005; Morrison, 2005). Dysregulated expression of the PcG protein BMI1 observed in leukemia is suggestive of failure in HSC self-renewal (Park et al., 2003; Raaphorst, 2005). Similar dysregulation of BMI1 also alters NSC self-renewal and proliferation (Molofsky et al., 2003, 2005).

Hypoacetylation of chromatin-associated proteins is a common phenomenon associated with tumorigenesis, and several pharmacological agents that inhibit deacetylation activity or promote acetyltransferase activity are in preclinical development (Cerny and Quesenberry, 2004). Of interest are the Sirtuins, a class of deacetylation enzymes originally identified in relation to the acetylation-mediated inhibition of Forkhead and p53-mediated apotosis (Langley et al., 2002). In mouse embryonic fibroblasts, human SIRT1 appears to mediate the deacetylation of histones localized to hypermethylated DNA, leading to the reactivation of heterochromatin-associated genes (Pruitt et al., 2006).

The exploitation of stem cell-associated properties such as lineage negativity and the presence of drug-resistance genes such as ABC transporters have already been used in the isolation of putative stem cells from tumors. It has been proposed that putative cancer stem cells may contain ABC transporters which explain the resistance of tumorigenic cells to conventional drug therapies (Jordan, 2005). The ABC transporters expressed by these cells were verapamil-sensitive and could export the externally introduced fluorescent dyes, Hoescht 33323 or Rhodamine 123. This dye exclusion property provides a strategy for cell separation of this "side-population" from dye-retaining cell types. Side-population analysis has been applied to isolate cells from the brain, muscle, skin, testis, and tumor tissues (Setoguchi et al., 2004). The properties of side-populations and how they relate somatic stem cells are beginning to be explored (Challen and Little, 2006).

Germ-line tumor cells, or EC cells, responsible for the formation of teratocarcinomas express certain genes associated with early blastocyst development and ESC pluripotency, and can give rise to mesodermal, endodermal and ectodermal lineages. While exhibiting substantial karyotypic abnormalities, EC cells were an interesting model to study the regulation of pre-implantation development before the isolation of ESCs. Although EC cells are a useful paradigm for understanding the biogenesis of a highly malignant tumor, they have provided a theoretical and technical framework for the subsequent culture and manipulation of ESCs derived from pre-implantation embryos (Martin and Evans, 1975; Andrews, 2002). The artificial selection of these cell populations for growth in vitro limited the extent to which they could be used to study cell differentiation (Andrews et al., 2005). However, comparisons of EC cells with ESCs have provided a useful framework for examining the mechanistic basis for cancerous progression from a self-renewing phenotype such as ESCs, to a dysregulated, aneuploid cell type largely restricted to self-renewal in the case of EC cells. Interestingly, human ESCs maintained for long periods in culture exhibit similar changes in ploidy to EC cells – duplications of chromosome 12 and 17, suggestive of tumorigenic potential (Sperger et al., 2003; Draper et al., 2004). These studies highlight the potential risks of prolonged human ESC culture, and recommend the use of low-passage ESCs for clinical research (Maitra et al., 2005; Plaia et al., 2005).

Directed Differentiation of Stem Cells

To realize the applications of stem cells as transplantable vectors for gene product delivery or as a source for tissue regeneration, reliable methods for directing their differentiation must be adequately defined. Given the teratoma-forming property of human ESCs in particular, the complete irreversible differentiation of somatic and embryonic stem cells into a particular, lineage-restricted cell phenotype is a key requirement for cell-based therapy. Molecular examination of stem cells would provide clues to understanding the transitions necessary and sufficient to direct stem cell differentiation.

The differentiation of stem cells has been informed by the molecular and cellular characterizations of developmental processes in invertebrate and vertebrate systems. Cell-extrinsic cues influences cell-intrinsic signaling and transcriptional activities, as observed in vertebrate cell fate specification (Tanabe and Jessell, 1996). High-throughput analyses of gene transcription in the study of cell differentiation during development have been useful in defining patterns of gene expression associated with neuronal specification. Investigations into the developing mouse neocortex that chart global patterns of rostral to caudal and dorsal to ventral gene expression has enabled the development of genome "maps" (Livesey et al., 2004; Sansom et al., 2005) that can provide insights into attempts to derive subtype-specific cortical neuronal populations from somatic or embryonic stem cells. Studies have also indicated that mouse ESCs which exhibit a default neural differentiation pathway, recapitulate appropriate temporal patterns of developmental gene transcription (Bouhon et al., 2005; Bouhon et al., 2006). Microarray platforms have facilitated the comparison of in vitro cell differentiation with developmental gene expression. This provides information to discriminate the genes associated with undifferentiated stem cells and their lineage-specified progeny (Yang et al., 2005).

Gene expression profiling during development has been useful in informing translational cell therapies. Mouse ESCs have been directed to acquire motor neuron identity by the sequential application of the signaling molecules retinoic acid (RA) and Sonic Hedgehog (SHH) along with the over-expression of specific transcription factors. The motor neurons were capable of appropriate integration into developing mouse embryos with axonal elaboration, muscle innervation, and the formation of functional synapses and electrophysiological properties (Wichterle et al., 2002; Miles et al., 2004). Dopaminergic neurons were also derived through the sequential application of morphogens, growth factors, and transcription factor over-expression (Lee et al., 2000; Kim et al., 2002; Kim et al., 2003). The molecular interactions necessary for SHH-dependent dopaminergic neuronal differentiation of mouse ESCs have recently been described (Andersson et al., 2006). Similar to mouse ESCs, human ESCs acquire pan-neuronal characteristics after differentiation (Itskovitz-Eldor et al., 2000; Carpenter et al., 2001; Reubinoff et al., 2001; Schuldiner et al., 2001). The mechanisms that underlie subtype-specific neuronal specification from human ESCs are less defined. However, mouse ESC experiments have informed of the conditions that could direct differentiation of human ESCs to dopaminergic neurons (Schulz et al., 2003; Schulz et al., 2004; Zeng et al., 2004), neural crest–derived peripheral neurons

(Pomp et al., 2005), motor neurons (Li et al., 2005), and retinal pigment epithelial cells (Klimanskaya et al., 2004).

Cell-extrinsic signaling cues that promote pan-neuronal differentiation include Noggin, an antagonist of extracellular bone morphogenic protein (BMP; Itsykson et al., 2005), and neurotrophins such as nerve growth factor, neurotrophin-3 (Levenberg et al., 2005) and GDNF (Buytaert-Hoefen et al., 2004). The derivation of oligodendrocyte-competent glial stem cells from mouse and human ESCs has also been demonstrated (Brustle et al., 1999; Faulkner and Keirstead, 2005; Keirstead, 2005). Human ESC-derived oligodendrocytes can induce remyelination after transplantation to demyelinated tissue, confirming functional competence (Glaser et al., 2005).

Conditions for ex vivo culture of somatic stem cells to promote self-renewal may also be improved through simulation of a stem cell "niche" signaling environment. Candidate factors-based immunohistochemistry and *in situ* hybridization methods have identified extracellular antagonists of the BMP pathway which includes Noggin and Neurogenesin that may act to eliminate the action of differentiation-inducing BMPs within the NSC niche environment (Lim et al., 2000; Ueki et al., 2003). Although HSCs are difficult to propagate in culture (Greenberger et al., 1979; Cohen et al., 1982; Boggs, 1999), bone marrow-derived stromal support cells which present within the HSC niche in vivo are known to provide soluble factors that promote self-renewal (Shadduck et al., 1983) and contact-mediated cues that promote differentiation (Verfaillie and Catanzaro, 1996).

Genetic manipulation provides another possible avenue for directed differentiation or improved stem cell isolation. To facilitate the activation of differentiation genes directly, retroviral and lentiviral mediated infections have been used for exogenous over-expression of genes in somatic and embryonic stem cells. The clinical application of gene therapy vectors in controlling stem cell function for the repair of monogenic disease is limited by the risk of insertional mutagenesis leading to oncogenic transformation. Over-expression or gene knockdown in somatic and embryonic stem cells can be utilized as a form of genetic manipulation (Lerou and Daley, 2005). For example, mouse and human ESC specification into atrial, ventricular or sinus node-like cell types have been observed after the introduction of a cardiomyocyte-specific promoter-driven antibiotic-resistance gene, followed by conventional antibiotic selection (Klug et al., 1996). However, the in vitro structural and functional properties of ESC-derived cardiomyocytes derived from gene perturbation were characteristic of embryonic (fetal) rather than adult cardiomyocytes, suggesting that further maturation of these cells would be required.

Promoters of genes active in specific stem cell populations can be conjugated to fluorescent reporter constructs that, when introduced, are transcribed in cells in which the promoter is active. The emitted fluorescence signal can be used to isolate these putative stem cells by FACS. For example, introduction of a reporter construct conjugated to the NSC-associated cytoskeletal protein NESTIN, into homogenized neural tissue has been used to enrich for NSCs (Keyoung et al., 2001). Although these studies establish that directed differentiation of stem cells is possible, there may be undesirable consequences in some instances. For instance, HSCs retain a capacity for stochastic differentiation which may be important for continuous maintenance of multiple lineages independent of environmental demand (Wagers et al., 2002). Consequently, artificially biasing stem cells toward one lineage by gene perturbation may be of corrective benefit, but can also potentially alter long-term homeostasis of the hematopoietic system.

Trans-differentiation, Metaplasia and Cell Fusion

Several studies have raised the possibility that somatic stem cells may transit between the "niches" of different tissues, and contribute to stem cell maintenance in multiple lineages. The ability of certain stem cells to exhibit broad lineage potential, and contribute to functional regeneration of multiple lineage-specific cell phenotypes, remains controversial (Blau et al., 2001; Morrison, 2001; Weissman et al., 2001). Of interest is the contribution of bone marrow-derived stem cell populations to non-hematopoietic and non-mesenchymal lineages. Adult skeletal muscle regeneration, while primarily mediated by local, tissue-resident stem cell populations, appears to incorporate bone marrow-derived cells after muscle injury (LaBarge and Blau, 2002). Contributions of circulating, bone marrow-derived stem cells to cardiomyocyte (Eisenberg et al., 2006), epithelial (Krause et al., 2001), and neural (Jiang et al., 2003) phenotypes have also been reported, although NSCs may not contribute to hematopoietic lineages (Massengale et al., 2005). A surprising, recent example of suggests bone marrow-derived stem cells retain germ-line potential and are capable of contributing to female oogenesis (Johnson et al., 2005). The manipulation and mobilization of endogenous bone marrow stem cell populations to initiate the repair of distant tissues is a tantalizing prospect (Tsai et al., 2002).

The mechanistic basis of stem cell plasticity remains unclear and its contribution to normal homeostasis is uncertain. Observations of the trans-differentiation phenomenon, or metaplasia, are not new. Indeed, while associated with tissue regeneration in anurans and urodeles, trans-differentiation is largely associated with disease or tissue injury in mammals (Slack and Tosh, 2001; Tosh and Slack, 2002; Masson et al., 2004). The assessment of the genome-level changes that correlate with the re-expression of genes associated with unexpected lineages may provide clues on the mechanisms by which cell identity is stabilized in the various somatic stem cell populations.

Alternatively, mobilized stem cells may contribute to tissue repair through fusion with resident differentiated cells and subsequent transfer of genetic material (Orlic, 2004). Cell fusion-mediated incorporation of circulatory stem cells has been observed in cerebellar Purkinje neurons (Alvarez-Dolado et al., 2003; Weimann et al., 2003), hepatocytes (Vassilopoulos and Russell, 2003; Vassilopoulos et al., 2003), and cardiac tissue (Terada et al., 2002; Ying et al., 2002). Cell fusion results in the formation of stable heterokaryons, causing the interchange of transcription and chromatin remodeling factors between distinct nuclei, or synkaryons, in which chromosome rearrangement and expulsion can enable the derivation of an integrated nucleus of apparently normal karyotype (Ogle et al., 2005). The observation of stem cell fusion has provoked interest in the mechanisms of cell fusion and its physiological role in endogenous tissue repair. For example, the fusion of ex vivo modified, autologous stem cells may represent an alternative to viral-mediated gene delivery for the manipulation of endogenous gene expression. It has been proposed that fusion events may also be operant in the development of cancers (Bjerkvig et al., 2005). The biological properties that enable cell fusion, as well as the susceptibility of certain somatic and embryonic stem cells to this process are poorly defined.

Investigations into cell fusion in vitro have also provided a useful tool in evaluating nuclear- or cytoplasm-associated factors that regulate stem cell function (Hakelien and Collas, 2002; Collas, 2003; Taranger et al., 2005). The creation of heterokaryons by chemical mediated fusion of pluripotent cells with differentiated cells has been used to assess global transcriptional and epigenetic changes (Tada et al., 2001; Cowan et al., 2005). Similar fusion mechanisms have been used across species barriers to address the role of fusion-mediated gene activation or de-repression. The cytoplasm of the *Xenopus laevis* oocyte represents an environment competent for altering gene transcription of a donor cell nucleus to resemble that of a pluripotent cell (Gurdon et al., 2003). The introduction of mammalian somatic cell nuclei into this environment results in the demethylation of the *OCT4* promoter (normally silent in the majority of somatic cell lineages) activating its transcription (Byrne et al., 2003; Simonsson and Gurdon, 2004, 2005).

In a landmark experiment, it was demonstrated that it is possible to reverse a stable differentiated state by carefully manipulating the levels of key gene transcripts in mouse somatic cells (Takahashi and Yamanaka, 2006). By over-expressing a specific subset of ESC-associated molecules selected on the basis of previous genomics profiling, pluripotent stem-like cells could be induced from mouse embryonic fibroblasts. It was further demonstrated that these karyotypically normal cells exhibit the morphology and growth properties of ESCs, including the ability to contribute to germ-line cells (Okita et al., 2007; Wernig et al., 2007). The ability of somatic cells to become transformed into pluripotent cells capable of differentiating into all cell types will usher in a new era in personalized regenerative medicine. While this technique of bioengineering stem cells from a patient's somatic cells has not yet been successful using human cells, there appears to be no insurmountable biological or ethical challenges preventing us from doing so.

Conclusion

The existence of stem cells during development and within adult human tissues highlights the nature of the cell genome as a dynamic entity, whose accessibility and transcriptional output is regulated in a temporal and spatial manner. Stem cells have shown considerable promise as vectors for gene therapy, and genetic manipulation of stem cell differentiation has enabled the enrichment and isolation of specific cell types for transplantation medicine or experimental research. We have also gained valuable insights into the core mechanism governing stem cell identity and function as a result of in vitro culture. It is widely anticipated that better understanding of the stem cell "epigenome" will further provide a more comprehensive understanding of stem cell properties, such as self-renewal, resistance to senescence, and multi-lineage differentiation. The link between somatic stem cells and cancer stem cells have sparked interest in how dysregulation of the stem cell identity and genome architecture is related to disease initiation and progression. This may provide insights into the development of novel treatment for cancer by examining how we can target these cancer stem cells. The molecular profiling of donor and recipient genomes can improve the efficiency and clinical outcomes for tissue transplantations. However, we note that considerable clinical, technical, and ethical barriers still remain for the implementation of stem cells as therapeutic agents.

Acknowledgment

We are grateful to the Agency for Science, Technology and Research (A*STAR) and Biomedical Research Council (BMRC) for funding. W.L.T. is supported by the A*STAR graduate scholarship. B.L. is partially supported by grants from NIH (DK47636 and AI54973).

References

Allegrucci, C, Denning, C, Priddle, H, and Young, L (2004). Stem-cell consequences of embryo epigenetic defects. *Lancet*, 364:206–208.

Alvarez-Buylla, A and Lim, DA (2004). For the long run: maintaining germinal niches in the adult brain. *Neuron*, 41:683–686.

Alvarez-Dolado, M, Pardal, R, Garcia-Verdugo, JM, Fike, JR, Lee, HO, Pfeffer, K, et al. (2003). Fusion of bone-marrow-derived cells with Purkinje neurons, cardiomyocytes and hepatocytes. *Nature*, 425:968–973.

Amit, M, Carpenter, MK, Inokuma, MS, Chiu, CP, Harris, CP, Waknitz, MA, et al. (2000). Clonally derived human embryonic stem cell lines maintain pluripotency and proliferative potential for prolonged periods of culture. *Dev Biol*, 227:271–278.

Andersson E, Tryggvason U, Deng Q, Friling S, Alekseenko Z, Robert B, et al. (2006). Identification of intrinsic determinants of midbrain dopamine neurons. *Cell*, 124:393–405.

Andrews, PW (2002). From teratocarcinomas to embryonic stem cells. *Philos Trans R Soc Lond B Biol Sci*, 357:405–417.

Andrews, PW, Goodfellow, PN, Shevinsky, LH, Bronson, DL, and Knowles, BB (1982). Cell-surface antigens of a clonal human embryonal carcinoma cell line: morphological and antigenic differentiation in culture. *Int J Cancer*, 29:523–531.

Andrews, PW, Matin, MM, Bahrami, AR, Damjanov, I, Gokhale, P, and Draper, JS (2005). Embryonic stem (ES) cells and embryonal carcinoma (EC) cells: opposite sides of the same coin. *Biochem Soc Trans*, 33:1526–1530.

Anisimov, SV, Tarasov, KV, Riordon, D, Wobus, AM, and Boheler, KR (2002). SAGE identification of differentiation responsive genes in P19 embryonic cells induced to form cardiomyocytes in vitro. *Mech Dev*, 117:25–74.

Anisimov, SV, Tarasov, KV, Tweedie, D, Stern, MD, Wobus, AM, and Boheler, KR (2002). SAGE identification of gene transcripts with profiles unique to pluripotent mouse R1 embryonic stem cells. *Genomics*, 79:169–176.

Arai, F, Hirao, A, and Suda, T (2005). Regulation of hematopoietic stem cells by the niche. *Trends Cardiovasc Med*, 15:75–79.

Baksh, D, Song, L, and Tuan, RS (2004). Adult mesenchymal stem cells: characterization, differentiation, and application in cell and gene therapy. *J Cell Mol Med*, 8:301–316.

Barclay, AN, Brown, MH, Law, SKA, McKnight, AJ, Tomlinson, MG, van der Merwe, PA. (1997) *The Leucocyte Antigen FactsBook (2nd Ed)*. Academic Press; London / San Diego, (1997).Barker, RA (2002). Repairing the brain in Parkinson's disease: where next? *Mov Disord*, 17:233–241.

Bartel, DP and Chen, CZ (2004). Micromanagers of gene expression: the potentially widespread influence of metazoan microRNAs. *Nat Rev Genet*, 5:396–400.

Baylin, SB and Herman, JG (2000). DNA hypermethylation in tumorigenesis: epigenetics joins genetics. *Trends Genet*, 16:168–174.

Behrstock, S, Ebert, A, McHugh, J, Vosberg, S, Moore, J, Schneider, B, et al. (2006). Human neural progenitors deliver glial cell line-derived neurotrophic factor to parkinsonian rodents and aged primates. *Gene Ther*, 13:379–388.

Behrstock, S and Svendsen, CN (2004). Combining growth factors, stem cells, and gene therapy for the aging brain. *Ann N Y Acad Sci*, 1019:5–14.

Bernstein, BE, Mikkelsen, TS, Xie, X, Kamal, M, Huebert, DJ, Cuff, J, et al. (2006). A bivalent chromatin structure marks key developmental genes in embryonic stem cells. *Cell*, 125:315–326.

Bird, A (2002). DNA methylation patterns and epigenetic memory. *Genes Dev*, 16:6–21.

Bjerkvig, R, Tysnes, BB, Aboody, KS, Najbauer, J, and Terzis, AJ (2005). Opinion: the origin of the cancer stem cell: current controversies and new insights. *Nat Rev Cancer*, 5:899–904.

Bjorklund, A, Dunnett, SB, Brundin, P, Stoessl, AJ, Freed, CR, Breeze, RE, et al. (2003). Neural transplantation for the treatment of Parkinson's disease. *Lancet Neurol*, 2:437–445.

Blais, A, and Dynlacht, BD (2005). Constructing transcriptional regulatory networks. *Genes Dev*, 19:1499–1511.

Blau, HM, Brazelton, TR, and Weimann, JM (2001). The evolving concept of a stem cell: entity or function? *Cell*, 105:829–841.

Boggs, SS (1999). The hematopoietic microenvironment: phylogeny and ontogeny of the hematopoietic microenvironment. *Hematol*, 4:31–44.

Bouhon IA, Joannides A, Kato H, Chandran S, Allen ND. (2006) Embryonic stem cell-derived neural progenitors display temporal restriction to neural patterning. *Stem Cells*. 24(8):1908–13.

Bouhon, IA, Kato, H, Chandran, S, and Allen, ND (2005). Neural differentiation of mouse embryonic stem cells in chemically defined medium. *Brain Res Bull*, 68:62–75.

Boyer, LA, Lee, TI, Cole, MF, Johnstone, SE, Levine, SS, Zucker, JP, et al. (2005). Core transcriptional regulatory circuitry in human embryonic stem cells. *Cell*, 122:947–956.

Boyer, LA, Plath, K, Zeitlinger, J, Brambrink, T, Medeiros, LA, Lee, TI, et al. (2006). Polycomb complexes repress developmental regulators in murine embryonic stem cells. *Nature*, 441:349–353.

Bradley, A (1990). Embryonic stem cells: proliferation and differentiation. *Curr Opin Cell Biol*, 2:1013–1017.

Bradley, A and Robertson, E (1986). Embryo-derived stem cells: a tool for elucidating the developmental genetics of the mouse. *Curr Top Dev Biol*, 20:357–371.

Bradley, JA, Bolton, EM, and Pedersen, RA (2002). Stem cell medicine encounters the immune system. *Nat Rev Immunol*, 2:859–871.

Brandenberger, R, Khrebtukova, I, Thies, RS, Miura, T, Jingli, C, Puri, R, et al. (2004). MPSS profiling of human embryonic stem cells. *BMC Dev Biol*, 4:10.

Brustle, O, Jones, KN, Learish, RD, Karram, K, Choudhary, K, Wiestler, OD, et al. (1999). Embryonic stem cell-derived glial precursors: a source of myelinating transplants. *Science*, 285:754–756.

Buszczak, M and Spradling, AC (2006). Searching chromatin for stem cell identity. *Cell*, 125:233–236.

Buytaert-Hoefen, KA, Alvarez, E, and Freed, CR (2004). Generation of tyrosine hydroxylase positive neurons from human embryonic stem cells after coculture with cellular substrates and exposure to GDNF. *Stem Cells*, 22:669–674.

Byrne, JA, Simonsson, S, Western, PS, and Gurdon, JB (2003). Nuclei of adult mammalian somatic cells are directly reprogrammed to oct-4 stem cell gene expression by amphibian oocytes. *Curr Biol*, 13:1206–1213.

Cai, J, Shin, S, Wright, L, Liu, Y, Zhou, D, Xue, H, et al. (2006) Massively parallel signature sequencing profiling of fetal human neural precursor cells. *Stem Cells Dev*, 15:232–244.

Cai, J, Weiss, ML, Rao, MS (2004). In search of "stemness." *Exp Hematol*, 32:585–598.

Cai, Y, Wu, P, Ozen, M, Yu, Y, Wang, J, Ittmann, M, et al. (2006). Gene expression profiling and analysis of signaling pathways involved in priming and differentiation of human neural stem cells. *Neuroscience*, 138:133–148.

Capela, A and Temple, S (2002). LeX/ssea-1 is expressed by adult mouse CNS stem cells, identifying them as nonependymal. *Neuron*, 35:865–875.

Carpenter, MK, Inokuma, MS, Denham, J, Mujtaba, T, Chiu, CP, and Rao, MS (2001). Enrichment of neurons and neural precursors from human embryonic stem cells. *Exp Neurol*, 172:383–397.

Cerny, J and Quesenberry, PJ (2004). Chromatin remodeling and stem cell theory of relativity. *J Cell Physiol*, 201:1–16.

Challen, GA and Little, MH (2006). A side order of stem cells: the SP phenotype. Stem *Cells* 24:3–12.

Chambers, I, Colby, D, Robertson, M, Nichols, J, Lee, S, Tweedie, S, et al. (2003). Functional expression cloning of NANOG, a pluripotency sustaining factor in embryonic stem cells. *Cell*, 113:643–655.

Charge, SB and Rudnicki, MA (2004). Cellular and molecular regulation of muscle regeneration. *Physiol Rev*, 84:209–238.

Chen, CZ, Li, L, Lodish, HF, and Bartel, DP (2004). MicroRNAs modulate hematopoietic lineage differentiation. *Science*, 303:83–86.

Chen, JC and Goldhamer, DJ (2003). Skeletal muscle stem cells. *Reprod Biol Endocrinol*, 1:101.

Cheng, LC, Tavazoie, M, and Doetsch, F (2005). Stem cells: from epigenetics to microRNAs. *Neuron*, 46:363–367.

Cobo, F, Stacey, GN, Hunt, C, Cabrera, C, Nieto, A, Montes, R, et al. (2005). Microbiological control in stem cell banks: approaches to standardisation. *Appl Microbiol Biotechnol*, 68:456–466.

Cohen, GI, Greenberger, JS, and Canellos, GP (1982). Effect of chemotherapy and irradiation on interactions between stromal and hemopoietic cells in vitro. *Scan Electron Microsc*,(Pt 1):359–365.

Collas, P (2003). Nuclear reprogramming in cell-free extracts. *Philos Trans R Soc Lond B Biol Sci*, 358:1389–1395.

Cosgrove, MS, Boeke, JD, and Wolberger, C (2004). Regulated nucleosome mobility and the histone code. *Nat Struct Mol Biol*, 11:1037–1043.

Cowan, CA, Atienza, J, Melton, DA, and Eggan, K (2005). Nuclear reprogramming of somatic cells after fusion with human embryonic stem cells. *Science*, 309:1369–1373.

da Rocha ST, and Ferguson-Smith, AC (2004). Genomic imprinting. *Curr Biol*, 14:R646–649.

Dawson, L, Bateman-House, AS, Mueller Agnew, D, Bok, H, Brock, DW, et al. (2003). Safety issues in cell-based intervention trials. *Fertil Steril*, 80:1077–1085.

Dillon, N (2004). Heterochromatin structure and function. *Biol Cell*, 96:631–637.

Dillon, N and Festenstein, R (2002). Unravelling heterochromatin: competition between positive and negative factors regulates accessibility. *Trends Genet*, 18:252–258.

Doetsch, F (2003). A niche for adult neural stem cells. *Curr Opin Genet Dev*, 13:543–550.

Doetsch, F, Caille, I, Lim, DA, Garcia-Verdugo, JM, and Alvarez-Buylla, A (1999). Subventricular zone astrocytes are neural stem cells in the adult mammalian brain. *Cell*, 97:703–716.

Draper, JS, Smith, K, Gokhale, P, Moore, HD, Maltby, E, Johnson, J, et al. (2004). Recurrent gain of chromosomes 17q and 12 in cultured human embryonic stem cells. *Nat Biotechnol*, 22:53–54.

Eisenberg, CA, Burch, JB, and Eisenberg, LM (2006). Bone marrow cells transdifferentiate to cardiomyocytes when introduced into the embryonic heart. *Stem Cells*, 24:1236–1245.

Eriksson, PS, Perfilieva, E, Bjork-Eriksson, T, Alborn, AM, Nordborg, C, Peterson, DA, et al. (1998). Neurogenesis in the adult human hippocampus. *Nat Med*, 4:1313–1317.

Esau, C, Kang, X, Peralta, E, Hanson, E, Marcusson, EG, Ravichandran, LV, et al. (2004). MicroRNA-143 regulates adipocyte differentiation. *J Biol Chem*, 279:52361–52365.

Evans, MJ and Kaufman, MH (1981). Establishment in culture of pluripotential cells from mouse embryos. *Nature*, 292:154–156.

Evsikov, AV and Solter, D (2003). Comment on "'Stemness': transcriptional profiling of embryonic and adult stem cells" and "a stem cell molecular signature." *Science*, 302:393; author reply 393.

Faden, RR, Dawson, L, Bateman-House, AS, Agnew, DM, Bok, H, Brock, DW, et al. (2003). Public stem cell banks: considerations of justice in stem cell research and therapy. *Hastings Cent Rep*, 33:13–27.

Fandrich, F, Lin, X, Chai, GX, Schulze, M, Ganten D, Bader M, et al. (2002) Preimplantation-stage stem cells induce long-term allogeneic graft acceptance without supplementary host conditioning. Nat Med. (2):171–8.

Faulkner J, and Keirstead, HS (2005). Human embryonic stem cell-derived oligodendrocyte progenitors for the treatment of spinal cord injury. *Transpl Immunol*, 15:131–142.

Fawcett, JW and Asher, RA (1999). The glial scar and central nervous system repair. *Brain Res Bull*, 49:377–391.

Feldman, N, Gerson, A, Fang, J, Li, E, Zhang, Y, Shinkai, Y, et al. (2006). G9a-mediated irreversible epigenetic inactivation of Oct-3/4 during early embryogenesis.*Nat Cell Biol*, 8:188–194.

Fortunel, NO, Otu, HH, Ng, HH, Chen, J, Mu, X, Chevassut, T, et al. (2003) Comment on "'Stemness': transcriptional profiling of embryonic and adult stem cells" and "a stem cell molecular signature." *Science*, 302:393; author reply 393.

Franklin, RJ (2002b). Why does remyelination fail in multiple sclerosis? *Nat Rev Neurosci*, 3:705–714.

Fuchs, E and Segre, JA (2000). Stem cells: a new lease on life. *Cell*, 100:143–155.

Futcher, B (2000). Microarrays and cell cycle transcription in yeast. *Curr Opin Cell Biol*, 12:710–715.

Gage, FH (2000). Mammalian neural stem cells. *Science*, 287:1433–1438.

Gearhart, J (2004). New human embryonic stem-cell lines–more is better. *N Engl J Med*, 350:1275–1276.

Geschwind, DH, Ou, J, Easterday, MC, Dougherty, JD, Jackson, RL, et al. (2001). A genetic analysis of neural progenitor differentiation. *Neuron*, 29:325–339.

Gill, SS, Patel, NK, Hotton, GR, O'Sullivan, K, McCarter, R, Bunnage, M, et al. (2003). Direct brain infusion of glial cell line-derived neurotrophic factor in Parkinson disease. *Nat Med*, 9:589–595.

Giraldez, AJ, Cinalli, RM, Glasner, ME, Enright, AJ, Thomson, JM, Baskerville, S, et al. (2005). MicroRNAs regulate brain morphogenesis in zebrafish. *Science*, 308:833–838.

Glaser, T, Perez-Bouza, A, Klein, K, and Brustle, O (2005). Generation of purified oligodendrocyte progenitors from embryonic stem cells. *Faseb J*, 19:112–114.

Greenberger, JS, Sakakeeny, M, and Parker, LM (1979). In vitro proliferation of hemopoietic stem cells in long-term marrow cultures: principles in mouse applied to man. *Exp Hematol*, 7 (Suppl. 5):135–148.

Gronthos, S, Zannettino, AC, Hay, SJ, Shi, S, Graves, SE, Kortesidis, A, et al. (2003). Molecular and cellular characterisation of highly purified stromal stem cells derived from human bone marrow. *J Cell Sci*, 116:1827–1835.

Gurdon. JB, Byrne, JA, and Simonsson, S (2003). Nuclear reprogramming and stem cell creation. *Proc Natl Acad Sci USA*, 100 (Suppl. 1):11819–11822.

Hakelien, AM and Collas, P (2002). Novel approaches to transdifferentiation. *Cloning Stem Cells*, 4:379–387.

Harrower, TP and Barker, RA (2004). Is there a future for neural transplantation? *BioDrugs*, 18:141–153.

Hoffman, LM and Carpenter, MK (2005). Characterization and culture of human embryonic stem cells. *Nat Biotechnol*, 23:699–708.

Honaramooz, A, Snedaker, A, Boiani, M, Scholer, H, Dobrinski, I, Schlatt, S (2002). Sperm from neonatal mammalian testes grafted in mice. *Nature*, 418:778–781.

Houbaviy, HB, Dennis, L, Jaenisch, R, and Sharp, PA (2005). Characterization of a highly variable eutherian microRNA gene. *Rna*, 11:1245–1257.

Houbaviy, HB, Murray, MF, and Sharp, PA (2003). Embryonic stem cell-specific MicroRNAs. *Dev Cell*, 5:351–358.

Ibrahim, A, Li, Y, Li, D, Raisman, G, El Masry, WS (2006). Olfactory ensheathing cells: ripples of an incoming tide? *Lancet Neurol*, 5:453–457.

Itskovitz-Eldor, J, Schuldiner, M, Karsenti, D, Eden, A, Yanuka, O, Amit, M, et al. (2000). Differentiation of human embryonic stem cells into embryoid bodies compromising the three embryonic germ layers. *Mol Med*, 6:88–95.

Itsykson, P, Ilouz, N, Turetsky, T, Goldstein, RS, Pera, MF, Fishbein, I, et al. (2005). Derivation of neural precursors from human embryonic stem cells in the presence of noggin. *Mol Cell Neurosci*, 30:24–36.

Ivanova, NB, Dimos, JT, Schaniel, C, Hackney, JA, Moore, KA, and Lemischka, IR (2002) A stem cell molecular signature. *Science*, 298:601–604.

Iwata, M, Awaya, N, Graf, L, Kahl, C, and Torok-Storb, B (2004). Human marrow stromal cells activate monocytes to secrete osteopontin, which down-regulates Notch1 gene expression in CD34+ cells. *Blood*, 103:4496–4502.

Jaenisch, R and Bird, A (2003). Epigenetic regulation of gene expression: how the genome integrates intrinsic and environmental signals. *Nat Genet*, 33 (Suppl.):245–254.

Jani, HR and Raisman, G (2004). Ensheathing cell cultures from the olfactory bulb and mucosa. *Glia*, 47:130–137.

Jenuwein, T and Allis, CD (2001). Translating the histone code. *Science*, 293:1074–1080.

Jiang, Y, Henderson, D, Blackstad, M, Chen, A, Miller, RF, and Verfaillie, CM (2003). Neuroectodermal differentiation from mouse multipotent adult progenitor cells. *Proc Natl Acad Sci USA*, 100 (Suppl. 1):11854–11860.

Johnson, J, Bagley, J, Skaznik-Wikiel, M, Lee, HJ, Adams, GB, Niikura, Y, et al. (2005). Oocyte generation in adult mammalian ovaries by putative germ cells in bone marrow and peripheral blood. *Cell*, 122:303–315.

Jordan, CT (2004). Cancer stem cell biology: from leukemia to solid tumors. *Curr Opin Cell Biol*, 16:708–712.

Jordan, CT (2005). Targeting the most critical cells: approaching leukemia therapy as a problem in stem cell biology. *Nat Clin Pract Oncol*, 2:224–225.

Kanellopoulou, C, Muljo, SA, Kung, AL, Ganesan, S, Drapkin, R, Jenuwein, T, et al. (2005). Dicer-deficient mouse embryonic stem cells are defective in differentiation and centromeric silencing. *Genes Dev*, 19:489–501.

Kannagi, R, Cochran, NA, Ishigami, F, Hakomori, S, Andrews, PW, Knowles, BB, Solter, D. (1983) Stage-specific embryonic antigens (SSEA-3 and -4) are epitopes of a unique globo-series ganglioside isolated from human teratocarcinoma cells. EMBO Journal 2(12):2355–2361.

Karsten, SL, Kudo, LC, and Geschwind, DH (2004). Microarray platforms: introduction and application to neurobiology. *Int Rev Neurobiol*, 60:1–23.

Karsten, SL, Kudo, LC, Jackson R, Sabatti, C, Kornblum, HI, and Geschwind, DH (2003). Global analysis of gene expression in neural progenitors reveals specific cell-cycle, signaling, and metabolic networks. *Dev Biol*, 261:165–182.

Katayama, Y, Battista, M, Kao, WM, Hidalgo, A, Peired, AJ, Thomas, SA, et al. (2006). Signals from the sympathetic nervous system regulate hematopoietic stem cell egress from bone marrow. *Cell*, 124:407–421.

Kehler, J, Hubner, K, Garrett, S, and Scholer, HR (2005). Generating oocytes and sperm from embryonic stem cells. *Semin Reprod Med*, 23:222–233.

Keirstead, HS (2005). Stem cells for the treatment of myelin loss. *Trends Neurosci*, 28:677–683.

Kempermann, G, Wiskott, L, and Gage, FH (2004). Functional significance of adult neurogenesis. *Curr Opin Neurobiol*, 14:186–191.

Kennison, JA (2004). Introduction to Trx-G and Pc-G genes. Methods *Enzymol*, 377:61–70.

Keyoung, HM, Roy, NS, Benraiss, A, Louissaint, A Jr., Suzuki, A, Hashimoto, M, et al. (2001). High-yield selection and extraction of two promoter-defined phenotypes of neural stem cells from the fetal human brain. *Nat Biotechnol*, 19:843–850.

Kiel, MJ, Yilmaz, OH, Iwashita, T, Yilmaz, OH, Terhorst, C, Morrison, SJ (2005). SLAM family receptors distinguish hematopoietic stem and progenitor cells and reveal endothelial niches for stem cells. *Cell*, 121:1109–1121.

Kim, JH, Auerbach, JM, Rodriguez-Gomez, JA, Velasco, I, Gavin, D, Lumelsky, N, et al. (2002). Dopamine neurons derived from embryonic stem cells function in an animal model of Parkinson's disease. *Nature*, 418:50–56.

Kim JY, Koh HC, Lee JY, Chang MY, Kim YC, Chung HY, et al. (2003). Dopaminergic neuronal differentiation from rat embryonic neural precursors by Nurr1 overexpression. *J Neurochem*, 85:1443–1454.

Kirmizis, A, Bartley, SM, Kuzmichev, A, Margueron, R, Reinberg, D, Green, R, et al. (2004). Silencing of human polycomb target genes is associated with methylation of histone H3 Lys 27. *Genes Dev*, 18:1592–1605.

Klein, SM, Behrstock, S, McHugh, J, Hoffmann, K, Wallace, K, Suzuki, M, et al. (2005). GDNF delivery using human neural progenitor cells in a rat model of ALS. *Hum Gene Ther*, 16:509–521.

Klimanskaya, I, Hipp, J, Rezai, KA, West, M, Atala, A, and Lanza, R (2004). Derivation and comparative assessment of retinal pigment epithelium from human embryonic stem cells using transcriptomics. *Cloning Stem Cells*, 6:217–245.

Klug, MG, Soonpaa, MH, Koh, GY, Field, LJ (1996). Genetically selected cardiomyocytes from differentiating embronic stem cells form stable intracardiac grafts. *J Clin Invest*, 98:216–224.

Krause, DS, Theise, ND, Collector, MI, Henegariu, O, Hwang, S, Gardner, R, et al. (2001). Multi-organ, multi-lineage engraftment by a single bone marrow-derived stem cell. *Cell*, 105:369–377.

LaBarge, MA and Blau, HM (2002). Biological progression from adult bone marrow to mononucleate muscle stem cell to multinucleate muscle fiber in response to injury. *Cell*, 111:589–601.

Lagasse, E, Shizuru, JA, Uchida, N, Tsukamoto, A, and Weissman, IL (2001). Toward regenerative medicine. *Immunity*, 14:425–436.

Laird, PW (2003). The power and the promise of DNA methylation markers. *Nat Rev Cancer*, 3:253–266.

Langley, E, Pearson, M, Faretta, M, Bauer, UM, Frye, RA, Minucci, S, et al. (2002). Human SIR2 deacetylates p53 and antagonizes PML/p53-induced cellular senescence. *Embo J*, 21:2383–2396.

Lapidot, T and Petit, I (2002). Current understanding of stem cell mobilization: the roles of chemokines, proteolytic enzymes, adhesion molecules, cytokines, and stromal cells. *Exp Hematol*, 30:973–981.

Lee, S, Zhou, G, Clark, T, Chen, J, Rowley, JD, and Wang, SM (2001). The pattern of gene expression in human CD15+ myeloid progenitor cells. *Proc Natl Acad Sci USA*, 98(6):3340–3345.

Lee, SH, Lumelsky, N, Studer, L, Auerbach, JM, and McKay, RD (2000). Efficient generation of midbrain and hindbrain neurons from mouse embryonic stem cells. *Nat Biotechnol*, 18:675–679.

Lee, TI, Jenner, RG, Boyer, LA, Guenther, MG, Levine, SS, Kumar, RM, et al. (2006). Control of developmental regulators by Polycomb in human embryonic stem cells. *Cell*, 125:301–313.

Lerou, PH and Daley, GQ (2005). Therapeutic potential of embryonic stem cells. *Blood Rev*, 19:321–331.

Levenberg, S, Burdick, JA, Kraehenbuehl, T, and Langer, R (2005). Neurotrophin-induced differentiation of human embryonic stem cells on three-dimensional polymeric scaffolds. *Tissue Eng*, 11:506–512.

Levine, SS, King, IF, and Kingston, RE (2004). Division of labor in polycomb group repression. *Trends Biochem Sci*, 29(9):478–485.

Li, E (2002). Chromatin modification and epigenetic reprogramming in mammalian development. *Nat Rev Genet*, 3:662–673.

Li, L and Xie, T (2005). Stem cell niche: structure and function. *Annu Rev Cell Dev Biol*, 21:605–631.

Li, XJ, Du, ZW, Zarnowska, ED, Pankratz, M, Hansen, LO, Pearce, RA, et al. (2005). Specification of motoneurons from human embryonic stem cells. *Nat Biotechnol*, 23:215–221.

Lim, DA, Tramontin, AD, Trevejo, JM, Herrera, DG, Garcia-Verdugo, JM, Alvarez-Buylla, A (2000). Noggin antagonizes BMP signaling to create a niche for adult neurogenesis. *Neuron*, 28:713–726.

Lim, LS, Loh, YH, Zhang, W, Li, Y, Chen, X, Wang, Y, Bakre, M, Ng, HH, and Stanton, LW. (2007). Zic3 is required for maintenance of pluripotency in embryonic stem cells. *Mol Biol Cell*, 18:1348–1358.

Lin, H (2002). The stem-cell niche theory: lessons from flies. *Nat Rev Genet*, 3:931–940.

Lindvall, O and Bjorklund, A (2004). Cell therapy in Parkinson's disease. *NeuroRx*, 1:382–393.

Liu, ET (2003). Molecular oncodiagnostics: where we are and where we need to go. *J Clin Oncol*, 21:2052–2055.

Liu, ET (2005). Genomic technologies and the interrogation of the transcriptome. *Mech Ageing Dev*, 126:153–159.

Liu, ET and Karuturi, KR (2004). Microarrays and clinical investigations. *N Engl J Med*, 350:1595–1597.

Liu, Y, Shin, S, Zeng, X, Zhan, M, Gonzalez, R, Mueller, FJ, et al. (2006). Genome wide profiling of human embryonic stem cells (hESCs), their derivatives and embryonal carcinoma cells to develop base profiles of U.S. Federal government approved hESC lines. *BMC Dev Biol*, 6:20.

Livesey, FJ, Young, TL, and Cepko, CL (2004). An analysis of the gene expression program of mammalian neural progenitor cells. *Proc Natl Acad Sci USA*, 101:1374–1379.

Loh, YH, Wu, Q, Chew, JL, Vega, VB, Zhang, W, Chen, X, et al. (2006). The OCT4 and NANOG transcription network regulates pluripotency in mouse embryonic stem cells. *Nat Genet*, 38:431–440.

Lu, J, Getz, G, Miska, EA, Alvarez-Saavedra, E, Lamb, J, Peck, D, et al. (2005). MicroRNA expression profiles classify human cancers. *Nature*, 435:834–838.

Lund, AH, and van Lohuizen, M (2004). Polycomb complexes and silencing mechanisms. *Curr Opin Cell Biol*, 16:239–246.

Maitra, A, Arking, DE, Shivapurkar, N, Ikeda, M, Stastny, V, Kassauei, K, et al. (2005). Genomic alterations in cultured human embryonic stem cells. *Nat Genet*, 37:1099–1103.

Mansfield, JH, Harfe, BD, Nissen, R, Obenauer, J, Srineel, J, Chaudhuri, A, et al. (2004). MicroRNA-responsive "sensor" transgenes uncover Hox-like and other developmentally regulated patterns of vertebrate microRNA expression. *Nat Genet*, 36:1079–1083.

Margueron, R, Trojer, P, and Reinberg, D (2005). The key to development: interpreting the histone code? *Curr Opin Genet Dev*, 15:163–176.

Martin, GR (1981). Isolation of a pluripotent cell line from early mouse embryos cultured in medium conditioned by teratocarcinoma stem cells. *Proc Natl Acad Sci USA*, 78:7634–7638.

Martin, MA and Bhatia, M (2005). Analysis of the human fetal liver hematopoietic microenvironment. *Stem Cells Dev*, 14:493–504.

Martin, GR and Evans, MJ (1975). Differentiation of clonal lines of teratocarcinoma cells: formation of embryoid bodies in vitro. *Proc Natl Acad Sci USA*, 72:1441–1445.

Massengale, M, Wagers, AJ, Vogel, H, and Weissman, IL (2005). Hematopoietic cells maintain hematopoietic fates upon entering the brain. *J Exp Med*, 201:1579–1589.

Masson, S, Harrison, DJ, Plevris, JN, and Newsome, PN (2004). Potential of hematopoietic stem cell therapy in hepatology: a critical review. *Stem Cells*, 22:897–907.

Mattick, JS and Makunin, IV (2005). Small regulatory RNAs in mammals. *Hum Mol Genet*, 14(Spec No 1):R121–R132.

McCulloch, EA and Till, JE (1964). Proliferation of Hemopoietic Colony-Forming Cells Transplanted into Irradiated Mice. *Radiat Res*, 22:383–397.

McCulloch, EA and Till, JE (2005). Perspectives on the properties of stem cells. *Nat Med*, 11:1026–1028.

McLaren, A (2002). Human embryonic stem cell lines: socio-legal concerns and therapeutic promise. *C R Biol*, 325:1009–1012.

McLean, DJ (2005). Spermatogonial stem cell transplantation and testicular function. *Cell Tissue Res*, 322:21–31.

McManus, MT (2003). MicroRNAs and cancer. *Semin Cancer Biol*, 13:253–258.

McNeish, J (2004). Embryonic stem cells in drug discovery. *Nat Rev Drug Discov*, 3:70–80.

Meyer, MJ and Nelson, LJ (2001). Respecting what we destroy. Reflections on human embryo research. *Hastings Cent Rep*, 31:16–23.

Mikkers, H and Frisen, J (2005). Deconstructing stemness. *EMBO J*,24:2715–2719.

Miles, GB, Yohn, DC, Wichterle, H, Jessell, TM, Rafuse, VF, and Brownstone, RM (2004). Functional properties of motoneurons derived from mouse embryonic stem cells. *J Neurosci*, 24:7848–7858.

Miska, EA (2005). How microRNAs control cell division, differentiation and death. *Curr Opin Genet Dev*, 15:563–568.

Mitsui, K, Tokuzawa, Y, Itoh, H, Segawa, K, Murakami, M, Takahashi, K, et al. (2003). The homeoprotein NANOG is required for maintenance of pluripotency in mouse epiblast and ES cells. *Cell*, 113:631–642.

Molofsky, AV, He, S, Bydon, M, Morrison, SJ, and Pardal, R (2005). Bmi-1 promotes neural stem cell self-renewal and neural development but not mouse growth and survival by repressing the p16Ink4a and p19Arf senescence pathways. *Genes Dev*, 19:1432–1437.

Molofsky, AV, Pardal, R, Iwashita, T, Park, IK, Clarke, MF, and Morrison, SJ (2003). Bmi-1 dependence distinguishes neural stem cell self-renewal from progenitor proliferation. *Nature*, 425:962–967.

Montarras, D, Morgan, J, Collins, C, Relaix, F, Zaffran, S, Cumano, A, et al. (2005). Direct isolation of satellite cells for skeletal muscle regeneration. Science 309:2064–2067.

Morrison, SJ (2001). Stem cell potential: can anything make anything? *Curr Biol*, 11:R7–9.

Morrison, SJ (2005). Cancer stem cells. *Clin Adv Hematol Oncol*, 3:171–172.

Mullighan, CG and Petersdorf, EW (2006). Genomic polymorphism and allogeneic hematopoietic transplantation outcome. *Biol Blood Marrow Transplant*, 12:19–27.

Murchison, EP and Hannon, GJ (2004). miRNAs on the move: miRNA biogenesis and the RNAi machinery. *Curr Opin Cell Biol*, 16:223–229.

Murchison, EP, Partridge, JF, Tam, OH, Cheloufi, S, and Hannon, GJ. (2005) Characterization of Dicer-deficient murine embryonic stem cells. *Proc Natl Acad Sci U S A*. 102(34):12135–12140.

Nelson, LJ and Meyer, MJ (2005). Confronting deep moral disagreement: the President's Council on Bioethics, moral status, and human embryos. *Am J Bioeth*, 5:33–42.

Ng, P, Wei, CL, Sung, WK, Chiu, KP, Lipovich, L, Ang, CC, et al. (2005). Gene identification signature (GIS) analysis for transcriptome characterization and genome annotation. *Nat Methods*, 2:105–111.

Nilsson, SK and Simmons, PJ (2004). Transplantable stem cells: home to specific niches. *Curr Opin Hematol*, 11:102–106.

Ogle, BM, Cascalho, M, and Platt, JL (2005). Biological implications of cell fusion. *Nat Rev Mol Cell Biol*, 6:567–575.

Ohlstein, B, Kai, T, Decotto, E, and Spradling, A (2004). The stem cell niche: theme and variations. *Curr Opin Cell Biol*, 16:693–699.

Orkin SH. (2005) Chipping away at the embryonic stem cell network. *Cell*. 122 (6):828–830.

Okita, K, Ichisaka, T, and Yamanaka, S (2007). Generation of germline-competent induced pluripotent stem cells. Nature.

Orlic, D (2004). The strength of plasticity: stem cells for cardiac repair. *Int J Cardiol*, 95 Suppl. 1):S16–19.

Ostenfeld, T and Svendsen, CN (2003). Recent advances in stem cell neurobiology. *Adv Tech Stand Neurosurg*, 28:3–89.

Palmer, TD, Schwartz, PH, Taupin, P, Kaspar, B, Stein, SA, and Gage, FH (2001). Cell culture. Progenitor cells from human brain after death. *Nature*, 411:42–43.

Pera, MF (2005). Stem cell culture, one step at a time. *Nat Methods*, 2:164–165.

Pera, MF, Reubinoff, B, and Trounson, A (2000). Human embryonic stem cells. *J Cell Sci*, 113 (Pt 1):5–10.

Pera, MF and Trounson, AO (2004). Human embryonic stem cells: prospects for development. *Development*, 131:5515–5525.

Petersdorf, EW and Deeg, HJ (1992). Diagnostic use of molecular probes before and after marrow transplantation. *Clin Lab Med*, 12:113–128.

Petersdorf, EW and Malkki, M (2006). Genetics of risk factors for graft-versus-host disease. *Semin Hematol*, 43:11–23.

Picard-Riera, N, Nait-Oumesmar, B, and Baron-Van Evercooren, A (2004). Endogenous adult neural stem cells: limits and potential to repair the injured central nervous system. *J Neurosci Res*, 76:223–231.

Plaia, TW, Josephson, R, Liu, Y, Zeng, X, Ording, C, Toumadje, A, et al. (2005). Characterization of a New NIH Registered Variant Human Embryonic Stem Cell Line BG01V: A Tool for Human Embryonic Stem Cell Research. *Stem Cells*, 24(3):531–546.

Pomp, O, Brokhman, I, Ben-Dor, I, Reubinoff, B, and Goldstein, RS (2005). Generation of peripheral sensory and sympathetic neurons and neural crest cells from human embryonic stem cells. *Stem Cells*, 23:923–930.

Potten, CS, Booth, C, Tudor, GL, Booth, D, Brady, G, Hurley, P, et al. (2003). Identification of a putative intestinal stem cell and early lineage marker; musashi-1. *Differentiation*, 71:28–41.

Potten, CS and Loeffler, M (1987). A comprehensive model of the crypts of the small intestine of the mouse provides insight into the mechanisms of cell migration and the proliferation hierarchy. *J Theor Biol*, 127:381–391.

Potten, CS and Loeffler, M (1990). Stem cells: attributes, cycles, spirals, pitfalls and uncertainties. Lessons for and from the crypt. *Development*, 110:1001–1020.

Poy, MN, Eliasson, L, Krutzfeldt, J, Kuwajima, S, Ma, X, Macdonald, PE, et al. (2004). A pancreatic islet-specific microRNA regulates insulin secretion. *Nature*, 432:226–230.

Prestoz, L, Relvas, JB, Hopkins, K, Patel, S, Sowinski, P, Price, J, et al. (2001) Association between integrin-dependent migration capacity of neural stem cells in vitro and anatomical repair following transplantation. *Mol Cell Neurosci*, 18:473–484.

Properzi, F and Fawcett, JW (2004). Proteoglycans and brain repair. *News Physiol Sci*, 19:33–38.

Pruitt, K, Zinn, RL, Ohm, JE, McGarvey, KM, Kang, SH, Watkins, DN, et al. (2006). Inhibition of SIRT1 reactivates silenced cancer genes without loss of promoter DNA hypermethylation. *PLoS Genet*, 2:e40.

Raaphorst, FM (2005). Deregulated expression of Polycomb-group oncogenes in human malignant lymphomas and epithelial tumors. *Hum Mol Genet*, 14 (Spec. No 1):R93–R100.

Raetz, EA and Moos, PJ (2004). Impact of microarray technology in clinical oncology. *Cancer Invest*, 22:312–320.

Raff, MC and Miller, RH, and Noble, M (1983). A glial progenitor cell that develops in vitro into an astrocyte or an oligodendrocyte depending on culture medium. *Nature*, 303:390–396.

Ramalho-Santos, M, Yoon, S, Matsuzaki, Y, Mulligan, RC, and Melton, DA (2002). "Stemness": transcriptional profiling of embryonic and adult stem cells. *Science*, 298:597–600.

Reubinoff, BE, Itsykson, P, Turetsky, T, Pera, MF, Reinhartz, E, Itzik, A, et al. (2001). Neural progenitors from human embryonic stem cells. *Nat Biotechnol*, 19:1134–1140.

Reya, T, Morrison, SJ, Clarke, MF, and Weissman, IL (2001). Stem cells, cancer, and cancer stem cells. *Nature*, 414:105–111.

Reynolds, BA and Rietze, RL (2005). Neural stem cells and neurospheres–re-evaluating the relationship. *Nat Methods*, 2:333–336.

Reynolds, BA and Weiss, S (1992). Generation of neurons and astrocytes from isolated cells of the adult mammalian central nervous system. *Science*, 255:1707–1710.

Richards, M, Tan, SP, Tan, JH, Chan, WK, and Bongso, A (2004). The transcriptome profile of human embryonic stem cells as defined by SAGE. *Stem Cells*, 22:51–64.

Ringrose, L and Paro, R (2004). Epigenetic regulation of cellular memory by the Polycomb and Trithorax group proteins. *Annu Rev Genet*, 38:413–443.

Robertson, KD (2002). DNA methylation and chromatin-unraveling the tangled web. *Oncogene*, 21:5361–5379.

Robertson, KD and Wolffe, AP (2000). DNA methylation in health and disease. *Nat Rev Genet*, 1:11–19.

Robson, P (2004). The maturing of the human embryonic stem cell transcriptome profile. *Trends Biotechnol*, 22:609–612.

Rodriguez, CI, Galan, A, Valbuena, D, and Simon, C (2006). Derivation of clinical-grade human embryonic stem cells. *Reprod Biomed Online*, 12:112–118.

Rugg-Gunn, PJ, Ferguson-Smith, AC, and Pedersen, RA (2005). Epigenetic status of human embryonic stem cells. *Nat Genet*, 37:585–587.

Saghatelyan, A, de Chevigny, A, Schachner, M, and Lledo, PM (2004). Tenascin-R mediates activity-dependent recruitment of neuroblasts in the adult mouse forebrain. *Nat Neurosci*, 7:347–356.

Sansom, SN, Hebert, JM, Thammongkol, U, Smith, J, Nisbet, G, Surani, MA, et al. (2005) Genomic characterisation of a Fgf-regulated gradient-based neocortical protomap. *Development*, 132:3947–3961.

Schuldiner, M, Eiges, R, Eden, A, Yanuka, O, Itskovitz-Eldor, J, Goldstein, RS, et al. (2001). Induced neuronal differentiation of human embryonic stem cells. *Brain Res*, 913:201–205.

Schulz, TC, Noggle, SA, Palmarini, GM, Weiler, DA, Lyons, IG, Pensa, KA, et al. (2004). Differentiation of human embryonic stem cells to dopaminergic neurons in serum-free suspension culture. *Stem Cells*, 22:1218–1238.

Schulz, TC, Palmarini, GM, Noggle, SA, Weiler, DA, Mitalipova, MM, and Condie, BG (2003). Directed neuronal differentiation of human embryonic stem cells. *BMC Neurosci*, 4:27.

Scolding, N (2001a). Use of stem cells in creation of embyros. *Lancet*, 358:2078.

Scolding, N (2001b). New cells from old. *Lancet*, 357:329–330.

Scolding, N (2005). Stem-cell therapy: hope and hype. *Lancet*, 365:2073–2075.

Seaberg, RM, van der Kooy, D (2003). Stem and progenitor cells: the premature desertion of rigorous definitions. *Trends Neurosci*, 26:125–131.

Seitz, H, Youngson, N, Lin, SP, Dalbert, S, Paulsen, M, Bachellerie, JP, et al. (2003). Imprinted microRNA genes transcribed antisense to a reciprocally imprinted retrotransposon-like gene. *Nat Genet*, 34:261–262.

Setoguchi, T, Taga, T, and Kondo, T (2004). Cancer stem cells persist in many cancer cell lines. *Cell Cycle*, 3:414–415.

Shadduck, RK, Waheed, A, Greenberger, JS, and Dexter, TM (1983). Production of colony stimulating factor in long-term bone marrow cultures. *J Cell Physiol*, 114:88–92.

Shamblott, MJ, Axelman, J, Littlefield, JW, Blumenthal, PD, Huggins, GR, Cui, Y, Cheng, L, and Gearhart JD. (2001) Human embryonic germ cell derivatives express a broad range of developmentally distinct markers and proliferate extensively in vitro. *Proc Natl Acad Sci U S A*. 2001 2;98(1):113–118.

Shamblott,MJ, Axelman, J, Wang, S, Bugg, EM, Littlefield, JW, Donovan, PJ, Blumenthal, PD, Huggins, GR, and Gearhart, JD. (1998) Derivation of pluripotent stem cells from cultured human primordial germ cells. *Proc Natl Acad Sci U S A*. 10;95(23):13726–13731.

Shaughnessy, JD Jr., and Barlogie, B (2003). Interpreting the molecular biology and clinical behavior of multiple myeloma in the context of global gene expression profiling. *Immunol Rev*, 194:140–163.

Shannon, MF, and Rao, S (2002). Transcription: Of chips and ChIPs. *Science*, 296:666–669.

Shi, Y, Lan, F, Matson, C, Mulligan, P, Whetstine, JR, Cole, PA, et al. (2004). Histone demethylation mediated by the nuclear amine oxidase homolog LSD1. *Cell*, 119:941–953.

Short, B, Brouard, N, Occhiodoro-Scott, T, Ramakrishnan, A, and Simmons, PJ (2003). Mesenchymal stem cells. *Arch Med Res*, 34:565–571.

Siminovitch, L, McCulloch, EA, and Till, JE (1963). The Distribution of Colony-Forming Cells among Spleen Colonies. *J Cell Physiol*, 62:327–336.

Simonsson, S and Gurdon, J (2004). DNA demethylation is necessary for the epigenetic reprogramming of somatic cell nuclei. *Nat Cell Biol*, 6:984–990.

Simonsson, S and Gurdon, JB (2005). Changing cell fate by nuclear reprogramming. *Cell*, Cycle 4:513–515.

Skuk, D (2004). Myoblast transplantation for inherited myopathies: a clinical approach. *Expert Opin Biol Ther*, 4:1871–1885.

Slack, JM and Tosh, D (2001). Transdifferentiation and metaplasia–switching cell types. *Curr Opin Genet Dev*, 11:581–586.

Solter, D, Knowles, BB. (1978) Monoclonal antibody defining a stage-specific mouse embryonic antigen (SSEA-1). Proc Natl Acad Sci U S A. 75 (11): 5565–5569.

Sperger, JM, Chen, X, Draper, JS, Antosiewicz, JE, Chon, CH, Jones, SB, et al. (2003). Gene expression patterns in human embryonic stem cells and human pluripotent germ cell tumors. *Proc Natl Acad Sci USA*, 100:13350–13355.

Spradling, A, Drummond-Barbosa, D, and Kai, T (2001). Stem cells find their niche. *Nature*, 414:98–104.

Stiles, CD (2003). Lost in space: misregulated positional cues create tripotent neural progenitors in cell culture. *Neuron*, 40:447–449.

Strahl, BD, and Allis, CD (2000) The language of covalent histone modifications. *Nature*. 403 (6765):41–45.

Sun, Y, Koo, S, White, N, Peralta, E, Esau, C, Dean, NM, et al. (2004). Development of a micro-array to detect human and mouse microRNAs and characterization of expression in human organs. *Nucleic Acids Res*, 32:e188.

Tada, M, Takahama, Y, Abe, K, Nakatsuji, N, and Tada, T (2001). Nuclear reprogramming of somatic cells by in vitro hybridization with ES cells. *Curr Biol*, 11:1553–1558.

Takahashi, K, and Yamanaka, S. (2006) Induction of pluripotent stem cells from mouse embryonic and adult fibroblast cultures by defined factors. *Cell*. 2006 Aug 25;126(4):663–676.

Takamizawa, J, Konishi, H, Yanagisawa, K, Tomida, S, Osada, H, Endoh, H, Harano T, et al. (2004) Reduced expression of the let-7 microRNAs in human lung cancers in association with shortened postoperative survival. *Cancer Res*, 64:3753–3756.

Tamaki, S, Eckert, K, He, D, Sutton, R, Doshe, M, Jain, G, et al. (2002). Engraftment of sorted/expanded human central nervous system stem cells from fetal brain. *J Neurosci Res*, 69:976–986.

Tanabe, Y and Jessell, TM (1996). Diversity and pattern in the developing spinal cord. *Science*, 274:1115–1123.

Taranger, CK, Noer, A, Sorensen, AL, Hakelien, AM, Boquest, AC, and Collas, P (2005). Induction of dedifferentiation, genomewide transcriptional programming, and epigenetic reprogramming by extracts of carcinoma and embryonic stem cells. *Mol Biol Cell*, 16:5719–5735.

Terada, N, Hamazaki, T, Oka, M, Hoki, M, Mastalerz, DM, Nakano, Y, et al. (2002) Bone marrow cells adopt the phenotype of other cells by spontaneous cell fusion. *Nature*, 416:542–545.

Terskikh, AV, Easterday, MC, Li, L, Hood, L, Kornblum, HI, Geschwind, DH, et al. (2001). From hematopoiesis to neuropoiesis: evidence of overlapping genetic programs. *Proc Natl Acad Sci USA*, 98:7934–7939.

Thomson, JA, Itskovitz-Eldor, J, Shapiro, SS, Waknitz, MA, Swiergiel, JJ, Marshall, VS, et al. (1998). Embryonic stem cell lines derived from human blastocysts. *Science*, 282:1145–1147.

Tosh, D and Slack, JM (2002). How cells change their phenotype. *Nat Rev Mol Cell Biol*, 3:187–194.

Tsai, RY Kittappa, R, and McKay, RD (2002). Plasticity, niches, and the use of stem cells. *Dev Cell*, 2:707–712.

Tsukada, Y, Fang, J, Erdjument-Bromage, H, Warren, ME, Borchers, CH, Tempst, P, et al. (2006). Histone demethylation by a family of JmjC domain-containing proteins. *Nature*, 439:811–816.

Tumbar, T, Guasch, G, Greco, V, Blanpain, C, Lowry, WE, Rendl, M, et al. (2004). Defining the epithelial stem cell niche in skin. *Science*, 303:359–363.

Ueki, T, Tanaka, M, Yamashita, K, Mikawa, S, Qiu, Z, Maragakis, NJ, et al. (2003). A novel secretory factor, Neurogenesin-1, provides neurogenic environmental cues for neural stem cells in the adult hippocampus. *J Neurosci*, 23:11732–11740.

Vassilopoulos, G and Russell, DW (2003). Cell fusion: an alternative to stem cell plasticity and its therapeutic implications. *Curr Opin Genet Dev*, 13:480–485.

Vassilopoulos, G Wang, PR and Russell, DW (2003). Transplanted bone marrow regenerates liver by cell fusion. *Nature*, 422:901–904.

Verfaillie, CM and Catanzaro, P (1996). Direct contact with stroma inhibits proliferation of human long-term culture initiating cells. *Leukemia*, 10:498–504.

Vire, E, Brenner, C, Deplus, R, Blanchon, L, Fraga, M, Didelot, C, et al. (2006). The Polycomb group protein EZH2 directly controls DNA methylation. *Nature*, 439:871–874.

Wagers, AJ, Christensen, JL, and Weissman, IL (2002). Cell fate determination from stem cells. *Gene Ther*, 9:606–612.

Wagner, W, Feldmann, RE, Seckinger, A, Maurer, MH, Wein, F, Blake, J, et al. (2006). The heterogeneity of human mesenchymal stem cell preparations--evidence from simultaneous analysis of proteomes and transcriptomes. *Exp Hematol*, 34:536–548.

Wagner, W, Saffrich, R, Wirkner, U, Eckstein, V, Blake, J, Ansorge, A, et al. (2005). Hematopoietic progenitor cells and cellular microenvironment: behavioral and molecular changes upon interaction. *Stem Cells*, 23:1180–1191.

Wei, CL, Miura, T, Robson, P, Lim, SK, Xu, XQ, Lee, MY, et al. (2005). Transcriptome profiling of human and murine ESCs identifies divergent paths required to maintain the stem cell state. *Stem Cells*, 23:166–185.

Weimann, JM, Johansson, CB, Trejo, A, and Blau, HM (2003). Stable reprogrammed heterokaryons form spontaneously in Purkinje neurons after bone marrow transplant. *Nat Cell Biol*, 5:959–966.

Weissman, IL (2000a). Translating stem and progenitor cell biology to the clinic: barriers and opportunities. *Science*, 287:1442–1446.

Weissman, IL (2000b). Stem cells: units of development, units of regeneration, and units in evolution. *Cell*, 100:157–168.

Weissman, IL, Anderson, DJ, and Gage, F (2001). Stem and progenitor cells: origins, phenotypes, lineage commitments, and transdifferentiations. *Annu Rev Cell Dev Biol*, 17:387–403.

West, JA and Daley, GQ (2004). In vitro gametogenesis from embryonic stem cells. *Curr Opin Cell Biol*, 16:688–692.

Wichterle, H, Lieberam, I, Porter, JA, and Jessell, TM (2002). Directed differentiation of embryonic stem cells into motor neurons. *Cell*, 110:385–397.

Wollert, KC and Drexler, H (2004). Cell therapy for acute myocardial infarction: where are we heading? *Nat Clin Pract Cardiovasc Med*, 1:61.

Wollert, KC and Drexler, H (2005). Clinical applications of stem cells for the heart. *Circ Res*, 96:151–163.

Wong, MH (2004). Regulation of intestinal stem cells. *J Investig Dermatol Symp Proc*, 9:224–228.

Wright, DE, Wagers, AJ, Gulati, AP, Johnson, FL, and Weissman, IL (2001). Physiological migration of hematopoietic stem and progenitor cells. *Science*, 294:1933–1936.

Yang, AX, Mejido, J, Luo, Y, Zeng, X, Schwartz, C, Wu, T, et al. (2005). Development of a focused microarray to assess human embryonic stem cell differentiation. *Stem Cells Dev*, 14:270–284.

Yekta, S, Shih, IH, and Bartel, DP (2004) MicroRNA-directed cleavage of HOXB8 mRNA. *Science*, 304:594–596.

Ying, QL, Nichols, J, Evans, EP, and Smith, AG (2002). Changing potency by spontaneous fusion. *Nature*, 416:545–548.

Yoo, CB and Jones, PA (2006). Epigenetic therapy of cancer: past, present and future. *Nat Rev Drug Discov*, 5:37–50.

Zaboikin, M, Srinivasakumar, N, and Schuening, F (2006). Gene therapy with drug resistance genes. *Cancer Gene Ther*, 13:335–345.

Zeng, X, Cai, J, Chen, J, Luo, Y, You, ZB, Fotter, E, et al. (2004). Dopaminergic differentiation of human embryonic stem cells. *Stem Cells*, 22:925–940.

Zhang, J, Tam, WL, Tong, GQ, Wu, Q, Chan, HY, Soh, BS, Lou, Y, Yang, J, Ma, Y, Chai, L, Ng, HH, Lufkin, T, Robson, P, and Lim B (2006). Sall4 modulates embryonic stem cell pluripotency and early embryonic development by the transcriptional regulation of Pou5f1. *Nat Cell Biol*, 8:1114–1123.

Zhou, G, Chen, J, Lee, S, Clark, T, Rowley, JD, and Wang, SM (2001). The pattern of gene expression in human CD34(+) stem/progenitor cells. *Proc Natl Acad Sci USA*, 98(24):13966–13971.

Part III

Health Genomics

38

Genomics and Global Health

Sir David Weatherall and Dhavendra Kumar

The provision of healthcare in the future offers completely different challenges to the developed, industrialized nations as compared to developing countries. For the former, the major problem is how to control the increasing costs of healthcare arising from the prevention and management of chronic disease in aged populations. The role of genetics and genomics in the prevention and management of these conditions is described in earlier chapters. Here, we focus on the quite different problems of the developing countries.

Although throughout the 1990s gross domestic product per head in the developing countries grew by 1.6% per year, and the proportion of those living on less than $1 per day fell from 29% to 23%, most of this progress occurred in Asia (Weatherall, 2003). In other regions, the level of poverty increased, and it is estimated that 150 million children in low- and middle-income economies are now suffering from malnutrition, and a similar number will be underweight by 2020, unless the situation improves. These problems are accentuated by epidemics of communicable diseases. Approximately 70% of the 40 million people affected by HIV/AIDS (human immunodeficiency virus/acquired immunodeficiency syndrome) live in countries with dysfunctional healthcare systems. Tuberculosis has reemerged, with 9 million new cases and 2 million deaths each year, and similar death rates are occurring from malaria. In all these diseases, the emergence of drug-resistant organisms is increasing. And, as countries pass through the epidemiological transition from infectious to noninfectious disease, many of them are already encountering major epidemics of the diseases of westernization; approximately 150 million people worldwide have type 2 diabetes, and this number is expected to double by 2025. In some of these populations, stroke and cardiovascular disease are already occurring at a higher frequency than in richer, developed countries.

Although the reasons for the in health between developing and developed nations are complex, reflecting poverty, war, natural disasters, and ineffective administration, a factor of undoubted importance is the lack of awareness and support on the part of developed countries. For example, less than 10% of global spending on medical research has been devoted to diseases that account for 90% of the global disease burden, and of the 1233 new drugs marketed between 1975 and 1999, only 13 were approved for tropical diseases (WHO, 2002a). At first sight, therefore, it is difficult to imagine how genetics could be of any value to the provision of healthcare in the developing countries.

The World Health Organization (WHO) published a report in 2002 entitled *Genomics and World Health* (WHO, 2002b). This study was based on wide-scale consultation with representatives from many developing countries and attempted to define the potential role for postgenome research for healthcare in the developing world. While this chapter is based on experiences in producing this report, it also summarizes progress since it was written.

The Scope of Genetics and Genomics for Disease Control in Developing Countries

An analysis of the potential of genomics for the future of global health requires a much broader view of the field than that needed for the better-defined diseases of the richer countries. In addition to the human genome, it encompasses the genomes of a wide variety of pathogens, including bacteria, viruses, and parasites. It also involves such information as is available about the genetic makeup of disease vectors and includes the genetic modification of crops.

It is beyond the scope of this chapter to deal with each of these aspects of genomics and global health detail. Readers who wish to obtain further information are referred to the WHO report and the reading list at the end of this chapter.

The Potential for Genomics to Improve the Health of Developing Countries

Although the applications of genetics and genomics to healthcare that are described elsewhere in this book may, ultimately, become applicable to most countries of the world, the most immediate applications are for the control of the common monogenic disorders and for the control and management of communicable diseases.

Monogenic Disease

Although there are some regional and ethnic variations, monogenic diseases occur in all ethnic groups, and the main problems globally relate to the development of adequate facilities and staffing to provide clinical genetics services. It is only in the case of the inherited disorders of hemoglobin that a genuine global problem in healthcare is emerging.

The important inherited hemoglobin disorders are summarized in Table 38–1. They are comprised of the α- and β-thalassemias

Table 38–1 Common Inherited Disorders of Hemoglobin

Thalassemias
α-thalassemia
α^0
α^+
β-thalassemia
β^0
β^+
$\delta\beta$-thalassemia
$\varepsilon\gamma\delta\beta$-thalassemia
Structural hemoglobin (Hb) variants
Hb S
Hb C
Hb E
Many other rare variants

and the common structural hemoglobin variants. Although hundreds of structural variants have been identified, only three, hemoglobins S, C, and E, occur at high frequencies. Hb S is associated with the sickling disorders, while Hb E, because it is synthesized at a reduced rate, behaves like a mild form of β-thalassemia; when it is inherited together with β-thalassemia, it produces a condition called Hb E β-thalassemia, which is one of the commonest severe forms of the disease in Asia. The molecular pathology, pathophysiology, and clinical features of these conditions are described elsewhere (Steinberg et al., 2001; Weatherall and Clegg, 2001a). Here, we focus on the health burden that they pose on developing countries and outline the approaches to their control.

Distribution and Frequency

Almost certainly as the result of heterozygote advantage against malaria, inherited hemoglobin disorders are the commonest monogenic diseases. Although these figures will undoubtedly have to be revised as better information is obtained, it has been estimated that 7% of the world population are carriers and that 3,00,000–5,00,000 babies with severe forms are born each year (WHO, 1987). Although they occur mainly in tropical regions, because of extensive population migrations, they are now encountered in most countries. Improvements in nutrition and public health in developing countries result in an epidemiological transition characterized by a fall of mortality in children under the age of 5 years. It follows that many of those with serious genetic diseases, who would formerly have died in infancy, are now surviving to present for diagnosis and management, a trend that is occurring throughout the Middle East, the Indian subcontinent, and many parts of Asia (Weatherall and Clegg, 2001b). Although many sub-Saharan African countries have yet to pass through this transition, the same phenomena is likely to occur in the future. In the light of these epidemiological and demographic changes, it is not surprising that hemoglobin disorders are presenting an increasing challenge for healthcare provision in developing countries.

The distribution of the thalassemias and the disorders due to Hb S are summarized in Figs. 38–1 and 38–2, and an outline of the global distribution of thalassemia and hemoglobin variants is summarized in Table 38–2. The thalassemias occur at a high frequency right across a broad tropical zone, ranging from Africa, through the Middle East and Indian subcontinent, to Southeast Asia. The sickling disorders are more restricted, occurring at an extremely high frequency in sub-Saharan Africa and in parts of the Middle East and Indian subcontinent, but they do not occur further east. On the other hand, Hb E reaches extremely high frequencies in many parts of Southeast Asia, but does not occur further west than the eastern half of the Indian subcontinent. The serious forms of α-thalassemia, the α^0 thalassemias, are restricted to parts of Southeast Asia and the Mediterranean Islands, while the milder forms, the α^+ thalassemias, occur at a very high frequency over much the same region as the β-thalassemias. Because of the very high frequencies of Hb E in many Asian countries (Table 38–2), the compound heterozygous state with β-thalassemia, Hb E β-thalassemia, occurs at a very high frequency; in some countries, such as Thailand, Bangladesh, Indonesia, and Myanmar, for example, it reaches a much higher frequency than the homozygous state for β-thalassemia, thalassemia major.

One of the major problems in determining the health burden caused by thalassemias and the sickling disorders is lack of adequate data about their true gene frequencies. The few studies that have attempted to micromap populations in Asia have shown that, even within small regions, there is a marked heterogeneity in the frequency of different forms of thalassemia (see Weatherall and Clegg, 2001a). For example, in an analysis of some 18 different island populations in Indonesia, it was found that there was quite remarkable heterogeneity in the frequency of both β-thalassemia and Hb E. Based on such incomplete data, a very approximate assessment of the numbers of births of children with serious hemoglobin disorders is shown in Figure 38–3.

Molecular Pathology

The α-thalassemias usually result from deletions of one or more of the α globin genes. Normally, individuals have two α genes per haploid genome designated $\alpha\alpha/\alpha\alpha$. The α^+ thalassemias result from a series of different-sized deletions of one of the pair of α globin genes ($-\alpha/\alpha\alpha$; $-\alpha/-\alpha$), while the α^0 thalassemias result from long deletions involving both α globin genes ($-/\alpha\alpha$; $-/-$). Some forms of α^+ thalassemia result from point mutations in the α globin genes ($\alpha^T\alpha/\alpha\alpha$; $\alpha^T\alpha/\alpha^T\alpha$). The resultant disorders are the direct result of gene dosage effect. The loss of a single α globin gene causes minimal hematological abnormalities. The loss of two α globin genes, either $-\alpha/-\alpha$ or $-/\alpha\alpha$, results in mild thalassemia traits; the loss of three genes ($-\alpha/-$) causes a variable hemolytic anemia called Hb H disease; while the loss of all four α genes ($-/-$) results in profound intrauterine hypoxia and stillbirth.

More than 150 different mutations have been described in patients with β-thalassemia. Unlike the α-thalassemias, major deletions are unusual. The bulk of β-thalassemia mutations are single base changes or small deletions or insertions at critical points along the gene. They include nonsense mutations, frameshift mutations, or mutations within introns or exons, or their junctions, that interfere with the mechanisms of splicing. In addition, there is a family of mutations that involve the promoter regions or adjacent areas. The clinical effect of these different mutations is variable. Although the frameshift and nonsense mutations result in absence of β chain production, some of the splice site mutations and particularly the promoter mutations may result in a much milder deficit of β chain production and hence in a milder phenotype.

Clinical Features

The homozygous states for β-thalassemias are usually associated with a transfusion-dependent disease, and children can only survive by receiving regular blood transfusion and expensive drug regimens to deal with the resulting accumulation of iron. The course and

Figure 38–1 World distribution of thalassemias.

management of the severe forms of Hb E β-thalassemia are similar. This condition, like many forms of thalassemia, shows remarkable phenotypic heterogeneity, even in the presence of identical β-thalassemia mutations. The homozygous states for α^0 thalassemia are associated with stillbirth, but there is also considerable maternal morbidity. Hb H disease has a variable phenotype; some patients have only moderate hemolytic anemia and splenomegaly, while others are transfusion dependent.

Although sickle cell anemia and its variants are not associated with such a severe degree of anemia, and show remarkable phenotypic heterogeneity for reasons that are not fully explained, there is an increased frequency of infection at all ages, and also a variable occurrence of acute complications called "crises." These may be associated with severe bone pain, or worse, priapism, sequestration of sickle cells in the lungs, stroke, marrow aplasia, and acute sequestration of red cells in the spleen. In addition, there is a variety of longer-term complications including aseptic necrosis of the femoral or humeral heads, pulmonary hypertension, renal failure, and gallstones. Thus, although these conditions do not cause as much strain on medical resources as the thalassemias, the management of the complications and the increasing use of blood transfusion to prevent them still causes a major strain on the provision of medical care (see Steinberg et al., 2001). The compound heterozygous states for the sickle cell gene and other hemoglobin variants produce pictures of varying severity. Hemoglobin SD disease and SE disease are similar to sickle cell anemia, whereas hemoglobin SC disease is milder, although it is associated with aseptic necrosis of the femoral heads and proliferative retinal disease.

The complex interactions between modifying genes and the environment that underlies the remarkable clinical diversity of the hemoglobin disorders is reviewed in detail by Steinberg et al. (2001) and Weatherall (2003).

Control and Management

Although progress in the management of hemoglobin disorders has been reasonably successful in the richer countries, this is not the case in the developing world, where the only children to receive adequate treatment are those whose parents can afford it. In Thailand, for example, where there are approximately three quarters of a million patients with severe forms of thalassemia, the mean age of survival for homozygous β-thalassemias is only 10 years, and that for patients with Hb E β-thalassemia, 30 years. There are very limited survival data for patients with sickle cell anemia in sub-Saharan Africa; in the United States, the median age for death in men is 42 years, and for women 48 years. Generally, sickle cell anemia has a milder phenotype in the Middle East and parts of the Indian subcontinent, and survival figures are much better than in Africa.

The major approach to the control of the thalassemias has been to develop screening and education with the offer of prenatal diagnosis, where acceptable. Programs of this kind, incorporating early prenatal diagnosis by chorion villus sampling and DNA analysis, have been applied successfully in Mediterranean countries and elsewhere, and are gradually being established in countries in the Middle East and in Southeast Asia. In Cyprus, Sardinia, and Italy, they have resulted in a major decline in the births of children with severe β-thalassemia. Although similar programs have been established in a few countries

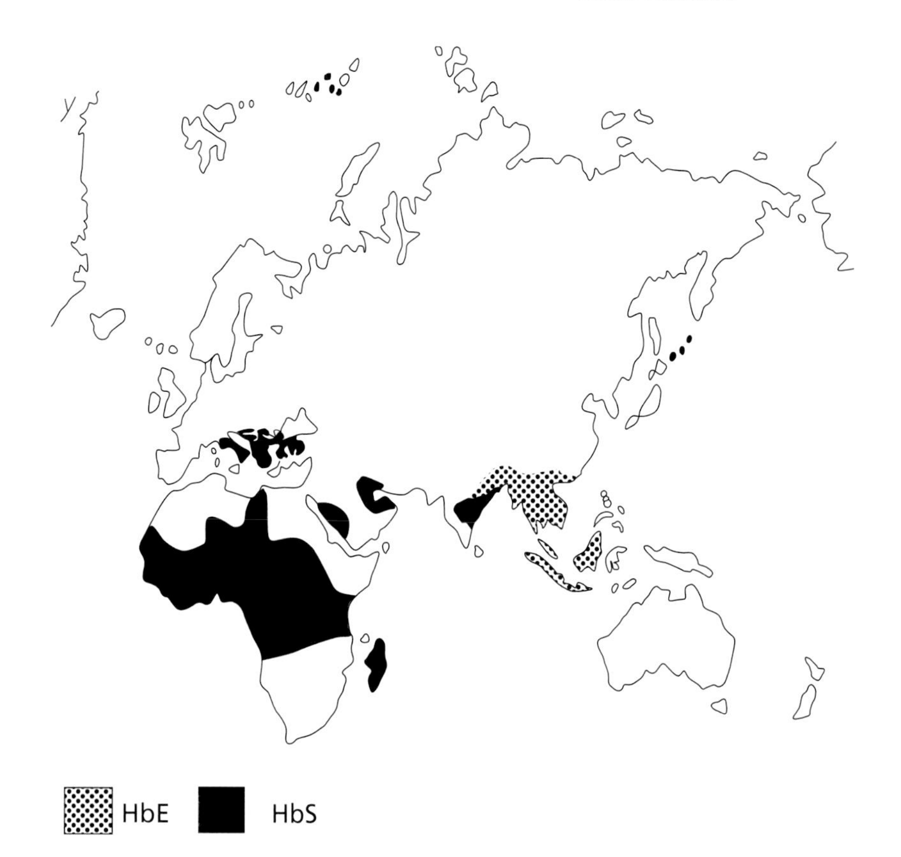

Figure 38–2 World distribution of Hb S and Hb E.

for the prenatal detection of sickle cell anemia, because of its phenotypic diversity they have been much less widely applied.

In many Asian countries, lack of financial support and facilities and, in some cases, religious belief, prenatal diagnosis programs are still not available. The importance of establishing community education programs before any form of screening or prenatal diagnosis is established cannot be overemphasized. In developed countries, and for those who can afford it in a few poor countries, bone marrow transplantation is now an option for those with a compatible donor. However, until more definitive forms of treatment by gene therapy become available for the hemoglobin disorders (even when they do, they may be extremely expensive), the most economic approach toward the control of thalassemia is likely to evolve around public education and screening, backed up, where acceptable, by prenatal diagnosis.

Table 38–2 Overall Distribution of Hemoglobin Disorders by WHO Region

WHO Region	Hb S	Hb C	Hb E	β Thal	α^0 Thal	α^+ Thal
AFR	1–38	0–21	0	0–12	0	10–50
AMR	1–20	0–10	0–20	0–3	0–5	0–40
EMR	0–60	0–3	0–2	2–18	0–2	1–60
EUR	0–30	0–5	0–20	0–19	1–2	0–12
SEAR	0–40	0	0–70	0–11	1–30	3–40
WPR	0	0	0	0–13	0	2–60

Source: Adapted from Weatherall and Clegg, 2001b.

Abbreviations: AFR, sub-Saharan Africa; AMR, the Americas; EMR, Eastern Mediterranean; EUR, Europe; SEAR, Southeast Asia; WPR, West Pacific.

Social Issues

Clearly, therefore, one of the major barriers to the more humane control of the thalassemias and sickling disorders in the developing countries is poverty. To make the case to their governments and to international health agencies that these conditions are causing a genuine public health problem, it is vital to obtain better information about the financial burden that they pose on health services. Although attempts have been made to cost the management of these diseases in different populations (Modell and Kuliev, 1993; Alwan and Modell, 1997), in order to argue the case for these genetic diseases compared with the major infectious killers of the developing world, it is necessary to try to assess their health burden in terms that can be compared with other diseases. Currently, the burden of disease is estimated in terms of disability-adjusted life years (DALYs) (Murray and Lopez, 1996). Essentially, rather than simply using mortality as a figure of health burden, DALYs include disability as well as death in attempting to estimate the total burden of disease in a particular community. Recently, an attempt has been made to calculate the health burden of the hemoglobin disorders in terms of DALYs (Weatherall et al., 2006). Because of doubts about both the frequency and costs of controlling and treating the hemoglobin disorders, these data must be viewed as very approximate, although the same can be said for many other diseases. Omitting α-thalassemia, the hemoglobin disorders account for approximately 3.9–10 million DALYs (a small figure compared to the estimated worldwide total of disease burden of 1537 million DALYs). Undoubtedly, they make a small contribution to total disease in Africa, where they are swamped by HIV, AIDS, and malaria. However, in other regions,

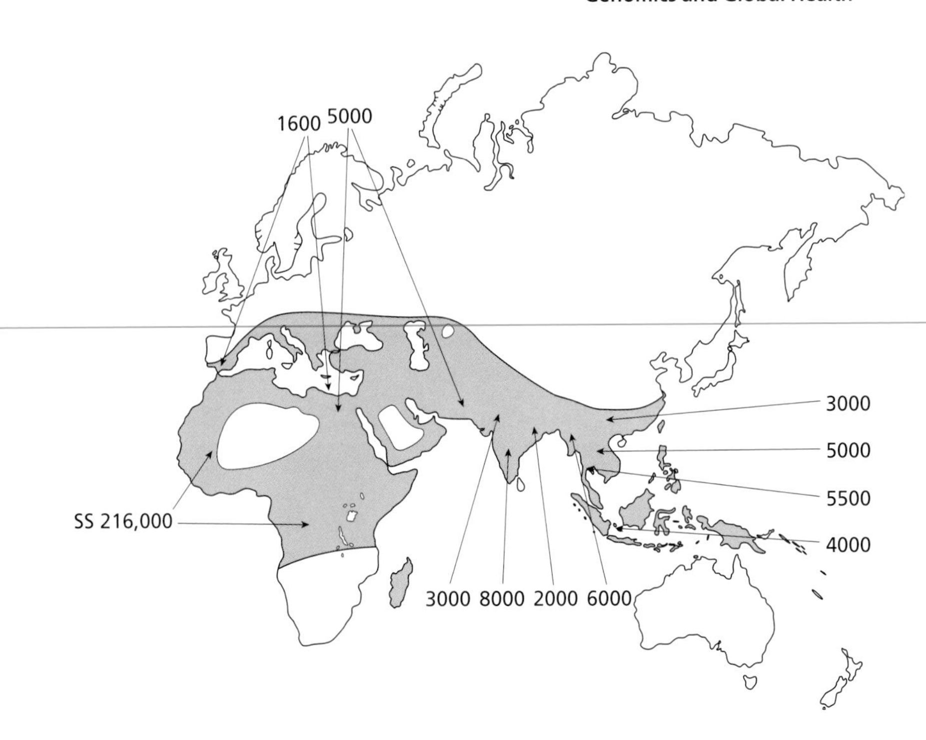

Figure 38–3 Approximate number of annual births of babies with β-thalassemia and sickle cell (SS) disease. (From Weatherall and Clegg, 2001a).

they appear to be much more significant. For example, the β-thalassemias account for between 1.61 and 3.45 million DALYs worldwide. Since they reach their highest frequency in Asia, and in East Asia and the Pacific and in South Asia HIV/AIDS accounts for some 3.1–7.4 million DALYs, respectively, it is clear that the thalassemias, at least in those regions, can be regarded as a major health burden.

These very preliminary studies indicate how important it will be in the future for clinical genetics to interact with public health and health economics in order to make the best possible case for the importance of monogenic disease as a public health burden, at least in the case of conditions that occur at a relatively high frequency in particular populations.

Communicable Diseases

One of the major hopes and predictions of the postgenome period was that new technology springing from this field would have a major effect on the prevention, diagnosis, and treatment of common communicable diseases. It should be remembered that, while major publicity is given to the "big 3" killers HIV/AIDS, tuberculosis, and malaria, there are many other communicable diseases that still cause large morbidity/mortality in developing countries. Many of these conditions are difficult to prevent, and treatment is still largely symptomatic. The widespread morbidity and mortality that occur regularly as epidemics of dengue surge across large regions of Asia and South America is but one example.

To put these problems into perspective, data collected in 1999 suggested that, of approximately 12 million childhood deaths from infection, approximately one-third were due to diseases for which vaccines or other preventative measures are already available. However, even if improved health facilities were to correct this disgraceful situation, there are still many communicable diseases for which there is no adequate approach to prevention or treatment. Moreover, as evidenced by the severe acute respiratory syndrome (SARS) epidemic, and recent concerns about the avian influenza virus, new epidemics will always be with us.

The role of genomics in the control of communicable disease includes the development of DNA diagnostics, the definition of targets for chemotherapy or vaccine development and vector control, and a better understanding of the factors that modify response to infection in human hosts.

Diagnostics

Remarkable progress has been made in sequencing the genomes of bacteria, viruses, and parasites, and it will not be long before the genome sequence of most of the major infectious agents is available. Rapid diagnostic methods are being developed based on the polymerase chain reaction (PCR) technique to identify pathogen sequences in blood or tissues. These approaches are being further refined for identifying organisms that exhibit drug resistance and also for subtyping many classes of bacteria and viruses. Although much remains to be known about the cost-effectiveness of these methods as compared to more conventional diagnostic procedures, at least some promising results have been obtained, particularly for the identification of organisms that are difficult to grow or when a very early diagnosis is required. Examples include the early identification and typing for dengue epidemics, the early diagnosis of leptospirosis, and the diagnosis of several parasitic diseases in which it is extremely difficult to grow the organism. These approaches have been shown to be cost-effective in the field (Harris and Tanner, 2000). This type of DNA technology is also being applied widely for the identification of new organisms, and is slowly gaining a place in monitoring vaccine trials (Felger et al., 2003). The extraordinary speed with which it was possible to identify a new coronavirus and its subtypes as the causative agent of SARS, and the way in which this information was applied epidemiologically to tracing the putative origins of the infection, is an example of the power of this approach (Ruan et al., 2003). PCR technology is also playing a valuable role in determining the activity of some virus infections, hepatitis C, for example.

Clearly, DNA-based diagnostics have considerable potential in the identification, subtyping, and resistance typing of a wide variety of infectious agents. When other methods are inadequate or too slow, they will undoubtedly find a place, but their more general role will depend on careful cost/benefit analyses compared to techniques that are more conventional. It is too early to anticipate their true place in global healthcare, but it is quite clear that the role of PCR and sequencing techniques is revolutionizing public health approaches to monitoring potential epidemics of viral infection.

Therapeutics and Vaccines

The use of genomic technology for the identification of potential drug targets and the development of vaccines has been much more difficult problems. Some promising drugs targets have been identified, and a wide variety of DNA-based vaccines, with or without adjuvants, are being developed.

One of the major problems in developing vaccines against parasites like those that cause malaria is their remarkable ability to change their antigenic makeup, a feature that is common to many infectious agents. In the case of malaria, the parasites that enter the body after a bite from a mosquito are called sporozoites. Hence, much effort has been directed at the development of vaccines based on a key antigen, the circumsporozoite protein (CSP) of *Plasmodium falciparum*. Efforts are also being directed at the development of vaccines that stimulate T-cell responses. Recombinant DNA technology has made it possible to produce plasmid DNA vaccines that can induce cell-mediated immunity. However, since on its own DNA stimulates only weak T-cell responses, vectors such as Yersinia, recombinant for DNA that encodes particular parasite proteins, are now being used, although, alone, these constructs produced unfocused T-cell responses. However, it has been found that priming with plasmid DNA and then boosting with the same DNA sequences in the vector generates a very strong T-cell response. Similarly, it has been found that the modified vaccinia virus Ankara, after attenuation by multiple in vitro serial passages, is extremely immunogenic for a malarial antigen. Similar approaches are being used to attempt to develop vaccines for tuberculosis and HIV. The problems involved are discussed by Targett (2005) and Kaufmann and McMichael (2005).

Pharmacogenetics

Pharmacogenetics is likely to play an increasingly important role in the control of some of the common diseases of developing countries. The prime example of the importance of this field in the rapidly changing patterns of drug resistance to common infectious agents is the current role of glucose-6-phosphate-dehydrogenase (G6PD) deficiency. Presumably, because it provides relative resistance to *P. falciparum* malaria, this polymorphism affects millions of individuals in countries in which malaria is or was a common killer. It results from many different mutations, which vary between ethnic groups and which are associated with different degrees of severity of the hemolytic anemia that may follow the ingestion of a wide variety of drugs and other agents (Luzzatto et al., 2001). For example, in African populations, the common mutation results in a self-limiting form of hemolysis, whereas there may be a much more severe degree of anemia in many other ethnic groups. Primaquine, which is the only agent that is currently available for the eradication of *P. vivax* malaria, now has to be given either in higher doses or for longer courses due to emerging parasite resistance. Thus, it will become increasingly important to screen patients for G6PD deficiency before treatment and to modify the dosage of the drug accordingly. Once the particular variety of G6PD deficiency has been established, there are simple, single-tube dye tests available for testing for G6PD levels, although even these are too expensive for routine use in many developing countries. Currently, a simple dipstick test is being developed.

It has been found that the frequency of a particular polymorphism of the gene *MDR1* is much more common in West Africa and African–American populations than those of European or Japanese background (Schaeffeler et al., 2001). This gene regulates the expression of P-glycoprotein, which is important in defense mechanisms against potentially toxic agents ingested in the diet. It has been suggested that this variant is common in Africa because it has offered a selective advantage against gastrointestinal tract infections. However, it appears that this gene reduces the efficacy of drugs such as protease inhibitors, which are now widely used for the treatment of HIV infections. No doubt, other examples of this type will appear.

Similarly, there have already been successes in defining some of the genes that are responsible for drug resistance in important pathogens, including those responsible for tuberculosis and malaria. For example, mutations in the chloroquine-resistance transporter of *P. falciparum*, encoded by the gene *pfcrt*, confer chloroquine resistance in laboratory strains of the parasite. Work in Malawi has shown that there is a stable relationship between rates of chloroquine-resistant genotypes and in vivo chloroquine resistance at sites where there are different population sizes, ethnic compositions, and levels of drug resistance and malaria transmission (Djimdé et al., 2001). This approach may have the potential for public health surveillance of antimalarial resistance and may permit comprehensive mapping of resistance at country and regional levels without the need to carry out numerous repeated longitudinal efficacy studies, particularly in parts of Africa where chloroquine is still used. Again, cost-benefit studies will have to be used to compare this genetic approach with other well-tried methods for identifying chloroquine resistance.

Vector Control

Clinical applications of genomics for the control of communicable disease are not restricted to infective agents. With the announcement of the complete sequence of the Anopheles mosquito genome, thoughts have turned to the possibility of genetically engineering this vector to make it unable to transmit the malarial parasite; a variety of vector/modification approaches are being explored (Land, 2003).

Variation in Susceptibility to Infection

A great deal is also being learned about genetic resistance to particular infections in human beings (Weatherall and Clegg, 2002), information that may become increasingly important when vaccines are submitted to clinical trial in populations with a high frequency of genetically resistant individuals, particularly if they are attenuating vaccines. In the case of malaria, as well as the well-known red cell polymorphisms discussed earlier in this chapter, there is a variety of blood group antigens, red cell enzymes, and genetic variations in the structure of membrane proteins that appear to have come under selection by resistance to malaria. Furthermore, systems other than the red cell have been modified in this way, including HLA/DR, tumor necrosis factor α, and a variety of other cytokines (Table 38–3). While some of these protective polymorphisms may play a small role in defense against infection, and all of them vary between ethnic groups, the effect of some of them is not trivial. Case–control studies in Africa and Papua New Guinea have shown that, as judged by strict WHO criteria for severe infection, the degree of protection provided by the sickle cell trait is approximately 70%–80%, while that by the milder forms of α-thalassemia approaches 60% (Hill et al., 1991; Allen et al., 1997), findings that

Table 38–3 Examples of Human Genes Involved in Varying Susceptibility to Communicable Disease

Disease	Genes Influencing Susceptibility
Malaria	α-globin; β-globin; Duffy chemokine receptor; G6PD; Blood group O; Erythrocyte band 3; HLA-B; HLA-DR; TNF; ICAM-1; Spectrin; Glycophorin A; Glycophorin B; CD36; NO Synthase; PK; CR1
Tuberculosis	HLA-A; HLA-DR; SLC11A1; VDR; IFNγR1
HIV/AIDS	CCR5; CCR2; IL-10
Leprosy	HLA-DR
Hepatitis-B	HLA-DR; IL-10
Acute bacterial infection	MBL-2; FCγR11-R; Sec 2

Abbreviations: CCR-5, chemokine receptor-5; CR1, complement receptor 1; FCγR11-R, receptor for constant region of immunoglobulin; G6PD, glucose-6-phosphate-deficiency; ICAM, intercellular adhesion network; IFNγR1, interferon-γ-receptor-1; IL-10, interleukin-10; MBL, mannose binding lectin; NO, nitric oxide; PK, pyruvate kinase (murine); Sec, secretor of blood group substance; SLC11A1, solute carrier family 11, member 1; TNF, tumor necros is factor; VDR, vitamin D receptor.

have been replicated in Africa recently (Williams et al., 2005). Knowledge of innate resistance to diseases such as malaria among different populations may be of considerable importance in designing vaccine trials in the future.

A wide range of resistance or susceptibility genes have been determined for bacterial and virus infections, including HIV, and genome searching in murine models has become a valuable approach to defining similar loci (Fortin et al., 2001). The recent identification of malaria resistance associated with mild pyruvate kinase deficiency in murine systems is an example of the kind of finding that may have application to human pathogen resistance (Min-Oo et al., 2003).

Some of the practical issues that will have to be addressed and overcome in the application of pharmacogenetics into clinical practice, both in developed and developing countries, have been reviewed recently (Royal Society, 2006).

Diseases of "Westernization"

As developing countries pass through the epidemiological transition and diseases due to poor living conditions, communicable diseases, and malnutrition gradually come under control, they tend to develop the same pattern of disease as richer countries. Indeed, coronary artery disease is now the commonest cause of death globally. This change in the pattern of illness is creating a major problem for many developing countries, since while they struggle to improve and maintain basic standards of nutrition and hygiene and to control the problems of communicable disease, they are also having to deal with this rapid increase in diseases of westernization (Unwin et al., 2001).

While many of the environmental triggers for these conditions may be the same as those in richer countries, the genetic factors that render populations more or less susceptible to these diseases may be different. For example, the complex interplay between genetic susceptibility, fetal and neonatal development, and environmental factors that are combining to produce an increasing global epidemic of type 2 diabetes may well reflect different patterns of genetic variability due to varying patterns of evolutionary adaptations in different ethnic groups. Similar principles may apply in the case of other common diseases, particularly cardiovascular disease and stroke. However, from the public health viewpoint, the very rapid increase of many of these diseases in the developing world consequent on changes in diet and lifestyle indicate that, regardless of the genetic component, the environment is playing a central role.

Undoubtedly, the prime example of these epidemiological transitions is the growing epidemic of type 2 diabetes, which is also an extremely important risk factor for vascular disease (see Chapter 13 on Diabetes Mellitus). It appears that the prevalence of this disease could be approximately 20%–70% in many populations and that the global figure may reach approximately 300 million people by 2020 (Zimmer et al., 2001). Although monogenic forms of the condition only account for 2%–3% of cases, twin studies suggest that there is a strong genetic component as well as a clear-cut environmental contribution to this disease. Although some of these findings may have to be revised in the light of recent information about the role of the fetal environment, there is no doubt that it will be extremely valuable to attempt to define the different genes that are involved in increased susceptibility to the rapidly changing environments and diets of the developing countries. In addition to providing information about the better management of this condition, it is possible that, by a better understanding of its genetic basis, it may be possible to focus public health measures for its control on subsets of the population.

Since type 2 diabetes is the subject of intense research by many academic and commercial organizations in the developed countries, this disease would seem to be an ideal subject for increasing interactions between the institutions of the rich and poor countries; both will have much to learn from each other. A similar approach could be of considerable value for the investigation of many other common communicable diseases (WHO, 2002a).

Ethical and Social Issues

The application of clinical genetics in developing countries raises a wide variety of social and ethical issues that are not encountered in developed countries (see also Chapter 40). While hitherto this field has been neglected, and it is vital that further research into the implications of genetic disease in the developing world is carried out, it is important that clinicians are aware of some of these problems before introducing even the most simple programs for the control of genetic disease in these countries.

From experiences in the thalassemia field, it is clear that it is vital to try to develop an extensive public education program before any form of genetic intervention is attempted. Even using the simplest diagrams, it is very difficult to explain the principles of Mendelian genetics to populations with limited education and whose beliefs about the nature of disease may be completely different to those of Western society. Misconceptions about how recessive diseases are inherited lead to a wide variety of social problems. Husbands tend to blame their wives for producing children with these conditions, marriages come under enormous stress, and there may be a high frequency of divorce and even suicide. It is also becoming clear that children with these diseases may suffer from a variety of social problems due to ignorance about the nature of genetic disease. Their parents may attempt to hide them away from any social contact, or they may be ostracized by their fellow students when they go to school.

Screening for genetic disease in the developing countries also raises a number of problems. In those populations in which there are arranged marriages, the discovery that a woman is a carrier for a serious genetic disorder may preclude her from finding a marriage partner. Although the pattern is changing, many of the world's great religions will still not accept termination of pregnancy for a serious genetic disorder, and it is difficult to develop prenatal diagnosis programs. In addition, in countries in which these programs are well developed, there are pressures on women who decide that they do not want to be involved in this procedure; their peers may look down on them for allowing children with serious genetic diseases to be born into the world.

These issues, and many more, underline how important it is to try to develop a broad-based education program before any attempts are made at screening or the development of control programs for genetic diseases in developing countries. Similarly, it is vital that these countries evolve ethics committees and other bodies for discussion of implications of introducing genetic control programs.

The Future

In its report *Genomics and World Health*, the WHO suggested that simple DNA technology should be introduced into the developing countries, at least for the control of the common hemoglobin disorders and, where cost-effective, for the diagnosis of communicable diseases that are difficult to identify by current techniques. How can this be achieved?

In the thalassemia field, one of the most effective approaches to disseminating simple DNA technology has been the establishment of North/South partnerships; that is, partnerships between centers in the rich and poor countries. Once several of the developing countries have evolved the expertise to develop national control programs for these diseases, the next step has been to try to evolve regional networks for the dissemination of knowledge about their control and management.

One of the major problems that impede the introduction of clinical genetics and modern methods of diagnosis and therapy of genetic disease into developing countries is lack of awareness of their importance on the part of both individual governments and international agencies. For example, although the thalassemias are causing an increasingly serious burden on healthcare services in Asian countries, only a handful of their governments have recognized that this problem exists and gone on to develop national control programs. Apart from poverty, one of the major reasons for this state of affairs is that genetic disease is not considered to be of importance compared with to major communicable disorders and conditions that arise from poverty and dysfunctional healthcare systems. However, particularly as countries pass through the epidemiological transition, this situation is changing and, as discussed earlier in this chapter, at least some monogenic diseases are starting to pose a major cause of ill health and an increasing burden on healthcare resources in developing countries. Until this situation is recognized by governments and the international health community, thousands of children with these inherited diseases will continue to die prematurely in many countries of the world.

Academia and Research Developments

The wider application of genomics for healthcare will require strong links between industry and academia. It is clear that certain areas of genomics will have major applications in the healthcare of developing countries, and thus will have significant commercial value. However, some of the developments in genomics could be less attractive to industry due to their limited commercial benefit. This work will need to be developed in the university sector and related research institutions. Hence, it will be important to foster strong international academic research partnerships between rich and poor countries (Weatherall, 2003).

An excellent example is the research field involving the thalassemias and other inherited hemoglobinopathies, where a relatively rapid transfer of DNA technology from the developed to the developing countries has been achieved through strong relationships between universities and academic researchers. A series of personal contacts, professional meetings, joint research initiatives, and exchange programs enabled this to be achieved. Leading universities and other organizations, including WHO, provided some support to this endeavor. However, there are some constraints to this collaboration. The main problem is that granting bodies that support research in developed countries, with some notable exceptions such as the UK's Wellcome Trust, sometimes are loathe to support research and development in developing countries. Even though some of these funding bodies are attempting to move in this direction, the only sustainable way forward will be if the governments of richer countries are persuaded of the value of sustainable collaborations between the North and South, and hence are able to direct at least some of their overseas aid programs in this direction.

If partnerships of this type are to increase, there will have to be major changes in the pattern of medical education, particularly in developed countries. There is very limited time allocated to global health in the curricula of undergraduate medical courses across the developed world. It is also important that universities and WHO take a lead in revising the curricula in light of rapid developments in genomics. In addition, student exchange programs between developed and developing countries should be encouraged, to increase awareness of health problems that are amenable to genomics-based healthcare provision. A major emphasis should be placed on the increasingly homogeneous distribution of illness among different countries in the future, and the potential importance of genomics for improving global health.

Information Technology and Bioinformatics

Increasingly sophisticated information technology and bioinformatics are essential in analyzing and interpreting the vast amount of data that are being generated in various sections of genomics-related research. The medical developments of genome projects depend heavily on this core facility. A great deal of this information is in the public domain and is freely available to those with the facilities and expertise to use it. For example, all the information derived from the International Human Genome Sequencing Consortium, the International SNP Map Working Group, the Pathogen Genome Project, the HapMap Project, and many similar research programs is made available immediately throughout the world. Furthermore, the majority of scientific journals insist on DNA data being deposited in publicly available databases. The complex algorithm programs directed at analyzing both the genome and the proteome are now freely available. Except for some restrictions, Celera allows scientists to view their draft human genome sequence data at no charge.

Effective and efficient use of all available "genome-based" information will require reliable and up-to-date information technology and computational facilities.

It is essential that developing countries acquire the necessary equipment and technical skills to take maximum benefit of various genomics-related resources. There is a shortage of skilled technicians and scientists in this field. Without these personnel, it will be impossible for developing countries to develop either strong research bases or establish collaborative industrial biotechnology programs.

Regional and International Collaboration

It is encouraging that several developing countries have established advanced biotechnology bases, with further expansion planned in the future. These developments are outlined in the report of the WHO (2002); the examples that follow provide some insights into the trend of developments of this kind.

The Asia–Pacific International Molecular Biology Network

The International Molecular Biology Network (IMBN) provides an excellent model of regional approach to capacity building. It was established in 1997 by a network of scientists from leading research institutions across Asia and the Pacific Rim to promote regional collaboration in the development of molecular biology and biotechnology, and benefit from the considerable potential of these technologies to improve the economic and health status of various countries in the region. The IMBN is structured in line with the European Molecular Biology Organization (EMBO). It strives to maintain and promote the development of excellence in molecular biology and biotechnology among scientists and institutions in the region, and to promote awareness of the importance of research among policy-makers.

The key objectives include:

1. Promoting study, research innovation, development, and dissemination of knowledge in molecular biology, genetic engineering, and directly related areas,
2. Establishing facilities and training programs to strengthen expertise and capacity for these disciplines,
3. Coordinating the conduct of research and development activities in laboratories designated by supporting institutions as Asia–Pacific International Molecular Biology Laboratories,
4. Cooperating with industry to identify areas of common interest.

The IMBN works with other international organizations in streamlining and coordinating activities targeted toward common goals and for minimizing the duplication of research time and resources. It receives funding from several supporting institutions across the region, and focuses its activities through designated participating centers of excellence.

Other Networks for Biotechnology Research

In addition to regional and international collaboration, some developing countries have made excellent progress in organizing and developing their own networks for advanced biotechnological research. A significant proportion of the technological and scientific resource is allocated for genomics-related research and development. Notable examples include Brazil, China, and India.

The strategies adopted by Brazil since 1997 have led to major advances in international genomics research. For example, a network of 200 researchers based at 30 institutions initiated a program to sequence the genome of the bacterium *Xylella fastidiosa*, a pathogen of citrus crops that is estimated to cost the Brazilian economy more than US$100 million each year in lost revenue. The program was funded by the State of Sao Paulo and managed by the Organization for Nucleotide Sequencing and Analysis (ONSA). The network sequenced 90% of the 3-million-base *Xylella* genome in less than 1 year, and published the full sequence soon after. This achievement was soon followed by the Cancer Genome Project, aimed at identifying expressed sequence tags (ESTs) associated with various cancers, especially those of the stomach, head, and neck, which are particularly prevalent in Brazil.

China has also made significant progress in organizing itself for genomics research and development. Investment by the Chinese Ministry of Science and Technology in establishing a center in Shanghai (Chinese National Human Genome Center; CHGC), and the Beijing Institute of Genomics (BGI), helped China to join the International Human Genome Sequencing Consortium, and played a key role in sequencing 1% of the 3 billion base pair sequence. This helped in developing technological capacity to undertake cutting-edge genomic science, and sparked the development of advanced bioinformatics infrastructure and supercomputing facilities needed to support this research.

The Government of India established the Department of Biotechnology in 1986, which attracts an annual budget of US$30 million. It supports both capacity building and frontline R&D programs in areas in which technical capacity exists but for which financial resources are lacking, encouraging collaborative links between academia and industry. In addition, it is responsible for developing regulatory standards for biotechnology research in the country. The biotechnology programs are being implemented through several autonomous and statutory institutions and universities that receive grants-in-aid.

There are several programs in human genetics and genomics currently supported by the Department of Biotechnology. These include clinical genetics research, genetic services, human genome diversity studies, gene therapy research, and a major functional genomics program at the Centre for Biochemical Technology in New Delhi. Recently, the Government of India has supported a major program in molecular medicine and genomics through the Indian Council of Medical Research (ICMR). Perhaps, the main strength of the Department of Biotechnology in India is the Biotechnology Information System Network that covers practically the whole country, comprising 10 distribution information centers and 46 subcenters around an apex center at the department headquarters. As a result of the network, a large number of public domain databases in the area of molecular biology and genetic engineering are now available to Indian scientists. India is now a major nodal point for various genomics-related databanks and networks (http://dbtindia.nic.in).

The population of India currently exceeds 1 billion people, comprising of 4693 communities with several thousand endogamous groups, 325 functioning languages and 25 scripts (Indian Genome Variation Consortium, 2005). Successive migrations and foreign invasions have contributed to enormous ethnic diversity and population heterogeneity. There are several small tribal founder population groups. In this context, the Government of India has funded a nationwide comprehensive genomics project that is aimed at setting up a database on the genomic variation. It is envisaged that this important resource will help in studying complex medical diseases and in pharmacogenomics research leading to the development of new drugs.

Six constituent laboratories of the Indian Council of Scientific Research (CSIR) have been commissioned for IGVdb project. The consortium aims to provide data on validated single nucleotide polymorphisms (SNPs), microsatellites, and short tandem repeats (STRs), along with gene duplications in over 1000 genes relevant to complex disorders. The aim is to estimate allele frequencies of novel as well as reported SNPs and microsatellites, to construct haplotypes, and to determine the extent of linkage disequilibrium within and across genes and across various population subgroups. The ultimate goal of the project is to create a DNA variation database of the people of India and make it available to researchers for understanding the molecular basis of disease predisposition and adverse drug response.

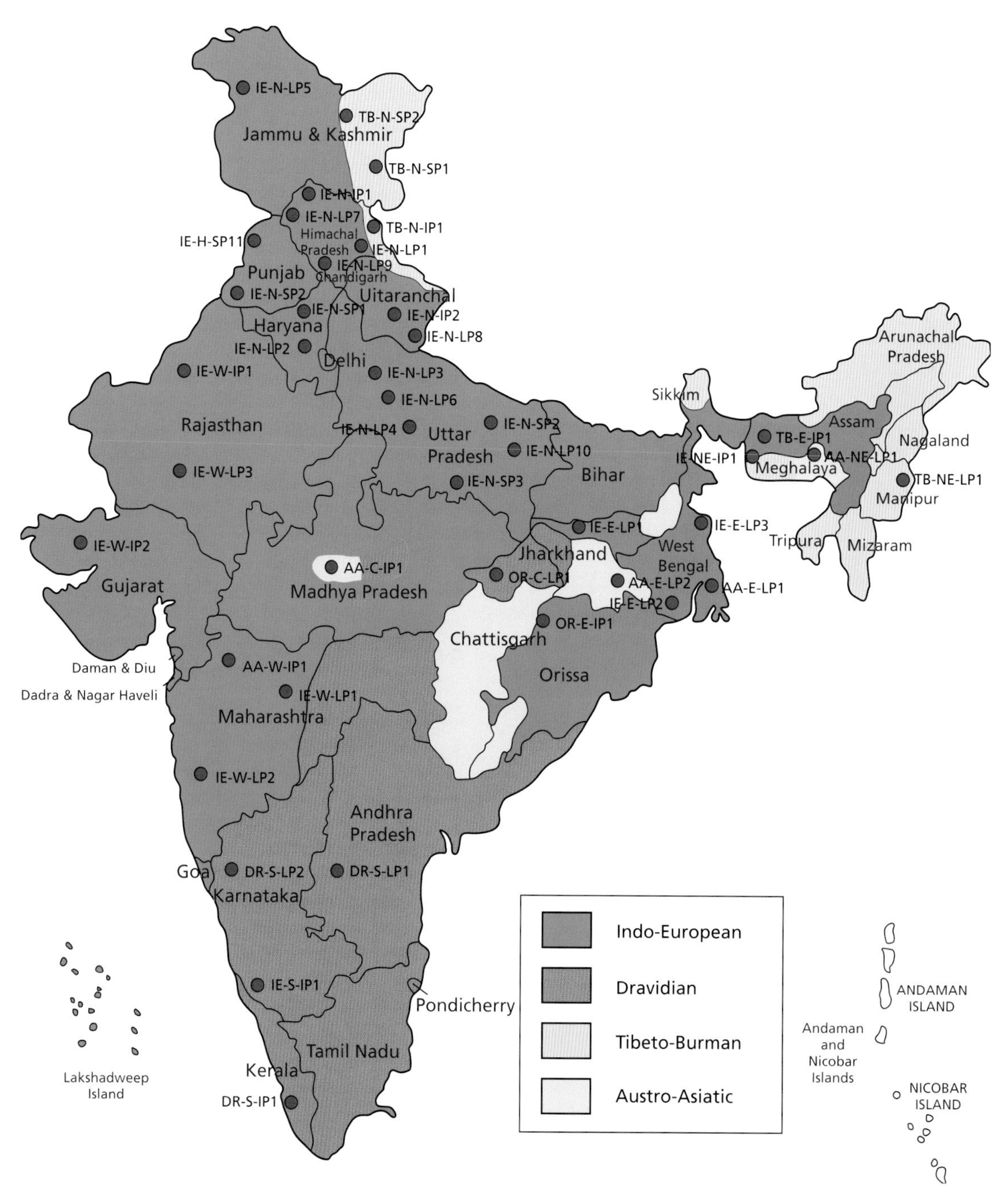

Figure 38–4 The political map of India showing the broad distribution of the four major language families [Indo-European (blue), Dravidian (brown), Tibeto-Burman (pink), and Austro-Asiatic (green)] selected for the Indian Genome Variation Project. Red circles indicate population studied, followed by symbols (see below) for the language family, geographic zones, and type of population admixture. Symbols: IE, Indo-European; DR, Dravidian; TB, Tibeto-Burman; AA, Austro-Asiatic; N, North; S, South; W, West; E, East; C, Central; NE, Northeast; LP, large endogamous; IP, isolated population; and SP, special population. (Reproduced from Indian Genome Variation Consortium, 2005 with permission from Springer, publishers for Human Genetics.)

The project involves a cohort of 15,000 individuals drawn from various subgroups of the Indian population, from all ethnic and tribal backgrounds (Fig. 38–4).

Postscript

Although progress may be slow, and will require a major move toward collaboration between molecular and cell biology, epidemiology, and public health, it is already clear that genomics is likely to contribute a great deal to global health in the future. It cannot continue to be ignored in global and national health planning in the future.

References

Allen, SJ, O'Donnell, A, Alexander, NDE, Alpers, MP, Peto, TEA, Clegg, JB et al. (1997). α^+-Thalassaemia protects children against disease due to malaria *and* other infections. *Proc Natl Acad Sci USA*, 94:14736–14741.

Alwan, A and Modell, B (1997). Community control of genetic and congenital disorders. *World Health Organization*, Eastern Mediterranean Region Office Technical Publication Series 24.

Djimde, A, Doumbo, OK, Steketee, RW, and Plowe, CV (2001). Application of a molecular marker for surveillance of chloroquine-resistant falciparum malaria. *Lancet*, 358, 890–891.

Felger, I, Genton, B, Smith, T, Tanner, M, and Beck, HP (2003). Molecular monitoring in malaria vaccine trials. *Trends Parasitol*, 19:60–63.

Fortin, A, Cardon, LR, Tam, M, Skamene, E, Stevenson, MM, and Gros, P (2001). Identification of a new malaria susceptibility locus (Char4) in recombinant congenic strains of mice. *Proc Natl Acad Sci USA*, 98:10793–10798.

Harris, E and Tanner, M (2000). Health technology transfer. *BMJ*, 321:817–820.

Hill, AVS, Allsopp, CEM, Kwiatkowski, D, Anstey, NM, Twunmasi, P, Rowe, PA, et al. (1991). Common west African HLA antigens are associated with protection from severe malaria. *Nature*, 352:595–600.

Indian Genome Variation Consortium (2005). The Indian genome variation database (IGVdb): a project overview. *Hum Genet*, 118:1–11.

Kaufmann, SH and McMichael, AJ (2005). Annulling a dangerous liaison: vaccination strategies against AIDS and tuberculosis. *Nat Med*, 11:S33–S44.

Land, KM (2003). The mosquito genome: perspectives and possibilities. *Trends Parasitol*, 19:103–105.

Luzzatto, L, Mehta, A, and Vulliamy, T (2001). Glucose 6-phosphate dehydrogenase. In CR Scriver, AL Beaudet, WS Sly, D Valle, B Childs, and B Vogelstein (eds.), *The Metabolic and Molecular Basis of Inherited Disease*. New York: McGraw Hill, pp. 4517–4554.

Min-Oo, G, Fortin, A, Tam, MF, Nantel, A, Stevenson, MM, and Gros, P (2003). Pyruvate kinase deficiency in mice protects against malaria. *Nat Genet*, 35:357–362.

Modell, B and Kuliev, AM (1993). A scientific basis for cost-benefit analysis of genetics services. *Trends Genet*, 9:46–52.

Murray, CJL and Lopez, AD (eds.) (1996). *The Global Burden of Disease: A Comprehensive Assessment of Mortality and Disability from Diseases, Injuries and Risk Factors in 1990 and Projected to 2020*. Cambridge, MA: Harvard University Press.

Royal Society (2006). *Personalised Medicine: Hopes and Realities*. London: Royal Society.

Ruan, YJ, Wei, CL, Ee, AL, Vega, VB, Thoreau, H, Su, ST, et al. (2003). Comparative full-length genome sequence analysis of 14 SARS coronavirus isolates and common mutations associated with putative origins of infection. *Lancet*, 361:1779–1785.

Schaeffeler, E, Eichelbaum, M, Brinkmann, U, Penger, A, Asante-Poku, S, Zanger, UM, et al. (2001). Frequency of C3435T polymorphism of MDR1 gene in African people. *Lancet*, 358:383–384.

Steinberg, MH, Forget, BG, Higgs, DR, and Nagel, RL (eds.) (2001). *Disorders of Hemoglobin*. New York: Cambridge University Press.

Targett, GA (2005). Malaria vaccines 1985–2005: a full circle? *Trends Parasitol*, 21:499–503.

Unwin N, Setel P, Rashid S, Mugusi F, Mbanya JC, Kitange H, et al. (2001). Noncommunicable diseases in sub-Saharan Africa: where do they feature in the health research agenda. *Bulletin of the World Health Organization*, 79:947–953.

Weatherall, DJ (2003). A New Year's resolution after a lost decade. *BMJ*, 327:1415–1416.

Weatherall, DJ, Akinyanju, O, Fucharoen, S, Olivieri, N, and Musgrove, P (2006). Inherited disorders of hemoglobin. In DJ Jamison et al. (eds.), *Disease Control Priorities in Developing Countries*, (2nd edn.) (pp. 663–694). Oxford University Press and World Bank.

Weatherall, DJ and Clegg, JB (2001a). Inherited haemoglobin disorders: an increasing global health problem. *Bulletin of the World Health Organisation*, 79:704–712.

Weatherall, DJ and Clegg, JB (2001b). *The Thalassaemia Syndromes*. Oxford: Blackwell Science.

Weatherall, DJ and Clegg, JB (2002). Genetic variability in response to infection. Malaria and after. *Genes Immun*, 3:331–337.

WHO (1987). Report of the V WHO Working Group on the Feasibility Study on Hereditary Disease Community Control Programmes, Heraklion, Crete, 24–25 October 1987 (WHO/HDP/WG/HA/89.2).

WHO (2002b). *Genomics and World Health*. Geneva: World Health Organization.

WHO (2002a). *Global Forum for Health Research. The 10/90 Report on Health Research 2001–2002*. WHO, Geneva.

Williams, TN, Wambua, S, Ugoga, S, Macaria, A, Mwacharo, JK, Newton, CR, et al. (2005) Both heterozygous and homozygous α^+ thalassemic protect against severe and fatal Plasmodium falciparum malaria on the coast of Kenya. *Blood*, 106:368–371.

Zimmer, P, Alberti, KGMM, and Shaw, J (2001). Global and social implications of the diabetes epidemic. *Nature*, 414:782–787.

39

Genetic Testing and Genomic Screening

Angus Clarke

Context

Twenty years ago, it became possible to isolate genes by first inferring their chromosomal location—the reverse approach to previous attempts at gene isolation. The recognition of widespread variation in the deoxyribonucleic acid (DNA) sequence of noncoding regions of the genome enabled the coinheritance of such variation with genetic diseases to be detected and measured, so that linkage could be established between the sites of variation and specific disease loci. This required (i) the preliminary work of gathering DNA samples and clinical information from many individuals in families segregating for Mendelian (monogenic) disorders, (ii) the ingenuity to manipulate the DNA with restriction enzymes and simple microbial vectors and then (iii) the ability to interpret the accumulated data in a statistically meaningful manner. Even when the protein encoded by a gene was completely unknown, it became possible to predict its amino acid sequence and then to recognize, isolate, and study the protein. This gave insight into completely novel cellular processes and greatly accelerated the pace of research in much of biology. Testing also become feasible for diagnostic and predictive purposes within many families affected by genetic diseases and some of the developmental disorders transmitted in the same fashion. These simply inherited conditions are, however, uncommon and together they account for only a small proportion of the overall burden of human disease.

Now, some 5 years after the success of the Human Genome Project in revealing the sequence of the human genome, there has been an associated transformation of the study of human disease and more generally of biology. The life sciences have become a quantitative and comparative discipline, and the new approaches emphasize the evolutionary interrelatedness of all life by common descent, as Darwin had forcefully pointed out a century earlier, as well as through our shared global environment. There has been a great acceleration in the generation of information on the genetic constitution of individuals and their tissues in health and disease, not just for the simply inherited disorders but also in the context of the common diseases of adult life in developed societies. There is an eager anticipation that this will lead to substantial, practical improvements in the provision of healthcare for these conditions.

Some of these anticipated benefits are beginning to arrive. In a few rare cases, gene product replacement can be made available, as in the inherited forms of leptin deficiency (Oral et al., 2002) and growth hormone deficiency. Microarray approaches to the study of the patterns of gene expression in tumors may give an indication of prognosis in breast cancer (Ahr et al., 2002; van't Veer et al., 2002) and searching fecal DNA for mutations in relevant genes (such as the *apc* gene) looks promising as a way of identifying asymptomatic individuals with colorectal malignancy (Traverso et al., 2002a,b). Rational therapies are also now being developed and translated into practice—as with the use of small molecules targeted against tumor-specific fusion proteins (e.g., the Philadelphia bcr-abl tyrosine kinase; Savage and Antman, 2002), and the generation of monoclonal antibodies against the disease-modifying protein products of tumor-amplified genes (Piccart-Gebhart et al., 2005)—although numerous practical, safety, and regulatory issues still remain to be addressed. The postgenome move into proteomics is also creating opportunities to classify tumors more sensitively, as with certain lung cancers (Yanagisawa et al., 2003) and leukemias (Pui and Evans, 2006), and this may lead in the longer term to the improved targeting of therapies and so to improved outcomes.

For the most part, however, our collective ability to understand the genetic basis of disease is outstripping our ability to intervene so as to treat a disease more effectively or to prevent it altogether. For the moment, perhaps a rather long interim "moment" of one or several decades, genetic testing of healthy individuals is more likely to identify those at risk of developing or transmitting a disorder rather than being of direct medical benefit to the individual. This chapter sets out one possible framework within which to consider and assess current practice and likely future developments.

Terminology

The term "genetic testing" can be used to refer to a wide range of activities. We must be clear as to the scope intended by this term and so we must distinguish between two pairs of words—genetic and genomic; testing and screening—and at least three different types of clinical investigation.

Genomics may be regarded as a comparative perspective taken to the study of genetics in general, whether different genes are being compared within an organism, or the "same" (i.e., the homologous) genes are being studied in different organisms, or properties of the whole genome are being considered collectively. Here, however, we will use "genetic" to refer to the testing of a single gene of major effect (a Mendelian gene locus), and "genomic" testing will refer to the investigation of the overall pattern of genetic variation within an individual that may be relevant to their potential development of a specific disease.

Similarly, "screening" may be regarded as the testing of a person (or of people) at more-or-less population risk of a disorder, whereas "testing" in the narrower sense refers to the examination or investigation of an individual in the context of their personal or family history of disease, that is, where there is a specific reason to believe that their individual chance of carrying a disease-associated genetic variant, or of developing a particular type of illness, may be different from the population average.

We use "genetic testing," therefore, to refer to the examination of one or several genetic loci in an individual because of their personal risk of a genetic disorder, usually because of their family history of disease. In contrast, "genetic screening," perhaps best prefaced by the word "population" for the sake of clarity, refers to the performance of a particular genetic investigation on a large group of individuals without preparatory sorting into risk categories. An individual who seeks access to over-the-counter genetic testing in a pharmacy or through the Internet may therefore be arranging a genetic test on himself or herself if he/she has a strong family history of a relevant disease, or they may be arranging a genetic screening test if there are (as far as they know) no particular features in their personal or family history suggesting an increased risk. It is therefore the context that distinguishes a specific test applied to one individual from a population screening test applied to others; there may be differences in the technology employed but they will reflect differences in the scale of the enterprise rather than the distinction between individual testing and population screening.

In addition to genetic testing of an individual in their family context, and the population genetic screening of individuals intended to identify those carrying a particular genetic variant predisposing to disease, there is a third type of genetic test in clinical use—the testing of a tissue sample from a patient with disease, where the genetic changes may be present only in some tissues rather than constitutionally, that is, in all tissues. This usually entails the testing of a malignant tumor to identify the genetic changes that have either led to the tumor or that may be useful in guiding treatment or predicting the likely response to treatment. Occasionally, the patient shows signs suggesting that they manifest a genetic disorder present in only some parts of the body—and investigations then focus on the possibility of somatic mosaicism (e.g., for neurofibromatosis type 2 or type 1).

Finally, there is a fourth category of testing to consider, "genomic screening," which term can be somewhat ambiguous. It can refer to the examination of the genome of one individual for evidence of variation that could be disease-related, such as alterations in the DNA sequence of genes or intergenic regions, or the variation in copy number of regions of the genome. It could also refer to the typing of many individuals for genetic variants, such as the patterns of specific single nucleotide polymorphisms (SNPs), already known from previous studies to be associated with disease or with response to therapy. The "screening" can then refer to the laboratory part of the process (testing across the whole genome) or to the human part of the process (testing many individuals at standard population risk of disease). In this chapter, we try to avoid the term "genomic screening" and refer instead to genome-based analysis or to population genetic screening, in the appropriate contexts.

There can also be some confusion as to the scope of the word "genetic" in the context of genetic testing. For our purposes, it refers to the information being generated rather than the mode of investigation. Hence, in the appropriate family context, a renal ultrasound scan can generate essentially predictive information about whether or not a young adult is likely to develop his family's polycystic kidney disease; a biochemical assay performed on a male infant at risk of Duchenne muscular dystrophy (DMD) will usually give very reliable predictive information as to whether he will go on to develop that disease. It can be seen, therefore, that genetic information can often be inferred from investigations that do not examine the genotype directly—the DNA or chromosomes—but look instead at the phenotype or some intermediate between the two.

A Preliminary Historical Perspective

The initial development of "reverse genetics" as outlined above depended on the accumulation of DNA samples and phenotypic data from family linkage studies for a wide range of single gene disorders. This stimulated the parallel development of both (1) the physical mapping and cloning of the human genome in a variety of vectors and of (2) the bioinformatics required to assess linkage, to recognize open reading frames and to predict protein sequences. Great as was the progress made in this way, it became clear that the recognition of relevant genes would proceed much more smoothly if the physical map of the human genome were tackled systematically rather than piecemeal, and by researchers interested in the genome as a whole rather than by those whose primary interest lay in the handful of particular genes they were seeking to identify, whose connection was their involvement in disease affecting one particular tissue rather than their being located close together on the same chromosome. This realization led to the Human Genome Project, which has succeeded in releasing several drafts of the human genome sequence and which, in effect, launched the field of "genomics"—characterizing different regions and properties of the human genome and comparing human genomes with those of other species.

The identification of Mendelian loci, at which mutations can be said to cause a disease, has greatly increased our understanding of many pathological processes. However, although mutations at these loci contribute heavily to the burden of those diseases that are clearly inherited within families, the genetic constitution of individuals at these loci makes relatively little direct contribution to the burden of the common, complex diseases of Western societies—especially coronary artery disease (CAD), hypertension, cerebrovascular disease, Alzheimer disease, colorectal cancer (CRC), breast cancer, depressive illness, schizophrenia, rheumatoid disease, and more. Many of these diseases cluster loosely in families without being inherited in any straightforward fashion. They are the common, degenerative disorders of our time, that arise from the continuous and complex processes of interaction between the individual's genome and their environment, from conception through to old age.

Genetic approaches to the study of these diseases have largely followed two courses. Those unusual families, in which such conditions do appear to be transmitted as a Mendelian trait and where the age at onset is usually younger than usual, have been studied by family linkage analysis and have often been found to be caused by mutations in single genes; they constitute the clearly Mendelian subsets concealed within the larger mass of those affected by such common diseases in the absence of a readily apparent inherited predisposition. These subgroups often comprise approximately 1 in 20 (5%) of the individuals affected by these diseases.

The remaining cases—the large majority, where there is no clear pattern of inheritance but where there may be some familial clustering—have been studied by the genomewide search for association between specific genetic variants and the occurrence of disease. Where familial clustering has been evident, investigators have looked for areas of the genome where the same variant is present in the (two or more) affected members of the same family—the particular variant (i.e., the particular allele at a polymorphic site) perhaps being specific to each extended family network, but with the researchers hoping that the same locus would be involved in different families. Investigators have searched for loci where alleles are shared within a family but not between families. In contrast, where the cases appear more often to be sporadic (or isolated), with less familial clustering apparent, investigators have searched for evidence of a more remote common ancestry through looking for the sharing of genetic variants between affected but apparently unrelated individuals from the same population, that is, they are searching for loci where affected individuals share the same allele, even though they are not known to be closely related. It is not expected in such studies that the polymorphism used to conduct the study, the particular SNP or microsatellite (CA repeat), will necessarily prove to be the causal factor contributing to disease susceptibility, but the hope is that the causally relevant genetic variation will lie in the vicinity of the implicated genetic variant—the polymorphism studied and the relevant causal factor having been kept together over time in linkage disequilibrium. It is then possible to use a knowledge of the nearby genes to devise hypotheses as to which might be associated with what is known of the pathophysiology of the disease in question. The investigators then assess the various *candidate loci* and look further at those where the association is most plausible.

The genetic variants studied have changed as the technology has developed. The first types of genomewide variation to be employed were the sequence variations associated with restriction enzyme recognition sites—the restriction fragment length polymorphism (RFLP)—and the minisatellite, but the potential for genome association studies with these was limited by their dependence on Southern blotting and hybridization with radioactive probes. The use of RFLPs was much more difficult than minisatellites, as a different probe was needed for each site, but even the use of minisatellites was rapidly superseded by the use of the more widely scattered microsatellites, mostly (CA) repeats, which can be investigated using methods based on the polymerase chain reaction (PCR). These approaches have in their turn been largely superseded by SNP-based approaches, because SNPs are more widely and evenly distributed than (CA) repeats and are simpler to assay.

The importance of these changes in methodology will become more apparent when we turn later to look at the causation of the common complex disorders. Here, we need to think through the relationship between the type of evidence produced by the different methodologies, the penetrance of any relevant genetic predisposition and population events that occurred at different periods of human history and prehistory. This means that apparently different conclusions, derived from studies employing different methodologies, are not necessarily contradictory but may rather provide important and complementary evidence about events at different stages of the human past that are all relevant to disease today.

High penetrance genetic variation will usually be identified by family-based linkage analyses in the context of an overtly Mendelian family history. This generalization will not be true of all diseases—such as Rett syndrome, a genetically lethal condition where most cases have occurred in association with a de novo mutation on the paternally-derived X chromosome—but will be applicable under most circumstances. When the penetrance is somewhat less, so that the relative risk of disease associated with different variants at the same site in the genome—under a given set of circumstances—is less striking, then the degree of familial clustering will be less and, very often, the origin of the genetic variation being examined will be more ancient. Another way of saying this is that genetic variation that is clearly disadvantageous tends to be of recent origin and is subject to strong negative selection, whereas variation that modifies risk to a lesser degree may be widespread because the relative advantage of different alleles may vary, depending upon environmental circumstances, or variation at other loci, or in the two different sexes, or because of linkage disequilibrium to a site of genetic variation that is itself subject to stronger selection (see discussion in Clarke, 2004).

The range of characteristics that are likely to have been subject to important selection in our collective history and prehistory of course includes physical fitness and fertility, natural selection for resistance to infectious disease, hardiness in the face of prevailing nutritional circumstances, and qualities such as intelligence and style of social interaction (e.g., cooperation versus competition/domination). We may think of epidemic infectious diseases such as syphilis, tuberculosis, and plague, nutritional challenges that may have contributed to the "thrifty" (and possibly diabetogenic) genotype, and variation in skin pigmentation as traits subject to powerful selective forces. The optimal degree of pigmentation would be set as a trade-off between sunburn and malignancy favoring more pigment and the need for activated vitamin D favoring less pigment, so as to avoid rickets and the perilous obstetric hazards of a rachitic pelvis. Interactions between environmental and genetic factors are likely to have been of great evolutionary importance, as with an interaction between vitamin D deficiency and vitamin D receptor genotype (Wilkinson et al., 2000). The consequences for human health today of our evolutionary history must inevitably be enormous but still remain unclear in detail. Just how the history of our immune system influences our bodies' responses to the social and material circumstances of life in developed countries today has still to be elucidated but will inevitably be relevant to the development of our contemporary degenerative disorders, especially those involving autoimmune phenomena. One factor implicated in resistance to infection is human population density, that has increased over several centuries, resuting in changes to the optimal genetic consitution (Xue et al., 2006). There will have been many such genetic changes in response to the varying patterns of infectious disease over human history and prehistory. In a similar way, our species' past nutritional circumstances will have consequences for our health today and the ways in which this is influenced by our contemporary diet.

In addition to natural selection, as Darwin pointed out, sexual selection is also likely to have been a major factor in shaping human evolution generally and variation between population groups in particular (Darwin 1874, 2003). Physical attractiveness will have been an important factor in generating and maintaining the physical and behavioral differences found between the sexes and may well have varied both between groups and within groups. Male competition and female choice are both likely to have been operating throughout human evolution—and to be doing so still. Between-group variation in physical features may have resulted in part from the failure of sexual "fashion" to favour any one particular physical appearance over the long term, whereas within-group variation may have developed in a frequency-dependent manner. Success may at times attend the extreme manifestation of a trait but, under other circumstances, extremes in a physical or behavioral trait can become a liability. Although the extreme in a desirable and attractive trait may lead to reproductive success, this may also attend the less common genotypes through their novelty value, which thereby maintains genetic variation in the population. This will be true for behaviors as well as physical traits—the attraction of wit and humour, for example, is well recognized, and sociobiology and its successor, evolutionary psychology, both elaborate on the success of two very different masculine behavior patterns, the macho "he-man" and the gentler, more caring "new man" strategies. The direction in which both natural and sexual selection take our species in the future remains, of course, obscure. The legacy of our evolutionary past, however, is that the environments in which our species' genome has been developed and moulded over many millennia resemble less and less the environments found on today's Earth; our past evolution has adapted us to past circumstances rather than those of modernity or postmodernity.

This lack of congruence between our innate biology and our contemporary social and physical environment will clearly have consequences for human health. Claims are made that we can now choose self-consciously to direct the evolution of our species, but social practices are likely to prove both more complex and less predictable than has been imagined. In attempting to select for property "p" we are likely, in fact, to find that we have selected instead, or additionally, for property "z." These attempts to control our own evolution seem likely to represent *hubris* and could never be more than partially successful. And this is to put to one side the vexed question of how "we" as a species would decide in advance what would count as a successful evolutionary manipulation; what social processes would operate to make or impose such collective decisions?

Medical genetics in the past has concerned itself with clearly disadvantageous mutations of recent historical origin, whereas the genomic approach to medicine now allows us to look at genetic variation of ancient origin and its relevance to the patterns of health and disease in whole populations rather than the ill-health of a small number of unfortunate families.

We now turn to look at what can be learned from the accumulated experience with family-based genetic testing that may be relevant to the application of genetics in the context of the common, complex diseases.

Genetic Testing

Family-based Genetic Testing

Genetic methods may, of course, be used to establish the diagnosis when an individual presents with clinical features likely to result from a genetic condition. Beyond that circumstance, however, there may be several different family contexts in which genetic testing of an individual may be considered as appropriate. In relation to their own health and healthcare, genetic testing is most likely to be appropriate where the person knows they are at risk of a disorder present in other members of the family but does not yet know if they will be affected; this amounts to predictive genetic testing.

Predictive genetic testing has been available for Huntington's disease (HD) for almost two decades and for several familial cancer syndromes for more than a decade. There have been numerous studies of the impact of testing on individuals and families, and of the circumstances in which testing can be clinically helpful or emotionally beneficial. One most important factor is whether the genetic test result makes a difference to the medical care of the individual, including surveillance for treatable complications of the disease. Where the test result does have implications for medical management, the health professionals involved will need to ensure that their client understands this. Indeed, professionals may wish frankly to recommend testing if this is clearly the best way to safeguard their patient's health and welfare—but they will wish to help the client make the best decision for their personal circumstances when the scope for medical intervention is less and it is the family and emotional factors that are more relevant. The usually "nondirective" ethos of genetic services will not be so appropriate where the decision about testing has clear implications for medical management (Elwyn et al., 2000). An example of such a context might arise with an adolescent or young adult at risk of the familial adenomatous polyposis (FAP) coli present in other family members. Because the prognosis in an affected individual is so poor in the absence of tumor surveillance, leading to colectomy in those at high risk of malignancy, it would be good practice for any health professional involved to recommend genetic testing. A similar situation arises with infants at risk of inherited retinoblastoma, where the benefits of genetic testing as an aid to the coordinated management of the at-risk child are so substantial that the only sensible course of action is to recommend it, so that frequent surveillance for tumors can be provided to those who are at high risk and can be avoided in those whose risk is not increased. Because of the implications of test results for the use of other medical resources, a number of perspectives unfamiliar to clinical geneticists may need to be introduced into decisions about which genetic testing services it would be appropriate to fund (Fulda and Lykens, 2006).

In the context of an individual at risk of a usually late-onset neurodegenerative disorder, such as HD, the situation is very different and will remain so until an effective treatment has been developed for symptomatic disease or effective, presymptomatic disease prevention has become possible. In this context, the genetic professionals, whether clinical geneticist or genetic counselor, will engage the at-risk individual who is seeking predictive genetic testing in an extended conversation. This process is designed to help the individual decide whether or not predictive testing would be helpful and, where the client chooses to have the test, to prepare for the full range of possible test results.

Even in such an apparently simple context, there are complexities to consider before proceeding with the test. Clients will be helped to think through how they, and others, would react to a favorable or an unfavorable result, or to a result of uncertain significance (an intermediate, "gray-zone" allele), or to a change of heart—a decision not to go ahead with testing. Not everyone wishes, or is able, to engage in discussions about such hypothetical scenarios (McAllister, 2002), but it is good practice to encourage such reflection (Sarangi et al., 2004, 2005). Having clear reasons for testing, and the willingness to engage in advance in reflection about the possible consequences of testing, may be associated with better outcomes after testing (Dudok de Wit et al., 1998; Decruyenaere et al., 2003).

There may be decisions to make about whom to tell about the risk, the test and the result, and there may be important potential consequences of testing for insurance or employment. The implications of testing for health or life insurance and for employment will clearly differ between countries, according to the legal context and the system of healthcare—and may lead to problems even within a European social democracy (Harper et al., 2004). There will also be consequences of testing for other individuals, including the person's partner, any surviving parents, and their siblings and children; there will be an impact on the whole family system (Sobel and Cowan, 2000). The most appropriate clinical approach to adopt has been refined, as experience has accumulated of predictive testing in this context of high risk without effective treatments (Harper, 1997; Kessler and Resta, 2000; Soldan et al., 2000; Decruyenaere et al., 2003). The early finding, however, was that there will be costs as well as benefits for those who are tested, whatever the result; this has not been contradicted by subsequent experience (Codori and Brandt, 1994).

One type of difficulty can emerge, even with a favorable result, if life-plans have been made and acted on and the result now undermines the basis of those decisions. Perhaps, a decision was made about marriage, children, or career that seemed safe in the context of genetic risk and years of life may now be viewed with regret. Difficulties may also arise if the results, whether "good" or "bad," conflict with a prior expectation of the person tested or their family. Individuals and families can have their own ideas about the pattern of inheritance that applies to *them*—perhaps it is the oldest girl in each generation who is affected, or the youngest boy, or the one with red hair. Such ideas, that can be referred to by the term "lay beliefs about inheritance," can lead to serious distortions in medical care; for example, women at risk of breast cancer but related to affected relatives through their father are less likely to seek risk assessment or additional surveillance because many families imagine that susceptibility to breast cancer has to be transmitted through the female line (Richards, 1996). Particular individuals may also be picked out by others in the family as destined to develop the family disease—they are "preselected"—and this phenomenon can contribute to complex processes of family psychodynamics (Kessler and Resta, 2000).

Experience has shown that those given an unfavorable predictive result for HD (and many other conditions) do often experience transient distress but, like those given a favorable result, this is followed by a fall in distress and anxiety over some months, associated with the loss of uncertainty, that returns to baseline levels by 1 year. Overall, the best predictor of posttest emotional state is the pretest state, and those whose motivation for testing is unfocused and nonspecific tend to fare worse than others (Broadstock et al., 2000; Decruyenaere et al., 2003). Those who show greater concern and distress during the counseling before the testing may be those who can engage constructively with their risk in advance and they may cope better with an unfavorable result (DudokdeWit et al., 1998). Those who proceed with testing for HD are a self-selected group (Codori et al., 1994) and results of these studies cannot be generalized to others—we could not draw any conclusions from these studies as to how those at low risk of a serious disorder would respond to unfavorable test results.

Using the framework of Burke and colleagues to evaluate genetic tests (Burke et al., 2001), through examining both the clinical validity of testing and the scope for useful medical interventions, we can turn to contexts where the risk level is high and there may be some medical or emotional benefits from testing. Testing for familial predisposition to breast and ovarian cancer at the *BRCA1* and *BRCA2* loci is in this category, with the benefits of prophylactic surgery in the prevention of malignancy now becoming better defined. In such contexts, the potential benefits of testing need to be weighed against the possible drawbacks jointly by the at-risk individual and their health professionals, and the contextual details of the individual case might then be crucial. One of the factors to be taken into account is the ability or willingness of the concerned individual to approach any surviving, affected relatives to see if they would be willing to have a genetic test, so that issues of family communication are central. Equally, although many women at risk of breast and ovarian cancer find testing helpful in both the practical and emotional domains, there are often real emotional barriers to their discussing test results with relatives. Indeed, given that the primary motivation of many is to provide information for their children, the unanticipated emotional and communication difficulties that arise for those who have undergone testing constitute an important phenomenon (Lim et al., 2004). Those already affected by cancer may have additional counseling and support needs, in part because of the additional implications for their own health (Hallowell et al., 2003).

The value of a test—in the sense of its clinical utility—can vary enormously with details of the family history and the availability for testing of other family members or of stored pathological specimens. If a family mutation can be identified in a sample from a definitely affected individual, then the interpretation to be made of the results of testing an at-risk relative can be defended with greater confidence than if no family mutation is known and only the at-risk individual can be tested. In the former case, a negative test result (the failure to find the family's mutation in one of the two *BRCA* loci) will give very substantial reassurance—although it will still leave the woman concerned at the population risk of developing a breast cancer. In the latter setting, finding no mutation in either of the two loci will provide some but much less reassurance as it is really of uncertain significance and therefore is rather unsatisfactory for all concerned. It is for this reason that some services only perform mutation searches in samples from a definitely affected individual and only when the chance of finding a *BRCA* gene mutation exceeds a given (arbitrary) threshold. Testing for the family's particular mutation in an individual at risk becomes possible only once it has been identified by a mutation search in the affected individual. This is one way of attempting to contain the costs of healthcare, although other models of service provision could well be defended and might permit at-risk women with fewer surviving female relatives to be given access to testing.

One approach that would maximize the cost-efficiency of mutation searching, which is much more expensive than testing for a known, familial mutation in large genes such as the *BRCA* loci and *APC*, would be to make mutation searching available to those affected individuals found to have a malignancy at a young age and who have a family history of relevant disease. This may be more cost-effective in the long term than a genetics service operating solely in response to concerns from unaffected, at-risk relatives. Proactively suggesting to such affected individuals that molecular testing might be appropriate would very substantially alter the emotional context for the patient. We already know that women affected by a breast cancer find that a positive genetic test result (a pathogenic mutation in a relevant gene) can cause more distress than they had anticipated (Dorval et al., 2000; Hallowell et al., 2002, 2003). There may be several reasons for this, including both the direct prognostic implications for themselves and the need for them now to consider the implications for others in their families, with the associated burden of having to raise this topic with their loved ones.

It is apparent that testing may have clear implications for the management of an individual's health care but that is often not the case. Where there is real uncertainty about the interpretation of a test result, as in many disease contexts, the decision about testing will be complex and personal (Evans et al., 2001). This arises, for example, where a positive (unfavorable) test result does not give an absolute yes/no answer but rather indicates an increased or decreased risk of developing the condition, as with testing for many inherited cancers, so that a negative (favorable) result leaves the person at population risk of disease rather than at zero risk. There may also be uncertainty about the way in which it might affect the individual or the age at which the malignancy or the illness might develop, even when it is highly likely to do so at some stage.

Where testing requires access to information about, and perhaps samples from, other family members, the decision to seek testing will for many be difficult and, even if they decide to go ahead, may be frustrated by a lack of cooperation by their relatives. Furthermore, the possible adverse consequences of unfavorable test results for insurance, and the possible distress or anger that may be anticipated in close family members and those whose risks will be modified by the result, may deter an individual from proceeding.

One context that has generated much debate among professionals, and also within families, is that of testing an individual at 1 in 4 (25%) prior risk of HD, when an unfavorable result on the individual implies an unfavorable predictive result for the person's parent, who was presumably at 1 in 2 (50%) prior risk but who may have decided not to have predictive testing and may resent the fact that they have, in effect, been tested without giving consent (Lindblad, 2001). Even apparently favorable results may have some difficult consequences, such as altered family relationships (Kessler and Resta, 2000) or an unanticipated reluctance to discontinue screening for disease (Michie et al., 2003).

In the future, with the development of technologies including microarrays and tandem mass spectroscopy (TMS), it may become possible for testing to cover so much of the genetic variation relevant to risk of a particular disease (e.g., breast and ovarian cancer) that an individual could be tested in their own right, without the need to access the results of tests performed on family members to interpret the individual's result. That would transform the situation for many with a family history of cancer into that now current in the context of HD, where the one category of mutation accounts for all cases of the disease; however, this still appears a distant prospect for susceptibility to breast and ovarian or colorectal cancer. Furthermore, even if all the relevant genetic factors were known, there would remain great uncertainty about the fate of each individual; the occurrence of second-hit mutations, and therefore of the anticipated malignancy, will inevitably be affected by purely random processes.

Diagnostic Genetic Testing

When an individual presents with disease that may have a genetic cause, whether or not there is a relevant family history, genetic investigations may very reasonably be initiated. Although it will often be prudent and helpful to explain to the patient, and perhaps their relatives, that the investigations may generate results with implications for others as well, the fact of these genetic implications—a potential impact on the immediate or even the extended family—will emerge once the diagnosis has been made, whatever technology is used to make the diagnosis. If a child has cystic fibrosis (CF) or DMD, for example, then there will be a risk of recurrence within the immediate family and there may be healthy carriers of the disease in the extended family too. This is true whether the diagnostic process entails DNA technology or more traditional methods (assay of sweat electrolytes in CF, or the histochemical or immunocytochemical examination of a muscle biopsy in DMD).

Similarly, if a healthy pregnant woman has renal cysts identified in herself during an ultrasound scan carried out on her fetus, then the possible implications for her own health, the fetus, and the family will clearly emerge without the need for any specifically *genetic* testing.

It is the clinical imperative of arriving at the correct diagnosis for a sick individual that drives these investigations—very appropriately—so that not only the patient but also other family members may need to come to terms simultaneously with both the serious nature of the established diagnosis and the fact of the genetic implications. This is unavoidable. It is important, however, to make sure that patients and families caught up in such a situation have ready access to timely and supportive clinical genetic assessment and genetic counseling as well as the best clinical care for established disease.

There are rather different considerations in the context of a child with developmental difficulties and dysmorphic features. In this setting, diagnostic investigations will usually include tests for various genetic conditions; this will come as no surprise to most parents. The problems that commonly arise in this context are threefold: (1) The failure to make a diagnosis that accounts for the child's difficulties, with resulting uncertainty for the prognosis, the risk of recurrence and the implications for other family members. This situation is slowly becoming less common as the diagnostic sensitivity of the relevant genetic investigations is improving, but it still applies in at least 40% or so of children with developmental delay. (2) The distress caused by the specific diagnostic label that is attached, either because it confirms the syndromic association of dysmorphic features with the developmental problems or because the parents suddenly come to appreciate that the prognosis for their child is substantially worse than they had previously accepted—perhaps, the child is unlikely ever to talk or to lead an independent life as an adult. (3) Distress and ambivalence about the risk of recurrence of the condition, in siblings or in the extended family, if this risk is significant. Many parents find that the prospect of terminating a pregnancy causes great distress, especially if it is associated in any way with the perceived devaluation of a much-loved child.

The Common Complex Diseases

What we refer to as the common complex diseases (CCDs) are not all distinct conditions but can be seen rather as symptom clusters with a variety of contributing causal factors and with contestable and somewhat arbitrary diagnostic criteria. Although this may be less true for the common cancers, such as breast cancer and CRC, and psychiatric disease may result from a different but equally unclear cluster of predispositions, this generalization does seem to apply to type 2 diabetes (T2D), CAD, hypertension, hypercholesterolemia, cerebrovascular disease, and Alzheimer disease. In this group of conditions, there is a health-disease continuum extending from clear normality through susceptibility and then minor clinical features to an unambiguous disease state. Furthermore, many of these complex disorders share at least some of the same causal factors, often grouped as the "metabolic syndrome," and the various associated degenerative pathologies often coexist in the same patient.

Interestingly, epidemiological data now strongly suggest that intrauterine experiences can influence the future health of the fetus as an adult individual—with poor fetal growth being associated with a higher incidence of these conditions in adult life. Although the causal processes involved are still being elucidated, the association between intrauterine growth retardation and susceptibility to these conditions later in life is well established and there are exciting indications that there is epigenetic modification of DNA and that paternal as well as maternal factors can be involved (Barker, 1992; Færgeman, 2003; Gluckman and Hanson, 2005; Pembrey et al., 2006). The interaction between maternal folate metabolism, fetal genotype, and the risk of malformation suggests that delicate, methylation-sensitive processes operate in a number of embryonic and fetal contexts (Gaspar et al., 2004).

The mechanisms through which environmental factors contribute to disease are also being elucidated through gene-based research. The way in which smoking interacts with genetic variants so as to produce its damage is being revealed. Smoking interacts with apolipoprotein E4 (apoE4) in increasing the risk of CAD (Humphries et al., 2001; Wang and Mahaney 2001), it interacts with folate and folate metabolism (*MTHFR* genotype) in causing colorectal polyps (Ulvik et al., 2001) and it interacts with at least two maternal gene loci in its effect on fetal growth (Wang et al., 2002).

Identifying the Mendelian Subsets: Taking a Family History

At present, the application of genetic testing to the common complex disorders is restricted in clinical practice to those families where there is a clear pattern of monogenic (Mendelian), usually autosomal dominant, disease susceptibility. The age at disease onset is often somewhat earlier than average in these families. Important Mendelian disease subsets are found in T2D (maturity-onset diabetes of the young—MODY), CAD and hypercholesterolemia (familial hypercholesterolemia—FH) and the common cancers of breast and bowel. The risk of an affected individual transmitting the susceptibility to their children is then 50%, so that the risk of a child going on to develop the condition at some stage is given as 50% of the lifetime penetrance. In the families showing predisposition to cancer, penetrance varies with the particular disease gene mutation but may be virtually 100% in FAP families, and perhaps 70%–80% for breast cancer in women carrying mutations at one of the *BRCA* gene loci, compared to the average lifetime risk for women of 8%–10%. In MODY, the prognosis and response to treatment vary with the particular gene involved and the risk to offspring is high—very much greater than the approximately 10% risk to the offspring of other parents with T2D. Mendelian subsets exist for the other complex disorders too, often with a relatively early age of onset—as is typical with familial Alzheimer disease.

We have already outlined the issues that arise in family-based genetic testing in these Mendelian subsets (vide supra); the most important question is then how a family with a Mendelian predisposition to disease comes to be recognized as such. The essential route to this is the recognition by family members or professionals that more individuals in the family have been affected than would be expected by chance, and often at a younger age than average. Once the suspicion arises, that a family has a Mendelian disease predisposition, then the health professional will want to obtain an accurate family tree and to consider molecular genetic investigation of an affected family member if there is an appreciable chance that there is a strong disease predisposition, and if the at-risk relatives express interest in predictive testing. The particular criteria used to decide whether or not to embark upon a molecular genetic mutation search vary between diseases and between families because of the practical, clinical consequences of finding a mutation as well as the costs of the laboratory work and the emotional impact on the family.

Criteria, such as the revised Amsterdam criteria in the context of CRC, have been drawn up to identify those families likely to harbor a germline (constitutional) mutation in a cancer-predisposing gene. This helps to focus genetic testing on families or tissue samples (e.g., tumor biopsies) where a result is more likely to be of value. Those at increased risk of CRC from hereditary nonpolypotic colon cancer (HNPCC) are offered colonoscopy at regular intervals to help diagnose any tumors before they progress to malignancy. Similar criteria exist to guide genetic testing in other disease contexts.

Professionals in primary healthcare may wish to identify those with Mendelian disorders because many of these are associated with increased risks of developing CCDs, so that there is real benefit to be had from surveillance and disease prevention among those affected by these individually uncommon but collectively important disorders (Scheuner et al., 2004). They may also wish to take a family history to assess risk of disease even in the absence of a recognized predisposition to disease or any concern raised by the patient but simply as a tool for identifying those at increased risk who may benefit from surveillance. This, then, is in effect a form of population screening with the risk of generating problems and the need for caution that inevitably accompany a proactive, population intervention even when it is introduced on a local scale.

Attempts were made to introduce population carrier screening for autosomal recessive disorders, especially CF, into primary care in the UK in the early 1990s and were not taken up with enthusiasm by either the public or the primary care professionals (Bekker et al., 1993; Payne et al., 1997). Inquiring about a family history of CAD has been more generally accepted by professionals as a guide to clinical decision making, at least in the UK (Summerton and Garrod, 1997), although it may not be so readily accepted by patients (Hutchison et al., 1998). There are likely to be at least two reasons for this acceptance of screening by professionals: (1) a family history of CAD has immediate relevance to the wellbeing of the practitioner's patients instead of only a long-term relevance to potential future patients and (2) it does not raise difficult and emotive topics such as prenatal diagnosis and the termination of wanted pregnancies, as inevitably occurs in the context of screening for carriers of recessively inherited disorders. With support, primary care professionals can learn to take more broadly based family history information (Rose et al., 1999) and this can be of particular value in relation to a family history of breast and colon cancer (Emery et al., 2001) in addition to cardiac disease. The reasons why primary healthcare professionals have often not embraced genetics, however, require serious consideration. Although educational initiatives on genetics in primary healthcare will be important in the long term, it will be unhelpful for some professionals to persist in making inflated claims about the imminent introduction of genetic testing into the community in advance of the reality. There is more than enough work to be done with the application of current knowledge into primary and secondary healthcare (Suther and Goodson, 2003); hyperbolic claims will lower the credibility of those who are speaking sense.

To consider the appropriateness of screening healthy individuals, those at population risk of specific disorders, we will set out one approach to the evaluation of screening programmes. We will then reflect on what can be learned from the early experiences with cholesterol screening, when that was introduced in advance of really effective treatments, because that may resemble the situation if genetic susceptibility screening is introduced ahead of therapeutic or preventive interventions of proven efficacy.

Criteria for Population Screening

The World Health Organization introduced a definition of screening in 1968 (Wilson and Jungner, 1968), and this has been usefully modified by the UK's National Screening Committee (NSC) (NSC, 2000) to frame an approach to the evaluation of potential screening tests. This is applicable in the context of genetic factors as in other contexts, although there may be different views on some issues in the context of genetic disease. Thus, if one outcome of antenatal screening for Down syndrome is that affected pregnancies are terminated, then not everyone would count this as a therapy or an improved outcome; this would be contentious.

The NSC definition of screening is, "a public health service in which members of a defined population, who do not necessarily perceive they are at risk of, or are already affected by, a disease or its complications, are asked a question or offered a test to identify those individuals who are more likely to be helped by further tests or treatment to reduce the risk of disease or its complications" (NSC, 2000). This definition has been helpful in that it leads directly into a set of criteria that a screening programme must meet for it to be deemed worthwhile. These criteria apply to the disease or condition, the test, the treatment or intervention available for the condition, and the programme of screening.

The condition must be an important health problem, its epidemiology and natural history must be understood and all other cost-effective approaches to primary prevention must have been implemented and found to be inadequate.

The screening test must be simple, safe, and precise and it must have been validated as a screening test so that the distribution of test values in the population is known, there must be an agreed cutoff level for the assay, the test must be acceptable to the population and there should be an agreed policy on further diagnostic investigation and the choices available to those who screen positive.

Early detection of the condition through screening must lead to improved outcomes through earlier application of an effective intervention. There should be agreed policies for the offer of appropriate treatment, and patient outcomes should already have been optimized through the introduction of effective clinical management at the time of disease presentation.

The overall programme of screening should, ideally, have demonstrated an improvement in mortality or morbidity in a randomized, controlled trial of screening. The complete screening programme must be acceptable to health professionals and the public, and the benefits of screening should outweigh the physical and psychological harm as well as the opportunity costs of screening (the health interventions that will not be implemented because the healthcare system is screening instead). There must be adequate staffing and facilities to implement the programme and a plan for managing and monitoring the programme and for quality assurance, and all other options for managing the condition must have been considered.

Although this framework for the assessment of possible screening programmes has been immensely helpful, it is still open to interrogation and the assumptions underlying it can be questioned. Just what counts as evidence may be open to dispute; the strong emphasis on quantifiable outcomes may lead to the neglect of ethical problems that arise for participants in the course of a screening programme, and some types of adverse outcomes for patients and families may be more readily apparent using qualitative research methods if the difficulties that arise are not immediately apparent from scores on a psychometric scale. From a different perspective, it may be argued that a collective healthcare system—whether a national scheme as in the UK, a social insurance model as in much of Europe, or a Health Maintenance Organization model as in the USA—will have an interest in containing the costs of healthcare, so that a cautious assessment of screening programmes may serve to delay the introduction of programmes that would benefit patients but would also consume scarce resources, that is, the criteria to be met by a screening programme may function as a barrier to implementation, in effect becoming a form of healthcare rationing. The debate as to how genetic services should be evaluated is important and ongoing (Goel, 2001; Wang et al., 2004; Sanderson et al., 2005).

Familial Hypercholesterolemia

This autosomal dominant condition affects approximately 1 in 500 of the population in the UK and many developed countries; it predisposes to CAD at an early age. The prevention of such excess and early CAD is feasible by treatment with diet and medication (the statin drugs, inhibitors of HMG CoA reductase) to achieve a reduction of serum cholesterol levels to within the normal range. Before such safe and effective treatment of hypercholesterolemia became available, measurement of serum cholesterol was often advocated to assess risk of CAD and that context—the awareness of risk without any very effective remedy—raised a number of issues.

It has long been recognized that families have their own understandings of how they come to be at risk of heart disease—acknowledging the effects of smoking, diet, and genetic factors in an intuitive fashion and without necessarily formulating a clear mechanism through which such factors could operate (Davison et al., 1992). Although identifying those at increased risk of disease can motivate some to comply with medical advice, it can also lead to paradoxical consequences such as inappropriate feelings of fatalism or of invulnerability, perhaps encouraging indulgence in harmful behaviors (Davison et al., 1989) and therefore leading on to unhelpful health consequences (Williams, 1988; Kinlay and Heller, 1990; and see discussion in Clarke 1995). Interestingly, CAD is often seen as a disease predominantly affecting men, so there can be a readiness to accept that men are at increased risk but a reluctance to acknowledge that women can also be affected (Emslie et al., 2001). This could, of course, be highly relevant to decisions about who seeks screening for disease risk, and who complies with behavioral recommendations or takes a prescribed risk-reducing medication (Senior et al., 2002).

Now that effective treatments are available for elevated serum cholesterol, will the introduction of genetic testing improve the management of at-risk individuals and families? There are many mutations recognized—predominantly in the *LDLR* gene but also in the *apoB* gene and some other loci—that are associated with hypercholesterolamia, so that the molecular diagnostic work involved in cascade testing within families is substantial; measurement of serum cholesterol remains the primary screening test and is also used in family cascade testing once an individual with FH has been ascertained. Nevertheless, molecular diagnostics improves the sensitivity of testing within families once the FH-associated mutation in each family has been identified (Heath et al., 2001; Umans-Eckenhausen et al., 2001). The question then arises as to whether establishing the diagnosis of FH in an individual through "traditional" biochemical measures of serum cholesterol or with molecular methods of mutation detection will have different consequences for patients' perceptions of the disease and for their motivation to comply with medical recommendations. There is some evidence from experience with newborn screening that a DNA-based diagnosis of FH can lead to a sense of fatalism (Senior et al., 1999). This area of research is currently being pursued vigorously, given its potential public health importance, but whether these behavioral considerations will influence the choice of screening or cascade-testing laboratory methods, however, is rather doubtful as the test sensitivity, specificity, and cost are likely to be decisive.

The general question of whether individuals with an increased but readily modifiable risk of disease should be identified through population screening or through family-based cascade testing is important, and will have to be kept under review as laboratory methods develop over time for each relevant disease (Krawczak et al., 2001). Family-based cascade testing for FH is certainly more cost-effective than population screening (Marks et al., 2002; Leren, 2004), but there are issues about the "best" (most effective and most ethical) way to achieve effective and efficient cascading through a family (Newson and Humphries, 2005). There are good reasons for primary care practitioners to remain alert to those at risk of FH—active, opportunistic case ascertainment—and to consider serum cholesterol screening in healthy young and middle-aged adults. Once one case in a family has been identified biochemically, cascade testing may then use molecular or biochemical methods or both. Clearly, it is important that cases of FH are identified so that they can be offered treatment for their susceptibility to CAD; this remains a Mendelian predisposition to cardiac disease, however, rather than a common, complex disorder. Research into the "complex genetic" contribution to CAD is not ready for general clinical application and preliminary findings may in fact have been made available too early, without adequate evidence to support the interpretation of results in practice (Humphries et al., 2004).

There are small but important Mendelian subsets concealed within many of the common, complex disorders. We have already referred to some of the familial cancer predispositions, including the *BRCA1* and *BRCA2* loci, in which mutations predispose to breast, ovarian, and other cancers, the *apc* locus associated with FAP coli and the mismatch repair loci associated with HNPCC. In current medical practice, the role of genetic testing for cancer susceptibility consists primarily of attempts to distinguish the approximately 5% of affected cases where there is a strong familial predisposition, with clear implications for the gene carrier and other members of the family, from the other approximately 95% where any inherited element in the predisposition is much weaker and the implications for others in the family are also correspondingly weaker. The small Mendelian subset among patients with diabetes mellitus consists largely of families in which the susceptibility is transmitted as an autosomal dominant trait, and there is in addition a group whose predisposition is mitochondrial. Clinical recognition of the Mendelian group can be helpful in management. The condition usually presents as the noninsulin-dependent or maturity-onset type of diabetes but often occurring at rather a young age; these features earn it the title, MODY. Knowledge of the diagnosis may be useful in ensuring that other affected cases in the family are recognized promptly, and families may seek genetic testing so as to resolve the uncertainty as to who will go on to develop the condition (Shepherd et al., 2000).

HFE Hemochromatosis

Genetic hemochromatosis presents an entirely different set of circumstances. Instead of a high-penetrance but uncommon, Mendelian subset of a common disorder, as with FH or MODY, hemochromatosis is a predisposition to excessive iron absorption, inherited as an autosomal recessive trait but of low penetrance and accounting for the large majority of noniatrogenic iron overload in many countries. This condition leads to overt disease in an uncertain but probably small or modest proportion of those with the predisposing genotype at the *HFE* locus on chromosome 6 (Beutler et al., 2002). Those with the predisposition, especially those homozygous for the predominant disease-associated allele, can be monitored for the accumulation of iron; removal of excess iron by phlebotomy is a cheap, safe, and effective means of preventing the serious health complications of the condition. How, then, should we decide whether to introduce population screening for this condition in countries where the frequency of the mutant allele is substantial—as in many Western countries including UK, Australia, and North America? In addition to problems relating to intellectual property rights, this question raises many practical and ethical issues that are intertwined and complex but instructive to consider.

One important reason why screening for hemochromatosis has not been widely introduced is that one of the NSC criteria for screening is not met: The natural history of homozygosity for the C282Y mutation at the *HFE* locus, and for the C282Y/H63D compound heterozygote state, remains imperfectly understood. As a consequence, the high carrier frequency for this autosomal recessive susceptibility state (10% in many White populations) makes genetic counselling problematic: What information should be provided to individuals with the relevant genotypes? How should healthy carriers or healthy homozygotes talk about the condition with their relatives? Should long-term hematological surveillance be recommended for a substantial proportion of the population, without much confidence that this would benefit more than a few?

Our collective approach to screening for hemochromatosis may have important repercussions for how we approach other genetic tests of disease susceptibility. First, we will have to watch the trend to redefine a "clinical disease" in terms of a "genetic constitution." It would be most unhelpful if the clinical disease "hemochromatosis" came to be redefined as synonymous with "homozygosity for the C282Y mutation at the *HFE* locus," although this verbal sleight could be used as a rhetorical device by those wishing to promote screening—the prevalence of this rare disease could then be given as 1 in 400! A redefinition of "CF" as "carrying mutations in both copies of the *CFTR* gene" would be similarly unhelpful in the context of newborn screening, where the goal is to identify those infants likely to benefit from early treatment rather than all those with mutations, who will include some with very mild and late-onset disease or even simply male infertility with no serious systemic illness.

How, then, do we decide what services to make available? Screening can be clearly beneficial for a few but does not (yet) fulfil the formal NSC criteria and so, in the UK, it is not provided through the National Health Service (NHS). Should we then present the information about screening to the public—or allow those with a commercial interest in screening to do so—and see what demand is generated? In effect, should we allow "autonomy" to trump other concerns, including "the public health," objective analysis of the evidence of costs and benefits of screening and equity of access to health services? Or should we aim for an appropriate, and hopefully benevolent, form of paternalism, even if this delays the introduction of possibly beneficial programmes until the evidence is clearer? Medical opinion in the UK is cautious (Haddow and Bradley, 1999; Dooley and Worwood, 2000), whereas in Australia, for example, it is more enthusiastic (Allen and Williamson, 2000; Nisselle et al., 2004). There is evidence from the UK that cascade screening is much more effective in families with a clinically affected index patient than when the index case is unaffected and has been identified through a screening test (McCune et al., 2003). This suggests that testing the general population will be much less efficient than family-cascade testing. This then raises a futher question: where testing is introduced privately, how should the costs of counseling, testing and surveillance for the extended family of those with a *HFE* gene mutation be met? Should monitoring the body iron store status of identified homozygotes be met through the NHS, that is, by general taxation, or should it be met by the commercial providers of the *HFE* test?

The ascertainment of neonates homozygous for the C282Y allele has been practised for some years in one region of France, along with family-cascade testing from these infants (Cadet et al., 2005). The rationale for this appears very weak as an approach to population screening; this has also been tried before in the contexts of FH (Senior et al., 1999) and a1-antitrypsin deficiency (McNeil et al., 1988) where the rationale for neonatal screening is somewhat stronger—the delay before intervention may be advised in the management of the individual child is much less—but neither of those programmes was continued.

The case of hemochromatosis raises many of the problems that we are likely to meet when it becomes possible to perform validated genetic tests giving information about low-penetrance disease susceptibility states. How can we reach agreement about when a test is ready for general clinical application outside the context of a research evaluation? The mere absence of continuing, overt psychological distress after testing for such a susceptibility in a highly selected group of research participants (Romero et al., 2005) is not at all sufficient to justify such testing (Wang et al., 2004; Sanderson et al., 2005). And how should the test be funded? How do we protect individuals from the effects of adverse discrimination in the face of such genetic tests? (Clayton, 2003) Who should pay for the subsequent additional health care measures, for the index case or their family, that are triggered by these test results?

Lessons from Newborn Screening Programmes

Population genetic screening programmes have been introduced for a variety of disorders. What can we learn from these programmes, that may be relevant to the introduction of screening for genetic susceptibility to the common, complex disorders? Although prenatal screening and carrier screening programmes are primarily concerned with reproduction and reproductive decisions, newborn screening is focused rather on the future health of the infant, as with genetic screening for susceptibility to disease.

Newborn screening programmes were established to identify those infants affected by disorders for which the treatment was effective if begun before a clinical diagnosis would usually be made, as with phenylketonuria (PKU) and congenital hypothyroidism. The benefits of screening for these conditions are undoubted, at least where the incidence of disease is appreciable. As other conditions have been included in the metabolic screen, however, the improved clinical outcomes resulting from screening have become less clear. It is possible for the introduction of new (even improved) laboratory techniques to drive alterations in the service provided to families. An enhanced sensitivity in detecting unusual metabolites, feasible with TMS, may lead to the recognition of milder cases of "disease" who will stand to gain less from early identification. This has happened in the context of medium chain acyl-CoA dehydrogenase (MCAD) deficiency, for example, potentially creating inappropriate anxiety in the parents of young children, who would be unlikely to come to any harm even if undetected. Similarly, changes in laboratory technique may lead to the early recognition of additional diseases where interventions are unlikely to lead to any substantial improvement in clinical outcomes, as with the broad range of organic acid and amino acid disorders that can be detected by TMS. We must recognize that such changes in technology have important consequences for families caught up in population screening programmes and must be introduced with caution.

Another learning point from newborn screening is that even minor differences in clinical procedures can make a substantial difference to how clients experience delivery of the service. For example, in the context of newborn screening for DMD, it is important to ensure that parents recognize that screening for this disease is not being offered in the expectation that early treatment will result in improved outcomes for the child but because the quality of the diagnostic process is improved and because early diagnosis makes reproductive choices available to the parents and the extended family in future pregnancies. There is a tendency for any offer of a health screening test to be routinized; while this may be acceptable in the context of newborn screening for PKU, it is unacceptable in newborn screening for this very different condition, DMD; although medical care can improve the quality and duration of life, there is currently no effective treatment to remedy this disorder. That the offer of testing is not a recommendation, and that the parents need to make their own decision, can be emphasized by modifications to the testing procedures. For example, the blood spot for DMD testing can be taken onto a different card from the one used to test for PKU and hypothyroidism. This minor change in practice serves to emphasize that this test is different from the others and this results in a somewhat lower rate of uptake of the test, hopefully reflecting a significant and useful effect on the quality of the informed consent for testing (Parsons et al., 2000; Parsons et al., 2005).

Rationale for Population Genetic Research into the CCDs

Population-based research into the CCDs aims to identify genetic variation that influences whether or not individuals will develop a particular disease and whether or not they will respond to specific treatments. Such research requires very substantial resources because large numbers of individuals have to be recruited to participate; demographic, environmental, and clinical data all need to be gathered and genomewide molecular analyses must be performed on biological samples collected from the participants. Furthermore, these collections of samples and data need to be maintained for decades so as to reveal those individuals who go on to develop disease and their responses to treatment. Finally, the mass of accumulated information needs to be analyzed so as to become interpretable first to the researcher and then to the working clinician.

How will this be of benefit to patients? The recognition of genetic variation related to the risk of disease may well identify genes whose corresponding proteins are involved in development of the disease—that is, in pathogenesis. This will improve our knowledge of disease mechanisms and open up new possibilities for interference in disease-related processes, identifying possible targets for the development of new, rational therapies as outlined above. We might also be able to identify those who are especially likely to develop disease by virtue of genetic or environmental factors. This could lead to interventions to reduce the risk of disease, by modifying environmental exposures, or influencing behavior or through the administration of novel preventive agents. This scenario—the "predict and prevent" paradigm of Baird (1990)—is of course still not with us except in the context of Mendelian factors. It will be worthwhile considering why this is so.

Many of the gene-based studies of the CCDs have so far given disappointing results in that reported associations have often been weak and have often not been replicated in other population groups or even in different samples from the same population. They have only occasionally led to the identification of clinically important genetic variation related to disease occurrence—as will be apparent from much of the content of this volume. One explanation for this is that the sample sizes achieved have so far been too small and that larger samples are needed—either prospective studies as in the large gene databanks of Iceland (deCODE), Estonia (the Estonian Genome Project) and UK (Biobank), or smaller and less costly case–control studies if interest is focussed on just one disease and if time and resources are limited. This perspective concedes that gene-disease associations are weak but still hopes that these associations, when pinned down by data-rich research, will prove to be clinically useful. Other explanations, however, are possible and plausible.

The first is that populations differ in their pattern of disease-related genetic variation because of their population history—their original pool of variation, their exposure to disease and climate, their nutritional circumstances, their contact with other population groups and, most importantly, their fluctuations in population size. With every substantial fall in population size, there is likely to be a shift in allele frequencies—especially likely when the population passes through a bottleneck, when this may happen in the absence of selection but purely from chance effects (genetic drift).

Then, there is the knowledge from other simpler species, such as *Drosophila*, even from inbred laboratory stocks, that variation can be maintained by disruptive selection as well as by heterozygote advantage. Genetic variation will persist in a population when different variants are favored at different stages of the life cycle, as with the factor V Leiden mutation that may be maintained through an advantage at implantation (Göpel et al., 2001), or in the two different sexes, or in different (e.g., fluctuating or otherwise changing) environments, as already discussed above (Vieira et al., 2000). Such complexities will inevitably be present in *Homo sapiens* to at least the extent found in *Drosophila* species, and will confound attempts to incorporate gene–environment interactions into the analyses. Discussions of human genetic polymorphism often refer to heterozygote advantage but in general pay much less attention to other important categories of selection that act to maintain variation. Gene–gene interactions can be viewed as generating another class of disruptive selection. Density-dependent selection operating in a context where population size fluctuates *for other reasons* is yet another important factor that will have led to genetic differences between and within populations that render gene–disease association studies more difficult to interpret. The very fact that environments do alter, through migration, climate change, population density and so on, will give an advantage to those organisms whose phenotype can be fine-tuned to the particular circumstances likely to be encountered—promoting the "predictive adaptive responses" mediated by epigenetic mechanisms, as discussed by Gluckman and Hanson (2005). This is conceptually akin to the notion that, when circumstances change to give an advantage to new mutations, selection will also and inevitably be operating in favor of increased mutation rates—the latter will hitchhike along with the success of the former.

These are reasons why the analysis of clinical and genetic data to give insights into the causation of the CCDs will be *very* complex and difficult. Just how complex and difficult remains to be seen; it will doubtless vary between diseases and between populations so that generalizations will be contentious and often inappropriate. This is not a reason for not setting out to perform such research but it is a reason for caution in making claims until the evidence is in place. A long-running programme of research by Charles Sing and colleagues, for example, has acknowledged these complexities and has set out to meet the challenges in a series of important contributions (Clark et al., 1998; Zerba et al., 2000 and many more).

There are two other strategies for gaining understanding of gene–environment interactions that may yield useful information more rapidly than the massive frontal assault strategy of gene databanks, although both approaches have their weaknesses. One is the use of case–control series in which environmental and genetic factors are compared between cases of disease and matched controls; the weakness is that the research is inevitably retrospective and the "matching" may be inadequate. The second has been termed "Mendelian randomization" by Davey Smith and Ebrahim (2003), and entails research conducted on those known to be at high risk of a disease because they have a Mendelian gene predisposing them to heart disease, cancer, or another CCD phenotype. Environmental and other genetic factors (apart from the implicated Mendelian gene) can be studied to give insight into other risk factors or preventive strategies that may be applicable in the general population but which will be demonstrated as effective (or otherwise) in a smaller study during a shorter period of time in the very high risk group than would be possible in the general population. An example may be the demonstrated interaction between the *CHEK2* gene and risk of breast cancer in those already at increased risk because of a family history of bilateral disease (Johnson et al., 2005), although of course the acceptability of inferring such broad conclusions from the (high risk Mendelian) few to the (general population) many may be challenged.

Haplotypes and Beyond

The generation of a map representing the allele frequencies and distribution of a large number of SNPs across the genome has been achieved after a concerted international effort. The International HapMap has created many opportunities to study associations of the SNPs with disease, with gene expression, with recombination and with linkage disequilibrium—giving insight to human origins, migration, and selection in the face of disease and other factors (International HapMap Consortium, 2005). This has confirmed that regions of linkage disequilibrium are found in areas of lower recombination, as would be expected from much previous work, but there are important limitations to this interpretation; fine-grained variation in recombination rates exist across the genome, even over short distances within genes (Crawford et al., 2004), and may be missed in studies of the whole genome. Such additional complexity casts doubt on some of the optimistic assessments of the (small) number of loci likely to be contributing to CCD susceptibility (Yang et al., 2005), along with the problems that arise from the use of models that ignore disruptive selection (e.g., from assuming simple multiplicative joint effects of the relevant loci). In short, several of the operating assumptions of the models commonly used in the analysis of SNP-disease associations are flawed (Terwilliger and Hiekkalinna, 2006).

One important factor that could confound disease-association studies is the existence of population "stratification" or "substructures," which could either conceal real effects or suggest the existence of spurious effects if the relationship between allele and disease is different within different groups because of other historical factors or other genetic or environmental mechanisms. The recognition of stratification is difficult and may be obscured despite studies being well designed (Freedman et al., 2004). This can be a problem even in relatively homogeneous populations such as Iceland (Helgason et al., 2005) or in carefully collected patient groups within a European country such as Germany (Berger et al., 2006).

The overall set of problems still to be resolved in the analysis and reporting of disease-association studies remains very substantial (Little et al., 2002). These include the essentially arbitrary assignment of the limits of statistical significance in such studies—often set, in effect, by the resources available to study possible associations in greater depth—and how to make allowances for the performance of multiple tests. This rather bleak view is not shared by all, and it may be that disease-association studies will produce biologically meaningful and clinically helpful results in a range of different contexts. Certainly, some problems will be more amenable to remedy than those discussed above, and these include publication bias and inadequate power (Colhoun et al., 2003). The need for larger studies and the avoidance of publication bias have both become apparent with the meta analysis of the angiotensin-converting enzyme in/del polymorphism and the frequency of CAD and responses to treatment (Bonnici et al., 2002). Similarly, limitations on the quality of data employed in these studies of the CCDs can be critical (Chang et al., 2006), and this will inevitably become a larger problem as the sample size and number of loci assayed both increase.

A new method relying on automated SNP analysis is the mapping of genetic variants that influence gene expression—able to identify local and remote effects, roughly equivalent to cis and trans effects (Cheung et al., 2005). This promises great insight into disease-related mechanisms. At the same time, however, the recognition of widespread and functionally important structural variation within the genome introduces a new order of complexity (Feuk et al., 2006). Extensive studies will be required to determine the relative antiquity of the SNPs and these microstructural and copy number variants. The interaction between short inversions, SNPs within the inversions, and the fine structure of recombination rates will be fascinating as it unfolds—but this is likely to set back the timing of our real understanding of SNP-disease associations. One clinically important example of this is the recognition that copy number of the *Fcgr3* gene influences susceptibility to glomerulonephritis (Aitman et al., 2006). The research community is girding its loins to tackle these new complexities (Thornton-Wells et al., 2004), including the challenges of epigenetics (Bjornsson et al., 2004) but it will take time for a comprehensive picture to emerge.

Population Genetic Structure—Models of Human History

The automated sequencing capacity of major laboratories has now been turned from the Human Genome Project itself to other studies, including the study of differences between human population groups. This is a sensitive area because "race," however defined, is a socially charged term and racial discrimination is widespread around the globe. The potential for misuse of genetic research data to bolster political campaigns seeking support for racist policies is all too real (Clarke, 1997) but, on the other hand, genetic differences between human populations are real and have important biological and health-related consequences.

As was found with older approaches, current methods confirm that the extent of genetic variation within population groups is much greater than variation between groups (Rosenberg et al., 2002). Allele frequencies are commonly graded along clines across a landmass where abrupt discontinuities in allele frequencies are unusual because they do not respect social or political frontiers. This can be summed up as the old slogan that race is a social construct and not a biological fact; this does not deny that there are systematic differences between population groups but asserts that there are no clear lines of demarcation (and no simple genetic tests) that would discriminate between the groups. Furthermore, many of the consequences of racial (ethnic or population) differences are socially mediated—and associated with inequalities in social class or of political or economic status, racial discrimination, and other social processes. Racial (or similar) categories have to be used in human genetic research so that we do not ignore important causes of ill health—whether the causal processes operate in a simple, biological fashion or through complex social processes—but it will be important not to reify such categories, treating them as if they indicated deeper or more important differences than is the case (Am Soc Human Genetics, 2005).

From our understanding of human evolution, and the way that different groups will have been exposed to different selective pressures or will have responded in different ways to similar pressures, it is clear that different populations may well have accumulated different sets of disease-related genetic variants. Disease-association studies may therefore give different results or the interactions between particular alleles and environmental or other genetic factors may differ from those found previously—as with susceptibility to leprosy (Malhotra et al., 2006) or circulating levels of angiotensin I-converting enzyme (McKenzie et al., 2001). The same variants may be present but the relationship between them and the phenotype will differ because of the multiple differences that have accumulated elsewhere in the genome. The pattern of disease susceptibility alleles in a small, inbred group separated only recently from the surrounding population, may be similar to that of the surrounding group (Newman et al., 2004). In contrast, when social and reproductive isolation have persisted for millennia, many disease-associated variants may accumulate. This will be especially true when the group has suffered a series of population expansions and bottlenecks, whether caused by malnutrition or pogrom, so that there have been repeated opportunities for stochastic effects to result in genetic drift—as with several Ashkenazi Jewish variants, for example, several lysosomal storage disoders and the common *LDLR* mutation causing FH (Durst et al., 2001).

When (if) our knowledge of genetics advances to the point where we can readily compute the interactions between multiple loci, then there would be no medical or biological point in considering race or population as a factor in genetic research or clinical medicine. Until then, however, "population group" may need to serve probabilistically (and cautiously) as a proxy for the full set of the relevant population-associated genetic differences (Foster and Sharp, 2004).

Our knowledge of human evolution is still developing fast, with estimates of the most recent common ancestor for all humankind varying. One extreme (and not very plausible) estimate is given as recent as "several thousand years ago" (Rohde et al., 2004). Studies of Y chromosome and mitochondrial sequences to track male and female lineages have given insight into human population movements, and use of autosomal and X chromosome sequences are adding to this picture (Xiao et al., 2004; Laan et al., 2005). The longer that human populations have had to diverge genetically, the more complex will be the small-scale differences between groups and individuals and the more complex will be the pattern of gene–disease association or gene–drug interaction. Although our knowledge of the genetic differences between population groups is still rudimentary, the marketing of drugs to access a particular "ethnic niche" is a possible strategy for corporations that have failed to justify the introduction of a product for general release (Kahn, 2004). Once we understand the factors that lead to different drug responses, however, these factors could be assessed specifically instead of using "race" as an imperfect proxy. For the interim, however, such approaches may prove helpful in the context of impoverished developing countries (Daar and Singer, 2005) whereas simultaneously being unacceptable within rich industrial countries where this ethnic marketing strategy is unlikely to be successful or even acceptable in the long term.Perhaps as important as any results from further population genetic researches, however, will be the results of research into epigenetics and other mechanisms mediating the long term, transgenerational consequences of adverse nutritional and social circumstances. The very high prevalence of T2D and CAD in certain indigenous population groups in Australia, the Pacific islands, and North and Central America can be interpreted in different ways. Attributing the different disease prevalence figures to the presence of specific genes in the different populations is likely to be far too simplistic, especially as the disease incidence figures have increased rapidly during the past half century. Neel's thrifty genotype model may be relevant here and so may Barker's fetal programming hypothesis, and Wilkinson's focus on the health consequences of inequality and persisting social injustice is a third perspective to consider (Wilkinson, 2005). Neel's thrifty genes have not been identified, however, and the mechanisms that mediate the ill effects of political inequity are unclear. Fetal programming looks more promising as an important factor that could, in part, be mediated by epigenetic effects; research programmes are actively investigating this area. The clear differences in public health implications of these three models of disease causation should be noted; these debates may have very real consequences for human welfare around the globe (McDermott, 1998).

Pharmacogenetics and Pharmacogenomics

Pharmacogenetics is the study of the genetic influences on how our bodies respond to exogenous chemicals, especially but not only prescribed medication, which may have immediate toxic or therapeutic effects or longer-term consequences. It has a long history (Nebert, 1997) but has been applied clinically in rather limited circumstances (Shah, 2004). Pharmacogenomics has a much wider compass, including all genetic variation relevant to causation or progression of a particular disease and that could be relevant to the design or development of new therapies. Pharmacogenetics then predicts the body's response to a medicine and helps to choose an appropriate treatment of *this* illness in *that* patient, while pharmacogenomics broadens out to become the study of genetic influences on the development and course of disease. Pharmacogenetics can be regarded as genetic testing for therapeutic guidance, whereas pharmacogenomics is a research activity to assist in the development of new therapies. Both might require molecular testing to identify which subcategory of a particular disease is present—which would be important for decisions to be made between the available treatments and for research into the basic mechanisms underlying the disease. The stated hope of gene-based research relating to pharmacology is that disease mechanisms will be better understood so that more effective treatments can be devised, so that prescribing (the choice of drug and the dosage) can be tailored to the individual's own constitution. In addition, it is hoped that adverse drug reactions (ADRs), that cause numerous and often serious health problems for those who are already sick, can be avoided. Drawing up "personal pharmacogenetic profiles" (Wolf et al., 2000) would mean that each individual could benefit from this knowledge every time they chose between over-the-counter medications.

Estimates of the likely timescale for developments in pharmacogenetics to work through to routine clinical practice in the CCDs are more sanguine—much longer—than in the past. The Royal Society, for example, has suggested this may take 15–20 years (Royal Society, 2005) although DNA diagnostics could be very helpful long before then, perhaps especially in developing countries, where testing for glucose-6-phosphate dehydrogenase (G6PD) deficiency, for example, or testing for infecting microorganisms could all be most valuable.

Enthusiasts such as Roses (2000), Goldstein et al., (2003) or Bell (2004) emphasise that using an individual's pattern of SNPs will indicate their risk of disease or of disease-related complications, so this will enable trials of treatment to be conducted with fewer participants than usual and in less time as those recruited would be at increased risk and would be expected to develop complications sooner than members of the general population. This approach would be similar to Davey Smith and Ebrahim's idea of Mendelian randomisation, as discussed above. One concern about this approach suggests that the drugs market will be fragmented by this strategy and that the industry will be able to identify or develop effective treatments for the larger groups but that small groups of patients will be neglected as being too small to make targeted drug development economically feasible.

There are a number of potential hazards in the development of pharmacogenetics, which have been carefully considered by the Nuffield Council on Bioethics (2003). These include the premature release of preliminary results from research studies to individuals (or to their insurance companies) when their interpretation would be unclear, the development of drugs just for those populations or population subgroups that are sufficiently large or wealthy to make these profitable or the possibility that drugs will be marketed globally when their efficacy has been determined in just one population. In addition, pharmacogenetic tests aimed at guiding therapy may give information about the likely response to treatment of a disease with the new drug, or the likely lack of response. The possibility that a test will generate prognostic information as well as therapeutic guidance should be considered and discussed in advance because of the potential personal impact and the possible consequences for insurance or employment.

Early steps have been taken toward elucidating the genetic factors influencing response to antidepressants (McMahon et al., 2006) or aspirin (Hankey and Eikelboom, 2006). Promising developments have occurred in the field of metabonomics that can be applied to predict the likely pattern of drug response in individual rats (Clayton et al., 2006). Although this work is far removed from application to humans, where there is a much greater degree of genetic and environmental variation than in the laboratory rat, this may in practice lead to a more readily applicable set of data than the patient's DNA sequence.

In future clinical practice, it is just as likely that the genetic variation to be studied will have arisen somatically, perhaps just in the diseased organ, as it is to be constitutional and readily detected in leukocytes. This idea is very familiar in the context of malignancy (vide supra) but may also apply for much less obviously genetic disorders. A recent example that has come to light is the association of atrial fibrillation with somatic changes in the Connexin 40 gene specifically in atrial tissue (Gollob et al., 2006).

Conclusion

Policy or What is to be done?

Current genetic services are focused on the clinical and laboratory needs of specific groups of patients—especially those affected by, or at risk of, Mendelian or chromosomal disorders, disorders of physical and mental development and reproductive loss. In the laboratory, molecular and cytogenetic methods are of great value in research in all areas of medicine and are already applied to the investigation of infectious diseases and the categorization of tumors. To the extent that genetics is entering the clinical arena of mainstream medicine, it is largely engaged in studies of the Mendelian genetic variants that cause diseases—the CCDs—that are not usually associated with a strong familial predisposition. The genetics involved is largely the application of molecular and statistical methods appropriate to traditional Mendelian conditions to reveal such predispositions that lie submerged or concealed within the much larger pool of the CCDs—the common cancers, hypertension, CAD, diabetes and Alzheimer disease.

All health professionals need to be aware of genetic services and when referral might be appropriate. They require an awareness that a family history of disease can indicate an increased risk of disease in their patients—and they must be willing to ask patients about their families. They need to know how to assess the family history and how to use criteria to identify those with a relevant family history who wold be likely to benefit from referral to genetic services. They need to understand the three principal modes of inheritance of Mendelian genes and they should be aware of mitochondrial inheritance and of the transmission of chromosomal rearrangements. They do not require much technical knowledge of molecular genetics.

When we shift our attention to the CCDs themselves, we must first recognize that the mechanisms that link genetic variation with the development of disease are still mostly obscure. There are many good reasons for anticipating that the links will be difficult to unravel and not especially important in ordinary clinical practice. Such details may be important to researchers, and could lead to new insights and then—if we are fortunate—to new therapies. But this research will not need to be understood by the average physician, let alone the average patient. The final common path of evidence-based medicine—the randomised controlled trial—will usually be the basis on which physicians will make their therapeutic decisions; some patients also want to consider the evidence and it is at that point that they will be able to find details as to the desired and undesired effects of treatment. How the treatment came to be devised—the underlying rationale—will not be of immediate relevance, merely of background interest. New treatments will be developed and become available in piecemeal fashion—as the evidence accumulates—and not in a predictable manner.

There will doubtless be examples where it turns out that the relationship between a disease and a specific genetic variant will be straightforward, although this will not happen very frequently as such cases will often already have been recognized as Mendelian predispositions to disease. There will need to be a great deal of research and development work done to establish those contexts where testing an individual for a genetic variant (or pattern of genetic variants) turns out to be clinically useful. Unless we are very fortunate as a species, and unless our population history is shorter and simpler than can reasonably be expected, then there will be a long road to travel before the clinical benefits of "therapeutic guidance" and "predict and prevent" are attained for the common, complex diseases—and, in many contexts, these happy outcomes will not be attained because the biology will turn out to be too complex. We will not know in advance where these happy accidents are waiting to be found. Some may regard this perspective as too bleak. It can be countered that it is simply realistic and it is better to be prepared for a hard journey than to be continually disappointed. The personality and experiences of researchers and commentators will doubtless contribute to their position on the optimism-realism spectrum, but it would be bad science to ignore the likely complexities of our species' evolutionary history for the sake of some short-lived optimism. Although research is well worthwhile (Smith et al., 2005), the potential rush to introduce tests of unproven clinical utility will need to be watched (Melzer and Zimmern, 2002; Burke and Zimmern, 2004; Javitt, 2006).

References

Ahr, A, Karn, T, Solbach, C, Seiter, T, Strebhardt, K, Holtrich, U, et al. (2002). Identification of high risk breast-cancer patients by gene expression profiling. *Lancet,* 359:131–132.

Aitman, TJ, Dong, R, Vyse, TJ, Norsworthy, PJ, Johnson, MD, Smith, J, et al. (2006). Copy number polymorphism in *Fcgr3* predisposes to glomerulonephritis in rats and humans. *Nature,* 439:851–855.

Allen, K and Williamson, R (2000). Screening for hereditary haemochromatosis should be implemented now. *BMJ,* 320:183–184.

American Society of Human Genetics (2005). Race, ethnicity and genetics working group. The use of racial, ethnic and ancestral categories in human genetics research. *Am J Hum Genet,* 77:519–532.

Baird, PA (1990). Genetics and health care. *Perspect Biol Med,* 33:203–213.

Barker, DJP (ed.) (1992). *Fetal and Infant Origins of Adult Disease.* London: British Medical Journal.

Bekker, H, Modell, M, Denniss, G, Silver, A, Mathew, C, Bobrow, M, et al. (1993). Uptake of cystic fibrosis testing in primary care: supply push or demand pull? *BMJ,* 306:1584–1586.

Bell, J (2004). Predicting disease using genomics. *Nature,* 429:453–456

Berger, M, Stassen, HH, Kohler, K, Krane, V, Monks, D, Wanner, C, et al. (2006). Hidden population substructures in an apparently homogeneous population bias association studies. *Eur J Hum Genet,* 14:236–244.

Beutler, E, Felitti, VJ, Koziol, JA, Ho, NJ, and Gelbart, T (2002). Penetrance of 845G→A(C282Y)*HFE* hereditary haemochromatosis mutation in the USA. *Lancet,* 359:211–218.

Bjornsson, HT, Fallin, MD, and Feinberg, AP (2004). An integrated epigenetic and genetic approach to common human disease. *Trends Genet,* 20:350–358

Bonnici, F, Keavney, B, Collins, R, and Danesh, J (2002). Angiotensin converting enzyme insertion or deletion polymorphism and coronary restenosis: meta-analysis of 16 studies. *BMJ,* 325:517–520.

Broadstock, M, Michie, S, and Marteau, T (2000). Psychological consequences of predictive genetic testing: a systematic review. *Eur J Hum Genet,* 8:731–738.

Burke, W, Pinsky, LE, and Press, NA (2001). Categorizing genetic tests to identify their ethical, legal and social implications. *Am J Med Genet,* 106:233–240.

Burke, W and Zimmern, RL (2004). Ensuring the appropriate use of genetic tests. *Nat Rev Genet,* 5:955–959.

Cadet, E, Capron, D, Gallet, M, Omanga-Leke, ML, Boutignon, H, Julier, C, et al. (2005). Reverse cascade screening of newborns for hereditary haemochromatosis: a model for other late onset diseases? *J Med Genet,* 42:390–395.

Chang, YP, Kim, JD, Schwander, K, Rao, DC, Miller, MB, Weder, AB, et al. (2006). The impact of data quality on the identification of complex disease genes: experience from the Family Blood Pressure Program. *Eur J Hum Genet,* 14:469–477.

Cheung, VG, Spielman, RS, Ewens, KG, Weber, TM, Morley, M, and Burdick, JT (2005). Mapping determinants of human gene expression by regional and genome-wide association. *Nature,* 437:1365–1369.

Clarke, A (1995). Population screening for genetic susceptibility to disease. *BMJ,* 311:35–38.

Clarke, A (1997). Limits to genetic research? Human diversity, intelligence and race. In PS Harper and A Clarke (eds), *Genetics, Society and Clinical Practice.* (Chapter 15, pp. 207–218). Oxford: Bios Scientific Publishers.

Clarke, A. (2004). On dissecting the genetic basis of behavior and intelligence. In D Rees and S Rose (eds), *The New Brain Sciences: Perils and Prospects.* (Chapter 11, pp.181–194). Cambridge: Cambridge University Press.

Clark, AG, Weiss, KM, Nickerson, DA, Taylor, SL, Buchanan, A, Stengard, J, et al. (1998). Haplotype structure and population genetic inferences from nucleotide-sequence variation in human lipoprotein lipase. *Am J Hum Genet,* 63:595–612.

Clayton, EW (2003). Ethical, legal and social implications of genomic medicine. *N Engl J Med,* 349(6):562–569.

Clayton, TA, Lindon, JC, Cloarec, O, Antti, H, Charuel, C, Hanton, G, et al. (2006). Pharmaco-metabonomic phenotyping and personalized drug treatment. *Nature,* 440:1073–1077.

Codori, AM and Brandt, J (1994). Psychological costs and benefits of predictive testing for Huntington's disease. *Am J Med Genet (Neuropsychiatr Genet),* 54:174–184.

Codori, AM, Hanson, R, and Brandt, J (1994). Self-selection in predictive testing for Huntington's disease. *Am J Med Genet (Neuropsychiatr Genet),* 54:167–173.

Colhoun, HM, McKeigue, PM, and Smith, GD. (2003). Problems of reporting genetic associations with complex outcomes. *Lancet,* 361:865–872.

Crawford, DC, Bhangale, T, Li, N, Hellenthal, G, Rieder, MJ, Nickerson, DA, et al. (2004). Evidence for substantial fine-scale variation in recombination rates across the human genome. *Nat Genet,* 36(7):700–706.

Daar, AS and Singer, PA (2005). Pharmacogenetics and geographical ancestry: implications for drug development and global health. *Nat Rev Genet,* 6:241–246.

Darwin, C (1874, 2003). *The Descent of Man and Selection in Relation to Sex* (2nd edn). John Murray, 1874 and Gibson Square Books, London, 2003.

Davison, C, Frankel, S, and Smith, GD (1989). Inheriting heart trouble. The relevance of common-sense ideas to preventive measures. *Health Educ Res,* 4:329–340.

Davison, C, Frankel, S, and Smith, GD (1992). The limits of lifestyle: re-assessing 'fatalism' in the popular culture of illness prevention. *Soc Sci Med,* 34(6):675–685.

Decruyenaere, M, Evers-Kiebooms, G, Cloostermans, T, Boogaerts, A, Demyttenaere, K, Dom, R, et al. (2003). Psychological distress in the 5-year period after predictive testing for Huntington's disease. *Eur J Hum Genet,* 11:30–38.

Dooley, J and Worwood, M (2000). Genetic haemochromatosis. A guideline compiled on behalf of the Clinical Task Force of the *British Committee for Standards in Haematology.* Abingdon, Oxfordshire 2000, Darwin Medical Communications Ltd. For and on behalf of the British Committee for Standards in Haematology

Dorval, M, Patenaude, AF, Schneider, KA, Kieffer, SA, DiGianni, L, Kalkbrenner, KJ, et al. (2000). Anticipated versus actual emotional reactions to disclosure of results of genetic tests for cancer susceptibility: findings from *p53* and *BRCA1* testing programs. *J Clin Oncol,* 18(10):2135–2142.

DudokdeWit, AC, Tibben, A, Duivenvoorden, HJ, Niermeijer, MF, Passchier, J, Trijsburg, RW (1998). Distress in individuals facing predictive DNA testing for autosomal dominant late-onset disorders: comparing questionnaire results with in-depth interviews. Rotterdam/Leiden Genetics Workgroup. *Am J Med Genet,* 75:62–74.

Durst, R, Colombo, R, Shpitzen, S, Avi, LB, Friedlander, Y, Wexler, R, et al. (2001). Recent origin and spread of a common Lithuanian mutation, G197del LDLR, causing familial hypercholesterolemia: positive selection is not always necessary to account for disease incidence among Ashkenazi Jews. *Am J Hum Genet,* 68:1172–1188.

Elwyn, G, Gray, J, and Clarke, A (2000). Shared decision making and non-directiveness in genetic counselling. *J Med Genet,* 37:135–138.

Emery, J, Lucassen, A, and Murphy, M (2001). Common hereditary cancers and implications for primary care. *Lancet,* 358:56–63.

Emslie, C, Hunt, K, and Watt, G (2001). Invisible women? The importance of gender in lay beliefs about heart problems. *Sociol Health Illn,* 23(2):203–233.

Evans, JP, Skrzynia, C, and Burke, W (2001). The complexities of predictive genetic testing. *BMJ,* 322:1052–1056.

Færgeman, O (2003). Coronary Artery Disease. Genes, Drugs and the Agricultural Connection. Amsterdam, Netherlands: Elsevier Science B.V.

Feuk, L, Carson, AR, and Scherer, SW (2006). Structural variation in the human genome. *Nat Rev Genet,* 7:85–97.

Freedman, ML, Reich, D, Penney, KL, McDonald, Mignault, AA, Patterson, N, et al. (2004). Assessing the impact of population stratification on genetic association studies. *Nat Genet,* 36(4):388–393.

Foster, MW and Sharp, RR (2004). Beyond race: towards a whole-genome perspective on human populations and genetic variation. *Nat Rev Genet,* 5:790–796.

Fulda, KG and Lykens, K (2006). Ethical issues in predictive genetic testing: a public health perspective. *J Med Ethics,* 32:143–147.

Gaspar, DA, Matioli, SR, Pavanello, R de C, Araujo, BC, Alonso, N, and Wyszynski D (2004). Maternal MTHFR interacts with the offspring's BCL3 genotypes, but not with TGFA, in increasing risk to nonsyndromic cleft lip with or without cleft palate. *Eur J Hum Genet,* 12:521–526.

Gluckman, P and Hanson, M (2005). *The Fetal Matrix. Evolution, Development and Disease.* Cambridge: Cambridge University Press.

Goel, V (2001). For Crossroads 99 Group. Appraising organised screening programmes for testing for genetic susceptibility to cancer. *BMJ,* 32:1174–1178.

Goldstein, DR, Tate, SK, and Sisodiya, SM (2003). Pharmacogenetics goes genomic. *Nat Rev Genet,* 4:937–947.

Gollob, MH, Jones, DL, Krahn, AD, Danis, LMLT, Gong, X-Q, Shao, Q, et al. (2006). Somatic mutations in the Connexin 40 gene (GJA5) in atrial fibrillation. *N Engl J Med,* 354:2677–2688.

Göpel, W, Ludwig, M, Junge, AK, Kohlmann, T, Diedrich, K, and Moller, J (2001). Selection pressure for the factor-V-Leiden mutation and embryo implantation. *Lancet,* 358:1238–1239.

Haddow, JE and Bradley, LA (1999). Hereditary haemochromatosis: to screen or not to screen. Conditions for screening are not yet fulfilled. *BMJ,* 319:531–532.

Hallowell, N, Foster, C, Ardern-Jones, A, Eeles, R, Murday, V, and Watson, M (2002). Genetic testing for women previously diagnosed with breast/ovarian cancer: examining the impact of *BRCA1* and *BRCA2* mutation searching. *Genet Test,* 6(2):79–87.

Hallowell, N, Foster, C, Eeles, R, Ardern-Jones, A, Murday, V, and Watson, M (2003). Balancing autonomy and responsibility: the ethics of generating and disclosing genetic information. *J Med Ethics,* 29:74–83.

Hankey, GJ and Eikelboom, JW (2006). Aspirin resistance. *Lancet,* 367:606–617.

Harper, PS (1997). Presymptomatic testing for late-onset genetic disorders. Lessons from Huntington's disease. In PS Harper and A Clarke (eds), *Genetics, Society and Clinical Practice* (Chapter 2, pp. 31–48). Oxford: Bios Scientific Publishers.

Harper, PS and Clarke, A (1997). *Genetics, Society and Clinical Practice.* Oxford: Bios Scientific Publishers.

Harper, PS, Gevers, S, de Wert, G, Creighton, S, Bombard, Y, and Hayden, MR. (2004). Genetic testing and Huntington's disease: issues of employment. *Lancet,* 3:249–259.

Heath, KE, Humphries, SE, Middleton-Price, and Boxer M (2001). A molecular genetic service for diagnosing individuals with familial hypercholesterolaemia (FH) in the United Kingdom. *Eur J Hum Genet,* 9:244–252.

Helgason, A, Yngvadottir, B, Hrafnkelsson, B, Gulcher, J, and Stefansson, K (2005). An Icelandic example of the impact of population structure on association studies. *Nat Genet,* 37:90–95.

Humphries, SE, Ridker, PM, and Talmud, PJ (2004). Genetic testing for CVD susceptibility: a useful clinical management tool or possible misinformation? Arterioscler Thromb Vasc Biol, 24:628–636.

Humphries, SE, Talmund, PJ, Hawe, E, Bolla, M, Day, INM, and Miller, GJ (2001). Apolipoprotein E4 and coronary heart disease in middle-aged men who smoke: a prospective study. *Lancet,* 358:115–119.

Hutchison, B, Birch, S, Evans, CE, Goldsmith, LJ, Markham, BA, Frank, J, et al. (1998). Screening for hypercholesterolaemia in primary care: randomised controlled trial of postal questionnaire appraising risk of coronary heart disease. *BMJ,* 316:1208–1213.

International HapMap Consortium (2005). A haplotype map of the human genome. *Nature,* 437:1299–1320.

Javitt, GH (2006). Policy implications of genetic testing: not just for geneticists anymore. *Adv Chronic Kidney Dis,* 13(2):178–182.

Johnson, N, Fletcher, O, Naceur-Lombardelli, C, Silva, IdS, Ashworth, A, and Peto J (2005). Interaction between CHEK2 1100delC and other low-penetrance breast-cancer susceptibility genes: a familial study. *Lancet,* 366:1554–1557.

Kahn, J (2004). How a drug becomes 'ethnic': law, commerce and the production of racial categories in medicine. *Yale J Health Policy, Law and ethics,* 4 (1):1–46.

Kessler, S, (ed.) and Resta, RG (2000). *Psyche and Helix. Psychological aspects of genetic counselling.* New York and Chichester, England: Wiley-Liss.

Kinlay, S and Heller, RF (1990). Effectiveness and hazards of case finding for a high cholesterol concentration. *BMJ,* 300:1545–1547.

Krawczak, M, Cooper, DN, and Schmidtke, J (2001). Estimating the efficacy and efficiency of cascade genetic screening. *Am J Hum Genet,* 69:361–370.

Laan, M, Wiebe, V, Khusnutdinova, E, Remm, M, and Paabo, S (2005). X-chromosome as a marker for population history: linkage disequilibrium and haplotype study in Eurasian populations. *Eur J Hum Genet,* 13:452–462.

Leren, TP (2004). Cascade genetic screening for familial hypercholesterolemia. *Clin Genet,* 66:483–487.

Lim, J, Macluran, M, Price, M, Bennett, B, Butow, P and the kConFab Psychosocial Group (2004). Short- and long-term impact of receiving genetic mutation results in women at increased risk for hereditary breast cancer. *J Genet Couns,* 13(2):115–133.

Lindblad, AN (2001). To test or not to test: an ethical conflict with presymptomatic testing of individuals at 25% risk for Huntington's disorder. *Clin Genet,* 60:442–446.

Little, J, Bradley, L, Bray, MS, Clyne, M, Dorman, J, Darrell, L, et al. (2002). Reporting, appraising and integrating data on genotype prevalence and gene-disease associations. *Am J Epidemiol,* 156(4):300–310.

Malhotra, D, Darvishi, K, Lohra, M, Kumar, H, Grover, C, Sood, S, et al. (2006). Association study of major risk single nucleotide polymorphisms in the common regulatory region of PARK2 and PACRG genes with leprosy in an Indian population. *Eur J Hum Genet,* 14:438–442.

Marks, D, Wonderling, D, Thorogood, M, Lambert, H, Humphries, SE, and Neil, HAW (2002). Cost effectiveness analysis of different approaches of screening for familial hypercholesterolaemia. *BMJ,* 324:1303–1308.

McAllister, M (2002). Predictive genetic testing and beyond: a theory of engagement. *J Health Psychol,* 7(5):491–508.
McCune, CA, Ravine, D, Worwood, M, Jackson, HA, Evans, HM, and Hutton, D (2003). Screening for hereditary haemochromatosis within families and beyond. *Lancet,* 362:1897–1898.
McDermott, R (1998). Ethics, epidemiology and the thrifty gene: biological determinism as a health hazard. *Soc Sci Med,* 47:1189–1195.
McKenzie, CA, Abecasis, GR, Keavney, B, Forrester, T, Ratcliffe, PJ, Julier, C, et al. (2001). Trans-ethnic fine mapping of a quantitative trait locus for circulating angiotensin I-converting enzyme (ACE). *Hum Mol Genet,* 10 (10):1077–1084.
McMahon, FJ, Buervenich, S, Charney, D, Lipsky, R, Rusdh, AJ, Wilson, AF, et al. (2006). Variation in the gene encoding serotonin 2A receptor is associated with outcome of antidepressant treatment. *Am J Hum Genet,* 78:804–814.
McNeil, TF, Sveger, T, and Thelin, T (1988). Psychosocial effects of screening for somatic risk: the Swedish α_1-antitrypsin experience. *Thorax,* 43:505–557.
Melzer, D and Zimmern, R (2002). Genetics and medicalisation. *BMJ,* 324:863–864.
Michie, S, Smith, JA, Senior, V, and Marteau, TM (2003). Understanding why negative genetic test results sometimes fail to reassure. *Am J Med Genet,* 119A:340–347.
National Screening Committee (2000). *Second Report of the National Screening Committee.* Health Departments of the United Kingdom, London.
Nebert, DW (1997). Polymorphisms in drug-metabolizing enzymes: what is their clinical relevance and why do they exist? *Am J Hum Genet* 60:265–271.
Neel, JV (1962). Diabetes mellitus: a "thrifty" genotype rendered detrimental by "progress"? *Am J Hum Genet,* 14:353–362.
Newman, DL, Hoffjan, S, Bourgain, C, Abney, M, Nicolae, RI, Profits, ET, et al. (2004). Are common disease susceptibility alleles the same in outbred and founder populations? *Eur J Hum Genet,* 12:584–590.
Newson, AJ and Humphries, SE (2005). Cascade testing in familial hypercholesterolaemia: how should family members be contacted? *Eur J Hum Genet,* 13:401–408.
Nisselle, AE, Delatatycki, MB, Collins, V, Metcalfe, S, Aitken, MA, du Sart, D, et al. (2004). Implementation of HaemScreen, a workplace-based genetic screening program for hemochromatosis. *Clin Genet,* 65:358–367.
Nuffield Council on Bioethics. (2003). *Pharmacogenetics: Ethical Issues.* London: Nuffield Council on Bioethics.
Oral, EA, Simha, V, Ruiz, E, Andewelt, A, Premkumar, A, Snell, P, et al. (2002). Leptin-replacement therapy for lipodystrophy. *N Engl J Med,* 346(8):570–578.
Parsons, EP, Clarke, AJ, Hood, K, Bradley, DM, and Bradley, DM (2000). Feasibility of a change in service delivery: the case of optional newborn screening for Duchenne muscular dystrophy. *Community Genet,* 3:17–23.
Parsons, EP, Moore, C, Israel, J, Hood, K, Clarke, AJ, and Bradley, DM (2005). Emphasizing parental choice on newborn screening. *Br J Midwifery,* 13:165–168.
Payne, Y, Williams, M, Cheadle, J, Stott, NCH, Rowlands, M, Shickle, D, et al. (1997). Carrier screening for cystic fibrosis in primary care: evaluation of a project in South Wales. *Clin Genet,* 51:153–163.
Pembrey, M, Bygren, LO, Kaati, G, Edvinsson, S, Northstone, K, Sjostrom, M, et al. (2006). Sex-specific, male-line transgenerational responses in humans. *Eur J Hum Genet,* 14:159–166.
Piccart-Gebhart, MJ, Procter, M, Leyland-Jones, B, Goldhirsh, A, Untch, M, Smith, I, et al. (2005). Trastuzumab after adjuvant chemotherapy in HER2-positive breast cancer. *N Engl J Med,* 353:1659–1651.
Pui, CH and Evans, WE (2006). Treatment of acute lymphoblastic leukemia. *N Engl J Med,* 354:166–178.
Richards, MPM (1996). Families, kinship and genetics. In T Marteau and M Richards (eds) *The Troubled Helix.* (Chapter 12, pp 249–273), Cambridge: Cambridge University Press.
Rohde, DLT, Olson, S, and Chang, JT (2004). Modelling the recent common ancestry of all living humans. *Nature,* 431:562–566.
Romero, LJ, Garry, PJ, Schuyler, M, Bennahum, DA, Qualls, C, Ballinger, L, et al. (2005). Emotional responses to APO E genotype disclosure for Alzheimer disease. *J Genet Couns,* 12(2):141–150.
Rose, P, Humm, E, Hey, K, Jones, L, and Huson, SM (1999). Family history taking and genetic counselling in primary care. *Fam Pract,* 16(1):78–83.
Rosenberg, NA, Pritchard, JK, Weber, JL, Cann, HM, Kidd, KK, Zhivotovsky, LA, et al. (2002). Genetic structure of human populations. *Science,* 298:2381–2385.
Roses, AD (2000). Pharmacogenetics and future drug development and delivery. *Lancet,* 355:1358–1361.
Royal Society (2005). Personalised medicines: hopes and realities. Royal Society Policy Document 18/05. London: The Royal Society.
Sanderson, S, Zimmern, R, Kroese, M, Higgins, J, Patch, C, and Emery, J (2005). How can the evaluation of genetic tests be enhanced? Lessons learned from the ACCE framework and evaluating genetic tests in the United Kingdom. *Genet Med,* 7(7):495–500.
Sarangi, S, Bennert, K, Howell, L, Clarke, A, Harper, P, and Gray, J (2004). Initiation of reflective frames in counselling for Huntington's disease predictive testing. *J Genet Couns,* 13:135–155.
Sarangi, S, Bennert, K, Howell, L, Clarke, A, Harper, P, and Gray J (2005). (Mis) alignments in counselling for Huntington's disease predictive testing: clients' responses to reflective frames. *J Genet Couns,* 14:29–42.
Savage, DG and Antman, KH (2002). Imatinib mesylate—A new oral targeted therapy. *N Engl J Med,* 346(9):683–693.
Scheuner, MT, Yoon, PW, and Khoury, MJ (2004). Contribution of Mendelian disorders to common chronic disease: opportunities for recognition, intervention and prevention. *Am J Med Genet,* 125C:50–65.
Senior, V, Marteau, TM, and Peters, TJ (1999). Will genetic testing for predisposition for disease result in fatalism? A qualitative study of parents' responses to neonatal screening for familial hypercholesterolaemia. *Soc Sci Med,* 48:1857–1860Senior, V, Smith, JA, Michie, S, and Marteau, TM (2002). Making sense of risk: an interpretative phenomenological analysis of vulnerability to heart disease. *J Health Psychol,* 7(2):157–168.
Shah, J (2004). Criteria influencing the clinical uptake of pharmacogenomic strategies. *BMJ,* 328:1482–1486.
Shepherd, M, Hattersley, AT, and Sparkes, AC (2000). Predictive genetic testing in diabetes: a case study of multiple perspectives. *Qual Health Res,* 10 (2):242–259.
Smith, GD and Ebrahim, S (2003). 'Mendelian randomization': can genetic epidemiology contribute to understanding environmental determinants of disease? *Int J Epidemiol,* 32:1–22.
Smith, DS, Ebrahaim, S, Lewis, S, Hansell, AL, Palmer, LJ, and Burton, PR (2005). Genetic epidemiology and public health: hope, hype and future prospects. *Lancet,* 366:1484–1498.
Sobel, SK and Cowan, DB (2000). Impact of genetic testing for Huntington disease on the family system. *Am J Med Genet,* 90:49–59.
Soldan, J, Street, E, Gray, J, Binedell, J, and Harper, PS (2000). Psychological model for presymptomatic test interviews: lessons learned from Huntington disease. *J Genet Couns,* 9(1):15–31.
Summerton, N and Garrood, PVA (1997). The family history in family practice: a questionnaire study. *Fam Pract,* 14(4):285–288.
Suther, S and Goodson, P (2003). Barriers to the provision of genetic services by primary care physicians: a systematic review of the literature. *Genet Med,* 5 (2):70–76.
Terwilliger, JD and Hiekkalinna, T (2006). An utter refutation of the 'fundamental theorem of the HapMap'. *Eur J Hum Genet,* 14:426–437.
Thornton-Wells, TA, Moore, JH, and Haines, JL (2004). Genetics, statistics and human disease: analytical retooling for complexity. *Trends Genet,* 20:640–647.
Traverso, G, Shuber, A, Olsson, L, Levin, B, Johnson, C, Hamilton, SR, et al. (2002a). Detection of proximal colorectal cancers through analysis of faecal DNA. *Lancet,* 359:403–404.
Traverso, G, Shuber, A, Levin, B, Johnson, C, Olsson, L, Schoetz, DJ, et al. (2002b). Detection of *APC* mutations in fecal DNA from patients with colorectal tumors. *N Engl J Med,* 346:311–320.
Ulvik, A, Evensen, ET, Lien, EA, Hoff, G, Vollset, SE, Majak, BM, et al. (2001). Smoking, Folate and Methylenetetrahydrofolate reductase status as interactive determinants of adenomatous and hyperplastic polyps of colorectum. *Am J Med Genet,* 101:246–254.
Umans-Eckenhausen, MAW, Defesche, JC, Sijbrands, EJG, Scheerder, RLJM, Kastelein, JJP.(2001). Review of first 5 years of screening for familial hypercholesterolaemia in the Netherlands. *Lancet,* 357:165–168.

van't Veer, LJ, Dai, H, Van de Vijver, MJ, He, YD, Hart, AAM, Mao, M et al. (2002). Gene expression profiling predicts clinical outcome of breast cancer. *Nature,* 415:530–536.

Vieira, C, Pasyukova, EG, Zeng, Zh-B, Hackett, JB, Lyman, RF, and Mackay, TFC (2000). Genotype-environment interaction for quantitative trait loci affecting life span in *Drosophila melanogaster*. *Genetics,* 154:213–227.

Wang, C, Gonzalez, R, and Merajver, SD (2004). Assessment of genetic testing and related counselling services: current research and future directions. *Soc Sci Med,* 58:1427–1442.

Wang, XL and Mahaney, MC (2001). Genotype-specific effects of smoking on risk of CHD. *Lancet,* 358:87–88.

Wang, X, Zuckerman, B, Pearson, C, Kaufman, G, Chen, C, Wang, G, et al. (2002). Maternal cigarette smoking, metabolic gene polymorphism, and infant birth weight. *JAMA,* 287:195–202.

Wilkinson, RG (2005). *The Impact of Inequality. How to Make Sick Societies Healthier.* London and New York: Routledge.

Wilkinson, RJ, Llewellyn, M, Toossi, Z, Patel, P, Pasvol, G, Lalvani, A et al. (2000). Influence of vitamin D deficiency and vitamin D receptor polymorphisms on tuberculosis among Gujarati Asians in west London: a case-control study. *Lancet,* 355:618–621.

Williams, RR (1998). Nature, nurture and family predisposition. *N Engl J Med,* 318:769–771.

Wilson, JMG and Junger, G (1968). *Principles and Practices of Screening for Disease.* Geneva: World Health Organization.

Wolf CR, Smith, G, and Smith, RL (2000). Pharmacogenetics. *BMJ,* 320:987–990.

Xiao, F-X, Yotova, V, Zietkiewicz, E, Lovell, A, Gehl, D, Bourgeois, S, et al. (2004). Human X-chromosomal lineages in Europe reveal Middle Eastern and Asiatic contacts. *Eur J Hum Genet,* 12:301–311.

Xue, Y, Daly, A, Yngvadottir, B, Liu, M, Coop, G, Kim, Y, et al. (2006). Spread of an inactive form of caspase-12 in humans is due to recent positive selection. *Am J Hum Genet,* 78:659–670.

Yanagisawa, K, Shyr, Y, Zu, BJ, Massion, PP, Larsen, PH, White, BC et al. (2003). Proteomic patterns of tumour subsets in non-small-cell lung cancer. *Lancet,* 362:433–439.

Yang, Q, Khoury, MJ, Friedman, JM, Little, J, and Flanders, WD (2005). How many genes underlie the occurrence of common complex diseases in the population? *Int J Epidemiol,* 34:1129–1137.

Zerba, KE, Ferrell, RE, and Sing, CF (2000). Complex adaptive systems and human health: the influence of common genotypes of the apolipoprotein E (*ApoE*) gene polymorphism and age on the relational order within a field of lipid metabolism traits. *Hum Genet,* 107:466–475.

40

Ethical, Legal, and Social Issues (ELSI)

Michael Parker

The increasing availability of high-throughput genomic technologies such as those developed for the Human Genome Project has led to qualitative and quantitative changes on a previously unparalleled scale in the processes of biological and biomedical science and to research endeavors: often of international, consortial and industrial dimensions. For example, concerted and coordinated efforts have led to successful completion of human haplotype mapping (www.hapmap.org) and to significant advances in malaria research (www.malariagen.net). In addition to the potential benefits of their use in medical research, genomic technologies, a rapidly changing technological playing-field, also have the potential to revolutionize medical practice more directly through the development of faster and more sensitive methods of clinical diagnosis and screening. Whilst inevitably bringing in their wake significant economic, social, educational, and workforce challenges for clinical laboratories and health service providers, such techniques are nevertheless likely to greatly enhance the clinical tools available in genetics and to bring significant benefits to patients and their families. This chapter explores some of the ethical and social dimensions of these developments in relation to the potential benefits, and the challenges in the use of genomics in research and clinical practice and identifying key questions relating to their governance.

Genomics in Medical Research and Clinical Practice

Until the later decades of the twentieth century, the identification and treatment of inherited disorders and the counseling of patients and family members about such disorders was based on a number of reasonably well-established clinical tools. These included the taking of family histories, the identification and delineation, through what is known as dysmorphology, of unusual clinical features (the phenotype) associated with inherited disorders (e.g., congenital skull malformations), and the use of biochemical and other clinical tests (e.g., electrocardiograms (ECGs) for the identification of unusual heart arrhythmias). Such clinical and research tools enabled health professionals to understand the inheritance patterns of a wide range of serious inherited disorders through the study of affected families, and to provide appropriate counseling, treatment, and reproductive advice to patients and their families. These important techniques continue to form the basis of much clinical practice in genetics, particularly in the identification of new cases of rare genetic disorders (www.orphanet.org).

The techniques of family history and dysmorphology have for a long time been supplemented by the use of microscopic techniques (cytogenetics) for the identification of deletions, additions or "translocations" of larger sections of deoxyribonucleic acid (DNA) at the chromosomal level. These chromosomal abnormalities are important clinically. Down syndrome, for example, which is caused by the presence of an additional copy of chromosome 21, can be identified through the use of such techniques, as can a number of important chromosomal abnormalities associated with learning disabilities and other disorders. One of the limitations of cytogenetics, however, is that it works at reasonably limited resolutions and is unable to pick up the very small genetic changes that are the cause of a great many, if not the majority, of rare inherited conditions. In recent years, however, it has become increasingly possible to test directly at the molecular level for changes associated with the absence or presence of the risk of developing serious inherited disorders such as breast cancer, cystic fibrosis, and Duchenne muscular dystrophy, and the number of molecular genetic tests available for clinical use, either through health service or in many cases research laboratories continues to rise (www.genetests.org; www.cmgs.co.uk). As a consequence, clinicians are increasingly well placed to provide patients and their families with accurate information about the extent to which they, and their siblings and descendents, are at risk of developing rare but serious inherited disorders. This approach to clinical genetics, based on a combination of family history-taking, delineation of the phenotype, supported by relevant clinical tests, laboratory genetic testing will inevitably continue to form the backbone of clinical practice in relation to rare and serious genetic disorders.

The vast majority of human disease burden, such as that arising from heart disease and common cancers, is however by definition not rare and not the result of unusual single genetic or chromosomal changes. Such common disorders are likely instead to be caused by combinations from a potentially vast range of very small disease-causing genetic changes (pathogenic mutations) or variations (genetic polymorphisms), each individually of small or negligible effect, combined with a number of significant environmental factors. Such complexity and diversity at the molecular level cannot be captured through the use of the "genetic" or "cytogenetic" techniques described earlier. One of the main advantages of high-throughput genomic technology, combined with sophisticated statistical analysis, is that it provides the possibility of systematic investigation of how such complex patterns of genetic variation across the whole human genome (hence the term "genomics") affect the ways in which individuals carrying those variants exhibit resistance or susceptibility to common diseases where the environmental component is high.

Following the completion of the Human Genome Project, it is now known that the human genome contains something in the order of 10 million single nucleotide polymorphisms (SNPs) (defined as single changes each of which having a frequency in the population of >1%). This means that SNPs are very common and their presence characterized by much diversity across populations. Interestingly, SNPs are not distributed evenly throughout the genome and tend to be clustered in groups (haplotypes) preserved in reproduction (through what is called "linkage disequilibrium") and to travel together through successive generations. Recent years have seen a great deal of progress on the basis of this relative stability, in describing the patterns of SNP (haplotype) diversity occurring in different populations (www.hapmap.org). The very existence of SNPs in combination with the patterns of their diversity, and its limits, makes it possible to study the variations of resistance and susceptibility to common diseases through the investigation of correlations (associations) between such SNPs or patterns of SNPs (haplotypes) and the occurrence of disease. Thus the existence of SNPs creates the basis for epidemiological research that seeks to establish the genetic features of common diseases by making it possible to screen very large numbers, often many millions, of genetic variations across whole human genomes in order to identify "natural variants" of genes and molecular pathways that correlate with and potentially influence disease susceptibility or resistance, and thereby improve the understanding of the molecular dimensions of disease.

Given the very large numbers of such genetic changes and the very small effects involved, the key to the availability and feasibility of this kind of research has been the development of high speed and, inevitably highly expensive, sequencing technology and associated medical informatics. Much of this was developed during the course of the Human Genome Project. As is always the case, the development of this new technology has changed the nature of research practice in a variety of interrelated ways. For example, because such technology is expensive and mostly beyond the budgets of individual research teams, the technology behind genomic research is increasingly tending to be owned and run by large international, commercially minded laboratories, to whom other researchers contract out, in order to get access to the most up-to-date technology. In addition to the costs and economies of scale involved in such research, the very large numbers (and diversity) of samples required tends to mean that such research is both collaborative and international in nature, with samples being collected from many countries and locations. Genomics has therefore inevitably led both to a globalization of biomedical science and in relation to sample collection and, in relation to its processing, increasing localization in a small number of high-tech, wealthy, and industrially minded genomic research/genotyping institutions.

Translational Research in Genomics

The development and adoption of genomic technologies, combined with their global and consortial implementation, is transforming biological science and clinical practice (and social processes) in a variety of complex, interrelated, and not easily predictable ways. Genomic research is making a significant contribution to biological science and to the understanding of the mechanisms involved in common complex disorders. In addition to their use in research, examples already exist of the use of high-throughput genomic technologies in the provision of health care, for example, the use of array technology in cancer genetics. It is for these reasons, in addition to the potential economic benefits of medical research, that large amounts of governmental and other resources are being put into this area.

The conceptualization of "translational" research projects, their funding and management, and the exploitation of their potential benefits do not, of course, take place in a vacuum. They are inevitably and unavoidably complex social, economic, and political processes in their own right. If the translation of data and methods from genomic research into clinical practice, and their use in the identification, prevention, and treatment of disease, are to be understood in socio-ethical terms, this suggests that what is to count as "successful" or "appropriate" translation will be contested. Important areas of contestation will include: what does it mean for science to be of "high quality"? How ought clinical services to be structured and funded? What forms of governance are appropriate for the regulation of genomic science, clinical practice and of the creation, management and use of databases and sample collections? How are processes of innovation in genomics best to be understood? These considerations and many others only serve to emphasize the importance of robust research on the social, ethical, and conceptual dimensions of innovation in science and clinical practice. The successful and appropriate development, implementation, and use of genomics is dependent upon a range of interrelated social, ethical, and regulatory factors, in addition to and including the successful completion of the science, and it will be important to consider such factors from an early stage as an integral part of the scientific process, including research design. The ethico-social dimensions of scientific developments in genomics are currently often only addressed as the science nears completion, or as tests are developed for clinical use; and this post hoc model of the role of ethical, social, legal, and economic considerations has the potential to lead, amongst other things, both to significant public distrust of science and of genomics, and to less effective health care interventions. Of equal importance is the fact that the failure to embed empirical social research and critical ethical analysis into science and technological innovation from the earliest stages has the potential to lead to the neglect of alternative, socially and ethically informed research ideas, concepts, and innovation. What this suggests is that a key factor in successful translational genomics ought to be the development of embedded multidisciplinary ethico-social research.

Ethical and Social Dimensions of Genomic Research

For obvious reasons, it is not possible or desirable in the context of a chapter such as this to second guess the outcomes of such research or to attempt to provide a definitive, once and for all, list of the "ethical and social implications" of genomic medicine in the abstract. This would go against the main thrust of this chapter, which is to call for open-ended, creative, and engaged "conversations" about such issues within science and clinical medicine themselves. Nevertheless, it is reasonable to expect that many of the following ethical and social issues are likely to continue to be central to discussion of the uses of genomics for some time to come.

Valid Consent

Whilst far from problematic, the concept of informed, or "valid," consent has come to be seen to be central to any consideration of the ethical aspects of biomedical research. The clarification of what is to count as an appropriate approach to consent, in the context of genomic epidemiology, is of vital importance for continued public trust in research that involves the collection and storage of very large numbers of biological samples combined with medical and non-medical information such as the information about the presence or absence of disease, its severity and duration, demographic information such as sex, age, ethnicity, and information about treatments and other interventions. It is widely accepted that the creation and future use of such databases must be based on valid consent, even if the concept of "valid consent" is itself contested.

Guidelines such as the Declaration of Helsinki suggest that, in order to be valid, the consent obtained for participation in medical research must be based on the provision of adequate information about the proposed research (WMA, 2000). That is, it must be *informed and understood.* It must also be *voluntarily* and *competently given.* There is the potential for each of these requirements to prove difficult to meet in practice in the context of genomics. The first of these requirements, that is, to be valid, consent to participate in research should be based on the provision of adequate information in ways that are understood, can be extremely demanding in practice, particularly where such samples are gathered in very diverse settings including remote locations in developing countries. Research ethics committees (RECs) require that those who provide samples and related information understand what participation in the research will mean for them (e.g., that blood will be taken), that they understand what will happen to the samples and information during and after the study, and that they are also provided with related information, such as what steps will be taken by the researchers to ensure anonymity of the samples and, where anonymity is not possible (e.g., where medical records and biological samples need to be actively linked), about what security measures are going to be put in place to ensure confidentiality and privacy. In some settings, information about the implications of genomic information for insurance or for employment will also be important. Conveying information of this diverse and complex kind presents major challenges, particularly when significant difficulty may be encountered even in conveying concepts such as that of a "gene" or of the difference between "clinical practice" and "research." The achievement of the conditions under which it is going to be possible to achieve informed consent will face three main challenges (Chokshi et al., 2006). The first of these will be that arising out of the linguistic and conceptual barriers presented by carrying out research in linguistically diverse settings. The second main challenge, arising out of the first, will be the development of appropriate educational models to help individuals, families, and communities to understand what is involved in participating in such research. The third, and again related, challenge will be finding ways of explaining the broader nature of the research and its potential benefits and harms for the individuals, families, and communities involved.

The second key element in valid consent, notably *voluntariness*, also presents challenges. Central among these, particularly in linguistically diverse community settings, will be the need to ensure that those who participate do not confuse research with clinical practice, the so-called "therapeutic misconception" (Molyneux et al., 2005). It is generally considered to be important that potential participants in research do not misunderstand it as clinical practice. Such misunderstanding is thought to be problematic for a number of reasons, some of which are linguistic and some of which are due to the fact that such research is often based in clinical settings and is carried out by those who are also responsible for the clinical care of the patient-participant. The key consideration here in relation to voluntariness is the extent to which potential participants feel able to choose not to participate. The perception that such research is part of clinical practice, combined with the perceived high social status of medical professionals, has the potential to make it difficult for potential participants to say no. Voluntariness can also be undermined by factors outside of the research itself such as communal or familial pressures to participate, or not to participate, in the research. A key perennial question for epidemiological researchers, and for regulators of such research, for example, in this respect continues to be how best to respond to culturally diverse approaches to the very idea of consent itself. In some settings, for example, recognized decision-making processes place more emphasis on the views of key social figures such as village elders, heads of "households" or "compounds" than on the individual. How ought researchers to go about both respecting local practices (including the diversity of such practices) and, at the same time, show appropriate respect for the wishes of individual people and concern for their protection?

The third key element in valid consent is *competence.* In order to be valid, consent must not only be informed and free from coercion, it must also be obtained from someone who is competent to provide it. This presents problems in genomics, as in other forms of medical research where such research is concerned with diseases affecting those who are not competent, such as is often, but not always, the case with very young children or with those affected by serious mental health problems. It will also potentially have implications for carrying out research on people who are very sick, for example, those affected by fever, and on those who are distressed, for example, the mothers of very sick children. Such research can be important, and vulnerable groups can, in some cases, be made more vulnerable still by the absence of research into their conditions. Notwithstanding the above, research on participants who do not reach the levels of competence required for valid consent raises important ethical and practical issues.

Implications for Communities and Populations

The requirement of valid consent in research is important for the protection of individual research participants, that is, those who will be most affected by participation in research. Like much other research, however, genomics has the potential to have implications for people other than those who actually consent to participate. The research methods associated with genomics, described earlier, involve the relating of common patterns of genetic diversity to the susceptibility to disease. The existence of such common patterns means that such diversity can also relate to differences between particular populations, for example, ethnic groups and this has the potential for groups as a whole to be identified, correctly or incorrectly, as more or less likely to be vulnerable to a particular disease such as malaria. Such information could potentially have significant implications, both for populations as a whole and for individuals other than those involved in the research (Nuffield Council of Bioethics, 2002). It might, for example, have implications for marriageability or perhaps employability of people identified as members of an "at risk" or "vulnerable" group. This raises the question of the extent to which such research and its social, population implications should be subject to more public, social processes of consent. It might be argued by some that the consideration of such issues ought to be the responsibility of research ethics committees. If this is to be so, it has implications for the training and representativeness nature of such committees, as well as for the deliberative processes adopted by them. It might however be felt, in some settings, that an REC is not the most appropriate mechanism and that some form of public engagement or deliberation is required.

Vulnerable Groups

It is evident that genomic research may have effects on communities and populations, in addition to those on individual groups, and I have suggested that this might, at least in some cases, mean that consent for the participation of individuals in genomic research (whether provided by individuals themselves or through some other, perhaps more culturally appropriate method) may be appropriately complemented by community discussions of the communal and perhaps ethnic dimensions of the research. Recognition of the populations/group effects of such research presents a number of potential ethical implications in addition to the question of whether group consent is required. The first of these is the practical one that the groups identified by genomic research (e.g., ethnic groups or groups of people sharing a condition) may not easily map on to existing or recognized decision-making or political groupings, and this may mean that the effect of the technology is both to *create* groups that might not have existed before and to lead to political and other changes. The second implication is that many of the groups, in addition to any vulnerability associated with the research, will be vulnerable in other ways. For example, the findings of the research may map onto groups such as people with mental health problems, the deaf, or onto ethnic groups already vulnerable to racism. The existing social disadvantages of such groups may be compounded by the findings of the research or by their already existing political isolation and relative powerlessness.

The Future Use of Samples

The notion of informed consent set out in the Declaration of Helsinki and other guidelines requires that research participants are "adequately" informed about the research. This raises the question of what is to count as "adequate." The concept of adequacy is problematic in all medical research, but in the context of research involving the creation of large genetic databases and the use of rapidly evolving technologies including, for example, the development of online publicly accessible, databases of genotype and phenotype information and the increasingly common requirement for rapid public release of data, this is particularly so (Gibbons et al., 2005). A key question here concerns the validity of what has come to be known as "broad" consent. Research ethics committees have tended to regard as adequate only the consent in which participants are informed about the specific uses to which their samples and information, for example, their medical records, will be put and when these uses, that is, the research, will be completed. This, however, is increasingly looking like an unrealistic model of consent in the context of very rapid change, and the development of available research technologies unforeseeable at the time of consent (Parker 2005).

Security, Confidentiality, and Data-sharing

Genomic research inevitably leads to the creation of large databases. Such databases include associated genotypical and phenotypical information (such as that about disease status, gender, age, treatments, and so on). These databases are also usually associated with sample collections, for example, blood or tissue. The creation, management, and use of such databases raise a number of important regulatory, ethical, and social considerations (Kaye et al., 2001). In addition to the questions around what constitutes valid consent, these will include important questions of security, confidentiality, and anonymity or de-identification. Whilst high standards of confidentiality and security are likely to be considered important, these will, in any consideration of appropriate governance, need to be set against the background of the limitations that would be placed on research by the inability to link samples, genotypes, phenotypical information, and information about treatments.

Related to this, and to the nature of broad consent, will be the considerations about the acceptability of third-party access to the data. It seems likely that some regulated third-party access is going to offer the potential of scientific and clinical benefits, for example, through the use of information in previously unforeseen research collaborations. A further set of considerations might arise in relation to the use of such data in research collaborations between commercial organizations such as pharmaceutical companies and university research groups. To what extent should such uses be permitted and how ought they to be governed?

Whilst not strictly questions of third-party access, an increasingly important set of ethico-regulatory issues arising in large-scale, expensive, international genomics research concerns the public release of de-identified genomic and phenotypical data via the Internet, as is being increasingly demanded by the major funders of such research. This raises important ethical issues around consent and about the use of archived samples for which consent was obtained before such public release was envisaged as well as issues of benefit-sharing.

Benefit-sharing

Genomic research is increasingly collaborative and international in nature. The collaborations involved often tend, however, to be of a type. In particular, it tends to be the case that expertise is localized: the technological/genotyping expertise based in high-tech medical research centers and separated from that associated with the collection of samples and phenotypes. This separation of expertise raises questions about how the credit for and the benefits of research, for example, intellectual property, are to be allocated and shared within such consortia. Similar questions arise around the release of data in its various forms to consortia members for analysis, and how this is to be managed in ways that both support the work of the consortia and maintain the research careers/intellectual property of the collaborators, particularly those in small, relatively poorly funded centers, such as those in developing countries.

The discussion of benefit-sharing in international research ethics has tended to concentrate on another aspect of benefit-sharing. That is, on the extent to which the benefits of the research should be shared with those who take part in the research as participants and their communities. This raises a number of ethical questions, such as do researchers have an obligation to share the benefits of their research with the individuals who participate as research subjects, or are these obligations better thought of as obligations to communities? Is it acceptable for researchers to provide benefits to communities, and perhaps even individuals, in ways that are not directly related to the nature of the research, for example, by offering to build a health center? (Benatar and Singer, 2000).

The Clinical Use of Genomics

There is much debate about the extent to which genomic research will lead to improvements in health care through the discovery of genetic variants associated with susceptibility to common complex disorders. Only time will tell. There seems no doubt, however, that genomics will lead to a much better understanding of the biological mechanisms at work in disease such as the immunological aspects of malaria, and that by so doing will contribute significantly to the development of vaccines and other treatments. In addition to the longer-term research benefits of genomics, this research has the potential to lead more directly to the benefits for clinical practice through the development of technology of use in diagnosis and screening. An example of this is the currently increasing clinical use of array technology to identify deletions, translocations, and so on in tumor samples. Whatever the other outcomes of genomic research, the introduction of high-throughput genomic technologies into clinical laboratories has the potential to revolutionize medicine in many ways through the development of very much more accurate and more rapid testing and diagnosis. Such innovation will have important implications for practice and for the health professional–patient relationship, and will inevitably generate new ethical and social issues in clinical practice such as those arising out of the increased likelihood of the identification of unexpected findings and its implications for consent.

Conclusions

The increasing availability of high-throughput genomic technologies has brought about significant qualitative and quantitative changes to research and practice of biomedical science. In addition to the benefits of their use in medical research, genomic technologies have the potential to revolutionize medical practice more directly through the development of faster and more sensitive methods of clinical diagnosis and screening. This chapter has sketched some of the ethical and social dimensions of developments in genomics, considering both the potential benefits and the challenges of the use of genomics in research and clinical practice and identifying key questions relating to its governance. The chapter has throughout emphasized the importance of placing science and technology in socio-ethical and global contexts and the related importance of embedded, reflexive, multidisciplinary research into the social, ethical, legal, and economic dimensions of innovation.

References

Benatar, S and Singer, P (2000). A new look at international research ethics. *Br Med J*, 321:824–826.

Chokshi, D, Thera, M, Parker, M, Diakite, M, Kwiatkowski, D, and Duombo, O (2006). Informed consent for genetic epidemiology in developing countries. *PLoS Med*, 2007, Vol. 4(4), pp. 636–641, e95.

Gibbons, SM, Helgason, HH, Kaye, J, Nõmper, A, and Wendel, L (2005). Lessons from European population genetic databases: comparing the law in Estonia, Iceland, Sweden and the United Kingdom. *Eur J Health Law*, 12(2):103–134.

Kaye, J (2001). Genetic research on the UK population—do new principles need to be developed. *Trends Mol Med*, 7(11):528–530.

Molyneux, CS, Peshu, N, and Marsh, K (2005). Trust and informed consent: insights from community members on the Kenyan coast. *Soc Sci Med*, 61:1463–1473.

Nuffield Council on Bioethics (2002). The ethics of research related to healthcare in developing countries, London.

Parker, M (2005). When is research on patient records without consent ethical? *Health Services Research and Policy*, 10(3):183–186.

World Medical Association (2000). Declaration of Helsinki: ethical principles for medical research involving human subjects. Fifty-second WMA General Assembly, Edinburgh, Scotland.

International human haplotype mapping (HapMap) project. Available: www.hapmap.org.

www.malariagen.net MalariaGen.

www.orphanet.org On-line resource on rare genetic disorders.

www.ncbi.nlm.nih.gov/entrez/query/OMIM On-line Mendelian inheritance in Man.

www.cmgs.co.uk Network of accredited UK molecular genetic laboratories.

www.genetests.org On-line resource on molecular genetic testing for research and clinical use.

41

The Regulation of Human Genomics Research

Jane Kaye

A key event in the regulation of medical research was the Nuremberg Trials that were held after World War II, which found Nazi physicians and researchers guilty of atrocities that were committed in the name of medical research. Since then, research ethics committees (RECs) and oversight mechanisms for medical research on human beings have become the accepted norm in most countries. The regulation of genomic research, like other subsets of medical research, is subject to the same regulatory controls that apply to medical research in general. Therefore, the focus of this chapter will be on the regulation of medical research on human beings, excluding the approval of medicine and medical devices, in order to discuss the ways in which these more general requirements apply to genomic research. This paper will also make reference to some of the specialist bodies that have been established as expert advisory agencies for genomics.

It will compare the regulatory systems in the UK and the USA in order to provide a general overview, while at the same time illustrating some of the significant differences that exist between them. These countries have quite different ways of regulating medical research on human beings. For example, the UK is increasingly influenced by European law and moves toward harmonization across Europe, whereas the USA has its own unique regulatory regime. In order to elucidate these differences, this chapter will compare the legal framework between these two jurisdictions, the institutions that have been established to oversee medical research on human beings and the powers that they have.

It is hoped that this chapter will highlight issues and difficulties faced by existing complex legal frameworks across the Western world. This review might be useful to readers from other countries where issues on the regulation for human genomics research are also actively debated and considered.

The Legal Framework for Medical Research

The Constitutional Structure

In order to understand the legal framework for medical research, it is important to understand the role of law and its origins in each jurisdiction. The legal frameworks in the UK and the USA have been developed through different processes. In the UK, there is a national parliament that makes law for the whole of England, although with devolution the Welsh and Scottish Parliaments increasingly have more powers to make laws specific to their jurisdiction. As a member of the European Union, the Parliament located in London has an obligation to implement directives that are passed by the European Parliament into UK law. Failure to do so will result in legal proceedings against the UK in the Court of Justice of the European Communities. However, each of the member states have its own "margin of appreciation" in the way that they implement and interpret European legislation, and this has led to discrepancies between the legislation that is found in each member state. The UK can also sign Conventions of the European Council, such as the Convention on Human Rights and Biomedicine 1997, but this is a voluntary act and is not compulsory, unlike the legal instruments of the European Union. The most significant legal instrument regarding medical research at the European level is the Convention on Human Rights and Biomedicine 1997. However, until now, the UK has refused to sign it, because of the provisions that relate to stem cell research.

In contrast, the US Congress (consisting of the House of Representatives and the Senate) and the Supreme Court are the highest law-making bodies in the USA. There is no obligation to sign or ratify the law of another authority, such as the United Nations or an international treaty, unless this has been a decision of the US Congress. The USA is a federal system, which means that the Congress and the state legislatures have different constitutional powers. For medical research, there is federal legislation, but the states also have considerable powers to legislate. The lack of a National Health Service (NHS) has also resulted in differences and anomalies in the provision of healthcare in different parts of the USA.

The Role of Statute

The USA has specially drafted legislation that covers the regulation of medical research, which applies across the USA because it is federal legislation. In 1991, the Department of Health and Human Services (HHS) issued the "Common Rule" (US Fed. Reg. 1991). The philosophical underpinnings of the Common Rule come from the Belmont Report—Ethical Principles and Guidelines for the Protection of Human Subjects. The Common Rule has a very narrow ambit, as it only applies to medical research that is conducted by a federal department or agency, or is federally funded. Research that comes under this scope must be reviewed and approved by the Institutional

Review Board (IRB) that operates in accordance with the requirements of this policy. However, under the Common Rule, the following research does not require IRB approval:

1. Research conducted in established or commonly accepted educational settings, involving normal educational practices;
2. Research involving the use of educational tests (cognitive, diagnostic, aptitude, achievement), survey procedures, interview procedures, or observation of public behavior, unless an individual can be identified or the information provided by the individual will incriminate or be detrimental to him or her;
3. Research involving the use of educational tests as long as the human subjects are elected or appointed public officials or candidates for public office; and federal statute(s) require(s) without exception that the confidentiality of the personally identifiable information will be maintained throughout the research and thereafter;
4. Research involving the collection or study of existing data, documents, records, pathological specimens, or diagnostic specimens, if these sources are publicly available or if the information is recorded by the investigator in such a manner that subjects cannot be identified, directly or through identifiers linked to the subjects;
5. Research and demonstration projects which are conducted by or subject to the approval of department or agency heads;
6. Taste and food quality evaluation and consumer acceptance studies. (US Fed. Reg. 1991).

There are a number of pieces of legislation that give power to supervisory bodies to act, which directly, or indirectly, have implications for research. Some examples are the Bayh–Doyle Act 1980 (Pub Law, 1980), which has had an effect on research as it requires that researchers who receive government funding have a legal obligation to protect their intellectual property in the event of commercialization. In addition, the Health Information Patient Privacy Act (HIPPA), which was introduced across the USA in 2003 to ensure the uniformity of privacy protections for patients across the country, has had a significant impact on research practice. The HIPPA regulations require that a patient authorize the use of their records for medical research purposes. However, this can be waived by an IRB, a privacy board, or a covered entity, such as a hospital. These "may give researchers access to medical records without IRB review or authorization by individual patients in two specific instances: preparing a research protocol (as long as access to medical records is needed for its preparation and no protected medical information is removed from the site), and performing research that concerns only people who have died" (Annas, 2002). However, there are a number of other pieces of legislation that have relevance for medical research. A search on the Library of Congress website (http://thomas.loc.gov/) revealed that there were in excess of 91 bills that directly related to medical research, and more than 300 had the words "medical" and "research" in them.

In contrast, the UK regulatory system for medical research does not have one main piece of legislation that provides a framework for the governance of all medical research. Animal research, on the other hand, is governed by the Animal (Scientific Procedures) Act 1986, which has its origins in a nineteenth century piece of legislation. Providing a statutory framework for an activity gives a mandate from a democratically elected body, but also is the opportunity to allocate powers of enforcement and accountability. Instead of a comprehensive piece of legislation, the law that applies to medical research on human beings in the UK is a complex patchwork of legislation, regulations, common law principles, guidelines, and codes of practice. At the level of statute, there are a number of general pieces of legislation that have clauses which make reference to medical research. The primary pieces of legislation that relate to medical research, or the use of personal information and human tissue are the Data Protection Act 1998, the Human Rights Act 1998, the Mental Capacity Act 2004, the Human Tissue Act 2004, the Freedom of Information Act 2000, the Human Fertilisation and Embryology Act 1990, the Access to Health Records Act 1990, the Health and Social Care Act 2001, the Computer Misuse Act 1990, and the Electronic Communications Act 2000.

Alongside this are regulations that also have a bearing on medical research, which are approved by the Secretary of State, who is, in this case, the Minister of Health. The most significant one is the Medicines for Human Use (Clinical Trials) Regulations 2004 that apply to clinical trials. These regulations are the only ones in the UK that detail the requirements for informed consent of medical research. These requirements are derived from the Declaration of Helsinki. The other important features of these regulations are the sections that relate to research ethics procedures and standards. These regulations are a direct implementation of the European Community Clinical Trials Directive 2001/20/EC (L 121/34 OJEC 1.5, 2001). This is an example of the increasing influence that the law of the European Community has on its member states.

The Role of Guidelines

There are also a number of key guidelines that apply to all employees in the NHS, which are written by the Department of Health and the National Institute for Health and Clinical Excellence (NICE), an independent organization which is responsible for providing national guidance. There is a comprehensive health system that is constantly being reorganized, and there is a move to more public–private partnerships. The NHS in the UK employs most of the medical professionals. However, not all medical research is carried out in the NHS, although this is one of the key access points for research participants and patients. There is a great deal of research carried out in universities and through the commercial sector. There are guidelines written by professional bodies such as the General Medical Council (the statutory body for the registration and accreditation of doctors), the British Medical Association, and the Royal Colleges. In addition, major funding bodies such as the Medical Research Council, Wellcome Trust, Action Research, Cancer Research UK, and the British Heart Foundation write guidelines or notes for practice. If a case is litigated, the courts will take into account professional guidelines, particularly the General Medical Council, as a basis for determining whether the appropriate standard of care has been followed. In the USA, there are similar bodies that write guidelines on medical research, but those that are binding on researchers are those of the National Institute of Health and other federally funded bodies.

The Role of the Courts

While the courts in the UK are prepared to accept guidelines as a means of assessing what the standard of care is, the court will then make the final decision as to what the appropriate standard is. In the vast majority of cases, the court will accept the medical standard of a body such as the General Medical Council unless "it is not capable of logical analysis." [*Bolitho v. City and Hackney Health Authority* (1998) AC 232]. The decisions of the courts have an effect on the manner in which medical research is conducted in the UK, although many of the decisions are based on negligence cases about medical treatment. There has only been one case concerning negligence in medical research, and that was *The Creutzfeldt–Jakob Disease*

Litigation, Plaintiffs v. United Kingdom Medical Research Council and another case (QB 54, BMLR 8). It is worth quoting from the judgment, as while the courts acknowledge that they should err on the side of caution in finding negligence, they are also prepared to impose a high standard of care on researchers, using the same legal standards that are applied in all negligence cases.

> The courts must be very cautious in condemning a clinical trial or therapeutic programme. Too ready a labelling of an act or omission as negligent by the courts could stultify progress in medical and scientific research and render eminent experts reluctant to serve on committees voluntarily. However, during the clinical trial of a new drug or form of treatment, and especially when the clinical trial is becoming a general therapeutic programme, all reasonably practicable steps should be taken to minimise dangers and side-effects. To discharge this duty, constant alert and inquiring evaluation of the trial or programme is required. I do not accept that a government department or a quasi-governmental agency such as the MRC can discharge this duty by a lower standard of care than a commercial pharmaceutical company... In my judgment, the same duty with the same standard of care is owed to all patients who are the subjects of clinical trials or new therapeutic programmes, whether the responsibility of a pharmaceutical company, government department or other agency.

The requirements for consent in the UK have been determined by the courts primarily through the common law doctrine of negligence and the medical professional's duty to warn. In contrast to the USA, Canada, and Australia, the test in the UK has been based on what a doctor would consider necessary to tell a patient, rather than what a reasonable patient would expect to know [*Sidaway v. Bethlem Royal Hospital Governors* (1985) AC 871 (HL)]. The recent decision in the case *Chester v. Afshar* [(2004) UKHL 41] suggests that the House of Lords may be moving more to a position where the information that is required to be given to the patient should be based on what a reasonable patient would want to know. This is more in line with other countries.

The courts in the USA have also been very influential in directing the standards that should be applied to medical research. As in the UK, the courts arbitrate on the difficult decisions and have established the requirements on issues such as consent, duty to warn, privacy, and anti-discrimination. Examples of important US decisions are *Washington University v. William J. Catalona* (Case No. 4:03 cv 01065 SNL) on the ownership of biological samples collected for research purposes, and the cases on genetic testing, such as *Pate v. Threlkel* [661 So2d. 278 (Fla 1995)] and *Safer v. Pack* [677 A. 2d 1188, 683 A 2d 1163 (NJ 1996)]. However, it is difficult to rely on the courts to establish a legal framework for research, as their law making is dependent on the cases that come before them and is an incremental extension of existing principles rather than a complete overhaul of a particular area. Therefore, the courts can explain and develop certain areas of law for guidance of researchers, but their role is not to develop completely new legal frameworks.

Regulatory Bodies and Their Powers

The Situation in the USA

The US Department of HHS funds two key organizations for the regulation of medical research on human beings. These are the Office for Human Research Protections (OHRP) and the National Institutes of Health (NIH). The NIH is responsible for research in general, whereas the OHRP has more responsibility for the compliance and accountability of the research process. The focus of both these institutions is on federally funded research, although their activities and guidelines have an effect on other bodies that carry out medical research in the USA.

The NIH is the primary federal agency in the USA for conducting and supporting medical research. The organization has its own research centers, which are leaders in their field, and also provides funding for research and guidance and information to researchers. It is a part of the US Department of HHS. The NIH Office of Biotechnology Activities (OBA) "monitors scientific progress in human genetics research in order to anticipate future developments, including ethical, legal, and social concerns in basic and clinical research involving Recombinant DNA, Genetic Technologies, and Xenotransplantation" (US Fed. Reg. 1991). The OBA is responsible for writing guidance, providing advice, keeping up-to-date with new developments, and maintaining a register of research activities. There are a number of expert advisory committees that develop guidance and feed back findings to the OBA and the NIH Director, such as the Recombinant DNA Advisory Committee (RAC) and the Secretary's Advisory Committee on Genetics, Health, and Society (http://www.nih.gov/oba/rac). "Compliance with the *NIH Guidelines* is mandatory for investigators at institutions receiving NIH funds for research involving recombinant DNA, they have become a universal standard for safe scientific practice in this area of research and are followed voluntarily by many companies and other institutions not otherwise subject to their requirements."

The OHRP "supports, strengthens, and provides leadership to the nation's system for protecting volunteers in research that is conducted or supported by the US Department of Health and Human Services (HHS)." The OHRP has a strong role in education and in providing guidance to individuals and institutions conducting human subject research. This involves organizing conferences, the development of resource materials, as well as quality improvement consultations.

As a part of its accountability strategy, it has a system of assurances where "universities, hospitals, and other research institutions in the USA and abroad have formal agreements with OHRP to comply with the regulations pertaining to human subject protections." It also runs an accreditation scheme for researchers in order to improve the ethos of research practice within an organization.

The OHRP is also responsible for maintaining a register of IRBs or Independent Ethics Committees (IECs) that approve federally funded research or research that comes under the Common Rule. The OHRP is responsible for investigating any allegations of non-compliance with the Common Rule and has the power to take action to protect human research subjects. The initial approach is to ask the institution to investigate, and then OHRP may conduct further investigations through on-site evaluations. The OHRP has the power to make recommendations about further training, education, and procedures. In extreme circumstances, it has the power to suspend research programs if they are proving to be unsafe, such as in the case of the death of Jesse Gelsinger in a gene therapy trial (George, et al., 2002). These powers of inspection and investigation ensure that there are protections throughout the research process.

In addition to these key institutions, there are also a number of organizations that have responsibility for the regulation of medical research on human beings; for example, the Environmental Protection Agency, the Centers for Disease Control and Prevention (CDC), the Food and Drug Administration, the Federal Trade Commission, and the Securities and Exchange Commission.

As with the RECs in the UK, it is the IRBs in the USA that are responsible for the approval of research projects and deal with the applications. Before any research commences, IRB or IEC approval must be obtained. The IRBs evaluate proposed research activities, making sure that the design of the study is consistent with sound scientific principles and ethical norms, such as informed consent. Applications can be rejected if they do not meet the criteria. The IRBs also have authority to seek a yearly review of the research. If the research involves vulnerable participants, reviews may be carried out more frequently. In addition, "the IRB may modify, suspend, or terminate approval of research that has been associated with serious harm to subjects or is not being conducted in accord with the NIH MPA or the IRBs decisions, stipulations, and requirements." (http://www.nih.gov/science/).

The Situation in the United Kingdom

In the UK, there is no single body that is responsible for oversight of the whole research process, which may be directly related to the fact that there is not a comprehensive legislative framework. The Department of Health has tried to remedy this with the "Research Framework for Health and Social Care," which "sets standards, details the responsibilities of the key people involved in research, outlines the delivery systems, and describes local and national monitoring systems" (George, et al., 2002). The National Research Ethics Service (NRES) is also responsible for providing guidance and for the system of accreditation for RECs. NRES does not have any formal compliance or inspection powers, though it is required to take out an audit of REC's in the UK on a regular basis. In the UK, there are in excess of 30 bodies that write guidelines which have relevance in medical research. The result is that there is a lot of activity at the front end of research, in the form of guidance and the approval of medical research by RECs, but little supervision after it has been approved and is underway.

There are only five bodies in the UK, the General Medical Council, the Information Commissioner, the Human Tissue Authority, the Human Fertilisation and Embryology Authority, and the NHS, that have any powers of enforcement and supervision. The organizations that have responsibility for licensing of collections of biological samples are the Human Tissue Authority, but there are a number of exemptions for research collections. The Information Commissioner has a light touch when it comes to supervision, as he will respond to complaints but does not have a system of regular inspection or oversight. The Human Fertilisation and Embryology Authority has a number of powers for inspection and enforcement ensuring compliance, but its scope is restricted to gametic materials and embryos (Human Fertilisation and Embryology Act 1990). The NHS as an employer also has professional supervisory procedures in place, such as Caldicott Guardians, which ensure that ethical practices are followed. In the UK, the Human Genetics Commission makes recommendations for change, but does not develop guidance.

RECs are the main decision makers in determining if, and how, medical research should proceed in the UK. All biomedical research projects must be subject to prior ethical review as well as technical scientific appraisal. RECs are the gatekeepers in deciding whether research will proceed, but they have no legal basis for their powers, and so their role is purely advisory. Apart from the clinical trials regulations, there is no statute that makes it mandatory for researchers to submit proposals to RECs for approval. However, it would be impossible to get access to patients through the NHS or to get the results of the research published without research ethics approval. Therefore, RECs do have considerable influence over the research that is carried out and the way that it is conducted, although there is little legal authority for their activities. The RECs have no authority to inspect research projects once they have commenced, and so that approval has become a hurdle that just needs to be passed. This is in contrast to the powers of the IRBs in the USA.

In addition to RECs, there may be a number of other bodies that require documents to be completed. For example, "similar information is required in differing formats for university registration (10 pages), REC approval (25 pages) and hospital approval (5 pages). Each form is independent of the others and the common data is not shared" (George, et al., 2002). RECs are concerned only with research that involves NHS patients, premises, staff, and tissue samples. If research does not fall within this ambit, "this does not mean that the project team can pursue their work unrestrained. It would be highly unusual if the study did not fall the remit of at least one (and possibly several) other regulatory bodies" (Shaw et al., 2005).

Conclusion

The result of the ad-hoc approach to governance in the UK is that the legal requirements are confusing, unduly complex, inconsistent, and at times incomplete. This can result in a lack of certainty for medical researchers as to the law that applies in a given situation (Ferguson, 2003) and a possibility that researchers may be at risk of legal liability. It also means that some types of research may be governed by different kinds of law—for example, clinical trials have a more comprehensive framework than other types of research. The lack of a coherent framework for all types of research based on accepted principles also makes it difficult to understand the full ramification of changes to the law.

New developments in the regulation of medical research have come about largely in response to different scandals. In the UK, for instance, the events at Alder Hey and Bristol led to the new Human Tissue Act 2004. Over the past few years, there have also been changes in the procedures for the approval of medical research on human beings. This change has led to a "layering" effect of oversight mechanisms and an increase in the requirements for research approval, which in turn has led to frustration on the part of researchers. As Shaw and colleagues (2005) state:

> In the UK, the rapid growth of systems and procedures for research management and governance has generated confusion and resentment in the research community. They bemoan the rising burden of paperwork, the curtailment of research freedom, expensive delays caused by lengthy application procedures, inconsistent decisions, and in some cases, the halting of entire research programmes by allegedly heavy handed but misinformed ethics committees.

In the USA, scandals have also led to changes in the regulatory system for medical research. For example, the Tuskegee Syphilis Study led to the enactment of the National Research Act of 1974 and the formation of the National Commission for the Protection of Human Subjects of Biomedical and Behavioral Research. Another example are the financial conflicts of interest in the early 1990s, which led to monographs providing guidance by the Association of American Medical Colleges and the Association of Academic Health Centers (Korn, 2000).

However, the essential difference between the USA and the UK is that there is a clear legislative basis for the regulation of research in the USA. Federal legislation has provided a framework that gives clarity and certainty as to how, and on what basis, medical research should proceed. The bodies that are involved in the supervision of research, such as the IRB and the OHRP, have considerable powers and legal authority to enforce decisions, as they are able to act when research projects are failing to meet required standards. This is in stark contrast to the UK system, where there are numerous bodies that write guidance, but only four that have compliance and enforcement powers. However, the weakness of the USA system is that it does not have universal application to all sectors where research is carried out.

The purpose of this chapter has been to give an overview of the legal framework for medical research in the UK and the USA, to identify the bodies that are responsible for oversight of research, and to illustrate the different powers that they exercise. In the space available, I have only been able to give a broad overview of the current regulatory framework for medical research in the two jurisdictions. It is evident that there are significant differences between the two systems, which reflect the societal concerns and the historical context of medical research in which they have been developed. There is scope for further comparison and review for the two jurisdictions in order to identify the best procedures for medical research on human beings, but also the implications that this may have for genomic research.

Bibliography

Animal (Scientific Procedures) Act 1986.

Annas, GJ (2002). Medical privacy and medical research—judging the new federal regulations. *N Engl J Med*, 346(3):216–220.

Ferguson, PR (2003). Legal and ethical aspects of clinical trials: the views of researchers. *Med Law Rev*, 11:48–66.

George, AJ, Gale, R, et al. (2002). Research governance at the crossroads. *Nat Med*, 8(2):99–101.

Governing Genetic Databases (UK) project (http://www.ggd.org.uk/).

Human Fertilisation and Embryology Act 1990.

Korn, D (2000). Conflicts of interest in biomedical research. *JAMA*, 284(17):2234–2237.

NIH website. Available: http://www.nih.gov/science/ [April 28, 2006].

Office of Human Research Protections (OHRP) website. Available: http://www.hhs.gov/ohrp/about/ohrpfactsheet.pdf [April 28, 2006].

L 121/34 *Official Journal of the European Communities*, May 1, 2001.

Public Law No. 96-517, 35 USC (1980).

Recombinant DNA Advisory Committee (RAC) website. Available: http://www4.od.nih.gov/oba/rac/aboutrdagt.htm [April 28, 2006].

Shaw, S, Boynton, PM, et al. (2005). Research governance: where did it come from, what does it mean? *J R Soc Med*, 98:496–502.

US Department of Health and Human Services. 45 CFR 46. Fed. Reg. 1991;56: 28012.

Glossary of Selected Terms and Phrases

Compiled by *Dhavendra Kumar*

Accession number: a unique number assigned to a nucleotide, protein, structure, or genome record by a sequence database builder.
Acrocentric: a chromosome having the centromere close to one end.
Algorithm: a step-by-step method for solving a computational problem.
Alignment: a one-to-one matching of two sequences so that each character in a pair of sequences is associated with a single character.
Allele: an alternative form of a gene at the same chromosomal locus.
Allelic heterogeneity: different alleles for one gene.
Alternative splicing: a regulatory mechanism by which variations in the incorporation of coding regions (*see Exon*) of the gene into messenger ribonucleic acid (mRNA) lead to the production of more than one related protein, or isoform.
Alu-**PCR**: a polymerase chain reaction (PCR) that uses an oligonucleotide primer with a sequence derived from the *Alu* repeat.
Alu **repeat (or sequence)**: one of a family of approximately 750,000 interspersed sequences in the human genome, which are thought to have originated from 7SL RNA gene.
Amino acid: a chemical subunit of a protein. Amino acids polymerize to form linear chains linked by peptide bonds called polypeptides. All proteins are made from 20 naturally occurring amino acids.
Amplification refractory mutation system (ARMS): an allele-specific PCR amplification reaction.
Amplimer: an oligonucleotide used as a primer of deoxyribonucleic acid (DNA) synthesis in the PCR.
Analytical specificity: the proportion of persons without a disease genotype who test negative.
Annealing: the association of complementary DNA (or RNA) strand to form the double-stranded structure.
Annotation: the descriptive text that accompanies a sequence in a database method.
Anonymous DNA: DNA not known to have a coding function.
Antibody: a protein produced by the immune system in response to an antigen (*see Antigen*). Antibodies bind to their target antigen to help the immune system destroy the foreign entity.
Anticipation: a phenomenon in which the age at onset of a disorder is reduced and/or severity of the phenotype is increased in successive generations.
Anticodon: the three bases of a transfer RNA (tRNA) molecule that form a complementary match to an mRNA codon and thus allow the tRNA to perform the key translation step in the process of information transfer from nucleic acid to protein.
Antigen: a molecule that is perceived by the immune system to be foreign.
Apoptosis: programmed cell death.
Autosome: any chromosome other than a sex chromosome (X or Y) and the mitochondrial chromosome.
Autozygosity: in an inbred person, homozygosity for alleles identical by descent.
Autozygosity mapping: a form of genetic mapping for autosomal recessive disorders in which affected individuals are expected to have two identical disease alleles by descent.
Bacterial artificial chromosome (BAC): DNA vectors into which large DNA fragments can be inserted and cloned in a bacterial host.
Bacteriophage: a virus, which infects bacterial cells, often referred to as *phage*.
Base: see Nucleotide base.
Bioinformatics: an applied computational system, which includes development and utilization of facilities to store, analyze, and interpret biological data.
Biotechnology: the industrial application of biological processes, particularly recombinant DNA technology and genetic engineering.
BLAST (Basic Local Alignment Search Tool): a fast database similarity search tool used by the National Center for Biotechnology Information (NCBI) that allows the world to search query sequences against the GeneBank database over the web.
Blastocyst: the mammalian embryo at the stage at which it is implanted into the wall of the uterus.
Bottleneck: a severe reduction in the number of individuals in a population, leading to a reduction in the genetic diversity of that population in later generations.
Candidate gene: any gene, which by virtue of a known property (function, expression pattern, chromosomal location, structural motif, etc.) is considered as a possible locus for a given disease.
Carrier: a person who carries an allele for a recessive disease (*see Heterozygote*) without the disease phenotype but can pass it on to the next generation.
Carrier testing: carried out to determine whether an individual carries one copy of an altered gene for a particular recessive disease.
Cell cycle: series of tightly regulated steps that a cell goes through from its creation to division to form two daughter cells.
"Central Dogma": a term proposed by Francis Crick in 1957—"DNA is transcribed into RNA which is translated into protein."
cDNA (complementary DNA): a piece of DNA copied in vitro from mRNA by a reverse transcription enzyme.
CentiMorgan (cM): a unit of genetic distance equivalent to 1% probability of recombination during meiosis. One centiMorgan is equivalent, on average, to a physical distance of approximately 1 megabase in the human genome.
Centromere: the constricted region near the center of a chromosome that has critical role in cell division.
Chimera: a hybrid, particularly a synthetic DNA molecule that is the result of ligation of DNA fragments that come from different organisms or an organism derived from more than one zygote.
Chromosome: subcellular structures, which contain and convey the genetic material of an organism.
Chromosome painting: fluorescent labeling of whole chromosomes by a fluorescence in situ hybridization (FISH) procedure in

which labeled probes each consist of complex mixture of different DNA sequences from a single chromosome.
Chromosome walking: a method of assembling clone contigs by using individual genomic DNA clones as hybridization probes for screening a genomic library.
Clinical genomics: the application of large-scale, high-throughput genomics technologies in clinical settings, such as clinical trials or primary care of patients. Examples include *pharmacogenomics* and *toxicogenomics*.
Clinical ontology: defined as hierarchies of concepts that apply to controlled syntax, database schema, semantic networks, or thesaurus. An ontological approach may be used to extract knowledge about disease progression, disease presentation, and comorbidities.
Clinical proteogenomics: applications of genomics, proteomics, small molecules, and informatics in delineation of the medical disorders. This includes identification of all genes, cataloguing the functional variation of their respective proteins thereby elucidating phenotypic differences that exhibit medical significance.
Clinical proteomics: translational application of new protein-based technologies in clinical medicine. For example, serum proteomic pattern diagnostics, a new concept in multiplexed biomarker analysis that may offer a new and exciting method for earlier and accurate disease prediction.
Clinomics: the application of oncogenomic research.
Clinical sensitivity: the proportion of persons with a disease phenotype who test positive.
Clinical specificity: the proportion of persons without a disease phenotype who test negative.
Clone: a line of cells derived from a single cell and therefore carrying identical genetic material.
Cloning vector: a DNA construct such as a plasmid, modified viral genome (bacteriophage or phage), or artificial chromosome that can be used to carry a gene or fragment of DNA for purposes of cloning (e.g., a bacterial, yeast, or mammalian cell).
Coagulation factors: various components of the blood coagulation system. The following factors have been identified: (Synonyms, which are or have been in use are included). Factor I (fibrinogen); Factor II (prothrombin); Factor III (thromboplastin, tissue factor); Factor IV (calcium); Factor V (labile factor); Factor VII (stable factor); Factor VIII [antihemophilic globulin (AHF), antihemophilic globulin (AHG), antihemophilic factor A Factor VIII: C]; Factor IX [plasma thromboplastin component (PTC), Christmas factor, antihemophilic factor B]; Factor X (Stuart factor, Prower factor, Stuart-Prower factor); Factor XI [plasma thromboplastin antecedent (PTA), antihemophilic factor C]; Factor XII (Hageman factor, surface factor, contact factor); Factor XIII [fibrin stabilizing factor (FSF), fibrin stabilizing enzyme, fibrinase; Other factors: [prekallikrein (Fletcher factor), and high molecular weight kininogen (Fizgerald)].
Coding DNA (sequence): the portion of a gene that is transcribed into mRNA.
Codon: a three-base sequence of DNA or RNA that specifies a single amino acid.
Companion diagnostics: the increasing trend for genetic diagnostics and therapeutics to become intertwined. These diagnostics identify the subset of patients who would benefit from a drug.
Comparative genomics: the comparison of genome structure and functional across different species in order to further understanding of biological mechanisms and evolutionary processes.
Comparative genome hybridization (CGH): use of competitive fluorescence in situ hybridization to detect chromosomal regions that are amplified or deleted, especially in tumors.
Complex diseases: diseases characterized by risk to relatives of an affected individual, which is greater than the incidence of the disorder in the population.
Complex trait: one, which is not strictly *Mendelian* (dominant, recessive, or sex linked) and may involve the interaction of two or more genes to produce a phenotype, or may involve *gene–environment* interactions.
Computational therapeutics: an emerging biomedical field concerned with the development of techniques for using software to collect, manipulate, and link biological and medical data from diverse sources. It also includes the use of such information in simulation models to make predictions or therapeutically relevant discoveries or advances.
Computer aided diagnosis (CAD): a general term used for a variety of artificial intelligence techniques applied to medical images. CAD methods are being rapidly developed at several academic and industry sites, particularly for large-scale breast, lung, and colon cancer screening studies. X-ray imaging for breast, lung, and colon cancer screening are good physical and clinical models for the development of CAD methods, related image database resources, and the development of common metrics and methods for evaluation. [*Large-scale screening applications include (a) improving the sensitivity of cancer detection, (b) reducing observer variation in image interpretation, (c) increasing the efficiency of reading large image arrays, (d) improving efficiency of screening by identifying suspect lesions or identifying normal images, and (e) facilitating remote reading by experts (e.g., telemammography)*].
Congenital: any trait, condition, or disorder that exists from birth.
Consanguinity: Marriage between two individuals having common ancestral parents, commonly between first cousins; an approved practice in some communities who share social, cultural, and religious beliefs. In genetic terms, two such individuals could be heterozygous by descent for an allele expressed as "*coefficient of relationship*," and any offspring could be therefore homozygous by descent for the same allele expressed as "*coefficient of inbreeding*."
Conservative mutation: a change in a DNA or RNA that leads to the replacement of one amino acid with a biochemically similar one.
Conserved sequence: a base sequence in a DNA molecule (or an amino acid sequence in a protein) that has remained essentially unchanged throughout evolution.
Constitutional mutation: a mutation, which is inherited and therefore present in all cells containing the relevant nucleic acid.
Contig: a consensus sequence generated from a set of overlapping sequence fragments that represent a large piece of DNA, usually a genomic region from a particular chromosome.
Copy number: the number of different copies of a particular DNA sequence in a genome.
CpG island: short stretch of DNA, often less than 1 kb, containing CpG dinucleotides, which are unmethylated and present at the expected frequency. CpG islands often occur at transcriptionally active DNA.
Cytoplasm: the internal matrix of a cell. The cytoplasm is the area between the outer periphery of a cell (the cell membrane) and the nucleus (in a eukaryotic cell).
Demographic transition: the change in the society from extreme poverty to a stronger economy, often associated by a transition in the pattern of diseases from malnutrition and infection to the intractable conditions of middle and old age, for example, cardiovascular disease, diabetes, and cancer.
Denaturation: dissociation of complementary strands to give single-stranded DNA and/or RNA.

Determinism (genetic): philosophical doctrine that human action is not free but determined by genetic factors.
Diploid: a genome (the total DNA content contained in each cell) that consists of two homologous copies of each chromosome.
Disease: a fluid concept influenced by societal and cultural attitudes that change with time and in response to new scientific and medical discoveries. The human genome sequence will dramatically alter how we define, prevent, and treat disease. Similar collection of symptoms and signs (*phenotype*) may have very different underlying genetic constitution (*genotype*). As genetic capabilities increase, additional tools will become available to subdivide disease designations that are clinically identical (*see Taxonomy of disease*).
Disease etiology: any factor or series of related events directly or indirectly causing a disease. For example, the genomics revolution has improved our understanding of disease determinants and provided a deeper understanding of molecular mechanisms and biological processes (*see Systems Biology*).
Disease expression: when a disease genotype is manifested in the phenotype.
Disease interventions: a term used in Genomics that refers to development of a new generation of therapeutics based on genes.
Disease management: a continuous, coordinated health care process that seeks to manage and improve the health status of a patient over the entire course of a disease. The term may also apply to a patient population. Disease management services include disease prevention efforts and as well as patient management.
Disease phenotype: includes disease related changes in tissues as judged by gross anatomical, histological, and molecular pathological changes. Gene and protein expression analysis and interpretation studies, particularly at the whole genome level are able to distinguish apparently similar phenotypes.
Divergence: the gradual acquisition of dissimilar characters by related organisms as they move away from a common point of origin (*see Sequence diversity*).
Diversity, genomic: the number of base differences between two genomes divided by the genome size.
DNA (deoxyribonucleic acid): the chemical that comprises the genetic material of all cellular organisms.
DNA cloning: replication of DNA sequences ligated into a suitable vector in an appropriate host organism (*see Cloning vector*).
DNA fingerprinting: use of hypervariable minisatllite probe (usually those developed by Alec Jeffreys) on a Southern blot to produce an individual-specific series of bands for identification of individuals or relationships.
DNA footprinting: a method of identifying and localizing DNA sequences within a DNA molecule that can specifically bind protein molecules that are protected against digestion by nucleases.
DNA library: a collection of cell clones containing different recombinant DNA clones.
DNA sequencing: technologies through which the order of base pairs in a DNA molecule can be determined.
Domain: A discrete portion of a protein with its own function. The combination of domains in a single protein determines its overall function.
Dominant: an allele (or the trait encoded by that allele), which produces its characteristic phenotype when present in the heterozygous form.
Dominant negative mutation: a mutation, which results in a mutant gene product, which can inhibit the function of the wild-type gene product in heterozygotes.
Dosage effect: the number of copies of a gene; variation in the number of copies can result in aberrant gene expression or associated with disease phenotype.
Dot-blot: a molecular hybridization method in which the target DNA is spotted on to a nitrocellulose or nylon membrane.
Drug design: development of new classes of medicines based on a reasoned approach using gene sequence and protein structure function information rather than the traditional trial and error method.
Drug interactions: refer to adverse drug interaction, drug- drug interaction, drug-laboratory interaction, drug-food interaction, etc. It is defined as an action of a drug on the effectiveness or toxicity of another drug.
Electronic Health Record (EHR): a real-time patient health record with access to evidence-based decision support tools that can be used to aid clinicians in decision-making, automating, and streamlining clinician's workflow, ensuring that all clinical information is communicated. It can also support the collection of data for uses other than clinical care, such as billing, quality management, outcome reporting, and public health disease surveillance and reporting.
Embryonic stem cells (ES cells): a cell line derived from undifferentiated, pluripotent cells from the embryo.
Emerging infectious diseases: refer to those infectious diseases whose incidence in humans have increased within the past two decades or threaten to increase in the near future. Emergence may also be used to describe the reappearance (or "re-emergence") of a known infection after a decline in incidence.
Enhancer: a regulatory DNA sequence that increases transcription of a gene. An enhancer can function in either orientation and it may be located up to several thousand base pairs upstream or down stream from the gene, it regulates.
ENTREZ: an online search and retrieval system that integrates information from databases at NCBI. These databases include nucleotide sequences, protein sequences, macromolecular structures, whole genomes, OMIM, and MEDLINE, through PubMed.
Environmental factors: may include chemical, dietary factors, infectious agents, physical, and social factors.
Enzyme: a protein, which acts as a biological catalyst that controls the rate of a biochemical reaction within a cell.
Epigenetic: a term describing nonmutational phenomenon, such as methylation and histone modification that modify the expression of a gene.
Euchromatin: the fraction of the nuclear genome, which contains transcriptionally active DNA and which, unlike *heterochromatin*, adopts a relatively extended conformation.
Eukaryote: an organism whose cells show internal compartmentalization in the form of membrane-bounded organelles (includes animals, plants, fungi, and algae).
Exon: the sections of a gene that code for all of its functional product. Eukaryotic genes may contain many exons interspersed with noncoding introns. An exon is represented in the mature mRNA product—the portions of an mRNA molecule that is left after all introns are spliced out, which serves as a template for protein synthesis.
Expression sequences tag (EST): partial or full complement DNA sequences, which can serve as markers for regions of the genome, which encode expressed products.
Family history: an essential tool in clinical genetics. Interpreting the family history can be complicated by many factors, including small families, incomplete or erroneous family histories,

consanguinity, variable penetrance, and the current lack of real understanding of the multiple genes involved in polygenic (complex) diseases.

FASTA format: a simple universal text format for storing DNA and protein sequences. The sequence begins with a ">" character followed by a single-line description (or header) followed by lines of sequence data.

Flow cytometry: the fractionation of chromosomes according to size and base composition in a fluorescence-activated chromosome (or cell) sorter.

Fluorescence in situ hybridization (FISH): a form of chromosome in situ hybridization in which nucleic acid probe is labeled by incorporation of a *fluorophore*, a chemical group that fluoresces when exposed to UV irradiation.

Founder effect: changes in allelic frequencies that occur when a small group is separated from a large population and establishes in a new location.

Frame-shift mutation: the addition or deletion of a number of DNA bases that is not a multiple of three, thus causing a shift in the reading frame of the gene. This shift leads to a change in the reading frame of all parts of a gene that are downstream from the mutation leading to a premature stop codon, and thus to a truncated protein product.

Functional genomics: the development and implementation of technologies to characterize the mechanisms through which genes and their products function and interact with each other and with the environment.

Gain-of-function mutation: a mutation that produces a protein that takes on a new or enhanced function.

Gene: the fundamental unit of heredity; in molecular terms, a gene comprises a length of DNA that encodes a functional product, which may be a polypeptide (a whole or constituent part of a protein or an enzyme) or a ribonucleic acid. It includes regions that precede and follow the coding region as well as introns and exons. The exact boundaries of a gene are often ill defined, since many promoter and enhancer regions dispersed over many kilobases may influence transcription.

Gene-based therapy: refers to all treatment regimens that employ or target genetic material. This includes (i) *transfection* (introducing cells whose genetic make- up is modified) (ii) *antisense* therapy, and (iii) *naked DNA* vaccination.

Gene expression: the process through which a gene is activated at a particular time and place so that its functional product is produced, that is, transcription into mRNA followed by translation into protein.

Gene expression profile: the pattern of changes in the expression of a specific set of genes that is relevant to a disease or treatment. The detection of this pattern depends upon the use of specific gene expression measurement technique.

Gene family: a group of closely related genes that make similar protein products.

Gene knockouts: a commonly used technique to demonstrate the phenotypic effects and/or variation related to a particular gene in a model organism, for example, in mouse (*see Knockout*); absence of many genes may have no apparent effect upon **phenotypes** (though stress situations may reveal specific susceptibilities). Other single knockouts may have a catastrophic effect upon the organism, or be lethal so that the organism cannot develop at all.

Gene regulatory network: a functional map of the relationships between a number of different genes and gene products (proteins), regulatory molecules, etc. that define the regulatory response of a cell with respect to a particular physiological function.

Gene therapy: a therapeutic medical procedure that involves either replacing/manipulating or supplementing nonfunctional genes with healthy genes. Gene therapy can be targeted to somatic (body) or germ (egg and sperm) cells. In *somatic gene therapy*, the recipient's genome is changed, but the change is not passed along to the next generation. In *germ-line gene therapy*, the parent's egg or sperm cells are changed with the goal of passing on the changes to their offspring.

Genetic architecture: refers to the full range of genetic effects on a trait. Genetic architecture is a moving target that changes according to gene and genotype frequencies, distributions of environmental factors, and such biological properties as age and sex.

Genetic code: the relationship between the order of nucleotide bases in the coding region of a gene and the order of amino acids in the polypeptide product. It is universal, triplet, nonoverlapping code such that each set of three bases (termed a codon) specifies which of the 20 amino acids is present in the polypeptide chain product of a particular position.

Genetic counseling: an important process for individuals and families who have a genetic disease or who are at risk for such a disease. Genetic counseling provides patients and other family members information about their condition and helps them make informed decisions.

Genetic determinism: the unsubstantiated theory that genetic factors determine a person's health, behavior, intelligence, or other complex attributes.

Genetic discrimination: unfavorable discrimination of an individual, a family, community, or an ethnic group on the basis of genetic information. Discrimination may include societal segregation, political persecution, opportunities for education and training, lack or restricted employment prospects, and adequate personal financial planning, for example, life insurance and mortgage.

Genetic engineering: the use of molecular biology techniques such as restriction enzymes, ligation, and cloning to transfer genes among organisms (*also known as recombinant DNA cloning*).

Genetic enhancement: the use of genetic methodologies to improve functional capacities of an organism rather than to treat disease.

Genetic epidemiology: a field of research in which correlations are sought between phenotypic trends and genetic variation across population groups.

Genetic map: a map showing the positions of genetic markers along the length of a chromosome relative to each other (genetic map) or in absolute distances from each other.

Genetic screening: testing a population group to identify a subset of individuals at high risk for having or transmitting a specific genetic disorder.

Genetic susceptibility: predisposition to a particular disease due to the presence of a specific allele or combination of alleles in an individual's genome.

Genetic test: an analysis performed on human DNA, RNA, genes, and/or chromosomes to detect heritable or acquired genotypes. A genetic test also is the analysis of human proteins and certain *metabolites*, which are predominantly used to detect heritable or acquired *genotypes, mutations*, or *phenotypes*.

Genetic testing: strictly refers to testing for a specific chromosomal abnormality or a DNA (nuclear or mitochondrial) mutation already known to exist in a family member. This includes diagnostic testing (postnatal or prenatal), presymptomatic or predictive genetic testing or for establishing the carrier status. The individual concerned should have been offered full information on all aspects of the genetic test through the process of "*nonjudgmental and nondirective*" genetic counseling. Most laboratories require a formal fully informed signed consent before carrying out the test. Genetic testing commonly involves DNA/RNA-based tests for single gene variants, complex genotypes, acquired mutations, and measures of gene expression. Epidemiologic studies are needed to establish clinical validity of each method to establish sensitivity, specificity, and predictive value.

Genetics: refers to the study of heredity, gene, and genetic material. In contrast to genomics, the genetics is traditionally related to lower-throughput, smaller-scale emphasis on single genes, rather than on studying structure, organization, and function of many genes.

Genome: the complete set of chromosomal and extra-chromosomal DNA/RNA of an organism, a cell, an organelle or a virus.

Genome annotation: the process through which landmarks in a genomic sequence are characterized using computational and other means; for example, genes are identified, predictions made as to the function of their products, their regulatory regions defined, and intergenic regions characterized (*see Annotation*).

Genome ontology: a standard set of consistent nomenclature systems that can be used to describe gene and protein functions in all organisms based on molecular function, biological process, and cellular location.

Genome project: the research and technology development effort aimed at mapping and sequencing the entire genome of human beings and other organisms.

Genomic drugs: drugs based on molecular targets; genomic knowledge of the genes involved in diseases, disease pathways, and drug-response.

Genomic instability: an increased tendency of the GENOME to acquire MUTATIONS when various processes involved in maintaining and replicating the genome are dysfunctional.

Genomic profiling: complete genomic sequence of an individual including the expression profile. This would be targeted to specific requirements, for example, most common complex diseases (diabetes, hypertension, and coronary heart disease).

Genomics: the study of the genome and its action. The term is commonly used to refer large-scale, high-throughput molecular analyses of multiple genes, gene products, or regions of genetic material (DNA and RNA). The term also includes the comparative aspect of genomes of various species, their evolution, and how they relate to each other (*see Comparative genomics*).

Genotype: the genetic constitution of an organism; commonly used in reference to a specific disease or trait.

GenPept: a comprehensive protein database that contains all of the translated regions of GenBank sequences.

Germ-line cells: a cell with a haploid chromosome content (also referred to as a gamete); in animals, sperm or egg, and in plants, pollen, or ovum.

Germline mosaic (germinal mosaic, gonadal mosaic, gonosomal mosaic): an individual who has a subset of germline cells carrying a mutation, which is not found in other germline cells.

Germline mutation: a gene change in the body's reproductive cells (egg or sperm) that becomes incorporated into the DNA of every cell in the body of offspring; germline mutations are passed on from parents to offspring, also called *hereditary mutation.*

Haploid: describing a cell (typically a gamete), which has only a single copy of each chromosome (i.e., 23 in man).

Haplotype: a series of closely linked loci on a particular chromosome, which tend to be inherited together as a block.

Heterozygosity: the presence of different alleles of a gene in one individual or in a population—a measure of genetic diversity.

Heterozygote: refers to a particular allele of a gene at a defined chromosome locus. A heterozygote has a different allelic form of the gene at each of the two homologous chromosomes.

Homology: similarity between two sequences due to their evolution from a common ancestor, often referred to as *homologs.*

Homozygote: refers to same allelic form of a gene on each of the two homologous chromosomes.

Human gene transfer: the process of transferring genetic material (DNA or RNA) into a person; an experimental therapeutic procedure to treat certain health problems by compensating for defective genes, producing a potentially therapeutic substance, or triggering the immune system to fight disease. This may help improve genetic disorders, particularly those conditions that result from inborn errors in a single gene (e.g., sickle cell anemia, hemophilia, and cystic fibrosis), and as well as with complex disorders, like cancer, heart disease, and certain infectious diseases, such as HIV/AIDS.

Human Genetics Commission: established in December 1999 in order to provide strategic advice to the UK Government on how new developments in human genetics will impact on people and on health care. It has a particular remit to advise on the social, ethical, and legal implications of these developments.(http://www.hgc.gov.uk/)

Human genome epidemiology: an evolving field of inquiry that uses systematic applications of epidemiologic methods and approaches in population based studies of the impact of human genetic variation on health and disease. Human genome epidemiology represents the intersection between genetic epidemiology and molecular epidemiology. The spectrum of topics addressed in human genome epidemiology range from basic to applied population based research.

Human Genome Project: a program to determine the sequence of the entire 3 billion bases of the human genome.

Identity by descent (IBD): alleles in an individual or in two people that are identical, because they have been inherited from the same common ancestor, as opposed to *identity by state* (*IBS*), which is coincidental possession of similar alleles in unrelated individuals. (*see Consanguinity*).

Immunogenomics: refers to the study of organization, function, and evolution of vertebrate defense genes, particularly those encoded by the major histocompatibility complex (MHC) and the leukocyte receptor complex (LRC). Both complexes form integral parts of the immune system. The MHC is the most important genetic region in relation to infection and common disease such as autoimmunity. Driven by pathogen variability, immune genes have become the most polymorphic loci known, with some genes having over 500 alleles. The main function of these genes is to provide protection against pathogens and they achieve this through complex pathways for antigen processing and presentation.

In situ hybridization: hybridization of a labeled nucleic acid to a target nucleic acid, which is typically immobilized on a microscopic slide, such as DNA of denatured metaphase chromosomes [as in *fluorescent in situ hybridization* (FISH)] or the RNA in a section of tissue [as in *tissue in situ hybridization* (TISH)].

In vitro: (Latin) literally "in glass," meaning outside of the organism in the laboratory, usually is a tissue culture.

In vivo: (Latin) literally "in life," meaning within a living organism.

Informatics: the study of the application of computer and statistical techniques to the management of information. In genome projects, informatics includes the development of methods to search databases quickly, to analyze DNA sequence information, and to predict protein sequence and structure from DNA sequence data.

Intron: a noncoding sequence within eukaryotic genes, which separates the exons (coding regions). Introns are spliced out of the messenger RNA molecule created from a gene after transcription, prior to protein translation (protein synthesis).
Isoforms/isozymes: alternative forms of protein/enzyme.
Knockout: a technique used primarily in mouse genetics to inactivate a particular gene in order to define its function.
Library: a collection of genomic or complementary DNA sequences from a particular organism that have been cloned in a vector and grown in an appropriate host organism (e.g., bacteria or yeast).
Ligase: an enzyme, which can use ATP to create phosphate bonds between the ends of two DNA fragments, effectively joining two DNA molecules into one.
Linkage: the phenomenon whereby pairs of genes, which are located in close proximity on the same chromosome tend to be coinherited.
Linkage analysis: a process of locating genes on the chromosome by measuring recombination rates between phenotypic and genetic markers (*see Lod score*)
Linkage disequilibrium: the nonrandom association in a population of alleles at nearby loci.
Locus: The specific site on a chromosome at which a particular gene or other DNA landmark is located.
Lod score: a measure of likelihood of genetic linkage between loci; a lod score greater than + 3 is often taken as evidence of linkage; one that is less than -2 is often taken as evidence against linkage.
Loss-of-function mutation: a mutation that decreases the production or function (or both) of the gene product.
Loss of heteorozygosity (LOH): loss of alleles on one chromosome detected by assaying for markers for which an individual is constitutionally heterozygous.
Lyonization: the process of random X chromosome inactivation in mammals.
Marker: a specific feature at an identified physical location on a chromosome, whose inheritance can be followed. The position of a gene implicated in a particular phenotypic effect can be defined through its linkage to such markers.
Meiosis: reductive cell division occurring exclusively in testis and ovary and resulting in the production of haploid cells, including sperm cells and egg cells.
Mendelian genetics: classical genetics that focuses on **monogenic** genes with high **penetrance**. The Mendelian genetics is a true **paradigm** and is used in discussing the mode of inheritance (*see Monogenic disease*).
Mendelian segregation: the process whereby individuals inherit and transmit to their offspring one out of the two alleles present in homologous chromosomes.
Messenger RNA (mRNA): RNA molecules that are synthesized from a DNA template in the nucleus (a gene) and transported to ribosomes in the cytoplasm where they serve as a template for the synthesis of protein (translation).
Microarrays-diagnostics: a rapidly developing tool increasingly used in pharmaceutical and genomics research and has the potential for applications in high-throughput diagnostic devices. Microarrays can be made of DNA sequences with known gene mutations, polymorphisms, and as well as selected protein molecules.
Microsatellite DNA: small array (often less than 0.1 kb) of short tandemly repeated DNA sequences.
Minisatellite DNA: an intermediate size array (often 0.1 to 20 kb long) of short tandemly repeated DNA sequences. *Hypervariable minisatellite* DNA is the basis of DNA fingerprinting and many VNTR markers.
Missense mutation: substitution of a single DNA base that results in a codon that specifies an alternative amino acid.
Mitochondria: cellular organelles present in eukaryotic organisms, which enable aerobic respiration and generate the energy to drive cellular processes. Each mitochondrion contains a small amount of circular DNA encoding a small number of genes (approximately 50).
Mitosis: cell division in somatic cells.
Model organism: an experimental organism in which a particular physiological process or disease has similar characteristics to the corresponding process in humans, permitting the investigation of the common underlying mechanisms.
Modifier gene: a gene whose expression can influence a phenotype resulting from mutation at another locus.
Molecular genetic screening: screening a section of the population known to be at a higher risk to be heterozygous for one of the mutations in the gene for a common autosomal recessive disease, for example, screening for cystic fibrosis in the North-European populations and beta-thalassaemia in the Mediterranean and Middle-East population groups.
Molecular genetic testing: molecular genetic testing for use in patient diagnosis, management, and genetic counseling; this is increasingly used in presymptomatic (predictive) genetic testing of "at-risk" family members using a previously known disease-causing mutation in the family.
Molecular hybridization: any process in which a probe DNA and a target DNA are denatured and allowed to reanneal under conditions, which encourage formation of heteroduplexes.
Mosaic: a genetic mosaic is an individual who has two or more genetically different cell lines derived from a single zygote.
Motif: a DNA-sequence pattern within a gene that, because of its similarity to sequences in other known genes, suggests a possible function of the gene, its protein products, or both.
Multifactorial disease: any disease or disorder caused by interaction of multiple genetic (polygenic) and environmental factors.
Multigene family: a set of evolutionary related loci within a genome, at least one of which can encode a functional product.
Mutation: a heritable alteration in the DNA sequence.
Natural selection: the process whereby some of the inherited genetic variation within a population will affect the ability of individuals to survive to reproduce (*fitness*).
Neutral mutation: a change or alteration in DNA sequence, which has no phenotypic effect (or has no effect on fitness).
Newborn screening: performed in newborns in state public health programs to detect certain genetic diseases for which early diagnosis and treatment are available.
Noncoding sequence: a region of DNA that is not translated into protein. Some noncoding sequences are regulatory portions of genes, others may serve structural purposes (telomeres and centromeres), while others may not have any function.
Nonconservative mutation: a change in the DNA or RNA sequence that leads to the replacement of one amino acid with a very dissimilar one.
Nonsense mutation: substitution of a single DNA base that leads in a stop codon, thus leading to the truncation of a protein.
Northern blot hybridization: a form of molecular hybridization in which target consists of RNA molecules that have been size fractioned by gel electrophoresis and subsequently transferred to a membrane.
Nucleosome: a structured unit of chromatin.
Nucleotide: a subunit of the DNA or RNA molecule. A nucleotide is a base molecule (adenine, cytosine, guanine, and thymine in the

case of DNA), linked to a sugar molecule (deoxyribose or ribose) and phosphate groups.

Nullizygous: lacking any copy of a gene or DNA sequence normally found in chromosomal DNA usually resulting from homozygous deletion in an autosome or from a single deletion in sex chromosomes in male.

OMIM: acronym for McKusick's Online Mendelian Inheritance in Man, a regularly updated electronic catalog of inherited human disorders and phenotypic traits accessible on NCBI network. Each entry is designated by a number (*MIM number*).

Oncogene: an acquired mutant form of a gene, which acts to transform a normal cell into a cancerous one.

Open reading frame (ORF): a significantly long sequence of DNA in which there are no termination codons. Each DNA duplex can have six reading frames, three for each single strand.

Ortholog: one set of homologous genes or proteins that perform similar functions in different species, that is, identical genes from different species, for example, *SRY* in humans and *Sry* in mice.

Paralog: similar genes (members of a gene family) or proteins (homologous) in a single species or different species that perform different functions.

Penetrance: the likelihood that a person carrying a particular mutant gene will have an altered phenotype (*see Phenotype*).

Pesudogene: a DNA sequence, which shows a high degree of sequence homology to a nonallelic functional gene but which is itself nonfunctional.

PFGE (pulse field gel eletrophoresis): a form of gel electrophoresis, which permits size fractionation of large DNA molecules.

Pharmacogenomics: the identification of the genes, which influence individual variation in the efficacy or toxicity of therapeutic agents, and the application of this information in clinical practice.

Phenotype: the clinical and/or any other manifestation or expression, such as a biochemical immunological alteration, of a specific gene or genes, environmental factors, or both.

Physical (gene) map: a map showing the absolute distances between genes.

PIC (polymorphism information content): a measure of informativeness of a genetic marker.

Plasmid: circular extra-chromosomal DNA molecules present in bacteria and yeast. Plasmids replicate autonomously each time a bacterium divides and are transmitted to the daughter cells. DNA segments are commonly cloned using plasmid vectors.

Point mutation: the substitution of a single DNA base in the normal DNA sequence.

Polygenic trait or character: a character or trait determined by the combined action of a number of loci, each with a small effect.

Polymerase chain reaction (PCR): **a** molecular biology technique developed in the mid-1980s through which specific DNA segments may be amplified selectively.

Polymorphism: the stable existence of two or more variant allelic forms of a gene within a particular population, or among different populations.

Positional cloning: the technique through which candidate genes are located in the genome through their coinheritance with linked markers. It allows to identify genes that lack information regarding the biochemical actions of their functional product.

Posttranscriptional modification: **a** series of steps through which protein molecules are biochemically modified within a cell following synthesis by translation of messenger RNA. A protein may undergo a complex series of modifications in different cellular compartments before its final functional form is produced.

Predictive testing: determines the probability that a healthy individual with or without a family history of a certain disease might develop that disease.

Predisposition, genetic: increased susceptibility to a particular disease due to the presence of one or more gene mutations, and/or a combination of alleles (haplotype), not necessarily abnormal, that is associated with an increased risk for the disease, and/or a family history that indicates an increased risk for the disease.

Predisposition test: a test for a genetic predisposition (incompletely **penetrant** conditions). Not all people with a positive test result will manifest the disease during their lifetimes.

Preimplantation genetic diagnosis (PIGD; PGD): used following in vitro fertilization to diagnose a genetic disease or condition in a preimplantation embryo.

Prenatal diagnosis: used to diagnose a genetic disease or condition in a developing fetus.

Presymptomatic test: predictive testing of individuals with a family history. Historically, the term has been used when testing for diseases or conditions such as Huntington's disease where the likelihood of developing the condition (known as **penetrance**) is very high in people with a positive test result.

Primer: a short nucleic acid sequence, often a synthetic oligonucleotide, which binds specifically to a single strand of a target nucleic acid sequence and initiates synthesis, using a suitable polymerase, of a complementary strand.

Probe: a DNA or RNA fragment, which has been labeled in some way, and used in a *molecular hybridization* assay to identify closely related DNA or RNA sequences.

Prokaryote: an organism or cell lacking a nucleus and other membrane bounded organelles. Bacteria are prokaryotic organisms.

Promoter: a combination of short sequence elements to which RNA polymerase binds in order to initiate transcription of a gene.

Protein: a protein is the biological effector molecule encoded by sequences of a gene. A protein molecule consists of one or more polypeptide chains of amino acid subunits. The functional action of a protein depends on its three-dimensional structure, which is determined by its amino acid composition.

Protein truncation test: a method of screening for chain-terminating mutations by artificially expressing a mutant allele in a coupled transcription-translation system.

Proteome: all of the proteins present in a cell or organism.

Proteomics: the development and application of techniques to investigate the protein products of the genome and how they interact to determine biological functions.

Proto-oncogene: a cellular gene, which when mutated is inappropriately expressed and becomes an oncogene.

Pseudoautosomal region (PAR): a region on the tips of mammalian X chromosomes, which is involved in recombination during male meiosis.

Recessive: an allele that has no phenotypic effect in the heterozygous state.

Recombinant DNA technology: the use of molecular biology techniques such as restriction enzymes, ligation, and cloning to transfer genes among organisms (*see Genetic engineering*).

Regulatory mutation: a mutation in a region of the genome that does not encode a protein but affects the expression of a gene.

Regulatory sequence: a DNA sequence to which specific proteins bind to activate or repress the expression of a gene.

Repeat sequences: a stretch of DNA bases that occurs in the genome in multiple identical or closely related copies.

Replication: a process by which a new DNA strand is synthesized by copying an existing strand, using it as a template for the addition of a complementary bases, catalyzed by a DNA polymerase enzyme.
Reproductive cloning: techniques aimed at the generation of an organism with an identical genome to an existing organism.
Restriction enzymes: a family of enzymes derived from bacterial that cut DNA at specific sequences of bases.
Restriction fragment length polymorphism (RFLP): a polymorphism due to difference in size of allelic restriction fragments as a result of restriction site polymorphism.
Ribonucleic acid (RNA): a single stranded nucleic acid molecule comprising a linear chain made up from four nucleotide subunits (A, C, G, and U). There are three types of RNA: messenger, transfer, and ribosomal.
Risk communication: an important aspect of genetic counseling, which involves pedigree analysis, interpretation of the inheritance pattern, genetic risk assessment, and explanation to the family member (or the family).
RT-PCR (reverse transcriptase-PCR): a PCR in which the target DNA is a cDNA copied by reverse transcriptase from an mRNA source.
Screening: carrying out of a test or tests, examination(s) or procedure(s) in order to expose undetected abnormalities, unrecognized (incipient) diseases, or defects: examples are early diagnosis of cancer using mass X-ray mammography for breast cancer and cervical smears for cancer of the cervix.
Segregation: the separation of chromosomes (and the alleles they carry) during meiosis; alleles on different chromosomes segregate randomly among the gametes (and the progeny).
Sensitivity (of a screening test): extent (usually expressed as a percentage) to which a method gives results that are free from false negatives; the fewer the false negatives, the greater the sensitivity. Quantitatively, sensitivity is the proportion of truly diseased persons in the screened population who are identified as diseased by the screening test.
Sex chromosome: the pair of chromosomes that determines the sex (gender) of an organism. In man, one X and one Y chromosomes constitute a male compared to two X chromosomes in a female.
Sex selection: preferential selection of the unborn child on the basis of the gender for social and cultural purposes. However, this may be acceptable for medical reasons, for example, to prevent the birth of a male assessed to be at risk for an X-linked recessive disease. For further information visit: http://www.bioethics.gov/topics/sex_index.html.
Shotgun sequencing: a cloning method in which total genomic DNA is randomly sheared and the fragments ligated into a cloning vector, also referred to as "shotgun" cloning.
Signal transduction: the molecular pathways through which a cell senses changes in its external environment and changes its gene expression patterns in response.
Silent mutation: substitution of a single DNA base that produces no change in the amino acid sequence of the encoded protein.
Single-nucleotide polymorphism (SNP): a common variant in the genome sequence; the human genome contains about10 million SNPs.
Somatic: all of the cells in the body, which are not gametes (germline).
Southern blot hybridization: a form of molecular hybridization in which the target nucleic acid consists of DNA molecules that have been size fractioned by gel electrophoresis and subsequently transferred to a nitrocellulose or nylon membrane.
Splicing: a process by which introns are removed from a messenger RNA prior to translation and the exons adjoined.
Stem cell: a cell, which has the potential to differentiate into a variety of different cell types depending on the environmental stimuli it receives.
Stop codon: a codon that leads to the termination of a protein rather than to the addition of an amino acid. The three stop codons are TGA, TAA, and TAG.
Synteny: a large group of genes that appear in the same order on the chromosomes of two different species.
Systems biology: refers to simultaneous measurement of thousands of molecular components (such as transcripts, proteins, and metabolites) and integrate these disparate data sets with clinical end points, in a biologically relevant manner; this model can be applied in understanding the etiology of disease.
Telomere: the natural end of the chromosome.
Therapeutic cloning: the generation and manipulation of stem cells with the objective of deriving cells of a particular organ or tissue to treat a disease.
Transcription: the process through which a gene is expressed to generate a complementary RNA molecule on a DNA template using RNA polymerase.
Transcription factor: a protein, which binds DNA at specific sequences and regulates the transcription of specific genes.
Transcriptome: the total messenger RNA expressed in a cell or tissue at a given point in time.
Transfection: a process by which new DNA is inserted in a eukaryotic cell allowing stable integration into the cell's genome.
Transformation: introduction of foreign DNA into a cell and expression of genes from the introduced DNA; this does not necessarily include integration into host cell genome.
Transgene: a gene from one source that has been incorporated into the genome of another organism.
Transgenic animal/plant: a fertile animal or plant that carries an introduced gene(s) in its germ-line.
Translation: a process through which a polypeptide chain of amino acid molecules is generated as directed by the sequence of a particular messenger RNA sequence.
Transposon: a mobile nucleic acid element.
Tumor suppressor gene: a gene, which serves to protect cells from entering a cancerous state; according to Knudson's "two-hit" hypothesis, both alleles of a particular tumor suppressor gene must acquire a mutation before the cell will enter a transformed cancerous state.
Unequal crossing over: recombination between nonallelic sequences on nonsister chromatids of homologous chromosomes.
Western blotting: a process in which proteins are size-fractioned in a polyacrylamide gel prior to transfer to a nitrocellulose membrane for probing with an antibody.
X-chromosome inactivation: random inactivation of one of the two X chromosomes in mammals by a specialized form of genetic imprinting (*see Lyonization*).
Yeast artificial chromosome (YAC): an artificial chromosome produced by combining large fragments of foreign DNA with small sequence elements necessary for chromosome function in yeast cells.
Yeast two-hybrid system: a genetic method for analyzing the interactions of proteins.
Zinc finger: a polypeptide motif, which is stabilized by binding a zinc atom and confers on proteins an ability to bind specifically to DNA sequences; commonly found in transcription factors.
Zoo blot: a Southern blot containing DNA samples from different species.

On-line Resources and Other Useful Contact Addresses

Useful Internet Genetic and Genomic Resources
All addresses have the prefix http://

Resource	Address
Genome Glossary	www.ornl.gov/sci/tecresources/Human_Genome/ www.genome.gov/glossary.cfm
Individual Chromosome Genome Map	www.ornl.gov/hgmis/posters/chromosome
National Center for Biotechnology Information	www.ncbi.nlm.nih.gov/
Medline/PubMed	www.ncbi.nlm.nih.gov/entrez/pubmed/
Mendelain Inheritance in Man	www.ncbi.nlm.nih.gov/entrez/OMIM/
Wellcome Trust Genome Resource	www.genome.wellcome.ac.uk/
The Sanger Center Genome Database	www.ensembl.org/index.html
Human Genome Organization (HUGO)	www.hugo-international.org/
Human Genome Database	www.hugo.gdb.org/
Genome Databases/GenBank	www.ncbi.nlm.nih.gov/databases
Celera Genomics	www.celera.com/
Pharmacogenomics/Genome Research	www.incyte.com/
Human-Mouse Homology Map	www.ncbi.nlm.nih.gov/Homology
The Jackson Laboratory/Mouse-Human Gene Map	www.informatics.jax.org
Human Genome Project/Wellcome Trust Sanger Center	www.sanger.ac.uk/HGP
The Golden Path Genome Viewer	www.genome.ucsc.edu
Cold Spring Harbor Laboratory/SNP Map	www.cshl.org/db/snp/map
University of Utah Genome Center/SNP Database	www.genome.utah.edu/genesnps
Human Gene Mutation Database	www.hgmd.cf.ac.uk
Gene Tests	www.genetests.org
Gene Clinics	www.geneclinics.org
Ethical, Legal and Social Implications of Human Genome Project (ELSI)	www.nhgri.nih.gov/ELSI
US Department of Defense Human Genome Project	www.er.doe.gov/
Human Genetics Commission (UK)	www.hgc.gov.uk/
UK Health Department Genetics Information	www.doh.gov.uk/genetics
American Society of Human Genetics	www.faseb.org/ashg/
British Society of Human Genetics	www.bshg.org.uk
European Society of Human Genetics	www.eshg.org/
American Society of gene therapy	www.asgt.org
Genomics at the FDA	www.fda.gov/cder/genomics/
Personalized Medicine Coalition	www.personalizedmedicinecoalition.org
National Center for Toxicological Research	www.fda.gov/nctr/
Office of Genomics and Disease Prevention, Center for Disease Control (CDC)	www.cdc.gov/genomics/
Society for Genomics Policy and Population health	www.sgpph.org

Index

Note: *t*= table, *f*= figure or illustration, *b*= box

D

E